Log-Gases and Random Matrices

The London Mathematical Society Monographs Series

The London Mathematical Society Monographs Series was established in 1968. Since that time it has published outstanding volumes that have been critically acclaimed by the mathematics community. The aim of this series is to publish authoritative accounts of current research in mathematics and high-quality expository works bringing the reader to the frontiers of research. Of particular interest are topics that have developed rapidly in the last ten years but that have reached a certain level of maturity. Clarity of exposition is important and each book should be accessible to those commencing work in its field.

The original series was founded in 1968 by the Society and Academic Press; the second series was launched by the Society and Oxford University Press in 1983. In January 2003, the Society and Princeton University Press united to expand the number of books published annually and to make the series more international in scope.

Vol. 34, *Log-Gases and Random Matrices*, by Peter J. Forrester
Vol. 33, *Prime-Detecting SievesI,* by Glyn Harman
Vol. 32, *The Geometry and Topology of Cȯxeter Groups*, by Michael W. Davis
Vol. 31, *Analysis of Heat Equations on Domains, by El Maati Ouhabaz*

Log-Gases and Random Matrices

P.J. Forrester

PRINCETON UNIVERSITY PRESS
PRINCETON AND OXFORD

Published by Princeton University Press,
41 William Street, Princeton, New Jersey 08540
In the United Kingdom: Princeton University Press,
6 Oxford Street, Woodstock, Oxfordshire OX20 1TW
press.princeton.edu

Library of Congress Cataloging-in-Publication Data

Forrester, Peter (Peter John)
Log-gases and random matrices / P.J. Forrester.
p. cm. -- (London Mathematical Society monographs)
ISBN 978-0-691-12829-0 (hardcover : alk. paper) 1. Random matrices.
2. Jacobi polynomials. 3. Integral theorems. I. Title.
QA188.F656 2010
519.2--dc22
2009053314

British Library Cataloging-in-Publication Data is available

The publisher would like to acknowledge the author of this volume for

providing the camera-ready copy from which this book was printed

Printed on acid-free paper. ∞

Printed in the United States of America

10 9 8 7 6 5 4 3 2 1

Preface

Often it is asked what makes a mathematical topic interesting. Some qualities which come to mind are usefulness, beauty, depth and fertility. Usefulness is usually measured by the utility of the topic outside mathematics. Beauty is an alluring quality of much of mathematics, with the caveat that it is often something only a trained eye can see. Depth comes via the linking together of multiple ideas and topics, often seemingly removed from the original context. And fertility means that with a reasonable effort there are new results, some useful, some with beauty, and a few maybe with depth, still awaiting to be found.

More than fifteen years ago I embarked on a project to write in monograph form a development of the theory of solvable log-gas systems in statistical mechanics. As a researcher in the field, I had personally witnessed and experienced some of the interesting qualities of this topic, and I was keen that these be recorded in a form which could serve as a reference for researchers in related fields. Little did I realize that in the ensuing years these related fields would be the subject of intense research activity, requiring a revision of both the focus of the book, and my own research directions, to properly reflect these developments.

Although my focus thus evolved away from the statistical mechanics of log-gas systems, this subject still proved to be a unifying theme in the presentation of the subject matter. And as a further give away as to my own research origins, there is a fairly strong flavor of the language of classical equilibrium statistical mechanics throughout, although a similar background of the reader can hardly be expected. More likely the motivation of the reader will come from the topics of random matrices, Painlevé systems, stochastic growth processes, or Jack polynomials. These are some of the the related fields referred to above, which have been the subject of much recent activity, and which promise to remain interesting topics into the future. Of these it is random matrices which appears along side log-gases as the unifying theme of the book. This marriage of topics has a fine historical pedigree, with the log-gas picture of eigenvalues of random matrices being used to great advantage in the pioneering work of Dyson [147].

While providing a directed logical framework, a development of the intersection between log-gases and random matrices necessarily excludes substantial portions of each of the topics taken separately. However, the latter is necessary in order to achieve a mostly self-contained presentation. Seeking the common intersection of two topics can then be seen as a way of achieving this in a fairly democratic manner. In addition there is intersection with a third topic at work, keeping a further bound on the content, but also being responsible for much of the richness of the mathematics. This third topic is integrable systems. In general the exact calculation of correlations and probability distributions for interacting statistical mechanical systems is an intractable problem; however, underlying integrable structures make log-gases and random matrices an exception. The development of this topic leads to the study of determinantal and Pfaffian processes and the corresponding orthogonal polynomials, as well as Painlevé systems and Jack polynomials.

The quality of usefulness marked the beginning of the study of random matrices and log-gases in mathematical physics. As already mentioned, log-gases were introduced as a tool by Dyson to study random matrices, or as expressed in [201], to liberate the mathematics where none yet exists. Random matrices themselves were introduced by Wigner as a model for the statistical properties of the highly excited energy levels of heavy nuclei. Many of the early works on this theme (up to 1965) are conveniently collected together in the work of Porter [447], along with an introductory review.

Long before their occurrence in physics, random matrices appeared in mathematics, especially in relation to the Haar measure on classical groups. Perhaps the first work of this type is due to Hurwitz, who computed the volume form of a general unitary matrix parametrized in terms of Euler angles [301]. The book of Weyl [540] contains the Haar volume form written in terms of eigenvalues and eigenvectors for the classical groups, and the book of Hua [300] inter-relates these forms to similar measures relating to spaces of Hermitian matrices. In mathematical statistics Wishart [547] gave the volume form of a rectangular matrix $\mathbf{X}$ in terms of the volume of the corresponding positive definite matrix $\mathbf{X}^T\mathbf{X}$. Two other early mathematical works of lasting importance to the field are those of Dixon [137] and Selberg [483], both of which relate to multidimensional integrals with integrands which can be interpreted as probability measures associated with random matrices.

The historical development of random matrices is well documented. Two recent informative accounts are [226], [75]. However, as already stated, the present work addresses only the intersection of the topics of log-gases, random matrices and integrable systems, and so a more extensive historical introduction beyond that already given does not serve as as an informative introduction to the content. Instead it is perhaps worth isolating some of our major topics, giving them some context and providing commentary on how they are to be developed.

Jacobians

All the works referenced above in relation to how random matrices appear in mathematics relate to Jacobians. To gain insight into the prevalance of Jacobians throughout random matrix theory, consider, for example, the problem of studying the eigenvalues of an $N \times N$ real symmetric random matrix, in the situation that the joint distribution on the space of the independent elements is given. The dimension of this space is $N(N+1)/2$. The eigenvalue/eigenvector decomposition provides a change of variables from the independent elements of the matrix to its N eigenvalues and $N(N-1)/2$ variables associated with its eigenvectors. A strategy then to study the eigenvalues is to perform this change of variables, and an essential ingredient for this task is the computation of the corresponding Jacobian.

In the case of real symmetric matrices, complex Hermitian and quaternion real Hermitian matrices, these Jacobians are computed in Chapter 1. Chapter 1 also contains the computation of the Jacobian for a change of variables from the independent elements of an $N \times N$ real symmetric tridiagonal matrix to its eigenvalues and a further $N-1$ independent variables relating to its eigenvectors, and a Jacobian relating to the Householder transformation. In Chapter 2 Jacobians are computed in relation to spaces of unitary matrices, including orthogonal and symplectic unitary matrices, which have dimensions $\mathrm{O}(N^2)$. Jacobians are also computed for the change of variables from the elements to the eigenvalues and variables relating to the eigenvectors for certain unitary and real orthogonal Hessenberg matrices. In these latter circumstances the underlying spaces are of dimension $\mathrm{O}(N)$. The singular value decomposition of rectangular matrices (or equivalently certain decompositions of positive definite matrices), the block decomposition of unitary matrices, and positive definite matrices formed from bidiagonal matrices are some of the settings which give rise to calculations of Jacobians undertaken in Chapter 3.

Jacobians of a different sought appear in Chapter 4. Here rational functions with random coefficients in their partial fraction expansion are encountered, and we seek to change variables from a description in terms of these coefficients to one in terms of the zeros. For this purpose use is make of tools already known from the computation of Jacobians in Chapters 1–3, in particular the calculus of wedge products, and also the classical Vandermonde and Cauchy determinants. In Chapter 11 Jacobians are encountered in the change of variables of differential operators given in terms of the elements of parameter-dependent random matrices, to the differential operators given in terms of corresponding eigenvalues and variables relating to the eigenvectors. Finally, in Chapter 15, a task similar to that addressed in Chapter 1 is undertaken, namely the change of variables from the description of $N \times N$ real, complex, or quaternion real matrices in terms of the independent elements, to one in terms of the eigenvalues (which are typically complex) and an appropriate number of other variables. Also computed are some Jacobians relating to the change of variables of a random polynomial from its coefficients to its zeros.

Determinantal point processes and orthogonal polynomials of one variable

A determinantal point process is a statistical system of many particles (points) in which the k-point correlation function is a $k \times k$ determinant for each k. The study of eigenvalues of random matrices with complex entries, and also of log-gas systems at the special coupling $\beta = 2$ (in the cases considered in this work, the former are mostly special cases of the latter, due to our subject matter being typically restricted to the intersection of the two fields) gives rise to determinantal point processes. Furthermore, the corresponding determinants are determined by just one quantity, referred to as the correlation kernel. To exhibit this fact an essential role is played by orthogonal polynomials. It turns out that in the cases of interest it is the classical orthogonal polynomials which are required. Because full information on the asymptotic properties of these polynomials is known in the existing literature, it is possible to proceed and calculate scaling limits.

A generalization of a determinant point process is a Pfaffian point process, in which the k-point correlation function is a $2k \times 2k$ Pfaffian (or equivalently a $k \times k$ quaternion determinant) for each k. The eigenvalues of matrix ensembles studied in Chapters 1–3 in which the matrices are diagonalized by real orthogonal or symplectic unitary matrices are examples of Pfaffian point processes. These eigenvalues can be interpreted in terms of log-gas systems at the particular coupling $\beta = 1$ and $\beta = 4$ respectively. In the theory of Pfaffian processes skew orthogonal polynomials play a role analogous to that played by orthogonal polynomials in the theory of determinantal processes. For the particular skew inner products encountered from the random matrix problems of Chapters 1–3, the required skew orthogonal polynomials can be expressed in terms of classical orthogonal polynomials, and moreover the elements of the Pfaffian are determined by a single 2×2 block, the elements of which can be expressed in a summed form suitable for asymptotic analysis. In Chapter 15 non-Hermitian Gaussian random matrices are studied, with real, complex, and real quaternion entries. The eigenvalues in the complex case form a determinantal point process, while in the other two cases a Pfaffian point process results.

The Selberg integral, Jack polynomials and generalized hypergeometric functions

Familiar in the theory of the Gauss hypergeometric function is the Euler integral, which has the feature that it can be evaluated in terms of gamma functions. The Selberg integral can be considered as an N-dimensional generalization of the Euler integral. In a random matrix context, it appears as the normalization of various ensembles considered in Chapters 1–3. In a log-gas context, it gives the partition function for general $\beta > 0$. When written in a trigonometric form, extra parameters can be interpreted as providing the full distribution of certain linear statistics in the circular β-ensemble. In the case $\beta = 2$, and in the limit $N \to \infty$, this ties in with the Fisher-Hartwig asymptotic formula from the theory of Toeplitz determinants, covered in Chapter 14.

One of the structures underlying the Selberg integral is a further multidimensional integral referred to as the Dixon-Anderson integral. Like the Selberg integral, it can arrived at by the consideration of a problem in random matrix theory, and it too can be evaluated in terms of gamma functions. The many free parameters in the Dixon-Anderson integral allow for an interpretation giving an inter-relation between the distribution of every second eigenvalue in classical matrix ensembles at $\beta = 1$, and the joint distribution of the eigenvalues for a related classical matrix ensemble at $\beta = 4$.

The integrand of the Selberg integral and its various limits is, up to normalization, the eigenvalue probability density function of the various classical β-ensembles given in Chapters 1–3. Theory linking the Selberg integral with the Dixon-Anderson integral can also be used to provide stochastic three-term recurrences (in the degree N) for the corresponding characteristic polynomials.

The integrand of the Euler integral is the weight function for the classical Jacobi polynomials (when defined on the interval $[0, 1]$). Likewise, the integrand of the Selberg integral, and its various limiting forms, can be used to define inner products which permit complete sets of orthogonal polynomials with special properties. The most fundamental are the Jack polynomials, which relate to the integrand of the Selberg integral in trigonometric form, specialized to correspond to the eigenvalue probability density function for the circular β-ensemble. Using the Jack polynomials as a basis, generalized classical Hermite, Laguerre and Jacobi polynomials, which are multivariable counterparts of the one-variable classical orthogonal polynomials of the same name, can be studied. Another viewpoint of the Jack polynomials is as the polynomial (in complex exponential variables) portion of the eigenfunctions for the Fokker-Planck operator of the Dyson Brownian motion model of the log-gas on a circle. This topic is developed in Chapter 11. A crucial feature is an algebraic theory of the Fokker-Planck operator, in which it is decomposed into fundamental commuting operators relating to the degenerate Hecke algebra of type A, and involving exchange operators. The presentation of the theory of Jack polynomials given in Chapter 12 begins from a study of the simultaneous nonsymmetric polynomial eigenfunctions of these commuting operators. The underlying degenerate Hecke algebra allows these polynomials to be constructed inductively using two fundamental operations (transposition and raising), and these operations allow for the explicit evaluation of associated scalar quantities such as various normalizations. The operations of symmetrization and antisymmetrization also play an important role.

It is well known that the Euler integral can be extended to provide the solutions of the Gauss hypergeometric differential equation. Likewise, weighting the Selberg integrand by an appropriate factor gives rise to multidimensional integrals which relate to multidimensional hypergeometric functions based on Jack polynomials. These are studied in Chapter 13. With the parameters specialized, these integrals can be interpreted as correlations for log-gas systems. Duality formulas, in which multidimensional integrals of this type are expressed as other multidimensional integrals, this time of dimension independent of N, provide the basis for the asymptotic analysis of the corresponding correlations for all even β at least. Furthermore, Jack polynomial theory can be used to compute the bulk dynamical two-point density-density correlation for the Dyson Brownian motion model perturbed from its equilibrium state for all values of rational values of β.

Painlevé transcendents

The Painlevé differential equations are a distinguished family of second order nonlinear equations. In applied mathematics they are perhaps best known for their role in soliton theory, and thus the study of integrable partial differential equations. Certain solutions of the Painlevé differential equations — the Painlevé transcendents — appear in the calculation of gap probabilities for classical random matrix systems corresponding to log-gas systems with $\beta = 1, 2$ and 4 (although the latter two are restricted to those instances in which their is an inter-relation with a $\beta = 2$ log-gas system; one way the latter comes about is by superimposing two $\beta = 1$ ensembles, and integrating over every second eigenvalue, while in the bulk the $\beta = 1$ gap probability is transformed by making use of an evenness symmetry). The viewpoint taken in Chapter 8 on these calculations is an algebraic theory of Painlevé systems based on a Hamiltonian formulation, due mainly to Okamoto, which has the feature of using the Toda lattice equation to inductively construct determinant solutions from a seed solution (the latter relating to an underlying linear second order equation). These determinants can be identified with the gap probabilities of certain log-gas systems at $\beta = 2$. Moreover (formal) scaling of the differential operators gives analogous characterization of the gap probabilities in various scaling limits. As a consequence of these characterizations, high precision calculation of the gap probabilities can be undertaken.

In Chapter 9 additional viewpoints on these results are considered. One is the study of function theoretic properties of the gap probabilities expressed as Fredholm determinants. Indeed, starting with the Fredholm determinant form seems necessary to account for the scaled limit rigorously. Instead of using function theoretic properties, this starting point can also be developed from a Riemann-Hilbert viewpoint, which in turn is closely related to studying isomonodromic deformations of linear second order differential equations. The Fredholm determinant evaluations allow the high precision calculations of the gap probabilities initiated from the Painlevé evaluations to be extended.

Macroscopic electrostatics and asymptotic formulas

Averaging a linear statistic against the eigenvalue spectrum of a random matrix gives a mean value proportional to N (the number of eigenvalues), but a variance of order unity. In applications, this effect shows itself in the study of the statistical properties of the conductance of a quantum wire, noted in Chapter 3. It can be anticipated, and a precise formula for the variance formulated, by hypothesizing that for large length scales the log-gas behaves like a macroscopic conductor, and then using linear response arguments based on the predictions of two-dimensional electrostatics. Moreover, this hypothesis leads to the prediction that the full distribution of the linear statistic will be a Gaussian. For some log-gas systems this has been rigorously established, one of these being that corresponding to the Szegö asymptotic formula from the theory of Toeplitz determinants.

One of the most basic predictions from macroscopic electrostatics is the leading form of the density profile for random matrix ensembles. It can be used too to predict the $O(1/N)$ correction to this form. Another application is to gap probabilities, for which the large gap size asymptotics can be predicted to the leading two orders, and at the soft edge the large deviation forms of the left and right tails can be computed. When available, the exact results agree with these predictions.

Non-intersecting paths and models in statistical mechanics

The generating function for non-intersecting paths on an acyclic directed graph is well known to be given in the form of a determinant. In a number of cases of interest this determinant can be evaluated, revealing that the joint probability density function for the paths possessing prescribed coordinates is of a log-gas form, with $\beta = 1$ or 2. Non-intersecting paths underly a number of statistical mechanical models, in particular the polynuclear growth model and the Hammersely model of directed percolation. To understand how this comes about requires a study of the Robinson-Schensted-Knuth correspondence from bijective combinatorics. This in turn leads naturally to the study of Schur polynomials, which are in fact examples of Jack polynomials. It is shown that fluctuations of the primary observable quantities in the polynuclear growth model and the Hammersely model of directed percolation (the height of the profile and length of the path, respectively) can be expressed as random matrix averages over the unitary group, and that these matrix averages can be rigorously analyzed in the appropriate scaling limits.

Various symmetrizations of the Hammersely model of directed percolation are particularly natural. Examples of these relate to averages over random matrices from the orthogonal and symplectic groups. Transformations of these averages to relate to gap probabilities in Laguerre random matrix ensembles with $\beta = 1$ and 4 allows the the rigorous analysis of the scaling limits.

Applications of random matrix theory

All the random matrix ensembles introduced in Chapters 1–3, for $\beta = 1, 2$ and 4 at least, can be associated with problems in quantum physics. The work of Wigner and Dyson relates the Gaussian ensembles to quantum Hamiltonians; the circular ensembles relate to scattering from a disordered cavity; Verbaarschot has given an interpretation of chiral random matrices in terms of the Dirac equation as it relates to QCD; and quantum transport problems lead to the Jacobi ensemble. For general values of $\beta > 0$ the eigenvalue p.d.f.'s of the β-ensembles appear as the ground state wave function of a class of quantum many-body problems with the $1/r^2$ pair potential. The eigenvalue p.d.f. for the complex random matrices of Chapter 15 has the interpretation as the absolute value squared of the ground state wave function for spinless fermions confined to a plane in the presence of a perpendicular magnetic field.

An application of the GOE to the statistics of high-dimensional random energy landscapes is given in Chapter 1. In Chapter 3 features of Wishart matrices and the Jacobi ensemble relating to multivariate statistics are discussed, as is the application of Wishart matrices to wireless communication, numerical analysis and quantum entanglement (the latter requires a further constraint on the trace). In Chapters 5 and 14 an account is given too of the application of both the GUE and CUE to the study of statistical properties of the zeros of the Riemann zeta function. The applications to statistical mechanics, as summarized under the previous heading, is given in Chapter 10.

It is clear from the above descriptions that the chapters have not been organized according to these headings. Instead the ordering has been determined by the desire to first define and motivate the various classical random matrix ensembles and their generalizations (for example, β extensions, minor processes), to give the mathematics leading to the determination of the corresponding eigenvalue p.d.f.'s, and to relate the latter to log-gases. This accounts for Chapters 1–4. Chapters 5–7 are about the calculation of correlations for the p.d.f's encountered in Chapters 1–4 when the former can be expressed in terms of determinants (Chapter 5) or Pfaffians (Chapter 6). Chapters 8 and 9 give the theory leading to the computation of gap probabilities and spacing distributions in some of the systems for which the correlations were computed in the previous three chapters. With knowledge of the evaluation of gap probabilities and related random matrix averages in terms of Painlevé transcendents thus established, we proceed in Chapter 10 to show how this can be put to use in the analysis of certain models in statistical mechanics relating to non-intersecting paths. The generalization of the Gaussian ensembles to be parameter dependent, or equivalently to have Brownian-motion valued entries, is introduced in Chapter 11, leading to the Calogero-Sutherland quantum many-body system and families of commuting operators. The polynomial eigenfunctions of these commuting operators are studied in Chapters 12 and 13, culminating in the computation of correlation functions for general β. Theory from Chapter 4 on the Dixon-Anderson integral again appears in Chapter 13, for its relevance to the computation of correlations for general β (or more precisely, for the inter-relations it provides), while theory from Chapters 1–3 relating to the β-ensembles is developed to give characterizations of the general β bulk and edge states in terms of stochastic differential equations. Continuing the general β theme, the study of fluctuation formulas is taken up in Chapter 14. The topic of the log-gas in a two-dimensional domain (which is in fact where my own studies began), and the corresponding random matrix ensembles in which the eigenvalues are complex, is the theme of final chapter of the book.

After this introduction to the content and organization, a few words about the presentation are appropriate. As already remarked it has been my desire to give enough detail so that the development is self-contained. A large portion of the necessary working is carried out in the body of the text, but use too has been made of an exercises format which is both more space efficient and less laborious. I have aimed to structure the exercises with sufficient intermediate results so that they can reasonably be worked through, without the need to consult the original references. Generally it has been my intention to keep the subject matter moving. Consequently, there are a small number of results requiring a technical working beyond the main stream of the book, which necessarily have been omitted.

I have been most fortunate to have had research fellowships from the Australian Research Council for the duration of this project. This has freed up time and energy for me to follow, and to be part of, many of the developments which have taken place since I began writing. Both being an active researcher in the field and following the developments have been necessary for writing this monograph. While rewarding, studying the research literature is often difficult and inefficient. My own learning was most efficient when studying instead monographs, in particular those of Gupta and Nagar [279], Haake [284], Hua [300], Mehta [398], Macdonald [376] and Muirhead [410]. Similarly it is my hope that this work will prove itself to be an efficient learning resource in preparation for future researches.

There are a number of individuals who have over the years lent their assistance to this project, both directly and indirectly. My wife Gail places value on the worth of such academic pursuits, and provided a home environment to make it possible. For getting me started in research, and teaching me some fundamentals, I thank R.J. Baxter, B. Jancovici and (the late) E.R. Smith. Collaborations with K. Aomoto, T.H. Baker, P. Desrosiers, N.E. Frankel, T. Nagao, E.M. Rains and N.S. Witte have been of great value. E. Dueñez provided some critical comments on my earlier writing on the circular ensembles which were of much help, and P. Sarnak saw enough potential in these earlier writings to recommend the work to Princeton University Press. Most recently F. Bornemann has provided me with high precision numerical data calculated from Fredholm determinants for use in Chapters 8 and 9, and A. Mays provided some help in relation to the proofreading.

Peter Forrester
Melbourne, Australia
January 2010

Contents

Chapter One

Gaussian matrix ensembles

The Gaussian ensembles are introduced as Hermitian matrices with independent elements distributed as Gaussians, and joint distribution of all independent elements invariant under conjugation by appropriate unitary matrices. The Hermitian matrices are divided into classes according to the elements being real, complex or real quaternion, and their invariance under conjugation by orthogonal, unitary, and unitary symplectic matrices, respectively. These invariances are intimately related to time reversal symmetry in quantum physics, and this in turn leads to the eigenvalues of the Gaussian ensembles being good models of the highly excited spectra of certain quantum systems. Calculation of the eigenvalue p.d.f.'s is essentially an exercise in change of variables, and to calculate the corresponding Jacobians both wedge products and metric forms are used. The p.d.f.'s coincide with the Boltzmann factor for a log-gas system at three special values of the inverse temperature $\beta = 1, 2$ and 4. Thus the eigenvalues behave as charged particles, all of like sign, which are in equilibrium. The Coulomb gas analogy, through the study of various integral equations, allows for the prediction of the leading asymptotic form of the eigenvalue density. After scaling, this leading asymptotic form is referred to as the Wigner semicircle law. The Wigner semicircle law is applied to the study of the statistics of critical points for a model of high-dimensional energy landscapes, and to relating matrix integrals to some combinatorial problems on the enumeration of maps. Conversely, the latter considerations also lead to the proof of the Wigner semicircle law in the case of the GUE. The shifted mean Gaussian ensembles are introduced, and it is shown how the Wigner semicircle law can be used to predict the condition for the separation of the largest eigenvalue. In the last section a family of random tridiagonal matrices, referred to as the Gaussian β-ensemble, are presented. These interpolate continuously between the eigenvalue p.d.f.'s of the Gaussian ensembles studied previously.

1.1 RANDOM REAL SYMMETRIC MATRICES

Quantum mechanics singles out three classes of random Hermitian matrices. We will begin our study by specifying one of these—Hermitian matrices with all entries real, or equivalently real symmetric matrices. The independent elements are taken to be distributed as independent Gaussians, but with the variance different for the diagonal and off-diagonal entries.

Definition 1.1.1 *A random real symmetric $N \times N$ matrix $\mathbf{X}$ is said to belong to the Gaussian orthogonal ensemble (GOE) if the diagonal and upper triangular elements are independently chosen with p.d.f.'s*

$$\frac{1}{\sqrt{2\pi}} e^{-x_{jj}^2/2} \quad \text{and} \quad \frac{1}{\sqrt{\pi}} e^{-x_{jk}^2},$$

respectively.

The p.d.f.'s of Definition 1.1.1 are examples of the normal (or Gaussian) distribution

$$\frac{1}{\sqrt{2\pi\sigma^2}} e^{-(x-\mu)^2/2\sigma^2},$$

denoted $\mathrm{N}[\mu, \sigma]$. With this notation, note that an equivalent construction of GOE matrices is to let $\mathbf{Y}$ be an $N \times N$ random matrix of independent standard Gaussians $\mathrm{N}[0, 1]$ and to form $\mathbf{X} = \frac{1}{2}(\mathbf{Y} + \mathbf{Y}^T)$.

The joint p.d.f. of all the independent elements is

$$P(\mathbf{X}) := \prod_{j=1}^{N} \frac{1}{\sqrt{2\pi}} e^{-x_{jj}^2/2} \prod_{1 \le j < k \le N} \frac{1}{\sqrt{\pi}} e^{-x_{jk}^2} = A_N \prod_{j,k=1}^{N} e^{-x_{jk}^2/2}$$
$$= A_N e^{-\sum_{j,k=1}^{N} x_{jk}^2/2} = A_N e^{-(1/2)\mathrm{Tr}\mathbf{X}^2}, \qquad (1.1)$$

where A_N is the normalization and Tr denotes the trace. This structure is behind the choices of the independent Gaussians in Definition 1.1.1. It provides the starting point to identify features of the GOE which make it relevant to quantum physics [447].

PROPOSITION 1.1.2 *Let $\mathbf{X}$ be a member of the GOE and let $\mathbf{R}$ be an $N \times N$ real orthogonal matrix. One has $P(\mathbf{R}^T\mathbf{X}\mathbf{R}) = P(\mathbf{X})$. Furthermore, the most general p.d.f. satisfying this equation which has the factorization property $P(\mathbf{X}) = \prod_{1\le j\le k\le N} f_{jk}(x_{jk})$ for f_{jk} differentiable is*

$$P(\mathbf{X}) = Ae^{-a\sum_{j,k=1}^{N}(x_{jk})^2 - b\sum_{j=1}^{N} x_{jj}} = Ae^{-a\mathrm{Tr}(\mathbf{X}^2) - b\mathrm{Tr}\mathbf{X}}.$$

Proof. See Exercises 1.1 q.1. □

PROPOSITION 1.1.3 *Define the entropy S of the joint p.d.f. P of the independent elements of $\mathbf{X}$ by $S[P] := -\int P \log P \, \mu(d\mathbf{X}) =: -\langle \log P \rangle_P$ where $\mu(d\mathbf{X}) := \prod_{1\le j\le k\le N} dx_{jk}$. Then P as given by (1.1) maximizes S subject to the constraint $\langle \mathrm{Tr}\mathbf{X}^2 \rangle_P = N^2$.*

Proof. Because of the constraint on the second moment, and the normalization constraint, we can write

$$S[P] = -\langle \log P \rangle_P - \lambda\Big(\langle \mathrm{Tr}\,\mathbf{X}^2 \rangle_P - N^2\Big) + (\log A + 1)\Big(\langle 1 \rangle_P - 1\Big),$$

where λ and $-(\log A + 1)$ are Lagrange multipliers. The condition for a maximum is $\delta S = 0$, where the variation is made with respect to P. This gives

$$-\log P - \lambda \mathrm{Tr}\mathbf{X}^2 + \log A = 0$$

and thus $P = Ae^{-\lambda \mathrm{Tr}\mathbf{X}^2}$. The value of λ is determined to be $\frac{1}{2}$ from the given constraint. □

From these properties an understanding of the applicability of the GOE in the study of quantum energy spectra can be obtained. However as a further prerequisite some theory from quantum mechanics is required [401], [284].

1.1.1 Time reversal in quantum systems

First it is necessary to understand the relevance of an $N \times N$ matrix to quantum energy spectra. A basic axiom of quantum mechanics says the energy spectrum of a quantum system is given by the eigenvalues of its (Hermitian) Hamiltonian operator H, the latter being in general infinite dimensional. Now, to model the discrete portion of the spectrum of a complicated quantum system, a reasonable approximation is to replace H by a finite-dimensional $N \times N$ Hermitian matrix, which has a discrete spectrum only.

Next we need to understand the significance of real symmetric matrices in quantum mechanics. This is related to the fact that in general the structure of a matrix modeling H is constrained by the symmetries of H.

DEFINITION 1.1.4 *A quantum Hamiltonian H is said to have a symmetry A if*

$$[H, A] = 0,$$

where $[\cdot,\cdot]$ denotes the commutator.

One basic symmetry of most quantum systems is time reversal.

DEFINITION 1.1.5 *A general time reversal operator T is any antiunitary operator, which means $T = UK$ where U is unitary and K is the complex conjugation operator.*

Hence we say a quantum system has a time reversal symmetry if the Hamiltonian commutes with an antiunitary operator.

Study of time reversal operators in the context of physical systems further restricts their form. For systems with an even number or no spin $\frac{1}{2}$ particles, it is required that

$$T^2 = 1,$$

while for a finite-dimensional system with an odd number of spin $\frac{1}{2}$ particles

$$T^2 = -1 \qquad \text{and} \qquad T = \mathbf{Z}_{2N} K,$$

where $\mathbf{Z}_{2N}$ is a $2N \times 2N$ block diagonal matrix with each 2×2 diagonal block given by

$$\begin{bmatrix} 0 & -1 \\ 1 & 0 \end{bmatrix} \tag{1.2}$$

(a tensor product formula for $\mathbf{Z}_{2N}$ is given in Exercises 1.1 q.2) which has the effect of reversing the spins. Real symmetric matrices arise in the former situation.

PROPOSITION 1.1.6 *Let H be a quantum Hamiltonian which is invariant with respect to a time reversal symmetry T, where T has the additional property $T^2 = 1$. Then H can always be given a T-invariant orthogonal basis, and with respect to this basis the (in general infinite) matrix representation of H is real.*

Proof. See Exercises 1.1 q.3. □

The above result tells us that a matrix chosen to model the discrete energy spectra of a quantum system with a time reversal symmetry T such that $T^2 = 1$ must be real symmetric. A further general property in quantum mechanics is that two operators related by a similarity transformation of unitary operators are equally valid descriptions of the operator, in that all observables are the same for both operators. A requirement of (1.1) is therefore that any two real symmetric matrices related by a similarity transformation of unitary matrices must have the same p.d.f. for the elements. For the two real symmetric matrices to be so related the unitary matrix must be real orthogonal (or i times a real orthogonal matrix; see Exercises 1.1 q.4). Thus this requirement is guaranteed by Proposition 1.1.2.

We are assuming no information on the Hamiltonian other than the time reversal symmetry. Proposition 1.1.3 says that the p.d.f. (1.1) is the most random subject to the given constraint, in that it maximizes the entropy.

These considerations thus show the applicability of the GOE in the study of quantum spectra. Explicitly, it is hypothesized that the statistical properties of the highly excited states of a complex quantum system with a time reversal symmetry $T^2 = 1$ coincide with the statistical properties of the bulk eigenvalues from large GOE matrices (see Section 7.1.1 for the notion of bulk eigenvalues). Here it is assumed that both spectra have been scaled (technically referred to as *unfolded*) so that the mean spacing is unity. The meaning of a complex quantum system requires further explanation. Wigner first made this hypothesis for the spectra of heavy nuclei in the 1950's. In 1984 Bohigas, Giannoni and Schmit made the same hypothesis for a single particle quantum billiard system, provided the underlying classical mechanics is chaotic and the system has a time reversal symmetry $T^2 = 1$. It is of interest to note that a GOE hypothesis also applies to eigenmodes of microwave cavities (this is not surprising as the Helmholtz equation is formally equivalent to the stationary Schrödinger equation), and also to the eigenmodes of systems governed by classical wave equations — vibrations of irregular shaped metal plates, electromechanical eigenmodes of aluminium and quartz blocks, among other examples. (For references to the original literature, and an extended discussion of GOE hypotheses, see [276].)

EXERCISES 1.1 1. The objective of this exercise is to prove Proposition 1.1.2.

(i) Note that the invariance $P(\mathbf{R}^T\mathbf{X}\mathbf{R}) = P(\mathbf{X})$ with $\mathbf{R}$ a permutation matrix requires that the distribution of all elements on the diagonal be equal, $f_{jj} = f$, and similarly the distribution of all elements on the off diagonal be equal, $f_{jk} = g$ $(j < k)$, for some f and g.

(ii) Choose

$$\mathbf{R} = \begin{bmatrix} 1 & \epsilon & 0 & \dots & 0 \\ -\epsilon & 1 & 0 & \dots & 0 \\ 0 & 0 & 1 & \dots & 0 \\ \vdots & \vdots & \vdots & \ddots & \vdots \\ 0 & 0 & 0 & \dots & 1 \end{bmatrix},$$

where $|\epsilon| \ll 1$. Ignoring terms $O(\epsilon^2)$, show that

$$\mathbf{R}^{-1}\mathbf{X}\mathbf{R} = \begin{bmatrix} x_{11} - 2\epsilon x_{12} & x_{12} + \epsilon(x_{11} - x_{22}) & x_{13} - \epsilon x_{23} & \dots & x_{1N} - \epsilon x_{2N} \\ * & x_{22} + 2\epsilon x_{12} & x_{23} + \epsilon x_{13} & \dots & x_{2N} + \epsilon x_{1N} \\ * & * & x_{33} & \dots & x_{3N} \\ \vdots & \vdots & \vdots & \ddots & \vdots \\ * & * & * & \dots & x_{NN} \end{bmatrix},$$

where the elements $*$ are such that the matrix is symmetric.

(iii) Use the result of (ii) to show that at first order in ϵ the requirement

$$\prod_{j=1}^{N} f(x_{jj}) \prod_{1 \le j < k \le N} g(x_{jk}) = \prod_{j=1}^{N} f(\tilde{x}_{jj}) \prod_{1 \le j < k \le N} g(\tilde{x}_{jk}),$$

where $\tilde{x}_{jk} := [\mathbf{R}^{-1}\mathbf{X}\mathbf{R}]_{jk}$ implies

$$\frac{(x_{11} - x_{22})g'(x_{12})}{g(x_{12})} - 2\frac{x_{12}f'(x_{11})}{f(x_{11})} + 2\frac{x_{12}f'(x_{22})}{f(x_{22})} - \sum_{j=3}^{N}\left(\frac{x_{2j}g'(x_{1j})}{g(x_{1j})} - \frac{x_{1j}g'(x_{2j})}{g(x_{2j})}\right) = 0,$$

which in turn, by separation of variables, implies

$$-\frac{f'(x_{11})}{f(x_{11})} + \frac{f'(x_{12})}{f(x_{12})} + \frac{(x_{11} - x_{22})g'(x_{12})}{2x_{12}g(x_{12})} = \gamma$$

for some constant γ.

(iv) By a further separation of variables in the last equation conclude

$$\frac{g'(x_{12})}{x_{12}g(x_{12})} = -b$$

for some constant b. Solve this differential equation.

(v) Note that the invariance $P(\mathbf{R}^T\mathbf{X}\mathbf{R}) = P(\mathbf{X})$ requires that P be a symmetric function of the eigenvalues, and thus a function of $\mathrm{Tr}(\mathbf{X}^k)$ $k = 1, 2, \dots$. Now combine the results of (i) and (iv) to deduce the result.

2. Let $\mathbf{A} = [a_{ij}]$ be a $p \times q$ matrix and $\mathbf{B} = [b_{i'j'}]$ be an $r \times s$ matrix. The tensor product, denoted $\mathbf{A} \otimes \mathbf{B}$, is the $pr \times qs$ matrix with elements

$$(\mathbf{A} \otimes \mathbf{B})_{ii',jj'} = a_{i,j} b_{i',j'},$$

and thus

$$\mathbf{A} \otimes \mathbf{B} = \begin{bmatrix} a_{11}\mathbf{B} & a_{12}\mathbf{B} & \cdots & a_{1q}\mathbf{B} \\ \vdots & \vdots & \ddots & \vdots \\ a_{p1}\mathbf{B} & a_{p2}\mathbf{B} & \cdots & a_{pq}\mathbf{B} \end{bmatrix}.$$

With $\mathbf{Z}_{2N}$ defined as above (1.2), show that

$$\mathbf{Z}_{2N} = \mathbf{1}_N \otimes \begin{bmatrix} 0 & -1 \\ 1 & 0 \end{bmatrix}. \tag{1.3}$$

3. [401] Let $\vec{\psi}_1 = \alpha_1 \vec{\phi}_1 + T(\alpha_1 \vec{\phi}_1)$, where α_1 is a scalar, $\vec{\psi}_1$ and $\vec{\phi}_1$ are vectors, T is anti-unitary and $T^2 = 1$. Note that $T\vec{\psi}_1 = \vec{\psi}_1$. Here Proposition 1.1.6 will be established.
   (i) From the antiunitarity property it follows that in general $\langle \vec{u} | T\vec{v} \rangle = \overline{\langle T\vec{u} | \vec{v} \rangle}$, where $\langle \cdot | \cdot \rangle$ denotes the inner product. Use this to show that $\langle \vec{u} | \vec{v} \rangle = \overline{\langle T\vec{u} | T\vec{v} \rangle}$.
   (ii) Suppose $\vec{\phi}_2$ is orthogonal to $\vec{\psi}_1$. Use (i) to show that $\vec{\psi}_2 := \alpha_2 \vec{\phi}_2 + T(\alpha_2 \vec{\phi}_2)$ is orthogonal to $\vec{\psi}_1$, and note how this construction can be used to create an orthogonal basis of vectors with the T-invariance property $T\vec{\psi}_n = \vec{\psi}_n$.
   (iii) Consider a Hamiltonian H which has symmetry T. Use the above properties of T to show that with respect to the basis $\{\vec{\psi}_n\}$ the matrix elements $\langle \vec{\psi}_m | H \vec{\psi}_n \rangle$ are real.

4. Let $\mathbf{X}$ be an arbitrary real symmetric $N \times N$ matrix and suppose $\mathbf{X}' = \mathbf{U}^{-1}\mathbf{X}\mathbf{U}$, where $\mathbf{U}$ is unitary and $\mathbf{X}'$ is real symmetric. Assume that the only symmetry of $\mathbf{X}$ and $\mathbf{X}'$ in general (other than some constant times the identity) is the time reversal operator T with $T^2 = 1$.
   (i) Deduce that $T\mathbf{U}T^{-1}\mathbf{U}^{-1}$ commutes with $\mathbf{X}$.
   (ii) Use (i) to show $T\mathbf{U} = c\mathbf{U}T$ and take the inverse of this equation to conclude $c = \pm 1$.
   (iii) Use (ii) and q.3(i) to show that with respect to the T invariant basis $\{\vec{\psi}_n\}$, $\langle \vec{\psi}_n | \mathbf{U} \vec{\psi}_m \rangle = c\overline{\langle \vec{\psi}_n | \mathbf{U} \vec{\psi}_m \rangle}$. Hence conclude that $\mathbf{U}$ has either real elements ($c = 1$) or pure imaginary elements ($c = -1$) and is thus either a real orthogonal matrix or i times a real orthogonal matrix.

1.2 THE EIGENVALUE P.D.F. FOR THE GOE

The p.d.f. for the elements of the matrices in the GOE is given by (1.1). We want to calculate the corresponding eigenvalue p.d.f. This was first accomplished as long ago as 1939 [299]. We will follow a more recent treatment [410].

The new variables and the final expression

The p.d.f. (1.1) has $N(N+1)/2$ independent variables, whereas there are only N eigenvalues, say, $\lambda_1 < \cdots < \lambda_N$. The remaining variables are linear combinations of the independent elements of the eigenvectors, denoted $p_1, \dots, p_{N(N-1)/2}$ say. Our task is to change variables

$$\exp\Big(-\frac{1}{2}\mathrm{Tr}(\mathbf{X}^2)\Big) \prod_{1 \le j \le k \le N} dx_{jk} = \exp\Big(-\frac{1}{2}\sum_{l=1}^{N} \lambda_l^2\Big) |J| \prod_{j=1}^{N} d\lambda_j \prod_{j=1}^{N(N-1)/2} dp_j,$$

where the Jacobian is given by

$$J := \det \begin{bmatrix} \frac{\partial x_{11}}{\partial \lambda_1} & \frac{\partial x_{12}}{\partial \lambda_1} & \cdots & \frac{\partial x_{NN}}{\partial \lambda_1} \\ \frac{\partial x_{11}}{\partial \lambda_2} & \frac{\partial x_{12}}{\partial \lambda_2} & \cdots & \frac{\partial x_{NN}}{\partial \lambda_2} \\ \vdots & \vdots & \ddots & \vdots \\ \frac{\partial x_{11}}{\partial p_{N(N-1)/2}} & \frac{\partial x_{12}}{\partial p_{N(N-1)/2}} & \cdots & \frac{\partial x_{NN}}{\partial p_{N(N-1)/2}} \end{bmatrix}.$$

Thus we must evaluate the Jacobian and then integrate over the variables $p_1, \dots, p_{N(N-1)/2}$ to obtain the eigenvalue p.d.f.

Below we will show that J factorizes,

$$J = \prod_{1\le j<k\le N} (\lambda_k - \lambda_j)\, f(p_1, \dots, p_{N(N-1)/2})$$

so the integration over the variables $p_1, \dots, p_{N(N-1)/2}$ only alters the normalization constant. Hence the final expression for the eigenvalue p.d.f. of the GOE is

$$\frac{1}{C_N} \exp\Big(- \frac{1}{2}\sum_{j=1}^N \lambda_j^2 \Big) \prod_{1\le j<k\le N} |\lambda_k - \lambda_j|, \tag{1.4}$$

where C_N is the normalization constant.

From the viewpoint of application to quantum mechanics, the important feature is the product of differences due to the Jacobian. It can be proved that the correlations are determined entirely by the product of differences, in the sense that the same so-called bulk correlations (see Section 7.1) result if the one body terms $e^{-\lambda^2/2}$ are replaced by some different functional forms $e^{-V(\lambda)/2}$, provided the local density is constant [125]. This feature is referred to as *universality* and gives rise to the notion [53] that the essential feature of a random matrix hypothesis applying to a quantum system is that the spectral correlations are geometrical, meaning that they are due to this Jacobian.

1.2.1 Wedge products

In the theory of multivariable calculus (see, e.g., [485]) the wedge product operation, which is linear and antisymmetric, is defined to give a signed volume element in the tangent space at a point in the manifold. However, for our purpose the latter concept plays no explicit role, and we can make do with the following definition.

DEFINITION 1.2.1 *With $du_i(j) := \delta_{i,j} du_i$ define*

$$du_1 \wedge \cdots \wedge du_N =: \bigwedge_{j=1}^N du_j := \det[du_i(j)]_{i,j=1,\dots,N}. \tag{1.5}$$

Note that it follows from (1.5) that

$$\int_\Omega f(u_1, \dots, u_N) du_1 \wedge \cdots \wedge du_N = \int_\Omega f(u_1, \dots, u_N) du_1 \cdots du_N,$$

since only the diagonal entries in the determinant are nonzero.

When changing variables from $\{u_1, \dots, u_N\}$ to $\{v_1, \dots, v_N\}$ the fundamental formula

$$du_i = \sum_{l=1}^N \frac{\partial u_i}{\partial v_l} dv_l$$

applies. Substituting this in (1.5), and noting the factorization

$$\Big[\sum_{l=1}^N \frac{\partial u_i}{\partial v_l} dv_l(j)\Big]_{i,j=1,\dots,N} = \Big[\frac{\partial u_i}{\partial v_j}\Big]_{i,j=1,\dots,N} \Big[dv_i(j)\Big]_{i,j=1,\dots,N}$$

shows

$$\bigwedge_{j=1}^N du_j = \det\Big[\frac{\partial u_i}{\partial v_j}\Big]_{i,j=1,\dots,N} \bigwedge_{j=1}^N dv_j. \tag{1.6}$$

The determinant in (1.6) is precisely the Jacobian for the change of variables. The practical use of calculating

Jacobians from this formula relies on an alternative way of calculating the l.h.s. of (1.6) in terms of $\{v_j\}$. For the problem at hand, this in turn is done by using the special feature that all the variables are connected by matrix relations. The following definitions are helpful.

DEFINITION 1.2.2 *For any $N \times N$ matrix $\mathbf{X} = [x_{jk}]$, the matrix of differentials is defined as*

$$d\mathbf{X} = \begin{bmatrix} dx_{11} & dx_{12} & \dots & dx_{1N} \\ dx_{21} & dx_{22} & \dots & dx_{2N} \\ \vdots & \vdots & \ddots & \vdots \\ dx_{N1} & dx_{N2} & \dots & dx_{NN} \end{bmatrix}.$$

With this definition the usual product rule for differentiation holds,

$$d(\mathbf{XY}) = d\mathbf{X}\,\mathbf{Y} + \mathbf{X}\,d\mathbf{Y}.$$

DEFINITION 1.2.3 *The symbol $(d\mathbf{X})$ denotes the wedge product of the independent elements of $d\mathbf{X}$. In particular, if $\mathbf{X}$ is a real symmetric matrix,*

$$(d\mathbf{X}) = \bigwedge_{1 \le j \le k \le N} dx_{jk},$$

while if $\mathbf{X} = [x_{jk} + iy_{jk}]_{j,k=1,\dots,N}$ is Hermitian ($x_{jk} = x_{kj}$, $y_{jk} = -y_{kj}$)

$$(d\mathbf{X}) = \bigwedge_{j=1}^{N} dx_{jj} \bigwedge_{1 \le j < k \le N} dx_{jk} dy_{jk}.$$

In integration formulas only the absolute value of the Jacobian occurring in the change of variables formula (1.6) is required, so consequently there is no need to strictly adhere to the ordering of wedge products specified in Definition 1.2.3 (according to the definition, reversing the order of two differentials changes the sign of the wedge product). Because of this, any overall factor of -1 will be ignored in subsequent formulas involving $(d\mathbf{X})$. With this convention $(d\mathbf{X})$ will be referred to as a *volume form*, or *volume measure.*

In preparation for the calculation of J, we note a result for the wedge product $(\mathbf{A}^T d\mathbf{M}\mathbf{A})$, where $\mathbf{A}$ is a real $N \times N$ matrix and $\mathbf{M}$ is a real symmetric $N \times N$ matrix [410].

PROPOSITION 1.2.4 *Let $\mathbf{A}$ and $\mathbf{M}$ be real $N \times N$ matrices, and suppose furthermore that $\mathbf{M}$ is symmetric. We have*

$$(\mathbf{A}^T d\mathbf{M}\mathbf{A}) = (\det \mathbf{A})^{N+1} (d\mathbf{M}).$$

Proof. We note from Definition 1.2.3 that

$$(\mathbf{A}^T d\mathbf{M}\mathbf{A}) = p(\mathbf{A})(d\mathbf{M}), \tag{1.7}$$

where p is a polynomial in the elements of $\mathbf{A}$. Furthermore, if $\mathbf{B}$ is also an $N \times N$ matrix, then

$$(\mathbf{B}^T \mathbf{A}^T d\mathbf{M}\mathbf{A}\mathbf{B}) = p(\mathbf{B})(\mathbf{A}^T d\mathbf{M}\mathbf{A}) = p(\mathbf{B})p(\mathbf{A})(d\mathbf{M}),$$

so we must have $p(\mathbf{AB}) = p(\mathbf{A})p(\mathbf{B})$, for arbitrary $\mathbf{A}$ and $\mathbf{B}$. But it is known [377] that the only polynomial in the matrix elements satisfying such a factorization is

$$p(\mathbf{A}) = (\det \mathbf{A})^k, \quad k \in \mathbb{Z}_{\ge 0}.$$

The value of k can be determined by making the special choice $\mathbf{A} = \mathrm{diag}(a, 1, \dots, 1)$ in (1.7).

For an alternative proof of this result, see Exercises 1.3 q.2. □

1.2.2 Calculation of the Jacobian

From Definition 1.2.3 and (1.6) we see

$$J \bigwedge_{i=1}^{N} d\lambda_i \bigwedge_{j=1}^{N(N-1)/2} dp_j = (d\mathbf{X}).$$

To calculate $(d\mathbf{X})$ in terms of the eigenvalues and eigenvectors we use the fact that all symmetric matrices are orthogonally diagonalizable [8] (see Exercises 1.9 q.3) to write

$$\mathbf{X} = \mathbf{R}\mathbf{L}\mathbf{R}^T. \tag{1.8}$$

Here $\mathbf{L}$ is a diagonal matrix consisting of the N eigenvalues of $\mathbf{X}$ and the columns of the real orthogonal matrix $\mathbf{R}$ consist of the corresponding normalized eigenvectors. Using the notation of Definition 1.2.2, the product rule for differentiation gives

$$d\mathbf{X} = d\mathbf{R}\,\mathbf{L}\mathbf{R}^T + \mathbf{R}d\mathbf{L}\,\mathbf{R}^T + \mathbf{R}\mathbf{L}d\mathbf{R}^T.$$

Rather than take the wedge product of both sides of this equation, it is simpler to premultiply by $\mathbf{R}^T$ and postmultiply by $\mathbf{R}$ to obtain

$$\begin{aligned}\mathbf{R}^T d\mathbf{X}\mathbf{R} &= \mathbf{R}^T d\mathbf{R}\,\mathbf{L} + \mathbf{L}d\mathbf{R}^T\mathbf{R} + d\mathbf{L} \\ &= \mathbf{R}^T d\mathbf{R}\,\mathbf{L} - \mathbf{L}\mathbf{R}^T d\mathbf{R} + d\mathbf{L},\end{aligned} \tag{1.9}$$

where to obtain the last line the formula $d\mathbf{R}^T\mathbf{R} = -\mathbf{R}^T d\mathbf{R}$ has been used (this follows from $\mathbf{R}\mathbf{R}^T = \mathbf{1}$).

According to Proposition 1.2.4

$$(\mathbf{R}^T d\mathbf{X}\mathbf{R}) = (\det \mathbf{R})^{N+1}(d\mathbf{X}). \tag{1.10}$$

But $\mathbf{R}$ is an orthogonal matrix and so $\det \mathbf{R} = \pm 1$. As already noted, since only the modulus of J occurs in the change of variables formula, this sign factor can be ignored.

The wedge product of the r.h.s. of (1.9) can be taken with the aid of the following result.

PROPOSITION 1.2.5 *With the notation $\vec{r_k} = (r_{1k}, r_{2k}, \dots, r_{Nk})^T$ for the kth column of $\mathbf{R}$, we have*

$$\mathbf{R}^T d\mathbf{R}\mathbf{L} - \mathbf{L}\mathbf{R}^T d\mathbf{R} + d\mathbf{L}$$
$$= \begin{bmatrix} d\lambda_1 & (\lambda_2-\lambda_1)\vec{r_1}^T\cdot d\vec{r_2} & \dots & (\lambda_N-\lambda_1)\vec{r_1}^T\cdot d\vec{r_N} \\ (\lambda_2-\lambda_1)\vec{r_1}^T\cdot d\vec{r_2} & d\lambda_2 & \dots & (\lambda_N-\lambda_2)\vec{r_2}^T\cdot d\vec{r_N} \\ \vdots & \vdots & \ddots & \vdots \\ (\lambda_N-\lambda_1)\vec{r_1}^T\cdot d\vec{r_N} & (\lambda_N-\lambda_2)\vec{r_2}^T\cdot d\vec{r_N} & \dots & d\lambda_N \end{bmatrix}.$$

Proof. This is obtained by explicitly forming the matrix products, and simplifying the resulting expression by noting from $d\mathbf{R}^T d\mathbf{R} = -\mathbf{R}^T d\mathbf{R}$ that $\mathbf{R}^T d\mathbf{R}$ is antisymmetric. □

From Proposition 1.2.5 and Definition 1.2.3, the wedge product of the r.h.s. of (1.9) can be written down (note in particular that the matrix in Proposition 1.2.5 is symmetric), whereas (1.10) gives the wedge product of the l.h.s. of (1.9). Equating these expressions gives

$$(d\mathbf{X}) = \prod_{1\le j<k\le N} (\lambda_k - \lambda_j) \bigwedge_{j=1}^{N} d\lambda_j \,(\mathbf{R}^T d\mathbf{R}). \tag{1.11}$$

The factorization property of the Jacobian between the eigenvalues and the variables involving the eigenvectors is evident and the expression (1.4) for the eigenvalue p.d.f. of the GOE follows. The p.d.f. for the components of the eigenvectors is calculated in Exercises 1.2 q.2.

1.2.3 Scaling of the Jacobian

Here we will show how the eigenvalue factor in the Jacobian can be deduced by considering a simple scaling property of the wedge product. Since there are $N(N+1)/2$ independent elements in $\mathbf{X}$, $(d\mathbf{X})$ consists of the product of $N(N+1)/2$ independent differentials. Thus if we multiply $\mathbf{X}$ by a scalar a, we have that $(d\,a\mathbf{X}) = a^{N(N+1)/2}(d\mathbf{X})$. On the other hand, with $\mathbf{X} = \mathbf{RLR}^T$, we know that $(d\mathbf{X})$ is a polynomial in $\lambda_1, \dots, \lambda_N$. Since $a\mathbf{X} = \mathbf{R}a\mathbf{LR}^T$, the scaling property of $(d\,a\mathbf{X})$ gives that in fact $(d\mathbf{X})$ is a homogeneous polynomial of degree $N(N-1)/2$ (here we have subtracted N from $N(N+1)/2$ to account for the scaling of the measure $d\lambda_1 \cdots d\lambda_N$). Furthermore, analysis of the 2×2 case reveals that the Jacobian must vanish linearly for $\lambda_j \to \lambda_k$ (see Exercises 1.2 q.3). Hence the polynomial factor is necessarily proportional to $\prod_{j<k}(\lambda_k - \lambda_j)$, in agreement with the above calculation.

1.2.4 Metric forms

Another approach to deriving (1.11) is through the use of a metric form defined on the space of symmetric matrices [300]. For an $N \times N$ real symmetric matrix $\mathbf{X}$, the metric form of the line element ds is specified by

$$(ds)^2 = \mathrm{Tr}(d\mathbf{X}d\mathbf{X}^T) = \sum_{j=1}^{N}(dx_{jj})^2 + 2\sum_{j<k}(dx_{jk})^2 \tag{1.12}$$

(of course $d\mathbf{X}^T = d\mathbf{X}$, but it is convenient to write as presented), and the volume measure is

$$(d\mathbf{X}) = \bigwedge_{j\le k} dx_{jk}.$$

If one now makes a change of variables, expressing the elements x_{jk} in terms of some new variables y_{jk} such that

$$(ds)^2 = \sum_{j=1}^{N}(h_{jj}dy_{jj})^2 + 2\sum_{j<k}(h_{jk}dy_{jk})^2, \tag{1.13}$$

where the h_{jk} typically depend on $\{y_{jk}\}$, the corresponding volume measure is

$$(d\mathbf{X}) = \Big(\bigwedge_{j\le k} h_{jk}\Big)(d\mathbf{Y}), \tag{1.14}$$

thus giving a change of variable formula for the volume measure.

More generally the metric forms method gives that if $(ds)^2$ is a symmetric quadratic form in some independent infinitesimals $\{dy_\mu\}$, so that

$$(ds)^2 = \sum_{\mu,\nu} g_{\mu,\nu}dy_\mu dy_\nu, \qquad g_{\mu,\nu} = g_{\nu,\mu}, \tag{1.15}$$

then the corresponding volume measure is

$$\Big(\det[g_{\mu,\nu}]\Big)^{1/2} \bigwedge_{\mu} dy_\mu. \tag{1.16}$$

Comparing (1.15) with (1.13), we see that there are only diagonal terms present in the formula for the line element. The determinant is then the product of the diagonal terms, which is consistent with (1.14).

We can apply this formalism by noting from (1.9) and Proposition 1.2.5 that

$$\mathrm{Tr}(d\mathbf{X}d\mathbf{X}^T) = 2\sum_{j<k}(\lambda_k - \lambda_j)^2(\vec{r}_j \cdot d\vec{r}_k)^2 + \sum_{j=1}^{N}(d\lambda_j)^2.$$

Application of (1.14) then reclaims (1.11).

EXERCISES 1.2 1. (i) Let $\mathbf{R}$ be a $N \times N$ real orthogonal matrix. Show that in general $\mathbf{R}$ has $N^2 - N(N-1)/2 - N$ independent elements.

(ii) Use (i) to show that the number of independent elements on the two sides of the equation $\mathbf{X} = \mathbf{RLR}^T$, where $\mathbf{X}$ is real symmetric and $\mathbf{L}$ is diagonal, are equal.

(iii) For $\vec{a}, \vec{b}$, $N \times 1$ real column vectors related by $\vec{a} = \mathbf{A}\vec{b}$ for some $N \times N$ matrix $\mathbf{A}$, show that

$$(d\vec{a}) = |\det \mathbf{A}|(d\vec{b}). \tag{1.17}$$

2. [284] Here the distribution of the components of the eigenvectors in the GOE is calculated.

(i) Note that for matrices in the GOE every eigenvector can be transformed by an arbitrary real orthogonal matrix, and still remain an eigenvector of a matrix in the GOE. Conclude from this that the only invariant of the eigenvectors is their norm, and so the joint distribution of the components $(u_1, \dots, u_N)$ is given by

$$\frac{1}{C}\delta\Big(1 - \sum_{p=1}^{N} u_p^2\Big),$$

where $C = 2\pi^{N/2}/\Gamma(N/2)$ represents the surface area of the unit $(N-1)$-sphere.

(ii) Show that the marginal joint distribution $p(u_1, \dots, u_n)$, obtained by integrating out the variables $u_{n+1}, \dots, u_N$, is given by

$$p(u_1, \dots, u_n) = \pi^{-n/2}\frac{\Gamma(N/2)}{\Gamma((N-n)/2)}\Big(1 - \sum_{p=1}^{n} u_p^2\Big)^{(N-n-2)/2}.$$

For this purpose write the delta function in (i) as a Fourier integral.

(iii) From (ii) show that for large N,

$$\frac{1}{N^{n/2}}p\Big(\frac{u_1}{\sqrt{N}}, \dots, \frac{u_n}{\sqrt{N}}\Big) \sim \Big(\frac{2}{\pi}\Big)^{n/2} e^{-\frac{1}{2}\sum_{p=1}^{n} u_p^2}. \tag{1.18}$$

(iv) Show that forming a vector $(u_1, \dots, u_N)$ in which each component has distribution $x_j/(x_1^2 + \cdots + x_N^2)^{1/2}$, with the x_js standard normal random variables, implies that the vector is uniformly distributed on the unit $(N-1)$-sphere and thus has joint density as in (i). Use this fact to rederive (1.18).

3. (i) For a general 2×2 real symmetric matrix

$$\mathbf{A} := \begin{bmatrix} a & b \\ b & c \end{bmatrix},$$

show that the (unordered) eigenvalues are given by

$$\lambda_\pm = \frac{1}{2}(a+c) \pm \frac{1}{2}\Big((a-c)^2 + 4b^2\Big)^{1/2}.$$

Note that the condition for a degenerate eigenvalue is $b = 0$ and $a = c$, and thus has codimension 2 in the space of matrix entries.

(ii) For the matrix in (i) parametrize the matrix of eigenvectors as

$$\mathbf{R} = \begin{bmatrix} \cos\theta & -\sin\theta \\ \sin\theta & \cos\theta \end{bmatrix}$$

and from the diagonalization equation $\mathbf{A} = \mathbf{R}\,\mathrm{diag}[\lambda_+, \lambda_1]\mathbf{R}^T$, read off that

$$a = \lambda_+ \cos^2\theta + \lambda_- \sin^2\theta, \quad b = (\lambda_+ - \lambda_-)\cos\theta\sin\theta, \quad c = \lambda_+ \sin^2\theta + \lambda_- \cos^2\theta.$$

(iii) Deduce from (ii) that

$$J := \begin{vmatrix} \frac{\partial a}{\partial \lambda_+} & \frac{\partial b}{\partial \lambda_+} & \frac{\partial c}{\partial \lambda_+} \\ \frac{\partial a}{\partial \lambda_-} & \frac{\partial b}{\partial \lambda_-} & \frac{\partial c}{\partial \lambda_-} \\ \frac{\partial a}{\partial \theta} & \frac{\partial b}{\partial \theta} & \frac{\partial c}{\partial \theta} \end{vmatrix} = (\lambda_+ - \lambda_-).$$

1.3 RANDOM COMPLEX HERMITIAN AND QUATERNION REAL HERMITIAN MATRICES

Since most physical systems possess a time reversal symmetry, the GOE correctly models statistical properties of the spectra of many quantum systems (recall the discussion at the end of Section 1.1). Nonetheless the considerations of time reversal symmetry of Section 1.1.1 indicate two further random matrix ensembles [149].

1.3.1 The Gaussian unitary ensemble

For a quantum system without time reversal symmetry the only constraint on the complex Hermitian matrix used to model the discrete portion of the energy spectrum is that two matrices related by a similarity transformation of unitary operators have the same joint p.d.f. for the elements. This requirement is satisfied by the following choice of matrix ensemble.

DEFINITION 1.3.1 *A random Hermitian $N \times N$ matrix $\mathbf{X}$ is said to belong to the Gaussian unitary ensemble (GUE) if the diagonal elements (which must be real) and the upper triangular elements $x_{jk} = u_{jk} + iv_{jk}$ are independently chosen with p.d.f.'s*

$$\frac{1}{\sqrt{\pi}} e^{-x_{jj}^2} \quad \text{and} \quad \frac{2}{\pi} e^{-2(u_{jk}^2 + v_{jk}^2)} = \frac{2}{\pi} e^{-2|x_{jk}|^2},$$

respectively. Equivalently, the diagonal entries have distribution $\mathrm{N}[0, 1/\sqrt{2}]$, while the upper triangular elements have distribution $\mathrm{N}[0, \frac{1}{2}] + i\mathrm{N}[0, \frac{1}{2}]$, and $\mathbf{X}$ can be specified in terms of the complex random matrix $\mathbf{Y}$ with entries independently chosen from $\mathrm{N}[0, 1/\sqrt{2}] + i\mathrm{N}[0, 1/\sqrt{2}]$, according to $\mathbf{X} = (\mathbf{Y} + \mathbf{Y}^\dagger)/2$.

From this definition the joint p.d.f. of all the independent elements is

$$P(\mathbf{X}) := \prod_{j=1}^{N} \frac{1}{\sqrt{\pi}} e^{-x_{jj}^2} \prod_{1 \le j < k \le N} \frac{2}{\pi} e^{-2|x_{jk}^2|} = A_N \prod_{j,k=1}^{N} e^{-|x_{jk}|^2} = A_N e^{-\mathrm{Tr}(\mathbf{X}^2)},$$

where A_N is the normalization. The invariance $P(\mathbf{U}^{-1}\mathbf{X}\mathbf{U}) = P(\mathbf{X})$ for any unitary matrix $\mathbf{U}$ follows immediately.

1.3.2 The Gaussian symplectic ensemble

In Section 1.1.1 it was remarked that in quantum systems with a time reversal symmetry T, either $T^2 = 1$ or $T^2 = -1$ with $T = \mathbf{Z}_{2N}K$. Consideration of the former case leads to real symmetric matrices. Here the latter possibility will be discussed.

Now, since T commutes with the $2N \times 2N$ matrix $\mathbf{X}$ modeling the Hamiltonian, $\mathbf{X}$ must in addition to being Hermitian have the property

$$\mathbf{X} = T\mathbf{X}T^{-1} = \mathbf{Z}_{2N}K\mathbf{X}K^{-1}\mathbf{Z}_{2N}^{-1} = \mathbf{Z}_{2N}K\mathbf{X}K\mathbf{Z}_{2N}^{-1} = \mathbf{Z}_{2N}\bar{\mathbf{X}}\mathbf{Z}_{2N}^{-1}. \tag{1.19}$$

Since $\mathbf{Z}_{2N}$ is block diagonal, with blocks (1.2), a $2N \times 2N$ matrix $\mathbf{X}$ with this property can be viewed as an $N \times N$ matrix with elements consisting of 2×2 blocks of the form

$$\begin{bmatrix} z & w \\ -\bar{w} & \bar{z} \end{bmatrix}, \tag{1.20}$$

where z and w are complex numbers. A 2×2 matrix of this form is said to be *real quaternion*. From an abstract perspective the quaternions are an algebra with elements of the form

$$a_0 + a_1 i + a_2 j + a_3 k, \qquad i^2 = j^2 = k^2 = -1, \;\; ijk = -1, \tag{1.21}$$

where $a_0, \dots, a_3$ are scalars. The basis elements $1, i, j, k$ can be realized as 2×2 matrices with complex elements given by

$$\mathbf{1} := \begin{bmatrix} 1 & 0 \\ 0 & 1 \end{bmatrix} \quad \mathbf{e}_1 := i\sigma_z = \begin{bmatrix} i & 0 \\ 0 & -i \end{bmatrix} \quad \mathbf{e}_2 := i\sigma_y = \begin{bmatrix} 0 & 1 \\ -1 & 0 \end{bmatrix} \quad \mathbf{e}_3 := i\sigma_x = \begin{bmatrix} 0 & i \\ i & 0 \end{bmatrix}, \tag{1.22}$$

respectively. Forming a general linear combination, consisting of real scalar multiples of these basis elements, gives the structure (1.20).

For future reference we note that with a real quaternion $\mathbf{q}$ written in the form $\mathbf{q} = c_0\mathbf{1} + c_1\mathbf{e}_1 + c_2\mathbf{e}_2 + c_3\mathbf{e}_3$ the dual, denoted $\bar{\mathbf{q}}$ or $\mathbf{q}^D$, is defined as

$$\bar{\mathbf{q}} = \mathbf{q}^D = c_0\mathbf{1} - c_1\mathbf{e}_1 - c_2\mathbf{e}_2 - c_3\mathbf{e}_3. \tag{1.23}$$

With this definition the dual of (1.20) is

$$\begin{bmatrix} \bar{z} & -w \\ \bar{w} & z \end{bmatrix}. \tag{1.24}$$

Furthermore, with $|\mathbf{q}|^2 := \bar{\mathbf{q}}\mathbf{q} = \mathbf{q}\bar{\mathbf{q}}$, we have $|\mathbf{q}|^2 = c_0^2 + c_1^2 + c_2^2 + c_3^2$, the relation $|\mathbf{q}_1\mathbf{q}_2| = |\mathbf{q}_1||\mathbf{q}_2|$ holds, and each nonzero $\mathbf{q}$ has a unique inverse, $\mathbf{q}^{-1} = \bar{\mathbf{q}}/|\mathbf{q}|^2$.

An $N \times N$ matrix with real quaternion elements is said to be *quaternion real*. This structure underlies the definition of the third and final ensemble of Gaussian random matrices as motivated by quantum physics.

DEFINITION 1.3.2 *A random Hermitian $N \times N$ matrix $\mathbf{X}$ with real quaternion elements is said to belong to the Gaussian symplectic ensemble (GSE) if the elements z_{jj} of each diagonal real quaternion (which must be real) are independently chosen with p.d.f.*

$$\sqrt{\frac{2}{\pi}} e^{-2z_{jj}^2}$$

(or equivalently have distribution $\mathrm{N}[0, 1/2]$*) while the upper triangular off-diagonal elements $z_{jk} = u_{jk} + iv_{jk}$ and $w_{jk} = u'_{jk} + iv'_{jk}$ are independently chosen with p.d.f.*

$$\frac{4}{\pi} e^{-4|z_{jk}|^2} \qquad \text{and} \qquad \frac{4}{\pi} e^{-4|w_{jk}|^2}$$

(or equivalently have distribution $\mathrm{N}[0, 1/2\sqrt{2}] + i\mathrm{N}[0, 1/2\sqrt{2}]$*). Thus $\mathbf{X} = (\mathbf{Y} + \mathbf{Y}^\dagger)/2$, where $\mathbf{Y}$ is an $N \times N$ random matrix of independent real quaternions with z and w in (1.20) having distribution $N[0, \frac{1}{2}] + iN[0, \frac{1}{2}]$.*

A fundamental property of quaternion real Hermitian matrices, which follows from the first equation in (1.19), is that their spectrum is doubly degenerate (see Exercises 1.3 q.1).

It follows from Definition 1.3.2 that the joint p.d.f. of all the independent elements of the GSE is given by

$$P(\mathbf{X}) = A_N e^{-2\mathrm{Tr}(\mathbf{X}^2)},$$

where A_N denotes the normalization and Tr denotes the trace with $\mathbf{X}^2$ regarded as a quaternion real matix (i.e., Tr($\mathbf{X}^2$) equals the sum of the scalar multiples of $\mathbf{1}_2$ on the diagonal of $\mathbf{X}^2$). This satisfies the general requirement of being invariant with respect to similarity transformations of appropriate unitary matrices. In fact the appropriate unitary matrices are those which under a similarity transformation map a quaternion real Hermitian matrix into another quaternion real Hermitian matrix. This subgroup of unitary matrices is specified by the following result.

PROPOSITION 1.3.3
(a) Let $\mathbf{X}$ be an arbitrary $N \times N$ Hermitian matrix with real quaternion elements, so that in general the only symmetry of $\mathbf{X}$ (other than some multiple of the identity) is the operator $T = \mathbf{Z}_{2N} K$. Then any unitary matrix $\mathbf{U}$ which under a similarity transformation maps $\mathbf{X}$ into another Hermitian matrix with real quaternion elements must commute or anticommute with T.
(b) A unitary matrix $\mathbf{U}$ which commutes with $T = \mathbf{Z}_{2N} K$ has the property

$$\mathbf{U}\mathbf{Z}_{2N}\mathbf{U}^T = \mathbf{Z}_{2N}, \tag{1.25}$$

which implies $\mathbf{U}$ is equivalent to a symplectic matrix, while a unitary matrix $\mathbf{U}$ which anticommutes with T has the property $-\mathbf{U}\mathbf{Z}_{2N}\mathbf{U}^T = \mathbf{Z}_{2N}$.

Proof. (a) Let $\mathbf{X}'$ be such that $\mathbf{U}^{-1}\mathbf{X}\mathbf{U} = \mathbf{X}'$. Since both $\mathbf{X}$ and $\mathbf{X}'$ are quaternion real, T commutes with both of these matrices. This implies $\mathbf{X}TUT^{-1} = TUT^{-1}\mathbf{X}'$. Comparing these two equations gives that $TUT^{-1}\mathbf{U}^{-1}$ commutes with $\mathbf{X}$. But the only operators which commute with $\mathbf{X}$ are T and some multiple of the identity, so the above combination of operators must equal one of these operators. We see that the first choice leads to $T = 1$, which is a contradiction, while the second gives $T\mathbf{U} = \pm \mathbf{U}T$ (regarding the signs, recall Exercises 1.1 q.4(ii)) as required.
(b) At the beginning of this subsection it was noted that any matrix, in this case $\mathbf{U}$, which commutes with T has the property $\mathbf{U} = \mathbf{Z}_{2N}\bar{\mathbf{U}}\mathbf{Z}_{2N}^{-1}$. Equation (1.25) follows after noting $\bar{\mathbf{U}} = (\mathbf{U}^{-1})^T$ and rearranging. To continue, recall that by definition a matrix is symplectic if

$$\mathbf{X}^T\mathbf{J}_{2N}\mathbf{X} = \mathbf{J}_{2N}, \qquad \mathbf{J}_{2N} := \begin{bmatrix} \mathbf{0}_N & \mathbf{1}_N \\ -\mathbf{1}_N & \mathbf{0}_N \end{bmatrix}. \tag{1.26}$$

If $\mathbf{X}$ is also unitary, this implies $\mathbf{X}$ has the block structure

$$\mathbf{X} = \begin{bmatrix} \mathbf{Z} & \mathbf{W} \\ -\bar{\mathbf{W}} & \bar{\mathbf{Z}} \end{bmatrix}$$

(cf. (1.20)). Now, the matrix $\mathbf{J}_{2N}$ is related to $\mathbf{Z}_{2N}$ by a similarity transformation $\mathbf{J}_{2N} = \mathbf{Q}^{-1}\mathbf{Z}_{2N}\mathbf{Q}$, where $\mathbf{Q}$ is a unitary matrix with elements ± 1 (there must therefore be exactly one nonzero element in each row/column). We thus conclude from (1.25) that $\mathbf{Q}^{-1}\mathbf{U}\mathbf{Q}$ is symplectic. The only difference in the anticommuting case is a minus sign, which gives the second result. □

1.3.3 The eigenvalue p.d.f.'s

The calculation of the eigenvalue p.d.f.'s from the joint p.d.f.'s for the elements of the GUE and GSE can be done in a similar way to that presented in Section 1.2 for the GOE. The required working is sketched in Exercises 1.3 q.3 and q.4, and the final results are summarized in the following, which for completeness also contains the eigenvalue p.d.f. for the GOE.

PROPOSITION 1.3.4 *Let $\mathbf{H}$ be a Hermitian matrix with real ($\beta = 1$), complex ($\beta = 2$), or real quaternion ($\beta = 4$) elements, and let $\mathbf{H}$ be decomposed in terms of its eigenvalues and eigenvectors via the formula $\mathbf{H} = \mathbf{U}\mathbf{L}\mathbf{U}^\dagger$, where $\mathbf{L}$ is a diagonal matrix consisting of the eigenvalues of $\mathbf{H}$, and $\mathbf{U}$ is a unitary matrix with real ($\beta = 1$), complex ($\beta = 2$) or real quaternion ($\beta = 4$) elements consisting of the corresponding eigenvectors. We have*

$$(d\mathbf{H}) = \prod_{1 \le j < k \le N} |\lambda_k - \lambda_j|^\beta \bigwedge_{j=1}^{N} d\lambda_j (\mathbf{U}^\dagger d\mathbf{U}), \tag{1.27}$$

and hence for an appropriate choice of the normalization $G_{\beta,N}$, which is given explicitly by (1.163),

$$\frac{1}{G_{\beta,N}} \exp\Big(-\frac{\beta}{2}\sum_{j=1}^{N}\lambda_j^2\Big) \prod_{1\le j<k\le N} |\lambda_k-\lambda_j|^\beta, \tag{1.28}$$

with $\beta = 1,2$ and 4 is the eigenvalue p.d.f. for the GOE, GUE and GSE, respectively.

We remark that for the decomposition $\mathbf{H} = \mathbf{U}\mathbf{L}\mathbf{U}^\dagger$ to be unique the eigenvalues must be ordered and the first component of the eigenvectors must be real and positive. Because (1.28) is a symmetric function of the eigenvalues, the ordering constraint can conveniently be removed, and the normalization appropriately adjusted. In particular, $G_{\beta,N}$ is the normalization without the ordering constraint.

1.3.4 Relationship to Lie algebras

The sets of matrices

$$\begin{aligned} gl(N,\mathbb{R}) &:= \{\text{all } N\times N \text{ real matrices}\},\\ gl(N,\mathbb{C}) &:= \{\text{all } N\times N \text{ complex matrices}\},\\ u^*(2N) &:= \{\text{all } N\times N \text{ real quaternion matrices}\} \end{aligned}$$

are each closed under commutation and so form matrix Lie algebras.

Now, in general a matrix can be decomposed as the sum of a Hermitian and an anti-Hermitian matrix. We see that the Hermitian component of the above Lie algebras corresponds to Hermitian matrices with real, complex and real quaternion elements, respectively. This is significant for a number of reasons. One is from a classification perspective. One can identify ten infinite families of matrix Lie algebras, in correspondence with the ten infinite families of symmetric spaces, as catalogued by Cartan [295]. We will see that each of the remaining seven cases also occurs in a basic quantum mechanics problem constrained by a global symmetry. Furthermore, the identification with matrix Lie algebras implies a one-to-one correspondence between the ten families of Hermitian matrices and the ten families of unitary matrices. This comes about because of the relationship between matrix Lie algebras and symmetric spaces. To each matrix Lie algebra there corresponds a noncompact and compact symmetric space, with the former being isomorphic to a certain set of Hermitian matrices, and the latter isomorphic to a certain set of unitary matrices. Some more details are given in Section 2.1.2. The isomorphism with symmetric spaces has the consequence that the eigenvalue-dependent portion of the Jacobian in $(d\mathbf{H})$ can be written in the form

$$\prod_{\vec{\alpha}\in R_+} |\langle\vec{\alpha},\vec{\lambda}\rangle|^{m_\alpha},$$

where $\vec{\alpha}$, an N component Euclidean vector, is a so-called root of the root system corresponding to the symmetric space, $\langle\cdot,\cdot\rangle$ is the dot product, R_+ is the set of positive roots, and m_α the multiplicity of $\vec{\alpha}$. This structure, in the case of the symmetric spaces corresponding to the classical groups, appears in the so called *Weyl integration formula* [540]. For the symmetric spaces corresponding to the Gaussian ensembles the positive roots are $\vec{e}_j - \vec{e}_k$ $(j<k)$ (root system of type A — see Section 4.7.2) with multiplicities $m_\alpha = \beta$, and this reclaims the eigenvalue-dependent portion of (1.27). However ,we will not pursue the derivation of these facts, which can be found in [295].

1.3.5 Octonions and the $N = 2, \beta = 8$ Gaussian ensemble

The p.d.f. (1.28) for $N = 2$, $\beta = 8$ can be realized as the eigenvalues of a random Hermitian matrix with real octonion elements. To see this, we must first revise aspects of the theory of real octonions [513]. The real octonions can be constructed out of the real quaternions. Let p_1, p_2, q_1, q_2 be abstract real quaternions, and thus linear combinations with real coefficients of $\{1, i, j, k\}$ as specified by (1.21). Let $\bar{q}$ denote the quaternionic dual defined by (1.23), and let l denote a quantity algebraically distinct from the real quaternions. The real octonion algebra then consists of elements of the form $a = p_1 + p_2 l$, $b = q_1 + q_2 l$, with addition and multiplication defined by

$$a + b = (p_1 + q_1) + (p_2 + q_2)l, \qquad ab = (p_1 q_1 - \bar{q}_2 p_2) + (q_2 p_1 + p_2 \bar{q}_1)l, \tag{1.29}$$

respectively. It follows that the real octonions are an eight-dimensional algebra with basis

$$1,\ e_1 := i,\ e_2 := j,\ e_3 := k,\ e_4 := l,\ e_5 := il,\ e_6 := jl,\ e_7 := kl.$$

In general

$$a(bc) \neq (ab)c$$

(for example with $a = e_5$, $b = e_6$, $c = e_7$ we have $a(bc) = -e_4$ and $(ab)c = e_4$), so unlike the real quaternions the real octonions are not associative. On the other hand, with $\bar{a} := \bar{p}_1 - p_2 l$, $\bar{p}_1$ denoting the quaternionic dual (1.23), we have $\overline{ab} = \bar{b}\bar{a}$ and thus with $|a| := \sqrt{a\bar{a}} = \sqrt{\bar{a}a}$,

$$|ab| = |a||b|. \tag{1.30}$$

Furthermore, with a general real octonion written as $a = a_0 + \sum_{j=1}^{7} a_j e_j$, we have

$$|a| = \sqrt{a_0^2 + a_1^2 + \cdots + a_7^2}, \tag{1.31}$$

and it is also true that each $a \neq 0$ has a unique inverse specified by

$$a^{-1} = \bar{a}/(\bar{a}a). \tag{1.32}$$

The properties (1.30)–(1.32) say that the real octonions are a normed division algebra. In fact a theorem of Hurwitz [301] says that up to isomorphisms, the only normed division algebras over the reals, with a unit element, are the reals, complex numbers, real quaternions and real octonions.

Because the real octonions are not associative, they cannot be represented as a matrix algebra. Nonetheless, the actions of right and left multiplication by a given real octonion a on a general real octonion x can be represented as a matrix. To specify these matrices, we first require the corresponding result for the real quaternions, which follows immediately from the explicit form of the multiplication rule.

PROPOSITION 1.3.5 *Let $x = x_0 + x_1 i + x_2 j + x_3 k$ be a real quaternion and let $\vec{x} = (x_0, x_1, x_2, x_3)^T$ denote the column vector formed from the coefficients. Then for $\vec{a}$ a real quaternion*

$$a\vec{x} = \phi(a)\vec{x}, \qquad x\vec{a} = \tau(a)\vec{x}$$

where

$$\phi(a) = \begin{bmatrix} a_0 & -a_1 & -a_2 & -a_3 \\ a_1 & a_0 & -a_3 & a_2 \\ a_2 & a_3 & a_0 & -a_1 \\ a_3 & -a_2 & a_1 & a_0 \end{bmatrix}, \quad \tau(a) = \mathbf{K}\phi^T(a)\mathbf{K} = \begin{bmatrix} a_0 & -a_1 & -a_2 & -a_3 \\ a_1 & a_0 & a_3 & -a_2 \\ a_2 & -a_3 & a_0 & a_1 \\ a_3 & a_2 & -a_1 & a_0 \end{bmatrix},$$

with $\mathbf{K} = \mathrm{diag}[1, -1, -1, -1]$.

Using Proposition 1.3.5, the corresponding result for the real octonions follows from the multiplication rule (1.29).

PROPOSITION 1.3.6 *Let $x = x_0 + \sum_{j=1}^{7} x_j e_j$ be a real octonion, and let $\vec{x} = (x_0, x_1, \dots, x_7)^T$ denote the column vector formed from the coefficients. Then with $a = a^{(1)} + a^{(2)} l$ a real octonion, and thus $a^{(1)}, a^{(2)}$*

real quaternions, and $\tilde{\mathbf{K}} := \mathrm{diag}[\mathbf{K}, \mathbf{1}_4]$ *we have*

$$a\vec{x} = \omega(a)\vec{x}, \qquad \vec{x}a = \nu(a)\vec{x},$$

where

$$\omega(a) = \begin{bmatrix} \phi(a^{(1)}) & -\tau(a^{(2)})\mathbf{K} \\ \phi(a_{(2)})\mathbf{K} & \tau(a^{(1)}) \end{bmatrix} = \begin{bmatrix} a_0 & -a_1 & -a_2 & -a_3 & -a_4 & -a_5 & -a_6 & -a_7 \\ a_1 & a_0 & -a_3 & a_2 & -a_5 & a_4 & a_7 & -a_6 \\ a_2 & a_3 & a_0 & -a_1 & -a_6 & -a_7 & a_4 & a_5 \\ a_3 & -a_2 & a_1 & a_0 & -a_7 & a_6 & -a_5 & a_4 \\ a_4 & a_5 & a_6 & a_7 & a_0 & -a_1 & -a_2 & -a_3 \\ a_5 & -a_4 & a_7 & -a_6 & a_1 & a_0 & a_3 & -a_2 \\ a_6 & -a_7 & -a_4 & a_5 & a_2 & -a_3 & a_0 & a_1 \\ a_7 & a_6 & -a_5 & -a_4 & a_3 & a_2 & -a_1 & a_0 \end{bmatrix},$$

$$\nu(a) = \tilde{\mathbf{K}}\omega^T(a)\tilde{\mathbf{K}}.$$

Consider now the 2×2 Hermitian matrix with real octonion entries

$$\mathbf{A} = \begin{bmatrix} a & b \\ \bar{b} & c \end{bmatrix}.$$

For $\mathbf{A}$ to be Hermitian, the elements a and c must in fact be real, and thus

$$\omega(\mathbf{A}) = \begin{bmatrix} a\mathbf{1}_8 & \omega(b) \\ \omega^T(b) & c\mathbf{1}_8 \end{bmatrix}. \tag{1.33}$$

Adding together appropriate (octonion) multiples of rows and columns shows that this matrix is similar to the matrix

$$\begin{bmatrix} a\mathbf{1}_8 & \mathbf{1}_4 \otimes \begin{bmatrix} b & 0 \\ 0 & \bar{b} \end{bmatrix} \\ \mathbf{1}_4 \otimes \begin{bmatrix} \bar{b} & 0 \\ 0 & b \end{bmatrix} & c\mathbf{1}_8 \end{bmatrix}$$

and thus the characteristic polynomial is given by

$$\det(\omega(\mathbf{A}) - \lambda\mathbf{1}_{16}) = ((a-\lambda)(c-\lambda) - b\bar{b})^8.$$

This shows that each eigenvalue is eightfold degenerate.

Regarding the eigenvectors, as the number of independent real elements in (1.33) is ten, and there are two distinct eigenvalues, there are a total of eight independent components. This implies that in the analogue of Proposition 1.2.5, exactly eight components are to be multiplied together in any one term, and consequently

$$(d\mathbf{A}) = (\lambda_1 - \lambda_2)^8 d\lambda_1 d\lambda_2 (\mathbf{U}^\dagger d\mathbf{U}). \tag{1.34}$$

Furthermore, choosing the elements a, c and the components b_j of $\omega(b)$ in (1.33) to have the Gaussian distributions

$$\frac{2}{\sqrt{\pi}} e^{-4a^2}, \quad \frac{2}{\sqrt{\pi}} e^{-4c^2}, \quad \sqrt{\frac{8}{\pi}} e^{-8b_j^2},$$

respectively, we have that the joint distribution of the independent elements is proportional to

$$e^{-\mathrm{Tr}((\omega(\mathbf{A}))^2)/2}.$$

This together with (1.34) implies that the eigenvalue p.d.f. is given by (1.28) with $N = 2$, $\beta = 8$.

EXERCISES 1.3 1. The aim of this exercise is to show that if a $2N \times 2N$ Hermitian matrix $\mathbf{X}$ commutes with the time reversal operator $T = \mathbf{Z}_{2N}K$, then the eigenvalues of $\mathbf{X}$ are doubly degenerate (this is known as Kramer's degeneracy).

(i) Suppose $\vec{\phi}$ is an eigenvector of $\mathbf{X}$ with eigenvalue λ. State why $T\vec{\phi}$ is also an eigenvector with eigenvalue λ.

(ii) Use the facts that T satisfies the formula of Exercises 1.1 q.3(i) and $T^2 = -1$ to show that $\langle \vec{\phi} | T\vec{\phi} \rangle = 0$ and hence deduce the desired result.

2. [389, p. 32] Let $\mathbf{A}$ and $\mathbf{M}$ be $N \times N$ matrices, where $\mathbf{A}$ is nonsingular. In this exercise it will be shown that for $\mathbf{A}$ real ($\beta = 1$), complex ($\beta = 2$) and real quaternion ($\beta = 4$), and $\mathbf{M}$ real symmetric ($\beta = 1$), Hermitian ($\beta = 2$) and quaternion real Hermitian ($\beta = 4$),

$$(\mathbf{A}^\dagger d\mathbf{M}\mathbf{A}) = \Big(\det(\mathbf{A}^\dagger \mathbf{A}) \Big)^{\beta(N-1)/2+1} (d\mathbf{M}), \tag{1.35}$$

up to a $\pm$ sign. In the case $\beta = 1$, this is the statement of Proposition 1.2.4. The idea is to decompose $\mathbf{A}$ in terms of elementary matrices $\mathbf{A} = \mathbf{E}_p \mathbf{E}_{p-1} \cdots \mathbf{E}_1$. Each elementary matrix is either a permutation matrix $\mathbf{E}^{(j \leftrightarrow k)}$ (the identity matrix with rows j and k interchanged), a matrix $\mathbf{E}^{(j \mapsto \alpha j)}$ which multiplies row j by the constant α with α real ($\beta = 1$), complex ($\beta = 2$) or real quaternion ($\beta = 4$) (the identity matrix with row j multiplied by α), or the matrix $\mathbf{E}^{(j \mapsto j+k)}$ which adds together two rows (the identity matrix with row k added to row j).

(i) Show by explicit calculation that for any matrix $\mathbf{X}$ of the same type as $\mathbf{M}$

$$(\mathbf{E}^{(j \leftrightarrow k)} d\mathbf{X} \mathbf{E}^{(j \leftrightarrow k)\dagger}) = (d\mathbf{X}), \qquad (\mathbf{E}^{(j \mapsto j+k)} d\mathbf{X} \mathbf{E}^{(j \mapsto j+k)\dagger}) = (d\mathbf{X}),$$

while, up to a $\pm$ sign,

$$(\mathbf{E}^{(j \mapsto \alpha j)} d\mathbf{X} \mathbf{E}^{(j \mapsto \alpha j)\dagger}) = |\alpha|^{\beta(N-1)+2} (d\mathbf{X}) = |\det \mathbf{E}^{(j \mapsto \alpha j)}|^{\beta(N-1)+2} (d\mathbf{X}).$$

(ii) Use the result of (i) to deduce the stated result.

For printing purposes, the symbol α^* rather than $\bar{\alpha}$ is used in the exercises below to denote the complex conjugate of α.

3. The aim of this exercise is to calculate the change of variables from the independent elements of a Hermitian matrix $\mathbf{X}$ to the eigenvalues $\lambda_1, \dots, \lambda_N$ and other independent variables.

(i) From the diagonalization formula $\mathbf{X} = \mathbf{U}\mathbf{L}\mathbf{U}^{-1}$, where $\mathbf{L} := \mathrm{diag}[\lambda_1, \dots, \lambda_N]$ and $\mathbf{U}$ is a unitary matrix with columns given by the eigenvectors of $\mathbf{X}$, show that

$$\mathbf{U}^{-1} d\mathbf{X} \mathbf{U} = \mathbf{U}^{-1} d\mathbf{U} \mathbf{L} - \mathbf{L}\mathbf{U}^{-1} d\mathbf{U} + d\mathbf{L}$$

and write down a formula for the Jacobian in terms of $(d\mathbf{X})$. Use the result of q.2 to show that the wedge product of the independent elements on the l.h.s. is equal to $(d\mathbf{X})$.

(ii) Show that $\mathbf{U}^{-1} d\mathbf{U} \mathbf{L} - \mathbf{L}\mathbf{U}^{-1} d\mathbf{U} + d\mathbf{L}$ equals

$$\begin{bmatrix} d\lambda_1 & (\lambda_2 - \lambda_1) \vec{u}_1^\dagger \cdot d\vec{u}_2 & \dots & (\lambda_N - \lambda_1) \vec{u}_1^\dagger \cdot d\vec{u}_N \\ (\lambda_2 - \lambda_1)(\vec{u}_1^\dagger \cdot d\vec{u}_2)^* & d\lambda_2 & \dots & (\lambda_N - \lambda_2) \vec{u}_2^\dagger \cdot d\vec{u}_N \\ \vdots & \vdots & \ddots & \vdots \\ (\lambda_N - \lambda_1)(\vec{u}_1^\dagger \cdot d\vec{u}_N)^* & (\lambda_N - \lambda_2)(\vec{u}_2^\dagger \cdot d\vec{u}_N)^* & \dots & d\lambda_N \end{bmatrix}.$$

(iii) Use the facts that $\vec{u}_j^\dagger \cdot d\vec{u}_k$ has independent real and imaginary parts and that only the elements on and above the diagonal are independent to conclude that the wedge product of the independent elements of the matrix in (ii) equals

$$\prod_{1 \le j < k \le N} (\lambda_k - \lambda_j)^2 \bigwedge_{j=1}^{N} d\lambda_j (\mathbf{U}^\dagger d\mathbf{U}).$$

(iv) Show that the factor dependent on the λ_j's is consistent with the form required by the scaling $\mathbf{X} \mapsto a\mathbf{X}$ (recall Section 1.2.3).

4. Here the objective is the same as in q.3 above, except $\mathbf{X}$ is now an $N \times N$ Hermitian matrix with real quaternion elements.

(i) From the diagonalization formula $\mathbf{X} = \mathbf{U}\mathbf{L}\mathbf{U}^{-1}$ where $\mathbf{L} = \mathrm{diag}[\lambda_1 \mathbf{1}_2, \dots, \lambda_N \mathbf{1}_2]$ and $\mathbf{U}$ is an $N \times N$ unitary matrix with real quaternion elements, write down the formulas analogous to those in q.3(i) and q.3(ii) above. To write down the analogue of (ii) use a matrix notation for the quaternion elements,

$$\vec{\mathbf{u}}_k = (\mathbf{u}_{1k}, \dots, \mathbf{u}_{Nk})^T, \quad \vec{\mathbf{u}}_j^\dagger \cdot \vec{\mathbf{u}}_k = \sum_{p=1}^{N} \mathbf{u}_{pj}^\dagger \mathbf{u}_{pk}.$$

(ii) Use the facts that $\vec{\mathbf{u}}_j^\dagger \cdot d\vec{\mathbf{u}}_k$ has four independent terms, corresponding to the real and imaginary parts of the two independent terms in each real quaternion element, to deduce the formula analogous to q.3(iii) above. Also repeat the scaling analysis of q.3(iv) above.

5. A Hermitian matrix with zero real part is antisymmetric.

(i) Show that the nonzero eigenvalues of antisymmetric Hermitian matrices come in $\pm$ pairs, λ_j and $-\lambda_j$, say, with corresponding eigenvectors $\vec{\phi}_j$ and $\vec{\phi}_j^*$, and that for N odd $\lambda = 0$ is an eigenvalue.

(ii) Use (i) to deduce that the equation of q.3(ii) holds with $\vec{u}_j = \vec{\phi}_j$, $\lambda_{N/2+j} = -\lambda_j$, $\vec{u}_{N/2+j} = \vec{u}_j^*$ $(j = 1, \dots, N/2)$ N even, and $\vec{u}_N^* = \vec{u}_N = \phi_0$, $\lambda_N = 0$, $\vec{u}_j = \vec{\phi}_j$, $\lambda_{(N-1)/2+j} = -\lambda_j$ $(j = 1, \dots, (N-1)/2)$ N odd. Use the fact that eigenvectors of a Hermitian matrix corresponding to distinct eigenvalues are orthogonal to deduce that $\vec{\phi}_0{}^\dagger \cdot d\vec{\phi}_j^* = 0$ $(j \neq 0)$, and note too that $\vec{\phi}_0^\dagger \cdot d\vec{\phi}_k$ and $\vec{\phi}_0^{*\dagger} \cdot d\vec{\phi}_j^*$ are not independent.

(iii) Use (ii) to show that for an antisymmetric Hermitian $N \times N$ matrix $\mathbf{H}^i$, diagonalized by $\mathbf{H}^i = \mathbf{U}\mathbf{L}\mathbf{U}^{-1}$,

$$(d\mathbf{H}^i) = \prod_{1 \le j < k \le N/2} (\lambda_j^2 - \lambda_k^2)^2 \bigwedge_{j=1}^{N/2} d\lambda_j \, (\mathbf{U}^\dagger d\mathbf{U}), \quad N \text{ even},$$

$$(d\mathbf{H}^i) = \prod_{j=1}^{(N-1)/2} \lambda_j^2 \prod_{1 \le j < k \le (N-1)/2} (\lambda_j^2 - \lambda_k^2)^2 \bigwedge_{j=1}^{(N-1)/2} d\lambda_j \, (\mathbf{U}^\dagger d\mathbf{U}), \quad N \text{ odd}.$$

(iv) Conclude from the result of (iii) that for a random antisymmetric Hermitian $N \times N$ matrix with upper triangular elements ix_{jk} chosen with p.d.f. $\sqrt{1/\pi} e^{-x_{jk}^2}$, the eigenvalue p.d.f. of the positive eigenvalues is equal to

$$\frac{1}{C_N} \prod_{j=1}^{N/2} e^{-\lambda_j^2} \prod_{1 \le j < k \le N/2} (\lambda_j^2 - \lambda_k^2)^2, \quad N \text{ even},$$

$$\frac{1}{C_N} \prod_{j=1}^{(N-1)/2} \lambda_j^2 e^{-\lambda_j^2} \prod_{1 \le j < k \le (N-1)/2} (\lambda_j^2 - \lambda_k^2)^2, \quad N \text{ odd},$$

where the normalizations C_N are given explicitly in (4.157) below.

6. (i) Let $\mathbf{Q}^r$ be a quaternion real Hermitian matrix in which all entries of each real quaternion are real, and let $\mathbf{H}$ be the Hermitian matrix formed by replacing each quaternion element (1.20) by the scalar $z + iw$. Show that $\mathbf{Q}^r$ and $\mathbf{H}$ have the same distinct eigenvalues, and that the eigenvalues of $\mathbf{Q}^r$ are doubly degenerate with eigenvectors of the form $\vec{\psi}^{(1)} = \begin{bmatrix} \phi_k^r \\ \phi_k^i \end{bmatrix}_{k=1,\dots,N}$ and $\vec{\psi}^{(2)} = \mathbf{Z}_{2N} \vec{\psi}^{(1)}$ where $\vec{\phi} = [\phi_k^r + i\phi_k^i]_{k=1,\dots,N}$ is an eigenvector of $\mathbf{H}$. Hence write down the eigenvalue p.d.f. of $\mathbf{Q}^r$.

(ii) Let $\mathbf{Q}$ be an $N \times N$ real quaternion Hermitian matrix. Proceed in a converse fashion to (i) to write down a $4N \times 4N$ real symmetric matrix $\mathbf{R}$ such that $\mathbf{Q}$ and $\mathbf{R}$ have the same distinct eigenvalues, and thus the same eigenvalue p.d.f. Also, relate the corresponding eigenvectors. Similarly, for $\mathbf{H}$ an $N \times N$ complex Hermitian matrix, replace each entry $x + iy$ by its 2×2 real matrix representation

$$\begin{bmatrix} x & y \\ -y & x \end{bmatrix} \tag{1.36}$$

to obtain a doubly degenerate $2N \times 2N$ matrix for which the distinct eigenvalues coincide with those of $\mathbf{H}$.

7. Consider a quaternion real Hermitian matrix $\mathbf{Q}^{\mathrm{i}}$ in which all entries of each real quaternion are pure imaginary so that $\mathbf{Q}^{\mathrm{i}}$ is antisymmetric.

(i) With the pair of eigenvectors corresponding to the doubly degenerate eigenvalues λ_j denoted by $\vec{\mathbf{u}}_j$ as in q.4, note from the theory of q.5(i) that $\vec{\mathbf{u}}_j^*$ is equal to the pair of eigenvectors corresponding to the doubly degenerate eigenvalue $-\lambda_j$.

(ii) Use the fact that eigenvectors of a Hermitian matrix corresponding to distinct eigenvalues are orthogonal to deduce that

$$\vec{\mathbf{u}}_j^\dagger \cdot \vec{\mathbf{u}}_j^* := \sum_{p=1}^N \mathbf{u}_{pj}^\dagger \mathbf{u}_{pj}^* = \sum_{p=1}^N \begin{bmatrix} 0 & -2\mathrm{Im}\,(z_{pj} w_{pj}) \\ -2\mathrm{Im}\,(z_{pj} w_{pj}) & 0 \end{bmatrix}, \quad \mathbf{u}_{jp} := \begin{bmatrix} z_{pj} & w_{pj} \\ -w_{pj}^* & z_{pj}^* \end{bmatrix},$$

and conclude from this that $\vec{\mathbf{u}}_j^\dagger \cdot d\vec{\mathbf{u}}_j^*$ has only one independent component. Note too that $\vec{\mathbf{u}}_j^\dagger \cdot d\vec{\mathbf{u}}_k$ and $\vec{\mathbf{u}}_j^{*\dagger} \cdot d\vec{\mathbf{u}}_k^*$ are not independent.

(iii) With the analogue of the equation of q.3(ii) in the quaternion case modified as in the first sentence of q.5(ii) (N even case), show from (i) that for an antisymmetric $N \times N$ quaternion real Hermitian matrix $\mathbf{Q}^{\mathrm{i}}$ diagonalized by $\mathbf{Q}^{\mathrm{i}} = \mathbf{U}\mathbf{L}\mathbf{U}^{-1}$,

$$(d\mathbf{Q}^{\mathrm{i}}) = \prod_{j=1}^{N/2} \lambda_j \prod_{1 \le j < k \le N/2} (\lambda_j^2 - \lambda_k^2)^4 \bigwedge_{j=1}^{N/2} d\lambda_j \, (\mathbf{U}^\dagger d\mathbf{U}), \quad N \text{ even},$$

$$(d\mathbf{Q}^{\mathrm{i}}) = \prod_{j=1}^{(N-1)/2} \lambda_j^5 \prod_{1 \le j < k \le (N-1)/2} (\lambda_j^2 - \lambda_k^2)^4 \bigwedge_{j=1}^{(N-1)/2} d\lambda_j \, (\mathbf{U}^\dagger d\mathbf{U}), \quad N \text{ odd}.$$

8. [146] Let $\mathbf{Q}$ be a quaternion real matrix with the property that $i\mathbf{Q}$ is Hermitian.

(i) Note that $\mathbf{Q}$ must anticommute with the time reversal operator $T = \mathbf{Z}_{2N} K$, and use this to show that if $|\phi\rangle$ is an eigenvector of $\mathbf{Q}$ with eigenvalue λ, then $T\vec{\phi}$ is an eigenvector with eigenvalue $-\lambda$.

(ii) Proceed as in q.4 to show that with $i\mathbf{Q}$ diagonalized by $i\mathbf{Q} = \mathbf{U} i\mathbf{L}\mathbf{U}^\dagger$, where

$$\mathbf{L} = \mathrm{diag}(\lambda_1, -\lambda_1, \dots, \lambda_N, -\lambda_N),$$

and $\mathbf{U}$ is unitary with real quaternion elements in which all elements are real,

$$(d\mathbf{Q}) = \prod_{j=1}^N (2\lambda_j)^2 \prod_{1 \le j < k \le N} (\lambda_k^2 - \lambda_j^2)^2 \bigwedge_{j=1}^N d\lambda_j (\mathbf{U}^\dagger d\mathbf{U}). \tag{1.37}$$

9. Let $G_{\beta,N}$ be the normalization in (1.28), which has the evaluation (1.163) below, and let

$$A_{\beta,N} = \left(\frac{\beta}{2\pi}\right)^{N/2} \left(\frac{\beta}{\pi}\right)^{\beta N(N-1)/4}$$

so that

$$A_{\beta,N} \int e^{-(\beta/2)\mathrm{Tr}(\mathbf{H}^2)} (d\mathbf{H}) = 1.$$

Deduce from this last equation and (1.27) that for $\beta = 1, 2$

$$\int (\mathbf{U}^\dagger d\mathbf{U}) = \frac{N!}{A_{\beta,N} G_{\beta,N}}, \tag{1.38}$$

where the first entry of each column of $\mathbf{U}$ is chosen to be real and positive.

10. Define the matrix

$$\tilde{\mathbf{H}} = \Big(\frac{c}{\operatorname{Tr} \mathbf{H}^2} \Big)^{1/2} \mathbf{H},$$

where $\mathbf{H}$ is a member of one of the Gaussian ensembles and $c > 0$ is a constant. By changing variables from the elements of $\mathbf{H}$ to the elements of $\tilde{\mathbf{H}}$, and $\operatorname{Tr} \mathbf{H}^2$, show that the eigenvalues $\{\tilde{\lambda}_j\}$ of $\tilde{\mathbf{H}}$ have distribution proportional to

$$\delta\Big(c - \sum_{j=1}^N \tilde{\lambda}_j^2\Big) \prod_{1 \le j < k \le N} |\tilde{\lambda}_k - \tilde{\lambda}_j|^\beta.$$

1.4 COULOMB GAS ANALOGY

The eigenvalue p.d.f. (1.28) can be identified with the Boltzmann factor of a particular log-gas, an observation which goes back to Dyson [146]. To appreciate this, we must revise some basic theory from statistical mechanics [390], and show how the Boltzmann factor of a log-gas, or more generally a one-component Coulomb system, is computed.

1.4.1 Boltzmann factors

The canonical formalism of statistical mechanics applies to any mechanical system of N particles free to move in a fixed domain Ω, in equilibrium at absolute temperature T. A fundamental postulate gives the p.d.f. for the event that the particles are at positions $\vec{r}_1, \dots, \vec{r}_N$ as

$$\frac{1}{\hat{Z}_N} e^{-\beta U(\vec{r}_1, \dots, \vec{r}_N)}.$$

Here $U(\vec{r}_1, \dots, \vec{r}_N)$ denotes the total potential energy of the system, $\beta := 1/(k_B T)$ (k_B is Boltzmann's constant), and the normalization $\hat{Z}_N$ is given by

$$\hat{Z}_N = \int_\Omega d\vec{r}_1 \cdots \int_\Omega d\vec{r}_N e^{-\beta U(\vec{r}_1, \dots, \vec{r}_N)}. \tag{1.39}$$

The term $e^{-\beta U(\vec{r}_1, \dots, \vec{r}_N)}$ is referred to as the *Boltzmann factor* and $\hat{Z}_N / N! =: Z_N$ is called the *(canonical) partition function.*

For log-potential Coulomb systems the potential energy U is calculated according to the laws of two-dimensional electrostatics, and Ω must be one- or two-dimensional. The particles can be thought of as infinitely long parallel charged lines, which are perpendicular to the confining domain. In a vacuum the electrostatic potential Φ at a point $\vec{r} = (x, y)$ due to a two-dimensional unit charge at $\vec{r}\,' = (x', y')$ is given by the solution of the *Poisson equation*

$$\nabla_{\vec{r}}^2 \Phi(\vec{r}, \vec{r}\,') = -2\pi \delta(\vec{r} - \vec{r}\,'), \tag{1.40}$$

where

$$\nabla_{\vec{r}}^2 := \frac{\partial^2}{\partial x^2} + \frac{\partial^2}{\partial y^2}.$$

It is straightforward to verify that the solution of the Poisson equation is (see Exercises 1.4 q.1)

$$\Phi(\vec{r},\vec{r}\,') = -\log(|\vec{r}-\vec{r}\,'|/l), \tag{1.41}$$

where l is some arbitrary length scale which will henceforth be set to unity.

A Coulomb system is said to consist of *one component* if all N particles are of like charge, q, say. To stop the particles from all repelling to the boundary, a neutralizing background charge density $-q\rho_b(\vec{r})$ is imposed, with the electroneutrality condition $\int_\Omega \rho_b(\vec{r})\,d\vec{r} = N$. The total potential energy U therefore consists of the sum of the electrostatic energy of the particle-particle interaction

$$U_1 := -q^2 \sum_{1\le j<k\le N} \log|\vec{r}_k - \vec{r}_j|,$$

the particle-background interaction

$$U_2 := q^2 \sum_{j=1}^N V(\vec{r}_j) \qquad \text{where} \qquad V(\vec{r}_j) := \int_\Omega \log|\vec{r}-\vec{r}_j|\,\rho_b(\vec{r})\,d\vec{r}, \tag{1.42}$$

and the background-background interaction

$$U_3 := -\frac{q^2}{2}\int_\Omega d\vec{r}\,'\,\rho_b(\vec{r}\,')\int_\Omega d\vec{r}\,\rho_b(\vec{r})\log|\vec{r}\,'-\vec{r}| = -\frac{q^2}{2}\int_\Omega \rho_b(\vec{r}\,')V(\vec{r}\,')\,d\vec{r}\,'. \tag{1.43}$$

The factor of $\frac{1}{2}$ in U_3 is included to compensate for the double counting of the potential energy implicit in the double integration.

From this expression for U we conclude that the Boltzmann factor of a one-component log-potential Coulomb system (log-gas) is of the form

$$e^{-\beta U_3}\prod_{l=1}^N e^{-\Gamma V(\vec{r}_l)}\prod_{1\le j<k\le N}|\vec{r}_k-\vec{r}_j|^\Gamma, \tag{1.44}$$

where $\Gamma := q^2/k_BT$. Furthermore, for a given geometry and background density the potentials $V(\vec{r})$ and U_3 can readily be evaluated. As an illustration, we have the following result.

PROPOSITION 1.4.1 *The Boltzmann factor of a one-component log-potential Coulomb system of N particles of charge $q = 1$, confined to a circle of radius R with a uniform neutralizing background, is given by*

$$R^{-N\beta/2}\prod_{1\le j<k\le N}|e^{i\theta_k}-e^{i\theta_j}|^\beta,$$

where the position of each particle has been specified in polar coordinates.

Proof. It is generally true that for two points $\vec{r}$ and $\vec{r}\,'$ in the plane $|\vec{r}-\vec{r}\,'| = |z-z'|$, where z and z' are the corresponding points in the complex plane. Hence, if $\vec{r}$ and $\vec{r}\,'$ are both on a circle of radius R with positions specified using polar coordinates, then $|\vec{r}-\vec{r}\,'| = R|e^{i\theta}-e^{i\theta'}|$. Use of this formula gives the required expression for the product over pairs in (1.44). It also allows the potential $V(\vec{r})$ to be written as

$$V(\vec{r}) = \frac{N}{2\pi R}\int_0^{2\pi}\log|Re^{i\theta'}-Re^{i\theta}|\,R d\theta' = N\log R + \frac{N}{2\pi}\int_0^{2\pi}\log|e^{i\theta'}-1|\,d\theta'.$$

But it is straightforward to show that the last integral vanishes (see Exercises 1.4 q.2), and so $V(\vec{r}) = N\log R$. Use of this result gives $U_3 = -\frac{q^2}{2}N^2\log R$. Substituting these evaluations in (1.44) and noting that since $q=1$, $\Gamma=\beta$ gives the desired expression for the Boltzmann factor. □

The Boltzmann factor, being proportional to the p.d.f. for the location of the particles, occurs in the definition of all statistical quantities associated with the equilibrium state. In particular the *canonical average* of any function $f(\vec{r}_1, \dots, \vec{r}_N)$ is given by

$$\langle f \rangle := \frac{1}{\hat{Z}_N} \int_\Omega d\vec{r}_1 \cdots \int_\Omega d\vec{r}_N \, f(\vec{r}_1, \dots, \vec{r}_N) e^{-\beta U(\vec{r}_1, \dots, \vec{r}_N)}. \tag{1.45}$$

With $f = \sum_{j=1}^N \delta(\vec{r} - \vec{r}_j)$ the canonical average is called the one-point correlation function, or particle density

$$\rho_{(1)}(\vec{r}) := \Big\langle \sum_{j=1}^N \delta(\vec{r} - \vec{r}_j) \Big\rangle = \frac{N}{\hat{Z}_N} \int_\Omega d\vec{r}_2 \cdots \int_\Omega d\vec{r}_N \, e^{-\beta U(\vec{r}, \vec{r}_2, \dots, \vec{r}_N)}, \tag{1.46}$$

where the equality is valid for a system of identical particles, and thus when the Boltzmann factor is a symmetric function of the particle coordinates.

1.4.2 The potential and calculation of $\rho_b(y)$

Comparison of (1.28) with (1.44) shows immediately that the eigenvalue p.d.f. is identical to the Boltzmann factor of a one-component log-potential Coulomb system confined to a line, with the position of the charged particles corresponding identically to the location of the eigenvalues. Furthermore, the background charge density $-q\rho_b(y)$ is such that

$$\frac{x^2}{2} + C = \int_{-\infty}^{\infty} \rho_b(y) \log|x - y| \, dy, \tag{1.47}$$

where C is a constant.

Note that it is not possible to satisfy (1.47) for $|x| \to \infty$, since in this limit the r.h.s. is to leading order $N \log|x|$ and is thus a different order from the l.h.s. Instead we seek to solve the integral equation for $\rho_b(y)$ with support on the finite interval $(-a, a)$ say, and x confined to the same interval. Then (1.47) reads

$$\frac{x^2}{2} + C = \int_{-a}^{a} \rho_b(y) \log|x - y| \, dy, \quad x \in (-a, a). \tag{1.48}$$

The solution of the equation can be computed exactly by the method of eigenfunction expansions (see, e.g., [448]).

Proposition 1.4.2 *Suppose all the eigenvalues $\{\lambda_n\}_{n=0,1,\dots}$ and corresponding normalized eigenfunctions $\{\phi_n\}_{n=0,1,\dots}$ of a linear operator A are known, all the eigenvalues are nonzero, and the eigenfunctions form a complete set. Then the operator equation $g = Af$, where g is given, has the solution*

$$f = \sum_{n=0}^{\infty} \frac{\langle g|\phi_n\rangle}{\lambda_n} \phi_n,$$

where $\langle \cdot|\cdot \rangle$ denotes the inner product.

Proof. Since the eigenfunctions form a complete set, $g = \sum_{n=0}^{\infty} \langle g|\phi_n\rangle \phi_n$ Also $f = \sum_{n=0}^{\infty} \langle f|\phi_n\rangle \phi_n$ and so $Af = \sum_{n=0}^{\infty} \langle f|\phi_n\rangle \lambda_n \phi_n$. The result follows by equating the coefficients of ϕ_n in the operator equation. □

To make use of this method, it is necessary to make the further change of variables

$$y = a\cos\theta, \qquad x = a\cos\sigma, \qquad \sin\theta \rho_b(a\cos\theta) =: a\phi(\theta)$$

so that (1.48) reads

$$-\frac{1}{4}\cos 2\sigma - \Big(\frac{1}{4} - \frac{N}{a^2}\log a + \frac{C}{a^2}\Big) = -\int_0^\pi \log|\cos\theta - \cos\sigma| \, \phi(\theta) \, d\theta. \tag{1.49}$$

Note that $\cos\theta - \cos\sigma = 2\sin(\sigma-\theta)/2\, \sin(\sigma+\theta)/2$. Since $2|\sin(\sigma-\theta)/2|$ gives the chord length for two points on the unit circle with angles σ and θ, the r.h.s. of (1.49) can be interpreted as giving (up to an additive constant) the electrostatic potential at the angle σ due to a charge density $\phi(\theta)$ and $\phi(2\pi-\theta)$ between 0 and π and π and 2π, respectively, on the unit circle.

The eigenvalues and eigenfunctions of the integral operator

$$A[\phi](\sigma) := -\int_0^\pi \log|\cos\theta - \cos\sigma|\phi(\theta)\, d\theta \tag{1.50}$$

are known (see Exercises 1.4 q.4). They are

$$\lambda_0 = \pi\log 2, \quad \phi_0(\theta) = \frac{1}{\pi^{1/2}}, \quad \lambda_n = \frac{\pi}{n}, \quad \phi_n(\theta) = \Big(\frac{2}{\pi}\Big)^{1/2}\cos n\theta \quad (n=1,2,\dots).$$

In terms of these eigenfunctions

$$-\frac{1}{4}\cos 2\sigma - \Big(\frac{1}{4} - \frac{N}{a^2}\log a + \frac{C}{a^2}\Big) = -\frac{1}{4}\Big(\frac{\pi}{2}\Big)^{\frac{1}{2}}\phi_2(\sigma) - \Big(\frac{1}{4} - \frac{N}{a^2}\log a + \frac{C}{a^2}\Big)\pi^{\frac{1}{2}}\phi_0(\sigma),$$

and so from the general formula of Proposition 1.9 the solution of the transformed integral equation is

$$\phi(\theta) = -\frac{1}{2\pi}(\cos 2\theta - 1) - \frac{1}{\pi\log 2}\Big(\frac{1}{4} - \frac{N}{a^2}\log a + \frac{C}{a^2} + \frac{1}{2}\log 2\Big),$$

where $1/(2\pi)$ has been added and subtracted for later convenience. Reverting back to the original variables we obtain the following result.

PROPOSITION 1.4.3 *The solution of the integral equation*

$$\frac{x^2}{2} + C = \int_{-a}^{a} \rho_b(y)\log|x-y|\, dy, \qquad -a \le x \le a,$$

is

$$\rho_b(y) = \frac{a}{\pi}\sqrt{1-\Big(\frac{y}{a}\Big)^2} - \frac{1}{\pi\log 2}\Big(\frac{1}{4} - \frac{N}{a^2}\log a + \frac{C}{a^2} + \frac{1}{2}\log 2\Big)\frac{a}{\sqrt{1-(y/a)^2}}.$$

We see that there are two drastically different classes of solution depending on the value of C. Unless we choose

$$C = N\log a - \frac{a^2}{4} - \frac{a^2}{2}\log 2 \tag{1.51}$$

the density profile $\rho_b(y)$ has an inverse square root singularity at $y = \pm a$. However, with C according to (1.51) the term proportional to $(1-(y/a)^2)^{-1/2}$ vanishes and a physically sensible result is obtained. Making this choice of C and fixing a by the neutrality condition $\int_{-a}^{a}\rho_b(y)\, dy = N$ gives the desired analogy between the Boltzmann factor of a one-component log-potential Coulomb system and the eigenvalue p.d.f.'s of Proposition 1.3.4.

PROPOSITION 1.4.4 *The Boltzmann factor of the one-component log-potential Coulomb system with particles of charge $q=1$ at $x_1,\dots,x_N$, confined to the interval $[-\sqrt{2N},\sqrt{2N}]$, with a neutralizing background charge density*

$$-\rho_b(y) = -\frac{\sqrt{2N}}{\pi}\sqrt{1-\frac{y^2}{2N}},$$

is

$$A\exp\Big(-\frac{\beta}{2}\sum_{j=1}^N x_j^2\Big)\prod_{1\le j<k\le N}|x_k-x_j|^\beta,\quad A=\exp\Big(-\frac{\beta N^2}{4}\log(N/2)+\frac{3\beta N^2}{8}\Big).$$

Proof. Apply the general formula (1.44) for the Boltzmann factor of a one-component log-potential Coulomb system with $\vec{r}_k=x_k$. From Proposition 1.4.3 and (1.51) with $a=\sqrt{2N}$

$$V(x)=\frac{x^2}{2}+4N\Big(\frac{1}{4}\log\sqrt{N/2}-\frac{1}{8}\Big),$$

$$U_3=-q^2\Big[\frac{N^2}{\pi}\int_{-1}^1 x^2\sqrt{1-x^2}\,dx+2N^2\Big(\frac{1}{4}\log\sqrt{N/2}-\frac{1}{8}\Big)\Big].$$

A simple change of variables $x=\cos\theta$ shows that the integral in the above equals $\pi/8$. The stated formula for the Boltzmann factor follows. □

Proposition 1.4.4 can be used to predict the eigenvalue density profile for Gaussian β-ensembles with eigenvalue p.d.f. (1.28). Physically, we expect that to leading order in N Coulomb systems are locally charge neutral, which for a one-component system implies that to leading order the particle density will equal the background density. For the log-potential system of Proposition 1.4.4 this gives the particle density as

$$\rho_{(1)}(y)=\frac{\sqrt{2N}}{\pi}\sqrt{1-\frac{y^2}{2N}}. \tag{1.52}$$

But the statistical properties of the log-potential system in Proposition 1.4.4 are identical to those of the eigenvalues of Gaussian random matrices, so we expect that the eigenvalue density profile will to leading order in N be given by this formula. Consequently we expect the so-called global density

$$\tilde{\rho}_{(1)}(Y):=\lim_{N\to\infty}\sqrt{2/N}\rho_{(1)}(\sqrt{2N}Y) \tag{1.53}$$

to obey the limit formula

$$\tilde{\rho}_{(1)}(x)=\begin{cases}\frac{2}{\pi}(1-x^2)^{1/2}, & |x|<1,\\ 0, & |x|\ge 1,\end{cases} \tag{1.54}$$

known as the *Wigner semicircle law*. The validity of this statement is known rigorously from [328]. Wigner's derivation, which is applicable to GUE matrices, is given in Exercises 1.6 q.1.

In Figure 1.1 we have plotted the empirical eigenvalue density for 1000 10×10 matrices from the GUE, using the variable $Y=y/\sqrt{2N}$. The accuracy of the Wigner semicircle law is evident.

1.4.3 The complex electric field and calculation of $\rho_b(y)$

The integral equation (1.48) is the special case $V(x)=x^2/2$ of the integral equation

$$V(x)+C=\int_{-a}^a\rho_b(y)\log|x-y|\,dy,\qquad x\in(-a,a). \tag{1.55}$$

For the log-gas at $\beta=2$ a rigorous derivation of this integral equation for the particle density is given in Exercises 14.4 q.4 below. We seek the solution such that $\rho_b(y)$ is bounded at $y=\pm a$ and normalized so that

$$\int_{-a}^a\rho_b(y)\,dy=N. \tag{1.56}$$

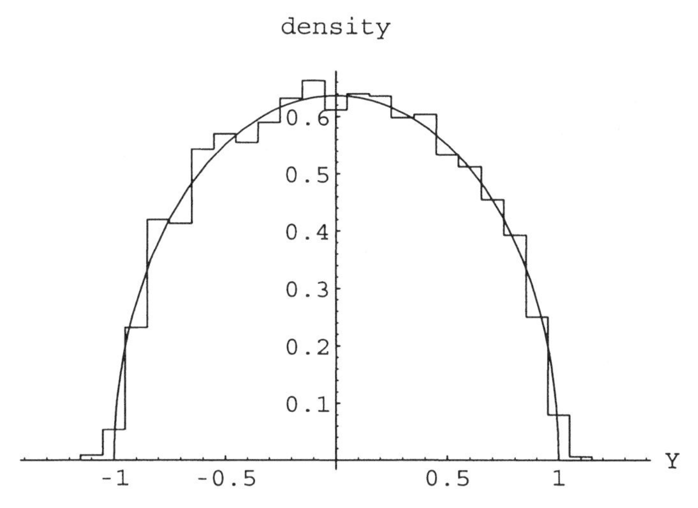

Figure 1.1 Empirical demonstration of the Wigner semicircle law for 10×10 matrices from the GUE.

If the primary concern is the calculation of $\rho_b(y)$ and not C, an alternative to the method of eigenfunctions used above is to introduce the complex electric field

$$E(z) := -\int_{-a}^{a} \frac{\rho_b(y)}{z-y}\, dy. \tag{1.57}$$

Note that for $z \notin [-a, a]$, $E(z)$ is analytic and has the asymptotic behavior

$$E(z) \underset{|z|\to\infty}{\sim} -\frac{N}{z}, \tag{1.58}$$

while for $z \sim \pm a$, by the assumption that $\rho_b(y)$ is bounded,

$$E(z) \underset{z\to\pm a}{\sim} \mathrm{O}(\log(z \mp a)). \tag{1.59}$$

Furthermore, if $\rho_b(y)$ can be analytically continued to a neighborhood of the interval $y \in (-a, a)$, it follows by deforming the path of integration in the neighborhood of $y = x$ and Cauchy's theorem that

$$E^+(x) - E^-(x) = 2\pi i \rho_b(x), \qquad x \in (-a, a), \tag{1.60}$$

where

$$E^{\pm}(x) = \lim_{\epsilon\to 0^+} E(x \pm i\epsilon).$$

Differentiating (1.55) shows

$$\mathrm{Re}\, E(x) = \frac{1}{2}(E^+(x) + E^-(x)) = -V'(x), \qquad \text{for} \quad x \in (-a, a). \tag{1.61}$$

The properties (1.58), (1.59) and (1.61) can be used to characterize $E(z)$, with $\rho_b(x)$ then computed from (1.60). For the quantity $W(z) := e^{E(z)}$ the properties (1.58), (1.59), (1.60), where $\rho_b(x)$ is given, specify a scalar *Riemann-Hilbert problem*.

Restricting attention to potentials $V(x)$ even in x, one can check that the function

$$E(z) = -\frac{1}{\pi}\sqrt{z^2 - a^2} \int_{-a}^{a} \frac{V'(t)}{(z-t)\sqrt{a^2 - t^2}}\, dt, \tag{1.62}$$

with a such that

$$\frac{1}{\pi}\int_{-a}^{a}\frac{tV'(t)}{\sqrt{a^2-t^2}}\,dt = N \tag{1.63}$$

has the properties (1.58), (1.59) and (1.61) and is thus the sought complex electric field. In particular, to verify (1.61), we note that (1.62) gives

$$E^{\pm}(x) = \mp\frac{i}{\pi}\sqrt{a^2-x^2}\lim_{\epsilon\to 0+}\int_{-a}^{a}\frac{V'(t)}{(x\pm i\epsilon-t)\sqrt{a^2-t^2}}\,dt \tag{1.64}$$

and then make use of Cauchy's theorem. Using this formula in (1.60) gives an explicit formula for $\rho_b(x)$ in terms of the potential V [412].

PROPOSITION 1.4.5 *In the case $V(x)$ even, the solution of the integral equation (1.55) with $\rho_b(y)$ bounded at $y=\pm a$ and normalized as in (1.56) is*

$$\rho_b(y) = \frac{1}{\pi^2}\sqrt{a^2-y^2}\int_{-a}^{a}\frac{V'(y)-V'(t)}{y-t}\frac{1}{\sqrt{a^2-t^2}}\,dt, \tag{1.65}$$

where a is specified by (1.63).

Proof. After substituting (1.64) in (1.60), we subtract an appropriate multiple of the identity

$$\lim_{\epsilon\to 0}\left(\int_{-a}^{a}\frac{1}{(x+i\epsilon-t)\sqrt{a^2-t^2}}\,dt - \int_{-a}^{a}\frac{1}{(x-i\epsilon-t)\sqrt{a^2-t^2}}\,dt\right) = 0 \tag{1.66}$$

from both sides to obtain

$$2\pi i\rho_b(x) = -\frac{2i}{\pi}\sqrt{a^2-x^2}\mathrm{Re}\lim_{\epsilon\to 0+}\int_{-a}^{a}\frac{V'(t)-V'(x)}{(x+i\epsilon-t)\sqrt{a^2-t^2}}\,dt.$$

The limit can be taken inside the integrand because the numerator vanishes for $x=t$, giving (1.65). □

In the special case $V(y)=y^2/2$, (1.65) gives

$$\rho_b(y) = \frac{a}{\pi}\sqrt{1-(y/a)^2},$$

while it follows from (1.63) that $a=\sqrt{2N}$, in agreement with (1.52). In this same special case the explicit form of the complex electric field can also be computed; this is done in Exercises 1.6 q.2. The generalization of (1.65) for $V(x)$ not necessarily even is

$$\rho_b(y) = \frac{1}{\pi^2}\sqrt{(y-a)(b-y)}\int_{a}^{b}\frac{V'(y)-V'(t)}{y-t}\frac{1}{\sqrt{(t-a)(b-t)}}\,dt, \tag{1.67}$$

where a and b are such that

$$\int_{a}^{b}\frac{V'(t)}{\sqrt{(t-a)(b-t)}}\,dt = 0, \qquad \frac{1}{\pi}\int_{a}^{b}\frac{tV'(t)}{\sqrt{(t-a)(b-t)}}\,dt = N, \tag{1.68}$$

as can be checked from a similar analysis.

It can happen that for certain $V(x)$ the solution (1.65) or (1.67) of (1.55) does not in fact correspond to the background density because it becomes negative within the interval $(-a,a)$. An example is the potential $V(x)=-cx^2+gx^4$ for c large enough. Formula (1.65) gives

$$\rho_b(y) = \frac{1}{\pi}(-2c+2ga^2+4gy^2)\sqrt{a^2-y^2}, \tag{1.69}$$

where, according to (1.63),

$$-ca^2 + \frac{3ga^4}{2} = N. \tag{1.70}$$

The solution (1.69) will take on negative values for some y whenever $c > ga^2$. According to (1.70), for this to happen it is sufficient that $c^2 > 2gN$. In such a circumstance, the original assumption that the support is on a single interval breaks down, and one must seek a solution supported on a double interval $(-a_2, -a_1) \cup (a_1, a_2)$.

EXERCISES 1.4 1. (i) By explicit differentiation show that $\Phi(\vec{r}, \vec{r}\,') = -\log\left(|\vec{r} - \vec{r}\,'|/l\right)$, satisfies the two-dimensional *Laplace equation* $\nabla_{\vec{r}}^2 \Phi(\vec{r}, \vec{r}\,') = 0$ for $\vec{r} \neq \vec{r}\,'$.

(ii) Use the divergence theorem in the plane

$$\int_{\mathcal{D}} \nabla^2 V(\vec{r})\, d\vec{r} = \int_{\mathcal{C}} \vec{n} \cdot \nabla V(\vec{r})\, d\vec{r}$$

with $V(\vec{r}) = \Phi(\vec{r}, \vec{r}\,')$, $\mathcal{D}$ a disk centered on $\vec{r}\,'$ and $\mathcal{C}$ the circle which is the boundary of the disk, to conclude

$$\int_{\mathcal{D}} \nabla_{\vec{r}}^2 \Phi(\vec{r}, \vec{r}\,')\, d\vec{r} = -2\pi.$$

Relate this result to the Poisson equation (1.40).

2. Use the power series expansion of $\log(1 - z)$ for $|z| < 1$ to show that for all $|\mu| < 1$,

$$\int_0^{2\pi} \log|1 - \mu e^{i\theta}|\, d\theta = 0.$$

Show that this integral is equal to $2\pi \log|\mu|$ for $|\mu| > 1$ by using the result for $|\mu| < 1$, and use the continuity of the integral as a function of μ to deduce its value for $|\mu| = 1$.

3. Suppose there are N mobile particles of charge q in a disk filled with a uniform neutralizing background $\rho_b = N/\pi R^2$. This specifies the two-dimensional one-component plasma confined to a disk.

(i) With the position of the particles specified in polar coordinates, use the integral evaluations of q.2 and the definition of $V(r)$ (1.42) to show

$$V(r) = \pi \rho_b (r^2/2 + R^2 \log R - R^2/2). \tag{1.71}$$

Write down the Poisson equation satisfied by $V(r)$.

(ii) Use this expression for $V(r)$ to calculate U_3 and thus show that the Boltzmann factor is equal to

$$e^{-\Gamma N^2((1/2)\log R - 3/8)} e^{-\pi \Gamma \rho_b \sum_{j=1}^N |\vec{r}_j|^2/2} \prod_{1 \le j < k \le N} |\vec{r}_k - \vec{r}_j|^\Gamma, \qquad \Gamma := q^2 \beta. \tag{1.72}$$

4. (i) Assuming the validity of the formula

$$\log|1 - ae^{ix}| = -\sum_{n=1}^\infty \frac{a^n \cos nx}{n}, \qquad 0 \le a < 1, \quad x \in \mathbb{R},$$

for $a = 1$ provided $x \neq 0 \bmod (2\pi)$, deduce that

$$\log|2\sin(x - t)/2| = -\sum_{n=1}^\infty \frac{\cos n(x - t)}{n}$$

for $x - t \neq 0 \bmod (2\pi)$, and write down a similar formula for $\log|2\sin(x + t)/2|$. Hence derive the cosine expansion

$$\log(2|\cos x - \cos t|) = -\sum_{n=1}^{\infty} \frac{2}{n} \cos nx \cos nt. \tag{1.73}$$

(ii) Use the above cosine expansion to verify that the eigenvalues and normalized eigenfunctions of the integral operator

$$A[\hat{\phi}](\sigma) := -\int_0^{\pi} \log|\cos\theta - \cos\sigma| \hat{\phi}(\theta)\, d\theta$$

are as specified below (1.50).

5. The objective of this exercise is to compute the background density and the Boltzmann factor for the one-component log-potential system confined to the interval $(-a, a)$ with unit charges and one-body potential

$$V(x) = \frac{x^2}{2} + g\frac{x^4}{N} + C. \tag{1.74}$$

This calculation is of interest in the graphical expansion of matrix integrals [98], [555], and will be used in this context in the next section.

(i) With $Y = \cos\theta$ verify that

$$\frac{\cos 4\theta - 1}{\sin\theta} = -8Y^2(1 - Y^2)^{1/2}.$$

(ii) Use the eigenfunction expansion method, the result of (i) and Proposition 1.4.3 to show that the solution of the integral equation

$$V(x) = \int_{-a}^{a} \rho_b(y) \log|x - y|\, dy, \qquad -a \le x \le a,$$

which is bounded at $y = \pm a$ is

$$\rho_b(y) = \frac{a}{\pi}\Big(1 + \frac{2ga^2}{N} + \frac{4g}{N}y^2\Big)\sqrt{1 - (y/a)^2},$$

provided

$$C = -a^2\left(\frac{1}{4} - \frac{N}{a^2}\log a + \frac{1}{2}\log 2 + \frac{3ga^2}{8N} + \frac{3ga^2}{2N}\log 2\right).$$

(iii) Use the neutrality condition to show

$$\frac{a^2}{2} + \frac{3ga^4}{2N} = N, \tag{1.75}$$

and use this in the formula for C to obtain the simplification

$$C = -\frac{a^2}{8} + N\log\frac{a}{2} - \frac{N}{4}.$$

(iv) Use the trigonometric Euler integral in Exercises 4.1 q.1(i) below and the neutrality condition to show

$$U_3 = -\frac{CN}{2} + \frac{a^4}{192} - \frac{a^2 N}{24} - \frac{N^2}{16},$$

and thus

$$\begin{aligned}(U_2 + U_3) &- (U_2 + U_3)|_{g=0} \\ &= \frac{g}{N}\sum_{j=1}^{N} x_j^4 - \frac{N^2}{2}\left[\frac{1}{24}\big((a/\sqrt{2N})^2 - 1\big)\big(9 - (a/\sqrt{2N})^2\big) - \log(a/\sqrt{2N})\right],\end{aligned} \tag{1.76}$$

where $(U_2 + U_3)|_{g=0}$ is as implicit in Proposition 1.4.4.

6. (i) For a general potential $u(x)$, use the eigenfunction expansion method to show that the solution $\rho_b(y)$ of the

integral equation

$$u(x) + C = \int_{-a}^{a} \rho_b(y) \log |x - y| \, dy, \quad x \in [-a, a], \tag{1.77}$$

which is bounded at $y = \pm a$ can be written

$$\rho_b(a \cos \theta) = -\frac{2}{a\pi^2 \sin \theta} \sum_{p=1}^{\infty} p \Big(\int_0^{\pi} u(a \cos \sigma) \cos p\sigma \, d\sigma \Big) \Big(\cos p\theta - 1 \Big).$$

(ii) For $u(x) = x^{2n}$, $n \in \mathbb{Z}^+$, use the integration formula

$$\int_0^{\pi} \cos^{2n} \sigma \cos 2p\sigma \, d\sigma = \frac{\pi}{2^{2n}} \binom{2n}{n+p},$$

verified using complex exponentials, and the transformation identity

$$\frac{1}{2n} \sum_{p=1}^{n} p \binom{2n}{n+p} \frac{1 - \cos 2p\theta}{1 - \cos^2 \theta} = \sum_{l=1}^{n} \binom{2(n-l)}{n-l} (2 \cos \theta)^{2(l-1)}$$

to show that [99]

$$\rho_b(x) = \frac{4n}{\pi} \Big(\frac{a}{2}\Big)^{2n-1} \Big(\sum_{l=1}^{n} \binom{2(n-l)}{n-l} \Big(\frac{2x}{a}\Big)^{2(l-1)} \Big) \sqrt{1 - \Big(\frac{x}{a}\Big)^2}.$$

Check that this is consistent with $\rho_b(y)$ in q.2(ii).

7. [124] The task of this exercise is to solve the integral equation

$$\frac{x^2}{2} + C = \int_{-b}^{a} \rho_b(y) \log |x - y| \, dy, \tag{1.78}$$

with $a = \sqrt{2N}s$, $s < 1$, subject to the neutrality constraint

$$\int_{-b}^{a} \rho_b(y) \, dy = N, \tag{1.79}$$

and to the constraint that $\rho_b(y)$ be bounded at $y = -b$.

(i) Change variables according to

$$y = Y + \frac{a-b}{2}, \qquad x = X + \frac{a-b}{2}$$

and then according to

$$Y = \frac{a+b}{2} \cos \theta, \quad Y = \frac{a+b}{2} \cos \sigma, \quad \sin \theta \rho_b \Big(\frac{a+b}{2} \cos \theta + \frac{a-b}{2} \Big) = \frac{a+b}{2} \phi(\theta),$$

to rewrite (1.78) to read

$$-\Big(\frac{2}{a+b}\Big)^2 \Big(\frac{(a+b)^2}{16} \cos 2\sigma - \frac{a^2 - b^2}{2} \cos \sigma + \frac{(a-b)^2}{8} + \frac{(a+b)^2}{16} + C - N \log \frac{a+b}{2} \Big)$$
$$= -\int_0^{\pi} \phi(\theta) \log | \cos \theta - \cos \phi | \, d\phi.$$

(ii) Use the method of derivation of Proposition 1.4.4 to show that only for the value

$$C = N \log \frac{l\sqrt{N}}{\sqrt{2}} - \frac{9Nl^2}{8} + 2Nls - Ns^2, \tag{1.80}$$

where $a+b=\sqrt{N}l$, does (1.78) permit a solution bounded at $y=-b$, and furthermore show that the latter has the explicit form

$$\rho_{\rm b}(y) = \frac{\sqrt{2N}}{\pi}(s-y/\sqrt{2N})^{1/2}(l-s+y/\sqrt{2N})^{1/2} + \sqrt{N}\frac{l-2s}{\sqrt{2}\pi}\Big(\frac{l-s+y/\sqrt{2N}}{s-y/\sqrt{2N}}\Big)^{1/2}. \qquad (1.81)$$

(iii) By making use of (4.2) below, show that the neutrality condition (1.79) gives

$$l = \frac{2}{3}(s+\sqrt{s^2+3}). \qquad (1.82)$$

8. In this exercise the location of the minimum of the function

$$H(x_1,\dots,x_N) := \frac{1}{2}\sum_{j=1}^N x_j^2 - \sum_{1\le j<k\le N}\log|x_k-x_j|$$

will be determined by following a calculation of Stieltjes [503]. This gives the equilibrium points of the system of Proposition 1.4.4.

(i) Show that H is convex by establishing that for $t_j\ne 0$ $(j=1,\dots,N)$, $\sum_{j,k=1}^N t_j t_k \frac{\partial^2 H}{\partial x_j x_k} > 0$, and conclude that H has a unique minimum.

(ii) Let $g(x)=\prod_{l=1}^N(x-x_l^{(0)})$. Show that the equations for the minimum $\partial H/\partial x_j = 0$ $(j=1,\dots,N)$ can be written

$$g''(x_j) - 2x_j g'(x_j) = 0 \quad (j=1,\dots,N).$$

(iii) Observe that the l.h.s. of the above equation is a polynomial of degree N which vanishes at the zeros of $g(x)$, and so must be proportional to $g(x)$, to deduce the d.e.

$$g''(x) - 2xg'(x) + 2Ng(x) = 0.$$

Hence show that the minimum of $H(x_1,\dots,x_N)$ occurs at the zeros of the Hermite polynomial $H_N(x)$.

1.5 HIGH-DIMENSIONAL RANDOM ENERGY LANDSCAPES

This and the next three sections all relate to the Wigner semicircle law for the eigenvalue density in the Gaussian ensembles. In this section it is the Wigner semicircle law in the case of the GOE which arises.

As we have seen, the GOE was formulated as a model of the eigenvalues of classically chaotic quantum Hamiltonians with a time reversal symmetry. Later, the GOE received prominence for its relevance to the study of the so-called replica trick in the theory of disordered systems [161]. As applied to random matrix theory, the replica trick corresponds to the identity

$$\Big\langle {\rm Tr}\,(\epsilon-\mathbf{H})^{-1}\Big\rangle_{\mathbf{H}\in{\rm GOE}} = \lim_{n\to 0}\frac{1}{n}\frac{\partial Z_n(\epsilon)}{\partial\epsilon}, \qquad Z_n(\epsilon) := \langle \det{}^n(\epsilon-\mathbf{H})\rangle_{\mathbf{H}\in{\rm GOE}}. \qquad (1.83)$$

In practice the difficulty with the implementation of (1.83) is that orthogonal polynomial methods to evaluate $Z_n(\epsilon)$ (see Chapter 5) require n to be a positive integer, and one is then faced with the problem of analytic continuation off the positive integers in order to take the limit.

More recently the GOE has been shown to be of relevance to another problem relating to the theory of disordered systems [248]. The problem is the computation of the distribution of the critical points for certain high-dimensional Gaussian random potentials (often referred to as landscapes). Specifically, consider the energy

$$\mathcal{H} := \frac{\mu}{2}\sum_{j=1}^N x_j^2 + V(x_1,\dots,x_N), \qquad (1.84)$$

where $\mu > 0$ and V is Gaussian distributed with zero mean and covariance

$$\langle V(\vec{x}_1)V(\vec{x}_2)\rangle = Nf\Big(\frac{1}{2N}(\vec{x}_1 - \vec{x}_2)^2\Big). \tag{1.85}$$

A critical point of $\mathcal{H}$ is characterized by the simultaneous stationarity conditions $\partial\mathcal{H}/\partial x_j = 0$ ($j = 1,\dots,N$). Let $\rho_{(1)}(\vec{x})$ be the density of critical points, so that $\mathcal{N}(D)$ — the expected number of critical points in the region D — is given by $\mathcal{N}(D) = \int_D \rho_{(1)}(\vec{x})\, d\vec{x}$. With $\{\vec{x}_k\}_{k=1,\dots,N^*}$ denoting the critical points, one has the change of variables type formula

$$\sum_{k=1}^{N^*} \delta(\vec{x} - \vec{x}_k) = \prod_{i=1}^{N} \delta\Big(\frac{\partial\mathcal{H}}{\partial x_i}\Big)\Big|\det\Big[\frac{\partial^2\mathcal{H}}{\partial x_i\partial x_j}\Big]_{i,j=1,\dots,N}\Big|, \tag{1.86}$$

and hence $\rho_{(1)}(\vec{x})$ can be computed as the ensemble average of the r.h.s. This form of $\rho_{(1)}(\vec{x})$ is referred to as the *generalized Kac-Rice formula* (see also (15.56) below).

PROPOSITION 1.5.1 *Let GOE$^\#$ refer to the GOE with matrices $\mathbf{X} \mapsto \sqrt{N/2f''(0)}\mathbf{X}$. We have*

$$\mathcal{N}(\mathbb{R}^N) = \mu^{-N}\sqrt{\frac{N}{2\pi}}\int_{-\infty}^{\infty} e^{-Nt^2/2}\Big\langle\Big|\det\Big((\mu + \sqrt{f''(0)}t)\mathbf{1}_N - \mathbf{X}\Big)\Big|\Big\rangle_{\mathbf{X}\in\mathrm{GOE}^\#} dt. \tag{1.87}$$

Proof. We begin with the formula implied by the sentence including (1.86). Recalling (1.84) this gives

$$\begin{aligned}\rho_{(1)}(\vec{x}) &= \Big\langle \prod_{i=1}^{N} \delta(\mu x_i + \partial_i V)|\det[\mu\delta_{j,k} + \partial_j\partial_k V]_{j,k=1,\dots,N}|\Big\rangle \\ &= \Big\langle \prod_{i=1}^{N} \delta(\mu x_i + \partial_i V)\Big\rangle\Big\langle|\det[\mu\delta_{j,k} + \partial_j\partial_k V]_{j,k=1,\dots,N}|\Big\rangle,\end{aligned} \tag{1.88}$$

where $\partial_i := \partial/\partial x_i$ and the second equality follows by noting that $\partial_i V$ and $\partial_j\partial_k V$ are statistically independent.

The Gaussian field formed by $\partial_i V$ has, according to (1.85), covariance $\langle\partial_j V \partial_k V\rangle = a^2\delta_{j,k}$, $a^2 := -f'(0)$. Hence, after making use of the Fourier integral representation of the delta function, one has

$$\Big\langle \prod_{i=1}^{N} \delta(\mu x_i + \partial_i V)\Big\rangle = \frac{1}{(\sqrt{2\pi a^2})^N} e^{-\mu^2\sum_{j=1}^N x_j^2/2a^2}$$

(see (1.93) below). The second average in (1.88) is independent of x_k, and so we can integrate over $D = \mathbb{R}^N$ to obtain

$$\mathcal{N}(\mathbb{R}^N) = \mu^{-N}\Big\langle\Big|\det[\mu\delta_{j,k} + \partial_j\partial_k V]_{j,k=1,\dots,N}\Big|\Big\rangle.$$

Set $H_{jk} := \partial_j\partial_k V$. It follows from (1.85) that

$$\langle H_{il}H_{jm}\rangle = \frac{f''(0)}{N}\Big(\delta_{ij}\delta_{lm} + \delta_{im}\delta_{lj} + \delta_{il}\delta_{jm}\Big). \tag{1.89}$$

Now let the diagonal elements H_{ii} and upper triangular elements H_{jk} ($j < k$) collectively be indexed H_μ, and form the vector $\vec{H} = (H_\mu)$. Being Gaussian variables, for some matrix $\mathbf{A}$ of appropriate size they have distribution proportional to $\exp(-\frac{1}{2}\vec{H}\mathbf{A}\vec{H})$. Furthermore $\mathbf{A}$ is completely determined by $\langle H_\mu H_\nu\rangle$ (see (1.95) below) with the task being to compute the inverse of the matrix of these averages. The final result can be written in a structured form, showing that $\mathbf{H} := [H_{jk}]$ is a real symmetric Gaussian random matrix with p.d.f. proportional to

$$\exp\Big(-\frac{N}{4f''(0)}\Big(\mathrm{Tr}\,\mathbf{H}^2 - \frac{1}{N+2}(\mathrm{Tr}\,\mathbf{H})^2\Big)\Big). \tag{1.90}$$

By completing the square in t we see that

$$\int_{-\infty}^{\infty} e^{-Nt^2/2} \exp\Big(-\frac{N}{4f''(0)}\Big(\mathrm{Tr}\,\big(\mathbf{H}-\sqrt{f''(0)}t\mathbf{1}_N\big)^2\Big)\Big)\,dt,$$

is proportional to (1.90) and (1.87) follows. (The N-dependent proportionality follows by requiring that $\mathcal{N}(\mathbb{R}^N) \to 1$ for $\mu \to \infty$.) □

With $J := \sqrt{f''(0)}$, changing variables $\mathbf{X} \mapsto J\sqrt{2/N}\mathbf{X}$ in the average of (1.87) gives

$$\begin{aligned}&(J\sqrt{2/N})^N\Big\langle\Big|\det\Big[\sqrt{N/2}((\mu/J)+t)\mathbf{1}_N-\mathbf{X}\Big]\Big|\Big\rangle_{\mathbf{X}\in\mathrm{GOE}}\\ &\quad=(J\sqrt{2/N})^N e^{N((\mu/J)+t)^2/4}\frac{G_{1,N+1}}{(N+1)G_{1,N}}\rho_{(1),N+1}(\sqrt{N/2}((\mu/J)+t))\end{aligned} \tag{1.91}$$

where $\rho_{(1),N+1}$ refers to the density in the GOE with $N+1$ eigenvalues, and $G_{1,N}$ is given by (1.163). The equality in (1.91) follows by writing the determinant as a product of eigenvalues, writing the average in terms of eigenvalues using (1.27), and recalling the formula for the density (one-particle correlation) (1.46). Substituting (1.91) in (1.87) shows

$$\mathcal{N}(\mathbb{R}^N)=\Big(\frac{J\sqrt{2}}{\mu\sqrt{N}}\Big)^N\Gamma\Big(\frac{N+1}{2}\Big)e^{N(\mu/J)^2/2}\sqrt{\frac{N}{\pi}}\int_{-\infty}^{\infty}e^{-N(t-(\mu/J))^2/4}\rho_{(1),N+1}(\sqrt{N/2}((\mu/J)+t))\,dt.$$

For large N, after making use of Stirling's formula

$$\Gamma(x+1)\sim(2\pi x)^{1/2}e^{x\log x-x}\qquad\text{as}\quad x\to\infty,\ \mathrm{Re}(x)>0, \tag{1.92}$$

and noting the delta function type behavior of the integral, we see that for the argument of $\rho_{(1),N+1}$ inside its support, and thus $\mu < J$,

$$\mathcal{N}(\mathbb{R}^N)\sim 2(2\pi)^{1/2}(J/\mu)^N e^{N(\mu/J)^2/2}e^{-N/2}\rho_{(1),N+1}(\sqrt{2N}(\mu/J)).$$

Making use now of (1.52), one obtains that for $\mu < J$

$$\Sigma(\mu):=\lim_{N\to\infty}\frac{1}{N}\log\mathcal{N}(\mathbb{R}^N)=\frac{1}{2}\Big(\frac{\mu^2}{J^2}-1\Big)-\log(\mu/J).$$

Note that $\Sigma(J) = 0$. In fact analysis of $\rho_{(1),N+1}(\sqrt{2N}X)$ for $|X| > 1$ undertaken in Exercises 14.4 q.5 below can be used to show that $\Sigma(\mu) = 0$ for $\mu > J$, and so the number of critical points undergoes a phase transition at $\mu = J$.

EXERCISES 1.5 1. (i) Let $\mathbf{A}$ be an $n \times n$ positive definite matrix. By changing variables $\vec{y} = \mathbf{A}^{1/2}\vec{x}$ and completing the square show

$$\begin{aligned}I_n[\mathbf{A},\vec{b}]&:=\int_{-\infty}^{\infty}dx_1\cdots\int_{-\infty}^{\infty}dx_n\exp\Big(-\frac{1}{2}\vec{x}^T\mathbf{A}\vec{x}+\vec{b}\cdot\vec{x}\Big)\\&=(2\pi)^{n/2}(\det\mathbf{A})^{-1/2}\exp\Big(\frac{1}{2}\vec{b}^T\mathbf{A}^{-1}\vec{b}\Big).\end{aligned} \tag{1.93}$$

(ii) Let

$$\langle f\rangle_{\mathbf{A}}=\frac{1}{I_n[\mathbf{A},\vec{0}]}\int_{-\infty}^{\infty}dx_1\cdots\int_{-\infty}^{\infty}dx_n\,f\exp\Big(-\frac{1}{2}\vec{x}^T\mathbf{A}\vec{x}\Big).$$

Use (1.93) and the method of derivation of (1.99) below to show that for l even

$$\langle x_{k_1}x_{k_2}\cdots x_{k_l}\rangle_{\mathbf{A}}=\sum_{\substack{\text{all possible}\\ \text{pairings of }(k_1\cdots k_l)}}\mathbf{A}^{-1}_{k_{p_1}k_{p_2}}\cdots\mathbf{A}^{-1}_{k_{p_{l-1}}k_{p_l}}, \tag{1.94}$$

while for l odd this average vanishes.

(iii) By choosing $l = 2$, deduce from (1.94) that

$$\mathbf{A}^{-1} = [\langle x_j x_k \rangle_{\mathbf{A}}]_{j,k=1,\dots,n},$$

which in words says that the covariance matrix associated with the average $\langle \cdot \rangle_{\mathbf{A}}$ is given by $\mathbf{A}^{-1}$.

(iv) Replace $\vec{b}$ by $i\vec{b}$ in (1.93), and integrate over $\vec{b}_{k+1}, \dots, \vec{b}_n$ $(k \le n)$ to deduce that

$$\begin{aligned}(2\pi)^{k/2}(\det \mathbf{A})^{-1/2} \int_{-\infty}^{\infty} db_{k+1} \cdots \int_{-\infty}^{\infty} db_n \exp\Big(- \frac{1}{2}\vec{b}^T \tilde{\mathbf{A}}^{-1}\vec{b}\Big) \\ = (2\pi)^{k/2}(\det \tilde{\mathbf{A}})^{-1/2} \exp\Big(- \frac{1}{2}\vec{b}^T \mathbf{A}^{-1}\vec{b}\Big)\Big|_{b_{k+1}=\cdots=b_n=0},\end{aligned}$$

where $\tilde{\mathbf{A}}$ is the $k \times k$ submatrix of $\mathbf{A}$ formed from the first k rows and columns.

(v) With $\vec{b}$ replaced by $i\vec{b}$, regard the r.h.s. of (1.93) as a p.d.f. for $\vec{b}$ (up to normalization), so that the covariance matrix is now

$$\mathbf{A} = [\langle b_j b_k \rangle_{\mathbf{A}^{-1}}]_{j,k=1,\dots,n}.$$

Show that under the linear change of variables $\vec{b} = \mathbf{L}\vec{c}$, the vector $\vec{c}$ has a Gaussian distribution

$$(2\pi)^{n/2}(\det \mathbf{B})^{-1/2} \exp\Big(- \frac{1}{2}\vec{c}^T \mathbf{B}^{-1}\vec{c}\Big), \quad \mathbf{B} = \langle c_j c_k \rangle_{\mathbf{A}^{-1}}. \tag{1.95}$$

1.6 MATRIX INTEGRALS AND COMBINATORICS

1.6.1 Combinatorics of $\langle \mathrm{Tr}(\mathbf{X}^{2k}) \rangle_{\mathrm{GUE}^*}$

In this section we put our knowledge of the asymptotic density for the GUE to use in the solution of a combinatorial problem. It has long been known [98] that the matrix integrals

$$\int f(\mathbf{X}) e^{-\mathrm{Tr}(\mathbf{X}^2)/2} (d\mathbf{X}),$$

for $\mathbf{X}$ a particular class of random matrices and suitable $f(\mathbf{X})$, have combinatorial significance in that they count certain diagrams embedded on surfaces according to their genus. Here, following [557], [239], we will detail such a combinatorial interpretation of the matrix integral

$$\frac{1}{C} \int \mathrm{Tr}(\mathbf{X}^{2k}) e^{-\mathrm{Tr}(\mathbf{X}^2)/2} (d\mathbf{X}) =: \langle \mathrm{Tr}(\mathbf{X}^{2k}) \rangle_{\mathrm{GUE}^*}, \tag{1.96}$$

where C is the normalization, and GUE* is identical to the GUE except that $\mathbf{X} \mapsto \mathbf{X}/\sqrt{2}$. By changing variables $\mathbf{X} = \mathbf{U}\mathbf{L}\mathbf{U}^{-1}$ for the eigenvalues and eigenvectors we see from the result of Exercises 1.3 q.3 that

$$\langle \mathrm{Tr}(\mathbf{X}^{2k}) \rangle_{\mathrm{GUE}^*} = \frac{1}{C} \int_{-\infty}^{\infty} d\lambda_1 \cdots \int_{-\infty}^{\infty} d\lambda_N \prod_{l=1}^{N} e^{-\lambda_l^2/2} \Big(\sum_{j=1}^{N} \lambda_j^{2k} \Big) \prod_{1 \le j < k \le N} |\lambda_k - \lambda_j|^2. \tag{1.97}$$

Changing variables $\lambda_l \mapsto \sqrt{2}\lambda_l$ we see from the definition (1.46) that in terms of the density $\rho_{(1)}(\lambda)$ for the GUE we have

$$\langle \mathrm{Tr}(\mathbf{X}^{2k}) \rangle_{\mathrm{GUE}^*} = 2^k \int_{-\infty}^{\infty} \lambda^{2k} \rho_{(1)}(\lambda)\, d\lambda. \tag{1.98}$$

However, it is not from (1.98) that the combinatorics arise; this comes from the evaluation of (1.96) as a Gaussian integral over the independent elements of the matrix $\mathbf{X}$. The latter task can be achieved by using a particular matrix version of *Wick's theorem*.

Proposition 1.6.1 *Let $\mathbf{X} = [z_{jk}]_{j,k=1,\dots,N}$, $z_{jk} = x_{jk} + iy_{jk}$ be Hermitian so that*

$$e^{-\mathrm{Tr}(\mathbf{X}^2)/2}(d\mathbf{X}) = e^{-\mathrm{Tr}(\mathbf{X}^2)/2} \prod_{j=1}^{N} dx_{jj} \prod_{1 \le j < k \le N} dx_{jk} dy_{jk}.$$

Let I be a finite ordered set of pairs of indices (j,k), $1 \le j,k \le N$, and let P denote a matching of the elements of I in pairs. Then we have

$$\Big\langle \prod_{(i,j)\in I} z_{ij} \Big\rangle_{\mathrm{GUE}^*} = \sum_{\substack{\text{pairings} \\ P \text{ of } I}} \prod_{(i,j),\,(k,l)} \langle z_{ij} z_{kl} \rangle_{\mathrm{GUE}^*}. \tag{1.99}$$

Proof. Introducing the Hermitian matrix $\mathbf{Y} = [w_{jk}]_{j,k=1,\dots,N}$ we observe that

$$\Big\langle \prod_{(i,j)\in I} z_{ij} \Big\rangle_{\mathrm{GUE}^*} = \Big(\prod_{(i,j)\in I} \frac{\partial}{\partial w_{ji}} \Big) \Big\langle e^{\mathrm{Tr}(\mathbf{YX})} \Big\rangle_{\mathrm{GUE}^*} \Big|_{\mathbf{Y}=0}. \tag{1.100}$$

In the integrand, writing

$$\mathrm{Tr}\,\mathbf{X}^2 - 2\mathrm{Tr}\,(\mathbf{YX}) = \mathrm{Tr}((\mathbf{X}-\mathbf{Y})^2) - \mathrm{Tr}\,\mathbf{Y}^2,$$

we see from the change of variables $\mathbf{X} \mapsto \mathbf{X} + \mathbf{Y}$ that

$$\Big\langle e^{\mathrm{Tr}(\mathbf{YX})} \Big\rangle_{\mathrm{GUE}^*} = e^{\mathrm{Tr}(\mathbf{Y}^2)/2} = \prod_{j=1}^{N} e^{w_{jj}^2/2} \prod_{1 \le j < k \le N} e^{w_{jk} w_{kj}}.$$

Thus (1.100) gives

$$\Big\langle \prod_{(i,j)\in I} z_{ij} \Big\rangle_{\mathrm{GUE}^*} = \sum_{\substack{\text{pairings} \\ P \text{ of } I}} \prod_{(i,j),\,(k,l)} \delta_{i,l}\delta_{j,k},$$

which reduces to (1.99) after noting

$$\langle z_{ij} z_{kl} \rangle_{\mathrm{GUE}^*} = \delta_{i,l}\delta_{j,k}. \tag{1.101}$$

□

Our task is to compute

$$\langle \mathrm{Tr}(\mathbf{X}^{2k}) \rangle_{\mathrm{GUE}^*} := \Big\langle \sum_{i_1,\dots,i_{2k}=1}^{N} z_{i_1 i_2} z_{i_2 i_3} \cdots z_{i_{2k-1} i_{2k}} z_{i_{2k} i_1} \Big\rangle_{\mathrm{GUE}^*}. \tag{1.102}$$

According to (1.99) we have

$$\langle \mathrm{Tr}(\mathbf{X}^{2k}) \rangle_{\mathrm{GUE}^*} = \sum_{i_1,\dots,i_{2k}=1}^{N} \sum_{\substack{\text{pairings } P \text{ of} \\ \{(i_1,i_2),(i_2,i_3),\dots,(i_{2k},i_1)\}}} \prod_{(j,j'),\,(l,l')} \langle z_{i_j i_{j'}} z_{i_l i_{l'}} \rangle_{\mathrm{GUE}^*}, \tag{1.103}$$

and (1.101) shows that various labels must coincide for a given term in this expression to be nonzero. For example, with $k = 4$ consider the particular term in (1.103)

$$\langle z_{i_1 i_2} z_{i_3 i_4} \rangle \langle z_{i_2 i_3} z_{i_8 i_1} \rangle \langle z_{i_4 i_5} z_{i_6 i_7} \rangle \langle z_{i_5 i_6} z_{i_7 i_8} \rangle = (\delta_{i_1,i_4}\delta_{i_2,i_3})(\delta_{i_2,i_1}\delta_{i_3,i_8})(\delta_{i_4,i_7}\delta_{i_5,i_6})(\delta_{i_5,i_8}\delta_{i_6,i_7}). \tag{1.104}$$

For this to be nonzero we must have $i_1 = i_2 = \cdots = i_8$, giving only one independent label. As another example, consider the term

$$\langle z_{i_1 i_2} z_{i_4 i_5} \rangle \langle z_{i_2 i_3} z_{i_3 i_4} \rangle \langle z_{i_5 i_6} z_{i_8 i_1} \rangle \langle z_{i_6 i_7} z_{i_7 i_8} \rangle = (\delta_{i_1,i_5}\delta_{i_2,i_4})(\delta_{i_2,i_4})(\delta_{i_5,i_1}\delta_{i_6,i_8})(\delta_{i_6,i_8}), \tag{1.105}$$

which is nonzero for $i_1 = i_5$, $i_2 = i_4$, $i_6 = i_8$, giving five independent labels, i_1, i_2, i_3, i_6, i_7, say.

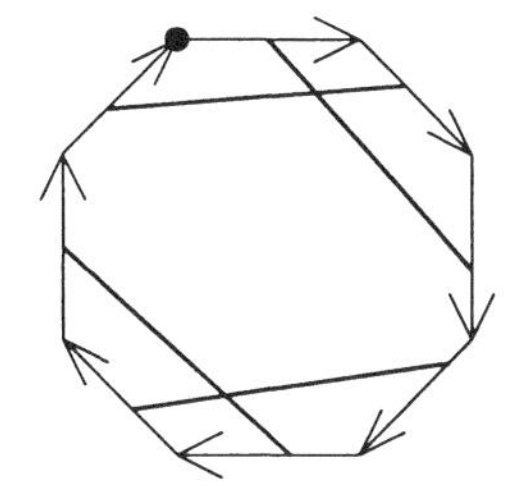
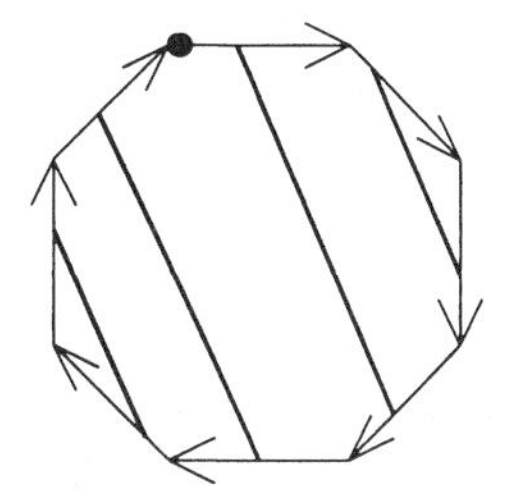

Figure 1.2 Graphical representation of the contributions (1.104) and (1.105). The heavy lines identify edges and the dot marks the location of the vertex labeled i_1, with the other vertices labeled clockwise.

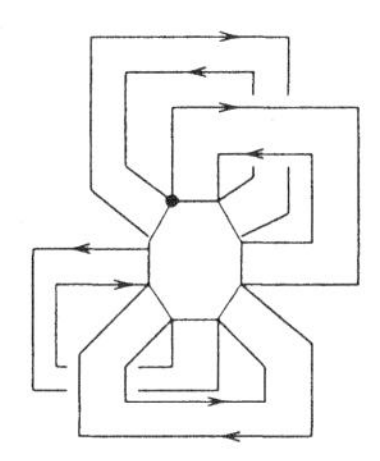
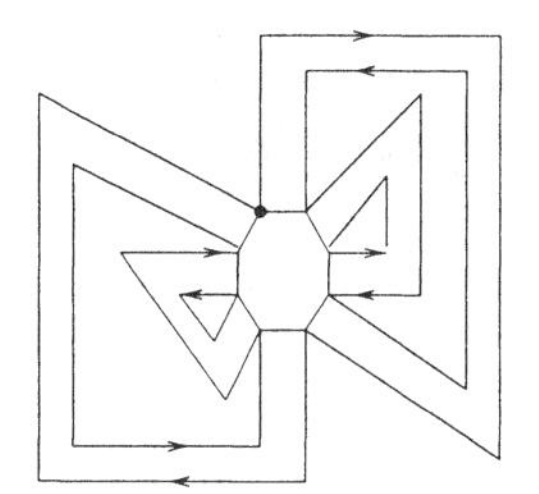

Figure 1.3 The dual graphical representation of Figure 1.2 for the contributions of (1.104) and (1.105). The dot marks the location of the vertex labeled i_1, with the other vertices labeled clockwise.

In general the nonzero terms in (1.103) can be represented graphically in two related ways, both of which involve a regular $2k$-gon, with the vertices labeled $i_1, \dots, i_{2k}$, and the edges oriented clockwise. One method to carry out the pairing between consecutive vertices (i_j, i_k) and consecutive vertices (i_l, i_m) is to join the corresponding edges on the $2k$-gon according to the rule that edges must be joined in opposite directions (see Figure 1.2 for this representation of (1.104) and (1.105)).

Another approach to carrying out the pairing is to draw a straight line segment perpendicular to and outward from the ends of each edge of the $2k$-gon. These segments are to be given the directions of out and in alternately around the $2k$-gon. Then each nonzero contribution to (1.103) can be represented by joining different pairs (i_j, i_k) and (i_l, i_m) of parallel straight line segments to form roadways (see Figure 1.3 for this representation of (1.104) and (1.105)). Note that the joining is such that edges of the roadways have definite directions.

The diagrams of Figure 1.2 can be catalogued according to the number ν of independent vertices after pairing. This of course is just the number of independent summation labels in (1.103) so we can write

$$\langle \mathrm{Tr}(\mathbf{X}^{2k}) \rangle_{\mathrm{GUE}^*} = \sum_{\nu=1}^{k+1} a_\nu(k) N^\nu, \tag{1.106}$$

where $a_\nu(k)$ denotes the number of different pairings which have ν vertices. On the other hand, the diagrams of Figure 1.3 are topological duals of Figure 1.2 with the ν independent vertices now ν independent faces. These can be determined by following the edge of a roadway and its continuation according to its direction, until arriving back at the starting point. The formal meaning of the faces is obtained by shrinking the width of the roadways in Figure 1.3 to single lines, at the same time as shrinking the $2k$-gon to a single vertex so the lines become loops, and then embedding the diagram on a closed surface as a *map*.

DEFINITION 1.6.2 *A map is a graph (a collection of vertices and edges) drawn on a closed surface such that the edges do not intersect and, if we cut the surface along the edges, a disjoint union of sets topologically*

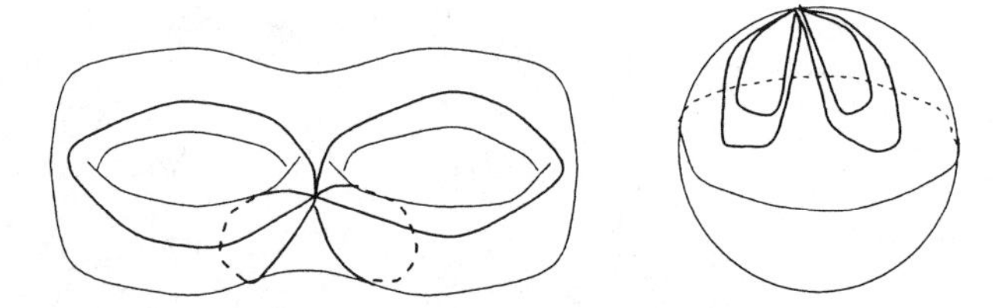

Figure 1.4 Embedding of the graphs of Figure 1.3 (after shrinking the k-gon to a single vertex, and the roadways to single lines) onto a closed surface to form a map with a single vertex.

equivalent to an open disk results. The number of such disks is by definition the number of faces of the map.

It is similarly the case that the number of independent vertices in the diagrams of Figure 1.2 can be specified in terms of the corresponding map.

The index ν in (1.106) determines the genus g (number of holes) of the closed surface. This follows from *Euler's relation*

$$2 - 2g = V - E + F, \tag{1.107}$$

where V denotes the number of vertices, E the number of edges and F the number of faces. In the diagrams of Figure 1.2 $V = \nu$, $F = 1$ and $E = k$, while in the diagrams of Figure 1.3 the roles of V and F are interchanged so that $V = 1$, $F = \nu$ and $E = k$. Either way (1.107) gives

$$\nu = k + 1 - 2g. \tag{1.108}$$

As shown in Figure 1.4, the diagrams of Figure 1.3 can be directly embedded on a surface of particular genus, thereby illustrating (1.108) (thus in the first case $k = 4$, $\nu = 1$, $g = 2$ while in the second case $k = 4$, $\nu = 5$, $g = 0$).

Using (1.108) in (1.106) gives

$$\langle \mathrm{Tr}(\mathbf{X}^{2k}) \rangle_{\mathrm{GUE}^*} = N^{k+1} \sum_{g=0}^{[k/2]} a_{k+1-2g}(k) N^{-2g}. \tag{1.109}$$

In particular

$$\lim_{N\to\infty} N^{-k-1} \langle \mathrm{Tr}(\mathbf{X}^{2k}) \rangle_{\mathrm{GUE}^*} = a_{k+1}(k) \tag{1.110}$$

where $a_{k+1}(k)$ denotes the number of matchings of the $2k$-gon which are planar (i.e. can be embedded on the surface of a sphere, which has $g = 0$). Substituting for $\langle \mathrm{Tr}(\mathbf{X}^{2k}) \rangle_{\mathrm{GUE}^*}$ using (1.97), and then substituting for $\rho_{(1)}(x)$ using (1.52), evaluating the integral using (4.3) below and simplifying the resulting gamma functions using the duplication formula

$$2^{2z-1}\Gamma(z)\Gamma(z + 1/2) = \pi^{1/2}\Gamma(2z), \tag{1.111}$$

one finds

$$a_{k+1}(k) = \frac{1}{k+1}\binom{2k}{k}. \tag{1.112}$$

This number is familiar in combinatorics and is called the kth *Catalan number.* In fact (1.112) can easily be derived without using (1.98), which has the significance of providing an alternative derivation of the Wigner semicircle law (1.52) for Hermitian matrices (see Exercises 1.6 q.1), one which applies to establishing the Wigner semicircle for a large class of symmetric random matrices with independent entries (see, e.g., [551]).

However, this is not the case for the coefficients $a_{k-1}(k), a_{k-3}(k), \dots$ for which the use of (1.98) is the most efficient. We will return to the evaluation of these numbers in Chapter 5 when the exact value of $\rho_{(1)}(\lambda)$ is available.

We remark that $\langle \mathrm{Tr}(\mathbf{X}^{2k}) \rangle_{\mathrm{GOE}}$ allows for a similar combinatorial description in terms of maps on surfaces, although now the surfaces may be nonorientable (corresponding to graphs with twisted ribbons) [348].

1.6.2 Combinatorics of the $\beta = 2$ partition function with a general power series potential

Closely related to the combinatorial interpretation of (1.96) is the combinatorial interpretation of

$$Z_N(\{g_j\}) := \Big\langle \prod_{l=1}^{N} e^{\sum_{j=1}^{\infty} g_j x_l^j / j N^{j/2-1}} \Big\rangle_{\mathrm{GUE}^*} \tag{1.113}$$

when expanded in a power series in $\{g_\imath\}$. For the latter, expanding the exponentials gives

$$Z_N(\{g_j\}) = \sum_{n_1, n_2, \dots = 0}^{\infty} \prod_{j=1}^{\infty} \frac{g_j^{n_j}}{j^{n_j} n_j! N^{n_j(j/2-1)}} \Big\langle \prod_{j=1}^{\infty} \Big(\sum_{l=1}^{N} x_l^j \Big)^{n_j} \Big\rangle_{\mathrm{GUE}^*}, \tag{1.114}$$

while

$$\Big\langle \prod_{j=1}^{\infty} \Big(\sum_{l=1}^{N} x_l^j \Big)^{n_j} \Big\rangle_{\mathrm{GUE}^*} = \Big\langle \prod_{j=1}^{\infty} \Big(\mathrm{Tr}\, \mathbf{X}^j \Big)^{n_j} \Big\rangle_{\mathrm{GUE}^*}. \tag{1.115}$$

From the discussion of the previous subsection we know how to give a combinatorial interpretation of (1.115) in the special case $n_j = 1\ (j = k)$, $n_j = 0\ (j \ne k)$. A natural generalization of this interpretation extends to the general case [557], [239].

Each factor of $\mathrm{Tr}\, \mathbf{X}^j$ is represented as a j-gon with vertices labeled $i_1, i_2, \dots, i_j$ clockwise, starting at a marked vertex. These labels on vertices extend to labels on pairs of oppositely directed roadway edges coming into and out of each vertex. Whereas in the case of a single factor of $\mathrm{Tr}\, \mathbf{X}^j$ the combinatorial interpretation of computing (1.115) via Wick's theorem involved connecting roadways within the single j-gon, the graphical representation of contributions to (1.115) is to connect roadways among or within any of the n_j j-gons ($j = 1, 2, \dots$). The resulting structure, referred to as a *labeled fatgraph*, has weight N^ν, where ν is the number of faces (which in turn is equal to the number of unpaired labels). The number of edges is equal to $\sum_{j=1}^{\infty} j n_j / 2$, which is required to be an integer, while the number of vertices — defined as the number of j-gons — is equal to $\sum_{j=1}^{\infty} n_j$. Recalling Euler's relation (1.107) we see that (1.114) can thus be written

$$Z_N(\{g_j\}) = \sum_{n_1, n_2, \dots = 0}^{\infty} \Big(\prod_{j=1}^{\infty} \frac{g_j^{n_j}}{j^{n_j} n_j!} \Big) \sum_g a_g(\{n_j\}) N^{2-2g},$$

where $a_g(\{n_j\})$ is the number of labeled graphs constructed out of n_j j-gons ($j = 1, 2, \dots$) which can be embedded on a surface of genus g.

The various j-gons in the labeled fatgraph will not in general be connected. However, as $Z_N(\{g_j\})$ is an exponential generating function for these quantities, it is a well-known fact that taking the logarithm restricts to connected components. Thus, denoting this restriction by an asterisk, we have

$$\log Z_N(\{g_j\}) = \sum_{n_1, n_2, \dots = 0}^{\infty} \Big(\prod_{j=1}^{\infty} \frac{g_j^{n_j}}{j^{n_j} n_j!} \Big) {\sum_g}^* a_g(\{n_j\}) N^{2-2g}. \tag{1.116}$$

Fatgraphs which are topologically equivalent define a class of maps Γ. For each class the maximum value of $a_g(\{n_j\})$ is $\prod_{j=1}^{\infty} j^{n_j} n_j!$ and furthermore $\prod_{j=1}^{\infty} j^{n_j} n_j! / a_g(\{n_j\})$ is an integer written as $|\mathrm{Aut}\, \Gamma|$. As the notation suggests, $|\mathrm{Aut}\, \Gamma|$ is in fact equal to the order of the group of automorphisms associated with Γ. This

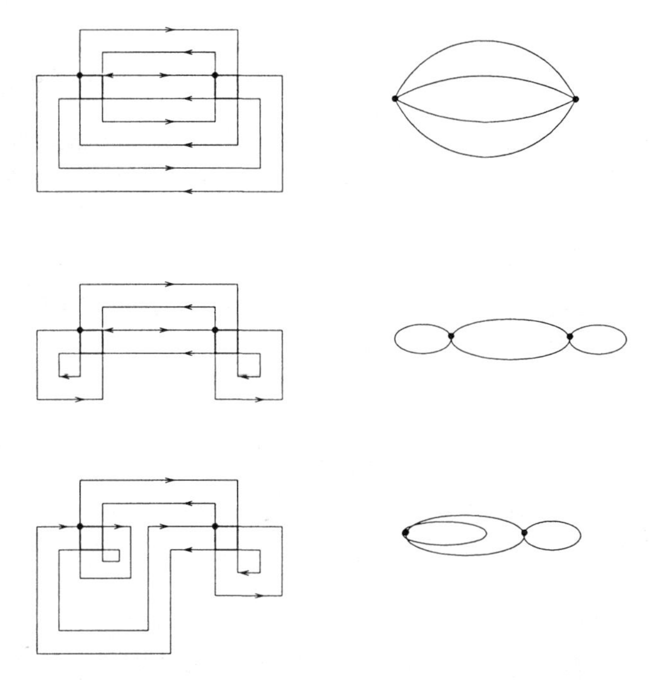

Figure 1.5 Three classes of fatgraphs can be constructed out of two 4-gons. An example from each class, together with the corresponding map is given. For the first class $|\mathrm{Aut}\,\Gamma| = 8$, while for the second and third classes $|\mathrm{Aut}\,\Gamma| = 2$.

can be specified as the number of equivalent labelings of the faces of Γ, which means the number of different labelings in the plane which result from topological transformations of the map on the closed surface.

In terms of $|\mathrm{Aut}\,\Gamma|$ (1.116) reads

$$\log Z_N(\{g_j\}) = \sum_{\text{connected }\Gamma} \frac{1}{|\mathrm{Aut}\,\Gamma|} N^{2-2g(\Gamma)} \prod_{j=1}^{\infty} g_j^{V_j(\Gamma)}, \tag{1.117}$$

where n_j in (1.116) has been written $V_j(\Gamma)$ in (1.117) to emphasize that it counts the number of vertices with coordination number j in the corresponding map. In particular

$$\lim_{N\to\infty} \frac{1}{N^2} \log Z_N(\{g_j\}) = \sum_{\substack{\text{connected }\Gamma \\ g(\Gamma)=0}} \frac{1}{|\mathrm{Aut}\,\Gamma|} \prod_{j=1}^{\infty} g_j^{V_j(\Gamma)}, \tag{1.118}$$

and thus we obtain a generating function for maps weighted according to the coordination number of the vertices. To illustrate (1.118), in Figure 1.5 we display the contributions to the coefficient of g_4^2 in terms of fatgraphs and the corresponding maps.

Suppose $g_j = 0$ for $j \neq 4$. From the definition (1.113) we have

$$\lim_{N\to\infty} \frac{1}{N^2} \log Z_N\Big(\{g_j\}\Big|_{g_j=0\ (j\neq 4)}\Big)$$
$$= \lim_{N\to\infty} \frac{1}{N^2} \log \Big\langle \prod_{l=1}^{N} e^{g_4 x_l^4/4N} \Big\rangle_{\mathrm{GUE}^*} = \lim_{N\to\infty} \frac{1}{N^2} \log \Big\langle \prod_{l=1}^{N} e^{g_4 x_l^4/N} \Big\rangle_{\mathrm{GUE}}. \tag{1.119}$$

To evaluate this limit we make use of the log-gas interpretation of the average as a ratio of configuration integrals relating to one-component log-potential systems with particular neutralizing background charge densities. The Boltzmann factor for these systems contains constant terms (i.e., terms independent of the particle coordinates) which are not present in (1.119). If these terms, A_N say, were included, its logarithm would then be expected to be proportional to N as the difference between two free energies is being calculated (see (4.160) below). Thus one expects

$$\lim_{N\to\infty} \frac{1}{N^2} \log Z_N\Big(\{g_j\}\Big|_{g_j=0\,(j\neq 4)}\Big) = \lim_{N\to\infty} \frac{1}{N^2} \log \frac{1}{A_N}.$$

The value of $1/A_N$ has been calculated in Exercises 1.4 q.5(iv). It is equal to the exponential of the x_j independent terms in (1.76) with g in (1.75) replaced by $-g_4$. This gives

$$\lim_{N\to\infty} \frac{1}{N^2} \log Z_N\Big(\{g_j\}\Big|_{g_j=0\,(j\neq 4)}\Big) = -\Big(\frac{1}{24}(u-1)(9-u) - \frac{1}{2}\log u\Big), \tag{1.120}$$

where u is defined in terms of g_4 as the solution of

$$u - 3g_4 u^2 = 1, \qquad u \to 1 \text{ as } g_4 \to 0. \tag{1.121}$$

According to the result of Exercises 1.6 q.1(iii),

$$u = -\sum_{k=0}^{\infty} \frac{1}{k+1}\binom{2k}{k}(3g_4)^k. \tag{1.122}$$

After substituting this in the r.h.s. of (1.120), and substituting (1.118) in the l.h.s, the following result is obtained [98].

PROPOSITION 1.6.3 *We have*

$$\sum_{\substack{\text{connected } \Gamma \\ g(\Gamma)=0}} \frac{1}{|\mathrm{Aut}\,\Gamma|} g_4^{V_4(\Gamma)} = \sum_{k=1}^{\infty} \frac{(2k-1)!}{k!(k+2)!}(3g_4)^k. \tag{1.123}$$

Proof. The remaining task is to expand the functions of u on the r.h.s. of (1.120) as power series in g_4. For the quadratic, this is immediate from (1.122) and (1.121). For $\log u$ this follows from the result of Exercises 1.6 q.3. □

Let us denote the coefficient of g_4^k in (1.123) by a_k, which then represents the number of (weighted) planar fatgraphs that can be constructed out of k 4-gons. Making use of Stirling's formula (1.92) shows

$$a_k \sim \frac{12^k}{k^{7/2}\pi}.$$

The particular value of the exponent of the algebraic term $k^{-7/2}$ has meaning in the conformal field theory associated with the graphical expansion [239].

If we cut an edge in any of the planar maps giving rise to (1.123), we obtain a planar fatgraph constructed from 4-gons, but now with two external legs in the same face. The external legs, when distinguished by different labelings, break the symmetry of the maps, so for all classes Γ_2 of such maps $|\mathrm{Aut}\,\Gamma_2| = 1$. Because the legs have been distinguished, and because there are twice as many edges as vertices, one sees [241]

$$\tilde{G} = 1 + 4g_4 \frac{\partial}{\partial g_4} G,$$

where $\tilde{G}$ denotes the generating function for the maps with external legs, and G denotes the l.h.s. of (1.123). Substituting the r.h.s. of (1.123) we see that the power series of $\tilde{G}$ has positive integer coefficients, as it must. In particular, the coefficient of g_4^2 is 9. One contribution results from the first map in Figure 1.5, while four result from each of the other two maps therein.

EXERCISES 1.6 1. Here the number $c_k := a_{k+1}(k)$ of diagrams which can be constructed from a $2k$-gon according to the prescription of Figure 1.2, and which contain no intersecting lines, will be computed directly.

(i) Suppose the lines from edge 1 join the lines from edge $2j$ ($j = 1, \dots, k$). Argue that inside these lines there can be c_{j-1} configurations of the allowed type, while there are c_{k-j} configurations of the allowed type joining the edges $2j+1, \dots, 2k$. Hence deduce that

$$c_k = \sum_{j=0}^{k-1} c_j c_{k-1-j}, \qquad c_0 = 1. \tag{1.124}$$

(ii) Verify that the Catalan numbers (1.112) solve this recurrence.

(iii) Introduce the generating function $C(t) = \sum_{k=0}^{\infty} c_k t^k$. Use the recurrence (1.124) to show that $C(t)$ satisfies the quadratic equation

$$C(t) = 1 + t(C(t))^2, \tag{1.125}$$

and consequently has the explicit form

$$tC(t) = \frac{1}{2}(1 - (1-4t)^{1/2}). \tag{1.126}$$

(iv) With the value of $a_{k+1}(k)$ known independently of the average in (1.110) according to the result of (ii), use (1.98) to deduce that the scaled density (1.53) is such that

$$2^{2k} \int_{-\infty}^{\infty} x^{2k} \tilde{\rho}_{(1)}(x)\, dx = \frac{1}{k+1}\binom{2k}{k} \tag{1.127}$$

while the odd moments vanish.

(v) A sufficient condition for a density function to be determined by its moments $\{c_0, c_1, c_2, \dots\}$ is that

$$\sum_{k=0}^{\infty} \frac{c_k t^k}{k!} \tag{1.128}$$

converges for some $t > 0$ (this implies the Fourier transform of the density function is analytic in the neighbourhood of the origin). Verify that this is the case for the moments in (iv). Now use the fact that (1.54) reproduces these values to conclude from (iv) that the Wigner semicircle law is valid.

2. (i) Consider the Gaussian β-ensemble p.d.f. in (1.160) below, scaled so that $\lambda_l \mapsto \sqrt{2\beta N}/J$, $J > 0$. Use the result of Proposition 1.4.4 to show that to leading order the density is then supported on the interval $[-J, J]$ and is given by

$$\frac{2N}{\pi J}\sqrt{1 - (y/J)^2}. \tag{1.129}$$

(ii) Use (1.55) with suitable $V(x)$ and a, to show that for $\rho_b(y)$ given by (1.129) and $z \in (-J, J)$,

$$\int_{-J}^{J} \frac{\rho_b(y)}{z-y}\, dy = \frac{2Nz}{J^2}. \tag{1.130}$$

Also, deduce from (1.127) that

$$\int_{-J}^{J} y^{2k} \rho_b(y)\, dy = \frac{2N}{J}(J/2)^{2k+1} c_k, \tag{1.131}$$

where c_k denotes the kth Catalan number (1.112). From this and the result of q.1(iii) deduce that for $|z| > J$,

$$\int_{-J}^{J} \frac{\rho_b(y)}{z-y}\, dy = \frac{2Nz}{J^2}\Big(1 - (1 - J^2/z^2)^{1/2}\Big). \tag{1.132}$$

3. [541] Let $f(z)$ and $\phi(z)$ be analytic in a neighborhood Ω of $z = a$. According to the Lagrange inversion formula, for t small enough that $|t\phi(z)| < |z - a|$, $z \in \Omega$, the equation $\zeta = a + t\phi(\zeta)$ has one solution in Ω, and furthermore

$$f(\zeta) = f(a) + \sum_{n=1}^{\infty} \frac{t^n}{n!} \frac{d^{n-1}}{da^{n-1}} \Big(f'(a)(\phi(a))^n \Big).$$

Use this formula to show that for x defined as the solution of the equation $x = 1 + yx^p$ with the property $x \to 1$ as $y \to 0$, one has

$$\log x = \sum_{k=1}^{\infty} \frac{(kp-1)!}{k!(kp-k)!} y^k.$$

1.7 CONVERGENCE

Consider for definiteness GUE matrices. As stated the Wigner semicircle law tells us the leading large N form of $\langle \frac{1}{N} \sum_{j=1}^{N} \sqrt{2N} \delta(\sqrt{2N}y - \lambda_j) \rangle_{\rm GUE}$. As the normalized empirical density integrated over an interval $[a, b]$ is the proportion of eigenvalues in that interval, $\#[a, b]$ say, equivalently the Wigner semicircle law tells us the expected value of this quantity when averaged over GUE matrices. Indeed, this was how Figure 1.1 was produced, with the theoretical means in each bin of the bar graph substituted by their empirical averages.

What if instead one considers $\#[a, b]$ for a sequence of single $n \times n$ matrices, $n = 1, 2, \dots$, each chosen from the GUE and with eigenvalues scaled $\lambda_j \mapsto \lambda_j / \sqrt{2N}$. Does the resulting sequence of values for $\#[a, b]$ converge to that predicted by the Wigner semicircle law? And what is the meaning of convergence in this setting? Regarding the latter point, two possibilities are convergence in probability, and almost sure convergence. Convergence in probability says that for a given $\epsilon > 0$, and sequence of single $n \times n$ GUE matrices $(n = 1, 2, \dots)$, $\Pr(|\mu_n - \mu| > \epsilon) \to 0$, where μ_n is the empirical value of $\#[a, b]$ for each matrix, and μ is the limiting ensemble average (the value implied by the Wigner semicircle law). Almost sure convergence says that the measure of the sequence of matrices for which $\mu_n \to \mu$ is equal to 1. A well-known consequence of the Borel-Cantelli lemma in probability theory (see, e.g., [66]) is that almost sure convergence is equivalent to the statement that for a given $\epsilon > 0$, $\sum_{n=1}^{\infty} \Pr(|\mu_n - \mu| > \epsilon) < \infty$. Note that a necessary condition for this is that $\Pr(|\mu_n - \mu| > \epsilon) \to 0$, and thus almost sure convergence implies convergence in probability. To estimate $\Pr(|\mu_n - \mu| > \epsilon)$, the *Chebyshev inequality* [66]

$$\Pr(|\mu_n - \mu| > \epsilon) \le \frac{\langle (\mu_n - \mu)^2 \rangle_{\rm GUE}}{\epsilon^2}$$

can be employed. Hence for convergence in probability, it is sufficient that $\langle (\mu_n - \mu)^2 \rangle_{\rm GUE} \to 0$ as $n \to \infty$, while for almost sure convergence, it is sufficient that $\sum_{n=1}^{\infty} \langle (\mu_n - \mu)^2 \rangle_{\rm GUE} < \infty$.

In Section 1.6.1 and Exercises 1.6 q.1 the Wigner semicircle law has been studied through its moments. We have shown that $\langle N^{-k-1} {\rm Tr}(\mathbf{X}^{2k}) \rangle_{\rm GUE^*} \to m_{2k}$ where m_{2k} is the corresponding moment of the Wigner semicircle law. To study convergence in probability and almost sure convergence one thus must study

$${\rm Var}(N^{-k-1}{\rm Tr}\mathbf{X}^{2k}) := \langle (N^{-k-1}{\rm Tr}\mathbf{X}^{2k})^2 \rangle_{\rm GUE^*} - \Big(\langle N^{-k-1}{\rm Tr}\mathbf{X}^{2k} \rangle_{\rm GUE^*} \Big)^2.$$

Now, analogous to (1.102) we have

$$\langle ({\rm Tr}\mathbf{X}^{2k})^2 \rangle_{\rm GUE^*} = \Big\langle \sum_{\substack{i_1, \dots, i_{2k} = 1 \\ j_1, \dots, j_{2k} = 1}} z_{i_1 i_2} \cdots z_{i_{2k} i_1} z_{j_1 j_2} \cdots z_{j_{2k} j_1} \Big\rangle.$$

Regarding $i_1, \ldots, i_{2k}$ as fixed, and taking into consideration (1.103), one sees [274]

$$\begin{aligned}\langle (\mathrm{Tr}\mathbf{X}^{2k})^2 \rangle_{\mathrm{GUE}^*} &= \Big\langle \sum_{i_1,\ldots,i_{2k}=1} z_{i_1 i_2} \cdots z_{i_{2k} i_1} \Big\rangle_{\mathrm{GUE}^*} \Big\langle \sum_{j_1,\ldots,j_{2k}=1} z_{j_1 j_2} \cdots z_{j_{2k} j_1} \Big\rangle_{\mathrm{GUE}^*} \Big(1 + \mathrm{O}\Big(\frac{1}{N^2}\Big)\Big) \\ &= \langle \mathrm{Tr}\mathbf{X}^{2k} \rangle^2_{\mathrm{GUE}^*} \Big(1 + \mathrm{O}\Big(\frac{1}{N^2}\Big)\Big).\end{aligned}$$

It follows from this that $\sum_{N=1}^{\infty} \mathrm{Var}(N^{-k-1}\mathrm{Tr}\mathbf{X}^{2k}) < \infty$, so we can conclude that almost sure convergence holds and so the Wigner semicircle law is the limiting density of all sequences of GUE matrices, up to a set of measure zero.

1.8 THE SHIFTED MEAN GAUSSIAN ENSEMBLES

The Gaussian orthogonal, unitary and symplectic ensembles have joint p.d.f. for the elements proportional to $\exp(-(\beta/2)\mathrm{Tr}\,\mathbf{H}^2)$. We know that this is equivalent to the independent entries in $\mathbf{H}$ having Gaussian distribution with mean zero and particular variance. It follows that for a fixed Hermitian matrix $\mathbf{H}_0$, a p.d.f. proportional to $\exp(-(\beta/2)\mathrm{Tr}\,(\mathbf{H}-\mathbf{H}_0)^2)$ specifies a Gaussian ensemble in which the mean of each element is equal to the corresponding element in $\mathbf{H}_0$. The simplest case is when all elements of $\mathbf{H}_0$ are constant, equal to c say. Then

$$\mathbf{H} = \mathbf{A} + c\vec{x}\vec{x}^T, \tag{1.133}$$

where $\vec{x}$ is a column vector with all entries equal to 1, and $\mathbf{A}$ is a member of the corresponding zero mean Gaussian ensemble. Thus in this case the shifted mean Gaussian ensembles correspond to a rank 1 perturbation of the original ensembles.

Consider for definiteness the GOE. Diagonalizing $\mathbf{A}$, $\mathbf{A} = \mathbf{OLO}^T$, $\mathbf{L} = \mathrm{diag}\,(a_1, \ldots, a_N)$, and writing $\mathbf{O}^T\vec{x} =: \vec{y}$ shows that from the viewpoint of the eigenvalues, the r.h.s. of (1.133) can be replaced by $\mathbf{L}+c\vec{y}\vec{y}^T$. We seek the eigenvalues of this matrix.

PROPOSITION 1.8.1 *The eigenvalues of the matrix*

$$\tilde{\mathbf{H}} := \mathrm{diag}(a_1, \ldots, a_N) + c\vec{y}\vec{y}^T$$

are given by the solutions of the equation

$$0 = 1 - c\sum_{i=1}^{N} \frac{y_i^2}{\lambda - a_i}. \tag{1.134}$$

Assuming the ordering $a_1 > \cdots > a_N$, and that $c > 0$, a corollary is that the eigenvalues satisfy the interlacing

$$\lambda_1 > a_1 > \lambda_2 > a_2 > \cdots > \lambda_N > a_N. \tag{1.135}$$

Proof. With $\tilde{\mathbf{A}} = \mathrm{diag}(a_1, \ldots, a_N)$ we have

$$\det(\mathbf{1}_N\lambda - \tilde{\mathbf{H}}) = \det(\mathbf{1}_N - \tilde{\mathbf{A}})\det(\mathbf{1}_N\lambda - c\vec{y}\vec{y}^T(\mathbf{1}_N\lambda - \tilde{\mathbf{A}})^{-1}). \tag{1.136}$$

But the matrix product in the second determinant has rank 1, and so

$$\det(\mathbf{1}_N - c\vec{y}\vec{y}^T(\mathbf{1}_N\lambda - \tilde{\mathbf{A}})^{-1}) = 1 - c\mathrm{Tr}(\vec{y}\vec{y}^T(\mathbf{1}_N\lambda - \tilde{\mathbf{A}})^{-1}) = 1 - c\sum_{i=1}^{N} \frac{y_i^2}{\lambda - a_i}. \tag{1.137}$$

The characteristic polynomial (1.136) vanishes at the zeros of this determinant, but not at the zeros of $\det(\mathbf{1}_N\lambda - \tilde{\mathbf{A}})$ due to the poles in (1.137), implying the first result. The interlacing condition can be seen by sketching a graph, and noting in the process that $cy_i^2 > 0$. □

The GUE and GSE lead to the same equation (1.134) but with y_i^2 replaced by $|y_i|^2 := \sum_{s=1}^{\beta}(y_i^{(s)})^2$, where the $y_i^{(s)}$ are the independent real parts of the complex and real quaternion entries, respectively. We remark too that in the case of the GSE the eigenvalues of $\mathbf{A}$ are doubly degenerate and the rank 1 perturbation leaves one copy of the eigenvalues unchanged; see the discussion about (4.20) below.

Of particular interest is the position of the largest eigenvalue of $\tilde{\mathbf{H}}$, and thus $\mathbf{H}$, as a function of c and N, which unlike the other eigenvalues is not trapped by the eigenvalues of $\mathbf{A}$. Following [335] this can analyzed by making use of the results of Exercises 1.6 q.2.

PROPOSITION 1.8.2 *Consider the Gaussian ensembles scaled so that the distribution of the elements is proportional to* $\exp(-(\beta N/J^2)\mathrm{Tr}\,(\mathbf{H}-\mathbf{H}_0)^2)$, *with* $\mathbf{H}_0$ *the constant matrix having all elements equal to* c/N, $c > 0$. *Suppose* N *is large. Then for* $2c > J$ *a single eigenvalue splits off from the bulk of the eigenvalues, these being supported on* $(-J, J)$, *and is located at*

$$\lambda = c + \frac{J^2}{4c}.$$

Proof. We seek a solution $\lambda > J$ of

$$1 = \frac{c}{N}\Big\langle \sum_{j=1}^{N} \frac{|y_j|^2}{\lambda - \lambda_j} \Big\rangle, \tag{1.138}$$

where each $|y_j|^2$ has mean unity, and $\{\lambda_j\}$ are the eigenvalues of a member of the specified Gaussian ensemble but with $\mathbf{H}_0 = \mathbf{0}$. We know that the density of the $\{\lambda_j\}$ is then given by the semicircle law (1.129). Hence for N large

$$\Big\langle \sum_{j=1}^{N} \frac{|y_j|^2}{\lambda - \lambda_j} \Big\rangle \sim \int_{-J}^{J} \frac{\rho_b(y)}{\lambda - y}\, dy = \frac{2N\lambda}{J^2}\Big(1 - \Big(1 - \frac{J^2}{\lambda^2}\Big)^{1/2}\Big),$$

where the first relation follows from the fact that the eigenvalues and eigenvectors are independently distributed and $\langle |y_j|^2 \rangle = 1$, while the equality, which requires that $\lambda > J$, follows from (1.132). Substituting this in (1.138) and solving for λ gives the stated result. □

1.9 GAUSSIAN β-ENSEMBLE

The p.d.f. (1.28) is realized by the eigenvalues of the GOE, GUE and GSE for the values of β equal to 1, 2 and 4 respectively. In this section a family of random tridiagonal matrices, referred to as the Gaussian β-ensemble, with (1.28) as their eigenvalue p.d.f. for general $\beta > 0$, will be studied. They can be motivated by the reduction of GOE or GUE matrices to tridiagonal form.

1.9.1 Householder transformations

A familiar technique in numerical linear algebra is the similarity transformation of a real symmetric matrix to tridiagonal form using a sequence of reflection matrices, referred to as *Householder transformations*. Explicitly, let $\mathbf{A}$ be a real symmetric matrix $[a_{ij}]_{i,j=1,\dots,N}$. Then one can construct a sequence of symmetric real orthogonal matrices $\mathbf{U}^{(1)}, \mathbf{U}^{(2)}, \dots, \mathbf{U}^{(N-2)}$ such that the transformed matrix

$$\mathbf{U}^{(N-2)}\mathbf{U}^{(N-3)}\cdots\mathbf{U}^{(1)}\mathbf{A}\mathbf{U}^{(1)}\mathbf{U}^{(2)}\cdots\mathbf{U}^{(N-2)} =: \mathbf{B}^{(N-2)} \tag{1.139}$$

is a symmetric tridiagonal matrix. These matrices have the structure

$$\mathbf{U}^{(j)} = \mathbf{1}_N - 2\vec{u}^{(j)}\vec{u}^{(j)T} = \begin{bmatrix} \mathbf{1}_j & \mathbf{0}_{j\times N-j} \\ \mathbf{0}_{N-j\times j} & \mathbf{V}_{N-j\times N-j} \end{bmatrix}, \tag{1.140}$$

where $\vec{u}^{(j)T}\vec{u}^{(j)} = 1$ and $\mathbf{V}_{N-j\times N-j}$ is symmetric real orthogonal. Geometrically $\mathbf{U}^{(j)}$ corresponds to a reflection in the hyperplane orthogonal to $\vec{u}^{(j)}$.

Consider first the construction of $\mathbf{U}^{(1)}$. Choosing the components $u_l^{(1)}$ of $\vec{u}^{(1)}$ as

$$u_1^{(1)} = 0, \qquad u_2^{(1)} = \Big[\frac{1}{2}\Big(1 - \frac{a_{12}}{\alpha}\Big)\Big]^{1/2}, \qquad u_l^{(1)} = -\frac{a_{1l}}{2\alpha u_2^{(1)}} \ (l \ge 3), \tag{1.141}$$

where $\alpha = (a_{12}^2 + \cdots + a_{1N}^2)^{1/2}$, we then have $\vec{u}^{(1)T}[a_{l1}]_{l=1,\dots,N} = (a_{12} - \alpha)/2u_2^{(1)}$. This in turn implies that

$$\mathbf{B}^{(1)} := \mathbf{U}^{(1)}\mathbf{A}\mathbf{U}^{(1)} \tag{1.142}$$

has

$$b_{11} = a_{11}, \qquad b_{12} = b_{21} = \alpha, \qquad b_{1k} = b_{k1} = 0 \ \ (k \ge 3)$$

and is thus tridiagonal with respect to the first row and column. The matrices $\mathbf{U}^{(j)}$, $j = 2, 3, \dots$ in order are now defined by the formulas (1.141), but with $u_1^{(j)} = u_2^{(j)} = \cdots = u_j^{(j)} = 0$, and the analogue of the entries a_{1l} replaced by the elements in the first row of the bottom right $(N - j + 1) \times (N - j + 1)$ submatrix of $\mathbf{B}^{(j-1)}$.

A number of works (see [157] and references therein) posed the question as to the form of $\mathbf{B}^{(N-2)}$ when $\mathbf{A}$ is a member of the GOE. It was found that like $\mathbf{A}$ itself, the elements of $\mathbf{B}^{(N-2)}$ are all independent (apart from the requirement that $\mathbf{B}^{(N-2)}$ be symmetric) with a distribution that can be calculated explicitly.

PROPOSITION 1.9.1 *Let* $\mathrm{N}[0,1]$ *refer to the standard normal distribution as defined below Definition 1.1.1, and let* $\tilde{\chi}_k$ *denote the square root of the gamma distribution* $\Gamma[k/2, 1]$*, the latter being specified by the p.d.f.* $(1/\Gamma(k/2))u^{k/2-1}e^{-u}$*,* $u > 0$*, and realized by the sum of the squares of* k *independent Gaussian distributions* $\mathrm{N}[0, 1/\sqrt{2}]$*. (The p.d.f. of* $\tilde{\chi}_k$ *is thus equal to* $(2/\Gamma(k/2))u^{k-1}e^{-u^2}$*,* $u > 0$*.) For* $\mathbf{A}$ *a member of the GOE, the tridiagonal matrix* $\mathbf{B}^{(N-2)}$ *obtained by successive Householder transformations is given by*

$$\begin{bmatrix} \mathrm{N}[0,1] & \tilde{\chi}_{N-1} & & & & \\ \tilde{\chi}_{N-1} & \mathrm{N}[0,1] & \tilde{\chi}_{N-2} & & & \\ & \tilde{\chi}_{N-2} & \mathrm{N}[0,1] & \tilde{\chi}_{N-3} & & \\ & \ddots & \ddots & \ddots & & \\ & & \tilde{\chi}_2 & \mathrm{N}[0,1] & \tilde{\chi}_1 \\ & & & \tilde{\chi}_1 & \mathrm{N}[0,1] \end{bmatrix}.$$

Proof. Let GOE_n denote the ensemble of $n \times n$ GOE matrices. From the Householder algorithm, the first row and column of $\mathbf{B}^{(N-2)}$ are he same as those of $\mathbf{B}^{(1)}$ in (1.142), and thus from (1.141) we have

$$b_{11}^{(N-2)} = \mathrm{N}[0,1], \qquad b_{12}^{(N-2)} = \tilde{\chi}_{N-1},$$

where use has been made of the assumption that $\mathbf{A}$ is a member of GOE_N, and the definition of $\tilde{\chi}_{N-1}^2$ as a sum of squares of Gaussians. To proceed further we must compute the distribution of the bottom $N-1 \times N-1$ block of $\mathbf{B}^{(1)}$. In general, denoting such a block of the matrix $\mathbf{X}$ by $\mathbf{X}_{N-1}$, it follows from (1.140) that $\mathbf{B}_{N-1}^{(1)} = \mathbf{V}_{N-1}\mathbf{A}_{N-1}\mathbf{V}_{N-1}$. Since the elements of the real orthogonal matrix $\mathbf{V}_{N-1}$ are independent of the elements of $\mathbf{A}_{N-1}$, which itself is a member of GOE_{N-1}, it follows immediately from the general invariance of the GOE under orthogonal transformations that $\mathbf{B}_{N-1}^{(1)}$ is also a member of GOE_{N-1}. Applying the Householder transformation to $\mathbf{B}_{N-1}^{(1)}$, we thus get

$$b_{22}^{(N-2)} = N[0,1], \qquad b_{23}^{(N-2)} = \tilde{\chi}_{N-2}.$$

Continuing inductively gives the stated result. □

1.9.2 Tridiagonal matrices

The result of Proposition 1.9.1 suggests investigating the Jacobian for the change of variables from a general real symmetric tridiagonal matrix

$$\mathbf{T} = \begin{bmatrix} a_n & b_{n-1} & & & & \\ b_{n-1} & a_{n-1} & b_{n-2} & & & \\ & b_{n-2} & a_{n-2} & b_{n-3} & & \\ & \ddots & \ddots & \ddots & & \\ & & b_2 & a_2 & b_1 \\ & & & b_1 & a_1 \end{bmatrix}, \tag{1.143}$$

to its eigenvalues and variables relating to its eigenvectors. First, for each eigenvalue λ_k and corresponding eigenvector $\vec{v}_k$, it is easy to see by direct substitution that once the first component $v_k^{(1)} =: q_k$ of $\vec{v}_k$ is specified, all other components can be expressed in terms of λ_k and the elements of $\mathbf{T}$. To make the eigendecomposition unique we specify that $q_k > 0$, and furthermore note that $\mathbf{T}$, being symmetric, can be orthogonally diagonalized, and so doing this we have

$$\sum_{k=1}^{n} q_k^2 = 1. \tag{1.144}$$

The Jacobian for the change of variables from

$$\vec{a} := (a_n, a_{n-1}, \dots, a_1), \qquad \vec{b} := (b_{n-1}, \dots, b_1), \tag{1.145}$$

to

$$\vec{\lambda} := (\lambda_1, \dots, \lambda_n), \qquad \vec{q} := (q_1, \dots, q_{n-1}) \tag{1.146}$$

can be calculated using the method of wedge products. However, one must first establish some auxiliary results.

PROPOSITION 1.9.2 *Let $(\mathbf{X})_{11}$ denote the top-left hand entry of the matrix $\mathbf{X}$. We have*

$$((\mathbf{T} - \lambda \mathbf{1})^{-1})_{11} = \sum_{j=1}^{n} \frac{q_j^2}{\lambda_j - \lambda}. \tag{1.147}$$

Also

$$\prod_{1 \le i < j \le n} (\lambda_i - \lambda_j)^2 = \frac{\prod_{i=1}^{n-1} b_i^{2i}}{\prod_{i=1}^{n} q_i^2}. \tag{1.148}$$

Proof. Now

$$((\mathbf{T} - \lambda \mathbf{1})^{-1})_{11} = \vec{e}_1 \cdot (\mathbf{T} - \lambda \mathbf{1})^{-1} \vec{e}_1,$$

where $\vec{e}_1 := (1, 0, \dots, 0)^T$. Since $\{\vec{v}_j\}$ is an orthonormal set,

$$\vec{e}_1 = \sum_{j=1}^{n} (\vec{e}_1 \cdot \vec{v}_j) \vec{v}_j = \sum_{j=1}^{n} q_j \vec{v}_j, \tag{1.149}$$

and substituting into the above equation gives (1.147). This derivation makes no explicit use of $\mathbf{T}$ being tridiagonal, rather just that (1.149) holds, for which it is sufficient $\mathbf{T}$ be real symmetric.

To derive (1.148) [140] we begin by recalling that in general for $\mathbf{X}$ an $n \times n$ nonsingular matrix,

$$(\mathbf{X}^{-1})_{11} = \frac{\det \mathbf{X}_{n-1}}{\det \mathbf{X}}, \tag{1.150}$$

where $\mathbf{X}_{n-1}$ denotes the bottom right $n-1 \times n-1$ submatrix of $\mathbf{X}$. Hence we can rewrite (1.147) to read

$$\frac{\prod_{i=1}^{n-1}(\lambda - \lambda_i^{(n-1)})}{\prod_{i=1}^{n}(\lambda - \lambda_i)} = \sum_{j=1}^{n} \frac{q_j^2}{\lambda - \lambda_j}, \tag{1.151}$$

where $\{\lambda_i^{(n-1)}\}$ denotes the eigenvalues of $\mathbf{X}_{n-1}$. It follows from this that

$$q_j^2 = \frac{P_{n-1}(\lambda_j)}{P_n'(\lambda_j)}, \qquad P_k(\lambda) := \prod_{i=1}^{k}(\lambda - \lambda_i^{(k)}), \tag{1.152}$$

where $P_k(\lambda)$ is the characteristic polynomial of the bottom right $k \times k$ submatrix of $\mathbf{T}$, say $\mathbf{T}_k$, and $\{\lambda_i^{(k)}\}$ the corresponding eigenvalues. Hence

$$\prod_{i=1}^{n} q_i^2 = \frac{\prod_{i=1}^{n} |P_{n-1}(\lambda_i)|}{\prod_{1 \le i < j \le n}(\lambda_i - \lambda_j)^2}. \tag{1.153}$$

Next, by expanding along the first row of $\lambda \mathbf{1}_k - \mathbf{T}_k$, one obtains the three-term recurrence

$$P_k(\lambda) = (\lambda - a_k)P_{k-1}(\lambda) - b_{k-1}^2 P_{k-2}(\lambda) \tag{1.154}$$

and it follows from this that

$$\prod_{i=1}^{k-1} |P_k(\lambda_i^{(k-1)})| = b_{k-1}^{2(k-1)} \prod_{i=1}^{k-1} |P_{k-2}(\lambda_i^{(k-1)})|.$$

Since

$$\prod_{i=1}^{k-1} |P_{k-2}(\lambda_i^{(k-1)})| = \prod_{i=1}^{k-1}\prod_{j=1}^{k-2} |\lambda_i^{(k-1)} - \lambda_j^{(k-2)}| = \prod_{j=1}^{k-2} |P_{k-1}(\lambda_j^{(k-2)})|, \tag{1.155}$$

this can be rewritten as

$$\prod_{i=1}^{k-1} |P_k(\lambda_i^{(k-1)})| = b_{k-1}^{2(k-1)} \prod_{j=1}^{k-2} |P_{k-1}(\lambda_j^{(k-2)})|,$$

and iteration shows

$$\prod_{i=1}^{n-1} |P_n(\lambda_i^{(n-1)})| = \prod_{i=1}^{n-1} b_i^{2i}.$$

Use of (1.155) with $k = n+1$ and substitution into (1.153) gives (1.148). □

PROPOSITION 1.9.3 *The Jacobian for the change of variables (1.145) to (1.146) can be written as*

$$\frac{1}{q_n} \frac{\prod_{i=1}^{n-1} b_i}{\prod_{i=1}^{n} q_i}. \tag{1.156}$$

Proof. [223] Rewriting (1.147) in the form

$$((\mathbf{1} - \lambda \mathbf{T})^{-1})_{11} = \sum_{j=1}^{n} \frac{q_j^2}{1 - \lambda \lambda_j} \tag{1.157}$$

and equating successive powers of λ on both sides gives

$$1=\sum_{j=1}^n q_j^2, \quad a_n=\sum_{j=1}^n q_j^2\lambda_j, \quad *+b_{n-1}^2=\sum_{j=1}^n q_j^2\lambda_j^2,$$
$$*+a_{n-1}b_{n-1}^2=\sum_{j=1}^n q_j^2\lambda_j^3, \quad *+b_{n-2}^2b_{n-1}^2=\sum_{j=1}^n q_j^2\lambda_j^4,$$
$$*+a_{n-2}b_{n-2}^2b_{n-1}^2=\sum_{j=1}^n q_j^2\lambda_j^5,\dots, \quad *+a_1b_1^2\cdots b_{n-2}^2b_{n-1}^2=\sum_{j=1}^n q_j^2\lambda_j^{2n-1},$$

where the $*$ denotes terms involving only variables already having appeared on the l.h.s. of preceding equations (thus the variables $a_n, b_{n-1}, a_{n-1}, b_{n-2},\dots$ occur in a triangular structure). The first of these equations implies

$$q_n dq_n = -\sum_{j=1}^{n-1} q_j dq_j. \tag{1.158}$$

Taking differentials of the remaining equations, substituting for $q_n dq_n$, and then taking wedge products of both sides (making use of the triangular structure on the l.h.s.) shows

$$\prod_{j=1}^{n-1} b_j^{4j-1} d\vec{a}\wedge d\vec{b} = q_n^2\prod_{j=1}^{n-1} q_j^3 \det\Big[[\lambda_k^j-\lambda_n^j]_{\substack{j=1,\dots,2n-1\\ k=1,\dots,n-1}}\,[j\lambda_k^{j-1}]_{\substack{j=1,\dots,2n-1\\ k=1,\dots,n}}\Big]d\vec{\lambda}\wedge d\vec{q},$$

where

$$d\vec{a}:=\bigwedge_{j=1}^n da_j,\quad d\vec{b}:=\bigwedge_{j=1}^{n-1} db_j,\quad d\vec{\lambda}:=\bigwedge_{j=1}^n d\lambda_j,\quad d\vec{q}:=\bigwedge_{j=1}^{n-1} dq_j.$$

By definition the Jacobian J is positive and such that $d\vec{a}\wedge d\vec{b} = \pm J d\vec{\lambda}\wedge d\vec{q}$ for some sign $\pm$. Making use of the determinant evaluation (1.175) below we thus read off that

$$J=\frac{1}{q_n}\frac{\prod_{j=1}^{n-1} b_j}{\prod_{j=1}^n q_j}\Big(\frac{\prod_{j=1}^n q_j^2}{\prod_{j=1}^{n-1} b_j^{2j}}\Big)^2\prod_{1\le j<k\le n}(\lambda_k-\lambda_j)^4.$$

Recalling (1.148) shows J is equal to (1.156). □

Using Proposition 1.9.3, the fact that the tridiagonal matrix of Proposition 1.9.1 has the same eigenvalue p.d.f. as GOE matrices can be reclaimed. Moreover, one can prescribe a tridiagonal matrix with eigenvalue p.d.f. (1.28) for general $\beta>0$ [140].

Proposition 1.9.4 *Let $\beta>0$ be fixed. In the notation of Proposition 1.9.1 define the Gaussian β-ensemble as the set of symmetric tridiagonal matrices*

$$\mathbf{T}_\beta := \begin{bmatrix} \mathrm{N}[0,1] & \tilde{\chi}_{(N-1)\beta} & & & & \\ \tilde{\chi}_{(N-1)\beta} & \mathrm{N}[0,1] & \tilde{\chi}_{(N-2)\beta} & & & \\ & \tilde{\chi}_{(N-2)\beta} & \mathrm{N}[0,1] & \tilde{\chi}_{(N-3)\beta} & & \\ & \ddots & \ddots & \ddots & & \\ & & \tilde{\chi}_{2\beta} & \mathrm{N}[0,1] & \tilde{\chi}_{\beta} \\ & & & \tilde{\chi}_{\beta} & \mathrm{N}[0,1] \end{bmatrix}. \tag{1.159}$$

The eigenvalues and first component of the eigenvectors (which form the vector $\vec{q}$) are independent, with the *distribution of the former given by*

$$\frac{1}{\tilde{G}_{\beta,N}} \prod_{l=1}^{N} e^{-\lambda_l^2/2} \prod_{1 \le j < k \le N} |\lambda_k - \lambda_j|^\beta \, d\vec{\lambda}, \qquad \tilde{G}_{\beta,N} = (2\pi)^{N/2} \prod_{j=0}^{N-1} \frac{\Gamma(1 + (j+1)\beta/2)}{\Gamma(1+\beta/2)}, \tag{1.160}$$

and the distribution of the latter given by

$$\frac{1}{c_{\beta,N} q_N} \prod_{i=1}^{N} q_i^{\beta-1} \, d\vec{q}, \qquad q_i > 0, \ \sum_{i=1}^{N} q_i^2 = 1, \quad \text{where} \quad c_{\beta,N} = \frac{\Gamma^N(\beta/2)}{2^{N-1}\Gamma(\beta N/2)}. \tag{1.161}$$

Proof. Denote the joint distribution of $\mathbf{T}_\beta$ by $P(\mathbf{T}_\beta)$. We have

$$\begin{aligned} P(\mathbf{T}_\beta)(d\mathbf{T}_\beta) &= \frac{2^{N-1}}{(2\pi)^{N/2}} \prod_{l=1}^{N-1} \frac{b_l^{\beta l - 1} e^{-b_l^2}}{\Gamma(\beta l/2)} \prod_{l=1}^{N} e^{-a_l^2/2} d\vec{a} \wedge d\vec{b} \\ &= \frac{2^{N-1}}{(2\pi)^{N/2}} \prod_{l=1}^{N-1} \frac{1}{\Gamma(\beta l/2)} \frac{1}{q_N} \frac{\prod_{l=1}^{N-1} b_l^{\beta l}}{\prod_{l=1}^{N} q_l} e^{-\mathrm{Tr}(\mathbf{T}_\beta^2)/2} d\vec{\lambda} \wedge d\vec{q}, \end{aligned}$$

where the second equality follows using (1.156). But

$$e^{-\mathrm{Tr}(\mathbf{T}_\beta^2)/2} = e^{-\sum_{j=1}^{N} \lambda_j^2/2}, \qquad \prod_{l=1}^{N-1} b_l^{\beta l} = \prod_{l=1}^{N} q_l^\beta \prod_{1 \le i < j \le N} |\lambda_j - \lambda_i|^\beta,$$

where the latter formula follows from (1.148), so indeed the dependence on $\vec{\lambda}$ and $\vec{q}$ factorizes into the functional forms specified in (1.160) and (1.161). The normalization for (1.161) follows from the Dirichlet integral [541]

$$\int_{\sum_{i=1}^{n+1} \rho_i = 1, \, \rho_i > 0} d\rho_1 \cdots d\rho_n \prod_{i=1}^{n+1} \rho_i^{s_i - 1} = \frac{\Gamma(s_1) \cdots \Gamma(s_{n+1})}{\Gamma(s_1 + \cdots + s_{n+1})} \tag{1.162}$$

with $n = N-1$, $s_i = \beta/2$ and the change of variables $\rho_i = q_i^2$. With this normalization specified, the value of $\tilde{G}_{\beta,N}$ follows (an extra factor of $N!$ is included to effectively remove the ordering on $\{\lambda_i\}$ implicit in the above working; recall the remark below Proposition 1.3.4). □

We remark that the evaluation of $\tilde{G}_{\beta,N}$ given in (1.160) implies, after a simple change of variables, that the normalization constant in (1.28) has the evaluation

$$G_{\beta,N} = \beta^{-N/2 - N\beta(N-1)/4} (2\pi)^{N/2} \prod_{j=0}^{N-1} \frac{\Gamma(1 + (j+1)\beta/2)}{\Gamma(1+\beta/2)}. \tag{1.163}$$

Another point of interest is that the recurrence (1.154) with

$$a_k \in \mathrm{N}[0,1], \qquad b_k^2 \in \Gamma[k\beta/2, 1] \tag{1.164}$$

can be used to generate the characteristic polynomial for a member of the Gaussian β-ensemble, so the p.d.f. (1.160) can be sampled by simply computing the zeros of this polynomial.

1.9.3 Sturm sequences

For tridiagonal matrices, the task of computing the cumulative microscopic eigenvalue density $N(\mu)$, that is, the number of eigenvalues less than μ, has a number of special features. This in turn follows from special features of the corresponding *Sturm sequences* [13].

DEFINITION 1.9.5 *Let $\mathbf{A}_n$ be a general $n \times n$ matrix, and let $\mathbf{A}_{n-k}$ $(k = 1, \ldots, n-1)$ denote the matrix* *obtained by deleting the first k rows and columns. Let $d_i := \det \mathbf{A}_i$ $(i = 1, \dots, n)$ and set $d_0 := 1$. The Sturm sequence refers to $(d_0, d_1, \dots, d_n)$.*

PROPOSITION 1.9.6 *Let $\mathbf{A}_n$ be a real symmetric matrix with no repeated eigenvalues and no zero eigenvalues, and similarly $\mathbf{A}_{n-k}$. The number of sign changes in the Sturm sequence (reading from right-to-left, say) is equal to the number of negative eigenvalues of $\mathbf{A}_n$.*

Proof. For a given $k = 2, \dots, n$ it is a fundamental result (see Exercises 4.2 q.2(iii) below) that the eigenvalues $\{a_i\}$ of $\mathbf{A}_k$ interlace the eigenvalues $\{\alpha_i\}$ of $\mathbf{A}_{k-1}$,

$$a_k < \alpha_{k-1} < a_{k-1} < \cdots < \alpha_1 < a_1.$$

We know too that the determinant is equal to the product of eigenvalues. Consequently, the number of negative eigenvalues of $\mathbf{A}_k$ equals the number of negative eigenvalues of $\mathbf{A}_{k-1}$, if d_k/d_{k-1} is positive, while we must add one if d_k/d_{k-1} is negative. Iteratively applying this for $k = n, \dots, 1$ gives the stated result. □

Applying Proposition 1.9.6 to the matrix $\mathbf{A}_n - \mu \mathbf{1}_n$ gives that $N(\mu)$ is equal to the number of sign changes in the Sturm sequence for $\mathbf{A}_n - \mu \mathbf{1}_n$. In the case that $\mathbf{A}_n$ is the tridiagonal matrix (1.143), one has that $d_k = (-1)^k P_k(\mu)$ as specified by (1.154), and using the recurrence (1.154) shows that $r_i := d_i/d_{i-1}$ can be specified by the recursive formula

$$r_i = \begin{cases} a_1 - \mu, & i = 1, \\ (a_i - \mu) - b_{i-1}^2/r_{i-1}, & i = 2, \dots, n. \end{cases} \tag{1.165}$$

As each sign change in the Sturm sequence $\{d_i\}$ corresponds to a negative in the ratio sequence $\{r_i\}$, we see that the number of negative values in $\{r_i\}$ equals $N(\mu)$. This latter result can be related to so-called *shooting eigenvectors*.

DEFINITION 1.9.7 *The vector $\vec{x}$ satisfying all but the first of the n linear equations implied by the matrix equation $(\mathbf{A}_n - \mu \mathbf{1}_n)\vec{x} = \vec{0}$, with $\vec{x} = (x_n, \dots, x_1)^T$ and x_1 given, is referred to as a shooting eigenvector. (Note that the first equation can only be satisfied as well if and only if μ is an eigenvalue.)*

For the tridiagonal matrix (1.143), and with x_{n+1} defined as the first component of $(\mathbf{A}_n - \mu \mathbf{1}_n)\vec{x}$, a recurrence for the ratio $s_i = x_i/x_{i-1}$, $i = 2, \dots, n+1$ is readily obtained, and comparison with (1.165) shows $s_i = -r_{i-1}/b_{i-1}$ (in the case $i = n+1$ this requires setting $b_n := 1$). Thus with each $b_i > 0$, the number of positive values in $\{s_i\}$ equals $N(\mu)$. This can equivalently be stated in terms of $\{x_i\}$.

PROPOSITION 1.9.8 *The number of sign changes in the shooting eigenvector $\vec{x}$ equals $n - N(\mu)$, which is the number of eigenvalues of $\mathbf{A}$ greater than μ.*

1.9.4 Prüfer phases

There is a parametrization, in terms of *Prüfer phases* and amplitudes, of the shooting vectors well suited to analysis of the large n limit of the bulk eigenvalues (see Section 13.6). To introduce the parametrization, first observe that the three-term recurrence satisfied by the shooting vector

$$b_j x_{j+1} + a_j x_j + b_{j-1} x_{j-1} = \mu x_j \qquad (j = 1, \dots, n;\ b_0 := 0, b_n := -1) \tag{1.166}$$

is equivalent to the matrix equation

$$\begin{bmatrix} (\mu - a_j)/b_j & -1/b_j \\ b_j & 0 \end{bmatrix} \begin{bmatrix} u_j \\ v_j \end{bmatrix} = \begin{bmatrix} u_{j+1} \\ v_{j+1} \end{bmatrix} \qquad (j = 1, \dots, n), \tag{1.167}$$

where

$$\begin{bmatrix} u_j \\ v_j \end{bmatrix} = \begin{bmatrix} 1 & 0 \\ 0 & b_{j-1} \end{bmatrix} \begin{bmatrix} x_j \\ x_{j-1} \end{bmatrix} \tag{1.168}$$

(note that the matrix in (1.167) has unit determinant and so as a transformation is volume preserving). Choosing the initial condition $u_1 = 1$, $v_1 = 0$ we see that

$$\begin{bmatrix} u_j \\ v_j \end{bmatrix} = T_j \begin{bmatrix} 1 \\ 0 \end{bmatrix},$$

where $T_j := V_{j-1} \cdots V_1$ is referred to as a transfer matrix

DEFINITION 1.9.9 *The Prüfer phases θ_j^μ and amplitudes $R_j^\mu > 0$ are such that*

$$\begin{bmatrix} u_j \\ v_j \end{bmatrix} = \begin{bmatrix} R_j^\mu \cos \theta_j^\mu \\ R_j^\mu \sin \theta_j^\mu \end{bmatrix}, \tag{1.169}$$

where $-\pi/2 < \theta_{j+1}^\mu - \theta_j^\mu < 3\pi/2$.

Note that it follows from (1.167) and (1.168) that $\{\theta_j^\mu\}$ satisfies the first order recurrence

$$b_j^2 \cot \theta_{j+1}^\mu = -\tan \theta_j^\mu + (\mu - a_j), \qquad \theta_1^\mu = 0. \tag{1.170}$$

A consequence is an identity which tells us that θ_j^μ is a decreasing function of μ (see also Exercises 1.9 q.5).

PROPOSITION 1.9.10 *We have*

$$(R_j^\mu)^2 \frac{\partial}{\partial \mu} \theta_j^\mu = -\sum_{l=1}^{j-1} u_l^2. \tag{1.171}$$

Proof. Differentiating (1.170) with respect to μ and making use of (1.168) and (1.169) gives the recurrence

$$(R_{j+1}^\mu)^2 \frac{\partial \theta_{j+1}^\mu}{\partial \mu} = (R_j^\mu)^2 \frac{\partial \theta_j^\mu}{\partial \mu} - u_j^2.$$

This together with the initial condition $\partial \theta_1^\mu / \partial \mu = 0$ implies (1.171). □

We are now in a position to relate θ_n^μ to $N(\mu)$ for the tridiagonal matrix (1.143) [326] . First note from the recurrence (1.166) that for $\mu \to \infty$, x_j is positive while $x_{j-1}/x_j \to 0$. Recalling (1.169), this implies $\lim_{\mu \to \infty} \theta_j^\mu = 0$. But it has just been shown that θ_j^μ is a decreasing function of μ. The facts that $x_{n+1} = u_{n+1} = R_{n+1}^\mu \cos \theta_{n+1}^\mu$ and that $x_{n+1} = 0$ if and only if μ is an eigenvalue then imply the kth largest eigenvalue λ_k of $\mathbf{T}$ is such that $\theta_{n+1}^{\lambda_k} = (\pi/2) + \pi(k-1)$, and moreover that θ_{n+1}^μ relates to the number of eigenvalues of $\mathbf{T}$ greater that μ, $n - N(\mu)$, according to

$$\Big| \frac{1}{\pi} \theta_{n+1}^\mu - (n - N(\mu)) \Big| \le \frac{1}{2}. \tag{1.172}$$

EXERCISES 1.9 1. The objective of this exercise is to derive the Vandermonde determinant evaluation

$$\det[x_j^{k-1}]_{j,k=1,\dots,N} := \begin{vmatrix} 1 & x_1 & x_1^2 & \cdots & x_1^{N-1} \\ 1 & x_2 & x_2^2 & \cdots & x_2^{N-1} \\ \vdots & \vdots & \vdots & \ddots & \vdots \\ 1 & x_N & x_N^2 & \cdots & x_N^{N-1} \end{vmatrix} = \prod_{1 \le j < k \le N} (x_k - x_j). \tag{1.173}$$

(i) Verify that both the determinant and product of differences are antisymmetric polynomials which are homogeneous of degree $\frac{1}{2}N(N-1)$ and hence must be proportional.

(ii) Show that the proportionality constant is unity by comparing the coefficients of the term $x_1^0 x_2^1 \dots x_N^{N-1}$ on both sides.

2. (i) In the Vandermonde determinant identity (1.173), replace N by pN. Subtract row one from row two, divide this row by $x_2 - x_1$ and take the limit $x_2 \to x_1$ by first differentiating the top and bottom lines with respect

to x_2. Next subtract the first and second row from the third, divide this row by $(x_3 - x_1)^2$ and take the limit $x_3 \longrightarrow x_1$ by differentiating top and bottom lines with respect to x_3 twice. Proceed in this fashion by subtracting rows $1, 2, \dots, j-1$ from row j ($j = 4, \dots, p$), dividing by $(x_j - x_1)^{j-1}$, and taking the limit $x_j \longrightarrow x_1$ by differentiating top and bottom lines with respect to x_j $j-1$ times. Repeat this procedure for successive blocks of p variables to deduce the confluent Vandermonde determinant identity

$$\det \begin{bmatrix} x_j^{k-1} \\ \binom{k-1}{1} x_j^{k-2} \\ \vdots \\ \binom{k-p+1}{p-1} x_j^{k-p} \end{bmatrix}_{\substack{j=1,\dots,N \\ k=1,\dots,pN}} = \prod_{1 \le j < k \le N} (x_k - x_j)^{p^2}. \tag{1.174}$$

(ii) Consider the identity (1.174) in the case $p = 2$. Take the transpose of the determinant, and rearrange columns so that it reads

$$(-1)^{N(N-1)/2} \det \left[[\lambda_k^{j-1}]_{\substack{j=1,\dots,2N \\ k=1,\dots,N}} [j\lambda_k^{j-1}]_{\substack{j=1,\dots,2N \\ k=1,\dots,N}} \right] = \prod_{1 \le j < k \le N} (\lambda_k - \lambda_j)^4.$$

Subtract column N from columns $1, \dots, k-1$, then expand by the first row to deduce that

$$(-1)^{(N-1)(N-2)/2} \det \left[[\lambda_k^j - \lambda_N^j]_{\substack{j=1,\dots,2N-1 \\ k=1,\dots,N-1}} [j\lambda_k^{j-1}]_{\substack{j=1,\dots,2N-1 \\ k=1,\dots,N}} \right] = \prod_{1 \le j < k \le N} (\lambda_k - \lambda_j)^4. \tag{1.175}$$

3. [546], [249] In this exercise the Householder transformation will be used to establish an identity of relevance to the Schur decomposition (15.3) below, and also to establish the diagonalization formula (1.8).

(i) Let $\vec{a}$ and $\vec{b}$ be unit vectors, and form the unit vector $\vec{v} = (\vec{a} + \vec{b})/|\vec{a} + \vec{b}|$. From the fact that $\vec{v}$ bisects the angle of $\vec{a}$ and $\vec{b}$, deduce from the geometrical interpretation of the Householder transformation

$$\mathbf{U}_N = \mathbf{1} - 2\vec{v}\vec{v}^T$$

as a reflection in the hyperplane orthogonal to $\vec{v}$ that $\mathbf{U}_N \vec{a} = -\vec{b}$, $\mathbf{U}_N \vec{b} = -\vec{a}$. Also derive these equations algebraically.

(ii) Let $\mathbf{A}_N$ be an $N \times N$ matrix and let λ be an eigenvalue of $\mathbf{A}_N$ with corresponding normalized eigenvector $\vec{w}$. Let $\vec{e}_1 := (1, 0, \dots, 0)^T$ be an $N \times 1$ elementary vector. In (i) set $\vec{a} = \vec{e}_1$, $\vec{b} = \vec{w}$, and use the formulas therein to deduce that

$$\mathbf{U}_N \mathbf{A}_N \mathbf{U}_N \vec{e}_1 = \lambda \vec{e}_1.$$

Hence conclude

$$\mathbf{U}_N \mathbf{A}_N \mathbf{U}_N = \begin{bmatrix} \lambda & \vec{\alpha}_{N-1}^T \\ \vec{0}_{N-1} & \mathbf{A}_{N-1} \end{bmatrix} \tag{1.176}$$

for some $1 \times (N-1)$ vector $\vec{\alpha}_{N-1}^T$ and $(N-1) \times (N-1)$ matrix $\mathbf{A}_{N-1}$.

(iii) Let $\mathbf{P}_N$ be a real orthogonal diagonal matrix (each diagonal entry ± 1). With $\mathbf{V}_N = \mathbf{P}_N \mathbf{U}_N$ note from (1.176) that

$$\mathbf{V}_N \mathbf{A}_N \mathbf{V}_N^T = \begin{bmatrix} \lambda & \vec{\beta}_{N-1}^T \\ \vec{0}_{N-1} & \tilde{\mathbf{A}}_{N-1} \end{bmatrix} \tag{1.177}$$

for some $\vec{\beta}_{N-1}^T$, $\tilde{\mathbf{A}}_{N-1}$. Now use a Householder transformation of the form

$$\begin{bmatrix} 1 & \vec{0}_{N-1}^T \\ \vec{0}_{N-1} & \mathbf{U}_{N-1} \end{bmatrix}$$

to reduce $\tilde{\mathbf{A}}_{N-1}$ to triangular form and proceed inductively to deduce that there exists a real orthogonal matrix $\mathbf{R}$ such that

$$\mathbf{R}\mathbf{A}_N \mathbf{R}^T = \mathbf{T}, \tag{1.178}$$

where $\mathbf{T}$ is upper triangular with diagonal entries equal to the eigenvalues of $\mathbf{A}_{N-1}$, and note that $\mathbf{R}$ is unique up to an overall sign of each column.

(iv) Show that (1.178) implies the diagonalization formula (1.8).

4. [156] In this exercise the change of variables implied by (1.176) will be used to derive a generalization of (1.11).

(i) From the decomposition (1.176) in the case $\mathbf{A}_N$ is symmetric so that $\vec{\alpha}_{N-1} = \vec{0}_{N-1}$, deduce that

$$\mathbf{U}_N d\mathbf{A}_N \mathbf{U}_N = \mathbf{U}_N d\mathbf{U}_N \begin{bmatrix} \lambda_1 & 0_{N-1}^T \\ \vec{0}_{N-1} & \mathbf{A}_{N-1} \end{bmatrix} - \begin{bmatrix} \lambda_1 & 0_{N-1}^T \\ \vec{0}_{N-1} & \mathbf{A}_{N-1} \end{bmatrix} \mathbf{U}_N d\mathbf{U}_N + \begin{bmatrix} d\lambda_1 & 0_{N-1}^T \\ \vec{0}_{N-1} & d\mathbf{A}_{N-1} \end{bmatrix},$$

where use is made of the fact that $\mathbf{U}_N d\mathbf{U}_N$ is antisymmetric. Make further use of this latter fact to show that it is permissible to write

$$\mathbf{U}_N d\mathbf{U}_N := \begin{bmatrix} 0 & -d\vec{s}_{N-1}^T \\ d\vec{s}_{N-1} & d\tilde{\mathbf{U}}_{N-1} \end{bmatrix}$$

and so obtain

$$\mathbf{U}_N d\mathbf{A}_N \mathbf{U}_N = \begin{bmatrix} 0 & d\vec{s}_{N-1}^T(\lambda_1 - \mathbf{A}_{N-1}) \\ (\lambda_1 - \mathbf{A}_{N-1}) d\vec{s}_{N-1} & d\tilde{\mathbf{U}}_{N-1}\mathbf{A}_{N-1} - \mathbf{A}_{N-1} d\tilde{\mathbf{U}}_{N-1} \end{bmatrix} + \begin{bmatrix} d\lambda_1 & 0_{N-1}^T \\ \vec{0}_{N-1} & d\mathbf{A}_{N-1} \end{bmatrix}.$$

(ii) Using (1.10) and (1.17) read off from the final equation in (i) that

$$(d\mathbf{A}_N) = |\det(\lambda_1 - \mathbf{A}_{N-1})| d\lambda_1 (d\vec{s}_{N-1})(d\mathbf{A}_{N-1}), \tag{1.179}$$

where use has been made of the fact that $d\tilde{\mathbf{U}}_{N-1}$ is a function of the components of $d\vec{s}_{N-1}$ and thus $d\vec{s}_{N-1} \wedge d\tilde{\mathbf{U}}_{N-1} = 0$.

(iii) Iterate (1.179) to obtain a result equivalent to (1.11).

5. (i) Use the fact that $r_i = -P_i(\mu)/P_{i-1}(\mu)$, to show that the Prüfer phase for $j = 2, \dots, n$ satisfies

$$\cot \theta_j^\mu = \frac{1}{(b_{j-1})^2} \frac{P_{j-1}(\mu)}{P_{j-2}(\mu)}. \tag{1.180}$$

(ii) Use the fact that $\{P_j(\mu)\}_{j=0,1,\dots}$ are, as a consequence of their obeying the three-term recurrence (1.154), a set of orthogonal polynomials with respect to an inner product defined by its moment (the so called Favard theorem, see, e.g., [384]), together with (5.13) below to show that $P_i'(\mu)P_{i-1}(\mu) - P_{i-1}'(\mu)P_i(\mu) > 0$. After differentiating (1.180) with respect to μ, use this fact to deduce $d\theta_j^\mu/d\mu < 0$.

Chapter Two

Circular ensembles

Invariance of the probability measure on the space of matrices under conjugation by the appropriate unitary matrices does not uniquely determine the Gaussian ensembles. This fact prompted Dyson to develop a theory of random unitary matrices with the same invariances as the Hermitian matrices used to model quantum Hamiltonians. In quantum mechanics, scattering matrices and Floquet operators are quantities which can be modeled by random unitary matrices. In the case of no time reversal symmetry, the unitary matrices have no further constraints and form the group $U(N)$. It is well known that the Haar measure is the unique uniform measure on $U(N)$. In the case of time reversal symmetry, the unitary matrices must be invariant under an appropriate transpose, and the required invariance properties of the Haar measure must be appropriately modified. The corresponding eigenvalue p.d.f.'s can be computed by using either the method of wedge products or metric forms. Another method is to map the unitary matrices to Hermitian matrices through a Cayley transform, and to make use of the known Jacobians for Hermitian matrices. The p.d.f.'s correspond to a log-gas system on a circle, again at three special values of the inverse temperature $\beta = 1, 2$ and 4. Although the Haar measure does not make explicit the distribution of the elements of a random unitary matrix, a suitable decomposition into certain elementary unitary matrices allows for distributions to be specified. It is further possible to average over the elements of a random unitary matrix. Unitary matrices are one of the three compact classical groups, the other two being the real orthogonal matrices and the unitary symplectic matrices. The eigenvalue p.d.f.'s corresponding to the uniform distribution (Haar measure) are calculated. In the last two sections unitary and real orthogonal Hessenberg matrices are used to provide generalizations of the eigenvalue p.d.f.'s obtained for special β previously in the chapter.

2.1 SCATTERING MATRICES AND FLOQUET OPERATORS

The three ensembles of random Hermitian matrices introduced at the beginning of Chapter 1 were motivated by their relevance as models of quantum Hamiltonians. Likewise, three ensembles of random unitary matrices can be isolated by considering settings in quantum physics, but now with the focus being on scattering matrices and evolution operators rather than Hamiltonians. The settings to be considered here are the scattering of plane waves within an irregular shaped cavity, and the evolution of periodic kicked quantum systems through their Floquet operators.

2.1.1 Random scattering matrices

We will consider first the random unitary matrices relating to the scattering of plane waves within an irregular shaped domain (cavity), or a cavity of arbitrary shape containing random scattering impurities (see, e.g., [53], [276]). The plane waves enter and leave the cavity through a lead (wave guide) which is assumed to permit N distinct plane wave states (channels). With the amplitudes of the N incoming plane wave states denoted by $\vec{I}$, and the amplitudes of the N outgoing states denoted by $\vec{O}$, the $N \times N$ scattering matrix is defined so that

$$\mathbf{S}\vec{I} = \vec{O}. \tag{2.1}$$

Flux conservation requires $|\vec{I}|^2 = |\vec{O}|^2$, and from this it follows that $\langle \mathbf{S}\vec{I}|\mathbf{S}\vec{I}\rangle = \langle \vec{I}|\vec{I}\rangle$ for arbitrary $\vec{I}$. This in turn implies $\mathbf{S}\mathbf{S}^\dagger = \mathbf{1}$, and thus $\mathbf{S}$ must be unitary. In fact, it is generally true (see, e.g., [401]) that $\mathbf{S}$ is a limiting form of the evolution operator

$$\mathbf{S} = \lim_{t_0 \to -\infty} \lim_{t \to \infty} U(t, t_0) \tag{2.2}$$

where, with $\mathcal{T}$ denoting time ordering,

$$U(t, t_0) := \mathcal{T} \exp\left(-\frac{i}{\hbar} \int_{t_0}^{t} H(t')\, dt' \right) \tag{2.3}$$

for some Hamiltonian H. Thus $\mathbf{S}$ is necessarily unitary. However, analogous to the theory of Section 1.1.1 for time-independent Hamiltonians, the structure of the $N \times N$ unitary matrix modeling $\mathbf{S}$ is constrained by the requirements of time reversal symmetry.

In Section 1.1.1 a time reversal symmetry T was defined by the statement that T is anti-unitary and $[H, T] = 0$. This definition assumed that the Hamiltonian was independent of time. For time-dependent Hamiltonians the following definition is used.

DEFINITION 2.1.1 *A time-dependent Hamiltonian $H(t)$ is said to have a time reversal symmetry T if T is anti-unitary and*

$$T^{-1} H(t) T = H(-t).$$

Notice that this reduces to the previous definition when H is independent of t.

From this definition the action of T on the evolution operator is easily deduced.

PROPOSITION 2.1.2 *If T is a time reversal symmetry of a time-dependent Hamiltonian H, then*

$$T^{-1} U(t, t_0) T = U(-t, -t_0). \tag{2.4}$$

Proof. As is well known (see, e.g., [401]), and can easily be checked from its definition in terms of a time ordered exponential, $U(t, t_0)$ satisfies the integral equation

$$U(t, t_0) = 1 + \frac{1}{i\hbar} \int_{t_0}^{t} H(t') U(t', t_0)\, dt'.$$

Thus

$$T^{-1} U(t, t_0) T = 1 - \frac{1}{i\hbar} \int_{t_0}^{t} T^{-1} H(t') U(t', t_0) T\, dt' = 1 - \frac{1}{i\hbar} \int_{t_0}^{t} H(-t') T^{-1} U(t', t_0) T\, dt',$$

where to obtain the first equality we used the fact that T^{-1} is anti-unitary, and to obtain the second equality we used Definition 2.1.1. Hence

$$T^{-1} U(-t, -t_0) T = 1 - \frac{1}{i\hbar} \int_{-t_0}^{-t} H(-t') T^{-1} U(t', -t_0) T\, dt' = 1 + \frac{1}{i\hbar} \int_{t_0}^{t} H(t') T^{-1} U(-t', -t_0) T\, dt'.$$

We therefore have that $U(t, t_0)$ and $T^{-1} U(-t, -t_0) T$ satisfy the same integral equation and are thus equal. □

We now assume that for an irregular shaped domain the statistical properties of the scattering matrix are determined solely by the global time reversal symmetry. To see what constraint a time reversal symmetry imposes on $\mathbf{S}$, suppose T is a time reversal symmetry of a time-dependent Hamiltonian H. By taking the limits $t_0 \to -\infty$, $t \to \infty$ in (2.4) and using (2.2) we conclude

$$T^{-1} \mathbf{S} T = \mathbf{S}^\dagger. \tag{2.5}$$

Consider now (2.5) with $T^2 = 1$. From Exercises 1.1 q.3, we know how to construct a T invariant basis,

$\{\psi_n\}$ say. Now

$$\begin{aligned}\langle\psi_m|\mathbf{S}\psi_n\rangle &= \overline{\langle T\psi_m|T\mathbf{S}\psi_n\rangle} \quad \text{(since } T \text{ is anti-unitary)}\\ &= \overline{\langle T\psi_m|\mathbf{S}^\dagger T\psi_n\rangle} \quad \text{(using (2.5) and } T^2 = 1)\\ &= \overline{\langle\psi_m|\mathbf{S}^\dagger\psi_n\rangle} \quad \text{(since } \{\psi_n\} \text{ is } T \text{ invariant)}\\ &= \langle\psi_n|\mathbf{S}\psi_m\rangle \quad \text{(from meaning of } \dagger \text{ and property of inner product).}\end{aligned}$$

Hence $\mathbf{S}$ must be symmetric.

Consider next (2.5) with $T = \mathbf{Z}_{2N}K$ and so $T^2 = -1$. Then Proposition 2.1.2 gives

$$\mathbf{S} = T\mathbf{S}^\dagger T^{-1} = \mathbf{Z}_{2N}K\mathbf{S}^\dagger K^{-1}\mathbf{Z}_{2N}^{-1} = \mathbf{Z}_{2N}K\mathbf{S}^\dagger K\mathbf{Z}_{2N}^{-1} = \mathbf{Z}_{2N}\mathbf{S}^T\mathbf{Z}_{2N}^{-1} =: \mathbf{S}^D. \tag{2.6}$$

This constraint is said to specify $\mathbf{S}$ as a *self-dual quaternion* matrix, and in terms of the 2×2 blocks in position (jk) and (kj) of $\mathbf{S}$ it requires

$$\begin{bmatrix} z_{jk}^{(1)} & z_{jk}^{(2)} \\ z_{jk}^{(3)} & z_{jk}^{(4)} \end{bmatrix} = \begin{bmatrix} z_{kj}^{(4)} & -z_{kj}^{(2)} \\ -z_{kj}^{(3)} & z_{kj}^{(1)} \end{bmatrix} \tag{2.7}$$

(cf. (1.24)).

As with Hamiltonians, two matrices $\mathbf{S}$ and $\mathbf{S}'$ modeling scattering matrices and related by a similarity transformation $\mathbf{S}' = \mathbf{U}^{-1}\mathbf{S}\mathbf{U}$ of unitary matrices are equally good physical descriptions, provided $\mathbf{S}$ and $\mathbf{S}'$ have the same symmetries. Let us investigate the restrictions on $\mathbf{U}$ implied by time reversal symmetry. Suppose first that the system has a time reversal symmetry with $T^2 = 1$. Then the matrix modeling the scattering matrix must be symmetric. The subgroup of unitary matrices which under a similarity transformation map $\mathbf{S}$ to another symmetric matrix is the real orthogonal matrices (or i times the real orthogonal matrices — see Exercises 2.1 q.2).

Next suppose the system has a time reversal symmetry with $T^2 = -1$ and $T = \mathbf{Z}_{2N}K$, so that the matrices modeling the scattering matrix must be self-dual quaternion. The subgroup of unitary matrices, which under a similarity transformation map self-dual quaternion matrices into self-dual quaternion matrices, is the symplectic equivalent matrices of Proposition 1.3.3 (see Exercises 2.1 q.3). Finally, if the system has no time reversal symmetry, $\mathbf{S}$ must be unitary but has no further constraints. Under similarity transformations all unitary matrices map $\mathbf{S}$ into another unitary matrix.

We remark that the element S_{nm} in $\mathbf{S}$ as specified by (2.1) has the property that its absolute value squared gives the probability that the state n scatters to the state m. Thus a quantity of interest is the mean value of $|S_{nm}^2|$ for $\mathbf{S}$ drawn from an ensemble of unitary matrices, appropriately constrained in the cases of a time reversal symmetry. This is computed in Section 2.3.2.

2.1.2 Random Floquet operators

We consider next evolution operators $U(t) := U(t, 0)$ as specified by (2.3) in the case that the Hamiltonian is periodic of period τ, $H(t) = H(t+\tau)$. Then for $n \in \mathbb{Z}$, $U(n\tau) = F^n$, where $F := U(\tau)$. The operator F is called the *Floquet operator*, which being a special value of the unitary operator U is itself unitary. To model F for periodic chaotic quantum systems such as kicked tops, random unitary matrices have been used (see, e.g., [284]). As in the above discussion of scattering matrices, the structure of a unitary matrix $\mathbf{F}$ modeling F is constrained by the requirement of time reversal symmetry.

Substituting $t = \tau$, $t_0 = 0$ in (2.4) implies $T^{-1}\mathbf{F}T = \mathbf{F}^{-1} = \mathbf{F}^\dagger$. This is identical to the constraint of scattering matrices (2.5). Thus for $T^2 = 1$, $\mathbf{F}$ must be a symmetric unitary matrix, while for $T^2 = -1$, $\mathbf{F}$ must be a self-dual quaternion unitary matrix.

EXERCISES 2.1 1. Show from the definitions that a Hermitian quaternion real matrix is self-dual.

2. Let T be a time reversal symmetry such that $T^2 = 1$. Suppose $\mathbf{S}$ and $\mathbf{S}'$ are two symmetric unitary matrices

related by a similarity transformation $\mathbf{S}' = \mathbf{U}^{-1}\mathbf{S}\mathbf{U}$, where $\mathbf{U}$ is a unitary matrix, which satisfy the time reversal symmetry constraint

$$\mathbf{S} = T\mathbf{S}^{\dagger}T^{-1} \qquad \text{and} \qquad \mathbf{S}' = T\mathbf{S}'^{\dagger}T^{-1}, \tag{2.8}$$

and suppose the only symmetry of $\mathbf{S}$ in general is some constant times the identity. Show that $\mathbf{U}$ is real orthogonal or i times a real orthogonal matrix.

3. Let T be a time reversal symmetry such that $T^2 = -1$. Follow the method of Proposition 1.3.3 to show that if $\mathbf{S}$ and $\mathbf{S}'$ are two self-dual quaternion matrices satisfying (2.8) and related by a similarity transformation as in q.2, and the only symmetry of $\mathbf{S}$ is in general some multiple of the identity, then $\mathbf{U}$ is a symplectic equivalent matrix or i times a symplectic equivalent matrix.

2.2 DEFINITIONS AND BASIC PROPERTIES

In keeping with the above findings on the classes of scattering matrices and Floquet operators consistent with time reversal symmetry, we seek a theory of symmetric, unitary, and self-dual quaternion random unitary matrices. This has been given by Dyson [146]. In contrast to the theory of Gaussian random matrices presented in the previous chapter, the theory was developed not from an explicit formula for the matrix elements of the random unitary matrices, but rather from the requirement that there is a uniform measure on each of these spaces. The definitions adopted were (at least implicitly) motivated by the relationship of the spaces of matrices to certain symmetric spaces. Some aspects of this are covered in Section 2.2.3. Here we adopt a simpler approach, by noting that in the case of unitary matrices with no further constraints, the construction of a uniform measure is a classical problem.

2.2.1 The circular unitary ensemble (CUE)

Among the spaces of symmetric, unitary and self-dual quaternion unitary matrices, the space of unitary matrices is special because it forms a group. Moreover, this group is compact. As such there are well established ways to specify and determine a volume form which defines a uniform measure on the space (see, e.g., [540]).

The volume form, or *Haar form* in keeping with the terminology of the corresponding measure — the *Haar measure* — is denoted $(d_H\mathbf{U})$ and is required to have the homogeneity property that, for any fixed unitary matrix $\mathbf{V}$,

$$(d_H\mathbf{U}\mathbf{V}) = (d_H\mathbf{V}\mathbf{U}) = (d_H\mathbf{U}). \tag{2.9}$$

In words this says the Haar form is invariant under left and right actions of the group, and in this sense the Haar form implies a uniform measure on the space of unitary matrices. It is a well-known theorem that the Haar form specified by (2.9) exists and is unique up to normalization for all compact groups.

PROPOSITION 2.2.1 *The Haar form for the unitary group is*

$$(d_H\mathbf{U}) = \frac{1}{C}(\mathbf{U}^{\dagger}d\mathbf{U}), \tag{2.10}$$

where C denotes the normalization.

Proof. According to the above discussion, we merely have to check (2.9). Now for fixed $\mathbf{V}$,

$$d\mathbf{U}\mathbf{V} = d\mathbf{U}\,\mathbf{V}, \qquad d\mathbf{V}\mathbf{U} = \mathbf{V}d\mathbf{U},$$

and thus (2.10) implies

$$(d_H\mathbf{U}\mathbf{V}) = \frac{1}{C}(\mathbf{V}^{\dagger}\mathbf{U}^{\dagger}d\mathbf{U}\,\mathbf{V}), \qquad (d_H\mathbf{V}\mathbf{U}) = \frac{1}{C}(\mathbf{U}^{\dagger}d\mathbf{U}).$$

Thus we are done if we can show

$$(\mathbf{V}^{\dagger}\mathbf{U}^{\dagger}d\mathbf{U}\,\mathbf{V}) = (\mathbf{U}^{\dagger}d\mathbf{U}), \tag{2.11}$$

which is in fact immediate from (1.35). □

The definition of an ensemble of random unitary matrices can now be made explicit.

DEFINITION 2.2.2 *The circular unitary ensemble (CUE) is the group of unitary matrices endowed with the volume form (2.10).*

Since $\mathbf{U}$ is unitary, a short calculation shows that the matrix of differentials $\mathbf{U}^\dagger d\mathbf{U}$ has the anti-Hermitian property $(\mathbf{U}^\dagger d\mathbf{U})^\dagger = -\mathbf{U}^\dagger d\mathbf{U}$. Thus we can write

$$\mathbf{U}^\dagger d\mathbf{U} = i\, d\mathbf{M}_2 \tag{2.12}$$

for some Hermitian matrix $\mathbf{M}_2$. There are no further constraints on the elements of $\mathbf{M}_2$ since $\mathbf{U}$ has the same number of independent elements as an arbitrary Hermitian matrix. Consequently computing the volume form $(\mathbf{U}^\dagger d\mathbf{U})$ for $\mathbf{U}$ unitary is equivalent to computing $(d\mathbf{M}_2)$ for $\mathbf{M}_2$ Hermitian, and thus the methods of the previous chapter are applicable.

2.2.2 The circular orthogonal and symplectic ensembles

The spaces of symmetric and self-dual quaternion unitary matrices do not form groups. Thus the invariance (2.9) expressing the homogeneity of the sought volume form must be modified. Consider first the case of symmetric unitary matrices. Two symmetric unitary matrices $\mathbf{S}_1$ and $\mathbf{S}_2$ can be related by $\mathbf{S}_2 = \mathbf{V}^T\mathbf{S}_1\mathbf{V}$ for some unitary matrix $\mathbf{V}$. This follows since each symmetric unitary matrix $\mathbf{S}$ can be written in the form

$$\mathbf{S} = \mathbf{U}^T\mathbf{U} \tag{2.13}$$

for $\mathbf{U}$ unitary; see the proof of Proposition 2.2.4 below. Thus for a uniform distribution we require that

$$(d_H\mathbf{S}) = (d_H\mathbf{V}^T\mathbf{S}\mathbf{V}) \tag{2.14}$$

for all unitary $\mathbf{V}$. Now, because the explicit formula (2.10) for the Haar volume form of the unitary group has both the left and right invariance properties (2.9), we see that the same formula (2.10) is consistent with (2.14). We also know from (2.11) that the underlying matrix of differentials, $\delta\mathbf{S}$ say, is arbitrary up to a similarity transformation with a unitary matrix. This arbitrariness can be used so that $\delta\mathbf{S}$ like $\mathbf{S}$ is symmetric, and thus analogous to (2.12) is of the form $i\,d\mathbf{M}_1$ for $\mathbf{M}_1$ real symmetric. The sought form is achieved by setting $\mathbf{V} = \mathbf{U}^\dagger$ in (2.11), where $\mathbf{U}$ is such that (2.13) holds, and so obtaining

$$\delta\mathbf{S} = (\mathbf{U}^T)^\dagger d\mathbf{S}\mathbf{U}^\dagger =: i\, d\mathbf{M}_1. \tag{2.15}$$

In the case of self-dual quaternion unitary matrices $\tilde{\mathbf{S}}$ the analogue of (2.13) is the decomposition

$$\tilde{\mathbf{S}} = \mathbf{Z}_{2N}^{-1}\mathbf{U}^T\mathbf{Z}_{2N}\mathbf{U} =: \mathbf{U}^D\mathbf{U} \tag{2.16}$$

for $\mathbf{U}$ a $2N \times 2N$ unitary matrix. It follows from this that two self-dual quaternion unitary matrices $\tilde{\mathbf{S}}_1$ and $\tilde{\mathbf{S}}_2$ can be related by $\tilde{\mathbf{S}}_2 = \mathbf{V}^D\tilde{\mathbf{S}}_1\mathbf{V}$ for some $2N \times 2N$ unitary matrix $\mathbf{V}$. Thus, in this case for a uniform distribution we require

$$(d_H\tilde{\mathbf{S}}) = (d_H\mathbf{V}^D\tilde{\mathbf{S}}\mathbf{V}).$$

Of course the Haar form (2.10) has this invariance property. As in the case of symmetric unitary matrices, we want to choose the arbitrary matrix in (2.11) so that the underlying matrix of differentials $\delta\tilde{\mathbf{S}}$ is itself self-dual (recall (2.6)), and thus analogous to (2.12) is of the form $i\,d\mathbf{M}_4$ for $\mathbf{M}_4$ a self-dual quaternion real matrix. Again as in the symmetric case, this is achieved by the choice $\mathbf{V} = \mathbf{U}^\dagger$, where $\mathbf{U}$ is such that (2.16) holds. We then have

$$\delta\tilde{\mathbf{S}} = (\mathbf{Z}_{2N}^{-1}\mathbf{U}^T\mathbf{Z}_{2N})^\dagger d\tilde{\mathbf{S}}\,\mathbf{U}^\dagger = (\mathbf{U}^D)^\dagger d\tilde{\mathbf{S}}\,\mathbf{U}^\dagger =: i\, d\mathbf{M}_4. \tag{2.17}$$

We are thus led to the following definition.

Definition 2.2.3 *The circular orthogonal ensemble (COE) is the space of symmetric unitary matrices endowed with the volume form corresponding to (2.15), and similarly the circular symplectic ensemble (CSE) is the space of self-dual quaternion unitary matrices endowed with the volume form corresponding to (2.17).*

2.2.3 Relationship to symmetric spaces

Notice from (2.15), (2.12) and (2.17) that the infinitesimal examples of real symmetric, Hermitian and self-dual real quaternion matrices occur in the volume forms of the COE, CUE and CSE respectively. As observed in Section 1.3.4, these infinitesimal generators are the Hermitian part of the matrix Lie algebras $gl(N,\mathbb{R})$, $gl(N,\mathbb{C})$ and $u^*(2N)$ respectively, and as such are invariant under the Hermitian conjugation mapping. The general theory of symmetric spaces [295] relates these Lie algebras and the Hermitian conjugation mapping to the compact quotient spaces

$$U(N)/O(N), \quad U(N)\times U(N)/U(N), \quad U(2N)/\mathrm{Sp}(2N)$$

respectively . Here

$$\begin{aligned} O(N) &:= \{N\times N\,\text{real unitary matrices}\} \\ U(N) &:= \{N\times N\,\text{complex unitary matrices}\} \\ \mathrm{Sp}(2N) &:= \{N\times N\,\text{real quaternion unitary matrices}\}, \end{aligned}$$

or equivalently $O(N)$ consists of the set of $N\times N$ real orthogonal matrices and $\mathrm{Sp}(2N)$ the set of $2N\times 2N$ unitary symplectic equivalent matrices. In fact, as observed by Dyson [150] these spaces are isomorphic to the space of symmetric unitary matrices, unitary matrices and symplectic self-dual unitary matrices, respectively. The demonstration of this fact exhibits some essential properties of the respective circular ensembles. For definiteness consider the quotient space $U(N)/O(N)$.

Proposition 2.2.4 *We have*

$$U(N)/O(N) \cong U_S(N),$$

where $U_S(N)$ denotes the set of $N\times N$ symmetric unitary matrices.

Proof. First we recall that in general for a group G and subgroup K, the quotient space $H = G/K$ is the set of equivalence classes with the equivalence relation $g_1 \sim g_2$ meaning that g_1 and g_2 are related by the transformation K (i.e., $g_1 = g_2 k$ for some $k\in K$).

We proceed by showing $\mathbf{g}_1 \sim \mathbf{g}_2$, $\mathbf{g}_1,\mathbf{g}_2 \in U(N)$, if and only if $\mathbf{g}_1\mathbf{g}_1^T = \mathbf{g}_2\mathbf{g}_2^T$. The only if direction is immediate. For the if direction we see that $\mathbf{g}_1\mathbf{g}_1^T = \mathbf{g}_2\mathbf{g}_2^T$ implies $(\mathbf{g}_2^{-1}\mathbf{g}_1)(\mathbf{g}_2^{-1}\mathbf{g}_1)^T = \mathbf{1}_N$. Thus $\mathbf{g}_1 = \mathbf{g}_2\mathbf{k}$ for some $\mathbf{k}\in O(N)$ as required. This demonstrates a one-to-one correspondence between elements of $U(N)/O(N)$ and elements of the form $\mathbf{g}\mathbf{g}^T$ in $U_S(N)$. It remains to check that all elements of $U_S(N)$ can be written in the form $\mathbf{g}\mathbf{g}^T$. For $\mathbf{h}\in U_S(N)$ there exists a matrix $\mathbf{k}\in O(N)$ such that $\mathbf{h} = \mathbf{k}\,\mathrm{diag}[e^{i\theta_j}]_{j=1,\dots,N}\mathbf{k}^T$, so choosing $\mathbf{g} = \mathbf{k}\,\mathrm{diag}[e^{i\theta_j/2}]_{j=1,\dots,N}$ expresses $\mathbf{h}$ in the desired form. □

One feature of the proof of Proposition 2.2.4 is the decomposition (2.13). Thus an involutive mapping $\phi = T$ (T denoting transpose) is identified which leaves the space K in the quotient G/K unchanged. It is the existence of the map ϕ which distinguishes the quotient space as a symmetric space. Another crucial aspect of the proof of Proposition 2.2.4 is the diagonalization formula

$$\mathbf{S} = \mathbf{R}\,\mathrm{diag}[e^{i\theta_j}]_{j=1,\dots,N}\mathbf{R}^T, \tag{2.18}$$

for some real orthogonal matrix $\mathbf{R}$. Comparison with (1.8) shows this to be the circular ensemble counterpart to the diagonalization formula for members of the GOE. As will be seen in the proof of Proposition 2.2.5, (2.18) is the key formula in the determination of the eigenvalue p.d.f.

The CSE allows for a description similar to that of the COE, with the role of (2.13) replaced by the decomposition formula (2.16). In particular this formula can be used to generate a member of the CSE from a member of the CUE. Also, the diagonalization formula for members of the CSE is

$$\mathbf{S} = \mathbf{B}\,\mathrm{diag}[e^{i\theta_j}\mathbf{1}_2]_{j=1,\dots,N}\mathbf{B}^D, \tag{2.19}$$

where $\mathbf{B}$ is symplectic equivalent, satisfying (1.25). This is the circular ensemble counterpart of the diagonalization formula for members of the GSE given in Exercises 1.3 q.4.

2.2.4 Eigenvalue p.d.f. for the circular ensembles

With each eigenvalue written $\lambda_j = e^{i\theta_j}$, the eigenvalue p.d.f.'s corresponding to the three circular ensembles can be calculated using the methods of Section 1.2.

PROPOSITION 2.2.5 *For an appropriate choice of normalization $C_{\beta,N}$, which is given explicitly in Proposition 2.8.7,*

$$\frac{1}{C_{\beta,N}} \prod_{1 \le j < k \le N} |e^{i\theta_k} - e^{i\theta_j}|^\beta, \qquad -\pi < \theta_l \le \pi \tag{2.20}$$

with $\beta = 1, 2$ and 4 is the eigenvalue p.d.f. for the COE, CUE and CSE, respectively.

Proof. Only the COE will be considered here; the remaining cases of the CUE and CSE are similar (the CUE is considered in Exercises 2.2 q.1). The starting point is the diagonalization formula (2.18). For notational convenience write $\boldsymbol{\Theta} = \mathrm{diag}[e^{i\theta_j}]_{j=1,\dots,N}$ in that formula. Analogous to (1.9), differentiation and minor manipulation gives

$$\mathbf{R}^T d\mathbf{S}\mathbf{R} = \mathbf{R}^T d\mathbf{R}\,\boldsymbol{\Theta} - \boldsymbol{\Theta}\mathbf{R}^T d\mathbf{R} + i\boldsymbol{\Theta} d\boldsymbol{\theta},$$

where θ is the diagonal matrix with entries θ_j ($j = 1, \dots, N$). But from (2.15)

$$\mathbf{R}^T d\mathbf{S}\mathbf{R} = i(\mathbf{U}\mathbf{R})^T d\mathbf{M}_1(\mathbf{U}\mathbf{R}),$$

where $\mathbf{U}$ is any unitary matrix such that $\mathbf{S} = \mathbf{U}^T\mathbf{U}$. In particular, with $\mathbf{U} = \boldsymbol{\Theta}^{1/2}\mathbf{R}^T$, where $\boldsymbol{\Theta}^{1/2}$ is the diagonal matrix with entries $e^{i\theta_j/2}$ ($j = 1, \dots, N$), comparison of the two equations for $\mathbf{R}^T d\mathbf{S}\mathbf{R}$ gives

$$\begin{aligned} d\mathbf{M}_1 &= d\boldsymbol{\theta} - i\boldsymbol{\Theta}^{-1/2}\mathbf{R}^T d\mathbf{R}\,\boldsymbol{\Theta}^{1/2} + i\boldsymbol{\Theta}^{1/2}\mathbf{R}^T d\mathbf{R}\,\boldsymbol{\Theta}^{-1/2} \\ &= \begin{bmatrix} d\theta_1 & 2\sin((\theta_2 - \theta_1)/2)\,\vec{r}_1 \cdot d\vec{r}_2 \dots & \dots & 2\sin((\theta_N - \theta_1)/2)\,\vec{r}_1 \cdot d\vec{r}_N \\ * & d\theta_2 & \dots & 2\sin((\theta_N - \theta_2)/2)\,\vec{r}_2 \cdot d\vec{r}_N \\ \vdots & \vdots & \ddots & \vdots \\ * & * & \dots & d\theta_N \end{bmatrix}, \end{aligned}$$

where the elements * are chosen so that the matrix is symmetric. Taking the wedge product of the independent elements, and noting that

$$2\sin((\theta_k - \theta_j)/2) = |e^{i\theta_k} - e^{i\theta_j}| \qquad \text{for } \theta_k > \theta_j \tag{2.21}$$

gives the result of the proposition for the COE.

□

Comparison of Proposition 2.2.5 with Proposition 1.4.1 shows immediately that the eigenvalue p.d.f. for the circular ensembles is directly proportional to the Boltzmann factor of the one-component log-potential Coulomb gas on a circle.

2.2.5 Antisymmetric unitary matrices

In Exercises 1.3 q.5 the eigenvalue p.d.f. of antisymmetric Gaussian random matrices with pure imaginary elements was calculated. If instead one considers antisymmetric unitary matrices, then the eigenvalue p.d.f. for the CUE, after a rescaling $2\theta_j \mapsto \theta_j$, is reclaimed. To understand this result, first note that for a unitary antisymmetric matrix $\mathbf{S}$, if λ is an eigenvalue then so is $-\lambda$. Because we must also have $|\lambda| = 1$ this implies that such matrices must be even dimensional. They can be constructed out of unitary matrices $\mathbf{U}$ by forming $\mathbf{U}^T\mathbf{Z}_{2N}\mathbf{U}$. A result of Hua [300] gives the decomposition

$$\mathbf{U}^T\mathbf{Z}_{2N}\mathbf{U} = \mathbf{R}\mathrm{diag}\left[\begin{bmatrix} 0 & e^{i\theta_1} \\ -e^{i\theta_1} & 0 \end{bmatrix}, \dots, \begin{bmatrix} 0 & e^{i\theta_N} \\ -e^{i\theta_N} & 0 \end{bmatrix}\right]\mathbf{R}^T,$$

where $\mathbf{R}$ is real orthogonal, and it follows from this that the eigenvalues of $\mathbf{U}^T\mathbf{Z}_{2N}\mathbf{U}$ are $\pm ie^{i\theta_j}$ $(j = 1, \dots, N)$. By defining $\delta\mathbf{S} = (\mathbf{U}^T\mathbf{Z}_{2N})^\dagger d\mathbf{S}\,\mathbf{U}^\dagger$ we can proceed as in Section 2.2.4 to deduce that the eigenvalue p.d.f. for the eigenvalues $e^{i\theta_j}$ $(j = 1, \dots, N)$ with $0 < \theta_j < \pi$ is proportional to

$$\prod_{1\le j<k\le N} |e^{2i\theta_k} - e^{2i\theta_j}|^2$$

and thus, after the replacements $2\theta_j \mapsto \theta_j$, formally equivalent to that of the CUE.

EXERCISES 2.2 1. Here the result of Proposition 2.2.5 in the case of CUE will be derived using the metric form approach of Section 1.2.4.

(i) Show from the diagonalization formula $\mathbf{U} = \mathbf{U}_2\Theta\mathbf{U}_2^\dagger$, where $\mathbf{U}_2$ is a unitary matrix with the first component of each of the entries in the first row chosen to be real and positive, that

$$\mathbf{U}_2^\dagger d\mathbf{U}\,\mathbf{U}_2 = \delta\mathbf{U}_2\Theta - \Theta\delta\mathbf{U}_2 + i\Theta d\theta,$$

where $\delta\mathbf{U}_2 := \mathbf{U}_2^\dagger d\mathbf{U}_2 = [\delta u_{2\,jk}]_{j,k=1,\dots,N}$.

(ii) Recalling (2.12), and using the fact that $\mathrm{Tr}((\delta\mathbf{U}_2\Theta - \Theta\delta\mathbf{U}_2)\Theta d\theta) = 0$, show from this that

$$\mathrm{Tr}(d\mathbf{M}_2 d\mathbf{M}_2^\dagger) = \sum_{\substack{j,k=1 \\ j\ne k}}^N |\delta u_{2\,jk}(e^{i\theta_j} - e^{i\theta_k})|^2 + \sum_{j=1}^N (d\theta_j)^2. \tag{2.22}$$

(iii) Read off from (ii) the result

$$(\mathbf{U}^\dagger d\mathbf{U}) = \prod_{1\le j<k\le N} |e^{i\theta_j} - e^{i\theta_k}|^2 \bigwedge_{j=1}^N d\theta_j (\mathbf{U}_2^\dagger d\mathbf{U}_2), \tag{2.23}$$

where $(\mathbf{U}_2^\dagger d\mathbf{U}_2) = \bigwedge_{j<k} \delta u_{2\,jk}^{\mathrm{r}} \delta u_{2\,jk}^{\mathrm{i}}$ and note that this is in agreement with Proposition 2.2.5 in the case $\beta = 2$.

2. [458] Suppose for a certain ensemble of $N \times N$ unitary matrices the eigenvalue p.d.f. is P_N. Let the p.d.f. of the p-th power of the eigenvalues in the same ensemble be denoted $P_N^{(p)}$, so that for any Laurent expandable function $f(e^{i\theta_1}, \dots, e^{i\theta_N})$,

$$\langle f(e^{ip\theta_1}, \dots, e^{ip\theta_N})\rangle_{P_N} = \langle f(e^{i\theta_1}, \dots, e^{i\theta_N})\rangle_{P_N^{(p)}}.$$

(i) Show that for P_N a Laurent polynomial in $\{e^{i\theta_j}\}$,

$$P_N^{(p)}(e^{i\theta_1}, \dots, e^{i\theta_N}) = \mathrm{LP}\Big(P_N(e^{i\theta_1/p}, \dots, e^{i\theta_N/p})\Big),$$

where LP denotes the Laurent polynomial portion of the function, i.e., only terms involving integer powers of the monomials $e^{i\theta_j}$.

(ii) For $U(N)$, note from Proposition 2.2.5 and the Vandermonde identity (1.173) that

$$P_N(e^{i\theta_1}, \dots, e^{i\theta_N}) = \frac{1}{(2\pi)^N N!}\mathrm{Asym}\Big(\sum_{Q\in S_N} \varepsilon(Q)\prod_{l=1}^N e^{i\theta_l(l-Q(l))}\Big),$$

where Asym denotes the operation of anti-symmetrization (see (4.135) below), S_N denotes the set of all permutations of $\{1,\dots,N\}$ and $\varepsilon(Q)$ denotes the corresponding parity.

(iii) For a given $0 \le \mu < p$, write $l = pl' + \mu + 1$, $0 \le l' < \lceil (N-\mu)/p \rceil$ and define $Q^{(\mu)}(l')$ by $Q(l) = pQ^{(\mu)}(l') + \mu + 1$ to show

$$\mathrm{LP}\Big(\sum_{Q\in S_N}\varepsilon(Q)\prod_{l=1}^{N} e^{i\theta_l(l-Q(l))/p}\Big) = \prod_{\mu=0}^{p-1}\Big(\sum_{Q^{(\mu)}\in S_{\lceil (N-\mu)/p\rceil}}\varepsilon(Q^{(\mu)})\prod_{l'=0}^{\lceil (N-\mu)/p\rceil -1} e^{i\theta_{pl'+\mu+1}(l'-Q^{(\mu)}(l'))}\Big).$$

(iv) Conclude from the result of (iii) that for $U(N)$, $P_N^{(p)}$ is proportional to the symmetrization of the product $\prod_{\mu=0}^{p-1} P_{\lceil (N-\mu)/p\rceil}$, and thus in particular that $P_N^{(p)}$ is the uniform measure for $p \ge N$.

2.3 THE ELEMENTS OF A RANDOM UNITARY MATRIX

2.3.1 Constructing a random unitary matrix

With members of the COE and CSE generated from a random unitary matrix according to (2.13) and (2.16), it is of some interest to consider the problem of specifying a random unitary matrix from a numerical generation perspective. In fact it is possible to give an explicit construction of the matrix elements of a member of the CUE [558], a fact which follows from a classical result of Hurwitz [302].

PROPOSITION 2.3.1 *Almost all $N \times N$ unitary matrices $\mathbf{U}$ have the unique decomposition*

$$\mathbf{U} = e^{i\alpha_0}\mathbf{U}_{N-1}\mathbf{U}_{N-2}\cdots\mathbf{U}_1$$

with

$$\begin{aligned}\mathbf{U}_k &= \mathbf{U}^{(k,k+1)}(\phi_{k,k+1},\psi_{k,k+1},0)\mathbf{U}^{(k,k+2)}(\phi_{k,k+2},\psi_{k,k+2},0)\\ &\quad\times\cdots\times\mathbf{U}^{(k,N-1)}(\phi_{k,N-1},\psi_{k,N-1},0)\mathbf{U}^{(k,N)}(\phi_{k,N},\psi_{k,N},\alpha_k),\end{aligned}$$

$$-\pi \le \alpha_k < \pi, \qquad -\pi \le \psi_{jk} < \pi, \qquad 0 \le \phi_{jk} \le \pi/2,$$

where the $N \times N$ unimodular matrices $\mathbf{U}^{(j,k)}(\phi,\psi,\alpha)$ are defined so that all diagonal elements are 1 except for the jth and kth which are equal to $\cos\phi\, e^{i\alpha}$ and $\cos\phi\, e^{-i\alpha}$ respectively, and all off-diagonal elements are zero except for the element in the jth row and kth column which is equal to $\sin\phi\, e^{i\psi}$, and the element in the kth row and jth column which is equal to $-\sin\phi\, e^{-i\psi}$ (the quantities ϕ and ψ are referred to as Euler angles). Furthermore

$$(\mathbf{U}^\dagger d\mathbf{U}) = \frac{1}{C}\bigwedge_{1\le j<k\le N} d[(\cos\phi_{jk})^{2(N-k+1)}]d\psi_{jk}\bigwedge_{j=0}^{N-1} d\alpha_j.$$

Proof. For the decomposition we essentially follow [411]. The strategy is first to choose the parameters $\phi_{1,j}$ and $\psi_{1,j}$ $(j=2,\dots,N)$ and α_1 such that

$$\mathbf{U}\mathbf{U}_1^\dagger = \begin{bmatrix} 1 & \mathbf{0}_{1\times N-1} \\ \mathbf{0}_{N-1\times 1} & \mathbf{V}\end{bmatrix}, \tag{2.24}$$

where $\mathbf{V}$ is an $(N-1)$-dimensional unitary matrix, and then to repeat the procedure to reduce $\mathbf{V}$ to an analogous form, and so on. To obtain (2.24), $\phi_{1,N}$, $\psi_{1,N}$ and α_1 are to be chosen so that $\mathbf{U}\mathbf{U}^{(1,N)\dagger}(\phi_{1,N},\psi_{1,N},\alpha_1)$ has the element 0 in position $(1,N)$ and element r in position $(1,1)$ for some $0<r<\infty$. A simple calculation using the definition of $\mathbf{U}^{(1,N)\dagger}$ shows this is achieved by requiring

$$\tan\phi_{1,N}\, e^{i(\psi_{1,N}-\alpha_1)} = \frac{u_{1N}}{u_{11}}, \qquad \frac{1}{\cos\phi_{1,N}} = \frac{r}{u_{11}e^{-i\alpha_1}}. \tag{2.25}$$

From the first equation in (2.25) $\phi_{1,N}$ is uniquely determined $(0 < \phi_{1,N} < \pi/2)$ as is $\psi_{1,N} - \alpha_1$ $(-\pi < \psi_{1,N} - \alpha_1 < \pi)$, while the second equation gives that $\alpha_1 = \arg(u_{11})$ $(-\pi \le \arg(u_{11}) < \pi)$ and $r = |u_{11}|/\cos\phi_{1,N}$.

Next $\phi_{1,N-1}$ and $\psi_{1,N-1}$ are chosen so that

$$\mathbf{U}\mathbf{U}^{(1,N)\dagger}(\phi_{1,N},\psi_{1,N},\alpha_1)\mathbf{U}^{(1,N-1)\dagger}(\phi_{1,N-1},\psi_{1,N-1},0) \tag{2.26}$$

has an element 0 in position $(1, N-1)$ and element r' in position $(1,1)$ for some $0 < r' < \infty$. This requires $\tan\phi_{1,N-1}\, e^{i\psi_{1,N-1}} = u_{1\,N-1}/r$ and $1/\cos\phi_{1,N-1} = r'/r$ which has a unique solution for $0 < \phi_{1,N-1} < \pi/2$, $-\pi \le \psi_{1,N-1} < \pi$ and $0 < r' < \infty$. Note from the definition of $\mathbf{U}^{(1,N-1)}$ that since $\mathbf{U}\mathbf{U}^{(1,N)\dagger}$ has element 0 in position $(1, N)$, (2.26) must also have element 0 in this position. Continuing this procedure gives that $\mathbf{U}\mathbf{U}_1^\dagger$ has all elements zero in the first row except that in position $(1,1)$ which is real positive. But $\mathbf{U}\mathbf{U}_1^\dagger$ is unitary so the element in position $(1,1)$ must equal unity and also the elements below this element in the first column must necessarily vanish.

To compute the volume form (2.10) we first note that the decomposition (2.24) gives

$$\mathbf{U}^\dagger d\mathbf{U} = \mathbf{U}_1^\dagger d\mathbf{U}_1 + \mathbf{U}_1^\dagger \begin{bmatrix} 0 & \mathbf{0}_{1\times N-1} \\ \mathbf{0}_{N-1\times 1} & \mathbf{V}^\dagger d\mathbf{V} \end{bmatrix} \mathbf{U}_1. \tag{2.27}$$

As the contribution to $(\mathbf{U}^\dagger d\mathbf{U})$ from $(\mathbf{U}_1^\dagger d\mathbf{U}_1)$ comes entirely from the first column, while the second term in (2.27) contains no such contribution, we obtain the factorization

$$(\mathbf{U}^\dagger d\mathbf{U}) = (\mathbf{U}_1^\dagger d\mathbf{U}_1)\Big(\mathbf{U}_1^\dagger \begin{bmatrix} 0 & \mathbf{0}_{1\times N-1} \\ \mathbf{0}_{N-1\times 1} & \mathbf{V}^\dagger d\mathbf{V} \end{bmatrix} \mathbf{U}_1\Big) = (\mathbf{U}_1^\dagger d\mathbf{U}_1)(\mathbf{V}^\dagger d\mathbf{V}),$$

where to obtain the second equality (1.35) has been used. It remains to compute $(\mathbf{U}_1^\dagger d\mathbf{U}_1)$, as the value of $(\mathbf{V}^\dagger d\mathbf{V})$ will then follow by induction. Now, since only the first column of $d\mathbf{U}_1$, $d\vec{u}_1^{(1)}$, say, is independent, we have

$$(\mathbf{U}_1^\dagger d\mathbf{U}_1) = (\mathbf{U}_1^\dagger d\vec{u}_1^{(1)}).$$

Here there are $2N-1$ independent variables $\alpha_1, \phi_{1,j}, \psi_{1,j}$ $(j = 2, \dots, N)$ which is in keeping with the first component of $\mathbf{U}_1^\dagger d\vec{u}^{(1)}$ being pure imaginary and the remaining component having independent real and imaginary parts.

Now, by definition, up to a sign

$$(\mathbf{U}_1^\dagger d\vec{u}^{(1)}) = \det \begin{bmatrix} \Big\{\mathbf{U}_1^\dagger \frac{\partial \vec{u}_1^{(1)}}{\partial\alpha}\Big\}_1^{\mathrm{i}} & \Big\{\mathbf{U}_1^\dagger \frac{\partial \vec{u}_1^{(1)}}{\partial\phi_{1,k}}\Big\}_1^{\mathrm{i}} & \Big\{\mathbf{U}_1^\dagger \frac{\partial \vec{u}_1^{(1)}}{\partial\psi_{1,k}}\Big\}_1^{\mathrm{i}} \\ \Big\{\mathbf{U}_1^\dagger \frac{\partial \vec{u}_1^{(1)}}{\partial\alpha}\Big\}_j^{\mathrm{r}} & \Big\{\mathbf{U}_1^\dagger \frac{\partial \vec{u}_1^{(1)}}{\partial\phi_{1,k}}\Big\}_j^{\mathrm{r}} & \Big\{\mathbf{U}_1^\dagger \frac{\partial \vec{u}_1^{(1)}}{\partial\psi_{1,k}}\Big\}_j^{\mathrm{r}} \\ \Big\{\mathbf{U}_1^\dagger \frac{\partial \vec{u}_1^{(1)}}{\partial\alpha}\Big\}_j^{\mathrm{i}} & \Big\{\mathbf{U}_1^\dagger \frac{\partial \vec{u}_1^{(1)}}{\partial\phi_{1,k}}\Big\}_j^{\mathrm{i}} & \Big\{\mathbf{U}_1^\dagger \frac{\partial \vec{u}_1^{(1)}}{\partial\psi_{1,k}}\Big\}_j^{\mathrm{i}} \end{bmatrix}_{j,k=2,\dots,N} d\alpha \bigwedge_{k=2}^N d\phi_{1,k} d\psi_{1,k}, \tag{2.28}$$

where $\{\ \}_j^{\mathrm{s}}$ denotes the real (s = r) or imaginary (s = i) part of the jth component. We observe that the matrix in (2.28) can be written in the factorized form, giving

$$\det\left(\begin{bmatrix} -\{\mathbf{U}_1\}_{1k}^{\mathrm{i}} & \{\mathbf{U}_1\}_{1k}^{\mathrm{r}} \\ \{\mathbf{U}_1\}_{jk}^{\mathrm{r}} & \{\mathbf{U}_1\}_{jk}^{\mathrm{i}} \\ -\{\mathbf{U}_1\}_{jk}^{\mathrm{i}} & \{\mathbf{U}_1\}_{jk}^{\mathrm{r}} \end{bmatrix}_{\substack{j=2,\dots,N \\ k=1,\dots,N}} \begin{bmatrix} \Big\{\frac{\partial \vec{u}_1^{(1)}}{\partial\alpha}\Big\}_j^{\mathrm{r}} & \Big\{\frac{\partial \vec{u}_1^{(1)}}{\partial\phi_{1,k}}\Big\}_j^{\mathrm{r}} & \Big\{\frac{\partial \vec{u}_1^{(1)}}{\partial\psi_{1,k}}\Big\}_j^{\mathrm{r}} \\ \Big\{\frac{\partial \vec{u}_1^{(1)}}{\partial\alpha}\Big\}_j^{\mathrm{i}} & \Big\{\frac{\partial \vec{u}_1^{(1)}}{\partial\phi_{1,k}}\Big\}_j^{\mathrm{i}} & \Big\{\frac{\partial \vec{u}_1^{(1)}}{\partial\psi_{1,k}}\Big\}_j^{\mathrm{i}} \end{bmatrix}_{\substack{j=1,\dots,N \\ k=2,\dots,N}}\right). \tag{2.29}$$

According to the Binet-Cauchy theorem [8] (see (6.88) below) the determinant of such a product of matrices is equal to the sum of the product of determinants of the matrix obtained by blocking out the lth column of the $2N-1 \times 2N$ matrix and the lth row of the $2N \times 2N-1$ matrix. From the explicit formulas

$$\begin{aligned} &u_{11} = \cos\phi_{1,2}\cdots\cos\phi_{1,N}e^{i\alpha_1}, \quad u_{21} = -\sin\phi_{1,2}e^{-i\psi_{1,2}}\cos\phi_{1,3}\cdots\cos\phi_{1,N}e^{i\alpha_1}, \\ &u_{31} = -\sin\phi_{1,3}e^{-i\psi_{1,3}}\cos\phi_{1,4}\cdots\cos\phi_{1,N}e^{i\alpha_1}, \quad \dots, \quad u_{N1} = -\sin\phi_{1,N}e^{-i\psi_{1,N}} \end{aligned} \tag{2.30}$$

we find

$$\det \begin{bmatrix} \left\{ \frac{\partial \vec{u}_1^{(1)}}{\partial \alpha} \right\}_j^{\rm r} & \left\{ \frac{\partial \vec{u}_1^{(1)}}{\partial \phi_{1,k}} \right\}_j^{\rm r} & \left\{ \frac{\partial \vec{u}_1^{(1)}}{\partial \psi_{1,k}} \right\}_j^{\rm r} \\ \left\{ \frac{\partial \vec{u}_1^{(1)}}{\partial \alpha} \right\}_j^{\rm i} & \left\{ \frac{\partial \vec{u}_1^{(1)}}{\partial \phi_{1,k}} \right\}_j^{\rm i} & \left\{ \frac{\partial \vec{u}_1^{(1)}}{\partial \psi_{1,k}} \right\}_j^{\rm i} \end{bmatrix}^{(l)}_{\substack{j=1,\dots,N \\ k=2,\dots,N}} = (-1)^{l-1} \{\mathbf{U}_1^\dagger\}^{\rm s}_{1\,[(l+1)/2]} \prod_{k=2}^N \sin \phi_{1,k} (\cos \phi_{1,k})^{2(N-k)+1}$$

where the superscript (l) on the l.h.s. denotes that the lth row of the matrix is to be deleted, and on the superscript s on r.h.s. denotes the real part for l odd and the imaginary part for l even. Thus (2.29) is equal to

$$\Big(\prod_{k=2}^N \sin \phi_{1,k} (\cos \phi_{1,k})^{2(N-k)+1} \Big) \sum_{l=1}^{2N} (-1)^{l-1} \{\mathbf{U}_1^\dagger\}^{\rm s}_{1\,[(l+1)/2]} \det \begin{bmatrix} -\{\mathbf{U}_1\}^{\rm i}_{1k} & \{\mathbf{U}_1\}^{\rm r}_{1k} \\ \{\mathbf{U}_1\}^{\rm r}_{jk} & \{\mathbf{U}_1\}^{\rm i}_{jk} \\ -\{\mathbf{U}_1\}^{\rm i}_{jk} & \{\mathbf{U}_1\}^{\rm r}_{jk} \end{bmatrix}^{(l)}_{\substack{j=2,\dots,N \\ k=1,\dots,N}}$$

where now the superscript (l) denotes that the lth column is to be deleted. But the sum over l is precisely the Laplace expansion of $\det \mathbf{U}_1^\dagger$ by the first row where the $N \times N$ complex matrix $\mathbf{U}_1^\dagger$ is written as a $2N \times 2N$ real orthogonal matrix by replacing each complex element by its 2×2 real matrix representation (1.36). Since a real orthogonal matrix has determinant equal to ± 1, we have that

$$(\mathbf{U}_1^\dagger d\vec{u}^{(1)}) = d\alpha \bigwedge_{k=2}^N \sin \phi_{1,k} (\cos \phi_{1,k})^{2(N-k)+1} d\phi_{1,k} d\psi_{1,k}$$

and the stated formula for the volume form now follows by induction. □

We remark that the precise detail of the decomposition presented in Proposition 2.3.1 differs from the one originally given by Hurwitz [302]. In Proposition 2.3.1 the matrix $\mathbf{U}\mathbf{U}_1^\dagger$ has the structure (2.24), whereas in the decomposition of [302] the analogous matrix product has the unit element in the bottom right corner. Further details are given in Exercises 2.3 q.1.

From Proposition 2.3.1, we see that with $\phi_{jk} = \arccos \xi_{jk}^{1/2(N-k+1)}$, $0 \le \xi_{jk} \le 1$, $(\mathbf{U}^\dagger d\mathbf{U})$ gives the uniform measure (in α_j, ψ_{jk} and ξ_{jk}) so that all matrices are equally probable, which is what is required for the CUE. Thus with α_j, ψ_{jk} and ξ_{jk} chosen at random with uniform density from their respective intervals, the unitary matrix $\mathbf{U}$ formed according to the decomposition in Proposition 2.3.1 will be a member of the CUE.

An essential ingredient in the Hurwitz construction is the decomposition (2.24). The analogous decomposition

$$\mathbf{U} = \mathbf{U}^{(1)} \begin{bmatrix} 1 & \mathbf{0}_{1 \times N-1} \\ \mathbf{0}_{N-1 \times 1} & \mathbf{V} \end{bmatrix} \tag{2.31}$$

has been used in [134] to give a probabilistic generation of $\mathbf{U}$. The idea is that $\mathbf{U}^{(1)}$ in (2.31) can be interpreted as an element of the quotient space $U(N)/U(N-1)$. The latter has the geometrical interpretation as the point on the complex $(N-1)$-sphere determining the axis of the lower-dimensional complex rotations specified by $U(N-1)$ (i.e., the image of the unit vector $\vec{e}_1$ under $U(N)$). Thus given $\mathbf{V} \in \mathrm{CUE}_{N-1}$ (CUE_n denoting the circular unitary ensemble of $n \times n$ random unitary matrices), $\mathbf{U} \in \mathrm{CUE}_N$ follows from (2.31) by choosing $\mathbf{U}^{(1)}$ as the matrix corresponding to mapping $\vec{e}_1$ to a random point $\vec{x}^{(1)}$ on the complex $(N-1)$-sphere. The first task is to generate $\vec{x}^{(1)}$. Analogous to the result of Exercises 1.2 q.2(iv) this can be achieved by forming the vector $\vec{z} = (z_1, \dots, z_N)$ in which each (complex) component has distribution $z_j/|\vec{z}|$, with each z_i a complex normal random variable with mean 0 and variance 1. Let $z_1 = |z_1| e^{i\theta_1}$, $0 \le \theta_1 < 2\pi$ and define $\vec{v}^{(1)} = \vec{x}_1^{(1)} + e^{i\theta_1} \vec{e}_1$, $\vec{u}^{(1)} = \vec{v}^{(1)}/|\vec{v}^{(1)}|$. Then it is straightforward to check that the matrix

$$\mathbf{U}^{(1)} = -e^{i\theta_1} (\mathbf{1}_N - 2\vec{u}^{(1)} (\vec{u}^{(1)*})^T),$$

where here $*$ denotes the complex conjugate, is unitary and has the sought property $\mathbf{U}^{(1)}\vec{e}_1 = \vec{x}^{(1)}$ (cf. Exercises 1.9 q.3).

We can now proceed inductively. Define $\vec{x}^{(j)} = \vec{z}^{(j)}/|\vec{z}^{(j)}|$ where $\vec{z}^{(j)} = (z_j, \dots, z_N)$ with each z_i a standard complex Gaussian chosen independently for a given j. Write $z_j = |z_j|e^{i\theta_j}, 0 \le \theta_j < 2\pi$, and define $\vec{v}^{(j)} = \vec{x}^{(j)} + e^{i\theta_j}\vec{e}_j$, and $\vec{u}^{(j)} = \vec{v}^{(j)}/|\vec{v}^{(j)}|$. From these quantities construct the unitary matrix

$$\mathbf{U}^{(j)} = \begin{bmatrix} \mathbf{1}_{j-1} & \mathbf{0}_{(j-1)\times(N-j+1)} \\ \mathbf{0}_{(N-j+1)\times(j-1)} & -e^{-i\theta_j}(\mathbf{1}_{N-j+1} - 2\vec{u}^{(j)}(\vec{u}^{(j)*})^T) \end{bmatrix}.$$

Then the matrix

$$\mathbf{U} = \mathbf{U}^{(1)}\mathbf{U}^{(2)}\cdots\mathbf{U}^{(N)}$$

gives a Haar distributed element of $U(N)$.

The diagonalization of a GUE matrix $\mathbf{X}$ also leads to the construction of a member of the CUE. To see this, we know from Exercises 1.3 q.3 that

$$(d\mathbf{X}) = \prod_{1\le j<k\le N} (\lambda_k - \lambda_j)^2 \bigwedge_{j=1}^{N} d\lambda_j\, (\mathbf{U}^\dagger d\mathbf{U}),$$

where the columns of the unitary matrix $\mathbf{U}$ consist of the normalized eigenvectors of $\mathbf{X}$, chosen, for example, so that their first element is real and positive. This subgroup of unitary matrices can be extended to the full group of unitary matrices by multiplying each eigenvector by an arbitrary phase $e^{i\alpha_j}, 0 \le \alpha_j < 2\pi$. Defining $\mathbf{U}$ out of these modified eigenvectors, the Haar form $(\mathbf{U}^\dagger d\mathbf{U})$ in the formula for $(d\mathbf{X})$ is unchanged since it is invariant under all unitary transformations, and thus a random unitary matrix results. A still more efficient way to calculate a member of the CUE via a decomposition of another random matrix is noted below the proof of Proposition 3.2.5.

2.3.2 Integration over the elements of a unitary matrix

The scattering matrix problem of Section 2.1.1 motivates the study of the matrix averages

$$Q^{a_1\alpha_1,\dots,a_l\alpha_l}_{b_1\beta_1,\dots,b_m\beta_m} := \langle (U_{b_1\beta_1}\cdots U_{b_m\beta_m})(\overline{U_{a_1\alpha_1}\cdots U_{a_l\alpha_l}})\rangle_{\mathbf{U}\in U(N)}. \tag{2.32}$$

We first make note of a number of general properties of this average [398].

PROPOSITION 2.3.2 *The average (2.32) is (a) nonzero only if $\{a_j\} = \{b_j\}$, and $\{\alpha_j\} = \{\beta_j\}$; (b) is unchanged by interchanging the labels of the rows or columns of $\mathbf{U}$; (c) is unchanged by permuting $a_1\alpha_1, \dots, a_l\alpha_l$ or $b_1\beta_1, \dots, b_m\beta_m$; (d) satisfies*

$$\sum_{n=1}^{N} Q^{n\alpha_1, a_2\alpha_2,\dots,a_m\alpha_m}_{n\beta_1, b_2\beta_2,\dots,b_m\beta_m} = \delta_{\alpha_1,\beta_1} Q^{a_2\alpha_2,\dots,a_m\alpha_m}_{b_2\beta_2,\dots,b_m\beta_m}.$$

Proof. Let $\mathbf{U}^{(0)} = \mathrm{diag}(e^{i\theta_1}, \dots, e^{i\theta_N})$. Since the Haar form is invariant under $\mathbf{U} \mapsto \mathbf{U}^{(0)}\mathbf{U}$, $\mathbf{U} \mapsto \mathbf{U}\mathbf{U}^{(0)}$ we must have

$$\begin{aligned} Q^{a_1\alpha_1,\dots,a_l\alpha_l}_{b_1\beta_1,\dots,b_m\beta_m} &= e^{i(\theta_{b_1}+\cdots+\theta_{b_m}) - i(\theta_{a_1}+\cdots+\theta_{a_l})} Q^{a_1\alpha_1,\dots,a_l\alpha_l}_{b_1\beta_1,\dots,b_m\beta_m} \\ &= Q^{a_1\alpha_1,\dots,a_l\alpha_l}_{b_1\beta_1,\dots,b_m\beta_m} e^{i(\theta_{\beta_1}+\cdots+\theta_{\beta_m}) - i(\theta_{\alpha_1}+\cdots+\theta_{\alpha_l})}, \end{aligned}$$

which implies (a). Property (b) follows from the invariance of the Haar form under the interchange of rows or columns, while (c) follows immediately from the definition (2.32). Finally, (d) follows from $\sum_{n=1}^{N} u_{n\alpha_1}\bar{u}_{n\beta_1} = \delta_{\alpha_1,\beta_1}$ which in turn characterizes $\mathbf{U}$ as a unitary matrix. □

The results of Proposition 2.3.2 can be used to calculate (2.32) for $l = m = 1$ and $l = m = 2$ [53].

PROPOSITION 2.3.3 *We have*

$$Q^{a_1\alpha_1}_{b_1\beta_1} = \frac{1}{N}\delta_{a_1,b_1}\delta_{\alpha_1,\beta_1}, \tag{2.33}$$

$$\begin{aligned} Q^{a_1\alpha_1,a_2\alpha_2}_{b_1\beta_1,b_2\beta_2} &= \frac{1}{N^2-1}\Big(\delta_{a_1,b_1}\delta_{\alpha_1,\beta_1}\delta_{a_2,b_2}\delta_{\alpha_2,\beta_2} + \delta_{a_1,b_2}\delta_{\alpha_1,\beta_2}\delta_{a_2,b_1}\delta_{\alpha_2,\beta_1}\Big) \\ &\quad - \frac{1}{N(N^2-1)}\Big(\delta_{a_1,b_1}\delta_{\alpha_1,\beta_2}\delta_{a_2,b_2}\delta_{\alpha_2,\beta_1} + \delta_{a_1,b_2}\delta_{\alpha_1,\beta_1}\delta_{a_2,b_1}\delta_{\alpha_2,\beta_2}\Big). \end{aligned} \tag{2.34}$$

Proof. The Kronecker deltas in (2.33) follow from (a), while (b) tells us that $Q^{a\alpha}_{a\alpha} = Q^{11}_{11}$, which when combined with (d) implies the nonzero value in (2.33) of $1/N$. In relation to (2.34), according to (a), (b), (c) it suffices to show that

$$Q^{11,11}_{11,11} = \frac{2}{N(N+1)}, \quad Q^{11,21}_{11,21} = \frac{1}{N(N+1)}, \quad Q^{13,24}_{13,24} = \frac{1}{N^2-1}, \quad Q^{23,14}_{13,24} = -\frac{1}{N(N^2-1)}. \tag{2.35}$$

The invariance of (2.32) under the mapping $\mathbf{U} \mapsto \mathbf{U}^{(0)}\mathbf{U}$ tells us that

$$Q^{11,11}_{11,11} = 2Q^{11,21}_{11,21}\sum_{m\neq n}|U^{(0)}_{1m}|^2|U^{(0)}_{1n}|^2 + Q^{11,11}_{11,11}\sum_{n=1}^{N}|U^{(0)}_{1n}|^4,$$

where use has been made of the facts that $Q^{m1,n1}_{m1,n1} = Q^{11,21}_{11,21}$ $(m\neq n)$, $Q^{n1,n1}_{n1,n1} = Q^{11,11}_{11,11}$. Using the unitarity of $\mathbf{U}^{(0)}$ it follows that

$$Q^{11,11}_{11,11} = 2Q^{11,21}_{11,21}.$$

On the other hand, (d) and (2.33) tell us that

$$Q^{11,11}_{11,11} + (N-1)Q^{11,21}_{11,21} = \frac{1}{N},$$

thereby implying the first two equations in (2.35). The derivation of the remaining equations is similar. □

It follows from (2.33) and (2.34) that

$$\begin{aligned} \langle |U_{nm}|^2\rangle_{\mathbf{U}\in U(N)} &= \frac{1}{N}, \\ \langle |U_{nm}|^2\rangle_{\mathbf{U}\in \mathrm{COE}} = \langle |(\mathbf{U}\mathbf{U}^T)_{nm}|^2\rangle_{\mathbf{U}\in U(N)} &= \frac{1+\delta_{n,m}}{N+1}, \\ \langle |U_{nm}|^2\rangle_{\mathbf{U}\in \mathrm{CSE}} = \langle |(\mathbf{U}\mathbf{U}^D)_{nm}|^2\rangle_{\mathbf{U}\in U(2N)} &= \frac{2-\delta_{n,m}}{2N-1} \end{aligned} \tag{2.36}$$

(in the final formula $\mathbf{U}\mathbf{U}^D$ is to be regarded a quaternion real matrix for purposes of computing the absolute values of the elements). For the diagonal entries $n = m$, the full distribution of $|U_{nm}|^2$ can in fact be computed in each case (see Exercises 3.8 q.3).

We remark that an evaluation formula for (2.32) in general, involving a double sum over permutations of N, with summand containing a further summation involving group characters (see Section 11.6.3) known as the Weingarten function, has been derived [114].

EXERCISES 2.3 1. Here the decomposition of Proposition 2.3.1 is modified to correspond with that of [302]. Let

$$\begin{aligned} \mathbf{E}_k := {} & \mathbf{U}^{(k,k+1)}(\phi_{k,k+1},\psi_{k,k+1},0)\mathbf{U}^{(k-1,k+1)}(\phi_{k-1,k+1},\psi_{k-1,k+1},0) \\ & \times\cdots\times \mathbf{U}^{(2,k+1)}(\phi_{2,k+1},\psi_{2,k+1},0)\mathbf{U}^{(1,k+1)}(\phi_{1,k+1},\psi_{1,k+1},\alpha_k). \end{aligned}$$

Modify the working of the proof of Proposition 2.3.1 to show that for appropriate choice of the parameters,

$$\mathbf{U}\mathbf{E}^\dagger_{N-1} = \begin{bmatrix} \mathbf{V} & \mathbf{0}_{N-1\times 1} \\ \mathbf{0}_{1\times N-1} & 1 \end{bmatrix},$$

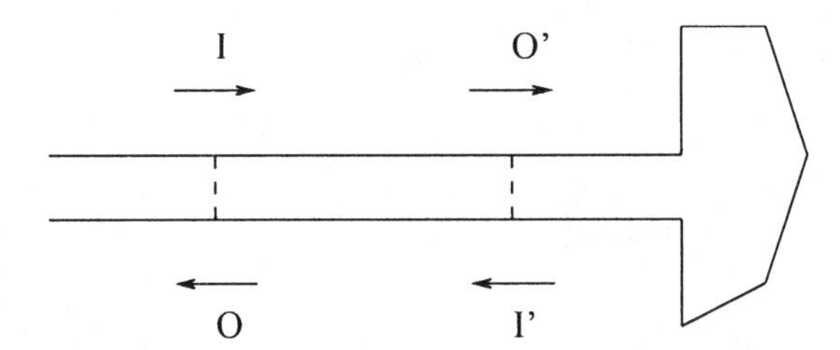

Figure 2.1 Schematic diagram of a lead with a tunneling barrier coupled to a chaotic cavity, indicating the notation for the various incoming and outgoing waves.

where $\mathbf{V}$ is an $(N-1)$-dimensional unitary matrix. Hence conclude that an arbitrary unitary matrix $\mathbf{U}$ can be decomposed as

$$\mathbf{U} = e^{i\alpha_0}\mathbf{E}_1\mathbf{E}_2\cdots\mathbf{E}_{N-1}.$$

2.4 POISSON KERNEL

The consideration of a more general scattering problem [101] than that of Section 2.1.1 leads to an ensemble of random unitary matrices S specified by the *Poisson kernel*

$$P(\mathbf{S}) = \frac{1}{C}\left(\frac{\det(\mathbf{1}-\bar{\mathbf{S}}^\dagger\bar{\mathbf{S}})}{|\det(\mathbf{1}-\bar{\mathbf{S}}^\dagger\mathbf{S})|^2}\right)^{\beta(N-1)/2+1}. \tag{2.37}$$

Here $\bar{\mathbf{S}} =: \int \mathbf{S}P(\mathbf{S})\,(d_H\mathbf{S})$ and thus denotes the mean of S. For the circular ensembles, $\bar{\mathbf{S}} = \mathbf{0}$ and (2.37) reduces to the uniform distribution. Generally (2.37) has the reproducing property [300]

$$f(\bar{\mathbf{S}}) = \int P(\mathbf{S})f(\mathbf{S})\,(d_H\mathbf{S}) \tag{2.38}$$

valid for any analytic function f of the matrix S. In the case $N=1$ (2.38) reads

$$f(z) = \frac{1}{2\pi}\int_0^{2\pi}\frac{(1-|z|^2)}{|1-\bar{z}e^{i\theta}|^2}f(e^{i\theta})\,d\theta, \tag{2.39}$$

which is the Poisson formula on a circle giving the value of a harmonic function f for $|z|<1$ in terms of its value on the unit circle.

The setting of the scattering problem is as in Figure 2.1. The lead contains a tunneling barrier, through which a portion of the incoming wave specified by the N-component vector $\vec{I}$ is transmitted and a portion is reflected. The scattering of the incoming wave through the tunneling barrier is described by the equation

$$\mathbf{S}_1\begin{bmatrix}\vec{I}\\ \vec{I'}\end{bmatrix} := \begin{bmatrix}\mathbf{r} & \mathbf{t}'\\ \mathbf{t} & \mathbf{r}'\end{bmatrix}\begin{bmatrix}\vec{I}\\ \vec{I'}\end{bmatrix} = \begin{bmatrix}\vec{O}\\ \vec{O'}\end{bmatrix}, \tag{2.40}$$

where $\mathbf{r}$ and $\mathbf{t}$ denote the $N\times N$ transmission and reflection matrix, respectively, for the incoming wave at the left of the barrier, and similarly $\mathbf{r}'$ and $\mathbf{t}'$ at the right of the barrier. As well as being unitary, the matrix $\mathbf{S}_1$ is required to be symmetric ($\beta=1$) and self-dual ($\beta=4$). Similarly, the scattering at the cavity entrance and the scattering in the total system, are described by the equations

$$\mathbf{S}_0\vec{O'} = \vec{I'}, \qquad \mathbf{S}\vec{I} = \vec{O}. \tag{2.41}$$

Straightforward manipulation of (2.40) and (2.41) shows that $\mathbf{S}$ is related to $\mathbf{S}_0$ by

$$\mathbf{S} = \mathbf{r} + \mathbf{t}'(\mathbf{1} - \mathbf{S}_0\mathbf{r}')^{-1}\mathbf{S}_0\mathbf{t}. \tag{2.42}$$

The Jacobian for the change of variables from $\mathbf{S}_0$ to $\mathbf{S}$ gives the Poisson kernel (2.37) for a certain choice of $\bar{\mathbf{S}}$ [300], [400].

PROPOSITION 2.4.1 *With* $\bar{\mathbf{S}} = \mathbf{r}$ *we have*

$$(d_H\mathbf{S}_0) = \left(\frac{\det(\mathbf{1} - \bar{\mathbf{S}}^\dagger\bar{\mathbf{S}})}{|\det(\mathbf{1} - \bar{\mathbf{S}}^\dagger\mathbf{S})|^2}\right)^{\beta(N-1)/2+1}(d_H\mathbf{S}) \tag{2.43}$$

and thus the Jacobian is proportional to the Poisson kernel (2.37).

Proof. First, one can check from the unitarity of $\mathbf{S}_1$ in (2.40) that

$$\mathbf{t} - \mathbf{r}'\mathbf{t}'^{-1}\mathbf{r} = (\mathbf{t}^\dagger)^{-1}, \quad \mathbf{t}^\dagger\mathbf{r}' = -\mathbf{r}^\dagger\mathbf{t}'.$$

These equations used in (2.42) imply

$$\mathbf{S}_0 = \mathbf{t}'^{-1}(\mathbf{S} - \mathbf{r})(\mathbf{1} - \mathbf{r}^\dagger\mathbf{S})^{-1}\mathbf{t}^\dagger. \tag{2.44}$$

Differentiating $\mathbf{X}\mathbf{X}^{-1} = \mathbf{1}$ gives $d\mathbf{X}^{-1} = -\mathbf{X}^{-1}d\mathbf{X}\mathbf{X}^{-1}$ and so we deduce from (2.44) that

$$d\mathbf{S}_0 = \mathbf{t}'^{-1}\Big(\mathbf{1} + (\mathbf{S} - \mathbf{r})(\mathbf{1} - \mathbf{r}^\dagger\mathbf{S})^{-1}\mathbf{r}^\dagger\Big)d\mathbf{S}(\mathbf{1} - \mathbf{r}^\dagger\mathbf{S})^{-1}\mathbf{t}^\dagger.$$

Use of the matrix identity

$$(\mathbf{1} - \mathbf{A}\mathbf{B})^{-1}\mathbf{A} = \mathbf{A}(\mathbf{1} - \mathbf{B}\mathbf{A})^{-1}, \tag{2.45}$$

and the formula $\mathbf{1} - \mathbf{r}\mathbf{r}^\dagger = \mathbf{t}'\mathbf{t}'^\dagger$ which follows from the unitarity of the scattering matrix in (2.40), shows that this can be rewritten

$$d\mathbf{S}_0 = \mathbf{t}'^\dagger(\mathbf{1} - \mathbf{S}\mathbf{r}^\dagger)^{-1}d\mathbf{S}(\mathbf{1} - \mathbf{r}^\dagger\mathbf{S})^{-1}\mathbf{t}^\dagger. \tag{2.46}$$

Let us suppose now that $\beta = 2$, so the volume form is given by (2.10). Writing $d\mathbf{S}$ as $\mathbf{S}\,\mathbf{S}^{-1}d\mathbf{S}$ in (2.46), using once more the identity (2.45) and multiplying both sides by $\mathbf{S}_0^\dagger$ shows

$$\mathbf{S}_0^\dagger d\mathbf{S}_0 = \mathbf{t}(\mathbf{1} - \mathbf{S}\mathbf{r}^\dagger)^{-1}\mathbf{S}d\mathbf{S}(\mathbf{1} - \mathbf{r}^\dagger\mathbf{S})^{-1}\mathbf{t}^\dagger. \tag{2.47}$$

The change of variables formula (1.35) in the case $\beta = 2$ applied to (2.47) gives (2.43) in the case $\beta = 2$, where use is also made of the fact that $\mathbf{1} - \mathbf{r}^\dagger\mathbf{r} = \mathbf{t}^\dagger\mathbf{t}$.

In the cases $\beta = 1$ and $\beta = 4$ we start again from (2.46). Recalling the formulas (2.15) and (2.17), we decompose $\mathbf{S}_0$ and $\mathbf{S}$ in terms of unitary matrices $\mathbf{U}_0$, $\mathbf{U}$, according to (2.13) and (2.16) as appropriate. Using the additional fact that $\mathbf{S}_1$ is symmetric (self-dual quaternion) for $\beta = 1$ ($\beta = 4$) and thus $\mathbf{r} = \mathbf{r}^T$ ($\mathbf{r} = \mathbf{r}^D$) we see that

$$\mathbf{U}_0 d\mathbf{S}_0\,\mathbf{U}_0^{-1} = \mathbf{B}^\dagger\mathbf{U}d\mathbf{S}\,\mathbf{U}^{-1}\mathbf{B}, \qquad \mathbf{B} := \mathbf{U}(\mathbf{1} - \mathbf{r}^\dagger\mathbf{S})^{-1}\mathbf{t}^\dagger\mathbf{U}_0^{-1}. \tag{2.48}$$

Furthermore, we can check that $\mathbf{B}$ has real elements for $\beta = 1$, complex elements for $\beta = 2$ and real quaternion elements for $\beta = 4$ (see Exercises 2.4 q.1). Applying the change of variables formula (1.35) in the cases $\beta = 1$ and $\beta = 4$ to (2.48) gives (2.43) in the cases $\beta = 1$ and $\beta = 4$. □

From (2.43) and the fact that $P(\mathbf{S}_0) = 1/C$ we have

$$P(\mathbf{S}_0)(d_H\mathbf{S}_0) = P(\mathbf{S})(d_H\mathbf{S}),$$

where $P(\mathbf{S})$ is given by (2.37). Thus the Poisson kernel with $\bar{\mathbf{S}} = \mathbf{r}$ represents the distribution of the $\mathbf{S}$ matrices for the lead with a tunneling barrier coupled to a chaotic cavity.

The eigenvalue p.d.f. corresponding to the Poisson kernel (2.37) can be specified in the special case $\bar{\mathbf{S}} = z\mathbf{1}_N$, when (2.37) is a function only of the eigenvalues $e^{i\theta_j}$. The eigenvalue p.d.f. is then obtained by multiplying the Poisson kernel by the eigenvalue p.d.f. (2.20) for the circular ensembles, and is thus given by

$$\frac{1}{\tilde{C}}\prod_{l=1}^{N}\frac{1}{|1-ze^{i\theta_l}|^{\beta(N-1)+2}}\prod_{1\le j<k\le N}|e^{i\theta_k}-e^{i\theta_j}|^{\beta}. \tag{2.49}$$

This is proportional to the Boltzmann factor of the one-component log-potential Coulomb gas on a circle, with an external charge of strength $-(N-1)+2/\beta$ at the point $1/z$ in the complex plane.

EXERCISES 2.4 1. [400] The objective of this exercise is to show that the matrix $\mathbf{B}$ in (2.48) has real elements for $\beta=1$ and real quaternion elements for $\beta=4$. This is equivalent to showing $\mathbf{B}^\dagger=\mathbf{B}^T$ and $\mathbf{B}^\dagger=\mathbf{B}^D$, respectively.

(i) Use (2.44) to show that the formula for $\mathbf{B}$ can be rewritten

$$\mathbf{B}=\mathbf{U}(\mathbf{S}-\mathbf{r})^{-1}\mathbf{t}'\mathbf{S}_0\mathbf{U}_0^{-1}.$$

(ii) Use the fact that $\mathbf{S}_0=\mathbf{U}_0^T\mathbf{U}_0$ $(\beta=1)$, $\mathbf{S}_0=\mathbf{U}_0^D\mathbf{U}_0$ $(\beta=4)$ to rewrite the formula in (i) as

$$\mathbf{B}=\mathbf{U}(\mathbf{S}-\mathbf{r})^{-1}\mathbf{t}'\mathbf{U}_0^T\ \ (\beta=1),\qquad \mathbf{B}=\mathbf{U}(\mathbf{S}-\mathbf{r})^{-1}\mathbf{t}'\mathbf{U}_0^D\ \ (\beta=4).$$

(iii) Similar to (ii) use the decompositions $\mathbf{S}=\mathbf{U}^T\mathbf{U}$ $(\beta=1)$, $\mathbf{S}=\mathbf{U}^D\mathbf{U}$ $(\beta=4)$ to show from the formula in (2.48) that

$$\mathbf{B}^\dagger=\mathbf{U}_0\mathbf{t}(\mathbf{S}-\mathbf{r})^{-1}\mathbf{U}^T\ \ (\beta=1),\qquad \mathbf{B}^\dagger=\mathbf{U}_0\mathbf{t}(\mathbf{S}-\mathbf{r})^{-1}\mathbf{U}^D\ \ (\beta=4).$$

(iv) Use the facts that $\mathbf{r}^T=\mathbf{r},\mathbf{t}^T=\mathbf{t}',\mathbf{S}^T=\mathbf{S}$ $(\beta=1)$, and $\mathbf{r}^D=\mathbf{r},\mathbf{t}^D=\mathbf{t}',\mathbf{S}^D=\mathbf{S}$ $(\beta=4)$ in the results of (iii) and compare with the results of (ii) to conclude that $\mathbf{B}^\dagger=\mathbf{B}^T$ $(\beta=1)$ and $\mathbf{B}^\dagger=\mathbf{B}^D$ $(\beta=4)$ as required.

2.5 CAUCHY ENSEMBLE

The mapping

$$e^{i\theta}=\frac{1+i\lambda}{1-i\lambda} \tag{2.50}$$

maps each point λ onto the real line to a point $e^{i\theta}$ $(-\pi<\theta<\pi)$ on the unit circle via a stereographic projection. The angle θ can be constructed by drawing the unit circle in the complex plane with center $(0,i)$ and projecting from the point $(0,2i)$ on the circle to the point λ on the real axis. The intersection of this line with the unit circle gives θ which is measured anticlockwise from the point $(0,0)$. Changing variables for each θ_j in (2.20) according to (2.50) gives

$$\prod_{1\le j<k\le N}|e^{i\theta_k}-e^{i\theta_j}|^{\beta}d\theta_1\cdots d\theta_N$$

$$=2^{N+\beta N(N-1)/2}\prod_{j=1}^{N}(1+\lambda_j^2)^{-\beta(N-1)/2-1}\prod_{1\le j<k\le N}|\lambda_k-\lambda_j|^{\beta}d\lambda_1\cdots d\lambda_N. \tag{2.51}$$

In orthogonal polynomial theory the function $w(x)=1/(1+x^2)^{\alpha}$ is referred to as the *Cauchy weight*. The result (2.51) motivates the definition of the Cauchy ensemble of random matrices via the eigenvalue p.d.f.

$$\frac{1}{C}\prod_{j=1}^{N}(1+\lambda_j^2)^{-\alpha}\prod_{1\le j<k\le N}|\lambda_k-\lambda_j|^{\beta}. \tag{2.52}$$

In the case $\alpha = \beta(N-1)/2+1$, $\beta = 1, 2$ and 4, (2.51) gives that this is realized by the stereographic projection of the eigenvalue p.d.f. of the circular ensembles onto the real line.

In fact Hermitian matrices $\mathbf{H}$ with the eigenvalue p.d.f. (2.52) for $\alpha = \beta(N-1)/2+1$, $\beta = 1, 2$ and 4, can be constructed out of their circular counterparts $\mathbf{U}$ by making the *Cayley transformation*

$$\mathbf{H} = i\frac{\mathbf{1}_N - \mathbf{U}}{\mathbf{1}_N + \mathbf{U}}. \tag{2.53}$$

To see this, note from the inverse of this relation,

$$\mathbf{U} = \frac{\mathbf{1}_N + i\mathbf{H}}{\mathbf{1}_N - i\mathbf{H}} \tag{2.54}$$

that the eigenvalues $e^{i\theta_j}$ of $\mathbf{U}$ are related to the eigenvalues λ_j of $\mathbf{H}$ by (2.50). It follows from (2.51) that the joint p.d.f. $P(\mathbf{H})$ of the matrices $\mathbf{H}$ can be specified by

$$P(\mathbf{H}) = \frac{1}{C}\Big(\det(\mathbf{1}_N + \mathbf{H}^2)\Big)^{-\beta(N-1)/2-1} \tag{2.55}$$

for $\mathbf{H}$ real symmetric ($\beta = 1$), complex Hermitian ($\beta = 2$) and self-dual quaternion real ($\beta = 4$).

The Cayley transformation can be used to compute the volume form $(\mathbf{U}^\dagger d\mathbf{U})$ in terms of the volume form $(d\mathbf{H})$. Since the volume form $(d\mathbf{H})$ has previously been computed in terms of the eigenvalues and eigenvectors of $\mathbf{H}$, and the eigenvalues of $\mathbf{H}$ and $\mathbf{U}$ are related by (2.50), this gives an alternative way to derive the result (2.20) [300]. The calculation begins by making use of the formula of Exercises 2.5 q.1 to deduce from (2.54) that

$$\mathbf{U}^\dagger d\mathbf{U} = 2i(\mathbf{1}_N + i\mathbf{H})^{-1} d\mathbf{H} (\mathbf{1}_N - i\mathbf{H})^{-1}. \tag{2.56}$$

For $\mathbf{U}$ symmetric (self-dual quaternion) it follows from (2.53) that $\mathbf{H}$ will be symmetric (self-dual quaternion), and thus so will be $(\mathbf{1}_N \pm i\mathbf{H})^{-1}$. Hence we can apply the result of Exercises 1.3 q.2 on the r.h.s. of (2.56) to deduce

$$(\delta\mathbf{U}) = 2^{N(\beta(N-1)/2+1)}\Big(\det(\mathbf{1}_N + \mathbf{H}^2)\Big)^{-\beta(N-1)/2-1}(d\mathbf{H}), \tag{2.57}$$

where $\beta = 1, 2$ or 4 for the elements of $\mathbf{H}$ real, complex or quaternion real, respectively. Changing variables to the eigenvalues and eigenvectors on the r.h.s. using the result of Proposition 1.3.4 and integrating out the eigenvector dependence gives the r.h.s. of (2.51), and thus the eigenvalue p.d.f. (2.20) after the change of variables (2.50).

For general β (2.52) represents the Boltzmann factor for a one-component log-gas on a line (the x-axis) subject to a one-body potential $\beta V(\lambda) = \alpha \log(1+\lambda^2)$, which corresponds to an external charge of strength $-2\alpha/\beta$ at the point $(0,1)$ in the xy-plane. In the case $2\alpha/\beta = N - 1 + 2/\beta$ when (2.51) holds, since the particle density in the circular ensemble is uniform (by symmetry), making the change of variables $\rho_{(1)}(\theta)\,d\theta = \rho_{(1)}(\lambda)\,d\lambda$ according to (2.50) shows the density in this particular case of the Cauchy ensemble is given by

$$\rho_{(1)}(\lambda) = \frac{N}{\pi(1+\lambda^2)}, \tag{2.58}$$

independent of β.

We conclude this section by noting some special properties of $P(\mathbf{H})$ as specified by (2.55) [300], [101].

Proposition 2.5.1 *Let $\mathbf{H}$ have the distribution (2.55). Then (a) $\mathbf{H}^{-1}$ is also distributed according to (2.55); (b) every $n \times n$ submatrix of $\mathbf{H}$, obtained by omitting $N-n$ rows and corresponding columns of $\mathbf{H}$, is distributed according to the ensemble (2.55) with N replaced by n.*

Proof. For the result (a), we simply change variables $\mathbf{Y} = \mathbf{H}^{-1}$. Since $\mathbf{Y}\mathbf{H} = \mathbf{1}_N$ we see that $d\mathbf{H} = -\mathbf{Y}^{-1} d\mathbf{Y}\,\mathbf{Y}^{-1}$.

Use of (1.35) then gives $(d\mathbf{H}) = (\det \mathbf{Y}^{-1})^{\beta(N-1)+2}(d\mathbf{Y})$, and so

$$P(\mathbf{Y}) \propto \det\left(\mathbf{1}_N + \mathbf{Y}^{-2}\right)^{-\beta(N-1)/2-1} (\det \mathbf{Y}^{-1})^{\beta(N-1)+2},$$

which after simple manipulation reduces to the r.h.s. of (2.55).

To obtain the result (b), suppose for definiteness that the final row and column of $\mathbf{H}$ are deleted, and write

$$\mathbf{H} = \begin{bmatrix} \mathbf{G} & \vec{y} \\ \vec{y}^\dagger & h_{NN} \end{bmatrix},$$

where $\mathbf{G}$ is the $(N-1) \times (N-1)$ submatrix, $\vec{y}$ is a vector of length $N-1$ with real ($\beta = 1$), complex ($\beta = 2$) or quaternion real ($\beta = 4$) elements, and h_{NN} is real. The task is to compute the distribution of $\mathbf{G}$ given that $\mathbf{H}$ is distributed according to (2.55).

From the identity

$$\begin{aligned} \det(\mathbf{1}_N + \mathbf{H}^2) &= \det(\mathbf{1}_{N-1} + \mathbf{G}^2 + \vec{y}\vec{y}^\dagger) \\ &\times \det \begin{bmatrix} (\mathbf{1}_{N-1} + \mathbf{G}^2 + \vec{y}\vec{y}^\dagger)^{-1} & \mathbf{0}_{(N-1)\times 1} \\ \mathbf{0}_{1\times(N-1)} & 1 \end{bmatrix} \det \begin{bmatrix} \mathbf{1}_{N-1} + \mathbf{G}^2 + \vec{y}\vec{y}^\dagger & \mathbf{G}\vec{y} + \vec{y}h_{NN} \\ \vec{y}^\dagger\mathbf{G} + \vec{y}^\dagger h_{NN} & 1 + h_{NN}^2 + \vec{y}^\dagger\vec{y} \end{bmatrix}, \end{aligned}$$

in which the product of the first two terms equals unity, we see that

$$\det(\mathbf{1}_N + \mathbf{H}^2) = \det(\mathbf{1}_{N-1} + \mathbf{G}^2 + \vec{y}\vec{y}^\dagger)(ah_{NN}^2 + 2bh_{NN} + c), \tag{2.59}$$

where

$$\begin{aligned} a &= 1 - \vec{y}^\dagger(\mathbf{1}_{N-1} + \mathbf{G}^2 + \vec{y}\vec{y}^\dagger)^{-1}\vec{y}, \\ 2b &= -\vec{y}^\dagger\mathbf{G}(\mathbf{1}_{N-1} + \mathbf{G}^2 + \vec{y}\vec{y}^\dagger)^{-1}\vec{y} - \vec{y}^\dagger(\mathbf{1}_{N-1} + \mathbf{G}^2 + \vec{y}\vec{y}^\dagger)^{-1}\mathbf{G}\vec{y}, \\ c &= 1 + \vec{y}^\dagger\vec{y} - \vec{y}^\dagger\mathbf{G}(\mathbf{1}_{N-1} + \mathbf{G}^2 + \vec{y}\vec{y}^\dagger)^{-1}\mathbf{G}\vec{y}. \end{aligned}$$

Now diagonalize $\mathbf{G}$ by $\mathbf{G} = \mathbf{U}\mathrm{diag}(\lambda_1, \dots, \lambda_{N-1})\mathbf{U}^\dagger$ and set

$$\mathbf{T} = \mathbf{U}\mathrm{diag}\left(\sqrt{1+\lambda_1^2}, \dots, \sqrt{1+\lambda_{N-1}^2}\right)\mathbf{U}^\dagger$$

so that $\mathbf{T}^\dagger = \mathbf{T}$ and $\mathbf{T}^2 = \mathbf{1}_{N-1} + \mathbf{G}^2$. Also, introduce the vector $\vec{x}$ according to $\vec{y}^\dagger = \vec{x}^\dagger\mathbf{T}$, and note that

$$(d\vec{y}) = (\det \mathbf{T})^\beta (d\vec{x}) = \left(\det(\mathbf{1}_{N-1} + \mathbf{G}^2)\right)^{\beta/2}(d\vec{x}).$$

These results show

$$\begin{aligned} P(\mathbf{G}) := \int P(\mathbf{H})(d\vec{y})dh_{NN} &= \det\left(\mathbf{1}_{N-1} + \mathbf{G}^2\right)^{-\beta(N-2)/2-1} \\ &\times \int (d\vec{x}) \det(\mathbf{1}_{N-1} + \vec{x}\vec{x}^\dagger)^{-\beta(N-1)/2-1} \int_{-\infty}^{\infty} dh_{NN}(ah_{NN}^2 + 2bh_{NN} + c)^{-\beta(N-1)/2-1} \end{aligned} \tag{2.60}$$

where in terms of the vector $\vec{x}$

$$a = 1 - \vec{x}^\dagger(\mathbf{1}_{N-1} + \vec{x}\vec{x}^\dagger)^{-1}\vec{x} = \frac{1}{1+\alpha}, \qquad \alpha := \vec{x}^\dagger\vec{x}$$

(the second equality follows by multiplying $(\mathbf{1}_{N-1} + \vec{x}\vec{x}^\dagger)^{-1}$ by $(\vec{x}\vec{x}^\dagger)(\vec{x}\vec{x}^\dagger)^{-1}$ and using the fact that $\vec{x}^\dagger\vec{x} =: \alpha$ is a scalar) and similarly

$$b = -\frac{\vec{x}^\dagger\mathbf{G}\vec{x}}{1+\alpha}, \qquad c = 1 + \alpha + \frac{(\vec{x}^\dagger\mathbf{G}\vec{x})^2}{1+\alpha}.$$

Now, changing variables $y = (a/\sqrt{ac - b^2})(h_{NN} + b/a)$ shows that in general

$$\int_{-\infty}^{\infty} \frac{dh_{NN}}{(ah_{NN}^2 + 2bh_{NN} + c)^p} = a^{p-1}(ac - b^2)^{1/2-p} \int_{-\infty}^{\infty} \frac{dy}{(1 + y^2)^p},$$

assuming $p > \frac{1}{2}$, $a > 0$, $ac - b^2 > 0$. But here we have that $ac - b^2 = 1$, while a is positive and independent of $\mathbf{G}$. Thus (2.60) gives

$$P(\mathbf{G}) \propto \det(\mathbf{1}_{N-1} + \mathbf{G}^2)^{-\beta(N-2)/2-1},$$

which is the required result. □

EXERCISES 2.5 1. The objective of this exercise is to derive the identity

$$\frac{d}{da}(1 - K)^{-1} = (1 - K)^{-1} \frac{dK}{da} (1 - K)^{-1},$$

where 1 denotes the identity operator and it is assumed the operator K is a smooth function of a.

(i) Write $(1 - K)^{-1}$ as a power series in K and use the differentiation formula

$$\frac{d}{da} K^n = \sum_{j=0}^{n-1} K^j \frac{dK}{da} K^{n-j-1}$$

to show

$$\frac{d}{da}(1 - K)^{-1} = \sum_{n=1}^{\infty} \sum_{j=0}^{n-1} K^j \frac{dK}{da} K^{n-j-1}.$$

(ii) Change summation labels $(n, j) \mapsto (l, p)$, where $l = j$ and $p = n - j - 1$, to obtain the stated result.

2.6 ORTHOGONAL AND SYMPLECTIC UNITARY RANDOM MATRICES

The compact classical groups [540] are the general unitary matrices $U(N)$, together with their restriction to real and real quaternion elements. Restricting the elements to be real gives the subgroup $O(N)$, while restricting the 2×2 sub-blocks in $2N \times 2N$ unitary matrices to be real quaternion gives matrices equivalent to the unitary symplectic matrices $\mathrm{Sp}(2N)$. While the set of all general unitary matrices coincides with the CUE, the real orthogonal matrices and unitary symplectic matrices do not make up the COE and CSE. Rather members of the COE are diagonalized by real orthogonal matrices, and members of the CSE are diagonalized by symplectic equivalent matrices. Nonetheless, as will be discussed in Chapters 7 and 10, there are applications requiring the eigenvalue distribution of all three classical groups, so we must address the problem of computing the eigenvalue p.d.f. for $O(N)$ and $\mathrm{Sp}(2N)$.

Orthogonal matrices can be divided into two classes, $O^+(N)$ and $O^-(N)$, according to their determinants equaling $+1$ or -1 respectively. For orthogonal matrices $\mathbf{R}$ in $O^+(N)$ it is possible to introduce the Cayley transformation

$$\mathbf{R} = \frac{\mathbf{1}_N + i\mathbf{A}}{\mathbf{1}_N - i\mathbf{A}} \tag{2.61}$$

for some antisymmetric Hermitian matrix $\mathbf{A}$ (both $\mathbf{R}$ and $\mathbf{A}$ have $\frac{1}{2}N(N-1)$ independent elements). It follows from (2.61), and the fact that the eigenvalues of $\mathbf{A}$ come in $\pm$ pairs, that the eigenvalues of $\mathbf{R}$ come in complex conjugate pairs $e^{\pm i\theta_j}$ (for N odd $+1$ is also an eigenvalue). Using (2.61) and proceeding as in Section 2.5 we find (see Exercises 2.6 q.1) the eigenvalue p.d.f. for matrices $\mathbf{R}$ is equal to

$$\frac{2}{(N/2)!(2\pi)^{N/2}} \prod_{1 \le j < k \le N/2} \Big(2(\cos\theta_k - \cos\theta_j)\Big)^2 \tag{2.62}$$

for N even and

$$\frac{1}{((N-1)/2)!(2\pi)^{(N-1)/2}} \prod_{j=1}^{(N-1)/2} 2(1-\cos\theta_j) \prod_{1\le j<k\le (N-1)/2} \Big(2(\cos\theta_k - \cos\theta_j)\Big)^2 \tag{2.63}$$

for N odd, where the normalizations follow from Exercises 5.5 q.4 below.

It is of interest to interpret the results (2.62) and (2.63) in the context of the calculation of Section 1.2.4. Consider for definiteness (2.62). For an orthogonal matrix there are $\frac{1}{2}N(N-1)$ independent elements. Now for each eigenvalue pair $e^{i\theta_j}, e^{-i\theta_j}$ there is an eigenvector pair $u_j, \bar{u}_j$. With Θ the diagonal matrix formed from $e^{i\theta_1}, e^{-i\theta_1}, \dots, e^{i\theta_{N/2}}, e^{-i\theta_{N/2}}$ the independent elements in $\delta\mathbf{U}_2$ are the strictly upper triangular entries, of which there are $\frac{1}{2}N(N-1)$, excluding the $N/2$ entries $(\delta\mathbf{U}_2)_{2j-1\,2j} = \mathbf{u}_j^T \cdot d\mathbf{u}_j = 0$, where the second equality follows from the fact that $\mathbf{U}_2$ is unitary with columns of the form $\mathbf{u}_j, \bar{\mathbf{u}}_j$. In the analogue of (2.22) the entries would otherwise contribute the factor $e^{i\theta_j} - e^{-i\theta_j}$, so the analogue of (2.23) reads

$$(\mathbf{R}^T d\mathbf{R}) = \prod_{1\le j<k\le N/2} |e^{i\theta_j} - e^{i\theta_k}|^2 |e^{i\theta_j} - e^{-i\theta_k}|^2 \bigwedge_{j=1}^{N/2} d\theta_j\, (\mathbf{U}_2^\dagger, d\mathbf{U}_2), \tag{2.64}$$

thus implying (2.62).

Consider now members of $O^-(N)$. In the case of N odd, the eigenvalues are as in the case $O^+(N)$ except that $\theta \mapsto \pi - \theta$ (this maps the eigenvalue $+1$ in $O^+(N)$ to an eigenvalue -1 in $O^-(N)$). Hence from (2.63) the eigenvalue p.d.f. is equal to

$$\frac{1}{((N-1)/2)!(2\pi)^{(N-1)/2}} \prod_{j=1}^{(N-1)/2} 2(1+\cos\theta_j) \prod_{1\le j<k\le (N-1)/2} (2(\cos\theta_k - \cos\theta_j))^2. \tag{2.65}$$

In the case of N even, members of $O^-(N)$ contain the pair of eigenvalues ± 1. Arguing as in the derivation of (2.64) shows the eigenvalue p.d.f. is equal to

$$\frac{1}{(N/2-1)!(2\pi)^{N/2-1}} \prod_{j=1}^{N/2-1} |1-e^{i\theta_j}|^2 |1+e^{i\theta_j}|^2 \prod_{1\le j<k\le N/2-1} |e^{i\theta_j} - e^{i\theta_k}|^2 |e^{i\theta_j} - e^{-i\theta_k}|^2. \tag{2.66}$$

where the normalization again follows from Exercises 5.5 q.4 below.

It remains to consider the case of unitary symplectic equivalent matrices, that is, $2N \times 2N$ unitary matrices satisfying the additional relation

$$\mathbf{U}\mathbf{Z}_{2N}\mathbf{U}^T = \mathbf{Z}_{2N} \tag{2.67}$$

constraining the elements to be real quaternion. The constraint (2.67) together with the unitarity requirement means that there are $N(2N+1)$ independent variables. We see from (2.67) that if $e^{i\theta}$ is an eigenvalue with eigenvector $\vec{v}$, then $e^{-i\theta}$ is an eigenvalue with eigenvector $\mathbf{Z}_{2N}\vec{v}$, thus implying that in the formalism of Section 1.2.4 there are $2N^2$ independent elements in $\delta\mathbf{U}_2$. With the eigenvalues ordered $e^{i\theta_1}, \dots, e^{i\theta_N}$, $e^{-i\theta_1}, \dots, e^{-i\theta_N}$, these can be taken as the real and imaginary parts of the elements $(\delta\mathbf{U}_2)_{jk}$ and $(\delta\mathbf{U}_2)_{j\,N+k}$ with $1 \le j < k \le N$ and $1 \le j \le k \le N$, respectively, thus giving for the analogue of (2.23) the formula

$$(\mathbf{U}^\dagger d\mathbf{U}) = \prod_{j=1}^{N} |e^{i\theta_j} - e^{-i\theta_j}|^2 \prod_{1\le j<k\le N} |e^{i\theta_j} - e^{i\theta_k}|^2 |1 - e^{i(\theta_j+\theta_k)}|^2 \bigwedge_{j=1}^{N} d\theta_j\, (\mathbf{U}_2^\dagger d\mathbf{U}_2). \tag{2.68}$$

Hence, with $0 \le \theta_j \le \pi$ $(j = 1, \dots, N)$ the corresponding eigenvalue p.d.f. is precisely the $N/2 - 1 \mapsto N$

case of (2.66), and thus equal to

$$\frac{1}{N!(2\pi)^N}\prod_{k=1}^{N}|e^{i\theta_k}-e^{-i\theta_k}|^2\prod_{1\le j<k\le N}|e^{i\theta_j}-e^{i\theta_k}|^2|1-e^{i(\theta_j+\theta_k)}|^2. \tag{2.69}$$

An alternative way to derive (2.69) is to note that unitary symplectic equivalent matrices can be constructed out of quaternion real matrices $\mathbf{Q}$ with the property that $i\mathbf{Q}$ is Hermitian via the Cayley transformation

$$\mathbf{U}=\frac{\mathbf{1}_{2N}-\mathbf{Q}}{\mathbf{1}_{2N}+\mathbf{Q}}. \tag{2.70}$$

Changing variables as in Section 2.5, deriving the analogue of (2.71) below and making use of the result of Exercises 1.3 q.8 gives (2.69).

The eigenvalue p.d.f.'s computed above for the orthogonal and symplectic groups of random matrices all have interpretations as the Boltzmann factor for certain log-gas systems. This is made explicit at the end of Section 2.9.

EXERCISES 2.6 1. The objective of this exercise is to derive the eigenvalue p.d.f.'s (2.62) and (2.63) for orthogonal matrices with determinant equal to $+1$.

(i) In general, for an antisymmetric Hermitian matrix $\mathbf{A}$,

$$(d\mathbf{A})=\prod_{j<k}dA_{jk}^{\mathrm{i}},$$

where the superscript i denotes the imaginary part (the real part is zero). From this, use the method of the proof of Proposition 1.2.5 to show that for a general real $N\times N$ matrix $\mathbf{B}$,

$$(\mathbf{B}^T d\mathbf{A}\,\mathbf{B})=\Big(\det(\mathbf{B}^T\mathbf{B})\Big)^{(N-1)/2}(d\mathbf{A}). \tag{2.71}$$

(ii) Deduce from (2.61) that

$$\mathbf{R}^T d\mathbf{R}=2i(\mathbf{1}+i\mathbf{A})^{-1}d\mathbf{A}^T(\mathbf{1}+i\mathbf{A}^T)^{-1}$$

(cf. (2.56)), and then use (2.71) to show from this that

$$(\mathbf{R}^T d\mathbf{R})=2^{N(N-1)/2}\Big(\det(\mathbf{1}+\mathbf{A}^2)\Big)^{-(N-1)/2}(d\mathbf{A}).$$

(iii) Use the result of (ii) together with the formulas of Exercises 1.3 q.5 to write down the eigenvalue p.d.f. for $(\mathbf{R}^T d\mathbf{R})$ in terms of the eigenvalues $\lambda_1,\dots,\lambda_{[N/2]}$ of $\mathbf{A}$. Change variables according to (2.61) to derive (2.62) and (2.63).

2.7 LOG-GAS SYSTEMS WITH PERIODIC BOUNDARY CONDITIONS

At the end of Section 2.2.4 the p.d.f. (2.20) was interpreted as the Boltzmann factor for a log-gas system on a circle. An equivalent interpretation is that the log-gas is defined on a line with periodic boundary conditions. To see this, suppose the line is in the x-direction and is of length L. To specify a two-dimensional Coulomb system in this setting, for the pair potential we seek the solution of the Poisson equation (1.40) subject to the semiperiodic boundary condition $\Phi((x+L,y),(x',y'))=\Phi((x,y),(x',y'))$. Using the further facts that the solution must depend only on $x-x'$ and $y-y'$, and that for $\vec{r}\sim\vec{r}\,'$, $\Phi(\vec{r},\vec{r}\,')\sim-\log|\vec{r}-\vec{r}\,'|$ (which is the solution in free boundary conditions (1.41) with $l=1$), we see from the theorem of elementary complex analysis asserting that the real part of an analytic function satisfies Laplace's equation that

$$\Phi(\vec{r},\vec{r}\,')=-\log\Big(|\sin(\pi(x-x'+i(y-y'))/L)|(L/\pi)\Big). \tag{2.72}$$

With the particles confined to the segment $[0, L]$ of the x-axis, this reduces to

$$-\log\Big(|\sin(\pi(x-x')/L)|\Big(\frac{L}{\pi}\Big)\Big) = -\log\Big(|e^{2\pi ix/L} - e^{2\pi ix'/L}|\Big(\frac{L}{2\pi}\Big)\Big),$$

thus revealing the equivalence to (1.41) with $\vec{r}, \vec{r}\,'$ confined to a circle.

In this section three different log-gas systems with periodic boundary conditions will be introduced. Two of these appear in different contexts in subsequent chapters; all three have solvability properties analogous to log-gas systems with unitary symmetry, as to be discussed in Chapter 5.

2.7.1 Transverse semiperiodic boundary conditions

Suppose that for a log-gas system interacting via the pair potential (2.72), instead of the particles being confined to the line segment $[0, L]$ in the x-direction, they are confined to the full line in the y-direction. Up to an additive constant, the pair potential is then $-\log|\sinh(\pi(y-y')/L)|$, and if the particles are restrained from repelling to infinity by an attractive harmonic potential, the Boltzmann factor is then of the form

$$\prod_{j=1}^{N} e^{-\beta c' y_j^2/2} \prod_{1\le j<k\le N} |\sinh(\pi(y_k - y_j)/L)|^\beta, \qquad -\infty < y_j < \infty. \tag{2.73}$$

We will see in Section 10.1.5 that (2.73) with $\beta = 1$ occurs in the theory of non-intersecting paths. In the case $\beta = 2$ the corresponding partition function occurs in Chern-Simon field theory [385, 514].

The quadratic term in (2.73) results from the particle-background interaction

$$c'y^2 + C = \int_{-\infty}^{\infty} \rho_b(x) \log|\sinh(\pi(y-x)/L)|\,dx. \tag{2.74}$$

Analogous to the situation with (1.47), this equation cannot be solved for $y \to \infty$, as each side is then of a different order. For x and y confined to a finite interval, reintroducing the variable $\lambda = e^{-2\pi y/L}$ the equation (2.74) can be solved according to the method of Section 1.4. However it is simpler to construct an asymptotic solution, valid for $N \to \infty$. This solution is obtained by noting $\log|\sinh(\pi(y-x)/L)| \sim \pi|y-x|/L$ for $|y-x| \to \infty$, and that $\pi|y-x|/L$ satisfies the one-dimensional Poisson equation $\frac{d^2}{dy^2}(\pi|y-x|/L) = (2\pi/L)\delta(x-y)$. Thus operating on both sides of (2.74) with $\frac{d^2}{dy^2}$ we obtain the asymptotic solution $\rho_b(x) = Lc'/\pi$, $|x| < N\pi/2Lc'$. As an application of this result, the principle of local charge neutrality for Coulomb systems implies that for fixed y and large N the leading order density of the quantities $\{y_j\}$ in (2.73) will be the constant Lc'/π for $|x| < N\pi/2Lc'$ and zero otherwise.

We remark that changing variables $x_j = e^{-2\pi(y_j - y_0)/L}$, for suitable y_0, shows that (2.73) can be written in the form

$$\prod_{l=1}^{N} w(x_l) \prod_{1\le j<k\le N} |x_k - x_j|^\beta, \tag{2.75}$$

with

$$w(x) = e^{-\beta c(\log x)^2}, \qquad x > 0. \tag{2.76}$$

Interpreted as an (abstract) eigenvalue p.d.f, this form has been used [107] in studies of conductance (see (3.97) below) in regimes for which there is an insulator to conductor transition.

2.7.2 Metal wall

Suppose that in addition to periodic boundary conditions in the x-direction, a perfect conductor occupies the region $y < 0$. This means that the pair potential satisfies Poisson's equation (1.40) subject to the boundary condition $\Phi((x,y),(x',y')) = 0$ for $y = 0$. According to the method of images the sought solution is

$$\Phi(\vec{r},\vec{r}') = -\log\left|\frac{\sin(\pi(x-x'+i(y-y'))/L)}{\sin(\pi(x-x'+i(y+y'))/L)}\right|,$$

which has the interpretation of there being an image charge of opposite sign created at $(x', -y')$.

Because of the image effect of the metal wall, from a log-gas perspective, it makes sense to impose a uniform background charge density $-\eta$, independent of the particle density. The Boltzmann factor for a system of N particles confined to the line $y = d$ is thus specified as follows.

PROPOSITION 2.7.1 *From the log-gas system near a metal wall as specified, mobile charges $q = 1$, the Boltzmann factor is*

$$e^{-\beta(\pi\eta^2 dL - 2\pi\eta N)}\left(\frac{\pi}{L\sinh 2\pi d/L}\right)^{N\beta/2}\prod_{1\le j<k\le N}\left|\frac{\sin(\pi(x_k-x_j)/L)}{\sin(\pi(x_k-x_j+2id)/L)}\right|^{\beta}. \tag{2.77}$$

Proof. The total potential energy consists of the particle-particle energy

$$U_1 = -\sum_{1\le j<k\le N}\log\left|\frac{\sin(\pi(x_k-x_j)/L)}{\sin(\pi(x_k-x_j+2id)/L)}\right|;$$

the self-energy U_1'

$$U_1' := \frac{1}{2}\sum_{j=1}^{N}\lim_{x'\to x_j}\Big(\Phi((x_j,d),(x',d)) - \log|x_j - x|\Big) = -\frac{N}{2}\log\left(\frac{\pi}{L\sinh 2\pi d/L}\right);$$

the particle-background energy

$$U_2 = \eta\sum_{j=1}^{N}\int_0^L \log\left|\frac{\sin(\pi x/L)}{\sin(\pi(x+2id)/L)}\right|dx = -2\pi d\eta N;$$

and the background-background energy

$$U_3 = \pi\eta^2 dL.$$

□

Note that for $d \to 0$ (2.77) becomes independent of $\{x_j\}$, while for $d \to 0$ the Boltzmann factor of Proposition 1.4.1 is reclaimed. Thus for general d there in an interpolation between a perfect gas and a one-component log-gas. For $\beta = 2$ we will see in Exercises 11.6 q.2 that the p.d.f. corresponding to (2.77) can be derived from a particular random matrix model.

2.7.3 Doubly periodic boundary conditions

The pair potential Φ with doubly periodic boundary conditions, period L and W, must obey the charge neutral Poisson equation

$$\frac{\partial^2\Phi}{\partial x^2} + \frac{\partial^2\Phi}{\partial y^2} = -2\pi\delta(x-x')\delta(y-y') + \frac{2\pi}{LW} \tag{2.78}$$

(note the integral of the right-hand side over the fundamental rectangle is 0; without this charge neutrality condition a doubly periodic solution is not possible). To solve (2.78) introduce the θ_1 function defined by

[541]

$$\theta_1(z;q) = -i \sum_{n=-\infty}^{\infty} (-1)^n q^{(n-1/2)^2} e^{2i(n-1/2)z}$$
$$= 2q^{1/4} \sin z \prod_{n=1}^{\infty} \left(1 - q^{2n} e^{2iz}\right) \left(1 - q^{2n} e^{-2iz}\right) \left(1 - q^{2n}\right). \tag{2.79}$$

With $q = e^{i\pi\tau}$, $\mathrm{Im}(\tau) > 0$ the facts that θ_1 is an entire function, $\theta_1(z;q) = 0$ if and only if $z = \pi m + \pi\tau n$, $m, n \in \mathbb{Z}$, and $\theta_1(z;q) \sim z\theta_1'(0;q)$ as $z \to 0$, tell us that

$$\tilde{\Phi}(z,z') := -\log\left(\frac{L|\theta_1(\pi(z-z')/L;q)|}{\pi\theta_1'(0;q)}\right), \quad q := e^{-\pi W/L} \tag{2.80}$$

satisfies the Poisson equation (1.40) for $0 \le x, x' < L$, $0 \le y, y' < W$, with the further specification that $\tilde{\Phi}(z,z') \sim -\log|z-z'|$ as $|z-z'| \to 0$. Since

$$\theta_1(z+\pi;q) = -\theta_1(z;q) \quad \text{and} \quad \theta_1(z+\pi\tau;q) = -q^{-1}e^{-2iz}\theta_1(z;q) \tag{2.81}$$

we see that (2.80) has the periodicity properties

$$\tilde{\Phi}((x+L,y),(x',y')) = \tilde{\Phi}((x,y),(x',y')),$$
$$\tilde{\Phi}((x,y+W),(x',y')) = -\frac{\pi}{L}(2y+W) + \tilde{\Phi}((x,y),(x',y')).$$

It follows that the potential

$$\Phi(z,z') := \frac{\pi y^2}{LW} + \tilde{\Phi}(z,z') \tag{2.82}$$

is doubly periodic, and satisfies the charge neutral Poisson equation (2.78).

For a system confined to the x-axis the quadratic term in (2.82) does not contribute, and the pair potential is given by (2.80). Using this, we find for the Boltzmann factor of a one-component system of N unit charges confined to the interval $[0, L]$, in the presence of a smeared out neutralizing background, the expression

$$\left(\frac{\pi\theta_1'(0;q)}{L}\right)^{N\beta/2} \left(q^{1/4}\prod_{n=1}^{\infty}(1-q^{2n})\right)^{-\beta N^2/2} \prod_{1\le j<k\le N} \left|\theta_1\left(\frac{\pi(x_k-x_j)}{L};q\right)\right|^{\beta}. \tag{2.83}$$

2.8 CIRCULAR β-ENSEMBLE

2.8.1 Hessenberg form

For non-Hermitian matrices, Householder similarity transformations analogous to (1.139) cannot give a tridiagonal form. Instead what results is a Hessenberg form, in which all entries are zero below the first sub-diagonal. By introducing into the conjugations a suitable diagonal unitary matrix, the entries on the first sub-diagonal can be chosen to be real and positive. Thus one can construct a sequence of unitary matrices $\mathbf{V}^{(1)}, \dots, \mathbf{V}^{(N-1)}$ and diagonal unitary matrices $\mathbf{D}^{(1)}, \dots, \mathbf{D}^{(N-1)}$ such that for a general complex matrix $\mathbf{W} = [w_{ij}]_{i,j=1,\dots,N}$,

$$\mathbf{Y}^{(N-1)} = \mathbf{D}^{(N-1)\dagger}\mathbf{V}^{(N-1)\dagger}\cdots\mathbf{D}^{(1)\dagger}\mathbf{V}^{(1)\dagger}\mathbf{W}\mathbf{V}^{(1)}\mathbf{D}^{(1)}\cdots\mathbf{V}^{(N-1)}\mathbf{D}^{(N-1)} \tag{2.84}$$

is of the sought Hessenberg form. The matrices $\mathbf{V}^{(j)}$ have the reflector structure of (1.140),

$$\mathbf{V}^{(j)} = \mathbf{1} - 2\vec{v}^{(j)}\vec{v}^{(j)\dagger} = \begin{bmatrix} \mathbf{1}_{j\times j} & \mathbf{0}_{j\times N-j} \\ \mathbf{0}_{N-j\times j} & \mathbf{U}_{N-j\times N-j} \end{bmatrix},$$

where $\vec{v}^{(j)\dagger}\vec{v}^{(j)} = 1$ and $\mathbf{U}_{N-j\times N-j}$ is unitary. The entries of $\vec{v}^{(j)}$ can be expressed iteratively, starting with $\mathbf{V}^{(1)}$. For this set

$$v_1^{(1)} = 0, \qquad v_2^{(1)} = \frac{\beta}{c}, \qquad v_l^{(1)} = -\frac{w_{l1}}{c} \quad (l \geq 3), \tag{2.85}$$

where, with $l := \sqrt{|w_{21}|^2 + \cdots + |w_{N1}|^2}$,

$$\beta = -\frac{w_{21}}{\sqrt{2}|w_{21}|}\Big(1 - \frac{|w_{21}|}{l}\Big)^{1/2}, \qquad c = \sqrt{2}l\Big(1 - \frac{|w_{21}|}{l}\Big)^{1/2}.$$

Noting that then $\vec{v}^{(1)\dagger}\vec{w}^{(1)} = -c/2$ shows $\tilde{\mathbf{Y}}^{(1)\dagger} := \mathbf{V}^{(1)\dagger}\mathbf{W}\mathbf{V}^{(1)}$ has

$$(\tilde{\mathbf{Y}}^{(1)\dagger})_{11} = w_{11}, \qquad (\tilde{\mathbf{Y}}^{(1)\dagger})_{21} = \frac{w_{21}}{|w_{21}|}l, \qquad (\tilde{\mathbf{Y}}^{(1)\dagger})_{l1} = 0 \ (l \geq 3). \tag{2.86}$$

This is the first column of a Hessenberg form, but with a complex subdiagonal entry. The phase $w_{21}/|w_{21}| = e^{i\phi}$ can be cancelled by forming the conjugation $\mathbf{Y}^{(1)} := \mathbf{D}^{(1)\dagger}\tilde{\mathbf{Y}}^{(1)}\mathbf{D}^{(1)}$, where

$$\mathbf{D}^{(1)} = \mathrm{diag}[1, e^{i\phi}, 1, \dots, 1]. \tag{2.87}$$

The matrices $\mathbf{V}^{(j)}$ $(j = 2, \dots, N-2)$, $\mathbf{D}^{(j)}$ $(j = 2, \dots, N-1)$ are defined by the formulas (2.85), (2.87), the former modified so that $v_1^{(j)} = \cdots = v_j^{(j)} = 0$ and the matrix elements w_{l1} replaced by those in the first column of the bottom right $N-j+1 \times N-j+1$ submatrix of $\mathbf{Y}^{(j-1)}$, and the latter modified so that the phase is in position $j+1$ on the diagonal. For $j = N-1$ we take $\mathbf{V}^{(N-1)} = \mathbf{1}$.

In the case that $\mathbf{W}$ is unitary, the Hessenberg form is special because the diagonal entries completely determine the matrix. Moreover, the remaining entries can be determined explicitly.

PROPOSITION 2.8.1 *Set $\alpha_{-1} := -1$, and suppose $|\alpha_{N-1}| = 1$. For $j = 0, \dots, N-2$ let α_j be complex numbers with $|\alpha_j| < 1$, and put $\rho_j = \sqrt{1 - |\alpha_j|^2}$. The Hessenberg matrix $\mathbf{H} = [H_{ij}]_{i,j=0,\dots,N-1}$ with diagonal entries $H_{ii} = -\alpha_{i-1}\bar{\alpha}_i$ and subdiagonal entries $H_{i+1,i} = \rho_i$ is unitary if*

$$H_{ij} = -\alpha_{i-1}\bar{\alpha}_j \prod_{l=i}^{j-1} \rho_l, \quad i < j.$$

Proof. One checks that

$$\sum_{i=0}^{N-1} |H_{ij}|^2 = 1, \qquad \sum_{i=0}^{N-1} H_{ij}\bar{H}_{ij'} = 0 \quad (j \neq j').$$

□

A special feature of the unitary Hessenberg matrix $\mathbf{H}$ as specified in Proposition 2.8.1 is a coupled recurrence satisfied by the characteristic polynomial [271].

PROPOSITION 2.8.2 *Let $\mathbf{H}_k$ denote the top $k \times k$ block of $\mathbf{H}$, and write $\chi_k(\lambda) = \det(\lambda\mathbf{1}_k - \mathbf{H}_k)$. We have*

$$\begin{aligned} \chi_k(\lambda) &= \lambda\chi_{k-1}(\lambda) - \bar{\alpha}_{k-1}\tilde{\chi}_{k-1}(\lambda), \\ \tilde{\chi}_k(\lambda) &= \tilde{\chi}_{k-1}(\lambda) - \lambda\alpha_{k-1}\chi_{k-1}(\lambda) \end{aligned} \tag{2.88}$$

$(k = 1, \dots, N)$, where $\chi_0(\lambda) = \tilde{\chi}_0(\lambda) = 1$. Furthermore $\tilde{\chi}_k(\lambda) = \lambda^k\bar{\chi}_k(1/\lambda)$ (here $\bar{\chi}_k$ denotes the polynomial χ_k with its coefficients replaced by their complex conjugates).

Note that with the replacements $\alpha_j \mapsto -\bar{\alpha}_j\alpha_{N-1}$ $(j = 0, \dots, N-2)$, the bottom $k \times k$ submatrix, after reflection in the anti-diagonal, becomes identical to the top $k \times k$ submatrix but with $\alpha_j \mapsto \alpha_{N-2-j}$

$(j = 0, \ldots, N-2)$. Hence the corresponding characteristic polynomial, $\chi_k^b(\lambda)$ say, satisfies the recurrences

$$\begin{aligned}\chi_k^b(\lambda) &= \lambda \chi_{k-1}^b(\lambda) + \bar{\alpha}_{N-1-k}\bar{\alpha}_{N-1}\tilde{\chi}_{k-1}^b(\lambda),\\ \tilde{\chi}_k^b(\lambda) &= \tilde{\chi}_{k-1}(\lambda) + \lambda \alpha_{k-1}\bar{\alpha}_{N-1-k}\alpha_{N-1}\tilde{\chi}_{k-1}^b(\lambda)\end{aligned} \tag{2.89}$$

$(k = 1, \ldots, N)$, where $\chi_0^b(\lambda) = \tilde{\chi}_0^b(\lambda) = 1$ and $\tilde{\chi}_k^b(\lambda) = \lambda^k \bar{\chi}_k^b(1/\lambda)$.

For random unitary matrices, the distribution of the parameters $\{\alpha_k\}$ occurring in Propositions 2.8.2 and (2.89) can be determined explicitly [356], thus allowing the corresponding eigenvalues to be calculated as the zeros of $\chi_N(\lambda)$, without the need to diagonalize a matrix. This requires introducing a particular distribution of complex numbers.

Definition 2.8.3 *For $\nu > 1$ and complex numbers $|z| < 1$, let Θ_ν denote the distribution of complex numbers with p.d.f.*

$$\frac{\nu - 1}{2\pi}(1 - |z|^2)^{(\nu-3)/2},$$

while for complex numbers $|z| = 1$ let Θ_1 denote the uniform distribution.

We remark that in the case $\nu > 1$, in terms of $z = re^{i\theta}$, $0 \le r < 1$, $0 \le \theta < 2\pi$, Θ_ν is such that θ has uniform distribution, while $s := r^2$ has distribution $\frac{\nu-1}{2}(1-s)^{(\nu-3)/2}$. Also, analogous to the result of Exercises 1.2 q.2(ii), one notes the marginal distribution of a single component of a vector on the complex $(N-1)$-sphere, chosen uniformly at random, or equivalently a single entry of a random element of $U(N)$, has distribution Θ_{2N-1}.

Proposition 2.8.4 *The unitary Hessenberg matrix, with positive elements on the subdiagonal, obtained by applying the augmented Householder transformations (2.84) to a random element of $U(N)$, has parameters $\{\alpha_{j-1}\}_{j=1,\ldots,N}$ distributed according to $\alpha_{N-j-1} \in \Theta_{2j+1}$ $(j = 0, \ldots, N-1)$.*

Proof. According to Proposition 2.8.1, $H_{00} = \bar{\alpha}_0$, while (2.86) gives that $H_{11} = w_{11}$, where $w_{11} = W_{11}$, $\mathbf{W} \in U(N)$. As remarked above, the latter has distribution Θ_{2N-1}. In the construction of $\mathbf{H}$, after determining the first column, we must apply the algorithm to the bottom right $N-1 \times N-1$ submatrix of $\mathbf{Y}^{(1)}$. In particular, we seek $(\mathbf{Y}^{(1)})_{22}$. The second column of $\mathbf{Y}^{(1)}$ can be regarded as a random point on the complex $(N-2)$-sphere, constrained to be orthogonal to the first column of $\mathbf{Y}^{(1)}$, and thus orthogonal to the vector $(w_{11}, \sqrt{1-|w_{11}|^2}, 0, \ldots, 0)$. This gives the structure

$$(\sqrt{1-|w_{11}|^2}z, -\bar{w}_{11}z, w_{32}, \ldots, w_{N2})^T,$$

where $(z, w_{32}, \ldots, w_{N2})$ is a random point on the complex $(N-2)$-sphere, telling us that $(\mathbf{Y}^{(1)})_{22} = -\bar{w}_{11}z = -\alpha_0 z$, $z \in \Theta_{2N-3}$. But $-\alpha_0\bar{\alpha}_1 = H_{11} = (\mathbf{Y}^{(1)})_{22}$ so $\alpha_1 \in \Theta_{2N-3}$.

In general, with the first $k-1$ columns in Hessenberg form, the k-th column of $\mathbf{Y}^{(k-1)}$ will have the structure

$$(\rho_0\rho_1\cdots\rho_{k-2}z, -\alpha_0\rho_1\cdots\rho_{k-2}z, -\alpha_1\rho_2\cdots\rho_{k-2}z, \ldots, -\alpha_{k-3}\rho_{k-2}z, -\alpha_{k-2}z, w_{k+1\,k}, \ldots, w_{Nk})^T$$

where $(z, w_{k+1\,k}, \ldots, w_{Nk})$ is a random point on the complex $(N-k)$-sphere, this being the most general unit vector orthogonal to the first $k-1$ columns. Consequently $(\mathbf{Y}^{(k-1)})_{kk} = -\alpha_{k-2}$, $z \in \Theta_{2N-2k+1}$. As $-\alpha_{k-2}\bar{\alpha}_{k-1} = H_{k-1\,k-1}$ it follows that $\alpha_{k-1} \in \Theta_{2N-2k+1}$ as required. □

The situation is now analogous to that which arose after obtaining the tridiagonal matrix of Proposition 1.9.1. Thus for the unitary Hessenberg $\mathbf{H}$ of Proposition 2.8.1 we would like to know how to directly change variables from the parameters $\vec{\alpha} = (\alpha_0, \ldots, \alpha_{N-1})$ to $\vec{\lambda} = (\lambda_1, \ldots, \lambda_N)$ and $\vec{q} = (q_1, \ldots, q_N)$. Here the $\lambda_j = e^{i\theta_j}$ are the eigenvalues of $\mathbf{H}$ and the q_i the modulus of the first component of the corresponding normalized eigenvectors. The latter must therefore satisfy (1.144). In preparation for the change of variables, we must establish the analogue of (1.148).

PROPOSITION 2.8.5 *For the Hessenberg matrix* $\mathbf{H}$ *of Proposition 2.8.1,*

$$\prod_{1\le i<j\le N} |\lambda_i - \lambda_j|^2 = \frac{\prod_{l=0}^{N-2}(1-|\alpha_l|^2)^{N-1-l}}{\prod_{j=1}^{N} q_j^2}. \tag{2.90}$$

Proof. See Exercises 2.8 q.1. □

PROPOSITION 2.8.6 *The Jacobian for the change of variables from* $\vec{\alpha}$ *to* $(\vec{\lambda}, \vec{q})$ *is equal to*

$$\frac{\prod_{i=0}^{N-2}(1-|\alpha_i|^2)}{q_N \prod_{i=1}^{N} q_i}. \tag{2.91}$$

Proof. [223] We follow the strategy of Proposition 1.9.3, starting with (1.157) ($\mathbf{T} \mapsto \mathbf{H}$ therein). Equating successive powers of λ on both sides (recalling the explicit form of the matrix elements from Proposition 2.8.1) gives

$$1 = \sum_{j=1}^{N} q_j^2, \qquad \bar{\alpha}_0 = \sum_{j=1}^{N} q_j^2 \lambda_j, \qquad * + \bar{\alpha}_1 \rho_0^2 = \sum_{j=1}^{N} q_j^2 \lambda_j^2,$$
$$* + \bar{\alpha}_2 \rho_0^2 \rho_1^2 = \sum_{j=1}^{N} q_j^2 \lambda_j^2, \quad \dots, \quad * + \bar{\alpha}_{N-1} \rho_0^2 \rho_1^2 \cdots \rho_{N-2}^2 = \sum_{j=1}^{N} q_j^2 \lambda_j^N, \tag{2.92}$$

where $*$ denotes terms involving only variables already having appeared on the l.h.s. of the preceding equations, thus implying a triangular structure. Recalling that α_j $(j = 0, \dots, N-2)$ have independent real and imaginary parts, while $\alpha_{N-1} := e^{i\phi}$, $\lambda_j := e^{i\theta_j}$ $(j = 1, \dots, N)$ have unit modulus, we see the number of equations can be made equal to the number of variables by using the first equation to eliminate q_N^2 in the subsequent equations, then forming the complex conjugate of all these equations but the last. Taking differentials, then wedge products, of both sides of these $2N-1$ equations shows

$$\rho_0^2 \rho_1^2 \cdots \rho_{N-2}^2 \prod_{l=1}^{N-2} \rho_{N-l-2}^{4l} d\vec{\alpha} \wedge d\phi$$
$$= q_N^2 \prod_{j=1}^{N-1} q_j^3 \left| \det \begin{bmatrix} \begin{bmatrix} \lambda_k^j - \lambda_N^j \\ \lambda_k^{-j} - \lambda_N^{-j} \end{bmatrix}_{j,k=1,\dots,N-1} & \begin{bmatrix} j\lambda_k^j \\ -j\lambda_k^{-j} \end{bmatrix}_{\substack{j=1,\dots,N-1 \\ k=1,\dots,N}} \\ [\lambda_k^N - \lambda_N^N]_{k=1,\dots,N-1} & [N\lambda_k^N]_{k=1,\dots,N} \end{bmatrix} \right| d\vec{\theta} \wedge d\vec{q}. \tag{2.93}$$

Now it is straightforward to check that the above determinant is a symmetric function of $\lambda_1, \dots, \lambda_N$, which is homogeneous of degree N. The highest negative power of λ_1 is $\lambda_1^{-(2N-3)}$, so it must be of the form

$$\frac{1}{\prod_{l=1}^{N} \lambda_l^{2N-3}} p(\lambda_1, \dots, \lambda_N),$$

where p is a symmetric polynomial of $\lambda_1, \dots, \lambda_N$ of degree $2N(N-1)$. We note too that the determinant vanishes when $\lambda_1 = \lambda_2$, as does its derivatives $(\lambda_1 \frac{\partial}{\partial \lambda_1})^j$ $(j = 1, 2, 3)$. The polynomial p must thus contain as a factor $\prod_{1\le j<k\le N}(\lambda_k - \lambda_j)^4$. But this is of degree $2N(N-1)$ so in fact the determinant must be proportional to

$$\frac{\prod_{1\le j<k\le N}(\lambda_k - \lambda_j)^4}{\prod_{l=1}^{N} \lambda_l^{2N-3}}. \tag{2.94}$$

In this expression the coefficient of $\prod_{l=1}^{N} \lambda_l^{4(l-1)-2N+3}$ is unity. In the determinant, let us add $(N-1)$ times the first column to the N-th column. Then we see that the coefficient of λ_1^{-2N+3} is given by the cofactor coming from multiplying together the $(2N-2, 1)$ and $(2N-4, N)$ elements. In the cofactor we add sgn $(2N-7))(N-2)$ times the first column to the $(N-1)$-st column (sgn$(x) := 1$ for $x > 0$, sgn$(x) := -1$ for $x < 0$). The coefficient of λ_1^{-2N+7} is given by the cofactor coming from multiplying together the $(2N-3, 1)$ and $(2N-5, N-1)$ elements. Proceeding in this manner we see that the coefficient of $\prod_{l=1}^N \lambda_l^{4(l-1)-2N+3}$ is $(-1)^{(N-1)(N-2)/2}$ in the determinant, so in fact the determinant is equal to (2.94) times this sign. Substituting in (2.93), recalling $|\lambda_l| = 1$ and making use of (2.90) gives (2.91). □

Knowledge of the Jacobian (2.93) allows a unitary Hessenberg matrix to be specified for which the eigenvalue p.d.f. is distributed as for the circular β-ensemble [356].

PROPOSITION 2.8.7 *Consider the matrix* $\mathbf{H}$ *of Proposition 2.8.1 with parameters* $\{\alpha_{j-1}\}_{j=1,\dots,N}$ *distributed according to*

$$\alpha_{N-j-1} \in \Theta_{\beta j+1} \qquad (j = 0, \dots, N-1).$$

The eigenvalues and first component of the eigenvectors (which form $\vec{q}$*) are independent, with the distribution of the eigenvalues given by*

$$\frac{1}{C_{\beta,N}} \prod_{1 \le j < k \le N} |e^{i\theta_k} - e^{i\theta_j}|^\beta \, d\vec{\theta}, \qquad C_{\beta,N} = (2\pi)^N \frac{\Gamma(\beta N/2 + 1)}{\Gamma(\beta/2 + 1)}$$

and the distribution of $\vec{q}$ *by (1.161).*

Proof. Denote the unitary Hessenberg matrix by $\mathbf{H}_\beta$, and its joint distribution by $P(\mathbf{H}_\beta)$. We have

$$\begin{aligned} P(\mathbf{H}_\beta)(d\mathbf{H}_\beta) &= \frac{1}{2\pi} \prod_{j=1}^{N-1} \frac{\beta j}{2\pi} (1 - |\alpha_{N-j-1}|^2)^{\beta j/2 - 1} d\vec{\alpha} \wedge d\vec{\phi} \\ &= \frac{\beta^{N-1}(N-1)!}{(2\pi)^N} \frac{1}{q_N \prod_{i=1}^N q_i} \prod_{l=0}^{N-2} (1 - |\alpha_l|^2)^{\beta(N-l-1)/2} d\vec{\theta} \wedge d\vec{q} \\ &= \frac{\beta^{N-1}(N-1)!}{(2\pi)^N} \frac{\pi_{i=1}^N q_i^{\beta-1}}{q_N} \prod_{1 \le j < k \le N} |e^{i\theta_j} - e^{i\theta_k}|^\beta d\vec{\theta} \wedge d\vec{q}, \end{aligned}$$

where the second equality uses (2.91) and the third (2.90). After recalling the normalization in (1.161), and dividing by $N!$ to effectively eliminate the ordering of the eigenvalues, we have the stated result. □

EXERCISES 2.8 1. The objective of this exercise is to derive (2.94).

(i) Use the analogue of (1.147) and follow the derivation of (1.152) to show

$$\prod_{i=1}^n q_i^2 = \frac{\prod_{i=1}^n |\chi_{n-1}^b(\lambda_i)| \Big|_{\alpha_j \mapsto -\bar{\alpha}_j \alpha_{n-1}}}{\prod_{1 \le i < j \le n} |\lambda_i - \lambda_j|^2},$$

thus reducing the task to that of showing

$$\prod_{i=1}^n |\chi_{n-1}^b(\lambda_i)| = \prod_{l=0}^{n-2} (1 - |\alpha_l|^2)^{n-1-l}. \tag{2.95}$$

(ii) With $\lambda_j^{(p)}$ denoting the jth zero of $\chi_p^b(\lambda)$, substitute $\lambda = 1/\lambda_j^{(k)}$ in (2.89) to show

$$\chi_k^b(1/\bar{\lambda}_j^{(k)}) = \frac{1}{\bar{\lambda}_j^{(k)}} (1 - |\alpha_{n-k-1}|^2) \chi_{k-1}^b(1/\bar{\lambda}_j^{(k)}),$$

and then introduce the factorizations

$$\chi_{k-1}^b(x) = \prod_{i=1}^{k-1} (x - \lambda_i^{(k-1)}), \qquad \tilde{\chi}_k^b(x) = \prod_{i=1}^k (1 - x\bar{\lambda}_i^{(k)})$$

to deduce

$$\prod_{i=1}^{k} \chi_k^b(1/\bar{\lambda}_i^{(k)}) = (1 - |\alpha_{n-k-1}|^2)^k \prod_{i=1}^{k} (1/\bar{\lambda}_i^{(k)})^k \prod_{j=1}^{k-1} \tilde{\chi}_{k-1}^b(\lambda_j^{(k-1)}).$$

Similarly derive the equation

$$\prod_{i=1}^{k} \tilde{\chi}_k^b(\lambda_i^{(k)}) = (1 - |\alpha_{n-k-1}|^2)^k \prod_{j=1}^{k-1} (\bar{\lambda}_j^{(k-1)})^{k-1} \chi_{k-1}^b(1/\bar{\lambda}_j^{(k-1)}).$$

(iii) Show that the final two equations of (ii) imply

$$\prod_{i=1}^{k} (\bar{\lambda}_i^{(k)})^k \chi_k^b(1/\bar{\lambda}_i^{(k)}) = \prod_{l=0}^{k-1} (1 - |\alpha_{n-l}|^2)^{l+1}.$$

Substitute for $\chi_k^b(1/\bar{\lambda}_i^{(k)})$ using the first equation in (ii) and set $k = n$ to deduce (2.95).

2.9 REAL ORTHOGONAL β-ENSEMBLE

The real orthogonal and symplectic unitary random matrices can be transformed to upper Hessenberg form. Consider in particular a member of $O^+(2N)$. Because the elements are real, the parameters $\{\alpha_j\}_{j=0,\dots,2N-1}$ are all real, and because the determinant is equal to $+1$, $\alpha_{2N-1} = -1$ (this can be seen by setting $\alpha_0 = \cdots = \alpha_{2N-2} = 0$). Following the strategy of the proof of Proposition 2.8.4 allows the distribution of the remaining α_j's to be determined explicitly [356].

DEFINITION 2.9.1 *The beta distribution on $(-1, 1)$, denoted $\tilde{B}[\alpha, \beta]$, is specified by the p.d.f.*

$$\frac{1}{2} \frac{\Gamma(\alpha + \beta)}{\Gamma(\alpha)\Gamma(\beta)} \Big(\frac{1-x}{2}\Big)^{\alpha-1} \Big(\frac{x+1}{2}\Big)^{\beta-1}.$$

PROPOSITION 2.9.2 *The real orthogonal Hessenberg matrix, with positive elements on the subdiagonal, obtained by applying the augmented Householder transformations (2.84) to a random element of $O^+(2N)$, has parameters $\{\alpha_j\}_{j=0,\dots,2n-2}$ distributed according to*

$$\alpha_j \in \tilde{B}\Big[\frac{2N-j-1}{2}, \frac{2N-j-1}{2}\Big].$$

This result has the immediate significance of implying that the eigenvalue p.d.f. for $O^+(2N)$ can be sampled by computing the zeros of the polynomial $\chi_{2n}(\lambda)$ as determined by (2.88) (because each χ_k is real, $\bar{\chi}_k(\lambda) = \lambda^k \chi_k(1/\lambda)$ and so only the first equation in (2.88) is required). Furthermore it suggests a β-generalization analogous to Proposition 2.8.7. First, the analogue of Propositions 2.8.5 and 2.8.6 must be noted. In preparation, we note that for a general real orthogonal upper Hessenberg $2N \times 2N$ matrix, there are $2N-1$ independent real parameters $\alpha_0, \dots, \alpha_{2N-2}$. In the corresponding eigendecomposition, there are N independent variables q_j $(j = 1, \dots, N)$ where $\frac{1}{2}q_j^2$ is the square of the first component of both the eigenvalues λ_j and $\bar{\lambda}_j$.

PROPOSITION 2.9.3 *For a $2N \times 2N$ real orthogonal Hessenberg matrix $\mathbf{H}$ of determinant $+1$, parametrized* *in terms of the real parameters* $\{\alpha_i\}_{i=0,\dots,2N-2}$, $|\alpha_i| < 1$, *we have*

$$\prod_{i=1}^{N} \Big|\lambda_i - \frac{1}{\lambda_i}\Big| \prod_{1\le i<j\le N} |\lambda_i - \lambda_j|^2 |\lambda_i - 1/\lambda_j|^2 = 2^N \frac{\prod_{l=0}^{2N-2}(1-\alpha_l^2)^{(2N-1-l)/2}}{\prod_{i=1}^N q_i^2}, \tag{2.96}$$

$$\prod_{j=1}^{N} |1-\lambda_j|^2 = 2\prod_{k=0}^{2N-2}(1-\alpha_k), \qquad \prod_{j=1}^{N}|1+\lambda_j|^2 = 2\prod_{k=0}^{2N-2}(1+(-1)^k\alpha_k). \tag{2.97}$$

Proof. Analogous to (1.157) we have

$$((I_{2N} - \lambda H)^{-1})_{11} = \frac{1}{2}\sum_{j=1}^{n} q_j^2\Big(\frac{1}{1-\lambda\lambda_j} + \frac{1}{1-\lambda\bar{\lambda}_j}\Big). \tag{2.98}$$

Since the r.h.s. is equal to $\chi_{2n-1}^b(1/\lambda)/\lambda\chi_{2n}^b(1/\lambda)$ it follows that

$$\Big|\frac{\chi_{2n-1}^b(\lambda_j)}{\chi_{2n}'(\lambda_j)}\Big| = \frac{1}{2}q_j^2 \qquad (j=1,\dots,2n),$$

where $\lambda_{j+N} = \bar{\lambda}_j$, $q_{j+N} = q_j$. Taking the product over $j = 1,\dots,2n$, making use of (2.95), then taking the square root gives (2.96). For the results (2.97), note

$$\prod_{j=1}^{n}|1-\lambda_j|^2 = \chi_{2n}(1), \qquad \prod_{j=1}^{n}|1+\lambda_j|^2 = \chi_{2n}(-1).$$

But from (2.88) $\chi_{k+1}(\lambda)|_{\lambda=\pm1} = (\lambda - \alpha_k\lambda^k)\chi_k(\lambda)|_{\lambda=\pm1}$. □

PROPOSITION 2.9.4 *For a real orthogonal upper Hessenberg* $2N \times 2N$ *matrix of determinant* $+1$, *the Jacobian for the change of variables from* $\vec{\alpha}$ *to* $(\vec{\theta},\vec{q})$ *is equal to*

$$\frac{2^{N-1}}{q_N\prod_{i=1}^N q_i}\frac{\prod_{l=0}^{2N-2}(1-|\alpha_l|^2)}{\prod_{k=0}^{2N-2}(1-\alpha_k)^{1/2}(1+(-1)^k\alpha_k)^{1/2}}. \tag{2.99}$$

Proof. Expanding (2.98) in powers of λ, analogous to (2.92) we obtain

$$1 = \sum_{j=1}^N q_j^2, \quad \alpha_0 = \frac{1}{2}\sum_{j=1}^N q_j^2(\lambda_j + \bar{\lambda}_j), \quad * + \alpha_1\rho_0^2 = \frac{1}{2}\sum_{j=1}^N q_j^2(\lambda_j^2 + \bar{\lambda}_j^2), \dots,$$

$$* + \alpha_{2N-2}\rho_0^2\cdots\rho_{2N-3}^2 = \frac{1}{2}\sum_{j=1}^N q_j^2(\lambda_j^{2N-1} + \bar{\lambda}_j^{2N-1}).$$

Using the first of these equations to substitute for q_N^2, taking differentials then wedge products of both sides, shows

$$\prod_{l=0}^{2N-3}\rho_l^{2(2N-2-l)}\,d\vec{\alpha}$$

$$= 2^{-N}q_N^2\prod_{j=1}^{N-1}q_j^3\Big|\det\Big[[\lambda_k^j + \lambda_k^{-j} - (\lambda_{\bar{k}}^j + \lambda_{\bar{k}}^{-j})]_{\substack{j=1,\dots,2N-1\\k=1,\dots,N-1}} \quad [j(\lambda_{\bar{k}}^j - \lambda_{\bar{k}}^{-j})]_{\substack{j=1,\dots,2N-1\\k=1,\dots,N}}\Big]\Big|\,d\vec{\theta}\wedge d\vec{q}.$$

The determinant is in fact equal to

$$\prod_{j=1}^N(\lambda_j - 1/\lambda_j)\prod_{1\le j<k\le N}(\lambda_k-\lambda_j)^2(1/\lambda_k - 1/\lambda_j)^2(\lambda_j - 1/\lambda_k)^2(1/\lambda_j - \lambda_k)^2.$$

To see this, note that it is a symmetric rational function of $\lambda_1, \dots, \lambda_N$, and is antisymmetric under the mapping $\lambda_i \mapsto 1/\lambda_i$ for any $i = 1, \dots, N$. It must thus be of the form

$$\prod_{j=1}^{N} (\lambda_j - 1/\lambda_j)\, q(\lambda_1, \dots, \lambda_N) \tag{2.100}$$

where q is symmetric and unchanged by the mapping $\lambda_i \mapsto 1/\lambda_i$. Noting too that the determinant vanishes when $\lambda_1 = \lambda_2$, we see that q must contain as a factor

$$\prod_{1 \le j < k \le N} (\lambda_k - \lambda_j)^2 (1/\lambda_k - 1/\lambda_j)^2 (\lambda_j - 1/\lambda_k)^2 (1/\lambda_j - \lambda_k)^2. \tag{2.101}$$

The highest order term in degree of (2.101) multiplied by the first factor in (2.100) is $\prod_{j=1}^{N} \lambda_j \prod_{1 \le j < k \le N} (\lambda_k - \lambda_j)^4$. The highest order term in degree in the determinant is

$$\det \Big[[\lambda_k^j - \lambda_N^j]_{\substack{j=1,\dots,2N-1 \\ k=1,\dots,N-1}} \, [j\lambda_k^j]_{\substack{j=1,\dots,2N-1 \\ k=1,\dots,N}} \Big].$$

According to (1.175), up to a sign this is equal to the same expression, so in fact q must equal (2.101) up to this sign. Using (2.96) to write the evaluation in terms of $\{\alpha_i\}$, $\{q_i\}$ gives the result. □

The above results allow a real orthogonal upper Hessenberg $2N \times 2N$ matrix with unit determinant, possessing an eigenvalue p.d.f. which β-generalizes that for $O^+(2N)$, to be specified [356].

PROPOSITION 2.9.5 *Consider the upper Hessenberg matrix of Proposition 2.9.2, but with*

$$\alpha_k \in \tilde{B}\Big[\frac{2N-k-2}{4}\beta + a + 1, \frac{2N-k-2}{4}\beta + b + 1\Big], \qquad k \text{ even},$$
$$\alpha_k \in \tilde{B}\Big[\frac{2N-k-3}{4}\beta + a + b + 1, \frac{2N-k-1}{4}\beta\Big], \qquad k \text{ odd},$$

$k = 0, \dots, 2N-2$. *The eigenvalues and first component of the eigenvectors (which form $\vec{q}$) are independent, with the distribution of the former given by*

$$\frac{1}{C_N(a,b;\beta)} \prod_{l=1}^{N} |1 - e^{i\theta_l}|^{2a+1} |1 + e^{i\theta_l}|^{2b+1} \prod_{1 \le j < k \le N} |e^{i\theta_j} - e^{i\theta_k}|^{\beta} |1 - e^{i(\theta_j + \theta_k)}|^{\beta}, \tag{2.102}$$

where the normalization is given by

$$\begin{aligned} C_N(a,b;\beta) &= 2^{(2a+2b+2)N + \beta N(N-1)} S_N(a,b,\beta/2), \\ S_N(\lambda_1, \lambda_2, \lambda) &= \prod_{j=0}^{N-1} \frac{\Gamma(\lambda_1 + 1 + j\lambda)\Gamma(\lambda_2 + 1 + j\lambda)\Gamma(1 + (j+1)\lambda)}{\Gamma(\lambda_1 + \lambda_2 + 2 + (N+j-1)\lambda)\Gamma(1+\lambda)}. \end{aligned} \tag{2.103}$$

Proof. With $\mathbf{H}_\beta$ denoting the Hessenberg matrix, we see

$$\begin{aligned} P(\mathbf{H}_\beta)(d\mathbf{H}_\beta) &= K_N(a,b;\beta) \\ &\times \prod_{k=0}^{2N-2} (1 - \alpha_k^2)^{-1+\beta(2N-k-1)/4} \prod_{k=0}^{2N-2} (1 - \alpha_k)^{a+1-\beta/4} (1 + (-1)^k \alpha_k)^{b+1-\beta/4} \, d\vec{\alpha}, \end{aligned}$$

where

$$K_N(a,b;\beta) = 2^{-(2N-1)} \prod_{\substack{k=0 \\ k \text{ even}}}^{2N-2} \frac{\Gamma((2N-k-2)\beta/2+a+b+2)2^{-(2N-k-2)\beta/2-a-b}}{\Gamma((2N-k-2)\beta/4+a+1)\Gamma((2N-k-2)\beta/4+b+1)}$$

$$\times \prod_{\substack{k=1 \\ k \text{ odd}}}^{2N-3} \frac{\Gamma((2N-k-2)\beta/2+a+b+2)2^{-(2N-k-2)\beta/2-a-b}}{\Gamma((2N-k-3)\beta/4+a+b+1)\Gamma((2N-k-1)\beta/4)}$$

$$= 2^{-\sigma} \frac{\prod_{p=N-1}^{2N-2} \Gamma(p\beta/2+a+b+2)}{\prod_{r=0}^{N-1} \Gamma(r\beta/2+a+1)\Gamma(r\beta/2+b+1) \prod_{s=0}^{N-2} \Gamma((s+1)\beta/2)},$$

with $\sigma := ((N-1)\beta/2+a+b+1)(2N-1)$. Changing variables using the results of Propositions 2.9.3 and 2.9.4 gives (2.102) with

$$\frac{1}{C_N(a,b;\beta)} = \frac{1}{N!} 2^{-N\beta/2-(a+b+1-\beta/2)} \frac{\Gamma^N(\beta/2)}{\Gamma(\beta N/2)} K_N(a,b;\beta),$$

and this can readily be written in the form (2.103). □

The p.d.f. (2.102) permits a log-gas interpretation.

PROPOSITION 2.9.6 *Consider a log-gas system of unit charges confined to a half circle $0 < \theta < \pi$. Let the coordinates of the particles be $\theta_1, \dots, \theta_N$ and suppose there are image charges at $-\theta_1, \dots, -\theta_N$. Suppose too that there are fixed particles at $\theta = 0, \pi$ of charge $(2a+1)/\beta - \frac{1}{2}$, $(2b+1)/\beta - \frac{1}{2}$, respectively. The Boltzmann factor is then proportional to (2.102).*

Proof. For the log-gas on a half circle with like image charges, the pair potential is

$$\Phi(\theta,\theta') = -\log|e^{i\theta} - e^{i\theta'}||1 - e^{i(\theta+\theta')}|.$$

The total potential energy then consists of the particle-particle energy

$$U_1 = -\sum_{1 \le j < k \le N} \log|e^{i\theta_j} - e^{i\theta_k}||1 - e^{i(\theta_j+\theta_k)}|,$$

the self energy

$$U_1' := \frac{1}{2}\sum_{j=1}^{N} \lim_{\theta \to \theta_j} \Big(\Phi(\theta,\theta_j) - \log|e^{i\theta} - e^{i\theta_j}|\Big) = -\frac{1}{2}\sum_{j=1}^{N} \log|1 - e^{i\theta_j}||1 + e^{i\theta_j}|$$

and the particle-fixed particle energy

$$U_f = -\Big(\frac{2a+1}{\beta} - \frac{1}{2}\Big)\sum_{j=1}^{N} \log|1 - e^{i\theta_j}| - \Big(\frac{2b+1}{\beta} - \frac{1}{2}\Big)\sum_{j=1}^{N} \log|1 + e^{i\theta_j}|.$$

Forming $e^{-\beta(U_1+U_1'+U_f)}$ gives (2.102). □

Chapter Three

Laguerre and Jacobi ensembles

A Hermitian random matrix $\mathbf{X}$ can be formed out of a rectangular Gaussian matrix in the top right block, and its Hermitian conjugate in the bottom left block, with zeros elsewhere. This structure, which defines the chiral ensembles, can be motivated by the consideration of Dirac operators in the context of quantum chromodynamics, and time reversal symmetry distinguishes the cases of real, complex and real quaternion elements. The positive eigenvalues of matrices from the chiral ensembles are the singular values of $\mathbf{X}$, or equivalently the nonzero eigenvalues of $\mathbf{X}^\dagger\mathbf{X}$. The ratio of the largest to smallest singular value is precisely the condition number of the linear system associated with $\mathbf{X}$. Eigenvalues of the matrix product $\mathbf{X}^\dagger\mathbf{X}$, in the case that $\mathbf{X}$ relates to a data matrix in multivariable statistics, are of basic importance to the method of principal component analysis. With $\mathbf{X}$ Gaussian, this matrix product is said to be a Wishart matrix. The eigenvalue p.d.f.'s for the chiral ensembles and Wishart matrices can be calculated using the method of wedge products, or metric forms. For certain values of the parameter α these same eigenvalue p.d.f.'s appear in the study of random matrix models of Hamiltonians for electron and hole wave functions at normal metal/superconductor junctions. The Gaussian ensembles of Chapter 1, the chiral ensembles, and these further random matrix models of Hamiltonians together form a list of ten, which can be identified with the ten infinite families of matrix Lie algebras. Also studied are Jacobi ensembles, defined by a family of eigenvalue p.d.f.'s with each eigenvalue supported on $[-1,1]$. Realizations for $\beta = 1,2$ and 4 in terms of Wishart matrices, unitary matrices corresponding to compact symmetric spaces, and singular values of block decompositions of unitary matrices are given. Motivated by an identity between canonical averages, a circular analogue of the Jacobi ensemble is defined. In the last three sections, random matrix realizations of the Laguerre, Jacobi and circular Jacobi ensembles are given for general $\beta > 0$.

3.1 CHIRAL RANDOM MATRICES

3.1.1 Random Dirac operators

A random matrix theory of Dirac operators, in the context of quantum chromodynamics (QCD), has been introduced by Verbaarschot [529] (for a review see [530]). More precisely, one is considering a massless Dirac particle coupled to a random gauge field, which has Hamiltonian $i\gamma_\mu(\partial_\mu + iA_\mu)$ (with summation over repeated indices implicit). As is well known, the γ-matrices $\gamma_1,\dots,\gamma_4$ anticommute with the matrix γ_5, so the nonzero eigenvalues of the massless Dirac operator occur in pairs $\pm\lambda$, corresponding to the eigenfunctions ψ and $\gamma_5\psi$. Now consider a basis, referred to as the *chiral* basis, which consists of eigenvectors of $i\gamma_5$. Since $(i\gamma_5)^2 = 1$ these eigenvectors have eigenvalue either $+1$ or -1, and so in forming a matrix representation of the Hamiltonian out of such a basis, the matrix elements between states with the same eigenvalue of γ_5 must vanish as γ_5 anticommutes with the Hamiltonian, leaving a block structure with nonzero elements in the upper-right and lower-left blocks only. Furthermore, the application to QCD requires that the Dirac operator has a given number, ν say, of zero eigenvalues (this determines the topological charge). A matrix structure consistent with these facts is given by the following result.

PROPOSITION 3.1.1 *The matrix*

$$\mathbf{H} = \begin{bmatrix} 0_{n\times n} & \mathbf{X} \\ \mathbf{X}^\dagger & 0_{m\times m} \end{bmatrix}, \tag{3.1}$$

where $\mathbf{X}$ *is an* $n \times m$ $(n \geq m)$ *matrix, has in general* $n-m$ *zero eigenvalues and the remaining eigenvalues given by* $\pm$ *the positive square roots of the eigenvalues of* $\mathbf{X}^\dagger \mathbf{X}$.

Proof. Write the eigenvalues of $\mathbf{H}$ in block form so that

$$\mathbf{H}\begin{bmatrix} \vec{\psi}_n \\ \vec{\phi}_m \end{bmatrix} = \lambda \begin{bmatrix} \vec{\psi}_n \\ \vec{\phi}_m \end{bmatrix}.$$

This is equivalent to the coupled equations

$$\mathbf{X}\vec{\phi}_m = \lambda\vec{\psi}_n, \qquad \mathbf{X}^\dagger\vec{\psi}_n = \lambda\vec{\phi}_m.$$

Replacing $\vec{\phi}_m$ by $-\vec{\phi}_m$ and λ by $-\lambda$ leaves these equations unchanged, showing that the eigenvalues come in $\pm$ pairs. Furthermore, the coupled equations imply $\mathbf{X}\mathbf{X}^\dagger\vec{\psi}_n = \lambda^2\vec{\psi}_n$. But the $n \times n$ matrix $\mathbf{X}\mathbf{X}^\dagger$ has rank m, and so has $n-m$ zero eigenvalues with the remaining eigenvalues equal to those of the full rank matrix $\mathbf{X}^\dagger\mathbf{X}$. □

Since this result holds independent of the details of $\mathbf{X}$ a random matrix hypothesis can be made: the statistical properties of the eigenvalues of the Dirac operator in the QCD problem will be the same as those of a generic matrix of the form (3.1), subject only to the constraints imposed by time reversal symmetry. Since $\mathbf{H}$ is Hermitian, the latter are the same as those of a nonrelativistic Hamiltonian. Thus, from Section 1.1, if the Dirac operator has a time reversal symmetry with $T^2 = 1$, then $\mathbf{X}$ can be chosen to have real elements, while if the time reversal symmetry is such that $T^2 = -1$ with $T = \mathbf{Z}_{2(n+m)}K$, then $\mathbf{X}$ must be quaternion real. A Gaussian distribution on the elements of $\mathbf{X}$ can be distinguished by the maximum entropy property of Proposition 1.1.3. Due to their origin in studying the Dirac equation with a chiral basis, random matrices of this type have become known as *chiral random matrices.*

DEFINITION 3.1.2 *Let* $\mathbf{X}$ *denote an* $n \times m$ $(n \geq m)$ *random matrix, and suppose the elements of* $\mathbf{X}$ *are determined by a parameter* $\beta = 1, 2$ *or* 4. *These elements are real, complex or real quaternion independent random variables with Gaussian densities*

$$\frac{1}{\sqrt{2\pi}}e^{-x_{jk}^2/2}, \quad \frac{1}{\pi}e^{-|z_{jk}|^2}, \quad \frac{2}{\pi}e^{-2|z_{jk}|^2} \quad \text{and} \quad \frac{2}{\pi}e^{-2|w_{jk}|^2}$$

in the three cases $\beta = 1, 2$ *and* 4 *respectively (recall from (1.20) that a real quaternion is specified by two complex numbers* z *and* w*). Use* $\mathbf{X}$ *to form the matrix* $\mathbf{H}$ *according to (3.1). The resulting ensembles of matrices are referred to as the chiral orthogonal ensemble* $(\beta = 1)$, *chiral unitary ensemble* $(\beta = 2)$ *and chiral symplectic ensemble* $(\beta = 4)$.

According to the singular value decomposition, any $n \times m$ $(n \geq m)$ matrix $\mathbf{X}$ can be written as

$$\mathbf{X} = \mathbf{U}\mathbf{\Lambda}\mathbf{V}^\dagger, \tag{3.2}$$

where $\mathbf{\Lambda}$ is an $n \times m$ diagonal matrix containing the m positive square roots of the eigenvalues of the matrix $\mathbf{X}^\dagger\mathbf{X}$, and $\mathbf{U}$ and $\mathbf{V}$ are $m \times m$ and $n \times n$ unitary matrices, respectively. This decomposition can be used to determine the eigenvalue p.d.f. of the chiral ensembles [529]. First, one notes that the constraint of real elements $(\beta = 1)$ and real quaternion elements $(\beta = 4)$, implies that

$$\mathbf{X} = \mathbf{O}_1\mathbf{\Lambda}\mathbf{O}_2^T, \qquad \mathbf{X} = \mathbf{B}_1\mathbf{\Lambda}\mathbf{Z}_{2n}\mathbf{B}_2^T\mathbf{Z}_{2n}^{-1} \tag{3.3}$$

for $\mathbf{O}_1, \mathbf{O}_2$ real orthogonal $(\beta = 1)$ and $\mathbf{B}_1, \mathbf{B}_2$ unitary symplectic equivalent $(\beta = 4)$, respectively. The metric forms approach of Section 1.2.4 can be used to calculate the eigenvalue p.d.f.

PROPOSITION 3.1.3 *The p.d.f. of the positive eigenvalues of the chiral ensembles is proportional to*

$$\prod_{j=1}^{m} \lambda_j^{\beta\alpha} e^{-\beta\lambda_j^2/2} \prod_{1\le j<k\le m} |\lambda_k^2 - \lambda_j^2|^\beta, \quad \alpha = n - m + 1 - \frac{1}{\beta}. \tag{3.4}$$

Proof. Consider first the real case ($\beta = 1$), for which the singular value decomposition is given by the first equation in (3.3). Since $\mathbf{X}$ has nm independent elements, and there are m nonzero elements in $\mathbf{\Lambda}$, the matrices $\mathbf{O}_1$ and $\mathbf{O}_2$ must together have a total of $m(n-1)$ independent elements (or combinations of elements). We begin by computing the differential of the singular value decomposition to obtain

$$\mathbf{O}_1^T d\mathbf{X}\mathbf{O}_2 = \delta\mathbf{O}_1\mathbf{\Lambda} + d\mathbf{\Lambda} - \mathbf{\Lambda}\delta\mathbf{O}_2,$$

where $\delta\mathbf{O} := \mathbf{O}^T d\mathbf{O}$. Substituting this result and its transpose in $\mathrm{Tr}(d\mathbf{X}d\mathbf{X}^T)$ and simplifying using the cyclic property of the trace and the identities

$$\mathbf{\Lambda} d\mathbf{\Lambda}^T - d\mathbf{\Lambda}\,\mathbf{\Lambda}^T = \mathbf{0}_{n\times n}, \qquad \mathbf{\Lambda}^T d\mathbf{\Lambda} - d\mathbf{\Lambda}^T\,\mathbf{\Lambda} = \mathbf{0}_{m\times m}$$

shows

$$\begin{aligned}\mathrm{Tr}(d\mathbf{X}d\mathbf{X}^T) &= \mathrm{Tr}\Big(- (\delta\mathbf{O}_1)^2\mathbf{\Lambda}\mathbf{\Lambda}^T - (\delta\mathbf{O}_2)^2\mathbf{\Lambda}^T\mathbf{\Lambda} + 2\mathbf{\Lambda}\delta\mathbf{O}_2\mathbf{\Lambda}^T\delta\mathbf{O}_1 + d\mathbf{\Lambda}d\mathbf{\Lambda}^T\Big) \\ &= \mathrm{Tr}\Big((\mathbf{\Lambda}\delta\mathbf{O}_2 - \delta\mathbf{O}_1\,\mathbf{\Lambda})(\mathbf{\Lambda}^T\delta\mathbf{O}_1 - \delta\mathbf{O}_2\,\mathbf{\Lambda}^T) + d\mathbf{\Lambda}d\mathbf{\Lambda}^T\Big).\end{aligned} \tag{3.5}$$

Writing this result in component form, using the fact that $\delta\mathbf{O}_1, \delta\mathbf{O}_2$ are real antisymmetric, gives

$$\begin{aligned}\mathrm{Tr}(d\mathbf{X}d\mathbf{X}^T) =& \sum_{1\le k<l\le m} \frac{1}{2}\Big([\delta\mathbf{O}_2]_{kl} - [\delta\mathbf{O}_1]_{kl}\Big)^2(\lambda_k+\lambda_l)^2 + \sum_{1\le k\le l\le m} \frac{1}{2}\Big([\delta\mathbf{O}_2]_{kl} + [\delta\mathbf{O}_1]_{kl}\Big)^2(\lambda_k-\lambda_l)^2 \\ &+ \sum_{k=1}^{m}\Big([\delta\mathbf{O}_2]_{kk} - [\delta\mathbf{O}_1]_{kk}\Big)^2\lambda_k^2 + \sum_{k=1}^{m}\sum_{l=m+1}^{n}(\delta\mathbf{O}_1)_{kl}^2\lambda_k^2 + \sum_{k=1}^{m}(d\lambda_k)^2.\end{aligned} \tag{3.6}$$

The fact that $\delta\mathbf{O}_1$ and $\delta\mathbf{O}_2$ are real antisymmetric also implies

$$[\delta\mathbf{O}_1]_{kk} = [\delta\mathbf{O}_2]_{kk} = 0, \tag{3.7}$$

and thus there are a total of $m(n-1)$ differentials in (3.6) involving elements of $\delta\mathbf{O}_1$ and $\delta\mathbf{O}_2$, which as noted above is the number required in the singular value decomposition to be contributed by $\mathbf{O}_1$ and $\mathbf{O}_2$. These differentials must therefore all be independent, so using (1.14) we read off the factors in (3.4) in the case $\beta = 1$ excluding the exponential (the terms involving the exponential result from the Gaussian measure).

Consider next the complex case ($\beta = 2$). The appropriate singular value decomposition is now given by (3.2). Writing $\mathbf{U} = \mathbf{U}_1$, $\mathbf{V} = \mathbf{U}_2$ in that formula and setting $\delta\mathbf{U}_i = \mathbf{U}_i^\dagger d\mathbf{U}_i$ we see that (3.6) is unchanged if we write $\delta\mathbf{O}_i \mapsto \delta\mathbf{U}_i$ and replace the squares of all quantities which are now complex by their absolute value squared. The formula for the Jacobian in (3.4) then follows by noting $[\delta\mathbf{U}_i]_{jk}$, $j<k$ has an independent real and imaginary part, while $[\delta\mathbf{U}_i]_{jj}$ has one independent (pure imaginary) component. In the quaternion case ($\beta = 4$) the matrix elements $[\delta\mathbf{U}_i]_{jk}$, $j<k$, have four independent components, while $[\delta\mathbf{U}_i]_{jj}$ has three independent components. □

3.1.2 Singular values and the condition number

A fundamental question in numerical analysis relates to the sensitivity of the computed solution to the input data. If an error of measure ϵ in the input produces an error of measure $c\epsilon$ in the computed solution, with c around unity, the problem is said to be well conditioned. However, if c is orders of magnitudes greater than unity, the input error is significantly amplified and the problem is said to be illconditioned. The proportionality c is termed the *condition number.*

To see how these ideas apply in a matrix setting, consider the task of computing the solution $\vec{x}$ to the $N\times N$

linear system

$$\mathbf{X}\vec{x} = \vec{b}. \tag{3.8}$$

Let $|\vec{x}|$ denote the usual (complex) Euclidean norm, and define the corresponding matrix norm (the so-called 2-norm) by $|\mathbf{X}| = \sup_{|\vec{x}|=1} |\mathbf{X}\vec{x}|$. Some immediate consequences of the definitions are the inequalities

$$|\mathbf{X}\vec{x}| \le |\mathbf{X}||\vec{x}|, \qquad |\mathbf{A}||\mathbf{B}| \le |\mathbf{A}\mathbf{B}|. \tag{3.9}$$

It is also true that $|\mathbf{X}^\dagger \mathbf{X}| = |\mathbf{X}|^2$. From this latter fact, by writing the vector in the definition of $|\mathbf{X}^\dagger \mathbf{X}|$ in terms of the eigenvectors of the Hermitian matrix $\mathbf{X}^\dagger \mathbf{X}$, it follows that

$$|\mathbf{X}| = \lambda_1^{1/2} = \mu_1, \tag{3.10}$$

where λ_1 is the largest eigenvalue of $\mathbf{X}^\dagger \mathbf{X}$ and μ_1 is the largest singular value of $\mathbf{X}$.

We seek to quantify the effect of perturbing either the matrix $\mathbf{X}$ or the vector $\vec{b}$ in (3.8). Suppose first that the matrix $\mathbf{X}$ is perturbed by the addition of $\delta\mathbf{X}$, and use the matrix norm to measure the size of the perturbation according to $\epsilon = |\delta\mathbf{X}|/|\mathbf{X}|$. With $\vec{y}$ the solution of the perturbed system

$$(\mathbf{X} + \delta\mathbf{X})\vec{y} = \vec{b}, \tag{3.11}$$

subtracting (3.11) from (3.8) and using the inequalities (3.9) show

$$\frac{|\vec{x} - \vec{y}|}{|\vec{y}|} \le \kappa(\mathbf{X})\epsilon, \qquad \kappa(\mathbf{X}) := |\mathbf{X}||\mathbf{X}^{-1}| = \frac{\mu_1}{\mu_N}. \tag{3.12}$$

Here, with μ_N denoting the smallest singular value of $\mathbf{X}$, the final equality follows from (3.10). Similarly, for a perturbation $\delta\vec{b}$ in the vector $\vec{b}$, defining $\epsilon = |\delta\vec{b}|/|\vec{b}|$, $\kappa(\mathbf{X})$ as in (3.12), and with $\vec{y}$ such that $\mathbf{X}\vec{y} = \vec{b} + \delta\vec{b}$, a straightforward calculation shows

$$\frac{|\vec{y} - \vec{x}|}{|\vec{x}|} \le \kappa(\mathbf{X})\epsilon. \tag{3.13}$$

For $\mathbf{X}$ a complex Gaussian matrix, and for $\mathbf{X}$ a real Gaussian matrix, the distribution of the condition number $\kappa(\mathbf{X})$ is calculated in Exercises 8.3 q.3 and Exercises 13.2 q.5, respectively.

3.1.3 Relationship to Lie algebras

The matrices of Definition 3.1.2 are the Hermitian part of the matrix Lie algebras

$$\mathrm{so}(p,q) = \left\{ \begin{bmatrix} \mathbf{X}_1 & \mathbf{X}_2 \\ \mathbf{X}_2^T & \mathbf{X}_3 \end{bmatrix}, \text{ all } \mathbf{X}_i \text{ real, } \mathbf{X}_1\ p \times p,\ \mathbf{X}_3\ q \times q, \text{ both skew symmetric} \right\},$$

$$u(p,q) = \left\{ \begin{bmatrix} \mathbf{Z}_1 & \mathbf{Z}_2 \\ \mathbf{Z}_2^\dagger & \mathbf{Z}_3 \end{bmatrix}, \text{ all } \mathbf{Z}_i \text{ complex, } \mathbf{Z}_1\ p \times p,\ \mathbf{Z}_3\ q \times q, \text{ both anti-Hermitian} \right\},$$

$$\mathrm{sp(p,q)} = \left\{ \begin{bmatrix} \mathbf{Q}_1 & \mathbf{Q}_2 \\ \mathbf{Q}_2^\dagger & \mathbf{Q}_3 \end{bmatrix}, \text{ all } \mathbf{Q}_i \text{ real quaternion, } \mathbf{Q}_1\ p \times p,\ \mathbf{Q}_3\ q \times q, \text{ both anti-Hermitian} \right\},$$

with $p = n$, $q = m$. In view of the relationship between the Lie algebras of Section 1.3.4 and the symmetric spaces of Section 2.2.3, we would expect that there is a corresponding theory of random unitary matrices. This is indeed the case, as will be considered in Section 3.7.

3.1.4 Coulomb gas analogy

Written in the form

$$\exp\Big(-\beta \sum_{j=1}^{m} \big(\lambda_j^2/2 - \alpha \log|\lambda_j| \big) - \beta \sum_{1 \le j < k \le m} \big(\log|\lambda_k - \lambda_j| + \log|\lambda_k + \lambda_j| \big) \Big) \tag{3.14}$$

we see that the eigenvalue p.d.f. (3.4) has the interpretation as the Boltzmann factor for a one-component log-potential system of unit charges on the half line $x > 0$, with image charges of like sign in the region $x < 0$. The charges are confined by a one-body harmonic potential as in the log-gas picture of the p.d.f. of Proposition 1.3.4, and there is also a fixed charge of strength α at the origin. Proceeding as in the proof of Proposition 1.4.4 (see Exercises 3.1 q.1) we can specify the background charge density in the Boltzmann factor of this log-potential system.

PROPOSITION 3.1.4 *The Boltzmann factor of the one-component log-potential Coulomb system with particles of unit charge at $x_1, \dots, x_N$, confined to the interval $[0, \sqrt{4N}]$, with image charges of the same sign in the interval $[-\sqrt{4N}, 0]$, a fixed charge of strength $\alpha - \frac{1}{2}$ at the origin, and a background charge density neutralizing the mobile charges*

$$-\rho_b(y) = -\frac{\sqrt{4N}}{\pi} \sqrt{1 - \frac{y^2}{4N}}$$

is proportional to

$$\prod_{j=1}^{N} |x_j|^{\beta\alpha} e^{-\beta x_j^2/2} \prod_{1 \le j < k \le N} |x_k^2 - x_j^2|^{\beta}.$$

In obtaining this result the fact that the finite portion of the self energy term $\frac{1}{2}\sum_{j=1}^{N} \log|x_j|$ must be included in the total energy has been used (recall the proof of Proposition 2.7.1). The background density again has a semicircle profile, although the independent particles are restricted to $x > 0$.

EXERCISES 3.1 1. For a one-component log-gas of unit charges, with image forces as in (3.14) and subject to a one-body potential $V(x)$, the background charge density $-\rho_b(x)$ satisfies the integral equation

$$V(x) + C = \int_0^a \rho_b(t) \log|x^2 - t^2| \, dt, \quad x \in (0, a),$$

where a is such that

$$\int_0^a \rho_b(t) \, dt = N.$$

Show that this is equivalent to the integral equation

$$V(x) + C = \int_{-a}^{a} \rho_b(|t|) \log|x - t| \, dt, \quad x \in (-a, a),$$

and so deduce the background density in Proposition 3.1.4 from that in Proposition 1.4.4.

2. [492] Consider the matrix $\mathbf{H}$ in (3.1) in the case $n = m$ with $\mathbf{X}$ a symmetric complex matrix.

   (i) Show that the singular value decomposition of $\mathbf{X}$ must be of the form $\mathbf{X} = \mathbf{U}\mathbf{\Lambda}\mathbf{U}^T$ for $\mathbf{U}$ unitary.

   (ii) Follow the proof of Proposition 3.1.3 to show

$$(d\mathbf{X}) = \prod_{1 \le j < k \le n} |\lambda_k^2 - \lambda_j^2| \prod_{j=1}^{n} \lambda_j \, d\lambda_j \, (\mathbf{U}^\dagger d\mathbf{U}).$$

3.2 WISHART MATRICES

3.2.1 Setting in multivariate statistics

In multivariate statistics there may be m variables y_k $(k = 1, \dots, m)$, with each variable measured n times. (It is typical in this setting to denote the number of different variables by p (for population) rather than m, but we will persist with m.) For example, y_k may represent the noise level at location k at 6 a.m. on a weekday. Measuring the noise levels at this time on n different weekdays gives vectors of data $\vec{y}_k := [y_k^{(j)}]_{j=1,\dots,n}$ for each variable y_k, which in turn can be used to define a data matrix $\mathbf{Y} := [\vec{y}_k]_{k=1,\dots,m}$. We remark that another convention has the vectors of data as row vectors, giving then our $\mathbf{Y}^T$ as the data matrix; see, e.g., [279]. The average of the measurements for variable y_k is

$$\bar{y}_k := \frac{1}{n}\sum_{j=1}^{n} y_k^{(j)}.$$

With the average data vector for variable y_k defined by $\vec{\bar{y}}_k = [\bar{y}_k]_{j=1,\dots,n}$, and the average data matrix by $\bar{\mathbf{Y}} := [\vec{\bar{y}}_k]_{k=1,\dots,m}$, the matrix product

$$\frac{1}{n^*}(\mathbf{Y} - \bar{\mathbf{Y}})^T(\mathbf{Y} - \bar{\mathbf{Y}}) = \Big[\frac{1}{n^*}\sum_{j=1}^{n}(y_{k_1}^{(j)} - \bar{y}_{k_1})(y_{k_2}^{(j)} - \bar{y}_{k_2})\Big]_{k_1,k_2=1,\dots,m},$$

with $n^* := n - 1$, then represents an empirical approximation to the covariance matrix

$$\Big[\Big\langle (y_{k_1} - \bar{y}_{k_1})(y_{k_2} - \bar{y}_{k_2})\Big\rangle\Big]_{k_1,k_2=1,\dots,m}$$

for the variables y_k.

The eigenvalue-eigenvector decomposition of $\mathbf{Y}^T\mathbf{Y}$ is of basic importance in the analysis of the covariance matrix. This comes under the name of principal component analysis (see, e.g., [410]), in which one considers the eigenvectors corresponding to the largest, second largest and subsequent eigenvalues as giving orthogonal linear combinations of variables which account for the successive maximum variations in the data. This then allows the statistically important linear combination of variables to be identified, and so effectively reduces the dimension of the problem.

Our interest is in the theoretical setting that the variables y_k are drawn from a multivariate Gaussian distribution with variance Σ and mean $\vec{\mu}$. Then a well-known result (see, e.g., [279]) gives that the distribution of $\mathbf{Y}^T\mathbf{Y}$ is the same as the distribution of

$$\mathbf{A} := \mathbf{X}^T\mathbf{X}, \tag{3.15}$$

where $\mathbf{X}$ is an $m \times n$ Gaussian matrix with elements of each row drawn from an m dimensional Gaussian distribution having variance Σ but mean zero. We define ensembles of matrices relating to the matrix structure (3.15) according to the following definition.

DEFINITION 3.2.1 *With $n \times m$ random matrices $\mathbf{X}$ specified as in Definition 3.1.2 in the three cases $\beta = 1, 2$ and 4, define the real ($\beta = 1$), complex ($\beta = 2$) and quaternion real ($\beta = 4$) Wishart ensembles as consisting of matrices $\mathbf{X}^\dagger\mathbf{X}$. The matrices $\mathbf{X}^\dagger\mathbf{X}$ are referred to as (uncorrelated) Wishart matrices (for the correlated case see Section 3.5).*

As already remarked, the square roots of the nonzero eigenvalues of (3.15) are the singular values of $\mathbf{X}$, or equivalently the positive eigenvalues of (3.1). The eigenvalue p.d.f. for matrices from the Wishart ensembles can therefore be written down from knowledge of the eigenvalue p.d.f. for the positive eigenvalues of the matrices $\mathbf{H}$ as given in Proposition 3.1.1, after the change of variables $\lambda_j^2 \mapsto \lambda_j$ in the latter.

PROPOSITION 3.2.2 *The eigenvalue p.d.f. for the real ($\beta = 1$), complex ($\beta = 2$) or quaternion real ($\beta = 4$) Wishart matrices is given by*

$$\frac{1}{W_{a\beta m}} \prod_{l=1}^{m} \lambda_l^{\beta a/2} e^{-\beta \lambda_l /2} \prod_{1 \le j < k \le m} |\lambda_k - \lambda_j|^{\beta}, \quad \lambda_l \ge 0, \tag{3.16}$$

where $a = n - m + 1 - 2/\beta$ and the normalization constant is given explicitly in (3.134) below.

Due to the one body factors of the form $\lambda^{\beta a/2} e^{-\beta\lambda/2}$ (3.16) is said to define the *Laguerre ensemble* of random matrices, or more accurately of eigenvalue p.d.f.'s. In particular, the case $\beta = 1$ is referred to as the Laguerre orthogonal ensemble (LOE), the case $\beta = 2$ is referred to as the Laguerre unitary ensemble (LUE), while the case $\beta = 4$ is referred to as the Laguerre symplectic ensemble (LSE). Note that the corresponding Wishart ensembles realize these eigenvalue p.d.f.'s. Changing variables $y^2 \mapsto y$, $x_j^2 \mapsto x_j$ in the result of Proposition 3.1.4 gives the log-gas analogy of the Laguerre ensemble p.d.f. (3.16).

PROPOSITION 3.2.3 *The Boltzmann factor of the one-component log-potential system with particles of unit charge at $x_1, \dots, x_N$, confined to the interval $[0, 4N]$ with a background charge density neutralizing the mobile charges*

$$-\rho_b(y) = -\frac{1}{2\pi y^{1/2}} (4N - y)^{1/2},$$

and with a fixed particle of charge $(a-1)/2 + 1/\beta$ at $x = 0$ is proportional to

$$\prod_{j=1}^{N} x_j^{\beta a/2} e^{-\beta x_j/2} \prod_{1 \le j < k \le N} |x_k - x_j|^{\beta}. \tag{3.17}$$

3.2.2 Wireless communication systems

At a theoretical level the problem of maximizing the information transfer between antennas and receivers in wireless communication systems relates to complex Wishart matrices [523]. Thus practical methods to achieve maximum rates [489] make use of multiple antennas (M_T, say) to transmit distinct bitstreams (signals), which are decoded by multiple receivers (M_R, say, with $M_R \ge M_T$). For the decoding to be possible in the case $M_R = M_T$ it is necessary that a distinct linear combination of the original signals be received at each antenna, the physical mechanism for which is scattering of the signal as it travels between antenna and receiver. This setting is described by the equation

$$\vec{y} = \mathbf{G}\vec{x} + \vec{z},$$

where the jth component of $\vec{y}$ is the received signal at receiver j, while the j-th component of $\vec{x}$ is the signal sent by transmitter j. The linear combination of transmitted signals received at each antenna is determined by the propagation matrix $\mathbf{G}$, while the vector $\vec{z}$ denotes the noise at the receivers due to the presence of extraneous signals.

The information capacity I measures the number of bits per second per frequency which can be transfered between input at the antenna and output at the receiver. In the case of a single antenna and receiver, this is given in terms of the signal to noise ratio $u := |x|/|z|$ by *Shannon's formula*

$$I = \log_2(1 + u).$$

In the case of multiple antennas and receivers Shannon's formula generalizes to

$$\tilde{I} = \mathrm{Tr}\, \log(\mathbf{1} + \tilde{u}\mathbf{G}^{\dagger}\mathbf{G}) = \sum_{j=1}^{M_T} \log(1 + \tilde{u}\lambda_j), \tag{3.18}$$

where $\tilde{I} := (\log 2)I$, $\tilde{u} := M_R\langle|\vec{x}|^2\rangle/M_T\langle|\vec{z}|^2\rangle$ and the λ_j denote the eigenvalues of $\mathbf{G}^\dagger\mathbf{G}$. Because the propagating signals are subject to random scattering, it is reasonable to model $\mathbf{G}$ as a random matrix with complex Gaussian entries. If furthermore there is no correlation between the random scattering of the components, we then have that $\mathbf{G}^\dagger\mathbf{G}$ is a complex Wishart matrix. With respect to the corresponding eigenvalues $\tilde{I}$ is then an example of a linear statistic (see Definition 14.3.1 below). In an appropriate scaled limit the distribution of general linear statistics are Gaussians with means given in terms of the eigenvalue density, and $\mathrm{O}(1)$ variances (see Sections 14.3 and 14.4).

3.2.3 The distribution of Wishart matrices

For $\mathbf{X}$ real in (3.15), the task of expressing the volume form $(d\mathbf{A})$ in terms of $(d\mathbf{X})$ was considered long ago by Wishart [547]. The result of this calculation, which we will carry out below in the complex case, can be used to rederive Proposition 3.2.2 for the eigenvalue p.d.f. of the matrices $\mathbf{A}$. We will follow the presentation in [410], generalizing the working to the complex case (only the real case is considered in [410]).

According to Definition 3.2.1, we are given that the joint probability distribution of the elements of the $n \times m$ complex matrix $\mathbf{X}$ is

$$\frac{1}{\pi^{nm}} \prod_{j=1}^{n} \prod_{k=1}^{m} e^{-|z_{jk}|^2} (d\mathbf{X}) \propto e^{-\mathrm{Tr}(\mathbf{X}^\dagger \mathbf{X})} (d\mathbf{X}). \tag{3.19}$$

With $\mathbf{A} := \mathbf{X}^\dagger\mathbf{X}$ the strategy is to first use the Gram-Schmidt orthogonalization procedure to write

$$\mathbf{X} = \mathbf{U}_1 \mathbf{T}, \tag{3.20}$$

where $\mathbf{U}_1$ is an $n \times m$ matrix such that $\mathbf{U}_1^\dagger \mathbf{U}_1 = \mathbf{1}_m$ and $\mathbf{T}$ is a $m \times m$ upper triangular matrix with diagonal entries real and positive. This allows $(d\mathbf{X})$ to be calculated in terms of $(d\mathbf{U}_1)$ and $(d\mathbf{T})$. Noting from (3.20) and the definition of $\mathbf{A}$ that $\mathbf{A} = \mathbf{T}^\dagger\mathbf{T}$ we then calculate $(d\mathbf{T})$ in terms of $(d\mathbf{A})$. Substituting the result of this second calculation into the first gives $(d\mathbf{X})$ in terms of $(d\mathbf{A})$ as required.

Proceeding as in the above outline, we must relate $(d\mathbf{T})$ and $(d\mathbf{A})$. This requires a preliminary result (cf. (1.17)).

PROPOSITION 3.2.4 *The Jacobian of the transformation $\vec{z} = \mathbf{A}\vec{w}$, where $\vec{w}$, $\vec{z}$ and $\mathbf{A}$ have complex entries, is $|\det \mathbf{A}|^2$.*

Proof. See Exercises 3.2 q.2. □

This result will be used to establish the relation between volume forms implied by (3.20).

PROPOSITION 3.2.5 *With $\mathbf{U}_1$ and $\mathbf{T}$ defined by (3.20) we have*

$$(d\mathbf{X}) = \prod_{j=1}^{m} t_{jj}^{2(n-j)+1} (d\mathbf{T})(\mathbf{U}_1^\dagger d\mathbf{U}_1),$$

where the t_{jj} are the diagonal entries of the matrix $\mathbf{T}$.

Proof. Since $\mathbf{X} = \mathbf{U}_1\mathbf{T}$ we have

$$d\mathbf{X} = d\mathbf{U}_1\,\mathbf{T} + \mathbf{U}_1 d\mathbf{T}.$$

Now extend the number of columns of $\mathbf{U}_1$ from m to n by defining an $(n-m) \times n$ matrix $\mathbf{U}_2$ such that $\mathbf{U} = [\mathbf{U}_1\mathbf{U}_2]$ and $\mathbf{U}^\dagger\mathbf{U} = \mathbf{1}$. Then

$$\mathbf{U}^\dagger d\mathbf{X} = \begin{bmatrix} \mathbf{U}_1^\dagger \\ \mathbf{U}_2^\dagger \end{bmatrix} d\mathbf{X} = \begin{bmatrix} \mathbf{U}_1^\dagger(d\mathbf{U}_1\,\mathbf{T} + \mathbf{U}_1 d\mathbf{T}) \\ \mathbf{U}_2^\dagger(d\mathbf{U}_1\,\mathbf{T} + \mathbf{U}_1 d\mathbf{T}) \end{bmatrix} = \begin{bmatrix} \mathbf{U}_1^\dagger d\mathbf{U}_1\,\mathbf{T} + d\mathbf{T} \\ \mathbf{U}_2^\dagger \mathbf{U}_1\,\mathbf{T} \end{bmatrix},$$

where to obtain the last equality the facts $\mathbf{U}_1^\dagger\mathbf{U}_1 = \mathbf{1}$ and $\mathbf{U}_2^\dagger\mathbf{U}_1 = \mathbf{0}$ have been used.

Consider the above equation. On the l.h.s., from Proposition 3.2.4 we have

$$(\mathbf{U}^\dagger d\mathbf{X}) = |\det \mathbf{U}^\dagger|^{2m} (d\mathbf{X}) = (d\mathbf{X}), \tag{3.21}$$

which is the l.h.s. of the assertion. On the r.h.s. consider the matrix product $\mathbf{U}_2^\dagger \mathbf{U}_1 \mathbf{T}$. The jth row is

$$(\vec{u}_j^\dagger d\vec{u}_1 \cdots \vec{u}_j^\dagger d\vec{u}_m)\mathbf{T} \qquad (m+1 \le j \le n).$$

Again using Proposition 3.2.4, the wedge product of the elements in this row is

$$|\det \mathbf{T}|^2 \bigwedge_{k=1}^m \vec{u}_j^\dagger \cdot d\vec{u}_k.$$

Hence, the wedge product of all the elements in the matrix product is

$$\bigwedge_{j=m+1}^n |\det \mathbf{T}|^2 \bigwedge_{k=1}^m \vec{u}_j^\dagger \cdot d\vec{u}_k = |\det \mathbf{T}|^{2(n-m)} \bigwedge_{j=m+1}^n \bigwedge_{k=1}^m \vec{u}_j^\dagger \cdot d\vec{u}_k.$$

It remains to consider the matrix $\mathbf{U}_1^\dagger d\mathbf{U}_1 \mathbf{T} + d\mathbf{T}$. From $\mathbf{U}_1^\dagger \mathbf{U}_1 = \mathbf{1}$ we see $\mathbf{U}_1^\dagger d\mathbf{U}_1 = -(\mathbf{U}_1^\dagger d\mathbf{U}_1)^\dagger$, so the matrix $\mathbf{U}_1^\dagger d\mathbf{U}_1$ is skew symmetric Hermitian. We therefore have

$$\mathbf{U}_1^\dagger d\mathbf{U}_1 \mathbf{T} = \begin{bmatrix} \vec{u}_1^\dagger \cdot d\vec{u}_1\, t_{11} & * & \dots & * \\ \vec{u}_2^\dagger \cdot d\vec{u}_1\, t_{11} & \vec{u}_2^\dagger \cdot d\vec{u}_2\, t_{22} & \dots & * \\ \vdots & \vdots & \ddots & \vdots \\ \vec{u}_m^\dagger \cdot d\vec{u}_1\, t_{11} & \vec{u}_m^\dagger \cdot d\vec{u}_2\, t_{22} + * & \dots & \vec{u}_m^\dagger \cdot d\vec{u}_m\, t_{mm} \end{bmatrix},$$

where the terms in which $\vec{u}_k^\dagger \cdot d\vec{u}_j$ and $(\vec{u}_k^\dagger \cdot d\vec{u}_j)^\dagger$ occur are ignored, as indicated by $*$, if $\vec{u}_k^\dagger \cdot d\vec{u}_j$ has already appeared in a previous column. We note that the diagonal terms $\vec{u}_j^\dagger \cdot d\vec{u}_j\, t_{jj}$ are pure imaginary.

If we now add $d\mathbf{T}$, we see that the wedge product of the resulting matrix is equal to

$$\prod_{i=1}^m t_{ii}^{2(m-i)+1} \bigwedge_{i=1}^m \bigwedge_{j=1}^m \vec{u}_j^\dagger \cdot d\vec{u}_i\, (d\mathbf{T}).$$

Hence the wedge product of all the elements on the r.h.s. is

$$|\det \mathbf{T}|^{2(n-m)} \prod_{i=1}^m t_{ii}^{2(m-i)+1} \bigwedge_{j=m+1}^n \bigwedge_{k=1}^m \vec{u}_j^\dagger \cdot d\vec{u}_k \bigwedge_{i=1}^m \bigwedge_{j=1}^m \vec{u}_j^\dagger \cdot d\vec{u}_i (d\mathbf{T}).$$

Since $\det \mathbf{T} = \prod_{i=1}^m t_{ii}$ and

$$\bigwedge_{j=m+1}^n \bigwedge_{k=1}^m \vec{u}_j^\dagger \cdot d\vec{u}_k \bigwedge_{i=1}^m \bigwedge_{j=1}^m \vec{u}_j^\dagger \cdot d\vec{u}_i = \bigwedge_{i=1}^m \bigwedge_{j=1}^n \vec{u}_j^\dagger \cdot d\vec{u}_i = (\mathbf{U}_1^\dagger d\mathbf{U}_1)$$

this is precisely the r.h.s. of the assertion. □

Note Proposition 3.2.5 gives that the elements of $\mathbf{T}$ are all independently distributed, while the dependence on $\mathbf{U}_1$ is of the form $(\mathbf{U}_1^\dagger d\mathbf{U}_1)$. Thus in the case $n = m$ the matrices $\mathbf{U}_1$ are Haar distributed, and so can be used to generate members of the CUE [157], [402].

The next result can be used to express $(d\mathbf{A})$ in terms of $(d\mathbf{T})$.

Proposition 3.2.6 *Let the $m \times m$ matrix $\mathbf{T}$ be as in (3.20), so that $\mathbf{A} = \mathbf{T}^\dagger \mathbf{T}$. We have*

$$(d\mathbf{A}) = 2^m \prod_{j=1}^m t_{jj}^{2m+1-2j} (d\mathbf{T}).$$

Proof. Since $\mathbf{A} = \mathbf{T}^\dagger \mathbf{T}$ we have $d\mathbf{A} = d\mathbf{T}^\dagger\, \mathbf{T} + \mathbf{T}^\dagger\, d\mathbf{T}$. Hence

$$d\mathbf{A} = \begin{bmatrix} 2t_{11}dt_{11} & t_{22}dt_{12} + * & \dots & t_{mm}dt_{1m} + * \\ * & 2t_{22}dt_{22} & \dots & t_{mm}dt_{2m} + * \\ \vdots & \vdots & \ddots & \vdots \\ * & * & * & 2t_{mm}dt_{mm} \end{bmatrix},$$

where * denotes differentials that have already appeared and thus do not contribute to the wedge product. Taking the wedge product of the above elements gives the stated result. □

Combining Propositions 3.2.5 and 3.2.6 we can obtain the p.d.f. of $\mathbf{A}$. This result was first given in [269], and is the complex analogue of the classical result obtained by Wishart [547] in the real case. We will include in the statement the analogous result for the real and quaternion real cases, the derivation of which is given in Exercises 3.2 q.4 and q.5

PROPOSITION 3.2.7 *Let $\mathbf{A}$ be a real ($\beta = 1$), complex ($\beta = 2$) or quaternion real ($\beta = 4$) Wishart matrix as defined in Definition 3.2.1. The p.d.f. of $\mathbf{A}$ is*

$$\frac{1}{\hat{C}_{\beta N}} e^{-(\beta/2)\mathrm{Tr}(\mathbf{A})} (\det \mathbf{A})^{(\beta/2)(n-m+1-2/\beta)}, \tag{3.22}$$

where $\hat{C}_{\beta N}$ is a normalization constant.

Proof. We are considering the case $\beta = 2$. Substituting the result of Proposition 3.2.6 in Proposition 3.2.5 gives

$$(d\mathbf{X}) = 2^{-m} (\det \mathbf{A})^{n-m} (d\mathbf{A})(\mathbf{U}_1^\dagger d\mathbf{U}_1), \tag{3.23}$$

where we have used the fact that $\det \mathbf{A} = (\det \mathbf{T})^2 = \prod_{j=1}^m t_{jj}^2$. Substituting this expression for $(d\mathbf{X})$ in the formula for the joint probability distribution of the elements of $\mathbf{X}$ (3.19), noting that $\mathrm{Tr}\mathbf{A} = \sum_{j,k=1}^m |x_{jk}|^2$, and integrating over the independent variables in $\mathbf{U}_1^\dagger d\mathbf{U}_1$ gives the stated result. □

In Exercises 3.2 q.6, a derivation of the $\mathbf{A}$-dependent portion of the Jacobian in (3.23), valid for all three β values, is given by exploiting functional equations.

The matrix $\mathbf{A}$ in Proposition 3.2.7 is Hermitian, with real ($\beta = 1$), complex ($\beta = 2$) or real quaternion elements ($\beta = 4$). Changing variables to the eigenvalues and eigenvectors according to the result of Proposition 1.3.4 gives the eigenvalue p.d.f. (3.16).

3.2.4 A matrix integral derivation

An alternative derivation [322] of Proposition 3.2.7 in the case $\beta = 2$ makes use of a matrix integral evaluation [247].

PROPOSITION 3.2.8 *Let*

$$I_{m,n}(\mathbf{Q}_m) := \int e^{\frac{i}{2}\mathrm{Tr}(\mathbf{H}_m \mathbf{Q}_m)} \Big(\det(\mathbf{H}_m - \mu \mathbf{1}_m) \Big)^{-n} (d\mathbf{H}_m), \tag{3.24}$$

where $\mathbf{H}_m$, $\mathbf{Q}_m$ are $m \times m$ Hermitian matrices, and suppose $n \ge m$, $Im\, \mu > 0$. For $\mathbf{Q}_m$ positive definite (i.e. all eigenvalues positive) one has

$$I_{m,n}(\mathbf{Q}_m) = \frac{2^m \pi^{m(m+1)/2} i^m (-1)^{m(m-1)/2}}{\prod_{j=n-m+1}^n \Gamma(j)} \Big(\det\Big(\frac{i}{2}\mathbf{Q}_m\Big)\Big)^{n-m} e^{\frac{i}{2}\mu \mathrm{Tr}(\mathbf{Q}_m)}$$

while $I_{m,n}(\mathbf{Q}_m)$ vanishes if $\mathbf{Q}_m$ has an eigenvalue less than or equal to zero.

Proof. We proceed in an analogous way to the proof of statement (b) in Proposition 2.5.1. Because the integral (3.24) is invariant under the transformation $\mathbf{H}_m \mapsto \mathbf{U}_m \mathbf{H}_m \mathbf{U}_m^{-1}$ for $\mathbf{U}_m$ unitary, $I_{m,n}(\mathbf{Q}_m)$ is a function of the eigenvalues of $\mathbf{Q}_m$ only and so we can take

$$\mathbf{Q}_m = \mathrm{diag}[q_1, \dots, q_m].$$

Now introduce the decomposition

$$\mathbf{H}_m = \begin{bmatrix} \mathbf{H}_{m-1} & \vec{h} \\ \vec{h}^\dagger & h_{mm} \end{bmatrix},$$

where $\vec{h}$ is a vector of length $m-1$ with complex entries, and proceed as in the derivation of (2.59) to show

$$\det(\mathbf{H}_m - \mu \mathbf{1}_m) = \det(\mathbf{H}_{m-1} - \mu \mathbf{1}_{m-1})(h_{mm} - a), \qquad a := \mu + \vec{h}^\dagger (\mathbf{H}_{m-1} - \mu \mathbf{1}_{m-1})^{-1} \vec{h}.$$

This shows

$$I_{m,n}(\mathbf{Q}_m) = \int (d\mathbf{H}_{m-1})\, e^{\frac{1}{2}\mathrm{Tr}(\mathbf{H}_{m-1}\mathbf{Q}_{m-1})} \times \Big(\det(\mathbf{H}_{m-1} - \mu \mathbf{1}_{m-1}) \Big)^{-n} \int (d\vec{h}) \int_{-\infty}^{\infty} dh_{mm}\, e^{\frac{1}{2} q_m h_{mm}} (h_{mm} - a)^{-n}.$$

Since by assumption $\mathrm{Im}(\mu) > 0$ we have $\mathrm{Im}(a) > 0$, which allows the integral over h_{mm} to be computed by closing the contour in the upper half-plane (for $q_m > 0$) to give

$$\int_{-\infty}^{\infty} \frac{e^{\frac{1}{2} q_m h_{mm}}}{(h_{mm} - a)^n}\, dh_{mm} = \frac{2\pi i}{\Gamma(n)} \Big(\frac{i q_m}{2} \Big)^{n-1} e^{\frac{1}{2} q_m a}.$$

On the other hand, closing the contour in the lower half-plane shows that for $q_m < 0$ the integral vanishes. The next task then is to evaluate

$$\int (d\vec{h})\, e^{\frac{1}{2} q_m a}. \tag{3.25}$$

Changing variables $\vec{h} = \mathbf{U}_{m-1}^{-1} \vec{w}$ where $\mathbf{U}_{m-1}$ is a unitary matrix such that $\mathbf{U}_{m-1}^{-1} \mathbf{H}_{m-1} \mathbf{U}_{m-1} = \mathrm{diag}[h_1, \dots, h_{m-1}]$ and applying Proposition 3.2.4 to deduce $(d\vec{h}) = (d\vec{w})$, separates (3.25) into $2(m-1)$ one-dimensional integrals,

$$\begin{aligned} \int (d\vec{h})\, e^{\frac{1}{2} q_m a} &= e^{\frac{1}{2} q_m \mu} \prod_{l=1}^{m-1} \int_{-\infty}^{\infty} dw_l^{\rm r}\, e^{\frac{i q_m}{2} (w_l^{\rm r})^2 (h_l - \mu)^{-1}} \int_{-\infty}^{\infty} dw_l^{\rm i}\, e^{\frac{i q_m}{2} (w_l^{\rm i})^2 (h_l - \mu)^{-1}} \\ &= e^{\frac{1}{2} q_m \mu} \prod_{l=1}^{m-1} \Big(-\frac{2\pi}{i q_m} \Big)(h_l - \mu) = e^{\frac{1}{2} q_m \mu} \Big(\frac{2\pi i}{q_m} \Big)^{m-1} \det\Big(\mathbf{H}_{m-1} - \mu \mathbf{1}_{m-1} \Big). \end{aligned}$$

Thus

$$I_{m,n}(\mathbf{Q}_m) = \frac{2\pi^m i (-1)^{m-1}}{\Gamma(n)} \Big(\frac{i q_m}{2} \Big)^{n-m} e^{\frac{1}{2} q_m \mu} I_{m-1,n-1}(\mathbf{Q}_{m-1}).$$

Iterating this and noting $I_{0,n'}(\mathbf{Q}_0) := 1$ gives the stated result. □

To make use of (3.24), note that with $\mathbf{X}$ an $n \times m$ complex Gaussian matrix,

$$P(\mathbf{X}) = \frac{1}{\pi^{mn}} e^{-\mathrm{Tr}(\mathbf{X}^\dagger \mathbf{X})},$$

and $\mathbf{A} = \mathbf{X}^\dagger \mathbf{X}$ we have

$$P(\mathbf{A}) = \int \delta(\mathbf{A} - \mathbf{X}^\dagger \mathbf{X}) P(\mathbf{X}) (d\mathbf{X}). \tag{3.26}$$

Here $\delta(\mathbf{A} - \mathbf{X}^\dagger \mathbf{X})$ is equal to the product of one-dimensional delta functions over the independent real and imaginary parts of $\mathbf{A}$. Writing each of these as a Fourier integral shows

$$\delta(\mathbf{A}-\mathbf{X}^\dagger\mathbf{X}) = \frac{1}{(2\pi)^{m^2}} \int e^{i\mathrm{Tr}(\mathbf{H}(\mathbf{A}-\mathbf{X}^\dagger\mathbf{X}))}(d\mathbf{H}), \tag{3.27}$$

where $\mathbf{H}$ is an $m \times m$ Hermitian matrix. Substituting this in (3.26) and noting that

$$\int e^{-\mathrm{Tr}(\mathbf{X}^\dagger\mathbf{X})} e^{-i\mathrm{Tr}(\mathbf{H}\mathbf{X}^\dagger\mathbf{X})}(d\mathbf{X}) = \pi^{mn}\Big(\det(\mathbf{1}+i\mathbf{H})\Big)^{-n},$$

which in turn is deduced from the fact that the integral is a function of the eigenvalues of $\mathbf{H}$ only and then separating the integration into a product of one-dimensional integrations, gives

$$P(\mathbf{A}) = \frac{\pi^{2mn}}{(2\pi)^{m^2}} \int \frac{e^{i\mathrm{Tr}(\mathbf{HA})}}{(\det(\mathbf{1}+i\mathbf{H}))^n}(d\mathbf{H}).$$

Now the result (3.24) can be applied, thus reclaiming (3.22).

EXERCISES 3.2 1. (i) Use the definition of the adjoint $\langle\vec{\phi}|\mathbf{X}^\dagger\vec{\psi}\rangle = \langle\mathbf{X}\vec{\phi}|\vec{\psi}\rangle$ and the property of the inner product $\langle\vec{\phi}|\vec{\phi}\rangle > 0$ for $\vec{\phi} \ne \vec{0}$ to show that the eigenvalues of $\mathbf{X}^\dagger\mathbf{X}$ are non-negative.

(ii) Consider the matrix product $\mathbf{X}^\dagger\mathbf{X}$ with $\mathbf{X}$ an $n \times m$ matrix with $n < m$. Show that $\mathbf{X}^\dagger\mathbf{X} = \mathbf{Y}^\dagger\mathbf{Y}$ where $\mathbf{Y}$ is an $m \times m$ matrix obtained from $\mathbf{X}$ by the addition of $m-n$ rows of zeros. Hence show that $\mathbf{X}^\dagger\mathbf{X}$ has $m-n$ zero eigenvalues.

(iii) By considering the corresponding characteristic polynomials, and making use of (5.26) below, show that in the setting of (ii) the nonzero eigenvalues of $\mathbf{X}^\dagger\mathbf{X}$ and $\mathbf{X}\mathbf{X}^\dagger$ are equal.

2. Here Proposition 3.2.4 will be established.

(i) With $\vec{z} = [x_j + iy_j]_{j=1,\dots,N}$, $\vec{w} = [u_j + iv_j]_{j=1,\dots,N}$ and $\mathbf{A} = [a_{jk} + ib_{jk}]_{j,k=1,\dots,N}$, show that the equation $d\vec{z} = \mathbf{A}d\vec{w}$ can be rewritten as the real matrix equation

$$\begin{bmatrix} [dx_j]_{j=1,\dots,N} \\ [dy_j]_{j=1,\dots,N} \end{bmatrix} = \begin{bmatrix} [a_{jk}]_{j,k=1,\dots,N} & -[b_{jk}]_{j,k=1,\dots,N} \\ [b_{jk}]_{j,k=1,\dots,N} & [a_{jk}]_{j,k=1,\dots,N} \end{bmatrix} \begin{bmatrix} [du_j]_{j=1,\dots,N} \\ [dv_j]_{j=1,\dots,N} \end{bmatrix}.$$

(ii) To evaluate the determinant of the $2N \times 2N$ matrix on the r.h.s. of the above equation, and thus the Jacobian, add i times the blocks in the bottom half to the blocks in the top half. Then subtract i times the blocks in the left half to the blocks in the right half so that the top right block is now the zero matrix.

3. (i) Show that the problem of calculating the eigenvalue p.d.f. of the Wishart matrices in the case $m = 1$ is equivalent to calculating the p.d.f. of $\sum_{j=1}^{\beta n} x_j^2$ where x_j has distribution $\sqrt{\beta/2\pi}e^{-\beta x_j^2/2}$.

(ii) Calculate the p.d.f. in (i), $p(\lambda)$ say, according to the formula

$$p(\lambda) = \Big(\frac{\beta}{2\pi}\Big)^{\beta n/2} \int_{-\infty}^{\infty} dx_1 e^{-\beta x_1^2/2} \cdots \int_{-\infty}^{\infty} dx_{\beta n} e^{-\beta x_{\beta n}^2/2} \delta\Big(\lambda - \sum_{j=1}^{\beta n} x_j^2\Big),$$

and thus reclaim Proposition 3.2.2 in the case $m = 1$.

4. [410] The objective of this exercise is to compute the eigenvalue p.d.f. for real Wishart matrices.

(i) Let $\mathbf{X}$ be a real $n \times m$ ($n \ge m$) matrix. Suppose the Gram-Schmidt orthogonalization procedure has been used to write $\mathbf{X} = \mathbf{RT}$ where $\mathbf{R}$ is an $n \times m$ real matrix such that $\mathbf{R}^T\mathbf{R} = \mathbf{1}_m$ and $\mathbf{T}$ is an upper triangular $m \times m$ real matrix with positive diagonal entries. Use the method of Proposition 3.2.5 to show that

$$(d\mathbf{X}) = \prod_{j=1}^m t_{jj}^{n-j}(d\mathbf{T})(\mathbf{R}^T d\mathbf{R}).$$

(ii) Let $\mathbf{A} = \mathbf{T}^T\mathbf{T}$ where $\mathbf{T}$ is as in (i). Use the method of the proof of Proposition 3.2.6 to show that

$$(d\mathbf{A}) = 2^m \prod_{j=1}^m t_{jj}^{m+1-j} (d\mathbf{T}).$$

(iii) Use the results of (i) and (ii) and Definition 3.2.1 to show that the p.d.f. of a real Wishart matrix is

$$\frac{1}{C} e^{-\frac{1}{2}\mathrm{Tr}(\mathbf{A})} (\det \mathbf{A})^{(n-m-1)/2},$$

where C is a normalization constant, and then use (1.11) to express this p.d.f. in terms of the eigenvalues of $\mathbf{A}$.

5. The objective of this exercise is to compute the eigenvalue p.d.f. for quaternion real Wishart matrices.

(i) Let $\mathbf{X}$ be an $n \times m$ $(n \ge m)$ matrix with real quaternion elements. Suppose the Gram-Schmidt orthogonalization procedure has been used to write $\mathbf{X} = \mathbf{UT}$ where $\mathbf{U}$ is an $n \times m$ matrix of real quaternions such that $\mathbf{U}^\dagger \mathbf{U} = \mathbf{1}$ and $\mathbf{T}$ is an upper triangular $m \times m$ matrix with diagonal entries positive real multiples of $\mathbf{1}_2$ and off-diagonal entries real quaternions. Use the method of the proof of Proposition 3.2.5 to show that

$$(d\mathbf{X}) = \prod_{j=1}^m t_{jj}^{4(n-j)+2} (d\mathbf{T})(\mathbf{U}^\dagger d\mathbf{U}).$$

(ii) Let $\mathbf{A} = \mathbf{T}^\dagger\mathbf{T}$ where $\mathbf{T}$ as in (i) above. Use the method of the proof of Proposition 3.2.6 to show that

$$(d\mathbf{A}) = 2^m \prod_{j=1}^m t_{jj}^{4(m-j)+1} (d\mathbf{T}).$$

(iii) Use the results of (i) and (ii) and Definition 3.2.1 to show that the p.d.f. of a quaternion real Wishart matrix is

$$\frac{1}{C} e^{-2\mathrm{Tr}(\mathbf{A})} (\det \mathbf{A})^{2(n-m+1/2)},$$

where C is a normalization constant and the operations Tr and det are not to include repeated eigenvalues, and then use (1.27) for $\beta = 4$ to express this p.d.f. in terms of the eigenvalues of $\mathbf{A}$.

6. [433] Let the p.d.f. of the $n \times m$ matrix $\mathbf{X}$ (with real ($\beta = 1$), complex ($\beta = 2$), real quaternion ($\beta = 4$) elements) be of the form $F(\mathbf{X}^\dagger\mathbf{X})$. The p.d.f. of the elements of $\mathbf{A} = \mathbf{X}^\dagger\mathbf{X}$ is then $h(\mathbf{A})F(\mathbf{A})$ for some h. The objective of this exercise is to determine h.

(i) Let $\mathbf{A} = \mathbf{B}^\dagger\mathbf{V}\mathbf{B}$, where $\mathbf{V}$ is positive definite. Making use of (1.35), show that the p.d.f. of $\mathbf{V}$ is then

$$F(\mathbf{B}^\dagger\mathbf{V}\mathbf{B})h(\mathbf{B}^\dagger\mathbf{V}\mathbf{B}) \det(\mathbf{B}^\dagger\mathbf{B})^{(\beta/2)(m-1+2/\beta)}. \tag{3.28}$$

(ii) Let $\mathbf{X} = \mathbf{YB}$, where $\mathbf{Y}$ is such that $\mathbf{V} = \mathbf{Y}^\dagger\mathbf{Y}$. Use an appropriate generalization of Proposition 3.2.4 to show that the p.d.f. of $\mathbf{Y}$ is $F(\mathbf{B}^\dagger\mathbf{Y}^\dagger\mathbf{Y}\mathbf{B})(\det \mathbf{B}^\dagger\mathbf{B})^{\beta n/2}$ and hence that of $\mathbf{V}$ is

$$F(\mathbf{B}^\dagger\mathbf{V}\mathbf{B})h(\mathbf{B}^\dagger\mathbf{B}) \det(\mathbf{B}^\dagger\mathbf{B})^{\beta n/2} h(\mathbf{V}). \tag{3.29}$$

(iii) Equate (3.28) and (3.29) with $\mathbf{V} = \mathbf{1}$ to deduce $h(\mathbf{B}^\dagger\mathbf{B}) = h(\mathbf{1})(\det \mathbf{B}^\dagger\mathbf{B})^{(\beta/2)(n-m+1-2/\beta)}$, and note that $h(\mathbf{1}) = c$ for some constant c to conclude

$$h(\mathbf{A}) \propto (\det \mathbf{A})^{(\beta/2)(n-m+1-2/\beta)}. \tag{3.30}$$

7. Let $\mathbf{Y}, \mathbf{Z}$ be $m \times m$ positive definite Hermitian matrices (denoted $\mathbf{Y} > \mathbf{0}, \mathbf{Z} > \mathbf{0}$) with real ($\beta = 1$), complex ($\beta = 2$) or real quaternion ($\beta = 4$) elements. By changing variables $\mathbf{X} = \mathbf{Z}^{1/2}\mathbf{Y}\mathbf{Z}^{1/2}$, making use of (1.35), show

$$\frac{1}{C} \int_{\mathbf{Y}>\mathbf{0}} e^{-(\beta/2)\mathrm{Tr}(\mathbf{YZ})} (\det \mathbf{Y})^{\beta(a-(m-1)-2/\beta)/2} (d\mathbf{Y}) = (\det \mathbf{Z})^{-\beta a/2}, \tag{3.31}$$

where C is such that both sides equal unity when $\mathbf{Z} = \mathbf{1}$.

3.3 FURTHER EXAMPLES OF THE LAGUERRE ENSEMBLE IN QUANTUM MECHANICS

The chiral ensemble, which from the viewpoint of its eigenvalue distribution is equivalent to the Laguerre ensemble, is motivated from a problem in quantum mechanics. Here some further problems in quantum mechanics are studied which lead to the Laguerre ensemble. As part of this four random Hamiltonians are isolated, which together with those forming the Gaussian and chiral ensembles, make up the ten Hermitian random matrix ensembles in correspondence with the ten matrix Lie algebras associated with infinite families of symmetric spaces.

3.3.1 Eigenvalue statistics of Wigner-Smith delay time matrix

In this subsection it will be shown, following [102], that for the quantum cavity problem in Section 2.1.1 the distribution of the scaled reciprocal eigenvalues of

$$\mathbf{Q}_E := -i\hbar \mathbf{S}^{-1/2} \frac{\partial \mathbf{S}}{\partial E} \mathbf{S}^{-1/2} \tag{3.32}$$

is given by the Laguerre ensemble (3.16) with $a = N$. Here E is the energy of the waves as they enter the cavity (for a long lead and fixed number of channels N the energy will to leading order be constant). The eigenvalues of $\mathbf{Q}_E$, denoted $\tau_1, \dots, \tau_N$ say, are referred to as the proper delay times, and are also the eigenvalues for the *Wigner-Smith matrix* $\mathbf{Q} = -i\hbar \mathbf{S}^{-1}\partial \mathbf{S}/\partial E$.

A precise formulation of the problem requires a Hamiltonian approach to the coupled lead-cavity system [531]. This can be done by introducing a basis of states $\{|a_i\rangle\}_{i=1,\dots,N}$ for the lead, and a basis of states $\{\mu_j\rangle\}_{j=1,\dots,M}$ for the cavity (typically $M \gg N$). The Hamiltonian is then defined as

$$\mathcal{H} = \sum_i |a_i\rangle E \langle a_i| + \sum_{j,j'} |\mu_j\rangle H_{jj'} \langle \nu_{j'}| + \sum_{j,i} \Big(|\mu_j\rangle W_{ji} \langle a_i| + |a_i\rangle \bar{W}_{ij} \langle \mu_i| \Big),$$

where the matrix elements $H_{jj'}$ form a random Hermitian $M \times M$ matrix $\mathbf{H}$ with real ($\beta = 1$), complex ($\beta = 2$) or quaternion real ($\beta = 4$) elements, while the coupling constants form a fixed $M \times N$ matrix $\mathbf{W}$. A key point is that for this $\mathcal{H}$ the corresponding $N \times N$ scattering matrix $\mathbf{S}$ can be calculated exactly as [380]

$$\mathbf{S} = \frac{\mathbf{1}_N + i\pi \mathbf{W}^\dagger (\mathbf{H} - E)^{-1} \mathbf{W}}{\mathbf{1}_N - i\pi \mathbf{W}^\dagger (\mathbf{H} - E)^{-1} \mathbf{W}}.$$

In the simplest case $\mathbf{W}$ can be taken as proportional to the identity,

$$\mathbf{W} = \frac{\sqrt{\Delta M}}{\pi} \mathbf{1}_{M \times N},$$

where Δ is the mean level spacing in the cavity. Furthermore it is assumed that the Hermitian matrix $\mathbf{H}$ is a Gaussian random matrix with

$$P(\mathbf{H}) \propto \exp\Big(-\beta\pi^2 \mathrm{Tr}\mathbf{H}^2 / 4\Delta^2 M \Big). \tag{3.33}$$

Writing

$$\mathbf{H} - E = \mathbf{U}_{M \times M} \mathrm{diag}[E_1 - E, \dots, E_M - E] \mathbf{U}^\dagger_{M \times M},$$

where the columns of $\mathbf{U}$ consist of the eigenvectors of $\mathbf{H}$, gives

$$\mathbf{S} = \frac{\mathbf{1}_N + i\mathbf{K}}{\mathbf{1}_N - i\mathbf{K}}, \quad \mathbf{K} = \frac{\Delta M}{\pi} \mathbf{U}_{N \times M} \mathrm{diag}\Big[(E_1 - E)^{-1}, \dots, (E_M - E)^{-1}\Big] (\mathbf{U}_{N \times M})^\dagger. \tag{3.34}$$

Here $\mathbf{U}_{N\times M}$ denotes the matrix $\mathbf{U}_{M\times M}$ with the last $M-N$ rows deleted. Since $\mathbf{H}$ has distribution (3.33) and $N \ll M$, it follows that the distribution of the submatrix $\mathbf{U}_{N\times N}$ is given by

$$P\Big(\frac{1}{\sqrt{M}}\mathbf{U}\Big) \propto \exp\Big(-\beta \mathrm{Tr}\, \mathbf{U}\mathbf{U}^\dagger/2\Big) \tag{3.35}$$

(cf. Exercises 1.3 q.2).

The formula (3.34), which expresses $\mathbf{S}$ as a function of E, can be used to compute the eigenvalue distribution of (3.32) [102]. But before undertaking this task we will present a matrix integration formula which is required in the course of the derivation.

PROPOSITION 3.3.1 *Let $\mathbf{X}$ be a $N \times N$ random matrix with real elements ($\beta = 1$), complex elements ($\beta = 2$) or real quaternion elements ($\beta = 4$), and let $\mathbf{B}$ be an $N \times N$ Hermitian random matrix which is real ($\beta = 1$), complex ($\beta = 2$) or quaternion real ($\beta = 4$). Furthermore suppose the joint distribution of $\mathbf{X}$ and $\mathbf{B}$ is proportional to*

$$e^{-\beta \mathrm{Tr}(\mathbf{X}\mathbf{X}^\dagger)/2}\delta(\mathbf{X}^{\dagger -1}\mathbf{B}\mathbf{X}^{-1}),$$

where δ denotes the Dirac delta function. Then the distribution of the eigenvalues $a_1, \dots, a_N$ of $\mathbf{A} = \mathbf{X}\mathbf{X}^\dagger$, defined up to a multiplicative constant as the eigenvalue dependent factor in

$$e^{-\beta \mathrm{Tr}(\mathbf{X}\mathbf{X}^\dagger)/2}\int \delta(\mathbf{X}^{\dagger -1}\mathbf{B}\mathbf{X}^{-1})(d\mathbf{B}), \tag{3.36}$$

is proportional to

$$\prod_{j=1}^N a_j^{\beta N/2} e^{-\beta a_j/2} \prod_{1\le j<k\le N} |a_k - a_j|^\beta. \tag{3.37}$$

Proof. To perform the integration over $\mathbf{B}$ in (3.36) we make the change of variables $\mathbf{X}^{\dagger -1}\mathbf{B}\mathbf{X}^{-1} = \mathbf{C}$ which in turn requires the wedge product formula (1.35),

$$(d\mathbf{B}) = (\mathbf{X}^\dagger d\mathbf{C}\mathbf{X}) = \Big(\det(\mathbf{X}\mathbf{X}^\dagger)\Big)^{\beta(N-1)/2+1}(d\mathbf{C}).$$

This shows that (3.36) reduces to

$$e^{-\beta \mathrm{Tr}(\mathbf{X}\mathbf{X}^\dagger)/2}\Big(\det(\mathbf{X}\mathbf{X}^\dagger)\Big)^{\beta(N-1)/2+1}. \tag{3.38}$$

But we know from our study of Wishart matrices that with $\mathbf{A} = \mathbf{X}\mathbf{X}^\dagger$, $(d\mathbf{X}) \propto (\det \mathbf{A})^{\beta/2-1}(d\mathbf{A})$. Using this in (3.38) to deduce the distribution of $\mathbf{A}$, and then changing variables to the eigenvalues and eigenvectors of $\mathbf{A}$ according to (1.27) gives the stated result. □

PROPOSITION 3.3.2 *Let the eigenvalues of the matrix $\mathbf{Q}_E$ (3.32) be denoted $\tau_1, \dots, \tau_N$, and write $\gamma_j = 1/\tau_j$. Then the p.d.f. of the γ_j is proportional to*

$$\prod_{l=1}^N \gamma_l^{\beta N/2} e^{-\beta \tau_H \gamma_l/2} \prod_{j<k} |\gamma_j - \gamma_k|^\beta,$$

where $\tau_H := 2\pi\hbar/\Delta$ and $\beta = 1, 2$ or 4 according to the scattering matrix $\mathbf{S}$ being a symmetric, unitary or self-dual quaternion random unitary matrix.

Derivation. Since the Haar form $(d_H\mathbf{S})$ has the invariance (2.11) it suffices to consider the neighborhood of $\mathbf{S} = -\mathbf{1}_N$. According to (3.34) this means that N eigenvalues (at least) of $\mathbf{H}$ are almost degenerate with E, so that $|E - E_\mu| \ll \Delta$, $\mu = 1, \dots, N$. Hence in fact only this special subclass of Gaussian random matrices make up $\mathbf{H}$. Ignoring the other

$M - N$ eigenvalues in (3.34), as well as the final $M - N$ components of each eigenvector, gives

$$\begin{aligned} \mathbf{S} &\approx \frac{\mathbf{1}_N + i\tilde{\mathbf{K}}}{\mathbf{1}_N - i\tilde{\mathbf{K}}}, \quad \tilde{\mathbf{K}} = \frac{\Delta M}{\pi}\mathbf{U}_{N\times N}\mathrm{diag}[(E_1 - E)^{-1}, \dots, (E_N - E)^{-1}]\mathbf{U}_{N\times N}^\dagger \\ &\approx -\mathbf{1}_N + 2i\tilde{\mathbf{K}}^{-1}, \quad \tilde{\mathbf{K}}^{-1} = \frac{\pi}{\Delta M}\mathbf{U}_{N\times N}\mathrm{diag}[E_1 - E, \dots, E_N - E]\mathbf{U}_{N\times N}^\dagger. \end{aligned} \tag{3.39}$$

Introducing the scaled matrix $\boldsymbol{\Psi} := M^{1/2}\mathbf{U}_{N\times N}$, we see from (3.35) that the joint distribution of the elements of $\boldsymbol{\Psi}$ is proportional to $\exp(-\beta \mathrm{Tr}\boldsymbol{\Psi}\boldsymbol{\Psi}^\dagger/2)$. This distribution is invariant under the transformation $\boldsymbol{\Psi} \mapsto \boldsymbol{\Psi}\mathbf{O}$, where $\mathbf{O}$ is an orthogonal ($\beta = 1$), unitary ($\beta = 2$) or symplectic ($\beta = 4$) $N \times N$ random unitary matrix with uniform measure. Defining $\tilde{\mathbf{H}} := \mathbf{O}\,\mathrm{diag}(E - E_1, \dots, E - E_N)\mathbf{O}^\dagger$, by the assumption that the $\{E_\mu\}$ are near degenerate eigenvalues from the Gaussian ensemble, we see from (3.33) that $\tilde{\mathbf{H}}$ has a uniform distribution near $\tilde{\mathbf{H}} = 0$. Furthermore, (3.39) then reads

$$\mathbf{S} = -\mathbf{1}_N + (i\tau_H/\hbar)\boldsymbol{\Psi}^{\dagger-1}\tilde{\mathbf{H}}\boldsymbol{\Psi}^{-1}.$$

Hence at $\mathbf{S} = -\mathbf{1}_N$ the distribution of $\boldsymbol{\Psi}\boldsymbol{\Psi}^\dagger$ is proportional to

$$\int e^{-\beta \mathrm{Tr}\boldsymbol{\Psi}\boldsymbol{\Psi}^\dagger/2}\delta(\boldsymbol{\Psi}^{\dagger-1}\tilde{\mathbf{H}}\boldsymbol{\Psi}^{-1})\,(d\tilde{\mathbf{H}}).$$

According to Proposition 3.3.1 the corresponding eigenvalue distribution of $\boldsymbol{\Psi}\boldsymbol{\Psi}^\dagger$ is given by (3.37). The stated result now follows since from (3.32) and (3.39), to leading order in M,

$$\mathbf{Q}_E = \tau_H \boldsymbol{\Psi}^{\dagger-1}\boldsymbol{\Psi}^{-1}$$

and so the eigenvalues of $\boldsymbol{\Psi}\boldsymbol{\Psi}^\dagger$ are equal to τ_H times the reciprocal of the eigenvalues of $\mathbf{Q}_E$.

3.3.2 Normal metal–superconductor junctions

In mesoscopic physics, a situation of interest is conductance through a normal metal–superconductor (NS) junction (see, e.g., [53]). At this junction the phenomenon of retroreflection (also known as Andreev reflection) can take place: an electron from the normal metal may be reflected at the junction as a hole with the same momentum but opposite velocity, and with the missing charge $2e$ absorbed as a Cooper pair by the superconducting condensate.

The theoretical description of the NS junction is via the (matrix) Bogoliubov-deGennes equation

$$\begin{bmatrix} \mathbf{h} & \boldsymbol{\Delta} \\ -\bar{\boldsymbol{\Delta}} & -\mathbf{h}^T \end{bmatrix} \begin{bmatrix} \vec{u} \\ \vec{v} \end{bmatrix} = \epsilon \begin{bmatrix} \vec{u} \\ \vec{v} \end{bmatrix}, \tag{3.40}$$

where $\boldsymbol{\Delta} = -\boldsymbol{\Delta}^T$ (this is a requirement of Fermi statistics) while $\mathbf{h} = \mathbf{h}^\dagger$, and both $\boldsymbol{\Delta}$ and $\mathbf{h}$ are $2N \times 2N$ matrices. The vectors $\vec{u}$ and $\vec{v}$ represent the electron and hole wavefunctions (their dimension is $2N$ because for each electron (hole) there are an up and down state so $\vec{u} = (\vec{u}_+, \vec{u}_-)$ and $\vec{v} = (\vec{v}_+, \vec{v}_-)$). Here, following [14], the constraints on the matrix elements of $\mathbf{h}$ and $\boldsymbol{\Delta}$ in (3.40) due to time reversal symmetry and spin rotation invariance will be considered.

It turns out that for this problem the appropriate time reversal operator T is such that $T^2 = -1$ and has the special structure

$$T = \mathbf{1}_2 \otimes \begin{bmatrix} \mathbf{0}_N & \mathbf{1}_N \\ -\mathbf{1}_N & \mathbf{0}_N \end{bmatrix} K, \tag{3.41}$$

where K denotes the complex conjugation operator. Thus

$$T \begin{bmatrix} \vec{u}_+ \\ \vec{u}_- \\ \vec{u}_+ \\ \vec{u}_- \end{bmatrix} = \begin{bmatrix} \vec{u}_-^* \\ -\vec{u}_+^* \\ \vec{u}_-^* \\ -\vec{u}_+^* \end{bmatrix},$$

where $*$ denotes complex conjugate. Also, spin-rotation invariance is the requirement that the Hamiltonian matrix in (3.40), $\mathcal{H}$ say, commutes with the matrix

$$\begin{bmatrix} \sigma_k \otimes \mathbf{1}_N & \mathbf{0}_{2N} \\ \mathbf{0}_{2N} & -\sigma_k^T \otimes \mathbf{1}_N \end{bmatrix} \tag{3.42}$$

for each $k = x, y, z$, where the σ_k are the Pauli matrices of (1.22). There are four cases to consider, depending on the presence or absence of time reversal symmetry and spin-rotation invariance.

(1) No time reversal symmetry or spin-rotation invariance.

We have the general structure exhibited in (3.40),

$$\mathcal{H} = \begin{bmatrix} \mathbf{A} & \mathbf{B} \\ -\bar{\mathbf{B}} & -\bar{\mathbf{A}} \end{bmatrix} \tag{3.43}$$

with $\mathbf{A} = \mathbf{A}^\dagger$, $\mathbf{B} = -\mathbf{B}^T$. Introducing the transformation $\mathcal{H} \mapsto \tilde{\mathcal{H}} := \mathbf{U}_0 \mathcal{H} \mathbf{U}_0^{-1}$ with

$$\mathbf{U}_0 = \frac{1}{\sqrt{2}} \begin{bmatrix} \mathbf{1}_{2N} & \mathbf{1}_{2N} \\ i\mathbf{1}_{2N} & -i\mathbf{1}_{2N} \end{bmatrix}$$

shows that $\tilde{\mathcal{H}}^* = -\tilde{\mathcal{H}} = \tilde{\mathcal{H}}^T$, and hence $\tilde{\mathcal{H}}$ is a Hermitian matrix with pure imaginary elements, or equivalently equal to i times a real antisymmetric matrix. The number of independent elements is equal to the number of independent elements in $\mathbf{A}$ and $\mathbf{B}$ together, and thus equal to $8N^2 - 2N$. This is the number of independent elements in a general $4N \times 4N$ antisymmetric Hermitian matrix, so we conclude $\tilde{H}$ belongs to this class.

(2) Time reversal symmetry, no spin-rotation invariance.

For $\mathcal{H}$ to commute with T we require that $\mathbf{h}$ and $\boldsymbol{\Delta}$ have the block structure

$$\mathbf{h} = \begin{bmatrix} \mathbf{h}_{++} & \mathbf{h}_{+-} \\ -\bar{\mathbf{h}}_{+-} & \mathbf{h}_{++}^T \end{bmatrix}, \qquad \boldsymbol{\Delta} = \begin{bmatrix} \boldsymbol{\Delta}_{++} & \boldsymbol{\Delta}_{+-} \\ -\boldsymbol{\Delta}_{+-}^T & -\boldsymbol{\Delta}_{++}^\dagger \end{bmatrix}.$$

Rearranging the blocks via an appropriate similarity transformation gives a matrix of the form (3.43) but with the elements of $\mathbf{A}$ and $\mathbf{B}$, now real quaternions, and thus consist of $N \times N$ lots of 2×2 blocks of the form (1.20). The matrices $\mathbf{A}$ and $\mathbf{B}$ must also be Hermitian and antisymmetric respectively as in (3.43). As in case (1) a further similarity transformation can be made to obtain an antisymmetric matrix, but now with a 2×2 sub-block structure of real quaternions in which the entries are pure imaginary.

Using a similarity transformation to rearrange appropriate rows and columns shows the latter can be written in the form

$$\begin{bmatrix} \mathbf{A} & \mathbf{B} \\ \mathbf{B} & -\mathbf{A} \end{bmatrix}, \qquad \mathbf{A} = -\mathbf{A}^T, \, \mathbf{B} = -\mathbf{B}^T, \tag{3.44}$$

with the elements of $\mathbf{A}$ and $\mathbf{B}$ pure imaginary. The eigenvalue problem for this matrix is equivalent to the eigenvalue problem for

$$\begin{bmatrix} \mathbf{0}_{2N} & \mathbf{W} \\ \mathbf{W}^\dagger & \mathbf{0}_{2N} \end{bmatrix}, \tag{3.45}$$

where $\mathbf{W}$ is a $2N \times 2N$ antisymmetric matrix with complex elements (see Exercises 3.3 q.1). Also shown in Exercises 3.3 q.1 is the fact that the eigenvalue equation for (3.45) is equivalent to the eigenvalue problem for

$$\begin{bmatrix} \mathbf{0}_{2N} & \mathbf{D} \\ \mathbf{D}^\dagger & \mathbf{0}_{2N} \end{bmatrix}, \tag{3.46}$$

where $\mathbf{D}$ is a $2N \times 2N$ self-dual matrix with complex elements.

(3) Spin-rotation invariance, no time reversal invariance.

The requirement that $\mathcal{H}$ commutes with the matrix (3.42) restricts $\mathbf{h}$ and $\mathbf{\Delta}$ to have the block structure

$$\mathbf{h} = \begin{bmatrix} \mathbf{h}_{++} & \mathbf{0}_N \\ \mathbf{0}_N & \mathbf{h}_{++} \end{bmatrix}, \qquad \mathbf{\Delta} = \begin{bmatrix} \mathbf{0}_N & \mathbf{\Delta}_{+-} \\ -\mathbf{\Delta}_{+-} & \mathbf{0}_N \end{bmatrix},$$

where $\mathbf{h}_{++} = \mathbf{h}_{++}^\dagger$ and $\mathbf{\Delta}_{+-} = \mathbf{\Delta}_{+-}^T$. Thus $\mathcal{H}$ consists of two commuting sub-blocks. Concentrating on one of these gives the structure

$$\mathcal{H}_r = \begin{bmatrix} \mathbf{A} & \mathbf{B} \\ \bar{\mathbf{B}} & -\bar{\mathbf{A}} \end{bmatrix} \tag{3.47}$$

with $\mathbf{A} = \mathbf{A}^\dagger, \mathbf{B} = \mathbf{B}^T$. Multiplying (3.47) by i and using a similarity transformation to rearrange appropriate rows and columns we see

$$i\mathcal{H}_r = \mathbf{Q}, \qquad \mathbf{Q} = -\mathbf{Q}^\dagger, \tag{3.48}$$

where $\mathbf{Q}$ is quaternion real.
(4) Spin-rotation invariance and time reversal invariance.
The constraints of (2) and (3) together imply that the structure of (3) holds with $\mathbf{h}_{++}$ and $\mathbf{\Delta}_{+-}$ real symmetric. Thus the reduced Hamiltonian (3.47) results with $\mathbf{A}$ and $\mathbf{B}$ real symmetric. We show in Exercises 3.3 q.2 that the eigenvalue equation for such matrices is equivalent to that for matrices of the form

$$\begin{bmatrix} \mathbf{0}_N & \mathbf{D} \\ \mathbf{D}^\dagger & \mathbf{0}_N \end{bmatrix}, \tag{3.49}$$

where $\mathbf{D}$ is symmetric with complex elements.

The matrices in each of the classes (1)–(4) have the property that if λ is an eigenvalue, then so is $-\lambda$. This can be seen from (3.40). Thus given the eigenvector as in (3.40) with eigenvalue ϵ, the eigenvector $\begin{bmatrix} \vec{v}^* \\ \vec{u}^* \end{bmatrix}$, also satisfies (3.40) but with eigenvalue $-\epsilon$, a fact which can be verified by taking the complex conjugate of the latter equation and noting $\mathbf{h} = \mathbf{h}^\dagger$. Since the matrices are Hermitian, they can all be diagonalized by a unitary matrix, where the elements of the unitary matrix are further constrained due to the special symmetries for each class. In fact these matrices have all appeared in earlier sections, and the Jacobian for the change of variables to the eigenvalues and eigenvectors has been computed.

In case (1), the class of matrices is equivalent to $4N \times 4N$ antisymmetric Hermitian matrices. From Exercises 1.3 q.5 the eigenvalue dependent portion of the Jacobian is

$$\prod_{1 \le j < k \le 2N} (\lambda_k^2 - \lambda_j^2)^2. \tag{3.50}$$

Case (2) led to $4N \times 4N$ antisymmetric Hermitian matrices with quaternion real elements (the entries of which are pure imaginary numbers). Since the matrix is Hermitian with quaternion real elements, each eigenvalue is doubly degenerate. From Exercises 1.3 q.7 we read off that the eigenvalue dependent portion of the Jacobian is

$$\prod_{j=1}^N |\lambda_j| \prod_{1 \le j < k \le N} (\lambda_k^2 - \lambda_j^2)^4. \tag{3.51}$$

The matrices of case (3) can be written in the form (3.48). We read off from Exercises 1.3 q.8 that the eigenvalue dependent portion of the Jacobian is

$$\prod_{j=1}^N \lambda_j^2 \prod_{1 \le j < k \le N} (\lambda_k^2 - \lambda_j^2)^2. \tag{3.52}$$

For the matrices of case (4) we read off from Exercises 3.1 q.8 that the eigenvalue-dependent portion of the

Jacobian is

$$\prod_{j=1}^{N} |\lambda_j| \prod_{1 \le j < k \le N} |\lambda_k^2 - \lambda_j^2|. \tag{3.53}$$

Choosing the independent real and imaginary parts of the random matrices $\mathbf{X}$ to be Gaussian random variables such that their joint distribution is proportional to $\exp(-\beta \mathbf{X}^2/2)$, where β is given by the exponent in the product of differences in each of (3.50)–(3.53), we see that the eigenvalue p.d.f.'s for the classes (1)–(4) are examples of the Laguerre eigenvalue p.d.f. written in the form of (3.4).

3.3.3 Relationship to Lie algebras

It is emphasized in [14] and [556] that the above classes of matrices are the Hermitian part of the matrix Lie algebras

$$\begin{aligned}
i \times \Big(\mathrm{so}(n, \mathbf{C})\Big) &:= \{ i \text{ times } n \times n \text{ skew symmetric complex matrices} \}, \\
i \times \Big(\mathrm{so}^*(2n)\Big) &:= \{ i \text{ times } n \times n \text{ skew symmetric real quaternion matrices} \}, \\
\mathrm{sp}(n, \mathbf{C}) &:= \left\{ \begin{bmatrix} \mathbf{Z}_1 & \mathbf{Z}_2 \\ \mathbf{Z}_3 & -\mathbf{Z}_1^T \end{bmatrix}, \text{ all } \mathbf{Z}_i \text{ complex, } \mathbf{Z}_2,\, \mathbf{Z}_3 \text{ symmetric} \right\}, \\
\mathrm{sp}(n, \mathbf{R}) &:= \left\{ \begin{bmatrix} \mathbf{X}_1 & \mathbf{X}_2 \\ \mathbf{X}_3 & -\mathbf{X}_1^T \end{bmatrix}, \text{ all } \mathbf{X}_i \text{ real, } \mathbf{X}_2,\, \mathbf{X}_3 \text{ symmetric} \right\},
\end{aligned}$$

and furthermore these matrix Lie algebras, together with those of Sections 1.3.4 and 3.1.3, complete Cartan's list [295] of the ten Lie algebras associated with the ten infinite families of symmetric spaces.

We summarize the matrix structures corresponding to matrix Lie algebras giving rise to p.d.f.'s of the form (3.4) in Table 3.1.

3.3.4 Entanglement of a random pure quantum state

The p.d.f. proportional to (3.16), with the further constraint that $\sum_{j=1}^{m} \lambda_j = 1$, occurs in studies of quantum entanglement (see, e.g., [56]). The setting is a finite-dimensional quantum system decomposed into distinct subsystems A and B of dimensions n and m, respectively (for convenience it is assumed that $n \ge m$). A state $|\psi\rangle$ of the composite system will have the decomposition

$$|\psi\rangle = \sum_{i=1}^{n} \sum_{j=1}^{m} x_{i,j} |a_i\rangle \otimes |b_j\rangle, \tag{3.54}$$

where $\{|a_i\rangle\}_{i=1,\dots,n}$, $\{|b_i\rangle\}_{i=1,\dots,m}$ are basis of the corresponding subsystems, and the $x_{i,j}$ are scalars.

The density matrix corresponding to $|\psi\rangle$ is the projection operator $\rho = |\psi\rangle\langle\psi|$. The reduced density matrix of the subsystem B is defined as the trace of ρ over the states of A,

$$\rho_B := \mathrm{Tr}_A[\rho] = \sum_{i=1}^{n} \langle a_i | \rho | a_i \rangle.$$

Substituting in this the definition of ρ, and then substituting (3.54), one obtains the form

$$\rho_B = \sum_{i,j=1}^{m} (\mathbf{X}^\dagger \mathbf{X})_{i,j} |b_i\rangle\langle b_j|$$

where $\mathbf{X} = [x_{i,j}]_{\substack{i=1,\dots,n \\ j=1,\dots,m}}$.

Matrix structure	Parameter values in (3.4)
$\begin{bmatrix} \mathbf{0}_{n\times n} & \mathbf{X}_{n\times m} \\ (\mathbf{X}_{n\times m})^\dagger & \mathbf{0}_{n\times n} \end{bmatrix}$ real elements ($\beta = 1$) complex elements ($\beta = 2$) real quaternion elements ($\beta = 4$)	$\beta = 1, 2$ or 4, $\alpha = n - m + 1 - 1/\beta$
$\begin{bmatrix} \mathbf{0}_{N\times N} & \mathbf{X}_{N\times N} \\ \mathbf{X}^\dagger_{N\times N} & \mathbf{0}_{N\times N} \end{bmatrix}$ $\mathbf{X}$ antisymmetric complex or $\mathbf{X}$ self-dual (N even)	These matrices are equivalent to antisymmetric matrices with pure imaginary real quaternion elements
$\mathbf{X}$ symmetric complex	$\beta = 1, \alpha = 1, m = N$
antisymmetric $N \times N$ matrices, $\mathbf{X}^T = -\mathbf{X}$ pure imaginary complex elements N odd : $m = (N-1)/2$, $\beta = 2$, $\alpha = 1$	N even : $m = N/2$, $\beta = 2$, $\alpha = 0$
pure imaginary real quaternion elements	N even: $m = N/2, \beta = 4, \alpha = 1/4$ N odd: $m = (N-1)/2, \beta = 4, \alpha = 5/4$
anti-Hermitian $N \times N$ matrices, $\mathbf{X}^\dagger = -\mathbf{X}$ real quaternion elements	$i\lambda_j \mapsto \lambda_j, \beta = 2, \alpha = 1, m = N$

Table 3.1 Matrix structures which give rise to the eigenvalue p.d.f. (3.4).

On the other hand, the diagonalization of $\mathbf{X}^\dagger\mathbf{X}$ gives the diagonal form $\rho_B = \sum_{i=1}^m \lambda_i |v_i^B\rangle$, where $\{|v_i^B\rangle\}_{i=1,\dots,m}$ are the normalized eigenvectors of $\mathbf{X}^\dagger\mathbf{X}$. Analogous reasoning gives $\rho_A = \sum_{i=1}^m \lambda_i |v_i^A\rangle$, where $\{|v_i^A\rangle\}_{i=1,\dots,m}$ are the normalized eigenvectors of $\mathbf{X}\mathbf{X}^\dagger$ corresponding to the nonzero eigenvalues. Now writing $|\psi\rangle$ as a linear combination of the product states $\{|v_i^A\rangle \otimes |v_j^B\rangle\}_{i,j=1,\dots,m}$ it follows from these formulas for the reduced density matrices that

$$|\psi\rangle = \sum_{i=1}^m \sqrt{\lambda_i} |v_i^A\rangle \otimes |v_i^B\rangle, \tag{3.55}$$

which is referred to as the *Schmidt decomposition.*

The coefficients $\{\sqrt{\lambda_i}\}$ can be used to quantify the degree of entanglement of the subsystems A and B via the *Shannon entropy* $S = -\sum_{i=1}^m \lambda_i \log \lambda_i$. One extreme is when $\lambda_1 = 1, \lambda_i = 0$ $(i = 2, \dots, m)$, in which case $|\psi\rangle$ is a direct product of states from subsystems A and B, and thus disentangled. In this case $S = 0$. Another extreme is when $\lambda_i = 1/m$ $(i = 1, \dots, m)$. This weights all states in (3.55) equally and gives $S = \log m$, which is the maximum allowed value.

Suppose the coefficients $\{x_{i,j}\}$ in (3.54) are independent Gaussians. If the subsystems A and B are invariant under time reversal symmetry with $T^2 = 1$ then these coefficients are real, while a time reversal symmetry with $T^2 = -1$ implies the coefficients are real quaternions. If there is no time reversal symmetry, the coefficients will be complex. In such a circumstance, $\mathbf{X}^\dagger\mathbf{X}$ is a Wishart matrix and the eigenvalues, if not further constrained, would have distribution (3.16). However, here there is a further constraint, as the normalization of $|\psi\rangle$ requires that $\sum_{i=1}^m \lambda_i = 1$, and thus the eigenvalue distribution is equal to

$$\frac{1}{\hat{W}_{a\beta m}} \prod_{l=1}^m \lambda_l^{\beta a/2} \prod_{1 \le j < k \le m} |\lambda_k - \lambda_j|^\beta \delta\Big(\sum_{j=1}^m \lambda_j - 1\Big), \quad \lambda_l \ge 0 \tag{3.56}$$

with $a = n - m + 1 - 2/\beta$. The explicit form of the normalization can be read off from (4.155) below. The eigenvalue p.d.f. (3.56) will be said to specify the *fixed trace Laguerre β-ensemble*. Another setting of this ensemble is given in Exercises 3.3 q.3, while in Exercises 3.3 q.4 the mean value of the Shannon entropy is computed.

EXERCISES 3.3 1. The objective of this exercise is to show that the eigenvalue problem for the matrix (3.44) is equivalent to that for the matrices (3.45) and (3.46).

(i) With the eigenvector for (3.44) corresponding to the eigenvalue λ denoted

$$\begin{bmatrix} \vec{u} + i\vec{v} \\ \vec{x} + i\vec{y} \end{bmatrix},$$

where $\vec{u}, \vec{v}, \vec{x}, \vec{y}$ are $2N$-component real vectors, equate real and imaginary parts in each block to obtain the coupled equations

$$i\mathbf{A}\vec{v} + i\mathbf{B}\vec{y} = \lambda\vec{u}, \quad -i\mathbf{A}\vec{u} - i\mathbf{B}\vec{x} = \lambda\vec{v}, \quad i\mathbf{B}\vec{v} - i\mathbf{A}\vec{y} = \lambda\vec{x}, \quad -i\mathbf{B}\vec{u} + i\mathbf{A}\vec{x} = \lambda\vec{y}.$$

(ii) Show that the coupled equations in (i) are equivalent to the eigenvalue equation

$$\begin{bmatrix} \mathbf{0}_{2N} & -i\mathbf{A} + \mathbf{B} \\ i\mathbf{A} + \mathbf{B} & \mathbf{0}_{2N} \end{bmatrix} \begin{bmatrix} \vec{u} + i\vec{x} \\ -\vec{v} + i\vec{y} \end{bmatrix} = \lambda \begin{bmatrix} \vec{u} + i\vec{x} \\ -\vec{v} + i\vec{y} \end{bmatrix}$$

and thus conclude that the eigenvalue problem for (3.44) is equivalent to that for (3.45).

(iii) Note that

$$\begin{bmatrix} \mathbf{Z}_{2N}^{-1} & \mathbf{0}_{2N} \\ \mathbf{0}_{2N} & \mathbf{1}_{2N} \end{bmatrix} \begin{bmatrix} \mathbf{0}_{2N} & -i\mathbf{A} + \mathbf{B} \\ i\mathbf{A} + \mathbf{B} & \mathbf{0}_{2N} \end{bmatrix} \begin{bmatrix} \mathbf{Z}_{2N} & \mathbf{0}_{2N} \\ \mathbf{0}_{2N} & \mathbf{1}_{2N} \end{bmatrix} = \begin{bmatrix} \mathbf{0}_{2N} & \mathbf{Z}_{2N}^{-1}(-i\mathbf{A} + \mathbf{B}) \\ (i\mathbf{A} + \mathbf{B})\mathbf{Z}_{2N} & \mathbf{0}_{2N} \end{bmatrix}$$

while

$$\left(\mathbf{Z}_{2N}^{-1}(-i\mathbf{A} + \mathbf{B})\right)^D = \mathbf{Z}_{2N}(-i\mathbf{A} + \mathbf{B})^T(\mathbf{Z}_{2N}^{-1})^T\mathbf{Z}_{2N}^{-1} = \mathbf{Z}_{2N}^{-1}(-i\mathbf{A} + \mathbf{B})$$

and conclude from this that the eigenvalue problem for (3.44) is equivalent to that for (3.46).

2. (i) Suppose $\mathbf{A}$ and $\mathbf{B}$ are real symmetric $N \times N$ matrices. Show that the real eigenvalue equation

$$\begin{bmatrix} \mathbf{A} & \mathbf{B} \\ \mathbf{B} & -\mathbf{A} \end{bmatrix} \begin{bmatrix} \vec{u} \\ \vec{v} \end{bmatrix} = \lambda \begin{bmatrix} \vec{u} \\ \vec{v} \end{bmatrix}$$

is equivalent to the complex equation $(\mathbf{A} + i\mathbf{B})(\vec{u} - i\vec{v}) = \lambda(\vec{u} + i\vec{v})$, and thus to the complex eigenvalue equation

$$\begin{bmatrix} \mathbf{0}_N & \mathbf{A} + i\mathbf{B} \\ \mathbf{A} - i\mathbf{B} & \mathbf{0}_N \end{bmatrix} \begin{bmatrix} \vec{u} + i\vec{v} \\ \vec{u} - i\vec{v} \end{bmatrix} = \lambda \begin{bmatrix} \vec{u} + i\vec{v} \\ \vec{u} - i\vec{v} \end{bmatrix}.$$

(ii) Conclude from (i) that the eigenvalue problem for the matrices of case (4) is equivalent to that for the matrices (3.49).

3. [279]; cf. Exercises 1.3 q.10. Let $\mathbf{A} = [a_{jk}]_{j,k=1,\dots,m}$ be a Wishart matrix, and thus have p.d.f. (3.22). The matrix $\mathbf{C} := [a_{jk}/(a_{jj}a_{kk})^{1/2}]_{j,k=1,\dots,m}$ is then called a correlation matrix, and it has the property $\mathrm{Tr}\,\mathbf{C} = m$. By noting that

$$\mathbf{A} = \mathrm{diag}(a_{11}^{1/2}, \dots, a_{mm}^{1/2})\mathbf{C}\mathrm{diag}(a_{11}^{1/2}, \dots, a_{mm}^{1/2})$$

show that the Jacobian for changing variables from $\mathbf{A}$ to $(a_{11}, \dots, a_{mm})$ and $\mathbf{C}$ is $\prod_{i=1}^m a_{ii}^{(m-1)/2}$. Use (3.22) to then show that the joint density of $a_{11}, \dots, a_{mm}$ and $\mathbf{C}$ is proportional to

$$\prod_{i=1}^m a_{ii}^{(m-1)/2-(\beta/2)(n-m+1-2/\beta)} (\det \mathbf{C})^{(\beta/2)(n-m+1-2/\beta)} \delta(c_1 + \cdots + c_m - m)$$

and read off from this that the p.d.f. of $\mathbf{C}$ is proportional to

$$(\det \mathbf{C})^{(\beta/2)(n-m+1-2/\beta)} \delta(c_1 + \cdots + c_m - m).$$

Hence conclude that the eigenvalues of $\mathbf{C}$ have distribution (3.56).

4. [560] This exercise develops some of the properties of the fixed trace Laguerre β-ensemble (3.56).

   (i) Denote the fixed trace Laguerre β-ensemble with one body weight x^a and $m = N$ by fLβE$_a$, and the corresponding Laguerre β-ensemble with one body weight $x^a e^{-x}$ by LβE$_a$. By writing the delta function constraint as a Fourier integral show that

$$\Big\langle \sum_{j=1}^N \lambda_j^p \Big\rangle_{\mathrm{fL}\beta\mathrm{E}_a} = \frac{(\beta N(N-1)/2 + N(a+1) - 1)!}{(\beta N(N-1)/2 + N(a+1) + p - 1)!} \Big\langle \sum_{j=1}^N \lambda_j^p \Big\rangle_{\mathrm{L}\beta\mathrm{E}_a}.$$

   (ii) Note that for large N the ratio of factorials in (i) has the large N form $(\beta N^2/2)^{-p}$, and conclude from this that for large N

$$\rho_{(1)}^{\mathrm{fL}\beta\mathrm{E}_a}(x) \sim (\beta N^2/2)\rho_{(1)}^{\mathrm{L}\beta\mathrm{E}_a}(\beta N^2 x/2) \sim \begin{cases} \dfrac{N}{2\pi x^{1/2}}(4N - xN^2)^{1/2}, & 0 < x < 4/N \\ 0, & x > 4/N. \end{cases}$$

   (iii) Use the result of (ii) to compute that for large N

$$\int_0^\infty x^\nu \rho_{(1)}^{\mathrm{fL}\beta\mathrm{E}_a}(x)\, dx \sim \Big(\frac{4}{N}\Big)^{\nu+1} \frac{N^2}{2\pi} \frac{\Gamma(\nu+1/2)\Gamma(3/2)}{\Gamma(\nu+2)}.$$

   Read off that the first moment equals unity, as required by the fixed trace condition. Also, by differentiating with respect to ν and making use of the dilogarithm function, show

$$-\Big\langle \sum_{j=1}^N \lambda_j \log \lambda_j \Big\rangle \sim \log N - \frac{1}{2} + \mathrm{O}\Big(\frac{\log N}{N}\Big).$$

3.4 THE EIGENVALUE DENSITY

3.4.1 Marčenko-Pastur law

Proposition 3.2.3 predicts that the eigenvalue density in the Laguerre ensemble is such that

$$\lim_{N\to\infty} 4\rho_{(1)}(4Ny) = \begin{cases} \frac{2}{\pi y^{1/2}}(1-y)^{1/2}, & 0 < y < 1, \\ 0, & y \ge 1. \end{cases} \tag{3.57}$$

This result in fact remains valid for a large class of Wishart type matrices, in which one relaxes the requirement that the elements of the matrix $\mathbf{X}$ be Gaussian, and is an example of the Marčenko-Pastur law [388]. Furthermore, it gives a good approximation for finite N, as illustrated in Figure 3.1 where the r.h.s. of (3.57) is compared to the empirical value of $\rho_{(1)}(Y)$, $Y = 4Ny$ for 500 10×10 complex Wishart matrices.

We know from Proposition 3.1.4 that in the case α fixed the support of the density corresponding to (3.14) with $m = N$ is to leading order in the interval $[0, \sqrt{4N}]$. Consequently, under the change of variable $y \mapsto \sqrt{N}y$, the support is in the interval $[0, 2]$. We see from (3.14) that if we also make the replacement $\alpha \mapsto \alpha N$, both the quadratic term and the log term will then be of the same order, so it is to be expected that the support of the density will be in some interval $[c, d]$, $c > 0$. According to Exercises 3.1 q.1, the integral equation satisfied by the background charge density for a general potential $V(x)$ in this setting is equivalent to

$$V(x) + C = \int_J \rho_b(t) \log|x - t|\, dt,$$

where $J = [-d, -c] \cup [c, d]$. For even potentials, the physically relevant solution of such "double cut" integral

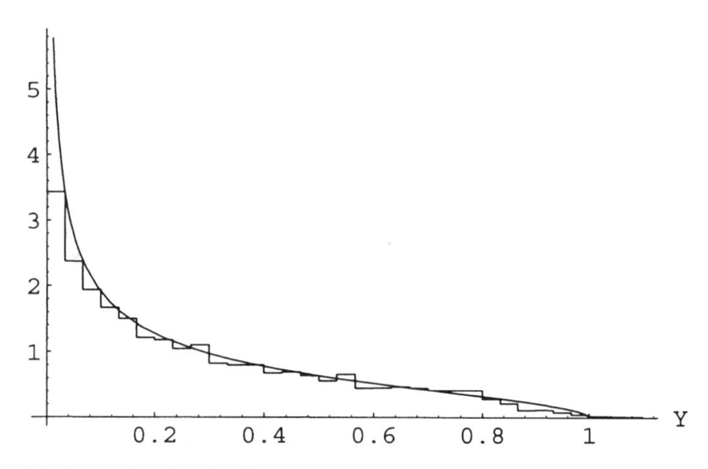

Figure 3.1 Empirical scaled eigenvalue density for 10×10 complex Wishart matrices compared with the theoretical scaled density (3.57).

equations can be deduced from the complex electric field approach of Section 1.4.3. One finds [345], [7]

$$\rho_b(x) = \frac{2}{\pi^2} \int_c^d \frac{xV'(t) - tV'(x)}{t^2 - x^2} \sqrt{\frac{(d^2 - x^2)(x^2 - c^2)}{(d^2 - t^2)(t^2 - c^2)}}\, dt \tag{3.58}$$

(cf. (1.65)), where c and d are determined by the conditions

$$\int_c^d \frac{V'(t)}{\sqrt{(d^2 - t^2)(t^2 - c^2)}}\, dt = 0, \qquad \int_c^d \rho_b(x)\, dx = N. \tag{3.59}$$

Specializing $V(x)$ to correspond to the chiral ensemble gives the following prediction for the eigenvalue density, known rigorously from [388] (see also next subsection).

PROPOSITION 3.4.1 *In the case*

$$V(x) = N\Big(\frac{x^2}{2} - \alpha \log |x|\Big), \tag{3.60}$$

which corresponds to the chiral ensemble with $m = N$, the change of scale $x \mapsto \sqrt{N}x$ and $\alpha \mapsto N\alpha$, the explicit form of (3.58) with the constraints (3.59) is

$$\rho_b(x) = \frac{N}{\pi} \frac{1}{x} \sqrt{(d^2 - x^2)(x^2 - c^2)}, \qquad c < x < d, \tag{3.61}$$

where $d - c = 2$, $cd = \alpha$.

Proof. With the substitution (3.60), we read off from (3.58) and (3.59) that

$$\rho_b(x) = \frac{2N}{\pi^2} \frac{\alpha}{x} \sqrt{R(x)} I_1, \quad I_2 - \alpha I_1 = 0, \quad \frac{\pi^2}{2\alpha} = I_1 I_3, \tag{3.62}$$

where $R(u) := (d^2 - u^2)(u^2 - c^2)$ and

$$I_1 := \int_c^d \frac{dt}{t\sqrt{R(t)}}, \qquad I_2 := \int_c^d \frac{t\, dt}{\sqrt{R(t)}}, \qquad I_3 := \int_c^d \frac{1}{t} \sqrt{R(t)}\, dt.$$

It turns out that after the change of variables $s = t^2$, the antiderivative of all these integrals can be computed exactly, leading to the results

$$I_1 = \frac{\pi}{2cd}, \qquad I_2 = \frac{\pi}{2}, \qquad I_3 = \frac{\pi}{4}(d-c)^2.$$

Substituting in (3.62) gives (3.61). □

3.4.2 Combinatorial derivation of the density

We have seen in Section 1.6 and the ensuing exercises how the density (1.52) can be deduced from a combinatorial interpretation of $\langle \mathrm{Tr}(\mathbf{X}^{2k}) \rangle_{\mathrm{GUE}}$. Here we will show how (3.61) can be deduced from consideration of $\langle \mathrm{Tr}(\mathbf{X}^\dagger \mathbf{X})^k \rangle_{\mathrm{cG}}$, where cG denotes that the average is over $n \times m$ complex Gaussian matrices $\mathbf{X}$ [535], [240].

Let I, I' be finite ordered sets of pairs of indices (j,k), $j \in [1,n]$, $k \in [1,m]$. Let P denote a matching between elements of I and elements of I'. Then analogous to (1.99) we have

$$\Big\langle \prod_{(i,j)\in I} \bar{z}_{i,j} \prod_{(i',j')\in I'} z_{i',j'} \Big\rangle_{\mathrm{cG}} = \sum_{\substack{\text{pairings } P \\ \text{between } I, I'}} \prod_{(i,j),(i',j')} \langle \bar{z}_{i,j} z_{i',j'} \rangle_{\mathrm{cG}}. \tag{3.63}$$

Noting

$$\mathrm{Tr}(\mathbf{X}^\dagger \mathbf{X})^k = \sum_{i_1,\dots,i_k=1}^{n} \sum_{j_1,\dots,j_k=1}^{m} \bar{z}_{i_1,j_k} \bar{z}_{i_2,j_1} \cdots \bar{z}_{i_k,j_{k-1}} z_{i_1,j_1} z_{i_2,j_2} \cdots z_{i_k,j_k}$$

it follows from (3.63) that

$$\Big\langle \mathrm{Tr}(\mathbf{X}^\dagger \mathbf{X})^k \Big\rangle_{\mathrm{cG}} = \sum_{i_1,\dots,i_k=1}^{n} \sum_{j_1,\dots,j_k=1}^{m} \sum_{\text{pairings}} \prod_{(l,l'),(m,m')} \langle \bar{z}_{l,m} z_{l',m'} \rangle_{\mathrm{cG}}, \tag{3.64}$$

where the pairings are between the sets

$$\{(i_1, j_k), (i_2, j_1), \dots, (i_k, j_{k-1})\} \quad \text{and} \quad \{(i_1, j_1), (i_2, j_2), \dots, (i_k, j_k)\}.$$

The pairing in (3.64) can be carried out graphically using a construction similar to that in Figure 1.3, where the vertices in the $2k$-gon correspond to $i_1, j_1, \dots, i_k, j_k$ in that order, and the odd (even) labeled edges must connect to odd (even) labeled edges. Each independent vertex corresponding to the is is to be weighted by n, while each independent vertex corresponding to the js is to be weighted by m. If the total number of independent vertices after pairing is ν, then analogous to (1.106) we have

$$\Big\langle \mathrm{Tr}(\mathbf{X}^\dagger \mathbf{X})^k \Big\rangle_{\mathrm{cG}} = \sum_{\nu=1}^{k+1} \sum_{\substack{i,j \geq 0 \\ i+j=\nu}} a_\nu(k,i,j) n^i m^j,$$

where $a_\nu(k,i,j)$ denotes the number of pairings which have i independent vertices corresponding to the is and j independent vertices corresponding to the js.

The equation (1.108) again allows the pairings to be catalogued according to the genus of the underlying map, and we have in particular

$$\lim_{m\to\infty} m^{-k-1} \Big\langle \mathrm{Tr}(\mathbf{X}^\dagger \mathbf{X})^k \Big\rangle_{\mathrm{cG}} = \sum_{i=1}^{k} a_{k+1}(k,i,k+1-i) \Big(\frac{n}{m}\Big)^i,$$

where $a_{k+1}(k,i,k+1-i)$ counts pairings corresponding to planar maps. Another interpretation of the meaning of $a_{k+1}(k,i,k+1-i)$ comes from the graphical picture of the pairings given in Figure 1.2, the planar version of which has no lines intersecting. The independent vertices corresponding to the is and js can be determined by the bi-coloring (white and gray, say) the interior of the $2k$-gon according to the rule that the colors alternate, and the region containing the marked vertex always has the same color (white, say). Then the number of regions colored white gives the value of the i index. Writing

$$A_k(p,q) := \sum_{i=1}^{k} a_{k+1}(k,i,k+1-i)p^i q^{k+1-i}, \tag{3.65}$$

it follows from a recursive argument analogous to the one leading to (1.124) that

$$A_{k+1}(p,q) = \sum_{j=0}^{k} A_j(q,p)A_{k-j}(p,q), \qquad A_0(p,q) = q.$$

Analogous to (1.125), introducing the generating function $\psi(p,q,t) = \sum_{k=1}^{\infty} A_k(p,q)t^k$ and noting the symmetry $\psi(p,q,t) = \psi(q,p,t)$, we thus have

$$\psi(p,q,t) = t(\psi(p,q,t)+p)(\psi(p,q,t)+q)$$

and so, with $u = pt$, $v = qt$,

$$t\psi(p,q,t) = \frac{1}{2}\Big(1 - u - v - (1 - 2(u+v) + (u-v)^2)^{1/2}\Big). \tag{3.66}$$

Applying the Lagrange inversion formula of Exercises 1.6 q.3 to the previous equation to expand ψ as a function of t allows $A_k(p,q)$ to be deduced, and expanding this as required by (3.65) shows

$$a_{k+1}(k,i,k+1-i) = \frac{1}{k}\binom{k}{i}\binom{k}{i-1}. \tag{3.67}$$

This is the *Narayana number* $N(k,i-1)$ [501].

It follows that the scaled density $\tilde{\rho}_{(1)}(x) := \lim_{m\to\infty} \rho_{(1)}(mx)|_{n/m=\alpha+1}$ is such that

$$\int_0^{\infty} x^{2k}\tilde{\rho}_{(1)}(x)\,dx = \sum_{i=1}^{k} \frac{1}{k}\binom{k}{i}\binom{k}{i-1}(\alpha+1)^i. \tag{3.68}$$

Using the general formula [270]

$$\int x^{m+1}\sqrt{R}\,dx = \frac{x^m R^{3/2}}{c(m+3)} - \frac{am}{c(m+4)}\int x^{m-1}\sqrt{R}\,dx - \frac{b(m+3/2)}{c(m+4)}\int x^m\sqrt{R}\,dx,$$

where $R = a + bx + cx^2$ we can verify that $\tilde{\rho}_{(1)}(x) = \rho_b(x)/N$ as specified by (3.61) is consistent with (3.68). Moreover, the moments (3.68) have the property (1.128) and thus uniquely determine $\tilde{\rho}_{(1)}(x)$.

Comparing (3.68) to (3.67) we see from (3.65) that

$$A_k(\alpha+1,1) = \int_c^d x^{2k}\tilde{\rho}_{(1)}(x)\,dx$$

and hence, for $|t|d^2 < 1$,

$$\int_c^d \frac{\tilde{\rho}_{(1)}(x)}{1-tx^2}\,dx = 1 + \psi(\alpha+1,1,t).$$

Making use now of (3.66) gives for the complex electric in the region $\mathrm{Re}(z) > d$,

$$2z\int_c^d \frac{\tilde{\rho}_{(1)}(x)}{z^2-x^2}\,dx = z\Big(1 - \frac{\alpha}{z^2} - \Big(1 - \frac{2(\alpha+2)}{z^2} + \frac{\alpha^2}{z^4}\Big)^{1/2}\Big). \tag{3.69}$$

We can check using the equations below (3.61) that the term in the square root is negative real for z real and between c and d only, so by analytic continuation this evaluation extends to all z outside this interval.

EXERCISES 3.4 1. With $a > 0$ and x_i positive, let

$$F(a; x_1, \dots, x_N) := \frac{1}{2}\sum_{j=1}^{N}(x_j - a\log x_j) - \sum_{1 \le j < k \le N} \log |x_k - x_j|.$$

By following the method of Exercises 1.4 q.8 show that the location of the minimum of F occurs at the zeros of the Laguerre polynomial $L_N^{(a-1)}(x)$. Conclude that for $N \to \infty$ the density of zeros must obey the limit law (3.57).

3.5 CORRELATED WISHART MATRICES

As formulated in Section 3.2.1 the $n \times m$ data matrix $\mathbf{X}$ in (3.15) has each row drawn from an m-dimensional Gaussian with mean zero and variance $\boldsymbol{\Sigma}$. Equivalently, the distribution of $\mathbf{X}$ is proportional to

$$\exp\Big(- \frac{1}{2}\mathrm{Tr}(\mathbf{X}^T\mathbf{X}\boldsymbol{\Sigma}^{-1})\Big). \tag{3.70}$$

By the analogue of (3.21), $(d\mathbf{X})$ is invariant under the transformation $\mathbf{X} \mapsto \mathbf{X}\mathbf{O}$ for $\mathbf{O}$ an $m \times m$ real orthogonal matrix. Hence $\boldsymbol{\Sigma}^{-1}$ in (3.70) can be replaced by its diagonal form, which in particular prompts us to refer to any situation in which $\boldsymbol{\Sigma}^{-1}$ is not the identity as correlated.

Application of Wishart matrices to wireless communications (recall Section 3.2.2), in which the analogue of the data matrix $\mathbf{X}$ is complex, suggests that (3.70) be generalized to

$$\exp\Big(- \mathrm{Tr}(\mathbf{X}^\dagger\boldsymbol{\Omega}^{-1}\mathbf{X}\boldsymbol{\Sigma}^{-1})\Big) \tag{3.71}$$

(see, e.g., [488]). Setting $\mathbf{Y} = \boldsymbol{\Omega}^{-1/2}\mathbf{X}\boldsymbol{\Sigma}^{-1/2}$ in this show that $\mathbf{Y}^\dagger\mathbf{Y}$ is an uncorrelated complex Wishart matrix. Here we will consider a special feature of the case that $\boldsymbol{\Sigma}^{-1} = \mathbf{1}_m$ while $\boldsymbol{\Omega}^{-1} = \mathrm{diag}((1)^{n-1}, b)$, where $(1)^{n-1}$ denotes 1 repeated $n-1$ times. Then

$$\mathbf{X}^\dagger\mathbf{X} = \mathbf{A} + b\vec{x}\vec{x}^\dagger, \tag{3.72}$$

where $\mathbf{A}$ is an uncorrelated complex Wishart matrix as specified in Definition 3.2.1 with $n \mapsto n-1$, while $\vec{x}$ is an m component standard complex Gaussian vector. By an analysis similar to that undertaken in Section 1.8, we can use knowledge of the leading asymptotic form of the eigenvalue density to analyze the behavior of the largest eigenvalue of (3.72) as a function of b.

We first note that working analogous to Proposition 1.8.1, and the paragraph preceding that result, shows that with the eigenvalues of $\mathbf{A}$ denoted $\{a_i\}$ and $\vec{x} = [x_i]$ the eigenvalues of (3.72) satisfy

$$1 = b\sum_{i=1}^{m}\frac{|x_i|^2}{\lambda - a_i}, \tag{3.73}$$

and are constrained by the inequalities (1.135). For convenience, let us scale $\mathbf{A}$ by $\mathbf{A}/m$. According to (3.57) the eigenvalue density $\rho_{(1)}(y)$ then has support on $(0,4)$, and is given explicitly by $\rho_{(1)}(y) = (m/\pi\sqrt{y})(1 - y/4)^{1/2}$, $0 < y < 4$. Thus with $b \mapsto b/m$ in (3.73), and each $|x_i|^2$ replaced by its mean value unity, (3.73) assumes the large m form

$$1 = b\int_0^4 \frac{\rho_{(1)}(y)}{\lambda - y}\,dy = \frac{2b}{\pi}\int_0^2 \frac{(1 - x^2/4)^{1/2}}{\lambda - x^2}\,dx,$$

where the second equality follows by changing variables $y = x^2$. Making use now of (1.131) and (1.126) allows this equation to be solved by λ, and an eigenvalue separation phenomenon analogous to that seen in Proposition 1.8.1 exhibited.

PROPOSITION 3.5.1 *For the rank 1 perturbation (3.72) with $b \mapsto b/m$ and $\mathbf{A}$ a diagonal matrix formed from the scaled eigenvalues λ_j/m of the Laguerre ensemble, the bulk of the spectrum is supported on $(0,4)$ and for $b > 2$ a single eigenvalue separates from the bulk and occurs at*

$$\lambda = \frac{b^2}{b-1}.$$

We remark that the same result holds if in (3.71) we set $\mathbf{\Omega}^{-1} = \mathbf{1}_n$ and $\mathbf{\Sigma}^{-1} = \mathrm{diag}((1)^{m-1}, b)$ [29]. To see this, we use the fact that the nonzero eigenvalues of $\mathbf{Y}^\dagger \mathbf{\Omega} \mathbf{Y}$ are the same as those for $\mathbf{\Omega} \mathbf{Y} \mathbf{Y}^\dagger$ (recall Exercises 3.2 q.1) and then interchange the role of $\mathbf{Y}$ and $\mathbf{Y}^\dagger$.

3.6 JACOBI ENSEMBLE AND WISHART MATRICES

3.6.1 Correlation coefficients

The Laguerre ensemble has been specified as the family of eigenvalue p.d.f.'s (3.16). Likewise, the Jacobi ensemble is defined as the family of eigenvalue p.d.f.'s

$$C_{ab\beta N}^{-1} \prod_{j=1}^{N} (1-y_j)^{a\beta/2} (1+y_j)^{b\beta/2} \prod_{1 \le j < k \le N} |y_k - y_j|^\beta, \qquad y_j \in [-1,1]. \tag{3.74}$$

Random matrices, constructed out of Wishart matrices, which give rise to (3.74) in the cases $\beta = 1, 2$ and 4 and for certain discrete a, b are given by the following result (see, e.g., [118]).

PROPOSITION 3.6.1 *Let $\mathbf{A} = \mathbf{a}^\dagger \mathbf{a}$, $\mathbf{B} = \mathbf{b}^\dagger \mathbf{b}$, where $\mathbf{a}$ and $\mathbf{b}$ are $n_1 \times m$ and $n_2 \times m$ random matrices, specified for $\beta = 1, 2$ and 4 as in Definition 3.1.2. The eigenvalues $x_1, \dots, x_m$ of the matrix $\mathbf{A}(\mathbf{A}+\mathbf{B})^{-1}$ (or equivalently the Hermitian matrix $(\mathbf{A}+\mathbf{B})^{-1/2}\mathbf{A}(\mathbf{A}+\mathbf{B})^{-1/2}$) have p.d.f. (3.74) with*

$$N = m, \quad y_j = 1 - 2x_j, \quad a = n_1 - m + 1 - 2/\beta, \quad b = n_2 - m + 1 - 2/\beta.$$

Proof. See Exercises 3.6 q.2. □

We remark that the matrix $\mathbf{A}(\mathbf{A}+\mathbf{B})^{-1}$ can equivalently be written as $\mathbf{X}^\dagger \mathbf{J} \mathbf{X} (\mathbf{X}^\dagger \mathbf{X})^{-1}$, where $\mathbf{X}$ is the $(n_1 + n_2) \times m$ matrix $[\mathbf{a}\ \mathbf{b}]$ and $\mathbf{J} := \mathrm{diag}((1)^{n_1}, (0)^{n_2})$.

One use of Proposition 3.6.1 relates to *correlation coefficients* in multivariate statistics. Correlation coefficients arise in canonical correlation analysis (see, e.g., [130]), for which the task is to form linear combinations within two sets of variables, $x_1, \dots, x_p$ and $y_1, \dots, y_q$, so that the correlation between the new variables in the different sets is maximized, while the new variables within each of the sets are uncorrelated. Now, associating with the variables $x_1, \dots, x_p$ the $n \times p$ matrix of observed values $\mathbf{X} = [x_k^{(j)}]_{\substack{j=1,\dots,n \\ k=1,\dots,p}}$, and with the variables $y_1, \dots, y_q$ the matrix of observed values $\mathbf{Y} = [y_k^{(j)}]_{\substack{j=1,\dots,n \\ k=1,\dots,q}}$, the empirical covariance matrix has the block form

$$\begin{bmatrix} \mathbf{X}^T\mathbf{X} & \mathbf{X}^T\mathbf{Y} \\ \mathbf{Y}^T\mathbf{X} & \mathbf{Y}^T\mathbf{Y} \end{bmatrix}.$$

The correlation coefficients are defined as the square of the eigenvalues of

$$(\mathbf{X}^T\mathbf{X})^{-1}\mathbf{X}^T\mathbf{Y}(\mathbf{Y}^T\mathbf{Y})^{-1}\mathbf{Y}^T\mathbf{X}. \tag{3.75}$$

The eigenvalue distribution of (3.75) can be computed by reducing the problem down to that of Proposition 3.6.1 [118].

PROPOSITION 3.6.2 *Let $\mathbf{X}$ be a $n \times p$ matrix, and $\mathbf{Y}$ be a $n \times q$ matrix, $q \ge p$, $n \ge p + q$, with Gaussian entries as specified in Definition 3.1.2 for $\beta = 1, 2$ and 4. Then the eigenvalues $x_1, \dots, x_p$ of the matrix*

(3.75) (with the transpose operation T replaced by the Hermitian conjugate operation $\dagger$ for $\beta = 2$ and 4) have the p.d.f. (3.74) with

$$N = p, \quad y_j = 1 - 2x_j, \quad a = q - p + 1 - 2/\beta, \quad b = n - q - p + 1 - 2/\beta.$$

Proof. The first step is to form an $n \times n$ orthogonal ($\beta = 1$), unitary ($\beta = 2$) or unitary symplectic ($\beta = 4$) matrix $\mathbf{U}$ such that its last $n - q$ rows are in the orthogonal complement of the space spanned by the columns of $\mathbf{Y}$, and thus

$$\mathbf{UY} = \begin{bmatrix} \mathbf{Y}_{1\,q \times q} \\ \mathbf{0}_{n-q \times q} \end{bmatrix},$$

where $\mathbf{Y}_1$ is nonsingular. Setting

$$\mathbf{UX} =: \mathbf{W}_{n \times p} =: \begin{bmatrix} \mathbf{a}_{q \times p} \\ \mathbf{b}_{(n-q) \times p} \end{bmatrix}$$

we see that we can write

$$\begin{aligned} \mathbf{X}^\dagger \mathbf{Y}(\mathbf{Y}^\dagger \mathbf{Y})^{-1} \mathbf{Y}^\dagger \mathbf{X} - x \mathbf{X}^\dagger \mathbf{X} &= \mathbf{W}^\dagger \begin{bmatrix} \mathbf{Y}_1 \\ \mathbf{0} \end{bmatrix} (\mathbf{Y}_1^\dagger \mathbf{Y}_1)^{-1} [\mathbf{Y}_1^\dagger \quad \mathbf{0}] \mathbf{W} - x \mathbf{W}^\dagger \mathbf{W} \\ &= \mathbf{W}^\dagger \begin{bmatrix} \mathbf{1}_q & \mathbf{0} \\ \mathbf{0} & \mathbf{0} \end{bmatrix} \mathbf{W} - x \mathbf{W}^\dagger \mathbf{W} \\ &= \mathbf{a}^\dagger \mathbf{a} - x(\mathbf{a}^\dagger \mathbf{a} + \mathbf{b}^\dagger \mathbf{b}). \end{aligned}$$

Now the eigenvalue condition for (3.75) can be written

$$\det(\mathbf{X}^\dagger \mathbf{Y}(\mathbf{Y}^\dagger \mathbf{Y})^{-1} \mathbf{Y}^\dagger \mathbf{X} - x \mathbf{X}^\dagger \mathbf{X}) = 0$$

so the above working shows that this is equivalent to

$$\det(\mathbf{a}^\dagger \mathbf{a} - x(\mathbf{a}^\dagger \mathbf{a} + \mathbf{b}^\dagger \mathbf{b})) = 0,$$

which is exactly the eigenvalue condition for the matrix $\mathbf{a}^\dagger \mathbf{a}(\mathbf{a}^\dagger \mathbf{a} + \mathbf{b}^\dagger \mathbf{b})^{-1}$. Furthermore, recalling the distribution of the elements of $\mathbf{X}$ is proportional to $\exp(-(\beta/2) \mathrm{Tr}\, \mathbf{X}^\dagger \mathbf{X})$, and noting that

$$\mathrm{Tr}(\mathbf{X}^\dagger \mathbf{X}) = \mathrm{Tr}(\mathbf{W}^\dagger \mathbf{W}) = \mathrm{Tr}(\mathbf{a}^\dagger \mathbf{a}) + \mathrm{Tr}(\mathbf{b}^\dagger \mathbf{b}),$$

and that the matrix $\mathbf{X}$ has the same number of elements as the matrices $\mathbf{a}$ and $\mathbf{b}$ together, we see the elements of $\mathbf{a}$ and $\mathbf{b}$ are independently distributed as specified in Definition 3.1.2 for $\beta = 1, 2$ and 4. The result now follows from Proposition 3.6.1. □

3.6.2 Coulomb gas analogy

The Jacobi ensemble p.d.f. (3.74) has a Coulomb gas analogy which can be determined by the change of variables $y = \cos \theta$.

PROPOSITION 3.6.3 *The Boltzmann factor of a one-component log-potential system with particles of unit charge at $y_1, \dots, y_N$, confined to the interval $[-1, 1]$ with a neutralizing background density*

$$-\rho_b(y) = -\frac{N}{\pi}(1 - y^2)^{-1/2}$$

and with a fixed particle of charge $(a-1)/2 + 1/\beta$ at $y = 1$ and another fixed particle of charge $(b-1)/2 + 1/\beta$ at $y = -1$, is proportional to the eigenvalue p.d.f. (3.74).

Proof. Changing variables $y_j = \cos\theta_j$, $0 < \theta_j < \pi$ shows

$$\prod_{j=1}^N (1-y_j)^{a\beta/2}(1+y_j)^{b\beta/2} \prod_{1\le j<k\le N} |y_k - y_j|^\beta \prod_{j=1}^N dy_j = \prod_{j=1}^N (2\sin^2\theta_j/2)^{a\beta/2}(2\cos^2\theta_j/2)^{b\beta/2}$$
$$\times \prod_{1<j<k\le N} |2\sin((\theta_j-\theta_k)/2)\sin((\theta_j+\theta_k)/2)|^\beta \prod_{j=1}^N \sin\theta_j d\theta_j. \tag{3.76}$$

The r.h.s. is proportional to (2.102) with the identifications

$$a\beta + 1 \mapsto 2a+1, \qquad b\beta+1 \mapsto 2b+1.$$

This explains the values of the fixed charges. As on the r.h.s. of (3.76) there is no one-body potential from any other source, to leading order the background charge density $-\rho_b(\theta)$ must satisfy

$$C = \int_0^\pi \rho_b(\theta)\log|\cos\phi - \cos\theta|\, d\theta.$$

The method of eigenfunction expansions detailed below Proposition 1.4.2 shows that the solution of this integral equation is $\rho_b(\theta)$ a constant. Since we require $\int_0^\pi \rho_b(\theta)\, d\theta = N$, we must have $\rho_b(\theta) = N/\pi$. Changing variables

$$\frac{N}{\pi}d\theta = \frac{N}{\pi}\frac{1}{\sqrt{1-y^2}}dy$$

gives the stated form of the background density. □

EXERCISES 3.6 1. Consider the Jacobi ensemble eigenvalue p.d.f. (3.74) with $a \mapsto Na$, $b \mapsto Nb$. The aim of this exercise is to use the formulas (1.67), (1.68) to deduce the background density in the corresponding log-gas, or equivalently the global form of the eigenvalue density.

(i) Substitute $V(x) = -N((a/2)\log(1-x) + (b/2)\log(1+x))$ in (1.67), (1.68) (after changing notation in these equations $a \mapsto c$, $b \mapsto d$) to deduce that

$$\rho_b(y) = \frac{aN}{\pi^2}\frac{\sqrt{(y-c)(d-y)}}{1-y^2}\int_c^d \frac{1}{1-t}\frac{1}{\sqrt{(t-c)(d-t)}}dt,$$

where c and d are determined by the equations

$$a\int_c^d \frac{dt}{(1-t)\sqrt{(t-c)(d-t)}} = b\int_c^d \frac{dt}{(1+t)\sqrt{(t-c)(d-t)}},$$
$$\frac{a}{2\pi}\int_c^d \frac{tdt}{(1-t)\sqrt{(t-c)(d-t)}} - \frac{b}{2\pi}\int_c^d \frac{tdt}{(1+t)\sqrt{(t-c)(d-t)}} = 1$$

(ii) [536], [113] Make use of tabulated integrals to simplify the formulas of (i) to read

$$\rho_b(y) = \frac{2+a+b}{2\pi}\frac{\sqrt{(y-c)(d-y)}}{1-y^2}, \tag{3.77}$$

where c and d are determined by

$$\frac{a}{\sqrt{(1-c)(1-d)}} = \frac{b}{\sqrt{(1+c)(1+d)}} = \frac{a+b+2}{2}.$$

(iii) Show that in the variable X, with $y = 1 - 2X^2$, the density (3.77) reads

$$\tilde{\rho}_b(X) = \frac{2+a+b}{\pi} \frac{\sqrt{(B^2 - X^2)(X^2 - A^2)}}{X(1 - X^2)}, \qquad X \in [A, B],$$

with

$$A^2 = \frac{1-d}{2} = \frac{1}{(a+b+2)^2} \Big((a+1)^{1/2}(a+b+1)^{1/2} - (b+1)^{1/2} \Big)^2,$$
$$B^2 = \frac{1-c}{2} = \frac{1}{(a+b+2)^2} \Big((a+1)^{1/2}(a+b+1)^{1/2} + (b+1)^{1/2} \Big)^2.$$

From [536] we know that for $\mathrm{Re}(w) > B^2$,

$$2 \int_A^B \frac{\tilde{\rho}_b(X)}{w - X^2} \, dX = -\frac{a}{w} + \frac{b}{1-w} - (a+b+2) \frac{((w - A^2)(w - B^2))^{1/2}}{w(1-w)}. \tag{3.78}$$

Use this to compute the first moment in X^2,

$$\int_A^B X^2 \tilde{\rho}_b(X) \, dX = \frac{a+1}{a+b+2}.$$

2. In this exercise Proposition 3.6.1 will be proved, following [410, Thm. 3.3.1] where the real case was considered. Let $\mathbf{A} + \mathbf{B} = \mathbf{T}^\dagger \mathbf{T}$ where $\mathbf{T}$ is an upper triangular $m \times m$ matrix with positive diagonal elements, and let $\mathbf{U}$ be the $m \times m$ matrix defined by $\mathbf{A} = \mathbf{T}^\dagger \mathbf{U} \mathbf{T}$.

   (i) Write down from (3.20), (3.23) and their analogue in the real and real quaternion cases the joint distribution of $\mathbf{A}$ and $\mathbf{B}$, and conclude from this that the joint distribution of $\mathbf{C} := \mathbf{A} + \mathbf{B}$ and $\mathbf{A}$ is proportional to

$$e^{-(\beta/2)\mathrm{Tr}(\mathbf{C})} (\det \mathbf{A})^{(\beta/2)(n_1 - m + \chi_\beta)} (\det(\mathbf{C} - \mathbf{A}))^{(\beta/2)(n_2 - m + \chi_\beta)} (d\mathbf{A}) \wedge (d\mathbf{C}),$$

   where $\chi_\beta := 1 - 2/\beta$.

   (ii) Note from (1.35) that $(\mathbf{T}^\dagger d\mathbf{U} \mathbf{T}) = (\det \mathbf{T}^\dagger \mathbf{T})^{\beta(m+1)/2} (d\mathbf{U})$ and use this to show

$$(d\mathbf{A}) \wedge (d\mathbf{C}) = (\det \mathbf{T}^\dagger \mathbf{T})^{\beta(m+1)/2} (d\mathbf{U}) \wedge (d(\mathbf{T}^\dagger \mathbf{T})).$$

   (iii) Write $\det \mathbf{A}$ and $\det(\mathbf{C} - \mathbf{A})$ in terms of $\det \mathbf{U}$ and $\det(\mathbf{1}_m - \mathbf{U})$, and use (ii) to show that the joint p.d.f. of $\mathbf{T}^\dagger \mathbf{T}$ and $\mathbf{U}$ is proportional to

$$e^{-(\beta/2)\mathrm{Tr}(\mathbf{T}^\dagger \mathbf{T})} \det(\mathbf{T}^\dagger \mathbf{T})^{(\beta/2)(n_1 + n_2 + 2\chi_\beta - m + 1)} (\det \mathbf{U})^{(\beta/2)(n_1 - m + \chi_\beta)} (\det(\mathbf{1}_m - \mathbf{U}))^{(\beta/2)(n_2 - m + \chi_\beta)}$$

   and deduce from this that the p.d.f. of $\mathbf{U}$ is proportional to

$$(\det \mathbf{U})^{(\beta/2)(n_1 - m + \chi_\beta)} (\det(\mathbf{1}_m - \mathbf{U}))^{(\beta/2)(n_2 - m + \chi_\beta)} \tag{3.79}$$

   while the p.d.f. of $\mathbf{C} = \mathbf{A} + \mathbf{B}$ is proportional to

$$e^{-(\beta/2)\mathrm{Tr}(\mathbf{C})} (\det \mathbf{C})^{(\beta/2)(n_1 + n_2 + 2\chi_\beta - m + 1)}.$$

   (iv) Note that $(\mathbf{T}^{-1} \mathbf{T}^\dagger)^{1/2} \mathbf{U} (\mathbf{T}^{-1} \mathbf{T}^\dagger)^{-1/2} = (\mathbf{A} + \mathbf{B})^{-1/2} \mathbf{A} (\mathbf{A} + \mathbf{B})^{-1/2}$, and use (1.35) to deduce that $\mathbf{U}$ has the same distribution as $(\mathbf{A} + \mathbf{B})^{-1/2} \mathbf{A} (\mathbf{A} + \mathbf{B})^{-1/2}$, which is thus given by (3.79). Use this result and Proposition 1.3.4 to deduce Proposition 3.6.1.

3. [279] Define $\mathbf{A}$ in terms of $\mathbf{a}$ according to Proposition 3.6.1. Let $\mathbf{X}$ be an $m \times N$ Gaussian matrix specified for $\beta = 1, 2$ and 4 as in Definition 3.1.2, and set

$$\mathbf{Y} = \mathbf{A}^{-1/2} \mathbf{X}. \tag{3.80}$$

The objective of this exercise is to show that the p.d.f. of $\mathbf{Y}$ is proportional to

$$\det(\mathbf{1} + \mathbf{Y}^\dagger \mathbf{Y})^{-(\beta/2)(n_1+N)}. \tag{3.81}$$

(i) With a specified in Proposition 3.6.1, show that the joint density of $\mathbf{A}$ and $\mathbf{X}$ is proportional to

$$(\det \mathbf{A})^{\beta a/2} e^{-(\beta/2)(\mathrm{Tr}\mathbf{A} + \mathrm{Tr}\mathbf{X}^\dagger \mathbf{X})}.$$

(ii) Use a generalization of Proposition 3.2.4 and the result of (i) to show that the change of variables (3.80) gives that the joint density of $\mathbf{A}$ and $\mathbf{Y}$ is proportional to

$$(\det \mathbf{A})^{\beta(N+a)/2} e^{-(\beta/2)\mathrm{Tr}(\mathbf{A}(\mathbf{1}+\mathbf{Y}^\dagger \mathbf{Y}))}.$$

(iii) Use (3.31) to integrate over $\mathbf{A} > 0$, and thus deduce the result.

4. Let $\mathbf{A}$, $\mathbf{B}$ be $N \times N$ Gaussian matrices with entries for $\beta = 1, 2$ and 4 as specified in Definition 3.1.2. Proceed as in q.3 to show that the distribution of $\mathbf{Y} = \mathbf{A}^{-1}\mathbf{B}$ is proportional to

$$\det(\mathbf{1} + \mathbf{Y}^\dagger \mathbf{Y})^{-\beta N}. \tag{3.82}$$

5. [508], [305] Use the method of Exercises 1.4 q.8 to show that for $a, b \ge 0$ the minimum of

$$-\frac{1}{2}\sum_{j=1}^N \Big(a \log(1 - y_j) + b \log(1 + y_j) \Big) - \sum_{1 \le j < k \le N} \log |y_j - y_k|, \quad |y_j| < 1 \ (j = 1, \dots, N)$$

occurs at the zeros of the Jacobi polynomial $P_N^{(a-1,b-1)}(x)$.

6. [2] Let $\mathbf{a}$, $\mathbf{b}$ be $n_1 \times m$, $n_2 \times m$ $(n_1, n_2 \ge m)$ Gaussian matrices with real ($\beta = 1$), complex ($\beta = 2$) or quaternion real ($\beta = 4$) entries. Use the results of Exercises 1.3 q.1(iv) and the remark below (2.31) to argue that the rectangular matrix

$$\begin{bmatrix} \mathbf{a} \\ \mathbf{b} \end{bmatrix} (\mathbf{a}^\dagger \mathbf{a} + \mathbf{b}^\dagger \mathbf{b})^{-1/2}$$

is the $(n_1 + n_2) \times m$ block of an $(n_1 + n_2) \times (n_1 + n_2)$ Haar distributed unitary matrix with real ($\beta = 1$), complex ($\beta = 2$) or quaternion real ($\beta = 4$) entries. Now make use of Proposition 3.6.1 to deduce that the top $n_1 \times m$ block of such random matrices has the square of its singular values as specified therein.

3.7 JACOBI ENSEMBLE AND SYMMETRIC SPACES

Sections 1.3.4, 3.1.3 and 3.3.3 contain the ten Lie algebras associated with the ten infinite families of symmetric spaces. The symmetric spaces corresponding to the matrix Lie algebras of Section 1.3.4 have been made explicit in Section 2.2.3, and have been shown to be isomorphic to the three circular ensembles. In this section the remaining seven symmetric spaces are given explicitly, as are the corresponding unitary matrices and their eigenvalue p.d.f.'s. After the change of variables $y_j = \cos\theta_j$, all the eigenvalue p.d.f.'s obtained are found to be examples of the Jacobi ensemble p.d.f. (3.74).

3.7.1 Symmetric spaces isomorphic to $O^+(N)$ and $\mathrm{Sp}(2N)$

The compact quotient spaces

$$O^+(N) \times O^+(N)/O^+(N), \qquad \mathrm{Sp}(2N) \times \mathrm{Sp}(2N)/\mathrm{Sp}(2N)$$

are symmetric spaces, isomorphic to $O^+(N)$ and $\mathrm{Sp}(2N)$ respectively. The Cayley transformations (2.61), (2.70) show that the corresponding spaces of Hermitian matrices are antisymmetric Hermitian matrices with pure imaginary elements, and quaternion real matrices with the property that when multiplied by i they are Hermitian. These in turn are the Hermitian part of the Lie algebras $i \times (\mathrm{so}(n, \mathbb{C}))$ and $\mathrm{sp}(n, \mathbb{C})$.

To relate the eigenvalues p.d.f.'s for $O(N)$ and $\mathrm{Sp}(2N)$ to the Jacobi ensemble p.d.f, we note that in (3.76) the latter has been related to the p.d.f. for the β generalization of the orthogonal and symplectic ensembles (2.102). Recalling then the results of Section 2.6 gives the precise form of this relationship.

PROPOSITION 3.7.1 *In terms of the variable $y_j = \cos\theta_j$, the eigenvalue p.d.f. for random real orthogonal matrices is given by the Jacobi ensemble (3.74) with $\beta = 2$, N replaced by N^* and*

$$(N^*, a, b) = \begin{cases} (N/2, -1/2, -1/2) & \textit{for matrices in } O^+(N),\ N \textit{ even}, \\ ((N-1)/2, 1/2, -1/2) & \textit{for matrices in } O^+(N),\ N \textit{ odd}, \\ ((N-1)/2, -1/2, 1/2) & \textit{for matrices in } O^-(N),\ N \textit{ odd}, \\ (N/2 - 1, 1/2, 1/2) & \textit{for matrices in } O^-(N),\ N \textit{ even}. \end{cases}$$

Similarly, in terms of the variable $y_j = \cos\theta_j$ the eigenvalue p.d.f. for $2N \times 2N$ random unitary matrices with each 2×2 sub-block a real quaternion element (unitary symplectic equivalent matrices) is given by the p.d.f. (3.74) with $\beta = 2$, N replaced by N^* and

$$(N^*, a, b) = (N, 1/2, 1/2).$$

3.7.2 Symmetric spaces corresponding to the chiral ensembles

Consider the block matrix (3.1). Use of the singular value decomposition (3.2) gives

$$\begin{bmatrix} \mathbf{0}_{m\times n} & \mathbf{V} \\ \mathbf{U}^\dagger & \mathbf{0}_{n\times m} \end{bmatrix} \begin{bmatrix} \mathbf{0}_{n\times n} & \mathbf{X} \\ \mathbf{X}^\dagger & \mathbf{0}_{m\times m} \end{bmatrix} \begin{bmatrix} \mathbf{0}_{n\times m} & \mathbf{U} \\ \mathbf{V}^\dagger & \mathbf{0}_{m\times n} \end{bmatrix} = \begin{bmatrix} \mathbf{0}_{m\times m} & \mathbf{\Lambda}^T \\ \mathbf{\Lambda} & \mathbf{0}_{n\times n} \end{bmatrix}. \tag{3.83}$$

The block matrix involving $\mathbf{V}$ and $\mathbf{U}^\dagger$ can be characterized as an $(n+m) \times (n+m)$ unitary matrix g such that

$$\mathbf{g}^\phi = \mathbf{g}^{-1}, \quad \text{where} \quad \mathbf{g}^\phi = \mathbf{I}'_{n,m}\mathbf{g}^\dagger\mathbf{I}'_{n,m} \tag{3.84}$$

with

$$\mathbf{I}'_{n,m} := \begin{bmatrix} \mathbf{1}_n & \mathbf{0}_{n\times m} \\ \mathbf{0}_{m\times n} & -\mathbf{1}_m \end{bmatrix}.$$

In the case that the matrix $\mathbf{X}$ in Proposition 3.1.1 has real elements or quaternion real elements, we know from (3.3) that the unitary matrix is constrained to be real symmetric and unitary symplectic equivalent respectively. Thus in these cases the second equation in (3.84) should be replaced by

$$\mathbf{g}^\phi = \mathbf{I}'_{n,m}\mathbf{g}^T\mathbf{I}'_{n,m}, \ \mathbf{g} \in O(n+m) \quad \text{and} \quad \mathbf{g}^\phi = \mathbf{I}'_{2n,2m}\mathbf{g}^D\mathbf{I}'_{2n,2m}, \ \mathbf{g} \in \mathrm{Sp}(2(n+m)). \tag{3.85}$$

The formulas (2.13) and (2.16) for matrices in the COE and CSE suggest considering the spaces of matrices of the form

$$\mathbf{S} = \mathbf{g}^\phi\mathbf{g} \tag{3.86}$$

for $\mathbf{g} \in O(n+m)$ with $\mathbf{g}^\phi$ as in the first formula of (3.85) ($\beta = 1$), for $\mathbf{g} \in U(n+m)$ with $\mathbf{g}^\phi$ as in (3.84) ($\beta = 2$), and for $\mathbf{g} \in \mathrm{Sp}(2(n+m))$ with $\mathbf{g}^\phi$ as in the second formula of (3.85) ($\beta = 4$). With these specifications of $\mathbf{g}^\phi$ the considerations of the proof of Proposition 2.2.4 shows that the matrices (3.86) are isomorphic to

$$O(n+m)/O(n) \times O(m), \quad U(n+m)/U(n) \times U(m), \quad \mathrm{Sp}(2(n+m))/\mathrm{Sp(2n)} \times \mathrm{Sp(2m)},$$

respectively.

A uniform measure on these symmetric spaces is achieved by the Haar form (2.10), but with the matrix of differentials modified using the invariance (2.11) to be unchanged by the mapping ϕ. From (2.15) and (2.17) we know we should choose

$$\delta \mathbf{S} = (\mathbf{g}^{\phi})^{\dagger} d\mathbf{S}\, \mathbf{g}^{\dagger} =: i d\mathbf{H} \tag{3.87}$$

where $\mathbf{H}$ is as in (3.1) with real elements ($\beta = 1$), complex elements ($\beta = 2$) and quaternion real elements ($\beta = 4$).

Whereas the matrix (3.1) has $n - m$ zero eigenvalues, the matrix $\mathbf{S}$ has $n - m$ unit eigenvalues. This can be seen from the Cayley transformation

$$\mathbf{S} = \frac{\mathbf{1}_{n+m} + i\mathbf{H}}{\mathbf{1}_{n+m} - i\mathbf{H}} \tag{3.88}$$

for a certain $\mathbf{H}$ of the form (3.1). Recalling that the nonzero eigenvalues of $\mathbf{H}$ occur in $\pm$ pairs, we see too that the non-unit eigenvalues of $\mathbf{S}$ come in complex conjugate pairs. Now, we can use the facts that $\mathbf{H}$ has the eigenvalue decomposition given in the proof of Proposition 3.1.1 and that the elements of the matrix $\mathbf{X}$ in (3.1) are independent to deduce from (3.88) that

$$(\delta \mathbf{S}) \propto \Big(\det(\mathbf{1}_{n+m} + \mathbf{H}^2) \Big)^{-\beta m/2} (d\mathbf{H})$$

(compare the exponent with that in (2.57)). Knowledge of the eigenvalue p.d.f (3.4) for $\{\mathbf{H}\}$ then implies that the eigenvalue p.d.f. for $\{\mathbf{S}\}$ is given by

$$\frac{1}{C} \prod_{j=1}^{m} \sin\theta_j (\cos^2\theta_j/2)^{\beta(1-2/\beta)/2} (\sin^2\theta_j/2)^{\beta(n-m+1-2/\beta)/2} \prod_{1\le j<k\le m} |\cos\theta_j - \cos\theta_k|^{\beta}. \tag{3.89}$$

The change of variables $y_j = \cos\theta_j$ shows (3.89) to be the special case $a = n - m + 1 - 2/\beta, b = 1 - 2/\beta$ of the Jacobi ensemble (3.74).

3.7.3 The symmetric spaces $\mathrm{Sp}(2N)/U(N)$, $O^+(2N)/U(N)$

Out of the ten sets of Hermitian matrices corresponding to infinite families of matrix Lie algebras, we have so far made a correspondence with eight sets of unitary matrices. It remains to identify the unitary matrices corresponding to Hermitian matrices of the form (3.46), and antisymmetric Hermitian matrices with quaternion real elements.

From the theory of Exercises 3.3 q.1 we know that matrices (3.46) satisfy the diagonalization type equation (3.83) with $m = n = N, \mathbf{V} = (\mathbf{U}^{-1})^T$. Now unitary matrices $\mathbf{g}$ of the form

$$\begin{bmatrix} \mathbf{U} & \mathbf{0}_N \\ \mathbf{0}_N & (\mathbf{U}^{-1})^T \end{bmatrix}$$

have the property (3.84) and furthermore satisfy (1.26) and so are examples of unitary symplectic matrices. From this observation it can be seen [139] that unitary matrices constructed according to the prescription (3.86) with $\mathbf{g} \in \mathrm{Sp}(2N)$ are isomorphic to the symmetric space $\mathrm{Sp}(2N)/U(N)$. In particular the equation (3.87) holds with $\mathbf{H}$ of the form (3.46). Use of the Cayley transformation (3.88), (2.57) with $\beta = 1$, together with the result (3.52) shows that the corresponding eigenvalue p.d.f. is proportional to

$$\prod_{j=1}^{N} \sin\theta_j \prod_{1\le j<k\le N} |\cos\theta_k - \cos\theta_j|. \tag{3.90}$$

Regarding the unitary matrices corresponding to antisymmetric Hermitian matrices with real quaternion elements, we first note that such matrices can be diagonalized by unitary symplectic equivalent matrices with real entries, and these in turn can be identified with matrices $U(N)$ (recall Exercises 1.3 q.6(ii)). This suggests considering the unitary matrices

$$\mathbf{S} = \mathbf{g}^D \mathbf{g}, \qquad \mathbf{g} \in O^+(2N), \tag{3.91}$$

which are isomorphic to the symmetric space $O^+(2N)/U(N)$ [139]. Note that then $(\mathbf{Z}_{2N}\mathbf{S})^T = -\mathbf{Z}_{2N}\mathbf{S}$, so $\mathbf{Z}_{2N}\mathbf{S}$ is a $2N \times 2N$ antisymmetric unitary matrix. For the set of matrices (3.91), (3.87) holds with $\mathbf{H}$ an antisymmetric Hermitian matrix with real quaternion elements. Use of the Cayley transformation (3.88), (2.57) with $\beta = 4$, together with the results of Exercises 1.3 q.7(iii), shows that the corresponding eigenvalue p.d.f. is proportional to

$$\prod_{j=1}^{N/2} \sin\theta_j \prod_{1 \le j < k \le N/2} |\cos\theta_j - \cos\theta_k|^4, \quad N \text{ even}, \tag{3.92}$$

$$\prod_{j=1}^{(N-1)/2} \sin\theta_j \sin^4\frac{\theta_j}{2} \prod_{1 \le j < k \le (N-1)/2} |\cos\theta_j - \cos\theta_k|^4, \quad N \text{ odd}. \tag{3.93}$$

In terms of the variable $y_j = \cos\theta_j$ the eigenvalue p.d.f.'s (3.90), (3.92) and (3.93) are of the form (3.74). Explicitly, with N replaced by N^* the parameters are

$$(N^*, a, b; \beta) = \begin{cases} (N, 0, 0; 1) \text{ for matrices in } \mathrm{Sp}(2N)/U(N), \\ (N/2, 0, 0; 4) \text{ for matrices in } O^+(2N)/U(N),\ N \text{ even}, \\ ((N-1)/2, 1/2, 0; 4) \text{ for matrices in } O^+(2N)/U(N),\ N \text{ odd}. \end{cases}$$

EXERCISES 3.7 1. [247] Let $\mathbf{Q}$ be a $2n \times 2n$ positive definite Hermitian matrix (i.e. all eigenvalues positive) and let $\mathbf{I}'_{n,n} = \mathrm{diag}[\mathbf{1}_n, -\mathbf{1}_n]$. The objective of this exercise is to note special properties of the diagonalization of the non-Hermitian matrix $\mathbf{QI}'_{n,n}$.

(i) With the diagonalization formula for $\mathbf{Q}$ reading

$$\mathbf{Q} = \mathbf{U}\mathrm{diag}[\lambda_1, \dots, \lambda_{2n}]\mathbf{U}^\dagger,$$

set $\mathbf{S} = \mathbf{U}\mathrm{diag}[\sqrt{\lambda_1}, \dots, \sqrt{\lambda_{2n}}]\mathbf{U}^\dagger$ so that $\mathbf{S}^2 = \mathbf{Q}$. Show that $\mathbf{QI}'_{n,n}$ is similar to $\mathbf{SI}'_{n,n}\mathbf{S}$, and thus conclude that the eigenvalues of $\mathbf{QI}'_{n,n}$ are all real.

(ii) Let $\mathbf{H}$ be a Hermitian $n \times n$ matrix, and let $\mathbf{T}$ be an arbitrary nonsingular $n \times n$ matrix. It can be shown [255] that the number of positive eigenvalues of $\mathbf{H}$ is the same as the number of positive eigenvalues of $\mathbf{T}^\dagger\mathbf{HT}$. Use this result and the result of (i) to show that exactly half the eigenvalues of $\mathbf{QI}'_{n,n}$ are positive.

(iii) With $\{\vec{v}_j\}_{j=1,\dots,2n}$ denoting the eigenvectors of $\mathbf{QI}'_{n,n}$ corresponding to the eigenvalues $\{\mu_j\}_{j=1,\dots,2n}$, assumed distinct and with the first n being positive, set $\mathbf{V} = [\vec{v}_1 \cdots \vec{v}_{2n}]$. Show that

$$\mathbf{V}^\dagger\mathbf{I}'_{n,n}\mathbf{V} = \mathrm{diag}[\mathrm{sgn}(\mu_1), \dots, \mathrm{sgn}(\mu_{2n})] = \mathbf{I}'_{n,n}. \tag{3.94}$$

(Matrices with this property form a group denoted $U(n,n)$.)

3.8 JACOBI ENSEMBLE AND QUANTUM CONDUCTANCE

The situation to be considered is a quasi one-dimensional conductor containing scattering impurities and having n available scattering channels (which are the different wavenumbers of the plane wave states) at the left-hand edge and m available scattering channels at the right-hand edge. The length of the sample is assumed to be much smaller than the coherence length, but much larger than the mean free path of the electrons, which is the criterion for the metallic phase. At each end of the conductor is an electron reservoir of different chemical potentials, which causes a current to flow. A quantity of interest is then the conductance of the sample.

3.8.1 Definitions

Before presenting the formula for the conductance, it is necessary to revise the theoretical description of the above setting. Similar to the case for the scattering problem involving a lead with a tunneling barrier considered in Section 2.4, fundamental quantities are the electron fluxes at the left- and right-hand edges of the conductor, which are specified by the $2n$ component vector $\begin{bmatrix} \vec{I} \\ \vec{O} \end{bmatrix}$ and the $2m$ component vector $\begin{bmatrix} \vec{I}' \\ \vec{O}' \end{bmatrix}$, where $\vec{I}\,(\vec{O})$ and $\vec{I}'\,(\vec{O}')$ denote the n and m component amplitudes of the plane wave states traveling into (out of) the left and right sides of the conductor, respectively. For definiteness we will suppose $n \ge m$. Flux conservation requires

$$|\vec{I}|^2 + |\vec{I}'|^2 = |\vec{O}|^2 + |\vec{O}'|^2.$$

By definition, the $(n+m)\times(n+m)$ scattering matrix $\mathbf{S}$ relates the flux traveling into the conductor to that traveling out,

$$\mathbf{S}\begin{bmatrix} \vec{I} \\ \vec{I}' \end{bmatrix} := \begin{bmatrix} \vec{O} \\ \vec{O}' \end{bmatrix}. \tag{3.95}$$

As done in Section 2.1.1 the flux conservation condition can be used to show that $\mathbf{S}$ must be unitary, and as in (2.40) the scattering matrix is further decomposed in terms of reflection and transmission matrices by

$$\mathbf{S} = \begin{bmatrix} \mathbf{r}_{n\times n} & \mathbf{t}'_{n\times m} \\ \mathbf{t}_{m\times n} & \mathbf{r}'_{m\times m} \end{bmatrix}. \tag{3.96}$$

We know from Section 2.1.2 that $\mathbf{S}$ must be symmetric if the system has a time reversal symmetry with $T^2 = 1$, and a self-dual quaternion matrix when there is a time reversal symmetry with $T^2 = -1, T = \mathbf{Z}_{2(n+m)}K$.

The conductance G is given in terms of the transmission matrix $\mathbf{t}$ (or $\mathbf{t}'$) by the so-called two-probe *Landauer formula*,

$$G/G_0 = \mathrm{Tr}(\mathbf{t}^\dagger\mathbf{t}) = \mathrm{Tr}(\mathbf{t}'^\dagger\mathbf{t}'), \tag{3.97}$$

where $G_0 = 2e^2/h$ is twice the fundamental quantum unit of conductance. Thus the quantities of interest are the eigenvalues of $\mathbf{t}^\dagger\mathbf{t}$ (or $\mathbf{t}'^\dagger\mathbf{t}'$). In fact the matrix $\mathbf{S}$ can be decomposed in a form which isolates the eigenvalues of $\mathbf{t}^\dagger\mathbf{t}$.

3.8.2 Distribution of singular values

Suppose for definiteness that there is no time reversal symmetry, so $\mathbf{S}$ is a general unitary matrix. We can then decompose each block of $\mathbf{S}$ as in the singular value decomposition (3.2),

$$\mathbf{r}_{n\times n} = \mathbf{U}_r\mathbf{\Lambda}_r\mathbf{V}_r^\dagger, \quad \mathbf{t}_{m\times n} = \mathbf{U}_t\mathbf{\Lambda}_t\mathbf{V}_t^\dagger, \text{ etc.}$$

where $\mathbf{U}_r, \mathbf{V}_r^\dagger, \dots$ are unitary matrices and $\mathbf{\Lambda}_r$ ($\mathbf{\Lambda}_t$) is a rectangular diagonal matrix consisting of the positive square roots of the eigenvalues of $\mathbf{r}^\dagger\mathbf{r}$ ($\mathbf{t}^\dagger\mathbf{t}$) (these eigenvalues are between 0 and 1 since $\mathbf{r}^\dagger\mathbf{r} + \mathbf{t}^\dagger\mathbf{t} = \mathbf{1}$) and $\mathbf{t}^\dagger\mathbf{t}$ must have $n-m$ zero eigenvalues since the rank of $\mathbf{t}$ is equal to m. The unitarity constraint inter-relates the matrices $\mathbf{U}_r, \mathbf{U}_{r'}, \dots$ (see Exercises 3.8 q.1) and implies the decomposition

$$\mathbf{S} = \begin{bmatrix} \mathbf{U}_r & \mathbf{0} \\ \mathbf{0} & \mathbf{U}_{r'} \end{bmatrix} \mathbf{L} \begin{bmatrix} \mathbf{V}_r^\dagger & \mathbf{0} \\ \mathbf{0} & \mathbf{V}_{r'}^\dagger \end{bmatrix}, \tag{3.98}$$

where

$$\mathbf{L} := \begin{bmatrix} \sqrt{\mathbf{1} - \mathbf{\Lambda}_t\mathbf{\Lambda}_t^T} & i\mathbf{\Lambda}_t \\ i\mathbf{\Lambda}_t^T & \sqrt{\mathbf{1} - \mathbf{\Lambda}_t^T\mathbf{\Lambda}_t} \end{bmatrix}. \tag{3.99}$$

In the case of a time reversal symmetry with $T^2 = 1$, $\mathbf{S}$ must be symmetric, and thus so must be the blocks of $\mathbf{r}_{n\times n}$ and $\mathbf{r}'_{m\times m}$ in (3.96). These square matrices then permit the singular value decompositions

$$\mathbf{r}_{n\times n} = \mathbf{U}_r\mathbf{\Lambda}_r\mathbf{U}_r^T, \qquad \mathbf{r}'_{m\times m} = \mathbf{U}_{r'}\mathbf{\Lambda}_{r'}\mathbf{U}_{r'}^T$$

and thus (3.99) holds with

$$\mathbf{V}_r^\dagger = \mathbf{U}_r^T, \qquad \mathbf{V}_{r'}^\dagger = \mathbf{U}_{r'}^T. \tag{3.100}$$

Similarly in the case of a time reversal symmetry with $T^2 = -1$, when $\mathbf{S}$ is self-dual quaternion, (3.99) holds with

$$\mathbf{V}_r^\dagger = \mathbf{U}_r^D, \qquad \mathbf{V}_{r'}^\dagger = \mathbf{U}_{r'}^D. \tag{3.101}$$

It turns out that the distribution of the elements of $\mathbf{\Lambda}_t$ in the decomposition (3.99) coincides exactly with the distribution of the elements of $\mathbf{\Lambda}$ in the singular value decompositions (3.2) and (3.3), which is given by (3.4) excluding the Gaussian factors [399], [53].

PROPOSITION 3.8.1 *Let $\mathbf{S}$ be a $(n+m)\times(n+m)$ matrix which is a member of the COE ($\beta = 1$), CUE ($\beta = 2$) or CSE ($\beta = 4$) (in the latter case each element is itself a 2×2 matrix). Decomposing $\mathbf{S}$ as in (3.99), the distribution of the nonzero elements of $\mathbf{\Lambda}_t$, which are equal to the square root of the nonzero eigenvalues of $\mathbf{t}^\dagger\mathbf{t}$, and thus in the range $[0,1]$, is proportional to*

$$\prod_{j=1}^m \lambda_j^{\beta\alpha} \prod_{1\le j<k\le m} |\lambda_k^2 - \lambda_j^2|^\beta, \qquad \alpha := n - m + 1 - 1/\beta. \tag{3.102}$$

Proof. Consider for definiteness the case $\beta = 2$, so that $\mathbf{S}$ is a random unitary $(n+m)\times(n+m)$ matrix, and thus has $(n+m)\times(n+m)$ independent elements. We proceed in an analogous way to the proof of Proposition 3.1.3, first computing the differential of (3.98) to obtain

$$\begin{bmatrix} \mathbf{U}_r^\dagger & \mathbf{0} \\ \mathbf{0} & \mathbf{U}_{r'}^\dagger \end{bmatrix} d\mathbf{S} \begin{bmatrix} \mathbf{V}_r & \mathbf{0} \\ \mathbf{0} & \mathbf{V}_{r'} \end{bmatrix} = \begin{bmatrix} \delta\mathbf{U}_r & \mathbf{0} \\ \mathbf{0} & \delta\mathbf{U}_{r'} \end{bmatrix} \mathbf{L} + d\mathbf{L} - \mathbf{L} \begin{bmatrix} \delta\mathbf{V}_r & \mathbf{0} \\ \mathbf{0} & \delta\mathbf{V}_{r'} \end{bmatrix}.$$

Using this to form $\mathrm{Tr}(d\mathbf{S}d\mathbf{S}^\dagger)$, simplifying using the fact that

$$\mathrm{Tr}\Big(\begin{bmatrix} \delta\mathbf{V}_r & \mathbf{0} \\ \mathbf{0} & \delta\mathbf{V}_{r'} \end{bmatrix} \mathbf{L}^\dagger d\mathbf{L}\Big) = 0,$$

and factoring the terms as in (3.5) but retaining the block matrix structure shows

$$\begin{aligned}
&\mathrm{Tr}(d\mathbf{S}d\mathbf{S}^\dagger) \\
&= \mathrm{Tr}\Big(\Big(\sqrt{1-\mathbf{\Lambda}_t\mathbf{\Lambda}_t^T}\delta\mathbf{V}_r - \delta\mathbf{U}_r\sqrt{1-\mathbf{\Lambda}_t\mathbf{\Lambda}_t^T}\Big)\Big(\sqrt{1-\mathbf{\Lambda}_t\mathbf{\Lambda}_t^T}\delta\mathbf{U}_r - \delta\mathbf{V}_r\sqrt{1-\mathbf{\Lambda}_t\mathbf{\Lambda}_t^T}\Big)\Big) \\
&+\mathrm{Tr}\Big((\mathbf{\Lambda}_t\delta\mathbf{V}_{r'} - \delta\mathbf{U}_r\mathbf{\Lambda}_t)(\mathbf{\Lambda}_t^T\delta\mathbf{U}_r - \delta\mathbf{V}_{r'}\mathbf{\Lambda}_t^T)\Big) + \mathrm{Tr}\Big((\mathbf{\Lambda}_t^T\delta\mathbf{V}_r - \delta\mathbf{U}_{r'}\mathbf{\Lambda}_t^T)(\mathbf{\Lambda}_t\delta\mathbf{U}_{r'} - \delta\mathbf{V}_r\mathbf{\Lambda}_t)\Big) \\
&+\mathrm{Tr}\Big(\Big(\sqrt{1-\mathbf{\Lambda}_t^T\mathbf{\Lambda}_t}\delta\mathbf{V}_{r'} - \delta\mathbf{U}_{r'}\sqrt{1-\mathbf{\Lambda}_t^T\mathbf{\Lambda}_t}\Big)\Big(\sqrt{1-\mathbf{\Lambda}_t^T\mathbf{\Lambda}_t}\delta\mathbf{U}_{r'} - \delta\mathbf{V}_{r'}\sqrt{1-\mathbf{\Lambda}_t^T\mathbf{\Lambda}_t}\Big)\Big) + \mathrm{Tr}(d\mathbf{L}d\mathbf{L}^\dagger).
\end{aligned}$$

With the diagonal elements of $\mathbf{\Lambda}\mathbf{\Lambda}^T$ denoted λ_j, $(j = 1, \dots, n)$, and thus $\lambda_{m+1} = \cdots = \lambda_n = 0$, and the five traces above denoted $T_1, \dots, T_5$, we see that in component form

$$
\begin{aligned}
T_1 = & \sum_{k=1}^{n} (1-\lambda_k^2)|(\delta \mathbf{V}_r)_{kk} - (\delta \mathbf{U}_r)_{kk}|^2 + \sum_{1\le k<l\le n} \frac{1}{2}\Big(\sqrt{1-\lambda_l^2}+\sqrt{1-\lambda_k^2}\Big)^2 |(\delta \mathbf{V}_r)_{lk} - (\delta \mathbf{U}_r)_{lk}|^2 \\
& + \Big(\sum_{1\le k<l\le m} + \sum_{k=1}^{m}\sum_{l=m+1}^{n} \Big) \frac{1}{2}\Big(\sqrt{1-\lambda_l^2}-\sqrt{1-\lambda_k^2}\Big)^2 |(\delta \mathbf{V}_r)_{lk} + (\delta \mathbf{U}_r)_{lk}|^2, \\
T_2 = & \sum_{k=1}^{m} \lambda_k^2 |(\delta \mathbf{V}_{r'})_{kk} - (\delta \mathbf{U}_r)_{kk}|^2 + \sum_{1\le k<l\le m} \Big\{ \frac{1}{2}(\lambda_l+\lambda_k)^2 |(\delta \mathbf{V}_{r'})_{lk} - (\delta \mathbf{U}_r)_{lk}|^2 \\
& + \frac{1}{2}(\lambda_l-\lambda_k)^2 |(\delta \mathbf{V}_{r'})_{lk} + (\delta \mathbf{U}_r)_{lk}|^2 \Big\} + \sum_{k=1}^{m}\sum_{l=m+1}^{n} \lambda_k^2 |(\delta \mathbf{U}_r)_{lk}|^2, \\
T_3 = & \sum_{k=1}^{m} \lambda_k^2 |(\delta \mathbf{V}_{r})_{kk} - (\delta \mathbf{U}_{r'})_{kk}|^2 + \sum_{1\le k<l\le m} \Big\{ \frac{1}{2}(\lambda_l+\lambda_k)^2 |(\delta \mathbf{V}_{r})_{lk} - (\delta \mathbf{U}_{r'})_{lk}|^2 \\
& + \frac{1}{2}(\lambda_l-\lambda_k)^2 |(\delta \mathbf{V}_{r})_{lk} + (\delta \mathbf{U}_{r'})_{lk}|^2 \Big\} + \sum_{k=1}^{m}\sum_{l=m+1}^{n} \lambda_k^2 |(\delta \mathbf{V}_r)_{lk}|^2, \\
T_4 = & \sum_{k=1}^{m} (1-\lambda_k^2)|(\delta \mathbf{V}_{r'})_{kk} - (\delta \mathbf{U}_{r'})_{kk}|^2 + \sum_{1\le k<l\le m} \frac{1}{2}\Big(\sqrt{1-\lambda_l^2}+\sqrt{1-\lambda_k^2}\Big)^2 |(\delta \mathbf{V}_{r'})_{lk} - (\delta \mathbf{U}_{r'})_{lk}|^2 \\
& + \frac{1}{2}\Big(\sqrt{1-\lambda_l^2}-\sqrt{1-\lambda_k^2}\Big)^2 |(\delta \mathbf{V}_{r'})_{lk} + (\delta \mathbf{U}_{r'})_{lk}|^2, \\
T_5 = & 2\sum_{j=1}^{m} \frac{(d\lambda_j)^2}{1-\lambda_j^2}.
\end{aligned}
$$

The task now is to identify the set of independent differentials, as unlike in (3.6) the differentials occurring in T_1–T_5 are not all independent. In fact the independent differentials are those in T_1, T_4, T_5 and the first term in T_3. Explicitly, these are the imaginary part of

$$
\begin{aligned}
(\delta \mathbf{V}_r)_{kk} - (\delta \mathbf{U}_r)_{kk}, &\quad 1 \le k \le n, \\
(\delta \mathbf{V}_r)_{kk} - (\delta \mathbf{U}_{r'})_{kk}, &\quad 1 \le k \le m, \\
(\delta \mathbf{V}_{r'})_{kk} - (\delta \mathbf{U}_{r'})_{kk}, &\quad 1 \le k \le m,
\end{aligned} \tag{3.103}
$$

and the real and imaginary parts of

$$
\begin{aligned}
(\delta \mathbf{V}_r)_{lk} - (\delta \mathbf{U}_r)_{lk}, &\quad 1 \le k < l \le n, \\
(\delta \mathbf{V}_r)_{lk} + (\delta \mathbf{U}_r)_{lk}, &\quad 1 \le k < l \le m, \quad 1 \le k \le m \ \& \ m+1 \le l \le n \\
(\delta \mathbf{V}_{r'})_{lk} - (\delta \mathbf{U}_{r'})_{lk}, &\quad 1 \le k < l \le m, \\
(\delta \mathbf{V}_{r'})_{lk} + (\delta \mathbf{U}_{r'})_{lk}, &\quad 1 \le k < l \le m,
\end{aligned} \tag{3.104}
$$

as well as

$$
d\lambda_j, \quad 1 \le j \le m.
$$

This is a total of $(n+m)^2$ terms as required by the number of independent terms in $\mathbf{S}$.

To compute the Jacobian we must make use of the general formula (1.15) rather than the simpler formula (1.14) because the metric form is not diagonal, as will become apparent in the subsequent calculations. Now, for $1 \le k < l \le m$ we see from (3.104) that the independent differentials can be taken as the real and imaginary parts of

$$
(\delta \mathbf{U}_r)_{lk}, \quad (\delta \mathbf{U}_{r'})_{lk}, \quad (\delta \mathbf{V}_r)_{lk}, \quad (\delta \mathbf{V}_{r'})_{lk}. \tag{3.105}
$$

With $z_1 := x_1 + iy_1$, $z_2 := x_2 + iy_2$, using the simple formula

$$|z_1 - z_2|^2 = x_1^2 + x_2^2 + y_1^2 + y_2^2 - 2x_1x_2 - 2y_1y_2, \tag{3.106}$$

we see from the expressions for T_1 through to T_4 that the contribution to the metric form from the differentials (3.105) is

$$\sum_{\substack{s=\text{ real part}\\ \text{imag. part}}} \sum_{1\le k<l\le m} \Big(2((\delta \mathbf{U}_r)_{lk}^{(s)})^2 + 2((\delta \mathbf{U}_{r'})_{lk}^{(s)})^2 + 2((\delta \mathbf{V}_r)_{lk}^{(s)})^2 + 2((\delta \mathbf{V}_{r'})_{lk}^{(s)})^2$$

$$+2a_{lk}\Big\{ (\delta \mathbf{U}_r)_{lk}^{(s)} (\delta \mathbf{V}_r)_{lk}^{(s)} + (\delta \mathbf{U}_{r'})_{lk}^{(s)} (\delta \mathbf{V}_{r'})_{lk}^{(s)} \Big\} + 2b_{lk}\Big\{ (\delta \mathbf{U}_r)_{lk}^{(s)} (\delta \mathbf{V}_{r'})_{lk}^{(s)} + (\delta \mathbf{V}_r)_{lk}^{(s)} (\delta \mathbf{U}_{r'})_{lk}^{(s)} \Big\}\Big),$$

with

$$a_{lk} = \frac{1}{2}\Big(\sqrt{1-\lambda_l^2} - \sqrt{1-\lambda_k^2}\Big)^2 - \frac{1}{2}\Big(\sqrt{1-\lambda_l^2} + \sqrt{1-\lambda_k^2}\Big)^2$$
$$b_{lk} = \frac{1}{2}(\lambda_l - \lambda_k)^2 - \frac{1}{2}(\lambda_l + \lambda_k)^2.$$

This contributes to $(\det[g_{jk}])^{1/2}$ in (1.15) the factor

$$\prod_{k<l}^{m} \begin{vmatrix} 2 & a_{lk} & 0 & b_{lk} \\ a_{lk} & 2 & b_{lk} & 0 \\ 0 & b_{lk} & 2 & a_{lk} \\ b_{lk} & 0 & a_{lk} & 2 \end{vmatrix} \propto \prod_{k<l}^{m} (\lambda_l^2 - \lambda_k^2)^2. \tag{3.107}$$

For $m+1 \le k < l \le n$, the coefficient of $|(\delta \mathbf{V}_r)_{lk} - (\delta \mathbf{U}_r)_{lk}|^2$ in T_1 is independent of the λs, so for the purposes of computing the p.d.f. of the latter these differentials can be ignored.

For $1 \le k \le m$, $m+1 \le l \le n$, we see from (3.104) that the independent differentials can be taken as the real and imaginary parts of

$$(\delta \mathbf{U}_r)_{lk}, \qquad (\delta \mathbf{V}_r)_{lk}.$$

Furthermore, use of (3.106) in the formula for T_1 shows that the contribution to the metric form from these differentials is

$$\sum_{\substack{s=\text{ real part}\\ \text{imag. part}}} \sum_{k=1}^{m} \sum_{l=m+1}^{n} \Big(2((\delta \mathbf{U}_r)_{lk}^{(s)})^2 + 2((\delta \mathbf{V}_r)_{lk}^{(s)})^2 + 2c_k (\delta \mathbf{U}_r)_{lk}^{(s)} (\delta \mathbf{V}_r)_{lk}^{(s)} \Big),$$

where

$$c_k = -2\sqrt{1-\lambda_k^2}.$$

This contributes to $(\det[g_{jk}])^{1/2}$ in (1.15) the factor

$$\prod_{k=1}^{m} \prod_{l=m+1}^{n} \begin{vmatrix} 2 & -2\sqrt{1-\lambda_k^2} \\ -2\sqrt{1-\lambda_k^2} & 2 \end{vmatrix} \propto \prod_{k=1}^{m} \lambda_k^{2(n-m)}. \tag{3.108}$$

It remains to calculate the contribution from the differentials (3.103). Because the coefficient of $|(\delta \mathbf{V}_r)_{kk} - (\delta \mathbf{U}_r)_{kk}|^2$ for $m+1 \le k \le n$ is independent of the λ's, we can restrict attention to $1 \le k \le m$ in all the differentials of (3.103). In this range of k values, the first term in T_2 couples together the three differentials in (3.103). Thus making use of a formula analogous to (3.106), and recalling that only the imaginary parts of (3.103) are nonzero, we have that the contribution to the metric form is

$$\sum_{k=1}^{m} (\delta \mathbf{X}_k)^2 + (\delta \mathbf{Y}_k)^2 + 2\lambda_k^2 (\delta \mathbf{Z}_k)^2 - 2\lambda_k^2 \delta \mathbf{X}_k \delta \mathbf{Z}_k - 2\lambda_k^2 \delta \mathbf{Y}_k \delta \mathbf{Z}_k + 2\lambda_k^2 \delta \mathbf{X}_k \delta \mathbf{Y}_k,$$

where

$$\delta \mathbf{X}_k := \mathrm{Im}\Big((\delta \mathbf{V}_r)_{kk} - (\delta \mathbf{U}_r)_{kk} \Big), \quad \delta \mathbf{Y}_k := \mathrm{Im}\Big((\delta \mathbf{V}_{r'})_{kk} - (\delta \mathbf{U}_{r'})_{kk} \Big), \quad \delta \mathbf{Z}_k := \mathrm{Im}\Big((\delta \mathbf{V}_r)_{kk} - (\delta \mathbf{U}_{r'})_{kk} \Big).$$

This contributes to $(\det[g_{jk}])^{1/2}$ in (1.15) the factor

$$\prod_{k=1}^m \begin{vmatrix} 1 & \lambda_k^2 & -\lambda_k^2 \\ \lambda_k^2 & 1 & -\lambda_k^2 \\ -\lambda_k^2 & -\lambda_k^2 & 2\lambda_k^2 \end{vmatrix}^{1/2} \propto \prod_{k=1}^m \lambda_k (1-\lambda_k^2)^{1/2}. \tag{3.109}$$

Finally, the contribution to $(\det[g_{jk}])^{1/2}$ in (1.15) from the coefficients of the terms $(d\lambda_j)^2$ in T_5 is

$$\prod_{k=1}^m \frac{1}{(1-\lambda_k^2)^{1/2}}. \tag{3.110}$$

Multiplying together (3.107)–(3.110) gives (3.102) in the case $\beta = 2$. □

Changing variables $\lambda_j^2 \mapsto x_j$ in (3.102), where the x_j are the eigenvalues $\mathbf{t}^\dagger \mathbf{t}$, gives a p.d.f. proportional to

$$\prod_{j=1}^m x_j^{\beta a/2} \prod_{1 \le j < k \le m} |x_k - x_j|^\beta, \qquad x_j \in [0,1], \tag{3.111}$$

where $a = n - m + 1 - 2/\beta$ as in Proposition 3.2.2. With $y_j := 1 - 2x_j$, the p.d.f. (3.111) is an example of the Jacobi ensemble (3.74).

3.8.3 Wishart matrix of a rectangular sub-block of $U(N)$

In the case $\beta = 2$ the distribution of the Wishart matrix of a general $n_1 \times n_2$ rectangular block of elements from $\mathbf{S}$ can be determined [536], [487], [197]. Due to the invariance of the underlying Haar measure for $U(N)$ with respect to interchanges of rows and columns, this block can be taken to be in the top left corner.

PROPOSITION 3.8.2 *Let $\mathbf{S} \in U(N)$, and let $\mathbf{A}$ denote the $n_1 \times n_2$ block of elements in the top left. For $n_1 \ge n_2$, $N \ge n_1 + n_2$, the distribution of $\mathbf{Y} := \mathbf{A}^\dagger \mathbf{A}$ is proportional to*

$$(\det \mathbf{Y})^{(n_1 - n_2)} \Big(\det(\mathbf{1}_{n_2} - \mathbf{Y}) \Big)^{(N - n_1 - n_2)}.$$

Proof. Let $\mathbf{C}$ denote the block of elements in the first n_1 rows and last $N - n_2$ rows not in $\mathbf{A}$. Then the unitarity of $\mathbf{S}$ implies

$$\mathbf{A}\mathbf{A}^\dagger + \mathbf{C}\mathbf{C}^\dagger = \mathbf{1}_{n_1}.$$

We follow [559] and regard this as a constraint in the space of general complex rectangular matrices $\mathbf{A}$, $\mathbf{C}$. Up to normalization the distribution of $\mathbf{A}$ is then given by

$$\int \delta(\mathbf{A}\mathbf{A}^\dagger + \mathbf{C}\mathbf{C}^\dagger - \mathbf{1}_{n_2})(d\mathbf{C}) \propto \int (d\mathbf{C}) \int e^{-i\mathrm{Tr}(\mathbf{H}(\mathbf{A}\mathbf{A}^\dagger + \mathbf{C}\mathbf{C}^\dagger - \mathbf{1}_{n_2}))} (d\mathbf{H}), \tag{3.112}$$

where the second expression follows upon using (3.27). Following [252], we would like now to interchange the order of integration, and integrate first over $\mathbf{C}$. To get an answer which is integrable around $\mathbf{H} = \mathbf{0}$, we replace $\mathbf{H}$ in the exponent by $\mathbf{H} - i\mu \mathbf{1}_m$, $\mu > 0$. Computing the Gaussian integrals by taking $\mathbf{H}$ to be diagonal then shows that up to a multiplicative constant (3.112) reduces to

$$\lim_{\mu \to 0^+} \int \Big(\det(\mathbf{H} - i\mu \mathbf{1}_m) \Big)^{-(N - n_2)} e^{i\mathrm{Tr}(\mathbf{H}(\mathbf{1}_{n_1} - \mathbf{A}\mathbf{A}^\dagger))} (d\mathbf{H}).$$

Evaluating this via the matrix integral (3.24) shows that the distribution of $\mathbf{A}$ is proportional to

$$(\det(\mathbf{1}_{n_1} - \mathbf{A}\mathbf{A}^\dagger))^{(N - n_1 - n_2)} = (\det(\mathbf{1}_{n_2} - \mathbf{A}^\dagger \mathbf{A}))^{(N - n_1 - n_2)}.$$

The result now follows from (3.23) with $\mathbf{X} \mapsto \mathbf{A}$, $\mathbf{A} \mapsto \mathbf{Y}$. □

The corresponding eigenvalue p.d.f. is an example of (3.74) with $y_j := 1 - 2x_j$, $a = n_1 - n_2$, $b = N - n_1 - n_2$ and $\beta = 2$.

The derivation of Proposition 3.8.2 can be extended to the cases in which $\mathbf{S} \in O(N)$, or $\mathbf{S} \in \mathrm{Sp}(2N)$ [197] (see also Exercises 3.8 q.2). In particular, with $\mathbf{A}$ denoting the $n_1 \times n_2$ block of elements in the top left corner, the distribution of $\mathbf{A}$ is found to be proportional to

$$(\det(\mathbf{1}_{n_2} - \mathbf{A}^\dagger \mathbf{A}))^{(\beta/2)(N-n_1-n_2+1-2/\beta)} \tag{3.113}$$

and thus the distribution of $\mathbf{Y} := \mathbf{A}^\dagger \mathbf{A}$ is proportional to

$$(\det \mathbf{Y})^{(\beta/2)(n_1-n_2+1-2/\beta)} \Big(\det(\mathbf{1}_{n_2} - \mathbf{Y}) \Big)^{(\beta/2)(N-n_1-n_2+1-2/\beta)}.$$

The eigenvalue p.d.f. of the latter is an example of (3.74) with $y_j := 1 - 2x_j$,

$$a = (\beta/2)(n_1 - n_2 + 1 - 2/\beta), \qquad b = (\beta/2)(N - n_1 - n_2 + 1 - 2/\beta).$$

EXERCISES 3.8 1. (i) With $\mathbf{S}$ given by (3.96) write down the six equations implied by the unitarity condition $\mathbf{S}\mathbf{S}^\dagger = \mathbf{S}^\dagger\mathbf{S} = \mathbf{1}_{2N}$.

(ii) With the singular value decompositions $\mathbf{r} = \mathbf{U}_r \mathbf{\Lambda}_r \mathbf{V}_r$ etc. show that the equations in (i) allow a solution implied by (3.99).

(iii) From (3.96) and the unitarity of $\mathbf{S}$ derive the equations $\mathbf{r}\mathbf{t}^\dagger = -\mathbf{t}'\mathbf{r}'^\dagger$ and $\mathbf{t}'^\dagger\mathbf{r} = -\mathbf{r}'^\dagger\mathbf{t}$ and thus deduce that $\mathrm{Tr}(\mathbf{t}\mathbf{t}^\dagger) = \mathrm{Tr}(\mathbf{t}'\mathbf{t}'^\dagger)$.

2. [504] This exercise considers properties of the transfer matrix associated with a scattering problem.

(i) Consider the quantum conductance problem described in terms of the scattering matrix $\mathbf{S}$ by (3.95), and suppose $n = m = N$. The $2N \times 2N$ transfer matrix $\tilde{\mathbf{M}}$ relates the states at the left-hand edge to those at the right hand edge,

$$\tilde{\mathbf{M}} \begin{bmatrix} \vec{I} \\ \vec{O} \end{bmatrix} = \begin{bmatrix} \vec{I}' \\ \vec{O}' \end{bmatrix}.$$

Show from the unitarity of $\mathbf{S}$ that in the notation below (3.94) $\tilde{\mathbf{M}} \in U(N,N)$, and use this to show that if λ is an eigenvalue of $\tilde{\mathbf{M}}$, then so is $1/\lambda$.

(ii) Use the decomposition (3.96) in the case $n = m = N$ to show

$$\tilde{\mathbf{M}} = \begin{bmatrix} (\mathbf{t}^\dagger)^{-1} & \mathbf{r}\mathbf{t}'^{-1} \\ -\mathbf{t}'^{-1}\mathbf{r} & \mathbf{t}'^{-1} \end{bmatrix}.$$

From this deduce that with the eigenvalues of $\mathbf{t}^\dagger\mathbf{t}$ denoted $T_1, \dots, T_N$, $\mathbf{\Lambda} = \mathrm{diag}(\lambda_1, \dots, \lambda_N)$, $\lambda_i := (1 - T_i)/T_i$, and $\mathbf{L}$ as in (3.99) $\tilde{\mathbf{M}}$ permits the parametrization

$$\tilde{\mathbf{M}} = \begin{bmatrix} \mathbf{V}_r^\dagger & \mathbf{0} \\ \mathbf{0} & \mathbf{V}_{r'} \end{bmatrix} \mathbf{L} \begin{bmatrix} \mathbf{U}_r & \mathbf{0} \\ \mathbf{0} & \mathbf{U}_{r'}^\dagger \end{bmatrix}, \tag{3.114}$$

and hence show

$$\Big(2 + \tilde{\mathbf{M}}\tilde{\mathbf{M}}^\dagger + (\tilde{\mathbf{M}}\tilde{\mathbf{M}}^\dagger)^{-1}\Big)^{-1} = \frac{1}{4} \begin{bmatrix} \mathbf{t}\mathbf{t}^\dagger & \mathbf{0} \\ \mathbf{0} & \mathbf{t}'^\dagger\mathbf{t}' \end{bmatrix}.$$

Conclude that, with the eigenvalues of $\mathbf{M}$ denoted $e^{\pm 2x_j}$ $(j = 1, \dots, N)$,

$$\frac{1}{\cosh^2 x_j} = T_j. \tag{3.115}$$

3. Read off from Proposition 3.8.1 that, for $N \times N$ matrices $\mathbf{U}$ from the COE ($\beta = 1$), CUE ($\beta = 2$), CSE ($\beta = 4$), the modulus squared $|U_{NN}|^2$ of the bottom right element has distribution proportional to

$$(1 - t)^{(\beta/2)(N-1-2/\beta)}, \qquad 0 \le t \le 1,$$

and thus, making use of (4.2) below, compute that

$$\langle |U_{NN}|^{2p}\rangle = \frac{\Gamma((\beta/2)(N-1)+1)\Gamma(p+1)}{\Gamma((\beta/2)(N-1)+p+1)}.$$

3.9 A CIRCULAR JACOBI ENSEMBLE

There is a mathematical transformation which relates the normalization of (3.74) for the Jacobi ensemble to the normalization of the p.d.f. in Proposition 2.2.5 for the circular ensemble, and more generally leads to the consideration of a circular Jacobi ensemble. The transformation is accomplished by the following general result [192].

PROPOSITION 3.9.1 *Let $f(t_1,\dots,t_N)$ be a Laurent polynomial in $t_1,\dots,t_N$. For $\mathrm{Re}(\epsilon)$ large enough so that the r.h.s. exists,*

$$\int_{-1/2}^{1/2} d\theta_1 \cdots \int_{-1/2}^{1/2} d\theta_N \prod_{l=1}^N e^{2\pi i\theta_l \epsilon} f(-e^{2\pi i\theta_1},\dots,-e^{2\pi i\theta_N})$$
$$= \Big(\frac{\sin \pi\epsilon}{\pi}\Big)^N \int_0^1 dt_1 \cdots \int_0^1 dt_N \prod_{l=1}^N t_l^{-1+\epsilon} f(t_1,\dots,t_N).$$

This result follows immediately from term-by-term integration of the Laurent polynomial on both sides. Note that for ϵ an integer the Fourier integral is equal to

$$\mathrm{CT}_{\{t_1,\dots,t_N\}} \prod_{l=1}^N t_l^\epsilon f(-t_1,\dots,-t_N)$$

where $\mathrm{CT}_{\{t_1,\dots,t_N\}}$ denotes the constant term, i.e. the term independent of $t_1,\dots,t_N$ in the Laurent expansion.

Writing $t_j = \frac{1}{2}(1-y_j)$ in (3.74) shows that

$$C_{ab\beta N} = 2^{N+(a+b)\beta N/2+\beta N(N-1)/2} S_N(a\beta/2, b\beta/2, \beta/2), \tag{3.116}$$

where

$$S_N(\lambda_1,\lambda_2,\lambda) := \int_0^1 dt_1 \cdots \int_0^1 dt_N \prod_{l=1}^N t_l^{\lambda_1}(1-t_l)^{\lambda_2} \prod_{1\le j<k\le N} |t_k - t_j|^{2\lambda}. \tag{3.117}$$

The integral S_N is known as the *Selberg integral*. The change of variables formula (3.76) gives the trigonometric form

$$2^{(2\lambda_1+2\lambda_2+2)N+2\lambda N(N-1)} S_N(\lambda_1,\lambda_2,\lambda) = \int_0^\pi d\theta_1 \cdots \int_0^\pi d\theta_N \prod_{l=1}^N |1-e^{i\theta_l}|^{2\lambda_1+1}$$
$$\times |1+e^{i\theta_l}|^{2\lambda_2+1} \prod_{1\le j<k\le N} |e^{i\theta_j} - e^{i\theta_k}|^{2\lambda} |1-e^{i(\theta_j+\theta_k)}|^{2\lambda}. \tag{3.118}$$

Comparison with (2.102) and (2.103) shows that S_N is thus evaluated by the product of gamma functions as specified in (2.103). This topic will be further developed in the next chapter. Our interest here is that S_N is of the class of integrals for which Proposition 3.9.1 applies. Thus with λ, λ_2 non-negative integers we have

$$S_N(\lambda_1,\lambda_2,\lambda) = (-1)^{N+N(N-1)\lambda/2} \Big(\frac{\pi}{\sin \pi b}\Big)^N M_N(a,b,\lambda), \tag{3.119}$$

where

$$M_N(a,b,\lambda)$$
$$:= \int_{-1/2}^{1/2} d\theta_1 \cdots \int_{-1/2}^{1/2} d\theta_N \prod_{l=1}^N e^{\pi i\theta_l(a-b)}|1+e^{2\pi i\theta_l}|^{a+b} \prod_{1\le j<k\le N} |e^{2\pi i\theta_k} - e^{2\pi i\theta_j}|^{2\lambda}, \quad (3.120)$$

with $-1-b-\lambda(N-1) = \lambda_1$ and $a+b = \lambda_2$. This trigonometric integral is often associated with Morris [407], although it was known to Selberg [227]. For a, b, λ non-negative integers it can be written in the equivalent form

$$M_N(a,b,\lambda) = \mathrm{CT}_{\{t_1,\dots,t_N\}} \prod_{l=1}^N (1-t_l)^a (1-\frac{1}{t_l})^b \prod_{1\le j<k\le N} \Big(1-\frac{t_j}{t_k}\Big)^\lambda \Big(1-\frac{t_k}{t_j}\Big)^\lambda. \quad (3.121)$$

In the special case $a = b$ the integrand of (3.120) is real, and after making the replacement $2\pi\theta_l \mapsto \theta_l$ reads

$$\prod_{l=1}^N |1+e^{i\theta_l}|^{2b} \prod_{1\le j<k\le N} |e^{i\theta_k} - e^{i\theta_j}|^{2\lambda}, \qquad -\pi < \theta_j \le \pi. \quad (3.122)$$

This is a generalization of the p.d.f. of Proposition 2.2.5 for the circular ensemble. When properly normalized, (3.122) will be referred to as defining the *circular Jacobi ensemble* [414]. In a random matrix context, (3.122) with $b = \lambda$, $2\lambda = \beta$ is the eigenvalue p.d.f. in the circular ensemble given that there is an eigenvalue at $\theta = \pi$. For general b it is said to describe a *spectrum singularity*. From the log-gas viewpoint, we see from Proposition 1.4.1 that (3.122) is proportional to the Boltzmann factor for a one-component log-potential system on a circle at coupling $\beta = 2\lambda$, with an impurity charge of strength $q' = b/\lambda$ fixed at the angle π on the boundary of the circle.

Applying the mapping (2.50) we see that

$$\prod_{l=1}^N |1+e^{i\theta_l}|^{2b} \prod_{1\le j<k\le N} |e^{i\theta_k} - e^{i\theta_j}|^{2\lambda} d\theta_1\cdots d\theta_N$$
$$= 2^{\lambda N(N-1)+N(1+2b)} \prod_{j=1}^N (1+\lambda_j^2)^{-\alpha} \prod_{1\le j<k\le N} |\lambda_k - \lambda_j|^{2\lambda} d\lambda_1\cdots d\lambda_N, \quad \alpha = \lambda(N-1)+1+b, \quad (3.123)$$

so the circular Jacobi ensemble is in fact equivalent to the Cauchy ensemble. The most general choice of parameters for which the integrand in (3.120) is real is $a = \bar{b} =: b_1 - ib_2$. We then have the (unnormalized) probability

$$\prod_{l=1}^N e^{b_2\theta_l}|1+e^{i\theta_l}|^{2b_1} \prod_{1\le j<k\le N} |e^{i\theta_k} - e^{i\theta_j}|^{2\lambda} d\theta_1\cdots d\theta_N = 2^{\lambda N(N-1)+N(1+2\mathrm{Re}(b))}$$
$$\times \prod_{j=1}^N \frac{1}{(1+i\lambda_j)^{b+1+\lambda(N-1)}(1-i\lambda_j)^{\bar{b}+1+\lambda(N-1)}} \prod_{1\le j<k\le N} |\lambda_k-\lambda_j|^{2\lambda} d\lambda_1\cdots d\lambda_N, \quad (3.124)$$

where the r.h.s. results from the mapping (2.50). This generalizes the circular Jacobi ensemble and Cauchy ensemble, respectively.

EXERCISES 3.9 1. [225] Show that for $q_1, q_2 > 0$ the minimum of

$$-q_1 \sum_{k=1}^{2N} \log|1 - e^{i\theta_k}| - q_2 \sum_{k=1}^{2N} \log|1 + e^{i\theta_k}| - \sum_{1 \le j < k \le 2N} \log|e^{i\theta_k} - e^{i\theta_j}|$$

subject to the requirement that

$$0 < \theta_j < \pi \ (j = 1, \dots, N), \qquad \pi < \theta_j < 2\pi \ (j = N+1, \dots, 2N)$$

occurs at the zeros of the Jacobi polynomial $P_N^{(q_1-1/2, q_2-1/2)}(\cos\theta)$, $0 < \theta < 2\pi$.

3.10 LAGUERRE β-ENSEMBLE

Analogous to the Gaussian β-ensemble of Section 1.9, it is possible to construct random tridiagonal matrices which have as their eigenvalue p.d.f. (3.16) for general $\beta > 0$ [140]. To gain insight into their form, consider the effect of the Householder reduction (1.139) on matrices $\mathbf{X}^T\mathbf{X}$, with $\mathbf{X}$ an $n \times m$ ($n \ge m$) rectangular matrix of standard Gaussians. For this purpose, one constructs two sequences of Householder reflector matrices (1.140) $\{\tilde{\mathbf{U}}^{(j)}\}_{j=1,\dots,m-2}$, $\{\mathbf{U}^{(j)}\}_{j=1,\dots,m-1}$, where each $\tilde{\mathbf{U}}^{(j)}$ is $n \times n$ and $\mathbf{U}^{(j)}$ is $m \times m$, such that

$$\mathbf{U}^{(m-2)T} \cdots \mathbf{U}^{(1)T}\mathbf{X}^T\tilde{\mathbf{U}}^{(1)T} \cdots \tilde{\mathbf{U}}^{(m-1)T} \tag{3.125}$$

is a bidiagonal matrix. This is done by first choosing $\tilde{\mathbf{U}}^{(1)}$ so that $\mathbf{X}^T\tilde{\mathbf{U}}^{(1)T}$ has zeros in the first row for $k \ge 2$. Because $\tilde{\mathbf{U}}^{(1)T}$ is a projection, the norm of the first row of $\mathbf{X}^T\tilde{\mathbf{U}}^{(1)T}$ must be conserved and so the distribution of $(\mathbf{X}^T\tilde{\mathbf{U}}^{(1)T})_{11}$ is χ_n^2, where χ_n^2 is the particular gamma distribution $\Gamma[n/2, 2]$ (cf. $\tilde{\chi}_k$ as specified in Proposition 1.9.1; by definition the gamma distribution $\Gamma[s,\sigma]$ has p.d.f. proportional to $(x/\sigma)^{s/2-1}e^{-x/\sigma}$). The elements of $\mathbf{U}^{(0)T}$ are independent of the rows $2, \dots, m-1$ in $\mathbf{X}^T$, so the distribution of the elements of these columns is unchanged. Now one constructs $\mathbf{U}^{(1)T}$ so that $\mathbf{U}^{(1)T}\mathbf{X}^T\tilde{\mathbf{U}}^{(1)T}$ has zeros in the first column for $j \ge 3$. Because of the structure of $\mathbf{U}^{(1)T}$ from (1.140) with $\mathbf{V}$ therein a projector, it leaves invariant the norm of the entries $2, \dots, m$ of the first column of $\mathbf{X}^T\tilde{\mathbf{U}}^{(1)T}$ (which as remarked above has the same distribution as the corresponding elements of $\mathbf{X}^T$), and consequently the distribution of $(\mathbf{U}^{(1)T}\mathbf{X}^T\tilde{\mathbf{U}}^{(1)T})_{12}$ is χ_{m-1}^2.

Proceeding inductively shows that the real Gaussian $m \times n$ matrix $\mathbf{X}^T$ is mapped via (3.125) to the bidiagonal matrix [486]

$$\mathbf{B}_1^T := \begin{bmatrix} \chi_n & & & \\ \chi_{m-1} & \chi_{n-1} & & \\ & \ddots & \ddots & \\ & & \chi_1 & \chi_{n-m+1} \end{bmatrix}. \tag{3.126}$$

It follows immediately that the $m \times m$ random tridiagonal matrix $\mathbf{B}_1^T\mathbf{B}_1$ has the same eigenvalue p.d.f. as real Wishart matrices, and is thus given by (3.16) with $\beta = 1$. As in going from Proposition 1.9.1 to Proposition 1.9.4 in the Gaussian case, there is a simple generalization of (3.126) which leads to the eigenvalue p.d.f. (3.16) for general $\beta > 0$.

PROPOSITION 3.10.1 *Define the real $m \times m$ bidiagonal random matrix*

$$\mathbf{B}_\beta^T := \begin{bmatrix} \chi_{\beta n} & & & \\ \chi_{\beta(m-1)} & \chi_{\beta(n-1)} & & \\ & \ddots & \ddots & \\ & & \chi_\beta & \chi_{\beta(n-m+1)} \end{bmatrix}, \tag{3.127}$$

where $n > m-1$ and n may be real, and set $\mathbf{T}_\beta := \mathbf{B}_\beta^T \mathbf{B}_\beta$. The eigenvalues and the first component of the eigenvectors (which form the vector $\vec{q}$) are independent, with the distribution of the former given by

$$\frac{1}{\widetilde{W}_{a,\beta,m}} \prod_{j=1}^m \lambda_j^{\beta a/2} e^{-\lambda_j/2} \prod_{1\le j<k\le m} |\lambda_k - \lambda_j|^\beta, \qquad \lambda_j \ge 0, \tag{3.128}$$

where $a = n - m + 1 - 2/\beta$ and

$$\widetilde{W}_{a,\beta,m} = 2^{m(a\beta/2+1+(m-1)\beta/2)} \prod_{j=1}^m \frac{\Gamma(1+\beta j/2)\Gamma(1+\beta a/2+\beta(j-1)/2)}{\Gamma(1+\beta/2)} \tag{3.129}$$

while the distribution of the latter is given by (1.161) with $N \mapsto m$.

Proof. Let us write

$$\mathbf{B}_\beta^T := \begin{bmatrix} x_n & & & \\ y_{m-1} & x_{n-1} & & \\ & \ddots & \ddots & \\ & & y_1 & x_{n-m+1} \end{bmatrix}, \quad \mathbf{T}_\beta := \begin{bmatrix} a_m & b_{m-1} & & & \\ b_{m-1} & a_{m-1} & b_{m-2} & & \\ \ddots & \ddots & \ddots & & \\ & & b_2 & a_2 & b_1 \\ & & & b_1 & a_1 \end{bmatrix}. \tag{3.130}$$

Now, by definition, the probability measure of the matrix $\mathbf{B}_\beta$ is

$$P(\mathbf{B}_\beta)(d\mathbf{B}_\beta) = c_{m,n} \prod_{i=0}^{m-1} x_{n-i}^{\beta(n-i)-1} e^{-x_{n-i}^2/2} \prod_{k=1}^{m-1} y_k^{\beta k-1} e^{-y_k^2/2} d\vec{x} \wedge d\vec{y}, \tag{3.131}$$

$$c_{m,n} := 2^{-(2m-1)} \prod_{j=1}^{m-1} \Gamma(j\beta/2) \prod_{k=1}^{m} \Gamma((n-k+1)\beta/2),$$

where $d\vec{x} := \bigwedge_{i=1}^n dx_{n-m+i}$, $d\vec{y} := \bigwedge_{i=1}^{m-1} dy_i$. Our first task is to introduce the variables of the matrix $\mathbf{T}_\beta$. From (3.130) we see

$$a_m = x_n^2, \quad a_i = y_i^2 + x_{n-m+i}^2, \quad b_i = y_i x_{n-m+i+1} \quad (i = m-1, m-2, \dots, 1). \tag{3.132}$$

To compute the Jacobian for the change of variables from $\{x_i, y_j\} \mapsto \{a_i, b_j\}$ we note from these equations that

$$\begin{aligned} da_m &= 2x_n dx_n, \\ da_i &= 2(x_{n-m+i} dx_{n-m+i} + y_i dy_i), \\ db_i &= x_{n-m+i+1} dy_i + y_i dx_{n-m+i+1} \ (i = m-1, \dots, 1). \end{aligned}$$

Taking the wedge product of both sides of these equations shows

$$d\vec{x} \wedge d\vec{y} = \Big(2^n x_{n-m+1} \prod_{i=0}^{m-2} x_{n-i}^2\Big)^{-1} d\vec{a} \wedge d\vec{b}, \tag{3.133}$$

where $d\vec{x}$, $d\vec{y}$ are as in (3.131) and $d\vec{a} := \bigwedge_{i=1}^m da_i$, $d\vec{b} := \bigwedge_{i=1}^{m-1} db_i$.

We now substitute (3.133) in (3.131), and then change variables $\{a_i, b_j\} \mapsto \{\lambda_i, q_j\}$ using Proposition 1.9.3 to deduce

$$(d\mathbf{B}_\beta) = 2^{-n} c_{m,n} e^{-\sum_{i=0}^{m-1} x_{n-i}^2/2} e^{-\sum_{i=1}^{m-1} y_i^2/2} \frac{\prod_{i=0}^{m-1} x_{n-i}^{\beta(n-i)-2} \prod_{i=1}^{m-1} y_i^{\beta i}}{q_m \prod_{i=1}^m q_i}.$$

Use of (1.148) and the fact that $b_i = y_i x_{n-m+i+1}$ shows

$$\prod_{1\le i<j\le m} |\lambda_i - \lambda_j| = \frac{\prod_{i=1}^{m-1} b_i^i}{\prod_{i=1}^{m} q_i} = \frac{\prod_{i=0}^{m-1} y_i^i x_{n-i}^{m-1-i}}{\prod_{i=1}^{m} q_i}$$

and thus we have

$$(d\mathbf{B}_\beta) = 2^{-n} c_{m,n} e^{-\sum_{i=0}^{m-1} x_{n-i}^2/2} e^{-\sum_{i=1}^{m-1} y_i^2/2} \times \prod_{1\le i<j\le m} |\lambda_i - \lambda_j|^\beta \frac{1}{q_m} \prod_{i=1}^{m} q_i^{\beta-1} \Big(\prod_{i=0}^{m-1} x_{n-i} \Big)^{\beta(n-m+1)-2} d\vec{q} \wedge d\vec{\lambda}.$$

Finally, we note that

$$\sum_{i=0}^{m-1} x_{n-i}^2 + \sum_{i=1}^{m-1} y_i^2 = \operatorname{Tr} \mathbf{T}_\beta = \sum_{i=1}^{m} \lambda_i, \qquad \prod_{i=0}^{m-1} x_{n-i}^2 = \det \mathbf{T}_\beta = \prod_{i=1}^{m} \lambda_i$$

and make use of the normalization in (1.161) to deduce the stated result. □

Note that the evaluation of the normalization (3.129) implies that the normalization $W_{a\beta m}$ in (3.16) has the evaluation

$$W_{a\beta m} = (\beta/2)^{-m(a\beta/2+1+(m-1)\beta/2} \prod_{j=0}^{m-1} \frac{\Gamma(1+\beta(j+1)/2)\Gamma(1+\beta a/2+\beta j/2)}{\Gamma(1+\beta/2)}. \tag{3.134}$$

Furthermore, the three-term recurrence (1.154) with $k = 1, 2, \dots, m$ and $\{a_k\}$, $\{b_k\}$ specified by (3.132) can be used to give a rapid computation of the characteristic polynomial $p_m(\lambda)$. By calculating the zeros we can then sample from (3.128), and the eigenvalues histogrammed according to the method of Section 1.9.3.

3.11 JACOBI β-ENSEMBLE

We have seen that the p.d.f.'s (1.160) and (3.128), specifying the Gaussian and Laguerre β-ensembles, respectively, can be realized by the eigenvalues of certain tridiagonal matrices. Equivalently, the characteristic polynomials for the p.d.f.'s can be generated from the three-term recurrence (1.154) with the a_k, b_k therein chosen according to certain distributions. For the Jacobi β-ensemble (3.74) the same is true [356]. This is a corollary of Proposition 2.9.4 for the real orthogonal β-ensemble.

PROPOSITION 3.11.1 *For $k = 0, \dots, 2n-2$, in terms of the notation of Definition 2.9.1, let*

$$\alpha_k \in \begin{cases} \tilde{B}\Big[\frac{2n-k-2}{4}\beta + a + 1, \frac{2n-k-2}{4}\beta + b + 1\Big], & k \text{ even}, \\ \tilde{B}\Big[\frac{2n-k-2}{4}\beta + a + 1, \frac{2n-k-2}{4}\beta\Big], & k \text{ odd}, \end{cases}$$

and set $\alpha_{2n-1} = \alpha_{-1} = -1$. The eigenvalues of the tridiagonal matrix (1.143) with

$$\begin{aligned} a_{n-k} &= (1-\alpha_{2k-1})\alpha_{2k} - (1+\alpha_{2k-1})\alpha_{2k-1}, \\ b_{n-k} &= \{(1-\alpha_{2k-1})(1-\alpha_{2k}^2)(1+\alpha_{2k+1})\}^{1/2} \end{aligned} \tag{3.135}$$

are distributed according to the Jacobi β-ensemble (3.74) with $a\beta/2 \mapsto a$, $b\beta/2 \mapsto b$, $N \mapsto n$.

Proof. With the change of variables

$$x_l = \cos\theta_l, \tag{3.136}$$

(2.102) becomes the Jacobi β-ensemble (3.74) with $a\beta/2 \mapsto a$, $b\beta/2 \mapsto b$. This transformation maps the eigenvalues from the unit half circle to the interval $(-1, 1)$ (recall (3.76)). A result of Szegö in orthogonal polynomial theory implies that for each real upper Hessenberg $2N \times 2N$ matrix with unit determinant, there is a tridiagonal matrix of the form (1.143) with eigenvalues related by (3.136). Moreover, the parameters of the tridiagonal matrix are related to those of the Hessenberg matrix by (3.135), which are known as the *Geronimus relations* [261]. Thus we see that Proposition 3.11.1 follows as a corollary of Proposition 2.9.5. □

EXERCISES 3.11 1. [159] Given $\Theta = (\theta_n, \dots, \theta_1)$, $\Phi = (\phi_{n-1}, \dots, \phi_1)$, define four $n \times n$ bidiagonal matrices $\mathbf{B}_{11}, \mathbf{B}_{12}, \mathbf{B}_{21}, \mathbf{B}_{22}$ by

$$\mathbf{B} = \begin{bmatrix} \mathbf{B}_{11} & \mathbf{B}_{12} \\ \mathbf{B}_{21} & \mathbf{B}_{22} \end{bmatrix}$$

$$= \left[\begin{array}{cccc|cccc} c_n & -s_n c'_{n-1} & & & s_n s'_{n-1} & & & \\ & c_{n-1} s'_{n-1} & \ddots & & c_{n-1} c'_{n-1} & s_{n-1} s'_{n-2} & & \\ & & \ddots & -s_2 c'_1 & & \ddots & \ddots & \\ & & & c_1 s'_1 & & & c_1 c'_1 & s_1 \\ \hline -s_n & -c_n c'_{n-1} & & & c_n s'_{n-1} & & & \\ & -s_{n-1} s'_{n-1} & \ddots & & -s_{n-1} c'_{n-1} & c_{n-1} s'_{n-2} & & \\ & & \ddots & -c_2 c'_1 & & \ddots & \ddots & \\ & & & -s_1 s'_1 & & & -s_1 c'_1 & c_1 \end{array}\right]$$

where $c_i = \cos\theta_i$, $s_i = \sin\theta_i$, $c'_i = \cos\phi_i$, $s'_i = \sin\theta_i$.

(i) Show that $\mathbf{B}$ is orthogonal.

(ii) Decompose $\mathbf{B}$ into the form (3.98) with $n = m$, write $\mathbf{\Gamma}_r = \mathrm{diag}(\sigma_1, \dots, \sigma_n)$ for the singular values of $\mathbf{B}_{11}^T \mathbf{B}_{11}$, and denote the first entries of the corresponding normalized eigenvectors (constrained to be positive) by $(v_1, \dots, v_n)$. Use the working of the proof of Proposition 3.10.1 to show that

$$\prod_{j=1}^{n} \Big(c_j^{\beta(j-1)+1} s_j^{\beta(j-1)} dc_j \Big) \prod_{j=1}^{n-1} \Big((c'_j)^{\beta j-1} (s'_j)^{\beta(j-1)+2} dc'_j \Big),$$
$$= \prod_{j<k} |\sigma_j^2 - \sigma_k^2|^{\beta} \Big(\prod_{j=1}^{n} \sigma_j \, d\sigma_j \Big) \Big(\prod_{j=1}^{n-1} v_j^{\beta-1} dv_j \Big).$$

(iii) With

$$\cos^2\theta_j \in \mathrm{B}[\beta(a+j)/2, \beta(b+j)/2] \qquad (j = n, \dots, 1)$$
$$\cos^2\phi_j \in \mathrm{B}[\beta j/2, \beta(a+b+1+j)/2] \qquad (j = n-1, \dots, 1),$$

use the above result to show that $\{\sigma_j^2\}$ are distributed according to the Jacobi β-ensemble

$$\frac{1}{C} \prod_{j=1}^{n} \lambda_j^{\beta(a+1)/2-1} (1-\lambda_j)^{\beta(b+1)/2-1} \prod_{1 \le j < k \le n} |\lambda_k - \lambda_j|^{\beta}.$$

3.12 CIRCULAR JACOBI β-ENSEMBLE

In Section 2.8 a random unitary Hessenberg matrix having eigenvalue p.d.f. realizing the circular β-ensemble was specified. Here, following [93], it will be shown how to modify that construction so the eigenvalue p.d.f. of the circular Jacobi β-ensemble (3.122) is realized. With $\{\alpha_j\}$ relating to a unitary Hessenberg matrix as in Proposition 2.8.1, we begin by introducing some related variables.

DEFINITION 3.12.1 *Set*

$$\gamma_k = \bar{\alpha}_k \prod_{j=0}^{k-1} \frac{1-\bar{\gamma}_j}{1-\gamma_j}, \qquad k = 0, \dots, N-1. \tag{3.137}$$

We now examine the significance of the quantities (3.137). In (2.88), for $k = 1, \dots, N-1$ introduce $\{z_l^{(k)}\}_{l=1,\dots,k}$ so that

$$b_k(\lambda) := \frac{\chi_k(\lambda)}{\tilde{\chi}_k(\lambda)} = \prod_{j=1}^{k} \frac{(\lambda - z_j^{(k)})}{(1-\lambda \bar{z}_j)}. \tag{3.138}$$

Using the definition in (3.138), it follows from (2.88) that

$$\gamma_k(\lambda) := \lambda - \frac{\chi_{k+1}(\lambda)}{\chi_k(\lambda)} = \frac{\bar{\alpha}_k}{b_k(\lambda)}. \tag{3.139}$$

It follows from this definition that

$$\chi_k(\lambda) = \prod_{j=0}^{k-1} (\lambda - \gamma_j(\lambda)) \quad (k = 1, \dots, N), \tag{3.140}$$

and this in turn used in (3.138) gives

$$\gamma_k(\lambda) = \bar{\alpha}_k \prod_{j=0}^{k-1} \frac{1 - \lambda \tilde{\gamma}_j(\lambda)}{\lambda - \gamma_j(\lambda)},$$

where $\tilde{\gamma}_j(\lambda) := \bar{\gamma}_j(1/\lambda)$. Comparison with (3.137) shows $\gamma_k = \gamma_k(1)$.

The γ_k variables will be chosen to be independently distributed in terms of a generalization of Θ_ν (recall Definition 2.8.3).

DEFINITION 3.12.2 *Let $\Theta_{\nu+1}^b$ denote the distribution of complex numbers $|z| < 1$ with p.d.f.*

$$\frac{(\Gamma(\nu/2+1+b))^2}{\pi \Gamma(\nu/2)\Gamma(\nu/2+1+2b)} (1-|z|^2)^{(\nu/2-1)} |1-z|^{2b}$$

while for complex numbers $|z| = 1$ let Θ_1^b denote the distribution

$$\frac{1}{2\pi} \frac{(\Gamma(1+b))^2}{\Gamma(1+2b)} |1-z|^{2b}$$

(note that $\Theta_\nu^b|_{b=0} = \Theta_\nu$).

PROPOSITION 3.12.3 *Let $\{\alpha_k\}_{k=0,\dots,N-1}$ specify a unitary Hessenberg matrix as in Proposition 2.8.1, and let these variables be determined by $\{\gamma_k\}$ according to (3.137), where γ_{N-j-1} is required to have distribution $\Theta_{\beta j/2}^b$. The unitary Hessenberg matrix then has its eigenvalue p.d.f. given by the p.d.f. (3.122) specifying the circular Jacobi β-ensemble. Furthermore, the first component of its eigenvectors are independently distributed as specified by (1.161).*

Proof. The measure associated with the joint distribution of $\{\gamma_{N-j-1}\}$ is proportional to

$$\prod_{j=0}^{N-2} (1 - |\gamma_{N-j-1}|^2)^{\beta j/2-1} \prod_{j=0}^{N-1} |1-\gamma_{N-j-1}|^{2b} d\vec{\gamma}. \tag{3.141}$$

Let us write $\gamma_k = r_k e^{i\theta_k}$, $(k = 0, \dots, N-2)$, $\gamma_{N-1} = e^{i\theta_{N-1}}$, $\alpha_k = r_k e^{i\psi_k}$, $(k = 0, \dots, N-2)$ and $\alpha_{N-1} =$

$e^{i\psi_{N-1}}$. From the triangular structure of (3.137), the Jacobian for the transformation

$$\{r_0,\dots,r_{N-2},\theta_0,\dots,\theta_{N-1}\} \mapsto \{r_0,\dots,r_{N-1},\psi_0,\dots,\psi_{N-1}\}$$

is seen to equal one. Hence (3.141) is equal to

$$\prod_{j=0}^{N-2}(1-|\alpha_{N-j-1}|^2)^{\beta j/2-1}\prod_{j=0}^{N-1}|1-\gamma_{N-j-1}|^{2b}\,d\vec{\alpha}, \tag{3.142}$$

where we have used the fact that $|\gamma_k| = |\alpha_k|$. Now, making use of (3.140) with $k = N$, $\lambda = 1$, and the fact that

$$\chi_N(1) = \det(\mathbf{1}_N - \mathbf{H}_\beta) = \prod_{j=1}^{N}(1-e^{i\theta_j}),$$

shows

$$\prod_{j=0}^{N-1}|1-\gamma_{N-j-1}|^{2b} = \prod_{j=1}^{N}|1-e^{i\theta_j}|^{2b}.$$

Furthermore, we can rewrite the remaining factors in (3.142) in terms of $\{\lambda_j\}$ and $\{q_i\}$ using the working of the proof of Proposition 2.8.7. □

Chapter Four

The Selberg integral

The normalization of the Jacobi β-ensemble, known as the Selberg integral, is studied in its own right. Four different derivations of the evaluation of the Selberg integral, including the original one of Selberg, are presented. Two of these derivations give extensions of the Selberg integral which are of use in the calculation of correlation functions considered in subsequent chapters. Furthermore, the derivation due to Anderson has an interpretation in terms of the eigenvalues of a random corank 1 projection of a fixed matrix and leads to a further random three-term recurrence for the characteristic polynomial of the Jacobi β-ensemble. Alternative constructions of the Gaussian and Laguerre β-ensembles also result. Underlying Anderson's derivation is a further integral evaluation—the Dixon-Anderson integral—which implies that the classical ensembles at $\beta = 1$ and $\beta = 4$ are related by integrating over every second eigenvalue in the former. This integral evaluation can be generalized, with the consequence that classical β-ensembles at $\beta = 2/(r+1)$, integrated over all eigenvalues not labeled by a multiple of $r+1$, give back classical β-ensembles with $\beta = 2(r+1)$. The Selberg integral can be used to prove Macdonald's constant term conjectures relating to root systems. Another application, involving explicitly the normalizations for the circular and Gaussian β-ensembles, is the computation of the free energy of the corresponding log-gas systems.

4.1 SELBERG'S DERIVATION

The Selberg integral refers to the N-dimensional integral (3.117)

$$S_N(\lambda_1, \lambda_2, \lambda) := \int_0^1 dt_1 \cdots \int_0^1 dt_N \prod_{l=1}^N t_l^{\lambda_1}(1-t_l)^{\lambda_2} \prod_{1 \le j < k \le N} |t_k - t_j|^{2\lambda}. \tag{4.1}$$

In the case $N = 1$, this integral is the Euler beta integral, and has the evaluation [541]

$$\int_0^1 x^{\lambda_1}(1-x)^{\lambda_2}\, dx = \frac{\Gamma(\lambda_1+1)\Gamma(\lambda_2+1)}{\Gamma(\lambda_1+\lambda_2+2)}. \tag{4.2}$$

For general N Selberg [483] evaluated (4.1) as the product of gamma functions (2.103),

$$S_N(\lambda_1, \lambda_2, \lambda) = \prod_{j=0}^{N-1} \frac{\Gamma(\lambda_1+1+j\lambda)\Gamma(\lambda_2+1+j\lambda)\Gamma(1+(j+1)\lambda)}{\Gamma(\lambda_1+\lambda_2+2+(N+j-1)\lambda)\Gamma(1+\lambda)}, \tag{4.3}$$

long before the development of the theory of Section 2.9. Note that (4.3) has various poles. Those with the largest real part depend on the sign of $\mathrm{Re}(\lambda)$, being at $\lambda_1 = -1$ or $-1-(N-1)\lambda$, $\lambda_2 = -1$ or $-1-(N-1)\lambda$ and $\lambda = -1/N$, which is in keeping with (4.1) defining a convergent integral for parameter values to the right of these in the appropriate complex planes.

According to (3.119), the evaluation of the Selberg integral for general N gives the evaluation of the Morris integral. Thus we find

$$M_N(a,b,\lambda) := \int_{-1/2}^{1/2} d\theta_1 \cdots \int_{-1/2}^{1/2} d\theta_N \prod_{l=1}^{N} e^{\pi i\theta_l(a-b)}|1+e^{2\pi i\theta_l}|^{a+b} \prod_{1\le j<k\le N} |e^{2\pi i\theta_k} - e^{2\pi i\theta_j}|^{2\lambda}$$
$$= \prod_{j=0}^{N-1} \frac{\Gamma(\lambda j + a + b + 1)\Gamma(\lambda(j+1)+1)}{\Gamma(\lambda j + a + 1)\Gamma(\lambda j + b + 1)\Gamma(1+\lambda)}, \tag{4.4}$$

where use has been made of the functional equation

$$\Gamma(z)\Gamma(1-z) = \frac{\pi}{\sin \pi z}. \tag{4.5}$$

This includes as special cases the normalization of the circular Jacobi ensemble (3.122) and the eigenvalue p.d.f. (2.20) of the circular ensemble. In fact, as will be demonstrated below, the normalization of the Hermite and Laguerre ensembles similarly follow as limiting cases of the Selberg integral.

Before showing how (4.3) can be used to calculate the normalizations of the various ensembles, we will consider a number of different derivations of this evaluation in addition to the one already given in Section 2.9, each being of independent interest. The first derivation to be presented is the original one of Selberg [483], but with the last step (analytic continuing off the integers) done differently [23].

To begin, note that for λ a positive integer the product of differences in (4.1) is a multidimensional polynomial of order $2\lambda(N-1)$ and so can be expanded

$$\prod_{1\le j<k\le N} |t_k - t_j|^{2\lambda} = \sum_{0\le n_1,\dots,n_N \le 2\lambda(N-1)} c_{n_1,\dots,n_N} t_1^{n_1} \cdots t_N^{n_N}. \tag{4.6}$$

Substituting this expansion in the definition of S_N, and use of the Euler beta integral evaluation (4.2) shows

$$S_N(\lambda_1,\lambda_2,\lambda) = \sum_{0\le n_1,\dots,n_N\le 2\lambda(N-1)} c_{n_1,\dots,n_N} \prod_{j=1}^{N} \frac{\Gamma(\lambda_1+1+n_j)\Gamma(\lambda_2+1)}{\Gamma(\lambda_1+\lambda_2+2+n_j)}. \tag{4.7}$$

The next stage of Selberg's derivation of (4.3) consists of evaluating (4.7) as a function of λ_1 and λ_2. For this purpose some properties of the integers $c_{n_1,\dots,n_N}$ occurring in (4.6) are required.

PROPOSITION 4.1.1 *The value of the coefficients $c_{n_1,\dots,n_N}$ in (4.6) is independent of the order of the integers $n_1,\dots,n_N$. For $0 \le n_1 \le n_2 \cdots \le n_N$ the nonzero values of $c_{n_1,\dots,n_N}$ occur when the integers $n_1,\dots,n_N$ satisfy the summation formula $\sum_{j=1}^N n_j = N(N-1)\lambda$ and the inequalities $(j-1)\lambda \le n_j$ and $n_j \le (N+j-2)\lambda$, $(j=1,\dots,N)$.*

Proof. The fact that the value of the coefficients $c_{n_1,\dots,n_N}$ is independent of the order of the integers $n_1,\dots,n_N$ follows because the function being expanded is symmetric. With $\Delta^{(N)}(t_1,\dots,t_N) := \prod_{1\le j<k\le N}(t_k - t_j)^2$, the summation formula follows from the fact that $\Delta^{(N)}$ is homogeneous of degree $N(N-1)$. From the summation formula and the assumed ordering we have $n_N \ge (N-1)\lambda$. This inequality, and the fact that $\Delta^{(N)}(t_1,\dots,t_N)$ is divisible by $\Delta^{(j)}(t_1,\dots,t_j)$ for each $j=1,\dots,N-1$, gives $n_j \ge (j-1)\lambda$, which is the first of the stated inequalities. To establish the second inequality observe

$$(\Delta^{(N)}(t_1,\dots,t_N))^\lambda = \prod_{j=1}^{N} t_j^{2(N-1)\lambda} (\Delta^{(N)}(1/t_1,\dots,1/t_N))^\lambda.$$

Now the allowed powers on the r.h.s, n_j' say with $0 \le n_1' \le n_2' \cdots \le n_N'$, are given in terms of the allowed powers on the l.h.s. by $n_j' = 2(N-1)\lambda - n_{N+1-j}$, and must satisfy the inequality derived above for n_j, $n_j' \ge (j-1)\lambda$. The second stated inequality now follows. □

PROPOSITION 4.1.2 *The expression (4.7) can be evaluated as a function of λ_1 and λ_2 to give*

$$S_N(\lambda_1,\lambda_2,\lambda) = c_N(\lambda) \prod_{j=0}^{N-1} \frac{\Gamma(\lambda_1+1+j\lambda)\Gamma(\lambda_2+1+j\lambda)}{\Gamma(\lambda_1+\lambda_2+2+(N+j-1)\lambda)},$$

where $c_N(\lambda)$ remains to be determined.

Proof. From the inequalities for n_j in Proposition 4.1.1 we see

$$\frac{\Gamma(1+\lambda_1+n_j)}{\Gamma(2+\lambda_1+\lambda_2+n_j)} = \frac{\Gamma(1+\lambda_1+(j-1)\lambda)}{\Gamma(2+\lambda_1+\lambda_2+(N+j-2)\lambda)} q_{n_j}(\lambda_1,\lambda_2),$$

where $q_{n_j}(\lambda_1,\lambda_2)$ is a polynomial in λ_1 and λ_2 of degree $(N+j-2)\lambda - n_j$ in λ_2. Thus

$$\prod_{j=1}^{N} \frac{\Gamma(1+\lambda_1+n_j)\Gamma(1+\lambda_2)}{\Gamma(2+\lambda_1+\lambda_2+n_j)} = Q(\lambda_1,\lambda_2) \prod_{j=1}^{N} \frac{\Gamma(1+\lambda_1+(j-1)\lambda)\Gamma(1+\lambda_2)}{\Gamma(2+\lambda_1+\lambda_2+(N+j-2)\lambda)},$$

where the polynomial $Q(\lambda_1,\lambda_2)$ has degree

$$\sum_{j=1}^{N} ((N+j-2)\lambda - n_j) = N(N-1)\lambda/2$$

in λ_2. The expression (4.7) is a linear combination of terms of this form and can therefore be written

$$\begin{aligned} S_N(\lambda_1,\lambda_2,\lambda) &= \tilde{Q}(\lambda_1,\lambda_2) \prod_{j=1}^{N} \frac{\Gamma(1+\lambda_1+(j-1)\lambda)\Gamma(1+\lambda_2)}{\Gamma(2+\lambda_1+\lambda_2+(N+j-2)\lambda)} \\ &= \frac{\tilde{Q}(\lambda_1,\lambda_2)}{R(\lambda_2)} \prod_{j=1}^{N} \frac{\Gamma(1+\lambda_1+(j-1)\lambda)\Gamma(1+\lambda_2+(j-1)\lambda)}{\Gamma(2+\lambda_1+\lambda_2+(N+j-2)\lambda)}, \end{aligned}$$

where $\tilde{Q}(\lambda_1,\lambda_2)$ is a polynomial of degree at most $N(N-1)\lambda/2$ and

$$R(\lambda_2) := \prod_{j=1}^{N} \frac{\Gamma(1+\lambda_2+(j-1)\lambda)}{\Gamma(1+\lambda_2)}$$

is a polynomial in λ_2 of degree $N(N-1)\lambda/2$. Since S_N is symmetric in λ_1 and λ_2 we must have

$$\frac{\tilde{Q}(\lambda_1,\lambda_2)}{R(\lambda_2)} = \frac{\tilde{Q}(\lambda_2,\lambda_1)}{R(\lambda_1)}.$$

Now the r.h.s. of this identity is a polynomial in λ_2, which implies $\tilde{Q}(\lambda_1,\lambda_2)$ must be divisible by $R(\lambda_2)$. But the maximum allowed degree in λ_2 of $\tilde{Q}(\lambda_1,\lambda_2)$ is the same as the degree of $R(\lambda_2)$, so the maximum degree must be attained and $\tilde{Q}(\lambda_1,\lambda_2)/R(\lambda_2)$ must be independent of λ_2. By symmetry the quotient must also be independent of λ_1, so is in fact only dependent on N and λ. □

To specify $c_N(\lambda)$ a limiting case of the Selberg integral can be used.

PROPOSITION 4.1.3 *We have*

$$\lim_{\lambda_1 \to -1^+} (1+\lambda_1) S_N(\lambda_1,\lambda_2,\lambda) = N S_{N-1}(2\lambda-1,\lambda_2,\lambda).$$

Proof. Since the integrand in (4.1) is symmetrical in $t_1,\dots,t_N$ we can write

$$S_N(\lambda_1,\lambda_2,\lambda) = N! \int_0^1 dt_N \int_{t_N}^1 dt_{N-1} \cdots \int_{t_2}^1 dt_1 \prod_{l=1}^{N} t_l^{\lambda_1}(1-t_l)^{\lambda_2} \prod_{1\le j<k\le N} |t_k - t_j|^{2\lambda}.$$

The stated result then follows after application of the general formula

$$\lim_{\lambda_1 \to -1^+} (1+\lambda_1) \int_0^1 t^{\lambda_1} f(t)\, dt = \lim_{\lambda_1 \to -1^+} (1+\lambda_1) \Big(f(0) \int_0^1 t^{\lambda_1}\, dt + \int_0^1 t^{\lambda_1} (f(t) - f(0))\, dt \Big) = f(0),$$

applicable whenever $(f(t) - f(0))/t$ is integrable near $t = 0$. □

Propositions 4.1.2 and 4.1.3 together give the recurrence relation

$$c_N(\lambda) = \frac{\Gamma(1+N\lambda)}{\Gamma(1+\lambda)} c_{N-1}(\lambda),$$

which, since $c_1(\lambda) = 1$, has solution

$$c_N(\lambda) = \prod_{j=0}^{N-1} \frac{\Gamma(\lambda(j+1)+1)}{\Gamma(\lambda+1)}.$$

Substituting this equation in Proposition 4.1.2 gives the evaluation (4.3) of the Selberg integral, proved for all non-negative integers λ. In fact the evaluation is valid for all λ for which (4.1) is defined. To show this *Carlson's theorem* (see, e.g., Titchmarsh [515]) can be used, although in [483] Selberg carries out this step from first principles.

PROPOSITION 4.1.4 *For* $\mathrm{Re}(z) \ge 0$ *suppose* $f(z)$ *is analytic and has the bound* $|f(z)| = \mathrm{O}(e^{\mu |z|})$, $\mu < \pi$. *Suppose also that* $f(z) = 0$ *for* $z \in \mathbb{Z}^+$. *Then* $f(z) = 0$ *identically.*

With $f(\lambda - 1)$ equal to the integral in (4.1) minus the gamma functions in (4.3), all the criteria of the theorem are satisfied (note in particular that both (4.1) and (4.3) are bounded by constants for $\mathrm{Re}(\lambda) \ge 1$), so indeed the evaluation (4.3), proved above for non-negative integer λ, is valid for all λ for which (4.1) is defined.

EXERCISES 4.1 1. (i) Change variables $x = \sin^2\theta$ in the Euler beta integral (4.2) to obtain the integration formula

$$\int_0^{\pi/2} \sin^{2\lambda_1+1}\theta \cos^{2\lambda_2+1}\theta\, d\theta = \frac{\Gamma(\lambda_1+1)\Gamma(\lambda_2+1)}{2\Gamma(\lambda_1+\lambda_2+2)}.$$

(ii) Use the change of variable in (i) to note that (3.118) can be written

$$S_N(\lambda_1, \lambda_2, \lambda) = \int_0^{\pi} d\theta_1 \cdots \int_0^{\pi} d\theta_N \prod_{j=1}^N \sin^{2\lambda_1+1}(\theta_j/2) \cos^{2\lambda_2+1}(\theta_j/2) \\ \times \prod_{1 \le j < k \le N} \Big| \sin((\theta_j - \theta_k)/2) \sin((\theta_j + \theta_k)/2) \Big|^{2\lambda},$$

which is known as the BC_N Selberg integral (see Section 4.7.2). Interpret this integral in terms of the partition function for a log-gas on a half-circle $0 < \theta_j < \pi$ with image charges of the same sign at $2\pi - \theta_j$ (recall the proof of Proposition 3.6.3).

2. Use the change of variables $t_l \mapsto 1/t_l$ in (4.1) to show

$$\int_1^{\infty} dt_1 \cdots \int_1^{\infty} dt_N \prod_{l=1}^N t_l^{\lambda_1} (t_l - 1)^{\lambda_2} \prod_{1 \le j < k \le N} |t_k - t_j|^{2\lambda} = S_N(-(2+\lambda_1+\lambda_2+2\lambda(N-1)), \lambda_2, \lambda). \quad (4.8)$$

3. (i) Make the change of variables $t_l = 1 - 1/(1+e^{-s_l})$ in (4.1) to obtain

$$S_N(\lambda_1, \lambda_2, \lambda) = 2^N \int_{-\infty}^{\infty} ds_1 \cdots \int_{-\infty}^{\infty} ds_N \prod_{l=1}^N \frac{e^{-(\lambda_1 - \lambda_2)s_l}}{(2\cosh s_l)^{\lambda_1+\lambda_2+2+2\lambda(N-1)}} \prod_{1 \le j < k \le N} \Big(2\sinh|s_k - s_j|\Big)^{2\lambda}.$$

(ii) Make the change of variables $t_l = 2/(1+\cosh s_l)$ to obtain

$$S_N(\lambda_1,\lambda_2,\lambda) = \int_{-\infty}^{\infty} ds_1 \cdots \int_{-\infty}^{\infty} ds_N \prod_{l=1}^{N} \frac{(\sinh^2 s_l)^{1/2+\lambda_2}}{(\cosh^2 s_l)^{\lambda_1+\lambda_2+3/2+2\lambda(N-1)}}$$
$$\times \prod_{1\le j<k\le N} \Big(\sinh|s_k - s_j| \sinh|s_k + s_j| \Big)^{2\lambda}.$$

4. (i) Use Liouville's theorem to show that with

$$f(u) = \Big(\frac{\theta_1(u;q)}{\theta_1(u+\pi/2;q)} \Big)^2$$

one has

$$f(u) - f(v) = \frac{(\theta_1(\pi/2;q))^2 \theta_1(u+v;q)\theta_1(u-v;q)}{(\theta_1(u+\pi/2;q)\theta_1(v+\pi/2;q))^2},$$
$$\frac{d}{du} f(u) = \theta_1'(0;q) \frac{(\theta_1(\pi/2;q))^2 \theta_1(2u;q)}{(\theta_1(u+\pi/2;q))^4}.$$

(ii) [459] Change variables $t_l = f(u_l)$ in (4.1) using the formulas of (i) to show

$$S_N(\lambda_1,\lambda_2,\lambda) = \frac{(\theta_1(\pi/2;q))^{2\lambda N(N-1)+2N(\lambda_2+1)}}{(\theta_1(3\pi/4;q))^{2\lambda_2}} \int_0^{\pi/4} du_1 \cdots \int_0^{\pi/4} du_N \prod_{l=1}^{N} \theta_1(2u_l;q)$$
$$\times \frac{(\theta_1(u_l;q))^{2\lambda_1} (\theta_1(\pi/4+u_l;q)\theta_1(\pi/4-u_l;q))^{\lambda_2}}{(\theta_1(u_l+\pi/2;q))^{4+4\lambda(N-1)+2(\lambda_1+\lambda_2)}} \prod_{1\le j<k\le N} |\theta_1(u_j+u_k;q)\theta_1(u_j-u_k;q)|^{2\lambda}.$$

Interpret the product over pairs in terms of charges interacting via the pair potential (2.80) in the interval $u_l \in [0,\pi/4]$, together with image charges of the same sign in $u_l \in [-\pi/4, 0]$.

5. [193] Suppose λ and λ_2 are positive integers.

   (i) Show from (4.3) that S_N is the reciprocal of a polynomial in λ_1.

   (ii) Show from (4.1) that the zeros of the polynomial in (i) occur at the negative of the powers of $t_1,\dots,t_N$ in the expansion of $\prod_{l=1}^N t_l(1-t_l)^{\lambda_2} \prod_{1\le j<k\le N} |t_k - t_j|^{2\lambda}$ and show that the order of the zeros is equal to the maximum number of times the corresponding power can occur in a single term (this property can be used to provide a direct evaluation of the Morris!integral).

6. [50] Suppose λ_1 red points, λ_2 blue points and one yellow point are placed uniformly at random on the unit interval, and thus each distinct ordering of the colors occurs with probability $1/\mathcal{N}$, where

$$\mathcal{N} = \frac{(\lambda_1+\lambda_2+1)!}{\lambda_1!\lambda_2!}.$$

Show that the particular configuration with the yellow point in $[t, t+dt]$ and all the red points are to its left, and all the blue points to its right occurs with probability $t^{\lambda_1}(1-t)^{\lambda_2} dt$. Use this to deduce (4.2) in the case $\lambda_1, \lambda_2 \in \mathbb{Z}_{\ge 0}$.

4.2 ANDERSON'S DERIVATION

4.2.1 A recurrence in N for S_N

In the study of some conjectured finite field analogues of the Selberg integral (see [17]), Anderson [16] developed a method of relating $S_{N+1}(\lambda_1,\lambda_2,\lambda)$ to $S_N(\lambda_1+\lambda,\lambda_2+\lambda,\lambda)$, and thereby evaluating the Selberg integral by iteration in N. The key integration formula developed for this purpose can in fact be found in the classical work of Dixon [137]. The integration formula can be derived as a corollary to a result relating to the density of the roots of a certain random rational function.

PROPOSITION 4.2.1 *Consider the random rational function*

$$R(\lambda) := \sum_{i=1}^{n} \frac{w_i}{a_i - \lambda}, \tag{4.9}$$

where the w_i are distributed according to the Dirichlet distribution

$$\frac{\Gamma(s_1 + \cdots + s_n)}{\Gamma(s_1)\cdots\Gamma(s_n)} \prod_{j=1}^{n} w_j^{s_j - 1}, \tag{4.10}$$

with $w_1, \dots, w_n > 0$ and $\sum_{j=1}^{n} w_j = 1$, and to be denoted $D_n[s_1, \dots, s_n]$ (note that only $n-1$ of the w_is are independent). The roots of $R(\lambda)$, denoted $\{\lambda_j\}$, have the p.d.f.

$$\frac{\Gamma(s_1 + \cdots + s_n)}{\Gamma(s_1)\cdots\Gamma(s_n)} \frac{\prod_{1\le j<k\le n-1}(\lambda_j - \lambda_k)}{\prod_{1\le j<k\le n}(a_j - a_k)^{s_j+s_k-1}} \prod_{j=1}^{n-1}\prod_{p=1}^{n} |\lambda_j - a_p|^{s_p - 1}, \tag{4.11}$$

to be referred to as the Dixon-Anderson density, where it is required that

$$a_1 > \lambda_1 > a_2 > \lambda_2 > \cdots > \lambda_{n-1} > a_n. \tag{4.12}$$

Proof. Considering the degree of the numerator when written with a common denominator shows that $R(\lambda)$ has exactly $n-1$ roots. That these are real and interlace as specified in (4.12) follows from a graphical argument, making essential use of each w_i being positive (recall the proof of Proposition 1.8.1) Hence, using too the fact that $\sum_{i=1}^{n} w_i = 1$, we see that (4.9) can be written

$$\sum_{i=1}^{n} \frac{w_i}{\lambda - a_i} = \frac{\prod_{l=1}^{n-1}(\lambda - \lambda_l)}{\prod_{l=1}^{n}(\lambda - a_l)},$$

where the $\{\lambda_j\}$ are real.

Computing the residue at $\lambda = a_j$ gives

$$w_j = \frac{\prod_{l=1}^{n-1}(a_j - \lambda_l)}{\prod_{l=1, l\ne j}^{n}(a_j - a_l)}. \tag{4.13}$$

The distribution of $\{w_j\}_{j=1,\dots,n-1}$ is given by (4.10). We want to change variables to $\{\lambda_j\}_{j=1,\dots,n-1}$. First, from (4.13), up to a sign

$$\bigwedge_{j=1}^{n-1} dw_j = \prod_{j=1}^{n-1} w_j \det\left[\frac{1}{a_j - \lambda_l}\right]_{j,l=1,\dots,n-1} \bigwedge_{j=1}^{n-1} d\lambda_j,$$

while the determinant can be evaluated according to the Cauchy double alternant identity (4.33) below. Thus we have that the Jacobian is equal to

$$\prod_{j=1}^{n-1} w_j \frac{\prod_{1\le j<k\le n-1}(a_j - a_k)(\lambda_j - \lambda_k)}{\prod_{j,k=1}^{n-1}|a_j - \lambda_k|}. \tag{4.14}$$

Multiplying (4.14) and (4.10), and substituting for w_j according to (4.13) gives (4.11). □

Because (4.11) is a p.d.f. in $\{\lambda_j\}$, integrating it over the region (4.12) must give unity. Rearranging and replacing n by $N+1$ gives the integral evaluation, to be referred to as the Dixon-Anderson integral,

$$\int_X d\lambda_1 \cdots d\lambda_N \prod_{1\le j<k\le N}(\lambda_j - \lambda_k) \prod_{j=1}^{N}\prod_{p=1}^{N+1}|\lambda_j - a_p|^{s_p - 1} = \frac{\prod_{i=1}^{N+1}\Gamma(s_i)}{\Gamma(\sum_{i=1}^{N+1} s_i)} \prod_{1\le j<k\le N+1}(a_j - a_k)^{s_j+s_k-1}, \tag{4.15}$$

where X is the domain of integration (4.12) with $n \mapsto N+1$. We remark that this has the interpretation as the partition function of a log-gas at $\beta = 1$ in which there are fixed particles of charge $s_p - 1$ at the points a_p $(p = 1, 2, \dots, N)$ interlaced as specified by (4.12). Dixon in fact gave an integration formula more general than (4.15). This is presented in Exercises 4.2 q.2.

We can use (4.15) to establish a recurrence formula in N for the Selberg integral (4.1).

PROPOSITION 4.2.2 *Let X'_{N+1} denote the region*

$$1 > x_1 > y_1 > x_2 > y_2 > \cdots > y_N > x_{N+1} > 0. \tag{4.16}$$

Consider

$$K(\lambda_1, \lambda_2, \lambda) := \int_{X'_{N+1}} dx_1 \cdots dx_{N+1} dy_1 \cdots dy_N \prod_{l=1}^{N+1} x_l^{\lambda_1} (1 - x_l)^{\lambda_2} \\ \times \prod_{i=1}^{N} \prod_{j=1}^{N+1} |y_i - x_j|^{\lambda - 1} \prod_{i<j}^{N} |y_i - y_j| \prod_{i<j}^{N+1} |x_i - x_j|. \tag{4.17}$$

By first integrating over $\{y_j\}$ we obtain

$$K(\lambda_1, \lambda_2, \lambda) = S_{N+1}(\lambda_1, \lambda_2, \lambda) \frac{(\Gamma(\lambda))^{N+1}}{(N+1)!\Gamma((N+1)\lambda)}$$

while first integrating over $\{x_l\}$ gives

$$K(\lambda_1, \lambda_2, \lambda) = S_N(\lambda_1 + \lambda, \lambda_2 + \lambda, \lambda) \frac{\Gamma(1+\lambda_1)\Gamma(1+\lambda_2)(\Gamma(\lambda))^N}{N!\Gamma(2 + \lambda_1 + \lambda_2 + N\lambda)}.$$

Proof. To integrate over $\{y_j\}$ we make use of (4.15) with $s_p = \lambda$ $(p = 1, \dots, N+1)$. For the integral over $\{x_l\}$, replace $N \mapsto N+1$ in (4.15) and set $a_1 = 1, a_{N+2} = 0, s_1 = \lambda_2 + 1, s_{N+2} = \lambda_1 + 1, s_p = \lambda$ $(p = 2, \dots, N+1)$.

Proposition 4.2.2 gives the recurrence formula

$$S_{N+1}(\lambda_1, \lambda_2, \lambda) = S_N(\lambda_1 + \lambda, \lambda_2 + \lambda, \lambda) \frac{(N+1)\Gamma((N+1)\lambda)\Gamma(1+\lambda_1)\Gamma(1+\lambda_2)}{\Gamma(\lambda)\Gamma(2+\lambda_1+\lambda_2+N\lambda)}.$$

After iteration with the initial condition $S_0 := 1$ the Selberg integral evaluation (4.3) is obtained. □

4.2.2 Relationship to an eigenvalue probability density function

The result of Proposition 4.2.1 can be used to provide the eigenvalue p.d.f. for matrices of the form

$$\mathbf{M} := \mathbf{\Pi A \Pi}, \qquad \mathbf{\Pi} := \mathbf{1} - \vec{x}\vec{x}^\dagger, \tag{4.18}$$

where $\mathbf{A}$ is a real symmetric, or complex Hermitian, matrix with eigenvalues $a_1 > a_2 > \cdots > a_n$ having corresponding multiplicities $m_1, m_2, \dots, m_n$ and $\vec{x}$ is a real or complex normalized Gaussian vector of the same number of rows as $\mathbf{A}$. The matrix $\mathbf{\Pi}$ is thus a projection onto the subspace orthogonal to the vector $\vec{x}$ (for this reason $\mathbf{\Pi}$ is said to have codimension 1), and the matrix $\mathbf{M}$ corresponds to $\mathbf{A}$ projected onto this subspace.

We begin by noting that the eigenvalues a_i of $\mathbf{A}$ must occur in $\mathbf{M}$ with multiplicity $m_i - 1$. This is a corollary of the following formula for the characteristic polynomial.

PROPOSITION 4.2.3 *With $\mathbf{M}$ defined by (4.18) we have*

$$\det(\mathbf{M} - \lambda\mathbf{1}) = -\lambda \det(\mathbf{A} - \lambda\mathbf{1}) \mathrm{Tr}((\mathbf{A} - \lambda\mathbf{1})^{-1} \vec{x}\vec{x}^\dagger). \tag{4.19}$$

Proof. Because $\mathbf{\Pi}$ is a projector, we can check that $[\mathbf{M}, \mathbf{A\Pi}] = 0$. The matrices $\mathbf{M}$ and $\mathbf{A\Pi}$ therefore have the same eigenvalues, so we have

$$\begin{aligned}\det(\mathbf{M}-\lambda\mathbf{1}) &= \det(\mathbf{A}\boldsymbol{\Pi}-\lambda\mathbf{1}) = \det(\mathbf{A}-\lambda\mathbf{1}-\mathbf{A}\vec{x}\vec{x}^{\,\dagger}) \\ &= \det(\mathbf{A}-\lambda\mathbf{1})\det(\mathbf{1}-(\mathbf{A}-\lambda\mathbf{1})^{-1}\mathbf{A}\vec{x}\vec{x}^{\,\dagger}).\end{aligned}$$

But the matrix in the second determinant of the final expression is of the form $\mathbf{1}+\mathbf{Y}$, where $\mathbf{Y}$ has rank 1. In this circumstance, $\det(\mathbf{1}+\mathbf{Y}) = 1 + \mathrm{Tr}\,\mathbf{Y}$ as used in (1.136). The result (4.19) now follows after straightforward manipulation, and making use of the fact that $\mathrm{Tr}(\vec{x}\vec{x}^{\,\dagger}) = 1$. □

According to (4.19) there are an eigenvalue $\lambda = 0$, and eigenvalues λ satisfying

$$\prod_{l=1}^{n}(a_l-\lambda)^{m_l}\sum_{i=1}^{n}\frac{\sum_{j=1}^{m_i}u_i^{(j)}}{a_i-\lambda} = 0, \tag{4.20}$$

where the $u_i^{(j)}$ denote the diagonal elements of $\vec{x}\vec{x}^{\dagger}$. The fact that $\mathbf{M}$ has eigenvalues a_i with multiplicity no less than $m_i - 1$ follows immediately from (4.20). Of interest is the distribution of the $n-1$ eigenvalues differing (in general) from the a_i and 0. According to (4.20) these are given by the zeros of the random rational function (4.9) with $w_i = \sum_{j=1}^{m_i} u_i^{(j)}$.

Now the sum of the squares of s independent real Gaussians with mean zero and standard deviation σ has distribution proportional to $x^{s/2-1}e^{-x/2\sigma^2}$ (recall Exercises 3.2 q.3). This is referred to as the gamma distribution and denoted $\Gamma[s/2, 2\sigma^2]$, as used in Proposition 1.9.1 and Section 3.10. Also relevant is the fact that a vector of appropriately normalized gamma distributed variables specifies the Dirichlet distribution.

PROPOSITION 4.2.4 *Let $X_1,\dots,X_n$ be independent random variables such that each X_j has distribution $\Gamma[s_j,\gamma]$, and let*

$$\rho_j := \frac{X_j}{X_1+\cdots+X_n} \quad \text{for} \quad j = 1,\dots,n.$$

Then the p.d.f. for $\vec{\rho} := (\rho_1,\dots,\rho_n)$ is given by the Dirichlet distribution (4.10).

It follows from the above theory that $\vec{w} := (w_1,\dots,w_n)$ has a Dirichlet distribution with

$$s_i = \beta m_i/2 \tag{4.21}$$

where $\beta = 1$ for $\vec{x}$ real and $\beta = 2$ for $\vec{x}$ complex. As a corollary of Proposition 4.2.1 we thus can compute the p.d.f. for the perturbed eigenvalues of $\mathbf{M}$ [28], [222].

PROPOSITION 4.2.5 *The eigenvalues of $\mathbf{M}$ in (4.18) differing from the eigenvalues of $\mathbf{A}$ and from 0, are given by the zeros of the random rational function (4.9) and have the p.d.f. (4.11) with the s_j specified by (4.21).*

4.2.3 Inter-relations between matrix ensembles with $\beta = 1$ and $\beta = 4$

The Dixon-Anderson integral (4.15) implies that integrating over the eigenvalue p.d.f. of certain matrix ensembles with $\beta = 1$, the eigenvalue p.d.f. of $\beta = 4$ matrix ensembles results [220], [222]. To state these inter-relations, let $\mathrm{OE}_N(e^{-V(x)})$ denote the matrix ensemble with orthogonal symmetry having eigenvalue p.d.f. proportional to

$$\prod_{l=1}^{N} e^{-V(x_l)} \prod_{1\le j<k\le N} |x_k - x_j|. \tag{4.22}$$

Similarly, let us denote by $\mathrm{SE}_N(e^{-4V(x)})$ the matrix ensemble with symplectic symmetry having eigenvalue p.d.f. proportional to

$$\prod_{l=1}^{N} e^{-4V(x_l)} \prod_{1\le j<k\le N} (x_k - x_j)^4. \tag{4.23}$$

With the eigenvalues in (4.22) assumed ordered according to

$$x_1 > x_2 > \cdots > x_N, \tag{4.24}$$

let $\mathrm{even}(\mathrm{OE}_N(f))$ denote the distribution of the even labeled coordinates in the ensemble $\mathrm{OE}_N(f)$.

PROPOSITION 4.2.6 *We have the inter-relations between matrix ensembles*

$$\mathrm{even}(\mathrm{OE}_{2N+1}(x^{(a-1)/2}(1-x)^{(b-1)/2})) = \mathrm{SE}_N(x^{a+1}(1-x)^{b+1}), \tag{4.25}$$
$$\mathrm{even}(\mathrm{OE}_{2N}((1-x)^{(b-1)/2})) = \mathrm{SE}_N((1-x)^{b+1}). \tag{4.26}$$

Proof. Consider first (4.25). In (4.15) set

$$N \mapsto N+1,\ a_1 = 1,\ a_{N+2} = 0,\ s_1 = (b+1)/2,\ s_{N+2} = (a+1)/2,\ s_j = 2\ (j = 2,\dots,N+1)$$

and multiply both sides by $\prod_{j=1}^{N+1} a_j^{(a-3)/2}(1-a_j)^{(b-3)/2}$. Noting that then

$$\prod_{j=1}^{N+1} a_j^{(a-3)/2}(1-a_j)^{(b-3)/2} \prod_{1\le j<k\le N+2} (a_j - a_k) \prod_{1\le j<k\le N+1} (\lambda_j - \lambda_k) \prod_{j=1}^{N+1}\prod_{p=1}^{N+2} |\lambda_j - a_p|^{s_p - 1}$$
$$\propto \mathrm{OE}_{2N+1}(x^{(a-1)/2}(1-x)^{(b-1)/2})$$

while

$$\prod_{j=1}^{N+1} a_j^{(a-3)/2}(1-a_j)^{(b-3)/2} \prod_{1\le j<k\le N+2} (a_j - a_k)^{s_j+s_k} \propto \mathrm{SE}_N(x^{a+1}(1-x)^{b+1}),$$

we see that (4.25) follows. To derive (4.26), in (4.15) set

$$a_1 = 1,\ s_1 = (b+1)/2,\ s_j = 2\ (j = 2,\dots,N)$$

and multiply both sides by $\prod_{j=1}^{N}(1-a_j)^{(b-3)/2}$. □

The change of variables $x_j \mapsto \frac{1}{2}(1 - x_j/L)$ in (4.25) with $a = b = L^2$ and $L \to \infty$ is the Gaussian limit, and one obtains

$$\mathrm{even}(\mathrm{OE}_{2N+1}(e^{-x^2/2})) = \mathrm{SE}_N(e^{-x^2}). \tag{4.27}$$

Note that (4.26) does not permit a Gaussian limit. On the other hand both (4.25) and (4.26) permit the Laguerre limit $x_j \mapsto x_j/L$, $b = L$, $L \to \infty$, giving

$$\mathrm{even}(\mathrm{OE}_{2N+1}(x^{(a-1)/2}e^{-x/2})) = \mathrm{SE}_N(x^{a+1}e^{-x}), \tag{4.28}$$
$$\mathrm{even}(\mathrm{OE}_{2N}(e^{-x/2})) = \mathrm{SE}_N(e^{-x}), \tag{4.29}$$

respectively. Furthermore, the method of proof of Proposition 4.2.6 applied to the integration formula following from (4.88) below being a p.d.f. in $\{x_j\}$ gives the inter-relation between Cauchy ensembles

$$\mathrm{even}(\mathrm{OE}_{2N+1}((1+ix)^{-(\alpha+1)/2}(1-ix)^{-(\bar\alpha+1)/2})) = \mathrm{SE}_N((1+ix)^{-(\alpha-1)}(1-ix)^{-(\bar\alpha-1)}).$$

With the notation $\mathrm{C}^b_{\beta,N}$ referring to the p.d.f. corresponding to (3.122) with $2\lambda \mapsto \beta$, applying the change of variables (2.50) to this gives

$$\mathrm{even}\ \mathrm{CE}^{(b-1)/2}_{1,2N+1} = \mathrm{CE}^{(b+1)}_{4,N}. \tag{4.30}$$

Finally, setting $b = 3$ in (4.30), noting that in general

$$\mathrm{CE}^{\beta}_{\beta,N} = \mathrm{CE}^{0}_{\beta,N+1}|_{\theta_{N+1}=2\pi} \tag{4.31}$$

and replacing $N \mapsto N-1$, we obtain [397]

$$\mathrm{alt}\,\mathrm{COE}_{2N} = \mathrm{CSE}_N, \tag{4.32}$$

where alt denotes the distribution of every second eigenvalue ordered around the circle.

EXERCISES 4.2

1. The objective of this exercise is to verify the Cauchy double alternant identity

$$\frac{\prod_{1\le j<k\le N}(x_k - x_j)(y_k - y_j)}{\prod_{j,k=1}^{N}(x_j - y_k)} = (-1)^{N(N-1)/2} \det\left[\frac{1}{x_j - y_k}\right]_{j,k=1,\dots,N}. \tag{4.33}$$

(i) For each row j, extract a factor $1/\prod_{k=1}^{N}(x_j - y_k)$, so that the r.h.s. of (4.33) reads

$$(-1)^{N(N-1)/2}\frac{1}{\prod_{j,k=1}^{N}(x_j - y_k)}\det\Big[\prod_{\substack{l=1\\ l\ne k}}^{N}(x_j - y_l)\Big]_{j,k=1,\dots,N}.$$

Argue that this determinant is a homogeneous polynomial of degree $N(N-1)/2$ in both sets of variables.

(ii) Note that the determinant in (i) vanishes when $x_j = x_{j'}$ or $y_k = y_{k'}$ and so contains a factor of the two difference products

$$\prod_{1\le j<k\le N}(x_k - x_j)(y_k - y_j).$$

Since this is of the required degree in each set of variables, conclude that the determinant is proportional to the two difference products. Determine the proportionality constant by setting $x_j = y_j$, and thus verify (4.33).

(iii) Replace y_j by $1/y_j$ in (4.33) to write the Cauchy double alternant identity in the form

$$\det\left[\frac{1}{1 - x_j y_k}\right]_{j,k=1,\dots,N} = \frac{\prod_{1\le j<k\le N}(x_k - x_j)(y_k - y_j)}{\prod_{j,k=1}^{N}(1 - x_j y_k)}. \tag{4.34}$$

(iv) The permanent of a square matrix $\mathbf{A} = [a_{jk}]_{j,k=1,\dots,N}$ is specified as the symmetric function of the elements

$$\mathrm{perm}\,\mathbf{A} = \sum_{\sigma\in S_N}\prod_{l=1}^{N} a_{l,\sigma(l)} \tag{4.35}$$

(cf. the formula (5.22) below for a determinant). By differentiating (4.34) with respect to $x_1, \dots, x_N$, deduce the Borchardt identity

$$\det\left[\frac{1}{(1 - x_j y_k)^2}\right]_{j,k=1,\dots,N} = \frac{\prod_{1\le j<k\le N}(x_k - x_j)(y_k - y_j)}{\prod_{j,k=1}^{N}(1 - x_j y_k)}\mathrm{perm}\left[\frac{1}{1 - x_j y_k}\right]_{j,k=1,\dots,N}. \tag{4.36}$$

2. [137] Let $\alpha_0, \dots, \alpha_n, \beta_0, \dots, \beta_m$ have positive real parts, and suppose $\sum_{i=0}^{n}\alpha_i = \sum_{j=0}^{m}\beta_j$. Using a method

analogous to that presented in the proof of Proposition 4.2.1, Dixon derived an integration identity equivalent to

$$\int_{R_n} dx_1 \cdots dx_n \prod_{i<j}^{n} |x_i - x_j| \prod_{i=1}^{n} \Big(\prod_{j=0}^{n} |a_j - x_i|^{\alpha_j - 1} \Big) \Big(\prod_{j=0}^{m} |b_j - x_i|^{-\beta_j} \Big)$$
$$= \prod_{0 \le i<j \le n} (a_i - a_j)^{(\alpha_i + \alpha_j - 1)} \prod_{0 \le i<j \le m} (b_i - b_j)^{1-\beta_i-\beta_j} \prod_{i=0}^{n} \prod_{j=0}^{m} |b_j - a_i|^{\alpha_i - \beta_j}$$
$$\times \frac{\prod_{j=0}^{n} \Gamma(\alpha_j)}{\prod_{i=0}^{m} \Gamma(\beta_i)} \int_{R'_m} dx_1 \cdots dx_m \prod_{i<j}^{m} |x_i - x_j| \prod_{i=1}^{m} \Big(\prod_{j=0}^{n} |a_j - x_i|^{-\alpha_j} \Big) \Big(\prod_{j=0}^{m} |b_j - x_i|^{\beta_j - 1} \Big), \quad (4.37)$$

where R_n and R'_m denote the regions

$$a_n \le x_n \le a_{n-1} \le \cdots \le a_1 \le x_1 \le a_0, \qquad b_m \le x_m \le b_{m-1} \le \cdots \le b_1 \le x_1 \le b_0,$$

respectively. Here we will give a derivation of (4.37) due to Rains [457].

(i) By taking $b_0 \to \infty$, show that in the case $m = 0$, (4.37) reduces to (4.15).

(ii) Verify that for $b_0 > b_1 > a_0 > a_1$

$$(b_0 - b_1) \int_{a_1}^{a_0} \frac{dx}{(b_0 - x)(b_1 - x)} = (a_0 - a_1) \int_{b_1}^{b_0} \frac{dx}{(x - a_0)(x - a_1)}$$

and from this conclude that for $b_0 > \cdots > b_N > a_0 > \cdots > a_N$,

$$\prod_{j=1}^{N} (b_0 - b_j) \det \Big[\int_{a_i}^{a_0} \frac{dx}{(b_0 - x)(b_j - x)} \Big]_{i,j=1,\dots,N} = \prod_{j=1}^{N} (a_0 - a_j) \det \Big[\int_{b_i}^{b_0} \frac{dx}{(x - a_0)(x - a_j)} \Big]_{i,j=1,\dots,N}.$$

(iii) Subtract row $(i-1)$ from row i $(i = N, \dots, 2)$ in order) then make use of the Cauchy double alternant determinant formula (4.33) to rewrite the final formula of (ii) as the integral identity

$$\prod_{0 \le i<j \le N} (b_i - b_j) \int_{R_N} dx_1 \cdots dx_N \frac{\prod_{1 \le i<j \le N} (x_i - x_j)}{\prod_{i=0}^{N} \prod_{j=1}^{N} (b_i - x_j)}$$
$$= \prod_{0 \le i<j \le N} (a_i - a_j) \int_{R'_N} dx_1 \cdots dx_N \frac{\prod_{1 \le i<j \le N} (x_i - x_j)}{\prod_{i=0}^{N} \prod_{j=1}^{N} (a_i - x_j)}.$$

Now write $N = \sum_{i=0}^{n} \alpha_i = \sum_{j=0}^{m} \beta_j$ where $\alpha_i, \beta_j \in \mathbb{Z}_{\ge 0}$. With $p^* := \sum_{i=0}^{p} \alpha_i$ $(p = 0, \dots, n)$, by taking the limit $a_{p^*+1}, \dots, a_{p^*+\alpha_{p+1}} \to a_{p^*+1}$ and then relabeling $a_{p^*+1} \mapsto a_p$, and performing an analogous merging procedure with the bs, deduce (4.37) in the case $\alpha_i, \beta_j \in \mathbb{Z}_{\ge 0}$.

3. [223] The objective of this exercise is to present circular analogues of aspects of Dixon's and Anderson's working. Let $q_i > 0$, $\sum_{j=1}^{n} q_j = 1$, $\lambda_j = e^{i\theta_j}$ $(\theta_1 < \theta_2 < \cdots < \theta_n)$, $t = e^{i\phi}$, $\lambda = e^{i\psi}$ and consider the rational function

$$C_n(\lambda) := t - (t-1)\lambda \sum_{j=1}^{n} \frac{q_j}{\lambda - \lambda_j}. \quad (4.38)$$

(i) Let $\vec{e}_1$ denote the $n \times 1$ unit vector $(1, 0, \dots, 0)^T$ as in Exercises 1.9 q.3, let $\mathbf{U}$ be an $n \times n$ unitary matrix with distinct eigenvalues $e^{i\theta_1}, \dots, e^{i\theta_n}$, and let $\tilde{\mathbf{U}}$ denote the matrix which results by multiplying the first row of $\mathbf{U}$ by t. Show that the eigenvalues of $\tilde{\mathbf{U}}$ are given by (4.38) with $q_j = |v_{1j}|^2$, where v_{1j} denotes the first entry of the eigenvector corresponding to $e^{i\theta_j}$.

(ii) Deduce from the result of (i) that $C_n(z)$ has exactly n zeros, each of unit modulus. With these zeros denoted

$e^{i\psi_j}$ $(j = 1, \dots, n)$, and after setting $\lambda = e^{i\psi}$ in (4.38) to rewrite it as

$$C_n(\lambda) = \frac{(t-1)}{2i}\Big(\cot\frac{\phi}{2} - \sum_{i=1}^n c_i \cot\frac{(\psi - \theta_i)}{2}\Big),$$

show that the zeros interlace in the sense

$$\theta_{i-1} < \psi_i < \theta_i \qquad (i = 1, \dots, n, \quad \theta_0 := \theta_n \bmod 2\pi). \tag{4.39}$$

(iii) Note from the result of (i) that we can write

$$C_n(\lambda) = \frac{\prod_{j=1}^n (\lambda - \tilde{\lambda}_j)}{\prod_{j=1}^n (\lambda - \lambda_j)}, \tag{4.40}$$

where $\tilde{\lambda}_j := e^{i\psi_j}$ and so deduce the formulas

$$-(t-1)\lambda_j q_j = \frac{\prod_{l=1}^n (\lambda_j - \tilde{\lambda}_l)}{\prod_{l=1, l\neq j}^n (\lambda_j - \lambda_l)} \qquad (j = 1, \dots, n) \tag{4.41}$$

as well as the constraint

$$\prod_{l=1}^n \tilde{\lambda}_l = t \prod_{l=1}^n \lambda_l.$$

(iv) Let $\prod_{j=1}^n (z - \tilde{\lambda}_j) =: \sum_{j=0}^n (-1)^{n-j} \tilde{F}_j z^j$. Use the formulas of (iii) to show

$$\Big| \det \Big[\frac{\partial q_j}{\partial \tilde{F}_l}\Big]_{j,l=1,\dots,n-1} \Big| = |1-t|^{-(n-1)} \frac{1}{\prod_{1\le j<k\le n} |\lambda_j - \lambda_k|},$$

and show also that with $\tilde{F}_n$ redefined as t,

$$\Big| \det \Big[\frac{\partial \tilde{F}_l}{\partial \tilde{\lambda}_j}\Big]_{j,l=1,\dots,n} \Big| = \prod_{1\le j<k\le n} |\tilde{\lambda}_k - \tilde{\lambda}_j|.$$

Thus conclude

$$d\vec{q} \wedge dt = J d\vec{\psi}, \qquad J = |1-t|^{-(n-1)} \prod_{1\le j<k\le n} \Big| \frac{\tilde{\lambda}_k - \tilde{\lambda}_j}{\lambda_k - \lambda_j} \Big|. \tag{4.42}$$

(v) Use the formulas of (ii) and the fact that each $q_j > 0$ to show

$$\Big(\prod_{j=1}^{n-1} q_j^{\alpha-1}\Big) q_n^{\alpha_0 - 1} = \frac{1}{|1-t|^{\alpha_0 + (n-1)\alpha - n}} \frac{\prod_{j=1}^{n-1}\prod_{l=1}^n |\lambda_j - \tilde{\lambda}_l|^{\alpha-1}}{\prod_{1\le j<l\le n-1} |\lambda_l - \lambda_j|^{2(\alpha-1)}} \frac{\prod_{l=1}^n |\lambda_n - \tilde{\lambda}_l|^{\alpha_0 - 1}}{\prod_{l=1}^{n-1} |\lambda_n - \lambda_l|^{\alpha_0 + \alpha - 2}}.$$

Note from this and the results of (iii) that if $(q_1, \dots, q_n)$ is distributed according to the Dirichlet distribution $D_n[(\alpha)^{n-1}, \alpha_0]$ (here the notation $(\alpha)^{n-1}$ denotes α repeated $n-1$ times) while t has p.d.f. specified by the measure

$$\frac{\Gamma^2(\frac{1}{2}(\alpha_0 + (n-1)\alpha + 1))}{2\pi\Gamma((n-1)\alpha + \alpha_0)} |1-t|^{\alpha_0 + (n-1)\alpha - 1} d\phi,$$

then the conditional p.d.f. of $\tilde{\lambda}_1 := e^{i\psi_1}, \dots, \tilde{\lambda}_n := e^{i\psi_n}$ given $\lambda_1 := e^{i\theta_1}, \dots, \lambda_{n-1} := e^{i\theta_{n-1}}, \lambda_n := 1$ is equal to

$$A \frac{\prod_{l=1}^n |1 - e^{i\psi_l}|^{\alpha_0 - 1}}{\prod_{l=1}^{n-1} |1 - e^{i\theta_l}|^{\alpha_0 + \alpha - 1}} \frac{\prod_{j=1}^{n-1}\prod_{l=1}^n |e^{i\theta_j} - e^{i\psi_l}|^{\alpha-1}}{\prod_{1\le j<k\le n-1} |e^{i\theta_k} - e^{i\theta_j}|^{2\alpha-1}} \prod_{1\le j<k\le n} |e^{i\psi_k} - e^{i\psi_j}|, \tag{4.43}$$

$$A := \frac{\Gamma^2(\frac{1}{2}(\alpha_0 + (n-1)\alpha + 1))}{2\pi(\Gamma(\alpha))^{n-1}\Gamma(\alpha_0)}.$$

(vi) With A as in (v) but with $\alpha = \lambda$, $\alpha_0 - 1 = a$, let

$$C_{\rm o}^{(n,n-1)}(\psi,\theta) = \frac{A(n-1)!}{(2\pi)^{n-1}M_{n-1}((a+a_1+\lambda)/2,(a+a_1+\lambda)/2,\lambda)}\prod_{l=1}^{n}|1-e^{i\psi_l}|^a$$
$$\times \prod_{1\le j<k\le n}|e^{i\psi_k}-e^{i\psi_j}|\prod_{l=1}^{n-1}|1-e^{i\theta_l}|^{a_1}\prod_{1\le j<k\le n-1}|e^{i\theta_k}-e^{i\theta_j}|\prod_{j=1}^{n-1}\prod_{l=1}^{n}|e^{i\theta_j}-e^{i\psi_l}|^{\lambda-1}.$$

Show that (4.43) being a conditional p.d.f. implies the integration formula

$$\int_R d\psi_1\cdots d\psi_n\, C_{\rm o}^{(n,n-1)}(\psi,\theta) = \frac{(n-1)!}{(2\pi)^{n-1}M_{n-1}((a+a_1+\lambda)/2,(a+a_1+\lambda)/2,\lambda)}$$
$$\times\prod_{l=1}^{n-1}|1-e^{i\theta_l}|^{a+a_1+\lambda}\prod_{1\le j<k\le n-1}|e^{i\theta_k}-e^{i\theta_j}|^{2\lambda},$$

where R denotes the region implied by (4.39) with $\theta_0 = 0$ and thus conclude $C_{\rm o}^{(n,n-1)}(\psi,\theta)$ defines a joint p.d.f.

(vii) Show that (4.30) can be derived directly from the result of (vi).

4.3 CONSEQUENCES FOR THE β-ENSEMBLES

The theory of the previous section relating to Anderson's derivation of the Selberg integral plays a fundamental role in the theory of the various β-ensembles. In particular, it allows a random three-term recurrence for the characteristic polynomial distinct from that which follows from the tridiagonal matrix characterization Proposition 3.11.1; it provides for inductive constructions of both the Gaussian β-ensemble and Laguerre β-ensemble; it allows too for a three-term recurrence of the characteristic polynomial of the Cauchy β-ensemble, and thus the circular Jacobi β-ensemble.

4.3.1 Random three-term recurrence for the Jacobi β-ensemble

Consideration of Anderson's derivation of the Selberg integral as implied by Proposition 4.2.2 leads one naturally to the joint p.d.f.

$$J_{\rm o}^{(n,n-1)}(x,y) := \frac{\Gamma(2+\alpha+\beta+(n-1)\lambda)}{\Gamma(1+\alpha)\Gamma(1+\beta)(\Gamma(\lambda))^{n-1}}\frac{1}{S_{n-1}(\alpha+\alpha_1+\lambda,\beta+\beta_1+\lambda,\lambda)}\prod_{i=1}^{n}x_i^{\alpha}(1-x_i)^{\beta}$$
$$\times\prod_{1\le i<j\le n}|x_j-x_i|\prod_{i=1}^{n-1}y_i^{\alpha_1}(1-y_i)^{\beta_1}\prod_{1\le i<j\le n-1}|y_j-y_i|\prod_{i=1}^{n}\prod_{j=1}^{n-1}|x_j-y_i|^{\lambda-1}. \quad (4.44)$$

Corresponding to the integrations noted in the proof of Proposition 4.2.2, this has the properties

$$\int_{X_n'} dx_1\cdots dx_n\, J_{\rm o}^{(n,n-1)}(x,y) = \frac{1}{S_{n-1}(\alpha+\alpha_1+\lambda,\beta+\beta_1+\lambda,\lambda)}\prod_{i=1}^{n-1}y_i^{\alpha+\alpha_1+\lambda}(1-y_i)^{\beta+\beta_1+\lambda}$$
$$\times\prod_{1\le i<j\le n-1}|y_i-y_j|^{2\lambda} =: J_{\rm o}^{(\cdot,n-1)}(y), \quad (4.45)$$

$$\int_{X_n'} dy_1 \cdots dy_{n-1}\, J_0^{(n,n-1)}(x,y)\Big|_{\alpha_1=\beta_1=0}$$
$$= \frac{1}{S_n(\alpha,\beta,\lambda)} \prod_{i=1}^n x_i^\alpha (1-x_i)^\beta \prod_{1\le i<j\le n} |x_i - x_j|^{2\lambda} =: J_0^{(n,\cdot)}(x), \tag{4.46}$$

where X_n' is specified by (4.16) with $n = N-1$. Note that (4.46) is the p.d.f. for the Jacobi 2λ-ensemble on the interval $[0,1]$. Here we will show that the relationship of the conditional p.d.f.'s

$$\frac{J_0^{(n,n-1)}(x,y)|_{\alpha_1=\beta_1=0}}{J_0^{(n,\cdot)}(x)}, \qquad \frac{J_0^{(n,n-1)}(x,y)}{J_0^{(\cdot,n-1)}(y)} \tag{4.47}$$

to random rational functions allows the characteristic polynomial of the Jacobi 2λ-ensemble to be generated by a random three-term recurrence [222]. The first of the ratios (4.47) is formally identical to (4.11) with $s_1 = \cdots = s_n = \lambda$ and thus gives the density of zeros of the random rational function (4.9) with $(w_1,\dots,w_n)$ distributed according to the Dirichlet distribution $D_n[(\lambda)^n]$, with the notation $(\lambda)^n$ denoting λ repeated n times. The second of the ratios (4.47) is formally identical to (4.11) with $n \mapsto n+1$ (this is to be done by starting the labelling at zero), $\lambda_0 = 1$, $\lambda_n = 0$, $s_0 = \beta+1$, $s_n = \alpha+1$ and $s_i = \lambda$ $(i = 1,\dots,n-1)$, and thus gives the density of zeros of the random rational function

$$\tilde{R}_{n+1}(x) := \frac{w_0}{x-1} + \frac{w_{n+1}}{x} + \sum_{i=1}^n \frac{w_i}{x - y_i}, \tag{4.48}$$

with $(w_0,\dots,w_n)$ distributed according to $D_{n+1}[\beta+1,(\lambda)^{n-1},\alpha+1]$.

Let us now seek the recursive construction of a polynomial, the zeros of which have joint density given by (4.46), the latter to be referred to as the *Selberg density*.

PROPOSITION 4.3.1 *Let $A_{-1}^\#(x) := 0$, $A_0^\#(x) := 1$ and define the polynomials $\{A_j^\#(x)\}_{j=1,\dots,n}$ by the requirement that their zeros $\{\lambda_l^{(j)}\}_{l=1,\dots,j}$ say have the p.d.f. $J_0^{(j,\cdot)}(x)|_{\substack{\alpha=\alpha^{(j)}\\ \beta=\beta^{(j)}}}$ with*

$$\alpha^{(j)} := (n-j)\lambda + \alpha_0, \qquad \beta^{(j)} := (n-j)\lambda + \beta_0$$

(in particular the zeros of $A_n^\#(x)$ have the p.d.f. $J_0^{(j,\cdot)}(x)|_{\substack{\alpha=\alpha_0\\ \beta=\beta_0}}$). These polynomials satisfy the random three-term recurrence

$$A_j^\#(x) = w_2^{(j)}(x-1)A_{j-1}^\#(x) + w_0^{(j)} x A_{j-1}^\#(x) + w_1^{(j)} x(x-1)A_{j-2}^\#(x) \tag{4.49}$$

where $(w_0^{(j)}, w_1^{(j)}, w_2^{(j)})$ is distributed according to $D_3[\beta^{(j)}+1,(j-1)\lambda,\alpha^{(j)}+1]$.

Proof. Consider the rational function (4.9) with $n = j-1$. Let $\{a_i\}$ have the p.d.f. $J_0^{(j-1,\cdot)}(x)|_{\substack{\alpha=\alpha^{(j-1)}\\ \beta=\beta^{(j-1)}}}$ so that $a_l = \lambda_l^{(j)}$, and let $\{w_i\}$ be chosen from the Dirichlet distribution $D_{j-1}[(\lambda)^{j-2},\lambda]$. Then it follows from Proposition 4.2.2 and the integration formula (4.46) that the roots are the zeros of $A_{j-2}^\#(x)$ and thus

$$\sum_{l=1}^{j-1} \frac{w_l}{x - \lambda_l^{(j-1)}} = \frac{A_{j-2}^\#(x)}{A_{j-1}^\#(x)}. \tag{4.50}$$

Consider next the rational function (4.48) with $n = j-1$. Again let $\{a_i\}$ have the p.d.f. $J_0^{(j-1,\cdot)}(x)|_{\substack{\alpha=\alpha^{(j-1)}\\ \beta=\beta^{(j-1)}}}$, and let the $w_i = \tilde{w}_i^{(j)}$ be chosen from the Dirichlet distribution $D_{j+1}[\beta^{(j)}+1,(k)^{j-1},\alpha^{(j)}+1]$. Similarly, it then follows from Proposition 4.2.2 and the integration formula (4.45) that the roots occur at the zeros of $A_j^{\#}(x)$ and thus

$$\frac{A_j^{\#}(x)}{x(x-1)A_{j-1}^{\#}(x)} = \frac{\tilde{w}_j^{(j)}}{x} + \frac{\tilde{w}_0^{(j)}}{x-1} + \sum_{l=1}^{j-1} \frac{\tilde{w}_l^{(j)}}{x-\lambda_l^{(j-1)}}. \tag{4.51}$$

According to Exercises 4.3 q.2(i) the marginal distribution of each w_i in (4.50) is $D_2[\lambda, (j-2)\lambda]$, while each $\tilde{w}_l^{(j)}$ $(l = 1, \dots, j-1)$ in (4.51) has distribution $D_2[\lambda, (j-1)\lambda + \alpha^{(j)} + \beta^{(j)} + 2]$. We similarly calculate that the marginal distribution of $w_1^{(j)}$ as specified in the statement of the proposition is $D_2[(j-1)\lambda; \alpha^{(j)} + \beta^{(j)} + 2]$. Using the result of Exercises 4.3 q.1 it follows that $w_1^{(j)} w_i$ has the same distribution as $\tilde{w}_i^{(j)}$. This implies that we can substitute $w_1^{(j)}$ times the l.h.s. of (4.50) for the sum over l in (4.51) to obtain

$$\frac{A_j^{\#}(x)}{x(x-1)A_{j-1}^{\#}(x)} = \frac{\tilde{w}_j^{(j)}}{x} + \frac{\tilde{w}_0^{(j)}}{x-1} + \frac{w_1^{(j)} A_{j-2}^{\#}(x)}{A_{j-1}^{\#}(x)}, \tag{4.52}$$

where it is required that $\tilde{w}_j^{(j)} + \tilde{w}_0^{(j)} + \tilde{w}_1^{(j)} = 1$.

Again using the result of Exercises 4.3 q.2(i), we see that the marginal distribution of $w_2^{(j)}$ agrees with that of $\tilde{w}_j^{(j)}$, and that the marginal distributions of $w_0^{(j)}$ and $\tilde{w}_0^{(j)}$ agree. Thus (4.52) is a rewrite of (4.49). □

The recurrence (4.49) does not have the structure (1.154) and so does not relate to a single tridiagonal matrix. Thus it is distinct from the recurrence implied by Proposition 3.11.1 which also generates the characteristic polynomial for the Jacobi β-ensemble. In fact (4.49) relates to the generalized eigenvalue problem for two tridiagonal matrices [553].

4.3.2 Gaussian β-ensemble

A limiting form of (4.9) relating to Gaussian ensembles is obtained by first specializing some of the parameters to obtain

$$\frac{w_0}{x-1} + \frac{w_{n+1}}{x} + \sum_{i=1}^{n} \frac{w_i}{x-y_i}, \tag{4.53}$$

and requiring $(w_0, \dots, w_{n+1})$ to be distributed according to $D_{n+2}[\alpha/2, s_1, \dots, s_n, \alpha/2]$. If we then write $\alpha = L^2$ and take $L \to \infty$, the marginal distribution of w_0 and w_{n+1} have the asymptotic form $\frac{1}{2} + \frac{1}{2L}\mathrm{N}[0,1]$ while the w_i $(i = 1, \dots, n)$ have to leading order the marginal distribution $\frac{1}{L^2}\Gamma[s_i, 1]$. Hence with $x \mapsto \frac{1}{2}(1 - \frac{X}{L})$, $y_i \mapsto \frac{1}{2}(1 - \frac{a_i}{L})$ we obtain from (4.53)

$$R_{n+1}^G(X) := X - \mu_0 - \sum_{j=1}^{n} \frac{\mu_j}{X - a_j}, \tag{4.54}$$

where μ_0 has distribution $\mathrm{N}[0,1]$ and μ_j distribution $\Gamma[s_i, 1]$ $(i = 1, \dots, n)$. The same change of variables and scaling in the specialization of (4.11) giving the distribution of the zeros of (4.48) specifies the distribution of the zeros of (4.54).

PROPOSITION 4.3.2 *The zeros of the random rational function (4.54) have the p.d.f.*

$$\frac{1}{\sqrt{2\pi}\Gamma(s_1)\cdots\Gamma(s_n)} \frac{\prod_{1\le j<k\le n+1}(\lambda_j - \lambda_k)}{\prod_{1\le j<k\le n}(a_j - a_k)^{s_j+s_k-1}} \prod_{j=1}^{n+1}\prod_{p=1}^{n} |\lambda_j - a_p|^{s_p-1} \exp\Big(-\frac{1}{2}\Big(\sum_{j=1}^{n+1}\lambda_j^2 - \sum_{j=1}^{n} a_j^2\Big)\Big), \tag{4.55}$$

where

$$\infty > \lambda_1 > a_1 > \lambda_2 > \cdots > a_n > \lambda_{n+1} > -\infty. \tag{4.56}$$

We can likewise take the Gaussian limit of the joint p.d.f. (4.44), the marginal distribution (4.46) and the random three-term recurrence (4.49), the latter reproducing (1.154) for the computation of the characteristic polynomial of the Gaussian β-ensemble. Thus in (4.44) let us change variables $x_i \mapsto (\frac{1}{2}-\frac{x_i}{2L})$, $y_i \mapsto (\frac{1}{2}-\frac{y_i}{2L})$, set $\alpha = \beta = aL^2$, $\alpha_1 = \beta_1 = a_1 L^2$ and take $L \to \infty$. We then obtain the joint p.d.f.

$$G_{\rm o}^{(n,n-1)}(x,y) := \left(\frac{a}{\pi}\right)^{1/2} \frac{(2a)^{(n-1)\lambda}}{(\Gamma(\lambda))^{n-1}} \frac{1}{m_{n-1}(\lambda;2(a+a_1))}$$
$$\times \prod_{i=1}^{n} e^{-ax_i^2} \prod_{1\le i<j\le n} |x_j - x_i| \prod_{i=1}^{n-1} e^{-a_1 y_i^2} \prod_{1\le i<j\le n-1} |y_j - y_i| \prod_{i=1}^{n}\prod_{j=1}^{n-1} |x_j - y_i|^{\lambda-1}, \tag{4.57}$$

where

$$m_n(\lambda;c) := \int_{-\infty}^{\infty} dx_1 \cdots \int_{-\infty}^{\infty} dx_n \, e^{-(c/2)\sum_{l=1}^n x_l^2} \prod_{1\le i<j\le n} |x_j - x_i|^{2\lambda} \tag{4.58}$$

and the x's and y's are interlaced according to

$$\infty > x_1 > y_1 > x_2 > y_2 > \cdots > y_{n-1} > x_n > -\infty. \tag{4.59}$$

Also, in this limit (4.45) and (4.46) read

$$\int_R dx_1 \cdots dx_n \, G_{\rm o}^{(n,n-1)}(x,y) \tag{4.60}$$
$$= \frac{1}{m_{n-1}(\lambda;2(a+a_1))} \prod_{i=1}^{n-1} e^{-(a+a_1)y_i^2} \prod_{1\le i<j\le n-1} |y_i - y_j|^{2\lambda} =: G_{\rm o}^{(\cdot,n-1)}(y), \tag{4.61}$$

$$\int_R dy_1 \cdots dy_{n-1} \, G_{\rm o}^{(n,n-1)}(x,y)\Big|_{a_1=0} = \frac{1}{m_n(\lambda;2a)} \prod_{i=1}^{n} e^{-ax_i^2} \prod_{1\le i<j\le n} |x_i - x_j|^{2\lambda} =: G_{\rm o}^{(n,\cdot)}(x), \tag{4.62}$$

where here R refers to (4.59). With $a = \frac{1}{2}$, $\lambda = \beta/2$ we recognize $G_0^{(n,\cdot)}$ as the p.d.f. of the Gaussian β-ensemble (1.160).

To generate a random polynomial with zero p.d.f. (4.62) we can take the Gaussian limit of the three-term recurrence (4.49). First we note that with $(w_0^{(j)}, w_1^{(j)}, w_2^{(j)})$ distributed as specified in Proposition 4.3.1, setting $\alpha_0 = \beta_0 = aL^2$ and taking $L \to \infty$, the marginal distributions of $w_0^{(j)}$ and $w_2^{(j)}$ have the asymptotic form $\frac{1}{2} + \frac{1}{2L}{\rm N}[0, \frac{1}{\sqrt{2a}}]$ while $w_1^{(j)}$ has the leading order marginal distribution $\frac{1}{L^2}\Gamma[(j-1)\lambda, \frac{1}{2a}]$ (cf. the statements above (4.54)). Thus by also writing $x \mapsto \frac{1}{2}(1-\frac{x}{L})$, $A_j^{\#}(x) \mapsto (-2L)^{-j}C_j^{\#}(x)$, we see that (4.49) reduces to

$$C_j^{\#}(x) = (x-r)C_{j-1}^{\#}(x) - s^{(j-1)}C_{j-2}^{\#}(x) \tag{4.63}$$

where r has distribution ${\rm N}[0, \frac{1}{\sqrt{2a}}]$ while $s^{(j-1)}$ has distribution $\Gamma[(j-1)\lambda, \frac{1}{2a}]$, and $C_{-1}^{\#}(x) = 0$, $C_0^{\#}(x) = 1$. Setting $2a = 1$, $\lambda = \beta/2$ we see that (4.63) coincides with (1.154), the a_k, b_k^2 therein specified by (1.164).

A different derivation of (4.63) is also possible. This has its origin in considering a sequence of random matrices defined recursively by [222], [200]

$$\mathbf{M}_n = \begin{bmatrix} {\rm diag}\,\mathbf{M}_{n-1} & \vec{b} \\ \vec{b}^T & c \end{bmatrix} \tag{4.64}$$

where $c, \vec{b}$ are chosen from suitable distributions. First we will show that if, for fixed n, the eigenvalue distribution of $\mathbf{M}_{n-1}$ is prescribed to equal $G_{\rm o}^{(\cdot,n-1)}(y)$ as specified in (4.60), then the joint distribution of the eigenvalues of $\mathbf{M}_n$ and $\mathbf{M}_{n-1}$ is given by (4.57). To see this, note that with the eigenvalues of $\mathbf{M}_{n-1}$ denoted $\{y_1, \dots, y_{n-1}\}$,

$$\det(\mathbf{1}_n x - \mathbf{M}_n) = \det(\mathbf{1}_{n-1} x - \mathbf{M}_{n-1})\Big(x - c - \sum_{j=1}^{n-1} \frac{b_j^2}{x - y_j}\Big). \tag{4.65}$$

Hence the eigenvalues of $\mathbf{M}_n$ are specified by the zeros of (4.54) with $\mu_0 = c$, $\mu_j = b_j^2$. Suppose now that c has distribution $\mathrm{N}[0, 1/\sqrt{2a}]$, while b_j^2 has distribution $\Gamma[\lambda, 1/2a]$. It then follows from Proposition 4.3.2 that the eigenvalue p.d.f. of $\mathbf{M}_n$, given the eigenvalues of $\mathbf{M}_{n-1}$, is equal to (4.55) with $n \mapsto n-1$, $s_i = \lambda$, $\lambda_j \mapsto \sqrt{2a}x_j$, $a_j \mapsto \sqrt{2a}a_j$, and multiplied by $(2a)^{n/2}$. In the case that the eigenvalue p.d.f. of $\mathbf{M}_{n-1}$ is given by $G_{\rm o}^{(\cdot, n-1)}(y)$, multiplication by this latter quantity shows that the joint eigenvalue p.d.f. is indeed equal to (4.57).

Similar working suffices to compute the eigenvalue p.d.f. of $\mathbf{M}_n$ defined recursively by (4.64) for $n = 1, 2, \dots$.

PROPOSITION 4.3.3 *Consider the sequence of random matrices (4.64) with c, b_j^2 having distribution* $\mathrm{N}[0,1]$, $\Gamma[\beta/2, 1]$. *The eigenvalue p.d.f. of $\mathbf{M}_n$ is equal to the p.d.f. (4.62) (with $a = \frac{1}{2}$, $\lambda = \beta/2$) specifying the Gaussian β-ensemble, and furthermore the characteristic polynomial satisfies the three-term recurrence (4.63) with $a = 1/2$, $\lambda = \beta/2$.*

Proof. With the eigenvalues of $\mathbf{M}_n, \mathbf{M}_{n-1}$ denoted $\{x_i\}, \{y_j\}$, let $G_{n-1}(\{x_i\}, \{y_j\})$ denote the conditional eigenvalue p.d.f. of $\mathbf{M}_n$. We know then that G_{n-1} is given by (4.55) with $n \mapsto n-1$, $s_i = \beta/2$. The eigenvalue p.d.f. of $\mathbf{M}_n$, $p_n(x_1, \dots, x_n)$ say, must satisfy the recurrence

$$p_n(x_1, \dots, x_n) = \int_R dy_1 \cdots dy_{n-1}\, G_{n-1}(\{x_i\}, \{y_j\}) p_{n-1}(y_1, \dots, y_{n-1}), \tag{4.66}$$

where R denotes the region (4.59) and $p_0 := 1$. The integration formula (4.62) shows that the solution of this recurrence is given by setting p_n equal to (4.62), with $a = 1/2$, $\lambda = \beta/2$, as required.

Let $q_n(x)$ denote the characteristic polynomial of $\mathbf{M}_n$. Then (4.65) gives

$$\frac{q_{n+1}(x)}{q_n(x)} = x - c - \sum_{j=1}^{n} \frac{b_j^2}{x - y_j}. \tag{4.67}$$

On the other hand, since (4.64) is a symmetric matrix, according to (1.149) we must have

$$\frac{q_{n-1}(x)}{q_n(x)} = \sum_{j=1}^{n} \frac{\mu_j^2}{x - y_j} \tag{4.68}$$

for μ_j^2 the square of the first component of the eigenvector corresponding to the eigenvalue y_i. Choosing

$$\mu_j^2 = b_i^2 \Big/ \sum_{j=1}^{n} b_j^2 \tag{4.69}$$

we can check from Proposition 4.2.5 and (4.11) that the implied joint distribution of $\{y_i\}$ and the zeros $\{x_i\}$ of $q_{n-1}(x)$ is the same as that implied by (4.66), thus justifying (4.69). Substituting (4.69) in (4.68), and noting that $\sum_{j=1}^{n} b_j^2$ has distribution $\Gamma[n\beta/2, 1]$ reclaims the three-term recurrence (4.63). □

Note that in the case $\beta = 1$ the construction (4.64) produces a matrix $\mathbf{M}_n \in \mathrm{GOE}_n$ by the appendage of an extra row and column to a matrix $\mathbf{M}_{n-1} \in \mathrm{GOE}_{n-1}$.

4.3.3 Laguerre β-ensemble

In the Dirichlet distribution (4.10), set $s_n = L/\sigma$ and scale $\rho_1, \rho_2, \dots, \rho_{n-1}$ so that $\rho_i = x_i/L$ ($i = 1, \dots, n-1$). Taking the limit $L \to \infty$ we see that (4.10) reduces to the product of independent gamma distributions

$$\frac{\sigma^{-(n-1)}}{\Gamma(s_1)\cdots\Gamma(s_{n-1})} \prod_{i=1}^{n-1} (x_i/\sigma)^{s_i-1} e^{-x_i/\sigma}, \qquad x_i > 0. \tag{4.70}$$

With the lower terminal in (4.9) set equal to $i = 0$ (or equivalently $n \mapsto n+1$), this result suggests scaling $w_i \mapsto w_i/L$, $a_i \mapsto a_i/L$ ($i = 1, \dots, n$), $\lambda \mapsto \lambda/L$ therein, setting $a_0 = 1$ and taking the limit $L \to \infty$. Noting that then $w_0 \to 1$, we see the limiting form of (4.9) reads

$$R^{\rm L}(\lambda) := 1 + \sum_{i=1}^{n} \frac{w_i}{a_i - \lambda}, \tag{4.71}$$

where each w_i is now distributed according to $\Gamma[s_i, \sigma]$. The distribution of the roots of (4.71) is given by the corresponding Laguerre type limiting form of the Dixon-Anderson density (4.11) with $n \mapsto n+1$.

PROPOSITION 4.3.4 *Consider the random rational function (4.71) with the coefficients w_i distributed according to $\Gamma[s_i, 1]$. This has exactly n roots, all of which are real, and for given $a_1, \dots, a_n$ these roots have the p.d.f.*

$$\frac{1}{\Gamma(s_1)\cdots\Gamma(s_n)} e^{-\sum_{j=1}^{n}(\lambda_j - a_j)} \prod_{1 \le i < j \le n} \frac{(\lambda_i - \lambda_j)}{(a_i - a_j)^{s_i + s_j - 1}} \prod_{i,j=1}^{n} |\lambda_i - a_j|^{s_j - 1}, \tag{4.72}$$

where

$$\lambda_1 > a_1 > \lambda_2 > a_2 > \cdots > \lambda_n > a_n. \tag{4.73}$$

The joint p.d.f. (4.44) has a well-defined Laguerre limit, specified by changing variables $x_i \mapsto x_i/L$, $y_i \mapsto y_i/L$, setting $\beta = L/b$, $\beta_1 = L/b_1$ and taking the limit $L \to \infty$. This gives

$$\begin{aligned} &L_{\rm o}^{(n,n-1)}(x,y) \\ &:= \frac{1}{\Gamma(1+\alpha)(\Gamma(k))^{n-1}} \frac{1}{\widetilde{W}_{n-1}(\alpha + \alpha_1 + \lambda, \lambda; bb_1/(b+b_1))} \prod_{i=1}^{n} x_i^{\alpha} e^{-x_i/b} \prod_{1 \le i < j \le n} |x_j - x_i| \\ &\quad \times \prod_{i=1}^{n-1} y_i^{\alpha_1} e^{-y_i/b_1} \prod_{1 \le i < j \le n-1} |y_j - y_i| \prod_{i=1}^{n} \prod_{j=1}^{n-1} |x_j - y_i|^{\lambda - 1}, \end{aligned} \tag{4.74}$$

where

$$\widetilde{W}_n(a, \lambda; b) = \int_0^{\infty} dx_1 \cdots \int_0^{\infty} dx_n \prod_{l=1}^{n} x_l^{a} e^{-x_l/b} \prod_{1 \le j < k \le n} |x_k - x_j|^{2\lambda}$$

and the xs and ys are interlaced according to

$$x_1 > y_1 > x_2 > y_2 > \cdots > y_{n-1} > x_n > 0. \tag{4.75}$$

We note too the Laguerre limit of the marginal distributions (4.45), (4.46),

$$\int_{\tilde R} dx_1 \cdots dx_n \, L_{\rm o}^{(n,n-1)}(x,y) = \frac{1}{\widetilde W_n(\alpha+\alpha_1+\lambda,\lambda; bb_1/(b+b_1))} \prod_{i=1}^{n-1} y_i^{\alpha+\alpha_1+\lambda} e^{-bb_1 y_i/(b+b_1)}$$
$$\times \prod_{1\le j<k\le n-1} |y_k - y_j|^{2\lambda} =: L_{\rm o}^{(\cdot,n-1)}(y), \tag{4.76}$$

$$\int_{\tilde R} dy_1 \cdots dy_{n-1} \, L_{\rm o}^{(n,n-1)}(x,y)\Big|_{\substack{\alpha_1=0\\ 1/b_1=0}} = \frac{1}{\widetilde W_n(\alpha,\lambda;b)} \prod_{i=1}^{n} x_i^{\alpha} e^{-x_i/b} \prod_{1\le j<k\le n} |x_k - x_j|^{2\lambda} =: L_{\rm o}^{(n,\cdot)}(x), \tag{4.77}$$

where $\tilde R$ refers to (4.75). With $b=2$, $\alpha=\beta a/2$ and $\lambda=\beta/2$ the latter is the p.d.f. specifying the Laguerre β-ensemble (3.128).

Let us also note the Laguerre limit of the random three-term recurrence (4.49). The Laguerre limit is obtained by scaling $x\mapsto x/L$, $w_1^{(j)} \mapsto v_1^{(j)}/L$, $w_2^{(j)} \mapsto v_2^{(j)}/L$, $w_0^{(j)}=1$, where the $v_1^{(j)}, v_2^{(j)}$ are distributed according to the gamma distributions $\Gamma[(j-1)\lambda, b]$, $\Gamma[(n-j)\lambda+\alpha_0+1, b]$, respectively. With $v_1^{(j)}, v_2^{(j)}$ so specified, and introducing the further scaling $A_j^{\#}(x) = L^{-j} B_j^{\#}(x)$ we see that the Laguerre limit of (4.49) reads

$$B_j^{\#}(x) = (x - v_2^{(j)}) B_{j-1}^{\#}(x) - x v_1^{(j)} B_{j-2}^{\#}(x). \tag{4.78}$$

This recurrence is to be solved subject to the initial conditions $B_{-1}^{\#}(x)=0$ and $B_0^{\#}(x)=1$. The zeros of $B_n^{\#}(x)$ are then distributed according to $L_{\rm o}^{(n,\cdot)}(x)|_{\alpha=\alpha_0}$. Note that (4.78) differs from the random three-term recurrence (1.154) with the substitutions (3.132). Such recurrences are in fact related to Laurent orthogonal polynomials [499].

In Proposition 4.3.3 the recursive construction of a sequence of $n\times n$ matrices, each with eigenvalue p.d.f. specifying a Gaussian β-ensemble, was noted. It is similarly possible to give a recursive construction of a sequence of matrices whose eigenvalues realize the Laguerre β-ensemble p.d.f. [199]. As noted below the proof of Proposition 4.3.3, the recursive construction in the Gaussian case can be motivated by the natural recursive structure of GOE matrices. In the case of the LOE, the eigenvalue p.d.f. (3.16) with $a=N-n-1$, and n nonzero eigenvalues, is realized by $\mathbf{X}_{(n)}^T \mathbf{X}_{(n)}$, where $\mathbf{X}_{(n)}$ is an $n\times N$ matrix of standard Gaussian entries. These matrices satisfy the recurrence

$$\mathbf{X}_{(n+1)}^T \mathbf{X}_{(n+1)} = \mathbf{X}_{(n)}^T \mathbf{X}_{(n)} + \vec x \vec x^T, \qquad n=0,1,\dots,N, \tag{4.79}$$

for $\vec x$ an $N\times 1$ Gaussian column vector. This suggests inductively defining a sequence of $N\times N$ positive definite matrices, indexed by the number of nonzero eigenvalues n, according to

$$\mathbf{A}_{(n+1)} = \mathrm{diag}\, \mathbf{A}_{(n)} + \vec x \vec x^T, \qquad n=0,1,\dots,N, \tag{4.80}$$

where $\mathrm{diag}\, \mathbf{A}_{(n)}$ refers to the diagonal form of $\mathbf{A}_{(n)}$, and $\mathbf{A}_{(0)} := [0]_{N\times N}$. By an appropriate choice of the distribution of the elements of the random vector $\vec x$, the eigenvalue p.d.f. of each $\mathbf{A}_{(n)}$ is given by a Laguerre β-ensemble.

PROPOSITION 4.3.5 *Let the jth entry x_j $(j=1,\dots,N)$ of the random vector $\vec x$ in (4.79) be such that x_j^2 has distribution $\Gamma[\beta/2,1]$. The eigenvalue p.d.f. for the nonzero eigenvalues of $\mathbf{A}_{(n)}$ is proportional to*

$$\prod_{l=1}^{n} \lambda_l^{(N-n+1)\beta/2-1} e^{-\lambda_l} \prod_{1\le j<k\le n} |\lambda_k - \lambda_j|^{\beta}. \tag{4.81}$$

Proof. We see from (4.79) that

$$\det(\lambda \mathbf{1}_N - \mathbf{A}_{(n+1)}) = \det(\lambda \mathbf{1}_N - \mathbf{A}_{(n)}) \det(\mathbf{1}_N - (\lambda \mathbf{1}_N - \mathbf{A}_{(n)})^{-1} \vec{x}\vec{x}^T)$$

and thus

$$\frac{\det(\lambda \mathbf{1}_N - \mathbf{A}_{(n+1)})}{\det(\lambda \mathbf{1}_N - \mathbf{A}_{(n)})} = 1 - \sum_{j=1}^n \frac{x_j^2}{\lambda - a_j} - \frac{\sum_{j=n+1}^N x_j^2}{\lambda}, \tag{4.82}$$

where $\{a_j\}_{j=1,\dots,n}$ denotes the eigenvalues of $\mathbf{A}_{(n)}$. The eigenvalues of $\mathbf{A}_{(n+1)}$ are thus given by the zeros of (4.82). Proposition 4.3.4 then tells us that the conditional p.d.f. for the eigenvalues of $\mathbf{A}_{(n+1)}$ is given by (4.72) with $n \mapsto n+1$, $s_1 = \cdots = s_n = \beta/2$, $s_{n+1} = (N-n)\beta/2$, $a_{n+1} = 0$. With $L_n(\{\lambda_j\}_{j=1,\dots,n+1}; \{a_j\}_{j=1,\dots,n})$ denoting this conditional p.d.f, the eigenvalue p.d.f. of $\mathbf{A}_{(n)}$ can then be specified as the solution of the recurrence (4.66), modified so that $n \mapsto n+1$, $\{x_j\} \mapsto \{\lambda_j\}$, $\{y_j\} \mapsto \{a_j\}$, $G_n \mapsto L_n$ and R equal to the region (4.73) with $n \mapsto n+1$. The integration formula (4.77) shows that (4.81) satisfies this recurrence. □

4.3.4 The circular Jacobi β-ensemble

As we have seen, the Dixon-Anderson density is intimately related to the Jacobi β-ensemble, and the limiting forms (4.55) and (4.72) relate to the Gaussian and Laguerre β-ensembles. In Exercises 4.2 q.3 a circular analogue of the Dixon-Anderson density was given. We will show in this section that a stereographic projection of the latter gives rise to a conditional p.d.f. which relates to the generalized Cauchy β-ensemble (3.124) [223]. This in turn can be used to derive a random three-term recurrence for the characteristic polynomial of the circular Jacobi β-ensemble.

We begin by equating (4.38) and (4.40) from Exercises 4.2 q.3, extending the lower terminals to zero and projecting the variables $\lambda, t, \lambda_j, \tilde{\lambda}_j$ with modulus unity onto the real line via the stereographic projections

$$\lambda = \frac{z-i}{z+i}, \quad t = \frac{c-i}{c+i}, \quad \lambda_j = \frac{y_j - i}{y_j + i} \ (j \neq 0), \quad \lambda_0 = 1, \quad \tilde{\lambda}_j = \frac{x_j - i}{x_j + i}. \tag{4.83}$$

Noting from (4.41) that

$$\frac{q_0}{c+i} = \frac{\prod_{l=1}^n (y_l + i)}{\prod_{l=0}^n (x_l + i)}, \tag{4.84}$$

we obtain

$$\frac{\prod_{j=0}^n (z - x_j)}{(z^2+1)\prod_{j=1}^n (z - y_j)} = \frac{z - c}{q_0(z^2+1)} - \sum_{j=1}^n \frac{(q_j/q_0)}{z - y_j}, \tag{4.85}$$

where

$$x_0 > y_1 > x_1 > \cdots > y_n > x_{n+1}. \tag{4.86}$$

Analogous to the result of Exercises 4.3 q.3(v), with $\{q_j\}$ specified to have a Dirichlet distribution, the distribution of the zeros of (4.85) can readily be determined by making use of (4.84), the formula

$$q_j = \frac{q_0}{|1 + iy_j|^2} \frac{\prod_{l=0}^n |y_j - x_l|}{\prod_{l=1, l\neq j}^n |y_j - y_l|}$$

and the Jacobian relation

$$d\vec{q} \wedge dc = \frac{J}{2}|1 + ic|^2 \prod_{j=0}^n \frac{2}{(1 + ix_j)(1 - ix_j)} d\vec{x},$$

where J is specified by (4.42), rewritten according to (4.83).

PROPOSITION 4.3.6 *Consider the rational function (4.85). Let $\{q_j\}_{j=0,\dots,n-1}$ have the Dirichlet distribu-*

tion

$$\frac{\Gamma(\sum_{j=0}^n d_j)}{\prod_{j=0}^n \Gamma(d_j)} \prod_{j=0}^n q_j^{d_j-1}, \qquad \sum_{j=0}^n q_j = 1, \tag{4.87}$$

where

$$\sum_{j=0}^n d_j + 1 = 2\mathrm{Re}\,\gamma.$$

Let c have the generalized Cauchy distribution

$$\frac{\Gamma(\gamma)\Gamma(\bar\gamma)}{\pi 2^{2(1-\mathrm{Re}\,\gamma)}\Gamma(2\mathrm{Re}\,\gamma - 1)}(1+ic)^{-\gamma}(1-ic)^{-\bar\gamma}.$$

The zeros of the random rational function (4.85) have the p.d.f.

$$\tilde A \prod_{j=0}^n (1+ix_j)^{-\gamma}(1-ix_j)^{-\bar\gamma} \prod_{j=1}^n (1+iy_j)^{\gamma-d_j}(1-iy_j)^{\bar\gamma-d_j}$$
$$\times \prod_{j=1}^n \prod_{l=0}^n |y_j - x_l|^{d_j-1} \prod_{1\le j<k\le n} |y_j-y_k|^{1-d_j-d_k} \prod_{0\le j<k\le n} |x_j - x_k|, \tag{4.88}$$

where

$$\tilde A = \frac{\Gamma(\gamma)\Gamma(\bar\gamma)}{\pi 2^{2(1-\mathrm{Re}\,\gamma)}} \frac{1}{\Gamma(2\mathrm{Re}\,\gamma - 1 - \sum_{i=1}^n d_i)\prod_{j=1}^n \Gamma(d_j)}$$

and the inequalities (4.86) are assumed.

It is now possible to derive the sought three-term recurrence.

PROPOSITION 4.3.7 *Let $p_n(z;\gamma;d)$ denote the monic random polynomial of degree n, specified by its zeros $\{x_j\}_{j=1,\dots,n}$ having p.d.f. proportional to*

$$\prod_{j=1}^n (1+ix_j)^{-\gamma}(1-ix_j)^{-\bar\gamma} \prod_{1\le j<k\le n} |x_j - x_k|^{2d} \tag{4.89}$$

(the value of the proportionality can be read off from (4.156) below). With $\mathrm{B}[\alpha,\beta]$ denoting the classical beta distribution on $[0,1]$ as specified in Exercises 4.3 q.1, let

$$b_n \in \mathrm{B}[2\mathrm{Re}\,\gamma + nd - 1, nd] \quad (n \ne 0), \;\; b_0 = 1,$$

and let c_n have the Cauchy distribution

$$\frac{\Gamma(\gamma+nd)\Gamma(\bar\gamma+nd)}{\pi 2^{2(1-nd-\mathrm{Re}\,\gamma)}\Gamma(2(\mathrm{Re}\,\gamma+nd)-1)}(1+ic)^{-(\gamma+nd)}(1-ic)^{-(\bar\gamma+nd)}.$$

We have

$$p_{n+1}(z;\gamma+nd;d)$$
$$= \frac{(z-c_n)}{b_n} p_n(z;\gamma+(n-1)d;d) + \Big(1-\frac{1}{b_n}\Big)(1+z^2)p_{n-1}(z;\gamma+(n-2)d;d), \tag{4.90}$$

where $p_0 = 1$.

Proof. According to (4.85)

$$\frac{p_{n+1}(z;\gamma;d)}{(z^2+1)p_n(z;\gamma-d;d)} = \frac{z-c}{q_0(z^2+1)} - \sum_{j=1}^{n} \frac{(q_j/q_0)}{z-y_j}. \tag{4.91}$$

To obtain a companion identity to (4.91), introduce the random rational function

$$\frac{\prod_{k=1}^{n-1}(z-u_k)}{\prod_{j=1}^{n}(z-y_j)} = \sum_{j=1}^{n} \frac{\mu_j}{z-y_j},$$

where $\{\mu_j\}$ have the Dirichlet distribution $D_n[(d)^n]$. The distribution of $\{u_k\}$ is then given by Proposition 4.2.1, appropriately specialized. With $\{y_j\}_{j=1,\dots,n}$ again chosen to have distribution given by the zeros of $p_n(z;\gamma-d;d)$, the marginal distribution of $\{u_k\}$ can be calculated by noting from (4.88) that

$$\begin{aligned}\tilde{A}_d \int_{R'} dx_0 \cdots dx_n \prod_{j=0}^{n}(1+ix_j)^{-\gamma}(1-ix_j)^{-\bar{\gamma}} \prod_{k=1}^{n}\prod_{l=0}^{n}|y_k-x_l|^{d-1} \prod_{0\le j<k\le n}|x_j-x_k| \\ = \prod_{j=1}^{n}(1+iy_j)^{-\gamma+d}(1-iy_j)^{-\bar{\gamma}+d} \prod_{1\le j<k\le n}|y_j-y_k|^{2d-1},\end{aligned} \tag{4.92}$$

where R' denotes the region (4.86) while $\tilde{A}_d := \tilde{A}|_{d_1=\cdots=d_n=d}$, and is thus equal to the zero distribution of $p_{n-1}(z;\gamma-2d;d)$. Consequently

$$\frac{p_{n-1}(z;\gamma-2d;d)}{p_n(z;\gamma-d;d)} = \sum_{j=1}^{n} \frac{\mu_j}{z-y_j}. \tag{4.93}$$

In (4.91) and (4.93) let us replace $\gamma \mapsto \gamma + nd$. From Exercises 4.3 q.2(i) we calculate that

$$q_j \in \mathrm{B}[d,(n-1)d] \ \ (j\neq 0), \quad q_0 \in \mathrm{B}[2\mathrm{Re}\,\gamma + nd - 1, nd], \quad \mu_j \in \mathrm{B}[d,(n-1)d]$$

subject to the constraints $\sum_{j=0}^{n} q_j = 1$, $\sum_{j=1}^{n}\mu_j = 1$. Taking this into consideration, and comparing the two identities, (4.90) results. □

To relate this to the circular Jacobi 2λ-ensemble (3.122), we know from (3.123) that with the change of variables $x_j = i(1-e^{i\theta_j})/(1+e^{i\theta_j})$ the p.d.f. (4.89) with $\gamma \mapsto \gamma + 2d$ (γ real) maps to (3.122) with $b = 2\gamma - 2$, $\lambda = d$. Thus the zeros of the polynomial $p_n(z;\gamma+(n-1)d;d)$, with γ real, $x_1,\dots,x_n$ say, under the inverse of this change of variables give for $\{\theta_j\}$ the distribution (3.122) with $b = 2\gamma-2$, $\lambda = d$. In particular this gives a recurrence for the characteristic polynomial of the circular β-ensemble distinct from (2.89).

EXERCISES 4.3 1. The objective of this exercise is to show that with the beta distribution defined as the two-dimensional Dirichlet distribution, $B[\alpha,\beta] = D_2[\alpha;\beta]$, and $x \in B[a+b,c]$, $y \in B[a,b]$, then $xy \in B[a,b+c]$.

(i) For $x \in B[c,d]$, $y \in B[a,b]$ note that the joint distribution of xy is proportional to

$$x^{c-1}(1-x)^{d-1}y^{a-1}(1-y)^{b-1},$$

and from this deduce that the joint distribution of $u := x$ and $z := xy$ is proportional to

$$u^{c-a-b}(1-u)^{d-1}z^{a-1}(u-z)^{b-1}.$$

(ii) In the case $c = a+b$ integrate over u to deduce that the p.d.f. of z is proportional to

$$z^{a-1}(1-z)^{b+d-1}.$$

2. (i) If $(w_0, \dots, w_n)$ is distributed according to $D_{n+1}[\alpha_0, \dots, \alpha_n]$, show that the marginal distribution of w_j $(j = 0, \dots, n-1)$ is given by $B[\alpha_j, \sum_{i=0, i\neq j}^n \alpha_i]$.

   (ii) In the setting of (i), show that the marginal distribution of $w_j + w_k$, $(j \neq k,\ j,k \le n)$ is $B[\alpha_j + \alpha_k, \sum_{i=0, i\neq j,k}^n \alpha_i]$.

3. (i) [45], [222] For $\mathbf{X} \in \mathrm{GUE}_{n+1}^*$ (the $*$ is used as in Section 1.6.1), and with $\mathbf{Y}$ denoting the top $n \times n$ principal minor of $\mathbf{X}$, let $\mathbf{U}$ be such that $\mathbf{U}^{-1}\mathbf{Y}\mathbf{U} = \mathrm{diag}(y_1, \dots, y_n) =: \tilde{\mathbf{Y}}$ where $\{y_j\}$ are the eigenvalues of $\mathbf{Y}$. Show that

$$\begin{bmatrix} \mathbf{U}^{-1} & \vec{0} \\ \vec{0}^T & 1 \end{bmatrix} \begin{bmatrix} \mathbf{Y} & \vec{x}_{n+1} \\ \vec{x}_{n+1}^T & x_{n+1,n+1} \end{bmatrix} \begin{bmatrix} \mathbf{U} & \vec{0} \\ \vec{0}^T & 1 \end{bmatrix} = \begin{bmatrix} \tilde{\mathbf{Y}} & \vec{y} \\ \vec{y}^\dagger & a \end{bmatrix},$$

where $\vec{y}$ is an n-component complex vector with Gaussian entries $\mathrm{N}[0, 1/\sqrt{2}] + i\mathrm{N}[0, 1/\sqrt{2}]$ and a has distribution N[0,1]. Use the result of Proposition 4.3.2 with $s_p = 1$ $(p = 1, \dots, n)$ to deduce that the joint distribution of the eigenvalues $\{x_j\}$ of $\mathbf{X}$, and the eigenvalues $\{y_j\}$ of its top $n \times n$ principal minor is proportional to

$$\prod_{l=1}^{n+1} e^{-x_l^2/2} \prod_{1\le j<k\le n+1} (x_j - x_k) \prod_{1\le j'<k'\le n} (y_{j'} - y_{k'}), \tag{4.94}$$

where

$$x_1 > y_1 > \cdots > x_n > y_n > x_{n+1}. \tag{4.95}$$

Deduce this same result from Proposition 4.2.5 by noting

$$\begin{bmatrix} \mathbf{Y} & \vec{0} \\ \vec{0}^T & 0 \end{bmatrix} = \mathrm{diag}(1, \dots, 1, 0)\mathbf{X}\mathrm{diag}(1, \dots, 1, 0) \sim \mathbf{\Pi}\,\mathrm{diag}\,\mathbf{X}\,\mathbf{\Pi},$$

where $\mathbf{\Pi}$ is a complex corank 1 random projection, and the symbol $\sim$ indicates the matrices in question are similar (i.e. have the same eigenvalues).

   (ii) [45] In the setting of (i), let the eigenvalues of the top $j \times j$ principal minor $(j = 1, \dots, n+1)$ be denoted $x_l^{(j)}$, $(l = 1, \dots, j)$. Extend the working therein to show that the joint p.d.f. of $\{x_l^{(j)}\}_{l=1,\dots,j}$ for $j = 1, \dots, n+1$ is proportional to

$$\prod_{l=1}^{n+1} e^{-(x_l^{(n+1)})^2/2} \prod_{1\le j<k\le n+1} (x_j^{(n+1)} - x_k^{(n+1)}) \prod_{j=1}^{n} \chi(x^{(j+1)} > x^{(j)}), \tag{4.96}$$

where

$$\chi(x^{(j+1)} > x^{(j)}) := \chi_{x_1^{(j+1)} > x_1^{(j)} > \cdots > x_j^{(j+1)} > x_j^{(j)} > x_{j+1}^{(j+1)}}.$$

   (iii) Use the characteristic equation for the eigenvalues of a general $(n+1) \times (n+1)$ Hermitian matrix $\mathbf{X}$, implied by the first displayed equation in (i) to deduce the interlacing property (4.95) with the eigenvalues of a principal minor (assuming the latter are distinct). Deduce the same result from the final displayed relation in (i).

4. (i) [222] Let $\mathbf{A} = \mathbf{X}^\dagger\mathbf{X}$, where $\mathbf{X}$ is an $n \times N$ $(n > N)$ complex Gaussian matrix such that $\mathbf{A}$ is a member of the LUE with $a = n - N$. Suppose b times a row of complex Gaussians is appended to $\mathbf{X}$ to form $\mathbf{Y}$, and consider $\mathbf{B} = \mathbf{Y}^\dagger\mathbf{Y}$. By noting that this is equivalent to the rank 1 perturbation $\mathbf{B} = \mathbf{A} + b\vec{x}^\dagger\vec{x}$ as considered in (3.72), use (3.73) and Proposition 4.3.4 to conclude that the joint distribution of the eigenvalues of $\mathbf{A}$ and $\mathbf{B}$ is

$$\prod_{l=1}^{N} a_l^a e^{-\sum_{j=1}^N a_j} e^{-(1/b)\sum_{j=1}^N (b_j - a_j)} \prod_{1\le j<k\le N} (b_j - b_k)(a_j - a_k), \tag{4.97}$$

where

$$b_1 > a_1 > b_2 > a_2 > \cdots > b_N > a_N. \tag{4.98}$$

   (ii) [215] Let $\mathbf{X}_{(N)}$ denote an $N \times N$ complex Gaussian matrix such that $\mathbf{X}_{(N)}^\dagger\mathbf{X}_{(N)}$ is a member of the LUE with $a = 0$. Let $\mathbf{X}_{(n)}$ denote the $n \times N$ matrix formed from the first n rows of $\mathbf{X}_{(N)}$. Let $x_j^{(n)}$

$(j = 1, \dots, n)$ denote the nonzero eigenvalues of $\mathbf{X}_{(n)}$. Proceed as in (i) to show that the joint p.d.f. of $\{x_j^{(n)}\}_{j=1,\dots,n}$ for $n = 1, \dots, N$ is equal to

$$\prod_{l=1}^{N} e^{-x_l^{(N)}} \prod_{1 \le j < k \le N} (x_j^{(N)} - x_k^{(N)}) \prod_{p=1}^{N-1} \chi(x^{(p+1)} > x^{(p)}), \tag{4.99}$$

where

$$\chi(x^{(p+1)} > x^{(p)}) := \chi_{x_1^{(p+1)} > x_1^{(p)} > \cdots > x_p^{(p+1)} > x_p^{(p)} > x_{p+1}^{(p+1)} > 0}. \tag{4.100}$$

5. [222] Let $\mathbf{A}$ be an $n \times n$ member of the GSE with variance such that the eigenvalue p.d.f. of the independent eigenvalues is proportional to

$$\prod_{l=1}^{n} e^{-y_l^2} \prod_{1 \le j < k \le n} (y_j - y_k)^4. \tag{4.101}$$

Regard $\mathbf{A}$ as a $4n \times 4n$ real matrix by replacing each complex entry in the real quaternion according to (1.36) so that $\mathbf{A}$ is now fourfold degenerate. Next border $\mathbf{A}$ by one extra row and column specified as $\sqrt{b}$ times a vector of standard Gaussians, obtaining eigenvalues $\{x_i\}$. Use the theory below (4.65) to conclude that the joint distribution of $\{x_i\}$ and $\{y_i\}$ is proportional to

$$\prod_{l=1}^{n+1} e^{-x_l^2/2b} \prod_{1 \le j < k \le n+1} (x_j - x_k) \prod_{l=1}^{n} e^{-y_l^2(1-1/2b)} \prod_{1 \le j < k \le n} (x_j - x_k) \prod_{j=1}^{n+1} \prod_{k=1}^{n} |x_j - y_k|, \tag{4.102}$$

where

$$x_1 > y_1 > x_2 > \cdots > y_n > x_{n+1}.$$

Note that (4.102) in the case $b = 1$ is identical to the eigenvalue p.d.f. for GOE_{2n+1}, and so conclude from (4.101) that

$$\mathrm{even}(\mathrm{GOE}_{2n+1}) = \mathrm{GSE}_n,$$

thus realizing (4.27).

4.4 GENERALIZATION OF THE DIXON-ANDERSON INTEGRAL

We know from Proposition 4.2.6 that the Dixon-Anderson integral (4.15) implies an inter-relation between the distribution of every second eigenvalue in a Jacobi orthogonal ensemble, and the eigenvalue PDF of a Jacobi symplectic ensemble. This integral can be generalized to give an integration formula which relates the distribution of every $(r+1)$th eigenvalue in a certain Jacobi ensemble with $\beta = 2/(r+1)$ to the eigenvalue PDF of another Jacobi ensemble with $\beta = 2(r+1)$ [199]. First the generalization of (4.15) will be presented.

PROPOSITION 4.4.1 *Let A_r denote the interlaced region*

$$a_j > \lambda_{r(j-1)+1} > \lambda_{r(j-1)+2} > \cdots > \lambda_{rj} > a_{j+1} \qquad (j = 1, \dots, n-1), \tag{4.103}$$

and let $\hat{C}$ be specified in terms of the Selberg integral by

$$\hat{C} = \prod_{l=1}^{n-1} \frac{1}{r!} S_r\Big(\Big(\sum_{p=1}^{l} s_p\Big) + 2(l-1)r/(r+1) - l, s_l - 1, 1/(r+1)\Big). \tag{4.104}$$

One has

$$\frac{1}{\hat{C}}\int_{A_r} d\lambda_1\cdots d\lambda_{r(n-1)} \prod_{1\le j<k\le r(n-1)} (\lambda_j-\lambda_k)^{2/(r+1)} \prod_{j=1}^{r(n-1)}\prod_{p=1}^{n}|\lambda_j-a_p|^{s_p-1} = \prod_{1\le j<k\le n}(a_j-a_k)^{r(s_j+s_k-2/(r+1))}. \tag{4.105}$$

Proof. Essential use will be made of the fact, to be established in Exercises 4.4 q.1, that the integral itself, $L_{r,n}(\{a_p\})$, say, interpreted as an analytic function of $\{a_p\}_{p=1,\dots,n}$, is symmetric in $\{(a_p,s_p)\}$ up to a phase in $\{s_p\}$. With this assumed, we begin by considering $L_{r,n}(\{a_p\})$ as an analytic function of a_1 in the appropriately cut complex a_1-plane. This function has singularities at $a_2,\dots,a_n$. In relation to the singularity at a_2, taking $a_1\mapsto a_2$ reveals that the singular behavior results from the integration of (4.103) for $j=1$ only, which can effectively be factorized from the rest of the integral. Simple rescaling of the integrand makes this singular behavior explicit, leading to the result

$$\begin{aligned} L_{r,n}(\{a_p\}) = {} & \Big(\prod_{p=3}^{n}(a_2-a_p)^{r(s_p-1)}\Big)(a_1-a_2)^{r(r-1)/(r+1)+r(s_1+s_2-1)}\frac{1}{r!}S_r(s_1-1,s_2,1/(r+1)) \\ & \times L_{r,n-1}(\{a_p\}_{p=2,\dots,n})|_{s_1\mapsto s_1+s_2+2/(r+1)-1}F(a_1-a_2;\{a_p\}_{p=2,\dots,n}), \end{aligned} \tag{4.106}$$

where $F(z;\{a_p\}_{p=2,\dots,n})$ is analytic in z and equal to unity at $z=0$. Using now the fact that the analytic continuation of $L_{r,n}(\{a_p\})$ is symmetric in $\{(a_p,s_p)\}$ gives the further factorization formula

$$L_{r,n}(\{a_p\}) = R_{r,n}(\{a_p\})G(\{a_p\}), \tag{4.107}$$

where $R_{r,n}$ denotes the r.h.s. of (4.105) and G is analytic in $\{a_p\}$ and symmetric in $\{(a_p,s_p)\}$.

To determine G, one can check that upon the replacements $a_p\mapsto ca_p$ both $L_{r,n}$ and $R_{r,n}$ are homogeneous in c of the same degree, and so G must be a constant. The constant $\hat{C}$ can be calculated according to the formula

$$\hat{C} = \lim_{a_1,\dots,a_n\to a} L_{r,n}(\{a_j\})/R_{r,n}(\{a_j\})$$

and this in turn can be computed in terms of the Selberg integral by repeated use of (4.106). □

To present the generalization of Proposition 4.2.6, notation extending that relating to (4.22) and (4.23) is required. Thus we denote by $\mathrm{ME}_{\beta,N}(g)$ the eigenvalue p.d.f. proportional to

$$\prod_{l=1}^{N} g(x_l) \prod_{1\le j<k\le N}|x_k-x_j|^\beta, \tag{4.108}$$

and suppose the eigenvalues are ordered as in (4.24). Furthermore, we denote by D_r the operation of integrating over all eigenvalues not labeled by a multiple of r (or equivalently observing only those eigenvalues labeled by a multiple of r).

PROPOSITION 4.4.2 *We have*

$$\mathrm{D}_{r+1}(\mathrm{ME}_{2/(r+1),(r+1)N+r}(x^a(1-x)^b)) = \mathrm{ME}_{2(r+1),N}(x^{(r+1)a+2r}(1-x)^{(r+1)b+2r}), \tag{4.109}$$

$$\mathrm{D}_{r+1}(\mathrm{ME}_{2/(r+1),(r+1)N}((1-x)^b)) = \mathrm{ME}_{2(r+1),N}((1-x)^{(r+1)b+2r}). \tag{4.110}$$

Proof. One specializes the parameters in (4.105), and multiplies by a suitable function of $\{a_j\}$ as in the proof of Proposition 4.2.6. □

Taking the Gaussian limit of (4.109) gives

$$\mathrm{D}_{r+1}(\mathrm{ME}_{2/(r+1),(r+1)N+r}(e^{-x^2})) = \mathrm{ME}_{2(r+1),N}(e^{-(r+1)x^2}), \tag{4.111}$$

which generalizes (4.27). Taking the Laguerre limit of (4.109) and (4.110) gives

$$\mathrm{D}_{r+1}(\mathrm{ME}_{2/(r+1),(r+1)N+r}(x^a e^{-x})) = \mathrm{ME}_{2(r+1),N}(x^{(r+1)a+2r}e^{-(r+1)x}), \tag{4.112}$$

$$\mathrm{D}_{r+1}(\mathrm{ME}_{2/(r+1),(r+1)N}(e^{-x})) = \mathrm{ME}_{2(r+1),N}(e^{-(r+1)x}), \tag{4.113}$$

which generalizes (4.28) and (4.29).

Let us specify $\mathrm{CE}^b_{\beta,N}$ as above (4.30). The circular analogue of (4.105) given in Exercises 4.4 q.3, with the parameters appropriately specialized, gives

$$\mathrm{D}_{r+1}(\mathrm{CE}^b_{2/(r+1),(r+1)N+r}) = \mathrm{CE}^{(r+1)b+r}_{2(r+1),N}. \tag{4.114}$$

Setting now $b = 1/(r+1)$ and recalling (4.31) gives

$$\mathrm{alt}_{r+1}(\mathrm{CE}^0_{2/(r+1),(r+1)N}) = \mathrm{CE}^0_{2(r+1),N}, \tag{4.115}$$

where alt_{r+1} denotes the distribution of $\theta_{r+1}, \theta_{2(r+1)}, \dots$ with θ_{r+1} regarded as the origin, generalizing (4.32).

EXERCISES 4.4 1. [199] Let $\mathcal{A} = (n_j)_{j=1,\dots,r+q}$, $n_j = 0$ or 1 be a sequence of r 0s or q 1s in a line with $1 \le q \le r$. Furthermore let

$$K(n_j) = \left(\begin{cases} 0, & n_j = 0 \\ \#0's \text{ to the right of } n_j, & n_j = 1 \end{cases} \right) = n_j \sum_{k=j+1}^{r+q} (1 - n_k),$$

and in terms of this define

$$K(\mathcal{A}) = \sum_{j=1}^{r+q} K(n_j). \tag{4.116}$$

The objective of this exercise is to show

$$\sum_{\mathcal{A}} e^{-2\pi i K(\mathcal{A})/(r+1)} = 0. \tag{4.117}$$

(i) Show that (4.117) is equivalent to the summation

$$\sum_{\mathcal{A}} e^{-2\pi i \sum_{j=1}^{r+q} n_j (r+q-j)/(r+1)} = 0. \tag{4.118}$$

(ii) Note that the l.h.s. of (4.118) is equal to the coefficient of z^q in $F(z)$, where

$$F(z) := \sum_{n_1,\dots,n_{r+q}=0,1} e^{-2\pi i \sum_{j=1}^{r+q} n_j (r+q-j)/(r+1)} z^{\sum_{j=1}^{r+q} n_j}.$$

Evaluate $F(z)$ as a product, simplifying the product by using the factorization formula

$$\prod_{l=1}^{N} (1 - z e^{2\pi i (l-1)/N}) = 1 - z^N, \tag{4.119}$$

and then expand the simplified product to show that $F(z)$ contains no term proportional to z^q, as required.

2. [199] Consider the integral in (4.105), $L_{r,n}(\{a_p\})$, with the factors $|\lambda_j - a_p|^{s_p - 1}$ therein replaced by power functions according to the formula implied by (4.124) below. Note that then, according to Cauchy's theorem, the positions of $\{a_p\}$ can be moved into the complex plane, with the value of the integral being independent of the contours, provided no contour crosses a branch of the power functions, and so corresponds to the analytic continuation. Here this analytic continuation is to be studied in the case that a_l and a_{l+1} swap places on the real axis for general $l = 1, \dots, n-1$. This is achieved by the deforming of contours as indicated in Figure 4.1.

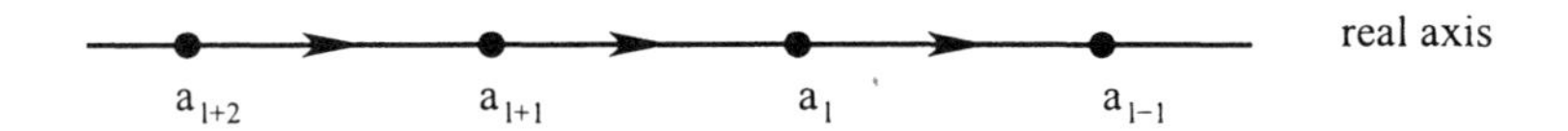

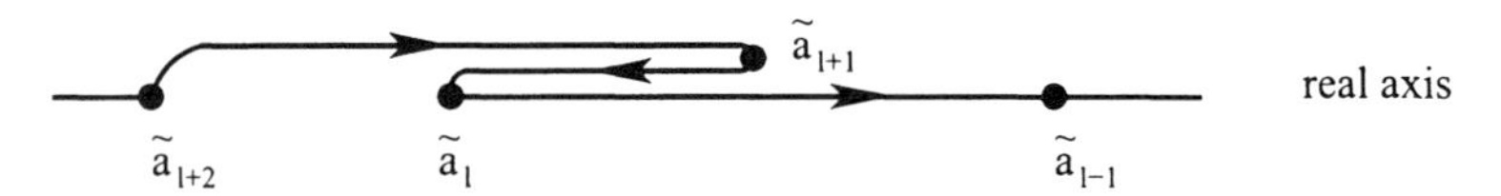

Figure 4.1 The contours from a_{l+2} to a_{l+1}, a_{l+1} to a_l, and a_l to a_{l-1} are deformed to the contours joining the corresponding tilded variables. Our interest is in the limit that $\tilde{a}_j = a_j$ $(j \neq l, l+1)$, $\tilde{a}_l = a_{l+1}$, $\tilde{a}_{l+1} = a_l$ and all contours in the second diagram run along the real axis. In the case $l = n-1$ the contour from $\tilde{a}_{l+2}$ to $\tilde{a}_{l+1}$ is to be deleted, while in the case $l = 1$ the contour from $\tilde{a}_l$ to $\tilde{a}_{l-1}$ is to be deleted.

(i) To study the integrand of (4.105) in the case of the second configuration of Figure 4.1, for notational convenience set $\lambda_{(j-1)r+\mu} = \lambda_j^{(\mu)}$ $(\mu = 1, \dots, r)$, and refer to these as species j. Moreover, in the second configuration let the integration variables be tilded so that $\lambda_j^{(\mu)} \mapsto \tilde{\lambda}_j^{(\mu)}$. From the Figure 4.1, note that between $\tilde{a}_l$ and $\tilde{a}_{l+1}$ there are r coordinates of species l, and for some $p, q \in \{0, \dots, r\}$ there are p coordinates of species $l-1$.

(ii) Next one would like to show that only configurations with $p = q = 0$ contribute, due to a cancellation effect. Suppose first that $p = 0$ while $q \neq 0$, and that furthermore the r species l variables are to the left of the q species $l-1$ variables in the interval $(\tilde{a}_l, \tilde{a}_{l+1})$. Inspection of (4.105) shows that interchanging the position of coordinates corresponding to different species leaves the magnitude of the integrand unchanged, but changes the phase by a factor of $e^{-2\pi i/(r+1)}$ for each interchange of a species $l-1$ and left neighbouring species l. Conclude from this that for a general ordering of the r species l variables, and q species $l-1$ variables, amongst a given set of $r+q$ positions in $(\tilde{a}_l, \tilde{a}_{l+1})$, the phase changes by $K(\mathcal{A})$ as specified by (4.116) and so according to (4.117) cancel when summed over all configurations. Make the same final conclusion in the cases $q = 0$ $(p \neq 0)$ and $p, q \neq 0$.

(iii) By comparing the contributing configurations for $p = q = 0$ with the allowed configurations before a_l and a_{l+1} were interchanged, together with the corresponding integrands, conclude

$$L_{r,n}(\{a_p\})\Big|_{\substack{a_l \leftrightarrow a_{l+1} \\ s_l \leftrightarrow s_{l+1}}} = e^{-\pi i r(s_l + s_{l+1} - 2/(r+1))} L_{r,n}(\{a_p\}),$$

which is the sought analytic continuation property.

3. Let R_r denote the interlaced region

$$\theta_{j-1} < \psi_{(r-1)j+1} < \psi_{(r-1)j+2} < \cdots < \psi_{rj} < \theta_j \qquad (j = 1, \dots, n)$$

with $\theta_0 := 0$, $\theta_n := 2\pi$, and define $\tilde{C}$ in terms of the constant $\hat{C}$ (4.104) and the Morris integral (4.4) according to

$$\tilde{C} = \hat{C}\Big|_{\{s_p\} \mapsto \{\alpha_p\}} M_r(a, a, 1/(r+1))$$

with $a := \frac{1}{2}(\sum_{p=1}^n \alpha_p + 2(n-1)r/(r+1) - n)$. The objective of this exercise is to show that

$$\begin{aligned} \frac{1}{\tilde{C}} \int_{R_r} d\psi_1 \cdots d\psi_{rn} \prod_{1 \le j < k \le rn} |e^{i\psi_k} - e^{i\psi_j}|^{2/(r+1)} \prod_{j=1}^{rn} \prod_{p=1}^{n} |e^{i\psi_j} - e^{i\theta_p}|^{\alpha_j - 1} \\ = \prod_{1 \le j < k \le n} |e^{i\theta_k} - e^{i\theta_j}|^{r(\alpha_j + \alpha_k - 2/(r+1))}. \end{aligned} \tag{4.120}$$

(i) Denote the l.h.s. of (4.120) by $Q_{r.n}(\{e^{i\theta_p}\})$. Use the method of q.2 to obtain the factorization

$$Q_{r,n}(\{w_p\}) = S_{r,n}(\{w_p\})\tilde{G}(\{w_p\}),$$

where

$$S_{r,n}(\{w_p\}) = \prod_{1\le j<k\le n} \Big(\Big(1-\frac{w_j}{w_k}\Big)\Big(1-\frac{w_k}{w_j}\Big)\Big)^{r((\alpha_j+\alpha_k)/2-1/(r+1))}$$

with $\tilde{G}(\{w_p\})$ a symmetric function of $\{(w_p,\alpha_p)\}$, analytic in $\{w_p\}$. Note furthermore that both sides are homogeneous of degree zero and so $\tilde{G}$ is a constant.

(ii) Show that $\tilde{G}$ is unity by computing

$$\lim_{\theta_1,\dots,\theta_{n-1}\to 0} \frac{Q_{r,n}(\{w_p\})}{S_{r,n}(\{w_p\})}.$$

4.5 DOTSENKO AND FATEEV'S DERIVATION

Dotsenko and Fateev [138] devised a contour integration method to calculate an integral more general than the Selberg integral, which occurred in their study of conformal field theory. Below we will detail a modification of their method as it applies to the original Selberg integral (4.1). The evaluation of the more general integral, which in fact will appear in the calculation of correlation functions in the one-component log-gas system given in Chapter 13, is the subject of Exercises 4.5 q.1.

4.5.1 Strategy

Let us suppose for definiteness that $0<\mathrm{Re}(\lambda)\ll 1$ and $-1<\mathrm{Re}(\lambda_2)\ll 0$. As a function of λ_1 the Selberg integral (4.1) is then analytic for $\mathrm{Re}(\lambda_1)>-1$. The contour integration technique of Dotsenko and Fateev gives the functional relationship

$$S_N(\lambda_1,\lambda_2,\lambda) = S_N(-\lambda_1-\lambda_2-2-2\lambda(N-1),\lambda_2,\lambda)\prod_{j=0}^{N-1}\frac{\sin\pi(\lambda_1+\lambda_2+2+(N-1+j)\lambda)}{\sin\pi(\lambda_1+1+j\lambda)}. \quad (4.121)$$

The r.h.s. is meromorphic for $\mathrm{Re}(\lambda_1)<-\epsilon$, $0<\epsilon\ll 1$, so this equation provides the analytic continuation of S_N as a function of λ_1 into the region $\mathrm{Re}(\lambda_1)\le -1$.

To establish the functional relationship (4.121), the integrals

$$S_{(p,N-p)}(\lambda_1,\lambda_2,\lambda) := \int_0^1 dt_1\cdots\int_0^1 dt_p\int_1^\infty dt_{p+1}\cdots\int_1^\infty dt_N\, f(t_1,\dots,t_N), \quad (4.122)$$

where

$$f(t_1,\dots,t_N) := \prod_{l=1}^N |t_l|^{\lambda_1}|1-t_l|^{\lambda_2}\prod_{1\le j<k\le N}|t_k-t_j|^{2\lambda} \quad (4.123)$$

are required. The integral $S_{(p,N-p)}$ is related to $S_{(p-1,N-p+1)}$, and so by iteration $S_{(N,0)}:=S_N$ is related to $S_{(0,N)}$. A change of variables then gives (4.121). From (4.121) it is possible to prove that S_N, when divided by (4.3) is a bounded analytic function of λ_1 in the finite complex λ_1 plane.

The next step in the proof is to show that both S_N and (4.3) have the same asymptotic behavior as $|\lambda_1|\to\infty$. This then gives, by Liouville's theorem, that both sides are the same function of λ_1. It remains to specify the undetermined function of λ_2 (which is immediate since S_N is symmetric in λ_1 and λ_2), λ and N. This problem has already been solved in Selberg's method of Section 4.1.

4.5.2 Details

The following relationship between $S_{(p,N-p)}$ and $S_{(p-1,N-p+1)}$ holds.

PROPOSITION 4.5.1 *One has*

$$S_{(p,N-p)}(\lambda_1,\lambda_2,\lambda) = \frac{p}{N-p+1}\frac{\sin\pi(N-p+1)\lambda\,\sin\pi(\lambda_1+\lambda_2+2+(N+p-2)\lambda)}{\sin\pi p\lambda\,\sin\pi(\lambda_1+1+(p-1)\lambda)}$$

$$\times S_{(p-1,N-p+1)}(\lambda_1,\lambda_2,\lambda).$$

Proof. The integral (4.122) is studied via a related contour integral in the variable u, with $\{t_j\}_{j=1,\dots,N}/\{t_p\}$ parameters such that

$$0<t_1<\cdots<t_{p-1}<1<t_{p+1}<\cdots<t_N.$$

The path of integration is the contour $\mathcal{C}$, specified as the path along the real axis on the upper half-plane side, with indentations around the points $0,1,t_1,\dots,t_{p-1},t_{p+1},\dots,t_N$ into the upper half-plane consisting of semicircles with vanishing small radii. The integrand is chosen as

$$u^{\lambda_1}(1-u)^{\lambda_2}\prod_{\substack{l=1\\ l\neq p}}^{N}(t_l-u)^{2\lambda}|t_l|^{\lambda_1}|1-t_l|^{\lambda_2},\quad \prod_{\substack{1\le j<k\le N\\ j,k\neq p}}|t_k-t_j|^{2\lambda}.$$

The multivalued power functions are defined in terms of the principal argument of the variable. Thus if $v=re^{i\theta}$ with $r>0,-\pi<\arg\theta\le\pi$ then $v^\mu:=r^\mu e^{i\theta\mu}$. Consequently

$$(t_l-u)^{2\lambda}=\begin{cases}|t_l-u|^{2\lambda}, & u<t_l,\\ e^{-2\pi i\lambda}|t_l-u|^{2\lambda}, & u>t_l.\end{cases}\tag{4.124}$$

With this definition the integrand as a function of u is analytic in the half-plane $\mathrm{Im}(u)>0$. With the parameters λ_2,λ obeying the restrictions given in the strategy subsection, and with $\lambda_1<0$, the integrand as a function of u decays sufficiently fast that the contour $\mathcal{C}$ can be closed in the upper half-plane without changing the value of the integral. But the integrand is analytic in the upper half-plane so we conclude by Cauchy's theorem that the integral over $\mathcal{C}$ vanishes.

On the other hand, let us evaluate the integral over $\mathcal{C}$ by parametrization. For the portion of $\mathcal{C}$ along $(-\infty,0)$, taking into consideration the definition of the power functions, gives

$$e^{\pi i\lambda_1}\int_{-\infty}^{0}dt_p\,f(t_1,\dots,t_N)$$

with f given by (4.123). Next consider the portion of $\mathcal{C}$ along $(0,1)$ in the limit that the radius of the half-circles around $t_1,\dots t_{p-1}$ shrink to zero. We obtain

$$\sum_{l=0}^{p-1}e^{-2\pi il\lambda}\int_{t_l}^{t_{l+1}}dt_p\,f(t_1,\dots,t_N),$$

where in the terminals $t_0:=0$ and $t_p:=1$. Finally, for the integral over $(1,\infty)$ we obtain

$$e^{-\pi i\lambda_2}\sum_{l=p}^{N}e^{-2\pi i(l-1)\lambda}\int_{t_l}^{t_{l+1}}dt_p\,f(t_1,\dots,t_N),$$

where in the terminals $t_p:=1$ and $t_{N+1}:=\infty$.

The sum of the above three terms equals zero. Multiplying through by $e^{-\pi i\lambda_1}$ and taking imaginary parts eliminates the first of these terms and gives

$$-\sum_{l=0}^{p-1}\sin\pi(\lambda_1+2l\lambda)\int_{t_l}^{t_{l+1}}dt_p\,f(t_1,\dots,t_N)=\sum_{l=p}^{N}\sin\pi(\lambda_1+\lambda_2+2(l-1)\lambda)\int_{t_l}^{t_{l+1}}dt_p\,f(t_1,\dots,t_N).$$

Now integrate over $0 < t_1 < \cdots < t_{p-1} < 1$ and $1 < t_{p+1} < \cdots < t_N < \infty$. Since the integrands above are symmetrical in the variables $t_1, \dots, t_N$ we see that within the individual sums the integrals are equal. Use of the definition (4.122) then gives

$$-\Big(\frac{1}{p}\sum_{l=0}^{p-1} \sin \pi(\lambda_1 + 2l\lambda)\Big) S_{(p,N-p)}(\lambda_1, \lambda_2, \lambda)$$
$$= \Big(\frac{1}{N-p+1}\sum_{l=p}^{N} \sin \pi(\lambda_1 + \lambda_2 + 2(l-1)\lambda)\Big) S_{(p-1,N-p+1)}(\lambda_1, \lambda_2, \lambda).$$

Simple manipulation to absorb the minus sign and use of the summation formula

$$\sin\theta + \sin(\theta + a) + \cdots + \sin(\theta + na) = \frac{\sin((n+1)a/2)}{\sin(a/2)} \sin(\theta + na/2)$$

gives the stated result.

□

The functional relation (4.121) follows from Proposition 4.5.1 by iteration, starting with $p = N$, then decreasing p down to $p = 1$, and the change of variables $t_j \mapsto 1/t_j$ $(j = 1, \dots, N)$ in $S_{(0,N)}$, making use of (4.8).

Let us now use the functional equation to prove that the Selberg integral (4.1), when divided by the product of gamma functions (4.3), is an analytic function of λ_1 in the finite complex λ_1-plane. Since S_N is analytic for $\mathrm{Re}(\lambda_1) > -1$, the functional equation gives that in general the only possible singularities of S_N as a function of λ_1 are simple poles at

$$\lambda_1 = -1 - j\lambda - k, \qquad j = 0, \dots, N-1, \; k = 0, 1, \dots \;.$$

On the other hand the product of gamma functions (4.3) has poles of the same order at precisely these points. Furthermore the product of gamma functions (4.3) has in general simple zeros at

$$\lambda_1 = -2 - \lambda_2 - \lambda(N - 1 + j) - k, \qquad j = 0, \dots, N-1, \quad k = 0, 1, \dots \;.$$

which from the functional equation (4.121) are also simple zeros of S_N. Thus the possible singularities of (4.1) divided by (4.3) all cancel out, leaving a function analytic in the finite complex λ_1-plane, as claimed.

It remains to establish that (4.1) and (4.3) are the same function of λ_1 for $|\lambda_1| \to \infty$. The asymptotic behavior of the Selberg integral can be calculated as follows.

PROPOSITION 4.5.2 *For* $\mathrm{Re}(\lambda_1) > -1$ *and* $|\lambda_1| \to \infty$

$$S_N(\lambda_1, \lambda_2, \lambda) \sim \lambda_1^{-N(1+\lambda_2)-N(N-1)\lambda} \int_0^\infty ds_1 \cdots \int_0^\infty ds_N \prod_{l=1}^N s_l^{\lambda_2} e^{-s_l} \prod_{1 \le j < k \le N} |s_k - s_j|^{2\lambda}.$$

Proof. Change variables $t_l = e^{-s_l/\lambda_1}$ in (4.1) to obtain

$$S_N(\lambda_1, \lambda_2, \lambda) = \lambda_1^{-N} \int_0^\infty ds_1 \cdots \int_0^\infty ds_N \prod_{l=1}^N e^{-s_l(1+1/\lambda_1)} (1 - e^{-s_l/\lambda_1})^{\lambda_2} \prod_{1 \le j < k \le N} |e^{-s_k/\lambda_1} - e^{-s_j/\lambda_1}|^{2\lambda}. \tag{4.125}$$

Expanding the exponentials e^{-s/λ_1} to first order in $1/\lambda_1$ gives the stated result. □

The functional equation (4.121) gives that this asymptotic behavior remains valid for all $-\pi < \arg(\lambda_1) < \pi$. On the other hand, using the asymptotic formula

$$\frac{\Gamma(x+a)}{\Gamma(x)} \sim x^a \qquad \text{as} \quad |x| \to \infty, \; \arg(x) \neq -\pi, \tag{4.126}$$

we see that the product of gamma functions in (4.3) has precisely the same asymptotic power law decay in λ_1.

We thus conclude that (4.1) divided by (4.3) is a bounded entire function of λ_1, and thus by Liouville's theorem is independent of λ_1. By symmetry the ratio must also be independent of λ_2. The remaining unspecified function of λ and N is calculated as in Selberg's derivation.

EXERCISES 4.5 1. The objective of this exercise is to prove the integration formula

$$\begin{aligned} &J_{(0,n)(m,0)}(\alpha,\beta;\rho) \\ &:= \prod_{i=1}^{n} \int_1^\infty dt_i \prod_{j=1}^{m} \int_0^1 d\tau_j f_{nm}(\{t_i\},\{\tau_j\},\alpha,\beta;\rho) \\ &= m!n!\rho^{2nm} \prod_{l=1}^{n} \frac{\Gamma(l\rho')}{\Gamma(\rho')} \prod_{j=1}^{m} \frac{\Gamma(j\rho-n)}{\Gamma(\rho)} \prod_{l=0}^{n-1} \frac{\Gamma(1+\beta'+l\rho')\Gamma(-1+2m-\alpha'-\beta'-(n-1+l)\rho')}{\Gamma(-\alpha'-l\rho')} \\ &\times \prod_{j=0}^{m-1} \frac{\Gamma(1-n+\alpha+j\rho)\Gamma(1-n+\beta+j\rho)}{\Gamma(2-n+\alpha+\beta+(m-1+j)\rho)}, \end{aligned}$$

where

$$f_{nm}(\{t_i\},\{\tau_j\},\alpha,\beta;\rho) := \left| \prod_{i=1}^{n} t_i^{\alpha'}(1-t_i)^{\beta'} \prod_{j=1}^{m} \tau_j^{\alpha}(1-\tau_j)^{\beta} \frac{\prod_{i<i'}(t_i-t_{i'})^{2\rho'} \prod_{j<j'}(\tau_j-\tau_{j'})^{2\rho}}{\prod_{i=1}^{n}\prod_{j=1}^{m}(\tau_j-t_i)^2} \right|,$$

and the parameters are subject to the relations

$$\alpha' = -\rho'\alpha \qquad \beta' = -\rho'\beta \qquad \rho' = 1/\rho.$$

This integral was first given explicitly in [192], but is implicit in the work of Dotsenko and Fateev [138]. Let

$$\begin{aligned} &J_{(p,n-p)(q,m-q)}(\alpha,\beta;\rho) \\ &:= \mathrm{P}\prod_{i=1}^{p} \int_0^1 dt_i \prod_{i'=p+1}^{n} \int_1^\infty dt_{i'} \prod_{j=1}^{q} \int_0^1 d\tau_j \prod_{j'=q+1}^{m} \int_1^\infty d\tau_{j'} f_{nm}(\{t_i\},\{\tau_j\},\alpha,\beta;\rho), \end{aligned}$$

where P denotes the principal part, and suppose

$$\rho = -\nu + i\mu \qquad \text{with } 0<|\nu|\ll 1,\ 0\ll\mu<1 \text{ and } |\rho|=1.$$

A. Analytic continuation of $J_{(n,0)(m,0)}$ in the complex α-plane

(i) By considering the behavior of

$$\tau_1^\alpha \int_0 dt_1 \ldots \int_0 dt_p \frac{1}{\prod_{i=1}^{p}(\tau_1-t_i)^2}$$

with $p=n$ for $\tau_1 \to 0$, conclude that, up to terms $\mathrm{O}(\nu)$, $J_{(n,0)(m,0)}$ is analytic for $\mathrm{Re}(\alpha) > n-1$. By considering the same behavior for general $p<n$ as well as the behavior of $t_n^{-\alpha/\rho}$ for $t_n \to \infty$ conclude that $J_{(p,n-p)(m,0)}$ is an analytic function of α for $p-1 < \mathrm{Re}(\alpha) < \infty$, $\mathrm{Im}(\alpha) \ge 0$.

(ii) Revise the method of the proof of Proposition 4.5.1 to conclude that the workings are unaffected by the introduction of the τ variables and thus

$$\begin{aligned} &J_{(p,n-p)(m,0)}(\alpha,\beta;\rho) \\ &= J_{(p-1,n-p+1)(m,0)}(\alpha,\beta;\rho) \frac{p}{n-p+1} \frac{\sin\pi(n-p+1)\rho' \sin\pi(\alpha'+\beta'+2+(n+p-2)\rho')}{\sin\pi p\rho' \sin\pi(\alpha'+1+(p-1)\rho')}. \end{aligned}$$

Iterate this equation to show

$$J_{(n,0)(m,0)}(\alpha,\beta;\rho) = J_{(0,n)(m,0)}(\alpha,\beta;\rho)\prod_{j=0}^{n-1}\frac{\sin\pi(\alpha'+\beta'+2+(n-1+j)\rho')}{\sin\pi(\alpha'+1+j\rho')}.$$

From (i) note that this equation provides the analytic continuation of $J_{(n,0)(m,0)}$ into the region $-1<\mathrm{Re}(\alpha)\le n-1$, $\mathrm{Im}(\alpha)\ge 0$.

(iii) Change variables $\tau_j \mapsto 1/\tau_j$ and $t_i \mapsto 1/t_i$ in the definition of $J_{(0,n)(0,m)}$ to show

$$J_{(0,n)(0,m)}(\alpha,\beta;\rho) = J_{(n,0)(m,0)}(-\alpha-\beta+2(n-1)-2\rho(m-1),\beta;\rho).$$

(iv) Proceed similarly as in the first part of (ii) to show

$$\begin{aligned}&J_{(0,n)(q,m-q)}(\alpha,\beta;\rho)\\&\quad= J_{(0,n)(q-1,m-q+1)}(\alpha,\beta;\rho)\frac{q}{m-q+1}\frac{\sin\pi(m-q+1)\rho\,\sin\pi(\alpha+\beta+2+(m+q-2)\rho)}{\sin\pi q\rho\,\sin\pi(\alpha+1+(q-1)\rho)}.\end{aligned}$$

By iterating this equation, substituting in the final equation of (ii), and using (iii), obtain the functional equation

$$\begin{aligned}&J_{(n,0)(m,0)}(\alpha,\beta;\rho)\\&\quad=\prod_{j=0}^{n-1}\frac{\sin\pi(\alpha'+\beta'+2+(n-1+j)\rho')}{\sin\pi(\alpha'+1+j\rho')}\prod_{j=0}^{m-1}\frac{\sin\pi(\alpha+\beta+2+(m-1+j)\rho)}{\sin\pi(\alpha+1+j\rho)}\\&\qquad\times J_{(n,0)(m,0)}(-\alpha-\beta+2(n-1)-2\rho(m-1),\beta;\rho).\end{aligned}$$

With $\mathrm{Re}(\beta)\approx n-1$, from (i) note that this equation provides the analytic continuation of $J_{(n,0)(m,0)}$ into the region $\mathrm{Re}(\alpha)<0$. With $\mathrm{Re}(\beta)\approx n-1$ and $\mathrm{Im}(\beta)>0$, from the final statement in (ii) note that the functional equation also provides the analytic continuation into the region $0\le\mathrm{Re}(\alpha)\le n-1$, $\mathrm{Im}(\alpha)<0$.

B. *Verification of a product of gamma functions ansatz for the evaluation of* $J_{(n,0)(m,0)}$ *as a function of* α *and* β

(i) Use the analytic property in A(i) and the analytic continuations in A(ii) and A(iv) to show that

$$\begin{aligned}J_{(n,0)(m,0)}(\alpha,\beta;\rho)&=a_{n,m}(\alpha,\beta,\rho)\prod_{l=0}^{n-1}\frac{\Gamma(1+\alpha'+l\rho')\Gamma(1+\beta'+l\rho')}{\Gamma(2-m+\alpha'+\beta'+(n-1+l)\rho')}\\&\quad\times\prod_{j=0}^{m-1}\frac{\Gamma(1-n+\alpha+j\rho)\Gamma(1-n+\beta+j\rho)}{\Gamma(2-n+\alpha+\beta+(m-1+j)\rho)},\end{aligned}$$

where $a_{n,m}(\alpha,\beta,\rho)$ is analytic in the finite complex α-plane.

(ii) Use the method of Proposition 4.5.2 to show that for $|\alpha|\to\infty$ with $\mathrm{Re}(\alpha)>n-1$,

$$J_{(n,0)(m,0)}(\alpha,\beta;\rho)\sim c\alpha^{2nm-n-m-n\beta'-m\beta-n(n-1)\rho'-m(m-1)\rho},$$

where c is nonzero and independent of α. Use the analytic continuation formulas to argue that this asymptotic behavior is valid for all $-\pi<\arg(\alpha)<\pi$, and show that the product of gamma functions has the same asymptotic behavior.

(iii) Apply Liouville's theorem, using the results of B(i), (ii), to show that $a_{n,m}(\alpha,\beta,\rho)$ is independent of α. Use the symmetry of $J_{(n,0)(m,0)}$ in α and β to then conclude that $a_{n,m}(\alpha,\beta,\rho)$ is independent of β also, and so $a_{n,m}(\alpha,\beta,\rho)=c_{n,m}(\rho)$.

C. *Calculation of the function* $c_{n,m}(\rho)$ *and the final evaluations*

(i) Use the method of Proposition 4.1.3 to obtain the formula

$$\lim_{\alpha\to-1^+}(\alpha+1)J_{(n,0)(m,0)}(\alpha,\beta;\rho)=mJ_{(n,0)(m-1,0)}(2\rho-1,\beta;\rho).$$

(ii) From C(i) and the equations in B(iii) and B(i) obtain the recurrence relation

$$c_{n,m}(\rho) = m\rho^{2n}\frac{\Gamma(m\rho - n)}{\Gamma(\rho)}c_{n,m-1}(\rho)$$

and thus show that

$$c_{n,m}(\rho) = m!\rho^{2nm}\prod_{i=1}^{m}\frac{\Gamma(i\rho - n)}{\Gamma(\rho)}c_{n,0}(\rho).$$

(iii) Note that $J_{(n,0)(0,0)}(\alpha, \beta; \rho) = S_n(\alpha', \beta', \rho')$ and thus conclude

$$c_{n,0}(\rho) = n!\prod_{i=1}^{n}\frac{\Gamma(i\rho')}{\Gamma(\rho')}.$$

(iv) Substitute the results of C(i) and C(ii) in the equation in B(iii) and then substitute that result in B(i) to evaluate $J_{(n,0)(m,0)}$. Substitute the evaluation of $J_{(n,0)(m,0)}$ in the second equation in A(ii) to obtain the desired evaluation of $J_{(0,n)(m,0)}$.

2. An integration formula of M. Riesz (see, e.g., [444]) states that for Re(α), Re(β), Re($d - \alpha - \beta$) > 0

$$\int_{\mathbb{R}^d}\frac{d\vec{\xi}^d}{|\vec{z}_1 - \vec{\xi}|^{d-\alpha}|\vec{z}_2 - \vec{\xi}|^{d-\beta}} = \frac{k_{\alpha,\beta}}{|\vec{z}_1 - \vec{z}_2|^{d-(\alpha+\beta)}},$$

where

$$k_{\alpha,\beta} = \frac{\pi^{d/2}\Gamma(\alpha/2)\Gamma(\beta/2)\Gamma((d-\alpha-\beta)/2)}{\Gamma((d-\alpha)/2)\Gamma((d-\beta)/2)\Gamma((\alpha+\beta)/2)}.$$

(i) For $d = 1$ use the Euler integral evaluation and (4.8) in the case $n = 1$ to reclaim this result.

(ii) The so-called complex Selberg integral

$$A_N(\alpha, \beta, \gamma) := \int_{\mathbb{R}^2} d\vec{r}_1 \cdots \int_{\mathbb{R}^2} d\vec{r}_N \prod_{l=1}^{N}|\vec{r}_l|^{2\alpha}|\vec{r}_l - \vec{e}|^{2\beta}\prod_{1\le j<k\le N}|\vec{r}_k - \vec{r}_j|^{2\gamma},$$

$\vec{e}$ an arbitrary unit vector in $\mathbb{R}^2$, has been considered by Aomoto [18] and Dotsenko-Fateev [138], who have obtained the factorization type formula

$$A_N(\alpha, \beta, \gamma) = \frac{\prod_{j=1}^{N}\sin\pi(\alpha + (j-1)\gamma/2)\sin\pi(\beta + (j-1)\gamma/2)\sin\pi j\gamma/2}{N!\prod_{j=1}^{N}\sin\pi(\alpha+\beta+(N+j-2)\gamma/2)\sin\pi\gamma/2}\Big(S_N(\alpha, \beta, \gamma/2)\Big)^2$$

where S_N denotes the Selberg integral (4.1). Show that in the case $N = 1$ this result implies the Riesz formula for $d = 2$.

4.6 AOMOTO'S DERIVATION

Our final derivation of the Selberg integral evaluation is due to Aomoto [19]. The objective of Aomoto's approach is to relate $S_N(\lambda_1, \lambda_2, \lambda)$ to $S_N(\lambda_1 + 1, \lambda_2, \lambda)$. This is done via the auxiliary functions

$$I_p^{(\alpha)}(x) := \int_0^1 dt_1 \cdots \int_0^1 dt_N\, F(t_1, \dots, t_N; \alpha, p, x), \tag{4.127}$$

where

$$F(t_1,\dots,t_N;\alpha,p,x) := (t_1-x)^\alpha\cdots(t_p-x)^\alpha(t_{p+1}-x)^{\alpha-1}\cdots(t_N-x)^{\alpha-1}$$
$$\times\prod_{l=1}^N t_l^{\lambda_1}(1-t_l)^{\lambda_2}\prod_{1\le j<k\le N}|t_k-t_j|^{2\lambda}$$

and

$$I_p^{(\alpha)}(x)[g] := \int_0^1 dt_1\cdots\int_0^1 dt_N\, g(t_1,\dots,t_N)F(t_1,\dots,t_N;\alpha,p,x).$$

Note that $I_p^{(\alpha)}(x)[1]=I_p^{(\alpha)}(x)$, $I_0^{(1)}(x)=S_N(\lambda_1,\lambda_2,\lambda)$ and $I_N^{(1)}(0)=S_N(\lambda_1+1,\lambda_2,\lambda)$.

4.6.1 Method

Aomoto's derivation makes use of the fundamental theorem of calculus. In particular the two equations

$$\int_0^1 dt_1\cdots\int_0^1 dt_N\,\frac{\partial}{\partial t_{p+1}}\big(t_{p+1}F(t_1,\dots,t_N;\alpha,p,x)\big)=0 \tag{4.128}$$

and

$$\int_0^1 dt_1\cdots\int_0^1 dt_N\,\frac{\partial}{\partial t_{p+1}}\big(t_{p+1}^2F(t_1,\dots,t_N;\alpha,p,x)\big)=0, \tag{4.129}$$

which follow from the fundamental theorem of calculus provided $t_{p+1}F$ vanishes at $t_{p+1}=0$ and 1 (i.e., provided $\lambda_1>-1$ and $\lambda_2>0$), are used.

After the partial differentiation is performed, the resulting terms are simplified by using symmetry properties of the integrands. With $x=0$ an appropriate linear combination of the two equations then gives the recurrence

$$I_{p+1}^{(1)}(0)=\frac{\lambda_1+1+(N-p-1)\lambda}{\lambda_1+\lambda_2+2+(2N-p-2)\lambda}I_p^{(1)}(0). \tag{4.130}$$

Iteration, and the relationship between between $I_0^{(1)}(0)$, $I_N^{(1)}(0)$ and the Selberg integral, give the recurrence

$$S_N(\lambda_1+1,\lambda_2,\lambda)=\prod_{p=0}^{N-1}\frac{\lambda_1+1+p\lambda}{\lambda_1+\lambda_2+2+(N-1+p)\lambda}S_N(\lambda_1,\lambda_2,\lambda). \tag{4.131}$$

4.6.2 Derivation of the recurrence

We will consider the simplification of (4.128) and (4.129) separately.

PROPOSITION 4.6.1 *We have*

$$\lambda_2 I_p^{(1)}\left[\frac{1}{1-t_{p+1}}\right]=(\lambda_1+\lambda_2+1+(N-p-1)\lambda)I_p^{(1)},$$

where $I_p^{(1)} =: I_p^{(1)}(0)$ *and* $I_p^{(1)}[g] := I_p^{(1)}(0)[g]$.

Proof. Setting $\alpha=1$ and performing the partial derivatives in (4.128) gives

$$(\lambda_1+\lambda_2+1)I_p^{(1)}-\lambda_2 I_p^{(1)}\left[\frac{1}{1-t_{p+1}}\right]+2\lambda\sum_{\substack{k=1\\k\ne p+1}}^N I_p^{(1)}\left[\frac{t_{p+1}}{t_{p+1}-t_k}\right]=0.$$

The summation can be simplified. For $k > p+1$ the integrand of $I_p^{(1)}$ is symmetrical in t_k and t_{p+1}. Thus

$$I_p^{(1)}\left[\frac{t_{p+1}}{t_{p+1}-t_k}\right] = I_p^{(1)}\left[\frac{t_k}{t_k-t_{p+1}}\right].$$

Taking the arithmetic mean of both sides gives

$$I_p^{(1)}\left[\frac{t_{p+1}}{t_{p+1}-t_k}\right] = \frac{1}{2}I_p^{(1)}\left[\frac{t_{p+1}}{t_{p+1}-t_k}+\frac{t_k}{t_k-t_{p+1}}\right] = \frac{1}{2}I_p^{(1)}.$$

For $k \le p$ the integrand of $I_p^{(1)}$ is symmetrical in t_k and t_p. Thus

$$I_p^{(1)}\left[\frac{t_{p+1}}{t_{p+1}-t_k}\right] = I_p^{(1)}\left[\frac{t_{p+1}}{t_{p+1}-t_p}\right] = I_{p-1}^{(1)}\left[\frac{t_p t_{p+1}}{t_{p+1}-t_p}\right].$$

But the integrand of $I_{p-1}^{(1)}$ is symmetrical in t_p and t_{p+1} so the integrand of $I_{p-1}^{(1)}\left[\frac{t_p t_{p+1}}{t_{p+1}-t_p}\right]$ is antisymmetrical under the interchange $t_p \leftrightarrow t_{p+1}$ and this quantity therefore vanishes. The summand has thus been simplified for all k, and the stated equation follows. □

PROPOSITION 4.6.2 *We have*

$$\lambda_2 I_p^{(1)}\left[\frac{1}{1-t_{p+1}}\right] = \lambda_2 I_p^{(1)} + \big(\lambda_1+\lambda_2+2+(2N-p-2)\lambda\big)I_{p+1}^{(1)}.$$

Proof. Setting $\alpha = 1$ and performing the partial derivative in (4.129) gives

$$(\lambda_1+2)I_p^{(1)}[t_{p+1}] - \lambda_2 I_p^{(1)}\left[\frac{t_{p+1}^2}{1-t_{p+1}}\right] + 2\lambda \sum_{\substack{k=1\\k\neq p+1}}^{N} I_p^{(1)}\left[\frac{t_{p+1}^2}{t_{p+1}-t_k}\right] = 0.$$

Now $I_p^{(1)}[t_{p+1}] = I_{p+1}^{(1)}$ and

$$I_p^{(1)}\left[\frac{t_{p+1}^2}{1-t_{p+1}}\right] = I_p^{(1)}\left[\frac{(t_{p+1}-1)(t_{p+1}+1)+1}{1-t_{p+1}}\right] = -I_p^{(1)} - I_{p+1}^{(1)} + I_p^{(1)}\left[\frac{1}{1-t_{p+1}}\right].$$

Furthermore, for $k > p+1$

$$I_p^{(1)}\left[\frac{t_{p+1}^2}{t_{p+1}-t_k}\right] = \frac{1}{2}I_p^{(1)}\left[\frac{t_{p+1}^2}{t_{p+1}-t_k}+\frac{t_k^2}{t_k-t_{p+1}}\right] = \frac{1}{2}I_p^{(1)}[t_{p+1}+t_k] = I_{p+1}^{(1)},$$

while for $k < p+1$

$$I_p^{(1)}\left[\frac{t_{p+1}^2}{t_{p+1}-t_k}\right] = I_{p+1}^{(1)}\left[\frac{t_{p+1}}{t_{p+1}-t_k}\right] = \frac{1}{2}I_{p+1}^{(1)}.$$

Substitution of these simplifications and rearrangement gives the stated equation. □

The recurrence (4.130) results from subtracting the equation in Proposition 4.6.1 from the equation in Proposition 4.6.2.

4.6.3 Solution of the recurrence

Iteration of (4.131) $M-1$ times to increase λ_1 at each step gives

$$S_N(\lambda_1+M,\lambda_2,\lambda) = S_N(\lambda_1,\lambda_2,\lambda)\prod_{p=0}^{N-1}\frac{\Gamma(\lambda_1+1+M+p\lambda)\Gamma(\lambda_1+\lambda_2+2+(N-1+p)\lambda)}{\Gamma(\lambda_1+1+p\lambda)\Gamma(\lambda_1+\lambda_2+2+M+(N-1+p)\lambda)}.$$

Taking the limit $M \to \infty$ on both sides using the asymptotic formula of Proposition 4.5.2 on the l.h.s. and the asymptotic formula (4.126) on the r.h.s. determines S_N as a function of λ_1,

$$S_N(\lambda_1, \lambda_2, \lambda) = A_N(\lambda_2, \lambda) \prod_{p=0}^{N-1} \frac{\Gamma(\lambda_1 + 1 + p\lambda)}{\Gamma(\lambda_1 + \lambda_2 + 2 + (N-1+p)\lambda)}.$$

But S_N is symmetrical in λ_1 and λ_2, so

$$S_N(\lambda_1, \lambda_2, \lambda) = c_N(\lambda) \prod_{p=0}^{N-1} \frac{\Gamma(\lambda_1 + 1 + p\lambda)\Gamma(\lambda_2 + 1 + p\lambda)}{\Gamma(\lambda_1 + \lambda_2 + 2 + (N-1+p)\lambda)}.$$

This is precisely the equation obtained using Selberg's derivation, so the function $c_N(\lambda)$ is specified as in Section 4.1, and the derivation of (4.3) completed.

4.6.4 A Fuchsian differential equation

With multivariable notation $t = \{t_1, \dots, t_N\}$, the elementary symmetric functions are defined by

$$e_r(t) = e_r(t_1, \dots, t_N) := \sum_{1 \le p_1 < \cdots < p_r \le N} t_{p_1} \cdots t_{p_r}. \tag{4.132}$$

In terms of this notation, one has that the multidimensional integral (4.127) is proportional to the $q = N$ case of the family of integrals

$$I_{p,q}^{(\alpha)}(x) := \int_0^x dt_1 \cdots \int_0^x dt_{N-q} \int_0^1 dt_{N-q+1} \cdots \int_0^1 dt_N \, G(t_1, \dots, t_N; \alpha, p, x),$$

where

$$G(t_1, \dots, t_N; \alpha, p, x) := \prod_{l=1}^{N} t_l^{\lambda_1} (1-t_l)^{\lambda_2} |x - t_l|^{\alpha - 1} \prod_{1 \le j < k \le N} |t_k - t_j|^{2\lambda} e_p(x - t_1, \dots, x - t_N).$$

It is shown in Exercises 4.6 q.1 that (4.127) satisfies the linear differential-difference equation

$$(N-p)E_p I_{p+1}^{(\alpha)}(x) = -(A_p x + B_p) I_p^{(\alpha)}(x) + x(x-1)\frac{d}{dx} I_p^{(\alpha)}(x) + D_p x(x-1) I_{p-1}^{(\alpha)}(x), \tag{4.133}$$

where

$$A_p = (N-p)(\lambda_1 + \lambda_2 + 2\lambda(N-p-1) + 2\alpha), \quad B_p = (p-N)(\lambda_1 + \alpha + \lambda(N-p-1)),$$

$$D_p = p(\lambda(N-p) + \alpha), \quad E_p = \lambda_1 + \lambda_2 + 1 + \lambda(2N-p-2) + \alpha,$$

and thus so does $I_{p,N}^{(\alpha)}(x)$. In fact the derivation of (4.133) shows that $I_{p,q}^{(\alpha)}(x)$ for general $q = 0, 1, \dots, N$ satisfies (4.133) provided $\mathrm{Re}(\alpha) > 0$.

The differential-difference system (4.133) is in fact the component form of a single first order matrix differential equation

$$\frac{d}{dx} \vec{I}^{(\alpha)}(x) = \frac{\mathbf{U}(x)}{x(x-1)} \vec{I}^{(\alpha)}(x) \tag{4.134}$$

where $\vec{I}^{(\alpha)}(x) = [I_p^{(\alpha)}(x)]_{p=0,\dots,N}$ and $\mathbf{U}(x)$ is the $(N+1) \times (N+1)$ tridiagonal matrix with diagonal $(a_0, \dots, a_N)$, leading upper diagonal $(b_0, \dots, b_{N-1})$, and leading lower diagonal $(c_1, \dots, c_N)$ where, with $A_p, \dots, E_p$ as in (4.133)

$$a_p = A_p x + B_p, \quad b_p = (N-p)E_p, \quad c_p = x(x-1)D_p.$$

The fact that $\{I_{p,q}^{(\alpha)}(x)\}_{p=0,\dots,N}$ satisfies (4.133) for each q means that the vector solution $\vec{I}^{(\alpha)}(x)$ in (4.134) can be replaced by the matrix solution $[I_{p,q}^{(\alpha)}(x)]_{p,q=0,\dots,N}$. Because the determinant of this matrix is nonzero (a fact which can be seen by studying the small x behavior of $I_{p,q}^{(\alpha)}(x)$ [122]), this is a fundamental matrix of solutions for the differential equation. We remark too that all singular points of (4.134) ($x = 0, 1, x, \infty$) are regular, making it an example of a *Fuchsian differential equation*.

EXERCISES 4.6 1. [188] With $I_p^{(\alpha)}(x)$ specified by (4.127), the objective of this exercise is to derive the differential-difference equation (4.133).

(i) Using some of the manipulations in the proof of Propositions 4.3.4 and 4.6.2 derive the formulas

$$I_p^{(\alpha)}[t_{p+1}] = I_{p+1}^{(\alpha)} + xI_p^{(\alpha)}, \quad \frac{d}{dx}I_p^{(\alpha)} = -p\alpha I_{p-1}^{(\alpha)} - (N-p)(\alpha-1)I_p^{(\alpha)}\left[\frac{1}{t_{p+1}-x}\right],$$

$$I_p^{(\alpha)}\left[\frac{t_{p+1}}{t_{p+1}-t_k}\right] = \begin{cases} -\frac{x}{2}I_{p-1}^{(\alpha)}, & k \le p, \\ \frac{1}{2}I_p^{(\alpha)}, & k > p+1, \end{cases} \qquad I_{p+1}^{(\alpha)}\left[\frac{t_{p+1}}{t_{p+1}-t_k}\right] = \begin{cases} \frac{1}{2}I_{p+1}^{(\alpha)}, & k < p+1, \\ I_{p+1}^{(\alpha)} + \frac{x}{2}I_p^{(\alpha)}, & k > p+1, \end{cases}$$

where here $I_p^{(\alpha)} := I_p^{(\alpha)}(x)$ and $I_p^{(\alpha)}[g] := I_p^{(\alpha)}(x)[g]$.

(ii) Use (i) and the method of the proof of Proposition 4.3.4 to show

$$\lambda_2 I_p^{(\alpha)}\left[\frac{1}{1-t_{p+1}}\right] = \big(\lambda_1+\lambda_2+\alpha+\lambda(N-p-1)\big)I_p^{(\alpha)} - \frac{x}{N-p}\frac{d}{dx}I_p^{(\alpha)} - xp\Big(\frac{\alpha}{N-p}+\lambda\Big)I_{p-1}^{(\alpha)}.$$

(iii) Use (i) and the method of the proof of Proposition 4.6.2 to show

$$\begin{aligned} -E_p I_{p+1}^{(\alpha)} &= \big((\lambda_1+2)x + \lambda_2(x+1) + 2x\lambda(N-p-1) + 2x(\alpha-1)\big)I_p^{(\alpha)} \\ &\quad - \frac{x^2}{N-p}\frac{d}{dx}I_p^{(\alpha)} - \Big(\lambda p x^2 + \frac{p\alpha x^2}{N-p}\Big)I_{p-1}^{(\alpha)} - \lambda_2 I_p^{(\alpha)}\left[\frac{1}{1-t_{p+1}}\right]. \end{aligned}$$

(iv) Combine the results of (ii) and (iii) to derive the stated recurrence.

2. [20] In this exercise a q-integral which reduces to the Selberg integral in the limit $q \to 1$ will be evaluated. The q-integral to be considered is

$$J(\alpha) := \int_{\langle \xi_F \rangle} \Phi(t)\,{}_1D(t)\tilde{w},$$

where

$$\Phi(t) := \prod_{j=1}^n t_j^{\alpha+(j-1)(1-2\gamma)}\frac{(qt_j;q)_\infty}{(q^\beta t_j;q)_\infty}\prod_{1\le i<j\le n}\frac{(q^{1-\gamma}t_j/t_i;q)_\infty}{(q^\gamma t_j/t_i;q)_\infty}, \qquad (z;q)_\infty := \prod_{j=0}^\infty (1-zq^j),$$

and

$$\tilde{w} := \frac{d_q t_1}{t_1}\cdots\frac{d_q t_n}{t_n}, \qquad {}_QD(t) := \prod_{1\le i<j\le n}(t_i - Qt_j).$$

The q-integral over $\langle \xi_F \rangle$ is defined as

$$\int_{\langle \xi_F \rangle} f(t)\tilde{w} := (1-q)^n \sum f(t_1,\dots,t_n),$$

where the sum is over all points such that

$$t_1 = q^{\nu_1},\ t_2/t_1 = q^{\nu_2}q^\gamma,\ t_3/t_2 = q^{\nu_3}q^\gamma, \dots, t_n/t_{n-1} = q^{\nu_n}q^\gamma$$

with each $\nu_j \in \mathbb{Z}_{\ge 0}$. In the limit $q \to 1^-$ $J(\alpha)$ tends to the Selberg integral (4.1) with $\lambda_1 = \alpha+n-2-(n-1)\gamma$, $\lambda_2 = \beta - 1$, $\lambda = \gamma$, and integration domain $1 \ge t_1 \ge t_2 \cdots \ge t_n \ge 0$.

The strategy is again to seek a recurrence relating $J(\alpha+1)$ to $J(\alpha)$. Using the fact that the $\alpha \to \infty$ behavior can be computed explicitly (this contrasts with the Selberg integral, for which a multiple integral is obtained), the recurrence uniquely specifies $J(\alpha)$.

(i) Show that for $\alpha \to \infty$, the behavior of $J(\alpha)$ is given by the term $t_1 = 1,\ t_2 = q^\gamma, \dots, t_n = q^{(n-1)\gamma}$ in its definition as a sum, and thus

$$J(\alpha) \sim q^{A_n} \prod_{j=1}^{n} \frac{\Gamma_q(\beta+(j-1)\gamma)\Gamma_q(j\gamma)(1-q)^{2(j-1)\gamma+\beta}}{\Gamma_q(\gamma)},$$

where $A_n = \sum_{j=1}^{n}(\alpha+(j-1)(1-2\gamma)+n-j)((j-1)\gamma)$ and $\Gamma_q(u) := (1-q)^{1-u}(q;q)_\infty/(q^u;q)_\infty$.

(ii) Define the q-shift operator of the i-th coordinate of $\phi(t)$ by

$$T_i\phi(t) = \phi(t_1, \dots, qt_i, \dots, t_n)$$

and define the covariant derivative by

$$\nabla_i\phi(t) := \phi(t) - \frac{T_i\Phi(t)}{\Phi(t)} T_i\phi(t).$$

By noting that $\Phi(t)$ vanishes at $t_1 = q^{-1}, t_{j+1}/t_j = q^{-1+\gamma}$ $(j = 1, \dots, n-1)$, which defines the boundary of $\langle \xi_F \rangle$, show from the definitions that

$$\int_{\langle \xi_F \rangle} \Phi(t)\nabla_i\phi(t)\tilde{w} = 0.$$

(iii) Let $\sigma \in S_n$ denote a permutation of indices,

$$\sigma f(t) = f(t_{\sigma(1)}, \dots, t_{\sigma(n)}).$$

Show that

$$\sigma\Phi(t) = U_\sigma(t)\Phi(t), \qquad \text{where} \quad U_\sigma(t) = \prod_{\substack{1\le i<j\le n \\ \sigma^{-1}(i)>\sigma^{-1}(j)}} \left(\frac{t_j}{t_i}\right)^{2\gamma-1} \frac{\theta(q^\gamma t_j/t_i;q)}{\theta(q^{1-\gamma}t_j/t_i;q)},$$

with $\theta(x;q) := (x;q)_\infty(q/x;q)_\infty(q;q)_\infty$, and use the property $\theta(qx) = -(1/x)\theta(x)$ to show that $T_iU_\sigma(t) = U_\sigma(t)$ for every i so that $U_\sigma(t)$ is a constant on $\langle \xi_F \rangle$. Use these properties and the result of (ii) to show

$$\int_{\langle \xi_F \rangle} \Phi(t)\sigma\nabla_i\phi(t)\tilde{w} = 0$$

and hence conclude that

$$\int_{\langle \xi_F \rangle} \Phi(t)\mathrm{Asym}(\nabla_i\phi(t))\tilde{w} = 0, \qquad \text{where} \quad \mathrm{Asym} f := \sum_{\sigma\in S_n} \mathrm{sgn}(\sigma)\sigma f. \tag{4.135}$$

(iv) Show that

$$\frac{T_1\Phi(t)}{\Phi(t)} = q^\alpha \frac{1-q^\beta t_1}{1-qt_1} \prod_{j=2}^{n} \frac{t_1 - q^{-\gamma}t_j}{t_1 - q^{\gamma-1}t_j},$$

and use this result to show that with

$$\phi(t) := (1-t_1)(t_2\cdots t_r) \prod_{1\le i<j\le n} (t_i - q^\gamma t_j)$$

we have $\nabla_1 \phi(t) = a - b - c + d$, where

$$a := t_2 \cdots t_r \prod_{1 \le i < j \le n} (t_i - q^\gamma t_j), \quad c := q^{\alpha+n-1} t_2 \cdots t_r \prod_{k=2}^n (t_1 - q^{-\gamma} t_k) \prod_{2 \le i < j \le n} (t_i - q^\gamma t_j),$$

$$b := t_1 \cdots t_r \prod_{1 \le i < j \le n} (t_i - q^\gamma t_j), \quad d := q^{\alpha+\beta+n-1} t_1 \cdots t_r \prod_{k=2}^n (t_1 - q^{-\gamma} t_k) \prod_{2 \le i < j \le n} (t_i - q^\gamma t_j).$$

Use (iii) to conclude

$$\int_{\langle \xi_F \rangle} \Phi(t) \mathrm{Asym}(a - b - c + d) \tilde{\omega} = 0.$$

(v) Kadell's lemma [336, lemma 4] states that for $M \subset \{1, \dots, n\}$

$$\mathrm{Asym}\Big(\prod_{j \in M} t_j \, {}_Q D(t) \Big) = Q^{a(M)} \frac{(Q;Q)_{|M|} (Q;Q)_{n-|M|}}{(1-Q)^n} e_{|M|}(t) \, {}_1 D(t),$$

where

$$a(M) = |\{(i,j) : 1 \le i < j \le n, \, i \notin M, j \in M\}|, \quad (Q;Q)_n := \prod_{j=1}^n (1 - Q^j)$$

and $e_r(t)$ denotes the elementary symmetric function (4.132).

Noting that by the interchanges $t_2 \leftrightarrow t_1, t_3 \leftrightarrow t_2, \dots, t_n \leftrightarrow t_{n-1}$,

$$\begin{aligned} \mathrm{Asym}(c) &= q^{\alpha+n-1} q^{-(n-1)\gamma} \mathrm{Asym}(t_1 \cdots t_{r-1} \, {}_{q^\gamma} D(t)), \\ \mathrm{Asym}(d) &= q^{\alpha+\beta+n-1} q^{-(n-1)\gamma} \mathrm{Asym}(t_1 \cdots t_{r-1} t_n \, {}_{q^\gamma} D(t)) \end{aligned}$$

use Kadell's lemma to show

$$\begin{aligned} \mathrm{Asym}(a - c) &= q^{\gamma(r-1)} \frac{(q^\gamma;q^\gamma)_{r-1} (q^\gamma;q^\gamma)_{n-r+1}}{(1-q^\gamma)^n} (1 - q^{\alpha+n-1-(n+r-2)\gamma}) e_{r-1}(t) \, {}_1 D(t), \\ \mathrm{Asym}(-b + d) &= -\frac{(q^\gamma;q^\gamma)_r (q^\gamma;q^\gamma)_{n-r}}{(1-q^\gamma)^n} (1 - q^{\alpha+\beta+n-1+(1-r)\gamma}) e_r(t) \, {}_1 D(t). \end{aligned}$$

Substitute in the result of (iv) to deduce the recurrence

$$\int_{\langle \xi_F \rangle} \Phi(t) e_r(t) \, {}_1 D(t) \tilde{\omega} = q^{1+(r-1)\gamma} \frac{1 - q^{n-r+1}}{1 - q^r} \frac{1 - q^{\alpha+n-1-(n+r-2)\gamma}}{1 - q^{\alpha+\beta+n+(1-r)\gamma}} \int_{\langle \xi_F \rangle} \Phi(t) e_{r-1}(t) \, {}_1 D(t) \tilde{\omega},$$

and iterate this recurrence to obtain the desired recurrence for $J(\alpha)$,

$$J(\alpha + 1) = \prod_{r=1}^n q^{(r-1)\gamma} \frac{1 - q^{\alpha+n-1-(n+r-2)\gamma}}{1 - q^{\alpha+\beta+n-1+(1-r)\gamma}} J(\alpha).$$

(vi) Iterate the recurrence above and make use of the functional equation

$$\Gamma_q(x+1) = [x]_q \Gamma_q(x), \qquad [x]_q := \frac{1 - q^x}{1 - q}, \tag{4.136}$$

to get a relationship between $J(\alpha + M)$ and $J(\alpha)$. Now use the formula

$$\frac{\Gamma_q(x+p)}{\Gamma_q(x)} \underset{x \to \infty}{\sim} ([x]_q)^p \tag{4.137}$$

to take the limit $M \to \infty$, and then use the asymptotic result in (i) to obtain the evaluation

$$J(\alpha) = q^{A_n} \prod_{j=1}^{n} \frac{\Gamma_q(\beta + (j-1)\gamma)\Gamma_q(\alpha + n - 1 - (n+j-2)\gamma)\Gamma_q(j\gamma)}{\Gamma_q(\gamma)\Gamma_q(\alpha+\beta+n-1-(n-j)\gamma)},$$

where A_n is specified in (i).

3. (i) In the case $\gamma \in \mathbb{Z}_{>0}$ show that the integrand in the q-integral defining $J(\alpha)$ in q.2 above is symmetric and thus deduce that in this case

$$J(\alpha) = \frac{1}{n!} \int_0^1 \cdots \int_0^1 \Phi(t)_1 D(t) \tilde{w}$$

where $\int_0^1 f(t) \frac{d_q t}{t} := (1-q) \sum_{j=0}^{\infty} f(q^j)$.

(ii) From (i) and Kadell's lemma (q.2(v) above) with $M = \emptyset$, show that in the case $\gamma \in \mathbb{Z}_{>0}$

$$J(\alpha) = \frac{1}{\Gamma_{q^\gamma}(n+1)} \int_0^1 \cdots \int_0^1 \Phi(t)\,_{q^\gamma} D(t) \tilde{w}.$$

(iii) [36] Assuming the identities

$$\int_0^1 t^{x+p-1} \frac{(qt;q)_\infty}{(q^y t;q)_\infty} d_q t = \frac{(q^x;q)_p}{(q^{x+y};q)_p} \frac{\Gamma_q(x)\Gamma_q(y)}{\Gamma_q(x+y)},$$

$$\mathrm{CT}\, (t;q)_a \Big(\frac{q}{t};q\Big)_b (q^{-(b+1)}t)^p = \frac{(q^{-b};q)_p}{(q^{a+1};q)_p} \frac{(q;q)_{a+b}}{(q;q)_a (q;q)_b}, \qquad (x;q)_a := \frac{(x;q)_\infty}{(xq^a;q)_\infty},$$

deduce that for any Laurent polynomial f

$$\begin{aligned} &\Big(\frac{\Gamma_q(x+y)}{\Gamma_q(x)\Gamma_q(y)} \Big)^N \int_0^1 d_q t_1 \cdots \int_0^1 d_q t_N \prod_{j=1}^N t_j^{x-1} \frac{(qt_j;q)_\infty}{(q^y t_j;q)_\infty} f(t_1, \dots, t_N) \\ &= \Big(\frac{(q;q)_a (q;q)_b}{(q;q)_{a+b}} \Big)^N \mathrm{CT} \prod_{j=1}^N (t_j;q)_a \Big(\frac{q}{t_j};q\Big)_b f(q^{-(b+1)}t_1, \dots, q^{-(b+1)}t_N), \end{aligned}$$

where $x = -b$ and $y = a + b + 1$.

(iv) Use (ii), (iii) and the evaluation of $J(\alpha)$ given in q.2 (vi) to deduce the q-Morris identity [552]

$$\mathrm{CT} \prod_{j=1}^N (x_j;q)_a \Big(\frac{q}{x_j};q\Big)_b \prod_{1 \le j < k \le N} \Big(\frac{x_j}{x_k};q\Big)_\lambda \Big(q\frac{x_k}{x_j};q\Big)_\lambda = \prod_{l=0}^{N-1} \frac{\Gamma_q(a+b+1+\lambda l)\Gamma_q(1+\lambda(l+1))}{\Gamma_q(a+1+\lambda l)\Gamma_q(b+1+\lambda l)\Gamma_q(1+\lambda)}.$$

(v) Take $a = b = 0$ in the q-Morris identity, simplify the r.h.s. to $\Gamma_q(\lambda N + 1)/(\Gamma_q(\lambda+1))^N$ and use the definition of $\Gamma_q(u)$ to take the limit $\lambda \to \infty$ and thus obtain the identity [374]

$$\mathrm{CT} \prod_{1 \le j < k \le N} \Big(\frac{x_j}{x_k};q\Big)_\infty \Big(q\frac{x_k}{x_j};q\Big)_\infty = \frac{1}{(q;q)_\infty^{N-1}}. \tag{4.138}$$

4.7 NORMALIZATION OF THE EIGENVALUE P.D.F.'S

The multidimensional integrals specifying the normalization of the eigenvalue p.d.f.'s for the Gaussian, circular, Laguerre and Jacobi random matrix ensembles can all be calculated as limiting cases of the Selberg integral, as we will now show. Furthermore, in the Gaussian and Laguerre cases the same limiting procedure applied to (4.49) gives a random three-term recurrence for the corresponding characteristic polynomials.

4.7.1 The various ensembles

Gaussian ensemble

From Proposition 1.3.4 the normalization $G_{\beta,N}$ for the eigenvalue p.d.f. of the Gaussian ensemble is given by

$$\begin{aligned} G_{\beta,N} &= \int_{-\infty}^{\infty} d\lambda_1 \cdots \int_{-\infty}^{\infty} d\lambda_N \prod_{l=1}^{N} e^{-\frac{\beta}{2}\lambda_l^2} \prod_{1\le j<k\le N} |\lambda_k - \lambda_j|^\beta \\ &= \beta^{-N/2-N\beta(N-1)/4} \int_{-\infty}^{\infty} dt_1 \cdots \int_{-\infty}^{\infty} dt_N \prod_{l=1}^{N} e^{-\frac{1}{2}t_l^2} \prod_{1\le j<k\le N} |t_k - t_j|^\beta. \end{aligned} \tag{4.139}$$

To express $G_{\beta,N}$ in terms of the Selberg integral (4.1), change variables in (4.1) $t_l \mapsto 1/2 - t_l/2L$ and put $2\lambda = \beta$ and $\lambda_1 = \lambda_2 = L^2/2$. Use of the elementary limit $\big(1 - (t_l/L)^2\big)^{L^2/2} \to e^{-t_l^2/2}$ as $L \to \infty$ shows

$$\lim_{L\to\infty} 2^{L^2}(2L)^{N+\beta N(N-1)/2} S_N(L^2/2, L^2/2, \beta/2) = \beta^{N/2+N\beta(N-1)/4} G_{\beta,N}.$$

The limit is computed using the evaluation (4.3) of S_N and Stirling's formula (1.92), and the following result obtained.

PROPOSITION 4.7.1 *For all β such that the integral is defined,*

$$\begin{aligned} &N! m_N(\beta/2) \\ &:= \int_{-\infty}^{\infty} dt_1 \cdots \int_{-\infty}^{\infty} dt_N \prod_{l=1}^{N} e^{-\frac{1}{2}t_l^2} \prod_{1\le j<k\le N} |t_k - t_j|^\beta = (2\pi)^{N/2} \prod_{j=0}^{N-1} \frac{\Gamma(1+(j+1)\beta/2)}{\Gamma(1+\beta/2)} \end{aligned} \tag{4.140}$$

(this integral is referred to as Mehta's integral [395]) and thus

$$G_{\beta,N} = \beta^{-N/2-N\beta(N-1)/4} (2\pi)^{N/2} \prod_{j=0}^{N-1} \frac{\Gamma(1+(j+1)\beta/2)}{\Gamma(1+\beta/2)}.$$

Note that these results have already been obtained in (1.160) and (1.163). A further direct evaluation is given in Exercises 4.7 q.2 using the p.d.f. (4.55).

Circular ensemble

Proposition 2.2.5 gives the normalization $C_{\beta,N}$ as

$$C_{\beta,N} := \int_{-\pi}^{\pi} d\theta_1 \cdots \int_{-\pi}^{\pi} d\theta_N \prod_{1\le j<k\le N} |e^{i\theta_k} - e^{i\theta_j}|^\beta, \tag{4.141}$$

which is referred to as Dyson's integral [146]. This is a special case of the trigonometric integral $M_N(a, b, \lambda)$ (3.120). Specifically $C_{\beta,N} = (2\pi)^N M_N(0, 0, \beta/2)$. From the evaluation (4.4), the following result is then obtained.

PROPOSITION 4.7.2 *For all β such that the integral $C_{\beta,N}$ is defined, we have*

$$C_{\beta,N} = (2\pi)^N \frac{\Gamma(1+\beta N/2)}{(\Gamma(1+\beta/2))^N}.$$

This result was obtained in Proposition 2.8.7 as a consequence of the realization of the circular β-ensemble in terms of unitary Hessenberg matrices.

Laguerre ensemble

Here, from Proposition 3.2.2, the normalization is given by

$$W_{a\beta N} := \int_0^\infty dx_1 \cdots \int_0^\infty dx_N \prod_{l=1}^N x_l^{\beta a/2} e^{-\beta x_l/2} \prod_{1 \le j < k \le N} |x_k - x_j|^\beta. \tag{4.142}$$

To compute this integral from the Selberg integral, in (4.1) change variables $t_l \mapsto x_l/L$ and put $2\lambda = \beta$, $\lambda_2 = \beta L/2$, $\lambda_1 = \beta a/2$. Using the elementary limit $\left(1 - \frac{t_l}{L}\right)^{\beta L/2} \to e^{-\beta t_l/2}$ as $L \to \infty$ then gives

$$\lim_{L \to \infty} L^{N + Na\beta/2 + \beta N(N-1)/2} S_N(\beta a/2, \beta L/2, \beta/2) = W_{a\beta N}.$$

Use of (4.3) and Stirling's formula (1.92) allows this limit, and thus $W_{a\beta N}$, to be evaluated.

PROPOSITION 4.7.3 *For all β such that the integral is defined we have*

$$W_{a\beta N} = (\beta/2)^{-N(a\beta/2 + 1 + (N-1)\beta/2)} \prod_{j=0}^{N-1} \frac{\Gamma(1 + (j+1)\beta/2)\Gamma(a\beta/2 + 1 + j\beta/2)}{\Gamma(1 + \beta/2)}.$$

This result has previously been obtained in (3.134).

Jacobi ensemble

According to (3.74), the normalization is given by

$$C_{ab\beta N} := \int_{-1}^1 dy_1 \cdots \int_{-1}^1 dy_N \prod_{l=1}^N (1 + y_l)^{a\beta/2} (1 - y_l)^{b\beta/2} \prod_{1 \le j < k \le N} |y_k - y_j|^\beta. \tag{4.143}$$

This integral is obtained from the Selberg integral by changing variables $t_l \mapsto (y_l + 1)/2$ and putting $2\lambda = \beta$, $\lambda_1 = a\beta/2$ and $\lambda_2 = b\beta/2$. Thus

$$C_{ab\beta N} = 2^{N + (a+b)\beta N/2 + \beta N(N-1)/2} S_N(a\beta/2, b\beta/2, \beta/2), \tag{4.144}$$

as already noted in (3.116).

Cauchy ensemble

The equation (3.123) shows that the Cauchy ensemble is equivalent to the circular Jacobi ensemble. The normalization of the latter is the special case $a = b$ of (4.4). Thus

$$\begin{aligned} \mathcal{N}_N^{(\mathrm{Cy})} &:= \int_{-\infty}^\infty d\lambda_1 \cdots \int_{-\infty}^\infty d\lambda_N \prod_{l=1}^N \frac{1}{(1 + \lambda_l^2)^\alpha} \prod_{1 \le j < k \le N} |\lambda_k - \lambda_j|^\beta \\ &= 2^{\beta N(N-1)/2 - 2(\alpha - 1)N} \pi^N M_N(\alpha - \beta(N-1)/2 - 1, \alpha - \beta(N-1)/2 - 1, \beta/2). \end{aligned} \tag{4.145}$$

The normalization of the generalized Cauchy ensemble in (3.124) is given in Exercises 4.7 q.4.

4.7.2 Root systems and the Selberg integral

The integrand of the Selberg integral exhibits structures indicative of a wider mathematical setting. To investigate further, a convenient starting point [24] is the evaluation of the Dyson integral (4.141) given by

Proposition 4.7.2, written as the constant term identity

$$\mathrm{CT} \prod_{j \neq k}^{N} (1 - e^{t_j - t_k})^{\lambda} = \prod_{j=1}^{N} \binom{j\lambda}{\lambda}. \tag{4.146}$$

Macdonald [375] observed that with each t_j regarded as the unit vector in the direction of x_j in $\mathbb{R}^N$, the vectors $t_j - t_k$ all lie in the $(N-1)$-dimensional real hyperplane $x_1 + \cdots + x_N = 0$. Moreover, the set of vectors $\{t_j - t_k, j \neq k\}$ has the property of being invariant with respect to reflections in any hyperplane orthogonal to a member of the set. In general, for a set of vectors E, and a nonzero vector $\vec{\alpha} \in E$, the orthogonal hyperplane is $P_{\vec{\alpha}} = \{\vec{\beta} \in E : \langle \vec{\beta}, \vec{\alpha} \rangle = 0\}$, while the reflection $\sigma_{\vec{\alpha}}$ in the orthogonal hyperplane of $\vec{\alpha}$ is defined by

$$\sigma_{\vec{\alpha}}(\vec{\gamma}) = \vec{\gamma} - \frac{2\langle \vec{\gamma}, \vec{\alpha} \rangle}{\langle \vec{\alpha}, \vec{\alpha} \rangle} \vec{\alpha}. \tag{4.147}$$

Note that this mapping sends $\vec{\alpha}$ to $-\vec{\alpha}$ and leaves invariant all vectors in $P_{\vec{\alpha}}$. The set of vectors also has two other properties which allow it to be identified as a *crystallographic root system.*

DEFINITION 4.7.4 *Let E be a subspace of $\mathbb{R}^N$ and let $\langle \cdot, \cdot \rangle$ denote the usual inner product (dot product). Then R is said to form a root system if*
a. R is finite, spans E and $\vec{0}$ is not in R;
b. If $\vec{\alpha} \in R$, the reflection $\sigma_{\vec{\alpha}}$ leaves R invariant.
The set R is said to form a crystallographic root system if it has the further property
c. If $\vec{\alpha}$, $\vec{\beta} \in R$, then $2\langle \vec{\beta}, \vec{\alpha} \rangle / \langle \vec{\alpha}, \vec{\alpha} \rangle$ is an integer.

In fact there are only a small number of crystallographic root systems. The root system under consideration is referred to as A_{N-1}, or as the root system of type A. There are four other infinite families of crystallographic root systems. These are denoted B_N, C_N, D_N and BC_N. BC_N is the set of vectors $\pm t_j$, $\pm 2t_j$, $\pm(t_j \pm t_k)$. The root systems B_N, C_N and D_N are subsets of BC_N: B_N is obtained by omitting the vectors $\pm 2t_j$, C_N by omitting the vectors $\pm t_j$, and D_N by omitting both these sets. The root system BC_N is said to be nonreduced because it is the union of B_N and C_N.

This suggests considering the constant term of a product of the type in (4.146) where the variables, interpreted as vectors, correspond to one of these other root systems. In the BC_N case one is thus led to consider

$$\begin{aligned} &\mathrm{CT} \prod_{j=1}^{N} (1 - e^{-t_j})^{\lambda_1} (1 - e^{t_j})^{\lambda_1} (1 - e^{-2t_j})^{\lambda_2} (1 - e^{2t_j})^{\lambda_2} \\ &\times \prod_{1 \le j < k \le N} (1 - e^{t_j - t_k})^{\lambda} (1 - e^{t_k - t_j})^{\lambda} (1 - e^{t_k + t_j})^{\lambda} (1 - e^{-(t_j + t_k)})^{\lambda}. \end{aligned} \tag{4.148}$$

Here different exponents have been chosen for the different classes of roots. This allows the constant terms corresponding to the type B, C and D roots to be obtained by setting λ_2, λ_1 and both λ_1, λ_2 equal to zero, respectively.

As a step toward evaluating the specialization of (4.148) for all exponents equal, Macdonald [375] noted that the integers j which occur in the binomial coefficients on the r.h.s. of (4.146) are related to the root system A_{N-1}: they are the degrees of the homogeneous polynomials in $x_1, \dots, x_N$ that are invariant under the Weyl group associated with the root system. For A_{N-1} the Weyl group is just the permutation group S_N, and consequently the invariant polynomials can be chosen as the (symmetric) power sums $\sum_{l=1}^{N} x_l^j$ $(j = 2, \dots, N)$ ($j = 1$ is not included because $\sum_{l=1}^{N} x_l = 0$). For all root systems except BC_N, the invariant polynomials can be defined similarly and their corresponding degrees calculated. In the case of BC_N a more general definition is needed [375]. Either way, one obtains integers d_j corresponding to the various root systems listed in Table 4.1.

root system	integers d_j	positive roots
A_{N-1}	$j\ (j=2,\dots,N)$	$\vec{e}_j - \vec{e}_k\ (1 \le j < k \le N)$
B_N	$2j\ (j=1,\dots,N)$	$\vec{e}_j\ (j=1,\dots,N),\ \vec{e}_j \pm \vec{e}_k\ (1 \le j < k \le N)$
C_N	$2j\ (j=1,\dots,N)$	$2\vec{e}_j\ (j=1,\dots,N),\ \vec{e}_j \pm \vec{e}_k\ (1 \le j < k \le N)$
D_N	N and $2j\ (j=1,\dots,N-1)$	$\vec{e}_j \pm \vec{e}_k\ (1 \le j < k \le N)$
BC_N	$(2j+2)\ (j=1,\dots,N)$	$\vec{e}_j, 2\vec{e}_j\ (j=1,\dots,N)\ \vec{e}_j \pm \vec{e}_k,\ (1 \le j < k \le N)$
E_6	2,5,6,8,9,12	
E_7	2,6,8,10,12,14,18	
E_8	2,8,12,14,18,20,24,30	
F_4	2,6,8,12	
G_2	2,6	

Table 4.1 Integers associated with root systems and the positive roots in the cases of the infinite families. The notation $\vec{e}_j$ denotes the jth elementary unit vector with N components.

For the root systems B_N, C_N, D_N and BC_N Macdonald conjectured that the corresponding product in (4.148) with all exponents equal has constant terms given by the r.h.s. of (4.146) but with j replaced by d_j as specified in Table 4.1. In fact, writing the constant term in (4.148) as a Fourier integral it is easily seen to be proportional to the Selberg integral written in the form noted in Exercises 4.1 q.1 (ii). The evaluation of the Selberg integral (4.3) shows that indeed the Macdonald conjecture is correct.

More generally Macdonald conjectured that for the reduced root systems

$$\mathrm{CT} \prod_{\vec{\alpha}\in R_+} \prod_{i=1}^{\lambda} (1 - q^{i-1} e^{-\vec{\alpha}})(1 - q^i e^{\vec{\alpha}}) = \prod_{j=1}^{N} \begin{bmatrix} d_j\lambda \\ \lambda \end{bmatrix}_q, \qquad \begin{bmatrix} a \\ b \end{bmatrix}_q := \frac{\Gamma_q(a+1)}{\Gamma_q(a-b+1)\Gamma_q(b+1)}, \tag{4.149}$$

where R_+ denotes the positive roots (defined as all $\vec{\alpha} \in R$ such that for a given $\vec{v} \in E$, $\langle \vec{\alpha}, \vec{v} \rangle > 0$, where it is assumed that $\vec{v}$ is chosen so that $\langle \vec{\alpha}, \vec{v} \rangle \neq 0$ for all $\vec{\alpha} \in R$), $e^{\vec{e}_j} := e^{t_j}$, and the d_j are given by the degrees of the independent homogeneous polynomials in the variables $x_1, \dots, x_N$ of the associated Weyl group. For the root system A_{N-1} this is a special case of the q-Morris identity of Exercises 4.6 q.3(iv). In the BC$_N$ case, both sides of (4.149) require modifying, giving a conjectured identity which was further generalized by Morris [407] to read

$$\mathrm{CT} \prod_{i=1}^{N} (x_i; q)_a (q x_i^{-1}; q)_a (q x_i^2; q^2)_b (q x_i^{-2}; q^2)_b$$
$$\times \prod_{1 \le j < k \le N} (x_j x_k; q)_\lambda (q x_j^{-1} x_k^{-1}; q)_\lambda (x_j x_k^{-1}; q)_\lambda (q x_j^{-1} x_k; q)_\lambda$$
$$= \prod_{j=0}^{N-1} \frac{(q;q)_{2a+2b+2\lambda j} (q;q)_{2b+2\lambda j} (q;q)_{\lambda(j+1)} (q^2;q^2)_{2b+\lambda j} (q^2;q^2)_{a+\lambda j}}{(q;q)_{a+2b+(N+j-1)\lambda} (q;q)_{2b+\lambda j} (q;q)_{a+\lambda j} (q;q)_\lambda (q^2;q^2)_{a+b+\lambda j} \cdot (q^2;q^2)_{b+\lambda j}} \tag{4.150}$$

(the case given in [375] corresponds to $a = b = \lambda$). This was proved by Gustafson [280] and Kadell [337], the former of whom showed it could be deduced from the more general identity

$$\begin{aligned} \mathrm{CT} \prod_{j=1}^{N} & \frac{(x_j^2;q)_\infty (x_j^{-2};q)_\infty}{\prod_{k=1}^4 (a_k x_j;q)_\infty (a_k x_j^{-1};q)_\infty} \\ & \times \prod_{1 \le j < k \le N} \frac{(x_j x_k^{-1};q)_\infty (x_j^{-1} x_k;q)_\infty (x_j x_k;q)_\infty (x_j^{-1} x_k^{-1};q)_\infty}{(b x_j x_k^{-1};q)_\infty (b x_j^{-1} x_k;q)_\infty (b x_j x_k;q)_\infty (b x_j^{-1} x_k^{-1};q)_\infty} \\ = 2^N N! & \prod_{j=1}^{N} \frac{(b;q)_\infty (b^{N+j-2} a_1 a_2 a_3 a_4;q)_\infty}{(b^j;q)_\infty (q;q)_\infty \prod_{1 \le \mu < \nu \le 4} (a_\mu a_\nu b^{j-1};q)_\infty} \end{aligned} \tag{4.151}$$

(see Exercises 4.7 q.5).

The conjecture (4.149), appropriately generalized in the case of BC_N, has been proved for all crystallographic root systems ($A_{N-1}, B_N, C_N, D_N, BC_N$ and the exceptional root systems E_8, E_7, E_6, F_4 and G_2) by Cherednik [108].

If the extra condition is imposed that the only multiples of $\vec{\alpha} \in R$ allowed are $\pm\vec{\alpha}$, then as already noted the root system BC_N is excluded, while the other root systems remain. For these remaining root systems, Macdonald conjectured that

$$\int_{-\infty}^{\infty} dt_1 e^{-t_1^2/2} \cdots \int_{-\infty}^{\infty} dt_N\, e^{-t_N^2/2} \Big| \prod_{\vec{\alpha} \in R_+} \frac{\langle \vec{\alpha}, \vec{t} \rangle}{\|\vec{\alpha}\|} \Big|^{2\gamma} = 2^{-\nu\gamma} (2\pi)^{N/2} \prod_{j=1}^{N} \frac{\Gamma(1+\gamma d_j)}{\Gamma(1+\gamma)}, \tag{4.152}$$

where the d_j are as above and ν is equal to the degree of $\prod_{\vec{\alpha} \in R_+} \langle \vec{\alpha}, \vec{t} \rangle$. In the type A case, this conjecture coincides with the Mehta integral of Proposition 4.7.1, while for the types B, C and D root systems it is seen to be a special case of $W_{a\beta N}$ as defined by (4.142) and evaluated by Proposition 4.7.3 (this identification requires the change of variable $x_j = y_j^2$).

EXERCISES 4.7 1. The objective of this exercise is to give Good's proof [268] of the identity

$$\mathrm{CT} \prod_{\substack{j,k=1 \\ j \ne k}}^{N} \left(1 - \frac{x_j}{x_k}\right)^{a_j} = \frac{(a_1 + \cdots + a_N)!}{a_1! \cdots a_N!},$$

$a_j \in \mathbb{Z}_{\ge 0}$, which was conjectured by Dyson [146]. Note that the case $a_1 = \cdots = a_N = \beta/2$ is equivalent to Proposition 4.7.2.

(i) In the Lagrange interpolation formula

$$f(x) = \sum_{j=1}^{N+1} f(x_j) \prod_{\substack{l=1 \\ l \ne j}}^{N+1} \frac{x - x_l}{x_j - x_l}, \tag{4.153}$$

for $f(x)$ a polynomial of degree $\le N$, put $N+1 \mapsto N$, $f(x) = 1$, and choose a suitable value of x to obtain the identity

$$\sum_{j=1}^{N} \prod_{\substack{l=1 \\ l \ne j}}^{N} \left(1 - \frac{x_j}{x_l}\right)^{-1} = 1.$$

(ii) Denoting the l.h.s. of the stated identity by $G_N(a_1, \dots, a_N)$, use the identity in (i) to show

$$G_N(a_1, \dots, a_N) = \sum_{j=1}^{N} G_N(a_1, \dots, a_{j-1}, a_j - 1, a_{j+1}, \dots, a_N).$$

(iii) By considering the occurrence of x_j in the product, argue that

$$G_N(a_1,\dots,a_{j-1},0,a_{j+1},\dots,a_N) = G_{N-1}(a_1,\dots,a_{j-1},a_{j+1},\dots,a_N).$$

Verify that the stated multinomial coefficient satisfies the recurrences in (ii) and (iii), together with the initial condition $G_N(0,\dots,0) = 1$, and conclude that since the constant term is uniquely specified by these conditions, the stated identity holds.

2. [166] In this exercise the Mehta integral will be evaluated in an analogous way to Anderson's derivation of the Selberg integral.

(i) Use the fact that (4.55) is a p.d.f. in $\{\lambda_j\}$ to deduce that

$$\int_R d\lambda_1\cdots d\lambda_{n+1}\, e^{-(1/2)\sum_{j=1}^{n+1}\lambda_j^2} \prod_{1\le j<k\le n+1}(\lambda_j-\lambda_k)\prod_{j=1}^{n+1}\prod_{p=1}^{n}|\lambda_j-a_p|^c$$
$$= \sqrt{2\pi}(\Gamma(c))^n e^{-(1/2)\sum_{j=1}^n a_j^2}\prod_{1\le j<k\le n}(a_j-a_k)^{2c-1},$$

where R is the region (4.56). Conclude from this that

$$\int_R da_1\cdots da_n d\lambda_1\cdots d\lambda_{n+1}\, e^{-(1/2)\sum_{j=1}^{n+1}\lambda_j^2}\prod_{1\le j<k\le n+1}(\lambda_j-\lambda_k)\prod_{1\le j<k\le n}(a_j-a_k)\prod_{j=1}^{n+1}\prod_{p=1}^{n}|\lambda_j-a_p|^c$$
$$= \sqrt{2\pi}(\Gamma(c))^n m_n(c).$$

(ii) Use (4.15) to compute the integral over $\{a_i\}$ first in the integral over $\{\lambda_i\}$, $\{a_i\}$ above, and thus conclude that it is also equal to

$$\frac{(\Gamma(c))^{n+1}}{\Gamma(c(n+1))}m_{n+1}(c).$$

(iii) Equate the expressions in (i) and (ii) and iterate the resulting difference equation, using the initial condition $m_0(c) = 1$ to rederive the evaluation (4.140).

3. [395] The objective of this exercise is to derive the integration formula

$$\int_\Delta dy_1\cdots dy_N\prod_{j=1}^N y_j^{a-1}\Big(1-\sum_{i=1}^N y_j\Big)^{b-1}\prod_{1\le j<k\le N}|y_k-y_j|^\beta$$
$$= \frac{\Gamma(b)}{\Gamma(b+aN+\beta N(N-1)/2)}\prod_{j=1}^N\frac{\Gamma(a+\beta(N-j)/2)\Gamma(1+\beta j/2)}{\Gamma(1+\beta/2)}, \tag{4.154}$$

where Δ is the region $y_j>0$ $(j=1,\dots,N)$, $\sum_{i=1}^N y_i\le 1$. In the case $\beta=0$ this relates to the Dirichlet distribution (4.10).

(i) Put $\beta a/2\mapsto a-1$, $\beta x_l/2\mapsto\lambda x_l$ in (4.142), move all factors of λ to the l.h.s, multiply both sides by $\lambda^{b-1}e^{-\lambda}$ and integrate over $\lambda\in[0,\infty)$ to obtain the evaluation of

$$\int_0^\infty dx_1\,x_1^{a-1}\cdots\int_0^\infty dx_N\,x_N^{a-1}\Big(1+\sum_{i=1}^N x_i\Big)^{-b}\prod_{1\le j<k\le N}|x_k-x_j|^\beta.$$

(ii) Let $x_j = y_j(1-\sum_{i=1}^N y_i)^{-1}$. Note that the range of integration $x_j\ge 0$ $(j=1,\dots,N)$ maps to the region Δ, and evaluate the Jacobian as

$$\det\Big[\frac{\partial x_j}{\partial y_k}\Big]_{j,k=1,\dots,N} = \Big(1-\sum_{i=1}^N y_i\Big)^{-(N+1)}.$$

(iii) Apply the change of variables in (ii) to the integral in (i) with $b \mapsto b - aN - \beta N(N-1)/2$ to obtain (4.154).

(iv) [560] Consider (4.154) in the case $b = 1$. Implement the constraint $0 \le \sum_{i=1}^N y_i \le 1$ by multiplying the integrand by $\delta(t - \sum_{i=1}^N y_i)$ and integrating over t from 0 to 1. By changing variables $y_i \mapsto t y_i$, show that the t dependence can be factorized, and hence obtain the integral evaluation

$$\int_{\mathbb{R}^N} dy_1 \cdots dy_N \prod_{j=1}^N y_j^{a-1} \prod_{1 \le j < k \le N} |y_k - y_j|^\beta \delta\Big(1 - \sum_{i=1}^N y_i\Big) = \frac{1}{\Gamma(aN + \beta N(N-1)/2)} \prod_{j=1}^N \frac{\Gamma(a + \beta(N-j)/2)\Gamma(1+\beta j/2)}{\Gamma(1+\beta/2)}. \tag{4.155}$$

4. (i) Use (3.124) and (4.4) to show

$$\int_{-\infty}^{\infty} d\lambda_1 \cdots \int_{-\infty}^{\infty} d\lambda_N \prod_{l=1}^N \frac{1}{(1+i\lambda_l)^b (1-i\lambda_l)^{\bar b}} \prod_{1 \le j<k \le N} |\lambda_k - \lambda_j|^\beta = 2^{\beta N(N-1)/2 - 2(\mathrm{Re}\, b - 1)N} \pi^N M_N(\bar b - \beta(N-1)/2 - 1, b - \beta(N-1)/2 - 1, \beta/2),$$

where M_N is given by (4.4). Writing $2\mathrm{Re}\, b = b + \bar b$, argue that b and $\bar b$ can be regarded as independent complex parameters.

(ii) Change variables $x_l = (a+b)\lambda_l/2 + i(a-b)/2$, and use the result of (i) to deduce

$$\int_{-\infty}^{\infty} dx_1 \cdots \int_{-\infty}^{\infty} dx_N \prod_{l=1}^N \frac{1}{(a+ix_l)^c (b-ix_l)^{\bar c}} \prod_{1\le j<k\le N} |x_k - x_j|^\beta = \frac{(2\pi)^N}{(a+b)^{(c+\bar c)N - \beta N(N-1)/2 - N}} M_N(\bar c - \beta(N-1)/2 - 1, c - \beta(N-1)/2 - 1, \beta/2) \tag{4.156}$$

(iii) Let the normalization in (3.4) be denoted $C_{\alpha\beta m}$. Use a simple change of variables to show

$$C_{\alpha\beta m} = 2^{-m} W_{\alpha - 1/\beta, \beta, m}$$

where $W_{\alpha\beta N}$ is the integral in (4.142) and thus evaluated by (3.134).

(iv) From the result of (iii) show that the normalizations in Exercises 1.3 q.5(iv) have the evaluations

$$C_N = \begin{cases} 2^{-N/2} \prod_{j=0}^{N/2-1} \Gamma(j+2)\Gamma(j+1/2), & N \text{ even}, \\ 2^{-(N-1)/2} \prod_{j=0}^{(N-1)/2-1} \Gamma(j+2)\Gamma(j+3/2), & N \text{ odd}, \end{cases} \tag{4.157}$$

5. (i) Verify that

$$(x^2;q)_\infty = (x;q)_\infty (xq^{1/2};q)_\infty (-x;q)_\infty (-xq^{1/2};q)_\infty$$

and thus show that with $b = q^\lambda$, $a_1 = -1$, $a_2 = q^a$, $a_3 = -a_4 = q^{1/2} q^{b/2}$ the l.h.s. of (4.150) reads

$$\mathrm{CT} \prod_{i=1}^N (x_i;q)_a (x_i^{-1};q)_a (x_i^2 q; q^2)_b (x^{-2} q; q^2)_b \times \prod_{1\le j<k\le N} (x_j x_k^{-1};q)_\lambda (x_j^{-1} x_k; q)_\lambda (x_j x_k;q)_\lambda (x_j^{-1} x_k^{-1};q)_\lambda.$$

(ii) In the special case $M = \emptyset$, Kadell's lemma (Exercises 4.6 q.2(v)) can be written

$$\sum_{w \in P} \prod_{1 \le i<j\le N} \frac{1 - tw(x_j/x_i)}{1 - w(x_j/x_i)} = \frac{(t;t)_N}{(1-t^N)}, \tag{4.158}$$

where P is the set of all permutations of $\{x_1, \dots, x_n\}$. Macdonald [374] has given an identity associated with root systems which in the type A case implies (4.158). In the type B case the identity implies

$$\sum_{w \in W} \prod_{l=1}^{N} \frac{1 - q^a w(x_l^{-1})}{1 - w(x_l^{-1})} \prod_{1 \le j < k \le N} \frac{(1 - q^\lambda w(x_j^{-1} x_k^{-1}))(1 - q^\lambda w(x_j^{-1} x_k))}{(1 - w(x_j^{-1} x_k^{-1}))(1 - w(x_j^{-1} x_k))}$$
$$= \frac{(q^{2a}; q^{2\lambda})_N}{(q^a; q^\lambda)_N} \frac{(q^\lambda; q^\lambda)_N}{(1 - q^\lambda)^N}, \tag{4.159}$$

where W denotes the set of all permutations of $\{x_1, \dots, x_n\}$ together with the inversions $x_j \mapsto x_j^{-1}$. Show that upon multiplication of the l.h.s. of (4.159) with the constant term in (i) above, the latter becomes equal to $2^N N!$ times the constant term in (4.150), while multiplying the r.h.s. of (4.159) by the r.h.s. of (4.151) with parameters as in (i) gives $2^N N!$ times the r.h.s. of (4.150).

4.8 FREE ENERGY

4.8.1 Analytic evaluation

The evaluation of the integral in Proposition 4.7.2 can be used to compute thermodynamic quantities associated with the one-component log-potential system on a circle of Proposition 1.4.1. This follows because the integral is proportional to the partition function, and this in turn is related to the free energy [390].

Thus in the canonical formalism of statistical mechanics, the total dimensionless free energy βF is given by

$$\beta F = -\log \frac{1}{N!} \hat{Z}_N \tag{4.160}$$

and the corresponding dimensionless free energy per particle βf is given by

$$\beta f = \lim_{\substack{N, |\Omega| \to \infty \\ N/|\Omega| = \rho}} \frac{1}{N} \beta F. \tag{4.161}$$

The limit $N, |\Omega| \to \infty$, $N/|\Omega| = \rho$ (fixed) is referred to as the *thermodynamic limit*. From βf other thermodynamic quantities of interest can be computed by differentiation. In particular

$$\frac{\partial \beta f}{\partial \beta} = \lim_{\substack{N, |\Omega| \to \infty \\ N/|\Omega| = \rho}} \Big(\frac{1}{N} \langle U \rangle \Big) =: u \tag{4.162}$$

gives the mean energy per particle. A further differentiation gives the *specific heat* C_V at constant volume (also referred to as the heat capacity),

$$C_V / k_B = -\frac{1}{\beta^2} \frac{\partial u}{\partial \beta}. \tag{4.163}$$

We see from (1.45) and (4.162) that in terms of averages

$$C_V / k_B = \frac{1}{\beta^2} \lim_{\substack{N, \Omega \to \infty \\ N/|\Omega| = \rho}} \frac{1}{N} \Big(\langle U^2 \rangle - \langle U \rangle^2 \Big), \tag{4.164}$$

so in particular C_V is non-negative.

Also, in thermodynamics, the pressure is related to the free energy by $dF = -P d|\Omega| - S dT$ (S denotes the entropy) so one has

$$P = -\Big(\frac{\partial F}{\partial |\Omega|} \Big)_T \sim \rho^2 \frac{\partial f}{\partial \rho}, \tag{4.165}$$

where the final expression holds in the thermodynamic limit.

Let's now apply the formalism to the log-gas on a circle.

PROPOSITION 4.8.1 *Let $N/2\pi R := \rho$ be fixed, and consider the one-component log-potential system on a circle with a uniform neutralizing background and unit charges. We have*

$$-\log \frac{1}{N!}\hat{Z}_N \sim N\beta f + \mathrm{O}(1),$$

where

$$\beta f = (1-\beta/2)\log 2\pi\rho - (\beta/2)\log(\beta/2) + \beta/2 + \log\Gamma(1+\beta/2) - \log 2\pi - 1. \tag{4.166}$$

Proof. This follows immediately from the form of the Boltzmann factor in Proposition 1.4.1, the evaluation of the partition function given by Proposition 4.7.2 and Stirling's formula (1.92). □

Since βf is known explicitly, the thermodynamic formula (4.162) gives for the mean energy per particle

$$u := \lim_{\substack{N,L\to\infty \\ N/L=\rho}} \frac{1}{N}\langle U\rangle = -\frac{1}{2}\log 2\pi\rho - \frac{1}{2}\log(\beta/2) + \frac{1}{2}\frac{\Gamma'(1+\beta/2)}{\Gamma(1+\beta/2)} \tag{4.167}$$

and use of (4.163) then gives for the specific heat per particle

$$C_V/k_B = -\frac{1}{\beta^2}\frac{\partial u}{\partial \beta} = -\frac{1}{\beta^2}\Big(-\frac{1}{2\beta} + \frac{1}{4}\frac{\Gamma''(1+\beta/2)\Gamma(1+\beta/2) - (\Gamma'(1+\beta/2))^2}{(\Gamma(1+\beta/2))^2}\Big). \tag{4.168}$$

By substituting (4.166) in (4.165) we have that the thermodynamic pressure is given by

$$\beta P = (1-\beta/2)\rho. \tag{4.169}$$

Two remarks are in order here. First, we see from Proposition 1.4.1 that the volume dependence factors out of the partition function, so (4.169) can be deduced without explicit evaluation of the latter. Second, for $\beta > 2$, the expression (4.169) becomes negative. The reason for this is that in performing the partial derivative to calculate the pressure, the background density as well as the particle density is being varied. Physically one requires that the background be maintained at constant density by some external constraint.

We note also that (4.166) can be expressed in terms of the excess quantity

$$\beta f^{\mathrm{ex}} := \beta f - \beta f_\infty, \tag{4.170}$$

where βf_∞ refers to the dimensionless free energy per particle for the perfect gas, $\beta f_\infty := \log\rho - 1$. We have

$$\beta f^{\mathrm{ex}} = -\frac{\beta}{2}\log\pi\rho + \beta g(\beta), \qquad g(\beta) = -\frac{1}{2}\log\beta + \frac{1}{2} + \frac{1}{\beta}\log\Gamma(1+\beta/2). \tag{4.171}$$

4.8.2 Low temperature limit

In the low temperature $\beta\to\infty$ limit, use of Stirling's formula (1.92) shows that the exact dimensionless free energy per particle (4.166) has the expansion

$$\beta f \sim (1-\beta/2)\log 2\pi\rho + \frac{1}{2}\log\frac{\beta}{4\pi} - 1 + \mathrm{O}\Big(\frac{1}{\beta}\Big). \tag{4.172}$$

This expansion can be reproduced by making a harmonic approximation in which the total energy is expanded to second order about the minimum energy configuration (which is attained at zero temperature), and the partition function approximated by

$$Z_N \underset{\beta\to\infty}{\approx} R^N e^{-\beta U_0}\prod_{l=1}^{N-1}\Big(\frac{2\pi}{\beta\lambda_l}\Big)^{1/2}, \tag{4.173}$$

where the λ_l denote the nonzero eigenvalues of $[H_{jk}] := [\partial^2 U/\partial\theta_j\partial\theta_k]$ evaluated at the minimum energy configuration (a zero eigenvalue occurs due to the rotational invariance of the ground state). The reasoning which gives rise to (4.173) is given in Exercises 4.8 q.2.

For the one-component log-gas on a circle, by symmetry the minimum energy configuration must be equally spaced points $\theta_j = \nu_0 + 2\pi(j-1)/N$ $(j = 1, \dots, N)$, where $0 \le \nu_0 < 2\pi/N$. For all ν_0

$$U_0 = (2\pi\rho)^{N\beta/2}, \quad H_{jj} = \frac{1}{4}\sum_{n=1}^{N-1} \frac{1}{\sin^2 \pi n/N}, \qquad H_{jk} = -\frac{1}{4\sin^2 \pi(j-k)/N} \quad (j \ne k),$$

where to calculate U_0 use has been made of Exercises 4.8 q.1. Since $[H_{jk}]$ is cyclic: $H_{jk} = H_{j-k\,0} = H_{j+N-k\,0}$ its eigenvectors $\vec{\psi}_k$ can be taken to be $\vec{\psi}_k = [\frac{1}{\sqrt{N}}e^{2\pi ijk/N}]_{j=0,\dots,N-1}$, and the corresponding eigenvalues computed as

$$\lambda_k = \frac{1}{4}\sum_{n=1}^{N-1} \frac{1-\cos 2\pi kn/N}{\sin^2 \pi n/N}.$$

For $N \to \infty$, k/N fixed

$$\lambda_k \sim \frac{N^2}{2\pi^2}\sum_{n=1}^{\infty} \frac{1-\cos 2\pi kn/N}{n^2} = \frac{N^2}{2}\Big(|k/N| - (k/N)^2\Big), \quad |k/N| < \frac{1}{2}, \tag{4.174}$$

(the equality can be verified by expressing $|y| - y^2$ as a Fourier cosine series) so (4.161) applied to (4.173) gives

$$\beta f \sim (1-\beta/2)\log 2\pi\rho + \frac{1}{2}\log\frac{\beta}{4\pi} + \int_0^{1/2} \log t(1-t)\, dt.$$

The integral equals -1, so comparison with (4.172) gives that the exact dimensionless free energy per particle is indeed reproduced up to $\mathrm{O}(1/\beta)$.

4.8.3 High temperature limit

In the high temperature limit $\beta \to 0$ the dimensionless free energy per particle (4.166) has the expansion

$$\beta f \sim (1-\beta/2)\log 2\pi\rho - \frac{\beta}{2}\log\frac{\beta}{2} + \frac{\beta}{2} + \frac{\beta}{2}\Gamma'(1) - \log 2\pi - 1 + \mathrm{O}(\beta^2). \tag{4.175}$$

To understand this requires introducing the two-particle correlation function $\rho_{(2)}(\vec{r}, \vec{r}\,')$, defined as the canonical average (1.45) with $f = \sum_{j\ne k=1}^{N} \delta(\vec{r}\,' - \vec{r}_j)\delta(\vec{r} - \vec{r}_k)$,

$$\begin{aligned}\rho_{(2)}(\vec{r}, \vec{r}\,') &:= \Big\langle \sum_{\substack{j,k=1\\ j\ne k}}^{N} \delta(\vec{r} - \vec{r}_j)\delta(\vec{r}\,' - \vec{r}_k)\Big\rangle \\ &= \frac{N(N-1)}{\hat{Z}_N}\int_\Omega d\vec{r}_3 \cdots \int_\Omega d\vec{r}_N\, e^{-\beta U(\vec{r}, \vec{r}\,', \vec{r}_3, \dots, \vec{r}_N)},\end{aligned} \tag{4.176}$$

where the last equality is valid for a system of identical particles. Introducing the truncated correlation

$$\rho^T_{(2)}(\vec{r}, \vec{r}\,') := \rho_{(2)}(\vec{r}, \vec{r}\,') - \rho_{(1)}(\vec{r})\rho_{(1)}(\vec{r}\,'), \tag{4.177}$$

which has the property of decaying for large separation between $\vec{r}$ and $\vec{r}'$, we can check from the definitions that for the log-gas

$$u := \lim_{\substack{N,L\to\infty \\ N/L=\rho}} \frac{1}{N}\langle U\rangle = -\frac{1}{\rho}\int_0^\infty \rho_{(2)}^T(x,0)\log x\,dx. \tag{4.178}$$

In (4.178) we make the weak coupling approximation to the truncated two-particle correlation function (4.177),

$$\rho_{(2)}^T(x,0) = -\frac{\rho\beta}{x}\int_0^\infty \frac{\sin k\rho x}{(k+\pi\beta)^2}\,dk \tag{4.179}$$

(the derivation of this result is given in Exercises 14.1 q.3). This gives

$$u = -\frac{1}{2}\log \pi\beta\rho + \frac{1}{\pi}\int_0^\infty \frac{\sin x}{x}\log x\,dx.$$

The integral is equal to $\frac{\pi}{2}\Gamma'(1)$, which can be deduced from the formula $\int_0^\infty t^{s-1}e^{-t}\,dt = \Gamma(s)$, so after taking the antiderivative with respect to β and adding βf for a perfect gas, $\beta f_\infty = \log 2\pi\rho - \log 2\pi - 1$, we see that (4.175) is reclaimed up to $\mathrm{O}(\beta^2)$.

EXERCISES 4.8

1. (i) Use the factorization formula (4.119) to show $\prod_{j=1}^{N-1}(1-e^{2\pi i j/N}) = N$.

   (ii) Use the result of (i) to show $\prod_{1\le j<k\le N}|e^{2\pi i k/N} - e^{2\pi i j/N}|^2 = N^N$.

2. For a one-component system defined on a line with a unique ground state configuration, the harmonic approximation to the partition function is

$$Z_N^{(\mathrm{h.a.})} = e^{-\beta U_0}\int_\Omega dx_1\cdots dx_N \exp\Big(-\frac{\beta}{2}\sum_{j,k=1}^N (x_j - x_j^{(0)})H_{jk}(x_k - x_k^{(0)})\Big),$$

   where H_{jk} is defined as below (4.173), Ω denotes the ordered integration domain $x_1 < x_2 < \cdots < x_N$ and the $x_j^{(0)}$ denote the equilibrium points. For $\beta\to\infty$ the significant contribution to the integral comes from $x_j \approx x_j^{(0)}$, so the domain of integration can be extended over $(-\infty,\infty)^N$ without affecting the asymptotic limit. By diagonalizing $[H_{jk}]$ obtain the formula

$$Z_N^{(\mathrm{h.a.})} = e^{-\beta U_0}\prod_{l=0}^{N-1}\Big(\frac{2\pi}{\beta\lambda_l}\Big)^{1/2},$$

   where the λ_l denote the eigenvalues of $[H_{jk}]$.

3. (i) Consider the integral equation (1.47). Show that if $\rho_b(y)$ is a solution then

$$\frac{x^2}{2l^2} + C + N\log l = \int_{-\infty}^\infty \rho_b(y/l)\log|x-y|\,dy.$$

   Use this result and Proposition 1.4.4 to show that for a one-component system of unit charges confined to the interval $[-\sqrt{2N}l, \sqrt{2N}l]$ with background charge density $-(\sqrt{2N}/l\pi)\sqrt{1-(y^2/2Nl^2)}$ the Boltzmann factor is

$$Al^{-\beta N/2}\exp\Big(-\frac{\beta}{2}\sum_{j=1}^N (x_j/l)^2\Big)\prod_{1\le j<k\le N}|(x_k - x_j)/l|^\beta,$$

   where A is given in Proposition 1.4.4.

   (ii) Choose $l = \sqrt{2N}/\pi\rho$ so that as $N\to\infty$ with y fixed the background charge density equals $-\rho$. Furthermore suppose that with the Boltzmann factor above, the particles can move anywhere on the real line. Use the definition of $G_{\beta,N}$ in (4.139) to show that the partition function for this system is given by

$$\frac{1}{N!} A l^{-\beta N/2+N} G_{\beta,N}.$$

(iii) Let c be a positive integer. Use the duplication formula

$$\Gamma(z)\Gamma(z+1/c)\cdots\Gamma(z+(c-1)/c) = c^{1/2-cz}(2\pi)^{(c-1)/2}\Gamma(cz) \tag{4.180}$$

to show that

$$\prod_{j=0}^{N-1}\Gamma(\alpha+1+jc) = c^{cN(N-1)/2+N(\alpha+1/2)}(2\pi)^{-N(c-1)/2}\prod_{p=1}^{c}\prod_{j=0}^{N-1}\Gamma\Big(\frac{\alpha+p}{c}+j\Big).$$

(iv) The Barnes G-function [44] is defined by

$$G(z+1) = (2\pi)^{z/2}\exp\left(-z/2-(\gamma+1)z^2/2\right)\prod_{k=1}^{\infty}\Big(1+\frac{z}{k}\Big)^k\exp(-z+z^2/k) \tag{4.181}$$

(γ denotes Euler's constant), which has the special values

$$G(1) = G(2) = G(3) = 1, \qquad G(1/2) = \pi^{-1/4}2^{1/24}e^{(3/2)\zeta'(-1)} = 0.6032442812\cdots$$

and satisfies the functional relation

$$G(z+1) = \Gamma(z)G(z). \tag{4.182}$$

Use (4.182) to rewrite the identity in (iii) as [184]

$$\prod_{j=0}^{N-1}\Gamma(\alpha+1+jc) = c^{cN(N-1)/2+N(\alpha+1/2)}(2\pi)^{-N(c-1)/2}\prod_{p=1}^{c}\frac{G(N+(\alpha+p)/c)}{G((\alpha+p)/c)}. \tag{4.183}$$

Use this formula, and the asymptotic expansions [44]

$$\log G(x+1) \underset{x\to\infty}{\sim} \frac{x^2}{2}\log x - \frac{3}{4}x^2 + \frac{x}{2}\log 2\pi - \frac{1}{12}\log x + \zeta'(-1) + \mathrm{o}(1) \tag{4.184}$$

and

$$\log\left(\frac{G(N+a+1)}{G(N+b+1)}\right) \underset{N\to\infty}{\sim} (b-a)N + \frac{a-b}{2}\log 2\pi + \left((a-b)N + \frac{a^2-b^2}{2}\right)\log N + \mathrm{o}(1) \tag{4.185}$$

to show that for β even

$$\begin{aligned}\log\prod_{j=1}^{N}\Gamma(1+\beta j/2) \sim \frac{\beta}{2}\Big(\frac{N^2}{2}\log\frac{N\beta}{2} - \frac{3}{4}N^2 + \frac{N}{2}\log N\Big) + \frac{N}{2}\log N \\ + \frac{N}{2}(1+\frac{\beta}{2})\log\frac{\beta}{2} + \frac{N}{2}\log 2\pi - \frac{1}{2}(1+\frac{\beta}{2})N + \mathrm{O}(\log N).\end{aligned} \tag{4.186}$$

(v) To establish the validity of (4.186) for general $\beta > 0$, with $c = \beta/2$ substitute the expansion

$$\log\Gamma(1+cj) = cj\log j + c(\log c - 1)j + \frac{1}{2}\log j + \log(2\pi c) + \mathrm{O}(1/j),$$

uniform in $j \geq 1$, and sum over j.

(vi) Use the above results to show that the dimensionless free energy per particle of the system in (i) with β even is given by (4.166) with ρ replaced by $\frac{1}{2}e^{1/2}\rho$.

4. [184] Let $\beta F(n,q)$ denote the dimensionless free energy of the one-component plasma on a circle of radius R

with n mobile unit charges, an impurity particle of charge q at $\theta = 0$, and a uniform neutralizing background. The dimensionless chemical potential $\beta\mu_n^*$ due to the introduction of the impurity charge is defined by

$$\beta\mu_n^* = \beta F(n,q) - \beta F(n,0).$$

(i) Use the result of Proposition 1.4.1 and (4.160) to show

$$e^{-\beta\mu_n^*} = R^{-q^2\beta/2} \frac{Z_n[|1+e^{i\theta}|^{q\beta}]}{Z_n[1]},$$

where

$$Z_n[g(\theta)] := \frac{1}{(2\pi)^n} \int_{-\pi}^{\pi} d\theta_1 \cdots \int_{-\pi}^{\pi} d\theta_n \prod_{l=1}^{n} g(\theta_l) \prod_{1 \le j < k \le n} |e^{i\theta_k} - e^{i\theta_j}|^\beta,$$

and use (4.4) to obtain the gamma function evaluation

$$e^{-\beta\mu_n^*} = R^{-q^2 c} \frac{f_n(2cq,c)}{f_n^2(cq,c)}, \qquad \text{where } f_n(\alpha,c) := \prod_{j=0}^{n-1} \frac{(\alpha+jc)!}{(jc)!}, \quad c := \beta/2. \tag{4.187}$$

(ii) Use (4.183) to show that for $c \in \mathbb{Z}^+$ we can write

$$f_n(\alpha,c) = c^{\alpha n} \prod_{p=0}^{c-1} \frac{G(n+(\alpha-p)/c+1)G(-p/c+1)}{G(n-p/c+1)G((\alpha-p)/c+1)}$$

and apply (4.185) to deduce that for large n

$$f_n(\alpha,c) \sim e^{(\alpha n \log n)} c^{\alpha n} e^{-\alpha n} n^{-(c-1)\alpha/2+\alpha^2/2c} \prod_{p=0}^{c-1} \frac{G(-p/c+1)}{G((\alpha-p)/c+1)}.$$

(iii) Suppose c is rational, $c = s/r$ for $s, r \in \mathbb{Z}^+$. Show from the definition that

$$f_{rn}(\alpha, s/r) = \prod_{\nu=0}^{r-1} \frac{f_n(\alpha+s\nu/r, s)}{f_n(s\nu/r, s)}$$

and use this together with the asymptotic expansion of (ii) to deduce that

$$\lim_{\substack{n\to\infty \\ n/2\pi R=\rho}} e^{-\beta\mu_n^*} = (2\pi\rho)^{q^2 c} r^{-(q^2-b^2)\beta/2}$$
$$\times \prod_{\nu=0}^{r-1}\prod_{p=0}^{s-1} \frac{G((q+b)/r+\nu/r-p/s+1)G^2((q-b)/r+\nu/r-p/s+1)}{G(2q/r+\nu/r-p/s+1)G(\nu/r-p/s+1)}. \tag{4.188}$$

Chapter Five

Correlation functions at $\beta = 2$

At the special coupling $\beta = 2$, the general n-particle correlation function for the one-component log-gas can be expressed as an $n \times n$ determinant involving orthogonal polynomials. A number of different viewpoints on this result are presented. These include integration formulas, functional differentiation and a formulation in terms of an integral equation. Yet another is to first deduce a determinant formula for the canonical average of what in the random matrix interpretation corresponds to ratios of characteristic polynomials. A general element of the $n \times n$ determinant, referred to as the correlation kernel, is independent of n. It is written as a sum of orthogonal polynomials, which can be summed using the Christoffel-Darboux formula. The resulting structure takes on particular significance in the study of spacing distributions, undertaken in Chapter 9. In the case of a general one-body potential with a power series expansion it is the integrability properties of the corresponding configuration integral which are of interest, and in particular its relation to the KP hierarchy from soliton theory. A further generalization isolates the product of two determinants formed out of linearly independent functions, and this leads to the study of multiple orthogonal polynomials. The topic of the final section is correlation functions for multicomponent systems, in which the joint p.d.f. is a product of determinants. As well as when all species consist of an equal number of particles, we treat too the case in which each species j consists of j particles. This latter case is applicable to the GUE minor process.

5.1 SUCCESSIVE INTEGRATIONS

5.1.1 n-point correlations and truncated correlations

The definition of the one-point and two-point correlations have been given in (1.46) and (4.176). More generally, the n-point (or n-particle) correlation function $\rho_{(n)}(\vec{r}_1, \dots, \vec{r}_n)$, is specified as a canonical average (1.45) according to

$$\begin{aligned} \rho_{(n)}(\vec{r}_1, \dots, \vec{r}_n) &:= \Big\langle \sum_{j_1 \neq \cdots \neq j_n} \prod_{l=1}^{n} \delta(\vec{r}_l - \vec{r}_{j_l}) \Big\rangle \\ &= \frac{N(N-1)\cdots(N-n+1)}{\hat{Z}_N} \int_\Omega d\vec{r}_{n+1} \cdots \int_\Omega d\vec{r}_N \, e^{-\beta U(\vec{r}_1, \dots, \vec{r}_N)}, \end{aligned} \tag{5.1}$$

where the equality is valid for a system of identical particles. We remark that the ratio

$$\rho_{(n)}(\vec{r}_1, \dots, \vec{r}_n) / \rho_{(n-1)}(\vec{r}_1, \dots, \vec{r}_{n-1})$$

can be interpreted as the density at point $\vec{r}_n$ given that there are particles at points $\vec{r}_1, \dots, \vec{r}_{n-1}$.

The correlation function $\rho_{(n)}$ does not decay for large separation between particles. However, by adding and subtracting appropriate combinations of $\rho_{(1)}, \dots, \rho_{(n-1)}$ to $\rho_{(n)}$ we can obtain a quantity, denoted $\rho^T_{(n)}$ and called the *truncated n-particle correlation function*, which will decay when two or more particles are at large separation. The truncated two-particle correlation is given by (4.177), while the truncated three-particle correlation is given by

$$\rho^T_{(3)}(\vec{r}_1, \vec{r}_2, \vec{r}_3) := \rho_{(3)}(\vec{r}_1, \vec{r}_2, \vec{r}_3) - \rho_{(1)}(\vec{r}_1)\rho_{(2)}(\vec{r}_2, \vec{r}_3) - \rho_{(1)}(\vec{r}_2)\rho_{(2)}(\vec{r}_1, \vec{r}_3) \\ -\rho_{(1)}(\vec{r}_3)\rho_{(2)}(\vec{r}_1, \vec{r}_2) + 2\rho_{(1)}(\vec{r}_1)\rho_{(1)}(\vec{r}_2)\rho_{(1)}(\vec{r}_3) \tag{5.2}$$

and in general,

$$\rho^T_{(n)}(\vec{r}_1, \dots, \vec{r}_n) := \sum_{m=1}^{n} \sum_{G} (-1)^{m-1}(m-1)! \prod_{j=1}^{m} \rho_{(|G_j|)}(\vec{r}_{g_j(1)}, \dots, \vec{r}_{g_j(|G_j|)}) \tag{5.3}$$

where the sum over G is over all subdivisions of $\{1, 2, \dots, n\}$ into m subsets $G_1, \dots, G_m$ with $G_j = \{g_j(1), \dots, g_j(|G_j|)\}$. For example, when $n = 3$ the $m = 1$ term corresponds to $G_1 = \{1, 2, 3\}$, the $m = 2$ term to $G_1 = \{1\}, G_2 = \{2, 3\}$ or $G_1 = \{2\}, G_2 = \{1, 3\}$ or $G_1 = \{3\}, G_2 = \{1, 2\}$, and the $m = 3$ term to $G_1 = \{1\}, G_2 = \{2\}, G_3 = \{3\}$ (rearrangements of the G_j are not considered distinct). The inverse of this formula is given in Exercises 5.1 q.3.

5.1.2 Orthogonal polynomials

The one-component log-gas confined to a line and subject to a one-body potential $V(x)$ with corresponding Boltzmann factor $w_\beta(x) := e^{-\beta V(x)}$ has itself a Boltzmann factor proportional to the p.d.f. $\mathrm{ME}_{\beta,N}(w_\beta(x))$ (recall (4.108)). The problem of computing the corresponding n-particle correlation for $\beta = 2$ is intimately related to the theory of orthogonal polynomials. We first note that the p.d.f. can be written in terms of orthogonal polynomials associated with $w_2(x)$.

PROPOSITION 5.1.1 *With $p_k(x)$ a polynomial of degree k which is furthermore monic (i.e. the coefficient of x^k is unity), let $\{p_k(x)\}_{k=0,1,2,\dots}$ be the orthogonal polynomials associated with the weight function $w_2(x)$,*

$$\int_{-\infty}^{\infty} w_2(x) p_j(x) p_k(x)\, dx =: (p_j, p_k)_2 = (p_j, p_j)_2\, \delta_{j,k}. \tag{5.4}$$

We have

$$\begin{aligned}
&\prod_{l=1}^{N} w_2(x_l) \prod_{1 \le j < k \le N} (x_k - x_j)^2 \\
&= \prod_{l=1}^{N} w_2(x_l) \Big(\det[p_{k-1}(x_j)]_{j,k=1,\dots,N} \Big)^2 \\
&= \prod_{l=1}^{N} (p_{l-1}, p_{l-1})_2 \det\Big[(w_2(x_j))^{1/2} \frac{p_{k-1}(x_j)}{(p_{k-1}, p_{k-1})_2} \Big]_{j,k=1,\dots,N} \det[(w_2(x_k))^{1/2} p_{j-1}(x_k)]_{j,k=1,\dots,N} \\
&= \prod_{l=1}^{N} (p_{l-1}, p_{l-1})_2 \det[K_N(x_j, x_k)]_{j,k=1,\dots,N},
\end{aligned} \tag{5.5}$$

where

$$K_N(x, y) := \Big(w_2(x) w_2(y) \Big)^{1/2} \sum_{\nu=0}^{N-1} \frac{p_\nu(x) p_\nu(y)}{(p_\nu, p_\nu)_2}. \tag{5.6}$$

Proof. The Vandermonde determinant identity (1.173) says

$$\det[x_j^{k-1}]_{j,k=1,\dots,N} = \prod_{1 \le j < k \le N} (x_k - x_j).$$

The first column in the determinant consists of all 1s and thus is equal to $[p_0(x_j)]_{j=1,\dots,N}$. Adding an appropriate multiple of this column to the 2nd column gives $[p_1(x_j)]_{j=1,\dots,N}$. Next we add appropriate multiples of columns 1 and 2 to column 3 to obtain $[p_2(x_j)]_{j=1,\dots,N}$. Proceeding in this fashion we obtain the first line of the stated identity. The second line follows by elementary manipulation of the first line and the general fact that $\det \mathbf{A} = \det \mathbf{A}^T$, while the final expression is obtained by matrix multiplication of the determinants in the second line.

□

The utility of the expression (5.5) lies in the properties

$$\int_{-\infty}^{\infty} K_N(x_1, y) K_N(y, x_2)\, dy = K_N(x_1, x_2) \quad \text{and} \quad \int_{-\infty}^{\infty} K_N(y, y)\, dy = N, \tag{5.7}$$

which in turn follow from the orthogonality of $\{p_k(x)\}$. Indeed combining (5.5) and (5.7) allows the n-particle correlation function to be computed [150].

PROPOSITION 5.1.2 *We have*

$$\rho_{(n)}(x_1, \dots, x_n) := \frac{N!}{(N-n)!} \prod_{l=1}^{n} w_2(x_l) \frac{\int_{-\infty}^{\infty} dx_{n+1}\, w_2(x_{n+1}) \cdots \int_{-\infty}^{\infty} dx_N\, w_2(x_N) \prod_{1 \le j < k \le N} (x_k - x_j)^2}{\int_{-\infty}^{\infty} dx_1\, w_2(x_1) \cdots \int_{-\infty}^{\infty} dx_N\, w_2(x_N) \prod_{1 \le j < k \le N} (x_k - x_j)^2}$$
$$= \det[K_N(x_j, x_k)]_{j,k=1,\dots,n}$$

where $K_N(x,y)$ is defined by (5.6). The corresponding truncated n-particle correlation function is given by

$$\rho^T_{(n)}(x_1, \dots, x_n) = (-1)^{n-1} \sum_{\substack{\text{cycles} \\ \text{length } n}} K_N(x_{i_1}, x_{i_2}) K_N(x_{i_2}, x_{i_3}) \cdots K_N(x_{i_n}, x_{i_1}),$$

where the sum is over all distinct cycles $i_1 \to i_2 \to \cdots \to i_n \to i_1$ of $\{1, \dots, n\}$ which are of length n.

Proof. From Proposition 5.1.1 the integral over x_N in both top and bottom line is equivalent to the $m = N$ case of

$$\int_{-\infty}^{\infty} \det[K_N(x_j, x_k)]_{j,k=1,\dots,m}\, dx_m.$$

To compute this integral we make a Laplace expansion of the determinant along the bottom row, and multiply each factor $K_N(x_m, x_k)$ $(k = 1, \dots, m-1)$ in the expansion into the $(m-1)$st column of the corresponding cofactor,

$$\det[K_N(x_j, x_k)]_{j,k=1,\dots,m}$$
$$= \sum_{k=1}^{m-1} (-1)^{m+k} \det\Big[K_N(x_j, x_l) \quad K_N(x_j, x_{l'}) \quad K_N(x_j, x_m) K_N(x_m, x_k)\Big]_{\substack{j=1,\dots,m-1 \\ l=1,\dots,k-1 \\ l'=k+1,\dots,m-1}}$$
$$+ K_N(x_m, x_m) \det[K_N(x_j, x_k)]_{j,k=1,\dots,m-1}.$$

Since all the dependence on x_m is in the last column of the determinants in the sum, the integration over x_m can be performed in each element of this column. Using (5.7) then gives

$$\int_{-\infty}^{\infty} \det[K_N(x_j, x_k)]_{j,k=1,\dots,m}\, dx_m = \sum_{k=1}^{m-1} (-1)^{m+k} \det\Big[K_N(x_j, x_l) \quad K_N(x_j, x_{l'}) \quad K_N(x_j, x_k)\Big]_{\substack{j=1,\dots,m-1 \\ l=1,\dots,k-1 \\ l'=k+1,\dots,m-1}}$$
$$+ N \det[K_N(x_j, x_k)]_{j,k=1,\dots,m-1}$$
$$= (-(m-1) + N) \det[K_N(x_j, x_k)]_{j,k=1,\dots,m-1},$$

where the last line follows by noting that interchanging columns in the sum over k so that the final column is moved to column k shows that all terms in the sum are the same and equal to $-\det[K_N(x_j, x_l)]_{j,l=1,\dots,m-1}$. Repeatedly applying this result, first with $m = N, N-1, \dots, n+1$, then with $m = N, N-1, \dots, 1$, allows both integrals in the definition of $\rho_{(n)}$ to be computed and the stated result follows.

To obtain the formula for the truncated correlation, we recall that in general [8]

$$\det[a_{jk}]_{j,k=1,\dots,n} = \sum_{P \in S_n} (-1)^{n-l} \prod_{1}^{l} a_{\alpha\beta} a_{\beta\gamma} \cdots a_{\delta\alpha}, \tag{5.8}$$

where the sum is over all permutations of $\{1, \dots, n\}$ consisting of l cycles of the form $(\alpha \to \beta \to \delta \to \cdots \to \delta \to \alpha)$. Comparing (5.8) with the final formula of Exercises 5.1 q.3 we see that in general if $\rho_{(n)}(x_1, \dots, x_n) = \det[a_{jk}]_{j,k=1,\dots,n}$ then

$$\rho^T_{(n)}(x_1, \dots, x_n) = (-1)^{n-1} \sum_{\substack{\text{cycles} \\ \text{length } n}} a_{i_1 i_2} a_{i_2 i_3} \cdots a_{i_n i_1}, \tag{5.9}$$

and hence the result. □

The summation $K_N(x, y)$ as defined in (5.6) can be performed explicitly, according to the *Christoffel-Darboux formula* [508].

PROPOSITION 5.1.3 *We have*

$$K_N(x, y) = \frac{(w_2(x) w_2(y))^{1/2}}{(p_{N-1}, p_{N-1})_2} \frac{p_N(x) p_{N-1}(y) - p_{N-1}(x) p_N(y)}{x - y}. \tag{5.10}$$

Proof. Our strategy is to evaluate

$$f(x, y) := (w_2(x) w_2(y))^{1/2} \int_{-\infty}^{\infty} dx_1 \cdots \int_{-\infty}^{\infty} dx_{N-1} \prod_{l=1}^{N-1} w_2(x_l) \prod_{1 \le j < k \le N-1} (x_k - x_j)^2 \prod_{j'=1}^{N-1} (x_{j'} - x)(x_{j'} - y)$$

in two different ways. From the first equality in (5.5) we have

$$\prod_{1 \le j < k \le N-1} (x_k - x_j) \prod_{j'=1}^{N-1} (x_{j'} - u) = \det \begin{bmatrix} p_{k-1}(u) \\ p_{k-1}(x_j) \end{bmatrix}_{\substack{j=1,\dots,N-1 \\ k=1,\dots,N}} \qquad (u = x, y).$$

Multiplication of the determinant with $u = x$ by the determinant with $u = y$ gives for the product of the integrand and $(w_2(x) w_2(y))^{1/2}$

$$\prod_{l=1}^{N} (p_{l-1}, p_{l-1})_2 \det \begin{bmatrix} K_N(x, y) & K_N(x, x_k) \\ K_N(x_j, y) & K_N(x_j, x_k) \end{bmatrix}_{j,k=1,\dots,N-1}.$$

Performing the integration as in the proof of Proposition 5.1.2 gives

$$f(x, y) = (N-1)! \prod_{l=1}^{N} (p_{l-1}, p_{l-1})_2 K_N(x, y).$$

On the other hand, in the integrand we can write

$$\prod_{1 \le j < k \le N-1} (x_k - x_j) \prod_{j'=1}^{N-1} (x_{j'} - x)(x_{j'} - y) = \frac{1}{x - y} \det \begin{bmatrix} p_{k-1}(y) \\ p_{k-1}(x) \\ p_{k-1}(x_j) \end{bmatrix}_{\substack{j=1,\dots,N-1 \\ k=1,\dots,N+1}}$$

and

$$\prod_{1 \le j < k \le N-1} (x_k - x_j) = \det[p_{k-1}(x_j)]_{j,k=1,\dots,N-1}. \tag{5.11}$$

Since both factors are antisymmetric in $x_1, \dots, x_{N-1}$, the second determinant can be replaced by its diagonal term in the integral, provided we multiply the integrand by $(N-1)!$. Thus

$$f(x,y) = (N-1)! \frac{(w_2(x)w_2(y))^{1/2}}{x-y} \int_{-\infty}^{\infty} dx_1 \cdots \int_{-\infty}^{\infty} dx_{N-1} \times \prod_{l=1}^{N-1} w_2(x_l) \det \begin{bmatrix} p_{k-1}(y) \\ p_{k-1}(x) \\ p_{j-1}(x_j)p_{k-1}(x_j) \end{bmatrix}_{\substack{j=1,\dots,N-1 \\ k=1,\dots,N+1}}, \tag{5.12}$$

where we have multiplied the jth diagonal term into row $j+2$ of the first determinant. The integration can now be performed row by row. For each row $j = 3, \dots, N+1$ only the term in column $j-2$ is nonzero (and equal to $(p_{j'-1}, p_{j'-1})_2$, $j' = j-2 = 1, \dots, N-1$). Expanding by these elements gives

$$f(x,y) = (N-1)! \prod_{j'=1}^{N-1} (p_{j'-1}, p_{j'-1})_2 \frac{(w_2(x)w_2(y))^{1/2}}{x-y} \Big(p_N(x)p_{N-1}(y) - p_{N-1}(x)p_N(y) \Big).$$

Equating the two expressions for $f(x,y)$ gives the required summation formula. □

The above proof is nonstandard; the conventional proof is given in Exercises 5.1 q.2. By taking the limit $y \to x$ in Proposition 5.1.3, the summation formula

$$K_N(x,x) := w_2(x) \sum_{\nu=0}^{N-1} \frac{(p_\nu(x))^2}{(p_\nu, p_\nu)_2} = \frac{w_2(x)}{(p_{N-1}, p_{N-1})_2} \Big(p'_N(x)p_{N-1}(x) - p'_{N-1}(x)p_N(x) \Big) \tag{5.13}$$

results (the dashes denote differentiation). It is used to evaluate the diagonal elements of the determinant in Proposition 5.1.2.

The second method of evaluation of $f(x,y)$ in the proof of Proposition 5.1.3 gives a multidimensional integral formula for $p_N(x)$ due originally to Heine [508], and this in turn leads to the evaluation of a related multidimensional integral.

PROPOSITION 5.1.4 *We have*

$$p_N(x) = \frac{1}{C} \int_{-\infty}^{\infty} dx_1 \cdots \int_{-\infty}^{\infty} dx_N \prod_{l=1}^{N} w_2(x_l)(x - x_l) \prod_{1 \le j < k \le N} (x_k - x_j)^2,$$

$$\int_{-\infty}^{\infty} \frac{p_N(x)}{y-x} w_2(x)\, dx = \frac{1}{(N+1)C} \int_{-\infty}^{\infty} dx_0 \cdots \int_{-\infty}^{\infty} dx_N \prod_{l=0}^{N} \frac{w_2(x_l)}{y - x_l} \prod_{0 \le j < k \le N} (x_k - x_j)^2,$$

where

$$C = \int_{-\infty}^{\infty} dx_1 \cdots \int_{-\infty}^{\infty} dx_N \prod_{l=1}^{N} w_2(x_l) \prod_{1 \le j < k \le N} (x_k - x_j)^2.$$

Proof. It remains to consider the second identity. From the first we see that

$$\int_{-\infty}^{\infty} \frac{p_N(x)}{y-x} w_2(x)\, dx = \frac{1}{C} \int_{-\infty}^{\infty} dx_0 \int_{-\infty}^{\infty} dx_1 \cdots \int_{-\infty}^{\infty} dx_N \prod_{l=0}^{N} w_2(x_l) \prod_{l'=1}^{N} \frac{1}{x_0 - x_{l'}} \frac{1}{y - x_0} \prod_{0 \le j < k \le N} (x_k - x_j)^2.$$

Now the integrand is symmetric apart from the factor

$$\prod_{l'=1}^{N} \frac{1}{x_0 - x_{l'}} \frac{1}{y - x_0} =: g_y(x_0; x_1, \dots, x_N).$$

Symmetrizing allows this factor to be replaced by

$$\frac{1}{N+1}\Big(g_y(x_0; x_1, \dots, x_N) + \sum_{j=1}^{N} g_y(x_0; x_1, \dots, x_N)\Big|_{x_0 \leftrightarrow x_j} \Big) = \frac{1}{N+1} \prod_{l=0}^{N} \frac{1}{y - x_l},$$

where the last equality can be verified by performing a partial fractions expansion of the r.h.s. regarded as a function of y. □

EXERCISES 5.1 1. Let $\{p_n(x)\}_{n=0,1,\dots}$ be the set of monic orthogonal polynomials with respect to the weight function $w_2(x)$, and write $C_n := (p_{n-1}, p_{n-1})_2/(p_{n-2}, p_{n-2})_2$. The objective of parts (i)–(iii) of this exercise is to establish the three-term recurrence

$$p_n(x) = (x + B_n)p_{n-1}(x) - C_n p_{n-2}(x) \tag{5.14}$$

for an appropriate B_n.

(i) Observe that $p_n(x) - xp_{n-1}(x)$ is a polynomial of degree $n-1$ and thus can be written as a linear combination of $p_0(x), \dots, p_{n-1}(x)$. Use the property of the inner product,

$$(xp_{n-1}, p_k)_2 = (p_{n-1}, xp_k)_2, \tag{5.15}$$

to show that the coefficients of $p_0(x), \dots, p_{n-3}(x)$ in this linear combination vanish.

(ii) Take the inner product of both sides of (5.14) with respect to p_{n-2} and use (5.15) to deduce the value of C_n.

(iii) By comparing coefficients of x^{n-1} on both sides of (5.14), show that in the Gaussian, Laguerre and Jacobi cases, respectively,

$$B_n^{(\mathrm{G})} = 0, \quad B_n^{(\mathrm{L})} = -(2n + a - 1), \quad B_n^{(\mathrm{J})} = \frac{a^2 - b^2}{(2n + a + b - 2)(2n + a + b)}. \tag{5.16}$$

(iv) Starting with the three-term recurrence (5.14) for $\{p_n(x)\}_{n=0,1,\dots}$ deduce that

$$\Big\{ \int_{-\infty}^{\infty} \frac{p_n(x')}{x - x'} w_2(x')\, dx' \Big\}_{n=0,1,\dots}$$

satisfies the same three-term recurrence. Hence conclude that the general solution of (5.14) is a linear combination of this sequence and $\{p_n(x)\}_{n=0,1,\dots}$.

(v) With $w_2(x) = e^{-2V(x)}$, by considering

$$\Big(\frac{d}{dx} p_n, p_{n-1}\Big)_2 \quad \text{and} \quad \Big(\frac{d}{dx} p_n, p_n\Big)_2$$

show that

$$2(V'(x)p_n, p_{n-1})_2 = n(p_{n-1}, p_{n-1})_2, \qquad (V'(x)p_n, p_n)_2 = 0. \tag{5.17}$$

(vi) For $V(x)$ an even polynomial of degree $2k$, show from the three-term recurrence (5.14) with $B_n = 0$ that the first equation in (5.17) implies a nonlinear recursion of order $2k-2$ for $\{C_n\}_{n=0,1,\dots}$. For the specific potential

$$V(x) = \frac{a_2}{2}x^2 + \frac{a_4}{4}x^4 + \frac{a_6}{6}x^6$$

obtain the explicit recurrence [308]

$$\frac{n}{2} = C_n\Big(a_2 + a_4(C_{n-1} + C_n + C_{n+1}) + a_6(C_{n-1} + C_n + C_{n+1})^2 \\ + a_6(C_{n-2}C_{n-1} - C_{n-1}C_{n+1} + C_{n+1}C_{n+2})\Big).$$

(vii) With $q_n(x) = p_n(x)/(p_n, p_n)_2^{1/2}$, $b_n = (p_{n+1}, p_{n+1})_2^{1/2}/(p_n, p_n)_2^{1/2}$, show from (5.14) that for suitable $\{a_j\}$,

$$b_{j-1}q_{j-1}(x) + a_j q_j(x) + b_j q_{j+1}(x) = x q_j(x), \qquad j = 0, 1, \dots,$$

where $q_{-1}(x) := 0$. Use this to deduce that for all real x $(q_0(x), q_1(x), \dots)^T$ is an eigenfunction of the infinite symmetric tridiagonal matrix (Jacobi matrix)

$$\mathbf{T} = \begin{bmatrix} a_0 & b_0 & 0 & \cdots \\ b_0 & a_1 & b_1 & \ddots \\ 0 & b_1 & a_2 & \ddots \\ \vdots & \ddots & \ddots & \ddots \end{bmatrix}$$

with eigenvalue x. If x is a zero of $p_n(x)$ show that $(q_0(x), \dots, q_{n-1}(x))^T$ is an eigenfunction of the top $n \times n$ sub-block of $\mathbf{T}$.

2. [508] Here the conventional proof of the Christoffel-Darboux formula (5.10) will be given.

(i) Use (5.14) to show that

$$p_{n+1}(x)p_n(y) - p_n(x)p_{n+1}(y) = (x - y)p_n(x)p_n(y) + C_{n+1}\Big(p_n(x)p_{n-1}(y) - p_{n-1}(x)p_n(y)\Big).$$

(ii) In the identity of (i) divide through by $(x - y)(p_n, p_n)_2$, substitute the value of C_{n+1} from q.1, and sum over n to obtain (5.10).

3. Define

$$u_n[a] = \int_\Omega d\vec{r}_1 \cdots \int_\Omega d\vec{r}_n \prod_{l=1}^n a(\vec{r}_l)\rho_{(n)}(\vec{r}_1, \dots, \vec{r}_n), \qquad v_n[a] = \int_\Omega d\vec{r}_1 \cdots \int_\Omega d\vec{r}_n \prod_{l=1}^n a(\vec{r}_l)\rho_{(n)}^T(\vec{r}_1, \dots, \vec{r}_n)$$

and introduce the generating functions

$$U[z; a] = 1 + \sum_{n=1}^\infty \frac{u_n[a]}{n!} z^n, \qquad V[z; a] = \sum_{n=1}^\infty \frac{v_n[a]}{n!} z^n.$$

(i) Show that

$$U[z; a] = \frac{1}{\hat{Z}_N} \int_\Omega d\vec{r}_1 \cdots \int_\Omega d\vec{r}_N \prod_{l=1}^N (1 + z a(\vec{r}_l)) e^{-\beta U(\vec{r}_1, \dots, \vec{r}_N)}.$$

(ii) Note from (5.3) that

$$v_n[a] = n! \sum_{m=1}^n \frac{(-1)^{m-1}}{m} \sum_{\substack{k_1, \dots, k_m \ge 1 \\ k_1 + \cdots + k_m = n}} \prod_{j=1}^m \frac{u_{k_j}[a]}{k_j!}.$$

(iii) Use the result of (ii) to show

$$V[z; a] = \log U[z; a],$$

and from this deduce that

$$u_n[a] = n! \sum_{m=1}^{n} \sum_{\substack{k_1,\dots,k_m \ge 1 \\ k_1+\cdots+k_m=n}} \frac{1}{m!} \frac{v_{k_1}[a]\cdots v_{k_m}[a]}{k_1!\cdots k_m!},$$

or equivalently

$$\rho_{(n)}(x_1,\dots,x_n) = \sum_{m=1}^{n} \sum_{G} \prod_{j=1}^{m} \rho^T_{(|G_j|)}(x_{g_j(1)},\dots,x_{g_j(|G_j|)}).$$

5.2 FUNCTIONAL DIFFERENTIATION AND INTEGRAL EQUATION APPROACHES

5.2.1 Functional differentiation method

An alternative method to derive Proposition 5.1.2 is to introduce a generalized partition function and make use of the functional differentiation formula

$$\frac{\delta}{\delta a(x)} \int_{-\infty}^{\infty} a(y) f(y)\, dy := f(x). \tag{5.18}$$

With $A[a]$ a linear functional of a, and $\delta_x(y) := \delta(x-y)$ the operation $\delta/\delta a(x)$ is the special case $\psi = \delta_x$ of the operation

$$D_\psi A[a] := \lim_{\epsilon\to 0} \frac{A[a+\epsilon\psi] - A[a]}{\epsilon}.$$

The generalized partition function $\hat{Z}_N[a]$ is defined as in (1.39) for the ordinary partition function, except that there is an arbitrary one body potential with Boltzmann factor $a(\vec{r})$ included in the integrand,

$$\hat{Z}_N[a] := \int_\Omega d\vec{r}_1\, a(\vec{r}_1) \cdots \int_\Omega d\vec{r}_N\, a(\vec{r}_N) e^{-\beta U(\vec{r}_1,\dots,\vec{r}_N)}. \tag{5.19}$$

It then follows from (5.1) and (5.18) that

$$\rho_{(n)}(\vec{r}_1,\dots,\vec{r}_n) = \frac{1}{\hat{Z}_N[1]} \frac{\delta^n}{\delta a(\vec{r}_1)\cdots\delta a(\vec{r}_n)} \hat{Z}_N[a]\Big|_{a=1}. \tag{5.20}$$

In the case of the log-gas at $\beta = 2$, the generalized partition function can be written as a determinant.

PROPOSITION 5.2.1 *For Ω equal to the real line and*

$$e^{-\beta U(x_1,\dots,x_N)} = \prod_{l=1}^{N} w_2(x_l) \prod_{1\le j<k\le N} (x_k - x_j)^2,$$

(5.19) can be written as

$$\hat{Z}_N[a] = N! \det \Big[\int_{-\infty}^{\infty} w_2(x) a(x) p_{j-1}(x) p_{k-1}(x)\, dx \Big]_{j,k=1,\dots,N}, \tag{5.21}$$

where the $p_j(x)$ are as in Proposition 5.1.1.

Proof. Begin by rewriting the Boltzmann factor according to the first equality in (5.5). Expanding the determinants using the general formula [8] (cf. (5.8))

$$\det[u_{j,k}]_{j,k=1,\dots,N} = \sum_{P\in S_N} \varepsilon(P) \prod_{l=1}^{N} u_{l,P(l)}, \tag{5.22}$$

where $\varepsilon(P)$ denotes the signature of the permutation, gives

$$e^{-\beta U(x_1,\dots,x_N)} = \prod_{l=1}^{N} w_2(x_l) \sum_{P\in S_N} \varepsilon(P) \sum_{Q\in S_N} \varepsilon(Q) \prod_{l=1}^{N} p_{P(l)-1}(x_l) p_{Q(l)-1}(x_l),$$

and this implies

$$\hat{Z}_N[a] = \sum_{P\in S_N} \varepsilon(P) \sum_{Q\in S_N} \varepsilon(Q) \prod_{l=1}^{N} \int_{-\infty}^{\infty} w_2(x) a(x) p_{P(l)-1}(x) p_{Q(l)-1}(x)\, dx.$$

The stated formula now follows from the general formula

$$\sum_{P\in S_N} \varepsilon(P) \sum_{Q\in S_N} \varepsilon(Q) \prod_{l=1}^{N} u_{P(l),Q(l)} = N! \det[u_{j,k}]_{j,k=1,\dots,N}, \tag{5.23}$$

which itself is a corollary of (5.22). □

Due to the orthogonality of $\{p_j(x)\}_{j=0,1,\dots}$ we see immediately from (5.21) that

$$\hat{Z}_N[a]\Big|_{a=1} = N! \prod_{l=1}^{N} (p_{l-1}, p_{l-1})_2. \tag{5.24}$$

In Exercises 5.4 q.1(i) below this same result is deduced from the proof of Proposition 5.1.2. We also see from (5.21) that the functional differentiations required according to (5.20) to compute $\rho_{(n)}$ can be performed row by row. For a nonzero contribution this operation must act on n distinct rows. Setting $a = 1$, the remaining $N - n$ rows unaffected by the functional differentiation are nonzero only in their diagonals. Expanding the determinant by these elements then gives

$$\rho_{(n)}(x_1, \dots, x_n) = \sum_{\substack{j_1,\dots,j_n=1 \\ j_1\neq \cdots \neq j_n}}^{N} \prod_{l=1}^{n} (p_{j_l-1}, p_{j_l-1})_2^{-1} \det[w_2(x_\mu) p_{j_\mu-1}(x_\mu) p_{j_\gamma-1}(x_\mu)]_{\mu,\gamma=1,\dots,n}.$$

Minor manipulation reclaims the formula of Proposition 5.1.2.

5.2.2 Integral equation approach

Here we will show how the formula for $\rho_{(n)}^T$ in Proposition 5.1.2 can be derived directly. We require the formula

$$\rho_{(n)}^T(x_1, \dots, x_n) = \frac{\delta^n}{\delta a(x_1) \cdots \delta a(x_n)} \log \hat{Z}_N[a]\Big|_{a=1} \tag{5.25}$$

(cf. (5.20)), which in turn follows immediately from the formulas of Exercises 5.1 q.3. Also required is a formula which rewrites the expression (5.21) for $\hat{Z}_N[a]$ in terms of an integral operator [259].

PROPOSITION 5.2.2 *We have*

$$\begin{aligned}
&\prod_{l=1}^{N} (p_{l-1}, p_{l-1})_2^{-1} \det\Big[\int_{-\infty}^{\infty} w_2(y) a(y) p_{j-1}(y) p_{k-1}(y)\, dy\Big]_{j,k=1,\dots,N} \\
&\quad = \det\Big[\delta_{j,k} + (p_{j-1}, p_{j-1})_2^{-1} \int_{-\infty}^{\infty} w_2(y)(a(y) - 1) p_{j-1}(y) p_{k-1}(y)\, dy\Big]_{j,k=1,\dots,N} \\
&\quad = \det(1 + K_a),
\end{aligned}$$

where 1 denotes the identity operator, K_a is the integral operator on $(-\infty, \infty)$ with kernel $(a(y)-1)K_N(x,y)$ and $K_N(x,y)$ is given by (5.6).

Proof. The first equality follows from the orthogonality of $\{p_j(y)\}$. To derive the second equality, consider the eigenvalue equation

$$\int_{-\infty}^{\infty} (1 - a(y)) K_N(x,y) f_l(y)\, dy = \lambda_l f_l(x).$$

Seeking solutions of the form $f_l(x) = (w_2(x))^{1/2} \sum_{j=0}^{\infty} c_{lj} p_j(x)$ and equating coefficients of $p_k(x)$ gives that nonzero eigenvalues can only occur if $c_{lj} = 0$ for $j \geq N$, in which case

$$\lambda_l c_{lk} = \frac{1}{(p_k, p_k)_2} \sum_{j=0}^{N-1} c_{lj} \int_{-\infty}^{\infty} w_2(y)(1 - a(y)) p_k(y) p_j(y)\, dy, \quad (l = 0, \dots, N-1).$$

This is equivalent to the eigenvalue problem for an $N \times N$ matrix, the corresponding characteristic polynomial being

$$\det \left[\lambda \delta_{j,k} + (p_{j-1}, p_{j-1})_2^{-1} \int_{-\infty}^{\infty} w_2(y)(a(y) - 1) p_{j-1}(y) p_{k-1}(y)\, dy \right]_{j,k=1,\dots,N} = \prod_{l=0}^{N-1} (\lambda + \lambda_l).$$

Setting $\lambda = 1$ gives the stated result.

An alternative approach [522] is to make use of the general operator identity

$$\det(1 + AB) = \det(1 + BA) \tag{5.26}$$

(see, e.g., [128], and for the special case that A and B are matrices see Exercises 5.2 q.2), which can be verified from the second equality in (5.27) below. Let A be the column vector-valued integral operator on $(-\infty, \infty)$ with kernel

$$[(w_2(y))^{1/2}(a(y) - 1) p_{j-1}(y)]_{j=1,\dots,N},$$

and let B be the operator which multiplies by the row vector

$$[(w_2(x))^{1/2} p_{k-1}(x)/(p_{k-1}, p_{k-1})_2]_{k=1,\dots,N}.$$

We can then identify the l.h.s. of the stated identity with $\det(1 + AB)$, and the r.h.s. by $\det(1 + BA)$. □

Substituting Proposition 5.2.2 in (5.21) shows

$$\log \frac{\hat{Z}_N[a]}{\hat{Z}_N[1]} = \log \det(1 + K_a) = \operatorname{Tr} \log(1 + K_a). \tag{5.27}$$

We remark that because of its appearance in the kernel of an integral operator, $K_N(x,y)$ is often referred to as a *correlation kernel*. Substituting in (5.25), and using the general formulas

$$\frac{\delta}{\delta a(x)} \log(1 + K_a) = \frac{1}{1 + K_a} \frac{\delta K_a}{\delta a(x)}, \quad \frac{\delta}{\delta a(x)} \frac{1}{1 + K_a} = -\frac{1}{1 + K_a} \frac{\delta K_a}{\delta a(x)} \frac{1}{1 + K_a} \tag{5.28}$$

(for the latter formula, see Exercises 2.5 q.1), together with the fact that $\frac{\delta K_a}{\delta a(x)}\big|_{a=1} = K_N(\cdot, x)$ while $K_a|_{a=1} = 0$, reclaims the formula for $\rho_{(n)}^T$ in Proposition 5.1.2. This working shows that generally if

$$\log \frac{\hat{Z}_N[a]}{\hat{Z}_N[1]} = \log \det(1 + K_a), \tag{5.29}$$

where K_a is an operator with kernel of the form $(a(y)+c)K(x,y)$, then

$$\rho^T_{(n)}(x_1,\dots,x_n) = (-1)^{n+1} \sum_{\substack{\text{cycles}\\ \text{length } n}} G(x_{i_1},x_{i_2})G(x_{i_2},x_{i_3})\cdots G(x_{i_1},x_{i_n})$$

$$G(x_i,x_{i'}) := \Big\langle x_i \Big| K(1+K_a)^{-1}\Big|_{a=1} \Big| x_{i'} \Big\rangle, \tag{5.30}$$

with $\langle x|A|y\rangle$ denoting the kernel of the integral operator A and K denoting the integral operator with kernel $K(x,y)$. The working around (5.9) in the proof of Proposition 5.1.2 then gives

$$\rho_{(n)}(x_1,\dots,x_n) = \det[G(x_j,x_k)]_{j,k=1,\dots,n}. \tag{5.31}$$

The identity of Proposition 5.2.2 can also be used to derive the formula of Proposition 5.1.2 for $\rho_{(n)}$ directly, without having first to deduce the formula for $\rho^T_{(n)}$. For this one requires the Fredholm expansion formula from the theory of integral equations [541].

PROPOSITION 5.2.3 *Let $K(x,y)$ be a continuous function of x,y in the interval $[a,b]$. Let K be the integral operator on $[a,b]$ with kernel $K(x,y)$. One has*

$$1 + \sum_{N=1}^\infty \frac{1}{N!} \int_a^b dx_1 \cdots \int_a^b dx_N \, \det[K(x_j,x_l)]_{j,l=1,\dots,N} = \det(1+K). \tag{5.32}$$

From the definition of K_a in Proposition 5.2.2, it follows from (5.32) that

$$\det(1+K_a) = 1 + \sum_{k=1}^\infty \int_{-\infty}^\infty dx_1\,(a(x_1)-1)\cdots\int_{-\infty}^\infty dx_k\,(a(x_k)-1)\det[K(x_\alpha,x_\beta)]_{\alpha,\beta=1,\dots,k}.$$

Substituting this for $Z_N[a]$ in (5.20) we see that the functional derivatives can be computed immediately to reclaim the sought determinant formula for $\rho_{(n)}$.

EXERCISES 5.2 1. Consider the integral operator on $[a_1,a_2]$ with kernel $K(x,y) = \sum_{j=0}^{N-1} \phi_j(x)\phi_j(y)$, where $\{\phi_j(x)\}_{j=0,1,\dots}$ are a complete set of real linearly independent functions. From the proof of Proposition 5.2.2 we know that this integral operator admits N eigenfunctions $\{\psi_j(x)\}_{j=0,\dots,N-1}$ of the form

$$\psi_j(x) = \sum_{k=0}^{N-1} c_{kj}\phi_k(x),$$

with corresponding eigenvalues $\{\lambda_k\}_{k=0,\dots,N-1}$ specified by

$$\sum_{j=0}^{N-1} g_{ij}c_{jk} = c_{ik}\lambda_k, \qquad g_{ij} := \int_{a_1}^{a_2} \phi_i(x)\phi_j(x)\,dx.$$

(i) Note from the above equation that the matrix $[c_{jk}]$ has columns which consist of the eigenvectors of $[g_{ij}]$.

(ii) Use (i) and the fact that since $K(x,y)$ is real and symmetric, its eigenfunctions are orthogonal and can be chosen to be real, to note that $[c_{jk}]$ can be chosen to be real orthogonal. From this conclude that

$$\sum_{j=0}^{N-1} \psi_j(x)\psi_j(y) = \sum_{j=0}^{N-1} \phi_j(x)\phi_j(y) =: K(x,y), \qquad \int_{a_1}^{a_2} \psi_j^2(x)\,dx = \lambda_j.$$

2. The aim of this exercise is to prove the identity

$$\det(\mathbf{1}_p + \mathbf{A}_{p\times q}\mathbf{B}_{q\times p}) = \det(\mathbf{1}_q + \mathbf{B}_{q\times p}\mathbf{A}_{p\times q}), \tag{5.33}$$

which is the special case of (5.26) in which the operators A and B are matrices.

(i) Note the factorizations

$$\begin{bmatrix} \mathbf{1}_p + \mathbf{A}_{p\times q}\mathbf{B}_{q\times p} & \mathbf{A}_{p\times q} \\ \mathbf{0}_{q\times p} & \mathbf{1}_q \end{bmatrix} = \begin{bmatrix} \mathbf{1}_p & \mathbf{A}_{p\times q} \\ -\mathbf{B}_{q\times p} & \mathbf{1}_q \end{bmatrix} \begin{bmatrix} \mathbf{1}_p & \mathbf{0}_{p\times q} \\ \mathbf{B}_{q\times p} & \mathbf{1}_q \end{bmatrix},$$

$$\begin{bmatrix} \mathbf{1}_p & \mathbf{A}_{p\times q} \\ \mathbf{0}_{q\times p} & \mathbf{1}_q + \mathbf{B}_{q\times p}\mathbf{A}_{p\times q} \end{bmatrix} = \begin{bmatrix} \mathbf{1}_p & \mathbf{0}_{p\times q} \\ \mathbf{B}_{q\times p} & \mathbf{1}_q \end{bmatrix} \begin{bmatrix} \mathbf{1}_p & \mathbf{A}_{p\times q} \\ -\mathbf{B}_{q\times p} & \mathbf{1}_q \end{bmatrix}.$$

(ii) Take the determinant of the identities in (i) to deduce (5.26).

5.3 RATIOS OF CHARACTERISTIC POLYNOMIALS

Let us denote by $\langle \cdot \rangle_{\mathrm{UE}_N(w_2)}$ the canonical average (recall (1.45)) for a general one-component log-gas at $\beta = 2$, or equivalently a matrix ensemble with unitary symmetry, so that

$$\langle f \rangle_{\mathrm{UE}_N(w_2)} := \frac{1}{\hat{Z}_N} \int_{-\infty}^{\infty} dx_1 \cdots \int_{-\infty}^{\infty} dx_N \, f(x_1, \dots, x_N) \prod_{l=1}^{N} w_2(x_l) \prod_{1 \le j < k \le N} (x_k - x_j)^2.$$

In this notation, the result of Proposition 5.1.4 evaluates

$$\Big\langle \prod_{l=1}^{N} \frac{\prod_{j=1}^{K}(u_j - x_l)}{\prod_{j=1}^{Q}(v_j - x_l)} \Big\rangle_{\mathrm{UE}_N(w_2)} =: A_{K,Q}(u,v) \tag{5.34}$$

in the special cases $(K,Q) = (1,0),\ (0,1)$. Because in the case of $w_2(x)$ classical $\{x_j\}$ can be interpreted as the eigenvalues of a random matrix $\mathbf{X}$, one has $\prod_{l=1}^{N}(u - x_l) = \det(u\mathbf{1} - \mathbf{X})$, and (5.34) is then the average of ratios of characteristic polynomials. Here (5.34) will be evaluated as a $(K+Q) \times (K+Q)$ determinant [252], [32]. The special case $K = Q$ can be used to reclaim Proposition 5.1.2.

PROPOSITION 5.3.1 *Let $\{p_j(x)\}_{j=0,1,\dots}$ denote the set of monic orthogonal polynomials with respect to the weight $w_2(x)$, and let*

$$h_k(x) := \int_{-\infty}^{\infty} \frac{p_k(t)}{x - t} w_2(t)\, dt, \qquad \Delta(\{y_j\}_{j=1,\dots,p}) := \Delta(y) = \prod_{1 \le j < k \le p} (y_k - y_j). \tag{5.35}$$

For $Q \le N$

$$A_{K,Q}(u,v) = \frac{(-1)^{Q(Q-1)/2}}{\prod_{j=N-Q}^{N-1}(p_j,p_j)_2 \Delta(u)\Delta(v)} \det \begin{bmatrix} h_{N-Q}(v_1) & \cdots & h_{N+K-1}(v_1) \\ \vdots & & \vdots \\ h_{N-Q}(v_Q) & \cdots & h_{N+K-1}(v_Q) \\ p_{N-Q}(u_1) & \cdots & p_{N+K-1}(u_1) \\ \vdots & & \vdots \\ p_{N-Q}(u_K) & \cdots & p_{N+K-1}(u_K) \end{bmatrix}. \tag{5.36}$$

Proof. We base our proof on workings in [116]. In the notation

$$D(\{a_j\}_{j=1,\dots,m}; \{b_k\}_{k=1,\dots,n} := D(a;b) = \prod_{j=1}^{m} \prod_{k=1}^{n} (b_k - a_j) \tag{5.37}$$

we have

$$A_{K,Q}(u,v) = \frac{1}{C} \int_{-\infty}^{\infty} dx_1 \cdots \int_{-\infty}^{\infty} dx_N \prod_{l=1}^{N} w_2(x_l) \frac{D(x;u)}{D(x;v)} (\Delta(x))^2. \tag{5.38}$$

With $a \sqcup b$ denoting the concatenation of the sequences (a_i) and (b_j) we have $\Delta(a \sqcup b) = \Delta(a)D(a;b)\Delta(b)$ and thus

$$\frac{D(x;u)}{D(x;v)}(\Delta(x))^2 = \frac{1}{\Delta(u)\Delta(v)} \frac{\Delta(x)\Delta(v)}{D(x;v)} \Delta(x \sqcup u).$$

The first line of (5.5) tells us that

$$\Delta(x \sqcup u) = \det \begin{bmatrix} [p_{k-1}(x_j)]_{\substack{j=1,\dots,N \\ k=1,\dots,N+K}} \\ [p_{k-1}(u_j)]_{\substack{j=1,\dots,K \\ k=1,\dots,N+K}} \end{bmatrix}, \tag{5.39}$$

while according to the identity (5.45) of Exercises 5.3 q.1 below, for $N \geq Q$

$$\frac{\Delta(x)\Delta(v)}{D(x;v)} = (-1)^{Q(Q-1)/2} \det \left[p_0(x_j)\, p_1(x_j) \cdots p_{N-Q-1}(x_j)\, \frac{1}{v_1 - x_j} \cdots \frac{1}{v_Q - x_j} \right]_{j=1,\dots,N}. \tag{5.40}$$

Let us now substitute (5.40) and (5.39) in (5.38). Because of the orthogonality of $\{p_{k-1}(x)\}_{k=1,2,\dots}$ with respect to $w_2(x)$ we see that we can replace (5.39) by

$$\det \begin{bmatrix} [p_{k-1}(x_j)]_{\substack{j=1,\dots,N \\ k=1,\dots,N+K}} & \\ [0]_{K\times(N-Q)} & [p_{k-1}(u_j)]_{\substack{j=1,\dots,K \\ k=N-Q+1,\dots,N+K}} \end{bmatrix} \tag{5.41}$$

without changing the value of the integral. And in this, the elements chosen from columns $1,\dots,N-Q$ must be from rows $1,\dots,N$ for a nonzero contribution. Let us denote the rows so selected by $1 \leq \sigma(1) < \sigma(2) < \cdots < \sigma(N-Q) \leq N$. For the remainder of the first N rows we use the labels $1 \leq \sigma(N-Q+1) < \cdots < \sigma(N+K) \leq N$ where all the $\sigma(k)$ must be distinct, and we denote the set of such σ's by $\Sigma_{N-Q,Q}$. By first expanding the determinant (5.41) by elements in row $\sigma(j)$ of column j $(j = 1,\dots,N-Q)$, then antisymmetrizing in the ordering of these $\sigma(j)$ we obtain

$$\sum_{\Sigma_{N-Q,Q}} \varepsilon(\sigma) \det[p_{j-1}(x_{\sigma(k)})]_{j,k=1,\dots,N-Q} \det \begin{bmatrix} [p_{k-1}(x_{\sigma(j)})]_{\substack{j=N-Q+1,\dots,N \\ k=N-Q+1,\dots,N+K}} \\ [p_{k-1}(u_j)]_{\substack{j=1,\dots,K \\ k=N-Q+1,\dots,N+K}} \end{bmatrix},$$

which is an example of a Laplace expansion by multiple columns [8]. In the integrand the first determinant in this expression can be replaced by $(N-Q)!\prod_{j=1}^{N-Q} p_{j-1}(x_{\sigma(j)})$. Multiplying this with (5.40) and integrating over $x_{\sigma(j)}$ $(j = 1,\dots,N-Q)$ row by row in the determinant, due to the orthogonality, the determinant can be expanded down columns $\sigma(j)$ to show that (5.38) is equal to

$$\frac{(-1)^{Q(Q-1)/2}(N-Q)!}{C\Delta(u)\Delta(v)} \prod_{j=1}^{N-Q} (p_{j-1}, p_{j-1})_2 \sum_{\Sigma_{N-Q,Q}} \int_{-\infty}^{\infty} dx_{\sigma(N-Q+1)}\, w_2(x_{\sigma(N-Q+1)}) \cdots$$
$$\times \int_{-\infty}^{\infty} dx_{\sigma(N)}\, w_2(x_{\sigma(N)}) \det \left[\frac{1}{v_i - x_{\sigma(j)}} \right]_{\substack{i=1,\dots,Q \\ j=N-Q+1,\dots,N}} \det \begin{bmatrix} [p_{k-1}(x_{\sigma(j)})]_{\substack{j=N-Q+1,\dots,N \\ k=N-Q,\dots,N+K}} \\ [p_{k-1}(u_j)]_{\substack{j=1,\dots,K \\ k=N-Q,\dots,N+K}} \end{bmatrix}.$$

In this expression the first determinant can be replaced by $Q!/\prod_{j=1}^{Q}(v_j - x_{\sigma(N-Q+j)})$, and the integrations carried out row by row. As the resulting expression is independent of the set $\Sigma_{N-Q,Q}$ the sum over this quantity contributes $\binom{N}{Q}$, while C is given by (5.24), we see that (5.36) results. □

We have the following general formula relating $A_{K,Q}(u,v)$ in the case $K = Q = k$ to the k-point correlation function [91].

PROPOSITION 5.3.2 *Let*

$$\operatorname*{Res}_{v=y} \int_{-\infty}^{\infty} \frac{f(t)}{v-t}\, dt := \lim_{\epsilon\to 0^+} \frac{1}{\pi} \mathrm{Im} \int_{-\infty}^{\infty} \frac{f(t)}{y - i\epsilon - t}\, dt = f(y). \tag{5.42}$$

We have

$$\rho_{(k)}(y_1, \dots, y_k) = \operatorname*{Res}_{v_1=y_1} \cdots \operatorname*{Res}_{v_k=y_k} \frac{\partial^k}{\partial u_1 \cdots \partial u_k} A_{k,k}(u,v)\Big|_{u=v}. \tag{5.43}$$

Proof. This follows upon noting

$$\frac{\partial^k}{\partial u_1 \cdots \partial u_k} \prod_{l=1}^N \prod_{j=1}^k \frac{u_j - x_l}{v_j - x_l}\Big|_{u=v} = \sum_{\substack{j_1,\dots,j_k=1 \\ j_1 \ne \cdots \ne j_k}}^N \frac{1}{\prod_{p=1}^k (u_p - x_{j_p})}.$$

□

It should be noted that (5.43) holds with $\mathrm{UE}_N(w_2)$ in the definition of $A_{K,Q}$ replaced by a general p.d.f. It is shown in Exercises 5.3 q.2 that with $K = Q$ (5.36) can be rewritten to read

$$\begin{aligned} A_{Q,Q}(u,v) &= \frac{(-1)^{Q(Q-1)/2}}{\Delta(u)\Delta^2(v)} \int_{-\infty}^{\infty} ds_1 \cdots \int_{-\infty}^{\infty} ds_Q\, \Delta(s) D(s;u) \\ &\times \prod_{l=1}^Q \frac{1}{v_l - s_l} \Big(\frac{w_2(s_l)}{w_2(u_l)} \Big)^{1/2} \det \Big[K_N(u_j, s_k) \Big]_{j,k=1,\dots,Q}. \end{aligned} \tag{5.44}$$

With $Q = k$, to get a nonzero contribution from the operation of differentiating with respect to $u_1, \dots, u_k$ and setting $u = v$ we see that only the factor $\prod_{j=1}^k (u_j - s_j)$ in $D(s;u)$ need be differentiated. Furthermore the operations Res are immediate, and the result of Proposition 5.1.2 follows.

EXERCISES 5.3 1. [48] In terms of the notation (5.35), (5.37), here the identity

$$\det \left[[a_j^{k-1}]_{\substack{j=1,\dots,M \\ k=1,\dots,M-N}} \Big[\frac{1}{a_j - b_k} \Big]_{\substack{j=1,\dots,M \\ k=1,\dots,N}} \right] = (-1)^{N(N-1)/2 + N(M-1)} \frac{\Delta(a)\Delta(b)}{D(b;a)}, \tag{5.45}$$

where it is assumed $N \le M$, will be derived.

(i) In the Cauchy double alternant identity (4.33) replace $N \mapsto M$. Take the limit $y_1 \to \infty$ to deduce (5.45) in the case $N = M - 1$.

(ii) Subtract the first column of the determinant obtained in (i) from the second column, then take the limit $y_2 \to \infty$ to obtain (5.45) in the case $N = M - 2$. Subtract the first and second columns of the corresponding determinant from the third, take the limit as $y_3 \to \infty$ to obtain (5.45) in the case $N = M - 3$, and proceed analogously in the general case.

2. [32] The objective of this exercise is to derive (5.44).

(i) Generalize the workings of the proofs of Propositions 5.1.2 and 5.1.3 to show

$$\Big\langle \prod_{j=1}^N \prod_{\mu=1}^m (u_\mu - x_j)(v_\mu - x_j) \Big\rangle_{\mathrm{UE}_N(w_2)} = \frac{\prod_{l=N}^{N+m-1} (p_{l-1}, p_{l-1})_2}{\Delta(u)\Delta(v)} \det \Big[\sum_{\nu=0}^{N+m-1} \frac{p_\nu(u_j) p_\nu(v_k)}{(p_\nu, p_\nu)_2} \Big]_{j,k=1,\dots,m}.$$

(ii) Rewrite the determinant in (5.36) in the form

$$\int_{-\infty}^{\infty} ds_1 \cdots \int_{-\infty}^{\infty} ds_Q \prod_{l=1}^Q \frac{w_2(s_l)}{v_l - s_l} \det \begin{bmatrix} p_{N-Q}(s_1) & \cdots & p_{N+K-1}(s_1) \\ \vdots & & \vdots \\ p_{N-Q}(s_Q) & \cdots & p_{N+K-1}(s_Q) \\ p_{N-Q}(u_1) & \cdots & p_{N+K-1}(u_1) \\ \vdots & & \vdots \\ p_{N-Q}(u_K) & \cdots & p_{N+K-1}(u_K) \end{bmatrix}.$$

Use (5.36) in the case $Q = 0$ to show that the determinant in this expression is equal to

$$\Delta(s \sqcup u) A_{Q+K,0}(s \sqcup u, \cdot)|_{N \mapsto N-Q}.$$

(iii) Substitute for $A_{Q+K,0}(s \sqcup u, \cdot)|_{N \mapsto N-Q}$ according to the result of (i), substitute this for the determinant above, and finally substitute in (5.36) with $K = Q$ to deduce (5.44).

5.4 THE CLASSICAL WEIGHTS

5.4.1 The Gaussian, Laguerre and Jacobi weights

From Proposition 1.3.4, (3.16) and (3.74), the one-body Boltzmann factor $w_2(x)$ for the Gaussian, Laguerre and Jacobi ensembles is given by

$$w_2^{(\mathrm{G})}(x) = e^{-x^2}, \quad -\infty < x < \infty, \qquad w_2^{(\mathrm{L})}(x) = x^a e^{-x}, \quad x > 0, \; a > -1$$

and

$$w_2^{(\mathrm{J})}(x) = (1-x)^a (1+x)^b, \quad -1 < x < 1, \; a, b > -1$$

respectively (in the latter two cases $w_2(x) = 0$ outside the specified intervals). The monic orthogonal polynomials associated with these weight functions are proportional to the Hermite, Laguerre and Jacobi classical polynomials respectively (see, e.g., [508]). Explicitly

$$\begin{aligned} p_n^{(\mathrm{G})}(x) &= 2^{-n} H_n(x), \\ p_n^{(\mathrm{L})}(x) &= (-1)^n n! L_n^a(x), \\ p_n^{(\mathrm{J})}(x) &= 2^n n! \frac{\Gamma(a+b+n+1)}{\Gamma(a+b+2n+1)} P_n^{(a,b)}(x) \end{aligned} \tag{5.46}$$

where

$$\begin{aligned} H_n(x) &= \sum_{m=0}^{[n/2]} (-1)^m 2^{n-m} \binom{n}{2m} \frac{(2m)!}{2^m m!} x^{n-2m}, \\ L_n^a(x) &= \sum_{m=0}^{n} (-1)^m \binom{n+a}{n-m} \frac{x^m}{m!}, \\ P_n^{(a,b)}(x) &= \binom{n+a}{n} \sum_{m=0}^{n} \frac{(n+a+b+1)_m}{(a+1)_m} \binom{n}{m} \left(\frac{x-1}{2}\right)^m \end{aligned} \tag{5.47}$$

(the notation $(u)_p$ is defined in (5.83) below). The corresponding normalizations are

$$\begin{aligned} (p_n, p_n)_2^{(\mathrm{G})} &= \pi^{1/2} 2^{-n} n!, \\ (p_n, p_n)_2^{(\mathrm{L})} &= \Gamma(n+1)\Gamma(a+n+1), \\ (p_n, p_n)_2^{(\mathrm{J})} &= 2^{a+b+1+2n} \frac{\Gamma(n+1)\Gamma(a+b+1+n)\Gamma(a+1+n)\Gamma(b+1+n)}{\Gamma(a+b+2n+1)\Gamma(a+b+2n+2)}. \end{aligned} \tag{5.48}$$

Using the above formulas in Proposition 5.1.2, with $K_N(x,y)$ evaluated according to Proposition 5.1.3 and (5.13), gives an explicit expression for the n-particle correlation in each of the ensembles.

5.4.2 Circular ensembles and the Cauchy weight

By making the transformation (2.50) we know from (2.51) that

$$\prod_{j=1}^{N} w_\beta(e^{i\theta_j}) \prod_{1\le j<k\le N} |e^{i\theta_k} - e^{i\theta_j}|^\beta d\theta_1 \cdots d\theta_N$$
$$= 2^{\beta N(N-1)/2+N} \prod_{j=1}^{N} w_\beta\Big(\frac{1+i\lambda_j}{1-i\lambda_j}\Big)(1+\lambda_j^2)^{-\beta(N-1)/2-1} \prod_{1\le j<k\le N} |\lambda_k - \lambda_j|^\beta d\lambda_1 \cdots d\lambda_N. \tag{5.49}$$

In the particular case

$$w_\beta(z) = |1+z|^{\beta a}, \qquad z := e^{i\theta} \tag{5.50}$$

this coincides with (3.123), so the effective real weight on the r.h.s. of (5.49) is then

$$w_\beta(\lambda) = (1+\lambda^2)^{-\alpha}, \quad \alpha = \beta(N+a-1)/2+1, \tag{5.51}$$

which specifies the Cauchy ensemble. Our present interest is in the case $\beta = 2$, when the Cauchy weight is $(1+x^2)^{-(N+a)}$. It is known [470], [464] (see also Exercises 5.4 q.2) that the orthogonal polynomials associated with this weight are the Jacobi polynomials

$$\{P_j^{(-a-N,-a-N)}(ix)\}_{0\le j<N+a} \tag{5.52}$$

(the bound on j is required because (5.51) has only a finite number of bounded moments).

From the final formula in (5.46) it follows that the monic polynomials corresponding to the set (5.52) are

$$p_n^{(\mathrm{Cy})}(x) := i^{-n} p_n^{(\mathrm{J})}(ix)\Big|_{\substack{a=b \\ a\mapsto -N-a}}. \tag{5.53}$$

For the normalization one has (see Exercises 5.4 q.2)

$$(p_n,p_n)_2^{(\mathrm{Cy})} = (-1)^{n-1}\tan\pi a\,(p_n,p_n)_2^{(\mathrm{J})}\Big|_{\substack{a=b \\ a\mapsto -N-a}}$$
$$= \pi 2^{-2(N-n)-2(a-1)} \frac{\Gamma(n+1)\Gamma(2(N+a)-2n)\Gamma(2(N+a)-1-2n)}{\Gamma(2(N+a)-n)(\Gamma(N+a-n))^2}. \tag{5.54}$$

The n-point correlation for the weight (5.51) on the real line thus follows immediately from Propositions 5.1.2 and 5.1.3. To deduce from this $\rho_{(n)}$ for the weight (5.50) we use the fact that for θ_j $(= 2\pi x_j/L)$ and λ_j related by (2.50),

$$\rho_{(n)}(x_1,\dots,x_n) = \prod_{j=1}^{n} \frac{4\pi}{|1+z_j|^2 L} \rho_{(n)}(\lambda_1,\dots,\lambda_n)\Big|_{\lambda_j = i(1-z_j)/(1+z_j)}$$
$$= \Big(\frac{\pi}{L}\Big)^n \prod_{j=1}^{n} \Big(\frac{1}{2}|1+z_j|\Big)^{2(N-1+a)} (1+z_j)^2$$
$$\times \det\left[\frac{p_N^{(\mathrm{Cy})}\big(i\frac{1-z_j}{1+z_j}\big)p_{N-1}^{(\mathrm{Cy})}\big(i\frac{1-z_k}{1+z_k}\big) - p_N^{(\mathrm{Cy})}\big(i\frac{1-z_k}{1+z_k}\big)p_{N-1}^{(\mathrm{Cy})}\big(i\frac{1-z_j}{1+z_j}\big)}{2i(p_{N-1},p_{N-1})_2^{(\mathrm{Cy})}(z_j - z_k)}\right]_{j,k=1,\dots,n} \tag{5.55}$$

5.4.3 Structural properties associated with the classical weights

The weight functions

$$w_2(x) =: e^{-2V(x)} = \begin{cases} e^{-x^2}, & \text{Hermite,} \\ x^a e^{-x} \ (x > 0), & \text{Laguerre,} \\ (1-x)^a(1+x)^b \ (-1 < x < 1), & \text{Jacobi,} \\ (1+x^2)^{-\alpha}, & \text{Cauchy,} \end{cases} \tag{5.56}$$

all have the property that their logarithmic derivative is a rational function. Thus, if we write

$$\frac{w_2'(x)}{w_2(x)} = -2V'(x) =: -\frac{g(x)}{f(x)}, \tag{5.57}$$

where f and g have no common factors, and $f > 0$, then

$$(f, g) = \begin{cases} (1, 2x), & \text{Hermite,} \\ (x, x-a), & \text{Laguerre,} \\ (1-x^2, (a-b)+(a+b)x), & \text{Jacobi,} \\ (1+x^2, 2\alpha x), & \text{Cauchy.} \end{cases} \tag{5.58}$$

Note that in (5.58)

$$\text{degree } f \leq 2, \qquad \text{degree g} \leq 1. \tag{5.59}$$

We will call any weight function with this property *classical*. An equivalent characterization of a classical polynomial, which has an immediate consequence regarding the Christoffel-Darboux sum, and future consequence in the construction of skew orthogonal polynomials to be undertaken in the next chapter, follows by consideration of the operator [4]

$$A := f\frac{d}{dx} + \left(\frac{f' - g}{2}\right) = \sqrt{e^{2V} f}\,\frac{d}{dx}\left(\sqrt{f e^{-2V}}\right). \tag{5.60}$$

First let us consider a general property of A.

PROPOSITION 5.4.1 *Assuming $e^{-2V(x)}$ vanishes at the endpoints of its support,*

$$(\phi, A\psi)_2 = -(A\phi, \psi)_2. \tag{5.61}$$

Proof. This follows immediately upon using the second form of A in (5.60) and integration by parts. □

Consider now the matrix $[(p_j, Ap_k)_2]_{j,k=0,\dots,N-1}$, where $\{p_j\}_{j=0,1,\dots}$ are the monic orthogonal polynomials associated with the weight function $e^{-2V(x)}$. By (5.61) this matrix is antisymmetric. Furthermore, the bound on the order of the polynomial pairs (f, g) given in (5.59) and the definition (5.60) give that

$$Ap_k(x) = -\frac{c_k}{(p_{k+1}, p_{k+1})_2} p_{k+1}(x) + \text{polynomials of lower degree.} \tag{5.62}$$

These two facts together imply that in the classical cases

$$\mathbf{A} := [(p_j, Ap_k)_2]_{j,k=0,\dots,N-1} = \begin{bmatrix} 0 & c_0 & 0 & 0 & \cdots & & \\ -c_0 & 0 & c_1 & 0 & \cdots & & \\ 0 & -c_1 & 0 & c_2 & \cdots & & \\ 0 & 0 & -c_2 & 0 & \cdots & & \\ \vdots & \vdots & \vdots & \vdots & \ddots & & \\ & & & & & 0 & c_{N-1} \\ & & & & & -c_{N-1} & 0 \end{bmatrix}, \tag{5.63}$$

or equivalently

$$A p_k(x) = -\frac{c_k}{(p_{k+1}, p_{k+1})_2} p_{k+1}(x) + \frac{c_{k-1}}{(p_{k-1}, p_{k-1})_2} p_{k-1}(x). \tag{5.64}$$

From the explicit forms (5.58), the definition (5.60) and the fact that $\{p_k\}$ are monic we find

$$\frac{c_k}{(p_{k+1}, p_{k+1})_2} = \begin{cases} 1, & \text{Hermite,} \\ \frac{1}{2}, & \text{Laguerre,} \\ \frac{1}{2}(2k+2+a+b), & \text{Jacobi,} \\ \alpha - 1 - k, & \text{Cauchy,} \end{cases} \tag{5.65}$$

with the explicit values of $(p_{k+1}, p_{k+1})_2$ given by (5.48) and (5.54).

Writing $\psi_k := e^{-V} p_k$, it follows from (5.64) and the three-term recurrence (5.14) that $\{\psi_k, \psi_{k-1}\}$ satisfies the first order system

$$f(x) \begin{bmatrix} \psi_k'(x) \\ \psi_{k-1}'(x) \end{bmatrix} = \begin{bmatrix} A_{11}(x) & A_{12}(x) \\ A_{21}(x) & A_{22}(x) \end{bmatrix} \begin{bmatrix} \psi_k(x) \\ \psi_{k-1}(x) \end{bmatrix} \tag{5.66}$$

with

$$\begin{aligned} A_{11}(x) &= -\Big(\frac{c_k}{(p_{k+1}, p_{k+1})_2} (x + B_{k+1}) + \frac{1}{2} f' \Big), \\ A_{12}(x) &= \frac{c_{k-1}}{(p_{k-1}, p_{k-1})_2} + \frac{c_k (p_k, p_k)_2}{(p_{k+1}, p_{k+1})_2 (p_{k-1}, p_{k-1})_2}, \\ A_{21}(x) &= \frac{c_{k-1}}{(p_k, p_k)_2} + \frac{c_{k-2}}{(p_{k-1}, p_{k-1})_2}, \\ A_{22}(x) &= -\frac{1}{2} f' + \frac{c_{k-2}}{(p_{k-1}, p_{k-1})_2} (x + B_k). \end{aligned}$$

Using (5.65), (5.48), (5.58) and (5.16) one obtains from this that in the Hermite case

$$A_{11}^{(\mathrm{G})}(x) = -A_{22}^{(\mathrm{G})}(x) = -x, \quad A_{12}^{(\mathrm{G})}(x) = k, \quad A_{21}^{(\mathrm{G})}(x) = 2; \tag{5.67}$$

in the Laguerre case

$$A_{11}^{(\mathrm{L})}(x) = -A_{22}^{(\mathrm{L})}(x) = -\frac{1}{2}(x - 2k - a), \quad A_{12}^{(\mathrm{L})}(x) = k(a+k), \quad A_{21}^{(\mathrm{L})}(x) = 1; \tag{5.68}$$

while in the Jacobi case

$$\begin{aligned} A_{11}^{(\mathrm{J})}(x) &= -A_{22}^{(\mathrm{J})}(x) = -\frac{1}{2}(2k+a+b)x + \frac{b^2 - a^2}{2(2k+a+b)}, \\ A_{12}^{(\mathrm{J})}(x) &= \frac{k(a+b+k)(a+k)(b+k)}{2(a+b+2k+2)(a+b+2k+1)(a+b+2k)}, \quad A_{21}^{(\mathrm{J})}(x) = 2k+a+b-1. \end{aligned} \tag{5.69}$$

A consequence of the matrix differential equation (5.66) is integral formulas for the Christoffel-Darboux sum (5.10) in the classical cases [519], [334], [221].

PROPOSITION 5.4.2 *Let $\psi_k := e^{-V} p_k$. Then in the Gaussian case*

$$K_N^{(\mathrm{G})}(x,y) = \frac{1}{(p_{N-1}, p_{N-1})_2^{(\mathrm{G})}} \int_0^\infty \Big(\psi_N^{(\mathrm{G})}(x+t) \psi_{N-1}^{(\mathrm{G})}(y+t) + \psi_{N-1}^{(\mathrm{G})}(x+t) \psi_N^{(\mathrm{G})}(y+t) \Big)\, dt; \tag{5.70}$$

in the Laguerre case

$$K_N^{(\mathrm{L})}(x,y) = -\frac{1}{2(p_{N-1},p_{N-1})_2^{(\mathrm{L})}} \int_0^1 \Big(\psi_N^{(\mathrm{L})}(xt)\psi_{N-1}^{(\mathrm{L})}(yt) + \psi_{N-1}^{(\mathrm{L})}(xt)\psi_N^{(\mathrm{L})}(yt)\Big)\, dt; \tag{5.71}$$

and in the Jacobi case

$$K_N^{(\mathrm{J})}(x,y) = \frac{(2N+a+b)}{2(p_{N-1},p_{N-1})_2^{(\mathrm{J})}} \int_0^1 \Big(\psi_N^{(\mathrm{J})}\Big(\frac{x+s}{1+xs}\Big)\psi_{N-1}^{(\mathrm{J})}\Big(\frac{y+s}{1+ys}\Big) + \psi_{N-1}^{(\mathrm{J})}\Big(\frac{x+s}{1+xs}\Big)\psi_N^{(\mathrm{J})}\Big(\frac{y+s}{1+ys}\Big)\Big)\frac{ds}{(1+xs)(1+ys)}. \tag{5.72}$$

Proof. In general

$$\begin{aligned}
&\Big(\frac{\partial}{\partial x}+\frac{\partial}{\partial y}\Big)\frac{\psi_N(x)\psi_{N-1}(y)-\psi_N(y)\psi_{N-1}(x)}{x-y} \\
&= \frac{1}{x-y}\Big(\frac{\partial}{\partial x}+\frac{\partial}{\partial y}\Big)\Big(\psi_N(x)\psi_{N-1}(y)-\psi_N(y)\psi_{N-1}(x)\Big) \\
&= \frac{1}{x-y}\Big(\frac{\partial}{\partial x}+\frac{\partial}{\partial y}\Big)\begin{bmatrix}\psi_N(x) & \psi_{N-1}(x)\end{bmatrix}\begin{bmatrix}0 & 1\\ -1 & 0\end{bmatrix}\begin{bmatrix}\psi_N(y)\\ \psi_{N-1}(y)\end{bmatrix}.
\end{aligned}$$

Now specializing to the Hermite case, we see that (5.66) can be used to compute the derivatives, thus implying that it is equal to

$$\begin{aligned}
&\begin{bmatrix}\psi_N^{(\mathrm{G})}(x) & \psi_{N-1}^{(\mathrm{G})}(x)\end{bmatrix}\begin{bmatrix} -\dfrac{A_{21}^{(\mathrm{G})}(x)-A_{21}^{(\mathrm{G})}(y)}{x-y} & \dfrac{A_{11}^{(\mathrm{G})}(x)+A_{22}^{(\mathrm{G})}(y)}{x-y} \\ -\dfrac{A_{22}^{(\mathrm{G})}(x)+A_{11}^{(\mathrm{G})}(y)}{x-y} & \dfrac{A_{12}^{(\mathrm{G})}(x)-A_{12}^{(\mathrm{G})}(y)}{x-y}\end{bmatrix}\begin{bmatrix}\psi_N^{(\mathrm{G})}(y)\\ \psi_{N-1}^{(\mathrm{G})}(y)\end{bmatrix} \\
&= \begin{bmatrix}\psi_N^{(\mathrm{G})}(x) & \psi_{N-1}^{(\mathrm{G})}(x)\end{bmatrix}\begin{bmatrix}0 & -1\\ -1 & 0\end{bmatrix}\begin{bmatrix}\psi_N^{(\mathrm{G})}(y)\\ \psi_{N-1}^{(\mathrm{G})}(y)\end{bmatrix} \\
&= -\Big(\psi_N^{(\mathrm{G})}(x)\psi_{N-1}^{(\mathrm{G})}(y)+\psi_{N-1}^{(\mathrm{G})}(x)\psi_N^{(\mathrm{G})}(y)\Big),
\end{aligned}$$

where the second line follows by using the explicit form of the matrix elements given in (5.67). Replace $x \mapsto x+t$, $y \mapsto y+t$ and note that $(\partial/\partial x + \partial/\partial y)$ can be replaced by $\partial/\partial t$. Integrating both sides from $t=0$ to ∞ gives (5.70). In the Laguerre case, one begins by making use of the general formula

$$\Big(x\frac{\partial}{\partial x}+y\frac{\partial}{\partial y}\Big)\frac{(xy)^{1/2}}{x-y}f = \frac{(xy)^{1/2}}{x-y}\Big(x\frac{\partial}{\partial x}+y\frac{\partial}{\partial y}\Big)f.$$

Proceeding as in the Gaussian case then gives

$$\Big(x\frac{\partial}{\partial x}+y\frac{\partial}{\partial y}\Big)(xy)^{1/2}K_N^{(\mathrm{L})}(x,y) = -\frac{(xy)^{1/2}}{2(p_{N-1},p_{N-1})_2^{(\mathrm{L})}}\Big(\psi_N^{(\mathrm{L})}(x)\psi_{N-1}^{(\mathrm{L})}(y)+\psi_{N-1}^{(\mathrm{L})}(x)\psi_N^{(\mathrm{L})}(y)\Big).$$

Changing variables $x=e^{-t}$, $y=e^{-s}$, and arguing as in the Gaussian case, we see from this that

$$\begin{aligned}
K_N^{(\mathrm{L})}(e^{-t},e^{-s}) &= -\frac{1}{2(p_{N-1},p_{N-1})_2^{(\mathrm{L})}} \\
&\quad\times\int_0^\infty e^{-u}\Big(\psi_N^{(\mathrm{L})}(e^{-t-u})\psi_{N-1}^{(\mathrm{L})}(e^{-s-u})+\psi_{N-1}^{(\mathrm{L})}(e^{-t-u})\psi_N^{(\mathrm{L})}(e^{-s-u})\Big)\,du,
\end{aligned}$$

which is equivalent to (5.71).

Finally, in the Jacobi case, we make use of the general formula

$$\Big((1-x^2)\frac{\partial}{\partial x} + (1-y^2)\frac{\partial}{\partial y}\Big)\frac{(1-x^2)^{1/2}(1-y^2)^{1/2}}{x-y}f = \frac{(1-x^2)^{1/2}(1-y^2)^{1/2}}{x-y}\Big((1-x^2)\frac{\partial}{\partial x} + (1-y^2)\frac{\partial}{\partial y}\Big)f,$$

and proceed as in the Hermite and Jacobi cases to deduce

$$\Big((1-x^2)\frac{\partial}{\partial x} + (1-y^2)\frac{\partial}{\partial y}\Big)(1-x^2)^{1/2}(1-y^2)^{1/2}K_N^{(\mathrm{J})}(x,y)$$
$$= -\frac{(2N+a+b)}{2}\frac{(1-x^2)^{1/2}(1-y^2)^{1/2}}{(p_{N-1},p_{N-1})_2^{(\mathrm{J})}}\Big(\psi_N^{(\mathrm{J})}(x)\psi_{N-1}^{(\mathrm{J})}(y) + \psi_{N-1}^{(\mathrm{J})}(x)\psi_N^{(\mathrm{J})}(y)\Big).$$

Changing variables $x = \tanh u$, $y = \tanh v$ we deduce from this that

$$\frac{1}{\cosh u \cosh v}K_N^{(\mathrm{J})}(\tanh u, \tanh v)$$
$$= \frac{(2N+a+b)}{2(p_{N-1},p_{N-1})_2^{(\mathrm{J})}}\int_0^\infty \frac{1}{\cosh(u+t)\cosh(v+t)}$$
$$\times\Big(\psi_N^{(\mathrm{J})}(\tanh(u+t))\psi_{N-1}^{(\mathrm{J})}(\tanh(v+t)) + \psi_{N-1}^{(\mathrm{J})}(\tanh(u+t))\psi_N^{(\mathrm{J})}(\tanh(v+t))\Big)\,dt.$$

Another change of variables gives (5.72). □

With $f(x)$ as in (5.58) and $x_0 = 0,1,0$ for the Hermite, Laguerre and Jacobi cases, the change of variables in the proof of Proposition 5.4.2 can be written $u(x) = \int_{x_0}^x dz/f(z)$, so that

$$(x(u), \sqrt{f(x(u))}) = \begin{cases} (u,1), & \text{Hermite,} \\ (e^u, e^{u/2}), & \text{Laguerre,} \\ (\tanh u, 1/\cosh u), & \text{Jacobi.} \end{cases}$$

Writing $\tilde{\psi}_k^{(\cdot)}(u) := \sqrt{f(x(u))}\psi_k^{(\cdot)}(x(u))$, and writing $\tilde{c}_k$ for (5.65), one sees that the results in the proof of Proposition 5.4.2 can be written in the unified form [332]

$$\sqrt{f(x(u))}\sqrt{f(x(v))}K_N^{(\cdot)}(x,y)$$
$$= \frac{\tilde{c}_{N-1}}{(p_{N-1},p_{N-1})_2^{(\cdot)}}\int_0^\infty \Big(\tilde{\psi}_N^{(\cdot)}(u+t)\tilde{\psi}_{N-1}^{(\cdot)}(v+t) + \tilde{\psi}_{N-1}^{(\cdot)}(u+t)\tilde{\psi}_N^{(\cdot)}(v+t)\Big)\,dt. \tag{5.73}$$

EXERCISES 5.4 1. (i) From the proof of Proposition 5.1.2 show that

$$\int_{-\infty}^\infty dx_1 w_2(x_1)\cdots\int_{-\infty}^\infty dx_N w_2(x_N)\prod_{1\le j<k\le N}(x_k-x_j)^2 = N!\prod_{k=1}^N (p_{k-1},p_{k-1})_2. \tag{5.74}$$

(ii) Use the integral evaluations of Propositions 4.7.1 and 4.7.3, (4.144) and (4.145) in (i) to provide alternative evaluations of $(p_n,p_n)_2^{(\mathrm{G})}$, $(p_n,p_n)_2^{(\mathrm{L})}$, $(p_n,p_n)^{(\mathrm{J})}$ and $(p_n,p_n)^{(\mathrm{Cy})}$.

(iii) Make use of the Vandermonde determinant (1.173), and (5.22), (5.23) to show

$$\int_{-\infty}^\infty dx_1\, w_2(x_1)\cdots\int_{-\infty}^\infty dx_N\, w_2(x_N)\prod_{1\le j<k\le N}(x_k-x_j)^2 = N!\det\Big[\int_{-\infty}^\infty w_2(x)x^{j+k-2}dx\Big]_{j,k=1,\dots,N}, \tag{5.75}$$

$$\int_{-L/2}^{L/2} dx_1\, w_2(z_1)\cdots\int_{-L/2}^{L/2} dx_N\, w_2(z_N)\prod_{1\le j<k\le N}|z_k-z_j|^2 = N!\det\Big[\int_{-L/2}^{L/2} w_2(z)z^{j-k}dx\Big]_{j,k=1,\dots,N}, \tag{5.76}$$

where in (5.76) $z = e^{2\pi i x/L}$.

(iv) Use the result of Proposition 2.2.5 in the case $\beta = 2$, together with the fact from Proposition 4.7.2 that $C_{2,N}$ therein equals $(2\pi)^N N!$ (or alternatively derive this using (5.76) with $L = 2\pi$, $w_2(z) = 1$), to show that the final identity in (iii) can be written

$$\Big\langle \prod_{l=1}^{N} w_2(e^{i\theta_l}) \Big\rangle_{U(N)} = \det \left[\frac{1}{2\pi} \int_{-\pi}^{\pi} w_2(e^{i\theta}) e^{i\theta(j-k)} \, d\theta \right]_{j,k=1,\dots,N}. \tag{5.77}$$

(v) Use the Vandermonde determinant formula (1.173) in the multidimensional integral formula of Proposition 5.1.4 for $p_N(x)$, together with elementary row operations in the resulting determinant to show

$$p_N(x) = \frac{N!}{C} \det[c_{j+k}x - c_{j+k+1}]_{j,k=0,\dots,N-1}, \qquad c_j := \int_{-\infty}^{\infty} w_2(x) x^j \, dx. \tag{5.78}$$

2. (i) Show from the Euler beta integral (4.2) that for $q \in \mathbb{Z}_{\geq 0}$

$$\int_{-1}^{1} \frac{x^{2q}}{(1-x^2)^\alpha} \, dx = (-1)^{q-1} \cot \pi\alpha \frac{\Gamma(1/2+q)\Gamma(\alpha-q-1/2)}{\Gamma(\alpha)}.$$

(ii) Make the change of variables $x = y/(1+y)$ in the Euler beta integral, then put $y \mapsto x^2$ to deduce that for $q \in \mathbb{Z}_{\geq 0}$

$$\int_{-\infty}^{\infty} \frac{x^{2q}}{(1+x^2)^\alpha} \, dx = \frac{\Gamma(1/2+q)\Gamma(\alpha-q-1/2)}{\Gamma(\alpha)}.$$

(iii) By comparing (i) and (ii) deduce that for f an analytic function such that the integrals are defined

$$\int_{-1}^{1} \frac{f(x)}{(1-x^2)^\alpha} \, dx = -\cot \pi\alpha \int_{-\infty}^{\infty} \frac{f(ix)}{(1+x^2)^\alpha} \, dx,$$

where the integrals are to be interpreted as their analytic continuation in α, and from this deduce (5.54).

3. [291] The objective of this exercise is to compute the integral in (1.97) and so extract information on the coefficients $a_{k+1-2g}(k)$ in (1.106).

(i) Note from Proposition 5.1.2, (5.46) and (5.48) that in the GUE (Hermite case)

$$\rho_{(1)}(\lambda) = e^{-\lambda^2} \sum_{\nu=0}^{N-1} \frac{(H_\nu(\lambda))^2}{\pi^{1/2} 2^\nu \nu!}.$$

(ii) Substitute the result of (i) in (1.97) and use the expansions

$$(H_\nu(\lambda))^2 = \sum_{p=0}^{\nu} e_{p,\nu} H_{2p}(\lambda), \quad e_{p,\nu} = \frac{2^{\nu-p}\nu!}{p!} \binom{\nu}{p},$$

$$\lambda^{2k} = \sum_{l=0}^{k} c_{l,2k} H_{2k-2l}(\lambda), \quad c_{l,2k} = \frac{(2k)!}{2^{2k} l! (2k-2l)!}$$

together with the orthogonality of $\{H_j(\lambda)\}_{j=0,1,\dots}$ to show that

$$\langle \mathrm{Tr}(\mathbf{X}^{2k}) \rangle_{\mathrm{GUE}^*} = (2k-1)!! \sum_{p=0}^{k} 2^p \binom{k}{p} \binom{N}{p+1}$$

and use this to rederive (1.112). Read off too the value of $a_{k-1}(k)$.

(iii) Substitute the result of (ii) in (1.106) to deduce the recurrence

$$a_{k+1-2g}(k) = \frac{4k-2}{k+1} a_{k-2g}(k-1) + \frac{(k-1)(2k-1)(2k-3)}{k+1} a_{k+1-2g}(k-2),$$

where $a_l(k) = 0$ for $l > k+1$ and $a_l(0) = \delta_{l,1}$.

5.5 CIRCULAR ENSEMBLES AND THE CLASSICAL GROUPS

5.5.1 Orthogonal polynomials on the unit circle

The formalism of Section 5.4.2 relating the circular ensemble to an ensemble on the real line has the virtue of unifying the treatment of the two classes of problems. This is particularly useful in the next chapter, when the correlations at $\beta = 1$ and 4 are calculated. However, restricting attention to $\beta = 2$, the circular ensembles can readily be studied without recourse to the mapping (2.50) [421].

Write $z_j := e^{2\pi i x_j/L}$, $-L/2 \le x_j \le L/2$, and let $\{p_k(z)\}_{k=0,1,\dots}$ be the family of monic orthogonal polynomials associated with the real, non-negative weight function $w_2(z)$ so that analogous to (5.4) we have

$$p_k(z) = z^k + \alpha_{k-1} z^{k-1} + \cdots + \alpha_0,$$

$$\int_{-L/2}^{L/2} w_2(z) p_j(z) \overline{p_k(z)}\, dx := (p_j, p_k)_2^{(\mathrm{C})} = (p_j, p_j)_2^{(\mathrm{C})} \delta_{j,k}. \tag{5.79}$$

A Gram-Schmidt construction of these polynomials from $\{z^k\}_{k=0,1,\dots}$ shows that each $p_j(z)$ has real coefficients, provided $w_2(z) = w_2(\bar{z})$, and thus provided w_2 is even in x. This condition will be assumed henceforth.

Proceeding as in the proof of (5.5) we see that

$$\prod_{l=1}^N w_2(z_l) \prod_{1 \le j < k \le N} |z_k - z_j|^2 = \prod_{j=1}^N (p_{j-1}, p_{j-1})_2^{(\mathrm{C})} \det[K_N(z_j, \bar{z}_k)]_{j,k=1,\dots,N},$$

$$K_N(z, \bar{u}) := \Big(w_2(z) w_2(u) \Big)^{1/2} \sum_{\nu=0}^{N-1} \frac{p_\nu(z) p_\nu(\bar{u})}{(p_\nu, p_\nu)_2^{(C)}}. \tag{5.80}$$

From this and (5.79), the proof of Proposition 5.1.2 shows that the corresponding correlation function is given by

$$\rho_{(n)}(x_1, \dots, x_n) = \det[K_N(z_j, \bar{z}_k)]_{j,k=1,\dots,n}. \tag{5.81}$$

Also, the function $K_N(z, \bar{u})$ can be summed explicitly via a result analogous to the Christoffel-Darboux formula (5.10) [508].

PROPOSITION 5.5.1 *For* $|u| = |z| = 1$ *we have*

$$K_N(z, \bar{u}) = \frac{(w_2(z) w_2(u))^{1/2}}{(p_N, p_N)_2^{(C)}} \frac{\bar{u}^N p_N(u) z^N p_N(\bar{z}) - p_N(\bar{u}) p_N(z)}{1 - \bar{u} z}. \tag{5.82}$$

A derivation of this result, modeled on the derivation of (5.10), is given in Exercises 5.5 q.1.

Consider now the weight function (5.50). In this case the polynomials $p_n(z)$ can be obtained explicitly, being polynomial examples of the hypergeometric function ${}_2F_1$. The latter is defined in general by the infinite series

$${}_2F_1(\alpha, \beta; \gamma; x) := \sum_{p=0}^\infty \frac{(\alpha)_p (\beta)_p}{(\gamma)_p p!} x^p, \qquad (u)_p := u(u+1) \cdots (u+p-1), \tag{5.83}$$

where unless the series terminates, it is required that $|x| < 1$ for absolute convergence of the series.

PROPOSITION 5.5.2 *With $w_2(z)$ given by (5.50) and $z = e^{2\pi i x/L}$, the monic polynomials*

$$p_n^{(\mathrm{CJ})}(z) := \frac{a}{a+n}\, {}_2F_1(-n, a+1; -n+1-a; z) \tag{5.84}$$

have the orthogonality property

$$\int_{-L/2}^{L/2} w_2(z) p_j(z) p_k(\bar{z})\, dx = (p_j, p_j)_2^{(\mathrm{CJ})} \delta_{j,k},$$

$$(p_n, p_n)_2^{(\mathrm{CJ})} = L \frac{\Gamma(n+2a+1)\Gamma(n+1)}{(\Gamma(n+a+1))^2}. \tag{5.85}$$

Proof. This is established in Exercises 5.5 q.3. □

5.5.2 The classical groups

We know from Section 2.6 that the eigenvalue p.d.f.'s for the orthogonal and symplectic groups, after the change of variables $\cos\theta_j = y_j$, coincide with the Jacobi ensemble in the case $\beta = 2$ and a, b appropriately chosen. Because the Jacobi polynomial in the variable $\cos\theta$ is simple for these cases, one finds simple expressions for K_N after substituting back the original variables.

PROPOSITION 5.5.3 *In terms of the original variables $\{\theta_i\}$, the k-point correlation for the eigenvalues of the classical groups on the upper half-circle $(0 < \theta_i < \pi)$ is given by*

$$\rho_{(k)}(\theta_1, \dots, \theta_k) = \det[K_N(\theta_j, \theta_l)]_{j,l=1,\dots,k}, \tag{5.86}$$

where

$$K_N^{O^+(2N)}(\theta, \theta') = \frac{1}{2\pi}\left(\frac{\sin((2N-1)(\theta-\theta')/2)}{\sin((\theta-\theta')/2)} + \frac{\sin((2N-1)(\theta+\theta')/2)}{\sin((\theta+\theta')/2)}\right),$$

$$K_N^{O^\pm(2N+1)}(\theta, \theta') = \frac{1}{2\pi}\left(\frac{\sin N(\theta-\theta')}{\sin((\theta-\theta')/2)} \mp \frac{\sin N(\theta+\theta')}{\sin((\theta+\theta')/2)}\right),$$

$$K_N^{\mathrm{Sp}(2N)}(\theta, \theta') = \frac{1}{2\pi}\left(\frac{\sin((2N+1)(\theta-\theta')/2)}{\sin((\theta-\theta')/2)} - \frac{\sin((2N+1)(\theta+\theta')/2)}{\sin((\theta+\theta')/2)}\right),$$

$$K_N^{O^-(2N+2)}(\theta, \theta') = K_N^{\mathrm{Sp}(2N)}(\theta, \theta').$$

Proof. Now $K_N(\theta, \theta')$ in (5.86) is related to $K_N^{(\mathrm{J})}(y, y')$ for the corresponding Jacobi ensemble by

$$K_N(\theta, \theta') = (\sin\theta \sin\theta')^{1/2} K_N^{(\mathrm{J})}(\cos\theta, \cos\theta').$$

For definiteness consider $O^+(2N)$. According to Proposition 3.7.1 and the Christoffel-Darboux sum (5.10) with $w_2(x) = (1-x^2)^{-1/2}$, $p_N(x) = p_N^{(\mathrm{J})}(x)|_{a=b=-1/2}$, we have

$$K_N^{O^+(2N)}(\theta, \theta') = \frac{1}{(p_{N-1}, p_{N-1})_2^{(\mathrm{J})}} \frac{p_N^{(\mathrm{J})}(\cos\theta) p_{N-1}^{(\mathrm{J})}(\cos\theta') - p_N^{(\mathrm{J})}(\cos\theta') p_{N-1}^{(\mathrm{J})}(\cos\theta)}{\cos\theta - \cos\theta'}\Big|_{a=b=-1/2}.$$

But from (5.46) and [508]

$$p_n^{(\mathrm{J})}(x)|_{a=b=-1/2} = 2^n n! \frac{\Gamma(n)}{\Gamma(2n)} P_n^{(-1/2,-1/2)}(x) = 2^{-(n-1)} T_n(x), \quad n \ge 1,$$

where $T_n(x)$ denotes the Chebyshev polynomial of the first kind. Hence after recalling (5.48) and using the identity

$T_n(\cos\theta) = \cos n\theta$ we have

$$K_N^{O^+(2N)}(\theta,\theta') = \frac{1}{\pi}\frac{\cos N\theta\cos(N-1)\theta' - \cos N\theta'\cos(N-1)\theta}{\cos\theta - \cos\theta'}.$$

Use of trigonometric identities reduces this to the stated form. The other cases can be derived similarly. An alternative approach is given in Exercises 5.5 q.4. □

The function K_N in (5.86) in the case of the CUE, or equivalently the classical group $U(N)$, also takes on a simple form. The CUE corresponds to the case $w_2(z) = 1$ in (5.79). The circular orthogonal polynomials for this weight are simply $p_j(z) = z^j$ with normalization $(p_j, p_j)_2^{(C)} = L$. With $L = 2\pi$ and so using angular variables θ, $-\pi < \theta \le \pi$, we see that (5.82) simplifies to give

$$K_N^{\rm CUE}(e^{i\theta}, e^{-i\theta'}) = \frac{1}{2\pi}e^{i(\theta-\theta')(N-1)/2}\frac{\sin(N(\theta-\theta')/2)}{\sin((\theta-\theta')/2)}. \tag{5.87}$$

EXERCISES 5.5 1. In this exercise (5.82) will be derived according to the strategy of the proof of Proposition 5.1.3. Let

$$f(z,\bar u) := (w_2(z)w_2(u))^{1/2}\int_{-L/2}^{L/2} dx_1\cdots\int_{-L/2}^{L/2} dx_{N-1}\prod_{l=1}^{N-1} w_2(z_l)\prod_{1\le j<k\le N-1}|z_k - z_j|^2\prod_{j=1}^{N-1}(z_j - z)(\bar z_j - \bar u). \tag{5.88}$$

(i) Proceed as in the proof of the first part of the proof of Proposition 5.1.3 to show that

$$f(z,\bar u) := (N-1)!\prod_{j=1}^{N}(p_{j-1}, p_{j-1})_2^{(\rm C)}\, K_N(z,\bar u). \tag{5.89}$$

(ii) Write $\prod_{j=1}^{N-1}(\bar z_j - \bar u)(z_j - z) = (-\bar u)^{N-1}\prod_{j=1}^{N-1}\bar z_j(z_j - u)(z_j - z)$, and proceed as in the second part of the proof of Proposition 5.1.3, appropriately modified by use of (5.22), to deduce that

$$\begin{aligned} f(z,\bar u) := (N-1)!\prod_{j=1}^{N}(p_{j-1}, p_{j-1})_2^{(\rm C)}\frac{(w_2(z)w_2(u))^{1/2}}{u - z}\bar u^{N-1} \\ \times\sum_{\nu=0}^{N-1}\frac{p_\nu(z)p_\nu(0)p_N(u) - p_\nu(u)p_\nu(0)p_N(z)}{(p_\nu, p_\nu)_2^{(\rm C)}}. \end{aligned} \tag{5.90}$$

(iii) Consider now $f(z,0)$. By writing $\prod_{j=1}^{N-1}\bar z_j(z_j - z) = (-z)^{N-1}\prod_{j=1}^{N-1}(\bar z_j - \bar z)$ and proceeding as in (ii) show

$$f(z,0) = (N-1)!\prod_{j=1}^{N-1}(p_{j-1}, p_{j-1})_2^{(\rm C)}(w_2(z)w_2(0))^{1/2}z^{N-1}p_{N-1}(\bar z),$$

and thus, comparing with (5.90), conclude

$$\sum_{\nu=0}^{N-1}\frac{p_\nu(z)p_\nu(0)}{(p_\nu, p_\nu)_2^{(\rm C)}} = z^{N-1}\frac{p_{N-1}(\bar z)}{(p_{N-1}, p_{N-1})_2^{(\rm C)}},$$

or equivalently

$$\sum_{\nu=0}^{N-1}\frac{p_\nu(z)p_\nu(0)}{(p_\nu, p_\nu)_2^{(\rm C)}} = \frac{z^N p_N(\bar z)}{(p_N, p_N)_2^{(\rm C)}} - \frac{p_N(z)p_N(0)}{(p_N, p_N)_2^{(\rm C)}}. \tag{5.91}$$

Now substitute (5.91) in (5.90) and compare with (5.89) to deduce the summation (5.82).

2. [508] Suppose $w_2(z) = |1 - \alpha z|^{-2}$, $|\alpha| < 1$, $z := e^{2\pi i x/L}$. Verify that $p_n(z) = z^n(1 - \alpha/z)$, $(n \ge 1)$, $p_0(z) = 1$ are monic orthogonal polynomials with respect to the inner product (5.79) with

$$(p_0, p_0)_2^{(\mathrm{C})} = \frac{L}{1-\alpha^2}, \qquad (p_n, p_n)_2^{(\mathrm{C})} = L, \ n \ge 1.$$

Use (5.79) to deduce from this that

$$\int_{-L/2}^{L/2} dx_1 \cdots \int_{-L/2}^{L/2} dx_N \prod_{l=1}^N |1 - \alpha z_l|^{-2} \prod_{1 \le j < k \le N} |z_k - z_j|^2 = \frac{L^N}{1 - \alpha^2}.$$

3. (i) Use Proposition 3.9.1 to obtain the identity

$$\frac{1}{C}\int_0^1 dt_1 \cdots \int_0^1 dt_N \prod_{j=1}^N t_j^{\lambda_1}(1-t_j)^{\lambda_2}(x - t_j) \prod_{j<k}(t_k - t_j)^2$$
$$= \frac{1}{C'}\int_{-1/2}^{1/2} d\theta_1 \cdots \int_{-1/2}^{1/2} d\theta_N \prod_{l=1}^N e^{2\pi i \theta_l \epsilon}|1 + e^{2\pi i \theta_l}|^{\lambda_2}(x + e^{2\pi i \theta_l}) \prod_{j<k} |e^{2\pi i \theta_k} - e^{2\pi i \theta_j}|^2,$$

where $\epsilon = \lambda_1 + \lambda_2/2 + N$ and C and C' are constants so that the coefficient of x^N on both sides is unity.

(ii) The monic orthogonal polynomials corresponding to the weight function $x^{\lambda_1}(1-x)^{\lambda_2}$, $(0 < x < 1)$ are proportional to

$$P_n^{(\lambda_1,\lambda_2)}(1 - 2x) = \binom{n + \lambda_1}{\lambda_1} {}_2F_1(-n, n + \lambda_1 + \lambda_2 + 1; \lambda_1 + 1; x). \tag{5.92}$$

Use this fact together with the result of (i) and the first identity of Proposition 5.1.4 to deduce the orthogonality property stated in Proposition 5.5.2.

(iii) Use the analogue of (5.74) for periodic weight functions, together with the Morris integral evaluation (4.4), to deduce the normalization in Proposition 5.5.2.

4. (i) Verify the so-called C and D type Vandermonde formulas [452]

$$\det[z_j^k - z_j^{-k}]_{j,k=1,\dots,n} = \prod_{j=1}^n (z_j - z_j^{-1}) \prod_{1 \le j < k \le n} (z_k - z_j)\Big(1 - \frac{1}{z_j z_k}\Big),$$
$$\det[z_j^{k-1} + z_j^{-(k-1)}]_{j,k=1,\dots,n} = 2 \prod_{1 \le j < k \le n} (z_k - z_j)\Big(1 - \frac{1}{z_j z_k}\Big).$$

(ii) By appropriate choice of z_j, rewrite the formulas of (i) to read

$$\det[\sin k\theta_j]_{j,k=1,\dots,n} = \prod_{j=1}^n \sin\theta_j \prod_{1 \le j < k \le n} (2\cos\theta_k - 2\cos\theta_j),$$
$$\det[\cos(k-1)\theta_j]_{j,k=1,\dots,n} = 2^{1-n} \prod_{1 \le j < k \le n} (2\cos\theta_k - 2\cos\theta_j).$$

Use the orthogonality of $\{\sin k\theta\}_{k=1,2,\dots}$ and $\{\cos(k-1)\theta\}_{k=1,2,\dots}$ on $[0,\pi]$ to deduce that

$$\int_0^\pi d\theta_1 \cdots \int_0^\pi d\theta_n \prod_{j=1}^n \sin^2\theta_j \prod_{1 \le j < k \le n} (\cos\theta_k - \cos\theta_j)^2 = n! 2^{-n^2}\pi^n,$$
$$\int_0^\pi d\theta_1 \cdots \int_0^\pi d\theta_n \prod_{1 \le j < k \le n} (\cos\theta_k - \cos\theta_j)^2 = n! 2^{-(n-1)^2}\pi^n.$$

Recalling (2.62) and (2.66), similarly show that for N even

$$\rho_{(n)}^{O^-(N)}(\theta_1,\dots,\theta_n) = \det\Big[\frac{2}{\pi}\sum_{l=1}^{N/2} \sin l\theta_j \sin l\theta_k\Big]_{j,k=1,\dots,n},$$

$$\rho_{(n)}^{O^+(N)}(\theta_1,\dots,\theta_n) = \det\Big[\frac{1}{\pi} + \frac{2}{\pi}\sum_{l=1}^{N/2} \cos l\theta_j \cos l\theta_k\Big]_{j,k=1,\dots,n}$$

and thus rederive the corresponding results in Proposition 5.5.3.

(iii) Starting with the B type Vandermonde formulas

$$\det[z_j^{k-1/2} - z_j^{-k+1/2}]_{j,k=1,\dots,n} = \prod_{j=1}^n (z_j^{1/2} - z_j^{-1/2}) \prod_{1\le j<k\le n} (z_k - z_j)\Big(1 - \frac{1}{z_j z_k}\Big),$$

$$\det[z_j^{k-1/2} + z_j^{-k+1/2}]_{j,k=1,\dots,n} = \prod_{j=1}^n (z_j^{1/2} + z_j^{-1/2}) \prod_{1\le j<k\le n} \Big(z_k - z_j\Big)\Big(1 - \frac{1}{z_j z_k}\Big)$$

show

$$\det[\sin(k - \tfrac{1}{2})\theta_j]_{j,k=1,\dots,n} = \prod_{j=1}^n \sin\frac{\theta_j}{2} \prod_{1\le j<k\le n} (2\cos\theta_k - 2\cos\theta_j),$$

$$\det[\cos(k - \tfrac{1}{2})\theta_j]_{j,k=1,\dots,n} = \prod_{j=1}^n \cos\frac{\theta_j}{2} \prod_{1\le j<k\le n} (2\cos\theta_k - 2\cos\theta_j).$$

Now use the orthogonality of $\{\sin(k - 1/2)\theta\}_{k=1,2,\dots}$ and $\{\cos(k - 1/2)\theta\}_{k=1,2,\dots}$ on $[0,\pi]$ to deduce that

$$\int_0^\pi d\theta_1 \cdots \int_0^\pi d\theta_n \prod_{j=1}^n \sin^2\frac{\theta_j}{2} \prod_{1\le j<k\le n} (\cos\theta_k - \cos\theta_j)^2$$

$$= \int_0^\pi d\theta_1 \cdots \int_0^\pi d\theta_n \prod_{j=1}^n \cos^2\frac{\theta_j}{2} \prod_{1\le j<k\le n} (\cos\theta_k - \cos\theta_j)^2 = n! 2^{-n^2}\pi^n.$$

Recalling (2.63), (2.65) use these results to show that for N odd

$$\rho_{(n)}^{O^+(N)}(\theta_1,\dots,\theta_n) = \det\Big[\frac{2}{\pi}\sum_{l=0}^{(N-1)/2-1} \sin(l + \tfrac{1}{2})\theta_j \sin(l + \tfrac{1}{2})\theta_k\Big]_{j,k=1,\dots,n},$$

$$\rho_{(n)}^{O^-(N)}(\theta_1,\dots,\theta_n) = \det\Big[\frac{2}{\pi}\sum_{l=0}^{(N-1)/2-1} \cos(l + \tfrac{1}{2})\theta_j \cos(l + \tfrac{1}{2})\theta_k\Big]_{j,k=1,\dots,n},$$

and thus rederive the corresponding results in Proposition 5.5.3.

5. The objective of this exercise is to derive the factorizations [545], [34]

$$\det[a_{i-j}]_{i,j=1,\dots,2N} = \det[a_{i-j} + a_{i+j-1}]_{i,j=1,\dots,N} \det[a_{i-j} - a_{i+j-1}]_{i,j=1,\dots,N}, \tag{5.93}$$

$$\det[a_{i-j}]_{i,j=1,\dots,2N+1} = \frac{1}{2}\det[a_{i-j} + a_{i+j-2}]_{i,j=1,\dots,N+1} \det[a_{i-j} - a_{i+j}]_{i,j=1,\dots,N}, \tag{5.94}$$

where it is required that $a_{-i} = a_i$.

(i) Consider the l.h.s. of (5.93). For each $i = 1,\dots,N$ replace each row i by row i minus row $2N+1-i$, then for each $j = 1,\dots,N$ replace column $2N+1-j$ by column $2N+1-j$ minus column j. Note that the determinant then has the structure

$$\begin{bmatrix} \mathbf{A} & 0_N \\ \mathbf{B} & \mathbf{C} \end{bmatrix}$$

for $\mathbf{A}$, $\mathbf{B}$, $\mathbf{C}$ $N \times N$ matrices and thus deduce the factorization as given by the r.h.s. of (5.93).

(ii) Show that an analogous procedure yields (5.94).

6. [34] Here the determinants in (5.93), (5.94) will be related to averages over the classical groups.

(i) Suppose $a(e^{i\theta}) = a(e^{-i\theta})$ and set

$$a_j = \frac{1}{2\pi} \int_{-\pi}^{\pi} a(e^{i\theta}) e^{-i\theta j}\, d\theta.$$

Note that

$$\det[a_{j-k} + a_{j+k-1}]_{j,k=1,\dots,N} = \frac{1}{N!} \frac{1}{(2\pi)^N} \int_0^{\pi} d\theta_1 \cdots \int_0^{\pi} d\theta_N \prod_{j=1}^N a(e^{i\theta_j}) \Big(\det[2\cos\theta_j(k-1/2)]_{j,k=1,\dots,N} \Big)^2$$

and obtain similar formulas for all the determinants on the r.h.s.'s of (5.93) and (5.94). Now make use of these formulas, substituting for the determinants according to formulas presented in q.4 and recognising the integrals as averages over orthogonal groups according to the results of Section 2.6, to deduce that

$$\det[a_{j-k} + a_{j+k-1}]_{j,k=1,\dots,N} = \Big\langle \prod_{j=1}^N a(e^{i\theta_j}) \Big\rangle_{\widetilde{O^-(2N+1)}},$$

$$\det[a_{j-k} - a_{j+k-1}]_{j,k=1,\dots,N} = \Big\langle \prod_{j=1}^N a(e^{i\theta_j}) \Big\rangle_{\widetilde{O^+(2N+1)}},$$

$$\frac{1}{2} \det[a_{j-k} + a_{j+k-2}]_{j,k=1,\dots,N} = \Big\langle \prod_{j=1}^N a(e^{i\theta_j}) \Big\rangle_{\widetilde{O^+(2N)}},$$

$$\det[a_{j-k} - a_{j+k}]_{j,k=1,\dots,N} = \Big\langle \prod_{j=1}^N a(e^{i\theta_j}) \Big\rangle_{\widetilde{O^-(2N+2)}},$$

where the symbol $\tilde{}$ denotes that only eigenvalues $0 < \theta_j < \pi$ are considered in the average.

(ii) Use the above results, together with (5.77), to show that (5.93) and (5.94) can be rewritten

$$\Big\langle \prod_{j=1}^{n} a(e^{i\theta_j}) \Big\rangle_{U(n)} = \Big\langle \prod_{j=1}^{[(n-1)/2]} a(e^{i\theta_j}) \Big\rangle_{\widetilde{O^-(n+1)}} \Big\langle \prod_{j=1}^{[n/2]} a(e^{i\theta_j}) \Big\rangle_{\widetilde{O^+(n+1)}} \tag{5.95}$$

for $n = 2N, 2N+1$ respectively, where again it is required that $a(e^{i\theta}) = a(e^{-i\theta})$.

5.6 LOG-GAS SYSTEMS WITH PERIODIC BOUNDARY CONDITIONS

5.6.1 Semiperiodic boundary conditions

In addition to the classical weight functions, there are a large number of weight functions for which the corresponding orthogonal polynomials are known explicitly. Here we will make use of one such case to specify the correlation functions corresponding to (2.73) with $\beta = 2$ [192]. First the change of variables

$e^{2\pi(y_j+\pi N/Lc')/L} = u_j$ shows that

$$\prod_{j=1}^N e^{-c'y_j^2} \prod_{1\le j<l\le N} |\sinh \pi(y_l - y_j)/L|^2 dy_1 \cdots dy_N$$
$$\propto \prod_{j=1}^N e^{-k^2 \log^2 u_j} \prod_{1\le j<l\le N} (u_l - u_j)^2 du_1 \cdots du_N, \tag{5.96}$$

where $k^2 = c'L^2/(2\pi)^2$. But for the weight function

$$w(u;q) := \pi^{-1/2} k e^{-k^2 \log^2 u}, \qquad q = e^{-1/(2k^2)},$$

it is known (see, e.g., [508]) that the Stieltjes-Wigert polynomials

$$S_l(u;q) := \frac{(-1)^l q^{l/2+1/4}}{\{(1-q)(1-q^2)\cdots(1-q^l)\}^{1/2}} \sum_{\nu=0}^{l} \begin{bmatrix} l \\ \nu \end{bmatrix}_q q^{\nu^2}(-q^{1/2}u)^\nu, \tag{5.97}$$

where

$$\begin{bmatrix} l \\ \nu \end{bmatrix}_q := \frac{(1-q^l)(1-q^{l-1})\cdots(1-q^{l-\nu+1})}{(1-q^\nu)(1-q^{\nu-1})\cdots(1-q)}$$

are the corresponding set of orthonormal polynomials. Thus, by making use of the Christoffel-Darboux formula, the correlation functions are specified in terms of this class of polynomial.

5.6.2 Metal wall

In statistical mechanics, in addition to the canonical formulation, in which there are a fixed number N of particles, there is also a grand canonical formulation, in which the number of particles is a variable controlled by the so called *fugacity* ζ. Thus the probability density of there being N (distinguishable) particles at $\vec{r}_1, \dots, \vec{r}_N$ is postulated to be

$$\frac{\zeta^N}{\Xi_\beta(\zeta)} e^{-\beta U(\vec{r}_1,\dots,\vec{r}_N)},$$

where

$$\Xi_\beta(\zeta) := \sum_{N=0}^\infty \frac{\zeta^N}{N!} \int_\Omega d\vec{r}_1 \cdots \int_\Omega d\vec{r}_N \, e^{-\beta U(\vec{r}_1,\dots,\vec{r}_N)}. \tag{5.98}$$

The latter is referred to as the *grand partition function*. Note that

$$\langle N \rangle = \zeta \frac{\partial}{\partial \zeta} \log \Xi_\beta(\zeta). \tag{5.99}$$

It turns out that the log-gas near a metal wall is best suited for study in this formulation, a fact which is in keeping with there being no need for a neutralizing background, due to the presence of image charges of opposite sign.

Set $\beta = 2$ in (2.77), and introduce the scaled fugacity

$$\xi = \frac{\pi e^{4\pi d\eta}}{L} \zeta. \tag{5.100}$$

The corresponding generalized grand partition function, defined as in (5.98) but with an additional factor of

$\prod_{l=1}^{N} a(\vec{r}_l)$ in the integrand (cf. (5.19)) is then given by

$$\Xi_2[a](\xi) = e^{-2\pi\eta^2 dL} \sum_{N=0}^{\infty} \frac{\xi^N}{N!} \int_0^L dx_1 \cdots \int_0^L dx_N \prod_{l=1}^{N} \frac{a(x_l)}{\sinh 2\pi d/L} \prod_{1 \le j < k \le N} \left| \frac{\sin \pi (x_k - x_j)/L}{\sin \pi (x_k - x_j + 2id)/L} \right|^2. \tag{5.101}$$

The product over pairs in (5.101) can be written in terms of the Cauchy double alternant (4.33). Thus setting $x_j = e^{2\pi i(x_j + id)/L}$, $y_j = e^{2\pi i(x_j - id)/L}$ in (4.33) shows

$$\prod_{l=1}^{N} \frac{1}{\sinh 2\pi d/L} \prod_{1 \le j < k \le N} \left| \frac{\sin \pi (x_k - x_j)/L}{\sin \pi (x_k - x_j + 2id)/L} \right|^2 = i^N \det \left[\frac{1}{\sin \pi (x_j - x_k + 2id)/L} \right]_{j,k=1,\dots,N}. \tag{5.102}$$

To proceed further, observe that (5.102) substituted into (5.101) gives an expression which has be summed according to the identity (5.32). Thus

$$\Xi_2[a](\xi) = e^{-2\pi\eta^2 dL} \det(1 + i\xi K), \tag{5.103}$$

where K is the integral operator on $[0, L]$ with kernel

$$K(x, y) = \frac{a(y)}{\sin \pi (x - y + 2id)/L}.$$

This shows that (5.29) is valid, so using (5.31) gives

$$\rho_{(n)}(x_1, \dots, x_n) = \det[G(x_j, x_k)]_{j,k=1,\dots,n},$$

where

$$G(x, y) = \langle x | i\xi K (1 + i\xi K)^{-1} |_{a=1} | y \rangle.$$

The integral operator G with kernel $G(x, y)$ must satisfy

$$G(1 + i\xi K) = i\xi K,$$

and consequently

$$G(x, y) + i\xi \int_0^L \frac{G(x, u)}{\sin \pi (u - y + 2id)/L} \, du = \frac{i\xi}{\sin \pi (x - y + 2id)/L}. \tag{5.104}$$

This integral equation can be used to determine $G(x, y)$ explicitly.

PROPOSITION 5.6.1 *We have*

$$G(x, y) = 2\xi \sum_{n=0}^{\infty} \frac{e^{-4\pi d(n+1/2)/L} e^{2\pi i(n+1/2)(x-y)/L}}{1 + 2\xi L e^{-4\pi d(n+1/2)/L}}. \tag{5.105}$$

Proof. Taking the complex conjugate of (5.104) shows $G(y, x) = \overline{G(x, y)}$. This together with the periodicity of the system suggests we seek a solution of the form

$$G(x, y) = \sum_{n=-\infty}^{\infty} g_n e^{2\pi i(n+1/2)(x-y)/L}.$$

Noting that for $d > 0$

$$\frac{1}{\sin \pi (z + 2id)/L} = -2i e^{-2\pi d/L} \sum_{n=0}^{\infty} e^{-4\pi dn/L} e^{2\pi i z(n+1/2)/L},$$

we can equate Fourier coefficients and so deduce (5.105). □

5.6.3 Doubly periodic boundary conditions

In the case $\beta = 2$ it is possible to calculate the n-point correlation corresponding to (2.83) by writing the product over theta functions in terms of a determinant [507], [182], [181]. In addition to $\theta_1(u;q)$ as specified by (2.79), also required is the further Jacobi theta function

$$\theta_3(u;q) := \sum_{n=-\infty}^{\infty} q^{n^2} e^{2iun}. \tag{5.106}$$

PROPOSITION 5.6.2 *Define the Jacobi theta functions θ_1 and θ_3 by (2.79) and (5.106) and let*

$$f_N(q) := N^{N/2} q^{-(N-1)(N-2)/24} (q^2;q^2)_\infty^{-(N-1)(N-2)/2}. \tag{5.107}$$

Then with $s = 3$ for N odd and $s = 1$ for N even we have

$$\int_0^1 \det\left[\theta_s(\pi(x_j + \alpha - k/N); q^{1/N})\right]_{j,k=1,\dots,N} d\alpha = f_N(q) \prod_{1 \le j < k \le N} \theta_1(\pi(x_k - x_j); q). \tag{5.108}$$

Proof. Only the case N odd will be considered; the N even case follows similarly. We proceed as in [182]. Both the l.h.s. and r.h.s. of the stated equation are antisymmetric functions of $x_1, \dots, x_N$ that vanish whenever two of the variables are equal. It thus suffices to check that both sides are the same function of x_1. This is done by studying the periodicity properties of both sides, and using Liouville's theorem.

It is straightforward to check that both sides are periodic under the mapping $x_1 \mapsto x_1 + 1$. Now write $q = e^{\pi i \tau}$, where $\mathrm{Im}(\tau) > 0$, and consider the effect of the mapping $x_1 \mapsto x_1 + \tau$. From (2.79) we see that the r.h.s. remains the same apart from a factor

$$q^{-(N-1)} e^{-2\pi i x_1 (N-1)} \prod_{k=2}^{N} e^{2\pi i x_k}.$$

On the l.h.s., since

$$\theta_3(\pi(x+\tau); q) = q^{-1} e^{-2\pi i x} \theta_3(\pi x; q) \tag{5.109}$$

the kth term of the first row of the determinant can be written

$$q^{-(N^2-1)/N} e^{-2\pi i (x_1 - \alpha - k/N)(N-1)} \theta_3(\pi(x_1 - \alpha + \tau/N - k/N); q^{1/N}),$$

while the kth term of the jth row $(j = 2, \dots, N)$ can be written

$$q^{1/N} e^{2\pi i (x_j - \alpha - k/N)} \theta_3(\pi(x_j - \alpha + \tau/N - k/N); q^{1/N}).$$

In the first row we note that $e^{2\pi i k(N-1)/N} = e^{-2\pi i k/N}$ and thus a common factor of $e^{-2\pi i k/N}$ can be removed from the kth column (the product over k of such factors equals 1). Furthermore, removing obvious common factors from each row of the determinant, the l.h.s. becomes

$$q^{-(N-1)} e^{-2\pi i x_1 (N-1)} \prod_{k=2}^{N} e^{2\pi i x_k} \int_0^1 \det[\theta_3(\pi(x_j - \alpha + \tau/N - k/N); q^{1/N})]_{j,k=1,\dots,N}\, d\alpha.$$

This integral is in fact the same as in the l.h.s. of the proposition, as the line integral along the path $\alpha + i\mathrm{Im}(\tau)/N$, $0 < \alpha < 1$ is the same as that along the same path $0 < \alpha < 1$ by Cauchy's theorem, and the periodicity of the integrand under $\alpha \mapsto \alpha + 1$. Thus under the mapping $x_1 \mapsto x_1 + \tau$ the l.h.s. and the r.h.s. have the same quasi-periodicity property.

Finally, consider the ratio r.h.s./l.h.s. From the above results, this is a doubly periodic function with periods 1 and τ. Furthermore, since the zeros of the l.h.s. are cancelled by the zeros of the r.h.s. (due to two rows of the determinant being equal), we have that r.h.s./l.h.s. is a doubly periodic entire function and is thus by Liouville's theorem independent of x_1. The result of the Proposition now follows, except for the evaluation of $f_N(q)$. This latter task is taken up in Exercises 5.6 q.1. □

The formula (5.108) can be generalized so that the variable α is not integrated over. In fact knowledge of

(5.108) is the key to deriving the generalization.

PROPOSITION 5.6.3 *Let $f_N(q)$ be given by (5.107). For N odd*

$$\det\Big[\theta_3\Big(\pi(x_j+\alpha-k/N);q^{1/N}\Big)\Big]_{j,k=1,\dots,N}$$
$$=\theta_3\Big(\pi\sum_{j=1}^N(x_j+\alpha);q\Big)f_N(q)\prod_{1\le j<k\le N}\theta_1\Big(\pi(x_k-x_j);q\Big), \tag{5.110}$$

while for N even

$$\det\Big[\theta_1\Big(\pi(x_j+\alpha-k/N);q^{1/N}\Big)\Big]_{j,k=1,\dots,N}$$
$$=\theta_4\Big(\pi\sum_{j=1}^N(x_j+\alpha);q\Big)f_N(q)\prod_{1\le j<k\le N}\theta_1\Big(\pi(x_k-x_j);q\Big). \tag{5.111}$$

In the latter formula the Jacobi theta function θ_4 is defined by

$$\theta_4(u;q)=\sum_{n=-\infty}^{\infty}(-1)^n q^{n^2}e^{2iun}. \tag{5.112}$$

Proof. We will give the details in the N even case [181]. Consider the integrand in (5.108) as a function of α. Since it is periodic of period $1/N$, it can be expanded in the form $\sum_{p=-\infty}^{\infty} I_p e^{2\pi i pN\alpha}$, where

$$I_p=\int_0^1 e^{-2\pi i pN\alpha}\det[\theta_1(\pi(x_j+\alpha-l/N);q^{1/N}]_{j,l=1,\dots,N}\,d\alpha.$$

Shifting the contours of integration from the unit interval to $\gamma-\tau p/N$ ($0<\gamma<1$, $q=e^{\pi i\tau}$) (this leaves the value of the integral unchanged due to the periodicity of the integrand and Cauchy's theorem) and using the second property in (2.81) gives

$$I_p=q^{p^2}e^{2\pi i p\sum_{j=1}^N x_j}I_0 \tag{5.113}$$

and thus

$$\det\Big[\theta_1(\pi(x_j+\alpha-l/N);q^{1/N}\Big]_{j,l=1,\dots,N}=\theta_3\Big(\pi\sum_{j=1}^N(x_j+\alpha);q\Big)I_0.$$

The final step is to substitute for I_0 using (5.108). □

Although the functions $\{\theta_s(\pi(x-k/N);q^{1/N})\}_{k=1,\dots,N}$ forming the columns in the determinant (5.110) and (5.111) are not an orthogonal set of functions, they can readily be transformed into one. For this, multiply both sides of (5.110) with $\alpha=0$ by

$$\det[e^{2\pi i lk/N}]_{\substack{l=1,\dots,N\\k=0,\dots,N-1}}=N^{N/2}i^{(N-1)(3N/2+1)}$$

and multiply both sides of (5.111) with $\alpha=-\pi\tau/2$ by

$$\det[e^{2\pi i l(k+1/2)/N}]_{\substack{l=1,\dots,N\\k=0,\dots,N-1}}=N^{N/2}i^{N+1}i^{(N-1)(3N/2+1)}$$

to obtain

$$\det[h_k(x_j)]_{j,k=1,\dots,N}=q^{-\sum_{m=0}^{N-1}m^2/N}N^{-N/2}i^{(N-1)(3N/2+1)}f_N(q)$$
$$\times\theta_s\Big(-\pi\sum_{j=1}^N x_j;q\Big)\prod_{1\le j<k\le N}\theta_1\left(-\pi(x_k-x_j);q\right). \tag{5.114}$$

where as in Proposition 5.6.2 $s = 3$ for N odd and $s = 1$ for N even. Here

$$h_k(x) := e^{-2\pi i(k-1)x}\theta_3(\pi(Nx - \tau(k-1)); q^N), \tag{5.115}$$

which for $k = 1, \dots, N$ form an orthogonal set with respect to the inner product $\langle f|g\rangle := \int_0^1 \bar{f}(x)g(x)\,dx$.

It follows from (5.114) that for both N even and N odd,

$$\Big(\prod_{1\le j<k\le N}\theta_1(\pi(x_k - x_j); q)\Big)^2 \propto \int_0^1 \det\Big[h_k(x_j + \alpha/N)\Big]_{j,k=1,\dots,N}\det\Big[\bar{h}_k(x_j+\alpha/N)\Big]_{j,k=1,\dots,N}\,d\alpha.$$

Proceeding now as in the proofs of Propositions 5.1.1 and 5.1.2, we have that for the ensemble (2.83) with $\beta = 2$,

$$\rho_{(n)}(x_1, \dots, x_n) = \int_0^1 \det\Big[\sum_{p=0}^{N-1} e^{-2\pi i p(x_j - x_k)/L}\theta_3(\pi(Nx_j/L - \tau(p-1) + \alpha); q^N) \times \frac{\theta_3(\pi(Nx_k/L + \tau(p-1) + \alpha); q^N)}{L\theta_3(2\pi\tau(p-1); q^{2N})}\Big]_{j,k=1,\dots,n} d\alpha, \tag{5.116}$$

where $q = e^{\pi i\tau}$, $\tau = iW/L$.

EXERCISES 5.6 1. (i) Use the infinite product expansion (2.79) to deduce that

$$\prod_{1\le j<k\le N}\Big(\frac{x_j}{x_k}; q^2\Big)_\infty\Big(q^2\frac{x_k}{x_j}; q^2\Big)_\infty$$
$$= i^{N(N-1)/2}q^{-N(N-1)/8}(q^2;q^2)_\infty^{-N(N-1)/2}\prod_{j=1}^N e^{\pi i(N-2j+1)u_j}\prod_{1\le j<k\le N}\theta_1(\pi(u_k - u_j); q),$$

where $x_j = e^{2\pi i u_j}$.

(ii) Suppose N is odd. Make use of (5.108) to show

$$f_N(q)\int_0^1 du_1\cdots\int_0^1 du_N\prod_{j=1}^N e^{\pi i(N-2j+1)u_j}\prod_{1\le j<k\le N}\theta_1(\pi(u_k-u_j);q)$$
$$= q^{\sum_{j=1}^N((N+1)/2-j)^2/N}\det\Big[e^{2\pi i k((N+1)/2-j)/N}\Big]_{j,k=1,\dots,N}$$
$$= q^{\sum_{j=1}^N((N+1)/2-j)^2/N}i^{-N(N-1)/2}N^{N/2},$$

and obtain an analogous formula for N even.

(iii) Read off from the identity (4.138) that

$$\int_0^1 du_1\cdots\int_0^1 du_N\prod_{1\le j<k\le N}\Big(\frac{x_j}{x_k};q^2\Big)_\infty\Big(q^2\frac{x_k}{x_j};q^2\Big)_\infty = \frac{1}{(q^2;q^2)_\infty^{N-1}},$$

and compare this to the evaluation of (ii), recalling the equation of (i), to deduce (5.107).

5.7 PARTITION FUNCTION IN THE CASE OF A GENERAL POTENTIAL

The $\beta = 2$ log-gas partition function

$$Z_n[\{t_i\}] := \frac{1}{n!}\int_{-\infty}^{\infty}dx_1\cdots\int_{-\infty}^{\infty}dx_n\prod_{l=1}^n e^{\sum_{j=1}^\infty t_j x_l^j}\prod_{1\le j<k\le n}(x_k - x_j)^2, \tag{5.117}$$

considered as a function of $\{t_i\}$, is intimately related to the theory of integrable systems [5], [362]. First we will demonstrate that Z_n satisfies the *Toda lattice equation*

$$\frac{\partial^2}{\partial t_1^2}\log Z_n = \frac{Z_{n+1}Z_{n-1}}{Z_n^2}. \tag{5.118}$$

Let $\{p_j(x;\{t_i\})\}_{j=0,1,\dots}$ be the monic orthogonal polynomials with respect to the weight function $e^{\sum_{i=1}^{\infty} t_i x^i}$ so that

$$\int_{-\infty}^{\infty} e^{\sum_{j=1}^{\infty} t_j x^j} p_j(x;\{t_i\})p_k(x;\{t_i\})\,dx =: (p_j,p_k) = h_j^2\delta_{j,k}. \tag{5.119}$$

According to (5.74) we have

$$Z_n[\{t_i\}] = \prod_{j=0}^{n-1} h_j^2. \tag{5.120}$$

Let $\tilde{p}(x;\{t_i\}) = \frac{1}{h_j}p_j(x;\{t_i\})$ be the corresponding orthonormal polynomials. Introduce the matrix

$$\mathbf{L} = \left[\frac{1}{h_j h_k}(xp_j,p_k)\right]_{j,k=0,1,\dots},$$

so that

$$\mathbf{L}(\tilde{p}_0,\tilde{p}_1,\dots)^T = x(\tilde{p}_0,\tilde{p}_1,\dots)^T. \tag{5.121}$$

It then follows from the three term recurrence (5.14) that this matrix has the symmetric tridiagonal structure

$$\mathbf{L} = \begin{bmatrix} \frac{1}{h_0^2}(xp_0,p_0) & \frac{h_1}{h_0} & & & \\ \frac{h_1}{h_0} & \frac{1}{h_1^2}(xp_1,p_1) & \frac{h_2}{h_1} & & \\ & \frac{h_2}{h_1} & \frac{1}{h_2^2}(xp_2,p_2) & \frac{h_3}{h_2} & \\ & \ddots & \ddots & \ddots & \end{bmatrix}.$$

Furthermore

$$(xp_n,p_n) = \frac{\partial}{\partial t_1}(p_n,p_n) - 2\left(\frac{\partial}{\partial t_1}p_n,p_n\right) = \frac{\partial}{\partial t_1}h_n^2,$$

where the second equality follows upon noting $\frac{\partial}{\partial t_1}p_n$ must be of degree $n-1$ since p_n is monic and thus the second inner product in the first equality vanishes.

The matrix $\mathbf{L}$ has the special property that its evolution with respect to $\{t_i\}$ can be determined explicitly.

PROPOSITION 5.7.1 *We have*

$$\frac{\partial \mathbf{L}}{\partial t_i} = \frac{1}{2}[(\mathbf{L}^i)_+ - (\mathbf{L}^i)_-, \mathbf{L}], \tag{5.122}$$

where the notation $(A)_+$ ($(A)_-$) denotes the strictly upper (lower) triangular portion of the matrix $\mathbf{A}$.

Proof. We must have

$$\frac{\partial}{\partial t_i}\tilde{p}_k(x) = \sum_{l=0}^{k} A_{kl}^{(i)}\tilde{p}_l(x)$$

for some $A_{kl}^{(i)}$ and thus

$$\frac{\partial}{\partial t_i}(\tilde{p}_0(x),\tilde{p}_1(x),\dots)^T = \mathbf{A}^{(i)}(\tilde{p}_0(x),\tilde{p}_1(x),\dots)^T, \tag{5.123}$$

where $\mathbf{A}^{(i)}$ is lower triangular. Now for $l < k$, by orthogonality

$$
\begin{aligned}
0 &= \frac{\partial}{\partial t_i} \int_{-\infty}^{\infty} \tilde{p}_k(x)\tilde{p}_l(x) e^{\sum_{i=1}^{\infty} t_i x^i} \, dx \\
&= \int_{-\infty}^{\infty} \frac{\partial \tilde{p}_k(x)}{\partial t_i} \tilde{p}_l(x) e^{\sum_{i=1}^{\infty} t_i x^i} \, dx + \int_{-\infty}^{\infty} x^i \tilde{p}_k(x)\tilde{p}_l(x) e^{\sum_{i=1}^{\infty} t_i x^i} \, dx,
\end{aligned} \tag{5.124}
$$

where to obtain the second equality use has been made of the fact $\frac{\partial \tilde{p}_l(x)}{\partial t_i}$ has degree l so the integral involving this term implied by the differentiation vanishes. It follows from this that

$$
(\mathbf{A}^{(i)})_- = -(\mathbf{L}^i)_-.
$$

Similarly, for $l = k$

$$
0 = \frac{\partial}{\partial t_i} \int_{-\infty}^{\infty} (\tilde{p}_k(x))^2 e^{\sum_{i=1}^{\infty} t_i x^i} \, dx = 2A_{kk}^{(i)} + (\mathbf{L}^i)_{kk}
$$

and so

$$
\mathbf{A}^{(i)} = -(\mathbf{L}^i)_- - \frac{1}{2}(\mathbf{L}^i)_0, \tag{5.125}
$$

where the notation $(\mathbf{B})_0$ denotes the diagonal part of the matrix $\mathbf{B}$. It follows immediately upon multiplying (5.123) by x that

$$
\frac{\partial}{\partial t_i}\Big(\mathbf{L}(\tilde{p}_0(x), \tilde{p}_1(x), \dots)^T\Big) = \mathbf{A}^{(i)}\mathbf{L}(\tilde{p}_0(x), \tilde{p}_1(x), \dots)^T
$$

and thus

$$
\frac{\partial}{\partial t_i}\mathbf{L} = [\mathbf{A}^{(i)}, \mathbf{L}] = [-(\mathbf{A}^{(i)})^T, \mathbf{L}],
$$

where the second equality follows from the first and the fact that $\frac{\partial \mathbf{L}/}{\partial t_i}$ is symmetric. Taking the arithmetic mean of the final two expressions gives the stated result. □

When $i = 1$, equating diagonal entries in (5.122) gives

$$
\frac{\partial}{\partial t_1}\frac{1}{h_n^2}\frac{\partial}{\partial t_1}h_n^2 = \Big(\frac{h_{n+1}}{h_n}\Big)^2 + \Big(\frac{h_{n-1}}{h_{n-2}}\Big)^2 - 2\Big(\frac{h_n}{h_{n-1}}\Big)^2, \qquad n \ge 2.
$$

Using (5.120) we see that this implies the Toda lattice equation (5.118).

The next integrability property of (5.117) to be discussed relates to the construction of annihilating operators $\{L_p\}_{p=0,1,\dots}$ which satisfy the Virasoro algebra

$$
[L_p, L_q] = (p - q)L_{p+q}. \tag{5.126}
$$

PROPOSITION 5.7.2 *The operators*

$$
L_p := \sum_{k=0}^{p} \frac{\partial}{\partial t_k}\frac{\partial}{\partial t_{p-k}} + \sum_{k=1}^{\infty} k t_k \frac{\partial}{\partial t_{p+k}}, \qquad \frac{\partial}{\partial t_0} := N \tag{5.127}
$$

have the property that

$$
L_p Z_N = 0 \tag{5.128}
$$

and furthermore satisfy (5.126).

Proof. As in Aomoto's approach to the Selberg integral discussed in Section 4.6, we start by making use of the fundamental theorem of calculus to deduce

$$\Big\langle \sum_{i=1}^N \frac{\partial}{\partial x_i} \frac{1}{z - x_i} \Big\rangle_{Z_N}$$
$$:= \frac{1}{N!} \int_{-\infty}^{\infty} dx_1 \cdots \int_{-\infty}^{\infty} dx_N \prod_{i=1}^N e^{\sum_{j=1}^\infty t_j x_i^j} \prod_{1 \le i < j \le N} (x_j - x_i)^2 \sum_{i=1}^N \frac{\partial}{\partial x_i} \frac{1}{z - x_i} = 0.$$

Integrating by parts, and using the simple identity

$$\sum_{i=1}^N \Big(\frac{1}{(z - x_i)^2} + 2 \sum_{\substack{k=1 \\ k \ne i}}^N \frac{1}{(x_i - x_k)(z - x_i)} \Big) = \sum_{i,j=1}^N \frac{1}{(z - x_i)(z - x_j)},$$

we have that

$$\Big\langle \sum_{i,j=1}^N \frac{1}{(z - x_i)(z - x_j)} + \sum_{i=1}^N \sum_{n=1}^\infty \frac{n t_n x_i^{n-1}}{z - x_i} \Big\rangle_{Z_N} = 0.$$

Equating coefficients of z^{-p} for $p > 1$ shows

$$\Big\langle \sum_{i,j=1}^N \sum_{k=0}^{p-2} x_i^k x_j^{p-2-k} + \sum_{i=1}^N \sum_{n=1}^\infty n t_n x_i^{n+p-2} \Big\rangle_{Z_N} = 0.$$

The l.h.s. of this expression is easily identified with $L_{p-2} Z_N$. Regarding (5.126), we note from the explicit form (5.127) that

$$[L_p, L_q] f(\{t_i\}) = \Big(2 \sum_{k=0}^p k \frac{\partial^2}{\partial t_{p-k} \partial t_{q+k}} - 2 \sum_{k=0}^q k \frac{\partial^2}{\partial t_{q-k} \partial t_{p+k}} + (p - q) \sum_{k=1}^\infty k t_k \frac{\partial}{\partial t_{p+q+k}} \Big) f(\{t_i\}).$$

□

We remark that the above proof shows (5.128), referred to as *Virasoro constraints*, remains valid for $p = -1$, in which case the first term in (5.127) is not present.

Our final point on integrability aspects of (5.117) relates to the *KP hierarchy* of equations. To specify these equations, introduce polynomials $s_j(\{t_i\})$ according to the expansion

$$e^{\sum_{j=1}^\infty t_j x^j} = \sum_{k=0}^\infty s_k(\{t_i\}) x^k, \tag{5.129}$$

so that

$$s_k(\{t_i\}) = \sum_{\substack{\nu_1, \nu_2, \dots \ge 0 \\ \nu_1 + 2\nu_2 + 3\nu_3 + \cdots = k}} \frac{t_1^{\nu_1} t_2^{\nu_2} t_3^{\nu_3} \cdots}{\nu_1! \nu_2! \nu_3! \cdots}.$$

For functions $f(\{t_i\}) =: f(t)$, $g(\{t_i\}) =: g(t)$ define the *Hirota symbol* by

$$A\Big(\Big\{\frac{\partial}{\partial t_i}\Big\}\Big) f \circ g := A\Big(\Big\{\frac{\partial}{\partial y_i}\Big\}\Big) f(t + y) g(t - y) \Big|_{y=0}. \tag{5.130}$$

In terms of this notation we will show that for $k = 1, 2, \dots$

$$\Big(s_{k+3}\Big(\Big\{\frac{1}{i} \frac{\partial}{\partial t_i}\Big\}\Big) - \frac{1}{2} \frac{\partial^2}{\partial t_1 \partial t_{k+2}} \Big) Z_N \circ Z_N = 0. \tag{5.131}$$

PROPOSITION 5.7.3 *We have*

$$[z^{-1}] \exp\Big(\sum_{i=1}^{\infty}(t_i - t'_i)z^i\Big) \exp\Big(-\sum_{i=1}^{\infty}\frac{1}{iz^i}\Big(\frac{\partial}{\partial t_i} - \frac{\partial}{\partial t'_i}\Big)\Big) Z_n[\{t'_i\}]Z_n[\{t_i\}] = 0, \tag{5.132}$$

where $[z^{-1}]$ denotes the coefficient of z^{-1} in the formal Laurent expansion.

Proof. Suppose $m < n$. Observe that

$$\begin{aligned}
&[z^{-1}] \exp\Big(\sum_{i=1}^{\infty}(t_i - t'_i)z^i\Big) p_n(z;\{t_i\}) \int_{-\infty}^{\infty} \frac{p_m(u;\{t'_i\})e^{\sum_{i=1}^{\infty} t'_i u^i}}{z-u}\, du \\
&= \int_{-\infty}^{\infty} p_m(u,\{t'\})e^{\sum_{i=1}^{\infty} t'_i u^i} [z^{-1}] \exp\Big(\sum_{i=1}^{\infty}(t_i - t'_i)z^i\Big) \frac{p_n(z;\{t_i\})}{z-u} du \\
&= \int_{-\infty}^{\infty} p_m(u;\{t'_i\})p_n(u;\{t_i\})e^{\sum_{i=1}^{\infty} t_i u^i}\, du = 0,
\end{aligned} \tag{5.133}$$

where to obtain the second equality use has been made of the general result that $[z^{-1}]f(z)/(z-u) = f(u)$ for $f(z)$ analytic, while the final equality uses the fact that $m < n$ and the orthogonality of $\{p_n(u;\{t_i\})\}_{n=0,1,\dots}$ with respect to $e^{\sum_{i=1}^{\infty} t_i u^i}$.

On the other hand, we know from Proposition 5.1.4 that

$$p_n(z;\{t_i\}) = \frac{1}{Z_n[\{t_i\}]}\Big\langle \prod_{j=1}^{n}(z - x_j)\Big\rangle_{Z_n[\{t_i\}]} = \frac{z^n}{Z_n[\{t_i\}]} Z_n\Big[\Big\{t_i - \frac{1}{iz^i}\Big\}\Big],$$

$$\int_{-\infty}^{\infty} \frac{p_m(z;\{t'_i\})e^{\sum_{i=1}^{\infty} t'_i u^i}}{z-u}\, du = \frac{1}{Z_m[\{t'_i\}]}\Big\langle \prod_{j=1}^{m+1}\frac{1}{(z-x_j)}\Big\rangle_{Z_{m+1}[\{t'_i\}]} = \frac{z^{-(m+1)}}{Z_m[\{t'_i\}]} Z_{m+1}\Big[\Big\{t'_i + \frac{1}{iz^i}\Big\}\Big].$$

Substituting these formulas in (5.133) with $n = N$, $m = N-1$ gives

$$[z^{-1}]e^{\sum_{i=1}^{\infty}(t_i - t'_i)z^i} Z_N\Big[\Big\{t'_i + \frac{1}{iz^i}\Big\}\Big] Z_N\Big[\Big\{t_i - \frac{1}{iz^i}\Big\}\Big] = 0. \tag{5.134}$$

But from the definitions

$$Z_N\Big[\Big\{t'_i + \frac{1}{iz^i}\Big\}\Big] = \exp\Big(\sum_{i=1}^{\infty}\frac{1}{iz^i}\frac{\partial}{\partial t'_i}\Big)Z_N[\{t'_i\}], \qquad Z_N\Big[\Big\{t_i - \frac{1}{iz^i}\Big\}\Big] = \exp\Big(-\sum_{i=1}^{\infty}\frac{1}{iz^i}\frac{\partial}{\partial t_i}\Big)Z_N[\{t_i\}],$$

which when substituted in (5.134) gives (5.132). □

Let us now show how (5.132) can be used to derive (5.131). First, we note from (5.129) that (5.132) can be written

$$\sum_{k=0}^{\infty} s_k(\{t_i - t'_i\}) s_{k+1}\Big(\Big\{\frac{1}{i}\frac{\partial}{\partial t'_i} - \frac{1}{i}\frac{\partial}{\partial t_i}\Big\}\Big) Z_N[\{t'_i\}]Z_N[\{t_i\}] = 0.$$

Introducing the new variables $u_j := \frac{1}{2}(t'_j - t_j)$, $w_j := \frac{1}{2}(t_j + t'_j)$ this reads

$$\sum_{k=0}^{\infty} s_k(\{-2u_i\}) s_{k+1}\Big(\Big\{\frac{1}{i}\frac{\partial}{\partial u_i}\Big\}\Big) Z_N[\{w_j + u_j\}]Z_N[\{w_j - u_j\}] = 0.$$

But

$$
\begin{aligned}
&s_{k+1}\Big(\Big\{\frac{1}{i}\frac{\partial}{\partial u_i}\Big\}\Big)Z_N[\{w_j+u_j\}]Z_N[\{w_j-u_j\}]\\
&\quad = s_{k+1}\Big(\Big\{\frac{1}{i}\frac{\partial}{\partial y_i}\Big\}\Big)Z_N[\{w_j+u_j+y_j\}]Z_N[\{w_j-u_j-y_j\}]\Big|_{y=0}\\
&\quad = s_{k+1}\Big(\Big\{\frac{1}{i}\frac{\partial}{\partial y_i}\Big\}\Big)\Big(\exp\sum_{j=1}^\infty u_j\frac{\partial}{\partial y_j}\Big)Z_N[\{w_j+y_j\}]Z_N[\{w_j-y_j\}]\Big|_{y=0}\\
&\quad = s_{k+1}\Big(\Big\{\frac{1}{i}\frac{\partial}{\partial w_i}\Big\}\Big)\Big(\exp\sum_{j=1}^\infty u_j\frac{\partial}{\partial w_j}\Big)Z_N[\{w_j\}]\circ Z_N[\{w_j\}]
\end{aligned}
$$

where to obtain the final line use has been made of the definition (5.130). Noting too that

$$
s_1\Big(\Big\{\frac{1}{i}\frac{\partial}{\partial w_i}\Big\}\Big)=\frac{\partial}{\partial w_1},\qquad s_k(\{-2u_i\})=-2u_k+\mathrm{O}(u^2),\ \ k\ge 2,
$$

and expanding in $\{u_i\}$ then shows

$$
\Big(\frac{\partial}{\partial w_1}+\sum_{k=1}^\infty(-2u_k+\mathrm{O}(u^2))\Big)s_{k+1}\Big(\Big\{\frac{1}{i}\frac{\partial}{\partial w_i}\Big\}\Big)\Big)\Big(1+\sum_{j=1}^\infty u_j\frac{\partial}{\partial w_j}+\mathrm{O}(u^2)\Big)Z_N[\{w_j\}]\circ Z_N[\{w_j\}]=0.
$$

Equating terms linear in $\{u_i\}$ we read off from this the equation (5.131) with k replaced by $k-2$, valid for $k=1,2,\dots$. The fact that

$$
\frac{\partial^{2p-1}}{\partial w_j^{2p-1}}Z_N[\{w_j\}]\circ Z_N[\{w_j\}]=0 \tag{5.135}
$$

for all $j,p\ge 1$ shows that this form of (5.131) is trivially satisfied for $k=1$ and 2, leaving the cases as stated.

Note the case $k=1$ of (5.131), after making use of (5.129), (5.130) and (5.135), can be written as

$$
\Big(\Big(\frac{\partial}{\partial t_1}\Big)^4+3\Big(\frac{\partial}{\partial t_2}\Big)^2-4\frac{\partial^2}{\partial t_1\partial t_3}\Big)\log Z_N+6\Big(\frac{\partial^2}{\partial t_1^2}\log Z_N\Big)^2=0. \tag{5.136}
$$

EXERCISES 5.7 1. With $\psi_j(x)=\frac{1}{h_j}e^{-V(x)/2}p_j(x)$, $V(x)=-\sum_{j=1}^\infty t_jx^j$ and the inner product $(f,g):=\int_{-\infty}^\infty f(x)g(x)\,dx$, introduce the matrix

$$
\mathbf{P}=\Big[\Big(\frac{d}{dx}\psi_j(x),\psi_k(x)\Big)\Big]_{j,k=0,1,\dots}.
$$

(i) Use integration by parts to conclude that $\mathbf{P}$ is antisymmetric. Note too that $[\mathbf{L},\mathbf{P}]=\mathbf{1}$, where $\mathbf{L}$ is as in (5.121).

(ii) Differentiate using the product rule, and use (5.121) to show

$$
\mathbf{P}+\frac{1}{2}\Big[\Big((V'(\mathbf{L})\vec{\psi}(x))_j,\psi_k(x)\Big)\Big]_{j,k=0,1,\dots}=\Big[\frac{1}{h_j}(p_j'(x)e^{-V(x)/2},\psi_k(x))\Big]_{j,k=0,1,\dots}.
$$

By noting $p_j'(x)e^{-V(x)/2}=h_{j-1}j\psi_{j-1}(x)+\sum_{l=0}^{j-2}\gamma_{jl}\psi_l(x)$ for some γ_{jl}, show that the entries in the matrix on the r.h.s. vanish for $k\ge j$. Now make use of the first result of (i) to conclude

$$
\mathbf{P}=-\frac{1}{2}\Big(V'(\mathbf{L})_+-V'(\mathbf{L})_-\Big),
$$

where $V'(\mathbf{L}):=[(V'(\mathbf{L})\vec{\psi}(x))_j,\psi_k(x))]_{j,k=0,1,\dots}$ and furthermore

$$
(V'(\mathbf{L}))_{jj}=0. \tag{5.137}
$$

(iii) By noting that the entries of the matrix on the r.h.s. of the first equation in (ii) is equal to jh_{j-1}/h_j for $(j,k) = (j, j-1)$, deduce that

$$(V'(\mathbf{L}))_{j,j-1} = jh_{j-1}/h_j. \tag{5.138}$$

Relate (5.137) and (5.138) to (5.17).

5.8 BIORTHOGONAL STRUCTURES

The first equality in (5.5) expresses the Boltzmann factor for the one-component log-gas at $\beta = 2$ in terms of a product of two determinants. With $\{\xi_j(x)\}_{j=1,2,\dots}$ and $\{\eta_j(x)\}_{j=1,2,\dots}$ separately a sequence of linear independent functions, the Boltzmann factor is a special case of the functional form

$$\frac{1}{C}\prod_{j=1}^{N} w_2(x_j)\det[\xi_j(x_k)]_{j,k=1,\dots,N}\det[\eta_j(x_k)]_{j,k=1,\dots,N} \tag{5.139}$$

(take $\xi_j(x) = \eta_j(x) = x^{j-1}$). Like (5.5), the n-particle correlation function for (5.139) can itself be written as a determinant [80]. In general a many particle system for which the n-point correlations are $n\times n$ determinants is said to be a *determinantal point process* [495].

PROPOSITION 5.8.1 *With $g_{jk} := (\eta_j, \xi_k)_2$, let $[g_{jk}]_{j,k=1,\dots,n}$ be invertible for each $n = 1, 2, \dots,$ define c_{jk} by*

$$[c_{jk}]_{j,k=1,\dots,N} = \Big([g_{jk}]_{j,k=1,\dots,N}\Big)^{-1}, \tag{5.140}$$

and set

$$K_N(x,y) = (w_2(x)w_2(y))^{1/2}\sum_{j,k=1}^{N} c_{jk}\xi_j(x)\eta_k(y). \tag{5.141}$$

We have

$$\rho_{(n)}(x_1,\dots,x_n) = \det[K_N(x_j,x_k)]_{j,k=1,\dots,n}. \tag{5.142}$$

Proof. Assume, deforming the ξ_j if necessary, that $[g_{jk}]$ can be decomposed as the product of a lower triangular and upper triangular matrix. It is then possible to construct functions

$$\zeta_i \in \mathrm{Span}(\xi_1,\dots,\xi_i), \qquad \psi_i \in \mathrm{Span}(\eta_1,\dots,\eta_i) \tag{5.143}$$

with the biorthogonality property

$$(\zeta_j, \psi_k)_2 = \delta_{j,k}. \tag{5.144}$$

According to the method of the proof of Proposition 5.1.2, we can deduce that (5.142) holds with

$$K_N(x,y) = (w_2(x)w_2(y))^{1/2}\sum_{l=1}^{N}\zeta_l(x)\psi_l(y). \tag{5.145}$$

This has the reproducing property

$$\int_{-\infty}^{\infty}(w_2(y))^{1/2}K_N(x,y)\xi_l(y)\,dy = (w_2(x))^{1/2}\xi_l(x) \qquad (l = 1,\dots,N). \tag{5.146}$$

On the other hand, because of (5.143) the decomposition (5.141) must hold for some c_{jk}, and so

$$\int_{-\infty}^{\infty}(w_2(y))^{1/2}K_N(x,y)\xi_l(y)\,dy = (w_2(x))^{1/2}\sum_{j,k=1}^{N} c_{jk}g_{kl}\xi_j(x). \tag{5.147}$$

Equating (5.146) and (5.147) gives (5.140). □

An example of the functional form (5.139), generalizing the $\beta = 2$ Boltzmann factor as given on the l.h.s. of (5.5), is

$$\frac{1}{C}\prod_{j=1}^{N} w_2(x_j)\frac{\det[e^{-a_j x_k}]_{j,k=1,\dots,N}}{\prod_{1\le j<k\le N}(a_j-a_k)}\prod_{1\le j<k\le N}(x_k-x_j). \tag{5.148}$$

Thus in the limit $a_j \to 0$ $(j = 1,\dots,N)$ (5.148) reduces to the l.h.s. of (5.5). We will see in Section 11.6.4 (see also Exercises 5.8 q.2(iv)) that with $w_2(x)$ given by either the Gaussian or Laguerre weight, (5.148) has an interpretation as an eigenvalue p.d.f. Our present interest is to show that for these particular classical weights the correlation kernel (5.141) can be written as a double contour integral. For this we first establish the following preliminary result.

PROPOSITION 5.8.2 *With*

$$w_2(x) = e^{-x^2}, \qquad \xi_k(x) = e^{-a_k^2/4 - a_k x}, \qquad \eta_j(x) = \frac{1}{\sqrt{\pi}}H_{j-1}(x) \tag{5.149}$$

we have

$$\sum_{k=1}^{N} c_{jk} z^{k-1} = \prod_{\substack{l=1\\ l\ne j}}^{N}\frac{(z+a_l)}{(a_l-a_j)}, \tag{5.150}$$

while with

$$w_2(x) = x^a e^{-x}, \qquad \xi_k(x) = (1+a_k)^{a+1}e^{-a_k x}, \qquad \eta_j(x) = \frac{x^{j-1}}{\Gamma(a+j)} \tag{5.151}$$

we have

$$\sum_{k=1}^{N} c_{jk} z^{-(k-1)} = \frac{(1+a_j)^{N-1}}{z^{N-1}}\prod_{\substack{l=1\\ l\ne j}}^{N}\frac{z-(1+a_l)}{a_j-a_l}. \tag{5.152}$$

Proof. In the case of (5.149)

$$g_{jk} = \frac{1}{\sqrt{\pi}}e^{-a_k^2/4}\int_{-\infty}^{\infty} e^{-x^2-a_k x}H_{j-1}(x)\,dx = (-a_k)^{j-1}, \tag{5.153}$$

where the second equality follows from a well-known classical identity [508]. Since by definition

$$\sum_{k=1}^{N} c_{jk} g_{kl} = \delta_{j,l} \tag{5.154}$$

it follows that $\sum_{k=1}^{N} c_{jk}(-a_l)^{k-1} = \delta_{j,l}$. Hence the l.h.s. of (5.150), which is a polynomial of degree $N-1$, vanishes for $z = -a_l$ $(l \ne j)$, and equals unity for $z = -a_j$, and thus must be given by the r.h.s. In the case of (5.151)

$$g_{jk} = \frac{(1+a_k)^{a+1}}{\Gamma(a+j)}\int_0^{\infty} x^{a+j-1}e^{-(1+a_k)x}\,dx = (1+a_k)^{-(j-1)}.$$

According to (5.154) $\sum_{k=1}^{N} c_{jk}(1+a_l)^{-(k-1)} = \delta_{j,l}$ and (5.152) follows. □

In addition to these formulas, also required are the integral identities

$$H_k(x) = \frac{2^k}{\sqrt{\pi}} \int_{-\infty}^{\infty} e^{-y^2} (x+iy)^k \, dy, \tag{5.155}$$

$$\frac{z^{a+j-1}}{\Gamma(a+j)} = \frac{1}{2\pi i} \int_{C_H} \frac{e^{wz}}{w^{a+j}} \, dw, \quad \mathrm{Re}(z) > 0, \tag{5.156}$$

where C_H is a simple contour starting at $-\infty + i\epsilon$ ($\epsilon > 0$), staying in the upper half-plane until it is fully in the right half-plane, turning around and staying in the lower half-plane until it reaches $-\infty - i\epsilon$ (Hankel loop).

PROPOSITION 5.8.3 *[554], [29] For the p.d.f. (5.148) with $w_2(x) = e^{-x^2}$ the n-point correlation $\rho_{(n)}$ is given by (5.142) with*

$$K_N(x,y) = -\int_{C_{\{a\}}} \frac{dv}{2\pi i} \int_{-i\infty}^{i\infty} \frac{dt}{2\pi i} \frac{e^{-(v^2-t^2)/4 - vx - ty}}{t+v} \prod_{l=1}^{N} \frac{(t+a_l)}{(a_l - v)}, \tag{5.157}$$

where $C_{\{a\}}$ is a simple contour which encircles $\{a_1, \dots, a_N\}$ anti-clockwise and stays entirely in the right half-plane.

For the p.d.f. (5.148) with $w_2(x) = x^a e^{-x}$, (5.142) holds with

$$K_N(x,y) = \int_{C_{\{a\}}} \frac{dv}{2\pi i} \int_{C_H} \frac{dw}{2\pi i} \frac{e^{-vx+wy}}{w-v} \Big(\frac{v}{w}\Big)^{a+N} \prod_{l=1}^{N} \frac{(w-(1+a_l))}{(v-(1+a_l))}, \tag{5.158}$$

where $C_{\{a\}}$ is as in (5.157) with the additional constraint that it does not intersect C_H.

Proof. Substituting (5.149) in (5.141) and using the integral representation (5.155) shows, after a simple change of variables in the latter,

$$K_N(x,y) = \frac{e^{-x^2/2+y^2/2}}{2\pi i} \sum_{j,k=1}^{N} c_{jk} e^{-a_j^2/4 - a_j x} \int_{-i\infty}^{i\infty} e^{t^2/4 - ty} t^{k-1} \, dt.$$

The sum over k can be performed according to (5.150) to give

$$K_N(x,y) = e^{-x^2/2+y^2/2} \sum_{j=1}^{N} e^{-a_j^2/4 - a_j x} \int_{-i\infty}^{i\infty} e^{t^2/4 - ty} \prod_{\substack{l=1 \\ l \ne j}}^{N} \frac{(t+a_l)}{(a_l - a_j)} \frac{dt}{2\pi i},$$

while the sum over j can be done by way of a contour integral to yield (5.157), but with an additional factor of $e^{-x^2/2+y^2/2}$. However this factor cancels out of the determinant and so can be ignored. The derivation of (5.158) is similar, using (5.156) and (5.152). □

Although (5.157) and (5.158) were calculated without requiring explicit knowledge of the corresponding biorthogonal polynomials (functions), the double integral representations are in fact intimately related to integral representations of the latter. In general so-called multiple orthogonal polynomials [22] allow one to construct functions from $\mathrm{span}\{e^{-b_i x} x^k\}_{\substack{k=0,\dots,j \\ i=1,\dots,D}}$, $\psi_j(x)$, say, and polynomials $\zeta_i(x)$ of degree i, which have the biorthogonality property (5.144).

To see this, let us suppose that of the parameters $\vec{a} = (a_1, \dots, a_N)$ only D are distinct, so that $\vec{a} = \vec{b}^{\vec{m}} := (b_1^{m_1}, \dots, b_D^{m_D})$, where $b_p^{m_p}$ means b_p repeated m_p times. Note that $\sum_{i=1}^{D} m_i = N$. Let $\vec{m}^* = (m_1^*, \dots, m_D^*)$ be such that $0 \le m_j^* \le m_j$ and set $|\vec{m}^*| = \sum_{p=1}^{D} m_p^*$. Construct a family of monic polynomials $P_{\vec{m}^*}(x)$ of degree $|\vec{m}^*|$, referred to as multiple orthogonal polynomials of type II, with the properties

$$\int_{-\infty}^{\infty} w_2(x) P_{\vec{m}^*}(x) x^j e^{-b_k x}\, dx = 0 \qquad (j = 0, \dots, m_k - 1) \tag{5.159}$$

for each $k = 1, \dots, D$. Furthermore, let $Q^{(j)}_{\vec{m}^*}(x)$ be a polynomial of degree $m_j^* - 1$, referred to as multiple orthogonal polynomials of type I, such that the function

$$Q_{\vec{m}^*}(x) := w_2(x) \sum_{j=1}^{D} Q^{(j)}_{\vec{m}^*}(x) e^{-b_j x}$$

satisfies

$$\int_{-\infty}^{\infty} w_2(x) Q_{\vec{m}^*}(x) x^j\, dx = \begin{cases} 0, & j = 0, \dots, |\vec{m}^*| - 2, \\ 1, & j = |\vec{m}^*| - 1. \end{cases} \tag{5.160}$$

Choose a sequence of multi-indices $\vec{m}^{*(0)}, \dots, \vec{m}^{*(N)}$ such that $|\vec{m}^{*(j)}| = j$ and $\vec{m}^{*(j)} \le \vec{m}^{*(j+1)}$ (this inequality must hold for each component), and define

$$\zeta_j(x) = P_{\vec{m}^{*(j)}}(x), \qquad \psi_j(x) = Q_{\vec{m}^{*(j+1)}}(x). \tag{5.161}$$

A direct calculation gives that the polynomials $\{\zeta_i(x)\}$ and the functions $\{\psi_i(x)\}$ have the biorthogonality property (5.144).

For the Hermite and Laguerre weights, the biorthogonal polynomials and functions can be determined explicitly for general indices $\vec{m}$ as certain contour integrals. Let us consider first the Laguerre case. To avoid complication with the contours, it will be assumed that the parameter a in the Laguerre weight is an integer.

Following [68], [132], let the multiple Laguerre function and polynomial of type I and II be specified by

$$\tilde{\mathcal{L}}^a_{\vec{m}}(x) = \sum_{i=1}^{D} e^{-b_i x} \tilde{\mathcal{L}}^a_{\vec{m},i}(x) = \int_{\mathcal{C}^{\{-1\}}_{\vec{b}}} \frac{e^{-xz}(1+z)^{|\vec{m}|+a-1}}{\prod_{i=1}^{D}(z-b_i)^{m_i}} \frac{dz}{2\pi i}, \qquad a \in \mathbb{Z}, \tag{5.162}$$

$$\mathcal{L}^a_{\vec{m}}(x) = \frac{(|\vec{m}|+a)!}{|\vec{m}|!} x^{-a} \int_{\mathcal{C}_{\{0\}}} \frac{e^{xw}}{w^{|\vec{m}|+a+1}} \prod_{i=1}^{D} (w - 1 - b_i)^{m_i} \frac{dw}{2\pi i}, \qquad a \in \mathbb{Z}. \tag{5.163}$$

In (5.162) $\mathcal{C}^{\{-1\}}_{\vec{b}}$ denotes a simple closed contour which encircles anticlockwise $b_1, \dots, b_D$ but not, for $|\vec{m}| + a - 1 \le -1$, the point -1 (otherwise this latter restriction is not necessary), while in (5.163) $\mathcal{C}_{\{0\}}$ is a simple closed contour which encircles the origin anticlockwise. These functions and polynomials have the property of reducing to the classical Laguerre polynomials in the limit $\vec{b} \to \vec{0}$.

PROPOSITION 5.8.4 *We have*

$$\lim_{\vec{b}\to\vec{0}} \tilde{\mathcal{L}}^a_{\vec{m}}(x) = L^a_{|\vec{m}|-1}(x), \qquad \lim_{\vec{b}\to\vec{0}} \mathcal{L}^a_{\vec{m}}(x) = L^a_{|\vec{m}|}(x).$$

Proof. The first formula follows from the definitions (5.162), upon comparison with the integral representation of the Laguerre polynomials

$$L^a_n(x) = \int_{\mathcal{C}_{\{0\}}} \frac{e^{-xw}}{w^{n+1}} (1+w)^{n+a} \frac{dw}{2\pi i}, \qquad a \in \mathbb{Z}. \tag{5.164}$$

For the second formula we modify (5.164) in accordance with the identity

$$n!(-x)^{-a} L^{-a}_n(x) = (n-a)! L^a_{n-a}(x), \qquad a \in \mathbb{Z}.$$

□

Furthermore, the Laguerre functions and polynomials satisfy orthogonality relations of the type required by (5.160) and (5.159).

PROPOSITION 5.8.5 *Let $a \in \mathbb{Z}_{\geq 0}$ and $b_1, \dots, b_D > -1$. We have*

$$\int_0^\infty e^{-x} x^{a+j} \tilde{\mathcal{L}}^a_{\vec{m}}(x)\, dx = \begin{cases} 0, & j = 0, \dots, |\vec{m}| - 2, \\ \frac{(-1)^{|\vec{m}|-1}(|\vec{m}|+a-1)!}{\prod_{i=1}^D (1+b_i)^{m_i}}, & j = |\vec{m}| - 1, \end{cases}$$

and

$$\int_0^\infty e^{-x-b_i x} x^{a+j} \mathcal{L}^a_{\vec{m}}(x)\, dx = 0 \qquad (i = 1, \dots, D)\ (j = 0, \dots, m_i - 1).$$

Thus, with $w_2(x) = x^a e^{-x}$, the multiple orthogonal polynomials and functions are, respectively,

$$P_{\vec{m}}(x) = \frac{(-1)^{|\vec{m}|}|\vec{m}|!}{\prod_{i=1}^D (1+b_i)^{m_i}} \mathcal{L}^a_{\vec{m}}(x), \qquad Q_{\vec{m}}(x) = \frac{(-1)^{|\vec{m}|-1} \prod_{i=1}^D (1+b_i)^{m_i}}{(|\vec{m}| + a - 1)!} \tilde{\mathcal{L}}^a_{\vec{m}}(x). \tag{5.165}$$

Proof. Substituting (5.162), then performing the x-integration first, shows

$$\int_0^\infty e^{-x} x^{a+j} \tilde{\mathcal{L}}^a_{\vec{m}}(x)\, dx = (a+j)! \int_{\mathcal{C}_{\vec{b}}^{\{-1\}}} \frac{(1+z)^{|\vec{m}|-j-2}}{\prod_{i=1}^D (z-b_i)^{m_i}} \frac{dz}{2\pi i}.$$

For $j = |\vec{m}| - 1$, use of the residue theorem applied to the region outside the contour (this is valid since the integrand decays sufficiently fast), in which there is a simple pole at $z = -1$, gives the stated result. For $j < |\vec{m}| - 1$ the point $z = -1$ is analytic and so the residue theorem gives that the integral vanishes. The same strategy suffices to establish the second orthogonality relation. □

With $m_i = 1$, $b_i = a_i$ $(i = 1, \dots, N)$, $D = N$ and $\vec{m}^{*(j)}$ having its first j components equal to 1, and the remaining $N - j$ components equal to 0, the multiple orthogonal functions (5.165) substituted in (5.161) give the correlation kernel in terms of biorthogonal functions according to (5.145).

Suppose all but r of the a_j are zero. By choosing these to be the final r of the a_js, but labeling them $a_1, \dots, a_r$ (note that (5.148) is symmetric in the a_j) we see that the first $N - r$ terms in the correlation kernel (5.145) are independent of the a_j. This structure can also be exhibited from the double contour integral form (5.158). First note that up to a factor which leaves (5.142) unchanged, (5.158) with $N - r$ a_j zero reads

$$\int_{\mathcal{C}_{\vec{a}}^{\{-1\}}} \frac{dv}{2\pi i} \int_{\mathcal{C}_{\{-1\}}} \frac{dw}{2\pi i} \frac{e^{-xv+yw}}{w-v} \Big(\frac{1+v}{1+w}\Big)^{N+a} \Big(\frac{w}{v}\Big)^{N-r} \prod_{i=1}^r \frac{w-a_i}{v-a_i} =: \mathcal{K}^a_N(x,y), \tag{5.166}$$

where the Hankel loop has been deformed to a closed contour upon the assumption that a is an integer. Because (5.148) reduces to the l.h.s. of (5.5) in the limit $\vec{a} \to 0$, (5.166) must, up to a factor which leaves (5.142) unchanged, be equal to the Laguerre case of (5.7). Tracing through the calculation of (5.166) for this factor shows

$$\mathcal{K}^a_N(x,y)\Big|_{\vec{a}=\vec{0}} = \sqrt{\frac{w_2(y)}{w_2(x)}} K_N^{(\mathrm{L})}(x,y), \qquad w_2(u) = u^a e^{-u}. \tag{5.167}$$

We can write (5.166) in terms of (5.167), and so called incomplete multiple Laguerre functions of type I and II (special cases of (5.162) and (5.163)),

$$\tilde{\Lambda}^{(j)}(x) = \int_{\mathcal{C}_{\{0,a_1,\dots,a_j\}}} \frac{e^{-xz}(1+z)^{N+a}}{z^{N-r}\prod_{k=1}^j (z-a_k)} \frac{dz}{2\pi i}, \quad \Lambda^{(j)}(x) = \int_{\mathcal{C}_{\{-1\}}} \frac{e^{xw} w^{N-r} \prod_{k=1}^{j-1}(w-a_k)}{(1+w)^{N+a}} \frac{dw}{2\pi i}. \tag{5.168}$$

PROPOSITION 5.8.6 *We have*

$$\mathcal{K}_N^a(x,y) = \mathcal{K}_{N-r}^{a+r}(x,y)\Big|_{\vec{a}=\vec{0}} + \sum_{i=1}^{r} \tilde{\Lambda}^{(i)}(x)\Lambda^{(i)}(y), \tag{5.169}$$

which will be referred to as the perturbed Laguerre kernel.

Proof. First one can use induction to check that

$$\frac{1}{w-z}\prod_{i=1}^{r}\frac{w-a_i}{z-a_i} = \frac{1}{w-z} + \sum_{i=1}^{r}\frac{\prod_{k=1}^{i-1}(w-a_k)}{\prod_{k=1}^{i}(z-a_k)}.$$

Substituting this in (5.166), the result follows from the appropriate definitions. □

The analogue of Proposition 5.8.6 in the Hermite case is given in the exercises below.

EXERCISES 5.8 1. Follow the method of proof of Proposition 5.1.2 to show that

$$\int_{-\infty}^{\infty} dx_1 \cdots \int_{-\infty}^{\infty} dx_N \prod_{l=1}^{N} w_2(x_l) \det[\xi_j(x_k)]_{j,k=1,\dots,N} \det[\eta_j(x_k)]_{j,k=1,\dots,N}$$

$$= N! \det\Big[\int_{-\infty}^{\infty} w_2(x)\xi_j(x)\eta_k(x)\,dx\Big]_{j,k=1,\dots,N}. \tag{5.170}$$

2. [132] Let $K_N(x,y)$ refer to (5.157).

(i) Use the fact that for $\vec{a} \to \vec{0}$, (5.148) reduces to the l.h.s. of (5.5), and follow the proof of Proposition 5.8.3 to conclude

$$K_N(x,y)\Big|_{\vec{a}=\vec{0}} = \Big(\frac{w_2(y)}{w_2(x)}\Big)^{1/2} K_N^{(\mathrm{H})}(x,y), \qquad w_2(u) = e^{-u^2},$$

where $K_N^{(\mathrm{H})}$ refers to (5.6) in the Hermite case.

(ii) Let

$$\tilde{\Gamma}^{(j)}(x) := \int_{\mathcal{C}_{\{0,a_1,\dots,a_j\}}} \frac{e^{-xz-z^2/4}}{z^{N-r}\prod_{k=1}^{j}(z-a_k)}\frac{dz}{2\pi i}, \Gamma^{(j)}(x) := \int_{-i\infty}^{i\infty} e^{xw+w^2/4} w^{N-r}\prod_{k=1}^{j-1}(w-a_k)\frac{dw}{2\pi i}. \tag{5.171}$$

Use the method of the proof of Proposition 5.8.6 to show that

$$K_N(x,y)\Big|_{a_i=0\ (i=r+1,\dots,N)} = K_{N-r}(x,y)\Big|_{\vec{a}=\vec{0}} + \sum_{j=1}^{r}\tilde{\Gamma}^{(j)}(x)\Gamma^{(j)}(y). \tag{5.172}$$

(iii) Use the integral representations of the Hermite polynomial

$$H_n(x) = 2^n n! \int_{\mathcal{C}_{\{0\}}} \frac{e^{xz-z^2/4}}{z^{n+1}}\frac{dz}{2\pi i} = \sqrt{\pi}e^{x^2}\int_{-i\infty}^{i\infty} z^n e^{-xz+z^2/4}\frac{dz}{2\pi i}$$

to show that

$$\Gamma^{(1)}(x) = \frac{1}{\sqrt{\pi}}e^{-x^2}H_{N-1}(-x), \qquad \tilde{\Gamma}^{(1)}(x) = \frac{e^{-xa_1-a_1^2/4}}{a_1^{N-1}} - \frac{1}{a_1}\sum_{p=0}^{N-2}\frac{1}{a_1^p}\frac{H_{N-2-p}(-x)}{2^{N-2-p}(N-2-p)!}.$$

(iv) The discussions around (11.1) and in Section 11.6.4 below tell us that (5.148) with $w_2(x) = e^{-x^2}$ is the eigenvalue p.d.f. for the matrix $\mathbf{H} = \mathbf{X} + \frac{1}{2}\mathbf{H}^{(0)}$. Here $\mathbf{X}$ is a member of the GUE while $\mathbf{H}^{(0)}$ is a fixed matrix with eigenvalues $\{-a_1,\dots,-a_N\}$, so that $\mathbf{H}$ is a shifted mean GUE matrix. Note that in the case all entries of $\mathbf{H}^{(0)}$ are equal to $2a/N$, and thus $\mathbf{H}$ equal to a GUE matrix where the Gaussian entries all have mean a/N, one then has that $a_1 = -2a$, $a_2 = \cdots = a_N = 0$. Use the results of (ii) and (iii) to show that for large a

$$\tilde{\Gamma}^{(1)}(x)\Gamma^{(1)}(x)\Big|_{a_1=-2a} \sim \frac{e^{-(x-a)^2}}{\sqrt{\pi}}$$

and interpret this in the context of the result of Proposition 1.8.2.

5.9 DETERMINANTAL k-COMPONENT SYSTEMS

5.9.1 All components having an equal number of particles

Consider the p.d.f. specified by (4.97) and (4.98). This may be viewed as a two-species particle system as determined by $\{a_i\}$ and $\{b_j\}$. Furthermore, the constraint (4.98) can be replaced by a determinant according to the following identity.

PROPOSITION 5.9.1 *Let $\chi_T = 1$ if T is true and $\chi_T = 0$ otherwise. Then for*

$$x_1 > \cdots > x_N, \qquad y_1 > \cdots > y_N, \tag{5.173}$$

we have

$$\det[\chi_{x_j - y_k > 0}]_{j,k=1,\dots,N} = \chi_{x_1 > y_1 > \cdots > x_N > y_N}. \tag{5.174}$$

Proof. For the ordering $x_1 > y_1 > \cdots > x_N > y_N$ the determinant is triangular with 1s down the diagonal, so (5.174) is correct in this case. All other orderings must have at least two xs (or two ys) in succession. The corresponding rows (or columns) in the determinant will then be equal so the determinant vanishes. □

Use of this together with the Vandermonde identity (1.173) shows the p.d.f. in question to be a particular 2-component case of the p.d.f. for a k-component system

$$\frac{1}{C}\det[\phi_i(x_j^{(1)})]_{i,j=1,\dots,N}\prod_{l=1}^{k-1}\det[W_l(x_i^{(l)}, x_j^{(l+1)})]_{i,j=1,\dots,N}\det[\psi_i(x_j^{(k)})]_{i,j=1,\dots,N}, \tag{5.175}$$

where the coordinates of species l are $\{x_j^{(l)}\}_{j=1,\dots,N}$. Following [88], to compute the corresponding correlation functions, a discretization $\mathcal{M}$ of the domain $[a,b]$, say to which the particles are confined will be introduced. Also, use will be made of theory relating to an abstract point process referred to as an L-ensemble.

DEFINITION 5.9.2 *With $\mathcal{M}$ a discrete set let $\mathbf{L} = [L(m_i, m_j]_{m_i, m_j \in \mathcal{M}}$. For $X \subset \mathcal{M}$, denote by $\mathbf{L}_X$ the symmetrically labeled submatrix $\mathbf{L}_X = [\mathbf{L}(x_i, x_j)]_{x_i, x_j \in X}$, and suppose $\det \mathbf{L}_X$ is non-negative. The L-ensemble is specified by setting*

$$\text{Prob}\,(X) = \frac{\det \mathbf{L}_X}{\det(\mathbf{1} + \mathbf{L})}.$$

Note that when confined to a lattice, according to (5.102) the log-gas near a metal wall for $\beta = 2$ is an example of an L-ensemble. Now, in a lattice setting correlations are in general specified by

$$\rho(Y) = \sum_{X \subset \mathcal{M}} \text{Prob}(X | Y \subset X).$$

For the L-ensemble, according to the derivation of (5.32)

$$\rho(Y) = \det(\mathbf{L}(\mathbf{1} + \mathbf{L})^{-1}|_Y).$$

Also required is the notion of a conditional L-ensemble.

DEFINITION 5.9.3 *Let $\mathcal{Y} \subset \mathcal{M}$. For $U \subset \mathcal{Y}$ define the conditional L-ensemble by setting*

$$\mathrm{Prob}(U) = \frac{\det \mathbf{L}_{U \cup \bar{\mathcal{Y}}}}{\det(\mathbf{1}_{\mathcal{Y}} + \mathbf{L})}, \tag{5.176}$$

where $\mathbf{1}_{\mathcal{Y}}$ is the $|\mathcal{M}| \times |\mathcal{M}|$ matrix with all but the diagonal entries corresponding to the sublattice $\mathcal{Y}$ equal to zero, and $\bar{\mathcal{Y}}$ denotes the complement of the set $\mathcal{Y}$. (This has the interpretation that the sites $\bar{\mathcal{Y}}$ are required to be fully occupied.)

PROPOSITION 5.9.4 *For the conditional L-ensemble $\rho(Y) = \det \mathbf{K}_Y$ with*

$$\mathbf{K} = \mathbf{1}_{\mathcal{Y}} - (\mathbf{1}_{\mathcal{Y}} + \mathbf{L})^{-1}|_{\mathcal{Y}}. \tag{5.177}$$

Proof. Use will be made of a result of Jacobi [8] which states that for $\mathbf{B} = \mathbf{A}^{-1}$,

$$\det \mathbf{B}_X = \det \mathbf{A}_{\bar{X}} / \det \mathbf{A}. \tag{5.178}$$

By expanding (5.177) about the diagonal we see that for $Y \subset \mathcal{Y}$

$$\det \mathbf{K}_Y = \sum_{X \subset Y} (-1)^{|X|} \det((\mathbf{1}_{\mathcal{Y}} + \mathbf{L})^{-1})_X = \sum_{\bar{X} \supset Y} (-1)^{|X|} \frac{\det(\mathbf{1}_{\mathcal{Y}} + \mathbf{L})_{\bar{X}}}{\det(\mathbf{1}_{\mathcal{Y}} + \mathbf{L})}$$
$$= \sum_{\bar{X} \supset \bar{Y}} (-1)^{|X|} (\text{Prob all points are in } \bar{X}) = \sum_{X \subset Y} (-1)^{|X|} (\text{Prob no points are in } X) = \rho(Y),$$

where the last equality follows from the inclusion–exclusion principle. □

Consider next k discretizations $\mathcal{M}_1, \dots, \mathcal{M}_k$ of the interval $[a, b]$, and let $x_i^{(j)} \in \mathcal{M}_j$ for each $j = 1, \dots, k$ and $i = 1, \dots, N$. A particular choice of $\{x_i^{(j)}\}$ is said to define a configuration. Suppose each configuration occurs with probability (5.175). Associate with $[\phi_i(x_j^{(1)})]_{i,j=1,\dots,N}$ the $N \times |\mathcal{M}_1|$ matrix $\mathbf{\Phi} = [\phi_i(x_j^{(1)}))]_{\substack{i=1,\dots,N \\ x_j \in \mathcal{M}_1}}$, and similarly define $|\mathcal{M}_l| \times |\mathcal{M}_{l+1}|$ matrices $\mathbf{W}_l$ ($l = 1, \dots, k-1$) and a $|\mathcal{M}_k| \times N$ matrix $\mathbf{\Psi}$. Then one sees from the Cauchy-Binet formula (6.88) below that $C = \det \mathbf{M}$, where

$$\mathbf{M} := \mathbf{\Phi} \mathbf{W}_1 \cdots \mathbf{W}_{k-1} \mathbf{\Psi}.$$

We seek the multispecies correlation functions associated with (5.175).

PROPOSITION 5.9.5 *Suppose* $\det \mathbf{M} \neq 0$. *Let $Y_i \subset \mathcal{M}_i$, and set*

$$\mathbf{W}_{[i,j)} = \begin{cases} \mathbf{W}_i \cdots \mathbf{W}_{j-1}, & i < j, \\ \mathbf{0}, & i \geq j. \end{cases} \tag{5.179}$$

We have

$$\rho(Y_1, \dots, Y_k) = \det[\mathbf{K}_{ij}|_{Y_i \times Y_j}]$$

where

$$\mathbf{K}_{ij} = \tilde{\mathbf{W}}_{[i,k)} \mathbf{\Psi} \mathbf{M}^{-1} \mathbf{\Phi} \tilde{\mathbf{W}}_{[1,j)} - \tilde{\mathbf{W}}_{[i,j)} \tag{5.180}$$

with $\tilde{\mathbf{W}}_{[1,1)} = \tilde{\mathbf{W}}_{[k,k)} = \mathbf{1}$, and $\tilde{\mathbf{W}}_{[i,j)} = \mathbf{W}_{[i,j)}$ otherwise.

Proof. Set $\mathcal{M} = \{1, \dots, N\} \cup \mathcal{M}_1 \cup \cdots \cup \mathcal{M}_k$ and consider the conditional L-ensemble on $\mathcal{M}$ with $\mathcal{Y} = \mathcal{M}_1 \cup \cdots \cup \mathcal{M}_k$

and

$$\mathbf{L} = \begin{bmatrix} 0 & \mathbf{\Phi} & 0 & 0 & \cdots & 0 \\ 0 & 0 & -\mathbf{W}_1 & 0 & \cdots & 0 \\ 0 & 0 & 0 & -\mathbf{W}_2 & \cdots & 0 \\ \vdots & \vdots & \vdots & \vdots & \vdots & \vdots \\ 0 & 0 & 0 & 0 & \cdots & -\mathbf{W}_{k-1} \\ \mathbf{\Psi} & 0 & 0 & 0 & \cdots & 0 \end{bmatrix}. \tag{5.181}$$

From this, symmetrically labeled submatrices must be formed which include the first N rows. But the only such submatrices with nonzero determinant are blocks of size $N \times N$, which reduces (5.176) to (5.175). The correlation functions are therefore specified by (5.177), which requires us to invert $\mathbf{1}_{\mathcal{Y}} + \mathbf{L}$. For this we use the general identity

$$\begin{bmatrix} \mathbf{A} & \mathbf{B} \\ \mathbf{C} & \mathbf{D} \end{bmatrix}^{-1} = \begin{bmatrix} -\mathbf{E}^{-1} & \mathbf{E}^{-1}\mathbf{B}\mathbf{D}^{-1} \\ \mathbf{D}^{-1}\mathbf{C}\mathbf{E}^{-1} & \mathbf{D}^{-1} - \mathbf{D}^{-1}\mathbf{C}\mathbf{E}^{-1}\mathbf{B}\mathbf{D}^{-1} \end{bmatrix}, \tag{5.182}$$

where $\mathbf{A}$ and $\mathbf{D}$ are square matrices not necessarily of the same dimension, and $\mathbf{E} := \mathbf{B}\mathbf{D}^{-1}\mathbf{C} - \mathbf{A}$. With $\mathbf{A} = [\mathbf{0}]_{N\times N}$, we use (5.182) with $\mathbf{D}$ replaced by $\mathbf{1} + \mathbf{D}$ and $\mathbf{D}$ the square matrix resulting from blocking out the first block row and column of $\mathbf{L}$, and $\mathbf{B}$ and $\mathbf{C}$ correspondingly defined to relate to $\mathbf{L}$. Then one has

$$(\mathbf{1} + \mathbf{D})^{-1} = \begin{bmatrix} \mathbf{1} & \mathbf{W}_{[1,2)} & \mathbf{W}_{[1,3)} & \cdots & \mathbf{W}_{[1,k)} \\ \mathbf{0} & \mathbf{1} & \mathbf{W}_{[2,3)} & \cdots & \mathbf{W}_{[2,k)} \\ \mathbf{0} & \mathbf{0} & \mathbf{1} & \cdots & \mathbf{W}_{[3,k)} \\ \vdots & \vdots & \vdots & \vdots & \vdots \\ \mathbf{0} & \mathbf{0} & \mathbf{0} & \cdots & \mathbf{1} \end{bmatrix} \tag{5.183}$$

and $\mathbf{E} = \mathbf{\Phi}\mathbf{W}_{[1,k)}\mathbf{\Psi} = \mathbf{M}$. Consequently (5.177) reduces to (5.180). □

We return now to the case $k = 2$ of (5.175) (with $x_i^{(1)} \mapsto x_i$, $x_i^{(2)} \mapsto y_i$ for notational convenience) and take a continuum limit in which the discretizations $\mathcal{M}_1$ and $\mathcal{M}_2$ are dense in $[a, b]$. For special choices of ϕ_i, W_l, ψ_i this was studied directly in [168], [416], while a direct approach to the general case can be found in [293]. The following result is then obtained [456].

PROPOSITION 5.9.6 *In the above specified setting*

$$\rho_{(k_1,k_2)}(x_1, \dots, x_{k_1}; y_1, \dots, y_{k_2}) = \det \begin{bmatrix} [K_{\rm oo}(x_j, x_l)]_{j,l=1,\dots,k_1} & [K_{\rm oe}(x_j, y_l)]_{\substack{j=1,\dots,k_1 \\ l=1,\dots,k_2}} \\ [K_{\rm eo}(y_j, x_l)]_{\substack{j=1,\dots,k_2 \\ l=1,\dots,k_1}} & [K_{\rm ee}(y_j, y_l)]_{j,l=1,\dots,k_2} \end{bmatrix}, \tag{5.184}$$

where

$$\begin{aligned} K_{\rm oo}(x, x') &= \sum_{j,k=1}^{N} \phi_j(x)\, M_{jk}^{-t} \int_a^b W(x', u)\psi_k(u)\, du, \\ K_{\rm oe}(x, y) &= \sum_{j,k=1}^{N} \phi_j(x)\, M_{jk}^{-t}\, \psi_k(y), \\ K_{\rm eo}(y, x) &= -W(x, y) + \sum_{j,k=1}^{N} \Big(\int_a^b W(u, y)\psi_j(u)\, du \Big) M_{jk}^{-t} \Big(\int_a^b W(x, v)\phi_k(v)\, dv \Big), \\ K_{\rm ee}(y, y') &= \sum_{j,k=1}^{N} \Big(\int_a^b W(v, y)\phi_j(v)\, dv \Big) M_{jk}^{-t}\, \psi_k(y'). \end{aligned} \tag{5.185}$$

Here $\mathbf{X}^{-t}$ denotes the operation of taking the transpose of the inverse of $\mathbf{X}$ and $[M_{jk}]$ is the matrix with entries

$$M_{jk} = \int_a^b dx\, \phi_j(x) \int_a^b dy\, \psi_k(y) W(x,y). \tag{5.186}$$

At this stage the formulas of Proposition 5.9.6 could be applied to the p.d.f. (4.97), with the main technical task being to choose $\{\phi_i\}$ and $\{\psi_i\}$ in the Vandermonde determinants so that they are biorthogonal with respect to the inner product (5.186). This is done in [221], but will not be reported on here. Rather these formulas will be applied in Section 11.6 to the computation of dynamical correlations.

5.9.2 Components with an unequal number of particles

In contrast to the p.d.f. (4.97), the two-component system with p.d.f. specified by (4.94) and (4.95) has one species consisting of $n+1$ particles, and the other species consisting of n particles. Extending (4.94) to the joint p.d.f. for the eigenvalues $\{x_j^{(k)}\}_{j=1,\dots,k}$ of nested $k \times k$ minors ($k = 1, \dots, n+1$) of $\mathbf{X} \in \mathrm{GUE}_N^*$ gives the form (4.99) in which component k consists of k particles. Making use of the identity (5.174) shows that (4.99) is of the determinantal form

$$\frac{1}{C} \prod_{l=0}^{N-1} \det[W_l(x_i^{(l)}, x_j^{(l+1)})]_{i,j=1,\dots,l+1} \det[\Psi_{i-1}^N(x_j^{(N)})]_{i,j=1,\dots,N} \tag{5.187}$$

with $x_{l+1}^{(l)} = -\infty$ and $W_0(x_1^{(0)}, x) = 1$. Analogous to the result of Proposition 5.9.5, the general r-point correlation of (5.187) can be expressed as an $r \times r$ determinant [84]. With $(a * b)(x,y) := \int_{-\infty}^{\infty} a(x,z) b(z,y)\, dz$, this requires the quantities

$$W^{(n_1,n_2)}(x,y) = \begin{cases} (W_{n_1} * \cdots * W_{n_2-1})(x,y), & n_1 < n_2, \\ 0, & n_1 \ge n_2, \end{cases} \tag{5.188}$$

and, for $n = 1, \dots, N-1$, $j = 1, \dots, N$,

$$\Psi_{n-j}^n(x) = (W^{(n,N)} * \Psi_{N-j}^N)(x). \tag{5.189}$$

Furthermore, define functions $\{\Phi_j^n(x)\}_{j=0,\dots,n-1}$, $n = 1, \dots, N$, constructed from

$$\mathrm{span}\{W_{i-1} * \tilde{W}^{(i,n)}(x_i^{(i-1)}, x)\}_{i=1,\dots,n},$$

where $\tilde{W}^{(i,m)} = W^{(i,m)}$ for $i < m$ and $W_{i-1} * \tilde{W}^{(i,m)} = 1$ for $i = m$, by the orthogonality requirement

$$\int_{-\infty}^{\infty} \Phi_j^n(x) \Psi_k^n(x)\, dx = \delta_{j,k}. \tag{5.190}$$

PROPOSITION 5.9.7 *Suppose*

$$\Phi_0^{n+1}(x) \propto W_n(x_{n+1}^{(n)}, x). \tag{5.191}$$

For the p.d.f. (5.187) the correlation between eigenvalues of species s_j at positions y_j ($j = 1, \dots, r$) has the determinantal form

$$\rho(\{(s_j, y_j)\}_{j=1,\dots,r}) = \det[K(s_j, y_j; s_k, y_k)]_{j,k=1,\dots,r}$$

with the kernel K given in terms of the quantities $W^{(n_1,n_2)}(x,y)$, $\Psi_j^n(x)$, $\Phi_j^n(x)$ specified above according to

$$K(s_j, y_j; s_k, y_k) = -W^{(s_j,s_k)}(y_j, y_k) + \sum_{l=1}^{s_k} \Psi_{s_j-l}^{s_j}(y_j) \Phi_{s_k-l}^{s_k}(y_k). \tag{5.192}$$

Proof. We follow [84], which in turn follows the method of the proof of Proposition 5.9.5. Before beginning this strategy proper, we consider the significance of (5.191). Introduce the $m \times m$ matrix $\mathbf{B}_m$ according to its elements

$$(\mathbf{B}_m)_{ij} = (W_{i-1} * \tilde{W}^{(i,m)} * \Psi^m_{m-j})(x_i^{(i-1)}). \tag{5.193}$$

Because of the discretization of the domain which is part of the method of the proof of Proposition 5.9.5, $*$ is now interpreted as a summation operation. From the definition (5.189) we see that for $1 \le i,j \le m$, $(\mathbf{B}_m)_{ij} = (\mathbf{B}_N)_{ij}$. This tells us that $\mathbf{B}_N$ is upper triangular if and only if

$$(\mathbf{B}_{m+1})_{m+1,j} = (W_m * \Psi^{m+1}_{m+1-j})(x^{(m)}_{m+1}) = c_m \delta_{j,m+1} \quad (c_m \ne 0).$$

Expanding W_m in terms of $\{\Phi_j^{m+1}\}_{j=0,\dots,m}$ and using (5.190) shows that this condition is equivalent to requiring $W_m(x^{(m)}_{m+1}, x) = c_m \Phi_0^{m+1}(x)$, which is the condition (5.191).

We now implement the method of the proof of Proposition 5.9.5. The support of each of the species (j) is discretized to domains $\mathcal{M}_j$. Set $\mathcal{M} = \{1,\dots,N\} \cup \mathcal{M}_1 \cup \cdots \cup \mathcal{M}_N$ and consider the conditional L-ensemble with $\mathcal{Y} = \mathcal{M}_1 \cup \cdots \cup \mathcal{M}_k$ and

$$\mathbf{L} = \begin{bmatrix} \mathbf{0} & \mathbf{E}_0 & \mathbf{E}_1 & \mathbf{E}_2 & \cdots & \mathbf{E}_{N-1} \\ \mathbf{0} & \mathbf{0} & -\mathbf{W}_1 & \mathbf{0} & \cdots & \mathbf{0} \\ \mathbf{0} & \mathbf{0} & \mathbf{0} & -\mathbf{W}_2 & \cdots & \mathbf{0} \\ \vdots & \vdots & \vdots & \vdots & \ddots & \vdots \\ \mathbf{\Psi} & \mathbf{0} & \mathbf{0} & \mathbf{0} & \cdots & \mathbf{0} \end{bmatrix},$$

where $\mathbf{\Psi}$ is the $|\mathcal{M}_N| \times N$ matrix, and $\mathbf{W}_l$ the $|\mathcal{M}_l| \times |\mathcal{M}_{l+1}|$ matrix

$$[\psi^N_{j-1}(x_i)]_{\substack{x_i \in \mathcal{M}_N \\ j=1,\dots,N}}, \qquad \mathbf{W}_l = [W_l(x_i^{(l)}, x_j^{(l+1)})]_{x_i^{(l)} \in \mathcal{M}_l, x_j^{(l+1)} \in \mathcal{M}_{l+1}},$$

respectively, while $\mathbf{E}_i$ is the $N \times |\mathcal{M}_{i+1}|$ matrix with entries in row $i+1$ given by $W_i(x^{(i)}_{i+1}, x_j^{(i+1)})$, $x_j^{(i+1)} \in \mathcal{M}_{i+1}$, and entries in all other rows zero. As with (5.181), from this choice of $\mathbf{L}$, symmetrically labeled submatrices must be formed which include the first N columns. The structure of $\mathbf{L}$ is such that the only submatrices of this type which have a nonzero determinant have determinant of the form (5.187).

The task now is to compute (5.177). Writing $\mathbf{L}$ in the form

$$\begin{bmatrix} \mathbf{0}_N & \mathbf{B} \\ \mathbf{C} & \mathbf{1} + \mathbf{D}_0 \end{bmatrix},$$

where $\mathbf{B} = [\mathbf{E}_0, \dots, \mathbf{E}_{N-1}]$, $\mathbf{C} = [\mathbf{0}, \dots, \mathbf{0}, \Psi]^T$, it follows from (5.182) that the inverse in (5.177) can be computed to give

$$\mathbf{K} = \mathbf{1} - \mathbf{D}^{-1} + \mathbf{D}^{-1}\mathbf{C}\mathbf{M}^{-1}\mathbf{B}\mathbf{D}^{-1},$$

where $\mathbf{D} := \mathbf{1} + \mathbf{D}_0$, $\mathbf{M} := \mathbf{B}\mathbf{D}^{-1}\mathbf{C}$. Now $\mathbf{D}^{-1}$ is given explicitly by the r.h.s. of (5.183) with $k = N$. This tells us that $\mathbf{1} - \mathbf{D}^{-1} = -[\mathbf{W}_{[i,j)}]$ (recall the notation (5.188)), and furthermore

$$\mathbf{D}^{-1}\mathbf{C} = [\mathbf{W}_{[1,N]}\mathbf{\Psi}, \dots, \mathbf{W}_{[N-1,N)}\mathbf{\Psi}, \mathbf{\Psi}]^T,$$

$$\mathbf{B}\mathbf{D}^{-1} = [\mathbf{E}_0, \mathbf{E}_0\mathbf{W}_{[1,2)} + \mathbf{E}_1, \dots, \sum_{k=1}^{N-1} \mathbf{E}_{k-1}\mathbf{W}_{[k,N)} + \mathbf{E}_{N-1}].$$

Hence the (n,m) block of $\mathbf{K}$ is equal to

$$-\mathbf{W}_{[n,m)} + \mathbf{W}_{[n,N)}\mathbf{\Psi}\mathbf{M}^{-1}\Big(\sum_{k=1}^{m-1} \mathbf{E}_{k-1}\mathbf{W}_{[k,m)} + \mathbf{E}_{m-1}\Big), \tag{5.194}$$

(here $\mathbf{W}_{[n,N)}$ is to be replaced by $\mathbf{1}$ for $n = N$) while

$$\mathbf{M} = \Big(\sum_{k=1}^{N-1} \mathbf{E}_{k-1} \mathbf{W}_{[k,N)} + \mathbf{E}_{N-1} \Big) \mathbf{\Psi}. \tag{5.195}$$

Furthermore, it follows from (5.179) and (5.188) that $(\mathbf{W}_{[n,m)})_{i,j} = W^{(n,m)}(x_i^{(n)}, x_j^{(m)})$, and this together with (5.189) shows $(\mathbf{W}_{[n,N)}\mathbf{\Psi})_{i,j} = \Psi^n_{n-j}(x_i^n)$. Similarly, from the definitions

$$\Big(\sum_{k=1}^{m-1} \mathbf{E}_{k-1} \mathbf{W}_{[k,m)} + \mathbf{E}_{m-1} \Big)_{i, x_j^{(m)}} = \begin{cases} (W_{i-1} * \tilde{W}^{(i,m)})(x_i^{(i-1)}, x_j^{(m)}), & 1 \le i \le m, \\ 0, & m+1 \le i \le N. \end{cases}$$

These facts applied to (5.195) give

$$\mathbf{M} = [W_{i-1} * \tilde{W}^{(i,N)} * \Psi^N_{N-j}(x_i^{(i-1)})]_{i,j=1,\dots,N}.$$

Next, define the matrix $\mathbf{B}_m$ such that

$$W_{i-1} * \tilde{W}^{(i,m)}(x_i^{(i-1)}, x) = \sum_{l=1}^{m} (\mathbf{B}_m)_{i,l} \Phi^m_{m-l}(x).$$

It follows from the orthogonality (5.190) that $\mathbf{B}_m$ is also given by (5.193) and so in particular $\mathbf{M} = \mathbf{B}_N$. And with the $N \times m$ matrix $\mathbf{\Phi}^m$ specified by

$$(\mathbf{\Phi}^m)_{i,j} = \begin{cases} \Phi^m_{m-i}(x_j^{(m)}), & 1 \le i \le m, \\ 0, & m+1 \le i \le N, \end{cases}$$

we have

$$\sum_{k=1}^{m-1} \mathbf{E}_{k-1} \mathbf{W}_{[k,m)} + \mathbf{E}_{m-1} = \begin{bmatrix} \mathbf{B}_m & \mathbf{0} \\ \mathbf{0} & \mathbf{0}_{(N-m)\times(N-m)} \end{bmatrix} \mathbf{\Phi}^m.$$

The considerations of the first paragraph tell us that this multiplied on the left by $\mathbf{M}^{-1}$ gives $\mathbf{\Phi}^m$, and consequently the elements of (5.194) are precisely those implied by (5.192). □

5.9.3 GUE minor process

As recalled at the beginning of the previous subsection, the joint p.d.f. for the eigenvalues $\{x_l^{(j)}\}_{l=1,\dots,j}$, $j = 1, 2, \dots$, of the top $j \times j$ sub-block of matrices from GUE* is given by (4.96). Changing scale $x_l^{(j)} \mapsto \sqrt{2} x_l^{(j)}$ this gives the same joint p.d.f. for matrices from the GUE. Here we seek the correlation function $\rho_{(r)}(\{(s_j, x_j)\}_{j=1,\dots,r})$ between eigenvalues of value x_j from the top $s_j \times s_j$ sub-block [333], [423]. In fact this can be obtained as an example of a result which holds for each of the Gaussian, Laguerre and Jacobi weights [215].

PROPOSITION 5.9.8 *Let $w(x)$ be one of the classical weight functions, Gaussian, Laguerre or Jacobi (with the latter defined on $(0,1)$ and so given by $x^a(1-x)^b$), and consider the joint p.d.f. proportional to*

$$\prod_{l=1}^{N} w(x_l^{(N)}) \prod_{1 \le j < k \le N} (x_j^{(N)} - x_k^{(N)}) \prod_{s=1}^{N-1} \chi(x^{(s+1)} > x^{(s)}). \tag{5.196}$$

We have

$$\rho_{(r)}((s_j, x_j)_{j=1,\dots,r}) = \det[f((s_j, x_j); (s_l, x_l))]_{j,l=1,\dots,r}, \tag{5.197}$$

where

$$f((s,x),(t,y)) = \begin{cases} \Big(w^{(s)}(x)w^{(t)}(y)\Big)^{1/2} \sum_{k=1}^{t} \frac{e_{s-k}}{e_{t-k}} \frac{p_{s-k}^{(s)}(x)p_{t-k}^{(t)}(y)}{\mathcal{N}_{t-k}^{(t)}}, & s \ge t, \\ -\Big(w^{(s)}(x)w^{(t)}(y)\Big)^{1/2} \sum_{k=-\infty}^{0} \frac{e_{s-k}}{e_{t-k}} \frac{p_{s-k}^{(s)}(x)p_{t-k}^{(t)}(y)}{\mathcal{N}_{t-k}^{(t)}}, & s < t. \end{cases} \tag{5.198}$$

Here $p_j(x)$ denotes the monic orthogonal polynomial, $\mathcal{N}_j := \int_{-\infty}^{\infty} w(x)(p_j(x))^2\,dx$ is the normalization while $e_j = (-1)^j, j!, 2^j j!$ in the three cases respectively. The superscripts (n) in $w^{(n)}(x)$, $p_j^{(n)}(x)$ and $\mathcal{N}_j^{(n)}$ indicate that $a \mapsto a+n$ (Laguerre case), $a \mapsto a+n$, $b \mapsto b+n$ (Jacobi case), while in the Gaussian case they have no effect.

Proof. After recalling (5.174), and introducing the convention that $x_{s+1}^{(s)} := -\infty$, we see that (5.196) can be written in the form of (5.187) with $W(x,y) = \chi_{y>x}$, $\psi_j^N(x) = w(x)c_j p_j(x)$ (c_j is a proportionality constant chosen for convenience). Substituting these in the definitions (5.188) and (5.189) gives

$$W^{(n_1,n_2)}(x,y) = \frac{1}{(n_2-n_1-1)!}\chi_{y>x}(y-x)^{n_2-n_1-1}, \tag{5.199}$$

$$\Psi_{n-j}^{n}(x) = \frac{c_{N-j}}{(N-n-1)!}\int_x^{\infty} w(y)p_{N-j}(y)(y-x)^{N-n-1}\,dy. \tag{5.200}$$

The Rodrigues formula for orthogonal polynomials gives

$$c_j p_j(y) = \frac{1}{e_j w(y)}\frac{d^j}{dy^j}\Big(w(y)(Q(y))^j\Big) =: \tilde{p}_j(y), \tag{5.201}$$

where $Q(y)$ equals $1, y, y(1-y)$ (thus in the notation of (5.57), $Q(y) = g(y)$) and the proportionality constant c_j has been chosen so the r.h.s. is equal to $H_j(y)$, $L_j^a(y)$, $P_j^{(a,b)}(1-2y)$ in the three cases, respectively.

Substituting (5.201) in (5.200) and integrating by parts shows that for $j \ge 0$ ($n \ne N$)

$$\Psi_j^n(x) = (-1)^{N-n}\frac{e_j}{e_{N-n+j}} w^{(n)}(x)\tilde{p}_j^{(n)}(x) \tag{5.202}$$

while for $j < 0$

$$\Psi_j^n(x) = \frac{(-1)^{N-n+j}}{e_{N-n+j}}\frac{1}{(-j-1)!}\int_x^{\infty}(y-x)^{-j-1}w(y)(Q(y))^{N-n+j}\,dy. \tag{5.203}$$

In view of (5.202), the requirement (5.190) is satisfied by

$$\Phi_j^n(x) = (-1)^{N-n}\frac{e_{N-n+j}}{e_j}\frac{1}{\tilde{\mathcal{N}}_j^{(n)}}\tilde{p}_j^{(n)}(x), \tag{5.204}$$

where $\tilde{\mathcal{N}}_j^{(n)} := \int_{-\infty}^{\infty} w^{(n)}(x)(\tilde{p}_j^{(n)}(x))^2\,dx$.

Note from (5.204) that $\phi_0^{n+1}(x)$ is a constant. Also, due to the convention that $x_{n+1}^{(n)} := -\infty$, we have $W(x_{n+1}^{(n)}, x) = \chi_{x>-\infty} = 1$. Thus $\phi_0^{n+1}(x) \propto W(x_{n+1}^{(n)}, x)$, which is the condition (5.191) in Proposition 5.9.7, and so the correlations are specified by the kernel (5.192). For $s \ge t$ it is immediate that

$$f((s,x),(t,y)) = \frac{a(s,x)}{a(t,y)}K(s,x;t,y), \tag{5.205}$$

with $a(s,x) = (-1)^s/(w^{(s)}(x))^{1/2}$. By writing $W^{(s,t)}(x,y)$ as a basis in $p_k^{(t)}(y)$ this same equation can be checked for $s < t$. Because the factor $a(s,x)/a(t,y)$ does not effect the determinant, (5.197) follows. □

Chapter Six

Correlation functions at $\beta = 1$ and 4

At $\beta = 1$ and 4, generalizations of the orthogonal polynomial technique used to calculate the correlation functions of the log-gas at $\beta = 2$ are possible. These generalizations involve introducing skew orthogonal polynomials and quaternion determinants, or equivalently Pfaffians. For classical one-body potentials, the required skew orthogonal polynomials can be written in terms of the corresponding orthogonal polynomials, and the correlation kernel can be expressed as the Christoffel-Darboux summation plus a correction involving these polynomials. To obtain the quaternion determinant form both integration formulas and the method of functional differentiation, known from the studies of the previous chapter, are used. For $\beta = 1$ the method of integration over alternate variables plays an important role. In Chapter 4 the Dixon-Anderson integral was used to show that the classical ensembles at $\beta = 1$ and $\beta = 4$ are related by integrating over every second eigenvalue in the former. Here some different perspectives on this result are given, and it is further shown that the random superposition of two classical $\beta = 1$ ensembles, and then integration over every second eigenvalue, gives back a classical $\beta = 2$ ensemble. Also studied is a two-component log-gas, which interpolates between the one-component log-gas on a circle with $\beta = 1$ and $\beta = 4$.

6.1 CORRELATION FUNCTIONS AT $\beta = 4$

6.1.1 Skew orthogonal polynomials and quaternion determinants

We begin with $\beta = 4$ rather than $\beta = 1$ because in the latter case there are complications due to the need to treat separately the cases N even and N odd. The first point to note is that whereas the calculation of the correlations at $\beta = 2$ requires orthogonal polynomials, the calculation of the n-particle correlation function at $\beta = 4$ requires so called *(monic) skew orthogonal polynomials.*

DEFINITION 6.1.1 *A skew symmetric inner product* $\langle \cdot | \cdot \rangle_s$ *has the property*

$$\langle f | g \rangle_s = -\langle g | f \rangle_s.$$

Monic skew orthogonal polynomials, $\{U_n(x)\}_{n=0,1,\dots}$, *are a family of monic polynomials which satisfy the skew orthogonality*

$$\langle U_{2m} | U_{2n+1} \rangle_s = -\langle U_{2n+1} | U_{2m} \rangle_s = u_m \delta_{m,n}, \qquad \langle U_{2m} | U_{2n} \rangle_s = \langle U_{2m+1} | U_{2n+1} \rangle_s = 0.$$

The quantity u_m *is referred to as the normalization.*

From Definition 6.1.1 we see that the skew orthogonality property still holds if we make the replacement

$$U_{2m+1}(x) \mapsto U_{2m+1}(x) + \gamma_{2m} U_{2m}(x) \tag{6.1}$$

for arbitrary γ_{2m}. A Gram-Schmidt type procedure shows that the monic skew orthogonal polynomials are unique up to this mapping. Note also that for a set of $2N$ skew orthogonal polynomials $\{U_n(x)\}_{n=0,\dots,2N-1}$, the skew orthogonality condition can be written in matrix form as

$$[\langle U_j | U_k \rangle_s]_{j,k=0,\dots,2N-1} = \begin{bmatrix} 0 & u_0 & & & & & & \\ -u_0 & 0 & & & & & & \\ & & 0 & u_1 & & & & \\ & & -u_1 & 0 & & & & \\ & & & & \ddots & & & \\ & & & & & & 0 & u_{N-1} \\ & & & & & & -u_{N-1} & 0 \end{bmatrix}, \tag{6.2}$$

where the entries not explicitly shown are all zero.

The Boltzmann factor for the particle-particle interaction at $\beta = 4$ can be expressed in a useful form in terms of the polynomials $\{Q_n(x)\}_{n=0,1,\dots}$ (with normalizations $\{q_m\}$), which are skew orthogonal with respect to the inner product

$$\begin{aligned} \langle f|g \rangle_4 &:= \frac{1}{2} \int_{-\infty}^{\infty} e^{-4V(x)} (f(x) g'(x) - f'(x) g(x)) dx \\ &= \frac{1}{2} \int_{-\infty}^{\infty} e^{-2V(x)} \Big(f(x) \frac{d}{dx} (e^{-2V(x)} g(x)) - g(x) \frac{d}{dx} (e^{-2V(x)} f(x)) \Big) dx, \end{aligned} \tag{6.3}$$

by use of quaternion determinants [150], or equivalently Pfaffians. The former will be considered first.

Definition 6.1.2 *The determinant* qdet *of an $N \times N$ self-dual quaternion matrix $\mathbf{Q}$ (recall (2.6)) is defined as*

$$\mathrm{qdet}\, \mathbf{Q} = \sum_{P \in S_N} (-1)^{N-l} \prod_1^l (\mathbf{q}_{ab} \mathbf{q}_{bc} \cdots \mathbf{q}_{da})^{(0)}. \tag{6.4}$$

Here the superscript (0) denotes the operation $\frac{1}{2}$Tr, or equivalently the scalar part (i.e., the number α_0 in the expansion $\mathbf{q} = \alpha_0 \mathbf{1} + \alpha_1 \mathbf{e}_1 + \alpha_2 \mathbf{e}_2 + \alpha_3 \mathbf{e}_3$; recall Section 1.3.2), P is any permutation of the indices $(1, \dots, N)$ consisting of l disjoint cycles of the form $(a \to b \to c \cdots \to d \to a)$ and $(-1)^{N-l}$ is equal to the parity of P.

Recall from Section 1.3.2 that a quaternion, denoted $\mathbf{q}_{\alpha\beta}$ in the above definition, can be regarded as a 2×2 matrix. It follows from (6.4) that qdet $\mathbf{Q}$ is unchanged if each element $\mathbf{q}_{\alpha\beta}$ is replaced by

$$\mathbf{A}^{-1} \mathbf{q}_{\alpha\beta} \mathbf{A}, \tag{6.5}$$

where $\mathbf{A}$ is an arbitrary non-singular 2×2 matrix.

In the case that each quaternion is a multiple of the identity, $\mathbf{q}_{\alpha\beta} = a_{\alpha\beta} \mathbf{1}_2$, the above definition gives qdet $\mathbf{Q} = \det[a_{\alpha\beta}]_{\alpha,\beta=1,\dots,N}$ as specified by (5.8). The motivation for choosing (6.4) as the definition of the determinant of a self-dual quaternion matrix is that the product over cycles occurs naturally in the calculation of correlation functions (recall Proposition 5.1.2). From a more abstract viewpoint, it is possible to define the determinant of a general quaternion matrix in a number of non-equivalent ways (the determinant function is first specified by some appropriate axioms generalizing the fundamental properties of det$\mathbf{A}$ for a complex matrix $\mathbf{A}$; see [151] and [25]). The definition (6.4) is due to Moore [406].

Suppose in (6.4) we insist that the order of the cycles in the product be the same for permutations which differ by the directions of some or all of the cycles. Now, for a particular cycle $a \to b \to c \cdots \to d \to a$, reversing its direction gives the cycle $a \to d \to \cdots \to c \to b \to a$ and because the matrix is assumed self-dual this gives the product

$$\mathbf{q}_{ad} \cdots \mathbf{q}_{cb} \mathbf{q}_{ba} = (\mathbf{q}_{ab} \mathbf{q}_{bc} \cdots \mathbf{q}_{da})^D,$$

where use has been made of the general property of the dual, $(\mathbf{AB})^D = \mathbf{B}^D \mathbf{A}^D$. With this convention of the order assumed, it follows that

$$\sum_{P\in S_N} (-1)^{N-l} \prod_1^l (\mathbf{q}_{ab}\mathbf{q}_{bc}\cdots\mathbf{q}_{da})$$

is a multiple of the identity, and we can write

$$\mathrm{qdet}\,\mathbf{Q} = \Big(\sum_{P\in S_N} (-1)^{N-l} \prod_1^l (\mathbf{q}_{ab}\mathbf{q}_{bc}\cdots\mathbf{q}_{da}) \Big)^{(0)}. \tag{6.6}$$

A crucial result is that $\mathrm{qdet}\,\mathbf{Q}$ can be expressed in terms of $\det\mathbf{Q}$ (in $\mathrm{qdet}\,\mathbf{Q}$, $\mathbf{Q}$ is to be regarded as an $N\times N$ matrix with quaternion elements, while in $\det\mathbf{Q}$, $\mathbf{Q}$ is to be regarded as a $2N\times 2N$ matrix with complex elements) [150].

PROPOSITION 6.1.3 *For a self-dual quaternion matrix we have*

$$\big(\mathrm{qdet}\,\mathbf{Q}\big)^2 = \det\mathbf{Q}. \tag{6.7}$$

Proof. Since (6.7) is a polynomial identity it suffices to consider the case that $\mathbf{Q}$ has real elements. Thus a typical element $\mathbf{q}_{jk}$ of $\mathbf{Q}$ has the form

$$\mathbf{q}_{jk} = \begin{bmatrix} a_{jk} & b_{jk} \\ -b_{jk} & a_{jk} \end{bmatrix}$$

for a_{jk}, b_{jk} real. But as noted in (1.36) such 2×2 matrices are isomorphic to complex numbers $z_{jk} := a_{jk} + ib_{jk}$. Substituting in (6.6) gives

$$\begin{aligned} \mathrm{qdet}\mathbf{Q} &= \sum_{P\in S_N} (-1)^{N-l} \prod_1^l z_{ab}z_{bc}\cdots z_{da} \\ &= \det\mathbf{Z}, \qquad \mathbf{Z} := [z_{jk}]_{j,k=1,\dots,N}. \end{aligned} \tag{6.8}$$

But the result of Exercises 3.1 q.2(ii) tells us that

$$(\det\mathbf{Z})^2 = \det\mathbf{Q} \tag{6.9}$$

(since $\mathbf{Q}$ is self-dual, $\mathbf{Z}$ is Hermitian, so there is no need to take the absolute value on the l.h.s. of (6.9)). □

At this stage we introduce Pfaffians.

DEFINITION 6.1.4 *Let $\mathbf{X} = [\alpha_{ij}]_{i,j=1,\dots,2N}$, where $\alpha_{ji} = -\alpha_{ij}$ so that $\mathbf{X}$ is an antisymmetric matrix of even degree. Then the Pfaffian of $\mathbf{X}$, denoted $\mathrm{Pf}\,\mathbf{X}$, is defined by*

$$\begin{aligned} \mathrm{Pf}\,\mathbf{X} &= \sum\nolimits^*_{P(2l)>P(2l-1)} \varepsilon(P)\alpha_{P(1)P(2)}\alpha_{P(3)P(4)}\cdots\alpha_{P(2N-1)P(2N)} \\ &= \frac{1}{2^N N!} \sum_{P\in S_{2N}} \varepsilon(P)\alpha_{P(1)P(2)}\alpha_{P(3)P(4)}\cdots\alpha_{P(2N-1)P(2N)}, \end{aligned} \tag{6.10}$$

where in the first summation the $*$ denotes that the sum is restricted to distinct terms only (i.e., only one term from the $N!$ ways of permuting the pairs of indices is to be included).

We note that for $\mathbf{Q}$ a self-dual quaternion matrix, $\mathbf{Q}\mathbf{Z}_{2N}^{-1}$ (where now $\mathbf{Q}$ is regarded as a $2N\times 2N$ matrix with complex elements) is an antisymmetric matrix, so its Pfaffian is well defined. In fact the Pfaffian of this antisymmetric matrix is equal to the quaternion determinant of $\mathbf{Q}$.

PROPOSITION 6.1.5 *For a self-dual quaternion matrix* $\mathbf{Q}$

$$\mathrm{qdet}\,\mathbf{Q} = \mathrm{Pf}\,\mathbf{Q}\mathbf{Z}_{2N}^{-1}. \tag{6.11}$$

Proof. For $\mathbf{X}$ an antisymmetric matrix, a classical result (see Exercises 6.1 q.1) says

$$(\mathrm{Pf}\,\mathbf{X})^2 = \det\mathbf{X}. \tag{6.12}$$

Noting $\det \mathbf{Z}_{2N}^{-1} = 1$, it follows that

$$(\mathrm{Pf}\,\mathbf{Q}\mathbf{Z}_{2N}^{-1})^2 = \det \mathbf{Q},$$

and comparing this with (6.7) we see (6.11) is correct up to a sign. Taking $\mathbf{Q}$ the identity verifies the sign. □

A useful corollary of (6.7) is the formula

$$\det \mathbf{Q} = (\det \mathbf{Q}\mathbf{Q}^D)^{1/2} = \mathrm{qdet}(\mathbf{Q}\mathbf{Q}^D), \tag{6.13}$$

valid for a general $2N \times 2N$ matrix $\mathbf{Q}$ with a positive determinant (the fact that $\mathrm{qdet}(\mathbf{Q}\mathbf{Q}^D)$ is positive follows from (6.8)). We can use (6.13) to express the Boltzmann factor for the log-gas with $\beta = 4$ in terms of the monic skew symmetric polynomials orthogonal with respect to the inner product (6.3) [381].

PROPOSITION 6.1.6 *Let*

$$\chi_k(x) := \begin{bmatrix} Q_{2k}(x) & Q_{2k+1}(x) \\ Q'_{2k}(x) & Q'_{2k+1}(x) \end{bmatrix}, \qquad \mathbf{C}(x) := \begin{bmatrix} 1 & 0 \\ 2V'(x) & 1 \end{bmatrix},$$

where $\{Q_n(x)\}$ *are monic skew orthogonal polynomials with respect to the inner product (6.3), with corresponding normalizations* $\{q_n\}$, *and let* $\chi_k^D(x)$ *denote the dual of* $\chi_k(x)$. *We have*

$$\begin{aligned}\prod_{l=1}^N e^{-4V(x_l)} \prod_{1\le j<k\le N} (x_k - x_j)^4 &= \prod_{l=1}^N e^{-4V(x_l)} \det \begin{bmatrix} x_j^{k-1} \\ (k-1)x_j^{k-2} \end{bmatrix}_{\substack{j=1,\dots,N \\ k=1,\dots,2N}} \\ &= \prod_{l=0}^{N-1} 2q_l \,\mathrm{qdet}[\mathbf{f}_4(x_j,x_k)]_{j,k=1,\dots,N},\end{aligned} \tag{6.14}$$

where

$$\mathbf{f}_4(x,y) := e^{-2(V(x)+V(y))} \sum_{k=0}^{N-1} \frac{1}{2q_k} \mathbf{C}(x)\chi_k(x)\chi_k^D(y)\mathbf{C}^D(y) = \begin{bmatrix} S_4(x,y) & I_4(x,y) \\ D_4(x,y) & S_4(y,x) \end{bmatrix}$$

with

$$\begin{aligned} S_4(x,y) &= \sum_{m=0}^{N-1} \frac{e^{-2V(x)}}{2q_m}\Big(Q_{2m}(x)\frac{d}{dy}\Big(e^{-2V(y)}Q_{2m+1}(y)\Big) - Q_{2m+1}(x)\frac{d}{dy}\Big(e^{-2V(y)}Q_{2m}(y)\Big)\Big), \\ I_4(x,y) &= -\int_x^y S_4(x,y')\,dy', \\ D_4(x,y) &= \frac{\partial}{\partial x} S_4(x,y). \end{aligned} \tag{6.15}$$

Note that the matrix $[\mathbf{f}_4(x_j,x_k)]_{j,k=1,\dots,N}$ *is self-dual.*

Proof. The first stated identity is the case $p = 2$ of (1.174). By adding suitable multiples of columns $1, 2, \dots, j-1$ to row j ($j = 2, \dots, 2N$ in order) and using the fact that the Q_n are monic we see that

$$\prod_{l=1}^{N} e^{-4V(x_l)} \det \begin{bmatrix} x_j^{k-1} \\ (k-1)x_j^{k-2} \end{bmatrix}_{\substack{j=1,\dots,N \\ k=1,\dots,2N}}$$
$$= \prod_{l=1}^{N} 2q_{l-1} \det \begin{bmatrix} \frac{e^{-2V(x_j)}}{\sqrt{2q_{k-1}}} Q_{2k-2}(x_j) & \frac{e^{-2V(x_j)}}{\sqrt{2q_{k-1}}} Q_{2k-1}(x_j) \\ \frac{e^{-2V(x_j)}}{\sqrt{2q_{k-1}}} Q'_{2k-2}(x_j) & \frac{e^{-2V(x_j)}}{\sqrt{2q_{k-1}}} Q'_{2k-1}(x_j) \end{bmatrix}_{j,k=1,\dots,N}. \tag{6.16}$$

Next multiply the matrix in (6.16) by $\mathrm{diag}[\mathbf{C}(x_j)]_{j=1,\dots,N}$ on the left. Since the determinant of this latter matrix is unity, (6.16) is unchanged. Now applying (6.13) gives the second stated result. □

It follows from the skew orthogonality property of $\{Q_n(x)\}_{n=0,1,\dots}$ and the definition of $\mathbf{f}_4$ involving $\chi_k(x)$ and $\chi_k^D(y)$ that

$$\int_{-\infty}^{\infty} \mathbf{f}_4(x,x)dx = N \begin{bmatrix} 1 & 0 \\ 0 & 1 \end{bmatrix} \quad \text{and} \quad \int_{-\infty}^{\infty} \mathbf{f}_4(x,y)\mathbf{f}_4(y,z)dy = \mathbf{f}_4(x,z). \tag{6.17}$$

Due to the similarities between (6.17) and (5.7) we might expect that the integrations required to compute the n-particle correlation can be carried out according to the method of the proof of Proposition 5.1.2. This would require making a Laplace expansion of qdet, which is possible [406], [151] but due to the noncommutativity of the quaternions there are some complicating factors. Alternatively, the integrations can be carried out directly, as we will now demonstrate [150], [381].

PROPOSITION 6.1.7 *We have*

$$\int_{-\infty}^{\infty} dx_1 \cdots \int_{-\infty}^{\infty} dx_N \prod_{l=1}^{N} e^{-4V(x_l)} \prod_{1 \le j < k \le N} (x_k - x_j)^4 = N! \prod_{k=0}^{N-1} 2q_k,$$
$$\rho_{(n)}(x_1, \dots, x_n) = \mathrm{qdet}[\mathbf{f}_4(x_j, x_k)]_{j,k=1,\dots,n},$$
$$\rho_{(n)}^T(x_1, \dots, x_n) = (-1)^{n-1} \sum_{\substack{\text{cycles} \\ \text{length } n}} \Big(\prod_{(i,i')} \mathbf{f}_4(x_i, x_{i'}) \Big)^{(0)},$$

where $\mathbf{f}_4$ *is defined in Proposition 6.1.6.*

Proof. Consider $\mathrm{qdet}[\mathbf{f}_4(x_j, x_k)]_{j,k=1,\dots,m}$. According to (6.6), this can be expanded as a sum over all permutations of $\{1, \dots, m\}$ with summand

$$(-1)^{m-l} \mathbf{S}_1 \mathbf{S}_2 \cdots \mathbf{S}_l, \tag{6.18}$$

where each $\mathbf{S}_p$ consists of a particular product of the $\mathbf{f}_4(x_j, x_k)$ in which the indices form a cycle. Let us compare this with the expansion of $\mathrm{qdet}[\mathbf{f}_4(x_j, x_k)]_{j,k=1,\dots,m-1}$. Analogous to (6.18), this can be written as a sum over all permutations of $\{1, \dots, m-1\}$ with summand

$$(-1)^{m-1-l'} \mathbf{A}_1 \cdots \mathbf{A}_{l'}, \tag{6.19}$$

where again each $\mathbf{A}_p$ consists of a particular product of the $\mathbf{f}_4(x_j, x_k)$ in which the indices form a cycle. Now, for each term (6.19) with $l' = l - 1$ there are m ways to extend it to form a product of cycles of the form (6.18). One of these is to introduce an extra identity cycle $m \mapsto m$ which gives

$$(-1)^{m-l} \mathbf{A}_1 \cdots \mathbf{A}_{l-1} \mathbf{f}_4(x_m, x_m). \tag{6.20}$$

The other $m-1$ possibilities are to introduce m within an existing cycle, which gives

$$(-1)^{m-l+1} \mathbf{A}_1 \cdots \mathbf{A}_j^{(m)} \cdots \mathbf{A}_{l-1}, \tag{6.21}$$

where $\mathbf{A}_j^{(m)} := \mathbf{f}_4(x_\alpha, x_\beta) \cdots \mathbf{f}_4(x_\delta, x_m)\mathbf{f}_4(x_m, x_\gamma) \cdots \mathbf{f}_4(x_\tau, x_\alpha)$. The sum over all permutations of $\{1, \dots, m-1\}$ with (6.20) and the $m-1$ terms of the form (6.21) as the summand gives $\mathrm{qdet}[\mathbf{f}_4(x_j, x_k)]_{j,k=1,\dots,m}$. Integrating (6.20) over x_m using (6.17) gives

$$N(-1)^{m-l}\mathbf{A}_1 \cdots \mathbf{A}_{l-1},$$

while integrating the sum of the $m-1$ terms of the form (6.21) over x_m gives

$$-(m-1)(-1)^{m-l}\mathbf{A}_1 \cdots \mathbf{A}_{l-1}. \tag{6.22}$$

According to (6.19) the sum total of these two terms is just $(N-(m-1))$ times a typical term in the expansion of $\mathrm{qdet}[\mathbf{f}_4(x_j, x_k)]_{j,k=1,\dots,m-1}$, and so we have

$$\int_{-\infty}^{\infty} \mathrm{qdet}[\mathbf{f}_4(x_j, x_k)]_{j,k=1,\dots,m}\, dx_m = (N-(m-1))\mathrm{qdet}[\mathbf{f}_4(x_j, x_k)]_{j,k=1,\dots,m-1}. \tag{6.23}$$

The first two results follow. The expression for $\rho_{(n)}^T$ follows from that for $\rho_{(n)}$ in an analogous way to the deduction of $\rho_{(n)}^T$ from $\rho_{(n)}$ in Proposition 5.1.2. □

The n-point correlation for the $\beta = 4$ log-gas system is thus given by an $n \times n$ quaternion determinant, or equivalently a $2n \times 2n$ Pfaffian. This will be shown also to be the case for $\beta = 1$ log-gas systems. In keeping with the definition of a determinantal point process, a many particle system for which the n-point correlation can be written as a $2n \times 2n$ Pfaffian with entries independent of n is said to be a *Pfaffian point process.*

6.1.2 Functional differentiation method

Historically (see [392]), the normalization integral and correlations in Proposition 6.1.7 were first calculated without making explicit use of quaternion determinants. The method used was to make use of the functional derivative (5.18) and Pfaffian (6.10) instead. However the implementation was such that only the one- and two-point correlations were obtained. More recently [522] it has been shown how to obtain the general n-point correlation via this method. Analogous to Proposition 5.2.1, the strategy is to write the generalized partition function in terms of the determinant of an integral operator. First the analogue of the l.h.s. of the identity of Proposition 5.2.1 is required.

PROPOSITION 6.1.8 *For Ω equal to the real line and*

$$e^{-\beta U(x_1,\dots,x_N)} = \prod_{l=1}^{N} e^{-4V(x_l)} \prod_{1 \le j < k \le N} (x_k - x_j)^4 \tag{6.24}$$

in (5.19) we have

$$\hat{Z}_N[a] = N!2^N \mathrm{Pf}[\beta_{jk}]_{j,k=1,\dots,2N} = N!2^N \Big(\det[\beta_{jk}]_{j,k=1,\dots,2N}\Big)^{1/2} \tag{6.25}$$

where

$$\begin{aligned}\beta_{jk} &:= \frac{1}{2}\int_{-\infty}^{\infty} e^{-4V(x)}a(x)\Big(Q_{j-1}(x)Q'_{k-1}(x) - Q_{k-1}(x)Q'_{j-1}(x)\Big)\, dx \\ &= \frac{1}{2}\int_{-\infty}^{\infty} e^{-2V(x)}a(x)\Big(Q_{j-1}(x)\frac{d}{dx}(e^{-2V(x)}Q_{k-1}(x)) - Q_{k-1}(x)\frac{d}{dx}(e^{-2V(x)}Q_{j-1}(x))\Big)\, dx.\end{aligned} \tag{6.26}$$

Proof. According to the first equality in (6.14), and (6.16), we have

$$\prod_{1\le j<k\le N}(x_k-x_j)^4=\det\begin{bmatrix} Q_{2k-2}(x_j) & Q_{2k-1}(x_j) \\ Q'_{2k-2}(x_j) & Q'_{2k-1}(x_j)\end{bmatrix}_{j,k=1,\dots,N}$$
$$=\sum_{P\in S_{2N}}\varepsilon(P)\prod_{l=1}^{N}Q_{P(2l-1)-1}(x_l)Q'_{P(2l)-1}(x_l).$$

The dependence on each integration variable is now decoupled, and this allows us to write

$$\hat{Z}_N[a]=\sum_{P\in S_{2N}}\varepsilon(P)\prod_{l=1}^{N}\alpha_{P(2l-1),P(2l)},\qquad \alpha_{j,k}:=\int_{-\infty}^{\infty}e^{-4V(x)}a(x)Q_{j-1}(x)Q'_{k-1}(x)\,dx.$$

Introducing the restriction $P(2l)>P(2l-1)$, which can be done if we include all interchanges of this type in the summand by writing $\alpha_{P(2l-1),P(2l)}\mapsto\alpha_{P(2l-1),P(2l)}-\alpha_{P(2l),P(2l-1)}$ gives

$$\hat{Z}_N[a]=2^N\sum_{\substack{P\in S_{2N}\\ P(2l)>P(2l-1)}}\varepsilon(P)\prod_{l=1}^{N}\beta_{P(2l-1),P(2l)}.$$

Comparing this expression with the definition (6.10) of the Pfaffian, and using (6.12), gives the stated result. □

The strategy now is to make use of (5.26) and so introduce the sought determinant of an integral operator.

PROPOSITION 6.1.9 *With* $\mathbf{f}_4$ *as in Proposition 6.1.6, the expression (6.25) for the generalized partition function can be rewritten as*

$$\hat{Z}_N[a]=N!2^N\prod_{l=0}^{N-1}q_l\Big(\det[\mathbf{1}_2+\mathbf{f}_4^T(\mathbf{a}-\mathbf{1}_2)]\Big)^{1/2},\tag{6.27}$$

where $\mathbf{f}_4^T(\mathbf{a}-\mathbf{1}_2)$ *is the* 2×2 *matrix integral operator with kernel* $\mathbf{f}_4^T(x,y)\mathrm{diag}[a(y)-1,a(y)-1]$.

Proof. With $\beta_{jk}=\beta_{jk}[a]$ write

$$[\beta_{jk}[a]]_{j,k=1,\dots,2N}=\Big[\beta_{jk}[1]+\beta_{jk}[a-1]\Big]_{j,k=1,\dots,2N}.$$

From the definition (6.26) of the β_{jk} and the skew orthogonality of $\{Q_j(x)\}$ we see that $[\beta_{jk}[1]]$ is antisymmetric and of the form (6.2). Taking out an appropriate factor from each row and interchanging rows $2j-1$ and $2j$ for each $j=1,\dots,N$ shows

$$\det[\beta_{jk}[a]]_{j,k=1,\dots,2N}=\prod_{j=0}^{N-1}q_j^2\det\left(\mathbf{1}_{2N}+\begin{bmatrix}-q_{j-1}^{-1}\beta_{2j,k}[a-1]\\ q_{j-1}^{-1}\beta_{2j-1,k}[a-1]\end{bmatrix}_{\substack{j=1,\dots,N\\ k=1,\dots,2N}}\right).\tag{6.28}$$

In preparation for using (5.26) we note that with

$$F_{2j-1}(x)=-e^{-2V(x)}Q_{2j-1}(x),\qquad F_{2j}(x)=e^{-2V(x)}Q_{2j-2}(x)$$

the integrand in the element (jk) of the final matrix in (6.28) consists of $(a(x)-1)/q_{[j/2]}$ times

$$F_j(x)\frac{d}{dx}\Big(e^{-2V(x)}Q_{k-1}(x)\Big)-e^{-2V(x)}Q_{k-1}(x)\frac{d}{dx}(F_j(x))$$
$$=\begin{bmatrix}F_j(x) & -\frac{d}{dx}(F_j(x))\end{bmatrix}\begin{bmatrix}\frac{d}{dx}(e^{-2V(x)}Q_{k-1}(x))\\ e^{-2V(x)}Q_{k-1}(x)\end{bmatrix}$$

This shows that with $\mathbf{A}$ the $2N \times 2$ matrix valued integral operator on $(-\infty, \infty)$ with kernel

$$(a(y)-1)\Big[\frac{1}{q_{[j/2]}}F_j(y) \qquad -\frac{1}{q_{[j/2]}}\frac{d}{dy}F_j(y)\Big],$$

and $\mathbf{B}$ the $2 \times 2N$ matrix multiplication operator

$$\left[\begin{array}{c}\frac{d}{dx}\Big(e^{-2V(x)}Q_{k-1}(x)\Big)\\ e^{-2V(x)}Q_{k-1}(x)\end{array}\right],$$

the determinant in (6.28) can be written $\det(\mathbf{1}_{2N} + \mathbf{AB})$. Now using (5.26) gives the stated result.

The analogous result in the case $\beta = 2$, Proposition 5.2.2, suggests an alternative strategy. This is to consider the 2×2 matrix integral/ eigenfunction equation

$$\int_{-\infty}^{\infty}\begin{bmatrix}S_4(x,y) & I_4(x,y)\\ D_4(x,y) & S_4(y,x)\end{bmatrix}\begin{bmatrix}\psi_l^{(1)}(y)\\ \psi_l^{(2)}(y)\end{bmatrix}(1-a(y))\,dy = \lambda_l\begin{bmatrix}c\psi_l^{(1)}(x)\\ \psi_l^{(2)}(x)\end{bmatrix}, \tag{6.29}$$

and to seek eigenfunctions of the form

$$\left[\begin{array}{c}\psi_l^{(1)}(y)\\ \psi_l^{(2)}(y)\end{array}\right] = \sum_{\nu=0}^{\infty} c_{l,\nu}\left[\begin{array}{c}e^{-2V(y)}Q_\nu(y)\\ \frac{d}{dy}(e^{-2V(y)}Q_\nu(y))\end{array}\right]. \tag{6.30}$$

Substituting (6.30) in (6.29) and equating coefficients of the vector in the νth term of (6.30) shows $c_{l,\nu} = 0$ for $\nu \ge 2N$ while for $\nu < 2N$

$$\lambda_l c_{l,\nu} = \frac{1}{q_{[(\nu+1)/2]}}\sum_{\nu'=0}^{2N-1} c_{l,\nu'}\beta_{\nu'+1,\nu+2}[1-a(y)].$$

This is the eigenvalue problem for the $2N \times 2N$ antisymmetric matrix

$$\Big[\frac{1}{q_{[(j-1)/2]}}\beta_{j,k}[a(y)-1]\Big]_{j,k=1,\dots,2N},$$

which can therefore be substituted for $\mathbf{f}_4^T(\mathbf{a} - \mathbf{1}_2)$ in (6.27), reclaiming (6.25). □

Substituting (6.27) in (5.25) and proceeding as in the working which led to (5.30) we find

$$\rho_{(n)}^T(x_1,\dots,x_n) = (-1)^{n+1}\frac{1}{2}\mathrm{Tr}\sum_{\substack{\text{cycles}\\ \text{length } n}}\prod_{(i,i')}^{n}\mathbf{f}_4(x_i,x_{i'}),$$

which is equivalent to the final formula in Proposition 6.1.7. More generally, we note that if $\mathbf{K}$ is a 2×2 matrix integral operator depending on $a(x)$ and

$$\hat{Z}_N[a] \propto \Big(\det(\mathbf{1}_2 + \mathbf{K})\Big)^{1/2}$$

we have

$$\rho_{(n)}^T(x_1,\dots,x_n) = (-1)^{n+1}\frac{1}{2}\mathrm{Tr}\sum_{\substack{\text{cycles}\\ \text{length } n}}\prod_{(i,i')}^{n}\mathbf{G}(x_i,x_{i'}),$$

$$\mathbf{G}(x_i,x_{i'}) = \Big\langle x_i\Big|\frac{\partial\mathbf{K}}{\partial a(x)}(\mathbf{1}+\mathbf{K})^{-1}\Big|_{a=1}\Big|x_{i'}\Big\rangle \tag{6.31}$$

(cf. (5.30)).

The formula (6.27) can also be used to compute $\rho_{(n)}$ directly. First we make use of (6.7) to note

$$\Big(\det[\mathbf{1}_2 + \mathbf{f}_4^T(\mathbf{a} - \mathbf{1}_2)]\Big)^{1/2} = \mathrm{qdet}\,[\mathbf{1}_2 + \mathbf{f}_4(\mathbf{a} - \mathbf{1}_2)].$$

But analogous to (5.32) we have for $\mathbf{K}_4$ a 2×2 matrix integral operator with kernel

$$\mathbf{K}_4(x,y) := \begin{bmatrix} K_4^{(1)}(x,y) & K_4^{(2)}(x,y) \\ K_4^{(3)}(x,y) & K_4^{(1)}(y,x) \end{bmatrix},$$

and $K_4^{(2)}$, $K_4^{(3)}$ antisymmetric with respect to interchange of x and y so that $[K_4(x_j,x_l)]_{j,l=1,\dots,k}$ is a self-dual quaternion matrix

$$\mathrm{qdet}\,[\mathbf{1}_2+\mathbf{K}_4(\mathbf{a}-\mathbf{1}_2)] = 1+\sum_{k=1}^\infty \frac{1}{k!}\int_{-\infty}^\infty dx_1\,(a(x_1)-1)\cdots\int_{-\infty}^\infty dx_k\,(a(x_k)-1)\,\mathrm{qdet}\,[\mathbf{K}_4(x_j,x_l)]_{j,l=1,\dots,k}. \tag{6.32}$$

From this the functional derivative required by (5.20) can be computed to give the second formula in Proposition 6.1.7.

EXERCISES 6.1 1. Let $\{\vec{x}_j\}_{j=1,\dots,2N}$ be $2N$ component vectors and let $\mathbf{A}$ be a $2N \times 2N$ antisymmetric matrix so that

$$\mathrm{Pf}[\vec{x}_j \cdot (\mathbf{A}\vec{x}_k)]_{j,k=1,\dots,2N} \tag{6.33}$$

is well defined.

(i) Let σ be a permutation of $\{1,2,\dots,2N\}$. Show from the definition (6.10) that

$$\mathrm{Pf}[\vec{x}_{\sigma(j)} \cdot (\mathbf{A}\vec{x}_{\sigma(k)})]_{j,k=1,\dots,2N} = \mathrm{sgn}(\sigma)\mathrm{Pf}[\vec{x}_j \cdot (\mathbf{A}\vec{x}_k)]_{j,k=1,\dots,2N}.$$

Note also from the definition (6.10) that (6.33) is a linear function in the components of $\{x_j\}$. Since the above equation shows the linear function is antisymmetric, conclude that for some f

$$\mathrm{Pf}[\vec{x}_j \cdot (\mathbf{A}\vec{x}_k)]_{j,k=1,\dots,2N} = f(\mathbf{A})\det[\vec{x}_1\ \vec{x}_2\ \cdots\ \vec{x}_{2N}].$$

By choosing $\vec{x}_j = \vec{e}_j$ (an elementary unit vector), show that

$$f(\mathbf{A}) = \mathrm{Pf}\,\mathbf{A}.$$

(ii) Let $\mathbf{A}'$ denote the $2N \times 2N$ antisymmetric matrix obtained from $\mathbf{A}$ by interchanging two rows and the corresponding two columns. Show from the first equation in (i) that

$$\mathrm{Pf}\,\mathbf{A}' = -\mathrm{Pf}\,\mathbf{A}. \tag{6.34}$$

(iii) Replace $\vec{x}_j$ by $\mathbf{B}\vec{e}_j$, for $\mathbf{B}$ a general $2N \times 2N$ matrix, in the second equation of (i) to conclude

$$\mathrm{Pf}(\mathbf{B}^T\mathbf{A}\mathbf{B}) = \det\mathbf{B}\,\mathrm{Pf}\,\mathbf{A}. \tag{6.35}$$

(iv) Set $\mathbf{A} = \mathbf{Z}_{2N}^{-1}$, and $\mathbf{X} = \mathbf{B}^T\mathbf{A}\mathbf{B}$ to deduce from (6.35) the result (6.12).

(v) By considering the coefficient of $\alpha_{r,2N}$ in (6.10), deduce the Laplace type expansion

$$\mathrm{Pf}\,\mathbf{A} = \sum_{r=1}^{2N-1}(-1)^{r+1}\alpha_{r,2N}\mathrm{Pf}_{r,2N}(\mathbf{A}), \tag{6.36}$$

where $\mathrm{Pf}_{r,2N}$ is the Pfaffian of the $(2N-2)\times(2N-2)$ antisymmetric matrix obtained by deleting rows and columns r, $2N$.

2. [456], [89] Let $\mathbf{B}$, $\mathbf{C}$ be $2J \times 2J$ and $2K \times 2K$ invertible antisymmetric matrices, and $\mathbf{A}$ be a matrix of size

$2J \times 2K$. The aim of this exercise is to show that

$$\frac{\mathrm{Pf}((\mathbf{C}^{-1})^T - \mathbf{A}^T\mathbf{B}\mathbf{A})}{\mathrm{Pf}((\mathbf{C}^{-1})^T)} = \frac{\mathrm{Pf}((\mathbf{B}^{-1})^T - \mathbf{A}\mathbf{C}\mathbf{A}^T)}{\mathrm{Pf}((\mathbf{B}^{-1})^T)}. \tag{6.37}$$

(i) By writing $\mathbf{A}_{p\times q} \mapsto \mathbf{A}_{p\times q}\mathbf{R}_{q\times q}$, $\mathbf{B}_{q\times p} \mapsto \mathbf{B}_{q\times p}\mathbf{M}_{p\times p}$, where $\mathbf{R}$ and $\mathbf{M}$ are assumed invertible, in (5.33) obtain the determinant identity

$$\det \mathbf{M}_{p\times p} \det(\mathbf{M}^{-1}_{p\times p} + \mathbf{A}_{p\times q}\mathbf{R}_{q\times q}\mathbf{B}_{q\times p}) = \det \mathbf{R}_{q\times q} \det(\mathbf{R}^{-1}_{q\times q} + \mathbf{B}_{q\times p}\mathbf{M}_{p\times p}\mathbf{A}_{p\times q}). \tag{6.38}$$

(ii) In (6.38), note that $(\mathbf{C}^{-1})^T = -\mathbf{C}^{-1}$, $(\mathbf{B}^{-1})^T = -\mathbf{B}^{-1}$, then square both sides, write in terms of determinants using (6.12), and verify the resulting identity using (6.37). This establishes (6.38) up to a sign. Show that the sign is correct by setting $\mathbf{A} = \mathbf{0}$.

3. [167, 263] The objective of this exercise is to derive the formulas

$$Q_{2n}(x) = \frac{1}{\hat{Z}_n}\int_{-\infty}^{\infty} dx_1\, e^{-4V(x_1)} \cdots \int_{-\infty}^{\infty} dx_n\, e^{-4V(x_n)} \prod_{l=1}^n (x - x_l)^2 \prod_{1\le j<k\le n} (x_k - x_j)^4,$$

$$Q_{2n+1}(x) = \frac{1}{\hat{Z}_n}\int_{-\infty}^{\infty} dx_1\, e^{-4V(x_1)} \cdots \int_{-\infty}^{\infty} dx_n\, e^{-4V(x_n)} \Big(x + 2\sum_{l=1}^n x_l\Big) \prod_{l=1}^n (x - x_l)^2 \prod_{1\le j<k\le n} (x_k - x_j)^4$$

(cf. first formula of Proposition 5.1.4).

(i) Show that

$$\prod_{l=1}^n (x - x_l)^2 \prod_{1\le j<k\le n} (x_k - x_j)^4$$
$$= \det \begin{bmatrix} x_j^{k-1} \\ (k-1)x_j^{k-2} \\ x^{k-1} \end{bmatrix}_{\substack{j=1,\dots,n \\ k=1,\dots,2n+1}} = \det \begin{bmatrix} Q_{k-1}(x_j) \\ Q'_{k-1}(x_j) \\ Q_{k-1}(x) \end{bmatrix}_{\substack{j=1,\dots,n \\ k=1,\dots,2n+1}},$$

$$\Big(x + 2\sum_{l=1}^n x_l\Big) \prod_{l=1}^n (x - x_l)^2 \prod_{1\le j<k\le n} (x_k - x_j)^4$$
$$= \det \begin{bmatrix} x_j^{k-1} & x_j^{2n+1} \\ (k-1)x_j^{k-2} & (2n+1)x_j^{2n} \\ x^{k-1} & x^{2n+1} \end{bmatrix}_{\substack{j=1,\dots,n-1 \\ k=1,\dots,2n}} = \det \begin{bmatrix} Q_{k-1}(x_j) & Q_{2n+1}(x_j) \\ Q'_{k-1}(x_j) & Q'_{2n+1}(x_j) \\ Q_{k-1}(x) & Q_{2n+1}(x) \end{bmatrix}_{\substack{j=1,\dots,n \\ k=1,\dots,2n+1}},$$

where in obtaining the final equality the arbitrary constant γ_{2m} in (6.1) is chosen so that the coefficient of x^{2m} in $U_{2m+1}(x)$ vanishes.

(ii) By proceeding as in the proof of Proposition 6.1.8, and using the notation therein, show that the first of the multiple integrals is equal to

$$2^n \sum_{\substack{P\in S_{2n+1} \\ P(2l)>P(2l-1)}} \varepsilon(P) \prod_{l=1}^n \beta_{P(2l-1),P(2l)}|_{a=1} Q_{P(2n+1)-1}(x),$$

and show that this formula also holds for the second integral provided $Q_{2n}(x)$ is replaced by $Q_{2n+1}(x)$ throughout. From the skew orthogonality property of $\{Q_l(x)\}_{l=0,1,\dots}$ note that $\beta_{P(2l-1),P(2l)}|_{a=1}$ equals $q_l\delta_{P(2l),P(2l+1)+1}$ and thus derive the stated formulas.

4. [217] Let $\langle\cdot|\cdot\rangle_s$ denote a general skew symmetric inner product. Let $\{p_j(x)\}_{j=0,1,\dots}$ denote a general family of

monic polynomials in which $p_j(x)$ is of degree j. With $J^{mn} := \langle p_m | p_n \rangle_s$ let

$$\mathcal{D}_n = \det[J^{2n-j\,2n-k}]_{j,k=1,\dots,2n}, \qquad \mathcal{E}_n = -\det\Big[[J^{2n-j\,2n+1}]_{j=0,\dots,2n}\ [J^{2n-j\,2n-k}]_{\substack{j=0,\dots,2n\\k=1,\dots,2n}}\Big].$$

Verify that the monic polynomials of degree $2n$, $2n+1$,

$$U_{2n}(x) = \mathcal{D}_n^{-1}\det\Big[[C_{2n-j}(x)]_{j=0,\dots,2n}\ [J^{2n-j\,2n-k}]_{\substack{j=0,\dots,2n\\k=1,\dots,2n}}\Big],$$

$$U_{2n+1}(x) = \mathcal{E}_n^{-1}\det\Big[[J^{2n+1-j\,2n+1}\ C_{2n+1-j}(x)]_{j=0,\dots,2n+1}\ [J^{2n+1-j\,2n+1-k}]_{\substack{j=0,\dots,2n+1\\k=2,\dots,2n+1}}\Big] + \gamma_{2n}U_{2n+1}(x)$$

satisfy the skew orthogonality properties of Definition 6.1.1 by showing $\langle U_{2n}|p_j\rangle_s = 0$, $\langle U_{2n+1}|p_j\rangle_s = 0$ $(j = 0,\dots,2n-1)$.

6.2 CONSTRUCTION OF THE SKEW ORTHOGONAL POLYNOMIALS AT $\beta = 4$

6.2.1 Special properties of the classical cases

Polynomials skew orthogonal with respect to (6.3) can, for the classical weight functions of Section 5.4.1, be calculated explicitly [419]. Here we will follow a subsequent treatment [3] which emphasizes the relationship with particular orthogonal polynomials and also inter-relates the construction of the skew orthogonal polynomials at $\beta = 4$ with the construction at $\beta = 1$ to be undertaken below. The same general formulas apply to the Cauchy weight (5.51).

In the theory of [3], it is convenient to work with the inner product

$$\begin{aligned}\langle f|g\rangle_4' &:= \frac{1}{2}\int_{-\infty}^{\infty} e^{-2V(x)}(f(x)g'(x) - f'(x)g(x))dx \\ &= \frac{1}{2}\int_{-\infty}^{\infty} e^{-V(x)}\Big(f(x)\frac{d}{dx}(e^{-V(x)}g(x)) - g(x)\frac{d}{dx}(e^{-V(x)}f(x))\Big)\,dx,\end{aligned} \tag{6.39}$$

which differs from (6.3) by the replacement $e^{-4V(x)} \mapsto e^{-2V(x)}$. For the classical weight functions (5.56), this simply corresponds to rescaling x and/or the parameters a, b and α. It turns out that the operator A (5.60) relates the inner product (6.39) to the $\beta = 2$ inner product in (5.4).

PROPOSITION 6.2.1 *Assuming $e^{-2V(x)}$ vanishes at the endpoints of its support, we have*

$$(\phi, A\psi)_2 = \langle\phi|\psi\rangle_4'\Big|_{2V(x)\mapsto 2V(x)-\log f(x)}. \tag{6.40}$$

Proof. It follows from the definition of A (5.60) and the general formula

$$\frac{d}{dx}F(x) = \int_{-\infty}^{\infty}\delta'(x-y)F(y)\,dy$$

that we have

$$A\psi[x] = \sqrt{e^{2V(x)}f(x)}\int_{-\infty}^{\infty}\delta'(x-y)\sqrt{f(y)e^{-2V(y)}}\psi(y)\,dy.$$

This gives

$$\begin{aligned}(\phi, A\psi)_2 &= \int_{-\infty}^{\infty} dx \int_{-\infty}^{\infty} dy \sqrt{f(x)e^{-2V(x)}}\phi(x)\delta'(x-y)\sqrt{f(y)e^{-2V(y)}}\psi(y) \\ &= -\int_{-\infty}^{\infty} \sqrt{f(x)e^{-2V(x)}}\psi(x)\Big(\frac{\partial}{\partial x}\sqrt{f(x)e^{-2V(x)}}\phi(x)\Big)dx \\ &= -\frac{1}{2}\int_{-\infty}^{\infty} \sqrt{f(x)e^{-2V(x)}}\psi(x)\Big(\frac{\partial}{\partial x}\sqrt{f(x)e^{-2V(x)}}\phi(x)\Big)\,dx \\ &\quad +\frac{1}{2}\int_{-\infty}^{\infty} \sqrt{f(x)e^{-2V(x)}}\phi(x)\Big(\frac{\partial}{\partial x}\sqrt{f(x)e^{-2V(x)}}\psi(x)\Big)dx \\ &= \langle\phi|\psi\rangle_4'\Big|_{2V\mapsto 2V-\log f}.\end{aligned}$$

□

The result of Proposition 6.2.1 can be used in conjunction with the simple structure exhibited by (5.63) to specify the classical skew orthogonal polynomials at $\beta = 4$ in terms of their orthogonal polynomial counterparts. Let $\{\tilde{Q}_j\}$ denote the skew orthogonal polynomials corresponding to the inner product (6.39), modified by the replacement

$$2V(x) \mapsto 2\tilde{V}_4(x) := 2V(x) - \log f(x) \tag{6.41}$$

and write $\{\tilde{q}_l\}$ for the corresponding normalizations. Next introduce the lower triangular transition matrix $\mathbf{X}^{(4)}$ such that

$$[\tilde{Q}_j(x)]_{j=0,\dots,N-1} = \mathbf{X}^{(4)}[p_j(x)]_{j=0,\dots,N-1}, \qquad X_{jj}^{(4)} = 1. \tag{6.42}$$

Use of (6.40) then gives

$$\tilde{\mathbf{q}} := \Big[\langle\tilde{Q}_j, \tilde{Q}_k\rangle_4'\Big|_{2V(x)\mapsto 2V(x)-\log f(x)}\Big]_{j,k=0,\dots,N-1} = \mathbf{X}^{(4)}\mathbf{A}\mathbf{X}^{(4)\,T},$$

where $\mathbf{A}$ is specified by (5.63), or equivalently

$$\mathbf{X}^{(4)-1}\tilde{\mathbf{q}}(\mathbf{X}^{(4)\,T})^{-1} = \mathbf{A}. \tag{6.43}$$

The equation (6.43) can be solved for the matrix $\mathbf{X}^{(4)-1} := [\tilde{\beta}_{jk}^{(4)}]_{j,k=0,\dots,N-1}$, which is lower triangular with 1s along the diagonal. This can be done by multiplying both sides of (6.43) by $\mathbf{X}^{(4)\,T}$ on the right, and then equating the strictly lower triangular parts of both sides. On the l.h.s. we have

$$\Big(\mathbf{X}^{(4)-1}\tilde{\mathbf{q}}\Big)_- = \begin{bmatrix} * & & & & & \\ -\tilde{q}_0 & * & & & & \\ -\tilde{q}_0\tilde{\beta}_{21}^{(4)} & \tilde{q}_0\tilde{\beta}_{20}^{(4)} & * & & & \\ -\tilde{q}_1\tilde{\beta}_{31}^{(4)} & \tilde{q}_1\tilde{\beta}_{30}^{(4)} & -\tilde{q}_1 & * & & \\ -\tilde{q}_1\tilde{\beta}_{41}^{(4)} & \tilde{q}_1\tilde{\beta}_{40}^{(4)} & -\tilde{q}_1\tilde{\beta}_{43}^{(4)} & \tilde{q}_1\tilde{\beta}_{42}^{(4)} & * & \\ \vdots & \vdots & \vdots & \vdots & & \ddots \end{bmatrix}_{2N\times 2N}, \tag{6.44}$$

where the subscript in $(\,)_-$ denotes the strictly lower triangular entries, while on the r.h.s.

$$\Big(\mathbf{A}\mathbf{X}^{(4)T}\Big)_- = (\mathbf{A})_-. \tag{6.45}$$

Equating (6.44) and (6.45) gives

$$\begin{aligned} c_{2p} &= \tilde{q}_p, & p &= 0,\dots,N-1, \\ \tilde{\beta}^{(4)}_{2p+1,j} &= 0, & j &= 0,\dots,2p-1, \\ \tilde{\beta}^{(4)}_{2p,j} &= 0, & j &= 0,\dots,2p-1,\ j \ne 2p-2, \\ \tilde{\beta}^{(4)}_{2p,2p-2} &= -c_{2p-1}/\tilde{q}_{p-1} = -c_{2p-1}/c_{2p-2} \end{aligned} \tag{6.46}$$

while $\tilde{\beta}^{(4)}_{2p+1,2p}$ is undetermined. Thus we have

$$\begin{aligned} p_{2j+1}(x) &= \tilde{Q}_{2j+1}(x) + \tilde{\beta}^{(4)}_{2j+1,2j}\tilde{Q}_{2j}(x), \\ p_{2j}(x) &= \tilde{Q}_{2j}(x) - \frac{c_{2j-1}}{c_{2j-2}}\tilde{Q}_{2j-2}(x), \\ \tilde{q}_p &= c_{2p}. \end{aligned} \tag{6.47}$$

The nonuniqueness of the skew orthogonal polynomials up to the transformation (6.1) implies we can choose $\tilde{\beta}^{(4)}_{2j+1,2j} = 0$ (this gives the simplest results). Doing this, and solving for $\tilde{Q}_{2j}(x)$ in the second equation of (6.47) gives

$$\begin{aligned} \tilde{Q}_{2j+1}(x) &= p_{2j+1}(x), \\ \tilde{Q}_{2j}(x) &= \Big(\prod_{p=0}^{j-1}\frac{c_{2p+1}}{c_{2p}}\Big)\sum_{l=0}^{j}\prod_{p=0}^{l-1}\frac{c_{2p}}{c_{2p+1}}p_{2l}(x), \\ \tilde{q}_p &= c_{2p}. \end{aligned} \tag{6.48}$$

Furthermore, it is possible to write $\tilde{Q}_{2j}(x)$ in terms of an indefinite integral involving $p_{2j+1}(x)$.

PROPOSITION 6.2.2 *With the notation (6.41) we have in the classical cases*

$$e^{-\tilde{V}_4(x)}\tilde{Q}_{2m}(x) = -\frac{\tilde{q}_m}{(p_{2m+1},p_{2m+1})_2}\int_x^\infty e^{-2V(t)+\tilde{V}_4(t)}p_{2m+1}(t)\,dt. \tag{6.49}$$

Proof. We can check from the second formula in (6.39) that

$$\frac{d}{dy}\Big(e^{-\tilde{V}_4(y)}\tilde{Q}_{2m}(y)\Big) = e^{\tilde{V}_4(y)}\langle\delta(x-y)|\tilde{Q}_{2m}(x)\rangle_4'\Big|_{V\mapsto\tilde{V}_4}.$$

But the completeness of $\{p_j(x)\}$ shows

$$\delta(x-y) = e^{-2V(y)}\sum_{n=0}^{\infty}\frac{p_n(x)p_n(y)}{(p_n,p_n)_2}. \tag{6.50}$$

Furthermore substituting for $p_n(x)$ by using the expansion

$$p_n(x) = \sum_{j=0}^{n}\tilde{\beta}^{(4)}_{nj}\tilde{Q}_j(x), \quad \tilde{\beta}^{(4)}_{nn} = 1, \tag{6.51}$$

we conclude from the skew orthogonality of $\{\tilde{Q}_j\}$ that

$$\frac{d}{dy}\Big(e^{-\tilde{V}_4(y)}\tilde{Q}_{2m}(y)\Big) = -\tilde{q}_m e^{\tilde{V}_4(y)-2V(y)}\sum_{\nu=2m+1}^{\infty}\frac{p_\nu(y)}{(p_\nu,p_\nu)_2}\tilde{\beta}^{(4)}_{\nu,2m+1}. \tag{6.52}$$

The stated formula now follows after noting from (6.46) that the only nonzero value of $\tilde{\beta}^{(4)}_{\nu,2m+1}$ is when $\nu = 2m+1$, in which case $\tilde{\beta}^{(4)}_{\nu,2m+1} = 1$. □

Strictly speaking, the derivation of (6.50) given in the above proof does not apply to the Cauchy weight: according to (5.52) the set of orthogonal polynomials is then finite and so (6.50) cannot be justified. However an alternative proof of (6.50) which uses only (6.48) and (5.64), and so is applicable to the Cauchy case, can be given. This is done by taking (6.52) as given and verifying it as an identity. In the classical cases the r.h.s. consists of a single term only. On the l.h.s. one substitutes for $\tilde{Q}_{2m}(x)$ using (6.48) and computes the derivative using (5.64); simplification yields the same single term as resulted on the r.h.s.

Combining Proposition 6.2.2 and (6.47) we thus have that the monic skew orthogonal polynomials and corresponding normalization with respect to the inner product (6.39), and with the classical weight functions

$$e^{-2\tilde{V}_4(x)} = \begin{cases} e^{-x^2}, & \text{Hermite}, \\ x^{a+1}e^{-x}, & \text{Laguerre}, \\ (1-x)^{a+1}(1+x)^{b+1}, & \text{Jacobi}, \\ (1+x^2)^{-(\alpha-1)}, & \text{Cauchy} \end{cases} \tag{6.53}$$

(recall (5.56), (5.58) and (6.41)) are given in terms of the corresponding monic orthogonal polynomials of Section 5.4.1 according to

$$\begin{aligned} \tilde{Q}_{2j+1}(x) &= p_{2j+1}(x), \\ \tilde{Q}_{2j}(x) &= -\frac{c_{2j}e^{\tilde{V}_4(x)}}{(p_{2j+1}, p_{2j+1})_2} \int_x^\infty e^{-2V(t)+\tilde{V}_4(t)} p_{2j+1}(t)\, dt, \\ \tilde{q}_j &= c_{2j}, \end{aligned} \tag{6.54}$$

where c_{2j} is specified by (5.65).

6.2.2 Summation formula

From Propositions 6.1.6 and 6.1.7 we have

$$\rho_{(n)}(x_1, \dots, x_n) = \text{qdet} \begin{bmatrix} S_4(x_j, x_k) & -\int_{x_j}^{x_k} S_4(x_j, y)\, dy \\ \frac{\partial}{\partial x_j} S_4(x_j, x_k) & S_4(x_k, x_j) \end{bmatrix}_{j,k=1,\dots,n}, \tag{6.55}$$

where S_4 is specified in (6.15). It is possible to rewrite S_4 in a form which separates the $\beta = 2$ correlation kernel (5.6).

PROPOSITION 6.2.3 *With $\tilde{S}_4(x,y)$ defined by $S_4(x,y)$ in (6.15) but modified so that $V \mapsto \frac{1}{2}\tilde{V}_4$ as specified by (6.41), $Q_k(x) \mapsto \tilde{Q}_k(x)$, and $\{p_j(x)\}_{j=0,1,\dots}$ the set (assumed complete) of monic orthogonal polynomials associated with the weight function $e^{-2V(x)}$ we have*

$$\tilde{S}_4(x,y) = \frac{1}{2} e^{-2V(y)-\tilde{V}_4(x)+\tilde{V}_4(y)} \Big(\sum_{n=0}^{2N-1} \frac{p_n(x)p_n(y)}{(p_n, p_n)_2} + \sum_{n=2N}^{\infty} \sum_{k=0}^{2N-1} \frac{p_n(y)}{(p_n, p_n)_2} \tilde{\beta}_{nk}^{(4)} \tilde{Q}_k(x) \Big). \tag{6.56}$$

Proof. The key to this summation is the formula (6.52) and its counterpart

$$\frac{d}{dx}\Big(e^{-\tilde{V}_4(x)}\tilde{Q}_{2m+1}(x)\Big) = \tilde{q}_m e^{\tilde{V}_4(x)-2V(x)} \sum_{\nu=2m}^{\infty} \frac{p_\nu(x)}{(p_\nu, p_\nu)_2} \tilde{\beta}_{\nu,2m}^{(4)}, \tag{6.57}$$

which is derived in an analogous manner. Substituting in the first equation of (6.15) (modified as specified) and performing minor manipulation gives (6.56). □

The first sum in (6.56) is evaluated by the Christoffel-Darboux formula (5.10), so it remains to simplify the second sum. Although this is complicated in general, in the classical cases simplification occurs. Thus choosing $\tilde{\beta}^{(4)}_{2j+1,2j} = 0$, as done in obtaining (6.48), we see from (6.46) that in the classical cases the only nonzero value of $\tilde{\beta}^{(4)}_{nk}$ for $n > k$ is

$$\tilde{\beta}^{(4)}_{2n,2n-2} = -\frac{c_{2n-1}}{c_{2n}},$$

and thus

$$\begin{aligned}\tilde{S}_4(x,y) = {} & \frac{1}{2} e^{-(V(y)-\tilde{V}_4(y))} e^{V(x)-\tilde{V}_4(x)} K_{2N}(x,y) \\ & + \frac{1}{2} e^{-\tilde{V}_4(x)+\tilde{V}_4(y)-2V(y)} \frac{c_{2N-1}}{c_{2N}} \frac{p_{2N}(y)}{(p_{2N},p_{2N})_2} \tilde{Q}_{2N-2}(x).\end{aligned} \tag{6.58}$$

Substituting for $\tilde{Q}_{2N-2}(x)$ according to (6.54) eliminates all reference to the skew orthogonal polynomials in the final summation formula [3].

PROPOSITION 6.2.4 *In the classical cases*

$$\begin{aligned}\tilde{S}_4(x,y) = {} & \frac{1}{2} e^{-(V(y)-\tilde{V}_4(y))} e^{V(x)-\tilde{V}_4(x)} K_{2N}(x,y) \\ & - \frac{e^{\tilde{V}_4(y)-2V(y)} c_{2N-1} p_{2N}(y)}{2(p_{2N},p_{2N})_2 (p_{2N-1},p_{2N-1})_2} \int_x^\infty e^{-2V(t)+\tilde{V}_4(t)} p_{2N-1}(t)\, dt.\end{aligned} \tag{6.59}$$

EXERCISES 6.2 1. Use the first equation in Proposition 6.1.7 and the definitions (4.139), (4.142) and (4.143) to note that

$$\frac{G_{4,N}}{G_{4,N-1}} = 2Nq^{(\mathrm{G})}_{N-1}, \quad \frac{W_{a,4,N}}{W_{a,4,N-1}} = 2Nq^{(\mathrm{L})}_{N-1}, \quad \frac{C_{a,b,4,N}}{C_{a,b,4,N-1}} = 2Nq^{(\mathrm{J})}_{N-1}, \quad \frac{\mathcal{N}^{(\mathrm{Cy})}_{\alpha,4,N}}{\mathcal{N}^{(\mathrm{Cy})}_{\alpha,4,N-1}} = 2Nq^{(\mathrm{Cy})}_{N-1}$$

for the Gaussian, Laguerre, Jacobi and Cauchy ensembles, respectively. From the explicit evaluations Proposition 4.7.1 and 4.7.3, and (4.144), (4.145), verify the corresponding formulas for q_n are consistent with (6.54).

2. [522] In this exercise the analogue of (5.140) for $S_4(x,y)$ will be derived.

   (i) In the formula (6.15) for S_4 write $e^{-2V(x)} Q_p(x) = \phi_p(x)$, and then put

$$\phi_p(x) = \sum_{j=0}^{2N-1} \alpha_{pj} \psi_j(x), \qquad \psi_j(x) = e^{-2V(x)} u_j(x),$$

   where the span of $\{u_j(x)\}_{j=0,1,\dots,2N-1}$ is equal to the space of polynomials of degree $\le 2N-1$. Show that

$$S_4(x,y) = \sum_{l_1=0}^{2N-1} \sum_{l_2=0}^{2N-1} \mu^{(4)}_{l_1 l_2} \psi_{l_1}(x) \psi'_{l_2}(y),$$

$$\mu^{(4)}_{l_1 l_2} := \sum_{m=0}^{N-1} \frac{1}{2q_m} (\alpha_{2m\, l_1} \alpha_{2m+1\, l_2} - \alpha_{2m\, l_2} \alpha_{2m+1\, l_1}).$$

   (ii) Writing $\alpha = [\alpha_{jk}]_{j,k=0,\dots,2N-1}$ and $\mathbf{M} = [\langle u_j | u_k \rangle_4]_{j,k=0,\dots,2N-1}$, check that

$$\alpha^T \mathbf{M} \alpha = \mathrm{diag}\Big(\begin{bmatrix} 0 & q_0 \\ -q_0 & 0 \end{bmatrix}, \dots, \begin{bmatrix} 0 & q_{N-1} \\ -q_{N-1} & 0 \end{bmatrix} \Big) \mathbf{M} \vec{\alpha}_{2k+1} = -q_k \vec{\alpha}_{2k},$$

   and from this show

$$[\mu^{(4)}_{l_1,l_2}]_{l_1,l_2=0,\dots,2N-1} = \mathbf{M}^{-1}.$$

3. [544] In this exercise S_4 will be related to K_N from $\beta = 2$ theory. Throughout we consider the original $\beta = 4$ problem modified so that $V(x) \mapsto V(x)/2$, $N \mapsto N/2$, and we denote $\bar{S}_4(x,y) = S_4(y,x)|_{\substack{N \mapsto N/2 \\ V \mapsto V/2}}$.

   (i) Define $K_N(x,y)$ as in (5.6) and set $w_2(x) = e^{-2V(x)}$. With $\bar{S}_4$, K denoting the integral operators with kernels $\bar{S}_4(x,y)$, $K_N(x,y)$, show using the form of $\bar{S}_4(x,y)$ implied by q.2(i) above that $\bar{S}_4 K = \bar{S}_4$.

   (ii) With $\mathbf{M}$ defined as in q.2(ii), and assuming $\{\psi_j(x)\}$ vanishes at the endpoints of the support of $e^{-2V(x)}$, show $m_{jk} = \int_{-\infty}^{\infty} \psi_j(x)\psi_k'(x)\,dx$, where here $\psi_j(x) := e^{-V(x)}u_j(x)$ (since $V(x) \mapsto V(x)/2$). Use this and the result of (i) to show $\bar{S}_4 K \psi_j' = \psi_j'$ and thus conclude that for $\bar{S}_4$ restricted to functions in $\mathcal{H} = \mathrm{Span}\{\psi_i\}_{i=0,\dots,2N-1}$ we have $\bar{S}_4|_{\mathcal{H}} = D(KD_{\mathcal{H}})^{-1}$, where D denotes differentiation, and $D_{\mathcal{H}}$ differentiation restricted to $\mathcal{H}$. Note also that $\bar{S}_4|_{\mathcal{H}^\perp} = 0$, where $\mathcal{H}^\perp$ denotes the orthogonal complement of $\mathcal{H}$ with respect to the inner product $(f,g) := \int_{-\infty}^{\infty} f(x)g(x)\,dx$.

   (iii) Let $\epsilon(x) = \frac{1}{2}\mathrm{sgn}(x)$, and denote by ϵ the integral operator with kernel $\epsilon(x-y)$. Verify that when restricted to functions in $\mathcal{H}$,

$$(KD_{\mathcal{H}})^{-1} = (I_{\mathcal{H}} - K\epsilon(I-K)D_{\mathcal{H}})^{-1}K\epsilon_{\mathcal{H}}.$$

   Substitute this in the formula obtained in (ii) for $\bar{S}_4|_{\mathcal{H}}$ to deduce

$$\bar{S}_4|_{\mathcal{H}} = I_{\mathcal{H}} + A_{\mathcal{H}}(I_{\mathcal{H}} - BA)^{-1}B|_{\mathcal{H}} = (I_{\mathcal{H}+D\mathcal{H}} - AB)^{-1}|_{\mathcal{H}}$$

   where $A = (I-K)D$, $B = K\epsilon$. Conclude that in general

$$\bar{S}_4 = (I_{\mathcal{H}+D\mathcal{H}} - (I-K)DK\epsilon)^{-1}K. \tag{6.60}$$

6.3 CORRELATION FUNCTIONS AT $\beta = 1$

The identity necessary to compute the correlations at $\beta = 1$ is more difficult to derive than the corresponding identities at $\beta = 2$ and 4, although the general strategy is to adopt an approach analogous to that used for $\beta = 4$. The starting point is to introduce appropriate skew orthogonal polynomials. We require the function $\mathrm{sgn}(x)$ where $\mathrm{sgn}(x) = 1$ $(x > 0)$, $\mathrm{sgn}(x) = -1$ $(x < 0)$ and $\mathrm{sgn}(0) = 0$.

DEFINITION 6.3.1 *Define a skew symmetric inner product $\langle\cdot|\cdot\rangle_1$ by*

$$\langle f|g\rangle_1 := \frac{1}{2}\int_{-\infty}^{\infty} dx\, e^{-V(x)}f(x)\int_{-\infty}^{\infty} dy\, e^{-V(y)}g(y)\mathrm{sgn}(y-x), \tag{6.61}$$

and let $\{R_n(x)\}_{n=0,1,\dots}$ be a corresponding family of monic skew orthogonal polynomials so that

$$\langle R_{2m}|R_{2n+1}\rangle_1 = -\langle R_{2n+1}|R_{2m}\rangle_1 = r_n\delta_{mn}, \quad \langle R_{2m}|R_{2n}\rangle_1 = \langle R_{2m+1}|R_{2n+1}\rangle_1 = 0.$$

The Boltzmann factor of the log-gas on a line at $\beta = 1$ can be expressed [381] as a quaternion determinant involving these polynomials, which in turn can be used to compute the n-particle correlation. The N even and N odd cases must be treated separately.

6.3.1 N even

Use will be made of the identity [123]

$$\prod_{1 \le j < k \le 2n} \mathrm{sgn}(x_k - x_j) = \mathrm{Pf}[\mathrm{sgn}(x_k - x_j)]_{j,k=1,\dots,2n}. \tag{6.62}$$

To see the validity of (6.62), we note from (6.10) that

$$\mathrm{Pf}[\mathrm{sgn}(x_k - x_j)]_{j,k=1,\dots,2n} = \sum\nolimits^{*}_{P(2l)>P(2l-1)} \varepsilon(P)\prod_{l=1}^{n}\mathrm{sgn}(x_{P(2l)} - x_{P(2l-1)}). \tag{6.63}$$

In the case that $x_1 < x_2 < \cdots < x_{2n}$ all terms in the product are positive and so the r.h.s. reads

$$\sum\nolimits^{*}_{P(2l)>P(2l-1)} \varepsilon(P).$$

But there are $(2n-1)!!$ terms in this sum, which can be constructed systematically from the identity permutation by successive elementary transpositions, each of which reverses the sign of the signature. Since $(2n-1)!!$ is odd the sum then adds to 1, verifying (6.62) in this case. For other orderings we note that the l.h.s. of (6.62) changes sign if x_i and x_{i+1} are interchanged. On the r.h.s. this interchange is equivalent to interchanging rows i and $i+1$, then columns i and $i+1$, which we know from (6.34) also changes the sign.

PROPOSITION 6.3.2 *Let N be even and suppose*

$$\Phi_k(x), \quad r_k\,(>0) \quad \text{and} \quad \mathbf{H} = [h(x_j,x_k)]_{j,k=1,\dots,N} \ (h(x_k,x_j) = -h(x_j,x_k))$$

are arbitrary, and let $R_k(x)$ be such that

$$\det[R_k(x_j)]_{j,k=1,\dots,N} = \prod_{1\le j<k\le N} (x_k - x_j) \tag{6.64}$$

(here the skew orthogonality property (6.61) is not assumed). Set

$$\phi_k(x) := \begin{bmatrix} \Phi_{2k}(x) & \Phi_{2k+1}(x) \\ R_{2k}(x)e^{-V(x)} & R_{2k+1}(x)e^{-V(x)} \end{bmatrix}.$$

We have

$$\prod_{j=1}^{N} e^{-V(x_j)} \prod_{1\le j<k\le N} (x_k - x_j) \operatorname{Pf}\mathbf{H} = \prod_{p=0}^{N/2-1} r_p \operatorname{qdet}[\mathbf{f}_1(x_j,x_k)]_{j,k=1,\dots,N}, \tag{6.65}$$

where $[\mathbf{f}_1(x_j,x_k)]$ is a self-dual quaternion matrix specified by

$$\mathbf{f}_1(x,y) := \sum_{k=0}^{N/2-1} \frac{1}{r_k}\phi_k(x)\phi_k^D(y) + \begin{bmatrix} 0 & h(x,y) \\ 0 & 0 \end{bmatrix} = \begin{bmatrix} S_1(x,y) & \tilde{I}_1(x,y) \\ D_1(x,y) & S_1(y,x) \end{bmatrix}$$

with

$$S_1(x,y) = \sum_{k=0}^{N/2-1} \frac{e^{-V(y)}}{r_k}\Big(\Phi_{2k}(x)R_{2k+1}(y) - \Phi_{2k+1}(x)R_{2k}(y)\Big),$$

$$D_1(x,y) = \sum_{k=0}^{N/2-1} \frac{e^{-(V(x)+V(y))}}{r_k}\Big(R_{2k}(x)R_{2k+1}(y) - R_{2k+1}(x)R_{2k}(y)\Big),$$

$$\tilde{I}_1(x,y) = \sum_{k=0}^{N/2-1} \frac{1}{r_k}\Big(\Phi_{2k+1}(x)\Phi_{2k}(y) - \Phi_{2k}(x)\Phi_{2k+1}(y)\Big) + h(x,y)$$
$$=: I_1(x,y) + h(x,y).$$

Proof. We use a method first introduced in the more general parameter-dependent problem [417]. Let

$$\mathbf{D}_1 := [D_1(x_j,x_k)]_{j,k=1,\dots,N}, \quad \mathbf{I}_1 := [I_1(x_j,x_k)]_{j,k=1,\dots,N}, \quad \mathbf{S}_1 := [S_1(x_j,x_k)]_{j,k=1,\dots,N}.$$

With this notation, we begin by noting from (6.64) that

$$\prod_{l=1}^{N} e^{-2V(x_l)} \prod_{1 \le j < k \le N} (x_k - x_j)^2 = \prod_{l=1}^{N} e^{-2V(x_l)} \det[R_{k-1}(x_j)]_{j,k=1,\dots,N} \det \begin{bmatrix} R_{2j-1}(x_k) \\ -R_{2j-2}(x_k) \end{bmatrix}_{\substack{j=1,\dots,N/2 \\ k=1,\dots,N}}$$
$$= \prod_{j=1}^{N/2} r_{j-1}^2 \det \mathbf{D}_1.$$

Since $\mathbf{D}_1$ is antisymmetric we can make use of (6.12) to conclude

$$\prod_{j=1}^{N} e^{-V(x_j)} \prod_{j<k} (x_k - x_j) = \prod_{j=1}^{N/2} r_{j-1} \mathrm{Pf}\, \mathbf{D}_1$$

(there is a choice of signs in taking the square root; that the correct sign has been taken can be checked by considering the construction of the term $x_N^{N-1} x_{N-1}^{N-2} \cdots x_2 x_1^0$ on the r.h.s.). It follows from this that

$$\prod_{j=1}^{N} e^{-V(x_j)} \prod_{j<k} (x_k - x_j) \mathrm{Pf}\, \mathbf{H} = (-1)^{N/2} \prod_{j=1}^{N/2} r_{j-1} \mathrm{Pf} \begin{bmatrix} \mathbf{D}_1 & \mathbf{0} \\ \mathbf{0} & -\mathbf{H} \end{bmatrix}. \tag{6.66}$$

To introduce the matrices $\mathbf{S}_1$ and $\mathbf{I}_1$ into (6.66) we introduce the nonunique (x-dependent) matrix $\alpha = [\alpha_{ij}]_{i,j=1,\dots,N}$ such that

$$\Phi_k(x_j) = -\sum_{i=1}^{N} \alpha_{ji} e^{-V(x_i)} R_k(x_i).$$

It then follows from the definitions that

$$\alpha \mathbf{D}_1 = -\mathbf{S}_1, \qquad \alpha \mathbf{D}_1 \alpha^T = -\mathbf{I}_1.$$

We make use of these relations by adding to the second block row of the matrix on the r.h.s. of (6.66) the first block row multiplied by α on the left, and adding to the new second block column the first block column multiplied by α^T on the right. This leaves the value of the Pfaffian unchanged and gives

$$\prod_{j=1}^{N} e^{-V(x_j)} \prod_{j<k} (x_k - x_j) \mathrm{Pf}\, \mathbf{H} = (-1)^{N/2} \prod_{j=1}^{N/2} r_{j-1} \mathrm{Pf} \begin{bmatrix} \mathbf{D}_1 & \mathbf{S}_1^T \\ -\mathbf{S}_1 & -\tilde{\mathbf{I}}_1 \end{bmatrix}.$$

Interchanging the first block row with the second and using (6.11) gives the sought result. □

The identity (6.62) implies that with

$$h(x,y) = \frac{1}{2} \mathrm{sgn}(y - x) \tag{6.67}$$

(the factor of $\frac{1}{2}$ is for later convenience), (6.65) reads

$$\prod_{j=1}^{N} e^{-V(x_j)} \prod_{j<k} |x_k - x_j| = \prod_{p=0}^{N/2-1} 2 r_p \, \mathrm{qdet}[\mathbf{f}_1(x_j, x_k)]_{j,k=1,\dots,N}. \tag{6.68}$$

The use of (6.68) is that with the $R_k(x)$ chosen to have the skew orthogonality property of Definition 6.3.1, the r_k chosen as the corresponding normalization and

$$\Phi_k(x) := \int_{-\infty}^{\infty} h(y,x) R_k(y) e^{-V(y)}\, dy, \tag{6.69}$$

the quantity $\mathbf{f}_1$ in Proposition 6.3.2 has integration properties analogous to those of $\mathbf{f}_4$ given in the previous section. One finds [381]

$$\int_{-\infty}^{\infty} \mathbf{f}_1(x,y)\mathbf{f}_1(y,z)dy = \begin{bmatrix} 0 & 0 \\ 0 & 1 \end{bmatrix} \mathbf{f}_1(x,z) + \mathbf{f}_1(x,z) \begin{bmatrix} 1 & 0 \\ 0 & 0 \end{bmatrix},$$
$$\int_{-\infty}^{\infty} \mathbf{f}_1(x,x)\,dx = N \begin{bmatrix} 1 & 0 \\ 0 & 1 \end{bmatrix} \tag{6.70}$$

(cf. (6.17)). These formulas, together with the method of the proof of Proposition 6.1.7, allow the integrations required to compute the n-particle correlations to be performed.

PROPOSITION 6.3.3 *Let $\{R_n(x)\}_{n=0,1,\dots}$, $\{r_n\}_{n=0,1,\dots}$ be as in Definition 6.3.1, $h(x,y)$ as in (6.67) and $\Phi_k(x)$ as in (6.69). In terms of these quantities specify $\mathbf{f}_1$ as in Proposition 6.3.2. Then for N even we have*

$$\int_{-\infty}^{\infty} dx_1 \cdots \int_{-\infty}^{\infty} dx_N \prod_{j=1}^{N} e^{-V(x_j)} \prod_{1\le j<k\le N} |x_k - x_j| = N! \prod_{k=0}^{N/2-1} 2r_k,$$
$$\rho_{(n)}(x_1,\dots,x_n) = \mathrm{qdet}[\mathbf{f}_1(x_j,x_k)]_{j,k=1,\dots,n},$$
$$\rho_{(n)}^T(x_1,\dots,x_n) = (-1)^{n-1} \sum_{\substack{\text{cycles} \\ \text{length } n}} \Big(\prod_{(i,i')} \mathbf{f}_1(x_i,x_{i'})\Big)^{(0)}.$$

Furthermore,

$$D_1(x,y) = \frac{\partial}{\partial x} S_1(x,y), \qquad I_1(x,y) = \frac{1}{2}\int_{-\infty}^{\infty} S_1(x,z)\mathrm{sgn}(z-y)dz = -\int_x^y S_1(x,z)\,dz.$$

Proof. We see that all equations in the proof of Proposition 6.1.7 up to (6.22) remain valid. For the replacement of (6.22), note that integrating $\mathbf{A}_j$, as defined below (6.21), over x_m gives

$$\mathbf{f}_1(x_\alpha,x_\beta)\cdots\begin{bmatrix} 0 & 0 \\ 0 & 1 \end{bmatrix}\mathbf{f}_1(x_\delta,x_\gamma)\cdots\mathbf{f}_1(x_\tau,x_\alpha) + \mathbf{f}_1(x_\alpha,x_\beta)\cdots\mathbf{f}_1(x_\delta,x_\gamma)\begin{bmatrix} 1 & 0 \\ 0 & 0 \end{bmatrix}\cdots\mathbf{f}_1(x_\tau,x_\alpha).$$

If the cycle $\mathbf{A}_j$ is of length p, then summing this expression over the p possible ways of inserting x_m into the cycle gives

$$\begin{bmatrix} 0 & 0 \\ 0 & 1 \end{bmatrix}\mathbf{A}_j + \mathbf{A}_j\begin{bmatrix} 1 & 0 \\ 0 & 0 \end{bmatrix} + (p-1)\mathbf{A}_j.$$

Now, in the definition (6.4) of qdet, only the scalar part of each cycle is required. Since in general Tr(**BC**) = Tr(**CB**), we see that the scalar part of the above expression is the same as the scalar part of $p\mathbf{A}_j$. Thus (6.22) in the proof of Proposition 6.1.7 remains valid, provided we take the scalar part. Consequently (6.23) again holds, which implies the stated results.

The formulas for D_1 and I_1 are a consequence of the definitions in Proposition 6.3.2 together with the explicit formulas (6.67) and (6.69). □

6.3.2 Integration over alternate variables

As with the corresponding results at $\beta = 4$, historically the normalization and one- and two-particle correlation functions in Proposition 6.3.3 were first evaluated in a way that did not make explicit use of quaternion determinants. Instead, use was make of functional differentiation, and an expression for the generalized partition function in terms of a Pfaffian obtained via the *method of integration over alternate variables* [392].

PROPOSITION 6.3.4 *Let N be even. For Ω equal to the real line and*

$$e^{-\beta U(x_1,\dots,x_N)} = \prod_{l=1}^{N} e^{-V(x_l)} \prod_{1\le j<k\le N} |x_k - x_j| \tag{6.71}$$

in (5.19) we have

$$\hat{Z}_N[a] = N!2^{N/2}\mathrm{Pf}[\gamma_{jk}]_{j,k=1,\dots,N} = N!2^{N/2}\Big(\det[\gamma_{jk}]_{j,k=1,\dots,N}\Big)^{1/2},$$
$$\gamma_{jk} := \frac{1}{2}\int_{-\infty}^{\infty} dx\, e^{-V(x)}a(x)R_{j-1}(x)\int_{-\infty}^{\infty} dy\, e^{-V(y)}a(y)R_{k-1}(y)\mathrm{sgn}(y-x). \tag{6.72}$$

Proof. Ordering the integration variables $-\infty < x_1 < \cdots < x_N < \infty$ gives

$$\begin{aligned}\hat{Z}_N[a] &= N!\int_{-\infty<x_1<\cdots<x_N<\infty} dx_1\cdots dx_N \prod_{l=1}^N e^{-V(x_l)}a(x_l)\prod_{1\le j<k\le N}(x_k-x_j)\\ &= N!\int_{-\infty<x_1<\cdots<x_N<\infty} dx_1\cdots dx_N \det[e^{-V(x_j)}a(x_j)R_{k-1}(x_j)]_{j,k=1,\dots,N},\end{aligned}$$

where to obtain the second equality use has been made of the Vandermonde identity (1.173) modified to read $\prod_{1\le j<k\le N}(x_k-x_j) = \det[R_{k-1}(x_j)]_{j,k=1,\dots,N}$. The integration over x_1 can be done by integrating the elements in the first row, thereby replacing these elements by $\int_{-\infty}^{x_2} e^{-V(x)}a(x)R_{k-1}(x)\,dx$. The first and second rows now depend on x_2, so we next perform the integration over x_3 in the third row. Since $x_2 < x_3 < x_4$ this gives $\int_{x_2}^{x_4} e^{-V(x)}a(x)R_{k-1}(x)\,dx$ for the elements of the third row. By adding the first row, these elements become $\int_{-\infty}^{x_4} e^{-V(x)}a(x)R_{k-1}(x)\,dx$, and so have the same form as the first row except they are functions of x_4. Proceeding similarly, we see after integrating over all the coordinates with odd labels that

$$\hat{Z}_N[a] = N!\int_{-\infty<x_2<x_4<\cdots<x_N<\infty} dx_2dx_4\cdots dx_N \det\begin{bmatrix}\int_{-\infty}^{x_{2j}} e^{-V(x)}a(x)R_{k-1}(x)\,dx\\ e^{-V(x_{2j})}a(x_{2j})R_{k-1}(x_{2j})\end{bmatrix}_{\substack{j=1,\dots,N/2\\ k=1,\dots,N}}.$$

The determinant is now symmetrical in $x_2, x_4, \dots, x_N$, so we can replace the domain of integration by the entire real line in each of the variables, provided we divide by $(N/2)!$. Expanding out the determinant and integrating over each x_{2l} gives

$$\hat{Z}_N[a] = \frac{N!}{(N/2)!}\sum_{P\in S_N}\varepsilon(P)\prod_{l=1}^{N/2}\mu_{P(2l-1),P(2l)},$$

where

$$\mu_{j,k} := \int_{-\infty}^{\infty} dx\, e^{-V(x)}a(x)R_{k-1}(x)\int_{-\infty}^{x} dy\, e^{-V(y)}a(y)R_{j-1}(y).$$

As in the proof of Proposition 6.1.8, if we restrict $P(2l) > P(2l-1)$, we can generate all other terms in the sum over permutations by replacing each $\mu_{P(2l-1),P(2l)}$ by

$$\gamma_{P(2l-1),P(2l)} := \mu_{P(2l-1),P(2l)} - \mu_{P(2l),P(2l-1)}.$$

Thus

$$\hat{Z}_N[a] = \frac{N!}{(N/2)!}\sum_{P(2l)>P(2l-1)}\varepsilon(P)\prod_{l=1}^{N/2}\gamma_{P(2l-1),P(2l)},$$

which after comparison with the definition (6.10) of a Pfaffian implies the stated result. □

An alternative way to derive (6.72) is to make use of (6.62) and (6.64). In fact as with Proposition 6.3.2 there is a general integration formula [123] for the product

$$\mathrm{Pf}[h(x_j,x_k)]_{j,k=1,\dots,2n}\det[R_{j-1}(x_k)]_{j,k=1,\dots,2n}.$$

PROPOSITION 6.3.5 *Let N be even. We have*

$$\int_{-\infty<x_1<\cdots<x_N<\infty} dx_1\cdots dx_N \prod_{l=1}^{N} e^{-V(x_l)}a(x_l)\,\det[R_{k-1}(x_j)]_{j,k=1}^N \mathrm{Pf}[h(x_j,x_k)]_{j,k=1}^N$$
$$= \frac{1}{N!}\int_{-\infty}^{\infty} dx_1\, e^{-V(x_1)}a(x_1)\cdots \int_{-\infty}^{\infty} dx_N\, e^{-V(x_N)}a(x_N)\,\det[R_{k-1}(x_j)]_{j,k=1}^N \mathrm{Pf}[h(x_j,x_k)]_{j,k=1}^N$$
$$= \mathrm{Pf}\Big[\int_{-\infty}^{\infty} dx\, e^{-V(x)}a(x)\int_{-\infty}^{\infty} dy\, e^{-V(y)}a(y)R_{j-1}(x)h(x,y)R_{k-1}(y)\Big]_{j,k=1}^{N}. \tag{6.73}$$

Proof. The integrand is symmetric in the xs, so the second multiple integral is equal to $N!$ times the first. To derive the final equality note that since both the determinant and Pfaffian are antisymmetric in the xs, we can write the second line as

$$\int_{-\infty}^{\infty} dx_1\, e^{-V(x_1)}a(x_1)\cdots \int_{-\infty}^{\infty} dx_N\, e^{-V(x_N)}a(x_N)\prod_{j=1}^{N} R_{j-1}(x_j)\mathrm{Pf}[h(x_j,x_k)]_{j,k=1,\dots,N}. \tag{6.74}$$

Recalling (6.10), this can be rewritten

$$\sum_{P(2l)>P(2l-1)} \varepsilon(P)\int_{-\infty}^{\infty} dx_1\, e^{-V(x_1)}a(x_1)\cdots \int_{-\infty}^{\infty} dx_N\, e^{-V(x_N)}a(x_N)\prod_{j=1}^{N} R_{j-1}(x_j)\prod_{l=1}^{N/2} h(x_{P(2l-1)},x_{P(2l)}). \tag{6.75}$$

But $\prod_{j=1}^N R_{j-1}(x_j) = \prod_{j=1}^N R_{P(j)-1}(x_{P(j)})$ so (6.75) is equal to

$$\sum_{P(2l)>P(2l-1)} \varepsilon(P)\int_{-\infty}^{\infty} dx_1\, e^{-V(x_1)}a(x_1)\cdots \int_{-\infty}^{\infty} dx_N\, e^{-V(x_N)}a(x_N)$$
$$\times\prod_{l=1}^{N/2} R_{P(2l-1)-1}(x_{P(2l-1)})h(x_{P(2l-1)},x_{P(2l)})R_{P(2l)-1}(x_{P(2l)}),$$

and upon noting, by changing the order of integration, that

$$\int_{-\infty}^{\infty} dx_1\, e^{-V(x_1)}a(x_1)\cdots \int_{-\infty}^{\infty} dx_N\, e^{-V(x_N)}a(x_N)$$
$$= \prod_{l=1}^{N/2}\Big(\int_{-\infty}^{\infty} dx_{P(2l-1)}\, e^{-V(x_{P(2l-1)})}a(x_{2l-1})\int_{-\infty}^{\infty} dx_{P(2l)}\, e^{-V(x_{P(2l)})}a(x_{2l})\Big),$$

this in turn reduces to

$$\sum_{P(2l)>P(2l-1)} \varepsilon(P)\prod_{l=1}^{N/2}\int_{-\infty}^{\infty} dx\, e^{-V(x)}a(x)\int_{-\infty}^{\infty} dy\, e^{-V(y)}a(y)R_{P(2l-1)-1}(x)h(x,y)R_{P(2l)-1}(y),$$

which from (6.10) we recognize as the r.h.s. of (6.73). □

As with the $\beta=4$ theory of Section 6.1.2 it is possible to obtain from (6.72) the n-point correlation of Proposition 6.3.3 [522].

PROPOSITION 6.3.6 *The expression (6.72) for the generalized partition function can be rewritten as*

$$\hat{Z}_N[a] = N!2^{N/2}\prod_{k=0}^{N/2-1} r_k\Big(\det[\mathbf{1}_2 + \mathbf{f}_1^T(\mathbf{a}-\mathbf{1}_2)]\Big)^{1/2}, \tag{6.76}$$

where $\mathbf{f}_1^T(\mathbf{a}-\mathbf{1}_2)$ is the matrix integral operator with kernel $\mathbf{f}_1^T(x,y)\mathrm{diag}[a(y)-1,a(y)-1]$.

Proof. In the definition of γ_{jk} write

$$a := f + 1, \quad \psi_j(x) := e^{-V(x)} R_{j-1}(x), \tag{6.77}$$

and denote by ϵ the integral operator with kernel $\frac{1}{2}\mathrm{sgn}\,(x-y)$ to obtain

$$\gamma_{jk} = \gamma_{jk}\Big|_{a=1} - \int_{-\infty}^{\infty} \Big(f(x)\psi_j(x)\epsilon\psi_k[x] - f(x)\psi_k(x)\epsilon\psi_j[x] - f(x)\psi_k(x)\epsilon(f\psi_j)[x] \Big)\, dx.$$

Because of the skew orthogonality of $\{R_j(x)\}_{j=0,1,\dots}$, $[\gamma_{jk}]|_{a=1}$ is skew symmetric with the structure of (6.2) (except that $N \mapsto N/2$). With

$$G_{2j-1}(x) := \psi_{2j}(x), \qquad G_{2j}(x) := -\psi_{2j-1}(x) \tag{6.78}$$

we see that factoring out $[\gamma_{jk}]|_{a=1}$, multiplying the even rows by -1, and interchanging odd and even rows gives

$$\det[\gamma_{jk}]_{j,k=0,\dots,N-1} = \prod_{j=1}^{N/2} r_{j-1}^2 \det\Big[\delta_{j,k} + \frac{1}{r_{[(j-1)/2]}} \\ \times \int_{-\infty}^{\infty} \Big(f(x)G_j(x)\epsilon\psi_k[x] - f(x)\psi_k(x)\epsilon G_j[x] - f(x)\psi_k(x)\epsilon(fG_j)[x] \Big)\, dx\Big]_{j,k=1,\dots,N}. \tag{6.79}$$

Now the determinant in (6.79) can be written as $\det[\mathbf{1}_N + \mathbf{AB}]$, where $\mathbf{A}$ is the $N \times 2$ matrix valued integral operator on $(-\infty, \infty)$ with kernel

$$\Big[-f(y)\epsilon G_j[y] - f(y)\epsilon(fG_j)[y] \qquad f(y)G_j[y] \Big]_{j=1,\dots,N},$$

while $\mathbf{B}$ is the operator which multiplies by the $2 \times N$ matrix

$$\begin{bmatrix} \psi_k(y) \\ \epsilon\psi_k[y] \end{bmatrix}_{k=1,\dots,N}.$$

To apply (5.26), we note that with these definitions of $\mathbf{A}$ and $\mathbf{B}$, the operator $\mathbf{1} + \mathbf{BA}$ is the 2×2 matrix integral operator

$$\begin{bmatrix} 1 - \sum_{j=1}^N \Big(\psi_j \otimes f\epsilon G_j + \psi_j \otimes f\epsilon(fG_j)\Big) & \sum_{j=1}^N \psi_j \otimes fG_j \\ -\sum_{j=1}^N \Big(\epsilon\psi_j \otimes f\epsilon G_j + \epsilon\psi_j \otimes f\epsilon(fG_j)\Big) & 1 + \sum_{j=1}^N \epsilon\psi_j \otimes fG_j \end{bmatrix}$$
$$= \begin{bmatrix} 1 - \sum_{j=1}^N \psi_j \otimes f\epsilon G_j & \sum_{j=1}^N \psi_j \otimes fG_j \\ -\sum_{j=1}^N \epsilon\psi_j \otimes f\epsilon G_j - \epsilon f & 1 + \sum_{j=1}^N \epsilon\psi_j \otimes fG_j \end{bmatrix} \begin{bmatrix} 1 & 0 \\ \epsilon f & 1 \end{bmatrix}, \tag{6.80}$$

where the notation $a \otimes b$ denotes the integral operator with kernel $a(x)b(y)$. The determinant of the last matrix equals 1, while recalling (6.78) and (6.77) we can identify the first matrix with the r.h.s. of (6.76). □

With (6.76) established, the formula in Proposition 6.3.3 for $\rho_{(n)}^T$ now follows as an example of (6.31), or from the analogue of (6.32).

6.3.3 N odd

The N odd analogue of Proposition 6.3.2 can be derived [150], [238], provided $\mathbf{H}$ is replaced by the $N + 1 \times N + 1$ antisymmetric matrix

$$\mathbf{X} := \begin{bmatrix} [h(x_j, x_k)]_{j,k=1,\dots,N} & [F(x_j)]_{j=1,\dots,N} \\ -[F(x_k)]_{k=1,\dots,N} & 0 \end{bmatrix}. \tag{6.81}$$

The relevance of this structure is that with $h(x,y)$ given by (6.67) and $F(x) = \frac{1}{2}$ we have

$$\mathrm{Pf}\,\mathbf{X} = 2^{-(N+1)/2} \prod_{1 \le j < k \le N} \mathrm{sgn}(x_k - x_j),$$

which follows from (6.62) by taking $x_{2n} \to \infty$.

PROPOSITION 6.3.7 *Let N be odd, and suppose*

$$\hat{\Phi}_k(x), \ \hat{r}_k \ (\neq 0), \ F(x) \text{ and } \mathbf{H} = [h(x_j, x_k)]_{j,k=1,\dots,N}, \ (h(x_k,x_j) = -h(x_j,x_k))$$

are arbitrary, and let $\hat{R}_k(x)$ be such that

$$\det[\hat{R}_k(x_j)]_{j,k=1,\dots,N} = \prod_{1 \le j < k \le N} (x_k - x_j).$$

Then, with $\mathbf{X}$ defined by (6.81)

$$\prod_{j=1}^{N} e^{-V(x_j)} \prod_{1 \le j < k \le N} (x_k - x_j) \mathrm{Pf}\,\mathbf{X} = \prod_{p=0}^{(N-1)/2} \hat{r}_p \,\mathrm{qdet}[\mathbf{f}_1^{\mathrm{odd}}(x_j, x_k)]_{j,k=1,\dots,N},$$

where

$$\mathbf{f}_1^{\mathrm{odd}}(x,y) := \begin{bmatrix} S_1^{\mathrm{odd}}(x,y) & \tilde{I}_1^{\mathrm{odd}}(x,y) \\ D_1^{\mathrm{odd}}(x,y) & S_1^{\mathrm{odd}}(y,x) \end{bmatrix}$$

with

$$S_1^{\mathrm{odd}}(x,y) = \sum_{k=0}^{(N-1)/2-1} \frac{e^{-V(y)}}{\hat{r}_k} \Big(\hat{\Phi}_{2k}(x) \hat{R}_{2k+1}(y) - \hat{\Phi}_{2k+1}(x) \hat{R}_{2k}(y) \Big) + \frac{e^{-V(y)}}{\hat{r}_{(N-1)/2}} F(x) \hat{R}_{N-1}(y),$$

$$D_1^{\mathrm{odd}}(x,y) = \sum_{k=0}^{(N-1)/2-1} \frac{e^{-V(x)} e^{-V(y)}}{\hat{r}_k} \Big(\hat{R}_{2k}(x) \hat{R}_{2k+1}(y) - \hat{R}_{2k+1}(x) \hat{R}_{2k}(y) \Big),$$

$$\tilde{I}_1^{\mathrm{odd}}(x,y) = \sum_{k=0}^{(N-1)/2-1} \frac{1}{\hat{r}_k} \Big(\hat{\Phi}_{2k+1}(x) \hat{\Phi}_{2k}(y) - \hat{\Phi}_{2k}(x) \hat{\Phi}_{2k+1}(y) \Big) + h(x,y)$$
$$+ \frac{1}{\hat{r}_{(N-1)/2}} (\hat{\Phi}_{N-1}(x) F(y) - F(x) \hat{\Phi}_{N-1}(y)).$$

Proof. As in the analogous result for N even, here we will use a method first introduced in the more general parameter-dependent problem [417]. The first step is to note from the defining property of $\{\hat{R}_k(x)\}$ that

$$\prod_{j=1}^{N} e^{-2V(x_j)} \prod_{1 \le j < k \le N} (x_k - x_j)^2$$
$$= \prod_{j=1}^{N} e^{-2V(x_j)} \det[\hat{R}_{k-1}(x_j)]_{j,k=1,\dots,N} \det \begin{bmatrix} \begin{bmatrix} \hat{R}_{2j-1}(x_k) \\ -\hat{R}_{2j-2}(x_k) \end{bmatrix}_{\substack{j=1,\dots,(N-1)/2 \\ k=1,\dots,N}} \\ [\hat{R}_{N-1}(x_k)]_{k=1,\dots,N} \end{bmatrix}$$
$$= \prod_{j=1}^{(N+1)/2} \hat{r}_{j-1}^2 \det[\mathbf{D}_1^{\mathrm{odd}} + \vec{r}\vec{r}^{\,T}],$$

where $\mathbf{D}_1^{\mathrm{odd}} = [D_1^{\mathrm{odd}}(x_j, x_k)]_{j,k=1,\dots,N}$, $\vec{r} := [e^{-V(x_j)} \hat{R}_{N-1}(x_j)/r_{(N-1)/2}]_{j=1,\dots,N}$. But in general for a column vector $\vec{v} = [v_j]_{j=1,\dots,N}$, and $N \times N$ matrix $\mathbf{A}$, it is easy to check that

$$\det[\mathbf{A} + \vec{v}\vec{v}^T] = \det \begin{bmatrix} \mathbf{A} & \vec{v} \\ -\vec{v}^T & 1 \end{bmatrix} \tag{6.82}$$

(use elementary column operations to make the terms in the final row, except that in the final column, zero). Now with $\mathbf{A} = \mathbf{D}_1^{\rm odd}$, $\mathbf{A}$ is an antisymmetric matrix of odd dimension and so $\det[\mathbf{A}] = 0$. This means the element 1 in the lower right-hand corner on the r.h.s. of the above expression can be replaced by 0, giving an antisymmetric matrix. Making use of (6.12) then shows

$$\prod_{j=1}^N e^{-V(x_j)} \prod_{j<k}(x_k - x_j) = \prod_{j=1}^{(N+1)/2} \hat{r}_{j-1} \, \mathrm{Pf} \begin{bmatrix} \mathbf{D}_1^{\rm odd} & \vec{r} \\ -\vec{r}^T & 0 \end{bmatrix}$$

and in particular, with $\vec{F} := [F(x_j)]_{j=1,\dots,N}$,

$$\begin{aligned} \prod_{j=1}^N e^{-V(x_j)} \prod_{j<k}(x_k - x_j) \, \mathrm{Pf}\, \mathbf{X} &= (-1)^{(N+1)/2} \prod_{j=1}^{(N+1)/2} \hat{r}_{j-1} \, \mathrm{Pf} \begin{bmatrix} \mathbf{D}_1^{\rm odd} & \vec{r} & & \\ -\vec{r}^T & 0 & & \\ & & -\mathbf{H} & -\vec{F} \\ & & \vec{F}^T & 0 \end{bmatrix} \\ &= (-1)^{(N+1)/2} \prod_{j=1}^{(N+1)/2} \hat{r}_{j-1} \, \mathrm{Pf} \begin{bmatrix} \mathbf{D}_1^{\rm odd} & \vec{r}\vec{F}^T \\ -\vec{F}\vec{r}^T & -\mathbf{H} \end{bmatrix}, \end{aligned} \tag{6.83}$$

where the validity of the second equality can be checked from the definition (6.10).

Consider now the quantities S_1, D_1 and I_1 as defined in Proposition 6.3.2, all with $N \mapsto N-1$ and N odd (note that then $D_1 = D_1^{\rm odd}$). In terms of these quantities define the matrices

$$\mathbf{D}_1 := [D_1(x_j, x_k)]_{j,k=1,\dots,N}, \quad \mathbf{I}_1 := [I_1(x_j, x_k)]_{j,k=1,\dots,N}, \quad \mathbf{S}_1 := [S_1(x_j, x_k)]_{j,k=1,\dots,N}.$$

To transform between these matrices we introduce the (nonunique x-dependent) matrix $\alpha = [\alpha_{ij}]_{i,j=1,\dots,N}$ such that

$$\Phi_k(x_j) = -\sum_{i=1}^N \alpha_{ji} e^{-V(x_i)} R_k(x_i).$$

It then follows from the definitions that

$$\alpha \mathbf{D}_1 = -\mathbf{S}_1, \qquad \alpha \mathbf{D}_1 \alpha^T = -\mathbf{I}_1.$$

We make use of these relations by multiplying the first block row on the r.h.s. of (6.83) by α on the left, and adding to the second block row, then multiplying the first block column by α^T on the right and adding the second block column, to obtain

$$\prod_{j=1}^N e^{-V(x_j)} \prod_{j<k}(x_k - x_j) \, \mathrm{Pf}\, \mathbf{X} = (-1)^{(N+1)/2} \prod_{j=1}^{(N+1)/2} \hat{r}_{j-1} \, \mathrm{Pf} \begin{bmatrix} \mathbf{D}_1^{\rm odd} & (\mathbf{S}_1^{\rm odd})^T \\ -\mathbf{S}_1^{\rm odd} & -\tilde{\mathbf{I}}_1^{\rm odd} \end{bmatrix}.$$

Interchanging the first block row with the second gives the stated result. □

By an appropriate choice of the arbitrary terms in Proposition 6.3.7 the analogue of Proposition 6.3.3 can be established.

PROPOSITION 6.3.8 *Let $\{R_n(x)\}_{n=0,1,\dots}$ and $\{r_n\}_{n=0,\dots,(N-1)/2-1}$ be as in Definition 6.3.1, and let*

$$\hat{r}_{(N-1)/2} := \int_{-\infty}^{\infty} e^{-V(x)} F(x) R_{N-1}(x)\, dx, \qquad \hat{r}_n := r_n \quad (n = 0, \dots, (N-1)/2 - 1),$$

$$\hat{R}_n(x) := R_n(x) - \hat{r}_{(N-1)/2}^{-1} \Big(\int_{-\infty}^{\infty} e^{-V(x')} F(x') R_n(x')\, dx' \Big) R_{N-1}(x) \quad (n = 0, \dots, N-2),$$

$$\hat{R}_{N-1}(x) := R_{N-1}(x),$$

$$\hat{\Phi}_n(x) := \int_{-\infty}^{\infty} e^{-V(y)} h(y,x) \hat{R}_n(y)\, dy \quad (n = 0, \dots, N-1),$$

$$h(x_j, x_k) = \frac{1}{2} \mathrm{sgn}(x_k - x_j), \qquad F(x) = \frac{1}{2}.$$

Then we have

$$\int_{-\infty}^{\infty} dx_1 \cdots \int_{-\infty}^{\infty} dx_N \prod_{j=1}^{N} e^{-V(x_j)} \prod_{1 \le j < k \le N} |x_k - x_j| = N! \prod_{k=0}^{(N-1)/2} 2\hat{r}_k,$$

$$\rho_{(n)}(x_1, \dots, x_n) = \mathrm{qdet}[\mathbf{f}_1^{\mathrm{odd}}(x_j, x_k)]_{j,k=1,\dots,n}.$$

Proof. As in the proof of Proposition 6.3.3, this is a consequence of the validity of the two formulas (6.70) (with $\mathbf{f}_1^{\mathrm{odd}}$ replacing $\mathbf{f}_1$). The formulas are checked from the skew orthogonality property of $\{R_j(x)\}$ and the properties

$$\int_{-\infty}^{\infty} e^{-V(x)} F(x) \hat{R}_n(x)\, dx = \hat{r}_{(N-1)/2} \delta_{n,N-1}, \qquad \int_{-\infty}^{\infty} e^{-V(x)} \hat{\Phi}_m(x) \hat{R}_n(x)\, dx = \langle R_m | R_n \rangle_1.$$

□

The qdet formula of Proposition 6.3.8 for $\rho_{(n)} =: \rho_{(n)}^N$ in the case N odd can also be deduced as a corollary of the corresponding result in Proposition 6.3.3 for N even [210]. To see how this comes about, suppose N is odd and consider the correlation $\rho_{(n+1)}^{N+1}$. According to the definition

$$\rho_{(n+1)}^{N+1}(x_1, \dots, x_{n+1}) = \frac{1}{\hat{Z}_{N+1}} \frac{(N+1)!}{(N+n)!} \prod_{l=1}^{n+1} w(x_l) \int_{-\infty}^{\infty} dx_{n+2} \cdots \int_{-\infty}^{\infty} dx_{N+1} \prod_{l=n+2}^{N+1} w_1(x_l) \prod_{j<k}^{N+1} |x_k - x_j|,$$

and consequently

$$\lim_{x_{n+1} \to \infty} (w(x_{n+1}))^{-1} (x_{n+1})^{-N} \rho_{(n+1)}^{N+1}(x_1, \dots, x_{n+1}) = (N+1) \frac{\hat{Z}_N}{\hat{Z}_{N+1}} \rho_{(n)}^N(x_1, \dots, x_n).$$

In the qdet expression for $\rho_{(n+1)}^{N+1}$ implied by Proposition 6.3.3, the only dependence on x_{n+1} is in the final row and column. The leading asymptotic form of the entries can readily be determined and the limit computed. The qdet expression of Proposition 6.3.8 results after appropriate elementary reduction of the limiting final row and column.

EXERCISES 6.3 1. The objective of this exercise is to apply the method of integration over alternate variables in the case N odd and thus show that then

$$\hat{Z}_N[a] = N! 2^{(N+1)/2} \mathrm{Pf} \begin{bmatrix} [\gamma_{jk}]_{j,k=1,\dots,N} & [\nu_j]_{j=1,\dots,N} \\ -[\nu_k]_{k=1,\dots,N} & 0 \end{bmatrix},$$

where γ_{jk} is as in Proposition 6.3.4 while

$$\nu_k := \frac{1}{2} \int_{-\infty}^{\infty} e^{-V(x)} a(x) R_{k-1}(x)\, dx.$$

(i) Proceed as in the proof of Proposition 6.3.4 to show

$$
\begin{aligned}
\hat{Z}_N[a] &= N! \int_{-\infty < x_2 < x_4 < \cdots < x_{N-1} < \infty} \prod_{l=1}^{(N-1)/2} dx_{2l}\, e^{-V(x_{2l})} a(x_{2l}) \\
&\quad \times \det \begin{bmatrix} \begin{bmatrix} \int_{-\infty}^{x_{2j}} e^{-V(x)} a(x) R_{k-1}(x)\, dx \\ R_{k-1}(x_{2j}) \end{bmatrix}_{\substack{j=1,\dots,(N-1)/2 \\ k=1,\dots,N}} \\ [\int_{-\infty}^{\infty} e^{-V(x)} a(x) R_{k-1}(x)\, dx]_{k=1,\dots,N} \end{bmatrix} \\
&= \frac{N!}{((N-1)/2)!} \sum_{P \in S_N} \varepsilon(P) \Big(\prod_{l=1}^{(N-1)/2} \mu_{P(2l-1),P(2l)} \Big) 2\nu_{P(N)} \\
&= \frac{N!}{((N-1)/2)!} 2^{(N+1)/2} \sum_{\substack{P(2l) > P(2l-1) \\ l=1,\dots,(N-1)/2}} \varepsilon(P) \Big(\prod_{l=1}^{(N-1)/2} \gamma_{P(2l-1),P(2l)} \Big) \nu_{P(N)}.
\end{aligned}
$$

(ii) Write $\nu_{P(N)} =: \nu_{P(N),N+1} = -\nu_{N+1,P(N)}$ in the above expression, and make use of (6.36) to identify this expression as proportional to the stated Pfaffian.

2. [123] Let $\mathbf{X}$ be given as in (6.81). Derive the analogue of (6.73) for N odd,

$$
\begin{aligned}
&\int_{-\infty}^{\infty} dx_1\, e^{-V(x_1)} a(x_1) \cdots \int_{-\infty}^{\infty} dx_N\, e^{-V(x_N)} a(x_N) \det[R_{k-1}(x_j)]_{j,k=1,\dots,N} \mathrm{Pf}\, \mathbf{X} \\
&\quad = N! \mathrm{Pf} \begin{bmatrix} [a_{jk}]_{j,k=1,\dots,N} & [b_j]_{j=1,\dots,N} \\ -[b_k]_{k=1,\dots,N} & 0 \end{bmatrix},
\end{aligned} \tag{6.84}
$$

where

$$
\begin{aligned}
a_{jk} &= \int_{-\infty}^{\infty} dx\, e^{-V(x)} a(x) \int_{-\infty}^{\infty} dy\, e^{-V(y)} a(y)\, R_{j-1}(x) h(x,y) R_{k-1}(y), \\
b_j &= \int_{-\infty}^{\infty} e^{-V(x)} a(x) F(x) R_{j-1}(x)\, dx.
\end{aligned}
$$

3. Use an argument similar to that used in the verification of (6.62) to show

$$
\mathrm{Pf}[\mathrm{sgn}(j-i) x_i y_j]_{i,j=1,\dots,2n} = \prod_{i=1}^{n} x_{2i-1} \prod_{j=1}^{n} y_{2j}. \tag{6.85}
$$

4. Assume that $N \le M$ are positive integers and N is even. Let $\mathbf{T} = [t_{ik}]_{\substack{i=1,\dots,N \\ k=1,\dots,M}}$ be an $N \times M$ matrix and $\mathbf{A} = [a_{kl}]_{k,l=1,\dots,M}$ be an $M \times M$ skew symmetric matrix. Then the minor summation formula states that [304]

$$
\sum_{J \subset \{1,\dots,M\},\, |J|=N} \mathrm{Pf}[\mathbf{A}_{J,J}] \det[\mathbf{T}_J] = \mathrm{Pf}[\mathbf{T}\mathbf{A}^T\mathbf{T}], \tag{6.86}
$$

where $\mathbf{A}_{J,J}$ denotes the $N \times N$ square matrix which results from restricting $\mathbf{A}$ to the rows and columns indexed by J, while $\mathbf{T}_J$ denotes the $N \times N$ square matrix which results by restricting the columns of $\mathbf{T}$ to J. The objective of this exercise is to prove (6.86) and derive as a consequence the Binet-Cauchy formula.

(i) [34] In (6.73) suppose the integration measure is uniform and discrete,

$$
\int_{-\infty}^{\infty} dx_j\, e^{-V(x_j)} a(x_j) \mapsto \sum_{n_j=1}^{M}, \tag{6.87}
$$

and write

$$
R_{j-1}(n_k) =: r_{jk}, \qquad h(n_j, n_k) =: a_{jk}.
$$

With the determinant in (6.73) replaced by its transpose, identify (6.73) with (6.86).

(ii) [303] Show that with $M = 2m$, $N = 2n$ and

$$\mathbf{A} = \begin{bmatrix} \mathbf{0}_m & \mathbf{1}_m \\ -\mathbf{1}_m & \mathbf{0}_m \end{bmatrix}, \qquad \mathbf{T} = \begin{bmatrix} \mathbf{X}_{m\times n} & \mathbf{0}_{m\times n} \\ \mathbf{0}_{m\times n} & \mathbf{Y}_{m\times n} \end{bmatrix},$$

one has

$$\mathrm{Pf}[\mathbf{A}_{J,J}] = \begin{cases} (-1)^{n(n-1)/2}, & \text{if } J = \{k_1 < \cdots < k_n < k_1 + m < \cdots < k_n + m\}, \\ 0, & \text{otherwise,} \end{cases}$$

and for J as in the first case

$$\det[\mathbf{T}_J] = \det[\mathbf{X}_K]\det[\mathbf{Y}_K],$$

where $K = \{k_1, \dots, k_m\}$.

(iii) Note that with $\mathbf{A}$ and $\mathbf{T}$ as in (ii),

$$\mathrm{Pf}[\mathbf{T}\mathbf{A}^T\mathbf{T}] = (-1)^{n(n-1)/2}\det[\mathbf{X}^T\mathbf{Y}]$$

and thus substitute this together with the results of (ii) in (6.86) to deduce

$$\sum_{K\subset\{1,\dots,m\},\,|K|=n} \det[\mathbf{X}_K]\det[\mathbf{Y}_K] = \det[\mathbf{X}^T\mathbf{Y}], \tag{6.88}$$

which is the Binet-Cauchy formula.

(iv) Relate (6.88) to the discretization (6.87), with $e^{-V(x)}a(x)$ replaced by $w_2(x)$, of (5.170).

5. [91] For eigenvalues $\{x_i\}_{i=1,\dots,N}$ let $P(y) := \prod_{i=1}^N (y - x_i)$ denote the characteristic polynomial and let $\mathrm{OE}_N(e^{-V(x)})$ be defined as above (4.22). Let $Z_N := \hat{Z}_N/N!$ denote the corresponding canonical partition function (recall (1.39)). One has the result that with $k+m$ even, $k - m = 2s$, $s > 1 - N$,

$$\Big\langle \frac{\prod_{i=1}^k P(\alpha_i)}{\prod_{i=1}^m P(\beta_i)} \Big\rangle_{\mathrm{OE}_{2N}(e^{-V(x)})} = \frac{Z_{2N+2s}}{Z_{2N}} \frac{D(\beta;\alpha)}{\Delta(-\alpha)\Delta(-\beta)} \mathrm{Pf}\,[W_N^{(1)}(\{\alpha,\beta\}|\{\alpha,\beta\})],$$

where use has been made of the notation (5.35), (5.37), and $W_N^{(1)}$ is a skew symmetric $(k+m)\times(k+m)$ matrix with rows and columns parametrized by $\{\alpha,\beta\}$ and with matrix elements

$$W_N^{(1)}(\alpha_i,\alpha_j) = \frac{Z_{2N+2s-2}}{Z_{2N+2s}}(\alpha_i - \alpha_j)\Big\langle P(\alpha_i)P(\alpha_j)\Big\rangle_{\mathrm{OE}_{2N+2s-2}(e^{-V(x)})},$$

$$W_N^{(1)}(\alpha_i,\beta_j) = \frac{1}{\alpha_i - \beta_j}\Big\langle \frac{P(\alpha_i)}{P(\beta_j)}\Big\rangle_{\mathrm{OE}_{2N+2s}(e^{-V(x)})},$$

$$W_N^{(1)}(\beta_i,\beta_j) = \frac{Z_{2N+2s+2}}{Z_{2N+2s}}(\beta_i - \beta_j)\Big\langle \frac{1}{P(\beta_i)P(\beta_j)}\Big\rangle_{\mathrm{OE}_{2N+2s+2}(e^{-V(x)})}.$$

(i) Use Proposition 5.3.2 to deduce from this that for $\mathrm{OE}_{2N}(e^{-V(x)})$,

$$\rho_{(n)}(x_1,\dots,x_n) = \mathrm{Pf}\,[K_{2N}^{(1)}(x_i,x_j)]_{i,j=1,\dots,n}$$

where $K_{2N}^{(1)}(x,y) = \begin{bmatrix} k_{11} & k_{12} \\ k_{21} & k_{22} \end{bmatrix}$ is the 2×2 skew symmetric matrix with elements

$$k_{11} = \frac{Z_{2N-2}}{Z_{2N}}(x-y)\langle P(x)P(y)\rangle_{\mathrm{OE}_{2N-2}(e^{-V(x)})}, \qquad k_{12} = \frac{1}{x-y}\operatorname*{Res}_{z=y}\Big\langle \frac{P(x)}{P(z)}\Big\rangle_{\mathrm{OE}_{2N}(e^{-V(x)})},$$

$$k_{21} = -k_{12}, \qquad k_{22} = \frac{Z_{2N+2}}{Z_{2N}}(x-y)\operatorname*{Res}_{z=y}\Big\langle \frac{1}{P(x)P(z)}\Big\rangle_{\mathrm{OE}_{2N+2}(e^{-V(x)})},$$

with use having been made of (5.42).

(ii) Use the method of integration over alternate variables to show that for N even

$$\langle P(x)P(y)\rangle_{\mathrm{OE}_{N-2}(e^{-V(x)})} = \frac{r_{N/2-1}}{x-y} D_1(x,y).$$

Adopting the strategy of the proof of Proposition 5.1.4, use this to deduce

$$\Big\langle \frac{P(x)}{P(y)} \Big\rangle_{\mathrm{OE}_N(e^{-V(x)})} = (x-y) \sum_{i=0}^{N/2-1} \frac{R_{2i+1}(x)h_{2i}(y) - R_{2i}(x)h_{2i+1}(y)}{r_i} + 1,$$

$$\Big\langle \frac{1}{P(x)P(y)} \Big\rangle_{\mathrm{OE}_{N+2}(e^{-V(x)})}$$
$$= \frac{1}{r_{N/2}} \frac{1}{x-y} \Big(\sum_{i=0}^{N/2-1} \frac{h_{2i+1}(x)h_{2i}(y) - h_{2i}(x)h_{2i+1}(y)}{r_i} + \Big\langle \frac{1}{x-t} \Big| \frac{1}{y-t} \Big\rangle_1 \Big),$$

where $h_j(x) := \langle R_j(t) | 1/(x-t) \rangle_1$.

(iii) Combine the results of (i) and (ii), and make use of (6.11) to reclaim the evaluation of $\rho_{(n)}$ given in Proposition 6.3.3.

6.4 CONSTRUCTION OF THE SKEW ORTHOGONAL POLYNOMIALS AND SUMMATION FORMULAS

6.4.1 Skew orthogonal polynomials

The operator (5.60) can also be used to facilitate the calculation of the monic skew orthogonal polynomials with respect to the inner product (6.61) in the classical cases. This follows because of a relationship between the inner products (6.61) and (5.4) involving the operator A [6].

PROPOSITION 6.4.1 *We have*

$$(\phi, A^{-1}\psi)_2 = -\langle \phi | \psi \rangle_1 \Big|_{V(x) \mapsto V(x) + \log f(x)}. \tag{6.89}$$

Proof. From (5.60) we see that

$$A^{-1} = (f(x)e^{-2V(x)})^{-1/2} \Big(\frac{d}{dx} \Big)^{-1} (e^{2V(x)} f(x))^{-1/2}$$

while it is easy to check that

$$\Big(\frac{d}{dx} \Big)^{-1} \psi(x) = \frac{1}{2} \int_{-\infty}^{\infty} \mathrm{sgn}(x-y) \psi(y)\, dy.$$

Thus

$$A^{-1}\psi[x] = (f(x)e^{-2V(x)})^{-1/2} \frac{1}{2} \int_{-\infty}^{\infty} e^{-V(y)} \mathrm{sgn}(x-y) (f(y))^{-1/2} \psi(y)\, dy$$

and the result is immediate. □

To make use of Proposition 6.4.1 we first note that it is possible to write the matrix

$$[(p_j, A^{-1}p_k)_2]_{j,k=0,\dots,N-1}$$

in terms of $\mathbf{A}$ as specified by (5.63). First note that with $\mathbf{D} := \mathrm{diag}((p_j,p_j)_2)_{j=0,\dots,N-1}$ we have

$$\Big([(p_j, A^{-1}p_k)_2]_{j,k=0,\dots,N-1} \mathbf{D}^{-1} \mathbf{A} \Big)_{jk} = -\sum_{l=0}^{N-1} \frac{(A^{-1}p_j, p_l)_2}{(p_l,p_l)_2} (p_l, Ap_k) \tag{6.90}$$

where use has been made of the first equation in Proposition 6.2.1. But according to (5.64)

$$\sum_{l=0}^{N-1} \frac{(p_l, Ap_k)_2}{(p_l, p_l)_2} p_l(x) = Ap_k(x) \quad \text{for} \quad k \neq N-1,$$

which when substituted back in (6.90) implies

$$[(p_j, A^{-1}p_k)_2]_{j,k=0,\dots,N-1} \mathbf{D}^{-1}\mathbf{A} = \mathbf{D}', \tag{6.91}$$

where $\mathbf{D}' = \mathbf{D} + \mathbf{C}_k$ with $\mathbf{C}_k$ having all entries zero except for the final column. These latter entries are easy to compute but it turns out their explicit form is not needed.

The formula (6.91) together with (6.89) allows the determination of the classical skew orthogonal polynomials $\tilde{R}_j(x)$ and corresponding normalizations $\tilde{r}_j$, where the tilde denotes that the underlying inner product is the modification of (6.61) specified by

$$V(x) \mapsto \tilde{V}_1(x) := V(x) + \log f(x). \tag{6.92}$$

Thus with $\mathbf{X}^{(1)} := [\alpha_{jk}]_{j,k=0,\dots,N-1}$ denoting the lower triangular transition matrix defined by

$$[\tilde{R}_j(x)]_{j=0,\dots,N-1} = \mathbf{X}^{(1)}[p_j(x)]_{j=0,\dots,N-1}, \qquad \alpha_{jj} = 1,$$

we have

$$\begin{aligned}
\tilde{\mathbf{r}} &:= \Big[\langle \tilde{R}_j, \tilde{R}_k\rangle_1\Big|_{V(x)\mapsto V(x)+\log f(x)}\Big]_{j,k=0,\dots,N-1} \\
&= \mathbf{X}^{(1)}\Big[\langle p_j, p_k\rangle_1\Big|_{V(x)\mapsto V(x)+\log f(x)}\Big]_{j,k=0,\dots,N-1} \mathbf{X}^{(1)T} \\
&= -\mathbf{X}^{(1)}[(p_j, A^{-1}p_k)_2]_{j,k=0,\dots,N-1}\mathbf{X}^{(1)T} = -\mathbf{X}^{(1)}\mathbf{D}'\mathbf{A}^{-1}\mathbf{D}\mathbf{X}^{(1)T},
\end{aligned} \tag{6.93}$$

where in obtaining the first equality of the final line (6.89) has been used, while in the final equality (6.91) has been used. Taking inverses of both sides of (6.93) gives the equivalent formula

$$\mathbf{X}^{(1)T}\tilde{\mathbf{r}}^{-1}\mathbf{X}^{(1)} = -\mathbf{D}^{-1}\mathbf{A}\mathbf{D}'^{-1}. \tag{6.94}$$

In the classical cases the equation (6.94) can be solved for $\mathbf{X}^{(1)}$. It is sufficient to premultiply both sides by $(\mathbf{X}^{(1)T})^{-1}$ and then equate the strictly lower triangular entries on both sides. Using (6.2) we see that on the l.h.s.

$$\Big(\tilde{\mathbf{r}}^{-1}\mathbf{X}^{(1)}\Big)_- = \begin{bmatrix}
* & & & & & & \\
1/\tilde{r}_0 & * & & & & & \\
-\alpha_{30}/\tilde{r}_1 & -\alpha_{31}/\tilde{r}_1 & * & & & & \\
\alpha_{20}/\tilde{r}_1 & \alpha_{21}/\tilde{r}_1 & 1/\tilde{r}_1 & * & & & \\
-\alpha_{50}/\tilde{r}_2 & -\alpha_{51}/\tilde{r}_2 & -\alpha_{52}/\tilde{r}_2 & -\alpha_{53}/\tilde{r}_2 & * & & \\
\alpha_{40}/\tilde{r}_2 & \alpha_{41}/\tilde{r}_2 & \alpha_{42}/\tilde{r}_2 & \alpha_{43}/\tilde{r}_2 & 1/\tilde{r}_2 & * & \\
\vdots & \vdots & \vdots & \vdots & \vdots & & \ddots
\end{bmatrix}_{2N\times 2N}. \tag{6.95}$$

On the r.h.s. the fact that $(\mathbf{X}^{(1)T})^{-1}$ is an upper triangular matrix with 1's along the diagonal and the explicit form (5.63) of $\mathbf{A}$ together show

$$-\Big((\mathbf{X}^{(1)T})^{-1}\mathbf{D}^{-1}\mathbf{A}\mathbf{D}'^{-1}\Big)_{-} = -\Big(\mathbf{D}^{-1}\mathbf{A}\mathbf{D}'^{-1}\Big)_{-} = \begin{bmatrix} * & & & & \\ \gamma_0 & * & & & \\ 0 & \gamma_1 & * & & \\ 0 & 0 & \gamma_2 & * & \\ \vdots & \vdots & \vdots & & \ddots \end{bmatrix}, \tag{6.96}$$

where

$$\gamma_j := c_j/(p_{j+1}, p_{j+1})_2 (p_j, p_j)_2. \tag{6.97}$$

Equating (6.95) and (6.96) gives

$$\begin{aligned} \gamma_{2p} &= 1/\tilde{r}_p \quad (p = 0, \dots, N-1), \\ \alpha_{2p,l} &= 0 \quad (l = 0, \dots, 2p-1), \\ \alpha_{2p+1,l} &= 0 \quad (l = 0, \dots, 2p-2), \\ \alpha_{2p+1,2p-1} &= -\gamma_{2p-1}\tilde{r}_p = -\gamma_{2p-1}/\gamma_{2p}, \end{aligned}$$

while $\tilde{\alpha}_{2p+1,2p}$ is left unspecified. Hence

$$\begin{aligned} \tilde{R}_{2j}(x) &= p_{2j}(x), \\ \tilde{R}_{2j+1}(x) &= p_{2j+1}(x) + \alpha_{2p+1,2p} p_{2j}(x) - \frac{\gamma_{2j-1}}{\gamma_{2j}} p_{2j-1}(x), \\ \tilde{r}_p &= 1/\gamma_{2p}. \end{aligned} \tag{6.98}$$

Note that the fact that $\alpha_{2p+1,2p}$ is arbitrary in the formula for $\tilde{R}_{2j+1}(x)$ is consistent with the skew orthogonal polynomials being nonunique up to the transformation (6.1). It is simplest to choose $\alpha_{2p+1,2p} = 0$. Then, in light of (6.97) and (5.65) all quantities in (6.98) are known explicitly. Note that with the choice $\alpha_{2p+1,2p} = 0$ it follows from (5.64) and (6.98) that

$$\tilde{R}_{2j+1}(x) = -\frac{1}{\gamma_{2j}(p_{2j}, p_{2j})_2} e^{\tilde{V}_1(x)} \frac{d}{dx}\Big(e^{-\tilde{V}_4(x)} p_{2j}(x)\Big). \tag{6.99}$$

In summary we thus have that skew orthogonal polynomials and corresponding normalizations with respect to the skew symmetric inner product (6.61), with weight functions

$$e^{-\tilde{V}_1(x)} = \begin{cases} e^{-x^2/2}, & \text{Gaussian}, \\ x^{(a-1)/2}e^{-x/2} \ (x > 0), & \text{Laguerre}, \\ (1-x)^{(a-1)/2}(1+x)^{(b-1)/2} \ (-1 < x < 1), & \text{Jacobi}, \\ (1+x^2)^{-(\alpha+1)/2}, & \text{Cauchy} \end{cases} \tag{6.100}$$

are given by (6.98) with $\alpha_{2p+1,2p} = 0$.

6.4.2 Summation formulas — N even

Analogous to the situation at $\beta = 4$ for the classical weight functions, the fundamental quantity $S_1(x, y)$, defined in Proposition 6.3.2, determining the correlation function at $\beta = 1$ can be summed explicitly. The details depend on the parity of N.

PROPOSITION 6.4.2 *Let N be even. Define $\tilde{S}_1(x, y)$ as in Proposition 6.3.2, but with $V \mapsto \tilde{V}_1$ and similarly define $\tilde{R}_k$, $\tilde{\Phi}_k$ and $\tilde{r}_k$. Let $\{p_j(x)\}_{j=0,1,\dots}$ be the set of monic orthogonal polynomials (assumed complete)* *associated with the weight function $e^{-2V(x)}$ and also write*

$$p_l(x) = \sum_{j=0}^{l} \tilde{\beta}_{lj}^{(1)} \tilde{R}_j(x), \quad \tilde{\beta}_{ll}^{(1)} = 1. \tag{6.101}$$

We have

$$\tilde{S}_1(x,y) = e^{-2V(x)+\tilde{V}_1(x)-\tilde{V}_1(y)} \Big(\sum_{n=0}^{N-1} \frac{p_n(x)p_n(y)}{(p_n,p_n)_2} + \sum_{n=N}^{\infty} \sum_{k=0}^{N-1} \frac{p_n(x)}{(p_n,p_n)_2} \tilde{\beta}_{nk}^{(1)} \tilde{R}_k(y) \Big). \tag{6.102}$$

Proof. From the definition (6.61) of the skew inner product at $\beta = 1$, and the definition in Proposition 6.3.2 of $\Phi_k(x)$, both modified as specified, we see that it is possible to write

$$\tilde{\Phi}_k(x') = e^{\tilde{V}_1(x')} \langle \tilde{R}_k(x) | \delta(x'-y) \rangle_1 \Big|_{V \mapsto \tilde{V}_1}.$$

Making use of (6.50) and substituting for $p_n(y)$ therein by (6.101) we see from the skew orthogonality of $\{\tilde{R}_k(y)\}$ that

$$\tilde{\Phi}_{2k}(x) = \tilde{r}_k e^{\tilde{V}_1(x)} e^{-2V(x)} \sum_{\nu=2k+1}^{\infty} \frac{p_\nu(x)}{(p_\nu,p_\nu)_2} \tilde{\beta}_{\nu\, 2k+1}^{(1)}, \tag{6.103}$$

$$\tilde{\Phi}_{2k+1}(x) = -\tilde{r}_k e^{\tilde{V}_1(x)} e^{-2V(x)} \sum_{\nu=2k}^{\infty} \frac{p_\nu(x)}{(p_\nu,p_\nu)_2} \tilde{\beta}_{\nu\, 2k}^{(1)}. \tag{6.104}$$

Substituting these results in the formula of Proposition 6.3.2 for $S_1(x,y)$ we obtain (6.102) after minor manipulation. □

The first sum in (6.102) is evaluated according to the Christoffel-Darboux formula (6.3). To evaluate the second sum requires knowledge of the transition coefficients $\tilde{\beta}_{lj}^{(1)}$. In the case of the classical polynomials these coefficients can be determined from (6.98). Setting $\alpha_{2p+1,2p} = 0$ and comparing with the definition (6.101) we see that

$$\begin{aligned} \tilde{\beta}_{2l,j}^{(1)} &= 0 \quad (j = 0,\dots,2l-1), \\ \tilde{\beta}_{2l+1,2j}^{(1)} &= 0 \quad (j = 0,\dots,l), \\ \tilde{\beta}_{2l+1,2j+1}^{(1)} &= \frac{\prod_{k=1}^{l} a_k}{\prod_{k=1}^{j} a_k}, \quad a_k := \frac{\gamma_{2k+1}}{\gamma_{2k}}. \end{aligned} \tag{6.105}$$

In fact we don't require the explicit form of the $\tilde{\beta}_{lj}^{(1)}$, but rather their factorization property

$$\tilde{\beta}_{nk}^{(1)} = \tilde{\beta}_{n,N-1}^{(1)} \tilde{\beta}_{N-1,k}^{(1)}, \quad n \geq N, \tag{6.106}$$

which is evident from the above formulas. Substituting (6.106) in the double summation in (6.102) and recalling (6.101) shows

$$\begin{aligned} \sum_{n=N}^{\infty} \sum_{k=0}^{N-1} \frac{p_n(x)}{(p_n,p_n)_2} \tilde{\beta}_{nk}^{(1)} \tilde{R}_k(y) &= \Big(\sum_{n=N}^{\infty} \frac{p_n(x)}{(p_n,p_n)_2} \tilde{\beta}_{n,N-1}^{(1)} \Big) p_{N-1}(y) \\ &= \Big(\frac{1}{\tilde{r}_{N/2-1}} e^{-\tilde{V}(x)+2V(x)} \tilde{\Phi}_{N-2}(x) - \frac{p_{N-1}(x)}{(p_{N-1},p_{N-1})_2} \Big) p_{N-1}(y), \end{aligned}$$

where the second equality follows from (6.103) and the fact that N is assumed even. Thus

$$\tilde{S}_1(x,y) = e^{-(V(x)-\tilde{V}_1(x))} e^{(V(y)-\tilde{V}_1(y))} K_{N-1}(x,y) + \gamma_{N-2} e^{-\tilde{V}_1(y)} \tilde{\Phi}_{N-2}(x) p_{N-1}(y).$$

Substituting for $\tilde{R}_{N-2}$ in the definition of Φ_{N-2} according to (6.98) allows all reference to the skew orthogonal polynomials to be eliminated in the final summation formula.

PROPOSITION 6.4.3 *For the classical cases*

$$\tilde{S}_1(x,y) = e^{-(V(x)-\tilde{V}_1(x))} e^{(V(y)-\tilde{V}_1(y))} K_{N-1}(x,y) \\ +\gamma_{N-2} e^{-\tilde{V}_1(y)} p_{N-1}(y) \frac{1}{2} \int_{-\infty}^{\infty} \mathrm{sgn}(x-t) p_{N-2}(t) e^{-\tilde{V}_1(t)}\, dt, \tag{6.107}$$

where γ_{N-2} is specified in terms of $(p_{N-2}, p_{N-2})_2$ by (6.97) and (5.65).

Because the completeness of the orthogonal polynomials has been assumed in the derivation of (6.107), and in the Cauchy case the polynomials are a finite set and so not complete, the validity of (6.107) in the Cauchy case remains to be established. In fact (6.107) can be verified directly in all the classical cases by making use of (5.64) and (6.99) in the formula of Proposition 6.3.2 for $\tilde{S}_1(x,y)$.

6.4.3 Summation formulas — N odd

In the case of N odd, we see from the definitions of Proposition 6.3.7 and Proposition 6.3.8 that

$$\tilde{S}_1^{\mathrm{odd}}(x,y) = \tilde{S}_1(x,y)\Big|_{N \mapsto N-1} + e^{-\tilde{V}(y)} \frac{\tilde{R}_{N-1}(y)}{2\tilde{s}_{N-1}} \\ + \frac{\tilde{\Phi}_{N-1}(x)}{\tilde{s}_{N-1}} \sum_{k=0}^{(N-1)/2-1} \frac{e^{-\tilde{V}_1(y)}}{\tilde{r}_k^{(1)}} \Big(- \tilde{s}_{2k} \tilde{R}_{2k+1}(y) + \tilde{s}_{2k+1} \tilde{R}_{2k}(y) \Big) \\ - \frac{\tilde{R}_{N-1}(y)}{\tilde{s}_{N-1}} \sum_{k=0}^{(N-1)/2-1} \frac{e^{-\tilde{V}_1(y)}}{\tilde{r}_k^{(1)}} \Big(- \tilde{s}_{2k} \tilde{\Phi}_{2k+1}(x) + \tilde{s}_{2k+1} \tilde{\Phi}_{2k}(x) \Big),$$

where

$$\tilde{s}_k := \frac{1}{2} \int_{-\infty}^{\infty} e^{-\tilde{V}_1(x)} \tilde{R}_k(x)\, dx. \tag{6.108}$$

The quantity $\tilde{S}_1(x,y)|_{N \mapsto N-1}$ is evaluated by (6.107). Furthermore, from the definitions we see that

$$\sum_{k=0}^{(N-1)/2-1} \frac{e^{-\tilde{V}_1(y)}}{\tilde{r}_k^{(1)}} \Big(- \tilde{s}_{2k} \tilde{R}_{2k+1}(y) + \tilde{s}_{2k+1} \tilde{R}_{2k}(y) \Big) \\ = - \lim_{x \to \infty} \tilde{S}_1(x,y)\Big|_{N \mapsto N-1} = -\gamma_{N-3} \tilde{s}_{N-3} e^{-\tilde{V}_1(y)} p_{N-2}(y), \tag{6.109}$$

where the second equality follows from (6.107), while

$$\sum_{k=0}^{(N-1)/2-1} \frac{1}{\tilde{r}_k^{(1)}} \Big(- \tilde{s}_{2k} \tilde{\Phi}_{2k+1}(x) + \tilde{s}_{2k+1} \tilde{\Phi}_{2k}(x) \Big) \\ = \frac{1}{2} \int_{-\infty}^{\infty} \mathrm{sgn}(x-y) \Big(- \lim_{x' \to \infty} \tilde{S}_1(x',y)\Big|_{N \mapsto N-1} \Big)\, dy = -\gamma_{N-3} \tilde{s}_{N-3} \tilde{\phi}_{N-2}(x), \tag{6.110}$$

where the second equality follows from (6.109) and

$$\tilde{\phi}_j(x) := \frac{1}{2} \int_{-\infty}^{\infty} e^{-\tilde{V}_1(y)} \mathrm{sgn}(x-y) p_j(y)\, dy. \tag{6.111}$$

Hence

$$\tilde{S}_1^{\rm odd}(x,y) = \tilde{S}_1(x,y)\Big|_{N\mapsto N-1} + e^{-\tilde{V}_1(y)}\frac{p_{N-1}(y)}{2\tilde{s}_{N-1}}$$
$$-\gamma_{N-3}\tilde{s}_{N-3}\frac{e^{-\tilde{V}_1(y)}}{\tilde{s}_{N-1}}\Big(\tilde{\phi}_{N-1}(x)p_{N-2}(y) - p_{N-1}(y)\tilde{\phi}_{N-2}(x)\Big), \tag{6.112}$$

where the fact that $\tilde{R}_{N-1}(y) = p_{N-1}(y)$, which follows from (6.98), and the fact that N is odd have been used.

For future reference we note that $\tilde{s}_n$ for n even can be expressed in terms of the γ_k. Thus it follows from (6.103), (6.105) and (6.98) that taking the limit $x \to \infty$ in the ratio $\tilde{\Phi}_n(x)/\tilde{\Phi}_{n-2}(x)$ gives

$$\frac{\tilde{s}_{n-2}}{\tilde{s}_n} = \frac{\gamma_{n-1}}{\gamma_{n-2}},$$

and thus

$$\tilde{s}_{2l} = \tilde{s}_0 \prod_{j=0}^{l-1} \frac{\gamma_{2j}}{\gamma_{2j+1}}.$$

Recalling (6.98) and (6.108) this is equivalent to the evaluation

$$\tilde{c}_{2l} = \tilde{c}_0 \prod_{j=0}^{l-1} \frac{\gamma_{2j}}{\gamma_{2j+1}}, \qquad \tilde{c}_k := \int_{-\infty}^{\infty} e^{-\tilde{V}_1(t)} p_k(t)\, dt. \tag{6.113}$$

We note also that taking the limit $x \to \infty$ in (6.104) and using (6.98) implies

$$\tilde{c}_{2l-1} = 0. \tag{6.114}$$

EXERCISES 6.4 1. [167, 263] Use the method of integration over alternate variables to show that

$$R_{2n}(x) = \frac{1}{\hat{Z}_{2n}} \int_{-\infty}^{\infty} dx_1\, e^{-V(x_1)} \cdots \int_{-\infty}^{\infty} dx_{2n}\, e^{-V(x_{2n})} \prod_{l=1}^{2n}(x-x_l) \prod_{1\le j<k\le 2n} |x_k - x_j|,$$
$$R_{2n+1}(x) = \frac{1}{\hat{Z}_{2n}} \int_{-\infty}^{\infty} dx_1\, e^{-V(x_1)} \cdots \int_{-\infty}^{\infty} dx_{2n}\, e^{-V(x_{2n})} \Big(x + \sum_{j=1}^{2n} x_j\Big)$$
$$\times \prod_{l=1}^{2n}(x-x_l) \prod_{1\le j<k\le 2n} |x_k - x_j|$$

(cf. Exercises 6.1 q.2), where in $R_{2n+1}(x)$ the arbitrary constant γ_{2m} in (6.1) is chosen so that the coefficient of x^{2n} vanishes. For this purpose, as well as the Vandermonde determinant identity, make use of the determinant formula

$$\det \begin{bmatrix} x_j^{k-1} & x_j^{2n+1} \\ x^{k-1} & x^{2n+1} \end{bmatrix}_{\substack{j=1,\dots,2n \\ k=1,\dots,2n}} = \Big(x + \sum_{j=1}^{2n} x_j\Big) \prod_{l=1}^{2n}(x-x_l) \prod_{1\le j<k\le 2n} |x_k - x_j|.$$

2. [544] In this exercise the $\beta = 1$ analogue of (6.60) is obtained.

(i) In an analogous notation to Exercises 6.2 q.3, show $S_1 K = S_1$. Use this result to show $S_1 K \epsilon \psi_i = \epsilon \psi_i$ and thus $S_1|_{\mathcal{H}} = \epsilon(K\epsilon|_{\mathcal{H}})^{-1}$.

(ii) Verify that the equations of Exercises 6.2 q.3(iii) hold with ϵ and D interchanged, and thus

$$S_1 = (I_{\mathcal{H}+\epsilon\mathcal{H}} - (I-K)\epsilon K D)^{-1} K.$$

6.5 ALTERNATE CORRELATIONS AT $\beta = 1$

We know from the inter-relations of Section 4.2.3, and also from the method of integration over alternate variables (recall Section 6.3.2), that at $\beta = 1$ a special role is played by every second eigenvalue in the sequence. With the coordinates ordered according to (4.24) (this is the reverse of the ordering convention used in Section 6.3.2), it is possible to calculate the (n_1, n_2)-point correlation function $\rho_{(n_1,n_2)}$, where n_1 coordinates are odd numbered and n_2 coordinates are even numbered [456]. We will see that in the classical cases the distribution of the even labeled coordinates can be identified with a $\beta = 4$ distribution, as is consistent with the results of Section 4.2.3.

6.5.1 N even

As usual in $\beta = 1$ theory, the N even and N odd cases must be treated separately. We begin assuming N to be even. Let us write $x_j = w_{2j-1}$, $y_j = w_{2j}$ $(j = 1, \dots, N/2)$ where it is assumed N is even. Let $\{R_j(w)\}_{j=0,1,\dots}$ be such that

$$\prod_{1 \le j < k \le N} (w_k - w_j) = \det[R_{j-1}(w_k)]_{j,k=1,\dots,N} = \det[R_{j-1}(x_k) \ \ R_{j-1}(y_k)]_{\substack{j=1,\dots,N \\ k=1,\dots,N/2}}.$$

Then it follows from Proposition 5.9.1 that

$$\chi_{x_1 > y_1 > \cdots > x_{N/2} > y_{N/2}} \prod_{j=1}^{N} e^{-V(w_j)} \prod_{1 \le j < k \le N} (w_j - w_k)$$
$$= \prod_{j=1}^{N/2} e^{-V(x_j) - V(y_j)} \det[R_{j-1}(x_k) \ \ R_{j-1}(y_k)]_{\substack{j=1,\dots,N \\ k=1,\dots,N/2}} \det[-\chi_{x_j - y_k > 0}]_{j,k=1,\dots,N/2}, \tag{6.115}$$

assuming the inequalities (5.173). On the l.h.s. of this identity is the Boltzmann factor for the ordered log-gas at $\beta = 1$. The r.h.s. naturally separates the even and odd labeled coordinates, and has the further significant feature of being a symmetric function of $\{x_i\}$ and $\{y_j\}$ separately. The identity (6.115) can be used to compute $\rho_{(n_1,n_2)}$.

PROPOSITION 6.5.1 *Let $\rho_{(n_1,n_2)}(x_1, \dots, x_{n_1}; y_1, \dots, y_{n_2})$ denote the (n_1, n_2)-point correlation function corresponding to (6.115) for n_1 particles with odd numbered w-coordinates and n_2 particles with even numbered w-coordinates. With x, x' odd numbered w-coordinates and y, y' even numbered w-coordinates let*

$$h_{oo}(x, x') = h_{ee}(y, y') = 0, \qquad h_{oe}(x, y) = -h_{eo}(y, x) = \frac{1}{2}\chi_{x-y>0},$$
$$\Phi_k^{o}(x) = \int_{-\infty}^{\infty} h_{eo}(y, x) R_k(y) e^{-V(y)}\, dy, \qquad \Phi_k^{e}(y) = \int_{-\infty}^{\infty} h_{oe}(x, y) R_k(x) e^{-V(x)}\, dx,$$

and let $\{R_k(t)\}$ denote the monic skew orthogonal polynomials of Definition 6.3.1, with corresponding normalizations $\{r_j\}$. Substitute these quantities in Proposition 6.3.2 and denote the corresponding expressions for S_1, D_1 and $\tilde{I}_1$ by the symbols S, D and $\tilde{I}$, respectively. Then we have

$$\rho_{(k_1,k_2)}(x_1, \dots, x_{k_1}; y_1, \dots, y_{k_2})$$
$$= \operatorname{qdet} \begin{bmatrix} [\mathbf{f}_{oo}(x_j, x_l)]_{j,l=1,\dots,k_1} & [\mathbf{f}_{oe}(x_j, y_l)]_{\substack{j=1,\dots,k_1 \\ l=1,\dots,k_2}} \\ [\mathbf{f}_{eo}(y_j, x_l)]_{\substack{j=1,\dots,k_2 \\ l=1,\dots,k_1}} & [\mathbf{f}_{ee}(y_j, y_l)]_{j,l=1,\dots,k_2} \end{bmatrix}, \tag{6.116}$$

where with $p, p' \in \{\mathrm{e}, \mathrm{o}\}$

$$\mathbf{f}_{pp'}(w,z) = \begin{bmatrix} S_{pp'}(w,z) & \tilde{I}_{pp'}(w,z) \\ D_{pp'}(w,z) & S_{pp'}(z,w) \end{bmatrix}. \tag{6.117}$$

Proof. This result can be deduced from Proposition 6.3.2. It is necessary first to replace each integration $\int dw$ over the appropriate w-variables in (6.115) by a Riemann sum approximation $\int d\mu(w)$. In the latter the lattice points which make up the domain of integration are chosen to be the set X if w is odd numbered (an x variable), and the set Y if w is even numbered (a y variable). These lattice points must interlace so that for coordinates x_j restricted to X and y_j restricted to Y $(j = 1, \dots, N/2)$ it's always possible to have $x_j > y_j$. Now consider

$$\prod_{j=1}^{N} e^{-V(w_j)} \det[R_{j-1}(w_k)]_{j,k=1,\dots,N} \mathrm{Pf}[-h(w_j, w_k)]_{j,k=1,\dots,N}. \tag{6.118}$$

It follows from the definition of $h(w,w') = h_{\cdot,\cdot}$ that (6.118) vanishes unless exactly half the w_k variables are on X and the other half are on Y. In the latter case (6.118) equals $2^{-N/2}$ times (6.115).

The expression (6.118) has a structure identical to the l.h.s. of (6.65), although the variables are confined to lattice points. We can therefore use Propositions 6.3.2 and 6.3.3 to write down the (n_1, n_2)-point correlation function, where n_1 coordinates are on X (odd numbered) and n_2 coordinates on Y (even numbered). The polynomials $R_k(z)$ must be skew orthogonal with respect to the inner product

$$\begin{aligned} \langle f, g \rangle &= \int_{X\cup Y} d\mu(w) \int_{X\cup Y} d\mu(z)\, e^{-V(w)-V(z)} f(w) g(z) h(w,z) \\ &= \int_X d\mu(x) \int_Y d\mu(y)\, e^{-V(x)-V(y)} \Big(f(x)g(y)\chi_{x-y>0} - f(y)g(x)\chi_{y-x>0} \Big), \end{aligned}$$

which in the continuum limit is identical to (6.61). □

Propositions 6.5.1 and 6.3.2 together give

$$\begin{aligned} S_{\mathrm{oo}}(x,x') &= \sum_{k=0}^{N/2-1} \frac{e^{-V(x')}}{2r_k} \Big(R_{2k+1}(x') \int_{-\infty}^{x} e^{-V(t)} R_{2k}(t)\, dt - R_{2k}(x') \int_{-\infty}^{x} e^{-V(t)} R_{2k+1}(t)\, dt \Big), \\ D_{\mathrm{oo}}(x,x') &= 2\frac{\partial}{\partial x} S_{\mathrm{oo}}(x,x'), \qquad \tilde{I}_{\mathrm{oo}}(x,x') = -\frac{1}{2}\int_x^{x'} S_{\mathrm{oo}}(x,t)\, dt, \\ S_{\mathrm{ee}}(y,y') &= -\sum_{k=0}^{N/2-1} \frac{e^{-V(y')}}{2r_k} \Big(R_{2k+1}(y') \int_y^{\infty} e^{-V(t)} R_{2k}(t)\, dt - R_{2k}(y') \int_y^{\infty} e^{-V(t)} R_{2k+1}(t)\, dt \Big), \\ D_{\mathrm{ee}}(y,y') &= 2\frac{\partial}{\partial y} S_{\mathrm{ee}}(y,y'), \qquad \tilde{I}_{\mathrm{ee}}(y,y') = -\frac{1}{2}\int_y^{y'} S_{\mathrm{ee}}(y,t)\, dt. \end{aligned} \tag{6.119}$$

Next we will show that in the classical cases $S_{\mathrm{ee}}(y,y')$ is closely related to $S_4(y,y')$. First we note, by comparing the definition of S_{ee} in (6.119) and the definition of S_1 in Proposition 6.3.2, with h, Φ therein given by (6.67), (6.69), that in general

$$S_{\mathrm{ee}}(y,y') = \frac{1}{2}\Big(S_1(y,y') - S_1(\infty, y') \Big) \tag{6.120}$$

(in the Jacobi case $S_1(\infty, y')$ means $S_1(1,y')$). With $e^{-V} \mapsto e^{-\tilde{V}_1}$ (recall (6.100)), let the corresponding summations in (6.120) be denoted $S_{\mathrm{ee}} \mapsto \tilde{S}_{\mathrm{ee}}$, $S_1 \mapsto \tilde{S}_1$. Now $\tilde{S}_1$ is summed in all the classical cases by (6.107). Furthermore, taking $x \to \infty$ in (6.107) we see

$$\tilde{S}_1(\infty, y') = \gamma_{N-2} e^{-\tilde{V}_1(y')} p_{N-1}(y') \int_{-\infty}^{\infty} p_{N-2}(t) e^{-\tilde{V}_1(t)}\, dt.$$

Thus we have

$$\begin{aligned}2\tilde{S}_{\rm ee}(y,y') = & e^{-(V(y)-\tilde{V}_1(y))}e^{V(y')-\tilde{V}_1(y')}K_{N-1}(y,y') \\ & -\gamma_{N-2}e^{-\tilde{V}_1(y')}p_{N-1}(y')\int_y^\infty p_{N-2}(t)e^{-\tilde{V}_1(t)}\,dt. \end{aligned} \qquad (6.121)$$

We remark that, after replacing (y, y') by (x, x') and $\int_y^\infty$ by $-\int_{-\infty}^x$, the same formula applies for $\tilde{S}_{\rm oo}(x, x')$.

The expression (6.121) has the same structure as the summation formula (6.56) for $\tilde{S}_4(x, y)$ (recall here that the tilde denotes the modified weight functions (6.53)) in the classical cases. Noting that in general $e^{-2V(t)+\tilde{V}_4(t)} = e^{-\tilde{V}_1(t)}$ we thus observe

$$\tilde{S}_4(y,y')\Big|_{N\mapsto N/2} = \tilde{S}_{\rm ee}(y,y')\Big|_{N\mapsto N+1}. \qquad (6.122)$$

We will see in the next subsection that the r.h.s. of (6.122) does indeed correspond to $\tilde{S}_{\rm ee}(y, y')$ for N odd, or equivalently that (6.121) remains valid for N odd. Recalling the analogous structure of (6.14) and the formula (6.116) for $\rho_{(0,n)}$ (which too remains valid for N odd), we therefore have the following result [220].

PROPOSITION 6.5.2 *Let* $\mathrm{OE}_N(e^{-V(x)})$, $\mathrm{SE}_N(e^{-4V(x)})$ *and the operation even be as in Section 4.2.3. We have*

$$\rho_{(n)}(y_1,\dots,y_n;\mathrm{SE}_{N/2}(e^{-2\tilde{V}_4(x)})) = \tilde{\rho}_{(n)}(y_1,y_2,\dots,y_n;\mathrm{even}(\mathrm{OE}_{N+1}(e^{-\tilde{V}_1(x)}))). \qquad (6.123)$$

The identity (6.123) with $n = N/2$ says that integrating the p.d.f. at $\beta = 1$, with weight function (6.100) and $N + 1$ particles, over every odd numbered coordinate gives the p.d.f. at $\beta = 4$ with weight function (6.53). Equivalently, in terms of the distributions,

$$\mathrm{even}(\mathrm{OE}_{N+1}(e^{-\tilde{V}_1(x)})) = \mathrm{SE}_{N/2}(e^{-2\tilde{V}_4(x)}). \qquad (6.124)$$

In the case $N \mapsto 2n$, this summarizes in one formula the results of Section 4.2.3 relating to eigenvalue p.d.f.'s on the real line. In fact a simple direct derivation of (6.124) based on (5.64) can be given (see Exercises 6.5 q.1).

The result (6.124) raises the question as to the relationship of $\rho_{(0,n)}$ to a $\beta = 4$ correlation in the cases when the weight function is not classical. It was proved in [220] that the only continuous weight functions for which (6.124) holds are the classical weight functions (or weight functions related to the classical weight functions by a fractional linear transformation). Another question of interest is the relationship of the distribution of every even labeled coordinate for OE_N, N even, to a $\beta = 4$ distribution. The following result was proved in [220].

PROPOSITION 6.5.3 *Let N be even. We have*

$$\mathrm{even}(\mathrm{OE}_N(f)) = \mathrm{SE}_{N/2}((g/f)^2) \qquad (6.125)$$

for

$$(f,g) = \begin{cases} (e^{-x/2}, e^{-x}), & x > 0, \\ ((1-x)^{(a-1)/2}, (1-x)^a), & 0 < x < 1, \end{cases} \qquad (6.126)$$

and up to linear fractional transformations these pairs of weights are unique.

It remains to consider the N odd case, and to verify that indeed (6.121) still holds true.

6.5.2 N odd

The computation of $\rho_{(n_1,n_2)}$ for N odd yields to a similar strategy as does the N even case. First one notes that in the limit $y_n \to -\infty$, (5.174) reads

$$\det \left[[\chi_{x_j > y_k}]_{\substack{j=1,\dots,N \\ k=1,\dots,N-1}} \; [1]_{j=1,\dots,N} \right] = \chi_{x_1 > y_1 > \cdots > y_{n-1} > x_n},$$

valid for

$$x_1 > \cdots > x_n, \qquad y_1 > \cdots > y_{n-1}.$$

Furthermore, with the latter ordering

$$\det \left[[\chi_{x_j > y_k}]_{\substack{j=1,\dots,N \\ k=1,\dots,N-1}} \; [1]_{j=1,\dots,N} \right]$$
$$= \mathrm{Pf} \begin{bmatrix} 0 & \chi_{x_1>y_1} & 0 & \chi_{x_1>y_2} & 0 & \cdots & \chi_{x_1>y_{N-1}} & 0 & 1 \\ -\chi_{x_1>y_1} & 0 & -\chi_{x_2>y_1} & 0 & -\chi_{x_3>y_1} & \cdots & 0 & -\chi_{x_N>y_1} & 1 \\ 0 & \chi_{x_2>y_1} & 0 & \chi_{x_2>y_2} & 0 & \cdots & \chi_{x_2>y_{N-1}} & 0 & 1 \\ -\chi_{x_1>y_2} & 0 & -\chi_{x_2>y_2} & 0 & -\chi_{x_3>y_2} & \cdots & 0 & -\chi_{x_N>y_2} & 1 \\ \vdots & & & & & & & & \vdots \end{bmatrix},$$

and this extends to the setting (6.118), thereby allowing Proposition 6.3.7 to be used in an analogous way to the use made of Proposition 6.3.2 in the derivation of Proposition 6.5.1.

PROPOSITION 6.5.4 *Let x, x' be odd numbered w-coordinates and y, y' be even numbered w-coordinates as in Proposition 6.5.1. Let*

$$h_{\rm oo}(x,x') = h_{\rm ee}(y,y') = 0, \qquad h_{\rm oe}(x,y) = -h_{\rm eo}(y,x) = \frac{1}{2}\chi_{x-y>0}, \quad F(x) = F(y) = \frac{1}{2},$$
$$\hat{\Phi}_k^{\rm o}(x) = \frac{1}{2}\int_{-\infty}^{\infty} h(y,x)\hat{R}_k(y)e^{-V(y)}\,dy, \qquad \hat{\Phi}_k^{\rm e}(y) = \frac{1}{2}\int_{-\infty}^{\infty} h(x,y)\hat{R}_k(x)e^{-V(x)}\,dx,$$

let $\{\hat{R}_k(t)\}$ be specified in terms of the monic skew orthogonal polynomials of Definition 6.3.1 as in Proposition 6.3.8, and similarly specify $\{\hat{r}_j\}$. Substitute these quantities in Proposition 6.3.7 and denote the corresponding expressions for $S_1^{\rm odd}$, $D_1^{\rm odd}$ and $\tilde{I}_1^{\rm odd}$ by the symbols $S^{\rm odd}$, $D^{\rm odd}$ and $\tilde{I}^{\rm odd}$, respectively. Then the formulas (6.116) and (6.117) again apply, where in the latter the superscript "odd" is to be attached to the matrix elements.

Comparing the above specification of $S^{\rm odd}$ with the formula for $S_1^{\rm odd}$ in Proposition 6.3.7 we see that

$$S_{\rm ee}^{\rm odd}(y,y') = \frac{1}{2}\Big(S_1^{\rm odd}(y,y') - S_1^{\rm odd}(\infty,y') \Big).$$

For the modified weight functions (6.100), indicated by the use of a tilde, we have the summation formula (6.107) of $\tilde{S}_1$. Substituting this in the above formula gives

$$2\tilde{S}_{\rm ee}^{\rm odd}(y,y') = 2\tilde{S}_{\rm ee}(y,y')\Big|_{N\mapsto N-1} + \gamma_{N-3}\tilde{s}_{N-3}\frac{e^{-\tilde{V}_1(y')}}{\tilde{s}_{N-1}}$$
$$\times \Big(p_{N-2}(y')\int_y^{\infty} e^{-\tilde{V}_1(t)}p_{N-1}(t)\,dt - p_{N-1}(y')\int_y^{\infty} e^{-\tilde{V}_1(t)}p_{N-2}(t)\,dt \Big), \quad (6.127)$$

where $\tilde{S}_{\rm ee}(y,y')$ is specified by (6.121). The formula (6.127) can be further simplified. The first step is to note from (6.103), (6.105) and (6.98) that taking the limit $x \to \infty$ in the ratio $\hat{\Phi}_{N-3}(x)/\hat{\Phi}_{N-1}(x) =$

$\tilde{\phi}_{N-3}(x)/\tilde{\phi}_{N-1}(x)$ implies

$$\gamma_{N-3}\frac{\tilde{s}_{N-3}}{\tilde{s}_{N-1}} = \gamma_{N-2}. \tag{6.128}$$

Substituting this in (6.127) and recalling (6.121) we see that

$$\tilde{S}^{\rm odd}_{\rm ee}(y,y') = \tilde{S}_{\rm ee}(y,y'), \tag{6.129}$$

provided

$$\gamma_{N-3}\tilde{\phi}_{N-3}(x) - \gamma_{N-2}\tilde{\phi}_{N-1}(x) = e^{-2V(x)+\tilde{V}_1(x)}\frac{p_{N-2}(x)}{(p_{N-2},p_{N-2})_2} \tag{6.130}$$

for N odd. This latter identity is verified by checking that both sides agree for $x \to \infty$ and that it reduces to (6.131) below upon differentiation.

EXERCISES 6.5 1. (i) Note from (5.60) and (5.64) that

$$e^{\tilde{V}_1(x)}\frac{d}{dx}\Big(e^{-\tilde{V}_4(x)}p_l(x)\Big) = -\frac{c_l}{(p_{l+1},p_{l+1})_2}p_{l+1}(x) + \frac{c_{l-1}}{(p_{l-1},p_{l-1})_2}p_{l-1}(x) \tag{6.131}$$

for $l = 0,1,\dots$, where $c_{-1} := 0$.

(ii) For N odd let

$$I(x_2,x_4,\dots,x_{N-1}) := \frac{1}{C}\Big(\prod_{l=1}^{(N+1)/2}\int_{x_{2l-2}}^{x_{2l}} dx_{2l-1}\Big)\prod_{j=1}^{N} e^{-\tilde{V}_1(x_j)}\prod_{1\le j<k\le N}|x_k - x_j|,$$

where $x_0 := -\infty$, $x_{N+1} := \infty$. With $x_1 < x_2 < \cdots < x_N$ use the Vandermonde formula (1.173) to write

$$\begin{aligned}
&\prod_{j<k}^{N}|x_k - x_j| \\
&= \prod_{l=1}^{N-1}\Big(-\frac{(p_l,p_l)_2}{c_{l-1}}\Big)\det\Big[[p_0(x_j)]_{j=1,\dots,N}\ \Big[-\frac{c_{k-1}}{(p_k,p_k)_2}p_k(x_j) + \frac{c_{k-2}}{(p_{k-2},p_{k-2})_2}p_{k-2}(x_j)\Big]_{\substack{j=1,\dots,N\\k=2,\dots,N}}\Big] \\
&= \prod_{l=1}^{N-1}\Big(-\frac{(p_l,p_l)_2}{c_{l-1}}\Big)\det\Big[[p_0(x_j)]_{j=1,\dots,N}\ \Big[e^{\tilde{V}_1(x_j)}\frac{d}{dx}\Big(e^{-\tilde{V}_4(x_j)}p_{k-1}(x_j)\Big)\Big]_{\substack{j=1,\dots,N\\k=2,\dots,N}}\Big],
\end{aligned}$$

where the second equality follows from (i). Substitute this in the integrand, then use the method of integration over alternate variables to show

$$\begin{aligned}
&I(x_2,x_4,\dots,x_{N-1}) \\
&= \frac{1}{C}\prod_{l=1}^{N-1}\Big(-\frac{(p_l,p_l)_2}{c_{l-1}}\Big) \\
&\quad\times\det\begin{bmatrix}\begin{bmatrix}\int_{-\infty}^{x_{2j}} e^{-\tilde{V}_1(x)}p_0(x)\,dx \\ e^{-\tilde{V}_1(x_{2j})}p_0(x_{2j})\end{bmatrix}_{j=1,\dots,(N-1)/2} & \begin{bmatrix} e^{-\tilde{V}_4(x_{2j})}p_{k-1}(x_{2j}) \\ \frac{d}{dx_{2j}}e^{-\tilde{V}_4(x_{2j})}p_{k-1}(x_{2j})\end{bmatrix}_{\substack{j=1,\dots,(N-1)/2\\k=2,\dots,N}} \\ \int_{-\infty}^{\infty} e^{-\tilde{V}_1(x)}p_0(x)\,dx & [0]_{k=2,\dots,N}\end{bmatrix} \\
&= \frac{1}{C}\prod_{l=1}^{N-1}\Big(-\frac{(p_l,p_l)_2}{c_{l-1}}\Big)\Big(\int_{-\infty}^{\infty} e^{-\tilde{V}_1(x)}p_0(x)\,dx\Big)\det\begin{bmatrix} e^{-\tilde{V}_4(x_{2j})}p_{k-1}(x_{2j}) \\ \frac{d}{dx_{2j}}e^{-\tilde{V}_4(x_{2j})}p_{k-1}(x_{2j})\end{bmatrix}_{\substack{j=1,\dots,(N-1)/2\\k=2,\dots,N}}.
\end{aligned}$$

(iii) Deduce from the final result in (ii) that

$$I(x_2, x_4, \ldots, x_{N-1}) \propto \prod_{l=1}^{(N-1)/2} e^{-2\tilde{V}_4(x_{2l})} \prod_{j<k}^{(N-1)/2} (x_{2k} - x_{2j})^4,$$

which with $N \mapsto N+1$ (and N now even) is the result (6.124).

2. [397] In this exercise it will be shown that integrating the distribution COE_{2N} over alternate angles $\theta_1, \theta_3, \ldots, \theta_{2N-1}$ in the region

$$\theta_{2j} < \theta_{2j+1} < \theta_{2j+2} \qquad (j = 0, \ldots, N-1), \tag{6.132}$$

where $\theta_0 := \theta_{2N} - 2\pi$ gives the distribution CSE_N, and thus rederiving (4.32).

(i) Use the identity

$$|e^{i\theta_k} - e^{i\theta_j}| = i^{-1} e^{-i(\theta_k + \theta_j)/2} (e^{i\theta_k} - e^{i\theta_j}), \qquad \theta_k > \theta_j$$

to show that the p.d.f. COE_{2N} is proportional to

$$\det[e^{ip\theta_j}]_{\substack{j=1,\ldots,2N \\ p=-(N-1/2),\ldots,N-1/2}}. \tag{6.133}$$

(ii) Integrate the determinant in (i) over the region (6.132) using the method of integration over alternate variables to obtain

$$\det \begin{bmatrix} (ip)^{-1}(e^{ip\theta_{2j}} + e^{ip\theta_{2N}}) \\ e^{ip\theta_{2j}} \end{bmatrix}_{\substack{j=1,\ldots,N \\ p=-(N-1/2),\ldots,N-1/2}}.$$

Now subtract $(ip)^{-1}$ times the final row from each of the odd numbered rows, then take out a factor $(ip)^{-1}$ from each column to show that the determinant is proportional to

$$\det \begin{bmatrix} e^{ip\theta_{2j}} \\ pe^{ip\theta_{2j}} \end{bmatrix}_{\substack{j=1,\ldots,N \\ p=-(N-1)/2,\ldots,(N-1)/2}} = \prod_{1 \le j < k \le N} |e^{i\theta_{2k}} - e^{i\theta_{2j}}|^4,$$

where the final equality can be deduced by taking a confluent limit in (6.133) (recall Exercises 1.9 q.2).

6.6 SUPERIMPOSED $\beta = 1$ SYSTEMS

The superimposed ensemble is formed out of a system of N_1 particles distributed according to (6.71) with $N \mapsto N_1$, and a distinct set of N_2 particles distributed according to (6.71) with $N \mapsto N_2$. A p.d.f. of N_1+N_2 coordinates $x_1, \ldots, x_{N_1+N_2}$ is formed by summing over the $\binom{N_1+N_2}{N_1}$ distinct ways of choosing N_1 coordinates out of $\{x_1, \ldots, x_{N_1+N_2}\}$ to be distributed according to (6.71) with $N \mapsto N_1$, and the remaining N_2 coordinates to be distributed according to (6.71) with $N \mapsto N_2$. Symbolically, one writes $\mathrm{OE}_{N_1}(e^{-V(x)}) \cup \mathrm{OE}_{N_2}(e^{-V(x)})$ for the resulting ensemble. In the cases $(N_1, N_2) = (N, N)$ or $(N, N+1)$ the p.d.f. has a special structure in which the dependence on the even and odd numbered coordinates factorizes [278], [220].

PROPOSITION 6.6.1 *Let $S = \{s_1, s_2, \ldots, s_l\}$, $s_1 > s_2 > \cdots > s_l \ge 1$ denote a set of positive integers and write*

$$\Delta(x_S) := \prod_{1 \le j < k \le l} (x_{s_j} - x_{s_k}).$$

Then we have

$$\sum_{\substack{S\subset\{1,\dots,2N\}\\|S|=N}} \Delta(x_S)\Delta(x_{\{1,\dots,2N\}-S}) = 2^N \Delta(x_{\{1,3,\dots,2N-1\}})\Delta(x_{\{2,4,\dots,2N\}}), \tag{6.134}$$

$$\sum_{\substack{S\subset\{1,\dots,2N+1\}\\|S|=N}} \Delta(x_S)\Delta(x_{\{1,\dots,2N+1\}-S}) = 2^N \Delta(x_{\{1,3,\dots,2N+1\}})\Delta(x_{\{2,4,\dots,2N\}}). \tag{6.135}$$

Proof. Consider the effect of exchanging x_i and x_{i+2} in a term of either equation. Since Δ is alternating, we see immediately that for $i, i+2 \in S$ or $i, i+2 \notin S$

$$\Delta(x_S)\Delta(x_{\{1,\dots,l\}-S}) \mapsto -\Delta(x_S)\Delta(x_{\{1,\dots,l\}-S}).$$

For other values of $i, i+2$ we see that every factor $x_k - x_j$, $k > j$, is taken to another such factor, except for the factor $x_{i+1} - x_i$ or $x_{i+2} - x_{i+1}$, whichever is present (exactly one of these factors must be present due to the assumption on $i, i+2$) which changes sign. This means that in fact each term in the sum is taken to the negative of a term in the sum. It follows from this that the sum is alternating under parity-preserving permutations; the degrees of each side imply the sum must therefore be proportional to the given r.h.s. The proportionality constant can be calculated by computing the coefficient of an appropriate monomial in the sum. □

According to the result (6.135), the distribution which results from superimposing a system with Boltzmann factor (6.71) with another system with Boltzmann factor (6.71) but with $N \mapsto N+1$ is proportional to

$$e^{-\sum_{j=1}^{2N+1} V(x_j)} \prod_{1\le j<k\le N+1} (x_{2j-1} - x_{2k-1}) \prod_{1\le j<k\le N} (x_{2j} - x_{2k}),$$

where $x_1 > x_2 > \cdots > x_{2N+1}$. The factorized form of this distribution allows the odd numbered coordinates (for example) to be integrated out, thus giving the distribution of the even numbered coordinates in the superimposed ensemble. Doing this, by making use of the Vandermonde formula and integrating the resulting determinant row by row, we see that the corresponding distribution is proportional to

$$e^{-\sum_{j=1}^{N} V(x_{2j})} \det\Big[\int_{x_{2j}}^{\infty} e^{-V(t)} t^{k-1}\, dt\Big]_{j,k=1,\dots,N+1} \prod_{1\le j<k\le N} (x_{2j} - x_{2k}) \tag{6.136}$$

where $x_{2N+2} := -\infty$. By making the replacements $e^{-V(x)} \mapsto e^{-\tilde V_1(x)}$ in this expression, and setting $e^{-\tilde V_1(x)}$ equal to one of the classical forms (6.100) the first determinant can be considerably simplified.

PROPOSITION 6.6.2 *The distribution (6.136), with weight function one of the modified classical forms, is proportional to*

$$e^{-2\sum_{j=1}^{N} V(x_{2j})} \prod_{1\le j<k\le N} (x_{2k} - x_{2j})^2, \tag{6.137}$$

where V and $\tilde V_1$ are related by (6.92). Equivalently, in the classical cases and with the coordinates ordered $x_1 > x_2 > \cdots > x_{2N+1}$

$$\text{even}\Big(\text{OE}_N(e^{-\tilde V_1(x)}) \cup \text{OE}_{N+1}(e^{-\tilde V_1(x)})\Big) = \text{UE}_N(e^{-2V(x)}). \tag{6.138}$$

Proof. With $\{p_j\}_{j=0,1,\dots}$ denoting the monic orthogonal polynomials corresponding to the weight $e^{-2V(x)}$, we first add together appropriate linear combinations of columns and thus obtain the identity

$$\begin{aligned}
&\det\Big[\int_{x_{2j}}^{\infty} e^{-\tilde V_1(t)} t^{k-1}\, dt\Big]_{j,k=1,\dots,N+1} \\
&= \det\Big[\int_{x_{2j}}^{\infty} e^{-\tilde V_1(t)}\, dt \quad \int_{x_{2j}}^{\infty} e^{-\tilde V_1(t)} \Big(p_l(t) + \mu_{l-2} p_{l-2}(t)\Big)\, dt\Big]_{\substack{j=1,\dots,N+1\\ l=1,\dots,N}}
\end{aligned} \tag{6.139}$$

where $p_{-1} := 0$. Now, for appropriate μ_{l-2}, the formulas (6.98) and (6.99) together imply

$$\int_{x_{2j}}^{\infty} e^{-\tilde{V}_1(t)}\Big(p_l(t) + \mu_{l-2}p_{l-2}(t)\Big)\,dt = \alpha_l e^{-\tilde{V}_4(x_{2j})}p_{l-1}(x_{2j}) \tag{6.140}$$

for some α_l. Recalling $x_{2N+2} := -\infty$ we see that the only nonzero term in the final row of the determinant on the r.h.s. of (6.139) is in the first column. Expanding by the first column we thus have that (6.139) is proportional to

$$\prod_{j=1}^{N} e^{-\tilde{V}_4(x_{2j})} \det\Big[p_{l-1}(x_{2j})\Big]_{j,l=1,\dots,N} = \prod_{j=1}^{N} e^{-\tilde{V}_4(x_{2j})} \prod_{1\le j<k\le N}(x_{2k} - x_{2j}),$$

which implies the stated formula. □

If we write (6.138) as

$$\text{even}\Big(\text{OE}_N(f) \cup \text{OE}_{N+1}(f)\Big) = \text{UE}_N(g) \qquad (f,g) = (e^{-\tilde{V}_1(x)}, e^{-2V(x)}), \tag{6.141}$$

then (6.124) reads

$$\text{even}\Big(\text{OE}_{N+1}(f)\Big) = \text{SE}_{N/2}((g/f)^2) \tag{6.142}$$

valid for N even. In fact it is proved in [220] that the first equation in (6.141) and the equation (6.142) are equivalent. Similarly, it is proved that equation (6.125) is equivalent to

$$\text{even}\Big(\text{OE}_N(f) \cup \text{OE}_N(f)\Big) = \text{UE}_N(g) \tag{6.143}$$

and thus holds when (f, g) are specified by (6.126).

We will see in the next chapter that the inter-relationships (6.123) and (6.138), and their circular ensemble counterparts (4.32), and (6.154) below, in turn imply inter-relationships between spacing probabilities for the couplings $\beta = 1, 2$ and 4.

6.6.1 Distribution of the odd numbered coordinates

In addition to the distribution on the l.h.s. of (6.138), it is also of interest to consider the distribution

$$\text{odd}\Big(\text{OE}_N(e^{-\tilde{V}_1(x)}) \cup \text{OE}_{N+1}(e^{-\tilde{V}_1(x)})\Big) \tag{6.144}$$

in the classical cases. In this distribution the particle closest to the edges of the support of the density is not integrated out. With the ordering of Proposition 6.6.2 this feature is also true of

$$\text{odd}\Big(\text{OE}_N(e^{-\tilde{V}_1(x)}) \cup \text{OE}_N(e^{-\tilde{V}_1(x)})\Big) \tag{6.145}$$

for the particle at the right edge and

$$\text{even}\Big(\text{OE}_N(e^{-\tilde{V}_1(x)}) \cup \text{OE}_N(e^{-\tilde{V}_1(x)})\Big) \tag{6.146}$$

for the particle at the left edge. Thus for purposes of computing the scaled limit of (6.144) at the right hand soft edge of the Gaussian ensemble it suffices to consider (6.145), while to compute the scaled limit of (6.144) at the hard edge of the Laguerre ensemble it suffices to consider (6.146). It turns out that the n-point correlation of (6.145) and (6.146) is simpler than that of (6.144).

According to (6.134) the distribution $\text{OE}_N(e^{-\tilde{V}_1(x)}) \cup \text{OE}_N(e^{-\tilde{V}_1(x)})$ is proportional to

$$e^{-\sum_{j=1}^{2N}\tilde{V}_1(x_j)} \prod_{1\le j<k\le N}(x_{2j-1} - x_{2k-1}) \prod_{1\le j<k\le N}(x_{2j} - x_{2k}), \tag{6.147}$$

where $x_1 > x_2 > \cdots > x_{2N}$. Consider now the distribution (6.145), obtained by integrating out the even numbered coordinates in (6.147). Analogous to (6.136) we see that the corresponding distribution is proportional to

$$e^{-\sum_{j=1}^N \tilde{V}_1(x_{2j-1})} \det \left[\int_{-\infty}^{x_{2j-1}} e^{-\tilde{V}_1(t)} t^{k-1} \, dt \right]_{j,k=1,\dots,N} \prod_{1 \le j < k \le N} (x_{2j-1} - x_{2k-1}). \tag{6.148}$$

With $\tilde{V}_1(x)$ one of the classical forms, the procedure of the proof of Proposition 6.6.2 shows that the determinant in (6.148) is in fact proportional to

$$\begin{aligned} & e^{-\sum_{j=1}^N \tilde{V}_4(x_{2j-1})} \det \left[[p_{k-1}(x_{2j-1})]_{\substack{j=1,\dots,N \\ k=1,\dots,N-1}} \left[e^{\tilde{V}_4(x_{2j-1})} \int_{-\infty}^{x_{2j-1}} e^{-\tilde{V}_1(x)} \, dx \right]_{j=1,\dots,N} \right] \\ & = e^{-\sum_{j=1}^N \tilde{V}_4(x_{2j-1})} \det \left[[p_{k-1}(x_{2j-1})]_{\substack{j=1,\dots,N \\ k=1,\dots,N-1}} [F_{N-1}(x_{2j-1})]_{j=1,\dots,N} \right] \end{aligned} \tag{6.149}$$

where

$$F_{N-1}(x) = \sum_{l=N-1}^{\infty} \frac{(p_l, f)_2}{(p_l, p_l)_2} p_l(x), \qquad f(x) := e^{\tilde{V}_4(x)} \int_{-\infty}^{x} e^{-\tilde{V}_1(t)} \, dt. \tag{6.150}$$

Now write

$$\prod_{1 \le j < k \le N} (x_{2k-1} - x_{2j-1}) = \det \left[p_{k-1}(x_{2j-1}) \right]_{j,k=1,\dots,N}$$

and substitute this together with (6.149) in (6.148) to obtain a p.d.f. of the form (5.139). Noting that

$$\int_{-\infty}^{\infty} e^{-2V(x)} p_j(x) F_{N-1}(x) \, dx = 0 \qquad (j = 0, \dots, N-2)$$

we see that furthermore the columns of the two determinants have the biorthogonality property (5.144) (up to normalization), and the correlation kernel is therefore given by (5.145).

PROPOSITION 6.6.3 *For the distribution (6.145) the n-point correlation is given by*

$$\rho_{(n)}^{\mathrm{odd}(\mathrm{OE}_N)^2}(x_1, \dots, x_n) = \prod_{j=1}^{n} e^{-2V(x_j)} \det \left[\sum_{l=0}^{N-2} \frac{p_l(x_j) p_l(x_k)}{(p_l, p_l)_2} + \frac{p_{N-1}(x_j) F_{N-1}(x_k)}{(p_{N-1}, F_{N-1})_2} \right]_{j,k=1,\dots,n}. \tag{6.151}$$

The formulas (6.103), (6.105) and (6.113) allow the coefficients $(p_l, f)_2$ in (6.150) to be computed explicitly.

PROPOSITION 6.6.4 *In the classical cases*

$$F_{N-1}(x) = \frac{1}{\gamma_0} \sum_{\nu=[(N-1)/2]}^{\infty} \frac{\prod_{l=1}^{\nu} (\gamma_{2l-1}/\gamma_{2l})}{(p_{2\nu+1}, p_{2\nu+1})_2} p_{2\nu+1}(x) + 2\tilde{c}_0^2 \sum_{l=[N/2]}^{\infty} \frac{\prod_{j=0}^{l-1} (\gamma_{2j}/\gamma_{2j+1})}{(p_{2l}, p_{2l})_2} p_{2l}(x), \tag{6.152}$$

where $\tilde{c}_0$ is defined by (6.113) and the γ_j by (6.97).

Proof. According to (6.113) and (6.111)

$$\int_{-\infty}^{x} e^{-\tilde{V}_1(t)} \, dt = \tilde{\phi}_0(x) + \tilde{c}_0. \tag{6.153}$$

Substituting in (6.150) and making use of (6.103), (6.105) and (6.113), (6.114) gives (6.152). □

EXERCISES 6.6 1. [148], [278] In this exercise the circular analogue of (6.138) will be derived.

(i) With S as in Proposition 6.6.1, let

$$\tilde{\Delta}(\theta_S) := \prod_{1 \le j < k \le l} \sin((\theta_{s_k} - \theta_{s_j})/2)$$

(cf. the definition of $\Delta(x_S)$ in Proposition 6.6.1). By writing $x_j = e^{i\theta_j}$ in (6.134) show

$$\sum_{\substack{S \subset \{1,\dots,2N\} \\ |S|=N}} \tilde{\Delta}(\theta_S)\tilde{\Delta}(\theta_{\{1,\dots,2N\}-S}) = 2^N \tilde{\Delta}(\theta_{\{1,3,\dots,2N-1\}})\tilde{\Delta}(\theta_{\{2,4,\dots,2N\}}).$$

(ii) With the operation alt defined in (4.32) q.3 show that analogous to (6.138)

$$\mathrm{alt}\Big(\mathrm{COE}_N \cup \mathrm{COE}_N\Big) = \mathrm{CUE}_N. \tag{6.154}$$

2. In the case $a = 0, b = 2$, recognize (4.97) as $\mathrm{LOE}_n|_{a=0} \cup \mathrm{LOE}_n|_{a=0}$. On the other hand, with $a = 0$ note that the definition of $\mathbf{A}$ in the derivation of (4.97) gives that $\{a_i\}$ has distribution $\mathrm{LUE}_n|_{a=0}$ and thus obtain a realization of (6.143) with $(f, g) = (e^{-x/2}, e^{-x})$.

6.7 A TWO-COMPONENT LOG-GAS WITH CHARGE RATIO 1:2

In this section a generalization of integration techniques introduced in Sections 6.1.2 and 6.3.2 will be used to calculate the free energy and correlation functions for the two-component log-gas with charge ratio 1:2, on a line with periodic boundary conditions, at the special coupling $\beta = 1$. In the limit that the concentration of the $+2$ charges goes to zero, this system reduces to the one-component log-gas at $\beta = 1$, while in the limit of zero concentration of the $+1$ charges, the one-component log-gas at $\beta = 4$ is reclaimed.

6.7.1 The generalized partition function

The two-component log-gas to be considered consists of N_1 particles of charge $q = 1$ and N_2 particles of charge $q = 2$, with coordinates $x_1, \dots, x_{N_1}$ and $x_{N_1+1}, \dots, x_{N_1+N_2}$, respectively, interacting via the logarithmic potential

$$\phi(x_i, x_j) = -q_i q_j \log \Big| (e^{2\pi i x_i/L} - e^{2\pi i x_j/L})(L/2\pi) \Big|, \tag{6.155}$$

on a circle of radius $L/2\pi$. The coordinates x_i and x_j are scaled angles, and the constant $L/2\pi$ is included in (6.155) so that $\phi(x_i, x_j) \sim -\log|x_i - x_j|$ for $|x_i - x_j| \to 0$. The circle carries a neutralizing background charge density $-\rho_b$, where

$$\rho_b := (N_1 + 2N_2)/L. \tag{6.156}$$

An equivalent interpretation of (6.155) is as the potential resulting from solving the two-dimensional Poisson equation in periodic boundary conditions (recall Section 2.7). The total energy of the system is calculated in a way analogous to that of the one-component system on a circle, detailed in the proof of Proposition 1.4.1. One finds

$$U = (2N_2 + N_1/2)\log(L/2\pi) - \sum_{1 \le j < k \le N_1+N_2} \log|e^{2\pi i x_j/L} - e^{2\pi i x_k/L}|^{q_j q_k}. \tag{6.157}$$

Thus the generalized partition function for this system at $\beta = 1$ is given by

$$Z_{N_1,N_2}[u,v] := \frac{C}{N_1!N_2!} \int_0^L dx_1 \cdots \int_0^L dx_{N_1+N_2} \prod_{l=1}^{N_1} u(x_l) \prod_{l=N_1+1}^{N_1+N_2} v(x_l)$$
$$\times \prod_{1 \le j < k \le N_1+N_2} |e^{2\pi i x_j/L} - e^{2\pi i x_k/L}|^{q_j q_k}, \tag{6.158}$$

where $C = (2\pi/L)^{(2N_2+N_1/2)}$.

Using a special case of the confluent alternant determinant in Exercises 1.9 q.2, together with some integration techniques from Sections 6.1.2 and 6.3.2, an expression for Z_{N_1,N_2} suitable for computing the partition function and correlation functions can be obtained.

PROPOSITION 6.7.1 *For N_1 even we have*

$$Z_{N_1,N_2}[u,v] = C[\zeta^{N_1/2}]\,\mathrm{Pf}[\beta_{-\tilde{j}\tilde{k}} - i\zeta\alpha_{-\tilde{j}\tilde{k}}]_{\tilde{j},\tilde{k}=-(N_1/2+N_2-1/2),\dots,N_1/2+N_2-1/2}$$
$$= C[\zeta^{N_1/2}]\Big(\pm\det[\beta_{\tilde{j}\tilde{k}} - i\zeta\alpha_{\tilde{j}\tilde{k}}]_{\tilde{j},\tilde{k}=-(N_1/2+N_2-1/2),\dots,N_1/2+N_2-1/2}\Big)^{1/2}, \tag{6.159}$$

where $[\zeta^{N_1/2}]f$ denotes the coefficient of $\zeta^{N_1/2}$ in the power series expansion of f, the tildes on $\tilde{j},\tilde{k}$ indicate they are half odd integers, the sign in the second equality is chosen so that when $u = v = 1$ the determinant is positive and

$$\alpha_{\tilde{j},\tilde{k}} := \int_0^L dx\, u(x) \int_0^L dy\, u(y)\mathrm{sgn}(x-y) e^{2\pi i \tilde{k} x/L} e^{-2\pi i \tilde{j} y/L},$$
$$\beta_{\tilde{j},\tilde{k}} := (\tilde{k} + \tilde{j}) \int_0^L v(x) e^{2\pi i(\tilde{k}-\tilde{j})x/L}\, dx. \tag{6.160}$$

Proof. Recalling (2.21), it follows that for $0 \le x_1 < x_2 \cdots < x_{N_1} \le L$,

$$\prod_{1 \le j<k \le N_1+N_2} |e^{2\pi i x_j/L} - e^{2\pi i x_k/L}|^{q_j q_k} = (-i)^{N_1/2} \prod_{l=1}^{N_1} e^{-\pi i x_l(N_1+2N_2-1)/L}$$
$$\times \prod_{l=N_1+1}^{N_1+N_2} e^{-2\pi i x_l(N_1+2N_2-2)/L} \prod_{1 \le j<k \le N_1+N_2} (e^{2\pi i x_k/L} - e^{2\pi i x_j/L})^{q_j q_k},$$

while according to Exercises 1.9 q.2 we have

$$\prod_{1 \le j<k \le N_1+N_2} (e^{2\pi i x_j/L} - e^{2\pi i x_k/L})^{q_j q_k} = \det \begin{bmatrix} [e^{2\pi i x_j(k-1)/L}]_{\substack{j=1,\dots,N_1 \\ k=1,\dots,N_1+2N_2}} \\ \begin{bmatrix} e^{2\pi i x_j(k-1)/L} \\ (k-1)e^{2\pi i x_j(k-2)/L} \end{bmatrix}_{\substack{j=N_1+1,\dots,N_1+N_2 \\ k=1,\dots,N_1+2N_2}} \end{bmatrix}.$$

Substituting these formulas in (6.158), applying the method of integration over alternate variables (recall Section 6.3.2) to integrate over $x_1, x_3, \dots, x_{N_1-1}$, and expanding the resulting determinant to integrate over all other variables gives

$$Z_{N_1,N_2}[u,v] = C(-i)^{N_1/2} \frac{1}{(N_1/2)!N_2!} \sum_{P \in S_{N_1+2N_2}} \varepsilon(P) \prod_{l=1}^{N_1/2} a_{P(2l-1),P(2l)} \prod_{l=N_1/2+1}^{N_1/2+N_2} b_{P(2l-1),P(2l)},$$

where

$$a_{j,k} := \int_0^L dx\, u(x) \int_0^x dy\, u(y) e^{2\pi i y(j-N_1/2-N_2-1/2)/L} e^{2\pi i x(k-N_1/2-N_2-1/2)/L},$$

$$b_{j,k} := (k-1)\int_0^L v(x) e^{2\pi i x(j+k-N_1-2N_1-1)/L}\, dx.$$

Now put $P(l) \mapsto N_1/2 + N_2 + 1/2 + \tilde{Q}(l)$, $\tilde{Q}(l) \in \{-(N_1/2+N_2-1/2), \dots, (N_1/2+N_2-1/2)\}$ (a set of half odd integers). Making the restriction $\tilde{Q}(2l) > \tilde{Q}(2l-1)$, as in the proof of Proposition 6.1.8 we then have

$$Z_{N_1,N_2}[u,v] = C(-i)^{N_1/2} \frac{1}{(N_1/2)!N_2!} \sum_{\tilde{Q}(2l)>\tilde{Q}(2l-1)} \varepsilon(\tilde{Q}) \prod_{l=1}^{N_1/2} \alpha_{-\tilde{Q}(2l-1),\tilde{Q}(2l)} \prod_{l=N_1/2+1}^{N_1/2+N_2} \beta_{-\tilde{Q}(2l-1),\tilde{Q}(2l)},$$

where α and β are defined as in the statement of the proposition. The first formula in (6.159) now results as a consequence of the definition (6.10) of a Pfaffian, while the second equality follows from (6.12) and interchanging rows in the matrix. □

With $u = v = 1$ we see from (6.160) that

$$\alpha_{\bar{j},\bar{k}} = -\frac{L^2}{\bar{j}\pi i}\delta_{\bar{j},\bar{k}}, \qquad \beta_{\bar{j},\bar{k}} = 2L\bar{j}\delta_{\bar{j},\bar{k}}, \tag{6.161}$$

and thus

$$Z_{N_1,N_2} = C\Big(\frac{L}{2\pi}\Big)^{N_1/2} (8L)^{N_1/2+N_2} \frac{(N_1/2+N_2)!}{(N_1+2N_2)!} [\zeta^{N_1/2}] \prod_{l=1}^{N_1/2+N_2} \Big(\zeta + \Big(l-\frac{1}{2}\Big)^2\Big). \tag{6.162}$$

6.7.2 Evaluation of the correlation functions

For the one-component log-gas the general n-particle correlation function has been denoted $\rho_{(n)}(x_1,\dots,x_n)$. For the present two-component log-gas we will denote the correlation function formed by specifying the coordinates of n_1 particles of charge $+1$ and n_2 particles of charge $+2$ by $\rho_{+1^{n_1},+2^{n_2}}$. Analogous to (5.20) these correlations can be calculated from the generalized partition function $Z_{N_1,N_2}[u,v]$ according to the formula

$$\rho_{+1^{n_1},+2^{n_2}}(x_1,\dots,x_{n_1};y_1,\dots,y_{n_2}) = \frac{1}{Z_{N_1,N_2}} \frac{\delta^{n_1+n_2}}{\delta u(x_1)\cdots\delta u(x_{n_1})\delta v(y_1)\cdots\delta v(y_{n_2})} Z_{N_1,N_2}[u,v]\Big|_{u=v=1}. \tag{6.163}$$

Our ability to compute (6.163) relies on a formula for $Z_{N_1,N_2}[u,v]$ analogous to (6.27) and (6.76).

PROPOSITION 6.7.2 *Let*

$$c_{\bar{l}} := (\beta_{\bar{l},\bar{l}} - i\zeta\alpha_{\bar{l},\bar{l}})\Big|_{u=v=1} = 2L\Big(\bar{l} + \frac{L\zeta}{2\pi\bar{l}}\Big),$$

and put

$$u(x) = 1 + f(x), \qquad v(x) = 1 + g(x). \tag{6.164}$$

Also let $e_{\bar{j}}$ denote the integral operator with kernel $e^{2\pi i\bar{j}y/L}$, and let ϵ be the integral operator on $[0, L]$ with kernel $\frac{1}{2}\mathrm{sgn}\,(x-y)$. Then we have

$$Z_{N_1,N_2}[1+f,1+g] = [\zeta^{N_1/2}]\tilde{Z}_{N_1,N_2}\Big(\det \mathbf{X}\Big)^{1/2}, \tag{6.165}$$

where $\tilde{Z}_{N_1,N_2}$ is given by (6.162) but with the operation $[\zeta^{N_1/2}]$ therein removed, and $\mathbf{X}$ is the integral operator with 4×4 matrix kernel

$$\Big[\delta_{j,k} + \sum_{\bar{l}} b_{j\bar{l}}\hat{a}_{\bar{l}k} - \delta_{j,2}\delta_{k,1}\frac{2\pi}{L}f\epsilon\Big]_{j,k=1,\dots,4},$$

$$[\hat{a}_{\bar{j}k}]_{(N_1+2N_2)\times 4} := \Big[\frac{\zeta L}{\pi c_{\bar{j}}\bar{j}}fe_{-\bar{j}} \quad -\frac{\zeta L}{i\pi c_{\bar{j}}}fe_{-\bar{j}} \quad \frac{\bar{j}}{c_{\bar{j}}}ge_{-\bar{j}} \quad \frac{1}{ic_{\bar{j}}}ge_{-\bar{j}}\Big]_{(N_1+2N_2)\times 4},$$

$$[b_{j\bar{k}}]_{4\times(N_1+2N_2)} := \begin{bmatrix} e_{\bar{k}} \\ \frac{1}{i\bar{k}}e_{\bar{k}} \\ e_{\bar{k}} \\ i\bar{k}e_{\bar{k}} \end{bmatrix}_{4\times(N_1+2N_2)}. \tag{6.166}$$

Proof. As in Propositions 6.3.6 and 6.1.9 the formula relies on the identity (5.26). The first step is to substitute (6.164) in (6.160) and simplify to obtain

$$\begin{aligned}\beta_{\bar{j},\bar{k}} - i\zeta\alpha_{\bar{j},\bar{k}} = c_{\bar{j}}\Big(\delta_{\bar{j},\bar{k}} &+ \frac{(\bar{k}+\bar{j})}{c_{\bar{j}}}\int_0^L g(x)e^{2\pi i(\bar{k}-\bar{j})x/L}\,dx + \frac{\zeta}{c_{\bar{j}}}\Big(\frac{L}{\pi\bar{j}}+\frac{L}{\pi\bar{k}}\Big)\int_0^L f(x)e^{2\pi i(\bar{k}-\bar{j})x/L}\,dx \\ &- \frac{2i\zeta}{c_{\bar{j}}}\int_0^L f(x)e^{2\pi i\bar{k}x/L}\epsilon(fe^{-2\pi iy\bar{j}/L})\,dx\Big).\end{aligned} \tag{6.167}$$

Factoring out $c_{\bar{j}}$, we see the determinant of (6.167) can be written in the form $\det(\mathbf{1} + \mathbf{AB})$, where $\mathbf{A}$ is the integral operator with $(N_1 + 2N_2) \times 4$ matrix kernel

$$[a_{\bar{j}k}] = [\hat{a}_{\bar{j}k}] - [\frac{2i\zeta}{c_{\bar{j}}}f\epsilon(fe_{-\bar{j}}) \quad 0 \quad 0 \quad 0],$$

with $\hat{a}_{\bar{j}k}$ specified by the first equation in (6.166), and $\mathbf{B}$ is the $4 \times (N_1 + 2N_2)$ matrix multiplication operator specified by the second equation in (6.166). On the other hand, $\mathbf{BA}$ is the integral operator with 4×4 matrix kernel

$$\Big[\sum_{\bar{l}} b_{j\bar{l}}a_{\bar{l}k}\Big]_{j,k=1,\dots,4}$$

and furthermore, analogous to (6.80), we have the factorization

$$\Big[\delta_{j,k} + \sum_{\bar{l}} b_{j\bar{l}}a_{\bar{l}k}\Big]_{j,k=1,\dots,4} = \Big[\delta_{j,k} + \sum_{\bar{l}} b_{j\bar{l}}\hat{a}_{\bar{l}k} - \delta_{j,2}\delta_{k,1}\frac{2\pi}{L}f\epsilon\Big]_{j,k=1,\dots,4}\begin{bmatrix} 1 & 0 & 0 & 0 \\ \frac{2\pi}{L}f\epsilon & 1 & 0 & 0 \\ 0 & 0 & 1 & 0 \\ 0 & 0 & 0 & 1 \end{bmatrix}.$$

Since the determinant of the final matrix is unity, we deduce (6.165) upon using the identity (5.26). □

Using (6.32) the functional derivatives required by (6.163) can be computed. First, for consistency with (6.27) and (6.76) we take the transpose of the matrix $\mathbf{X}$ in (6.165), and also introduce the notation

$$S_{\bar{j}}(x) = \sum_{\bar{l}=-(N_1/2+N_2-1/2)}^{N_1/2+N_2-1/2} \frac{e^{2\pi ix\bar{l}/L}}{c_{\bar{l}}\bar{l}^{\bar{j}}}.$$

Then, making use of (6.6), we find

$$\frac{\delta^{n_1+n_2}}{\delta u(x_1)\cdots\delta u(x_{n_1})\delta v(y_1)\cdots\delta v(y_{n_2})} Z_{N_1,N_2}[u,v]\Big|_{u=v=1}$$
$$= [\zeta^{N_1/2}]\Big(\prod_{l=1}^{N_2+N_1/2} c_l\Big)\,\mathrm{qdet}\begin{bmatrix} [\mathbf{a}(x_j-x_k)]_{j,k=1,\dots,n_1} & [\mathbf{b}(x_j-y_k)]_{\substack{j=1,\dots,n_1\\ k=1,\dots,n_2}} \\ [\mathbf{c}(y_j-x_k)]_{\substack{j=1,\dots,n_2\\ k=1,\dots,n_1}} & [\mathbf{d}(y_j-y_k)]_{j,k=1,\dots,n_2} \end{bmatrix}, \tag{6.168}$$

where

$$\mathbf{a}(x) = \frac{\zeta L}{\pi}\begin{bmatrix} S_1(x) & -\frac{\pi^2}{L^2\zeta}\mathrm{sgn}(x) - iS_2(x) \\ iS_0(x) & S_1(x) \end{bmatrix}, \quad \mathbf{b}(x) = \sqrt{\frac{\zeta L}{\pi}}\begin{bmatrix} S_1(x) & -iS_0(x) \\ iS_0(x) & S_{-1}(x) \end{bmatrix}$$

$$\mathbf{c}(x) = \sqrt{\frac{\zeta L}{\pi}}\begin{bmatrix} S_{-1}(x) & -iS_0(x) \\ iS_0(x) & S_1(x) \end{bmatrix}, \quad \mathbf{d}(x) = \begin{bmatrix} S_{-1}(x) & -iS_{-2}(x) \\ iS_0(x) & S_{-1}(x) \end{bmatrix}.$$

This, together with (6.162), explicitly determines $\rho_{+1^{n_1},+2^{n_2}}$.

EXERCISES 6.7 1. In the case $N_2 = 1$, use the Morris integral evaluation (4.4) to evaluate $Z_{N_1,N_2}[1,1]$ as specified by (6.158), and verify that this agrees with the value given by (6.162).

Chapter Seven

Scaled limits at $\beta = 1, 2$ and 4

In the previous two chapters the general n-point correlations for log-gas systems at $\beta = 1, 2$ and 4 have been evaluated as determinants or quaternion determinants with entries given in terms of certain correlation kernels. In the classical cases the explicit form of the latter can readily be analyzed in certain scaling limits. Our main technique is to make use of known asymptotic expansions of the classical polynomials. The bulk, soft edge and hard edge scaling limits all refer to moving the origin to the respective portions of the support of the density, choosing the length scale to be of the order of the mean inter-particle spacing, then taking $N \to \infty$. Knowledge of the scaled form of the bulk two-, three- and four-point correlations is used to obtain the first order expansion in $\beta - \beta_0$, $\beta_0 = 1, 2$ and 4 for the two-point correlations. The density itself is also analyzed in terms of a global coordinate, which maps the support of the density to a finite interval, and this allows corrections to the Wigner and Marčenko-Pastur laws to be determined. In the case of the GUE minor process, when the minors differ by $\mathrm{O}(N^{2/3})$ in size, a soft edge correlation which occurs in the Dyson Brownian motion model of the GUE is obtained. We also compute the free energy and bulk correlations for the log-gas system with charge ratio 1:2. This requires the use of a particular local limit theorem relating to the coefficients of a polynomial given in factorized form. Asymptotic expansions of the various two-point scaled correlations are determined, revealing some simple expressions which are later explained from a macroscopic viewpoint in Chapter 14. The correlations in the bulk at $\beta = 2$ have been conjectured to be identical to the correlations between the large Riemann zeros in the theory of prime numbers; an introduction to this topic is given within this chapter.

7.1 SCALED LIMITS AT $\beta = 2$ — GAUSSIAN ENSEMBLES

7.1.1 Bulk correlations

Proposition 1.4.4 predicts the semicircle law for the global density of the Gaussian ensembles. It follows that with the change of scale $x_j \mapsto \pi\rho x_j/\sqrt{2N}$, the bulk density (i.e., density in the neighborhood of the origin) approaches ρ as $N \to \infty$. With this scale, the corresponding Christoffel-Darboux sum, multiplied by the scale factor $\pi\rho/\sqrt{2N}$, tends to a well defined limit as $N \to \infty$.

Proposition 7.1.1 *Denote by $K_N^{(\mathrm{G})}(x,y)$ (5.6) with $p_n \mapsto p_n^{(\mathrm{G})}$, $w_2 \mapsto w_2^{(\mathrm{G})}(x)$. We have*

$$K^{\mathrm{bulk}}(x,y) := \lim_{N\to\infty} \frac{\pi\rho}{\sqrt{2N}} K_N^{(\mathrm{G})}\Big(\frac{\pi\rho x}{\sqrt{2N}}, \frac{\pi\rho y}{\sqrt{2N}}\Big) = \frac{\sin \pi\rho(x-y)}{\pi(x-y)}$$

and thus for the Gaussian ensemble with $\beta = 2$

$$\rho_{(n)}^{\mathrm{bulk}}(x_1,\dots,x_n) := \lim_{N\to\infty} \Big(\frac{\pi\rho}{\sqrt{2N}}\Big)^n \rho_{(n)}\Big(\frac{\pi\rho x_1}{\sqrt{2N}},\dots,\frac{\pi\rho x_n}{\sqrt{2N}}\Big) = \rho^n \det\Big[\frac{\sin \pi\rho(x_j - x_k)}{\pi\rho(x_j - x_k)}\Big]_{j,k=1,\dots,n},$$

where the diagonal elements in the determinant are equal to unity. The corresponding scaled truncated n-

particle correlation function is given by

$$\rho_{(n)}^{T\,\mathrm{bulk}}(x_1,\dots,x_n) = (-1)^{n-1}\rho^n \sum_{\substack{\text{cycles}\\ \text{length } n}} \frac{\sin \pi\rho(x_{i_1}-x_{i_2})}{\pi\rho(x_{i_1}-x_{i_2})} \cdots \frac{\sin \pi\rho(x_{i_n}-x_{i_1})}{\pi\rho(x_{i_n}-x_{i_1})},$$

where the sum is over all distinct cycles $i_1 \to i_2 \to \cdots \to i_n \to i_1$ *of* $\{1,\dots,n\}$ *which are of length* n.

Proof. The first result follows immediately from the asymptotic formula [508]

$$\frac{\Gamma(n/2+1)}{\Gamma(n+1)} e^{-x^2/2} H_n(x) = \cos(\sqrt{2n+1}x - n\pi/2) + \mathrm{O}(n^{-1/2}), \tag{7.1}$$

together with Stirling's formula, applied to the r.h.s. of the Christoffel-Darboux formula (5.10) (a simple trigonometric formula is also required). The second result follows from the first and Proposition 5.1.2. The formula for the truncated correlation follows from (5.9). □

Letting $\rho_{(2)}^{\mathrm{bulk}}(x,y)$ now denote the bulk scaled two-point correlation, Proposition 7.1.1 gives

$$\rho_{(2)}^{\mathrm{bulk}}(x,y) = \rho^2 - \rho^2 \left(\frac{\sin \pi\rho(x-y)}{\pi\rho(x-y)} \right)^2. \tag{7.2}$$

We know that for the Gaussian ensemble with $\beta = 2$, $\rho_{(2)}(x,y)$ can be realized by calculating the two-point correlation function for the eigenvalues of random Hermitian matrices. Of course the limit $N \to \infty$ is not accessible, so we consider instead a finite value ($N = 15$) and calculate $\rho_{(2)}$ empirically using the middle (8th) eigenvalue as the origin, and scaling the mean spacing between the middle eigenvalue and its two neighbors to unity (i.e., $\rho = 1$). To compute $\rho_{(2)}$ empirically the quantities $p(n;s)$ ($n = 0,\dots,3$) — the p.d.f.'s for the event that there are exactly n eigenvalues a distance s from the middle eigenvalue — are computed from the eigenvalues of 5,000 computer-generated random Hermitian matrices, and added together according to the second formula of (8.18) below. The results of the empirical calculation are compared with the theoretical prediction in Figure 7.1.

A straightforward calculation using (7.2) allows the dimensionless Fourier transform

$$\tilde{S}(k) := 1 + \frac{1}{\rho}\int_{-\infty}^{\infty} \rho_{(2)}^T(x,0) e^{i\rho x k}\, dx, \tag{7.3}$$

referred to as the dimensionless *structure function*, to be calculated as

$$\tilde{S}(k) = \begin{cases} \frac{|k|}{2\pi}, & |k| < 2\pi, \\ 1, & |k| \ge 2\pi. \end{cases} \tag{7.4}$$

7.1.2 Perturbation about $\beta = 2$ in the bulk

Knowledge of the higher order correlation function at $\beta = 2$ can be used to compute the first order correction, in powers of $(\beta - 2)$, to $\rho_{(2)}(0,x)$ at $\beta = 2$ in the bulk [207]. Explicitly, introducing the dimensionless correlation

$$g_{(n)}(x_1,\dots,x_n) := \rho_{(n)}(x_1,\dots,x_n)/\rho^n$$

we can use knowledge of $g_{(n)}(x_1,\dots,x_n)$ for $n = 2,3$ and 4 at $\beta = \beta_0$ to expand $g_{(2)}(x_1,x_2) = g_{(2)}(x_1,x_2;\beta)$ about $\beta = \beta_0$ (here $\beta_0 = 2$) to first order in $\beta - \beta_0$. Beginning with the definition of $g_{(2)}(x_1,x_2;\beta)$ in the finite one-component log-gas on a circle, expanding the integrand in the numerator and denominator about β_0 to first order, then making use again of the definition of the correlations gives that in

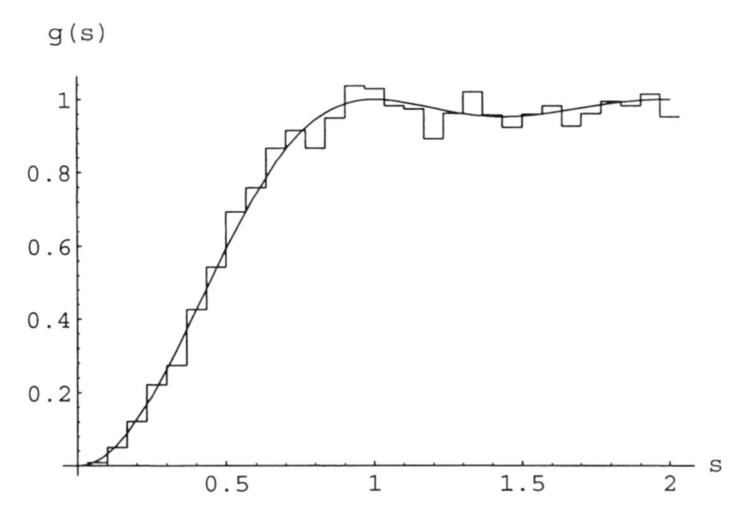

Figure 7.1 Comparison between the theoretical prediction for the two-point correlation function $g(s) := \rho_{(2)}(s,0)/\rho^2$ for infinite GUE matrices in the bulk, scaled so that the density is unity, and the empirical two-point correlation function for 5,000 computer-generated 15×15 matrices from the GUE.

the bulk limit, up to terms $O((\beta - \beta_0)^2)$ [313]

$$
\begin{aligned}
g_{(2)}(x_1, x_2; \beta) = g_{(2)}(x_1, x_2) + (\beta - \beta_0)\Big\{ & - g_{(2)}(x_1, x_2)\Phi(x_1, x_2) \\
& -2\rho \int_{-\infty}^{\infty} \Big(g_{(2)}(x_1, x_2, x_3) - g_{(2)}(x_1, x_2) \Big) \Phi(x_1, x_3)\, dx_3 \\
& -\frac{1}{2}\rho^2 \int_{-\infty}^{\infty}\int_{-\infty}^{\infty} \Big(g_{(2)}(x_1, x_2, x_3, x_4) - g_{(2)}(x_1, x_2) g_{(2)}(x_3, x_4) - g_{(2)}(x_1, x_2, x_3) \\
& \qquad - g_{(2)}(x_1, x_2, x_4) + 2 g_{(2)}(x_1, x_2) \Big) \Phi(x_3, x_4)\, dx_3 dx_4 \Big\},
\end{aligned} \tag{7.5}
$$

where $\Phi(x_1, x_2) := -\log|x_1 - x_2|$ and the correlations on the right hand side are evaluated at $\beta = \beta_0$.

For $\beta_0 = 2$ in (7.5), substituting for the correlations using Proposition 7.1.1 and simplifying by expanding out the determinants shows that up to terms $O((\beta - 2)^2)$

$$
\begin{aligned}
g_{(2)}(x_1, x_2; \beta) \; & = 1 - \Big(P_2(x_1, x_2) \Big)^2 + (\beta - 2)\Big\{ - (1 - (P_2(x_1, x_2))^2)\Phi(x_1, x_2) \\
& -2 \int_{-\infty}^{\infty} \Big(- (P_2(x_2, x_3))^2 - (P_2(x_1, x_3))^2 + 2P_2(x_1, x_2)P_2(x_2, x_3)P_2(x_3, x_1) \Big) \Phi(x_1, x_3)\, dx_3 \\
& -\frac{1}{2} \int_{-\infty}^{\infty}\int_{-\infty}^{\infty} \Big(4P_2(x_1, x_3)P_2(x_3, x_4)P_2(x_4, x_1) - 4P_2(x_1, x_2)P_2(x_2, x_3)P_2(x_3, x_4)P_2(x_4, x_1) \\
& -2P_2(x_1, x_3)P_2(x_3, x_2)P_2(x_2, x_4)P_2(x_4, x_1) + 2\Big(P_2(x_1, x_3)\Big)^2 \Big(P_2(x_2, x_4)\Big)^2 \Big) \Phi(x_3, x_4)\, dx_3 dx_4 \Big\}
\end{aligned} \tag{7.6}
$$

where

$$
P_2(x, y) := \frac{\sin \pi\rho(x - y)}{\pi\rho(x - y)}.
$$

It turns out that these integrals can be computed exactly in terms of elementary functions, together with the sine and cosine integrals

$$\mathrm{si}\,(x) := -\int_x^\infty \frac{\sin t}{t}\,dt, \qquad \mathrm{ci}\,(x) := -\int_x^\infty \frac{\cos t}{t}\,dt,$$

with the final result being

$$\begin{aligned} g_{(2)}(0,x;\beta) = 1 &- \Big(\frac{\sin \pi\rho x}{\pi\rho x}\Big)^2 \\ &+(\beta-2)\Big\{\frac{1}{2}\Big(\frac{\sin\pi\rho x}{\pi\rho x}\Big)^2 - \frac{\sin 2\pi\rho x}{2\pi\rho x} + \mathrm{ci}\,(2\pi\rho x) \\ &+\frac{1}{2(\pi\rho x)^2}\Big((\log 2\pi\rho|x| + C)\cos 2\pi\rho x - \mathrm{ci}\,(2\pi\rho x)\Big)\Big\} + \mathrm{O}((\beta-2)^2) \end{aligned} \tag{7.7}$$

(here C denotes Euler's constant). Moreover, the corresponding dimensionless structure function is given in terms of elementary functions only,

$$\tilde{S}(k;\beta) = \begin{cases} \frac{|k|}{2\pi} + (\beta-2)\Big\{\frac{1}{2}\log\Big(1-\frac{k^2}{(2\pi)^2}\Big) + \frac{|k|}{4\pi}\log\frac{2\pi+|k|}{2\pi-|k|} - \frac{|k|}{4\pi}\Big\}, & |k|<2\pi, \\ 1+(\beta-2)\Big\{\frac{1}{2}\log\frac{|k|+2\pi}{|k|-2\pi} + \frac{|k|}{4\pi}\log\Big(1-\frac{(2\pi)^2}{k^2}\Big) - \frac{\pi}{|k|}\Big\}, & |k|>2\pi, \end{cases} \tag{7.8}$$

up to terms $\mathrm{O}((\beta-2)^2)$.

7.1.3 Soft edge scaling limit

The edges of the spectrum are predicted by Proposition 1.4.4 to occur at $\pm\sqrt{2N}$. Since the eigenvalues are not confined by a wall at these points but have a nonzero density on either side, each edge is referred to as a *soft edge*. For x in the neighborhood of the right edge ($\sqrt{2N}$) we have the large N asymptotic formula [508]

$$\exp(-x^2/2)H_N(x) = \pi^{1/4}2^{N/2+1/4}(N!)^{1/2}N^{-1/12}\Big(\pi\mathrm{Ai}(t) + \mathrm{O}(N^{-2/3})\Big), \tag{7.9}$$

where $x = (2N)^{1/2} + 2^{-1/2}N^{-1/6}t$ and with $\mathrm{Ai}(x)$ denoting the Airy function. The Airy function in turn can be specified by the integral representation

$$\mathrm{Ai}(x) = \int_{\mathcal{A}} e^{-xv+v^3/3}\frac{dv}{2\pi i}, \tag{7.10}$$

where the contour starts at $e^{-\pi i/3}\infty$ and finishes at $e^{\pi i/3}\infty$, following the corresponding rays asymptotically, staying in the sector $-\pi/3 < \arg z < \pi/3$. The expansion (7.9) suggests that in order to evaluate the Christoffel-Darboux sum in the neighborhood of $x = \sqrt{2N}$ we should make the change of variables

$$x = (2N)^{1/2} + \frac{X}{2^{1/2}N^{1/6}}, \qquad y = (2N)^{1/2} + \frac{Y}{2^{1/2}N^{1/6}} \tag{7.11}$$

(the factors of $1/2^{1/2}$ are chosen for later convenience). Using the above asymptotic formula and Stirling's formula, the Christoffel-Darboux sum is then readily evaluated [189].

PROPOSITION 7.1.2 *We have*

$$\begin{aligned} K^{\mathrm{soft}}(X,Y) &:= \lim_{N\to\infty}\frac{1}{2^{1/2}N^{1/6}}K_N^{(\mathrm{G})}\Big((2N)^{1/2}+\frac{X}{2^{1/2}N^{1/6}}, (2N)^{1/2}+\frac{Y}{2^{1/2}N^{1/6}}\Big) \\ &= \frac{\mathrm{Ai}(X)\mathrm{Ai}'(Y) - \mathrm{Ai}(Y)\mathrm{Ai}'(X)}{X-Y}, \end{aligned} \tag{7.12}$$

where $K_N^{(\mathrm{G})}(x,y)$ *is given by Proposition 5.1.3 with* $p_N(x) = p_N^{(\mathrm{G})}(x)$.

We see from the formula of Proposition 5.1.2 that to specify $\rho_{(n)}$ we must also calculate $K^{\rm soft}(X,X)$. This can be deduced from (7.12) by taking the limit $Y \to X$ and simplifying using the fact that $\mathrm{Ai}(x)$ satisfies the differential equation $y''(x) = xy(x)$. This gives

$$K^{\rm soft}(X,X) = -X(\mathrm{Ai}(X))^2 + (\mathrm{Ai}'(X))^2. \tag{7.13}$$

7.1.4 Soft edge scaling of the perturbed Hermite kernel

The perturbed Hermite kernel is given by (5.172), and thus consists of the unperturbed kernel, which has the soft edge scaling (7.12), plus a sum of r correction terms. The latter depend on parameters $a_1, \dots, a_r$, which to give a well-defined soft edge limit must be replaced by the parameters $s_1, \dots, s_r$ according to the scaling [442], [132]

$$a_k = -\sqrt{2N} + \sqrt{2}N^{1/6} s_k.$$

To specify the limiting functional form, we introduce a class of incomplete multiple Airy functions

$$\begin{aligned} \tilde{\mathrm{Ai}}^{(j)}(x) &:= \int_{\mathcal{A}_{\{s_1,\dots,s_j\}}} \frac{e^{-xv+v^3/3}}{\prod_{k=1}^j (v-s_k)} \frac{dv}{2\pi i}, \\ \mathrm{Ai}^{(j)}(x) &:= (-1)^j \int_{\mathcal{A}} e^{-xv+v^3/3} \prod_{k=1}^{j-1} (v+s_k) \frac{dv}{2\pi i}. \end{aligned} \tag{7.14}$$

Here $\mathcal{A}_{\{s_1,\dots,s_i\}}$ is a contour which starts at $e^{-\pi i/3}\infty$ and finishes at $e^{\pi i/3}\infty$, following the corresponding rays asymptotically, staying in the sector $\arg z > \pi/3$, $\arg z < -\pi/3$, and crossing the real axis to the left of $\{s_k\}$. The contour $\mathcal{A}$ is defined as in (7.10).

PROPOSITION 7.1.3 *For large N*

$$\begin{aligned} e^{-x^2/2}\tilde{\Gamma}^{(j)}(x)\Big|_{\substack{x=\sqrt{2N}+X/2^{1/2}N^{1/6} \\ a_k=\sqrt{2N}(-1+s_k/N^{1/3})}} &= (-1)^{N+r+1} \frac{N^{(j-1)/3}}{(2N)^{(N+j-r-1)/2}} \tilde{\mathrm{Ai}}^{(j)}(x)\Big(1+\mathrm{O}(N^{-1/3})\Big), \\ e^{x^2/2}\Gamma^{(j)}(x)\Big|_{\substack{x=\sqrt{2N}+X/2^{1/2}N^{1/6} \\ a_k=\sqrt{2N}(-1+s_k/N^{1/3})}} &= (-1)^{N+r+1} \frac{(2N)^{(N+j-r)/2}}{N^{j/3}} \mathrm{Ai}^{(j)}(x)\Big(1+\mathrm{O}(N^{-1/3})\Big). \end{aligned} \tag{7.15}$$

Proof. Consider the first formula. In the definition (5.171) make the change of variables $z \mapsto \sqrt{2N}(-1+w/N^{1/3})$. This shows

$$\begin{aligned} & e^{-x^2/2}\tilde{\Gamma}^{(j)}(x)\Big|_{\substack{x=\sqrt{2N}+X/2^{1/2}N^{1/6} \\ a_k=\sqrt{2N}(-1+s_k/N^{1/3})}} \\ &= (-1)^{N-r} \frac{N^{(j-1)/3}}{(\sqrt{2N})^{N-r+j-1}} \int_{\mathcal{C}_{\{N^{1/3},s_1,\dots,s_r\}}} \frac{e^{-wX} e^{-N^{2/3}w} e^{-w^2 N^{1/3}/2}}{(1-w/N^{1/3})^{N-r} \prod_{k=1}^j (w-s_k)} \frac{dw}{2\pi i}. \end{aligned}$$

But for large N

$$\frac{e^{-N^{2/3}w} e^{-w^2 N^{1/3}/2}}{(1-w/N^{1/3})^{N-r}} \sim e^{w^3/3}(1+\mathrm{O}(N^{-1/3})),$$

and the method of steepest descent says we must deform the contour to rays such that the exponent is minimized. This occurs along the rays $\arg w = \pm\pi/3$, giving the first formula in (7.15). The second asymptotic formula is derived similarly. □

It follows immediately from these asymptotic formulas, and (7.12), that with the scaled variables X, Y,

$\{s_k\}$ specified by

$$x = \sqrt{2N} + X/2^{1/2}N^{1/6}, \quad y = \sqrt{2N} + Y/2^{1/2}N^{1/6}, \quad a_k = \sqrt{2N}(-1 + s_k/N^{1/3}), \tag{7.16}$$

the perturbed Hermite kernel (5.172) has the soft edge scaled form

$$\lim_{N\to\infty} \frac{1}{\sqrt{2}N^{1/6}} e^{-x^2/2+y^2/2} K_N(x,y) = K^{\rm soft}(X,Y) + \sum_{j=1}^{r} \tilde{\rm Ai}^{(j)}(X){\rm Ai}^{(j)}(Y). \tag{7.17}$$

We see from (7.10) and (7.14) that

$${\rm Ai}^{(1)}(x) = {\rm Ai}(x), \qquad \tilde{\rm Ai}^{(1)}(x) = \int_{-\infty}^{x} e^{s_1(x-t)} {\rm Ai}(t)\, dt,$$

so in the case $r = 1$ this simplifies to read

$$K^{\rm soft}(X,Y) + {\rm Ai}(Y) \int_{-\infty}^{X} e^{s_1(X-t)} {\rm Ai}(t)\, dt. \tag{7.18}$$

7.1.5 Soft edge of GUE minor process

The correlations for the GUE minor process are given by (5.198), with all quantities referring to the Gaussian case. They permit a soft edge scaling, in which the species are separated by $O(N^{2/3})$ [215]. The limiting correlation is the so-called dynamical extension of the Airy kernel, specified by

$$\rho_{(n)}^{\rm soft}((\tau_1, y_1), \dots, (\tau_n, y_n)) = \det[K^{\rm soft}((\tau_j, y_j), (\tau_k, y_k))]_{j,k=1,\dots,n}, \tag{7.19}$$

where

$$K^{\rm soft}((\tau_x, x), (\tau_y, y)) = \begin{cases} \displaystyle\int_0^\infty e^{-(\tau_y-\tau_x)u} {\rm Ai}(x+u){\rm Ai}(y+u)\, du, & \tau_y \ge \tau_x, \\ \displaystyle -\int_{-\infty}^0 e^{-(\tau_y-\tau_x)u} {\rm Ai}(x+u){\rm Ai}(y+u)\, du, & \tau_y < \tau_x \end{cases}$$

(see Section 11.7 below).

PROPOSITION 7.1.4 *In the Gaussian case of (5.198), introduce the scalings*

$$s_i = N - 2c_i N^{2/3}, \qquad y_i = (2s_i)^{1/2} + \frac{Y_i}{\sqrt{2}s_i^{1/6}}.$$

One has

$$\lim_{N\to\infty} \Big(\frac{1}{\sqrt{2}N^{1/6}}\Big)^r \rho_{(r)}(\{(s_j, y_j)\}_{j=1,\dots,r}) = \det[K^{\rm soft}((c_j, X_j), (c_k, X_k))]_{j,k=1,\dots,r}. \tag{7.20}$$

Proof. Substituting in (5.198) the appropriate Gaussian quantities we obtain, for $c_l \ge c_j$,

$$K(s_j, y_j; s_l, y_l) = \frac{e^{-y_j^2}}{\sqrt{\pi}} \sum_{k=1}^{s_l} \frac{1}{2^{s_l-k}(s_l-k)!} H_{s_j-k}(y_j) H_{s_l-k}(y_l), \tag{7.21}$$

while for $c_l < c_j$ the r.h.s. is to be modified by multiplying by -1 and changing the summation to $k \in \mathbb{Z}_{\le 0}$. Only the former case will be considered explicitly, as the latter is essentially the same. For the analysis of the sum (7.21) we substitute (7.9) to obtain

$$K(s_j, y_j; s_l, y_l) \sim e^{-N^{1/3}(Y_j - Y_l)} 2^{-(c_j - c_l)N^{2/3}} 2^{1/2} N^{-1/6}$$
$$\times \sum_{k=1}^{N} \Big(\frac{(N - 2c_j N^{2/3} - k)!}{(N - 2c_l N^{2/3} - k)!} \Big)^{1/2} \mathrm{Ai}(Y_j + 2k/(2N)^{1/3}) \mathrm{Ai}(Y_l + 2k/(2N)^{1/3}). \quad (7.22)$$

Noting that the leading order contribution to the summation comes from k of order $N^{1/3}$, then using Stirling's formula to simplify the ratio of factorials in this regime, we can recognize the sum as the Riemann sum approximation to the first integral in (7.20). □

7.1.6 Global limit of density

The scaled global density (1.53) was computed in Exercises 1.6 q.1. Here, following [257], we will show how this same result can be derived from the formula $\rho_{(1)}(x) = K_N(x, x)$, where $K_N(x, x)$ is given by (5.13) with the quantities therein specified in the Gaussian case of Section 5.4.1, but with x scaled $x \mapsto \sqrt{2N} x$, so the weight becomes e^{-2Nx^2}. With this scaling the relevant monic orthogonal polynomials are

$$\tilde{p}_n^{(\mathrm{G})}(x) = 2^{-3n/2} N^{-n/2} H_n(\sqrt{2N} x).$$

Making use of the differentiation formula for the Hermite polynomials we have that

$$\frac{d}{dx} \tilde{p}_n^{(\mathrm{G})}(x) = n \tilde{p}_{n-1}^{(\mathrm{G})}(x),$$

and thus

$$K_N^{(\mathrm{G})}(x, x) = \frac{e^{-2Nx^2}}{(\tilde{p}_{N-1}^{(\mathrm{G})}, \tilde{p}_{N-1}^{(\mathrm{G})})_2} \Big(N (\tilde{p}_{N-1}^{(\mathrm{G})}(x))^2 - (N-1) \tilde{p}_{N-2}^{(\mathrm{G})}(x) \tilde{p}_N^{(\mathrm{G})}(x) \Big). \quad (7.23)$$

The asymptotic form of $H_n(\sqrt{2N} x)$ for n near N is given by the Plancheral-Rotach formula [508]. It tells us that for $x \in (-1, 1)$

$$\tilde{p}_{N+m}^{(\mathrm{G})}(x) = \sqrt{\frac{2}{\pi}} \frac{2^{-m}}{(1 - x^2)^{1/4}} (\tilde{p}_N^{(\mathrm{G})}, \tilde{p}_N^{(\mathrm{G})})_2 \, e^{Nx^2} \cos\Big(f_N(x) - m \arccos x \Big) \Big(1 + \mathrm{O}\Big(\frac{1}{N}\Big) \Big),$$
$$f_N(x) := N x \sqrt{1 - x^2} - \Big(N + \frac{1}{2} \Big) \arccos x + \frac{\pi}{4}. \quad (7.24)$$

Substituting this in (7.23), with $N \mapsto N + 1$ therein, and making use of the trigonometric identity

$$\cos^2 a - \cos(a - b) \cos(a + b) = 1 - \cos^2 b,$$

we reclaim (1.54).

Asymptotic analysis can also be carried out on integral representations of (7.23) [341], [258]. This allows correction terms to the Wigner semicircle law to be obtained. In particular, with $|x| < 1$ fixed, one finds

$$\sqrt{2/N} \rho_{(1)}(\sqrt{2N} x) = \rho^{\mathrm{W}}(x) - \frac{2 \cos(2N\pi P(x))}{\pi^3 (\rho^{\mathrm{W}}(x))^2} \frac{1}{N} + \mathrm{O}\Big(\frac{1}{N^2}\Big), \quad (7.25)$$

where

$$\rho^{\mathrm{W}}(x) = \frac{2}{\pi} \sqrt{1 - x^2}, \qquad P(x) = 1 + \frac{x}{2} \rho^{\mathrm{W}}(x) - \frac{1}{\pi} \arccos x. \quad (7.26)$$

The analysis of [341] also applies at the soft edge, where it gives

$$\frac{1}{2^{1/2}N^{1/6}}\rho_{(1)}((2N)^{1/2}+\frac{X}{2^{1/2}N^{1/6}})$$
$$=\rho_{(1)}^{\mathrm{soft},2}(X)-\frac{1}{20}\Big(3X^2(\mathrm{Ai}(X))^2-2X(\mathrm{Ai}'(X))^2-3\mathrm{Ai}(X)\mathrm{Ai}'(X)\Big)\frac{1}{N^{2/3}}+\mathrm{O}\Big(\frac{1}{N}\Big) \tag{7.27}$$

with $\rho_{(1)}^{\mathrm{soft},2}(X)$ given by (7.13) (the 2 in the superscript soft,2 indicates $\beta=2$; below we will encounter $\rho_{(1)}^{\mathrm{soft},\beta}(X)$ for other values of β).

EXERCISES 7.1 1. Show from (7.8) that

$$\frac{\partial\tilde{S}(k;\beta)}{\partial\beta}\Big|_{\beta=2}=-\frac{|k|}{4\pi}+\sum_{j=1}^{\infty}\frac{1}{2j(2j-1)}\Big(\frac{|k|}{2\pi}\Big)^{2j},\qquad |k|<2\pi. \tag{7.28}$$

7.2 SCALED LIMITS AT $\beta=2$ — LAGUERRE AND JACOBI ENSEMBLES

7.2.1 Laguerre ensemble — hard edge

The global density for the Laguerre ensemble is predicted by Proposition 3.2.3. In particular the density of the large eigenvalues is identical to the density of the large eigenvalues in the Gaussian ensemble. In contrast, the small eigenvalues in the Laguerre ensemble bunch together as the origin is approached, giving rise to the $x^{-1/2}$ divergence. Due to the hard wall at $x=0$ in the Laguerre ensemble this is referred to as the *hard edge* of the spectrum, whereas the boundary of the support of the large eigenvalues is a soft edge, as there is no wall in the statistical mechanics interpretation.

To compute the correlation functions in the neighborhood of the origin, we first change scales so that the average inter-particle (inter-eigenvalue) spacing is O(1) in the $N\to\infty$ limit. From the large N asymptotic formula [508, p.199]

$$e^{-x/2}x^{a/2}L_N^a(x)\sim M^{-a/2}\frac{\Gamma(N+a+1)}{N!}J_a(2(Mx)^{1/2})+\mathrm{O}(N^{a/2-3/4}),\quad a>-1,$$
$$\sim N^{a/2}J_a(2(Nx)^{1/2}), \tag{7.29}$$

where $M=N+(a+1)/2$ and $J_a(z)$ denotes the Bessel function, we see that the appropriate choice of scale is provided by the change of variable

$$x=\frac{X}{4N} \tag{7.30}$$

(the factor of $\frac{1}{4}$ is chosen for convenience). With this scale, by using the above asymptotic formula and Stirling's formula, the Christoffel-Darboux summation is readily evaluated in the $N\to\infty$ limit [189].

PROPOSITION 7.2.1 *We have*

$$K_a^{\mathrm{hard}}(X,Y):=\lim_{N\to\infty}\frac{1}{4N}K_N^{(\mathrm{L})}\Big(\frac{X}{4N},\frac{Y}{4N}\Big)$$
$$=\frac{J_a(X^{1/2})Y^{1/2}J_a'(Y^{1/2})-X^{1/2}J_a'(X^{1/2})J_a(Y^{1/2})}{2(X-Y)}, \tag{7.31}$$

where $K_N^{(\mathrm{L})}(x,y)$ is given by Proposition 5.1.3 with $p_N(x)=p_N^{(\mathrm{L})}(x)$.

By taking the limit $Y\to X$ in the above formula and using the Bessel function identities

$$uJ_\alpha'(u)=\alpha J_\alpha(u)-uJ_{\alpha+1}(u),\quad J_\alpha'(u)=J_{\alpha-1}(u)-\frac{\alpha}{u}J_\alpha(u), \tag{7.32}$$

we see from (7.31) that

$$K_a^{\rm hard}(X,X) = \frac{1}{4}\Big((J_a(X^{1/2}))^2 - J_{a+1}(X^{1/2})J_{a-1}(X^{1/2}) \Big). \tag{7.33}$$

The Bessel function identities also show that (7.31) can be rewritten

$$K_a^{\rm hard}(X,Y) = \frac{X^{1/2}J_{a+1}(X^{1/2})J_a(Y^{1/2}) - Y^{1/2}J_{a+1}(Y^{1/2})J_a(X^{1/2})}{2(X-Y)}. \tag{7.34}$$

7.2.2 Laguerre ensemble — soft edge

It was remarked above that at the soft edge the global density in the Laguerre ensemble is the same as the global density at the spectrum edge in the Gaussian ensemble. In fact, use of the asymptotic expansion [508, p.201]

$$e^{-x/2}L_n^a(x) = (-1)^n 2^{-a-1/3} n^{-1/3}\Big({\rm Ai}(t) + {\rm O}(n^{-2/3}) \Big), \tag{7.35}$$

where $x = 4n + 2a + 2 + 2(2n)^{1/3}t$, shows that

$$\lim_{N\to\infty} 2(2N)^{1/3} K_N^{({\rm L})}(4N + 2(2N)^{1/3}X, 4N + 2(2N)^{1/3}Y) = K^{\rm soft}(X,Y). \tag{7.36}$$

Thus, with this scaling of the coordinates, all correlation functions are those of the Gaussian ensemble at the soft edge. As for the Gaussian kernel, correction terms to the l.h.s. of (7.36) regarded as a function of N can be computed using integral representations of $K_N^{(L)}$. For the corresponding density, one finds [341], [258]

$$2(2N)^{1/3}\rho_{(1)}^{({\rm L})}(4N + 2(2N)^{1/3}X) \sim \rho_{(1)}^{{\rm soft},2}(X) + \frac{a}{(2N)^{1/3}}\Big({\rm Ai}(X)\Big)^2 + {\rm O}(N^{-2/3}). \tag{7.37}$$

7.2.3 Laguerre ensemble — global density

The result (3.57) for the global density can be derived by computing the large N limit of $K_N^{({\rm L})}(4Ny, 4Ny)$. One approach is to make use of the differentiation formula for the Laguerre polynomials, as given in (7.130) below, to write the formula for $K_N^{({\rm L})}$ in terms of $\{L_n^\alpha\}$ for certain n and α, and then to make use of the Laguerre analogue of the Plancheral-Rotach formula (see, e.g., [202])

$$\begin{aligned} x^{a/2}e^{-x/2}L_{n+m}^a(x)|_{x=4nX} &= (-1)^{n+m}(2\pi\sqrt{X(1-X)})^{-1/2}n^{a/2-1/2}\Big(g_{m,n}^{({\rm L})}(X) + {\rm O}\Big(\frac{1}{n}\Big)\Big), \\ g_{m,n}^{({\rm L})}(X) &= \sin\Big(2n(\sqrt{X(1-X)} - \arccos\sqrt{X}) - (2m+a+1)\arccos\sqrt{X} + 3\pi/4\Big). \end{aligned} \tag{7.38}$$

An alternative approach is to make use of a contour integral form of $K_N^{({\rm L})}$. This has the advantage of allowing corrections to (3.57) to be computed, giving the expansion up to terms ${\rm O}(N^{-2})$ [258]

$$4\rho_{(1)}(4NX) \sim \rho^{\rm MP}(X) - \Big(\frac{\cos((2N+a)\pi P_{\rm MP}(X) - a\pi(1 + X\rho^{\rm MP}(X)))}{4\pi X(1-X)} - \frac{a}{2\pi\sqrt{X(1-X)}}\Big)\frac{1}{N}, \tag{7.39}$$

where

$$\rho^{\rm MP}(x) := \frac{2}{\pi}\sqrt{\frac{1-x}{x}}, \qquad P_{\rm MP}(x) := 1 + x\rho^{\rm MP}(x) - \frac{2}{\pi}\arccos\sqrt{x},$$

valid for $0 < X < 1$.

7.2.4 Hard edge scaling of the perturbed Laguerre kernel

Analogous to the situation with the perturbed Hermite kernel, its Laguerre counterpart (5.169) admits a scaling at both the soft and hard edges [29], [132]. At the soft edge the eigenvalues $\{x_j\}_{j=1,\dots,k}$ and parameters $\{a_j\}_{j=1,\dots,r}$ are to be scaled according to

$$x_j = 4N + 2(2N)^{1/3}X_j, \qquad a_j = -\frac{1}{2} + \frac{s_j}{2(2N)^{1/3}}, \tag{7.40}$$

with the result

$$\lim_{N\to\infty} (2(2N)^{1/3})^k \rho_{(k)}(x_1,\dots,x_k) = \det\Big[K^{\rm soft}(X_\alpha, X_\beta) + \sum_{l=1}^{r} \tilde{\rm Ai}^{(l)}(X_\alpha){\rm Ai}^{(l)}(X_\beta)\Big]_{\alpha,\beta=1,\dots,k}, \tag{7.41}$$

as found in (7.17). We will concentrate on the hard edge scaling. The limiting functional form involves incomplete multiple Bessel functions

$$\tilde{J}^{(p)}(x) = \int_{\mathcal{C}_{\{0,s_1,\dots,s_p\}}} \frac{e^{-xz+1/4z} z^{a+r}}{\prod_{k=1}^{p}(z-s_k)} \frac{dz}{2\pi i}, \quad J^{(p)}(x) = \int_{\mathcal{C}_{\{0\}}} \frac{e^{xw-1/4w}\prod_{k=1}^{p-1}(w-s_k)}{w^{a+r}} \frac{dw}{2\pi i}.$$

Explicitly, the following result holds [132].

PROPOSITION 7.2.2 *For large N, with $a_k = 4Ns_k$ $(k=1,\dots,r)$*

$$(4N)^{p-a-r-1}\tilde{\Lambda}^{(i)}\Big(\frac{X}{4N}\Big) = \tilde{J}^{(p)}(X) + {\rm O}\Big(\frac{1}{N}\Big), \qquad (4N)^{a+r-p}\Lambda^{(i)}\Big(\frac{X}{4N}\Big) = J^{(p)}(X) + {\rm O}\Big(\frac{1}{N}\Big). \tag{7.42}$$

Consequently

$$\lim_{N\to\infty} \frac{1}{4N} \mathcal{K}_N^a\Big(\frac{X}{4N}, \frac{Y}{4N}\Big)\Big|_{a_k = 4Ns_k} = \Big(\frac{Y}{X}\Big)^{(a+r)/2} K^{\rm hard}_{a+r}(X,Y) + \sum_{p=1}^{r} \tilde{J}^{(p)}(X) J^{(p)}(Y). \tag{7.43}$$

Proof. The first term in (7.43) follows immediately from (5.169), (5.167) and (7.31), while the sum in (7.43) follows from (5.169) and (7.42). To derive (7.42), substitute $a_k = 4Ns_k$ in (5.168), change variables $z \mapsto 4Nz$, $w \mapsto 4Nw$, scale the contours and make use of the elementary limit $(1+u/N)^N \to e^u$. □

From the integral representation of the Bessel function

$$J_\alpha(x) = \int_{\mathcal{C}_{\{0\}}} \frac{e^{x(z-z^{-1})/2}}{z^{\alpha+1}} \frac{dz}{2\pi i},$$

together with the formula $J_{-\alpha}(x) = (-1)^\alpha J_\alpha(x)$, both valid for $\alpha \in \mathbb{Z}$, we can check from the definitions that

$$\begin{aligned}
J^{(1)}(x) &= (4x)^{(a+r-1)/2} J_{a+r-1}(\sqrt{x}),\\
\tilde{J}^{(1)}(x) &= -2^{-(a+r+1)} \int_{-\infty}^{x} t^{-(a+r+1)/2} e^{s_1(t-x)} J_{a+r+1}(\sqrt{t})\, dt\\
&= (4x)^{-(a+r)/2} J_{a+r}(\sqrt{x}) - s_1 2^{-(a+r)} \int_{-\infty}^{x} t^{-(a+r)/2} e^{s_1(t-x)} t^{-(a+r)/2} J_{a+r}(\sqrt{t})\, dt.
\end{aligned}$$

Here the final line follows on integrating by parts, making use of the identity

$$u^{-\alpha-1}J_{\alpha+1}(u) = -\frac{1}{u}\frac{d}{du}(u^{-\alpha}J_\alpha(u)).$$

Furthermore, making use of the form (7.34), together with the Bessel function three-term recurrence

$$tJ_{\alpha+2}(t) = 2(\alpha+1)J_{\alpha+1}(t) - tJ_\alpha(t), \tag{7.44}$$

it is straightforward to verify that

$$\Big(\frac{Y}{X}\Big)^{1/2} K_{a+1}^{\rm hard}(X,Y) = K_a^{\rm hard}(X,Y) - \frac{1}{2} X^{-1/2} J_{a+1}(X^{1/2}) J_a(Y^{1/2}).$$

As a consequence of these facts, in the case $r = 1$ (7.43) can be written in terms of Bessel functions according to

$$\Big(\frac{Y}{X}\Big)^{a/2} \Big(K_a^{\rm hard}(X,Y) - \frac{s_1}{2} J_a(\sqrt{Y}) X^{a/2} \int_{-\infty}^X e^{s_1(t-X)} t^{-(a+1)/2} J_{a+1}(\sqrt{t})\, dt \Big).$$

7.2.5 Jacobi ensemble

For the Jacobi ensemble, Proposition 3.6.3 predicts the same global eigenvalue density in the neighborhood of the edges $x = \pm 1$ as for the Laguerre ensemble in the neighborhood of the edge $x = 0$. From the Jacobi polynomial large N asymptotic formula [508, p.197]

$$P_N^{(a,b)}\Big(1 - \frac{x}{2N^2}\Big) \sim \Big(\frac{x^{1/2}}{2N}\Big)^{-a} J_a(x^{1/2}) \tag{7.45}$$

we see that by making the shift of origin and change of scale $x = 1 - \frac{X}{2N^2}$ the scaled Christoffel-Darboux summation can be evaluated in the $N \to \infty$ limit to give the same expression as in Proposition 7.2.1. Thus the scaled correlations in the neighborhood of the edges $x = 0$ and $x = 1$ of the Laguerre and Jacobi ensembles, respectively, are the same. This is to be expected as in the neighborhood of these points the corresponding Boltzmann factors are proportional.

7.2.6 Circular Jacobi ensemble — spectrum singularity

Consider the n-point correlation function (5.55). Our interest is in the neighborhood of the point $z = -1$, which is analyzed by writing $x_j \mapsto x_j + L/2$ or equivalently $z_j \mapsto -z_j$. In preparation for taking the thermodynamic limit, we make use of the formula

$$P_n^{(\alpha,\beta)}(x) = \frac{\Gamma(\beta+1+2n+\alpha)}{n!\Gamma(\beta+1+n+\alpha)} \Big(\frac{x-1}{2}\Big)^n {}_2F_1\Big(-n, -n-\alpha; -2n-\alpha-\beta; -\frac{2}{x-1}\Big),$$

where ${}_2F_1$ denotes the Gauss hypergeometric function (5.83), to write

$$\begin{aligned} p_N^{(\rm CJ)}\Big(i\frac{1+z}{1-z}\Big) &= i^{-N}\Big(-\frac{2}{1-z}\Big)^N {}_2F_1(-N, a; 2a; 1-z), \\ p_{N-1}^{(\rm CJ)}\Big(i\frac{1+z}{1-z}\Big) &= i^{-(N-1)}\Big(-\frac{2}{1-z}\Big)^{N-1} {}_2F_1(-(N-1), a+1; 2a+2; 1-z). \end{aligned} \tag{7.46}$$

To take the thermodynamic limit we use the formulas

$$\begin{aligned} &\lim_{n\to\infty} {}_2F_1(-n, b; c; t/n) = {}_1F_1(b; c; -t), \quad {}_1F_1(b; c; -t) := \sum_{n=0}^\infty \frac{(b)_n}{(c)_n n!}(-t)^n, \\ &{}_1F_1(a; 2a; 2ix) = \Gamma\Big(a+\frac{1}{2}\Big)\Big(\frac{x}{2}\Big)^{-(a-1/2)} e^{ix} J_{a-1/2}(x), \end{aligned} \tag{7.47}$$

which give

$$\begin{aligned} p_N^{(\rm CJ)}\Big(i\frac{1+z_j}{1-z_j}\Big)\Big|_{\alpha=N+a} &\sim \frac{(-1)^N}{(\sin \pi x/L)^N}\Gamma\Big(a+\frac{1}{2}\Big)(\pi\rho x/2)^{-(a-1/2)} J_{a-1/2}(\pi\rho x), \\ p_{N-1}^{(\rm CJ)}\Big(i\frac{1+z_j}{1-z_j}\Big)\Big|_{\alpha=N+a} &\sim \frac{(-1)^{N-1}}{(\sin \pi x/L)^{N-1}}\Gamma\Big(a+\frac{3}{2}\Big)(\pi\rho x/2)^{-(a+1/2)} J_{a+1/2}(\pi\rho x). \end{aligned}$$

Using (5.48) to evaluate the normalization in (5.55) and using the duplication formula (4.180) we find that in the thermodynamic limit [418]

$$\rho_{(n)}^{\rm s.s.}(x_1,\dots,x_n) = \det\Big[K^{\rm s.s.}(x_j,x_k)\Big]_{j,k=1,\dots,n},$$

$$K^{\rm s.s.}(x,y) := (\pi\rho x)^{1/2}(\pi\rho y)^{1/2}\frac{\Big(J_{a+1/2}(\pi\rho x)J_{a-1/2}(\pi\rho y) - J_{a+1/2}(\pi\rho y)J_{a-1/2}(\pi\rho x)\Big)}{2(x-y)}, \tag{7.48}$$

where $\rho := N/L$, which is valid for all values of $x_1,\dots,x_n$ in the case a a non-negative integer, while it is valid for $x_1,\dots,x_n$ all positive otherwise. In the limit $y \to x$ we have

$$K^{\rm s.s.}(x,x) = \frac{(\pi\rho)^2 x}{2}\Big((J_{a-1/2}(\pi\rho x))^2 + (J_{a+1/2}(\pi\rho x))^2 - \frac{2a}{\pi\rho x}J_{a-1/2}(\pi\rho x)J_{a+1/2}(\pi\rho x)\Big). \tag{7.49}$$

Using the Bessel function formulas

$$J_{1/2}(x) = \Big(\frac{2}{\pi x}\Big)^{1/2}\sin x, \qquad J_{-1/2}(x) = \Big(\frac{2}{\pi x}\Big)^{1/2}\cos x, \tag{7.50}$$

together with the addition formula for the sine function, we see that

$$K^{\rm s.s.}(x,y)\Big|_{a=0} = \frac{\sin\pi\rho(x-y)}{\pi(x-y)} =: K^{\rm bulk}(x,y).$$

This is expected because with $a=0$ the circular Jacobi ensemble corresponds to Dyson's circular ensemble, and the latter is locally identical to the bulk of the Gaussian ensemble. Another relationship between (7.48) and the bulk correlation of Proposition 7.1.1 is given in Exercises 7.2 q.3.

7.2.7 The classical groups

There are three possible scalings of k-point correlations for the classical groups specified by Proposition 5.5.3 and (5.87). One is the bulk scaling with unit density. For the CUE this is achieved by $\theta \mapsto 2\pi x/N$, while for the cases of Proposition 5.5.3 one requires $\theta \mapsto \pi/2 + \pi x/N$. A simple calculation gives $\rho_{(k)}^{\rm bulk}$ as specified by the result of Proposition 7.1.1 with $\rho = 1$ as the scaled k-point correlation.

In addition to the bulk scaling, there are two possible edge scalings—when there is an eigenvalue at the endpoint, and when there is not. Typical of the former case is $O^+(2N+1)$ in the neighborhood of $\theta = 0$. A straightforward calculation shows

$$\begin{aligned}\rho_{(k)}^{+}(x_1,\dots,x_k) &:= \lim_{N\to\infty}\Big(\frac{\pi}{N}\Big)^k \rho_{(k)}^{O^+(2N+1)}(\pi x_1/N,\dots,\pi x_k/N)\\ &= \det\Big[\frac{\sin\pi(x_j-x_l)}{\pi(x_j-x_l)} - \frac{\sin\pi(x_j+x_l)}{\pi(x_j+x_l)}\Big]_{j,l=1,\dots,k}.\end{aligned} \tag{7.51}$$

Typical of the latter is $O^-(2N+1)$ in the neighborhood of $\theta = 0$. It follows from Proposition 5.5.3 that

$$\begin{aligned}\rho_{(k)}^{-}(x_1,\dots,x_k) &:= \lim_{N\to\infty}\Big(\frac{\pi}{N}\Big)^k \rho_{(k)}^{O^-(2N+1)}(\pi x_1/N,\dots,\pi x_k/N)\\ &= \det\Big[\frac{\sin\pi(x_j-x_l)}{\pi(x_j-x_l)} + \frac{\sin\pi(x_j+x_l)}{\pi(x_j+x_l)}\Big]_{j,l=1,\dots,k}.\end{aligned} \tag{7.52}$$

The two edge scalings can be related to the hard edge result (7.31). Thus we see that with

$$K_a^{\rm ch}(X,Y) := 2\pi^2(XY)^{1/2}K_a^{\rm hard}((\pi X)^2,(\pi Y)^2) \tag{7.53}$$

determining the edge correlations in the ensemble (3.4) with $\beta = 2$ and the bulk density unity,

$$K^{\rm ch}_{\pm 1/2}(X,Y) = \frac{\sin \pi(X-Y)}{\pi(X-Y)} \pm \frac{\sin \pi(X+Y)}{\pi(X+Y)}, \tag{7.54}$$

where use has been made of (7.50).

EXERCISES 7.2 1. Use the asymptotic formulas (7.9) and (7.29) to deduce from (5.70) and (5.71) that

$$K^{\rm soft}(X,Y) = \int_0^\infty {\rm Ai}(X+t){\rm Ai}(Y+t)\,dt, \qquad K^{\rm hard}_a(X,Y) = \frac{1}{4}\int_0^1 J_a(\sqrt{Xt})J_a(\sqrt{Yt})\,dt.$$

Note too the integral formula

$$K^{\rm bulk}(x,y) = \frac{\rho}{2}\int_{-1}^1 e^{\pi i \rho(x-y)t}\,dt.$$

2. The objective of this exercise is to show the connection between the quantities $K^{\rm soft}(X,Y)$, $K^{\rm hard}_a(X,Y)$, $K^{\rm s.s.}(X,Y)$ and $K^{\rm bulk}(X,Y)$.

(i) Use the asymptotic expansions (7.69) and (7.74) below to deduce that

$$\lim_{c\to\infty} \int_c^{c+\pi\rho x/\sqrt{c}} \rho^{\rm soft}_{(1)}(-X)\,dX = \rho x, \qquad \lim_{c\to\infty} \int_c^{c+2\pi\rho x\sqrt{c}} \rho^{\rm hard}_{(1)}(X)\,dX = \rho x$$

so the densities are asymptotically constant in the variable x.

(ii) Use the asymptotic expansions (7.68) and (7.73) to show

$$\begin{aligned} &\lim_{c\to\infty} \frac{\pi\rho}{\sqrt{c}} K^{\rm soft}(-(c+\pi\rho x/\sqrt{c}), -(c+\pi\rho y/\sqrt{c})) \\ &\quad = \lim_{c\to\infty} 2\pi\rho\sqrt{c}K^{\rm hard}_a(c+2\pi\rho x\sqrt{c}, c+2\pi\rho y\sqrt{c}) = K^{\rm bulk}(x,y). \end{aligned}$$

(iii) Use the asymptotic expansion [539]

$$J_a(x) \sim \Big(\frac{2}{a}\Big)^{1/3} {\rm Ai}\Big(\frac{2^{1/3}(a-x)}{x^{1/3}}\Big), \tag{7.55}$$

valid for a and x large such that the argument of the Airy function is order one, to show that

$$\lim_{a\to\infty} 2a(a/2)^{1/3} K^{\rm hard}_a(a^2 - 2a(a/2)^{1/3}x, a^2 - 2a(a/2)^{1/3}y) = K^{\rm soft}(x,y).$$

(iv) With $\phi(x) := \sqrt{x/2}J_{a+1/2}(x)$, show by using the Bessel function identities (7.32) that for $\rho = 1/\pi$ (7.48) can be rewritten

$$K^{\rm s.s.}(x,y) = \frac{\phi(x)y\phi'(y) - \phi(y)x\phi'(x)}{x-y}.$$

(v) Use the asymptotic expansion (7.55) in the result of (iv) to show that

$$\lim_{a\to\infty} (a/2)^{1/3} K^{\rm s.s.}\Big(a - (a/2)^{1/3}x, a - (a/2)^{1/3}y\Big) = K^{\rm soft}(x,y).$$

(vi) Consider the chiral ensemble (3.4) with $\beta = 2$ and $\alpha = a + \frac{1}{2}$ (note that this is defined on $\mathbb{R}^+$). Let $\{\tilde{p}_j(x)\}$ be orthonormal polynomials with respect to the weight function $w_2(x) = |x|^{2a+1}e^{-x^2}$, $x \in \mathbb{R}$. Show that

$$\begin{aligned} K_N(x,y) &= (w_2(x)w_2(y))^{1/2} \sum_{\nu=0}^{N-1} \tilde{p}_{2\nu}(x)\tilde{p}_{2\nu}(y) \\ &= (w_2(x)w_2(y))^{1/2} \sum_{\nu=0}^{2N-1} \Big(\tilde{p}_\nu(x)\tilde{p}_\nu(y) + \tilde{p}_\nu(x)\tilde{p}_\nu(-y)\Big), \end{aligned}$$

where the second line follows from the parity of the $\tilde{p}_\nu(x)$. Conclude from this that

$$2(XY)^{1/2}K^{\rm hard}(X^2,Y^2) = \Big(K^{\rm s.s.}(X,Y) + K^{\rm s.s.}(X,-Y)\Big)\Big|_{\substack{a\mapsto a+1/2\\ \rho=1/\pi}}.$$

(vii) With $\{\tilde{p}_j(x)\}$ as in (vi), note that $x\tilde{p}_{2\mu-1}(x) = \tilde{p}_{2\mu}(x)|_{a\mapsto a-1}$ and use this to show

$$K^{\rm s.s.}(X,Y)\Big|_{\rho=1/\pi} = (XY)^{1/2}\Big(K^{\rm hard}(X^2,Y^2)\Big|_{a\mapsto a-1/2} + K^{\rm hard}(X^2,Y^2)\Big|_{a\mapsto a+1/2}\Big).$$

3. Argue that

$$\rho^{\rm bulk}_{(n+1)}(x_1,\dots,x_n,0)/\rho = \rho^{\rm s.s.}_{(n)}(x_1,\dots,x_n)\Big|_{a=1}, \tag{7.56}$$

where $\rho^{\rm bulk}_{(n+1)}$ refers to the correlation function in Proposition 7.1.1 while $\rho^{\rm s.s.}_{(n)}$ refers to the correlation function (7.48) for the spectrum singularity. Use the trigonometric formula for $J_{1/2}(x)$ in (7.50), an analogous formula for $J_{3/2}(x)$ deducible from (7.50) and the three term recurrence (7.44) together with elementary row operations in the determinant formula for the l.h.s. to check this directly.

4. [215] In this exercise the soft edge limit of the Laguerre case of (5.198) will be analyzed. Explicitly, it will be shown that with the scalings

$$s_i = N - \tilde{s}_i,\ \ \tilde{s}_i := 2c_i(2N)^{2/3}, \qquad y_i = 4s_i + 2(a+N-s_i) + 2(2N)^{1/3}Y_i,$$

one has

$$\lim_{N\to\infty}\Big(2(2N)^{2/3}\Big)^r \rho_{(r)}(\{(s_j,y_j)\}_{j=1,\dots,r}) = \det[K^{\rm soft}((c_j,Y_j),(c_k,Y_k))]_{j,k=1,\dots,r}, \tag{7.57}$$

where $K^{\rm soft}$ is specified by (7.19).

(i) Substitute in (5.198) the appropriate Laguerre quantities to obtain, for $\tilde{s}_j \le \tilde{s}_l$,

$$K(N-\tilde{s}_j,y_j;N-\tilde{s}_l,y_l) = y_j^{a+\tilde{s}_j}e^{-y_j}\sum_{k=1}^{N-\tilde{s}_l}\frac{\Gamma(N-\tilde{s}_j-k+1)}{\Gamma(N-k+a+1)}L^{(a+\tilde{s}_j)}_{N-\tilde{s}_j-k}(y_j)L^{(a+\tilde{s}_l)}_{N-\tilde{s}_l-k}(y_l), \tag{7.58}$$

and note that for $\tilde{s}_j > \tilde{s}_l$ the r.h.s. is to be modified by multiplying by -1 and changing the summation to $k\in\mathbb{Z}_{\le 0}$.

(ii) Use the generalization of (7.35) applicable for $a = {\rm o}(n)$ [334],

$$x^{a/2}e^{-x/2}L^a_{n-k}(x) = (-1)^{n-k}(2n)^{-1/3}\sqrt{(n-k+a)!/(n-k)!}\times\Big({\rm Ai}\Big(X+\frac{2k}{(2n)^{1/3}}\Big) + {\rm O}(n^{-2/3})\begin{cases}{\rm O}(e^{-k/n^{1/3}}), & k\ge 0\\ {\rm O}(1), & k<0\end{cases}\Big) \tag{7.59}$$

with $n = N - \tilde{s}_i$ in (7.58) to show that for large N

$$K(N-\tilde{s}_j,y_j;N-\tilde{s}_l,y_l) \sim e^{-(2N)^{1/3}(Y_j-Y_l)}(2N)^{-2/3}\times\sum_{k=1}^{N-\tilde{s}_l}\Big(\frac{(N-\tilde{s}_j-k)!}{(N-\tilde{s}_l-k)!}\Big)^{1/2}{\rm Ai}\Big(Y_j+\frac{2k}{(2N)^{1/3}}\Big){\rm Ai}\Big(Y_l+\frac{2k}{(2N)^{1/3}}\Big).$$

Proceed now as in the analysis of (7.22) to deduce (7.57).

7.3 LOG-GAS SYSTEMS WITH PERIODIC BOUNDARY CONDITIONS

7.3.1 Semiperiodic boundary conditions

For the p.d.f. (5.96) the relationship with the Stieltjes-Wigert polynomials $S_n(y;q)$ shows that

$$\rho_{(n)}(y_1,\dots,y_n) = \det[L_N(y_j,y_k)]_{j,k=1,\dots,n}, \tag{7.60}$$

where, with $q = e^{-2\pi^2/L^2c'}$, $u = e^{c'(y+\pi N/Lc')}$ and a_n denoting the coefficient of y^n in $S_n(y;q)$,

$$L_N(y,y') = \frac{2\pi}{L}\Big(uu'w(u;q)w(u';q)\Big)^{1/2}\frac{a_{N-1}}{a_N}\Big(\frac{S_N(y;q)S_{N-1}(y';q) - S_N(y';q)S_{N-1}(y;q)}{y-y'}\Big). \tag{7.61}$$

The evaluation of the $N \to \infty$ limit (all other parameters fixed) is given by the following result [191].

Proposition 7.3.1 *We have*

$$\begin{aligned} L(y,y') := \lim_{N\to\infty} L_N(y,y') &= \Big(\frac{c'}{\pi}\Big)^{1/2}\frac{q^{-1/8}}{(q;q)^3_\infty}\Big(\frac{\ell(y;q)\ell(-y';q) - \ell(y';q)\ell(-y;q)}{2\sinh\pi(y-y')/L}\Big) \\ &= \frac{1}{L}\frac{1}{\theta_1'(0;(q')^2)}\Big(\frac{\hat\ell(y;q)\hat\ell(-y';q) - \hat\ell(y';q)\hat\ell(-y;q)}{2\sinh\pi(y-y')/L}\Big), \end{aligned}$$

where

$$\ell(y;q) := e^{\xi\pi y/2L}\sum_{\nu=-\infty}^{\infty}(-1)^\nu q^{(\nu+1/4-c'Ly/2\pi)^2},$$

$$\hat\ell(y;q) := e^{-\xi\pi y/2L}\theta_1\Big(\pi\Big(\frac{1}{4}+\frac{c'Ly}{2\pi}\Big);q'\Big),$$

with θ_1 denoting the Jacobi theta function (2.79), $q' = e^{-c'L^2/2}$ and $\xi = 1$ for N even and $\xi = -1$ for N odd.

Proof. Reading off the value of a_N from (5.97) shows that $a_{N-1}/a_N \sim q^{-2N}$ for $N \to \infty$. This estimate, together with the fact that

$$\begin{bmatrix} N-1 \\ \nu \end{bmatrix}_q \sim (1-q^{N-\nu})\begin{bmatrix} N \\ \nu \end{bmatrix}_q$$

shows

$$\begin{aligned} \frac{a_{N-1}}{a_N}S_N(u;q)S_{N-1}(u';q) \sim -&\frac{q^{-N}}{(1-q)\cdots(1-q^{N-1})}\sum_{\nu=0}^{N}\begin{bmatrix} N \\ \nu \end{bmatrix}_q q^{\nu^2+\nu/2}(-u)^\nu \\ &\times\sum_{\nu'=0}^{N}\begin{bmatrix} N \\ \nu' \end{bmatrix}_q(1-q^{N-\nu'})q^{\nu'^2+\nu'/2}(-u')^{\nu'}. \end{aligned}$$

Since we have to subtract the same term with u and u' interchanged we see that the factor $(1-q^{N-\nu'})$ in the sum over ν' can be replaced by $-q^{N-\nu'}$. Doing this, and completing the square in ν and ν' gives exponents $(\nu - N/2 + 1/4 - c'Ly/2\pi)^2$ and $(\nu' - N/2 - 1/4 - c'Ly/2\pi)^2$. The remaining step is to change the summation variable $\nu - [N/2] \mapsto \nu$, and similarly for ν', and note that

$$\begin{bmatrix} N \\ \nu + [N/2] \end{bmatrix}_q \sim \frac{1}{(q;q)_\infty}.$$

This gives the first expression for the limiting value of $L(y,y')$. The second expression follows by applying the conjugate modulus transformation from the theory of the Jacobi theta functions [541]. □

Here the density is a periodic function, period $1/\eta$, $\eta = c'L/2\pi$, so the system is in a crystalline state. This is to be expected, because as noted in Section 2.7.1 the pair potential is asymptotically that of the one-dimensional Coulomb system ($|x|$ potential), which is crystalline for all couplings [367]. The dependence on the parity of N is also understood as a consequence of the ordered state, and symmetry of the system about the origin: for N odd $y = 0$ is a maximum of the density while for N even $y = 0$ is a minimum.

7.3.2 Metal wall

Reinstating the fugacity (5.100) into (5.105) we see that the summation therein is a Riemann sum approximation to an integral which becomes exact for $L \to \infty$. Thus

$$\tilde{G}(x,y) := \lim_{L\to\infty} G(x,y) = 2\pi\zeta e^{4\pi d\eta} \int_0^\infty \frac{e^{-4\pi dt + 2\pi i t(x-y)}}{1 + 2\pi\zeta e^{4\pi d\eta} e^{-4\pi dt}}\, dt. \tag{7.62}$$

With $x = y$ this gives the density as a transcendental function of the fugacity. For large d (7.62) simplifies to

$$\tilde{G}(x,y)\Big|_{d\to\infty} = \int_0^1 e^{2\pi i t(x-y)}\, dt = \eta e^{\pi i \eta(x-y)} \frac{\sin \pi\eta(x-y)}{\pi\eta(x-y)},$$

reclaiming the bulk scaling result of Proposition 7.1.1 with $\rho = \eta$, as expected.

7.3.3 Doubly periodic boundary conditions

With $N, L \to \infty$ and $N/L = \rho$ fixed the summation defining the correlation kernel in (5.116) tends to a Riemann integral, giving that

$$\rho_{(n)}(x_1, \dots, x_n) = \rho^n \int_0^1 \det \Big[H(\rho x_j + \alpha, \rho x_k + \alpha) \Big]_{j,k=1,\dots,n} d\alpha, \tag{7.63}$$

where, with $\tilde{q} = e^{-\pi W \rho}$,

$$H(X,Y) = \int_0^1 e^{-2\pi i (X-Y)} \frac{\theta_3(\pi(X - iW\rho u); \tilde{q})\theta_3(\pi(Y + iW\rho u); \tilde{q})}{\theta_3(2\pi i W\rho u; \tilde{q}^2)}\, du. \tag{7.64}$$

As is to be expected from the underlying pair potentials, the state defined by the correlation kernel of Proposition 7.3.1, with $c'L/2pi = \rho$ and $L = W$ is closely related to the state defined by (7.63) [191]. In fact, use of residue calculus in (7.64) shows that then $H(x,y) = L(x,y)$, telling us in turn that the correlations for the doubly periodic (DP) system are related to the semiperiodic/free boundary condition (SP) system by averaging over a period,

$$\rho_{(n)}^{\rm DP}(y_1, \dots, y_n) = \rho \int_0^{1/\rho} \rho_{(n)}^{\rm SP}(y_1 + \alpha, \dots, y_n + \alpha)\, d\alpha.$$

7.4 ASYMPTOTIC BEHAVIOR OF THE ONE- AND TWO-POINT FUNCTIONS AT $\beta = 2$

In this section the large separation behavior of the exact truncated two-particle correlation functions, and the behavior of the density at large distances from the edge will be computed. Throughout, the notation $\rho_{(1)}$ and $\rho_{(2)}$ will refer to the appropriate scaled values.

7.4.1 Bulk

It follows from (7.2) that

$$\rho_{(2)}^T(x,0) = -\frac{1}{2(\pi x)^2} + \frac{\cos 2\pi\rho x}{2(\pi x)^2}. \tag{7.65}$$

This exhibits that the leading non-oscillatory form is $-1/2(\pi x)^2$, which is in keeping with the $|k| \to 0$ form of (7.4). For small x (7.2) gives

$$\rho_{(2)}(x,0) = \rho^2 \Big(\frac{(\pi \rho x)^2}{3} - \frac{2(\pi \rho x)^4}{45} - \cdots \Big). \tag{7.66}$$

The fact that the leading term is proportional to x^2 follows from the fact that the Boltzmann factor vanishes as $|x-y|^2$ for particles at positions x and y, in the limit $x \to y$.

7.4.2 Soft edge

For the scaled one- and two-point functions near the soft edge, Propositions 5.1.2 and 7.1.2 give

$$\rho_{(1)}(X) = K^{\rm soft}(X,X), \qquad \rho_{(2)}^T(X,Y) = -(K^{\rm soft}(X,Y))^2, \tag{7.67}$$

where $K^{\rm soft}(X,X)$ and $K^{\rm soft}(X,Y)$ are given by (7.13) and (7.12), respectively. From the asymptotic expansion [136]

$$\mathrm{Ai}(-x) \underset{x\to\infty}{\sim} \frac{1}{\pi^{1/2} x^{1/4}} \cos(2x^{3/2}/3 - \pi/4) \tag{7.68}$$

we deduce that [189], [318]

$$\rho_{(1)}(-X) \underset{X\to\infty}{\sim} \frac{\sqrt{X}}{\pi} - \frac{\cos(4X^{3/2}/3)}{4\pi X} + \mathrm{O}\Big(\frac{1}{X^{5/2}}\Big), \tag{7.69}$$

$$\rho_{(2)}^T(-X,-Y) \underset{X,Y\to\infty}{\sim} -\frac{1}{4\pi^2\sqrt{XY}} \frac{X+Y}{(X-Y)^2}, \tag{7.70}$$

where in (7.70) only non-oscillatory terms have been written down. The asymptotic expansion

$$\mathrm{Ai}(x) \underset{x\to\infty}{\sim} \frac{e^{-2x^{3/2}/3}}{2\sqrt{\pi} x^{1/4}} \tag{7.71}$$

tells us

$$\log \rho_{(1)}(X) \underset{X\to\infty}{\sim} -\frac{4}{3} X^{3/2}. \tag{7.72}$$

It is possible to exhibit a matching between the asymptotic expansion (7.69), and the global asymptotic expansion (7.25) expanded in the neighborhood of the edge (see Exercises 7.4 q.1).

7.4.3 Hard edge

Here, according to Propositions 5.1.2 and 7.2.1

$$\rho_{(1)}(X) = K_a^{\rm hard}(X,X), \qquad \rho_{(2)}^T(X,Y) = -(K_a^{\rm hard}(X,Y))^2,$$

where $K_a^{\rm hard}(X,X)$ and $K_a^{\rm hard}(X,Y)$ are given by (7.33) and (7.31) respectively. From the asymptotic formula [165]

$$J_\alpha(x) \underset{x\to\infty}{\sim} \Big(\frac{2}{\pi x}\Big)^{1/2} \cos(x - \pi\alpha/2 - \pi/4) \tag{7.73}$$

we find [189], [51] for the leading non-oscillatory behavior

$$\rho_{(1)}(X) \underset{X\to\infty}{\sim} \frac{1}{2\pi X^{1/2}}, \tag{7.74}$$

$$\rho_{(2)}^T(X,Y) \underset{X,Y\to\infty}{\sim} -\frac{1}{4\pi^2\sqrt{XY}} \frac{X+Y}{(X-Y)^2}. \tag{7.75}$$

Note in particular that (7.75) is identical to the behavior (7.70) found for the same quantity at the soft edge.

It is also of interest to consider the hard edge correlations in the squared variables of (7.53). Then we obtain the asymptotic behavior

$$\rho_{(1)}(X) \underset{X\to\infty}{\sim} 1, \qquad \rho^T_{(2)}(X,Y) \underset{X,Y\to\infty}{\sim} -\frac{1}{2\pi^2(X-Y)^2}.$$

The second of these is the non-oscillatory bulk behavior seen from (7.65).

7.4.4 Spectrum singularity

In the neighborhood of a spectrum singularity the correlation functions are given by (7.48). In particular

$$\rho_{(1)}(x) = K^{\text{s.s.}}(x,x) \underset{x\to\infty}{\sim} \rho + \rho\frac{a^2(a^2-3)}{4(\pi\rho x)^2}, \tag{7.76}$$

where only non-oscillatory terms have been written down. Also

$$\rho^T_{(2)}(x,y) = -\Big(K^{\text{s.s.}}(x,y)\Big)^2 \underset{x,y\to\infty}{\sim} -\frac{1}{2\pi^2(x-y)^2}, \tag{7.77}$$

where again oscillatory terms have been ignored. The latter is the non-oscillatory bulk behavior seen from (7.65).

7.4.5 Semiperiodic boundary conditions

We have seen that the state of the system in this case is crystalline, and the correlations are given in terms of Jacobi theta functions. Simplified results can be obtained by supposing $L \gg 1$, and then averaging over the period of the density oscillations. First, since $\theta_1(u;q) \sim 2q^{1/4}\sin u$, (7.60) and Proposition 7.3.1 give that for $L \gg 1$

$$\begin{aligned}\rho^T_{(2)}(y,y') \sim & -\frac{1}{2L^2\sinh^2\pi(y-y')/L}\Big(2\sin^2\Big(\frac{\pi}{4}+\pi\eta y\Big)\sin^2\Big(\frac{\pi}{4}+\pi\eta y'\Big)e^{-\pi(y-y')/L} \\ & +2\sin^2\Big(\frac{\pi}{4}+\pi\eta y'\Big)\sin^2\Big(\frac{\pi}{4}+\pi\eta y\Big)e^{\pi(y-y')/L} - \cos 2\pi\eta y\cos 2\pi\eta y'\Big).\end{aligned}$$

Now introducing the smoothed truncated correlation

$$\bar{\rho}^T_{(2)}(y,0) = \eta\int_0^{1/\eta}\rho^T_{(2)}(y+y',y')\,dy'$$

and substituting the above asymptotic form we obtain [208]

$$\bar{\rho}^T_{(2)}(y,y') \sim -\frac{\cosh(\pi(y-y')/L)}{2\sinh^2(\pi(y-y')/L)}. \tag{7.78}$$

Again with $L \gg 1$, a different translationally invariant approximation can be obtained directly from the kernel of Proposition 7.3.1, rather than first computing the two-point function. Thus by approximating the factor of $e^{-\xi\pi y/2L}$ in $\hat{l}(y;q)$ by unity one obtains the kernel

$$\tilde{L}(y,y') := \frac{\sin\pi\eta(x-y)}{L\sinh\pi(x-y)/L}, \tag{7.79}$$

which is a functional form first obtained in a random matrix calculation in [413]. Note that (7.79) reduces to the bulk scaling sine kernel of Proposition 7.1.1 in the limit $L\to\infty$. For small L, only the values $x\approx y$ are important and so only the diagonal terms in the determinant of Proposition 5.1.2 contribute, which is Poisson behavior. Thus the kernel interpolates between these limiting cases.

7.4.6 Metal wall

Integration by parts of the limiting kernel (5.105) gives that for $|x-y| \to \infty$,

$$\tilde{G}(x,y) \sim \frac{\zeta e^{4\pi d\eta}}{1+2\pi\zeta e^{4\pi d\eta}} \frac{1}{i(x-y)}.$$

Thus the corresponding decay of $\rho_{(2)}(x,y)$ is an inverse square, with a negative coefficient and no oscillatory term.

EXERCISES 7.4 1. [258] Here a type of connection formula between the bulk and soft edge densities in the scaled GUE will be exhibited.

(i) Note that with respect to the scaled Gaussian weight function $w_2(x) = e^{-2Nx^2}$, the asymptotic formula (7.25) reads

$$\rho_{(1)}(x) = \rho^{\rm W}(x) - \frac{2\cos(2N\pi P(x))}{\pi^3 \rho_{\rm W}^2(x)} \frac{1}{N} + {\rm O}\Big(\frac{1}{N^2}\Big)$$

while (7.27) gives

$$\frac{N^{1/3}}{2}\rho_{(1)}\Big(1+\frac{\xi}{2N^{2/3}}\Big) = \rho_{(1)}^{\rm soft,2}(\xi) + {\rm O}\Big(\frac{1}{N^{2/3}}\Big).$$

(ii) Show that substituting $x = 1 + \xi/2N^{2/3}$ in the first expression for $\rho_{(1)}(x)$ in (i), multiplying by $N^{1/3}$ and expanding in N keeping terms O(1) only gives the first two terms of the $\xi \to -\infty$ form of the second expression, as implied by (7.69).

7.5 BULK SCALING AND THE ZEROS OF THE RIEMANN ZETA FUNCTION

7.5.1 Montgomery pair correlation conjecture

From Chapter 1 we know that the eigenvalues of a chaotic quantum system without time reversal symmetry will have, after appropriate rescaling, the same statistical properties as the eigenvalues from the (infinite) GUE. In this section, following mainly [353], [74], we will discuss the occurrence of the GUE distributions in the study of the statistical properties of the Riemann zeros of large modulus. The Riemann zeros are the zeros of the analytic continuation of the Riemann zeta function

$$\zeta(s) = \sum_{n=1}^\infty \frac{1}{n^s}, \qquad {\rm Re}(s) > 1,$$

on the critical line ${\rm Re}(s) = \frac{1}{2}$. Indeed, it has been noted [62] that there are formal similarities between a certain representation of the density of zeros of $\zeta(s)$ on the critical line (Proposition 7.5.4 below), and the Gutzwiller trace formula for the density of states in a chaotic quantum system without time reversal symmetry, so from this viewpoint the occurrence of the GUE distributions is not unexpected. On the other hand we know that in the large N bulk scaling the GUE distributions coincide with those for the CUE or equivalently $U(N)$. In fact, as will be seen subsequently, it is $U(N)$ eigenvalues which more accurately models the Riemann zeros.

First we recall that $\zeta(s)$ is closely related to the prime numbers, as is seen from a product formula due to Euler.

PROPOSITION 7.5.1 *For* ${\rm Re}(s) > 1$,

$$\zeta(s) = \prod_p (1-p^{-s})^{-1},$$

where the product is over all primes p (= 2, 3, 5,...).

Proof. Subtracting $2^{-s}\zeta(s)$ from the definition of $\zeta(s)$ gives

$$\zeta(s)(1-2^{-s}) = \sum_{\substack{n=1 \\ n\neq 2m, m\in\mathbb{Z}^+}}^{\infty} \frac{1}{n^s}.$$

Similarly, subtracting from this series 3^{-s} times the series gives

$$\zeta(s)(1-2^{-s})(1-3^{-s}) = \sum_{n=1}^{\infty}{}^{*} \frac{1}{n^s} = 1^{-s} + 5^{-s} + 7^{-s} + \cdots,$$

where the * indicates all terms n which are multiples of 2 or 3 are to be omitted. Subtracting all the primes to the power of $-s$ in this fashion gives

$$\zeta(s)\prod_p (1-p^{-s}) = 1^{-s} = 1$$

as required. □

Our interest is in the statistical properties of the zeros of $\zeta(s)$ in the complex plane. The famous *Riemann hypothesis* asserts that all the zeros of $\zeta(s)$ with nonzero imaginary part lie on the so-called critical line $\mathrm{Re}(s) = \frac{1}{2}$. As noted above such zeros are called the Riemann zeros. The Riemann hypothesis dates back to 1859. Over a century later, beginning with a conjecture of Montgomery [405] for the pair correlation function $\langle d(E)d(E+\varepsilon)\rangle$, $d(E) := \sum_{j=1}^{\infty}\delta(E-E_j)$ (E_j denotes the (positive) imaginary part of the jth zero), the statistical properties of the Riemann zeros with large imaginary part have been the subject of study.

The Montgomery conjecture states that

$$1 + S(\hat{\varepsilon}) := \lim_{\substack{E\to\infty \\ \hat{\varepsilon}\text{ fixed}}} \frac{\langle d(E)d(E+\varepsilon)\rangle}{\langle d(E)\rangle\langle d(E+\varepsilon)\rangle} = \delta(\hat{\varepsilon}) + 1 - \frac{\sin^2\pi\hat{\varepsilon}}{(\pi\hat{\varepsilon})^2}, \tag{7.80}$$

where the averages $\langle\cdot\rangle$ are over a region $[E, E+\Delta E]$ such that $1 \ll \Delta E \ll E$, and $\hat{\varepsilon} = \varepsilon/\bar{d}(E)$ with $\bar{d}(E)$ denoting the mean spacing between zeros at point E on the critical line. Since in general $S(x) = \delta(x) + \rho_{(2)}^T(x,0)/\rho^2$, we see from (7.65) that (7.80) is identical to $S(x)$ for the GUE in the bulk limit. Indeed the so called *Montgomery-Odlyzko* law [352] asserts that after scaling the large Riemann zeros have the same distribution as the bulk eigenvalues for large GUE matrices. This conjecture has supporting evidence from large scale numerical computations [426], involving the accurate evaluation of the zero number of order 10^{20} along the critical line, and over 10^7 of its neighbors. Moreover for test functions $f(x)$ with Fourier transform $\hat{f}(k)$ supported on $|k| < 2\pi$, Montgomery rigorously proved that

$$\int_{-2\pi}^{2\pi} \hat{S}(k)\hat{f}(k)\,dk = \frac{1}{2\pi}\int_{-2\pi}^{2\pi} |k|\hat{f}(k)\,dk, \tag{7.81}$$

where $\hat{S}$ denotes the Fourier transform of S, which is consistent with (7.80) since $\hat{S}(k)$ is given by (7.4) (note that $\hat{S}(k) = \tilde{S}(k)$ for $\rho = 1$). This rigorous result has been extended to higher order correlations in [472]. By abandoning rigor, the general n-point correlation function for the limiting zeros along the critical line can be calculated exactly [73], and the result of Proposition 7.1.1 for the GUE obtained. Here we will give the latter argument in the simplest case ($n = 2$), and so derive (7.80).

7.5.2 Rigorous theory

Since Proposition 7.5.1 gives that there are no zeros of $\zeta(s)$ for $\mathrm{Re}(s) > 1$, to study the zeros it is necessary to analytically continue $\zeta(s)$. For this purpose a contour integral representation can be used. Furthermore, this representation can be used to establish a functional equation satisfied by $\zeta(s)$.

PROPOSITION 7.5.2 *We have*

$$\zeta(s) = -\frac{\Gamma(1-s)}{2\pi i} \int_C \frac{(-z)^{s-1}}{e^z - 1} dz$$

where C is any simple contour starting at $z = \infty + i\mu$ ($\mu > 0$), looping around $z = 0$, and finishing at $z = \infty - i\mu$ without crossing the real axis, and $|\arg(-z)| < \pi$. Furthermore

$$\zeta(s) = 2^s \pi^{s-1} \sin(\pi s/2) \Gamma(1-s) \zeta(1-s).$$

Proof. For $\mathrm{Re}(s) > 1$, the first result follows by writing on the portion of C with $\mathrm{Re}(z) > 0$

$$\frac{1}{e^z - 1} = \sum_{k=1}^{\infty} e^{-kz},$$

noting that the contribution from other portions of C can be made arbitrarily small, then changing variables $kz \mapsto z$ and integrating term by term using the formula [541]

$$\frac{1}{\Gamma(1-s)} = -\frac{1}{2\pi i} \int_C (-z)^{s-1} e^{-z} dz$$

(cf. (5.156)). For $\mathrm{Re}(s) \le 1$, the integral representation is the analytic continuation. The second result follows by supposing $\mathrm{Re}(s) < 0$ and expanding C to a circle of infinite radius about the origin (the resulting contour integral vanishes), and calculating the original contour integral as $2\pi i$ times the sum of the residues of the poles at $z = \pm 2\pi i, \pm 4\pi i, \dots$. □

It follows from Proposition 7.5.2 that the only singularity of $\zeta(s)$ is a simple pole at $s = 1$, that there are zeros (*trivial zeros*) at $s = -2, -4, \dots$ and that the only other zeros of $\zeta(s)$ must occur in the *critical strip* $0 \le s \le 1$. To investigate other zeros, Riemann introduced the entire function

$$\xi(t) := \frac{1}{2} s(s-1)\zeta(s)\Gamma(s/2)\pi^{-s/2}, \qquad s := \frac{1}{2} - it. \tag{7.82}$$

The Riemann hypothesis is equivalent to the statement that all zeros of $\xi(t)$ are on the real t-axis. Use of the functional equation for $\zeta(s)$, the functional equation (4.5) for the gamma function and the duplication formula (1.111) shows that

$$\xi(t) = \xi(-t), \quad \text{and thus} \quad \bar{\xi}(t) = \xi(t).$$

This last equation implies $\xi(t)$ is real for real t. Using $\xi(t)$, the number of zeros $N(E)$ of $\zeta(s)$ in the critical strip with $0 \le \mathrm{Im}(s) < E$ can be computed (see, e.g., [516]).

PROPOSITION 7.5.3 *We have*

$$N(E) = \bar{N}(E) + N_{\mathrm{osc}}(E),$$

where

$$\bar{N}(E) := 1 - \frac{1}{2\pi} E \log \pi - \frac{1}{\pi} \mathrm{Im}\, \log \Gamma\left(\frac{1}{4} - \frac{iE}{2}\right), \quad N_{\mathrm{osc}}(E) := -\frac{1}{\pi} \mathrm{Im}\, \log \zeta\left(\frac{1}{2} - iE\right).$$

Proof. In terms of $\xi(t)$, $N(E)$ is equal to the number of zeros of $\xi(t)$ in the region $-\frac{1}{2} \le \mathrm{Im}(t) \le \frac{1}{2}, 0 \le \mathrm{Re}(t) < E$. According to the residue theorem, this is given by

$$N(E) = \frac{1}{2} \frac{1}{2\pi i} \int_{C'} \frac{\xi'(t)}{\xi(t)} dt = \frac{1}{4\pi} \mathrm{Im} \int_{C'} \frac{\xi'(t)}{\xi(t)} dt,$$

where C' is the rectangular contour with sides parallel to the real and imaginary t-axes, which passes through the points $\pm E$ and $\pm i(\frac{1}{2} + \alpha)$, with $\alpha > 0$. It is assumed that E is not a zero of $\xi(t)$. The factor of $1/2$ comes from the fact that $\xi(t)$ is even. Since $\xi(t)$ is even and $\xi(t) = \bar{\xi}(t)$, the integral over C' can be replaced by 4 times the integral over C_1, where C_1 is the contour over the portion of C' in the upper right quadrant of the complex t plane. This gives

$$N(E) = \frac{1}{\pi}\Delta \arg \xi(t),$$

where $\Delta \arg\xi(t)$ refers to the change in argument of $\xi(t)$ in going from $t = E$ to $t = i(\frac{}{/2} + \alpha)$ along C_1. But at $t = i(\frac{1}{2} + \alpha)$, $\xi(t) > 0$ and thus

$$N(E) = -\frac{1}{\pi} \arg \xi(E) = -\frac{1}{\pi} \text{Im } \log \xi(E).$$

The result now follows by substituting (7.82). □

7.5.3 Heuristic argument

The above theory is entirely rigorous. However, to achieve our goal of establishing (7.80), it is necessary to proceed heuristically. We begin with a formula for the density of zeros, and assume the validity of the Riemann hypothesis, which gives that all zeros of $\xi(t)$ are real.

PROPOSITION 7.5.4 *Let $E_1,\ E_2, \dots$ denote the positive zeros of $\xi(E)$. We have*

$$d(E) := \sum_{j=1}^{\infty} \delta(E - E_j) = \frac{d}{dE} N(E) = \bar{d}(E) + d_{\rm osc}(E),$$

where

$$\bar{d}(E) \sim \frac{1}{2\pi} \log \frac{E}{2\pi}, \qquad d_{\rm osc}(E) = -\frac{1}{\pi} \sum_p \sum_{k=1}^{\infty} \frac{\log p}{p^{k/2}} \cos(Ek \log p).$$

Derivation. To derive the formula for $\bar{d}(E)$, which is rigorous, we simply use Stirling's formula in the formula for $\bar{N}(E)$ in Proposition 7.5.3 and differentiate. To derive the formula for $d_{\rm osc}(E)$, which is heuristic, we substitute the Euler product for $\zeta(\frac{1}{2} - iE)$ in the formula for $N_{\rm osc}(E)$ in Proposition 7.5.3 (note this is not justified since the product only converges for $\text{Re}(s) > 1$). Expanding $\log(1 - p^{-s})$ and differentiating gives the stated result.

In an obvious notation, this result gives

$$N_{\rm osc}(E) = -\frac{1}{\pi} \sum_p \sum_{k=1}^{\infty} \frac{\exp(-\frac{k}{2} \log p)}{k} \sin(Ek \log p). \tag{7.83}$$

It should be remarked [63], [64] that in the study of classically chaotic quantum systems, the so-called *trace formula* gives a formula of the same structure for the oscillating part of the quantum spectrum, as calculated from the classical data. Thus, after a simplifying approximation, one has for $\hbar \to 0$

$$N_{\rm osc}(E) \sim \frac{1}{\pi} \sum_{\tilde{p}} \sum_{k=1}^{\infty} \frac{\exp(-\frac{k}{2}\lambda_{\tilde{p}} T_{\tilde{p}})}{k} \sin\Big(\frac{k}{\hbar} S_{\tilde{p}}(E) - \frac{\pi k}{2} \mu_{\tilde{p}}\Big). \tag{7.84}$$

Here $\tilde{p}$ labels primitive periodic orbits (i.e., orbits traversed once), k labels their repetitions, $S_{\tilde{p}}(E)$ is the action of the primitive orbit with the property that the period of the latter is given by $T_{\tilde{p}} = \partial S_{\tilde{p}}/\partial E$, $\lambda_{\tilde{p}}$ is the instability exponent and $\mu_{\tilde{p}}$ is the Maslov phase. The formulas (7.83) and (7.84) coincide (apart from an overall minus sign; see [69] for a discussion of the possible significance of this) with the primes as primitive periodic orbits of period $\log p$, while $\lambda_{\tilde{p}} = 1$ and $\mu_{\tilde{p}} = 0$. Also, because time reversal symmetry would require all non-self-retracing orbits to have a partner, the fact that the prefactor in (7.83) is $1/\pi$ rather than $2/\pi$ implies the analogous quantum system does not possess time reversal symmetry.

Note that if we average $d(E)$ over some region $[E, E + \Delta E]$ with $1 \ll \Delta E \ll E$, Proposition 7.5.4 gives $\langle d(E) \rangle = \bar{d}(E)$. Use of Proposition 7.5.4 also allows $\langle d(E) d(E + \varepsilon) \rangle$, where $\langle \cdot \rangle$ denotes this same average, to be reduced into a form suitable for further analysis.

PROPOSITION 7.5.5 *For large E we have*

$$\begin{aligned}\langle d(E)d(E+\varepsilon)\rangle &\sim \langle \bar{d}(E)\bar{d}(E+\varepsilon)\rangle + \langle d_{\rm osc}(E)d_{\rm osc}(E+\varepsilon)\rangle \\ &= \bar{d}(E)\bar{d}(E+\varepsilon) + \langle d_{\rm osc}(E)d_{\rm osc}(E+\varepsilon)\rangle_{\rm diag} + \langle d_{\rm osc}(E)d_{\rm osc}(E+\varepsilon)\rangle_{\rm off},\end{aligned}$$

where

$$\langle d_{\rm osc}(E)d_{\rm osc}(E+\varepsilon)\rangle_{\rm diag} \sim \frac{1}{2\pi^2}{\rm Re}\sum_p \frac{\log^2 p}{p} e^{-i\varepsilon \log p},$$

$$\langle d_{\rm osc}(E)d_{\rm osc}(E+\varepsilon)\rangle_{\rm off} \sim \frac{1}{\pi^2}{\rm Re}\sum_p \frac{\log^2 p}{p} e^{-i\varepsilon \log p}\Big\langle \sum_{\substack{h \\ h\ll p}} \cos Eh/p \Big\rangle,$$

p denoting a prime, and where h is such that $p+h$ is prime.

Derivation. The first asymptotic equality follows by substituting the formula of Proposition 7.5.4 for $d(E)$ and $d(E+\varepsilon)$, and noting that due to the averaging $\langle \bar{d}(E)d_{\rm osc}(E+\varepsilon)\rangle \sim 0$. The averaging does not affect $\bar{d}(E)\bar{d}(E+\varepsilon)$ so the first term of the equality follows. This leaves the final asymptotic expressions. Now

$$\langle d_{\rm osc}(E)d_{\rm osc}(E+\varepsilon)\rangle = \frac{1}{\pi^2}\Big\langle \sum_{p,p'}\sum_{k,k'=1}^{\infty} \frac{\log p \log p'}{p^{k/2}p'^{k'/2}} \cos(Ek\log p)\cos((E+\varepsilon)k'\log p')\Big\rangle.$$

Due to the averaging, we can replace the cosine terms by

$$\frac{1}{2}\cos(Ek\log p - (E+\varepsilon)k'\log p')$$

(the term involving $(1/2)\cos(Ek\log p + (E+\varepsilon)k'\log p')$ will vanish). Furthermore, again due to the averaging, only terms with $k\log p \sim k'\log p'$ are relevant in the limit $E\to\infty$. Now for either $k>1$ or $k'>1$ the sum over these contributions converges, so we expect the sum to be dominated by the terms with $k=k'=1$, which formally diverges. Thus

$$\langle d_{\rm osc}(E)d_{\rm osc}(E+\varepsilon)\rangle \sim \frac{1}{2\pi^2}{\rm Re}\Big\langle \sum_{p,p'} \frac{\log p \log p'}{p^{1/2}p'^{1/2}} e^{iE(\log p - \log p') - i\varepsilon \log p'}\Big\rangle.$$

Now split the double sum into two parts $\sum_{p,p'} = \sum_{p=p'} + \sum_{p\ne p'}$. The first sum gives the first of the last two equations in the proposition. To obtain the last equation, make the ordering $p<p'$ (and thus multiply the sum by 2) and notice that the leading order contribution comes from primes $p'\sim p$. Writing $p' = p+h$, expanding the logarithm to first order in h/p, and noting that for $h \ll p$, $e^{-iEh/p} \sim \cos Eh/p$, gives the final stated equation.

A straightforward (heuristic) analysis gives the limiting value of the term $\langle\cdot\rangle_{\rm diag}$ in Proposition 7.5.5 (a rigorous analysis can also be given; in fact the result (7.81) was obtained in such a way). This requires use of the *prime number theorem*, which asserts that to leading order the number of prime numbers less than X is given by $X/\log X$.

PROPOSITION 7.5.6 *We have*

$$\lim_{\substack{E,\varepsilon\to\infty \\ \varepsilon/\bar{d}(E)=\hat{\varepsilon}}} \frac{1}{(\bar{d}(E))^2}\langle d_{\rm osc}(E)d_{\rm osc}(E+\varepsilon)\rangle_{\rm diag} \sim -\frac{1}{2(\pi\hat{\varepsilon})^2}.$$

Derivation. The basic idea is to replace the sum in the formula $\langle\cdot\rangle_{\rm diag}$ given in Proposition 7.5.5 by an integral. According to the prime number theorem, to leading order the density of primes is $1/\log X$, and so

$$\sum_p \frac{\log^2 p}{p} e^{-i\varepsilon\log p} \sim \int_1^\infty \frac{\log X}{X} e^{-i\varepsilon \log X}\, dX = (2\pi)^2 \int_0^\infty \tau\, e^{-2\pi i\tau\varepsilon}\, d\tau,$$

where the last equation follows from the change of variables $\log X = 2\pi\tau$. Finally note that for $\mathrm{Re}(z) > 0$

$$\int_0^\infty \tau e^{-\tau z}\, d\tau = \frac{1}{z^2}\int_0^\infty \tau e^{-\tau}\, d\tau = \frac{1}{z^2}, \tag{7.85}$$

and using this as an analytic continuation to evaluate the integral.

To mimic the strategy used to derive Proposition 7.5.6 requires an asymptotic formula not for the number of primes less than X, but for the more complicated quantity $\pi_m(X)$, the number of primes p not exceeding X and such that $(p-m)$ is a prime. For this one uses the Hardy-Littlewood conjecture [290], which asserts

$$\pi_m(X) \sim \alpha(m)\frac{X}{\log^2 X},$$

where $\alpha(m)$ (which was given explicitly by Hardy and Littlewood) has after averaging the large m behavior [353]

$$\langle \alpha(m) \rangle \sim 1 - \frac{1}{2m}. \tag{7.86}$$

PROPOSITION 7.5.7 *We have*

$$\lim_{\substack{E,\varepsilon\to\infty \\ \varepsilon/\bar{d}(E)=\hat{\varepsilon}}} \frac{1}{(\bar{d}(E))^2}\langle d_{\mathrm{osc}}(E)d_{\mathrm{osc}}(E+\varepsilon)\rangle_{\mathrm{off}} \sim \frac{\cos 2\pi\hat{\varepsilon}}{2(\pi\hat{\varepsilon})^2}.$$

Derivation. The first step is to use the Hardy-Littlewood conjecture to replace the sums over p and h in the expression of Proposition 7.5.5 for $\langle\cdot\rangle_{\mathrm{off}}$ by the corresponding density $\alpha(k)/\log^2 p$ to give

$$\langle d_{\mathrm{osc}}(E)d_{\mathrm{osc}}(E+\varepsilon)\rangle_{\mathrm{off}} \sim \frac{1}{\pi^2}\mathrm{Re}\int_1^\infty \frac{dp}{p}e^{-i\varepsilon\log p}\int_{l(p)}^\infty dh\, \langle\alpha(h)\rangle \cos Eh/p.$$

Here $l = l(p) = \mathrm{O}(\log p)$, and the upper terminal of integration in the second integral has been set equal to infinity as only the region $h \ll p$ contributes to the leading order.

Now make two changes of variables: $w = \log p/(2\pi\bar{d}(E))$ and $y = (2\pi)^w E^{1-w} h$. Since $2\pi\bar{d}(E) \sim \log(E/2\pi)$ we have $p \sim (E/2\pi)^w$ and $y \sim Eh/p$. Thus the change of variables gives

$$\begin{aligned}\langle d_{\mathrm{osc}}(E)d_{\mathrm{osc}}(E+\varepsilon)\rangle_{\mathrm{off}} \sim \frac{2\bar{d}(E)}{\pi}\mathrm{Re}\int_0^\infty dw\, \exp(-2\pi i\bar{d}(E)\varepsilon w)\frac{E^{w-1}}{(2\pi)^w} \\ \times \int_{l(2\pi)^w E^{1-w}}^\infty dy \left\langle \alpha\left(\frac{yE^{w-1}}{(2\pi)^w}\right)\right\rangle \cos y.\end{aligned}$$

Consider the case $w < 1$. When $E \to \infty$, $E^{1-w} \to \infty$ so the second integral vanishes. When $w > 1$, $E^{1-w} \to 0$ and $E^{w-1} \to \infty$ so we can use (7.86) to conclude

$$\left\langle \alpha\left(\frac{yE^{w-1}}{(2\pi)^w}\right)\right\rangle \sim 1 - \frac{(2\pi)^w}{2yE^{w-1}}.$$

Multiplying by $\cos y$ and integrating, we take the first term to integrate to zero, while the second term is to leading order given by the small y behavior

$$\int_{l(2\pi)^w E^{1-w}}^\infty \frac{\cos y}{y}\, dy \sim (w-1)\log E \sim (w-1)2\pi\bar{d}(E).$$

Thus

$$\langle d_{\mathrm{osc}}(E)d_{\mathrm{osc}}(E+\varepsilon)\rangle \sim -\frac{\bar{d}(E)}{\pi}\mathrm{Re}\int_1^\infty e^{-2\pi i\bar{d}(E)\varepsilon w}(w-1)(2\pi\bar{d}(E))\, dw.$$

Evaluating the integral by changing variables $w - 1 \mapsto w$ and using (7.85) establishes the result.

Combining the results of Propositions 7.5.5–7.5.7 gives the statement (7.80) of the Montgomery conjecture for $\hat{\varepsilon} \neq 0$ (the delta function at $\hat{\varepsilon} = 0$ follows from the definition).

The above heuristic analysis can also be carried out without taking the limit $E \to \infty$, although E is still assumed to be large. Then it is found that [64], [81]

$$\langle d_{\rm osc}(E) d_{\rm osc}(E+\varepsilon)\rangle_{\rm diag} = -\frac{1}{4\pi^2}\frac{\partial^2}{\partial\varepsilon^2}\log\Big(|\zeta(1+i\varepsilon)|^2\Big|\exp\sum_p\sum_{m=1}^\infty \frac{1-m}{m^2p^m}e^{im\log p\varepsilon}\Big|^2\Big),$$

$$\langle d_{\rm osc}(E) d_{\rm osc}(E+\varepsilon)\rangle_{\rm off} = \frac{1}{2\pi^2}|\zeta(1+i\varepsilon)|^2{\rm Re}\Big(e^{2\pi i\bar{d}(E)\varepsilon}\prod_p\big(1-\frac{(1-p^{i\varepsilon})^2}{p-1}\big)\Big).$$

Expanding as a function of ε gives [72]

$$\langle d_{\rm osc}(E) d_{\rm osc}(E+\varepsilon)\rangle_{\rm diag} = -\frac{1}{2\pi^2\varepsilon^2} - \frac{(\gamma_0^2+2\gamma_1+c_0)}{2\pi^2} + {\rm O}(\varepsilon^2),$$

$$\langle d_{\rm osc}(E) d_{\rm osc}(E+\varepsilon)\rangle_{\rm off} = \frac{1}{2\pi^2}{\rm Re}\Big(\Big(\frac{1}{\varepsilon^2}+(\gamma_0^2+2\gamma_1+c_0)+iQ+{\rm O}(\varepsilon^2)\Big)e^{2\pi i\bar{d}(E)\varepsilon}\Big),$$

where γ_0, γ_1 are specified by

$$\zeta(1+x) = \frac{1}{x} + \sum_{n=0}^\infty \frac{(-1)^n}{n!}\gamma_n x^n$$

while

$$c_0 := \sum_p (\log p)^2 \sum_{n=1}^\infty \frac{(n-1)}{p^n}, \qquad Q := \sum_p \frac{\log^3 p}{(p-1)^2}.$$

With $\hat{\varepsilon} := \varepsilon/\bar{d}(E)$ this implies that for $E \to \infty$

$$\frac{1}{(\bar{d}(E))^2}\langle d_{\rm osc}(E) d_{\rm osc}(E+\varepsilon)\rangle = 1 - \frac{\sin^2\pi\hat{\varepsilon}}{(\pi\hat{\varepsilon})^2} - \frac{\Lambda}{(\pi\bar{d}(E))^2}\sin^2(\pi\hat{\varepsilon}) + {\rm O}\Big(\frac{1}{(\bar{d}(E))^3}\Big), \tag{7.87}$$

where $\Lambda := \gamma_0^2 + 2\gamma_1 + c_0 = 1.57314...$

On the other hand (5.87) and (5.86) give that for matrices from $U(N)$, with lengths scaled so that $\rho = 1$,

$$\rho_{(2)}(x, 0) = 1 - \frac{\sin^2\pi x}{(\pi x)^2} - \frac{1}{3N^2}\sin^2\pi x + {\rm O}\Big(\frac{1}{N^4}\Big). \tag{7.88}$$

Comparison of (7.87) and (7.88) gives agreement of the first three terms provided one makes the identification

$$N = \frac{\pi\bar{d}(E)}{\sqrt{3\Lambda}} = \frac{1}{\sqrt{12\Lambda}}\log\Big(\frac{E}{2\pi}\Big). \tag{7.89}$$

It should be noted that the $U(N)$ type correction term in (7.87) can be seen in Odlyzko's numerical data [72].

7.5.4 L functions

The Riemann zeta function is an explicit example of a general class of functions referred to as L-functions. Such functions are Dirichlet series

$$L(s) = \sum_{n=1}^\infty \frac{a_n}{n^s} \tag{7.90}$$

with the a_n such that the series converges for ${\rm Re}(s) > 1$, and with three additional properties. Briefly, these relate to the analytic continuation, functional equation and Euler product, which are all required to be similar to that for $\zeta(s)$. The generalized Riemann hypothesis asserts that all complex zeros of an L-function lie on the critical line $\mathrm{Re}(s) = \frac{1}{2}$. Furthermore there is analytic [472] and numerical [471] evidence to suggest that the statistics of these zeros for large imaginary part agrees with the statistics of eigenvalues from large $U(N)$ matrices.

In addition to this, Katz and Sarnak [352], [351] have related the statistical properties of the zeros on the critical line nearest the real axis for families of L-functions to the eigenvalues of large random matrices from the classical groups near $\theta = 0$. One example of a family of L-functions are those associated with the quadratic Dirichlet characters, $a_n = \chi_d(n)$ in (7.90), for all allowed d. Here d is the fundamental discriminant, which can be either positive or negative. It has the further property of being square free and congruent to $1 \mod 4$ if it is odd, and of being 4 times a square free integer congruent to 2 or $3 \mod 4$ if it is even. The so called character $\chi_d(n)$ only takes on values $\pm 1, 0$ and has the periodicity property $\chi_d(n + |d|) = \chi_d(n)$, with $|d|$ being referred to as the conductor.

Consider now the zeros on the critical line closest to the real axis, $1/2 + i\gamma^{(n)}$ with $\cdots \gamma_d^{(-2)} < \gamma_d^{(-1)} < 0 < \gamma_d^{(1)} < \gamma_d^{(2)} < \cdots$. Define the scaled variable $\alpha_d^{(j)} \log(|d|/2\pi)$, and define an ensemble by allowing d to vary and averaging. Analytic [352], [351] and numerical [471] evidence suggests that the distribution of the resulting zeros is that of the eigenvalues of matrices from the classical group $\mathrm{Sp}(2N)$ for N large, about $\theta = 0$. Moreover, an analogous result is expected to hold true for general families of L-functions, but with the correspondence to L-functions now relating to one of the classical groups $U(N), O^+(2N), O^-(2N+1)$ or $\mathrm{Sp}(2N)$. We refer to [403], [115] as starting points for studies into this topic.

7.6 SCALED LIMITS AT $\beta = 4$ — GAUSSIAN ENSEMBLE

7.6.1 The bulk

In the Gaussian ensemble $e^{-4V(x)} = e^{-2\tilde{V}_4(\sqrt{2}x)}$, which implies

$$S_4(x, y) = \sqrt{2}\tilde{S}_4(\sqrt{2}x, \sqrt{2}y). \tag{7.91}$$

In the bulk we must scale x and y as in Proposition 7.1.1. It turns out that then only the Christoffel-Darboux term in (6.59) contributes to the leading order asymptotics.

Proposition 7.6.1 *For the Gaussian ensemble with $\beta = 4$*

$$\lim_{N\to\infty} \Big(\frac{\pi\rho}{\sqrt{2N}}\Big)^n \rho_{(n)}\Big(\frac{\pi\rho x_1}{\sqrt{2N}}, \dots, \frac{\pi\rho x_n}{\sqrt{2N}}\Big)$$
$$= \rho^n \,\mathrm{qdet} \begin{bmatrix} \dfrac{\sin 2\pi\rho(x_j - x_k)}{2\pi\rho(x_j - x_k)} & \dfrac{1}{2\pi\rho}\displaystyle\int_0^{2\pi\rho(x_j - x_k)} \frac{\sin x}{x}\, dx \\ \dfrac{\partial}{\partial x_j}\dfrac{\sin 2\pi\rho(x_j - x_k)}{2\pi\rho(x_j - x_k)} & \dfrac{\sin 2\pi\rho(x_j - x_k)}{2\pi\rho(x_j - x_k)} \end{bmatrix}.$$

Proof. In view of the above remarks, the only point which requires checking is that the second term in the expression (6.59) for $\tilde{S}_4(x, y)$ does not contribute to

$$\lim_{N\to\infty} (\pi\rho/\sqrt{2N}) S_4(\pi\rho x/\sqrt{2N}, \pi\rho y/\sqrt{2N}).$$

This is a consequence of the $N \to \infty$ asymptotic estimate [420] $\int_x^\infty e^{-t^2/2} H_{2N-1}(t)dt = \mathrm{O}(2^{2N}(N-1)!)$ together with (7.1). □

7.6.2 Properties of the two-point function

With $\rho_{(2)}^{\rm bulk}(x,y)$ now denoting the scaled two-point correlation function in the bulk, Proposition 7.6.1 gives

$$\rho_{(2)}^{T\rm bulk}(x,y) = -\rho^2 \Big(\frac{\sin 2\pi\rho(x-y)}{2\pi\rho(x-y)} \Big)^2 + \rho^2 \Big(\frac{1}{2\pi\rho} \frac{\partial}{\partial x} \frac{\sin 2\pi\rho(x-y)}{2\pi\rho(x-y)} \Big) \int_0^{2\pi\rho(x-y)} \frac{\sin t}{t}\, dt. \tag{7.92}$$

It follows from this that for small x

$$\rho_{(2)}^{\rm bulk}(x,0) = \rho^2 \Big(\frac{(2\pi\rho x)^4}{135} - \frac{2(2\pi\rho x)^6}{4725} + \frac{2(2\pi\rho x)^8}{165375} + {\rm O}((\rho x)^{10}) \Big), \tag{7.93}$$

while for large x,

$$\begin{aligned}\rho_{(2)}^{T\,\rm bulk}(x,0) = \rho^2 \Big(& \frac{\pi}{2} \frac{\cos 2\pi\rho x}{2\pi\rho x} - \frac{1}{(2\pi\rho x)^2} + \frac{3}{2(2\pi\rho x)^4} \Big(1 + {\rm O}\Big(\frac{1}{(\rho x)^2}\Big)\Big) \\ & - \frac{\pi}{2} \frac{\sin 2\pi\rho x}{(2\pi\rho x)^2} + \frac{\cos 4\pi\rho x}{2(2\pi\rho x)^4} \Big(1 + {\rm O}\Big(\frac{1}{(\rho x)^2}\Big)\Big) + {\rm O}\Big(\frac{1}{(\rho x)^5}\Big)\Big),\end{aligned} \tag{7.94}$$

where the terms ${\rm O}(1/x^2)$ do not contain any oscillatory factors. Also, making use of the 2×2 determinant form of $\rho_{(2)}^T$ implied by Proposition 7.6.1, the dimensionless Fourier transform (7.3) can be computed as (see Exercises 7.6 q.1)

$$\widetilde{S}(k) = \begin{cases} \frac{|k|}{4\pi} - \frac{|k|}{8\pi} \log|1 - \frac{|k|}{2\pi}|, & |k| \le 4\pi, \\ 1, & |k| \ge 4\pi. \end{cases} \tag{7.95}$$

Note in particular the small $|k|$ behavior

$$\widetilde{S}(k) \sim \frac{|k|}{4\pi} + \frac{k^2}{16\pi^2} + {\rm O}(|k|^3). \tag{7.96}$$

7.6.3 Perturbation about $\beta = 4$ in the bulk

The expansion (7.5) tells us the first order correction in $(\beta - \beta_0)$ to the two-point correlation function at coupling β_0 in terms of higher order correlations at β_0. As was the case at $\beta_0 = 2$, these higher order correlations are known for $\beta_0 = 4$. Also, according to (6.4) the expansion of a quaternion determinant is formally the same as that of an ordinary determinant allowing (7.5) in the case of $\beta_0 = 4$ to be written in a form analogous to (7.6) and computed exactly. Moreover the Fourier transform of this first order correction can be computed exactly in terms of elementary functions together with the dilogarithm

$${\rm dilog}(x) = \int_1^x \frac{\log t}{1-t}\, dt.$$

It has a different analytic form for each of the regions $|k| < 2\pi$, $2\pi < |k| < 4\pi$, $|k| > 4\pi$. For $|k| < 2\pi$, the exact expression (which is rather lengthy) implies the expansion [207]

$$\begin{aligned}\frac{\partial \widetilde{S}(k;\beta)}{\partial \beta}\Big|_{\beta=4} = & -\frac{|k|}{16\pi} + \frac{|k|^3}{256\pi^3} + \frac{5k^4}{3072\pi^4} + \frac{3|k|^5}{4096\pi^5} + \frac{27k^6}{81920\pi^6} \\ & + \frac{37|k|^7}{245760\pi^7} + \frac{1273k^8}{18350080\pi^8} + \frac{887|k|^9}{27525120\pi^9} + \frac{4423k^{10}}{293601280\pi^{10}} \\ & + \frac{1949|k|^{11}}{275251200\pi^{11}} + \cdots.\end{aligned} \tag{7.97}$$

7.6.4 Global limit of the density

The global density $\tilde{\rho}_{(1)}(x)$ for general Gaussian β-ensembles with eigenvalue p.d.f. (1.28) is defined by (1.53). We read off from (6.55) and (7.91) that for $\beta = 4$

$$\sqrt{2/N}\rho_{(1)}(\sqrt{2N}x) = (2/\sqrt{N})\tilde{S}_4(2\sqrt{N}x, 2\sqrt{N}x).$$

To compute the large N form of this expression, we see from (6.59) that in addition to the global limit of K_{2N}, which is already known from (7.25), we must also compute the global limit of $\int_x^\infty e^{-t^2/2}H_{2N-1}(t)\,dt$. For this, use of the integral evaluation implied by (7.146) below gives

$$\int_x^\infty e^{-t^2/2}H_{2N-1}(t)\,dt = \sqrt{\frac{\pi}{2}}\frac{(2N-1)!}{(N-1/2)!} - \int_0^x e^{-t^2/2}H_{2N-1}(t)\,dt.$$

The global limit of the integral on the r.h.s. can now be analyzed using the Plancheral-Rotach formula (7.24), and the following result obtained [341], [202].

PROPOSITION 7.6.2 *Let* $-1 < x < 1$, *set* $A_N(x) := \cos(2Nx\sqrt{1-x^2} + (2N + \frac{1}{2})\arcsin x - N\pi)$ *and let* $\rho^{\rm W}(x)$ *denote the Wigner semicircle density. One has*

$$\sqrt{2/N}\rho_{(1)}(\sqrt{2N}x) = \rho^{\rm W}(x) - \Big(\frac{1}{\sqrt{2\pi N}} + \frac{(-1)^N}{2\pi N}\Big)\frac{\cos A_N(x)}{(1-x^2)^{1/4}} + \frac{1}{4\pi N}\frac{1}{\sqrt{1-x^2}} + {\rm O}(N^{-3/2}). \quad (7.98)$$

7.6.5 Soft edge

Here we want to study (7.91) in the scaled limit with x and y given by (7.11), and thus to compute

$$S_4^{\rm soft}(X,Y) := \lim_{N\to\infty}\frac{1}{2^{1/2}N^{1/6}}S_4\Big((2N)^{1/2} + \frac{X}{2^{1/2}N^{1/6}}, (2N)^{1/2} + \frac{Y}{2^{1/2}N^{1/6}}\Big).$$

Now, use of the asymptotic formula (7.68) shows that

$$\begin{aligned}\lim_{N\to\infty}\frac{1}{2^{1/2}N^{1/6}}K_{2N}^{(\rm G)}\Big(2^{1/2}((2N)^{1/2} + \frac{X}{2^{1/2}N^{1/6}}), 2^{1/2}((2N)^{1/2} + \frac{Y}{2^{1/2}N^{1/6}})\Big)\\ = 2^{-1/3}K^{\rm soft}(2^{2/3}X, 2^{2/3}Y),\end{aligned}$$

where $K^{\rm soft}$ is given by (7.12). Writing out the explicit form of the second term in (6.59), substituting in (7.91) and introducing the scaled variables gives

$$-\frac{2^{-2N-1/2}}{\sqrt{N}\Gamma(2N)}e^{-y^2}H_{2N}(\sqrt{2}y)\int_x^\infty e^{-t^2}H_{2N-1}(\sqrt{2}t)\,dt\Big|_{\substack{x\mapsto(2N)^{1/2}+X/2^{1/2}N^{1/6}\\ y\mapsto(2N)^{1/2}+Y/2^{1/2}N^{1/6}}} =: A_4(X,Y).$$

Use of the asymptotic expansion (7.9) shows

$$\lim_{N\to\infty}\frac{1}{2^{1/2}N^{1/6}}A_4(X,Y) = -\frac{1}{2^{2/3}}{\rm Ai}(2^{2/3}Y)\int_X^\infty {\rm Ai}(2^{2/3}v)\,dv.$$

Hence

$$S_4^{\rm soft}(X,Y) = \frac{1}{2^{1/3}}K^{\rm soft}(2^{2/3}X, 2^{2/3}Y) - \frac{1}{2^{2/3}}{\rm Ai}(2^{2/3}Y)\int_X^\infty {\rm Ai}(2^{2/3}v)\,dv. \quad (7.99)$$

The density is obtained by setting $X = Y$ in (7.99). According to (7.68) the second term decays as $X \to -\infty$. This implies

$$\rho_{(1)}^{\rm soft}(-X) \underset{X\to\infty}{\sim} \frac{1}{2^{1/3}}K^{\rm soft}(-2^{2/3}X, -2^{2/3}X) \underset{X\to\infty}{\sim} \frac{\sqrt{X}}{\pi}, \quad (7.100)$$

which is the same leading order behavior as that exhibited at $\beta = 2$ (recall (7.69)). Another feature of interest relating to the density is the correction terms to the soft edge scaled form. According to (7.27) the leading correction to the first term in (6.59) is $O(N^{-2/3})$. The correction to the second term is found by making use of (7.9). It is $O(N^{-1/3})$, and gives the expansion [202]

$$\frac{1}{2^{1/2}N^{1/6}}\rho_{(1)}\Big((2N)^{1/2}+\frac{X}{2^{1/2}N^{1/6}}\Big) \sim \rho_{(1)}^{\rm soft,4}(X)+\frac{1}{2^{4/3}(2N)^{1/3}}\Big(\mathrm{Ai}(2^{2/3}X)\Big)^2+O(N^{-2/3}). \quad (7.101)$$

From Proposition 6.1.7 we have that the truncated two-particle correlation is given by

$$\rho_{(2)}^{T\rm soft}(X,Y) = -S_4^{\rm soft}(X,Y)S_4^{\rm soft}(Y,X) - \Big(\frac{\partial}{\partial X}S_4^{\rm soft}(X,Y)\Big)\int_X^Y S_4^{\rm soft}(X,Y')\,dY'. \quad (7.102)$$

We can check from (7.68) that only $K^{\rm soft}$ in the formula (7.99) for $S_4^{\rm soft}$ contributes to the leading non-oscillatory behavior of $\rho_{(2)}^T(X,Y)$ for $X,Y\to-\infty$. One finds

$$\rho_{(2)}^{T\rm soft}(-X,-Y) \underset{X,Y\to\infty}{\sim} -\frac{1}{8\pi^2\sqrt{XY}}\frac{X+Y}{(X-Y)^2}, \quad (7.103)$$

which is exactly one half of the asymptotic expression (7.70) found for the same quantity at $\beta = 2$.

EXERCISES 7.6 1. (i) Let

$$\mathrm{FT}\,f = \hat{f}(k) := \int_{-\infty}^{\infty} f(x)e^{ikx}\,dx.$$

Verify that

$$\int_{-\infty}^{\infty} f(y-x)g(x)e^{ikx}\,dx = \frac{1}{2\pi}\int_{-\infty}^{\infty} e^{-ily}\hat{f}(l-k)\hat{g}(l)\,dl. \quad (7.104)$$

(ii) Let

$$\mathbf{f}_4(x) := \begin{bmatrix} \dfrac{\sin 2\pi x}{2\pi x} & \displaystyle\int_0^x \frac{\sin 2\pi t}{2\pi t}\,dt \\ \dfrac{\partial}{\partial x}\dfrac{\sin 2\pi x}{2\pi x} & \dfrac{\sin 2\pi x}{2\pi x} \end{bmatrix}.$$

Use the formula

$$\mathrm{FT}\,\frac{\sin 2\pi x}{2\pi x} = \begin{cases} \frac{1}{2}, & |k|<2\pi, \\ 0, & |k|>2\pi \end{cases}$$

to check that

$$\mathrm{FT}\,\mathbf{f}_4(x) = \begin{bmatrix} \dfrac{1}{2} & \dfrac{i}{2k} \\ -\dfrac{ik}{2} & \dfrac{1}{2} \end{bmatrix} \quad (|k|<2\pi), \qquad \mathrm{FT}\,\mathbf{f}_4(x) = \begin{bmatrix} 0 & 0 \\ 0 & 0 \end{bmatrix} \quad (|k|>2\pi).$$

(iii) Use the result (7.104) with $y=0$, together with the result of (ii) to show

$$\mathrm{FT}\Big(\mathbf{f}_4(x)\mathbf{f}_4(-x)\Big) = \begin{cases} \begin{bmatrix} 1-\dfrac{|k|}{4\pi}+\dfrac{|k|}{8\pi}\log\Big|1-\dfrac{|k|}{4\pi}\Big| & 0 \\ 0 & 1-\dfrac{|k|}{4\pi}+\dfrac{|k|}{8\pi}\log\Big|1-\dfrac{|k|}{2\pi}\Big| \end{bmatrix}, & |k|<4\pi, \\ \begin{bmatrix} 0 & 0 \\ 0 & 0 \end{bmatrix}, & |k|>4\pi \end{cases} \quad (7.105)$$

and thus deduce (7.95).

2. [221] Use the integral form of the kernels given in Exercises 7.1 q.1 to show that for scaled = bulk or soft

$$\tilde{S}_4^{\rm scaled}(X,Y) = \frac{1}{2}K^{\rm scaled}(X,Y) + \frac{1}{2}\frac{\partial}{\partial Y}\int_X^{\infty} K^{\rm scaled}(t,Y)\,dt, \quad (7.106)$$

where $\tilde{S}_4^{\rm bulk}(x,y) = S_4^{\rm bulk}(x/2,y/2)$, $\tilde{S}_4^{\rm soft}(x,y) = 2^{1/3}S^{\rm soft}(x/2^{2/3},y/2^{2/3})$ and thus

$$\begin{aligned} \frac{\partial}{\partial X}\tilde{S}_4^{\rm scaled}(X,Y) &= \frac{1}{2}\Big(\frac{\partial}{\partial X} - \frac{\partial}{\partial Y}\Big)K^{\rm scaled}(X,Y), \\ -\int_X^Y \tilde{S}_4(X,u)\,du &= \frac{1}{2}\Big(\int_Y^\infty K^{\rm scaled}(X,t)\,dt - \int_X^\infty K^{\rm scaled}(Y,t)\,dt\Big). \end{aligned} \tag{7.107}$$

7.7 SCALED LIMITS AT $\beta = 4$ — LAGUERRE AND JACOBI ENSEMBLES

7.7.1 Hard edge

In the Laguerre case

$$e^{-4V(x)} \propto e^{-2\tilde{V}_4(2x)}\Big|_{a\mapsto 2a-1}$$

and so

$$S_4(x,y) = 2\tilde{S}_4(2x,2y)\Big|_{a\mapsto 2a-1}. \tag{7.108}$$

Furthermore, the explicit form of $\tilde{S}_4$ is, from (6.59),

$$\tilde{S}_4(x,y) = \frac{1}{2}\Big(\frac{x}{y}\Big)^{1/2} K_{2N}^{(\rm L)}(x,y) + \frac{(2N)!}{4\Gamma(a+2N)}y^{(a-1)/2}e^{-y/2}L_{2N}^a(y)\int_x^\infty t^{(a-1)/2}e^{-t/2}L_{2N-1}^a(t)\,dt. \tag{7.109}$$

The scaling at the hard edge is as specified in Proposition 7.2.1, so we seek to compute

$$S_4^{\rm hard}(X,Y) := \lim_{N\to\infty}\frac{1}{4N}S_4\Big(\frac{X}{4N},\frac{Y}{4N}\Big).$$

Now, according to Proposition 7.2.1

$$\lim_{N\to\infty}\frac{1}{4N}K_{2N}^{(\rm L)}\Big(\frac{X}{2N},\frac{Y}{2N}\Big) = 2K^{\rm hard}(4X,4Y),$$

while the asymptotic form of the second term in (7.109) is deduced from (7.29), giving in total

$$S_4^{\rm hard}(X,Y) = 2\Big(\frac{X}{Y}\Big)^{1/2} K^{\rm hard}(4X,4Y)\Big|_{a\mapsto 2a-1} + \frac{J_{2a-1}(2Y^{1/2})}{2Y^{1/2}}\int_{X^{1/2}}^\infty J_{2a-1}(2t)\,dt. \tag{7.110}$$

Bessel function identities applied to this give the alternative form

$$S_4^{\rm hard}(X,Y) = 2K^{\rm hard}(4X,4Y)\Big|_{a\mapsto 2a} - \frac{J_{2a-1}(2Y^{1/2})}{2Y^{1/2}}\int_0^{X^{1/2}} J_{2a+1}(2t)\,dt \tag{7.111}$$

(see also Exercises 7.7 q.3). Note that in the case $a=0$ the integral in (7.111) can be evaluated, and thus we have the simplified result

$$S_4^{\rm hard}(X,Y)\Big|_{a=0} = 2K^{\rm hard}(4X,4Y)\Big|_{a=0} - \frac{1}{4Y^{1/2}}J_1(2Y^{1/2})\Big(J_0(2X^{1/2})-1\Big). \tag{7.112}$$

In the case of the Jacobi ensemble a similar calculation shows

$$\lim_{N\to\infty}\frac{1}{2N^2}\tilde{S}_4^{(\rm J)}\Big(1-\frac{X}{2N^2},1-\frac{Y}{2N^2}\Big) = \lim_{N\to\infty}\frac{1}{4N}\tilde{S}_4^{(\rm L)}\Big(\frac{X}{4N},\frac{Y}{4N}\Big),$$

which is to be expected in accordance with the remarks following (7.45).

7.7.2 Asymptotics of the one- and two-point functions

From Proposition 6.1.7 the density in the scaled limit near the hard edge is given by

$$\begin{aligned}\rho_{(1)}^{\rm hard}(X) &= S_4^{\rm hard}(X,X) \\ &= \frac{1}{2}\Big((J_{2a+1}(2X^{1/2}))^2 - J_{2a+2}(2X^{1/2})J_{2a-2}(2X^{1/2})\Big) \\ &\quad + \frac{1}{2X^{1/2}} J_{2a-1}(2X^{1/2}) \int_{X^{1/2}}^\infty J_{2a-1}(2s)\, ds, \end{aligned} \tag{7.113}$$

where the second equality follows from (7.111) and (7.33). For large X, use of (7.73) shows that the leading order decay of $S_4^{\rm hard}(X,X)$ is given by the leading order behavior of $2K^{\rm hard}(4X,4X)$. Use of (7.74) then gives

$$\rho_{(1)}^{\rm hard}(X) \underset{X\to\infty}{\sim} \frac{1}{2\pi X^{1/2}}, \tag{7.114}$$

which is identical to the behavior exhibited at $\beta = 2$.

The truncated two-particle correlation is given by (7.102) with $S_4^{\rm soft}$ replaced by $S_4^{\rm hard}$. For the leading asymptotics, we again see from (7.73) that the leading behavior of $S_4^{\rm hard}(X,Y)$ is given by the leading behavior of $2K^{\rm hard}(4X,4Y)$. Substituting this in the modified form of (7.102) shows that

$$\rho_{(2)}^{T\,\rm hard}(X,Y) \underset{X,Y\to\infty}{\sim} -\frac{1}{8\pi^2\sqrt{XY}}\frac{X+Y}{(X-Y)^2} \tag{7.115}$$

independent of a. This is the same behavior as found in (7.103) for the soft edge at $\beta = 4$.

7.7.3 Laguerre ensemble — global and soft edge densities

For finite N the density is given by (7.108) with $x = y$. The global density is then four times this expression with $x \mapsto 4NX$ (recall (3.57)). The first term in (7.109) is the Christoffel-Darboux kernel, which has been analyzed in this limit to give (7.39). For the second term, use of (7.131) below allows the integral over (x,∞) to be replaced by minus the same integral over $(0,x)$. This latter integral can be analyzed using the Plancheral-Rotach type asymptotic expansion (7.38), with the final result being [202]

$$4\rho_{(1)}(4NX) \sim \rho^{\rm MP}(X) - \frac{1}{2(\pi N)^{1/2}}\frac{g_{0,2N}^{(L)}(X)|_{a\mapsto 2a-1}}{X^{3/4}(1-X)^{1/4}} + \frac{a}{2\pi N\sqrt{X(1-X)}} + {\rm o}(N^{-1}), \tag{7.116}$$

where $g_{0,2N}^{(L)}(X)$ is specified by (7.38), and $\rho^{\rm MP}(X)$ is as in (7.39).

At the soft edge, the expansion of the first term in (7.109) is given by (7.37), while that of the second term is determined by making use of (7.35). With $\rho_{(1)}^{\rm soft}(X)$ as in (7.101), one finds [202]

$$\begin{aligned} &2(2N)^{1/3}\rho_{(1)}(4N + 2(2N)^{1/3}X) \\ &\quad \sim \rho_{(1)}^{\rm soft,4}(X) + \frac{a}{2N^{1/3}}\Big((\mathrm{Ai}(2^{2/3}X))^2 + \mathrm{Ai}'(2^{2/3}X)\int_{2^{2/3}X}^\infty \mathrm{Ai}(t)\,dt\Big) + {\rm O}(N^{-2/3}). \end{aligned} \tag{7.117}$$

7.7.4 Circular Jacobi ensemble — spectrum singularity

We recall from (3.123) that to study the log-gas on a circle with weight (5.50), it is sufficient to study the log-gas on a line with Cauchy weight as specified by (5.56). According to (3.123) the parameter α in (5.56) is dependent on N, β as well as the parameter a in (5.50). For purposes of using the general formula (6.59), it is convenient to define α independent of β, by setting

$$e^{-2V(x)} = \frac{1}{(1+x^2)^\alpha}, \qquad \alpha = N + a, \tag{7.118}$$

which according to (3.123) is indeed the correct choice at $\beta = 2$. Then (6.53) gives

$$e^{-2\tilde{V}_4(x)} = \frac{1}{(1+x^2)^{\alpha-1}}, \tag{7.119}$$

and comparison with (3.123) shows that the correct weight function for the Cauchy ensemble equivalent to the circular Jacobi ensemble at $\beta = 4$ is $e^{-2\tilde{V}_4(x)}\big|_{\substack{\alpha \mapsto 2\alpha \\ \alpha = N+a}}$. This means that the quantity $\tilde{S}_4(x,y)\big|_{\substack{\alpha \mapsto 2\alpha \\ \alpha = N+a}}$ gives the sought n-point correlation at $\beta = 4$.

Now, with the weight function (7.118), use of (7.119) and (5.65) in (6.59) shows

$$\begin{aligned}\tilde{S}_4(x,y) = &\frac{1}{2}\Big(\frac{1+x^2}{1+y^2}\Big)^{1/2} K_{2N}^{(\mathrm{C})}(x,y) \\ &- \frac{(1+y^2)^{-(\alpha+1)/2}}{2(p_{2N-1}, p_{2N-1})_2}(\alpha - 2N) p_{2N}(y) \int_x^\infty \frac{p_{2N-1}(t)}{(1+t^2)^{(\alpha+1)/2}}\, dt,\end{aligned} \tag{7.120}$$

where

$$K_{2N}^{(\mathrm{CJ})}(x,y) := \frac{1}{(1+x^2)^{\alpha/2}(1+y^2)^{\alpha/2}} \frac{p_{2N}^{(\mathrm{C})}(x) p_{2N-1}^{(\mathrm{C})}(y) - p_{2N}^{(\mathrm{C})}(y) p_{2N-1}^{(\mathrm{C})}(x)}{x-y}.$$

As in the analysis leading to (7.48), our task is to compute the thermodynamic limit of $\tilde{S}_4$ with

$$x \mapsto i\frac{1+z}{1-z}, \quad z := e^{2\pi i x/L} \qquad y \mapsto i\frac{1+w}{1-w}, \quad w = e^{2\pi i y/L}. \tag{7.121}$$

Analogous to (7.46) we have

$$\begin{aligned} p_{2N}^{(\mathrm{CJ})}\Big(i\frac{1+z}{1-z}\Big)\Big|_{\substack{\alpha \mapsto 2\alpha \\ \alpha = N+a}} &= i^{-2N}\Big(-\frac{2}{1-z}\Big)^{2N} {}_2F_1(-2N, 2a; 4a; 1-z), \\ p_{2N-1}^{(\mathrm{C})}\Big(i\frac{1+z}{1-z}\Big)\Big|_{\substack{\alpha \mapsto 2\alpha \\ \alpha = N+a}} &= i^{-(2N-1)}\Big(-\frac{2}{1-z}\Big)^{2N-1} {}_2F_1(-(2N-1), 2a+1; 4a+2; 1-z), \end{aligned} \tag{7.122}$$

while $(p_{2N-1}, p_{2N-1})_2^{(\mathrm{C})}$ is specified by (5.54). Making use of the asymptotic formula (7.47) shows

$$K_{2N}^{(\mathrm{CJ})}\Big(i\frac{1+z}{1-z}, i\frac{1+w}{1-w}\Big)\Big|_{\substack{\alpha \mapsto 2\alpha \\ \alpha = N+a}} \sim \frac{2\pi\rho|xy|}{N} K^{\mathrm{s.s.}}(2x, 2y)\Big|_{a \mapsto 2a}, \tag{7.123}$$

where $K^{\mathrm{s.s.}}$ is specified by (7.48).

Regarding the second line in (7.120), making the substitutions (7.121) and (7.122), a similar analysis which leads to (7.123) shows that the leading asymptotic form is

$$-\frac{a}{N}(\pi\rho y)^{3/2} J_{2a-1/2}(2\pi\rho y) \int_0^{\pi\rho x} s^{-1/2} J_{2a+1/2}(2s)\, ds.$$

Hence

$$\tilde{S}_4\Big(i\frac{1+z}{1-z}, i\frac{1+w}{1-w}\Big)\Big|_{\substack{\alpha \mapsto 2\alpha \\ \alpha = N+a}} \sim \frac{\pi\rho y^2}{N} S_4^{\mathrm{s.s.}}(x,y), \tag{7.124}$$

$$S_4^{\mathrm{s.s.}}(x,y) := K^{\mathrm{s.s.}}(2x, 2y)\Big|_{a \mapsto 2a} - a\pi\rho \frac{J_{2a-1/2}(2\pi\rho y)}{(\pi\rho y)^{1/2}} \int_0^{\pi\rho x} s^{-1/2} J_{2a+1/2}(2s)\, ds. \tag{7.125}$$

The result (7.124) can be used to deduce the asymptotic behavior of $\tilde{I}_4$ and $\tilde{D}_4$ as specified by (6.15). One finds

$$\tilde{I}_4\Big(i\frac{1+z}{1-z}, i\frac{1+w}{1-w}\Big) \sim -\int_x^y S_4^{\rm s.s.}(x,u)\,du, \qquad \tilde{D}_4\Big(i\frac{1+z}{1-z}, i\frac{1+w}{1-w}\Big) \sim \Big(\frac{\pi\rho x y}{N}\Big)^2 \frac{\partial}{\partial x} S_4^{\rm s.s.}(x,y).$$

We substitute these asymptotic forms in the expression of (6.14) for $\mathbf{f}_4$, form qdet according to Proposition 6.1.7, and extract common factors from the odd numbered columns and even numbered rows which give an overall contribution

$$\prod_{j=1}^n \frac{\pi\rho}{N} x_j^2. \tag{7.126}$$

This procedure gives the leading asymptotic form in the thermodynamic limit of $\rho_{(n)}$ as specified by the first equality in (5.55). The factor (7.126) cancels with the prefactors on the r.h.s. of the first equality in (5.55), leaving the final expression [213]

$$\rho_{(n)}^{\rm s.s.}(x_1,\dots,x_n) = {\rm qdet}\begin{bmatrix} S_4^{\rm s.s.}(x_j,x_k) & -\int_{x_j}^{x_k} S_4^{\rm s.s.}(x_j,u)\,du \\ \frac{\partial}{\partial x_j} S_4^{\rm s.s.}(x_j,x_k) & S_4^{\rm s.s.}(x_k,x_j) \end{bmatrix}_{j,k=1,\dots,n}. \tag{7.127}$$

When $a = 0$ the circular Jacobi ensemble reduces to the circular ensemble of Dyson. In this case (7.127) should agree with the scaled n-point correlation given in Proposition 7.6.1 for the Gaussian ensemble in the bulk. Indeed, we see from (7.125), (7.48) and (7.50) that

$$S_4^{\rm s.s.}(x,y)\Big|_{a=0} = \rho\frac{\sin 2\pi\rho(x-y)}{2\pi\rho(x-y)},$$

which when substituted in (7.127) gives the result of Proposition 7.6.1.

Another case of interest is $a = 1$. Then the formula (7.56) must apply, with the l.h.s. given by Proposition 7.6.1, and the r.h.s. by (7.127). In particular, taking $n = 1$ we must have

$$S_4^{\rm s.s.}(x,0)\Big|_{a=1} = \rho - \rho\Big(\frac{\sin 2\pi\rho x}{2\pi\rho x}\Big)^2 + \rho\Big(\frac{1}{2\pi\rho}\frac{\partial}{\partial x}\frac{\sin 2\pi\rho x}{2\pi\rho x}\Big)\int_0^{2\pi\rho x}\frac{\sin t}{t}\,dt. \tag{7.128}$$

This can be checked using (7.125) and (7.49).

EXERCISES 7.7 1. Analogous to Exercises 7.1 q.2, verify that

$$\begin{aligned}
&\lim_{\alpha\to\infty}\frac{\pi\rho}{\sqrt{\alpha}} S_4^{\rm soft}(-(\alpha+\pi\rho x/\sqrt{\alpha}), -(\alpha+\pi\rho y/\sqrt{\alpha})) \\
&\quad = \lim_{a\to\infty} 2\pi\rho\sqrt{a} S_4^{\rm hard}(a+2\pi\rho x\sqrt{a}, a+2\pi\rho y\sqrt{a}) = S_4^{\rm bulk}(x,y), \\
&\lim_{a\to\infty} 2a(a/2)^{1/3} S_4^{\rm hard}(a^2-2a(a/2)^{1/3}x, a^2-2a(a/2)^{1/3}y) = S_4^{\rm soft}(x,y),
\end{aligned}$$

and with $\rho = 1/\pi$

$$\lim_{a\to\infty}(a/2)^{1/3} S_4^{\rm s.s.}(a-(a/2)^{1/3}x, a-(a/2)^{1/3}y) = S_4^{\rm soft}(x,y).$$

2. [3] The objective of this exercise is to verify that (7.109) can be rewritten as [544], [216]

$$\begin{aligned}
2\tilde{S}_4^{(\rm L)}(x,y) &= K_{2N}^{(\rm L)}(x,y)\Big|_{a\mapsto a+1} \\
&+\frac{(2N)!\,y^{(a+1)/2}e^{-y/2}}{2\Gamma(2N+a+1)}\frac{L_{2N}^{a+1}(y)-L_{2N-1}^{a+1}(y)}{y}\int_0^x t^{(a+1)/2}e^{-t/2}\frac{d}{dt}L_{2N}^{a+1}(t)\,dt.
\end{aligned} \tag{7.129}$$

The general strategy is to compare coefficients of $y^{(a-1)/2}e^{-y/2}L_k^{a+1}(y)$, $k = 0,\dots,2N$ in (7.109) and (7.129).

This requires using the Laguerre polynomial identities

$$\begin{aligned} L_n^{a-1}(x) &= L_n^a(x) - L_{n-1}^a(x), \\ \frac{d}{dx}L_n^a(x) &= -L_{n-1}^{a+1}(x), \\ xL_{n-1}^{a+1}(x) &= -nL_n^{a+1}(x) + (2n+a)L_{n-1}^{a+1}(x) - (n+a)L_{n-2}^{a+1}(x), \end{aligned} \tag{7.130}$$

and the definite integral (see (6.113) and (6.114))

$$\int_0^\infty x^{a/2}e^{-x/2}L_n^{a+1}(x)\,dx = \begin{cases} \frac{\Gamma((n+3)/2))\Gamma(a+n+2)}{2^{a/2-1}\Gamma(n+2)\Gamma((n+a+3)/2)}, & n \text{ even}, \\ 0, & n \text{ odd}. \end{cases} \tag{7.131}$$

(i) Use the first and third identities in (7.130) to show that the coefficient of $y^{(a-1)/2}e^{-y/2}L_{2N}^a(y)$ in (7.129) is

$$\frac{1}{2}x^{(a+1)/2}e^{-x/2}\frac{((2N-1)!)^2L_{2N-1}^{a+1}(x)}{\Gamma(2N)\Gamma(a+1+2N)}(-2N)+\frac{(2N)!}{4\Gamma(2N+a+1)}\int_0^x t^{(a+1)/2}e^{-t/2}\Big(\frac{d}{dt}L_{2N}^{a+1}(t)\Big)\,dt,$$

and read off that the same coefficient in (7.109) is

$$\frac{(2N)!}{4\Gamma(a+2N)}\int_x^\infty t^{(a+1)/2-1}e^{-t/2}L_{2N-1}^a(t)\,dt.$$

Use the definite integral (7.131) to show that both these expressions agree when $x = 0$. Then use the identities (7.130) to show that the two expressions have the same derivative.

(ii) Proceed as in (i) to show that for $k < 2N$, the coefficient of $y^{(a-1)/2}e^{-y/2}L_k^a(y)$ in (7.129) is

$$\frac{k!}{2\Gamma(a+1+k)}x^{(a+1)/2}e^{-x/2}\Big(L_{k+1}^{a+1}(x) - L_k^{a+1}(x)\Big)$$

while in (7.109) the coefficient is

$$\frac{k!}{2\Gamma(a+1+k)}x^{(a+1)/2}e^{-x/2}L_k^a(x),$$

and note from the first identity in (7.130) that these expressions are equal.

(iii) Use (7.129) to show that an alternative expression for (7.110) is (7.111).

7.8 SCALED LIMITS AT $\beta = 1$ — GAUSSIAN ENSEMBLE

7.8.1 The bulk

In the Gaussian ensemble $\tilde{V}_1 = V$ and so $\tilde{S}_1(x,y) = S_1(x,y)$. Thus from (6.107) we read off that for N even

$$\tilde{S}_1^{(G)}(x,y) = K_{N-1}^{(G)}(x,y) + \frac{e^{-y^2/2}H_{N-1}(y)}{2^N\pi^{1/2}(N-2)!}\int_{-\infty}^\infty \operatorname{sgn}(x-t)e^{-t^2/2}H_{N-2}(t)\,dt. \tag{7.132}$$

Proceeding as in the proof of Proposition 7.6.1 we can obtain from this formula the scaling limit of the n-point correlation function in the bulk.

PROPOSITION 7.8.1 *For the Gaussian ensemble at $\beta = 1$ we have*

$$\lim_{N\to\infty}\Big(\frac{\pi\rho}{\sqrt{N}}\Big)^n \rho_{(n)}\Big(\frac{\pi\rho x_1}{\sqrt{N}},\dots,\frac{\pi\rho x_n}{\sqrt{N}}\Big)$$
$$= \rho^n \,\mathrm{qdet}\begin{bmatrix} \dfrac{\sin\pi\rho(x_j-x_k)}{\pi\rho(x_j-x_k)} & \dfrac{1}{\pi\rho}\displaystyle\int_0^{\pi\rho(x_j-x_k)}\frac{\sin t}{t}dt - \frac{1}{2\rho}\mathrm{sgn}(x_j-x_k) \\ \dfrac{\partial}{\partial x_j}\dfrac{\sin\pi\rho(x_j-x_k)}{\pi\rho(x_j-x_k)} & \dfrac{\sin\pi\rho(x_j-x_k)}{\pi\rho(x_j-x_k)} \end{bmatrix}_{j,k=1,\dots,n}.$$

7.8.2 Properties of the two-point function

The truncated two-particle correlation, according to Proposition 7.8.1, is given by

$$\rho_{(2)}^T(x,0) = \rho^2\Big(-\frac{\sin^2\pi\rho x}{(\pi\rho x)^2} + \frac{1}{\pi\rho}\Big(\frac{d}{dx}\frac{\sin\pi\rho x}{\pi\rho x}\Big)\Big(-\frac{\pi}{2}\mathrm{sgn}\,x + \int_0^{\pi\rho x}\frac{\sin t}{t}dt\Big)\Big). \tag{7.133}$$

This implies the small x expansion

$$\rho_{(2)}(x,0) = \rho^2\Big(\frac{\pi}{6}|\pi\rho x| - \frac{\pi}{60}|\pi\rho x|^3 + \frac{1}{135}(\pi\rho x)^4 - \cdots\Big) \tag{7.134}$$

and the large x asymptotic expansion

$$\rho_{(2)}^T(x,0) = \rho^2\Big(-\frac{1}{(\pi\rho x)^2} + \frac{3}{2(\pi\rho x)^4}\Big(1+\mathrm{O}\Big(\frac{1}{x^2}\Big)\Big) + \frac{\cos 2\pi\rho x}{2(\pi\rho x)^4}\Big(1+\mathrm{O}\Big(\frac{1}{x^2}\Big)\Big)\Big), \tag{7.135}$$

where the terms $\mathrm{O}(1/x^2)$ do not contain any oscillatory factors. Analogous to (7.95), the corresponding dimensionless Fourier transform can be computed in terms of elementary functions [395],

$$\tilde{S}(k) = \begin{cases} \frac{|k|}{\pi} - \frac{|k|}{2\pi}\log\Big(1+\frac{|k|}{\pi}\Big), & |k| \le 2\pi, \\ 2 - \frac{|k|}{2\pi}\log\frac{|k|/\pi+1}{|k|/\pi-1}, & |k| \ge 2\pi, \end{cases} \tag{7.136}$$

which exhibits the small $|k|$ expansion

$$\tilde{S}(k) \sim \frac{|k|}{\pi} - \frac{k^2}{2\pi^2} + \mathrm{O}(k^3). \tag{7.137}$$

7.8.3 Perturbation about $\beta = 1$ in the bulk

As in the cases of $\beta_0 = 2$ and $\beta_0 = 4$, the explicit form of the n-point correlation ($n = 2, 3, 4$) can be used to compute the first order perturbation of the dimensionless two-point correlation about $\beta_0 = 1$, or equivalently the first order perturbation of the dimensionless structure function. However, as far as the $|k| < \pi$ form of the latter goes this is superfluous, as an integral formula obtained for $\tilde{S}(k;\beta)$ in Section 13.7.4 below shows that if we set

$$f(k;\beta) := \frac{\pi\beta}{|k|}S(k;\beta), \quad 0 < k < \min(2\pi, \pi\beta) \tag{7.138}$$

and define f for $k < 0$ by analytic continuation, then

$$f(k;\beta) = f\Big(-\frac{2k}{\beta};\frac{4}{\beta}\Big). \tag{7.139}$$

Thus if we write

$$\frac{\pi\beta}{|k|}\tilde{S}(k;\beta) = 1 + \sum_{j=1}^{\infty} A_j(\beta/2)\Big(\frac{|k|}{\pi\beta}\Big)^j, \quad |k| < \min(2\pi, \pi\beta) \tag{7.140}$$

the coefficients $A_j(\beta/2)$ must satisfy the functional relation

$$A_j(1/x) = (-x)^{-j} A_j(x). \tag{7.141}$$

In particular, knowing the expansion (7.97) of $\partial\tilde{S}(k;\beta)/\partial\beta|_{\beta=4}$ we can use this to deduce the corresponding expansion of $\partial\tilde{S}(k;\beta)/\partial\beta|_{\beta=1}$.

7.8.4 Global limit of the density

Suppose for definiteness that N is even, and write

$$\Phi_N(x) = 2^{-N}\int_0^x e^{-t^2/2} H_N(t)\, dt. \tag{7.142}$$

Using this in (7.132) gives, after simple manipulation,

$$\rho_{(1)}(x) = K_N^{(\mathrm{G})}(x,x) - \frac{2^{-(N-1)}e^{-x^2}}{\sqrt{\pi}(N-1)!}(H_{N-1}(x))^2 + \frac{e^{-x^2/2}}{2\sqrt{\pi}} H_{N-1}(x)\Phi_{N-2}(x). \tag{7.143}$$

The global asymptotic expansion of $K_N^{(\mathrm{G})}(x,x)$ is given by (7.25), while that of the other terms can be deduced by making use of (7.24). These together show [341]

$$\sqrt{2/N}\rho_{(1)}(\sqrt{2N}X) = \rho^{\mathrm{W}}(X) - \frac{1}{2\pi N\sqrt{1-X^2}} + \mathrm{O}(N^{-2}). \tag{7.144}$$

Inspection of (7.25), (7.98) and (7.144) shows that with oscillatory terms ignored they can be combined into the single formula

$$\sqrt{2/N}\rho_{(1)}(\sqrt{2N}X) \sim \rho^{\mathrm{W}}(X) - \frac{1}{\pi N}\Big(\frac{1}{\beta} - \frac{1}{2}\Big)\frac{1}{\sqrt{1-X^2}}, \qquad -1 < X < 1. \tag{7.145}$$

Macroscopic arguments will be used in Section 14.2 below to derive this formula for general β, and to extend it to all real $X \neq 1$.

7.8.5 Soft edge

The task here is to evaluate the scaled limit of (7.132) with coordinates (7.11). For the first term the limit is equal to $K^{\mathrm{soft}}(X,Y)$ as specified by (7.12). To compute the scaled limit of the second term we make use of (6.113) to rewrite (7.142) as

$$\Phi_N(x) = \sqrt{\frac{\pi}{2}}\frac{N!}{2^N(N/2)!} - 2^{-N}\int_x^\infty e^{-t^2/2} H_N(t)\, dt, \tag{7.146}$$

and then use the asymptotic expansion (7.9). This gives

$$\lim_{N\to\infty} \frac{1}{2^{1/2}N^{1/6}} e^{-y^2/2} \frac{1}{2\sqrt{\pi}} H_{N-1}(y)\Phi_{N-2}(x)\Big|_{\substack{x\mapsto\sqrt{2N}+X/2^{1/2}N^{1/6}\\ y\mapsto\sqrt{2N}+Y/2^{1/2}N^{1/6}}} = \frac{1}{2}\mathrm{Ai}(Y)\Big(1 - \int_X^\infty \mathrm{Ai}(t)\,dt\Big),$$

and thus, in an obvious notation

$$S_1^{\mathrm{soft}}(X,Y) = K^{\mathrm{soft}}(X,Y) + \frac{1}{2}\mathrm{Ai}(Y)\Big(1 - \int_X^\infty \mathrm{Ai}(t)\,dt\Big). \tag{7.147}$$

The density is given by (7.147) with $X = Y$. Since then the second term decreases as $X \to -\infty$, we see that

$$\rho_{(1)}(-X) \underset{X\to\infty}{\sim} K^{\rm soft}(-X,-X) \underset{X\to\infty}{\sim} \frac{\sqrt{X}}{\pi}, \tag{7.148}$$

which is identical to the leading asymptotic behavior of the soft edge density at $\beta = 2$ and $\beta = 4$. The asymptotic expansion of the density at the soft edge can be computed starting from (7.143). We know from (7.27) that the corrections to the first term therein are $O(N^{-2/3})$. Using (7.146) and (7.9) shows the remaining terms have corrections $O(N^{-1/3})$, giving the result [202]

$$\frac{1}{2^{1/2}N^{1/6}}\rho_{(1)}\Big((2N)^{1/2} + \frac{X}{2^{1/2}N^{1/6}}\Big) \sim \rho_{(1)}^{{\rm soft},1}(X) + \frac{1}{2N^{1/3}}{\rm Ai}'(X)\Big(1 - \int_X^\infty {\rm Ai}(t)\,dt\Big) + O(N^{-2/3}). \tag{7.149}$$

From Propositions 6.3.2 and 6.3.3 the truncated two-particle correlation is given by

$$\begin{aligned}\rho_{(2)}^T(X,Y) = &-S_1^{\rm soft}(X,Y)S_1^{\rm soft}(Y,X) + \frac{1}{2}\frac{\partial}{\partial X}S_1^{\rm soft}(X,Y) \\ &\times\Big(\int_{-\infty}^{\infty} S_1^{\rm soft}(X,z){\rm sgn}(z-Y)\,dz - {\rm sgn}(X-Y)\Big).\end{aligned} \tag{7.150}$$

Use of (7.68) shows that only $K^{\rm soft}(X,Y)$ in the formula (7.147) contributes to the leading non-oscillatory behavior of (7.150), which one calculates as

$$\rho_{(2)}^T(-X,-Y) \underset{X,Y\to\infty}{\sim} -\frac{1}{2\pi^2\sqrt{XY}}\frac{X+Y}{(X-Y)^2}. \tag{7.151}$$

Note that this is the same asymptotic behavior found in (7.70) and (7.103) provided we replace the factor of two in the denominator by 2β.

EXERCISES 7.8

1. From the definition (7.138) read off from (7.136) and (7.95) that

$$f(k;1) = 1 - \frac{1}{2}\log\Big(1 + \frac{k}{\pi}\Big), \qquad f(k;4) = 1 - \frac{1}{2}\log\Big(1 - \frac{k}{2\pi}\Big) \tag{7.152}$$

and thus illustrate (7.139) for $\beta = 1$.

2. [221] Use the integral form of the kernel given in Exercises 7.1 q.1 to show that for scaled = bulk or soft

$$S_1^{\rm scaled}(X,Y) = \frac{1}{2}K^{\rm scaled}(X,Y) - \frac{1}{2}\frac{\partial}{\partial Y}\int_{-\infty}^X K^{\rm scaled}(t,Y)\,dt \tag{7.153}$$

(cf. (7.106)) and thus the equations (7.107) hold with $\tilde S_4^{\rm scaled}$ replaced by $S_1^{\rm scaled}$.

7.9 SCALED LIMITS AT $\beta = 1$ — LAGUERRE AND JACOBI ENSEMBLES

7.9.1 Hard edge

In the Laguerre case at $\beta = 1$ we have $V(x) = \tilde V_1(x)\Big|_{a\mapsto a+1}$ and so

$$S_1(x,y) = \tilde S_1(x,y)\Big|_{a\mapsto a+1}.$$

Making use of (6.107) we thus have that for N even

$$\begin{aligned} S_1(x,y) &= \Big(\frac{x}{y}\Big)^{1/2} K^{(\mathrm{L})}_{N-1}(x,y)\Big|_{a\mapsto a+1} \\ &\quad - \frac{(N-1)!}{4\Gamma(a+N)} y^{a/2} e^{-y/2} L^{a+1}_{N-1}(y) \int_0^\infty \operatorname{sgn}(x-u) L^{a+1}_{N-2}(u) u^{a/2} e^{-u/2}\, du \\ &= \Big(\frac{x}{y}\Big)^{1/2} K^{(\mathrm{L})}_{N-1}(x,y)\Big|_{a\mapsto a+1} - \frac{(N-1)!}{4\Gamma(a+N)} y^{a/2} e^{-y/2} L^{a+1}_{N-1}(y) \\ &\quad \times \Big(2\int_0^x L^{a+1}_{N-2}(u) u^{a/2} e^{-u/2}\, du - \frac{\Gamma((N+1)/2)\Gamma(a+N)}{2^{a/2-1}\Gamma(N)\Gamma((N+1+a)/2)} \Big), \end{aligned} \tag{7.154}$$

where the second equality follows from (7.131). Making use of (7.29) then shows

$$\begin{aligned} S_1^{\mathrm{hard}}(X,Y) &:= \lim_{N\to\infty} \frac{1}{4N} S_1\Big(\frac{X}{4N}, \frac{Y}{4N}\Big) \\ &= \Big(\frac{X}{Y}\Big)^{1/2} K^{\mathrm{hard}}(X,Y)\Big|_{a\mapsto a+1} + \frac{J_{a+1}(Y^{1/2})}{4Y^{1/2}} \Big(1 - \int_0^{X^{1/2}} J_{a+1}(v)\, dv\Big), \end{aligned} \tag{7.155}$$

where K^{hard} is given by (7.31). Bessel function identities applied to this expression give the alternative form [216]

$$S_1^{\mathrm{hard}}(X,Y) = K^{\mathrm{hard}}(X,Y) + \frac{J_{a+1}(Y^{1/2})}{4Y^{1/2}} \Big(1 - \int_0^{X^{1/2}} J_{a-1}(v)\, dv\Big) \tag{7.156}$$

(see also Exercises 7.9 q.1). Both (7.155) and (7.156) have the feature that in the case $a = 0$ the integrals therein can be evaluated explicitly. Doing this in (7.156) gives [418]

$$S_1^{\mathrm{hard}}(X,Y)\Big|_{a=0} = K^{\mathrm{hard}}(X,Y)\Big|_{a=0} - \frac{1}{4\sqrt{Y}} J_0(\sqrt{X}) J_1(\sqrt{Y}). \tag{7.157}$$

As in the $\beta = 4$ theory, for the Jacobi ensemble it can be shown that [415]

$$\lim_{N\to\infty} \frac{1}{2N^2} S_1^{(\mathrm{J})}\Big(1 - \frac{X}{2N^2}, 1 - \frac{Y}{2N^2}\Big) = \lim_{N\to\infty} \frac{1}{4N} S_1^{(\mathrm{L})}\Big(\frac{X}{4N}, \frac{Y}{4N}\Big).$$

7.9.2 Asymptotics of the one- and two-point functions

From Propositions 6.3.3, 6.3.2 and (7.33), (7.155), we have that for general $a > -1$ the scaled density at a point X from the hard edge is given by

$$\begin{aligned} \rho_{(1)}^{\mathrm{hard}}(X) &= \frac{1}{4}\Big((J_{a+1}(X^{1/2}))^2 - J_{a+2}(X^{1/2}) J_a(X^{1/2}) \Big) \\ &\quad - \frac{J_{a+1}(\sqrt{X})}{4\sqrt{X}} \Big(\int_0^{\sqrt{X}} J_{a+1}(u)\, du - 1 \Big). \end{aligned} \tag{7.158}$$

The asymptotic expansion (7.73) allows the leading large X non-oscillatory behavior of (7.158) to be computed, analogous to the determination of the asymptotics of (7.113). Again the result (7.74) is obtained. This suggests a universality property: in general the large X behavior of $\rho_{(1)}(X)$ near the hard edge will be given by (7.74), independent of the value of β and a.

The truncated two-particle correlation function is given by the formula (7.150) with S_1^{soft} replaced by S_1^{hard} and the lower terminal of integration $-\infty$ replaced by 0. From the fact that the large X, Y asymptotics of S_1^{hard} come from $K^{\mathrm{hard}}(X,Y)$, the (modified) formula (7.150) shows that the asymptotics (7.151) persists independent of a.

7.9.3 Laguerre ensemble — global and soft edge densities

For finite N (assumed even) the density is given by setting $x = y$ in (7.154). Following the strategy detailed in Section 7.7.3 for the $\beta = 4$ Laguerre ensemble density, this expression can be analyzed globally by setting $x = 4NX$, and at the soft edge by setting $x = 4N + 2(2N)^{1/3}X$. The final results are [341], [202]

$$4\rho_{(1)}(4NX) \sim \rho^{\rm MP}(X) + \frac{1}{2\pi N}\frac{a}{\sqrt{X(1-X)}} + {\rm o}(N^{-1}), \tag{7.159}$$

$$\begin{aligned} &2(2N)^{1/3}\rho_{(1)}(4N + 2(2N)^{1/3}X) \\ &\quad \sim \rho_{(1)}^{\rm soft,1}(X) - \frac{a}{2(2N)^{1/3}}\Big({\rm Ai}'(X)(1 - \int_X^\infty {\rm Ai}(s)\,ds) - ({\rm Ai}(X))^2\Big) + {\rm O}(N^{-2/3}). \end{aligned} \tag{7.160}$$

Comparison of these formulas with their Laguerre counterparts at $\beta = 2$ and 4 reveals some common features. Consider first the global density. Ignoring oscillatory terms, inspection of (7.39), (7.116) and (7.159) shows that they all satisfy

$$4\rho_{(1)}(4NX) \sim \rho^{\rm MP}(X) + \frac{a}{2\pi N\sqrt{X(1-X)}} + {\rm o}(N^{-1}), \qquad 0 < X < 1. \tag{7.161}$$

As discussed in Section 14.2 below, this is closely related to the expansion (7.145). In relation to the expansion of the density at the soft edge, we see from (7.37), (7.117) and (7.160), together with the explicit functional forms of $\rho_{(1)}^{\rm soft,\beta}$ that they all satisfy

$$2(2N)^{1/3}\rho_{(1)}(4N + 2(2N)^{1/3}X) \sim \rho_{(1)}^{\rm soft,\beta}(X) - \frac{a}{(2N)^{1/3}}\frac{d}{dX}\rho_{(1)}^{\rm soft,\beta}(X). \tag{7.162}$$

This is in distinction to the analogous expansions for the Gaussian ensembles at the soft edge, which reveal no such structure.

7.9.4 Circular Jacobi ensemble — spectrum singularity

Considerations analogous to those of Section 7.7.4 show that the correct weight function for the Cauchy ensemble equivalent to the circular Jacobi ensemble is

$$e^{-\tilde V_1(x)} = \frac{1}{(1+x^2)^{(\alpha+1)/2}}, \qquad \alpha = N + a. \tag{7.163}$$

Thus the n-point correlation can be computed from the quantity $\tilde S_1(x,y)|_{N+\alpha=a}$ with $e^{-2V(x)}$ given by (7.118). Use of (7.118), (7.163) and (5.65) in (6.107) shows this is given by

$$\begin{aligned} \tilde S_1(x,y)\Big|_{N+a=\alpha} &= \Big(\frac{1+x^2}{1+y^2}\Big)^{1/2} K_{N-1}^{(\rm C)}(x,y) \\ &\quad + \frac{(a+1)p_{N-1}^{(\rm C)}(y)}{(p_{N-2},p_{N-2})_2^{(\rm C)}(1+y^2)^{(\alpha+1)/2}}\int_{-\infty}^\infty \frac{{\rm sgn}(x-t)p_{N-2}^{(\rm C)}(t)}{(1+t^2)^{(\alpha+1)/2}}\,dt. \end{aligned} \tag{7.164}$$

Regarding the thermodynamic limit, proceeding as in the derivation of (7.123) we readily find

$$K_{N-1}^{(\rm C)}\Big(i\frac{1+z}{1-z}, i\frac{1+w}{1-w}\Big)\Big|_{\alpha=N+a} \sim \frac{\pi\rho xy}{N}K_{\rm s.s.}(x,y)\Big|_{a\mapsto a+1}.$$

The asymptotic analysis of the second line in (7.164) is complicated by the region of integration being the whole real line. Analogous to the manipulations in (7.146) and (7.154), with $x \mapsto -i(1+z)/(1-z) =$

$-\cot \pi x/L$ we write

$$\int_{-\infty}^{\infty} \frac{\mathrm{sgn}(x-t)p_{N-2}^{(\mathrm{C})}(t)}{(1+t^2)^{(N+a+1)/2}}\,dt\Big|_{x\mapsto -\cot \pi x/L} = -\int_{-\infty}^{\infty} \frac{\mathrm{sgn}(x-t)p_{N-2}^{(\mathrm{C})}(t)}{(1+t^2)^{(N+a+1)/2}}\,dt\Big|_{x\mapsto \cot \pi x/L}$$

$$= -\int_{-\infty}^{\infty} \frac{p_{N-2}^{(\mathrm{C})}(t)}{(1+t^2)^{(N+a+1)/2}}\,dt + 2\int_{\cot \pi x/L}^{\infty} \frac{p_{N-2}^{(\mathrm{C})}(t)}{(1+t^2)^{(N+a+1)/2}}\,dt. \tag{7.165}$$

The definite integral is evaluated according to the general formula (6.113), which after recalling (6.97) and (5.65) gives

$$\int_{-\infty}^{\infty} \frac{p_{N-2}^{(\mathrm{C})}(t)}{(1+t^2)^{(N+a+1)/2}}\,dt = \Big(\int_{-\infty}^{\infty} \frac{dt}{(1+t^2)^{(N+a+1)/2}}\Big) \prod_{j=0}^{N/2-2} \frac{N+a-1-2j}{N+a-2j} \frac{(p_{2j+1},p_{2j+1})_2^{(\mathrm{C})}}{(p_{2j},p_{2j})_2^{(\mathrm{C})}}$$
$$= 2^{a+2}N^{-a-2}\frac{\Gamma(a/2+1)\Gamma(a+5/2)}{\Gamma(a/2+3/2)}, \tag{7.166}$$

where to obtain the first equality use has been made of (5.54), while the duplication formula (1.111) has been used in obtaining the final line.

The second integral in (7.165), along with the terms outside the integral in (7.164), are analyzed as in Section 7.2.6. Combining results, and substituting in (7.164) we find the asymptotic behavior

$$\tilde{S}_1\Big(i\frac{1+z}{1-z}, i\frac{1+w}{1-w}\Big)\Big|_{\alpha=N+a} \sim \frac{\pi\rho y^2}{N} S_1^{\mathrm{s.s.}}(x,y), \tag{7.167}$$

$$S_1^{\mathrm{s.s.}}(x,y) = K_{\mathrm{s.s.}}(x,y)\Big|_{a\mapsto a+1} + \frac{\pi\rho(a+1)\Gamma(a/2+1)}{2^{3/2}\Gamma(a/2+3/2)}(\pi\rho y)^{-1/2}J_{a+1/2}(\pi\rho y)$$
$$\times\Big(1 - 2^{1/2}\frac{\Gamma(a/2+3/2)}{\Gamma(a/2+1)}\int_0^{\pi\rho x} s^{-1/2}J_{a+3/2}(s)\,ds\Big). \tag{7.168}$$

Substituting (7.167) in the formulas of Proposition 6.3.2 for D_1 and $\tilde{I}_1$ and substituting this n-point function in (5.55), after cancellation of the prefactor in the first equality of (5.55), we see that in the thermodynamic limit the n-point correlation about the spectrum singularity at $\beta = 1$ is given by [213]

$$\rho_{(n)}(x_1,\dots,x_n) = \mathrm{qdet}\begin{bmatrix} S_1^{\mathrm{s.s.}}(x_j,x_k) & -\int_{x_j}^{x_k} S_1^{\mathrm{s.s.}}(x_j,u)\,du - \frac{1}{2}\mathrm{sgn}(x_j-x_k) \\ \frac{\partial}{\partial x_j} S_1^{\mathrm{s.s.}}(x_j,x_k) & S_1^{\mathrm{s.s.}}(x_k,x_j) \end{bmatrix}_{j,k=1,\dots,n}. \tag{7.169}$$

Analogous to the corresponding correlation at $\beta = 4$ (7.127), we can check from (7.168) that with $a = 0$ the result of Proposition 7.8.1 is obtained, while with $a = 1$ the analogue of (7.128) holds.

7.9.5 Superimposed $\beta = 1$ systems

The formula (6.152) can be used to compute the scaled limit of (6.151) at the soft edge of the Gaussian ensemble. Proceeding as in Sections 7.6.5 and 7.8.5 one finds [195]

$$\rho_{(n)}^{\mathrm{odd(OEsoft)}^2}(x_1,\dots,x_n)$$
$$:= \lim_{N\to\infty}\Big(\frac{1}{2^{1/2}N^{1/6}}\Big)^n \rho_{(n)}^{\mathrm{odd(GOE}_N)^2}\Big(\sqrt{2N}+\frac{x_1}{2^{1/2}N^{1/6}},\dots,\sqrt{2N}+\frac{x_n}{2^{1/2}N^{1/6}}\Big)$$
$$= \det\Big[K^{\mathrm{soft}}(x_j,x_k) + \mathrm{Ai}(x_j)\int_0^{\infty}\mathrm{Ai}(x_k-v)\,dv\Big]_{j,k=1,\dots,n}. \tag{7.170}$$

Note that this corresponds to the kernel (7.18) with $s_1 = 0$.

Consideration of the distribution (6.146) only requires minor modification to the above working. In particular we see the formula (6.149) applies but with

$$f(x) = e^{\tilde{V}_4(x)} \int_x^\infty e^{-\tilde{V}_1(t)}\, dt$$

in (6.150). This means that formula (6.151) formally applies for

$$\rho_{(n)}^{\mathrm{even}(\mathrm{OE}_N)^2}(x_1, \dots, x_n), \tag{7.171}$$

the only difference being that the first summation in (6.152) must be multiplied by a minus sign, which in turn follows from the fact that

$$\int_x^\infty e^{-\tilde{V}_1(t)}\, dt = -\tilde{\phi}_0(x) + \tilde{c}_0$$

(cf. (6.153)). This modified form of (6.152) can then be used to compute the scaled limit of (7.171) at the hard edge of the superimposed Laguerre orthogonal ensembles. Proceeding as in Sections 7.7.1 and 7.9 one finds [195]

$$\begin{aligned}\rho_{(n)}^{\mathrm{odd}(\mathrm{OEhard})^2}(x_1, \dots, x_n) &:= \lim_{N\to\infty} \Big(\frac{1}{4N}\Big)^n \rho_{(n)}^{\mathrm{even}(\mathrm{LOE}_N|_{a\mapsto a-1})^2}\Big(\frac{x_1}{4N}, \dots, \frac{x_n}{4N}\Big) \\ &= \det\Big[K^{\mathrm{hard}}(x_j, x_k) + \frac{J_a(\sqrt{x_j})}{2\sqrt{x_k}} \int_{\sqrt{x_k}}^\infty J_a(t)\, dt\Big]_{j,k=1,\dots,n} \end{aligned} \tag{7.172}$$

(the notation odd(OE)^2 on the l.h.s. refers to the labeling $x_1 < x_2 < \cdots$ which is natural at the hard edge in the scaled limit).

EXERCISES 7.9 1. Using Exercises 7.7 q.2 as a guide, verify that (7.154) can be rewritten as [544], [216]

$$\begin{aligned}S_1^{(\mathrm{L})}(x,y) = K_N^{(\mathrm{L})}(x,y) &+ \frac{N!}{4\Gamma(N+a)} y^{a/2} e^{-y/2} \\ &\times L_{N-1}^{a+1}(y) \int_0^\infty \mathrm{sgn}(x-u) u^{a/2-1} e^{-u/2} L_N^{a-1}(u)\, du.\end{aligned}$$

Deduce from this (7.156).

7.10 TWO-COMPONENT LOG-GAS WITH CHARGE RATIO 1:2

7.10.1 A local limit theorem and the evaluation of the free energy

To compute the free energy per particle from the closed form expression (6.162) for the partition function requires the asymptotic behavior of the coefficient of $\zeta^{N_1/2}$ in the polynomial

$$g_{N_1/2+N_2}(\zeta) := \prod_{l=1}^{N_1/2+N_2} \Big(\zeta + (l - \frac{1}{2})^2\Big). \tag{7.173}$$

This can be deduced as a special case of a local limit theorem of Bender [55], which relates to the asymptotic behavior of the coefficients of general polynomials given in a factorized form.

PROPOSITION 7.10.1 [55, Th. 2] *Let*

$$P_n(x) = \sum_{k=0}^n a_n(k) x^k = a_n(n) \prod_{j=1}^n (x + r_n(j))$$

be a polynomial in x whose roots are all real and nonpositive. Associate with $P_n(x)$ the normalized double sequence

$$p_n(k) := \frac{a_n(k)}{P_n(1)}.$$

Then with the mean and variance given by

$$\mu_n = \sum_{j=1}^{n} \frac{1}{1 + r_n(j)}, \qquad \sigma_n^2 = \sum_{j=1}^{n} \frac{r_n(j)}{(1 + r_n(j))^2},$$

the $a_n(k)$ satisfy a local limit theorem

$$\lim_{n\to\infty} \sup_{x\in(-\infty,\infty)} \left| \sigma_n p_n([\sigma_n x + \mu_n]) - \frac{1}{\sqrt{2\pi}} e^{-x^2/2} \right| = 0 \tag{7.174}$$

provided $\sigma_n \to \infty$ as $n \to \infty$. (The square brackets in $[\sigma_n x + \mu_n]$ denote the integer part.)

When $x = 0$, (7.174) gives

$$a_n([\mu_n]) \sim \frac{P_n(1)}{\sigma_n \sqrt{2\pi}}, \tag{7.175}$$

provided $\sigma_n \to \infty$ as $n \to \infty$. This formula can be used to deduce the asymptotics of $[\zeta^{N_1/2}]g_{N_1/2+N_2}(\zeta)$.

PROPOSITION 7.10.2 *We have*

$$[\zeta^{N_1/2}]g_{N_1/2+N_2}(\zeta) \sim \frac{|\Gamma(i\alpha + N_1/2 + N_2 + \frac{1}{2})|^2 \cosh \pi\alpha}{\pi \alpha^{N_1} \sqrt{2\pi\sigma^2}} \tag{7.176}$$

where

$$\alpha = \frac{N_2 + N_1/2}{\nu}, \tag{7.177}$$

with ν specified as the solution of the equation

$$\frac{N_1}{N_1 + 2N_2} = \frac{\arctan \nu}{\nu}, \tag{7.178}$$

and

$$\sigma^2 \sim \frac{N_2 + N_1/2}{2} \Big(\frac{\arctan \nu}{\nu} - \frac{1}{1+\nu^2} \Big).$$

Proof. To use (7.175) to compute the coefficient of $\zeta^{N_1/2}$ in $g_{N_1/2+N_2}(\zeta)$ requires that we scale ζ by a suitable function of $N_1/2 + N_2$ so that $\mu_{N_1/2+N_2} = N_1/2$. Thus we choose $P_n(x) = \alpha^{-N_1} g_n(\alpha^2 x)$ in Proposition 7.10.1, where α is to be so determined. From the formula for μ_n we can check that for large $n = N_1/2 + N_2$, with α given by (7.177)

$$\mu_{N_1/2+N_2} \sim \alpha \arctan \nu$$

which implies $\mu_{N_1/2+N_2} = N_1/2$ provided (7.178) is satisfied. The specification of σ^2 follows similarly to the formula for μ_n, and we note that indeed $\sigma^2 \to \infty$ as $N_1/2 + N_2 \to \infty$, which is required for the validity of (7.175). The stated asymptotic formula now follows from (7.175) after noting from (7.173) that

$$g_{N_1/2+N_2}(\alpha^2 x) = |\Gamma(i\alpha\sqrt{x} + N_2 + N_1/2 + \frac{1}{2})|^2 \frac{\cosh \pi\alpha\sqrt{x}}{\pi},$$

where use has been make of (4.5). □

Use of Stirling's formula in (7.176), together with the definitions (7.177) and (7.178) allows (7.176) to be simplified to read

$$[\zeta^{N_1/2}]g_{N_1/2+N_2}(\zeta) \sim \exp\Big(N_2 \log(1 + 1/\nu^2) + 2N_2 \log(N_2 + N_1/2) + (N_1/2)\log(1+\nu^2) - 2N_2 + \mathrm{O}(1)\Big). \tag{7.179}$$

Substituting this result in (6.162), and making a further use of Stirling's formula specifies the asymptotics of Z_{N_1,N_2}. Substituting the resulting expression in the general formula (4.160) gives the free energy per particle [179].

PROPOSITION 7.10.3 *Let*

$$x_1 := \lim_{N_1,N_2\to\infty} \frac{N_1}{N_1+N_2}, \qquad x_2 := \lim_{N_1,N_2\to\infty} \frac{N_2}{N_1+N_2}, \qquad \rho_b = \lim_{N_1,N_2,L\to\infty} \frac{N_1+2N_2}{L}$$

and specify ν by (7.178). Then

$$-\log Z_{N_1,N_2} \sim (N_1+N_2)\beta f + \mathrm{O}(1),$$

where

$$\beta f = \frac{1}{2}x_1 \log \frac{\rho_b}{4(1+\nu^2)} - x_2 \log\Big(\rho_b(2\pi)^2\Big(1+\frac{1}{\nu^2}\Big)\Big) - \frac{x_1}{2} + x_2. \tag{7.180}$$

The quantities x_1 and x_2 are referred to as concentrations. Note that in the limit $x_1 \to 1$, so that $\nu \to 0$, $\rho_b \to \rho$ and $x_2 \to 0$,

$$\beta f \sim \frac{1}{2}\log\frac{\rho}{4} - \frac{1}{2},$$

which agrees with the $\beta = 1$ case of (4.166) for the free energy per particle for the one-component log-gas at $\beta = 1$. Similarly, in the limit $x_2 \to 1$, so that $\nu \to \infty$, $\rho_b \to 2\rho$ and $x_1 \to 0$,

$$\beta f \sim -\log\Big(2\rho(2\pi)^2\Big) + 1,$$

which agrees with the $\beta = 4$ case of (4.166) for the free energy per particle of the one-component log-gas.

7.10.2 The correlation functions

The local limit theorem Proposition 7.10.1 can also be used to compute the thermodynamic limit of the correlation function (6.163). According to (6.163) and (6.168) it suffices to compute the asymptotics of

$$[\zeta^{N_1/2-q}] \prod_{\substack{l=1 \\ l\neq l_1,\dots,l_p}}^{N_1/2+N_2} (\zeta + (l-\frac{1}{2})^2) =: [\zeta^{N_1/2-q}] \frac{g_{N_1/2+N_2}(\zeta)}{\prod_{i=1}^p(\zeta + (l_i - 1/2)^2)}$$

for p and q fixed. Application of Proposition 7.10.1 gives

$$[\zeta^{N_1/2-q}] \prod_{\substack{l=1 \\ l\neq l_1,\dots,l_p}}^{N_1/2+N_2} (\zeta + (l-\frac{1}{2})^2) \sim \frac{\alpha^{2q}}{\prod_{i=1}^p(\alpha^2 + (l_i - \frac{1}{2})^2)} [\zeta^{N_1/2}]\, g_{N_1/2+N_2}(\zeta), \tag{7.181}$$

where α is given by (7.177). As a result, with

$$s_j(x) := \int_0^1 \frac{t^j \sin \pi\rho_b x t}{t^2 + 1/\nu^2}\,dt, \qquad c_j(x) := \int_0^1 \frac{t^j \cos \pi\rho_b x t}{t^2+1/\nu^2}\,dt, \tag{7.182}$$

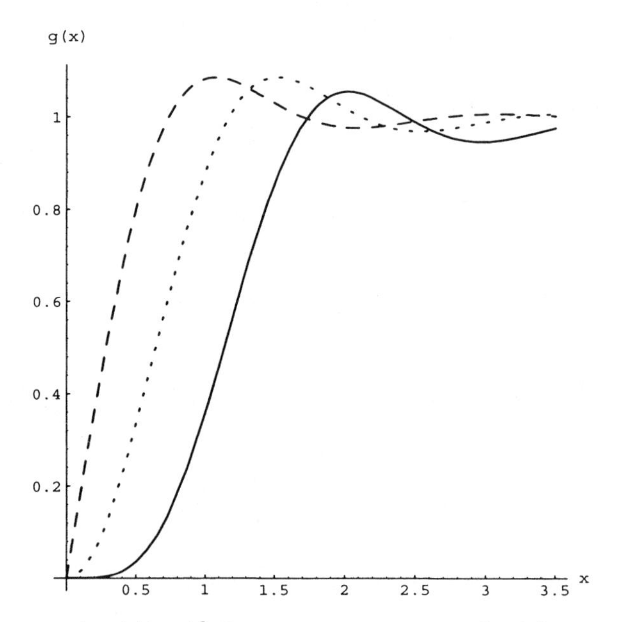

Figure 7.2 Plot of $g(x) = \rho_{+1,+1}(x,0)/(\rho_{+1})^2$ (first dashed curve), $\rho_{+1,+2}(x,0)/\rho_{+1}\rho_{+2}$ (second dashed curve) and $\rho_{+2,+2}(x,0)/(\rho_{+2})^2$ in the case $\rho_{+1} = \rho_{+2} = 1/3$, $\rho_b = 1$. Note the different position of the first maximum in each case.

we obtain from (6.163) and (6.168) the limiting formula

$$\begin{aligned}&\rho_{+1^{n_1},+2^{n_2}}(x_1,\dots,x_{n_1};y_1,\dots,y_{n_2})\\&= \mathrm{qdet}\begin{bmatrix}[\mathbf{A}(x_j-x_k)]_{j,k=1,\dots,n_1} & [\mathbf{B}(x_j-y_k)]_{\substack{j=1,\dots,n_1\\k=1,\dots,n_2}}\\ [\mathbf{C}(y_j-x_k)]_{\substack{j=1,\dots,n_2\\k=1,\dots,n_1}} & [\mathbf{D}(y_j-y_k)]_{j,k=1,\dots,n_2}\end{bmatrix},\end{aligned} \tag{7.183}$$

where

$$\mathbf{A}(x) = \frac{\rho_b}{\nu^2}\begin{bmatrix} c_0(x) & -\frac{\pi}{2}\nu^2\mathrm{sgn}(x) + s_{-1}(x)\\ -s_1(x) & c_0(x)\end{bmatrix}, \quad \mathbf{B}(x) = \frac{\rho_b}{\nu\sqrt{2}}\begin{bmatrix} c_0(x) & s_1(x)\\ -s_1(x) & c_2(x)\end{bmatrix},$$

$$\mathbf{C}(x) = \frac{\rho_b}{\nu\sqrt{2}}\begin{bmatrix} c_2(x) & s_1(x)\\ -s_1(x) & c_0(x)\end{bmatrix}, \quad \mathbf{D}(x) = \frac{\rho_b}{2}\begin{bmatrix} c_2(x) & s_3(x)\\ -s_1(x) & c_2(x)\end{bmatrix}.$$

Note that in the limit $\nu \to 0$ this expression reduces to the result of Proposition 7.8.1 for the correlations at $\beta = 1$, while it in the limit $\nu \to \infty$ reduces to the result of Proposition 7.6.1 for the correlations at $\beta = 4$.

The determinant form of (7.183) allows the general truncated correlation to be written down. Thus we have

$$\rho_{+1^{n_1},+2^{n_2}}(x_1,\dots,x_{n_1};y_1,\dots,y_{n_2}) = (-1)^{n_1+n_2-1}\sum_{\substack{\text{cycles}\\\text{length } n_1+n_2}}\prod_{(\imath,\imath')}\Big(\mathbf{f}(\vec{r}_\imath,\vec{r}_{\imath'})\Big)^{(0)}, \tag{7.184}$$

where $\mathbf{f}(\vec{r}_\imath,\vec{r}_{\imath'}) := \mathbf{A}(x_\imath - x_{\imath'})$, $i,i' \in [1,n_1]$; $\mathbf{f}(\vec{r}_\imath,\vec{r}_{\imath'}) := \mathbf{B}(x_\imath - y_{i'-n_1})$, $i \in [1,n_1]$, $i' \in [n_1+1, n_1+n_2]$; $\mathbf{f}(\vec{r}_\imath,\vec{r}_{\imath'}) := \mathbf{C}(y_{\imath-n_1} - x_{\imath'})$, $i \in [n_1+1, n_1+n_2]$, $i' \in [1,n_1]$; $\mathbf{f}(\vec{r}_\imath,\vec{r}_{\imath'}) := \mathbf{D}(y_{i-n_1} - y_{i'-n_1})$, $i,i' \in [n_1+1, n_1+n_2]$. This of course agrees with the results for the truncated correlations for the one-component log-gas at $\beta = 1$ and $\beta = 4$, as given by Propositions 7.8.1 and Propositions 7.6.1, respectively, in the appropriate limits.

For the two-particle correlation functions (7.183) and (7.184) give [179]

$$\rho^T_{+1,+1}(x) = -\frac{\rho_b^2}{\nu^4}\Big((c_0(x))^2 + s_1(x)s_{-1}(x)\Big) + \frac{\pi\rho_b^2}{2\nu^2}s_1(x),$$
$$\rho^T_{+1,+2}(x) = -\frac{\rho_b^2}{2\nu^2}\Big((s_1(x))^2 + c_0(x)c_2(x)\Big),$$
$$\rho^T_{+2,+2}(x) = -\frac{\rho_b^2}{4}\Big((c_2(x))^2 + s_1(x)s_3(x)\Big). \tag{7.185}$$

Plots of these quantities for a particular value of the parameter ν are given in Figure 7.2. Note that the value of the first maximum is characteristic of local charge neutrality.

EXERCISES 7.10 1. Show that to leading order the large x expansions of (7.185) are

$$\rho^T_{+1,+1}(x) \underset{x\to\infty}{\sim} -\frac{\rho_b^2}{(1+\nu^2)^2}\frac{1}{(\pi x\rho_b)^2}, \qquad \rho^T_{+1,+2}(x) \underset{x\to\infty}{\sim} -\frac{\rho_b^2\nu^2}{2(1+\nu^2)^2}\frac{1}{(\pi x\rho_b)^2},$$
$$\rho^T_{+2,+2}(x) \underset{x\to\infty}{\sim} -\frac{\rho_b^2\nu^4}{4(1+\nu^2)^2}\frac{1}{(\pi x\rho_b)^2}. \tag{7.186}$$

Chapter Eight

Eigenvalue probabilities — Painlevé systems approach

The generating function for the probability that there are exactly n eigenvalues in an interval J of a classical matrix ensemble with unitary symmetry can, for certain J, be identified with the τ-function of a Painlevé system. In particular, this is possible whenever J is a single interval containing an endpoint of the support of the density. This allows the distribution of certain eigenvalue probabilities relating to the largest and smallest eigenvalue, and the bulk spacing, to be characterized in terms of the solution of nonlinear equations of Painlevé type. A practical consequence is the rapid computation of the power series expansions of the spacing distribution, and the high precision numerical tabulations which follow from these. To obtain the Painlevé nonlinear equations we give a self-contained development of the Hamiltonian formulation of the Painlevé theory, which makes essential use of Bäcklund transformations and the Toda lattice equation. In the case of $J = [-a, a]$ and the weight being classical and even, the gap probability for $\beta = 1$ can be related to a $\beta = 2$ gap probability. This allows for a Painlevé characterization of the gap probability of the COE and its bulk scaling limit. Inter-relationships between the COE, CUE and CSE known from earlier chapters then imply the analogous result for the CSE. In the last section, the theory of orthogonal polynomial systems on the unit circle is used to characterize some random matrix averages over $U(N)$ in terms of discrete Painlevé equations.

8.1 DEFINITIONS

From experimental data of energy spectra, it is a simple matter to construct the histogram corresponding to the p.d.f. for the spacing between consecutive energy levels. For the spectra of heavy nuclei, this was first done in the late 1950s [447]. In accordance with our discussion in Chapter 1, the theoretical p.d.f. should be identical to that of the bulk eigenvalues in the GOE, appropriately scaled. Here we will discuss the mathematical theory of Painlevé systems as it relates to the calculation of such eigenvalue spacing distributions.

In general, for a continuous, one-dimensional statistical mechanical system the p.d.f. for the spacing distribution is simply related to the probability of a particle-free interval (sometimes called the gap, or hole, probability).

DEFINITION 8.1.1 *We denote by $E_{N,\beta}(n; J)$ the probability that there are exactly n particles in the interval J of a continuous, one-dimensional statistical mechanical system at inverse temperature β with N particles. With $J = (a_-, a_+)$ we denote by $p_{N,\beta}(n; J)$ the p.d.f. for the event that given there is a particle at a_-, there is a particle at a_+, with exactly n particles in between. Similarly, we use the notation $p_{N,\beta}(n; a_\pm)$ to denote the p.d.f. for the event that there is a particle at $a_\pm$ and exactly n particles in total to the right (left).*

PROPOSITION 8.1.2 *Define the generating functions*

$$E_{N,\beta}(J;\xi) := \frac{1}{\hat{Z}_N}\int_{-\infty}^{\infty} dx_1 \cdots \int_{-\infty}^{\infty} dx_N \prod_{l=1}^{N}(1-\xi\chi_J^{(l)})\, e^{-\beta U(x_1,\dots,x_N)},$$

$$p_{N,\beta}(J;\xi) := \frac{N(N-1)}{\rho_{(1)}(a_1)\hat{Z}_N}\int_{-\infty}^{\infty} dx_3 \cdots \int_{-\infty}^{\infty} dx_N \prod_{l=3}^{N}(1-\xi\chi_J^{(l)}) e^{-\beta U(a_-,a_+,x_3,\dots,x_N)},$$

$$p_{N,\beta}(a_\pm;\xi) := \frac{N}{\hat{Z}_N}\int_{-\infty}^{\infty} dx_2 \cdots \int_{-\infty}^{\infty} dx_N \prod_{l=2}^{N}(1-\xi\chi_J^{(l)}) e^{-\beta U(a_\pm,x_2,\dots,x_N)},$$

where $\chi_J^{(l)} = 1$ if $x_l \in J$ and $\chi_J^{(l)} = 0$ otherwise, in the second formula $J = (a_-, a_+)$, and in the third formula $J = (a_+,\infty)$ or $(-\infty, a_-)$. For a continuous one-dimensional statistical mechanical system of N particles, we have

$$E_{N,\beta}(n;J) = \frac{(-1)^n}{n!}\frac{\partial^n}{\partial\xi^n}E_{N,\beta}(J;\xi)\Big|_{\xi=1}, \tag{8.1}$$

$$\begin{aligned} p_{N,\beta}(n;J) &= \frac{(-1)^n}{n!}\frac{\partial^n}{\partial\xi^n}p_{N,\beta}(J;\xi)\Big|_{\xi=1} \\ &= -\frac{1}{\rho_{(1)}(a_-)}\frac{\partial^2}{\partial a_-\partial a_+}E_{N,\beta}(n;J) + 2p_{N,\beta}(n-1;J) - p_{N,\beta}(n-2;J), \end{aligned} \tag{8.2}$$

$$p_{N,\beta}(n;a_\pm) = \frac{(-1)^n}{n!}\frac{\partial^n}{\partial\xi^n}p_{N,\beta}(a_\pm;\xi)\Big|_{\xi=1} = \pm\frac{d}{da}E_{N,\beta}(n;J) + p_{N,\beta}(n-1;J), \tag{8.3}$$

where $p_{N,\beta}(k;J) = 0$ for $k<0$ and in (8.2) $J = (a_-, a_+)$, while in (8.3) $J = (a_+,\infty)$ or $(-\infty, a_-)$.

Proof. The first equalities in (8.1)–(8.3) follow from the definitions of the generating functions and the defining equations

$$E_{N,\beta}(n;J) = \frac{1}{\hat{Z}_N}\binom{N}{n}\int_J dx_1 \cdots \int_J dx_n \int_{\bar{J}} dx_{n+1} \cdots \int_{\bar{J}} dx_N\, e^{-\beta U(x_1,\dots,x_N)},$$

$$\begin{aligned} p_{N,\beta}(n;J) = &\frac{1}{\rho_{(1)}(a_-)}\frac{1}{\hat{Z}_N}\frac{N!}{n!(N-n-2)!}\int_J dx_3 \cdots \int_J dx_{n+2}\int_{\bar{J}} dx_{n+3}\cdots\int_{\bar{J}} dx_N \\ &\times e^{-\beta U(a_-,a_+,x_3,\dots,x_N)}, \end{aligned}$$

$$p_{N,\beta}(n;a_\pm) = \frac{1}{\hat{Z}_N}\frac{N!}{n!(N-n-1)!}\int_J dx_2 \cdots \int_J dx_{n+1}\int_{\bar{J}} dx_{n+2}\cdots\int_{\bar{J}} dx_N\, e^{-\beta U(a_\pm,x_2,\dots,x_N)},$$

where $\bar{J} := (-\infty,\infty) - J$. For the second equalities in (8.2) and (8.3), we note that it follows from the first equalities that

$$E_{N,\beta}(J;\xi) = \sum_{n=0}^{\infty}(1-\xi)^n E_{N,\beta}(n;J), \qquad p_{N,\beta}(J;\xi) = \sum_{n=0}^{\infty}(1-\xi)^n p_{N,\beta}(n;J),$$

$$p_{N,\beta}(a_\pm;\xi) = \sum_{n=0}^{\infty}(1-\xi)^n p_{N,\beta}(n;a_\pm), \tag{8.4}$$

while inspection of the definitions of the generating functions shows

$$p_{N,\beta}(J;\xi) = -\frac{1}{\xi^2\rho_{(1)}(a_-)}\frac{\partial^2}{\partial a_-\partial a_+}E_{N,\beta}(J;\xi), \qquad p_{N,\beta}(a_\pm;\xi) = \pm\frac{1}{\xi}\frac{d}{da_\pm}E_{N,\beta}(J;\xi). \tag{8.5}$$

Substituting (8.4) in (8.5) and equating powers of $(1-\xi)$ gives the second equalities. □

In the case of a matrix ensemble with unitary symmetry specified by the eigenvalue p.d.f.

$$\frac{1}{C}\prod_{j=1}^{N} g(x_j) \prod_{1\le j<k\le N} (x_k - x_j)^2 \tag{8.6}$$

(as in Section 5.3 this is to be denoted $\mathrm{UE}_N(g)$), the generating functions $E_{N,2}(J;\xi)$, in the case $J = (-\infty, a_-)$ or (a_+, ∞), and $p_{N,2}(a_\pm;\xi)$, are the particular cases $\mu = 0$, $\mu = 2$ (the latter after multiplication by a suitable constant times $g(a_\pm)$) of

$$E_{N,2}^{(\mu)}(J;\xi;g) := \Big\langle \prod_{l=1}^{N}(1-\xi\chi_J^{(l)})|a_\pm - x_l|^\mu \Big\rangle_{\mathrm{UE}_N(g)}. \tag{8.7}$$

Specifically

$$E_{N,2}(J;\xi) = E_{N,2}^{(0)}(J;\xi;g), \quad p_{N+1,2}(a_\pm;\xi) = \frac{\hat{Z}_N}{\hat{Z}_{N+1}} g(a_\pm) E_{N,2}^{(2)}(J;\xi;g). \tag{8.8}$$

Following [228], [230], [233], the theme of this chapter is to relate the general average (8.7) for classical weights g to the theory of Painlevé systems. Consequently it will be possible to characterize such averages as the solution of certain nonlinear equations.

We will see that the Painlevé systems are Hamiltonian systems associated with the Painlevé equations. There are six Painlevé equations. These are special second order nonlinear equations, labelled PI–PVI, given explicitly by

$$\begin{aligned}
\mathrm{PI}\ y'' &= 6y^2 + t, \\
\mathrm{PII}\ y'' &= 2y^3 + ty + \alpha, \\
\mathrm{PIII}\ y'' &= \frac{1}{y}(y')^2 - \frac{1}{t}y' + \gamma y^3 + \frac{1}{t}(\alpha y^2 + \beta) + \frac{\delta}{y}, \\
\mathrm{PIV}\ y'' &= \frac{1}{2y}(y')^2 + \frac{3}{2}y^3 + 4ty^2 + 2(t^2 - \alpha)y + \frac{\beta}{y}, \\
\mathrm{PV}\ y'' &= \Big(\frac{1}{2y} + \frac{1}{y-1}\Big)(y')^2 - \frac{1}{t}y' + \frac{(y-1)^2}{t^2}\Big(\alpha y + \frac{\beta}{y}\Big) + \frac{\gamma y}{t} + \frac{\delta y(y+1)}{y-1}, \\
\mathrm{PVI}\ y'' &= \frac{1}{2}\Big(\frac{1}{y} + \frac{1}{y-1} + \frac{1}{y-t}\Big)(y')^2 - \Big(\frac{1}{t} + \frac{1}{t-1} + \frac{1}{y-t}\Big)y' \\
&\quad + \frac{y(y-1)(y-t)}{t^2(t-1)^2}\Big(\alpha + \frac{\beta t}{y^2} + \frac{\gamma(t-1)}{(y-1)^2} + \frac{\delta t(t-1)}{(y-t)^2}\Big).
\end{aligned} \tag{8.9}$$

Following [310], let us make a few remarks as to their historical context. Leading up to the work of Painlevé and the isolation of the equations (8.9) were the studies of Fuchs and Poincaré in the late nineteenth century on first order differential equations of the form

$$P(y', y, t) = 0, \tag{8.10}$$

where P is a polynomial in y', y with coefficients meromorphic in t. In general, the solution of such equations have *moveable singularities*, by which one means singularities which depend on the initial condition. Of interest is to distinguish equations with moveable poles from those with moveable essential singularities. Thus for example

$$\frac{dy}{dt} = y^2 + 1 \tag{8.11}$$

has the general solution $y = \tan(t + c)$, where c determines the initial condition. Hence in this case all the singularities are moveable (first order) poles. In contrast the equation

$$\frac{dy}{dt} = \frac{1}{\alpha y^{\alpha-1}}, \qquad \alpha = 2, 3, \dots$$

has the general solution $y = (t-c)^{1/\alpha}$, for which the singularity is a movable branch point (essential singularity). Fuchs and Poincaré were able to classify all equations of the form (8.10) which are free from movable essential singularities — they must be reducible (i.e., equivalent after an analytic change of variables, or linear fractional transformation) to either the differential equation of the Weierstrass $\mathcal{P}$-function,

$$\Big(\frac{dy}{dt}\Big)^2 = 4y^3 - g_2 y - g_3, \tag{8.12}$$

or of the Riccati equation

$$\frac{dy}{dt} = a(t)y^2 + b(t)y + c(t), \tag{8.13}$$

where a, b, c are analytic in t (note that (8.11) is of the latter form).

Painlevé then took up the problem of classifying second order differential equations of the form

$$y'' = R(y', y, t), \tag{8.14}$$

where R is a rational function in all arguments, which are free from movable essential singularities. In general nonlinear equations free from movable essential singularities are now said to possess the *Painlevé property.* It was shown that the only equations of the form (8.14) with the Painlevé property which could not be reduced to the first order equations (8.12), (8.13) or to a linear differential equation are the equations (8.9).

In studying (8.7) we will encounter transcendents associated with PII–PVI. In fact we will typically not encounter these nonlinear equations directly, but rather the so-called Jimbo-Miwa-Okamoto σ-form of the Painlevé equations,

$$\begin{aligned}
&\sigma\mathrm{PII}\ (\sigma_{II}'')^2 + 4\sigma_{II}'\Big((\sigma_{II}')^2 - t\sigma_{II}' + \sigma_{II}\Big) - a^2 = 0,\\
&\sigma\mathrm{PIII}'\ (t\sigma_{III'}'')^2 - v_1 v_2 (\sigma_{III'}')^2 + \sigma_{III'}'(4\sigma_{III'}' - 1)(\sigma_{III'} - t\sigma_{III'}') - \frac{1}{4^3}(v_1 - v_2)^2 = 0,\\
&\sigma\mathrm{PIV}\ (\sigma_{IV}'')^2 - 4(t\sigma_{IV}' - \sigma_{IV})^2 + 4\sigma_{IV}'(\sigma_{IV}' + 2\alpha_1)(\sigma_{IV}' - 2\alpha_2) = 0,\\
&\sigma\mathrm{PV}\ (t\sigma_V'')^2 - \Big(\sigma_V - t\sigma_V' + 2(\sigma_V')^2 + (\nu_0 + \nu_1 + \nu_2 + \nu_3)\sigma_V'\Big)^2\\
&\qquad +4(\nu_0 + \sigma_V')(\nu_1 + \sigma_V')(\nu_2 + \sigma_V')(\nu_3 + \sigma_V') = 0,\\
&\sigma\mathrm{PVI}\ \sigma_{VI}'\Big(t(1-t)\sigma_{VI}''\Big)^2 + \Big(\sigma_{VI}'(2\sigma_{VI} - (2t-1)\sigma_{VI}') + v_1 v_2 v_3 v_4\Big)^2 = \prod_{k=1}^4 (\sigma_{VI}' + v_k^2).
\end{aligned} \tag{8.15}$$

We will see in the next section that the σ-form is the differential equation satisfied by certain auxiliary Hamiltonians in the Hamiltonian formulation of PII–PVI. (Analogous to the origin of the Painlevé equations (8.9), they also occur in classifying second order second degree equations with the Painlevé property [120].)

EXERCISES 8.1 1. (i) Use the second equality for $p_{N,\beta}(n; J)$ in (8.2) to deduce that

$$p_{N,\beta}(n; J) = -\frac{1}{\rho_{(1)}(a_-)} \frac{\partial^2}{\partial a_- \partial a_+} \sum_{j=0}^n (n - j + 1) E_{N,\beta}(j; J) \tag{8.16}$$

and show that in a translationally invariant state, $\rho_{(1)}(a_1) = \rho$, with $|a_1 - a_2| = s$ it reduces to

$$p_{N,\beta}(n;(0,s)) = \frac{1}{\rho}\frac{d^2}{ds^2}\sum_{j=0}^{n}(n-j+1)E_{N,\beta}(j;(0,s)). \tag{8.17}$$

(ii) Use the second and third formulas in (8.4), together with the definitions of $p_{N,\beta}$ and $\rho_{(1)}^N, \rho_{(2)}^N$ (in the latter the superscript N denotes a finite system of N particles) to deduce the sum rules

$$\sum_{n=0}^{N-1} p_{N,\beta}(n;a) = \rho_{(1)}^N(a), \qquad \sum_{n=0}^{N-2} p_{N,\beta}(n;(a_1,a_2)) = \frac{\rho_{(2)}^N(a_1,a_2)}{\rho_{(1)}^N(a_1)}. \tag{8.18}$$

2. For a perfect gas on a line, show that in the thermodynamic limit

$$E_\beta((a_1,a_2);\xi) = e^{-\xi\rho|a_1-a_2|}$$

and hence evaluate $E_\beta(n;(a_1,a_2))$ and $p_\beta(n;(a_1,a_2))$.

3. [447] (Wigner surmise) The p.d.f. for the Gaussian β-ensemble in the case $N = 2$ is

$$p(x_1,x_2) = \frac{1}{G_2}e^{-\beta c(x_1^2+x_2^2)/2}(x_2-x_1)^\beta, \qquad G_2 = 2^\beta \pi^{1/2}(\beta c)^{-\beta/2-1}\Gamma((\beta+1)/2), \tag{8.19}$$

where the extra factor c is a scale factor, and the eigenvalues are ordered so that $x_2 > x_1$. Define the averaged spacing p.d.f. by

$$p_\beta^{\rm W}(s) := \int_{-\infty}^{\infty} p(x,x+s)\,dx$$

with the value of c in (8.19) chosen so that $\int_0^\infty s p_\beta^{\rm W}(s)\,ds = 1$ (mean spacing equals unity).

(i) For general c show that

$$p_\beta^{\rm W}(s) = \frac{1}{\tilde{G}_2}s^\beta e^{-c\beta s^2/4}, \qquad \tilde{G}_2 = \Big(\frac{\beta c}{\pi}\Big)^{1/2} G_2.$$

(ii) Use the constraint that the mean spacing be unity to show

$$\frac{c\beta}{4} = \bigg(\frac{\Gamma(\frac{\beta}{2}+1)}{\Gamma(\frac{\beta}{2}+\frac{1}{2})}\bigg)^2$$

and thus in particular

$$p_1^{\rm W}(s) = \frac{\pi s}{2}e^{-\pi s^2/4}, \qquad p_2^{\rm W}(s) = \frac{32 s^2}{\pi^2}e^{-4s^2/\pi}, \qquad p_4^{\rm W}(s) = \frac{2^{18}s^4}{3^6\pi^3}e^{-64s^2/9\pi}. \tag{8.20}$$

(iii) [65] Show from (8.16) in the case $n = 0$ that the gap probability associated with $p_1^{\rm W}$ is $E_1^{\rm W}(0;s) = \int_s^\infty e^{-\pi t^2/4}\,dt$. Use this and your answer to q.2 to show that for a system of unit density, consisting of a fraction c_1 of particles which are a perfect gas (PG), and fraction $1 - c_1$ of particles obeying the $\beta = 1$ Wigner surmise, the probability that the interval $(0, s)$ is free of particles is given by

$$E_1^{\rm PG}(0;c_1 s)E_1^{\rm W}(0;(1-c_1)s) = e^{-c_1 s}\int_{(1-c_1)s}^\infty e^{-\pi t^2/4}\,dt.$$

4. [330] Suppose that in a general continuous one-dimensional statistical mechanical system of N particles, the right most particle is labeled x_1, the second rightmost x_2 and so on. Set $a_0 = \infty$, suppose $a_1 > a_2 > \cdots$ and put $I_j = (a_j, a_{j-1})$. Denote by $E_{N,\beta}(\{(n_r, I_r)\}_{r=1,\dots,l})$ the probability that exactly n_r of the particles are in I_r

$(r = 1, \dots, l)$. Show that

$$\Pr(x_1 < a_1, \dots, x_l < a_l) = \sum_{(n_1,\dots,n_l)\in\mathcal{L}_l} E_{N,\beta}(\{(n_r, I_r)\}_{r=1,\dots,l}),$$

where

$$\mathcal{L}_l = \{(n_1, \dots, n_l) \in \mathbb{Z}^l_{\ge 0} : \sum_{j=1}^{r} n_j \le r - 1 \; (r = 1, \dots, l)\}.$$

5. Show that with the change of variables $t \mapsto (is+1)/2$, $\sigma_{VI}(t) \mapsto i\tilde{\sigma}_{VI}(s)/2$, the σPVI equation in (8.15) reads

$$\tilde{\sigma}'_{VI}\Big((1+s^2)\tilde{\sigma}''_{VI}\Big)^2 + 4\Big(\tilde{\sigma}'_{VI}(\tilde{\sigma}_{VI} - s\tilde{\sigma}'_{VI}) - iv_1v_2v_3v_4\Big)^2 + 4\prod_{k=1}^{4}(\tilde{\sigma}'_{VI} + v_k^2) = 0. \tag{8.21}$$

(This will be referred to as the $\tilde{\sigma}$PVI equation.)

6. [120] In the general theory of Painlevé equations it is known that any differential equation of the form

$$\begin{aligned}(y'')^2 = -\frac{4}{g^2(t)}\Big\{&c_1(ty'-y)^3 + c_2y'(ty'-y)^2 + c_3(y')^2(ty'-y) \\ &+ c_4(y')^3 + c_5(ty'-y)^2 + c_6y'(ty'-y) + c_7(y')^2 + c_8(ty'-y) + c_9y' + c_{10}\Big\}\end{aligned} \tag{8.22}$$

where $g(t) := c_1t^3 + c_2t^2 + c_3t + c_4$, referred to as the *master Painlevé equation*, is integrable in terms of Painlevé transcendents. Show that only four of the ten parameters in (8.22) are essential because the equation retains its form under the gauge transformations

$$\bar{t} = \frac{at+b}{ct+d}, \qquad \bar{y} = \frac{hy+kt+m}{ct+d}.$$

8.2 HAMILTONIAN FORMULATION OF THE PAINLEVÉ THEORY

8.2.1 The auxiliary Hamiltonian

In the Hamiltonian approach to the Painlevé equations PII–PVI, one presents a Hamiltonian $H = H(p, q, t; \vec{v})$, where the components of $\vec{v}$ are parameters, such that after eliminating p in the Hamilton equations

$$q' = \frac{\partial H}{\partial p}, \qquad p' = -\frac{\partial H}{\partial q}, \tag{8.23}$$

q' and p' denoting derivatives with respect to t, the equation in q is the appropriate Painlevé equation. This was first achieved by Malmquist in 1922 [383]. However the consequences of the Hamiltonian formulation to be presented below, in particular the Bäcklund transformations (transformations of p and q which conserve the Hamiltonian structure) and the associated sequences of special solutions, were not explored until the work of Okamoto in the 1980s [428].

The forms of the Hamiltonians given in the work of Okamoto, after some renaming of the parameters, are

$$\begin{aligned}
H_{II} &= -\frac{1}{2}(2q^2 - p + t)p - \frac{v_1 - v_2}{2}q, \\
tH_{III'} &= q^2p^2 - (q^2 + v_1 q - t)p + \frac{1}{2}(v_1 + v_2)q, \\
H_{IV} &= (2p - q - 2t)pq - 2(v_1 - v_2)p + (v_3 - v_2)q, \\
tH_V &= q(q-1)^2p^2 - \{(v_1 - v_2)(q-1)^2 - 2(v_1 + v_2)q(q-1) + tq\}p \\
&\quad + (v_3 - v_2)(v_4 - v_2)(q-1), \\
t(t-1)H_{VI} &= q(q-1)(q-t)p^2 - \Big((v_3 + v_4)(q-1)(q-t) + (v_3 - v_4)q(q-t) \\
&\quad - (v_1 + v_2)q(q-1)\Big)p + (v_3 - v_1)(v_3 - v_2)(q-t),
\end{aligned} \tag{8.24}$$

where the parameters herein relate to those in (8.9) according to

PII $\quad v_1 + v_2 = 0, \quad \alpha = v_1 - \frac{1}{2},$

PIII$'$ $\quad \alpha = -4v_2, \quad \beta = 4(v_1 + 1), \quad \gamma = 4, \quad \delta = -4,$

PIV $\quad v_1 + v_2 + v_3 = 0, \quad \alpha = 1 + 2v_3 - v_1 - v_2, \quad \beta = -2\alpha_1^2,$

PV $\quad v_1 + v_2 + v_3 + v_4 = 0,\ \alpha = \frac{1}{2}(v_3 - v_4)^2,\ \beta = -\frac{1}{2}(v_1 - v_2)^2,\ \gamma = 2v_1 + 2v_2 - 1,\ \delta = -\frac{1}{2},$

PVI $\quad \alpha = \frac{1}{2}(v_1 - v_2)^2,\ \beta = -\frac{1}{2}(v_3 + v_4)^2,\ \gamma = \frac{1}{2}(v_3 - v_4)^2,\ \delta = \frac{1}{2}(1 - (1 - v_1 - v_2)^2).$

These Hamiltonians can be systematically derived from the isomonodromy deformation theory associated with the Painlevé equations [323], [427] (see also Section 9.9.2), although the details are somewhat complicated and will not be presented here. In the Hamiltonian $H_{III'}$, q satisfies the differential equation

$$y'' = \frac{1}{y}(y')^2 - \frac{1}{t}y' + \frac{\alpha y^3}{4t^2} + \frac{1}{4t^2}(\beta y^2 + \gamma t) + \frac{\delta}{4y} \tag{8.25}$$

for suitable parameters $\alpha, \dots, \delta$. This equation, referred to as PIII$'$, reduces to PIII in (8.9) upon making the replacements $t \mapsto t^2$, $y \mapsto ty$.

A feature of the Hamiltonians in (8.24) is that they involve parameters $\vec{v} = (v_1, v_2, \dots)$ instead of the parameters $\alpha, \beta, \dots$ in the original Painlevé equations (8.9). An advantage of introducing the parameters $\vec{v}$ is that the Hamiltonians display certain symmetries in these variables. These symmetries are revealed by the second order second degree differential equation satisfied by simple modifications of the Hamiltonians.

PROPOSITION 8.2.1 *Define the auxiliary Hamiltonians*

$$\begin{aligned}
h_{II}(t) &= H_{II}, \\
h_{III'}(t) &= tH_{III'} + \frac{1}{4}v_1^2 - \frac{1}{2}t, \\
h_{IV}(t) &= H_{IV} - 2v_2 t, \\
h_V(t) &= tH_V + (v_3 - v_2)(v_4 - v_2) - v_2 t - 2v_2^2, \\
h_{VI}(t) &= t(t-1)H_{VI} + e_2[-v_1, -v_2, v_3]t, -\frac{1}{2}e_2[-v_1, -v_2, v_3, v_4],
\end{aligned}$$

where $e_p[a_1, \dots, a_s] := \sum_{1 \le j_1 < \cdots < j_p \le s} a_{j_1} a_{j_2} \cdots a_{j_p}$. *These auxiliary Hamiltonians satisfy the nonlinear equations*

$$
\begin{aligned}
&(h''_{II})^2 + 4(h'_{II})^3 + 2h'_{II}(th'_{II} - h_{II}) - \frac{1}{4}\Big(\frac{v_1 - v_2}{2}\Big)^2 = 0,\\
&(th''_{III'})^2 + v_1 v_2 h'_{III'} - (4(h'_{III'})^2 - 1)(h_{III'} - th'_{III'}) - \frac{1}{4}\Big(v_1^2 + v_2^2\Big) = 0,\\
&(h''_{IV})^2 - 4(th'_{IV} - h_{IV})^2 + 4\prod_{k=1}^{3}(h'_{IV} + 2v_k) = 0,\\
&(th''_V)^2 - (h_V - th'_V + 2(h'_V)^2)^2 + 4\prod_{k=1}^{4}(h'_V + v_k) = 0,\\
&h'_{VI}(t(1-t)h''_{VI})^2 + (h'_{VI}\{2h_{VI} - (2t-1)h'_{VI}\} + v_1 v_2 v_3 v_4)^2 = \prod_{k=1}^{4}(h'_{VI} + v_k^2). \qquad (8.26)
\end{aligned}
$$

Proof. We will give the derivation of the differential equation for h_{VI} only; the derivations in the other cases are similar. Using the general fact that for Hamiltonians $H = H(p, q, t)$,

$$\frac{dH}{dt} = \frac{\partial H}{\partial t}$$

(a consequence of the Hamilton equations (8.23)), we see that

$$h'_{VI} = -q(q-1)p^2 + (v_3(2q-1) - v_4)p - v_3^2. \qquad (8.27)$$

This can be rewritten

$$q(q-1)(h'_{VI} + v_3^2) = -(q(q-1)p)^2 + (v_3(2q-1) - v_4)q(q-1)p, \qquad (8.28)$$

which shows a differential equation for h_{VI} will follow if we can express q and $q(q-1)p$ in terms of h_{VI} and its derivatives. To do this we note from (8.27) and the definition of h_{VI} that

$$h_{VI} - th'_{VI} = q(-h'_{VI} + e_2[-v_1, -v_2, v_3]) + (v_1 + v_2)q(q-1)p - \frac{1}{2}e_2[-v_1, -v_2, v_3, v_4],$$

and this differentiated, making use of the Hamilton equations, implies

$$
\begin{aligned}
t(t-1)h''_{VI} = {}& 2q(e_1[-v_1, -v_2, v_3]h'_{VI} - e_3[-v_1, -v_2, v_3]) - 2q(q-1)p(h'_{VI} - v_1 v_2)\\
&- e_1[-v_1, -v_2, v_3, v_4]h'_{VI} + e_3[-v_1, -v_2, v_3, v_4].
\end{aligned}
$$

These last two equations are linear in q and $q(q-1)p$. Solving for them and substituting in (8.28) gives the equation for h_{VI} in (8.26). □

The equations (8.26) are simply related to the σ-form of the Painlevé equations (8.15). Thus substituting

$$
\begin{aligned}
\sigma_{II}(t) &= -2^{1/3}h_{II}(-2^{1/3}t)\Big|_{(v_1, v_2) = (a, -a)},\\
\sigma_{III'}(t) &= -h_{III'}(t/4) + \frac{t}{8} + \frac{v_1 v_2}{4},\\
\sigma_{IV}(t) &= \Big(h_{IV}(t) + 2v_2 t\Big)\Big|_{(1+v_3-v_1, v_1-v_2, v_2-v_3) = (\alpha_0, \alpha_1, \alpha_2)},\\
\sigma_V(t) &= h_V(t) + v_2 t + 2v_2^2, \qquad \nu_{j-1} = v_j - v_2 \quad (j = 1, \dots, 4),\\
\sigma_{VI}(t) &= h_{VI}(t) \qquad (8.29)
\end{aligned}
$$

in (8.15) gives (8.26).

8.2.2 Affine Weyl group symmetries

Inspection of (8.26) shows the differential equations are invariant with respect to certain simple mappings of the parameters, and thus so are the auxiliary Hamiltonians. The generators of the mappings can be taken to be

$$\begin{aligned}
h_{II} &: s_1(v_1,v_2) = (v_2,v_1), \\
h_{III'} &: s_1(v_1,v_2) = (v_2,v_1), \quad s_2(v_1,v_2,t) = (v_1,-v_2,-t), \\
h_{IV} &: s_1(v_1,v_2,v_3) = (v_2,v_1,v_3), \quad s_2(v_1,v_2,v_3) = (v_1,v_3,v_2), \\
h_V &: s_1(v_1,v_2,v_3,v_4) = (v_2,v_1,v_3,v_4), \quad s_2(v_1,v_2,v_3,v_4) = (v_1,v_3,v_2,v_4), \\
&\quad s_3(v_1,v_2,v_3,v_4) = (v_1,v_2,v_4,v_3), \\
h_{VI} &: s_1(v_1,v_2,v_3,v_4) = (v_2,v_1,v_3,v_4), \quad s_2(v_1,v_2,v_3,v_4) = (v_1,v_3,v_2,v_4), \\
&\quad s_3(v_1,v_2,v_3,v_4) = (v_1,v_2,v_4,v_3), \; s_4(v_1,v_2,v_3,v_4) = (-v_2,-v_1,v_3,v_4).
\end{aligned} \tag{8.30}$$

Geometrically the above mappings are reflections in the hyperplane perpendicular to the following vectors,

$$\begin{aligned}
h_{II} &: \vec{\alpha}_1 := \vec{e}_1 - \vec{e}_2, \\
h_{III'} &: \vec{\alpha}_1 := \vec{e}_1 - \vec{e}_2, \quad \vec{\alpha}_2 := \vec{e}_2, \\
h_{IV} &: \vec{\alpha}_1 := \vec{e}_1 - \vec{e}_2, \quad \vec{\alpha}_2 := \vec{e}_2 - \vec{e}_3, \\
h_V &: \vec{\alpha}_1 := \vec{e}_1 - \vec{e}_2, \quad \vec{\alpha}_2 := \vec{e}_2 - \vec{e}_3, \quad \vec{\alpha}_3 = \vec{e}_3 - \vec{e}_4, \\
h_{VI} &: \vec{\alpha}_1 := \vec{e}_1 - \vec{e}_2, \quad \vec{\alpha}_2 := \vec{e}_2 - \vec{e}_3, \quad \vec{\alpha}_3 = \vec{e}_3 - \vec{e}_4, \quad \vec{\alpha}_4 = \vec{e}_3 + \vec{e}_4.
\end{aligned} \tag{8.31}$$

These vectors form a *base* for the root systems A_1, B_2, A_2, A_3, D_4, respectively.

DEFINITION 8.2.2 *Let R be a root system in a vector space E. We say that $\Delta = \{\vec{\alpha}_1,\dots,\vec{\alpha}_l\}$, $\vec{\alpha}_i \in R$ $(i = 1,\dots,l)$ is a base for R if Δ is a basis for E with the further requirement that for each $\vec{\alpha} \in R$, in the expansion $\vec{\alpha} = \sum_{i=1}^{l} \zeta_i \vec{\alpha}_i$ the coefficients ζ_i are integers which are either all non-negative or all non-positive. The former are called the positive roots, while the latter are the negative roots (this latter definition is consistent with that below (4.149), with the choice $\vec{v} = \sum_{i=1}^{l} \vec{\alpha}_i$, say).*

Next we note a further reflection operator closely related to the root system structure revealed in (8.31),

$$\begin{aligned}
A_1 &: \quad s_0(v_1,v_2) = (2+v_2,-2+v_1), \\
B_2 &: \quad s_0(v_1,v_2) = (-1-v_2,-1-v_1), \\
A_2 &: \quad s_0(v_1,v_2,v_3) = (1+v_3,v_2,-1+v_1), \\
A_3 &: \quad s_0(v_1,v_2,v_3,v_4) = (1+v_4,v_2,v_3,v_1-1), \\
D_4 &: \quad s_0(v_1,v_2,v_3,v_4) = (1-v_2,1-v_1,v_3,v_4).
\end{aligned} \tag{8.32}$$

To see the geometric significance of these operators requires the notion of the *highest root.*

DEFINITION 8.2.3 *With a positive root defined in Definition 8.2.2, one defines a partial order $<$ on R by the statement that $\vec{\alpha} < \vec{\beta}$ if $\vec{\beta} - \vec{\alpha}$ is a sum of positive roots. The maximal root with respect to this ordering is called the highest root.*

From this definition the highest roots are

$$\vec{\alpha}_0 = \begin{cases} \vec{e}_1 - \vec{e}_l & \text{for } A_{l-1} \\ \vec{e}_1 + \vec{e}_2 & \text{for } B_2, D_4. \end{cases}$$

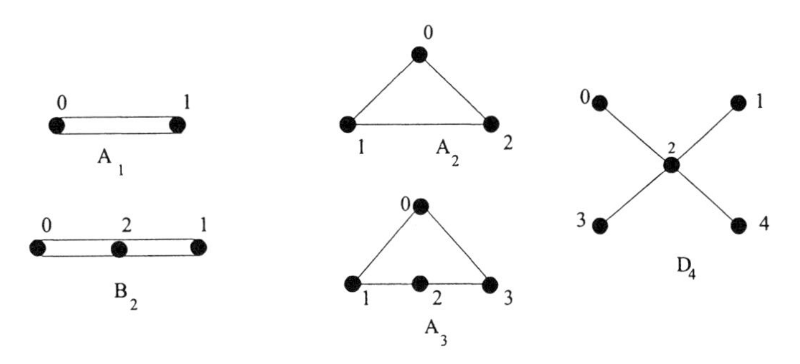

Figure 8.1 Dynkin diagrams for affine root systems relevant to Painlevé systems.

With reflections in the hyperplane $\{\vec{x} : \vec{\phi} \cdot \vec{x} = c\}$ defined by

$$\sigma_{c,\vec{\phi}}(\vec{\gamma}) = \vec{\gamma} - \frac{2\vec{\gamma} \cdot \vec{\phi}}{\vec{\phi} \cdot \vec{\phi}} \vec{\phi} + \frac{2c\vec{\phi}}{\vec{\phi} \cdot \vec{\phi}}$$

(cf. (4.147)), the mappings (8.32) then correspond to reflections in the hyperplanes $\{\vec{x} : \vec{\phi} \cdot \vec{x} = 2\}$ for A_1, $\{\vec{x} : \vec{\phi} \cdot \vec{x} = -1\}$ for B_2, $\{\vec{x} : \langle\vec{\phi}, \vec{x}\rangle = 1\}$ for A_2, A_3 and D_4, where $\vec{\phi} = \vec{\alpha}_0$. The mappings (8.32) together with (8.30) generate the so-called *affine Weyl group* corresponding to the respective root system.

The algebra specifying the affine Weyl groups for the root systems A_l, B_l, D_l are given by the Coxeter relations

$$(s_i s_j)^{m_{ij}} = 1, \qquad 0 \le i, j \le l. \tag{8.33}$$

The exponents m_{ij} are positive integers which can be read off from the Dynkin diagrams associated with the affine root systems as presented in Figure 8.1. The Dynkin diagrams are constructed by the rule that if $\vec{\alpha}_i$ and $\vec{\alpha}_j$ are perpendicular (i.e., have dot product equal to zero) then the vertices i and j are not connected; if the angle between $\vec{\alpha}_i$ and $\vec{\alpha}_j$ is $\pi/3$ then the vertices i and j are connected by a single line; if the angle between $\vec{\alpha}_i$ and $\vec{\alpha}_j$ is $\pi/4$ then the vertices i and j are connected by a double line. The root system A_1 is special because $\vec{\alpha}_0$ and $\vec{\alpha}_1$ are the same. In this case the two vertices are to be connected by a double line. To construct the m_{ij} in (8.33) from the Dynkin diagram, the rule is that $m_{ii} = 1$, $m_{ij} = 2$ if the corresponding vertices in the Dynkin diagram are not joined, $m_{ij} = 3$ if they are joined by one line and $m_{ij} = 4$ if they are joined by two lines. Related to the matrix $[m_{ij}]$ is the Cartan matrix $[a_{ij}]$. It is constructed from the rule that $a_{ii} = -2$, $a_{ij} = 0, -1, -2$ according to the vertices i and j being connected by zero, one or two lines, respectively. The explicit forms of $[m_{ij}]$ and $[a_{ij}]$ are listed in Table 8.1.

The introduction of the Cartan matrix allows the action in (8.30) and (8.32) to be written in a succinct way. For this purpose we replace the parameters $v_1, v_2, \dots$ by $\alpha_0, \alpha_1, \dots$ according to

$$\begin{aligned}
&A_1 : \alpha_0 = 1 + (v_2 - v_1)/2, \quad \alpha_1 = (v_1 - v_2)/2, \\
&B_2 : \alpha_0 = -\frac{1}{2}(1 + v_1 + v_2), \quad \alpha_1 = v_1 - v_2, \quad \alpha_2 = v_2, \\
&A_2 : \alpha_0 = 1 + v_3 - v_1, \quad \alpha_1 = v_1 - v_2, \quad \alpha_2 = v_2 - v_3, \\
&A_3 : \alpha_0 = 1 + v_4 - v_1, \quad \alpha_1 = v_1 - v_2, \quad \alpha_2 = v_2 - v_3, \quad \alpha_3 = v_3 - v_4, \\
&D_4 : \alpha_0 = 1 - v_1 - v_2, \quad \alpha_1 = v_1 - v_2, \quad \alpha_2 = v_2 - v_3, \quad \alpha_3 = v_3 - v_4, \quad \alpha_4 = v_3 + v_4.
\end{aligned} \tag{8.34}$$

Note that for $j = 1, 2, \dots$, $\alpha_j = 0$ defines the hyperplane $\vec{\alpha}_j \cdot \vec{v} = 0$, while $\alpha_0 = 0$ defines the hyperplane $\vec{\alpha}_0 \cdot \vec{v} = c$ ($c = 2$ for A_1, $c = -1$ for B_2, $c = 1$ for A_2, A_3, D_4). These are the hyperplanes with respect to which s_j and s_0 act as reflection operators. From this geometrical interpretation, it follows that $s_i \alpha_j$ must be

Root system	$[m_{ij}]$	Cartan matrix
A_1	$\begin{bmatrix} 1 & 4 \\ 4 & 1 \end{bmatrix}$	$\begin{bmatrix} 2 & -2 \\ -2 & 2 \end{bmatrix}$
B_2	$\begin{bmatrix} 1 & 2 & 4 \\ 2 & 1 & 4 \\ 4 & 4 & 1 \end{bmatrix}$	$\begin{bmatrix} 2 & 0 & -2 \\ 0 & 2 & -2 \\ -2 & -2 & 2 \end{bmatrix}$
A_2	$\begin{bmatrix} 1 & 3 & 3 \\ 3 & 1 & 3 \\ 3 & 3 & 1 \end{bmatrix}$	$\begin{bmatrix} 2 & -1 & -1 \\ -1 & 2 & -1 \\ -1 & -1 & 2 \end{bmatrix}$
A_3	$\begin{bmatrix} 1 & 3 & 2 & 3 \\ 3 & 1 & 3 & 2 \\ 2 & 3 & 1 & 3 \\ 3 & 2 & 3 & 1 \end{bmatrix}$	$\begin{bmatrix} 2 & -1 & 0 & -1 \\ -1 & 2 & -1 & 0 \\ 0 & -1 & 2 & -1 \\ -1 & 0 & -1 & 2 \end{bmatrix}$
D_4	$\begin{bmatrix} 1 & 2 & 3 & 2 & 2 \\ 2 & 1 & 3 & 2 & 2 \\ 3 & 3 & 1 & 3 & 3 \\ 2 & 2 & 3 & 1 & 2 \\ 2 & 2 & 3 & 2 & 1 \end{bmatrix}$	$\begin{bmatrix} 2 & 0 & -1 & 0 & 0 \\ 0 & 2 & -1 & 0 & 0 \\ -1 & -1 & 2 & -1 & -1 \\ 0 & 0 & -1 & 2 & 0 \\ 0 & 0 & -1 & 0 & 2 \end{bmatrix}$

Table 8.1 Specification of $[m_{ij}]$ and $[a_{ij}]$.

a linear combination of α_j and α_i, which we can check is given by the explicit formula

$$s_i \alpha_j = \alpha_j - \alpha_i a_{ij}, \tag{8.35}$$

where $[a_{ij}]$ is the corresponding Cartan matrix.

The introduction of the variables (8.34) leads to another elementary mapping which turns out to correspond to a Bäcklund transformation of the Painlevé system [424], [339]. This mapping relabels the α_j in a way which conserves the relative geometry of the root system. Equivalently, these mappings correspond to automorphisms of the Dynkin diagram which conserve the relative structure. Thus one introduces the operators π, r_1, r_2 such that

$$\begin{aligned} A_l &: \pi \alpha_i = \alpha_{i+1} \pi \quad i = 0, \dots, l \ (\alpha_{l+1} := \alpha_0), \\ D_4 &: r_1(\alpha_0, \alpha_1, \alpha_2, \alpha_3, \alpha_4) = (\alpha_1, \alpha_0, \alpha_2, \alpha_4, \alpha_3), \\ &\quad r_2(\alpha_0, \alpha_1, \alpha_2, \alpha_3, \alpha_4) = (\alpha_3, \alpha_4, \alpha_2, \alpha_0, \alpha_1). \end{aligned} \tag{8.36}$$

Here the B_2 affine root system is excluded, the reason being that the natural diagram automorphism $r(\alpha_0, \alpha_1, \alpha_2) = (\alpha_1, \alpha_0, \alpha_2)$ is incompatible with the constraint $-1 = 2\alpha_0 + \alpha_1 + 2\alpha_2$. The operators in (8.36) obey the algebraic relations

$$\begin{aligned} A_l &: \pi^l = 1, \qquad \pi s_j = s_{j+1} \pi \quad (s_l := s_0), \\ D_4 &: r_1^2 = r_2^2 = 1, \quad r_1 s_2 = s_2 r_1, \quad r_2 s_2 = s_2 r_2, \\ &\quad s_1 = r_1 s_0 r_1, \quad s_4 = r_1 s_3 r_1, \quad s_3 = r_2 s_0 r_2, \quad s_4 = r_2 s_1 r_2. \end{aligned}$$

8.2.3 Bäcklund transformations

For each of the operators, $\mathcal{O}$, say, introduced above the mappings $\bar{\alpha}_i := \mathcal{O}\alpha_i$ can be combined with particular birational mappings $\bar{p} = \mathcal{O}p$, $\bar{q} = \mathcal{O}q$ such that the Hamilton equations (8.23) remain valid in the variables

PII	α_0	α_1	p	q
s_0	$-\alpha_0$	$\alpha_1 + 2\alpha_0$	$p + \dfrac{4\alpha_0 q}{f} + \dfrac{2\alpha_0^2}{f^2}$	$q + \dfrac{\alpha_0}{f}$
s_1	$\alpha_0 + 2\alpha_1$	$-\alpha_1$	p	$q + \dfrac{\alpha_1}{p}$
π	α_1	α_0	$-f$	$-q$

Table 8.2 Bäcklund transformations for PII. Here $\alpha_0 + \alpha_1 = 1$, $f = p - 2q^2 - t$.

PIII′	α_0	α_1	α_2	p	q	t
s_0	$-\alpha_0$	α_1	$\alpha_2 + 2\alpha_0$	$\frac{q}{t}\left[q(p-1) - \frac{\alpha_1}{2}\right] + 1$	$-\frac{t}{q}$	t
s_1	α_0	$-\alpha_1$	$\alpha_2 + 2\alpha_1$	p	$q - \dfrac{\alpha_1}{2(p-1)}$	t
s_2	$\alpha_0 + 2\alpha_2$	$\alpha_1 + 2\alpha_2$	$-\alpha_2$	$1 - p$	$-q$	$-t$

Table 8.3 Bäcklund transformations for PIII′. Here $2\alpha_0 + \alpha_1 + 2\alpha_2 = -1$.

$\bar{p}, \bar{q}, \bar{H} := \mathcal{O}H$. Birational mappings which conserve the Hamilton equations are called *Bäcklund transformations.* A systematic way to construct $\bar{p}$, $\bar{q}$, valid in all cases except PIII′, has been given by Noumi and Yamada [424]. However, to proceed along those lines would require introducing a number of new concepts, which we refrain from doing here. Instead we will be content with a statement of the final results.

Before doing so, some further theory relating to the Hamiltonian of the PV system must be introduced. Thus it turns out [339] that the natural variables in relation to Bäcklund transformations for PV are not those in the Hamiltonian tH_V of (8.24), but rather a transformation of these variables. For ease of notation, let us first put $p \mapsto P$, $q \mapsto Q$, $tH_V \mapsto tK_V$ in the formula for tH_V in (8.24) so that it reads

$$tK_V = Q(Q-1)^2P^2 - \{(v_1-v_2)(Q-1)^2 - 2(v_1+v_2)Q(Q-1)+tQ\}P + (v_3-v_2)(v_4-v_2)(Q-1). \quad (8.37)$$

Introducing the new variables p and q (not to be confused with p and q in the definition of tH_V) in (8.37) via [538]

$$(q-1)(Q-1) = 1, \qquad (q-1)p + (Q-1)P = v_3 - v_2 \quad (8.38)$$

shows

$$tK_V = tH_{V^*} + (v_3 - v_2)(v_1 - v_4), \quad (8.39)$$

where

$$tH_{V^*} := q(q-1)p(p+t) - (v_1 - v_2 + v_3 - v_4)pq + (v_1 - v_2)p + (v_2 - v_3)tq. \quad (8.40)$$

Here $1 + 1/(q+1)$ satisfies the PV equation. The Hamiltonian (8.40) and more particularly the associated variables p and q have simple images under Bäcklund transformations. With the PV system so modified, to form what will be referred to as the PV* system, the Bäcklund transformations of PII–PVI are given by Tables 8.2–8.6.

Using the Bäcklund transformation tables, the action of the elementary operators on the Hamiltonians are

PIV	α_0	α_1	α_2	p	q
s_0	$-\alpha_0$	$\alpha_0+\alpha_1$	$\alpha_0+\alpha_2$	$p+\dfrac{\alpha_0}{f}$	$q-\dfrac{2\alpha_0}{f}$
s_1	$\alpha_0+\alpha_1$	$-\alpha_1$	$\alpha_2+\alpha_1$	$p-\dfrac{\alpha_1}{q}$	q
s_2	$\alpha_0+\alpha_2$	$\alpha_1+\alpha_2$	$-\alpha_2$	p	$q+\dfrac{\alpha_2}{p}$
π	α_1	α_2	α_0	$-\frac{1}{2}f$	$-2p$

Table 8.4 Bäcklund transformations for PIV. Here $\alpha_0+\alpha_1+\alpha_2=1$, $f=p-q-2t$.

PV*	α_0	α_1	α_2	α_3	p	q
s_0	$-\alpha_0$	$\alpha_1+\alpha_0$	α_2	$\alpha_3+\alpha_0$	p	$q+\frac{\alpha_0}{p+t}$
s_1	$\alpha_0+\alpha_1$	$-\alpha_1$	$\alpha_2+\alpha_1$	α_3	$p-\frac{\alpha_1}{q}$	q
s_2	α_0	$\alpha_1+\alpha_2$	$-\alpha_2$	$\alpha_3+\alpha_2$	p	$q+\frac{\alpha_2}{p}$
s_3	$\alpha_0+\alpha_3$	α_1	$\alpha_2+\alpha_3$	$-\alpha_3$	$p-\dfrac{\alpha_3}{q-1}$	q
π	α_1	α_2	α_3	α_0	$t(q-1)$	$-\frac{p}{t}$

Table 8.5 Bäcklund transformations for PV*. Here $\alpha_0+\alpha_1+\alpha_2+\alpha_3=1$.

PVI	α_0	α_1	α_2	α_3	α_4	p	q
s_0	$-\alpha_0$	α_1	$\alpha_2+\alpha_0$	α_3	α_4	$p-\dfrac{\alpha_0}{q-t}$	q
s_1	α_0	$-\alpha_1$	$\alpha_2+\alpha_1$	α_3	α_4	p	q
s_2	$\alpha_0+\alpha_2$	$\alpha_1+\alpha_2$	$-\alpha_2$	$\alpha_3+\alpha_2$	$\alpha_4+\alpha_2$	p	$q+\dfrac{\alpha_2}{p}$
s_3	α_0	α_1	$\alpha_2+\alpha_3$	$-\alpha_3$	α_4	$p-\dfrac{\alpha_3}{q-1}$	q
s_4	α_0	α_1	$\alpha_2+\alpha_4$	α_3	$-\alpha_4$	$p-\dfrac{\alpha_4}{q}$	q
r_1	α_1	α_0	α_2	α_4	α_3	$-\dfrac{p(q-t)^2+\alpha_2(q-t)}{t(t-1)}$	$\left(\dfrac{q-1}{q-t}\right)t$
r_2	α_3	α_4	α_2	α_0	α_1	$-\dfrac{q}{t}(qp+\alpha_2)$	$\dfrac{t}{q}$

Table 8.6 Bäcklund transformations for PVI. Here $\alpha_0+\alpha_1+2\alpha_2+\alpha_3+\alpha_4=1$.

Operator	Definition	$T(\alpha_0, \alpha_1, \dots)$	$T\vec{v}$
T_{II}	πs_1	$(\alpha_0 + 1, \alpha_1 - 1)$	$(v_1 - 1, v_2 + 1)$
$T_{III'}$	$s_0 s_2 s_1 s_2$	$(\alpha_0 - 1, \alpha_1, \alpha_2 + 1)$	$(v_1 + 1, v_2 + 1)$
T_{IV}	$\pi^{-1} s_1 s_2$	$(\alpha_0 + 1, \alpha_1, \alpha_2 - 1)$	$(v_1 - 1/3, v_2 - 1/3, v_3 + 2/3)$
T_V	$\pi^{-1} s_1 s_2 s_3$	$(\alpha_0 + 1, \alpha_1, \alpha_2, \alpha_3 - 1)$	$(v_1 - \frac{1}{4}, v_2 - \frac{1}{4}, v_3 - \frac{1}{4}, v_4 + \frac{3}{4})$
T_{VI}	$r_1 s_0 s_1 s_2 s_3 s_4 s_2$	$(\alpha_0 + 1, \alpha_1 + 1, \alpha_2 - 1, \alpha_3, \alpha_4)$	$(v_1, v_2 - 1, v_3, v_4)$

Table 8.7 Shift operators with the property $TH = H|_{T(\alpha_0,\alpha_1,\dots)}$

computed as

$$
\begin{aligned}
\mathrm{PII}:\ & s_0 H_{II} = H_{II} + \frac{\alpha_0}{f},\ \ s_1 H_{II} = H_{II},\ \ \pi H_{II} = H_{II} + q, \\
\mathrm{PIII}':\ & s_0(tH_{III'}) = tH_{III'} - q(p-1) + \frac{1}{2}\alpha_1\Big(\frac{1}{2} - \alpha_0\Big), \\
& s_1(tH_{III'}) = tH_{III'} - \frac{1}{2}\Big(\alpha_0 + \frac{1}{2}\Big)\alpha_1,\ \ s_2(tH_{III'}) = tH_{III'} - t - (\alpha_1 + \alpha_2)q, \\
\mathrm{PIV}:\ & s_0 H_{IV} = H_{IV} - \frac{2\alpha_0}{f},\ \ s_1 H_{IV} = H_{IV} + 2\alpha_1 t,\ \ s_2 H_{IV} = H_{IV} - 2\alpha_2 t, \\
& \pi H_{IV} = H_{IV} + 2p - 2\alpha_2 t, \\
\mathrm{PV}^*:\ & s_0(tH_{V^*}) = tH_{V^*} + \alpha_0 \frac{t}{p+t} + \alpha_0(\alpha_2 - 1),\ \ s_1(tH_{V^*}) = tH_{V^*} + \alpha_1 t + \alpha_1\alpha_3, \\
& s_2(tH_{V^*}) = tH_{V^*} - \alpha_2 t + \alpha_2(\alpha_0 - 1),\ \ s_3(tH_{V^*}) = tH_{V^*} + \alpha_1\alpha_3, \\
& \pi(tH_{V^*}) = tH_{V^*} + (q-1)p - \alpha_2 t, \\
\mathrm{PVI}:\ & s_0 t(t-1)H_{VI} = t(t-1)H_{VI} - \alpha_0\frac{t(t-1)}{q-t} + \alpha_0(\alpha_3 - 1)t + \alpha_0(\alpha_4 - 1)(t-1), \\
& s_1 t(t-1)H_{VI} = t(t-1)H_{VI}, \\
& s_2 t(t-1)H_{VI} = t(t-1)H_{VI} + \alpha_2(1 + \alpha_1 - \alpha_0)t - \alpha_2(\alpha_1 + \alpha_2 + \alpha_3), \\
& s_3 t(t-1)H_{VI} = t(t-1)H_{VI} - \alpha_3(1 - \alpha_0)t, \\
& s_4 t(t-1)H_{VI} = t(t-1)H_{VI} - \alpha_4(1 - \alpha_0)(t-1), \\
& r_1 t(t-1)H_{VI} = t(t-1)H_{VI} - q(q-1)p - \alpha_2 q + \alpha_2(\alpha_1 - \alpha_0)t + \alpha_2(\alpha_0 + \alpha_2 + \alpha_4), \\
& r_2 t(t-1)H_{VI} = t(t-1)H_{VI} + (1-t)qp + \alpha_2(\alpha_0 + \alpha_2 + \alpha_4)(1-t).
\end{aligned}
\tag{8.41}
$$

By composing suitable combinations of the elementary operators in the Bäcklund transformation tables, infinite order shift operators T can be constructed, with action on the parameters $\alpha_0, \alpha_1, \dots$ which changes their value by ± 1 or 0. Certain of the shift operators T have the further property that

$$
TH := H\Big|_{\substack{T(\alpha_0,\alpha_1,\dots)\\ Tp,Tq}} = H\Big|_{T(\alpha_0,\alpha_1,\dots)}, \tag{8.42}
$$

so that the action of T effectively leaves p and q unchanged. Specifically, one can check from (8.41), making use too of (8.39) in the PV case and the Bäcklund transformation tables, that the operators in Table 8.7, amongst other examples, have this property.

Painlevé system	Restriction on the α_j	Restriction on the v_j
PII	$\alpha_1 = 0$	$v_1 - v_2 = 0$
PIII$'$	$\alpha_1 + 2\alpha_2 = 0$	$v_1 + v_2 = 0$
PIV	$\alpha_2 = 0$	$v_2 - v_3 = 0$
PV	$\alpha_2 + \alpha_3 = 0$	$v_2 - v_4 = 0$
PVI	$\alpha_2 = 0$	$v_2 - v_3 = 0$

Table 8.8 Given these restrictions on the parameters, the Painlevé system permits a solution with $p = 0$, $H = 0$.

Let us write

$$T^n H =: H[n], \qquad T^n h =: h[n]. \tag{8.43}$$

Then it follows from (8.42) that $h[n]$ satisfies the appropriate equation of (8.26) with parameters $T^n\vec{v}$. On the other hand, if we introduce the τ-*function* by

$$H = \frac{d}{dt}\log\tau, \qquad H[n] = \frac{d}{dt}\log\tau[n], \tag{8.44}$$

it turns out that (8.42) together with the explicit form of Tp and Tq deduced from the Bäcklund transformation tables allows $\tau[n]$, multiplied by an appropriate elementary function, to be identified as an integral of the form (8.7). In this way the integrals (8.7) can be characterized as solutions of nonlinear equations of the type (8.15).

8.2.4 Classical solutions

When $N = 0$ it is natural to assign the averages (8.7) the value unity. To relate the Painlevé systems to the averages in this case we must show that for a specialization of the parameters it is permissable to choose $H[0] = 0$, which from the definition (8.44) permits $\tau[0] = 1$ as required. In fact inspection of (8.24) shows that each of the Painlevé systems PII–PVI permits a solution with

$$H = 0, \qquad p = 0, \tag{8.45}$$

provided the parameters are specified as in Table 8.8.

Next we want to show that $\tau[1]$, in the cases PIV, PV* and PVI, is equal to a particular average (8.7) in the case $N = 1$. Now, from (8.44) we seek $H[1] := TH[0]$. Using (8.42), (8.24), (8.45) and Tables 8.7, 8.8, with $\vec{v}[0] =: \vec{v}^{(0)}$ we note that

$$\begin{aligned} &H_{II}[1] = q_{II}[0], \quad tH_{III'}[1] = q_{III'}[0], \quad H_{IV}[1] = q_{IV}[0], \\ &tH_V[1] = (v_3^{(0)} - v_2^{(0)})(q_V[0] - 1), \quad t(t-1)H_{VI}[1] = (v_3^{(0)} - v_1^{(0)})(q_{VI}[0] - t). \end{aligned} \tag{8.46}$$

On the other hand $q[0]$ satisfies the Hamilton equation

$$q'[0] = \frac{\partial H}{\partial p}\Big|_{p=0,\vec{v}=\vec{v}[0]}$$

with parameters $\vec{v}[0]$ specified by the constraints in Table 8.8. This is a first order nonlinear differential equation in $q[0]$, which is seen to be of the Riccati type (8.13). First substituting for $q[0]$ according to (8.46), then substituting for $H[1]$ according to (8.44) gives a second order differential equation in $\tau[1]$, which furthermore in the cases of PIV, PV and PVI permits integral solutions of the type (8.7) with $N = 1$.

PROPOSITION 8.2.4 *Let $\vec{v}[0]$ be specified by the constraint in Table 8.8 in each case, and let $T\vec{v}[0]$ then be*

Painlevé system	Differential equation
PII	Airy differential equation $u'' + \frac{1}{2}tu = 0$
PIII$'$	Bessel differential equation $tu'' + (v_1^{(0)} + 1)u' - \frac{1}{4}u = 0, \quad u(t) = \tau_{III'}[1](t/4)$
PIV	Hermite-Weber equation $u'' + 2tu' + 2(v_1^{(0)} - v_2^{(0)})u = 0$
PV	Confluent hypergeometric differential equation $tu'' + (v_3^{(0)} - v_1^{(0)} + 1 + t)u' + (v_3^{(0)} - v_2^{(0)})u = 0$
PVI	Gauss hypergeometric differential equation $t(1-t)u'' + (c - (a+b+1)tu' - abu = 0$ $a = v_3^{(0)} - v_1^{(0)}, \; b = 1 - v_1^{(0)} - v_3^{(0)}, \; c = 1 - v_1^{(0)} + v_4^{(0)}$

Table 8.9 Classical differential equations satisfied by $\tau[1] := T\tau[0]$.

specified by Table 8.7. With $\tau[1]$ *specified by*

$$H\Big|_{T\vec{v}[0]} = \frac{d}{dt}\log\tau[1],$$

$\tau[1]$ *satisfies the linear differential equations of Table 8.9. For parameter values such that the integral converges, the general solution of the Hermite-Weber equation is proportional to*

$$\Big(\int_{-\infty}^{\infty} - \xi\int_t^{\infty}\Big)|t-u|^{-(v_1^{(0)} - v_2^{(0)})}e^{-u^2}\,du; \tag{8.47}$$

for the confluent hypergeometric equation it is proportional to

$$\Big(\int_0^{\infty} - \xi\int_0^{1}\Big)e^{-tu}(1-u)^{v_2 - v_1}u^{v_3 - v_2 - 1}\,du; \tag{8.48}$$

and for the Gauss hypergeometric equation it is proportional to

$$\Big(\int_0^{1} - \xi\int_0^{t}\Big)u^{a-c}(1-u)^{c-b-1}|t-u|^{-a}\,du \tag{8.49}$$

or

$$\Big(\int_{-\infty}^{\infty} - \xi\int_s^{\infty}\Big)(1+iu)^{a-c}(1-iu)^{c-b-1}|s-u|^{-a}\,du, \quad t = \frac{1+is}{2}. \tag{8.50}$$

Proof. It remains to consider the derivation of the integral solutions. For definiteness, consider the case of the Gauss hypergeometric differential equation. We use the fact [541] that in general

$$\int_r^s u^{a-c}(1-u)^{c-b-1}(t-u)^{-a}\,du,$$

where r and s are any of the values $0, 1, t, \nu\infty$ ($|\nu| = 1$) with the parameters such that the integrand vanishes at these points, satisfies this differential equation. Forming from this a linear combination with $(r,s) = (0,1)$ and $(r,s) = (0,t)$ gives (8.49). Similarly, choosing $(r,s) = (-\infty,\infty)$ and (t,∞) and suitably deforming these contours in the complex plane gives (8.50). □

Painlevé system	$a(t)$	$g[n]$
PII	1	$\tau_{II}[n]$
PIII′	t	$(t^{n^2/2}\tau_{III'}[n])\vert_{t\mapsto t/4}$
PIV	1	$e^{-t^2(\alpha_2^{(0)}-n)}\tau_{IV}[n]$
PV*	t	$t^{n^2/2}e^{(-\alpha_2^{(0)}-\alpha_3^{(0)}+n)t}\tau_{V^*}[n]$
PVI	$t(t-1)$	$(t(t-1))^{(n-\alpha_2^{(0)})(n+\alpha_0^{(0)}-1)}\tau_{VI}[n]$

Table 8.10 Specifications for the Toda lattice equation (8.51).

It follows from Proposition 8.2.4 that there are solutions of the PIV, PV and PVI systems such that

$$E_{1,2}^{(\mu)}((t,\infty);\xi;e^{-x^2}) = \tau_{IV}(t)\Big|_{\alpha_0=\mu+2,\alpha_1=-\mu,\alpha_2=-1},$$

$$E_{1,2}^{(\mu)}((0,t);\xi;x^a e^{-x}) = t^{a+\mu+1}\tau_V(t)\Big|_{\substack{\nu_0=-\mu\ \nu_1=0,\\ \nu_2=a+1,\nu_3=1}}, \quad \nu_{j-1} := v_j - v_2,$$

$$E_{1,2}^{(\mu)}((0,t);\xi;x^a(1-x)^b) = \tau_{VI}(t)\Big|_{\substack{v_1=(a+b)/2+1+\mu,v_2=(a+b)/2,\\ v_3=(a+b)/2+1,v_4=(b-a)/2}},$$

$$E_{1,2}^{(\mu)}((t,\infty);\xi;(1+x^2)^{-\alpha}) = \tau_{VI}\Big(\frac{1+it}{2}\Big)\Big|_{\substack{v_1=\mu+1-\alpha,v_2=-\alpha,\\ v_3=1-\alpha,v_4=0}}.$$

8.2.5 Toda lattice equation

With the cases $n=0$ and $n=1$ now settled, the key ingredient in relating $\tau[n]$ to the n-dimensional integrals implied by (8.7) for general $n \ge 2$ is the Toda lattice equation (cf. (5.118))

$$\delta^2 \log g[n] = \frac{g[n+1]g[n-1]}{g^2[n]}, \qquad \delta := a(t)\frac{d}{dt} \tag{8.51}$$

satisfied by the product of $\tau[n]$ and an elementary function.

PROPOSITION 8.2.5 *The τ-functions satisfy the Toda lattice equation (8.51) with $g[n]$ and $a(t)$ therein as specified by Table 8.10.*

Proof. We will explicitly consider the PV case only, although the same strategy suffices in all cases [339]. In this case it is convenient to work with the PV* Hamiltonian (8.40). According to (8.39) and (8.44) the τ-function of the latter, τ_{V^*} say, is related to the τ-function for (8.37) by

$$\tau_{V^*}[n] = t^{\alpha_2^{(0)}(1-\alpha_0^{(0)}-n)}\tau_V[n]. \tag{8.52}$$

From the definitions

$$t\frac{d}{dt}\log\frac{\tau_{V^*}[n-1]\tau_{V^*}[n+1]}{\tau_{V^*}^2[n]} = \Big(T_V(tH_{V^*}[n]) - tH_{V^*}[n]\Big) - \Big(tH_{V^*}[n] - T_V^{-1}(tH_{V^*}[n])\Big). \tag{8.53}$$

On the other hand, it follows from the property (8.42), Table 8.7 and (8.40) that

$$T_V(tH_{V^*}[n]) - tH_{V^*}[n] = q[n]p[n]$$

and thus the r.h.s. of (8.53) is equal to

$$q[n]p[n] - (T_V^{-1}q[n])(T_V^{-1}p[n]).$$

With T_V^{-1} given in terms of the elementary operators by Table 8.7, use of the Bäcklund transformation tables for PV*, together with the Hamilton equations for H_{V^*}, allows this term to be identified with

$$\begin{aligned} &t\frac{d}{dt}\log\Big(q[n](q[n]-1)p[n]+\alpha_2^{(0)}q[n]+(-\alpha_2^{(0)}-\alpha_3^{(0)}+n)t\Big) \\ &\quad = t\frac{d}{dt}\log\Big(\frac{d}{dt}tH_{V^*}[n]+(-\alpha_2^{(0)}-\alpha_3^{(0)}+n)t\Big) = t\frac{d}{dt}\log\frac{d}{dt}t\frac{d}{dt}\log\Big(e^{(-\alpha_2^{(0)}-\alpha_3^{(0)}+n)t}\tau_{V^*}[n]\Big). \end{aligned} \tag{8.54}$$

Equating (8.54) to the l.h.s. of (8.53) implies (8.51) with the $a(t)$ and $g(t)$ as in Table 8.10. □

Let us suppose the parameters are initially constrained as in Table 8.8. We know that then $\tau[0]=1$ in all cases. Inspection of Table 8.10 shows that we also have $g[0]=1$ in all cases. This is significant because an identity of Sylvester (see Exercises 8.2 q.1) gives that if $g[0]=1$, then the general solution of the Toda lattice equation (8.51) is given by

$$g[n]=\det\Big[\delta^{j+k}g[1]\Big]_{j,k=0,\dots,n-1}. \tag{8.55}$$

It therefore remains to show that with $g[1]$ given by the elementary function of t specified in Table 8.10, multiplied by the appropriate integral formula (8.47)–(8.50), substitution in (8.55) leads to n-dimensional integrals of the form (8.7).

PROPOSITION 8.2.6 *There are solutions of the PIV, PV and PVI systems such that*

$$\begin{aligned} E_{N,2}^{(\mu)}((t,\infty);\xi;e^{-x^2}) &= \tau_{IV}(t)\Big|_{\alpha_0=1+\mu+N,\alpha_1=-\mu,\alpha_2=-N}, \\ E_{N,2}^{(\mu)}((0,t);\xi;x^a e^{-x}) &= t^{(a+\mu)N+N^2}\tau_V(t)\Big|_{\substack{\nu_0=-\mu,\nu_1=0,\\ \nu_2=a+N,\nu_3=N}}, \quad \nu_{j-1}:=v_j-v_2, \\ E_{N,2}^{(\mu)}((0,t);\xi;x^a(1-x)^b) &= \tau_{VI}(t)\Big|_{\substack{v_1=(a+b)/2+N+\mu,v_2=(a+b)/2,\\ v_3=(a+b)/2+N,v_4=(b-a)/2}}, \\ E_{N,2}^{(\mu)}((t,\infty);\xi;(1+x^2)^{-\alpha}) &= \tau_{VI}\Big(\frac{1+it}{2}\Big)\Big|_{\substack{v_1=\mu+N-\alpha,v_2=-\alpha,\\ v_3=N-\alpha.v_4=0}}. \end{aligned}$$

Proof. For definiteness, let us consider the PV case. Then, according to Table 8.10, (8.52), (8.55) and (8.48) we have

$$\begin{aligned} \tau_V[n] &= t^{-n^2/2}e^{-nt}\det\Big[\Big(t\frac{d}{dt}\Big)^{j+k}g[1]\Big]_{j,k=0,\dots,n-1}, \\ g[1] &= t^{1/2}e^t\Big(\int_0^\infty - \xi\int_0^1\Big)e^{-tu}(1-u)^{v_2-v_1}u^{v_3-v_2-1}\,du \\ &= t^{1/2}\Big(\int_{-1}^\infty - \xi\int_{-1}^0\Big)e^{-tu}(-u)^{v_2-v_1}(u+1)^{v_3-v_2-1}\,du. \end{aligned}$$

Now a general property of the Toda lattice equation (8.51) with $a(t)=t$ is that if $\{g[n]\}_{n=0,1,\dots}$ is a solution with $g[0]=1$, then $\{t^{n\kappa}g[n]\}$ is also a solution which is given by the determinant formula (8.55) with $g[1]\mapsto t^\kappa g[1]$. With

$$F(a,c;t):=\Big(\int_{-1}^\infty - \xi\int_{-1}^0\Big)e^{-tu}(-u)^{a-1}(u+1)^{c-a-1}\,du, \tag{8.56}$$

where $a=v_2-v_1+1$, $c=v_3-v_1+1$, choosing $\kappa=-1/2$ shows

$$t^{-n/2}\tau_V[n]=t^{-n^2/2}e^{-nt}\det\Big[\Big(t\frac{d}{dt}\Big)^{j+k}F(a,c;t)\Big]_{j,k=0,\dots,n-1}. \tag{8.57}$$

Starting with this formula, our strategy is to use elementary row and column operations to eliminate the operator δ^{j+k}, $\delta:=t\frac{d}{dt}$.

Now

$$\begin{aligned}\delta F(a,c;t) &= \Big(\int_{-1}^{\infty} - \xi \int_{-1}^{0}\Big)\Big(t\frac{d}{dt}e^{-tu}\Big)(-u)^{a-1}(u+1)^{c-a-1}\,du \\ &= -\Big(\int_{-1}^{\infty} - \xi \int_{-1}^{0}\Big)\Big(\frac{d}{du}e^{-tu}\Big)(-u)^{a}(u+1)^{c-a-1}\,du \\ &= -aF(a,c;t) + (c-a-1)\Big(\int_{-1}^{\infty} - \xi \int_{-1}^{0}\Big)e^{-tu}(-u)^{a}(u+1)^{c-a-2}\,du \\ &= -aF(a,c;t) + (c-a-1)F(a+1,c;t), \end{aligned} \tag{8.58}$$

using integration by parts. This identity can be used to eliminate the operation δ^k from the r.h.s. of (8.57). Thus substituting (8.58) in column k, and adding a times column $k-1$, $(k = n, n-1, \dots, 2)$ in that order shows

$$\det[\delta^{j+k}F(a,c;t)]_{j,k=0,\dots,n-1} \propto \det\Big[\delta^j F(a,c;t) \quad \delta^{j+k-1}F(a+1,c;t)\Big]_{\substack{j=0,\dots,n-1\\ k=1,\dots,n-1}}.$$

Repeating this procedure on column k $(k = n-1, n-2, \dots, k')$ for each of $k' = 2, 3, \dots, n-2$ in that order shows

$$\det[\delta^{j+k}F(a,c;t)]_{j,k=0,\dots,n-1} \propto \det[\delta^j F(a+k,c;t)]_{j,k=0,\dots,n-1}. \tag{8.59}$$

Analogous to (8.58) we have

$$\frac{d}{dt}F(a,c;t) = -\Big(\int_{-1}^{\infty} - \xi \int_{-1}^{0}\Big)ue^{-tu}(-u)^{a-1}(u+1)^{c-a-1}\,du = -F(a,c+1;t) + F(a,c;t).$$

Substituting this in row j of (8.59) $(j = n-1, n-2, \dots, 1$ in this order) and subtracting row $j-1$ shows

$$\det[\delta^{j+k}F(a,c;t)]_{j,k=0,\dots,n-1} \propto \det\begin{bmatrix} F(a+k,c;t) \\ \delta^{j-1}tF(a+k,c+1;t)\end{bmatrix}_{\substack{j=1,\dots,n-1\\ k=0,\dots,n-1}}.$$

Repeating this procedure on row j, $j = n-1, n-2, \dots, j'$, for each of $j' = 2, 3, \dots, n-2$ in that order then gives

$$\det[\delta^{j+k}F(a,c;t)]_{j,k=0,\dots,n-1} \propto \det[t^j F(a+k,c+j;t)]_{j,k=0,\dots,n-1}.$$

Substituting this in (8.57) we thus have

$$\tau_V[n] \propto e^{-nt}\det[F(a+k,c+j;t)]_{j,k=0,\dots,n-1}. \tag{8.60}$$

We next substitute the integral form of F from (8.56) in (8.60). This shows

$$\begin{aligned}\tau_V[n] \propto e^{-nt}&\Big(\int_{-1}^{\infty} - \xi \int_{-1}^{0}\Big)du_1\, e^{-tu_1}(-u_1)^{a-1}(u_1+1)^{c-a-n}\cdots\Big(\int_{-1}^{\infty} - \xi \int_{-1}^{0}\Big)du_n \\ &\times e^{-tu_n}(-u_n)^{a-1}(u_n+1)^{c-a-n}\det[(-u_{j+1})^k(u_{j+1}+1)^{n+j-k-1}]_{j,k=0,\dots,n-1}.\end{aligned} \tag{8.61}$$

But using elementary column operations it follows that the determinant in this expression is equal to

$$\prod_{j=0}^{n-1}(1-u_{j+1})^j \det[u_{j+1}^k]_{j,k=0,\dots,n-1}. \tag{8.62}$$

All factors in the integrand of (8.61), apart from (8.62), are symmetric in $\{u_j\}$ so we can symmetrize (8.62) and then substitute it in (8.61). Noting that

$$\mathrm{Sym}\Big(\prod_{j=0}^{n-1}(1-u_{j+1})^j\det[u_{j+1}^k]_{j,k=0,\dots,n-1}\Big) = \pm\prod_{j<k}(u_j-u_k)^2$$

for some sign $\pm$, substituting this for the determinant in (8.61), and changing variables $u_j \mapsto u_j - 1$, then $u_j \mapsto u_j/t$, gives the sought identification between $\tau_V(t)$ and $E_{N,2}^{(\mu)}((0,t);\xi, x^a e^{-x})$. □

EXERCISES 8.2 1. [527] The objective of this exercise is to show that (8.55) satisfies the Toda lattice equation (8.51).

(i) Let $D = \det[a_{ij}]_{i,j=1,\dots,n}$, and define

$$D_{ij} := \frac{\partial D}{\partial a_{ij}}, \qquad D_{ij,pq} := \frac{\partial^2 D}{\partial a_{ip} \partial a_{jq}}.$$

Note that the Laplace expansion by row i gives

$$\sum_{j=1}^{n} a_{ij} D_{ij} = D, \qquad \sum_{j=1}^{n} a_{ij} D_{kj} = 0 \quad (k \neq i). \tag{8.63}$$

(Hint: the second equation corresponds to the Laplace expansion when two rows of $[a_{ij}]$ are equal.)

(ii) Let $'$ denote any derivative operation. By making use of the identity

$$\sum_{s=1}^{n} a'_{rs} \frac{D_{is}}{D} = -\sum_{s=1}^{n} a_{rs} \Big(\frac{D_{is}}{D}\Big)',$$

which follows from (8.63), and (8.63) itself, verify that

$$-\sum_{r=1}^{n}\sum_{s=1}^{n} a'_{rs} \frac{D_{is} D_{rj}}{D^2} = \Big(\frac{D_{ij}}{D}\Big)'.$$

Choose $' = \frac{\partial}{\partial a_{pq}}$ to deduce from this that

$$\frac{\partial}{\partial a_{pq}} \Big(\frac{D_{ij}}{D}\Big) = -\frac{1}{D^2} D_{iq} D_{pj}.$$

On the other hand, show directly from the definitions that

$$\frac{\partial}{\partial a_{pq}} \Big(\frac{D_{ij}}{D}\Big) = \frac{1}{D^2}(D D_{ip,jq} - D_{ij} D_{pq})$$

and thus conclude

$$\det \begin{bmatrix} D_{ij} & D_{iq} \\ D_{pj} & D_{pq} \end{bmatrix} = D D_{ip,jq}. \tag{8.64}$$

(iii) Let $D^{(n)} := \det[\delta_s^i \delta_t^j g(s,t)]_{i,j=0,\dots,n-1}$, where $\delta_u := a(u)\frac{d}{du}$, and define $D_{ij}^{(n)}$ and $D_{ij,pq}^{(n)}$ as in (i) with D therein replaced by $D^{(n)}$. Note that

$$\begin{aligned} \delta_s D^{(n)} &= -D_{nm}^{(n+1)}, \quad & \delta_t D^{(n)} &= -D_{mn}^{(n+1)}, \quad & \delta_s\delta_t D^{(n)} = D_{nn}^{(n+1)}, \\ D^{(n)} &= D_{m,m}^{(n+1)}, & D^{(n-1)} &= D_{nm,nm}^{(n+1)}, \end{aligned}$$

where $m = n+1$. Substitute these formulas in (8.64) with D replaced by $D^{(n+1)}$, $i = j = n$, $p = q = m$ to deduce that

$$\det \begin{bmatrix} \delta_s\delta_t D^{(n)} & \delta_s D^{(n)} \\ \delta_t D^{(n)} & D^{(n)} \end{bmatrix} = D^{(n+1)} D^{(n-1)},$$

and note that this can be rewritten to read

$$\delta_s \delta_t \log D^{(n)} = \frac{D^{(n+1)} D^{(n-1)}}{(D^{(n)})^2}. \tag{8.65}$$

(iv) Let $g(s,t) = g(s-t,0) =: g(u)$ so that $D^{(n)} = (-1)^{n(n-1)/2} \det[\delta_u^{i+j} g(u)]_{i,j=0,\dots,n-1}$. Use this in (8.65) to deduce (8.55) satisfies (8.51).

2. [230] In this exercise a determinant and integral solution of the PIII′ system will be presented.

(i) Note that $u(t) = t^{-\mu/2} I_\mu(\sqrt{t})$, $\mu = v_1^{(0)}$, is a solution of the Bessel differential equation in Table 8.9.

(ii) Conclude from (i), Table 8.10 and (8.55) that

$$t^{n^2/2} \tau_{III'}[n]\Big|_{t\mapsto t/4} = \det\Big[\Big(t\frac{d}{dt}\Big)^{j+k} t^{(1-\mu)/2} I_\mu(\sqrt{t})\Big]_{j,k=0,\dots,n-1},$$

and upon using the theory of the third sentence of the proof of Proposition 8.2.6 show that this can be rewritten as

$$t^{n(\mu-1)/2}(t^{n^2/2}\tau_{III'}[n])\Big|_{t\mapsto t/4} = \det\Big[\Big(t\frac{d}{dt}\Big)^{j+k} I_\mu(\sqrt{t})\Big]_{j,k=0,\dots,n-1}.$$

(iii) In the determinant change variables $t = s^2$ and make use of the Bessel function identities

$$s\frac{d}{ds} I_\mu(s) = s I_{\mu+1}(s) + \mu I_\mu(s),$$
$$s\frac{d}{ds}(s^j I_{\mu+j}(s)) = s^{j+1} I_{\mu+j-1}(s) - \mu s^j I_{\mu+j}(s)$$

together with elementary row and column operations to conclude that with $v_1^{(0)} = -v_2^{(0)} = \mu$

$$\tau_{III'}[n](t/4) = t^{-n\mu/2} \det[I_{\mu+j-k}(\sqrt{t}]_{j,k=1,\dots,n}. \tag{8.66}$$

(iv) Use the integral representation

$$I_\nu(s) := \frac{1}{2\pi}\int_{-\pi}^{\pi} e^{-i\nu\theta} e^{s\cos\theta}\, d\theta, \qquad \nu\in\mathbb{Z} \tag{8.67}$$

together with (5.76) to deduce from (8.66) that for $\mu \in \mathbb{Z}$

$$\begin{aligned}\tau_{III'}[n](t/4) &= \frac{t^{-n\mu/2}}{n!(2\pi)^n}\int_{-\pi}^{\pi} d\theta_1\, e^{\sqrt{t}\cos\theta_1 - i\mu\theta_1}\cdots\int_{-\pi}^{\pi} d\theta_n\, e^{\sqrt{t}\cos\theta_n - i\mu\theta_n}\prod_{1\le j<k\le n}|e^{i\theta_j}-e^{i\theta_k}|^2\\ &= t^{-n\mu/2}\Big\langle \prod_{j=1}^n e^{\sqrt{t}\cos\theta_j - i\mu\theta_j}\Big\rangle_{\mathrm{CUE}_n}.\end{aligned} \tag{8.68}$$

3. [230] The objective of this exercise is to derive a τ-function sequence for the PV^* system which involves an average over CUE_n analogous to that in (8.68), and has the feature of being singular in terms of the coordinates of the PV system. Write

$$T_V^n H_0^{V^*} = H_n^{V^*}, \qquad T_V^n \vec{v}^{(0)} = (v_1^{(0)} - n/4, v_2^{(0)} - n/4, v_3^{(0)} - n/4, v_4^{(0)} + 3n/4)$$

and define $\tau_n^{V^*}$ (up to normalization) by the requirement that

$$H_n^{V^*} = \frac{d}{dt}\log \tau_n^{V^*}.$$

(i) Show that with $q_0 = 1$, $v_3^{(0)} - v_4^{(0)} = 0$, the Hamiltonian $tH_0^{V^*}$ reduces to $tH_0^{V^*} = (v_1 - v_3)t$. Note that the transformation formulas (8.38) between $\{p,q\}$ and $\{P,Q\}$ are singular at $q=1$. Show too that

$$\tau_0 = e^{(v_1^{(0)} - v_3^{(0)})t}, \quad p_0 = t\frac{d}{dt}\log\tau_1^{V^*} + (v_3^{(0)} - v_1^{(0)})t,$$

Function	σ Painlevé equation	Parameters
$U_N^{\rm G}(t;\mu;\xi)$	σPIV	$\alpha_1=-\mu,\quad \alpha_2=-N$
$U_N^{\rm L}(t;a,\mu;\xi)$	σPV	$\nu_0=0,\ \nu_1=-\mu,\ \nu_2=N+a,\ \nu_3=N$
$U_N^{\rm J}(t;a,b,\mu;\xi)$	σPVI	$v_1=\frac{1}{2}(a+b)+N+\mu,\quad v_2=\frac{1}{2}(a+b)$ $v_3=\frac{1}{2}(a+b)+N,\quad v_4=\frac{1}{2}(b-a)$
$U_N^{\rm Cy}(t;\alpha,\mu;\xi)$	$\tilde\sigma$PVI	$v_1=\mu+N-\alpha,\quad v_2=-\alpha$ $v_3=N-\alpha,\quad v_4=0$

Table 8.11 The differential equations σPIV–σPVI are given in (8.15), while $\tilde\sigma$PVI is given in (8.21).

where $e^{-(v_1^{(0)}-v_3^{(0)})t}\tau_1^{V^*}$ satisfies the confluent hypergeometric differential equation

$$ty''+(v_1^{(0)}-v_2^{(0)}+1+t)y'+(v_1^{(0)}-v_3^{(0)})y=0.$$

(ii) Consider the solution of the above equation analytic at the origin,

$$e^{-(v_1^{(0)}-v_3^{(0)})t}\tau_1^{V^*}={}_1F_1(v_1^{(0)}-v_3^{(0)},v_1^{(0)}-v_2^{(0)}+1;-t).$$

Write this in an integral form by verifying, using (4.4) in the case $N=1$, that with $z=e^{i\theta}$

$${}_1F_1(-\nu,\mu+1;-t)=\frac{\Gamma(\mu+1)\Gamma(\nu+1)}{\Gamma(\mu+\nu+1)}\frac{1}{2\pi}\int_{-\pi}^{\pi}z^{(\mu-\nu)/2}|1+z|^{\mu+\nu}e^{tz}\,d\theta.$$

(iii) Proceed as in the proof of Proposition 8.2.6 to deduce from (ii) that with $v_1^{(0)}-v_3^{(0)}=-\nu$, $v_1^{(0)}-v_2^{(0)}=\mu$

$$\tau_n^{V^*}=e^{-\nu t}\Big(\frac{\Gamma(\mu+1)}{\Gamma(\mu+\nu+1)}\Big)^n\prod_{l=0}^{n-1}\Gamma(\nu+l+1)\Big\langle\prod_{l=1}^{n}z_l^{(\mu-\nu)/2}|1+z_l|^{\mu+\nu}e^{tz_l}\Big\rangle_{{\rm CUE}_n}. \tag{8.69}$$

8.3 σ-FORM PAINLEVÉ EQUATION CHARACTERIZATIONS

8.3.1 Finite N ensembles

In Proposition 8.2.6 the averages (8.7) in the four classical cases have been identified as τ-functions of certain Painlevé systems. By the definition (8.44), the logarithmic derivative of the τ-function is the Hamiltonian, and we know from Proposition 8.2.1 that for each Hamiltonian there is an auxiliary Hamiltonian, which from (8.29) is simply related to a σ-function satisfying the corresponding second order second degree differential equation in (8.15). Thus after multiplying (8.7) in the classical cases by an appropriate function of N, $a(t)$ (as specified in Table 8.10) times its logarithmic derivative satisfies a Painlevé equation in σ-form. Provided a boundary condition can be specified, we can therefore characterize the averages (8.7) in the classical cases in terms of particular solutions of the equations (8.15).

PROPOSITION 8.3.1 *Let $U_N^{\rm G}(t;\mu;\xi)$, $U_N^{\rm L}(t;a,\mu;\xi)$, $U_N^{\rm J}(t;a,b,\mu;\xi)$, $U_N^{\rm Cy}(t;\alpha,\mu;\xi)$ satisfy the differential equations as specified in Table 8.11, and write*

$$F_N^{\rm J}(t;a,b,\mu;\xi):=U_N^{\rm J}(t;a,b,\mu;\xi)-e_2[-v_1,-v_2,v_3]t+\frac{1}{2}e_2[-v_1,-v_2,v_3,v_4], \tag{8.70}$$

where e_2 is specified as in Proposition 8.2.1. Also let $L_N(a):=W_{a,2,N}$ as used in (4.142), $J_n(a,b)=$

$S_n(a,b,1)$ where S_n denotes the Selberg integral (4.1) and $M_n(a) := M_n(a,a,1)$ as specified by (4.4).

In the notation (8.7) we have

$$\begin{aligned}E_{N,2}^{(\mu)}((s,\infty);\xi;e^{-x^2}) &= s^{\mu N}\exp\Big(-\int_s^\infty \Big(U_N^{\rm G}(t;\mu;\xi)-\frac{N\mu}{t}\Big)\,dt\Big),\\ E_{N,2}^{(\mu)}((0,s);\xi;x^a e^{-x}) &= \frac{L_N(a+\mu)}{L_N(a)}\exp\int_0^s\Big(U_N^{\rm L}(t;a,\mu;\xi)+\mu N\Big)\frac{dt}{t},\\ E_{N,2}^{(\mu)}((0,s);\xi;x^a(1-x)^b) &= \frac{J_N(a+\mu,b)}{J_N(a,b)}\exp\int_0^s F_N^{\rm J}(t;a,b,\mu;\xi)\,\frac{dt}{t(t-1)},\\ E_{N,2}^{(\mu)}((s,\infty);\xi;(1+x^2)^{-\alpha}) &= (1+s^2)^{N\mu/2}\\ &\quad\times\exp\Big(-\int_s^\infty\Big(U_N^{\rm Cy}(t;\alpha,\mu;\xi)-te_2[-v_1,-v_2,v_3]-tN\mu\Big)\frac{dt}{1+t^2}\Big).\end{aligned} \tag{8.71}$$

Also, with

$$\begin{aligned}&E_{N,2}^{(\mu)}(\pi-x,\pi);\xi;|1+e^{i\theta}|^{2\omega}\\ &:= \frac{1}{C}\int_{-\pi}^{\pi}d\theta_1\cdots\int_{-\pi}^{\pi}d\theta_N\prod_{l=1}^N(1-\xi\chi_{\pi-x,\pi}^{(l)})|e^{-ix}-e^{i\theta_l}|^{2\mu}|1+e^{i\theta_l}|^{2\omega}\prod_{1\le j<k\le N}|e^{i\theta_k}-e^{i\theta_j}|^2\\ &=: \Big\langle\prod_{l=1}^N(1-\xi\chi_{(\pi-x,\pi)}^{(l)})|e^{-ix}-e^{i\theta_l}|^{2\mu}\Big\rangle_{\rm cJUE}\end{aligned}$$

we have

$$\begin{aligned}E_{N,2}^{(\mu)}((\pi-x,\pi);\xi;|1+e^{i\theta}|^{2\omega}) = \frac{M_N(\omega+\mu)}{M_N(\omega)}\exp\Big(-\int_0^{x/2}\Big(U_N^{\rm Cy}(\cot\phi;N+\omega+\mu/2,\mu;\xi)\\ -(e_2[-v_1,-v_2,v_3]\Big|_{\alpha=N+\omega+\mu/2}+N\mu)\cot\phi\Big)d\phi\Big).\end{aligned} \tag{8.72}$$

Proof. In addition to the reasoning noted above the statement of the proposition, one must also consider the limiting behavior of the particular $E_{N,2}^{(\mu)}(J;\xi;g)$ as $|J|\to 0$. The evaluation (8.72) follows from the final equation in (8.71) by the mapping (3.123) between the Cauchy ensemble and the circular Jacobi ensemble. □

We are still faced with the task of specifying the boundary conditions associated with the solutions of the nonlinear equations in Proposition 8.3.1. For general parameters this can be done by analyzing the $|J|\to 0$ limit of the individual elements in the determinant form of the $E_{N,2}^{(\mu)}$ [235]. For $\mu=0$, a different approach is possible, for then we have the general expansion

$$E_{N,2}^{(0)}(J;\xi;g(x)) = 1-\xi\int_J\rho_{(1)}^N(t)\,dt+\frac{\xi^2}{2!}\int_J dt_1\int_J dt_2\,\rho_{(2)}^N(t_1,t_2)+\cdots, \tag{8.73}$$

which is a consequence of (9.1) below. It implies

$$\log E_{N,2}^{(0)}(J;\xi;g(x)) \underset{|J|\to 0}{\sim} -\xi\int_J\rho_{(1)}^N(t)\,dt. \tag{8.74}$$

Small s expansion — Jacobi ensemble

The formulas of Proposition 8.3.1 can readily be used to generate power series expansions in the Laguerre, Jacobi and circular Jacobi cases. By use of the formulas (8.8), (8.1) and (8.2) the power series expansions of the corresponding probabilities $E_{N,2}(n;J)$ and p.d.f.'s $p_{N,2}(n;a)$ then follow. The results of Proposition 8.3.1

also allow the latter to be numerically tabulated or statistical quantities such as the variance to be computed. An example of the latter type of calculation is given in Section 8.3.4. Here the Painlevé characterization will be used to compute terms in the power series expansion of $E^{\rm J}_{N,2}((0,s);\xi)$ for general a,b (the superscript J denotes Jacobi).

We know from Proposition 8.3.1 that $U^{\rm J}_N(t;a,b,0;\xi)$ satisfies the σPVI equation in (8.8) with $v_1=v_3=N+(a+b)/2$, $v_2=(a+b)/2$, $v_4=(b-a)/2$. With $F^{\rm J}_N$ specified by (8.70), let us put

$$f^{\rm J}_N(t;a,b;\xi) := F^{\rm J}_N(t;a,b,0;\xi). \tag{8.75}$$

Substituting for $U^{\rm J}_N$ in terms of $f^{\rm J}_N$ in the σPVI equation, using the fact that in the case of interest $v_1=v_3$, shows $f^{\rm J}_N$ satisfies the equation

$$\begin{aligned}(t(1-t)f'')^2 - 4t(1-t)(f')^3 + 4(1-2t)(f')^2 f + 4f'f^2 - 4f^2v_1^2 \\ +(f')^2\Big(4tv_1^2(1-t) - (v_2-v_4)^2 - 4tv_2v_4\Big) + 4ff'(-v_1^2+2tv_1^2+v_2v_4) = 0.\end{aligned} \tag{8.76}$$

Furthermore, from (8.73) we know that this equation is to be solved subject to the boundary condition

$$f(t) \underset{t\to0^+}{\sim} -\xi t\rho^{\rm J}_{(1),N}(t) \underset{t\to0^+}{\sim} -\xi c^{\rm J}_N(a,b)t^{a+1}, \qquad c^{\rm J}_N(a,b) = \frac{\Gamma(N+a+b+1)\Gamma(N+a+1)}{\Gamma(a+1)\Gamma(a+2)\Gamma(N)\Gamma(N+b)},$$

where the second asymptotic equality follows by noting from the definitions that

$$\rho^{\rm J}_{(1),N}(t) \underset{t\to0^+}{\sim} t^a N\frac{S_{N-1}(a+2,b,1)}{S_N(a,b,1)},$$

S_n denoting the Selberg integral. Substituting this boundary condition in (8.76) generates the small t expansion

$$\begin{aligned}-f(t) \underset{t\to0^+}{\sim} &\ \xi c^{\rm J}_N(a,b)t^{a+1}\Big(1+\frac{a(1-b)-2N(a+b+N)}{2+a}t+{\rm O}(t^2)\Big) \\ &+\frac{(\xi c^{\rm J}_N(a,b))^2}{1+a}t^{2a+2}\Big(1+\frac{2+6a+3a^2-(3+2a)(2N^2+2N(a+b)+ab)}{(2+a)^2}t+{\rm O}(t^2)\Big) \\ &+\frac{(\xi c^{\rm J}_N(a,b))^3}{(1+a)^2}t^{3a+3}\Big(1+\frac{4+10a+5a^2-(4+3a)(2N^2+2N(a+b)+ab)}{(2+a)^2}t+{\rm O}(t^2)\Big) \\ &+\cdots.\end{aligned}$$

Substituting the above expansion (with the explicit form of the terms ${\rm O}(t^2)$ included) in the formula for $E^{\rm J}_{N,2}((0,s);\xi)$, which we read off from the $\mu=0$ case of the appropriate formula in Proposition 8.3.1, one finds after integrating and exponentiating that

$$\begin{aligned}E^{\rm J}_{N,2}((0,s);\xi) = 1 &- \frac{\xi c^{\rm J}_N(a,b)}{1+a}s^{a+1}\Big(1-\frac{1+a}{(2+a)^2}(-2-2a+ab+2aN+2bN+2N^2)s+{\rm O}(s^2)\Big) \\ &+\frac{(\xi c^{\rm J}_N(a,b))^2(N-1)(N+a+1)(N+b-1)(N+a+b+1)}{(a+2)^4(a^2+4a+3)^2}s^{2a+4}(1+{\rm O}(s))+\cdots.\end{aligned} \tag{8.77}$$

We remark that one can anticipate from the expansion (8.73), together with the small $t_1,\dots,t_n$ expansion

$$\rho^N_{(n)}(t_1,\dots,t_n) \sim C\prod_{j=1}^n t_j^a \prod_{1\le j<k\le n}(t_k-t_j)^2$$

for some C independent of the t_j, that the leading term in s accompanying the power ξ^k will be proportional to s^{ka+k^2} in agreement with (8.77).

Small s expansions — the classical groups

The CUE, or equivalently the classical group $U(N)$, is the special case $\omega = 0$ of the cJUE. Furthermore, the case $\omega = 1$ of the cJUE corresponds to the CUE with an eigenvalue fixed at $\theta = \pi$. For definiteness we will consider the generating function

$$E_{N,2}^{\rm cJ}((\pi - x, \pi); \xi)\Big|_{\omega=0} := E_{N,2}^{(\mu)}((\pi - x, \pi); \xi; |1 + e^{i\theta}|^{2\omega})\Big|_{\mu=\omega=0}.$$

This is characterized by $U_N^{\rm Cy}(t; N, 0; \xi)$, which according to Proposition 8.3.1 satisfies the $\tilde{\sigma}$PVI equation with $v_1 = v_3 = v_4 = 0$, $v_2 = -N$. The general expansion (8.73) shows the corresponding boundary condition is $U_N^{\rm Cy}(t; N, 0; \xi) \sim c$ as $t \to \infty$, where $c = \xi N/\pi$, and this substituted into the $\tilde{\sigma}$PVI equation generates the large t expansion

$$U_N^{\rm Cy}(t; N, 0; \xi) \underset{t\to\infty}{\sim} c + \frac{c^2}{t} + \frac{c^3}{t^2} + \frac{c^2(-2 - N^2 + 9c^2)}{9t^3} + {\rm O}\Big(\frac{1}{t^4}\Big). \tag{8.78}$$

Keeping terms through to ${\rm O}(1/t^8)$, substituting in the formula for $E_{N,2}^{\rm cJ}((\pi - x, \pi); \xi)|_{\omega=0}$ which we read off from (8.72), expanding the integral to ${\rm O}(x^9)$ then exponentiating, again keeping terms to ${\rm O}(x^9)$ one finds

$$\begin{aligned} E_{N,2}^{\rm CUE}((0, 2x); \xi) = {} & 1 - cx + \frac{(N^2 - 1)c^2x^4}{36} - \frac{(N^2 - 1)(2N^2 - 3)}{1350}c^2x^6 \\ & + \frac{(N^2 - 1)(N^2 - 2)(3N^2 - 5)}{52920}c^2x^8 - \frac{(N^2 - 4)(N^2 - 1)^2}{291600}c^3x^9 + {\rm O}(x^{10}). \end{aligned} \tag{8.79}$$

The results of Section 2.6 show that the eigenvalue p.d.f. for the classical groups $O^{\pm}(N)$ and ${\rm Sp}(2N)$, after the change of variables $x_j = \cos^2\theta_j/2$, corresponds to the eigenvalue p.d.f. for the JUE with weight $x^a(1 - x)^b$, $0 < x < 1$, and $a, b = \pm 1/2$. For example,

$$\begin{aligned} E_2^{O^-(2N+1)}((0, \phi); \xi) &= E_{N,2}^{\rm J}((0, \sin^2\phi/2); \xi)\Big|_{\substack{a=-1/2 \\ b=1/2}}, \\ E_2^{O^+(2N+1)}((0, \phi); \xi) &= E_{N,2}^{\rm J}((0, \sin^2\phi/2); \xi)\Big|_{\substack{a=1/2 \\ b=-1/2}}. \end{aligned} \tag{8.80}$$

Proceeding as in the derivation of (8.77) one can show

$$\begin{aligned} E_2^{O^-(2N+1)}((0, x); \xi) = {} & 1 - \tilde{c}x + \frac{(4N^2 - 1)}{36}\tilde{c}x^3 - \frac{(48N^4 - 40N^2 + 7)}{3600}\tilde{c}x^5 \\ & + \frac{(4N^4 - 5N^2 + 1)}{2025}\tilde{c}^2x^6 + \frac{(192N^6 - 336N^4 + 196N^2 - 31)}{211680}\tilde{c}x^7 \\ & - \frac{(48N^6 - 112N^4 + 77N^2 - 13)}{198450}\tilde{c}^2x^8 + {\rm O}(x^9), \\ E_2^{O^+(2N+1)}((0, x); \xi) = {} & 1 - \frac{(4N^2 - 1)}{36}\tilde{c}x^3 + \frac{(4N^2 - 1)(12N^2 - 7)}{3600}\tilde{c}x^5 \\ & - \frac{(4N^2 - 1)(48N^4 - 72N^2 + 31)}{211680}\tilde{c}x^7 + {\rm O}(x^9), \end{aligned} \tag{8.81}$$

where $\tilde{c} := 2N\xi/\pi =: 2c$.

8.3.2 Soft edge

Scaled limits — methodology

Painlevé himself noted that upon a suitable limiting procedure, the Painlevé transcendents degenerate downwards (see, e.g., [310]). Thus there is a limit transition from PVI to PIII′ for example, but not the other way around. This can be seen at the level of the corresponding differential equations, where changing variables as appropriate for the limit and equating the highest order terms exhibits the differential equation of the limiting Painlevé transcendent. It is this general approach we will take in characterizing the scaled limits of the finite N quantities $E_{N,2}$, known in terms of σ Painlevé transcendents from Proposition 8.3.1.

We know from Section 7.1 that with the replacement (7.11) the Gaussian ensemble scales to the soft edge. From this it can be rigorously established that

$$\lim_{N\to\infty} E_{N,2}^{(0)}\Big(((2N)^{1/2}+t/2^{1/2}N^{1/6},\infty);\xi;e^{-x^2}\Big),$$

which is the limiting generating function for the probability that the interval (t,∞) at the soft edge contains n eigenvalues, is well defined (see Section 9.1 below). Furthermore, with suitable C independent of t

$$\lim_{N\to\infty} Ce^{-t^2}E_{N,2}^{(2)}((t,\infty);\xi;e^{-x^2})\Big|_{t\mapsto(2N)^{1/2}+t/2^{1/2}N^{1/6}}$$

is the generating function for the p.d.f. of the event that the interval (t,∞) at the soft edge contains n eigenvalues and there is an eigenvalue at t. The fact that these scaled averages are well defined suggests that for general μ, again with suitable C independent of t,

$$\lim_{N\to\infty} Ce^{-\mu t^2/2}E_{N,2}^{(\mu)}((t,\infty);\xi;e^{-x^2})\Big|_{t\mapsto(2N)^{1/2}+t/2^{1/2}N^{1/6}}$$

will be well defined. Recalling Proposition 8.3.1 and taking the logarithmic derivative, this in turn suggests that

$$u^{\rm s}(t;\mu;\xi):=\lim_{N\to\infty}\frac{1}{\sqrt{2}N^{1/6}}\Big(-\mu t+U_N^{\rm G}(t;\mu;\xi)\Big)\Big|_{t\mapsto(2N)^{1/2}+t/2^{1/2}N^{1/6}} \tag{8.82}$$

will also be well defined.

PROPOSITION 8.3.2 *[519], [228] The function (8.82) satisfies the σPII equation in (8.15) with $a=\mu$. For $\mu=0$ the boundary condition is*

$$u^{\rm s}(t;0;\xi)\underset{t\to\infty}{\sim}\xi\rho_{(1)}^{\rm soft}(t). \tag{8.83}$$

We have

$$\begin{aligned} E_2^{\rm soft}((t,\infty);\xi) &= \exp\Big(-\int_t^\infty u^{\rm s}(s;0;\xi)\,ds\Big), \\ p_2^{\rm soft}(t;\xi) &= \rho_{(1)}^{\rm soft}(t)\exp\Big(-\int_t^\infty\Big(u^{\rm s}(s;2;\xi)-\rho_{(1)}^{\rm soft}{}'(s)/\rho_{(1)}^{\rm soft}(s)\Big)\,ds\Big), \end{aligned} \tag{8.84}$$

where $u^{\rm s}(s;0;\xi)$ and $u^{\rm s}(s;2;\xi)$ are related by

$$u^{\rm s}(t;2;\xi)=\frac{d}{dt}\log u^{\rm s}(t;0;\xi)+u^{\rm s}(t;0;\xi). \tag{8.85}$$

Proof. The fact that (8.82) satisfies the σPII equation with $a=\mu$ is deduced by recalling from Proposition 8.3.1 that $U_N^{\rm G}(t;\mu;\xi)$ satisfies the σPIV equation in (8.15) with $\alpha_1=-\mu$, $\alpha_2=-N$, replacing $U_N^{\rm G}$ by $\tilde U_N^{\rm G}+\mu t$, changing variables $t=(2N)^{1/2}+s/2^{1/2}N^{1/6}$, then replacing $\tilde U_N^{\rm G}$ by $\sqrt{2}N^{1/6}u^{\rm s}(s;\mu;\xi)$ and taking the limit $N\to\infty$. One sees that the σPIV equation degenerates to σPII.

For the boundary conditions in the case $\mu = 0$ one uses (8.73), while the inter-relation (8.85) follows from the fact that

$$p_2^{\rm soft}(t;\xi) = \frac{1}{\xi}\frac{d}{dt}E_2^{\rm soft}((t,\infty);\xi), \tag{8.86}$$

which in turn is an example of (8.5).

As noted in the text above the statement of the proposition we have

$$E_2^{\rm soft}((t,\infty);\xi) = \lim_{N\to\infty} E_{N,2}^{(0)}\Big(((2N)^{1/2} + t/2^{1/2}N^{1/6},\infty);\xi;e^{-x^2}\Big)$$

and

$$p_2^{\rm soft}((t,\infty);\xi) = \lim_{N\to\infty} Ce^{-t^2}E_{N,2}^{(2)}((t,\infty);\xi;e^{-x^2})\Big|_{t\mapsto(2N)^{1/2}+t/2^{1/2}N^{1/6}}.$$

Recalling the formula of Proposition 8.3.1 for $E_{N,2}(\mu;(t,\infty);\xi;e^{-x^2})$ immediately gives the first formula in (8.84). For the second formula we must take into consideration the asymptotic form (8.83) to ensure that the integrand is integrable at infinity, and fix the undetermined proportionality constant by requiring that $p_2^{\rm soft}((t,\infty);\xi) \underset{t\to\infty}{\sim} \rho_{(1)}^{\rm soft}(t)$. □

The formulas (8.84) can be used to compute a tabulation of $p_2^{\rm soft}(n;t)$, provided n is small. However we defer this task until Section 9.4 when we have available an even simpler Painlevé transcendent evaluation of $E_2^{\rm soft}((t,\infty);\xi)$.

8.3.3 Hard edge

With the variables scaled as specified by (7.30), the Laguerre ensemble scales to the hard edge, and we expect

$$E_2^{{\rm hard}\,(\mu)}((0,t);\xi;a) := \lim_{N\to\infty}\Big(\frac{L_N(a)}{L_N(a+\mu)}E_{N,2}^{(\mu)}\Big(\Big(0,\frac{t}{4N}\Big);\xi;x^a e^{-x}\Big)\Big) \tag{8.87}$$

(here $L_N(a) = W_{a,2,N}$ is as used in (8.71); note that the constants are chosen so that at $t = 0$ it equals 1) to be well defined. Recalling the second formula in (8.71) it follows that

$$E_2^{{\rm hard}\,(\mu)}((0,t);\xi;a) = \exp\int_0^t u^{\rm h}(s;a,\mu;\xi)\,\frac{ds}{s}, \tag{8.88}$$

where

$$u^{\rm h}(t;a,\mu;\xi) := \lim_{N\to\infty}\Big(U_N^{\rm L}\Big(\frac{t}{4N};a,\mu;\xi\Big) + \mu N\Big). \tag{8.89}$$

In this limit the σPV equation characterizing $U_N^{\rm L}$ degenerates to a variant of the σPIII$'$ equation.

PROPOSITION 8.3.3 *[520], [230] The function $u^{\rm h}$ specified by (8.89) satisfies the differential equation*

$$(tu'')^2 - (\mu+a)^2(u')^2 - u'(4u'+1)(u-tu') - \frac{\mu(\mu+a)}{2}u' - \frac{\mu^2}{4^2} = 0 \tag{8.90}$$

and consequently

$$u^{\rm h}(t;a,\mu;\xi) = -\Big(\sigma_{III'}(t) + \mu(\mu+a)/2\Big), \tag{8.91}$$

where $\sigma_{III'}$ satisfies the σPIII' equation in (8.15) with parameters $v_1 = a+\mu$, $v_2 = a-\mu$. For $\mu = 0$ the boundary condition is

$$u^{\rm h}(t;a,0;\xi) = -\sigma_{III'}(t) \underset{t\to0^+}{\sim} -\xi t\rho_{(1)}^{\rm hard}(t) \underset{t\to0^+}{\sim} -\xi\frac{t^{1+a}}{2^{2+2a}\Gamma(1+a)\Gamma(2+a)}. \tag{8.92}$$

We have

$$E_2^{\rm hard}((0,t);\xi;a) = \exp\Big(- \int_0^t \sigma_{III'}(s)\Big|_{\mu=0} \frac{ds}{s}\Big),$$
$$p_2^{\rm hard}((0,t);\xi;a) = \frac{t^a}{2^{2a+2}\Gamma(a+1)\Gamma(a+2)} \exp\Big(- \int_0^t \Big(\sigma_{III'}(s)\Big|_{\mu=2} + a + 2\Big) \frac{ds}{s}\Big), \tag{8.93}$$

where $\sigma_{III'}(s)|_{\mu=0}$ *and* $\sigma_{III'}(s)|_{\mu=2}$ *are related by*

$$\sigma_{III'}(t)|_{\mu=2} = \sigma_{III'}(t)|_{\mu=0} - 1 - t\frac{\sigma'_{III'}(t)|_{\mu=0}}{\sigma_{III'}(t)|_{\mu=0}}. \tag{8.94}$$

Proof. Making the replacement $\sigma_V \mapsto \sigma_V - N\mu$ in the σPV equation of (8.15), changing variables $t \mapsto t/4N$, $\sigma_V(t/4N) \mapsto u(t)$ and equating terms of order N^2 (which is the leading order) on both sides gives (8.90).

The $\mu = 0$ boundary condition follows from the general formula (8.73), while (8.94) follows by noting

$$p_2^{\rm hard}((0,t);\xi;a) = -\frac{1}{\xi}\frac{d}{dt} E_2^{\rm hard}((0,t);\xi;a). \tag{8.95}$$

The first formula in (8.93) follows immediately from the fact that

$$E_2^{\rm hard}((0,t);\xi;a) = E_2^{{\rm hard}(0)}((0,t);\xi;a),$$

together with the evaluation of $E_2^{{\rm hard}(0)}$ implied by (8.88) and (8.91). For the second formula note from (8.8) and (8.87) that

$$p_2^{\rm hard}((0,t);\xi;a) = t^a \lim_{N\to\infty} \frac{L_N(a+2)}{L_{N+1}(a)} \frac{1}{(4N)^a} E_{N,2}^{(2)}\Big(\Big(0,\frac{t}{4N}\Big);\xi;x^a e^{-x}\Big),$$

then make use of (8.88) and (8.91). □

The case $a \in \mathbb{Z}_{\ge 0}$ of $E_2^{{\rm hard}(\mu)}(s;\xi=1;a)$ is special, because then an a-dimensional determinant expression can be found. This is a consequence of the identity (8.119) in Exercises 8.3 q.1(v). First we note that a simple change of variables shows that for $J = (0,t)$ or $J = (t,\infty)$

$$E_{N,2}^{(\mu)}(J,\xi=0;x^a e^{-x})\Big|_{t\mapsto -t} = e^{Nt} E_{N,2}^{(a)}((0,t),\xi=1;x^\mu e^{-x}).$$

We then apply (8.119) below to conclude that for $a \in \mathbb{Z}_{\ge 0}$

$$E_{N,2}^{(a)}((0,t),\xi=1;x^\mu e^{-x}) \propto e^{-Nt}\Big\langle \prod_{l=1}^a (1+z_l)^\mu (1+1/z_l)^N e^{tz_l}\Big\rangle_{{\rm CUE}_a}.$$

The hard edge limit (8.87) can be taken on the r.h.s. by regarding each integration over z_l as a contour integral over $|z_l| = 1$, then shifting the contour of integration to $|z_l| = 2N/t^{1/2}$. This shows that for $a, \mu \in \mathbb{Z}_{\ge 0}$

$$E_2^{{\rm hard}\,(\mu)}((0,s);\xi=1;a) = a! \prod_{j=1}^a \frac{(j+\mu-1)!}{j!} e^{-s/4}\Big(\frac{2}{s^{1/2}}\Big)^{a\mu} \Big\langle \prod_{l=1}^a z_l^\mu e^{\sqrt{s}(z_l+z_l^{-1})/2}\Big\rangle_{{\rm CUE}_a} \tag{8.96}$$

(cf. (8.68)), where the proportionality constant has been specified by the requirement that both sides equal unity as $s \to 0$; see (13.27) below. Making use of the integral representation (8.67) and the identity (5.77) shows that this can equivalently be written as the determinant formula

$$E_2^{{\rm hard}\,(\mu)}((0,s);\xi=1;a) = a! \prod_{j=1}^a \frac{(j+\mu-1)!}{j!} e^{-s/4}\Big(\frac{2}{s^{1/2}}\Big)^{a\mu} \det\Big[I_{\mu+j-k}(\sqrt{s})\Big]_{j,k=1,\dots,a}, \tag{8.97}$$

where the restriction $\mu \in \mathbb{Z}_{\ge 0}$ present in (8.96) can be removed. It follows from this that for $a \in \mathbb{Z}_{\ge 0}$ we have [204]

$$E_2^{\rm hard}(n=0;t;a) = e^{-t/4}\det[I_{j-k}(\sqrt{t})]_{j,k=1,\dots,a},$$
$$p_2^{\rm hard}(n=0;t;a) = \frac{1}{4}e^{-t/4}\det[I_{2+j-k}(\sqrt{t})]_{j,k=1,\dots,a}, \tag{8.98}$$

where the latter formula follows from the fact that $p_2^{\rm hard}(0;t;a)$ is proportional to $E_2^{\rm hard(2)}(s;1;a)$, and fixing the proportionality constant by requiring that the small t behavior be as displayed in (8.93).

We remark that integration techniques similar to those of the second part of Proposition 5.1.3 allow us to give alternative derivations of the identities (8.98), and also to derive that for $a \in \mathbb{Z}_{\geq 0}$ [204]

$$p_2^{\rm hard}(n=1;t;a) = \int_0^t p_2^{(1,2)}(u,t)\,du, \tag{8.99}$$

where $p_2^{(1,2)}(t_1,t_2)$ — the joint distribution of the smallest and second smallest eigenvalue — has the determinant form

$$p_2^{\rm hard\,(1,2)}(t_1,t_2) = 2^{-4}e^{-t_2/4}(t_2/t_1)^a\det\begin{bmatrix} [I_{j-k+2}(t_2^{1/2})]_{\substack{j=1,\dots,a\\k=1,\dots,a+2}} \\ [(\frac{t_2-t_1}{t_2})^{(k-j)/2}I_{j-k+2}(\sqrt{t_2-t_1})]_{\substack{j=1,2\\k=1,\dots,a+2}}\end{bmatrix}. \tag{8.100}$$

8.3.4 High precision numerical computation

One use of the Painlevé transcendent evaluations of the spacing distributions is that it allows for their high precision numerical computation, provided the number of eigenvalues conditioned to be in the spacing is small. As an example we will consider the scaled distribution of the first and second eigenvalues as measured anticlockwise from $\theta = 0$, for the orthogonal groups $O^+(2N+1)$ and $O^-(2N+1)$. These are in fact related to hard edge distributions. Thus we see from (8.80), and the discussion around (7.45) relating to the hard edge scaling of the Jacobi ensemble, that

$$\lim_{N\to\infty} E_2^{O^-(2N+1)}((0,\phi);\xi)\Big|_{\phi=\pi s/N} := E_2^{O^-}((0,s);\xi) = E_2^{\rm hard}((0,(\pi s)^2);-\frac{1}{2};\xi),$$
$$\lim_{N\to\infty} E_2^{O^+(2N+1)}((0,\phi);\xi)\Big|_{\phi=\pi s/N} := E_2^{O^+}((0,s);\xi) = E_2^{\rm hard}((0,(\pi s)^2);\frac{1}{2};\xi). \tag{8.101}$$

According to (8.88) and Proposition 8.3.3

$$E_2^{O^\mp}((0,s);\xi) = \exp\Big(-\int_0^{(\pi s)^2} v\Big(t;\pm\frac{1}{2};\xi\Big)\frac{dt}{t}\Big), \tag{8.102}$$

where $v(t;\pm\frac{1}{2};\xi)$ satisfies the equation

$$(tv'')^2 - \frac{1}{4}(v')^2 + v'(4v'-1)(v-tv') = 0, \tag{8.103}$$

with the transcendents $v(t;a;\xi) =: v(t)$ distinguished by the boundary conditions,

$$v(t) \underset{t\to0^+}{\sim} \frac{\xi t^{1/2}}{\pi}\quad (a=-\frac{1}{2}), \qquad v(t) \underset{t\to0^+}{\sim} \frac{\xi t^{3/2}}{3\pi}\quad (a=\frac{1}{2}). \tag{8.104}$$

Noting from (8.95) that

$$p_2^{O^\pm}((0,s);\xi) = -\frac{1}{\xi}\frac{d}{ds}E^{O^\pm}((0,s);\xi) \tag{8.105}$$

we want to use (8.102) to compute

$$p_2^{O^\pm}(n;s) = \frac{(-1)^n}{n!}\frac{\partial^n}{\partial \xi^n} p_2^{O^\pm}((0,s);\xi)\Big|_{\xi=1}. \tag{8.106}$$

This will be done by computing power series solutions of (8.103) about the origin.

For this purpose, we begin by noting that a crucial feature of the boundary conditions (8.104) is that (8.103) generates unique power series expansions of $v(t)$ in terms of $t^{1/2}$,

$$v(t) = ct^{a+1} + \sum_{p=0}^\infty c_p t^{a+(p+3)/2}. \tag{8.107}$$

In particular direct substitution of (8.107) (with a specified) into (8.103) allows a general coefficient c_p to be expressed in terms of a polynomial function of $\{c, c_1, \dots, c_{p-1}\}$, which in turn can be written as an explicit system of equations. These equations are well suited to solving via computer algebra, and in this way we can obtain several hundred coefficients, all of which are polynomials in ξ. It turns out that for $a = -\frac{1}{2}$, c_p is of degree $p+2$ in ξ, while for $a = \frac{1}{2}$, c_p is of degree $p/3$ (approximately) in ξ. For this reason it is more difficult to compute c_p in the former case than the latter, and 436 and 700 coefficients were calculated respectively. Let this number be denoted by M. We next substitute the truncated series in (8.105) and compute the first $M+2$ terms of the power series of the exponential. Working in integer arithmetic, this gives the exact form of each of these $M+2$ terms.

Theoretically the fact of $E_2^{\rm hard}((0,(\pi s)^2);\xi)$ being a τ-function implies it is an entire function of s. Because a large number of terms in its power series expansion have been computed, high precision computation of $p_2^{O^\pm}(n;s)$ is possible. We want to use this to compute the moments $\int_0^\infty s^q p_2^{O^\pm}(n;s)\,ds$. Since $p_2^{O^\pm}(n;s)$ is given as a truncated power series, we compute term by term $\int_0^\alpha s^q p_2^{O^\pm}(n;s)\,ds$, where α is a cutoff tuned to get a value for the normalization $\int_0^\infty p_2^{O^\pm}(n;s)\,ds$ as close as possible to, but not exceeding, 1. For general n the number of accurate digits can be estimated by reducing the value of the degree M in the truncation of the power series. For $p_2^{O^+}(n;s)$ with $n > 5$ and $p_2^{O^-}(n;s)$ with $n > 4$ the power series are not large enough to get useful information. For smaller values of n, high precision evaluations of the moments are obtained. From these we compute for the mean, variance and higher order statistical quantities

$$\gamma_1 := \frac{\langle (X - \langle X\rangle)^3\rangle}{({\rm Var}\,X)^{3/2}}, \qquad \gamma_2 := \frac{\langle (X - \langle X\rangle)^4\rangle}{({\rm Var}\,X)^{2}} - 3, \tag{8.108}$$

referred to as the skewness and (excess) kurtosis, respectively, the values as listed in Table 8.12.

Table 8.12 has been extended to values of n beyond those accessible using the power series. This has been possible due to a computational procedure based on the Fredholm determinant formula (9.81) below (see Section 9.2 for the required methodology).

8.3.5 Spectrum singularity and bulk

The circular Jacobi average of Proposition 8.3.1 is expected to have a well-defined scaled limit after the replacement $\phi \mapsto X/N$ and multiplication by a suitable constant (i.e., term independent of X). To avoid this latter problem, one considers the scaled logarithmic derivative

$$u(X;\omega,\mu;\xi) := \lim_{N\to\infty} X \frac{d}{dX} \log E_{N,2}^{(\mu)}\Big(\Big(\pi - \frac{X}{N}, \pi\Big); \xi; |1 + e^{i\theta}|^{2\omega}\Big). \tag{8.109}$$

It follows from (8.72) that

$$\begin{aligned} &-2\frac{d}{d\phi} \log E_{N,2}^{(\mu)}\Big((\pi - \phi, \pi); \xi; |1 + e^{2\pi i\theta}|^{2\omega}\Big) \\ &\quad = U_N^{\rm Cy}(\cot\phi; N + \omega + \mu/2; \tilde{\xi}) - (e_2[-v_1, -v_2, v_3] + N\mu)\cot\phi, \end{aligned}$$

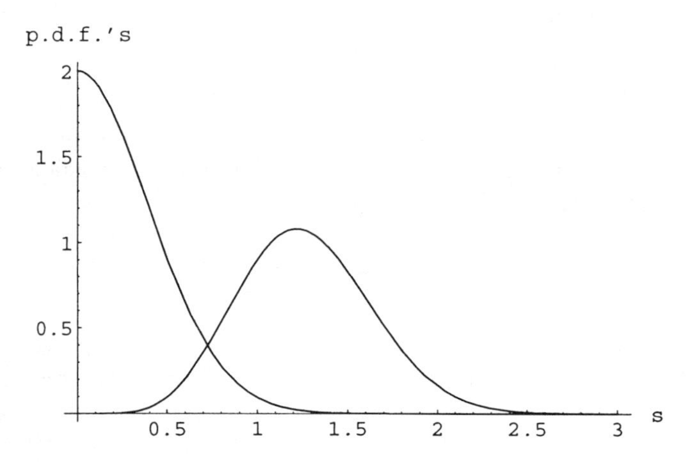

Figure 8.2 Plot of $p_2^{O^-}(n;t)$ for $n = 0$ (leftmost curve) and $n = 1$.

p.d.f.	mean	variance	skewness	kurtosis
$p_2^{O^-}(0;s)$	0.32138 26639	0.06016 64291	1.03522 47759	1.01497 61256
$p_2^{O^-}(1;s)$	1.27026 53928	0.13347 95698	0.27104 49106	−0.03359 26930
$p_2^{O^-}(2;s)$	2.26131 89835	0.16459 27733	0.14246 20157	−0.03907 67354
$p_2^{O^-}(3;s)$	3.25782 76925	0.18375 20687	0.09485 95413	−0.03218 00266
$p_2^{O^-}(4;s)$	4.25597 96097	0.19759 69896	0.07058 23711	−0.02733 17251
$p_2^{O^-}(5;s)$	5.25483 70464	0.20844 38843	0.05597 43856	−0.02404 98007
$p_2^{O^-}(6;s)$	6.25406 10064	0.21736 42245	0.04625 77696	−0.02171 86929
$p_2^{O^-}(7;s)$	7.25349 95584	0.22494 12584	0.03934 55728	−0.01998 19739
$p_2^{O^-}(8;s)$	8.25307 45177	0.23152 77403	0.03418 57830	−0.01863 62447
$p_2^{O^-}(9;s)$	9.25274 15598	0.23735 33656	0.03019 20747	−0.01756 00843
$p_2^{O^+}(0;s)$	0.78271 57582	0.10658 96782	0.45629 16499	0.05534 84703
$p_2^{O^+}(1;s)$	1.76454 67894	0.15135 50698	0.18800 06679	−0.04163 48906
$p_2^{O^+}(2;s)$	2.75925 65680	0.17507 73887	0.11409 64039	−0.03542 19954
$p_2^{O^+}(3;s)$	3.75678 01658	0.19114 91594	0.08100 77394	−0.02951 00917
$p_2^{O^+}(4;s)$	4.75534 80131	0.20331 19066	0.06246 53950	−02554 06905
$p_2^{O^+}(5;s)$	5.75441 51870	0.21310 11204	0.05067 01710	−0.02279 24207
$p_2^{O^+}(6;s)$	6.75375 94329	0.22129 48617	0.04253 19349	−0.02079 12811
$p_2^{O^+}(7;s)$	7.75327 32947	0.22834 18447	0.03659 05822	−0.01926 91981
$p_2^{O^+}(8;s)$	8.75289 85058	0.23452 44998	0.03206 90553	−0.01807 00081
$p_2^{O^+}(9;s)$	9.75260 07410	0.24003 21874	0.02851 66965	−0.01709 81199

Table 8.12 Statistical properties of $p_2^{O^\mp}(n;t)$ for various n. In the cases of $p_2^{O^-}(n;t)$, $n \geq 2$, and $p_2^{O^+}(n;t)$, $n \geq 3$, these are from [79].

where $U_N^{\rm Cy}(t;\alpha,\mu;\xi)$ satisfies the $\tilde{\sigma}$PVI equation (8.21) with parameters as noted in Table 8.11. Changing variables and equating terms of leading order in N shows that (8.109) satisfies the σPV equation [230].

PROPOSITION 8.3.4 *We have that*

$$h(t) := u(it;\omega,\mu;\xi) + \omega\mu$$

satisfies the σPV equation in (8.15) with

$$\nu_0 = 0,\ \ \nu_1 = \mu,\ \ \nu_2 = \omega + \mu/2,\ \ \nu_3 = -\omega + \mu/2,\ \ h(-it)|_{\mu=0} \underset{t\to 0^+}{\sim} -\xi t \rho_{(1)}^{\rm s.s.}(t).$$

Furthermore

$$\begin{aligned} E_2^{\rm s.s.}((0,t);\xi) &= \exp \int_0^{2\pi t} h(-is)|_{\mu=0}\, \frac{ds}{s}, \\ p_2^{\rm s.s.}((0,t);\xi) &= (2\pi t)^{2\omega} \frac{\Gamma^2(\omega+1)}{\Gamma(2\omega+1)\Gamma(2\omega+2)} \exp \int_0^{2\pi t} \Big(h(-is)|_{\mu=2} - 2\omega \Big) \frac{ds}{s}, \end{aligned} \tag{8.110}$$

where $h(u)|_{\mu=0}$ and $h(u)|_{\mu=2}$ are related by

$$h(u)|_{\mu=2} = -1 + h(u)|_{\mu=0} + u \frac{h'(u)|_{\mu=0}}{h(u)|_{\mu=0}}.$$

Proof. The formulas (8.110) follow from (8.109) and the respective finite N formulas in Proposition 8.3.1. The boundary condition for $\mu = 0$ follows from the general formula (8.74) and the inter-relation from the fact that

$$p_2^{\rm s.s.}((0,t);\xi) = -\frac{1}{\xi}\frac{d}{dt} E_2^{\rm s.s.}((0,t);\xi).$$

□

From the definitions

$$E_2^{\rm s.s.}(n;(0,t))\Big|_{\omega=0} = E_2^{\rm bulk}(n;t), \qquad p_2^{\rm s.s.}(n;(0,t))\Big|_{\omega=1} = p_2^{\rm bulk}(n;t),$$

where the bulk state has $\rho = 1$. Writing $\sigma(s;\xi) := h(-2is)|_{\mu=0}$, $u(s;\xi) := h(-is)|_{\mu=2} - 2$, it follows from Proposition 8.3.4 that [324], [233]

$$\begin{aligned} E_2^{\rm bulk}((0,t);\xi) &= \exp \int_0^{\pi t} \frac{\sigma(s;\xi)}{s}\, ds, \\ p_2^{\rm bulk}((0,t);\xi) &= \frac{\pi^2}{3} t^2 \exp \int_0^{2\pi t} u(s;\xi)\, \frac{ds}{s} \end{aligned} \tag{8.111}$$

where σ and u satisfy the equations

$$\begin{aligned} &(s\sigma'')^2 + 4(s\sigma' - \sigma)\Big(s\sigma' - \sigma + (\sigma')^2\Big) = 0, \\ &(su'')^2 + (u - su')(u - su' + 4 - 4(u')^2) - 16(u')^2 = 0 \end{aligned} \tag{8.112}$$

with boundary conditions

$$\begin{aligned} \sigma(s;\xi) &\underset{s\to 0}{\sim} -\frac{\xi}{\pi}s - \frac{\xi^2}{\pi^2}s^2, \\ u(s;\xi) &\underset{s\to 0}{\sim} -\frac{1}{15}s^2 + {\rm O}(s^4) - \frac{\xi}{8640\pi}\Big(s^5 + {\rm O}(s^7)\Big). \end{aligned} \tag{8.113}$$

From (8.111)–(8.113), proceeding as in the derivation of (8.79) we obtain the power series expansion for the

p.d.f.	mean	variance	skewness	kurtosis
$p_2^{\rm bulk}(0;s)$	1	0.17999 38776	0.49706 36204	0.12669 98480
$p_2^{\rm bulk}(1;s)$	2	0.24897 77536	0.24167 43158	−0.01494 23984
$p_2^{\rm bulk}(2;s)$	3	0.29016 98290	0.15542 00591	−0.02317 40428
$p_2^{\rm bulk}(3;s)$	4	0.31944 35563	0.11334 61773	−0.02150 23114
$p_2^{\rm bulk}(4;s)$	5	0.34214 08054	0.08871 43069	−0.01914 18388
$p_2^{\rm bulk}(5;s)$	6	0.36067 45961	0.07263 43907	−0.01714 28515
$p_2^{\rm bulk}(5;s)$	7	0.37633 63928	0.06135 08835	−0.01555 25979
$p_2^{\rm bulk}(7;s)$	8	0.38989 74631	0.05301 56552	−0.01428 79010
$p_2^{\rm bulk}(8;s)$	9	0.40185 51105	0.04661 73337	−0.01326 81121
$p_2^{\rm bulk}(9;s)$	10	0.41254 86854	0.04155 73856	−0.01243 20513

Table 8.13 Statistical properties of $p_2^{\rm bulk}(n;s)$ for various n. In the cases $n \ge 3$ these are from [79].

generating function of (8.111)

$$E_2^{\rm bulk}((0,t);\xi) = 1 - \xi t + \frac{\xi^2\pi^2t^4}{36} - \frac{\xi^2\pi^4t^6}{675} + \frac{\xi^2\pi^6t^8}{17640} - \frac{\xi^3\pi^6t^9}{291600} - \frac{\xi^2\pi^8t^{10}}{637875} + \frac{\xi^3\pi^8t^{11}}{4961250} + \cdots \quad (8.114)$$

which we observe is consistent with (8.79). Consequently, making use also of (8.17), and reinstating a general density ρ, we have

$$\begin{aligned}\frac{1}{\rho}p_2^{\rm bulk}(0;t) &= \frac{1}{\rho^2}\frac{d^2E_2^{\rm bulk}(t;\xi)|_{\xi=1}}{dt^2} \\ &= \frac{\pi^2(\rho t)^2}{3} - \frac{2\pi^4(\rho t)^4}{45} + \frac{\pi^6(\rho t)^6}{315} - \frac{\pi^6(\rho t)^7}{4050} - \frac{2\pi^8(\rho t)^8}{14175} + \frac{11\pi^8(\rho t)^9}{496125} + \cdots, \\ \frac{1}{\rho}p_2^{\rm bulk}(1;t) &= \frac{1}{\rho^2}\Big(2\frac{d^2E_2^{\rm bulk}(t;\xi)|_{\xi=1}}{dt^2} - \frac{d^2}{dt^2}\Big(\frac{\partial}{\partial\xi}E_2^{\rm bulk}(t;\xi)\Big)\Big|_{\xi=1}\Big) \\ &= \frac{\pi^6(\rho t)^7}{4050} - \frac{11\pi^8(\rho t)^9}{496125} + \cdots \end{aligned} \quad (8.115)$$

The numerical evaluation of $p_2^{\rm bulk}(n;t)$ for small n and the corresponding moments can be carried out using the expression for $p_2^{\rm bulk}((0,t);\xi)$ in (8.111) according to the method detailed in Section 8.3.4. Results of such a calculation, obtained from computing 700 terms in the power series expansion of $E_2^{\rm bulk}((0,t);\xi)$, are given in Table 8.13. As to be discussed in Section 9.2, for general $n \ge 0$ a numerical procedure based on the Fredholm determinant formula (11.26) can be used to provide accurate determination of the $p_2^{\rm bulk}(n;t)$ [79]. This has allowed our table to extended beyond the cases accessible from the power series.

EXERCISES 8.3 1. [230] In this exercise averages over the classical group will be related to averages over the JUE, and so given evaluations in terms of Painlevé transcedents.

(i) Show from the definitions that

$$E_{N,2}^{(\mu)}((t,\infty);\xi;x^ae^{-x}) = (1-\xi)^N E_{N,2}^{(\mu)}\Big((0,t);\frac{\xi}{\xi-1};x^ae^{-x}\Big). \quad (8.116)$$

Conclude from Proposition 8.3.1 that

$$V_N^{\rm L}(t;a,\mu;\xi) := t\frac{d}{dt}\log E_{N,2}^{(\mu)}((t,\infty);\xi;x^ae^{-x}) - \mu N,$$

like $U_N^{\rm L}$, satisfies the σPV equation in (8.15) with $\nu_0 = 0$, $\nu_1 = -\mu$, $\nu_2 = N + a$, $\nu_3 = N$.

(ii) Show from the definitions that

$$E_{N,2}^{(\mu)}((t,\infty);1;x^a e^{-x}) \propto t^{(a+\mu)N+N^2} \Big\langle e^{-t\sum_{j=1}^N x_j} \Big\rangle_{\mathrm{JUE}_N|_{b\mapsto\mu}},$$

where here JUE_N refers to the ensemble (8.6) with weight $x^a(1-x)^b$, $0<x<1$. Use the result of (i) to then obtain the evaluation

$$\Big\langle e^{-t\sum_{j=1}^N x_j} \Big\rangle_{\mathrm{JUE}_N|_{b\mapsto\mu}} = \exp \int_0^t \frac{V_N^{\mathrm{L}}(s,a,\mu;1) - aN - N^2}{s}\, ds$$

and deduce from Proposition 8.2.6 that there is a solution of the PV system such that

$$\Big\langle e^{-t\sum_{j=1}^N x_j} \Big\rangle_{\mathrm{JUE}_N|_{b\mapsto\mu}} = \tau_V(t)\Big|_{\substack{\nu_0=-\mu,\nu_1=0\\ \nu_2=a+N,\nu_3=N}} \qquad (\nu_{j-1} = v_j - v_2).$$

(iii) With (N^*,a,b) specified as in Proposition 3.7.1 note from the results therein that for $G = O^+(N), O^-(N)$ or $\mathrm{Sp}(2N)$,

$$\langle e^{t\mathrm{Tr}(\mathrm{U})}\rangle_{\mathrm{U}\in G} = e^{-2tN^*} e^{\chi_{N^*}t} \Big\langle e^{4t\sum_{j=1}^{N^*}\lambda_j} \Big\rangle_{\mathrm{JUE}_{N^*}}, \tag{8.117}$$

where $\chi_{N^*}=0$ for $G=\mathrm{Sp}(2N)$ and $O^+(N)$, $O^-(N)$ (N even), while $\chi_{N^*}=\pm1$ for $G=O^+(N)$, $O^-(N)$ (N odd), respectively. Use the result of (ii) to deduce from this that

$$e^{\chi_{N^*}t}\langle e^{-t\mathrm{Tr}(\mathrm{U})}\rangle_{\mathrm{U}\in G} = e^{2tN^*} \exp \int_0^{4t} \frac{V_{N^*}^{\mathrm{L}}(s;a,b;1) - aN^* - N^{*2}}{s}\, ds,$$

where $V_{N^*}^{\mathrm{L}}$ is specified as in (i) together with the boundary condition

$$V_N^{\mathrm{L}}(t;a,b;1) \underset{t\to0^+}{\sim} aN + N^2 + \mathrm{O}(t).$$

(iv) Use Proposition 3.9.1 and Carlson's theorem, Proposition 4.1.4, to deduce that

$$\Big\langle e^{t\sum_{j=1}^N x_j} \Big\rangle_{\mathrm{JUE}_N}\Big|_{b\mapsto\mu} \propto \Big\langle \prod_{l=1}^N e^{\pi i x_l(a'-b')} |1+e^{2\pi i x_l}|^{a'+b'} e^{-te^{2\pi i x_l}} \Big\rangle_{\mathrm{CUE}_N}, \tag{8.118}$$

where $a' = N+a+\mu$, $b' = -(N+a)$ and thus from the result of (ii) conclude

$$\Big\langle \prod_{l=1}^N e^{\pi i x_l(a'-b')} |1+e^{2\pi i x_l}|^{a'+b'} e^{te^{2\pi i x_l}} \Big\rangle_{\mathrm{CUE}_N} = \frac{M_N(a',b',1)}{M_N(0,0,1)} \exp \int_0^t \frac{V_N^{\mathrm{L}}(s,a,\mu;1) - aN - N^2}{s}\, ds \Big|_{\substack{a=-(N+b'),\\ \mu=a'+b'}}$$

where M_N denotes the Morris integral.

(v) For $a',b' \in \mathbb{Z}_{\ge0}$, use the identity

$$z_l^{(a'-b')/2}|1+z_l|^{a'+b'} = (1+z_l)^{a'}(1+1/z_l)^{b'}, \qquad z_l := e^{2\pi i x_l}$$

(which has implicitly been used previously in going from (3.120) to (3.121)) to deduce that the CUE_N average in (iv) is a polynomial of degree $b'N$ in t, and combine (i) and (iv) to deduce that

$$t\frac{d}{dt}\log t^{-b'N} \Big\langle \prod_{l=1}^N e^{\pi i x_l(a'-b')} |1+e^{2\pi i x_l}|^{a'+b'} e^{te^{-2\pi i x_l}} \Big\rangle_{\mathrm{CUE}_N}$$

satisfies the σPV equation in (8.15) with $\{\nu_0,\nu_1,\nu_2,\nu_3\} = \{0, a'+b', b', -N\}$ (this also requires using the fact that if $\sigma(t)$ satisfies the σPV equation with parameters $\{\nu_0, \nu_1, \nu_2, \nu_3\}$, then $\sigma(-t)$ satisfies the same equation with parameters $\{-\nu_0, -\nu_1, -\nu_2, -\nu_3\}$). Show that this result could also be deduced from the result of Exercises 8.2 q.3(iii). Note from the definitions that with $\mu \in \mathbb{Z}_{\geq 0}$, $J = (t, \infty)$ or $(0, t)$, $E_{N,2}^{(\mu)}(J, 0; x^a e^{-x})$ is a polynomial of degree μN. Recalling too the characterization of $E_{N,2}^{(\mu)}$ from (i), combine these facts to conclude that for $\mu \in \mathbb{Z}_{\geq 0}$

$$E_{N,2}^{(\mu)}(J, 0; x^a e^{-x}) \propto \Big\langle \prod_{l=1}^{\mu} (1 + z_l)^a (1 + 1/z_l)^N e^{-tz_l} \Big\rangle_{\mathrm{CUE}_\mu} \tag{8.119}$$

(note the dual role played by N and μ on the two sides of this identity).

2. [233] The aim of this exercise is to derive a duality formula between certain averages over the CUE.

   (i) Show that analogous to (8.116),

$$E_{N,2}^{(\mu)}((t,1); \xi; x^a(1-x)^b) = (1-\xi)^N (-1)^{\mu N} E_{N,2}^{(\mu)}\Big((0,t), \frac{\xi}{\xi - 1}; x^a(1-x)^b\Big).$$

   Conclude from Proposition 8.3.1 that, like $U_N^{\mathrm{J}}(t; a, b, \mu; \xi)$,

$$V_N^{\mathrm{J}}(t; a, b, \mu; \xi) := e_2^{(1)} t - \frac{1}{2} e_2^{(2)} + t(t-1) \frac{d}{dt} \log E_{N,2}^{(\mu)}((t,1); \xi; x^a(1-x)^b),$$

   where $e_2^{(1)} := e_2[-v_1, -v_2, v_3]$ and $e_2^{(2)} := e_2[-v_1, -v_2, v_3, v_4]$, satisfies the σPVI equation with

$$v_1 = \frac{1}{2}(a+b) + N + \mu, \quad v_2 = \frac{1}{2}(a+b), \quad v_3 = \frac{1}{2}(a+b) + N, \quad v_4 = \frac{1}{2}(b-a).$$

   (ii) Use a simple change of variables to show

$$E_{N,2}^{(\mu)}((t,1); \xi = 1; x^a(1-x)^b) = t^{N(a+\mu+N)} \Big\langle \prod_{l=1}^{N} (1 - tx_l)^b \Big\rangle_{\mathrm{JUE}_N|_{b \mapsto \mu}}$$

   and note too that $N(a + \mu + N) = (v_3 - v_2)(v_1 - v_4)$ to conclude from (i) that

$$\Big\langle \prod_{l=1}^{N} (1 - tx_l)^b \Big\rangle_{\mathrm{JUE}_N|_{b \mapsto \mu}}$$
$$= \exp \int_0^t \Big((v_3 - v_2)(v_4 - v_1)(t-1) - e_2^{(1)} t + \frac{1}{2} e_2^{(2)} + V_N^{\mathrm{J}}(t; a, b, \mu; \xi = 1) \Big) \frac{dt}{t(t-1)},$$

   where JUE_N is defined on $0 < x < 1$ as in q.1(iii) above.

   (iii) Let $\hat{v} = (v_3, v_4, -v_2, -v_1)$. Note that the σPVI equation is unchanged by the mapping $v \mapsto \hat{v}$. Show too that

$$e_2^{(1)} t - \frac{1}{2} e_2^{(2)} - (t-1)(v_3 - v_2)(v_1 - v_4) = \hat{e}_2^{(1)} t - \frac{1}{2} \hat{e}_2^{(2)}$$

   and thus conclude that the result of (ii) can be rewritten

$$\Big\langle \prod_{l=1}^{N} (1 - tx_l)^b \Big\rangle_{\mathrm{JUE}_N|_{b \mapsto \mu}} = \exp \int_0^t \Big(- \hat{e}_2^{(1)} t + \frac{1}{2} \hat{e}_2^{(2)} + \hat{V}_N(t; a, b, \mu; \xi = 1) \Big) \frac{dt}{t(t-1)},$$

   where $\hat{V}_N$ satisfies the same σPVI equation as V_N.

   (iv) As in the derivation of (8.118), use Proposition 3.9.1 and Carlson's theorem to show

$$\Big\langle \prod_{l=1}^{N} (1 - tx_l)^\mu \Big\rangle_{\mathrm{JUE}_N} = \frac{M_N(0,0,1)}{M_N(a', b', 1)} \Big\langle \prod_{l=1}^{N} z_l^{(a'-b')/2} |1 + z_l|^{a'+b'} (1 + tz_l)^\mu \Big\rangle_{\mathrm{CUE}_N},$$

where $a' = N + a + \mu$, $b' = -(N+a)$. Use (iii) to conclude

$$\Big\langle \prod_{l=1}^N z_l^{(a'-b')/2} |1+z_l|^{a'+b'} (1+tz_l)^\mu \Big\rangle_{\mathrm{CUE}_N} = \frac{M_N(a',b',1)}{M_N(0,0,1)} \exp \int_0^t \Big(-e_2^{(1)}t + \frac{1}{2}e_2^{(2)} + V_N^{\mathrm{J}}(t; -N-b', \mu, a'+b'; 1) \Big) \frac{dt}{t(t-1)},$$

where here V_N^{J} satisfies the σPVI equation with

$$v_1 = a' + \frac{N+b'+\mu}{2}, \quad v_2 = -\frac{N+b'-\mu}{2}, \quad v_3 = \frac{N-b'+\mu}{2}, \quad v_4 = \frac{N+b'+\mu}{2}$$

and $e_2^{(1)}, e_2^{(2)}$ relate to these vs. Note too from Proposition 8.2.6 that this CUE_N is a τ-function for the same PVI system.

(v) Note that under the mappings

$$\mu \leftrightarrow N, \quad a' \mapsto a' + b', \quad b' \mapsto -a'$$

the v parameters in (iv) map to $\bar{v} := (v_2, v_1, v_4, -v_3)$, while under this mapping together with the mapping $t \mapsto 1-t$ the σPVI equation is unchanged. Noting too that

$$\bar{e}_2^{(1)}(1-t) - \frac{1}{2}\bar{e}_2^{(2)} = -\Big(e_2^{(1)}t - \frac{1}{2}e_2^{(2)}\Big),$$

argue as in the derivation of (8.119) to conclude that for $\mu \in \mathbb{Z}_{\ge 0}$

$$\begin{aligned} \Big\langle \prod_{l=1}^N z_l^{(a'-b')/2} |1+z_l|^{a'+b'} (1+tz_l)^\mu \Big\rangle_{\mathrm{CUE}_N} &\propto \Big\langle \prod_{l=1}^\mu z_l^{(a'+2b')/2} |1+z_l|^{a'} (1+(1-t)z_l)^N \Big\rangle_{\mathrm{CUE}_\mu} \\ &\propto \Big\langle \prod_{l=1}^\mu (1-(1-t)x_l)^N \Big\rangle_{\mathrm{JUE}_\mu} \Big|_{\substack{a=b'-\mu \\ b=a'}} . \end{aligned} \tag{8.120}$$

3. [154] According to (8.98) the p.d.f. for the scaled smallest eigenvalue $\tilde{\lambda}_{\min} = \lambda_{\min}/4N$ for matrices $\mathbf{X}^\dagger \mathbf{X}$, with $\mathbf{X}$ an $N \times N$ square complex Gaussian, N large, takes the simple form $\frac{1}{4}e^{-t/4}$. Use (3.12), together with the subsequent discussion, to note that for $N \to \infty$ the scaled condition number $\tilde{\kappa}(\mathbf{X}) := \kappa(\mathbf{X})/4N$ is such that

$$\tilde{\kappa}(\mathbf{X}) = (1/\tilde{\kappa}_{\min})^{1/2}. \tag{8.121}$$

Hence conclude that the p.d.f. of $\tilde{\kappa}(\mathbf{X})$ is equal to

$$\frac{e^{-1/4y^2}}{2y^3}, \quad y > 0.$$

8.4 THE CASES $\beta = 1$ AND 4 — CIRCULAR ENSEMBLES AND BULK

8.4.1 Inter-relationships between gap probabilities with unitary symmetry

In general the generating function $E_{N,2}^{(0)}(J;\xi;g)$, in the case the weight $g(x)$ is even and interval J is symmetrical about the origin, can be written as a product of generating functions with different J, g and N for which the new g have support for $x > 0$ [197]. In the case of the Cauchy ensemble related to the CUE via a stereographic projection, this leads to a factorization in terms of generating functions for gap probabilities in ensembles of real orthogonal matrices. In the bulk limit this same factorization gives the generating function for the bulk gap probability as the product of two generating functions for the gap probability at the hard edge with parameters $a = \pm\frac{1}{2}$. We will see subsequently that these factorizations are intimately related to inter-relationships between gap probabilities in the case of unitary symmetry, with gap probabilities in the cases of orthogonal and symplectic symmetry.

PROPOSITION 8.4.1 *Let*

$$J_{\{t_{2i},t_{2i+1}\}} := \Big(\bigcup_{i=0}^{p-1}(t_{2i},t_{2i+1})\Big)\cup\Big(\bigcup_{i=0}^{p-1}(-t_{2i+1},-t_{2i})\Big) \tag{8.122}$$

with $t_{2i+1} > t_{2i} \ge 0$ and each (t_{2i},t_{2i+1}) disjoint so that $J_{\{t_{2i},t_{2i+1}\}}$ consists of p disjoint intervals on the positive half line and these same intervals reflected through the origin onto the negative half line. Suppose also that $g(x)$ is even. Then we have

$$E_{N,2}^{(0)}(J_{\{t_{2i},t_{2i+1}\}};\xi;g(x)) = E_{[(N+1)/2],2}^{(0)}(J^+_{\{t_{2i}^2,t_{2i+1}^2\}};\xi;x^{-1/2}g(x^{1/2})\chi_{x>0}) \times E_{[N/2],2}^{(0)}(J^+_{\{t_{2i}^2,t_{2i+1}^2\}};\xi;x^{1/2}g(x^{1/2})\chi_{x>0}), \tag{8.123}$$

where $J^+_{\{t_{2i}^2,t_{2i+1}^2\}} := \cup_{i=0}^{p-1}(t_{2i}^2,t_{2i+1}^2)$.

Proof. By definition

$$E_{N,2}^{(0)}(J_{\{t_{2i},t_{2i+1}\}};\xi;g(x)) = \frac{1}{C}\Big(\int_{-\infty}^{\infty} - \xi\int_{J_{\{t_{2i},t_{2i+1}\}}}\Big)dx_1\cdots\Big(\int_{-\infty}^{\infty} - \xi\int_{J_{\{t_{2i},t_{2i+1}\}}}\Big)dx_N \times\prod_{l=1}^N g(x_l)\prod_{1\le j<k\le N}|x_k-x_j|^2, \tag{8.124}$$

where C is independent of ξ and such that $E_{N,2}^{(0)}(J_{\{t_{2i},t_{2i+1}\}};\xi;g(x))|_{\xi=0} = 1$. Using (5.75) allows this to be rewritten

$$E_{N,2}^{(0)}(J_{\{t_{2i},t_{2i+1}\}};\xi;g(x)) = \frac{N!}{C}\det\Big[\Big(\int_{-\infty}^{\infty} - \xi\int_{J_{\{t_{2i},t_{2i+1}\}}}\Big)g(x)x^{j+k}\,dx\Big]_{j,k=0,\dots,N-1}. \tag{8.125}$$

Due to both $J_{\{t_{2i},t_{2i+1}\}}$ and $g(x)$ being even about the origin, we see the elements in (8.125) vanish whenever j and k have different parity. For $p=1,2$, introducing

$$\begin{aligned}\mathbf{A}_p &:= \Big[\Big(\int_{-\infty}^{\infty} - \xi\int_{J_{\{t_{2i},t_{2i+1}\}}}\Big)g(x)x^{2(j+k)}\,dx\Big]_{j,k=0,\dots,N_p-1} \\ &= \Big[\Big(\int_0^{\infty} - \xi\int_{J^+_{\{t_{2i}^2,t_{2i+1}^2\}}}\Big)g(y^{1/2})y^{j+k-1/2}\,dy\Big]_{j,k=0,\dots,N_p-1},\end{aligned}$$

where $N_1 := [(N+1)/2]$, $N_2 := [N/2]$, we see by rearranging rows and columns that

$$E_{N,2}^{(0)}(J_{\{t_{2i},t_{2i+1}\}};\xi;g(x)) = \frac{N!}{C}\det\begin{bmatrix}\mathbf{A}_1 & 0\\ 0 & \mathbf{A}_2\end{bmatrix} = \frac{N!}{C}\det\mathbf{A}_1\det\mathbf{A}_2.$$

Use of (8.125) on the r.h.s, and noting that both sides must equal unity when $\xi = 0$ then implies (8.123). □

Let us consider the special case $J=(-t,t)$, $g(x)=(1+x^2)^{-N}$ of (8.123). We know from (2.51) that the l.h.s. then maps to the CUE upon a stereographic projection,

$$E_{N,2}^{(0)}\Big(\Big(-\tan\frac{\theta}{2},\tan\frac{\theta}{2}\Big);\xi;(1+x^2)^{-N}\Big) = E_{N,2}^{\rm CUE}((-\theta,\theta);\xi). \tag{8.126}$$

Applying (8.123) we therefore have [34], [197]

$$
\begin{aligned}
E_{N,2}^{\rm CUE}((-\theta,\theta);\xi) &= E_{[(N+1)/2],2}^{(0)}\Big(\Big(0,\tan^2\frac{\theta}{2}\Big);\xi;x^{-1/2}(1+x)^{-N}\chi_{x>0}\Big) \\
&\quad\times E_{[N/2],2}^{(0)}\Big(\Big(0,\tan^2\frac{\theta}{2}\Big);\xi;x^{1/2}(1+x)^{-N}\chi_{x>0}\Big) \\
&= E_{[(N+1)/2],2}^{\rm J}\Big(\Big(0,\sin^2\frac{\theta}{2}\Big);\xi\Big)\Big|_{a=-1/2,b=b^*} E_{[N/2],2}^{\rm J}\Big((0,\sin^2\frac{\theta}{2});\xi\Big)\Big|_{a=1/2,b=-b^*} \\
&= E^{O^-(2[(N+1)/2]+1)}((0,\theta);\xi)E^{O^+(2[N/2]+1)}((0,\theta);\xi),
\end{aligned}
\tag{8.127}
$$

where $b^*=\frac{1}{2}$ (N even), $b^*=-\frac{1}{2}$ (N odd), and the second equality follows from the first by a change of variables $y=x/(1+x)$, and the third equality from (8.80).

For definiteness consider the case $N\mapsto 2N$ even. Choosing the second and third equalities (8.127) reads

$$
\begin{aligned}
E_{2N,2}^{\rm CUE}((-\theta,\theta);\xi) &= E_{N,2}^{\rm J}\Big(\Big(0,\sin^2\frac{\theta}{2}\Big);\xi\Big)\Big|_{\substack{a=-1/2\\ b=1/2}} E_{N,2}^{\rm J}\Big(\Big(0,\sin^2\frac{\theta}{2}\Big);\xi\Big)\Big|_{\substack{a=1/2\\ b=-1/2}} \\
&= E_2^{O^-(2N+1)}((0,\theta);\xi)E_2^{O^+(2N+1)}((0,\theta);\xi).
\end{aligned}
\tag{8.128}
$$

(An alternative way to derive this formula is to use the identity (5.95).) We remark too that one can check that this is consistent with the power series expansions (8.79), (8.81). In relation to the first equality in (8.128), we know from Proposition 8.3.1 that $E_{N,2}^{\rm J}((0,s);\xi)$ for general a,b is characterized by the quantity $f_N^{\rm J}$ (8.75), and furthermore with $a=-b$ the parameters are $v_1=v_3=N$, $v_2=0$, $v_4=b$. With these parameters we see that the differential equation (8.76) determining $f_N^{\rm J}$ is invariant with respect to the mapping $b\mapsto -b$. Thus the $f_N^{\rm J}$ characterizing the two terms on the r.h.s. of the first equality in (8.128) satisfies the same differential equation, differing only in the boundary condition.

The generating function for the bulk gap probabilities can be obtained from (8.127) by scaling $\theta\mapsto 2\pi s/N$ and taking $N\to\infty$. Recalling from Section 7.2 that $X\mapsto X/2N^2$ gives the hard edge scaling in the Jacobi ensemble we see

$$
\begin{aligned}
E_2^{\rm bulk}((-s,s);\xi) &= E_2^{\rm hard}((0,\pi^2 s^2);\xi)\Big|_{a=-1/2} E_2^{\rm hard}((0,\pi^2 s^2);\xi)\Big|_{a=1/2} \\
&= E_2^{O^-}((0,s);\xi)E_2^{O^+}((0,s);\xi),
\end{aligned}
\tag{8.129}
$$

where on the l.h.s. the bulk density has been set to unity.

8.4.2 Gap probabilities with orthogonal symmetry and an evenness symmetry

Analogous to the definition (8.7), let us define

$$
E_{N,1}(J;\xi;f) := \Big\langle \prod_{l=1}^N (1-\xi\chi_J^{(l)})\Big\rangle_{{\rm OE}_N(f)}. \tag{8.130}
$$

For f a classical weight (6.100) and furthermore even (this then excludes the Laguerre case, and restricts the Jacobi case to $a=b$), and with $J=(-t,t)$, it is possible to relate (8.130) to an average over a matrix ensemble with unitary symmetry [197].

PROPOSITION 8.4.2 *Let $f=e^{-\tilde V_1(x)}$ be as specified above, let $e^{-2V(x)}$ correspond to the different cases of $e^{-\tilde V_1(x)}$ as specified in (5.56) and suppose N is even. We have*

$$
E_{N,1}((-t,t);\xi;e^{-\tilde V_1(x)})\Big|_{\xi=1} = E_{N/2,2}^{(0)}((0,t^2);\xi;y^{-1/2}e^{-2V(y^{1/2})}\chi_{y>0})\Big|_{\xi=1}. \tag{8.131}
$$

Proof. According to the method of integration over alternate variables (recall Section 6.3.2)

$$E_{N,1}((-t,t);\xi;f) = \frac{N!}{C}\det[A_{j,k}]^{1/2}_{j,k=1,\dots,N}, \tag{8.132}$$

where

$$A_{j,k} = \frac{1}{2}\Big(\int_{-\infty}^{\infty} - \xi\int_{-t}^{t}\Big)dx\, f(x)R_{j-1}(x)\Big(\int_{-\infty}^{\infty} - \xi\int_{-t}^{t}\Big)dy\, f(y)R_{k-1}(y)\mathrm{sgn}(y-x) \tag{8.133}$$

and C is independent of t. Here $\{R_j(x)\}_{j=0,1,\dots}$ is an arbitrary set of monic polynomials. Let us choose $R_j(x)$ even (odd) for j even (odd). Then

$$A_{2j,2k} = A_{2j-1,2k-1} = 0 \qquad \text{for } j,k = 1,\dots,N/2$$

so every alternate element in the matrix $[A_{j,k}]$ is zero. Interchanging rows and columns so that the zero elements are all in the top left and bottom right block and noting $A_{2k,2j-1} = -A_{2j-1,2k}$ shows

$$E_{N,1}((-t,t);\xi;f) = \frac{N!}{C}\det[A_{2j-1,2k}]_{j,k=1,\dots,N/2}. \tag{8.134}$$

We now make use of the assumption that f is an even classical weight, $f = e^{-\tilde V_1(x)}$. Let $\{p_j(x)\}_{j=0,1,\dots}$ denote the monic orthogonal polynomials corresponding to the even weight functions $e^{-2V(x)}$ in (5.56). Then $p_j(x)$ is even (odd) for j even (odd). Furthermore, from (6.98) and (6.99),

$$\tilde R_{2j}(x) = p_{2j}(x), \qquad \tilde R_{2j+1}(x) = -\frac{1}{\gamma_{2j}(p_{2j},p_{2j})_2}e^{\tilde V_1(x)}\frac{d}{dx}\Big(e^{-\tilde V_4(x)}p_{2j}(x)\Big), \tag{8.135}$$

where $e^{-\tilde V_4(x)}$ is specified by (6.53), have the skew orthogonality property

$$\frac{1}{2}\int_{-\infty}^{\infty} dx\, e^{-\tilde V_1(x)}\tilde R_{2j-2}(x)\int_{-\infty}^{\infty} dy\, e^{-\tilde V_1(y)}\tilde R_{2k-1}(y)\,\mathrm{sgn}(y-x) = \frac{1}{\gamma_{2k-2}}\delta_{j,k}. \tag{8.136}$$

Choosing $R_j(x) = \tilde R_j(x)$, $j = 0,1,\dots$ in (8.133), using (8.136) and integrating by parts shows

$$\begin{aligned} A_{2j-1,2k} = {} & \frac{1}{\gamma_{2k-2}}\delta_{j,k} - \frac{(2\xi-\xi^2)}{\gamma_{2k-2}(p_{2k-2},p_{2k-2})_2}\int_{-t}^{t} e^{-V_2(x)}p_{2j-2}(x)p_{2k-2}(x)\,dx \\ & + \frac{(\xi-\xi^2)e^{-\tilde V_4(t)}p_{2k-2}(t)}{\gamma_{2k-2}(p_{2k-2},p_{2k-2})_2}\int_{-t}^{t} e^{-\tilde V_1(x)}p_{2j-2}(x)dx. \end{aligned}$$

It follows from this that

$$E_{N,1}((-t,t);1;e^{-\tilde V_1(x)}) = \frac{1}{C}\det\Big[\Big(\int_{-\infty}^{\infty} - \int_{-t}^{t}\Big)e^{-V_2(x)}p_{2j-2}(x)p_{2k-2}(x)\,dx\Big]_{j,k=1,\dots,N/2} \tag{8.137}$$

for some C independent of t. Repeating the workings of the proof of Proposition 5.2.1, but in the reverse order, shows that the r.h.s. of (8.137) is equal to

$$\frac{(N/2)!}{C}\Big(\int_{-\infty}^{\infty} - \int_{-t}^{t}\Big)dx_1\, e^{-V_2(x_1)}\cdots\Big(\int_{-\infty}^{\infty} - \int_{-t}^{t}\Big)dx_{N/2}\, e^{-V_2(x_{N/2})}\prod_{1\le j<k\le N/2}(x_k^2 - x_j^2)^2.$$

Changing variables $x_j^2 = y_j$ and substituting back in (8.137) gives (8.131). □

The most interesting case of (8.131) is when $e^{-\tilde V_1(x)} = (1+x^2)^{-(N+1)/2}$. Recalling from (2.51) that this special case of the Cauchy orthogonal ensemble is related to the COE via a stereographic projection, and thus in particular

$$E_{N,1}\Big(\Big(-\tan\frac{\theta}{2},\tan\frac{\theta}{2}\Big);\xi;(1+x^2)^{-(N+1)/2}\Big)\Big|_{\xi=1} = E_{N,1}^{\rm COE}()(-\theta,\theta);\xi)\Big|_{\xi=1},$$

(8.131) then implies

$$\begin{aligned} E_{2N,1}^{\rm COE}((-\theta,\theta);\xi)\Big|_{\xi=1} &= E_{N,2}^{(0)}\Big(\Big(0,\tan^2\frac{\theta}{2}\Big);\xi;y^{-1/2}(1+y)^{-2N}\chi_{y>0}\Big)\Big|_{\xi=1} \\ &= E_{N,2}^{\rm J}\Big(0,\Big(0,\sin^2\frac{\theta}{2}\Big)\Big)\Big|_{\substack{a=-1/2\\b=1/2}} = E_2^{O^-(2N+1)}(0;(0,\theta)), \end{aligned} \tag{8.138}$$

where the second and third equalities follow as in (8.127). Thus the probability of there being no eigenvalues in the interval $(-\theta,\theta)$ for the COE — a $\beta=1$ quantity — is equal to the probability of there being no eigenvalues in $(0,\theta)$ for matrices from $O^-(2N+1)$ — a $\beta=2$ quantity. The evaluation of the latter in terms of a σPVI transcendent has been discussed in the last subsection of Section 8.3.1.

As with going from (8.127) to (8.129), we can take the bulk limit (bulk density set equal to unity) in (8.138) to obtain

$$E_1^{\rm bulk}(0;(-s,s)) = E_2^{\rm hard}(0;(0,\pi^2s^2))\Big|_{a=-1/2} = E_2^{O^-}(0;s). \tag{8.139}$$

This characterization and the results of Section 8.3.4 give the expansion

$$E_1^{\rm bulk}(0;(0,s)) = 1-\rho s+\frac{\pi^2(\rho s)^3}{36}-\frac{\pi^4(\rho s)^5}{1200}+\frac{\pi^4(\rho s)^6}{8100}+\frac{\pi^6(\rho s)^7}{70560}-\frac{\pi^6(\rho s)^8}{264600}+{\rm O}(s^9). \tag{8.140}$$

Note that this could have been obtained from the first series expansion in (8.81), by replacing $N\mapsto N/2$, $x\mapsto\rho\pi s/N$, setting $\xi=1$ and taking $N\to\infty$, in agreement with (8.138). It follows from (8.140) that

$$\begin{aligned} \frac{1}{\rho}p_1^{\rm bulk}(0;s) &= \frac{1}{\rho^2}\frac{d^2}{ds^2}E_1^{\rm bulk}(0;(0,s)) \\ &= \frac{\pi^2(\rho s)}{6}-\frac{\pi^4(\rho s)^3}{60}+\frac{\pi^4(\rho s)^4}{270}+\frac{\pi^6(\rho s)^5}{1680}-\frac{\pi^6(\rho s)^6}{4725}+{\rm O}(s^7). \end{aligned} \tag{8.141}$$

Substituting the Painlevé evaluation (8.102) of (8.139) in the first equality of (8.141) gives a Painlevé evaluation of $p_1^{\rm bulk}(0;s)$. This can be simplified if we first note from (8.93) and (8.95) that

$$\frac{d}{dt}\exp\Big(-\int_0^{(\pi t)^2}v\Big(s;-\frac{1}{2};\xi=1\Big)\frac{ds}{s}\Big) = -\exp\Big(-\int_0^{(\pi t)^2}\Big(\sigma_{III'}(s)\Big|_{\substack{\mu=2\\a=-1/2}}+\frac{3}{2}\Big)\frac{ds}{s}\Big),$$

and the following result deduced [229].

PROPOSITION 8.4.3 *We have*

$$p_1^{\rm bulk}(0;s) = \frac{2\tilde u((\pi s/2)^2)}{s}\exp\Big(-\int_0^{(\pi s/2)^2}\frac{\tilde u(t)}{t}\,dt\Big),$$

where $\tilde u$ satisfies the nonlinear equation

$$s^2(\tilde u'')^2 = (4(\tilde u')^2-\tilde u')(s\tilde u'-\tilde u)+\frac{9}{4}(\tilde u')^2-\frac{3}{2}\tilde u'+\frac{1}{4}$$

subject to the boundary condition

$$\tilde u(s)\underset{s\to0^+}{\sim}\frac{s}{3}-\frac{s^2}{45}+\frac{8s^{5/2}}{135\pi}.$$

Although this result is of theoretical interest as an exact form of the Wigner surmise (8.20), for purposes of numerical computations we can make use of the fact that the moments of $p_2^{O^-}(s;\xi)|_{\xi=1}$ are known from

p.d.f.	mean	variance	skewness	kurtosis
$p_1^{\rm bulk}(0;s)$	1	0.28553 06557	0.68718 99889	0.37123 80638
$p_1^{\rm bulk}(1;s)$	2	0.41639 36889	0.34939 68438	0.02858 27332
$p_1^{\rm bulk}(2;s)$	3	0.49745 52604	0.22741 44134	−0.01329 56588
$p_1^{\rm bulk}(3;s)$	4	0.55564 24180	0.16645 68639	−0.01994 68028
$p_1^{\rm bulk}(4;s)$	5	0.60091 83521	0.13042 07251	−0.02007 29233
$p_1^{\rm bulk}(5;s)$	6	0.63794 46245	0.10679 47124	−0.01884 07449
$p_1^{\rm bulk}(6;s)$	7	0.66925 53948	0.09018 32871	−0.01743 19487
$p_1^{\rm bulk}(7;s)$	8	0.69637 60657	0.07790 15490	−0.01613 54800
$p_1^{\rm bulk}(8;s)$	9	0.72029 45046	0.06847 07897	−0.01500 75200
$p_1^{\rm bulk}(9;s)$	10	0.74168 65573	0.06101 25387	−0.01404 07984

Table 8.14 Statistical properties of $p_1^{\rm bulk}(n;s)$ for various values of n. In the cases $n \ge 4$ these are from [79].

Section 8.3.4. Thus we see from (8.139) and the first equality in (8.141) that for $p \ge 1$

$$\int_0^\infty s^p p_1^{\rm bulk}(0;s)\,ds = p2^{p-1}\int_0^\infty s^{p-1} p_2^{O^-}(0;s)\,ds, \tag{8.142}$$

where we have set $\rho = 1$. The results of Table 8.12 then give the results listed in Table 8.14 for the case $n = 0$.

The result of Proposition 8.4.2 leading to (8.138) applies only in the case $\xi = 1$. Thus at this stage we are still to obtain results on the generating function $E_{N,1}^{\rm COE}((-\theta,\theta);\xi)$ for general ξ. A step in this direction is obtained by considering the implication of the identity (6.154) [146], [396].

PROPOSITION 8.4.4 *We have the inter-relationship*

$$\begin{aligned} E_{N,2}^{\rm CUE}(n;(-\theta,\theta)) = \sum_{l=0}^{n} \Big(E_{N,1}^{\rm COE}(2(n-l);(-\theta,\theta)) + E_{N,1}^{\rm COE}(2(n-l)-1;(-\theta,\theta))\Big) \\ \times\Big(E_{N,1}^{\rm COE}(2l;(-\theta,\theta)) + E_{N,1}^{\rm COE}(2l+1;(-\theta,\theta))\Big), \end{aligned} \tag{8.143}$$

where $E_{N,1}^{\rm COE}(-1;(-\theta,\theta)) := 0$, *and* $E_{N,p}(n;(-\theta,\theta)) = 0$ *for* $n > N$. *Equivalently, in terms of the generating functions*

$$\begin{aligned} E_{N,2}^{\rm CUE}((-\theta,\theta);\xi) &:= \sum_{n=0}^\infty (1-\xi)^n E_{N,2}^{\rm CUE}(n;(-\theta,\theta)), \\ E_{N,1}^{\pm}((-\theta,\theta);\xi) &:= \sum_{n=0}^\infty (1-\xi)^n \Big(E_{N,1}^{\rm COE}(2n;(-\theta,\theta)) + E_{N,1}^{\rm COE}(2n\pm 1;(-\theta,\theta))\Big) \end{aligned} \tag{8.144}$$

we have

$$E_{N,2}^{\rm CUE}((-\theta,\theta);\xi) = E_{N,1}^{-}((-\theta,\theta);\xi)E_{N,1}^{+}((-\theta,\theta);\xi). \tag{8.145}$$

Proof. To derive (8.143), we suppose there are n eigenvalues in the interval $(0,s)$ of the CUE, and ask what (6.154) says about the corresponding number of eigenvalues in $(-\theta,\theta)$ for the individual COEs in the ensemble COE ∪ COE. Let us suppose there are an even number $2(n-l)$ $(0 \le l \le n)$ in $(-\theta,\theta)$ from one of the individual COEs in COE ∪ COE. Because every second eigenvalue is integrated over in the operation alt, this means there must be either $2l$ or $2l \pm 1$ (these latter two possibilities occurring with probability $\frac{1}{2}$) eigenvalues from the other COE to leave n eigenvalues in $(0,s)$.

Thus we obtain the probability

$$E_{N,1}^{\rm COE}(2(n-l);(-\theta,\theta))\Big(E_{N,1}^{\rm COE}(2l;(-\theta,\theta))+\frac{1}{2}\Big(E_{N,1}^{\rm COE}(2l+1;(-\theta,\theta))+E_{N,1}^{\rm COE}(2l-1;(-\theta,\theta))\Big)\Big), \tag{8.146}$$

where $E_{N,1}^{\rm COE}(-1;(-\theta,\theta)) := 0$. Similarly, if there is an odd number $2(n-l)-1$ $(-1\le l\le n)$ in $(-\theta,\theta)$ from this same COE, there must be either $2l+1$, or $2l, 2l+2$ (these latter two possibilities both occurring with probability $\frac{1}{2}$), so we obtain the probability

$$E_{N,1}^{\rm COE}(2(n-l)-1;(-\theta,\theta))\Big(E_{N,1}^{\rm COE}(2l+1;(-\theta,\theta))+\frac{1}{2}\Big(E_{N,1}^{\rm COE}(2l;(-\theta,\theta))+E_{N,1}^{\rm COE}(2l+2;(-\theta,\theta))\Big)\Big). \tag{8.147}$$

Adding together (8.146) and (8.147) and summing over l gives (8.143). With (8.143) established, multiplying both sides by $(1-\xi)^n$ and summing over n, (8.145) follows immediately. □

In addition to the factorization (8.145), we have previously encountered the factorization (8.127). At $\xi=1$ we can see that these are in fact the very same factorizations. Thus according to (8.138) we have

$$E_{N,2}^{\rm J}((0,\sin^2\frac{\theta}{2});\xi)\Big|_{\substack{a=-b=-1/2\\ \xi=1}} = E_{2N,1}^{\rm COE}(0;(-\theta,\theta)) = E_{2N,1}^{-}((-\theta,\theta);\xi)\Big|_{\xi=1}, \tag{8.148}$$

where the second equality follows from the definition (8.144). Equivalently, the equation

$$E_{2N,1}^{\pm}((-\theta,\theta);\xi) = E_2^{O^{\pm}(2N+1)}((0,\theta);\xi) \tag{8.149}$$

holds for $\xi=1$, and is consistent with both factorizations (8.128) and (8.145) for general ξ. In fact methods based on expansions in terms of the eigenvalues and eigenvectors of underlying Fredholm integral operators (see Section 9.6.1) can be used to establish that (8.149) holds for general ξ in the case of $E_{2N,1}^{+}$. The factorization (8.128) then establishes (8.149) for $E_{2N,1}^{-}$. With $E_{2N,1}^{\pm}$ thus evaluated, we substitute in (8.144), and use this equation to deduce, with $\bar\xi := 2\xi-\xi^2$ and thus $1-\bar\xi=(1-\xi)^2$, that

$$E_{2N,1}^{\rm COE}((-\theta,\theta);\xi) = \frac{(1-\xi)E_2^{O^+(2N+1)}((-\theta,\theta);\bar\xi)+E_2^{O^-(2N+1)}((-\theta,\theta);\bar\xi)}{2-\xi}. \tag{8.150}$$

In the bulk scaling limit (bulk density unity) (8.149) reads

$$E_1^{\rm bulk\pm}((-s,s);\xi) = E_2^{O^{\pm}(2N+1)}((0,s);\xi), \tag{8.151}$$

where

$$E_1^{\rm bulk\pm}((-s,s);\xi) := \sum_{n=0}^{\infty}(1-\xi)^n\Big(E_1^{\rm bulk}(2n;(-s,s))+E_1^{\rm bulk}(2n\pm1;(-s,s))\Big),$$

while (8.150) reads

$$E_1^{\rm bulk}((-s,s);\xi) = \frac{(1-\xi)E_2^{O^+}((0,s);\bar\xi)+E_2^{O^-}((0,s);\bar\xi)}{2-\xi}. \tag{8.152}$$

This latter formula is equivalent to the equation

$$E_1^{\rm bulk}(n;2s) = (-1)^n\Big(\sum_{l=0}^{[n/2]}E_2^{O^-}(l;s) - \sum_{l=0}^{[(n-1)/2]}E_2^{O^+}(l;s)\Big). \tag{8.153}$$

Noting from (8.105) that

$$p_2^{O^{\pm}}(n;s) = -\frac{d}{ds}\sum_{p=0}^{n}E_2^{O^{\pm}}(p;s)$$

this implies the simple formula

$$\frac{d}{ds}E_1^{\rm bulk}(n;2s) = (-1)^n\Big(p_2^{O^+}([(n-1)/2];s) - p_2^{O^-}([n/2];s)\Big).$$

But according to (8.2)

$$p_1^{\rm bulk}(n;s) = \frac{d^2}{ds^2}E_1^{\rm bulk}(n;s) + 2p_1^{\rm bulk}(n-1;s) - p_1^{\rm bulk}(n-2;s), \tag{8.154}$$

where $p_1^{\rm bulk}(k;s) = 0$ for $k < 0$, so in fact the following recurrence for $p_1^{\rm bulk}(n;s)$ involving $p_2^{O^\pm}$ holds.

PROPOSITION 8.4.5 *We have*

$$\begin{aligned} p_1^{\rm bulk}(n;s) = & \frac{(-1)^n}{2}\frac{d}{ds}\Big(p_2^{O^+}([(n-1)/2];s/2) - p_2^{O^-}([n/2];s/2)\Big) \\ & +2p_1^{\rm bulk}(n-1;s) - p_1^{\rm bulk}(n-2;s), \end{aligned} \tag{8.155}$$

where $p_1^{\rm bulk}(k;s) := 0$ *for* $k < 0$.

Using (8.155) and knowledge of the moments of $p_2^{O^\pm}(k;s)$ from Table 8.12 we then obtain the statistical characterizations listed in Table 8.14 in the cases $n = 1, 2, 3$.

8.4.3 A relationship between gap probabilities with orthogonal and symplectic symmetry

It turns out that knowledge of the COE gap probabilities is sufficient for the determination of the CSE gap probabilities. This is a consequence of (4.32) [397].

PROPOSITION 8.4.6 *We have*

$$E_{N,4}^{\rm CSE}(j;(-\theta,\theta)) = E_{2N,1}^{\rm COE}(2j;(-\theta,\theta)) + \frac{1}{2}\Big(E_{2N,1}^{\rm COE}(2j-1;(-\theta,\theta)) + E_{2N,1}^{\rm COE}(2j+1;(-\theta,\theta))\Big) \tag{8.156}$$

or equivalently

$$p_{N,4}^{\rm CSE}(n;(-\theta,\theta)) = p_{2N,1}^{\rm COE}(2n+1;(-\theta,\theta)). \tag{8.157}$$

Proof. Multiplying both sides of (8.156) by $n - j + 1$, summing over j from 0 to n and making use of (8.17) shows that (8.156) is equivalent to (8.157). To derive (8.157) we note that with eigenvalues fixed at $-\theta, \theta$, the only way these eigenvalues can remain unaffected in the even operation of (4.32), and furthermore this operation leave n eigenvalues inside $(-\theta,\theta)$, is that there originally be $2n+1$ eigenvalues inside $(-\theta,\theta)$ ($n+1$ of which are integrated over in the even operation). According to (4.32) the eigenvalue p.d.f. for COE_{2N} reduces to that for CSE_N upon the even operation, so (8.157) follows. □

Recalling the definition (8.149), it follows from (8.156) and (8.149) that

$$E_{N,4}^{\rm CSE}((-\theta,\theta);\xi) = \frac{1}{2}\Big(E_2^{O^+(2N+1)}((0,\theta);\xi) + E_2^{O^-(2N+1)}((0,\theta);\xi)\Big). \tag{8.158}$$

In the bulk limit ($\rho = 1$) this reads

$$E_4^{\rm bulk}((0,s);\xi) = \frac{1}{2}\Big(E_2^{O^+}((0,s);\xi) + E_2^{O^-}((0,s);\xi)\Big), \tag{8.159}$$

while the bulk limit of (8.157) gives

$$p_4^{\rm bulk}(n;s) = 2p_1^{\rm bulk}(2n+1;2s). \tag{8.160}$$

Note that the latter implies

$$\int_0^\infty s^p p_4^{\rm bulk}(n;s)\,ds = 2^{-p}\int_0^\infty s^p p_1^{\rm bulk}(2n+1;s)\,ds. \tag{8.161}$$

EXERCISES 8.4 1. [458] Let $\mathbf{U}$ be a $N \times N$ unitary matrix, and replace each entry $x+iy$, $x,y \in \mathbb{R}$ by its 2×2 real matrix representation (1.36) to obtain a $2N \times 2N$ real orthogonal matrix to be denoted $\mathrm{Re}(\mathbf{U})$.

(i) Show that if the eigenvalues of $\mathbf{U}$ are $e^{i\theta_j}$, $(j = 1,\dots,N)$, then the eigenvalues of $\mathrm{Re}(\mathbf{U})$ are $e^{\pm i\theta_j}$, $(j = 1,\dots,N)$.

(ii) With $a(e^{i\theta}) = a(e^{-i\theta})$ note that

$$\Big\langle \prod_{j=1}^{N} a(e^{i\theta_j}) \Big\rangle_{U(N)} = \Big\langle \prod_{j=1}^{N} a(e^{i\theta_j}) \Big\rangle_{\mathrm{Re}(U(N))},$$

where on the r.h.s. the average is over one of each of the eigenvalues $e^{\pm i\theta_j}$ $(j = 1,\dots,N)$. Substitute (5.95) for the l.h.s. to deduce

$$\mathrm{Ev}(\mathrm{Re}(\mathbf{U})) = \mathrm{Ev}(O^+(N+1)) \oplus \mathrm{Ev}(O^-(N+1)),$$

where Ev denotes the eigenvalue distribution of the eigenvalues $e^{i\theta}$, $0 < \theta < \pi$ of the corresponding matrix ensembles.

2. The generalized Hermite ensemble is specified by the weight function $w_2(x) = |x|^{2a} e^{-x^2}$, $x \in (-\infty, \infty)$. Under the scaling $x \mapsto x/\sqrt{2N}$ the correlations coincide with those for the spectrum singularity (7.48) with $\rho = 1/\pi$ [418], and in particular

$$\lim_{N\to\infty} E_{N,2}\Big(\Big(-\frac{t}{\sqrt{2N}}, \frac{t}{\sqrt{2N}}\Big); \xi; |x|^{2a} e^{-x^2}\Big) = E_2^{\mathrm{s.s.}}((-t,t);\xi). \tag{8.162}$$

(i) Show that for $a \neq 0$ the logarithmic derivative of the generalized Hermite weight has the structure (5.57) with degree $f = 2$, and so in general this does not define a classical weight.

(ii) Apply (8.123) with $J_{\{t_{2i},t_{2i+1}\}} = (-t,t)$, $g(x) = |x|^{2a} e^{-x^2}$ and take the scaling limit (8.162) to conclude

$$E_2^{\mathrm{s.s.}}((-t,t);\xi)\Big|_{\rho=1/\pi} = E_2^{\mathrm{hard}}((0,t^2);\xi)\Big|_{a\mapsto a-1/2} E_2^{\mathrm{hard}}((0,t^2);\xi)\Big|_{a\mapsto a+1/2}. \tag{8.163}$$

3. (i) Consider the circular β-ensemble. Show from the definitions that for $x_1,\dots,x_n$ small

$$\rho_{(n)}(x_1,\dots,x_n) \sim \frac{N!}{(N-n)!L^n}\Big(\frac{2\pi}{L}\Big)^{\beta n(n-1)/2} \prod_{1\le j<k\le n} |x_k - x_j|^\beta \frac{M_{N-n}(n\beta/2, n\beta/2, \beta/2)}{M_N(0,0,\beta/2)},$$

where M_k is specified by (4.4). Hence deduce that in the bulk scaling limit

$$\rho_{(n)}(x_1,\dots,x_n) \sim \rho^n c_n(\beta)(2\pi)^{\beta n(n-1)/2} \prod_{1\le j<k\le n} |\rho x_k - \rho x_j|^\beta,$$

where

$$c_n(\beta) = (\beta/2)^{\beta n(n-1)/2}((\beta/2)!)^n \prod_{k=0}^{n-1} \frac{\Gamma(\beta k/2+1)}{\Gamma(\beta(n+k)/2+1)}. \tag{8.164}$$

(ii) Show from the definitions that for small s

$$p_\beta^{\mathrm{bulk}}(n;s) \sim \frac{1}{\rho n!}\int_0^s dx_1 \cdots \int_0^s dx_n\, \rho_{(n+2)}(0,s,x_1,\dots,x_n)$$

and use the result of (i) to deduce from this that

$$\frac{1}{\rho} p_\beta^{\mathrm{bulk}}(n;s) \sim S_n(\beta,\beta,\beta/2) c_{n+2}(\beta)(2\pi)^{\beta(n+2)(n+1)/2}(\rho s)^{n+\beta(n+2)(n+1)/2}, \tag{8.165}$$

where S_n is specified by (4.3). Verify that the leading terms in (8.115) and (8.141) exhibit this behavior.

(iii) [1] Set $\rho = 1$. In keeping with the Wigner surmise of Exercises 8.1 q.3, and the result of the above exercise, make an ansatz for $p_\beta^{\rm bulk}(n;s)$ of the form

$$p_\beta^{\rm W}(n;s) := As^\gamma e^{-(Bs)^2}, \qquad \gamma = n + \beta(n+2)(n+1)/2.$$

Requiring that $\int_0^\infty p_\beta^{\rm W}(n;s)\,ds = 1$, $\int_0^\infty s p_\beta^{\rm W}(n;s)\,ds = n+1$, show that

$$A = \frac{2}{(n+1)^{\gamma+1}} \frac{(\Gamma(\gamma/2+1))^{\gamma+1}}{(\Gamma((\gamma+1)/2)^{\gamma+2}}, \qquad B = \frac{\Gamma(\gamma/2+1)}{(n+1)\Gamma((\gamma+1)/2)}.$$

4. [148] Use the identity (8.157) and the second sum rule in (8.18) to deduce

$$\sum_{n=0}^{N-2} p_{2N,1}^{\rm COE}(2n+1;(0,\alpha)) = \frac{1}{2\rho}\rho_{(2)}^{{\rm CSE}_N}(0,\alpha),$$

$$\sum_{n=0}^{N-1} p_{2N,1}^{\rm COE}(2n;(0,\alpha)) = \frac{1}{\rho}\Big(\rho_{(2)}^{{\rm COE}_{2N}}(0,\alpha) - \frac{1}{2}\rho_{(2)}^{{\rm CSE}_{2N}}(0,\alpha)\Big),$$

where $\rho = N/\pi$.

8.5 DISCRETE PAINLEVÉ EQUATIONS

Amongst the random matrix averages characterizable from the results of this chapter as the solution of certain Painlevé equations in σ-form are the $U(N)$ (or equivalently CUE$_N$) averages

$$\tau^{III'}[N](t;\mu) := \Big\langle \prod_{l=1}^N z_l^\mu e^{\frac{1}{2}\sqrt{t}(z_l + z_l^{-1})} \Big\rangle_{U(N)}, \tag{8.166}$$

$$\tau^{V}[N](t;\mu,\nu) := \Big\langle \prod_{l=1}^N (1+z_l)^\mu (1+1/z_l)^\nu e^{tz_l} \Big\rangle_{U(N)}, \tag{8.167}$$

$$\tau^{VI}[N](t;\mu,w_1,w_2;\xi) := \Big\langle \prod_{l=1}^N (1-\xi\chi_{(\pi-\phi,\pi)}^{(l)}) e^{w_2\theta_l} |1+z_l|^{2w_1} \Big(\frac{1}{tz_l}\Big)^\mu (1+tz_l)^{2\mu} \Big\rangle_{U(N)}, \tag{8.168}$$

where in (8.168) $t = e^{i\phi}$. It will be shown in this section that these averages have the further property of satisfying a recurrence in N. The recurrences involve auxiliary quantities which themselves satisfy so-called discrete Painlevé equations. The explicit form of the latter of interest to us are the alternate discrete Painlevé II equation [422]

$$\frac{1}{2}\frac{v_1^{(0)} + v_2^{(0)} + 2 + 2n}{q_n q_{n+1} + t} + \frac{1}{2}\frac{v_1^{(0)} + v_2^{(0)} + 2n}{q_{n-1}q_n + t} = \frac{1}{q_n} - \frac{q_n}{t} + \frac{v_2^{(0)} + n}{t}, \tag{8.169}$$

the discrete Painlevé IV recurrences [461]

$$\begin{aligned} x_n + x_{n-1} &= \frac{t}{y_n} + \frac{v_3^{(0)} - v_4^{(0)} - n}{1 - y_n}, \\ y_n y_{n+1} &= t\frac{x_n - \frac{1}{2}(v_1^{(0)} + v_2^{(0)}) + 1 + v_4^{(0)} + n}{x_n^2 - \frac{1}{4}(v_2^{(0)} - v_1^{(0)})^2}, \end{aligned} \tag{8.170}$$

and the discrete Painlevé V recurrences [478]

$$g_{n+1}g_n = \frac{t}{t-1}\frac{(f_n+1+n-\alpha_2^{(0)})(f_n+1+n-\alpha_2^{(0)}-\alpha_4^{(0)})}{f_n(f_n+\alpha_3^{(0)})},$$
$$f_n+f_{n-1} = -\alpha_3^{(0)}+\frac{\alpha_1^{(0)}+n}{g_n-1}+\frac{(\alpha_0^{(0)}+n)t}{t(g_n-1)-g_n}. \tag{8.171}$$

The alternate discrete Painlevé II equation is satisfied by

$$q_n^{III'} := T_{III'}^n q^{III'}\Big|_{\vec{v}=\vec{v}^{(0)}}, \tag{8.172}$$

the discrete Painlevé IV recurrences by

$$x_n = (p_n^{V^*}+t)q_n^{V^*} - \frac{1}{2}(v_2^{(0)}-v_1^{(0)}),$$
$$y_n = \frac{1}{q_n^{V^*}}, \tag{8.173}$$

where

$$p_n^{V^*} := T_V^n p^{V^*}\Big|_{\vec{v}=\vec{v}^{(0)}}, \qquad q_n^{V^*} := T_V^n q^{V^*}\Big|_{\vec{v}=\vec{v}^{(0)}},$$

and the discrete Painlevé V recurrences by

$$g_n := \frac{q_n^{VI}}{q_n^{VI}-1},$$
$$f_n := q_n^{VI}(q_n^{VI}-1)p_n^{VI} + (1+n-\alpha_2^{(0)}-\alpha_4^{(0)})(q_n^{VI}-1) - \alpha_3^{(0)}q_n^{VI} - (\alpha_0^{(0)}+n)\frac{q_n^{VI}(q_n^{VI}-1)}{q_n^{VI}-t}, \tag{8.174}$$

where

$$p_n^{VI} := T_{VI}^n p^{VI}\Big|_{\vec{\alpha}=\vec{\alpha}^{(0)}}, \qquad q_n^{VI} := T_{VI}^n q^{VI}\Big|_{\vec{\alpha}=\vec{\alpha}^{(0)}}.$$

It is simplest to show that (8.172) satisfies (8.169). The required strategy is common to the results relating to (8.170) and (8.171), but avoids some of the more complicated algebraic manipulation required in these latter cases.

PROPOSITION 8.5.1 *The nth shift of the general Painlevé III′ transcendent $q^{III'}|_{\vec{v}=\vec{v}^{(0)}}$ by the Schlesinger operator $T_{III'}$, $q_n^{III'}$, satisfies the alternate discrete Painlevé II equation (8.169).*

Proof. For notational convenience, let us write $q_n^{III'} =: q_n$. Also define $p_n^{III'} := T_{III'}^n p^{III'}|_{\vec{v}=\vec{v}^{(0)}}$ and write $p_n^{III'} =: p_n$. Now from the definition of $T_{III'}$ in Table 8.7, the actions in Table 8.3 can be used to express $q_{n+1} = T_{III'}q_n$, $q_{n-1} = T_{III'}^{-1}q_n$, $p_{n+1} = T_{III'}p_n$, $p_{n-1} = T_{III'}^{-1}p_n$ each in terms of p_n, q_n. Explicitly

$$q_{n+1} = -\frac{t}{q_n} + \frac{1}{2}\frac{(v_1^{(0)}+v_2^{(0)}+2+2n)t}{q_n[q_n(p_n-1)-(v_1^{(0)}-v_2^{(0)})/2]+t}, \tag{8.175}$$

$$p_{n+1} = \frac{q_n}{t}[q_n(p_n-1)-\frac{1}{2}(v_1^{(0)}-v_2^{(0)})]+1, \tag{8.176}$$

$$q_{n-1} = \frac{t}{(v_1^{(0)}+v_2^{(0)}+2n)/2p_n - q_n}, \tag{8.177}$$

Painlevé system	$F(p,q)$
PII	$\frac{\partial}{\partial t} H_{II} = -\frac{1}{2}p$
PIII$'$	$\frac{\partial}{\partial t} t H_{III'} = p$
PIV	$\frac{\partial}{\partial t} H_{IV} = -2pq$
PV*	$\frac{\partial}{\partial t}(tH_{V^*} + (v_4^{(0)} - v_2^{(0)} + n)t) = q(q-1)p + (v_2^{(0)} - v_3^{(0)})q + (v_4^{(0)} - v_2^{(0)} + n)$
PVI	$\frac{\partial}{\partial t}(t(t-1)H_{VI}) + (v_3^{(0)} - v_2^{(0)} + N)(-v_1^{(0)} - v_2^{(0)} + N)$ $= q(1-q)p^2 + 2v_3^{(0)}pq - (v_3^{(0)} + v_4^{(0)})p + (v_3^{(0)} - v_2^{(0)} + N)(-v_2^{(0)} - v_3^{(0)} + N)$

Table 8.15 Specification of $F(p,q)$ in (8.179).

while the formula for p_{n-1} is not needed. The recurrences (8.175) and (8.176) imply

$$q_{n+1} + \frac{t}{q_n} = \frac{\frac{1}{2}(v_1^{(0)} + v_2^{(0)} + 2 + 2n)}{p_{n+1}}. \tag{8.178}$$

Solving (8.177) for p_n and substituting in (8.178) gives (8.169). □

Next we will relate the variables p_n, q_n to the particular ratio of τ-functions $\tau[n+1]\tau[n-1]/\tau^2[n]$ for each of the Painlevé systems PII–PVI. For this we must decide on a normalization of $\tau[n]$ — we suppose $\tau[n]$ is given by $g[n]$ as in Table 8.10, and thus $g[n]$ satisfies the Toda lattice equation (8.51) (the key point here is that the possible proportionality constant on the r.h.s. of (8.51), permitted from the derivation, is chosen to be unity). Then from the derivation of Proposition 8.2.5 we can read off the following results.

Proposition 8.5.2 *With $\{\tau[n]\}_{n=0,1,\dots}$ for each of the Painlevé systems normalized as specified, and $\tau[n] := \tau[n](t/4)$ in the PIII$'$ case,*

$$\frac{\tau[n+1]\tau[n-1]}{\tau^2[n]} = F(p_n, q_n), \tag{8.179}$$

where $F(p,q)$ is specified by Table 8.15 for each of the Painlevé systems.

In the PIII$'$ case the appropriately normalized τ-function sequences is given by (8.68). Making use of Propositions 8.5.2 and 8.5.1 the following results are almost immediate [232].

Proposition 8.5.3 *Let $\tau^{III'}[N] = \tau^{III'}[N](t;\mu)$ as given by (8.166) for $\mu \in \mathbb{Z}$, and by*

$$\tau^{III'}[N](t;\mu) = \det[I_{\mu+j-k}(\sqrt{t})]_{j,k=1,\dots,N} \tag{8.180}$$

for general μ (recall (8.97)), and let p_N, q_N denote the variables in the corresponding Hamiltonian. The sequences $\{\tau^{III'}[N]\}_{N=0,1,\dots}$, $\{p_N\}_{N=0,1,\dots}$, $\{q_N\}_{N=0,1,\dots}$ satisfy the coupled recurrences

$$\begin{aligned}
\frac{\tau^{III'}[N+1]\tau^{III'}[N-1]}{(\tau^{III'}[N])^2}\Big|_{t\mapsto 4t} &= p_N \qquad (N=1,2,\dots), \\
p_{N+1} &= \frac{q_N^2}{t}(p_N - 1) - \frac{\mu q_N}{t} + 1 \qquad (N=0,1,\dots), \\
q_{N+1} &= -\frac{t}{q_N} + \frac{(1+N)t}{q_N(q_N(p_N-1)-\mu)+t} \qquad (N=0,1,\dots),
\end{aligned} \tag{8.181}$$

subject to the initial conditions

$$p_0 = 0, \qquad q_0 = t\frac{d}{dt}\log t^{-\mu/2} I_\mu(\sqrt{t}),$$
$$\tau^{III'}[0] = 1, \qquad \tau^{III'}[1] = I_\mu(\sqrt{t}).$$

8.6 ORTHOGONAL POLYNOMIAL APPROACH

Another approach to deriving recurrences for the averages (8.166)–(8.168) is via the theory of (bi)orthogonal polynomials on the unit circle [234], [236]. For this some aspects of the latter theory must first be developed. With $z = e^{i\theta}$ let

$$w(z) = \sum_{k=-\infty}^{\infty} w_k z^k, \qquad w_k = \frac{1}{2\pi}\int_{-\pi}^{\pi} w(z) e^{-ik\theta}\, d\theta. \tag{8.182}$$

Introduce the Toeplitz determinant, or equivalently $U(N)$ average,

$$I_N^0[w] := \det[w_{j-k}]_{j,k=1,\dots,N} = \Big\langle \prod_{l=1}^{N} w(z_l) \Big\rangle_{U(N)} \tag{8.183}$$

(for the equality recall (5.77)). If $I_N^0[w]$ is nonzero for each $N = 1, 2, \dots$, for otherwise general complex $w(z)$ one can construct a system of biorthogonal polynomials $\{\phi_n(z)\}_{n=0,1,\dots}$ and $\{\bar{\phi}_n(z)\}_{n=0,1,\dots}$ (here $\bar{\phi}$ does not in general stand for the complex conjugate of ϕ) such that

$$\frac{1}{2\pi}\int_{-\pi}^{\pi} \tilde{w}(z)\phi_m(z)\bar{\phi}_n(1/z)\, d\theta = \delta_{m,n}, \qquad \tilde{w}(z) = \frac{w(z)}{w_0}. \tag{8.184}$$

In the case that $w(z)$ is real for $|z| = 1$ ($w_k = w_{-k}$) the biorthogonal system reduces to an orthogonal system because then $\bar{\phi}_n = \phi_n$, however this will not be the case in general. Some fundamental quantities related to the biorthogonal system are identified in the following definition.

DEFINITION 8.6.1 *Introduce notation for the various coefficients in $\phi_n(z)$ and $\bar{\phi}_n(z)$ according to*

$$\phi_n(z) = \kappa_n z^n + l_n z^{n-1} + m_n z^{n-2} + \cdots + \phi_n(0),$$
$$\bar{\phi}_n(z) = \kappa_n z^n + \bar{l}_n z^{n-1} + \bar{m}_n z^{n-2} + \cdots + \bar{\phi}_n(0),$$

where $\bar{l}_n, \bar{m}_n, \bar{\phi}_n(0)$ do not in general denote the complex conjugates of $l_n, m_n, \phi_n(0)$, and κ_n is real and positive. Introduce the reciprocal polynomials by

$$\phi_n^*(z) = z^n \bar{\phi}_n(1/z), \qquad \bar{\phi}_n^*(z) = z^n \phi_n(1/z)$$

(cf. the definition of $\tilde{\chi}_k(\lambda)$ in (2.88)) and define the so called reflection coefficients

$$r_n = \frac{\phi_n(0)}{\kappa_n}, \qquad \bar{r}_n = \frac{\bar{\phi}_n(0)}{\kappa_n}.$$

Next we make note of a number of key formulas associated with the biorthogonal system and the quantities of Definition 8.6.1.

PROPOSITION 8.6.2 *We have*

$$\sum_{j=0}^{N-1} \phi_j(z)\bar{\phi}_j(\zeta) = \frac{\phi_N^*(z)\bar{\phi}_N^*(\zeta) - \phi_N(z)\bar{\phi}_N(\zeta)}{1 - z\zeta},$$
$$\left(\frac{\kappa_{N-1}}{\kappa_N}\right)^2 = 1 - r_N\bar{r}_N,$$
$$\kappa_n z\phi_n(z) = \kappa_{n+1}\phi_{n+1}(z) - \phi_{n+1}(0)\phi_{n+1}^*(z),$$
$$\kappa_n\phi_{n+1}(z) = \kappa_{n+1}z\phi_n(z) + \phi_{n+1}(0)\phi_n^*(z),$$
$$\kappa_n\phi_n(0)\phi_{n+1}(z) + \kappa_{n-1}\phi_{n+1}(0)z\phi_{n-1}(z) = (\kappa_n\phi_{n+1}(0) + \kappa_{n+1}\phi_n(0)z)\phi_n(z),$$
$$\frac{l_n}{\kappa_n} = \sum_{j=0}^{n-1} \bar{r}_j r_{j+1}. \tag{8.185}$$

Furthermore the third, fourth and fifth equations remain valid if ϕ is formally replaced by $\bar{\phi}$.

Proof. Let us refer to the equations (8.185) as (i)–(vi) in order. The summation (i) is analogous to and consistent with the circular form of the Christoffel-Darboux formula (5.82). Like that formula, it can be derived by following the strategy of Exercises 5.1 q.6. Setting $z = \zeta = 0$ in (i) implies that

$$\sum_{j=0}^{N} \phi_j(0)\bar{\phi}_j(0) = \kappa_N^2.$$

Subtracting this from its counterpart with $N \mapsto N-1$ and recalling the definitions of $r_N, \bar{r}_N$ gives (ii). With $N = n+1$, multiplication of both sides of (i) with $1 - z\zeta$ and comparison of the coefficient of ζ^{n+1} in (i) implies (iii), while its analogue with ϕ formally replaced by $\bar{\phi}$ follows by comparison of the coefficient of z^{n+1}. Replacing z by $1/z$ and recalling the various definitions this latter equation can be written

$$\kappa_n\phi_n^*(z) = \kappa_{n+1}\phi_{n+1}^*(z) - \bar{\phi}_{n+1}(0)\phi_{n+1}(z).$$

Substituting for $\phi_{n+1}^*(z)$ using (iii) and simplifying using (ii) implies (iv). Replacing n by $n-1$ in (iii) and using (iv) to eliminate $\phi_n^*(z)$ then gives (v). To derive (vi), first equate coefficients of z^n in (v). This gives

$$\kappa_n\phi_n(0)l_{n+1} + \kappa_{n-1}^2\phi_{n+1}(0) = \kappa_n^2\phi_{n+1}(0) + \kappa_{n+1}\phi_n(0)l_n.$$

Making use of (ii) it follows from this that

$$\frac{l_{n+1}}{\kappa_{n+1}} = \frac{l_n}{\kappa_n} + \bar{r}_n r_{n+1}$$

which is equivalent to (vi). □

The second and sixth equations of (8.185) can be used to derive a recurrence system for $\{\tau^{III'}[N]\}_{N=0,1,\dots}$ which is in fact equivalent to the one of Proposition 8.5.3. First we note that following the strategy of the derivation of (5.74) shows

$$I_N^0[w] = w_0^N \prod_{j=0}^{N-1} \kappa_j^{-2},$$

which allows the second equation in (8.185) to be rewritten

$$\frac{I_{N+1}^0[w]I_{N-1}^0[w]}{(I_N^0[w])^2} = 1 - r_N\bar{r}_N. \tag{8.186}$$

Now, according to (8.166), with

$$w(z) = z^{\mu}e^{\frac{1}{2}\sqrt{t}(z+z^{-1})} \tag{8.187}$$

we have

$$I_N^0[w] = \tau^{III'}[N](t;\mu) \tag{8.188}$$

so the left hand side of (8.186) coincides with the left hand side of (8.179). Thus, whereas in proceeding from (8.179) the strategy was to use the fact that p_n and q_n satisfy recurrences to determine $\{\tau^{III'}[N]\}_{N=0,1,\dots}$ by recurrence, our present task is to determine recurrences for r_N and $\bar{r}_N$.

PROPOSITION 8.6.3 *For the weight (8.187) the sequences $\{r_N\}_{N=0,1,\dots}$, $\{\bar{r}_N\}_{N=0,1,\dots}$ satisfy the coupled recurrences*

$$\frac{1}{2}\sqrt{t}(r_{N+1}\bar{r}_N + r_N\bar{r}_{N-1}) + N\frac{r_N\bar{r}_N}{1-r_N\bar{r}_N} - \mu = 0,$$
$$\frac{1}{2}\sqrt{t}(\bar{r}_{N+1}r_N + \bar{r}_N r_{N-1}) + N\frac{r_N\bar{r}_N}{1-r_N\bar{r}_N} + \mu = 0$$

subject to the initial conditions

$$r_0 = \bar{r}_0 = 1, \qquad r_1 = -\frac{I_{\mu+1}(\sqrt{t})}{I_\mu(\sqrt{t})}, \qquad \bar{r}_1 = -\frac{I_{\mu-1}(\sqrt{t})}{I_\mu(\sqrt{t})}.$$

Proof. To avoid minor complications we suppose here that μ is an integer (if μ is not an integer the contour of integration implied by (8.182) must be modified not to cross the cut along the negative real axis, about which $w(z)$ is discontinuous). Following a method of Laguerre and Freud (see [244]) we consider two different ways to evaluate the integral

$$J_1 = \frac{1}{2\pi}\int_{-\pi}^{\pi} z^2 w'(z)\phi_N(z)\bar{\phi}(1/z)\,d\theta = \int_C z^2 w'(z)\phi_N(z)\bar{\phi}_{N+1}(1/z)\frac{dz}{2\pi i z}, \tag{8.189}$$

where C is the unit circle in the complex plane. Integrating by parts gives

$$J_1 = -\int_C z w(z)\phi_N(z)\bar{\phi}_{N+1}(1/z)\frac{dz}{2\pi i z} - \int_C z^2 w(z)\phi_N'(z)\bar{\phi}_{N+1}(1/z)\frac{dz}{2\pi i z}$$
$$+ \int_C z^2 w(z)\phi_N(z)\bar{\phi}_{N+1}'(1/z)\frac{dz}{2\pi i z}.$$

But

$$z\phi_N(z) = \frac{\kappa_N}{\kappa_{N+1}}\phi_{N+1}(z) + \pi_N(z), \qquad z^2\phi_N'(z) = \frac{N\kappa_N}{\kappa_{N+1}}\phi_{N+1}(z) + \pi_N(z),$$
$$\bar{\phi}_{N+1}'(1/z) = \frac{(N+1)\kappa_{N+1}}{\kappa_N}\bar{\phi}_N(1/z) \tag{8.190}$$

for some polynomials $\pi_N(z)$ of degree N so the biorthogonality relation (8.184) implies

$$J_1 = -\frac{\kappa_N}{\kappa_{N+1}}(N+1) + \frac{\kappa_{N+1}}{\kappa_N}(N+1).$$

Alternatively, differentiating (8.187) shows (8.189) can be rewritten

$$J_1 = \int_C z^2\Big(\frac{\mu}{z} + \frac{1}{2}\sqrt{t}\Big(1 - \frac{1}{z^2}\Big)\Big)w(z)\phi_N(z)\bar{\phi}_{N+1}(1/z)\frac{dz}{2\pi i z}. \tag{8.191}$$

Using the first equation in (8.190) together with

$$z^2\phi_N(z) = \frac{\kappa_N}{\kappa_{N+2}}\phi_{N+2}(z) + \frac{1}{\kappa_{N+1}}\Big(l_N - \frac{\kappa_N l_{N+2}}{\kappa_{N+2}}\Big)\phi_{N+1}(z) + \pi_N(z),$$

it follows from the biorthogonality relation (8.184) that

$$J_1 = \mu \frac{\kappa_N}{\kappa_{N+1}} + \frac{1}{2}\sqrt{t}\Big(\frac{l_N}{\kappa_{N+1}} - \frac{l_{N+2}}{\kappa_{N+2}} \frac{\kappa_N}{\kappa_{N+1}} \Big).$$

Equating the two expressions for J_1 and eliminating l_N using the final equation of (8.185) we deduce the first recurrence. To deduce the second recurrence we apply an analogous strategy to

$$J_2 = \int_C w'(z)\phi_{N+1}(z)\bar{\phi}_N(1/z)\frac{dz}{2\pi i z}.$$

□

In the case $\mu = 0$ the weight (8.187) is real. Then $\bar{r}_n = r_n$ and the coupled recurrences of Proposition 8.6.3 reduce to the single recurrence

$$-\frac{2N}{\sqrt{t}}\frac{r_N}{1-r_N^2} = r_{N+1} + r_{N-1} \tag{8.192}$$

subject to the initial conditions

$$r_0 = 1, \qquad r_1 = -\frac{I_1(\sqrt{t})}{I_0(\sqrt{t})}.$$

Different derivations of this result can be found in [28], [82].

The relationship between $r_N, \bar{r}_N$ in Proposition 8.6.3 and p_N, q_N in Proposition 8.5.3 is easily deduced. First note that subtracting the two equations in Proposition 8.5.3 shows

$$\frac{1}{2}\sqrt{t}(L_{N+1} + L_N) + 2\mu = 0, \qquad L_N := \bar{r}_N r_{N-1} - r_N \bar{r}_{N-1}, \tag{8.193}$$

and the initial conditions therein imply this to be solved subject to the initial condition

$$L_1 = -\frac{2\mu}{\sqrt{t}}.$$

By inspection we see that the solution of (8.193) satisfying this initial condition is the constant $L_N = -2\mu/\sqrt{t}$, $N = 1, 2, \dots$ and thus

$$\frac{1}{2}\sqrt{t}(\bar{r}_{N+1} r_N - r_{N+1}\bar{r}_N) + \mu = 0. \tag{8.194}$$

PROPOSITION 8.6.4 *We have*

$$p_N = 1 - r_N \bar{r}_N \big|_{t\mapsto 4t}, \qquad q_N = -\sqrt{t}\frac{r_{N+1}}{r_N}\Big|_{t\mapsto 4t}.$$

Proof. The equation for p_N follows immediately upon comparing the first equation of (8.181) with (8.186) and recalling (8.188). Substituting this equation for p_N in the second equation of (8.181) we see that (8.194) results if q_N is as specified. □

For the PV τ-function sequence relating to (8.167), the formulas of Proposition 8.6.2 again yield coupled recurrences for $\{r_N\}_{N=0,1,\dots}$, $\{\bar{r}_N\}_{N=0,1,\dots}$, which in turn can be related to the recurrences obtained from the method of Section 8.5 [234]. In the case of the PVI τ-function sequence relating to (8.168), the formulas (8.185) are inadequate and a more sophisticated formulation based on the Riemann-Hilbert characterization of orthogonal polynomials on the unit circle is called for [236].

EXERCISES 8.6 1. (i) Follow the derivation of (5.78) to show

$$\phi_n(z) = \frac{\kappa_n}{I_n^0[w]} \det[c_{j-k}z - c_{j-k-1}]_{j,k=1,\dots,n}, \qquad \bar{\phi}_n(z) = \frac{\kappa_n}{I_n^0[w]} \det[c_{j-k}z - c_{j-k+1}]_{j,k=1,\dots,n}.$$

(ii) Set $z = 0$ in the above formulas to deduce

$$r_n = (-1)^n \frac{I_n^1[w]}{I_n^0[w]}, \qquad \bar{r}_n = (-1)^n \frac{I_n^{-1}[w]}{I_n^0[w]},$$

where $I_n^\mu[w] := I_n^0[z^\mu w]$.

2. [306] The objective of this exercise is to show that for $w(z)$ given by (8.187) with $\mu = 0$

$$\phi_n'(z) = \frac{\kappa_{n-1}}{\kappa_n}\Big(n + \frac{\sqrt{t}}{2z} + \frac{\sqrt{t}}{2}\frac{\kappa_{n-1}\phi_{n-1}(0)}{\kappa_n\phi_n(0)} - \frac{t}{2}\frac{\phi_{n+1}(0)\phi_n(0)}{\kappa_{n+1}\kappa_n}\Big)\phi_{n-1}(z) \\ - \frac{\sqrt{t}}{2z}\frac{\kappa_{n-1}}{\kappa_n}\frac{\phi_{n-1}(0)}{\phi_n(0)}\phi_n(z). \tag{8.195}$$

(i) Let $v(z) = -\log w(z)$. Expand $\phi_n'(z)$ as a series in $\{\phi_k(z)\}$ using (8.184), noting that because $w(e^{i\theta})$ is real $\bar{\phi}_k = \phi_k$, to deduce

$$\phi_n'(z) = \phi_{n-1}(z)\int_C \Big(\frac{1}{\zeta}\phi_{n-1}(1/\zeta)\phi_k(1/\zeta) + \frac{1}{\zeta^2}\phi_{n-1}'(1/\zeta)\Big)\phi_n(\zeta)\tilde{w}(\zeta)\,\frac{d\zeta}{2\pi i\zeta} \\ + \sum_{k=0}^{n-1}\phi_k(z)\int_C v'(\zeta)\phi_k(1/\zeta)\phi_n(\zeta)\tilde{w}(\zeta)\,\frac{d\zeta}{2\pi i\zeta},$$

where use is to be made of the fact that for $p_k(z)$ a polynomial of degree k and $k < n$, $\int_C \phi_n(\zeta)p_k(1/\zeta)\frac{d\zeta}{\zeta} = 0$.

(ii) Now make use of the structure of $\phi_n(z)$ in Definition 8.6.1 and the first equation in Proposition 8.6.2 to deduce the differential relation

$$\phi_n'(z) = n\frac{\kappa_{n-1}}{\kappa_n}\phi_{n-1}(z) + \phi_n^*(z)\int_C \frac{v'(z) - v'(\zeta)}{z - \zeta}\phi_n(\zeta)\phi_n^*(1/\zeta)\tilde{w}(\zeta)\,\frac{d\zeta}{2\pi i} \\ - \phi_n(z)\int_C \frac{v'(z) - v'(\zeta)}{z - \zeta}\phi_n(\zeta)\phi(1/\zeta)\tilde{w}(\zeta)\,\frac{d\zeta}{2\pi i}. \tag{8.196}$$

(iii) Note from the explicit form of $w(z)$ that

$$\frac{v'(z) - v'(\zeta)}{z - \zeta} = -\frac{\sqrt{t}}{2}\Big(\frac{1}{z\zeta^2} + \frac{1}{z^2\zeta}\Big).$$

Substitute this in (8.196) and simplify by noting from Definition 8.6.1

$$\zeta\Big(\phi_n(\zeta) - \frac{\kappa_n}{\phi(0)}\phi_n^*(\zeta)\Big) = -\frac{\kappa_{n-1}}{\kappa_n}\frac{\phi_{n-1}(0)}{\phi_n(0)}\phi_n(\zeta) + r_{n-1}(z),$$

$$\zeta\frac{\phi_n^*(\zeta)}{\phi(0)} = \frac{\phi_{n+1}(\zeta)}{\kappa_{n+1}} + \Big(\frac{\kappa_n l_n - \kappa_{n-1}l_{n-1}}{\kappa_n(\phi_n(0))^2} - \frac{l_{n+1}}{\kappa_{n+1}\kappa_n}\Big)\phi_n(\zeta) + r_{n-1}(z),$$

where $r_{n-1}(z)$ is some polynomial of degree $n - 1$, to deduce the sought equation.

Chapter Nine

Eigenvalue probabilities — Fredholm determinant approach

The theme of characterizing eigenvalue probabilities as solutions of nonlinear equations is continued, this time using various function theoretic and integrable systems aspects of Fredholm determinants, an approach complimentary to that of Painlevé systems. The starting point is a Fredholm determinant formula for the generating function of the gap probability in the case of matrix ensembles with unitary symmetry. Fredholm determinants with analytic kernels are well suited to numerical evaluation, and this allows tables of statistical properties of various distributions obtained in the previous chapter using power series to be extended. Function theoretic properties of the Fredholm determinants are used to directly derive Painlevé transcendent evaluations of such gap probabilities in the scaled limit. In the cases of the soft and hard edges the evaluations involve Painlevé transcendents distinct from those obtained in the previous chapter. Analysis of the eigenvalues and eigenvectors of the integral operators corresponding to the Fredholm determinants leads to asymptotic formulas for $E_2^{\rm scale}(n;J)/E_2^{\rm scale}(0;J)$ as $|J| \to \infty$ in the cases that scale equals bulk, soft and hard. The Painlevé transcendent evaluations can be used to determine the $|J| \to \infty$ behavior of $E_2^{\rm scale}(0;J)$, provided the leading form is first deduced from other considerations. In the case of $J = [-a,a]$ and the weight being even, an analysis of the eigenvalues and eigenvectors of the even part of the corresponding Christoffel-Darboux kernel leads to a generating function identity complementary to that obtained for the same quantity in the previous chapter, which justifies the Painlevé transcendent evaluation of $E_{N,1}^{\rm COE}(n;(-\theta,\theta))$ given therein. Painlevé transcendent evaluations of $E_\beta^{\rm scale}(n;J)$ for scale = soft and hard are obtained by making use of superimposed $\beta = 1$ ensembles, as considered in Chapter 8, and inter-relations of $\beta = 1$ and $\beta = 4$ ensembles known from Chapter 6. Integrable systems methods — relating to the Riemann-Hilbert problem and the isomonodromic deformation of linear second order differential equations, and the KP equation and Virasoro constraints — are applied to the $\beta = 2$ soft edge gap probability and $\beta = 2$ Gaussian ensemble right edge gap probability, respectively.

9.1 FREDHOLM DETERMINANTS

Underlying the Fredholm determinant approach to gap probabilities in the case of matrix ensembles with unitary symmetry is a general expression for the generating function of the gap probability as a sum over the corresponding k-point correlation functions.

PROPOSITION 9.1.1 *For a continuous one-dimensional statistical mechanical system of N particles*

$$E_{N,\beta}((a_1,a_2);\xi) = 1 + \sum_{n=1}^{N} \frac{(-\xi)^n}{n!} \int_{a_1}^{a_2} dx_1 \cdots \int_{a_1}^{a_2} dx_n \, \rho_{(n)}^N(x_1,\dots,x_n), \tag{9.1}$$

where $\rho_{(n)}^N$ denotes the n-particle correlation function defined by (5.1).

Proof. Comparing the definition of $E_{N,\beta}((s_N(a_1), s_N(a_2));\xi)$ from Proposition 8.1.2 with the formula for the quantity $U[a;z]$ in Exercises 5.1 q.3(i), we see that they coincide provided we set $a(x) = 1$, $x \in (a_1,a_2)$, $a(x) = 0$ otherwise and $z = -\xi$. Thus the expansion (9.1) follows from the definition of $U[a;z]$ given in Exercises 5.1 q.3. □

In the general theory of infinite point processes, a relevant concern is conditions on the correlations $\rho_{(n)}^{\rm scaled}$

which are sufficient for the point process to be uniquely determined by its correlations. One such condition is that [372]

$$\int_R dx_1 \cdots \int_R dx_n \, \rho_{(n)}^{\rm scaled}(x_1,\dots,x_n) = {\rm o}(n!) \tag{9.2}$$

for $R = \bigcup_{j=1}^{l}(a_j,b_j)$ any domain which contains a finite number of particles ($\int_R \rho_{(1)}(x)\,dx < \infty$). The point process being uniquely determined by its correlations has the consequence that the joint distribution of the number of particles in R is uniquely determined by its moments, and that the analogue of (9.1) holds for the corresponding generating function. Now, it follows from the definitions that

$$\int_R dx_1 \cdots \int_R dx_n \, \rho_{(n)}^{\rm scaled}(x_1,\dots,x_n) = \Big\langle \#R(\#R-1)\cdots(\#R-n)\Big\rangle,$$

where $\#R$ denotes the number of particles in the interval R and is thus a so-called factorial moment, from which the usual moments can be calculated by linear combinations. One sees from this that if upon the linear change of scale $s_N(x) = \alpha_N + \beta_N x$

$$\begin{aligned}\lim_{N\to\infty} |\beta_N|^n \int_{s_N(R)} dx_1 \cdots \int_{s_N(R)} dx_n \, \rho_{(n)}^{N}(s_N(x_1),\dots,s_N(x_n)) \\ = \int_R dx_1 \cdots \int_R dx_n \, \rho_{(n)}^{\rm scaled}(x_1,\dots,x_n),\end{aligned} \tag{9.3}$$

then the joint distribution of the number of particles in $s_N(R)$ for the sequence of point processes specified by $\{\rho_{(n)}^N\}$ converges to the joint distribution of the number of particles in R for the process specified by $\{\rho_{(n)}^{\rm scaled}\}$ [496]. In particular, the following result holds.

PROPOSITION 9.1.2 *The convergence (9.3), together with the bound (9.2), implies the convergence*

$$\begin{aligned}E_\beta((a_1,a_2);\xi) &:= \lim_{N\to\infty} E_{N,\beta}((s_N(a_1),s_N(a_2));\xi) \\ &= 1 + \sum_{n=1}^{\infty} \frac{(-\xi)^n}{n!} \int_{a_1}^{a_2} dx_1 \cdots \int_{a_1}^{a_2} dx_n \, \rho_{(n)}^{\rm scaled}(x_1,\dots,x_n).\end{aligned} \tag{9.4}$$

Before proceeding to study the limit in (9.4), we remark that according to Proposition 8.1.2, the parameter ξ in (9.1) and (9.4) is an auxiliary quantity used to form a generating function. However, a physical interpretation of ξ is also possible [76]. The idea is to suppose that eigenvalues (particles) are independently deleted from samples with probability $(1-\xi)$. Theoretically, this has the simple effect of multiplying each n-point correlation by ξ^n. It thus follows that for $0 < \xi < 1$, (9.1) and (9.4) can be interpreted as gap probabilities for the corresponding diluted systems.

Returning now to the limit in (9.4), we will first show that for the scaled classical random matrix ensembles with unitary symmetry, the convergence (9.3) can be established. This follows from the asymptotic form of the corresponding Christoffel-Darboux kernels.

PROPOSITION 9.1.3 *In the notations of Chapters 5 and 7, we have*

$$\frac{\pi\rho}{\sqrt{2N}} K_N^{(\mathrm{G})}\Big(\frac{\pi\rho x}{\sqrt{2N}}, \frac{\pi\rho y}{\sqrt{2N}}\Big) = K^{\mathrm{bulk}}(x,y) + \mathrm{O}\Big(\frac{1}{\sqrt{N}}\Big)\mathrm{O}(1), \tag{9.5}$$

$$\frac{1}{2^{1/2}N^{1/6}} K_N^{(\mathrm{G})}\Big((2N)^{1/2} + \frac{X}{2^{1/2}N^{1/6}}, (2N)^{1/2} + \frac{Y}{2^{1/2}N^{1/6}}\Big) = K^{\mathrm{soft}}(X,Y) + \mathrm{O}\Big(\frac{1}{N^{2/3}}\Big)\mathrm{O}(e^{-X-Y}), \tag{9.6}$$

$$\frac{1}{4N} K_N^{(\mathrm{L})}\Big(\frac{X}{4N}, \frac{Y}{4N}\Big) = K^{\mathrm{hard}}(X,Y) + \mathrm{O}\Big(\frac{1}{N}\Big)\mathrm{O}(1), \tag{9.7}$$

$$\frac{2\pi i}{L} K_N^{(\mathrm{CJ})}(e^{2\pi i x/L}, e^{2\pi i y/L}) = K^{\mathrm{s.s.}}(x,y) + \mathrm{O}\Big(\frac{1}{N}\Big)\mathrm{O}(1), \tag{9.8}$$

where in the remainder terms the dependence on N is uniform, as is the dependence on x, y in a finite interval for (9.5), $X, Y \in [w, \infty)$, $w \in \mathbb{R}$, for (9.6), $X, Y \in [0, w]$, $w \in \mathbb{R}^+$ for (9.7) and $x, y \in [0, L]$ for (9.8).

Proof. In Chapter 7 we have used asymptotic expansions to show that the pointwise limit of the l.h.s. of the above expressions is given by the first term on the r.h.s. To obtain the correction term to the pointwise limit we make use of the correction term in the asymptotic expansion of the corresponding orthogonal polynomials. Consider (9.5). We know from [508] that the correction term in (7.1) also has a uniform dependence on $\sqrt{2nx}$, in that

$$\frac{\Gamma(n/2+1)}{\Gamma(n+1)} e^{-x^2/2} H_n(x) = \cos(\sqrt{2n+1}x - n\pi/2) + \mathrm{O}(n^{-1/2})\mathrm{O}(e^{\pm i\sqrt{2n}x}). \tag{9.9}$$

Of course for x real the term $\mathrm{O}(e^{\pm i\sqrt{2n}x})$ can be replaced by a constant; however we also have use for (9.9) in the case of x complex. Using (9.9) to substitute for the denominator in (5.10), and making use too of the normalization (5.48) and Stirling's formula we immediately deduce (9.5), although at this stage we have to assume $|x-y|$ is bounded away from zero to be sure of the error term. To see that this restriction is not necessary, we take advantage of $K_N^{(\mathrm{G})}(x,y)$ being an analytic function of y and write

$$\frac{\pi\rho}{\sqrt{2N}} K_N^{(\mathrm{G})}\Big(\frac{\pi\rho x}{\sqrt{2N}}, \frac{\pi\rho y}{\sqrt{2N}}\Big) = \frac{\pi\rho}{\sqrt{2N}}\frac{1}{2\pi}\int_0^{2\pi} K_N^{(\mathrm{G})}\Big(\frac{\pi\rho x}{\sqrt{2N}}, \frac{\pi\rho y}{\sqrt{2N}} + Re^{it}\Big)dt$$

for arbitrary $R > 0$. Choosing $R = \pi\rho/\sqrt{2N}$ we see from (9.9) that the leading term is an integral representation of $K^{\mathrm{bulk}}(x,y)$, while the remainder term is bounded by $\mathrm{O}(N^{-1/2})$ coming from the numerator of the definition of $K_N^{(\mathrm{G})}$, times the maximum of the scaled denominator

$$\frac{1}{|y - x + e^{it}|}.$$

For $|y-x| \ll 1$ the latter is bounded, thus demonstrating the validity of (9.5) in this domain.

The same strategy works in the other cases, once the correction term in the analogue of (9.9) is known. For (9.6) we make use of the asymptotic expansion (7.9) refined [435] so that the correction term $\mathrm{O}(N^{-2/3})$ is replaced by the uniform bound $\mathrm{O}(N^{-2/3})\mathrm{O}(e^{-t})$ (cf. (7.59)), while for (9.7) we replace the correction term $\mathrm{O}(N^{a/2-3/4})$ in (7.29) by [508]

$$\begin{cases} x^{5/4}\mathrm{O}(N^{a/2-3/4}), & cN^{-1} \le x \le w, \\ x^{a/2+2}\mathrm{O}(N^{a}), & 0 < x \le cN^{-1}, \end{cases}$$

where $c, w > 0$ are arbitrary. To analyze the remainder term for the Christoffel-Darboux sum in the case of the circular Jacobi ensemble we make use of the integral representation

$$\frac{\Gamma(a+b+1)}{\Gamma(a+1)\Gamma(b+1)}\int_{-1/2}^{1/2} e^{\pi i x(a-b)}|1+e^{2\pi i x}|^{a+b}(1+te^{2\pi i x})^r\,dx = {}_2F_1(-r,-b;a+1;t), \tag{9.10}$$

which can be deduced by expanding $(1+te^{2\pi i x})^r$ according to the binomial theorem and integrating term by term using

(4.4) in the case $N = 1$. This shows

$$ {}_2F_1(-n, -b; a+1; t/n) = {}_1F_1(-b; a+1; -t)\Big(1 + \mathrm{O}\Big(\frac{1}{n}\Big)\mathrm{O}(t^2)\Big). $$

□

It remains to establish the bound (9.2). We make use of the fact that $\rho_{(n)}$ is non-negative and, according to Proposition 5.1.2, equal to the determinant of a symmetric matrix. For the determinant of such matrices we have the following inequality, equivalent to a result of Hadamard given in (9.12) below.

PROPOSITION 9.1.4 *For* $\mathbf{A} = [a_{ij}]_{i,j=1,\dots,n}$ *non-negative and symmetric,*

$$ \det \mathbf{A} \le \prod_{i=1}^{n} a_{ii}. \tag{9.11} $$

Proof. [393] Consider first the case $a_{ii} = 0$ for some $i = 1, \dots, n$. Because a matrix is non-negative if and only if its principal minors are non-negative, it follows that $a_{ii}a_{jj} - |a_{ij}|^2 \ge 0$ for each $j \ne i$ and thus all elements in the ith row of $\mathbf{A}$ must vanish, so $\det \mathbf{A} = 0$, in agreement with (9.11). Hence it suffices now to consider the case that $\mathbf{A}$ is positive definite.

We begin by performing the Laplace expansion

$$ \det \mathbf{A} = a_{11} \det \mathbf{A}_1 - \det \begin{bmatrix} 0 & \vec{\alpha}^\dagger \\ \vec{\alpha} & \mathbf{A}_1 \end{bmatrix}, $$

where $\vec{\alpha} = [a_{k1}]_{k=2,\dots,n}$ and $\mathbf{A}_1 = [a_{ij}]_{i,j=2,\dots,n}$. Analogous to a manipulation performed in the proof of Proposition 2.5.1 we have

$$ \begin{aligned} \det \begin{bmatrix} 0 & \vec{\alpha}^\dagger \\ \vec{\alpha} & \mathbf{A}_1 \end{bmatrix} &= \det \mathbf{A}_1 \det \begin{bmatrix} 1 & \mathbf{0}_{(N-1)\times 1} \\ \mathbf{0}_{1\times(N-1)} & \mathbf{A}_1^{-1} \end{bmatrix} \det \begin{bmatrix} 0 & \vec{\alpha}^\dagger \\ \vec{\alpha} & \mathbf{A}_1 \end{bmatrix} \\ &= (\det \mathbf{A}_1)(\vec{\alpha}^\dagger \mathbf{A}_1^{-1} \vec{\alpha}) > 0, \end{aligned} $$

where the inequality follows because $\mathbf{A}_1$ is a principal minor of $\mathbf{A}$ and thus is itself positive definite, and this implies $\mathbf{A}_1^{-1}$ is positive definite. Hence

$$ \det \mathbf{A} < a_{11} \det \mathbf{A}_1, $$

and the result now follows by induction. □

Note that writing $\mathbf{A} = \mathbf{B}^\dagger \mathbf{B}$ for $\mathbf{B}$ an $n \times n$ square matrix shows (9.11) is equivalent to the inequality

$$ |\det \mathbf{B}|^2 \le \prod_{i=1}^{n} \Big(\sum_{j=1}^{n} |b_{ij}|^2 \Big), \tag{9.12} $$

which is Hadamard's result [541].

In light of the determinant structure of $\rho_{(n)}$ noted above Proposition 9.1.4, and the fact that the diagonal entries of the determinant equal the one-body density, the bound (9.11) tells us that

$$ \rho_{(n)}(x_1, \dots, x_n) \le \prod_{l=1}^{n} \rho_{(1)}(x_l). \tag{9.13} $$

Assuming the region R is such that the integral of the one-body density over R is finite (i.e., contains a finite number of eigenvalues) we see the integral in (9.2) in fact only grow exponentially fast in n. All criteria of Proposition 9.1.2 are therefore met and so it follows that for the scaled classical random matrix ensemble with unitary symmetry

$$E_2^{\rm scaled}(J;\xi) = 1 + \sum_{n=1}^{\infty} \frac{(-\xi)^n}{n!} \int_J dx_1 \cdots \int_J dx_n \det \left[K^{\rm scaled}(x_j, x_k) \right]_{j,k=1,\dots,n}, \tag{9.14}$$

where scale = bulk, soft, hard or s.s. and $J = (-t,t)$, (t,∞), $(0,t)$, $(-t,t)$, respectively. We recognise this sum as that occurring in the expansion (5.32) of the determinant of a Fredholm integral operator. Hence

$$E_2^{\rm scaled}(J;\xi) = \det(1 - \xi K_J^{\rm scaled}), \tag{9.15}$$

where

$$(1 - \xi K_J^{\rm scale}) f := f(x) - \xi \int_J K^{\rm scaled}(x,y) f(y)\, dy. \tag{9.16}$$

With (9.14) as the starting point, and with scaled = bulk, $J = (0,t)$ Jimbo, Miwa, Môri and Sato [324] obtained the Painlevé transcendent evaluation of (9.15) implied by the first equation of (8.111). Furthermore, this work also identified the Fredholm determinant in (9.15) as a τ-function associated with the integrable structure of a particular monodromy preserving deformation of a linear differential equation. In this chapter we will present both the function theoretic properties of (9.15) which lead to its Painlevé transcendent evaluations, as well as aspects of the underlying integrable systems theory.

EXERCISES 9.1 1. [90] Let $J_{(p),I}(x_1,\dots,x_p)$ denote the p-point correlation for the event that there are exactly n particles in the interval I, and these are at the positions $x_1,\dots,x_p$. The quantity $J_{(p),I}$ is referred to as a *Janossy density.*

(i) Consider the p.d.f. (5.139) and let $\zeta_j, \tilde{\zeta}_j$ be linear combinations of $\{\xi_l\}_{l=1,\dots,j}$, and let $\psi_k, \tilde{\psi}_k$ be linear combinations of $\{\eta_l\}_{l=1,\dots,k}$. With

$$\int_{-\infty}^{\infty} w_2(x) \zeta_j(x) \psi_k(x)\, dx = \delta_{j,k}, \qquad \Big(\int_{-\infty}^{\infty} - \int_I \Big) w_2(x) \tilde{\zeta}_j(x) \tilde{\psi}_k(x)\, dx = \delta_{j,k}$$

let $K_N(x,y) = \sum_{l=0}^{N-1} \zeta_l(x)\psi_l(y)$ and $\tilde{K}_N(x,y) = \sum_{l=0}^{N-1} \tilde{\zeta}_l(x)\tilde{\psi}_l(y)$. Show

$$J_{(p),I}(x_1,\dots,x_p) = \det(1 - K_{N,I}) \det \left[\tilde{K}_N(x_i, x_j) \right]_{i,j=1,\dots,p}, \tag{9.17}$$

where $K_{N,I}$ is the integral operator on I with kernel $K_N(x,y)$.

(ii) Verify that the Laguerre polynomials $\{L_k^0(x-t)\}_{k=0,1,\dots}$ are orthogonal on the interval (t,∞) with respect to the weight e^{-x}, and use this together with the result of (i) to show that with this choice of w_2 and $\xi_l = \eta_l = x^{l-1}$

$$\tilde{K}_N(x,y) = K_N^{(\rm L)}(x-t, y-t) \Big|_{a=0},$$

where $K_N^{(\rm L)}$ is the Christoffel-Darboux kernel in the Laguerre case with $a=0$. Substitute this in (9.17), and relate its hard edge scaled form with $p=2$, $t=t_2$, $x_1=t_1$, $x_2=t_2$, to the formula (8.100) with $a=0$.

2. (i) By making use of Propositions 5.2.1 and 5.2.2 with $a(x) = 1 - \xi\chi_{x\in J}$, deduce that

$$E_{N,2}(J;\xi;w_2(x)) = \det(1-K),$$

where K is the integral operator on J with kernel $\xi K_N(x,y)$, as is consistent with the formula (9.15).

(ii) Note from (6.27) and (6.76) with $a(x)$ as in (i) that for $\beta = 1,4$

$$\Big(E_{N,\beta}(J;\xi;e^{-\beta V(x)}) \Big)^2 = \det[\mathbf{1}_2 - \xi \mathbf{f}_\beta \boldsymbol{\chi}_J], \tag{9.18}$$

where $\boldsymbol{\chi}_J := {\rm diag}[\chi_J, \chi_J]$, and χ_J is the indicator function for J.

3. Consider a two-species system, species a and b, with coordinates $\{a_j\}$ and $\{b_j\}$. Let

$$\rho_{(n_1,n_2)}(\{a_j\}_{j=1,\dots,n_1};\{b_j\}_{j=1,\dots,n_2})$$

denote the corresponding (n_1,n_2)-point correlation, and introduce the generating function

$$E_N(J_a,J_b;\xi_x,\xi_y) := \Big\langle \prod_{l=1}^{N}(1-\xi_a\chi_{J_a}^{(l,a)})(1-\xi_b\chi_{J_b}^{(l,b)})\Big\rangle$$

(cf. the case $\mu=0$ of (8.7)). Show that

$$E_N(J_a,J_b;\xi_a,\xi_b) = 1 + \sum_{\substack{n_1,n_2=0\\(n_1,n_2)\neq(0,0)}}^{N} \frac{(-\xi_a)^{n_1}(-\xi_b)^{n_2}}{n_1!n_2!}$$
$$\times \int_{J_a} da_1\cdots\int_{J_a} da_{n_1}\int_{J_b} db_1\cdots\int_{J_b} db_{n_2}\,\rho_{((n_1,n_2))}(\{a_j\}_{j=1,\dots,n_1};\{b_j\}_{j=1,\dots,n_2}).$$

In the cases that the correlations have a determinant structure

$$\rho_{((n_1,n_2))}(\{a_j\}_{j=1,\dots,n_1};\{b_j\}_{j=1,\dots,n_2}) = \det\begin{bmatrix}[K_{aa}(a_j,a_k)]_{j,k=1,\dots,n_1} & [K_{ab}(a_j,b_k)]_{\substack{j=1,\dots,n_1\\k=1,\dots,n_2}}\\ [K_{ba}(b_j,a_k)]_{\substack{j=1,\dots,n_2\\k=1,\dots,n_1}} & [K_{bb}(b_j,b_k)]_{j,k=1,\dots,n_2}\end{bmatrix}$$

deduce from this that

$$E_N(J_a,J_b;\xi_a,\xi_b) = \det(1-K), \tag{9.19}$$

where K is the 2×2 matrix integral operator on $\mathbb{R}$ with kernel

$$\mathbf{K}(x,y) = \begin{bmatrix}\chi_{x\in J_a}\xi_a K_{aa}(x,y)\chi_{y\in J_a} & \chi_{x\in J_a}\xi_a K_{ab}(x,y)\chi_{y\in J_b}\\ \chi_{y\in J_b}\xi_b K_{ba}(y,x)\chi_{x\in J_a} & \chi_{y\in J_b}\xi_b K_{bb}(y,y)\chi_{y\in J_b}\end{bmatrix}.$$

9.2 NUMERICAL COMPUTATIONS USING FREDHOLM DETERMINANTS

The sine kernel refers to the kernel

$$K^{\rm bulk}(x,y) := \frac{\sin\pi\rho(x-y)}{\pi(x-y)}, \tag{9.20}$$

and the task is to compute

$$E_2^{\rm bulk}((0,s);\xi) = \det(1-\xi K^{\rm bulk}_{(0,s)}). \tag{9.21}$$

Let us first address this problem in a numerical sense.

In Section 8.3.5 the Painlevé transcendent formula (8.111) was used to generate the first 700 terms in its power series, and this in turn was used to compute Table 8.13, giving statistical properties of $\{p_2^{\rm bulk}(n;s)\}$ for n up to 4. Due to the computational expense incurred in computing the power series, to get more terms is difficult, and this restriction in turn implies a loss of accuracy in the determination of $p_2^{\rm bulk}(n;s)$ for higher values of n (recall from (8.165) that the leading term in the power series of $p_2^{\rm bulk}(n;s)$ is proportional to $s^{n+(n+2)(n+1)}$). It turns out that a numerical scheme based on (9.21) can overcome this problem [79].

The first tabulations of spacing distributions were based on Fredholm determinants [259]. Thus the Fredholm determinant formula (9.81) below for $E_2^{O^-}((0,s);\xi=1)$ was used to obtain a graphically accurate determination of the bulk spacing distribution $p_1^{\rm bulk}(0;s)$ (recall (8.139)). This was done by first developing the theory of Section 9.6.1 below, and thereby determining the eigenvalues of the Fredholm integral operator in terms of prolate spheroidal functions. By making use of tables of these functions, $E_1^{\rm bulk}(0;s)$ was computed, and a numerical differentiation scheme then used to compute $p_1^{\rm bulk}(0;s)$ according to (8.154). The same strategy was employed in [392] to tabulate $p_2^{\rm bulk}(0;s)$ (see [425] for a discussion of the source of inaccuracies in this table). Subsequently, as reproduced in [395], the prolate spheroidal functions were used to plot $\{E_\beta^{\rm bulk}(n;s)\}$, $\beta = 1,2$ for successive values of n up to 10.

Recently Bornemann [78], [79] has shown how the Fredholm determinant formulas can be used to compute spacing distributions such as $\{p_2^{\rm bulk}(n;s)\}$ for successive values of n up to 14 (and beyond if required), to machine precision of 15-digit accuracy. The strategy relates to the approximation of the integral in definition (9.16) by a sum, according to an m-point quadrature rule of order m (i.e., a rule replacing the integral by a weighted sum of m terms, such that it is exact for a polynomial of degree $m-1$) with positive weights w_j,

$$\int_J K(x,y)f(y)\,dy \approx \sum_{k=1}^m w_j K(x,y_k)f(y_k).$$

The integral operator eigenvalue equation $\xi K_J \psi = \lambda\psi$ is correspondingly replaced by the system of m linear equations

$$\xi \sum_{j=1}^m w_j K(y_j,y_k)\psi(y_k) = \lambda\psi(y_j) \qquad (k=1,\dots,m),$$

or equivalently (with $\psi \mapsto w_j^{-1/2}\psi$)

$$\xi \sum_{j=1}^m w_j^{1/2} K(y_j,y_k) w_k^{1/2}\psi(y_k) = \lambda\psi(y_j).$$

The characteristic polynomial of the symmetric matrix $[\xi w_j^{1/2}K(y_j,y_k)w_k^{1/2}]_{j,k=1,\dots,m}$ at $\lambda = 1$ is the Fredholm determinant of $1-\xi K_J$ in this approximation, and so

$$\det(1-\xi K_J) \approx \det[\delta_{j,k} - \xi w_j^{1/2}K(y_j,y_k)w_k^{1/2}]_{j,k=1,\dots,m}. \tag{9.22}$$

It is shown in [78] that with the kernel $K(x,y)$ analytic in a complex neighbourhood of (a,b), this approximation has error $\mathrm{O}(\rho^{-m})$ $(\rho > 1)$ and thus converges exponentially fast. In practice this means doubling m will typically double the number of correct digits. The derivative in s, required by (8.115) and analogous formulas, is carried out by converting the numerical tabulation to an interpolation in Chebyshev points.

According to the formulas of Proposition 8.1.2, to compute higher order spacing distributions from the generating function, derivatives in ξ of (9.22) are required. This being an entire function of ξ (it is a polynomial), one can use the formula for the k-th derivative

$$f^{(k)}(z) = \frac{k!}{2\pi r^k}\int_0^{2\pi} e^{-ik\theta} f(z+re^{i\theta})\,d\theta, \qquad r>0,$$

with the trapezoidal rule approximation, which is known to converge exponentially fast for periodic analytic integrands. For the bulk spacings the choice $r=1$ was empirically determined to be the most numerically stable in this regard.

9.3 THE SINE KERNEL

We now take up the problem of using (9.21) to obtain the characterization in terms of the σPV equation (8.112). Key equations from [324] for this purpose were isolated by Tracy and Widom [517], following on from the works [394], [152]. These equations relate to a number of quantities associated with the integral operator in (9.15). One of these quantities is the resolvent kernel $R(x,y)$, specified by

$$\xi K_J(1-\xi K_J)^{-1} \doteq R(x,y), \tag{9.23}$$

where the notation $\doteq$ denotes that the r.h.s. is the kernel of the integral operator on the l.h.s. Our interest is in the case that J is a single interval, $J=[a_1,a_2]$ say. Closely related to (9.23), and used in the definition of the other quantities entering the coupled equations, is $\rho(x,y)$ specified by

$$(1-\xi K_J)^{-1} \doteq \rho(x,y) \tag{9.24}$$

(note that $\rho(x,y)=\delta(x-y)+R(x,y)$). Both (9.23) and (9.24) have meaning for general kernels; however, the other quantities are special to the Christoffel-Darboux type structure of $K^{\rm scale}(x,y)$.

DEFINITION 9.3.1 *Let $\phi(x)$, $\psi(x)$ be such that*

$$\xi K^{\rm scaled}(x,y) := \frac{\phi(x)\psi(y)-\phi(y)\psi(x)}{x-y}. \tag{9.25}$$

In terms of $\phi(x)$, $\psi(x)$ and (9.24) introduce the quantities $Q(x)$, q_j, $P(x)$, p_j according to

$$Q(x) := (1-\xi K_J)^{-1}\phi := \int_{a_1}^{a_2} \rho(x,y)\phi(y)dy, \quad q_j = Q(a_j) := \lim_{\substack{x\to a_j \\ x\in(a_1,a_2)}} Q(x),$$

$$P(x) := (1-\xi K_J)^{-1}\psi := \int_{a_1}^{a_2} \rho(x,y)\psi(y)dy, \quad p_j = P(a_j) := \lim_{\substack{x\to a_j \\ x\in(a_1,a_2)}} P(x).$$

The equations of [517] relating to $E_2^{\rm bulk}((0,t);\xi)$ can now be stated.

PROPOSITION 9.3.2 *For (9.14) with*

$$K^{\rm scaled}(x,y) = K^{\rm bulk}(x,y)\Big|_{\rho=1/\pi},$$

and in the notation of Definition 9.3.1 with $a_1=-t, a_2=t$, we have

$$\text{(i) } R(-t,t)=\frac{qp}{t}, \quad \text{(ii) } R(t,t)=p^2+q^2-\frac{2q^2p^2}{t}, \quad \text{(iii) } \frac{d}{dt}R(t,t)=2(R(-t,t))^2,$$
$$\text{(iv) } \frac{dq}{dt}=p-\frac{2q^2p}{t}, \quad \text{(v) } \frac{dp}{dt}=-q+\frac{2qp^2}{t},$$

where $p:=p_2$ and $q:=q_2$.

From these coupled equations a single differential equation for $\sigma(t) := -tR(t/2,t/2)$ can be obtained, and knowledge of $\sigma(t)$ allows $E_2((0,t);\xi)$ to be computed. For this we require the general formula

$$\frac{\partial}{\partial a_j}\log\det(1-\xi K_J) = (-1)^{j-1}R(a_j,a_j) \qquad (j=1,2) \tag{9.26}$$

(see Exercises 9.3 q.1). It follows from this that

$$-2R(t,t) = \frac{d}{dt}\log E_2((-t,t);\xi), \tag{9.27}$$

which in turn implies

$$E_2((0,t);\xi) = E_2((-t/2,t/2);\xi) = \exp\int_0^{\pi\rho t}\frac{\sigma(u)}{u}\,du. \tag{9.28}$$

On the r.h.s. of (9.28) t has been replaced by $\pi\rho t$ to account for a general value of ρ (this is valid because ρ is the only length scale in the problem).

PROPOSITION 9.3.3 *The quantity $\sigma(t) := -tR(t/2, t/2)$ satisfies the first of the differential equations in (8.112) (which is the σPV equation of (8.15) with $\nu_0 = \nu_1 = \nu_2 = \nu_3 = 0$ and $t \mapsto -2is$) subject to the first of the boundary conditions in (8.113).*

Proof. From equations (i), (ii), (iv) and (v) of Proposition 9.3.2 it is straightforward to deduce the equations

$$\frac{d}{dt}(tR(-t,t)) = p^2 - q^2 \quad \text{and} \quad \frac{d}{dt}(tR(t,t)) = p^2 + q^2.$$

Squaring these equations and subtracting gives

$$\left(\frac{d}{dt}(tR(-t,t))\right)^2 - \left(\frac{d}{dt}(tR(t,t))\right)^2 = -4p^2q^2 = -4t^2(R(-t,t))^2,$$

where to obtain the last equality equation (i) has again been used. The stated equation now follows by using the equations

$$4t^2(R(-t,t))^2 = 2t\frac{d}{dt}(tR(t,t)) - 2tR(t,t) \quad \text{and} \quad 4tR(-t,t)\frac{d}{dt}(tR(-t,t)) = t\frac{d^2}{dt^2}(tR(t,t)),$$

which are consequences of equation (iii), to eliminate $tR(-t,t)$ and its derivative. The boundary condition follows from the leading small t computation of the $n = 1$ and $n = 2$ terms in (9.4) with $K(x,y)$ given by $K^{\rm bulk}(x,y)$ and ρ therein set equal to $\frac{1}{\pi}$. □

9.3.1 Derivation of the coupled equations

Two general operator identities are used throughout the derivation.

PROPOSITION 9.3.4 *For any operators L and K,*

$$[L, (1-K)^{-1}] = (1-K)^{-1}[L,K](1-K)^{-1}, \tag{9.29}$$

and for any operator K which can be differentiated with respect to a parameter a,

$$\frac{d}{da}(1-K)^{-1} = (1-K)^{-1}\frac{dK}{da}(1-K)^{-1}. \tag{9.30}$$

Proof. The first identity is verified by letting both sides act on an arbitrary function f, expanding the commutators, and comparing the resulting expressions. The second identity has been derived in Exercises 2.5 q.1. □

The following simple lemma is also required.

PROPOSITION 9.3.5 *Let M denote multiplication by the independent variable, that is, $Mf(x) = xf(x)$. Then for an integral operator L_J with kernel $L(x,y)$ supported on the interval $J = [a_1, a_2]$,*

$$[M, L_J] \doteq (x-y)L(x,y).$$

Proof. Since

$$(ML_J - L_JM)f := x\int_{a_1}^{a_2} L(x,y)f(y)dy - \int_{a_1}^{a_2} L(x,y)yf(y)dy = \int_{a_1}^{a_2}(x-y)L(x,y)f(y)dy,$$

the result follows. □

After these preliminaries, we begin the derivation proper with an expression for the resolvent kernel [307].

PROPOSITION 9.3.6 *Let $R(x,y)$ be defined as in (9.23). We have*

$$R(x,y) = \frac{Q(x)P(y) - P(x)Q(y)}{x-y}, \tag{9.31}$$

where Q and P are specified in Definition 9.3.1, and thus

$$R(x,x) = -Q(x)P'(x) + P(x)Q'(x). \tag{9.32}$$

Proof. We will compute (9.29) with $L = M$ in two different ways. By Proposition 9.3.5 and (9.25),

$$[M, \xi K_J] \doteq \phi(x)\psi(y) - \phi(y)\psi(x).$$

Thus $[M, \xi K_J] = A_1 - A_2$ where $A_1 \doteq \phi(x)\psi(y)$, $A_2 \doteq \phi(y)\psi(x)$, so we have

$$\begin{aligned}
&[M, \xi K_J](1 - \xi K_J)^{-1} f \\
&= A_1(1-\xi K_J)^{-1} f - A_2(1-\xi K_J)^{-1} f \\
&= \phi(x) \int_{a_1}^{a_2} dy\, \psi(y) \int_{a_1}^{a_2} \rho(y,y') f(y') dy' - \psi(x) \int_{a_1}^{a_2} dy\, \phi(y) \int_{a_1}^{a_2} \rho(y,y') f(y') dy' \\
&= \phi(x) \int_{a_1}^{a_2} dy\, f(y) \int_{a_1}^{a_2} dy' \psi(y') \rho(y,y') - \psi(x) \int_{a_1}^{a_2} dy\, f(y) \int_{a_1}^{a_2} dy' \phi(y') \rho(y,y'),
\end{aligned}$$

where we have used the symmetry $\rho(y,y') = \rho(y',y)$ which follows since $K(x,y) = K(y,x)$. Hence, using Definition 9.3.1, we have

$$[M, \xi K_J](1 - \xi K_J)^{-1} f = \phi(x) \int_{a_1}^{a_2} f(y) P(y)\, dy - \psi(x) \int_{a_1}^{a_2} f(y) Q(y)\, dy,$$

and so

$$[M, \xi K_J](1 - \xi K_J)^{-1} \doteq \phi(x) P(y) - \psi(x) Q(y),$$

which gives

$$(1 - \xi K_J)^{-1} [M, \xi K_J](1 - \xi K_J)^{-1} \doteq Q(x) P(y) - P(x) Q(y).$$

On the other hand, for the l.h.s of (9.29), we have from Proposition 9.3.5

$$[M, (1 - \xi K_J)^{-1}] \doteq (x-y)\rho(x,y) = (x-y) R(x,y).$$

Equating the above two equations gives the stated result. □

Equation (i) of Proposition 9.3.2 can now be derived by using (9.31). To see this, we note that with $a_1 = -t$, $a_2 = t$, from the facts that $\phi(x)$ is odd, $\psi(x)$ is even and $\rho(x,y) = \rho(-x,-y)$ (which follows since $K(x,y) = K(-x,-y)$), it follows from Definition 9.3.1 that

$$q_2 = -q_1 = q, \qquad p_2 = p_1 = p, \tag{9.33}$$

so substitution gives the desired equation.

Whereas Proposition 9.3.6 holds for general ϕ and ψ, the final conclusion of the next result uses a special property of the particular choice

$$\phi(x) = \sqrt{\frac{\xi}{\pi}} \sin \pi \rho x, \qquad \psi(x) = \sqrt{\frac{\xi}{\pi}} \cos \pi \rho x, \tag{9.34}$$

which gives the kernel (9.20). This result is presented in preparation for the computation of $P'(x)$ and $Q'(x)$ which occur in the expression (9.32) for $R(x,x)$.

PROPOSITION 9.3.7 *Let D denote the operator for differentiation with respect to the independent variable. We have*

$$[D, K_J] \doteq (D_x + D_y) K(x,y) - \delta^-(y - a_2) K(x,y) + \delta^+(y - a_1) K(x,y),$$

where

$$\int_{a_1}^{a_2} \delta^-(y - a_2) f(y) dy := f(a_2) \quad \text{and} \quad \int_{a_1}^{a_2} \delta^+(y - a_1) f(y) dy := f(a_1),$$

and consequently, in the cases that $(D_x + D_y)K(x,y) = 0$,

$$[D, (1-\xi K_J)^{-1}] \doteq R(x,a_1)\rho(a_1,y) - R(x,a_2)\rho(a_2,y).$$

Proof. We have

$$\begin{aligned}[D,K_J]f :=& \frac{\partial}{\partial x}\int_{a_1}^{a_2} K(x,y)f(y)\,dy - \int_{a_1}^{a_2} K(x,y)\frac{\partial}{\partial y}f(y)\,dy \\ :=& \int_{a_1}^{a_2}\frac{\partial}{\partial x}K(x,y)f(y)\,dy - \Big(f(a_2)K(x,a_2) - f(a_1)K(x,a_1)\Big) + \int_{a_1}^{a_2} f(y)\frac{\partial}{\partial y}K(x,y)\,dy,\end{aligned}$$

which gives the first result. Now suppose $(D_x + D_y)K(x,y) = 0$. The first result then gives

$$\begin{aligned}[D,K_J](1-\xi K_J)^{-1}f =& \int_{a_1}^{a_2} dy\,K(x,y)(\delta^+(y-a_1) - \delta^-(y-a_2))\int_{a_1}^{a_2} dy'\,\rho(y,y')f(y') \\ =& K(x,a_1)\int_{a_1}^{a_2}\rho(a_1,y')f(y')dy' - K(x,a_2)\int_{a_1}^{a_2}\rho(a_2,y')f(y')dy'.\end{aligned}$$

The second result now follows upon applying (9.29) with $L = D$ and recalling $(1-\xi K_J)^{-1}\xi K_J = R$. □

The derivatives $P'(x)$ and $Q'(x)$, and consequently $R(x,x)$, can now be computed in terms of quantities in Definition 9.3.1.

PROPOSITION 9.3.8 *With* $K(x,y)$ *given by (9.20) we have*

$$\begin{aligned}Q'(x) =& \pi\rho P(x) + R(x,a_1)q_1 - R(x,a_2)q_2, \\ P'(x) =& -\pi\rho Q(x) + R(x,a_1)p_1 - R(x,a_2)p_2\end{aligned}$$

and consequently

$$R(a_j,a_j) = \pi\rho(p_j^2 + q_j^2) - \frac{(q_2p_1 - p_2q_1)^2}{a_2 - a_1} \qquad (j = 1,2).$$

Proof. From Definition 9.3.1 we have

$$Q'(x) = D(1-\xi K_J)^{-1}\phi = (1-\xi K_J)^{-1}D\phi + [D,(1-\xi K_J)^{-1}]\phi.$$

The stated formula for $Q'(x)$ now follows from the fact that $D\phi = \pi\rho\psi$, Proposition 9.3.7, (9.23) and Definition 9.3.1. The formula for $P'(x)$ is derived similarly, and the formula for $R(a_j,a_j)$ is derived by substituting for $Q'(x)$ and $P'(x)$ in (9.32), and using (9.31) and Definition 9.3.1. □

Equation (ii) of Proposition 9.3.2 follows from the formula above for $R(a_j,a_j)$ with $j = 2$, $a_2 = t$ and the substitutions (9.33). For the derivation of equation (iii), a preliminary result is needed.

PROPOSITION 9.3.9 *We have* $\frac{\partial}{\partial a_j}R(x,y) = (-1)^j R(x,a_j)R(a_j,y)$ *and so with* $(D_x + D_y)K(x,y) = 0$,

$$\left(\frac{\partial}{\partial a_j} + \frac{\partial}{\partial x} + \frac{\partial}{\partial y}\right)R(x,y) = \begin{cases} -R(x,a_2)R(a_2,y), & j = 1, \\ R(x,a_1)R(a_1,y), & j = 2, \end{cases}$$

for $x, y \in (a_1,a_2)$.

Proof. We have

$$\frac{\partial}{\partial a_j}R(x,y) = \frac{\partial}{\partial a_j}\rho(x,y)$$

so we consider (9.30) with $a = a_j$. Now

$$\frac{\partial}{\partial a_j}K_J f = \frac{\partial}{\partial a_j}\int_{a_1}^{a_2} K(x,y)f(y)dy = (-1)^j K(x,a_j)f(a_j)$$

and so

$$\frac{\partial}{\partial a_j} K_J \doteq (-1)^j K(x, a_j)\delta^{\mp}(y - a_j) \quad (j = 1, 2).$$

This gives

$$\frac{\partial}{\partial a_j} \rho(x, y) = (-1)^j R(x, a_j)\rho(a_j, y) \tag{9.35}$$

and the first result follows since $\rho(a_j, y) = R(a_j, y)$ for $y \in (a_1, a_2)$. The second result follows from the first result, the fact that for $x, y \in (a_1, a_2)$

$$[D, (1 - \xi K_J)^{-1}] \doteq \left(\frac{\partial}{\partial x} + \frac{\partial}{\partial y}\right) R(x, y)$$

(this can be verified by using integration by parts) and the final equation in Proposition 9.3.7. □

Equation (iii) of Proposition 9.3.2 can now be derived by first noting

$$\frac{\partial}{\partial a_j} R(a_j, a_j) = \left(\frac{\partial}{\partial a_j} + \frac{\partial}{\partial x} + \frac{\partial}{\partial y}\right) R(x, y)\bigg|_{x=y=a_j} = (-1)^j (R(a_1, a_2))^2, \tag{9.36}$$

where the final equality follows from Proposition 9.3.9. This gives

$$\frac{d}{dt} R(t, t) = \left(\frac{\partial}{\partial a_2} R(a_2, a_2) - \frac{\partial}{\partial a_1} R(a_2, a_2)\right)\bigg|_{a_2 = -a_1 = t} = 2(R(-t, t))^2 \tag{9.37}$$

as required.

For the derivation of the final two equations in Proposition 9.3.2, the formulas

$$\frac{\partial q_j}{\partial a_j} = \left(\frac{\partial}{\partial x} + \frac{\partial}{\partial a_j}\right) Q(x)\bigg|_{x=a_j} \quad \text{and} \quad \frac{\partial p_j}{\partial a_j} = \left(\frac{\partial}{\partial x} + \frac{\partial}{\partial a_j}\right) P(x)\bigg|_{x=a_j} \tag{9.38}$$

are used. The partial derivatives with respect to x are given by Proposition 9.3.8. The derivatives with respect to a_j are given by the following result, which is valid for all kernels of the form (9.25).

PROPOSITION 9.3.10 *We have*

$$\frac{\partial}{\partial a_j} Q(x) = (-1)^j R(x, a_j) q_j \quad \textit{and} \quad \frac{\partial}{\partial a_j} P(x) = (-1)^j R(x, a_j) p_j.$$

Proof. Now

$$\frac{\partial}{\partial a_j} Q(x) := \frac{\partial}{\partial a_j} (1 - K)^{-1} \phi.$$

But from (9.35)

$$\frac{\partial}{\partial a_j} (1 - \xi K_J)^{-1} \doteq (-1)^j R(x, a_j) \rho(a_j, y),$$

and the first result follows from Definition 9.3.1. The second result follows similarly. □

The final two equations of Proposition 9.3.2 now follow by noting

$$\frac{dq_j}{dt} = \frac{\partial q_j}{\partial a_2} - \frac{\partial q_j}{\partial a_1},$$

and similarly for $\frac{dp_j}{dt}$, substituting the results of Propositions 9.3.10 and 9.3.8 into (9.38) to compute one of the derivatives, and using Proposition 9.3.10 to compute the other derivative.

EXERCISES 9.3 1. The objective of this exercise is to derive the formula (9.26). For an integral operator K_J on $J = [a_1, a_2]$ with kernel $K(x, y)$ we have $\operatorname{Tr} K_J := \int_{a_1}^{a_2} K(x, x)\, dx$. Use the formula $\log \det A = \operatorname{Tr} \log A$, and the power series expansion of $\log(1-\xi K_J)$ to show that

$$\begin{aligned}
&\frac{\partial}{\partial a_j}\det\log(1-\xi K_J)\\
&=-\frac{\partial}{\partial a_j}\sum_{n=1}^{\infty}\frac{\xi^n}{n}\int_{a_1}^{a_2}dx_1\cdots\int_{a_1}^{a_2}dx_n\,K(x_1,x_2)K(x_2,x_3)\cdots K(x_{n-1},x_n)K(x_n,x_1)\\
&=(-1)^{j-1}\sum_{n=1}^{\infty}\xi^n\int_{a_1}^{a_2}dx_2\cdots\int_{a_1}^{a_2}dx_n\,K(a_j,x_2)K(x_2,x_3)\cdots K(x_{n-1},x_n)K(x_n,a_j).
\end{aligned}$$

Identify the final expression with $(-1)^{j-1}R(a_j,a_j)$ as required by (9.26).

2. [324], [517] Here the multiple gap probability will be exhibited as an integrable system.

   (i) Let $E_2(0;J)$ denote the probability that there are no particles in some collection of intervals J. Show that (9.15) still holds, with the domain of integration of the integral operator now J.

   (ii) In the case $J=\bigcup_{j=1}^{m}(a_{2j-1},a_{2j})$ modify the proofs of Propositions 9.3.6–9.3.10 to show that (with $\rho=1/\pi$)

$$R(a_i,a_i)=p_i^2+q_i^2+\sum_{\substack{k=1\\k\neq i}}^{2m}(-1)^k\frac{(q_ip_k-p_iq_k)^2}{a_i-a_k},$$

$$\frac{\partial q_i}{\partial a_i}=p_i-\sum_{\substack{k=1\\k\neq i}}^{2m}(-1)^kR(a_i,a_k)q_k\quad\text{and}\quad\frac{\partial p_i}{\partial a_i}=-q_i-\sum_{\substack{k=1\\k\neq i}}^{2m}(-1)^kR(a_i,a_k)p_k,$$

$$\frac{\partial q_j}{\partial a_k}=(-1)^kR(a_j,a_k)q_k\quad\text{and}\quad\frac{\partial p_j}{\partial a_k}=(-1)^kR(a_j,a_k)p_k$$

   with

$$R(a_j,a_k)=\frac{q_jp_k-p_jq_k}{a_j-a_k}q_k.$$

   (iii) Introduce the notation

$$q_{2j}=-x_{2j},\quad q_{2j+1}=x_{2j+1},\quad p_{2j}=-iy_{2j},\quad p_{2j+1}=y_{2j+1},$$

$$\omega(a)=d_a\log(1-K)=-\sum_{i=1}^{2m}(-1)^iR(a_i,a_i)da_i=\sum_{i=1}^{2m}G_ida_i,$$

$$G_i=x_i^2+y_i^2-\sum_{\substack{k=1\\k\neq i}}^{2m}\frac{(x_iy_k-y_ix_k)^2}{a_i-a_k},$$

   and define the canonical symplectic structure by

$$\{x_j,x_k\}=0,\qquad\{y_j,y_k\}=0,\qquad\{x_j,y_k\}=\frac{1}{2}\delta_{j,k}.$$

   Verify that the two pairs of equations in (ii) are equivalent to the Hamilton equations

$$d_ax_j=\{x_j,\omega(a)\}\quad\text{and}\quad d_ay_j=\{y_j,\omega(a)\}.$$

   Furthermore, verify that the G_is are in involution so that

$$\{G_i,G_j\}=0.$$

9.4 THE AIRY KERNEL

9.4.1 Painlevé II equation

The Airy kernel refers to the kernel

$$K^{\rm soft}(X,Y) := \frac{{\rm Ai}(X){\rm Ai}'(Y) - {\rm Ai}(Y){\rm Ai}'(X)}{X-Y}, \tag{9.39}$$

and the task is to compute

$$E_2^{\rm soft}((s,\infty);\xi) = \det(1 - \xi K^{\rm soft}_{(s,\infty)}). \tag{9.40}$$

In [518] it is emphasized that a key ingredient in the derivation of differential equations associated with kernels of the form (9.25) is that ϕ and ψ are related by the coupled first order differential equations

$$\begin{aligned} m(x)\phi'(x) &= A(x)\phi(x) + B(x)\psi(x),\\ m(x)\psi'(x) &= -C(x)\phi(x) - A(x)\psi(x), \end{aligned} \tag{9.41}$$

where m, A, B and C are polynomials. For the Airy kernel (9.39), $\phi(x) = \sqrt{\xi}{\rm Ai}(x)$ and $\psi(x) = \sqrt{\xi}{\rm Ai}'(x)$, and since ${\rm Ai}''(x) = x{\rm Ai}(x)$ we see that the equations (9.41) are satisfied with

$$m(x) = 1, \quad A(x) = 0, \quad B(x) = 1, \quad C(x) = -x. \tag{9.42}$$

Using (9.41), the working of Section 9.3 can be modified to derive coupled nonlinear equations for quantities associated with the Fredholm determinant of the Airy kernel on the interval $(a_1, a_2) = (s, \infty)$ [519]. The equations involve two quantities in addition to those of Definition 9.3.1.

Definition 9.4.1 *Let ϕ, ψ, P and Q be as defined in Definition 9.3.1. In terms of these quantities we write*

$$u := \langle \phi | Q \rangle \qquad \text{and} \qquad v := \langle \phi | P \rangle = \langle \psi | Q \rangle,$$

where $\langle f | g \rangle := \int_{a_1}^{a_2} f(y) g(y) dy$.

Proposition 9.4.2 *For the Airy kernel (9.39), in the notation of Definitions 9.3.1 and 9.4.1 with $a_1 = s$, $a_2 = \infty$, $p := p_1$, $q := q_1$ and $R := R(s,s)$, we have*

$$\begin{aligned} &q' = p - qu, \quad p' = sq + pu - 2qv, \quad R = pq' - qp',\\ &R' = -q^2, \qquad u' = -q^2, \qquad\qquad v' = -pq. \end{aligned}$$

The derivation of these equations will be given in the next subsection. Presently we use of them to derive a single differential equation for $R =: R(s;\xi)$. From this the generating function $E_2^{\rm soft}((s,\infty);\xi)$ that there are no eigenvalues in the interval (s, ∞) can be calculated, since from the general formula (9.26) we have

$$R(s;\xi) = \frac{d}{ds} \log E_2((s,\infty);\xi)$$

and so

$$E_2^{\rm soft}((s,\infty);\xi) = \exp\left(- \int_s^\infty R(t;\xi) dt \right). \tag{9.43}$$

In (8.84) of Proposition 8.3.2 the same formula was obtained, but with $R(t;\xi)$ replaced by $u^s(t;0;\xi)$. Indeed it follows from Proposition 9.4.2 that $R(t;\xi)$ and $u^s(t;0;\xi)$ are the same quantity.

Proposition 9.4.3 *As with $u^s(t;0;\xi)$ specified in Proposition 8.3.2, the quantity R of Proposition 9.4.2 satisfies the differential equation*

$$(R'')^2 + 4R'\Big((R')^2 - sR' + R\Big) = 0, \tag{9.44}$$

and this equation is to be solved subject to the boundary condition $R(t;\xi) \underset{t\to\infty}{\sim} \xi\rho_{(1)}^{\rm soft}(t)$.

Proof. Let us refer to the equations of Proposition 9.4.2 according to their order of presentation. Differentiating the first of the stated equations above, and substituting for the first derivatives using the first, second and fifth equations gives

$$q'' = sq + q^3 + q(u^2 - 2v).$$

But from the first, fifth and sixth equations we see $qq' = uu' - v'$ and thus $q^2 = u^2 - 2v$ (the constant of integration is zero since all terms vanish as $s \to \infty$). Substituting for $u^2 - 2v$ gives

$$q'' = sq + 2q^3, \tag{9.45}$$

which is the special case $\alpha = 0$ of the Painlevé II differential equation as listed in (8.9). Using the fourth equation it can be verified from this that

$$R = q'^2 - sq^2 - q^4.$$

The stated differential equation for R can now be verified using this equation together with further use of the fourth equation. □

The derivation of (9.44) implies a formula for $E_2^{\rm soft}((s,\infty);\xi)$ involving $q = q(t;\xi)$, which according to (9.45) is a Painlevé II transcendent, in contrast to $R(t;\xi)$ which is a σPII transcendent. Thus, making use of the fourth equation in Proposition 9.4.2 and integrating by parts in (9.43), one obtains [519]

$$E_2^{\rm soft}((s,\infty);\xi) = \exp\Big(- \int_s^\infty (t-s) q^2(t;\xi)\, dt \Big), \tag{9.46}$$

where q is the solution of the Painlevé II equation (9.45) satisfying the condition

$$q(s;\xi) \sim \sqrt{\xi}{\rm Ai}(s) \quad \text{as} \quad s \to \infty. \tag{9.47}$$

We remark that it has been proved [294] that (9.45) has a unique solution subject to (9.46) with $\xi = 1$. Furthermore, this solution exhibits the asymptotic behavior

$$q(s;1) \underset{s\to-\infty}{\sim} \sqrt{-s/2} \tag{9.48}$$

and consequently it follows from (9.46) that [519]

$$E_2^{\rm soft}(0;(s,\infty)) \underset{s\to-\infty}{\sim} e^{s^3/12}. \tag{9.49}$$

9.4.2 Numerical computation using the PII equation

The evaluation (9.46) is well suited to the numerical computation of $p_2^{\rm soft}(n;t)$, provided n is small [451]. The strategy is first to calculate the asymptotic expansion of $q(s) := q(s;\xi)$ in (9.45) about $s = \infty$, and to use this to obtain an accurate evaluation of $q(s_0)$ and $q'(s_0)$ for some particular s_0. These values are then used to obtain a power series expansion of $q(s)$ about $s = s_0$, which in turn provides accurate evaluations of $q(s_1)$ and $q'(s_1)$ with $s_1 < s_0$, and the procedure continues (cf. Section 8.3.4).

Proposition 9.4.4 *Write for the $s \to \infty$ asymptotic expansion of the Airy function*

$${\rm Ai}(s) \sim \frac{e^{-(2/3)s^{3/2}}}{2\sqrt{\pi}s^{1/4}} \sum_{n=0}^\infty \frac{(-1)^n}{(\frac{2}{3}s^{3/2})^n}\alpha_n$$

so that [435] $\alpha_n = (6n-1)(6n-5)\alpha_{n-1}/72n$, $\alpha_0 = 1$. *Let* $\alpha_n^{(3)} = \sum_{0\le k\le l\le n} \alpha_{n-l}\alpha_{l-k}\alpha_k$ *and specify* $\{a_n\}_{n=0,1,\dots}$ *by the recurrence*

$$a_n = \alpha_n^{(3)} + \frac{3}{4}na_{n-1} - \frac{1}{8}\Big(n - \frac{1}{6}\Big)\Big(n - \frac{5}{6}\Big)a_{n-2} \tag{9.50}$$

subject to the initial conditions $a_{-2} = a_{-1} = 0$. *We have*

$$q(s;\xi) \sim \sqrt{\xi}\mathrm{Ai}(s) + \xi^{3/2}\frac{e^{-2s^{3/2}}}{32\pi^{3/2}s^{7/4}}\sum_{k=0}^{\infty}\frac{(-1)^k a_k}{(\frac{2}{3}s^{3/2})^k}. \tag{9.51}$$

Proof. Put $q = \sqrt{\xi}\mathrm{Ai}(s) + \xi^{3/2}Q(s)$, where $|Q(s)| \ll \mathrm{Ai}(s)$ for $s \to \infty$. Then

$$Q'' \sim sQ + 2(\mathrm{Ai}(s))^3. \tag{9.52}$$

Since

$$(\mathrm{Ai}(s))^3 \sim \frac{e^{-2s^{3/2}}}{8\pi^{3/2}s^{3/4}}\sum_{n=0}^{\infty}\frac{(-1)^n}{(\frac{2}{3}s^{3/2})^n}\alpha_n^{(3)},$$

we see by substituting

$$Q(s) = \frac{e^{-2s^{3/2}}}{32\pi^{3/2}s^{7/4}}\sum_{k=0}^{\infty}\frac{(-1)^k a_k}{(\frac{2}{3}s^{3/2})^k}$$

in (9.52) and equating like terms that (9.50) results. □

It is empirically observed in [451] that the sum over k in (9.51) is optimally truncated at $k \approx \frac{4}{3}s_0^{3/2}$ for given $s_0 \gg 0$. This then gives accurate numerical evaluations of $q(s_0;\xi)$ and $q'(s_0;\xi)$. For a general s_0, knowledge of $c_0 := q(s_0;\xi)$ and $c_1 := q'(s_0;\xi)$ allows the power series expansion

$$q(s;\xi) = \sum_{l=0}^{\infty} c_l(s-s_0)^l \tag{9.53}$$

to be computed by recurrence.

PROPOSITION 9.4.5 *Let* $c_n^{(k)} := \sum_{j=0}^{n} c_{n-j}c_j^{(k-1)}$ *be the coefficients in the power series expansion of* $(q(s;\xi))^k$ *about* s_0 *(note* $c_n^{(1)} = c_n$*). We have*

$$c_{n+2} = \frac{2c_n^{(3)} + s_0c_n + c_{n-1}}{(n+2)(n+1)}. \tag{9.54}$$

Proof. This follows by direct substitution of (9.53) in (9.45), and equating like powers of $(s-s_0)$. □

According to (8.86), (8.3) and (9.46), with $u(t) := q(t;1)$, $v(t) := \frac{\partial}{\partial\xi}q(t;\xi)|_{\xi=1}$, we have

$$p_2^{\mathrm{soft}}(0;s) = \frac{d}{ds}\exp\Big(-\int_s^{\infty}(t-s)u^2(t)\,dt\Big),$$
$$p_2^{\mathrm{soft}}(1;s) = p_2^{\mathrm{soft}}(0;s) - 2\frac{d}{ds}\Big(\int_s^{\infty}(t-s)u(t)v(t)\,dt\Big)\exp\Big(-\int_s^{\infty}(t-s)u^2(t)\,dt\Big).$$

As it is not practical to leave ξ as a variable in the iteration of (9.54) for general $c_0 = q(s_0;\xi)$ and $c_1 = q'(s_0;\xi)$, we calculate the power series for $v(s)$ about s_0 from knowledge of the power series of $u(s)$ and the initial values $v(s_0)$, $v'(s_0)$, by noting from (9.45) that

$$v'' = (s+6u^2)v.$$

The moments calculated from this procedure [449] are presented in Table 9.1. Numerical computation of the Fredholm determinant (9.40) according to the method of Section 9.2 allows for the high precision computation of $\{p_2^{\mathrm{soft}}(n;t)\}$ beyond $n=0$ and 1 [79], and these too are presented in Table 9.1.

p.d.f.	mean	variance	skewness	kurtosis
$p_2^{\rm soft}(0;t)$	−1.77108 68074	0.81319 47928	0.22408 42036	0.09344 80876
$p_2^{\rm soft}(1;t)$	−3.67543 72971	0.54054 50473	0.12502 70941	0.02173 96385
$p_2^{\rm soft}(2;t)$	−5.17132 31745	0.43348 13326	0.08880 80227	0.00509 66000
$p_2^{\rm soft}(3;t)$	−6.47453 77733	0.37213 08147	0.06970 92726	−0.00114 15160
$p_2^{\rm soft}(4;t)$	−7.65724 22912	0.33101 06544	0.05777 55438	−0.00405 83706
$p_2^{\rm soft}(5;t)$	−8.75452 24419	0.30094 94654	0.04955 14791	−0.00559 9855_

Table 9.1 Statistical properties of $p_2^{\rm soft}(n;t)$ for various n.

9.4.3 Further theory

In this subsection the theory contained in [517] will be used to derive the equations of Proposition 9.4.2. Now, examination of the theory of Section 9.3.1 shows that for any integral operator on the interval (a_1, a_2), with kernel K_J of the form (9.25), there are certain inter-relationships between the quantities of Definitions 9.3.1 and 9.4.1 which are always valid. These are referred to as the *universal equations.*

Proposition 9.4.6 *For $j, k = 1, 2$ we have*

$$\begin{aligned}
&\text{(i)}\ R(a_j,a_k) = \frac{q_j p_k - p_j q_k}{a_j - a_k}\ (j\neq k), && \text{(ii)}\ \frac{\partial}{\partial a_j}\log\det(1-\xi K_J) = (-1)^{j-1}R(a_j,a_j),\\
&\text{(iii)}\ \frac{\partial q_j}{\partial a_k} = (-1)^k R(a_j,a_k) q_k\ (j\neq k), && \text{(iv)}\ \frac{\partial p_j}{\partial a_k} = (-1)^k R(a_j,a_k) p_k\ (j\neq k),\\
&\text{(v)}\ \frac{\partial u}{\partial a_k} = (-1)^k q_k^2, && \text{(vi)}\ \frac{\partial v}{\partial a_k} = (-1)^k p_k q_k,\\
&\text{(vii)}\ \frac{\partial w}{\partial a_k} = (-1)^k p_k^2,
\end{aligned} \tag{9.55}$$

where $j,k = 1,2$; R, a_j, p_j, q_j and K_J are as in Definition 9.3.1; u and v are as in Definition 9.4.1, and $w := \langle\psi|P\rangle := \int_{a_1}^{a_2} P(y)\psi(y)\,dy$ with P and ψ as defined in Definition 9.3.1.

Proof. Equation (i) follows from Proposition 9.3.6, equation (ii) is just (9.26), while (iii) and (iv) follow from Proposition 9.3.10. To derive (v), we note that

$$\frac{\partial u}{\partial a_k} = \frac{\partial}{\partial a_k}\int_{a_1}^{a_2} Q(y)\phi(y)\,dy = (-1)^k\phi(a_k)q_k + \int_{a_1}^{a_2}\phi(y)\frac{\partial}{\partial a_k}Q(y)\,dy.$$

Substituting the value of $\partial Q(y)/\partial a_k$ from Proposition 9.3.10 and then noting that

$$\int_{a_1}^{a_2} R(y,a_k)\phi(y)\,dy = \lim_{\substack{x\to a_k\\ x\in(a_1,a_2)}}\int_{a_1}^{a_2}\Big(\rho(y,x)-\delta(y-x)\Big)\phi(y)\,dy = Q(a_k)-\phi(a_k)$$

gives the stated result. Equations (vi) and (vii) are derived similarly. □

In addition to the universal equations of Proposition 9.4.6 there are equations which depend on the specific form of ψ and ϕ. Let us take up the task of deriving these equations in the case

$$m(x) = 1, \quad A(x) = \alpha_0 + \alpha_1 x, \quad B(x) = \beta_0 + \beta_1 x, \quad C(x) = \gamma_0 + \gamma_1 x. \tag{9.56}$$

First we seek formulas for $P'(x)$ and $Q'(x)$ analogous to those given in Proposition 9.3.8 for the sine kernel. For this purpose the result analogous to Proposition 9.3.7 is required.

PROPOSITION 9.4.7 *With D denoting the operator for differentiation with respect to the independent variable, we have*

$$[D,(1-\xi K_J)^{-1}] \doteq \alpha_1\Big(Q(x)P(y)+P(x)Q(y)\Big)+\beta_1 P(x)P(y)+\gamma_1 Q(x)Q(y)-\sum_{k=1}^{2}(-1)^k R(x,a_k)\rho(a_k,y).$$

Proof. From (9.41) and (9.56) it follows that

$$\begin{aligned}(D_x+D_y)\xi K(x,y) &= \frac{A(x)-A(y)}{x-y}\Big(\phi(x)\psi(y)+\psi(x)\phi(y)\Big)+\frac{B(x)-B(y)}{x-y}\psi(x)\psi(y)+\frac{C(x)-C(y)}{x-y}\phi(x)\phi(y)\\ &= \alpha_1\Big(\phi(x)\psi(y)+\psi(x)\phi(y)\Big)+\beta_1\psi(x)\psi(y)+\gamma_1\phi(x)\phi(y).\end{aligned}$$

The result follows from this according to the proof of Proposition 9.3.7. □

The differentiation formulas for $P(x)$ and $Q(x)$ can now be obtained by following the method of the proof of Proposition 9.3.8.

PROPOSITION 9.4.8 *We have*

$$\begin{aligned}Q'(x) &= \alpha_0 Q(x)+\alpha_1 Q_1(x)+(\alpha_1 v+\gamma_1 u)Q(x)+\beta_0 P(x)+\beta_1 P_1(x)\\ &\quad +(\alpha_1 u+\beta_1 v)P(x)-\sum_{k=1}^{2}(-1)^k R(x,a_k)q_k,\end{aligned}$$

$$\begin{aligned}P'(x) &= -\gamma_0 Q(x)-\gamma_1 Q_1(x)+(\alpha_1 w+\gamma_1 v)Q(x)-\alpha_0 P(x)-\alpha_1 P_1(x)\\ &\quad +(\alpha_1 v+\beta_1 w)P(x)-\sum_{k=1}^{2}(-1)^k R(x,a_k)p_k,\end{aligned}$$

where $Q_1(x) := \int_{a_1}^{a_2} y\rho(x,y)\phi(y)\,dy$ and $P_1(x) := \int_{a_1}^{a_2} y\rho(x,y)\psi(y)\,dy$.

The quantities $Q_1(x)$ and $P_1(x)$ can be written in terms of $Q(x)$, $P(x)$, u, v and w.

PROPOSITION 9.4.9 *We have*

$$Q_1(x)=xQ(x)-\Big(vQ(x)-uP(x)\Big)\quad\text{and}\quad P_1(x)=xP(x)-\Big(wQ(x)-vP(x)\Big).$$

Proof. From the proof of Proposition 9.3.6 we have

$$[M,(1-\xi K_J)^{-1}]\doteq Q(x)P(y)-P(x)Q(y).$$

Applying this to ϕ and ψ and rearranging gives the two stated formulas. □

Substituting the results of Proposition 9.4.9 in Proposition 9.4.8, then substituting the resulting equations in the formula (9.32) for $R(x,x)$ and taking the limit as $x\to a_j$ gives the following formula to supplement the equations of Proposition 9.4.6.

PROPOSITION 9.4.10 *We have*

$$\begin{aligned}R(a_j,a_j) &= q_j^2(\gamma_0+\gamma_1 a_j-2\gamma_1 v-2\alpha_1 w)+2p_jq_j(\gamma_1 u-\alpha_0+\alpha_1 a_j-\beta_1 w)\\ &\quad +p_j^2(\beta_0+\beta_1 a_j+2\alpha_1 u+2\beta_1 v)+\sum_{\substack{k=1\\k\neq j}}^{2}(-1)^k R(a_j,a_k)(q_jp_k-p_jq_k).\end{aligned}$$

According to (9.38) and Proposition 9.3.10, the formulas for the derivatives of $Q(x)$ and $P(x)$ give us formulas for $\frac{\partial q_j}{\partial a_j}$ and $\frac{\partial p_j}{\partial a_j}$.

PROPOSITION 9.4.11 *We have*

$$\frac{\partial q_j}{\partial a_j} = q_j(\alpha_0 + \alpha_1 a_j + \gamma_1 u - \beta_1 w) + p_j(\beta_0 + \beta_1 a_j + 2\alpha_1 u + 2\beta_1 v) - \sum_{\substack{k=1 \\ k \neq j}}^{2} (-1)^k R(a_j, a_k) q_k,$$

$$\frac{\partial p_j}{\partial a_j} = p_j(-\alpha_0 - \alpha_1 a_j - \gamma_1 u + \beta_1 w) + q_j(-\gamma_0 - \gamma_1 a_j + 2\gamma_1 v + 2\alpha_1 w) - \sum_{\substack{k=1 \\ k \neq j}}^{2} (-1)^k R(a_j, a_k) p_k.$$

As our final equation we give the analogue of the second equality in (9.36), which follows from the first equality in (9.36), Proposition 9.4.7, the final statement in the proof of Proposition 9.3.9 and the first statement of Proposition 9.3.9.

PROPOSITION 9.4.12 *We have*

$$\frac{\partial}{\partial a_j} R(a_j, a_j) = -\sum_{\substack{k=1 \\ k \neq j}}^{2} (-1)^k \Big(R(a_j, a_k) \Big)^2 + 2\alpha_1 q_j p_j + \beta_1 p_j^2 + \gamma_1 q_j^2.$$

Derivation of the d.e.'s for the Airy kernel

For the Airy kernel, from (9.42) and (9.56),

$$\alpha_0 = \alpha_1 = 0, \quad \beta_0 = 1, \ \beta_1 = 0, \quad \gamma_0 = 0, \ \gamma_1 = -1. \tag{9.57}$$

Also $(a_1, a_2) = (s, \infty)$, which means that $p_2 = q_2 = R(a_1, a_2) = 0$ and p_1, q_1 depend only on s. The first two equations of Proposition 9.4.2 can thus be seen to follow immediately from Proposition 9.4.11, the third equation from Proposition 9.4.10 and the first two equations, the fourth equation from Proposition 9.4.12, and the fifth and sixth equations from equations (v) and (vi) respectively of Proposition 9.4.7.

EXERCISES 9.4 1. [231] The aim of this exercise is to use the two different evaluations of the soft edge gap probability to deduce an identity between Painlevé II transcendents.

(i) Use the first equation in (8.29), the definition of h_{II} in Proposition 8.2.1 and the first of the Hamilton equations in (8.23) to show that

$$\sigma_{II}'(t) = -\frac{1}{2^{1/3}} \Big(q' + q^2 + \frac{t}{2} \Big) \Big|_{t \mapsto -2^{1/3} t}.$$

Write $q = q_\alpha(t)$ where α is the parameter in the PII equation of (8.9) to conclude from Proposition 8.3.3 and (9.43) that

$$\Big(u^s(t; 0; \xi) \Big)' = R'(t; \xi) = -\frac{1}{2^{1/3}} \Big(q_{-1/2}'(t) + q_{-1/2}^2(t) + \frac{t}{2} \Big) \Big|_{t \mapsto -2^{1/3} t}. \tag{9.58}$$

(ii) Use the fourth equation in Proposition 9.4.2 and (9.45) to deduce from (i) that

$$q_0^2(t) = \frac{1}{2^{1/3}} \Big(q_{-1/2}'(t) + q_{-1/2}^2(t) + \frac{t}{2} \Big) \Big|_{t \mapsto -2^{1/3} t}. \tag{9.59}$$

(In fact an identity of Gambier [254] gives

$$-\epsilon 2^{1/3} q_0^2(-2^{-1/3} t) = \frac{d}{dt} q_{\epsilon/2}(t) - \epsilon q_{\epsilon/2}^2(t) - \frac{\epsilon}{2} t$$

valid for both $\epsilon = \pm$.)

9.5 BESSEL KERNELS

As seen in Chapter 7, there are two kernels which qualify for the title of a Bessel kernel,

$$K^{\rm hard}(X,Y) := \frac{J_a(X^{1/2})Y^{1/2}J_a'(Y^{1/2}) - X^{1/2}J_a'(X^{1/2})J_a(Y^{1/2})}{2(X-Y)} \tag{9.60}$$

$$K^{\rm s.s.}(x,y) := (\pi\rho x)^{1/2}(\pi\rho y)^{1/2}\frac{(J_{a+1/2}(\pi\rho x)J_{a-1/2}(\pi\rho y) - J_{a+1/2}(\pi\rho y)J_{a-1/2}(\pi\rho x))}{2(x-y)}. \tag{9.61}$$

We will treat each separately.

9.5.1 Spacings at the hard edge

Referring to the integral operator on $J = (0,t)$ with kernel $K^{\rm hard}(x,y)$ and resolvent kernel $R(x,y)$, it follows from (9.26) that

$$E_2^{\rm hard}((0,s);\xi) =: E_2^{\rm hard}((0,s);\xi;a) = \exp\Big(-\int_0^s R(t;\xi)\,dt\Big), \tag{9.62}$$

where $R(t;\xi) := R(t,t)$. A theory analogous to that of the sine and Airy kernels can be developed to rederive the characterization of $R(t;\xi)$ as a particular σPIII$'$ transcendent implied by (8.88) and Proposition 8.3.3. Furthermore, this development leads to an alternative formula expressing $E_2^{\rm hard}((0,s);\xi)$ in terms of a particular transformed Painlevé V transcendent [520].

For the Bessel kernel (9.60), $\phi(x) = \sqrt{\xi}J_a(\sqrt{x})$ and $\psi(x) = x\phi'(x)$. From the d.e. satisfied by the Bessel function it follows that (9.41) holds with $m(x) = x$, $A(x) = 0$, $B(x) = 1$ and $C(x) = \frac{1}{4}(x - a^2)$. Thus $A(x)$, $B(x)$ and $C(x)$ are of the form assumed in (9.56) with

$$\alpha_0 = \alpha_1 = 0, \qquad \beta_0 = 1, \beta_1 = 0, \qquad \gamma_0 = -\frac{a^2}{4}, \gamma_1 = \frac{1}{4} \tag{9.63}$$

but now $m(x) = x$. In this case the kernel specific formulas based on Proposition 9.4.7 require modification as the commutator $[D,(1-\xi K_J)^{-1}]$ no longer has a simple kernel. Instead it is the commutator $[MD,(1-\xi K_J)^{-1}]$ which is used to give the desired formulas.

PROPOSITION 9.5.1 *We have*

$$[MD,(1-\xi K_J)^{-1}] \doteq \alpha_1\Big(Q(x)P(y) + P(x)Q(y)\Big) + \beta_1 P(x)P(y) + \gamma_1 Q(x)Q(y) - \sum_{k=1}^{2}(-1)^k a_k R(x,a_k)\rho(a_k,y).$$

Proof. Analogous to the calculation of $[D,K]$ in the proof of Proposition 9.3.7 we find

$$[MD,K] \doteq (xD_x + yD_y + 1)K(x,y) - a_2\delta^-(y-a_2)K(x,y) + a_1\delta^+(y-a_1)K(x,y).$$

But from (9.41) and (9.56) we see that $(xD_x + yD_y + 1)K(x,y)$ is given by the r.h.s. of the equation in the proof of Proposition 9.4.7. The stated result now follows from the workings of the proof of the final identity in Proposition 9.3.7. □

Due to the similarity of this result with the result of Proposition 9.4.7, by inspecting the role of the commutator $[D,K]$ in the derivations of the formulas for $R(a_j,a_j)$ and similar given in the previous section, we see that these formulas require only minor modification.

PROPOSITION 9.5.2 *In the case $m(x) = x$ the formulas of Propositions 9.4.10 and 9.4.12 for $R(a_j, a_j)$ and $\partial R(a_j, a_j)/\partial a_j$ hold provided a factor of a_j is inserted on the l.h.s. in front of $R(a_j, a_j)$ and a factor of a_k is inserted in the summand on the r.h.s.. Similarly, the formulas of Proposition 9.4.11 for $\partial p_j/\partial a_j$ and $\partial p_j/\partial a_j$ hold provided a factor of a_j is inserted on the l.h.s. and a factor of a_k is inserted in the summand on the r.h.s.*

The analogue of the equations of Proposition 9.4.6 obtained for the Airy kernel can now be obtained.

PROPOSITION 9.5.3 *Consider the Bessel kernel (9.60). In the notation of Definitions 9.3.1 and 9.4.1 with $a_1 = 0$, $a_2 = s$, $p := p_2$, $q := q_2$ and $R := R(s, s)$ we have*

$$\begin{aligned} sq' &= p + \tfrac{1}{4}qu, \quad sp' = \tfrac{1}{4}(a^2 - s)q + \tfrac{1}{2}qv - \tfrac{1}{4}pu, \quad (sR)' = \tfrac{1}{4}q^2, \\ u' &= q^2, \qquad\qquad v' = pq. \end{aligned} \tag{9.64}$$

Proof. The first and second equations follow from Proposition 9.4.11 with $j = 2$, modified according to Proposition 9.5.2 and with the substitutions (9.63). The third equation follows from Proposition 9.4.12, while the fourth and fifth equations follow immediately from equations (v) and (vi) of Proposition 9.4.6. □

Making use of the equations (9.64), a single non-linear equation can be derived for the quantity R.

PROPOSITION 9.5.4 *With $R(t; \xi)$, set $\sigma(t; \xi) = -tR$. As with $u^h(t; a, 0; \xi)$ in (8.88), $\sigma(t)$ satisfies the differential equation (8.90) with $\mu = 0$, subject to the boundary condition (8.92).*

Proof. A procedure similar to that used in the derivation of (9.44) can be followed. We begin by applying $s\frac{d}{ds}$ to both sides of the first equation in (9.64), and eliminating the first derivatives by using the second, fourth and fifth equations. This gives

$$s(sq')' = \frac{1}{4}(a^2 - s)q + \frac{1}{16}(u^2 + 8v)q + \frac{1}{4}sq^3. \tag{9.65}$$

Next, by multiplying the first equation in (9.64) by $2q$ and using the fourth and fifth equations we see

$$q^2 + 2sqq' = 2v' + \frac{1}{2}uu' + u', \quad \text{which implies} \quad sq^2 = 2v + \frac{1}{4}u^2 + u.$$

Substituting for $u^2 + 8v$ thus allows (9.65) to be rewritten as

$$s(sq')' = \frac{1}{4}(a^2 - s)q - \frac{1}{4}uq + \frac{1}{2}sq^3. \tag{9.66}$$

Multiplying this equation by q', adding $-\frac{1}{2}(sR)'$ to the l.h.s. and $-\frac{1}{8}q^2$ to the r.h.s. (this is permitted by the third equation of (9.64)) and antidifferentiating shows

$$-sR + (sq')^2 = \frac{1}{4}(a^2 - s - u)q^2 + \frac{1}{4}sq^4.$$

But from the third and fourth equations $u = 4sR$, so

$$(q^2 - 1)sR = \frac{1}{4}(a^2 - s)q^2 + \frac{1}{4}sq^4 - (sq')^2. \tag{9.67}$$

Use of this equation, together with the third equation of (9.64), gives the stated differential equation for $-tR$. □

The above working shows that the quantity q can also be characterized as the solution of a non-linear equation. First note substituting $sR = u/4$ in (9.67) gives an equation expressing $(q^2 - 1)u$ in terms of q. Therefore after multiplying (9.66) by $(1 - q^2)$ we can substitute for the term involving $(q^2 - 1)u$ on the r.h.s. to obtain the differential equation

$$s(q^2-1)(sq')' = q(sq')^2 + \frac{1}{4}(s-a^2)q + \frac{1}{4}sq^3(q^2-2). \tag{9.68}$$

The corresponding boundary condition, which follows from the boundary condition for $\sigma(s;\xi) := -sR$ in Proposition 9.5.4 and the third equation in (9.64), is

$$q(s;\xi) \underset{s\to 0^+}{\sim} \frac{\sqrt{\xi}}{2^a\Gamma(1+a)} s^{a/2}. \tag{9.69}$$

The equation (9.68) is related to a Painlevé equation by a fractional linear transformation. Thus making the transformation

$$q(s) = \frac{1+y(x)}{1-y(x)}, \qquad s = x^2, \tag{9.70}$$

one finds that $y(x)$ satisfies the Painlevé V equation in (8.9) with $\alpha = -\beta = a^2/8$, $\gamma = 0$ and $\delta = -2$. Furthermore, making use of the third equation in (9.64) we deduce from (9.62) that

$$E_2^{\rm hard}((0,s);\xi;a) = \exp\Big(-\frac{1}{4}\int_0^s (\log\frac{s}{t}) q^2(t;\xi)\,dt\Big), \tag{9.71}$$

which shows that $q(t;\xi)$ determines $E_2^{\rm hard}((0,s);\xi)$.

9.5.2 Symmetrical gap about the spectrum singularity

In Section 8.3.5 the generating function $E_2^{\rm s.s.}((0,t);\xi)$ for n eigenvalues in the interval $(0,t)$ with a spectrum singularity at the origin was computed. In this section we will compute the generating function $E_2^{\rm s.s.}((-t,t);\xi)$ for $E_2^{\rm s.s.}(n;\xi)$, the probability that there are exactly n eigenvalues in the symmetric interval $(-t,t)$ about the spectrum singularity at the origin.

Let $p_{2,\rm n.n.}^{\rm bulk}(0;t)$ denote the probability density function for the spacing between nearest neighbour eigenvalues in the bulk. This is distinct from $p_2^{\rm bulk}(0;t)$ which represents the p.d.f. for the spacing between consecutive eigenvalues in the bulk, and thus the spacing distribution of the average of the spacing between the left neighbor and the right neighbor, only one of which being the nearest neighbor. It follows from the definitions that

$$p_{2,\rm n.n.}^{\rm bulk}(0;t) = -\frac{d}{dt}E_2^{\rm s.s.}(0;(-t,t))\Big|_{a=1}.$$

We consider the integral operator in (9.15) with $K(x,y) = K^{\rm s.s.}(x,y)|_{\rho=1/\pi}$. With $R(x,y)$ denoting the kernel of the corresponding resolvent operator and $R(t;\xi) := R(t,t)$, it follows from (9.26) that

$$E_2^{\rm s.s.}(0;(-t,t)) = \exp\Big(-2\int_0^{\pi\rho t} R(s;\xi)\,ds\Big). \tag{9.72}$$

Here, in the terminal of the integral t has been replaced by $\pi\rho t$ to reinstate the general density, and use has been made of the fact that $R(s;\xi)$ is even in s. According to (9.61), the kernel $K^{\rm s.s.}(x,y)|_{\rho=1/\pi}$ is of the type (9.25) with

$$\phi(x) = \sqrt{\frac{\xi x}{2}}J_{a+1/2}(x), \qquad \psi(x) = \sqrt{\frac{\xi x}{2}}J_{a-1/2}(x), \tag{9.73}$$

and the equations (9.41) hold with

$$m(x) = x, \quad \alpha_0 = -a,\ \alpha_1 = 0, \quad \beta_0 = 0,\ \beta_1 = 1, \quad \gamma_0 = 0,\ \gamma_1 = 1. \tag{9.74}$$

In relation to this kernel, the following coupled equations can be derived [218].

PROPOSITION 9.5.5 *In the notation of Definitions 9.3.1 and 9.4.1 with $a_1 = -t$, $a_2 = t$, $p := p_2$, $q := q_2$, $R := R(t,t)$, relating to the kernel (9.61) with $\rho = 1/\pi$ and $a \in \mathbb{Z}_{\geq 0}$, we have*

(i) $tR = 2(-a+u-w)pq + t(p^2+q^2) + 2(pq)^2$ (ii) $tq' = (-a+u-w)q + tp$,
(iii) $tp' = -tq - (-a+u-w)p$, (iv) $(tR)' = p^2 + q^2$,
(v) $u' = 2q^2$, (vi) $w' = 2p^2$.

Proof. For $a \in \mathbb{Z}_{\geq 0}$, we have from (9.73) that

$$\phi(-x) = (-1)^{a-1}\phi(x), \qquad \psi(-x) = (-1)^a \psi(x),$$

which gives $K^{\rm s.s.}(x,y) = K^{\rm s.s.}(-x,-y)$ and thus $\rho(x,y) = \rho(-x,-y)$. From Definition 9.3.1 we then see that

$$q_1 = (-1)^{a-1} q_2 = (-1)^{a-1} q, \quad p_1 = (-1)^a p_2 = (-1)^a p, \quad v = 0. \tag{9.75}$$

Equation (i) can now be derived from (9.4.10) with $j = 2$, modified according to Proposition 9.5.2, by substituting (9.75) and (9.74). Equations (ii) and (iii) follow from Proposition 9.4.11, modified according to Proposition 9.5.2, after noting from equation (i) of Proposition 9.4.6 and (9.75) that

$$R(-t,t) = (-1)^a \frac{pq}{t}.$$

Equation (iv) follows from Proposition 9.4.11, modified according to Proposition 9.5.2, now with use being made of the general formula

$$\frac{d}{dt} f(-t,t) = \Big(\frac{\partial}{\partial a_2} - \frac{\partial}{\partial a_1}\Big) f(a_1,a_2)\Big|_{a_2=-a_1=t}. \tag{9.76}$$

Equations (v) and (vi) result from the universal equations (v) and (vii) of Proposition 9.4.6, together with (9.76). □

From the equations of Proposition 9.5.5, a procedure similar to that used to derive Proposition 9.3.3 from Proposition 9.3.2 allows $\sigma(s;\xi) = -sR(s/2;\xi)$ to be specified as the solution of a certain nonlinear equation.

PROPOSITION 9.5.6 *The quantity $\sigma(s;\xi) := -sR(s/2;\xi)$, where $R(s;\xi)$ occurs in (9.72), satisfies the differential equation*

$$(s\sigma'')^2 + 4(-a^2 + s\sigma' - \sigma)\Big((\sigma')^2 - \Big(a - \sqrt{(a^2 - s\sigma' + \sigma)}\Big)^2\Big) = 0 \tag{9.77}$$

subject to the boundary condition

$$\sigma(s;\xi) \sim -\xi \frac{2(s/4)^{2a+1}}{\Gamma(1/2+a)\Gamma(3/2+a)} \quad \text{as} \quad s \to 0. \tag{9.78}$$

Proof. Consider the equations of Proposition 9.5.5. Multiply (ii) by p, multiply (iii) by q, add and use (v) to obtain

$$(pq)' = p^2 - q^2 = \frac{1}{2}(w' - u').$$

Antidifferentiating gives

$$pq = \frac{1}{2}(w - u)$$

which together with (iv) allows (i) to be rewritten as

$$tR = -2a(pq) - 2(pq)^2 + t(tR)'. \tag{9.79}$$

Another equation relating tR to pq is obtained by squaring (iv), squaring the formula $(pq)' = p^2 - q^2$ obtained above, and subtracting. This gives

$$((pq)')^2 - ((tR)')^2 = -4(pq)^2. \tag{9.80}$$

Solving (9.79) for pq (it follows from a small t expansion that the negative square root is to be taken) and $(pq)'$, substituting in (9.80) and introducing the notation $\sigma(2t) := -2tR$ gives (9.77). The boundary condition follows from the fact that $R(t;\xi) \sim \xi K^{\rm s.s.}(t,t)\Big|_{\rho=1/\pi}$ as $t \to 0$, and the corresponding behavior of $K^{\rm s.s.}(t,t)$ deduced from (7.49). □

The second order second degree equation (9.77) involves a square root and so is not a σ-form Painlevé equation (8.15). However (8.15) lists σPIII′; if instead one considers the Hamiltonian theory of PIII rather than PIII′, the equation (9.77) can be recognized in this context and $\sigma(s;\xi)$ identified as an auxiliary Painlevé III Hamiltonian (see Exercises 9.5 q.1).

EXERCISES 9.5 1. [548] The Hamiltonian for the PIII system, in contrast to the PIII′ system (recall Section 8.2.1), is given by [429]

$$tH_{III} = 2q^2p^2 - \Big(2tq^2 + (2v_1+1)q - 2t\Big)p + (v_1+v_2)tq,$$

where the parameters v_1, v_2 relate to the parameters in the PIII equation of (8.9) by

$$\alpha = -4v_2, \quad \beta = 4(v_1+1), \quad \gamma = 4, \quad \delta = -4.$$

(i) With

$$h_{III}(t) := tH_{III} + \frac{1}{8}(2v_1+1)^2$$

show that

$$h'_{III} = -2q^2 + 2p + (v_1+v_2)q, \qquad 8(h_{III} - th'_{III}) = (4pq - 2v_1 - 1)^2.$$

Differentiate the latter and use the Hamilton equations (8.23) to show

$$-th''_{III} = (4pq - 2v_1 - 1)(2pq^2 + 2p - (v_1+v_2)q).$$

(ii) Use the results of (i) to verify that for $\epsilon = \pm 1$

$$\begin{aligned}(th''_{III})^2 = {}& [2(h_{III} - th'_{III})]\Big(4(h'_{III})^2 + 16[2(h_{III} - th'_{III})] \\ & - 16\epsilon(v_2 - v_1 - 1)\sqrt{2(h_{III} - th'_{III})} - 16\Big(v_2 - \frac{1}{2}\Big)\Big(v_1 + \frac{1}{2}\Big)\Big).\end{aligned}$$

(iii) For $\epsilon = \pm$, show that with

$$v_1 = -\epsilon a - \frac{1}{2}, \qquad v_2 = \epsilon a + \frac{1}{2}, \qquad \sigma(t) = 2h_{III}\Big(\frac{t}{4i}\Big) - a^2$$

the equation in (ii) reduces to (9.77).

2. Make use of (7.53), (8.101), (9.15) and (9.61) to show

$$E_2^{O^\pm}(s;\xi) = \det\Big(1 - \xi K^{\rm ch}_{\pm 1/2}\Big|_{(0,s)}\Big), \tag{9.81}$$

where $K^{\rm ch}_{\pm 1/2}|_{(0,s)}$ is the integral operator on $(0,s)$ with kernel (7.54).

9.6 EIGENVALUE EXPANSIONS FOR GAP PROBABILITIES

9.6.1 Commuting differential operators and $E_2(n;J)$

We have seen that the logarithmic derivatives of the Fredholm determinants (9.15) satisfy second order nonlinear differential equations of Painlevé type. In this section we will show that the eigenvalues of the integral operators $K_J^{\rm scale}$ can be obtained from the eigenvalues of certain second order linear differential operators. As a consequence, the asymptotic form of the eigenvalues for large values of the parameter determining J

can be determined, and this in turn allows the asymptotic form of $E_2^{\rm scale}(n;J)/E_2^{\rm scale}(0;J)$ — the ratio of the probability that there are n eigenvalues in J to the probability that there are no eigenvalues in J — to be computed [517].

To appreciate this last point, we first remark that for each of the three cases under consideration (no direct analysis of scale = s.s. is considered) $K_J^{\rm scale}$ has distinct eigenvalues which can be labeled by a non-negative integer and so ordered

$$E_2^{\rm scale}(n;J) = \prod_{l=0}^{\infty}(1-\xi\lambda_l). \tag{9.82}$$

Use of this formula in (8.1) gives

$$\frac{E_2^{\rm scale}(n;J)}{E_2^{\rm scale}(0;J)} = \sum_{0\le j_1<\cdots<j_n} \frac{\lambda_{j_1}\cdots\lambda_{j_n}}{(1-\lambda_{j_1})\cdots(1-\lambda_{j_n})}. \tag{9.83}$$

The utility of this equation lies in the fact that for large values of the parameter specifying J the eigenvalues decrease exponentially fast allowing the corresponding asymptotic form of the l.h.s. to be read off from the first term of the multiple sum, $(j_1,\dots,j_n)=(0,1,\dots,n-1)$.

Some qualitative features of the λ_l can readily be deduced. First we note an upper bound. For this we begin by observing that as $|J|\to 0$, the eigenvalues of $1-\xi K_J$ must tend to 1, and thus each λ_l must tend to zero. These eigenvalues are a continuous function of the parameter determining J, so the fact that $E_2^{\rm scale}(J;1) = E_2^{\rm scale}(0;J)$ is a strictly positive quantity then implies $1-\lambda_l>0$ for each l and for all values of the parameter specifying J.

A lower bound on the λ_l can also be deduced. For this each integral operator $K_J^{\rm scale}$, scale = bulk, soft, hard, is written as the square of another integral operator, thereby implying $\lambda_l\ge 0$ for each l. Thus the following results hold [259], [519], [520].

PROPOSITION 9.6.1 *Let $\tilde K_{(-1,1)}^{\rm bulk}$ be the integral operator on $(-1,1)$ defined by*

$$\tilde K_{(-1,1)}^{\rm bulk}[f] = \int_{-1}^{1} \frac{\sin(t(x-y))}{\pi(x-y)} f(y)\,dy.$$

We have

$$\tilde K_{(-1,1)}^{\rm bulk} = (V_{(-1,1)}^{\rm bulk})^\dagger V_{(-1,1)}^{\rm bulk}, \qquad V_{(-1,1)}^{\rm bulk}[f] = \int_{-1}^{1}\Big(\frac{t}{2\pi}\Big)^{1/2} e^{ itxy} f(y)\,dy.$$

Let $\tilde K_{(0,\infty)}^{\rm soft}$ be the integral operator on $(0,\infty)$ defined by

$$\tilde K_{(0,\infty)}^{\rm soft}[f] = \int_0^\infty dy \int_0^\infty du\, {\rm Ai}(x+u+t){\rm Ai}(y+u+t) f(y).$$

We have

$$\tilde K_{(0,\infty)}^{\rm soft} = (V_{(0,\infty)}^{\rm soft})^2, \qquad V_{(0,\infty)}^{\rm soft}[f] = \int_0^\infty {\rm Ai}(x+u+s) f(u)\,du.$$

Let $\tilde K_{(0,1)}^{\rm hard}$ be the integral operator on $(0,1)$ defined by

$$\tilde K_{(0,1)}^{\rm hard}[f] = \frac{t}{4}\int_0^1 dy \int_0^1 du\, J_a(\sqrt{txu})J_a(\sqrt{tyu}) f(y).$$

We have

$$\tilde K_{(0,1)}^{\rm hard} = (V_{(0,1)}^{\rm hard})^2, \qquad V_{(0,1)}^{\rm hard}[f] = \frac{\sqrt{t}}{2}\int_0^1 J_a(\sqrt{txu}) f(u)\,du.$$

Proof. In general, if $V_J \doteq V(x,y)$, then $V_J^\dagger V_J \doteq \int_J \overline{V(u,x)} V(u,y)\, du$, so the results are immediate. □

The operators $\tilde{K}^{\rm bulk}_{(-1,1)}$, $\tilde{K}^{\rm soft}_{(0,\infty)}$, $\tilde{K}^{\rm hard}_{(0,1)}$ result from $K^{\rm bulk}_{(-t,t)}|_{\rho=1/\pi}$, $K^{\rm soft}_{(t,\infty)}$, $K^{\rm hard}_{(0,t)}$ respectively, with the kernels written in integral form as given in Exercises 7.1 q.1, after a simple change of variable to scale the t-dependence from the terminals of integration, a transformation which leaves unchanged the eigenvalues.

The decompositions of Proposition 9.6.1 lead to the characterizations of the eigenvalues in (9.15) in terms of the eigenvalues of differential operators. This comes about because for each of the integral operators V_J, one can construct a second order linear differential operator of the form

$$L = \frac{d}{dx}\alpha(x)\frac{d}{dx} + \beta(x) \tag{9.84}$$

which commutes with V_J.

PROPOSITION 9.6.2 *Let $V_{[a_1,a_2]} \doteq V(x,y)$ and let L be given by (9.84). If $\alpha(a_1) = \alpha(a_2) = 0$ and*

$$\Big(\frac{\partial}{\partial x}\alpha(x)\frac{\partial}{\partial x} + \beta(x)\Big)V(x,y) = \Big(\frac{\partial}{\partial y}\alpha(y)\frac{\partial}{\partial y} + \beta(y)\Big)V(x,y), \tag{9.85}$$

then L commutes with $V_{[a_1,a_2]}$.

Proof. We have

$$\begin{aligned} LV_{[a_1,a_2]}[f] &= \int_{a_1}^{a_2} \Big(\frac{\partial}{\partial x}\alpha(x)\frac{\partial}{\partial x} + \beta(x)\Big)V(x,y)f(y)\,dy \\ &= \int_{a_1}^{a_2} \Big(\frac{\partial}{\partial y}\alpha(y)\frac{\partial}{\partial y} + \beta(y)\Big)V(x,y)f(y)\,dy \\ &= -\int_{a_1}^{a_2} \Big(\frac{\partial}{\partial y}V(x,y)\Big)\alpha(y)\frac{\partial}{\partial y}f(y)\,dy + \int_{a_1}^{a_2} \beta(y)V(x,y)f(y)\,dy \\ &= \int_{a_1}^{a_2} V(x,y)\Big(\frac{\partial}{\partial y}\alpha(y)\frac{\partial}{\partial y} + \beta(y)\Big)f(y)\,dy, \end{aligned}$$

where the second equality follows from (9.85), while the third and fourth equalities follow by integration by parts and use of the assumption that $\alpha(y)$ vanishes at $y = a_1, a_2$. □

One can verify that the assumption on $\alpha(x)$ and condition (9.85) are satisfied with

$$\begin{aligned} &\alpha^{\rm bulk}(x) = x^2 - 1, &&\beta^{\rm bulk}(x) = t^2x^2, \\ &\alpha^{\rm soft}(x) = x, &&\beta^{\rm soft}(x) = -(x+t)x, \\ &\alpha^{\rm hard}(x) = x(1-x), &&\beta^{\rm hard}(x) = -\tfrac{a^2}{4x} - \tfrac{tx}{4}. \end{aligned}$$

Consider first the bulk case. Then the corresponding eigenfunctions $S_n(x;t) =: S_n(x)$ say, are known as the *prolate spheroidal functions.* They form an orthonormal set on $(-1,1)$ and are entire functions of x, which are real for real x (in the limit $t \to 0^+$ they reduce to the Legendre polynomials, up to normalization). For n even $S_n(x)$ is an even function of x, while for n odd, $S_n(x)$ is an odd function of x.

Because $L^{\rm bulk}$ commutes with $V^{\rm bulk}_{(-1,1)}$ it follows that the $S_n(x) =: S_n(x;t)$ defined as the eigenfunctions of $L^{\rm bulk}$ are also the eigenfunctions of $V^{\rm bulk}_{(-1,1)}$. Using the parity property of the $S_n(x)$ gives that the corresponding eigenvalue equation can be written

$$\begin{aligned} &\Big(\frac{t}{2\pi}\Big)^{1/2} \int_{-1}^{1} \cos xyt\, S_n(y;t)\,dy = \mu_n(t)S_n(x;t), \quad n \text{ even}, \\ &\Big(\frac{t}{2\pi}\Big)^{1/2} i \int_{-1}^{1} \sin xyt\, S_n(y;t)\,dy = \mu_n(t)S_n(x;t), \quad n \text{ odd}, \end{aligned}$$

where the $\mu_n(t)$ are the eigenvalues. The fact that $S_n(x)$ is real for real x implies that $\mu_n(t)$ is real for n even and is pure imaginary for n odd. It also shows that the $S_n(x)$ are the eigenfunctions of $V^\dagger$, with eigenvalues $\bar{\mu}_n(t)$, and thus of $\tilde{K}^{\rm bulk}_{(-1,1)} = (V^{\rm bulk}_{(-1,1)})^\dagger V^{\rm bulk}_{(-1,1)}$ with corresponding eigenvalues $|\mu_n(t)|^2 =: \lambda_n(t)$.

The computation of the large t asymptotics of the $\lambda_n(t)$ makes use of the *Hellmann-Feynman* formula [296], [174], which gives a relation between $\lambda_n(t)$ and the eigenfunction $S_n(x;t)$ evaluated at the special point $x = 1$.

PROPOSITION 9.6.3 *Let $K(x,y;t)$ be a symmetric (in x and y) kernel on (a_1,a_2) depending smoothly on t. Let $g_j(x;t)$ be an eigenfunction, normalized so that*

$$\int_{a_1}^{a_2} (g_j(x;t))^2 dx = 1,$$

with corresponding eigenvalue $\lambda_j(t)$. We have

$$\lambda_j'(t) = \int_{a_1}^{a_2} dx\, g_j(x;t) \int_{a_1}^{a_2} dy\, g_j(y;t) \frac{\partial}{\partial t} K(x,y;t).$$

In particular, for the operator $V^{\rm bulk}_{(-1,1)}$ of Proposition 9.6.2 and with $a_1 = -1$, $a_2 = 1$, this gives

$$\mu_j'(t) = \frac{\mu_j(t)}{t} S_j^2(1;t) \quad \textit{or equivalently} \quad \lambda_j'(t) = 2\frac{\lambda_j(t)}{t} S_j^2(1;t).$$

Proof. For the first statement, we follow [519]. Differentiating the eigenvalue equation for $g_j(x;t)$ gives

$$\int_{a_1}^{a_2} \frac{\partial}{\partial t} K(x,y;t) g_j(y;t)\, dy + \int_{a_1}^{a_2} K(x,y;t) \frac{\partial}{\partial t} g_j(y;t) dy = \lambda_j'(t) g_j(x;t) + \lambda_j(t) \frac{\partial}{\partial t} g_j(x;t).$$

Multiplying both sides by $g_j(x;t)$, integrating over x, and using the normalization, the eigenvalue equation and the symmetry in x and y of K gives the desired result. The first equation of the second result follows simply from the first result by noting that

$$\frac{\partial}{\partial t} e^{itxy} = \frac{x}{t}\frac{\partial}{\partial x} e^{itxy},$$

and using integration by parts. The second part follows from the first part and the fact that $\lambda_j = (-1)^j \mu_j^2$. □

Applying the WKB method of asymptotic analysis to be the eigenvalue equation for the differential operator L, it was shown by Fuchs [245] (after the identification $S_j(x;t) = a^{-1/2} f_j(ax;t)$, $a = \sqrt{t}$, where f_j is the eigenfunction studied in [245]) that the eigenfunctions have the asymptotic behavior

$$\pm S_j(1;t) = 2^{(3j+2)/2} \pi^{1/4} (j!)^{-1/2} t^{(j+1/2)/2} e^{-t} \mu_j(t)(1 + {\rm o}(1))$$

for j fixed and $t \to \infty$. Substituting in the final equation of Proposition 9.6.3 gives

$$\frac{1}{\lambda_j^2(t)} \lambda_j'(t) = \frac{\pi^{1/2}}{j!} 8^{j+1} t^{j+1/2} e^{-2t} (1 + {\rm o}(1)),$$

and this, when integrated from t' to ∞, implies

$$\frac{1}{\lambda_j(t')} - 1 \sim \frac{\pi^{1/2}}{j!} 8^{j+1} \int_{t'}^\infty t^{j+1/2} e^{-2t}\, dt \sim \frac{4\pi^{1/2}}{j!} 8^j t'^{j+1/2} e^{-2t'}$$

(here use has been made of the fact that $\lambda_j(t) \sim 1$ as $t \to \infty$). Thus one has the following result [245].

PROPOSITION 9.6.4 *The eigenvalues $\lambda_j(t)$ of the integral operator $\tilde{K}^{\rm bulk}_{(-1,1)}$ have the asymptotic behavior*

$$1 - \lambda_j(t) \sim 4\pi^{1/2} 8^j (j!)^{-1} t^{j+1/2} e^{-2t} \tag{9.86}$$

for j fixed and $t \to \infty$.

In the soft and hard edge cases the differential operator implied by (9.86) does not lead to special functions previously studied from this viewpoint. Nonetheless an analysis parallel to that sketched above leading to (9.86) can be carried out to obtain the following results [519], [520].

PROPOSITION 9.6.5 *Let i be fixed. One has*

$$1 - \lambda_i^{\rm soft} \underset{t\to-\infty}{\sim} \frac{\sqrt{\pi}}{i!} 2^{5i+3} \Big(-\frac{t}{2} \Big)^{3i/2+3/4} \exp\Big(-\frac{8}{3}\Big(-\frac{t}{2}\Big)^{3/2}\Big) \tag{9.87}$$

$$1 - \lambda_i^{\rm hard} \underset{t\to\infty}{\sim} \frac{2\pi}{\Gamma(a+i+1)i!} t^{i+(a+1)/2} e^{-2\sqrt{t}} 2^{4i+2a+2}. \tag{9.88}$$

Substituting (9.86), (9.87), (9.88) for the first term of (9.83) gives the sought asymptotic forms for the l.h.s. of the latter.

PROPOSITION 9.6.6 *Let $G(x)$ denote the Barnes G-function. For n fixed*

$$\begin{aligned} \frac{E_2^{\rm bulk}(n;t)}{E_2^{\rm bulk}(0;t)} &\underset{t\to\infty}{\sim} G(n+1)\pi^{-n/2} 2^{-n^2-n/2} (\pi\rho t)^{-n^2/2} e^{n\pi\rho t}, \\ \frac{E_2^{\rm soft}(n;(t,\infty))}{E_2^{\rm soft}(0;(t,\infty))} &\underset{t\to-\infty}{\sim} \frac{G(n+1)}{\pi^{n/2} 2^{(5n^2+n)/2}} (-t/2)^{-3n^2/4} \exp\Big(\frac{8n}{3}\Big(-\frac{t}{2}\Big)^{3/2}\Big), \\ \frac{E_2^{\rm hard}(n;(0,t);a)}{E_2^{\rm hard}(0;(0,t);a)} &\underset{t\to\infty}{\sim} \frac{G(a+n+1)G(n+1)}{G(a+1)} \pi^{-n} 2^{-n(2n+2a+1)} t^{-n^2/2-an/2} e^{2n\sqrt{t}}, \end{aligned} \tag{9.89}$$

where in the bulk case the setting of considering the interval $J=(-t,t)$ and $\rho = 1/\pi$ has been changed to the interval $J=(0,t)$ with a general ρ by scaling t.

9.6.2 Further asymptotics

The results of Proposition 9.6.6 can be supplemented by specifying the asymptotic behavior of $E_2^{\rm bulk}(0;t)$, $E_2^{\rm soft}(0,(t,\infty))$ and $E_2^{\rm hard}(0;(0,t))$ as the size of the interval gets large. This in turn is related to the asymptotic form of the Painlevé transcendents specifying these quantities.

A log-gas argument, given in Section 14.6, predicts that for some unspecified case dependent constants c,

$$\log E_2^{\rm bulk}(0;t) \underset{t\to\infty}{\sim} -ct^2, \quad \log E_2^{\rm soft}(0,(t,\infty)) \underset{t\to-\infty}{\sim} ct^3, \quad \log E_2^{\rm hard}(0,(0,t)) \underset{t\to\infty}{\sim} -ct. \tag{9.90}$$

Rigorous determination of the first of these behaviors, which also specifies the proportionality constant as $c = \pi^2/8$ (assuming $\rho = 1$), can be found in [543], [129]. The second result is just (9.49) and so for this $c = 1/12$. The behaviors (9.90) imply specific leading behaviors of the corresponding Painlevé transcendents, and with this established the differential equations generate unique asymptotic expansions. The latter allow (9.90) to be extended [517], [519], [520].

PROPOSITION 9.6.7 *We have*

$$\begin{aligned} E_2^{\rm bulk}(0;t) &\underset{t\to\infty}{\sim} c_{\rm bulk} \frac{e^{-(\pi\rho t)^2/8}}{(\pi\rho t)^{1/4}} \Big(1 - \frac{1}{8(\pi\rho t)^2} + {\rm O}((\pi\rho t)^{-4})\Big), \\ E_2^{\rm soft}(0;(t,\infty)) &\underset{t\to-\infty}{\sim} \frac{c_{\rm soft}}{(|t|)^{1/8}} e^{-|t|^3/12} \Big(1 + \frac{3}{2^6|t|^3} + \frac{2025}{2^{13}t^6} + {\rm O}(t^{-9})\Big), \\ E_2^{\rm hard}(0;(0,t);a) &\underset{t\to\infty}{\sim} c_{\rm hard} \frac{e^{-t/4+a\sqrt{t}}}{t^{a^2/4}} \Big(1 + \frac{a}{8t^{1/2}} + \frac{9a^2}{128t} + {\rm O}\Big(\frac{1}{t^{3/2}}\Big)\Big), \end{aligned} \tag{9.91}$$

where $c_{\rm bulk}, c_{\rm soft}, c_{\rm hard}$ have the values

$$c_{\rm bulk} = 2^{1/6} e^{3\zeta'(-1)}, \quad c_{\rm soft} = 2^{1/24} e^{\zeta'(-1)}, \quad c_{\rm hard} = \frac{G(1+a)}{(2\pi)^{a/2}}. \tag{9.92}$$

Proof. It follows from (8.1) and the first equation in (8.111) that

$$\log E_2^{\rm bulk}(0;t) = \int_0^{\pi t} \frac{\sigma(s;1)}{s}\,ds, \tag{9.93}$$

where σ satisfies the first equation in (8.112). To reproduce the leading behavior (9.90) with c as specified it must be that

$$\sigma(s;1) \underset{s\to\infty}{\sim} -\frac{s^2}{4}.$$

In fact the differential equation satisfied by σ has a unique solution of the form

$$\sigma(s;1) = a_2 s^2 + a_0 + a_{-2}s^{-2} + \cdots \qquad (a_2 \ne 0)$$

which from the knowledge that $a_2 = -\frac{1}{4}$ is calculated as

$$\sigma(s;1) = -\frac{s^2}{4} - \frac{1}{4} - \frac{1}{4s^2} - \frac{5}{2s^4} - \frac{131}{2s^6} + {\rm O}\Big(\frac{1}{s^8}\Big).$$

Substituting in (9.93) and reinstating the general density ρ gives the first expansion in (9.91). The value of the undetermined constant $\tau_{\rm bulk}$ can be deduced from a result in the theory of Toeplitz determinants (see Exercise 9.6 q.1). A rigorous determination is also known [363], [163], [127].

An analogous strategy applies in the soft and hard edge cases. Thus according to (9.43) and the first equation in (8.93)

$$\log E_2^{\rm soft}(0;(t,\infty)) = -\int_t^\infty R(s;1)\,ds, \qquad \log E_2^{\rm hard}(0;(0,t)) = -\int_0^t \sigma_{III'}(s)\Big|_{\substack{\mu=0\\ \xi=1}} \frac{ds}{s}, \tag{9.94}$$

where $R(s;1)$ satisfies the differential equation (9.44) and $\sigma_{III'}(s)$ satisfies the σPIII$'$ equation in (8.15) with $v_1 = v_2 = a$. The first of these differential equations has a unique solution of the form $\sum_{j=0}^\infty a_{2-j}s^{2-j}$, which calculation shows is given by

$$R(s) \sim \frac{s^2}{4} - \frac{1}{8s} + \frac{9}{64s^4} + {\rm O}\Big(\frac{1}{s^7}\Big),$$

while the second has a unique solution of the form $\sigma(s) = c_1 s + c_{1/2}s^{1/2} + c_0 + c_{-1/2}s^{-1/2} + \cdots$ provided the sign in $\sqrt{a^2} = \pm a$ is fixed. Choosing the minus sign (this makes sense physically from the resulting asymptotic expression for $\log E_2(0;(0,s))$) allows the expansion

$$\sigma(s) = \frac{s}{4} - \frac{a}{2}s^{1/2} + \frac{a^2}{4} + \frac{a}{16}s^{-1/2} + \frac{a^2}{16}s^{-1} + \cdots \tag{9.95}$$

to be generated. Substituting as appropriate in (9.94) gives the final two expansions in (9.91), up to the constants.

In the soft edge case the constant follows from the result [126], [30] that for $c < 0$

$$\int_c^\infty R(y;1)\,dy + \int_{-\infty}^c \Big(R(y;1) - \frac{1}{4}y^2 + \frac{1}{8y}\Big)dy = -\frac{1}{24}\log 2 - \zeta'(-1) + \frac{1}{12}|c|^3 + \frac{1}{8}\log|c|.$$

For $a \in \mathbb{Z}_{\ge 0}$, the constant in the hard edge case can be deduced from the integral formula (8.96), as is done for a more general integral formula in (13.52) below. It's validity for $-1 < a < 1$ has been proved in [162]. □

The expansions (9.90) and (9.92) in the hard edge case with $a = \pm\frac{1}{2}$ allow us to deduce the large t expansions of $E_\beta^{\rm bulk}(n;t)$ for $\beta = 1$ and 4 [49]. First the case $n = 0$ will be considered. According to (8.139), $E_1^{\rm bulk}(0;s) = E_2^{\rm hard}(0;(0,(\pi s/2)^2))|_{a=-1/2}$, where on the l.h.s. the density has been set equal to 1. It thus follows from the final formula in (9.91) that

$$E_1^{\rm bulk}(0;t) \underset{t\to\infty}{\sim} (2\pi)^{1/4} G(1/2) \frac{e^{-(\pi\rho t)^2/16 - \pi\rho t/4}}{(\pi\rho t/2)^{1/8}}\Big(1 + {\rm O}\Big(\frac{1}{t}\Big)\Big). \tag{9.96}$$

Furthermore, it follows from (8.159) that

$$E_4^{\rm bulk}(0;t) \underset{t\to\infty}{\sim} \frac{1}{2} E_2^{\rm hard}(0;(0,(\pi t)^2))|_{a=1/2}. \tag{9.97}$$

The final formula in (9.91) then gives

$$E_4^{\rm bulk}(0;t) \underset{t\to\infty}{\sim} \frac{G(3/2)}{2(2\pi)^{1/4}} \frac{e^{-(\pi\rho t)^2/4+\pi\rho t/2}}{(\pi\rho t)^{1/8}} \Big(1+{\rm O}\Big(\frac{1}{t}\Big)\Big). \tag{9.98}$$

The asymptotic formulas for $E_\beta^{\rm bulk}(0;t)$, $\beta=1,2,4$ are all consistent with the form

$$E_\beta^{\rm bulk}(0;t) \underset{t\to\infty}{\sim} c_\beta \frac{e^{-\beta(\pi\rho t)^2/16+(\beta/2-1)\pi\rho t/2}}{(\pi\rho t)^{(1-(2/\beta)(1-\beta/2)^2)/4}}. \tag{9.99}$$

Valko and Virág [525] have used the stochastic sine equation (13.179) below to prove this form for general $\beta>0$.

In relation to the general n case, the formula (8.153) tells us that

$$E_1^{\rm bulk}(n;t) \underset{t\to\infty}{\sim} \begin{cases} E_2^{\rm hard}(n/2;(\pi\rho t/2)^2)|_{a=-1/2}, & n \text{ even} \\ E_2^{\rm hard}((n-1)/2;(\pi\rho t/2)^2)|_{a=1/2}, & n \text{ odd}. \end{cases}$$

Making use of both (9.90) and (9.92) it follows from this that

$$\frac{E_1^{\rm bulk}(n;t)}{E_1^{\rm bulk}(0;t)} \underset{t\to\infty}{\sim} \frac{G(n/2+1/2)G(n/2+1)}{G(1/2)} \pi^{-n/2} 2^{-n^2/2} (\pi\rho t/2)^{-n(n-1)/4} e^{n\pi\rho t/2}. \tag{9.100}$$

Similarly, from (8.159) we have

$$E_4^{\rm bulk}(n;t) \underset{t\to\infty}{\sim} \frac{1}{2} E_2^{\rm hard}(n;(\pi\rho t)^2)\Big|_{a=1/2}. \tag{9.101}$$

Recalling (9.97) and making use of (9.90) this gives

$$\frac{E_4^{\rm bulk}(n;t)}{E_4^{\rm bulk}(0;t)} \underset{t\to\infty}{\sim} \frac{G(n+3/2)G(n+1)}{G(3/2)} \pi^{-n} 2^{-2n(n+1)} (\pi\rho t)^{-n^2-n/2} e^{2n\pi\rho t}. \tag{9.102}$$

We see from Proposition 9.6.6, (9.100) and (9.102) that the asymptotic formulas for $E_\beta^{\rm bulk}(n;t)/E_\beta^{\rm bulk}(0;t)$, $\beta=1,2,4$ are all consistent with the form

$$\frac{E_\beta^{\rm bulk}(n;t)}{E_\beta^{\rm bulk}(0;t)} \underset{t\to\infty}{\sim} c_{\beta,n} \frac{e^{\beta n\pi\rho t/2}}{(\pi\rho t)^{\beta n^2/4+(\beta/2-1)n/2}}, \tag{9.103}$$

which can in fact be obtained from a log-gas analysis (see Section 14.6.2).

9.6.3 Some gap probabilities at $\beta=1$ with an evenness symmetry

In Section 8.4.2 the equation (8.149) relating gap probabilities for the COE ($\beta=1$ quantities) to gap probabilities in the JUE ($\beta=2$ quantities) was shown to hold for $\xi=1$. Here (8.149), along with some companion identities, will be derived for general ξ via a calculation which uses evenness symmetry as well as the eigenvalues and eigenvectors of an underlying Fredholm integral operator [395], [197].

First we will derive a determinant formula for the generating function

$$E_{2N,1}((-t,t);\xi;w_1) := \frac{1}{\hat{Z}_{2N}} \int_{-\infty}^{\infty} dx_1 \cdots \int_{-\infty}^{\infty} dx_{2N} \prod_{l=1}^{2N} w_1(x_l)(1-\xi\chi_{(-t,t)}^{(l)}) \prod_{1\le j<k\le 2N} |x_k-x_j| \tag{9.104}$$

in the cases that $w_1(x)$ is an even classical weight function, and thus according to (6.100) one of

$$w_1(x) = \begin{cases} e^{-x^2/2}, & \text{Gaussian}, \\ (1-x^2)^{(a-1)/2} \ (-1 < x < 1), & \text{symmetric Jacobi}, \\ (1+x^2)^{-(\alpha+1)/2}, & \text{Cauchy}. \end{cases} \tag{9.105}$$

For this purpose use will be made of the polynomials $\{p_j(x)\}_{j=0,1,\dots}$ orthogonal with respect to the corresponding classical weight $w_2(x)$ in (5.56), as well as of the eigenvalues and eigenvectors of the integral operator on $(-t,t)$ with kernel

$$K^+(x,y) := (w_2(x)w_2(y))^{1/2} \sum_{l=0}^{N-1} \frac{p_{2l}(x)p_{2l}(y)}{(p_{2l},p_{2l})_2}.$$

According to the proof of Proposition 5.2.2 there are exactly N eigenfunctions. These are of the form $(w_2(x))^{1/2} q_{2j}^{(2N)}(x,t)$, where each $q_{2j}^{(2N)}$ an even polynomial of degree $2N$ having the structure

$$q_{2j}^{(2N)}(x;t) = \sum_{l=0}^{N-1} c_{jl} \frac{1}{\sqrt{(p_{2l},p_{2l})_2}} p_{2l}(x), \tag{9.106}$$

with

$$[c_{jl}]_{j,l=0,\dots,N-1} =: \mathbf{C} \tag{9.107}$$

a real orthogonal matrix. Furthermore

$$\int_{-t}^{t} w_2(x) q_{2j}^{(2N)}(x;t) q_{2k}^{(2N)}(x;t)\, dx = \nu_{2j}^{(2N)}(t)\delta_{j,k}, \tag{9.108}$$

where $\{\nu_{2j}^{(2N)}(t)\}_{j=0,1,\dots,N-1}$ are the eigenvalues of the integral operator. The $\{q_{2j}^{(2N)}(x,t)\}_{j=0,\dots,N-1}$ also satisfy the orthonormality condition

$$(q_{2j}^{(2N)}(x,t), q_{2k}^{(2N)}(x;t))_2 = \delta_{j,k}, \qquad j,k = 0,\dots,N-1, \tag{9.109}$$

as can be seen from the orthogonality of $\{p_{2l}(x)\}_{l=0,\dots,N-1}$ with respect to this inner product and the structure (9.106).

PROPOSITION 9.6.8 *Suppose $w_1(x)$ is given by one of (9.105), and $w_2(x)$ by its companion in (5.56). Let $\{\nu_{2j}^{(2N)}(t)\}_{j=0,1,\dots,N-1}$ and $\{q_{2j}^{(2N)}(x;t)\}_{j=0,\dots,N-1}$ be specified as in the above text. We have*

$$E_{2N,1}((-t,t);\xi;w_1) = \det \Big[A_{2j-1,2k} \Big]_{j,k=1,\dots,N}, \tag{9.110}$$

where

$$\begin{aligned} A_{2j-1,2k} = {} & \delta_{j,k}\Big(1 - (2\xi - \xi^2)\nu_{2(j-1)}^{(2N)}(t)\Big) \\ & + (\xi - \xi^2)\frac{w_2(t)}{w_1(t)} q_{2k-2}^{(2N)}(t) \int_{-t}^{t} w_1(x) q_{2j-2}^{(2N)}(x;t)\, dx. \end{aligned} \tag{9.111}$$

Proof. Let $\{R_j(x)\}_{j=0,1,\dots}$ be the skew orthogonal polynomials with respect to the skew inner product (6.61), setting $e^{-V(x)} = w_1(x)$ therein, and let $\{p_j(x)\}_{j=0,1,\dots}$ be the monic orthogonal polynomials with respect to $w_2(x)$. According to (6.98), (6.99), (6.97) and (5.65)

$$R_{2j}(x) = p_{2j}(x), \quad R_{2j+1}(x) = -\frac{a_{2j}}{w_1(x)}\frac{d}{dx}\Big(\frac{w_2(x)}{w_1(x)} p_{2j}(x)\Big), \quad r_j = \frac{(p_{2j},p_{2j})_2}{a_{2j}}, \tag{9.112}$$

where

$$a_k = \begin{cases} 1, & \text{Gaussian}, \\ k+1+a, & \text{symmetric Jacobi}, \\ 2N+a-1-k, & \text{Cauchy}. \end{cases}$$

Substituting (9.112) in the Vandermonde type identity

$$\prod_{1 \le j < k \le 2N} (x_k - x_j) = \det[R_{j-1}(x_k)]_{j,k=1,\dots,2N}$$

(cf. (9.71)) we see that

$$\prod_{j=1}^{2N} w_1(x_j) \prod_{1 \le j < k \le 2N} |x_k - x_j|$$
$$= \prod_{j=0}^{N-1} (-a_{2j}) \det \begin{bmatrix} w_1(x_k) p_{2j}(x_k) \\ \frac{d}{dx_k}\Big(\frac{w_2(x_k)}{w_1(x_k)} p_{2j}(x_k)\Big) \end{bmatrix}_{\substack{j=0,\dots,N-1 \\ k=1,\dots,2N}} \prod_{1 \le j < k \le 2N} \mathrm{sgn}(x_k - x_j). \tag{9.113}$$

This identity holds true with the replacements

$$a_{2j} \mapsto r_{2j}, \qquad p_{2j}(x_k) \mapsto q_{2j}^{(N)}(x_k;t) \tag{9.114}$$

on the r.h.s. To see this first rearrange rows in the determinant so they contain in order

$$w_1(x_k)p_0(x_k), \dots, w_1(x_k)p_{N-2}(x_k), \frac{d}{dx_k}\Big(\frac{w_2(x_k)}{w_1(x_k)} p_0(x_k)\Big), \dots, \frac{d}{dx_k}\Big(\frac{w_2(x_k)}{w_1(x_k)} p_{N-2}(x_k)\Big).$$

Then multiplying both sides of (9.113) by

$$\det \begin{bmatrix} \mathbf{DC} & \mathbf{0}_N \\ \mathbf{0}_N & \mathbf{DC} \end{bmatrix},$$

where $\mathbf{C}$ is given by (9.107) and

$$\mathbf{D} := \mathrm{diag}\Big(\frac{1}{\sqrt{(p_0, p_0)_2}}, \dots, \frac{1}{\sqrt{(p_{N-1}, p_{N-1})_2}} \Big),$$

gives (9.113) back but with the substitutions (9.114).

Recalling the identity (6.62) we see the r.h.s. of (9.113) with the substitutions (9.114) can be written

$$\prod_{j=0}^{N-1} r_j \det[\phi_{j-1}(x_k)]_{j,k=1,\dots,2N} \mathrm{Pf}[h(x_k, x_j)]_{j,k=1,\dots,2N}$$

with

$$\phi_{2j}(x) = w_1(x) q_{2j}^{(2N)}(x;t), \quad \phi_{2j+1}(x) = -\frac{d}{dx}\Big(\frac{w_2(x)}{w_1(x)} q_{2j}^{(2N)}(x;t)\Big), \quad h(x,y) = \mathrm{sgn}(y-x). \tag{9.115}$$

Substituting this for the integrand of (9.104) and making use of the general integration formula (6.73) and the identity (6.12) shows that

$$E_{2N,1}((-t,t);\xi;w_1) = \frac{2^N 2N!}{\hat{Z}_{2N}} \prod_{j=0}^{N-1} r_j \det\Big[A_{j,k}\Big]^{1/2}_{j,k=1,\dots,2N}, \tag{9.116}$$

where

$$A_{j,k} = \frac{1}{2}\Big(\int_{-\infty}^{\infty} - \xi \int_{-t}^{t}\Big) dx \Big(\int_{-\infty}^{\infty} - \xi \int_{-t}^{t}\Big) dy \, \phi_{j-1}(x)\phi_{k-1}(y) \mathrm{sgn}(y-x). \tag{9.117}$$

Because each of $w_1(x), w_2(x), q_{2j}^{(2N)}(x;t)$ is even we see from (9.115) that $\phi_{2j}(x)$ is even while $\phi_{2j+1}(x)$ is odd. It then follows from the definition (9.117) that

$$A_{2j,2k} = A_{2j-1,2k-1} = 0 \qquad \text{for} \quad j,k = 1,\dots,\frac{N}{2}.$$

Interchanging rows and columns so that the zero elements are all in the top left and bottom right blocks, and noting $A_{2k,2j-1} = -A_{2j-1,2k}$ gives the stated result (9.110) (here $\hat{Z}_{2N}$ in (9.116) has been eliminated by noting that both sides must equal unity when $\xi = 0$). □

The generating function $E^{+}_{2N,1}$ in (8.149) is by definition a generating function for the sum of probabilities $E^{\rm COE}_{2N,1}(2n;(-\theta,\theta)) + E^{\rm COE}_{2N,1}(2n-1;(-\theta,\theta))$. Such probabilities can be computed from the generating function (9.104) according to the general formula (8.1). Applying this formula to (9.110) allows the following results to be derived.

PROPOSITION 9.6.9 *In the setting of Proposition 9.6.8, with $q_j(x;t) =: q_j(x)$, we have*

$$\begin{aligned} &E_{2N,1}(2n;(-t,t);w_1) \\ &\quad = \prod_{j=0}^{N-1}(1-\nu_{2j}^{(2N)}(t)) \sum_{0\le j_1<j_2<\cdots\le j_n\le N-1} \frac{\nu_{2j_1}^{(2N)}(t)\cdots\nu_{2j_n}^{(2N)}(t)}{(1-\nu_{2j_1}^{(2N)}(t))\cdots(1-\nu_{2j_n}^{(2N)}(t))} \\ &\quad \times\Big(1-\frac{w_2(t)}{w_1(t)}\sum_{\alpha=1}^{n}\frac{q_{2j_\alpha}^{(2N)}(t)}{\nu_{2j_\alpha}^{(2N)}(t)}\int_{-t}^{t} w_1(x)q_{2j_\alpha}^{(2N)}(x)\,dx\Big), \end{aligned} \tag{9.118}$$

$$\begin{aligned} &E_{2N,1}(2n-1;(-t,t);w_1) \\ &\quad = \prod_{j=0}^{N-1}(1-\nu_{2j}^{(2N)}(t)) \sum_{0\le j_1<j_2<\cdots\le j_n\le N-1} \frac{\nu_{2j_1}^{(2N)}(t)\cdots\nu_{2j_n}^{(2N)}(t)}{(1-\nu_{2j_1}^{(2N)}(t))\cdots(1-\nu_{2j_n}^{(2N)}(t))} \\ &\quad \times\frac{w_2(t)}{w_1(t)}\sum_{\alpha=1}^{n}\frac{q_{2j_\alpha}^{(2N)}(t)}{\nu_{2j_\alpha}^{(2N)}(t)}\int_{-t}^{t} w_1(x)q_{2j_\alpha}^{(2N)}(x)\,dx. \end{aligned} \tag{9.119}$$

Thus

$$\begin{aligned} &E_{2N,1}(2n;(-t,t);w_1) + E_{2N,1}(2n-1;(-t,t);w_1) \\ &\quad = \prod_{j=0}^{N-1}(1-\nu_{2j}^{(2N)}(t)) \sum_{0\le j_1<j_2<\cdots\le j_n\le N-1} \frac{\nu_{2j_1}^{(2N)}(t)\cdots\nu_{2j_n}^{(2N)}(t)}{(1-\nu_{2j_1}^{(2N)}(t))\cdots(1-\nu_{2j_n}^{(2N)}(t))} \\ &\quad = E_{N,2}(n;(0,t^2);x^{-1/2}w_2(x^{1/2})\chi_{x>0}). \end{aligned} \tag{9.120}$$

Proof. The general formula (8.1) tells us that to compute (9.118) we must differentiate (9.116) $2n$ times with respect to ξ and then set $\xi = 1$. Now each term in the determinant of (9.116) is quadratic in ξ. If we were to differentiate two distinct rows once with respect to ξ and then set $\xi = 1$, the two rows would have the form

$$-\frac{w_2(t)}{w_1(t)}q_{2k-2}^{(2N)}(t)\int_{-t}^{t} w_1(x)q_{2j-2}^{(2N)}(x)\,dx$$

and thus be proportional to each other, so the determinant will vanish. Hence for a nonzero contribution no two distinct rows are to be differentiated just once. Since a total of $2n$ differentiations are required, it follows that n distinct rows $j_1,\dots,j_n$ say must be differentiated twice. We see immediately from (9.117) that upon setting $\xi = 1$ the remaining rows are only nonzero on the diagonal. Expanding by these elements, and multiplying by the combinatorial factor $(2n-1)!!$

to account for the number of ways of forming n pairs from $2n$ rows shows

$$E_{2N,1}(2n;(-t,t);w_1) = \prod_{j=0}^{N-1}(1-\nu_{2j}^{(2N)}(t)) \sum_{0\le j_1<j_2<\cdots\le j_n\le N-1} \frac{\nu_{2j_1}^{(2N)}(t)\cdots\nu_{2j_n}^{(2N)}(t)}{(1-\nu_{2j_1}^{(2N)}(t))\cdots(1-\nu_{2j_n}^{(2N)}(t))}$$
$$\times \det\Big[\delta_{j_\alpha,j_\beta} - \frac{w_2(t)}{w_1(t)}\frac{q_{2j_\beta}^{(2N)}(t)}{\nu_{2j_\beta}^{(2N)}(t)}\int_{-t}^{t} w_1(x)q_{2j_\alpha}^{(2N)}(x)\,dx\Big]_{\alpha,\beta=1,\dots,n}.$$

The determinant in this expression is a special case of the general determinant

$$\det[\delta_{i,j} - x_i y_j]_{i,j=1,\dots,n} = 1 - \sum_{j=1}^{n} x_j y_j \tag{9.121}$$

(whenever $\mathbf{A}$ has rank 1, $\det(\mathbf{1}-\mathbf{A}) = 1 - \operatorname{Tr}\mathbf{A}$), as used in (1.137), so (9.118) results.

The only difference in the derivation of (9.119) is that one row is differentiated with respect to ξ exactly once, while $n-1$ rows are differentiated twice. This gives

$$E_{2N,1}(2n-1;(-t,t);w_1) = -\prod_{j=0}^{N-1}(1-\nu_{2j}^{(2N)}(t)) \sum_{0\le j_1<j_2<\cdots\le j_n\le N-1} \frac{\nu_{2j_1}^{(2N)}(t)\cdots\nu_{2j_n}^{(2N)}(t)}{(1-\nu_{2j_1}^{(2N)}(t))\cdots(1-\nu_{2j_n}^{(2N)}(t))}$$
$$\times \frac{d}{d\epsilon}\det\Big[\delta_{j_\alpha,j_\beta} - \epsilon\frac{w_2(t)}{w_1(t)}\frac{q_{2j_\beta}^{(2N)}(t)}{\nu_{2j_\beta}^{(2N)}}\int_{-t}^{t} w_1(x)q_{2j_\alpha}^{(2N)}(x)\,dx\Big]_{\alpha,\beta=1,\dots,n}\Big|_{\epsilon=1}.$$

The determinant can again be calculated using (9.121), and (9.119) results.

It remains to justify the second equality in (9.120). Now application of the theory of Exercises 9.6 q.3 tells us that

$$\prod_{j=0}^{N-1}(1-\xi\nu_{2j}^{(2N)}(t)) = E_{N,2}((0,t^2);\xi;x^{-1/2}w_2(x^{1/2})\chi_{x>0})$$

so we can identify the r.h.s. of the first equality in (9.120) as

$$\frac{(-1)^n}{n!}\frac{\partial^n}{\partial\xi^n}E_{N,2}((0,t^2);\xi;x^{-1/2}w_2(x^{1/2})\chi_{x>0})\Big|_{\xi=1}.$$

Application of (8.1) then gives the second equality in (9.120). □

To apply (9.120) to the COE, note that analogous to (8.126) it follows from (2.51) that

$$E_{N,1}^{\rm COE}((-\theta,\theta);\xi) = E_{N,1}^{(0)}\Big(\Big(-\tan\frac{\theta}{2},\tan\frac{\theta}{2}\Big);\xi;(1+x^2)^{-(N+1)/2}\Big),$$

or equivalently

$$E_{N,1}^{\rm COE}\Big(n;(-\theta,\theta)\Big) = E_{N,1}\Big(n;\Big(-\tan\frac{\theta}{2},\tan\frac{\theta}{2}\Big);(1+x^2)^{-(N+1)/2}\Big). \tag{9.122}$$

Because $w_1(x) = (1+x^2)^{-(N+1)/2}$ is an example of the Cauchy weight in (9.105) and the interval $(-\tan\frac{\theta}{2},\tan\frac{\theta}{2})$ is symmetric about the origin, for N even (9.120) applies and we conclude

$$\begin{aligned}E_{2N,1}^{\rm COE}(2n;(-\theta,\theta)) &+ E_{2N,1}^{\rm COE}(2n-1;(-\theta,\theta))\\ &= E_{N,2}\Big(n;\Big(0,\tan^2\frac{\theta}{2}\Big);x^{-1/2}(1+x)^{-(N+1)/2}\chi_{x>0}\Big)\\ &= E_{N,2}\Big(n;\Big(0,\sin^2\frac{\theta}{2}\Big);x^{-1/2}(1-x)^{1/2}\chi_{0<x<1}\Big)\end{aligned}$$

where the second equality follows by a change of variables. This equation is equivalent to (8.149) in the case of $E^+_{2N,1}$.

EXERCISES 9.6 1. [219] Consider the one-component log-gas at $\beta = 2$ on a circle with circumference length L. Suppose that the circle is divided into M equally spaced lattice points, and that the N particles ($N \le M$) are constrained to lie on these points.

(i) Use the method of the proof of Proposition 5.2.1 and the orthonormality relation

$$\frac{1}{M} \sum_{n=0}^{M-1} e^{2\pi i n(k-j)/M} = \delta_{k,j}$$

for $|k - j| < M, k, j \in \mathbb{Z}$ to show that the n-particle correlation function is given by

$$\rho_{(n)}(l_1, \dots, l_n) = \det \left[\frac{\sin \pi N(l_j - l_k)/M}{L \sin \pi (l_j - l_k)/M} \right]_{j,k=1,\dots,n}.$$

(ii) Modify the derivation of the formula (9.1) for $E_\beta(0; (a_1, a_2))$ to show that in general for a one-dimensional lattice gas excluded from the lattice sites $l = m_1, m_1 + 1, \dots, m_2$

$$E_\beta(0; [m_1, m_2]) = 1 + \sum_{n=1}^{N} \frac{(-1)^n}{n!} \left(\frac{L}{M} \right)^n \sum_{l_1=m_1}^{m_2} \cdots \sum_{l_n=m_1}^{m_2} \rho_{(n)}(l_1, \dots, l_n),$$

and hence conclude that for the present system

$$E_2(0; [m_1, m_2]) = \det \left[\delta_{j,k} - \frac{\sin \pi N(j-k)/M}{M \sin \pi (j-k)/M} \right]_{j,k=m_1,\dots,m_2}.$$

With $m := m_2 - m_1$, in the thermodynamic limit $N, M, L \to \infty$, $\tau := L/M$ and $\rho := N/L$ fixed, rewrite this result to read

$$E_2(0; m) = \det \left[\frac{1}{2\pi} \int_{\pi\tau\rho}^{2\pi - \pi\tau\rho} e^{i(j-k)\theta} \, d\theta \right]_{j,k=1,\dots,m}.$$

(iii) Widom [542] has proved the following theorem regarding the large m asymptotic behavior of Toeplitz determinants of the type occurring in (ii): Let $D_m[f] := \det[\frac{1}{2\pi} \int_0^{2\pi} f(\theta) e^{i(j-k)\theta} \, d\theta]_{j,k=1,\dots,m}$, where f satisfies $f(\theta) = f(2\pi - \theta)$, is supported on a closed arc $\alpha \le \theta \le 2\pi - \alpha$ and is positive along this arc. Then (assuming some mild restrictions on the derivative of f), as $m \to \infty$

$$D_m[f] \sim \exp\left(m^2 \log \cos \frac{\alpha}{2} + mG[f] - \frac{1}{4} \log \left(m \sin \frac{\alpha}{2} \right) + 3\zeta'(-1) \right) 2^{1/12} T[F], \tag{9.123}$$

where $F(\theta) = f(2 \cos^{-1}(\cos \alpha/2 \cos \theta))$, and if $\log g(\theta) = \sum_{k=-\infty}^{\infty} g_k e^{ik\theta}$, then $G[g] := e^{g_0}$ and $T[g] := e^{\frac{1}{4} \sum_{k=1}^{\infty} k g_k g_{-k}}$. Use this theorem in the final formula of (ii) above to show that for large m

$$E_2(0; m) \sim \exp\left(m^2 \log \cos \frac{\pi\rho\tau}{2} - \frac{1}{4} \log \left(m \sin \frac{\pi\rho\tau}{2} \right) + \frac{1}{12} \log 2 + 3\zeta'(-1) \right)$$

and thus deduce the first equation in (9.91) by taking the limit $\tau \to 0$ with $m\tau = x$. (For a rigorous justification of this argument, see [363].)

2. Use the identity (8.162) and the third asymptotic expansion in (9.91) to deduce that

$$E_2^{\rm s.s.}(0; (-t, t)) \Big|_{\rho = 1/\pi} \underset{t \to \infty}{\sim} {\rm Chard}|_{a-1/2} {\rm Chard}|_{a+1/2} \frac{e^{-t^2/2 + 2at}}{t^{a^2+1/4}} \left(1 + \frac{a}{4t} + \cdots \right).$$

Show that in the case $a = 0$ this is consistent with the first asymptotic expansion in (9.91) (this requires the special value of $G\left(\frac{1}{2}\right)$ given above (4.182)).

3. [197] The objective of this exercise is to interpret the factorization (8.123) in terms of the eigenvalues of the underlying integral operator. In particular we aim to show that if in (8.123)

$$E_{N,2}^{(0)}(J_{\{t_{2i},t_{2i+1}\}};\xi;g(x)) = \prod_{j=0}^{N-1}(1-\xi\lambda_j(J_{\{t_{2i},t_{2i+1}\}})),$$

where $\{\lambda_j(J_{\{t_{2i},t_{2i+1}\}})\}$ are the eigenvalues of the integral operator on $J_{\{t_{2i},t_{2i+1}\}}$ with kernel

$$K_N(x,y) = (g(x)g(y))^{1/2}\sum_{l=0}^{N-1}\frac{p_l(x)p_l(y)}{(p_l,p_l)_2}$$

(here $\{p_l(x)\}$ are the monic orthogonal polynomials corresponding to the weight $g(x)$), then

$$E_{[(N+1)/2],2}^{(0)}(J^+_{\{t_{2i}^2,t_{2i+1}^2\}};\xi;x^{-1/2}g(x^{1/2})\chi_{x>0}) = \prod_{\substack{j=0\\ j\text{ even}}}^{N-1}(1-\xi\lambda_j(J_{\{t_{2i},t_{2i+1}\}})),$$

$$E_{[N/2],2}^{(0)}(J^+_{\{t_{2i}^2,t_{2i+1}^2\}};\xi;x^{1/2}g(x^{1/2})\chi_{x>0}) = \prod_{\substack{j=0\\ j\text{ odd}}}^{N-1}(1-\xi\lambda_j(J_{\{t_{2i},t_{2i+1}\}})).$$

(i) Set $N_1 = [(N+1)/2]$. Deduce from the proof of Proposition 8.4.1 that

$$\begin{aligned}
&E_{N_1,2}^{(0)}(0;J^+_{\{t_{2i}^2,t_{2i+1}^2\}};\xi;x^{-1/2}g(x^{1/2})\chi_{x>0})\\
&= \frac{1}{C}\Big(\int_{-\infty}^{\infty}-\xi\int_{J_{\{t_{2i},t_{2i+1}\}}}\Big)dx_1\,w_2(x_1)\cdots\Big(\int_{-\infty}^{\infty}-\xi\int_{J_{\{t_{2i},t_{2i+1}\}}}\Big)dx_{N_1}\,w_2(x_{N_1})\\
&\quad\times\prod_{1\le j<k\le N_1}(x_k^2-x_j^2)^2.
\end{aligned}$$

(ii) Let $\rho^+_{(k)}(x_1,\dots,x_k)$ denote the k-point correlation function associated with the p.d.f.

$$\frac{1}{C}\prod_{l=1}^{N_1}g(x_l)\prod_{1\le j<k\le N_1}(x_k^2-x_j^2)^2. \tag{9.124}$$

Proceed as in the derivation of Propositions 5.1.1 and 5.1.2 to show

$$\rho^+_{(k)}(x_1,\dots,x_k) = \det\Big[K_N^+(x_\alpha,x_\beta)\Big]_{\alpha,\beta=1,\dots,k}, \qquad K_N^+(x,y) := (g(x)g(y))^{1/2}\sum_{l=0}^{N_1}\frac{p_{2l}(x)p_{2l}(y)}{(p_{2l},p_{2l})_2}$$

and use the equality between (9.14) and (9.15) to conclude from this that

$$E_{[(N+1)/2],2}^{(0)}(J^+_{\{t_{2i}^2,t_{2i+1}^2\}};\xi;x^{-1/2}g(x^{1/2})\chi_{x>0}) = \det(1-\xi K_N^+),$$

where K_N^+ is the integral operator on $J_{\{t_{2i},t_{2i+1}\}}$ with kernel $K_N^+(x,y)$. Now use the fact that the eigenfunctions of the latter are the even eigenfunctions of the integral operator on $J_{\{t_{2i},t_{2i+1}\}}$ with kernel $K_N(x,y)$ to deduce the first of the sought results. Apply a similar strategy to derive the second result.

9.7 THE PROBABILITIES $E_\beta^{\rm soft}(n;(s,\infty))$ FOR $\beta = 1,4$

In the bulk the probability $E_1^{\rm bulk}(0;(0,s))$ was evaluated via the mapping (8.139) to a gap probability for a $\beta = 2$ quantity. For the probability $E_1^{\rm soft}(0;(s,\infty))$ an analogous result is the formula

$$\Big(E_1^{\rm soft}(0;(s,\infty))\Big)^2 = E_1^{{\rm odd(soft)}^2}(0;(s,\infty)), \tag{9.125}$$

where $E^{{\rm odd(soft)}^2}$ refers to the system with k-point correlation (7.170). Since the latter is a $k \times k$ determinant, (9.15) gives

$$\Big(E_1^{\rm soft}(0;(s,\infty))\Big)^2 = \det\Big(1 - (K^{\rm soft} + A \otimes B)\Big), \tag{9.126}$$

where $K^{\rm soft}$ is the integral operator on (s,∞) with kernel (9.39) while A is the operator which multiplies by ${\rm Ai}(x)$, while B is the integral operator with kernel $\int_0^\infty {\rm Ai}(y-v)\,dv$. (Strictly speaking (9.126) is ill-defined due to B not decaying for $y \to -\infty$; this can be overcome by redefining A and B in such a way that the corresponding determinant is unchanged. However this technicality does not effect the working formula (9.128) below.) Using (9.126), $E_1^{\rm soft}(0;(s,\infty))$ can be evaluated in terms of the Painlevé II transcendent $q(s)$ occurring in the expression (9.46) for $E_2^{\rm soft}(0;(s,\infty))$ [521], [195].

PROPOSITION 9.7.1 *For the infinite GOE at the soft edge, scaled as in (7.11), and with q the Painlevé II transcendent satisfying (9.45) subject to the boundary condition (9.47) with $\xi = 1$,*

$$\Big(E_1^{\rm soft}(0;(s,\infty))\Big)^2 = E_2^{\rm soft}(0;(s,\infty))\exp\Big(-\int_s^\infty q(x)\,dx\Big). \tag{9.127}$$

Proof. Factoring out $\det(1 - K^{\rm soft})$ from (9.126) and making use of (9.15) shows

$$\begin{aligned}\Big(E_1^{\rm soft}(0;(s,\infty))\Big)^2 &= E_2^{\rm soft}(0;(s,\infty))\det\Big(1 - (1-K^{\rm soft})^{-1}A\otimes B\Big)\\ &= E_2^{\rm soft}(0;(s,\infty))\Big(1 - \int_s^\infty (1-K^{\rm soft})^{-1}A[y]B(y)\,dy\Big),\end{aligned} \tag{9.128}$$

where the second equality follows from the fact that $(1-K^{\rm soft})^{-1}A[y]$ is the eigenfunction of the operator $(1-K^{\rm soft})^{-1}A\otimes B$, so the eigenvalue is

$$\int_s^\infty (1-K^{\rm soft})^{-1}A[y]B(y)\,dy.$$

Introducing the notation of Section 9.4, we put $\phi(x) = A(x) = {\rm Ai}(x)$ and $Q(x) = (1-K^{\rm soft})^{-1}A[x]$ so that

$$\int_s^\infty (1-K^{\rm soft})^{-1}A[y]B(y)\,dy = \int_s^\infty dy\,Q(y)\int_{-\infty}^y dv\,\phi(v) =: u_\epsilon. \tag{9.129}$$

Note from (9.128) that with the notation (9.129) we have

$$\Big(E_1^{\rm soft}(0;(s,\infty))\Big)^2 = E_2^{\rm soft}(0;(0;(s,\infty))(1-u_\epsilon). \tag{9.130}$$

The strategy now is to derive coupled differential equations for u_ϵ and the quantity

$$q_\epsilon := \int_s^\infty dy\,\rho(s,y)\int_{-\infty}^y dv\,\phi(v). \tag{9.131}$$

According to Proposition 9.3.10 we have

$$\frac{\partial}{\partial s}Q(y) = -qR(y,s) = -q\Big(-\delta^+(y-s) + \rho(s,y)\Big). \tag{9.132}$$

p.d.f.	mean	variance	skewness	kurtosis
$p_1^{\rm soft}(0;t)$	−1.20653 35745	1.60778 10345	0.29346 45240	0.16524 29384
$p_1^{\rm soft}(1;t)$	−3.26242 79028	1.03544 74415	0.16550 94943	0.04919 51565
$p_1^{\rm soft}(2;t)$	−4.82163 02757	0.82239 01151	0.11762 14761	0.01977 46604
$p_1^{\rm soft}(3;t)$	−6.16203 99636	0.70315 81054	0.09232 83954	0.00816 06305
$p_1^{\rm soft}(4;t)$	−7.37011 47042	0.62425 23679	0.07653 98210	0.00245 40580
$p_1^{\rm soft}(5;t)$	−8.48621 83723	0.56700 71487	0.06567 07705	−0.00073 42515

Table 9.2 Statistical properties of $p_1^{\rm soft}(n;t)$ for various n. Note from (9.142) that $\tilde{p}_4^{\rm soft}(n;t) = p_1^{\rm soft}(2n+1;t)$, so the second, fourth and sixth lines give the statistical properties of $\{\tilde{p}_4^{\rm soft}(n;t)\}_{n=0,1,2}$.

Making use of this formula, and recalling $q := Q(s)$, shows

$$\frac{d}{ds}u_\epsilon = -q\int_{-\infty}^{s}\phi(v)\,dv + \int_s^\infty dy\Big(\frac{\partial}{\partial s}Q(y)\Big)\int_{-\infty}^{y} dv\,\phi(v) = -qq_\epsilon. \tag{9.133}$$

To derive the corresponding formula for the derivative of q_ϵ we will make use of the formula

$$\Big(\frac{\partial}{\partial s}+\frac{\partial}{\partial x}+\frac{\partial}{\partial y}\Big)\rho(x,y) = -Q(x)Q(y), \tag{9.134}$$

which is a consequence of Proposition 9.3.7 with the substitutions (9.57) and (9.35). This shows

$$\begin{aligned}\frac{d}{ds}q_\epsilon &= -\int_s^\infty dy\,\frac{\partial}{\partial y}\rho(s,y)\int_{-\infty}^{y} dv\,\phi(v) - q\int_s^\infty dy\,Q(y)\int_{-\infty}^{y} dv\,\phi(v)\\ &= \int_s^\infty \rho(s,y)\phi(y)\,dy - qu_\epsilon \;=\; q(1-u_\epsilon).\end{aligned} \tag{9.135}$$

As q is known, the system of equations (9.133) and (9.135) fully determines u_ϵ and q_ϵ once boundary conditions are specified. Now $Q(y)$ is smooth, so we see from (9.129) that

$$u_\epsilon \to 0 \tag{9.136}$$

as $s\to\infty$. On the other hand $\rho(s,y) = \delta^+(s-y) + R(s,y)$, where $R(s,y)$ is smooth, so for $s\to\infty$

$$q_\epsilon \sim \int_{-\infty}^\infty \phi(v)\,dv = \int_{-\infty}^\infty {\rm Ai}(v)\,dv = 1. \tag{9.137}$$

The unique solution of the coupled equations (9.133) and (9.135) satisfying (9.136) and (9.137) is then verified to be

$$u_\epsilon = 1 - e^{-\mu}, \qquad q_\epsilon = e^{-\mu}, \tag{9.138}$$

where

$$\mu := \int_s^\infty q(x)\,dx.$$

Substituting the evaluation of $u_\epsilon^{\rm s}$ from (9.138) in (9.130) gives (9.127), as required. □

Recalling (9.46) and the analogue of (8.86) for $\beta = 1, \zeta = 1$, it follows from (9.127) that

$$p_1^{\rm soft}(0;t) = \frac{d}{dt}\exp\Big(-\frac{1}{2}\int_t^\infty (x-t)q^2(x)\,dx - \frac{1}{2}\int_t^\infty q(x)\,dx\Big), \tag{9.139}$$

where q is as in (9.127). Using this formula, statistical characterizations of $p_1^{\rm soft}(0;t)$ have been computed [449] as in Table 9.2.

To compute $E_1^{\rm soft}(n;(s,\infty))$ for $n \ge 1$, and to compute

$$\tilde{E}_4^{\rm soft}(n;(s,\infty)) := \lim_{N\to\infty} E_4\Big(n;\Big((2N)^{1/2} + \frac{s}{2^{1/2}N^{1/6}},\infty\Big);e^{-x^2};N/2\Big),$$

(this particular definition is made to make connection with (6.124)) we first make note of some inter-relationships between $\{E_\beta^{\rm soft}(n;(s,\infty))\}_{n=0,1,\dots}$ for $\beta = 1,2,4$.

PROPOSITION 9.7.2 *We have*

$$E_2^{\rm soft}(n;(s,\infty)) = \sum_{l=0}^{2n+1} E_1^{\rm soft}(2n+1-l;(s,\infty))\Big(E_1^{\rm soft}(l;(s,\infty)) + E_1^{\rm soft}(l-1;(s,\infty))\Big) \tag{9.140}$$

and

$$\tilde{E}_4^{\rm soft}(n;(s,\infty)) = E_1^{\rm soft}(2n;(s,\infty)) + E_1^{\rm soft}(2n+1;(s,\infty)). \tag{9.141}$$

The latter equation is equivalent to

$$\tilde{p}_4^{\rm soft}(n;(s,\infty)) = p_1^{\rm soft}(2n+1;(s,\infty)). \tag{9.142}$$

Proof. The identity (9.140) is deduced from (6.138) in an analogous way to how (8.143) is deduced from (6.154). The identity (9.141) follows from (6.142) by an argument similar to that used to deduce Proposition 8.4.6. □

We remark that with the notation $\operatorname*{odd}_{1-\xi} f$ denoting the odd powers in $1-\xi$ of the expansion of $f(\xi)$ about $\xi = 1$, and with

$$\bar{\xi} := 2\xi - \xi^2 \tag{9.143}$$

so that $1 - \bar{\xi} = (1-\xi)^2$, it follows from (9.140) that in relation to the corresponding generating functions

$$(1-\xi)E_2^{\rm soft}((s,\infty);\bar{\xi}) = \operatorname*{odd}_{1-\xi}\Big[\Big(E_1^{\rm soft}((s,\infty);\xi)\Big)^2(2-\xi)\Big]. \tag{9.144}$$

Now, it follows from (9.140) with $n = 0$ that

$$E_1^{\rm soft}(1;(s,\infty)) = \frac{1}{2}\bigg(E_1^{\rm soft}(0;(s,\infty)) + \frac{E_2^{\rm soft}(0;(s,\infty))}{E_1^{\rm soft}(0;(s,\infty))}\bigg). \tag{9.145}$$

Making use of the evaluation (9.127) shows

$$\Big(E_1^{\rm soft}(1;(s,\infty))\Big)^2 = E_2^{\rm soft}(0;(s,\infty))\cosh^2\Big(\frac{1}{2}\int_s^\infty q(x)\,dx\Big), \tag{9.146}$$

while it follows from this, (9.46) and (8.3) that

$$p_1^{\rm soft}(1;t) = p_2^{\rm soft}(0;t) + \frac{d}{dt}\Big(\exp\Big(-\frac{1}{2}\int_t^\infty (x-t)q^2(x)\,dx\Big)\cosh\Big(\frac{1}{2}\int_t^\infty q(x)\,dx\Big)\Big). \tag{9.147}$$

The formula (9.147) was used in the computation of the corresponding statistical data of Table 9.2. However, the equation (9.140) is not sufficient to compute $E_1^{\rm soft}(n;(s,\infty))$ for $n \ge 2$. It must be supplemented by the following [198].

PROPOSITION 9.7.3 *We have*

$$E_1^{\rm odd(soft)^2}(n;(s,\infty)) = \sum_{l=0}^{2n} E_1^{\rm soft}(2n-l;(s,\infty))\Big(E_1^{\rm soft}(l;(s,\infty)) + E_1^{\rm soft}(l-1;(s,\infty))\Big), \tag{9.148}$$

which in terms of generating functions and with $\bar{\xi}$ specified by (9.143) reads

$$E_1^{\mathrm{odd(soft)}^2}((s,\infty);\bar{\xi}) = \underset{1-\xi}{\mathrm{even}}\left[\left(E_1^{\mathrm{soft}}((s,\infty);\xi)\right)^2(2-\xi)\right]. \tag{9.149}$$

Furthermore, with $q(x;\xi)$ as in (9.46),

$$E_1^{\mathrm{odd(soft)}^2}((s,\infty);\xi) = E_2^{\mathrm{soft}}((s,\infty);\xi)\Big(\cosh\int_s^\infty q(x;\xi)\,dx - \sqrt{\xi}\sinh\int_s^\infty q(x;\xi)\,dx\Big). \tag{9.150}$$

Proof. The identity (9.148), which for $n=0$ reduces to (9.125), follows from the construction of the scaled ensemble odd(soft)^2 as a superposition of two $\beta=1$ soft edge ensembles.

In relation to (9.150), because the k-point correlation is given by the determinant formula (7.170), application of (9.15) gives

$$E_1^{\mathrm{odd(soft)}^2}((s,\infty);\bar{\xi}) = \det(1-\xi K_{(s,\infty)}^{\mathrm{soft}})\Big(1-\xi\int_s^\infty (1-\xi K^{\mathrm{soft}})^{-1}A[y]B(y)\,dy\Big).$$

But $\det(1-\xi K_{(s,\infty)}^{\mathrm{soft}}) = E_2^{\mathrm{soft}}((s,\infty);\xi)$, so our task is to show that

$$1-\xi\int_s^\infty (1-\xi K^{\mathrm{soft}})^{-1}A[y]B(y)\,dy = \cosh\int_s^\infty q(x;\xi)\,dx - \sqrt{\xi}\sinh\int_s^\infty q(x;\xi)\,dx.$$

This can be done by modifying the working of the proof of Proposition 9.7.1.

First introduce the notation

$$\phi^{\mathrm{s}}(x) = \sqrt{\xi}\mathrm{Ai}(x), \qquad Q^{\mathrm{s}}(x) = (1-\xi K^{\mathrm{soft}})^{-1}\phi^{\mathrm{s}}(x)$$

so that

$$\xi\int_s^\infty (1-\xi K^{\mathrm{soft}})^{-1}A[y]B(y)\,dy = \int_s^\infty dy\,Q^{\mathrm{s}}(x)\int_{-\infty}^y \phi^{\mathrm{s}}(v)\,dv =: u_\epsilon^{\mathrm{s}}$$

(cf. (9.129)), and similarly introduce the ξ generalization of (9.131)

$$q_\epsilon^{\mathrm{s}} := \int_s^\infty dy\,\rho(s,y)\int_{-\infty}^y dv\,\phi(v),$$

where $\rho(s,y)$ is the kernel of the integral operator $(1-\xi K^{\mathrm{soft}})^{-1}$. The workings of the proof of Proposition 9.7.1 tell us that u_ϵ^{s} and q_ϵ^{s} satisfy the coupled equations

$$\frac{du_\epsilon^{\mathrm{s}}}{ds} = -q(s;\xi)q_\epsilon^{\mathrm{s}}, \qquad \frac{dq_\epsilon^{\mathrm{s}}}{ds} = q(s;\xi)(1-u_\epsilon^{\mathrm{s}})$$

subject to the boundary conditions

$$u_\epsilon^{\mathrm{s}} \to 0, \qquad q_\epsilon^{\mathrm{s}} \to \sqrt{\xi} \quad \text{as} \quad s\to\infty.$$

This system has solution

$$q_\epsilon^{\mathrm{s}} = \sqrt{\xi}\cosh\int_s^\infty q(x;\xi)\,dx - \sinh\int_s^\infty q(x;\xi)\,dx, \qquad u_\epsilon^{\mathrm{s}} = 1-\cosh\int_s^\infty q(x;\xi)\,dx - \sqrt{\xi}\sinh\int_s^\infty q(x;\xi)\,dx.$$

The latter is precisely the sought identity. □

Adding together (9.140) and (9.148) gives the Painlevé transcendent evaluation of the generating function $E_1^{\mathrm{soft}}((s,\infty);\xi)$ [135], [198].

PROPOSITION 9.7.4 *We have*

$$\left(E_1^{\mathrm{soft}}((s,\infty);\xi)\right)^2 = E_2^{\mathrm{soft}}((s,\infty);\bar{\xi})\frac{\xi-1-\cosh\mu(s,\bar{\xi})+\sqrt{\bar{\xi}}\sinh\mu(s,\bar{\xi})}{\xi-2}, \tag{9.151}$$

where $\bar{\xi}$ is given by (9.143) and

$$\mu(s,\bar{\xi}) := \int_s^\infty q(t;\bar{\xi})\,dt. \tag{9.152}$$

The expression (9.151) allows $\{E_1^{\rm soft}(n;(s,\infty))\}$ and thus $\{p_1^{\rm soft}(n;(s,\infty))\}$ to be computed for small n according to the method detailed in Section 9.4.2, although the need to solve coupled equations makes this a difficult exercise, even for $n=2$. Fortunately the generating function $E_1^{\rm soft}((s,\infty);\xi)$ can be expressed in terms of Fredholm determinants, and the method of Section 9.2 employed. Thus it turns out that $\mu(s,\xi)$ as specified by (9.152), is intimately related to the operator $V_{(0,\infty)}^{\rm soft}$ introduced in Proposition 9.6.1. Explicitly [132]

$$\exp\Big(-\mu(s;\xi)\Big) = \frac{\det(1-\sqrt{\xi}V_{(0,\infty)}^{\rm soft})}{\det(1+\sqrt{\xi}V_{(0,\infty)}^{\rm soft})}. \tag{9.153}$$

This is to be substituted in (9.150), together with the formula

$$E_2^{\rm soft}((s,\infty);\xi) = \det(1-\sqrt{\xi}V_{(0,\infty)}^{\rm soft})\det(1+\sqrt{\xi}V_{(0,\infty)}^{\rm soft}), \tag{9.154}$$

which follows from (9.40) and Proposition 9.6.1. Computing appropriate derivates with respect to ξ then gives the l.h.s. of (9.148), and this in turn allows for the recursive computation of $\{E_1^{\rm soft}(n,(s,\infty))\}_{n=0,1,\dots}$ in terms of Fredholm determinants. The simplest case is $n=0$, which shows $E_1^{\rm soft}(0,(s,\infty))$ admits the simple Fredholm form [480], [173]

$$E_1^{\rm soft}(0,(s,\infty)) = \det(1-V_{(0,\infty)}^{\rm soft}). \tag{9.155}$$

Statistical properties of $\{p_1^{\rm soft}(n;(s,\infty))\}$ for $n=2,\dots,5$ computed using this formalism [79] are given in Table 9.2.

We now turn our attention to the $s\to-\infty$ asymptotic behavior of $E_\beta^{\rm soft}(0;(s,\infty))$ for $\beta=1$ and 4, known from (9.91) in the case $\beta=2$. According to (9.127), (9.141) and (9.146), in addition to the expansion (9.91) this requires knowledge of the $s\to-\infty$ expansion of $\int_s^\infty q(x)\,dx$. On this point it is known that [30]

$$\exp\Big(-\frac{1}{2}\int_s^\infty q(x)\,dx\Big) \underset{s\to-\infty}{\sim} 2^{-1/4}e^{-|s|^{3/2}/3\sqrt{2}}\Big(1+{\rm O}(1/s^4)\Big).$$

Noting from Section 7.6.5 that $E_4^{\rm soft}(n;(s,\infty)) = \tilde{E}_4^{\rm soft}(n;(2^{2/3}s,\infty))$, the following result then follows [30].

PROPOSITION 9.7.5 *For $\beta=1$ or 4 one has*

$$E_\beta^{\rm soft}(0;(s,\infty)) \underset{s\to-\infty}{\sim} \tau_\beta \frac{e^{-\beta|s|^3/24+(\beta/2-1)(\sqrt{2}/3)|s|^{3/2}}}{|s|^{(1/8)(1-(2/\beta)(1-\beta/2)^2)}}, \tag{9.156}$$

where $\tau_1 = 2^{-11/48}e^{\frac{1}{2}\zeta'(-1)}$, $\tau_4 = 2^{-37/48}e^{\frac{1}{2}\zeta'(-1)}$. Moreover, comparison with (9.91) shows this same asymptotic form to be valid for $\beta=2$.

Theory related to the characterization of the distribution of the largest eigenvalue in terms of (13.186) below allows the leading behavior exhibited in (9.156) to be rigorously established for general $\beta>0$ [463].

EXERCISES 9.7 1. [231] In this exercise $E_1^{\rm soft}$ and $\tilde{E}_4^{\rm soft}$ will be identified as τ-functions of Painlevé systems.

(i) Use (9.46) to show that (9.127) can be written

$$E_1^{\rm soft}(0;(s,\infty)) = \exp\Big(-\frac{1}{2}\int_s^\infty (t-s)(q^2(t)-q'(t))\,dt\Big).$$

(ii) With u^s defined as in Proposition 8.3.2, generalize the derivation of (9.58) to show

$$-2^{-1/3}\Big(u^s(-2^{-1/3}t;\mu;\xi)\Big)' = -\frac{1}{2}\Big(q'_{\mu-1/2}(t) + q^2_{\mu-1/2}(t) + \frac{t}{2}\Big).$$

(iii) Recalling that $q(t) = q_0(t)$, as well as the definition of σ_{II} $(= u^s)$ from (8.29) and Proposition 8.2.1, and noting that the differential equation for q (9.45) is unchanged by the negation $q \mapsto -q$, show from results (i) and (ii) that we can write

$$E_1^{\rm soft}(0;(s,\infty)) = \exp\Big(-\int_s^\infty \Big(H_{II}(t)\Big|_{\alpha=0} + \frac{t^2}{8}\Big)dt\Big),$$

where $H_{II}(t)|_{\alpha=0} + \frac{t^2}{8} \sim \frac{1}{2}{\rm Ai}(t)$ for $t \to \infty$. Thus, defining the auxiliary PII Hamiltonian and corresponding τ-function by

$$\tilde{H}_{II} := H_{II} + \frac{t^2}{8}, \qquad \tilde{\tau}_{II}(t;\alpha) = \frac{d}{dt}\log \tilde{H}_{II},$$

conclude

$$E_1^{\rm soft}(0;(s,\infty)) = \tilde{\tau}_{II}^{+}(s;0), \tag{9.157}$$

where the superscript $+$ indicates that $\tilde{H}_{II}$ is subject to the boundary condition $\tilde{H}_{II}(t) \sim \frac{1}{2}{\rm Ai}(t)$.

(iv) By writing $q \mapsto -q$ in the above working show that

$$\exp\Big(-\frac{1}{2}\int_s^\infty (t-s)(q^2(t) + q'(t))\,dt\Big) = \tilde{\tau}_{II}^{-}(s;0),$$

where the superscript $-$ indicates that $\tilde{H}_{II}$ is subject to the boundary condition $\tilde{H}_{II}(t) \sim -\frac{1}{2}{\rm Ai}(t)$. Use this result in (9.46) to deduce that

$$E_2^{\rm soft}(0;(s,\infty)) = \tilde{\tau}_{II}^{+}(s;0)\tilde{\tau}_{II}^{-}(s;0). \tag{9.158}$$

Substitute (9.157) and (9.158) in (9.141) with $n = 0$ to show

$$\tilde{E}_4^{\rm soft}(0;(s,\infty)) = \frac{1}{2}\Big(\tilde{\tau}_{II}^{+}(s;0) + \tilde{\tau}_{II}^{-}(s;0)\Big). \tag{9.159}$$

9.8 THE PROBABILITIES $E_\beta^{\rm hard}(n;(0,s);a)$ FOR $\beta = 1, 4$

Considerations similar to those which yielded the evaluation of $E_\beta^{\rm soft}(n;(s,\infty))$ for $\beta = 1$ and 4 suffice for the evaluation of $E_\beta^{\rm hard}$ [197]. For $\beta = 2$ the latter is specified by (8.87) with $\mu = 0$. For $\beta = 1$ and 4, so as to relate to (6.124), we define

$$\begin{aligned} E_1^{\rm hard}((0,s);\xi;a) &:= \lim_{N\to\infty} E_1((0,s/4N);\xi;x^a e^{-x/2};N), \\ \tilde{E}_4^{\rm hard}((0,s);\xi;a) &:= \lim_{N\to\infty} E_4((0,s/4N);\xi;x^a e^{-x};N/2). \end{aligned}$$

The first step is to make note of the analogue of Proposition 9.7.2, derived as for the results therein using (6.138) and (6.142) with $f = x^{(a-1)/2}e^{-x/2}$, $g = x^a e^{-x}$.

PROPOSITION 9.8.1 *We have*

$$\begin{aligned} E_2^{\rm hard}(n;(0,s);a) = \sum_{l=0}^{2n+1} & E_1^{\rm hard}(2n+1-l;(0,s);(a-1)/2) \\ & \times \Big(E_1^{\rm hard}(l;(0,s);(a-1)/2) + E_1^{\rm hard}(l-1;(0,s);(a-1)/2)\Big) \end{aligned} \tag{9.160}$$

and

$$\tilde{E}_4^{\rm hard}(n;(0,s);a+1) = E_1^{\rm hard}(2n;(0,s);(a-1)/2) + E_1^{\rm hard}(2n+1;(0,s);(a-1)/2). \qquad (9.161)$$

The first is equivalent to the generating function identity

$$(1-\xi)E_2^{\rm hard}((0,s);\bar{\xi};a) = \underset{1-\xi}{\rm odd}\Big[\Big(E_1^{\rm hard}((0,s);\xi;(a-1)/2)\Big)^2(2-\xi)\Big] \qquad (9.162)$$

and the second to the relation

$$\tilde{p}_4^{\rm hard}(n;(s,\infty);a+1) = p_1^{\rm hard}(2n+1;(s,\infty);(a-1)/2). \qquad (9.163)$$

The second step is to formulate the analogue of Proposition 9.7.3. For this let odd(hard)^2 refer to the ensemble with k-point correlation (7.172), and let the corresponding generating function for k eigenvalues in $(0,s)$ be denoted $E^{{\rm odd(hard)}^2}((0,s);\xi;(a-1)/2)$.

PROPOSITION 9.8.2 *We have*

$$E_1^{{\rm odd(hard)}^2}(n;(0,s);(a-1)/2) = \sum_{l=0}^{2n} E_1^{\rm hard}(2n-l;(0,s);(a-1)/2)$$
$$\times\Big(E_1^{\rm hard}(l;(0,s);(a-1)/2) + E_1^{\rm hard}(l-1;(0,s);(a-1)/2)\Big), \qquad (9.164)$$

which equivalently in terms of generating functions reads

$$E_2^{{\rm odd(hard)}^2}((0,s);\bar{\xi};(a-1)/2) = \underset{1-\xi}{\rm even}\Big[\Big(E_1^{\rm hard}((0,s);\xi;(a-1)/2)\Big)^2(2-\xi)\Big], \qquad (9.165)$$

where $\bar{\xi}$ is given by (9.143). Furthermore

$$E_2^{{\rm odd(hard)}^2}((0,s);\bar{\xi};(a-1)/2) = E_2^{\rm hard}((0,s);\bar{\xi};a)\Big(\cosh\tilde{\mu}(s;\bar{\xi};a) - \sqrt{\xi}\sinh\tilde{\mu}(s;\bar{\xi};a)\Big), \quad (9.166)$$

where with $q(t;\xi)$ as in (9.71)

$$\tilde{\mu}(s;a;\bar{\xi}) := \frac{1}{2}\int_0^s \frac{q(t;\bar{\xi})}{\sqrt{t}}\,dt.$$

Proof. The relation (9.164) follows from the definition of the ensemble odd(hard)^2 as a superposition, while (9.165) follows from (9.164) by multiplication by $(1-\bar{\xi})^{2n}$ and summation over n. To deduce (9.166), we note from (7.172) and (9.15) that

$$E_1^{{\rm odd(hard)}^2}(0;(0,s);\xi;(a-1)/2) = \det\Big(1-\xi(K^{\rm hard}+C\otimes D)\Big), \qquad (9.167)$$

where $K^{\rm hard}$ is the operator on $(0,s)$ with kernel (9.60), C is the operator which multiplies by $J_a(\sqrt{y})$, and D is the integral operator with kernel

$$\frac{1}{2\sqrt{y}}\int_{\sqrt{y}}^\infty J_a(t)\,dt. \qquad (9.168)$$

We see from (9.167) that (9.166) is equivalent to the identity

$$1-\xi\int_0^s [(1-\xi K^{\rm hard})^{-1}C](y)D(y)\,dy = \cosh\tilde{\mu}(s;\xi;a) - \sqrt{\xi}\sinh\tilde{\mu}(s;\xi;a). \qquad (9.169)$$

With

$$\phi^{\rm h}(x) = \sqrt{\xi}J_a(\sqrt{x}), \qquad Q^{\rm h}(x) = [(1-\xi K^{\rm hard})^{-1}\phi^{\rm h}](x),$$

after changing variables $t=\sqrt{u}$ in (9.168) we have

$$\xi\int_0^s (1-\xi K^{\rm hard})^{-1}C[y]D(y)\,dy = \frac{1}{4}\int_0^s dy\,Q^{\rm h}(y)\frac{1}{\sqrt{y}}\int_y^\infty du\,\frac{1}{\sqrt{u}}\phi^{\rm h}(u) =: u_\epsilon^{\rm h}, \qquad (9.170)$$

and analogous to (9.131) we also introduce

$$q_\epsilon^{\rm h} := \int_0^\infty dy\, \rho^{\rm h}(s,y) \int_{-\infty}^y du\, \frac{1}{\sqrt{u}} \phi^{\rm h}(u), \tag{9.171}$$

where $\rho^{\rm h}(s,y)$ denotes the kernel of the integral operator $(1 - \xi K^{\rm hard})^{-1}$. Now Proposition 9.3.10 gives

$$\frac{\partial}{\partial s} Q^{\rm h}(y) = q(s;\xi)\Big(- \delta^+(y-s) + \rho^{\rm h}(s,y)\Big)$$

(cf. (9.132)). Using this formula and recalling $q(s;\xi) := Q(s)$ we see from (9.170) that

$$\frac{d}{ds} u_\epsilon^{\rm h} = \frac{1}{4} q(s;\xi) q_\epsilon^{\rm h}. \tag{9.172}$$

The corresponding formula for the derivative of $q_\epsilon^{\rm h}$ is derived by making use of the formula

$$x \frac{\partial}{\partial x} \rho^{\rm h}(x,y) + s \frac{\partial}{\partial s} \rho^{\rm h}(x,y) = - \frac{\partial}{\partial y} \Big(y \rho^{\rm h}(x,y) \Big) + \frac{1}{4} Q^{\rm h}(x) Q^{\rm h}(y)$$

(cf. (9.134)) which is a consequence of Proposition 9.5.1, (9.63) and (9.35). This shows

$$\begin{aligned} s \frac{dq_\epsilon^{\rm h}}{ds} &= - \int_0^s dy \left(\frac{d}{dy} (y \rho^{\rm h}(s,y)) \right) \frac{1}{\sqrt{y}} \int_y^\infty du\, \frac{1}{\sqrt{u}} \phi^{\rm h}(u) + q(s;\xi) u_\epsilon^{\rm h} \\ &= - \frac{1}{2} \int_0^s dy\, \rho^{\rm h}(s,y) \frac{1}{\sqrt{y}} \int_y^\infty du\, \frac{1}{\sqrt{u}} \phi^{\rm h}(u) - \int_0^s dy\, \rho^{\rm h}(s,y) \phi^{\rm h}(y) + q(s;\xi) u_\epsilon^{\rm h} \\ &= - \frac{1}{2} q_\epsilon^{\rm h} - q(s;\xi)(1 - u_\epsilon^{\rm h}). \end{aligned} \tag{9.173}$$

The coupled equations (9.172) and (9.173) must be solved subject to the $s \to 0$ boundary conditions

$$u_\epsilon^{\rm h} \sim 0, \quad \sqrt{s} q_\epsilon^{\rm h} \sim \int_0^\infty \frac{1}{\sqrt{u}} \phi^{\rm h}(u)\, du = 2\sqrt{\xi} \int_0^\infty J_a(v)\, dv = 2\sqrt{\xi}. \tag{9.174}$$

The occurrence of $\sqrt{s} q_\epsilon$ in (9.174) suggests we introduce $\bar{q}_\epsilon^{\rm h} := \sqrt{s} q_\epsilon^{\rm h}$ in (9.172) and (9.173). Doing this gives the system of equations

$$\sqrt{s} \frac{du_\epsilon^{\rm h}}{ds} = \frac{1}{4} q(s;\xi) \bar{q}_\epsilon^{\rm h}, \qquad \sqrt{s} (\bar{q}_\epsilon^{\rm h})' = -q(s;\xi)(1 - u_\epsilon^{\rm h}). \tag{9.175}$$

The solution satisfying (9.174) can be checked to be

$$\begin{aligned} u_\epsilon^{\rm h} &= 2\Big(\sqrt{\xi} \cosh \tilde{\mu}(s;a;\xi) - \sinh \tilde{\mu}(s;a;\xi) \Big), \\ \bar{q}_\epsilon^{\rm h} &= 1 - \cosh \tilde{\mu}(s;a;\xi) + \sqrt{\xi} \sinh \tilde{\mu}(s;a;\xi), \end{aligned}$$

thus verifying (9.169). □

Adding together (9.165) and (9.162), and using (9.166) gives the analogue of (9.151) for the hard edge,

$$\Big(E_1^{\rm hard}((0,s);\xi;\frac{a-1}{2}) \Big)^2 = E_2^{\rm hard}((0,s);\bar{\xi};a) \frac{\xi - 1 - \cosh \tilde{\mu}(s;a;\bar{\xi}) + \sqrt{\bar{\xi}} \sinh \tilde{\mu}(s;a;\bar{\xi})}{\xi - 2}. \tag{9.176}$$

Setting $\xi = 1$ this reads

$$\Big(E_1^{\rm hard}\Big(0;(0,s);\frac{a-1}{2}\Big) \Big)^2 = E_2^{\rm hard}(0;(0,s);a) \exp\Big(- \tilde{\mu}(s;\xi=1;a) \Big). \tag{9.177}$$

Analogous to (9.153), one can show that [132]

$$\exp\Big(- \tilde{\mu}(s;\xi;a)\Big) = \frac{\det(1 - \sqrt{\xi} V^{\rm hard}_{(0,1)})}{\det(1 + \sqrt{\xi} V^{\rm hard}_{(0,1)})}. \tag{9.178}$$

This substituted in (9.176) implies

$$E_1^{\rm hard}\Big(0;(0,s);\frac{a-1}{2}\Big) = \det(1 - V^{\rm hard}_{(0,1)}), \tag{9.179}$$

where use is also required of the fact that

$$E_2^{\rm hard}(0;(0,s);a) = \det(1 - \sqrt{\xi} V^{\rm hard}_{(0,1)}) \det(1 + \sqrt{\xi} V^{\rm hard}_{(0,1)}) \tag{9.180}$$

(cf. (9.154)), which in turn follows from (9.15) and Proposition 9.6.1. Substituting (9.178) and (9.179) in (9.164), and computing appropriate derivatives with respect to ξ, we see that the l.h.s. of (9.166) can be expressed in terms of Fredholm determinants. This allows for a recursive computation of $\{E_1^{\rm hard}(n;(0,s);(a-1)/2)\}$ in terms of Fredholm determinants.

We remark that in the case $(a-1)/2 = m \in \mathbb{Z}^+$ the probability $E_1^{\rm hard}(0;(0,s);m)$ can be written as an m-dimensional integral (see Exercises 9.8 q.1). From this we can deduce the large s asymptotic expansion

$$E_1^{\rm hard}(0;(0,s);m) \sim \frac{G(3/2)}{G(m+3/2)} \frac{G(2m+2)}{G(m+2)} \frac{e^{-s/8 + m\sqrt{s}}}{(2\sqrt{s})^{m(m+1/2)}}. \tag{9.181}$$

We will see in Chapter 12 that an m-dimensional integral evaluation of $E_\beta^{\rm hard}(0;(0,s);m)$ can also be obtained for $m \in \mathbb{Z}^+$ and general $\beta > 0$. The corresponding asymptotic expansion is given in (13.52).

EXERCISES 9.8 1. [233], [231] The objective of this exercise is to derive the identity

$$\begin{aligned} E_1^{\rm hard}(0;(0,s^2);m) &= e^{-s^2/8 + ms} \Big\langle e^{-2s\sum_{j=1}^m x_j} \Big\rangle_{{\rm JUE}_m}\Big|_{a=b=1/2} \\ &= e^{-s^2/8} \Big\langle e^{s \sum_{j=1}^m \cos\theta_j} \Big\rangle_{Sp(2m)}, \end{aligned} \tag{9.182}$$

where the second equality follows from (8.117).

(i) Use (9.71) to show that (9.177) can be written

$$E_1^{\rm hard}\Big(0;(0,s);\frac{a-1}{2}\Big) = \exp\Big(- \frac{1}{8} \int_0^s \Big((\log \frac{s}{t}) q^2(t) + 2\frac{q(t)}{\sqrt{t}}\Big) dt\Big). \tag{9.183}$$

(ii) Recalling the $s \to 0^+$ behavior (9.69) verify that

$$\int_0^s \frac{q(t)}{\sqrt{t}}\, dt = \int_0^s \Big(\log s - \log t\Big) \frac{d}{dt}(\sqrt{t} q(t))\, dt$$

and use this in (i) to show

$$E_1^{\rm hard}\Big(0;(0,s);\frac{a-1}{2}\Big) = \exp\Big(- \frac{1}{4} \int_0^{\sqrt{s}} \Big(\log \frac{s}{t}\Big) \Big(x\frac{dq}{dx} + q + xq^2\Big)\, dx\Big), \quad t = x^2.$$

(iii) Introduce the function $y(x)$ according to (9.70) with $s = t$. Verify that

$$x\frac{dq}{dx} + q + xq^2 = \frac{1}{(1-y)^2}\Big(2x\frac{dy}{dx} - y^2 + 4xy + 1\Big) + x.$$

(iv) In the Hamiltonian H_V of (8.24) make the substitutions

$$q \mapsto y, \ p \mapsto z, \ t \mapsto \eta x, \ H_V \mapsto \frac{1}{\eta}\tilde{H}_V$$

(cf. (9.145)). Use the relationship between the PV parameters listed below (8.24), and the fact that the general P_V equation with $\delta \neq 0$ can be reduced to the case with $\delta = -\frac{1}{2}$ by the mapping $t \mapsto \sqrt{-2\delta t}$, to show that with y the PV transcendent in (9.70) we must set

$$\eta = 2, \qquad v_2 = -v_3 = -\frac{1}{4}(a+1), \qquad v_1 = -v_4 = \frac{1}{4}(a-1). \tag{9.184}$$

With $\tilde{H}_V$ so specified, make use of the first of the corresponding Hamilton equations together with the result of (iii) to show

$$x\frac{dq}{dx} + q + xq^2 = -2\Big(\frac{a-1}{2} - \frac{x}{2} + \frac{d}{dx}(x\tilde{H}_V)\Big).$$

(v) Substitute the result of (iv) in (ii), and use the relationship between H_V and σ_V as specified by (8.29) and Proposition 8.2.1 to conclude that

$$\tilde{\sigma}_V := x\frac{d}{dx}\log E_1^{\rm hard}\Big(0;(0,x^2);\frac{a-1}{2}\Big) + \frac{1}{4}x^2 - \frac{(a-1)}{2}x + \frac{a(a-1)}{4} \tag{9.185}$$

satisfies the σPV equation in (8.15) with parameters (9.184) and $t \mapsto 2x$.

(vi) From the results of Exercises 8.3 q.1(i)&(ii) note that

$$x\frac{d}{dx}\log\Big\langle e^{-2x\sum_{j=1}^m x_j}\Big\rangle_{{\rm JUE}_m|_{a=b=1/2}} + m\Big(m+\frac{1}{2}\Big) \tag{9.186}$$

satisfies the same equation as that satisfied by (9.185) provided $(a-1)/2 = m \in \mathbb{Z}^+$. By verifying that (9.185) and (9.186) have the same small x expansion (for this use the matrix average form in (9.182) together with a result from Exercises 10.7 q.2 below), deduce from this (9.182).

2. [233], [231] The objective of this exercise is to show that

$$\tilde{E}_4^{\rm hard}(0;(0,s^2);2l) = e^{-s^2/8}\Big\langle e^{(s/2){\rm Tr\,U}}\Big\rangle_{{\rm U}\in O(2l)} \tag{9.187}$$

(cf. (9.182)).

(i) Use (9.160) and (9.161) with $n=0$ to show

$$\tilde{E}_4^{\rm hard}(0;(0,s);a+1) = \frac{1}{2}\Big(E_1^{\rm hard}\Big(0;(0,s);\Big(\frac{a-1}{2}\Big)\Big) + \frac{E_2^{\rm hard}(0;(0,s);a)}{E_1^{\rm hard}(0;(0,s);(a-1)/2)}\Big).$$

(ii) Use the relationship between the eigenvalue p.d.f. for ${\rm Sp}(2m)$ and $O^-(2m+2)$ noted above (2.69), together with the second formula in (9.182), to note that we can write

$$E_1^{\rm hard}(0;(0,s);l-1) = e^{-s^2/8}\Big\langle e^{s{\rm Tr\,U}}\Big\rangle_{{\rm U}\in O^-(2l)}.$$

(iii) Note from (9.71) and (9.183) that

$$\frac{E_2^{\rm hard}(0;(0,s);a)}{E_1^{\rm hard}(0;(0,s);(a-1)/2)} = \exp\Big(-\frac{1}{8}\int_0^s\Big(\Big(\log\frac{s}{t}\Big)q^2(t) - 2\frac{q(t)}{\sqrt{t}}\Big)dt\Big),$$

and use the workings of q.1 to deduce that the conclusion of q.1(v) holds with $E_1^{\rm hard}$ therein replaced by the l.h.s. of this equation, with $s \mapsto x^2$.

(iv) Use the results of Exercises 8.3 q.1(i)&(ii) to note that

$$x\frac{d}{dx}\log\Big\langle e^{-2x\sum_{j=1}^m x_j}\Big\rangle_{{\rm JUE}_{m+1}|_{a=b=1/2}} + m\Big(m+\frac{1}{2}\Big)$$

satisfies the same equation as (9.185), provided $(a-1)/2 = m \in \mathbb{Z}^+$. Use (8.117) to rewrite this as

$$x\frac{d}{dx}\log e^{-(m+1)x}\Big\langle e^{(s/2)\mathrm{Tr}\,\mathbf{U}}\Big\rangle_{\mathbf{U}\in O^+(2m+2)} + m(m+1/2)$$

and show from this (making use of a result from Exercises 10.7 q.2 below) that the small x expansion is the same as that of (9.185). Conclude that

$$\frac{E_2^{\rm hard}(0;(0,s);2l-1)}{E_1^{\rm hard}(0;(0,s);l-1)} = e^{-s^2/8}\Big\langle e^{(s/2)\mathrm{Tr}\,\mathbf{U}}\Big\rangle_{\mathbf{U}\in O^+(2l)}.$$

Substitute this, together with the result of (ii), in the result of (i) with $a \mapsto 2l-1$ to deduce (9.187).

9.9 RIEMANN-HILBERT VIEWPOINT

We turn our attention from the function theoretic properties of the Fredholm determinant (9.15) to aspects of the underlying integrable systems theory. A crucial first step [307] is to associate with integral operators having kernels of the form (9.25) a Riemann-Hilbert problem (recall Section 1.4.3).

9.9.1 Integrable kernels

Generalize the kernel (9.25) to

$$K(\lambda,\mu) = \frac{\vec{f}^T(\lambda)\vec{g}(\mu)}{\lambda-\mu}, \qquad \vec{f} = \begin{bmatrix} f_1 \\ f_2 \end{bmatrix},\ \vec{g} = \begin{bmatrix} g_1 \\ g_2 \end{bmatrix}, \tag{9.188}$$

where it is required that

$$\vec{f}^T(\lambda)\vec{g}(\lambda) = 0. \tag{9.189}$$

Note that this includes the Christoffel-Darboux structure (9.25). The proof of Proposition 9.3.6 again shows

$$(1-K_J)^{-1} = 1 + R_J$$

with the kernel of R_J given by

$$R(\lambda,\mu) = \frac{\vec{F}^T(\lambda)\vec{G}(\mu)}{\lambda-\mu}, \qquad F_j = (1-K_J)^{-1}f_j,\ G_j = (1-K_J)^{-1}g_j\ \ (j=1,2). \tag{9.190}$$

Introduce the 2×2 matrix

$$\mathbf{Y}(\lambda) := \mathbf{1}_2 - \int_{a_1}^{a_2} \frac{\vec{F}(\mu)\vec{g}^T(\mu)}{\mu-\lambda}\,d\mu. \tag{9.191}$$

We see immediately that

$$\mathbf{Y}(\lambda) \in \mathrm{Reg}(\mathbb{C}\backslash[a_1,a_2]), \tag{9.192}$$

where $\mathrm{Reg}(X)$ denotes the space of analytic function on the set X, and we see too that

$$\mathbf{Y}(\infty) = \mathbf{1}_2. \tag{9.193}$$

In traversing the interval $[a_1,a_2]$ in the complex plane from a_1 to a_2, let the l.h.s. of the interval be denoted $+$, and the r.h.s. $-$. For $\lambda\in(a_1,a_2)$, let

$$\mathbf{Y}_\pm(\lambda) := \lim_{\substack{\lambda'\to\lambda \\ \lambda'\in(\pm)\,\mathrm{side}}} \mathbf{Y}(\lambda'). \tag{9.194}$$

The function $\mathbf{Y}(\lambda)$ is discontinuous across (a_1, a_2), with $\mathbf{Y}_-(\lambda)$ being related to $\mathbf{Y}_+(\lambda)$ by the so-called jump matrix $\mathbf{H}(\lambda)$ specified in the following result.

PROPOSITION 9.9.1 *We have*

$$\mathbf{Y}_-(\lambda) = \mathbf{Y}_+(\lambda)\mathbf{H}(\lambda), \tag{9.195}$$

where $\mathbf{H}(\lambda)$ *is the* 2×2 *matrix specified by*

$$\mathbf{H}(\lambda) = \mathbf{1}_2 + 2\pi i \vec{f}(\lambda)\vec{g}^T(\lambda). \tag{9.196}$$

Also, for $\lambda \in (a_1, a_2)$

$$\vec{F}(\lambda) = \mathbf{Y}_+(\lambda)\vec{f}(\lambda) = \mathbf{Y}_-(\lambda)\vec{f}(\lambda), \quad \vec{G}(\lambda) = (\mathbf{Y}_+^T(\lambda))^{-1}\vec{g}(\lambda) = (\mathbf{Y}^T{}_-(\lambda))^{-1}\vec{g}(\lambda). \tag{9.197}$$

Proof. We will first derive (9.197). Let $\lambda \in (a_1, a_2)$ and consider

$$\mathbf{Y}_\pm(\lambda)\vec{f}(\lambda) := \vec{f}(\lambda) - \lim_{\substack{\lambda' \to \lambda \\ \lambda' \in (\pm)\,\mathrm{side}}} \int_{a_1}^{a_2} \frac{\vec{F}(\mu)\vec{g}^T(\mu)\vec{f}(\lambda)}{\mu - \lambda'} d\mu = \vec{f}(\lambda) - \int_{a_1}^{a_2} \frac{\vec{F}(\mu)\vec{g}^T(\mu)\vec{f}(\lambda)}{\mu - \lambda} d\mu,$$

where the second equality follows by noting that when $\mu = \lambda$, the scalar factor

$$\vec{g}^T(\mu)\vec{f}(\lambda) = \vec{f}^T(\lambda)\vec{g}(\mu) \tag{9.198}$$

in the numerator vanishes (recall (9.189)), and thus the integral is now continuous across $[a_1, a_2]$. Use of (9.198) for general λ, μ shows

$$\mathbf{Y}_\pm(\lambda)\vec{f}(\lambda) = \vec{f}(\lambda) + \int_{a_1}^{a_2} \frac{\vec{f}^T(\lambda)\vec{g}(\mu)}{\lambda - \mu} \vec{F}(\mu)\, d\mu = \vec{f}(\lambda) + K_J(F)[\lambda] \; = \; \vec{F}(\lambda),$$

where the final equality follows by making use of the second equation in (9.190). The derivation of the second equation in (9.197) is analogous.

Consider now (9.195). From the definition (9.194) and Cauchy's residue theorem

$$\mathbf{Y}_+(\lambda) - \mathbf{Y}_-(\lambda) = -\int_{C_\lambda} \frac{\vec{F}(\mu)\vec{g}^T(\mu)}{\mu - \lambda} d\mu = -2\pi i \vec{F}(\lambda)\vec{g}^T(\lambda),$$

where C_λ is a circle in the complex μ-plane about the point $\mu = \lambda$, traversed anticlockwise. Substituting (9.199) for $\vec{F}(\lambda)$ in this equation gives (9.195). □

The function $\mathbf{Y}(\lambda)$, specified only by the analyticity condition (9.192), normalization (9.193) and jump condition (9.195) (and thus jump matrix $\mathbf{H}$) is said to solve the normalized Riemann-Hilbert problem, RHP($[a_1, a_2], \mathbf{H}$). The solution given by (9.191) is unique (see Exercises 9.9 q.1). An important property relates to its determinant.

PROPOSITION 9.9.2 *With* $\mathbf{Y}(\lambda)$ *specified by (9.191),*

$$\det \mathbf{Y}(\lambda) = 1. \tag{9.199}$$

Proof. According to (9.189), the matrix $2\pi i \vec{f}(\lambda)\vec{g}^T(\lambda) =: i\mathbf{X}$ is nilpotent ($\mathbf{X}^2 = \mathbf{0}$) and thus $\det \mathbf{H}(\lambda) = \det[\mathbf{1}_2 + i\mathbf{X}] = 1$ since the eigenvalues of $\mathbf{1}_2 + i\mathbf{X}$ must equal unity. Consequently, from (9.195)

$$\det \mathbf{Y}_-(\lambda) = \det \mathbf{Y}_+(\lambda), \tag{9.200}$$

and so $\det \mathbf{Y}(\lambda)$ is analytic for $\lambda \in (a_1, a_2)$. Furthermore $\mathbf{Y}(\lambda)$ is analytic for $\lambda \in \mathbb{C}\backslash[a_1, a_2]$ and thus $\det \mathbf{Y}(\lambda)$ is analytic in this region. At $\lambda = a_1, a_2$ we see from (9.191) that $\mathbf{Y}(\lambda)$ has a logarithmic singularity. But since $\det \mathbf{Y}(\lambda)$ is analytic for all $\lambda \neq a_1, a_2$, this singularity must cancel out of $\det \mathbf{Y}(\lambda)$, or else $\det \mathbf{Y}(\lambda)$ would have a branch cut.

Consequently, $\det \mathbf{Y}(\lambda)$ is analytic in the finite complex λ-plane. From (9.193), $\det \mathbf{Y}(\lambda) \to 1$ as $|\lambda| \to \infty$. Thus (9.199) follows by Liouville's theorem. □

9.9.2 Schlesinger equations

We now specialize to kernels of the form (9.25), and suppose ψ and ϕ therein furthermore obey the coupled equations (9.41). Note that the latter can be written in the form

$$\frac{d}{dx}\begin{bmatrix} \psi(x) \\ \phi(x) \end{bmatrix} = \mathbf{A}_0(x)\begin{bmatrix} \psi(x) \\ \phi(x) \end{bmatrix}, \qquad \mathbf{A}_0(x) = \begin{bmatrix} a(x) & b(x) \\ c(x) & -a(x) \end{bmatrix}, \tag{9.201}$$

where $a(x), b(x), c(x)$ are rational functions of x. From the general theory of linear differential equations, (9.201) can be extended to the matrix differential equation

$$\frac{d}{dx}\mathbf{\Phi}(x) = \mathbf{A}_0(x)\mathbf{\Phi}(x), \qquad \mathbf{\Phi}(x) = \begin{bmatrix} \psi(x) & \tilde{\psi}(x) \\ \phi(x) & \tilde{\phi}(x) \end{bmatrix}, \tag{9.202}$$

where the two columns of $\mathbf{\Phi}(x)$ are linearly independent. We see from the fact that (9.202) has the formal solution

$$\mathbf{\Phi}(x) = \mathbf{C}_0 \exp \int^x \mathbf{A}_0(y)\, dy,$$

$\mathbf{C}_0$ a constant matrix, and the fact that $\mathbf{A}_0$ is traceless, that $\tilde{\psi}$ and $\tilde{\phi}$ in (9.202) can be chosen so that

$$\det \mathbf{\Phi}(x) = 1. \tag{9.203}$$

Following [346], [83], our objective is to show that the matrix product

$$\mathbf{\Psi}(x) := \mathbf{Y}(x)\mathbf{\Phi}(x), \tag{9.204}$$

where $\mathbf{Y}(x)$ is specified by (9.191) with

$$\vec{f}(x) = \begin{bmatrix} \psi(x) \\ \phi(x) \end{bmatrix}, \qquad \vec{g}(x) = \begin{bmatrix} -\phi(x) \\ \psi(x) \end{bmatrix} \tag{9.205}$$

satisfies some special partial differential equations. With the choice (9.205), (i6.6e) reduces to $\xi K^{\rm scale}(x,y)$ as specified by (9.25). Note that for general $\vec{f}, \vec{g}$, it follows from (9.199) and (9.204) that

$$\det \mathbf{\Psi}(x) = 1. \tag{9.206}$$

PROPOSITION 9.9.3 *With $\mathbf{H}(x)$ specified by (9.196), and $\vec{f}, \vec{g}$ therein specified by (9.205), we have*

$$\begin{aligned} \mathbf{H}(x) &= \begin{bmatrix} 1 - 2\pi i\psi(x)\phi(x) & 2\pi i\phi^2(x) \\ -2\pi i\phi^2(x) & 1 + 2\pi i\psi(x)\phi(x) \end{bmatrix} \\ &= \mathbf{\Phi}(x)\begin{bmatrix} 1 & 2\pi i \\ 0 & 1 \end{bmatrix}\mathbf{\Phi}^{-1}(x). \end{aligned} \tag{9.207}$$

Proof. This can be verified by a direct calculation. □

It follows from (9.207) and the jump relation (9.195) that the matrix $\mathbf{\Psi}(x)$ specified by (9.204) satisfies the jump relation

$$\mathbf{\Psi}_-(x) = \mathbf{\Psi}_+(x)\begin{bmatrix} 1 & 2\pi i \\ 0 & 1 \end{bmatrix}, \qquad x \in (a_1, a_2). \tag{9.208}$$

For $|x| \to \infty$, (9.191) and (9.204) show

$$\mathbf{\Psi}(x) \sim \Big(\mathbf{1}_2 + \mathrm{O}\Big(\frac{1}{x}\Big)\Big)\mathbf{\Phi}(x).$$

In relation to the neighborhood of the endpoints $x \sim a_k$ $(k = 1, 2)$, introduce

$$\overset{\circ}{\mathbf{\Psi}}(x) := \begin{bmatrix} 1 & (-1)^{k-1}\log(x - a_k) \\ 0 & 1 \end{bmatrix} =: (x - a_k)^{(-1)^{k-1}\begin{bmatrix} 0 & 1 \\ 0 & 0 \end{bmatrix}},$$

$$\hat{\mathbf{\Psi}}(x) := \mathbf{\Psi}_-(x)\overset{\circ}{\mathbf{\Psi}}{}_-^{-1}(x) = \mathbf{\Psi}_+(x)\overset{\circ}{\mathbf{\Psi}}{}_+^{-1}(x),$$

where the final equality follows on use of (9.208) and shows that $\hat{\mathbf{\Psi}}(x)$ is analytic in the neighborhood of a_k. Then the facts that $\mathbf{Y}(x)$ and thus $\mathbf{\Psi}(x)$ has a logarithmic singularity at a_k, that $\det \mathbf{Y}(x) = 1$, and the jump condition (9.208) imply that in the neighborhood of the endpoints the singular behavior factorizes as

$$\mathbf{\Psi}(x) = \hat{\mathbf{\Psi}}(x)\overset{\circ}{\mathbf{\Psi}}(x). \tag{9.209}$$

PROPOSITION 9.9.4 *We have*

$$\frac{\partial \mathbf{\Psi}(x)}{\partial x} = \mathbf{A}(x)\mathbf{\Psi}(x), \tag{9.210}$$

where

$$\mathbf{A}(x) = \hat{\mathbf{A}}_0(x) + \sum_{k=1}^{2} \frac{\mathbf{A}_k}{x - a_k}, \tag{9.211}$$

$$\mathbf{A}_k := (-1)^{k-1}\hat{\mathbf{\Psi}}(a_k)\begin{bmatrix} 0 & 1 \\ 0 & 0 \end{bmatrix}\hat{\mathbf{\Psi}}^{-1}(a_k) \tag{9.212}$$

and the matrix $\hat{\mathbf{A}}_0(x)$ is analytic in the neighborhood of a_k. Furthermore

$$\frac{\partial \mathbf{\Psi}}{\partial a_k} = \mathbf{U}_k(x)\mathbf{\Psi}, \qquad \mathbf{U}_k(x) = -\frac{\mathbf{A}_k}{x - a_k}. \tag{9.213}$$

Proof. According to (9.204) and (9.202)

$$\frac{\partial \mathbf{\Psi}}{\partial x}\mathbf{\Psi}^{-1} = \frac{\partial \mathbf{Y}}{\partial x}\mathbf{Y}^{-1} + \mathbf{Y}(x)\mathbf{A}_0(x)\mathbf{Y}^{-1}(x). \tag{9.214}$$

This formula tells us that the possible singularities of $\mathbf{A}(x)$ in (9.210) occur at the end points a_1, a_2, along the branch $x \in (a_1, a_2)$ of $\mathbf{Y}$, or at the poles, $\{b_j\}$, say, of $\mathbf{A}_0(x)$. In fact there is no branch cut along $x \in (a_1, a_2)$. To see this note that the fact that the jump matrix in (9.208) is a constant implies

$$\frac{\partial \mathbf{\Psi}_+}{\partial x}\mathbf{\Psi}_+^{-1} = \frac{\partial \mathbf{\Psi}_-}{\partial x}\mathbf{\Psi}_-^{-1}.$$

To study the neighborhood of $x = a_1, a_2$ we note from (9.209) that

$$\frac{\partial \mathbf{\Psi}}{\partial x}\mathbf{\Psi}^{-1} = \frac{(-1)^{k-1}}{x - a_k}\hat{\mathbf{\Psi}}(x)\begin{bmatrix} 0 & 1 \\ 0 & 0 \end{bmatrix}\hat{\mathbf{\Psi}}^{-1}(x) + \frac{\partial \hat{\mathbf{\Psi}}}{\partial x}\hat{\mathbf{\Psi}}^{-1}.$$

Since $\hat{\mathbf{\Psi}}$ is analytic in the neighborhood of each a_k, the expansion (9.211) follows. The equation (9.213) is a simple consequence of (9.209). □

We remark that it follows from (9.212) that

$$\mathbf{A}_1^2 = \mathbf{A}_2^2 = \mathbf{0} \tag{9.215}$$

and in particular $\mathrm{Tr}\mathbf{A}_1 = \mathrm{Tr}\mathbf{A}_2 = 0$. Also, because according to (9.193) $\mathbf{Y}$ is asymptotically equal to the identity, it follows from (9.214) and (9.212) that

$$\hat{\mathbf{A}}_0(x) \underset{|x|\to\infty}{\sim} \mathbf{Y}(x)\mathbf{A}_0(x)\mathbf{Y}^{-1}(x) + \mathrm{O}\Big(\frac{1}{x}\Big) \underset{|x|\to\infty}{\sim} \mathbf{A}_0(x). \tag{9.216}$$

Together the formulas (9.210) and (9.213) form a *Lax pair.* Taking the partial derivative with respect to a_k of (9.210), and the partial derivative of (9.213) with respect to x, equating and making further use of (9.210) and (9.213) shows

$$\frac{\partial \mathbf{U}_k}{\partial x} - \frac{\partial \mathbf{A}}{\partial a_k} = [\mathbf{A}, \mathbf{U}_k]. \tag{9.217}$$

Furthermore, for $j \neq k$, it follows by taking the partial derivative of (9.213) with respect to a_j that

$$\frac{\partial \mathbf{U}_k}{\partial a_j} - \frac{\partial \mathbf{U}_j}{\partial a_k} = [\mathbf{U}_j, \mathbf{U}_k]. \tag{9.218}$$

The equations (9.217) and (9.218) themselves imply a pair of partial differential equations for the matrices $\mathbf{A}_k$, known as the *Schlesinger equations.*

PROPOSITION 9.9.5 *We have*

$$\frac{\partial \mathbf{A}_k}{\partial a_j} = \frac{[\mathbf{A}_j, \mathbf{A}_k]}{a_j - a_k} \qquad (j \neq k),$$
$$\frac{\partial \mathbf{A}_j}{\partial a_j} = \sum_{\substack{l=1 \\ l \neq j}}^{2} \frac{[\mathbf{A}_j, \mathbf{A}_l]}{a_l - a_j} - [\mathbf{A}_j, \hat{\mathbf{A}}_0(a_j)]. \tag{9.219}$$

Proof. The first equation follows by substituting the definition of $\mathbf{U}_k$ from (9.213) in (9.218) and equating coefficients of $1/x$ and $1/x^2$ on both sides. The second follows by substituting (9.211) and the definition of $\mathbf{U}_k$ from (9.213) in (9.217), interchanging $k \leftrightarrow j$, and equating residues at $x = a_j$. □

Historically (see, e.g., [310]), the Schlesinger equations arose in the study of the *isomonodromy deformation* of linear matrix differential equations of the form (9.210). Now, in general the effect of the poles at a_1, a_2 in (9.211), and the poles at infinity revealed by the change of variable $x \mapsto 1/x$, is that the solution $\mathbf{\Psi}(x)$ at these singularities of $\mathbf{A}(x)$ has a branch point. Thus following the solution $\mathbf{\Psi}(x)$ about a closed contour encircling any one of these points will not give back $\mathbf{\Psi}(x)$ itself but rather some linear combination of the fundamental solutions,

$$\mathbf{\Psi}(x)\Big|_{x=u+\epsilon e^{2\pi i}} = \mathbf{\Psi}(x)\Big|_{x=u+\epsilon} \mathbf{X}_u, \quad 0 < \epsilon \ll 1, \ \ u = a_1, a_2, \infty. \tag{9.220}$$

The matrices $\mathbf{X}_{a_1}$, $\mathbf{X}_{a_2}$, $\mathbf{X}_\infty$ are called *monodromy matrices.* In the special circumstance that the monodromy matrices are independent of a_1 and a_2, the differential equation (9.210) is said to be *monodromy preserving.* In the case that $\hat{\mathbf{A}}_0(x) = \mathbf{0}$ in the definition (9.211) of $\mathbf{A}(x)$, it was shown by Schlesinger that (9.210) being monodromy preserving implies (9.213) and thus the equations (9.219). A derivation of this result in given in Exercises 9.9 q.1. The converse is also true. In the case that $\hat{\mathbf{A}}_0(x) \neq \mathbf{0}$, the analogous result is due to Jimbo, Miwa and Ueno [325].

Associated with the Schlesinger equations is the one-form

$$\omega = \Big(\frac{\mathrm{Tr}(\mathbf{A}_1\mathbf{A}_2)}{a_1 - a_2} + \mathrm{Tr}(\mathbf{A}_1\hat{\mathbf{A}}_0)\Big)\Big|_{x=a_1} da_1 + \Big(\frac{\mathrm{Tr}(\mathbf{A}_2\mathbf{A}_1)}{a_2 - a_1} + \mathrm{Tr}(\mathbf{A}_2\hat{\mathbf{A}}_0)\Big)\Big|_{x=a_2} da_2$$
$$=: f_1(a_1, a_2)da_1 + f_2(a_1, a_2)da_2. \tag{9.221}$$

Its significance is that with

$$d\omega := \frac{\partial}{\partial a_2} f_1(a_1,a_2) da_1 \wedge da_2 + \frac{\partial}{\partial a_1} f_2(a_1,a_2) da_2 \wedge da_1$$

the equations (9.219) imply $d\omega = 0$ and thus ω is closed. The general Stokes theorem then implies it is possible to write

$$\omega = d \log \tau \tag{9.222}$$

for some τ. It's explicit form is given by the following result.

PROPOSITION 9.9.6 *With τ defined implicitly in terms of ω by (9.222), which in turn is defined in terms of quantities associated with the kernel (9.188) by (9.221), we have that up to a multiplicative constant*

$$\tau = \det(1 - K_J). \tag{9.223}$$

Proof. Following [83], according to (9.23) and (9.26)

$$\begin{aligned} d \log \det(1 - K_J) &:= \Big(\frac{\partial}{\partial a_1} \log \det(1 - K_J)\Big) da_1 + \Big(\frac{\partial}{\partial a_2} \log \det(1 - K_J)\Big) da_2 \\ &= R(a_1,a_1) da_1 + R(a_2,a_2) da_2. \end{aligned}$$

Comparing with (9.221) implies we must show

$$R(a_1,a_1) = \frac{\mathrm{Tr}(\mathbf{A}_1 \mathbf{A}_2)}{a_1 - a_2} + \mathrm{Tr}(\mathbf{A}_1 \hat{\mathbf{A}}_0)\Big|_{x=a_1} \tag{9.224}$$

and the same equation with the indices 1 and 2 reversed. Now we know from (9.190) that

$$R(a_1,a_1) = \lim_{x,y \to a_1} \frac{\vec{F}^T(x)\vec{G}(y)}{x - y} = \lim_{x,y \to a_1} \frac{\vec{G}^T(y)\vec{F}(x)}{x - y}.$$

Making use of (9.197) and (9.204) then shows

$$\begin{aligned} R(a_1,a_1) &= \lim_{x,y \to a_1} \frac{\vec{g}^T(y)\mathbf{Y}_+^{-1}(y)\mathbf{Y}_-(x)\vec{f}(x)}{x - y} \\ &= \lim_{x,y \to a_1} \frac{\vec{g}^T(y)\mathbf{\Phi}(y)\mathbf{\Psi}_+^{-1}(y)\mathbf{\Psi}_-(x)\mathbf{\Phi}^{-1}(x)\vec{f}(x)}{x - y}. \end{aligned}$$

From the definitions (9.202), (9.205)

$$\mathbf{\Phi}^{-1}\vec{f} = \begin{bmatrix} 1 \\ 0 \end{bmatrix} =: \hat{e}_1, \qquad \vec{g}^T \mathbf{\Phi} = [0 \ \ 1] =: \hat{e}_2^T,$$

and so

$$R(a_1,a_1) = \lim_{x,y \to a_1} \frac{\hat{e}_2^T \mathbf{\Psi}_+^{-1}(y)\mathbf{\Psi}_-(x)\hat{e}_1}{x - y} = \hat{e}_2^T \mathbf{\Psi}_+^{-1}(x)\mathbf{\Psi}_-'(x)\hat{e}_1\Big|_{x=a_1},$$

where the final equality follows by l'Hôpital's rule. Making use of (9.207) and (9.210) then shows

$$
\begin{aligned}
R(a_1,a_1) &= \hat{e}_2^T \begin{bmatrix} 1 & -2\pi i \\ 0 & 1 \end{bmatrix} \boldsymbol{\Psi}_-^{-1}(x)\Big(\hat{\mathbf{A}}_0 + \sum_{k=1}^{2} \frac{\mathbf{A}_k}{x-a_k}\Big)\boldsymbol{\Psi}_-(x)\hat{e}_1\Big|_{x=a_1} \\
&= \hat{e}_2^T \boldsymbol{\Psi}_-^{-1}(x)\Big(\hat{\mathbf{A}}_0 + \sum_{k=1}^{2} \frac{\mathbf{A}_k}{x-a_k}\Big)\boldsymbol{\Psi}_-(x)\hat{e}_1\Big|_{x=a_1} \\
&= \mathrm{Tr}\Big(\begin{bmatrix} 0 & 1 \\ 0 & 0 \end{bmatrix} \boldsymbol{\Psi}_-(x)^{-1}\hat{\mathbf{A}}_0\boldsymbol{\Psi}_-(x)\Big)\Big|_{x=a_1} \\
&\quad + \sum_{k=1}^{2} \mathrm{Tr}\Big(\begin{bmatrix} 0 & 1 \\ 0 & 0 \end{bmatrix} \boldsymbol{\Psi}_-(x)^{-1}\frac{\mathbf{A}_k}{x-a_k}\boldsymbol{\Psi}_-(x)\Big)\Big|_{x=a_1},
\end{aligned}
$$

where the second equality follows on multiplying out the first two terms, while the third equality uses the simple identity

$$
\hat{e}_2^T \mathbf{X} \hat{e}_1 = \mathrm{Tr}\Big(\begin{bmatrix} 0 & 1 \\ 0 & 0 \end{bmatrix} \mathbf{X}\Big)
$$

valid for any 2×2 matrix $\mathbf{X}$. Recalling (9.209), and noting that $\overset{\circ}{\boldsymbol{\Psi}}(x)$ therein is upper triangular and $\hat{\psi}(x)$ is analytic at $x = a_1$ allows this last expression to be rewritten as

$$
\begin{aligned}
R(a_1,a_1) &= \mathrm{Tr}\Big(\begin{bmatrix} 0 & 1 \\ 0 & 0 \end{bmatrix} \hat{\boldsymbol{\Psi}}(x)^{-1}\hat{\mathbf{A}}_0\hat{\boldsymbol{\Psi}}(x)\Big)\Big|_{x=a_1} \\
&\quad + \sum_{k=1}^{2} \mathrm{Tr}\Big(\begin{bmatrix} 0 & 1 \\ 0 & 0 \end{bmatrix} \hat{\boldsymbol{\Psi}}(x)^{-1}\frac{\mathbf{A}_k}{x-a_k}\hat{\boldsymbol{\Psi}}(x)\Big)\Big|_{x=a_1}.
\end{aligned}
$$

Recalling the definition (9.212) of $\mathbf{A}_1$ we see that

$$
\begin{aligned}
\mathrm{Tr}\Big(\begin{bmatrix} 0 & 1 \\ 0 & 0 \end{bmatrix} \hat{\boldsymbol{\Psi}}(x)^{-1}\hat{\mathbf{A}}_0\hat{\boldsymbol{\Psi}}(x)\Big)\Big|_{x=a_1} &= \mathrm{Tr}(\mathbf{A}_1\hat{\mathbf{A}}_0), \\
\mathrm{Tr}\Big(\begin{bmatrix} 0 & 1 \\ 0 & 0 \end{bmatrix} \hat{\boldsymbol{\Psi}}(x)^{-1}\frac{\mathbf{A}_2}{x-a_2}\hat{\boldsymbol{\Psi}}(x)\Big)\Big|_{x=a_1} &= \frac{\mathrm{Tr}(\mathbf{A}_1\mathbf{A}_2)}{a_1-a_2},
\end{aligned}
$$

making use too of the facts that $\hat{\psi}(x)$ is analytic in the neighborhood of a_1 and so can be expanded in a power series and that $\mathrm{Tr}\,\mathbf{A}_1^2 = 0$ shows

$$
\mathrm{Tr}\Big(\begin{bmatrix} 0 & 1 \\ 0 & 0 \end{bmatrix} \hat{\boldsymbol{\Psi}}(x)^{-1}\frac{\mathbf{A}_1}{x-a_1}\hat{\boldsymbol{\Psi}}(x)\Big) = \mathrm{O}((x-a_1)),
$$

telling us that this vanishes at $x = a_1$. Thus (9.224) results. The equation for $R(a_2, a_2)$ follows similarly. □

9.9.3 Application to the Airy kernel

We know from (9.41) and (9.42) that in the case of the Airy kernel (9.39), and thus $\psi(x) = \mathrm{Ai}(x)$, $\phi(x) = \mathrm{Ai}'(x)$ in (9.201), the corresponding matrix $\mathbf{A}_0(x)$ has the explicit form

$$
\mathbf{A}_0(x) = \begin{bmatrix} 0 & x \\ 1 & 0 \end{bmatrix}. \tag{9.225}
$$

For the Airy kernel we consider $J = (s, \infty)$ and thus $a_1 = s$, $a_2 = \infty$, so in this case (9.211) reads

$$
\mathbf{A}(x) = \hat{\mathbf{A}}_0(x) + \frac{\mathbf{A}_1}{x-s}.
$$

We note that it follows from (9.212) that $\mathbf{A}_1$ has the structure

$$\mathbf{A}_1 = \begin{bmatrix} \alpha & -\alpha^2/\beta \\ \beta & -\alpha \end{bmatrix}. \tag{9.226}$$

Also it follows by substituting (9.225) in the first relation of (9.216) and equating terms proportional to x and proportional to unity, using the fact that $\mathbf{Y}(x)$ is to leading order equal to the unit matrix and has unit determinant, that $\hat{\mathbf{A}}_0(x)$ has the structure

$$\hat{\mathbf{A}}_0(x) = \begin{bmatrix} \mu & x+\nu \\ 1 & -\mu \end{bmatrix}. \tag{9.227}$$

In (9.226) and (9.227), α, β, μ, ν depend on s but not x. The compatibility condition (9.217) can be used to derive inter-relations between these quantities.

PROPOSITION 9.9.7 *We have*

$$\begin{aligned} \frac{d\mu}{ds} &= \beta, & \frac{d\nu}{ds} &= -2\alpha, \\ \frac{d\alpha}{ds} &= \frac{\alpha^2}{\beta} + (s+\nu)\beta, & \frac{d\beta}{ds} &= 2(\alpha - \mu\beta). \end{aligned} \tag{9.228}$$

As a consequence

$$\mu = 2\mu\alpha - \frac{\alpha^2}{\beta} + (s+\nu)\beta, \qquad \beta = -(\nu + \mu^2). \tag{9.229}$$

Proof. With $(a_1, a_2) = (s, \infty)$ the compatibility condition (9.217) takes on the simplified form

$$(x-s)\frac{\partial}{\partial s}\hat{\mathbf{A}}_0(x) + \frac{d}{ds}\mathbf{A}_1 = [\hat{\mathbf{A}}_0(x), \mathbf{A}_1].$$

With $\hat{\mathbf{A}}_0(x)$ given by (9.227) and $\mathbf{A}_1$ by (9.226), this expression is linear in x. Equating terms proportional to x gives

$$\frac{d}{ds}\begin{bmatrix} \mu & \nu \\ 1 & -\mu \end{bmatrix} = \left[\begin{bmatrix} 0 & 1 \\ 0 & 0 \end{bmatrix}, \begin{bmatrix} \alpha & -\alpha^2/\beta \\ \beta & -\alpha \end{bmatrix}\right],$$

which implies the first two differential equations. Equating terms independent of x gives

$$-s\frac{d}{ds}\begin{bmatrix} \mu & \nu \\ 1 & -\mu \end{bmatrix} + \frac{d}{ds}\begin{bmatrix} \alpha & -\alpha^2/\beta \\ \beta & -\alpha \end{bmatrix} = \left[\begin{bmatrix} \mu & \nu \\ 1 & -\mu \end{bmatrix}, \begin{bmatrix} \alpha & -\alpha^2/\beta \\ \beta & -\alpha \end{bmatrix}\right],$$

and from this the second two differential equations follow. Alternatively these two differential equations can be derived from the second equation in (9.219), appropriately specialized. The stated relations (9.229) can be verified by differentiating and using (9.228). □

Using Proposition 9.9.7 we can readily obtain a characterization of the specialization of the resolvent kernel $R(s) := R(s,s)$.

PROPOSITION 9.9.8 *We have*

$$R(s) = \mu, \qquad \frac{dR(s)}{ds} = \beta,$$

where β satisfies the differential equation

$$\frac{d^2\beta}{ds^2} - \frac{1}{2\beta}\left(\frac{d\beta}{ds}\right)^2 - 2s\beta + 4\beta^2 = 0,$$

which is a particular case of what is referred to in the literature as the Painlevé XXXIV equation. Setting

$\beta = -q^2$ the latter reduces to the particular Painlevé II equation (9.45), and thus the fourth equation of Proposition 9.4.2 is reclaimed.

Proof. The relation (9.224), together with the parametrizations (9.226) and (9.227) tell us that for the Airy kernel

$$R(s) = \mathrm{Tr}\Big(\begin{bmatrix} \alpha & -\alpha^2/\beta \\ \beta & -\alpha \end{bmatrix} \begin{bmatrix} \mu & s+\nu \\ 1 & -\mu \end{bmatrix} \Big) = 2\mu\alpha - \frac{\alpha^2}{\beta} + (s+\nu)\beta = \mu,$$

where the final equality follows from the first equation in (9.229). The stated equation for the derivative of R now follows from the first differential equation (9.228). The differential equation for β can be verified by differentiating the formula for $d\beta/ds$ in (9.228), making use of the other differential equations therein, substituting for $2\alpha^2/\beta - 4\alpha\mu$ in the resulting expression in terms of $(d\beta/ds)^2$, and making use of the second equation in (9.229). □

EXERCISES 9.9 1. The objective of this exercise is to show that if the $n \times n$ matrix differential equation

$$\frac{\partial}{\partial x}\mathbf{Y} = \mathbf{A}\mathbf{Y}, \qquad \mathbf{A} := \sum_{\nu=1}^{n} \frac{\mathbf{A}_\nu}{x - a_\nu}, \tag{9.230}$$

where the $\mathbf{A}_\nu$ are diagonalizable and $\mathbf{A}_\infty := -\sum_{\nu=1}^{n} \mathbf{A}_\nu$ is diagonal and has monodromy and connection matrices independent of the a_ν, then $\mathbf{Y}$ also satisfies

$$\frac{\partial}{\partial a_\nu}\mathbf{Y} = -\frac{\mathbf{A}_\nu}{x - a_\nu}\mathbf{Y}. \tag{9.231}$$

For this we require the readily verifiable facts that with the diagonal form of $\mathbf{A}_\nu$ given by $\mathbf{A}_0^{(\nu)} = \mathbf{G}_\nu^{-1}\mathbf{A}_\nu\mathbf{G}_\nu$, in the neighborhood of a_ν the solution of (9.230) permits the expansion

$$\mathbf{Y}^{(\nu)}(x) = \mathbf{G}_\nu \Big(\mathbf{1}_2 + \sum_{j=1}^{\infty} \mathbf{Y}_j^{\nu} (x - a_\nu)^j \Big)(x - a_\nu)^{\mathbf{A}_0^{(\nu)}}, \tag{9.232}$$

while in the neighborhood of infinity the solution is

$$\mathbf{Y}^{(\infty)}(x) = \Big(\mathbf{1}_2 + \sum_{j=1}^{\infty} \mathbf{Y}_j^{(\infty)} x^{-j} \Big) x^{-\mathbf{A}_\infty}. \tag{9.233}$$

The connection matrices $\mathbf{C}_\nu$ are defined so that

$$\mathbf{Y}(x) = \mathbf{Y}^{\infty}(x), \qquad \mathbf{Y}(x) = \mathbf{Y}^{(\nu)}(x)\mathbf{C}_\nu. \tag{9.234}$$

(i) With $d\mathbf{X} := \sum_{\nu=1}^{n} \frac{\partial}{\partial a_\nu}\mathbf{X} da_\nu$, set $\mathbf{\Omega}(x,a) := d\mathbf{Y}\,\mathbf{Y}^{-1}$. Defining the monodromy matrices $\mathbf{M}_\nu$ by the property

$$\mathbf{Y}\Big|_{x=u+\epsilon e^{2\pi i}} = \mathbf{Y}\Big|_{x=u+\epsilon} \mathbf{X}_u, \quad 0 < \epsilon \ll 1, \ u = a_1, \dots, a_n, \infty$$

(cf. (9.220)), use the fact the monodromy matrices are independent of the a_ν to show that $\mathbf{\Omega}(x,a)$ is free of branch points.

(ii) Use (9.232) and (9.234) to show that for each $\nu = 1, \dots, n$

$$d\mathbf{Y} \sim \mathbf{G}_\nu \mathbf{A}_0^{(\nu)} \frac{da_\nu}{x - a_\nu}(x - a_\nu)^{\mathbf{A}_0^{(\nu)}}\mathbf{C}_\nu, \quad \mathbf{Y}^{-1} \sim \mathbf{C}_\nu^{-1}(x - a_\nu)^{-\mathbf{A}_0^{(\nu)}}\mathbf{G}_\nu^{-1}$$

and thus

$$\mathbf{\Omega} = -\mathbf{G}_\nu \frac{\mathbf{A}_0^{(\nu)}}{x - a_\nu}\mathbf{G}_\nu^{-1} da_\nu + \mathrm{O}(1).$$

Show too that $\mathbf{\Omega} = \mathrm{O}\Big(\frac{1}{x}\Big)$ as $|x| \to \infty$. Use Liouville's theorem to conclude

$$\mathbf{\Omega} = -\sum_{\nu=1}^{n} \frac{\mathbf{A}_\nu}{x - a_\nu} da_\nu,$$

and thus the equations (9.231) must hold.

9.10 NONLINEAR EQUATIONS FROM THE VIRASORO CONSTRAINTS

In Section 5.7 special integrability properties of the $\beta = 2$ log-gas configuration integral (5.117) were considered. Consideration of such properties of the more general integral

$$Z_n[\{t_i\}; R] := \frac{1}{n!} \int_R dx_1 \cdots \int_R dx_n \prod_{l=1}^{n} e^{-V(x_l)} e^{\sum_{j=1}^\infty t_j x_l^j} \prod_{1 \le j < k \le n} (x_k - x_j)^2, \tag{9.235}$$

where

$$R = \bigcup_{i=1}^{r} [c_{2i-1}, c_{2i}] \subset \text{support}(e^{-V(x)}),$$

provides a further method to deduce nonlinear equations for gap probabilities [526]. The key ingredients for this purpose are the partial differential equation (5.136), which remains valid with Z_n as specified by (5.117) replaced by Z_n as specified by (9.235), and the appropriate generalization of the Virasoro constraints (5.128). It thus remains to consider the latter.

PROPOSITION 9.10.1 *Let Z_n be given by (9.235), and suppose for $x \in$ support$(e^{-V(x)})$*

$$V'(x) = \frac{\sum_{i=0}^\infty b_i x^i}{\sum_{i=0}^\infty a_i x^i} =: \frac{g(x)}{f(x)}. \tag{9.236}$$

With

$$\mathcal{B}_p := \sum_{i=1}^{2r} c_i^{p+1} f(c_i) \frac{\partial}{\partial c_i}$$

and L_p as specified by (5.127) we have

$$\Big(-\mathcal{B}_p + \sum_{i=0}^{\infty} \Big(a_i L_{p+i} - b_i \frac{\partial}{\partial t_{p+i+1}} \Big) \Big) Z_N = 0, \qquad p \ge -1. \tag{9.237}$$

Proof. Proceeding as in the proof of Proposition 5.7.2, we consider

$$\Big\langle \sum_{i=1}^{N} \frac{\partial}{\partial x_i} \frac{f(x_i)}{z - x_i} \Big\rangle_{Z_N}$$

$$:= \frac{1}{N!} \int_E dx_1 \cdots \int_E dx_N \prod_{k=1}^{N} e^{-V(x_k)} e^{\sum_{j=1}^\infty t_j x_k^j} \prod_{1 \le j < k \le N} (x_k - x_j)^2 \sum_{i=1}^{N} \frac{\partial}{\partial x_i} \frac{f(x_i)}{z - x_i}.$$

Integration by parts shows

$$
\begin{aligned}
&-\sum_{k=1}^{r}\Big(\frac{f(c_{2k})N}{z-c_{2k}}e^{-V(c_{2k})+\sum_{j=1}^{\infty}t_j(c_{2k})^j}\\
&\quad\times\int_E dx_1\cdots\int_E dx_{N-1}\prod_{k=1}^{N-1}e^{-V(x_k)}e^{\sum_{j=1}^{\infty}t_jx_k^j}\prod_{1\le j<k\le N-1}(x_k-x_j)^2\prod_{j=1}^{N-1}(c_{2k}-x_j)^2\\
&\quad-\frac{f(c_{2k-1})N}{z-c_{2k-1}}e^{-V(c_{2k-1})+\sum_{j=1}^{\infty}t_j(c_{2k-1})^j}\\
&\quad\times\int_E dx_1\cdots\int_E dx_{N-1}\prod_{k=1}^{N-1}e^{-V(x_k)}e^{\sum_{j=1}^{\infty}t_jx_k^j}\prod_{1\le j<k\le N-1}(x_k-x_j)^2\prod_{j=1}^{N-1}(c_{2k-1}-x_j)^2\Big)\\
&\quad-\Big\langle\sum_{i=1}^{N}\frac{f(x_i)V'(x_i)}{z-x_i}\Big\rangle_{Z_N}\\
&\quad+\Big\langle\sum_{i=1}^{N}\Big(\frac{f(x_i)}{(z-x_i)^2}+2\sum_{\substack{k=1\\k\neq i}}^{N}\frac{f(x_i)}{(x_i-x_k)(z-x_i)}\Big)\Big\rangle_{Z_N}+\Big\langle\sum_{i=1}^{N}\frac{f'(x_i)}{z-x_i}\Big\rangle_{Z_N}\\
&\quad+\Big\langle\sum_{i=1}^{N}f(x_i)\sum_{n=1}^{\infty}\frac{nt_nx_i^{n-1}}{z-x_i}\Big\rangle_{Z_N}=0
\end{aligned}
\tag{9.238}
$$

The sum over k from 1 to r in this expression can be recognized as

$$
\sum_{k=1}^{2r}\frac{f(c_k)}{z-c_k}\frac{\partial}{\partial c_k}Z_N \tag{9.239}
$$

and we see from (9.236) that

$$
\Big\langle\sum_{i=1}^{N}\frac{f(x_i)V'(x_i)}{z-x_i}\Big\rangle_{Z_N}=\Big\langle\sum_{i=1}^{N}\frac{g(x_i)}{z-x_i}\Big\rangle_{Z_N}. \tag{9.240}
$$

Using the simple identity

$$
\frac{1}{(x_i-x_k)(z-x_i)}=\Big(\frac{1}{x_i-x_k}+\frac{1}{z-x_i}\Big)\frac{1}{z-x_k}
$$

it is easy to see that

$$
\begin{aligned}
&\Big\langle\sum_{i=1}^{N}\Big(\frac{f(x_i)}{(z-x_i)^2}+2\sum_{\substack{k=1\\k\neq i}}^{N}\frac{f(x_i)}{(x_i-x_k)(z-x_i)}\Big)\Big\rangle_{Z_N}+\Big\langle\sum_{i=1}^{N}\frac{f'(x_i)}{z-x_i}\Big\rangle_{Z_N}\\
&=\Big\langle\sum_{i,k=1}^{N}\frac{f(x_i)-f(x_k)}{(x_i-x_k)(z-x_k)}\Big\rangle_{Z_N}+\Big\langle\sum_{i,k=1}^{N}\frac{f(x_i)}{(z-x_i)(z-x_k)}\Big\rangle_{Z_N}.
\end{aligned}
\tag{9.241}
$$

Substituting (9.239), (9.240), (9.241) in (9.238) shows

$$
\begin{aligned}
&-\sum_{k=1}^{2r}\frac{f(c_k)}{z-c_k}\frac{\partial}{\partial c_k}Z_N-\Big\langle\sum_{i=1}^{N}\frac{g(x_i)}{z-x_i}\Big\rangle_{Z_N}+\Big\langle\sum_{i,k=1}^{N}\frac{f(x_i)-f(x_k)}{(x_i-x_k)(z-x_k)}\Big\rangle_{Z_N}\\
&+\Big\langle\sum_{i,k=1}^{N}\frac{f(x_i)}{(z-x_i)(z-x_k)}\Big\rangle_{Z_N}+\Big\langle\sum_{i=1}^{N}f(x_i)\sum_{n=1}^{\infty}\frac{nt_nx_i^{n-1}}{z-x_i}\Big\rangle_{Z_N}=0,
\end{aligned}
\tag{9.242}
$$

and equating coefficients of z^{-p} for $p > 1$ gives

$$-\sum_{k=1}^{2r} f(c_k)c_k^{p-1}\frac{\partial}{\partial c_k}Z_N - \Big\langle \sum_{i=1}^{N} x_i^{p-1}g(x_i)\Big\rangle_{Z_N} + \Big\langle \sum_{i,k=1}^{N} \frac{f(x_i)-f(x_k)}{(x_i-x_k)}x_k^{p-1}\Big\rangle_{Z_N}$$
$$+\Big\langle \sum_{i,j=1}^{N} f(x_i)\sum_{k=0}^{p-2} x_i^k x_j^{p-2-k}\Big\rangle_{Z_N} + \Big\langle \sum_{i=1}^{N} f(x_i)\sum_{n=1}^{\infty} nt_n x_i^{n+p-2}\Big\rangle_{Z_N} = 0.$$

Substituting the power series expansions (9.236) shows that we can write

$$\begin{aligned}
\Big\langle \sum_{i=1}^{N} x_i^{p-1}g(x_i)\Big\rangle_{Z_N} &= \sum_{i=0}^{\infty} b_i \frac{\partial}{\partial t_{p+i-1}} Z_N,\\
\Big\langle \sum_{i,j=1}^{N} f(x_i)\sum_{k=0}^{p-2} x_i^k x_j^{p-2-k}\Big\rangle_{Z_N} &= \sum_{i=0}^{\infty} a_i \sum_{k=0}^{p-2} \frac{\partial}{\partial t_{k+i}}\frac{\partial}{\partial t_{p-2-k}} Z_N,\\
\Big\langle \sum_{i=1}^{N} f(x_i)\sum_{n=1}^{\infty} nt_n x_i^{n+p-2}\Big\rangle_{Z_N} &= \sum_{i=0}^{\infty} a_i \sum_{k=1}^{\infty} kt_k \frac{\partial}{\partial t_{p+k-2}} Z_N,
\end{aligned}$$

while using too the simple expansion

$$\frac{a^n-b^n}{a-b} = \sum_{p=0}^{n-1} a^{n-1-p}b^p$$

shows

$$\Big\langle \sum_{i,k=1}^{N} \frac{f(x_i)-f(x_k)}{(x_i-x_k)}x_k^{p-1}\Big\rangle_{Z_N} = \sum_{i=0}^{\infty} a_i \sum_{k=0}^{i-1} \frac{\partial}{\partial t_k}\frac{\partial}{\partial t_{p+i-2}} Z_N.$$

Noting that

$$\sum_{k=0}^{p-2} \frac{\partial}{\partial t_{k+i}}\frac{\partial}{\partial t_{p-2-k}} + \sum_{k=0}^{i-1} \frac{\partial}{\partial t_k}\frac{\partial}{\partial t_{p+i-2}} = \sum_{k=0}^{p-2+i} \frac{\partial}{\partial t_k}\frac{\partial}{\partial t_{p+i-2-k}}$$

we obtain (9.237) with p replaced by $p-2$, $p > 1$. Equating coefficients of z^{-1} in (9.242) gives formally the same results, thus establishing (9.237) in the remaining case, $p = -1$. □

Gap probabilities for the GUE

Following [526], we will use the result of Proposition 9.10.1 to specify a partial differential equation for

$$F := \log Z_n\Big|_{t_1=t_2=\cdots=0,\, V(x)=x^2} \tag{9.243}$$

in the variables $\{c_1,\dots,c_{2r}\}$. First we note that with $V(x) = x^2$, (9.236) holds with

$$a_0 = 1,\ a_1 = a_2 = \cdots = 0,\ b_0 = 0,\ b_1 = 2,\ b_2 = b_3 = \cdots = 0.$$

The first three Virasoro constraints (9.237) can then be written

$$\begin{aligned}
-\mathcal{B}_{-1}\log Z_n &= \Big(2\frac{\partial}{\partial t_1} - \sum_{i=2}^{\infty} it_i\frac{\partial}{\partial t_{i-1}}\Big)\log Z_n - nt_1,\\
-\mathcal{B}_0\log Z_n &= \Big(2\frac{\partial}{\partial t_2} - \sum_{i=1}^{\infty} it_i\frac{\partial}{\partial t_i}\Big)\log Z_n - n^2,\\
-\mathcal{B}_1\log Z_n &= \Big(2\frac{\partial}{\partial t_3} - 2n\frac{\partial}{\partial t_1} - \sum_{i=1}^{\infty} it_i\frac{\partial}{\partial t_{i+1}}\Big)\log Z_n.
\end{aligned} \tag{9.244}$$

Using the equations (9.244) and the KP equation (5.136), the following partial differential equation for (9.243) can be derived.

PROPOSITION 9.10.2 *We have*

$$\Big(\mathcal{B}_{-1}^4 + (8n + 6\mathcal{B}_{-1}^2 F)\mathcal{B}_{-1}^2 + 4(3\mathcal{B}_0^2 - 4\mathcal{B}_{-1}\mathcal{B}_1 + 6\mathcal{B}_0)\Big)F = 0. \tag{9.245}$$

Proof. We want to substitute for the partial derivatives in (5.136), which all involve t-variables, partial derivatives involving $c_1, \dots, c_{2r}$. For this purpose we make use of (9.244).

It is easy to see from the first equation in (9.244) that

$$\frac{\partial^2}{\partial t_1^2} \log Z_n\Big|_{t_1=t_2=\cdots=0} = \frac{\mathcal{B}_{-1}^2 F}{4} + \frac{n}{2}, \qquad \frac{\partial^4}{\partial t_1^4} \log Z_n\Big|_{t_1=t_2=\cdots=0} = \frac{\mathcal{B}_{-1}^4 F}{16}. \tag{9.246}$$

The second equation in (9.244) gives

$$-\mathcal{B}_0 F = 2\frac{\partial}{\partial t_2} \log Z_n\Big|_{t_1=t_2=\cdots=0} - n^2,$$

$$\mathcal{B}_0^2 F = 4\Big(\frac{\partial^2}{\partial t_2^2} - \frac{\partial}{\partial t_2}\Big) \log Z_n\Big|_{t_1=t_2=\cdots=0}$$

and thus we have

$$\frac{\partial^2}{\partial t_2^2} \log Z_n\Big|_{t_1=t_2=\cdots=0} = \Big(\frac{\mathcal{B}_0^2}{4} - \frac{\mathcal{B}_0}{2}\Big)F + \frac{n^2}{2}. \tag{9.247}$$

A similar calculation using all three equations in (9.244) shows

$$\frac{\partial^2}{\partial t_1 \partial t_3} \log Z_n\Big|_{t_1=t_2=\cdots=0} = \frac{\mathcal{B}_{-1}}{4}(\mathcal{B}_1 + n\mathcal{B}_{-1})F - \frac{3\mathcal{B}_0 F}{4} + \frac{3n^2}{4}. \tag{9.248}$$

Substituting (9.246)–(9.248) in (5.136) gives the stated result. □

Consider the special case $r = 1$, $(c_1, c_2) = (-\infty, x)$. Then in (9.245) we can make the replacements

$$\mathcal{B}_{-1} \mapsto \frac{d}{dx}, \quad \mathcal{B}_0 \mapsto x\frac{d}{dx}, \quad \mathcal{B}_1 \mapsto x^2\frac{d}{dx}$$

and after simplification deduce the nonlinear differential equation

$$\frac{d^3}{dx^3}F + 6\Big(\frac{d}{dx}F\Big)^2 + (-4x^2 + 8n)\frac{d}{dx}F + 4xF = 0. \tag{9.249}$$

On the other hand in this case $F = U_N^G(x;\mu;\xi)\Big|_{\substack{\mu=0\\ \xi=1}}$ (recall Proposition 8.3.1), and thus F must satisfy the σPIV equation in (8.15) with $\alpha_1 = 0$ and $\alpha_2 = -N$. In fact results of Cosgrove [119], [120], reducing a class of third order nonlinear differential equations to the equations (8.15), can be used to give a direct reduction of (9.249) to the sought σPIV equation [526].

EXERCISES 9.10 1. [526] Write

$$Z_{n,\beta}[\{t_i\}; R] := \frac{1}{n!}\int_R dx_1 \cdots \int_R dx_n \prod_{l=1}^n e^{-V(x_l)} e^{\sum_{j=1}^\infty t_j x_l^j} \prod_{1\le j<k\le n} |x_k - x_j|^\beta$$

(cf. (9.235)). The aim of this exercise is to show that (9.237) generalizes to read

$$\Big(-\mathcal{B}_p + \sum_{i=0}^\infty \Big(a_i L_{p+i}^{(\beta)} - b_i \frac{\partial}{\partial t_{p+i+1}}\Big)\Big) Z_{N,\beta} = 0, \qquad p \ge -1.$$

where

$$L_p^{(\beta)} := \frac{\beta}{2} \sum_{k=0}^{p} \frac{\partial}{\partial t_k} \frac{\partial}{\partial t_{p-k}} + \sum_{k=1}^{\infty} k t_k \frac{\partial}{\partial t_{p+k}} + \Big(1 - \frac{\beta}{2}\Big)(p+1)\frac{\partial}{\partial t_p}.$$

(i) Modify the working of the proof of Proposition 9.10.1 to show that (9.241) applies with $\mathcal{Z}_N$ replaced by $\mathcal{Z}_{N,\beta}$ provided

$$\Big\langle \sum_{i,k=1}^{N} \frac{f(x_i) - f(x_k)}{(x_i - x_k)(z - x_k)} \Big\rangle_{\mathcal{Z}_N} + \Big\langle \sum_{i,k=1}^{N} \frac{f(x_i)}{(z - x_i)(z - x_k)} \Big\rangle_{\mathcal{Z}_N}$$

is replaced by

$$\frac{\beta}{2}\Big(\Big\langle \sum_{i,k=1}^{N} \frac{f(x_i) - f(x_k)}{(x_i - x_k)(z - x_k)} \Big\rangle_{\mathcal{Z}_N} + \Big\langle \sum_{i,k=1}^{N} \frac{f(x_i)}{(z - x_i)(z - x_k)} \Big\rangle_{\mathcal{Z}_N}\Big)$$
$$+ \Big(1 - \frac{\beta}{2}\Big)\Big(\Big\langle \sum_{i=1}^{N} \frac{f(x_i)}{(z - x_i)^2} \Big\rangle + \Big\langle \sum_{i=1}^{N} \frac{f'(x_i)}{z - x_i} \Big\rangle\Big).$$

(ii) Show that the coefficient of z^{-p}, $p \ge 1$, in the final line above is equal to

$$(1 - \beta/2)\Big(\sum_{k=0}^{\infty} a_k (p - 1 + k) \frac{\partial}{\partial t_{k+p-2}}\Big) \mathcal{Z}_{N,\beta}$$

and use this together with the working below (9.241) to deduce the sought result.

Chapter Ten

Lattice paths and growth models

Nonintersecting lattice paths are the space-time trajectories of random walkers on a one-dimensional lattice. In some circumstances the probability distribution for the final position of the paths can be written in the form of the Boltzmann factor of a $\beta = 1$ or 2 log-gas in which the particles are confined to lattice sites, or as an average over the unitary or symplectic groups. An important role is played by the Schur polynomials, which can be defined combinatorially in terms of weighted lattice paths. Tiling of a hexagon by three species of rhombi is equivalent to some nonintersecting lattice path configurations. By analyzing the vertical rhombi, a multi-species system referred to as a bead process is obtained; the p.d.f. for any one of these species have the form of the JUE, while a further scaled limit gives the joint p.d.f. for the GUE minor process. The Robinson-Schensted-Knuth correspondence from bijective combinatorics leads to a statistical model—the discrete polynuclear growth model—which can be specified in terms of nonintersecting paths. The maximum height can be given as a matrix integral over the unitary group, while imposing certain symmetry constraints on the underlying matrix of nucleation events gives rise to matrix integrals over the symplectic and orthogonal groups. An alternative interpretation of the polynuclear growth model is as a model of directed last passage percolation. This permits a limit to the Hammersley process—a last passage percolation model associated with points distributed in a unit square according to a Poisson rate. The Hammersley process in turn relates to the length of the largest increasing subsequence of a random permutation, and results from random matrix theory relating to spacing distributions are used to compute the corresponding scaled distribution.

10.1 COUNTING FORMULAS FOR DIRECTED NONINTERSECTING PATHS

10.1.1 Nonintersecting ld/rd lattice paths

Consider the square lattice, and mark in points along the x-axis at $l_1^{(0)} > l_2^{(0)} > \cdots > l_p^{(0)}$ where all $l_j^{(0)}$ are required to have the same parity (i.e. be all even or all odd). For each $j = 1, \dots, p$ move a copy of the point $(l_j^{(0)}, 0)$ to $(l_j^{(0)} - 1, 1)$ or $(l_j^{(0)} + 1, 1)$ according to the rule that all the new points are distinct, and draw a left or right sloping segment connecting the point and its copy. Weight each line segment $w_1^{\mp}$ according to it sloping to the left (right). We build up p weighted nonintersecting paths each of N segments (also referred to as steps) by repeating this procedure a total of N times, with the weight for going from points along $y = j-1$ to points along $y = j$ given by $w_j^{\mp}$ for left (right) diagonal segments (abbreviated ld, rd, respectively). An example is given in Figure 10.1. The lattice paths can be considered as the space-time trajectories of random walkers on a one-dimensional lattice. At each time interval the random walkers can move one step to the left or one step to the right subject to the constraint that no two walkers can occupy the same site. In this picture the model under consideration is referred to as the *lock step model of vicious walkers* [175].

Let us denote the x-coordinates of the final positions by $\{l_i\}_{i=1,\dots,p}$, $l_1 > l_2 > \cdots > l_p$. We seek the total weight $G_N^{\mathrm{ld/rd}}(\vec{l}^{(0)}; \vec{l})$ of all allowed nonintersecting ld/rd lattice paths of N segments starting at $\{l_i^{(0)}\}_{i=1,\dots,p}$, $l_1^{(0)} > l_2^{(0)} > \cdots > l_p^{(0)}$, and finishing at $\{l_i\}_{i=1,\dots,p}$. This can be determined as a special case of a theorem due to Linström, Gessel and Viennot (see, e.g., [475]) on computing the generating function for the total weight of a general class of nonintersecting lattice paths.

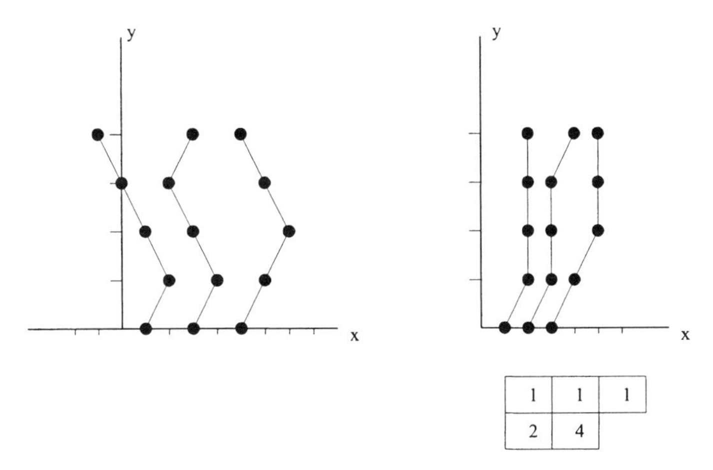

Figure 10.1 A configuration of 3 nonintersecting left diagonal/right diagonal lattice paths of 4 steps starting two units apart (leftmost diagram). An equivalent configuration of up/right diagonal lattice paths starting one unit apart is given in the rightmost diagram, and the paths are also encoded as a semi-standard tableau, with the kth column corresponding to the kth lattice path from the right. The total weight of the right steps in the configuration can be read off from the semi-standard tableau as $w_1^{+\#1's} \cdots w_p^{+\#p's}$.

Proposition 10.1.1 *Let D denote a directed acyclic graph (i.e., no loops), and let $\vec{u} = (u_1, \dots, u_r)$, $\vec{v} = (v_1, \dots, v_r)$ be sets of vertices in D so that in connecting $\{u_i\}$ to $\{v_i\}$ by nonintersecting paths along edges of the graph, the only possibility is to connect u_i to v_i for each $i = 1, 2, \dots, r$. Let each edge of the graph be weighted, let $h(u, v)$ denote the total weight of all single paths from u to v, and let $H(\{u\}, \{v\})$ denote the total weight of all nonintersecting paths starting at $\{u\}$ and finishing at $\{v\}$. Then*

$$H(\{u\}, \{v\}) = \det[h(u_i, v_j)]_{i,j=1,\dots,r}. \tag{10.1}$$

Proof. Following [502], we note that by definition

$$\det[h(u_i, v_j)]_{i,j=1,\dots,r} = \sum_{\sigma \in S_r} \operatorname{sgn}(\sigma) h(u_1, v_{\sigma(1)}) \cdots h(u_r, v_{\sigma(r)}) \tag{10.2}$$

(recall (5.22)). With P_i denoting a path from u_i to $v_{\sigma(i)}$, and $w(P_i) := h(u_i, v_{\sigma(i)})$ its corresponding weight, this expansion can be viewed as a generating function for $(r+1)$-tuples $(\sigma, P_1, \dots, P_r)$, assigned weight $\operatorname{sgn}(\sigma) w(P_1) \cdots w(P_r)$. We want to show that in the sum all such weights of intersecting paths cancel in pairs, while for the nonintersecting paths $\operatorname{sgn}(\sigma) = 1$.

Choose a total ordering of the vertices so that $u_1 < \cdots < u_r < v_1 < \cdots < v_r$. Consider a $(\sigma, P_1, \dots, P_r)$ with at least one pair of intersecting paths, and let v be the smallest vertex in the total ordering such that the paths intersect. Let i and j be the smallest indices such that paths P_i and P_j pass through v, and introduce the notation $P_k(\to v)$ and $P_k(v \to)$ to denote the subpaths of P_k from u_k to v and v to $v_{\sigma(k)}$. Next define $(\sigma', P_1', \dots, P_r')$ such that $P_l' = P_l$ for $l \neq i, j$ and

$$P_i' = P_i(\to v) P_j(v \to), \qquad P_j' = P_j(\to v) P_i(v \to), \qquad \sigma' = \sigma \circ (i, j)$$

(see Figure 10.2). We note that the set of edges of $(P_1', \dots, P_r')$ is identical to the set of edges of $(P_1, \dots, P_r)$ and thus $w(P_1) \cdots w(P_r) = w(P_1') \cdots w(P_r')$, while $\operatorname{sgn}(\sigma') = -\operatorname{sgn}(\sigma)$. Consequently

$$\operatorname{sgn}(\sigma) w(P_1) \cdots w(P_r) + \operatorname{sgn}(\sigma') w(P_1') \cdots w(P_r') = 0. \tag{10.3}$$

Since the graph D is assumed acyclic, this new set of paths has the same set of intersection points as before (as illustrated in [502], if D has cycles an intersection between two paths may be mapped to a self-intersection, thus violating this property). Hence this construction is an involution and so gives a unique pairing of intersecting paths each with the property (10.3), demonstrating that their contribution cancels in (10.2).

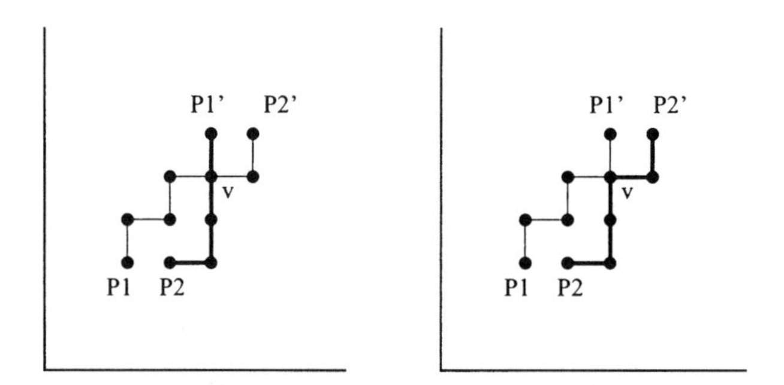

Figure 10.2 The intersecting lattice paths from P_1 to P_1' and P_2 to P_2' in the second diagram are constructed from the intersecting lattice paths from P_1 to P_2' and P_2 to P_1' according to the prescription in the proof of Proposition 10.1.1.

It remains to show that the nonintersecting paths have $\operatorname{sgn}(\sigma) = 1$. In fact it follows from the assumption on $\{u\}$ and $\{v\}$ that the only time nonintersecting paths occur is when $\sigma = I$ (the identity), so this property is immediate. □

To apply Proposition 10.1.1 to nonintersecting ld/rd lattice paths, note that the ld and rd segments form the edges of a directed graph, and that the generating function (total weight) for single lattice paths from $l^{(0)}$ to l is

$$\frac{1}{2\pi}\int_{-\pi}^{\pi}\prod_{j=1}^{N}(w_j^- e^{-i\theta} + w_j^+ e^{i\theta})e^{-i(l-l^{(0)})\theta}\, d\theta =: g_N^{\mathrm{ld/rd}}(l^{(0)}; l). \tag{10.4}$$

Thus

$$G_N^{\mathrm{ld/rd}}(\vec{l}^{(0)}; \vec{l}) = \det\left[g_N^{\mathrm{ld/rd}}(l_j^{(0)}, l_k)\right]_{j,k=1,\dots,p}. \tag{10.5}$$

The case that the initial and final sites are all odd,

$$\vec{l}^{(0)} = (2p-1, 2p-3, \dots, 1), \qquad \vec{l} = (2l_1 - 1, \dots, 2l_p - 1), \tag{10.6}$$

or all even, so that the initial points are the minimum allowed distance apart (2 lattice sites), has some special features. First it is easy to see that then each configuration of nonintersecting lattice paths is equivalent to a configuration of nonintersecting lattice paths starting one site apart with segments either vertical or diagonal and to the right (see Figure 10.1).

Let $G_N^{\mathrm{u/rd}}$ denote the total weight of the up/right diagonal (abbreviated u/rd) nonintersecting paths in the case that each rd path segment at step j is weighted q_j, and each vertical path segment is weighted unity. From Proposition 10.1.1, for general initial conditions

$$G_N^{\mathrm{u/rd}}(\vec{m}^{(0)}; \vec{m}) = \det\left[g_N^{\mathrm{u/rd}}(m_j^{(0)}, m_k)\right]_{j,k=1,\dots,p}, \tag{10.7}$$

where, with e_j denoting the elementary symmetric functions (polynomials) as defined in (4.132) ($e_n := 0$ for $n \in \mathbb{Z}^-$),

$$\begin{aligned} g_N^{\mathrm{u/rd}}(m^{(0)}; m) &= \frac{1}{2\pi}\int_{-\pi}^{\pi}\prod_{j=1}^{N}(1 + q_j e^{i\theta})e^{-i(m-m^{(0)})\theta}\, d\theta \\ &= e_{m-m^{(0)}}(q_1, \dots, q_N). \end{aligned} \tag{10.8}$$

In the case that the paths start one unit apart, the number of distinct u/rd nonintersecting paths can be written in terms of the components of $\vec{m}$ in a simple form [365].

Proposition 10.1.2 *Let $\vec{m}^{(0)} = (p-1, p-2, \dots, 0)$ and set $q_j = 1$ $(j = 1, \dots, p)$. We have*

$$G_N^{\mathrm{u/rd}}(\vec{m}^{(0)}; \vec{m}) = \prod_{i=1}^{p} \frac{(N+p-i)!}{m_i!(p+N-1-m_i)!} \prod_{1 \le j < k \le p} (m_j - m_k). \tag{10.9}$$

Proof. With $q_j = 1$ $(j = 1, \dots, p)$ we see from (10.8) that

$$g_N^{\mathrm{u/rd}}(m^{(0)}; m) = \binom{N}{m - m^{(0)}}. \tag{10.10}$$

Substituting in (10.7) and extracting a common factor from each column shows

$$\begin{aligned} & G_N^{\mathrm{u/rd}}(\vec{m}^{(0)}; \vec{m}) \\ & = \prod_{i=1}^{p} \frac{N!}{m_i!(p+N-1-m_i)!} \det \Big[(m_k - p + j + 1)_{p-j} (N - m_k + p - j + 1)_{j-1} \Big]_{j,k=1,\dots,p}, \end{aligned} \tag{10.11}$$

where $(a)_j$ is specified in (5.83). But in general [364]

$$\det \Big[(x_j + a_{k+1}) \cdots (x_j + a_n)(x_j + b_2) \cdots (x_j + b_k) \Big]_{j,k=1,\dots,n} = \prod_{1 \le j < k \le n} (x_k - x_j) \prod_{2 \le j \le k \le n} (a_k - b_j). \tag{10.12}$$

This can be seen by subtracting columns $j+1$ from columns j $(j = 1, \dots, n-1)$ in order, and extracting the common factors of $(b_n - a_n), (b_{n-1} - a_{n-1}), \dots, (b_2 - a_2)$ which result in the new columns $j = 1, \dots, n-1$. Repeating this procedure starting from column 2 then column 3 etc. reduces the determinant to

$$\prod_{2 \le j \le k \le n} (b_j - a_k) \det \Big[(x_j + a_{k+1}) \cdots (x_j + a_n) \Big]_{j,k=1,\dots,n}.$$

The argument used to evaluate the Vandermonde determinant (1.173) shows that the last determinant is equal to $\prod_{j<k}^{n} (x_j - x_k)$ and thus (10.12) follows. The transpose of the determinant in (10.11) is the special case (up to a sign) of (10.12) with

$$n = p, \quad x_j = m_j - p, \quad a_k = k, \quad b_k = k - 1 - N - p$$

and (10.9) results. □

We note that (10.9) has the form of a Boltzmann factor of a one-component log-gas at $\beta = 1$ with its domain the integer lattice points $-N \le l \le N + p - 1$, subject to a one-body potential with Boltzmann factor $w(l)$ proportional to $1/(N+p-1-l)!(N+l)!$. Each configuration of these new u/rd nonintersecting paths is equivalent to a *semi-standard tableau*.

Definition 10.1.3 *Let $\lambda_1 \ge \lambda_2 \ge \cdots \ge \lambda_n \ge 0$ be non-negative integers. Refer to this ordered set of non-negative integers by λ, which in turn is called a partition. The diagram of a partition is constructed by imagining a blank matrix grid, then for each row $j = 1, \dots, n$ inserting in this grid λ_j boxes of unit dimension drawn side-by-side starting from column 1. The conjugate partition λ' is defined as a diagram by interchanging the rows and columns of the diagram of λ. A semi-standard tableau of shape λ and content N is then a filling of this diagram with the numbers $\{1, 2, \dots, N\}$ such that the numbers weakly increase along any row, and strictly increase down any column.*

Given a u/rd nonintersecting lattice path configuration, a semi-standard tableau can be constructed by labeling each right diagonal segment between $y = j - 1$ and $y = j$ by the number j. The numbers of the rightmost lattice path form the first column of the tableau, the numbers of the second rightmost lattice path form the second column and so on. Here the length of each column k of the tableau $(k = 1, \dots, p)$ is equal to $m_{p-k+1} - (p - k + 1)$. An example of the construction is given in Figure 10.1. Of fundamental importance is a particular weighted sum over all semi-standard tableaux of a given shape.

DEFINITION 10.1.4 *For semi-standard tableaux of a given shape λ, and content N, define the total weight by*

$$s_\lambda(q_1,\dots,q_N) := \sum_{\substack{\text{semi standard tableaux}\\ \text{shape } \lambda, \text{ content } N}} q_1^{\#1's} q_2^{\#2's} \cdots q_N^{\#N's}. \tag{10.13}$$

This is referred to as the Schur polynomial.

Let $G_N^{\mathrm{u/rd}}$ be as in (10.7). From the above discussion it follows that

$$G_N^{\mathrm{u/rd}}((p-1,\dots,0);(\lambda_1'+p-1,\dots,\lambda_p')) = s_\lambda(q_1,\dots,q_N). \tag{10.14}$$

Combining (10.14) and (10.7) and recalling (10.8) shows

$$s_\lambda(q_1,\dots,q_N) = \det\left[e_{\lambda_k'+j-k}\right]_{j,k=1,\dots,p}, \tag{10.15}$$

where $p=\lambda_1$. Observe from (10.15) that s_λ is a symmetric function of $q_1,\dots,q_N$, a property which is not obvious from the definition (10.13).

A recurrence implied by the definition (10.13) can be used to verify an expression for the Schur polynomials as the ratio of two determinants.

PROPOSITION 10.1.5 *We have*

$$s_\lambda(q_1,\dots,q_N) = \frac{\det[q_j^{N-k+\lambda_k}]_{j,k=1,\dots,N}}{\det[q_j^{N-k}]_{j,k=1,\dots,N}}. \tag{10.16}$$

Proof. First note from the definition (10.13) that, with $*$ denoting the region

$$\lambda_1 \ge \mu_1 \ge \lambda_2 \ge \mu_2 \ge \cdots \ge \lambda_{N-1} \ge \mu_{N-1} \ge \lambda_N, \tag{10.17}$$

the Schur polynomials satisfy the recurrence

$$s_\lambda(q_1,\dots,q_N) = \sum\nolimits_\mu^* s_\mu(q_1,\dots,q_{N-1}) q_N^{|\lambda|-|\mu|}. \tag{10.18}$$

Subject to the initial condition $s_\lambda = 1$ for $\lambda=\emptyset$, this uniquely determines the Schur polynomials.

Let us denote the r.h.s. of (10.16) by $r_\lambda(q_1,\dots,q_N)$. Following [453], our objective is to show that r_λ satisfies the same recurrence and initial condition. The initial condition is immediate. To derive the recurrence, we begin by setting $q_N=1$ in the determinants and subtracting the last row (which now consists of 1's in each entry) from all the other rows. Dividing the jth row by q_j-1 $(j=1,\dots,N-1)$, and expanding each term as a power series shows

$$r_\lambda(q_1,\dots,q_{N-1},1) = \frac{\det\begin{bmatrix} \sum_{l=0}^{N-k+\lambda_k-1} q_j^l \\ 1 \end{bmatrix}_{\substack{j=1,\dots,N-1\\ k=1,\dots,N}}}{\det\begin{bmatrix} \sum_{l=0}^{N-k-1} q_j^l \\ 1 \end{bmatrix}_{\substack{j=1,\dots,N-1\\ k=1,\dots,N}}}. \tag{10.19}$$

Next in each determinant, subtract column k from column $(k-1)$, $k=2,\dots,N$ in order. The last row then has only its final element nonzero. Expanding by this element shows that (10.19) reduces to the ratio of $(N-1)$-dimensional determinants

$$\frac{\det\left[\sum_{l=\lambda_{k+1}}^{\lambda_k} q_j^{l+N-k-1}\right]_{j,k=1,\dots,N-1}}{\det\left[q_j^k\right]_{j,k=1,\dots,N-1}} = \sum\nolimits^* r_\mu(q_1,\dots,q_{N-1}). \tag{10.20}$$

Because the determinant in (10.16) is homogeneous of degree $|\lambda| := \sum_{l=1}^N \lambda_l$ it follows from the equality between (10.19) and (10.16) that r_λ satisfies the recurrence (10.18).

□

A fundamental property of the Schur polynomials relates to their orthonormality with respect to averaging over the unitary group, or equivalently over the circular unitary ensemble,

$$\Big\langle s_\lambda(e^{-i\theta_1},\dots,e^{-i\theta_N}) s_\kappa(e^{i\theta_1},\dots,e^{i\theta_N}) \Big\rangle_{U(N)} = \delta_{\lambda,\kappa}. \tag{10.21}$$

This property is simple to verify from the determinant formula in (10.16). It can also be understood from the fact that the Schur polynomials are eigenfunctions of a certain differential operator which is self-adjoint with respect to the inner product $\langle \bar{f} g\rangle_{U(N)}$ (see Section 12.1.2).

A further corollary of (10.16), together with the Vandermonde determinant evaluation (1.173), is the specialization formula

$$s_\lambda(1,q,\dots,q^{N-1}) = q^{\sum_{j=1}^N (j-1)\lambda_j} \prod_{1\le j<k\le N} \frac{1-q^{\lambda_j-\lambda_k+k-j}}{1-q^{k-j}}. \tag{10.22}$$

In the limit $q \to 1$ this yields the evaluation formula

$$s_\lambda(x_1,\dots,x_N)\Big|_{x_1=\cdots=x_N=1} := s_\lambda((1)^N) = \prod_{1\le j<k\le N} \frac{\lambda_j-\lambda_k+k-j}{k-j}. \tag{10.23}$$

Given a semi-standard tableau, a nonintersecting lattice path configuration can also be constructed where each row (rather than column) corresponds to a lattice path. This will be referred to as the *conjugate* configuration. Thus if there are p^* rows, one marks points along $y=1$ at $x=0,1,\dots,p^*-1$. The numbers in row j correspond to a horizontal right segment of lattice path number p^*-j at $y=j$ which is weighted by q_j, while otherwise the lattice path segments are up. If the length of row j in the tableau is equal to λ_j, then the final x-coordinate of walker j is λ_j+p^*-j (see Figure 10.3). Let the total weight of such paths be denoted $G_N^{\rm u/rh}$. From the definition (10.13) we have

$$G_N^{\rm u/rh}((p^*-1,p^*-2,\dots,0);(\lambda_1+p^*-1,\dots,\lambda_{p^*})) = s_\lambda(q_1,\dots,q_N). \tag{10.24}$$

We can use (10.24) to obtain a different expression for $G_N^{\rm u/rh}$ and thus the Schur polynomials. For this, introduce the complete symmetric functions (polynomials) h_j in the variables $q_1,\dots,q_n$ by

$$h_j = \sum_{1\le k_1\le\cdots\le k_j\le n} q_{k_1}\cdots q_{k_j} = \sum_{\substack{l_1,\dots,l_n\ge 0\\ l_1+\cdots+l_n=j}} q_1^{l_1}q_2^{l_2}\cdots q_n^{l_n} \qquad (j\ge 1),$$

$h_0=1$, $h_{-j}=0$. These polynomials are generated by

$$\prod_{j=1}^n \frac{1}{1-q_j u} = \sum_{j=0}^\infty h_j u^j. \tag{10.25}$$

Noting that

$$\frac{1}{1-q_j e^{i\theta}} = 1 + \sum_{l=1}^\infty q_j^l e^{il\theta}$$

and making use of (10.25) we see that the generating function for a single u/rh path is

$$g_N^{\rm u/rh}(r^{(0)};r) = \frac{1}{2\pi}\int_{-\pi}^{\pi} \prod_{l=1}^N \frac{1}{1-q_l e^{i\theta}} e^{-i(r-r^{(0)})\theta}\, d\theta = h_{r-r^{(0)}}. \tag{10.26}$$

Substituting this in (10.1) and comparing with (10.24) shows

$$s_\lambda(q_1,\dots,q_N) = \det\Big[h_{\lambda_k+j-k}\Big]_{j,k=1,\dots,p^*}, \tag{10.27}$$

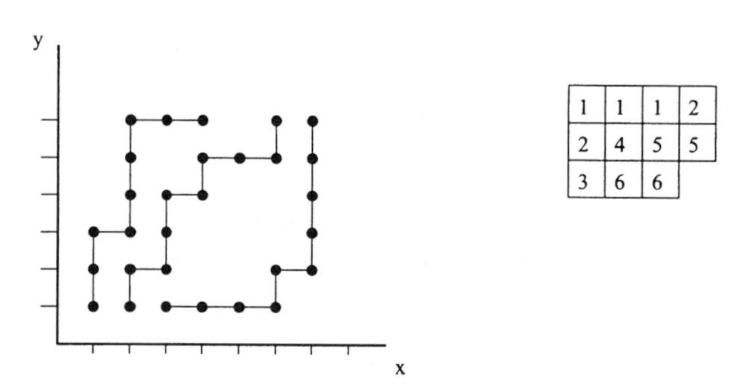

Figure 10.3 A configuration of 3 u/rh nonintersecting lattice paths starting one unit apart with horizontal segments in any of 6 consecutive levels. The level of each right horizontal step of the jth walker from the right is recorded in the jth row to obtain a semi-standard tableau.

where $p^* = \mu_1'$.

Another consequence of (10.24), and the evaluation formula (10.23), is a counting formula for the number of u/rh nonintersecting paths starting one unit apart and finishing at arbitrary points $l_1, \dots, l_{p^*}$ [282].

Proposition 10.1.6 *Let $\vec{l}^{(0)} = (p^* - 1, p^* - 2, \dots, 0)$, $\vec{l} = (l_1, \dots, l_{p^*})$. Let $G_N^{\rm u/rh}(\vec{l}^{(0)}; \vec{l})$ denote the number of u/rh paths from $\vec{l}^{(0)}$ at $y = 1$ to $\vec{l}$ at $y = N$. We have*

$$G_N^{\rm u/rh}(\vec{l}^{(0)}; \vec{l}) = \prod_{i=1}^{p^*} \frac{(N - p^* + l_i)!}{l_i!(N - p^* + i - 1)!} \prod_{1 \le j < k \le p^*} (l_j - l_k). \tag{10.28}$$

Proof. With $q_1 = \cdots = q_N = 1$ in (10.24), substitute (10.23), set $\lambda_j = l_j + j - p^*$ $(j = 1, \dots, p^*)$, $\lambda_j = 0$ $(j = p^* + 1, \dots, N)$, and simplify. □

10.1.2 Nonintersecting lattice paths near a wall

In the nonintersecting ld/rd lattice path model of the previous section, suppose the paths are conditioned so that their x-coordinate is always positive. The y-axis can then be thought of as an infinitely repelling wall. Consider a single weighted ld/rd lattice path of N segments starting at $l^{(0)}$ and finishing at l with both $l^{(0)}, l > 0$. We know the total weight of such paths is given by $g_N^{\rm ld/rd}$ as specified in (10.4). Consider now

$$g_N^{\rm w,ld/rd}(l^{(0)}, l) := g_N^{\rm ld/rd}(l^{(0)}, l) - g_N^{\rm ld/rd}(l^{(0)}, -l). \tag{10.29}$$

To go from $l^{(0)}$ to $-l$ the path must cross the y-axis at least once. If we reflect the portion of the lattice path from the first crossing to $-l$ about the y-axis we obtain a lattice path from $l^{(0)}$ to l which touches or crosses the y-axis (see Figure 10.4). This lattice path will be in $g_N^{\rm ld/rd}$ and thus will cancel out of (10.29), leaving only the contribution from lattice paths strictly to the right of the y-axis.

The formula (10.1) shows that forming a determinant out of (10.29) gives the total weight of multiple nonintersecting paths that are restricted to the region $x > 0$ [180]. Furthermore, in the special case that the walkers start at the minimum allowed distance from the wall (1 lattice site) and the minimum distance from each other (2 lattice sites), the determinant can be evaluated [365].

Proposition 10.1.7 *Define $g_N^{\rm w,ld/rd}$ by (10.29) and let $G_N^{\rm w,ld/rd}(l_1^{(0)}, \dots, l_p^{(0)}; l_1, \dots, l_p)$ denote the total*

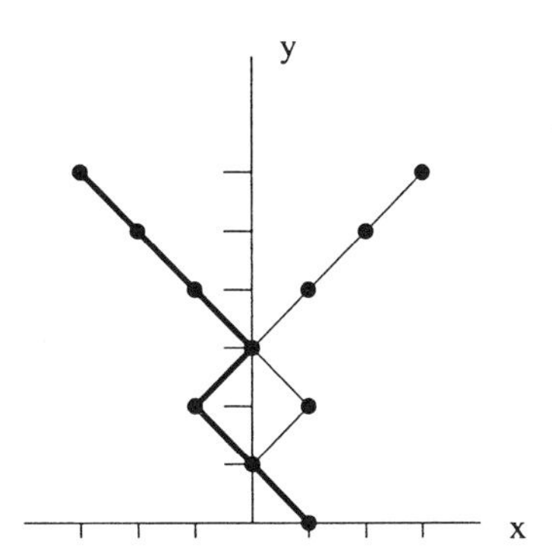

Figure 10.4 In bold is drawn an ld/rd lattice path starting at $x = l^{(0)} = 1$ and finishing at $x = -l = -3$ after 6 steps, while the union of this path from $x = 1$ to its first contact with the y-axis and the reflection of the remaining portion of the path in the y-axis gives a path starting at $x = l^{(0)} = 1$ and finishing at $x = l = 3$.

weight of the lattice paths. We have

$$G_N^{\mathrm{w,ld/rd}}(l_1^{(0)},\dots,l_p^{(0)};l_1,\dots,l_p) = \det\Big[g_N^{\mathrm{w,ld/rd}}(l_j^{(0)},l_k)\Big]_{j,k=1,\dots,p}, \tag{10.30}$$

and it follows from this that

$$\begin{aligned} &G_{2N}^{\mathrm{w,ld/rd}}(\{l_j^{(0)} = 2(p-j)+1\}_{j=1,\dots,p}; 2l_1-1,\dots,2l_p-1)\Big|_{\substack{w_k^{\pm}=1\\(k=1,\dots,2N)}} \\ &\quad = \prod_{i=1}^{p} \frac{(2l_i-1)(2N+2i-2)!}{(N+l_i+p-1)!(N-l_i+p)!} \prod_{1\le i<j\le p}(l_i-l_j)(l_i+l_j-1). \end{aligned} \tag{10.31}$$

Proof. As already commented (10.30) follows from (10.1). Regarding (10.31), we make use of (10.10) then proceed as in the derivation of (10.11) to deduce

$$\begin{aligned} &G_{2N}^{\mathrm{w,ld/rd}}(\{l_j^{(0)} = 2(p-j)+1\}_{j=1,\dots,p}; 2l_1-1,\dots,2l_p-1)\Big|_{\substack{w_k^{\pm}=1\\(k=1,\dots,2N)}} \\ &\quad = \prod_{i=1}^{p} \frac{(2N)!}{(N+p-l_i)!(N+p-1+l_i)!} \det\Big[(N+p-j+2-l_k)_{j-1}(N-p+j)_{2p-j} \\ &\qquad\qquad -(N+p-j+1+l_k)_{j-1}(N-p+j+1-l_k)_{2p-j}\Big]_{j,k=1,\dots,p}. \end{aligned} \tag{10.32}$$

We observe that this determinant with $j \mapsto p-j+1$ is the special case of the determinant

$$\begin{aligned} &W_p(\{x_k\}_{k=1,\dots,p},\{a_j\}_{j=2,\dots,p},\{b_j\}_{j=-p+1,\dots,p-1}) \\ &\quad := \det\Big[(a_{j+1}-x_k)(a_{j+2}-x_k)\cdots(a_p-x_k)(b_{-j+1}+x_k)(b_{-j+2}+x_k)\cdots(b_{p-1}+x_k) - *\Big]_{j,k=1,\dots,p} \end{aligned} \tag{10.33}$$

where the * denotes the same term but with $x_k \mapsto -x_k$, obtained by setting

$$x_k = l_k - \frac{1}{2}, \quad a_j = N + j - \frac{1}{2}, \quad b_j = N + j + \frac{1}{2}. \tag{10.34}$$

But in general (see Exercises 10.1 q.3)

$$\begin{aligned} &W_p(\{x_k\}_{k=1,\dots,p},\{a_j\}_{j=2,\dots,p},\{b_j\}_{j=-p+1,\dots,p-1}) \\ &\quad = 2^p \prod_{l=1}^{p} x_l \prod_{1\le j<k\le p}(x_k-x_j)(x_k+x_j)\prod_{l=2}^{p}\prod_{j=l}^{p}(b_{-j-1+l}+a_j)\prod_{0\le j<k\le p-1}(b_j+b_k). \end{aligned} \tag{10.35}$$

Substituting (10.34) we obtain (10.31) after some minor simplification. □

The nonintersecting w,ld/rd lattice paths can be mapped to lattice paths restricted to $x > 0$ in which the allowed steps are u/rd for odd numbered steps and u/ld for even numbered steps. The configurations of the latter are denoted w,u/rd(o)/ld(e), where (e) denotes (e)ven while (o) denotes (o)dd. The steps up are to be weighted unity, the steps rd(o) weighted w_j^+ for step number $2j-1$, while the steps ld(e) weighted w_j^- for step number $2j$. For the initial conditions, if the original x coordinates in the w,ld/rd case are specified by $2l_1^{(0)}-1, 2l_2^{(0)}-1, \dots, 2l_p^{(0)}-1$ say, then the original x-coordinates of the case of the w,u/rd(o)/ld(e) lattice paths are given by $l_1^{(0)}, l_2^{(0)}, \dots, l_p^{(0)}$. At odd numbered steps the ld segments then map to up segments with weight 1 and the rd segments remain unaltered, while at even numbered steps the rd segments map to up segments with weight 1 and the ld segments remain unaltered (cf. Exercises 10.1 q.2). It is easy to see that if the final x-coordinates after $2N$ steps in the w,ld/rd case are $2l_1-1, \dots, 2l_p-1$ then the final x-coordinates in the case of the w,u/rd(o)/ld(e) lattice paths are given by $m_j = l_j$ $(j=1,\dots,p)$. Thus in particular

$$\begin{aligned} & G_{2N}^{\mathrm{w,u/rd(o)/ld(e)}}(r_1^{(0)},\dots,r_p^{(0)};r_1,\dots,r_p) \\ & \quad = G_{2N}^{\mathrm{w,ld/rd}}(2r_1^{(0)}-1,\dots,2r_p^{(0)}-1;2r_1-1,\dots,2r_p-1)\Big|_{\substack{w_{2j-1}^-=w_{2j}^+=1 \\ (j=1,\dots,N)}}. \end{aligned} \tag{10.36}$$

For the initial spacings 1 unit apart, the w,rd(o)/ld(e) lattice paths (or equivalently the w,u/rd/ld lattice paths) can be encoded as *oscillating tableaux*. The latter are a sequence of semi-standard tableaux, one for each $y = 1, 2, \dots, 2N$ for paths of $2N$ segments, in which an rd(o) segment of the kth lattice path from the right between $y = 2j-2$ and $y = 2j-1$ is recorded in column k by a box labeled j, while an ld(e) segment of the same lattice path in going between $y = 2j-1$ and $y = 2j$ is recorded by removing the bottom box from column k (see Figure 10.5). Analogous to the situation with semi-standard tableaux noted in Section 10.1.1, by reading the oscillating tableaux along rows rather than down columns we can associate a conjugate set of lattice paths. These lattice paths start one unit from the wall and one unit from each other, and for odd (even) y-coordinates can move an arbitrary number of units to the right (left), or one unit up, subject to the wall constraint that $x > 0$. Such paths are denoted w,u/rh(o)/lh(e).

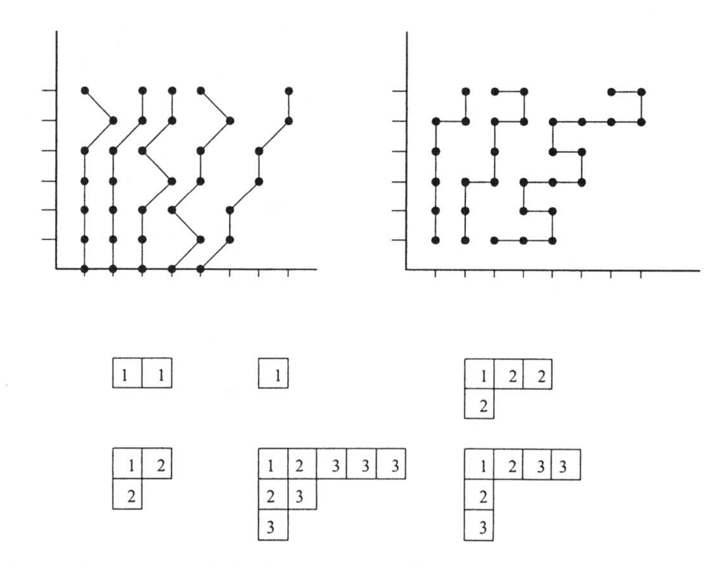

Figure 10.5 Illustration of the correspondence between w,u/rd/ld lattice paths, their conjugate w,u/rh(o)/lh(e) lattice paths and oscillating tableaux.

10.1.3 Single move ld/rd lattice paths

As an extension of the ld/rd nonintersecting paths model of Section 10.1.1, allow in addition up steps to get ld/rd/u lattice paths. At step j weight the ld/rd segments by w_j^- and w_j^+, respectively, and the up steps by unity. The generating function for a single path of N segments is

$$g_N^{\mathrm{ld/rd/u}}(l^{(0)};l) = \frac{1}{2\pi}\int_{-\pi}^{\pi}\prod_{k=1}^{N}(1+w_k^+e^{i\theta}+w_k^-e^{-i\theta})e^{-i\theta(l-l^{(0)})}\,d\theta. \tag{10.37}$$

Application of Proposition 10.1.1 on the directed lattice formed by left and right sloping diagonal lines as well as vertical lines, gives that the generating function for multiple ld/rd/u paths is equal to

$$\det\left[g_N^{\mathrm{ld/rd/u}}(l_j^{(0)};l_k)\right]_{j,k=1,\dots,l}. \tag{10.38}$$

Suppose now that at each step j exactly one segment is ld or rd, while all others are up. Denote such lattice paths by s,ld/rd ("s" is for single). In the picture of evolution of random walkers on a one-dimensional lattice, this means that at each time step exactly one walker moves to the left or right while all other walkers remain stationary, specifying the so-called *random turns model of vicious walkers* [175]. We remark that in the version of this model in which only steps in one direction (say to the right) are allowed, and in which the walkers are initially one unit apart, the configurations can be encoded as a *standard tableau* (see Figure 10.6). A standard tableau is defined as for a semi-standard tableau, except that the content of the tableau must equal the number of boxes, and so each box must be labeled by a different number.

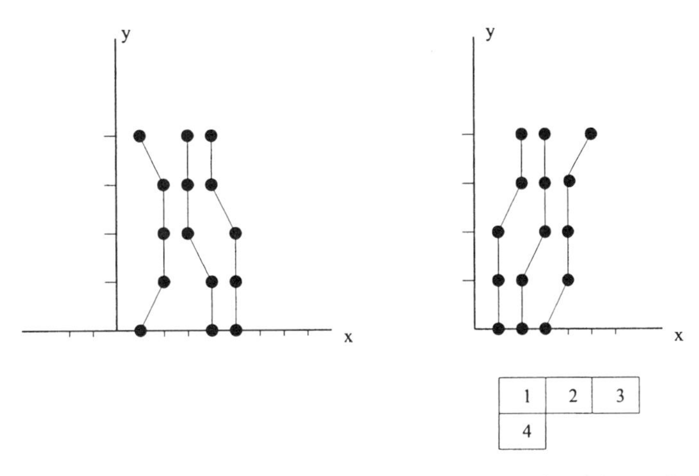

Figure 10.6 Two configurations of 3 nonintersecting single move ld/rd lattice paths of 4 steps. In the second configuration the paths are initially one unit apart and only rd segments are allowed. In this case the paths can be encoded as a standard tableau.

The total weight of p nonintersecting s,ld/rd paths of N steps is the term in (10.38) linear in each of w_k^+, w_k^- $(k=1,\dots,N)$. Hence we read off from (10.37) and (10.38) the following result [183].

PROPOSITION 10.1.8 *We have*

$$G_N^{\mathrm{s,ld/rd}}(\vec{l}^{(0)};\vec{l}) = \frac{1}{(2\pi)^p}\int_{-\pi}^{\pi}d\theta_1\cdots\int_{-\pi}^{\pi}d\theta_p \times\prod_{l=1}^{N}\sum_{k=1}^{p}\left(w_k^+e^{i\theta_j}+w_k^-e^{-i\theta_j}\right)\det\left[e^{-i(l_\alpha-l_\beta^{(0)})\theta_\alpha}\right]_{\alpha,\beta=1,\dots,p}. \tag{10.39}$$

The w,ld/rd lattice paths of Section 10.1.2 can be similarly modified to accommodate the single move condition [41].

PROPOSITION 10.1.9 *Let $G_N^{\mathrm{s,w,ld/rd}}(\vec{l}^{(0)};\vec{l})$ denote the total weight of p nonintersecting single move ld/rd paths restricted to $x > 0$. We have*

$$G_N^{\mathrm{s,w,ld/rd}}(\vec{l}^{(0)};\vec{l}) = \frac{1}{(2\pi)^p}\int_{-\pi}^{\pi} d\theta_1 \cdots \int_{-\pi}^{\pi} d\theta_p \prod_{k=1}^{N}\Big(\sum_{j=1}^{p} w_k^+ e^{i\theta_j} + w_k^- e^{-i\theta_j}\Big)$$
$$\times \det\Big[e^{-i(l_\alpha - l_\beta^{(0)})\theta_\alpha} - e^{-i(l_\alpha + l_\beta^{(0)})\theta_\alpha}\Big]_{\alpha,\beta=1,\dots,p}. \tag{10.40}$$

10.1.4 Nonintersecting paths on a triangular lattice

Following [330], consider a triangular lattice with sites formed from the integer span of the vectors $\vec{u}_1 = (-1,-1)$ and $\vec{u}_2 = (-1,1)$, and with edges formed by connecting nearest neighbour points in the directions of $\vec{u}_1, \vec{u}_2$ and $\vec{u}_1 + \vec{u}_2$ (see Figure 10.7). Let us use the basis $U = \{\vec{u}_1, \vec{u}_2\}$ to specify a general site via the coordinate $(a,b)_U := a\vec{u}_1 + b\vec{u}_2$. We define a directed path on this lattice by traversing the edges in their positive direction only. With reference to the xy-coordinate system, such paths consist of left diagonal (ld), negative right diagonal ($-$rd) and left horizontal steps. Let each step in the direction of $\vec{u}_1$ or $\vec{u}_2$ be weighted w, while the steps in the direction of $\vec{u}_1 + \vec{u}_2$ are to be weighted unity. Let $g^{\mathrm{ld}/-\mathrm{rd/lh}}((0,0)_U,(n,m)_U)$ denote the weight of all allowed lattice paths starting at the origin and finishing at $(n,m)_U$.

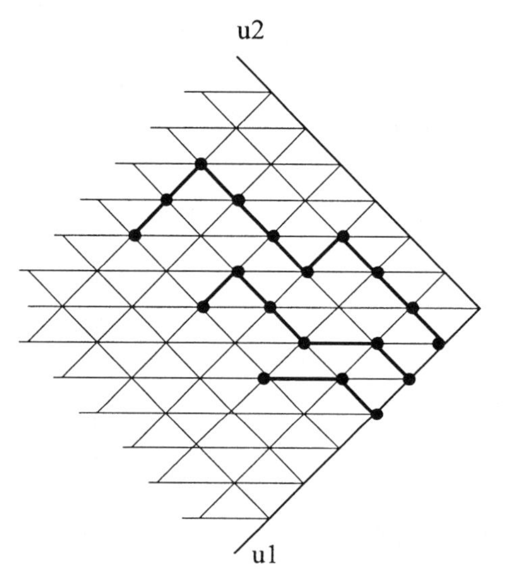

Figure 10.7 Axes in the directions of the vectors $\vec{u}_1 = (-1,-1)$ and $\vec{u}_2 = (-1,1)$ provide a convenient coordinate system for the triangular lattice formed by the integer span of these vectors, and connecting points as described in the text. Drawn on this lattice is a family of 3 nonintersecting rd/$-$ld/rh lattice paths, starting at $(1,0)_U, (2,0)_U, (3,0)_U$ and finishing at $(4,h_j)_U$ for particular h_j.

PROPOSITION 10.1.10 *We have*

$$g^{\mathrm{ld}/-\mathrm{rd/lh}}((0,0)_U,(n,m)_U) = \frac{w^{n+m}}{n!}\sum_{c=0}^{\infty}(m-c+1)_n \frac{n!}{(n-c)!c!} w^{-2c}. \tag{10.41}$$

Proof. Let a, b, c denote the number of steps in the direction of $\vec{u}_1, \vec{u}_2, \vec{u}_1 + \vec{u}_2$, respectively. Then $n = a+c$, $m = b+c$, while the number of paths with a given number of steps a, b, c is equal to

$$\frac{(a+b+c)!}{a!b!c!} = \frac{(n+m-c)!}{(n-c)!(m-c)!c!}$$

each of which is to be weighted by

$$w^{a+b} = w^{n+m-2c}.$$

Multiplying together the right-hand sides of the two displayed equations and summing over c gives (10.41). □

Knowledge of (10.41) allows the total weight of p nonintersecting paths starting at $(p+1-j, 0)_U$ ($j = 1, \dots, p$) and finishing at $(p, h_j)_U$ with $0 \le h_1 < \cdots < h_p$ to be computed.

PROPOSITION 10.1.11 *Let $G^{\mathrm{ld}/-\mathrm{rd}/\mathrm{rh}}(\{(p+1-j,0)_U\}_{j=1,\dots,p}; \{(p,h_j)_U\}_{j=1,\dots,p})$ denote the sought total weight. We have*

$$G^{\mathrm{ld}/-\mathrm{rd}/\mathrm{rh}}(\{(p+1-j,0)_U\}_{j=1,\dots,p}; \{(p,h_j)_U\}_{j=1,\dots,p})$$
$$= \det\Big[g^{\mathrm{ld}/-\mathrm{rd}/\mathrm{lh}}((0,0)_U, (j-1, h_k)_U)\Big]_{j,k=1,\dots,p} = \prod_{j=1}^{p} w^{h_j} \frac{(1+w^2)^{j-1}}{w^{j-1}(j-1)!} \prod_{1 \le j < k \le p} (h_k - h_j).$$

Proof. The first equality follows immediately from the fact that the paths are directed and Proposition 10.1.1. For the second equality, we note that with (10.41) substituted, the determinant is equal to

$$\prod_{j=1}^{p} \frac{w^{j-1+h_j}}{(j-1)!} \sum_{c_1,\dots,c_p=0}^{\infty} \prod_{i=1}^{p} \frac{(i-c_i)_{c_i}}{c_i! w^{2c_i}} \det\Big[(h_j - c_i + 1)_{i-1}\Big]_{i,j=1,\dots,p}.$$

Elementary row operations reduce the determinant in this expression to the Vandermonde determinant (1.173) in the variables $\{h_1, \dots, h_p\}$ independent of the c_i, and so the multiple summation is a product of p independent sums of the form

$$\sum_{c=0}^{\infty} (i-c)_c \frac{1}{c! w^{2c}} = \sum_{c=0}^{i-1} \binom{i-1}{c} \frac{1}{w^{2c}} = \Big(1 + \frac{1}{w^2}\Big)^{i-1}.$$

□

10.1.5 Continuous nonintersecting paths

Consider the generating function (10.4) for a single ld/rd lattice path in the case $w_j^- = w_j^+ = 1/2$, and suppose for definiteness that N, l, $l^{(0)}$ are all even, which is accomplished by writing $N \mapsto 2n$, $l \mapsto 2l$, $l^{(0)} \mapsto 2l^{(0)}$. We then have

$$g_{2N}^{\mathrm{ld}/\mathrm{rd}}(2l^{(0)}, 2l) = 2^{-2N} \binom{2N}{N + l - l^{(0)}}, \tag{10.42}$$

which also have the interpretation as being the probability that a walker reaches site $2l$ after $2N$ steps, starting at site $l^{(0)}$ and given that at each step the walker goes to the right or to the left each with probability $\frac{1}{2}$. The corresponding mean square displacement is

$$\sum_{l=-N}^{N} (2l)^2 g_{2N}^{\mathrm{ld}/\mathrm{rd}}(0, 2l) = 2N.$$

In the random walk picture, suppose now that the spacing between lattice sites is τ, and consider a continuum limit in which

$$\tau \to 0, \ \ N, l, l^{(0)} \to \infty, \ \ 2l\tau \to x, \ \ 2l^{(0)}\tau \to x^{(0)}, \ \ 2N\tau^2 \to Dt.$$

Then, using Stirling's formula, one obtains from (10.42) the limiting p.d.f.

$$\frac{1}{2\tau} g_{2N}^{\mathrm{ld/rd}}(2l^{(0)}, 2l) \to \Big(\frac{1}{2\pi Dt}\Big)^{1/2} e^{-(x-x^{(0)})^2/2Dt} =: g_t^{(B)}(x^{(0)}, x), \tag{10.43}$$

which has mean square displacement Dt. Note that $g_t^{(\mathrm{B})}$ (B for Brownian) is the unique solution of the diffusion equation

$$\frac{1}{D}\frac{\partial u}{\partial t} = \frac{1}{2}\frac{\partial^2 u}{\partial x^2}$$

subject to the initial condition $u|_{t=0} = \delta(x - x^{(0)})$. It follows from (10.43) and (10.5) that the p.d.f. for p nonintersecting Brownian walkers, or equivalently for continuous nonintersecting paths in the xt-plane, is given by the Karlin-MacGregor formula [347]

$$\begin{aligned} G_t^{(\mathrm{B})}(\vec{x}^{(0)}; \vec{x}) &= \det[g_t^{(\mathrm{B})}(x_j^{(0)}, x_k)]_{j,k=1,\dots,p} \\ &= \Big(\frac{1}{2\pi Dt}\Big)^{p/2} e^{-\sum_{j=1}^p (x_j^2 + (x_j^{(0)})^2)/2Dt} \det[e^{x_j^{(0)} x_k/Dt}]_{j,k=1,\dots,p}. \end{aligned} \tag{10.44}$$

Setting $x_j^{(0)} = (j-1)a$ (equally spaced initial condition) and using the Vandermonde formula (1.173) shows that $G_t^{(\mathrm{B})}(\{(j-1)a\}; \vec{x})$ has the functional form (2.73) with $\beta = 1$.

Suppose now that the positions $\vec{x}^{(0)}$ at time t are prescribed, and the nonintersecting condition is required up to time T. For a given starting position $\vec{y}$ define the average over positions time t later as

$$\mathcal{N}_t(\vec{y}) = \int_{x_N > \cdots > x_1} G_t^{(\mathrm{B})}(\vec{y}; \vec{x})\, d\vec{x}.$$

For $0 < t_0 < t < T$ the p.d.f. for the particles arriving at positions $\vec{x}$ at time t is then [349]

$$\frac{G_{t-t_0}^{(\mathrm{B})}(\vec{x}^{(0)}; \vec{x}) \mathcal{N}_{T-t}(\vec{x})}{\mathcal{N}_{T-t_0}(\vec{x}^{(0)})}, \tag{10.45}$$

as the numerator is the p.d.f. for the event that there is a transition from $\vec{x}^{(0)}$ to $\vec{x}$ in time $t-t_0$, and that following this the particles continue to obey the nonintersecting condition, while the denominator is the p.d.f. for the nonintersecting condition on the prescribed $\vec{x}^{(0)}$. The following relationship to the GUE eigenvalue p.d.f. can now be deduced.

PROPOSITION 10.1.12 *For p continuous nonintersecting Brownian walkers, starting at the origin and with the nonintersecting condition required for all times, the p.d.f. for the event that they arrive at $\vec{x}$ after time t is*

$$\frac{1}{(2\pi)^{p/2}\prod_{l=1}^{p-1} l!}\Big(\frac{1}{Dt}\Big)^{p^2/2} e^{-\sum_{j=1}^p x_j^2/2Dt} \prod_{1 \le j < k \le p} (x_k - x_j)^2. \tag{10.46}$$

Proof. Since the determinant in the final expression of (10.44) vanishes for any of $\{x_j^{(0)}\}$ equal or $\{x_j\}$ equal, it follows that for large t

$$\det[e^{x_j^{(0)} x_k/Dt}]_{j,k=1,\dots,p} \sim \frac{1}{\prod_{l=1}^{p-1} l!}\Big(\frac{1}{Dt}\Big)^{p(p-1)/2} \prod_{1 \le j < k \le p} (x_k - x_j)(x_k^{(0)} - x_j^{(0)}). \tag{10.47}$$

Consequently, in the limit $T \to \infty$ (10.45) reduces to

$$\frac{G_{t-t_0}^{(\mathrm{B})}(\vec{x}^{(0)}; \vec{x}) \prod_{1 \le j<k \le p}(x_k - x_j)}{\prod_{1 \le j<k \le p}(x_k^{(0)} - x_j^{(0)})}.$$

Substituting (10.44), the limits $\vec{x}^{(0)} \to \vec{0}$ can now be taken by further use of (10.47). □

The continuous limit of nonintersecting paths near a wall is considered in Exercises 10.1 q.6.

EXERCISES 10.1 1. Consider $2N$-step ld/rd lattice paths with

$$w_{2j-1}^- = 1, \quad w_{2j-1}^+ = a_j, \quad w_{2j}^+ = 1, \quad w_{2j}^- = b_j \qquad (j = 1, \dots, N).$$

(i) Show that the generating function for a single path is

$$g_{2N}^{\rm ld/rd}(2l^{(0)}, 2l) = \frac{1}{2\pi}\int_{-\pi}^{\pi} \prod_{j=1}^{N}(1 + a_j e^{i\theta})(1 + b_j e^{-i\theta}) e^{-i(l-l^{(0)})\theta}\, d\theta.$$

(ii) Make use of (10.1) and (5.77) to deduce that for p nonintersecting paths

$$G_{2N}^{\rm ld/rd}(\{l_j^{(0)} = 2(p-j)\}_{j=1,\dots,p}; \{l_j = 2(p-j) + 2r\}_{j=1,\dots,p})$$
$$= \Big\langle \prod_{j=1}^{N}\prod_{k=1}^{p} e^{-2i\theta_k r}(1 + a_j e^{i\theta_k})(1 + b_j e^{-i\theta_k})\Big\rangle_{U(p)}. \tag{10.48}$$

(iii) Use (3.121) and (4.4) to deduce from (10.48) that when $a_j = b_j = 1$ $(j = 1, \dots, N)$ the number of nonintersecting paths of the type in (ii) is equal to

$$\prod_{i=1}^{p} \frac{(2N + p - i)!(i-1)!}{(N - r + p - i)!(N + r + i - 1)!}$$

and show that this same formula can be deduced from (10.9).

2. (i) Make use of q1.(i), (10.30), (10.29) and (10.36) to show that with rd(o) steps weighted $a_1, \dots, a_N$ and ld(e) steps weighted $b_1, \dots, b_N$,

$$G_{2N}^{\rm w,u/rd(o)/ld(e)}(\{r_j^{(0)} = p - j + 1\}_{j=1,\dots,p}; r_1, \dots, r_p)$$
$$= \frac{1}{(2\pi)^p}\int_{-\pi}^{\pi} d\theta_1 \cdots \int_{-\pi}^{\pi} d\theta_p$$
$$\times \prod_{l=1}^{p}\Big(\prod_{j=1}^{N}(1 + a_j e^{i\theta_l})(1 + b_j e^{-i\theta_l})\Big) e^{-i(p-l+1)\theta_l} \det\Big[e^{ir_\alpha\theta_\beta} - e^{-ir_\alpha\theta_\beta}\Big]_{\alpha,\beta=1,\dots,p}.$$

(ii) Suppose $\{a_j\} = \{b_j\}$ (in any order), so that

$$\prod_{l=1}^{p}\prod_{j=1}^{N}(1 + a_j e^{i\theta_l})(1 + b_j e^{-i\theta_l}) = \prod_{l=1}^{p}\prod_{j=1}^{N}|1 + a_j e^{i\theta_l}|^2. \tag{10.49}$$

Note that in general for an integral over an interval symmetric about the origin the replacement $\theta_l \mapsto -\theta_l$ leaves the integral unchanged, as does the relabeling $\theta_j \leftrightarrow \theta_k$ for a multiple integral. For the integral in (ii) with the substitution (10.49), use the fact that the integrand changes sign under such mappings to deduce

$$G_{2N}^{\rm w,u/rd(o)/ld(e)}(\{r_j^{(0)} = p - j + 1\}_{j=1,\dots,p}; r_1, \dots, r_p)$$
$$= \frac{1}{(2\pi)^p}\frac{1}{2^p p!}\int_{-\pi}^{\pi} d\theta_1 \cdots \int_{-\pi}^{\pi} d\theta_p$$
$$\times \prod_{l=1}^{p}\prod_{j=1}^{N}|1 + a_j e^{i\theta_l}|^2 \Big(\sum_{w\in W}\varepsilon(w)\prod_{l=1}^{p} e^{-i(p-l+1)w(\theta_l)}\Big) \det\Big[e^{ir_\alpha\theta_\beta} - e^{-ir_\alpha\theta_\beta}\Big]_{\alpha,\beta=1,\dots,p},$$

where W denotes the set of all permutations of $\{\theta_1, \dots, \theta_p\}$ together with negations, while $\varepsilon(w)$ denotes the parity of the sum of the number of interchanges and the number of negations.

(iii) Note that

$$\sum_{w\in W}\varepsilon(w)\prod_{l=1}^{p}e^{-i(p-l+1)w(\theta_l)}=\sum_{\sigma\in S_p}\varepsilon(\sigma)\prod_{l=1}^{p}\Big(e^{-i(p-l+1)\sigma(\theta_l)}-e^{i(p-l+1)\sigma(\theta_l)}\Big)$$
$$=\det\Big[e^{-i(p-\alpha+1)\theta_\beta}-e^{i(p-\alpha+1)\theta_\beta}\Big]_{\alpha,\beta=1,\dots,p},$$

and thus conclude

$$G_{2N}^{\mathrm{w,u/rd(o)/ld(e)}}(\{r_j^{(0)}=p-j+1\}_{j=1,\dots,p};r_1,\dots,r_p)$$
$$=\frac{1}{(2\pi)^p}\frac{1}{2^p p!}\int_{-\pi}^{\pi}d\theta_1\cdots\int_{-\pi}^{\pi}d\theta_p\prod_{l=1}^{p}\prod_{j=1}^{N}|1+a_je^{i\theta_l}|^2$$
$$\times\det\Big[e^{-i(p-\alpha+1)\theta_\beta}-e^{i(p-\alpha+1)\theta_\beta}\Big]_{\alpha,\beta=1,\dots,p}\det\Big[e^{ir_\alpha\theta_\beta}-e^{-ir_\alpha\theta_\beta}\Big]_{\alpha,\beta=1,\dots,p}. \quad (10.50)$$

(iv) Make use of the type C Vandermonde formula in Exercises 5.5 q.4, and the explicit form of the eigenvalue p.d.f. for Sp($2p$) (2.69), to deduce from (10.50) that

$$G_{2N}^{\mathrm{w,u/rd(o)/ld(e)}}(\{r_j^{(0)}=p-j+1\}_{j=1,\dots,p};\{r_j=p-j+1\}_{j=1,\dots,p})=\Big\langle\prod_{l=1}^{p}\prod_{j=1}^{N}|1+a_je^{i\theta_l}|^2\Big\rangle_{\mathrm{Sp}(2p)}. \quad (10.51)$$

3. The objective of this exercise is to prove (10.35).

(i) In (10.33) add row $j-1$ to row j for $j=p,p-1,\dots,2$ in order to deduce that

$$W_p(\{x_k\}_{k=1,\dots,p},\{a_j\}_{j=2,\dots,p},\{b_j\}_{j=-p+1,\dots,p-1})=\prod_{j=2}^{p}(b_{-j+1}+a_j)$$
$$\times\det\begin{bmatrix}(a_2-x_k)\cdots(a_p-x_k)(b_0+x_k)\cdots(b_{p-1}+x_k)-(x_k\mapsto -x_k)\\ \Big[(a_{j+1}-x_k)\cdots(a_p-x_k)(b_{-j+2}+x_k)\cdots(b_{p-1}+x_k)-(x_k\mapsto -x_k)\Big]_{j=2,\dots,p}\end{bmatrix}_{k=1,\dots,p},$$

and show that repeated use of this strategy yields

$$W_p(\{x_k\}_{k=1,\dots,p},\{a_j\}_{j=2,\dots,p},\{b_j\}_{j=-p+1,\dots,p-1})$$
$$=\prod_{l=2}^{p}\prod_{j=l}^{p}(b_{-j-1+l}+a_j)X_p(\{x_k\}_{k=1,\dots,p},\{a_j\}_{j=2,\dots,p},\{b_j\}_{j=0,\dots,p-1}),$$

where

$$X_p(\{x_k\}_{k=1,\dots,p},\{a_j\}_{j=2,\dots,p},\{b_j\}_{j=0,\dots,p-1})$$
$$:=\det\Big[(a_{j+1}-x_k)\cdots(a_p-x_k)(b_0+x_k)\cdots(b_{p-1}+x_k)-(x_k\mapsto -x_k)\Big]_{j,k=1,\dots,p} \quad (10.52)$$

(ii) Note that the determinant X_p is homogeneous of degree $\frac{3}{2}p^2-\frac{1}{2}p$, that it vanishes for $x_j=x_k,-x_k$ ($k\neq j$) and for $x_j=-x_j$, and is antisymmetric in the xs and symmetric in the bs. Thus conclude that

$$X_p(\{x_k\}_{k=1,\dots,p},\{a_j\}_{j=2,\dots,p},\{b_j\}_{j=0,\dots,p-1})$$
$$=\prod_{l=1}^{p}x_l\prod_{1\le j<k\le p}(x_k-x_j)(x_k+x_j)f(\{x_j\}_{j=1,\dots,p},\{b_j\}_{j=0,\dots,p-1}), \quad (10.53)$$

where f is homogeneous of degree $\frac{1}{2}p(p-1)$ and symmetric in $\{x_j\}_{j=1,\dots,p}$, $\{b_j\}_{j=0,\dots,p-1}$, so our remaining task is to show that

$$f(\{x_j\}_{j=1,\dots,p},\{b_j\}_{j=0,\dots,p-1}) = 2^p \prod_{0\le j<k\le p-1}(b_j+b_k). \tag{10.54}$$

(iii) Observe that (10.54) would follow if we could show that the coefficient of

$$x_1^{2p-1}x_2^{2p-3}\cdots x_p b_0^{p-1}b_1^{p-2}\cdots b_{p-1}^0$$

in X_p was equal to $(-1)^{(p-1)(p-2)/2}2^p$. For this purpose, note that the only term in (10.52) containing a factor of x_1^{2p-1} is the top left entry, and expand by the corresponding coefficient in this entry to show

$$\begin{aligned}
&[x_1^{2p-1}]X_p(\{x_k\}_{k=1,\dots,p},\{a_j\}_{j=2,\dots,p},\{b_j\}_{j=0,\dots,p-1})\\
&\quad = 2(-1)^{p-1}\det\Big[(a_{j+1}-x_k)\cdots(a_p-x_k)(b_0+x_k)\cdots(b_{p-1}+x_k)-(x_k\mapsto -x_k)\Big]_{j,k=2,\dots,p}\\
&\quad = 2(-1)^{p-1}X_{p-1}(\{x_k\}_{k=2,\dots,p},\{a_j\}_{j=3,\dots,p},\{b_j\}_{j=0,\dots,p-1}).
\end{aligned}$$

Note that here we have a $(p-1)\times(p-1)$ determinant in which each entry is linear in b_0, so the coefficient of b_0^{p-1} is equal to this determinant with (b_0+x_k) and (b_0-x_k) removed from the entries, and thus

$$\begin{aligned}
&[b_0^{p-1}x_1^{2p-1}]X_p(\{x_k\}_{k=1,\dots,p},\{a_j\}_{j=2,\dots,p},\{b_j\}_{j=0,\dots,p-1})\\
&\quad = 2(-1)^{p-1}X_{p-1}(\{x_k\}_{k=2,\dots,p},\{a_j\}_{j=3,\dots,p},\{b_j\}_{j=1,\dots,p-1}).
\end{aligned}$$

By iterating this deduce the sought result.

4. Let $\vec{m}^{(0)} = (p-1,p-2,\dots,0)$ and $\vec{m} = (\lambda_1'+p-1,\dots,\lambda_p')$.

(i) Note that (1.173) can be written

$$\mathrm{Asym}\,(x_1^{n-1}x_2^{n-2}\cdots x_n^0) = \prod_{1\le j<k\le n}(x_j-x_k),$$

where Asym is specified by (4.135).

(ii) Use the result of (i), together with (10.7), the first equality in (10.8), and the formulas (10.16), (10.15) and (5.77) to show

$$s_\mu(q_1,\dots,q_n) = \Big\langle \prod_{l=1}^p\prod_{j=1}^n(1+q_je^{i\theta_l})s_{\mu'}(e^{-i\theta_1},\dots,e^{-i\theta_p})\Big\rangle_{U(p)}. \tag{10.55}$$

(iii) Note from (10.21) that (10.55) is equivalent to the so called *dual Cauchy identity*

$$\prod_{l=1}^p\prod_{j=1}^n(1+q_jz_l) = \sum_{\mu:\mu_1\le p} s_\mu(q_1,\dots,q_n)s_{\mu'}(z_1,\dots,z_p). \tag{10.56}$$

5. (i) Starting with (10.24), proceed as in the derivation of (10.55) to show

$$s_\mu(q_1,\dots,q_n) = \Big\langle \prod_{l=1}^{p^*}\prod_{j=1}^n\frac{1}{1-q_je^{i\theta_l}}s_\mu(e^{-i\theta_1},\dots,e^{-i\theta_{p^*}})\Big\rangle_{U(p^*)}, \tag{10.57}$$

where $p^* = \mu_1'$.

(ii) Note from (10.21) that (10.57) is equivalent to the so called *Cauchy identity*

$$\prod_{l=1}^p\prod_{j=1}^n\frac{1}{1-q_jz_l} = \sum_\mu s_\mu(q_1,\dots,q_n)s_\mu(z_1,\dots,z_p). \tag{10.58}$$

6. [180], [350] In this exercise the continuous limit of noninteracting paths near a wall is studied.

   (i) Let the continuous generating function be denoted $G_t^{(\mathrm{wB})}(\vec{y}^{(0)};\vec{y})$. Proceed as in the derivation of (10.44) to show

$$G_t^{(\mathrm{wB})}(\vec{y}^{(0)};\vec{y}) = \Big(\frac{1}{2\pi Dt}\Big)^{p/2} e^{-\sum_{j=1}^p (y_j^2+(y_j^{(0)})^2)/2Dt} \det\Big[e^{y_j^{(0)}y_k/Dt} - e^{-y_j^{(0)}y_k/Dt}\Big]_{j,k=1,\dots,p}.$$

   Now set $y_j^{(0)} = ja$ (equally spaced initial condition) and use the type C Vandermonde formula from Exercises 5.5 q.4 to show

$$G_t^{(\mathrm{wB})}(\{ja\}_{j=1,\dots,p};\vec{y}) \propto e^{-\sum_{j=1}^p y_j^2/2Dt} \prod_{j=1}^p \sinh\frac{ay_j}{Dt} \prod_{1\le j<k\le p} \sinh\frac{a(y_k-y_j)}{2Dt} \sinh\frac{a(y_k+y_j)}{2Dt}.$$

   Interpret this as a log-gas system at $\beta = 1$ on the half-line $y > 0$ interacting via the pair potential (2.72) with image charges of like sign in $y < 0$, and with an appropriate neutralizing background.

   (ii) For p continuous nonintersecting Brownian walkers, starting at the origin and confined to the half line $y > 0$ for $t > 0$, and with the nonintersecting condition required for all times, show the p.d.f. for the event that they arrive at $\vec{y}$ after time t is proportional to

$$\Big(\frac{1}{Dt}\Big)^{p^2+p/2} e^{-\sum_{j=1}^p y_j^2/2Dt} \prod_{l=1}^p y_l^2 \prod_{1\le j<k\le p} (y_k^2-y_j^2)^2.$$

   Recognize this as the eigenvalue p.d.f. for $(2p+1)\times(2p+1)$ antisymmetric Gaussian unitary matrices (recall Exercises 1.3 q.5).

10.2 DIMERS AND TILINGS

In this section it will be shown how some statistical mechanical models of dimers and tilings relate to nonintersecting paths.

10.2.1 Dimers

Consider the brickwork lattice (which is equivalent to the hexagonal lattice) with an odd number of rows and impose periodic boundary conditions in the y-direction by identifying the bottom and top rows. Suppose that on each vertical edge a hard rod (dimer) is placed, thus covering all the lattice sites without overlap. Let us suppose now that along the bottom row a number of dimers are placed across nonconsecutive horizontal edges. For this to be done without overlap, but still cover all lattice sites, for each horizontal dimer a vertical dimer sharing the same site must be removed. This dimer is to be replaced by a horizontal dimer in the next row: this horizontal dimer can in general lie either to the left or to the right of the vacated site, but is constrained not to intersect another horizontal dimer. Because these new horizontal dimers will overlap vertical dimers going between the second and third rows, the procedure of removing vertical dimers and replacing them with non-overlapping horizontal dimers lying either to the left or to the right of the vacated site is repeated. Continuing in this fashion leads to the same number of horizontal dimers being introduced on the final row as were introduced on the first row, which by the assumption of periodic boundary conditions must occur at the same positions as on the first row.

Each matching horizontal dimer in the bottom and top rows can be connected by a ld/rd path which starts at an odd numbered column half a lattice spacing below the bottom row and finishes at the same column one half a lattice spacing below the top row (remember the top and bottom rows are identified), with each segment passing through a horizontal dimer. Because dimers do not intersect this gives a family of nonintersecting ld/rd paths. The periodic boundary conditions with an odd number of rows imply the initial positions equal to the final positions, and it has already been noted that the former are restricted to odd numbered columns. The paths have the additional constraint of lying between the first and last column inclusive (see Figure 10.8). Conversely, any such family of nonintersecting ld/rd paths determines a dimer configuration on the brickwork lattice.

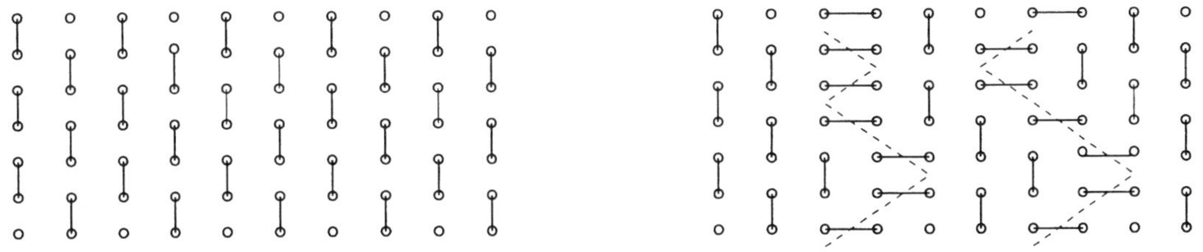

Figure 10.8 The configuration of all vertical dimers on the brickwork lattice with periodic boundary conditions in the vertical direction (leftmost diagram), and a configuration in which two horizontal dimers are introduced into the bottom row, giving rise to nonintersecting ld/rd paths.

With M columns and $2N+1$ rows, and each horizontal dimer weighted w, the generalized partition function relating to summing over all weighted configurations is given by

$$Z_{2N,M}(w^{2N})[a] = 1 + \sum_{p=1}^{M^*} w^{2Np} \sum_{1 \le n_1 < \cdots < n_p \le M^*} \prod_{l=1}^{p} a(2n_l - 1)$$
$$\times G_{2N}^{\mathrm{ww,ld/rd}}(\{2n_j - 1\}_{j=1,\dots,p}; \{2n_j - 1\}_{j=1,\dots,p}). \qquad (10.59)$$

Here $M^* = [(M-1)/2]$ and $G_{2N}^{\mathrm{ww,ld/rd}}(\vec{l}^{(0)}; \vec{l})$ denotes the number of ways p nonintersecting paths, constrained to have x-coordinates in $\{1, \dots, M-1\}$ (or equivalently confined to integer sites between infinitely repelling walls at $x=0$ and $x=M$), and taking N steps each of which is ld/rd. The initial positions along $y=0$ are at $\vec{l}^{(0)}$, and the final positions along $y=2N$ are at $\vec{l}$. It is possible to write (10.59) as a determinant.

PROPOSITION 10.2.1 *We have*

$$Z_{2N,M}(w^{2N})[a] = \det\Big(\mathbf{1}_{M^*} + [a(2j-1)w^{2N} g_{2N}^{\mathrm{ww;ld/rd}}(2j-1, 2l-1)]_{j,l=1,\dots,M^*}\Big), \qquad (10.60)$$

where

$$g_{2N}^{\mathrm{ww;ld/rd}}(l^{(0)}, l) = \frac{2}{M} \sum_{k=1}^{M-1} \Big(2\cos\frac{\pi k}{M}\Big)^N \sin\frac{\pi l k}{M} \sin\frac{\pi l^{(0)} k}{M}. \qquad (10.61)$$

Proof. Applying (10.1) gives

$$G_{2N}^{\mathrm{ww,ld/rd}}(\vec{l}^{(0)}; \vec{l}) = \det\Big[g_{2N}^{\mathrm{ww,ld/rd}}(l_j^{(0)}, l_k)\Big]_{j,k=1,\dots,N},$$

where $g_{2N}^{\mathrm{ww,ld/rd}}(l^{(0)}, l)$ is the number of single lattice paths from $l^{(0)}$ to l. Hence

$$Z_{2N,M}(w^{2N})[a] = 1 + \sum_{p=1}^{M^*} \sum_{1 \le n_1 < \cdots < n_p \le M^*} \det[a_{n_j,n_k}]_{j,k=1,\dots,p}, \qquad (10.62)$$

where

$$a_{n_j,n_k} = a(2n_j - 1)w^{2N} g_{2N}^{\mathrm{ww,ld/rd}}(2n_j - 1, 2n_k - 1).$$

Now, by discretizing (5.32) we see that the r.h.s. of (10.62) can be summed to give

$$Z_{2N,M}(w^{2N})[a] = \det\Big(\mathbf{1}_{M^*} + [a_{j,l}]_{j,l=1,\dots,M^*}\Big)$$

which is (10.60).

To obtain (10.61) we first generalize the reasoning leading to (10.29) to conclude

$$g_{2N}^{\mathrm{ww,ld/rd}}(l_j^{(0)}; l_k) = \sum_{k=-\infty}^{\infty} \Big(g_{2N}^{\mathrm{ld/rd}}(l^{(0)}, l + 2kM) - g_{2N}^{\mathrm{ld/rd}}(l^{(0)}, -l - 2kM) \Big).$$

Substituting (10.4), performing the sum over k using the Poisson summation formula, and simplifying gives (10.61). □

The determinant formula (10.60) implies the correlations are given by (5.31) and thus

$$\rho_{(k)}(2n_1 - 1, \dots, 2n_k - 1) = \det[G(2n_j - 1, 2n_l - 1)]_{j,l=1,\dots,k},$$

where, with $\mathbf{K} = [g_N^{\mathrm{ww,ld/rd}}(2j-1, 2l-1)]_{j,l=1,\dots,M^*}$,

$$G(2n_j - 1, 2n_l - 1) = \Big(w^{2N}\mathbf{K}(1 + w^{2N}\mathbf{K})^{-1} \Big)_{2n_j-1, 2n_l-1}.$$

To make this latter formula explicit we note that $\mathbf{G} := [G(2j-1, 2l-1)]_{j,l=1,\dots,M^*}$ must satisfy $\mathbf{G}(\mathbf{1} + w^{2N}\mathbf{K}) = w^{2N}\mathbf{K}$ and then proceed as in the derivation of (5.105) to deduce that [330]

$$G(2n_j - 1, 2n_l - 1) = \frac{2}{M^*} \sum_{k=1}^{M^*} \sin \frac{\pi(2n_j - 1)k}{M} \sin \frac{\pi(2n_l - 1)k}{M} \frac{(2w\cos \pi k/M)^{2N}}{1 + (2w\cos \pi k/M)^{2N}}. \tag{10.63}$$

In Exercises 10.2 q.1 it is shown that in an appropriate large M, N limit this becomes the sine kernel.

10.2.2 Tiling of a hexagon

Nonintersecting ld/rd paths initially equally spaced 2 units apart, and with final points also equally spaced 2 units apart, are equivalent to certain three-species parallelogram tilings of a hexagon, the shape of the latter being determined by the parameters associated with the paths. Thus suppose there are a paths, which each consist of $b+c$ steps, and the x-coordinates of the final points are displaced by $c-b$ lattice sites to the right of the initial points. Then it is easy to see that the paths are bounded by a hexagon of side lengths $2a$, $\sqrt{2}b$, $\sqrt{2}c$, $2a$, $\sqrt{2}b$, $\sqrt{2}c$, and the paths uniquely determine a tiling of the hexagon (see Figure 10.9). The ld segments correspond to the left (right) sloping parallelograms, sloping sides of length $\sqrt{2}$, horizontal sides length 2. Changing the angles in the hexagon so that they are all 120°, and changing the side length $2a \mapsto \sqrt{2}a$ maps this three-species parallelogram tiling into a three-species rhombi tiling. The rhombi tiling can be viewed as the three dimensional diagram of stacked cubes which weakly decrease in height going down columns and across rows of the base, with row 1 counted from the back and column 1 from the left. The heights then specify an array of non-negative integers

$$\{a_{ij}\}: \quad a_{i,j} \ge a_{i+1,j}, \; a_{i,j} \ge a_{i,j+1},$$

referred to as *plane partitions*. Note that plane partitions are a variation on semi-standard tableau in which the entries now weakly decrease along rows and down columns of the diagram of a partition. The number of paths is equal to the maximum number of columns a, while the number of steps to the left in any one path is equal to the maximum number of rows b, and the number of steps to the right determines the maximum height.

The correspondence between nonintersecting paths and tilings can be used to enumerate the latter.

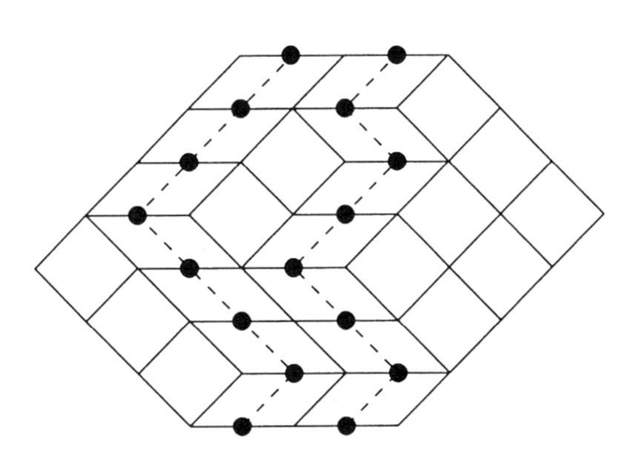

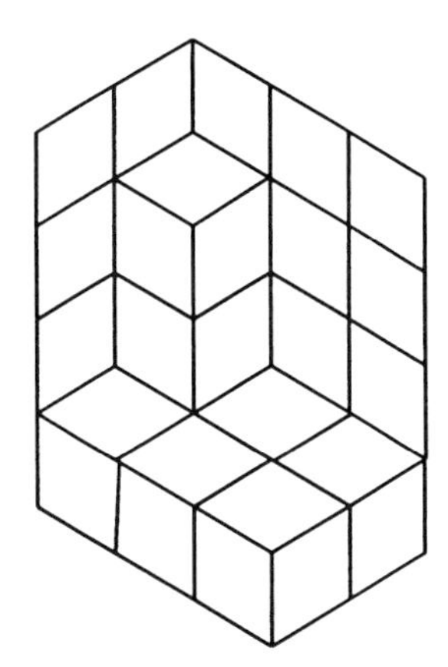

Figure 10.9 Nonintersecting ld/rd paths with initial and final positions 2 units apart are equivalent to a three-species parallelogram tiling of a hexagon involving left-sloping parallelograms (ld segment), right sloping parallelograms (rd segment) and vertical-axis parallelograms. This can be deformed into a rhombi tiling of a hexagon. The rhombi tiling in turn can be viewed as the diagram of a plane partition. Here the rows are counted from the top, and the columns are counted from the right.

PROPOSITION 10.2.2 *Consider a hexagon of the second type in Figure 10.9, with dimensions $a \times b \times c$. The number of distinct rhombi tilings is equal to*

$$\prod_{i=1}^{a} \frac{(a+b+c-i)!(i-1)!}{(a+b-i)!(c+i-1)!}.$$

Proof. This follows by setting $N = b + c$, $p = a$, $m_j = p + c - b - j$ in (10.9). □

The dimer model of Section 10.2.1 can be formulated as a tiling problem by placing over each vertical dimer a 1×2 rectangle, and placing over each horizontal dimer a 2×1 rectangle. It is also true that the rhombi tiling of the hexagon is equivalent to a dimer covering of the underlying hexagonal lattice — place a dimer of unit length (a unit being the side length of a rhombus) symmetrically on the major axis of each rhombus.

Let us return now to the setting of the first of the diagrams in Figure 10.9. We see that for each level $y = 1, 2, \dots$, complementary to the x-coordinates of the paths are the x-coordinates of the center point of the vertical-axis parallelograms (see Figure 10.10). Regarding these coordinates as specifying a configuration of a certain lattice gas confined to parallel lines at $y = 1, 2, \dots$, the number of particles on each line is determined only by the total number of steps in the corresponding lattice paths $(b + c)$, and the displacement $c - b$ of the x-coordinates of the starting positions relative to the final positions; the number of lattice paths plays no role. Thus for $y = 1, 2, \dots, b$ there are $1, 2, \dots, b$ particles $\{x_i^{(y)}\}_{i=1,\dots,y}$; for $y = b+1, \dots, c$ there are b particles $\{x_i^{(y)}\}_{i=1,\dots,b}$; and for $y = c+1, \dots, b+c-y$ there are $b-1, b-2, \dots, 1$ particles $\{x_i^{(y)}\}_{i=1,\dots,b+c-y}$. This configuration of particles exhibits the interlacings

$$\begin{aligned} &x_1^{(y+1)} > x_1^{(y)} > x_2^{(y+1)} > \cdots > x_y^{(y)} > x_{y+1}^{(y+1)} \qquad (y = 1, \dots, b-1), \\ &x_1^{(y)} > x_1^{(y+1)} > \cdots > x_y^{(y)} > x_y^{(y+1)} \qquad (y = b, \dots, c-1), \\ &x_1^{(y)} > x_1^{(y+1)} > x_2^{(y)} > \cdots > x_{b+c-y-1}^{(y+1)} > x_{b+c-y}^{(y)} \qquad (y = c, \dots, b+c-1). \end{aligned} \tag{10.64}$$

Furthermore, they must stay within the bounds of the hexagon, and are restricted to odd (even) numbered sites for y odd (even).

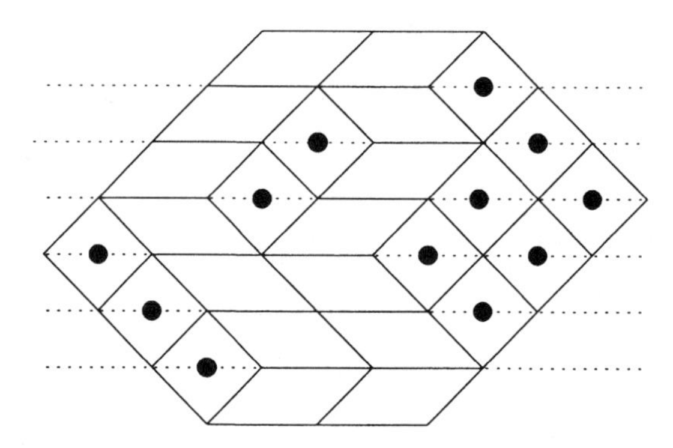

Figure 10.10 Interlacing particle system which results by associating particles with the center point of the vertical-axis parallelograms of a tiling.

Suppose now that the parallelograms are rescaled so that the leftmost extremity (which occurs on the line $y = b$) is at $x = 0$, while the rightmost extremity (which occurs on the line $y = 0$) is at $x = 1$. By taking the number of lattice paths and thus rhombi to infinity, a continuous multi-species particle system is obtained (the different lines are the different species), which is specified by the interlacings (10.64), supplemented by the requirement that all coordinates are bound between 0 and 1. This multi-species system is referred to as a *bead process* [94], [177]. One point of interest is the distribution of particles on a given line r.

PROPOSITION 10.2.3 *The p.d.f. on line r is proportional to*

$$\begin{cases} \prod_{j=1}^{r} (x_j^{(r)})^{c-r}(1-x_j^{(r)})^{b-r} \prod_{1\le j<k\le r} (x_j^{(r)} - x_k^{(r)})^2, & 1 \le r \le b, \\ \prod_{j=1}^{r} (x_j^{(r)})^{c-r}(1-x_j^{(r)})^{r-b} \prod_{1\le j<k\le b} (x_j^{(r)} - x_k^{(r)})^2, & b < r \le c, \\ \prod_{j=1}^{r} (x_j^{(r)})^{r-c}(1-x_j^{(r)})^{r-b} \prod_{1\le j<k\le b+c-r} (x_j^{(r)} - x_k^{(r)})^2, & c < r \le b+c-1. \end{cases} \tag{10.65}$$

Proof. We integrate over all particles on lines $1, \dots, r-1$ in order, and further integrate over all particles on lines $b+c-1, \dots, r+1$ in order. The particle coordinate on line 1, $x_1^{(1)}$ is to be integrated between $x_2^{(2)}$ and $x_1^{(2)}$, giving $x_1^{(2)} - x_2^{(2)}$. This function now must be integrated over $x_1^{(3)} < x_1^{(2)} < x_2^{(3)} < x_3^{(3)}$. For this purpose use is made of the Vandermonde determinant expansion (1.173) to write

$$x_1^{(2)} - x_2^{(2)} = \det \begin{bmatrix} 1 & x_2^{(2)} \\ 1 & x_1^{(2)} \end{bmatrix}.$$

The integration can be done row-by-row to give

$$\frac{1}{2}(x_1^{(3)} - x_2^{(3)})(x_1^{(3)} - x_3^{(3)})(x_2^{(3)} - x_3^{(3)}) = \frac{1}{2} \det \begin{bmatrix} 1 & x_3^{(3)} & (x_3^{(3)})^2 \\ 1 & x_2^{(3)} & (x_2^{(3)})^2 \\ 1 & x_1^{(3)} & (x_1^{(3)})^2 \end{bmatrix},$$

where the equality follows by further use of (1.173). This must now be integrated over the first region in (10.64) with $y = 4$, a task which can again be done row-by-row to give

$$\frac{1}{3!2!} \prod_{1\le j<k\le 4} (x_j^{(4)} - x_k^{(4)}) = \det[(x_{5-j}^{(4)})^{k-1}]_{j,k=1,\dots,4}.$$

Proceeding in this fashion we see that for $r \le b$, integrating over all particles on lines $1, \dots, r-1$ will give

$$\prod_{s=1}^{r-1} \frac{1}{s!} \prod_{1 \le j < k \le r} (x_j^{(r)} - x_k^{(r)}). \tag{10.66}$$

For $r > b$, we must continue to integrate (10.66) over the second of the regions in (10.64), noting too that all variables are between 0 and 1. In the case $y = b$ the above method gives an expression proportional to

$$C \prod_{j=1}^{b} (1 - x_j^{(b+1)}) \prod_{1 \le j < k \le b} (x_j^{(b+1)} - x_k^{(b+1)}) = \det \Big[(1 - x_{b+2-j}^{(b+1)})^k \Big]_{j,k=1,\dots,b+1}.$$

Continuing this working shows that for $b < r \le c$, integrating over all particles on lines $1, \dots, r-1$ will give

$$\prod_{j=1}^{b} (1 - x_j^{(r)})^{r-b} \prod_{1 \le j < k \le b} (x_j^{(r)} - x_k^{(r)}), \tag{10.67}$$

up to a proportionality constant.

For $r > c$ we must continue to integrate (10.67) over the third of the regions in (10.64), supplemented by the constraint that all variables are between 0 and 1. To do this we can formally continue to integrate over the inequalities of the second of the regions in (10.64), then take the limit $x_c^{(r)} \to 0$, $x_{c-1}^{(r-1)} \to 0$ (after first dividing by $x_{c-1}^{(r-1)}$), then take the limit $x_{c-2}^{(r-2)} \to 0$ (after first dividing by $(x_{c-2}^{(r-2)})^2$) etc. to so obtain

$$\prod_{j=1}^{b+c-r} (1 - x_j^{(r)})^{r-b} (x_j^{(r)})^{r-c} \prod_{1 \le j < k \le b+c-r} (x_j^{(r)} - x_k^{(r)}),$$

up to a proportionality constant.

The results from integrating over all particles on lines $b+c-1, \dots, r+1$ in order are formally the same, after the identification $x_j^{(r)} \mapsto 1 - x_j^{(b+c-r)}$, and the stated result thus follows. □

We note that each of the forms in (10.65) is an example of a Jacobi unitary ensemble with weight defined on $[0,1]$.

Another point of interest is the joint distribution of the lines $1, \dots, r$ in the case that $c, b \to \infty$ and the coordinates $\{x_j^{(r)}\}$ are appropriately scaled.

PROPOSITION 10.2.4 *Consider the bead process with the coordinates scaled* $x_j^{(r)} \mapsto \frac{1}{2}(1 + y_j^{(r)}/\sqrt{b})$ *and* $b, c \to \infty$ *with* $c/b \to 1$. *The joint distribution of* $\bigcup_{s=1}^{r} \{y_j^{(s)}\}_{j=1,\dots,s}$ *is proportional to*

$$\prod_{l=1}^{r} e^{-(y_l^{(r)})^2} \prod_{1 \le j < k \le r} (y_j^{(r)} - y_k^{(r)}) \prod_{s=1}^{r-1} \chi_{y_1^{(s+1)} > y_1^{(s)} > y_2^{(s+1)} > \cdots > y_s^{(s)} > y_{s+1}^{(s+1)}}. \tag{10.68}$$

Proof. The joint p.d.f of the first r lines is obtained by integrating over all lines from $y = b+c-1$ down to $y = r+1$ inclusive. From the workings of the proof of Proposition 10.2.3 we know this is proportional to

$$\prod_{j=1}^{r} (1 - x_j^{(r)})^{b-r} (x_j^{(r)})^{c-r} \prod_{1 \le j < k \le r} (x_j^{(r)} - x_k^{(r)}) \prod_{s=1}^{r-1} \chi_{x_1^{(s+1)} > x_1^{(s)} > x_2^{(s+1)} > \cdots > x_s^{(s)} > x_{s+1}^{(s+1)}}.$$

Under the stated limiting conditions, this give (10.68). □

One recognizes (10.68) as the eigenvalue p.d.f. (5.196) (in the case $w(x) = e^{-x^2}$) of the GUE minor process [333].

10.2.3 Tiling of the Aztec diamond

Another tiling problem which gives rise to nonintersecting lattice paths is domino tiling of the so-called *Aztec diamond* lattice. The latter can be defined as the union of all vertices and edges which lie on lattice squares $[m, m+1] \times [l, l+1]$, $(m, l \in \mathbb{Z})$ within the diamond shaped region $\{(x, y) : |x| + |y| \le n+1\}$. A domino tiling is a covering of this lattice by 2×1 and 1×2 rectangles whose corners lie on the lattice points. To associate the tiling with lattice paths, color the squares in the Aztec diamond alternating white-black, choosing the left-most square in the top half as white, and the left-most square in the bottom half as black. This gives a checkerboard pattern in which the squares alternate between white-black and black-white down any column or across any row. Then for a horizontal domino which covers a white-black (black-white) pair of squares, reading left to right, no (horizontal) segment of a path is marked; for a vertical domino which covers a white-black (black-white) pair of squares, reading top to bottom, an ld ($-$rd) segment of a path is marked. As seen in Figure 10.11 (the B denotes an underlying black square), the tiling then maps uniquely to certain returning nonintersecting rd/$-$ld/rh paths (recall Section 10.1.4).

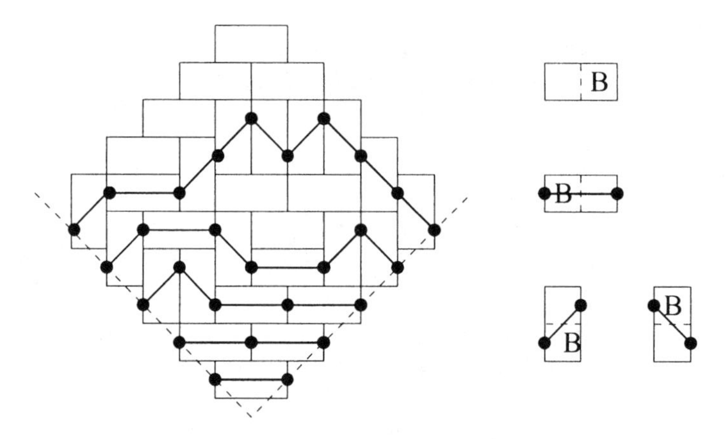

Figure 10.11 Mapping between a domino tiling of the Aztec diamond and nonintersecting rd/$-$ld/rh paths.

From the correspondence the total number of distinct tilings can be calculated [164].

Proposition 10.2.5 *The number of domino tilings of the Aztec diamond lattice bounded by* $\{(x, y) : |x| + |y| \le n+1\}$ *is equal to* $2^{n(n+1)/2}$.

Proof. This follows by setting $h_k = k - 1$, $p = n + 1$, $w = 1$ in the formula of Proposition 10.1.11. □

Exercises 10.2 1. [330] In this exercise a grand canonical ensemble of ww,ld/rd lattice paths is studied.

(i) In (10.63) write $M \mapsto 2M + 1$ and note that then

$$\lim_{M \to \infty} G(M + x, M + y) = \int_0^1 \cos \pi t(x - y) \frac{(2w \cos \pi t/2)^{2N}}{1 + (2w \cos \pi t/2)^{2N}} \, dt.$$

(ii) With $2w > 1$ and $\theta_0 := \frac{2}{\pi} \arccos \frac{1}{2w}$, take the $N \to \infty$ limit of the result in (i) to obtain as the limiting kernel

$$\int_0^{\theta_0} \cos \pi t(x - y) \, dt = \frac{\sin \pi \theta_0 (x - y)}{\pi (x - y)}.$$

(iii) Verify that for $j, k = 1, \dots, [(M-1)/2]$

$$\sum_{l=1}^{[(M-1)/2]} g_N^{\mathrm{ww,ld/rd}}(2j - 1, 2l - 1) \sin \frac{\pi (2l - 1)k}{M} = \Big(2 \cos \frac{\pi k}{M}\Big)^N \sin \frac{\pi (2j - 1)k}{M}.$$

Conclude that the eigenvalues of

$$[g_N^{\rm ww,ld/rd}(2j-1,2l-1)]_{j,l=1,\dots,[(M-1)/2]}$$

are given by $\{(2\cos\frac{\pi k}{M})^N\}_{k=1,\dots,[(M-1)/2]}$ and use this to obtain the evaluation

$$Z_{N,M}(\zeta)[1] = \prod_{k=1}^{[(M-1)/2]} \Big(1+\zeta\Big(2\cos\frac{\pi k}{M}\Big)^N\Big). \tag{10.69}$$

10.3 DISCRETE POLYNUCLEAR GROWTH MODEL

10.3.1 Robinson-Schensted-Knuth correspondence

A u/rh and u/lh pair of nonintersecting lattice paths, initially equally spaced, can be put into one-to-one correspondence with a matrix of integers, a result (with a pair of paths recorded as a pair of semi-standard tableaux) usually referred to as the *Robinson-Schensted-Knuth* (RSK) correspondence [246]. The non-negative integer matrix can itself be thought of as recording events in a discrete space and time growth process [329].

Thus consider an $n\times n$ square non-negative integer matrix $\mathbf{X} = [x_{i,j}]_{i,j=1,\dots,n}$ with rows numbered from the bottom, rotated 45° anticlockwise. Label the horizontal rows of the rotated matrix by $t=1,2,\dots,2n-1$ and the vertical columns by $x=0,\pm1,\dots,\pm(n-1)$, where $x=0$ corresponds to the diagonal of the original matrix (see Figure 10.12). The entries $x_{i,j}$ in the matrix for successive t values ($t=i+j-1$) are heights of "nucleation events" — columns of unit width and height $x_{i,j}$ centered about the corresponding x-coordinate which are placed on top of the profile formed by earlier nucleation events and their growth. Thus at $t=1$ there is a nucleation event at $x=0$ which consists of a column of width 1 and height $x_{1,1}$ marked on the line at $y=0$ in the xy-plane. In general, as $t\mapsto t+1$ the profile of all nucleation events so far recorded is to "grow" one unit in the $-x$ direction and one unit in the $+x$ direction. Thus in going from $t=1$ to $t=2$ the nucleation event centered at $x=0$ of height $x_{1,1}$ now has width 3 units. On top of this profile, centered at $x=-1$ and $x=1$ nucleation events of unit width and height $x_{2,1}, x_{1,2}$, respectively are then drawn. In now going from $t=2$ to $t=3$ this new profile is to grow one unit to the left and one unit to the right. In so doing we see that an overlap of width one unit and height $\min(x_{2,1},x_{1,2})$ will occur. This overlap is ignored in the first diagram (profile on $y=0$), and recorded instead as a profile on the line immediately below (here $y=-1$). The process is repeated with these rules until the nucleation event of height $x_{n,n}$ at $t=2n-1$ has been recorded above $x=0$ on the first diagram.

The boundary of each profile forms a pair of lattice paths — one to the left and one to the right of $x=0$. The lattice paths for $x<0$ start at $x=-(2n-\frac{3}{2})$ and go either up (in integer amounts with each unit regarded as a step) or to the right (in steps of two units) until they reach $x=-\frac{1}{2}$, while the lattice paths for $x>0$ start at $x=(2n-3/2)$ and go either up or right until they reach $x=\frac{1}{2}$. In the y-direction the paths start one unit apart at $y=0,\dots,-(n-1)$ (see Figure 10.12).

Conversely, each such family of nonintersecting paths corresponds to a unique $n\times n$ non-negative integer matrix, which can be constructed by reversing the above procedure, so we thus have a bijection between paths and matrices. The paths in the pair are separately equivalent to the lattice path type of Figure 10.3 and thus to a semi-standard tableau, so the bijection is also between non-negative integer matrices and pairs of semi-standard tableaux of the same shape (the latter constraint comes about because the pairs of paths have the same finishing points).

The above correspondence between integer matrices and pairs of nonintersecting lattice paths can be extended to a correspondence between weighted integer matrices and pairs of weighted nonintersecting lattice paths. For this each entry $x_{i,j}$ in the integer matrix is weighted by $(1-a_ib_j)(a_ib_j)^{x_{i,j}}$, with the factor $(1-a_ib_j)$ representing a normalization allowing the weighting to be interpreted as a probability from a ge-

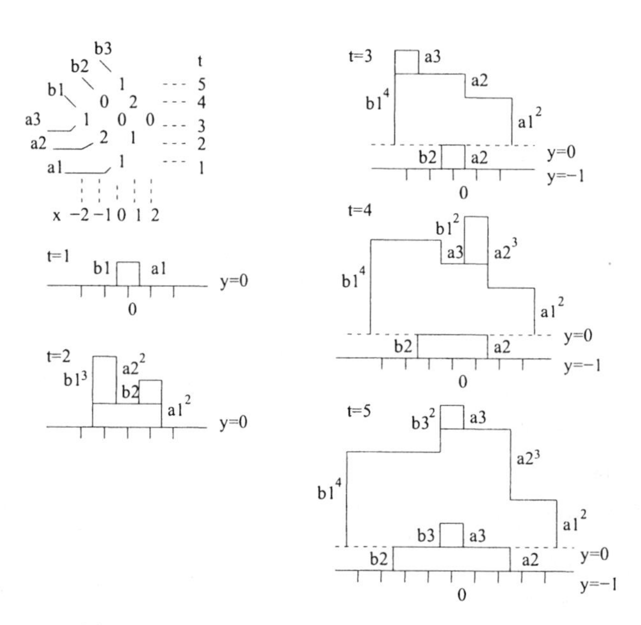

Figure 10.12 Mapping from a weighted non-negative integer matrix to a pair of weighted nonintersecting lattice paths.

ometric distribution. Since each entry $x_{i,j}$ corresponds to a nucleation event which adds a unit column of height $x_{i,j}$ to the top profile, the weight can be recorded on the profile by weighting the left side of the new column by $b_j^{x_{i,j}}$ and the right side by $a_i^{x_{i,j}}$ (we take $\prod_{i,j=1}^n (1 - a_i b_j)$ as an overall normalization factor). If columns should overlap as the profiles grow, the resulting column recorded on the profile below is to carry a left side and a right side weight which are exactly equal to those erased in the overlap, thus conserving the total vertical weight in the direction of any side. Crucially the rules of the growth process ensure that at each time step the weight in the direction of any one side is proportional to one particular b_j (left sides) or one particular a_i (right sides). At the end of the process this means that each vertical step at $x = -(2n + \frac{1}{2} - 2j)$ is weighted by b_j, and each vertical step at $x = 2n + \frac{1}{2} - 2j$ by a_j (see Figure 10.12).

The profile with endpoints at $y = -l + 1$ ($l = 1, \dots, n$) will be referred to as the level-l path, and its evolution forms a growth process known as the *discrete polynuclear growth model* [450], [330]. Of interest is the statistical properties of the profile, in particular its maximum displacement μ_1. To study this quantity the maximum displacements μ_l of the level-l paths for each $l = 1, 2, \dots, n$ are relevant.

The nonintersecting condition implies $\mu_1 \ge \mu_2 \ge \cdots \ge \mu_n$ so $\mu = (\mu_1, \mu_2, \dots, \mu_n)$ forms a partition. We know from (10.24) that with each of the vertical steps at $x = -(2n + \frac{1}{2} - 2j)$ weighted by b_j, the total weight of all nonintersecting u/rh paths (with horizontal steps two units) initially equally spaced at $y = 0, \dots, -(n-1)$ along $x = -(2n - \frac{3}{2})$ and finishing at $y = \mu_1, \mu_2 - 1, \dots, \mu_n - (n-1)$ along $x = -1/2$ is given by the Schur polynomial $s_\mu(b_1, b_2, \dots, b_n)$. Similarly, with vertical steps at $x = (2n + \frac{1}{2} - 2j)$ each weighted by a_j, the total weight of all nonintersecting u/lh (with horizontal steps two units) initially equally spaced at $y = 0, \dots, -(n-1)$ along $x = 2n - \frac{3}{2}$ and finishing at $y = \mu_1, \mu_2 - 1, \dots, \mu_n - (n-1)$ along $x = \frac{1}{2}$ is given by the Schur polynomial $s_\mu(a_1, a_2, \dots, a_n)$. It follows that if the non-negative integers $x_{i,j}$ in the matrix $\mathbf{X}$ are independent geometric random variables with parameter $a_i b_j$ and thus

$$\Pr(x_{i,j} = k) = (1 - a_i b_j)(a_i b_j)^k, \tag{10.70}$$

then the probability that such an integer matrix $\mathbf{X}$ corresponds to a pair of nonintersecting lattice paths with maximum heights μ is given by

$$\prod_{i,j=1}^{n} (1 - a_i b_j) s_\mu(a_1, a_2, \dots, a_n) s_\mu(b_1, b_2, \dots, b_n). \tag{10.71}$$

A fundamental statistical quantity relating to the maximum height $\mu_1 =: h^\Box$ of the level-1 path, namely its cumulative probability density, follows from this by summing over $\mu : \mu_1 \le l$. Thus

$$\Pr(h^\Box \le l) = \prod_{i,j=1}^{n} (1 - a_i b_j) \sum_{\mu:\, \mu_1 \le l} s_\mu(a_1, a_2, \dots, a_n) s_\mu(b_1, b_2, \dots, b_n). \tag{10.72}$$

We remark that since the Schur polynomials are symmetric functions this probability is symmetric in $\{a_i\}$ and $\{b_j\}$, a feature which is not obvious from the definitions. We note too that the normalization condition for this probability implies the Cauchy identity (10.58).

From the combinatorial definition of the Schur polynomial in (10.16), the sum over μ in (10.72) can be regarded as a sum over all pairs of weighted semi-standard tableaux of shape μ, content n and with row length no greater than l. Now we know from Figure 10.1 that such tableaux uniquely code weighted ld/rd nonintersecting lattice paths initially two sites apart consisting of at most l paths and n steps. This can be extended to exactly l paths by appending a suitable number of paths starting immediately to the left of those already present and which move only ld (such paths are not recorded as part of the tableau). A pair of tableaux of the same shape, content n and row length no greater than l, then corresponds to l returning ld/rd nonintersecting lattice paths of $2n$ steps in which both the initial and final spacings are two sites apart. This viewpoint allows $\Pr(h^\Box \le l)$ to be written in terms of a random matrix average [34].

PROPOSITION 10.3.1 *We have*

$$\Pr(h^\Box \le l) = \prod_{i,j=1}^{n} (1 - a_i b_j) \Big\langle \prod_{j=1}^{n} \prod_{k=1}^{l} (1 + a_j e^{-i\theta_k})(1 + b_j e^{i\theta_k}) \Big\rangle_{\mathrm{CUE}_l}. \tag{10.73}$$

Proof. From the returning paths interpretation,

$$\Pr(h^\Box \le l) = \prod_{i,j=1}^{n} (1 - a_i b_j) G_{2n}^{\mathrm{ld/rd}}(\{l_j^{(0)} = 2(l-j)+1\}_{j=1,\dots,l}; \{l_k = 2(l-k)+1\}_{k=1,\dots,l}), \tag{10.74}$$

where the weights in $G_{2n}^{\mathrm{ld/rd}}$ are specified by $w_s^- = a_s, w_s^+ = 1\ (s = 1, \dots, n), w_s^+ = b_s, w_s^- = 1\ (s = n+1, \dots, 2n)$. The result now follows from (10.48). □

10.3.2 Joint probabilities

Consider the measure on the space of non-negative integer matrices implied by (10.70). Then, as discussed in Section 10.3.1, the probability an $n_1 \times n_2$ non-negative integer matrix maps to a pair of semi-standard tableaux with shape μ, one of content n_1, the other of content n_2, under the RSK correspondence is equal to

$$\prod_{i=1}^{n_1} \prod_{j=1}^{n_2} (1 - a_i b_j) s_\mu(a_1, \dots, a_{n_1}) s_\mu(b_1, \dots, b_{n_2}) \tag{10.75}$$

(cf. (10.71)). Note that this vanishes for $\ell(\mu) > \min(n_1, n_2)$.

Next we seek the joint probability that an $n_1 \times (n_2 + 1)$ non-negative integer matrix with measure implied by (10.70) corresponds to a pair of semi-standard tableaux with shape μ, contents n_1 and $n_2 + 1$, and that the $n_1 \times n_2$ bottom left sub-block corresponds to a pair of semi-standard tableaux with shape κ and contents n_1 and n_2. According to the remark below (10.75), for the probabilities to be nonzero one requires $\ell(\mu) \le \min(n_1, n_2 + 1)$ and $\ell(\kappa) \le \min(n_1, n_2)$.

The growth model picture of the RSK correspondence tells us that μ_l equals the height of the level-l path at $x = \pm 1/2$, while κ_l equals the heights of the level-l path at $x = -\frac{3}{2}$. The inequalities (10.104) imply that

$$\mu_1 \geq \kappa_1 \geq \mu_2 \geq \kappa_2 \geq \cdots \geq \kappa_n \geq \mu_n, \tag{10.76}$$

where $n = \min(n_1, n_2 + 1)$. The nonintersecting paths corresponding to these heights have total weights

$$s_\mu(a_1, \ldots, a_{n_1}) \qquad \text{and} \qquad s_\kappa(b_1, \ldots, b_{n_2})$$

to the right and left, respectively. The path segments connecting the level-l path at $x = -\frac{3}{2}$ to $x = -\frac{1}{2}$ have weight $b_{n_2+1}^{\mu_l - \kappa_l}$. Noting too that the normalization for the paths is $\prod_{j=1}^{n_1} \prod_{k=1}^{n_2+1} (1 - a_j b_k)$, the following result is obtained [222].

PROPOSITION 10.3.2 *Consider an $n_1 \times (n_2+1)$ non-negative integer matrix with entries chosen according to (10.70). The joint probability that this matrix maps under the RSK correspondence to a tableau of shape μ, and that the $n_1 \times n_2$ matrix obtained by deleting the final column maps to a tableaux of shape κ is equal to*

$$\prod_{j=1}^{n_1} \prod_{k=1}^{n_2+1} (1 - a_j b_k) s_\mu(a_1, \ldots, a_{n_1}) s_\kappa(b_1, \ldots, b_{n_2}) b_{n_2+1}^{\sum_{l=1}^{n} (\mu_l - \kappa_l)},$$

where $n = \min(n_1, n_2 + 1)$, supported on partitions interlaced according to (10.76) and with $\ell(\mu) \leq n_1$, $\ell(\kappa) \leq n_2$.

Suppose now that under the RSK correspondence with weights chosen according to (10.70), the pair of semi-standard tableau corresponding to a particular $n_1 \times n_2$ matrix ($n_1 > n_2$) has shape κ. It then follows from Proposition 10.3.2 and (10.71) that the probability an $n_1 \times (n_2 + 1)$ matrix, obtained by adding an extra column to the existing matrix, of having shape μ is

$$P(\mu, \kappa) = \chi(\mu > \kappa) \prod_{i=1}^{n_1} (1 - a_i b_{n_2+1}) \frac{s_\mu(a_1, \ldots, a_{n_1})}{s_\kappa(a_1, \ldots, a_{n_1})} b_{n_2+1}^{\sum_{j=1}^{n_2} (\mu_j - \kappa_j) + \mu_{n_2+1}}, \tag{10.77}$$

where

$$\chi(\mu > \kappa) = \chi(\mu_1 \geq \kappa_1 \geq \mu_2 \geq \cdots \geq \mu_{n_2} \geq \kappa_{n_2} \geq \mu_{n_2+1} \geq 0).$$

This in turn allows for an extension of Proposition 10.3.2.

PROPOSITION 10.3.3 *Consider an $n_1 \times p$ ($n_1 \geq p$) non-negative integer matrix with entries chosen according to (10.70). The joint probability that under the RSK correspondence this matrix is such that the principal $n_1 \times s$ submatrices ($s = 0, \ldots, p$) corresponds to pairs of semi-standard tableaux with shape $\mu^{(s)}$ (note that $\mu_i^{(s)} = 0$ for $i > s$) is equal to*

$$\prod_{i=1}^{n_1} \prod_{j=1}^{p} (1 - a_i b_j) s_{\mu^{(p)}}(a_1, \ldots, a_{n_1}) \prod_{s=1}^{p} b_s^{\sum_{j=1}^{s-1} (\mu_j^{(s)} - \mu_j^{(s-1)}) + \mu_p^{(s)}} \chi(\mu^{(s)} > \mu^{(s-1)}). \tag{10.78}$$

Proof. Since in the case $n = 0$ (10.71) can be taken as equal to unity, the sought joint probability is obtained from (10.77) by forming the product $\prod_{s=0}^{p-1} P(\mu^{(s+1)}, \mu^{(s)})$. □

10.3.3 The exponential limit and continuous RSK

With $a_i = e^{-\alpha_i/L}$, $b_i = e^{-\beta_i/L}$ and the height $x_{i,j}$ at each site scaled by $x_{i,j} \mapsto L x_{i,j}$ we see that the discrete geometrical distribution (10.70) becomes the exponential distribution

$$\Pr(x_{i,j} \in [y, y + dy]) = (\alpha_i + \beta_j) e^{-(\alpha_i + \beta_j) y} dy, \quad y \geq 0. \tag{10.79}$$

This distribution can be used to define a probability measure on $n \times n$ matrices with non-negative real numbers as entries. The RSK correspondence gives a bijective mapping with a pair of u/rh nonintersecting lattice paths, each starting two units apart along $y = 0$ and finishing at $y_1 > y_2 > \cdots > y_n > 0$ along $x = 0$, in which the up steps are continuous and weighted by an exponential variable proportional to the increment. Explicitly, with the entries of $\mathbf{X}$ distributed according to (10.79), the RSK mapping requires that at $x = -(2n + \frac{1}{2} - 2j)$ ($x = 2n + \frac{1}{2} - 2j$) the up increment v of each level-l path with $l \le j$ is to be weighted $e^{-\beta_j v}$ ($e^{-\alpha_j v}$), and there is an overall weighting of the paths by the normalization $\prod_{i,j=1}^n (\alpha_i + \beta_j)$.

The total weight of a single path with vertical increments of length v_j at $x = -(2n + \frac{1}{2} - 2j)$ ($j = l, \dots, n$), weighted by $e^{-\beta_j v_j}$ with prescribed total displacement $\sum_{j=l}^n v_j = y$ is given by

$$\int_0^\infty dv_l\, e^{-\beta_l v_l} \cdots \int_0^\infty dv_n\, e^{-\beta_n v_n}\, \delta\Big(y - \sum_{j=l}^n v_j\Big) = \sum_{j=l}^n \frac{e^{-\beta_j y}}{\prod_{\mu=l \atop \mu \ne j}^n (\beta_j - \beta_\mu)} =: u_l(\{\beta_j\}_{j=l,\dots,n}; y). \quad (10.80)$$

A (semi-)continuous analogue of Proposition 10.1.1 then gives that the corresponding total weight of the set of continuous nonintersecting paths for $x < 0$ is $\det[u_l(\{\beta_j\}_{j=l,\dots,n}; y_k)]_{k,l=1,\dots,n}$. Multiplying this by the weight of the paths for $x > 0$, and the normalization, we see that the continuous analogue of (10.71) is [222], [87]

$$\begin{aligned}\prod_{i,j=1}^n (\alpha_i + \beta_j) \det[u_l(\{\alpha_j\}_{j=l,\dots,n}; y_k)]_{k,l=1,\dots,n} \det[u_l(\{\beta_j\}_{j=l,\dots,n}; y_k)]_{k,l=1,\dots,n} \\ = \frac{\prod_{i,j=1}^n (\alpha_i + \beta_j)}{\prod_{i<j}^n (\alpha_j - \alpha_i)(\beta_j - \beta_i)} \det[e^{\alpha_j y_k}]_{j,k=1,\dots,n} \det[e^{\beta_j y_k}]_{j,k=1,\dots,n}. \quad (10.81)\end{aligned}$$

In the special case

$$\alpha_i = a + (i-1)c, \qquad \beta_j = \tilde{a} + (j-1)\tilde{c} \quad (10.82)$$

(10.80) can be simplified to read

$$u_l(\{\alpha_j\}_{j=l,\dots,n}; y_k) = \frac{e^{-(a+(l-1)c)y_k}}{c^{n-l}(n-l)!}(1 - e^{-cy_k})^{n-l}, \quad (10.83)$$

and similarly for $u_l(\{\beta_j\}_{j=l,\dots,n}; y_k)$. Substituting in (10.81) and making use of the Vandermonde determinant identity (1.173) we obtain the p.d.f.

$$\begin{aligned}\Big(\prod_{i,j=1}^n (a + \tilde{a} + (i-1)c + (j-1)\tilde{c})\Big)\Big(\prod_{j=1}^n \frac{1}{(c\tilde{c})^{j-1}\Gamma^2(j)}\Big) \\ \times e^{-(a+\tilde{a})\sum_{j=1}^n y_j} \prod_{1 \le i < j \le n} (e^{-cy_j} - e^{-cy_i})(e^{-\tilde{c}y_j} - e^{-\tilde{c}y_i}). \quad (10.84)\end{aligned}$$

In the case $c = \tilde{c}$, after the change of variables and replacement of parameters $e^{-cy_j} \mapsto y_{n+1-j}$, $(a + \tilde{a}) \mapsto c(\alpha + 1)$, (10.84) reduces to the JUE supported on $(0, 1)$ with $a = \alpha, b = 0$, while with $a + \tilde{a} = 1$ and $c = \tilde{c} \to 0$ it reduces to the LUE with $a = 0$. Thus, with $E(0; J; \mathrm{ME})$ denoting the probability of having no eigenvalues in the interval J of the matrix ensemble ME, in the case $c = \tilde{c}$ we see that

$$\Pr\Big(l_1 \le \frac{1}{c} \log \frac{1}{s}\Big) = E(0; (0, s); \mathrm{JUE}_n|_{a=\alpha \atop b=0}),$$

where the JUE is supported on $(0, 1)$ and $0 < s < 1$. Similarly, in the case $c = \tilde{c} \to 0$,

$$\Pr(l_1 \le s) = E(0; (s, \infty); \mathrm{LUE}_n|_{a=0}).$$

In Exercises 10.3 q.1 it is shown how a modification of (10.81) leads to more general examples of the JUE and LUE.

The p.d.f. (10.84) can also be obtained from (10.71), by first specializing the parameters

$$(a_1,\dots,a_N)=(z_1,z_1t_1,z_1t_1^2,\dots,z_1t_1^{N-1}),\quad (b_1,\dots,b_N)=(z_2,z_2t_2,z_2t_2^2,\dots,z_2t_2^{N-1}),$$

then simplifying the Schur polynomials according to (10.22). After setting

$$t_1=e^{-c/L},\ t_2=e^{-\tilde{c}/L},\ z_1=e^{-\tilde{a}/L},\ (\mu_i+N-i)/L=y_i$$

and taking the limit as $L\to\infty$ we see that (10.71) multiplied by L^{2N} tends to (10.84). This is to be expected as then the geometric distribution reduces to (10.79) with α_i and β_j therein equal to (10.82).

Continuous RSK limits of joint probabilities can also be taken. In particular, substituting in (10.78) $a_i=b_j=e^{-1/2L}$, writing $(\mu_i^{(s)}+s-i)/L=x_i^{(s)}$ and taking $L\to\infty$ (note that then $\Pr(x_{i,j}\in[x,x+dx])=e^{-x}dx$) gives, upon use of (10.23), a functional form known from the study of the recursive construction of Wishart matrices in Exercises 4.3 q.4. Explicitly, the joint p.d.f. for $\{x_i^{(s)}\}_{i=1,\dots,s}$, $s=1,\dots,p$ is proportional to

$$\prod_{l=1}^{p}(x_l^{(p)})^{n_1-p}e^{-x_l^{(p)}}\prod_{1\le i<j\le p}(x_i^{(p)}-x_j^{(p)})\prod_{s=1}^{p-1}\chi(x^{(s+1)}>x^{(s)}),\tag{10.85}$$

where $\chi(x^{(s+1)}>x^{(s)})$ is specified as in (4.100).

In the growth model picture the variables $x_s^{(s)}$ correspond to the heights $\lambda_1(n_1,s)$ on the outer profile. It has been shown in Proposition 7.1.4 that with the horizontal distance along the profile $\mathrm{O}(N^{2/3})$, the correlation between the variables $x_j^{(s)}$, appropriately scaled, is given by the dynamical extension of the Airy kernel (7.19). It thus follows that the distribution function for two heights along the profile with this separation have distribution in terms of the dynamical Airy kernel as implied by the result of Exercises 9.1 q.3.

10.3.4 Correlation functions and the Borodin-Okounkov identity

The joint p.d.f. (10.71) for the heights in the polynuclear growth model is, in view of the determinant formula (10.16) for the Schur polynomials, of the form (5.139) so the theory of Section 5.8 applies in relation to the calculation of the correlation functions. In fact, by using the determinant formula (10.27) instead of (10.16), a double contour form of the corresponding correlation kernel in the limit $n\to\infty$ is possible [430].

PROPOSITION 10.3.4 *With μ a partition, write $\mu_k-(k-1)=:n_k$, and consider the joint p.d.f. on $\vec{n}$ implied by (10.71). For $n\to\infty$ we have*

$$\rho_{(l)}(n_1,\dots,n_l)=\det[K(n_\alpha,n_\beta)]_{\alpha,\beta=1,\dots,l},\tag{10.86}$$

where

$$K(k,l)=\int_{C_{r_1}}\frac{d\alpha}{2\pi i\alpha^k}\int_{C_{r_2}}\frac{d\beta}{2\pi i\beta^{-l+1}}\frac{1}{\alpha-\beta}\frac{H(1/\beta;\{b_i\})H(\alpha;\{a_i\})}{H(\beta;\{a_i\})H(1/\alpha;\{b_i\})}.\tag{10.87}$$

Here $r_1>1>r_2>0$ and C_r denotes a circle about the origin of radius r, while

$$H(u;\{q_i\}):=\prod_{j=1}^{\infty}\frac{1}{1-q_ju}.$$

Proof. In the following some formal manipulations on semi-infinite Toeplitz determinants are carried out; for their justification we refer to [456], [331].

In terms of $\{n_k\}$, after use of (10.27) the joint p.d.f. (10.71) reads

$$\prod_{i,j=1}^{\infty}(1-a_ib_j)\det\Big[h_{n_k+j-1}(\{a_i\})\Big]_{j,k=1,\dots,n}\det\Big[h_{n_k+j-1}(\{b_i\})\Big]_{j,k=1,\dots,n}.$$

As this is a symmetric function in $\{n_k\}$ which vanishes if $n_j=n_k$ $(j\neq k)$, we can relax the ordering constraint $n_1>n_2>\cdots$ implied by the definition of $\{n_k\}$. According to the result of Proposition 5.8.1 with $\xi_j(n)=h_{n+j-1}(\{a_i\})$, $\eta_j(n)=h_{n+j-1}(\{b_i\})$, for $n\to\infty$ the l-point correlation is given by (10.86) with

$$K(p,q)=\sum_{j,k=1}^{\infty}c_{jk}h_{p+j-1}(\{a_i\})h_{q+j-1}(\{b_i\}), \tag{10.88}$$

where $[c_{jk}]_{j,k=1,2,\dots}=([\sum_{l=-\infty}^{\infty}\xi_j(l)\eta_k(l)]_{j,k=1,2,\dots})^{-1}$.

In preparation for computing the matrix inverse, introduce the notation for a semi-infinite Toeplitz matrix

$$T[f(z)]=\Big[[z^{j-k}]f(z)\Big]_{j,k=1,2,\dots} \tag{10.89}$$

for $f(z)$ a Laurent expandable function. We then have

$$\sum_{l=-\infty}^{\infty}\xi_j(l)\eta_k(l)=\sum_{l=1}^{\infty}h_{l-j}(\{a_i\})h_{l-k}(\{b_i\})=\Big(T\Big[H(1/z;\{a_i\})\Big]T\Big[H(z;\{b_i\})\Big]\Big)_{jk}, \tag{10.90}$$

where in obtaining the first equality the change of variables $l\mapsto l-k-j$ has been performed and the fact that the summation vanishes for $l<1$ used, while for the second equality use has been made of (10.25). Because (10.90) is a decomposition into the product of an upper triangular and a lower triangular Toeplitz matrix, and for such matrices in general the inverse is a Toeplitz matrix of the same type but with reciprocal generating function, it follows

$$[c_{jk}]_{j,k=1,2,\dots}=T\Big[\frac{1}{H(z;\{b_i\})}\Big]T\Big[\frac{1}{H(1/z;\{a_i\})}\Big]=\Big[\sum_{l=1}^{\infty}e_{j-l}(\{b_i\})e_{k-l}(\{a_i\})\Big]_{j,k=1,2,\dots},$$

where $\{e_j\}$ denote the elementary symmetric functions (4.132). This result in turn allows us to compute that

$$\sum_{j,k=1}^{\infty}c_{jk}\beta^k\alpha^{-j}=\frac{\beta/\alpha}{1-\beta/\alpha}\frac{1}{H(1/\alpha;\{b_i\})H(\beta;\{a_i\})}, \tag{10.91}$$

assuming $0<|\beta|<1<|\alpha|$.

Consider now the generating function

$$\sum_{k,l=-\infty}^{\infty}\alpha^k\beta^{-l}K(k,l).$$

Substituting (10.88) and making use of (10.91) shows

$$\sum_{k,l=-\infty}^{\infty}\alpha^k\beta^{-l}K(k,l)=\frac{1}{1-\beta/\alpha}\frac{H(1/\beta;\{b_i\})H(\alpha;\{a_i\})}{H(\beta;\{a_i\})H(1/\alpha;\{b_i\})}. \tag{10.92}$$

This is equivalent to the sought result (10.87). □

Suppose the coordinates $\{n_j\}$ are restricted to be less than or equal to l. According to the general theory of Section 9.1 we have that

$$\lim_{n\to\infty}\Pr(h^{\square}\le l)=\det(1-K_J), \tag{10.93}$$

where K_J is the Wiener-Hopf operator on $J=\{l+1,l+2,\dots,\}$ with kernel $K(j,k)$. On the other hand

(10.73) gives

$$\lim_{n\to\infty} \Pr(h^{\square} \le l) = \prod_{i,j=1}^{\infty} (1 - a_i b_j) \Big\langle \prod_{j=1}^{\infty} \prod_{k=1}^{l} (1 + a_j e^{-i\theta_k})(1 + b_j e^{i\theta_k}) \Big\rangle_{\mathrm{CUE}_l}. \tag{10.94}$$

Equate (10.93) and (10.94) (with $\theta_k \mapsto -\theta_k$), and substitute

$$\sum_{l=1}^{\infty} a_l^k = k c_k, \qquad \sum_{l=1}^{\infty} b_l^k = k c_{-k}$$

for the corresponding power sum symmetric functions. Noting that with these substitutions, and $u_j = a_j$, $u_j = b_j$, respectively, we have

$$\prod_{j=1}^{\infty} (1 - u_j z) = \exp\Big(- \sum_{k=1}^{\infty} c_{\mp k} z^k \Big), \tag{10.95}$$

and recalling the general identity (5.32) gives an identity expressing a general Toeplitz determinant in terms of a Fredholm determinant.

PROPOSITION 10.3.5 *Let $D_n[f]$ be specified as in Proposition 14.4.1 below. With $c(\theta) := \sum_{n=-\infty}^{\infty} c_n e^{in\theta}$ we have*

$$D_n[e^{c(\theta)}] = e^{n c_0 + \sum_{p=1}^{\infty} p c_{-p} c_p} \det(1 - K_J), \tag{10.96}$$

where K_J is the Wiener-Hopf operator on $[n+1, n+2, \dots)$ with kernel

$$K(k,l) = \int_{\mathcal{C}_{r_1}} \frac{d\alpha}{2\pi i \alpha^k} \int_{\mathcal{C}_{r_2}} \frac{d\beta}{2\pi i \beta^{-l+1}} \frac{1}{\alpha - \beta} \exp\Big(\sum_{k=1}^{\infty} c_k (\beta^k - \alpha^k) + c_{-k} (\alpha^{-k} - \beta^{-k}) \Big).$$

A reformulation of (10.96), and an alternative derivation, are given in Exercises 10.3 q.2.

This result was given in the context of random matrix theory by Borodin and Okounkov [86]. Later it was realized that the same identity had occurred in the context of Toeplitz determinant theory in an earlier study by Geronimo and Case [260]. In this context, observe that (10.96) effectively gives higher order corrections to the Szegö formula (14.71) below.

EXERCISES 10.3 1. [222] In this exercise a generalization of (10.84) will be derived.

(i) In the case that $\mathbf{X}$ is an $n \times m$ ($n \ge m$) matrix of non-negative real numbers distributed according to (10.79), by taking $b_{m+1} = \cdots = b_n = 0$ in (10.71), and noting that we must then have $\ell(\mu) \le m$ for this to be nonzero, argue that the continuous RSK correspondence maps the distribution on $\mathbf{X}$ to

$$\prod_{i=1}^{n} \prod_{j=1}^{m} (\alpha_i + \beta_j) \det[u_l(\{\alpha_j\}_{j=l,\dots,n}; y_k)]_{k,l=1,\dots,m} \det[u_l(\{\beta_j\}_{j=l,\dots,m}; y_k)]_{k,l=1,\dots,m},$$

where u_l is specified by (10.80).

(ii) Make use of (10.83) (appropriately modified) to show that with $\alpha_i = a + (i-1)c$, $\beta_j = \tilde{a} + (j-1)\tilde{c}$ this simplifies to read

$$\Big(\prod_{i=1}^{n} \prod_{j=1}^{m} (a + \tilde{a} + (i-1)c + (j-1)\tilde{c}) \Big) \Big(\prod_{j=1}^{m} \frac{1}{(c\tilde{c})^{j-1} \Gamma(j) \Gamma(n-m+j)} \Big) \frac{1}{c^{(n-m)m}}$$
$$\times \prod_{k=1}^{m} (1 - e^{-c y_k})^{n-m} e^{-(a+\tilde{a}) \sum_{j=1}^{m} y_j} \prod_{1 \le i < j \le m} (e^{-c y_j} - e^{-c y_i})(e^{-\tilde{c} y_j} - e^{-\tilde{c} y_i}). \tag{10.97}$$

Relate this to the JUE and the LUE, as done for (10.84).

2. [47], [92] The aim of this exercise is to give a reformulation and alternative derivation of (10.96).

(i) Let $K = [K(l,k)]_{k,l=1,\dots}$ be the transpose of the semi-infinite matrix with entries as specified by (10.87). With $\phi(z) := 1/(H(1/z;\{a_j\})H(z;\{b_j\}))$ show that (10.96) can alternatively be written

$$D_n[\phi(e^{i\theta})] = D_\infty[\phi(e^{i\theta})]\det(1 - Q_n K Q_n), \tag{10.98}$$

where Q_n is the semi-infinite diagonal matrix differing from the identity by the deletion of the first n elements.

(ii) With $\phi_-(z) := 1/H(1/z;\{a_j\})$, $\phi_+(z) := 1/H(z;\{b_j\})$ so that $\phi(z) = \phi_-(z)\phi_+(z)$, verify from (10.92) that

$$K(k,l) = \sum_{p=1}^{\infty}(\phi_-/\phi_+)_{p+l-1}(\phi_+/\phi_-)_{-p-k+1}$$

where we have made use of the notation $f_j = [z^j]f(z)$. After introducing the further notation

$$H(a) = [a_{j+k-1}]_{j,k=1,2,\dots} \qquad H(\tilde{a}) = [a_{-j-k+1}]_{j,k=1,2,\dots}, \tag{10.99}$$

conclude that K as specified in (i) can be written

$$K = H(\phi_-/\phi_+)H(\widetilde{\phi_+/\phi_-}). \tag{10.100}$$

(iii) In terms of the notation in (ii), together with $T(a) = [a_{j-k}]_{j,k=1,2,\dots}$, by verifying the formula

$$(ab)_l = \sum_{p=-\infty}^{\infty} a_p b_{l-p}$$

show that

$$T(ab) = T(a)T(b) + H(a)H(\tilde{b}). \tag{10.101}$$

(iv) Let $\mathbf{A}$ be an $N \times N$ invertible matrix, let $\mathbf{P}_{n,N}$ be the diagonal matrix with the first n elements 1 and all other entries 0, and set $\mathbf{Q}_{n,N} = \mathbf{1}_N - \mathbf{P}_{n,N}$. Then, as already noted in (5.178) but in a different notation, a theorem of Jacobi gives [8]

$$\det \mathbf{P}_{n,N}\mathbf{A}^{-1}\mathbf{P}_{n,N} = \frac{\det \mathbf{Q}_{n,N}\mathbf{A}\mathbf{Q}_{n,N}}{\det \mathbf{A}}. \tag{10.102}$$

(note too that (1.150) corresponds to the case $n = 1$). Under general conditions this remains valid for $N \to \infty$ and thus semi-infinite matrices. By noting from (10.101) that with $A = 1 - K$, K as in (10.100),

$$\begin{aligned} P_{n,\infty}A^{-1}P_{n,\infty} &= P_{n,\infty}T(\phi_-/\phi_+)T(\phi_+/\phi_-)P_{n,\infty} \\ &= P_{n,\infty}T(\phi_+^{-1})T(\phi)T(\phi_-^{-1})P_{n,\infty} \\ &= T_n(\phi_+^{-1})T_n(\phi)T_n(\phi_-^{-1}), \end{aligned}$$

where $T_n := P_{n,\infty}TP_{n,\infty}$, deduce (10.98).

10.4 FURTHER INTERPRETATIONS AND VARIANTS OF THE RSK CORRESPONDENCE

10.4.1 Stochastic recurrences

The nonintersecting paths underlying the discrete polynuclear growth process can be generated by recurrences which couple together the displacements along level-l and level-$(l-1)$ at varying positions along the paths. Furthermore, in the case of the top path, this recurrence allows for the derivation of some inter-relations between various growth models, and also for some different interpretations of the process relating to this path.

Let $\mathbf{X}_{n_1,n_2}$ denote the first n_1 rows and n_2 columns of the integer matrix $\mathbf{X}$. Extend this to a square matrix $\tilde{\mathbf{X}}_{n_1,n_2}$ of dimension $\max(n_1,n_2) \times \max(n_1,n_2)$ by adjoining rows of zeros to the top (or columns of zeros to the right as appropriate), then apply the growth algorithm and record $\lambda_1(n_1,n_2)$, the maximum displacement of the corresponding level-1 path.

Consideration of the rules of the growth process leads to an interpretation of $\lambda_l(n_1,n_2)$ for $n_1 = n$ or $n_2 = n$ as displacements in the paths relating to $\mathbf{X}$ itself. Thus one has that $\lambda_l(n,j)$ gives the height of the level-l path at $x = -(2n + \frac{1}{2} - 2j)$, while $\lambda_l(i,n)$ gives the height of the level-l path at $x = (2n + \frac{1}{2} - 2i)$. One immediate implication is the summations

$$\sum_{l=1}^{n} \Big(\lambda_l(n,j) - \lambda_l(n,j-1) \Big) = \sum_{i=1}^{n} x_{i,j}, \qquad \sum_{l=1}^{n} \Big(\lambda_l(i,n) - \lambda_l(i-1,n) \Big) = \sum_{j=1}^{n} x_{i,j}, \tag{10.103}$$

where $\lambda_l(0,j) = \lambda_l(i,0)$. These equate the increments at a given position to the sum of the nucleation events contributing to those increments.

For the paths contributing to $\tilde{\mathbf{X}}_{n_1,n_2}$, the same considerations give that $\lambda_l(n^*,j)$ is equal to the height of the level-l paths at $x = -(2n^* + \frac{1}{2} - 2j)$, while $\lambda_l(i,n^*)$ is equal to the height of the level-l path at $x = (2n + \frac{1}{2} - 2i)$, where $n^* = \max(n_1,n_2)$. One notes too that $\lambda_l(n^*,j) = \lambda_l(n_1,j)$, $\lambda_l(i,n^*) = \lambda_l(i,n_2)$. Further relevant points are that the $\lambda_l(n_1 - 1, n_2 - 1)$ occur as the displacement of the level-l path at the origin for times $t = 2n^* - 2$ (second-last step), and that

$$\lambda_l(i,j) \ge \lambda_l(i,j-1) \ge \lambda_{l+1}(i,j), \qquad \lambda_l(i,j) \ge \lambda_l(i-1,j) \ge \lambda_{l+1}(i,j). \tag{10.104}$$

These considerations, together with the rules of the growth process, imply recurrences for $\{\lambda_l(i,j)\}$. Thus the rules of the growth process give that $\lambda_1(n_1,n_2)$ is equal to the maximum of the heights in level-1 at $x = -1$ and $x = 1$, plus the nucleation height x_{n_1,n_2}, and so

$$\lambda_1(n_1,n_2) = \max \Big(\lambda_1(n_1,n_2-1), \lambda_1(n_1-1,n_2) \Big) + x_{n_1,n_2}, \tag{10.105}$$

where $\lambda_1(0,j) = \lambda_1(i,0) = 0$. Similarly, the maximum displacement $\lambda_l(n_1,n_2)$ of the level-l path ($l = 2,\dots,n$) satisfies

$$\lambda_l(n_1,n_2) = \max \Big(\lambda_l(n_1,n_2-1), \lambda_l(n_1-1,n_2) \Big) + x^{(l-1)}_{n_1,n_2}, \tag{10.106}$$

where $x^{(l-1)}_{n_1,n_2}$ is the height of an overlap event (if any) which occurs in the growth of the nucleation events x_{n_1,n_2-1} or x_{n_1-1,n_2} and/or corresponding plateaux. This latter height is equal to the minimum of the height in level-$(l-1)$ at $x = \pm 1$, with the height at the origin in the second-last step subtracted so that

$$x^{(l-1)}_{n_1,n_2} = \min \Big(\lambda_{l-1}(n_1,n_2-1), \lambda_{l-1}(n_1-1,n_2) \Big) - \lambda_{l-1}(n_1-1,n_2-1).$$

Substituting this in (10.106) gives that for $l > 1$

$$\begin{aligned} \lambda_l(n_1,n_2) = {} & \max \Big(\lambda_l(n_1,n_2-1), \lambda_l(n_1-1,n_2) \Big) \\ & + \min \Big(\lambda_{l-1}(n_1,n_2-1), \lambda_{l-1}(n_1-1,n_2) \Big) - \lambda_{l-1}(n_1-1,n_2-1). \end{aligned} \tag{10.107}$$

10.4.2 Last passage percolation and increasing subsequences

The process interpretation of the mapping from the non-negative integer matrix $\mathbf{X}$ to nonintersecting paths can alternatively be viewed as a certain directed last passage percolation [329]. Let us regard each element $x_{i,j}$ of the matrix $\mathbf{X}$ as a waiting time associated with the lattice site (i,j) $(1 \le i,j \le n)$. The directed last passage percolation is defined by forming a u/rh lattice path from the site (1,1) to the site (n_1,n_2), such that it maximizes the sum of the waiting times (in general such a path will not be unique). Thus the quantity of interest is

$$L_{\mathbf{X}}^{(1)}(n_1,n_2) := \max \sum_{(1,1)\,\mathrm{u/rh}\,(n_1,n_2)} x_{i,j}, \tag{10.108}$$

which is referred to as the *last passage time*. From this definition we see that

$$L_{\mathbf{X}}^{(1)}(n_1,n_2) = \max\Big(L_{\mathbf{X}}^{(1)}(n_1,n_2-1), L_{\mathbf{X}}^{(1)}(n_1-1,n_2)\Big) + x_{n_1,n_2} \tag{10.109}$$

with $L_{\mathbf{X}}^{(1)}(0,k) = L_{\mathbf{X}}^{(1)}(j,0) := 0$. This is identical to the recurrence (10.105) with

$$L_{\mathbf{X}}^{(1)}(n_1,n_2) = \lambda_1(n_1,n_2) \tag{10.110}$$

for all $1 \le n_1, n_2 \le n$, so the last passage time is equal to the maximum height in the growth process applied to $\tilde{\mathbf{X}}_{n_1,n_2}$.

The definition (10.108) of the last passage time can be generalized to a quantity $L_{\mathbf{X}}^{(l)}(n_1,n_2)$ which relates to the level-p heights $\lambda_p(n_1,n_2)$ for each $p = 1,\dots,l$. Thus let $(\mathrm{rd}^*)^l$ denote the set of l disjoint rd* lattice paths, the latter defined as either a single point, or points connected by segments formed out of arbitrary positive integer multiples of steps to the right and steps up in the rectangle $1 \le i \le n_1$, $1 \le j \le n_2$ (the steps are said to be disjoint if they connect no common lattice sites). In terms of this set generalize (10.108) to

$$L_{\mathbf{X}}^{(l)}(n_1,n_2) = \max \sum_{(\mathrm{rd}^*)^l} x_{i,j}. \tag{10.111}$$

These quantities obey a recurrence generalizing (10.109),

$$\begin{aligned} L_{\mathbf{X}}^{(l)}(n_1,n_2) + L_{\mathbf{X}}^{(l-1)}(n_1-1,n_2-1) = \max\Big(&L_{\mathbf{X}}^{(l-1)}(n_1-1,n_2) + L_{\mathbf{X}}^{(l)}(n_1,n_2-1), \\ &L_{\mathbf{X}}^{(l-1)}(n_1,n_2-1) + L_{\mathbf{X}}^{(l)}(n_1-1,n_2)\Big) + x_{n_1,n_2}, \end{aligned} \tag{10.112}$$

where $L_{\mathbf{X}}^{(0)}(n,m) := 0$, which can be proved by consideration of the union of the paths realizing $L_{\mathbf{X}}^{(l-1)}$ and $L_{\mathbf{X}}^{(l)}$ in each term [457]. Noting that in general

$$\begin{aligned} &\max(a_{l-1}+b_l, b_{l-1}+a_l) - \max(a_{l-2}+b_{l-1}, b_{l-2}+a_{l-1}) \\ &\quad = \max(b_l - b_{l-1}, a_l - a_{l-1}) + \min(a_{l-1}-a_{l-2}, b_{l-1}-b_{l-2}) \end{aligned} \tag{10.113}$$

we see by subtracting from the recurrence (10.112) the same recurrence with $l \mapsto l-1$, that the recurrence (10.107) results with the identification

$$\lambda_l(n_1,n_2) = L_{\mathbf{X}}^{(l)}(n_1,n_2) - L_{\mathbf{X}}^{(l-1)}(n_1,n_2),$$

or equivalently

$$L_{\mathbf{X}}^{(l)}(n_1,n_2) = \sum_{p=1}^{l} \lambda_p(n_1,n_2). \tag{10.114}$$

The non-negative matrix $\mathbf{X}$ can be presented as a two-line array, and $L_{\mathbf{X}}^{(l)}(n_1,n_2)$ interpreted in terms of weakly increasing subsequences (see Figure 10.13). In fact it is this viewpoint which is most prevalent in the combinatorics literature, being the one taken in the pioneering works of Schensted and Greene [481], [273]. A two-line array is constructed from the $n_1 \times n_2$ matrix $\mathbf{X}_{n_1,n_2}$ of non-negative integers by writing down in sequence, reading along the first row of the matrix, then the second row and so on (recall that our convention is to count rows from the bottom), $x_{i,j}$ copies of the ordered pair $\binom{i}{j}$ (see Figure 10.13 for an example). By construction the entries in the top row of the two-line array are all weakly increasing, but not so

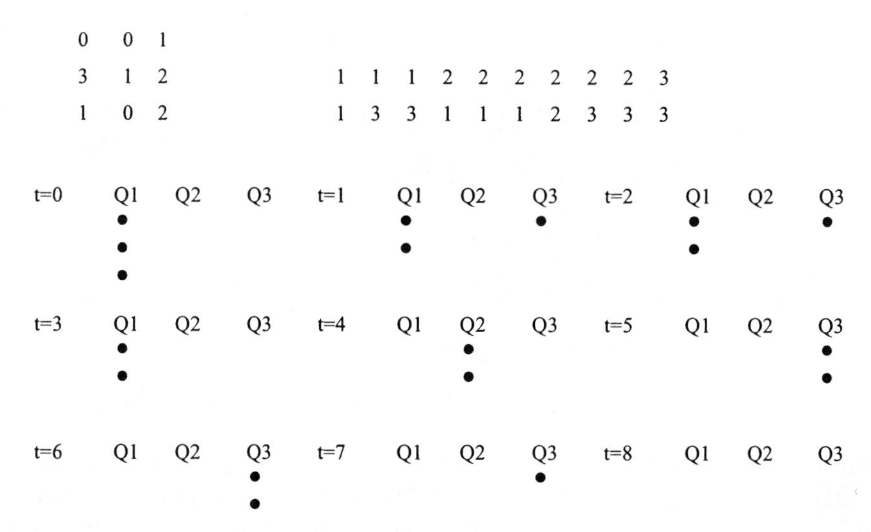

Figure 10.13 A 3×3 non-negative integer matrix, with the corresponding queueing process and two-line array. Note that $L_{\mathbf{X}}^{(1)}(3,3) = 8$, corresponding to the path through the elements $\binom{1}{1}, \binom{2}{1}, \binom{2}{2}, \binom{2}{3}, \binom{3}{3}$, while the copies of these ordered pairs forms an increasing subsequence of length 8 in the bottom line of the two-line array.

in general the bottom row. The maximum of the lengths of subsequences in the bottom row which are weakly increasing (i.e., are of the form $l_{j_1} \le l_{j_2} \le \cdots \le l_{j_m}$) is readily seen to be given by the formula (10.108) specifying $L_{\mathbf{X}}^{(1)}(n_1, n_2)$. In this context (10.110), with $\lambda_1(n_1, n_2)$ interpreted as the length of the first row of the corresponding semi-standard tableau, is due to Schensted. Furthermore, the maximum of the lengths of l disjoint subsequences in the bottom row which are all weakly increasing is the same thing as $L_{\mathbf{X}}^{(l)}(n_1, n_2)$ and in this context (10.114) is due to Greene.

10.4.3 Queues

Focusing attention on $L_{\mathbf{X}}^{(1)}(n_1, n_2)$ only, a number of further interpretations of the non-negative integer elements of $\mathbf{X}$ and the corresponding meaning of the last passage time are possible. One such interpretation is to view the $x_{i,j}$ as service times in a queueing process [28]. Specifically, consider an infinite number of queues Q_j $(j = 1, 2, \dots)$ which initially have n jobs labeled $i = 1, 2, \dots$ in queue Q_1, and zero jobs in all other queues, and suppose that immediately after service in Q_j, each job moves to Q_{j+1} (see Figure 10.13 for an example). Then with $x_{i,j}$ denoting the time it takes server j to process job i (once it reaches the server), we see that the time $T(i, j)$ it takes job i to leave queue Q_j satisfies the recurrence (10.105) and is thus equal to $\lambda_1(i, j)$.

Theory related to the queueing problem when all the service times are chosen from the same distribution allows for a characterization of $\{\lambda_1(i, N)\}_{i=1,\dots,n}$. Thus one has the general result that with the service times all chosen from a distribution of mean μ and variance σ^2, the limiting scaled exit times

$$D_i := \lim_{N\to\infty} \frac{T(i, N) - \mu N}{\sigma\sqrt{N}} \tag{10.115}$$

of job i from queue N have joint p.d.f. specified by [267]

$$D_i = \sup_{0=t_0<t_1<\cdots<t_i=1} \Big(\sum_{l=0}^{i-1} (B_l(t_{l+1}) - B_l(t_l)) \Big), \quad i = 1, \dots, n. \tag{10.116}$$

In (10.116) each B_l is an independent standard Brownian motion.

This general result is independent of the specific distribution from which the $x_{i,j}$ are being sampled. In the case that the distribution is the exponential with mean unity, we know from (10.85) the joint distribution of the variables $\{x_l^{(s)}\}_{l=1,\dots,s}$, $s = 1, \dots, p$, having the interpretation as the maximum heights in the corresponding sequence of the semicontinuous polynuclear growth models. These heights are such that they relate to the exit times in the queueing model by $T(i, N) = x_i^{(i)}$, provided we set $n_1 = N$. Thus if in (10.85) with $n_1 = N$ we introduce the scaled variables $d_l^{(s)} := (x_l^{(s)} - N)/\sqrt{N}$ as suggested by (10.115) (in particular, then $d_s^{(s)} = D_s$) we obtain for the joint p.d.f. of $\{d_l^{(s)}\}_{l=1,\dots,p}$

$$\frac{1}{C} \prod_{l=1}^{p} e^{-(x_l^{(p)})^2/2} \prod_{1 \le i < j \le p} (x_i^{(p)} - x_j^{(p)}) \prod_{s=2}^{p} \chi(x^{(s)} > x^{(s-1)})$$

for some normalization C. We recognize as the joint p.d.f. (4.96) for the eigenvalues of successive minors of a random matrix from GUE$_p^*$. In this interpretation $x_s^{(s)}$ corresponds to the largest eigenvalue of the $s \times s$ minor and so the following result holds true [45].

Proposition 10.4.1 *The joint distribution of $\{D_1, \dots, D_n\}$ as specified by (10.116) is equal to the joint distribution of $\{X_1, \dots, X_n\}$, where X_j denotes the largest eigenvalue in the top $j \times j$ sub-block of a random matrix chosen from GUE$_n^*$.*

10.4.4 Corner growth model

We consider next a directed growth process (the *corner growth model*) associated with $\mathbf{X}$, distinct from the polynuclear growth model already considered. For this define a subset of entries of $\mathbf{X}$ by

$$A(t) := \Big\{ (m, n) : \max \sum_{(1,1)\mathrm{u/rh}(m,n)} (x_{ij} + 1) \le t \Big\}. \tag{10.117}$$

Comparing with the definition of $L_{\mathbf{X}}^{(1)}(m, n)$, we see that

$$\max \sum_{(1,1)\mathrm{u/rh}(m,n)} (x_{ij} + 1) = L_{\mathbf{X}}^{(1)}(m, n) + (m + n - 1) =: \tilde{L}_{\mathbf{X}}(m, n). \tag{10.118}$$

Because the sum is over all paths $(1,1)\mathrm{u/rh}(m,n)$ and we must take the maximum, it follows that if $(m^*, n^*) \in A(t)$, then $(i, j) \in A(t)$ for all $1 \le i \le m^*$, $1 \le j \le n^*$, so $A(t)$ forms a staircase-shaped region in the square $1 \le m, n \le N$. The boundary of $A(t)$ can be extended from the x-axis to the y-axis by a single u/lh lattice path enclosing a region referred to as the diagram of $A(t)$. Incrementing t by 1 unit, we see that $A(t + 1)$ is related to $A(t)$ by the addition of at most one lattice site to each column of the latter, and for all columns but the first this site must have its left neighbor in $A(t)$ (the left neighbor will then be referred to as occupied). Thus if

$$\tilde{L}_{\mathbf{X}}(m, n) = k + 1 \tag{10.119}$$

then the lattice site (m, n) is added to the diagram of $A(k)$ and so becomes part of the diagram of $A(k + 1)$. We say this event happens at time $t = k + 1$.

Let us now focus attention on the probability of the event (10.119). For $m + n > k + 2$ we see from

(10.118) that

$$\Pr(\tilde{L}_{\mathbf{X}}(m,n) = k+1) = 0.$$

For $m+n \le k+2$, (10.118) and the recurrence (10.109) shows

$$\begin{aligned}\Pr(\tilde{L}_{\mathbf{X}}(m,n) = k+1) \\ = \sum_{j=0}^{k+2-m-n} \Pr(x_{mn} = j)\Pr\Big(\max(\tilde{L}_{\mathbf{X}}(m,n-1),(\tilde{L}_{\mathbf{X}}(m-1,n)) = k-j\Big).\end{aligned} \qquad (10.120)$$

On the other hand consider a growth process in which $A(k+1)$ is created from $A(k)$ by first adding to columns according to the rule that for all columns but the first its left neighbour is in $A(k)$, then for each such site (i,j) so added, accepting with probability $1-q_{ij}$ and rejecting with probability q_{ij}. For such a process we see that

$$\begin{aligned}\Pr(\tilde{L}_{\mathbf{X}}(m,n) = k+1) \\ = \sum_{j=0}^{k+2-m-n} (1-q_{mn})q_{mn}^j \Pr\Big(\max(\tilde{L}_{\mathbf{X}}(m,n-1),\tilde{L}_{\mathbf{X}}(m-1,n)) = k-j\Big)\end{aligned} \qquad (10.121)$$

(the significance here of the operation max$(\cdot)$ is that it ensures that both $(m,n-1)$ and $(m-1,n)$ are in $A(k)$, while the factor $(1-q_{mn})q_{mn}^j$ represents the probability that site (m,n) was sampled a total of j times before being accepted). Comparing (10.120) and (10.121) we see that they are identical provided

$$\Pr(x_{mn} = j) = (1-q_{mn})q_{mn}^j. \qquad (10.122)$$

This condition is fulfilled by the weighting (10.70) of $\mathbf{X}$ if we choose

$$q_{mn} = a_m b_n. \qquad (10.123)$$

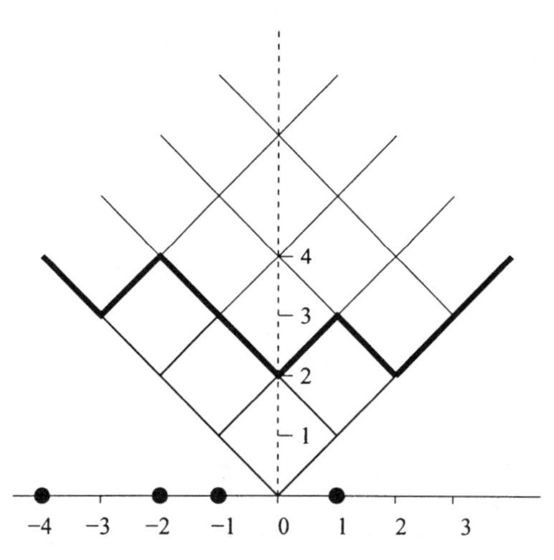

Figure 10.14 The square lattice $1 \le m,n \le N$ rotated by 45° anticlockwise and stretched by a factor of $\sqrt{2}$. On the rotated lattice squares are placed in positions stable with respect to motion under gravity according to the rules of the growth process. The top endpoint of each left diagonal segment of the boundary of the accepted placed squares, extended to run along the rotated axes, can be projected onto the x-axis to give a particle configuration.

10.4.5 Totally asymmetric simple exclusion process

The growth process relating to $A(t)$ is itself related to certain stochastic dynamics of particles on a line [330]. To see this rotate the diagram of $A(t)$ as indicated in Figure 10.14. Instead of thinking of the growth process as adding lattice points, we now think in terms of squares being added. From the shape of the boundary of the accepted squares, a particle configuration on the integer lattice of the x-axis is defined according to the rule specified in Figure 10.14. It follows that in a time interval $t \mapsto t+1$ of the stochastic dynamics, only particles with their right neighbouring site vacant can move, and they must either move to that lattice site or stay where they are. The probability of these events is the same as the probability of accepting or rejecting the corresponding square in the growth process. According to the prescriptions (10.122) and (10.123) these probabilities are in general lattice site dependent. But in the mapping to the particle configurations all points in the rotated lattice with $i-j$ constant are mapped to the same particle coordinate. For (10.123) then to be constant and also to be less than 1 we must take $q_{ij} = q$ so that it is independent of the lattice site. Thus particles with right neighbor vacant move to that site with probability $1-q$, and stay where they are with probability q. This stochastic dynamics is referred to as the *totally asymmetric simple exclusion process.* Note that here $\tilde{L}_{\mathbf{X}}(m,n) = k+1$ means the particle initially at position $i = -(n-1)$ has moved m steps at time $t = k+1$.

10.4.6 The dual RSK correspondence

The correspondence between non-negative integer matrices and pairs of nonintersecting u/rd paths has a variant, equivalent to the dual Robinson-Schensted-Knuth correspondence [246], in which a matrix of 0s and 1s is put into one-to-one correspondence with u/rh nonintersecting lattice paths paired with ld/lh nonintersecting paths. Again the matrix itself can be thought of as recording events in a discrete space and time growth process.

Let $\mathbf{X} = [x_{i,j}]_{\substack{i=1,\dots,m \\ j=1,\dots,n}}$ be an $m \times n$ matrix with rows counted from the bottom in which each entry $x_{i,j}$ is either 0 or 1. Here $x_{i,j} = 1$ represents a nucleation event (a unit square) which is positioned above the segment $x = j-1$ to $x = j$, on top of earlier nucleation events and their growth. These nucleation events occur at successive time intervals $t = 1, 2, \dots, n$ corresponding to the 1st, 2nd and subsequent rows of the matrix $\mathbf{X}$. Thus at $t = 1$ there are nucleation events between $x = [j_1 - 1, j_1], [j_2 - 1, j_2], \dots, [j_p - 1, j_p]$ where $x_{1j_1}, \dots, x_{1j_p}$ are the nonzero entries of the first row of $\mathbf{X}$. These are marked on the line $y = 0$ as unit squares. In this growth process, as $t \mapsto t+1$ the profile of all nucleation events so far recorded grows to the right (but not to the left) until they join up with the neighboring nucleation event on the right. If there is no right neighboring nucleation event, and this nucleation event has not yet grown, it is to grow to $x = n+1$, while the rightmost profile from the growth of all nucleation events at earlier times is to grow one unit to the right. The shape of the rightmost unit square after growth is to be modified by the removal of the upper triangular half. Furthermore, the meeting of nucleation events in this top profile (which occurs at $j_2, \dots, j_p$ in going from $t = 1$ to $t = 2$) is to be recorded on the line $y = -1$ as new nucleation events drawn immediately to the right of these points (thus along the segments $[j_2, j_2+1], \dots, [j_p, j_p+1]$ in going from $t = 1$ to $t = 2$), which themselves grow according to the specified rule and give rise to nucleation events on the line $y = -2$ and so on.

After the growth and creation of new nucleation events in lower levels at time $t+1$, new nucleation events are placed on the 1st level in correspondence with the nonzero entries of the $(t+1)$st row from the bottom of the matrix $\mathbf{X}$, and this completes the process at that time. This procedure is to stop after time $m+l$ in level l $(l = 1, \dots, m+1)$, this being the maximum time for which new nucleation events can be created and then grow once. The growth profile at each level is of the form of a u/rh path from $x = 0$ to $x = n-1$ and a ld/lh path from $x = n+m$ to $x = n+1$ (see Figure 10.15). In level-l these pairs of paths start and finish at $y = -(l-1)$, have the same maximum height μ_l units, say, above this line, and do not intersect with the path above, so that $\mu = (\mu_1, \mu_2, \dots, \mu_n)$ forms a partition.

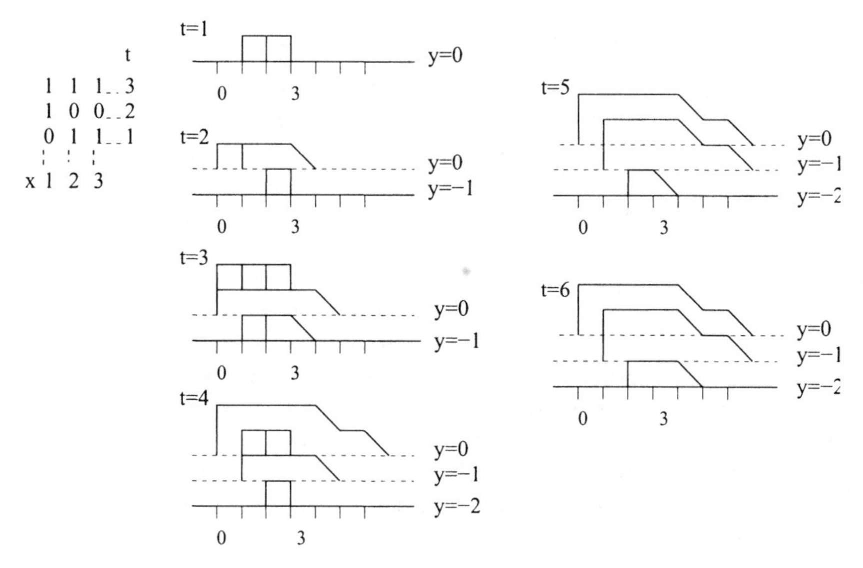

Figure 10.15 Mapping from a 0,1 matrix to a pair of nonintersecting lattice paths.

It follows that with each entry of **X** weighted according to the probability

$$\Pr(x_{ij} = k) = \frac{(a_i b_j)^k}{1 + a_i b_j}, \qquad k = 0, 1, \tag{10.124}$$

to obtain a one-to-one correspondence between weighted 0,1 matrices and weighted nonintersecting lattice paths, in the latter the displacements $\mu(j, l)$ should be weighted $b_j^{\mu(j,l)}$ and $\nu(i, l)$ should be weighted $a_i^{\nu(i,l)}$.

It follows from (10.24) that the total weight of all nonintersecting u/rh paths initially equally spaced at $y = 0, \dots, -(n-1)$ along $x = 0$, finishing at $y = \mu_j - (j-1)$ $(j = 1, \dots, n)$ along $x = n - 1$, with up steps at $x = j - 1$ weighted b_j is given by $s_\mu(b_1, b_2, \dots, b_n)$. Similarly, the total weight of all nonintersecting ld/lh paths initially equally spaced at $y = 0, \dots, -(n-1)$ along $x = n + m$, finishing at $y = \mu_j - (j-1)$ $(j = 1, \dots, n)$ along $x = n + 1$, with no more than one up step at $x = n + m + 1 - i$ (the latter weighted by a_i) is given by $s_{\mu'}(a_1, a_2, \dots, a_m)$. Here μ' denotes the conjugate partition (recall Definition 10.1.3). Hence the probability that a 0,1 matrix with entries distributed according to (10.124) corresponds to a pair of such lattice paths is given by

$$\prod_{i=1}^{m} \prod_{j=1}^{n} (1 + a_i b_j)^{-1} s_{\mu'}(a_1, \dots, a_m) s_\mu(b_1, \dots, b_n) \tag{10.125}$$

(cf. (10.71)). Note that the normalization condition for this probability implies the dual Cauchy identity (10.56). Let $\Pr(h^{0,1} \le l)$ denote the probability that the maximum height in the level-1 profile is less than or equal to l. According to (10.125) and the fact that μ_1 therein corresponds to the maximum height, we have

$$\Pr(h^{0,1} \le l) = \prod_{i=1}^{m} \prod_{j=1}^{n} (1 + a_i b_j)^{-1} \sum_{\mu_1 \le l} s_{\mu'}(a_1, \dots, a_m) s_\mu(b_1, \dots, b_n). \tag{10.126}$$

We know that the Schur polynomial $s_{\mu'}$ can be interpreted in terms of nonintersecting u/rh paths by reading along rows (there being μ_1 rows in μ' and thus $\mu_1 \le l$ paths), while s_μ can be interpreted in terms of u/rd paths by reading down columns (in μ there are μ_1 columns and thus again $\mu_1 \le l$ paths). The first set of paths consist of m steps, and the second consists of n steps, with the final positions matching. Let

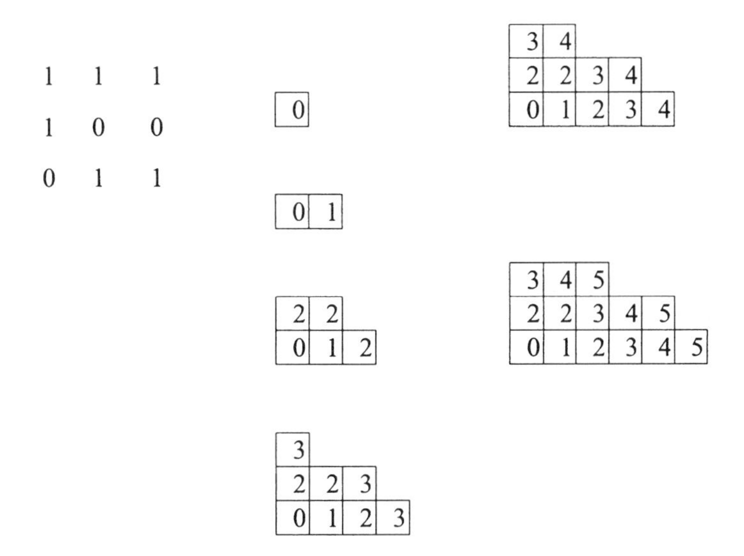

Figure 10.16 Oriented digital boiling evolution corresponding to the same 0,1 matrix as in Figure 10.15.

$G_{m,n}^{\mathrm{u/rh,u/rd}}(\vec{l}^{(0)};\vec{l}^{(0)})$, $\vec{l}^{(0)} := (l-1, l-2, \dots, 0)$, denote the total weight of this joining of paths, starting and finishing at $\vec{l}^{(0)}$, with rh steps weighted by $\{a_j\}$ and rd steps by $\{b_j\}$. We then have

$$\Pr(h^{0,1} \le l) = \prod_{i=1}^{m}\prod_{j=1}^{n}(1+a_i b_j)^{-1} G_{m,n}^{\mathrm{u/rh,u/rd}}(\vec{l}^{(0)};\vec{l}^{(0)}).$$

On the other hand, the generating function for a single path of this type is

$$g_{m,n}^{\mathrm{u/rh,u/rd}}(l^{(0)};l) = \frac{1}{2\pi}\int_{-\pi}^{\pi}\prod_{j=1}^{m}(1+a_j e^{i\theta})\prod_{k=1}^{n}(1-b_k e^{-i\theta})^{-1} e^{-i\theta(l-l^{(0)})}\,d\theta.$$

Making use of (10.1) and (5.77) allows $G_{m,n}^{\mathrm{u/rh,u/rd}}(\vec{l}^{(0)};\vec{l}^{(0)})$ to be written as an average over $U(l)$, giving [34]

$$\Pr(h^{0,1} \le l) = \prod_{i=1}^{m}\prod_{j=1}^{n}(1+a_i b_j)^{-1}\Big\langle\Big(\prod_{j=1}^{m}\prod_{k=1}^{l}(1+a_j e^{i\theta_k})\Big)\Big(\prod_{j=1}^{n}\prod_{k=1}^{l}(1-b_j e^{-i\theta_k})\Big)^{-1}\Big\rangle_{U(l)}.$$

The top path in the above dual RSK growth process allows for an alternative description, known as *oriented digital boiling* [272]. To begin, a unit square is placed with its bottom left corner at $x=0$, $y=-1$. At time $t=1$ a unit square is added flush right, and a unit square is also added above the existing cube with probability p. Generally, in going from time t to time $t+1$, all plateaux grow one unit to the right. Furthermore, above all existing squares still part of the outer profile, a square is added to each independently with probability p. A 0,1 matrix can be associated with the addition of a square, with a 1 (0) placed in column j and row i (rows counted from the bottom) if a square is added (not added) to column j of the growth process at time $i-j+1$. If the square is added deterministically, either a 0 or a 1 may be recorded. Conversely, given a 0,1 matrix, a sequence of oriented digital boiling configuration can be constructed (see Figure 10.16 for an example).

EXERCISES 10.4 (i) Argue that

$$h_1^{0,1}(n_1,n_2) = \max\Big(h_1^{0,1}(n_1,n_2-1), h_1^{0,1}(n_1-1,n_2)+x_{n_1,n_2}\Big).$$

(ii) Show that $L_{\mathbf{X}}^{0,1}(n_1,n_2) := \sum x_{i,j}$, where the sum is over all u/rd paths from the bottom row (row 1) to the top row (row n_1), satisfies the same recurrence and thus conclude that $h_1^{0,1}(n_1,n_2) = L_{\mathbf{X}}^{0,1}(n_1,n_2)$.

10.5 SYMMETRIZED GROWTH MODELS

10.5.1 Matrices symmetric about the antidiagonal

In general the antidiagonal of an $n \times n$ matrix $\mathbf{X}$ refers to the elements $x_{i,j}$ with $i = n+1-j$. Reflecting the entries about the antidiagonal gives the matrix $\mathbf{X}^R := [x_{n+1-j,n+1-i}]_{i,j=1,\dots,n}$. Represented geometrically, because we have adopted the convention of numbering rows from the bottom, this reflection is about the top left to bottom right diagonal.

Our interest here is in the Robinson-Schensted-Knuth correspondence applied to non-negative integer matrices with the symmetry $\mathbf{X} = \mathbf{X}^R$. For this use will be made of two theoretical results, which will be stated without proof. First we require a relation between the paths resulting from the RSK mapping applied to $\mathbf{X}^R$ to those resulting from applying the RSK mapping to $\mathbf{X}$ in the generic case that $\mathbf{X} \neq \mathbf{X}^R$ (see, e.g., [475]).

PROPOSITION 10.5.1 *Let (Q^R, P^R) and (P, Q) denote the pairs of tableaux which result from applying the RSK mapping to $\mathbf{X}^R$ and $\mathbf{X}$ respectively. We have*

$$\#j\,s \text{ in } P = \#n+1-j\,s \text{ in } P^R \tag{10.127}$$

and similarly for Q and Q^R. The tableau P^R is referred to as the Schützenberger dual of the tableau P and can be constructed by the evacuation algorithm (see Exercises 10.5 q.2).

Note that (10.127) is a simple consequence of (10.103).

It follows from Proposition 10.5.1 that if $\mathbf{X} = \mathbf{X}^R$ and thus $\mathbf{X}$ is symmetric about the antidiagonal, the RSK correspondence gives a bijection with a pair of tableaux (P, P^R). Because P^R can be constructed from P, we have that in fact there is a bijection between weighted versions of each. For this we choose the weight of P^R to equal the weight of P by weighting each occurrence of j in P as well as each occurrence of $n+1-j$ in P^R by a_j. This weighting of the tableaux is equivalent in the path picture to weighting the up steps at both $x = 2n + \frac{1}{2} - 2i$ and $x = -2i + \frac{3}{2}$ by a_i. With the maximum displacement of the level-l path μ_l, the total weight of such paths is equal to $s_\mu(a_1^2,\dots,a_n^2)$. Also, from Figure 10.12 such weighting of the paths results by choosing the weights $x_{i,j}$ of the matrix $\mathbf{X}$ to be proportional to $(a_i a_{n+1-j})^{x_{i,j}}$. But to obtain a probabilistic setting we should weight only the sites $i \le n+1-j$, with the value of $x_{i,j}$ for $i < n+1-j$ fixed by symmetry. This can be achieved without affecting the weights of the pairs of paths by simply squaring the weights at sites $i < n+1-j$ and readjusting the normalizations. With $a_i = \sqrt{q_i}$, we therefore choose

$$\begin{aligned}\Pr(x_{i,j} = k) &= (1 - q_i q_{n+1-j})(q_i q_{n+1-j})^k, \quad i < n+1-j,\\ \Pr(x_{i,n+1-i} = k) &= (1 - q_i) q_i^k.\end{aligned} \tag{10.128}$$

From the above discussion we then have the following result.

PROPOSITION 10.5.2 *Consider a non-negative integer matrix $\mathbf{X}$, symmetric about the antidiagonal with independent entries chosen according to (10.128). The probability that under the RSK correspondence $\mathbf{X}$ maps to a pair of paths with up steps weighted proportional to $\sqrt{q_i}$ for $x = 2n + \frac{1}{2} - 2i$ and $x = -2i + \frac{3}{2}$, and with final displacements μ, is equal to*

$$\prod_{i=1}^{n}(1 - q_i) \prod_{1 \le i < j \le n} (1 - q_i q_j) s_\mu(q_1,\dots,q_n). \tag{10.129}$$

The second property of the RSK correspondence when $[x_{i,j}]$ is symmetric about the antidiagonal to be used but not proved is [246]

$$\#\{x_{i,n+1-i} : x_{i,n+1-i} \text{ odd}\} = \#\{\mu_j : \mu_j \text{ odd}\} = \sum_{j=1}^{n} (-1)^{j-1} \mu_j'. \tag{10.130}$$

Hence if we generalize the second probability in (10.128) to read

$$\Pr(x_{i,n+1-i} = k) = \frac{(1-q_i^2)}{1+\beta q_i} \beta^{k \bmod 2} q_i^k, \tag{10.131}$$

then we have the corresponding generalization of (10.129) [34],

$$\prod_{i=1}^{n} \frac{(1-q_i^2)}{1+\beta q_i} \prod_{1 \le i < j \le n} (1 - q_i q_j) \beta^{\sum_{j=1}^{n} (-1)^{j-1} \mu_j'} s_\mu(q_1, \dots, q_n). \tag{10.132}$$

Writing $h^{\boxslash}$ for the maximum height in the growth model picture of this generalized model, it follows from (10.132) that we have

$$\Pr(h^{\boxslash} \le l) = \prod_{i=1}^{n} \frac{1-q_i^2}{1+\beta q_i} \prod_{1 \le i < j \le n} (1 - q_i q_j) \sum_{\mu : \mu_1 \le l} \beta^{\sum_{j=1}^{n} (-1)^{j-1} \mu_j'} s_\mu(q_1, \dots, q_n). \tag{10.133}$$

It is possible to write (10.133) as a random matrix average analogous to (10.73) [34].

PROPOSITION 10.5.3 *Let $\langle \cdot \rangle_{\mathrm{Sp}(2l)}$ denote an average with respect to the eigenvalue p.d.f. (2.69) for random matrices from the classical group Sp$(2l)$. We have*

$$\Pr(h^{\boxslash} \le 2l) = \prod_{i=1}^{n} \frac{1-q_i^2}{1+\beta q_i} \prod_{1 \le i < j \le n} (1 - q_i q_j) \Big\langle \prod_{k=1}^{l} \Big(\frac{1}{|1 - \beta e^{-i\theta_k}|^2} \prod_{j=1}^{n} |1 + q_j e^{i\theta_k}|^2 \Big) \Big\rangle_{\mathrm{Sp}(2l)}, \tag{10.134}$$

$$\Pr(h^{\boxslash} \le 2l+1) = \prod_{i=1}^{n} (1 - q_i^2) \prod_{1 \le i < j \le n} (1 - q_i q_j) \Big\langle \prod_{k=1}^{l} \prod_{j=1}^{n} |1 + q_j e^{i\theta_k}|^2 \Big\rangle_{\mathrm{Sp}(2l)}. \tag{10.135}$$

Proof. We follow [224]. According to (10.56) we can write

$$\prod_{k=1}^{l} \prod_{j=1}^{n} |1 + q_j e^{i\theta_k}|^2 = \sum_{\mu : \mu_1 \le 2l} s_\mu(q_1, \dots, q_n) s_{\mu'}(e^{i\theta_1}, e^{-i\theta_1}, \dots, e^{i\theta_l}, e^{-i\theta_l}). \tag{10.136}$$

Comparing (10.134) with this substitution to (10.133) shows the former is equivalent to the integration formula

$$\Big\langle \prod_{k=1}^{l} \frac{1}{|1 - \beta e^{-i\theta_k}|^2} s_\rho(e^{i\theta_1}, e^{-i\theta_1}, \dots, e^{i\theta_l}, e^{-i\theta_l}) \Big\rangle_{\mathrm{Sp}(2l)} = \beta^{\sum_{j=1}^{2l} (-1)^{j-1} \rho_j}. \tag{10.137}$$

Recalling now the eigenvalue p.d.f. for Sp$(2l)$ (2.69), noting that the integrand is unchanged by $\theta_l \mapsto -\theta_l$ and making use of (10.13) shows the l.h.s. of (10.137) is equal to

$$\frac{1}{(2\pi)^l 2^l l!} \int_{-\pi}^{\pi} d\theta_1 \cdots \int_{-\pi}^{\pi} d\theta_l \prod_{k=1}^{l} \frac{(e^{i\theta_k} - e^{-i\theta_k})}{|1 - \beta e^{-i\theta_k}|^2} \det \begin{bmatrix} e^{i\theta_j (\rho_{2l-k+1} + k - 1)} \\ e^{-i\theta_j (\rho_{2l-k+1} + k - 1)} \end{bmatrix}_{\substack{j=1,\dots,l \\ k=1,\dots,2l}}. \tag{10.138}$$

Proceeding as in the proof of Proposition 6.1.8 reduces this to

$$\frac{1}{2^l} \mathrm{Pf} \Big[\frac{1}{2\pi} \int_{-\pi}^{\pi} \frac{e^{i\theta} - e^{-i\theta}}{|1 - \beta e^{-i\theta}|^2} (e^{i\theta(\rho_j - j - \rho_k + k)} - e^{-i\theta(\rho_j - j - \rho_k + k)}) d\theta \Big]_{j,k=1,\dots,2l}, \tag{10.139}$$

where as the final step the order of all rows and columns of the Pfaffian have been reversed. Suppose in the integral of (10.139) that $k > j$ so that $\rho_j - \rho_k + k - j > 0$. The integrand is readily computed, showing that (10.139) is equal to

$$\beta^{-l}\mathrm{Pf}\Big[\mathrm{sgn}(k-j)\beta^{|\rho_j-\rho_k+k-j|}\Big]_{j,k=1,\dots,2l}. \tag{10.140}$$

The above Pfaffian can be evaluated by setting $x_j = \rho_j - j$, $f(x_j) = \beta^{x_j}$ in the general formula

$$\mathrm{Pf}\Big[\Big(\frac{f(x_j)}{f(x_k)}\Big)^{\mathrm{sgn}(x_j-x_k)}\mathrm{sgn}(x_j-x_k)\Big]_{j,k=1,\dots,2l} = \prod_{j=1}^{l}\frac{f(x_{Q(2j-1)})}{f(x_{Q(2j)})}\varepsilon(Q), \tag{10.141}$$

where the permutation Q is such that

$$x_{Q(2j-1)} > x_{Q(2j)}, \qquad Q(2j) > Q(2j-1) \ (j = 1,\dots,l)$$

to give the r.h.s. of (10.137). To see the validity of (10.141), note that of the $(2l-1)!!$ permutations in the sum contributing to the Pfaffian, only one term contributes to the r.h.s. with the rest cancelling in pairs.

It remains to verify (10.135). Use of an appropriate modification of (10.136) shows that this is equivalent to the integration formula

$$\Big\langle s_\rho(e^{i\theta_1}, e^{-i\theta_1}, \dots, e^{i\theta_l}, e^{-i\theta_l}, \beta)\Big\rangle_{\mathrm{Sp}(2l)} = \beta^{\sum_{j=1}^{2l+1}(-1)^{j-1}\rho_j}. \tag{10.142}$$

Proceeding as in the derivation of (10.138) and (10.139) shows that the l.h.s. is equal to

$$\frac{1}{(2\pi)^l 2^l l!}\int_{-\pi}^{\pi} d\theta_1 \cdots \int_{-\pi}^{\pi} d\theta_l \prod_{k=1}^{l}\frac{(e^{i\theta_k}-e^{-i\theta_k})}{|1-\beta e^{-i\theta_k}|^2}\det\begin{bmatrix} e^{i\theta_j(\rho_{2l-k+2}+k-1)} \\ e^{-i\theta_j(\rho_{2l-k+2}+k-1)} \\ \beta^{\rho_{2l-k+2}+k-1}\end{bmatrix}_{\substack{j=1,\dots,l \\ k=1,\dots,2l+1}}$$

$$= \frac{1}{2^l}\mathrm{Pf}\begin{bmatrix} \mathbf{A}_{(2l+1)\times(2l+1)} & [\beta^{\rho_j+2l+1-j}]_{j=1,\dots,2l+1} \\ [\beta^{\rho_k+2l+1-k}]_{k=1,\dots,2l+1} & 0 \end{bmatrix},$$

where

$$\mathbf{A}_{(2l+1)\times(2l+1)} := \Big[\frac{1}{2\pi}\int_{-\pi}^{\pi}\frac{e^{i\theta}-e^{-i\theta}}{|1-\beta e^{-i\theta}|^2}(e^{i\theta(\rho_j-j-\rho_k+k)} - e^{-i\theta(\rho_j-j-\rho_k+k)})d\theta\Big]_{j,k=1,\dots,2l+1}.$$

The integral can be computed to give for the l.h.s.

$$\beta^{-(l+1)}\mathrm{Pf}\begin{bmatrix} [\mathrm{sgn}(k-j)\beta^{|\rho_j-\rho_k+k-j|}]_{j,k=1,\dots,2l+1} & [\beta^{\rho_j+2l+1-j}]_{j=1,\dots,2l} \\ [-\beta^{\rho_k+2l+1-k}]_{k=1,\dots,2l} & 0 \end{bmatrix}.$$

This is precisely the same as (10.140) with $l \mapsto l+1$, $\rho_{2l+2} = 0$, and so reduces to the r.h.s. of (10.142). □

We see from the exact results of Proposition 10.5.3 that

$$\Pr(h^{\boxslash} \le 2l+1) = \Pr(h^{\boxslash} \le 2l)|_{\beta=0}. \tag{10.143}$$

To understand this result, note that for matrices symmetric about the antidiagonal, the path $(0,0)\mathrm{u/rh}(n,n)$ which maximizes $\sum_{(0,0)\mathrm{u/rh}(n,n)} x_{i,j}$ can likewise be chosen to be symmetric about the antidiagonal. This means that for $h^{\boxslash}$ to be odd, the value of $x_{i,n+1-i}$ must be odd, as all values of $x_{i,j}$ off the antidiagonal contributing to $h^{\boxslash}$ occur in pairs. Moreover (10.131) gives

$$\Pr(x_{i,n+1-i} = 2l'+1) + \Pr(x_{i,n+1-i} = 2l') = \Pr(x_{i,n+1-i} = 2l')\Big|_{\beta=0},$$

whereas $\Pr(x_{i,n+1-i} = 2l'-1)|_{\beta=0} = 0$, so pairing the odd values on the diagonal with the even values in this way we reclaim the setting of $\Pr(h^{\boxslash} \le 2l)|_{\beta=0}$ as required by (10.143).

Matrices zero above the antidiagonal

Consider a general $n \times n$ non-negative integer matrix $\mathbf{X} = [x_{i,j}]_{i,j=1,\dots,n}$ and suppose $x_{i,j} = 0$ for $i > n+1-j$. For $i \le n+1-j$ let $x_{i,j}$ occur with the geometric probability (10.70). Applying the RSK correspondence in the nonintersecting paths formulation but stopping at $t=n$ rather than $t=2n-1$ gives a bijection between $\mathbf{X}$ and weighted w,u/rh(o)/lh(e) nonintersecting lattice paths as specified in Section 10.1.2, with the spacing between levels one unit (see Figure 10.17). As usual $L^{(1)}_{\mathbf{X}}(n,n)$ is equal to the maximum of the heights in the level-1 path, $\tilde{h}$, say. On the other hand $\tilde{h}$ is equal to the number of nonintersecting w,u/rd(o)/ld(e) lattice paths dual to the w,u/rh(o)/lh(e) lattice paths, where in both cases the paths start and finish on the same level, with the interspacing between levels one unit. Hence

$$\Pr(\tilde{h} \le l) = \prod_{1 \le i \le j \le n} (1 - a_i b_{n+1-j}) G_{2n}^{\mathrm{w,u/rd(o)/ld(e)}}(\{r_j^{(0)} = l - j + 1\}_{j=1,\dots,l}; \{r_j = l - j + 1\}_{j=1,\dots,l}).$$

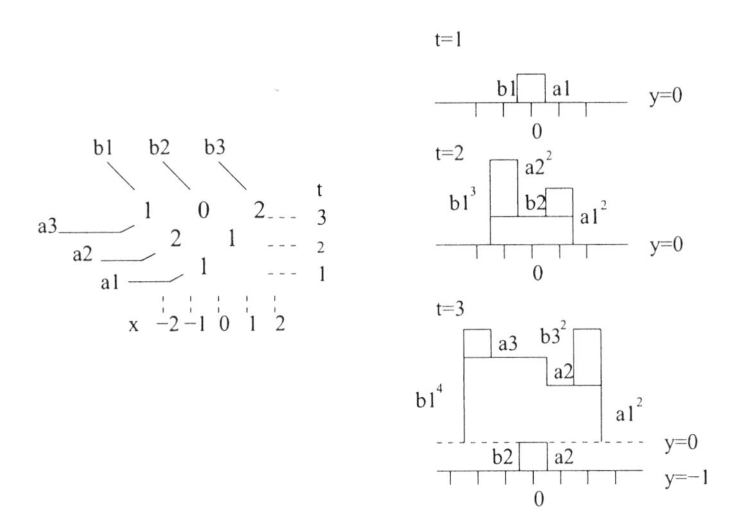

Figure 10.17 Mapping from a weighted non-negative integer matrix with entries zero above the antidiagonal (this portion of the matrix is not shown) and weighted nonintersecting lattice paths. The latter, upon rotation by 90° clockwise, can be regarded as w,u/rh(o)/lh(e) type paths.

Let us now set $b_i = a_{n+1-i}$, $a_i = q_i$ $(i = 1,\dots,n)$ so that for sites (i,j) with $i < n+1-j$ the probabilities are the same as (10.128) determining $h^{\boxslash}$. Making use of (10.51) shows

$$\Pr(\tilde{h} \le l)\Big|_{\substack{b_i = a_{n+1-i} \\ a_i = q_i}} = \prod_{1 \le i \le j \le n} (1 - q_i q_j) \Big\langle \prod_{k=1}^{l} \prod_{j=1}^{n} |1 + q_j e^{i\theta_k}|^2 \Big\rangle_{\mathrm{Sp}(2l)}.$$

Comparing with (10.134) we thus have

$$\Pr(\tilde{h} \le l) = \Pr(h^{\boxslash} \le 2l)\Big|_{\beta=0}. \tag{10.144}$$

To anticipate this, we note from (10.131) that

$$\Pr(x_{i,n+1-i} = k)\Big|_{\beta=0} = \begin{cases} (1-q_i^2)q_i^k, & k \text{ even}, \\ 0, & k \text{ odd}. \end{cases} \tag{10.145}$$

Also, as remarked in the discussion of (10.143), in the case $[x_{i,j}]$ is symmetric about the antidiagonal

$$\max \sum_{(1,1)\mathrm{u/rh}(n,n)} x_{i,j} = \max \sum_{(1,1)\mathrm{u/rh}(l,n+1-l)} \chi_{i,j} x_{i,j}, \tag{10.146}$$

where $(l, n+1-l)$ is any point on the antidiagonal while $\chi_{i,j} = 1$ for points on the antidiagonal and $\chi_{i,j} = 2$ otherwise. Because according to (10.145), in the case $\beta = 0$ the points on the antidiagonal are restricted to even values, we can redefine $\chi_{i,j}$ in (10.146) to equal 2 for all points without changing $\Pr(h^{\boxslash} \le 2l)\Big|_{\beta=0}$, provided we redefine the probability (10.145) so that

$$\Pr(x_{i,n+1-i} = k)\Big|_{\beta=0} = (1-q_i^2)q_i^{2k}, \qquad k = 0, 1, \dots$$

With this redefinition we see that the probabilities making up $\Pr(\tilde{h} \le l)$ in (10.144) are precisely the same as those specifying $\Pr(h^{\boxslash} \le 2l)|_{\beta=0}$, and furthermore $h^{\boxslash} = 2\tilde{h}$, which together imply (10.144).

10.5.2 Matrices symmetric about the diagonal

The rules of the growth process tell us that if the non-negative integer matrix $\mathbf{X} = [x_{i,j}]_{i,j=1,\dots,n}$ maps to the pair of tableaux (P, Q), then the transposed matrix $\mathbf{X}^T := [x_{j,i}]_{i,j=1,\dots,n}$ maps to (Q, P). Hence the Robinson-Schensted-Knuth correspondence when applied to non-negative integer matrices symmetric about the diagonal gives a bijection with a single semi-standard tableau, or equivalently a single set of u/rh lattice paths, since then $P = Q$. To obtain a bijection between weighted symmetric matrices and a weighted set of u/rh lattice paths the simplest situation is to require that the sets of paths have the same weighting, and then to weight a single set by the square of the original weighting. We see from Figure 10.12 that this will happen if we restrict the weights so that $a_i = b_i$ $(i = 1, \dots, n)$, and thus weight each entry $x_{i,j}$ by $(1-a_ia_j)(a_ia_j)^{x_{i,j}}$. But as in (10.70) and (10.128) we would like to relate these weights to probabilities, for which purpose we should weight only the sites $i \le j$, with the value of $x_{i,j}$ for $i > j$ fixed by symmetry. This can be done by squaring the weights at sites $i < j$ and readjusting the normalizations. The following analogue of Proposition 10.5.2 is then obtained [329].

Proposition 10.5.4 *Consider a non-negative integer matrix* $\mathbf{X}$*, symmetric about the diagonal with independent entries chosen according to*

$$\Pr(x_{i,j} = k) = (1-q_iq_j)(q_iq_j)^k, \ i < j, \qquad \Pr(x_{i,i} = k) = (1-q_i)q_i^k. \tag{10.147}$$

The probability that under the RSK correspondence $\mathbf{X}$ *maps to a pair of paths with up steps weighted proportional to* $\sqrt{q_i}$ *for* $x = \pm(2n + \frac{1}{2} - 2i)$*, and with final displacements* μ*, is equal to*

$$\prod_{i=1}^{n}(1-q_i) \prod_{1 \le i < j \le n} (1-q_iq_j) s_\mu(q_1, \dots, q_n). \tag{10.148}$$

The probability on the diagonal entries in (10.147) can be generalized [34]. This is possible because in the RSK mapping for a symmetric matrix

$$\sum_{j=1}^{n} x_{j,j} = \sum_{j=1}^{n} (-1)^{j-1}\mu_j \tag{10.149}$$

(see Exercises 10.5 q.1). Thus if the probability in question is generalized to read

$$\Pr(x_{i,i} = k) = (1-\alpha q_i)(\alpha q_i)^k, \tag{10.150}$$

(10.148) should correspondingly be generalized to

$$\prod_{i=1}^{n}(1-\alpha q_i)\prod_{1\le i<j\le n}(1-q_iq_j)\alpha^{\sum_{j=1}^{n}(-1)^{j-1}\mu_j}s_\mu(q_1,\dots,q_n). \tag{10.151}$$

With $\mathbf{X}$ symmetric about the diagonal, let us denote the maximum height of the level-1 path by $h^{\boxslash}$. Noting that

$$\sum_{j=1}^{n}(-1)^{j-1}\mu_j = \#(\text{columns of odd length in }\mu)=\sum_{k=1}^{n}\mu'_k \bmod 2, \tag{10.152}$$

where $l=\mu_1$, we see from (10.151) that

$$\Pr(h^{\boxslash}\le l)=\prod_{i=1}^{n}(1-\alpha q_i)\prod_{1\le i<j\le n}(1-q_iq_j)\sum_{\mu:\mu_1\le l}\alpha^{\sum_{k=1}^{n}\mu'_k \bmod 2}s_\mu(q_1,\dots,q_n). \tag{10.153}$$

As with (10.133) this can be written in terms of an average over a classical group.

PROPOSITION 10.5.5 *We have*

$$\Pr(h^{\boxslash}\le l)=\prod_{i=1}^{n}(1-\alpha q_i)\prod_{1\le i<j\le n}(1-q_iq_j)\Big\langle \det(\mathbf{1}_l+\alpha\mathbf{U})\prod_{j=1}^{n}\det(\mathbf{1}_l+q_j\mathbf{U})\Big\rangle_{\mathbf{U}\in O(l)}. \tag{10.154}$$

Proof. As with the proof of Proposition 10.5.3, we follow [224]. Use of (10.56) in (10.154) and comparison with (10.153) shows that (10.154) is equivalent to the matrix integral evaluation

$$\Big\langle \det(\mathbf{1}_l+\alpha\mathbf{U})s_\rho(\mathbf{U})\Big\rangle_{\mathbf{U}\in O(l)}=\alpha^{\sum_{j=1}^{l}\rho_j \bmod 2}, \tag{10.155}$$

where $s_\rho(\mathbf{U})$ denotes the Schur polynomial as a function of all the eigenvalues of $\mathbf{U}$.

For definiteness, consider the l even case, $l\mapsto 2l$, and consider separately the components $O^{\pm}(2l)$ of $O(2l)$. Recalling the eigenvalue p.d.f. for $O^{+}(2l)$ (2.62), and proceeding as in the derivation of (10.140) shows

$$\langle \det(\mathbf{1}_l+\alpha\mathbf{U})s_\rho(\mathbf{U})\rangle_{\mathbf{U}\in O^{+}(2l)}=2^{1-l}\mathrm{Pf}[a_{jk}]_{j,k=1,\dots,2l} \tag{10.156}$$

where

$$\begin{aligned}a_{jk}&=\Big((1+\alpha^2)\delta_{(\rho_j-j)-(\rho_k-k),\mathrm{odd}}+2\alpha\delta_{(\rho_j-j)-(\rho_k-k),\mathrm{even}}\Big)\mathrm{sgn}(k-j)\\&=\Big(\frac{1}{2}(1+\alpha)^2-\frac{1}{2}(1-\alpha)^2(-1)^{(\rho_j-j)-(\rho_k-k)}\Big)\mathrm{sgn}(k-j).\end{aligned}$$

The task is therefore to compute the Pfaffian of the matrix with these entries. For this one uses the identity [502]

$$\mathrm{Pf}(\mathbf{A}+\mathbf{B})=\sum_{\substack{S\subseteq\{1,2,\dots,2l\}\\|S|\ \mathrm{even}}}(-1)^{\sum_{j\in S}j-|S|/2}\mathrm{Pf}_S(\mathbf{A})\mathrm{Pf}_{\bar S}(\mathbf{B}), \tag{10.157}$$

where $\mathrm{Pf}_S(\mathbf{A})$ denotes the Pfaffian of $\mathbf{A}$ restricted to rows and columns specified by the index set S, and similarly $\mathrm{Pf}_{\bar S}(\mathbf{B})$ ($\bar S$ denotes the complement of S). With

$$\mathbf{A}=\Big[\frac{1}{2}(1+\alpha)^2\mathrm{sgn}(k-j)\Big]_{j,k=1,\dots,2l},\qquad \mathbf{B}=\Big[-\frac{1}{2}(1-\alpha)^2(-1)^{(\rho_j-j)-(\rho_k-k)}\mathrm{sgn}(k-j)\Big]_{j,k=1,\dots,2l},$$

from the simple identities

$$\mathrm{Pf}[\mathrm{sgn}(k-j)]=1,\qquad \mathrm{Pf}[a_{j,k}(-1)^{(\rho_j-j)-(\rho_k-k)}]=(-1)^{\sum(\rho_j-j)}\mathrm{Pf}[a_{j,k}],$$

we see that

$$\mathrm{Pf}_S\mathbf{A} = 2^{-|S|/2}(1+\alpha)^{|S|}, \qquad \mathrm{Pf}_{\bar{S}}\mathbf{B} = (-2)^{-|\bar{S}|/2}(1-\alpha)^{2l-|S|}(-1)^{\sum_{j\in\bar{S}}\rho_j - j}.$$

Application of (10.157) gives

$$2^{1-l}\mathrm{Pf}(\mathbf{A}+\mathbf{B}) = 2 \sum_{\substack{S\subseteq\{1,2,\dots,2l\}\\ |S|\ \mathrm{even}}} \Big(\frac{1+\alpha}{2}\Big)^{|S|}\Big(\frac{1-\alpha}{2}\Big)^{2l-|S|}(-1)^{\sum_{j\in\bar{S}}\rho_j}. \tag{10.158}$$

But in general

$$\sum_{\substack{S\subseteq\{1,2,\dots,2l\}\\ |S|\ \mathrm{even}}} x^{|S|}y^{2l-|S|}(-1)^{\sum_{j\in\bar{S}}\rho_j}$$
$$= \frac{1}{2}\Big(\sum_{S\subseteq\{1,2,\dots,2l\}} x^{|S|}y^{2l-|S|}(-1)^{\sum_{j\in\bar{S}}\rho_j} + \sum_{S\subseteq\{1,2,\dots,2l\}} x^{|S|}(-y)^{2l-|S|}(-1)^{\sum_{j\in\bar{S}}\rho_j}\Big)$$
$$= \frac{1}{2}\Big(\prod_{j=1}^{2l}(x+(-1)^{\rho_j}y) + \prod_{j=1}^{2l}(x-(-1)^{\rho_j}y)\Big).$$

Using this result to evaluate (10.158) and substituting in (10.156) gives the matrix integral evaluation

$$\langle \det(\mathbf{1}_l+\alpha\mathbf{U})s_\rho(\mathbf{U})\rangle_{\mathbf{U}\in O^+(2l)} = \alpha^{\sum_{j=1}^{2l}\rho_j \bmod 2} + \alpha^{\sum_{j=1}^{2l}(\rho_j+1)\bmod 2}. \tag{10.159}$$

It remains to compute the corresponding formula for the average over $O^-(2l)$. The analogue of (10.156) in this case is

$$\langle \det(\mathbf{1}_l+\alpha\mathbf{U})s_\rho(\mathbf{U})\rangle_{U\in O^-(2l)} = \frac{(1-\alpha^2)}{2^{l-1}}[\zeta]\mathrm{Pf}[a_{jk}+\zeta b_{jk}]_{j,k=1,\dots,2l} \tag{10.160}$$

(cf. (6.159)), where a_{jk} is as in (10.156) while $b_{jk} = (-1)^{\rho_k-k} - (-1)^{\rho_j-j}$. Observing

$$[b_{jk}] = \vec{u}\vec{w}^T - \vec{w}\vec{u}^T, \qquad \vec{u} = [1]_{j=1,\dots,2l}, \ \vec{w} = [(-1)^{\rho_j-j}]_{j=1,\dots,2l} \tag{10.161}$$

shows that $[b_{jk}]$ has rank 2. It follows that the Pfaffian in (10.160) is linear in ζ, and so the r.h.s. of (10.160) can be rewritten

$$\frac{(1-\alpha^2)}{2^{l-1}}\frac{1}{\zeta}\Big(\mathrm{Pf}\big[[a_{jk}]+\zeta[b_{jk}]\big] - \mathrm{Pf}[a_{jk}]\Big). \tag{10.162}$$

With γ, ζ_1, ζ_2 arbitrary nonzero constants, the structure (10.161), and applying elementary row and column operations, verifies that this in turn can be rewritten

$$\frac{(1-\alpha^2)}{2^{l-1}}\frac{1}{\zeta_1\zeta_2}\left(\mathrm{Pf}\begin{bmatrix}[a_{jk}] & \zeta_1\vec{w} & \zeta_2\vec{u}\\ -\zeta_1\vec{w}^T & 0 & \gamma\\ -\zeta_2\vec{u}^T & -\gamma & 0\end{bmatrix} - \gamma\mathrm{Pf}[a_{jk}]\right). \tag{10.163}$$

Setting $\zeta_1 = \frac{1}{2}(1-\alpha)^2$, $\zeta_2 = (1+\alpha)^2$, adding one half of the final row/column to the second-last row/column, subtracting the second-last row/column from the final row column, and setting $\gamma = (1+\alpha^2)$ allows (10.163) to be recognised as

$$\frac{2^{1-l}}{1-\alpha^2}\Big(\mathrm{Pf}[a_{jk}]_{2(l+1)\times 2(l+1)}\Big|_{\rho_{2l+1}=\rho_{2l+2}=0} - (1+\alpha^2)\mathrm{Pf}[a_{jk}]_{2l\times 2l}\Big). \tag{10.164}$$

Comparing (10.156) and (10.159) tells us that

$$\mathrm{Pf}[a_{jk}]_{2l\times 2l} = 2^{l-1}\Big(\alpha^{\sum_{j=1}^{2l}\rho_j \bmod 2} + \alpha^{\sum_{j=1}^{2l}(\rho_j+1)\bmod 2}\Big).$$

Substituting in (10.164) and simplifying implies the matrix integral evaluation

$$\langle \det(\mathbf{1}_l+\alpha\mathbf{U})s_\rho(\mathbf{U})\rangle_{\mathbf{U}\in O^-(2l)} = \alpha^{\sum_{j=1}^{2l}\rho_j \bmod 2} - \alpha^{\sum_{j=1}^{2l}(\rho_j+1)\bmod 2}. \tag{10.165}$$

As the matrix integrals over both components $O^+(2l)$ and $O^-(2l)$ have now been evaluated, it only remains to take the arithmetic mean of these to obtain (10.155). □

We remark that the Pieri formula (12.190) below enables the general α case of (10.155) to be deduced from the result for $\alpha = 0$.

EXERCISES 10.5 1. [360], [221] The objective of this exercise is to derive (10.149).

(i) Note that for the nonintersecting paths interpretation of the RSK correspondence, $[x_{i,j}]_{i,j=1,\dots,n}$ symmetric about $i = j$ implies $\lambda_l(j,i) = \lambda_l(i,j)$, and use this in (10.105) and (10.107) to conclude

$$\begin{aligned}\lambda_1(i,i) &= \lambda_1(i,i-1) + x_{i,i},\\ \lambda_l(i,i) &= \lambda_l(i,i-1) + \lambda_{l-1}(i,i-1) - \lambda_{l-1}(i-1,i-1), \quad l > 1.\end{aligned}$$

(ii) Form appropriate linear combinations of the equations in (i), making use of the fact that $\lambda_l(l,l-1) = 0$, to deduce the equation

$$x_{i,i} = \sum_{l=1}^{i}(-1)^{l-1}\lambda_l(i,i) - \sum_{l=1}^{i-1}(-1)^{l-1}\lambda_l(i-1,i-1).$$

Sum this over i from 1 to n and recall that $\lambda_l(n,n) = \mu_l$ to deduce (10.149).

2. Schützenberger's evacuation operation on a semi-standard tableau P can be described as follows (see, e.g., [246]). Remove the number n_0, say, of the first square in the first row of P, thus creating a blank square. Interchange the place of the blank square and its neighbor immediately to the right, or immediately below, whichever is smaller (if both neighbors have the same number, choose the one below). Repeat this procedure until the blank square is at the end of a row or the end of a column. Then the square in this position of P^R is numbered $n + 1 - n_0$. The blank square is removed from P and the procedure is repeated to complete the filling of P^R.

(i) Consider the pair of tableaux (P, Q) equivalent to the pair of lattice paths in Figure 10.12. Note that since the matrix $\mathbf{X}$ therein is such that $\mathbf{X} = \mathbf{X}^R$, we must have $P^R = Q$.

(ii) Show that the result of (i) is consistent with the construction of P^R from the evacuation operation.

10.6 THE HAMMERSLEY PROCESS

Closely related to the last passage percolation model associated with the non-negative matrix $\mathbf{X}$ is a last passage percolation model associated with random points in the unit square, known as the *Hammersley process.* The latter is defined by first marking in the unit square points uniformly at random according to a Poisson rate with intensity z^2, so that the probability the square contains N points is equal to $z^{2N}e^{-z^2}/N!$. From the points one forms a continuous path by joining points with straight line segments of positive slope, and this path is extended to begin at $(0,0)$ and finish at $(1,1)$ by adding an extra segment at both ends. With the length of the extended path defined as the number of points it contains, the stochastic variable $l^\square$ is defined as the maximum of the lengths of all possible extended paths (see Figure 10.18). To relate the Hammersley process so defined to the last passage percolation model of Section 10.4.2, consider an $n \times n$ matrix $\mathbf{X}$ with entries chosen according to the geometric distribution (10.70). Set $a_i = b_j = z/n$. Then to leading order in $1/n^2$ the probability that $x_{ij} = 1$ is equal to z^2/n^2 while to the same order the probability that $x_{ij} > 1$ is zero. If we now think of the lattice sites (i,j) scaled to the points $(i/n, j/n)$ then in the limit $n \to \infty$ we see that a Poisson process with intensity z^2 in the unit square is generated for the distribution of matrix elements with value unity. This latter point follows from the general fact that the Poisson process in question can be realized as the $M \to \infty$ limit of the discrete process of dividing the unit square up into a regular $M \times M$ grid and marking a point randomly within each subsquare with probability z^2/M^2.

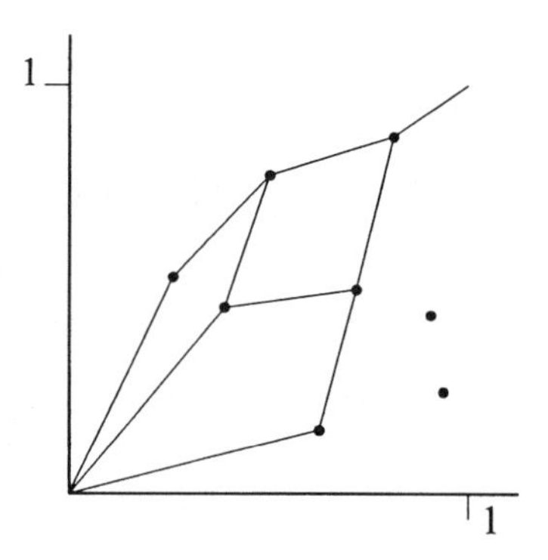

Figure 10.18 Eight points in the unit square, and the extended directed paths of maximum length. Since the number of points in these paths is 3, here $l_n^\square = 3$.

Furthermore, one has that in this limit the last passage time $L_{\mathbf{X}}^{(1)}(n,n)$ as defined by (10.108) coincides with the definition of $l^\square$ given above. As a consequence the cumulative probability distribution for $l^\square$ can be deduced from knowledge of the formula for the last passage time (denoted $h^\square$) in Proposition 10.3.1.

PROPOSITION 10.6.1 *For the Hammersley process*

$$\Pr(l^\square \le l) = e^{-z^2} \Big\langle e^{2z\sum_{j=1}^{l} \cos\theta_j} \Big\rangle_{U(l)}. \tag{10.166}$$

Proof. The relationship between the Hammersley process and the last passage percolation model of Section 10.4.2 tells us that

$$\Pr(l^\square \le l) = \lim_{n\to\infty} \Pr(h^\square \le l)\Big|_{a_i=b_j=z/n}.$$

Computing the limit in (10.73) gives (10.166). □

10.6.1 Relationship to random permutations

Any particular realization of the Hammersley process containing N points gives a geometrical construction of a random permutation of $\{1,2,\dots,N\}$. To see this label the x-coordinates of the points by $0 < x_1 < \cdots < x_N < 1$ and similarly the y-coordinates by $0 < y_1 < \cdots < y_N < 1$. Each point will then have a coordinate of the form $(x_j, y_{P(j)})$, where $\{P(1),\dots,P(N)\}$ is a permutation of $\{1,2,\dots,N\}$. Because the N points are distributed at random with uniform distribution, the permutation is also random with uniform distribution. In addition, the quantity $l^\square$ has an interpretation in terms of the permutation. Thus the analogue of a continuous path consisting of segments of positive slope is a subsequence $1 \le j_1 < j_2 < \cdots < j_r \le N$ such that $P(j_1) < P(j_2) < \cdots < P(j_r)$, which is referred to as an increasing subsequence. The length of an increasing subsequence is defined as the value r. We then see from the definitions that the maximum length of all increasing subsequences of P coincides with $l^\square$.

10.6.2 Droplet PNG model

Associated with the non-negative matrix $\mathbf{X}$, both a last passage percolation model and a discrete polynuclear growth model have been identified. Likewise, associated with random points in the unit square are a last passage percolation model (the Hammersley process) and a particular polynuclear growth model known as the droplet PNG model [450].

To define the latter consider the xt-half-plane $t > 0$. Let this half plane be filled with points uniformly at random and such that the mean density is unity. Analogous to the entries x_{ij} of the matrix $\mathbf{X}$ these points are to be thought of as seeds for nucleation events of layered growth of unit height, although now the nucleation event grows continuously to the left and right with unit velocity and is created with zero width. In the droplet model, at $(x, t) = (0, 0)$ a single layer, taken to be at height zero, starts spreading with unit velocity to the left and to the right. All nucleation events and thus subsequent layers are constrained to occur above this initial layer. As the initial layer grows with unit speed, only nucleation events bounded by the "lightcone" axis $v = (t + x)/\sqrt{2}$, $u = (t - x)/\sqrt{2}$ are created at a time that their position coordinate makes contact with the ground layer or its growth. The nucleation events (x_i, t_i) inside the lightcone create the beginning of a portion of a layer of unit height on top of the ground layer, or existing layers, at position x_i. The layers are formed by the growth of the nucleation events with unit velocity to the left and to the right; if two growing portions of a layer collide, then growth at that point ceases and the two portions become one, growing only at the end points of this one portion (see Figure 10.19 for an example). Of interest are the statistical properties of the height at the origin after this growth process has been underway for time $t = T$.

Figure 10.19 Example of the plateau profile at the time of four successive nucleation events, including the initial event (which is labeled the 0th event and its plateau the 0th level). Note that between the second and third nucleation events, two plateaux on the first level have coalesced.

To analyze this quantity, the first observation is that only those nucleation events in the region $[u = 0, u = T/\sqrt{2}] \times [v = 0, v = T/\sqrt{2}]$ of the lightcone can contribute to the height at $x = 0$ up to time $t = T$. Suppose in a realization of the nucleation events there are N points in this region. For a Poisson process of unit density this occurs with probability $z^{2N} e^{-z^2}/N!$, where $z^2 = T^2/2$ is the area of the region. One then uses the construction of the previous subsection to associate with the configuration of points a permutation P, and furthermore marks in the world lines of the growth of the nucleation events (see Figure 10.20). The world lines show clearly the layered structure of the growth, and in particular the height at the origin after time T. To relate this height to a property of P, we first note a construction which determines the layer in which each particular nucleation event occurs. This can be done by partitioning the permutation into decreasing subsequences using the leftmost digits at all times. The jth such decreasing subsequence tells us the coordinates of the nucleation events, projected onto the line $x = -t$, which lie in the jth layer in the growth process. For example, in Figure 10.20 the permutation is 5374162, and the decreasing subsequences formed from the leftmost digits are $(531)(742)(6)$. As is shown in Exercises 10.6 q.1, it is generally true that the number of decreasing subsequences of this type is equal to the length of the longest increasing subsequence of the same permutation. Thus studying the height at the origin in the PNG model after time T is equivalent to studying the maximum path length in the Hammersley process with intensity $z^2 = T^2/2$.

10.6.3 Permutation matrices and increasing subsequences

We have seen that each configuration of N points in a realization of the Hammersley process is equivalent to a random permutation of $\{1, 2, \dots, N\}$, read off from the labels of the points $(x_j, y_{P(j)})$, $j = 1, \dots, N$. Also associated with the labels is a permutation matrix defined so that the entry $(j, P(j))$ of row j is equal to unity, while all other entries in the row are equal to zero. The RSK correspondence applied to the permutation matrix maps to a pair of u/rh and u/lh lattice paths with the constraint that for each allowed position of the up

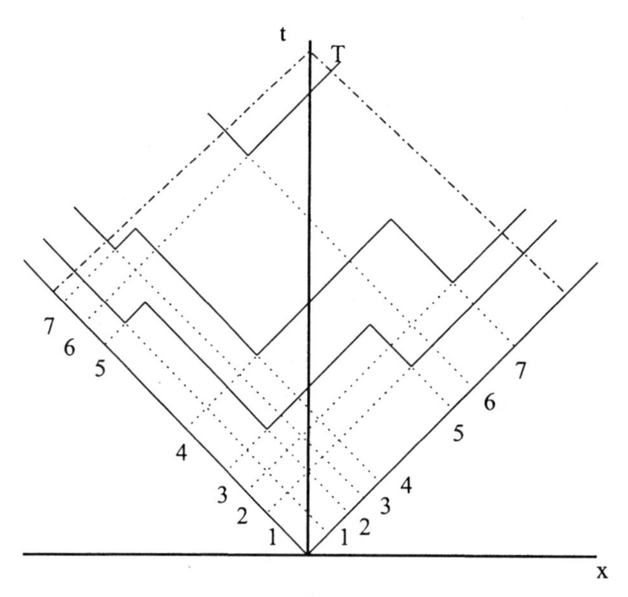

Figure 10.20 World lines for the endpoints of the plateaux. Only nucleation events inside the square-shaped region including the lines $x = \pm t$, and the lines from $t = T$ to these lines affect the height at the origin. The nucleation points occur at v-shaped configurations, while the inverted v part of the world lines correspond to the joining of the plateaux originating from two different nucleation events. The labeling on the lines $x = t$ and $x = -t$ allow the world lines to be identified uniquely with a permutation [532].

steps (recall Figure 10.12) there is a total of 1 such step only. Thus there is a correspondence with a pair of single move u/rh and u/lh lattice paths initially equally spaced, with a total of n moves each, or equivalently with a pair of standard tableaux having the same shape and each of content n (recall Section 10.1.3).

From the discussion of the paragraph below (10.114) we know that if the pair of tableaux have shape κ, then κ_1 is equal to the longest increasing subsequence length of the permutation, while $\sum_{i=1}^{l} \kappa_i$ is equal to the maximum of the lengths of l disjoint increasing subsequences. Denoting by $l_n^\square$ the longest increasing subsequence length, it follows that for a random permutation with uniform measure

$$\Pr(l_n^\square \le l) = \frac{1}{n!}\Big(\#\,(\text{of pairs of standard tableaux with content } n,\ \kappa_1 \le l)\Big). \tag{10.167}$$

This can be written as a random matrix average [262], [455].

Proposition 10.6.2 *We have*

$$\Pr(l_n^\square \le l) = \frac{1}{n!}\Big\langle \Big| \sum_{j=1}^{l} e^{i\theta_j} \Big|^{2n} \Big\rangle_{U(l)} = \frac{n!}{(2n)!}\Big\langle \Big(\sum_{j=1}^{l} 2\cos\theta_j \Big)^{2n} \Big\rangle_{U(l)}. \tag{10.168}$$

Proof. We note that

$$f_n^\lambda := \#(\text{standard tableaux of shape } \lambda \text{ and content } n) \tag{10.169}$$

occurs as a particular coefficient in the monomial expansion of the Schur polynomial $s_\lambda(w_1, \dots, w_n)$. Thus we recall from (10.16) that s_λ is defined as a weighted sum over semi-standard tableau of shape λ and content n with each tableau weighted by $\prod_{j=1}^{n} w_j^{\#j}$. Since a standard tableau is a special case of a semi-standard tableau in which each number $j = 1, \dots, n$ occurs exactly once, we see immediately that

$$[w_1 w_2 \cdots w_n] s_\lambda(w_1, \dots, w_n) = f_n^\lambda, \tag{10.170}$$

where $[w_1 w_2 \cdots w_n]$ denotes the coefficient of $w_1 w_2 \cdots w_n$. Hence

$$\Pr(l_n^{\square} \le l) = \frac{1}{n!}[a_1 \cdots a_n b_1 \cdots b_n] \sum_{\kappa:\kappa_1 \le l} s_\kappa(a_1, a_2, \dots, a_n) s_\kappa(b_1, b_2, \dots, b_n). \tag{10.171}$$

But according to (10.72) and (10.73)

$$\sum_{\kappa:\kappa_1 \le l} s_\kappa(a_1, a_2, \dots, a_n) s_\kappa(b_1, b_2, \dots, b_n) = \Big\langle \prod_{j=1}^{n} \prod_{k=1}^{l} (1 + a_j e^{-i\theta_k})(1 + b_j e^{i\theta_k}) \Big\rangle_{\mathrm{CUE}_l}.$$

Reading off the coefficients of $a_1 \cdots a_n b_1 \cdots b_n$ in the random matrix average gives the first equality in (10.168).
The second equality follows by noting

$$\Big(\sum_{j=1}^{l} 2\cos\theta_j\Big)^{2n} = \sum_{j=0}^{2n} \binom{2n}{j} \Big(\sum_{k=1}^{l} e^{i\theta_k}\Big)^j \Big(\sum_{k=1}^{l} e^{-i\theta_k}\Big)^{2n-j}, \tag{10.172}$$

then observing that only the term $j = n$ contributes to the $U(l)$ average. □

We note that the result (10.168) could have been derived from the result (10.166) for the Hammersley process. Thus from the definitions, $\Pr(l_N^{\square} \le l)$ and $\Pr(l^{\square} \le l)$ are related by

$$\Pr(l^{\square} \le l) = e^{-z^2} \sum_{N=0}^{\infty} \frac{z^{2N}}{N!} \Pr(l_N^{\square} \le l). \tag{10.173}$$

Expanding the random matrix average in (10.166) in a power series in z and using this relation leads to the second equality in (10.168).

10.6.4 Random words

A permutation can be regarded as a bijective mapping from $\{1, \dots, n\}$ to $\{1, \dots, n\}$. If we consider instead functions which map from $\{1, \dots, n\}$ to $\{1, \dots, k\}$, then presenting the function as a two-line array specifies a so-called *word* of length n from an alphabet of k letters. Presenting the two-line array for a word as a non-negative integer matrix gives a $0, 1$ matrix of dimension $n \times k$, constrained so that each row contains exactly a single 1. Applying the mapping of Figure 10.15 with this constraint gives a bijection with u/rh nonintersecting lattice paths making k steps paired with ld/lh nonintersecting lattice paths making n steps (as for general 0,1 matrices), but with the latter lattice paths constrained so that at any one step the number of ld steps is exactly one. In terms of tableaux we thus have a bijection with a semi-standard tableau of content k, and a standard tableau of the same shape also of content n.

Denoting by $l_{n,k}^{\mathrm{word}}$ the longest increasing subsequence length, it follows that for a random word

$$\Pr(l_{n,k}^{\mathrm{word}} \le l) = k^{-n} \sum_{\mu:\mu_1 \le l} \#(\text{ semi-standard tableaux, shape } \mu, \text{ content } k) f_n^{\mu}.$$

Use of (10.13) and (10.169) then shows

$$\begin{aligned}
\Pr(l_{n,k}^{\mathrm{word}} \le l) &= k^{-n}[b_1 \cdots b_n] \sum_{\mu:\mu_1 \le l} s_\mu(a_1, \dots, a_k) s_\mu(b_1, \dots, b_n)\Big|_{a_1 = \cdots = a_k = 1} \\
&= k^{-n}[b_1 \cdots b_n] \Big\langle \prod_{p=1}^{l} \Big(\prod_{r=1}^{k} (1 + a_r e^{i\theta_p}) \Big) \Big(\prod_{s=1}^{n} (1 + b_s e^{-i\theta_p}) \Big) \Big\rangle_{U(l)} \Big|_{a_1 = \cdots = a_k = 1} \\
&= k^{-n} \Big\langle \prod_{j=1}^{l} (1 + e^{i\theta_j})^k \Big(\sum_{j=1}^{l} e^{-i\theta_j} \Big)^n \Big\rangle_{U(l)},
\end{aligned} \tag{10.174}$$

where the second equality follows from the equality between (10.72) and (10.73).

EXERCISES 10.6 1. Note from the discussion of Section 10.1.3 that, with $\vec{l}^{(0)} = (l-1, l-2, \dots, 0)$ and $w_l^- = 0$, $w_l^+ = 1$ $(l = 1, \dots, n)$, $w_l^+ = 0$, $w_l^- = 1$ $(l = n+1, \dots, 2n)$,

$$\# \, (\text{of pairs of standard tableaux with content } n, \, \kappa_1 \le l)) = G_{2n}^{\mathrm{s,ld/rd}}(\vec{l}^{(0)}; \vec{l}^{(0)}).$$

Substitute this in (10.167), and make use of (10.39) to reclaim the first equality in (10.168).

2. [133] Consider a permutation π of $\{1, 2, \dots, N\}$, and let $\ell(\pi)$ denote the maximum length of the increasing subsequences.

   (i) Show that the minimum number of single integer moves required to return π back to the identity is $N - \ell(\pi)$.

   (ii) Suppose the integers in the permutation, reading from left to right, are sorted into piles according to the rules that low numbers are put on top of high numbers on the left-most possible pile, and if this is not possible, a new pile is started to the right of existing piles. Show that the number of piles is equal to $\ell(\pi)$.

10.7 SYMMETRIZED PERMUTATION MATRICES

10.7.1 Longest increasing and decreasing subsequence lengths of a random involution

A permutation matrix $\mathbf{X}$ is a real orthogonal matrix and thus has the property $\mathbf{X}^{-1} = \mathbf{X}^T$ (remember that we are counting the rows from the bottom, so the operation T corresponds to reflection about the bottom left to top right diagonal). Thus permutation matrices with the symmetry $\mathbf{X} = \mathbf{X}^T$ have the property $\mathbf{X}^2 = \mathbf{1}$ and therefore correspond to involutions. Further, the number of 1s on the diagonal corresponds to the number of fixed points of the involution. Our interest is in the cumulative distribution for $l_{n,k}^{\boxslash}$, the longest increasing subsequence length of an involution of $\{1, \dots, n\}$ consisting of k 2-cycles.

One approach to computing the cumulative distribution for $l_{n,k}^{\boxslash}$ is to consider the symmetrized Hammersley model underlying random involutions. First one notes that a random permutation matrix with the symmetry $\mathbf{X} = \mathbf{X}^T$ and $n - 2k$ points on the diagonal corresponds to n random points in the unit square symmetric about the diagonal $y = x$ and with $n - 2k$ of the n points on this line. A Poissonized version of this setting is to break the unit square into $M \times M$ equal subsquares, and to break the diagonal up into M equal segments. In each subsquare below the diagonal (together with its image above the diagonal) a point is marked with probability z^2/M^2, while a point is marked on each segment of the diagonal with probability $\alpha z/M$. The probability that there are a total of n points in the square with $n - 2k$ of the points on the diagonal is then given by the coefficient of $w^n \gamma^{n-2k}$ in

$$\Big(1 - \frac{\alpha z}{M} + \frac{\alpha z w \gamma}{M}\Big)^M \Big(1 - \frac{z^2}{M^2} + \frac{z^2 w^2}{M^2}\Big)^{M^2/2}.$$

Taking the limit $M \to \infty$ it follows from this that the probability is equal to

$$[w^n \gamma^{n-2k}] e^{\alpha z \gamma w + z^2 w^2/2} e^{-z^2/2 - \alpha z} = e^{-z^2/2 - \alpha z} \frac{z^n}{n!} \alpha^{n-2k} s_{n,k},$$

where

$$s_{n,k} = \binom{n}{2k} \frac{(2k)!}{2^k k!}$$

is the number of involutions consisting of k 2-cycles and $n - 2k$ fixed points. With $l^{\boxslash}$ denoting the maximum of the lengths of the paths consisting of positive sloping segments, it then follows that

$$\Pr(l^{\boxslash} \le l) = e^{-\alpha z - z^2/2} \sum_{n=0}^{\infty} \frac{z^n}{n!} \sum_{k=0}^{[n/2]} \alpha^{n-2k} s_{n,k} \Pr(l_{n,k}^{\boxslash} \le l). \tag{10.175}$$

The significance of the relation (10.175) is that the symmetrized Hammersley model can be realized as a limiting case of the growth process associated with matrices symmetric about the diagonal discussed in Section 10.5.2. We see from the first formula in (10.147), and (10.150), that the required limit is to set $q_i = z/n$ for each $i = 1, \dots, n$ and take $n \to \infty$. From the evaluation of the cumulative distribution for $h^{\boxslash}$ given in Proposition 10.5.5, we can then compute the cumulative distribution for $l^{\boxslash}$ according to

$$\Pr(l^{\boxslash} \le l) = \lim_{n \to \infty} \Pr(h^{\boxslash} \le l)\Big|_{q_i = z/n} \tag{10.176}$$

to obtain the following result [34].

Proposition 10.7.1 *We have*

$$\Pr(l^{\boxslash} \le l) = e^{-\alpha z} e^{-z^2/2} \Big\langle \det(\mathbf{1}_l + \alpha \mathbf{U}) e^{z \mathrm{Tr}\, \mathbf{U}} \Big\rangle_{\mathbf{U} \in \mathrm{O}(l)}. \tag{10.177}$$

Rotating all the elements of a permutation matrix with the symmetry $\mathbf{X} = \mathbf{X}^T$ by 90° anticlockwise gives a permutation matrix with the symmetry $\mathbf{X} = \mathbf{X}^R$. In the two line array presentation of an involution, this is equivalent to reversing the ordering of the bottom line, and so interchanging increasing subsequences with decreasing subsequences. Thus the maximum length of the increasing subsequences in the permutation corresponding to the rotated matrix $l_{n,k}^{\boxbslash}$ say, is equal to the maximum length of the decreasing subsequences in the underlying involution. As with (10.175), this is related to the maximum of the length of paths consisting of positive sloping segments in the corresponding Hammersley process by

$$\Pr(l^{\boxbslash} \le l) = e^{-\beta z - z^2/2} \sum_{n=0}^{\infty} \frac{z^n}{n!} \sum_{k=0}^{[n/2]} \beta^{n-2k} s_{n,k} \Pr(l_{n,k}^{\boxbslash} \le l), \tag{10.178}$$

where now the antidiagonal $y = 1 - x$ $(0 < x < 1)$ contains points with a Poisson intensity βz. The analogue of the formula (10.176), together with the results of Proposition 10.5.3, then give the following random matrix form [34].

Proposition 10.7.2 *We have*

$$\Pr(l^{\boxbslash} \le 2l) = e^{-\beta z - z^2/2} \Big\langle \prod_{k=1}^{l} \frac{1}{|1 - \beta e^{-i\theta_k}|^2} e^{z \sum_{j=1}^{l} 2\cos\theta_j} \Big\rangle_{\mathrm{Sp}(2l)}, \tag{10.179}$$

$$\Pr(l^{\boxbslash} \le 2l + 1) = e^{-z^2/2} \Big\langle e^{z \sum_{j=1}^{l} 2\cos\theta_j} \Big\rangle_{\mathrm{Sp}(2l)}. \tag{10.180}$$

10.7.2 Relationship to the flat PNG model

The Hammersley process symmetric about the antidiagonal, but with no points thereon, is relevant to the study of the PNG model in which growth is from a flat substrate [450]. Here, in distinction to the case of the droplet PNG model, growth from the nucleation events (which again occur at random with unit mean in space and time) is no longer restricted to happen above an initial plateau, but rather may occur over the length of the whole x-axis. The profile will then on average be flat and the statistical properties of the height fluctuation independent of the position, which can therefore without loss of generality be studied at $x = 0$. For nucleation events to affect the height at the origin after time $\sqrt{2}T$ they must occur at positions x and times t such that $|x| < \sqrt{2}T - t$. This is a triangular shaped region, with vertices $(x, t) = (-\sqrt{2}T, 0)$, $(\sqrt{2}T, 0)$, $(0, \sqrt{2}T)$ in the xt-plane, in which nucleation events occur uniformly at random with unit density. From the rules of the PNG model one sees that the height at the origin is equal to the maximum number of points in an upward directed path formed from the nucleation events, which starts along the line $(x, 0)$ and finishes at the point $(0, \sqrt{2}T)$.

The triangular shaped region of points can be extended to a square shaped region by reflecting it about the x-axis. The length of the longest upwards-directed path from $(0, -\sqrt{2}T)$ to $(0, \sqrt{2}T)$ within the square is the same thing as the length of the longest up/right path in a symmetrized Hammersley process. The symmetrization is with respect to the antidiagonal, which itself contains no points. The height h at the origin of the PNG model is equal to half the length of the longest up/right path in the Hammersley process, and thus the cumulative distribution $\Pr(h \le l)$ is given by (10.179) with $z = T$ and $\beta = 0$.

EXERCISES 10.7

1. [246] The objective of this exercise is to derive the explicit formula

$$f_n^\lambda = n! \frac{\prod_{1 \le i < j \le p} (l_i - l_j)}{\prod_{i=1}^p l_i!}, \qquad l_i := \lambda_i + p - i, \tag{10.181}$$

where $p := \mu_1'$ and $|\lambda| = n$.

(i) Argue that

$$f_n^\lambda = \sum_{i=1}^n f_{n-1}^{\lambda_{(i)}}, \tag{10.182}$$

where $\lambda_{(i)} := (\lambda_1, \dots, \lambda_i - 1, \dots, \lambda_n)$ and $f_{n-1}^{\lambda_{(i)}} := 0$ for $\lambda_{(i)}$ not a partition, and note that this formula together with the initial condition $f_0^{0^n} = 1$ specifies f_n^λ by induction.

(ii) With $\Delta := \prod_{i<j}(l_i - l_j)$ show that for the r.h.s. of (10.181) to satisfy the formula of (i) we require

$$n\Delta(l_1, \dots, l_p) = \sum_{i=1}^p l_i \Delta(l_1, \dots, l_i - 1, \dots, l_p).$$

Verify this formula by establishing the identity

$$\sum_{i=1}^p x_i \Delta(x_1, \dots, x_i + t, \dots, x_p) = \Big(x_1 + \dots + x_p + \binom{p}{2} t\Big) \Delta(x_1, \dots, x_p).$$

(First observe that the l.h.s. is anti-symmetric in the x_i, and hence has as a factor $\Delta(x_1, \dots, x_p)$. From this conclude the expression must be a linear function of t.)

2. [7] In this exercise the combinatorial setting will be used to determine the asymptotic behavior of some matrix integrals.

(i) Consider (10.178) in the case $\beta = 0$. Noting that

$$\Pr(l_{2n,n}^{\boxslash} \le 2l) = 1 \ (n \le l), \qquad \Pr(l_{2(l+1),l+1}^{\boxslash} \le 2l) = 1 - \frac{1}{s_{2(l+1),l+1}},$$

show

$$\Pr(l^{\boxslash} \le 2l)\Big|_{\beta=0} = 1 - \frac{z^{2l+2}}{(2l+2)!} + \mathrm{O}(z^{2l+4}).$$

Substitute this in (10.179) with $\beta = 0$ to conclude

$$\Big\langle e^{z \sum_{j=1}^l 2\cos\theta_j} \Big\rangle_{\mathrm{Sp}(2l)} = e^{z^2/2 - z^{2l+2}/(2l+2)! + \mathrm{O}(z^{2l+4})}.$$

(ii) Repeat the considerations of (i), applied to (10.175) and (10.177) with $\alpha = 0$, to show

$$\langle e^{z\mathrm{Tr}(\mathrm{U})} \rangle_{\mathrm{U} \in \mathrm{O}(2l)} = e^{z^2/2 - z^{2l+2}/(2l+2)! + \mathrm{O}(z^{2l+4})}.$$

(iii) Use the fact that the eigenvalues $0 < \theta < \pi$ for $O^-(2l+2)$ have the same p.d.f. as the eigenvalues in $0 < \theta < \pi$ for $\mathrm{Sp}(2l)$ to deduce from the results of (i) and (ii) that

$$\langle e^{z\mathrm{Tr}(\mathrm{U})} \rangle_{\mathrm{U} \in \mathrm{O}^\pm(2l)} = e^{z^2/2 \pm z^{2l}/(2l)! + \mathrm{O}(z^{2l+2})}.$$

10.8 GAP PROBABILITIES AND SCALED LIMITS

10.8.1 Growth in the square

Consider the probability (10.73) relating to the maximum height in the polynuclear growth model with $a_1 = \cdots = a_n = \sqrt{q}$, $b_1 = \cdots = b_m = \sqrt{q}$, $b_{m+1} = \cdots = b_n = 0$. In this case

$$\Pr(h^{\square} \le l) = (1-q)^{nm} \Big\langle \prod_{k=1}^{l} (1 + q e^{i\theta_k})^m (1 + e^{-i\theta_k})^n \Big\rangle_{U(l)}.$$

Let us now write $q = e^{-1/L}$, $l \mapsto Lt$ and take the limit $L \to \infty$ (this is the Laguerre case of the exponential limit discussed in Section 10.3.3). The result of Exercises 10.3 q.1 implies [329]

$$\begin{aligned} &\lim_{L\to\infty} \Pr(h^{\square} \le Lt) \\ &= \frac{1}{\prod_{j=1}^{m} \Gamma(j)\Gamma(n-m+j)} \int_0^t dx_1 \cdots \int_0^t dx_m \prod_{l=1}^{m} x_l^{n-m} e^{-x_l} \prod_{1 \le j < k \le m} (x_k - x_j)^2 \\ &= E_{m,2}(0; (t,\infty); x^{n-m} e^{-x}) \end{aligned} \tag{10.183}$$

where the final equality follows from the definition. In words this says that the limiting probability is equal to the probability of no eigenvalues in the interval (t, ∞) (the soft edge) of the Laguerre unitary ensemble $\mathrm{UE}_m(x^{n-m} e^{-x})$.

Consider now the Poisson points limit, in which the cumulative distribution for the maximum height is given by (10.166). The CUE_l average in this expression is a special case of (8.96), so we see that

$$\Pr(l^{\square} \le l) = E_2^{\mathrm{hard}}(0; (0, 4z^2); l), \tag{10.184}$$

where use is made of (8.87). Hence $\Pr(l^{\square} \le l)$ is equal to the probability of no eigenvalues at the hard edge of the LUE with $a = l$.

10.8.2 Growth symmetric about the antidiagonal

For the polynuclear growth model with the underlying integer matrix symmetric about the antidiagonal, and with only even entries allowed on the diagonal, the maximum height $h^{\boxslash}$ is given by (10.134) with $\beta = 0$. In the case $q_1 = \cdots = q_n = q$ this reads

$$\Pr(h^{\boxslash} \le 2l) = (1-q^2)^{n(n+1)/2} \Big\langle \det(\mathbf{1} + q\mathbf{U})^n \Big\rangle_{\mathbf{U} \in Sp(2l)}. \tag{10.185}$$

We consider the particular exponential limit obtained by writing $q = e^{-1/2L}$, $2l \mapsto Lt$ and taking $L \to \infty$. From the appropriate limit of (10.133) we see that then [27]

$$\begin{aligned} \lim_{L\to\infty} \Pr(h^{\boxslash} \le Lt) &= \frac{1}{W_{0,1,n}} \int_0^t dx_1 \cdots \int_0^t dx_n \prod_{l=1}^{n} e^{-x_l/2} \prod_{1 \le j < k \le n} |x_k - x_j| \\ &= E_{n,1}(0; (t,\infty); e^{-x/2}), \end{aligned} \tag{10.186}$$

and thus the limiting probability is equal to the probability that there are no eigenvalues in the soft edge interval (t, ∞) of LOE_n with $a = 0$. Although this probability is a $\beta = 1$ quantity, it can be written as a determinant [233].

PROPOSITION 10.8.1 *In addition to (10.186) we also have*

$$\lim_{L\to\infty} \Pr(h^{\boxslash} \le Lt) = \frac{(2\pi)^{n/2}}{G(n+1)} (t/4)^{n^2/2} e^{-nt/4} \det[I_{j-k+1/2}(t/4)]_{j,k=1,\dots,n}. \tag{10.187}$$

Proof. According to Exercises 8.3 q.3 it follows from (10.185) that

$$\Pr(h^{\boxslash} \le 2l) = (1-q^2)^{n(n+1)/2}(4q)^{nl}\Big\langle \prod_{j=1}^{l}\Big(\frac{1}{4q}(1-q)^2+\lambda_j\Big)^n\Big\rangle_{\mathrm{JUE}_l}\Big|_{a=b=1/2}. \tag{10.188}$$

Now, simple manipulation of (8.120) shows that for the JUE_l supported on $(0,1)$ with general a,b

$$\Big\langle \prod_{j=1}^{l}(u+x_j)^n\Big\rangle_{\mathrm{JUE}_l} = \frac{n!(1+u)^{nl}}{M_n(b+l,a+l,1)}\Big\langle \prod_{j=1}^{n} z_j^{(b-a-l)/2}|1+z_j|^{a+b+l}\Big(\frac{u}{1+u}+z_j\Big)^l\Big\rangle_{\mathrm{CUE}_n}, \tag{10.189}$$

where the explicit form of the proportionality follows by consideration of the $u\to\infty$ limit. According to (5.76), with $z=e^{i\theta}$ the CUE_n average can be written as the Toeplitz determinant

$$\det\Big[\frac{1}{2\pi}\int_{-\pi}^{\pi} z^{(b-a-l)/2+(j-k)}|1+z|^{a+b+l}\Big(\frac{u}{1+u}+z\Big)^l\,d\theta\Big]_{j,k=1,\dots,n}.$$

The integral in the determinant can be evaluated in terms of the Gauss hypergeometric function (expand $(u/(1+u)+z)^l$ by the binomial theorem and make use of (4.4) in the case $N=1$) to give

$$\det\Big[\frac{\Gamma(a+b+l+1)}{\Gamma(a-(j-k)+1)\Gamma(b+l+j-k+1)}\,{}_2F_1\Big(-l,-b-l-(j-k);a+1-(j-k);\frac{u}{1+u}\Big)\Big]_{j,k=1,\dots,n}.$$

With $q=e^{-t/4l}$ and $u=(1-q)^2/4q$, for $l\to\infty$ we have $u\to(t/8l)^2$. Making use of the series (5.83) and the identity

$${}_0F_1(-;1+\mu;s^2/4)=\Gamma(\mu+1)\Big(\frac{2}{s}\Big)^{\mu}I_\mu(s), \tag{10.190}$$

we see that the limiting form of the general entry of the determinant is

$$(t/8l)^{(j-k)-a}I_{a-(j-k)}(t/4)$$

and thus to leading order the determinant is equal to

$$(t/8l)^{-(an)}\det[I_{a-(j-k)}(t/4)]_{j,k=1,\dots,n}. \tag{10.191}$$

Also, noting from (4.4) and (4.182) that

$$M_n(a,b,1)=\frac{G(n+1+a+b)}{G(1+a+b)}\frac{G(1+a)}{G(n+1+a)}\frac{G(1+b)}{G(n+1+b)}G(n+2),$$

we compute using (4.185) the asymptotic expansion

$$M_n(a+l,b+l,1)\underset{l\to\infty}{\sim} l^{-n^2/2}(2\pi)^{-n/2}2^{2nl+n^2/2+n(a+b)}G(n+2). \tag{10.192}$$

Substituting (10.191) and (10.192) in (10.189) and substituting the result in (10.188) implies (10.187). □

Consider now the Poisson points limit of this model, in which the cumulative distribution of the maximum height is given by (10.179) with $\beta=0$. Recalling (9.182) we see that

$$\Pr(l^{\boxslash}\le 2l)=E_1^{\mathrm{hard}}(0;(0,(2z)^2);l). \tag{10.193}$$

Thus $\Pr(l^{\boxslash}\le 2l)$ is equal to the probability of no eigenvalues in the scaled hard edge interval $(0,(2z)^2)$ of a matrix ensemble with orthogonal symmetry and singularity proportional to x^l near the hard edge.

10.8.3 Growth symmetric about the diagonal

In the case of the polynuclear growth model with the underlying integer matrix symmetric about the diagonal, and all entries on the diagonal itself set equal to zero, the distribution of the maximum height $h^{\boxslash}$ is specified by (10.154) with $\alpha = 0$. With $q_1 = \cdots = q_n = q$ this can be written

$$\Pr(h^{\boxslash} \le l) = \frac{1}{2}(1-q^2)^{n(n-1)/2}\Big(\langle \det(\mathbf{1} + q\mathbf{U})^n \rangle_{\mathbf{U} \in O^+(l)} + \langle \det(\mathbf{1} + q\mathbf{U})^n \rangle_{\mathbf{U} \in O^-(l)} \Big). \tag{10.194}$$

In the RSK mapping between integer matrices symmetric about the diagonal and semi-standard tableaux, it follows from (10.149) that with all the diagonal elements set equal to zero, the tableaux must have row lengths equal in pairs, $\mu_1 = \mu_2, \mu_3 = \mu_4, \dots$, so only $\mu_1, \mu_3, \dots$ are independent. Suppose for convenience that n is even. We see from (10.151) and (10.23) that with $q_1 = \cdots = q_n = q$ and $\alpha = 0$, $h_j = \mu_j + (n-j)$ the joint p.d.f. for $h^{\boxslash} := h_1$ and $h_3, \dots, h_{n-1}$ is given by

$$q^{-(n-1)n/2} \frac{(1-q^2)^{n(n-1)/2}}{\prod_{l=1}^{n-1} l!} q^{\sum_{j=1}^{n/2} h_{2j-1}} \\ \times \prod_{1 \le i < j \le n/2} (h_{2i-1} - h_{2j-1})^2 (h_{2i-1} - h_{2j-1} + 1)(h_{2i-1} - h_{2j-1} - 1),$$

where $h_j = \mu_j + (n-j)$ and so $h_1 > h_3 > \cdots > h_{n-1} \ge 0$. It follows from this that in the particular exponential limit obtained by writing $q = e^{-1/L}$, $l \mapsto Lt$ and taking $L \to \infty$ [27],

$$\begin{aligned} \lim_{L \to \infty} \Pr(h^{\boxslash} \le Lt) &= \frac{1}{(n/2)! \prod_{l=1}^{n-1} l!} \int_0^t dx_1 \cdots \int_0^t dx_{n/2} \prod_{j=1}^{n/2} e^{-x_j} \prod_{1 \le j < k \le n/2} (x_k - x_j)^4 \\ &= E_{n/2,4}(0, (t, \infty); e^{-x}). \end{aligned} \tag{10.195}$$

Hence here the limiting probability is equal to the probability that there are no eigenvalues in the soft edge interval (t, ∞) of the $\mathrm{LSE}_{n/2}$ with $a = 0$. As with the gap probability in (10.185), manipulation of (10.194) implies that it has a determinantal form [233].

PROPOSITION 10.8.2 *We have*

$$\lim_{L \to \infty} \Pr(h^{\boxslash} \le Lt) \Big|_{q_1 = \cdots = q_n = e^{-1/L}} = \frac{1}{2} \frac{(2\pi)^{n/2}}{G(n+1)} \\ \times (t/4)^{n^2/2} e^{-nt/4} \Big(\det[I_{j-k+1/2}(t/4)]_{j,k=1,\dots,n} + \det[I_{j-k-1/2}(t/4)]_{j,k=1,\dots,n} \Big). \tag{10.196}$$

It remains to consider the Poisson points limit of this model, in which the cumulative distribution of the maximum height is given by (10.177) with $\alpha = 0$. Recalling (9.187) we see that

$$\Pr(l^{\boxslash} \le 2l) = \tilde{E}_4^{\mathrm{hard}}(0; (0, (2z)^2); 2l). \tag{10.197}$$

Thus $\Pr(l^{\boxslash} \le 2l)$ is equal to the probability of no eigenvalues in the scaled hard edge interval $(0, (2z)^2)$ of a matrix ensemble with symplectic symmetry and singularity proportional to x^{2l} near the hard edge.

10.8.4 Scaled limit

In the setup of the Hammersley process there are on average z^2 points in the unit square. As $z \to \infty$ the length of the longest up/right path $l^{\square}$ will thus diverge. Of interest is the statistical properties of $l^{\square}$ in this limit. In particular, what is the mean of $l^{\square}$ and can the fluctuations about the mean be quantified? Some insight into these questions is provided by the identity (10.184). Thus we know from Exercises 7.1 q.1(iii) that the hard edge with singularity proportional to x^a scales to the soft edge upon change of variables $x \mapsto a^2 - 2a(a/2)^{1/3}x$. This then suggests that to obtain a well-defined scaled limit we should introduce a scaled variable y such that $4z^2 = l^2 - 2l(l/2)^{1/3}y$ or equivalently $y = (l-2z)/(l/2)^{1/3}$. Indeed, at a formal level it follows from Exercises 7.1 q.2(iii) that

$$\lim_{l\to\infty} E_2^{\rm hard}(0;(0,l^2-2l(l/2)^{1/3}y);l) = E_2^{\rm soft}(0;(y,\infty)), \tag{10.198}$$

and thus

$$\lim_{z\to\infty} \Pr\Big(\frac{l^\square - 2z}{z^{1/3}} \le y\Big) = E_2^{\rm soft}(0;(y,\infty)). \tag{10.199}$$

One reads off that for $z \to \infty$ the mean of $l^\square$ is to leading order equal to $2z$, and the fluctuations about the mean are proportional to $z^{1/3}$, with the distribution of the scaled quantity $(l^\square - 2z)/z^{1/3}$ identical to that of the largest eigenvalue at the soft edge of a scaled matrix ensemble with unitary symmetry.

The limit formula (10.199) was first derived and proved by Baik, Deift and Johansson [31]. Here we will present a proof based on the hard to soft edge limiting transition at $\beta = 2$ [85]. The key result is a variation of (10.198).

PROPOSITION 10.8.3 *Let $Q_l(y) := (l - (l/2)^{1/3}y)^2$. We have*

$$\lim_{l\to\infty} E_2^{\rm hard}(0;(0,Q_l(y));l) = E_2^{\rm soft}(0;(y,\infty)). \tag{10.200}$$

Proof. Starting with the formula

$$E_2^{\rm hard}(0,(0,l^2-2l(l/2)^{1/3}y);l) = 1 + \sum_{n=1}^\infty \frac{(-1)^n}{n!} \int_0^y dx_1 \cdots \int_0^y dx_n \det\Big[K_l^{\rm hard}(x_j,x_k)\Big]_{j,k=1,\dots,n},$$

which is a consequence of (9.4) and Propositions 5.1.2 and 7.2.1, we adopt a strategy analogous to the proofs of Propositions 9.1.2 and 9.1.3. Thus we focus attention on the $l \to \infty$ form of

$$(Q_l'(x)Q_l'(y))^{1/2} K_l^{\rm hard}(Q_l(x),Q_l(y)).$$

According to (7.31) and the definition of Q_l, we require asymptotic estimates of

$$J_l(l-(l/2)^{1/3}y), \qquad J_l'(l-(l/2)^{1/3}y).$$

The uniform large ν asymptotic expansion of $J_\nu(\nu z)$ valid for all $z \in \mathbb{C}$, $\arg z \ne \pi$ obtained by Olver [435] can be used to show that

$$J_\nu(\nu - w(\nu/2)^{1/3}) \sim \Big(\frac{2}{\nu}\Big)^{1/3} {\rm Ai}(w) + {\rm O}(\nu^{-1}){\rm O}(e^{-w}),$$
$$-\Big(\frac{\nu}{2}\Big) J_\nu'(\nu - w(\nu/2)^{1/3}) \sim \Big(\frac{2}{\nu}\Big)^{1/3} {\rm Ai}'(w) + {\rm O}(\nu^{-2/3}){\rm O}(e^{-w}).$$

Substituting in (7.31) shows that for $|x-y|$ bounded away from zero at least

$$(Q_l'(x)Q_l'(y))^{1/2} K_2^{\rm hard}(Q_l(x),Q_l(y)) = K_2^{\rm soft}(x,y) + {\rm O}(l^{-1/3}){\rm O}(e^{-x-y}).$$

The validity of this estimate for $|x-y| \to 0$, and thus all $x, y > 0$ can be established by the same technique as used in the proof of Proposition 9.1.3. It follows from this that

$$|Q_l'(x_1)\cdots Q_l'(x_n)|\rho_{(n)}^{\rm hard}(Q_l(x_1),\dots,Q_l(x_n))|_{a=l} = \rho_{(n)}^{\rm soft}(x_1,\dots,x_n) + R_n^l(x_1,\dots,x_n), \tag{10.201}$$

where $R_n^l \to 0$ as $l \to \infty$ uniformly on $x_j \in [y,\infty)$ $(j = 1,\dots,n)$. Thus the analogue of the property (9.3) relating to Proposition 9.1.1 is valid. Hence the conclusion of the latter applies, telling us that

$$\lim_{l\to\infty} E_2^{\rm hard}(Q_l(y);l) = 1 + \sum_{n=1}^\infty \frac{(-1)^n}{n!} \int_y^\infty dx_1 \cdots \int_y^\infty dx_n\, \rho_{(n)}^{\rm soft}(x_1,\dots,x_n),$$

which recalling (9.4) is (10.200). □

We can deduce from Proposition 10.8.3 the following limit theorem, which includes (10.198).

PROPOSITION 10.8.4 *We have*

$$\lim_{l\to\infty} E_2^{\rm hard}(0;(0,l^2-2l(l/2)^{1/3}y+{\rm O}(l));l) = E_2^{\rm soft}(0;(y,\infty)). \tag{10.202}$$

Proof. Since $Q_l(y) = l^2 - 2l(l/2)^{1/3}y + {\rm O}(l^{2/3})$, we can rewrite the l.h.s. of (10.202) as

$$\lim_{l\to\infty} E_2^{\rm hard}(0;(0,Q_l(y)+{\rm O}(l));l)$$

which in turn can be written as

$$\lim_{l\to\infty} E_2^{\rm hard}(0;(0,Q_l(y));l) + \lim_{l\to\infty} E_2^{\rm hard}(0;(Q_l(y),Q_l(y)+{\rm O}(l));l).$$

From Proposition 10.8.3 we know that the first term is equal to $E_2^{\rm soft}(0;(y,\infty))$. For the second term, the working of the proof of Proposition 10.8.3 gives that it equals

$$\lim_{l\to\infty} E_2^{\rm soft}(0;(y-{\rm O}(l^{-1/3}),y))$$

and thus vanishes. □

Closely related to the scaled form (10.199) for the cumulative distribution of $l^\square$ is an analogous limit formula for $l_N^\square$,

$$\lim_{N\to\infty} {\rm Pr}\Big(\frac{l_N^\square - 2\sqrt{N}}{N^{1/6}} \le y\Big) = E_2^{\rm soft}(0;(y,\infty)). \tag{10.203}$$

This result, quantifying the limiting distribution of the largest increasing subsequence length of a large random permutation, was derived in [31] as a corollary of (10.199). Some of the working is given in Exercises 10.8.

The analogue of (10.201) can readily be established for $\beta = 1$ and 4. This allows us to deduce from (10.193) and (10.197) the limit formulas [35]

$$\lim_{z\to\infty} {\rm Pr}\Big(\frac{l^{\boxbslash} - 2z}{z^{1/3}} \le y\Big) = E_1^{\rm soft}(0;(y,\infty)), \qquad \lim_{z\to\infty} {\rm Pr}\Big(\frac{l^{\boxslash} - 2z}{z^{1/3}} \le y\Big) = \tilde{E}_4^{\rm soft}(0;(y,\infty)),$$

and these in turn tell us that in relation to increasing subsequences

$$\lim_{k\to\infty} {\rm Pr}\Big(\frac{l_{2k,k}^{\boxbslash} - 2\sqrt{2k}}{(2k)^{1/6}} \le y\Big) = E_1^{\rm soft}(0;(y,\infty)), \qquad \lim_{k\to\infty} {\rm Pr}\Big(\frac{l_{2k,k}^{\boxslash} - 2\sqrt{2k}}{(2k)^{1/6}} \le y\Big) = \tilde{E}_4^{\rm soft}(0;(y,\infty)).$$

EXERCISES 10.8 1. [328] Let the sequence $\{q_n\}_{n=0,1,\dots}$ satisfy the bounds $0 \le q_n \le 1$ and be monotonically decreasing so that $q_n \ge q_{n+1}$. Let

$$\phi(\xi) := e^{-\xi}\sum_{n=0}^\infty q_n \frac{\xi^n}{n!}$$

and for given $d > 0$ write

$$\mu_n^{(d)} = n + (2\sqrt{d+1}+1)\sqrt{n\log n}, \qquad \nu_n^{(d)} = n - (2\sqrt{d+1}+1)\sqrt{n\log n}.$$

The objective of this exercise is to show that

$$\phi(\mu_n^{(d)}) - Cn^{-d} \le q_n \le \phi(\nu_n^{(d)}) + Cn^{-d} \tag{10.204}$$

for all $n \ge n_0$, where C is some positive constant.

(i) Suppose $\xi > 0$. With $w_n(\xi) := e^{-\xi}\xi^n/n!$, and $f(x) = x \log x + 1 - x$, use Stirling's formula to show

$$w_n(\xi) \le C \exp(-\xi f(n/\xi)).$$

(ii) Note that for $0 \le x \le 2$, $f(x) \ge (x-1)^2/4$ and use this in the result of (i) to deduce

$$\sum_{n=0}^{[\xi - 2\sqrt{d+1}\sqrt{\xi \log \xi}]} w_n(\xi) \le \frac{C}{\xi^d}, \qquad \sum_{n=[\xi+2\sqrt{d+1}\sqrt{\xi \log \xi}]}^{[2\xi]} w_n(\xi) \le \frac{C}{\xi^d}.$$

Also, use the fact that $f(x) \ge x/10$ for $x \ge 2$ to show that for $\xi > \xi_0 > 0$

$$\sum_{n=[2\xi]}^{\infty} w_n(\xi) \le C(\xi_0) e^{-\xi/5}.$$

Combine these inequalities to conclude that for ξ large enough

$$\sum_{n=0}^{\infty} w_n(\xi) - \sum_{n:|n-\xi| \le 2\sqrt{d+1}\sqrt{\xi \log \xi}} w_n(\xi) \le \frac{C}{\xi^d}.$$

(iii) Use the assumption that $0 \le q_n \le 1$ to deduce from the final result of (ii) that

$$\phi(\xi) - \sum_{n:|n-\xi| \le 2\sqrt{d+1}\sqrt{\xi \log \xi}} q_n w_n(\xi) \le \frac{C}{\xi^d}.$$

Use the assumption that $q_n \ge q_{n+1}$ and the fact that $\sum_{n=0}^{\infty} w_n(\xi) = 1$ to show

$$\sum_{n:|n-\xi| \le 2\sqrt{d+1}\sqrt{\xi \log \xi}} q_n w_n(\xi) \le q_{\xi - 2\sqrt{d+1}\sqrt{\xi \log \xi}},$$

and similarly show

$$\sum_{n:|n-\xi| \le 2\sqrt{d+1}\sqrt{\xi \log \xi}} q_n w_n(\xi) \ge q_{[\xi+2\sqrt{d+1}\sqrt{\xi \log \xi}]} - \frac{C}{\xi^d}.$$

By combining the above three inequalities, deduce (10.204).

2. [31] In (10.173) replace l by $2z + z^{1/3}y$ so that it reads

$$\Pr\Big(\frac{l^{\square} - 2z}{z^{1/3}} \le y\Big) = e^{-z^2} \sum_{N=0}^{\infty} \frac{z^{2N}}{N!} \Pr\Big(\frac{l_N^{\square} - 2z}{z^{1/3}} \le y\Big).$$

Now apply (10.204) with $q_n \mapsto q_n(\xi)$, $\xi = z^2$ to deduce from (10.199) the limit formula (10.203).

10.9 HAMMERSLEY PROCESS WITH SOURCES ON THE BOUNDARY

In the droplet PNG model nucleation events occur above the plateau formed by an initial nucleation event at $(x,t) = (0,0)$. An extension of this model is to allow for growth of the droplet due to nucleation events forming on the left and right boundaries of the droplet. In particular, suppose nucleation events are created at the rate $\alpha_- dt$ on the left boundary, and at the rate $\alpha_+ dt$ on the right boundary. An equivalent viewpoint is that surrounding the initial nucleation event at $(x,t) = (0,0)$ are two staircase structures of growing plateaux (downward sloping for $x < 0$, upward sloping for $x > 0$) with vertical increments of one unit having intensity α_- for $x < 0$ and α_+ for $x > 0$. The arrival of a layer of the staircase structure at the boundary of the droplet realizes the creation of nucleation events thereon. It turns out that this particular extension of the droplet PNG model is relevant to the study of the KPZ (Kardar-Parisi-Zhang) universality class of critical phenomena [451].

To analyze the height at the origin in this model after time $t = \sqrt{2}T$ we note that it is equivalent to the length $L(T, \alpha_+, \alpha_-)$ of the longest up/right path in a Hammersley process with $z^2 = T^2$, and furthermore in which there are points on the x-axis (y-axis) of the unit square with Poisson distribution of intensity α_+ (α_-). The cumulative distribution of the longest path length $L(T, \alpha_+, \alpha_-)$ can be expressed as a random matrix average by a limiting procedure applied to the discrete PNG model of Section 10.3.1. For this purpose we extend the $n \times n$ non-negative integer matrix $[x_{ij}]_{i,j=1,\dots,n}$ to an $(n+1) \times (n+1)$ matrix $[x_{ij}]_{i,j=0,\dots,n}$, but again with all elements chosen according to the geometric distribution (10.70). As in the discussion at the beginning of Section 10.6, to recover the setting of the Hammersley model in the region of the square away from the boundary we set $a_i = b_j = T/n$ $(i, j = 1, \dots, n)$ and take $n \to \infty$. The boundary distributions in this limit are obtained by setting $a_0 = \alpha_+$, $b_0 = \alpha_-$. However this procedure leaves at the origin a non-negative integer geometric random variable with parameter $\alpha_+\alpha_-$ (this distribution will be denoted $g(\alpha_+\alpha_-)$). Denoting by $L^+(T, \alpha_+, \alpha_-)$ the longest up/right path in this process, we see it is related to $L(T, \alpha_+, \alpha_-)$ by the simple relation

$$L^+(T, \alpha_+, \alpha_-) = L(T, \alpha_+, \alpha_-) + \chi, \qquad \chi \in g(\alpha_+\alpha_-). \tag{10.205}$$

Furthermore, the limiting procedure applied to (10.73) (appropriately extended to include the variables a_0 and b_0) gives

$$\Pr(L^+(T, \alpha_+, \alpha_-) \le l) = (1 - \alpha_+\alpha_-) e^{-(\alpha_+ + \alpha_-)T - T^2} \tilde{D}_l, \tag{10.206}$$

where

$$\tilde{D}_l := \Big\langle \prod_{j=1}^{l} (1 + \alpha_+ e^{i\theta_j})(1 + \alpha_- e^{-i\theta_j}) e^{2T \sum_{j=1}^{l} \cos\theta_j} \Big\rangle_{U(l)}. \tag{10.207}$$

The formulas (10.205) and (10.206) provide us with a formula for the cumulative distribution of $L(T, \alpha_+, \alpha_-)$.

PROPOSITION 10.9.1 *We have*

$$\Pr(L(T, \alpha_+, \alpha_-) \le l) = e^{-(\alpha_+ + \alpha_-)T - T^2} (\tilde{D}_l - \alpha_+\alpha_- \tilde{D}_{l-1}). \tag{10.208}$$

Proof. Introducing the generating functions

$$Q(x) = \sum_{l=0}^{\infty} \Pr(L(T, \alpha_+, \alpha_-) \le l) x^l, \qquad Q^+(x) = \sum_{l=0}^{\infty} \Pr(L^+(T, \alpha_+, \alpha_-) \le l) x^l,$$

we see from (10.205) that

$$\begin{aligned} Q^+(x) &= \sum_{l=0}^{\infty} x^l \sum_{k=0}^{l} \Pr(L(T, \alpha_+, \alpha_-) \le l-k) \Pr(\chi = k) \\ &= (1 - \alpha_+\alpha_-) \sum_{l=0}^{\infty} x^l \sum_{k=0}^{l} \Pr(L(T, \alpha_+, \alpha_-) \le l-k)(\alpha_+\alpha_-)^k = \frac{1 - \alpha_+\alpha_-}{1 - x\alpha_+\alpha_-} Q(x), \end{aligned}$$

where the final equality follows by writing $x^l = x^{l-k} x^k$ and summing independently over $l-k$ and k. Multiplying both sides of this identity by $1 - x\alpha_+\alpha_-$ and equating like powers of x gives (10.208). □

The quantity $\tilde{D}_l$ in (10.208) with $\alpha_+ = 1/\alpha_- = \alpha$ can be expressed in terms of

$$D_l = \tilde{D}_l \Big|_{\alpha_+ = \alpha_- = 0} = \Big\langle e^{2T \sum_{j=1}^{l} \cos\theta_j} \Big\rangle_{U(l)}$$

familiar from (10.166), and monic orthogonal polynomials $\{\pi_j(e^{i\theta})\}_{j=0,1,\dots}$ with respect to the weight $e^{2T\cos\theta}$,

$$\frac{1}{2\pi}\int_{-\pi}^{\pi} \pi_j(e^{i\theta})\overline{\pi_k(e^{i\theta})} e^{2T\cos\theta}\, d\theta = \frac{1}{\kappa_j^2}\delta_{j,k}.$$

PROPOSITION 10.9.2 *We have*

$$\tilde{D}_l\Big|_{\alpha_+=1/\alpha_-=\alpha} = \Big((1-l)\pi_l(-\alpha)\pi_l(-\alpha^{-1}) - \alpha\pi_l'(-\alpha)\pi_l(-\alpha^{-1}) - \alpha^{-1}\pi_l(-\alpha)\pi_l'(-\alpha^{-1})\Big) D_l. \quad (10.209)$$

Proof. Let $\pi_n^*(z) := z^n\pi_n(1/z)$ (cf. Definition 8.6.1). It follows from the working of Exercises 5.1 q.6(i) (for the first equality) and the first, third and fourth equations in Proposition 8.6.2 (for the second equality) that

$$\frac{\Big\langle \prod_{j=1}^n w(\theta_j)(e^{i\theta_j}-x)(e^{-i\theta_j}-y)\Big\rangle_{U(n)}}{\Big\langle \prod_{j=1}^n w(\theta_j)\Big\rangle_{U(n)}} = \frac{1}{\kappa_n^2}\sum_{k=0}^n \kappa_k^2\pi_k(x)\pi_k(y) = \frac{\pi_n^*(x)\pi_n^*(y) - xy\pi_n(x)\pi_n(y)}{1-xy}. \quad (10.210)$$

Taking the limit $y \to 1/x = -1/\alpha$ in the second equality gives the stated result. □

Our interest is in the scaled limit of (10.208) with $\alpha_+ = 1/\alpha_- = \alpha$ for

$$l = [2T + T^{1/3}s], \qquad \alpha = 1 - y/T^{1/3}, \qquad T \to \infty.$$

According to (10.166), (10.199) and (9.46)

$$\lim_{T\to\infty} e^{-T^2} D_{[2T+T^{1/3}s]}(t) = E_2^{\rm soft}(0,(s,\infty)) = \exp\Big(-\int_s^\infty (t-s)q^2(t)\,dt\Big), \quad (10.211)$$

where $q(t)$ is the Painlevé II transcendent satisfying (9.45) subject to the boundary condition (9.47) with $\xi = 1$. The remaining task then is to analyze the polynomials $\pi_l(-\alpha)$ etc. in (10.209).

A rigorous treatment of this problem has been given by Baik and Rains [35], [33] using Riemann-Hilbert methods. To avoid technicalities we will proceed instead via a heuristic analysis [451], [196]. First we note from the third and fourth equation in Proposition 8.6.2 that

$$\pi_{n+1}(z) = z\pi_n(z) + r_{n+1}\pi_n^*(z), \qquad \pi_{n+1}^*(z) = r_{n+1}z\pi_n(z) + \pi_n^*(z). \quad (10.212)$$

Introduce

$$R_n(T) := (-1)^{n-1}r_n(T), \quad P_n(\alpha) = e^{-T\alpha}\pi_n^*(-\alpha), \quad Q_n(\alpha) = -e^{-T\alpha}(-1)^n\pi_n(-\alpha). \quad (10.213)$$

Then we see that the equations (10.212) are consistent with the asymptotic forms

$$R_{[2T+T^{1/3}s]}(T) \sim T^{-1/3}u(s), \quad P_{[2T+T^{1/3}s]}(1-y/T^{1/3}) \sim a(s,y), \quad Q_{[2t+T^{1/3}s]}(1-y/T^{1/3}) \sim b(s,y), \quad (10.214)$$

in which u, a, b are coupled by the partial differential equations

$$\frac{\partial a}{\partial s} = ub, \qquad \frac{\partial b}{\partial s} = ua - yb. \quad (10.215)$$

Moreover, in terms of the limiting functions a and b we see from the first equality in (10.210) that

$$\begin{aligned} g(s,y) &:= \lim_{T\to\infty} e^{-(\alpha_++\alpha_-)T}\tilde{D}_{[2T+T^{1/3}s]}(T)\Big|_{\alpha_+=1/\alpha_-=1-y/T^{1/3}} \Big/ D_{[2T+T^{1/3}s]}(T) \\ &= \int_{-\infty}^s a(u,y)a(u,-y)\,du = \int_{-\infty}^s b(u,y)b(u,-y)\,du, \end{aligned} \quad (10.216)$$

while (10.209) gives

$$g(s,y) = (s - y^2)a(s,y)a(s,-y) + a(s,-y)\frac{\partial}{\partial y}a(s,y) - a(s,y)\frac{\partial}{\partial y}a(s,-y). \tag{10.217}$$

Knowledge of $g(s,y)$ is sufficient for the scaled form of the cumulative distribution of $L(t,\alpha_+,\alpha_-)$, since from (10.208) and (10.211) it follows that

$$\begin{aligned}\tilde{F}_y(s) &:= \lim_{T\to\infty} \Pr\Big(\frac{L(T,1-y/T^{1/3},1+y/T^{1/3}) - 2\sqrt{T}}{T^{1/6}} \le s\Big)\\ &= \frac{\partial}{\partial s}\Big(g(s,y)E_2^{\rm soft}(0,(s,\infty))\Big).\end{aligned} \tag{10.218}$$

From the definition (10.210) of ϕ_n^* and the definitions (10.213) and (10.214), we see that a and b are related by

$$a(s,y) = -b(s,-y)e^{\frac{1}{3}y^3 - sy}. \tag{10.219}$$

The task then is to determine $a(s,y)$ or $b(s,y)$. In fact it is more convenient to obtain a system of equations determining both these quantities. The equations (10.215) give their dependence on s, but involve the unspecified limiting function $u(s)$. Thus it remains to determine $u(s)$, partial differential equations for the dependence of a and b on y, and boundary conditions for the partial differential equations. First we will determine equations for the dependence of a and b on y.

PROPOSITION 10.9.3 *We have*

$$\pi_n'(z) = \Big(\frac{n}{z} + \frac{t}{z^2} - \frac{r_{n+1}r_n t}{z}\Big)\pi_n(z) + \Big(\frac{r_{n+1}t}{z} - \frac{r_n t}{z^2}\Big)\pi_n^*(z), \tag{10.220}$$

$$\pi_n^{*\prime}(z) = \Big(-\frac{r_{n+1}t}{z} + r_n t\Big)\pi_n(z) + \Big(-t + \frac{r_{n+1}r_n t}{z}\Big)\pi_n^*(z). \tag{10.221}$$

Proof. The first equation follows immediately from (8.196) with $t \mapsto (2T)^2$, after making use of the second and third equations in Proposition 8.6.2. The second equation is derived similarly, after deriving an equation for $\pi_n^{*\prime}(z)$ analogous to (8.196) [232]. □

Formal substitution of the asymptotic forms (10.214) in the equations of Proposition 10.9.3 gives the partial differential equations

$$\frac{\partial a}{\partial y} = u^2 a - (u' + yu)b, \qquad \frac{\partial b}{\partial y} = (u' - yu)a + (y^2 - s - u^2)b. \tag{10.222}$$

To determine u we read off from (8.192) with $\sqrt{t} \mapsto 2T$ that r_n satisfies the second order difference equation

$$-\frac{n}{T}\frac{r_n}{1-r_n^2} = r_{n+1} + r_{n-1} \qquad \text{with} \qquad r_0 = 1, \quad r_1 = -\frac{I_1(2T)}{I_0(2T)}, \tag{10.223}$$

which we know is a transformed version of a particular discrete PII equation. With $R_n(t)$ as specified in (10.213), and the scaled form of this quantity as specified by (10.214), this difference equation becomes the differential equation

$$\frac{d^2u}{ds^2} = su + 2u^3. \tag{10.224}$$

We recognize (10.224) as the differential equation (9.45) satisfied by the Painlevé II transcendent q in (10.211). Furthermore, to be compatible with the first of the initial conditions in (10.223), we require for

$s = -2T^{2/3}$ and $T \to \infty$ that $T^{-1/3}u(s) \to -1$, or equivalently

$$u(s) \underset{s\to-\infty}{\sim} -\sqrt{-s/2}.$$

Up to the minus sign, this is precisely the boundary condition (9.48), and we know from [294] that (10.224) has the unique solution with this property,

$$u(s) = -q(s). \tag{10.225}$$

In relation to the boundary condition, we note from (10.219) that $a(s,0) = -b(s,0)$. Using this in (10.215) with $y = 0$ and u given by (10.225) shows

$$a(s,0) = A\exp\Big(-\int_s^\infty q(t)\,dt\Big).$$

To determine the constant A we note from (10.227) below that $a(s,0) \to 1$ as $s \to \infty$ and thus

$$a(s,0) = -b(s,0) = \exp\Big(-\int_s^\infty q(t)\,dt\Big). \tag{10.226}$$

With all quantities now completely determined, the results can be summarized as follows [35], [33].

PROPOSITION 10.9.4 *Denote by $-q(t)$ the Painlevé II transcendent satisfying (10.211), subject to the boundary condition (9.47) with $\xi = 1$, and define $\tilde{F}_y(s)$ as the scaled limit in (10.218). We have*

$$\tilde{F}_y(s) = \frac{\partial}{\partial s}\Big(g(s,y)\exp\Big(-\int_s^\infty (t-s)q^2(t)\,dt\Big)\Big),$$

where $g(s,y)$ is given in terms of $a(s,y)$ by (10.217). The function $a(s,y)$ in turn is specified as the solution of the coupled partial differential equations

$$\frac{\partial a}{\partial s} = qb, \qquad \frac{\partial b}{\partial s} = qa - yb,$$
$$\frac{\partial a}{\partial y} = q^2 a - (q' + yq)b, \qquad \frac{\partial b}{\partial y} = (q' - yq)a + (y^2 - s - q^2)b,$$

subject to the boundary condition (10.226).

EXERCISES 10.9 1. (i) For a general real weight $w(\theta)$ show

$$\frac{\langle \prod_{j=1}^n w(\theta_j)(1 - xe^{i\theta_j})\rangle_{U(n)}}{\langle \prod_{j=1}^n w(\theta_j)\rangle_{U(n)}} = \pi_n^*(x).$$

(ii) In the case of the weight $w(\theta) = e^{2T\cos\theta}$, use the result of (i) and (10.206) to show that

$$\pi_n^*(-x) = \frac{\tilde{D}_n|_{\alpha_+=x,\ \alpha_-=0}}{\tilde{D}_n|_{\alpha_+=\alpha_-=0}} = \frac{e^{xT}\Pr(L^+(T,x,0)\le n)}{\Pr(L^+(T,0,0)\le n)}.$$

From this conclude $\pi_n^*(-x) \to e^{xT}$ as $n \to \infty$ and thus

$$a(s,y) \to 1 \qquad \text{as} \quad s \to \infty. \tag{10.227}$$

Chapter Eleven

The Calogero–Sutherland model

Consideration of shifted mean parameter-dependent Gaussian random matrices, or equivalently Hermitian matrices with entries undergoing Brownian motion, leads to the Dyson Brownian motion model of the one-component log-gas. In the classical cases, a similarity transformation of the corresponding Fokker-Planck operator gives the Schrödinger operator for the Calogero-Sutherland model, which is the name given to the quantum many-body system for particles interacting on a line or a circle via the $1/r^2$ pair potential. By generalizing these Schrödinger operators to include exchange terms, decompositions into more elementary operators can be exhibited, and these operators can be used to establish integrability. For the coupling $\beta = 2$ the pair potential term in the Schrödinger operator is not present, giving rise to a free Fermi system and allowing the corresponding Green function to be expressed as a determinant. In the Gaussian and cases, this determinant form can be related to certain matrix integrals, including that due to Harish-Chandra, and Itzykson and Zuber. The determinant form also allows for the calculation of dynamical correlation functions, by applying formulas worked out in Chapter 5. These are analyzed in various scaled limits.

11.1 SHIFTED MEAN PARAMETER-DEPENDENT GAUSSIAN RANDOM MATRICES

In some energy spectra problems for chaotic quantum systems there is a parameter which varies the spectrum continuously. A well-known example, highlighted in [284], is the spectrum of the hydrogen atom as a function of the magnetic field strength. One approach to modeling such systems is to consider a random matrix $\mathbf{H}$ with distribution interpolating between a fixed matrix $\mathbf{H}_0$ (when $\tau = 0$) and one of the Gaussian ensembles of Chapter 1 (when $\tau = \infty$).

DEFINITION 11.1.1 *With $\mathbf{H}$ real symmetric ($\beta = 1$), Hermitian ($\beta = 2$), or self-dual quaternion real ($\beta = 4$), the parameter-dependent Gaussian ensembles are defined to have the joint p.d.f. for the independent elements*

$$\begin{aligned} P_\tau(\mathbf{H}^{(0)}; \mathbf{H}) &= \frac{1}{C} \exp\Big(-\beta \sum_{j,k=1}^{N} |H_{jk} - e^{-\tau} H_{jk}^{(0)}|^2 / 2|1 - e^{-2\tau}| \Big) \\ &= \frac{1}{C} \exp\Big(-\beta \mathrm{Tr}(\mathbf{H} - e^{-\tau}\mathbf{H}^{(0)})^2 / 2|1 - e^{-2\tau}| \Big), \end{aligned}$$

where C is the normalization and $\mathbf{H}^{(0)}$ is a fixed Hermitian matrix.

Comparing Definition 11.1.1 with the definitions given in Chapter 1 of the parameter-independent Gaussian ensembles we see that instead of the elements being chosen with mean zero and fixed standard deviation, they now have a mean determined by $\mathbf{H}^{(0)}$ (recall Section 1.8) and a parameter-dependent variance. For $\mathbf{H}^{(0)}$ fixed, we see by changing variables that

$$\mathbf{H} = |1 - e^{-2\tau}|^{1/2} \mathbf{X} + e^{-\tau} \mathbf{H}^{(0)}, \tag{11.1}$$

where $\mathbf{X}$ a member of the Gaussian β-ensemble ($\beta = 1, 2, 4$), and thus $\mathbf{H}$ is equal to the sum of a random and deterministic matrix. This setting is sometimes referred to as a random matrix model with a source [95].

In applications $\mathbf{H}^{(0)}$ may itself be a random matrix. For example, let $\beta = 2$ and suppose the joint p.d.f. for the elements at $\tau = 0$ is that of the GOE,

$$P_0(\mathbf{H}^{(0)}) = \prod_{j=1}^{N} \Big(\frac{1}{2\pi}\Big)^{1/2} e^{-(H_{jj}^{(0)})^2/2} \prod_{j<k} \frac{1}{\pi^{1/2}} e^{-(H_{jk}^{(0)})^2}$$

(recall Definition 1.1.1). The joint distribution of the elements of $\mathbf{H}$, $P_\tau(\mathbf{H})$ say, is obtained by integrating over $H_{jj}^{(0)}$ and $H_{jk}^{(0)}$,

$$\begin{aligned} P_\tau(\mathbf{H}) &:= \int P_\tau(\mathbf{H}^{(0)};\mathbf{H}) P_0(\mathbf{H}^{(0)})\,(d\mathbf{H}^{(0)}) \\ &= \prod_{j=1}^{N} \frac{e^{-H_{jj}^2/(1+e^{-2\tau})}}{\sqrt{\pi(1+e^{-2\tau})}} \prod_{j<k} \frac{2}{\pi\sqrt{1-e^{-4\tau}}} e^{-2(\mathrm{Re}\, H_{jk})^2/(1+e^{-2\tau}) - 2(\mathrm{Im}\, H_{jk})^2/(1-e^{-2\tau})}. \end{aligned}$$

This expression shows that $\mathbf{H}$ can be decomposed as a sum of a real symmetric matrix $\mathbf{X}^{(1)}$, and a real antisymmetric matrix $\mathbf{X}^{(2)}$,

$$\mathbf{H} = \Big(\frac{1+e^{-2\tau}}{2}\Big)^{1/2} \mathbf{X}^{(1)} + i\Big(\frac{1-e^{-2\tau}}{2}\Big)^{1/2} \mathbf{X}^{(2)}. \tag{11.2}$$

The diagonal and off-diagonal elements of $\mathbf{X}^{(1)}$ have independent Gaussian distributions with mean zero and variance 1 and $\frac{1}{2}$ respectively, while the off-diagonal elements of $\mathbf{X}^{(2)}$ have mean zero and variance $\frac{1}{2}$.

As emphasized in [284], the precise dependence on τ in Definition 11.1.1 has been chosen so that the joint p.d.f. satisfies the multidimensional Fokker-Planck equation of the Ornstein-Uhlenbeck process, or Brownian motion in a harmonic potential,

$$\frac{\partial P_\tau}{\partial \tau} = \sum_{\mu} \Big(\frac{\partial}{\partial H_\mu}(H_\mu P_\tau) + \frac{1}{\beta} D_\mu \frac{\partial^2 P_\tau}{\partial H_\mu^2} \Big). \tag{11.3}$$

Here the label μ ranges over the independent elements, including both the real and imaginary parts of the off-diagonal elements if they are complex, and $D_\mu = 1$ for the diagonal elements and $D_\mu = \frac{1}{2}$ for the off-diagonal elements. This equation can be verified from the factorization property in Definition 11.1.1, the fact that (11.3) is valid for the p.d.f. of each particular μ, and the product rule for differentiation.

Analogous to the studies of Chapters 1–3, let us take up the problem of calculating the eigenvalue p.d.f. of $\mathbf{H}$ from the distribution of the elements of P_τ. Our task is to compute an integral involving the variables associated with the eigenvectors of $\mathbf{H}$. Recalling (1.27) we see the eigenvalue p.d.f. for a particular $\mathbf{H}^{(0)}$ is given by

$$\prod_{1 \le j<k \le N} |\lambda_k - \lambda_j|^\beta \int P_\tau(\mathbf{H}^{(0)};\mathbf{H})(\mathbf{U}^\dagger d\mathbf{U}), \tag{11.4}$$

where $\mathbf{H} = \mathbf{U}\mathbf{L}\mathbf{U}^\dagger$ and $\mathbf{U}$ is real orthogonal, unitary or unitary symplectic for $\beta = 1, 2$ or 4, respectively.

As written it appears the p.d.f. (11.4) is a function of $\mathbf{H}^{(0)}$ and the eigenvalues of $\mathbf{H}$. In fact (11.4) depends only on the eigenvalues of both matrices. This follows by making the change of variables $\mathbf{U} \mapsto \mathbf{U}^{(0)}\mathbf{U}$, where $\mathbf{H}^{(0)} = \mathbf{U}^{(0)}\mathbf{L}^{(0)}\mathbf{U}^{(0)\dagger}$, and it is assumed $\mathbf{H}^{(0)}$ is diagonalized by the same set, or a subset, of the unitary matrices which diagonalize $\mathbf{H}$. From the cyclic property of the trace we check that then $P_\tau(\mathbf{H}^{(0)};\mathbf{H}) \mapsto P_\tau(\mathbf{L}^{(0)};\mathbf{H})$, while it is a fundamental property of the volume form $(\mathbf{U}^\dagger d\mathbf{U})$ that it is unchanged by such a transformation (recall (2.9)), thus showing that there is indeed no dependence on $\mathbf{U}^{(0)}$.

Next we want to change our viewpoint and regard $\mathbf{H}$ in (11.4) as fixed, in the sense that the unitary matrices $\mathbf{U}$ being integrated over are independent of $\mathbf{H}$. This can be achieved by changing variables $\mathbf{U} \mapsto \mathbf{U}\mathbf{V}$ and defining $\mathbf{V}\mathbf{L}\mathbf{V}^\dagger$ to be the fixed matrix $\mathbf{H}$. Doing this, noting again that $(\mathbf{U}^\dagger d\mathbf{U})$ is unchanged by this mapping, and interchanging the names of $\mathbf{U}$ and $\mathbf{V}$ for consistency with (11.4), we conclude that (11.4) is equal to

$$\prod_{1\le j<k\le N} |\lambda_k - \lambda_j|^\beta \int P_\tau(\mathbf{H}^{(0)}; \mathbf{VHV}^\dagger)\,(\mathbf{V}^\dagger d\mathbf{V}). \tag{11.5}$$

Finally, with the assumption $\{\mathbf{V}\}$ is sufficient to diagonalise $\mathbf{H}^{(0)}$, suppose the eigenvalue p.d.f. of $\mathbf{H}^{(0)}$ is $p_0(\vec{\lambda}^{(0)})$. Then we have that the eigenvalue p.d.f. of $\mathbf{H}$ averaged over p_0, $p_\tau(\vec{\lambda})$ say, is given by

$$\begin{aligned} p_\tau(\vec{\lambda}) = &\prod_{1\le j<k\le N} |\lambda_k - \lambda_j|^\beta \int_{-\infty}^{\infty} d\lambda_1^{(0)} \cdots \int_{-\infty}^{\infty} d\lambda_N^{(0)}\, p_0(\vec{\lambda}^{(0)}) \\ &\times \int P_\tau(\mathbf{H}^{(0)}; \mathbf{VHV}^\dagger)\,(\mathbf{V}^\dagger d\mathbf{V}). \end{aligned} \tag{11.6}$$

The p.d.f. (11.6) can be specified as the solution of a certain transformed Fokker-Planck equation. Essential to the analysis is the fact that since $P_\tau(\mathbf{H}^{(0)}; \mathbf{H})$ satisfies the Fokker-Planck equation (11.3), the integral in (11.6) satisfies the same equation, provided we change variables in the Fokker-Planck equation from the elements of $\mathbf{H}$ to the eigenvalues and variables dependent on the eigenvectors. This latter task is accomplished starting with the formula (1.12) for the metric form (and its analogue in the complex Hermitian and quaternion real cases) and using formulas from tensor calculus, as we will proceed to demonstrate.

For the metric form we have

$$(ds)^2 = \mathrm{Tr}(d\mathbf{H} d\mathbf{H}^\dagger) = \sum_{\mu,\nu} g_{\mu\nu} dH_\mu dH_\nu, \tag{11.7}$$

where $g_{\mu\nu} = \delta_{\mu,\nu}$ for the diagonal elements and $g_{\mu\nu} = 2\delta_{\mu,\nu}$ for the off-diagonal elements. In general the Laplacian associated with a metric form of the type (11.7) is given by (see, e.g., [410])

$$\nabla^2 = \sum_{\mu,\nu} \frac{1}{\sqrt{\det \mathbf{g}}} \frac{\partial}{\partial H_\mu} \sqrt{\det \mathbf{g}} (\mathbf{g}^{-1})_{\mu,\nu} \frac{\partial}{\partial H_\nu}. \tag{11.8}$$

With $\mathbf{g}$ diagonal we see

$$\nabla^2 = \sum_\mu D_\mu \frac{\partial^2}{\partial H_\mu^2}, \tag{11.9}$$

which is precisely the second order operator in (11.3). Also

$$\sum_\mu \frac{\partial}{\partial H_\mu}(H_\mu P_\tau) = \sum_\mu H_\mu \frac{\partial}{\partial H_\mu} P_\tau + \Big(N + \frac{\beta}{2} N(N-1)\Big) P_\tau, \tag{11.10}$$

so the task is to change variables in (11.9) and (11.10).

With $\mathbf{H} = \mathbf{ULU}^\dagger$, $\mathbf{U}$ specified as in (11.4) we have

$$\mathbf{U}^\dagger d\mathbf{H}\mathbf{U} = \mathbf{U}^\dagger d\mathbf{U}\mathbf{L} + d\mathbf{L} - \mathbf{L}\mathbf{U}^\dagger d\mathbf{U},$$

which when substituted in (11.7) gives

$$(ds)^2 = \sum_{j=1}^N (d\lambda_j)^2 + \sum_{\substack{j,k=1 \\ j\ne k}}^N |\lambda_k - \lambda_j|^2 \sum_{s=1}^\beta (\delta u_{jk}^{(s)})^2, \tag{11.11}$$

where $\delta u_{jk}^{(s)}$ denotes the real and imaginary components in $\mathbf{U}^\dagger d\mathbf{U} := [\delta u_{j,k}]$. This metric form is of the type (1.13) with

$$y_{jj} = \lambda_j, \quad h_{jj} = 1, \quad dy_{jk}^{(s)} = \delta u_{jk}^{(s)}, \quad h_{jk}^{(s)} = |\lambda_k - \lambda_j|. \tag{11.12}$$

Thus if we write

$$(ds)^2 = \sum_{\mu,\nu} g_{\mu\nu} dy_\mu dy_\nu$$

then $\mathbf{g} := [g_{\mu\nu}]$ is diagonal with

$$\sqrt{\det \mathbf{g}} = \prod_{j<k} |\lambda_k - \lambda_j|^\beta =: J \quad \text{and} \quad g_{\mu\mu} = 1 \ (\mu = (jj)).$$

Use of these facts in (11.8) (with $\{H_\nu\}$ replaced by $\{y_\nu\}$) gives that in terms of $\{\lambda_j\}$ and variables relating to $\mathbf{U}$

$$\nabla^2 = \frac{1}{J} \sum_{j=1}^N \frac{\partial}{\partial \lambda_j} \Big(J \frac{\partial}{\partial \lambda_j} \Big) + O_{\mathbf{U}}, \tag{11.13}$$

where the operator $O_{\mathbf{U}}$ involves derivatives with respect to variables relating to $\mathbf{U}$.

Analogous to (11.8) we have

$$\sum_\mu H_\mu \frac{\partial}{\partial H_\mu} = \sum_{\mu,\nu} H_\mu (\mathbf{g}^{-1})_{\mu,\nu} \frac{\partial}{\partial H_\nu}.$$

Changing variables as implied by (11.12) gives

$$\sum_\nu H_\nu \frac{\partial}{\partial H_\nu} = \sum_{j=1}^N \lambda_j \frac{\partial}{\partial \lambda_j} + O'_{\mathbf{U}}, \tag{11.14}$$

where $O'_{\mathbf{U}}$ involves derivatives with respect to variables relating to $\mathbf{U}$.

PROPOSITION 11.1.2 *We have*

$$\frac{\partial p_\tau}{\partial \tau} = \mathcal{L} p_\tau, \qquad \mathcal{L} := \frac{1}{\beta} \sum_{j=1}^N \frac{\partial^2}{\partial \lambda_j^2} + \sum_{j=1}^N \frac{\partial}{\partial \lambda_j} \Big(\lambda_j - \sum_{\substack{k=1 \\ k \ne j}}^N \frac{1}{\lambda_j - \lambda_k} \Big),$$

subject to the initial condition that $p_0(\lambda_1, \dots, \lambda_N)$ *is specified.*

Proof. To obtain the p.d.e, we note that since $P_\tau(\mathbf{H}^{(0)}; \mathbf{H})$ satisfies the Fokker-Planck equation (11.3) it follows immediately that the final integral in (11.6), $I := I(\mathbf{H}; \tau)$, say, satisfies the same equation, provided we change variables in the Fokker-Planck equation from the elements of $\mathbf{H}$ to the eigenvalues and variables dependent on the eigenvectors. The change of variables is accomplished by using the formulas (11.9), (11.10), (11.13) and (11.14), which give

$$\frac{\partial I}{\partial \tau} = \mathcal{L}_0 I + \Big(\beta N(N-1)/2\Big) I, \qquad \mathcal{L}_0 := \frac{1}{\beta J} \sum_{j=1}^N \frac{\partial}{\partial \lambda_j} \Big(J \frac{\partial}{\partial \lambda_j} \Big) + \sum_{j=1}^N \frac{\partial}{\partial \lambda_j} \lambda_j.$$

Substituting $I = J^{-1} p$ and expanding the derivatives using the product rule gives

$$\frac{\partial p_\tau}{\partial \tau} = \frac{1}{\beta} \sum_{j=1}^N \Big(\frac{\partial^2 p_\tau}{\partial \lambda_j^2} + p_\tau \frac{\partial J}{\partial \lambda_j} \frac{\partial J^{-1}}{\partial \lambda_j} + J \frac{\partial p_\tau}{\partial \lambda_j} \frac{\partial J^{-1}}{\partial \lambda_j} + p_\tau J \frac{\partial^2 J^{-1}}{\partial \lambda_j^2} \Big)$$
$$+ J p_\tau \sum_{j=1}^N \lambda_j \frac{\partial}{\partial \lambda_j} J^{-1} + \sum_{j=1}^N \frac{\partial}{\partial \lambda_j} \lambda_j p_\tau + \Big(\beta N(N-1)/2\Big) p_\tau.$$

This can be simplified by noting

$$J\frac{\partial J^{-1}}{\partial\lambda_j} = -\frac{\partial}{\partial\lambda_j}\log J \qquad \text{and} \qquad \sum_{j=1}^N \lambda_j \frac{\partial}{\partial\lambda_j}\log J = \beta \sum_{\substack{j,k=1\\ j\neq k}}^N \frac{\lambda_j}{\lambda_j - \lambda_k} = \beta N(N-1)/2,$$

where the last equality follows by interchanging j and k in the sum and taking the arithmetic mean. Thus

$$\frac{\partial p_\tau}{\partial\tau} = \frac{1}{\beta}\sum_{j=1}^N \left(\frac{\partial^2 p_\tau}{\partial\lambda_j^2} - p_\tau \frac{\partial^2}{\partial\lambda_j^2}\log J - \frac{\partial p_\tau}{\partial\lambda_j}\frac{\partial}{\partial\lambda_j}\log J + \frac{\partial}{\partial\lambda_j}\lambda_j p_\tau\right)$$

which, after computing $\partial \log J/\partial\lambda_j$, is the stated p.d.e.

The initial condition follows from the $\tau \to 0$ behavior of $P_\tau(\mathbf{H}^{(0)};\mathbf{H})$: it is nonzero only if the elements and thus the eigenvalues of $\mathbf{H}$ and $\mathbf{H}^{(0)}$ coincide. □

The connection between log-potential systems and the eigenvalue p.d.f. specified by the solution of a p.d.e. in Proposition 11.1.2 comes from the classification of the p.d.e. [469]. It is the Fokker-Planck equation for the log-potential system of Proposition 1.4.4. To see this we recall that for an interacting N-particle system with a general potential energy W, executing overdamped Brownian motion in a fictitious viscous fluid with friction coefficient γ at inverse temperature β, the evolution of the p.d.f. $p_\tau(\lambda_1,\dots,\lambda_N)$ for the location of the N particles at the points $\lambda_1,\dots,\lambda_N$ is given by the solution of the Fokker-Planck equation

$$\gamma\frac{\partial p_\tau}{\partial\tau} = \mathcal{L}p_\tau \qquad \text{where} \quad \mathcal{L} = \sum_{j=1}^N \frac{\partial}{\partial\lambda_j}\left(\frac{\partial W}{\partial\lambda_j} + \beta^{-1}\frac{\partial}{\partial\lambda_j}\right). \tag{11.15}$$

For the system of Proposition 1.4.4

$$W = W^{(\mathrm{H})} := -\sum_{1\le j<k\le N}\log|\lambda_j - \lambda_k| + \frac{1}{2}\sum_{j=1}^N \lambda_j^2 \quad \text{and so} \quad \frac{\partial W}{\partial\lambda_j} = -\sum_{\substack{k=1\\ k\neq j}}^N \frac{1}{\lambda_j - \lambda_k} + \lambda_j, \tag{11.16}$$

and (11.15) becomes the evolution equation of Proposition 11.1.2.

We remark that in small time intervals $\delta\tau := \tau - \tau_0$ the displacements $\lambda_j - \lambda_j^{(0)} := \lambda_j(\tau) - \lambda_j^{(0)}(\tau_0)$ as described by (11.15) are Gaussian random variables, with mean and variance

$$\langle(\lambda_j - \lambda_j^{(0)})\rangle = -\frac{\partial W}{\partial\lambda_j}\delta\tau, \qquad \langle(\lambda_j - \lambda_j^{(0)})^2\rangle = \frac{2}{\beta}\delta\tau, \tag{11.17}$$

respectively, while to leading order in $\delta\tau$ the correlation $\langle(\lambda_j - \lambda_j^{(0)})(\lambda_k - \lambda_k^{(0)})\rangle$ vanishes, as do all correlations involving three or more displacements. In fact these criteria uniquely specify the equation (11.15), and have been used by Dyson [145] to provide a different method of establishing Proposition 11.1.2. We will present this method in Exercises 11.1 q.3.

The equations (11.17) can be used to recast the Fokker-Planck equation (11.15) as the set of coupled Langevin equations

$$\gamma\frac{d\lambda_j(\tau)}{d\tau} = -\frac{\partial W}{\partial\lambda_j} + \mathcal{F}_j(\tau) \qquad (j = 1,\dots,N). \tag{11.18}$$

Here $\mathcal{F}_j(\tau)$ denotes a random force related to standard Brownian motion $B_j(\tau)$ by

$$\mathcal{F}_j(\tau) = \sqrt{\frac{2\gamma}{\beta}}B_j'(\tau). \tag{11.19}$$

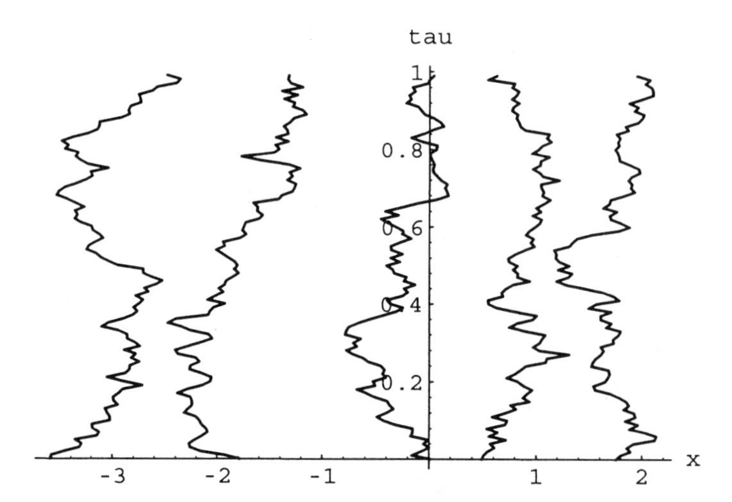

Figure 11.1 Brownian paths of the eigenvalue motion of a parameter-dependent Gaussian random matrix.

Thus $\mathcal{F}_j(\tau)$ is a Gaussian random variable with zero mean and variance given by

$$\langle \mathcal{F}_i(\tau)\mathcal{F}_j(\tau')\rangle = \frac{2\gamma}{\beta}\delta_{i,j}\delta(\tau-\tau').$$

Another general property of the Fokker-Planck equation (11.15) is that as $\tau \to \infty$, p_τ becomes proportional to $e^{-\beta W}$ (see Exercises 11.1 q.1 (iii)). Now with W as in (11.16), $e^{-\beta W}$ is proportional to the p.d.f. for the eigenvalues of the Gaussian β-ensemble of Chapter 1, and so the eigenvalue p.d.f. of the parameter-dependent Gaussian random matrices converges to the Gaussian β-ensemble in the limit $\tau \to \infty$. For $\beta = 1, 2$ and 4 this of course must be the case, as for $\tau \to \infty$ the joint p.d.f. of the elements of the parameter-dependent Gaussian random matrices coincides with the Gaussian ensembles of Chapter 1.

Using Definition 11.1.1, a realization of the Brownian motion paths described by the Fokker-Planck equation in Proposition 11.1.2 can be constructed. Thus we choose a matrix $\mathbf{H}^{(0)}$ from the GUE, say. The eigenvalues of this matrix give the initial positions. Then, for a particular increment of the parameter τ, $\delta\tau$ say ($\delta\tau \ll 1$), we construct a matrix according to Definition 11.1.1 (take $\beta = 2$ for definiteness): the diagonal and upper triangular off-diagonal elements (the latter have an independent real and imaginary part) are chosen with p.d.f.'s

$$\Big(\frac{1}{\pi(1-e^{-2\delta\tau})}\Big)^{1/2} e^{-(H_{jj}-H_{jj}^{(0)})^2/(1-e^{-2\delta\tau})} \quad \text{and} \quad \frac{2}{\pi(1-e^{-2\delta\tau})}e^{-2|H_{jk}-H_{jk}^{(0)}|^2/(1-e^{-2\delta\tau})}.$$

The eigenvalues of this matrix give the new positions. This procedure is continually repeated, with the matrix $\mathbf{H}^{(0)}$ given by the matrix computed in the previous step. Non-intersecting Brownian paths corresponding to the eigenvalues computed in this way are shown in Figure 11.1. This numerical procedure draws attention to the alternative viewpoint on parameter-dependent random matrices implied by the working of Exercises 11.1 q.3, namely, that the entries themselves of the matrix $\mathbf{H}$ can be regarded as Brownian paths.

EXERCISES 11.1 1. [469] Consider the Fokker-Planck operator (11.15).

(i) Show that as an operator identity

$$\mathcal{L}e^{-\beta W} = \frac{1}{\beta}\sum_{j=1}^N \frac{\partial}{\partial\lambda_j}e^{-\beta W}\frac{\partial}{\partial\lambda_j} = e^{-\beta W}\sum_{j=1}^N\Big(\frac{1}{\beta}\frac{\partial^2}{\partial\lambda_j^2} - \frac{\partial W}{\partial\lambda_j}\frac{\partial}{\partial\lambda_j}\Big),$$

and hence conclude $p = e^{-\beta W}$ is, up to normalization, the steady state solution of the Fokker-Planck equation.

(ii) Use the first equality in (i) to show that $e^{\beta W/2}\mathcal{L}e^{-\beta W/2}$ is a Hermitian operator, and use the second equality to obtain the explicit formula

$$e^{\beta W/2}\mathcal{L}e^{-\beta W/2} = \sum_{j=1}^{N}\Big(\frac{1}{\beta}\frac{\partial^2}{\partial\lambda_j^2} - \frac{\beta}{4}\Big(\frac{\partial W}{\partial \lambda_j}\Big)^2 + \frac{1}{2}\frac{\partial^2 W}{\partial\lambda_j^2}\Big).$$

(iii) Use (ii) to check the identity

$$e^{\beta W/2}\mathcal{L}e^{-\beta W/2} = -\beta^{-1}\sum_{j=1}^{N}\Pi_j^\dagger\Pi_j \quad \text{where} \quad \Pi_j := \frac{1}{i}\frac{\partial}{\partial\lambda_j} - \frac{i\beta}{2}\frac{\partial W}{\partial\lambda_j},$$

show from this that the eigenvalues of $\mathcal{L}$ are ≤ 0, and thus conclude that as $\tau\to\infty$ p is proportional to the steady state solution $e^{-\beta W}$. Also check that $\Pi_j e^{-\beta W/2} = 0$.

2. Let $\{|n\rangle\}_{n=0,1,\dots}$ be the complete set of eigenfunctions of a Hermitian operator A with corresponding eigenvalues $\{\lambda_n\}_{n=0,1,\dots}$, all assumed distinct. Let $\{|n\rangle^{(\epsilon)}\}_{n=0,1,\dots}$ be the complete set of eigenfunctions of the perturbed operator $A+\epsilon B$ with corresponding eigenvalues $\{\lambda_n^{(\epsilon)}\}_{n=0,1,\dots}$. Write

$$|n\rangle^{(\epsilon)} = |n\rangle + \epsilon|n\rangle_1 + \epsilon^2|n\rangle_2 + \cdots, \qquad \lambda_n^{(\epsilon)} = \lambda_n + \epsilon\lambda_{n,1} + \epsilon^2\lambda_{n,2} + \cdots,$$

and choose the normalization of $|n\rangle^{(\epsilon)}$ so that ${}^{(\epsilon)}\langle n|n\rangle^{(\epsilon)} = 1$ for all ϵ.

(i) Use the normalization condition to deduce ${}_j\langle n|n\rangle_j = 0$ $(j = 1,2,\dots)$.

(ii) Equate coefficients of ϵ and ϵ^2 in the eigenvalue equation for $|n\rangle^{(\epsilon)}$ and use (i) to show that $\lambda_{n,1} = \langle n|B|n\rangle$ and $\lambda_{n,2} = {}_1\langle n|B|n\rangle_1$.

(iii) In the equation $(A-\lambda_n)|n\rangle_1 + (B-\lambda_{n,1})|n\rangle = 0$ obtained by equating coefficients of ϵ in the eigenvalue equation for $|n\rangle^{(\epsilon)}$, note from (ii) that $(B-\lambda_{n,1})|n\rangle = \sum_{j\ne n}\langle j|B|n\rangle|j\rangle$ and operate on both sides with $(A-\lambda_n)^{-1}$ to express $|n\rangle_1$ in terms of $\{|j\rangle\}$ to obtain the formula

$$\lambda_n^{(\epsilon)} - \lambda_n = \langle n|\epsilon B|n\rangle - \sum_{j\ne n}\frac{\langle n|\epsilon B|j\rangle\langle j|\epsilon B|n\rangle}{\lambda_j - \lambda_n} + \mathrm{O}(\epsilon^3). \tag{11.20}$$

3. [145] The following results were used by Dyson to derive Proposition 11.1.2. To begin, replace τ by $\delta\tau := \tau - \tau_0$ in P_τ as given by Definition 11.1.1 and write

$$\langle f\rangle := \int f P_{\delta\tau}(\mathbf{H}^{(0)};\mathbf{H})\,(d\mathbf{H}).$$

(i) Show that as $\delta\tau\to 0$

$$\langle (H_\mu - H_\mu^{(0)})\rangle = -\delta\tau H_\mu^{(0)}, \qquad \langle |H_\mu - H_\mu^{(0)}|^2\rangle = \delta\tau\begin{cases} 2/\beta, & \mu = (jj), \\ 1 & \text{otherwise}, \end{cases}$$

while $\langle (H_\mu - H_\mu^{(0)})(\bar H_\nu - \bar H_\nu^{(0)})\rangle$ vanishes at order $\delta\tau$ as do all higher order correlations.

(ii) Suppose that with $\tau = \tau_0$, $\mathbf{H}$ is diagonal and thus $H_{jj}^{(0)} = \lambda_j^{(0)}$, $H_{jk}^{(0)} = 0$ $(j\ne k)$. Deduce from the formula in q.2(iii) that to second order in $\delta\tau$

$$\lambda_j - \lambda_j^{(0)} = H_{jj} - H_{jj}^{(0)} + \sum_{k\ne j}^{N}\frac{|(H_{jk} - H_{jk}^{(0)}|^2}{\lambda_j^{(0)} - \lambda_k^{(0)}}.$$

Use the results of (i) and the perturbation expansion to show that as $\tau \to \tau_0$

$$\langle \lambda_j - \lambda_j^{(0)} \rangle = -\delta\tau \Big(\lambda_j - \sum_{\substack{k=1 \\ k \ne j}}^{N} \frac{1}{\lambda_j - \lambda_k} \Big) = -\delta\tau \frac{\partial W^{(\mathrm{H})}}{\partial \lambda_j} \quad \text{and} \quad \langle (\lambda_j - \lambda_j^{(0)})^2 \rangle = \delta\tau \frac{2}{\beta},$$

which are the equations (11.17).

4. [534], [251] Consider the parameter-dependent Gaussian ensemble (11.1) with $\mathbf{H}^{(0)}$ also a member of the Gaussian β-ensemble and set $e^{-\tau} = \cos\phi$. Note that upon an appropriate modification, (11.20) implies

$$K_n := \frac{\partial^2 \lambda_n^{(\phi)}}{\partial \phi^2}\Big|_{\phi=0} = -\lambda_n + 2 \sum_{j \ne n} \frac{|\langle n|\mathbf{X}|j\rangle|^2}{\lambda_n - \lambda_j}.$$

The quantity K_n is referred to as the level curvature.

(i) Show that for the Gaussian β-ensembles ($\beta = 1, 2, 4$), averaging over $\mathbf{X}$ allows $|\langle n|\mathbf{X}|j\rangle|^2$ to be replaced by $\frac{1}{2}$ (i.e. the sum of the variances of the independent real numbers making up each of the off-diagonal elements equals $\frac{1}{2}$ in each case). As in the proof of Proposition 1.8.2 note that for large N

$$\sum_{j \ne n} \frac{1}{\lambda_n - \lambda_j} \sim \int_{-\sqrt{2N}}^{\sqrt{2N}} \frac{\rho_b(y)}{\lambda_n - y}\, dy = \lambda_n,$$

where the equality follows from (1.130) with $J = \sqrt{2N}$. Hence conclude $\langle K_n \rangle = 0$.

(ii) The distribution of level curvatures at $\lambda_n = 0$ is defined as

$$\mathcal{P}(K) = \frac{1}{N\rho_{(1)}(0)} \Big\langle \sum_{n=1}^{N} \delta(K - K_n)\delta(\lambda_n) \Big\rangle_{\mathbf{X}, \mathbf{H}^{(0)}}.$$

Use the Fourier integral form of $\delta(K - K_n)$ to perform the average over $\mathbf{X}$ and so obtain (after implementing too the effect of $\delta(\lambda_n)$)

$$\mathcal{P}(K) = \frac{1}{N\rho_{(1)}(0)} \frac{1}{2\pi} \int_{-\infty}^{\infty} e^{-i\omega K} \Big\langle \prod_{j=2}^{N} \frac{|\lambda_j|^\beta}{(1 + 2i\omega/\beta\lambda_j)^{\beta/2}} \Big\rangle_{\mathrm{E}_{\beta,N-1}(e^{-\beta\lambda^2/2})} d\omega.$$

Remark: In the large N limit this has been shown to be proportional to $1/(K^2 + 2N)^{\beta/2+1}$.

11.2 OTHER PARAMETER-DEPENDENT ENSEMBLES

11.2.1 Circular ensembles

A parameter-dependent extension of the circular ensembles presented in Chapter 2 has been given by Dyson [145]. First it is necessary to extend the defining equation in Definitions 2.2.2 and 2.2.3 so that $d\mathbf{S}$ is unitary at second order,

$$\mathbf{S} + d\mathbf{S} = \mathbf{V}\Big(\mathbf{1} + i\, d\mathbf{M} - \frac{1}{2}(d\mathbf{M})^2\Big)\mathbf{U} \tag{11.21}$$

($\mathbf{V} = \mathbf{U}^T$ for $\beta = 1$, $\mathbf{V} = \mathbf{U}^D$ for $\beta = 4$). A parameter $\delta\tau$ is introduced into the entries of the matrix of differentials $d\mathbf{M}$ by requiring

$$\frac{1}{\delta\tau}\langle \delta M_\mu^s \rangle = 0, \qquad \frac{1}{\delta\tau}\langle (\delta M_\mu^s)^2 \rangle = \begin{cases} 2/\beta, & \mu = (jj), \\ 1/\beta, & \text{otherwise}, \end{cases} \tag{11.22}$$

where δM_μ^s denotes the parameter-dependent version of dM_μ^s, while all other moments vanish to the same order.

Proceeding as in Exercises 11.1 q.3 (ii) we now suppose that for a particular parameter τ, $\mathbf{U} = \mathbf{V} = \mathrm{diag}[e^{i\theta_j/2}]$, so that $\mathbf{S}$ is diagonal. Then the perturbation expansion of Exercises 11.1 q.2(iii) applied to the r.h.s. of (11.21) gives that up to second order in dM_{jk}^s (s labels the independent real component(s) of the entries)

$$\begin{aligned} e^{i(\theta_j+\delta\theta_j)} &= e^{i\theta_j}\Big(1 + i\,dM_{jj} - \frac{1}{2}\sum_{k,s}(dM_{jk}^s)^2 + \sum_{k\neq j}\frac{e^{i\theta_k}\sum_s(dM_{jk}^s)^2}{e^{i\theta_k}-e^{i\theta_j}}\Big) \\ &= e^{i\theta_j}\Big(1 + i\,dM_{jj} - \frac{1}{2}(dM_{jj})^2 - \frac{i}{2}\sum_{k\neq j}\sum_s (dM_{jk}^s)^2\cot(\theta_k-\theta_j)/2\Big). \end{aligned} \tag{11.23}$$

Expanding the l.h.s. to second order in $\delta\theta_j$, equating real and imaginary parts on both sides and averaging over dM_{jj} and dM_{jk}^s, we see using (11.22) that

$$\langle \delta\theta_j \rangle = -\frac{\partial W}{\partial \theta_j}\delta\tau \quad \text{and} \quad \langle (\delta\theta_j)^2 \rangle = \frac{2}{\beta}\delta\tau,$$

where

$$W = W^{(\mathrm{C})} := -\sum_{j<k}\log|e^{i\theta_k} - e^{i\theta_j}|, \tag{11.24}$$

which are the equations (11.17) and thus equivalent to the Fokker-Planck equation (11.15) (with λ_j replaced by θ_j and $W = W^{(\mathrm{C})}$). The eigenvalues for matrices from the parameter-dependent circular ensemble can therefore be regarded as particles in a one-component log-gas on a circle undergoing overdamped Brownian motion.

An alternative approach is to begin with the equation (2.12) (for $\beta = 2$) and (2.15), (2.17) (for $\beta = 1, 4$) with the matrix elements of $d\mathbf{M}$ therein, dM_μ^s, say, replaced by δM_μ^s as specified by (11.22). The distribution of the elements of $\mathbf{M}$ then satisfy the Fokker-Planck equation

$$\frac{\partial P}{\partial \tau} = \frac{1}{\beta}\sum_{\mu,s} D_\mu \frac{\partial^2 P}{\partial (M_\mu^s)^2} \tag{11.25}$$

(cf. (11.3)). According to the theory given by (11.7)–(11.9) the r.h.s. is equal to $\frac{1}{\beta}\nabla^2$, and ∇^2 can in turn be calculated by changing variables in (11.8), using the metric form expansion (2.22) as in going from (11.11) to (11.13). This reclaims the Fokker-Planck equation (11.15) with $W = W^{(\mathrm{C})}$, $\lambda_j = \theta_j$.

Notice that the perturbative expansion (11.23) can be interpreted as originating from the equation

$$\mathbf{S}(\tau + d\tau) = \mathbf{S}(\tau)e^{i\,d\mathbf{M}}. \tag{11.26}$$

This can be used to approximately compute members of the parameter-dependent circular ensembles.

11.2.2 Wishart matrices

Parameter-dependent Wishart matrices are constructed from parameter-dependent Gaussian random matrices in an analogous fashion to their parameter independent counterparts. Thus if $\mathbf{X}$ is an $n \times m$ parameter-dependent Gaussian random matrix with joint p.d.f. for the elements given by

$$P_\tau(\mathbf{X}^{(0)};\mathbf{X}) := \frac{1}{C} \exp\Big(-\beta \sum_{j=1}^{n} \sum_{k=1}^{m} |X_{jk} - e^{-\tau} X_{jk}^{(0)}|^2 / 2|1 - e^{-2\tau}| \Big)$$
$$= \frac{1}{C} \exp\Big(-\beta \mathrm{Tr}((\mathbf{X} - e^{-\tau}\mathbf{X}^{(0)})^\dagger (\mathbf{X} - e^{-\tau}\mathbf{X}^{(0)}))/2|1 - e^{-2\tau}| \Big),$$

then $\mathbf{A} = \mathbf{X}^\dagger \mathbf{X}$ defines a parameter-dependent Wishart matrix. As with parameter-dependent Gaussian random matrices, the corresponding eigenvalue p.d.f. can be computed in at least two different ways: via the method of the proof of Proposition 11.1.2 [311] or via Dyson's method of Exercises 11.1 q.3. Here the former method will be outlined (for the latter method see Exercises 11.2 q.1).

The starting point is to decompose $\mathbf{X}$ according to its singular value decomposition $\mathbf{X} = \mathbf{U}\mathbf{L}\mathbf{V}^\dagger$ ($\mathbf{L} = \mathrm{diag}(x_1, \dots, x_m)$), where $\{x_j^2\}$ are the eigenvalues of $\mathbf{A}$. We want to substitute this in P_τ and integrate out the dependence on $\mathbf{U}$ and $\mathbf{V}$ to obtain the p.d.f. of $\{x_j^2\}$. Recalling from (3.4) the eigenvalue portion of the Jacobian, analogous to (11.4) this operation is equivalently given by

$$J \int (\mathbf{U}^\dagger d\mathbf{U}) \int (\mathbf{V}^\dagger d\mathbf{V}) P_\tau(\mathbf{X}^{(0)}; \mathbf{U}\mathbf{X}\mathbf{V}^\dagger), \tag{11.27}$$

where $J \propto \prod_{j=1}^m x_j^{\beta a + 1} \prod_{1 \le j < k \le m} |x_k^2 - x_j^2|^\beta$, $a = n - m + 1 - 2/\beta$.

Assuming $\mathbf{X}^{(0)}$ can be decomposed by the same group of matrices as $\mathbf{X}$, that is, $\{\mathbf{U}\}$, $\{\mathbf{V}\}$, or a subgroup of these matrices, it follows from the cyclic property of the trace in the definition of P_τ that (11.27) depends only on $\{x_j^2\}$ and $\{(x_j^{(0)})^2\}$. In this circumstance, with $p_0((x_1^{(0)})^2, \dots, (x_m^{(0)})^2)$ denoting the p.d.f. for $\{(x_j^{(0)})^2\}$, the p.d.f. for $\{x_j^2\}$ corresponding to the eigenvalues of $\mathbf{A} = \mathbf{X}^\dagger\mathbf{X}$ averaged over $\mathbf{X}^{(0)}$ is given by

$$p_\tau(x_1^2, \dots, x_m^2) = J \int_0^\infty d(x_1^{(0)})^2 \cdots \int_0^\infty d(x_m^{(0)})^2 \, p_0((x_1^{(0)})^2, \dots, (x_m^{(0)})^2)$$
$$\times \int (\mathbf{U}^\dagger d\mathbf{U}) \int (\mathbf{V}^\dagger d\mathbf{V}) P_\tau(\mathbf{X}^{(0)}; \mathbf{U}\mathbf{X}\mathbf{V}^\dagger). \tag{11.28}$$

By repeating the proof of Proposition 11.1.2 with J now given as below (11.27), we find that (11.28) can be characterized as the solution of the Fokker-Planck equation describing the overdamped Brownian dynamics of the log-gas with potential energy

$$W = W^{(\mathrm{L})} := \frac{1}{2} \sum_{j=1}^N x_j^2 - \frac{a'}{2} \sum_{j=1}^N \log x_j^2 - \sum_{1 \le j < k \le N} \log |x_k^2 - x_j^2|, \tag{11.29}$$

where $a' = a + 1/\beta$ and $N = m$.

11.2.3 Jacobi ensembles

We have seen in Section 3.7 that the Jacobi ensemble can be realized by the p.d.f. for the singular values which form the matrix $\mathbf{\Lambda}_t$ in the decomposition (3.99). As $\mathbf{S}$ in (3.99) is unitary (with the additional constraint of being symmetric for $\beta = 1$, and self-dual quaternion for $\beta = 4$), a parameter-dependent extension is possible via the evolution equation (11.26) [238]. The corresponding evolution equation for the singular values can be calculated via the method used for the parameter-dependent circular ensembles, as given in the paragraph including (11.25). This in turn relies on the validity of (11.13), for which (up to a proportionality constant) one requires that $h_{jj} = c$, c a constant, in the analogue of (1.13), as holds true in (11.12). Now, in terms of the variables λ_j, we see from the expression for T_5 in the proof of Proposition 3.8.1 that $h_{jj} = 2/(1 - \lambda_j^2)$. However, if we write $\lambda_j = \sin\phi_j$ and consider $\{\phi_j\}$ as the variables we then obtain $h_{jj} = 2$. The theory of Section 11.1 then tells us that $\{\phi_j\}$ evolves according to the Fokker-Planck equation (11.15) with λ_j replaced by ϕ_j and W such that $e^{-\beta W} = J$, J being the Jacobian specified in Proposition 3.8.1, with the change of variables $\lambda_j = \sin\phi_j$. Specifically we have

$$W = W^{(\mathrm{J})} := -\frac{a'}{2}\sum_{j=1}^{N}\log\sin^2\phi_j - \frac{b'}{2}\sum_{j=1}^{N}\log\cos^2\phi_j - \sum_{1\le j<k\le N}\log|\sin^2\phi_j - \sin^2\phi_k| \tag{11.30}$$

for certain a', b'.

11.2.4 DMPK equation

Important in the theory of conductance in disordered wires is a parameter-dependent extension of the transfer matrix decomposition (3.114). As these matrices have the property (3.94), one sees that $d\tilde{\mathbf{M}}^\dagger I'_{n,n}\tilde{\mathbf{M}} = -\tilde{\mathbf{M}}^\dagger I'_{n,n}d\tilde{\mathbf{M}}$ and consequently $\tilde{\mathbf{M}}^\dagger I'_{n,n}d\tilde{\mathbf{M}} = i\,d\mathbf{M}$ for some $2n\times 2n$ Hermitian matrix $\mathbf{M}$. This tells us that the appropriate metric form is $\mathrm{Tr}((d\mathbf{M})^2)$, which together with the decomposition (3.114) can be used in the evolution equation (11.26) to obtain an evolution equation for the variables λ_i in $\mathbf{\Lambda} = \mathrm{diag}(\lambda_1,\dots,\lambda_N)$ of (3.114). As in the above discussion of Jacobi ensembles, we must first make a change of variables in order that $h_{jj} = c$ (constant) in the analogue of (1.13). This is achieved by setting $\lambda_j = \sinh^2 x_j$, and considering $\{x_j\}$ as the independent variables. We conclude that $\{x_j\}$ evolves according to the Fokker-Planck equation (11.15) with λ_j replaced by x_j and $e^{-\beta W} = J$. The Jacobian is first computed in the variable λ_j according to the calculation of Proposition 3.8.1, and then the change of variables to $\{x_j\}$ is made. All three values $\beta = 1,2,4$ can be considered by imposing the constraints (3.100), (3.101). One finds (see e.g. [53])

$$W = W^{(\mathrm{D})} := -\frac{1}{2\beta}\sum_{j=1}^{N}\log|\sinh 2x_j| - \sum_{1\le j<k\le N}\log|\sinh^2 x_k - \sinh^2 x_j|. \tag{11.31}$$

In the variables $\{\lambda_i\}$ this particular Fokker-Planck equation is named the Dorokhov-Mello-Pereyra-Kumar (DMPK) equation.

EXERCISES 11.2 1. [11] In this exercise the Fokker-Planck equation for Wishart matrices will be rederived using the method of Dyson. The eigenvalues of $\mathbf{X}^\dagger\mathbf{X}$ ($\mathbf{X}^{(0)\dagger}\mathbf{X}^{(0)}$) are denoted by $x_1^2,\dots,x_m^2$ ($(x_1^{(0)})^2,\dots,(x_m^{(0)})^2$) and $\mathbf{X} = [x_{jk}]$, $\mathbf{X}^{(0)} = [x_{jk}^{(0)}]$, $\delta x_j := x_j - x_j^{(0)}$, and $\delta x_{jk} := x_{jk} - x_{jk}^{(0)}$. Furthermore, we suppose that $\mathbf{X}^{(0)\dagger}\mathbf{X}^{(0)}$ is diagonal, and consider a small increment in parameter $\delta\tau$.

(i) From the p.d.f. for the elements of $\mathbf{X}$ show that $\langle\delta x_{jk}\rangle = -x_{jk}^{(0)}\delta\tau$ and $\langle(\delta x_{jk}^s)^2\rangle = \frac{2}{\beta}\delta\tau$, while all other moments vanish to the same order in $\delta\tau$. Also, note from the assumption regarding $\mathbf{X}^{(0)\dagger}\mathbf{X}^{(0)}$ that $(x_j^{(0)})^2 = \sum_{k=1}^n |x_{kj}^{(0)}|^2 := \sum_{k=1}^n\sum_{s=1}^\beta (x_{kj}^{(0)s})^2$.

(ii) Use the perturbation formula of Exercises 11.1 q.2 with $\epsilon B = \mathbf{X}^\dagger\mathbf{X} - \mathbf{X}^{(0)\dagger}\mathbf{X}^{(0)}$ to show that to second order in ϵ

$$x_j^2 - (x_j^{(0)})^2 = \sum_{k=1}^n\Big(|\delta x_{kj}|^2 + \delta\bar{x}_{kj}\,x_{kj}^{(0)} + \delta x_{kj}\,\bar{x}_{kj}^{(0)}\Big) + \sum_{\substack{k=1\\ \ne j}}^n \frac{|\sum_{l=1}^m(\delta\bar{x}_{lj}\delta x_{lk} + \delta\bar{x}_{lj}\,x_{lk}^{(0)} + \bar{x}_{lj}^{(0)}\delta x_{lk})|^2}{x_j^2 - x_k^2}.$$

(iii) Average over $\{\delta x_{jk}\}$ in (ii) using the formulas of (i) to deduce that to order $\delta\tau$

$$\langle(\delta x_j)^2\rangle + 2x_j\langle\delta x_j\rangle = \Big(-2x_j^2 + 2n + 2\sum_{\substack{k=1\\ \ne j}}^m \frac{x_j^2 + x_k^2}{x_j^2 - x_k^2}\Big)\delta\tau.$$

(iv) Add $(x_j^{(0)})^2$ to both sides of the equation in (ii), take the square root, expand the r.h.s, and average over

$\{\delta x_{jk}\}$ to deduce that

$$\langle \delta x_j \rangle = \frac{1}{2x_j}\Big(-2x_j^2 + 2n - \frac{2}{\beta} + 2\sum_{\substack{k=1 \\ \neq j}}^{m} \frac{x_j^2 + x_k^2}{x_j^2 - x_k^2} \Big) \delta\tau.$$

(v) From the results of (iii) and (iv) deduce the equations (11.17) with W given by (11.29).

2. [106] In the theory of radial stochastic Loewner evolution (radial SLE), the conformal mapping associated with N such processes, $G_t^{(N)}(z)$ satisfies the differential equation

$$\frac{dG_t^{(N)}}{dt} = -G_t^{(N)} \sum_{j=1}^{N} \frac{G_t^{(N)} + e^{i\theta_j(t)}}{G_t^{(N)} - e^{i\theta_j(t)}},$$

where

$$\frac{d\theta_j(t)}{dt} = \sum_{\substack{k=1 \\ k\neq j}}^{N} \cot \frac{\theta_j(t) - \theta_k(t)}{2} + \sqrt{\kappa} B_j'(t). \tag{11.32}$$

By comparing with (11.18), identify (11.32) as equivalent to the Fokker-Planck equation (11.15) in the case W is given by (11.24), $\gamma = \frac{1}{2}$ and $\kappa = 4/\beta$.

11.3 THE CALOGERO-SUTHERLAND QUANTUM SYSTEMS

11.3.1 The Schrödinger operators

In Exercises 11.1 q.1(i) we have seen that the Fokker-Planck operator of the type (11.15) can always be transformed into a Hermitian operator. When W corresponds to the parameter-dependent random matrix ensembles, the transformation can be written

$$-e^{\beta W/2} \mathcal{L} e^{-\beta W/2} = (H - E_0)/\beta, \tag{11.33}$$

where H is the Schrödinger operator for a quantum many-body system in which the particles interact via a $1/r^2$ pair potential (all three-body terms which occur in general cancel). In the Gaussian and circular case, these quantum many-body systems were first studied by Calogero [105] and Sutherland [506]. The name Calogero-Sutherland model has now been given to a class of quantum many-body systems, including those originally studied by Calogero and Sutherland, which are related to root systems for classical reflection groups [434] (see Section 11.5.5). We will consider a subclass of these models for which the Schrödinger operator has a potential energy consisting of one- and two-body terms only,

$$H = -\sum_{j=1}^{N} \frac{\partial^2}{\partial x_j^2} + \sum_{j=1}^{N} V_1(x_j) + \sum_{1\le j<k\le N} V_2(x_j, x_k), \tag{11.34}$$

and for which the ground state ψ_0 factorizes into the form

$$\psi_0 = \prod_{j=1}^{N} f_1(x_j) \prod_{1\le j<k\le N} f_2(x_j, x_k). \tag{11.35}$$

Note that all Schrödinger operators derived from (11.33) with W consisting of one- and two-body potentials only must have the property (11.35) since, as follows from Exercises 11.1 q.1, ψ_0 is proportional to $e^{-\beta W/2}$.

PROPOSITION 11.3.1 *Consider the Fokker-Planck operator (11.15) with λ_j replaced by x_j and W given* *by (11.16). The operator identity (11.33) holds with*

$$H = H^{(\mathrm{H})} := -\sum_{j=1}^{N} \frac{\partial^2}{\partial x_j^2} + \frac{\beta^2}{4} \sum_{j=1}^{N} x_j^2 + \beta(\beta/2 - 1) \sum_{1 \le j < k \le N} \frac{1}{(x_j - x_k)^2}$$

and $E_0 = \frac{N\beta}{2} + \frac{\beta^2}{4} N(N-1)$.

Proof. We use the general formula of Exercises 11.1 q.1(ii). For the given W

$$\sum_{j=1}^{N} \frac{\partial^2 W}{\partial x_j^2} = N + \sum_{j=1}^{N} \sum_{\substack{k=1 \\ k \ne j}}^{N} \frac{1}{(x_j - x_k)^2}$$

and

$$\sum_{j=1}^{N} \Big(\frac{\partial W}{\partial x_j} \Big)^2 = \sum_{j=1}^{N} x_j^2 + \sum_{\substack{j,k_1,k_2=1 \\ k_1,k_2 \ne j}}^{N} \frac{1}{(x_j - x_{k_1})} \frac{1}{(x_j - x_{k_2})} - 2 \sum_{\substack{j,k=1 \\ k \ne j}}^{N} \frac{x_j}{x_j - x_k}.$$

The last sum occurred in the proof of Proposition 11.1.2. It is equal to $N(N-1)/2$. Also, in the triple sum all terms with $k_1 \ne k_2$ cancel (these are the apparent three-body terms), due to the identity

$$\frac{1}{(a-b)(a-c)} + \frac{1}{(b-a)(b-c)} + \frac{1}{(c-a)(c-b)} = 0. \tag{11.36}$$

Thus

$$\sum_{j=1}^{N} \Big(\frac{\partial W}{\partial x_j} \Big)^2 = \sum_{j=1}^{N} x_j^2 + \sum_{\substack{j,k=1 \\ k \ne j}}^{N} \frac{1}{(x_k - x_j)^2} - N(N-1).$$

Substituting in the general formula gives the stated Schrödinger operator. □

For W corresponding to the other parameter-dependent random matrix ensembles, a similar result holds.

PROPOSITION 11.3.2 *With W given by (11.24), (11.29), (11.30) and (11.31), the operator identity (11.33) holds with*

$$H = H^{(\mathrm{C})} := -\sum_{j=1}^{N} \frac{\partial^2}{\partial x_j^2} + \beta(\beta/2 - 1) \Big(\frac{\pi}{L} \Big)^2 \sum_{1 \le j < k \le N} \frac{1}{\sin^2(\pi(x_j - x_k)/L)},$$

$$H = H^{(\mathrm{L})} := -\sum_{j=1}^{N} \frac{\partial^2}{\partial x_j^2} + \sum_{j=1}^{N} \Big(\frac{\beta a'}{2} \Big(\frac{\beta a'}{2} - 1 \Big) \frac{1}{x_j^2} - \frac{\beta^2}{4} x_j^2 \Big)$$
$$+ \beta(\beta/2 - 1) \sum_{1 \le j < k \le N} \Big(\frac{1}{(x_k - x_j)^2} + \frac{1}{(x_k + x_j)^2} \Big),$$

$$H = H^{(\mathrm{J})} := -\sum_{j=1}^{N} \frac{\partial^2}{\partial \phi_j^2} + \sum_{j=1}^{N} \Big(\frac{a'\beta}{2} \Big(\frac{a'\beta}{2} - 1 \Big) \frac{1}{\sin^2 \phi_j} + \frac{b'\beta}{2} \Big(\frac{b'\beta}{2} - 1 \Big) \frac{1}{\cos^2 \phi_j} \Big)$$
$$+ \beta(\beta/2 - 1) \sum_{1 \le j < k \le N} \Big(\frac{1}{\sin^2(\phi_j - \phi_k)} + \frac{1}{\sin^2(\phi_j + \phi_k)} \Big),$$

$$H = H^{(\mathrm{D})} := -\sum_{j=1}^{N} \Big(\frac{\partial^2}{\partial x_j^2} + \frac{1}{\sinh^2 2x_j} \Big)$$
$$+ \beta(\beta/2 - 1) \sum_{1 \le j < k \le N} \Big(\frac{1}{\sinh^2(x_j - x_k)} + \frac{1}{\sinh^2(x_j + x_k)} \Big),$$

where $0 \le x_j \le L$, $x_j > 0$, $0 < \phi_j < \pi$, $x_j > 0$ *and*

$$
\begin{aligned}
E = E_0^{(\mathrm{C})} &:= \beta^2 \Big(\frac{\pi}{L}\Big)^2 \frac{N(N^2-1)}{12}, \\
E = E_0^{(\mathrm{L})} &:= \frac{\beta^2 N}{2}(a' + N - 1) + \frac{\beta N}{2}, \\
E = E_0^{(\mathrm{J})} &:= \frac{\beta^2 N}{4}\Big(a'^2 + b'^2 + (N-1)(a'+b') + 2a'b' + 4(N-1)(N-2)\Big) + \beta N(N-1), \\
E = E_0^{(\mathrm{D})} &:= -N - 2\beta N(N-1) - \frac{\beta^2}{3} N(N-1)(N-2),
\end{aligned}
$$

respectively.

Proof. The derivation follows the same procedure as the proof of Proposition 11.3.1, with (11.36) being replaced by the identities

$$
\begin{aligned}
\cot(a-b)\cot(a-c) + \cot(b-a)\cot(b-c) + \cot(c-a)\cot(c-b) &= -1, \\
\frac{a}{(a-b)(a-c)} + \frac{b}{(b-a)(b-c)} + \frac{c}{(c-a)(c-b)} &= 0, \\
\frac{a(1-a)}{(a-b)(a-c)} + \frac{b(1-b)}{(b-a)(b-c)} + \frac{c(1-c)}{(c-a)(c-b)} &= -1
\end{aligned}
\tag{11.37}
$$

(the latter applies to both $W^{(\mathrm{J})}$ and $W^{(\mathrm{D})}$). □

11.3.2 The Green function and eigenfunctions

Note that in general the substitutions (11.33) and

$$p_\tau = e^{E_0 \tau/\beta} e^{-\beta W/2} \psi_\tau \tag{11.38}$$

in the Fokker-Planck equation (11.15) with $\gamma = 1$ give the imaginary time Schrödinger equation

$$-\beta \frac{\partial}{\partial \tau} \psi_\tau = H \psi_\tau. \tag{11.39}$$

Since (11.39) is linear, to calculate its solution for general initial conditions it suffices to obtain the solution when the precise particle positions are known initially,

$$\psi_\tau(\vec{x})\Big|_{\tau=0} = \prod_{l=1}^{N} \delta(x_l - x_l^{(0)}) \qquad (x_1^{(0)} < \cdots < x_N^{(0)}). \tag{11.40}$$

With this initial condition we say that $\psi_\tau =: G_\tau^{\mathrm{S}}(\vec{x}^{(0)}; \vec{x})$ is the *Green function* solution of the equation (11.39). Because of the ordering constraint in (11.40), G_τ^{S} has the $\tau = 0$ normalization

$$\int_A dx_1 \cdots dx_N \, G_\tau^{\mathrm{S}}(\vec{x}^{(0)}; \vec{x})|_{\tau=0} = 1, \tag{11.41}$$

where A is the region $-\infty < x_1 < x_2 < \cdots < x_N < \infty$. Note that it follows from (11.38) that the Green function solution of the Fokker-Planck equation (11.15), G_τ^{FP} say, is given in terms of G_τ^{S} by

$$G_\tau^{\mathrm{FP}}(\vec{x}^{(0)}; \vec{x}) = e^{E_0 \tau/\beta} \frac{e^{-\beta W(x_1, \dots, x_N)/2}}{e^{-\beta W(x_1^{(0)}, \dots, x_N^{(0)})/2}} G_\tau^{\mathrm{S}}(\vec{x}^{(0)}; \vec{x}). \tag{11.42}$$

The Green function solution of (11.39) can be expressed in terms of the eigenfunctions and eigenvalues of the Schrödinger operator H. Thus, according to the method of separation of variables, if H has a complete set of normalized eigenfunctions $\{\psi_\kappa\}$ with corresponding eigenvalues $\{E_\kappa\}$ (in general the label κ will be an N-tuple of integers), then

$$G_\tau^{\rm S}(\vec{x}^{(0)};\vec{x}) = \sum_\kappa \psi_\kappa(\vec{x}^{(0)})\psi_\kappa(\vec{x})e^{-E_\kappa \tau/\beta}. \tag{11.43}$$

At this stage the Hamiltonian $H^{(\rm D)}$ is distinguished from the other Hamiltonians in that it does not support bound states. This means the complete set of eigenfunctions are labeled by continuous quantum numbers. For the other Hamiltonians there is a complete set of bound states, and the label κ is an N-tuple of integers. Only these cases will be considered in the following.

It turns out that the study of the eigenfunctions has its natural setting in generalizations of H which involve exchange terms. These generalizations also play a crucial role in the study of the integrability of H.

DEFINITION 11.3.3 *A Schrödinger operator H of the form (11.34) is said to be integrable if there exists a set of N algebraically independent commuting operators $\{X_i\}_{i=1,\dots,N}$ such that*

$$[H, X_i] = 0, \qquad i = 1,\dots,N.$$

The theme of integrability will be addressed in the next two sections. The commuting operators which emerge will be used to study the eigenfunctions of the Schrödinger operators in subsequent chapters.

EXERCISES 11.3 1. [505] The objective of this exercise is to generalize the Schrödinger equation $H^{(\rm C)}\psi_0^{(\rm C)} = E_0^{(\rm C)}\psi_0^{(\rm C)}$. In particular, we seek $W(x)$ even and periodic of period L such that the Schrödinger operator

$$H = -\sum_{j=1}^N \frac{\partial^2}{\partial x_j^2} + \sum_{1\le j<k\le N} W(x_k - x_j) \tag{11.44}$$

has a ground state of the form

$$\psi_0 = \prod_{1\le j<k\le N} u(x_k - x_j) \tag{11.45}$$

for $u(x)$ even or odd.

(i) Substitute (11.45) in (11.44) to deduce that

$$H\psi_0 = \sum_{1\le j<k\le N}\Big(W(x_k - x_j) - 2\phi'(x_k - x_j)\Big)\psi_0 - \sum_{j=1}^N\Big(\sum_{\substack{k=1\\k\ne j}}^N \phi(x_j - x_k)\Big)^2\psi_0,$$

where $\phi(x) := u'(x)/u(x)$, and thus ϕ is odd.

(ii) Note from the equation of (i) that if there is an even function f such that

$$\sum_{j=1}^N\Big(\sum_{\substack{k=1\\k\ne j}}^N \phi(x_j - x_k)\Big)^2 = \sum_{j=1}^N\sum_{\substack{k=1\\k\ne j}}^N\Big(\phi^2(x_j - x_k) - (N-2)f(x_j - x_k)\Big), \tag{11.46}$$

then ψ_0 as specified by (11.45) is the ground state of (11.44) with

$$W(x) = 2\Big(\phi'(x) + \phi^2(x) - (N-2)f(x)\Big) + \frac{2E_0}{N(N-1)}.$$

(iii) Show that (11.46) holds if, for $x + y + z = 0$,

$$\phi(x)\phi(y) + \phi(y)\phi(z) + \phi(z)\phi(x) = f(x) + f(y) + f(z).$$

(iv) With

$$\phi(x) = \alpha\zeta(x), \qquad \zeta(x) := \frac{\pi}{L}\frac{\theta_1'(\pi x/L;q)}{\theta_1(\pi x/L;q)},$$

where $\theta_1(z;q)$ is specified by (2.79), use the identity [541]

$$(\zeta(x)+\zeta(y)+\zeta(z))^2 = -(\zeta'(x)+\zeta'(y)+\zeta'(z))$$

valid for $x+y+z=0$ to show that with

$$W(x) = (2\alpha - N + 2)\zeta'(x) + (2\alpha^2 - N + 2)\zeta^2(x) + \frac{2E_0}{N(N-1)}$$

the ground state of (11.44) is given by (11.45) with

$$u(x) = (\theta_1(\pi x/L;q))^{\alpha}.$$

(With $\alpha = \beta/2$, the absolute value squared of (11.45) is therefore proportional to (2.83).)

(v) Reclaim the equation $H^{(\mathrm{C})}\psi_0^{(\mathrm{C})} = E_0^{(\mathrm{C})}\psi_0^{(\mathrm{C})}$ by taking the $q \to 0$ limit.

2. [77] With $a_1 < a_2 < \cdots < a_n$ and $g_j > 0$ $(j = 1,\dots,n)$ consider the function of x specified by

$$y = x - \sum_{l=1}^{n} \frac{g_l}{x - a_l}$$

(cf. (4.54)).

(i) By sketching a graph, show that for each value of y there are $n+1$ values of x, one in each of the intervals (a_{l-1}, a_l), $(l = 1,\dots,n)$ with $a_0 := -\infty$, $a_{n+1} := \infty$. Let these values of x be denoted $x^{(l)}, l = 0,\dots,n$.

(ii) Deduce from (i) that

$$\int_{-\infty}^{\infty} f(y)\,dx = \int_{-\infty}^{\infty} f(y)\Big(\sum_{l=0}^{n} 1/(dy/dx^{(l)})\Big)\,dy,$$

valid or general f such that the integral is well defined.

(iii) Let C_R denote the circle of radius R, centered at 0. By evaluating $\lim_{R\to\infty}(1/2\pi i)\int_{C_R}\frac{dx/}{y}$ first as a contour integral, then by the residue theorem, deduce that

$$\sum_{l=0}^{n} 1/(dy/dx^{(l)}) = 1 \tag{11.47}$$

and hence conclude

$$\int_{-\infty}^{\infty} f(y)\,dx = \int_{-\infty}^{\infty} f(y)\,dy. \tag{11.48}$$

(iv) Use the result of (iii) in the case $n = 1$, $a_1 = 0$ to show

$$\int_{-\infty}^{\infty} e^{-(x^2+g^2/x^2)}\,dx = \sqrt{\pi}e^{-2g}, \qquad g > 0. \tag{11.49}$$

3. [21] In this exercise a derivation of the integral evaluation

$$\int_{-\infty}^{\infty} dx_1 \cdots \int_{-\infty}^{\infty} dx_N \exp\Big(-\beta\Big(\frac{1}{2}\sum_{j=1}^{N} x_j^2 + \sum_{1\le j<k\le N}\frac{g^2}{(x_k - x_j)^2}\Big)\Big) = \Big(\frac{2\pi}{\beta}\Big)^{N/2} e^{-\beta g N(N-1)/2}, \tag{11.50}$$

valid for $g > 0$, which can be regarded as a multidimensional generalization of the working of q.2 will be outlined. This integral evaluation was first derived by taking the classical limit of the quantum partition functions associated with $H^{(\mathrm{H})}$ [253].

(i) Consider the integral in (11.50) for $N = 2$. Show that after an appropriate change of variables, (11.49) can be used to obtain its evaluation.

(ii) With

$$w_j = x_j - g \sum_{k=1, k\neq j}^{N} \frac{1}{x_j - x_k} \tag{11.51}$$

show that

$$\sum_{j=1}^{N} w_j^2 = \sum_{j=1}^{N} x_j^2 + 2g^2 \sum_{j<k}^{N} \frac{1}{(x_k - x_j)^2} - N(N-1)g.$$

(iii) The denominators in (11.51), when set equal to zero, form a hyperplane arrangement, which partitions space into chambers Δ. Analogous to the result of q.2(i) there is exactly one solution of (11.51) for each chamber Δ, denoted $\vec{x}^{\Delta}$. Further, analogous to (11.47) it can be proved that

$$\sum_{\text{chambers } \Delta} \frac{1}{J(\Delta)} = 1, \qquad J(\Delta) := \det \left[\frac{\partial w_j}{\partial x_k} \right] \Big|_{\vec{x}=\vec{x}^{\Delta}}.$$

Use this to deduce that

$$\int_{-\infty}^{\infty} dx_1 \cdots \int_{-\infty}^{\infty} dx_N \, f(w_1, \dots, w_N) = \int_{-\infty}^{\infty} dx_1 \cdots \int_{-\infty}^{\infty} dx_N \, f(x_1, \dots, x_N)$$

(cf. (11.48)), valid for any f such that the integral is well defined. With an appropriate choice of f, use this result and the result of (ii) to reclaim (11.50).

11.4 THE SCHRÖDINGER OPERATORS WITH EXCHANGE TERMS

11.4.1 Definitions

Let s_{jk} denote the operator which acts on the coordinates $x_1, \dots, x_N$ of a function f by interchanging x_j and x_k. In investigating the integrability of $H^{(\mathrm{H})}$ and $H^{(\mathrm{C})}$ as defined in Propositions 11.3.1 and 11.3.2 respectively, Polychronakos [446] introduced more general Schrödinger operators $H^{(\mathrm{H,Ex})}$ and $H^{(\mathrm{C,Ex})}$ in which the factor $(\beta/2 - 1)$ outside the sum over pairs is replaced by $(\beta/2 - s_{jk})$ and placed inside the sum over pairs. Explicitly

$$H^{(\mathrm{C,Ex})} = -\sum_{j=1}^{N} \frac{\partial^2}{\partial x_j^2} + \beta \Big(\frac{\pi}{L}\Big)^2 \sum_{1 \le j < k \le N} \frac{\beta/2 - s_{jk}}{\sin^2(\pi(x_j - x_k)/L)}, \qquad 0 \le x_j \le L, \tag{11.52}$$

$$H^{(\mathrm{H,Ex})} := -\sum_{j=1}^{N} \frac{\partial^2}{\partial x_j^2} + \frac{\beta^2}{4} \sum_{j=1}^{N} x_j^2 + \beta \sum_{1 \le j < k \le N} \frac{\beta/2 - s_{jk}}{(x_j - x_k)^2} \tag{11.53}$$

(here and below in similar terms written as fractions, the operation in the numerator $(\beta/2 - s_{jk})$ is to be carried out first). The Schrödinger operators $H^{(\mathrm{L})}$ and $H^{(\mathrm{J})}$ can be similarly generalized to the operators $H^{(\mathrm{L,Ex})}$ and $H^{(\mathrm{J,Ex})}$ with exchange terms, although the procedure is complicated by the underlying root system being of B and BC type rather than the simpler A type which corresponds to $H^{(\mathrm{H})}$ and $H^{(\mathrm{C})}$. The complication means that in addition to the exchange operator s_{jk} it is also necessary to introduce the change of sign operator, σ_j say, which replaces the coordinate x_j by $-x_j$. Note that the operators s_{jk} and σ_j make up the type B Weyl group (recall Section 4.7.2). In terms of these operators one defines

$$\begin{aligned} H^{(\mathrm{L,Ex})} := &-\sum_{j=1}^{N} \frac{\partial^2}{\partial x_j^2} + \frac{\beta^2}{4} \sum_{j=1}^{N} x_j^2 + \frac{\beta a'}{2} \sum_{j=1}^{N} \frac{\beta a'/2 - \sigma_j}{x_j^2} \\ &+ \beta \sum_{1 \le j < k \le N} \Big(\frac{\beta/2 - s_{jk}}{(x_j - x_k)^2} + \frac{\beta/2 - \sigma_j \sigma_k s_{jk}}{(x_j + x_k)^2} \Big), \qquad x_j > 0, \end{aligned} \tag{11.54}$$

$$H^{(\mathrm{J,Ex})} := -\sum_{j=1}^{N} \frac{\partial^2}{\partial \phi_j^2} + \frac{\beta a'}{2} \sum_{j=1}^{N} \frac{\beta a'/2 - \sigma_j}{\sin^2 \phi_j} + \frac{\beta b'}{2} \sum_{j=1}^{N} \frac{\beta b'/2 - \sigma_j}{\cos^2 \phi_j}$$
$$+\beta \sum_{1 \le j < k \le N} \Big(\frac{\beta/2 - s_{jk}}{\sin^2(\phi_j - \phi_k)} + \frac{\beta/2 - \sigma_j \sigma_k s_{jk}}{\sin^2(\phi_j + \phi_k)} \Big), \quad 0 < \phi_j < \pi. \tag{11.55}$$

From the definition we see that the operator s_{jk} acts like the identity on symmetric functions. Thus when restricted to act on symmetric functions (which must also be even in the cases of (11.54) and (11.55)) the operators $H^{(\cdot,\mathrm{Ex})}$ reduce to their original form $H^{(\cdot)}$. One consequence is that $e^{-\beta W^{(\cdot)}/2}$, which we know from Exercises 11.1 q.1 is the symmetric ground state wave function of $H^{(\cdot)}$, remains the symmetric ground state wave function of $H^{(\cdot,\mathrm{Ex})}$. Another is that a decomposition of $H^{(\cdot,\mathrm{Ex})}$ implies a decomposition of $H^{(\cdot)}$ after restricting the former to act on symmetric functions. In subsequent sections, for each $H^{(\cdot,\mathrm{Ex})}$, a set of N commuting operators will be identified which establishes integrability and provides a simple decomposition of the Schrödinger operator; by the above remarks these properties are established for $H^{(\cdot)}$ after restricting the operators to act on symmetric functions.

11.4.2 The type A Dunkl operator

A few years before Polychronakos introduced the operators $H^{(\cdot,\mathrm{Ex})}$, Dunkl [141] introduced a class of differential operators with exchange terms. These came about in the context of studying multivariable polynomials and integral transforms associated with weight functions of the form of the integrand in (4.152). It turns out that these same operators play an essential role in establishing the integrability of $H^{(\cdot,\mathrm{Ex})}$.

DEFINITION 11.4.1 *The type A Dunkl operators are defined by*

$$d_j := \frac{\partial}{\partial z_j} + \frac{1}{\alpha} \sum_{\substack{k=1 \\ k \ne j}}^{N} \frac{1 - s_{jk}}{z_j - z_k}, \qquad j = 1, \dots, N.$$

When acting on homogeneous polynomials, the Dunkl operators lower the degree by 1. Another fundamental property is that they mutually commute.

PROPOSITION 11.4.2 *For all* $1 \le j, k \le N$ *we have*

$$[d_j, d_k] = 0.$$

Proof. From the definitions of the d_j and s_{jk} we see

$$[d_j, d_k] = \frac{1}{\alpha^2} \sum_{\substack{l=1 \\ l \ne j,k}}^{N} \Big(\frac{s_{jl}s_{kl} - s_{kl}s_{jl}}{(z_j - z_l)(z_k - z_l)} + \frac{s_{jl}s_{kj} - s_{kj}s_{jl}}{(z_k - z_j)(z_j - z_l)} + \frac{s_{jk}s_{kl} - s_{kl}s_{jk}}{(z_j - z_k)(z_k - z_l)} \Big).$$

But for distinct indices

$$s_{jk}s_{kl} = s_{lj}s_{jk} = s_{kl}s_{lj} \tag{11.56}$$

(the indices are cycles of (jkl) with the middle index repeated twice), so we can write

$$[d_j, d_k] = \frac{1}{\alpha^2} \sum_{\substack{l=1 \\ l \ne j,k}}^{N} \Big(-\frac{1}{(z_j - z_l)(z_k - z_l)} + \frac{1}{(z_k - z_j)(z_j - z_l)} + \frac{1}{(z_j - z_k)(z_k - z_l)} \Big) (s_{jl}s_{lk} - s_{kl}s_{lj})$$

and according to the identity (11.36) the terms in the large brackets cancel. □

The Dunkl operators can be used to define a set of commuting degree preserving operators ξ_i, and these operators in turn can be used to decompose the operator $H^{(\mathrm{C,Ex})}$ [58].

DEFINITION 11.4.3 *The Cherednik operators ξ_i are defined in terms of the Dunkl operators d_i by*

$$\begin{aligned}\xi_i &:= \alpha z_i d_i + 1 - N + \sum_{p=i+1}^{N} s_{ip} \\ &= \alpha z_i \frac{\partial}{\partial z_i} + \sum_{p<i} \frac{z_i}{z_i - z_p}(1 - s_{ip}) + \sum_{p>i} \frac{z_p}{z_i - z_p}(1 - s_{ip}) + 1 - i,\end{aligned} \tag{11.57}$$

where $i = 1, \dots, N$.

PROPOSITION 11.4.4 *For all $1 \le i, j \le N$ we have*

$$[\xi_i, \xi_j] = 0.$$

Proof. From Proposition 11.4.2 and the second and third members of the readily verified relations

$$\begin{array}{llll}\text{(i)} & [d_i, z_i] = 1 + \frac{1}{\alpha}\sum_{p\ne i} s_{ip} & \text{(ii)} & d_i s_{ip} = s_{ip} d_p, \\ \text{(iii)} & [d_i, z_j] = -\frac{1}{\alpha} s_{ij} \quad (i \ne j) & \text{(iv)} & [d_i, s_{jp}] = 0 \quad (i \ne j, p),\end{array} \tag{11.58}$$

we see that

$$[z_i d_i, z_j d_j] = -\alpha (z_i d_i - z_j d_j) s_{ij}.$$

Also, the second and fourth relations give

$$\Big[z_i d_i, \alpha \sum_{p=j+1}^{N} s_{jp}\Big] + \Big[\sum_{p=i+1}^{N} s_{ip}, \alpha z_j d_j\Big] = \alpha (z_i d_i - z_j d_j) s_{ij}$$

while for $i > j$ (and similarly $i < j$)

$$\Big[\sum_{p=i+1}^{N} s_{ip}, \sum_{p=j+1}^{N} s_{jp}\Big] = \sum_{p=i+1}^{N} \Big([s_{ip}, s_{jp}] + [s_{ip}, s_{ji}]\Big) = 0,$$

where to obtain the last equality (11.56) has been used. □

PROPOSITION 11.4.5 *With $\alpha = 2/\beta$ and $z_j := e^{2\pi i x_j / L}$ we have*

$$H^{(\mathrm{C,Ex})} = \Big(\frac{2\pi}{L}\Big)^2 \frac{1}{\alpha^2} \sum_{j=1}^{N} \Big(e^{-\beta W^{(\mathrm{C})}/2}\Big(\xi_j + \frac{N-1}{2}\Big) e^{\beta W^{(\mathrm{C})}/2}\Big)^2.$$

Proof. Writing (11.52) in terms of α and the z_j gives

$$H^{(\mathrm{C,Ex})} = \Big(\frac{2\pi}{L}\Big)^2 \Big(\sum_{j=1}^{N}\Big(z_j \frac{\partial}{\partial z_j}\Big)^2 - \frac{1}{\alpha^2} \sum_{\substack{j,l=1 \\ j\ne l}}^{N} \frac{z_j z_l}{(z_j - z_l)^2} + \frac{1}{\alpha} \sum_{\substack{j,l=1 \\ j\ne l}}^{N} \frac{z_j z_l}{(z_j - z_l)^2} s_{jl}\Big). \tag{11.59}$$

On the other hand, we see from (11.24) with $\theta_j = 2\pi x_j / L$ that $e^{-\beta W^{(\mathrm{C})}/2}$ is proportional to

$$\prod_{j=1}^{N} z_j^{-(N-1)/2\alpha} \prod_{j<k} (z_k - z_j)^{1/\alpha}, \tag{11.60}$$

so Definition 11.4.3 gives

$$e^{-\beta W^{(\mathrm{C})}/2}\Big(\xi_j + \frac{N-1}{2}\Big) e^{\beta W^{(\mathrm{C})}/2} = \alpha z_j \frac{\partial}{\partial z_j} - \sum_{p<j} z_j \frac{s_{jp}}{z_j - z_p} - \sum_{p>j} z_p \frac{s_{jp}}{z_j - z_p}. \tag{11.61}$$

Summing over j, dividing by α and squaring we see that the term independent of α in the resulting expression is

$$\Big(\sum_{j=1}^N z_j \frac{\partial}{\partial z_j}\Big)^2,$$

and, after some straightforward manipulation, that the term proportional to $1/\alpha$ is equal to the term required by (11.59). Furthermore, use of (11.36) and (11.56) shows that the term proportional to $1/\alpha^2$ is also as required by (11.59). □

11.4.3 Integrability of $H^{(\mathrm{C,Ex})}$ and $H^{(\mathrm{C})}$

Consider the operator

$$X^{(\mathrm{C})}(u) := \prod_{j=1}^N (1 + u\xi_j) = \sum_{p=0}^N u^p X_p^{(\mathrm{C})}, \qquad X_p^{(\mathrm{C})} := \sum_{1 \le i_1 < \cdots < i_p \le N} \xi_{i_1} \cdots \xi_{i_p}. \tag{11.62}$$

According to Propositions 11.4.4 and 11.4.5, $e^{-\beta W^{(\mathrm{C})}/2} X^{(\mathrm{C})}(u) e^{\beta W^{(\mathrm{C})}/2}$ commutes with $H^{(\mathrm{C,Ex})}$. Thus the set of N algebraically independent operators $e^{-\beta W^{(\mathrm{C})}/2} X_p^{(\mathrm{C})}(u) e^{\beta W^{(\mathrm{C})}/2}$, $p = 1, \dots, N$, are mutually commuting, and commute with $H^{(\mathrm{C,Ex})}$ which is itself a linear combination of $X_2^{(\mathrm{C})}$, $X_1^{(\mathrm{C})}$ and a constant. This establishes the integrability of $H^{(\mathrm{C,Ex})}$.

Also, we know that when restricted to act on symmetric functions $H^{(\mathrm{C,Ex})}$ reduces to $H^{(\mathrm{C})}$. It follows that the integrability property of $H^{(\mathrm{C,Ex})}$ carries over to $H^{(\mathrm{C})}$ with the set of algebraically independent, mutually commuting operators $e^{-\beta W^{(\mathrm{C})}/2} X_p^{(\mathrm{C})}(u) e^{\beta W^{(\mathrm{C})}/2}$ restricted to act on symmetric functions.

11.5 THE OPERATORS $H^{(\mathrm{H,Ex})}$, $H^{(\mathrm{L,Ex})}$ AND $H^{(\mathrm{J,Ex})}$

11.5.1 Definitions of the transformed operators

An alternative way to view Proposition 11.4.5 and the results of Section 11.4.3 is as a decomposition of the transformed Schrödinger operator

$$\begin{aligned}
\tilde{H}^{(\mathrm{C,Ex})} &:= \Big(\frac{L}{2\pi}\Big)^2 e^{\beta W^{(\mathrm{C})}/2} (H^{(\mathrm{C,Ex})} - E_0^{(\mathrm{C})}) e^{-\beta W^{(\mathrm{C})}/2} \\
&= \sum_{j=1}^N \Big(z_j \frac{\partial}{\partial z_j}\Big)^2 + \frac{N-1}{\alpha} \sum_{j=1}^N z_j \frac{\partial}{\partial z_j} \\
&\quad + \frac{2}{\alpha} \sum_{1 \le j < k \le N} \frac{z_j z_k}{z_j - z_k} \Big(\Big(\frac{\partial}{\partial z_j} - \frac{\partial}{\partial z_k} \Big) - \frac{1 - s_{jk}}{z_j - z_k} \Big),
\end{aligned} \tag{11.63}$$

where $z_j := e^{2\pi i x_j / L}$ and $\alpha := 2/\beta$, in terms of a set of mutually commuting operators. The analogous transformation in the Hermite, Laguerre and Jacobi cases gives

$$\begin{aligned}
\tilde{H}^{(\mathrm{H,Ex})} &:= -\frac{2}{\beta} e^{\beta W^{(\mathrm{H})}/2} (H^{(\mathrm{H,Ex})} - E_0^{(\mathrm{H})}) e^{-\beta W^{(\mathrm{H})}/2} \\
&= \sum_{j=1}^N \Big(\frac{\partial^2}{\partial y_j^2} - 2y_j \frac{\partial}{\partial y_j} \Big) + \frac{2}{\alpha} \sum_{j<k} \frac{1}{y_j - y_k} \Big(\Big(\frac{\partial}{\partial y_j} - \frac{\partial}{\partial y_k} \Big) - \frac{1 - s_{jk}}{y_j - y_k} \Big),
\end{aligned} \tag{11.64}$$

$$\begin{aligned}\tilde{H}^{(\mathrm{L,Ex})} &:= -\frac{1}{2\beta}e^{\beta W^{(\mathrm{L})}/2}(H^{(\mathrm{L,Ex})} - E_0^{(\mathrm{L})})e^{-\beta W^{(\mathrm{L})}/2} \\ &= \frac{1}{4}\sum_{j=1}^{N}\Big(\frac{\partial^2}{\partial y_j^2} + -2y_j\frac{\partial}{\partial y_j} + (2a+1)\frac{1}{y_j}\frac{\partial}{\partial y_j}\Big) \\ &\quad + \frac{1}{\alpha}\sum_{j<k}\frac{1}{y_j^2 - y_k^2}\Big(\Big(y_j\frac{\partial}{\partial y_j} - y_k\frac{\partial}{\partial y_k}\Big) - \frac{y_j^2 + y_k^2}{y_j^2 - y_k^2}(1 - s_{jk})\Big),\end{aligned} \tag{11.65}$$

$$\begin{aligned}\tilde{H}^{(\mathrm{J,Ex})} &:= \frac{1}{4}e^{\beta W^{(\mathrm{J})}/2}(H^{(\mathrm{J,Ex})} - E_0^{(\mathrm{J})})e^{-\beta W^{(\mathrm{J})}/2} \\ &= \sum_{j=1}^{N}\Big(z_j\frac{\partial}{\partial z_j}\Big)^2 + \Big(\Big(a+\frac{1}{2}\Big)\frac{z_j+1}{z_j-1} + \Big(b+\frac{1}{2}\Big)\frac{z_j-1}{z_j+1} + \frac{N-1}{\alpha}\Big)z_j\frac{\partial}{\partial z_j} \\ &\quad + \frac{2}{\alpha}\sum_{1\le j<k\le N}\frac{z_j z_k}{z_j - z_k}\Big(\Big(\frac{\partial}{\partial z_j} - \frac{\partial}{\partial z_k}\Big) - \frac{1-s_{jk}}{z_j - z_k}\Big) \\ &\quad + \frac{2}{\alpha}\sum_{1\le j<k\le N}\frac{1}{z_j z_k - 1}\Big(\Big(z_j\frac{\partial}{\partial z_j} - z_k\frac{\partial}{\partial z_k}\Big) - \frac{z_j z_k(1-s_{jk})}{z_j z_k - 1}\Big),\end{aligned} \tag{11.66}$$

where

$$a := (a'\beta - 1)/2, \qquad b := (b'\beta - 1)/2.$$

To obtain (11.64) and (11.65) the change of variables $y_j = \sqrt{\beta/2}x_j$ has been made, and for (11.65) the restriction to functions even in y_j has been made, while for (11.66) the change of variables $z_j = e^{2i\phi_j}$ has been made along with the restriction to eigenfunctions unchanged by the mapping $z_j \mapsto 1/z_j$.

11.5.2 The operators h_i

Associated with the type A Dunkl operators d_i is the generalized Laplacian

$$\Delta_A := \sum_{i=1}^{N} d_i^2. \tag{11.67}$$

The explicit expansion of Δ_A is closely related to (11.64) [141].

Proposition 11.5.1 *With the type A Dunkl operators d_i defined in terms of the variables $\{y_j\}$ we have*

$$\Delta_A = \sum_{j=1}^{N}\frac{\partial^2}{\partial y_j^2} + \frac{2}{\alpha}\sum_{j<k}\frac{1}{y_j - y_k}\Big(\Big(\frac{\partial}{\partial y_j} - \frac{\partial}{\partial y_k}\Big) - \frac{1-s_{jk}}{y_j - y_k}\Big). \tag{11.68}$$

Proof. This follows from Definition 11.4.1 of the d_i, the relations (11.56) and the identity (11.36) (the latter two results are used to show that the apparent term proportional to $1/\alpha^2$ vanishes). □

Comparison between (11.64) and (11.68) shows

$$\tilde{H}^{(\mathrm{H,Ex})} = \Delta_A - 2\sum_{j=1}^{N} y_j\frac{\partial}{\partial y_j} = \Delta_A - \frac{2}{\alpha}\sum_{j=1}^{N}\xi_j - \frac{1}{\alpha}N(N-1) \tag{11.69}$$

where the second equality follows from Definition 11.4.3 of ξ_j (written in terms of $\{y_j\}$), or equivalently

$$\tilde{H}^{(\mathrm{H,Ex})} = -\frac{2}{\alpha}\sum_{i=1}^{N} h_i - \frac{1}{\alpha}N(N-1), \qquad h_i := \xi_i - \frac{\alpha}{2}d_i^2. \tag{11.70}$$

Like the result of Proposition 11.4.5 this gives a decomposition of the corresponding Schrödinger operator in terms of simpler operators, although the simpler operators occur only to the first power in (11.70) whereas they are squared in Proposition 11.4.5. From the results of Section 11.4.3 we see that the operator

$$X^{(\mathrm{H})}(u) := \prod_{j=1}^{N}(1+uh_j) = \sum_{p=0}^{N} u^p X_p^{(\mathrm{H})}, \qquad X_p^{(\mathrm{H})} := \sum_{1\le i_1<\cdots<i_p\le N} h_{i_1}\cdots h_{i_p} \tag{11.71}$$

can be used to establish integrability, provided we can show that the $\{h_j\}$ mutually commute. This latter task can be readily accomplished, once note is taken of some commutation formulas.

PROPOSITION 11.5.2 *We have*

$$[\xi_j, d_i] = \begin{cases} d_i s_{ij}, & i<j, \\ d_j s_{ij}, & i>j, \\ -\alpha d_j - \sum_{p<j} s_{jp} d_j - \sum_{p>j} d_j s_{jp} & i=j. \end{cases}$$

Proof. From the definition of ξ_j in terms of d_j we see that

$$[\xi_j, d_i] = \frac{1}{\alpha}[y_j d_j, d_i] + \sum_{p=j+1}^{N}[s_{jp}, d_i]. \tag{11.72}$$

But in general

$$[AB, C] = A[B,C] + [A,C]B, \tag{11.73}$$

which together with Proposition 11.4.2 implies

$$[y_j d_j, d_i] = [y_j, d_i] d_j.$$

Using this equation and (11.58) in (11.72) gives the stated result. □

PROPOSITION 11.5.3 *The operators h_i, $(i = 1, \ldots, N)$ as defined by (11.70) mutually commute.*

Proof. We have

$$[h_i, h_j] = [\xi_i, \xi_j] - \frac{\alpha}{2}[\xi_i, d_j^2] - \frac{\alpha}{2}[d_i^2, \xi_j] + \frac{\alpha^2}{4}[d_i^2, d_j^2].$$

According to Propositions 11.4.5 and 11.4.2 the first and last commutators vanish, while making use of (11.73) and Proposition 11.5.2 shows that the sum of the second and third commutators vanishes too. □

Another way of showing that $\{h_i\}$ commute is first to note from the definition (11.70) and Definition 11.4.3 that we can write

$$h_i = \alpha\Big(y_i - \frac{d_i}{2}\Big)d_i + 1 - N + \sum_{p=i+1}^{N} s_{ip}, \tag{11.74}$$

and then note that the algebra (11.58), supplemented by the relation $[y_i, y_j] = 0$, is unchanged by the mapping

$$\phi^{(A)}(y_i) = y_i - \frac{d_i}{2}, \quad \phi^{(A)}(d_i) = d_i, \quad \phi^{(A)}(s_{ij}) = s_{ij}, \tag{11.75}$$

where $\phi^{(A)}$ is multiplicative (i.e., $\phi^{(A)}(uv) = \phi^{(A)}(u)\phi^{(A)}(v)$). Since by these rules

$$\phi^{(A)}(\xi_i) = h_i, \tag{11.76}$$

we see by applying $\phi^{(A)}$ to the commutation relation $[\xi_i, \xi_j] = 0$ that $[h_i, h_j] = 0$ as required.

It follows immediately from Proposition 11.5.3 that $\{X_p^{(\mathrm{H})}\}_{p=1,\dots,N}$ as defined by (11.71) mutually commute and commute with $\tilde{H}^{(\mathrm{H,Ex})}$, thus establishing the integrability of the latter operator. These operators restricted to act on symmetric functions provide a mutually commuting set of algebraically independent operators which commute with $\tilde{H}^{(\mathrm{H})}$, and establish integrability in this case as well.

11.5.3 The operators l_i

The study of the Schrödinger operator $\tilde{H}^{(\mathrm{L,Ex})}$ requires the Dunkl operator of type B.

Definition 11.5.4 *For $i = 1, \dots, N$ the Dunkl operators for the root system of type B are defined by*

$$d_i^{(B)} := \frac{\partial}{\partial y_i} + \frac{1}{\alpha} \sum_{p \neq i} \Big(\frac{1 - s_{ip}}{y_i - y_p} + \frac{1 - \sigma_i \sigma_p s_{ip}}{y_i + y_p} \Big) + \frac{a + 1/2}{y_i} (1 - \sigma_i).$$

A direct calculation using (11.56) and the second of the identities (11.37) shows that the $d_i^{(B)}$ commute,

$$[d_i^{(B)}, d_j^{(B)}] = 0. \tag{11.77}$$

Analogous to the result of Proposition 11.5.1, the corresponding generalized Laplacian is closely related to the operator $\tilde{H}^{(\mathrm{L,Ex})}$. Thus a direct calculation, making use of the same identities required to establish (11.77), shows

$$\Delta_B := \sum_{i=1}^N (d_i^{(B)})^2 = 4\Big(\tilde{H}^{(\mathrm{L,Ex})} + \frac{1}{2} \sum_{j=1}^N y_j \frac{\partial}{\partial y_j} \Big), \tag{11.78}$$

provided Δ_B is restricted to act on functions even in each y_j. Rewriting $\sum_{j=1}^N y_j \frac{\partial}{\partial y_j}$ in terms of $\{\hat{\xi}_i\}$, where $\hat{\xi}_i$ denotes the Cherednik operators of Definition 11.4.3 with the change of variables $z_j = y_j^2$ (the operators $\hat{\xi}_i$ rather than ξ_i are needed for latter purposes), we see from this that

$$\tilde{H}^{(\mathrm{L,Ex})} = -\frac{1}{2\alpha} \sum_{i=1}^N l_i - \frac{1}{2\alpha} N(N-1), \qquad l_i := \hat{\xi}_i - \frac{\alpha}{4} (d_i^{(B)})^2. \tag{11.79}$$

The algebra isomorphism (11.75) can be extended to the type B case, and from this the fact that the $\{l_i\}$ commute can be deduced. First we note from the definitions that

$$\hat{\xi}_i = \Big(\frac{\alpha}{2} y_i d_i^{(B)} + 1 - N + \frac{1}{2} \sum_{p=i+1}^N (s_{ip} + \sigma_i \sigma_p s_{ip}) \Big) \Big|_{\substack{\text{even}\\ \text{functions}}},$$

which in turn shows that l_i as specified by (11.79) can be rewritten

$$l_i = \Big(\frac{\alpha}{4} (2y_i - d_i^{(B)}) d_i^{(B)} + 1 - N + \frac{1}{2} \sum_{p=i+1}^N (s_{ip} + \sigma_i \sigma_p s_{ip}) \Big) \Big|_{\substack{\text{even}\\ \text{functions}}}. \tag{11.80}$$

Now the algebra satisfied by $\{y_i, d_i^{(B)}, s_{ij}, \sigma_i\}$ is noted in (11.88) below. We see that the mapping

$$\phi^{(B)}(y_i) = y_i - \frac{d_i^B}{2}, \quad \phi^{(B)}(d_i^B) = d_i^B, \quad \phi^{(B)}(s_{ij}) = s_{ij}, \quad \phi^{(B)}(\sigma_i) = \sigma_i, \tag{11.81}$$

with $\phi^{(B)}$ multiplicative is an algebra isomorphism of (11.88), and also

$$\phi^{(B)}(\hat{\xi}_i) = l_i. \tag{11.82}$$

Thus it follows by applying $\phi^{(B)}$ to the commutation relation $[\hat{\xi}_i, \hat{\xi}_j] = 0$ that $[l_i, l_j] = 0$, which is the sought result.

From the decomposition of $\tilde{H}^{(\mathrm{L,Ex})}$ in (11.79), and the commutivity of $\{l_i\}$, we see that the operator

$$X^{(\mathrm{L})}(u) := \prod_{j=1}^{N}(1 + ul_j) = \sum_{p=0}^{N} u^p X_p^{(\mathrm{L})}, \qquad X_p^{(\mathrm{L})} := \sum_{1 \le i_1 < \cdots < i_p \le N} l_{i_1} \cdots l_{i_p}, \tag{11.83}$$

when restricted to act on functions even in $y_1, \dots, y_N$ commutes with $\tilde{H}^{(\mathrm{L,Ex})}$ (which itself was defined with the restriction that it acts on functions even in $y_1, \dots, y_N$). By also restricting $X^{(\mathrm{L})}(u)$ to act on symmetric functions, the fact that the decomposition (11.79) when restricted to symmetric functions gives $\tilde{H}^{(\mathrm{L})}$ shows that $X^{(\mathrm{L})}(u)$ commutes with $\tilde{H}^{(\mathrm{L})}$, thus establishing integrability in both cases.

11.5.4 The operators $\xi_j^{(BC)}$

The analogue of the Cherednik operator of Definition 11.4.3 for the BC root system is [60], [297]

$$\xi_j^{(BC)} = \frac{1}{\alpha}\xi_j - \frac{1}{\alpha}\sum_{\substack{k=1 \\ \neq j}}^{N} \frac{1 - \sigma_j \sigma_k s_{jk}}{1 - z_j z_k} - \Big(a + \frac{1}{2}\Big)\frac{1 - \sigma_j}{1 - z_j} - \Big(b + \frac{1}{2}\Big)\frac{1 - \sigma_j}{1 + z_j} + \frac{1}{2}(a + b + 1). \tag{11.84}$$

A direct calculation using the final identity of (11.37), and (11.56), shows that

$$\tilde{H}^{(\mathrm{J,Ex})} = \sum_{j=1}^{N} \Big(\xi_j^{(BC)}\Big)^2 - \frac{1}{4}E_0^{(\mathrm{J})}, \tag{11.85}$$

provided the action of each $(\xi_j^{(BC)})^2$ is restricted to functions which are even in the z_i and unchanged by the mapping $z_i \mapsto 1/z_i$ $(i = 1, \dots, N)$. This latter point indicates an essential difference with the decomposition of Proposition 11.4.5 which is otherwise very similar: the operator $(\xi_j^{(BC)})^2$ must be considered as a single entity which cannot be further decomposed as the product $\xi_j^{(BC)}\xi_j^{(BC)}$. Consequently, the analogue of the commuting operator (11.62) is

$$X^{(\mathrm{J})}(u) := \prod_{j=1}^{N}(1+u(\xi_j^{(BC)})^2) = \sum_{p=0}^{N} u^p X_p^{(\mathrm{J})}, \qquad X_p^{(\mathrm{J})} := \sum_{1 \le i_1 < \cdots < i_p \le N} (\xi_{i_1}^{(BC)})^2 \cdots (\xi_{i_p}^{(BC)})^2, \tag{11.86}$$

where the action is restricted to functions which are even in the z_i and unchanged by the mapping $z_i \mapsto 1/z_i$ $(i = 1, \dots, N)$. By further restricting the action to functions which are symmetric in the z_i we see that $\{X_p^{(\mathrm{J})}\}_{p=1,\dots,N}$ commute with the operator $\tilde{H}^{(\mathrm{J})}$ and thus establish its integrability.

11.5.5 Root system structure

The Schrödinger operators (11.52) and (11.55) both describe quantum many-body systems on a circle. They can be cast in the common form

$$H^{(\cdot,Ex)} = -\sum_{j=1}^{N} \frac{\partial^2}{\partial x_j^2} + \Big(\frac{\pi}{L}\Big)^2 \sum_{\vec{\alpha} \in \Delta_+} \frac{k_\alpha(k_\alpha - s_\alpha)\vec{\alpha} \cdot \vec{\alpha}}{(z^{\alpha/2} - z^{-\alpha/2})^2}, \tag{11.87}$$

where $z_j := e^{2\pi i x_j/L}$ and $z^\alpha := z_1^{\alpha_1} \cdots z_N^{\alpha_N}$ for $\vec{\alpha}$ equal to the N-component vector $(\alpha_1, \dots, \alpha_N)$. In (11.87) Δ_+ denotes the positive roots of the underlying root system (A type for $H^{(\mathrm{C,Ex})}$ and BC type for $H^{(\mathrm{J,Ex})}$), which are listed in Table 4.1. The operator s_α denotes the reflection in the subspaces perpendicular to α. These operators are specified in terms of the exchange and negation operators for the root systems A_{N-1}

and BC_N and thus by s_{jk} and $s_{jk}, \sigma_j \sigma_k s_{jk}, \sigma_j$, respectively. Finally, the k_α are required to have the property that $k_\alpha = k_{w(\alpha)}$ for all $w \in W$ (the Weyl group generated by the reflections s_α). In the A type case this means k_α is constant. Choosing $k_\alpha = \beta/2$, we reclaim (11.52).

EXERCISES 11.5 1. [39] Show that the analogue of the equations (11.58) for the operators $d_i^{(B)}$ are the equations

$$[d_i^{(B)}, y_j] = \delta_{ij}\Big(1 + \frac{1}{\alpha}\sum_{k \neq i}(s_{ik} + \sigma_i \sigma_j s_{ij}) + (2a+1)\sigma_j\Big) - \frac{1}{\alpha}(1 - \delta_{ij})(s_{ij} - \sigma_i \sigma_k s_{ik}),$$

$$\begin{aligned} s_{ij} d_j^{(B)} &= d_i^{(B)}, & s_{ij} d_k^{(B)} &= d_k^{(B)} s_{ij} \quad (k \neq i, j), \\ \sigma_j d_j^{(B)} &= -d_i^{(B)} \sigma_j, & \sigma_j d_k^{(B)} &= d_k^{(B)} \sigma_j \quad (k \neq j). \end{aligned} \tag{11.88}$$

2. [39] This exercise relates to some algebraic properties of $\{\xi_i\}$.

(i) Use Definition 11.4.3 of the ξ_i and the formulas (11.58) to verify that

$$\xi_i s_i - s_i \xi_{i+1} = 1, \quad \xi_{i+1} s_i - s_i \xi_i = -1, \quad [\xi_i, s_j] = 0 \ (j \neq i-1, i). \tag{11.89}$$

Remark: These relations, together with

$$[\xi_i, \xi_j] = 0$$

and

$$s_j^2 = 1, \quad s_j s_{j+1} s_j = s_{j+1} s_j s_{j+1}, \quad [s_i, s_j] = 0 \ (|i - j| \geq 2) \tag{11.90}$$

define the degenerate type A Hecke algebra. Note that the second relation in (11.89) follows from the first by pre- and post-multiplying by s_i.

(i) Use Definition 11.4.3 of the ξ_i, and the commutator relations (11.58) to show

$$[\xi_i, z_j] = \begin{cases} -z_j s_{ij}, & i < j, \\ -z_i s_{ij}, & i > j, \\ \alpha z_i + \sum_{p > i} z_p s_{ip} + \sum_{p < i} z_i s_{ip}, & i = j \end{cases}$$

(cf. Proposition 11.5.2).

(ii) Show from (i) that

$$[s_i, (\xi_i + \xi_{i+1})] = 0, \qquad [s_i, \xi_i \xi_{i+1}] = 0$$

and from these equations, together with the last equation of (i), conclude that

$$[s_i, f(\xi_1, \dots, \xi_N)] = 0$$

for any symmetric function f.

(iii) Use the algebra isomorphism (11.75) and (11.76) to verify that the equations of (i) remain valid with the ξ_i replaced by the h_i.

3. [36] Here some changes of variables for $\tilde{H}^{(L)}$ and $\tilde{H}^{(J)}$ are noted.

(i) Let $\tilde{H}^{(\mathrm{L})}$ refer to (11.54) when acting on symmetric functions, so that $s_{jk} = 1$. Replace y_j in favor of u_j according to $u_j = y_j^2$ to show that in terms of u_j

$$\tilde{H}^{(\mathrm{L})} = \sum_{j=1}^{N}\Big(u_j \frac{\partial^2}{\partial u_j^2} + (a - u_j + 1)\frac{\partial}{\partial u_j} + \frac{2}{\alpha}\sum_{\substack{k=1 \\ \neq j}}^{N} \frac{u_j}{u_j - u_k}\frac{\partial}{\partial u_j}\Big).$$

(ii) Let $\tilde{H}^{(J)}$ refer to (11.55) when acting on symmetric functions, so that $s_{jk} = 1$. Replace z_j in favor of y_j

according to $y_j = \sin^2 \phi_j$ to show that in terms of y_j

$$\tilde{H}^{(\mathrm{J})} = \sum_{j=1}^{N} \Big(y_j(1-y_j)\frac{\partial^2}{\partial y_j^2} + (a+1-y_j(a+b+2))\frac{\partial}{\partial y_j} + \frac{2}{\alpha}\sum_{\substack{k=1\\ \neq j}}^{N} \frac{y_j(1-y_j)}{y_j-y_k}\frac{\partial}{\partial y_j}\Big). \tag{11.91}$$

11.6 DYNAMICAL CORRELATIONS FOR $\beta = 2$

11.6.1 Formulation

The Schrödinger operators of Proposition 11.3.1 and 11.3.2, related to the corresponding Fokker-Planck operators by the transformation (11.33), have the property that at $\beta = 2$ the two-body term vanishes identically. The Schrödinger operators then describe free particles in an external potential, permitting computations of dynamical correlation functions not possible for the other couplings.

In relation to the dynamical correlations, we recall from Section 11.3.2 that the Green functions G_τ^{FP} is the solution of the Fokker-Planck equation (11.15) which satisfies the delta function initial condition as in (11.40). The p.d.f. for the event that the particles are at positions $\vec{x}^{(1)}$ for parameter value τ_1, and at positions $\vec{x}^{(2)}$ for parameter value τ_2, is calculated from the Green functions according to

$$p_{\tau_1,\tau_2}(\vec{x}^{(1)};\vec{x}^{(2)}) = \frac{1}{N!}\int_{-\infty}^{\infty} dx_1^{(0)} \cdots \int_{-\infty}^{\infty} dx_N^{(0)}\, p_0(\vec{x}^{(0)}) G_{\tau_1}^{\mathrm{FP}}(\vec{x}^{(0)};\vec{x}^{(1)}) G_{\tau_2-\tau_1}^{\mathrm{FP}}(\vec{x}^{(1)};\vec{x}^{(2)}), \tag{11.92}$$

where p_0 is the prescribed initial p.d.f. in terms of the variables $\vec{x}^{(0)}$. The corresponding dynamical correlation $\rho_{(m_1,m_2)}(\{x_j^{(1)}\}_{j=1,\dots,m_1};\tau_1;\{x_j^{(2)}\}_{j=1,\dots,m_2};\tau_2)$ is then calculated from the formula

$$\rho_{(m_1,m_2)}(\{x_j^{(1)}\}_{j=1,\dots,m_1};\tau_1;\{x_j^{(2)}\}_{j=1,\dots,m_2};\tau_2) := \frac{N!}{(N-m_1)!}\frac{N!}{(N-m_2)!}$$
$$\times \int_{-\infty}^{\infty} dx_{m_1+1}^{(1)} \cdots \int_{-\infty}^{\infty} dx_N^{(1)} \int_{-\infty}^{\infty} dx_{m_2+1}^{(2)} \cdots \int_{-\infty}^{\infty} dx_N^{(2)}\, p_{\tau_1,\tau_2}(\vec{x}^{(1)};\vec{x}^{(2)}). \tag{11.93}$$

At $\beta = 2$ we read off from Propositions 11.3.1 and 11.3.2 that

$$H^{(\mathrm{H})} = -\sum_{j=1}^{N}\frac{\partial^2}{\partial x_j^2} + \sum_{j=1}^{N} x_j^2,$$
$$H^{(\mathrm{L})} = -\sum_{j=1}^{N}\frac{\partial^2}{\partial x_j^2} + \sum_{j=1}^{N}\Big(\frac{a'(a'-1)}{x_j^2} + x_j^2\Big),$$
$$H^{(\mathrm{J})} = -\sum_{j=1}^{N}\frac{\partial^2}{\partial x_j^2} + \sum_{j=1}^{N}\Big(\frac{a'(a'-1)}{\sin^2 x_j} + \frac{b'(b'-1)}{\cos^2 x_j}\Big), \tag{11.94}$$

which all describe independent particles in an external one-body potential. The same is true for $H^{(\mathrm{D})}$, although as already remarked that case does not fit into the scheme below since there are no bound states. However, the fact that in the original Fokker-Planck description the particles cannot overlap due to the repulsive two-body interaction means the particles are to be regarded as fermions and thus an antisymmetric Green function is sought.

Now in general for Schrödinger operators of the form (11.94), the free fermion Green function solution G_τ^{S} is given in terms of the Green function solution with $N = 1$, $g_\tau^{\mathrm{S}}(x_1, x_2)$, say, by

$$G_\tau^{\mathrm{S}}(\vec{x};\vec{x}') = \det[g_\tau^{\mathrm{S}}(x_j, x_k')]_{j,k=1,\dots,N}. \tag{11.95}$$

Furthermore the method of separation of variables gives that for $N = 1$ the Green function solution of the imaginary time Schrödinger equation (11.39) with $\beta = 2$ can be written in terms of the eigenfunctions $\{\Phi_n\}_{n=0,1,\dots}$ and corresponding eigenvalues $\{\nu_n\}_{n=0,1,\dots}$ according to

$$g_\tau^{\rm S}(u^{(0)}, u) = \sum_{j=0}^\infty \frac{\Phi_j(u^{(0)})\Phi_j(u)}{\langle \Phi_j | \Phi_j \rangle} e^{-\gamma_j \tau}, \quad \gamma_j = \frac{\nu_j}{2}$$

where $\langle f | g \rangle := \int_I f(u) g(u)\, du$.

For the particular operators (11.94) the eigenfunctions can be expressed in terms of the classical Hermite, Laguerre and Jacobi polynomials. A unified presentation of the three cases is possible by introducing the new variable $y_j = x_j$ in $H^{(\rm H)}$, $y_j = x_j^2$ in $H^{(\rm L)}$ and $\frac{1}{2}(1 - y_j) = \sin^2 x_j$ in $H^{(\rm J)}$, and defining the transformed single-particle Green function g_τ by

$$g_\tau(y, y') dy = g_\tau^{\rm S}(x, x') dx.$$

Recalling (11.42) we then find that

$$G_\tau^{\rm FP}(\vec{x}; \vec{x}') dx_1 \cdots dx_N = G_\tau(\vec{y}; \vec{y}\,') dy_1 \cdots dy_N, \tag{11.96}$$

where

$$G_\tau(\vec{y}, \vec{y}\,') = e^{E_0 \tau/2} \prod_{j=1}^N \sqrt{\frac{w_2(y_j)}{w_2(y_j')}} \prod_{1 \le j < k \le N} \frac{(y_k - y_j)}{(y_k' - y_j')} \det[g_\tau(y_j, y_k')]_{j,k=1,\dots,N} \tag{11.97}$$

with

$$g_\tau(y, y') = \sqrt{w_2(y) w_2(y')} \sum_{j=0}^\infty \frac{p_j(y) p_j(y')}{(p_j, p_j)_2} e^{-\gamma_j \tau}. \tag{11.98}$$

Here the weight function is one of the classical forms of Section 5.4.1, the corresponding monic polynomials are given by (5.46), while

$$\gamma_n = \begin{cases} (n + \frac{1}{2}), & \text{Hermite}, \\ 2(n + (a+1)/2), & \text{Laguerre}, \\ 2(n + (a+b+1)/2)^2, & \text{Jacobi}, \end{cases} \tag{11.99}$$

where $a := a' - \frac{1}{2}$, $b := b' - \frac{1}{2}$.

To proceed further the initial condition p_0 in (11.92) must be specified. In the example of the parameter-dependent random matrix given by (11.2), the appropriate choice would be the GOE eigenvalue p.d.f. This leads to a quaternion determinant formula for (11.93) [216], but the details will not be pursued. Instead we consider the simpler situation that $P_0(\mathbf{H})$ is the joint p.d.f. for elements of the GUE,

$$p_{\tau_1, \tau_2}(\vec{x}^{(1)}; \vec{x}^{(2)}) = G_{\tau_2 - \tau_1}^{\rm FP}(\vec{x}^{(1)}; \vec{x}^{(2)}).$$

The latter in turn is the special case $\tilde{w}_2(x) = w_2(x)$ of choosing in (11.92)

$$p_0(\vec{x}^{(0)}) = \frac{1}{C} \prod_{l=1}^N \tilde{w}_2(x_l) \prod_{1 \le j < k \le N} (x_k - x_j)^2. \tag{11.100}$$

11.6.2 Relationship to matrix integrals

Consider the Fokker-Planck equation (11.15) with W given by (11.16). For $\beta = 1, 2, 4$ the Green function solution is given by (11.5) and thus

$$G_\tau^{\rm FP}(\vec{\lambda}^{(0)}; \vec{\lambda}) \propto \prod_{1 \le j < k \le N} |\lambda_k - \lambda_j|^\beta e^{-\tilde{\beta}\sum_{j=1}^N \lambda_j^2 - \tilde{\beta}t^2 \sum_{j=1}^N (\lambda_j^{(0)})^2} \int e^{2\tilde{\beta}t{\rm Tr}(\mathbf{VHV}^\dagger \mathbf{H}^{(0)})} (\mathbf{V}^\dagger d\mathbf{V}), \quad (11.101)$$

where $\tilde{\beta} := \beta/2|1 - e^{-2\tau}|$, $t := e^{-\tau}$ and $\mathbf{V}$ is real orthogonal, unitary or unitary symplectic for $\beta = 1, 2$ or 4, respectively.

On the other hand, for $\beta = 2$, this Green function is also given by the Hermite case of (11.96). Comparing the two expressions, and making use of the classical summation formula [508]

$$\sum_{k=0}^\infty \frac{H_k(w)H_k(z)}{k!2^k\sqrt{\pi}} t^k = \frac{1}{\sqrt{\pi}}(1-t^2)^{-1/2} e^{-t^2(z^2+w^2)/(1-t^2)} e^{2wzt/(1-t^2)}, \quad |t| < 1, \quad (11.102)$$

gives the following matrix integral formula, due to Harish-Chandra [292] and Itzykson and Zuber [309].

PROPOSITION 11.6.1 *Let $(\mathbf{U}^\dagger d\mathbf{U})$ denote the Haar volume form for $N \times N$ unitary matrices, normalized so that $\int (\mathbf{U}^\dagger d\mathbf{U}) = 1$. Let $\mathbf{A}$ and $\mathbf{B}$ be $N \times N$ Hermitian matrices with eigenvalues $\{a_j\}$ and $\{b_j\}$, respectively. We have*

$$\int e^{{\rm Tr}(\mathbf{UAU}^\dagger \mathbf{B})} (\mathbf{U}^\dagger d\mathbf{U}) = \prod_{j=1}^{N-1} j! \frac{\det[e^{a_j b_k}]_{j,k=1,\dots,N}}{\prod_{1 \le j < k \le N}(a_k - a_j)(b_k - b_j)}. \quad (11.103)$$

Proof. Summing (11.98) in the Gaussian case according to (11.102) and substituting in (11.97) shows

$$G_\tau^{\rm FP}(\vec{\lambda}^{(0)}; \vec{\lambda}) \propto \prod_{1 \le j < k \le N} \frac{(\lambda_k - \lambda_j)}{(\lambda_k^{(0)} - \lambda_j^{(0)})} e^{-\sum_{j=1}^N \lambda_j^2} e^{-\tilde{\beta}t^2 \sum_{j=1}^N (\lambda_j^2 + (\lambda_j^{(0)})^2)} \det\Big[e^{2\tilde{\beta}t\lambda_j \lambda_k^{(0)}} \Big]_{j,k=1,\dots,N}. \quad (11.104)$$

Comparing this with (11.101), and setting $2\tilde{\beta}t\lambda_j = a_j$, $\lambda_j^{(0)} = b_j$, gives (11.103) up to the proportionality constant. To verify the latter is given correctly, take the limit $a_1, \dots, a_N \to 0$ and make use of the Vandermonde determinant formula (1.173) to show that both sides are equal to unity. □

Consider next the Fokker-Planck equation (11.15) with W given by (11.29). Then the Green function solution is given by (11.27) and thus

$$G_\tau^{\rm FP}(\vec{x}^{(0)}; \vec{x}) \propto \prod_{j=1}^N x_j^{\beta a+1} \prod_{1 \le j < k \le N} |x_k^2 - x_j^2|^\beta e^{-\tilde{\beta}\sum_{j=1}^N x_j^2 - \tilde{\beta}t^2 \sum_{j=1}^N (x_j^{(0)})^2}$$
$$\times \int (\mathbf{U}^\dagger d\mathbf{U}) \int (\mathbf{V}^\dagger d\mathbf{V}) e^{\tilde{\beta}t{\rm Tr}(\mathbf{VX}^\dagger \mathbf{U}^\dagger \mathbf{X}^{(0)} + \mathbf{X}^{(0)\dagger} \mathbf{UXV}^\dagger)}, \quad (11.105)$$

where $\tilde{\beta}, t$ are as in (11.101)

For $\beta = 2$ this Green function is also given by the Laguerre case of the l.h.s. of (11.96). Making use of the classical summation formula [508]

$$\sum_{n=0}^\infty \frac{n!}{(a+1)_n} L_n^a(x) L_n^a(y) t^n = (1-t)^{-a-1} \exp\Big(-\frac{t}{1-t}(x+y) \Big) {}_0F_1\Big(a+1; \frac{xyt}{(1-t)^2} \Big) \quad (11.106)$$

we obtain the alternative formula

$$G_\tau^{\rm FP}(\vec{x}^{(0)};\vec{x}) \propto \prod_{j=1}^N x_j^{2a+1} e^{-x_j^2} \prod_{1\le j<k\le N} \frac{(x_k^2-x_j^2)}{((x_k^{(0)})^2-(x_j^{(0)})^2)} e^{-\tilde{\beta}t^2\sum_{j=1}^N (x_j^2+(x_j^{(0)})^2)}$$
$$\times \det\Big[{}_0F_1(a+1;(x_j x_k^{(0)} t\tilde{\beta})^2)\Big]_{j,k=1,\dots,N}. \tag{11.107}$$

Comparing (11.105) in the case $\beta=2$ with (11.107), and interchanging the role of $\mathbf{U}$ and $\mathbf{V}$ for convenience, the following analogue of (11.103) is obtained [277], [311].

Proposition 11.6.2 *Let $(\mathbf{U}^\dagger d\mathbf{U})$, $(\mathbf{V}^\dagger d\mathbf{V})$ denote the normalized Haar volume for $N\times N$, $n\times n$ unitary matrices with $n\ge N$. Let $\mathbf{X}$ and $\mathbf{Y}$ be $n\times N$ complex matrices, with the eigenvalues of $\mathbf{X}^\dagger\mathbf{X}$, $\mathbf{Y}^\dagger\mathbf{Y}$ given by $x_1^2,\dots,x_N^2$ and $y_1^2,\dots,y_N^2$, respectively. We have*

$$\int (\mathbf{U}^\dagger d\mathbf{U}) \int (\mathbf{V}^\dagger d\mathbf{V}) e^{\mathrm{Tr}(\mathbf{U}\mathbf{X}^\dagger\mathbf{V}^\dagger\mathbf{Y}+\mathbf{Y}^\dagger\mathbf{V}\mathbf{X}\mathbf{U}^\dagger)}$$
$$= \prod_{j=0}^{N-1}(n-N+1)_j j!\, \frac{\det[{}_0F_1(n-N+1;(x_j y_k)^2]_{j,k=1,\dots,N}}{\prod_{1\le j<k\le N}(y_k^2-y_j^2)(x_k^2-x_j^2)}. \tag{11.108}$$

Here the normalization has been fixed by requiring that both sides equal unity in the limit $x_1,\dots,x_N\to 0$. We note that the ${}_0F_1$ function can be replaced in favor of the Bessel function of pure imaginary argument by using the identity (10.190).

11.6.3 Character expansions

A major aspect of the theory of the classical group $U(N)$, and consequently of group integrals over the corresponding Haar measure, is that of its representations. This theory is detailed in many texts (see e.g. [540]). Following [487] we will present, without further explanation, a number of key notions and formulas which together allow for the computation of matrix integrals of the type (11.103) and (11.108). A key role is played by the Schur polynomials in their determinant form (10.16).

First, irreducible representations of $U(N)$ and more generally $\mathrm{Gl}(N)$ (the group of $N\times N$ invertible complex matrices) are labeled by partitions μ, and they have dimension expressible in terms of Schur polynomials by

$$s_\mu((1)^N) = \frac{1}{h_\mu}\prod_{j=1}^N \frac{\Gamma(N-j+1+\kappa_j)}{\Gamma(N-j+1)}, \tag{11.109}$$

where the equality follows from (12.105) below and h_μ is specified by (12.58) below with $\alpha=1$. Second, if $\mathbf{X}\in Gl(N)$ has eigenvalues $\lambda_1,\dots,\lambda_N$, then its representations $\mathbf{X}^{(\mu)}$ have the property

$$\chi_\mu(\mathbf{X}) := \mathrm{Tr}(\mathbf{X}^{(\mu)}) = s_\mu(\lambda_1,\dots,\lambda_N) \tag{11.110}$$

independent of the basis. The quantity $\chi_\mu(\mathbf{X})$ is a group character. Third, for $\mathbf{U}\in U(N)$, the elements of $\mathbf{U}^{(\mu)}$ have the integration property

$$\int U_{ij}^{(\mu)}\bar{U}_{kl}^{(\rho)}(\mathbf{U}^\dagger d\mathbf{U}) = \frac{1}{s_\mu((1)^N)}\delta_{\mu,\rho}\delta_{i,k}\delta_{j,l} \tag{11.111}$$

(cf. the first result in Proposition 2.3.3), where $(\mathbf{U}^\dagger d\mathbf{U})$ denotes the normalized Haar volume form.

We will make use of these results to deduce the following matrix integral, which generalizes (11.103) [438].

Proposition 11.6.3 *Let $\mathbf{A},\mathbf{B}$ be $N\times N$ complex matrices with eigenvalues $\{a_1,\dots,a_N\}$, $\{b_1,\dots,b_N\}$.*

We have

$$\int \det(\mathbf{1}_N - \mathbf{A}\mathbf{U}^\dagger\mathbf{B}\mathbf{U})^{-r}(\mathbf{U}^\dagger d\mathbf{U}) = \prod_{k=1}^{N-1} \frac{k!}{(r-N+1)_k} \frac{\det[(1-a_jb_k)^{N-1-r}]_{j,k=1,\dots,N}}{\prod_{1\le j<k\le N}(a_k-a_j)(b_k-b_j)}. \tag{11.112}$$

Proof. Denote the integral by $I(\mathbf{A},\mathbf{B})$. According to the generalized binomial theorem, (12.133) below in the case $\alpha = 1$,

$$\frac{1}{(\det(\mathbf{1}-\mathbf{X}))^r} = \sum_\kappa \prod_{j=1}^N \frac{\Gamma(r-j+1+\kappa_j)}{\Gamma(r-j+1)} \frac{s_\kappa(\mathbf{X})}{h_\kappa}.$$

Substituting this in the integrand and using (11.110) shows

$$I(\mathbf{A},\mathbf{B}) = \sum_\kappa \frac{1}{h_\kappa} \prod_{j=1}^N \frac{\Gamma(r-j+1+\kappa_j)}{\Gamma(r-j+1)} \int \mathrm{Tr}(\mathbf{A}^{(\kappa)}\mathbf{U}^{(\kappa)\dagger}\mathbf{B}^{(\kappa)}\mathbf{U}^{(\kappa)})(\mathbf{U}^\dagger d\mathbf{U}).$$

The integral is quadratic in the entries of the unitary matrix representation and so can be computed using (11.111) to give

$$I(\mathbf{A},\mathbf{B}) = \sum_\kappa \frac{1}{h_\kappa} \prod_{j=1}^N \frac{\Gamma(r-j+1+\kappa_j)}{\Gamma(r-j+1)} \frac{1}{s_\kappa((1)^N)} \chi_\kappa(\mathbf{A})\chi_\kappa(\mathbf{B}). \tag{11.113}$$

Next, substitute for $s_\kappa((1)^N)$ using (11.109) and substitute the determinant form of the Schur polynomials (10.16). Finally, write $k_j := \kappa_j - j + N$ so that $0 \le k_N < k_{N-1} < \cdots < k_1$ and apply the general formula (5.170) to obtain

$$I(\mathbf{A},\mathbf{B}) = \frac{1}{\Delta(a)\Delta(b)} \prod_{j=1}^N \frac{\Gamma(N-j+1)}{\Gamma(r-j+1)} \det\Big[\sum_{p=0}^\infty \frac{\Gamma(r+1-N+p)}{\Gamma(1+p)}(a_jb_k)^p\Big]_{j,k=1,\dots,N},$$

where here $\Delta(a) := \prod_{1\le j<k\le N}(a_k - a_j)$ and similarly $\Delta(b)$. The evaluation (11.112) now follows upon application of the classical binomial theorem. □

11.6.4 Matrix integrals and biorthogonal ensembles

The matrix integrals (11.103), (11.108) and (11.112) all lead to biorthogonal ensembles. For definiteness, consider the Harish-Chandra/Itzykson-Zuber integral (11.103). As already noted, it implies for the Green function solution of the Fokker-Planck equation (11.15) with W given by (11.16) and $\beta = 2$ the explicit form (11.104). This Green function is equivalent to the eigenvalue p.d.f. of the complex parameter-dependent random matrix specified by Definition 11.1.1. Its functional form is the biorthogonal ensemble (5.148) with $w_2(x) = e^{-x^2}$.

An alternative way to state this result is in terms of the shifted mean GUE, with distribution of elements proportional to $\exp(-(\mathbf{H} - \mathbf{H}^{(0)})^2)$. Thus we have that the p.d.f. for the eigenvalues $\{x_j\}$ of $\mathbf{H}$, given the eigenvalues $\{y_j\}$ of $\mathbf{H}^{(0)}$, is equal to (5.148) with $w_2(x) = e^{-x^2}$ and $a_j = -2y_j$. In the case that $a_j = 0$ ($j = r+1,\dots,N$) the matrix $\mathbf{H}^{(0)}$ has rank r, and the corresponding correlation kernel is given by (5.172). We know that the correlation kernel exhibits a soft edge scaling limit on introducing the scaled variables (7.16), and in particular setting $y_k = \sqrt{N/2} - (N^{1/6}/\sqrt{2})s_k$ for the nonzero eigenvalues of $\mathbf{H}^{(0)}$. The significance of this latter value can be seen from Proposition 1.8.2, which tells us that in the case $r = 1$ a single eigenvalue will separate from the Wigner semicircle for $y_1 > \sqrt{N/2}$, thereby giving $y_1 = \sqrt{N/2}$ as the critical value of this effect.

Another application [256] of (11.103) leading to the biorthogonal ensemble (5.148) is the computation of the eigenvalue p.d.f. of complex correlated Wishart matrices $\mathbf{X}^\dagger\mathbf{X}$, where $\mathbf{X}$ is a $n \times N$ ($n \ge N$) complex Gaussian matrix with joint p.d.f.

$$\frac{1}{C} e^{-\mathrm{Tr}(\mathbf{X}^\dagger\mathbf{X}\mathbf{\Sigma}^{-1})}$$

(cf. (3.71)). Writing $\mathbf{\Sigma}^{-1} = \mathbf{1}_N + \mathbf{B}$, we see that the joint p.d.f. of the elements of $\mathbf{A} = \mathbf{X}^\dagger \mathbf{X}$ is proportional to

$$(\det \mathbf{A})^{n-N} e^{-\mathrm{Tr}(\mathbf{A})} e^{-\mathrm{Tr}(\mathbf{AB})},$$

where use has been made of (3.23). Now changing variables to the eigenvalues and eigenvectors of $\mathbf{A}$ according to (1.27) with $\beta = 2$ shows that the eigenvalue p.d.f. of $\mathbf{A}$ is proportional to

$$\prod_{l=1}^{N} \lambda_l^{n-N} e^{-\lambda_l} \prod_{1 \le j < k \le N} (\lambda_k - \lambda_j)^2 \int e^{-\mathrm{Tr}(\mathbf{U}^\dagger \mathbf{AUB})} (\mathbf{U}^\dagger d\mathbf{U}). \tag{11.114}$$

Applying (11.103) gives (5.148) with $w_2(x) = x^{n-N} e^{-x}$.

Let the eigenvalues of $\mathbf{B}$ be denoted $\{b_i\}$, and suppose $b_1 = \cdots = b_{N-r} = 0$ (in this situation the covariance matrix is said to be spiked [334], [29]). The corresponding correlation kernel is then given by (5.169) with $a := n - N$, and the nonzero eigenvalues of $\mathbf{B}$ denoted $a_1, \dots, a_r$. We know from (7.40), (7.41) that this correlation kernel permits a soft edge scaling, where in particular it is required that $a_j = -\frac{1}{2} + s_j/2N^{1/3}$, or equivalently that the eigenvalues of the covariance matrix $\mathbf{\Sigma}$ different from unity are of the form $2 + 2s_j/N^{1/3}$. To understand the significance of this value, specialize to the case $n = N, r = 1$. For $\mathbf{Y}$ an $n \times n$ matrix of standard complex Gaussians, the matrix $\mathbf{X}^\dagger \mathbf{X} = \mathbf{\Sigma}^{1/2} \mathbf{Y}^\dagger \mathbf{Y} \mathbf{\Sigma}^{1/2}$ has the first column of $\mathbf{X}$ having variance unity. Noting that the eigenvalues of $\mathbf{X}^\dagger \mathbf{X}$ are the same as those for $\mathbf{X}\mathbf{X}^\dagger$, from this latter matrix we see the problem can be regarded as a rank 1 perturbation of the type (3.72) with $b = 2 + 2s_j/N^{1/3}$. According to Proposition 3.5.1 the value $b = 2$ is critical for the largest eigenvalue to separate from the bulk.

11.6.5 Continuous nonintersecting paths

In the setting of nonintersecting paths (recall Section 10.1.5) it is natural to prescribe not just the initial but also the final configuration. Take $\vec{x}^{(0)}$ to be the initial configuration at time $t = 0$, and $\vec{x}^{(m+1)}$ to be the final configuration at time $t = 2T$. With $G_t^{(\mathrm{B})}$ given by (10.44), the p.d.f. for arriving at $\vec{x}^{(s)}$ in time $t = t_s$, $s = 1, \dots, m$, where $0 < t_1 < \cdots t_m < 2T$, is then

$$\frac{\prod_{l=1}^{m+1} G_{t_l - t_{l-1}}^{(\mathrm{B})} (\vec{x}^{(l-1)}; \vec{x}^{(l)})}{G_{2T}^{(\mathrm{B})} (\vec{x}^{(0)}; \vec{x}^{(m+1)})}. \tag{11.115}$$

The simplest situation is that when $\vec{x}^{(0)} = \vec{x}^{(m+1)} = \vec{0}$. According to (10.47), for $\vec{x}$ and/or $\vec{y}$ approaching $\vec{0}$,

$$G_t^{(\mathrm{B})}(\vec{x}; \vec{y}) \sim \frac{1}{(2\pi)^{p/2} \prod_{l=1}^{p-1} l!} \Big(\frac{1}{t}\Big)^{p^2/2} e^{-\sum_{j=1}^{p} (x_j^2 + y_j^2)/2t} \prod_{1 \le j < k \le p} (x_j - x_k)(y_j - y_k).$$

This shows that for $\vec{x}^{(0)}, \vec{x}^{(m+1)} \to \vec{0}$, (11.115) reduces to

$$\frac{1}{(2\pi)^{p/2} \prod_{l=1}^{p-1} l!} \Big(\frac{2T}{t_1(2T - t_m)}\Big)^{p^2/2} e^{-\sum_{j=1}^{p} (x_j^{(1)})^2/2t_1} e^{-\sum_{j=1}^{p} (x_j^{(m)})^2/2(2T - t_m)}$$
$$\times \prod_{1 \le j < k < p} (x_j^{(1)} - x_k^{(1)})(x_j^{(m)} - x_k^{(m)}) \prod_{l=1}^{m-1} G_{t_{l+1} - t_l}^{(\mathrm{B})} (\vec{x}^{(l)}; \vec{x}^{(l+1)}). \tag{11.116}$$

Prescribed initial and final positions can also be realized in the setting of parameter-dependent random matrices. Thus it was remarked at the end of Section 11.1 that the latter can be regarded as random matrices in which the entries are Brownian paths. More explicitly, the Brownian paths are those determined by the Ornstein-Uhlenbeck process corresponding to the Fokker-Planck equation (11.3). If instead we define the parameter-dependent ensemble to have joint p.d.f. for the elements, P_t say, to be proportional to

$$\exp\Big(- \beta \mathrm{Tr}\,(\mathbf{H} - \mathbf{H}^{(0)})^2/4t\Big), \tag{11.117}$$

then P_t satisfies the multidimensional heat equation

$$\frac{\partial P_t}{\partial t} = \frac{1}{\beta} \sum_\mu D_\mu \frac{\partial^2 P_t}{\partial H_\mu^2},$$

where D_μ is as in (11.3). The entries are now thus free Brownian motion, beginning at the entries of $\mathbf{H}^{(0)}$, and so this latter matrix determines the eigenvalues of $\mathbf{H}$ at $t = 0$. In particular, if we choose $\mathbf{H}^{(0)} = \mathbf{0}$, then the Brownian motions corresponding to the entries begin at the origin, and the eigenvalues of $\mathbf{H}$ are all initially zero. If the Brownian paths specifying the elements are also constrained to return to the origin at $t = 2T$ (and thus be Brownian bridges), then we have a situation in which both the initial and final positions of the eigenvalues are prescribed as equaling $\vec{0}$. In the case $\beta = 2$, when the matrices in (11.117) are complex Hermitian, the corresponding p.d.f. for the eigenvalues equaling $\vec{\lambda}^{(s)}$ at times t_s $(s = 1, \dots, m)$ with $0 < t_1 < \cdots < t_m < 2T$ is in fact precisely (11.116). To see this, we read off from the limit $\lambda, \tau \to 0$, $\lambda/\sqrt{\tau}$ fixed of (11.104) that the Green function G_t in this case is related to the Green function $G_t^{(\mathrm{B})}$ (10.44) for the nonintersecting paths problem by

$$G_t(\vec{\lambda}^{(0)}; \vec{\lambda}) = \prod_{1 \le j < k \le N} \frac{(\lambda_k - \lambda_j)}{(\lambda_k^{(0)} - \lambda_j^{(0)})} G_t^{(\mathrm{B})}(\vec{\lambda}^{(0)}; \vec{\lambda}).$$

But the ratio of products of differences cancels out of (11.115), giving the sought equivalence.

11.6.6 Correlations and determinant formulas

The first task in evaluating the dynamical correlation (11.93) is to compute (11.92) in a convenient form. Throughout we will work with the variables y_j used in (11.97) and consider the initial condition (11.100).

PROPOSITION 11.6.4 *Let $\{u_j(z;\tau)\}_{j=0,1,\dots}$, $\{v_j(z;\tau)\}_{j=0,1,\dots}$ be a set of monic polynomials with u_j, v_j of degree j. For the p.d.f. (11.92) with Green function (11.97) and initial condition (11.100) we have*

$$p_{\tau_1,\tau_2}(\vec{y}^{(1)}; \vec{y}^{(2)}) = \frac{1}{C} \prod_{j=1}^N \sqrt{w_2(y_j^{(2)})} \det\Big[v_{j-1}(y_k^{(2)}; \tau_2)\Big]_{j,k=1,\dots,N} \det[g_{\tau_2 - \tau_1}(y_l^{(1)}, y_j^{(2)})]_{j,l=1,\dots,N}$$
$$\times \prod_{j=1}^N \sqrt{w_2(y_j^{(1)})} \det\Big[\tilde{u}_{j-1}(y_k^{(1)}; \tau_1)\Big]_{j,k=1,\dots,N}, \tag{11.118}$$

where

$$\tilde{u}_{j-1}(y;\tau) := \frac{1}{\sqrt{w_2(y)}} \int_I \frac{\tilde{w}_2(z)}{\sqrt{w_2(z)}} u_{j-1}(z;\tau) g_\tau(z,y)\, dz. \tag{11.119}$$

Proof. From the definition

$$p_{\tau_1,\tau_2}(\vec{y}^{(1)}; \vec{y}^{(2)}) = \frac{1}{C} \prod_{j=1}^N \sqrt{w_2(y_j^{(2)})} \prod_{j>l}^N (y_j^{(2)} - y_l^{(2)}) \det[g_{\tau_2-\tau_1}(y_l^{(1)}, y_j^{(2)})]_{j,l=1,\dots,N}$$
$$\times \int_I dy_1^{(0)} \cdots \int_I dy_N^{(0)} \prod_{l=1}^N \frac{\tilde{w}_2(y_l^{(0)})}{\sqrt{w_2(y_l^{(0)})}} \prod_{j<k} (y_k^{(0)} - y_j^{(0)}) \det[g_{\tau_1}(y_k^{(0)}, y_j^{(1)})]_{j,k=1,\dots,N} \tag{11.120}$$

for some normalization C. Write

$$\prod_{j<k}(y_k^{(0)} - y_j^{(0)}) = \det[u_{j-1}(y_k^{(0)};\tau_1)]_{j,k=1,\dots,N}, \tag{11.121}$$

so that there are two determinant factors in the integrand. Since both determinants are antisymmetric in $\{y_j^{(0)}\}_{j=0,\dots,N-1}$ we can replace (11.121) by $N!\prod_{j=1}^N u_{j-1}(y_j^{(0)};\tau_1)$ in the integrand. Integrating row-by-row in the remaining determinant shows that the last line in (11.120) can be written

$$N!\Big(\prod_{l=1}^N \sqrt{w_2(y_l^{(1)})}\Big)\det\Big[\tilde{u}_{j-1}(y_k^{(1)};\tau_1)\Big]_{j,k=1,\dots,N},$$

and the result follows. □

We recognize the structure (11.118) as an example of the p.d.f. (5.175) with $k = 2$. Consequently the correlations are given by Proposition 5.9.6. The double summations therein reduces to a single summation if we choose the polynomials $\{u_j(x;\tau)\}$ and $\{v_j(x;\tau)\}$ so that they have the biorthogonality property

$$\int_I dy\,\sqrt{w_2(y)}\Big(\int_I dz\,\frac{\tilde{w}_2(z)}{\sqrt{w_2(z)}}u_{j-1}(z;\tau_1)g_{\tau_1}(z;y)\Big)v_{k-1}(y;\tau_2) = h_{j-1}(\tau_2)\delta_{j,k}. \tag{11.122}$$

At $\tau_1 = \tau_2 = 0$ the l.h.s. of (11.122) reduces to

$$\int_I \tilde{w}_2(y)u_{j-1}(y;0)v_{k-1}(y;0)\,dy.$$

Thus if $\{p_l^{(0)}(y)\}_{l=0,1,\dots}$ denotes the monic orthogonal polynomials with respect to the weight function $\tilde{w}_2(y)$, then we must have

$$u_{j-1}(y;0) = p_{j-1}^{(0)}(y), \qquad v_{k-1}(y;0) = p_{k-1}^{(0)}(y).$$

Now let $\{p_l(y)\}_{l=0,1,\dots}$ denote the orthogonal polynomials with respect to the weight function $w_2(y)$ as in (11.98), and write

$$p_{k-1}^{(0)}(y) = \sum_{l=0}^{k-1}\tilde{\alpha}_{k-1\,l}\,p_l(y), \qquad \tilde{\alpha}_{k-1\,k-1} = 1. \tag{11.123}$$

A direct calculation verifies the following result.

PROPOSITION 11.6.5 *The monic polynomials*

$$u_{j-1}(z;\tau) = p_{j-1}^{(0)}(z), \qquad v_{k-1}(y;\tau) = e^{-\gamma_{k-1}\tau}\sum_{l=0}^{k-1}\tilde{\alpha}_{k-1\,l}\,p_l(y)e^{\gamma_l\tau}$$

with normalization $h_{j-1}(\tau) = e^{-\gamma_{j-1}\tau}(p_{j-1}^{(0)}, p_{j-1}^{(0)})_{\tilde{w}_2}$ have the biorthogonality property (11.122).

Note that in the special case $\tilde{w}_2(y) = w_2(y)$ Proposition 11.6.5 gives

$$u_j(z;\tau) = p_j(z), \qquad v_k(z;\tau) = p_k(z), \tag{11.124}$$

with normalization $h_j(\tau) = e^{-\gamma_j\tau}(p_j, p_j)_2$. Application of Proposition 5.9.6 gives the following formula for the dynamical correlations.

PROPOSITION 11.6.6 *For the p.d.f. (11.118)*

$$\rho_{(m_1,m_2)}(y_1^{(1)},\dots,y_{m_1}^{(1)};\tau_1;y_1^{(2)},\dots,y_{m_2}^{(2)};\tau_2) = \det\begin{bmatrix}[\sigma_{jk}^{11}]_{j,k=1,\dots,m_1} & [\sigma_{jk}^{12}]_{\substack{j=1,\dots,m_1\\k=1,\dots,m_2}}\\ [\tilde{\sigma}_{jk}^{21}]_{\substack{j=1,\dots,m_2\\k=1,\dots,m_1}} & [\sigma_{jk}^{22}]_{j,k=1,\dots,m_2}\end{bmatrix}, \tag{11.125}$$

where

$$\sigma_{jk}^{mn} = \sqrt{w_2(y_j^{(m)})w_2(y_k^{(n)})} \sum_{l=1}^{N} \frac{\tilde{u}_{l-1}(y_j^{(m)};\tau_m)v_{l-1}(y_k^{(n)};\tau_n)}{h_{l-1}(\tau_n)},$$

$$\tilde{\sigma}_{jk}^{mn} = \sigma_{jk}^{mn} - g_{\tau_m-\tau_n}(y_j^{(m)}|y_k^{(n)}) \quad (\tau_m > \tau_n)$$

$$= -\sqrt{w_2(y_j^{(m)})w_2(y_k^{(n)})} \sum_{l=N}^{\infty} \frac{\tilde{u}_l(y_j^{(m)};\tau_m)v_l(y_k^{(n)};\tau_n)}{h_l(\tau_n)}. \tag{11.126}$$

Proof. It follows from the biorthogonality (11.122) and the definition (11.119) that an arbitrary function $f(x)$ can be expanded

$$f(x) = \sum_{l=0}^{\infty} \frac{\int_I w_2(t)f(t)\tilde{u}_l(t;\tau)\,dt}{h_l(\tau)} v_l(x;\tau).$$

Choosing $f(x) = g_{\tau'}(x,x')/\sqrt{w_2(x)}$ and noting from Proposition 11.6.5 that $u_{j-1}(z;\tau)$ in (11.119) is independent of τ, we see from this that

$$g_{\tau'}(y_j^{(m)},y_k^{(n)}) = \sqrt{w_2(y_j^{(m)})w_2(y_k^{(n)})} \sum_{l=0}^{\infty} \frac{\tilde{u}_l(y_j^{(m)};\tau+\tau')v_l(y_k^{(n)};\tau)}{h_l(\tau)}.$$

Choosing $\tau = \tau_n$, $\tau' = \tau_m - \tau_n$ $(\tau_m > \tau_n)$, verifies the final equality in (11.126), and allows the integrals in the formulas of Proposition 5.9.6 for σ_{jk}^{mn} to be evaluated to give the first equality. □

We remark that $p_{\tau_1,\tau_2}(\vec{x};\vec{y})$ effectively specifies a two-species system. As such the formula (9.19) applies for the corresponding gap probability generating function.

EXERCISES 11.6 1. (i) [508] Consider the second order differential equation

$$\alpha(z)\frac{d^2p_n(z)}{dz^2} + \beta(z)\frac{dp_n(z)}{dz} = \lambda_n p_n(z). \tag{11.127}$$

Let $x = x(z)$ be such that $x'(z) = 1/\sqrt{\alpha(z)}$, and put

$$\psi_0(x) = e^{-w(z(x))} \qquad \text{with } w' = \frac{\alpha' - 2\beta}{4\alpha}.$$

Show that the functions $\psi_n(x) = \psi_0(z)p_n(z(x))$ then solve the Schrödinger equation

$$\Big(-\frac{d^2}{dx^2} + V(x)\Big)\psi_n(x) = -\lambda_n\psi_n(x)$$

where

$$V(x) = v(z(x)), \qquad v = \frac{(2\beta-\alpha')(2\beta-3\alpha')}{16\alpha} - \frac{1}{4}\alpha' + \frac{1}{2}\beta'.$$

(ii) [285] The Hermite, Laguerre and Jacobi polynomials satisfy (11.127) with

$$(\alpha(z),\beta(z),\gamma_n) = \begin{cases} (1,-2z,-2n), & \text{Hermite,} \\ 4(z,(a+1-z),-n), & \text{Laguerre,} \\ 4(1-z^2, b-a-(a+b+2)z, -n(n+a+b+1)), & \text{Jacobi} \end{cases}$$

(the factors of 4 in the Laguerre and Jacobi cases are chosen for convenience). Show that the theory of (i) is consistent with the results stated in the paragraph including (11.96).

2. [409] The following p.d.f. can be interpreted as specifying a preferred basis.

(i) Let $\mathbf{U}$ be a member of the CUE, and define a probability density function for Hermitian matrices $\mathbf{H}$ by requiring that

$$\Pr(\mathbf{H}) \propto e^{-\operatorname{Tr}\mathbf{H}^2} e^{-b\operatorname{Tr}([\mathbf{U},\mathbf{H}][\mathbf{U},\mathbf{H}]^\dagger)}. \tag{11.128}$$

Integrate over $\mathbf{U}$ according to (11.103) to deduce that the corresponding eigenvalue p.d.f. is proportional to

$$\det\Big[e^{-(b+1/2)(x_j^2+x_k^2)+2bx_jx_k}\Big]_{j,k=1,\dots,N}.$$

(ii) Regard (11.128) as specifying a probability distribution on unitary matrices $\mathbf{U}$ rather than Hermitian matrices $\mathbf{H}$. By integrating over $\mathbf{H}$ show that the corresponding eigenvalue p.d.f. is proportional to

$$\prod_{1\le j<k\le N}\frac{\sin^2(\theta_k-\theta_j)/2}{1+4b\sin^2(\theta_k-\theta_j)/2}.$$

3. (i) Set $c=\sqrt{t(2T-t)/T}$. In the case $m=1$, note that (11.116) with $x_j^{(1)} =: x_j$ can be written

$$\frac{2^{p^2/2}c^{-p}}{(2\pi)^{p/2}\prod_{l=1}^{p-1}l!}e^{-\sum_{j=1}^p(x_j/c)^2}\prod_{1\le j<k\le p}\Big(\frac{x_k}{c}-\frac{x_j}{c}\Big)^2$$

and check from Proposition 4.7.1 that it is correctly normalized with respect to the domain $x_1>\cdots>x_p$.

(ii) Make use of (11.102) to show that for general p, (11.116) can be written

$$\begin{aligned}&\frac{2^{-p(p-1)/2}}{\pi^{p/2}\prod_{l=1}^{p-1}l!}\Big(\frac{T}{t_1(2T-t_m)}\Big)^{p(p-1)/2}(c_1c_m)^{p(p-1)/2}\prod_{l=1}^m c_l^{-p}\\ &\times\prod_{s=1}^m e^{-\sum_{j=1}^p(x_j^{(s)}/c_s)^2}\det[H_{j-1}(x_k^{(1)}/c_1)]_{j,k=1,\dots,p}\det[H_{j-1}(x_k^{(m)}/c_m)]_{j,k=1,\dots,p}\\ &\times\det\Big[\sum_{n=0}^\infty\frac{e^{-n(\tau_{s+1}-\tau_s)}}{\sqrt{\pi}2^n n!}H_n(x_j^{(s+1)}/c_{s+1})H_n(x_k^{(s)}/c_s)\Big]_{j,k=1,\dots,p},\end{aligned}$$

where $c_j:=\sqrt{t_j(2T-t_j)/T}$, $\tau_s:=-\frac{1}{2}\log(2T-t_s)/t_s$.

(iii) Use the continuum version of Proposition 5.9.5 to obtain from the result of (ii) the dynamical correlation function

$$\rho_{(r)}((x_1,\tau_1),\dots,(x_r,\tau_r))=\prod_{j=1}^r c_j^{-1}e^{-\sum_{j=1}^r(x_j/c_j)^2}\det[\tilde K((x_j,\tau_j),(x_k,\tau_k))]_{j,k=1,\dots,r}$$

where

$$\tilde K((x,s),(y,t))=\begin{cases}\displaystyle\sum_{k=0}^{p-1}\frac{e^{k(t-s)}}{\sqrt{\pi}2^k k!}H_k(y/c_y)H_k(x/c_x), & t\ge s,\\ \displaystyle-\sum_{k=p}^{\infty}\frac{e^{k(t-s)}}{\sqrt{\pi}2^k k!}H_k(y/c_y)H_k(x/c_x), & t<s.\end{cases}$$

4. Let $(\mathbf{U}^\dagger d\mathbf{U})$ denote the normalized Haar volume form for $N\times N$ unitary matrices, and let $\mathbf{A}$, $\mathbf{B}$ be general $N\times N$ complex matrices. The method of Section 11.6.3 has been used to derive the matrix integral evaluation [482]

$$\begin{aligned}&\int(\det\mathbf{U})^\nu e^{(1/2)(\mathbf{AU}+\mathbf{BU}^\dagger)}(\mathbf{U}^\dagger d\mathbf{U})\\ &=2^{N(N-1)/2}\prod_{k=1}^{N-1}k!\Big(\frac{\det\mathbf{B}}{\det\mathbf{A}}\Big)^{\nu/2}\frac{\det[\mu_i^{j-1}I_{\nu+j-1}(\mu_i)]_{i,j=1,\dots,N}}{\prod_{i<j}^N(\mu_i^2-\mu_j^2)},\end{aligned}$$

where $\nu \in \mathbb{Z}$ and $\{\mu_i^2\}$ are the eigenvalues of $\mathbf{AB}$. By taking the limits $\mu_2 \to \mu_1 = \sqrt{t}$, $\mu_3 \to \mu_2$, etc., in succession deduce from this (8.180).

11.7 SCALED LIMITS

With the $\beta = 2$ initial state (11.100) assumed, let us suppose further that $\tilde{w}_2(y) = w_2(y)$ so that this state corresponds to the equilibrium state. In this case (11.124) gives that the in general biorthogonal, parameter-dependent polynomials $\{u_j(x;\tau)\}$, $\{v_j(x;\tau)\}$ reduce to orthogonal, parameter-independent polynomials, and we see from (11.126) that

$$\sigma_{jk}^{mn} = \sqrt{w_2(y_j^{(m)})w_2(y_k^{(n)})}\sum_{l=1}^{N} \frac{p_{l-1}(y_j^{(m)})p_{l-1}(y_k^{(n)})}{(p_{l-1},p_{l-1})_2} e^{\gamma_{l-1}(\tau_n-\tau_m)},$$

$$\tilde{\sigma}_{jk}^{mn} = -\sqrt{w_2(y_j^{(m)})w_2(y_k^{(n)})}\sum_{l=N+1}^{\infty} \frac{p_{l-1}(y_j^{(m)})p_{l-1}(y_k^{(n)})}{(p_{l-1},p_{l-1})_2} e^{\gamma_{l-1}(\tau_n-\tau_m)}. \tag{11.129}$$

The scaled form of these quantities can be computed in the bulk and at the hard and soft edges [441], [378], [216].

11.7.1 Bulk

Consider the Gaussian weight $w_2(x) = e^{-x^2}$. The monic orthogonal polynomials are given by the polynomials $p_n^{(G)}(x)$ in (5.46), their corresponding normalization is given in (5.48) and the constant γ_n is given in (11.99). We see from (11.129) that in the case $m = n$, σ_{jk}^{mn} is precisely the function $K_N(y_j^{(m)}, y_k^{(m)})$ of Proposition 5.1.3 and is thus summed by the Christoffel-Darboux formula (5.10). Its bulk scaled limit is given in Proposition 7.1.1. Thus with $\sigma_{jk}^{mn} := \sigma(y_j^{(m)}, y_k^{(n)}; \tau_m - \tau_n)$ we have

$$\lim_{N\to\infty} \frac{\pi\rho}{\sqrt{2N}}\sigma\Big(\frac{\pi\rho y_j^{(m)}}{\sqrt{2N}}, \frac{\pi\rho y_k^{(m)}}{\sqrt{2N}}; 0\Big) = \frac{\sin \pi\rho(y_j^{(m)} - y_k^{(m)})}{\pi(y_j^{(m)} - y_k^{(m)})}.$$

In the case $m \neq n$ the Christoffel-Darboux formula cannot be used. Instead we proceed as in the asymptotic analysis of the summations in (6.152), which involves substituting the asymptotic form (7.1). One finds

$$\lim_{N\to\infty} \frac{\pi\rho}{\sqrt{2N}}\sigma\Big(\frac{\pi\rho y_j^{(m)}}{\sqrt{2N}}, \frac{\pi\rho y_k^{(n)}}{\sqrt{2N}}; \frac{(\pi\rho)^2}{2N}(\tau_m - \tau_n)\Big)$$
$$=: S_2^{\rm bulk}(y_j^{(m)}, y_k^{(n)}; \tau_m - \tau_n) = \rho\int_0^1 e^{(\tau_n-\tau_m)(\pi\rho u)^2/2}\cos \pi u\rho(y_j^{(m)} - y_k^{(n)})\,du. \tag{11.130}$$

Note that τ has been scaled by $1/N$ to obtain a well-defined result. Similarly, now assuming $\tau_n < \tau_m$, we have

$$\lim_{N\to\infty} \frac{\pi\rho}{\sqrt{2N}}\tilde{\sigma}\Big(\frac{\pi\rho y_j^{(m)}}{\sqrt{2N}}, \frac{\pi\rho y_k^{(n)}}{\sqrt{2N}}; \frac{(\pi\rho)^2}{2N}(\tau_m - \tau_n)\Big)$$
$$=: \tilde{S}_2^{\rm bulk}(y_j^{(m)}, y_k^{(n)}; \tau_m - \tau_n) = -\rho\int_1^\infty e^{(\tau_n-\tau_m)(\pi\rho u)^2/2}\cos \pi u\rho(y_j^{(m)} - y_k^{(n)})\,du. \tag{11.131}$$

Recalling (11.125) we therefore have

$$\lim_{N\to\infty}\Big(\frac{\pi\rho}{\sqrt{2N}}\Big)^{m_1+m_2}\rho_{(m_1,m_2)}\Big(\frac{\pi\rho y_1^{(1)}}{\sqrt{2N}},\dots,\frac{\pi\rho y_{m_1}^{(1)}}{\sqrt{2N}};\tau_1;\frac{\pi\rho y_1^{(2)}}{\sqrt{2N}},\dots,\frac{\pi\rho y_{m_2}^{(2)}}{\sqrt{2N}};\tau_2\Big)\Big|_{\substack{\tau_1\mapsto(\pi\rho)^2\tau_1/2N\\ \tau_2\mapsto(\pi\rho)^2\tau_2/2N}}$$
$$=:\rho^{\rm bulk}_{(m_1,m_2)}(y_1^{(1)},\dots,y_{m_1}^{(1)};\tau_1;y_1^{(2)},\dots,y_{m_2}^{(2)};\tau_2)$$
$$=\det\begin{bmatrix}[S_2^{\rm bulk}(y_j^{(1)},y_k^{(1)};0)]_{m_1\times m_1} & [S_2^{\rm bulk}(y_j^{(1)},y_k^{(2)};\tau_1-\tau_2)]_{m_1\times m_2}\\ [\tilde S_2^{\rm bulk}(y_j^{(2)},y_k^{(1)};\tau_2-\tau_1)]_{m_2\times m_1} & [S_2^{\rm bulk}(y_j^{(2)},y_k^{(2)};0)]_{m_2\times m_2}\end{bmatrix}. \tag{11.132}$$

11.7.2 Soft edge

At the soft edge the appropriate scaling is

$$y\mapsto\sqrt{2N}+\frac{y}{2^{1/2}N^{1/6}},\qquad \tau\mapsto\frac{\tau}{N^{1/3}},$$

with the scaling of y thus being the same as in the static theory (recall (7.11)). Using (7.9) one finds, for $\tau_n\ge\tau_m$

$$\lim_{N\to\infty}\frac{e^{-N^{2/3}(\tau_m-\tau_n)}}{2^{1/2}N^{1/6}}\sigma\Big(\sqrt{2N}+\frac{y_j^{(m)}}{2^{1/2}N^{1/6}},\sqrt{2N}+\frac{y_k^{(n)}}{2^{1/2}N^{1/6}};\frac{1}{N^{1/3}}(\tau_m-\tau_n)\Big)$$
$$=:S_2^{\rm soft}(y_j^{(m)},y_k^{(n)};\tau_m-\tau_n)=\int_0^\infty {\rm Ai}\,(y_j^{(m)}+v){\rm Ai}(y_k^{(n)}+v)e^{-v(\tau_n-\tau_m)}\,dv, \tag{11.133}$$

while for $\tau_n<\tau_m$

$$\lim_{N\to\infty}\frac{e^{-N^{2/3}(\tau_m-\tau_n)}}{2^{1/2}N^{1/6}}\tilde\sigma\Big(\sqrt{2N}+\frac{y_j^{(m)}}{2^{1/2}N^{1/6}},\sqrt{2N}+\frac{y_k^{(n)}}{2^{1/2}N^{1/6}};\frac{1}{N^{1/3}}(\tau_m-\tau_n)\Big)$$
$$=:\tilde S_2^{\rm soft}(y_j^{(m)},y_k^{(n)};\tau_m-\tau_n)=-\int_{-\infty}^0 {\rm Ai}\,(y_j^{(m)}+v){\rm Ai}(y_k^{(n)}+v)e^{-v(\tau_n-\tau_m)}\,dv. \tag{11.134}$$

Note that in the case $\tau_n=\tau_m$, (11.133) must reduce to the soft edge scaled form of the Gaussian kernel (7.12), a fact which is seen from the first formula of Exercises 7.1 q.1(ii).

Setting

$$\rho^{\rm soft}_{(m_1,m_2)}(y_1^{(1)},\dots,y_{m_1}^{(1)};\tau_1;y_1^{(2)},\dots,y_{m_2}^{(2)};\tau_2)$$
$$:=\lim_{N\to\infty}\Big(\frac{1}{2^{1/2}N^{1/6}}\Big)^{m_1+m_2}\rho_{(m_1,m_2)}\Big(\sqrt{2N}+\frac{y_1^{(1)}}{2^{1/2}N^{1/6}},\dots,\sqrt{2N}+\frac{y_{m_1}^{(1)}}{2^{1/2}N^{1/6}};\frac{\tau_1}{N^{1/3}};$$
$$\sqrt{2N}+\frac{y_1^{(2)}}{2^{1/2}N^{1/6}},\dots,\sqrt{2N}+\frac{y_{m_2}^{(2)}}{2^{1/2}N^{1/6}};\frac{\tau_2}{N^{1/3}}\Big) \tag{11.135}$$

and recalling (11.125) it follows that $\rho^{\rm soft}_{(m_1,m_2)}$ is given by the determinant in (11.132), but with $S_2^{\rm bulk}$ and $\tilde S_2^{\rm bulk}$ replaced by $S_2^{\rm soft}$ and $\tilde S_2^{\rm soft}$ respectively.

11.7.3 Hard edge

To study the hard edge we use the Laguerre weight $w_2(x)=x^ae^{-x}$, $x>0$, and the corresponding monic orthogonal polynomials from (5.46). The appropriate scaling is

$$y\mapsto\frac{y}{4N},\qquad \tau\mapsto\frac{\tau}{2N},$$

where again the scaling of y is chosen to coincide with the static theory (recall (7.30)).

Use of (7.29) gives

$$\lim_{N\to\infty}\frac{1}{4N}\sigma\Big(\frac{y_j^{(m)}}{4N},\frac{y_k^{(n)}}{4N};\frac{\tau_m-\tau_n}{2N}\Big)$$
$$:= S_2^{\rm hard}(y_j^{(m)},y_k^{(n)};\tau_m-\tau_n) = \frac{1}{4}\int_0^1 J_a\Big(\sqrt{uy_j^{(m)}}\Big)J_a\Big(\sqrt{uy_k^{(n)}}\Big)e^{-u(\tau_m-\tau_n)}\,du \qquad (11.136)$$

and, for $\tau_n<\tau_m$,

$$\lim_{N\to\infty}\frac{1}{4N}\tilde\sigma\Big(\frac{y_j^{(m)}}{4N},\frac{y_k^{(n)}}{4N};\frac{\tau_m-\tau_n}{2N}\Big)$$
$$:= \tilde S_2^{\rm hard}(y_j^{(m)},y_k^{(n)};\tau_m-\tau_n) = -\frac{1}{4}\int_1^\infty J_a\Big(\sqrt{uy_j^{(m)}}\Big)J_a\Big(\sqrt{uy_k^{(n)}}\Big)e^{-u(\tau_m-\tau_n)}\,du. \qquad (11.137)$$

Setting

$$\rho^{\rm hard}_{(m_1,m_2)}(y_1^{(1)},\dots,y_{m_1}^{(1)};\tau_1;y_1^{(2)},\dots,y_{m_2}^{(2)};\tau_2)$$
$$=\lim_{N\to\infty}\Big(\frac{1}{4N}\Big)^{m_1+m_2}\rho_{(m_1,m_2)}\Big(\frac{y_1^{(1)}}{4N},\dots,\frac{y_{m_1}^{(1)}}{4N};\frac{\tau_1}{2N}\,\frac{y_1^{(2)}}{4N},\dots\frac{y_{m_2}^{(2)}}{4N};\frac{\tau_2}{2N}\Big) \qquad (11.138)$$

and recalling (11.125) we thus have that $\rho^{\rm hard}_{(m_1,m_2)}$ is given by the determinant in (11.132) with $S_2^{\rm bulk}$ and $\tilde S_2^{\rm bulk}$ replaced by $S_2^{\rm hard}$ and $\tilde S_2^{\rm hard}$, respectively.

EXERCISES 11.7 1. (i) With

$$f_{jk}^{mn} := \sigma_{jk}^{mn}\ \ ((m,n)\ne(2,1)),\qquad f_{jk}^{21}=\tilde\sigma_{jk}^{21},$$

where σ_{jk}^{mn} and $\tilde\sigma_{jk}^{21}$ are specified by (11.126), use (11.122) to check that

$$\int_I f_{jl}^{m1}f_{lk}^{1p}\,dy_l = f_{jk}^{mp}\xi^1_{mp},\qquad \int_I f_{jl}^{m2}f_{lk}^{2p}\,dy_l = f_{jk}^{mp}\xi^2_{mp}, \qquad (11.139)$$

where $m,p=1$ or 2 and

$$\xi^1=\begin{bmatrix}1&1\\0&0\end{bmatrix},\qquad \xi^2=\begin{bmatrix}0&1\\0&1\end{bmatrix}.$$

(ii) Using the orthogonality relations (11.139), adopt the method of integration of Proposition 5.1.2 to show that

$$\int_I\det\begin{bmatrix}[\sigma^{11}_{jk}]_{j,k=1,\dots,\nu} & [\sigma^{12}_{jk}]_{\substack{j=1,\dots,\nu\\k=1,\dots,\mu}}\\ [\tilde\sigma^{21}_{jk}]_{\substack{j=1,\dots,\mu\\k=1,\dots,\nu}} & [\sigma^{22}_{jk}]_{j,k=1,\dots,\mu}\end{bmatrix}dy_\mu^{(2)}$$
$$=(N-\mu+1)\det\begin{bmatrix}[\sigma^{11}_{jk}]_{j,k=1,\dots,\nu} & [\sigma^{12}_{jk}]_{\substack{j=1,\dots,\nu\\k=1,\dots,\mu-1}}\\ [\tilde\sigma^{21}_{jk}]_{\substack{j=1,\dots,\mu-1\\k=1,\dots,\nu}} & [\sigma^{22}_{jk}]_{j,k=1,\dots,\mu-1}\end{bmatrix}$$

and

$$\int_I\det\begin{bmatrix}[\sigma^{11}_{jk}]_{j,k=1,\dots,\nu} & [\sigma^{12}_{jk}]_{\substack{j=1,\dots,\nu\\k=1,\dots,\mu}}\\ [\tilde\sigma^{21}_{jk}]_{\substack{j=1,\dots,\mu\\k=1,\dots,\nu}} & [\sigma^{22}_{jk}]_{j,k=1,\dots,\mu}\end{bmatrix}dy_\nu^{(1)}$$
$$=(N-\nu+1)\det\begin{bmatrix}[\sigma^{11}_{jk}]_{j,k=1,\dots,\nu-1} & [\sigma^{12}_{jk}]_{\substack{j=1,\dots,\nu-1\\k=1,\dots,\mu}}\\ [\tilde\sigma^{21}_{jk}]_{\substack{j=1,\dots,\mu\\k=1,\dots,\nu-1}} & [\sigma^{22}_{jk}]_{j,k=1,\dots,\mu}\end{bmatrix}.$$

Explain how these results are consistent with (11.125).

Chapter Twelve

Jack polynomials

We have seen that the calculation of dynamical correlation functions for the Brownian evolution of the log-gas requires knowledge of the Green function solution of the Fokker-Planck equation. It has also been noted that the Green function can be expressed in terms of the eigenvalues and eigenfunctions of the corresponding Schrödinger operator. In the cases of interest, these eigenfunctions factorize into a product of the ground state wave function times a multivariable polynomial. The most fundamental case is the Schrödinger operator $H^{(\mathrm{C},\mathrm{Ex})}$. The polynomial part of the eigenfunctions are then termed the nonsymmetric Jack polynomials, and they form the natural starting point from which to investigate the properties of their symmetric counterparts, which are the polynomial part of the eigenfunctions of $H^{(\mathrm{C})}$. Here a self-contained development of the theory of Jack polynomials is undertaken. We discuss orthogonality properties, with respect to both a pairing inner product and an integral inner product, and compute the associated normalization constants. The pairing inner product relates to the Cauchy product expansion, and this combined with the integral inner product allows a generalization of the Selberg integral, in which the Jack polynomial appears as a factor in the integrand, to be evaluated. Interpolation Jack polynomials, which can be defined by their vanishing properties, are introduced for their use in the derivation of the branching coefficients in the expansion of the product of a symmetric Jack polynomial and an elementary symmetric function (Pieri formulas), and their relation to generalized binomial coefficients.

12.1 NONSYMMETRIC JACK POLYNOMIALS

12.1.1 Construction as eigenfunctions

Consider the transformed operator $\tilde{H}^{(\mathrm{C},\mathrm{Ex})}$ (11.63). According to Proposition 11.4.5 this operator has the decomposition

$$\tilde{H}^{(\mathrm{C},\mathrm{Ex})} = \frac{1}{\alpha^2}\sum_{j=1}^{N}\Big(\xi_j + \frac{N-1}{2}\Big)^2 - \Big(\frac{L}{2\pi}\Big)^2 E_0^{(\mathrm{C})}.$$

The fact that the ξ_j all commute suggests that we seek simultaneous eigenfunctions of the ξ_j. Such eigenfunctions (if they exist) will then be eigenfunctions of $\tilde{H}^{(\mathrm{C},\mathrm{Ex})}$. We will show not only that such eigenfunctions exist, but that they form a complete set, being in one-to-one correspondence with the monomials $z_1^{\eta_1} z_2^{\eta_2}\cdots z_N^{\eta_N} =: z^\eta$ ($\eta_j \in \mathbb{Z}_{\ge 0}$; η is said to form a *composition* of non-negative integers, and the η_j form the *parts*). The mechanism behind this result is that the ξ_j have a triangular action on the monomial z^η. To be more precise, we must first introduce a partial ordering for compositions.

DEFINITION 12.1.1 *Let η^+ denote the particular reordering of the parts of the composition*

$$(\eta_1, \eta_2, \dots, \eta_N) \mapsto (\eta_{P(1)}, \eta_{P(2)}, \dots, \eta_{P(N)})$$

such that $\eta_{P(1)} \ge \eta_{P(2)} \ge \cdots \ge \eta_{P(N)}$ (note that the associated permutation P need not be unique). We know from Definition 10.1.3 that in general an N-tuple of non-negative integers so ordered is referred to as *a partition. The modulus of a composition η, or its associated partition η^+, is defined as*

$$|\eta| := \sum_{j=1}^{N} \eta_j.$$

Suppose $|\eta| = |\nu|$ for compositions $\eta \neq \nu$. In such cases the partial ordering $<$, known as the dominance ordering, is defined by the statement that $\nu < \eta$ if

$$\sum_{j=1}^{p} \nu_j \leq \sum_{j=1}^{p} \eta_j \tag{12.1}$$

for each $p = 1, \dots, N$. A further partial ordering $\prec$, referred to as the Bruhat ordering, is defined on compositions by the statement that $\nu \prec \eta$ if $\nu^+ < \eta^+$, or in the case $\nu^+ = \eta^+$, if $\nu = \prod_{l=1}^{r} s_{i_l j_l}\eta$, where $\eta_{i_l} > \eta_{j_l}$, $i_l < j_l$.

We emphasize that both $<$ and $\prec$ are partial orderings only. For example, the partitions $(4,1,1)$ and $(3,3,0)$ are incomparable according to the requirement (12.1). We will see that $\xi_j z^\eta$ gives monomials for which the corresponding exponents are all comparable with η in the partial ordering $\prec$, and are in fact smaller than η in this ordering.

PROPOSITION 12.1.2 *Let*

$$\eta(j,k,r) = \begin{cases} (\eta_1, \dots, \eta_j + r, \dots, \eta_k - r, \dots, \eta_N), & 1 \leq r \leq \eta_k - \eta_j - 1, \quad \eta_k > \eta_j, \\ (\eta_1, \dots, \eta_j - r, \dots, \eta_k + r, \dots, \eta_N), & 1 \leq r \leq \eta_j - \eta_k, \quad \eta_j > \eta_k, \end{cases}$$

and

$$\bar{\eta}_j = \alpha\eta_j - \#\{k < j | \eta_k \geq \eta_j\} - \#\{k > j | \eta_k > \eta_j\}. \tag{12.2}$$

For each $i = 1, \dots, N$ we have

$$\xi_i z^\eta = \bar{\eta}_i z^\eta + \sum_{p<i} \operatorname{sgn}(\eta_i - \eta_p) \sum_r z^{\eta(p,i,r)} + \sum_{p>i} \operatorname{sgn}(\eta_i - \eta_p) \sum_r z^{\eta(i,p,r)}, \tag{12.3}$$

where the first sum over r is from $1 \leq r \leq \eta_i - \eta_p - 1$ ($\eta_i > \eta_p$) and $1 \leq r \leq \eta_p - \eta_i$ ($\eta_p > \eta_i$), while the second sum over r has the same constraints with η_i and η_p interchanged. The exponents $\eta(j,k,r)$ occurring on the r.h.s. of (12.3) are smaller in the partial ordering $\prec$ than η.

Proof. The identity

$$\frac{a^n b^m - b^n a^m}{a - b} = a^{n-1}b^m + a^{n-2}b^{m+1} + \cdots + a^m b^{n-1}, \quad n > m, \tag{12.4}$$

shows that

$$\frac{z_i}{z_i - z_p}(1 - s_{ip}) z_i^{\eta_i} z_p^{\eta_p} = \begin{cases} (z_i^{\eta_i} z_p^{\eta_p} + z_i^{\eta_i - 1} z_p^{\eta_p + 1} + \cdots + z_p^{\eta_p+1} z_i^{\eta_i - 1}), & \eta_i > \eta_p, \\ -(z_p^{\eta_p - 1} z_i^{\eta_i + 1} + z_p^{\eta_p - 2} z_i^{\eta_i + 2} + \cdots + z_p^{\eta_i} z_i^{\eta_p}), & \eta_p > \eta_i, \end{cases}$$

$$\frac{z_p}{z_i - z_p}(1 - s_{ip}) z_i^{\eta_i} z_p^{\eta_p} = \begin{cases} -(z_p^{\eta_p} z_i^{\eta_i} + z_p^{\eta_p - 1} z_i^{\eta_i + 1} + \cdots + z_p^{\eta_i + 1} z_i^{\eta_p - 1}), & \eta_p > \eta_i, \\ (z_i^{\eta_i - 1} z_p^{\eta_p + 1} + z_i^{\eta_i - 2} z_p^{\eta_p + 2} + \cdots + z_i^{\eta_p} z_p^{\eta_i}), & \eta_i > \eta_p. \end{cases}$$

According to Definition 11.4.3 of the ξ_i, these identities suffice to give the action of ξ_i on z^η, and they indeed imply the stated formulas (to obtain $\bar{\eta}_i$ in the form stated requires using the identity $\#\{k < j | \eta_j > \eta_k\} = j - 1 - \#\{k < j | \eta_k \geq \eta_j\}$).

Finally, note that in the definition of $\eta(j,k,r)$ for $r \neq \eta_j - \eta_k$ we have $\eta(j,k,r) < \eta$, while for $r = \eta_j - \eta_k$, $\eta(j,k,r) \prec \eta$ (the latter follows because then $\eta(j,k,r) = s_{jk}\eta$ with $\eta_j > \eta_k$, $j < k$). □

In general a triangular action on an ordered set of basis vectors $\phi_0 < \phi_1 < \phi_2 < \cdots$, $A\phi_j = \alpha_j \phi_j +$

$\sum_{l=0}^{j-1} \tilde{c}_{jl}\phi_l$, allows for a systematic construction of eigenfunctions of A,

$$\psi_j = \phi_j + \sum_{l=0}^{j-1} c_{jl}\phi_l$$

with eigenvalue α_j, provided $\{\alpha_j\}$ are distinct. For each individual ξ_i, although there is a triangular action, the eigenvalues are not distinct. Thus instead of considering a particular ξ_i, we consider their generating function as specified by the operator

$$X^{(\mathrm{C})}(\{u_l\}) := \prod_{l=1}^{N}(1 + u_l\xi_l) \tag{12.5}$$

which generalizes (11.62). Proposition 12.1.2 gives that this operator also has a triangular structure when acting on z^η, showing

$$\prod_{l=1}^{N}(1 + u_l\bar{\eta}_l)z^\eta + \text{lower order terms}, \tag{12.6}$$

where the lower order terms are with respect to the partial ordering $\prec$. Thus $X^{(\mathrm{C})}(\{u_l\})$ has eigenfunctions of the form

$$E_\eta(z) := E_\eta(z;\alpha) := E_\eta(z_1,\dots,z_N;\alpha) = z^\eta + \sum_{\nu\prec\eta} a_{\eta\nu}z^\nu, \tag{12.7}$$

with corresponding eigenvalues $\prod_{j=1}^{N}(1 + u_j\bar{\eta}_j)$, which according to (12.2) are distinct for distinct compositions. Due to this latter fact the construction of the eigenfunctions from their triangular structure is now unique. However, it is not immediately clear that (12.7) is independent of $\{u_l\}$. To see that indeed this must be the case, we can use the fact that the ξ_j commute to act on the eigenvalue equation

$$X^{(\mathrm{C})}(\{u_l\})E_\eta(z) = \prod_{l=1}^{N}(1 + u_l\bar{\eta}_l)E_\eta(z) \tag{12.8}$$

with the operator $X^{(\mathrm{C})}(\{v_l\})$, $\{v_l\} \neq \{u_l\}$. This leads to the conclusion that $X^{(\mathrm{C})}(\{v_l\})E_\eta(z)$ is an eigenfunction of $X^{(\mathrm{C})}(\{u_l\})$ with eigenvalue $\prod_{l=1}^{N}(1 + u_l\bar{\eta}_l)$. But the unique (up to normalization) polynomial with this property is $E_\eta(z)$, so $X^{(\mathrm{C})}(\{v_l\})E_\eta(z)$ must be proportional to $E_\eta(z)$. Of course eigenfunctions of $X^{(\mathrm{C})}(\{v_l\})$ cannot depend on $\{u_l\}$ so $E_\eta(z)$ must in fact be independent of $\{u_l\}$.

Equating coefficients of u_j on both sides of the eigenvalue equation (12.8), we see that

$$\xi_j E_\eta(z) = \bar{\eta}_j E_\eta(z) \tag{12.9}$$

for $j = 1,\dots,N$ so that the E_η are simultaneous eigenfunctions of each ξ_j. The fact that $X^{(\mathrm{C})}(\{u_l\})$ has a unique eigenfunction of the form (12.7) implies that E_η is the unique polynomial of the form (12.7) which is a simultaneous eigenfunction of $\{\xi_j\}$. The E_η are referred to as the *nonsymmetric Jack polynomials* [436], [58].

In the limit $\alpha \to \infty$, we see by including only the terms in ξ_j and $\bar{\eta}_j$ proportional to α that the eigenvalue equation (12.9) reduces to

$$z_j\frac{\partial}{\partial z_j}E_\eta(z) = \eta_j E_\eta(z).$$

Thus we see that

$$E_\eta(z;\infty) = z^\eta. \tag{12.10}$$

12.1.2 Orthogonality and a multidimensional integral

The Schrödinger operator $H^{(\mathrm{C,Ex})}$ as specified by (11.52) is self-adjoint with respect to the inner product

$$\langle f|g\rangle := \int_{-L/2}^{L/2} dx_1 \cdots \int_{-L/2}^{L/2} dx_N \, \overline{f(z_1,\dots,z_N)} g(z_1,\dots,z_N), \tag{12.11}$$

where $z_j := e^{2\pi i x_j/L}$, so that

$$\langle f|H^{(\mathrm{C,Ex})}g\rangle = \langle H^{(\mathrm{C,Ex})}f|g\rangle.$$

From the relationship (11.63) between $H^{(\mathrm{C,Ex})}$ and $\tilde{H}^{(\mathrm{C,Ex})}$, it follows that $\tilde{H}^{(\mathrm{C,Ex})}$ is self-adjoint with respect to the inner product

$$\langle f|g\rangle^{(\mathrm{C})} := \int_{-L/2}^{L/2} dx_1 \cdots \int_{-L/2}^{L/2} dx_N \prod_{1\le j<k\le N} |z_k - z_j|^{2/\alpha} \overline{f(z_1,\dots,z_N)} g(z_1,\dots,z_N) \tag{12.12}$$

(recall that $2/\alpha := \beta$). In light of the decomposition of Proposition 11.4.5 we might also suspect that the Cherednik operator ξ_i is self-adjoint with respect to (12.12). In fact this is indeed the case.

PROPOSITION 12.1.3 *For each $i = 1,\dots,N$ we have*

$$\langle f|\xi_i g\rangle^{(\mathrm{C})} = \langle \xi_i f|g\rangle^{(\mathrm{C})}.$$

Proof. By writing $F = e^{-\beta W^{(\mathrm{C})}/2} f$ and $G = e^{-\beta W^{(\mathrm{C})}/2} g$ we see that the stated formula is equivalent to the statement that

$$\langle F|(e^{-\beta W^{(\mathrm{C})}/2}\xi_i e^{\beta W^{(\mathrm{C})}/2})G\rangle = \langle (e^{-\beta W^{(\mathrm{C})}/2}\xi_i e^{\beta W^{(\mathrm{C})}/2})F|G\rangle,$$

where $\langle\cdot|\cdot\rangle$ refers to the inner product (12.11). The validity of this latter formula is seen from the explicit expression (11.61) for $e^{-\beta W^{(\mathrm{C})}/2}\xi_i e^{\beta W^{(\mathrm{C})}/2}$. □

An immediate consequence of Proposition 12.1.3 is that the operator $X^{(\mathrm{C})}(\{u_l\})$ is self-adjoint with respect to (12.12). Now we know that the nonsymmetric Jack polynomials E_η are the polynomial eigenfunctions of $X^{(\mathrm{C})}(\{u_l\})$, and that the corresponding eigenvalue is distinct. Thus we can apply the elementary theorem, stating that eigenfunctions of a Hermitian (self-adjoint) operator with distinct eigenvalues are orthogonal, to conclude the following result.

PROPOSITION 12.1.4 *For compositions $\eta \ne \nu$,*

$$\langle E_\eta|E_\nu\rangle^{(\mathrm{C})} = 0.$$

We note that Proposition 12.1.4 together with the structure (12.7) implies that it is possible to construct $\{E_\eta\}$ via the Gram-Schmidt orthogonalization procedure from $\{z^\eta\}$ with the partial order $\prec$. Because $\{z^\eta\}$ is a basis for analytic functions in N variables, it follows that so is $\{E_\eta\}$.

12.1.3 Orthogonality and a product expansion

The pairing inner product

$$\langle E_\eta, E_\nu\rangle_q := u_\eta \delta_{\eta,\nu}, \tag{12.13}$$

for suitable normalization u_η, is of importance in the subsequent development of the general theory. Motivating the definition of (12.13) and in particular the choice of u_η is the expansion of the function

$$\Omega(x,y) := \prod_{i=1}^N \frac{1}{(1-x_i y_i)} \prod_{j,k=1}^N \frac{1}{(1-x_j y_k)^{1/\alpha}} \tag{12.14}$$

in a basis of $\{E_\eta(x)\}$ [476].

PROPOSITION 12.1.5 *For some coefficients $u_\eta \neq 0$ we have*

$$\Omega(x,y) = \sum_\eta \frac{1}{u_\eta} E_\eta(x) E_\eta(y). \tag{12.15}$$

Proof. Since $\{E_\eta\}$ is a basis for analytic functions of N variables and $\Omega(\alpha x, y/\alpha) = \Omega(x,y)$ we must have

$$\Omega = \sum_\eta E_\eta(x) U_\eta(y) \tag{12.16}$$

for some homogeneous polynomials $U_\eta(y)$ of degree $|\eta|$. With $\xi_i^{(x)}$ ($\xi_i^{(y)}$) denoting that ξ_i is written in terms of $x_1,\dots,x_N$ ($y_1,\dots,y_N$), we will show

$$\xi_i^{(x)}\Omega = \xi_i^{(y)}\Omega \qquad (i=1,\dots,N). \tag{12.17}$$

Now, using (12.17) in (12.16) gives $\xi_i^{(y)} U_\eta(y) = \bar{\eta}_i U_\eta(y)$ for each $i = 1,\dots,N$, so we conclude that $U_\eta(y)$ is a multiple of $E_\eta(y)$, and thus (12.15) holds.

It remains to derive (12.17). Since Ω is symmetric under interchange of x_i and y_i it suffices to show that $\frac{1}{\Omega}\xi_i^{(x)}\Omega$ is symmetric under interchange of x_i and y_i. Now, from the definitions

$$\frac{1}{\Omega}\alpha x_i \frac{\partial}{\partial x_i}\Omega = \frac{\alpha x_i y_i}{1 - x_i y_i} - N + \sum_{j=1}^N \frac{1}{1 - x_i y_j}, \qquad \frac{1}{\Omega}\frac{1 - s_{ij}}{x_i - x_j}\Omega = \frac{y_i - y_j}{(1 - x_i y_j)(1 - x_j y_i)},$$

which allows the explicit form of $\frac{1}{\Omega}\xi_i^{(x)}\Omega$ to be written down. Straightforward manipulation then exhibits the sought symmetry property. □

With the u_η occurring in (12.15) being chosen as the u_η in (12.13), another basis fundamental to the nonsymmetric Jack polynomials can be identified [139].

DEFINITION 12.1.6 *With $\Omega(x,y)$ as specified in (12.14), for each composition η define a polynomial q_η homogeneous of degree $|\eta|$ by*

$$\Omega(x,y) = \sum_\eta q_\eta(x) y^\eta. \tag{12.18}$$

The monomial y^η by definition has the factorization $y_1^{\eta_1}\cdots y_N^{\eta_N}$. Similarly, defining $q_k(x_i;x)$ by the expansion

$$\frac{1}{(1 - x_i y_i)} \prod_{j=1}^N \frac{1}{(1 - x_j y_i)^{1/\alpha}} = \sum_{k=0}^\infty q_k(x_i;x) y_i^k, \tag{12.19}$$

we see that

$$q_\eta(x) = q_{\eta_1}(x_1;x) q_{\eta_2}(x_2;x) \cdots q_{\eta_N}(x_N;x). \tag{12.20}$$

Note from (12.19) that the simplest nontrivial example of the elementary polynomials $q_k(x_i;x)$ is

$$q_1(x_i;x) = x_i + \frac{1}{\alpha}\sum_{j=1}^N x_j. \tag{12.21}$$

We saw in Proposition 12.1.2 that $\xi_j^{(x)} x^\eta$ gives a series in $\{x^\nu\}$ with $\nu \preceq \eta$. It turns out that $\xi_j^{(x)} q_\eta(x)$ gives a series in $\{q_\nu(x)\}$ with $\eta \preceq \nu$, implying an upper triangular structure, whereas the former structure is lower triangular. To demonstrate this, we note from the definition (12.18) that

$$\xi_i^{(x)}\Omega(x,y) = \sum_\eta (\xi_i^{(x)} q_\eta(x)) y^\eta. \tag{12.22}$$

But making use of (12.17) in (12.18) we also have

$$\xi_i^{(x)}\Omega(x,y) = \sum_\eta q_\eta(x)\xi_i^{(y)} y^\eta.$$

Substituting (12.3) in this and rearranging the sum over η to be of the form in (12.22) we deduce the following analogue of Proposition 12.1.2, which exhibits the claimed triangular structure.

PROPOSITION 12.1.7 *We have*

$$\xi_i^{(x)} q_\eta(x) = \bar{\eta}_i q_\eta(x) + \sum_{p<i} \mathrm{sgn}(\eta_i - \eta_p) \sum_r q_{\eta^*(p,i,r)} + \sum_{p>i} \mathrm{sgn}(\eta_i - \eta_p) \sum_r q_{\eta^*(i,p,r)}, \tag{12.23}$$

where

$$\eta^*(j,k,r) = \begin{cases} (\eta_1,\dots,\eta_j - r,\dots,\eta_k + r,\dots,\eta_N), & 1 \le r \le \eta_k - \eta_j - 1, \quad \eta_k > \eta_j, \\ (\eta_1,\dots,\eta_j + r,\dots,\eta_k - r,\dots,\eta_N), & 1 \le r \le \eta_j - \eta_k, \quad \eta_j > \eta_k, \end{cases}$$

and $q_{\eta^*(j,k,r)} = 0$ *if* η^* *contains a negative part.*

The largest composition with respect to the ordering $\prec$ and such that $|\eta| = p$ is $\eta^\# := (p,0,\dots,0)$. For this composition (12.23) gives $\xi_i^{(x)} q_{\eta^\#}(x) = \bar{\eta}_i q_{\eta^\#}(x)$ for each $i = 1,\dots,N$ so $q_{\eta^\#}(x) \propto E_{\eta^\#}(x)$. For general η we can make use of the upper triangular structure and the operator (12.5) to systematically construct simultaneous eigenfunctions of the $\xi_i^{(x)}$, which must have the same upper triangular structure. On the other hand we know these eigenfunctions to be the Jack polynomials. Thus we obtain the analogue of (12.7) expressing the Jack polynomials as a series in $\{q_\nu(x)\}$,

$$E_\eta(x) = c_\eta q_\eta(x) + \sum_{\eta \prec \nu} \tilde{a}_{\eta\nu} q_\nu(x), \tag{12.24}$$

for some c_η. We will show subsequently that c_η is precisely the normalization u_η in (12.13).

Let us now proceed to relate $\{q_\eta\}$ to the pairing inner product (12.13).

PROPOSITION 12.1.8 *Let* $\{v_\eta\}$ *and* $\{w_\eta\}$ *be sets of homogeneous multivariable polynomials such that*

$$\Omega(x,y) = \sum_\eta v_\eta(x) w_\eta(y) \tag{12.25}$$

with $\Omega(x,y)$ *specified by (12.14) and suppose that* $\{w_\eta(y)\}$ *is a basis. Then* $\{v_\eta(x)\}$ *is a basis and furthermore*

$$\langle v_\eta, w_\nu \rangle_q = \delta_{\eta,\nu}. \tag{12.26}$$

Proof. Since $\{E_\gamma\}$ is a basis we can write

$$v_\eta = \sum_\gamma a_{\eta\gamma} E_\gamma / u_\gamma \quad \text{and} \quad w_\nu = \sum_\gamma b_{\nu\gamma} E_\gamma.$$

Then from the definition (12.13) we have

$$\langle v_\eta, w_\nu \rangle_q = \sum_\gamma a_{\eta\gamma} b_{\nu\gamma}. \tag{12.27}$$

On the other hand, equating (12.15) and (12.25) with the above expansions substituted in the latter implies

$$\sum_\gamma a_{\gamma\eta} b_{\gamma\nu} = \delta_{\eta,\nu}. \tag{12.28}$$

Consider compositions of modulus p, and with a total order on such compositions define $\mathbf{A} = [a_{\gamma\eta}]$, $\mathbf{B} = [b_{\gamma\nu}]$. Then (12.28) is equivalent to the matrix equation $\mathbf{A}^T\mathbf{B} = \mathbf{1}$. Since $\{v_\eta\}$ is a basis the matrix $\mathbf{A}$ is invertible so it follows from this that

$$\mathbf{B} = (\mathbf{A}^T)^{-1}; \tag{12.29}$$

this in turn tells us that $\mathbf{B}$ is invertible and thus $\{w_\nu\}$ is a basis. Also, in this notation the r.h.s. of (12.27) reads $[\mathbf{A}\mathbf{B}^T]_{\eta\nu}$. According to (12.29) this is equal to $[\mathbf{1}]_{\eta\nu}$ and (12.26) follows. □

With $v_\eta(x) = q_\eta(x)$ and $w_\eta(y) = y^\eta$ all the assumptions of Proposition 12.1.8 are met, so we conclude

$$\langle q_\eta, x^\nu \rangle_q = \delta_{\eta,\nu}. \tag{12.30}$$

This relation may well be taken as the definition of $\langle \cdot, \cdot \rangle_q$, with (12.13) then a corollary, although then one has to work harder to first establish that $\{q_\eta\}$ form a basis [139].

As a consequence of (12.30) the quantity c_η in (12.24) can be determined. Thus substituting (12.24) for E_η and (12.7) for E_ν in (12.13) it follows from (12.30) that $c_\eta = u_\eta$. Hence (12.24) can be written

$$E_\eta(x) = u_\eta q_\eta(x) + \sum_{\eta \prec \nu} \tilde{a}_{\eta\nu} q_\nu(x). \tag{12.31}$$

We would like to express the quantity u_η occurring in (12.13), (12.15) and now (12.31) as an explicit function of the parts of η. However, this is not possible until some further theory, introduced in the next section, is developed.

EXERCISES 12.1

1. Show that $z^p E_\eta(z) = E_{\eta+p^N}(z)$, where $z^p := z_1^p \cdots z_N^p$ and

$$\eta + p^N := (\eta_1 + p, \eta_2 + p, \dots, \eta_N + p).$$

(It is assumed that $p \geq -\min_j \eta_j$; however, more generally this equation can be used to give meaning to E_η in the case that η has negative parts.)

2. Show that

$$E_\eta\Big(\frac{1}{z}\Big) = E_{-\eta^R}(z^R),$$

where

$$z^R := (z_N, z_{N-1}, \dots, z_1), \qquad \eta^R := (\eta_N, \eta_{N-1}, \dots, \eta_1) \tag{12.32}$$

and $E_{-\eta^R}$ is interpreted as remarked in q.1. (Hint: Make the change of variables $z_i \mapsto 1/z_i$ in the definition of ξ_i, and relate the resulting operator to $\xi\,|_{z\mapsto z^R}$.)

3. Show from Definition 11.4.3 that $\xi_i^{(N)}|_{z_N=0} = \xi_i^{(N-1)}$, where $\xi_i = \xi_i^{(n)}$ indicates that the variables in the definition of ξ_i are $z_1, \dots, z_n$. Use this to deduce that for $\eta_N = 0$,

$$E_\eta(z_1, \dots, z_{N-1}, 0) = E_\eta(z_1, \dots, z_{N-1}).$$

4. Let κ denote a partition, and let κ' denote its conjugate (recall Definition 10.1.3)κ' Show that

$$\sum_{i=1}^N (i-1)\kappa_i = \sum_{i=1}^N \kappa_i'(\kappa_i' - 1)/2. \tag{12.33}$$

5. (i) Verify that

$$(\overline{s_i\eta})_i = \begin{cases} \bar{\eta}_{i+1}, & \eta_i \neq \eta_{i+1}, \\ \bar{\eta}_i, & \eta_i = \eta_{i+1}, \end{cases} \qquad (\overline{s_i\eta})_{i+1} = \begin{cases} \bar{\eta}_i, & \eta_i \neq \eta_{i+1}, \\ \bar{\eta}_{i+1}, & \eta_i = \eta_{i+1}, \end{cases} \qquad (\overline{s_i\eta})_j = \bar{\eta}_j \quad (j \neq i, i+1)$$

to deduce that $\{\bar{\eta}_j\}_{j=1,\dots,N} = \{(\overline{P\eta})_j\}_{j=1,\dots,N}$, where $P\eta$ denotes some permutation of η.

(ii) Deduce from (i) that the eigenvalue of $X^{(\mathrm{C})}(u) := X^{(\mathrm{C})}(\{u_j\})|_{u_j=u}$ corresponding to the eigenfunction

$E_\eta(z)$ is the same for all compositions η giving rise to the same partition η^+, and has the explicit form

$$\prod_{j=1}^N (1 + u(\overline{\eta^+})_j) = \prod_{j=1}^N (1 + u(\alpha\eta_j^+ - (j-1))).$$

(iii) Show that the same features hold true of the operator $\tilde{H}^{(\mathrm{C,Ex})}$, with the eigenvalue corresponding to $E_\eta(z)$ being given by

$$e(\eta^+;\alpha) := 2b((\eta^+)') + \Big(\frac{N-1}{\alpha} + 1\Big)|\eta| - \frac{2}{\alpha}b(\eta^+), \quad b(\kappa) := \sum_{j=1}^N (j-1)\kappa_j, \tag{12.34}$$

where use has been made of the formula of (12.33).

12.2 RECURRENCE RELATIONS

12.2.1 Elementary recurrences

It turns out that all the nonsymmetric Jack polynomials can, in principle, be calculated recursively by just two elementary operations. The first is the elementary transposition $s_i := s_{i\,i+1}$, which interchanges the ith and $(i+1)$th coordinates [436].

PROPOSITION 12.2.1 *Let $\bar{\delta}_{i,\eta} := \bar{\eta}_i - \bar{\eta}_{i+1}$. We have*

$$s_i E_\eta(z) = \begin{cases} \frac{1}{\bar{\delta}_{i,\eta}} E_\eta(z) + \Big(1 - \frac{1}{\bar{\delta}^2_{i,\eta}}\Big) E_{s_i\eta}(z), & \eta_i > \eta_{i+1} \\ E_\eta(z), & \eta_i = \eta_{i+1} \\ \frac{1}{\bar{\delta}_{i,\eta}} E_\eta(z) + E_{s_i\eta}(z), & \eta_i < \eta_{i+1} \end{cases}$$

where s_i acts on compositions by interchanging the ith and $(i+1)$th part.

Proof. The derivation requires the three relations (11.89). Let $E := \bar{\delta}_{i,\eta} s_i E_\eta - E_\eta$. Then making use of the first of the relations in (11.89) shows $\xi_i E = \bar{\eta}_{i+1} E$, the second relation gives $\xi_{i+1} E = \bar{\eta}_i E$ and the final relation gives $\xi_j E = \bar{\eta}_j E$ $(j \ne i, i+1)$. Hence E must be proportional to $E_{s_i\eta}$. For $\eta_i < \eta_{i+1}$ it follows from the structure (12.7) that the coefficient of $z^{s_i\eta}$ in E is $\bar{\delta}_{i,\eta}$, while for $\eta_i = \eta_{i+1}$ the coefficient is $-\bar{\delta}_{i,\eta}^{-1}$, thus implying the stated formula in these cases. The stated formula for $\eta_i > \eta_{i+1}$ follows from the formula for $\eta_i < \eta_{i+1}$ by replacing η by $s_i\eta$, acting on both sides with s_i and then once again using the formula in its original form. □

The second elementary operation is specified by a raising-type operator Φ [359].

DEFINITION 12.2.2 *When acting on functions define*

$$\Phi := z_N s_{N-1} s_{N-2} \cdots s_1$$

so that

$$\Phi f(z_1, \dots, z_N) = z_N f(z_N, z_1, \dots, z_{N-1}).$$

Also define Φ to act on compositions by

$$\Phi\eta := (\eta_2, \dots, \eta_N, \eta_1 + 1).$$

PROPOSITION 12.2.3 *The operator Φ obeys the relations*

$$\xi_i \Phi = \Phi \xi_{i+1} \quad (i = 1, \dots, N-1), \qquad \xi_N \Phi = \Phi(\xi_1 + \alpha)$$

η	$E_\eta(z)$
0^N	1
$0^{N-1}1$	z_N
$0^{N-2}10$	$z_{N-1} + \frac{1}{\alpha+N-1} z_N$
$0^{N-3}10^2$	$z_{N-2} + \frac{1}{\alpha+N-2}(z_{N-1} + z_N)$
$\vdots$	$\vdots$
10^{N-1}	$z_1 + \frac{1}{\alpha+1}(z_2 + z_3 + \cdots + z_N)$
$0^{N-2}1^2$	$z_N z_{N-1}$
$0^{N-3}101$	$z_N(z_{N-2} + \frac{1}{\alpha+N-1} z_{N-1})$
$0^{N-4}10^21$	$z_N(z_{N-3} + \frac{1}{\alpha+N-2}(z_{N-2} + z_{N-1}))$
$\vdots$	$\vdots$
$10^{N-2}1$	$z_N(z_1 + \frac{1}{\alpha+2}(z_2 + \cdots + z_{N-1}))$
$0^{N-1}2$	$z_N(z_N + \frac{1}{\alpha+1}(z_1 + \cdots + z_{N-1}))$

Table 12.1 Recursive generation of some nonsymmetric Jack polynomials

and consequently

$$\Phi E_\eta(z) = E_{\Phi\eta}(z).$$

Proof. The first relation follows from the facts that

$$\begin{aligned} z_i \frac{\partial}{\partial z_i}\Phi &= \Phi z_{i+1}\frac{\partial}{\partial z_{i+1}}, \\ \frac{z_i}{z_i - z_p}(1 - s_{ip})\Phi &= \Phi \frac{z_{i+1}}{z_{i+1} - z_{p+1}}(1 - s_{i+1\,p+1}), \\ \frac{z_p}{z_i - z_p}(1 - s_{ip})\Phi &= \Phi \frac{z_{p+1}}{z_{p+1} - z_{i+1}}(1 - s_{i+1\,p+1}), \quad p \neq N, \\ \frac{z_N}{z_i - z_N}(1 - s_{iN})\Phi &= \Phi \frac{z_1}{z_{i+1} - z_1}(1 - s_{i+1\,1}) \end{aligned}$$

and the definition of ξ_i, while the second relation is a consequence of the same equations with the first replaced by

$$z_N \frac{\partial}{\partial z_N}\Phi = \Phi z_1 \frac{\partial}{\partial z_1} + \Phi.$$

To derive the final statement, we note from the first two relations that

$$\xi_i(\Phi E_\eta) = \bar{\eta}_{i+1}(\Phi E_\eta) \ (i = 1, \dots, N-1), \qquad \xi_N(\Phi E_\eta) = (\bar{\eta}_1 + \alpha)(\Phi E_\eta),$$

which imply ΦE_η is proportional to $E_{\Phi\eta}$. That the proportionality constant is unity follows from comparison of the term proportional to $z^{\Phi\eta}$ on both sides. □

The recursive generation of some particular E_η obtained using the above formulas is given in Table 12.1. Note that only the case $\eta_i < \eta_{i+1}$ of the action of s_i is required.

Finally we will discuss a lowering-type operator $\hat{\Phi}$ which is the counterpart to the raising-type operator Φ [39].

DEFINITION 12.2.4 *Define $\hat{\Phi}$ when acting on functions of $z_1, \dots, z_N$ by*

$$\hat{\Phi} = d_1 s_1 s_2 \cdots s_{N-1} = s_1 \cdots s_{i-1} d_i s_i \cdots s_{N-1},$$

where d_i denotes the type A Dunkl operator, and when acting on compositions by

$$\hat{\Phi}\eta = (\eta_N - 1, \eta_1, \dots, \eta_{N-1}).$$

A result analogous to Proposition 12.2.3 for Φ holds for $\hat{\Phi}$.

PROPOSITION 12.2.5 *The operator $\hat{\Phi}$ obeys the relations*

$$\xi_i \hat{\Phi} = \hat{\Phi}\xi_{i-1} \ (i = 2, \dots, N) \qquad \xi_1 \hat{\Phi} = \hat{\Phi}(\xi_N - \alpha)$$

and consequently

$$\hat{\Phi} E_\eta = \frac{\bar{\eta}_N + N - 1}{\alpha} E_{\hat{\Phi}\eta} = \frac{1}{\alpha} \frac{d'_\eta}{d'_{\hat{\phi}\eta}} E_{\hat{\Phi}\eta}.$$

Proof. Using the formulas of (11.89) we see that

$$\begin{aligned} \xi_j \, s_1 s_2 \cdots s_{N-1} &= s_1 s_2 \cdots s_{j-2} \, \xi_j \, s_{j-1} s_j \cdots s_{N-1} \\ &= s_1 s_2 \cdots s_{j-2} \, (s_{j-1}\xi_{j-1} - 1) \, s_j \cdots s_{N-1} \\ &= s_1 s_2 \cdots s_{N-1} \xi_{j-1} - s_1 s_2 \cdots s_{j-1} \, s_j \cdots s_{N-1}, \end{aligned}$$

which together with Proposition 11.5.2 gives

$$\begin{aligned} \xi_j \, \hat{\Phi} &= \xi_j \, d_1 \, s_1 s_2 \cdots s_{N-1} = d_1(\xi_j + s_{1j}) s_1 s_2 \cdots s_{N-1} \\ &= d_1 \left(s_1 s_2 \cdots s_{N-1} \xi_{j-1} - s_1 s_2 \cdots s_{j-2} \, s_j \cdots s_{N-1} + s_{1j} \, s_1 s_2 \cdots s_{N-1} \right). \end{aligned}$$

Since the last two terms cancel, the first relation follows. For the second relation, use of the formulas of (11.89) shows

$$\xi_1 \, s_1 s_2 \cdots s_{N-1} = s_1 s_2 \cdots s_{N-1} \, \xi_N + \sum_{j=1}^{N-1} s_1 \cdots s_{j-1} \, s_{j+1} \cdots s_{N-1},$$

and this together with Proposition 11.5.2 gives

$$\begin{aligned} \xi_1 \, \hat{\Phi} &= d_1 \left(\xi_1 - \alpha - \sum_{p>1} s_{1p} \right) s_1 s_2 \cdots s_{N-1} \\ &= d_1 \left(s_1 s_2 \cdots s_{N-1} \, (\xi_N - \alpha) + \sum_{j=1}^{N-1} s_1 \cdots s_{j-1} \, s_{j+1} \cdots s_{N-1} - \sum_{p>1} s_{1p} \, s_1 s_2 \cdots s_{N-1} \right) \\ &= \hat{\Phi}(\xi_N - \alpha), \end{aligned}$$

where the last equality follows after noting that the last two terms in the line above cancel.

The two relations together imply that $\hat{\Phi} E_\eta$ is a constant multiple of $E_{\hat{\Phi}\eta}$. To determine this constant, note that the leading term in $E_{\hat{\Phi}\eta}$ is $z^{\hat{\Phi}\eta}$. To determine the coefficient of this term in $\hat{\Phi} E_\eta$, writing $\hat{\Phi} = s_1 \cdots s_{N-1} d_N$ shows that the coefficient is equal to that of $z_1^{\eta_1} \cdots z_{N-1}^{\eta_{N-1}} z_N^{\eta_N - 1}$ in $d_N E_\eta$. Recalling that

$$\xi_N = \alpha z_N d_N + 1 - N \tag{12.35}$$

then gives the coefficient as $(\bar{\eta}_N + N - 1)/\alpha$ as required. □

As remarked, the operators s_i and Φ generate the E_η recursively. For some scalar operators A, it is possible to compute $A(E_{s_i \eta})$ and $A(E_{\Phi\eta})$ and so then to compute $A(E_\eta)$ recursively. In the next section three examples of this procedure will be given: the first will provide the evaluation of $E_\eta((1)^N)$ (the notation $(1)^N$ means $z_1 = \cdots = z_N = 1$) [476]; in the second the normalization of the E_η with respect to the inner product (12.12) will be established [39]; in the third the normalization u_η occurring in (12.13), (12.15) and (12.31) will be evaluated [144].

12.3 APPLICATION OF THE RECURRENCES

12.3.1 The specialization formula

The evaluation of $E_\eta((1)^N)$ is in terms of the quantities

$$d_\eta := \prod_{(i,j)\in\eta} \Big(\alpha(a(i,j)+1) + l(i,j) + 1\Big), \quad e_\eta := \prod_{(i,j)\in\eta} \Big(\alpha(a'(i,j)+1) + N - l'(i,j)\Big). \tag{12.36}$$

For completeness we also list at this stage similar quantities which are required subsequently:

$$d'_\eta := \prod_{(i,j)\in\eta} \Big(\alpha(a(i,j)+1) + l(i,j)\Big), \quad e'_\eta := \prod_{(i,j)\in\eta} \Big(\alpha(a'(i,j)+1) + N - 1 - l'(i,j)\Big). \tag{12.37}$$

Here the notation $(i,j) \in \eta$ refers to the *diagram* of the composition η, in which each part η_i becomes the nodes (i,j), $1 \le j \le \eta_i$ on a square lattice labeled as is conventional for a matrix (cf. Definition 10.1.3) The quantities $a(i,j)$, $a'(i,j)$, $l(i,j)$ and $l'(i,j)$ are specified by the following definition [476].

DEFINITION 12.3.1 *Referring to the diagram of a composition η, let $a(i,j) := \eta_i - j$ denote its arm length at node (i,j), and let $a'(i,j) := j-1$ denote its coarm length. Also define the leg length $l(i,j) = ul(i,j) + ll(i,j)$ and coleg length $l'(i,j) = ul'(i,j) + ll'(i,j)$, by setting*

$$\begin{aligned} ll(i,j) &= \#\{k > i \mid j \le \eta_k \le \eta_i\}, & ul(i,j) &= \#\{k < i \mid j \le \eta_k + 1 \le \eta_i\} \\ ll'(i,j) &= \#\{k > i \mid \eta_k > \eta_i\}, & ul'(i,j) &= \#\{k < i \mid \eta_k \ge \eta_i\} \end{aligned}$$

(note that $ll'(i,j)$ and $ul'(i,j)$ are independent of j).

Use will be made of the quantities (12.36) and (12.37) with η replaced by $s_i\eta$ and $\Phi\eta$. In particular, we require the readily verified formulas [476]

$$e_{s_i\eta} = e_\eta, \quad e'_{s_i\eta} = e'_\eta, \quad \frac{d_{s_i\eta}}{d_\eta} = \begin{cases} \frac{\bar\delta_{i,\eta}+1}{\bar\delta_{i,\eta}}, & \eta_i > \eta_{i+1}, \\ \frac{\bar\delta_{i,\eta}}{\bar\delta_{i,\eta}-1}, & \eta_i < \eta_{i+1}, \end{cases} \quad \frac{d'_{s_i\eta}}{d'_\eta} = \begin{cases} \frac{\bar\delta_{i,\eta}}{\bar\delta_{i,\eta}-1}, & \eta_i > \eta_{i+1}, \\ \frac{\bar\delta_{i,\eta}+1}{\bar\delta_{i,\eta}}, & \eta_i < \eta_{i+1}, \end{cases}$$

$$\frac{d_{\Phi\eta}}{d_\eta} = \frac{e_{\Phi\eta}}{e_\eta} = \bar\eta_1 + \alpha + N, \qquad \frac{d'_{\Phi\eta}}{d'_\eta} = \frac{e'_{\Phi\eta}}{e'_\eta} = \bar\eta_1 + \alpha + N - 1,$$

$$\frac{d'_\eta}{d'_{\hat\Phi\eta}} = \bar\eta_N + N - 1, \qquad \frac{d_\eta}{d_{\hat\Phi\eta}} = \bar\eta_N + N. \tag{12.38}$$

PROPOSITION 12.3.2 *We have*

$$E_\eta((1)^N) = \frac{e_\eta}{d_\eta}. \tag{12.39}$$

Proof. Starting with $\eta = 0^N$, all η can be generated from the action of Φ and s_i on η, and furthermore the latter can be restricted to the case $\eta_i < \eta_{i+1}$. Thus it suffices to show $E_{\Phi\eta}((1)^N)$ and $E_{s_i\eta}((1)^N)$ (the latter with $\eta_i < \eta_{i+1}$) are given correctly by (12.39), which is certainly true for $\eta = 0^N$. Now according to Proposition 12.2.5, $E_{\Phi\eta}((1)^N) =$

$\Phi E_\eta((1)^N) = E_\eta((1)^N)$. On the other hand, from (12.38) we also have $e_{\Phi\eta}/d_{\Phi\eta} = e_\eta/d_\eta$, which is in agreement with (12.39). Furthermore, according to Proposition 12.2.1,

$$E_{s_i\eta}((1)^N) = \frac{\bar{\delta}_{i,\eta} - 1}{\bar{\delta}_{i,\eta}} E_\eta((1)^N), \quad \eta_i < \eta_{i+1}$$

while (12.38) gives

$$\frac{e_{s_i\eta}}{d_{s_i\eta}} = \frac{\bar{\delta}_{i,\eta} - 1}{\bar{\delta}_{i,\eta}} \frac{e_\eta}{d_\eta},$$

again in agreement with (12.39). □

12.3.2 The normalization integral

Next we will turn our attention to the evaluation of the normalization integral

$$\mathcal{N}_\eta^{(C)} := \langle E_\eta | E_\eta \rangle^{(C)} := \int_{-L/2}^{L/2} dx_1 \cdots \int_{-L/2}^{L/2} dx_N \prod_{1 \le j < k \le N} |z_k - z_j|^{2/\alpha} E_\eta(\bar{z}) E_\eta(z), \tag{12.40}$$

where $z_j = e^{2\pi i x_j / L}$.

PROPOSITION 12.3.3 *We have*

$$\frac{\mathcal{N}_\eta^{(C)}}{\mathcal{N}_{0^N}^{(C)}} = \frac{d'_\eta e_\eta}{d_\eta e'_\eta}. \tag{12.41}$$

Proof. As in the proof of Proposition 12.3.2, it suffices to show $\langle E_{\Phi\eta} | E_{\Phi\eta} \rangle^{(C)}$ and $\langle E_{s_i\eta} | E_{s_i\eta} \rangle^{(C)}$ are correctly given by (12.41). Since the transpositions s_i are self-adjoint with respect to the inner product (12.12), we see from Definition 12.2.2 that in general $\langle \Phi f | \Phi g \rangle^{(C)} = \langle f | g \rangle^{(C)}$, so Φ is an isometry. Using the final equation of Proposition 12.2.5 we thus have

$$\langle E_{\Phi\eta} | E_{\Phi\eta} \rangle^{(C)} = \langle \Phi E_\eta | \Phi E_\eta \rangle^{(C)} = \langle E_\eta | E_\eta \rangle^{(C)}.$$

Using (12.38) shows that the r.h.s. of (12.41) satisfies the same recurrence. Consider next $\mathcal{N}_\eta^{(C)}$ with η replaced by $s_i\eta$. In the case $\eta_i < \eta_{i+1}$, Proposition 12.2.1 gives $E_{s_i\eta} = s_i E_\eta - \bar{\delta}_{i,\eta}^{-1} E_\eta$, so we have

$$\begin{aligned}
\langle E_{s_i\eta} | E_{s_i\eta} \rangle^{(C)} &= \langle (s_i - \bar{\delta}_{i,\eta}^{-1}) E_\eta | (s_i - \bar{\delta}_{i,\eta}^{-1}) E_\eta \rangle^{(C)} \\
&= (1 + \bar{\delta}_{i,\eta}^{-2}) \langle E_\eta | E_\eta \rangle^{(C)} - 2\bar{\delta}_{i,\eta}^{-1} \langle E_\eta | s_i E_\eta \rangle^{(C)} \\
&= (1 - \bar{\delta}_{i,\eta}^{-2}) \langle E_\eta | E_\eta \rangle^{(C)},
\end{aligned} \tag{12.42}$$

where the final equality follows by substituting for $s_i E_\eta$ according to Proposition 12.2.1 in the previous line, and using the fact that for $\eta_i \ne \eta_{i+1}$, E_η and $E_{s_i\eta}$ are orthogonal. On the other hand (12.38) shows that the r.h.s. of (12.41) satisfies the same recurrences under the action of s_i. □

12.3.3 The normalization u_η

To calculate the normalization u_η in (12.13), a strategy similar to that used in the proof of Proposition 12.3.3 can be adopted, except that direct use of the lowering type operator $\hat{\Phi}$ rather than the raising type operator Φ is made.

PROPOSITION 12.3.4 *We have*

$$u_\eta = \frac{d'_\eta}{d_\eta}. \tag{12.43}$$

Proof. Again we seek to express $\langle E_{s_i\eta}, E_{s_i\eta} \rangle_q$ and $\langle E_{\Phi\eta}, E_{\Phi\eta} \rangle_q$ in terms of $\langle E_\eta, E_\eta \rangle_q$ and show that the equations are consistent with (12.43). Equation (12.42) makes use of the orthogonality of $\{E_\nu\}$ and self-adjointness of s_i, but no other detail of the inner product, so again we have

$$\langle E_{s_i\eta}, E_{s_i\eta}\rangle_q = (1 - \bar{\delta}_{i,\eta}^{-2})\langle E_\eta, E_\eta\rangle_q. \tag{12.44}$$

This is indeed consistent with (12.43).

For the second sought recurrence, following [144], note from (12.30) and (12.20) that in general

$$\langle q_1(x_i;x)f, x_i g\rangle_q = \langle f, g\rangle_q = \langle \sigma f, \sigma g\rangle_q,$$

where $q_1(x_i;x)$ is given by (12.21) and σ is any permutation of N symbols. Recalling Definition 12.2.4 of $\hat{\Phi}$ we thus have

$$\langle E_{\hat{\Phi}\eta}, E_{\hat{\Phi}\eta}\rangle_q = \langle d_N E_\eta, d_N E_\eta\rangle_q = \langle q_1(x_N;x) d_N E_\eta, x_N d_N E_\eta\rangle_q = \frac{1}{\alpha}(\bar{\eta}_N - 1 + N)\langle q_1(x_N;x) d_N E_\eta, E_\eta\rangle_q, \tag{12.45}$$

where the final equality follows by making use of (12.35). To proceed further, we note from Definition 11.4.1 of d_N and the explicit form (12.21) of $q_1(x_N;x)$ that

$$d_N q_1(x_N;x) = d_N x_N + \frac{1}{\alpha}.$$

From this and (12.35) we see that

$$d_N q_1(x_N;x) d_N E_\eta = \frac{1}{\alpha}(\bar{\eta}_N + N) d_N E_\eta.$$

Thus $q_1(x_N;x) d_N E_\eta - \frac{1}{\alpha}(\bar{\eta}_N + N)E_\eta$ is annihilated by d_N, and consequently from (12.23)

$$q_N(x_N;x) d_N E_\eta = \frac{1}{\alpha}(\bar{\eta}_N + N)E_\eta + \sum_{\nu:|\nu|=|\eta|,\ \nu_N=0} c_{\eta\nu} q_\nu.$$

Substituting this in (12.45) and using the structure formula (12.24), together with the orthogonality (12.30), show that for $\eta_N > 0$

$$\langle \hat{\Phi} E_\eta, \hat{\Phi} E_\eta\rangle_q = \frac{1}{\alpha^2}(\bar{\eta}_N + N)(\bar{\eta}_N - 1 + N)\langle E_\eta, E_\eta\rangle_q = \frac{1}{\alpha^2}\frac{d'_\eta d_\eta}{d'_{\hat{\Phi}\eta} d_{\hat{\Phi}\eta}}\langle E_\eta, E_\eta\rangle_q,$$

where the second equality follows from (12.38). Substituting the result of Proposition 12.2.5 for the $\hat{\Phi}E_\eta$, then replacing η by $\Phi\eta$ (this is always possible since we are requiring $\eta_N > 0$) shows

$$\langle E_{\Phi\eta}, E_{\Phi\eta}\rangle_q = \frac{d_\eta d'_{\Phi\eta}}{d_{\Phi\eta} d'_\eta}\langle E_\eta, E_\eta\rangle_q,$$

which like (12.44) is consistent with (12.43). □

12.4 A GENERALIZED BINOMIAL THEOREM AND AN INTEGRATION FORMULA

12.4.1 The generalized binomial theorem

The classical binomial theorem refers to the power series expansion

$$\frac{1}{(1-x)^r} = \sum_{n=0}^{\infty} \frac{(r)_n}{n!} x^n, \quad |x| < 1,$$

where $(r)_n$ is specified as in (5.83). Here the binomial theorem will be generalized so as to express $\prod_{j=1}^N (1 - x_j)^{-r}$ as a series in $\{E_\eta\}$. This will be done by making use of the expansion formula (12.15) as well as the specialization formula (12.39) [39]. Replacing the quantity $(r)_n$ will be the generalized Pochhammer symbol

$$[u]^{(\alpha)}_{\eta^+} := \prod_{j=1}^N \frac{\Gamma(u - \frac{1}{\alpha}(j-1) + \eta_j^+)}{\Gamma(u - \frac{1}{\alpha}(j-1))}. \tag{12.46}$$

Note from the definitions (12.36) and (12.37) that e_η and e'_η are expressible in terms of the generalized Pochhammer symbol. Thus

$$e_\eta = \alpha^{|\eta|}[1 + N/\alpha]^{(\alpha)}_{\eta^+}, \qquad e'_\eta = \alpha^{|\eta|}[1 + (N-1)/\alpha]^{(\alpha)}_{\eta^+}. \tag{12.47}$$

PROPOSITION 12.4.1 *With u_η as in (12.15), and thus given explicitly by (12.43), we have*

$$\prod_{j=1}^N \frac{1}{(1-x_j)^r} = \sum_\eta \frac{\alpha^{|\eta|}[r]^{(\alpha)}_{\eta^+}}{u_\eta d_\eta} E_\eta(x) = \sum_\eta \frac{\alpha^{|\eta|}[r]^{(\alpha)}_{\eta^+}}{d'_\eta} E_\eta(x). \tag{12.48}$$

Proof. We begin by replacing N by kN, $kN \in \mathbb{Z}^+$ in (12.15), then setting $y_1, \dots, y_{kN}$ equal to 1 and $x_{N+1}, \dots, x_{kN}$ equal to 0. Noting that u_η as defined below (12.15) and d_η as defined by (12.36) are independent of N, and making use of the formula (12.47) for e_η gives the formula

$$\prod_{j=1}^N \frac{1}{(1-x_j)^{kN/\alpha+1}} = \sum_\eta \frac{e_\eta|_{N\mapsto kN}}{u_\eta d_\eta} E_\eta(x) = \sum_\eta \frac{\alpha^{|\eta|}[1 + kN/\alpha]^{(\alpha)}_{\eta^+}}{u_\eta d_\eta} E_\eta(x).$$

Now, from the definition (12.46) we see that each coefficient of $E_\eta(x)$ is a polynomial in kN/α. Also, expanding the l.h.s. as a power series gives that the same must be true here too. It follows that since both sides are equal for each $kN \in \mathbb{Z}^+$, they must in fact be equal for all (complex) values $kN/\alpha =: r - 1$. □

12.4.2 A generalization of the Morris constant term identity and the Selberg integral

Since $\{E_\eta\}$ is an orthogonal basis for analytic functions f with respect to the scalar product $\langle\cdot|\cdot\rangle^{(C)}$, it follows that a general analytic function f has the expansion

$$f(x_1, \dots, x_N) = \sum_\eta \frac{\langle E_\eta|f\rangle^{(C)}}{\mathcal{N}^{(C)}_\eta} E_\eta(x). \tag{12.49}$$

Setting $f = \prod_{j=1}^N (1-x_j)^{-r}$ and comparing with (12.48) shows that we can evaluate the inner product $\langle E_\eta|f\rangle^{(C)}$ in this case. In fact manipulation of the inner product provides us with a generalization of the Morris constant term identity (4.4) [39].

PROPOSITION 12.4.2 *Let $\widetilde{\Delta}(x) := \prod_{j\neq k}(1 - x_j/x_k)^{1/\alpha}$. For $a, b, 1/\alpha \in \mathbb{Z}_{\geq 0}$ we have*

$$\frac{\mathrm{CT}\left(\prod_{i=1}^N (1-x_i)^a (1 - \frac{1}{x_i})^b E_\eta(x)\widetilde{\Delta}(x)\right)}{\mathrm{CT}\left(\prod_{i=1}^N (1-x_i)^a (1 - \frac{1}{x_i})^b \widetilde{\Delta}(x)\right)} = E_\eta((1)^N) \frac{[-b]^{(\alpha)}_{\eta^+}}{[1 + a + (N-1)/\alpha]^{(\alpha)}_{\eta^+}}. \tag{12.50}$$

Proof. First note that for (Laurent) polynomials f and g and $1/\alpha \in \mathbb{Z}_{\geq 0}$ we can write (12.12) as

$$\langle f|g\rangle^{(C)} = L^N \mathrm{CT}\Big(f\Big(\frac{1}{x_1}, \dots, \frac{1}{x_N}\Big) g(x_1, \dots, x_N)\widetilde{\Delta}(x)\Big).$$

Thus for $r \in \mathbb{Z}_{\leq 0}$, comparison of (12.49) with $f = \prod_{j=1}^N (1-x_j)^{-r}$ and (12.48) give

$$\mathrm{CT}\left(\prod_{i=1}^N (1 - x_i^{-1})^{-r} E_\eta(x)\widetilde{\Delta}(x)\right) = \frac{\alpha^{|\eta|}[r]^{(\alpha)}_{\eta^+}}{d'_\eta} \mathrm{CT}\left(E_\eta(x^{-1})E_\eta(x)\widetilde{\Delta}(x)\right).$$

Now make the replacement $\eta \mapsto \eta + a$ and rewrite $E_{\eta+a}$ according to the formula of Exercises 12.1 q.1. Also, set $r = -a - b$ and note

$$x_i^a\Big(1 - \frac{1}{x_i}\Big)^{a+b} = (-1)^a(1 - x_i)^a\Big(1 - \frac{1}{x_i}\Big)^b,$$

to conclude that

$$\mathrm{CT}\Big(\prod_{i=1}^N (1-x_i)^a\Big(1-\frac{1}{x_i}\Big)^b E_\eta(x)\tilde{\Delta}(x)\Big) = (-1)^{aN}\alpha^{|\eta|}[-a-b]^{(\alpha)}_{\eta^+ + a^N}\frac{1}{d'_{\eta+a^N}}\,\mathrm{CT}\Big(E_\eta(x^{-1})E_\eta(x)\tilde{\Delta}(x)\Big). \tag{12.51}$$

Expressing the l.h.s. of this in the form of the l.h.s. of (12.50) we have

$$\frac{\mathrm{CT}\Big(\prod_{i=1}^N(1-x_i)^a(1-\frac{1}{x_i})^b E_\eta(x)\tilde{\Delta}(x)\Big)}{\mathrm{CT}\Big(\prod_{i=1}^N(1-x_i)^a(1-\frac{1}{x_i})^b\tilde{\Delta}(x)\Big)} = \frac{\alpha^{|\eta|}[-a-b]^{(\alpha)}_{\eta^+ + a^N}d'_{a^N}}{[-a-b]^{(\alpha)}_{a^N}d'_{\eta+a^N}}\frac{\mathcal{N}_\eta^{(C)}}{\mathcal{N}_{0^N}^{(C)}}, \tag{12.52}$$

where $\mathcal{N}_\eta^{(C)}$ is defined by (12.40) and the ratio $\mathcal{N}_\eta^{(C)}/\mathcal{N}_{0^N}^{(C)}$ evaluated by (12.41). To simplify the r.h.s. of this expression, we first note from the definition (12.46) that

$$\frac{[-a-b]^{(\alpha)}_{\eta^+ + a^N}}{[-a-b]^{(\alpha)}_{a^N}} = [-b]^{(\alpha)}_{\eta^+}. \tag{12.53}$$

To deduce the dependence of $d'_{\eta+a^N}$ on a we note from the formula of Exercises 12.1 q.1 and (12.39) that

$$\frac{d_{\eta+a^N}}{d_\eta} = \frac{e_{\eta+a^N}}{e_\eta}. \tag{12.54}$$

The formula of Exercises 12.1 q.1 also tells us that $\mathcal{N}^{(C)}_{\eta+a^N}/\mathcal{N}^{(C)}_\eta = 1$, so making use of (12.54) in (12.41) shows

$$\frac{d'_{\eta+a^N}}{d'_\eta} = \frac{e'_{\eta+a^N}}{e'_\eta} = \alpha^{aN}\frac{[1+a+(N-1)/\alpha]_{\eta^+}}{[1+(N-1)/\alpha]_{\eta^+}}\prod_{j=1}^N\frac{\Gamma(1+a+(j-1)/\alpha)}{\Gamma(1+(j-1)/\alpha)}$$

and thus

$$\frac{d'_{\eta+a^N}}{d'_{a^N}} = \frac{[1+a+(N-1)/\alpha]_{\eta^+}}{[1+(N-1)/\alpha]_{\eta^+}}d'_\eta. \tag{12.55}$$

Substituting (12.53) and (12.55) in (12.52) and making further use of (12.47) and (12.41) give the sought result. □

With the notation (4.4), (12.50) can be rewritten as the trigonometric integral

$$\frac{1}{M_N(a,b;1/\alpha)}\int_{-1/2}^{1/2}dx_1\cdots\int_{-1/2}^{1/2}dx_N\prod_{l=1}^N e^{\pi i x_l(a-b)}|1+e^{2\pi i x_l}|^{a+b}E_\eta(-e^{2\pi i x_1},\dots,-e^{2\pi i x_N})$$
$$\times\prod_{1\le j<k\le N}|e^{2\pi i x_k}-e^{2\pi i x_j}|^{2/\alpha} = E_\eta((1)^N)\frac{[-b]^{(\alpha)}_{\eta^+}}{[1+a+(N-1)/\alpha]^{(\alpha)}_{\eta^+}}. \tag{12.56}$$

In this form the restriction to $a, b, 1/\alpha \in \mathbb{Z}_{\ge 0}$ required in (12.50) can be lifted. Furthermore, Proposition

3.9.1 can be used to convert (12.56) to the generalized Selberg integral

$$\frac{1}{S_N(\lambda_1,\lambda_2,1/\alpha)}\int_0^1 dt_1\cdots\int_0^1 dt_N\prod_{l=1}^N t_l^{\lambda_1}(1-t_l)^{\lambda_2}E_\eta(t_1,\dots,t_N)\prod_{1\le j<k\le N}|t_j-t_k|^{2/\alpha}$$
$$= E_\eta((1)^N)\frac{[\lambda_1+(N-1)/\alpha+1]^{(\alpha)}_{\eta^+}}{[\lambda_1+\lambda_2+2(N-1)/\alpha+2]^{(\alpha)}_{\eta^+}}. \qquad (12.57)$$

EXERCISES 12.4 1. Argue that $E_\eta((-1)^N)=(-1)^{|\eta|}E_\eta((1)^N)$ is the limiting value of the l.h.s. of (12.56) for $a=b$ and $a\to\infty$. Show that this is consistent with the asymptotic form of the r.h.s.

2. Let κ' denote the partition conjugate to κ (recall Definition 10.1.3). Show that

$$[a]^{(\alpha)}_\kappa=(-\alpha)^{-|\kappa|}[-\alpha a]^{(1/\alpha)}_{\kappa'}.$$

3. [342] Let κ denote a partition. Define d'_κ as in (12.37) and set

$$h_\kappa:=\prod_{(i,j)\in\kappa}\Big(\alpha a(i,j)+l(i,j)+1\Big). \qquad (12.58)$$

In this exercise the formulas

$$h_\kappa=\frac{\alpha^{|\kappa|}[N/\alpha]^{(\alpha)}_\kappa}{f^{1/\alpha}_N(\kappa)},\qquad f^\lambda_n(\kappa):=\prod_{1\le i<j\le n}\frac{((j-i)\lambda+\kappa_i-\kappa_j)_\lambda}{((j-i)\lambda)_\lambda}, \qquad (12.59)$$

$$d'_\kappa=\frac{\alpha^{|\kappa|}[(N-1)/\alpha+1]^{(\alpha)}_\kappa}{\bar f^{1/\alpha}_N(\kappa)},\qquad \bar f^\lambda_n(\kappa):=\prod_{1\le i<j\le n}\frac{(1-\lambda+(j-i)\lambda+\kappa_i-\kappa_j)_\lambda}{(1-\lambda+(j-i)\lambda)_\lambda}, \qquad (12.60)$$

will be established by induction. Consider for definiteness the first identity. Note that it is true for $\kappa=0^N$.

(i) Assume the identity is true for a particular partition $\kappa=(\kappa_1,\dots,\kappa_N)$. Verify that the identity is then true for the partition $\kappa^{(i)}:=(\kappa_1,\dots,\kappa_{i-1},\kappa_i+1,\kappa_{i+1},\dots,\kappa_N)$, provided

$$\prod_{j=1}^{\kappa_i}\frac{\kappa'_j-i+1+(\kappa_i+1-j)\alpha}{\kappa'_j-i+1+(\kappa_i-j)\alpha}\prod_{j=i+1}^N\frac{(j-i)/\alpha+\kappa_i-\kappa_j+1/\alpha}{(j-i)/\alpha+\kappa_i-\kappa_j}=N-(i-1)+\kappa_i\alpha.$$

(ii) Verify the identity in (i) by defining $i\le j_1<j_2\cdots<j_i\le N$ such that

$$\kappa_{i+1},\dots,\kappa_{j_1}=\kappa_i,\quad \kappa_{j_1+1},\dots,\kappa_{j_2}=\kappa_i-1,\quad \dots\quad ,\kappa_{j_i+1},\dots,\kappa_N=0,$$

and expressing κ'_j in terms of $j_1,\dots,j_i$.

12.5 INTERPOLATION NONSYMMETRIC JACK POLYNOMIALS

12.5.1 Definitions and fundamental properties

Of importance in the development of aspects of the theory of nonsymmetric Jack polynomials which will arise later, in particular generalized binomial coefficients, and also in the derivation of Pieri type formulas, is the notion of an *interpolation nonsymmetric Jack polynomial* (also referred to as a shifted nonsymmetric Jack polynomial) [432], [358], [477]. In one variable the Jack polynomials are simply the monomials $\{x^p\}$. The interpolation Jack polynomials in one variable generalize the monomial x^p to the polynomial

$$(-1)^p(-x)_p=x(x-1)\cdots(x-p+1). \qquad (12.61)$$

This can be characterized by being the unique (up to normalization) polynomial of degree at most p which vanishes at $x = 0, 1, \dots, p-1$ and is nonzero at $x = p$. Likewise, in the N-variable case the interpolation nonsymmetric Jack polynomials are defined by a vanishing condition, which is based on the following result [358], [477].

PROPOSITION 12.5.1 *For a given composition ν, define $\bar{\nu}_j$ by (12.2), and let $\{F(\bar{\nu})\}$ be a given set of complex numbers indexed by $\bar{\nu}$. There exists a unique polynomial f in N variables of degree at most d such that for all $|\nu| \le d$*

$$f(\bar{\nu}_1/\alpha, \dots, \bar{\nu}_N/\alpha) = F(\bar{\nu}_1, \dots, \bar{\nu}_N). \tag{12.62}$$

Proof. Since for each distinct ν there is a distinct value of $\bar{\nu}$, (12.62) gives the same number of linear equations for the coefficients a_ν in the expansion

$$f(x_1, \dots, x_N) = \sum_{\nu: |\nu| \le d} a_\nu x^\nu$$

as there are distinct coefficients. Hence if a solution to the equations exists, then it must be unique. To show existence, we will construct f of the form

$$f(x_1, \dots, x_N) = g(x_1, \dots, x_{N-1}) + \Big(x_N + \frac{N-1}{\alpha}\Big) h(x_N - 1, x_1, \dots, x_{N-1}) \tag{12.63}$$

by induction on N and d. The validity for $N = 1$, d arbitrary follows from the fundamental theorem of algebra, while in the case $d = 0$, N arbitrary, there is nothing to prove. Consider first the set of compositions such that $|\nu| \le d$ and with $\nu_N = 0$. From (12.2) we then have $\bar{\nu}_N = -(N-1)$ and so

$$f(x_1, \dots, x_{N-1}, \bar{\nu}_N/\alpha) = g(x_1, \dots, x_{N-1}). \tag{12.64}$$

Since g depends only on the variables $x_1, \dots, x_{N-1}$, the induction hypothesis gives that with $x_j = \bar{\nu}_j/\alpha$, f can take on prescribed values.

Consider next the cases that $\nu_N \ne 0$. We want to find a function h such that

$$f(\bar{\nu}_1/\alpha, \dots, \bar{\nu}_N/\alpha) - g(\bar{\nu}_1/\alpha, \dots, \bar{\nu}_{N-1}/\alpha) = \frac{1}{\alpha}(\bar{\nu}_N + N - 1) h(\bar{\nu}_N/\alpha - 1, \bar{\nu}_1/\alpha, \dots, \bar{\nu}_{N-1}/\alpha)$$

and h can take on prescribed values at these points. The essential point here is that the composition $\hat{\Phi}\nu$ (recall Definition 12.2.2) which has degree $|\nu| - 1$ is such that $(\overline{\hat{\Phi}\nu})_1 = \bar{\nu}_N - \alpha$, $(\overline{\hat{\Phi}\nu})_j = \bar{\nu}_{j-1}$, $j \ge 1$. Since $\bar{\nu}_N + (N-1) \ne 0$ for $\nu_N \ne 0$, it follows by induction in d that there is a polynomial h which takes on prescribed values at $\overline{\hat{\Phi}\nu}/\alpha$. □

It follows from Proposition 12.5.1 that there is a unique polynomial (up to normalization), $E^*_\eta(x)$, say, of degree $\le |\eta|$, which has the *vanishing property*

$$E^*_\eta(\bar{\nu}/\alpha) := E^*_\eta(\bar{\nu}_1/\alpha, \dots, \bar{\nu}_N/\alpha) = 0, \quad |\nu| \le |\eta| \ (\nu \ne \eta), \tag{12.65}$$

while

$$E^*_\eta(\bar{\eta}/\alpha) \ne 0. \tag{12.66}$$

In the next proposition it will be shown that the coefficient of x^η in $E^*_\eta(x)$ is nonzero. Choosing the coefficient to be unity gives

$$E^*_\eta(x) = x^\eta + \sum_{\substack{|\nu| \le |\eta| \\ \nu \ne \eta}} a_{\eta\nu} x^\nu. \tag{12.67}$$

Note in particular that $\{E^*_\eta(x)\}$ forms a basis for analytic functions. With this normalization, the polynomial $E^*_\eta(x)$ is referred to as the interpolation nonsymmetric Jack polynomial (its relationship to the nonsymmetric Jack polynomial will become evident as the theory is developed).

PROPOSITION 12.5.2 *The polynomial E^*_η contains the monomial x^η with a nonzero coefficient. Furthermore, with*

$$\Delta^{(N)} f(x_1, \dots, x_N) := f(x_N - 1, x_1, \dots, x_{N-1})$$

we have

$$\Big(x_N + \frac{N-1}{\alpha}\Big)\Delta^{(N)} E^*_\eta(x) = E^*_{\Phi\eta}(x), \tag{12.68}$$

where $\Phi\eta := (\eta_2, \dots, \eta_N, \eta_1 + 1)$ as in Definition 12.2.2.

Proof. Analogous to the decomposition (12.63) write

$$E^*_\eta(x_1, \dots, x_N) = g(x_1, \dots, x_{N-1}) + \Big(x_N + \frac{N-1}{\alpha}\Big) h(x_N - 1, x_1, \dots, x_{N-1}). \tag{12.69}$$

For the first statement we will proceed by induction on N and $|\eta|$. In the case $\eta_N = 0$, (12.64) gives

$$g(x_1, \dots, x_{N-1}) = E^*_{\eta_\#}(x_1, \dots, x_{N-1}, \bar{\eta}_N/\alpha) \propto E^*_{\eta_\#}(x_1, \dots, x_{N-1}), \tag{12.70}$$

where $\eta_\# := (\eta_1, \dots, \eta_{N-1})$ and the validity of the final proportionality follows from (12.65) and (12.66). Thus g contains $x^{\eta_\#}$ by the induction hypothesis in N. For $\eta_N \neq 0$, from the proof of Proposition 12.5.1 we can take $g = 0$ and

$$h(x_1, \dots, x_N) = E^*_{\hat{\Phi}\eta}(x_1, \dots, x_N). \tag{12.71}$$

But $E^*_{\hat{\Phi}\eta}$ contains $x^{\hat{\Phi}\eta}$ by the induction hypothesis in $|\eta|$, and so it follows from (12.69) that E^*_η contains x^η.
To derive (12.68), substitute (12.71) in (12.69) with $g = 0$ and replace η by $\Phi\eta$. □

A consequence of (12.68) is an evaluation formula for $E^*_\eta(x)$ at the special point $x = \bar{\eta}/\alpha$.

PROPOSITION 12.5.3 *We have*

$$E^*_\eta(\bar{\eta}/\alpha) = \alpha^{-|\eta|} d'_\eta. \tag{12.72}$$

Proof. Noting from (12.67) that the proportionality constant in (12.70) is unity, we see that if $\eta_N = 0$, then

$$E^*_\eta(\bar{\eta}/\alpha) = E^*_{\eta_\#}(\bar{\eta}_\#/\alpha),$$

where $\eta_\# := (\eta_1, \dots, \eta_{N-1})$. The r.h.s. of (12.72) has this property, which provides us with an induction in N. For induction in $|\eta|$, we note from (12.68) that

$$\Big(\frac{\bar{\eta}_1 + \alpha + N - 1}{\alpha}\Big) E^*_\eta(\bar{\eta}/\alpha) = E^*_{\Phi\eta}(\overline{\Phi\eta}/\alpha),$$

where we have used the facts $(\overline{\Phi\eta})_j = \eta_{j+1}$ $(j = 1, \dots, N-1)$, $(\overline{\Phi\eta})_N = \bar{\eta}_1 + \alpha$. We see from (12.38) that the r.h.s. of (12.72) satisfies this same recurrence. □

To proceed further we require the fact that the E^*_η are eigenfunctions of the family of operators [358]

$$\tilde{\Xi}_i := x_i - \sigma_i \cdots \sigma_{N-1}\Big(x_N + \frac{N-1}{\alpha}\Big)\Delta^{(N)}\sigma_1 \cdots \sigma_{i-1}, \tag{12.73}$$

where

$$\sigma_i := s_i + \frac{1}{\alpha}\frac{1 - s_i}{x_i - x_{i+1}} \qquad (i = 1, \dots, N-1). \tag{12.74}$$

Note that in the case $N = 1$ we have

$$\tilde{\Xi}_1 = x_1(1 - \Delta^{(1)}), \tag{12.75}$$

where $\Delta^{(1)} f(x_1) := f(x_1 - 1)$, which is indeed an eigenoperator for (12.61). A feature of (12.75), deduced by noting that

$$\Delta^{(1)} = \sum_{j=0}^{\infty} \frac{(-1)^j}{j!} \frac{d^j}{dx_1^j}, \tag{12.76}$$

is that it consists of a degree preserving term equivalent to $x_1 \frac{d}{dx_1}$, plus degree lowering terms. Similarly, inspection of (12.73) using (12.76) and the fact that

$$\Delta^{(N)} = s_{N-1} s_{N-2} \cdots s_1 \Delta^{(1)}, \quad \Delta^{(1)} f(x_1, x_2, \dots, x_N) := f(x_1 - 1, x_2, \dots, x_N) \tag{12.77}$$

for general N shows

$$\tilde{\Xi}_i = \frac{1}{\alpha} \xi_i + \text{degree-lowering terms.} \tag{12.78}$$

If $\{\tilde{\Xi}_i\}$ are eigenoperators, this means we can replace (12.67) by the more refined statement

$$E^*_\eta(x) = E_\eta(x) + \sum_{|\nu|<|\eta|} \tilde{a}_{\eta\nu} E_\nu(x). \tag{12.79}$$

To establish that (12.73) is indeed an eigenoperator, it is first necessary to specify the structure of $\sigma_i f(x)|_{x=\bar{\nu}/\alpha}$. From the definition (12.74) we see

$$\sigma_i f(x)\Big|_{x=\bar{\nu}/\alpha} = \frac{1}{\bar{\nu}_i - \bar{\nu}_{i+1}} f(\bar{\nu}/\alpha) + \frac{\bar{\nu}_i - \bar{\nu}_{i+1} - 1}{\bar{\nu}_i - \bar{\nu}_{i+1}} f(s_i \bar{\nu}/\alpha).$$

Now, from Exercises 12.1 q.5(i) we have that for $\nu_i \neq \nu_{i+1}$, $s_i \bar{\nu} = \overline{s_i \nu}$, while for $\nu_i = \nu_{i+1}$, $\bar{\nu}_i - \bar{\nu}_{i+1} = 1$, so we conclude that in all cases

$$\sigma_i f(\bar{\nu}/\alpha) = a f(\bar{\nu}/\alpha) + b f(\overline{s_i \nu}/\alpha) \tag{12.80}$$

for a and b independent of f.

PROPOSITION 12.5.4 *We have*

$$\tilde{\Xi}_i E^*_\eta = \frac{1}{\alpha} \bar{\eta}_i E^*_\eta. \tag{12.81}$$

Proof. The E^*_η are uniquely characterized, up to normalization, by the vanishing condition (12.65), together with the property (12.66). Now, we see from (12.80) and the action of $\Delta^{(N)}$ that

$$\sigma_i \cdots \sigma_{N-1} \Big(x_N + \frac{N-1}{\alpha} \Big) \Delta^{(N)} \sigma_1 \cdots \sigma_{i-1} E^*_\eta(x) \Big|_{x=\bar{\nu}/\alpha}$$

consists of a linear combination of $E^*_\eta(\bar{\rho}/\alpha)$ for compositions ρ such that $|\rho| = |\nu| - 1$. But for such compositions $E^*_\eta(\bar{\rho}/\alpha) = 0$. Recalling the definition (12.73), this means $\tilde{\Xi}_i E^*_\eta(\bar{\nu}/\alpha) = 0$ for $|\nu| \le |\eta|$, $\nu \neq \eta$, while the x_i term in (12.73) gives $\tilde{\Xi}_i E^*_\eta(\bar{\eta}/\alpha) = \frac{1}{\alpha} \bar{\eta}_i E^*_\eta(\bar{\eta}/\alpha)$, thus establishing (12.81). □

12.5.2 Algebra isomorphism

The eigenvalue equations (12.81) and the fact that $\{E^*_\eta(x)\}$ form a basis for analytic functions imply

$$[\tilde{\Xi}_i, \tilde{\Xi}_j] = 0. \tag{12.82}$$

A direct calculation using the definition (12.74) shows that like $\{s_i\}$, the $\{\sigma_i\}$ satisfy the algebra

$$\sigma_j^2 = 1, \quad \sigma_j \sigma_{j+1} \sigma_j = \sigma_{j+1} \sigma_j \sigma_{j+1}, \quad [\sigma_i, \sigma_j] = 0 \ (|i - j| \ge 2). \tag{12.83}$$

In fact $\{\sigma_i, \alpha\tilde{\Xi}_j\}$ form a representation of the degenerate type A Hecke algebra specified in Exercises 11.5 q.2. To see this, first note that

$$x_i\sigma_i = \sigma_i x_{i+1} + \frac{1}{\alpha}, \qquad [x_i, \sigma_j] = 0 \ (j \neq i-1, i). \tag{12.84}$$

Now introduce the operator [358]

$$Z_i := x_i - \tilde{\Xi}_i = \sigma_i \cdots \sigma_{N-1}\Big(x_N + \frac{N-1}{\alpha}\Big)\Delta^{(N)}\sigma_1 \cdots \sigma_{i-1}. \tag{12.85}$$

We see immediately that

$$Z_i\sigma_i = \sigma_i Z_{i+1}. \tag{12.86}$$

Also, noting that

$$\Delta^{(N)}\sigma_j = \sigma_{j-1}\Delta^{(N)}$$

and using the second relation in (12.84) we see

$$Z_i\sigma_j = \sigma_j Z_i \qquad (j \neq i-1, i). \tag{12.87}$$

Combining (12.87), (12.86) and (12.84), we have from the definition of Z_i that

$$(\alpha\tilde{\Xi}_i)\sigma_i - \sigma_i(\alpha\tilde{\Xi}_{i+1}) = 1, \qquad [\alpha\tilde{\Xi}_i, \sigma_j] = 0 \ (j \neq i-1, i), \tag{12.88}$$

which is precisely the relations (11.89), thus completing (with (12.82) and (12.83)) the defining relations of the degenerate type A Hecke algebra. An immediate consequence is the analogue of Proposition 12.2.1.

PROPOSITION 12.5.5 *We have*

$$\sigma_i E^*_\eta = \begin{cases} \frac{1}{\delta_{i,\eta}}E^*_\eta + \Big(1 - \frac{1}{\delta^2_{i,\eta}}\Big)E^*_{s_i\eta}, & \eta_i > \eta_{i+1}, \\ E^*_\eta, & \eta_i = \eta_{i+1}, \\ \frac{1}{\delta_{i,\eta}}E^*_\eta + E^*_{s_i\eta}, & \eta_i < \eta_{i+1}. \end{cases}$$

Let ψ be the operator which has the property

$$\psi E_\eta = E^*_\eta\psi$$

of mapping the nonsymmetric Jack polynomials to the interpolation nonsymmetric Jack polynomials. It follows immediately from (12.81) that

$$\psi^{-1}\tilde{\Xi}_i\psi E_\eta = \frac{1}{\alpha}\bar{\eta}_i E_\eta$$

and thus

$$\psi\xi_i = \tilde{\Xi}_i\psi.$$

Also, Propositions 12.2.1 and 12.5.5 together imply

$$\psi s_i = \sigma_i\psi.$$

It is also possible to determine explicitly the action of ψ on x_i [358], which turns out to involve the operator Z_i (12.85). Now, from its definition, Z_i acts on E^*_η by

$$Z_i E^*_\eta = (x_i - \bar{\eta}_i/\alpha)E^*_\eta. \tag{12.89}$$

Because the r.h.s. vanishes for all $x = \nu/\alpha$, $|\nu| \le |\eta|$, and has degree $|\eta| + 1$ we must have

$$Z_i E^*_\eta = \sum_{\nu:|\nu|=|\eta|+1} a_{\eta\nu} E^*_\nu. \tag{12.90}$$

Equating the leading degree terms on the r.h.s.'s of (12.89) and (12.90) gives

$$x_i E_\eta = \sum_{\nu:|\nu|=|\eta|+1} a_{\eta\nu} E_\nu,$$

and this equation together with (12.90) implies

$$\psi x_i = Z_i \psi \tag{12.91}$$

which is the sought relation. Since $\{x_i\}$ commute, it follows immediately from this that

$$[Z_i, Z_j] = 0,$$

while another consequence is the Rodrigues type formula

$$E_\eta(Z_1, \dots, Z_N) \cdot 1 = E^*_\eta(x_1, \dots, x_N). \tag{12.92}$$

12.5.3 Extra vanishing condition

By definition $E^*_\eta(\bar{\nu}/\alpha) = 0$ for $|\nu| \le |\eta|$, $\nu \ne \eta$. In fact the eigen-equation (12.81) implies there are further conditions under which E^*_η must vanish [358]. This requires introducing a further partial order on compositions.

DEFINITION 12.5.6 *Let ν, η be compositions. We write $\nu \preceq \eta$ if there exists a permutation π such that $\nu_i < \eta_{\pi(i)}$ for $i < \pi(i)$ and $\nu_i \le \eta_{\pi(i)}$ for $i \ge \pi(i)$.*

Note that for ν and η partitions the statement $\nu \preceq \eta$ is equivalent to $\nu \subseteq \eta$ (inclusion of diagrams). However for compositions, although $\nu \subseteq \eta$ implies $\nu \preceq \eta$, the converse is not true in general. The significance of the partial order $\preceq$ lies with the following result.

PROPOSITION 12.5.7 *For a composition η let I_S be the set of functions which vanish at $\bar{S}_\eta := \{\bar{\nu}/\alpha : \eta \not\preceq \nu\}$. Then $\tilde{\Xi}_i(I_S) \subseteq I_S$.*

Proof. For $\bar{\nu}/\alpha \in \bar{S}_\eta$ we have to show that $\tilde{\Xi}_i f(\bar{\nu}/\alpha) = 0$. We see from (12.73) and (12.74) that $\tilde{\Xi}_i f(\bar{\nu}/\alpha)$ is a linear combination of $f(\bar{\nu}/\alpha)$ and

$$\tilde{\sigma}_i \cdots \tilde{\sigma}_{N-1} \Delta^{(N)} \tilde{\sigma}_1 \cdots \tilde{\sigma}_{i-1} f(x)\Big|_{x=\bar{\nu}/\alpha},$$

where each $\tilde{\sigma}_i$ is either 1 or s_i. This is equal to $f(\bar{\rho}/\alpha)$, where for some subset $J = \{j_1, \dots, j_r\}$ of $\{1, \dots, N\}$ with $j_1 < \cdots < j_r$

$$\nu_{j_i} = \rho_{j_{i+1}} \ (i = 1, \dots, r-1), \quad \nu_{j_r} = \rho_{j_1} + 1, \quad \nu_j = \rho_j \ (j \notin J).$$

From Definition 12.5.6 we see that $\rho \preceq \nu$. Since $\nu \preceq \eta$ we have $\rho \preceq \eta$. Thus $\bar{\rho}/\alpha \in \bar{S}_\eta$ and so $f(\bar{\rho}/\alpha) = 0$, which in turn implies $\tilde{\Xi}_i f(\bar{\nu}/\alpha) = 0$. □

The extra vanishing property can now be established.

PROPOSITION 12.5.8 *For compositions η, ν such that $\eta \not\preceq \nu$ we have $E^*_\eta(\bar{\nu}/\alpha) = 0$.*

Proof. Consider the set I_S of Proposition 12.5.7, and choose a member of this set, h_η say, which does not vanish at η. Writing

$$h_\eta(x) = \sum_\rho a_\rho E^*_\rho(x)$$

and noting from Proposition 12.5.7 that $\tilde{\Xi}_i h_\eta \in I_S$ we must have that

$$E^*_\rho(\bar{\nu}/\alpha) = 0 \quad \text{for all } \bar{\nu}/\alpha \in \bar{S}_\eta. \tag{12.93}$$

The nonvanishing condition gives that for some ρ (ρ' say) $E^*_{\rho'}(\bar{\eta}/\alpha) \neq 0$ and so from (12.65) $|\eta| \geq |\rho'|$. But $E^*_{\rho'}(\bar{\rho}'/\alpha) \neq 0$ by (12.66) and so $\bar{\rho}'/\alpha \notin \bar{S}_\eta$, which means $\eta \preceq \rho'$. The conditions $|\eta| \geq |\rho'|$ and $\eta \preceq \rho'$ together imply $\rho' = \eta$, so we conclude from (12.93) that $E^*_\eta(\bar{\nu}/\alpha) = 0$ for all $\eta \npreceq \nu$. □

12.6 THE SYMMETRIC JACK POLYNOMIALS

12.6.1 The symmetrization operator

We know that when restricted to act on symmetric functions, the Schrödinger operator $H^{(\mathrm{C,Ex})}$ (11.52) reduces to the Schrödinger operator $H^{(\mathrm{C})}$ of Proposition 11.3.2. Furthermore, transforming the operator $H^{(\mathrm{C})}$ by factoring out the ground state gives the operator

$$\begin{aligned}\tilde{H}^{(\mathrm{C})} &:= \Big(\frac{L}{2\pi}\Big)^2 e^{\beta W^{(\mathrm{C})}/2}(H^{(\mathrm{C})} - E_0^{(\mathrm{C})})e^{-\beta W^{(\mathrm{C})}/2} \\ &= \sum_{j=1}^N \Big(z_j \frac{\partial}{\partial z_j}\Big)^2 + \frac{N-1}{\alpha}\sum_{j=1}^N z_j \frac{\partial}{\partial z_j} + \frac{2}{\alpha} \sum_{1 \le j < k \le N} \frac{z_j z_k}{z_j - z_k}\Big(\frac{\partial}{\partial z_j} - \frac{\partial}{\partial z_k}\Big),\end{aligned} \tag{12.94}$$

where $z_j := e^{2\pi i x_j/L}$. Comparison with (11.63) shows that the transformed Schrödinger operator $H^{(\mathrm{C,Ex})}$ when restricted to act on symmetric functions reduces to (12.94). Now, as seen in (12.34), the eigenvalue of $H^{(\mathrm{C,Ex})}$ for the eigenfunction E_η depends only on the underlying partition η^+. Thus the linear combination

$$\sum_{\eta:\eta^+=\kappa} \tilde{a}_\eta E_\eta(z), \tag{12.95}$$

where κ denotes a partition, is also an eigenfunction of $H^{(\mathrm{C,Ex})}$. In fact it is possible to choose the coefficients $\tilde{a}_\eta$ such that the linear combination is a symmetric function of z, thereby constructing a symmetric polynomial eigenfunction of $\tilde{H}^{(\mathrm{C})}$ for each partition κ.

PROPOSITION 12.6.1 *The polynomial*

$$P_\kappa(z) = P_\kappa(z;\alpha) := d'_\kappa \sum_{\eta:\eta^+=\kappa} \frac{1}{d'_\eta} E_\eta(z) \tag{12.96}$$

is symmetric in $z_1, \dots, z_N$, and an eigenfunction of (12.94), and is referred to as the symmetric Jack polynomial. Furthermore, it has the structure

$$P_\kappa(z) = m_\kappa + \sum_{\sigma<\kappa} b_{\kappa\sigma} m_\sigma, \tag{12.97}$$

where m_κ denotes the monomial symmetric function in the variables $z_1, \dots, z_N$ associated with the partition κ (for example, with $\kappa = 21$ and $N = 2$, $m_{21} = z_1^2 z_2 + z_1 z_2^2$), $<$ denotes the dominance ordering for partitions, and the coefficients $b_{\kappa\sigma}$ are independent of N.

Proof. The special structure is a consequence of the special structure of the nonsymmetric Jack polynomials exhibited by (12.7). The fact that the coefficient of m_κ in (12.96) is unity is seen by noting that the only polynomial in $\{E_\eta\}_{\eta^+=\kappa}$ containing the monomial z^κ is E_κ itself (a consequence of (12.7)) and this occurs in (12.96) with coefficient unity. The coefficients $b_{\kappa\sigma}$ are independent of N as a consequence of the stability property of Exercises 12.1 q.3. It remains to check that the r.h.s. of (12.96) is symmetric, which is equivalent to showing that it is unchanged by the action of the elementary transposition s_i $(i = 1, \dots, N-1)$. This can be done by writing

$$\sum_{\eta:\eta^+=\kappa} \frac{1}{d'_\eta} E_\eta(z) = \sum_{\substack{\eta:\eta^+=\kappa \\ \eta_i \le \eta_{i+1}}} \chi_{\eta_i\eta_{i+1}} \Big(\frac{1}{d'_\eta} E_\eta(z) + \frac{1}{d'_{s_i\eta}} E_{s_i\eta}(z) \Big)$$

with $\chi_{\eta_i\eta_{i+1}} = 1/2$ $(\eta_i = \eta_{i+1})$, $\chi_{\eta_i\eta_{i+1}} = 1$ otherwise, then using Proposition 12.2.1 to compute the action of s_i on the r.h.s, which shows that in fact the individual terms in this grouping remain unchanged. □

We remark that in the case $N = 2$, $P_\kappa(z;\alpha)$ can be expressed in terms of the Jacobi polynomials $P^{(\gamma,\gamma)}_{\kappa_1-\kappa_2}(x)$, $\gamma = (\alpha-1)/2$ and x related to z_1, z_2 (see Exercises 12.6 q.6).

In Exercises 12.1 q.5(ii), it was noted for the operator $X^{(\mathrm{C})}(u)$ (as with $\tilde{H}^{(\mathrm{C,Ex})}$), the eigenvalue corresponding to the eigenfunction E_η depends only on η^+. Hence any linear combination of the form (12.95) is also an eigenfunction of $X^{(\mathrm{C})}(u)$. In particular P_κ is an eigenfunction satisfying the eigenvalue equation

$$X^{(\mathrm{C})}(u) P_\kappa(z) = \Big(\prod_{j=1}^N (1 + u(\alpha\kappa_j - (N-j))) \Big) P_\kappa(z)$$

and is in fact the unique eigenfunction of $X^{(\mathrm{C})}(u)$ with the structure (12.97). To see this latter point, consider $X^{(\mathrm{C})}(u) z^\eta$, as implied by (12.6). The action of $X^{(\mathrm{C})}(u)$ on m_κ can be deduced by symmetrizing both sides, as specified by the operation

$$\mathrm{Sym}\, f(x_1, \dots, x_N) := \sum_{P \in S_N} f(x_{P(1)}, \dots, x_{P(N)})$$

(cf. (4.135)). Since Sym can be constructed out of elementary transpositions s_i, and $X^{(\mathrm{C})}(u)$ is symmetric in the $\{\xi_i\}$, it follows from Exercises 11.5 q.2(ii) that Sym commutes with $X^{(\mathrm{C})}(u)$. This shows

$$X^{(\mathrm{C})}(u) m_\kappa = \Big(\prod_{j=1}^N (1 + u(\alpha\kappa_j - (N-j))) \Big) m_\kappa + \sum_{\sigma < \kappa} a_{\kappa\sigma} m_\sigma \tag{12.98}$$

for some coefficients $a_{\kappa\sigma}$. Because of this triangular structure, and the fact that the eigenvalue $\prod_{j=1}^N (1 + u(\alpha\kappa_j - (N-j)))$ is distinct for distinct partitions κ, the uniqueness follows.

The Sym operator can be used to give a different formula to (12.96) relating $P_\kappa(z)$ to $\{E_\eta(z)\}$. Now, from (12.8) and Exercises 12.1 q.5(ii) we have

$$X^{(\mathrm{C})}(u) E_\eta(z) = \Big(\prod_{j=1}^N (1 + u(\alpha\kappa_j - (N-j))) \Big) E_\eta(z).$$

By applying Sym to both sides, and using the fact that Sym commutes with $X^{(\mathrm{C})}(u)$, we thus have

$$X^{(\mathrm{C})}(u) \mathrm{Sym}\, E_\eta(z) = \Big(\prod_{j=1}^N (1 + u(\alpha\kappa_j - (N-j))) \Big) \mathrm{Sym}\, E_\eta(z). \tag{12.99}$$

Since, from the structure formula (12.7) Sym $E_\eta(z)$ must, up to normalization, have the structure (12.97), we conclude from this that

$$\mathrm{Sym}\, E_\eta(z) = a_\eta P_{\eta^+}(z) \tag{12.100}$$

for some constant a_η. Substituting $z = (1)^N$ shows

$$a_\eta = \frac{N! E_\eta((1)^N)}{P_{\eta^+}((1)^N)}. \tag{12.101}$$

The explicit value of $E_\eta((1)^N)$ is known from (12.39), while the explicit value of $P_{\eta^+}((1)^N)$ can in fact be deduced from (12.101) by making use of the additional fact that a_η can be independently computed for the composition η^{+R} (recall (12.32)). Thus the structure (12.7) shows the only term in $E_{\eta^{+R}}(z)$ which when symmetrized contributes to m_{η^+} in $P_{\eta^+}(z)$ is $z^{\eta^{+R}}$, and has coefficient unity. It follows from (12.100) that

$$\mathrm{Sym}\, z^{\eta^{+R}} = a_{\eta^R} m_{\eta^+},$$

and so

$$a_{\eta^{+R}} = \prod_{j=0}^{\eta_1^+} f_j!, \tag{12.102}$$

where f_j denotes the *frequency* of the integer j in η^+ (for example, if $\eta^+ = 211100$ then $f_0 = 2$, $f_1 = 3$, $f_2 = 1$).

Equating (12.101) with (12.102) in the case $\eta = \kappa$ for κ a partition shows

$$P_\kappa((1)^N) = \frac{N!}{\prod_{j=0}^{\eta_1^+} f_j!} \frac{e_\kappa}{d_{\kappa^R}}, \tag{12.103}$$

where use has also been made of (12.39) and the fact, which follows from (12.47), that $e_{\kappa^R} = e_\kappa$. Simplification of the r.h.s. of (12.103) gives the sought analogue of (12.39) in the case of the symmetric Jack polynomial [500].

PROPOSITION 12.6.2 *Let h_κ be given by (12.58) and let*

$$b_\kappa := \prod_{(i,j)\in\kappa} \Big(\alpha a'(i,j) + N - l'(i,j)\Big) = \alpha^{|\kappa|}[N/\alpha]_\kappa^{(\alpha)}. \tag{12.104}$$

We have

$$P_\kappa((1)^N) = \frac{b_\kappa}{h_\kappa}. \tag{12.105}$$

Proof. From the formulas (12.47) and (12.104) we see that

$$\frac{e_\kappa}{b_\kappa} = \frac{1}{N!}\prod_{j=1}^{N}(\alpha\kappa_j + N - j + 1),$$

so it suffices to show

$$\frac{h_\kappa}{\prod_{j=0}^{\kappa_1} f_j!} = \frac{d_{\kappa^R}}{\prod_{j=1}^{N}(\alpha\kappa_j + N - j + 1)}. \tag{12.106}$$

This is done in Exercises 12.6 q.1. □

12.6.2 Orthogonality with respect to $\langle\cdot|\cdot\rangle^{(\mathrm{C})}$

The orthogonality of $\{E_\eta\}$ with respect to (12.12), together with the expansion formula (12.96) implies that $\{P_\kappa\}$ are also orthogonal with respect to (12.12). This can also be deduced from the fact that the eigenoperator $X^{(\mathrm{C})}(u)$ for $\{P_\kappa\}$ is self-adjoint with respect to (12.12) and has distinct eigenvalues (recall the argument leading to Proposition 12.1.4). Note also that the orthogonality together with the structure (12.97) implies that $\{P_\kappa\}$ can be constructed from $\{m_\kappa\}$ with the partial ordering $<$ according to the Gram-Schmidt procedure.

The normalization of P_κ with respect to (12.12) can be deduced from the knowledge of the normalization of E_η (12.41), together with the formulas (12.96), (12.100) and (12.102) [376], [40].

PROPOSITION 12.6.3 *Let $\mathcal{N}_\kappa^{(\mathrm{C})} := \langle P_\kappa | P_\kappa \rangle^{(\mathrm{C})}$. We have*

$$\frac{\mathcal{N}_\kappa^{(\mathrm{C})}}{\mathcal{N}_{0^N}^{(\mathrm{C})}} = \frac{N!}{\prod_{j=0}^{\kappa_1} f_j!} \frac{d'_\kappa e_\kappa}{d_{\kappa^R} e'_\kappa} = \frac{b_\kappa d'_\kappa}{e'_\kappa h_\kappa},$$

where d_κ, d'_κ, e_κ, e'_κ, h_κ, b_κ are as defined by (12.36), (12.37), (12.58), (12.104).

Proof. According to (12.96), the orthogonality of $\{E_\eta\}$ with respect to (12.12) and the normalization formula (12.41) we see that

$$\langle P_\kappa | E_{\kappa^R} \rangle^{(\mathrm{C})} = \frac{d'_\kappa}{d'_{\kappa^R}} \langle E_{\kappa^R} | E_{\kappa^R} \rangle^{(\mathrm{C})} = \frac{d'_\kappa e_\kappa}{d_{\kappa^R} e'_\kappa} \mathcal{N}_0^{(\mathrm{C})}.$$

On the other hand, from the fact that the weight functions in $\langle \cdot | \cdot \rangle^{(\mathrm{C})}$ is symmetric, as is P_κ, we have

$$\langle P_\kappa | E_{\kappa^R} \rangle^{(\mathrm{C})} = \frac{1}{N!} \langle P_\kappa | \mathrm{Sym}\, E_{\kappa^R} \rangle^{(\mathrm{C})} = \frac{\prod_{j=0}^{\kappa_1} f_j!}{N!} \mathcal{N}_\kappa^{(\mathrm{C})},$$

where the final equality follows from (12.100) and (12.102). The first stated formula for $\mathcal{N}_\kappa^{(\mathrm{C})} / \mathcal{N}_{0^N}^{(\mathrm{C})}$ follows. For the second equality we substitute for $N! e_\kappa / \prod_{j=0}^{\kappa_1} f_j! d_{\kappa^R}$ using the equality between (12.103) and (12.105). □

12.6.3 Orthogonality, a Cauchy-type product and the antisymmetric Jack polynomials

In Section 12.1.3 the product $\Omega(x, y)$ was introduced, as were the associated polynomials $\{q_\eta\}$ and the inner product $\langle \cdot, \cdot \rangle_q$ for which $\{E_\eta\}$ forms an orthogonal set. An analogous theory can be developed for the symmetric Jack polynomials [376], although the path we will take is to first study the antisymmetric Jack polynomials [40].

In general the antisymmetrization operation Asym is specified as in (4.135). Since it is always possible to write

$$\mathrm{Asym}\, f(z_1, \dots, z_N) = \Delta(z) g(z_1, \dots, z_N), \qquad \Delta(z) := \prod_{1 \le j < k \le N} (z_j - z_k), \tag{12.107}$$

where g is symmetric, we see that for f analytic it is possible to expand $\mathrm{Asym}\, f$ in terms of $\{\Delta m_\kappa\}$. Now with

$$\delta := (N-1, N-2, \dots, 0) \tag{12.108}$$

and κ a partition we have

$$\mathrm{Asym}\, z^{\delta+\kappa} = \Delta(z)(m_\kappa + \text{terms smaller with respect to } <),$$

so proceeding as in the derivation of (12.98) we can conclude

$$X^{(\mathrm{C})}(u) \Delta(z) m_\kappa = \Big(\prod_{j=1}^N (1 + u(\alpha(\kappa_j + N - j) - (N-j))) \Big) \Delta(z) m_\kappa + \sum_{\sigma < \kappa} \alpha'_{\kappa\sigma} \Delta(z) m_\sigma$$

for some constants $\alpha'_{\kappa\sigma}$. By the argument used below (12.98), this implies that $X^{(\mathrm{C})}(u)$ possesses unique antisymmetric eigenfunctions of the form

$$S_{\kappa+\delta}(z) = S_{\kappa+\delta}(z; \alpha) := \Delta(z) \Big(m_\kappa + \sum_{\sigma < \kappa} a'_{\kappa\sigma} m_\sigma \Big), \tag{12.109}$$

and furthermore, if ρ has distinct parts,

$$\mathrm{Asym}\, E_\rho(z) = c_\rho S_{\rho^+}(z; \alpha), \quad \rho^+ = \kappa + \delta \tag{12.110}$$

for some proportionality constant c_ρ (in the cases that ρ has repeated parts, $\mathrm{Asym}\, E_\rho = 0$; see Exercises 12.6

q.2).

In preparation for computing the explicit value of c_ρ, we first make note of the expansion of the antisymmetric Jack polynomial $S_{\rho^+}(z)$ in terms of $\{E_\eta\}$.

PROPOSITION 12.6.4 *Let the partition μ have all parts distinct. We have*

$$S_\mu(z) = \frac{1}{d_\mu} \sum_{\sigma \in S_N} \varepsilon(\sigma) d_{\sigma(\mu)} E_{\sigma(\mu)}(z). \tag{12.111}$$

Proof. As noted in the proof of Proposition 12.6.1, the only polynomial in $\{E_{\sigma(\mu)}\}_{\sigma \in S_N}$ containing the monomial z^μ is E_μ itself. This corresponds to the term $\sigma = \mathrm{id}$ (the identity), and gives that the coefficient of z^μ is unity as required by (12.109). It thus remains to check that the r.h.s. of (12.111) is antisymmetric. For this we write (12.111) as

$$S_\kappa(z) = \frac{1}{d_\mu} \sum_{\substack{\sigma \in S_N \\ \sigma(\mu)_i \le \sigma(\mu)_{i+1}}} \varepsilon(\sigma) \Big(d_{\sigma(\mu)} E_{\sigma(\mu)}(z) - d_{s_i \sigma(\mu)} E_{s_i \sigma(\mu)}(z) \Big)$$

for a given i, $(1 \le i \le N-1)$, and proceed as in the proof of Proposition 12.6.1. □

Also required is the analogue of (12.102), whereby we independently determine c_ρ in the special case $\rho = (\kappa + \delta)^R$. From the structure (12.7) we have that the only term in $E_{(\kappa+\delta)^R}(z)$ which contributes to the term $\Delta(z) m_\kappa$ in the expansion (12.109) of $S_{\kappa+\delta}(z)$ is $z^{(\kappa+\delta)^R}$. We therefore have that

$$\mathrm{Asym}\, z^{(\kappa+\delta)^R} = c_{(\kappa+\delta)^R} \Delta(z) m_\kappa$$

and so

$$c_{(\kappa+\delta)^R} = (-1)^N. \tag{12.112}$$

Using (12.112) and (12.111) the value of c_ρ for general ρ can be determined.

PROPOSITION 12.6.5 *We have*

$$c_\rho = \varepsilon(\sigma) \frac{d'_\rho}{d'_{\rho^{+R}}}, \tag{12.113}$$

where σ is the permutation such that $\rho = \sigma(\rho^+)$.

Proof. Following [211], we introduce the polynomial

$$G(x,y) = \sum_{\rho : \rho^+ = \kappa} \frac{d_\rho}{d'_\rho} E_\rho(x) E_\rho(y). \tag{12.114}$$

We can check from Proposition 12.2.1 and the relations (12.38) that

$$s_i^{(x)} G(x,y) = s_i^{(y)} G(x,y), \qquad i = 1, \dots, N-1$$

and consequently

$$\mathrm{Asym}^{(x)} G(x,y) = \mathrm{Asym}^{(y)} G(x,y). \tag{12.115}$$

Applying (12.115)–(12.114) and using (12.110) shows

$$S_{\rho^+}(x) \sum_{\rho : \rho^+ = \kappa} \frac{d_\rho}{d'_\rho} c_\rho E_\rho(y) = S_{\rho^+}(y) \sum_{\rho : \rho^+ = \kappa} \frac{d_\rho}{d'_\rho} c_\rho E_\rho(x).$$

It follows from this that

$$S_{\rho^+}(x) = \tilde{c}_{\rho^+} \sum_{\rho : \rho^+ = \kappa} \frac{d_\rho}{d'_\rho} c_\rho E_\rho(x) \tag{12.116}$$

for some proportionality constant $\tilde{c}_{\rho^+}$. Comparing with (12.111) shows

$$\tilde{c}_{\rho^+} \frac{d_\rho}{d'_\rho} c_\rho = \varepsilon(\sigma) \frac{d_\rho}{d_{\rho^+}}, \tag{12.117}$$

where σ is such that $\rho = \sigma(\rho^+)$. Setting $\rho = \rho^{+R}$ and making use of (12.112) evaluates $\tilde{c}_{\rho^+}$ as

$$\tilde{c}_{\rho^+} = \frac{d'_{\rho^{+R}}}{d_{\rho^+}},$$

and substituting this in turn in (12.117) we deduce (12.113). □

An immediate consequence of the $\{S_{\kappa+\delta}\}$ being eigenfunctions of the operator $X^{(\mathrm{C})}(u)$ with distinct eigenvalues is that the antisymmetric Jack polynomials are orthogonal with respect to the inner product (12.12). Noting that

$$\prod_{j<k} |z_k - z_j|^{2/\alpha} |\Delta(z)|^2 = \prod_{j<k} |z_k - z_j|^{2(\alpha+1)/\alpha},$$

we have that the polynomials $S_{\kappa+\delta}(z)/\Delta(z)$ can be constructed by orthogonalizing $\{m_\kappa\}$ in the order implied by the partial order $<$, with respect to the inner product

$$\langle \cdot | \cdot \rangle^{(\mathrm{C})} \Big|_{\alpha \mapsto \alpha/(1+\alpha)}.$$

But we already know that the unique symmetric polynomials with this property are the symmetric Jack polynomials $P_\kappa^{(\alpha/(1+\alpha))}(z)$, so we have

$$S_{\kappa+\delta}(z;\alpha) = \Delta(z) P_\kappa(z; \alpha/(1+\alpha)), \tag{12.118}$$

and consequently with $\rho^+ = \kappa + \delta$

$$\mathrm{Asym}\, E_\rho(z) = c_\rho \Delta(z) P_\kappa(z; \alpha/(1+\alpha)). \tag{12.119}$$

In the limit $\alpha \to \infty$, substituting (12.10) in (12.119) with $\rho = (\delta + \kappa)^R$ and making use of the Vandermonde determinant formula (1.173) shows

$$P_\kappa(z; 1) = \frac{\det[z_j^{k-1+\kappa_{N+1-k}}]_{j,k=1,\dots,N}}{\det[z_j^{k-1}]_{j,k=1,\dots,N}}. \tag{12.120}$$

This is the ratio of alternants formula (10.16) for the Schur polynomials in the N variables $z_1, \dots, z_N$, and thus

$$P_\kappa(z; 1) = s_\kappa(z). \tag{12.121}$$

The formula (12.119) can be used to obtain the analogue of (12.15) for the symmetric Jack polynomials, referred to as a Cauchy type expansion [500] ((10.58) is the special case $\alpha = 1$).

PROPOSITION 12.6.6 *Let*

$$\Pi^{(\alpha)}(x, y) := \prod_{i,j=1}^{N} \frac{1}{(1 - x_i y_j)^{1/\alpha}}. \tag{12.122}$$

We have

$$\Pi^{(\alpha)}(x, y) = \sum_\kappa \frac{h_\kappa}{d'_\kappa} P_\kappa(x) P_\kappa(y). \tag{12.123}$$

Proof. Suppose we can show that

$$\Pi^{(\alpha)}(x,y) = \sum_\kappa \frac{1}{v_\kappa} P_\kappa(x) P_\kappa(y) \tag{12.124}$$

for some nonzero v_κ. Then to determine v_κ, we proceed as in the derivation of (12.48) to deduce from (12.124), (12.105) and (12.104) that

$$\prod_{j=1}^N \frac{1}{(1-x_j)^r} = \sum_\kappa \frac{\alpha^{|\kappa|}[r]_\kappa^{(\alpha)}}{v_\kappa h_\kappa} P_\kappa(x). \tag{12.125}$$

Now substituting (12.96) for P_κ and comparing with (12.48) gives

$$v_\kappa = \frac{d'_\kappa}{h_\kappa} \tag{12.126}$$

as required. It thus remains to establish (12.124). For this we begin by applying $\mathrm{Asym}^{(x)}$ to both sides of (12.14). On the l.h.s. we make use of the Cauchy double alternant formula (4.34) to note that

$$\mathrm{Asym}^{(x)} \frac{1}{\prod_{j=1}^N (1-x_j y_j)} = \frac{\Delta(x)\Delta(y)}{\prod_{j,k=1}^N (1-x_j y_k)}, \tag{12.127}$$

and so obtain

$$\mathrm{Asym}^{(x)} \Omega(x,y) = \frac{\Delta(x)\Delta(y)}{\prod_{j,k=1}^N (1-x_j y_k)^{(\alpha+1)/\alpha}}, \tag{12.128}$$

while on the r.h.s. we use (12.119). This gives

$$\Delta(x)\Delta(y)\Pi^{(\alpha/(1+\alpha))}(x,y) = \Delta(x) \sum_\rho{}^{*} \frac{c_\rho}{u_\rho} P_\kappa(x;\alpha/(\alpha+1)) E_\rho(y;\alpha), \tag{12.129}$$

where the $*$ denotes that the sum is restricted to ρ with distinct parts and $\rho^+ = \kappa + \delta$. Applying $\mathrm{Asym}^{(y)}$ to both sides of this identity we see that the l.h.s. is simply multiplied by $N!$, while on the r.h.s. we again use (12.119). Cancelling $\Delta(x)\Delta(y)$ from both sides and summing over permutations of ρ which give the same κ, and finally replacing $\alpha/(1+\alpha)$ by α gives the sought formula. □

The structure of (12.123) together with the result of Proposition 12.1.8 suggests we define a pairing inner product for symmetric functions by

$$\langle P_\kappa, P_\mu \rangle_g := \frac{d'_\kappa}{h_\kappa} \delta_{\kappa,\mu}. \tag{12.130}$$

Introducing the symmetric homogeneous polynomials g_κ by

$$\Pi^{(\alpha)}(x,y) = \sum_\kappa g_\kappa(x) m_\kappa(y) \tag{12.131}$$

allows us to deduce that $\{g_\kappa\}$ are a basis for symmetric functions and furthermore

$$\langle g_\kappa, m_\mu \rangle_g = \delta_{\kappa,\mu}. \tag{12.132}$$

We remark that there is an alternative way to expand the product (12.122), in terms of the power sum basis for symmetric functions. This is covered in Exercises 12.6 q.5, and leads to an alternative characterization of the scalar product $\langle \cdot, \cdot \rangle_g$.

The formula

$$\prod_{j=1}^N \frac{1}{(1-x_j)^r} = \sum_\kappa \frac{\alpha^{|\kappa|}[r]_\kappa^{(\alpha)}}{d'_\kappa} P_\kappa(x), \tag{12.133}$$

obtained by substituting (12.126) in (12.125), is a more compact way to write the generalized binomial theorem (12.48). A useful corollary of (12.133) follows by making the replacements $x_j \mapsto x_j/r$ and taking the limit $r \to \infty$. Thus, since

$$\lim_{r\to\infty} r^{-|\kappa|}[r]_\kappa^{(\alpha)} = 1, \qquad \lim_{r\to\infty} (1 - x_j/r)^{-r} = e^{x_j}$$

we obtain

$$e^{(x_1+\cdots+x_N)} = \sum_\kappa \frac{\alpha^{|\kappa|}}{d'_\kappa} P_\kappa(x;\alpha). \tag{12.134}$$

Equating terms homogeneous of degree p on both sides shows as a further corollary,

$$(x_1 + \cdots + x_N)^p = p! \sum_{|\kappa|=p} \frac{\alpha^{|\kappa|}}{d'_\kappa} P_\kappa(x;\alpha). \tag{12.135}$$

The result (12.135) together with the expansion formula (12.123) allows a property of P_κ first noted in the work of Stanley [500] to be deduced.

PROPOSITION 12.6.7 *Suppose $|\kappa| \le N$, and denote by $[x_1 \cdots x_k]f(x)$ the coefficient of $x_1 \cdots x_k$ in f. One has*

$$[x_1 \cdots x_{|\kappa|}]P_\kappa(x) = \frac{|\kappa|!}{h_\kappa}.$$

Proof. Since

$$\Pi^{(\alpha)}(x,y) = \prod_{i,j=1}^N \Big(1 + \frac{1}{\alpha} x_i y_j + \mathrm{O}((x_i y_j)^2)\Big)$$

we see that

$$[x_1 \cdots x_k]\Pi^{(\alpha)}(x,y) = \frac{1}{\alpha^k}(y_1 + \cdots + y_N)^k = |\kappa|! \sum_{|\kappa|=k} \frac{1}{d'_\kappa} P_\kappa(y;\alpha),$$

where to obtain the second equality (12.135) has been used. But according to (12.123)

$$[x_1 \cdots x_k]\Pi^{(\alpha)}(x,y) = \sum_\kappa [x_1 \cdots x_k]P_\kappa(x;\alpha)\, \frac{h_\kappa}{d'_\kappa} P_\kappa(y;\alpha).$$

Equating coefficients of $\frac{1}{d'_\kappa} P_\kappa^{(\alpha)}(y)$ in the above two equations gives the stated result. □

12.6.4 Shift operators

The Cherednik operators (11.57) can be used to define operators G_+, G_- with an action on symmetric Jack polynomials which shifts the value of $1/\alpha$ by unity,

$$G_- P_{\kappa+\delta}(z;\alpha) = c_\kappa^- P_\kappa(z;\alpha/(1+\alpha)), \qquad G_+ P_\kappa(z;\alpha/(1+\alpha)) = c_\kappa^+ P_{\kappa+\delta}(z;\alpha), \tag{12.136}$$

for some proportionality constants $c_\kappa^\pm$ [437], [344]. Use of (12.118) shows that equivalently

$$G_- P_{\kappa+\delta}(z;\alpha) = c_\kappa^- \Delta^{-1}(z) S_{\kappa+\delta}(z;\alpha), \qquad G_+(\Delta^{-1}(z) S_{\kappa+\delta}(z;\alpha)) = c_\kappa^+ P_{\kappa+\delta}(z;\alpha),$$

thereby exhibiting that the $G_\pm$ map between the symmetric and antisymmetric Jack polynomials (the latter divided by $\Delta(z)$) of the same index and parameter.

We will see that the operators $G_\pm$ can be constructed out of the operators

$$\tilde{G}_\pm := \prod_{1\le j<k\le N} (\xi_k - \xi_j \pm 1).$$

To see the significance of these operators, we note from the results of Exercises 11.5 q.2(ii) that

$$(\xi_{i+1} - \xi_i \mp 1)s_i = -s_i(\xi_{i+1} - \xi_i \pm 1 - 2s_i), \quad (\xi_j - \xi_{i+1} - 1)s_i = -s_i(\xi_j - \xi_i + 1), \tag{12.137}$$

where the latter equation is valid for $j \neq i, i+1$. It follows immediately that $\tilde{G}_+$ maps antisymmetric functions to symmetric functions, while $\tilde{G}_-$ maps symmetric functions to antisymmetric functions. Also of importance is the action of $\tilde{G}_-$ on antisymmetric functions. In this regard, the relations (12.137) allow us to establish, via an induction in N, that

$$\Big(1 + \prod_{l=1}^{N-1} \prod_{j=1}^{l} s_j\Big)\tilde{G}_- A(z) = 2\tilde{G}_+ A(z) \tag{12.138}$$

for general antisymmetric functions A.

From (12.138) an adjoint type property involving $\tilde{G}_\pm$ and relating inner products (12.12) with $1/\alpha$ shifted by unity can be established.

PROPOSITION 12.6.8 *Write $\langle f|g\rangle^{(C)} = \langle f|g\rangle_\alpha^{(C)}$, and let $\Delta(z)$ be given by (12.107). For f, g symmetric functions*

$$\langle \Delta^{-1}(\bar{z})\tilde{G}_- f|g\rangle^{(C)}_{\alpha/(\alpha+1)} = \langle f|\tilde{G}_+(\Delta(z)g)\rangle^{(C)}_\alpha. \tag{12.139}$$

Proof. From the definition of $\langle f|g\rangle^{(C)}$,

$$\langle \Delta^{-1}(\bar{z})\tilde{G}_- f|g\rangle^{(C)}_{\alpha/(\alpha+1)} = \langle \tilde{G}_- f|\Delta(z)g\rangle^{(C)}_\alpha = \langle f|\tilde{G}_-(\Delta(z)g)\rangle^{(C)}_\alpha, \tag{12.140}$$

where the second equality follows from the fact that $\tilde{G}_-$ is self-adjoint with respect to $\langle\cdot|\cdot\rangle^{(C)}_\alpha$. Changing the names of the integration variables $z_j \mapsto z_{N+1-j}$ $(j = 1, \dots, N)$ in the integral specifying the inner product does not change its value. Doing this to the final inner product in (12.140) and then taking the arithmetic mean with the final inner product unaltered shows

$$\langle f|\tilde{G}_-(\Delta(z)g)\rangle^{(C)}_\alpha = \frac{1}{2}\Big\langle f\Big|\Big(1 + \prod_{l=1}^{N-1}\prod_{j=1}^{l} s_j\Big)\tilde{G}_-(\Delta(z)g)\Big\rangle^{(C)}_\alpha.$$

The sought identity (12.139) now follows upon use of (12.138). □

The property (12.139) allows the operators $G_\pm$ (12.136) to be made explicit.

PROPOSITION 12.6.9 *Choosing*

$$G_+ = \alpha^{-N(N-1)/2}\tilde{G}_+\Delta(z), \qquad G_- = \alpha^{-N(N-1)/2}(\Delta(z))^{-1}\tilde{G}_- \tag{12.141}$$

the actions (12.136) hold with

$$c^\pm_\kappa = a^\pm_\kappa((\alpha+1)/\alpha), \qquad a^\pm_\kappa(\beta) := \prod_{j<k}(\lambda_j - \lambda_k \mp 1 + (k - j \pm 1)\beta).$$

Proof. Only the case of G_- will be considered; similar working establishes the results for G_+.

The fact that

$$\Delta(z)P_\kappa(z;\alpha) = P_{\kappa+\delta}(z;\alpha) + \sum_{\mu<\kappa+\delta} a_{\kappa\mu}P_\mu(z;\alpha)$$

together with $\tilde{G}_+$ being an eigenoperator of $P_{\kappa+\delta}(z;\alpha)$ show that

$$c^+_\kappa = \alpha^{-N(N-1)/2} \prod_{1\le j<k\le N} (\overline{\kappa+\delta}_j - \overline{\kappa+\delta}_k + 1).$$

Recalling that for a partition ρ, $\bar{\rho}_j = \alpha\rho_j - (j-1)$, the stated formula for c^+_κ follows.

To establish that G_- has the sought action, we note that in terms of $G_\pm$ (12.139) reads

$$\langle G_- f | g \rangle^{(\mathrm{C})}_{\alpha/(\alpha+1)} = \langle f | G_+ g \rangle^{(\mathrm{C})}_\alpha .$$

Choose $g = P_\kappa(z; \alpha/(1+\alpha))$ and $f = m_\mu$ with $\mu < \kappa + \delta$. Noting that then

$$G_- f = P_{\mu-\delta}(z; \alpha/(\alpha+1)) + \sum_{\rho < \mu - \delta} a_{\mu\rho} P_\rho(z; \alpha/(\alpha+1))$$

for $\mu - \delta$ a partition, $G_- f = 0$ otherwise, it follows that $\langle m_\mu | G_+ g \rangle^{(\mathrm{C})}_\alpha = 0$ for all $\mu < \kappa + \delta$, and hence $G_+ g$ is proportional to $P_{\kappa+\delta}(z; \alpha)$.

□

The actions (12.136) together with the adjoint type property (12.141) can be used to deduce a formula for the normalization $\langle P_\kappa | P_\kappa \rangle^{(\mathrm{C})}_\alpha$ in the case that $1/\alpha$ is a positive integer. Thus we see that

$$\langle P_\kappa(\cdot, 1+1/\alpha) | P_\kappa(\cdot, 1+1/\alpha) \rangle^{(\mathrm{C})}_{1/(1+1/\alpha)} = \frac{a^-_\kappa(1+1/\alpha)}{a^+_\kappa(1+1/\alpha)} \langle P_{\kappa+\delta}(\cdot, 1/\alpha) | P_{\kappa+\delta}(\cdot, 1/\alpha) \rangle^{(\mathrm{C})}_\alpha$$

which upon iteration gives

$$\langle P_\kappa(\cdot, k+1/\alpha) | P_\kappa(\cdot, k+1/\alpha) \rangle^{(C)}_{1/(k+1/\alpha)} = \prod_{j=0}^{k-1} \frac{a^+_{\kappa+jk}((k-j)+1/\alpha)}{a^-_{\kappa+jk}((k-j)+1/\alpha)} \langle P_{\kappa+k\delta}(\cdot, 1/\alpha) | P_{\kappa+k\delta}(\cdot, 1/\alpha) \rangle^{(C)}_\alpha .$$

Now set $1/\alpha = 0$. Since $P_\mu(z; 0) = m_\mu$, we see that the inner product on the r.h.s. is equal to $L^N N!$ and thus an explicit formula for $\langle P_\kappa(\cdot, k) | P_\kappa(\cdot, k) \rangle^{(\mathrm{C})}_k$ is obtained. This can be checked to agree with the formula in Proposition 12.6.3 upon making use of (12.59) and (12.60).

12.6.5 Integration formulas

By summing (12.56) over η, $\eta^+ = \kappa$, according to (12.96) we obtain the trigonometric integral [192]

$$\frac{1}{M_N(a,b;1/\alpha)} \int_{-1/2}^{1/2} dx_1 \cdots \int_{-1/2}^{1/2} dx_N \prod_{l=1}^N e^{\pi i x_l (a-b)} |1 + e^{2\pi i x_l}|^{a+b} P_\kappa(-e^{2\pi i x_1}, \dots, -e^{2\pi i x_N})$$
$$\times \prod_{1 \le j < k \le N} |e^{2\pi i x_k} - e^{2\pi i x_j}|^{2/\alpha} = P_\kappa((1)^N) \frac{[-b]^{(\alpha)}_\kappa}{[1 + a + (N-1)/\alpha]^{(\alpha)}_\kappa}. \tag{12.142}$$

Similarly, we obtain from (12.57) the generalized Selberg integral [386], [338], [342]

$$\frac{1}{S_N(\lambda_1, \lambda_2; 1/\alpha)} \int_0^1 dt_1 \cdots \int_0^1 dt_N \prod_{l=1}^N t_l^{\lambda_1} (1-t_l)^{\lambda_2} P_\kappa(t_1, \dots, t_N) \prod_{j<k} |t_j - t_k|^{2/\alpha}$$
$$= P_\kappa((1)^N) \frac{[\lambda_1 + (N-1)/\alpha + 1]^{(\alpha)}_\kappa}{[\lambda_1 + \lambda_2 + 2(N-1)/\alpha + 2]^{(\alpha)}_\kappa}. \tag{12.143}$$

We recall that M_N and S_N are given explicitly in terms of gamma functions by (4.4) and (4.3) respectively. On the other hand, substituting for P_κ using (12.100) shows that (12.142) implies (12.56) and (12.143) implies (12.57), so the nonsymmetric and symmetric versions are equivalent. A corollary of (12.142), which is of particular use in the calculation of dynamical correlation functions, can be deduced as a consequence of the following result.

PROPOSITION 12.6.10 *Let $f(z_1,\dots,z_N)$ be symmetric in $z_1,\dots,z_N$ ($z_j := e^{2\pi i x_j}$), periodic in x_j of period 1 and homogeneous of integer order k ($k \neq 0$). Let $u_\epsilon(x_l)$ have the small ϵ expansion*

$$u_\epsilon(x_l) = 1 + \epsilon a(x_l) + \mathrm{o}(\epsilon).$$

Then we have

$$\begin{aligned}\lim_{\epsilon\to 0}\frac{1}{\epsilon}\int_{-1/2}^{1/2} dx_1 \cdots \int_{-1/2}^{1/2} dx_N \prod_{l=1}^N u_\epsilon(x_l) f(z_1,\dots,z_N) \\ = N\Big(\int_{-1/2}^{1/2} z_1^k a(x_1)dx_1\Big)\int_0^1 dx_2 \cdots \int_0^1 dx_N\, f(1,z_2,\dots,z_N).\end{aligned}$$

Proof. For small ϵ,

$$\prod_{l=1}^N u_\epsilon(x_l) \sim 1 + \epsilon \sum_{l=1}^N a(x_l)$$

and thus, since f is assumed homogeneous of nonzero integer order,

$$\int_{-1/2}^{1/2} dx_1 \cdots \int_{-1/2}^{1/2} dx_N \prod_{l=1}^N u_\epsilon(x_l) f(z_1,\dots,z_N) \sim \epsilon \int_{-1/2}^{1/2} dx_1 \cdots \int_{-1/2}^{1/2} dx_N \sum_{l=1}^N a(x_l) f(z_1,\dots,z_N).$$

The stated result now follows by using the assumption that f is symmetric to replace $\sum_{l=1}^N a(x_l)$ in the integrand by $Na(x_1)$, then using the assumption that f is periodic to replace z_j by $z_1 z_j$ ($j = 2,\dots,N$) and finally the fact that f is homogeneous of order k to write

$$f(z_1, z_1 z_2,\dots,z_1 z_N) = z_1^k f(1,z_2,\dots,z_N).$$

□

PROPOSITION 12.6.11 *We have*

$$\begin{aligned}N\int_0^1 dx_2 \cdots \int_0^1 dx_N \prod_{l=2}^N |1 - e^{2\pi i x_l}|^{2/\alpha} P_\kappa(1, e^{2\pi i x_2},\dots,e^{2\pi i x_N}) \\ \times \prod_{2\le j<k\le N} |e^{2\pi i x_k} - e^{2\pi i x_j}|^{2/\alpha} = P_\kappa((1)^N)|\kappa|(\kappa_1 - 1)!\frac{(N/\alpha)!}{(1/\alpha)!^N}\frac{[0]'^{(\alpha)}_\kappa}{[1+(N-1)/\alpha]^{(\alpha)}_\kappa},\end{aligned} \tag{12.144}$$

where the dash on $[0]'^{(\alpha)}_\kappa$ means that the $j = 1$ term in its definition (12.46) is to be omitted.

Proof. This is deduced from (12.142) by setting $a = -b = \epsilon$ and taking the limit $\epsilon \to 0$. To do this on the l.h.s. we use Proposition 12.6.10 with $u_\epsilon(x) = z^\epsilon$ and

$$f(z_1,\dots,z_N) = P_\kappa(-z_1,\dots,-z_N) \prod_{1\le j<k\le N} |z_k - z_j|^{2/\alpha},$$

and use the fact that

$$\int_{-1/2}^{1/2} e^{2\pi i x|\kappa|} x\, dx = \frac{(-1)^{|\kappa|}}{2\pi i|\kappa|}.$$

On the r.h.s. this limit is taken directly, using the formula

$$\lim_{\epsilon\to 0}\frac{1}{\epsilon}[\epsilon]^{(\alpha)}_\kappa = (\kappa_1 - 1)![0]'^{(\alpha)}_\kappa.$$

□

We remark that it follows from Proposition 12.6.11 that one has the expansion [287]

$$p_k := z_1^k + \cdots + z_N^k = \sum_{|\kappa|=k} \frac{\beta_\kappa}{\langle P_\kappa | P_\kappa \rangle^{(\mathrm{C})}} P_\kappa(z_1, \ldots, z_N) \tag{12.145}$$

where β_κ is equal to the r.h.s. of (12.144) (see Exercises 12.6 q.7).

Another class of integration formulas involving Jack polynomials generalize the vanishing implied by (10.137) in the limit $\beta \to 0$, (10.155) and (10.159) in the limit $\alpha \to 0$, and the counterparts of these latter two integrals relating to $O^\pm(2l+1)$. Recalling the eigenvalue p.d.f.'s for $\mathrm{Sp}(2l)$ and $O^\pm(l)$ from Section 2.6, these vanishings can be seen to be the case $\alpha = 1$ of the following integrals [460], which we state without proof.

PROPOSITION 12.6.12 *Let*

$$\Delta_\pm := \prod_{1 \le j < k \le l} (e^{i\theta_k} - e^{i\theta_j})(1 - e^{i\theta_j} e^{i\theta_k}),$$

write $(z^\pm, w^\pm, \ldots)$ *as an abbreviation for* $(z, z^{-1}, w, w^{-1}, \ldots)$, *and denote by* $\ell(\kappa)$ *the number of nonzero parts of* κ. *We have the generalized* $\mathrm{Sp}(2l)$ *vanishing integral*

$$\int_0^\pi d\theta_1 \cdots \int_0^\pi d\theta_l \, P_\kappa(e^{\pm i\theta_1}, \ldots, e^{\pm i\theta_l}) \prod_{j=1}^l |1 - e^{i\theta_j}|^{2/\alpha} |1 + e^{i\theta_j}|^{2/\alpha} |\Delta_\pm|^{2/\alpha} = 0$$

unless $\ell(\kappa) \le 2l$ *and each* $\kappa_1 = \kappa_2$, $\kappa_3 = \kappa_4$, $\ldots$*; the generalized* $O^+(2l)$ *vanishing integral*

$$\int_0^\pi d\theta_1 \cdots \int_0^\pi d\theta_l \, P_\kappa(e^{\pm i\theta_1}, \ldots, e^{\pm i\theta_l}) \prod_{j=1}^l |1 - e^{i\theta_j}|^{1/\alpha - 1} |1 + e^{i\theta_j}|^{1/\alpha - 1} |\Delta_\pm|^{2/\alpha} = 0$$

unless $\ell(\kappa) \le 2l$ *and each* κ_i *has the same parity; the generalized* $O^\mp(2l+1)$ *vanishing integral*

$$\int_0^\pi d\theta_1 \cdots \int_0^\pi d\theta_l \, P_\kappa(e^{\pm i\theta_1}, \ldots, e^{\pm i\theta_l}, 1) \prod_{j=1}^l |1 - e^{i\theta_j}|^{3/\alpha - 1} |1 + e^{i\theta_j}|^{1/\alpha - 1} |\Delta_\pm|^{2/\alpha} = 0$$

unless $\ell(\kappa) \le 2l + 1$ *and each* κ_i *has the same parity; the generalized* $O^-(2l)$ *vanishing integral*

$$\int_0^\pi d\theta_1 \cdots \int_0^\pi d\theta_{l-1} \, P_\kappa(e^{\pm i\theta_1}, \ldots, e^{\pm i\theta_{l-1}}, 1, -1) \prod_{j=1}^{l-1} |1 - e^{i\theta_j}|^{3/\alpha - 1} |1 + e^{i\theta_j}|^{3/\alpha - 1} |\Delta_\pm|^{2/\alpha} = 0$$

unless $\ell(\kappa) \le 2l$ *and each* κ_i *has the same parity. Moreover, with* $\kappa_{2i-1} = \mu_i$ $(i = 1, \ldots, l)$ *the normalized value of non-zero cases of the first integral equals*

$$\prod_{(i,j) \in \mu} \frac{(j - 1 + (2l + 2 - 2i)/\alpha)(\mu_i - j + 1 + (2\mu_j' - 2i + 1)/\alpha)}{(j + (2l + 1 - 2i)/\alpha)(\mu_i - j + (2\mu_j' - 2i + 2)/\alpha)},$$

while with $\kappa_i = 2\mu_i$ $(i = 1, \ldots, 2l)$ *the arithmetic mean of the normalized value of second and fourth integral in the nonzero cases equals*

$$\prod_{(i,j) \in \mu} \frac{(2j - 2 + (2l + 1 - i)/\alpha)(2\mu_i - 2j + 1 + (\mu_j' - i)/\alpha)}{(2j - 1 + (2l - i)/\alpha)(2\mu_i - 2j + (\mu_j' - i + 1)/\alpha)}.$$

EXERCISES 12.6 1. [40] The objective of this exercise is to verify the formula (12.106).

(i) Let $\ell(\kappa)$ be as in Proposition 12.6.12. Check that $f_0! = \prod_{j=\ell(\kappa)+1}^N (N - j + 1)$.

(ii) Show from the definitions that the nodes at the end of each row of the diagram of κ contribute $\prod_{i=1}^{\kappa_1} f_i!$ to h_κ, while the nodes at the beginning of each row of κ^R contribute $\prod_{j=1}^{\ell(\kappa)}(\alpha\kappa_j + N - j + 1)$ to d_{κ^R}.

(iii) Check that the node $(i', j') \in \kappa$ (not at the end of a row) in the ith row of the jth block (the jth block contains parts equal to j) has the same leg length as the node $(i'', j'') \in \kappa^R$ in the ith row of the jth block, shifted one column to the right. As these nodes have arm lengths differing by one, conclude they give a contribution to h_κ and d_{κ^R}, respectively, which is equal.

2. (i) For ρ with repeated parts, use the structure (12.7) to deduce that the coefficient of $\Delta(z)m_\kappa(z)$ in Asym E_ρ, $\kappa = \rho - \delta$ must vanish, and thus Asym E_ρ is proportional to $S_\mu^{(\alpha)}(z)$ with $\mu < \kappa$. Note that this is inconsistent with (12.111) unless the proportionality constant is zero, and thus Asym $E_\rho = 0$.

(ii) Substitute the r.h.s. of (12.123) with $\alpha \mapsto \alpha/(\alpha+1)$ in (12.129), rewrite $\Delta(y)P_\kappa(y;\alpha/(\alpha+1))$ therein using (12.118), and equate coefficients of $P_{\eta^+}(x;\alpha/(\alpha+1))E_\rho(y;\alpha)$ on both sides to deduce that

$$c_\rho = \varepsilon(\sigma)\frac{d'_\rho(\alpha)}{d_{\rho^+}(\alpha)}\frac{h_{\eta^+}(\alpha/(\alpha+1))}{d'_{\eta^+}(\alpha/(\alpha+1))}.$$

Remark: This can be further simplified (see [40]) to reclaim the formula (12.113).

3. [376] The power sum symmetric functions are given by

$$p_\kappa = p_{\kappa_1}p_{\kappa_2}\cdots p_{\kappa_N} = p_1^{f_1}p_2^{f_2}\cdots p_{\kappa_1}^{f_{\kappa_1}}, \qquad p_j = p_j(x) = \sum_{m=1}^N x_m^j \tag{12.146}$$

(f_j is used in the same sense as in (12.102)). They form a basis for symmetric functions in an infinite number of variables, i.e. in the limit $N \to \infty$ (for fixed N, $\{p_\kappa\}$ are not linearly independent).

(i) Show from the definition of p_j in an infinite number of variables $x = \{x_i\}_{i=1,2,\dots}$ that

$$\sum_{j=1}^\infty p_j(x)t^{j-1} = \sum_{k=1}^\infty \frac{d}{dt}\log\frac{1}{1-x_kt},$$

and deduce from this that

$$\prod_{k=1}^\infty \frac{1}{(1-x_kt)^{1/\alpha}} = \exp\Big(\sum_{j=1}^\infty \frac{p_j(x)t^j}{\alpha j}\Big).$$

(ii) Replace t by y_l in the above identity and take the product over l to conclude

$$\prod_{k,l=1}^\infty \frac{1}{(1-x_ky_l)^{1/\alpha}} = \sum_\kappa \frac{\alpha^{-\ell(\kappa)}}{z_\kappa}p_\kappa(x)p_\kappa(y), \qquad z_\kappa := \prod_{l=1}^{\kappa_1} l^{f_l} f_l!.$$

(iii) Use the result of (ii) together with Proposition 12.1.8 to deduce

$$\langle p_\kappa, p_\sigma\rangle_g = \alpha^{\ell(\kappa)}z_\kappa\delta_{\kappa,\sigma},$$

where $\langle\cdot,\cdot\rangle_g$ is specified by (12.130).

4. [376], [500] This exercise relates to expansion formulas for $P_\kappa(x)$.

(i) Check from the definition (12.34) that $e(\mu;\alpha) < e(\kappa;\alpha)$ for $\mu < \kappa$.

(ii) Use (12.94) together with (12.4) to show that

$$\tilde H^{(C)}m_\kappa = e(\kappa;\alpha)m_\kappa + \frac{1}{\alpha}\sum_{1\le j<k\le N}(\kappa_j - \kappa_k)\sum_{r=1}^{[(\kappa_j-\kappa_k)/2]} m_{\kappa(j,k,r)},$$

where $\kappa(j,k,r)$ is the partition obtained by rearranging the composition

$$(\kappa_1,\dots,\kappa_j - r,\dots,\kappa_k + r,\dots,\kappa_N).$$

(iii) Use the results of (i) and (ii) to show that $\tilde{H}^{(\mathrm{C})}$ has unique eigenfunctions of the form (12.97), and thus the P_κ can be characterized uniquely as the eigenfunctions of $\tilde{H}^{(\mathrm{C})}$ which are of the form (12.97).

(iv) Use these results to show

$$P_{21^k0^{N-k-1}}(z) = m_{21^k0^{N-k-1}}(z) + \frac{(k+1)(k+2)}{k+1+\alpha} m_{1^{k+2}0^{N-k-2}}(z). \qquad (12.147)$$

(v) Suppose $N > \ell(\kappa)$. Introduce the notation

$$\kappa^* := (\kappa_2,\dots,\kappa_N,0), \quad x^* := (x_2,\dots,x_N),$$

and write

$$\tilde{H}^{(\mathrm{C})} m_\kappa(x) = \sum_{\mu\le\kappa} b_{\kappa\mu} m_\mu(x), \quad \tilde{H}^{(\mathrm{C})*} m_{\kappa^*}(x^*) = \sum_{\mu^*\le\kappa^*} b_{\kappa^*\mu^*} m_{\mu^*}(x^*). \qquad (12.148)$$

Use the result of (ii) to show that

$$b_{\kappa\mu} = b_{\kappa^*\mu^*} \quad \text{for} \quad \mu < \kappa, \qquad b_{\kappa\kappa} - b_{\mu\mu} = b_{\kappa^*\kappa^*} - b_{\mu^*\mu^*}.$$

From this deduce that if

$$P_\kappa(x) = \sum_{\mu\le\kappa} a_{\kappa\mu} m_\mu \quad \text{then} \quad P_{\kappa^*}(x^*) = \sum_{\mu^*\le\kappa^*} a_{\kappa^*\mu^*} m_{\mu^*}.$$

(vi) Use (v) to show

$$[x_1^{\kappa_1}]\, P_\kappa(x) = P_{\kappa^*}(x^*), \qquad (12.149)$$

where the notation $[x_1^{\kappa_1}]$ denotes the coefficient of $x_1^{\kappa_1}$.

(vii) Let $<^R$ denote the reverse lexicographical ordering $\lambda < \kappa$ iff $\lambda_i = \kappa_i$ $(i = 1,\dots,j-1)$ and $\lambda_j < \kappa_j$ for some j. Note that this total ordering is compatible with the dominance ordering (12.1). Let $\kappa^{(1)} <^R \kappa^{(2)} <^R \cdots <^R \kappa^{(n)} = \kappa$ denote the order of all partitions less than κ with respect to $<^R$, and having equal modulus. Define $b_{\kappa\mu}$ as in (12.148), with $b_{\kappa\kappa} := e(\kappa)$, and $\mathcal{E}_\lambda := \prod_{i=1}^{n-1}(e(\lambda) - e(\lambda^{(i)}))$. A result of [370] gives the Hessenberg determinant formula

$$P_\lambda(x) = \frac{1}{\mathcal{E}_\lambda} \det \begin{bmatrix} m_{\lambda^{(1)}} & e(\lambda^{(1)}) - e(\lambda) & 0 & 0 & \cdots & 0 \\ m_{\lambda^{(2)}} & b_{\lambda^{(2)}\lambda^{(1)}} & e(\lambda^{(2)}) - e(\lambda) & 0 & \cdots & 0 \\ \vdots & \vdots & & \ddots & & \vdots \\ m_{\lambda^{(n-1)}} & b_{\lambda^{(n-1)}\lambda^{(1)}} & b_{\lambda^{(n-1)}\lambda^{(2)}} & & \cdots & e(\lambda^{(n-1)}) - e(\lambda) \\ m_\lambda & b_{\lambda\lambda^{(1)}} & b_{\lambda\lambda^{(2)}} & & \cdots & b_{\lambda\lambda^{(n-1)}} \end{bmatrix}. \qquad (12.150)$$

Use this to rederive (12.147).

5. [437] Here a factorization property of $P_{\kappa+2\alpha\delta}(z;1/(-\alpha+1/2))$, $\alpha \in \mathbb{Z}^+$ is deduced.

(i) Recalling the notation (12.34) and (12.107), start from the eigenvalue equation

$$H^{(C)}\Big(|\Delta(z)|^\alpha P_\kappa(z;1/\alpha)\Big) = \Big(E_0^{(C)} + \Big(\frac{2\pi}{L}\Big)^2 e(\kappa;\alpha)\Big)|\Delta(z)|^\alpha P_\kappa(z;1/\alpha),$$

then replace α by $1-\alpha$ and make use of the fact that $H^{(C)}|_{\alpha\mapsto 1-\alpha} = H^{(C)}$ and (12.94) to deduce

$$\tilde{H}^{(C)}\Big(|\Delta(z)|^{1-2\alpha} P_\kappa(z;1/(1-\alpha))\Big) = e(\kappa;1-\alpha)|\Delta(z)|^{1-2\alpha} P_\kappa(z;1/(1-\alpha)).$$

(ii) Next replace α by $-\alpha+\frac{1}{2}$, κ by $\kappa+\alpha(N-1)$, and use the fact that $z^p P_\kappa(z) = P_{\kappa+p}(z)$ (recall Exercises 12.1 q.1) to show that for α a non-negative integer

$$\tilde{H}^{(C)}\Big|_{\alpha\mapsto-\alpha+1/2}\Big((\Delta(z))^{2\alpha}P_\kappa(z;1/(\alpha+1/2))\Big) = e\Big(\kappa+\alpha(N-1);\alpha+\frac{1}{2}\Big)(\Delta(z))^{2\alpha}P_\kappa(z;1/(\alpha+1/2)).$$

With δ as in (12.108), deduce from this that

$$(\Delta(z))^{2\alpha}P_\kappa^{(1/(\alpha+1/2))}(z) = P_{\kappa+2\alpha\delta}^{(1/(-\alpha+1/2))}(z). \tag{12.151}$$

(iii) Let $\alpha = -(k+1)/(r-1)$ with $k+1, r-1$ positive, relatively prime integers. For κ a partition of N parts and $s \in \mathbb{Z}^+$, let $N-\ell(\kappa) = (k+1)s-1 =: n_0$ and let the smallest nonzero part of κ equal $(r-1)s+1$. Specifically, κ is specified in terms of its frequencies $[f_0 f_1 f_2 \cdots]$ according to

$$\kappa := \kappa(k,r,s) = [n_0 0^{s(r-1)} k 0^{r-1} k 0^{r-1} k \cdots].$$

A conjecture of [61] (consistent in the case $s=1$ with properties of Jack polynomials with negative, rational α known from [171]) gives

$$P_{\kappa(k,r,s)}(z_1,\dots,z_N)\Big|_{z_1=\cdots=z_{(k+1)s-1}=z} = \prod_{i=s(k+1)}^{N}(z-z_i)^{(r-1)s+1}P_{\kappa(k,r,1)}(z_{s(k+1)},z_{s(k+1)+1},\dots,z_N).$$

Show that with $k=s=1$ this is equivalent to the $\kappa=0^N$ case of (12.151).

6. [550] The objective of this exercise is to derive the formula

$$P_{\kappa_1\kappa_2}(z_1,z_2;2/\alpha) = \frac{2^{\kappa_1-\kappa_2}}{a_{\kappa_1-\kappa_2}}(z_1z_2)^{(\kappa_1+\kappa_2)/2}P_{\kappa_1-\kappa_2}^{(\gamma,\gamma)}\Big(\frac{1}{2}(z_1+z_2)(z_1z_2)^{-1/2}\Big), \tag{12.152}$$

where $P_n^{(\alpha,\beta)}(x)$ denotes the Jacobi polynomial (5.47), $\gamma := (\alpha-1)/2$ and

$$a_n := \binom{n+\gamma}{n}\frac{(n+2\gamma+1)_n}{(\gamma+1)_n}2^{-n}.$$

(i) Let $\epsilon_{\kappa_1,\kappa_2} = \kappa_1^2+\kappa_2^2-\alpha\kappa_2$ and

$$L := \Big(z_1\frac{\partial}{\partial z_1}\Big)^2 + \Big(z_2\frac{\partial}{\partial z_2}\Big)^2 + \alpha\frac{z_1z_2}{z_1-z_2}\Big(\frac{\partial}{\partial z_1}-\frac{\partial}{\partial z_2}\Big).$$

Use the result of q.4 (iii) above to show that

$$(L-\epsilon_{\kappa_1,\kappa_2})P_{\kappa_1\kappa_2}(z_1,z_2;2/\alpha) = 0.$$

(ii) Show that with the change of variables $u = \frac{1}{2}(z_1+z_2)(z_1z_2)^{-1/2}$, $v = (z_1z_2)^{1/2}$, the operator L reads

$$L = \frac{1}{2}\Big((u^2-1)\frac{\partial^2}{\partial u^2} + v^2\frac{\partial^2}{\partial v^2} + u\frac{\partial}{\partial u} + v\frac{\partial}{\partial v}\Big) + \frac{\alpha}{2}\Big(u\frac{\partial}{\partial u} - v\frac{\partial}{\partial v}\Big).$$

(iii) With the ansatz $P_{\kappa_1\kappa_2}(z_1,z_2;2/\alpha) = v^{\kappa_1+\kappa_2}f(u)$ in the equation of (i), use the result of (ii) to show that f satisfies the equation

$$\Big((u^2-1)\frac{\partial^2}{\partial u^2} + (1+\alpha)u\frac{\partial}{\partial u} - (\kappa_1-\kappa_2)(\kappa_1-\kappa_2-\alpha)\Big)f(u) = 0.$$

Since this is the differential equation satisfied by the Jacobi polynomial $P_{\kappa_1-\kappa_2}^{(\gamma,\gamma)}(u)$, $\gamma = (\alpha-1)/2$, conclude that (12.152) is correct up to the proportionality constant. Use the expansion (5.47) to choose the proportionality constant so the coefficient of $z_1^{\kappa_1}z_2^{\kappa_2}$ on the r.h.s. of (12.152) is unity.

(iv) Note from (12.118) and (12.152) that in the case $N=2$ the antisymmetric Jack polynomial has the explicit form

$$S_{\kappa_1\kappa_2}(z_1,z_2;2/\alpha) = (z_1-z_2)\frac{2^{\kappa_1-\kappa_2}}{a_{\kappa_1-\kappa_2}}(z_1z_2)^{(\kappa_1+\kappa_2)/2}P^{(\gamma',\gamma')}_{\kappa_1-\kappa_2}\Big(\frac{1}{2}(z_1+z_2)(z_1z_2)^{-1/2}\Big), \quad \gamma' := \frac{\alpha+1}{2},$$

where a_n is defined as in (12.152) except that γ is replaced by γ'.

7. In this exercise the expansion (12.145) will be derived.

(i) Note from the orthonormality and completeness of $\{P_\kappa\}$ that

$$\beta_\kappa = \int_{-1/2}^{1/2} dx_1 \cdots \int_{-1/2}^{1/2} dx_N\, (z_1^{-k}+\cdots+z_N^{-k})P_\kappa(z_1,\dots,z_N)\prod_{j<k}^{N}|z_k-z_j|^{2/\alpha}.$$

(ii) Use the symmetry of the integrand to replace $\sum_{p=1}^N z_p^{-k}$ by Nz_1^{-k}. Now change variables $z_j \mapsto z_1z_j$ $(j=2,\dots,N)$ to deduce that β_k is equal to the l.h.s. of (12.144).

8. Apply the limiting procedure used to deduce Proposition 4.7.3 to show from (12.143) that

$$\frac{1}{\tilde W_{\lambda_1,\alpha,N}}\int_0^\infty dt_1\cdots\int_0^\infty dt_N \prod_{l=1}^N t_l^{\lambda_1}e^{-t_l}P_\kappa(t_1,\dots,t_N)\prod_{j<k}|t_k-t_j|^{2/\alpha} = P_\kappa((1)^N)[\lambda_1+(N-1)/\alpha+1]_\kappa^{(\alpha)}, \tag{12.153}$$

where

$$\tilde W_{\lambda_1,\alpha,N} = \int_0^\infty dt_1\cdots\int_0^\infty dt_N \prod_{l=1}^N t_l^{\lambda_1}e^{-t_l}\prod_{j<k}|t_k-t_j|^{2/\alpha}.$$

12.7 INTERPOLATION SYMMETRIC JACK POLYNOMIALS

With a_η as in (12.100), let us introduce the interpolation symmetric Jack polynomials $P^*_\kappa(x)$ according to

$$\mathrm{Sym}\, E^*_\eta(x) = a_\eta P^*_\kappa(x), \qquad \kappa = \eta^+. \tag{12.154}$$

Suppose η itself is a partition, and let ν be a composition such that $\eta \not\leq \nu^+$ or equivalently $\eta \not\subseteq \nu^+$. It follows from Definition 12.5.6 that then $\eta \not\leq \rho(\nu^+)$ for any permutation ρ, and the extra vanishing condition Proposition 12.5.8 gives $E^*_\eta(\rho(\nu^+)/\alpha)=0$. Thus with $\nu^+=\mu$ we have from (12.154) that

$$P^*_\kappa(\bar\mu/\alpha) = 0 \qquad \text{for } \kappa \not\subseteq \mu \tag{12.155}$$

and in particular

$$P^*_\kappa(\bar\mu/\alpha) = 0 \qquad \text{for } |\mu|\le|\kappa|,\ (\mu\ne\kappa). \tag{12.156}$$

The proof of Proposition 12.5.1 shows there is a unique (up to normalization) polynomial with the property (12.156) together with the requirement that $P^*_\kappa(\bar\kappa/\alpha)\ne 0$, and this can be used to characterize the interpolation symmetric Jack polynomials, with (12.155) an extra vanishing condition.

It follows from (12.79) that

$$P^*_\kappa(x) = P_\kappa(x) + \sum_{|\mu|<|\kappa|} c_{\kappa\mu}P_\mu(x) \tag{12.157}$$

for some coefficients $\{c_{\kappa\mu}\}$. Also, proceeding as in the proof of Proposition 12.6.1, making use of Proposition 12.5.5 rather than Proposition 12.2.1, we see that analogous to (12.96)

$$P^*_\kappa(x) = d'_\kappa \sum_{\eta:\eta^+=\kappa}\frac{1}{d'_\eta}E^*_\eta(x).$$

From this and the vanishing property (12.65) we see that

$$P^*_\kappa(\bar{\kappa}/\alpha) = E^*_\kappa(\bar{\kappa}/\alpha) = \alpha^{-|\kappa|} d'_\kappa, \tag{12.158}$$

where the final equality follows from (12.72).

Next we turn our attention to characterizing $\{P^*_\kappa(x)\}$ as eigenfunctions and deriving a special form of the eigenoperator. Regarding the former, because of the algebra isomorphism of Section 12.5.2, the working leading to the result of Exercises 11.5 q.3(iii) allows us to conclude

$$[s_i, f(\hat{\Xi}_1, \dots, \hat{\Xi}_N)] = 0.$$

Consequently, as in the derivation of (12.99) we conclude

$$\prod_{j=1}^N (1 + u\hat{\Xi}_j) P^*_\kappa(x) = \prod_{j=1}^N (1 + u(\kappa_j - (N-j)/\alpha)) P^*_\kappa(x). \tag{12.159}$$

Next we seek a special form of this eigenoperator, which has significance in the development of the theory of the symmetric Jack polynomials. This is the determinant formula [359]

$$\prod_{j=1}^N (1 + u\hat{\Xi}_j)\Big|_{\substack{\text{symmetric} \\ \text{functions}}} = \frac{1}{\Delta(x)} \det \Big[(x_i u + 1)(x_i + 1/\alpha)^{N-j} - u x_i^{N-j+1} \Delta_i \Big]_{i,j=1,\dots,N}, \tag{12.160}$$

where $\Delta(x)$ is specified by (12.107) and

$$\Delta_i f(x_1, \dots, x_N) = f(x_1, \dots, x_{i-1}, x_i - 1, x_{i+1}, \dots, x_N) =: f(x - \varepsilon_i). \tag{12.161}$$

To establish this formula, we first show that the r.h.s. of (12.160) maps $\mathrm{Sym}(x^\kappa)$ for κ a partition to a polynomial of degree less than or equal to $|\kappa|$.

Proposition 12.7.1 *Let*

$$\Delta_{ij} := (x_i u + 1)(x_i + 1/\alpha)^{N-j} - u x_i^{N-j+1} \Delta_i.$$

With s_κ denoting the Schur polynomial, we have

$$\frac{1}{\Delta(x)} \det[\Delta_{ij}]_{i,j=1,\dots,N} \mathrm{Sym}(x^\kappa)$$
$$= \prod_{j=1}^N (1 + u(\kappa_j - (N-j)/\alpha)) s_\kappa(x_1, \dots, x_N) + \sum_{\substack{|\mu| \le |\kappa| \\ \mu \ne \kappa}} c_{\kappa\mu} s_\mu(x_1, \dots, x_N). \tag{12.162}$$

Proof. Making use of (10.16) shows (12.162) can be rewritten

$$\det[\Delta_{ij}]_{i,j=1,\dots,N} \mathrm{Sym}(x^\kappa) = \prod_{j=1}^N (1 + u(\kappa_j - (N-j)/\alpha)) \mathrm{Asym}(x^{\kappa+\delta_N}) + \sum_{\substack{|\mu| \le |\kappa| \\ \mu \ne \kappa}} c_{\kappa\mu} \mathrm{Asym}(x^{\mu+\delta_N}), \tag{12.163}$$

where $\delta =: \delta_N$ is specified by (12.108). Since in general an operator of the form $\det[A_j[x_i]]_{i,j=1,\dots,N}$ where $A_j[x_i]$ acts only on the x_i variable maps symmetric functions to antisymmetric functions, it suffices to check that the action of $\det[\Delta_{ij}]_{i,j=1,\dots,N}$ on the monomial x^η ($\eta^+ = \kappa$) is consistent with (12.163). For this task, expanding $(x_i + 1/\alpha)^{N-j}$ by the binomial theorem shows

$$\Delta_{ij} = ((N-j)u/\alpha + 1) x_i^{N-j} + u x_i^{N-j+1}(1 - \Delta_i) + \mathrm{O}(x_i^{N-j-1}),$$

and thus

$$\Delta_{ij} x_i^{\eta_i} = (1 + u(\eta_i + (N-j)/\alpha)) x_i^{\eta_i + N - j} + \mathrm{O}(x_i^{\eta_i + N - j - 1}). \tag{12.164}$$

Taking the determinant indeed gives a structure consistent with (12.163). □

With the property (12.162) established, we verify (12.160) by checking that the r.h.s. is an eigenoperator for the $\{P^*_\kappa\}$. For this, let $I \subseteq \{1, \dots, N\}$, and with ε_i as in (12.161) put $\varepsilon_I := \sum_{i\in I} \varepsilon_i$, $\Delta_I f := (\prod_{i\in I} \Delta_i) f = f(x - \varepsilon_I)$. Also, introduce the functions $\phi_I(x) := \det[c^I_{ij}(x)]_{i,j=1,\dots,N}$ where

$$c^I_{ij}(x) = \begin{cases} x_i^{N-j+1} & \text{for } i \in I, \\ (x_i + 1/\alpha)^{N-j} & \text{for } i \notin I. \end{cases} \tag{12.165}$$

PROPOSITION 12.7.2 *Let μ be a partition and δ_N be as in (12.144). If $\mu - \varepsilon_I$ is not a partition then $\phi_I(\mu + \delta_N/\alpha) = 0$.*

Proof. Consider first the case that $\mu_N = 0$ and $N \in I$. Then with $x = \mu + \delta_N/\alpha$, $x_N = 0$ and thus the final row of $\phi_I(\mu + \delta_N/\alpha)$ vanishes. The only other instance for which $\mu - \varepsilon_I$ is not a partition is when there is an $i < N$ such that $i \in I$, $i + 1 \notin I$ and $\mu_i = \mu_{i+1}$. Then with $x = \mu + \delta_N/\alpha$, $x_i = x_{i+1} + 1/\alpha$ and according to the definition (12.165), $[c^I_{ij}(\mu + \delta_N/\alpha]$ has two rows the same. □

PROPOSITION 12.7.3 *For each partition κ,*

$$\frac{1}{\Delta(x)} \det[\Delta_{ij}]_{i,j=1,\dots,N} P^*_\kappa(x) = \prod_{j=1}^N (1 + u(\kappa_j - (N-j)/\alpha)) P^*_\kappa(x).$$

Proof. The value of the eigenvalue follows from (12.162). To check that $P^*_\kappa(x)$ is the corresponding eigenfunction we use the characterization (12.156). Put $D := \det[\Delta_{ij}]_{i,j=1,\dots,N}$ and expand

$$D = \sum_I d_I \Delta_I \tag{12.166}$$

with

$$d_I = \det[d^I_{ij}], \qquad d^I_{ij} = \begin{cases} -u x_i^{N-j+1} & \text{for } i \in I, \\ (x_i u + 1)(x_i + 1/\alpha)^{N-j} & \text{for } i \notin I. \end{cases}$$

Since d_I has ϕ_I as a factor, Proposition 12.7.2 holds. To make use of this fact, we note (12.166) gives

$$D P^*_\kappa(\mu + \delta_N/\alpha) = \sum_I d_I|_{x=\mu+\delta_N/\alpha} P^*_\kappa(\mu - \epsilon_I + \delta_N/\alpha). \tag{12.167}$$

For $\mu - \varepsilon_I$ a partition the factor $P^*_\kappa(\mu - \varepsilon_I + \delta_N/\alpha)$ vanishes according to (12.156), while for $\mu - \varepsilon_I$ not a partition the factor $d_I|_{x=\mu+\delta_N/\alpha}$ vanishes according to Proposition 12.7.2. Hence all terms in (12.167) vanish so we conclude that $(\Delta(x))^{-1} D P^*_\kappa(x)$ is proportional to $P^*_\kappa(x)$ as required. □

12.7.1 Isomorphism between P_κ and P^*_κ

Let Ψ be the invertible linear map with the action

$$\Psi P_\kappa = P^*_\kappa \tag{12.168}$$

for all partitions κ. Closely related to Ψ is the operator [359]

$$\mathcal{E} := \frac{1}{\Delta(x)} \det \left[(x_i + 1/\alpha)^{N-j} + u x_i^{N-j+1} \Delta_i \right]_{i,j=1,\dots,N}$$

(cf. definition of Δ_{ij} in Proposition 12.7.1). Expanding

$$\mathcal{E} = 1 + u\mathcal{E}_1 + \cdots + u^N \mathcal{E}_N$$

we can show that $\mathcal{E}_k$ maps P^*_κ to the space spanned by $\{P^*_\lambda\}$ with $|\lambda| = |\kappa| + k$.

Proposition 12.7.4 *Let $e_k(x)$ denote the elementary symmetric functions (4.132) in $x_1, \dots, x_N$ of degree k. We have*

$$\mathcal{E}_k P^*_\kappa(x) = \sum_{\lambda: |\lambda| = |\kappa| + k} c_{\kappa\lambda} P^*_\lambda(x), \tag{12.169}$$

where the $c_{\kappa\lambda}$ are such that

$$\sum_{\lambda: |\lambda| = |\kappa| + k} c_{\kappa\lambda} P_\lambda(x) = e_k(x) P_\kappa(x). \tag{12.170}$$

Proof. For (12.169) it suffices to check that $\mathcal{E}_k P^*_\kappa(x)$ vanishes at $x = \mu + \delta_N/\alpha$ for all $|\mu| < |\kappa| + k$. With ϕ_I defined above (12.165), we see

$$\mathcal{E}_k = \frac{1}{\Delta(x)} \sum_{|I|=k} \phi_I \Delta_I. \tag{12.171}$$

Now arguing as in the proof of Proposition 12.7.3 gives (12.169). The result (12.170) follows by recalling the structure (12.157), and noting that the degree preserving action of Δ_I is just the identity while the highest degree terms in $(\Delta(x))^{-1}\phi_I$ are equal to $(\Delta(x))^{-1} \det[\bar{c}^I_{ij}(x)]$ where

$$\bar{c}^I_{ij}(x) = \begin{cases} x_i^{N-j+1} & \text{for } i \in I, \\ x_i^{N-j} & \text{for } i \notin I. \end{cases}$$

This in turn simplifies to $\prod_{i \in I} x_i$, and tells us that the highest degree action of $\mathcal{E}_k$ is simply multiplication by $e_k(x)$. □

It follows immediately from (12.170) and (12.169) that

$$\Psi(e_k P_\kappa) = \mathcal{E}_k \Psi(P_\kappa), \tag{12.172}$$

and this in turn implies

$$\Psi(e_j^m e_k^n) = \Psi(e_k^n e_j^m) = \mathcal{E}_j^m \mathcal{E}_k^n \cdot 1 = \mathcal{E}_k^n \mathcal{E}_j^m \cdot 1, \qquad n, m \in \mathbb{Z}_{\geq 0},$$

or equivalently for analytic f symmetric in $x_1, \dots, x_N$

$$\Psi f = f(\mathcal{E}_1, \dots, \mathcal{E}_N) \cdot 1. \tag{12.173}$$

12.7.2 Binomial coefficients

The classical binomial coefficients are related to the polynomials (12.61), $f_p(x)$, say, by

$$\binom{l}{p} = \frac{f_p(l)}{f_p(p)}. \tag{12.174}$$

They furthermore occur in the expansion

$$(1+x)^l = \sum_{p=0}^{l} \binom{l}{p} x^p, \tag{12.175}$$

which may alternatively be used to define the binomial coefficients, with (12.174) then a corollary. From the latter viewpoint, generalized symmetric binomial coefficients can then be defined by

$$\frac{P_\kappa(1+x)}{P_\kappa((1)^N)} = \sum_{\mu: |\mu| \leq |\kappa|} \binom{\kappa}{\mu} \frac{P_\mu(x)}{P_\mu((1)^N)}. \tag{12.176}$$

Then, in keeping with (12.174), the generalized symmetric binomial coefficients can be expressed in terms of the interpolation symmetric Jack polynomials [431].

PROPOSITION 12.7.5 *We have*

$$\binom{\kappa}{\mu} = \frac{P^*_\mu(\bar\kappa/\alpha)}{P^*_\mu(\bar\mu/\alpha)}. \tag{12.177}$$

Proof. Let $D_\mu(\{\xi_i\})$ be a symmetric polynomial of degree $\le |\mu|$ in the Cherednik operators $\{\xi_i\}$. Then, according to Taylor's theorem, and making use too of the structure (12.97), we have that for some particular polynomials D_μ

$$P_\kappa(1+x) = \sum_{\mu:|\mu|\le|\kappa|} D_\mu(\{\xi_i\})P_\kappa|_{x=(1)^N} P_\mu(x).$$

Since $D_\mu(\{\xi_i\})P_\kappa|_{x=(1)^N} = D_\mu(\{\bar\kappa_i\})P_\kappa((1)^N)$, this reads

$$P_\kappa(1+x) = P_\kappa((1)^N) \sum_{\mu:|\mu|\le|\kappa|} D_\mu(\{\bar\kappa_i\})P_\mu(x). \tag{12.178}$$

Now $D_\mu(\{\bar\kappa_i\})$ is a polynomial in $\{\kappa_i\}$ of degree $\le |\mu|$ with the additional features that

$$D_\mu(\{\bar\kappa_i\}) = \begin{cases} 0, & |\kappa| \le |\mu| \ (\kappa \ne \mu), \\ 1/P_\mu((1)^N), & \kappa = \mu. \end{cases}$$

But the only polynomial with these properties is $P^*_\mu(\bar\kappa/\alpha)/(P^*_\mu(\mu/\alpha)P_\mu((1)^N))$. Substituting in (12.178) and comparing with (12.176) gives the result. □

It follows from (12.155) that (12.177) is nonzero only if $\mu \subseteq \kappa$. Hence (12.176) can be refined to read

$$\frac{P_\kappa(1+x)}{P_\kappa((1)^N)} = \sum_{\mu\subseteq\kappa} \binom{\kappa}{\mu} \frac{P_\mu(x)}{P_\mu((1)^N)}. \tag{12.179}$$

EXERCISES 12.7 1. Use the fact that for $\kappa = p^N$, $P_\kappa(x) = x_1^p \cdots x_N^p$, and the binomial theorem (12.133) with $r = -p$ to show from (12.179) that for $\kappa = p^N$,

$$\binom{\kappa}{\mu} = \frac{\alpha^{|\mu|}[-p]^{(\alpha)}_\mu}{d'_\mu}. \tag{12.180}$$

12.8 PIERI FORMULAS

12.8.1 The Macdonald automorphism ω_α

Macdonald [376] introduced the operator ω_α according to its action on the power sum basis,

$$\omega_\alpha p_\kappa := \alpha^{\ell(\kappa)} p_\kappa, \tag{12.181}$$

where $\ell(\kappa)$ denotes the length of κ as in Exercises 12.6 q.1. The action of this operator on the Cauchy-type product (12.122) and the symmetric Jack polynomials turns out to be very simple, leading to a useful variant of the identity (12.123).

Consider first the product (12.122). From the identity of Exercises 12.6 q.3(ii) we see that

$$\omega^{(y)}_{-\alpha} \prod_{k,l=1}^\infty \frac{1}{(1-x_k y_l)^{1/\alpha}} = \sum_\kappa \frac{(-1)^{\ell(\kappa)}}{z_\kappa} p_\kappa(x) p_\kappa(y) = \prod_{k,l=1}^\infty (1 - x_k y_l).$$

This result can be used to deduce the action of $\omega_{-\alpha}$ on P_κ. Setting $y_l = 0$ for $l > N$ we can substitute

(12.123) in the l.h.s to conclude

$$\sum_{\kappa} \frac{h_\kappa}{d'_\kappa} P_\kappa(x;\alpha)\Big(\omega_{-\alpha} P_\kappa(y;\alpha)\Big) = \prod_{k,l=1}^{N} (1 - x_k y_l). \tag{12.182}$$

Further progress relies on the following identity for the action of $\tilde{H}^{(\mathrm{C})}$ on $\prod_{k,l=1}^{N}(1 - x_k y_l)$.

PROPOSITION 12.8.1 *With $\tilde{E}_1$ given by*

$$\tilde{E}_1 = \sum_{j=1}^{N} x_j \frac{\partial}{\partial x_j} \tag{12.183}$$

we have

$$\Big(\tilde{H}^{(\mathrm{C})(x)} - (\frac{N-1}{\alpha}+1)\tilde{E}_1^{(x)}\Big) \prod_{k,l=1}^{N} (1 - x_k y_l) = -\frac{1}{\alpha}\Big(\tilde{H}^{(\mathrm{C})(y)} - (\frac{N-1}{\alpha}+1)\tilde{E}_1^{(y)}\Big)\Big|_{\alpha\mapsto 1/\alpha} \prod_{k,l=1}^{N} (1 - x_k y_l). \tag{12.184}$$

Proof. From the definition (12.94) of $\tilde{H}^{(\mathrm{C})}$, we see that there are terms on both sides independent of α, and terms proportional to $1/\alpha$. The equality of these terms can be checked directly. For example, with $\Lambda := \prod_{k,l=1}^{N}(1 - x_k y_l)$, the terms independent of α on the l.h.s. are

$$\Lambda \sum_{j=1}^{N} \sum_{\substack{l,l'=1 \\ l\neq l'}}^{N} \frac{x_j^2 y_l y_{l'}}{(1 - x_j y_l)(1 - x_j y_{l'})}, \tag{12.185}$$

while the terms independent of α on the r.h.s. are

$$2\Lambda \sum_{\substack{j,k=1 \\ j\neq k}}^{N} \frac{y_j y_k}{y_j - y_k} \sum_{l=1}^{N} \frac{x_l}{1 - y_j x_l}.$$

The final multiple summation can be rewritten by interchanging the summation labels j and k, then taking the arithmetic mean with the original form. This gives agreement with the term (12.185). Similar manipulation gives equality between the terms proportional to $1/\alpha$. □

If we substitute for $\prod_{k,l=1}^{N}(1 - x_k y_l)$ in the l.h.s. of (12.184) according to the formula of (12.182), the action of $\tilde{H}^{(\mathrm{C})(x)} - ((N-1)/\alpha + 1)\tilde{E}_1^{(x)}$ can be computed from the fact that it is an eigenoperator for the symmetric Jack polynomials $P_\kappa(x;\alpha)$ with eigenvalue $2b(\kappa') - \frac{2}{\alpha}b(\kappa)$ (recall (12.34)). Making the same substitution on the r.h.s. and equating coefficients of $P_\kappa^{(\alpha)}(x)$ we see that $\omega_{-\alpha}P_\kappa^{(\alpha)}(y)$ satisfies the eigenvalue equation

$$\Big(\tilde{H}^{(\mathrm{C})(y)} - \Big(\frac{N-1}{\alpha}+1\Big)\tilde{E}_1^{(y)}\Big)\Big|_{\alpha\mapsto 1/\alpha} \omega_{-\alpha} P_\kappa(y;\alpha) = \Big(2b(\kappa') - \frac{2}{\alpha}b(\kappa)\Big)\omega_{-\alpha} P_\kappa(y;\alpha).$$

Using the fact that $P_\kappa(x;\alpha)$ has leading term $m_\kappa(x)$, we see from (12.182) that $(h_\kappa/d'_\kappa)\omega_{-\alpha}P_\kappa(y;\alpha)$ must have leading term $(-1)^{|\kappa|}m_{\kappa'}(y)$. But the unique polynomial satisfying the eigenvalue equation with this property is $(-1)^{|\kappa|}P_{\kappa'}(y;\alpha)$, so we have the following result.

PROPOSITION 12.8.2 *The Macdonald automorphism (12.181) acts on the symmetric Jack polynomials according to*

$$\omega_{-\alpha} P_\kappa(x;\alpha) = (-1)^{|\kappa|} \frac{d'_\kappa}{h_\kappa} P_{\kappa'}(x;1/\alpha) \tag{12.186}$$

and consequently

$$\prod_{k,l=1}^{N} (1 - x_k y_l) = \sum_{\kappa} (-1)^{|\kappa|} P_\kappa(x;\alpha) P_{\kappa'}(y;1/\alpha). \tag{12.187}$$

The identity (12.187) is referred to as the dual Cauchy product type expansion ((10.56) is the case $\alpha = 1$).

12.8.2 The Pieri and combinatorial formulas

According to (12.121) the Schur polynomials are the special case $\alpha = 1$ of the symmetric Jack polynomials $P_\kappa(x; 1/\alpha)$. Earlier, in (10.16) we defined the Schur polynomials in terms of semi-standard tableaux. Here the analogue of this combinatorial formula will be derived in the Jack case. This will be done by first providing the Jack analogue of the *Pieri formula* for Schur polynomials. The latter expresses $e_m(x) s_\kappa(x)$, where $e_m(x)$ denotes the mth elementary symmetric function (recall (4.132)), in terms of $\{s_\mu(x)\}$. Its Jack analogue is the expansion of $e_m(x) P_\kappa(x;\alpha)$ in terms of $\{P_\mu(x;\alpha)\}$.

The Pieri formula for Schur polynomials is easy to deduce by making use of the ratio of alternants formula (12.120), but before doing so it is necessary to introduce a definition relating to diagrams of partitions.

DEFINITION 12.8.3 *Let κ and λ be partitions described by their diagrams, and suppose $\kappa \subset \lambda$. The skew diagram λ/κ consists of those boxes of λ which are not in κ. A skew diagram is said to be a vertical (horizontal) m-strip if λ/κ consists of m boxes, all of which are in distinct rows (columns).*

PROPOSITION 12.8.4 *We have*

$$e_m(x) s_\kappa(x) = \sum_{\substack{\lambda \\ \lambda/\kappa \text{ a vertical } m \text{ strip}}} s_\lambda(x). \tag{12.188}$$

Proof. From (12.120) and the mth elementary function written in the form

$$e_m(x) = \sum_{\substack{M \in \{1,\dots,N\} \\ |M|=m}} \prod_{i \in M} x_i$$

we have

$$\begin{aligned} e_m(x) s_\kappa(x) &= \frac{1}{\Delta(x)} \text{Asym}\left(e_m(x) x^{\delta+\kappa} \right) \\ &= \frac{1}{\Delta(x)} \sum_{\substack{M \subseteq \{1,\dots,N\} \\ |M|=m}} \text{Asym}(x^{\delta+\kappa+\chi_M}) = \sum_{\substack{M \subseteq \{1,\dots,N\} \\ |M|=m}} s_{\kappa+\chi_M}(x), \end{aligned} \tag{12.189}$$

where $\chi_M = (\chi(i \in M))_{i=1,\dots,N}$ with $\chi(i \in M) = 1$ if $i \in M$ and $\chi(i \in M) = 0$ otherwise. In other words $\kappa + \chi_M$ is obtained from κ by adding 1 to κ_i for each $i \in M$. Now if $\kappa_j = \kappa_{j+1}$ and $\chi(j \in M) = 0$, $\chi(j+1 \in M) = 1$ then $\kappa + \chi_M$ is not a partition. However, we see from (12.120) that then two columns in the definition of $s_{\kappa+\chi_M}(x)$ are equal and so in such cases $s_{\kappa+\chi_M}(x) = 0$. Hence all nonzero contributions to (12.189) are of the form $\lambda = \kappa + \chi_M$ for λ/κ a vertical m-strip and (12.188) results. □

The general structure of (12.188) persists in the Jack case [500], [376].

PROPOSITION 12.8.5 *For suitable coefficients $\psi'_{\lambda/\kappa}$,*

$$e_m(x) P_\kappa(x) = \sum_{\substack{\lambda \\ \lambda/\kappa \text{ a vertical } m \text{ strip}}} \psi'_{\lambda/\kappa} P_\lambda(x). \tag{12.190}$$

Proof. We must have

$$e_m(x)P_\kappa(x) = \sum_{\lambda:|\lambda|=|\kappa|+m} c_{\kappa,\lambda} P_\lambda(x). \tag{12.191}$$

By considering the leading term in the monomial basis on the l.h.s. of (12.191) we must have $\lambda \le \kappa + 1^m 0^{N-m}$. Applying (12.186) to both sides of (12.191), then replacing α by $1/\alpha$ gives

$$P_{(m)}(x)P_{\kappa'}(x) = \sum_{\lambda:|\lambda|=|\kappa|+m} \tilde{c}_{\kappa,\lambda} P_{\lambda'}(x) \tag{12.192}$$

for suitable $\tilde{c}_{\kappa,\lambda}$ proportional to $c_{\kappa,\lambda}$. Consideration of the leading term on the l.h.s. of (12.192) shows $\lambda' \le \kappa' + m^1 0^{N-1}$. This combined with the inequality $\lambda \le \kappa + 1^m 0^{N-m}$ just derived implies

$$\ell(\lambda) = \ell(\kappa) \quad \text{or} \quad \ell(\lambda) = \ell(\kappa) + 1. \tag{12.193}$$

We will now show that these two classes of permitted terms in (12.191) have the further requirement that λ/κ is a vertical m-strip.

Setting all terms but $x_1, \dots, x_{\ell(\kappa)}$ in (12.192) equal to zero eliminates the terms with $\ell(\lambda) = \ell(\kappa) + 1$ and gives

$$P_{m^1 0^{N-1}}(x_1, \dots, x_{\ell(\kappa)}) P_{\kappa'}(x_1, \dots, x_{\ell(\kappa)}) = \sum_{\substack{\lambda:|\lambda|=|\kappa|+m \\ \ell(\lambda)=\ell(\kappa)}} \tilde{c}_{\kappa,\lambda} P_{\lambda'}(x_1, \dots, x_{\ell(\kappa)}). \tag{12.194}$$

We now proceed by induction. Suppose (12.194), with the sum restricted to λ/κ a vertical m-strip, is true for partitions $\hat{\kappa}$ with $|\hat{\kappa}| < |\kappa|$ and $\ell(\hat{\kappa}) \le \ell(\kappa)$ (this is certainly the case for $\hat{\kappa} = 0^{\ell(\kappa)}$). Noting that

$$P_{\kappa'}(x_1, \dots, x_{\ell(\kappa)}) = x_1 \cdots x_{\ell(\kappa)} P_{\kappa'-1}(x_1, \dots, x_{\ell(\kappa)})$$

and applying the induction hypothesis to $P_{(m)}P_{\kappa'-1}$ gives that the sum in (12.194) can be restricted to λ/κ a vertical m-strip for all κ.

It remains to obtain the same result for partitions λ such that $\ell(\lambda) = \ell(\kappa) + 1$ in (12.191) or (12.192). Such partitions can be isolated in (12.191) by equating coefficients of $x_1^{\kappa_1+1}$ on both sides. Making use of (12.149) with $x^* = (x_2, \dots, x_N)$, $N \ge \ell(\kappa) + 1$, shows that (12.191) then reduces to

$$e_{m-1}(x^*)P_{\kappa^*}(x^*) = \sum_{\substack{\lambda:|\lambda|=|\kappa|+m \\ \ell(\lambda)=\ell(\kappa)+1}} c_{\kappa,\lambda} P_{\lambda^*}(x^*).$$

After application of (12.186) this reads

$$P_{m-1\,0^{N-1}}(x^*)P_{\kappa'-1}(x^*) = \sum_{\substack{\lambda:|\lambda|=|\kappa|+m \\ \ell(\lambda)=\ell(\kappa)+1}} \tilde{c}_{\kappa,\lambda} P_{\lambda'-1}(x^*).$$

The same argument as used below (12.194) now gives that λ can be restricted to λ/κ a vertical m-strip, and (12.190) follows. □

The Jack polynomial generalization of (12.188) can be determined by using the theory of the interpolation symmetric Jack polynomials [359].

PROPOSITION 12.8.6 *For each lattice point (i,j) of the diagram of a partition λ, define*

$$\begin{aligned} c_\lambda(i,j) &:= \alpha(\lambda_i - j) + (\lambda'_j - i + 1) = \alpha a(i,j) + \ell(i,j) + 1, \\ c'_\lambda(i,j) &:= \alpha(\lambda_i - j + 1) + (\lambda'_j - i) = \alpha(a(i,j) + 1) + \ell(i,j). \end{aligned}$$

In terms of these quantities, for $\kappa \subset \lambda$ define

$$\psi'_{\lambda/\kappa} = \prod_{(i,j)} \frac{c_\lambda(i,j)/c'_\lambda(i,j)}{c_\kappa(i,j)/c'_\kappa(i,j)}, \tag{12.195}$$

where the product is taken over all pairs $(i,j) \in \lambda$ such that $\kappa_i = \lambda_i$ and $\kappa'_j < \lambda'_j$ (i.e., rows such that λ/κ is empty, and columns such that λ/κ is non-empty). We have

$$e_m(x)P_\kappa(x) = \sum_{\substack{\lambda \\ \lambda/\kappa \text{ a vertical } m \text{ strip}}} \psi'_{\lambda/\kappa} P_\lambda(x). \tag{12.196}$$

Proof. We seek the coefficients $a_{\kappa\lambda}$, λ/κ a vertical m-strip, such that

$$e_m(x)P_\kappa(x) = \sum_{\lambda:|\lambda|=|\kappa|+m} a_{\kappa\lambda} P_\lambda(x).$$

From (12.172) this can equivalently be written

$$\mathcal{E}_m P^*_\kappa(x) = \sum_{\lambda:|\lambda|=|\kappa|+m} a_{\kappa\lambda} P^*_\lambda(x).$$

From the vanishing property (12.156) a particular coefficient $a_{\kappa\lambda}$ in this equation can be isolated by evaluation at $x = \bar{\lambda}/\alpha$. Thus, with $I \subset \{1,\dots,N\}$,

$$\begin{aligned} a_{\kappa\lambda} &= \frac{1}{P^*_\lambda(\bar{\lambda}/\alpha)} \mathcal{E}_m P^*_\kappa(x)\Big|_{x=\bar{\lambda}/\alpha} = \frac{1}{P^*_\lambda(\bar{\lambda}/\alpha)} \Big(\frac{1}{\Delta(x)} \sum_{|I|=m} \phi_I(x) \Delta_I P^*_\kappa(x) \Big)\Big|_{x=\bar{\lambda}/\alpha} \\ &= \frac{1}{P^*_\lambda(\bar{\lambda}/\alpha)\Delta(\bar{\lambda}/\alpha)} \phi_{\chi_M}(\bar{\lambda}/\alpha) P^*_\kappa(\bar{\kappa}/\alpha), \end{aligned} \tag{12.197}$$

where the second equality follows from (12.171), and the third equality (in which M is such that $\lambda = \kappa + \chi_M$) makes further use of the vanishing property, together with the definition of Δ_I given above (12.165).

According to (12.158) and the definition (12.37)

$$\frac{P^*_\kappa(\bar{\kappa}/\alpha)}{P^*_\lambda(\bar{\lambda}/\alpha)} = \alpha^m \frac{d'_\kappa}{d'_\lambda} = \alpha^m \frac{\prod_{(i,j)\in\kappa} c'_\kappa(i,j)}{\prod_{(i,j)\in\lambda} c'_\lambda(i,j)}. \tag{12.198}$$

We write $\bar{M} := \{i \notin M\}$, $J := \{\lambda_i | i \in M\}$, $\bar{J} = \{\lambda_i | i \notin M\}$. From the definitions we see that unless $i \in M, j \notin J$, $c'_\lambda(i,j+1) = c'_\kappa(i,j)$ while unless $i \in \bar{M}$, $j \notin \bar{J}$, $c'_\lambda(i,j) = c'_\kappa(i,j)$. Thus (12.198) can be written

$$\frac{P^*_\kappa(\bar{\kappa}/\alpha)}{P^*_\lambda(\bar{\lambda}/\alpha)} = \alpha^m \prod_{i\in M} \frac{1}{c'_\lambda(i,1)} \prod_{i\in M, j\in \bar{J}} \frac{c'_\kappa(i,j)}{c'_\lambda(i,j+1)} \prod_{i\in\bar{M}, j\in J} \frac{c'_\kappa(i,j)}{c'_\lambda(i,j)}. \tag{12.199}$$

Also, from the definitions,

$$\Delta(\bar{\lambda}/\alpha) = \prod_{i<j} \Big((\lambda_i + (N-i)/\alpha) - (\lambda_j + (N-j)/\alpha) \Big)$$

while it follows from the Vandermonde formula (1.173) and (12.165) that

$$\phi_{\chi_M}(\bar{\lambda}/\alpha) = \prod_{i\in M} (\lambda_i + (N-i)/\alpha) \prod_{i<k} \Big((\lambda_i + (N - i + \chi_{i\in\bar{M}})/\alpha) - (\lambda_k + (N - k + \chi_{k\in\bar{M}})/\alpha) \Big).$$

To relate these products to those in (12.195), we note that for $j \in \bar{J}$,

$$\{k \in \bar{M} | \lambda_k = j\} = \{\lambda'_{j+1} + 1, \lambda'_{j+1} + 2, \dots, \kappa'_j\},$$

while if $j = 0$ this same formula holds with $\kappa'_0 := N$. Similarly, for $j \in J$

$$\{i \in M | \lambda_i = j\} = \{\kappa'_j + 1, \kappa'_j + 2, \dots, \lambda'_j\}.$$

Using these formulas we can rewrite the products to obtain

$$\frac{\phi_{\chi_M}(\bar{\lambda}/\alpha)}{\Delta(\bar{\lambda}/\alpha)} = \alpha^{-m} \prod_{i\in M} c'_\lambda(i,1) \prod_{i\in M, j\in \bar{J}} \frac{c'_\lambda(i,j+1)}{c'_\kappa(i,j)} \prod_{i\in \bar{M}, j\in J} \frac{c_\lambda(i,j)}{c_\kappa(i,j)}. \tag{12.200}$$

Multiplying together (12.199) and (12.200), and substituting in (12.197) gives (12.195). □

Combinatorial formula

A consequence of the dual Cauchy expansion (12.187), together with the Pieri formula (12.190), is a formula expressing $P_\kappa(x)$ as a series in $\{x_N^n\}_{n=0,\dots,\kappa_1}$ [343], [376].

PROPOSITION 12.8.7 *With $\psi'_{\lambda/\kappa} = \psi'_{\lambda/\kappa}(\alpha)$ specified by (12.195), define*

$$\psi_{\kappa/\mu} =: \psi_{\kappa/\mu}(\alpha) = \psi'_{\kappa'/\mu'}(1/\alpha). \tag{12.201}$$

We have

$$P_\kappa(x_1,\dots,x_N) = \sum_{r=0}^{\kappa_1} x_N^r \sum_{\substack{\mu \\ \kappa/\mu \text{ a horizontal } r \text{ strip}}} \psi_{\kappa/\mu} P_\mu(x_1,\dots,x_{N-1}). \tag{12.202}$$

Proof. We must have

$$P_\kappa(x_1,\dots,x_N;\alpha) = \sum_{r=0}^{\kappa_1} x_N^r \sum_{\substack{\mu \\ |\kappa|=|\mu|+r}} a_{\kappa\mu} P_\mu(x_1,\dots,x_{N-1};\alpha)$$

for some $a_{\kappa\mu}$. Substituting in the r.h.s. of (12.187) gives

$$\sum_\kappa (-1)^{|\kappa|} P_{\kappa'}(y;1/\alpha) \sum_{r=0}^{\kappa_1} x_N^r \sum_{\substack{\mu \\ |\kappa|=|\mu|+r}} a_{\kappa\mu} P_\mu(x_1,\dots,x_{N-1};\alpha). \tag{12.203}$$

On the l.h.s. of (12.187), we make use of (12.196) and lower dimensional cases of the r.h.s. by writing

$$\begin{aligned}\prod_{k,l=1}^{N} (1-x_k y_l) &= \prod_{j=1}^{N} (1-x_N y_j) \prod_{k=2,l=1}^{N} (1-x_k y_l) \\ &= \Big(\sum_{j=1}^{N} (-x_N)^j e_j(y)\Big) \sum_\mu (-1)^{|\mu|} P_{\mu'}(y;1/\alpha) P_\mu(x_1,\dots,x_{N-1};\alpha) \\ &= \sum_\mu (-1)^{|\mu|} P_\mu(x_1,\dots,x_{N-1};\alpha) \sum_{j=1}^{N} (-x_N)^j \sum_{\substack{\kappa \\ \kappa'/\mu' \text{ a vertical } j \text{ strip}}} \psi'_{\kappa'/\mu'}(1/\alpha) P_{\kappa'}(y;1/\alpha).\end{aligned} \tag{12.204}$$

Equating coefficients of $P_{\kappa'}(y;1/\alpha)P_\mu(x_1,\dots,x_{N-1};\alpha)$ in (12.203) and (12.204) gives $a_{\kappa\mu} = \psi'_{\kappa'/\mu'}(1/\alpha)$ for κ'/μ' a vertical r-strip, or equivalently κ/μ a horizontal r-strip, and $a_{\kappa\mu} = 0$ otherwise, as required by (12.202). □

The combinatorial significance of (12.202) comes about because it can be expressed as a sum over semi-standard tableaux. To see this let $x = \{x_1,\dots,x_N\}$ be elementary weights and in terms of these quantities define the weight $W_1(T)$ of a semi-standard tableau T of content N as

$$W_x(T) = x_1^{\#1's} x_2^{\#2's} \cdots x_N^{\#N's}. \tag{12.205}$$

Furthermore, let $p_{i,j}$ denote the number entered in the diagram of κ at site (i,j). Define the partitions $\kappa^{(h)}$ in terms of their diagram by

$$\kappa^{(h)} = \{(i,j) \in \kappa \,|\, p_{i,j} \le h\}.$$

With $\alpha_h := \#h$'s we have that $\kappa^{(h)}/\kappa^{(h-1)}$ is a horizontal α_h-strip, and T can be uniquely specified in terms of the sequence of partitions $\kappa^{(0)}, \kappa^{(1)}, \dots, \kappa^{(N)}$ (note $\kappa^{(N)} = \kappa$).

Consider now the general structure

$$g_\kappa(x_1, \dots x_N) := \sum_{\substack{T \\ \text{shape } \kappa}} \prod_{h=1}^{N} c(\kappa^{(h)}/\kappa^{(h-1)}) W_x(T).$$

This satisfies the recurrence relation

$$\begin{aligned} g_\kappa(x_1, \dots, x_N) &= \sum_{r=0}^{N} x_N^r \sum_{\substack{\mu \\ \kappa/\mu \text{ a horizontal } r \text{ strip}}} \sum_{\substack{T \\ \text{shape } \mu}} c(\kappa/\mu) \prod_{h=1}^{N-1} c(\kappa^{(h)}/\kappa^{(h-1)}) W_{x_1,\dots,x_{N-1}}(T) \\ &= \sum_{r=0}^{N} x_N^r \sum_{\substack{\mu \\ \kappa/\mu \text{ a horizontal } r \text{ strip}}} c(\kappa/\mu) g_\mu(x_1, \dots, x_{N-1}), \end{aligned} \tag{12.206}$$

where in obtaining the final line the fact that $\kappa^{(h)} = \mu^{(h)}$ for $h = 0, 1, \dots, N-1$ has been used. Comparing (12.206) with (12.202) gives the sought combinatorial formula for the Jack polynomials [500].

PROPOSITION 12.8.8 *With $\psi_{\kappa/\mu} = \psi_{\kappa/\mu}(\alpha)$ specified by (12.201) and $W_x(T)$ by (12.205), we have*

$$P_\kappa(x_1, \dots, x_N; \alpha) = \sum_{\substack{T \\ \text{shape } \kappa}} \prod_{h=1}^{N} \psi_{\kappa^{(h)}/\kappa^{(h-1)}}(\alpha) W_x(T). \tag{12.207}$$

Note that when $\alpha = 1$, (12.207) reduces to the combinatorial definition of the Schur polynomial in (10.16), and furthermore (12.202) reduces to the recurrence (10.18).

12.8.3 An average over the Dixon-Anderson density

The binomial expansion (12.176) can be used in the recurrence (12.202) to deduce an identity which in turn can be used to compute the average of a Jack polynomial with respect to the Dixon-Anderson density (4.11) [431], [369], [361].

PROPOSITION 12.8.9 *We have*

$$P_\mu^*(\bar\lambda/\alpha) = \frac{[N/\alpha]_\mu^{(\alpha)}}{[(N-1)/\alpha]_\mu^{(\alpha)}} \sum{}^{*} \psi_{\lambda/\nu} \frac{P_\nu((1)^{N-1})}{P_\lambda((1)^N)} P_\mu^*(\bar\nu/\alpha), \tag{12.208}$$

*where the sum * is over partitions ν of $N-1$ parts, such that*

$$\lambda_1 \ge \nu_1 \ge \lambda_2 \ge \nu_2 \ge \cdots \ge \nu_{N-1} \ge \lambda_N. \tag{12.209}$$

Proof. Setting $x_N = 1$ in (12.202) shows

$$P_\lambda(x_1, \dots, x_{N-1}, 1) = \sum{}^{*} \psi_{\lambda/\nu} P_\nu(x_1, \dots, x_{N-1}).$$

Next set $x_j = 1 + y_j$ $(j = 1, \dots, N-1)$, make use of the binomial expansion (12.176) on both sides, and equate coefficients of $P_\mu(y_1, \dots, y_{N-1})$ to deduce

$$P_\lambda((1)^N)\binom{\lambda}{\mu} = \frac{P_\mu((1)^N)}{P_\mu((1)^{N-1})} \sum{}^{*} \psi_{\lambda/\nu} P_\nu((1)^{N-1})\binom{\nu}{\mu}.$$

The result (12.208) follows from this upon use of (12.177) and (12.105). □

Suppose λ in (12.208) is replaced by ax, where a is a scalar such that each component of ax is a non-negative integer. Taking the limit $a \to \infty$ allows the component of x to take on continuous values, and we have $P^*_\mu(\bar{\lambda}/\alpha) \sim P^*_\mu(ax)$, while (12.157) gives $\lim_{a\to\infty} P^*_\mu(ax)/a^{|\mu|} = P_\mu(x)$. Applying the same scaling to the r.h.s. of (12.208), replacing $\nu \mapsto ay$, and making use of (12.105), (12.59) and (12.211) below gives the evaluation of the average of a Jack polynomial with respect to the Dixon-Anderson density (4.11).

PROPOSITION 12.8.10 *Let* $\mathrm{DA}_N(1/\alpha)$ *refer the Dixon-Anderson density (4.11) with* $n \mapsto N$, $s_1, \dots, s_n = 1/\alpha$, $\{\lambda_j\} \mapsto \{y_j\}$, $\{a_j\} \mapsto \{x_j\}$. *For* $\ell(\mu) \le N - 1$ *we have*

$$P_\mu(x) = \frac{[N/\alpha]^{(\alpha)}_\mu}{[(N-1)/\alpha]^{(\alpha)}_\mu} \langle P_\mu(y) \rangle_{\mathrm{DA}_N(1/\alpha)}. \tag{12.210}$$

Note that with $x_N = 0$ (12.210) can be interpreted as an integral eigenoperator equation for Jack polynomials in $N-1$ variables.

EXERCISES 12.8

1. [376] Let

$$g^k_n(\lambda, \nu) = \prod_{1 \le i \le j \le n} \frac{(\nu_i - \lambda_{j+1} + k(j-i) + 1)_{k-1} (\lambda_i - \nu_j + k(j-i) + 1)_{k-1}}{(\lambda_i - \lambda_{j+1} + k(j-i) + 1)_{k-1} (\nu_i - \nu_j + k(j-i) + 1)_{k-1}}.$$

Use an induction procedure similar to that used in Exercises 12.4 q.3 to show that with $\psi_{\kappa/\mu}$ specified according to (12.201) and (12.195),

$$\psi_{\kappa/\mu} = g^{1/\alpha}_N(\kappa, \mu). \tag{12.211}$$

2. [343] In this exercise a recursion for $\mathcal{N}^{(\mathrm{C})}_\kappa$ is established.

   (i) Show from the definition (4.132) that

   $$e_m\Big(\frac{1}{t_1}, \dots, \frac{1}{t_N}\Big) = \prod_{i=1}^N t_i^{-1} e_{N-m}(t_1, \dots, t_N),$$

   and show that analogous to the formula of Exercises 12.1 q.1, $z^p P_\kappa(z) = P_{\kappa + p^N}(z)$.

   (ii) From the results of (i) deduce that

   $$e_m(t) P_{\kappa - \chi_M}(t) P_\kappa(1/t) = P_{\kappa + \chi_{\bar{M}}}(t) e_{N-m}(1/t) P_\kappa(1/t),$$

   where χ_M is as in (12.189).

   (iii) Expand $e_m(t) P_{\kappa-\chi_M}(t)$ and $e_{N-m}(1/t) P_\kappa(1/t)$ according to (12.196), set $t_j = e^{2\pi i x_j}$ and take the inner product (12.12) of both sides to deduce

   $$\mathcal{N}^{(\mathrm{C})}_\kappa = \frac{\psi'_{(\kappa+\chi_{\bar{M}})/\kappa}}{\psi'_{\kappa/(\kappa-\chi_M)}} \mathcal{N}^{(\mathrm{C})}_{\kappa-\chi_M}.$$

   Check that this is consistent with the result of Proposition 12.6.3.

3. [537] Let $K(\lambda_1, \lambda_2, \lambda)[P_\mu(y)]$ be defined as in (4.17) but with $P_\mu(y)$ included in the integrand. By making use of (12.210) obtain the analogue of the first formula for K in Proposition 4.2.2, and use the same argument as that

used in the proof of Proposition 4.2.2 to derive the second formula for K therein. Hence conclude

$$\frac{S_{N+1}(\lambda_1, \lambda_2, 1/\alpha)[P_\nu]}{S_{N+1}(\lambda_1, \lambda_2, 1/\alpha)} = \frac{[(N+1)/\alpha]_\mu^{(\alpha)}}{[N/\alpha]_\mu^{(\alpha)}} \frac{S_N(\lambda_1 + 1/\alpha, \lambda_2 + 1/\alpha, 1/\alpha)[P_\mu]}{S_N(\lambda_1 + 1/\alpha, \lambda_2 + 1/\alpha, 1/\alpha)}, \qquad \ell(\mu) \le N.$$

Now use the symmetric Jack analogue of Exercises 12.1 q.1 to show that with $\ell(\mu) \le N+1$ this implies

$$\frac{S_{N+1}(\lambda_1 - \mu_{N+1}, \lambda_2, 1/\alpha)[P_{\nu - \mu_{N+1}}]}{S_{N+1}(\lambda_1, \lambda_2, 1/\alpha)} = \frac{[(N+1)/\alpha]_{\mu-\mu_{N+1}}^{(\alpha)}}{[N/\alpha]_{\mu-\mu_{N+1}}^{(\alpha)}} \frac{S_N(\lambda_1 + 1/\alpha, \lambda_2 + 1/\alpha, 1/\alpha)[P_{\mu - \mu_{N+1}}]}{S_N(\lambda_1 + 1/\alpha, \lambda_2 + 1/\alpha, 1/\alpha)}.$$

Making use of (12.105), (12.59), show that this can be iterated to reclaim (12.143).

4. Replace y_l by $1/y_l$ and make use of the results of Exercises 12.1 q.1 and q.2 (as they apply to symmetric Jack polynomials) to rewrite (12.187) to read

$$\prod_{k,l=1}^{N} (x_k - y_l) = \sum_\kappa (-1)^{|\mu|} P_\kappa(x;\alpha) P_\mu(y; 1/\alpha), \tag{12.212}$$

where $\mu = N^N - \kappa'^R$.

Chapter Thirteen

Correlations for general β

Our ability to calculate correlations for log-gas systems beyond the special random matrix couplings $\beta = 1, 2$ and 4 relies on Jack polynomial theory. One application of the theory is to particular generalizations of the Selberg integral—the so-called Selberg correlation integrals—which allow the exact calculation of the two particle correlation for the log-gas on a circle at even β. Another is to the exact calculation of the density-density correlation in the case of two different initial conditions: the first corresponding to a perfect gas and the second, which is restricted to β rational, to the equilibrium state. In the latter case, with $\beta = p/q$ in reduced form, the correlation is expressed as a $(p+q)$-dimensional integral, with p integrals on $[0,1]$ and q integrals on $[1,\infty)$, and is closely related to the Dotsenko-Fateev integral from Chapter 4. A consequence is a functional equation relating the static structure functions at β and $4/\beta$. These applied studies motivate further development of Jack polynomial theory, in particular to generalized hypergeometric functions, generalized classical polynomials and zonal polynomials. Two other topics relating to general β are also given in this chapter. One is the inter-relations between spacing distributions, linking $\beta = 2(r+1)$ to $\beta = 2/(r+1)$ $(r \in \mathbb{Z}^+)$, which follow as a consequence of the generalization of the Dixon-Anderson integral given in Chapter 4; the other is the stochastic differential equations specifying the bulk and edge distributions.

13.1 HYPERGEOMETRIC FUNCTIONS AND SELBERG CORRELATION INTEGRALS

13.1.1 Generalized hypergeometric functions

The Cauchy-type identities (12.123) and (12.187), used in conjunction with the generalized Selberg integral (12.143), give rise to a further class of generalized Selberg integrals (the Selberg correlation integrals [19], [342], [185]) of direct relevance to the calculation of correlation functions in the log-gas, and displays their relationship to generalized hypergeometric functions. First the latter quantities will be specified [342], [373], [550].

DEFINITION 13.1.1 *Define the renormalized symmetric Jack polynomial $C_\kappa^{(\alpha)}(x)$ by*

$$C_\kappa^{(\alpha)}(x) := \frac{\alpha^{|\kappa|}|\kappa|!}{d'_\kappa} P_\kappa(x;\alpha) \tag{13.1}$$

and let $[u]_\kappa^{(\alpha)}$ denote the generalized Pochhammer symbol as defined by (12.46). In terms of these quantities, the generalized hypergeometric functions ${}_pF_q^{(\alpha)}$ are specified by

$${}_pF_q^{(\alpha)}(a_1,\dots,a_p;b_1,\dots,b_q;x_1,\dots,x_m) := \sum_\kappa \frac{1}{|\kappa|!} \frac{[a_1]_\kappa^{(\alpha)}\cdots[a_p]_\kappa^{(\alpha)}}{[b_1]_\kappa^{(\alpha)}\cdots[b_q]_\kappa^{(\alpha)}} C_\kappa^{(\alpha)}(x_1,\dots,x_m). \tag{13.2}$$

Since in the one-variable case we have $\kappa = k$, $C_k^{(\alpha)}(x) = x^k$ and $[u]_k^{(\alpha)} = (u)_k$, we see that with $m = 1$ the generalized hypergeometric function ${}_pF_q^{(\alpha)}$ reduces to the classical hypergeometric function, specified in the case $p = 2$, $q = 1$ by (5.83). We note that for general number of variables, in the cases $p = q = 0$ and

$p=1, q=0$, (13.2) can be expressed in terms of elementary functions which are independent of α. Thus it follows from (12.134) that

$$ {}_0F_0^{(\alpha)}(x_1,\dots,x_m) = e^{x_1+\cdots+x_m} \tag{13.3} $$

while the generalized binomial theorem (12.133) implies

$$ {}_1F_0^{(\alpha)}(a;x_1,\dots,x_m) = \prod_{j=1}^m (1-x_j)^{-a}. \tag{13.4} $$

In fact these results are not independent. Indeed (13.3) can be deduced as a limiting case of (13.4) by making use of the general formula

$$ \begin{aligned} &\lim_{a_p\to\infty} {}_pF_q^{(\alpha)}(a_1,\dots,a_p;b_1,\dots,b_q;x_1/a_p,\dots,x_m/a_p) \\ &\quad = {}_{p-1}F_q^{(\alpha)}(a_1,\dots,a_{p-1};b_1,\dots,b_q;x_1,\dots,x_m), \end{aligned} \tag{13.5} $$

which follows from the explicit form (12.46) of $[a_p]_\kappa^{(\alpha)}$ and the fact that $C_\kappa^{(\alpha)}(x)$ is homogeneous of degree $|\kappa|$.

Next we will show that it is the generalized hypergeometric function ${}_2F_1^{(\alpha)}$ that is related to Selberg correlation integrals.

Proposition 13.1.2 *We have*

$$ \begin{aligned} &\frac{1}{M_N(a,b;1/\alpha)} \int_{-1/2}^{1/2} dx_1 \cdots \int_{-1/2}^{1/2} dx_N \prod_{l=1}^N \Big(e^{\pi i x_l (a-b)} |1+e^{2\pi i x_l}|^{a+b} \prod_{l'=1}^m (1+t_{l'}e^{2\pi i x_l}) \Big) \\ &\times \prod_{j<k} |e^{2\pi i x_k} - e^{2\pi i x_j}|^{2/\alpha} = {}_2F_1^{(1/\alpha)}(-N,\alpha b; -(N-1)-\alpha(1+a); t_1,\dots,t_m), \end{aligned} \tag{13.6} $$

$$ \begin{aligned} &\frac{1}{S_N(\lambda_1,\lambda_2;1/\alpha)} \int_0^1 dx_1 \cdots \int_0^1 dx_N \prod_{l=1}^N \Big(x_l^{\lambda_1}(1-x_l)^{\lambda_2} \prod_{l'=1}^m (1-t_{l'}x_l) \Big) \prod_{j<k} |x_j-x_k|^{2/\alpha} \\ &\quad = {}_2F_1^{(1/\alpha)}(-N, -(N-1)-\alpha(\lambda_1+1); -2(N-1)-\alpha(\lambda_1+\lambda_2+2); t_1,\dots,t_m). \end{aligned} \tag{13.7} $$

Proof. We will give only the derivation of (13.6); (13.7) follows from similar reasoning. Consider the r.h.s. of (12.142). According to (12.105),

$$ P_\kappa((1)^N) = \frac{b_\kappa}{h_\kappa} = \frac{\alpha^{|\kappa|}[N/\alpha]_\kappa^{(\alpha)}}{h_\kappa}, \tag{13.8} $$

where the second equality follows form (12.104). Next we write all the terms on the r.h.s. in terms of the conjugate partition κ'. From the definitions (12.58) and (12.37) and the facts that for $\kappa \mapsto \kappa'$, $a(i,j) \leftrightarrow l(i,j)$, we see that $h_{\kappa'} = \alpha^{|\kappa|} d'_\kappa|_{\alpha\mapsto 1/\alpha}$. This together with the formula of Exercises 12.4 q.2 shows that the r.h.s. can be written

$$ \frac{(-1)^{|\kappa|}\alpha^{-|\kappa|}[-N]_{\kappa'}^{(1/\alpha)}[\alpha b]_{\kappa'}^{(1/\alpha)}}{d_{\kappa'}|_{\alpha\mapsto 1/\alpha}[-(N-1)-\alpha(1+a)]_{\kappa'}^{(1/\alpha)}}. $$

With this substitution, now multiply both sides of (12.142) by $(-1)^{|\kappa|}P_{\kappa'}(t_1,\dots,t_m;1/\alpha)$ and make use of (12.187) on the l.h.s. and the formulas (13.1) and (13.2) (the latter with $p=2$, $q=1$ and κ replaced by κ') on the r.h.s. □

Also of interest are Selberg correlation integrals based on the Cauchy type formula (12.123).

PROPOSITION 13.1.3 *We have*

$$\frac{1}{M_N(a,b,1/\alpha)} \int_{-1/2}^{1/2} dx_1 \cdots \int_{-1/2}^{1/2} dx_N \prod_{l=1}^{N} \Big(e^{\pi i x_l (a-b)} |1 + e^{2\pi i x_l}|^{a+b} \prod_{l'=1}^{m} (1 + t_{l'} e^{2\pi i x_l})^{-1/\alpha} \Big)$$
$$\times \prod_{j<k} |e^{2\pi i x_k} - e^{2\pi i x_j}|^{2/\alpha} = {}_2F_1^{(\alpha)}\Big(\frac{N}{\alpha}, -b; \frac{1}{\alpha}(N-1) + a + 1; t_1, \dots, t_m\Big), \tag{13.9}$$

$$\frac{1}{S_N(\lambda_1,\lambda_2,1/\alpha)} \int_0^1 dx_1 \cdots \int_0^1 dx_N \prod_{l=1}^{N} \Big(x_l^{\lambda_1} (1 - x_l)^{\lambda_2} \prod_{l'=1}^{m} (1 - t_{l'} x_l)^{-1/\alpha} \Big) \prod_{j<k} |x_j - x_k|^{2/\alpha}$$
$$= {}_2F_1^{(\alpha)}\Big(\frac{N}{\alpha}, \frac{1}{\alpha}(N-1) + \lambda_1 + 1; \frac{2}{\alpha}(N-1) + \lambda_1 + \lambda_2 + 2; t_1, \dots, t_m\Big). \tag{13.10}$$

Proof. These formulas follow immediately upon multiplying (12.142) and (12.143) by

$$\frac{h_\kappa}{d'_\kappa} P_\kappa^{(\alpha)}(t_1, \dots, t_m)$$

and summing over κ, making use of (12.123) on the l.h.s. and (13.8), (13.1), (13.2) on the r.h.s. □

Finally we will note the analogous multidimensional integral formulas which result from the generalized binomial theorem (12.133).

PROPOSITION 13.1.4 *We have*

$$\frac{1}{M_N(a,b,1/\alpha)} \int_{-1/2}^{1/2} dx_1 \cdots \int_{-1/2}^{1/2} dx_N \prod_{l=1}^{N} e^{\pi i x_l (a-b)} |1 + e^{2\pi i x_l}|^{a+b} (1 + t e^{2\pi i x_l})^{-r}$$
$$\times \prod_{1 \le j < k \le N} |e^{2\pi i x_k} - e^{2\pi i x_j}|^{2/\alpha} = {}_2F_1^{(\alpha)}\Big(r, -b; \frac{1}{\alpha}(N-1) + a + 1; (t)^N\Big), \tag{13.11}$$

$$\frac{1}{S_N(\lambda_1,\lambda_2,1/\alpha)} \int_0^1 dx_1 \cdots \int_0^1 dx_N \prod_{l=1}^{N} x_l^{\lambda_1} (1 - x_l)^{\lambda_2} (1 - t x_l)^{-r} \prod_{j<k} |x_j - x_k|^{2/\alpha}$$
$$= {}_2F_1^{(\alpha)}\Big(r, \frac{1}{\alpha}(N-1) + \lambda_1 + 1; \frac{2}{\alpha}(N-1) + \lambda_1 + \lambda_2 + 2; (t)^N\Big). \tag{13.12}$$

Proof. These formulas follow immediately from (12.142) and (12.143) upon multiplying by $\alpha^{|\kappa|}[r]_\kappa^{(\alpha)}/d'_\kappa$ and summing over κ using (12.133) on the l.h.s. and (13.8), (13.1), (13.2) on the r.h.s. □

An immediate consequence of Proposition 13.1.4 is that it provides a multidimensional generalization of the classical Gauss summation formula for ${}_2F_1$ at $t = 1$. Thus by observing that with $t = 1$ the integral in the second formula becomes an example of the Selberg integral (4.1) we see that [550]

$${}_2F_1^{(\alpha)}\Big(r, \frac{1}{\alpha}(N-1) + \lambda_1 + 1; \frac{2}{\alpha}(N-1) + \lambda_1 + \lambda_2 + 2; (1)^N\Big) = \frac{S_N(\lambda_1, \lambda_2 - r, 1/\alpha)}{S_N(\lambda_1, \lambda_2, 1/\alpha)}, \tag{13.13}$$

or equivalently, after making use of (4.3),

$${}_2F_1^{(\alpha)}(a, b; c; (1)^N) = \prod_{j=0}^{N-1} \frac{\Gamma(c - j/\alpha)\Gamma(c - a - b - j/\alpha)}{\Gamma(c - a - j/\alpha)\Gamma(c - b - j/\alpha)}. \tag{13.14}$$

13.1.2 Holonomic partial differential equations

In general a differential equation (or set of differential equations) is said to be *holonomic* if it generates a unique (up to normalization) solution of a given type. A classical example is the differential equation for the hypergeometric function ${}_2F_1(a, b; c; t)$,

$$t(1-t)\frac{d^2F}{dt^2} + \Big(c - (a+b+1)t\Big)\frac{dF}{dt} - abF = 0.$$

Seeking a power series solution of this equation gives a recurrence relation for the coefficients, the solution to which gives (5.83). It turns out [342] (see also [550], [373]) that this property extends to (13.2) with $p = 2$, $q = 1$ and general m.

PROPOSITION 13.1.5 *For $c - \frac{1}{\alpha}(i-1) \notin \mathbb{Z}_{\le 0}$, $i = 1, \dots, m$, each of the partial differential equations*

$$t_p(1-t_p)\frac{\partial^2 F}{\partial t_p^2} + \Big(c - \frac{1}{\alpha}(m-1) - (a+b+1-\frac{1}{\alpha}(m-1))t_p\Big)\frac{\partial F}{\partial t_p} - abF$$
$$+\frac{1}{\alpha}\sum_{\substack{j=1\\ j\neq p}}^{m}\frac{1}{t_p - t_j}\Big(t_p(1-t_p)\frac{\partial F}{\partial t_p} - t_j(1-t_j)\frac{\partial F}{\partial t_j}\Big) = 0, \qquad (13.15)$$

$p = 1, \dots, m$ has the same symmetric formal power series solution equal to unity at the origin. This solution can be calculated by recurrence and is given by the series (13.2) with $p = 2$, $q = 1$, which is analytic in the neighbourhood of the origin for $|t_j| < 1$, $j = 1, \dots, m$.

The proof of Proposition 13.1.5, which is a generalization of the proof given by Muirhead [410] in the case $\alpha = 2$, first establishes the uniqueness property by transforming the equations in terms of elementary symmetric functions and establishing a recurrence relation for the coefficient. Next it is shown that there is a unique series solution of the equations (13.15) summed over p from 1 to m of the form

$$F = \sum_{\kappa} \alpha_\kappa C_\kappa^{(\alpha)}(t_1, \dots, t_m),$$

provided the coefficients α_κ are independent of m. This is done by establishing a recurrence equation for the α_κ. Finally the recurrence equation is solved to give (13.2) in the case $p = 2$, $q = 1$. In Exercises 13.1 q.1 it will be shown directly that the Selberg correlation integral (13.7) satisfies the set of partial differential equations (13.15).

In the remainder of this subsection, some consequences of Proposition 13.1.5 will be discussed by way of transformation formulas for the generalized hypergeometric function ${}_2F_1^{(\alpha)}$. First we note the analogue of the classical Kummer relations [550].

PROPOSITION 13.1.6 *We have*

$$\begin{aligned}{}_2F_1^{(\alpha)}(a, b; c; t_1, \dots, t_m) &= \prod_{j=1}^{m}(1-t_j)^{-a}\,{}_2F_1^{(\alpha)}\Big(a, c-b; c; -\frac{t_1}{1-t_1}, \dots, -\frac{t_m}{1-t_m}\Big)\\ &= \prod_{j=1}^{m}(1-t_j)^{c-a-b}\,{}_2F_1^{(\alpha)}(c-a, c-b; c; t_1, \dots, t_m).\end{aligned}$$

Proof. These formulas are established by verifying that the functions on the r.h.s. satisfy the p.d.e.'s (13.15) for which use is made of the particular form of the p.d.e.'s satisfied by the ${}_2F_1^{(\alpha)}$ functions on the r.h.s. Because the functions of the r.h.s. are analytic at the origin, according to Proposition 13.1.5 the fact that they satisfy the p.d.e.'s (13.15) implies that they are equal to the function ${}_2F_1^{(\alpha)}(a, b; c; t_1, \dots, t_m)$. □

Replacing t_j by t_j/b ($j = 1, \dots, m$) in the second equality of Proposition 13.1.6, taking $b \to -\infty$ and using (13.5) gives a generalization of the second Kummer formula,

$$ {}_1F_1^{(\alpha)}(a;c;t_1,\dots,t_m) = \prod_{j=1}^m e^{t_j}\, {}_1F_1^{(\alpha)}(c-a;c;-t_1,\dots,-t_m). \tag{13.16}$$

When a or $b \in \mathbb{Z}_{\le 0}$ for some $1 \le i \le m$, the series (13.2) defining ${}_2F_1^{(\alpha)}$ terminates. In such cases the method of proof of Proposition 13.1.6 shows that a further transformation formula holds.

PROPOSITION 13.1.7 *Subject to the above noted condition for the termination of the series,*

$$ {}_2F_1^{(\alpha)}(a,b;c;t_1,\dots,t_m) = \frac{{}_2F_1^{(\alpha)}(a,b;a+b+1+(m-1)/\alpha-c;1-t_1,\dots,1-t_m)}{{}_2F_1^{(\alpha)}(a,b;a+b+1+(m-1)/\alpha-c;(1)^m)}$$

where the denominator on the r.h.s. can be evaluated according to (13.13).

13.1.3 Hypergeometric functions of two sets of variables

The generalized hypergeometric function (13.2) can be obtained as a special case of a more general class of generalized hypergeometric functions based on two sets of variables.

DEFINITION 13.1.8 *With $E_\eta(x) := E_\eta(x_1,\dots,x_n;\alpha)$ denoting the nonsymmetric Jack polynomial of n variables, let*

$$ {}_p\mathcal{K}_q(a_1,\dots,a_p;b_1,\dots,b_q;x_1,\dots,x_n;y_1,\dots,y_n) := \sum_\eta \frac{[a_1]_{\eta^+}^{(\alpha)}\cdots[a_p]_{\eta^+}^{(\alpha)}}{[b_1]_{\eta^+}^{(\alpha)}\cdots[b_q]_{\eta^+}^{(\alpha)}} \frac{\alpha^{|\eta|}}{d'_\eta} \frac{E_\eta(x)E_\eta(y)}{E_\eta(1^n)}. \tag{13.17}$$

Setting $y_1 = \cdots = y_n = 1$ in (13.17), using (12.96) and comparing with the definition (13.2) we see that

$$ {}_p\mathcal{K}_q(a_1,\dots,a_p;b_1,\dots,b_q;x_1,\dots,x_n;(1)^n) = {}_pF_q^{(\alpha)}(a_1,\dots,a_p;b_1,\dots,b_q;x_1,\dots,x_n). \tag{13.18}$$

Furthermore, by symmetrizing the function ${}_p\mathcal{K}_q$ in the variables $x_1,\dots,x_n$ (or $y_1,\dots,y_n$) a hypergeometric function of two sets of variables results which contains the symmetric Jack polynomials in its definition. To show this requires first establishing a transformation property of (13.17) under the action of the elementary transposition operator $s_i^{(x)}$ [39].

PROPOSITION 13.1.9 *Let $F = \sum_\eta A_\eta E_\eta(x)E_\eta(y)$. Then $s_i^{(x)}F = s_i^{(y)}F$ if and only if the coefficients A_η satisfy*

$$ A_{s_i\eta} = \Big(1 - \frac{1}{\bar\delta_{i,\eta}^2}\Big)A_\eta, \qquad \eta_i > \eta_{i+1},$$

or equivalently

$$ A_{s_i\eta} = \Big(1 - \frac{1}{\bar\delta_{i,\eta}^2}\Big)^{-1}A_\eta, \qquad \eta_i < \eta_{i+1}.$$

Proof. This follows from the formulas for s_iE_η given in Proposition 12.2.1. □

Making use of (12.38) we see that the conditions of Proposition 13.1.9 are met by the definition (13.17) of ${}_p\mathcal{K}_q$, so we have

$$\begin{aligned} &s_i^{(x)}\, {}_p\mathcal{K}_q(a_1,\dots,a_p;b_1,\dots,b_q;x_1,\dots,x_n;y_1,\dots,y_n) \\ &\quad = s_i^{(y)}\, {}_p\mathcal{K}_q(a_1,\dots,a_p;b_1,\dots,b_q;x_1,\dots,x_n;y_1,\dots,y_n). \end{aligned} \tag{13.19}$$

This result in turn can be used to determine the form of ${}_p\mathcal{K}_q$ under the action of $\mathrm{Sym}^{(x)}$.

PROPOSITION 13.1.10 *Let*

$$\begin{aligned}&{}_p\mathcal{F}_q^{(\alpha)}(a_1,\dots,a_p;b_1,\dots,b_q;x_1,\dots,x_n;y_1,\dots,y_n)\\&:=\sum_\kappa \frac{1}{|\kappa|!}\frac{[a_1]_\kappa^{(\alpha)}\cdots[a_p]_\kappa^{(\alpha)}}{[b_1]_\kappa^{(\alpha)}\cdots[b_q]_\kappa^{(\alpha)}}\frac{C_\kappa^{(\alpha)}(x_1,\dots,x_n)C_\kappa^{(\alpha)}(y_1,\dots,y_n)}{C_\kappa^{(\alpha)}((1)^n)}.\end{aligned}\tag{13.20}$$

We have

$$\begin{aligned}&\mathrm{Sym}^{(x)}\,{}_p\mathcal{K}_q(a_1,\dots,a_p;b_1,\dots,b_q;x_1,\dots,x_n;y_1,\dots,y_n)\\&=n!\,{}_p\mathcal{F}_q(a_1,\dots,a_p;b_1,\dots,b_q;x_1,\dots,x_n;y_1,\dots,y_n).\end{aligned}\tag{13.21}$$

Proof. For convenience we first abbreviate (13.17) by writing

$$ {}_p\mathcal{K}_q=\sum_\eta A_\eta E_\eta(x)E_\eta(y).$$

Making use of (12.100) gives

$$\mathrm{Sym}^{(x)}\,{}_p\mathcal{K}_q=\sum_\eta A_\eta a_\eta P_{\eta^+}(x)E_\eta(y).\tag{13.22}$$

Suppose we now make a further application of $\mathrm{Sym}^{(x)}$. On the l.h.s. of the above formula a factor of $n!$ results. However, on the r.h.s. we can use the formula (13.19) to replace $\mathrm{Sym}^{(x)}$ with $\mathrm{Sym}^{(y)}$ and compute its action again using (12.100) to conclude

$$\mathrm{Sym}^{(x)}\,{}_p\mathcal{K}_q=\frac{1}{n!}\sum_\eta A_\eta a_\eta^2 P_{\eta^+}(x)P_{\eta^+}(y).$$

Recalling (12.101) and making use of (12.96) in the case $z=(1)^n$ gives the sought result. □

The hypergeometric function of two sets of variables can be used to provide an integral representation of ${}_2F_1^{(\alpha)}(a,b;c;t_1,\dots,t_N)$ [550].

PROPOSITION 13.1.11 *We have*

$$\begin{aligned}&{}_2F_1^{(\alpha)}\Big(a,\frac{1}{\alpha}(N-1)+\lambda_1+1;\frac{2}{\alpha}(N-1)+\lambda_1+\lambda_2+2;t_1,\dots,t_N\Big)=\frac{1}{S_N(\lambda_1,\lambda_2,1/\alpha)}\\&\times\int_0^1dx_1\cdots\int_0^1dx_N\,{}_1\mathcal{F}_0^{(\alpha)}(a;t_1,\dots,t_N;x_1,\dots,x_N)\prod_{l=1}^N x_l^{\lambda_1}(1-x_l)^{\lambda_2}\prod_{1\le j<k\le N}|x_k-x_j|^{2/\alpha}.\end{aligned}\tag{13.23}$$

Proof. This follows by multiplying both sides of (12.143) by

$$\frac{[a]_\kappa^{(\alpha)}}{|\kappa|!}\frac{C_\kappa^{(\alpha)}(t)}{P_\kappa((1)^N)}$$

and summing over N. □

Since, from the definition (13.20), together with the formula (13.4), we have

$${}_1\mathcal{F}_0^{(\alpha)}(a;(t)^N;x_1,\dots,x_N)={}_1F_0^{(\alpha)}(a;tx_1,\dots,tx_N)=\prod_{j=1}^N(1-tx_j)^{-a},$$

and we see that (13.12) follows as a special case of the formula (13.23).

EXERCISES 13.1 1. [342] The objective of this exercise is to show that the Selberg correlation integral in (13.7) satisfies the partial differential equations (13.15) for suitable values of the parameters. For this purpose let

$$S_{N,m}[f] := \int_{[0,1]^N} dx_1 \dots dx_N\, f \prod_{j=1}^N \Big(x_j^{\lambda_1} (1-x_j)^{\lambda_2} \prod_{k=1}^m (x_j - t_k) \Big) \prod_{1\le j<k\le N} |x_k - x_j|^{2\lambda},$$

where f may be an operator acting on all terms to the right.

(i) Use the three equations

$$0 = S_{N,m}\Big[\sum_{j=1}^N \frac{\partial}{\partial x_j}\Big],\quad 0 = S_{N,m}\Big[\sum_{j=1}^N \frac{\partial}{\partial x_j} x_j\Big],\quad 0 = S_{N,m}\Big[\sum_{j=1}^N \frac{\partial}{\partial x_j}\frac{1}{x_j - t_p}\Big]$$

($p = 1, \dots, m$) (cf. (4.128) and (4.129)), which follow from the fundamental theorem of calculus, to show that for $\lambda_1, \lambda_2 > 0$

$$0 = \lambda_1 S_{N,m}\Big[\sum_{j=1}^N \frac{1}{x_j}\Big] - \lambda_2 S_{N,m}\Big[\sum_{j=1}^N \frac{1}{1-x_j}\Big] + S_{N,m}\Big[\sum_{j=1}^N\sum_{k=1}^m \frac{1}{x_j - t_k}\Big],$$

$$0 = N(1+\lambda_1+\lambda_2+m+(N-1)\lambda) S_{N,m}[1] - \lambda_2 S_{N,m}\Big[\sum_{j=1}^N \frac{1}{1-x_j}\Big] + S_{N,m}\Big[\sum_{j=1}^N\sum_{k=1}^m \frac{t_k}{x_j - t_k}\Big],$$

and

$$0 = -2\lambda S_{N,m}\Big[\sum_{1\le j<k\le N} \frac{1}{(x_j-t_p)(x_k-t_p)}\Big] + \frac{\lambda_1}{t_p} S_{N,m}\Big[\sum_{j=1}^N \Big(\frac{1}{x_j - t_p} - \frac{1}{x_j}\Big)\Big]$$
$$- \frac{\lambda_2}{1-t_p} S_{N,m}\Big[\sum_{j=1}^N \Big(\frac{1}{x_j-t_p} + \frac{1}{1-x_j}\Big)\Big] + \sum_{\substack{l=1\\ l\ne p}}^m \frac{1}{t_p - t_l} S_{N,m}\Big[\sum_{j=1}^N \Big(\frac{1}{x_j - t_p} - \frac{1}{x_j - t_l}\Big)\Big].$$

(ii) By regarding

$$\lambda_1 S_{N,m}\Big[\sum_{j=1}^N \frac{1}{x_j}\Big] \quad \text{and} \quad \lambda_2 S_{N,m}\Big[\sum_{j=1}^N \frac{1}{1-x_j}\Big]$$

as unknowns in the first two equations of (i) obtain the equations

$$\lambda_1 S_{N,m}\Big[\sum_{j=1}^N \frac{1}{x_j}\Big] = N(1+\lambda_1+\lambda_2+m+(N-1)\lambda) S_{N,m}[1] - S_{N,m}\Big[\sum_{j=1}^N\sum_{k=1}^m \frac{1-t_k}{x_j - t_k}\Big],$$

$$\lambda_2 S_{N,m}\Big[\sum_{j=1}^N \frac{1}{1-x_j}\Big] = N(1+\lambda_1+\lambda_2+m+(N-1)\lambda) S_{N,m}[1] + S_{N,m}\Big[\sum_{j=1}^N\sum_{k=1}^m \frac{t_k}{x_j - t_k}\Big],$$

and substitute these equations in the third equation of (i) to show

$$0 = -2\lambda S_{N,m}\Big[\sum_{1\le j<k\le N} \frac{1}{(x_j - t_p)(x_k - t_p)}\Big]$$
$$+\Big(\frac{\lambda_1}{t_p} - \frac{\lambda_2}{1-t_p}\Big) S_{N,m}\Big[\sum_{j=1}^{N} \frac{1}{x_j - t_p}\Big] - \frac{1}{t_p}\Big(N(1+\lambda_1+\lambda_2+m+(N-1)\lambda)S_{N,m}[1]$$
$$-S_{N,m}\Big[\sum_{j=1}^{N}\sum_{k=1}^{m} \frac{1-t_k}{x_j - t_k}\Big]\Big) - \frac{1}{1-t_p}\Big(N(1+\lambda_1+\lambda_2+m+(N-1)\lambda)S_{N,m}[1]$$
$$+S_{N,m}\Big[\sum_{j=1}^{N}\sum_{k=1}^{m} \frac{t_k}{x_j - t_k}\Big]\Big) + S_{N,m}\Big[\sum_{\substack{l=1\\ l\ne p}}^{m} \frac{1}{t_p - t_l}\sum_{j=1}^{N} \frac{1}{x_j - t_p} - \frac{1}{x_j - t_l}\Big].$$

(iii) Show from the definition of $S_{N,m} := S_{N,m}[1]$ that

$$\frac{\partial S_{N,m}}{\partial t_p} = -S_{N,m}\Big[\sum_{j=1}^{N} \frac{1}{x_j - t_p}\Big],$$
$$\frac{\partial^2 S_{N,m}}{\partial t_p^2} = 2S_{N,m}\Big[\sum_{1\le j<k\le N} \frac{1}{(x_j - t_p)(x_k - t_p)}\Big]$$

and substitute in the final equation of (ii) to obtain the p.d.e. (13.15) with

$$\alpha \mapsto \lambda, \quad a = -N, \quad b = (\lambda_1+\lambda_2+m+1)/\lambda + N - 1, \quad c = (\lambda_1+m)/\lambda, \quad F = S_{N,m}. \tag{13.24}$$

(iv) Write the general solution F of (13.15) as

$$F = \prod_{j=1}^{m} t_j^{-a}\tilde{F}(1/t_1,\dots,1/t_m),$$

and show that $\tilde{F}(t_1,\dots,t_m)$ satisfies the same p.d.e.'s with

$$b \mapsto a - c + 1 + (m-1)/\alpha, \qquad c \mapsto a - b + 1 + (m-1)/\alpha.$$

(v) Use the results of (iii) and (iv) to show that the Selberg correlation integral in (13.7) satisfies the system of equations (13.15) with

$$\alpha \mapsto \frac{1}{\alpha}, \quad a = -N, \quad b = -(N-1) - \alpha(\lambda_1+1), \quad c = -2(N-1) - \alpha(\lambda_1+\lambda_2+2).$$

2. (i) Use the result of q.1(iii) and Proposition 13.1.5 to show

$$\frac{1}{S_N(\lambda_1+m,\lambda_2,\lambda)}\int_{[0,1]^N} dx_1\cdots dx_N \prod_{j=1}^{N}\Big(x_j^{\lambda_1}(1-x_j)^{\lambda_2}\prod_{k=1}^{m}(x_j - t_k)\Big)\prod_{1\le j<k\le N}|x_k - x_j|^{2\lambda}$$
$$= {}_2F_1^{(\lambda)}(-N, (\lambda_1+\lambda_2+m+1)/\lambda + N - 1, (\lambda_1+m)/\lambda; t_1,\dots,t_m). \tag{13.25}$$

Use this in the case $m = 1$ together with (5.92) to show

$$\Big\langle \prod_{j=1}^{N}(t - x_j)\Big\rangle_{\mathrm{ME}_{2\lambda,N}(x^{\lambda_1}(1-x)^{\lambda_2})} \propto P_N^{(\alpha,\beta)}(1-2t)$$

with $\alpha = (\lambda_1+1)/\lambda - 1$, $\beta = (\lambda_2+1)/\lambda - 1$.

(ii) Replace the $w_i^{(j)}$, $(i = 0, 1, 2)$ in (4.49) by their mean values and verify that the resulting recurrence is satisfied by $A_j^{\#}(x) = c_j P_j^{(\hat{\alpha},\hat{\beta})}(1 - 2x)$ where $\hat{\alpha} = (\alpha^{(j)} + 1)/\lambda - 1$, $\hat{\beta} = (\beta^{(j)} + 1)/\lambda - 1$ and c_j is chosen such that $A_j^{\#}(x)$ is monic. Note that this is consistent with the result of (i).

(iii) Show that in the case $\lambda = 1$, the result of (i) is consistent with the first formula in Proposition 5.1.4.

(iv) Use the first formula in Exercises 6.4 q.1 to show that the result of (i) with $\lambda = \frac{1}{2}$ is consistent with the first formula in (6.98).

3. Set $t_p \mapsto t_p/b$ in (13.15) and take the limit $b \to \infty$ to conclude that ${}_1F_1^{(\alpha)}(a; c; t)$ satisfies each of the p.d.e.'s

$$t_p \frac{\partial^2 F}{\partial t_p^2} + \Big(c - \frac{1}{\alpha}(m-1) - t_p\Big)\frac{\partial F}{\partial t_p} - aF + \frac{1}{\alpha}\sum_{\substack{j=1 \\ j\neq p}}^{m} \frac{1}{t_p - t_j}\Big(t_p\frac{\partial F}{\partial t_p} - t_j\frac{\partial F}{\partial t_j}\Big) = 0, \tag{13.26}$$

and that these uniquely determine this function.

4. (i) By making the replacement $t \mapsto t/r$ and taking the limit $r \to \infty$, show from (13.11) that

$$\frac{1}{M_N(a,b,1/\alpha)} \int_{-1/2}^{1/2} dx_1 \cdots \int_{-1/2}^{1/2} dx_N \prod_{l=1}^{N} e^{\pi i x_l(a-b)} |1+e^{2\pi i x_l}|^{a+b} e^{-te^{2\pi i x_l}} \prod_{j<k} |e^{2\pi i x_k} - e^{2\pi i x_j}|^{2/\alpha}$$
$$= {}_1F_1^{(\alpha)}(-b; a+1+(N-1)/\alpha; (t)^N).$$

(ii) Consider the case $a, b, 2/\alpha \in \mathbb{Z}_{\geq 0}$ when the integrand in (i) is analytic. Argue that in this case it is possible to replace $e^{2\pi i x_j}$ by $Re^{2\pi i x_j}$, $R > 0$. Now make the replacement $t \mapsto -t/b$, put $R = bt^{-1/2}$ and take the limit $b \to \infty$ to show that [187]

$${}_0F_1^{(\alpha)}(c + (N-1)/\alpha; (t)^N) = B_N(c,\alpha)\Big(\frac{1}{t}\Big)^{(c-1)N/2}$$
$$\times \Big(\frac{1}{2\pi}\Big)^N \int_{[-\pi,\pi]^N} \prod_{j=1}^{N} e^{2t^{1/2}\cos\theta_j} e^{i(c-1)\theta_j} \prod_{1\leq j<k\leq N} |e^{i\theta_k} - e^{i\theta_j}|^{2/\alpha} d\theta_1 \cdots d\theta_N, \tag{13.27}$$

where $c \in \mathbb{Z}_{\geq 1}$ and

$$B_N(c,\alpha) = \prod_{j=1}^{N} \frac{\Gamma(1+1/\alpha)\Gamma(c+(j-1)/\alpha)}{\Gamma(1+j/\alpha)}. \tag{13.28}$$

Argue that this is valid for general α.

5. [38] The objective of this exercise is to deduce the identity

$$e^{p_1(x)}{}_0\mathcal{K}_0(x;y) = {}_0\mathcal{K}_0(x;y+1), \tag{13.29}$$

where $p_1(x)$ is specified by (12.146).

(i) Let $\xi_i^{(y)}$ refer to (11.57) with $z \mapsto y$. By noting $\xi_i^{(y+1)} = \xi_i^{(y)} + \alpha d_i^{(y)}$, and using the fact that $\xi_i^{(y)}{}_0\mathcal{K}_0(x;y) = \xi_i^{(x)}{}_0\mathcal{K}_0(x;y)$ together with the second identity in (13.86) below, deduce that

$$\xi_i^{(y+1)} e^{p_1(x)}{}_0\mathcal{K}_0(x;y) = \xi_i^{(x)}\Big(e^{p_1(x)}{}_0\mathcal{K}_0(x;y)\Big).$$

(ii) Note from the definition (13.17) that

$$e^{p_1(x)}{}_0\mathcal{K}_0(x;y) = \sum_\eta \alpha^{|\eta|}\frac{d_\eta}{d'_\eta e_\eta} U_\eta(y) E_\eta(x)$$

for some polynomial $U_\eta(y)$ with leading term $E_\eta(y)$. Now use the result from (i) to deduce

$$\xi_i^{(y+1)} U_\eta(y) = \bar{\eta}_i U_\eta(y),$$

and thus conclude $U_\eta(y)$ is equal to $E_\eta(y+1)$, as required by (13.29).

(iii) Analogous to (12.176) the nonsymmetric binomial coefficients can be defined by

$$\frac{E_\eta(1+z)}{E_\eta((1)^N)} = \sum_{\nu:|\nu|\le|\eta|} \binom{\eta}{\nu} \frac{E_\nu(z)}{E_\nu((1)^N)}. \tag{13.30}$$

Use this to substitute for the $E_\eta(y+1)$ which occur on the r.h.s. of (13.29), and equate coefficients of $E_\eta(y)$ on both sides to obtain

$$e^{p_1(x)} E_\eta(x) \alpha^{|\eta|} \frac{1}{d'_\eta} = \sum_\nu \alpha^{|\nu|} \frac{1}{d'_\nu} \binom{\nu}{\eta} E_\nu(x). \tag{13.31}$$

6. [342], [38] Here the symmetric analogues of q.5 are considered.

(i) Symmetrize (13.29) and so obtain

$$e^{p_1(x)} {}_0\mathcal{F}_0^{(\alpha)}(x;y) = {}_0\mathcal{F}_0^{(\alpha)}(x;y+1).$$

(ii) Substitute for $P_\kappa(y+1)$ in the series expansion of the r.h.s. of the identity in (i) using (12.179) and equate coefficients of $P_\kappa(y)$ to deduce the symmetric analogue of (13.31)

$$e^{p_1(x)} C_\kappa^{(\alpha)}(x) = \sum_\mu \binom{\mu}{\kappa} \frac{|\kappa|!}{|\mu|!} C_\kappa^{(\alpha)}(x).$$

(iii) By equating terms of degree $|\kappa|+1$ on both sides of the identity of (ii) note that

$$p_1(x) P_\kappa(x) = \alpha \sum_{\mu:|\mu|=|\kappa|+1} \binom{\mu}{\kappa} \frac{d'_\kappa}{d'_\mu} P_\mu(x).$$

By comparing this with (12.196) in the case $m=1$ read off that for $|\mu| = |\kappa|+1$,

$$\binom{\mu}{\kappa} = \frac{1}{\alpha} \frac{d'_\mu}{d'_\kappa} \psi_{\mu/\kappa}.$$

13.2 CORRELATIONS AT EVEN β

13.2.1 Circular β-ensembles

The generalizations of the Selberg integral presented in Section 13.1.1 allow the general n-point correlation for the log-gas on a circle at even β, or equivalently the circular β-ensemble, to be expressed as a particular generalized hypergeometric function ${}_2F_1^{(\beta/2)}$ in $\beta(n-1)/2$ variables. Suppose for convenience that there are $N+n$ particles in the system. According to the results of Proposition 1.4.1 for the Boltzmann factor, and Proposition 4.7.2 for the normalization, the general formula (5.1) gives

$$\rho_{(n)}(r_1,\dots,r_n) = \frac{(N+1)_n}{L^n} \frac{((\beta/2)!)^{N+n}}{(\beta(N+n)/2)!} \prod_{1\le j<k\le n} |e^{2\pi i r_k/L} - e^{2\pi i r_j/L}|^\beta I_{N,n}(\beta; r_1,\dots,r_n), \tag{13.32}$$

where

$$I_{N,n}(\beta; r_1,\dots,r_n) := \int_{[0,1]^N} dx_1 \cdots dx_N \prod_{j=1}^N \prod_{k=1}^n |1 - e^{2\pi i (x_j - r_k/L)}|^\beta \prod_{1\le j<k\le N} |e^{2\pi i x_k} - e^{2\pi i x_j}|^\beta, \tag{13.33}$$

and we have scaled the angles $\theta_j \mapsto 2\pi x_j/L$. In the case β even, the Selberg correlation integral (13.6) can be shown to include (13.33), which thus implies a formula for $\rho_{(n)}$ in terms of the generalized hypergeometric function ${}_2F_1^{(\beta/2)}$ [185].

PROPOSITION 13.2.1 *For the n-point correlation (13.32) with β even we have*

$$\rho_{(n)}(r_1,\dots,r_n) = \frac{(N+1)_n}{L^n}\frac{((\beta/2)!)^{N+n}}{(\beta(N+n)/2)!}\prod_{1\le j<k\le n}|e^{2\pi i r_k/L}-e^{2\pi i r_j/L}|^\beta M_N(n\beta/2,n\beta/2,\beta/2)$$
$$\times\prod_{k=2}^n e^{\pi i N\beta(r_k-r_1)/L}\,{}_2F_1^{(\beta/2)}(-N,n;2n;1-t_1,\dots,1-t_{(n-1)\beta}), \tag{13.34}$$

where M_N is specified by (4.4),

$$t_k := e^{-2\pi i(r_j-r_1)/L}, \qquad k=1+(j-2)\beta,\dots,(j-1)\beta\ (j=2,\dots,n)$$

(i.e., $e^{-2\pi i(r_j-r_1)/L}$ for each $j=2,\dots,n$ is repeated β times).

Proof. First change variables $x_j\mapsto x_j+r_1/L+\frac{1}{2}$ in (13.33) to obtain

$$I_{N,n}(\beta;r_1,\dots,r_n) = \int_{[-1/2,1/2]^N} dx_1\cdots dx_N\prod_{j=1}^N\Big(|1+e^{2\pi i x_j}|^\beta\prod_{k=2}^n|1+e^{2\pi i(x_j+(r_1-r_k)/L)}|^\beta\Big)$$
$$\times\prod_{j<k}|e^{2\pi i x_k}-e^{2\pi i x_j}|^\beta.$$

Now, for β even,

$$|1+e^{2\pi i(x_j+(r_1-r_k)/L)}|^\beta = e^{-\pi i\beta(x_j+(r_1-r_k)/L)}(1+e^{2\pi i(x_j+(r_1-r_k)/L)})^\beta,$$

which shows

$$I_{N,n}(\beta;r_1,\dots,r_n) = \prod_{k=2}^n e^{-\pi i\beta N(r_1-r_k)/L}\int_{[-1/2,1/2]^N} dx_1\cdots dx_N\prod_{j=1}^N\Big(e^{-\pi i\beta(n-1)x_j}|1+e^{2\pi i x_j}|^\beta$$
$$\times\prod_{k=2}^n(1+e^{2\pi i(x_j+(r_1-r_k)/L)})^\beta\Big)\prod_{j<k}|e^{2\pi i x_k}-e^{2\pi i x_j}|^\beta.$$

Comparison with (13.6) shows that this integral is of the type therein with

$$m=(n-1)\beta,\quad a-b=-\beta(n-1),\quad a+b=\beta,\quad 2/\alpha=\beta$$

and the t_k as specified in the statement of the proposition. Thus (13.6) shows that $I_{N,n}$ is proportional to

$${}_2F_1^{(\beta/2)}(-N,n;n-2/\beta-N-1;t_1,\dots,t_{(n-1)\beta}).$$

Use of Proposition 13.1.7 gives the form stated in the proposition; the proportionality constant then follows since ${}_2F_1^{(\beta/2)}$ at the origin is unity, while at the origin $I_{N,n}=M_N(n\beta/2,n\beta/2,\beta/2)$. □

According to (13.23) the generalized hypergeometric function in m variables has an m-dimensional integral representation. However, the kernel of this integral representation is itself a generalized hypergeometric function of two sets of variables, which in general does not further simplify. An exception is the case in which all the m variables are equal. Then the particular generalized hypergeometric function of two sets of variables can be summed using the generalized binomial theorem, and the integral representation (13.12) involving only elementary functions holds. Proposition 13.2.1 shows this is precisely the case of the two-point correlation function. Thus we have the following β-dimensional integral representation [190].

PROPOSITION 13.2.2 *For the two-point correlation, (13.32) with $n = 2$, and β even,*

$$\rho_{(2)}(r_1, r_2) = \frac{(N+2)(N+1)}{L^2} \frac{(\beta N/2)!((\beta/2)!)^{N+2}}{(\beta(N+2)/2)!} \frac{M_N(n\beta/2, n\beta/2, \beta/2)}{S_\beta(1-2/\beta, 1-2/\beta, 2/\beta)}$$
$$\times (2\sin\pi(r_1 - r_2)/L)^\beta e^{-\pi i \beta N(r_1-r_2)/L} \int_{[0,1]^\beta} du_1 \cdots du_\beta$$
$$\times \prod_{j=1}^{\beta} (1 - (1 - e^{2\pi i(r_1-r_2)/L}) u_j)^N u_j^{-1+2/\beta} (1-u_j)^{-1+2/\beta} \prod_{j<k} |u_k - u_j|^{4/\beta}.$$

13.2.2 The bulk limit

By using Proposition 13.2.1, the evaluation of $\rho_{(n)}$ in the bulk limit follows immediately from (13.5) and the formulas (4.183), (4.185) and (4.180).

PROPOSITION 13.2.3 *For the n-point correlation (13.32) with β even we have*

$$\rho_{(n)}^{\text{bulk}}(r_1, \dots, r_n) := \lim_{\substack{N,L\to\infty \\ N/L=\rho}} \rho_{(n)}(r_1, \dots, r_n)$$
$$= \rho^n c_n(\beta) \prod_{1 \le j < k \le n} |2\pi\rho(r_k - r_j)|^\beta \prod_{k=2}^{n} e^{\pi i \rho \beta (r_k - r_1)}$$
$$\times {}_1F_1^{(\beta/2)}(n, 2n; -2\pi i\rho(r_2 - r_1), \dots, -2\pi i \rho(r_n - r_1)),$$

where in the argument of ${}_1F_1^{(\beta/2)}$ each $-2\pi i\rho(r_j - r_1)$ $(j = 2, \dots, n)$ occurs β times, and

$$c_n(\beta) = (\beta/2)^{\beta n(n-1)/2} ((\beta/2)!)^n \prod_{k=0}^{n-1} \frac{\Gamma(\beta k/2 + 1)}{\Gamma(\beta(n+k)/2 + 1)}.$$

In the case of the two-point function we have, from Proposition 13.2.2, the integral representation

$$\rho_{(2)}^{\text{bulk}}(r_1, r_2) = \rho^2 (\beta/2)^\beta \frac{((\beta/2)!)^3}{\beta!(3\beta/2)!} \frac{e^{-\pi i \beta\rho(r_1-r_2)} (2\pi\rho(r_1 - r_2))^\beta}{S_\beta(-1+2/\beta, -1+2/\beta, 2/\beta)} \int_{[0,1]^\beta} du_1 \cdots du_\beta$$
$$\times \prod_{j=1}^{\beta} e^{2\pi i \rho(r_1-r_2)u_j} u_j^{-1+2/\beta} (1-u_j)^{-1+2/\beta} \prod_{j<k} |u_k - u_j|^{4/\beta}. \tag{13.35}$$

13.2.3 Asymptotics

The integral representation (13.35) is well suited to study the large $|r_1 - r_2| =: r$ asymptotics [188]. Being a Fourier transform on a finite (multidimensional) interval, its major contributions will come from the neighborhood of the end points. We partition the integration variables into two sets, and expand the integrand about zero in the first set and about unity in the second set of variables. To leading order this procedure gives for the expansion of the integral

$$\frac{e^{\pi i \beta \rho r}}{(2\pi\rho r)^\beta} 2 \sum_{l=0}^{\beta/2} \binom{\beta}{\beta/2 - l} \frac{\cos 2\pi l \rho r}{(2\pi\rho r)^{4l^2/\beta}} J_{\beta/2-l,\beta}(-i) J_{\beta/2+l,\beta}(i), \tag{13.36}$$

where

$$J_{n,\beta}(z) := \int_{(0,\infty)^n} \prod_{j=1}^{n} t_j^{-1+2/\beta} e^{-zt_j} \prod_{1\le j<k\le n} |t_k - t_j|^{4/\beta} \, dt_1 \cdots dt_n. \tag{13.37}$$

The integral (13.37) is convergent for $\mathrm{Re}(z) > 0$, but the formula (13.36) requires its value with $\mathrm{Re}(z) = 0$. However, a simple change of variables $zt_j \mapsto s_j$ shows that

$$J_{n,\beta}(z) = \frac{1}{z^{2n^2/\beta}} J_{n,\beta}(1),$$

which is an analytic function for all $z \neq 0$. Furthermore, according to Proposition 4.7.3 we have the explicit evaluation

$$J_{n,\beta}(1) = \prod_{j=1}^{n} \frac{\Gamma(1+2j/\beta)\Gamma(2j/\beta)}{\Gamma(1+2/\beta)},$$

so all terms in (13.36) are known explicitly.

The next order oscillatory terms are calculated by including in the expansion of the integrand about the endpoints the leading order correction terms. For each $l = 0, \dots, \beta/2$, writing $t_j = 1 - s_j$ ($j = \beta/2 - l + 1, \dots, \beta$), and noting

$$\prod_{j=1}^{\beta/2-l} (1-t_j)^{-1+2/\beta} \prod_{j'=\beta/2-l+1}^{\beta} (1-s_{j'})^{-1+2/\beta} \prod_{a=1}^{\beta/2-l} \prod_{l'=\beta/2-l+1}^{\beta} (1-t_a-s_{l'})^{4/\beta}$$

$$\sim 1 - \Big(1 + \frac{4}{\beta}l + \frac{2}{\beta}\Big) \sum_{j=1}^{\beta/2-l} t_j - \Big(1 - \frac{4}{\beta}l + \frac{2}{\beta}\Big) \sum_{j=\beta/2-l+1}^{\beta} s_j, \tag{13.38}$$

shows that we must evaluate

$$J_{n,\beta}(z)\Big[\sum_{j=1}^{n} t_j\Big] := \int_{(0,\infty)^n} \Big(\sum_{j=1}^{n} t_j\Big) \prod_{j=1}^{n} t_j^{-1+2/\beta} e^{-zt_j} \prod_{1\le j<k\le n} |t_k - t_j|^{4/\beta} \, dt_1 \cdots dt_n.$$

This can be evaluated explicitly for we see immediately that

$$J_{n,\beta}(z)\Big[\sum_{j=1}^{n} t_j\Big] = -\frac{d}{dz} J_{n,\beta}(z) = \frac{(2n^2/\beta)}{z^{2n^2/\beta+1}} J_{n,\beta}(1). \tag{13.39}$$

Consequently, the first order correction to (13.36) is given by

$$\frac{e^{\pi i\beta\rho r}}{(2\pi\rho r)^{\beta}} 2 \sum_{l=0}^{\beta/2} \binom{\beta}{\beta/2 - l} \frac{\sin 2\pi l\rho r}{(2\pi\rho r)^{4l^2/\beta}} b_l J_{\beta/2-l,\beta}(-i) J_{\beta/2+l,\beta}(i), \quad b_l = -\frac{8l}{\beta}\Big(1 - \frac{2l^2}{\beta}\Big). \tag{13.40}$$

The $l = 0$ term in (13.40), which in (13.36) gives the leading order nonoscillatory term (and thus must equal ρ^2), vanishes. To get the next order non-oscillatory term we must therefore expand the case $l = 0$ (equal numbers of variables expanded about the end points) to second order. This is done in Exercises 13.2 q.2. Putting together the above results we thus obtain the desired asymptotic expansion.

PROPOSITION 13.2.4 *The two-point correlation function (13.35) for the log-gas at even β has the large r asymptotic expansion*

$$\frac{1}{\rho^2}\rho_{(2)}(r,0) \sim 1 - \frac{1}{\beta(\pi\rho r)^2} + \mathrm{O}\Big(\frac{1}{r^4}\Big) + 2\sum_{p=1}^{\beta/2} \frac{a_p}{(2\pi\rho r)^{4p^2/\beta}} \Big(\cos 2\pi p\rho r + b_p \frac{\sin 2\pi p\rho r}{2\pi\rho r} + \mathrm{O}\Big(\frac{\cos 2\pi p\rho r}{r^2}\Big)\Big),$$

where

$$a_p = \left(\frac{\prod_{j=1}^{p} \Gamma(2j/\beta+1)}{\prod_{j=1}^{p-1} \Gamma(-2j/\beta+1)} \right)^2$$

and b_p is as in (13.40).

For $\beta = 2$ this agrees with (7.65), and for $\beta = 4$ it agrees with (7.94).

13.2.4 Jacobi β-ensemble

Here we will consider the Jacobi β-ensemble on the interval $[0,1]$, which is obtained from (3.74) by the change of variables $x_j = (1-y_j)/2$, and furthermore we will set $a\beta/2 = \lambda_1$ and $b\beta/2 = \lambda_2$. It follows from (13.25) that for the Jacobi β-ensemble in question, with $N+n$ particles and β even, the n-point correlation is given by

$$\begin{aligned}
\rho_{(n)}(r_1,\dots,r_n) &:= \frac{(N+n)_n}{S_{N+n}(\lambda_1,\lambda_2,\beta/2)} \prod_{k=1}^{n} r_k^{\lambda_1}(1-r_k)^{\lambda_2} \prod_{j<k}^{n} |r_k - r_j|^\beta \\
&\times \int_{[0,1]^N} dx_1 \dots dx_N \prod_{j=1}^{N} \Big(x_j^{\lambda_1}(1-x_j)^{\lambda_2} \prod_{k=1}^{n} (x_j - r_k)^\beta \Big) \prod_{1\le j<k\le N} |x_k - x_j|^\beta \\
&= (N+n)_n \frac{S_N(\lambda_1 + n\beta, \lambda_2, \beta/2)}{S_{N+n}(\lambda_1,\lambda_2,\beta/2)} \prod_{k=1}^{n} t_k^{\lambda_1}(1-t_k)^{\lambda_2} \prod_{j<k}^{n} |t_k - t_j|^\beta \\
&\times {}_2F_1^{(\beta/2)}(-N, 2(\lambda_1+\lambda_2+m+1)/\beta + N - 1; 2(\lambda_1+m)/\beta; t_1,\dots,t_{\beta n})\Big|_{\{t_j\}\mapsto\{r_j\}},
\end{aligned} \tag{13.41}$$

where $m = \beta n$ and

$$t_k = r_j \quad \text{for} \quad k = 1 + (j-1)\beta, \dots, j\beta \ (j = 1,\dots,n). \tag{13.42}$$

In the special case $n = 1$ the arguments of the generalized hypergeometric function in (13.41) are all equal, and we have available β-dimensional integral representations by way of (13.11) and (13.12). However the parameters in the hypergeometric function are such that these integrals are singular. This can be overcome by using the first of the Kummer type relations in Proposition 13.1.6 to note that the hypergeometric function can be transformed to read

$$(1-r)^{\beta N}\, {}_2F_1^{(\beta/2)}\Big(-N, 2(\lambda_1+1)/\beta - N + 1; 2\lambda_1/\beta + 2; -\frac{t_1}{1-t_1}, \dots, -\frac{t_\beta}{1-t_\beta} \Big)\Big|_{t_1=\cdots=t_\beta=r}.$$

Now we can make use of (13.11) to conclude

$$\begin{aligned}
\rho_{(1)}(r) &= (N+1) \frac{S_N(\lambda_1+\beta,\lambda_2,\beta/2)}{S_{N+1}(\lambda_1,\lambda_2,\beta/2)} \frac{r^{\lambda_1}(1-r)^{\lambda_2+\beta N}}{M_\beta(2(\lambda_1+1)/\beta - 1, 2(\lambda_2+1)/\beta + N - 1; 2/\beta)} \\
&\times \int_{-1/2}^{1/2} dx_1 \cdots \int_{-1/2}^{1/2} dx_\beta \prod_{l=1}^{\beta} e^{\pi i x_l (2(\lambda_1-\lambda_2)/\beta)} |1 + e^{2\pi i x_l}|^{2(\lambda_1+\lambda_2+2)/\beta + N - 2} \\
&\times \Big(e^{-\pi i x_l} - \frac{r}{1-r} e^{\pi i x_l} \Big)^N \prod_{1\le j<k\le\beta} |e^{2\pi i x_k} - e^{2\pi i x_j}|^{4/\beta}.
\end{aligned} \tag{13.43}$$

13.2.5 Laguerre β-ensemble

The limiting procedure specified in Section 4.7.1 converting the Selberg integral into the normalization for the Laguerre ensemble allows the Selberg correlation integral to be similarly transformed.

Proposition 13.2.5 *We have*

$$\frac{1}{W_{a+2m/\beta,\beta,N}} \int_0^\infty dx_1 \cdots \int_0^\infty dx_N \prod_{j=1}^N \Big(x_j^{a\beta/2} e^{-\beta x_j/2} \prod_{k=1}^m (x_j - t_k) \Big) \prod_{1 \le j<k \le N} |x_k - x_j|^\beta$$
$$= {}_1F_1^{(\beta/2)}(-N; a + 2m/\beta; t_1, \dots, t_m). \tag{13.44}$$

Proof. In (13.25) set

$$x_j \mapsto x_j/L, \ t_k \mapsto t_k/L, \ \lambda_2 \mapsto \beta L/2, \ \lambda_1 = a\beta/2, \ \lambda = \beta/2$$

and taking the limit $L \to \infty$, using (13.5) on the r.h.s, gives (13.44) up to a proportionality constant. The latter is fixed by requiring that both sides agree at $t_1 = \cdots = t_m = 0$. □

We can use this result to express the general n-point correlation function for the Laguerre ensemble, β even, in terms of the generalized hypergeometric function ${}_1F_1^{(\beta/2)}$. In a system of $N+n$ particles, this correlation is defined by the N-dimensional integral

$$\rho_{(n)}(r_1, \dots, r_n) = \frac{(N+n)_n}{W_{a,\beta,N+n}} \Big(\prod_{j=1}^n r_j^{a\beta/2} e^{-\beta r_j/2} \Big) \prod_{1 \le j<k \le n} |r_k - r_j|^\beta$$
$$\times \int_{(0,\infty)^N} dx_1 \cdots dx_N \prod_{j=1}^N \Big(x_j^{a\beta/2} e^{-\beta x_j/2} \prod_{k=1}^n |r_k - x_j|^\beta \Big) \prod_{j<k} |x_k - x_j|^\beta.$$

For β even this is a special case of the integral in (13.44) with $m = \beta n$ and $\{t_k\}$ related to $\{r_j\}$ by (13.42), and thus

$$\rho_{(n)}(r_1, \dots, r_n) = \frac{(N+n)_n}{W_{a+2n/\beta,\beta,N+n}} W_{a,\beta,N} \Big(\prod_{j=1}^n r_j^{a\beta/2} e^{-\beta r_j/2} \Big) \prod_{1 \le j<k \le n} |r_k - r_j|^\beta$$
$$\times {}_1F_1^{(\beta/2)}(-N; a+2n; t_1, \dots, t_{\beta n}) \Big|_{\{t_j\} \mapsto \{r_k\}}. \tag{13.45}$$

We remark that in the case $n = 1$ the function ${}_1F_1^{(\beta/2)}$ has the β-dimensional integral representation of Exercises 13.1 q.4(i), and thus

$$\rho_{(1)}(r) = (N+1) \frac{W_{a,\beta,N}}{W_{a+2/\beta,\beta,N+1}} \frac{r^{a\beta/2} e^{-\beta r/2}}{M_\beta(2/\beta - 1 + a, N, \beta/2)} \int_{-1/2}^{1/2} dx_1 \cdots \int_{-1/2}^{1/2} dx_\beta$$
$$\times \prod_{l=1}^\beta e^{\pi i x_l (2/\beta - 1 + a - N)} |1 + e^{2\pi i x_l}|^{N+2/\beta-1+a} e^{-r e^{2\pi i x_l}} \prod_{j<k} |e^{2\pi i x_k} - e^{2\pi i x_j}|^{4/\beta}. \tag{13.46}$$

According to (7.30) hard edge scaling requires we compute the limit

$$\rho_{(n)}^{\rm hard}(X_1, \dots, X_n) = \lim_{N \to \infty} \Big(\frac{1}{4N} \Big)^n \rho_{(n)} \Big(\frac{X_1}{4N}, \dots, \frac{X_n}{4N} \Big).$$

Substituting (13.46) and using the confluence (13.5) and the explicit evaluation of $W_{a,\beta,N}$ given in Proposition 4.7.3 shows that for β even

$$\rho_{(n)}^{\rm hard}(X_1,\dots,X_n) = A_n(\beta)\prod_{j=1}^n X_j^{\beta a/2} \prod_{1\le j<k\le n} |X_k - X_j|^\beta \\ \times {}_0F_1^{(\beta/2)}(a+2n; Y_1,\dots,Y_{n\beta})\Big|_{\{Y_j\}\mapsto\{-X_j/4\}}, \tag{13.47}$$

where

$$A_n(\beta) = 2^{-n(2+a\beta+\beta(n-1))}(\beta/2)^{n(1+a\beta+\beta(n-1))}\frac{(\Gamma(1+\beta/2))^n}{\prod_{j=1}^{2n}\Gamma(1+a\beta/2+\beta(j-1)/2)}.$$

For the density ($n=1$), and with $a = c - 2/\beta$, c a positive integer, the generalized hypergeometric function ${}_0F_1^{(\beta/2)}$ can be written as a β-dimensional integral according to (13.27) to give

$$\rho_{(1)}(X) = a(c,\beta)X^{\beta/2-1}\int_{[-\pi,\pi]^\beta}\prod_{j=1}^\beta e^{iX^{1/2}\cos\theta_j}e^{i(c-1)\theta_j}\prod_{1\le j<k\le\beta}|e^{i\theta_k}-e^{i\theta_j}|^{4/\beta}d\theta_1\cdots d\theta_\beta, \tag{13.48}$$

where

$$a(c,\beta) = (-1)^{(c-1)\beta/2}(2\pi)^{-\beta}\frac{1}{2}\Big(\frac{\beta}{4}\Big)^\beta\frac{(\Gamma(1+2/\beta))^\beta}{\Gamma(\beta)}.$$

The correlation integrals (13.44) can also be used to give expressions for some spacing distributions at the hard edge in the Laguerre ensemble. Now, by definition

$$\begin{aligned} E_{N,\beta}(0;(0,s);x^ae^{-\beta x/2}) &= \frac{1}{W_{2a/\beta,\beta,N}}\int_s^\infty dx_1\cdots\int_s^\infty dx_N\prod_{j=1}^N e^{-\beta x_j/2}x_j^a\prod_{j<k}|x_k-x_j|^\beta \\ &= \frac{e^{-N\beta s/2}}{W_{2a/\beta,\beta,N}}\int_0^\infty dx_1\cdots\int_0^\infty dx_N\prod_{j=1}^N e^{-\beta x_j/2}(x_j+s)^a\prod_{j<k}|x_k-x_j|^\beta, \end{aligned} \tag{13.49}$$

where the second equality follows by the change of variables $x_j\mapsto x_j+s$. From this, we can calculate the distribution of the smallest eigenvalue according to

$$\begin{aligned} p_\beta^{(N)}(0;s;a) &= -\frac{d}{ds}E_{N,\beta}(0;(0,s);x^ae^{-\beta x/2}) \\ &= \frac{Ne^{-N\beta s/2}}{W_{2a/\beta,\beta,N}}s^a\int_0^\infty dx_1\cdots\int_0^\infty dx_{N-1}\prod_{j=1}^{N-1}x_j^\beta e^{-\beta x_j/2}(x_j+s)^a\prod_{j<k}|x_k-x_j|^\beta, \end{aligned} \tag{13.50}$$

where the second equality follows by differentiating the first equality in (13.49) and then changing variables $x_j\mapsto x_j+s$.

We see that for $a\in\mathbb{Z}_{\ge0}$, the final integrals in both (13.49) and (13.50) are examples of the correlation integral (13.44) and thus can be expressed in terms of the hypergeometric function ${}_1F_1^{(\beta/2)}$.

PROPOSITION 13.2.6 *For $a\in\mathbb{Z}_{\ge0}$ we have*

$$\begin{aligned} E_{N,\beta}(0;(0,s);x^ae^{-\beta x/2}) &= e^{-\beta Ns/2}{}_1F_1^{(\beta/2)}(-N;2a/\beta;(-s)^a), \\ p_\beta^{(N)}(0;s;a) &= Ns^ae^{-\beta Ns/2}\frac{W_{2a/\beta+2,\beta,N-1}}{W_{a,\beta,N}}{}_1F_1^{(\beta/2)}(-N+1;2a/\beta+2;(-s)^a). \end{aligned}$$

From the above formulas, and with the aid of the confluent limit (13.5), we can calculate the scaled limits

$$E_\beta^{\rm hard}(0;(0,s);a) := \lim_{N\to\infty} E_{N,\beta}(0;(0,s/4N);x^a e^{-\beta x/2}),$$

$$p_\beta^{\rm hard}(0;s;a) := \lim_{N\to\infty} \frac{1}{4N} p_\beta^{(N)}(0;(0,s/4N);a).$$

PROPOSITION 13.2.7 *For $a \in \mathbb{Z}_{\ge 0}$ we have*

$$E_\beta^{\rm hard}(0;(0,s);a) = e^{-\beta s/8} {}_0F_1^{(\beta/2)}(2a/\beta;(s/4)^a),$$

$$p_\beta^{\rm hard}(0;s;a) = A_{a,\beta} s^a e^{-\beta s/8} {}_0F_1^{(\beta/2)}(2a/\beta+2;;(s/4)^a),$$

where

$$A_{a,\beta} = 4^{-(a+1)}(\beta/2)^{2a+1} \frac{\Gamma(1+\beta/2)}{\Gamma(1+a)\Gamma(1+a+\beta/2)}.$$

We know from Exercises 13.1 q.4 that both ${}_1F_1^{(\beta/2)}$ and ${}_0F_1^{(\beta/2)}$ occurring in Propositions 13.2.6 and 13.2.7 have a dimensional integral representations. In particular, this gives for the probability that there are no eigenvalues in the interval $(0,s)$ in the scaled limit the formula

$$E_\beta^{\rm hard}(0;(0,s);a) = e^{-\beta s/8} B_a(2/\beta,\beta/2) \Big(\frac{1}{s}\Big)^{(-1+2/\beta)a/2} \Big(\frac{1}{2\pi}\Big)^a \times \int_{[-\pi,\pi]^a} \prod_{j=1}^a e^{s^{1/2}\cos\theta_j} e^{i(-1+2/\beta)\theta_j} \prod_{1\le j<k\le a} |e^{i\theta_k} - e^{i\theta_j}|^{4/\beta} \, d\theta_1 \cdots d\theta_a, \tag{13.51}$$

where B_a is specified by (13.28).

An immediate application of this formula is the calculation of the large s asymptotic expansion. Now, in this limit the maximum of the factors $e^{s^{1/2}\cos\theta_j}$ in the integrand occur at $\theta_j = 0$. Expanding to leading order about these points gives

$$E_\beta^{\rm hard}(0;(0,s);a) \sim e^{-\beta s/8 + a s^{1/2}} \Big(\frac{1}{s}\Big)^{a(a+1)/2\beta - a/4} \tau_{a,\beta} \Big(1 + {\rm O}\Big(\frac{1}{s^{1/2}}\Big)\Big), \tag{13.52}$$

where

$$\tau_{a,\beta} = 2^{(2/\beta-1)a} \Big(\frac{1}{2\pi}\Big)^a B_a(2/\beta,\beta/2) \int_{(-\infty,\infty)^a} \prod_{j=1}^a e^{-t_j^2/2} \prod_{1\le j<k\le m} |t_k - t_j|^{4/\beta} \, dt_1 \cdots dt_a.$$

The latter integral is evaluated explicitly in Proposition 4.7.1, which together with the explicit form (13.28) for B_m gives the simplification

$$\tau_{a,\beta} = 2^{(2/\beta-1)a} \Big(\frac{1}{2\pi}\Big)^{a/2} \prod_{j=1}^a \Gamma(2j/\beta).$$

In the case $\beta = 2$ this gives $\tau_{a,\beta} = (2\pi)^{-a/2} G(a+1)$, which verifies the conjectured asymptotic form in (9.92) for $a \in \mathbb{Z}_{\ge 0}$.

13.2.6 Gaussian β-ensemble

We know from Section 4.7.1 that changing variables and replacing according to

$$x_j \mapsto \frac{1}{2}\Big(1 - \sqrt{\frac{\beta}{2}\frac{x_j}{L}}\Big), \quad t_j \mapsto \frac{1}{2}\Big(1 - \sqrt{\frac{\beta}{2}\frac{t_j}{L}}\Big), \quad \lambda_1 = \lambda_2 = L^2, \tag{13.53}$$

and taking the limit $L \to \infty$, the Selberg correlation integral of Exercises 13.1 q.1 becomes proportional to

$$I_{m,N}(t_1,\dots,t_m;\beta) := \int_{(-\infty,\infty)^N} dx_1 \cdots dx_N \prod_{j=1}^{N} e^{-\beta x_j^2/2} \prod_{k=1}^{m} (t_k - x_j) \prod_{j<k} |x_k - x_j|^\beta. \tag{13.54}$$

However, unlike in (13.44) the corresponding generalized hypergeometric function does not reduce to another generalized hypergeometric function. Nonetheless (13.54) satisfies analogous p.d.e.'s. To see this note that the substitutions (13.53) in the p.d.e.'s (13.15) with parameters (13.24) satisfied by the Selberg correlation integral tell us that for each $p = 1,\dots,m$,

$$\Big(\frac{\partial^2}{\partial t_p^2} - 2t_p \frac{\partial}{\partial t_p} + 2N + \frac{2}{\alpha} \sum_{\substack{l=1 \\ l\neq p}}^{N} \frac{1}{t_p - t_l} \Big(\frac{\partial}{\partial t_p} - \frac{\partial}{\partial t_l} \Big) \Big) I_{m,N} = 0, \tag{13.55}$$

where $\alpha = \beta/2$. Summing up the equations (13.55) and comparing with (11.53) we see that the resulting equation can be written

$$\tilde{H}_m^{(\mathrm{H})} I_m = -2NmI_{m,N} \tag{13.56}$$

with $\tilde{H}_m^{(\mathrm{H})}$ equal to the operator $\tilde{H}^{(\mathrm{H,Ex})}$ with m coordinates and restricted to symmetric functions.

The logical development of the theory at this stage is to consider the polynomial eigenfunctions of $\tilde{H}^{(\mathrm{H,Ex})}$. This will be carried out in the next section. Here we list some results as relevant for present purposes.

In the case $N = 1$ the monic polynomial eigenfunctions of $\tilde{H}^{(\mathrm{H,Ex})}$ are the Hermite polynomials $2^{-k}H_k(x)$, which have eigenvalue $-2k$, and these polynomials have the integral representation (5.156). Likewise, for general N, $\tilde{H}^{(\mathrm{H,Ex})}$ permits polynomial eigenfunctions $E_\eta^{(\mathrm{H})}(x)$, referred to as *nonsymmetric Hermite polynomials* with the structure

$$E_\eta^{(\mathrm{H})}(x) = E_\eta(x) + \sum_{|\nu|<|\eta|} a_{\eta\nu} E_\nu(x), \tag{13.57}$$

having eigenvalue $-2|\eta|$ and admitting the integral representation

$$E_\eta^{(\mathrm{H})}(x) = \frac{e^{x_1^2+\cdots+x_N^2}}{\mathcal{N}_0^{(\mathrm{H})}} \int_{(-\infty,\infty)^N} {}_0\mathcal{K}_0(2y;-ix) E_\eta(iy)\, d\mu^{(\mathrm{H})}(y), \tag{13.58}$$

where

$$d\mu^{(\mathrm{H})}(y) := e^{-(y_1^2+\cdots+y_N^2)} \prod_{1\le j<k\le N} |y_k - y_j|^{2/\alpha} dy_1 \cdots dy_N,$$

$$\mathcal{N}_0^{(\mathrm{H})} := \int_{(-\infty,\infty)^N} d\mu^{(\mathrm{H})}(y) \tag{13.59}$$

(note that $\mathcal{N}_0^{(\mathrm{H})}$ is essentially the Mehta integral (4.140); it is given in terms of gamma functions by (13.80) below) and ${}_0\mathcal{K}_0$ is specified by (13.17).

Now (13.54) has the structure

$$I_{m,N}(t_1,\dots,t_m;\beta) = (t_1\cdots t_m)^N + \text{lower terms}. \tag{13.60}$$

But as

$$(t_1\cdots t_m)^N = E_{N^m}(t_1,\dots,t_m),$$

and as this is symmetric, it follows from (13.19) and (13.58) that $E_{N^m}^{(\mathrm{H})}$ is symmetric and has the structure

(13.60). Furthermore $E^{(\rm H)}_{N^m}$ is an eigenfunction of $\tilde{H}^{(\rm H)}_m$ with eigenvalue $-2|\eta| = -2Nm$, so we conclude

$$I_{m,N}(t_1,\dots,t_m;\beta) = E^{(\rm H)}_{N^m}(t_1,\dots,t_m)\Big|_{\alpha=\beta/2}. \tag{13.61}$$

Analogous to (13.41) we have that for the Gaussian β-ensemble with $N+n$ particles and β even the n-point correlation function is given in terms of $I_{m,N}$ and thus $E^{(\rm H)}_{N^m}$ by

$$\begin{aligned}\rho_{(n)}(r_1,\dots,r_n) &= (N+n)_n \frac{G_{\beta,N}}{G_{\beta,N+n}} I_{\beta n,N}(\{t_j\}\mapsto\{r_k\};\beta) \\ &= (N+n)_n \frac{G_{\beta,N}}{G_{\beta,N+n}} E^{(\rm H)}_{N^{\beta n}}(\{r_k\})\Big|_{\alpha=\beta/2},\end{aligned} \tag{13.62}$$

where $G_{\beta,N}$ is the normalization in (1.28). In the special case $n=1$, according to (13.42) all the arguments in $E^{(\rm H)}_{N^{\beta n}}(\{r_k\})$ are equal. Noting from (13.18) and (13.3) that

$${}_0\mathcal{K}_0(2y_1,\dots,2y_m;(-it)^m) = {}_0F_0^{(\alpha)}(-2iy_1t,\dots,-2iy_mt) = e^{-2it\sum_{j=1}^m y_j} \tag{13.63}$$

one can then deduce from (13.58) the integral representation

$$\begin{aligned}\rho_{(1)}(r) = (N+1)\frac{G_{\beta,N}}{G_{\beta,N+1}}\frac{e^{-\beta r^2/2}}{\mathcal{N}_0^{(\rm H)}\big|_{\substack{N\mapsto\beta\\ \alpha\mapsto\beta/2}}} \\ \times \int_{(-\infty,\infty)^\beta} du_1\cdots du_\beta \prod_{j=1}^{\beta}(iu_j+r)^N e^{-u_j^2}\prod_{1\le j<k\le\beta}|u_k-u_j|^{4/\beta}.\end{aligned} \tag{13.64}$$

The form (13.64) is suitable for asymptotic analysis [132]. One finds for the global density

$$\begin{aligned}\sqrt{\frac{2}{N}}\rho_{(1)}(\sqrt{2N}x) \sim \rho^{\rm W}(x) - \frac{2}{\pi}\frac{\Gamma(1+2/\beta)}{(\pi\rho^{\rm W}(x))^{6/\beta-1}}\frac{1}{N^{2/\beta}} \\ \times\cos\Big(2\pi NP(x) + (1-2/\beta)\mathrm{Arcsin}\,x\Big) + \mathrm{O}\Big(\min\Big(\frac{1}{N},\frac{1}{N^{8/\beta}}\Big)\Big),\end{aligned} \tag{13.65}$$

where $\rho^{\rm W}(x)$ and $P(x)$ are as in (7.26), while at the soft edge

$$\frac{1}{\sqrt{2}N^{1/6}}\rho_{(1)}\Big(\sqrt{2N}+\frac{x}{\sqrt{2}N^{1/6}}\Big) \sim \frac{\Gamma(1+\beta/2)}{2\pi}\Big(\frac{4\pi}{\beta}\Big)^{\beta/2}\prod_{j=1}^{\beta}\frac{\Gamma(1+2/\beta)}{\Gamma(1+2j/\beta)}K_{\beta,\beta}(x) + \mathrm{O}(N^{-1/3}), \tag{13.66}$$

where

$$K_{n,\beta}(x) := -\frac{1}{(2\pi i)^n}\int_{-i\infty}^{i\infty}dv_1\cdots\int_{-i\infty}^{i\infty}dv_n\prod_{j=1}^{n}e^{v_j^3/3-xv_j}\prod_{1\le k<l\le n}|v_k-v_l|^{4/\beta}. \tag{13.67}$$

We remark that the expansion (13.65) is consistent with the previously computed global expansions of the Gaussian ensemble density at $\beta=2$ and 4 given by (7.25) and (7.98) respectively.

Let us denote the $N\to\infty$ limit of (13.66) by $\rho^{{\rm soft},\beta}_{(1)}(x)$. Asymptotic analysis of the multiple integral

(13.67) shows [132]

$$\rho_{(1)}^{\mathrm{soft},\beta}(x) \underset{x\to\infty}{\sim} \frac{1}{2\pi}\frac{\Gamma(1+\beta/2)}{(4\beta)^{\beta/2}}\frac{e^{-2\beta x^{3/2}/3}}{x^{3\beta/4-1/2}} + \mathrm{O}\Big(\frac{1}{x^{3\beta/4+1}}\Big),$$
$$\rho_{(1)}^{\mathrm{soft},\beta}(x) \underset{x\to-\infty}{\sim} \frac{\sqrt{|x|}}{\pi} - \frac{\Gamma(1+\beta/2)}{2^{6/\beta-1}|x|^{3/\beta-1/2}}\cos\Big(\frac{4}{3}|x|^{3/2} - \frac{\pi}{2}\Big(1-\frac{2}{\beta}\Big)\Big) + \mathrm{O}\Big(\frac{1}{|x|^{5/2}}, \frac{1}{|x|^{6/\beta-1/2}}\Big). \tag{13.68}$$

The latter result is consistent with the expansions (7.69), (7.100) given previously for $\beta = 2, 4$, while the former is consistent with the expansion (7.72) obtained previously for $\beta = 2$.

EXERCISES 13.2 1. Use the formula (13.16) to verify that the expression for $\rho_{(n)}$ in Proposition 13.2.3 is real. Similarly, use the first transformation formula in Proposition 13.1.6 to verify that the expression for $\rho_{(n)}$ in Proposition 13.2.1 is real.

2. (i) Suppose that the integrand in (13.35) has $\beta/2$ variables expanded about $t_j = 0$ and $\beta/2$ variables expanded about $t_{j'} = 1$. Show that the next order term in the expansion (13.38) is then

$$\Big(1+\frac{4}{\beta^2}\Big)\Big(\sum_{j=1}^{\beta/2} t_j\Big)\Big(\sum_{j=\beta/2+1}^{\beta} s_j\Big) + \frac{1}{2}\Big(1+\frac{2}{\beta}\Big)^2\Big(\sum_{j=1}^{\beta/2} t_j\Big)^2 + \frac{1}{2}\Big(1+\frac{2}{\beta}\Big)^2\Big(\sum_{j=\beta/2+1}^{\beta} s_j\Big)^2$$
$$- \frac{(1+2/\beta)}{2}\sum_{j=1}^{\beta/2} t_j^2 - \frac{(1+2/\beta)}{2}\sum_{j=\beta/2+1}^{\beta} s_j^2.$$

(ii) Use the result (13.39), and its derivative, to evaluate $J_{n,\beta}(z)[\sum_{j=1}^n t_j]$ and $J_{n,\beta}(z)[(\sum_{j=1}^n t_j)^2]$. Also, note from (12.147) that

$$\sum_{j=1}^n t_j^2 := m_{20^{n-1}} = P_{20^{n-1}}(t_1,\dots,t_n) - \frac{2}{1+\alpha}P_{110^{n-2}}(t_1,\dots,t_n),$$

and use this together with (12.153) to deduce that

$$J_{\beta/2,\beta}(z)\Big[\sum_{j=1}^n t_j^2\Big] = \frac{1}{z^2}\Big(\frac{-2/\beta+2+3\beta/2}{1+2/\beta}\Big)J_{n,\beta}(z).$$

(iii) Use the results of (ii) to show that the total contribution of the second order terms in (i) to the asymptotic expansion of (13.35) is

$$-\frac{\rho^2}{\beta(\pi\rho r)^2}.$$

3. [39] The nonsymmetric Hermite polynomials $E_\eta^{(\mathrm{H})}$ can be specified as the polynomials with the structure (13.57) which satisfy the eigenfunction equation $h_i E_\eta^{(\mathrm{H})} = \bar{\eta}_i E_\eta^{(\mathrm{H})}$ for each $i = 1,\dots,N$. The nonsymmetric Laguerre polynomials $E_\eta^{(\mathrm{L})}$ are even functions of the variables $y = (y_1,\dots,y_N)$, and this is indicated by writing $E_\eta^{(\mathrm{L})}(y^2)$. They can be specified as the polynomials with the structure (13.57) (with $x \mapsto y^2$) which satisfy the eigenfunction equation $l_i E_\eta^{(\mathrm{L})}(y^2) = \bar{\eta}_i E_\eta^{(\mathrm{L})}(y^2)$ for each $i = 1,\dots,N$. In this exercise Rodrigues-type formulas for these polynomials will be derived.

(i) Apply the map $\phi^{(A)}$ specified by (11.75), (11.76) to the eigenvalue equation (12.9), to show

$$h_i E_\eta\Big(y - \frac{1}{2}d\Big)\cdot 1 = \bar{\eta}_i E_\eta\Big(y-\frac{1}{2}d\Big)\cdot 1$$

where $E_\eta(y - \frac{1}{2}d)$ denotes the polynomial $E_\eta(y)$ with each variable y_i replaced by $y_i - \frac{1}{2}d_i$. Deduce from this that

$$E_\eta^{(\mathrm{H})}(y) = E_\eta\Big(y - \frac{1}{2}d\Big)\cdot 1 \tag{13.69}$$

(cf. (12.92)).

(ii) Use the map $\phi^{(B)}$ specified by (11.81), (11.82) to show

$$E_\eta^{(\mathrm{L})}(y^2) = E_\eta\Big(\Big(y - \frac{1}{2}d^{(B)}\Big)^2\Big)\cdot 1. \tag{13.70}$$

4. [187] Use the fact that the integrand in (13.48) is periodic to translate the domain of integration to $[-\pi+\epsilon, \pi+\epsilon]$, $0 < \epsilon \ll 1$. By then expanding $\beta/2$ of the variables about $\theta = 0$ and $\beta/2$ variables about $\theta = \pi$, deduce that for $X \to \infty$

$$\rho_{(1)}(X) \sim |a(c,\beta)| X^{\beta/2-1} 2^\beta \binom{\beta}{\beta/2} \Big| \int_{(-\infty,\infty)^{\beta/2}} e^{iX^{1/2}(\theta_1^2+\cdots+\theta_{\beta/2}^2)/2} \prod_{1\le j<k\le\beta/2} |\theta_k - \theta_j|^{4/\beta} d\theta_1\cdots d\theta_{\beta/2}\Big|^2.$$

Change variables $iX^{1/2}\theta_k^2 \mapsto -\lambda_k^2$ and use the evaluation (1.160) to deduce from this the explicit asymptotic form

$$\rho_{(1)}(X) \sim \frac{1}{2\pi X^{1/2}}.$$

Note that this is consistent with (7.74) and (7.114).

5. (i) By an appropriate change of variables show that

$$\begin{aligned} &E_{N,\beta}(0; (0,t); x^a(1-x)^b\chi_{0<x<1}) \\ &= (1-t)^{b+N+\beta N(N-1)/2}\frac{1}{C}\int_0^1 dx_1\cdots\int_0^1 dx_N \prod_{l=1}^N \Big(x_l + \frac{t}{1-t}\Big)^a (1-x_l)^b \prod_{1\le j<k\le N} |x_k - x_j|^\beta, \end{aligned}$$

where C is such that the r.h.s. equals unity at $t = 0$. With the hard edge scaling specified by $t = X/4N^2$, as is consistent with the scale identified in (7.45) after a change of variables appropriate for the Jacobi ensemble defined on $(0,1)$, use this result with $a = 0$ to reclaim the result of Proposition 13.2.7 in the case $a = 0$.

(ii) Make use of (13.10) in the case $m = 1$, $\lambda_1 = \lambda_2 = 0$, $\alpha = 2$ to deduce that

$$\begin{aligned} &\int_0^1 dx_1\cdots\int_0^1 dx_N \prod_{l=1}^N \Big(x_l + \frac{t}{1-t}\Big)^{-1/2} \prod_{1\le j<k\le N} |x_k - x_j| \\ &\propto \Big(\frac{1}{t} - 1\Big)^{-N/2} {}_2F_1\Big(\frac{N}{2}, \frac{N+1}{2}; N+1; 1-\frac{1}{t}\Big) \\ &= (1-t)^{-N/2} {}_2F_1\Big(\frac{N}{2}, \frac{N+1}{2}; N+1; 1-t\Big) \end{aligned}$$

where the equality follows upon use of the first of the Kummer relations in Proposition 13.1.6.

(iii) By use of the connection formula [541]

$${}_2F_1(a,b;c;z) = A\,{}_2F_1(a,b;a+b-c;1-z) + B\,{}_2F_1(c-a,c-b;c-a-b+1;1-z),$$

$$A := \frac{\Gamma(c)\Gamma(c-a-b)}{\Gamma(c-a)\Gamma(c-b)}, \qquad B := \frac{\Gamma(c)\Gamma(a+b-c)}{\Gamma(a)\Gamma(b)}$$

deduce that

$$\lim_{N\to\infty} {}_2F_1\Big(\frac{N}{2}, \frac{N+1}{2}; N+1; -\frac{X}{4N^2}\Big) \propto {}_0F_1\Big(\frac{1}{2}; \frac{X}{16}\Big) - \frac{X^{1/2}}{4}\,{}_0F_1\Big(\frac{3}{2}; \frac{X}{16}\Big) = e^{\sqrt{X}/4},$$

where the equality follows from (10.190) and (7.50).

(iv) [154] Use the results of (i)–(iii) to show

$$E_1^{\rm hard}(0;(0,X);a=-1/2)=e^{-X/4+\sqrt{X}/4}$$

and hence by following the reasoning of Exercises 8.3 q.4 conclude that the p.d.f. for $\tilde{\kappa}(\mathbf{A})$ in the case of real Gaussian matrices is equal to

$$\frac{y+2}{8y^3}e^{-1/(2y)-1/(8y^2)}$$

(use must be made too of Proposition 3.2.2).

6. (i) Suppose $\mu\in\mathbb{Z}^+$. By manipulating the integrands as implied by the average, making use of Proposition 13.1.4, and applying Proposition 13.1.7, show that

$$\Big\langle\prod_{l=1}^N|z-e^{i\theta_l}|^{2\mu}\Big\rangle_{{\rm CE}^0_{\beta,N}}\propto{}_2F_1^{(2/\beta)}(-\mu,-\mu;-2\mu;(1-|z|^2)^N).$$

Now apply Proposition 13.1.2 to conclude

$$\Big\langle\prod_{l=1}^N|z-e^{i\theta_l}|^{2\mu}\Big\rangle_{{\rm CE}^0_{\beta,N}}\propto\Big\langle\prod_{l=1}^{\mu}\Big(1-(1-|z|^2)x_l\Big)^N\Big\rangle_{{\rm ME}_{4/\beta,\mu}(x^\alpha(1-x)^\alpha)},\tag{13.71}$$

where $\alpha=2/\beta-1$.

(ii) Suppose $\mu\in\mathbb{Z}^+$. By making use of (13.12), then transforming according to the theory of Exercises 13.1 q.1 (iv), and finally applying the first of the Kummer type transformations of Proposition 13.1.6, show that

$$\Big\langle\prod_{l=1}^N(x-x_l)^{2\mu}\Big\rangle_{{\rm ME}_{\beta,N}(x^a(1-x)^b)}\propto(1-x)^{2\mu N}{}_2F_1^{(2/\beta)}(-2\mu,1+(N-1)\beta/2+b;-2\mu-a;(x/(x-1))^N).$$

Now make use of (13.6) to conclude this average is proportional to

$$\Big\langle\prod_{l=1}^{2\mu}e^{i\theta_l(a-b)/\beta}|1+e^{i\theta_l}|^{2(a+b+2)/\beta+N-2}((1-x)e^{-i\theta_l/2}-xe^{i\theta_l/2})^N\Big\rangle_{{\rm CE}^0_{4/\beta,2\mu}}.\tag{13.72}$$

13.3 GENERALIZED CLASSICAL POLYNOMIALS

The nonsymmetric Hermite polynomials $E_\eta^{(\rm H)}$ have been introduced in Section 13.2.6, while the nonsymmetric Laguerre polynomials have appeared in Exercises 13.2 q.3. Here, following [38], the properties of both will be further developed. A brief development of the theory of the generalized Jacobi polynomials is also given.

13.3.1 Generalized Hermite polynomials

In Section 13.2.6 the nonsymmetric Hermite polynomials $E_\eta^{(\rm H)}$ were defined as the eigenfunctions of $\tilde{H}^{(\rm H,Ex)}$ with the structure (13.57). In Exercises 13.2 q.3 it was remarked that they can alternatively be specified as simultaneous eigenfunctions of $\{h_i\}$ with corresponding eigenvalues $\bar{\eta}_i$. To see the consistency between these two specifications, note from the definition (11.70) that h_i consists of ξ_i, a degree-preserving operator, and d_i^2 which lowers the degree by 2. Thus the h_i have the action

$$h_iE_\eta(y)=\bar{\eta}_iE_\eta(y)+\sum_{|\nu|<|\eta|}\tilde{b}_{\eta\nu}E_\nu(y).$$

Proceeding now as in the working which leads to (12.9), we can use this to conclude the existence of a simultaneous set of polynomial eigenfunctions of the operator

$$X^{(\mathrm{H})}(\{u\}) := \prod_{l=1}^{N}(1 + u_l h_l) \tag{13.73}$$

of the form (13.57), with $\{a_{\eta\,\nu}\}$ therein independent of $\{u_l\}$. The relation (11.70) then gives that the $E_\eta^{(\mathrm{H})}$ are also polynomial eigenfunctions of $H^{(\mathrm{H},\mathrm{Ex})}$ with eigenvalue $-2|\eta|$.

Orthogonality

The polynomials $\{E_\eta^{(\mathrm{H})}\}$ are orthogonal with respect to the inner product

$$\langle f|g\rangle^{(\mathrm{H})} := \int_{(-\infty,\infty)^N} f(y)g(y)\, d\mu^{(\mathrm{H})}(y), \tag{13.74}$$

where $d\mu^{(\mathrm{H})}(y)$ is given by (13.59), so that

$$\langle E_\nu^{(\mathrm{H})}|E_\eta^{(\mathrm{H})}\rangle^{(\mathrm{H})} = \mathcal{N}_\eta^{(\mathrm{H})}\delta_{\nu,\eta}. \tag{13.75}$$

This fact relies on the eigenoperator (13.73) separating the eigenvalues and being self-adjoint with respect to (13.74). The self-adjointness property of $X^{(\mathrm{H})}(u)$ in turn follows from the fact that each $h_\imath$ is self-adjoint with respect to (13.74).

PROPOSITION 13.3.1 *We have*

$$\langle f|d_i g\rangle^{(\mathrm{H})} = \langle (2y_i - d_i)f|g\rangle^{(\mathrm{H})}$$

and thus

$$\langle f|h_i g\rangle^{(\mathrm{H})} = \langle h_i f|g\rangle^{(\mathrm{H})}.$$

Proof. The first equation is verified by direct substitution of Definition 11.4.1 and integration by parts. Denote the adjoint just determined by d_i^* so that $d_i^* = 2y_i - d_i$. Then according to (11.76) we can write

$$h_i = \frac{\alpha}{2} d_i^* d_i + 1 - N + \sum_{p=i+1}^{N} s_{ip}$$

which immediately implies the second result. □

Recurrences

We know that all the nonsymmetric Jack polynomials can be generated from the constant polynomial $E_{0^N}(z) := 1$ by application of just two fundamental operators: the transposition s_i and the raising type operator Φ. Also the explicit form of the lowering operator $\hat{\Phi}$ was identified. The analogous operators for the $E_\eta^{(\mathrm{H})}(y)$ can be deduced by simply applying the map $\phi^{(A)}$ as specified by (11.75) to the corresponding formulas for $E_\eta(z)$ in Propositions 12.2.1, 12.2.3 and 12.2.5, and making use of (11.76) and (13.69).

PROPOSITION 13.3.2 *We have*

$$s_i E_\eta^{(\mathrm{H})}(y) = \begin{cases} \frac{1}{\delta_{i,\eta}} E_\eta^{(\mathrm{H})}(y) + \left(1 - \frac{1}{\delta_{i,\eta}^2}\right) E_{s_i\eta}^{(\mathrm{H})}(y), & \eta_i > \eta_{i+1}, \\ E_\eta^{(\mathrm{H})}(y), & \eta_i = \eta_{i+1}, \\ \frac{1}{\delta_{i,\eta}} E_\eta^{(\mathrm{H})}(y) + E_{s_i\eta}^{(\mathrm{H})}(y), & \eta_i < \eta_{i+1}, \end{cases} \tag{13.76}$$

$$\hat{\Phi}^* E_\eta^{(\mathrm{H})}(y) = 2 E_{\Phi\eta}^{(\mathrm{H})}(y), \quad \hat{\Phi}^* := s_{N-1} \cdots s_1 d_1^* = s_{N-1} \cdots s_1 (2y_1 - d_1) \tag{13.77}$$

$$\hat{\Phi} E_\eta^{(\mathrm{H})}(y) = \frac{1}{\alpha} \frac{d'_\eta}{d'_{\Phi\eta}} E_{\Phi\eta}^{(\mathrm{H})}(y), \tag{13.78}$$

where $\hat{\Phi}^*$ is the adjoint of $\hat{\Phi}$ with respect to the inner product (13.74).

Analogous to the determination of the normalization $\mathcal{N}_\eta^{(\mathrm{C})}$ using the operators s_i and Φ given in the proof of Proposition 12.3.3, the normalization of $E_\eta^{(\mathrm{H})}$ with respect to the inner product (13.74) can be determined using the operators s_i and $\hat{\Phi}^*$.

PROPOSITION 13.3.3 *We have*

$$\mathcal{N}_\eta^{(\mathrm{H})} := \langle E_\eta^{(\mathrm{H})} | E_\eta^{(\mathrm{H})} \rangle^{(\mathrm{H})} = \frac{1}{(2\alpha)^{|\eta|}} \frac{d'_\eta e_\eta}{d_\eta} \mathcal{N}_0^{(\mathrm{H})}, \tag{13.79}$$

where

$$\mathcal{N}_0^{(\mathrm{H})} := \mathcal{N}_{0^N}^{(\mathrm{H})} = \langle 1 | 1 \rangle^{(\mathrm{H})} = 2^{-N(N-1)/2\alpha} \pi^{N/2} \prod_{j=0}^{N-1} \frac{\Gamma(1 + (j+1)/\alpha)}{\Gamma(1 + 1/\alpha)}. \tag{13.80}$$

Proof. Because the formula (13.76) is formally the same as in Proposition 12.2.1, we have analogous to (12.42)

$$\langle E_{s_i\eta}^{(\mathrm{H})} | E_{s_i\eta}^{(\mathrm{H})} \rangle^{(\mathrm{H})} = (1 - \bar{\delta}_{i,\eta}^{-2}) \langle E_\eta^{(\mathrm{H})} | E_\eta^{(\mathrm{H})} \rangle^{(\mathrm{H})}. \tag{13.81}$$

Furthermore, taking the inner product of both sides of (13.77) with $E_{\Phi\eta}^{(\mathrm{H})}$ and dividing by 2 gives

$$\langle E_{\Phi\eta}^{(\mathrm{H})} | E_{\Phi\eta}^{(\mathrm{H})} \rangle^{(\mathrm{H})} = \frac{1}{2} \langle E_{\Phi\eta}^{(\mathrm{H})} | \hat{\Phi}^* E_\eta^{(\mathrm{H})} \rangle^{(\mathrm{H})} = \frac{1}{2} \langle \hat{\Phi} E_{\Phi\eta}^{(\mathrm{H})} | E_\eta^{(\mathrm{H})} \rangle^{(\mathrm{H})}. \tag{13.82}$$

But a simple change of variables in (13.78) gives

$$\hat{\Phi} E_{\Phi\eta}^{(\mathrm{H})} = \frac{1}{\alpha} \frac{d'_{\Phi\eta}}{d'_\eta} E_\eta^{(\mathrm{H})},$$

so we have

$$\langle E_{\Phi\eta}^{(\mathrm{H})} | E_{\Phi\eta}^{(\mathrm{H})} \rangle^{(\mathrm{H})} = \frac{1}{2\alpha} \frac{d'_{\Phi\eta}}{d'_\eta} \langle E_\eta^{(\mathrm{H})} | E_\eta^{(\mathrm{H})} \rangle^{(\mathrm{H})}. \tag{13.83}$$

The formulas (13.83) and (13.82) completely determine $\langle E_\eta^{(\mathrm{H})}, | E_\eta^{(\mathrm{H})} \rangle^{(\mathrm{H})}$ starting from $\mathcal{N}_{0^N}^{(\mathrm{H})} = \langle 1 | 1 \rangle^{(\mathrm{H})}$, the value of which is given by Proposition 4.7.1. It follows immediately from (12.38) that the r.h.s. of (13.79) obeys the same recurrences, implying the result. □

Generating function

The classical Hermite polynomials can be defined by the generating function

$$\sum_{k=0}^{\infty} \frac{H_k(x) y^k}{k!} = e^{2xy} e^{-y^2}. \tag{13.84}$$

A similar result holds for the nonsymmetric Hermite polynomials, with the role of the exponential e^{2xy} played by the *type A Dunkl kernel* $\mathcal{K}_A(2x;y)$, which is specified in terms of the hypergeometric function (13.17) by

$$\mathcal{K}_A(x;y) = {}_0\mathcal{K}_0(x;y) = \sum_\eta \alpha^{|\eta|} \frac{d_\eta}{d'_\eta e_\eta} E_\eta(x) E_\eta(y) \tag{13.85}$$

(here (12.39) has been substituted for $E_\eta((1)^N)$). We first require some additional properties of $\mathcal{K}_A(2x;y)$.

PROPOSITION 13.3.4 *We have*

$$\hat{\Phi}^{(x)}\, \mathcal{K}_A(x;y) = \Phi^{(y)}\, \mathcal{K}_A(x;y), \qquad d_i^{(x)}\, \mathcal{K}_A(x;y) = y_i\, \mathcal{K}_A(x;y). \tag{13.86}$$

Proof. The first equality follows by applying $\hat{\Phi}^{(x)}$ term by term to the series definition (13.17) using Proposition 12.2.5, writing $\nu = \hat{\Phi}\eta$ in the summation, then using the final formula of Proposition 12.2.3 to rewrite $E_{\Phi\nu}(y)$ in terms of $\Phi^{(y)}\, E_\nu(y)$. The second result follows by writing y_i and $d_i^{(x)}$ in terms of $\Phi^{(y)}$ and $\hat{\Phi}^{(x)}$, respectively, and using the first equality. □

With $\tilde{E}_1$ specified by (12.183) and Δ_A by (11.67), we see from (11.68) that the Hamiltonian (11.64) can be written as

$$\tilde{H}^{(\mathrm{H,Ex})} = \Delta_A - 2\tilde{E}_1. \tag{13.87}$$

Now from the definition (11.67) of Δ_A, it follows immediately from the second equality in (13.86) that

$$\Delta_A^{(x)} \mathcal{K}_A(2x;y) = 4p_2(y)\mathcal{K}_A(2x;y) \tag{13.88}$$

(recall the meaning of $p_2(y)$ as specified in (12.146)). Also, the fact that the E_η are homogeneous of degree $|\eta|$ implies

$$\tilde{E}_1^{(y)} E_\eta(y) = |\eta| E_\eta(y), \tag{13.89}$$

where $\tilde{E}_1^{(y)}$ is given by (12.183) with $\{x\} \mapsto \{y\}$. The generating function can now be established.

PROPOSITION 13.3.5 *We have*

$$\sum_\eta \frac{(2\alpha)^{|\eta|} d_\eta}{e_\eta d'_\eta} E_\eta^{(\mathrm{H})}(x) E_\eta(y) = \mathcal{K}_A(2x;y) e^{-p_2(y)}. \tag{13.90}$$

Proof. From the definition (13.85) we see that

$$\mathcal{K}_A(2x;y) e^{-p_2(y)} = \sum_\eta \frac{(2\alpha)^{|\eta|} d_\eta}{e_\eta d'_\eta} Q_\eta(x) E_\eta(y),$$

where $Q_\eta(x)$ is a polynomial with leading term $E_\eta(x)$. Applying $\tilde{E}_1^{(y)}$ to both sides using (13.89) and the formula

$$\begin{aligned} \tilde{E}_1^{(y)} \left(\mathcal{K}_A(2x;y) e^{-p_2(y)} \right) &= e^{-p_2(y)} \tilde{E}_1^{(y)} \mathcal{K}_A(2x;y) - 2p_2(y) \mathcal{K}_A(2x;y) e^{-p_2(y)} \\ &= e^{-p_2(y)} \tilde{E}_1^{(x)} \mathcal{K}_A(2x;y) - \frac{1}{2} \Delta_A^{(x)} \mathcal{K}_A(2x;y), \end{aligned}$$

where use has been made of (13.88), shows

$$\left(\tilde{E}_1^{(x)} - \frac{1}{2} \Delta_A^{(x)} \right) e^{-p_2(y)} \mathcal{K}_A(2x;y) = \sum_\eta \frac{(2\alpha)^{|\eta|} d_\eta}{e_\eta d'_\eta} |\eta| Q_\eta(x) E_\eta(y).$$

Substituting (13.85) on the l.h.s, recalling (13.87) and equating $E_\eta(y)$ on both sides shows

$$\tilde{H}^{(\mathrm{H,Ex})} Q_\eta(x) = -2|\eta| Q_\eta(x).$$

The unique polynomial eigenfunction solution of this equation with leading term E_η is $E_\eta^{(\mathrm{H})}$, and the stated formula follows. □

By combining the generating function with a corollary of (13.88), an exponential operator can be identified which maps the Jack polynomials to the nonsymmetric Hermite polynomials. First, note that (13.88) implies

$$\Big(\frac{1}{4}\Delta_A^{(x)}\Big)^k \mathcal{K}_A(2x;y) = \Big(p_2(y)\Big)^k \mathcal{K}_A(2x;y),$$

which after multiplication by $(-1)^k/k!$ and summing over k gives

$$\exp\Big(-\frac{1}{4}\Delta_A^{(x)}\Big)\mathcal{K}_A(2x;y) = e^{-p_2(y)}\mathcal{K}_A(2x;y).$$

Using the generating function (13.90) on the r.h.s. and the series formula (13.85) on the l.h.s, and equating coefficients of $E_\eta(y)$ on both sides shows that $\exp(-\frac{1}{4}\Delta_A)$ has the desired action,

$$\exp\Big(-\frac{1}{4}\Delta_A\Big)E_\eta(x) = E_\eta^{(\mathrm{H})}(x). \tag{13.91}$$

Now $E_\eta(x)$ is an eigenfunction of ξ_j with eigenvalue $\bar\eta_j$, while $E_\eta^{(\mathrm{H})}(x)$ is an eigenfunction of h_j with the same eigenvalue. Hence (13.91) remains valid with the replacements $E_\eta(x) \mapsto \xi_j E_\eta(x)$ and $E_\eta^{(\mathrm{H})}(x) \mapsto h_j \exp(-\frac{1}{4}\Delta_A)E_\eta(x)$. This shows that we have the intertwining relation

$$\exp\Big(-\frac{1}{4}\Delta_A\Big)\xi_j = h_j \exp\Big(-\frac{1}{4}\Delta_A\Big). \tag{13.92}$$

Integration formulas

The Dunkl kernel $\mathcal{K}_A(x,y)$ occurs as the kernel in an integral transform that relates the nonsymmetric Jack and Hermite polynomials. One of these is (13.58), while its inverse and the transform underlying these results are given by the following.

PROPOSITION 13.3.6 *Let $d\mu^{(\mathrm{H})}(y)$ be given by (13.59), and let $\mathcal{N}_{0N}^{(\mathrm{H})}$ be as in (13.79). Then we have*

$$\int_{(-\infty,\infty)^N} \mathcal{K}_A(2y;z)\mathcal{K}_A(2y;w)\,d\mu^{(\mathrm{H})}(y) = \mathcal{N}_{0N}^{(\mathrm{H})} e^{p_2(w)+p_2(z)}\mathcal{K}_A(2z;w), \tag{13.93}$$

$$\int_{(-\infty,\infty)^N} \mathcal{K}_A(2y;z)E_\eta^{(\mathrm{H})}(y)\,d\mu^{(\mathrm{H})}(y) = \mathcal{N}_{0N}^{(\mathrm{H})} e^{p_2(z)}E_\eta(z). \tag{13.94}$$

Proof. To derive the first formula, multiply both sides by $e^{-p_2(w)-p_2(z)}$, use the generating function (13.90) twice on the l.h.s., and then use the orthogonality of $\{E_\eta^{(\mathrm{H})}(y)\}$ with respect to (13.74) to compute the integral. The resulting sum is identified as proportional to $\mathcal{K}_A(2z;w)$. The second formula follows from the first after multiplying by $e^{-p_2(w)}$, using the generating function on the l.h.s., and equating coefficients of $E_\eta(w)$ on both sides. We remark that (13.58) is derived from (13.93) similarly, by replacing z by iz and w by $-iw$, then using the generating function on the r.h.s., and equating coefficients of $E_\eta(w)$. □

We remark that in the case $N=1$ (13.94) reduces to the classical identity (5.153).

Symmetric Hermite polynomials

The symmetric Hermite polynomials $P_\kappa^{(\mathrm{H})}(y;\alpha) := P_\kappa^{(\mathrm{H})}(y)$ can be defined as the eigenfunctions of the operator (11.71) with the structure

$$P_\kappa^{(\mathrm{H})}(y) = P_\kappa(y) + \sum_{|\mu|<|\kappa|} a_{\kappa\mu}P_\mu(y). \tag{13.95}$$

Analogous to the case of the symmetric Jack polynomials (recall Section 12.6), the linear combination

$$\sum_{\eta:\eta^+=\kappa} \tilde{a}_\eta E_\eta^{(\mathrm{H})}(y)$$

is an eigenfunction of $X^{(\mathrm{H})}(u)$. Furthermore, according to (12.96) and (13.57), choosing $\tilde{a}_\eta = d'_{\eta^+}/d_\eta$ gives the leading term equal to $P_\kappa(y)$ as required by (13.95). Because the leading term is symmetric, the polynomial itself must be symmetric ($X^{(\mathrm{H})}(u)$ maps symmetric functions to symmetric functions), so we must have

$$P_\kappa^{(\mathrm{H})}(y) = d'_{\eta^+} \sum_{|\eta|:\eta^+=\kappa} \frac{1}{d'_\eta} E_\eta^{(\mathrm{H})}(y). \tag{13.96}$$

Alternatively, analogous to (12.100), the symmetric Hermite polynomials can be constructed from their nonsymmetric counterparts by symmetrization. Thus

$$P_\kappa^{(\mathrm{H})}(y) = \frac{1}{a_\eta} \mathrm{Sym}\, E_\eta^{(\mathrm{H})}(y), \tag{13.97}$$

where $\kappa = \eta^+$ and a_η is given by (12.101). Applying Sym to both sides of the generating function (13.90), making use of Proposition 13.1.10, (12.100) and (13.96), we obtain the generating function for the symmetric Hermite polynomials,

$$\sum_{|\kappa|} \frac{(2\alpha)^{|\kappa|}}{d'_\kappa} \frac{P_\kappa^{(\mathrm{H})}(x) P_\kappa(y)}{P_\kappa((1)^N)} = {}_0\mathcal{F}_0^{(\alpha)}(2x; y) e^{-p_2(y)}. \tag{13.98}$$

Furthermore, applying Sym to both sides of (13.91) shows that

$$\exp\Big(-\frac{1}{4}D_0\Big) P_\kappa(y) = P_\kappa^{(\mathrm{H})}(y). \tag{13.99}$$

Here D_0 is the generalized Laplacian Δ_A restricted to symmetric functions, which from (11.68) has the explicit form

$$D_0 = \sum_{j=1}^N \frac{\partial^2}{\partial y_j^2} + \frac{2}{\alpha} \sum_{j<k} \frac{1}{y_j - y_k}\Big(\frac{\partial}{\partial y_j} - \frac{\partial}{\partial y_k}\Big).$$

The integration formulas of Proposition 13.3.6 can also be symmetrized. Consider in particular the formula (13.58). Applying $\mathrm{Sym}^{(z)}$ to both sides, making use of (13.97) and Proposition 13.1.10 gives

$$N! \int_{(-\infty,\infty)^N} {}_0\mathcal{F}_0^{(\alpha)}(2y; -iz) E_\eta(iy)\, d\mu^{(\mathrm{H})}(y) = \mathcal{N}_{0^N}^{(\mathrm{H})} e^{-p_2(z)} a_\eta P_{\eta^+}^{(\mathrm{H})}(z).$$

But ${}_0\mathcal{F}_0^{(\alpha)}(2y; -iz)$ is symmetric in $\{y_i\}$ so

$$\begin{aligned} N! \int_{(-\infty,\infty)^N} {}_0\mathcal{F}_0^{(\alpha)}(2y; -iz) E_\eta(iy)\, d\mu^{(\mathrm{H})}(y) &= \int_{(-\infty,\infty)^N} {}_0\mathcal{F}_0^{(\alpha)}(2y; -iz) \mathrm{Sym} E_\eta(iy)\, d\mu^{(\mathrm{H})}(y) \\ &= a_\eta \int_{(-\infty,\infty)^N} {}_0\mathcal{F}_0^{(\alpha)}(2y; -iz) P_{\eta^+}(iy)\, d\mu^{(\mathrm{H})}(y). \end{aligned}$$

Hence the symmetric analogue of (13.58) is

$$\int_{(-\infty,\infty)^N} {}_0\mathcal{F}_0^{(\alpha)}(2y; -iz) P_\kappa(iy)\, d\mu^{(\mathrm{H})}(y) = \mathcal{N}_{0^N}^{(\mathrm{H})} e^{-p_2(z)} P_\kappa^{(\mathrm{H})}(z). \tag{13.100}$$

The formulas (13.96) and (13.97) together allow the norm of the $P_\kappa^{(\mathrm{H})}$ with respect to (13.74) to be computed from knowledge of the nonsymmetric norm (13.79).

PROPOSITION 13.3.7 *We have*

$$\langle P_\kappa^{(\mathrm{H})} | P_\kappa^{(\mathrm{H})} \rangle^{(\mathrm{H})} = (2\alpha)^{|\kappa|} \frac{P_\kappa((1)^N)}{d'_\kappa} \mathcal{N}_{0^N}^{(\mathrm{H})}.$$

Proof. This is derived in a similar fashion to the norm of the symmetric Jack polynomials as given in the proof of Proposition 12.3.3. □

An alternative derivation of this result using the generating function (13.98) is given in Exercises 13.3 q.1.

13.3.2 Generalized Laguerre polynomials

The nonsymmetric generalized Laguerre polynomials, to be denoted $E_\eta^{(\mathrm{L})}$, are simultaneous eigenfunctions of the $\{l_i\}$. Analogous to the theory of the nonsymmetric Hermite polynomials, the fact that l_i consists of $\hat{\xi}_i$ plus a degree lowering operator, together with the fact that the $\{l_i\}$ commute, implies the simultaneous eigenfunctions have the structure

$$E_\eta^{(\mathrm{L})}(y^2) = E_\eta(y^2) + \sum_{|\nu|<|\eta|} c_{\eta\nu} E_\nu(y^2), \tag{13.101}$$

and the corresponding eigenvalue is $\bar{\eta}_i$. The relationship (11.79) shows that each $E_\eta^{(\mathrm{L})}(y^2)$ is also a polynomial eigenfunction of $\tilde{H}^{(\mathrm{L,Ex})}$ with corresponding eigenvalue $-|\eta|$.

Orthogonality

Since the Schrödinger operator $H^{(\mathrm{L,Ex})}$ (11.54) is Hermitian, the definition of $\tilde{H}^{(\mathrm{L,Ex})}$ together with the above facts tells us immediately that for compositions ν and η such that $|\nu| \neq |\eta|$ (at least),

$$\langle E_\nu^{(\mathrm{L})} | E_\eta^{(\mathrm{L})} \rangle^{(\mathrm{L})} = 0 \tag{13.102}$$

where

$$\langle f | g \rangle^{(\mathrm{L})} := \int_{(-\infty,\infty)^N} f(y_1^2, \dots, y_N^2) g(y_1^2, \dots, y_N^2) \, d\mu^{(\mathrm{L})}(y^2), \tag{13.103}$$

$$d\mu^{(\mathrm{L})}(y^2) := \prod_{l=1}^N |y_l|^{2a+1} e^{-y_l^2} \prod_{1 \le j < k \le N} |y_k^2 - y_j^2|^{2/\alpha} dy_1 \cdots dy_N. \tag{13.104}$$

In fact the orthogonality (13.102) holds for general $\nu \neq \eta$. This in turn follows from the fact that the l_i, as well as being eigenvalues of the $\{E_\eta^{(\mathrm{L})}(y^2)\}$ which separate the eigenvalues, are self-adjoint with respect to (13.103).

PROPOSITION 13.3.8 *We have*

$$\langle f | \hat{d}_i g \rangle^{(\mathrm{L})} = \langle (1 - \frac{a}{y_i^2} - \hat{d}_i) f | g \rangle^{(\mathrm{L})}, \tag{13.105}$$

$$\langle f | l_i g \rangle^{(\mathrm{L})} = \langle l_i f | g \rangle^{(\mathrm{L})}. \tag{13.106}$$

Proof. The result (13.105) follows from (13.103) and Definition 11.5.4 after integrating by parts (the fact that f and g are even in (13.103) is not used in the derivation). Hence with $d_i^{(B)*}$ denoting the adjoint of $d_i^{(B)}$ with respect to (13.103) we have $d_i^{(B)*} = 2y_i - d_i^{(B)}$. Substituting this in (11.80) shows

$$l_i = \Big(\frac{\alpha}{4} d_i^{(B)*} d_i^{(B)} + 1 - N + \frac{1}{2} \sum_{p=i+1}^{N} (s_{ip} + \sigma_i \sigma_p s_{ip})\Big)\Big|_{\substack{\text{even} \\ \text{functions}}},$$

and thus l_i is self-adjoint with respect to (13.103), which is the statement (13.106). □

Recurrences

If we make the E_η in Proposition 12.2.1 a function of y_i^2 and then apply the map $\phi^{(B)}$ specified by (11.81) and make use of (13.70) we obtain the recurrences

$$s_i E_\eta^{(\mathrm{L})}(y^2) = \begin{cases} \frac{1}{\delta_{i,\eta}} E_\eta^{(\mathrm{L})}(y^2) + \Big(1 - \frac{1}{\delta_{i,\eta}^2}\Big) E_{s_i\eta}^{(\mathrm{L})}(y^2), & \eta_i > \eta_{i+1}, \\ E_\eta^{(\mathrm{L})}(y^2), & \eta_i = \eta_{i+1}, \\ \frac{1}{\delta_{i,\eta}} E_\eta^{(\mathrm{H})}(y^2) + E_{s_i\eta}^{(\mathrm{L})}(y^2), & \eta_i < \eta_{i+1}. \end{cases} \tag{13.107}$$

As in the theory of the nonsymmetric Jack and Hermite polynomials, the operator s_i is one of the two fundamental operators needed to generate the $E_\eta^{(\mathrm{L})}$ by recurrence. The other is a raising type operator. In preparation for defining the raising type operator, and its lowering type companion, we introduce the raising type operator for $E_\eta(y^2)$, $\Psi := y_N^2 s_{N-1} \cdots s_2 s_1$ so that

$$\Psi\, E_\eta(y^2) = E_{\Phi\eta}(y^2) \tag{13.108}$$

(cf. the final equation in Proposition 12.2.3). One might also suspect we should consider the operator $\hat{\Phi}$ of Definition 12.2.4 with the change of variables $z_j = y_j^2$. However this is not the case; instead the lowering type operator is defined by

$$\hat{\Psi} := B_1 s_1 s_2 \cdots s_{N-1},$$

where $B_i := \frac{1}{4}(d_i^{(B)})^2$.

PROPOSITION 13.3.9 *With $q := 1 + (N-1)/\alpha$ we have*

$$\begin{aligned} \hat{\xi}_j\, \hat{\Psi} &= \hat{\Psi}\, \hat{\xi}_{j-1}, \qquad 2 \le j \le n, \\ \hat{\xi}_1\, \hat{\Psi} &= \hat{\Psi}\, (\hat{\xi}_N - \alpha), \\ \hat{\Psi}\, E_\eta(y^2) &= \frac{1}{\alpha} \frac{[a+q]_\eta}{[a+q]_{\hat{\Phi}\eta}} \frac{d'_\eta}{d'_{\hat{\Phi}\eta}} E_{\hat{\Phi}\eta}(y^2). \end{aligned} \tag{13.109}$$

Proof. Let $\hat{d}_i$ denote the type A Dunkl operator of Definition 11.4.1 with the change of variables $z_j = y_j^2$ $(j = 1, \dots, N)$ so that

$$\hat{d}_i = \frac{1}{2y_i} \frac{\partial}{\partial y_i} + \frac{1}{\alpha} \sum_{p \ne i} \frac{1 - s_{ip}}{y_i^2 - y_p^2}.$$

Similarly define the Cherednik operator with the same change of variables

$$\hat{\xi}_j := \alpha y_j^2 \hat{d}_j + (1 - N) + \sum_{p>j} s_{jp}.$$

One can verify directly that

$$[\hat{\xi}_j, B_i] = \begin{cases} B_i s_{ij}, & i < j, \\ B_j s_{ij}, & i > j, \\ -\alpha B_j - \sum_{p<j} s_{jp} B_j - \sum_{p>j} B_j s_{jp}, & i = j. \end{cases}$$

The first two results of (13.109) follow from this commutator formula in a manner analogous to the deduction of the first two results of Proposition 12.2.5 from Proposition 11.5.2. The final result, up to the proportionality constant c_η, say, follows from the first two results. Examination of the leading term gives

$$c_\eta = (a + q - 1 + \bar{\eta}_N/\alpha)\Big(\frac{\bar{\eta}_N + N - 1}{\alpha}\Big),$$

which can be identified with the stated form. □

Consider now the action of $\phi^{(B)}$ on Ψ and $\hat{\Psi}$. From the definitions we see

$$\phi^{(B)}\hat{\Psi} = \hat{\Psi},$$

$$\phi^{(B)}\Psi = \Big(y_N - \frac{1}{2}d_i^{(B)}\Big)^2 s_{N-1}\cdots s_2 s_1 \;=\; s_{N-1}\cdots s_2 s_1\Big(y_N - \frac{1}{2}d_i^{(B)}\Big)^2 \;=\; \hat{\Psi}^*,$$

where $\hat{\Psi}^*$ denotes the adjoint of $\hat{\Psi}$ with respect to (13.103). Thus, applying the map $\phi^{(B)}$ to (13.108) and (13.109) we obtain the sought raising and lowering formulas for the $E_\eta^{(\mathrm{L})}$.

PROPOSITION 13.3.10 *We have*

$$\hat{\Psi}\, E_\eta^{(\mathrm{L})}(y^2) = \frac{1}{\alpha}\frac{[a+q]_\eta}{[a+q]_{\hat{\Phi}\eta}}\frac{d'_\eta}{d'_{\hat{\Phi}\eta}}\, E_{\hat{\Phi}\eta}^{(\mathrm{L})}(y^2),$$

$$\hat{\Psi}^*\, E_\eta^{(\mathrm{L})}(y^2) = E_{\Phi\eta}^{(\mathrm{L})}(y^2).$$

The action of $\hat{\Psi}^*$ given by the above result allows $\langle E_{\Phi\eta}^{(\mathrm{L})}, |E_{\Phi\eta}^{(\mathrm{L})}\rangle^{(\mathrm{L})}$ to be related to $\langle E_\eta^{(\mathrm{L})}, |E_\eta^{(\mathrm{L})}\rangle^{(\mathrm{L})}$ as in the derivation of (13.83). Furthermore, the formulas (13.107) and the orthogonality of the $\{E_\eta^{(\mathrm{L})}\}$ with respect to (13.103) implies the formula corresponding to (13.81) holds in the Laguerre case. Proceeding as in the proof of Proposition 13.3.3 allows the normalization of the nonsymmetric Laguerre polynomials to be computed.

PROPOSITION 13.3.11 *With $q = 1 + (N-1)/\alpha$ as in Proposition 13.3.9, we have*

$$\mathcal{N}_\eta^{(\mathrm{L})} := \langle E_\eta^{(\mathrm{L})}|E_\eta^{(\mathrm{L})}\rangle^{(\mathrm{L})} = \frac{[a+q]_\eta}{\alpha^{|\eta|}}\frac{d'_\eta e_\eta}{d_\eta}\,\mathcal{N}_0^{(\mathrm{L})},$$

where

$$\mathcal{N}_0^{(\mathrm{L})} := \mathcal{N}_{0^N}^{(\mathrm{L})} = \langle 1|1\rangle^{(\mathrm{L})} = \alpha^{(1-N-(N-1)^2/\alpha)}\prod_{j=0}^{N-1}\frac{\Gamma(1+(j+1)/\alpha)\Gamma(a+1+j/\alpha)}{\Gamma(1+1/\alpha)}.$$

Generating function

The classical Laguerre polynomials can be specified by the generating function

$$e^{z^2}{}_0F_1(a+1;-(xz)^2) = \sum_{k=0}^{\infty}\frac{L_k^a(x^2)z^{2k}}{(a+1)_k}. \tag{13.110}$$

In the case of the $E_\eta^{(\mathrm{L})}$, an analogous formula holds, with the hypergeometric function ${}_0F_1(a+1;-(xz)^2)$ replaced by

$$\mathcal{K}_B(x^2;z^2) := {}_0\mathcal{K}_1(a+q;x^2;z^2) := \sum_\eta \frac{\alpha^{|\eta|}}{[a+q]_\eta}\frac{d_\eta}{d'_\eta e_\eta}E_\eta(x^2)E_\eta(z^2). \tag{13.111}$$

Before presenting the generating function, we make note of some fundamental properties of $\mathcal{K}_B$, which are analogous to the properties of $\mathcal{K}_A$ presented in Proposition 13.3.4.

PROPOSITION 13.3.12 *We have*

$$\hat{\Psi}^{(x)}\,\mathcal{K}_B(x^2;y^2) = \Psi^{(y)}\,\mathcal{K}_B(x^2;y^2),$$
$$B_i^{(x)}\,\mathcal{K}_B(x^2;y^2) = y_i^2\,\mathcal{K}_B(x^2;y^2). \tag{13.112}$$

Proof. The first equality follows by applying $\hat{\Psi}^{(x)}$ term-by-term to the series definition (13.111) using the final equation in Proposition 13.3.9, writing $\nu = \hat{\Phi}\eta$ in the summation, then using (13.108) to rewrite $E_{\Phi\nu}(y^2)$ in terms of $\Psi^{(y)} E_\nu(y^2)$. The second result follows by writing y_i and $B_i^{(x)}$ in terms of $\Psi^{(y)}$ and $\hat{\Psi}^{(x)}$, respectively, and using the first result. □

In further preparation for deriving the generating function, we note with the notation (11.78) the operator $\tilde{H}^{(\mathrm{L,Ex})}$ can be rewritten to read

$$\tilde{H}^{(\mathrm{L,Ex})} = \Delta_B - 2\sum_{l=1}^N x_l \frac{\partial}{\partial x_l}. \tag{13.113}$$

Furthermore, we note the second equality in Proposition 13.3.12 implies

$$\Delta_B\,\mathcal{K}_B(x^2;y^2) = 4p_1(x^2)\mathcal{K}_B(x^2;y^2) \tag{13.114}$$

(p_1 is as specified in (12.146)). These facts allow us to proceed as in the derivation of (13.90) to establish the following formula.

PROPOSITION 13.3.13 *We have*

$$\sum_\eta \frac{(-\alpha)^{|\eta|}}{[a+q]_\eta}\frac{d_\eta}{d'_\eta e_\eta} E_\eta^{(\mathrm{L})}(x)E_\eta(z) = \mathcal{K}_B(x;-z)e^{p_1(z)}. \tag{13.115}$$

In Exercises 13.3 q.6, the generating function (13.115) is used to give the explicit form of the coefficients $c_{\eta\nu}$ in (13.101) in terms of the generalized binomial coefficients.

Analogous to the exponential operator formula (13.91) we can use (13.114) to deduce from the generating function (13.115) an exponential operator formula expressing $E_\eta^{(\mathrm{L})}(x^2)$ in terms of $E_\eta(x^2)$. Thus we have

$$\exp\Big(-\frac{1}{4}\Delta_B\Big)E_\eta(x^2) = E_\eta^{(\mathrm{L})}(x^2). \tag{13.116}$$

Similarly, analogous to (13.92) this in turn implies the intertwining relation

$$\exp\Big(-\frac{1}{4}\Delta_B\Big)\hat{\xi}_j = l_j \exp\Big(-\frac{1}{4}\Delta_B\Big). \tag{13.117}$$

Integration formulas

The strategy used to derive the integration formulas of Proposition 13.3.6, with the generating function (13.115) taking the role of the generating function (13.90), implies analogous integration formulas in the Laguerre case.

PROPOSITION 13.3.14 *We have*

$$\int_{[0,\infty)^N} \mathcal{K}_B(x;-z_a)\mathcal{K}_B(x;-z_b)\,d\mu^{(\mathrm{L})}(x^2) = \mathcal{N}_{0^N}^{(\mathrm{L})} e^{-p_1(z_a)}e^{-p_1(z_b)}\mathcal{K}_B(z_a;z_b),$$
$$\int_{[0,\infty)^N} \mathcal{K}_B(x;-z_a)E_\eta^{(\mathrm{L})}(x)\,d\mu^{(\mathrm{L})}(x^2) = \mathcal{N}_{0^N}^{(\mathrm{L})} e^{-p_1(z_a)}E_\eta(-z_a),$$
$$\int_{[0,\infty)^N} \mathcal{K}_B(x;-z_a)E_\eta(-x)\,d\mu^{(\mathrm{L})}(x^2) = \mathcal{N}_{0^N}^{(\mathrm{L})} e^{-p_1(z_a)}E_\eta^{(\mathrm{L})}(z_a). \tag{13.118}$$

Symmetric Laguerre polynomials

The theory of symmetric Laguerre polynomials can be developed from the nonsymmetric Laguerre polynomials in an analogous fashion to the development of the symmetric Hermite polynomials. Thus the symmetric Laguerre polynomials can be defined as the eigenfunctions of (11.54) with the structure

$$P_\kappa^{(\mathrm{L})}(y^2) = P_\kappa(y^2) + \sum_{|\mu|<|\kappa|} \tilde{a}_{\kappa\mu} P_\mu(y^2). \tag{13.119}$$

Furthermore, they can be expressed in terms of the nonsymmetric Laguerre polynomials by either of the formulas

$$P_\kappa^{(\mathrm{L})}(y) = d'_{\eta^+} \sum_{\eta:\eta^+=\kappa} \frac{1}{d'_\eta} E_\eta^{(\mathrm{L})}(y) \tag{13.120}$$

or

$$P_\kappa^{(\mathrm{L})}(y) = \frac{1}{a_\eta} \mathrm{Sym} E_\eta^{(\mathrm{L})}(y), \tag{13.121}$$

where a_η is specified by (12.101).

Applying Sym to both sides of (13.116) shows

$$\exp\Big(-\frac{1}{4}\Delta_B\Big|_{\substack{\text{symmetric}\\ \text{functions}}}\Big) P_\kappa(y^2) = P_\kappa^{(\mathrm{L})}(y^2). \tag{13.122}$$

Changing variables $y^2 \mapsto x$, and noting that

$$\frac{1}{4}\Delta_B\Big|_{\substack{\text{symmetric}\\ \text{functions}}} = D_1 + (a+1)\sum_{j=1}^N \frac{\partial}{\partial x_j},$$

where

$$D_1 := \sum_{j=1}^N x_j \frac{\partial^2}{\partial x_j^2} + \frac{2}{\alpha}\sum_{j\neq k} \frac{x_j}{x_j - x_k}\frac{\partial}{\partial x_j},$$

we see that (13.122) can equivalently be written

$$\exp\Big(-\Big(D_1 + (a+1)\sum_{l=1}^N \frac{\partial}{\partial x_l}\Big)\Big) P_\kappa(x) = P_\kappa^{(\mathrm{L})}(x). \tag{13.123}$$

The Sym operator can also be applied to both sides of the generating function (13.115) to give that

$$\sum_\kappa \frac{(-\alpha)^{|\kappa|}}{[a+q]_\kappa} d'_\kappa \frac{P_\kappa^{(\mathrm{L})}(x) P_\kappa(z)}{P_\kappa((1)^N)} = {}_0\mathcal{F}_1(a+q; x; -z) e^{p_1(z)}. \tag{13.124}$$

Also, similar to the derivation of the norm of the symmetric Hermite polynomials, Proposition 13.3.7, the formulas (13.120) and (13.121) allow the norm of the symmetric Laguerre polynomials to be computed from knowledge of the nonsymmetric norm Proposition 13.3.11.

PROPOSITION 13.3.15 *We have*

$$\langle P_\kappa^{(\mathrm{L})} | P_\kappa^{(\mathrm{L})} \rangle^{(\mathrm{L})} = |\kappa|! \mathcal{N}_0^{(\mathrm{L})} C_\kappa^{(\alpha)}((1)^N) [a+q]_\kappa^{(\alpha)},$$

where $C_\kappa^{(\alpha)}$ is specified by (13.1).

13.3.3 Generalized Jacobi polynomials

Let η be an N-tuple of non-negative integers, let ε be an N-tuple with each entry $+1$ or -1, and define $\varepsilon\eta$ as the N-tuple formed from ε and η by multiplication of their respective parts. A direct calculation using (11.84) shows that for suitable $e^{(\mathrm{J})}_{j,\varepsilon\eta}$, $c_{\varepsilon\eta,\varepsilon'\eta'}$

$$\xi_j^{BC} z^{\varepsilon\eta} = e^{(\mathrm{J})}_{j,\varepsilon\eta} z^{\varepsilon\eta} + \sum_{\eta'<\eta} \sum_{\varepsilon'} c_{\varepsilon\eta,\varepsilon'\eta'} z^{\varepsilon'\eta'}.$$

From this action we can make use of the operator (11.86) to conclude that $\tilde{H}^{(\mathrm{J,Ex})}$ permits a complete set of nonsymmetric eigenfunctions

$$E^{(\mathrm{J})}_\eta(\sin^2\phi) = E_\eta(\sin^2\phi) + \sum_{|\nu|<|\eta|} a_{\nu\eta} E_\nu(\sin^2\phi). \tag{13.125}$$

In the variable $y = \sin^2\phi$, the $E^{(\mathrm{J})}_\eta(y)$ are polynomials, referred to as the nonsymmetric Jacobi polynomials.

Another direct calculation shows $\xi_j^{(BC)}$ is self-adjoint with respect to the inner product

$$\langle f|g\rangle^{(\mathrm{J})}_\phi := \int_0^{\pi/2} d\phi_1 \cdots \int_0^{\pi/2} d\phi_N \, |\psi_0^{(\mathrm{J})}|^2 f(\bar{z}) g(z), \tag{13.126}$$

where $z := e^{2i\phi}$ and

$$\psi_0^{(\mathrm{J})} = \prod_{j=1}^N z_j^{-(N-1)/\alpha-(a+b+1)/2} (z_j-1)^{a+1/2}(z_j+1)^{b+1/2} \prod_{j<k} (z_k - z_j)^{1/\alpha}(1-z_jz_k)^{1/\alpha}. \tag{13.127}$$

To see this, one makes use of the operator identity

$$\psi_0^{(\mathrm{J})} \xi_j^{(BC)} (\psi_0^{(\mathrm{J})})^{-1} = z_j \frac{\partial}{\partial z_j} - \frac{1}{\alpha}\Big(\sum_{l<j} \frac{z_l}{z_j - z_l} s_{lj} + \sum_{l>j} \frac{z_j}{z_j - z_l} s_{lj}\Big)$$
$$+ \frac{1}{\alpha} \sum_{\substack{k=1 \\ \ne j}}^N \frac{\sigma_j \sigma_k s_{jk}}{1 - z_j z_k} + \Big(a + \frac{1}{2}\Big)\frac{\sigma_j}{1 - z_j} + \Big(b + \frac{1}{2}\Big)\frac{\sigma_j}{1 + z_j}.$$

As a consequence, $\{E^{(\mathrm{J})}_\eta(\sin^2\phi)\}$ is orthogonal with respect to the inner product (13.126). Changing variables $y_j = \sin^2\phi_j$ shows the nonsymmetric Jacobi polynomials $\{E^{(\mathrm{J})}_\eta(y)\}$ are orthogonal with respect to the inner product

$$\langle f|g\rangle^{(\mathrm{J})}_y := \int_0^1 dy_1 \cdots \int_0^1 dy_N \prod_{l=1}^N y_l^a (1-y_l)^b \prod_{j<k} |y_k - y_j|^{2/\alpha} f(y) g(y). \tag{13.128}$$

Symmetrizing $E^{(\mathrm{J})}_\eta(y)$ gives the corresponding symmetric Jacobi polynomial

$$P^{(\mathrm{J})}_\kappa(y) = \frac{1}{a_\eta} \mathrm{Sym}\, E^{(\mathrm{J})}_\eta(y),$$

where $\kappa = \eta^+$ and a_η is given by (12.101). The symmetric Jacobi polynomials form a complete set of symmetric polynomial eigenfunctions of $\tilde{H}^{(\mathrm{J})}$ (11.91), with corresponding eigenvalues

$$e^{(\mathrm{J})}_\kappa := \sum_{j=1}^N \kappa_j(\kappa_j + 2(N-j)/\alpha + a + b + 1). \tag{13.129}$$

They appear in a generalization of the dual Cauchy product (12.212) [484] ,[404].

PROPOSITION 13.3.16 *With $\mu := N^N - \kappa'$ one has*

$$\prod_{k,l=1}^{N} (x_k - y_l) = \sum_{\kappa} (-1)^{|\mu|} P_\kappa^{(\mathrm{J})}(x;\alpha) P_\mu^{(\mathrm{J})}(y;1/\alpha). \tag{13.130}$$

Proof. First, by a direct calculation, we can check that

$$-\tilde{H}^{(\mathrm{J})x} \prod_{k,l=1}^{N} (x_k - y_l) = \frac{1}{\alpha}\Big(\tilde{H}_*^{(\mathrm{J})y} - e_{(N)^N *}^{(\mathrm{J})}\Big) \prod_{k,l=1}^{N} (x_k - y_l), \tag{13.131}$$

where $\tilde{H}^{(\mathrm{J})x}$ refers to (11.91) with $y \mapsto x$, while $\tilde{H}_*^{(\mathrm{J})y}$, $e_{\kappa *}^{(\mathrm{J})}$ refer to (11.91), (13.129) with parameters

$$\alpha \mapsto 1/\alpha, \qquad a \mapsto a\alpha + \alpha - 1, \qquad b \mapsto b\alpha + \alpha - 1.$$

Expanding

$$\prod_{k,l=1}^{N} (x_k - y_l) = \sum_{\kappa \subseteq (N)^N} f_\kappa(y) P_\kappa^{(\mathrm{J})}(x), \tag{13.132}$$

we have from (12.212) that f_κ is of the form

$$f_\kappa(y) = (-1)^{|\mu|}\Big(P_\kappa(y;1/\alpha) + \sum_{\rho<\mu} c_{\kappa\rho} P_\rho(y;1/\alpha)\Big), \tag{13.133}$$

where $\mu := N^N - \kappa'$. Now, substituting (13.132) in (13.131) we obtain

$$-\sum_{\kappa \subseteq N^N} \Big(e_\kappa^{(\mathrm{J})} - \frac{1}{\alpha} e_{N^N *}^{(\mathrm{J})}\Big) f_\kappa(y) P_\kappa^{(\mathrm{J})}(x) = \frac{1}{\alpha} \sum_{\kappa \subseteq N^N} \tilde{H}_*^{(\mathrm{J})y} f_\kappa(y) P_\kappa^{(\mathrm{J})}(x).$$

But from the explicit formula (13.129) we can make use of (12.33) to check that

$$-\alpha\Big(e_\kappa^{(\mathrm{J})} - \frac{1}{\alpha} e_{N^N *}^{(\mathrm{J})}\Big) = e_{\mu *}^{(\mathrm{J})}.$$

As the unique eigenfunction of $\tilde{H}_*^{(\mathrm{J})y}$ with the structure (13.133), eigenvalue $e_{\mu *}^{(\mathrm{J})}$, is

$$f_\kappa(y) = (-1)^{|\mu|} P_\kappa^{(\mathrm{J})}(y;1/\alpha)$$

(13.130) follows. □

EXERCISES 13.3 1. Following the method of [36] an alternative derivation of Proposition 13.3.7 will be given.

(i) By multiplying both sides of (13.98) by $P_\kappa^{(\mathrm{H})}(y)$ and integrating with respect to the measure $d\mu^{(\mathrm{H})}(y)$ deduce that with $\mathcal{N}_\kappa^{(\mathrm{H})} := \langle P_\kappa^{(\mathrm{H})} | P_\kappa^{(\mathrm{H})} \rangle^{(\mathrm{H})}$,

$$\mathcal{N}_\kappa^{(\mathrm{H})} \frac{(2\alpha)^{|\kappa|}}{d'_\kappa P_\kappa((1)^N)} P_\kappa(z) = e^{-p_2(z)} \int_{(-\infty,\infty)^N} {}_0\mathcal{F}_0^{(\alpha)}(2y;z) P_\kappa^{(\mathrm{H})}(y)\, d\mu^{(\mathrm{H})}(y).$$

(ii) Set $z_1 = \cdots = z_N = c$ in the formula of (i), make use of (13.3) and complete the square to show

$$\mathcal{N}_\kappa^{(\mathrm{H})} \frac{(2\alpha)^{|\kappa|}}{d'_\kappa P_\kappa((1)^N)} P_\kappa((c)^N) = \int_{(-\infty,\infty)^N} e^{-p_2(y)} \prod_{j<k} |y_k - y_j|^{2/\alpha} P_\kappa^{(\mathrm{H})}(y+c)\, dy_1 \cdots dy_N.$$

Now take the limit $c \to \infty$, noting from (13.95) that $\lim_{c\to\infty} P_\kappa^{(\mathrm{H})}(y+c)/P_\kappa(c) = 1$ to deduce the formula for $\mathcal{N}_\kappa^{(\mathrm{H})}$ given in Proposition 13.3.7.

2. Use the integral representation (13.58), the definition (13.85) and the integration formula (13.93) for $|t| < 1$ to show that

$$\sum_\eta \frac{1}{\mathcal{N}_\eta^{(\mathrm{H})}} E_\eta^{(\mathrm{H})}(w) E_\eta^{(\mathrm{H})}(z) t^{|\eta|}$$
$$= \frac{1}{\mathcal{N}_{0^N}^{(\mathrm{H})}} (1-t^2)^{-Nq/2} \exp\Big(-\frac{t^2}{(1-t^2)}(p_2(w)+p_2(z))\Big) \mathcal{K}_A\Big(\frac{2wt}{(1-t^2)^{1/2}}; \frac{z}{(1-t^2)^{1/2}}\Big), \quad (13.134)$$

where $q := 1 + (N-1)/\alpha$. Similarly, make use of the first and third formulas in (13.118) and (13.111) to provide a verification of the formula

$$\sum_\eta \frac{1}{\mathcal{N}_\eta^{(\mathrm{L})}} E_\eta^{(\mathrm{L})}(x) E_\eta^{(\mathrm{L})}(y) t^{|\eta|}$$
$$= \frac{1}{\mathcal{N}_{0^N}^{(\mathrm{L})}} (1-t)^{-N(a+q)} \exp\Big(-\frac{t}{1-t}(p_1(x)+p_1(y))\Big) \mathcal{K}_B\Big(\frac{y}{1-t}; \frac{tx}{1-t}\Big), \quad (13.135)$$

valid for $|t| < 1$.

3. (i) Use the formula (13.91) in (13.58) to deduce that for f analytic such that all terms converge

$$\int_{(-\infty,\infty)^N} \mathcal{K}_A(2y; -iz) f(iy)\, d\mu^{(\mathrm{H})}(y) = \mathcal{N}_{0^N}^{(\mathrm{H})} e^{-p_2(z)} e^{-\Delta_A/4} f(z). \quad (13.136)$$

(ii) Use this result to show that if we define the integral transform (a generalized Fourier transform, or *Dunkl transform*) by

$$F(z) = \frac{e^{p_2(z)}}{\mathcal{N}_{0^N}^{(\mathrm{H})}} \int_{(-\infty,\infty)^N} \mathcal{K}_A(2y; -iz) f(iy)\, d\mu^{(\mathrm{H})}(y)$$

then the inverse transform is

$$f(z) = \frac{e^{-p_2(z)}}{\mathcal{N}_{0^N}^{(\mathrm{H})}} \int_{(-\infty,\infty)^N} \mathcal{K}_A(2y; z) F(y)\, d\mu^{(\mathrm{H})}(y).$$

4. [142] With d_i denoting the type A Dunkl operator, and p and q homogeneous polynomials of N variables, define the pairing

$$[p,q]_A := p(d)q\Big|_{x=0}, \quad (13.137)$$

where $p(d)$ is the operator obtained from $p(x)$ by replacing each x_i by d_i.

(i) Note that for p and q of different degrees, $[p,q]_A = 0$.

(ii) By forming an appropriate linear combination of the second formula in (13.86) note that

$$E_\nu(d^{(x)}) \mathcal{K}_A(x;y) = E_\nu(y) \mathcal{K}_A(x;y).$$

Now set $x = 0$ and equate coefficients of $E_\nu(y)$ to deduce that [39]

$$[E_\nu, E_\eta]_A = \frac{1}{\alpha^{|\eta|}} \frac{d'_\eta e_\eta}{d_\eta} \delta_{\nu,\eta}. \quad (13.138)$$

(iii) Compare (13.138) and (13.75) to show

$$[E_\nu, E_\eta]_A = \frac{2^{|\eta|}}{\mathcal{N}_{0^N}^{(\mathrm{H})}} \langle E_\nu^{(\mathrm{H})} | E_\eta^{(\mathrm{H})} \rangle^{(\mathrm{H})}.$$

Now make use of (13.91) to deduce

$$[p,q]_A = \frac{2^{|\eta|}}{\mathcal{N}^{(\mathrm{H})}_{0^N}} \langle e^{-\Delta_A/4} p | e^{-\Delta_A/4} q \rangle^{(\mathrm{H})}.$$

5. With the symmetric Hermite polynomials characterized by the structure (13.95) and orthogonality

$$\langle P^{(\mathrm{H})}_\kappa | P^{(\mathrm{H})}_\mu \rangle^{(\mathrm{H})} \propto \delta_{\kappa,\mu},$$

and the symmetric Laguerre polynomials characterized by the structure (13.119) and orthogonality

$$\langle P^{(\mathrm{L})}_\kappa | P^{(\mathrm{L})}_\mu \rangle^{(\mathrm{L})} \propto \delta_{\kappa,\mu},$$

use the characterization of the symmetric Jacobi polynomials as having the structure (13.125) and orthogonality $\langle P^{(\mathrm{J})}_\kappa | P^{(\mathrm{J})}_\mu \rangle^{(\mathrm{J})}_y \propto \delta_{\kappa,\mu}$ to deduce the limit relations

$$\lim_{L\to\infty} L^{|\kappa|} P^{(\mathrm{J})}_\kappa(x/L)\Big|_{b\mapsto L} = P^{(\mathrm{L})}_\kappa(x), \qquad \lim_{L\to\infty} (-2L)^{|\kappa|} P^{(\mathrm{J})}_\kappa\Big(\frac{1}{2}(1-x/L)\Big)\Big|_{\substack{a\mapsto L^2\\ b\mapsto L^2}} = P^{(\mathrm{H})}_\kappa(x).$$

6. The objective of this exercise is to derive the formula

$$E^{(\mathrm{L})}_\eta(x) = \frac{(-1)^{|\eta|}[a+q]_\eta e_\eta}{d_\eta} \sum_{\nu\preceq\eta} \frac{(-1)^{|\nu|}}{[a+q]_\nu} \frac{d_\nu}{e_\nu} \binom{\eta}{\nu} E_\nu(x), \tag{13.139}$$

giving the explicit expansion of the nonsymmetric Laguerre polynomials in terms of the nonsymmetric Jack polynomials.

(i) Use (13.31) to obtain the formula

$$\mathcal{K}_B(x;-z)e^{p_1(z)} = \sum_\eta \frac{(-1)^{|\eta|}}{[a+q]_\eta} \frac{d_\eta}{e_\eta} E_\eta(x) \sum_\nu \frac{\alpha^{|\nu|}}{d'_\nu} \binom{\nu}{\eta} E_\nu(z).$$

(ii) Substitute the result of (i) in (13.115) and equate coefficients of $E_\eta(z)$ to deduce (13.139).

(iii) Symmetrize (13.139) to show

$$P^{(\mathrm{L})}_\kappa(x) = (-1)^{|\kappa|} P^{(\alpha)}_\kappa((1)^N)[a+q]^{(\alpha)}_\kappa \sum_{\sigma\subseteq|\kappa|} \binom{\kappa}{\sigma} \frac{(-1)^{|\sigma|} P_\sigma(x)}{[a+q]^{(\alpha)}_\sigma P_\sigma((1)^N)}.$$

(iv) Use (12.180) together with (12.105) to deduce from (iii) that in the case $\kappa = p^N$

$$P^{(\mathrm{L})}_\kappa(x) = (-1)^{|\kappa|} P_\kappa((1)^N)[a+q]^{(\alpha)}_\kappa \, {}_1F^{(\alpha)}_1(-N;a+q;x).$$

Note also that with $\eta = p^N$, $E^{(\mathrm{L})}_\eta(x) = P^{(\mathrm{L})}_\eta(x)$, so $E^{(\mathrm{L})}_\eta(x)$ is also thus related to ${}_1F^{(\alpha)}_1$.

13.4 GREEN FUNCTIONS AND ZONAL POLYNOMIALS

13.4.1 Hermite case

The symmetric Hermite polynomials $P^{(\mathrm{H})}_\kappa$ (13.97) are a complete set of polynomial eigenfunctions of $H^{(\mathrm{H,Ex})}$ with eigenvalue $-2|\kappa|$. Recalling the transformation and change of variables which gave rise to the eigenoperator (11.64) for $\{P^{(\mathrm{H})}_\kappa\}$, we see that the complete set of symmetric eigenfunctions of $H = H^{(\mathrm{H})}$ is

$$\Big\{ e^{-\beta W^{(\mathrm{H})}(x)/2} P^{(\mathrm{H})}_\kappa(\sqrt{\beta/2}x) \Big\}_\kappa, \tag{13.140}$$

where here and in the remainder of the subsection $P_\kappa^{(\mathrm{H})}(u) = P_\kappa^{(\mathrm{H})}(u;2/\beta)$, with corresponding eigenvalues $E_0 + \beta|\kappa|$. Denoting $\mathcal{N}_\kappa^{(\mathrm{H})}|_{\alpha=2/\beta} := \langle P_\kappa^{(\mathrm{H})} | P_\kappa^{(\mathrm{H})} \rangle^{(\mathrm{H})}|_{\alpha=2/\beta}$, a simple change of variables gives $(\sqrt{\beta/2})^{-N-\beta N(N-1)/2} \mathcal{N}_\kappa^{(\mathrm{H})}|_{\alpha=2/\beta}$ for the normalization of (13.140).

The formula (11.43) shows the corresponding Green function is specified by

$$G_\tau^{\mathrm{S(H)}}(\vec{x}^{(0)};\vec{x}) = (\sqrt{\beta/2})^{N+\beta N(N-1)/2} e^{-E_0\tau/\beta} e^{-\beta W^{(\mathrm{H})}(x^{(0)})/2} e^{-\beta W^{(\mathrm{H})}(x)/2} \times \sum_\kappa \frac{P_\kappa^{(\mathrm{H})}(\sqrt{\beta/2}x^{(0)}) P_\kappa^{(\mathrm{H})}(\sqrt{\beta/2}x)}{\mathcal{N}_\kappa^{(\mathrm{H})}|_{\alpha=2/\beta}} e^{-|\kappa|\tau}. \tag{13.141}$$

With $G_\tau^{\mathrm{S(H)}}$ known, the general formula (11.42) gives the Green function for the corresponding Fokker-Planck operator as

$$G_\tau^{\mathrm{FP(H)}}(\vec{x}^{(0)};\vec{x}) = (\sqrt{\beta/2})^{N+\beta N(N-1)/2} e^{-\beta W^{(\mathrm{H})}(x)} \sum_\kappa \frac{P_\kappa^{(\mathrm{H})}(\sqrt{\beta/2}x^{(0)}) P_\kappa^{(\mathrm{H})}(\sqrt{\beta/2}x)}{\mathcal{N}_\kappa^{(\mathrm{H})}|_{\alpha=2/\beta}} e^{-|\kappa|\tau}. \tag{13.142}$$

Symmetrizing (13.134) allows the multidimensional analogue of the classical summation formula (11.102) to be deduced,

$$\sum_\kappa \frac{P_\kappa^{(\mathrm{H})}(x) P_\kappa^{(\mathrm{H})}(y) t^{|\kappa|}}{\mathcal{N}_\kappa^{(\mathrm{H})}|_{\alpha=2/\beta}} = \frac{1}{\mathcal{N}_0^{(\mathrm{H})}|_{\alpha=2/\beta}} (1-t^2)^{-N(1+\beta(N-1)/2)/2} \exp\Big(-\frac{t^2}{1-t^2}\sum_{j=1}^N (x_j^2+y_j^2)\Big) \times {}_0\mathcal{F}_0^{(2/\beta)}\Big(\frac{2xt}{(1-t^2)^{1/2}}; \frac{y}{(1-t^2)^{1/2}}\Big), \quad |t|<1. \tag{13.143}$$

Recalling (13.20), substituting (13.143) in (13.142) gives $G_{\mathrm{FP}}^{(\mathrm{H})}$ as a sum over Jack polynomials.

13.4.2 Laguerre case

The symmetric Laguerre polynomials $P_\kappa^{(\mathrm{L})}(y^2)$ (13.119) are a complete set of symmetric polynomial eigenfunctions of $\tilde{H}^{(\mathrm{L,Ex})}$ with eigenvalue $-|\kappa|$, and so the complete set of symmetric eigenfunctions of $H^{(\mathrm{L})}$ is

$$\Big\{ e^{-\beta W^{(\mathrm{L})}(x)/2} P_\kappa^{(\mathrm{L})}(\beta x^2/2)\Big|_{\substack{a\mapsto \beta a'/2-1/2\\ \alpha\mapsto 2/\beta}} \Big\}_\kappa$$

with eigenvalue $E_\kappa^{(\mathrm{L})} = E_0^{(\mathrm{L})} + 2\beta|\kappa|$. Denoting $\mathcal{N}_\kappa^{\mathrm{L}}|_{\alpha=2/\beta} := \langle P_\kappa^{\mathrm{L}} | P_\kappa^{\mathrm{L}} \rangle^{(\mathrm{L})}|_{\alpha=2/\beta}$, and proceeding as in the derivation of (13.142) shows

$$G_\tau^{\mathrm{FP(L)}}(\vec{x}^{(0)};\vec{x}) = (\beta/2)^{N(1+\beta(N+a'-1))/2} e^{-\beta W^{(\mathrm{L})}(x)} \times \sum_\kappa \frac{P_\kappa^{(\mathrm{L})}(\beta(x^{(0)})^2/2) P_\kappa^{(\mathrm{L})}(\beta x^2/2)}{\mathcal{N}_\kappa^{(\mathrm{L})}} e^{-2|\kappa|\tau}\Big|_{\substack{a\mapsto\beta a'/2-1/2\\ \alpha=2/\beta}}. \tag{13.144}$$

In the classical case $N=1$, this can be summed according to (11.106). By symmetrizing (13.135) one obtains the multivariable generalization [36]

$$\sum_\kappa \frac{P_\kappa^{(\mathrm{L})}(x) P_\kappa^{(\mathrm{L})}(y) t^{|\kappa|}}{\mathcal{N}_\kappa^{(\mathrm{L})}}\Big|_{\alpha=2/\beta} = \frac{1}{\mathcal{N}_{0^N}^{(\mathrm{L})}|_{\alpha=2/\beta}} (1-t)^{-N(a+q)} \exp\Big(-\frac{t}{1-t}\sum_{j=1}^N (x_j+y_j)\Big) \times {}_0\mathcal{F}_1^{(2/\beta)}\Big(a+q; \frac{xt}{1-t}; \frac{y}{1-t}\Big), \quad |t|<1, \tag{13.145}$$

where $q = 1+\beta(N-1)/2$, thus giving an alternative form for the r.h.s. of (13.144).

13.4.3 Zonal polynomials

We know from (11.101) that for $\beta = 1, 2, 4$ the Green function $G_\tau^{\rm FP(H)}$ is given in terms of a matrix integral. Similarly (11.105) gives $G_\tau^{\rm FP(L)}$ for $\beta = 1, 2, 4$ in terms of a matrix integral. Comparing with the expressions for the Green functions as implied by (13.142), (13.143) and (13.144), (13.145) the following matrix integral evaluations result.

PROPOSITION 13.4.1 *Let $(\mathbf{U}^\dagger d\mathbf{U})$, $(\mathbf{V}^\dagger d\mathbf{V})$ denote the normalized Haar volume form for $N \times N$, $n \times n$ matrices with $n \ge N$. Let $\mathbf{H}$, $\mathbf{H}^{(0)}$ be as in (11.101) and let $\mathbf{X}$, $\mathbf{Y}$ be $n \times N$ matrices with real ($\beta = 1$), complex ($\beta = 2$) or real quaternion ($\beta = 4$) elements. We have*

$$\int e^{\mathrm{Tr}(\mathbf{H}^{(0)}\mathbf{U}^\dagger\mathbf{H}\mathbf{U})}(\mathbf{U}^\dagger d\mathbf{U}) = {}_0\mathcal{F}_0^{(2/\beta)}(\lambda^{(0)};\lambda), \tag{13.146}$$

$$\int (\mathbf{U}^\dagger d\mathbf{U}) \int (\mathbf{V}^\dagger d\mathbf{V})\, e^{\mathrm{Tr}(\mathbf{U}\mathbf{X}^\dagger\mathbf{V}^\dagger\mathbf{Y}+\mathbf{Y}^\dagger\mathbf{V}\mathbf{X}\mathbf{U}^\dagger)} = {}_0\mathcal{F}_1^{(2/\beta)}(\beta n/2; x^2; y^2), \tag{13.147}$$

where $\{\lambda_j^{(0)}\}$, $\{\lambda_j\}$ are the eigenvalues of $\mathbf{H}^{(0)}$, $\mathbf{H}$ and $\{x_j^2\}$, $\{y_j^2\}$ are the eigenvalues of $\mathbf{X}^\dagger\mathbf{X}$, $\mathbf{Y}^\dagger\mathbf{Y}$.

In the case $\beta = 2$ these are the same matrix integrals evaluated by (11.103) and (11.108), respectively.

Underlying the integration formulas of Proposition 13.4.1 are further integration formulas directly involving $\{C_\kappa^{(2/\beta)}\}$ in the integrand. These relate to a fundamental theory for the $\{C_\kappa^{(2/\beta)}\}$ as so-called *zonal polynomials* associated with the symmetric spaces $\mathrm{gl}(N,\mathbb{R})/O(N)$ ($\beta = 1$), $\mathrm{gl}(N,\mathbb{C})/U(N)$ ($\beta = 2$) and $u^*(2N)/\mathrm{Sp}(2N)$ ($\beta = 4$) [376]. However a development of this theory is outside the scope of the present work. Instead we will be content with noting some key integration formulas from the theory.

PROPOSITION 13.4.2 *Let $\mathbf{U}$ be as in (13.146). Let $\mathbf{A}$, $\mathbf{B}$ be $N \times N$ matrices with real, complex and real quaternion entries for $\beta = 1, 2, 4$, respectively. We have*

$$\int C_\kappa^{(2/\beta)}(\mathbf{A}\mathbf{U}^\dagger\mathbf{B}\mathbf{U})\,(\mathbf{U}^\dagger d\mathbf{U}) = \frac{C_\kappa^{(2/\beta)}(\mathbf{A})C_\kappa^{(2/\beta)}(\mathbf{B})}{C_\kappa^{(2/\beta)}((1)^N)}, \tag{13.148}$$

where $C_\kappa^{(2/\beta)}(\mathbf{A}\mathbf{U}^\dagger\mathbf{B}\mathbf{U})$ is defined as $C_\kappa^{(2/\beta)}(y)$ with $y_1, \dots, y_N$ denoting the eigenvalues of $\mathbf{A}\mathbf{U}^\dagger\mathbf{B}\mathbf{U}$, and similarly the meaning of $C_\kappa^{(2/\beta)}(\mathbf{A})$, $C_\kappa^{(2/\beta)}(\mathbf{B})$.

In fact this integration formula can be taken to be the defining property of the corresponding zonal polynomials, up to normalization. Note that summing (13.148) over κ using (13.2), (13.3), (13.20) reclaims (13.146).

Consider now (13.147) in the case $n = N$, $\mathbf{Y} = \mathbf{1}$. Noting that the integrand is then independent of $\mathbf{V}$, and recalling from Section 1.3.2 that in the case $\beta = 4$ Tr in (13.147) refers to the matrices regarded as having real quaternion rather than scalar elements, we obtain

$$\langle e^{\mathrm{Tr}(\mathbf{X}\mathbf{O})}\rangle_{\mathbf{O}\in O(N)} = {}_0F_1^{(2)}(N/2; \mathbf{X}\mathbf{X}^T/4), \tag{13.149}$$

$$\langle e^{\mathrm{Tr}(\mathbf{X}\mathbf{U})}e^{\mathrm{Tr}(\mathbf{U}^\dagger\mathbf{X}^\dagger)}\rangle_{\mathbf{U}\in U(N)} = {}_0F_1^{(1)}(N; \mathbf{X}\mathbf{X}^\dagger), \tag{13.150}$$

$$\langle e^{\mathrm{Tr}(\mathbf{X}\mathbf{S})}\rangle_{\mathbf{S}\in \mathrm{Sp}(2N)} = {}_0F_1^{(1/2)}(2N; \mathbf{X}\mathbf{X}^\dagger). \tag{13.151}$$

In (13.151) Tr now refers to the matrices regarded as having complex entries and use has been made of the fact that for $\mathbf{X}$ real quaternion $\mathrm{Tr}(\mathbf{X}\mathbf{S}) = \mathrm{Tr}(\mathbf{X}\mathbf{S})^\dagger$.

It turns out that underlying (13.149)–(13.151) are group integrals transforming Schur polynomials to zonal polynomials [312], [376], [454].

PROPOSITION 13.4.3 *One has*

$$\langle s_\lambda(\mathbf{AO})\rangle_{\mathbf{O}\in O(N)} = \begin{cases} \dfrac{C_\kappa^{(2)}(\mathbf{AA}^T)}{C_\kappa^{(2)}((1)^N)}, & \lambda = 2\kappa, \\ 0 & \text{otherwise}, \end{cases} \tag{13.152}$$

$$\langle s_\lambda(\mathbf{AU}) s_\kappa(\mathbf{U}^\dagger \mathbf{A}^\dagger)\rangle_{\mathbf{U}\in U(N)} = \delta_{\lambda,\kappa} \frac{C_\kappa^{(1)}(\mathbf{AA}^\dagger)}{C_\kappa^{(1)}((1)^N)}, \tag{13.153}$$

$$\langle s_\lambda(\mathbf{AS})\rangle_{\mathbf{S}\in \mathrm{Sp}(2N)} = \begin{cases} \dfrac{C_\kappa^{(1/2)}(\mathbf{AA}^\dagger)}{C_\kappa^{(1/2)}((1)^N)}, & \lambda = \kappa^2, \\ 0 & \text{otherwise}, \end{cases} \tag{13.154}$$

where in (13.152) the partition 2κ is the partition obtained by doubling each part of κ, while in (13.154), κ^2 is the partition obtained by repeating each part of κ twice.

Note that with $\mathbf{A} = \mathbf{1}$ (13.154) reduces to (10.137) with $\beta = 0$, while (13.152) reduces to (10.155) with $\alpha = 0$. To deduce (13.149)–(13.151) from the results of Proposition 13.4.3 first note from (13.1) and (13.8) that for general α

$$C_\kappa^{(\alpha)}((1)^N) = |\kappa|! \frac{\alpha^{2|\kappa|}[N/\alpha]_\kappa^{(\alpha)}}{h_\kappa d'_\kappa}. \tag{13.155}$$

In (13.152) and (13.154) multiply both sides by $1/d'_\lambda|_{\alpha=1}$ and use (13.3) on the l.h.s. On the r.h.s. use (13.155), (13.2) and the identities

$$\frac{(h_\kappa d'_\kappa)|_{\alpha=2}}{d'_{2\kappa}|_{\alpha=1}} = 1, \qquad \frac{2^{2|\kappa|}(h_\kappa d'_\kappa)|_{\alpha=1/2}}{d'_{\kappa^2}|_{\alpha=1}} = 1 \tag{13.156}$$

to obtain (13.149) and (13.151). For the identity (13.150), first multiply both sides of (13.153) by $1/d'_\kappa|_{\alpha=1}$, and simplify by noting that for $\alpha = 1$, $d'_\kappa = h_\kappa$.

13.4.4 Duality formulas

The calculations of Section 13.2 are based on expressions for the correlations in terms of multidimensional integrals whose dimension is independent of N. As the correlations themselves are multidimensional integrals whose dimension is proportional to N, a feature of Jack polynomial theory which gives identities between multidimensional integrals is being put to use (see also Exercises 13.2 q.6). Here, following [131], such a duality type formula will be developed for states in the Dyson Brownian motion model of the log-gas in a harmonic well with general initial conditions.

PROPOSITION 13.4.4 *We have*

$$e^{p_2(x)} \Big\langle {}_0\mathcal{F}_0^{(\alpha)}(2y; -ix) \prod_{j=1}^n \prod_{k=1}^N (s_j - \sqrt{\alpha} y_k) \Big\rangle_{\mathrm{ME}_{2/\alpha,N}(e^{-y^2})}$$
$$= (-i)^{nN} e^{p_2(s)} \Big\langle {}_0\mathcal{F}_0^{(1/\alpha)}(2y; -is) \prod_{j=1}^n \prod_{k=1}^N (y_j + \sqrt{\alpha} x_k) \Big\rangle_{\mathrm{ME}_{2/\alpha.n}(e^{-y^2})}. \tag{13.157}$$

Proof. In the case that f is a symmetric function, (13.136) reads

$$\int_{(-\infty,\infty)^N} {}_0\mathcal{F}_0^{(\alpha)}(2y; -iz) f(iy)\, d\mu^{(\mathrm{H})}(y) = \mathcal{N}_0^{(\mathrm{H})} e^{-p_2(z)} e^{-\Delta_A/4} f(z),$$

or equivalently

$$e^{p_2(x)}\Big\langle {}_0\mathcal{F}_0^{(\alpha)}(2y;-ix)f(iy)\Big\rangle_{\mathrm{ME}_{2/\alpha,N}(e^{-y^2})} = e^{-\Delta_A/4}f(x). \tag{13.158}$$

Let $\Delta_A^{(p,s)}$ be specified by (11.68) with $N \mapsto p$, $y \mapsto s$ and $1 - s_{jk}$ replaced by zero. A direct calculation shows

$$\Delta_A^{(N,y)}\prod_{j=1}^{n}\prod_{k=1}^{N}(s_j - i\sqrt{\alpha}y_k) = \Delta_A^{(n,s)}\Big|_{\alpha\mapsto 1/\alpha}\prod_{j=1}^{n}\prod_{k=1}^{N}(s_j - i\sqrt{\alpha}y_k)$$

and consequently

$$e^{-\Delta_A^{(N,y)}/4}\prod_{j=1}^{n}\prod_{k=1}^{N}(s_j - i\sqrt{\alpha}y_k) = e^{-\Delta_A^{(n,s)}|_{\alpha\mapsto 1/\alpha}/4}\prod_{j=1}^{n}\prod_{k=1}^{N}(s_j - i\sqrt{\alpha}y_k). \tag{13.159}$$

Suppose now that we choose

$$f(y) = \prod_{j=1}^{n}\prod_{k=1}^{N}(s_j - i\sqrt{\alpha}y_k) \tag{13.160}$$

in (13.158). Making use of (13.159) shows

$$e^{p_2(x)}\Big\langle {}_0\mathcal{F}_0^{(\alpha)}(2y;-ix)f(iy)\Big\rangle_{\mathrm{ME}_{2/\alpha,N}(e^{-y^2})} = e^{-\Delta_A^{(n,s)}|_{\alpha\mapsto 1/\alpha}/4}\prod_{j=1}^{n}\prod_{k=1}^{N}(s_j - i\sqrt{\alpha}x_k). \tag{13.161}$$

To deduce (13.157) we make a further application of (13.158), this time with $N \mapsto n$, $\alpha \mapsto 1/\alpha$, $x = s$, $y = x$ and now $f(u)$ given by the r.h.s. of (13.160) with $y \mapsto x$. Substituting for the r.h.s. of (13.161) the l.h.s. of (13.158) so modified gives (13.157). □

Consider (13.157) with $s = (c)^n$, $x = (0)^N$. Making use of (13.63) with ${}_0\mathcal{K}_0$ replaced by ${}_0\mathcal{F}_0^{(\alpha)}$ we see that (13.157) simplifies to involve only elementary functions in the average. Completing the square gives

$$\Big\langle \prod_{j=1}^{N}(c - \sqrt{\alpha}y_j)^n\Big\rangle_{\mathrm{ME}_{2/\alpha,N}(e^{-y^2})} = \Big\langle \prod_{j=1}^{n}(c - iy_j)^N\Big\rangle_{\mathrm{ME}_{2\alpha,n}(e^{-y^2})}. \tag{13.162}$$

This is a generalization of the duality formula used to derive (13.64).

In the special cases $2\alpha = 1, 2$ and 4 we can substitute (13.146) in (13.157) to obtain the matrix integrals duality [96], [97]

$$e^{\mathrm{Tr}\,\mathbf{X}^2}\Big\langle e^{-2i\mathrm{Tr}(\mathbf{XY})}\prod_{j=1}^{n}\det(s_j - \sqrt{\alpha}\mathbf{Y})\Big\rangle_{\mathbf{Y}\in\mathrm{ME}_{2/\alpha,N}(e^{-y^2})}$$
$$= (-i)^{nN}e^{\mathrm{Tr}\,\mathbf{S}^2}\Big\langle e^{-2i\mathrm{Tr}(\mathbf{SY})}\prod_{k=1}^{N}\det(\mathbf{Y} + \sqrt{\alpha}x_k)\Big\rangle_{\mathbf{Y}\in\mathrm{ME}_{2\alpha,n}(e^{-y^2})},$$

where now $\mathrm{ME}_{\beta,m}$ refers to the appropriate ensemble of Gaussian random matrices.

EXERCISES 13.4 1. [509], [287], [454] For $\beta = 1, 2, 4$, let $\mathbf{X}, \mathbf{A}, \mathbf{B}$ be $N \times N$ random matrices with standard Gaussian real, complex, real quaternion entries respectively. Define

$$\langle f(\mathbf{X})\rangle_{\mathbf{X}} := \frac{1}{C}\int_{\mathbf{X}} e^{-\mathrm{Tr}(\mathbf{XX}^\dagger)}f(\mathbf{X})(d\mathbf{X}),$$

where C is such that the r.h.s. equals unity for $f = 1$.

(i) By noting that the distribution of $\mathbf{X}$ is invariant under the mapping $\mathbf{X} \mapsto \mathbf{UX}$ for $\mathbf{U} \in O(N), U(N), \mathrm{Sp}(2N)$, respectively, conclude

$$\langle f(\mathbf{AXBX}^\dagger)\rangle_{\mathbf{X}} = \langle\langle f(\mathbf{AUXBX}^\dagger\mathbf{U}^\dagger)\rangle_{\mathbf{U}}\rangle_{\mathbf{X}}. \tag{13.163}$$

Similarly, use the invariance under $\mathbf{X} \mapsto \mathbf{XU}$ to conclude

$$\langle f(\mathbf{AXBX}^\dagger)\rangle_{\mathbf{X}} = \langle\langle f(\mathbf{AXUBU}^\dagger\mathbf{X}^\dagger)\rangle_{\mathbf{U}}\rangle_{\mathbf{X}}. \tag{13.164}$$

(ii) Choose $f = C_\kappa^{(2/\beta)}$ in (13.163) to deduce from (13.148) that

$$\langle C_\kappa^{(2/\beta)}(\mathbf{AXBX}^\dagger)\rangle_{\mathbf{X}} = \frac{C_\kappa^{(2/\beta)}(\mathbf{A})}{C_\kappa^{(2/\beta)}((1)^N)}\langle C_\kappa^{(2/\beta)}(\mathbf{XBX}^\dagger)\rangle_{\mathbf{X}}.$$

(iii) Use (13.164) and (13.148) to evaluate the average on the r.h.s. of the identity in (ii) and obtain

$$\langle C_\kappa^{(2/\beta)}(\mathbf{AXBX}^\dagger)\rangle_{\mathbf{X}} = \frac{C_\kappa^{(2/\beta)}(\mathbf{A})C_\kappa^{(2/\beta)}(\mathbf{B})}{(C_\kappa^{(2/\beta)}((1)^N))^2}\langle C_\kappa^{(2/\beta)}(\mathbf{XX}^\dagger)\rangle_{\mathbf{X}}. \tag{13.165}$$

(iv) Note from (13.152) that with $\mathbf{X}$ an $N \times N$ random matrix of standard real Gaussians

$$\langle s_{2\kappa}(\mathbf{AX})\rangle_{\mathbf{X}} = \langle\langle s_{2\kappa}(\mathbf{AXO})\rangle_{\mathbf{O}}\rangle_{\mathbf{X}} = \frac{1}{C_\kappa^{(2)}((1)^N)}\langle C_\kappa^{(2)}(\mathbf{A}^T\mathbf{AXX}^T)\rangle_{\mathbf{X}}.$$

Now use (13.165) in the case $\beta = 1$ to deduce from this that

$$\langle s_{2\kappa}(\mathbf{AX})\rangle_{\mathbf{X}} = \frac{C_\kappa^{(2)}(\mathbf{AA}^T)}{(C_\kappa^{(2)}((1)^N))^2}\langle C_\kappa^{(2)}(\mathbf{XX}^T)\rangle_{\mathbf{X}}.$$

(v) With $\mathbf{X}$ an $N \times N$ random matrix of standard complex Gaussians, show from (13.153) and (13.165) in the case $\beta = 2$ that

$$\langle s_\lambda(\mathbf{AX})s_\kappa(\mathbf{X}^\dagger\mathbf{A}^\dagger)\rangle_{\mathbf{X}} = \delta_{\lambda,\kappa}\frac{C_\kappa^{(1)}(\mathbf{AA}^\dagger)}{(C_\kappa^{(1)}((1)^N))^2}\langle C_\kappa^{(1)}(\mathbf{XX}^\dagger)\rangle_{\mathbf{X}}.$$

(vi) With $\mathbf{X}$ an $N \times N$ random matrix of standard real quaternions, show from (13.154) and (13.165) in the case $\beta = 4$ that

$$\langle s_{\kappa^2}(\mathbf{AX})\rangle_{\mathbf{X}} = \frac{C_\kappa^{(1/2)}(\mathbf{AA}^\dagger)}{(C_\kappa^{(1/2)}((1)^N))^2}\langle C_\kappa^{(1/2)}(\mathbf{XX}^\dagger)\rangle_{\mathbf{X}}.$$

2. [249] In this exercise, the simplest case ($p = 1$) of the matrix integral duality formula

$$\langle|\det(z\mathbf{1}_N - \mathbf{AU})|^{2p}\rangle_{\mathbf{U}\in U(N)} \propto \int_0^\infty dt_1 \cdots \int_0^\infty dt_p \prod_{l=1}^p \frac{\det(|z|^2\mathbf{1}_N + t_l\mathbf{AA}^\dagger)}{(1+t_l)^{N+2p}} \prod_{1\le j<k\le p}|t_k - t_j|^2 \tag{13.166}$$

will be established.

(i) With $s_\kappa(\mathbf{X})$ denoting the Schur polynomial in the eigenvalues of $\mathbf{X}$, note from the dual Cauchy identity (10.56) that

$$\det(\mathbf{1}_N + q\mathbf{X}) = \sum_{r=0}^N q^r s_{1^r}(\mathbf{X}). \tag{13.167}$$

(ii) Use (13.167) together with the integration formula (13.153), to show

$$\langle|\det(z\mathbf{1}_N - \mathbf{AU})|^2\rangle_{\mathbf{U}\in U(N)} = |z|^{2N}\sum_{r=0}^N |z|^{-2r}\frac{C_{1^r}^{(1)}(\mathbf{AA}^\dagger)}{C_{1^r}^{(1)}((1)^N)}.$$

(iii) Note from the definition (13.1) and (12.105) that

$$\frac{C_{1^r}^{(1)}(\mathbf{A}\mathbf{A}^\dagger)}{C_{1^r}^{(1)}((1)^N)} = \binom{N}{r}^{-1} s_{1^r}(\mathbf{A}\mathbf{A}^\dagger),$$

and change variables $x \mapsto t/(1-t)$ in the Euler beta integral (4.2) to show

$$\binom{N}{r}^{-1} = (N+1)\int_0^\infty \frac{t^r}{(1+t)^{N+2}}\,dt.$$

Substitute this in the previous formula, substitute the result in the formula of (ii), and make a further use of (13.167) to obtain (13.166) in the case $p=1$.

13.5 INTER-RELATIONS FOR SPACING DISTRIBUTIONS

A generalization of the Dixon-Anderson integral was used in Section 4.4 to show that the p.d.f. of every $(r+1)$th eigenvalue in certain β-ensembles with $\beta = 2/(r+1)$ is equal to the p.d.f. of another β-ensemble with $\beta = 2(r+1)$. In the case $r=1$ this has been used in Sections 8.4.3, 9.7 and 9.8 to relate gap probabilities in the bulk, and at the soft and hard edges, respectively, of matrix ensembles with orthogonal and symplectic symmetry. In this section the inter-relations of Section 4.4 will be used for general $r \in \mathbb{Z}^+$ to deduce the corresponding inter-relations between spacing distributions, both in the case of finite N ensembles and in scaled limits [199]. First some notation is required.

Definition 13.5.1 *Suppose the eigenvalues in the matrix ensemble are ordered. Let the p.d.f. for the $(k+1)$th largest eigenvalue be denoted $p^{\max}(k;s;\mathrm{ME}_{\beta,N})$. In the case that the support of the spectrum is restricted to $x>0$ let the p.d.f. for the $(k+1)$th smallest eigenvalue be denoted $p^{\min}(k;s;\mathrm{ME}_{\beta,N})$. In the circular β-ensemble $\mathrm{CE}_{\beta,N}$ let $p^{\rm spacing}(k;s;\mathrm{CE}_{\beta,N})$ denote the p.d.f. for eigenvalues which are $(k+1)$th neighbors.*

With these definitions, we read off from (4.112)–(4.115) the following inter-relations.

Proposition 13.5.2 *One has*

$$\begin{aligned}
p^{\max}((r+1)k+r;s;\mathrm{ME}_{2/(r+1),(r+1)N+r}(e^{-x})) &= p^{\max}(k;s;\mathrm{ME}_{2(r+1),N}(e^{-(r+1)x})),\\
p^{\min}((r+1)k+r;s;\mathrm{ME}_{2/(r+1),(r+1)N+r}(x^a e^{-x})) &= p^{\min}(k;s;\mathrm{ME}_{2(r+1),N}(x^{(r+1)a+2r}e^{-(r+1)x})),\\
p^{\min}(((r+1)k+r;s;\mathrm{CE}^b_{2/(r+1),(r+1)N+r}) &= p^{\min}(k;s;\mathrm{CE}^{(r+1)b+2r}_{2(r+1),N}),\\
p^{\rm spacing}((r+1)k+r;s;\mathrm{CE}^0_{2/(r+1),(r+1)N}) &= p^{\rm spacing}(k;s;\mathrm{CE}^0_{2(r+1),N}).
\end{aligned}$$

We know from Section 7.2 that the soft and hard edge scalings of $\mathrm{ME}_{\beta,N}(x^a e^{-cx})$ are given by

$$x \mapsto \frac{\beta}{2c}(4N + 2(2N)^{1/3}s_\beta x), \qquad x \mapsto \frac{\beta}{2c}\frac{x}{4N\tilde{s}_\beta}, \tag{13.168}$$

respectively, where s_β, $\tilde{s}_\beta$ are arbitrary length scales. Thus we can calculate the soft and hard edge distributions according to

$$\begin{aligned}
p_\beta^{\rm soft}(k;s) &:= \lim_{N\to\infty}\frac{\beta}{c}(2N)^{1/3}s_\beta p^{\max}(k;s;\mathrm{ME}_{\beta,N}(e^{-cx})),\\
p_\beta^{\rm hard}(k;s;a) &:= \lim_{N\to\infty}\frac{\beta}{2c}\frac{1}{4N\tilde{s}_\beta}p^{\min}(k;s;\mathrm{ME}_{\beta,N}(x^a e^{-cx})).
\end{aligned}$$

In the ensemble $\mathrm{CE}^b_{\beta,N}$ the mean spacing between eigenvalues is $2\pi/N$ and so the appropriate scaled distributions are

$$p_\beta^{\rm bulk,s.s.}(k;s;b) := \lim_{N\to\infty} \frac{2\pi}{N} p^{\min}(k;2\pi s/N;{\rm CE}^b_{\beta,N}),$$

$$p_\beta^{\rm bulk,sp.}(k;s) := \lim_{N\to\infty} \frac{2\pi}{N} p^{\rm spacing}(k;2\pi s/N;{\rm CE}^0_{\beta,N}).$$

These limits can be taken in the results of Proposition 13.5.2 to obtain generalizations of the inter-relations (9.141), (9.163) and (8.160).

PROPOSITION 13.5.3 *Let the scales in (13.168) be chosen so that* $s_{2/(r+1)} = (r+1)^{2/3} s_{2(r+1)}$ *and* $\tilde{s}_{2/(r+1)}(r+1)^2 = \tilde{s}_{2(r+1)}$. *One has*

$$p^{\rm soft}_{2/(r+1)}((r+1)k+r;s) = p^{\rm soft}_{2(r+1)}(k;s),$$
$$p^{\rm hard}_{2/(r+1)}((r+1)k+r;s;a) = p^{\rm hard}_{2(r+1)}(k;s;(r+1)a+2r),$$
$$(r+1)p^{\rm bulk,s.s.}_{2/(r+1)}((r+1)k+r;(r+1)s;b) = p^{\rm bulk,s.s.}_{2(r+1)}(k;s;(r+1)b+2r),$$
$$(r+1)p^{\rm bulk,sp.}_{2/(r+1)}((r+1)k+r;(r+1)s) = p^{\rm bulk,sp.}_{2(r+1)}(k;s).$$

EXERCISES 13.5 *1.* [199] According to (8.17)

$$p_\beta^{\rm bulk,sp.}(k;s) = \frac{d^2}{ds^2}\sum_{j=0}^{k}(k-j+1)E_\beta^{\rm bulk}(j;s).$$

Argue that for s large the term $j=k$ on the r.h.s. will dominate, and then use the final equation in Proposition 13.5.3 to deduce

$$(r+1)E^{\rm bulk}_{2/(r+1)}((r+1)k;(r+1)s) \sim E^{\rm bulk}_{2(r+1)}(k;s).$$

Show that this is consistent with (9.99) and (9.103).

13.6 STOCHASTIC DIFFERENTIAL EQUATIONS

13.6.1 The bulk

The random recurrences obtained in Chapters 1–3 for characteristic polynomials of the various β-ensembles are well suited to obtaining a characterization of the number of eigenvalues in a certain interval, and the spacing between eigenvalues, in terms of stochastic differential equations. We consider first the circular β-ensemble, which for finite N has the property that the eigenvalues are equal to the zeros of $\chi_N(\lambda)$ as specified by the coupled system (2.89). For the purpose of studying the $N\to\infty$ bulk scaling limit of the circular β-ensemble, this suggests seeking the scaled, limiting form of these equations. To do this we don't work with $\chi_k(\lambda)$ itself, since for $|\lambda|=1$ it takes values in a two-dimensional region of the complex plane. Instead, following [357] we work with

$$B_k(\lambda) := \frac{\lambda\chi_{k-1}(\lambda)}{\tilde{\chi}_{k-1}(\lambda)} \tag{13.169}$$

(cf. (3.138)) which has the property that $|B_k(\lambda)| = 1$ for $|\lambda|=1$. The fact that the zeros of $\tilde{\chi}_{k-1}(\lambda)$, $k=2,\dots,N-1$, are all outside the unit circle, together with the relationship between χ_{k-1} and $\tilde{\chi}_{k-1}$, shows $B_k(\lambda)$ is a finite *Blanske product* and thus has the further property $|B_k(\lambda)|<1$ for $|\lambda|<1$. The (relative) Prüfer phase (recall Section 1.9.4) can be used to parametrize $B_k(\lambda)$ for $|\lambda|=1$.

PROPOSITION 13.6.1 *Write*

$$\frac{B_k(e^{i\theta})}{B_k(1)} = e^{i\psi_k(\theta)} \tag{13.170}$$

so that $\psi_k(\theta)$ denotes the relative Prüfer phase. For $k = 0, \dots, N-1$ we have that $\psi_k(\theta)$ is an increasing function of θ, and satisfies the recurrence

$$\psi_{k+1}(\theta) = \psi_k(\theta) + \theta + 2\mathrm{Im}\log\Big(\frac{1-\gamma_k}{1-\gamma_k e^{i\psi_k(\theta)}}\Big), \qquad \psi_0 = \theta, \tag{13.171}$$

where $\gamma_k := B_k(1)\alpha_k$ and the logarithm is defined by its power series.

Proof. We see from the coupled recurrences (2.89) that $\{B_k\}$ satisfies the first order recurrence

$$B_{k+1}(\lambda) = \lambda B_k(\lambda)\frac{1-\bar{\alpha}_k B_k(\lambda)}{1-\alpha_k B_k(\lambda)}, \qquad B_0(\lambda) = \lambda. \tag{13.172}$$

It follows from this and (13.170) that

$$e^{i\psi_{k+1}(\theta)} = e^{i(\theta+\psi_k(\theta))}\frac{1-\gamma_k}{1-\bar{\gamma}_k}\frac{1-\bar{\gamma}_k e^{-i\psi_k(\theta)}}{1-\gamma_k e^{-i\psi_k(\theta)}}.$$

The recurrence (13.171) now follows by noting that for general $z \ne 0$, $z/\bar{z} = \exp(2i\mathrm{Im}\,\log z)$.

To show ψ_k is an increasing function of θ we use the fact that $|B_k(z)|$ is less than one, equal to one and greater than 1 for $|z| < 1$, $|z| = 1$, $|z| > 1$, respectively, to conclude

$$\frac{\partial |B_k(z)|}{\partial r} > 0 \tag{13.173}$$

for $r := |z| = 1$. Writing $B_k(z) = |B_k(z)|\exp i\psi_k((r,\theta))$ (and in particular $\psi_k((1,\theta)) = \psi_k(\theta)$), we see from the radial Cauchy-Riemann equation that

$$\frac{\partial}{\partial r}\Big(|B_k(z)|e^{i\psi_k((r,\theta))}\Big)\Big|_{r=1} = -i\frac{\partial}{\partial\theta}e^{i\psi_k(\theta)}.$$

Minor manipulation and use of (13.173) shows $\partial\psi_k(\theta)/\partial\theta > 0$ as required. □

According to (13.172), $B_N(\lambda) = 0$ when $B_{N-1}(\lambda) = 1/\bar{\alpha}_{N-1}$, with the latter a variable uniform on the unit circle. Thus, with $\bar{\alpha}_{N-1} = e^{-i\eta}$, the eigenvalues can be characterized as the values of the relative Prüfer phases

$$\psi_{N-1}(\theta) = 2\pi k + \eta, \qquad k \in \mathbb{Z}_{\ge 0}. \tag{13.174}$$

We seek the bulk scaling of (13.171), in which $\theta \mapsto 2\pi x/N$ so that in the variable x the mean spacing is unity. It may be expected that $\psi_k(2\pi x/N) \sim \Phi(t;x)$ where $t = k/N$, and consequently

$$\frac{\partial\Phi(t;x)}{\partial t} = 2\pi x + 2\mathrm{Im}\Big((e^{i\Phi(t;x)} - 1)A(t)\Big), \qquad \Phi(0;x) = 0 \tag{13.175}$$

(for this type of scaling of the difference equations for polynomials orthogonal on the unit circle the resulting differential equations are referred to as a *Krein system*). Here $A(t)$ is the process obtained from the $N \to \infty$ form of $N\gamma_k$ (or equivalently $N\alpha_k$, due to the rotational invariance of the latter), as seen by expanding the logarithm to first order. Recalling Proposition 2.8.4, and noting from Definition 2.8.3 that $\langle r^2\rangle_{\Theta_\nu} = 2/(\nu+1)$, shows that the process is consistent with

$$A(t) = \frac{1}{(\beta t)^{1/2}}(B_1'(t) + iB_2'(t)) \tag{13.176}$$

for B_1, B_2 two independent standard Brownian paths. A rigorous demonstration of (13.175), with $A(t)$ given by (13.176), as the scaling limit of (13.171) is given in [357]. The criteria (13.174) for an eigenvalue before scaling implies the criteria in relation to (13.175) that

$$\Phi(1;x) = 2\pi k + \eta, \qquad k \in \mathbb{Z}_{\geq 0},$$

where $0 \leq \eta \leq 2\pi$ is chosen uniformly at random and then fixed.

At $\beta = 1, 2$ and 4, the study of the scaled correlation functions undertaken in Chapter 7 reveals that the process defined by the scaled $N \to \infty$ eigenvalues in the circular β-ensemble is the same as that defined by the bulk scaled eigenvalues of the Gaussian β-ensemble. The characterization in terms of a stochastic differential equation allows for this coincidence to be verified for general $\beta > 0$. Moreover, study of the Gaussian β-ensemble from this perspective leads to further insights into the stochastic differential equation [524]. Such a study is based on the matrix recurrence (1.167), and the parametrization in terms of Prüfer phases (1.169). The analysis is complicated by the corresponding recurrence (1.170) being nonlinear. This complication is overcome by viewing the matrix recurrence as specifying a distance preserving mapping in hyperbolic geometry. This in turn leads to a construction of the corresponding point process in terms of the *Brownian carousel.*

DEFINITION 13.6.2 *Consider the Poincaré disk model of the hyperbolic plane (see Section 15.7.2 below). Let $b = b(t)$ be a path in the disk, z a point on the boundary and $f : \mathbb{R}^+ \mapsto \mathbb{R}^+$ an integrable function. Rotate z about center $b(t)$ at angular speed λf, and count the integer-valued total windings $N(\lambda)$ of z about $b(t)$. The hyperbolic carousel point process is the point process corresponding to $N(\lambda)$. The Brownian carousel point process is the hyperbolic carousel point process with b equal to hyperbolic Brownian motion B_H.*

Workings from [524] give that with the point written as $z = e^{i\gamma_\lambda(t)}$, the angle γ_λ satisfies the differential equation

$$\frac{d\gamma_\lambda}{dt} = \lambda f \frac{|e^{i\gamma_\lambda} - b|^2}{1 - |b|^2} \tag{13.177}$$

(the final factor can be recognized as proportional to the reciprocal of the Poisson kernel ; recall (2.39)). Consider the hyperbolic angle $\alpha_\lambda(t)$ determined by $z_0 := e^{i\gamma_\lambda(0)}$, $b(t)$ and $e^{i\gamma_\lambda(t)}$. With the Möbius transformation of the unit disk taking z_0 to 1, and w to 0 denoted $\mathcal{T}(z_0, w; z)$, so that

$$\mathcal{T}(z_0, w; z) = \frac{z - w}{1 - \bar{w}z}\frac{1 - \bar{w}z}{z_0 - w},$$

this angle is specified by

$$\alpha_\lambda(0) = 0, \qquad e^{i\alpha_\lambda(t)} = \mathcal{T}(e^{i\gamma_\lambda(0)}, b(t); e^{i\gamma_\lambda(t)}). \tag{13.178}$$

It is shown in [524] (by applying *Ito's formula*) that in the Brownian carousel ($b = B_H$) the equations (13.177) and (13.178) together imply $\alpha_\lambda(t)$ satisfies the *stochastic sine equation*

$$d\alpha_\lambda = \lambda f dt + \mathrm{Re}((e^{-i\alpha_\lambda(t)} - 1)(dB_1 + i\, dB_2)), \qquad \alpha_\lambda(0) = 0 \tag{13.179}$$

(a coupled one-parameter family of stochastic differential equations — the same B_1, B_2 must be used for each λ), where B_1 and B_2 are as in (13.176). For fixed λ the simplification $\mathrm{Im}(e^{-i\alpha_\lambda(t)/2}(dB_1 + i\, dB_2)) = dB$, where B is one-dimensional Brownian motion, is valid so it follows from (13.179) that in this circumstance

$$d\alpha_\lambda = \lambda f dt + 2\sin(\alpha_\lambda/2) dB. \tag{13.180}$$

The term involving Brownian motion in (13.180) vanishes when $\alpha_\lambda = 2\pi k$, $k \in \mathbb{Z}$, and this together with the fact that $\alpha_\lambda(t)$ has a limit for $t \to \infty$ implies $\alpha_\lambda(\infty)$ is an integer multiple of 2π (in the case of (13.179) this statement holds almost surely: as a function of λ, $\alpha_\lambda(\infty)$ jumps between successive integer values).

Suppose that in (13.179)

$$f(t) = (\beta/4)e^{-\beta t/4}. \tag{13.181}$$

Changing variables $t = -(4/\beta)\log s$ $(0 < s \le 1)$, $\alpha_\lambda(t) = \Phi_\lambda(s)$ shows

$$d\Phi_\lambda(s) = \lambda ds + \frac{4}{\sqrt{\beta s}} \sin\Phi_\lambda(s)\, \mathrm{Im}\Big(e^{-i\Phi_\lambda(s)/2}(dB_1(s) + i\, dB_2(s))\Big), \qquad \Phi_\lambda(1) = 0, \tag{13.182}$$

where to change variables in $B(t)$ we have used the realization $dB(t) = G\sqrt{dt}$ with G a standard normal (see, e.g., [157]). Compare now (13.182) with (13.175). Minor manipulation shows the two equations to be identical, with $s \leftrightarrow t$, $\Phi_\lambda(s) \leftrightarrow \Phi(t;x)$, $\lambda \leftrightarrow 2\pi x$. However the boundary equations are different, with (13.175) prescribing Φ at $t = 0$, while (13.182) prescribes Φ at $s = 1$. Despite this difference in boundary conditions, it is indeed the case that with the choice of f (13.181) the Brownian carousel point process, measured in units of 2π, coincides with the point process for eigenvalues in the bulk of the Gaussian β-ensemble (mean density unity). Moreover α_λ can then be constructed as a relative Prüfer phase, and the distribution of $\alpha_\lambda(\infty)/2\pi$ as specified by (13.180) is equal to the distribution of the number of eigenvalues in $[0,\lambda]$ of the limiting matrix ensemble.

13.6.2 Soft edge

We turn our attention now to the characterization of the scaled eigenvalues at the soft edge of the Gaussian β-ensemble in terms of a stochastic operator [158], [463]. We begin by studying the limit $\beta \to \infty$.

PROPOSITION 13.6.3 *Let $\mathbf{T}_\beta = [t_{ij}]_{i,j=1,\dots,N}$ be the random tridiagonal matrix specified in (1.159). Let $\mathbf{\Delta}_N$ be the $N \times N$ symmetric tridiagonal matrix with -1s above and below the diagonal, and 2s on the diagonal, and let $\tilde{\mathbf{J}}_N$ be the $N \times N$ matrix with nonzero entries on the subdiagonal below the diagonal only, these being $1, 2, \dots, N-1$ in order. With*

$$\mathbf{D} = \mathrm{diag}\Big((N/2)^{-(i-1)/2} \prod_{k=1}^{i-1} \frac{1}{\sqrt{\beta}} t_{k,k+1}\Big)_{i=1,\dots,N}$$

one has

$$\lim_{\beta\to\infty} \frac{1}{\sqrt{\beta}} \mathbf{D}\mathbf{T}_\beta \mathbf{D}^{-1} - \sqrt{2N}\mathbf{1}_N = -\frac{1}{\sqrt{2}N^{1/6}}\Big(N^{2/3}\mathbf{\Delta}_N + N^{-1/3}\tilde{\mathbf{J}}_N\Big), \tag{13.183}$$

where the equality is in values of the distribution. Furthermore, the eigenvectors of (13.183) are given by $\mathbf{D}\vec{x}$ where $\vec{x} = [x_j]_{j=1,\dots,N}$ with

$$x_{N-n} = \Big(\frac{1}{\sqrt{\pi}n!2^n}\Big)^{1/2} e^{-\lambda^2/2} H_n(\lambda).$$

Here $H_n(\lambda)$ denotes the Hermite polynomial of degree n, and λ is required to be such that $H_N(\lambda) = 0$.

Proof. Since $\lim_{\beta\to\infty} \frac{1}{\sqrt{\beta}} t_{k,k+1} = \lim_{\beta\to\infty} \frac{1}{\sqrt{\beta}} \tilde{X}_{(n-k)\beta} = \frac{1}{\sqrt{\beta}}\sqrt{n-k}$ is an equality in values of the distribution, we see that for large β

$$\lim_{\beta\to\infty} \frac{1}{\sqrt{\beta}} \mathbf{T}_\beta = \frac{1}{\sqrt{2}} \begin{bmatrix} 0 & \sqrt{N-1} & & & & \\ \sqrt{N-1} & 0 & \sqrt{N-2} & & & \\ & \sqrt{N-2} & 0 & \sqrt{N-3} & & \\ & & \ddots & \ddots & \ddots & \\ & & & \sqrt{2} & 0 & 1 \\ & & & & 1 & 0 \end{bmatrix}$$

and the result (13.183) now follows by direct calculation. The statement in regards to the eigenvectors can be verified from the three-term recurrence of Exercises 5.1 q.1(vii). □

The significance of the r.h.s. of (13.183) is that it is a discretization, lattice spacing $h = N^{-1/3}$, of the Airy operator

$$-\frac{d^2}{dx^2} + x. \tag{13.184}$$

With the boundary condition that the eigenfunctions vanish at $x = 0$ and decay to zero as $x \to \infty$, so as to be consistent with the form of the eigenfunctions for finite N, the Airy operator has eigenvalues given by the zeros of the Airy function. This leads to the conclusion that for $\beta \to \infty$, and in the soft edge scaling, the eigenvalues of the Gaussian β-ensemble are given by the zeros of the Airy function. For finite N we know from Exercises 1.4 q.8 that for $\beta \to \infty$ the eigenvalues of the Gaussian β-ensemble crystallize at the zeros of the Hermite polynomial. The asymptotic formula (7.9) tells us that at the soft edge these scale to the zeros of the Airy function, in agreement with this conclusion.

These findings suggest studying the l.h.s. of (13.183) for fixed β. Then one has [158]

$$\frac{1}{\sqrt{\beta}}\mathbf{D}\mathbf{T}_\beta\mathbf{D}^{-1} - \sqrt{2N}I_N = -\frac{1}{\sqrt{2}N^{1/6}}\Big(N^{2/3}\mathbf{\Delta}_N + N^{-1/3}\tilde{\mathbf{J}}_N + \frac{2}{\sqrt{\beta}}\mathbf{W}\Big), \tag{13.185}$$

where $\mathbf{W}$ is the bidiagonal random matrix

$$\mathbf{W} = -\frac{N^{1/6}}{\sqrt{2}}\begin{bmatrix} N[0,1] & & & & \\ b_{(N-1)\beta} & N[0,1] & & & \\ & b_{(N-2)\beta} & N[0,1] & & \\ & & \ddots & \ddots & \\ & & & b_\beta & N[0,1] \end{bmatrix},$$

with $b_{(N-j)\beta} = (2\tilde{\chi}^2_{(N-j)\beta} - (N-j)\beta)/\sqrt{2\beta N}$. We can check that $b_{(N-j)\beta}$ has mean zero and variance $1 - j/N$, so that each element of $\mathbf{W}$ has mean zero, and to leading order for j fixed and N large has standard deviation $N^{1/6}$. This is consistent with a discretization, lattice spacing $h = N^{-1/3}$, of a Brownian motion process which has mean zero and standard deviation $\sqrt{h}$ over an interval $(x, x+h]$. Recalling that (13.183) is a discretization of (13.184), we are therefore led to suspect that (13.185) is a discretization of the stochastic Airy operator

$$-\frac{d^2}{dx^2} + x + \frac{2}{\sqrt{\beta}}B'(x), \tag{13.186}$$

where $B(x)$ denotes standard Brownian motion. Furthermore, this suggests the distribution of the eigenvalues, calculated when the eigenfunctions are required to vanish at $x = 0$ and decay to zero as $x \to \infty$, will coincide with the distribution of the scaled eigenvalues at the soft edge of the Gaussian β-ensemble. In the work [463] these statements have been made rigorous.

13.6.3 Hard edge

The eigenvalues at the hard edge of the Laguerre β-ensemble also allow for a description in terms of a stochastic operator [158], [462]. To begin, define $\tilde{\mathbf{B}}_\beta$ to be the random bidiagonal matrix (3.127) with $n - m =: \tilde{a}$ and $m \mapsto n$. Let $\mathbf{A} = [(-1)^i\delta_{i,n+1-j}]_{i,j=1,\dots,n}$, and then form

$$\mathbf{L}_\beta := \frac{1}{\sqrt{\beta}}\mathbf{A}\tilde{\mathbf{B}}_\beta\mathbf{A} = \frac{1}{\sqrt{\beta}}\begin{bmatrix} \chi_{(\tilde{a}+1)\beta} & -\chi_\beta & & & \\ & \chi_{(\tilde{a}+2)\beta} & -\chi_{2\beta} & & \\ & & \ddots & \ddots & \\ & & & \chi_{(\tilde{a}+n-1)\beta} & -\chi_{(n-2)\beta} \\ & & & & \chi_{(\tilde{a}+n)\beta} \end{bmatrix}.$$

In the limit $\beta \to \infty$ this reduces to the fixed matrix

$$\mathbf{L}_\infty = \begin{bmatrix} \sqrt{\tilde{a}+1} & -\sqrt{1} & & & \\ & \sqrt{\tilde{a}+2} & -\sqrt{2} & & \\ & & \ddots & \ddots & \\ & & & \sqrt{\tilde{a}+n-1} & -\sqrt{n-1} \\ & & & & \sqrt{\tilde{a}+n} \end{bmatrix}, \tag{13.187}$$

which can readily be decomposed as the finite difference scheme for a differential operator [158].

PROPOSITION 13.6.4 *Let $h = 1/(2n+\tilde{a}+1)$, $x_i = h(\tilde{a}+i)$; ∇ equal the $n \times n$ upper biadiagonal matrix with -1s down the diagonal and 1s along the next diagonal; $\mathbf{S}$ equal half the upper bidiagonal matrix with all entries on the diagonal and next diagonal equal to 1; and $\mathbf{E}$ equal the $n \times n$ upper biadiagonal matrix specified by*

$$(\mathbf{E})_{i,i} = \frac{1}{h}\sqrt{x_{2i}+h\tilde{a}} - \frac{1}{h}\sqrt{x_{2i}} - \frac{\tilde{a}}{2}x_{2i}^{-1/2}, \quad (\mathbf{E})_{i,i+1} = -\frac{1}{h}\sqrt{x_{2i}-h\tilde{a}} + \frac{1}{h}\sqrt{x_{2i}} - \frac{\tilde{a}}{2}x_{2i}^{-1/2}.$$

We have

$$\sqrt{\frac{2}{h}}\mathbf{L}_\infty = -2\,\mathrm{diag}(\sqrt{x_2}, \sqrt{x_4}, \dots, \sqrt{x_{2n}})\Big(\frac{1}{2h}\nabla\Big) + \tilde{a}\,\mathrm{diag}\Big(\frac{1}{\sqrt{x_2}}, \frac{1}{\sqrt{x_4}}, \dots, \frac{1}{\sqrt{x_{2n}}}\Big)\mathbf{S} + \mathbf{E}. \tag{13.188}$$

A Taylor expansion shows that the entries of $\mathbf{E}$ are $\mathrm{O}(h)$, and so $\mathbf{E}$ can be ignored in the limit $n \to \infty$. The other matrices can be viewed as a discretization of the operator

$$\mathcal{J}_{\tilde{a}}^\infty := -2\sqrt{x}\frac{d}{dx} + \frac{\tilde{a}}{\sqrt{x}}, \qquad x \in [0,1] \tag{13.189}$$

acting on square integrable functions with the properties that $v(1) = 0$ (this accounts for the final row of (13.187) having only one entry) and $(\mathcal{J}_{\tilde{a}}^\infty v)(0) = 0$ (a consequence of the entries in the first row of (13.187) being $\mathrm{O}(1)$).

Now matrices realizing the Laguerre β-ensemble specified by the p.d.f. (3.17) with $N \mapsto n$ are given the product $\frac{1}{\beta}\mathbf{B}_\beta\mathbf{B}_\beta^T$, where $\mathbf{B}_\beta$ is specified by (3.127) with $a = n-m+1-2/\beta = \tilde{a}+1-2/\beta$ and $m \mapsto n$. To obtain the hard edge scaling, this product must be multiplied by $4n \sim 2(2n+\tilde{a}+1)$, giving a matrix product similar to $\frac{2}{h}\mathbf{L}_\beta\mathbf{L}_\beta^T$ provided $a = \tilde{a}+1-2/\beta$. Hence, the hard edge scaled eigenvalues of the $\beta \to \infty$ limit of the Laguerre β-ensemble must be given by the eigenvalues of the operator $\mathcal{J}_{\tilde{a}}^\infty(\mathcal{J}_{\tilde{a}}^\infty)^*$, where

$$(\mathcal{J}_{\tilde{a}}^\infty)^* = 2\sqrt{x}\frac{d}{dx} + \frac{\tilde{a}+1}{\sqrt{x}}$$

is the adjoint of $\mathcal{J}_{\tilde{a}}^\infty$ and the boundary conditions noted below (13.189) are enforced. One can check that the eigenfunctions of this operator are the Bessel functions $J_{\tilde{a}+1}(\sigma_k\sqrt{x})$ where σ_k denotes the kth zero of $J_{\tilde{a}}(x)$, and the corresponding eigenvalues are σ_k^2. This is in keeping with the result of Exercises 4.6 q.1 and the asympototic formula (7.29).

The study of (13.188) for fixed β [158] introduces a noise term into the discretization, and suggests that $\sqrt{\frac{2}{h}}\mathbf{L}_\beta$ is a discretization of the *stochastic Bessel operator*

$$\mathcal{J}_{\tilde{a}}^\beta = -2\sqrt{x}\frac{d}{dx} + \frac{\tilde{a}}{\sqrt{x}} + \frac{2}{\sqrt{\beta}}B'(x), \tag{13.190}$$

where $B(x)$ denotes standard Brownian motion. The corresponding eigenfunctions must vanish at $x = 1$, and so too must the action of (13.190) on the eigenfunctions at $x = 0$. Thus the hard edge scaled eigenvalues in the Laguerre β-ensemble specified by the p.d.f. (3.17) with $N \mapsto n$ must therefore be given by the eigenvalues of the operator $\mathcal{J}_{\tilde{a}}^\beta(\mathcal{J}_{\tilde{a}}^\beta)^*$, a fact proved rigorously in [462].

EXERCISES 13.6 1. [357] Consider the coupled differential equations (Dirac-Krein system)

$$\left(\begin{bmatrix} b(x) & -\dfrac{d}{dx} - a(x) \\ \frac{d}{dx} - a(x) & -b(x) \end{bmatrix} - E\right)\begin{bmatrix} p(x;E) \\ q(x;E) \end{bmatrix} = 0$$

subject to the initial conditions $p(0;E) = 0$, $q(0;E) = 1$.

(i) By writing

$$\begin{bmatrix} p(x;E) \\ q(x;E) \end{bmatrix} = \begin{bmatrix} r(x;E)\sin\theta(x;E) \\ r(x;E)\cos\theta(x;E) \end{bmatrix}, \qquad \theta(0;E) = 0$$

show that

$$\frac{d}{dx}\theta(x;E) = E + \mathrm{Im}\Big(e^{2i\theta(x;E)}(a(x) + ib(x))\Big).$$

(ii) Suppose $a(x) + ib(x)$ is a random complex function, unchanged by complex rotations so that

$$e^{2i\theta(x;E_0)}(a(x) + ib(x)) = a(x) + ib(x).$$

Under this circumstance, deduce from the differential equation in (i) that

$$\frac{d}{dx}\Psi(x;E) = 2E + \mathrm{Im}\Big((e^{i\Psi(x;E)} - 1)(a(x) + ib(x))\Big),$$

where $\Psi(x;E) = 2(\theta(x;E) - \theta(x;0))$, and after an appropriate choice of $a(x)$, $b(x)$ and identification of Ψ, x and E recognize this as the stochastic equation (13.175).

13.7 DYNAMICAL CORRELATIONS IN THE CIRCULAR β ENSEMBLE

13.7.1 Eigenfunction expansion of the density-density correlation

Consider the setting of dynamical correlations relating the initial state to a state with parameter value τ. Our interest is in the truncation of the two-point function $\rho_{(1,1)}(x;0;y;\tau) =: \rho_{(1,1)}(x,y;\tau)$, denoted $\rho^T_{(1,1)}$, obtained by subtracting the product of dynamical one-point functions $\rho_{(1)}(x;0)\rho_{(1)}(y;\tau)$. Making use of (11.92) and (11.93) we therefore have

$$\begin{aligned}
\rho^T_{(1,1)}(x,y;\tau) = N^2 &\int_{-\infty}^{\infty} dx_2^{(0)} \cdots \int_{-\infty}^{\infty} dx_N^{(0)}\, p_0(x, x_2^{(0)}, \dots, x_N^{(0)}) \int_{-\infty}^{\infty} dx_2^{(1)} \cdots \int_{-\infty}^{\infty} dx_N^{(1)} \\
&\times G_\tau^{\mathrm{FP}}(y, x_2^{(1)}, \dots, x_N^{(1)} | x, x_2^{(0)}, \dots, x_N^{(0)}) - \rho_{(1)}(x;0)\rho_{(1)}(y;\tau) \\
= &\int_{-\infty}^{\infty} dx_1^{(0)} \cdots \int_{-\infty}^{\infty} dx_N^{(0)} \Big(\sum_{j=1}^N \delta(x_j^{(0)} - x)\Big) p_0(\vec{x}^{(0)}) \\
&\times \int_{-\infty}^{\infty} dx_1^{(1)} \cdots \int_{-\infty}^{\infty} dx_N^{(1)} \Big(\sum_{j=1}^N \delta(x_j^{(1)} - y)\Big) \\
&\times G_\tau^{\mathrm{FP}}(\vec{x}^{(0)}; \vec{x}^{(1)}) - \rho_{(1)}(x;0)\rho_{(1)}(y;\tau).
\end{aligned} \tag{13.191}$$

For $G_\tau^{\rm FP}$ we substitute (11.42) with $G_\tau^{\rm S}$ given in terms of the eigenfunctions ψ_κ of the corresponding Schrödinger operator by (11.43). This gives

$$\begin{aligned}\rho_{(1,1)}^T&(x,y;\tau)\\ &= \sum_\kappa \frac{\Big\langle e^{-\beta W/2}\Big(\sum_{j=1}^N \delta(x_j^{(1)}-y)\Big)\Big|\psi_\kappa\Big\rangle\Big\langle\psi_\kappa\Big|p_0 e^{\beta W/2}\Big(\sum_{j=1}^N \delta(x_j^{(1)}-x)\Big)\Big\rangle}{\langle\psi_\kappa|\psi_\kappa\rangle} e^{-(E_\kappa-E_0)\tau/\beta}\\ &\quad -\rho_{(1)}(x;0)\rho_{(1)}(y;\tau),\end{aligned} \tag{13.192}$$

where

$$\langle f|g\rangle := \int_{-\infty}^{\infty} dx_1 \cdots \int_{-\infty}^{\infty} dx_N\, f(x_1,\dots,x_N)g(x_1,\dots,x_N). \tag{13.193}$$

The formula (11.43) for the Green function, and consequently the formula (13.192) for $\rho_{(1,1)}^T$, assumed we are choosing a real basis of eigenfunctions. We know from Proposition 1.1.6 that this is always possible for a real Hamiltonian operator. However, in studying the Dyson Brownian motion model on the circle the corresponding Schrödinger operator $H^{({\rm C})}$ has been transformed to the operator $\tilde H^{({\rm C})}$ (12.94) involving the complex coordinates $z_j = e^{2\pi i x_j/L}$. In this situation the appropriate inner product is that specified by (12.11) (cf. (13.193)). Correspondingly (11.43) must be modified to read

$$G_\tau^{\rm S}(\vec{x};\vec{x}^{(0)}) = \sum_\kappa \overline{\psi_\kappa(z_1^{(0)},\dots,z_N^{(0)})}\psi_\kappa(z_1,\dots,z_N)e^{-E_\kappa\tau/\beta}. \tag{13.194}$$

Now we know from Proposition 12.6.1 that the eigenfunctions of $H^{({\rm C})}$, which after factorization of the ground state are analytic in the $\{z_j\}$, are given by

$$\psi_\kappa(z_1,\dots,z_N) = \frac{1}{C}e^{-\beta W^{({\rm C})}(x_1,\dots,x_N)/2}P_\kappa(z_1,\dots,z_N;2/\beta), \tag{13.195}$$

where P_κ denotes the symmetric Jack polynomial and C the normalization. We remark that in terms of the inner product (12.11) the normalization is given by

$$C^2 = \langle e^{-\beta W^{({\rm C})}/2}P_\kappa | e^{-\beta W^{({\rm C})}/2}P_\kappa\rangle. \tag{13.196}$$

According to the result of Exercises 12.1 q.5(iii) the corresponding eigenvalue is

$$E_\kappa = E_0 + \Big(\frac{2\pi}{L}\Big)^2 e(\kappa;2/\beta). \tag{13.197}$$

However, the functions (13.195) are not a complete set with respect to the space of Laurent expandable functions in the $\{z_j\}$ as required in the derivation of (13.194). To extend the set we note that, analogous to the result of Exercises 12.1 q.1,

$$z_1^{-l}\cdots z_N^{-l}P_\kappa(z_1,\dots,z_N;2/\beta) = P_{\kappa-l^N}(z_1,\dots,z_N;2/\beta), \tag{13.198}$$

$\kappa_N - l^N \ge 0$, and moreover the l.h.s. of (13.198) is an eigenfunction of (12.94) with eigenvalue $e(\kappa-l;\alpha)$ for all $l\in\mathbb{Z}$. Hence a complete set of eigenfunctions is given by

$$\{e^{-\beta W^{({\rm C})}/2}P_\kappa(z;2/\beta)\}_\kappa \cup \{e^{-\beta W^{({\rm C})}/2}z^{-l}P_\kappa(z;2/\beta)\}_{\substack{\kappa:\kappa_N=0,\\ l=1,2,\dots}} \tag{13.199}$$

where the κ are partitions. Equivalently, if we replace κ by $\kappa_\pm$, where $\kappa_\pm$ denotes an ordered set of integers (not necessarily non-negative), and use (13.198) to define $P_{\kappa_\pm}(z)$ in this setting, we see that

$$\{e^{-\beta W^{({\rm C})}/2}P_{\kappa_\pm}(z;2/\beta)\}_{\kappa_\pm} \tag{13.200}$$

suffices for the complete set of eigenfunctions.

Let us adopt the latter viewpoint. We then have

$$G_\tau^S(\vec{x}^{(0)};\vec{x}) = e^{-E_0\tau/\beta} \sum_{\kappa_\pm} \frac{P_{\kappa_\pm}(z^{(0)};2/\beta)\overline{P_{\kappa_\pm}(z;2/\beta)}}{\mathcal{N}_{\kappa_\pm}^{(\mathrm{C})}|_{\alpha=2/\beta}} e^{-e(\kappa_\pm;2/\beta)\tau/\beta}.$$

Using this formula in place of (11.43), and making use of the inner product (12.12) we see from (13.191) that

$$\begin{aligned}
&\rho_{(1,1)}^T(x,y;\tau) \\
&= \sum_{\kappa_\pm \ne 0^N} \frac{\langle \sum_{j=1}^N \delta(x_j^{(1)}-y) | P_{\kappa_\pm}(z^{(1)};2/\beta)\rangle^{(\mathrm{C})} \langle P_{\kappa_\pm}(z^{(0)};2/\beta) | p_0 e^{\beta W^{(\mathrm{C})}} \sum_{j=1}^N \delta(x_j^{(0)}-x)\rangle^{(\mathrm{C})}}{\mathcal{N}_{\kappa_\pm}^{(\mathrm{C})}} \\
&\quad \times e^{-\tau e(\kappa_\pm;2/\beta)/\beta} \\
&= \sum_{\kappa_\pm : \kappa_\pm \ne 0^N} \frac{\langle \sum_{j=1}^N \delta(x_j^{(1)}) | P_{\kappa_\pm}(z^{(1)};2/\beta)\rangle^{(\mathrm{C})} \langle P_{\kappa_\pm}(z^{(0)};2/\beta) | p_0 e^{\beta W^{(\mathrm{C})}} \sum_{j=1}^N \delta(x_j^{(0)})\rangle^{(\mathrm{C})}}{\mathcal{N}_{\kappa_\pm}^{(\mathrm{C})}} \\
&\quad \times e^{2\pi i |\kappa_\pm|(y-x)/L} e^{-\tau e(\kappa_\pm;2/\beta)/\beta}.
\end{aligned} \tag{13.201}$$

In obtaining the first equality use has been made of the fact that $\rho_{(1)}(y;\tau) = N/L$ for all τ, while the second equality follows from changing variables in the inner product and using the fact that P_κ is homogeneous of degree $|\kappa|$.

The normalization $\mathcal{N}_\kappa^{(\mathrm{C})}$ for a partition is given by Proposition 12.6.3. Making use of the formulas (12.105) and (12.47) we see that this can be written

$$\mathcal{N}_\kappa^{(\mathrm{C})} = \mathcal{N}_{0^N}^{(\mathrm{C})} \frac{(\beta/2)^{|\kappa|} d'_\kappa P_\kappa^{(2/\beta)}((1)^N)}{[\beta(N-1)/2+1]_\kappa^{(2/\beta)}} \tag{13.202}$$

while Proposition 4.6.2 gives

$$\mathcal{N}_{0^N}^{(\mathrm{C})} = L^N \frac{\Gamma(\beta N/2+1)}{(\Gamma(\beta/2+1))^N}. \tag{13.203}$$

Suppose $\kappa_\pm := \kappa - l^N$ is such that $\kappa_N - l < 0$. We then note from (13.198) that $\mathcal{N}_{\kappa-l^N}^{(\mathrm{C})} = \mathcal{N}_\kappa^{(\mathrm{C})}$ and make use of (13.202).

Consider next the inner product $\langle \sum_{j=1}^N \delta(x_j^{(1)}) | P_{\kappa_\pm}(z^{(1)};2/\beta)\rangle^{(\mathrm{C})}$ in (13.201). For κ a partition the integration formula (12.144) gives

$$\Big\langle \sum_{j=1}^N \delta(x_j^{(1)}) \Big| P_\kappa(z^{(1)};2/\beta) \Big\rangle^{(\mathrm{C})} = \mathcal{N}_{0^N}^{(\mathrm{C})} \frac{|\kappa|(\kappa_1-1)!}{L} P_\kappa^{(2/\beta)}((1)^N) \frac{\prod_{j=2}^{l(\kappa)}(-\frac{\beta}{2}(j-1))_{\kappa_j}}{[\beta(N-1)/2+1]_\kappa^{(2/\beta)}}. \tag{13.204}$$

In the case that P_κ is replaced by $P_{\kappa_\pm}$, it turns out that this inner product vanishes unless $\kappa_\pm$ has either all positive parts (which is the case (13.204)) or all negative parts (which we will see can be related to (13.204)).

PROPOSITION 13.7.1 *Suppose κ is a partition with $\kappa_N = 0$ and $l \in \mathbb{Z}_{>0}$, and for η a composition, let η^+*

denote the associated partition, as in Definition 12.1.1. We have

$$\Big\langle \sum_{j=1}^N \delta(x_j) \Big| z^{-l} P_\kappa(z;2/\beta) \Big\rangle^{(\mathrm{C})}$$
$$= L^{N-1} N \int_{-1/2}^{1/2} dx_2\, e^{-2\pi i l x_2} \cdots \int_{-1/2}^{1/2} dx_N\, e^{-2\pi i l x_N} P_\kappa(1, e^{2\pi i x_2}, \dots, e^{2\pi i x_N})$$
$$\times \prod_{j=2}^N |1 - e^{2\pi i x_j}|^\beta \prod_{2 \le j < k \le N} |e^{2\pi i x_k} - e^{2\pi i x_j}|^\beta$$
$$= \begin{cases} \langle \sum_{j=1}^N \delta(x_j) | P_{(l-\kappa)^+}(1/z;2/\beta)\rangle^{(\mathrm{C})}, & \kappa_1 \le l, \\ 0 & \text{otherwise.} \end{cases}$$

Proof. The symmetric analogue of the formula of Exercises 12.1 q.2 reads

$$z^{-l} P_\kappa(z;2/\beta) = P_{\kappa - l^N}(z;2/\beta) = P_{(l^N - \kappa)^+}(1/z;2/\beta),$$

which immediately implies the stated result for $\kappa_1 \le l$. To show that the inner product vanishes for $\kappa_1 > l$, we proceed as in the proof of Proposition 12.3.2, this time setting $a = -b = -l + \epsilon$ in (12.142). According to Proposition 12.6.10, (4.4) and (12.46) we then have

$$\Big\langle \sum_{j=1}^N \delta(x_j) \Big| z^{-l} P_\kappa(z;2/\beta) \Big\rangle^{(\mathrm{C})} = L^{N-1} N(|\kappa| - Nl) P_\kappa((1)^N;2/\beta)$$
$$\times \lim_{\epsilon \to 0} \frac{1}{\epsilon} \prod_{j=1}^N \frac{\Gamma(1 + \beta(j-1)/2)\Gamma(\beta j/2 + 1)}{\Gamma(1 - l + \epsilon + \beta(N-j)/2 + \kappa_j)\Gamma(1 + l - \epsilon + \beta(N-j)/2 - \kappa_j)\Gamma(1+\lambda)}.$$

The product contains the terms

$$\frac{1}{\Gamma(1 - l + \epsilon + \kappa_N)\Gamma(1 + l - \epsilon - \kappa_1)}.$$

With $\kappa_N = 0$ and $l \in \mathbb{Z}_{>0}$, $1/\Gamma(1-l+\epsilon+\kappa_N)$ is proportional to ϵ for small ϵ, and similarly for $\kappa_1 > l$, $1/\Gamma(1+l-\epsilon-\kappa_1)$ is also proportional to ϵ for small ϵ, thus implying the result. □

The result of Proposition 13.7.1 implies the sum in (13.201) can be reduced to a sum over partitions only, provided twice the real part is taken. Thus

$$\rho^T_{(1,1)}(x,y;\tau) = 2\mathrm{Re} \sum_{\substack{\kappa \\ \kappa \ne 0^N}} \frac{1}{\mathcal{N}_\kappa^{(\mathrm{C})}} \Big\langle \sum_{j=1}^N \delta(x_j^{(1)}) \Big| P_\kappa(z^{(1)};2/\beta) \Big\rangle^{(\mathrm{C})}$$
$$\times \Big\langle P_\kappa(z^{(0)};2/\beta) \Big| p_0 e^{\beta W^{(\mathrm{C})}} \sum_{j=1}^N \delta(x_j^{(0)}) \Big\rangle^{(\mathrm{C})} e^{2\pi i |\kappa|(y-x)/L} e^{-\tau e(\kappa;2/\beta)/\beta}. \quad (13.205)$$

It remains to evaluate the inner product $\langle P_\kappa(z^{(0)};2/\beta) | p_0 e^{\beta W^{(\mathrm{C})}} \sum_{j=1}^N \delta(x_j^{(0)})\rangle^{(\mathrm{C})}$ in (13.205). In the case $p_0 = e^{-\beta W^{(\mathrm{C})}} / \mathcal{N}_{0^N}^{(\mathrm{C})}$ (initial state equals equilibrium state),

$$\Big\langle P_\kappa(z^{(0)};2/\beta) \Big| p_0 e^{\beta W^{(\mathrm{C})}} \sum_{j=1}^N \delta(x_j^{(0)}) \Big\rangle^{(\mathrm{C})} = \frac{1}{\mathcal{N}_{0^N}^{(\mathrm{C})}} \Big\langle \sum_{j=1}^N \delta(x_j^{(0)}) \Big| P_\kappa(1/z^{(0)};2/\beta) \Big\rangle^{(\mathrm{C})} \quad (13.206)$$

and thus is evaluated according to (13.204). In the case $p_0 = 1/L^N$ (initial state a perfect gas)

$$\begin{aligned}\Big\langle P_\kappa(z^{(0)};2/\beta)\Big|p_0 e^{\beta W^{(C)}} \sum_{j=1}^N \delta(x_j^{(0)})\Big\rangle^{(C)} &= \frac{N}{L}\int_{-1/2}^{1/2} dx_2^{(0)} \cdots \int_{-1/2}^{1/2} dx_N^{(0)}\, P_\kappa(1/z^{(0)};2/\beta) \\ &= \frac{N}{L} P_\kappa(1/z_1^{(0)},0,\dots,0) \\ &= \begin{cases} \frac{N}{L}(z_1^{(0)})^{-\kappa_1}, & \kappa = (\kappa_1,0,\dots,0), \\ 0 & \text{otherwise.}\end{cases} \end{aligned} \tag{13.207}$$

Consideration of these facts gives the following formulas for $\rho_{(1,1)}^T$ in the finite system [212].

PROPOSITION 13.7.2 *For the perfect gas initial condition*

$$\rho_{(1,1)}^T(x,y;\tau) = \frac{2N}{L^2}\mathrm{Re}\sum_{\kappa_1=1}^\infty e^{2\pi i\kappa_1(y-x)/L} e^{-(2\pi)^2\tau(\kappa_1^2+\beta(N-1)\kappa_1/2)/\beta L^2}, \tag{13.208}$$

while for the case that the initial state is equal to the equilibrium state

$$\begin{aligned}\rho_{(1,1)}^T(x,y;\tau) = &\frac{2}{L^2}\mathrm{Re}\sum_{\kappa\neq 0}|\kappa|^2(\kappa_1-1)!^2 f_N^{2/\beta}(\kappa) f_N^{2/\beta}(\kappa)\prod_{j=1}^N \frac{(\Gamma(1+\beta(j-1)/2))^4}{(\Gamma(1+\beta(N-j)/2+\kappa_j))^2} \\ &\times \prod_{j=2}^N \frac{1}{(\Gamma(1+\beta(j-1)/2-\kappa_j))^2}\exp\Big(2\pi i(y-x)|\kappa|/L - e(\kappa;2/\beta)\tau/\beta\Big)\end{aligned} \tag{13.209}$$

Proof. In deriving (13.208) we use the result (13.207) in (13.205), together with the facts that for $(\kappa_1, 0, \dots, 0)$, $(\beta/2)^{|\kappa|}d'_\kappa = 1$, which allows the ratio of (13.204) and (13.202) to be simplified. In deriving (13.209), we substitute (13.204), (13.206) and (13.202) in (13.205), and use the results

$$\Big(\prod_{j=2}^{l(\kappa)}\Big(-\frac{\beta}{2}(j-1)\Big)_{\kappa_j}\Big)^2 = \Big(\prod_{j=2}^N \frac{\Gamma(1+\beta(j-1)/2)}{\Gamma(1+\beta(j-1)/2-\kappa_j)}\Big)^2,$$

$$P_\kappa^{(2/\beta)}((1)^N) = \frac{b_\kappa}{h_\kappa}\Big|_{\alpha\mapsto 2/\beta} = f_N^{\beta/2}(\kappa), \tag{13.210}$$

where $f_N^{\beta/2}(\kappa)$ is specified by (12.59), as well as the formula (12.60) for d'_κ. □

13.7.2 Bulk limit

The formula (13.208) for $\rho_{(1,1)}^T$ in the case of perfect gas initial conditions is in the form of a Riemann sum approximation to an integral, and the following result is immediate.

PROPOSITION 13.7.3 *In the bulk limit $N, L \to \infty$, $N/L = \rho$ (constant),*

$$\rho_{(1,1)}^{T\text{bulk}}(x,y;\tau) = 2\rho^2\int_0^\infty e^{-(2\pi\rho)^2(u^2+\beta u/2)\tau/\beta}\cos(2\pi\rho(x-y)u)\,du. \tag{13.211}$$

In contrast, the formula (13.209) for $\rho_{(1,1)}^T$ in the case that the initial state is equal to the equilibrium state requires further analysis before its limiting value can be computed. One essential difficulty is that in (13.209) the sum is over all partitions κ, and thus N independent quantities, whereas in (13.208) only the largest part κ_1 enters into the summation. This feature would appear to make the problem of computing the thermodynamic limit intractable. However, if we restrict attention to the case β rational a significant simplification takes place.

PROPOSITION 13.7.4 *Let β be rational and write $\beta/2 = p/q$ where p and q are relatively prime. Then the summand in (13.209) is nonzero if and only if κ is of the form*

$$\kappa = (\alpha_1, \dots, \alpha_q, \underbrace{p, \dots, p}_{\beta_1\ p's}, \dots, \underbrace{1, \dots, 1}_{\beta_p\ 1's}, 0, \dots, 0), \tag{13.212}$$

where $\alpha_1 \ge \alpha_2 \ge \cdots \ge \alpha_q$ and $q + \sum_{j=1}^p \beta_j \le N$.

Proof. The factor $(\prod_{j=2}^N \Gamma(1+\beta(j-1)/2-\kappa_j))^2$ in (13.209) is non-zero if and only if $\beta(j-1)/2-\kappa_j \notin \{-1,-2,\dots\}$ for each $j = 1, 2, \dots, N$. With $\beta/2 = p/q$ and $j = q+1$ this means $p - \kappa_{q+1} \notin \{-1,-2,\dots\}$ and thus $\kappa_{q+1} \le p$ as required. □

As a consequence of (13.7.4), for β rational the summation in (13.209) can be taken over the $p+q$ coordinates α_i $(i = 1, \dots, q)$ and β_j $(j = 1, \dots, p)$. In fact it is more convenient to replace the coordinates β_j by γ_j $(j = 1, \dots, p)$ defined so that $\kappa_{q+\gamma_a+k} = p - a$, $k = 1, \dots, \gamma_{a+1} - \gamma_a$ $(\gamma_0 := 0, \gamma_{p+1} := N - q)$.

PROPOSITION 13.7.5 *For $\beta/2 = p/q$ and κ as specified above and large N, $\{\alpha_j\}, \{\gamma_j\}$ we have $f_N^{\beta/2}(\kappa) = ABC/D$, where*

$$A := \prod_{1 \le i < j \le q} \Big((j-i)\beta/2 + \alpha_i - \alpha_j\Big)_{\beta/2} \sim \prod_{1 \le i < j \le q} (\alpha_i - \alpha_j)^{\beta/2};$$

$$B := \prod_{i=1}^{q} \prod_{j=q+1}^{N} \Big((j-i)\beta/2 + \alpha_i - \kappa_j\Big)_{\beta/2} \sim b_1 b_2 \prod_{i=1}^{q} \prod_{a=1}^{p} \frac{1}{\gamma_a \beta/2 + \alpha_i}$$

with

$$b_1 = \prod_{i=1}^{q} \Gamma((N-i+1)\beta/2 + \alpha_i), \quad b_2 = \frac{1}{\prod_{i=1}^{q} \Gamma((q-i+1)\beta/2 + \alpha_i - p)};$$

and

$$C := \prod_{q+1 \le i < j \le N} \Big((j-i)\beta/2 + \kappa_i - \kappa_j\Big)_{\beta/2} \sim \Big(\Gamma(\beta/2)\Big)^{-N+q} (\beta/2)^{-\sum_{j=1}^{p} \gamma_j} c_1 c_2 \prod_{a=1}^{p-1} \prod_{a'=a+1}^{p} (\gamma_{a'} - \gamma_a)^{2/\beta}$$

with

$$c_1 := \prod_{a=0}^{p} \prod_{i=q+\gamma_a+1}^{q+\gamma_{a+1}} \Gamma((N+1-i)\beta/2 + p - a), \quad c_2 := \prod_{a'=1}^{p} \frac{1}{\Gamma(\gamma_{a'} + 1 + 2(a'-1)/\beta)};$$

$$D := \prod_{1 \le i < j \le N} \Big((j-i)\beta/2\Big)_{\beta/2} = \Big(\frac{1}{\Gamma(\beta/2)}\Big)^{N} \prod_{j=1}^{N} \Gamma(j\beta/2).$$

Proof. The decomposition $f_N^{\beta/2}(\kappa) = ABC/D$ is immediate from the definitions (12.59) and (13.212). For the asymptotic form of A we have

$$\prod_{1 \le i < j \le q} \Big((j-i)\beta/2 + \alpha_i - \alpha_j\Big)_{\beta/2} = \prod_{1 \le i < j \le q} \frac{\Gamma((j-i)\beta/2 + \alpha_i - \alpha_j + \beta/2)}{\Gamma((j-i)\beta/2 + \alpha_i - \alpha_j)}$$
$$\sim \prod_{1 \le i < j \le q} \Big((j-i)\beta/2 + \alpha_i - \alpha_j\Big)^{\beta/2} \sim \prod_{1 \le i < j \le q} (\alpha_i - \alpha_j)^{\beta/2},$$

where the first asymptotic expression follows upon using (4.126).

For the asymptotic form of B we write

$$\prod_{j=q+1}^{N} = \prod_{a=0}^{p} \prod_{j=1+q+\gamma_a}^{\gamma_{a+1}+q}$$

and note

$$\prod_{a=0}^{p} \prod_{j=1+q+\gamma_a}^{\gamma_{a+1}+q} \frac{\Gamma((j-i)\beta/2+\alpha_i-(p-a)+\beta/2)}{\Gamma((j-i)\beta/2+\alpha_i-(p-a))}$$
$$= \prod_{a=0}^{p} \frac{\Gamma((\gamma_{a+1}+q-i)\beta/2+\alpha_i-(p-a)+\beta/2)}{\Gamma((\gamma_a+1+q-i)\beta/2+\alpha_i-(p-a))}$$
$$= \frac{\Gamma((N-i)\beta/2+\alpha_i+\beta/2)}{\Gamma((1+q-i)\beta/2+\alpha_i-p)} \prod_{a=1}^{p} \frac{\Gamma((\gamma_a+q-i)\beta/2+\alpha_i-(p-a)-1+\beta/2)}{\Gamma((\gamma_a+1+q-i)\beta/2+\alpha_i-(p-a))}$$
$$= \frac{\Gamma((N-i)\beta/2+\alpha_i+\beta/2)}{\Gamma((1+q-i)\beta/2+\alpha_i-p)} \prod_{a=1}^{p} \frac{1}{(\gamma_a+q-i)\beta/2+\alpha_i-(p-a)-1-\beta/2},$$

where in obtaining the second equality the facts $\gamma_0 := 0$, $\gamma_{p+1} := N-q$ have been used. The result now follows since $\gamma_a, \alpha_i \gg 1$ while the remaining quantities in the product are fixed.

For the asymptotic form of C note

$$\prod_{q+1 \le i < j \le N} = \Big(\prod_{a=0}^{p-1} \prod_{i=q+\gamma_a+1}^{q+\gamma_{a+1}} \prod_{a'=a+1}^{p} \prod_{j=q+\gamma_{a'}+a}^{q+\gamma_{a'}+1} \Big) \Big(\prod_{a=0}^{p} \prod_{i=q+\gamma_a+1}^{\min(q+\gamma_{a+1}, N-1)} \prod_{j=i+1}^{q+\gamma_{a+1}} \Big).$$

Proceeding as in the simplification of B we see that for the range of indices implied by the first grouping of products

$$\prod_{a=0}^{p-1} \prod_{i=q+\gamma_a+1}^{q+\gamma_{a+1}} \prod_{a'=a+1}^{p} \prod_{j=q+\gamma_{a'}+1}^{q+\gamma_{a'}+1} \Big((j-i)\beta/2+\kappa_i-\kappa_j \Big)_{\beta/2}$$
$$= \prod_{a=0}^{p-1} \prod_{i=q+\gamma_a+1}^{q+\gamma_{a+1}} \frac{\Gamma((N+1-i)\beta/2+p-a)}{\Gamma((q+\gamma_{a+1}+1-i)\beta/2)} \prod_{a'=a+1}^{p} \frac{1}{(q+\gamma_{a'}+1-i)\beta/2+a'-a-1}.$$

Now

$$\prod_{i=q+\gamma_a+1}^{q+\gamma_{a+1}} \frac{1}{(q+\gamma_{a'}+1-i)\beta/2+a'-a-1} = \Big(\frac{2}{\beta}\Big)^{\gamma_{a+1}-\gamma_a} \frac{\Gamma(\gamma_{a'}-\gamma_{a+1}+1+2(a'-a-1)/\beta)}{\Gamma(\gamma_{a'}-\gamma_a+1+2(a'-a-1)/\beta)},$$

while

$$\prod_{a=0}^{p-1} \prod_{a'=a+1}^{p} \Big(\frac{2}{\beta}\Big)^{\gamma_{a+1}-\gamma_a} \frac{\Gamma(\gamma_{a'}-\gamma_{a+1}+1+2(a'-a-1)/\beta)}{\Gamma(\gamma_{a'}-\gamma_a+1+2(a'-a-1)/\beta)}$$
$$= \Big(\prod_{a=0}^{p-1} \Big(\frac{2}{\beta}\Big)^{(\gamma_{a+1}-\gamma_a)(p-a)} \Big)$$
$$\times \prod_{a=1}^{p-1} \prod_{a'=a+1}^{p} \frac{\Gamma(\gamma_{a'}-\gamma_a+1+2(a'-a)/\beta)}{\Gamma(\gamma_{a'}-\gamma_a+1+2(a'-a-1)/\beta)} \Big(\prod_{a'=1}^{p} \frac{1}{\Gamma(\gamma_{a'}+1+2(a'-1)/\beta)} \Big)$$
$$\sim \Big(\frac{2}{\beta}\Big)^{N-q} \Big(\prod_{a'=1}^{p} \frac{1}{\Gamma(\gamma_{a'}+1+2(a'-1)/\beta)} \Big) \prod_{a=0}^{p-1} \prod_{a'=a+1}^{p} (\gamma_{a'}-\gamma_a)^{2/\beta}.$$

Finally, for the range of indices implied by the second grouping of products

$$\prod_{a=0}^{p} \prod_{i=q+\gamma_a+1}^{\min(q+\gamma_{a+1},N-1)} \prod_{j=i+1}^{q+\gamma_{a+1}} \frac{\Gamma((j-i+1)\beta/2+\kappa_i-\kappa_j)}{\Gamma((j-i)\beta/2+\kappa_i-\kappa_j)}$$
$$= \prod_{a=0}^{p} \prod_{i=q+\gamma_a+1}^{\min(q+\gamma_{a+1},N-1)} \prod_{j=i+1}^{q+\gamma_{a+1}} \frac{\Gamma((j-i+1)\beta/2)}{\Gamma((j-i)\beta/2)}$$
$$= \Big(\frac{1}{\Gamma(\beta/2)}\Big)^{N-1-q} \prod_{a=0}^{p} \prod_{i=q+\gamma_a+1}^{\min(q+\gamma_{a+1},N-1)} \Gamma((q+\gamma_{a+1}-i+1)\beta/2)$$

and

$$\prod_{a=0}^{p-1} \prod_{i=q+\gamma_a+1}^{q+\gamma_{a+1}} \frac{\Gamma((N+1-i)\beta/2+p-a)}{\Gamma((q+\gamma_{a+1}+1-i)\beta/2)} \Big(\prod_{a=0}^{p} \prod_{i=q+\gamma_a+1}^{\min(q+\gamma_{a+1},N-1)} \Gamma((q+\gamma_{a+1}-i+1)\beta/2)\Big)$$
$$= \frac{1}{\Gamma(\beta/2)} \prod_{a=0}^{p} \prod_{i=q+\gamma_a+1}^{q+\gamma_{a+1}} \Gamma((N+1-i)\beta/2+p-a).$$

□

A similar calculation gives the following result.

PROPOSITION 13.7.6 *For $\beta/2 = p/q$ and κ as specified by (13.212), and large N, $\{\alpha_j\}$, $\{\gamma_j\}$ we have $\bar{f}_N^{\beta/2}(\kappa) = A'B'C'/D'$ where*

$$A' := \prod_{1\le i<j\le q} \Big((j-i)\beta/2+\alpha_i-\alpha_j+1-\beta/2\Big)_{\beta/2} \sim \prod_{1\le i<j\le q} (\alpha_i-\alpha_j)^{\beta/2};$$

$$B' := \prod_{i=1}^{q} \prod_{j=q+1}^{N} \Big((j-i)\beta/2+\alpha_i-\kappa_j+1-\beta/2\Big)_{\beta/2} \sim b_1'b_2' \prod_{i=1}^{q} \prod_{a=1}^{p} \frac{1}{\gamma_a\beta/2+\alpha_i}$$

with

$$b_1' = \prod_{i=1}^{q} \Gamma((N-i+1)\beta/2+\alpha_i+1-\beta/2), \quad b_2' = \frac{1}{\prod_{i=1}^{q} \Gamma((q-i+1)\beta/2+\alpha_i-p+1-\beta/2)};$$

$$C' := \prod_{q+1\le i<j\le N} \Big((j-i)\beta/2+\kappa_i-\kappa_j+1-\beta/2\Big)_{\beta/2}$$
$$\sim \Big(\Gamma(2/\beta)\Big)^p (\beta/2)^{-\sum_{j=1}^p \gamma_j} c_1'c_2' \prod_{a=1}^{p-1} \prod_{a'=a+1}^{p} (\gamma_{a'}-\gamma_a)^{2/\beta}$$

with

$$c_1' := \prod_{a=0}^{p} \prod_{i=q+\gamma_a+1}^{q+\gamma_{a+1}} \Gamma((N+1-i)\beta/2+p-a+1-\beta/2), \quad c_2' := \prod_{a'=1}^{p} \frac{1}{\Gamma(\gamma_{a'}+2a'/\beta)};$$

and

$$D' := \prod_{i\le i<j\le N} \Big((j-i)\beta/2+1-\beta/2\Big)_{\beta/2} = \prod_{j=1}^{N} \Gamma(\beta(j-1)/2+1).$$

Next we combine terms and products involving N from Propositions 13.7.5 and 13.7.6 with the products in (13.209).

PROPOSITION 13.7.7 *With the notation introduced in Propositions 13.7.5 and 13.7.6 we have*

$$b_1 b_1' c_1 c_1' \Gamma^2(\kappa_1) \prod_{j=1}^{N} \Big(\Gamma(1+(N-j)\beta/2+\kappa_j) \Big)^{-2}$$
$$\sim N^{q(\beta/2-1)} \prod_{j=1}^{q} \Big(\beta/2 + \frac{\alpha_j}{N} \Big)^{\beta/2-1} (\beta/2)^{-N+1+q-p} \frac{\Gamma((N-q)\beta/2+p)}{\Gamma(N-q+2p/\beta)} \prod_{a=1}^{p} (N-\gamma_a)^{2/\beta-1},$$

$$b_2 b_2' c_2 c_2' \prod_{j=2}^{N} \Big(\Gamma(1+(j-1)\beta/2-\kappa_j) \Big)^{-2} \prod_{j=1}^{N} \Big(\Gamma(1+(j-1)\beta/2) \Big)^2$$
$$\sim \prod_{j=2}^{q} \Big(\frac{1}{\Gamma(-(j-1)\beta/2)} \Big)^2 \prod_{j=1}^{q} \alpha_j^{\beta/2-1} \prod_{a=1}^{p} \Big(\Gamma(2a/\beta) \Big)^{-2} (\beta/2)^{2\sum_{j=1}^{p}\gamma_j} \prod_{a=1}^{p} \gamma_a^{2/\beta-1}$$

and

$$\frac{1}{DD'} \prod_{j=1}^{N} (\Gamma(1+\beta(j-1)/2)^2 = (\Gamma(\beta/2))^N (\beta/2)^{N-1} \frac{\Gamma(N)}{\Gamma(\beta N/2)}.$$

Proof. To derive the first formula we begin by noting

$$b_1 b_1' \prod_{j=1}^{q} \Big(\Gamma(1+(N-j)\beta/2+\kappa_j) \Big)^{-2} = \prod_{j=1}^{q} \frac{|\Gamma((N-j+1)\beta/2+\alpha_j)}{\Gamma(1+(N_j)\beta/2+\alpha_j)} \sim N^{q(\beta/2-1)} \prod_{j=1}^{q} \Big(\beta/2 + \frac{\alpha_j}{N} \Big)^{\beta/2-1},$$

where in the final line use has been made of (4.126). Also, using the fact that

$$\prod_{j=q+1}^{N} \Big(\Gamma(1+(N-j)\beta/2+\kappa_j) \Big)^{-2} = \prod_{a=0}^{p} \prod_{i=q+\gamma_a+1}^{q+\gamma_{a+1}} \Big(\Gamma(1+(N-j)\beta/2+p-a) \Big)^{-2}$$

we have

$$c_1 c_1' \prod_{j=q+1}^{N} \Big(\Gamma(1+(N-j)\beta/2+\kappa_j) \Big)^{-2}$$
$$= \prod_{a=0}^{p} \prod_{i=q+\gamma_a+1}^{q+\gamma_{a+1}} \frac{\Gamma((N+1-i)\beta/2+p-a)}{\Gamma(1+(N-i)\beta/2+p-a)}$$
$$= \lim_{\epsilon\to 0} \prod_{a=0}^{p} \prod_{i=q+\gamma_a+1}^{q+\gamma_{a+1}} \frac{\Gamma((N+1-i)\beta/2+p-a)}{((N-i)\beta/2+p-a+\epsilon)\Gamma(1+(N-i)\beta/2+p-a+\epsilon)}$$
$$= (\beta/2)^{-(N-q)} \lim_{\epsilon\to 0} \prod_{a=0}^{p} \frac{\Gamma((N-q-\gamma_{a+1})+2(p-a+\epsilon)/\beta)}{\Gamma((N-q-\gamma_a)+2(p-a)/\beta)} \frac{\Gamma((N-q-\gamma_a)+2(p-a)/\beta)}{\Gamma((N-q-\gamma_{a+1})+2(p-a+\epsilon)/\beta)}$$
$$\sim (\beta/2)^{-N+1+q-p} \frac{\Gamma((N-q)\beta/2+p)}{\Gamma((N-q)+2p/\beta)} \prod_{a=1}^{p} (N-\gamma_a)^{2/\beta-1},$$

where in the second equality the quantity ϵ is introduced because $(N-i)\beta/2+p-a=0$ for $a=p, i=q+\gamma_{p+1}:=N$. These results imply the first stated result.

To derive the second result we begin by noting

$$\prod_{j=2}^{q}\Big(\Gamma(1+(j-1)\beta/2-\kappa_j)\Big)^{-2}=\prod_{j=2}^{q}\Big(\frac{\sin\pi((j-1)\beta/2)}{\pi}\Big)^2\prod_{j=2}^{q}\Big(\Gamma(\kappa_j-(j-1)\beta/2)\Big)^2.$$

Hence

$$\begin{aligned}
b_2b_2'\Gamma^2(\kappa_1)&\prod_{j=2}^{q}\Big(\Gamma(1+(j-1)\beta/2-\kappa_j)\Big)^{-2}\prod_{j=1}^{q}\Big(\Gamma(1+(j-1)\beta/2)\Big)^2\\
&=\prod_{j=2}^{q}\Big(\frac{\sin\pi((j-1)\beta/2)}{\pi}\Big)^2\prod_{j=1}^{q}\frac{\Gamma(\alpha_j-(j-1)\beta/2)}{\Gamma(\alpha_j+1-j\beta/2)}\prod_{j=1}^{q}\Big(\Gamma(1+(j-1)\beta/2)\Big)^2\\
&\sim\prod_{j=2}^{q}\Big(\frac{\sin\pi((j-1)\beta/2)}{\pi}\Big)^2\prod_{j=1}^{q}\alpha_j^{\beta/2-1}\prod_{j=1}^{q}\Big(\Gamma(1+(j-1)\beta/2)\Big)^2=\prod_{j=1}^{q}\alpha_j^{\beta/2-1}\prod_{j=1}^{q}\Big(\frac{1}{\Gamma(-(j-1)\beta/2)}\Big)^2.
\end{aligned}$$

Next we note

$$\begin{aligned}
&\prod_{j=q+1}^{N}\frac{\Gamma(1+(j-1)\beta/2)}{\Gamma(1+(j-1)\beta/2-\kappa_j)}\\
&\quad=\prod_{a=0}^{p}\prod_{j=q+\gamma_a+1}^{q+\gamma_{a+1}}\frac{\Gamma(1+(j-1)\beta/2)}{\Gamma(1+(j-1)\beta/2-(p-a))}=(\beta/2)^{\sum_{j=1}^{p}\gamma_j}\prod_{i=1}^{p}\frac{1}{\Gamma(2i/\beta)}\prod_{a=1}^{p}\Gamma(\gamma_a+2a/\beta),
\end{aligned}$$

which shows

$$c_2c_2'\prod_{j=q+1}^{N}\Big(\frac{\Gamma(1+(j-1)\beta/2)}{\Gamma(1+(j-1)\beta/2-\kappa_j)}\Big)^2\sim(\beta/2)^{2\sum_{j=1}^{p}\gamma_j}\Big(\prod_{i=1}^{p}\frac{1}{\Gamma(2i/\beta)}\Big)^2\prod_{a=1}^{p}\gamma_a^{2/\beta-1}.$$

□

It remains to specify e_κ and $|\kappa|$ in terms of the coordinates $\{\alpha_j\}$, $\{\gamma_a\}$.

PROPOSITION 13.7.8 *For the partitions (13.212) we have*

$$\beta e_\kappa\sim\Big(\frac{2\pi}{L}\Big)^2\Big(\sum_{j=1}^{q}(\alpha_j^2+(\beta/2)N\alpha_j)+\sum_{a=1}^{p}(\beta/2)(N\gamma_a-\gamma_a^2)\Big)$$

and $|\kappa|=\sum_{j=1}^{q}\alpha_j+\sum_{a=1}^{p}\gamma_a$.

Proof. From the definition of $\{\alpha_j\}$ and $\{\gamma_a\}$ we have

$$|\kappa|=\sum_{j=1}^{q}\alpha_j+\sum_{a=0}^{p}(p-a)(\gamma_{a+1}-\gamma_a)$$

with $\gamma_0:=0$. Simplification of the last series gives the stated result for $|\kappa|$. For βe_κ, from (12.34) we have

$$\begin{aligned}
\beta e_\kappa&=\Big(\frac{2\pi}{L}\Big)^2\sum_{j=1}^{q}\Big(\alpha_j^2+(\beta/2)\alpha_j(N+1-2q)\Big)\\
&\quad+\Big(\frac{2\pi}{L}\Big)^2\sum_{a=0}^{p}\Big((p-a)^2(\gamma_{a+1}-\gamma_a)+(\beta/2)(p-a)(\gamma_{a+1}-\gamma_a)(N-\gamma_{a+1}-\gamma_a-2q)\Big).
\end{aligned}$$

We see that the leading order terms in the summand are proportional to N^2. Ignoring terms of a lesser order gives

$$\beta e_\kappa \sim \Big(\frac{2\pi}{L}\Big)^2 \sum_{j=1}^{q} (\alpha_j^2 + (\beta/2)\alpha_j N) + \Big(\frac{2\pi}{L}\Big)^2 \sum_{a=0}^{p} (\beta/2)(p-a)(\gamma_{a+1}-\gamma_a)(N-\gamma_{a+1}-\gamma_a),$$

which we identify with the stated result. □

Substituting the results of Propositions 13.7.5 to 13.7.8 in (13.209) shows that for large L and N, $\rho^T_{(1,1)}(x,y;\tau)$ is the Riemann approximation to a multidimensional integral in the variables $x_j := \alpha_j/N$ ($x_j \ge 0$) and $y_j := \gamma_j/N$ ($1 \ge y_j \ge 0$) [283].

PROPOSITION 13.7.9 *For $\beta/2 := \lambda = p/q$ (p and q relatively prime) we have*

$$\rho^{T\mathrm{bulk}}_{(1,1)}(x,y;\tau) = C_{p,q}(\lambda) \prod_{i=1}^{q} \int_0^\infty dx_i \prod_{j=1}^{p} \int_0^1 dy_j \, Q^2_{p,q} F(p,q,\lambda|\{x_i,y_j\}) \cos(Q_{p,q}(x-y)) \exp(-E_{p,q}\tau/2\lambda), \tag{13.213}$$

where the momentum Q and the energy E variables are given by

$$Q_{p,q} := 2\pi\rho\Big(\sum_{i=1}^{q} x_i + \sum_{j=1}^{p} y_j\Big), \quad E_{p,q} := (2\pi\rho)^2\Big(\sum_{i=1}^{q} \epsilon_P(x_i) + \sum_{j=1}^{p} \epsilon_H(y_j)\Big)$$

with $\epsilon_P(x) = x(x+\lambda)$ and $\epsilon_H(y) = \lambda y(1-y)$, the so called form factor F is given by

$$F(p,q,\lambda|\{x_i,y_j\}) = \frac{\prod_{i<i'} |x_i - x_{i'}|^{2\lambda} \prod_{j<j'} |y_j - y_{j'}|^{2/\lambda}}{\prod_{i=1}^{q} (\epsilon_P(x_i))^{1-\lambda} \prod_{j=1}^{p} (\epsilon_H(y_j))^{1-1/\lambda} \prod_{i=1}^{q} \prod_{j=1}^{p} (x_i + \lambda y_j)^2} \tag{13.214}$$

and the normalization is given by

$$C_{p,q}(\lambda) = \frac{\lambda^{2p(q-1)}\Gamma^2(p)}{2\pi^2 p! q!} \frac{\Gamma^q(\lambda)\Gamma^p(1/\lambda)}{\prod_{i=1}^{q} \Gamma^2(p-\lambda(i-1)) \prod_{j=1}^{p} \Gamma^2(1-(j-1)/\lambda)}. \tag{13.215}$$

Proof. Substituting the results of Propositions 13.7.5 – 13.7.8 in (13.209) and minor manipulation gives

$$\begin{aligned} \rho^{T\mathrm{bulk}}_{(1,1)}(x,y;\tau) &\sim \frac{1}{2\pi^2}\Big(\prod_{j=2}^{q} \frac{1}{\Gamma(-(j-1)\beta/2)}\Big)^2 \Big(\prod_{a=1}^{p} \frac{1}{\Gamma(2a/\beta)}\Big)^2 (\Gamma(2/\beta))^p ((\Gamma(\beta/2))^q \\ &\times N^{-(p+q)} \sum_\kappa{}^{*} Q^2_{p,q} \prod_{a=1}^{p} (1-v_a)^{2/\beta-1} (\beta v_a/2)^{2/\beta-1} \prod_{i=1}^{q} u_i^{\beta/2-1} (\beta/2+u_i)^{\beta/2-1} \\ &\times \prod_{1\le i<j\le q} (u_j - u_i)^\beta \prod_{1\le a<a'\le p} (v_{a'} - v_a)^{4/\beta} \cos((x-y)Q_{p,q}) e^{-E_{p,q}\tau/\beta}, \end{aligned} \tag{13.216}$$

where $v_a := \gamma_a/N$, $u_i := \alpha_i/N$, while $Q_{p,q}$ and $E_{p,q}$ are as in (13.213) and the asterisk denotes the summation is over the partitions (13.212). Because the summand depends only on the scaled variables $\{v_a\}$, $\{u_i\}$ and further is symmetrical in these sets of variables, we have that for large N

$$N^{-(p+q)} \sum_\kappa{}^{*} \sim \frac{1}{p!q!} \int_0^\infty du_1 \cdots \int_0^\infty du_q \int_0^1 dv_1 \cdots \int_0^1 dv_p. \tag{13.217}$$

The remaining task is to show the constant prefactor in (13.216), multiplied by $1/p!q!$ from (13.217), can be written in the form (13.215). Now

$$\prod_{i=2}^{q}\Gamma(p-\beta(i-1)/2)$$
$$=\prod_{i=2}^{q}(\beta/2)^p\Big(2(p-1)/\beta-(i-1)\Big)\Big(2(p-2)/\beta-(i-1)\Big)\cdots\Big(-(i-1)\Big)\prod_{i=2}^{q}\Gamma(-\beta(i-1)/2)$$
$$=(\beta/2)^{p(q-1)}(-1)^{q-1}(q-1)!\prod_{j=2}^{p}\frac{\Gamma(2(j-1)/\beta-1)}{\Gamma(2(j-1)/\beta-q)}\prod_{i=2}^{q}\Gamma(-\beta(i-1)/2),$$

while

$$\prod_{j=2}^{p}\frac{\Gamma(2(j-1)/\beta-1)}{\Gamma(2(j-1)/\beta-q)}=\prod_{j=2}^{p}\Gamma(2(j-1)/\beta)\prod_{j=1}^{p-1}\frac{1}{\Gamma(1-2j/\beta)}.$$

Thus

$$\frac{1}{\prod_{j=2}^{q}\Gamma(-\beta(j-1)/2)}=(\beta/2)^{p(q-1)}\frac{\prod_{j=2}^{p}\Gamma(2(j-1)/\beta)}{\Gamma(p-\beta(i-1)/2)}\frac{(q-1)!}{\prod_{j=1}^{p-1}\Gamma(1-2j/\beta)},$$

which when substituted in the prefactor of (13.216) gives the form (13.215). □

13.7.3 Asymptotic expansions

Consider first the formula (13.211) for $\rho_{(1,1)}^{T\text{bulk}}(x,y;\tau)$ in the case of perfect gas initial conditions. Rewriting the cosine term as a complex exponential and linearizing the exponent about $s=0$ shows that for large x and τ

$$\rho_{(1,1)}^{T\text{bulk}}(x,0;\tau)\sim 2\rho^2\text{Re}\left(\frac{1}{\tau(2\pi\rho)^2/2-2\pi i\rho x}\right).$$

Consider next the formula (13.213) for $\rho_{(1,1)}^{T\text{bulk}}(x,y;\tau)$ in the case that the initial state equals the final state. Here, analogous to the procedure of Section 13.2.3, the large x and τ asymptotic expansion of $\rho_{(1,1)}^{T\text{bulk}}(x,y;\tau)$ is obtained by expanding the integrand in (13.213) about $x_i=0$ for all $i=1,\dots,q$ and either $y_j=0$ or $y_j=1$. Choosing to expand about $y_j=0$ for $j=1,\dots,p-m$ and $y_j=1$ for $j=p-m+1,\dots,m$ (this is one of $\binom{p}{m}$ equivalent choices of partitioning $\{y_j\}$ into $p-m$ variables to be expanded about 1 and m variables to be expanded about 0), and writing $w_j=1-y_{j-(p-m)}$ $(j=p-m+1,\dots,p)$ gives

$$Q_{p,q}=2\pi\rho m+2\pi\rho\Big(\sum_{i=1}^{q}x_i+\sum_{j=1}^{p-m}y_j-\sum_{j=1}^{m}w_j\Big),$$
$$E_{p,q}\sim(2\pi\rho)^2\Big(\sum_{i=1}^{q}\lambda x_i+\sum_{j=1}^{p-m}\lambda y_j+\sum_{j=1}^{m}\lambda w_j\Big),$$
$$F\sim\binom{p}{m}\lambda^{-2mq}\frac{\prod_{i<i'}^{q}|x_i-x_{i'}|^{2\lambda}\prod_{j<j'}^{p-m}|y_j-y_{j'}|^{2/\lambda}\prod_{k<k'}^{m}|w_k-w_{k'}|^{2/\lambda}}{\prod_{i=1}^{q}x_i^{1-\lambda}\prod_{j=1}^{p-m}y_j^{1-1/\lambda}\prod_{k=1}^{m}w_k^{1-1/\lambda}}.$$

Extending the integration interval in the variables y_j $(j=1,\dots,p-m)$ and w_j $(j=1,\dots,m)$ to $(0,\infty)$, and changing variables

$$2\pi\rho(-ix+\pi\rho\tau)x_i\mapsto x_i,\quad 2\pi\rho(-ix+\pi\rho\tau)y_j\mapsto y_j,\quad 2\pi\rho(ix+\pi\rho\tau)w_j\mapsto w_j,$$

shows that to leading order in terms proportional to $\cos 2\pi m\rho x$ $(m = 0, \dots, p)$,

$$\begin{aligned}
&\rho_{(1,1)}^{T\mathrm{bulk}}(x, 0; \tau) \\
&\sim C_{p,q}(\lambda)(2\pi\rho)^2 \Big(\frac{1}{2} \Big(\frac{1}{(-i(2\pi\rho)x + (2\pi\rho)^2\tau/2)^2} + \frac{1}{(i(2\pi\rho)x + (2\pi\rho)^2\tau/2)^2} \Big) A(q, p, \lambda) \\
&\quad + \sum_{m=1}^{p} \binom{p}{m} \lambda^{-2mq} m^2 \cos 2\pi m\rho x \Big(\frac{1}{|i(2\pi\rho)x + (2\pi\rho)^2\tau/2|^2} \Big)^{m^2/\lambda} B(q, p-m, \lambda) C(m, \lambda) \Big),
\end{aligned} \tag{13.218}$$

where

$$\begin{aligned}
&A(q, p, \lambda) \\
&= \int_0^\infty dx_1\, x_1^{\lambda-1} e^{-x_1} \cdots \int_0^\infty dx_q\, x_q^{\lambda-1} e^{-x_q} \int_0^\infty dy_1\, y_1^{1/\lambda-1} e^{-y_1} \cdots \\
&\quad \times \int_0^\infty dy_p\, y_p^{1/\lambda-1} e^{-y_p} \Big(\sum_{i=1}^q x_i + \sum_{j=1}^p y_j \Big)^2 \frac{\prod_{i<i'}^q |x_i - x_{i'}|^{2\lambda} \prod_{j<j'}^p |y_j - y_{j'}|^{2/\lambda}}{\prod_{i=1}^q \prod_{j=1}^p (x_i + \lambda y_j)^2},
\end{aligned} \tag{13.219}$$

$$\begin{aligned}
&B(q, p-m, \lambda) \\
&= \int_0^\infty dx_1\, x_1^{\lambda-1} e^{-x_1} \cdots \int_0^\infty dx_q\, x_q^{\lambda-1} e^{-x_q} \int_0^\infty dy_1\, y_1^{1/\lambda-1} e^{-y_1} \cdots \\
&\quad \times \int_0^\infty dy_p\, y_{p-m}^{1/\lambda-1} e^{-y_{p-m}} \frac{\prod_{i<i'}^q |x_i - x_{i'}|^{2\lambda} \prod_{j<j'}^{p-m} |y_j - y_{j'}|^{2/\lambda}}{\prod_{i=1}^q \prod_{j=1}^{p-m} (x_i + \lambda y_j)^2},
\end{aligned} \tag{13.220}$$

$$C(m, \lambda) = \int_0^\infty dw_1\, w_1^{1/\lambda-1} e^{-w_1} \cdots \int_0^\infty dw_m\, w_m^{1/\lambda-1} e^{-w_m} \prod_{j<j'}^{m} |w_j - w_{j'}|^{2/\lambda}. \tag{13.221}$$

All the above integrals can be evaluated in terms of the following limiting form of the Dotsenko-Fateev integral from Exercises 4.5 q.1 [237].

PROPOSITION 13.7.10 *Let*

$$I_{n,m}(z; \beta, \rho) := \frac{1}{m!n!} \prod_{i=1}^n \int_0^\infty dr_i\, r_i^{-\beta/\rho} e^{-z r_i} \prod_{j=1}^m \int_0^\infty d\eta_j\, \eta_j^{\beta} e^{-z\eta_j} \frac{\prod_{i<i'} |r_i - r_{i'}|^{2/\rho} \prod_{j<j'} |\eta_{j'} - \eta_j|^{2\rho}}{\prod_{i=1}^n \prod_{j=1}^m (\eta_j + \rho r_i)^2}. \tag{13.222}$$

Then we have the evaluation

$$I_{n,m}(z; \beta, \rho) = z^{-c_{nm}(\beta,\rho)} \prod_{l=1}^n \frac{\Gamma(l/\rho)}{\Gamma(1/\rho)} \prod_{j=1}^m \frac{\Gamma(j\rho - n)}{\Gamma(\rho)} \prod_{l=0}^{n-1} \Gamma(1 - \beta/\rho + l/\rho) \prod_{j=0}^{m-1} \Gamma(1 - n + \beta + j\rho), \tag{13.223}$$

where

$$c_{nm}(\beta, \rho) = m + n + m\beta - n\beta/\rho + m(m-1)\rho + n(n-1)/\rho - 2mn.$$

Proof. Consider $J_{(0,n)(m,0)}(\alpha, \beta, \rho)$ as defined in Exercises 4.5 q.1. Setting $t_i = \exp(s_i/\alpha)$ and $\tau_j = \exp(-\xi_j/\alpha)$ after eliminating the parameters α', β', ρ' in favor of α, β, ρ, and then considering the limit $\alpha \to \infty$ shows

$$\begin{aligned}
f_{nm}(\{e^{s_i/\rho}\}, \{e^{-\xi_j/\rho}\}; \alpha, \beta, \rho) &\underset{\alpha\to\infty}{\sim} \prod_{i=1}^n (s_i/\alpha)^{-\beta/\rho} e^{-s_i/\rho} \prod_{j=1}^m (\xi_j/\alpha)^{\beta} e^{-\xi_j} \\
&\quad \times \frac{\prod_{i<i'} \alpha^{-2/\rho}(s_i - s_{i'})^{2/\rho} \prod_{j<j'} \alpha^{-2\rho}(\xi_{j'} - \xi_j)^{2\rho}}{\prod_{i=1}^n \prod_{j=1}^m \alpha^{-2mn}(s_i + \xi_j)^2}.
\end{aligned}$$

Noting also that

$$\int_1^\infty dt_\imath \underset{\alpha\to\infty}{\sim} \frac{1}{\alpha}\int_0^\infty ds_\imath, \qquad \int_0^1 d\tau_\jmath \underset{\alpha\to\infty}{\sim} \frac{1}{\alpha}\int_0^\infty d\xi_\jmath,$$

we see that

$$J_{(0,n)(m,0)}(\alpha,\beta,\rho) \underset{\alpha\to\infty}{\sim} \alpha^{-c_{nm}(\beta,\rho)} \prod_{\imath=1}^n \int_0^\infty ds_\imath\, s_\imath^{-\beta/\rho} e^{-s_\imath/\rho} \prod_{\jmath=1}^\infty d\xi_\jmath\, \xi^\beta e^{-\xi_\jmath}$$
$$\times \frac{\prod_{i<\imath'} |s_\imath - s_{\imath'}|^{2/\rho} \prod_{\jmath<\jmath'} |\xi_{\jmath'} - \xi_\jmath|^{2\rho}}{\prod_{\imath=1}^n \prod_{\jmath=1}^m (s_\imath + \xi_\jmath)^2}. \tag{13.224}$$

On the other hand, using the relation (4.126), we see from the evaluation formula in Exercises 4.5 q.1 that

$$J_{(0,n)(m,0)}(\alpha,\beta,\rho) \underset{\alpha\to\infty}{\sim} \alpha^{-c_{nm}(\beta,\rho)} \rho^{(1-\beta/\rho+(n-1)/\rho)n} \prod_{l=1}^n \frac{\Gamma(l/\rho)}{\Gamma(1/\rho)} \prod_{j=1}^m \frac{\Gamma(j\rho-n)}{\Gamma(\rho)}$$
$$\times \prod_{l=0}^{n-1} \Gamma(1-\beta/\rho+l/\rho) \prod_{j=0}^{m-1} \Gamma(1-n+\beta+j\rho). \tag{13.225}$$

After equating (13.224) with (13.225), and introducing the common scaling factor z via the change of integration variables $s_\imath/\rho = zr_\imath$, $\xi_j = z\eta_j$, the integration formula (13.223) results. □

Comparison of (13.219), (13.220) and (13.221) with (13.222) shows

$$\begin{aligned}
A(q,p,\lambda) &= q!p! \lim_{\epsilon\to 0} \frac{\partial^2}{\partial z^2} I_{p,q}(z;(p+\epsilon)/q-1,(p+\epsilon)/q)\Big|_{z=1} \\
&= \frac{q!p!}{\lambda\Gamma^q(\lambda)} \prod_{l=1}^p \frac{\Gamma^2(l/\lambda)}{\Gamma(1/\lambda)} \prod_{j=1}^{q-1} \Gamma^2(j\lambda-p), \\
B(q,p-m,\lambda) &= q!(p-m)! I_{p-m,q}(1;\lambda-1;\lambda) \\
&= q!(p-m)! \prod_{l=1}^{p-m} \frac{\Gamma^2(l/\lambda)}{\Gamma(1/\lambda)} \prod_{j=1}^q \frac{\Gamma^2(j\lambda-(p-m))}{\Gamma(\lambda)}, \\
C(m,\lambda) &= m! I_{0,m}(1;1/\lambda-1,1/\lambda),
\end{aligned}$$

where in the formula for $A(q,p,\lambda)$ the auxiliary variable ϵ is introduced because $I_{p,q}(z;p/q-1,p/q)$ before the double differentiation is singular. Substituting the above results in (13.218), and manipulating (13.215) to write it in the form

$$C_{p,q}(\lambda) = \frac{1}{2\pi^2 p!q!} \frac{\Gamma^q(\lambda)\Gamma^p(1/\lambda)}{\prod_{j=1}^p \Gamma^2(j/\lambda) \prod_{i=1}^{q-1} \Gamma^2(i\lambda-p)}$$

shows that

$$\rho_{(1,1)}^{T\mathrm{bulk}}(x,0;\tau) \sim -\frac{\rho^2}{\lambda}\left(\frac{1}{(2\pi\rho x + i(2\pi\rho)^2\tau/2)^2} + \frac{1}{(2\pi\rho x - i(2\pi\rho)^2\tau/2)^2}\right)$$
$$+ \sum_{m=1}^p a_m \frac{\cos 2\pi m\rho x}{|2\pi\rho x + i(2\pi\rho)^2\tau/2|^{4m^2/\beta}}, \tag{13.226}$$

$$a_m = 2m^2\lambda^{-2mq} \frac{\prod_{j=1}^m \Gamma^2(j/\lambda)}{\prod_{l=p-m+1}^p \Gamma^2(l/\lambda)} \frac{\prod_{j=1}^q \Gamma^2(j\lambda-p+m)}{\prod_{i=1}^{q-1} \Gamma^2(i\lambda-p)},$$

where as in Proposition 13.7.9 $\lambda = \beta/2 = p/q$. It is straightforward to check that in the special case

$\beta/2 \in \mathbb{Z}_{>0}$, so that for $q = 1$ this expansion agrees with the asymptotic expansion of Proposition 13.2.4.

We turn our attention now to asymptotics associated with the dynamical structure function, defined by

$$\hat{S}(k;\tau) := \int_{-\infty}^{\infty} \rho_{(1,1)}^{T\mathrm{bulk}}(x,0;\tau) e^{ikx}\, dx. \tag{13.227}$$

For the case that the initial state equals the equilibrium state, the non-oscillatory term of the asymptotic behavior (13.226) implies that for k small

$$\hat{S}(k;\tau) \sim \frac{|k|}{\pi\beta} e^{-\pi\rho|k|\tau}. \tag{13.228}$$

This can be derived directly from the integral formula (13.213) substituted in (13.227) (see Exercises 13.7 q.1). An analogous calculation [206] also allows the next leading term of the small k expansion for $\tau = 0$ to be computed, giving the result

$$\hat{S}(k;0) \sim \frac{|k|}{\pi\beta} + \frac{1}{2\pi\rho}\Big(\frac{1}{\pi\beta}\Big)\Big(\frac{\beta-2}{\beta}\Big)k^2. \tag{13.229}$$

In (14.15) below the same expansion will be deduced from physical principles.

13.7.4 Analytic properties of the static structure function

The formula (13.213) in the case $\tau = 0$ allows the structure function for the static log-gas to be computed. Thus, in the notation of that formula,

$$\begin{aligned}\hat{S}(k) =: \hat{S}(k;\beta) &= \int_{-\infty}^{\infty} \rho_{(1,1)}^{T\mathrm{bulk}}(x,0;0) e^{ikx}\, dx \\ &= \pi C_{p,q}(\lambda) \prod_{i=1}^{q} \int_0^\infty dx_i \prod_{j=1}^{p} \int_0^1 dy_j\, Q_{p,q}^2 F(q,p,\lambda|\{x_i,y_j\})\, \delta(k - Q_{p,q}). \end{aligned} \tag{13.230}$$

For values of $|k|$ small enough, (13.230) can be used to relate $\hat{S}(k;\beta)$ to an analytic function [207].

PROPOSITION 13.7.11 *Set $\rho = 1$ and suppose $|k| < 2\pi$. Then*

$$\hat{S}(k;\beta) = \frac{|k|}{\pi\beta} f(|k|;\beta), \tag{13.231}$$

where $f(k;\beta)$ is analytic in the interval $|k| < \min(2\pi, \pi\beta)$, and is given explicitly by

$$f(k;\beta) = 2\pi^2\lambda C_{p,q}(\lambda) \prod_{i=1}^{q} \int_0^\infty dx_i \prod_{j=1}^{p} \int_0^\infty dy_j\, Q_{p,q}^2 \hat{F}(q,p,\lambda|\{x_i,y_j\};k)\, \delta(1 - Q_{p,q}|_{\rho=1}), \tag{13.232}$$

with $\lambda = \beta/2$ and

$$\begin{aligned}\hat{F}(q,p,\lambda|\{x_i,y_j\};k) &= \frac{1}{\prod_{i=1}^{q}(x_i(1+kx_i/\lambda))^{1-\lambda}\prod_{j=1}^{p}(y_j(1-ky_j))^{1-1/\lambda}} \\ &\quad\times \frac{\prod_{i<i'}|x_i - x_{i'}|^{2\lambda}\prod_{j<j'}|y_j - y_{j'}|^{2/\lambda}}{\prod_{i=1}^{q}\prod_{j=1}^{p}(x_i + \lambda y_j)^2}. \end{aligned} \tag{13.233}$$

Proof. Note that in (13.230) the integration variables are all positive, and because of the delta function are restricted to the hyperplane

$$\sum_{i=1}^{q} x_i + \sum_{j=1}^{p} y_j = \frac{|k|}{2\pi}.$$

We see immediately from these constraints that the restriction $y_j < 1$ in the domain of integration is redundant for $|k| < 2\pi$. Assuming this condition the integration over y_j can be extended to the region $(0, \infty)$. The change of variables $x_i \mapsto |k|x_i$ and $y_j \mapsto |k|y_j$ then shows that for $|k| < 2\pi$ the formula (13.231) is valid. The analytic properties of $f(k;\beta)$ are evident by inspection. □

An immediate consequence of (13.232) is a functional equation satisfied by $f(k;\beta)$.

PROPOSITION 13.7.12 *The function $f(k;\beta)$, related to $\hat{S}(k;\beta)$ by (13.231), satisfies the functional equation (7.139).*

Proof. The integral in (13.232) is unchanged by the mapping $\lambda \mapsto 1/\lambda$ (and thus $p \leftrightarrow q$) followed by $k \mapsto -k/\lambda$. The precise functional equation (7.139) then follows provided we can show that

$$C_{p,q}(\lambda) = \lambda^{2pq-2} C_{q,p}(1/\lambda),$$

which indeed readily follows from the definition of $C_{p,q}(\lambda)$ in (13.215). □

As noted in (7.152), the functional equation (7.139) can be illustrated using the evaluations (7.136) and (7.95) of $\hat{S}(k;\beta)$ for $\beta = 1$ and 4, respectively in terms of elementary functions. Furthermore, in relation to the small $|k|$ expansion of $\hat{S}(k;\beta)$, in the notation of (7.141) the functional equation (7.139) implies the functional relation (7.141). The simplest structure of the coefficients $A_j(x)$ allowing (7.141) is $A_j(x) = p_j(x)$, where $p_j(x)$ is a polynomial of degree j. The functional relation (7.141) can be stated as requiring

$$p_j(x) = \sum_{l=0}^{j} a_{j,l} x^l, \qquad a_{j,l} = a_{j,j-l} \quad (j \text{ even}), \tag{13.234}$$

$$p_j(x) = (x-1) \sum_{l=0}^{j-1} \tilde{a}_{j,l} x^l, \qquad \tilde{a}_{j,l} = \tilde{a}_{j,j-1-l} \quad (j \text{ odd}). \tag{13.235}$$

In fact the structure of (13.234) can be refined by noting from the exact result (7.4) that

$$f(k;2) = 1. \tag{13.236}$$

This implies

$$p_j(x) = (x-1)^2 \sum_{l=0}^{j-2} b_{j,l} x^l, \qquad b_{j,l} = b_{j,j-2-l} \quad (j \text{ even}). \tag{13.237}$$

Assuming the validity of the assumption that $A_j(x)$ is a polynomial of degree j, the exact results giving $\widetilde{S}_\beta(k)$ as an elementary function of k for various β allow the polynomials $p_j(x)$ for j up to nine to be completely specified [207], thus extending the expansion (13.229) up to order k^{10}.

EXERCISES 13.7 1. [206] The aim of this exercise is to derive (13.228).

(i) Suppose $\rho = 1$. Show that the formula (13.231) holds for $\hat{S}(k;\tau)$ provided a factor $\exp(-\hat{E}_{p,q,k}\tau/2\lambda)$ is included in the integrand of the formula (13.232) defining $f(k;\beta)$, where

$$\hat{E}_{p,q,k} = (2\pi)^2 k^2 \Big(\sum_{i=1}^{q} x_i^2 - \lambda \sum_{j=1}^{p} y_j^2 \Big).$$

(ii) Note that in the limit $k \to 0$ the quantity $f(k;\beta)$ thus modified is independent of k and thus

$$\hat{S}(k;\tau) \sim \pi |k| C_{p,q}(\lambda) I(\lambda) e^{-|k|\pi\rho\tau}, \tag{13.238}$$

where

$$I(\lambda) = \prod_{i=1}^{q} \int_0^\infty dx_i \prod_{j=1}^{p} \int_0^\infty dy_j \, Q_{p,q}^2 G(q,p,\lambda | \{x_i, y_j\}; k) \, \delta(1 - Q_{p,q}),$$

$$G(q,p,\lambda | \{x_i, y_j\}; k) = \frac{1}{\prod_{i=1}^{q} x_i^{1-\lambda} \prod_{j=1}^{p} \lambda y_j^{1-1/\lambda}} \frac{\prod_{i<i'} |x_i - x_{i'}|^{2\lambda} \prod_{j<j'} |y_j - y_{j'}|^{2/\lambda}}{\prod_{i=1}^{q} \prod_{j=1}^{p} (x_i + \lambda y_j)^2}.$$

(iii) To evaluate $I(\lambda)$ rewrite it as

$$I(\lambda) = \lim_{\epsilon \to 0+} \frac{1}{2\pi} \int_{-\infty}^{\infty} du \, e^{iu} \prod_{i=1}^{q} \int_0^\infty dx_i \prod_{j=1}^{p} \int_0^\infty dy_j \, Q_{p,q}^2 G(q,p,\lambda | \{x_i, y_j\}; k) e^{-iQ_{p,q} u} e^{-\epsilon Q_{p,q}},$$

and change variables $x_i \mapsto x_i/(\epsilon + iu)$, $y_j \mapsto y_j/(\epsilon + iu)$ to obtain

$$I(\lambda) = \Big(\lim_{\epsilon \to 0+} \frac{1}{2\pi} \int_{-\infty}^{\infty} \frac{e^{iu}}{(\epsilon + iu)^2} du \Big) \prod_{i=1}^{q} \int_0^\infty dx_i \prod_{j=1}^{p} \int_0^\infty dy_j \, Q_{p,q}^2 G(q,p,\lambda | \{x_i, y_j\}; k) e^{-Q_{p,q}}.$$

Note that the first integral on the r.h.s. is unity, while the second integral is $(2\pi\rho)^2$ times the integral $A(q,p,\lambda)$ defined by (13.219). Use the fact that $(2\pi\rho)^2 A(q,p,\lambda) = 1/(C_{p,q}(\lambda)\pi^2\beta)$, which follows from the workings leading to obtaining the first term in (13.226) from (13.218), to deduce (13.228) from (13.238).

2. (i) Note from (13.230) that for $p = q = 1$ and thus $\beta = 2$

$$\hat{S}(k) = \rho \int_0^\infty dx \int_0^1 dy \, \delta(|k|/2\pi\rho - (x + y))$$

and from this reclaim the result (7.4).

(ii) Consider (13.230) with $p = 1$, $q = 2$ so that $\beta = 1$. By changing variables

$$y = \frac{1}{2}(1 - Y), \qquad x_1 = \frac{1}{4}(X_1 - 1), \qquad x_2 = \frac{1}{4}(X_2 - 1)$$

deduce that

$$\hat{S}(k) = \frac{\rho}{4} \int_1^\infty dX_1 \int_1^\infty dX_2 \int_{-1}^1 dY \, (\frac{X_1 + X_2}{2} - Y)^2 \frac{|X_1 - X_2|(1 - Y^2)}{(X_1^2 - 1)^{1/2}(X_2^2 - 1)^{1/2}} \times \frac{\delta(|k|/\pi\rho - ((X_1 + X_2)/2 - Y))}{(X_1 - Y)^2 (X_2 - Y)^2}. \tag{13.239}$$

(iii) Now introduce the change of variables

$$Y = v, \qquad X_1 + X_2 = 2u_1 u_2, \qquad X_1 X_2 = u_1^2 + u_2^2 - 1.$$

Show that, for $X_1 > X_2$, the final equation can be replaced by $X_1 - X_2 = (4u_1^2 u_2^2 - 2u_1^2 - 2u_2^2 + 2)^{1/2}$, and hence deduce that the Jacobian for the change of variables from $\{X_1, X_2\}$ to $\{u_1, u_2\}$ is equal to

$$\frac{2}{X_1 - X_2}(u_1^2 - u_2^2) = \frac{2}{(X_1 - X_2)}(X_1^2 - 1)^{1/2}(X_2^2 - 1)^{1/2}.$$

Use this to rewrite (13.239) as

$$\hat{S}(k) = \rho \int_1^\infty du_1 \int_1^\infty du_2 \int_{-1}^1 dv \, (u_1 u_2 - v)^2 (1 - v^2) \frac{\delta(|k|/\pi\rho - u_1 u_2 + v)}{(u_1^2 + u_2^2 - 1 - 2v u_1 u_2 + v^2)^2}.$$

(This form of $\hat{S}(k)$ for $\beta = 1$ first appeared in [160], where it was further evaluated to reclaim (7.136).)

(iv) Let $E_{p,q}$ be as specified in Proposition 13.7.9. Show that after the change of variables introduced in (ii) and (iii)

$$E_{1,2} = \frac{\pi^2 \rho^2}{2}(2u_1^2 u_2^2 - u_1^2 - u_2^2 + 1 - v^2),$$

and hence deduce from (13.213) that for $\beta = 1$ [490]

$$\rho_{(1,1)}^{T\text{bulk}}(x, y; \tau) = \rho^2 \int_1^\infty du_1 \int_1^\infty du_2 \int_{-1}^1 dv \, \frac{(u_1 u_2 - v)^2 (1 - v^2)}{(u_1^2 + u_2^2 - 1 - 2v u_1 u_2 + v^2)^2}$$
$$\times \cos(\pi\rho(x - y)(u_1 u_2 - v)) \exp\Big(- (\tau/2)(\pi\rho)^2 (2u_1^2 u_2^2 - u_1^2 - u_2^2 + 1 - v^2)\Big).$$

Chapter Fourteen

Fluctuation formulas and universal behavior of correlations

In previous chapters, asymptotic properties of log-gas systems relating to the two-point correlation, density profiles at the edge and spacing distributions were calculated exactly at the random matrix couplings $\beta = 1, 2$ and 4. Here macroscopic physical characterizations of the log-gas will be used to predict extensions of the asymptotic forms to the general β case, and furthermore to study fluctuation formulas for certain statistics. As an example, the log-gas has the physical property that it will perfectly screen an external charge density in the long wavelength limit, and this can be used to predict universal asymptotic forms for the two-point correlation. These universal forms can in turn be used to predict that the variance of a linear statistic will be of order unity, and furthermore to give closed forms for this quantity. Macroscopic electrostatics combined with a linear response relation implies that the p.d.f. for such linear statistics will be Gaussian. It is shown that this latter fact ties in with the Szegö and Fisher-Hartwig asymptotic formulas from the theory of Toeplitz determinants. In the second to last section, macroscopic arguments are used to predict asymptotic properties of the gap and spacing probabilities $E_\beta(n; s)$ and $P_\beta(n; s)$, while in the last section the current correlation is introduced and is used to study the dynamical structure function.

14.1 PERFECT SCREENING

In Section 1.4 a macroscopic physical characterization of one-component log-gas systems was used to predict the leading behavior of the particle density. Thus it was hypothesized that, to leading order in the number of particles N, the charge density of the mobile particles will exactly cancel the fixed background density, which is of opposite sign. The reasoning for this is that the absence of cancellation would imply the total charge density in some macroscopic region to be nonzero. An electric field would then be created, and the state would no longer be in equilibrium.

A refinement of this physical reasoning is to consider the response of a log-gas system to a periodic perturbation in the background density. If the wavelength of this perturbation is large compared to the inter-particle spacing, the above argument says that for an equilibrium state to be obtained the mobile particles must adjust their density to exactly cancel the charge density of the perturbation. A charged system with this property is said to be in a *conductive phase*. We will show how it can be used together with a linear response argument to deduce the small $|k|$ behavior of the structure function $\hat{S}(k)$, or equivalently the long distance non-oscillatory behavior of the truncated two-particle correlation, for the log-gas on a line [314].

First the linear response formula must be introduced. It states that for any observable A in a statistical mechanical system, the change in its mean value due to a perturbation δU in the total energy U is to leading order in δU given by

$$\langle A \rangle_\epsilon - \langle A \rangle_0 = -\beta \langle A \delta U \rangle_0^T, \tag{14.1}$$

where the subscripts ϵ and 0 denote the presence and absence, respectively, of the perturbation in the average. It is simple to derive (14.1) in the canonical ensemble from the meaning of the averages,

$$\langle A \rangle_\epsilon := \frac{1}{Z_\epsilon} \int_\Omega d\vec{x}_1 \cdots \int_\Omega d\vec{x}_N \, A e^{-\beta(U+\delta U)} \text{ with } Z_\epsilon := \int_\Omega d\vec{x}_1 \cdots \int_\Omega d\vec{x}_N \, e^{-\beta(U+\delta U)}$$

and $\langle A\delta U\rangle_0^T := \langle A\delta U\rangle_0 - \langle A\rangle_0\langle\delta U\rangle_0$.

Now we will apply (14.1) to the characterization of a conductive phase. For this suppose the system is perturbed by an external charge density ϵe^{ikx} (physically this can be thought of as a pair of external charge densities $(\epsilon \cos kx, \epsilon \sin kx)$). The total energy of the system is then perturbed by an amount

$$\delta U = -\epsilon \sum_{j=1}^{N} \int_{-\infty}^{\infty} \log|x - x_j|\, e^{ikx}\, dx = \epsilon \frac{\pi}{|k|} \hat{n}_{(1)}(k), \quad \hat{n}_{(1)}(k) := \int_{-\infty}^{\infty} n_{(1)}(x) e^{ikx}\, dx = \sum_{j=1}^{N} e^{ikx_j}, \tag{14.2}$$

where $n_{(1)}(x) = \sum_{j=1}^{N} \delta(x - x_j)$ denotes the microscopic particle density and the result

$$-\int_{-\infty}^{\infty} \log|x|\, e^{ikx}\, dx = \frac{\pi}{|k|} \tag{14.3}$$

from the theory of generalized functions has been used. Take for the observable A the charge density at the point x, which for a one-component system with unit charges is equal to the microscopic particle density $n_{(1)}(x)$. (Strictly speaking the microscopic charge density $c_{(1)}(x)$ is equal to $n_{(1)}(x) - 1$ for a system consisting of unit positive charges and a unit density neutralizing negative background, but the constant -1 plays no role in the subsequent analysis.) Then (14.1) reads

$$\langle n_{(1)}(x)\rangle_\epsilon - \langle n_{(1)}(x)\rangle_0 = -\beta\epsilon \frac{\pi}{|k|} \langle n_{(1)}(x) \hat{n}_{(1)}(k)\rangle_0^T. \tag{14.4}$$

Consider first the l.h.s. of (14.4). According to the characterization of a conductive phase, as $|k| \to 0$ the system responds so that its charge density is equal and opposite to that of the external charge density, and thus

$$\langle n_{(1)}(x)\rangle_\epsilon - \langle n_{(1)}(x)\rangle_0 \underset{|k|\to 0}{\sim} -\epsilon e^{ikx}. \tag{14.5}$$

Consider now the r.h.s. of (14.4). Using the translational invariance of the system in the bulk we have

$$\langle n_{(1)}(x) \hat{n}_{(1)}(k)\rangle_0^T = e^{ikx} \hat{S}(k), \tag{14.6}$$

where

$$\hat{S}(k) := \rho \tilde{S}(k/\rho), \tag{14.7}$$

with $\tilde{S}(k)$ denoting the dimensionless structure function (7.3). Hence (14.4) implies

$$\hat{S}(k) \underset{|k|\to 0}{\sim} \frac{|k|}{\pi\beta}, \tag{14.8}$$

or equivalently, by taking the inverse transformation,

$$\rho_{(2)}^T(x, x') \underset{|x-x'|\to\infty}{\sim} -\frac{1}{\beta\pi^2 (x - x')^2}, \tag{14.9}$$

for the leading non-oscillatory term. The exact results (7.136), (7.4), (7.95) and (7.135), (7.65), (7.94) for $\tilde{S}(k)$ and $\rho_{(2)}^T(x, 0)$ at $\beta = 1, 2$ and 4 in the bulk exhibit the behaviors (14.8) and (14.9) respectively, as do the exact results (13.228) and (13.226) for general rational β. Also, the perturbation results (7.28) and (7.97) are consistent with (14.8).

Setting $k = 0$ in (14.8) gives $\hat{S}(k) = 0$ or equivalently

$$\int_{-\infty}^{\infty} \rho_{(2)}^T(x, 0)\, dx = -\rho, \tag{14.10}$$

which can be interpreted as saying that the screening cloud about a fixed internal charge has total charge equal and opposite to that of the internal charge. It is known as a *charge sum rule*, and is expected to hold true independent of the phase of the Coulomb system, being necessary for thermodynamic stability [386]. In this regard, note that the formula $\hat{S}(k) = 0$ can be deduced from (14.4) without the assumption (14.5), by simply multiplying both sides by $|k|$ and setting $k = 0$. Internal screening should also be valid in a semi-infinite system, where it implies

$$\int_0^\infty \rho_{(2)}^T(y_1, y) \, dy = -\rho_{(1)}(y_1), \tag{14.11}$$

and in the case that a multiple number n of internal charges are fixed, which requires

$$\int_I \rho_{(n+1)}^T(y_1, y_2, \dots, y_n, y) \, dy = -n\rho_{(n)}^T(y_1, \dots, y_n), \tag{14.12}$$

valid for an infinite or semi-infinite interval I. A derivation of (14.12) for I equal to the real line is given in Exercises 14.1 q.1, while the case of I semi-infinite is the subject of Exercises 14.1 q.2.

The exact results for $\rho_{(n)}^T$ in Chapter 7 all satisfy (14.12). At $\beta = 2$ this follows from the fact $\rho_{(n)}^T$ is of the form

$$\rho_{(n)}^T(x_1, \dots, x_n) = (-1)^{n-1} \sum_{\substack{\text{cycles} \\ \text{length } n}} K(x_{i_1}, x_{i_2}) \cdots K(x_{i_n}, x_{i_1}),$$

where $K(x, y)$ has the property

$$\int_I K(x, y) K(y, z) \, dy = K(x, z),$$

these being limiting forms of the second equation in Proposition 5.1.2 and the first equation in (5.7) respectively. Similarly, at $\beta = 1$ and 4 the validity of (14.12) follows from the structure of $\rho_{(n)}^T$ in Propositions 6.1.7 and 6.3.3, together with the fact that the second equation in (6.17) and the first equation in (6.70) remain valid in the thermodynamic limit.

14.1.1 Second order correction

A refinement of the linear response argument leading to (14.8) can be given [315] which predicts the term of order k^2 in the small $|k|$ expansion of $\hat{S}(k)$. This refinement requires introducing into the r.h.s. of (14.5) a wave number-dependent factor $A(k)$, where $A(k) \to 1$ as $|k| \to 0$. Thus, for $|k|$ small, we have that the net charge in the system after the perturbation is $\epsilon e^{ikx}(1 - A(k))$. This creates an electric potential

$$\phi(x) = -\epsilon(1 - A(k)) \int_{-\infty}^{\infty} \log|x - y| e^{iky} \, dy = \frac{\pi\epsilon}{|k|}(1 - A(k)) e^{ikx}$$

and thus a corresponding electric field

$$E_x = -\frac{\partial \phi(x)}{\partial x} = -i\pi\epsilon \, \mathrm{sgn}(k) \, (1 - A(k)) e^{ikx}. \tag{14.13}$$

Now the existence of this electric field means that there is a force density $\rho_{(1)}(x) E_x$ at point x in the system. Physically, this force density must be balanced for the system to be stable. The balancing force is provided by the gradient of the pressure fluctuation,

$$\frac{\partial}{\partial x}\Big(\delta P(x)\Big) = \frac{\partial}{\partial x}\Big(\frac{\partial P}{\partial \rho} \delta\rho_{(1)}(x)\Big) = -\frac{\partial}{\partial x}\Big(\frac{\partial P}{\partial \rho} \epsilon A(k) e^{ikx}\Big) = -ik\frac{\partial P}{\partial \rho} \epsilon A(k) e^{ikx}, \tag{14.14}$$

so equating ρ times (14.13) with (14.14) gives

$$A(k) = \Big(1 + \frac{|k|}{\pi\rho} \frac{\partial P}{\partial \rho}\Big)^{-1} \sim 1 - \frac{|k|}{\pi\rho} \frac{\partial P}{\partial \rho}.$$

Note that in deriving (14.14) the fact that for a one-component system the charge and particle densities are proportional has been used. Since an extra factor $A(k)$ on the r.h.s. of (14.5) carries through to an extra factor $A(k)$ on the r.h.s. of (14.8), after recalling (4.169) we thus have

$$\hat{S}(k) \underset{|k|\to 0}{\sim} \frac{|k|}{\pi\beta} - \frac{1}{\rho}\Big(\frac{|k|}{\pi\beta}\Big)^2(1-\beta/2). \tag{14.15}$$

Inspection of the exact results (7.4), (7.96), and (7.137), and recalling (14.7), gives agreement with (14.15). Similarly the perturbation results (7.28) and (7.97) agree with (14.15).

14.1.2 Multicomponent systems

In general for a multicomponent Coulomb system, the *charge-charge correlation* $C_{(2)}(\vec{r},\vec{r}\,')$ is defined by

$$C_{(2)}(\vec{r},\vec{r}\,') = \langle c_{(1)}(\vec{r})c_{(1)}(\vec{r}\,')\rangle - \langle c_{(1)}(\vec{r})\rangle\langle c_{(1)}(\vec{r}\,')\rangle, \tag{14.16}$$

where $c_{(1)}(\vec{r})$ denotes the microscopic charge density

$$c_{(1)}(\vec{r}) := \sum_{j=1}^{N} q_j\delta(\vec{r}-\vec{r}_j) \tag{14.17}$$

with q_j denoting the charge of the jth particle. For a p-component system this can be written in terms of the truncated two-particle correlations and the particle densities as

$$C_{(2)}(\vec{r},\vec{r}\,') = \sum_{\alpha,\gamma=1}^{p} q_\alpha q_\gamma \rho_{(2)}^T(\vec{r}^{\,(\alpha)},\vec{r}^{\,(\gamma)})\Big|_{\substack{\vec{r}^{\,(\alpha)}=\vec{r}\\ \vec{r}^{\,(\gamma)}=\vec{r}\,'}} + \sum_{\alpha=1}^{p} q_\alpha^2\delta(\vec{r}^{\,(\alpha)}-\vec{r}\,')\rho_{(1)}(\vec{r}^{\,(\alpha)})\Big|_{\vec{r}^{\,(\alpha)}=\vec{r}}, \tag{14.18}$$

where here the labels α and γ specify the species.

For a multicomponent log-gas on a line, the perfect screening argument of Section 14.1 can be used to predict the leading order non-oscillatory decay of $C_{(2)}$. Thus, repeating that argument with the observable A chosen to be the microscopic charge density (14.17), we conclude that

$$C_{(2)}(x,x') \underset{|x-x'|\to\infty}{\sim} -\frac{1}{\beta\pi^2(x-x')^2} \tag{14.19}$$

(cf. (14.9)). Now, for the two-component log-gas with charge ratio 1:2, (14.18) gives

$$C_{(2)}(x,0) = \rho_{+1,+1}^T(x,0) + 4\rho_{+1,+2}^T(x,0) + 4\rho_{+2,+2}^T(x,0) + \delta(x)(\rho_{+1}(x) + 4\rho_{+2}(x)).$$

Use of (7.186) shows that for $\beta = 1$

$$C_{(2)}(x,0) \underset{x\to\infty}{\sim} -\frac{1}{\pi^2x^2},$$

in agreement with the prediction (14.19).

In a p-component system we can also consider the consequence of perfect screening of internal charges. Analogous to the equations of Exercises 14.1 q.1 we have

$$\int_I \langle n_{(1)}(y_1^{(\alpha_1)})\cdots n_{(1)}(y_n^{(\alpha_n)})c_{(1)}(x)\rangle_0^T\,dx = 0,$$

where

$$\langle n_{(1)}(y_1^{(\alpha_1)})\cdots n_{(1)}(y_n^{(\alpha_n)})c_{(1)}(x)\rangle_0^T = \sum_{\alpha=1}^p q_\alpha \Big(\rho_{(n+1)}(y_1^{(\alpha_1)},\dots,y_n^{(\alpha_n)};x^{(\alpha)})\Big|_{x^{(\alpha)}=x}$$
$$+\Big(\sum_{j=1}^n \delta_{\alpha,\alpha_j}\delta(x-y_j^{(\alpha_j)})\Big)\rho_{(n)}(y_1^{(\alpha_1)},\dots,y_n^{(\alpha_n)}) - \rho_{(1)}(x^{(\alpha)})\Big|_{x^{(\alpha)}=x}\rho_{(n)}(y_1^{(\alpha_1)},\dots,y_n^{(\alpha_n)})\Big),$$

and as in (14.18) the label α specifies the species. Inductive use of this equation gives

$$\sum_{\alpha=1}^p q_\alpha \int_I \rho_{(n+1)}^T(y_1^{(\alpha_1)},\dots,y_n^{(\alpha_n)},x^{(\alpha)})\,dx^{(\alpha)} = -\Big(\sum_{j=1}^n q_{\alpha_j}\Big)\rho_{(n)}^T(y_1^{(\alpha_1)},\dots,y_n^{(\alpha_n)}) \tag{14.20}$$

(cf. (14.12)). The exact results (7.185) satisfy this requirement in the case $n=1$.

EXERCISES 14.1 1. (i) Suppose the observable $n_{(1)}(x)$ in (14.4) is replaced by $n_{(1)}(y_1)\cdots n_{(1)}(y_n)$. By taking the limit $k\to 0$ in the resulting equation, deduce that

$$\int_I \langle n_{(1)}(y_1)\cdots n_{(1)}(y_n)n_{(1)}(y)\rangle_0^T\,dy = 0, \tag{14.21}$$

where I denotes the real line and

$$\begin{aligned}&\langle n_{(1)}(y_1)\cdots n_{(1)}(y_n)n_{(1)}(y)\rangle_0^T\\ &:= \langle n_{(1)}(y_1)\cdots n_{(1)}(y_n)n_{(1)}(y)\rangle_0 - \langle n_{(1)}(y_1)\cdots n_{(1)}(y_n)\rangle_0\langle n_{(1)}(y)\rangle_0\\ &= \rho_{(n+1)}(y_1,\dots,y_n,y) + \Big(\sum_{j=1}^n \delta(y-y_j)\Big)\rho_{(n)}(y_1,\dots,y_n) - \rho_{(n)}(y_1,\dots,y_n)\rho_{(1)}(y).\end{aligned}$$

Use this equation inductively to deduce (14.12).

(ii) Show from the definitions that (14.21) holds in the finite system, irrespective of the system being a log-gas.

2. For the log-gas on a half-line, let the system be perturbed by an external charge of strength ϵ at the point x, so that the energy is perturbed by an amount $\delta U = -\epsilon\int_I \log|x-y|n_{(1)}(y)\,dy$. Take for the observable $n_{(1)}(y_1)\cdots n_{(1)}(y_n)$, so the linear response relation (14.1) then reads

$$\langle n_{(1)}(y_1)\cdots n_{(1)}(y_n)\rangle_\epsilon - \langle n_{(1)}(y_1)\cdots n_{(1)}(y_n)\rangle_0 = -\epsilon\beta\int_I \log|x-y|\langle n_{(1)}(y_1)\cdots n_{(1)}(y_n)n_{(1)}(y)\rangle_0^T\,dy.$$

(i) By taking the limit $x\to\infty$, deduce (14.21).

(ii) In the case $n=1$ (one-component system), the observable corresponds to the charge density. Perfect screening of the external charge density requires that

$$\int_I \Big(\langle n_{(1)}(y_1)\rangle_\epsilon - \langle n_{(1)}(y_1)\rangle_0\Big)\,dy_1 = -\epsilon.$$

Substitute this in the linear response relation to deduce that

$$1 = \beta\int_I dy_1\int_I dy\,\log|y_1-y|C_{(2)}(y_1,y), \tag{14.22}$$

where $C_{(2)}(y_1,y)$ corresponds to the one-component case of the charge-charge correlation (14.18).

3. Let $\tilde S(k)$ denote the dimensionless structure function (7.3), and let $\tilde v(k)$ denote the dimensionless Fourier transform of the pair potential $v(x)$,

$$\tilde v(k) = \rho\int_{-\infty}^{\infty} v(x)e^{ik\rho x}\,dx.$$

For long-range potentials $v(x)$ such that $\tilde{v}(k) \to \infty$ as $k \to 0$, the weak coupling ($\beta \to 0$) approximation from the theory of classical fluids [289] says that

$$\tilde{S}(k) - 1 = -\frac{\beta \tilde{v}(k)}{1 + \beta \tilde{v}(k)}$$

and thus

$$\rho_{(2)}^T(x, 0) = -\frac{\rho^2}{2\pi} \int_{-\infty}^{\infty} \frac{\beta \tilde{v}(k)}{1 + \beta \tilde{v}(k)} e^{-i\rho x k}\, dk.$$

(i) By rearranging and taking the inverse Fourier transform, show that the approximation for $\tilde{S}(k)$ is equivalent to the integral equation

$$\frac{1}{\rho} \rho_{(2)}^T(y, 0) = -\beta \int_{-\infty}^{\infty} v(|y - x|) \Big(\rho_{(2)}^T(x, 0) + \rho \delta(x) \Big)\, dx.$$

Show that this equation can be derived from the linear response relation (14.1) with δU the energy due to an external charge placed at the origin, A chosen as the microscopic density at the point y and $\langle A \rangle_\epsilon - \langle A \rangle_0$ set equal to $\rho_{(2)}^T(y, 0)/\rho$.

(ii) For the log-gas show that the formula for $\tilde{S}(k)$ reproduces (14.8), and use the formula for $\rho_{(2)}^T(x, 0)$ to derive (4.179).

(iii) Use the formula for $\rho_{(2)}^T(x, 0)$ and (4.178) to deduce that the mean energy per particle is given by

$$u = -\frac{1}{2\beta} \lim_{x \to 0^+} \Big(\beta v(x) + \frac{1}{\rho^2} \rho_{(2)}^T(x, 0) \Big).$$

14.2 MACROSCOPIC BALANCE AND DENSITY

In this section the global asymptotic expansions of the Gaussian and Laguerre densities (7.162) and (7.161) will be considered from a macroscopic viewpoint. Consider first the Gaussian case. In Section 1.4 the macroscopic balance equation (1.48) was used to deduce the Wigner semicircle form for the density. Here we want to refine (1.48) to allow a correction to the Wigner semicircle form to be computed. The basic hypothesis is that the electrostatic energy on the r.h.s. should be corrected by subtracting the free energy. According to (4.166), with ρ therein interpreted as $\rho_{(1)}(x)$, this modifies (1.48) so it reads

$$\frac{x^2}{2} + C = \int_{-a}^{a} \rho_{(1)}(y) \log |x - y|\, dy + \Big(\frac{1}{2} - \frac{1}{\beta} \Big) \log \rho_{(1)}(x), \qquad x \in (-a, a) \tag{14.23}$$

for some constant C. We know that to leading order the density is supported on $(-\sqrt{2N}, \sqrt{2N})$ so that $a = \sqrt{2N}$.

If we introduce the global density $\sigma(X) := \sqrt{2/N} \rho_{(1)}(\sqrt{2N} X)$, normalized so that

$$\int_{-1}^{1} \sigma(X)\, dX = 1, \tag{14.24}$$

the balance equation (14.23) reads

$$x^2 + C = \int_{-1}^{1} \sigma(y) \log |x - y|\, dy + \frac{1}{N} \Big(\frac{1}{2} - \frac{1}{\beta} \Big) \log \sigma(x), \tag{14.25}$$

again for some constant C. This integral equation can be solved to give the sought correction to $\rho^{\mathrm{W}}(y)$.

PROPOSITION 14.2.1 *The integral equation (14.25) permits the solution*

$$\sigma(y) = \rho^{\mathrm{W}}(y) + \frac{\mu^{\mathrm{G}}(y)}{N}, \tag{14.26}$$

where $\rho^{\mathrm{W}}(y)$ is the Wigner semicircle form as specified in (7.26) while

$$\mu^{\mathrm{G}}(y) = \Big(\frac{1}{\beta} - \frac{1}{2}\Big)\Big(\frac{1}{2}(\delta(y-1) + \delta(y+1)) - \frac{1}{\pi}\frac{1}{\sqrt{1-y^2}}\Big). \tag{14.27}$$

Note that this is consistent with (7.162).

Proof. Substituting (14.26) in (14.25) and equating terms $\mathrm{O}(1/N)$ gives

$$C = \int_{-1}^{1} \mu^{\mathrm{G}}(y) \log|x-y|\, dy + \Big(\frac{1}{2} - \frac{1}{\beta}\Big) \log \rho^{\mathrm{W}}(x), \qquad x \in (-1,1), \tag{14.28}$$

which must be solved subject to the constraint

$$\int_{-1}^{1} \mu^{\mathrm{G}}(y)\, dy = 0, \tag{14.29}$$

deduced by substituting (14.26) in (14.24).

Differentiating (14.28) gives

$$0 = \mathrm{PV} \int_{-1}^{1} \frac{\mu^{\mathrm{G}}(y)}{x-y}\, dy - \Big(\frac{1}{2} - \frac{1}{\beta}\Big)\frac{x}{1-x^2},$$

where PV denotes the principal value. The validity of (14.27) can be verified by making use of the fact that [448]

$$\mathrm{PV} \int_{-1}^{1} \frac{1}{x-y}\frac{1}{\sqrt{1-y^2}}\, dy = 0, \qquad x \in (-1,1)$$

(cf. (1.66)) and recalling the requirement (14.29). □

To study the expansion of the global density in the Laguerre case (7.161) we begin by recalling the result of Proposition 3.1.4. This allows the general chiral ensemble p.d.f.

$$\frac{1}{C}\prod_{l=1}^{N} e^{-\beta V(x_l)} x_l^{\beta/2} \prod_{1\le j<k\le N} |x_k^2 - x_j^2|^{\beta} \qquad (x_l > 0) \tag{14.30}$$

to be interpreted as a log-gas confined to $x>0$, with image charges of the same sign in $x<0$, and subject to a one-body potential $V(x)$. It has been noted in Exercises 3.1 q.9 that to leading order in N the background density satisfies the same integral equation (1.55) as the background density for the log-gas on a line with confining potential $V(x)$. Analogous reasoning gives that with

$$V(x) = \frac{x^2}{2} - a\log|x| \tag{14.31}$$

the global density $\sigma(X) := 4\rho(4NX)$, which is normalized so that $\int_0^1 \sigma(X)\, dX = 1$ satisfies (14.25) with x^2 on the l.h.s. replaced by $2x^2 - \frac{a}{2N}\log|x|$. Seeking a solution of the form (14.26) with $\rho^{\mathrm{W}}(y)$ replaced by $2\rho^{\mathrm{W}}(y)$ and $\mu^{\mathrm{G}}(y)$ by $\mu^{\mathrm{ch}}(y)$ gives (14.27) with $\mu^{\mathrm{G}}(y) \mapsto \mu^{\mathrm{ch}}(y) - \frac{a}{2}\delta(y)$. Taking into consideration the normalization requirement we thus obtain

$$\mu^{\mathrm{ch}}(y) = \frac{a}{\pi\sqrt{1-y^2}} - \frac{a}{2}\delta(y) + \Big(\frac{1}{\beta} - \frac{1}{2}\Big)\Big(\delta(y-1) - \frac{1}{\pi}\frac{1}{\sqrt{1-y^2}}\Big). \tag{14.32}$$

On the other hand, changing variables $x_l \mapsto x_l^2$ in the Laguerre β-ensemble $\mathrm{ME}_{\beta,N}(x^{\beta(a+(1-1/\beta)/2}e^{-\beta x/2})$ gives (14.30) with $V(x)$ specified by (14.31). It thus follows that with $\mu^{\mathrm{L}}(X)$ the $\mathrm{O}(1/N)$ correction to the density in the Laguerre β-ensemble with one-body terms $x^{\beta a}e^{-\beta x/2}$ one has

$$2X\mu^{\rm L}(X^2)|_{a\mapsto a+(1-1/\beta)/2} = \mu^{\rm ch}(X). \tag{14.33}$$

This is indeed consistent with (7.161).

EXERCISES 14.2 1. Appropriately refine the argument of the proof of Proposition 3.6.3 to conclude that for the Jacobi β-ensemble specified therein

$$N^{-1}\rho_{(1)}(y) \sim \frac{1}{\pi(1-y^2)^{1/2}} - \frac{1}{N}\Big(\Big(\frac{a-1}{2}+\frac{1}{\beta}\Big)\delta(y-1)+\Big(\frac{b-1}{2}+\frac{1}{\beta}\Big)\delta(y+1)-\frac{(2/\beta-1+(a+b)/2)}{\pi(1-y^2)^{1/2}}\Big). \tag{14.34}$$

By changing variables $y=\cos\theta$ deduce that for the β-ensemble on the half-circle specified by the r.h.s. of (3.76),

$$N^{-1}\rho_{(1)}(\theta) \sim \frac{1}{\pi} - \frac{1}{N}\Big(\Big(\frac{a-1}{2}+\frac{1}{\beta}\Big)\delta(\theta)+\Big(\frac{b-1}{2}+\frac{1}{\beta}\Big)\delta(\theta-\pi)-\frac{(2/\beta-1+(a+b)/2)}{\pi}\Big). \tag{14.35}$$

14.3 VARIANCE OF A LINEAR STATISTIC

One application of the formula (14.9) relates to the calculation of the variance of a *linear statistic* in the infinite density limit for a log-gas on the line. Before discussing this and similar applications, some preliminary remarks and theory relating to linear statistics and their variance are required.

DEFINITION 14.3.1 *A linear statistic of a one-component system is any function A that can be written in the form $A=\sum_{j=1}^N a(\vec{r}_j)$ where $\vec{r}_j$ denotes the particle coordinates.*

In random matrix theory, an example of a linear statistic is the conductance

$$G/G_0 = \sum_{j=1}^N \lambda_j, \tag{14.36}$$

where $\{\lambda_j\}$ are the eigenvalues of $\mathbf{t}^\dagger\mathbf{t}$ (recall Section 3.8). An observable quantity associated with (14.36) is the mean square fluctuation (i.e., variance)

$$\mathrm{Var}(G/G_0) := \langle (G/G_0-\langle G/G_0\rangle)^2\rangle = \langle (G/G_0)^2\rangle - \langle G/G_0\rangle^2. \tag{14.37}$$

For a general linear statistic the variance can be expressed in terms of the truncated two-particle correlation function.

PROPOSITION 14.3.2 *For a one-component system of N particles confined to an interval I*

$$\mathrm{Var}(A) = \int_I dx \int_I dx' a(x)a(x')\Big(\rho_{(2)}^T(x,x')+\rho_{(1)}(x)\delta(x-x')\Big). \tag{14.38}$$

Proof. We have

$$\langle A^2\rangle = \int_I dx\, a(x)\int_I dx' a(x')\Big\langle \sum_{\substack{j,j'=1\\ j\neq j'}}^N \delta(x-x_j)\delta(x'-x_{j'})\Big\rangle + \int_I a^2(x)\Big\langle\sum_{j=1}^N\delta(x-x_j)\Big\rangle dx$$
$$= \int_I dx\, a(x)\int_I dx' a(x')\big(\rho_{(2)}(x,x')+\delta(x-x')\rho_{(1)}(x)\big)$$

and

$$\langle A\rangle = \int_I a(x)\rho_{(1)}(x)\,dx,$$

so the result follows from the definition of $\mathrm{Var}(A)$ and $\rho_{(2)}^T$. □

Experiments (see e.g., [53]) have shown that (14.37) is of order unity, independent of N. An important feature in explaining this effect is that the density of the eigenvalues of $\mathbf{t}^\dagger \mathbf{t}$, which take values between 0 and 1, is infinite in the large N limit. In fact in the infinite density limit similar $\mathrm{O}(1)$ fluctuation formulas hold for the one-component log-gas confined to an interval I finite or otherwise. In particular, due to the analogy between the eigenvalues of the Wishart matrix and the log-gas confined to the half-line, the log-gas results can be used to calculate (14.37).

Other linear statistics also show themselves in the theory of quantum conductance [53]. One is the shot noise power P, which characterizes time-dependent fluctuations in the current due to the discreteness of the electron charge. With λ_j, G_0 as in (14.36), and V the voltage difference between two electron reservoirs,

$$P = P_0 \sum_{j=1}^{N} \lambda_j (1 - \lambda_j), \qquad P_0 = 2eVG_0. \tag{14.39}$$

Another relates to the conductance $G_{\rm NS}$ of a normal metal-superconductor junction (recall Section 3.3.2), which is given by

$$G_{\rm NS} = \frac{4e^2}{h} \sum_{j=1}^{N} \frac{\lambda_j^2}{(2 - \lambda_j)^2}. \tag{14.40}$$

From Proposition 14.3.2, to evaluate $\mathrm{Var}(A)$ in the infinite density limit we require the infinite density limit of $\rho_{(2)}^T(x, x')$ smoothed over the microscopic interparticle spacing. The smoothing kills all terms with period proportional to the inter-particle spacing — these terms do not contribute to $\mathrm{Var}(A)$ in the infinite density limit as they integrate to zero. Since in a one-component system of point particles the only length scale is the interparticle spacing $1/\rho$, we see that the asymptotic formula (14.9) must also apply for $|x - x'|$ fixed and $\rho \to \infty$ and thus the sought asymptotic form is already known in the case that the domain is the real line. Also of interest in the study of $\mathrm{Var}(A)$ is the infinite density limit of $\rho_{(2)}^T(x, x')$ for the log-gas confined to different one-dimensional domains I, in particular the circle, finite interval and semi-infinite interval. The linear response argument of the previous section cannot readily be adapted to these cases. However, Jancovici [316] has shown how a different linear response argument can be formulated which is also applicable in the more general settings.

In the method of Jancovici, the one-dimensional domain I is regarded as embedded in two dimensions, and it is hypothesized that in the infinite density limit a log-gas confined to the domain I behaves as a conductor obeying the laws of two-dimensional electrostatics. The system is perturbed by adding an external charge δq at the point $\vec{r}'$. The total energy of the system is then perturbed by an amount $\delta U = \delta q \Phi(\vec{r}')$, where $\Phi(\vec{r}')$ is the electrostatic potential at $\vec{r}'$ due to the charge density induced on I. The observable A is chosen to be the potential $\Phi(\vec{r})$ at another point $\vec{r}$ due to the charges in I ($\Phi(\vec{r}) = -\sum_{j=1}^N \log|\vec{r} - x_j|$) so that the linear response formula reads

$$\langle \Phi(\vec{r}) \rangle_\epsilon - \langle \Phi(\vec{r}) \rangle_0 = -\beta \delta q \langle \Phi(\vec{r}) \Phi(\vec{r}') \rangle_0^T.$$

But the assumption of the applicability of two-dimensional electrostatics gives that in the infinite density limit the response of the system is such that

$$\langle \Phi(\vec{r}) \rangle_\epsilon - \langle \Phi(\vec{r}) \rangle_0 = \phi(\vec{r}),$$

where $\phi(\vec{r})$ denotes the macroscopic potential at $\vec{r}$ due to the charge induced on I by the charge at $\vec{r}'$ ($\phi(\vec{r})$ does not include the potential due to the charge at $\vec{r}'$ itself). Thus

$$\phi(\vec{r}) = -\beta \delta q \langle \Phi(\vec{r}) \Phi(\vec{r}') \rangle_0^T. \tag{14.41}$$

Suppose for definiteness that the domain I is some segment (finite or infinite) of the real line. To relate the r.h.s of (14.41) to $\rho_{(2)}^T(x, x')$ we note that for a one-component system with unit charges, $\rho_{(2)}^T(x, x') =$

$\langle\sigma(x)\sigma(x')\rangle_0^T$, where $\sigma(x)$ denotes the charge density, and it is assumed $x \neq x'$. But from 2d electrostatics $E^+(x) - E^-(x) = 2\pi\sigma(x)$, where $E^+(x) - E^-(x)$ is the discontinuity of the electric field perpendicular to the point x on I (cf. Section 1.4.3), so we have

$$\rho_{(2)}^T(x, x') = \frac{1}{(2\pi)^2}\Big\langle(E^+(x) - E^-(x))(E^+(x') - E^-(x'))\Big\rangle_0^T. \tag{14.42}$$

Now the electric fields and their corresponding averages can be calculated from the potential $\Phi(\vec{r})$ occurring in (14.41) according to

$$E^\pm(x) = \lim_{y\to 0^\pm}\frac{\partial}{\partial y}\Phi(\vec{r}), \quad \langle E^r(x)E^s(x')\rangle_0^T = \lim_{y\to 0^\pm}\lim_{y'\to 0^\pm}\frac{\partial^2}{\partial y\partial y'}\langle\Phi(\vec{r})\Phi(\vec{r}\,')\rangle_0^T, \tag{14.43}$$

where $r, s = \pm$. Substituting (14.41) in this last equation, and making use of the symmetry relation

$$\langle E^r(x)E^s(x')\rangle_0^T = \langle E^{-r}(x)E^{-s}(x')\rangle_0^T$$

allows (14.42) to be rewritten as

$$\rho_{(2)}^T(x, x') = \frac{2}{(2\pi)^2}\Big(\lim_{y\to 0^+}\lim_{y'\to 0^+}\frac{\partial^2}{\partial y\partial y'}\Big(-\frac{\phi(\vec{r})}{\beta\delta q}\Big) - \lim_{y\to 0^+}\lim_{y'\to 0^-}\frac{\partial^2}{\partial y\partial y'}\Big(-\frac{\phi(\vec{r})}{\beta\delta q}\Big)\Big). \tag{14.44}$$

At this stage the details of the calculation depend on the particular interval I. To illustrate the details, suppose I is the whole real line, and let us use (14.43) to rederive (14.9).

PROPOSITION 14.3.3 *For the one-component log-gas on a line*

$$\lim_{\rho\to\infty}\rho\int_x^{x+1/\rho}\rho_{(2)}^T(y, x')dy = -\frac{1}{\beta\pi^2(x - x')^2} = -\frac{1}{\beta\pi^2}\frac{\partial^2}{\partial x\partial x'}\log|x - x'| \tag{14.45}$$

or equivalently, for a fixed density ρ, the leading non-oscillatory term in the large $|x - x'|$ expansion of $\rho_{(2)}^T(x, x')$ is given by the r.h.s of (14.45).

Derivation. In using macroscopic electrostatics we are implicitly assuming the quantities such as $\sigma(x)$ and $\rho_{(2)}^T(x, x')$ are averaged on a microscopic scale so (14.44) will yield the average quantity in (14.45). To calculate $\phi(\vec{r})$ suppose for definiteness the perturbing charge is in the upper half plane $y' > 0$. Then for $y > 0$, from the method of images

$$\phi(\vec{r}) - \delta q\log|z - z'| = -\delta q(\log|z - z'| - \log|z - \bar{z}'|) \tag{14.46}$$

(recall that $\phi(\vec{r})$ does not include the potential due to the charge at $\vec{r}\,'$, which is the second term on the l.h.s.), where $z = x + iy$, $z' = x' + iy'$, while for $y < 0$,

$$\phi(\vec{r}) - \delta q\log|z - z'| = 0. \tag{14.47}$$

An elementary calculation according to (14.44) yields the first equality in (14.45).

The prediction of Proposition 14.3.3 that the leading non-oscillatory term in the large $|x - x'|$ expansion of $\rho_{(2)}^T(x, x')$ is equal to $-1/(\beta\pi^2(x - x')^2)$ has already been obtained in (14.9), and checked against exact results.

To apply this method to other domains I we see the main task is to specify the potential $\phi(\vec{r})$. There are two cases to consider: I infinite or I finite. For I infinite the system can support a charge excess by redistributing charges at infinity, while for I finite the situation is that of a grounded conductor which carries no nett charge. From the viewpoint of two-dimensional electrostatics, in the former situation we require $\phi(\vec{r}) - \delta q\log|z - z'|$ to vanish for $\vec{r} \in I$ independent of the value of $\vec{r}\,'$, while in the latter situation we only require this quantity to vanish for both $\vec{r}, \vec{r}\,' \in I$. Note that in the latter situation we also require $\phi(\vec{r}) \to 0$ as $|\vec{r}| \to \infty$ since the total charge on the conductor is zero. In all cases $\phi(\vec{r})$ must satisfy Laplace's equation and be symmetric in $\vec{r}$ and $\vec{r}\,'$ unless these points are separated by a closed domain I.

We will consider the different domains of particular interest separately.

Circle

For $|z'| > 1$, we can verify that

$$\phi(\vec{r}) - \delta q \log|z - z'| = \begin{cases} -\delta q\Big(\log\left(|z - z'|/|1/\bar{z} - z'|\right) + \log|z'|\Big), & |z| > 1, \\ -\delta q \log|z'|, & |z| < 1, \end{cases} \tag{14.48}$$

is the required potential, as it satisfies all the above requirements. In particular, the "constant" $-\delta q \log|z'|$ for $|z| < 1$ is required by continuity of the two solutions at $|z| = 1$. For $|z'| < 1$ a similar verification shows that

$$\phi(\vec{r}) - \delta q \log|z - z'| = \begin{cases} -\delta q \log|z|, & |z| > 1, \\ -\delta q \log\left(|z - z'|/|1 - \bar{z}z'|\right), & |z| < 1 \end{cases} \tag{14.49}$$

is the required potential.

Finite Interval $(-1, 1)$

Here $\phi(\vec{r})$ is obtained from the solution for $\phi(\vec{r})$ above for the circle geometry in the case $|z'|, |z| > 1$ via the conformal mapping $2w = z + 1/z$, which maps the exterior of the unit disk to the plane with a cut from -1 to 1. We thus simply substitute $z = w + (w^2 - 1)^{1/2}$, $z' = w' + (w'^2 - 1)^{1/2}$ $(w = re^{i\theta}, w' = r'e^{i\theta'})$, where the principal branch of the square root is to be taken, in (14.48). The resulting expression is valid for all $\vec{r}$ and $\vec{r}\,'$.

Semi-infinite Interval $(0, \infty)$

Here we begin with the infinite line potential (14.46) and map the infinite line to the half-line $x > 0$ by the conformal mapping $w = z^2$. Thus the required potential is obtained by substituting $z = w^{1/2} = r^{1/2}e^{i\theta/2}$ and $z' = w'^{1/2} = r'^{1/2}e^{i\theta'/2}$ in (14.46). The resulting expression is valid for all $\vec{r}$ and $\vec{r}\,'$.

From these explicit evaluations of $\phi(\vec{r})$ we can use (14.44) (appropriately modified in the case of a circle) to evaluate $\rho_{(2)}^T(x, x')$ in the infinite density limit. One finds [52], [192]

$$\begin{aligned} \lim_{N\to\infty} \frac{N}{2\pi} \int_{\theta'}^{\theta'+2\pi/N} \rho_{(2)}^T(\phi, \theta)\, d\phi &= -\frac{1}{\beta(2\pi)^2 \sin^2(\theta - \theta')/2}, \\ &= -\frac{1}{\beta\pi^2} \frac{\partial^2}{\partial\theta\partial\theta'} \log|\sin(\theta - \theta')/2|, \end{aligned} \tag{14.50}$$

$$\begin{aligned} \lim_{N\to\infty} \frac{N}{2} \int_{x'}^{x'+2/N} \rho_{(2)}^T(y, x)\, dy &= -\frac{1}{\beta\pi^2 (x - x')^2} \frac{1 - xx'}{[(1 - x^2)(1 - x'^2)]^{1/2}}, \\ &= -\frac{1}{\beta\pi^2} \frac{1}{(1 - x^2)^{1/2}} \frac{\partial^2}{\partial x \partial x'} \Big((1 - x'^2)^{1/2} \log|x - x'| \Big), \end{aligned} \tag{14.51}$$

$$\begin{aligned} \lim_{\rho\to\infty} \rho \int_x^{x+1/\rho} \rho_{(2)}^T(y, x')\, dy &= -\frac{1}{2\pi\beta\sqrt{xx'}} \frac{x + x'}{(x - x')^2} \\ &= -\frac{1}{\beta\pi^2} \frac{\partial^2}{\partial x \partial x'} \log\left| \frac{\sqrt{x} - \sqrt{x'}}{\sqrt{x} + \sqrt{x'}} \right|, \end{aligned} \tag{14.52}$$

for the unit circle, interval $(-1,1)$ and semi-infinite line respectively. Note that (14.52) will also give the leading non-oscillatory decay of $\rho_{(2)}^T(x,x')$ for ρ fixed and $x,x' \to \infty$ (cf. the statement of Proposition 14.3.3). Note too that the soft and hard edge asymptotic formulas obtained in Chapter 7 for $\rho_{(2)}^T(x,y)$ are all in agreement with the first equality in (14.52).

Now, denoting the partial derivative representations in Proposition 14.3.3 and (14.50)–(14.52) by $K_2(x,x')$ we see that in each case

$$\int_I K_2(x,x')dx' = 0 \tag{14.53}$$

provided we take the differentiation with respect to x outside the integral. On the other hand, without this interchange of limits, there is a non-integrable singularity at $x = x'$ in $K_2(x,x')$. Since (14.53) is precisely the condition (14.10) for the perfect screening of an internal charge, this suggests we substitute $K_2(x,x')$ for $\rho_{(2)}^T(x,x') + \rho_{(1)}(x)\delta(x-x')$ in the formula of Proposition 14.3.2 and interchange the order of integration with respect to x and differentiation with respect to x'. This gives

$$\begin{aligned}\mathrm{Var}(A) &= -\frac{1}{\beta\pi^2}\int_{-\infty}^{\infty} dx\, a'(x)\int_{-\infty}^{\infty} dy\, a'(y)\log|x-y| \\ &= \frac{1}{\beta 2\pi^2}\int_{-\infty}^{\infty} |k||\tilde{a}(k)|^2\,dk,\end{aligned} \tag{14.54}$$

$$\begin{aligned}\mathrm{Var}(A) &= -\frac{1}{\beta\pi^2}\int_{-\pi}^{\pi} d\theta\, a'(\theta)\int_{-\pi}^{\pi} d\phi\, a'(\phi)\log|\sin(\theta-\phi)/2| \\ &= \frac{4}{\beta}\sum_{n=1}^{\infty} n a_n a_{-n}, \quad a(\theta) = \sum_{n=-\infty}^{\infty} a_n e^{in\theta},\end{aligned} \tag{14.55}$$

$$\begin{aligned}\mathrm{Var}(A) &= \frac{1}{\beta\pi^2}\int_{-1}^{1} dx \frac{a(x)}{(1-x^2)^{1/2}}\int_{-1}^{1} dy \frac{a'(y)(1-y^2)^{1/2}}{x-y} \\ &= \frac{2}{\beta}\sum_{n=1}^{\infty} n c_n^2, \qquad a(\cos\theta) = c_0 + 2\sum_{n=1}^{\infty} c_n \cos n\theta,\end{aligned} \tag{14.56}$$

$$\begin{aligned}\mathrm{Var}(A) &= -\frac{1}{\beta\pi^2}\int_0^{\infty} dx\, a'(x)\int_0^{\infty} dx'\, a'(y)\log\left|\frac{\sqrt{x}-\sqrt{y}}{\sqrt{x}+\sqrt{y}}\right| \\ &= \frac{1}{\beta 2\pi^2}\int_{-\infty}^{\infty} |\hat{a}(k)|^2 k\tanh(\pi k)\,dk, \quad \hat{a}(k) = \int_{-\infty}^{\infty} e^{ikx} a(e^x)\,dx\end{aligned} \tag{14.57}$$

for the infinite line, unit circle, interval $(-1,1)$ and semi-infinite line, respectively, where $a'(u)$ denotes the derivative of $a(u)$. In deriving the second equality in (14.54) the Fourier transform (14.3) has been used, while the second equality in (14.55) makes use of the Fourier expansion

$$\log|\sin(\theta-\theta')/2| = \sum_{p=-\infty}^{\infty} \alpha_p e^{ip(\theta-\theta')}, \quad \alpha_p = \frac{1}{2\pi}\int_{-\pi}^{\pi} \log|\sin(\theta/2)|e^{-ip\theta}d\theta = -\frac{1}{2|p|}(p\neq 0). \tag{14.58}$$

The second equality in (14.56) follows after the change of variables $x = \cos\theta$, $y = \cos\phi$ and use of the cosine expansion (1.73). Also, the second equality in (14.57) makes use of the Fourier transform [165]

$$-\log\left|\tanh\frac{\pi y}{4a}\right| = \frac{1}{2}\int_{-\infty}^{\infty} \frac{\tanh ax}{x} e^{ixy}\,dx$$

in the case $a = \pi$.

From (14.36) and (14.37) the formula (14.56) with $a(x) = (1+x)/2$ (this is the linear function on $(-1,1)$

that takes on values 0 and 1 at the endpoints), and thus $c_1 = 1/4$, $c_n = 0$ $(n > 1)$, gives for the variance of the dimensionless conductance in the quantum transport problem [51]

$$\mathrm{Var}(G/G_0) = \frac{1}{8\beta}.$$

14.3.1 Alternative calculation of the smoothed two-point density correlation

The linear response formula (14.1) can be used in a different way [51] to that detailed in the previous section to give an alternative derivation of the formulas (14.45) and (14.50)–(14.52). The idea is to perturb the system by an arbitrary one-body potential so that $\delta U = \sum_{j=1}^{N} u(x_j)$. The observable A is taken to be the microscopic density at a point x' in the system so the linear response relation (14.1) then reads

$$\langle n_{(1)}(x')\rangle_\epsilon - \langle n_{(1)}(x')\rangle_0 = -\beta \int_I u(x)\Big(\rho_{(2)}^T(x,x') + \rho_{(1)}(x')\delta(x-x')\Big)\,dx, \tag{14.59}$$

where we have used the definition of $\rho_{(2)}^T$ to rewrite the r.h.s. of (14.1). Our task is to compute the l.h.s. of (14.59) for arbitrary u and then to calculate $\rho_{(2)}^T$ by functional differentiation,

$$\frac{\delta}{\delta u(x)}\Big(\langle n_{(1)}(x')\rangle_\epsilon - \langle n_{(1)}(x')\rangle_0\Big) = -\beta\Big(\rho_{(2)}^T(x,x') + \rho_{(1)}(x')\delta(x-x')\Big). \tag{14.60}$$

Note that (14.60) is exact if we set $u(x) = 0$ after the functional differentiation and thus is valid with this qualification independent of the applicability of the linear response relation. We can calculate the l.h.s. of (14.59) by viewing $u(x)$ as the potential energy of an external charge density. As in Section 1.4 it is hypothesized that the density difference is determined by macroscopic electrostatics and thus satisfies

$$-\int_I \log|x-x'|\Big(\big\langle n_{(1)}(x')\big\rangle_\epsilon - \big\langle n_{(1)}(x')\big\rangle_0\Big)\,dx' = u(x) + C, \tag{14.61}$$

where the constant C is determined by the particle conservation condition

$$\int_I \Big(\big\langle n_{(1)}(x')\big\rangle_\epsilon - \big\langle n_{(1)}(x')\big\rangle_0\Big)\,dx' = 0.$$

The equation (14.61) for I an infinite line is solved using Fourier integrals, for I a circle it is solved using Fourier sums, and for I a finite or semi-infinite interval it can be solved using the eigenfunction method of Proposition 1.4.2. We will illustrate the details of the method in the case of a circle (the finite and semi-infinite intervals are covered in Exercises 14.3).

PROPOSITION 14.3.4 *The solution of the integral equation*

$$\int_{-\pi}^{\pi} \log|\sin(\theta-\theta')/2|\Delta\rho(\theta')\,d\theta' = u(\theta) + C,$$

where C is chosen so that $\int_{-\pi}^{\pi}\Delta\rho(\theta')\,d\theta' = 0$ and $u(\theta) = \sum_{p=-\infty}^{\infty} u_p e^{ip\theta}$ is given by the Fourier series

$$\Delta\rho(\theta) = -\frac{1}{\pi}\sum_{p=-\infty}^{\infty} |p| u_p e^{ip\theta}.$$

Proof. Substituting for $\log|\sin(\theta-\theta')/2|$ in the integral equation using (14.58) gives

$$\sum_{p=-\infty}^{\infty} \alpha_p e^{ip\theta}\int_{-\pi}^{\pi}\Delta\rho(\theta')e^{-ip\theta'}\,d\theta' = \sum_{p=-\infty}^{\infty} u_p e^{ip\theta} + C,$$

where α_p is defined in (14.58). Equating coefficients of $e^{ip\theta}$ gives the value of the Fourier coefficients of $\Delta\rho(\theta)$, and the stated result follows. □

Since $u_p = \frac{1}{2\pi}\int_{-\pi}^{\pi} u(\theta) e^{-ip\theta}\, d\theta$ we see from Proposition 14.3.4 that

$$\frac{\delta}{\delta u(\theta)} \Delta\rho(\theta') = -\frac{1}{2\pi^2} \sum_{p=-\infty}^{\infty} |p| e^{ip(\theta'-\theta)} = \frac{1}{\pi^2} \frac{\partial^2}{\partial\theta \partial\theta'} \log|\sin(\theta-\theta')/2|, \tag{14.62}$$

where to obtain the last equality we have used (14.58). Substituting (14.62) in (14.60) reclaims the infinite density formula (14.50) for the two point correlation on a circle.

EXERCISES 14.3 1. The objective of this exercise is to solve the integral equation

$$\int_{-1}^{1} \log|x-x'| \Delta\rho(x')\, dx' = u(x) + C,$$

where C is chosen so that $\int_{-1}^{1} \Delta\rho(x')\, dx' = 0$, and to use this solution to rederive (14.51) for the case of the interval $(-1, 1)$.

(i) Substitute $x = \cos\theta$, $x' = \cos\sigma$, then use the eigenfunction method of Section 1.5 to show that

$$\Delta\rho(\cos\theta) = -\frac{2}{\pi^2 \sin\theta} \sum_{p=1}^{\infty} p \Big(\int_0^{\pi} u(\cos\sigma) \cos p\sigma \, d\sigma \Big) \cos p\theta.$$

(ii) From this result and the final expression of Exercises 1.5 q.1 show that

$$\frac{\delta\Delta\rho(x)}{\delta u(x')} = \frac{1}{\pi^2(1-x^2)^{1/2}(1-x'^2)^{1/2}} \Big((1-x'^2)^{1/2} \frac{\partial}{\partial x'} \Big)^2 \log|x-x'|$$

and thus reclaim (14.51).

2. In this exercise the integral equation

$$\int_{0}^{\infty} \log|x-x'| \Delta\rho(x')\, dx' = u(x) + C,$$

with C chosen so that $\int_0^{\infty} \Delta\rho(x')\, dx' = 0$, is solved and the solution is used to rederive (14.52).

(i) Use the formal identity

$$\log|x^2 - t^2| = -2 \int_0^{\infty} \cos kx \, \cos kt \, \frac{dk}{k}$$

to show that

$$\Delta\rho(y^2) = -\frac{1}{\pi^2 y} \int_0^{\infty} dk\, k \Big(\int_0^{\infty} dx\, u(x^2) \cos kx \Big) \cos ky.$$

(ii) Use the formulas in (i) to show that

$$\frac{\delta\Delta\rho(x)}{\delta u(x')} = -\frac{1}{\pi^2 x'^{1/2}} \frac{\partial}{\partial x} \Big(x^{1/2} \frac{\partial}{\partial x} \Big) \log|x-x'|$$

and thus reclaim (14.50) in the case of the semi-infinite interval.

3. For two statistics A and B, the covariance is defined by

$$\mathrm{Cov}(A,B) := \langle (A - \langle A \rangle)(B - \langle B \rangle) \rangle = \langle AB \rangle - \langle A \rangle \langle B \rangle,$$

while the correlation coefficient is defined by

$$\mathrm{Cor}(A,B) := \frac{\mathrm{Cov(A,B)}}{(\mathrm{Var}(A)\mathrm{Var}(B))^{1/2}}.$$

Use the Schwarz inequality to show

$$|\mathrm{Cor}(A,B)| \le 1.$$

4. Use (14.39) and (14.56) with $a(x) = (1-x^2)/4$ to show that

$$\mathrm{Var}\,(P/P_0) = \frac{1}{64\beta}.$$

14.4 GAUSSIAN FLUCTUATIONS OF A LINEAR STATISTIC

14.4.1 The infinite density limit

After having evaluated the variance of the linear statistic A, it is natural to seek the full distribution $P_A(t)$ of A. For a general statistic A this is defined as

$$P_A(t) = \frac{1}{\hat{Z}} \int_I dx_1 \cdots \int_I dx_N \, \delta(t-A) e^{-\beta U} \tag{14.63}$$

(cf. the equation of Exercises 3.2 q.3(ii)) and represents the p.d.f. for the event that $A = t$. The evaluation of $P(t)$ for a log-gas in the infinite density limit can be deduced by a linear response argument [445], [316].

To do this, suppose $\delta U = \sum_{j=1}^N a(x_j)$ in (14.1), and take for the observable A therein (not to be confused with the linear statistic A) the microscopic density $n_{(1)}(x')$. This gives

$$\langle n_{(1)}(x')\rangle_\epsilon - \langle \rho(x')\rangle_0 = -\beta \langle n_{(1)}(x') A\rangle_0^T. \tag{14.64}$$

Now terms ignored on the r.h.s. of (14.64) are quadratic and higher in $a(x)$. But on the l.h.s, the formula (14.61) with $u(x)$ replaced by $a(x)$ says that in the infinite density limit $\langle \rho_{(1)}(x')\rangle_\epsilon - \langle \rho_{(1)}(x')\rangle_0$ is linearly related to $a(x)$. This implies (14.64) must be exact in the infinite density limit. Multiplying by $a(x')$ and integrating over x' then shows, at the same level of physical argument, that the relation

$$\langle A\rangle_\epsilon - \langle A\rangle_0 = -\beta\langle A^2\rangle_0^T \tag{14.65}$$

is exact in the infinite density limit. Replacing A by $-ikA/\beta$ and introducing the infinite density $\rho \to \infty$ limiting form of $P_A(t)$, $P_A^{(\infty)}(t)$ say, we see that (14.65) can be rewritten as

$$\frac{d}{dk} \log \int_{-\infty}^{\infty} e^{ik(t-\langle A\rangle_0)} P_A^{(\infty)}(t)\, dt = -k\langle A^2\rangle_0^T. \tag{14.66}$$

After integrating with respect to k and exponentiating this implies

$$\int_{-\infty}^{\infty} e^{ik(t-\langle A\rangle_0)} P_A^{(\infty)}(t)\, dt := \lim_{\rho\to\infty} \Big\langle \prod_{l=1}^N e^{ik(a(x_l)-\langle a\rangle)} \Big\rangle = e^{-k^2\langle A^2\rangle_0^T/2}. \tag{14.67}$$

The distribution $P_A^{(\infty)}(t-\langle A\rangle_0)$ is thus exactly given by the Gaussian

$$P_A^{(\infty)}(t-\langle A\rangle_0) = \Big(\frac{1}{2\pi\sigma^2}\Big)^{1/2} e^{-t^2/2\sigma^2}, \qquad \sigma^2 = \mathrm{Var}(A) = \langle A^2\rangle_0^T. \tag{14.68}$$

Note that for the particular choices of I discussed in the previous two sections $\mathrm{Var}(A)$ is order one and known explicitly, so $P_A^{(\infty)}(t-\langle A\rangle_0)$ is completely specified.

The exactness of (14.65) also has consequence for the higher order correlations $\rho_{(n)}^T(x_1,\dots,x_n)$ implying they must vanish in the infinite density limit for $x_1,\dots,x_n$ distinct [445]. This can be seen from the cumulant expansion

$$\log \int_{-\infty}^{\infty} e^{ikt} P_A(t)\, dt = \sum_{n=1}^{\infty} \frac{(ik)^n}{n!} M_n(a), \tag{14.69}$$

where

$$M_n(a) := \int_I dx_1\, a(x_1) \cdots \int_I dx_n\, a(x_n)\, \rho_{(n)}^T(x_1|x_2|\dots|x_n). \tag{14.70}$$

In (14.70) $\rho_{(n)}^T(x_1|x_2|\dots|x_n)$ is the truncated n-particle correlation function which includes the contribution of coincident points,

$$\begin{aligned}
\rho_{(2)}^T(x_1|x_2) &= \rho_{(2)}^T(x_1, x_2) + \delta(x_1 - x_2)\rho_{(1)}(x_1), \\
\rho_{(3)}^T(x_1|x_2|x_3) &= \rho_{(3)}^T(x_1, x_2, x_3) + \delta(x_1 - x_2)\rho_{(2)}^T(x_2, x_3) + \delta(x_1 - x_3)\rho_{(2)}^T(x_1, x_2) \\
&\quad + \delta(x_2 - x_3)\rho_{(2)}^T(x_1, x_3) + \delta(x_1 - x_2)\delta(x_2 - x_3)\rho_{(1)}(x_1),
\end{aligned}$$

and so on. Substituting (14.69) in (14.66) we see that $M_n(A) = 0$ for $n \ge 3$. Since A is arbitrary, this implies $\rho_{(n)}^T(x_1|x_2|\dots|x_n) = 0$. But for $x_1, \dots, x_n$ distinct, $\rho_{(n)}^T(x_1|x_2|\dots|x_n) = \rho_{(n)}^T(x_1, \dots, x_n)$, so indeed $\rho_{(n)}^T(x_1, \dots, x_n)$, $n \ge 3$, must vanish for noncoincident points.

14.4.2 Szegö's asymptotic formula

The result (14.68) can be used both to anticipate Szëgo's asymptotic formula from the theory of Toeplitz determinants, and to interpret it as a fluctuation formula. We recall that an $N \times N$ Toeplitz determinant is characterized by each entry in row j, column k depending on the difference $j - k$. Thus all entries are constant along each diagonal parallel to the main diagonal. Szegö's asymptotic formula gives the large N asymptotic behavior of a general class of Toeplitz determinants [327].

PROPOSITION 14.4.1 *Let $e^{a(\theta)}$ and $a(\theta)$ have Fourier decompositions*

$$e^{a(\theta)} = \sum_{p=-\infty}^{\infty} \alpha_p e^{ip\theta}, \quad a(\theta) = \sum_{p=-\infty}^{\infty} a_p e^{ip\theta}, \qquad -\pi \le \theta \le \pi,$$

respectively, where it is required $\sum_{p=-\infty}^{\infty} |p||a_p|^2 < \infty$. One has

$$D_N[e^{a(\theta)}] := \det[\alpha_{j-k}]_{j,k=1,\dots,N} \underset{N\to\infty}{\sim} \exp\Big(N a_0 + \sum_{p=1}^{\infty} p a_p a_{-p} + \mathrm{o}(1)\Big). \tag{14.71}$$

We remark that the Borodin-Okounkov identity (10.96) shows that the correction terms to (14.71) are in general exponentially small.

To anticipate (14.71), we note from (5.77) that

$$D_N[e^{a(\theta)}] = \Big\langle \prod_{l=1}^{N} e^{a(\theta_l)} \Big\rangle_{U(N)}, \tag{14.72}$$

where $\langle \cdot \rangle_{U(N)}$ can be interpreted as the canonical average for the one-component log-gas on a circle at $\beta = 2$. Hence

$$D_N[e^{ika(\theta)}] = \hat{P}_A(k),$$

where $\hat{P}_A(k)$ is the Fourier transform of the probability distribution $P_A(t)$ for the linear statistic $A =$

$\sum_{j=1}^N a(\theta_j)$ in the log-gas. According to (14.67) and (14.55)

$$\hat{P}_A(k) = \exp\Big(Nika_0 - k^2 \sum_{p=1}^\infty pa_p a_{-p} + \mathrm{o}(1)\Big), \tag{14.73}$$

which is the sought result.

A consequence of (14.73) is that Szegö's asymptotic formula can be interpreted as a Gaussian fluctuation formula [327]. In Exercises 14.4 q.3 below, it will be shown that (14.73) can be derived using the continuous approximation applied directly to the potential energy of the Boltzmann factor in (14.72). This procedure has been adapted by Johansson [328] to rigorously prove (14.73), and more generally (14.68) in the case of the log-gas on the unit circle for all general $\beta > 0$.

The general formula (14.67) can be used to give the analogue of (14.71), interpreted as an asymptotic formula for the $U(N)$ average (14.72), for averages over the classical groups $O(N)$ and $\mathrm{Sp}(2N)$. With $a(\theta) \mapsto a(\cos\theta)$ the variance is given by (14.56). This follows from Proposition 3.7.1 which relates the eigenvalue p.d.f.'s for the classical groups to particular Jacobi unitary ensembles upon changing variables $y_j = \cos\theta_j$.

Let G be one of the classical groups $O^\pm(N), \mathrm{Sp}(2N)$, and let $\tilde{G}$ denote the eigenvalue p.d.f. of the eigenvalues with angles $\theta \in (0,\pi)$. The formulas (14.67), (14.56) and (14.35) give that

$$\Big\langle \prod_{l=1}^{N^*} e^{a(\cos\theta_l)} \Big\rangle_{\tilde{G}} \underset{N\to\infty}{\sim} \exp\Big(N^* c_0 + a\sum_{n=1}^\infty c_n + b\sum_{n=1}^\infty (-1)^n c_n + \frac{1}{2}\sum_{p=1}^\infty pc_p^2\Big), \tag{14.74}$$

where $a(\cos\theta) = c_0 + 2\sum_{n=1}^\infty c_n \cos n\theta$ and N^*, a, b are as in Proposition 3.7.1. This is in agreement with known rigorous results (see, e.g., citeJo97, [46]).

14.4.3 Asymptotics of a class of Hankel determinants

The first identity in Exercises 5.1 q.3(iii) tells us that

$$\begin{aligned} &\tilde{D}_N(e^{-NV(x)+a(x)}) \\ &:= \frac{1}{N!}\int_{-\infty}^\infty dx_1\, e^{-NV(x_1)} \cdots \int_{-\infty}^\infty dx_N\, e^{-NV(x_N)} \prod_{l=1}^N e^{a(x_l)} \prod_{1\le j<k\le N} (x_k - x_j)^2 \\ &= \det[a_{j+k}]_{j,k=0,\dots,N-1}, \qquad a_p = \int_{-\infty}^\infty x^p e^{a(x)-NV(x)}\, dx, \end{aligned} \tag{14.75}$$

where $a_p = \int_{-\infty}^\infty x^p e^{a(x)-NV(x)}\,dx$, thus relating the multiple integral to a Hankel determinant (by definition the latter have entries which depend on the sum of the row index j and the column index k). The following asymptotic formula for these integrals is known [328].

PROPOSITION 14.4.2 *Let $V(x)$ be an even degree polynomial which is independent of N with positive leading coefficient and no real zeros. Let $\rho_{(1)}(x)$ be the scaled density in the log-gas system corresponding to $\tilde{D}_N(e^{-NV(x)})$, supported on $[c_1, c_2]$ and normalized so that*

$$\int_{c_1}^{c_2} \rho_{(1)}(x)\,dx = 1.$$

For sufficiently well behaved functions $a(x)$ one has

$$\frac{\tilde{D}_N(e^{-NV(x)+a(x)})}{\tilde{D}_N(e^{-NV(x)})} \underset{N\to\infty}{\sim} \exp\Big(N\int_{c_1}^{c_2} a(x)\rho_{(1)}(x)\,dx\Big)$$
$$\times \exp\Big(\frac{1}{4\pi^2}\int_{c_1}^{c_2} dx\,\frac{a(x)}{\sqrt{(x-c_1)(c_2-x)}}\int_{c_1}^{c_2} dy\,\frac{a'(y)\sqrt{(y-c_1)(c_2-y)}}{x-y}\Big). \tag{14.76}$$

A key identity used in the derivation of (14.76) is given in Exercises 14.4 q.4(iii).

We recognize (14.76) as an example of (14.67) with $\beta = 2$, $a(x) \mapsto -ia(x)$ and $\langle A^2\rangle_0^T = \mathrm{Var}(A)$ given by (14.56) (in the latter the kernel must be translated so that the endpoints are at $x = c_1, c_2$). The result of Proposition 14.4.2 holds with the product of differences squared in (14.75) raised to the power of $\beta/2$, provided a factor of $2/\beta$ multiplies the argument of the second exponential in (14.76), and furthermore the r.h.s. is multiplied by the additional factor $\exp(\int_c^d a(y)\mu(y)\,dy)$, where $\mu(y)$ is the $\mathrm{O}(1/N)$ correction to the density (recall (14.26)).

14.4.4 Gaussian fluctuations of the spacing distribution

Consider a continuous one-dimensional statistical mechanical system in the bulk in a fluid state with unit density. Let $F_\beta(s)$ denote the probability that the spacing, to the right say, of a given particle is greater that s. Analogous to the formula (9.4) we can write

$$F_\beta(s) = \sum_{l=1}^{\infty} \frac{(-1)^{l-1}}{(l-1)!}\int_0^s dx_2 \cdots \int_0^s dx_l\, \rho_{(l)}(0, x_2, \dots, x_l). \tag{14.77}$$

Let $\eta(L,s)$ denote the number of particles in an interval of length L which have their right-neighbor spacing greater than s. Clearly the expected value of $\eta(L,s)$ is $LF_\beta(s)$. For the log-gas at $\beta = 2$, Soshnikov [494] has considered the distribution of $\eta(L,s)$, and proved that it has a Gaussian form as $L \to \infty$.

The analysis of $\eta(L,s)$ is complicated by the fact that it is not a linear statistic. Consequently the formulas (14.65) and (14.69) do not hold, so other methods of analysis are necessary. The approach taken in [494] is to prove

$$\Big\langle\Big((\eta(L,s) - \langle\eta(L,s)\rangle)/L^{1/2}\Big)^p\Big\rangle = \begin{cases} (p-1)!!(b(s))^{p/2} + \mathrm{o}(1), & p \text{ even} \\ \mathrm{o}(1), & p \text{ odd} \end{cases} \tag{14.78}$$

where $b(s)$ is the variance of the scaled statistic $(\eta(L,s) - \langle\eta(L,s)\rangle)/L^{1/2}$ and $(2n-1)!! := (2n-1)(2n-3)\cdots 3\cdot 1$. These moments uniquely characterize a Gaussian distribution. Note that the variance of $\eta(L,s)$ itself is proportional to L, which is the same behavior exhibited by a gas with short-range interactions.

We will not attempt to give details of the working which leads to (14.78) for general p, but we will consider the details of the case $p = 2$ and thus calculate the variance explicitly. First, some notation is required. Denote by $\xi(x_i, dx_i, s)$ the indicator for the event that there is an eigenvalue in $[x_i, x_i + dx_i]$ and no eigenvalue in $[x_i + dx_i, x_i + s]$. Then

$$\eta(L,s) = \int_0^L \xi(x, dx, s) \quad \text{and} \quad \langle\xi(x_i, dx_i, s)\rangle = F_\beta(s)dx_i \tag{14.79}$$

while for $|x_1 - x_2| < s$ we must have

$$\langle\xi(x_1, dx_1, s)\xi(x_2, dx_2, s)\rangle = 0. \tag{14.80}$$

For $|x_1 - x_2| > s$ the following result holds.

PROPOSITION 14.4.3 *Consider a fluid state in the thermodynamic limit. Let $x_1, \dots, x_{m+2}$ be particle* *coordinates with* $x_3, \dots, x_{m+2}$ *restricted to the intervals* $[x_1, x_1+s] \cup [x_2, x_2+s]$. *Suppose in a particular configuration of the particles*

$$x_{i_1}, \dots, x_{i_k} \in [x_1, x_1+s] \quad \text{and} \quad x_{j_1}, \dots, x_{j_{m+2-k}} \in [x_2, x_2+s],$$

and define the correlation function $\rho_{m+2,2}$ *by*

$$\rho_{m+2,2}(x_1, \dots, x_{m+2}) = \rho_{m+2}(x_1, \dots, x_{m+2}) - \rho_k(x_1, x_{i_1}, \dots, x_{i_k})\rho_{2+m-k}(x_2, x_{j_1}, \dots, x_{j_{2+m-k}}).$$

Then for $|x_1 - x_2| > s$ *we have*

$$\operatorname{Cov}\Big(\xi(x_1, dx_1, s)\xi(x_2, dx_2, s)\Big) = g(x_1, x_2)dx_1 dx_2, \tag{14.81}$$

where

$$g(x_1, x_2) = \Big(\sum_{m=0}^{\infty} \frac{(-1)^m}{m!} \int_{I^m} \rho_{m+2,2}(0, |x_1 - x_2|, x_3, \dots, x_{m+2}) dx_3 \cdots dx_{m+2}\Big)$$

with $I = [0, s] \cup [|x_1 - x_2|, |x_1 - x_2| + s]$.

Proof. Analogous to (14.77), for $|x_1 - x_2| > s$ we can write

$$\langle \xi(x_1, dx_1, s)\xi(x_2, dx_2, s)\rangle = \Big(\sum_{m=0}^{\infty} \frac{(-1)^m}{m!} \int_{(I')^m} \rho_{m+2}(x_1, x_2, x_3, \dots, x_{m+2}) dx_3 \cdots dx_{m+2}\Big) dx_1 dx_2,$$

where $I' = [x_1, x_1+s] \cup [x_2, x_2+s]$ for $|x_1 - x_2| < s$. Introducing the correlation function $\rho_{m+2,2}$ this reads

$$\begin{aligned}
&\langle \xi(x_1, dx_1, s)\xi(x_2, dx_2, s)\rangle \\
&= \Big(\sum_{m=0}^{\infty} \frac{(-1)^m}{m!} \int_{(I')^m} \rho_{m+2,2}(x_1, x_2, x_3, \dots, x_{m+2}) dx_3 \cdots dx_{m+2}\Big) dx_1 dx_2 \\
&+ \Big(\sum_{l=0}^{\infty} \frac{(-1)^l}{l!} \int_{[x_1, x_1+s]^l} \rho_{l+1}(x_1, x_3, \dots, x_{l+2}) dx_3 \cdots dx_{l+2}\Big) \\
&\times \Big(\sum_{l=0}^{\infty} \frac{(-1)^l}{l!} \int_{[x_2, x_2+s]^l} \rho_{l+1}(x_2, x_3, \dots, x_{l+2}) dx_3 \cdots dx_{l+2}\Big) dx_1 dx_2.
\end{aligned}$$

The final term is precisely $\langle \xi(x_1, dx_1, s)\rangle\langle \xi(x_2, dx_2, s)\rangle$, so subtracting this term from both sides gives the l.h.s. of (14.81) (recall the definition of covariance given in Exercises 14.3 q.3). The r.h.s. is obtained after using translation invariance to shift the interval I' to I. □

The necessary ingredients are now available to compute the variance.

PROPOSITION 14.4.4 *For* $L \to \infty$ *we have*

$$\operatorname{Var}(\eta(L, s)) \sim Lb(s), \quad b(s) = \int_{|x|>s} g(s, x)\, dx - 2sF_\beta^2(s) + F_\beta(s).$$

Proof. By definition, and making use of (14.79),

$$\begin{aligned}
\operatorname{Var}(\eta(L, s)) = &\Big\langle \int_0^L \Big(\xi(x_1, dx_1, s) - F_\beta(s)dx_1\Big) \int_0^L \Big(\xi(x_2, dx_2, s) - F_\beta(s)dx_2\Big)\Big\rangle_{x_1 \neq x_2} \\
&+ \Big\langle \int_0^L \Big(\xi(x_1, dx_1, s) - F_\beta(s)dx_1\Big)^2\Big\rangle.
\end{aligned}$$

In the integrand of the final average, since $(\xi(x_1, dx_1, s))^2 = \xi(x_1, dx_1, s)$ we see that this term of the integrand is of order dx_1, while all other terms are of order $(dx_1)^2$ and so do not contribute. Computing the average of $\xi(x_1, dx_1, s)$

using (14.79) shows that this term equals $LF_\beta(s)$. To compute the first average, we need to consider separately the cases $|x_1 - x_2| > s$ and $|x_1 - x_2| < s$. For $|x_1 - x_2| > s$ the average of the terms in the integrand is precisely equal to $\mathrm{Cov}(\xi(x_1, dx_1, s)\xi(x_2, dx_2, s))$ and so is given by (14.81). For $|x_1 - x_2| < s$, (14.80) applies, so after using the second formula in (14.79), the average of the integrand of the first term in this region reduces to $-(F_\beta(s))^2 dx_1 dx_2$. Thus we have

$$\mathrm{Var}(\eta(L,s)) = \int_0^L dx_1 \int_{\substack{0\\|x_1-x_2|>s}}^L dx_2\, g(x_1,x_2) - \int_0^L dx_1 \int_{\substack{0\\|x_1-x_2|<s}}^L dx_2 (F_\beta(s))^2 + LF_\beta(s),$$

from which the stated result follows. □

One notes that no specific properties of the log-gas at $\beta = 2$ were used in the derivation of Proposition 14.4.4, rather, this result for $\mathrm{Var}(\eta(L,s))$ is universal for general statistical mechanical systems in one-dimension.

EXERCISES 14.4 1. [37] Suppose $e^{-\beta U}$ in (14.67) is given by the eigenvalue p.d.f. of Proposition 1.3.4 for the Gaussian ensemble and $a(x) = x$. Also, make the change of variables $x_j \to \sqrt{2N} x_j$ so that the support of the density is $(-1, 1)$.

(i) For β even complete the square to show that independent of N, $\hat{P}(k) = e^{-k^2/4\beta}$, and so

$$P(u) = \Big(\frac{\beta}{\pi}\Big)^{1/2} e^{-\beta u^2}.$$

Use Carlson's theorem (recall Proposition 4.1.4) to argue that this result must remain true for all $\beta > 0$.

(ii) Show that the formula for $P(u)$ in (i) is in agreement with (14.68) by evaluating (14.56) with $a(x) = x$.

2. Consider the one-component log-gas at $\beta = 2$ subject to a one-body potential with Boltzmann factor $e^{u(\theta)}$. Make the replacement $a(\theta) \mapsto u(\theta) + ika(\theta)$ in the Szegö formula (14.71) to deduce that (14.73) again holds true.

3. (i) Let

$$e^{-\beta W[a]} := e^{\beta \sum_{j<k} \log |e^{i\theta_j} - e^{i\theta_k}| + ik \sum_{j=1}^N a(\theta_j)}.$$

Show that for large N the maximum of the exponent occurs when

$$\frac{\beta}{2} \int_{-\pi}^{\pi} \Delta\rho(\theta) \frac{\cos((\phi-\theta)/2)}{\sin((\phi-\theta)/2)} d\theta + ika'(\phi) = 0, \qquad \phi \in [-\pi, \pi],$$

where $\rho_{(1)}(\theta) = 1 + \Delta\rho(\theta)/N$ is the density of points $\{\theta_j\}$ and solve this equation for $\Delta\rho(\theta)$ using the result of Proposition 14.3.4.

(ii) Note that for large N

$$e^{-\beta(U[a]-U[0])} \sim \exp\Big(\frac{ikN}{2\pi}\int_{-\pi}^{\pi} a(\theta) d\theta + \frac{ik}{2\pi}\int_{-\pi}^{\pi} \Delta\rho(\theta) a(\theta) d\theta + \frac{\beta}{8\pi^2}\int_{-\pi}^{\pi} d\theta\, \Delta\rho(\theta) \int_{-\pi}^{\pi} d\phi\, \Delta\rho(\phi) \log|e^{i\theta} - e^{i\phi}|\Big).$$

Substitute for $\Delta\rho(\theta)$ using the result of (i) and simplify the integrals making use of the second equality in (14.55) to reclaim the r.h.s. of (14.73) in the case $\beta = 2$.

4. [328] Generalize the definition (14.75) to read

$$\tilde{D}_N(e^{-NV(x)+a(x)})[f] := \frac{1}{N!}\int_{-\infty}^{\infty} dx_1\, e^{-NV(x_1)} \cdots \int_{-\infty}^{\infty} dx_N\, e^{-NV(x_N)} \prod_{j=1}^N e^{a(x_j)} \prod_{1 \le j < k \le N} (x_k - x_j)^2 f.$$

(i) Let $\phi(x)$ be differentiable for all $x \in \mathbb{R}$ and suppose $\phi'(x)$ is bounded at infinity. Make the change of variables $x_j = y_j + \epsilon\phi(y_j)$, $0 < \epsilon \ll 1$, in the above definition of $\tilde{D}_N$ with $f = 1$ and expand to leading order in ϵ to conclude

$$-N\tilde{D}_N(e^{-NV(x)+a(x)})\Big[\sum_{j=1}^N \phi(x_j)V'(x_j)\Big] + \tilde{D}_N(e^{-NV(x)+a(x)})\Big[\sum_{j=1}^N \phi(x_j)a'(x_j)\Big]$$

$$+2\tilde{D}_N(e^{-NV(x)+a(x)})\Big[\sum_{1\le j<k\le N} \frac{\phi(x_k)-\phi(x_j)}{x_k - x_j}\Big] + \tilde{D}_N(e^{-NV(x)+a(x)})\Big[\sum_{j=1}^N \phi'(x_j)\Big] = 0.$$

(ii) Show that this same equation can be obtained by integrating by parts

$$\tilde{D}_N(e^{-NV(x)+a(x)})\Big[\sum_{j=1}^N \frac{d}{dx_j}\phi(x_j)\Big]$$

as in Aomoto's method of Section 4.6.

(iii) Choose $\phi(x) = \frac{1}{z-x}$ to deduce from the equation in (i) that

$$N\tilde{D}_N(e^{-NV(x)+a(x)})\Big[\sum_{j=1}^N \frac{V'(z)-V'(x_j)}{z-x_j}\Big] - NV'(z)\tilde{D}_N(e^{-NV(x)+a(x)})\Big[\sum_{j=1}^N \frac{1}{z-x_j}\Big]$$

$$+\Big(\tilde{D}_N(e^{-NV(x)+a(x)})\Big[\sum_{j=1}^N \frac{1}{z-x_j}\Big]\Big)^2 + R = 0,$$

where

$$R = 2\tilde{D}_N(e^{-NV(x)+a(x)})\Big[\sum_{1\le j<k\le N} \frac{1}{(z-x_k)(z-x_j)}\Big] - \Big(\tilde{D}_N(e^{-NV(x)+a(x)})\Big[\sum_{j=1}^N \frac{1}{z-x_j}\Big]\Big)^2$$

$$+\tilde{D}_N(e^{-NV(x)+a(x)})\Big[\sum_{j=1}^N \frac{a'(x_j)}{z-x_j}\Big] + \tilde{D}_N(e^{-NV(x)+a(x)})\Big[\sum_{j=1}^N \frac{1}{(z-x_j)^2}\Big].$$

(iv) In general it is known that

$$\frac{1}{N^2}\bigg(\frac{\tilde{D}_N(e^{-NV(x)+a(x)})[2\sum_{1\le j<k\le N} u(x_j)u(x_k)]}{\tilde{D}_N(e^{-NV(x)+a(x)})[1]} - \Big(\frac{\tilde{D}_N(e^{-NV(x)+a(x)})[\sum_{j=1}^N u(x_j)]}{\tilde{D}_N(e^{-NV(x)+a(x)})[1]}\Big)^2\bigg) \to 0$$

as $N \to \infty$ while

$$\lim_{N\to\infty} \frac{1}{N}\frac{\tilde{D}_N(e^{-NV(x)+a(x)})[\sum_{j=1}^N u(x_j)]}{\tilde{D}_N(e^{-NV(x)+a(x)})[1]} = \int_{-\infty}^{\infty} u(x)\rho_{(1)}(x)\,dx,$$

where $\rho_{(1)}(x)$ is the limiting normalized particle density for the statistical mechanical system with Boltzmann factor implied by $\tilde{D}_N(e^{-NV(x)+a(x)})[1]$. Use these facts to deduce from the equation in (iii) that

$$\int_{-\infty}^{\infty} \frac{V'(z)-V'(x)}{z-x}\rho_{(1)}(x)\,dx - V'(z)\int_{-\infty}^{\infty}\frac{1}{z-x}\rho_{(1)}(x)\,dx + \Big(\int_{-\infty}^{\infty}\frac{1}{z-x}\rho_{(1)}(x)\,dx\Big)^2 = 0,$$

and hence conclude from this that for $z \to$ support $\rho(x)$,

$$\int_{-\infty}^{\infty}\frac{1}{z-x}\rho_{(1)}(x)\,dx \sim V'(z).$$

Identify this as the fundamental equation in the complex electric field approach of Section 1.4.3 to the calculation of $\rho_b(x)$.

5. (i) Let $\rho^{\rm W}(x)$ be as in (7.26). Verify that for $s > 1$

$$\int_{-1}^{1} \log |s - x| \rho^{\rm W}(x)\, dx = s^2 - s\sqrt{s^2-1} - \log(2(s - \sqrt{s^2-1})) - \frac{1}{2}$$

by first showing that both sides tend to $\log s + {\rm O}(1/s)$ as $s \to \infty$, and then verifying that the derivative is consistent with (1.132).

(ii) By making use of the leading asymptotics in (14.76), which remains valid for general $\beta > 0$ (the leading asymptotics is independent of β), deduce that for $s > 1$ and large N [248]

$$e^{-Ns^2} \Big\langle \prod_{l=1}^{N} |s - x_l| \Big\rangle_{{\rm OE}_N(e^{-Nx^2})} \sim \exp\Big(- N\Big(s\sqrt{s^2-1} + \log(2(s - \sqrt{s^2-1})) + \frac{1}{2}\Big) + {\rm O}(1)\Big).$$

Conclude from this that for the GOE

$$\rho_{(1),N+1}(\sqrt{2N}s) \sim \exp\Big(- N\Big(s\sqrt{s^2-1} + \log(s - \sqrt{s^2-1})\Big) + {\rm O}(1)\Big), \tag{14.82}$$

valid for $s > 1$.

6. Let $\tilde{\rho}_{(1)}(x) = \rho_b(x)/N$, where $\rho_b(x)$ is the density (3.61) for the scaled chiral ensemble. Furthermore let

$$\rho^{\rm L}_{(1)}(y) = \frac{1}{2\sqrt{y}} \tilde{\rho}_{(1)}(\sqrt{y})$$

and thus $\rho^{\rm L}_{(1)}(y)$ is the normalized density in the Laguerre ensemble scaled so that $m = N$, $\lambda \mapsto N\lambda$, $a = N\alpha$ with N large.

(i) Verify from (3.69) that for $x \notin [c^2, d^2]$

$$2\int_{c^2}^{d^2} \frac{\rho^{\rm L}_{(1)}(y)}{x - y}\, dy = 1 - \frac{\alpha}{x} - \Big(1 - \frac{2(\alpha+2)}{x} + \frac{\alpha^2}{x^2}\Big)^{1/2},$$

and deduce from this that

$$2\int_{c^2}^{d^2} \rho^{\rm L}_{(1)}(y) \log|x - y|\, dy = x - \alpha - u + \alpha \log\Big|\frac{\alpha(\alpha + u - x) - 2x}{2x^2}\Big| + (2+\alpha)\log\Big|\frac{u + x - 2 - \alpha}{2}\Big|,$$

where $u := ((x - \alpha)^2 - 4x)^{1/2}$.

(ii) Assuming the validity of the large N formula

$$\Big\langle \prod_{l=1}^{N} |x - \lambda_l|^\beta \Big\rangle_{{\rm ME}_{\beta,N}(\lambda^{\beta N\alpha/2} e^{-\beta N\lambda/2})} \sim \exp\Big(\beta N \int_{c^2}^{d^2} \log|x - y| \rho^{\rm L}_{(1)}(y)\, dy\Big)$$

for $x \notin [c^2, d^2]$, read off from (i) that

$$x^{\beta N\alpha/2} e^{-\beta N x/2} \Big\langle \prod_{l=1}^{N} |x - \lambda_l|^\beta \Big\rangle_{{\rm ME}_{\beta,N}(\lambda^{\beta N\alpha/2} e^{-\beta N\lambda/2})}$$
$$\sim \exp\Big(\frac{\beta N}{2}\Big(- u + \alpha\log\Big|\frac{\alpha(\alpha + u - x) - 2x}{2x}\Big| + (2+\alpha)\log\Big|\frac{u + x - 2 - \alpha}{2}\Big|\Big)\Big).$$

Hence conclude that for $x \notin [c^2, d^2]$

$$\rho^{\rm L}_{(1),N+1}(x) \sim \exp\Big(\frac{\beta N}{2}\Big(-u + \alpha\log\Big|\frac{\alpha(\alpha + u - x) - 2x}{2(\alpha+1)^{1/2}x}\Big| + (2+\alpha)\log\Big|\frac{u + x - 2 - \alpha}{2(\alpha+1)^{1/2}}\Big|\Big)\Big). \tag{14.83}$$

14.5 CHARGE AND POTENTIAL FLUCTUATIONS

14.5.1 Charge fluctuations

In the bulk of the one-component log-gas, by setting $a(x) = \chi_{[-l/2,l/2]}$, where $\chi_{[-l/2,l/2]} = 1$ if $x \in [-l/2,l/2]$ and zero otherwise, we have that the linear statistic $A := \sum_{j=1}^N (a(x_j) - \langle a \rangle)$ represents the fluctuation in the number of particles, or equivalently charge, in the interval $[-l/2,l/2]$. Since $\partial a(x)/\partial x = \delta(x-l)$, (14.54) gives that the variance diverges in the infinite density limit. In the log-gas on a line with fixed density (which we take equal to unity, for convenience), it follows that the variance of A diverges as $l \to \infty$. Here we want to determine the asymptotic form of this divergence.

Rewriting the formula (14.38) for Var(A) by use of the Fourier transform gives

$$\mathrm{Var}(A) = \frac{1}{2\pi}\int_{-\infty}^{\infty} |\hat{a}(k)|^2 \hat{S}(k)dk = \frac{4l}{\pi}\int_0^{\infty} \frac{\sin^2(y/2)}{y^2}\hat{S}(y/l)dy, \tag{14.84}$$

where $\hat{S}(k)$ denotes the structure function (14.7). The asymptotics can be determined by breaking the range of integration into the intervals [0,1], $[1,l]$ and $[l,\infty)$, and using (14.8) in the first two intervals. One finds [146]

$$\mathrm{Var}(A) \sim \frac{2}{\pi^2\beta}\log l + B_\beta, \tag{14.85}$$

where

$$B_\beta = \frac{2}{\pi^2\beta}C + \frac{2}{\pi}\int_0^1 \frac{1}{y^2}\Big(\hat{S}(y) - \frac{y}{\pi\beta}\Big)\,dy + \frac{2}{\pi}\int_1^\infty \frac{1}{y^2}\hat{S}(y)dy, \tag{14.86}$$

with C denoting Euler's constant. In deriving (14.86) use has been made of the identity

$$\int_0^1 \frac{(1-\cos y)}{y}\,dy - \int_1^\infty \frac{\cos y}{y}dy = C.$$

For $\beta = 1, 2$ and 4, $\hat{S}(k)$ is known in terms of elementary functions according to (7.136), (7.4) and (7.95), respectively (from (14.7), for $\rho = 1$, $\hat{S}(k) = \tilde{S}(k)$). The integrals in (14.86) can then be computed explicitly to give

$$\begin{aligned}
B_1 &= \frac{2}{\pi^2}C + \frac{2}{\pi^2}\Big(1+\log 2\pi\Big) - \frac{1}{4},\\
B_2 &= \frac{1}{\pi^2}C + \frac{1}{\pi^2}\Big(1+\log 2\pi\Big),\\
B_4 &= \frac{1}{2\pi^2}C + \frac{1}{2\pi^2}\Big(1+\log 4\pi\Big) + \frac{1}{4}.
\end{aligned} \tag{14.87}$$

We remark that the leading behavior exhibited by (14.85) is consistent with that obtained by substituting the r.h.s. of (14.45) for $\rho_{(2)}^T(x,x') + \rho_{(1)}(x)\delta(x-x')$ in (14.38) with $a(x) = \chi_{[-l/2,l/2]}$ and ignoring the singular terms in the ensuing evaluation of the integral. This same heuristic procedure predicts the leading behavior

$$\mathrm{Var}(A) \sim \frac{1}{2\pi^2\beta}\log l \tag{14.88}$$

for the charge fluctuation in the interval $[0,l]$ of the log-gas on the half-line, a result which can be proved rigorously for the hard edge at $\beta = 2$ [498]. The fact that the fluctuations in the number of particles in an interval of size l is proportional to $\log l$ contrasts to the situation with an integrable (at infinity) pair potential $V(x)$. In such systems $\hat{S}(0) \neq 0$ and the formula (14.84) gives that the variance is proportional to l.

Also of interest is the probability distribution (14.63) associated with charge fluctuations. We expect the linear response relation (14.1) to again be exact in the "macroscopic" limit, which here is $l \to \infty$, provided the ratio $\Big(\sum_{j=1}^N (a(x_j) - \langle a \rangle)\Big)^2 / \mathrm{Var}(A)$ is finite. For the infinite system with a finite density this implies

$$\Big(\frac{2}{\pi^2\beta}\log l\Big)^{1/2} P\Big(\Big(\frac{2}{\pi^2\beta}\log l\Big)^{1/2} u\Big) \sim \frac{1}{(2\pi)^{1/2}} e^{-u^2/2} \quad \mathrm{as} || \to \infty. \tag{14.89}$$

For $\beta = 1, 2$ and 4 this has been rigorously established by Costin and Lebowitz [121]. The proof in the case $\beta = 2$ has been extended in [495], [465] to apply to all determinantal point processes (recall Section 5.8).

PROPOSITION 14.5.1 *Let I be an interval in a determinantal point process such that $A := \sum_{j=1}^N \chi_{x_j \in I} \to \infty$ as $N \to \infty$. Then for $N \to \infty$, $(A - \langle A \rangle)/\sqrt{\mathrm{Var}\, A}$ has Gaussian distribution with zero mean and unit variance.*

Proof. Setting $a(x) = \chi_{x \in I}$ in the first result of Exercises 5.1 q.3(iii) allows (14.69) to be rewritten

$$\log \int_{-\infty}^{\infty} e^{ikt} P(t)\, dt = \sum_{n=1}^{\infty} \frac{1}{n!} (e^{ik} - 1)^n \bar{U}_n, \quad \bar{U}_n := \int_I dx_1 \cdots \int_I dx_n\, \rho_{(n)}^T(x_1, \dots, x_n).$$

Putting $ik = \log(1 + x)$ and equating with the r.h.s. of (14.69) shows that for $k > 2$,

$$\bar{U}_k = M_k - \sum_{j=2}^{k-1} b_{kj} M_j - (-1)^k (k-1)! M_1 \tag{14.90}$$

for some coefficients b_{kj}. On the other hand, because we are considering a determinantal point process, $\rho_{(n)}^T$ has the form given in Proposition 5.1.2 and consequently

$$\begin{aligned} \bar{U}_k &= (-1)^{k-1}(k-1)! \int_I dx_1 \cdots \int_I dx_k\, K_N(x_1 - x_2) K_N(x_2 - x_3) \cdots K_N(x_k - x_1) \\ &= (-1)^{k-1}(k-1)! \mathrm{Tr}\, \mathbf{K}_N^k, \end{aligned}$$

where $\mathbf{K}_N$ is the integral operator on I with kernel $K_N(x, y)$ (cf. Exercises 9.3 q.1). Since $M_1 = \mathrm{Tr}\, \mathbf{K}$ this allows (14.90) to be rewritten

$$M_k = (-1)^k (k-1)! \mathrm{Tr}(\mathbf{K}_N - \mathbf{K}_N^k) + \sum_{j=2}^{k-1} b_{kj} M_j. \tag{14.91}$$

But

$$\mathrm{Tr}(\mathbf{K}_N - \mathbf{K}_N^k) = \sum_{j=1}^{k-1} \mathrm{Tr}(\mathbf{K}_N^j - \mathbf{K}_N^{j+1}) < (k-1)\mathrm{Tr}(\mathbf{K}_N - \mathbf{K}_N^2) = (k-1)\mathrm{Var}\, A, \tag{14.92}$$

where the inequality follows from the fact that the eigenvalues of $\mathbf{K}_N$ must satisfy the inequality $0 \le \lambda_l < 1$ (recall Section 9.6.1). Substituting (14.92) in (14.91) and proceeding inductively gives $M_k = \mathrm{O}(\mathrm{Var}\, A)$ for $k > 2$. Since by assumption $\mathrm{Var}\, A \to \infty$ as $N \to \infty$, it follows that

$$\frac{1}{(\mathrm{Var}\, A)^k} M_k \to 0, \qquad k \ge 3,$$

which implies the stated result. □

A proof of (14.89) for general β has been given in [355], by making use of the theory of Section 2.8.

14.5.2 Potential fluctuations

Another linear statistic giving rise to a logarithmic divergence of the variance is the potential at the origin $a(x) = -\log|x|$. First, note from (14.54) and (14.3) that the variance of $A = -\sum_{j=1}^N \log x_j$ diverges in the infinite density limit. In the case of the semi-infinite or infinite line this divergence must persist at finite density due to the equivalence with the infinite density limit for the logarithmic potential. We should therefore consider the asymptotic form of the divergence in the variance of A in a finite system with fixed density as a function of the size of the system.

Suppose in particular the log-gas is confined to a circle of circumference length L, and consider the linear statistic $A = -\sum_{j=1}^N \log|2\sin(\pi x_j/L)|$, $-L/2 < x_j \le L/2$, which corresponds to the potential at the origin. Using (14.38) and (14.58) we have

$$\mathrm{Var}(A) = \frac{L}{4} \sum_{\substack{n=-\infty \\ n\neq 0}}^{\infty} \frac{1}{n^2} r_n,$$

where the Fourier coefficients r_n are specified by $\rho_{(2)}^T(x-x') + \delta(x-x') = \frac{1}{L}\sum_{n=-\infty}^{\infty} r_n e^{2\pi i(x-x')n/L}$. But analogous to (14.8) we expect $r_n \sim (2\pi/L)(|n|/\pi\beta)$ for $0 \le |n| \le \mathrm{O}(L)$, which gives

$$\mathrm{Var}(A) \sim \frac{1}{\beta}\log L \quad \text{as } L \to \infty. \tag{14.93}$$

Furthermore, analogous to (14.89) we expect the corresponding probability distribution to be asymptotically Gaussian,

$$\Big(\frac{1}{\beta}\log L\Big)^{1/2} P\Big(\Big(\frac{1}{\beta}\log L\Big)^{1/2} u\Big) \sim \frac{1}{(2\pi)^{1/2}} e^{-u^2/2} \quad \text{as } L \to \infty. \tag{14.94}$$

A statistic closely related to the potential at the origin is the fluctuation of the average (angular) displacement from its mean, $A = \frac{\pi}{L}\sum_{l=1}^N x_l$, or equivalently the fluctuation in the scaled dipole moment. Thus comparing the Fourier expansions

$$-\log|2\sin\frac{\pi x}{L}| = \sum_{m=1}^{\infty} \frac{\cos 2\pi m x/L}{m} \qquad \text{and} \qquad \frac{\pi x}{L} = \sum_{m=1}^{\infty} \frac{\sin 2\pi m x/L}{m} \tag{14.95}$$

we see that the variance will again exhibit the behavior (14.93), and the probability distribution the behavior (14.94).

Whereas the potential fluctuations are infinite in the infinite system at finite density, the fluctuation of the potential difference

$$A(x) := -\Big(\sum_{j=1}^N \log|x - x_j| - \sum_{j=1}^N \log|x_j|\Big) \tag{14.96}$$

is a well-behaved quantity. Thus by use of the formula (14.3) in (14.38) we have

$$\mathrm{Var}(A(x)) = 2\pi \int_{-\infty}^{\infty} \frac{\sin^2(kx/2)}{k^2} \hat{S}(k)dk, \tag{14.97}$$

which, on identifying ρx with ρl, is precisely π^2 times (14.84). This quantity therefore diverges in the infinite density limit, or equivalently as $x \to \infty$, exhibiting the behavior

$$\mathrm{Var}(A(x)) \underset{x\to\infty}{\sim} \frac{2}{\beta}\log x + \pi^2 B_\beta, \tag{14.98}$$

and analogous to (14.89) we expect

$$\Big(\frac{2}{\beta}\Big)^{1/2} P\Big(\Big(\frac{2}{\beta}\Big)^{1/2} u\Big) \sim \frac{1}{(2\pi)^{1/2}} e^{-u^2/2} \qquad \text{as } x \to \infty. \tag{14.99}$$

14.5.3 Fisher-Hartwig asymptotics

The prediction (14.94) for the probability distribution of the linear statistics in the log-gas on a circle with $a(x) = -\log 2|\sin \pi x/L|$ and $a(x) = \frac{\pi}{L}x$ can be proved [37] at $\beta = 2$ using a generalization of the Szegö asymptotic formula (14.71). This generalization was conjectured by Fisher and Hartwig [176], and has subsequently been proved to varying degrees of generality in relation to the number of singularities R and constraints on $\{a_r, b_r\}$; see, e.g., [43].

Proposition 14.5.2 *Consider the Toeplitz determinant in Proposition 14.4.1 with*

$$\begin{aligned} a(\theta) &= g(\theta) - i\sum_{r=1}^{R} b_r \arg e^{i(\theta_r+\pi-\theta)} + \sum_{r=1}^{R} a_r \log|2 - 2\cos(\theta-\theta_r)| \\ &= g(\theta) + \sum_{r=1}^{R}\Big((a_r+b_r)\log(1+e^{i(\theta-\theta_r-\pi)}) + (a_r-b_r)\log(1+e^{i(\theta_r+\pi-\theta)})\Big), \end{aligned}$$

where $-\pi < \arg z \le \pi$, and assume $g(\theta) = \sum_{p=-\infty}^{\infty} g_p e^{ip\theta}$, where $\sum_{p=-\infty}^{\infty} |p||g_p|^2 < \infty$. Then for $|\mathrm{Re}(a_r)| < \frac{1}{2}$, $|\mathrm{Re}(b_r)| < \frac{1}{2}$ or $\mathrm{Re}(a_r) > -\frac{1}{2}$, $\mathrm{Re}(b_r) = 0$ (and possibly in other regions; although see Exercises 14.5 q.2)

$$D_N[e^{a(\theta)}] \sim e^{g_0 N} e^{\sum_{r=1}^{R}(a_r^2-b_r^2)\log N} E, \tag{14.100}$$

where E is independent of N. To specify E, write $g(\theta) - g_0 = g_+(\theta) + g_-(\theta)$, where $g_+(\theta) = \sum_{p=1}^{\infty} g_p e^{ip\theta}$ and $g_-(\theta) = \sum_{p=-\infty}^{-1} g_p e^{ip\theta}$. Then

$$\begin{aligned} E &= e^{\sum_{k=1}^{\infty} k g_k g_{-k}} \prod_{r=1}^{R} e^{-(a_r+b_r)g_-(\theta_r)} e^{-(a_r-b_r)g_+(\theta_r)} \\ &\times \prod_{1\le r\ne s\le R}\Big(1-e^{i(\theta_s-\theta_r)}\Big)^{-(a_r+b_r)(a_s-b_s)} \prod_{r=1}^{R}\frac{G(1+a_r+b_r)G(1+a_r-b_r)}{G(1+2a_r)}, \end{aligned}$$

where G is the Barnes G-function.

The functional form of (14.100), together with the structure of the constant E, can be anticipated by suitably truncating the logarithms in the definition of $a(\theta)$, and applying Szegö's asymptotic formula (14.71). The necessary working is given in Exercises 14.5 q.3.

Consider now the application of Proposition 14.5.2 to the p.d.f. of $A = -\sum_{j=1}^{N}\log 2|\sin \pi x_j/L|$ with $\beta = 2$. From the first line of (14.67) and the definition of D_N in (14.71) we have

$$\hat{P}(k) = D_N\Big[(2-2\cos\theta)^{ik/2}\Big].$$

The Fisher-Hartwig formula (14.100) with $g(\theta) = 0, R = 1, b_1 = 0, a_1 = ik/2$ and $\theta_1 = 0$ gives that

$$\hat{P}(k) \underset{N\to\infty}{\sim} e^{-(k^2/4)\log N}, \tag{14.101}$$

which is equivalent to (14.94) with $\beta = 2$, $L \sim N$. Similarly, with $A := \frac{\pi}{L}\sum_{l=1}^{N} x_l$ we have that at $\beta = 2$

$$\hat{P}(k) = D_N[e^{ik\theta/2}].$$

The Fisher-Hartwig formula (14.100), this time with $g(\theta) = 0, R = 1, a_1 = 0, b_1 = ik/2$ and $\theta_1 = -\pi$, again implies the asymptotic behavior (14.101), as expected.

For the log-gas on a circle at general rational β, to prove the Gaussian behavior (14.101) for the linear statistics $a(x) = -\log 2|\sin \pi x/L|$ and $a(x) = \pi x/L$ it suffices to compute the asymptotic form of the average

$$\left\langle \prod_{l=1}^{N} e^{ib\theta_l} |2 + 2\cos\theta_l|^a \right\rangle_{\mathrm{C}\beta\mathrm{E}_N}, \tag{14.102}$$

where $\mathrm{C}\beta\mathrm{E}_N$ refers to the circular β-ensemble as specified by the p.d.f. (2.20). In fact (14.102) is given in terms of gamma functions according to the Morris integral (4.4). In terms of the quantity $f_n(\alpha, c)$ defined by (4.187) we see that

$$\left\langle \prod_{l=1}^{N} e^{ib\theta_l} |2 + 2\cos\theta_l|^a \right\rangle_{\mathrm{C}\beta\mathrm{E}_N} = \frac{f_N(2a, c)}{f_N(a+b, c) f_N(a-b, c)}, \qquad c := \beta/2. \tag{14.103}$$

Proceeding now as in the derivation of (4.188) provides the sought asymptotic form for β rational.

PROPOSITION 14.5.3 *Let $\beta/2 = s/r$, $s, r \in \mathbb{Z}_{\geq 0}$. We have*

$$\left\langle \prod_{l=1}^{N} e^{ib\theta_l} |2 + 2\cos\theta_l|^a \right\rangle_{\mathrm{C}\beta\mathrm{E}_N} \underset{N\to\infty}{\sim} N^{(2/\beta)(a^2-b^2)} A_{a,b}^{(\beta)}, \tag{14.104}$$

where

$$A_{a,b}^{(\beta)} := r^{-(2/\beta)(a^2-b^2)} \prod_{\nu=0}^{r-1} \prod_{p=0}^{s-1} \frac{G((a+b)/s + \nu/r - p/s + 1) G((a-b)/s + \nu/r - p/s + 1)}{G(2a/s + \nu/r - p/s + 1) G(\nu/r - p/s + 1)}. \tag{14.105}$$

It follows that

$$\left\langle \prod_{l=1}^{N} e^{ik\theta_l/2} \right\rangle_{\mathrm{C}\beta\mathrm{E}_N} \sim \left\langle \prod_{l=1}^{N} |2 - 2\cos\theta_l|^{ik/2} \right\rangle_{\mathrm{C}\beta\mathrm{E}_N} \sim e^{-(k^2/2\beta)\log N},$$

thus establishing the Gaussian behavior (14.101) for the respective linear statistics [37].

The result (14.104) can also be used to formulate a β-generalization of the Fisher-Hartwig formula (14.100). Thus for general rational β, in the notation of Propositions 14.5.2 and 14.5.3 one conjectures [201]

$$\left\langle \prod_{l=1}^{N} e^{a(\theta_l)} \right\rangle_{\mathrm{C}\beta\mathrm{E}_N} \underset{N\to\infty}{\sim} e^{g_0 N} e^{(2/\beta)\sum_{r=1}^{R}(a_r^2 - b_r^2)\log N} E^{(\beta)},$$

where

$$E^{(\beta)} = e^{(2/\beta)\sum_{k=1}^{\infty} k g_k g_{-k}} \prod_{r=1}^{R} e^{-(2/\beta)(a_r+b_r) g_-(\theta_r)} e^{-(2/\beta)(a_r-b_r) g_+(\theta_r)}$$

$$\times \prod_{1\leq r\neq s\leq R} \left(1 - e^{i(\theta_s-\theta_r)}\right)^{-(2/\beta)(a_r+b_r)(a_s-b_s)} \prod_{j=1}^{R} A_{a_j,b_j}^{(\beta)}.$$

14.5.4 Value distribution of characteristic polynomials for $U(N)$

Closely related to the results of Section 14.5.2 and 14.5.3 are statistical properties of the characteristic polynomial $\Lambda(z) := \prod_{l=1}^{N}(\lambda_l - z)$ for the classical group $U(N)$. Consider first the distribution of the values of $\log \Lambda(z)$ for z on the unit circle. Because of rotational invariance this point can be chosen to be $z = -1$. We seek the joint distribution of the real and imaginary parts,

$$P(s,t) = \left\langle \delta(s - \mathrm{Re}\,\log \Lambda(-1)) \delta(t - \mathrm{Im}\,\log \Lambda(-1)) \right\rangle_{U(N)}. \tag{14.106}$$

With the eigenvalues given by $\{e^{i\theta_j}\}_{j=1,\dots,N}$, $-\pi < \theta_j \le \pi$,

$$\mathrm{Re}\,\log\Lambda(-1) = \sum_{j=1}^N \log|1+e^{i\theta_j}|, \qquad \mathrm{Im}\,\log\Lambda(-1) = \frac{1}{2}\sum_{j=1}^N \theta_j$$

and so

$$\hat{P}(k,l) := \int_{-\infty}^{\infty} ds\, e^{iks} \int_{-\infty}^{\infty} dt\, e^{ilt} P(s,t) = \Big\langle \prod_{j=1}^N e^{il\theta_j/2}|1+2\cos\theta_j|^{ik/2}\Big\rangle_{U(N)}.$$

This is precisely the average (14.102) in the case $\beta = 2$ with $b = l/2$, $a = ik/2$, and so has the same evaluation as specified by (14.103). According to (14.104), for $N \to \infty$

$$\hat{P}(k,l) \sim e^{-\log N(k^2+l^2)/4}$$

(for a derivation of this result using the theory of Section 2.8, see [473]), so the following result holds true [354].

PROPOSITION 14.5.4 *Introduce the scaled logarithm of the characteristic polynomial,*

$$\log\tilde{\Lambda}(z) = \Big(\frac{2}{\log N}\Big)^{1/2}\log\Lambda(z),$$

and denote the joint distribution of the real and imaginary parts of $\log\tilde{\Lambda}(-1)$ *by* $\tilde{P}(s,t)$. *We have*

$$\lim_{N\to\infty} \tilde{P}(s,t) = \frac{1}{2\pi}e^{-(s^2+t^2)/2},$$

and thus the limiting distribution of the values of $\log\tilde{\Lambda}(-1)$ *is the complex Gaussian* $\mathrm{N}[0,1] + i\mathrm{N}[0,1]$.

The *Keating-Snaith hypothesis* [354] relates this result to the value distribution of the Riemann zeta function on the critical line. The former is an extension of the Montgomery-Odlyzko law (recall Section 7.5.1), and relates the value distribution of $\zeta(\frac{1}{2}+it)$ to the value distribution of $\Lambda(z)$. In particular it is hypothesized that for large t the value distribution of $\log\zeta(\frac{1}{2}+it)$ is well described by the value distribution of $\log\Lambda(z)$ for $N \times N$ unitary random matrices with $N = \log t/2\pi$ (this ensures that the mean spacing between zeros and between eigenvalues is asymptotically equal, but on the other hand is not the identification used in (7.89)). Evidence for this hypothesis is a result of Selberg [516] which gives that for any rectangle $B \in \mathbb{C}$

$$\lim_{T\to\infty}\frac{1}{T}\left|\left\{t : T \le t \le 2T,\ \frac{\log\zeta(1/2+it)}{\sqrt{\frac{1}{2}\log\log T}} \in B\right\}\right| = \frac{1}{2}\iint_B e^{-(x^2+y^2)/2}\,dxdy, \qquad (14.107)$$

thus exhibiting the complex Gaussian behavior for the logarithm of the characteristic polynomials seen in Proposition 14.5.4 for $U(N)$, with the identification

$$N = \log T. \qquad (14.108)$$

The value distribution of $|\Lambda(z)|$ itself is also of interest. Here the corresponding Fourier transform does not lead to a tractable integral. Instead one observes that for $p(s)$ a distribution with support $s > 0$, knowledge of the Mellin transform (complex moments)

$$m(\lambda) := \int_0^\infty s^{\lambda-1}p(s)\,ds$$

as a function in the complex plane gives

$$p(s) = \frac{1}{2\pi i}\int_{c-i\infty}^{c+i\infty} s^{-x}m(x)\,dx$$

via the inverse Mellin transform. Now, for random matrices from $U(N)$, and with $p(s)$ denoting the distribution of $|\Lambda(-1)|^2$,

$$m(a+1) = \Big\langle \prod_{l=1}^{N} |2 + 2\cos\theta_l|^a \Big\rangle_{U(N)}, \tag{14.109}$$

and this in turn has an explicit product of gamma function evaluation which follows from (14.103) (for a discussion of computing the corresponding Mellin transform, see [471]). The asymptotic formula (14.104) with $\beta = 2$, $r = s = 1$, $b = 0$ shows that for $N \to \infty$

$$N^{-a^2} m(a+1) \sim \frac{G^2(a+1)}{G(2a+1)}. \tag{14.110}$$

An application to zeta function theory of this result has been given by Keating and Snaith [354], in keeping with their hypothesis relating to (14.107). Thus, after recalling the identification (14.108), one anticipates that the $U(N)$ result (14.110) should relate to the large T value of

$$\frac{1}{(\log T)^{a^2}} \frac{1}{T} \int_0^T \Big| \zeta\Big(\frac{1}{2} + it\Big)\Big|^{2a} dt. \tag{14.111}$$

In zeta function theory this was conjectured to have the structure $f(a)A(a)$, where $A(a)$ is the number theoretic quantity

$$A(a) = \prod_{\text{primes } p} \Big(\Big(1 - \frac{1}{p}\Big)^{\lambda^2} \sum_{m=0}^{\infty} \Big(\frac{\Gamma(a+m)}{m!\Gamma(a)}\Big)^2 p^{-m} \Big)$$

and

$$f(1) = 1, \quad f(2) = \frac{1}{12}, \quad f(3) = \frac{42}{9!}, \quad f(4) = \frac{24024}{16!}$$

(the first two such values are known rigorously), with the other values of $f(a)$ unknown. Indeed, these values are precisely those given by the r.h.s. of (14.110). Thus one is led to interpret this term as due to $U(N)$ like properties of the Riemann zeros, leading to the conjecture that for general a

$$f(a) = \frac{G^2(a+1)}{G(2a+1)}.$$

EXERCISES 14.5 1. (i) Note from the results of Exercises 5.1 q.3(i) and (ii) that

$$\Big\langle \prod_{l=1}^{N} e^{ib\theta_l} |2 + 2\cos\theta_l|^a (1 + xe^{-i\theta_l}) \Big\rangle_{\mathrm{CUE}_N} = \frac{M_N(a+b, a-b, 1)}{M_N(0,0,1)}\, {}_2F_1(-N, a+b; b-a-N; x).$$

(ii) Deduce from this the limiting behavior

$$\lim_{N\to\infty} \frac{M_N(0,0,1)}{M_N(a+b, a-b, 1)} \Big\langle \prod_{l=1}^{N} e^{ib\theta_l} |2 + 2\cos\theta_l|^a (1 + xe^{-i\theta_l}) \Big\rangle_{\mathrm{CUE}_N} = (1+x)^{-(a+b)}$$

and show that this is consistent with the prediction of the Fisher-Hartwig formula (14.100).

2. (i) Note that with $a(\theta)$ of Proposition 14.5.2 specified by $g(\theta) = 0$, $R = 2$, $\theta_1 = \phi_1$, $\theta_2 = \phi_2$, $a_1 = b_1 = \frac{1}{2}$, $a_2 = -b_2 = \frac{1}{2}$ we have

$$D_N[e^{a(\theta)}] = e^{iN(\phi_2 - \phi_1)} \Big\langle \prod_{l=1}^{N} (e^{i\theta_l} - e^{i\phi_1})(e^{-i\theta_l} - e^{-i\phi_2}) \Big\rangle_{\mathrm{CUE}_N}.$$

(ii) Use (5.89) and (5.87) to deduce from this that

$$D_N[e^{a(\theta)}] = e^{iN(\phi_2-\phi_1)/2}\frac{\sin((N+1)(\phi_2-\phi_1)/2))}{\sin((\phi_2-\phi_1)/2)}.$$

As this does not exhibit the $N \to \infty$ behavior (14.100), conclude that the region $|\mathrm{Re}(a_r)| < \frac{1}{2}$, $|\mathrm{Re}(b_r)| < \frac{1}{2}$ sufficient for the validity of the latter cannot in general be enlarged in all parameters.

3. [332] The objective of this exercise is to use the Szegö asymptotic formula (14.71) to anticipate features of the Fisher-Hartwig asymptotic formula (14.100).

(i) In the second expression of Proposition 14.5.2 for $a(\theta)$, replace the logarithms by their corresponding power series truncated at the Nth term, and read off that the Fourier coefficients of $a(\theta)$ are then given by

$$a_p = \begin{cases} g_p - \frac{1}{p}\sum_{r=1}^R (a_r+b_r)e^{-i\theta_r p}\chi_{p\le N}, & p \ge 1, \\ g_0, & p=0, \\ g_p - \frac{1}{|p|}\sum_{r=1}^R (a_r-b_r)e^{i\theta_r p}\chi_{|p|\le N}, & p \le -1. \end{cases}$$

(ii) Show from (i) that for N large

$$\sum_{p=1}^\infty p a_p a_{-p} \sim \sum_{p=1}^\infty p g_p g_{-p} - \sum_{r=1}^R (a_r - b_r) g_+(\theta_r) - \sum_{r=1}^R (a_r+b_r) g_-(\theta_r)$$
$$- \sum_{\substack{r,r'=1\\ r\ne r'}}^R (a_r+b_r)(a_{r'}-b_{r'})\log(1-e^{i(\theta_{r'}-\theta_r)}) + \log N \sum_{r=1}^R (a_r^2 - b_r^2).$$

Substitute this in (14.71) to reproduce (14.100) up to the product over the Barnes G-function in E.

4. [368] The aim of this exercise is to determine the distribution of the form factor for the log-gas on a circle. For this purpose, with $k = p/N, p \in \mathbb{Z}$ consider

$$\Big\langle \prod_{l=1}^N e^{i(u/\sqrt{N})\cos Nk\theta_l + i(v/\sqrt{N})\sin Nk\theta_l} \Big\rangle_{\mathrm{C}\beta\mathrm{E}_N}.$$

(i) Use the cumulant expansion (14.69), (14.70) to deduce that the logarithm of this average has the large N form

$$1 - \frac{u^2+v^2}{4}\int_{-\pi}^{\pi} \rho_{(2)}^T(\theta,0)\cos Nk\theta\, d\theta + \mathrm{O}\Big(\frac{1}{\sqrt{N}}\Big).$$

(ii) Conclude that the joint distribution of

$$(A,B) = \Big(\frac{1}{\sqrt{N}}\sum_{j=1}^N \cos Nk\theta_j, \frac{1}{\sqrt{N}}\sum_{j=1}^N \sin Nk\theta_j\Big)$$

is for large N the Gaussian

$$e^{-(A^2+B^2)/2\sigma^2}, \qquad \sigma^2 = \frac{1}{2}\int_{-\pi}^{\pi} \rho_{(2)}^T(\theta,0)\cos Nk\theta\, d\theta.$$

(iii) In the notation of (ii) note that

$$A^2+B^2 = \frac{1}{N}\Big|\sum_{j=1}^N e^{ikN\theta_j}\Big|^2$$

so that for $k \ne 0$

$$\langle (A^2+B^2)\rangle_{\mathrm{C}\beta\mathrm{E}_N} = \hat{S}(Nk) = \int_{-\pi}^{\pi} \rho_{(2)}^T(\theta,0)\cos Nk\theta\, d\theta.$$

Conclude from the result of (ii) that the distribution of $s = A^2 + B^2$ is, for large N, the exponential distribution with mean $\hat{S}(Nk)$.

5. [117, 235] With $U \in U(N)$, eigenvalues $\{e^{i\theta_l}\}_{l=1,\dots,N}$, introduce the characteristic polynomial as above (14.106), and in terms of this define

$$\mathcal{Z}(s) = e^{-\pi i N/2} e^{-i\sum_{n=1}^{N}\theta_n/2} z^{-N/2}\Lambda(z).$$

(note that $Z(s)$ is real for $|s| = 1$). Results from [117] give that for $k \in \mathbb{Z}^+$

$$N^{-k^2-2k}\langle|\mathcal{Z}'(1)|^{2k}\rangle_{U\in U(N)} \underset{N\to\infty}{\sim} \tilde{b}_k, \tag{14.112}$$

where, with $[x^j]f(x)$ denoting the coefficient of x^j in the power series expansion of $f(x)$,

$$\tilde{b}_k = (-1)^{k(k+1)/2}(2k)![x^{2k}]\Big(e^{-x/2}x^{-k^2/2}\det[I_{\alpha+\beta-1}(2\sqrt{x})]_{\alpha,\beta=1,\dots,k}\Big)$$

(cf. (14.110)).

(i) Interchange row β with row $k-\beta+1$ in the above determinant, then compare with (8.97) to conclude that

$$\tilde{b}_k = \frac{(-1)^k}{A(k,k)}(2k)![x^{2k}]\Big(e^{x/2}\tilde{E}_2^{\text{hard}(k)}((0,4x);\xi = 1;k)\Big), \qquad A(a,\mu) := a!\prod_{j=1}^{a}\frac{(j+\mu-1)!}{j!}.$$

(ii) Make use of (8.88) and Proposition 8.3.3 to conclude from (i) that

$$\tilde{b}_k = \frac{(-1)^k}{A(k,k)}(2k)![x^{2k}]\exp\Big(-\int_0^{4x}(\eta(s)+k^2)\frac{ds}{s}\Big)$$

where $\eta(s)$ satisfies the differential equation

$$(s\eta'')^2 + 4\Big((\eta')^2 - \frac{1}{64}\Big)(\eta - s\eta') - \frac{k^2}{4^2} = 0$$

subject to the requirement that it is even in s with $\eta(0) = -k^2$.

14.6 ASYMPTOTIC PROPERTIES OF $E_\beta(n;J)$ AND $P_\beta(n;J)$

14.6.1 Large s behavior of $E_\beta(0;s)$

In (9.90) results of a Coulomb gas argument predicting that for an eigenvalue free region J, $\log E_2(0;J)$ is to leading order in $|J|$ proportional to $|J|^3$, $|J|$ and $|J|^2$ at the soft edge, hard edge and in the bulk, respectively, were required. Here these results will be derived by combining macroscopic electrostatics with a scaling argument [189].

One considers the setting of the one-component log-gas confined to the half-line $x > 0$ with background charge density given by the power law $\rho_b(x) = -Ax^\mu$. The basic hypothesis is that for large s

$$E_\beta(0;s) \sim e^{-\beta E(s)}, \tag{14.113}$$

where $E(s)$ is the electrostatic energy due to excluding the mobile positive charge from the interval $(0,s)$. Note that $e^{-\beta E(s)}$ is the Boltzmann factor for the configuration with energy $E(s)$.

Now, let $\rho_{(1)}(x)$ denote the charge density of the positive charge ($\rho(x) = 0$ for $x \in [0,s]$), and let $\hat{\rho}(x)$ denote the total charge, $\hat{\rho}(x) = \rho(x) + \rho_b(x)$. The latter is constrained by the condition of global charge neutrality

$$\int_0^\infty \hat{\rho}(x)\,dx = 0. \tag{14.114}$$

With $\phi(x) = -\int_{-\infty}^{\infty} \hat{\rho}(y) \log|x-y|\, dy$ one then has

$$E(s) = \frac{1}{2} \int_0^\infty \hat{\rho}(x)\phi(x)\, dx. \tag{14.115}$$

For the region (s, ∞), where there are mobile charges, the potential is a constant which can be taken to be zero. Noting that for $x \in [0, s]$, $\hat{\rho}(x) = \rho_b(x) = -Ax^\mu$, and making use of (14.114) we see that this latter condition is consistent with the homogeniety property

$$\hat{\rho}(sx) = s^\mu \hat{\rho}(x). \tag{14.116}$$

Substituting in (14.115) gives

$$E(s) = Cs^{2u+2}, \qquad C = E(1) \tag{14.117}$$

and hence (14.113) gives the prediction

$$E_\beta(0; s) \sim e^{-\beta C s^{2\mu+2}}. \tag{14.118}$$

With $\mu = \frac{1}{2}, -\frac{1}{2}$ and 0 so as to give the known asymptotic form of the density at the soft edge, hard edge and in the bulk, respectively, found in Chapter 7, this gives the corresponding leading behavior of $E_2(0; s)$ as quoted in (9.90). We remark that the stochastic differential equation characterizations (13.186) and (13.180) of the β-ensembles at the soft edge and the bulk can be used to prove (14.118) in those cases, and furthermore to give the value of C [463], [524], [462]. As already noted, this method can also be used to deduce higher order terms, giving for example the large s expansion in the bulk (9.99).

14.6.2 Large s behavior of $E_\beta^{\rm bulk}(n; s)$

In the infinite log-gas with a constant background charge density $-\rho$, the leading large s asymptotics of $E_\beta^{\rm bulk}(n; s)$—the probability that an interval of length s contains exactly n eigenvalues—can be determined using an extension of the macroscopic electrostatics argument used above [152], [178]. As in (14.113) the basic hypothesis is that for large s

$$E_\beta^{\rm bulk}(n; s) \sim e^{-\beta \delta F}, \tag{14.119}$$

where here δF is the change in energy caused by changing the particle density so that the interval of length s contains n particles.

Analogous to the hypothesis used in Section 14.2, the change in energy δF is taken to consist of two parts — an electrostatic energy V_1 and a free energy V_2. These are calculated from the one-body density $\rho_{(1)}(x)$ according to

$$V_1 = \frac{1}{2} \int_{-\infty}^\infty (\rho_{(1)}(x) - \rho)\phi(x)\, dx, \tag{14.120}$$

where $\phi(x) = -\int_{-\infty}^\infty (\rho_{(1)}(y) - \rho) \log|x-y|\, dy$, and

$$V_2 = \int_{-\infty}^\infty \rho_{(1)}(x) \left(f_\beta[\rho_{(1)}(x)] - f_\beta[\rho] \right) dx. \tag{14.121}$$

Since, according to Proposition 4.8.1, the free energy per particle f_β is such that $f_\beta[\rho_{(1)}(x)] - f_\beta[\rho] = \left(\frac{1}{\beta} - \frac{1}{2}\right) \log(\rho_{(1)}(x)/\rho)$, we have

$$V_2 = \left(\frac{1}{\beta} - \frac{1}{2}\right) \int_{-\infty}^\infty \rho_{(1)}(x) \log(\rho_{(1)}(x)/\rho)\, dx. \tag{14.122}$$

The potential $\phi(x)$ and density $\rho_{(1)}(x)$ are calculated via two-dimensional macroscopic electrostatics. For

$n < \rho s$ we suppose the n particles are confined to an interval $(-b, b)$ and within that interval the system behaves like a conductor so $\phi(x) = -c$ (constant). The quantities b and c are to be calculated. The remaining particles are confined to an open-ended region $(-\infty, -t)$ and (t, ∞), where $2t = s$. These regions are conductors with $\phi(x) = 0$. The remaining intervals, $(-t, -b)$ and (b, t), contain no mobile particles and so behave as an insulator on which ϕ will vary continuously between 0 and $-c$. The density $\rho_{(1)}(x)$ is further constrained by the conditions

$$\int_{-b}^{b} \rho_{(1)}(x)\, dx = n, \qquad \int_{-\infty}^{\infty} (\rho_{(1)}(x) - \rho) dx = 0. \tag{14.123}$$

Substituting the constraints on $\phi(x)$ in (14.120), together with the condition $\rho_{(1)}(x) = 0$ for $b < |x| < t$ and the first condition in (14.123), then integrating by parts gives

$$V_1 = -\frac{cn}{2} + \rho \int_b^t x \frac{d\phi(x)}{dx} dx. \tag{14.124}$$

To determine $d\phi/dx$ we introduce the complex electric field $E(z) = -\partial\phi(z)/\partial x + i\partial\phi(z)/\partial y$, so that $d\phi/dx = -\mathrm{Re}E(x)$ (recall Section 1.4.3). Now $E(z)$ is required to be an analytic function of z in the upper half-plane (its value in the lower half-plane is given by symmetry), with real part which vanishes on the conducting regions of the real line. Furthermore, $E(z)$ must have branch points at $z = \pm t, \pm b$, it must vanish as $|z| \to \infty$ (since there is no net charge) and as $t \to \infty$ with $b = 0$ it must equal $\pi i \rho$ (since then the real line is an insulator with uniform charge density ρ and so $\phi(z) = \pi y$; recall Section 2.7.1). The unique function with these properties is

$$E(z) = \pi\rho \left[i - \left(\frac{z^2 - b^2}{t^2 - z^2} \right)^{1/2} \right],$$

where the square root is chosen to be positive real on the real axis between $(-t, -b)$ and (b, t). Taking the real part and changing sign gives that for $x \in (b, t)$

$$\frac{d\phi}{dx} = \pi\rho \left(\frac{x^2 - b^2}{t^2 - x^2} \right)^{1/2}, \tag{14.125}$$

while the density in the interval $(-b, b)$ is given by

$$\rho_{(1)}(x) = -\frac{1}{2\pi i} \left(E^+(x) - E^-(x) \right) = \rho \left(\frac{b^2 - x^2}{t^2 - x^2} \right)^{1/2}. \tag{14.126}$$

Note that b is specified by the first equation in (14.123) with the substitution of (14.126). An integral identity in [270] allows the resulting equation to be rewritten

$$n = 2\rho t \Big(E(k') - k^2 K(k') \Big), \tag{14.127}$$

where K and E are the complete elliptic integrals of the first and second kinds and $k^2 = 1 - k'^2$ with $k' = b/t$.

The explicit formula (14.125) allows the evaluation of (14.124),

$$V_1 = -\frac{cn}{2} + \frac{\pi^2 \rho^2}{4} (t^2 - b^2). \tag{14.128}$$

Also, integrating (14.125) from b to t and recalling that $\phi(t) = 0$, $\phi(b) = -c$, gives

$$c = \pi\rho \int_b^t \left(\frac{x^2 - b^2}{t^2 - x^2} \right)^{1/2} dx = \pi\rho t \Big(E'(k') - k'^2 K'(k') \Big). \tag{14.129}$$

The integral in (14.121) defining V_2 can also be calculated exactly [152].

Proposition 14.6.1 *One has*

$$V_2 = \Big(\frac{1}{\beta} - \frac{1}{2}\Big)c,$$

where $c = -\phi(b) = -\phi(x)$, $x \in [-b, b]$.

Proof. Define a function $F(z)$, analytic in the upper half-plane, by

$$F(z) = -\frac{1}{2\pi i}(E(z) - E(\bar{z})) + \frac{i}{2\pi}(E(z) + \overline{E(z)}) = \rho + \frac{i}{\pi}E(z)$$

(on the real axis $F(x) = \rho_{(1)}(x) - \frac{i}{\pi}\frac{d\phi}{dx}$). Since $\log(F(z)/\rho)$ is also analytic in the upper half-plane, and decays of order $1/|z|^2$ as $|z| \to \infty$, the residue theorem gives

$$\int_{-\infty}^{\infty} F(x)\log(F(x)/\rho)dx := \int_{-\infty}^{\infty}\Big(\rho_{(1)}(x) - \frac{i}{\pi}\frac{d\phi}{dx}\Big)\log\Big(\Big(\rho_{(1)}(x) - \frac{i}{\pi}\frac{d\phi}{dx}\Big)/\rho\Big)dx = 0. \qquad (14.130)$$

Now for $x \notin (-t, -b), (b, t)$ we know that $d\phi/dx = 0$ while for x in these intervals $\rho_{(1)}(x) = 0$, $d\phi/dx$ is positive and $\log\big(\rho_{(1)}(x) - \frac{i}{\pi}\frac{d\phi}{dx}\big) = \log\big(\frac{1}{\pi}\frac{d\phi}{dx}\big) - i\pi/2$. Taking the real part of (14.130) we therefore have

$$\int_{-\infty}^{\infty}\rho_{(1)}(x)\log(\rho_{(1)}(x)/\rho)dx = \frac{1}{2}\Big(\int_{-t}^{-b} + \int_{b}^{t}\Big)\frac{d\phi}{dx}dx = -\frac{1}{2}(\phi(-b) + \phi(b)),$$

and the stated result follows upon recalling (14.122). □

We see that V_1 and V_2 are completely specified by (14.128), (14.129) and Proposition 14.6.1. Hence $E_\beta^{\rm bulk}(n; s)$ is completely specified — all that remains is to calculate the large s behavior of $\log E_\beta(n; s)$. For this purpose, note that substituting (14.126) in the first equation of (14.123) and expanding for $t \gg b$ shows

$$b \sim \Big(\frac{2nt}{\pi\rho}\Big)^{1/2}. \qquad (14.131)$$

Since $k' = b/t$, we see that $(k')^2 \sim (2n/\pi\rho t)$. From the known asymptotic expansions for $E'(k')$ and $K'(k')$ as $k' \to 0$, (14.129) then gives that

$$c \sim \pi\rho t - \frac{n}{2}\Big(\log\frac{8\pi\rho t}{n} + 1\Big).$$

Substituting this result in the expressions (14.128) and Proposition 14.6.1 for V_1 and V_2, then substituting the sum for δF in (14.119) shows

$$\log E_\beta^{\rm bulk}(n; 2t) \underset{\substack{t,n\to\infty\\ t\gg n}}{\sim} -\beta\frac{(\pi\rho t)^2}{4} + \Big(\beta n + \frac{\beta}{2} - 1\Big)\pi\rho t + \frac{n}{2}\Big(1 - \frac{\beta}{2} - \frac{\beta n}{2}\Big)\Big(\log\frac{8\pi\rho t}{n} + 1\Big), \qquad (14.132)$$

and this in turn implies

$$\frac{E_\beta^{\rm bulk}(n; 2t)}{E_\beta^{\rm bulk}(0; 2t)} \underset{\substack{t,n\to\infty\\ t\gg n}}{\sim} \tilde{c}_{\beta,N}\frac{e^{\beta n\pi\rho t}}{(\pi\rho t)^{\beta n^2/4 + (\beta/2-1)n/2}} \qquad (14.133)$$

for some $\tilde{c}_{\beta,N}$. We recognize this latter form as precisely the conjectured asymptotic result (9.103) for the same ratio with n fixed, and we recognize (14.132) with $n = 0$ as agreeing with the first two terms of (9.99).

Also of interest is the case $0 \ll \rho s - n \ll \rho s$ because of its relationship to charge fluctuations, and in particular the formula (14.89). This case is considered in Exercises 14.6 q.1.

14.6.3 $E_{\beta,N}(0;J)$ and $p_{\beta,N}(0;J)$ for large $N, |J|$

Log-gas arguments can also be used to study the asymptotics of gap probabilities and spacing distributions when N is large but finite. Here one is typically concerned with *large deviations*, when the size of the gap or spacing (measured in appropriate units) depends on N. Moreover, a rigorous justification of the log-gas heuristics can in some circumstances be given [329].

The simplest situation of this type is the gap probability $E_{\beta,N}(0;(-\alpha,\alpha))$ in the finite N circular β-ensemble. Here $0<\alpha<\pi$ is an angle on the circle, and so on average the interval $(-\alpha,\alpha)$ would contain $\alpha N/\pi$ eigenvalues if not constrained to be a gap. According to the hypothesis (14.113)

$$E_{\beta,N}(0;(-\alpha,\alpha)) \sim \exp\Big(\frac{\beta}{2}\int_\alpha^{2\pi-\alpha} d\theta_1\,\hat\rho(\theta_1)\int_\alpha^{2\pi-\alpha} d\theta_2\,\hat\rho(\theta_2)\log|e^{i\theta_1}-e^{i\theta_2}|\Big), \tag{14.134}$$

where $\hat\rho(\theta) := \rho_{(1)}(\theta) - N/2\pi$. An electrostatics evaluation of $\hat\rho(\theta)$ and the corresponding electrostatic energy is known [147], but we can in fact bypass such a calculation. The reason is that through the Toeplitz determinant asymptotic formula (9.123) we have knowledge of the electrostatic energy in the case $\beta=2$. Since the electrostatic energy is independent of β this gives

$$E_{\beta,N}(0;(-\alpha,\alpha)) \sim \exp\Big(\frac{\beta}{2}N^2\log\cos\frac{\alpha}{2}\Big). \tag{14.135}$$

The next situation to be considered is the right tail large deviation form of the distribution of the largest eigenvalue in the Gaussian β-ensemble, scaled so that the leading order support is the interval $(-1,1)$. One sees from (8.73) and (8.86) that in the large s asymptotic regime

$$p_{\beta,N}(0;(s,\infty)) \sim \rho_{(1),N}(s). \tag{14.136}$$

With $s>1$, and thus outside of the interval of leading support, the density is known from (14.82) in the case $\beta=1$, and so we read off that for general $\beta>0$

$$p_{\beta,N}(0;(s,\infty)) \sim \exp\Big(-\beta N\Big(s\sqrt{s^2-1}+\log(s-\sqrt{s^2-1})\Big)\Big). \tag{14.137}$$

In the case of the scaled Laguerre ensemble the corresponding asymptotic form of the distribution of the largest eigenvalue outside the leading support is similarly given by (14.83).

Finally we consider the left tail large deviation form of the distribution of the largest eigenvalue in the Gaussian β-ensemble, again scaled so that the leading order support is the interval $(-1,1)$. This is calculated from the log-gas formula

$$p_{\beta,N}(0;(s,\infty)) \sim e^{-\beta(U(s)-U(1))}, \qquad s<1,$$

$U(s)$ is the leading order in N portion of the total potential energy for the scaled Gaussian β-ensemble constrained so that the eigenvalues are restricted to be less than s. The working of Exercises 1.4 q.4 gives that such a constraint corresponds to a background density (1.81), and we see from (1.78) that to leading order in N

$$U(s) = \frac{1}{4}\int_{\sqrt{2N}(s-l)}^{\sqrt{2N}s} x^2\rho_b(x)\,dx - \frac{CN}{2},$$

where C is given by (1.80). Computing the integral making use of (4.2) then gives that for large N and $s<1$ [124]

$$p_{\beta,N}(0;(s,\infty)) \sim \exp\Big(-\beta N^2\Big(\frac{2s^2}{3}-\frac{s^4}{27}-\frac{5}{18}s\sqrt{3+s^2}-\frac{1}{27}s^3\sqrt{3+s^2}-\frac{1}{2}\log\frac{s+\sqrt{s^2+3}}{3}\Big)\Big). \tag{14.138}$$

14.6.4 Covariance and variance of spacing distributions in the bulk

For a one-dimensional system, the covariance of particular spacing configurations can be related to the variance of related configurations [100].

PROPOSITION 14.6.2 *In the bulk of a one-dimensional system, choose a particular particle at position x_i, and denote the positions of the successive particles to the right by $x_{i+1}, x_{i+2}, \dots$. Then, independent of i,*

$$p_\beta(n;s) = \langle \delta(s_i(n) - s) \rangle, \quad s_i(n) := x_{i+n+1} - x_i \tag{14.139}$$

($s_i(n)$ will be referred to as the nth order spacing), while

$$\begin{aligned}\mathrm{Cov}(s_i(n), s_{i+r+1}(n')) = &\mathrm{Var}(s_i(|r-n|-1)) + \mathrm{Var}(s_i(r+n'+1)) \\ &-\mathrm{Var}(s_i(|r+n'-n+1|-1)) - \mathrm{Var}(s_i(r))\end{aligned} \tag{14.140}$$

with the convention that $s_i(-1) = 0$.

Proof. The first formula is essentially the definition of $p_\beta(n;s)$. For the second formula, we note from the identity

$$(z_1 - z_2)(z_3 - z_4) = \frac{1}{2}\Big((z_1 - z_4)^2 + (z_2 - z_3)^2 - (z_1 - z_3)^2 - (z_2 - z_4)^2\Big)$$

that in general

$$\mathrm{Cov}(z_1 - z_2, z_3 - z_4) = \frac{1}{2}\Big(\mathrm{Var}(z_1 - z_4) + \mathrm{Var}(z_2 - z_3) - \mathrm{Var}(z_1 - z_3) - \mathrm{Var}(z_2 - z_4)\Big).$$

The covariance formula (14.140) now follows from the difference formula (14.139) for $s_i(n)$. □

A noteworthy special case of (14.140) is $n = n' = 0$, which gives the covariance of two nearest neighbor spacings (to the right) with r particles in between as

$$\mathrm{Cov}(s_i(0), s_{i+r+1}(0)) = \frac{1}{2}\Big(\mathrm{Var}(s_i(r+1)) - 2\mathrm{Var}(s_i(r)) + \mathrm{Var}(s_i(r-1))\Big) \tag{14.141}$$

(an alternative derivation of this result is given in Exercises 14.6 q.2). Another special case of interest is $r = n = n'$. This represents the covariance of successive nth order spacings, and (14.140) then reads

$$\mathrm{Cov}(s_i(n), s_{i+n+1}(n)) = \frac{1}{2}\Big(\mathrm{Var}(s_i(2n+1)) - 2\mathrm{Var}(s_i(n))\Big). \tag{14.142}$$

To make use of the above formulas the value of $\mathrm{Var}(s_i(n))$ is required. For the random matrix couplings $\beta = 1, 2, 4$, and for small values of n, these can be read off from Tables 8.13, 8.14 and the inter-relation (8.161). The corresponding covariances are listed in Table 14.1. In general, since $s_i(n)$ is not a linear statistic, $\mathrm{Var}(s_i(n))$ is not easily accessible. However, physically one would expect $\mathrm{Var}(s_i(n))$ to be asymptotically equal to $\mathrm{Var}(\chi_{[0,n+1]})$ (here we are assuming unit density), where the latter is the variance in the number of particles in an interval of length $n+1$. Using the explicit asymptotic formula (14.85), we thus expect for the one-component log-gas that

$$\mathrm{Var}(s_i(n)) \underset{n\to\infty}{\sim} \frac{2}{\pi^2 \beta} \log n + \mathrm{O}(1). \tag{14.143}$$

Substituting in (14.141) then gives

$$\mathrm{Cov}(s_i(0), s_{i+r+1}(0)) \underset{r\to\infty}{\sim} -\frac{1}{\pi^2 \beta} \frac{1}{r^2}, \tag{14.144}$$

while substituting in (14.142) shows

$$\mathrm{Cov}(s_i(n), s_{i+n+1}(n)) \underset{n\to\infty}{\sim} -\frac{1}{\pi^2 \beta} \log n. \tag{14.145}$$

	$A_n(1)$	$A_n(2)$	$A_n(4)$	$B_n(1)$	$B_n(2)$
$n=0$	-0.077333811	-0.05550500	-0.034643119	-0.077333811	-0.05550500
$n=1$	-0.024900730	-0.013895003	-0.007118315	-0.138572479	-0.089255975
$n=2$	-0.011437207	-0.005959174	-0.002983845	-0.1784829481	-0.10983253094
$n=3$	-0.006455611	-0.003288239	-0.001640118	-0.20745438515	-0.12449482474
$n=4$	-0.004124830	-0.0020817292		-0.23007507345	-0.13586646270
$n=5$	-0.002857751	-0.001435997			
$n=6$	-0.002095049	-0.0010503631			
$n=7$	-0.001601116	-0.0008017114			
$n=8$	-0.0012631931	-0.0006320362			

Table 14.1 Tabulation of $A_n(\beta) := \mathrm{Cov}(s_i(0), s_{i+n+1}(0))$, $B_n(\beta) := \mathrm{Cov}(s_i(n), s_{i+n+1}(n))$ for small values of n and the random matrix couplings $\beta = 1, 2$ and 4. Note that in general $A_0(\beta) = B_0(\beta)$.

Even with $r = 3$ the formula (14.144) accurately approximates the exact value given in Table 14.1. More generally we would expect the full distribution of $s_i(n)$ to be asymptotically equal to that of $\chi_{[0,n+1]}$. It should therefore obey the Gaussian law (14.89). This trend is already seen in the small n data of Tables 8.13, 8.14, in that the skewness and kurtosis is typically decreasing as n increases.

Further remarks can be made in relation to both (14.144) and (14.145). Consider first (14.145). Together with (14.143) it implies

$$\mathrm{Cov}\Big(s_{i-n-1}(n), s_i(n)\Big) + \mathrm{Var}(s_i(n)) + \mathrm{Cov}\Big(s_i(n), s_{i+n+1}(n)\Big) \sim \mathrm{O}(1). \tag{14.146}$$

This is essentially a charge neutrality result, which says that the charge excess in the interval containing the first $n+1$ particles will be compensated by the excess of opposite sign in the two neighboring intervals of $n+1$ particles.

Regarding (14.144), we note that the asymptotic behavior is identical to the leading non-oscillatory behavior of the charge-charge correlation as given by (14.9). The physical reasoning which led to (14.143) is consistent with this result. An analogous result is that

$$\mathrm{Cov}(\chi_{[0,1]}, \chi_{[n,n+1]}) = \frac{1}{2}\Big(\mathrm{Var}(\chi_{[0,n+1]}) - 2\mathrm{Var}(\chi_{[0,n]}) - \mathrm{Var}(\chi_{[0,n-1]})\Big),$$

which according to (14.85) also exhibits the asymptotic behavior (14.144).

Analytic information is also available on $\mathrm{Var}(s_i(n))$ for finite n. In particular, French et al. [243] (see also Exercises 14.6 q.3) have refined the physical relationship between $\mathrm{Var}(s_i(n))$ and $\mathrm{Var}(\chi_{[0,n+1]})$ by compensating for the fact that fixed particles are present at the endpoints in the definition of $s_i(n)$ but not in $\chi_{[0,n+1]}$. Consequently they have deduced the formula

$$\mathrm{Var}(s_i(n)) \approx \mathrm{Var}(\chi_{[0,n+1]}) - \frac{1}{6}. \tag{14.147}$$

The accuracy of this formula for the log-gas at $\beta = 1, 2$ and 4, and small values of n, can be tested by reading off from Tables 8.13, 8.14 and the inter-relation (8.161), and comparing with the corresponding value of $\mathrm{Var}(\chi_{[0,n+1]})$ deduced from (14.84). The results displayed in Table 14.2 demonstrate an accuracy of up to 1 part in 10^4. However for larger n the discrepancy increases; for example we have $c_9(2) = 0.1585$.

Related to the topics of this section is the large k form of $p_\beta^{\mathrm{soft}}(k; s)$, the latter being the distribution of the

n	$c_n(1)$	$c_n(2)$	$c_n(4)$
0	0.1606	0.1640	0.1545
1	0.1670	0.1664	0.1603
2	0.1674	0.1665	0.1622
3	0.1672	0.1664	0.1631
4	0.1644	0.1663	0.1635

Table 14.2 Tabulation of $c_n(\beta) := \mathrm{Var}(\chi_{[0,n+1)} - \mathrm{Var}(s_i(n))$ for the random matrix couplings $\beta = 1, 2$ and 4. The approximation (14.147) predicts $c_n(\beta) \approx 1/6$.

kth largest eigenvalue at the soft edge. To leading order the mean μ_k of this distribution must satisfy

$$k \sim \int_{-\infty}^{\mu_k} \rho_{(1)}^{\rm soft}(X)\,dX \sim \frac{2}{3\pi}|\mu_k|^{3/2},$$

where the second asymptotic form follows from (13.68). It is proved in [281], for $\beta = 2$, that with σ_k^2 proportional to $(\log k)/(\beta k^{2/3})$, as $k \to \infty$

$$(X_k + |\mu_k|)/\sigma_k \sim \mathrm{N}[0,1].$$

This is consistent with the data of Tables 9.1 and 9.2.

14.6.5 The Δ_3 statistic

In the bulk of the spectrum, after appropriate scaling, the mean eigenvalue spacing is a constant $1/\rho$. Thus the mean number of eigenvalues $\langle n(x)\rangle$ from some (arbitrary) origin to a point x in the spectrum increases linearly with x. The actual number of eigenvalues between 0 and x, which is a staircase function that jumps one unit at the position of each eigenvalue, will deviate from a straight line. On the other hand, there will be a unique line of best fit according to the criterion of least square deviation. For such a straight line within an interval of length ℓ, Dyson and Mehta [153] have introduced this deviation as a statistic, denoted Δ_3 (the subscript 3 occurs because two similar statistics, denoted Δ_1 and Δ_2, were also introduced), characterizing the eigenvalue spectrum. Explicitly

$$\Delta_3 := \min_{\mathrm{A,B}} \frac{1}{\ell}\int_{-\ell/2}^{\ell/2} (n(y) - Ay - B)^2\,dy, \tag{14.148}$$

where $n(y)$ measures the number of eigenvalues from some arbitrary point a large distance from the interval $(-\ell/2, \ell/2)$. Equating the partial derivatives with respect to A and B in (14.148) to zero to calculate the minimum shows

$$A = \frac{12}{l^3}\int_{-\ell/2}^{\ell/2} y n(y)\,dy, \qquad B = \frac{1}{l}\int_{-\ell/2}^{\ell/2} n(y)\,dy.$$

Minor manipulation then gives

$$\begin{aligned}\Delta_3 &= \frac{1}{\ell}\int_{-l/2}^{l/2} du \int_{-\ell/2}^{\ell/2} dv \left(\delta(v-u)n(u)n(v) - \frac{12}{l^3}uvn(u)n(v) - \frac{1}{\ell}n(u)n(v)\right) \\ &= \frac{1}{2\ell^2}\int_{-\ell/2}^{\ell/2} du \int_{-\ell/2}^{\ell/2} dv \left(1 + \frac{12uv}{l^2}\right)\Big(n(u) - n(v)\Big)^2.\end{aligned}$$

Thus if we let $N(|u-v|)$ denote the variance of the number of particles in the interval $|u-v|$, then by taking the ensemble average we see that

$$\langle \Delta_3 \rangle = \frac{1}{2\ell^2} \int_{-\ell/2}^{\ell/2} du \int_{-\ell/2}^{\ell/2} dv \left(1 + \frac{12uv}{l^2}\right) N(|u-v|).$$

Finally, changing variables $u - v = y$, $u + v = z$ and performing the integration over z give [440]

$$\langle \Delta_3 \rangle = \frac{2}{\ell} \int_0^\ell \left(\left(\frac{y}{\ell}\right)^3 - 2\frac{y}{\ell} + 1 \right) N(y)\, dy. \tag{14.149}$$

Substituting the asymptotic expansion (14.85) (which assumes $\rho = 1$) in this expression shows that for large ℓ

$$\langle \Delta_3 \rangle \sim \frac{1}{\pi^2 \beta} \log \ell - \frac{9}{4\pi^2 \beta} + \frac{B_\beta}{2}.$$

EXERCISES 14.6 1. [152], [178] In this exercise the asymptotic form of $E_\beta(n;s)$ will be computed in the case $0 \ll s - n \ll s$ using the theory which led to (14.132).

(i) For $0 \ll \rho s - n \ll \rho s$ $(s = 2t)$, $k' = b/t \sim 1$. Thus use known expansions of $E(k)$ and $K(k)$ for k near 0 and 1 in the appropriate formulas given in the text to deduce that

$$\rho b \sim \frac{\rho s}{2} + \frac{n - \rho s}{\log(\rho s/(n - \rho s))}, \qquad c \sim \pi^2 \frac{(\rho s - n)}{2 \log(\rho s/(n - \rho s))}.$$

(ii) Substitute the results of (i) in (14.128) and Proposition 14.6.1 to deduce from (14.119) that in the region in question

$$E_\beta(n;s) \sim \exp\left(- \frac{\pi^2 \beta}{4 \log \rho s} (\rho s - n)^2 \right),$$

and relate this to (14.89).

2. (i) Use the formula

$$p_\beta(n;s) = \frac{d^2}{ds^2} \sum_{j=0}^{n} (n - j + 1) E_\beta(j;s),$$

which follows from (8.16), to show that

$$\int_0^\infty s^2 p_\beta(n;s)\, ds = 2 \sum_{j=0}^{n} (n - j + 1) I_\beta(j), \qquad I_\beta(p) := \int_0^\infty E_\beta(p;s)\, ds.$$

(ii) Use the identities

$$(s_i(n))^2 = \Big(\sum_{j=0}^{n} s_{i+j}(0) \Big)^2 = \sum_{j=0}^{n} \Big(s_{i+j}(0) \Big)^2 + 2 \sum_{0 \le j < k \le n} s_{i+j}(0) s_{i+k}(0)$$

to deduce that

$$\int_0^\infty s^2 p_\beta(n;s)\, ds = (n+1) \langle s_i(0) \rangle^2 + 2 \sum_{j=1}^{n} (n - j + 1) \langle s_i(0) s_{i+j}(0) \rangle \tag{14.150}$$

and compare this with the result of (i) for $n = 0, 1, 2, \ldots$ successively to deduce

$$\langle s_i^2(0) \rangle = 2 I_\beta(0), \qquad \langle s_i(0) s_{i+j}(0) \rangle = I_\beta(j) \ (j \ge 1).$$

Rewrite (14.150) in the form

$$\mathrm{Var}\Big(s_i(n)\Big) = (n+1)\mathrm{Var}\Big(s_i(0)\Big) + 2\sum_{j=1}^{n} j\mathrm{Cov}\Big(s_i(0), s_{i+n+1-j}(0)\Big)$$

and use this inductively to deduce (14.141).

3. [243] Let $n(x)$ represent the microscopic number of particles between 0 and x, where the origin 0 is chosen in the bulk. Then $n(x)$ is a staircase function, increasing by 1 each time x is increased past a particle coordinate. Let $n^{(C)}(x)$ denote some continuous approximation to $n(x)$, and let $\epsilon(x)$ denote the correction at a particle coordinate x_i, so that $n^{(C)}(x_i) = n(x_i) + \epsilon(x_i)$ with $|\epsilon(x_i)| \le \frac{1}{2}$. Assuming, as is reasonable, that $\epsilon(x)$ is uniformly distributed and $n(x)$ and $\epsilon(x)$ are uncorrelated, show that

$$\Big\langle (n^{(C)}(x_i) - n^{(C)}(x_j)) \Big\rangle = \Big\langle (n(x_i) - n(x_j)) \Big\rangle + \frac{1}{6}.$$

(This result forms the basis for the approximate formula (14.147).)

4. [124] Show that for $s \to 1^-$, the r.h.s. of (14.138) is to leading order equal to $e^{-\beta N^2 u^3/3}$, where $u = 1 - s$. Interpret $\sqrt{2N}u$ as the displacement $\sqrt{2N} - x$ from the mean position of the largest eigenvalue of the Gaussian β-ensemble without scaling, then write $\sqrt{2N}u = X/\sqrt{2}N^{1/6}$ as is consistent with the soft edge scaling (7.11) to reclaim the leading order term of (9.156).

5. [533], [497] Consider eigenvalues from CUE$_N$. Let Z_N be the minimum of all the N spacings between consecutive eigenvalues. It is known that

$$\lim_{N\to\infty} \Pr\Big(Z_N \Big(\frac{N^4}{72\pi}\Big)^{1/3} > x \Big) = e^{-x^3}.$$

The aim of this exercise is to give an heuristic prediction of this result.

(i) Let $\hat{Z}_N(x)$ denote the random variable for the number of consecutive spacings less than x. Note that $\Pr(\hat{Z}_N(x) = 0) = \Pr(Z_N > x)$. Integrate the leading term of the expansion (8.165) with $\rho = N/2\pi$ to deduce that as $x \to 0$, the probability that a single spacing is less than x is equal to

$$c_\beta (Nx)^{1+\beta}, \qquad c_\beta = \frac{(\beta/2)^\beta((\beta/2)!)^3}{2\pi(\beta+1)!(3\beta/2)!}.$$

(ii) From (i), by assuming that to some degree of approximation such spacings are independent, conclude

$$\Pr(\hat{Z}_N(x) = 0) \approx (1 - c_\beta(Nx)^{1+\beta})^N \approx e^{-c_\beta N (Nx)^{1+\beta}}.$$

Now scale x and set $\beta = 2$ to obtain the result.

6. [169] Consider a sequence of $2N+1$ particles on a circle of radius $(2N+1)/2\pi$, label them by their scaled angles

$$N + \frac{1}{2} > x_n > \cdots x_1 > x_0 > x_{-1} > \cdots > -\Big(N + \frac{1}{2}\Big)$$

and consider that statistic $\delta_n := x_n - x_0 - n$. For $p = \pm 1, \pm 2, \dots, \pm N$ define the corresponding power spectrum

$$\mathcal{P}(p) = \frac{1}{2N+1}\Big| \sum_{n=-N}^{N} \delta_n e^{-2\pi i p n/(2N+1)} \Big|^2.$$

As a long wavelength ($|p| \ll N$, $N \gg 1$), continuum approximation write $\delta_n \approx \int_0^x (n_{(1)}(y) - y)\, dy$, where $n_{(1)}(y)$ is the microscopic density, so that

$$\begin{aligned}\mathcal{P}(p) &\approx \frac{1}{2N}\Big| \int_{-N}^{N} \Big(\int_0^x (n_{(1)}(y) - y)\, dy \Big) e^{-2\pi i k x}\, dx \Big|^2 \\ &\approx \frac{1}{2N}\Big(\frac{1}{2\pi k}\Big)^2 \Big| \int_{-N}^{N} n_{(1)}(x) e^{-2\pi i k x}\, dx \Big|^2,\end{aligned}$$

where the second line follows by integration by parts and the assumption that N is large. Deduce from this that

$$\langle \mathcal{P}(p) \rangle \sim \Big(\frac{1}{2\pi k}\Big)^2 \hat{S}(2\pi k),$$

where $\hat{S}$ refers to the structure function. In the case of the log-gas, note from (14.8) that this implies

$$\langle \mathcal{P}(p) \rangle \sim \frac{1}{2\pi^2 \beta |k|},$$

thus exhibiting $1/f$ noise.

14.7 DYNAMICAL CORRELATIONS

14.7.1 Definition

In dynamical many-body systems, the current correlations are fundamental quantities closely related to the density correlations. In particular the macroscopic one-body current $J(x;\tau)$ is related to the one-body dynamical particle density $\rho_{(1)}(x;\tau)$ via the continuity equation

$$\frac{\partial}{\partial \tau} \rho_{(1)}(x;\tau) = -\frac{\partial}{\partial x} J(x;\tau). \tag{14.151}$$

The macroscopic one-body current is usually defined as the averaged value of the classical microscopic current

$$j_\tau(x) = \sum_{j=1}^N \frac{dx_j(\tau)}{d\tau} \delta(x - x_j(\tau)). \tag{14.152}$$

The latter satisfies

$$\frac{\partial}{\partial \tau} n_\tau(x) = -\frac{\partial}{\partial x} j_\tau(x), \qquad n_\tau(x) := \sum_{j=1}^N \delta(x - x_j(\tau)), \tag{14.153}$$

and thus the continuity equation is true at a microscopic level. However, in the Fokker-Planck description of Brownian motion the classical microscopic current has no immediate meaning because the velocities do not explicitly occur in (11.15).

One way to deduce the analogue of (14.152) is to insist on the applicability of the microscopic continuity equation. First note that for an observable $A_{\tau_1} = A(\{x_j^{(1)}\})$ measured at parameter value τ_1, and an observable $B_{\tau_2} = B(\{x_j^{(2)}\})$ measured at parameter value τ_2, the average of the product $A_{\tau_1} B_{\tau_2}$ is given by

$$\begin{aligned}\langle A_{\tau_1} B_{\tau_2} \rangle = & \frac{1}{N!} \int_{-\infty}^{\infty} dx_1^{(0)} \cdots \int_{-\infty}^{\infty} dx_N^{(0)}\, p_0(\vec{x}^{(0)}) \int_{-\infty}^{\infty} dx_1^{(1)} \cdots \int_{-\infty}^{\infty} dx_N^{(1)}\, A_{\tau_1} \\ & \times G_{\tau_1}^{\rm FP}(\vec{x}^{(0)}; \vec{x}^{(1)}) \int_{-\infty}^{\infty} dx_1^{(2)} \cdots \int_{-\infty}^{\infty} dx_N^{(2)}\, B_{\tau_2} G_{\tau_2 - \tau_1}^{\rm FP}(\vec{x}^{(1)}; \vec{x}^{(2)}).\end{aligned} \tag{14.154}$$

The delta function initial condition of the Green function and the structure of (11.15) show

$$G_{\tau_1}(\vec{x}^{(0)}; \vec{x}^{(1)}) = e^{\mathcal{L}\tau} \prod_{l=1}^N \delta(x_l^{(1)} - x_l^{(0)}), \tag{14.155}$$

where it is understood that $\mathcal{L}$ acts on $\{x_l^{(1)}\}$, and in (11.15) we have set $\gamma = 1$. Substituting this formula in

(14.154) allows the integration over $\{x_l^{(0)}\}$ to be carried out. Then substituting (14.155) with $\{x_l^{(0)}\}, \{x_l^{(0)}\}$ replaced by $\{x_l^{(1)}\}, \{x_l^{(2)}\}$ in the resulting expression and integrating over $\{x_l^{(1)}\}$ we obtain

$$\langle A_{\tau_1} B_{\tau_2} \rangle = \int_{-\infty}^{\infty} dx_1^{(2)} \cdots \int_{-\infty}^{\infty} dx_N^{(2)} \, B_{\tau_2} e^{\mathcal{L}(\tau_2-\tau_1)/\gamma} A_{\tau_1} e^{\mathcal{L}\tau_1/\gamma} p_0(\vec{x}^{(2)}). \tag{14.156}$$

Next we note from Exercises 11.1 q.1 (ii) and (11.15) with $\gamma = 1$ that

$$e^{\mathcal{L}\tau} = e^{-\beta W/2} e^{-\tau \sum_{j=1}^N \Pi_j^\dagger \Pi_j / \beta\gamma} e^{\beta W/2}$$

so we can rewrite (14.156) as

$$\langle A_{\tau_1} B_{\tau_2} \rangle = \int_{-\infty}^{\infty} dx_1^{(2)} \cdots \int_{-\infty}^{\infty} dx_N^{(2)} \, e^{-\beta W/2} B(\tau_2) A(\tau_1) p_0(\vec{x}^{(2)}) e^{\beta W/2}, \tag{14.157}$$

where

$$A(\tau) := e^{\tau \sum_{j=1}^N \Pi_j^\dagger \Pi_j/\beta\gamma} A_\tau e^{-\tau \sum_{j=1}^N \Pi_j^\dagger \Pi_j/\beta\gamma} \tag{14.158}$$

and similarly the definition of $B(\tau)$.

The equation (14.158) with $A_\tau = n_\tau(x) = \sum_{j=1}^N \delta(x - x_j(\tau))$ provides a definition of the dynamical microscopic density $n(x;\tau)$ in operator form. Substituting in (14.151) then allows the sought formula for the microscopic current to be obtained.

Proposition 14.7.1 *Let $n(x;\tau)$ be defined by (14.158) with $A_\tau = n_\tau(x) = \sum_{j=1}^N \delta(x - x_j(\tau))$, and similarly define $j(x;\tau)$ with $A_\tau = j_\tau(x)$, $j_\tau(x)$ to be determined. Then for the continuity equation (14.151) (which now refers to microscopic quantities) to be satisfied we require*

$$j_\tau(x) = -\frac{i}{\gamma\beta} \sum_{j=1}^N \Big(\frac{1}{i} \frac{\partial}{\partial x_j} \delta(x - x_j(\tau)) + \delta(x - x_j(\tau)) \frac{1}{i} \frac{\partial}{\partial x_j} \Big). \tag{14.159}$$

Proof. With $n(x;\tau)$ and n_τ as specified it follows from (14.158) that

$$\frac{\partial n(x;\tau)}{\partial \tau} = \frac{1}{\gamma\beta} e^{\tau \sum_{j=1}^N \Pi_j^\dagger \Pi_j/\beta\gamma} \Big[\sum_{j=1}^N \Pi_j^\dagger \Pi_j, n_\tau \Big] e^{-\tau \sum_{j=1}^N \Pi_j^\dagger \Pi_j/\beta\gamma}.$$

Comparison with (14.151) and recalling the definition of j_τ then shows

$$\frac{\partial}{\partial x} j_\tau(x) = -\frac{1}{\gamma\beta} \Big[\sum_{j=1}^N \Pi^\dagger \Pi_j, n_\tau \Big].$$

The stated result now follows from the fact that

$$\sum_{j=1}^N \Pi^\dagger \Pi_j = \sum_{j=1}^N \frac{\partial^2}{\partial x_j^2} + V(x_1, \dots, x_N)$$

for some V. □

14.7.2 Hydrodynamic limit

Physical principles involving the continuity equation (14.153) can be used to predict the small k form of the dynamical structure function as given by (13.228) [54]. For a single particle moving in a viscous medium ($\gamma = 1$) Newton's law of motion gives

$$m \frac{dv(\tau)}{d\tau} = -v(\tau) + \mathcal{F},$$

where $\mathcal{F}$ is the applied force. Hence in an equilibrium situation $v(\tau) = \mathcal{F}$. The many-body analogue of this latter equation is

$$j(x;\tau) = \mathcal{F}(x;\tau),$$

where $\mathcal{F}(x;\tau)$ now refers to the macroscopic force density. For the log-gas, in the long wavelength regime, the force density will to leading order be of electrostatic origin, implying [145]

$$j(x;\tau) = -n_\tau(x)\frac{\partial}{\partial x}\Big(V(x) - \int_{-\infty}^{\infty} \rho_{(1)}(x';\tau)\log|x-x'|\Big)\,dx'. \tag{14.160}$$

To proceed further one takes the partial derivative with respect to x on both sides, and substitutes for $\partial j(x;\tau)/\partial x$ on the l.h.s. using the continuity equation (14.153). Next the resulting equation is linearized by writing

$$\rho_{(1)}(x';\tau) = \rho_{(1)}(x) + \delta n_\tau(x)$$

(recall that (14.160) refers to the long wavelength regime so the microscopic quantity $n_\tau(x)$ is essentially smoothed) and terms of order $(\delta n_\tau(x))^2$ are ignored. Using the equilibrium condition

$$\frac{\partial}{\partial x}\Big(V(x) - \rho\int_{-\infty}^{\infty} \rho_{(1)}(x)\log|x-x'|\,dx'\Big) = 0,$$

the linearized equation then reads

$$\begin{aligned}\frac{\partial \delta n_\tau(x)}{\delta\tau} &= -\rho\frac{\partial^2}{\partial x^2}\int_{-\infty}^{\infty} \delta n_\tau(x')\log|x-x'|\,dx' \\ &= -\rho\frac{\partial}{\partial x}\int_{-\infty}^{\infty}\Big(\frac{\partial}{\partial x'}\delta n_\tau(x')\Big)\log|x-x'|\,dx'.\end{aligned} \tag{14.161}$$

Introducing Fourier transforms and recalling (14.3) this gives

$$\frac{\partial \delta\hat{n}_\tau(k)}{\partial\tau} = -\rho\pi|k|\delta\hat{n}_\tau(k).$$

Thus $\delta\hat{n}_\tau(k) = \delta\hat{n}_\tau(0)e^{-\pi\rho|k|\tau}$, or equivalently

$$\hat{n}_\tau(k) = \hat{n}_\tau(0)e^{-\pi\rho|k|\tau}. \tag{14.162}$$

But for a system confined to a region of length L, $\hat{S}(k;\tau) = \frac{1}{L}\langle\hat{n}_0(k)\hat{n}_\tau(k)\rangle$. Substituting (14.162) predicts that for the general β log-gas, in the long wavelength $k \to 0$ limit,

$$\hat{S}(k;\tau) \sim \hat{S}(k;0)e^{-\pi\rho|k|\tau}, \tag{14.163}$$

which when combined with (14.8) implies (13.228).

Chapter Fifteen

The two-dimensional one-component plasma

The two-dimensional one-component plasma (2dOCP) consists of log-potential charges of the same sign in a two-dimensional domain which contains a smeared out neutralizing background, and so is the two-dimensional version of the one-component log-gas. Although only one value of the coupling allows an exact solution, there are a number of different two-dimensional geometries and boundary conditions for which this exact solution is possible. Here the exact solutions for disk, sphere and antisphere geometries are considered, as well as the exact solution for metallic and Neumann boundary conditions. The first three of these allow for interpretations as eigenvalue p.d.f.'s, and as the modulus squared of the many-body wave function formed by free fermions confined to these surfaces in the presence of a particular magnetic field. Also associated with these three geometries are the zeros of three families of random polynomials, although the correlations are given not by determinants, but rather by permanents. For the 2dOCP at general coupling, macroscopic arguments of the type used in Chapter 14 imply a number of sum rules and asymptotic formulas, which can be illustrated on the exact results. The fast decay of the correlations in the bulk is responsible for sum rules which have no one-component log-gas analogues. In the last section, a classification scheme for random matrix ensembles with complex eigenvalues is considered.

15.1 COMPLEX GAUSSIAN RANDOM MATRICES AND POLYNOMIALS

15.1.1 Eigenvalues of complex random matrices

The Boltzmann factor for the 2dOCP in a disk has been calculated in Exercises 1.4 q.3 as proportional to

$$\prod_{j=1}^{N} e^{-\pi\rho\Gamma|\vec{r}_j|^2/2} \prod_{1\le j<k\le N} |\vec{r}_k - \vec{r}_j|^\Gamma \tag{15.1}$$

(recall $\Gamma := q^2\beta$, where q denotes the magnitude of the charge; in two dimensions it is conventional to keep Γ as the coupling in the Boltzmann factor even if we take $q = 1$). For the special coupling $\Gamma = 2$, this same p.d.f. occurs as the eigenvalue p.d.f. for complex Gaussian random matrices.

Proposition 15.1.1 *Consider an $N \times N$ random matrix $\mathbf{X}$ in which the elements $u_{jk} + iv_{jk}$ are independently distributed with p.d.f.*

$$\frac{1}{\pi} e^{-|u_{jk}|^2 - |v_{jk}|^2}. \tag{15.2}$$

(The set of such matrices is said to define the complex Ginibre ensemble, after [264].) The corresponding eigenvalue p.d.f. for the (complex) eigenvalues $z_j = x_j + iy_j$ is proportional to (15.1) with $\vec{r}_j = (x_j, y_j)$, $\Gamma = 2$ and $\rho = 1/\pi$.

Proof. We follow [395, Appendix 35] (see also [266]), as refined by [298], and begin by writing $\mathbf{X}$ in terms of its Schur decomposition,

$$\mathbf{X} = \mathbf{U}\mathbf{T}\mathbf{U}^{-1}, \tag{15.3}$$

where $\mathbf{U}$ is a unitary matrix which is unique only up to the phase of each column, and $\mathbf{T}$ is a triangular matrix with all elements below the diagonal zero and the diagonal elements equal to the eigenvalues. The Schur decomposition (15.3) follows by iterating the complex form of (1.176), obtained by Householder transformations (see also Exercises 15.1 q.2). The number of independent real parts of $\mathbf{X}$ is $2N^2$, while the number in $\mathbf{T}$ is N^2+N. Consistent with this is that making a specific choice of the phase of each column of $\mathbf{U}$ gives the number of independent real variables associated with $\mathbf{U}$ as N^2-N. Now

$$\mathbf{U}^\dagger d\mathbf{X}\mathbf{U} = d\mathbf{T} + \mathbf{U}^\dagger d\mathbf{U}\mathbf{T} - \mathbf{T}\mathbf{U}^\dagger d\mathbf{U}, \tag{15.4}$$

and $(\mathbf{U}^\dagger d\mathbf{X}\mathbf{U}) = (d\mathbf{X})$ (this latter equation follows from Proposition 3.2.4). With $d\mathbf{V} := \mathbf{U}^\dagger d\mathbf{U}$, the element in position (jk) of the matrix on the r.h.s. of (15.4) is

$$dT_{jk} + \sum_{l\le k} dV_{jl}T_{lk} - \sum_{j\le l} T_{jl}dV_{lk}. \tag{15.5}$$

For $j>k$, we rewrite (15.5) to read

$$(T_{kk}-T_{jj})dV_{jk} + \Big(\sum_{l<k} dV_{jl}T_{lk} - \sum_{j<l} T_{jl}dV_{lk}\Big) \tag{15.6}$$

while for $j\le k$ we rewrite it as

$$dT_{jk} + T_{jk}(dV_{jj} - dV_{kk}) + \Big(\sum_{\substack{l=1\\ l\ne j}}^{k} dV_{jl}T_{lk} - \sum_{\substack{l=j\\ l\ne k}}^{N} T_{jl}dV_{lk}\Big). \tag{15.7}$$

With the wedge product computed in the order of indices $(N\,1)\ ((N-1)\,1)\dots(1\,1)\ (N\,2)\dots(1\,2)$, and so on, after recalling that $dV_{jk} = -d\bar{V}_{kj}$, we see that the differentials are introduced for the first time from the terms outside the bracketed summations in (15.6) and (15.7). Hence the bracketed terms do not contribute to the wedge product, which, after recalling $T_{jj}=z_j$, is therefore equal to

$$\prod_{j<k}|z_j-z_k|^2 \wedge_j dz_j^{\mathrm{r}}dz_j^{\mathrm{i}} \wedge_{j<k} dV_{jk}^{\mathrm{r}}dV_{jk}^{\mathrm{i}}(dT_{jk}+T_{jk}(dV_{jj}-dV_{kk}))^{\mathrm{r}}(dT_{jk}+T_{jk}(dV_{jj}-dV_{kk}))^{\mathrm{i}}, \tag{15.8}$$

where as used previously the superscripts r and i denote real and imaginary parts, respectively. But dV_{jj} must be expressible in terms of $\{dV_{jk}\}_{j<k}$ by the count on the number of independent real variables associated with $\mathbf{U}$. Hence $(d\mathbf{V}) := \wedge_{j<k}dV_{jk}^{\mathrm{r}}dV_{jk}^{\mathrm{i}}$, when wedged with dV_{jj} must give zero and so (15.8) reduces to

$$\prod_{j<k}|z_j-z_k|^2 \wedge_j dz_j^{\mathrm{r}}dz_j^{\mathrm{i}}(d\mathbf{V}) \wedge_{j<k} dT_{jk}^{\mathrm{r}}dT_{jk}^{\mathrm{i}}. \tag{15.9}$$

Finally, the joint distribution of the elements is proportional to $e^{-\mathrm{Tr}(\mathbf{X}\mathbf{X}^\dagger)} = e^{-\sum_{j=1}^N |z_j|^2 - \sum_{j<k}|T_{jk}|^2}$, and so

$$e^{-\mathrm{Tr}(\mathbf{X}\mathbf{X}^\dagger)}(d\mathbf{X}) = e^{-\sum_{j=1}^N |z_j|^2}\prod_{j<k}|z_k-z_j|^2 \wedge_j dz_j^{\mathrm{r}}dz_j^{\mathrm{i}}\, e^{-\sum_{j<k}|T_{jk}|^2} \bigwedge_{j<k} dT_{jk}^{\mathrm{r}}dT_{jk}^{\mathrm{i}}(d\mathbf{V}).$$

The dependence on the eigenvalues thus factorizes from the dependence on the other variables, and is indeed proportional to (15.1) with $\Gamma=2$ and $\rho=1/\pi$. □

One immediate implication of Proposition 15.1.1 is in relation to the density of eigenvalues of complex Gaussian random matrices. According to Exercises 1.4 q.3 the Boltzmann factor (15.1) results from the 2dOCP in a disk of radius R with a uniform background of charge density $-\rho$ $(=-1/\pi$ in the eigenvalue analogy). By the principle of local charge neutrality for Coulomb systems, this implies that the eigenvalues will uniformly occupy a disk in the complex plane of radius $R=\sqrt{N}$. The validity of this prediction, which will subsequently be established analytically, is supported by displaying graphically the numerical computation of the eigenvalues of a complex Gaussian random matrix (see Figure 15.1). More generally, eigenvalues of random matrices with i.i.d. elements (from a real or complex distribution) of zero mean and

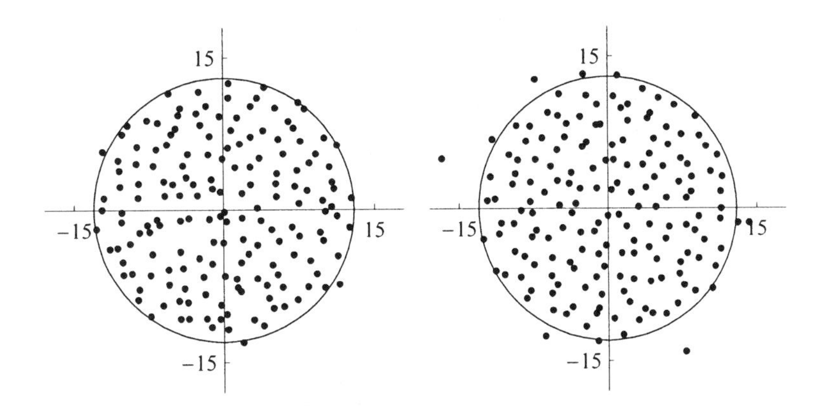

Figure 15.1 Eigenvalues in the complex plane of a 169×169 random matrix with complex Gaussian elements (left most plot) and zeros of a random polynomial with complex coefficients a_j chosen from the Gaussian distribution with zero mean and variance $1/j!$. Note that the edge of the leading order support $|z| < 13$ (drawn as a circle) is sharper in the former case.

finite variance are to leading order supported uniformly in a disk about the origin in the complex plane, a result known as the *circle law* [265], [26], [510].

15.1.2 Zeros of complex random polynomials

The Jacobian $\prod_{j<k} |z_k - z_j|^2$ which occurs in the proof of Proposition 15.1.1 for the change of variables from the elements to the eigenvalues and associated variables of a complex matrix also occurs in the change of variables from the complex coefficients a_j of the polynomial

$$p(z) = a_0 + a_1 z + a_2 z^2 + \cdots + a_N z^N \tag{15.10}$$

to the variables $\{a_N, z_1, \dots, z_N\}$, where the z_j are the zeros of $p(z)$ [286].

PROPOSITION 15.1.2 *With (15.10) factorized as* $p(z) = a_N \prod_{j=1}^N (z - z_j)$ *we have*

$$da_0 \wedge d\bar{a}_0 \wedge \cdots \wedge da_{N-1} \wedge d\bar{a}_{N-1} = J dz_1 \wedge d\bar{z}_1 \wedge \cdots \wedge dz_N \wedge d\bar{z}_N,$$

where

$$J = |a_N|^{2N} \prod_{1 \le j < k \le N} |z_k - z_j|^2.$$

Proof. The coefficients are related to the zeros by the formula

$$a_j / a_N = \sum_{1 \le i_1 \le \cdots \le i_{N-j} \le N} z_{i_1} z_{i_2} \cdots z_{i_{N-j}} =: e_{N-j} \tag{15.11}$$

(for the definition, recall (4.132)), while the Jacobian is given by

$$J = \det \begin{bmatrix} \frac{\partial}{\partial z_j} a_{k-1} & \frac{\partial}{\partial \bar{z}_j} a_{k-1} \\ \frac{\partial}{\partial z_j} \bar{a}_{k-1} & \frac{\partial}{\partial \bar{z}_j} \bar{a}_{k-1} \end{bmatrix}_{j,k=1,\dots,N}.$$

Substituting (15.11) in the Jacobian shows that J is a homogeneous polynomial in $\{z_j\}$ and $\{\bar{z}_j\}$ of degree $\frac{1}{2}N(N-1)$. It vanishes whenever $z_j = z_k$ or $\bar{z}_j = \bar{z}_k$ for any $j \ne k$ and thus contains $\prod_{j<k}(z_k - z_j)(\bar{z}_k - \bar{z}_j)$ as a factor. But this is homogeneous of the required degree, and so is equal to J, up to a proportionality constant. That the proportionality constant equals unity follows by examination of the diagonal term in J. □

Consider now the case that the coefficients a_j in (15.10) are complex Gaussian random variables with mean zero and standard deviation σ_j, so that the measure associated with the joint distribution of the elements is

$$\prod_{j=0}^{N} \frac{1}{2\pi\sigma_j^2} \exp\Big(- \sum_{j=0}^{N} |a_j|^2/2\sigma_j^2 \Big) \prod_{j=0}^{N} da_j^{\rm r} da_j^{\rm i}. \tag{15.12}$$

Changing variables according to Proposition 15.1.2, this becomes

$$\prod_{j=0}^{N} \frac{1}{2\pi\sigma_j^2} \exp\Big(- \sum_{j=0}^{N} |e_j|^2/2\sigma_{N-j}^2 \Big) \prod_{1 \le j < k \le N} |z_k - z_j|^2 \Big(\prod_{j=1}^{N} dx_j dy_j \Big) da_N^{\rm r} da_N^{\rm i} \tag{15.13}$$

where $e_0 := 1$ and $x_j + iy_j =: z_j$. Integrating over $a_N^{\rm r}$ and $a_N^{\rm i}$ shows that the p.d.f. for the roots of $p(z)$ is given by [70]

$$\pi N! \prod_{l=0}^{N} \frac{1}{\pi\sigma_l^2} \frac{\prod_{1 \le j<k \le N} |z_k - z_j|^2}{(\sum_{j=0}^{N} |e_j|^2/\sigma_{N-j}^2)^N}, \tag{15.14}$$

where the e_j are given in terms of the $\{z_j\}$ by (15.11).

Unlike the numerator in (15.14), which represents the Boltzmann factor for particle-particle interaction in the 2dOCP with $\Gamma = 2$, the denominator has no simple plasma interpretation as all particle coordinates are coupled. Nonetheless, we will see in Section 15.3.3 that there are some similarities with the plasma system. In particular, choosing $\sigma_j^2 = 1/j!$ gives a uniform density of zeros in the complex plane with support (to leading order) in a disk of radius $\sqrt{N}$.

EXERCISES 15.1 1. [250] In this exercise we will consider a random matrix $\mathbf{J}$ of the form $\mathbf{J} = \mathbf{H} + iv\mathbf{A}$, where $\mathbf{H}$ and $\mathbf{A}$ are Gaussian Hermitian random matrices with joint p.d.f.'s for the elements proportional to $\exp(-\frac{1}{1+\tau}\mathrm{Tr}\mathbf{X}^2)$ $(\mathbf{X} = \mathbf{H}, \mathbf{A})$, where $\tau = (1 - v^2)/(1 + v^2)$.

(i) Verify the formulas

$$\mathrm{Tr}\,\mathbf{H}^2 = \frac{1}{2}\Big(\mathrm{Tr}(\mathbf{J}\mathbf{J}^\dagger) + \mathrm{Re}\,\mathrm{Tr}(\mathbf{J}^2)\Big) \ \text{ and } \ \mathrm{Tr}\,\mathbf{A}^2 = \frac{1}{2v^2}\Big(\mathrm{Tr}(\mathbf{J}\mathbf{J}^\dagger) - \mathrm{Re}\,\mathrm{Tr}(\mathbf{J}^2)\Big)$$

and thus show that the joint p.d.f. for the elements of $\mathbf{J}$, obtained by changing variables in the joint distributions for the elements of $\mathbf{H}$ and $v\mathbf{A}$, is proportional to

$$\exp\Big(- \frac{1}{1-\tau^2}\mathrm{Tr}(\mathbf{J}\mathbf{J}^\dagger - \tau \mathrm{Re}\,\mathbf{J}^2)\Big). \tag{15.15}$$

(ii) Decomposing $\mathbf{J}$ in terms of its Schur decomposition (15.3), note that the above joint distribution becomes

$$\exp\Big(- \frac{1}{1-\tau^2}\Big(\sum_{j=1}^{N} |z_j|^2 - \tau \mathrm{Re} \sum_{j=1}^{N} z_j^2 + \sum_{j<k} |T_{jk}|^2\Big)\Big),$$

and hence deduce that the eigenvalue p.d.f. is proportional to

$$\exp\Big(- \frac{1}{1-\tau^2}\sum_{j=1}^{N}\Big(|z_j|^2 - \frac{\tau}{2}(z_j^2 + \bar{z}_j^2)\Big)\Big) \prod_{1 \le j<k \le N} |z_k - z_j|^2. \tag{15.16}$$

2. [249] Let $\mathbf{X}_N$ be an $N \times N$ complex matrix, and let z be an eigenvalue with normalized eigenvector $\vec{x}$ chosen with $x_1 > 0$. Define $\mathbf{U} = \mathbf{1}_N - 2\vec{v}\vec{v}^\dagger$, where $\vec{v} = (\vec{x} + \vec{e}_1)/|\vec{x} + \vec{e}_1|$. The working of Exercises 1.9 q.3 gives

$$\mathbf{X}_N = \mathbf{U} \begin{bmatrix} z & \vec{w}_{N-1}^\dagger \\ \vec{0}_{N-1} & \mathbf{X}_{N-1} \end{bmatrix} \mathbf{U}$$

for some $N-1\times 1$ vector $\vec{w}$ and $N-1\times N-1$ matrix $\mathbf{X}_{N-1}$.

(i) Repeat the working which led to (1.179) to show

$$(d\mathbf{X}_N) = |\det(z\mathbf{1}_{N-1} - \mathbf{X}_{N-1})|^2 dz^{\mathrm{r}} dz^{\mathrm{i}} (d\vec{w}_{N-1})(d\vec{s}_{N-1})(d\mathbf{X}_{N-1}),$$

where $\vec{s}_{N-1}$ is the first column of $\mathbf{U}d\mathbf{U}$ with the first entry removed.

(ii) Write $\vec{v} = (v_1, \vec{q})$. The definition of $\vec{v}$ implies $\frac{1}{2} \le v_1^2 \le 1$ and thus $0 \le |\vec{q}|^2 \le \frac{1}{2}$. Note that

$$\mathbf{U} = \begin{bmatrix} 2|\vec{q}|^2 - 1 & -2v_1\vec{q}^T \\ -2v_1\vec{q} & \mathbf{1}_{N-1} - 2\vec{q}\vec{q}^\dagger \end{bmatrix},$$

and from this show

$$(d\vec{s}_{N-1}) = (a\mathbf{1}_{N-1} + b\vec{q}\vec{q}^\dagger)(d\vec{q}) + (2a+b)\vec{q}\vec{q}^T(d\vec{q}^*)$$

where here $*$ denotes complex conjugate and

$$a = -2\sqrt{1-|\vec{q}|^2}, \qquad b = \frac{1-2|\vec{q}|^2}{\sqrt{1-|\vec{q}|^2}}.$$

(iii) Deduce from (ii) that

$$(d\vec{s}_{N-1}) = \det(\mathbf{1}_{2N-2} + \mathbf{L})(d\vec{q}),$$

where

$$\mathbf{L} = \begin{bmatrix} b\vec{q}\vec{q}^\dagger & (2a+b)\vec{q}\vec{q}^T \\ (2a+b)\vec{q}^*\vec{q}^\dagger & b\vec{q}^*\vec{q}^T \end{bmatrix} = \begin{bmatrix} b\vec{q} & (2a+b)\vec{q} \\ (2a+b)\vec{q}^* & b\vec{q}^* \end{bmatrix} \begin{bmatrix} \vec{q}^\dagger & \vec{0}^T \\ \vec{0}^T & b\vec{q}^T \end{bmatrix}.$$

Now make use of (5.33) to obtain the evaluation

$$\det(\mathbf{1}_{2N-2} + \mathbf{L}) = 2^{2N-2}(1-|\vec{q}|^2)^{N-2}(1-2|\vec{q}|^2).$$

3. [156] Use Householder reduction to Hessenberg form, as specified by (2.84), to show that complex Gaussian random matrices with entries independently distributed according to (15.2) are equivalent under unitary conjugation to the Hessenberg matrix $[a_{ij}]_{i,j=1,\dots,N}$ where all entries are again independently distributed. Explicitly, show that for $i \le j$, a_{ij} is again distributed according to (15.2), while $a_{i+1,i}$ has distribution $\tilde{\chi}_{2(N-i)}$ (recall Proposition 1.9.1).

4. In this exercise the problem of finding the zeros of a general degree N polynomial

$$p(x) = x^N + a_{N-1}x^{N-1} + \cdots + a_1 x + a_0$$

is shown to be equivalent to finding the eigenvalues of the *companion matrix*

$$C_N := \begin{bmatrix} 0 & 0 & \cdots & \cdots & 0 & -a_0 \\ 1 & 0 & \cdots & \cdots & 0 & -a_1 \\ 0 & 1 & \cdots & \cdots & 0 & -a_2 \\ \vdots & \vdots & & \ddots & \vdots & \\ 0 & 0 & \cdots & \cdots & 1 & -a_{N-1} \end{bmatrix}.$$

(i) Let the zeros of $p(x)$ be $\{\lambda_j\}_{j=1,\dots,N}$. Show that

$$C_N^T[\lambda_j^p]_{p=0,\dots,N-1} = \lambda_j[\lambda_j^p]_{p=0,\dots,N-1}.$$

(ii) Conclude from (i) and the Vandermonde determinant formula (1.173) that if the zeros of $p(x)$ are all distinct, C_N^T and thus C_N is diagonalizable and has eigenvalues $\lambda_1, \dots, \lambda_N$.

15.2 QUANTUM PARTICLES IN A MAGNETIC FIELD

As well as an interpretation as an eigenvalue p.d.f, the Boltzmann factor (15.1) with $\Gamma = 2$ occurs as the modulus squared of the exact ground state wave function for a certain quantum many-body system. This is the quantum many-body system consisting of N non-interacting fermions confined to a plane with a perpendicular magnetic field. All the fermions (electrons) are assumed to have their spin magnetic moment frozen along the direction of the magnetic field, thus implying that the spin of the electrons can be ignored. The demonstration of the analogy requires first revising the quantum mechanics of a single particle in a magnetic field (see, e.g., [112]).

15.2.1 Single particle wave function

Consider the setting of a quantum particle confined to the xy-plane. Suppose the particle has mass m and charge $-e$, and is subject to a perpendicular magnetic field $\vec{B} = B\hat{\mathbf{z}}$, $B > 0$. The Hamiltonian is then

$$H := \frac{1}{2m}\Big(-i\hbar\nabla + \frac{e}{c}\vec{A}\Big)^2 = \frac{1}{2m}\vec{\Pi}^2 = \frac{1}{2}\hbar w_c(a^\dagger a + a a^\dagger), \tag{15.17}$$

where

$$\Pi_x = -i\hbar\frac{\partial}{\partial x} + \frac{e}{c}A_x, \quad \Pi_y = -i\hbar\frac{\partial}{\partial y} + \frac{e}{c}A_y,$$

$$a^\dagger = \frac{l}{\sqrt{2}\hbar}(\Pi_x + i\Pi_y), \quad a = \frac{l}{\sqrt{2}\hbar}(\Pi_x - i\Pi_y), \tag{15.18}$$

$w_c := eB/mc$ (c denotes the speed of light) is called the cyclotron frequency, $l := \sqrt{\hbar c/eB}$ is called the magnetic length, and the vector potential $\vec{A}$ must satisfy

$$\nabla \times \vec{A} = B\hat{\mathbf{z}}. \tag{15.19}$$

Using (15.19) we can check from the definitions (15.18) that the commutation relation $[\Pi_x, \Pi_y] = -i\hbar^2/l^2$ holds, and from this we can check that

$$[a, a^\dagger] = 1. \tag{15.20}$$

Use of (15.20) in (15.17) gives that the Hamiltonian can be written in the harmonic oscillator-like form

$$H = \hbar w_c\Big(a^\dagger a + \frac{1}{2}\Big) = \hbar w_c\Big(a a^\dagger - \frac{1}{2}\Big). \tag{15.21}$$

Thus there are eigenstates $|n\rangle$ of H with energy $E_n = (n + \frac{1}{2})\hbar w_c$, $(n = 0, 1, 2, \dots)$, referred to as *Landau levels*, which are given in terms of the ground state $|0\rangle$ by

$$|n\rangle = \frac{(a^\dagger)^n}{\sqrt{n!}}|0\rangle \tag{15.22}$$

with the state $|0\rangle$ specified by $a|0\rangle = 0$.

Unlike for the harmonic oscillator, these states do not form a complete set (this is not surprising as here the system is two-dimensional). In fact, by studying the classical problem (see Exercises 15.2 q.1) and using the correspondence principle, we can interpret H as proportional to the square of the radius of the cyclotron orbit, which gives that the eigenstates $|n\rangle$ have cyclotron radius $\sqrt{2n+1}l$. The classical theory also tells us that once the cyclotron orbit radius is fixed, the only remaining degree of freedom is the center of the orbit. Applying the correspondence principle to the formulas of Exercises 15.2 q.1 for the classical orbit center gives the quantum center of orbit operators

$$X = x - \frac{l^2}{\hbar}\Pi_y, \quad Y = y + \frac{l^2}{\hbar}\Pi_x. \tag{15.23}$$

Using (15.19) and the explicit formulas for Π_x and Π_y in (15.18) gives the commutation relation $[X, Y] = il^2$. This indicates that we should consider the operator $X^2 + Y^2$, for then by defining

$$b^\dagger = \frac{1}{\sqrt{2}l}(X - iY), \quad b = \frac{1}{\sqrt{2}l}(X + iY) \tag{15.24}$$

we see that $[b, b^\dagger] = 1$ and furthermore

$$X^2 + Y^2 = l^2(bb^\dagger + b^\dagger b) = 2l^2\Big(b^\dagger b + \frac{1}{2}\Big).$$

Thus this operator also has a harmonic oscillator form, and so has eigenstates $|m\rangle$ with eigenvalue $R_m^2 = (2m+1)l^2$, $m = 0, 1, 2, \ldots$, which are given in terms of the state $|0\rangle$ (defined as the solution of $b|0\rangle = 0$) by

$$|m\rangle = \frac{(b^\dagger)^m}{\sqrt{m!}}|0\rangle. \tag{15.25}$$

A direct calculation from (15.18) and (15.24) shows that $a, a^\dagger$ commutes with $b, b^\dagger$, and so simultaneous eigenstates of H and $X^2 + Y^2$ are permitted. Thus for each state (15.22) of specific energy, the operator $(b^\dagger)^m$ creates orthogonal states

$$|n, m\rangle = \frac{(a^\dagger)^n (b^\dagger)^m}{\sqrt{n!m!}}|0, 0\rangle$$

with the same energy $E_n = (n + \frac{1}{2})\hbar\omega_c$ and eigenvalue of $X^2 + Y^2$ equal to $(2m+1)l^2$.

15.2.2 Many particle ground state

The states $|m\rangle =: \psi_m(\vec{r})$ given by (15.25) are orthogonal states with the minimal allowed energy $\frac{1}{2}\hbar\omega_c$ (i.e., the lowest Landau level) and their interpretation is that they have definite values of the distance from the origin to the center of their cyclotron orbit, which increases with m. The densest N-particle ground state ψ, in which the particles are fermions but otherwise non-interacting, is therefore obtained by constructing a Slater determinant from the states $\psi_0(\vec{r}), \ldots, \psi_{N-1}(\vec{r})$,

$$\psi(\vec{r}_1, \ldots, \vec{r}_N) = \frac{1}{\sqrt{N!}}\det[\psi_{j-1}(\vec{r}_k)]_{j,k=1,\ldots,N}. \tag{15.26}$$

The immediate task is thus to evaluate the single-particle states (15.25) which from (15.23), (15.24) and (15.18) depend on the particular choice of A_x and A_y. Here we will consider the *symmetric gauge*

$$\vec{A} = \frac{B}{2}(-y\hat{\mathbf{x}} + x\hat{\mathbf{y}}). \tag{15.27}$$

In the gauge (15.27), with $z = x + iy$, $\frac{\partial}{\partial z} := \frac{1}{2}(\frac{\partial}{\partial x} - i\frac{\partial}{\partial y})$ and $\frac{\partial}{\partial \bar{z}} = \frac{1}{2}(\frac{\partial}{\partial x} + i\frac{\partial}{\partial y})$ we can write

$$a = -\frac{i}{\sqrt{2}}\left(\frac{1}{2l}\bar{z} + 2l\frac{\partial}{\partial z}\right), \quad b = \frac{1}{\sqrt{2}}\left(\frac{1}{2l}z + 2l\frac{\partial}{\partial \bar{z}}\right). \tag{15.28}$$

Using in (15.21) the explicit form of a from (15.28) and the corresponding form of $a^\dagger$, we see that H can be written

$$H = \frac{\hbar\omega_c}{2}\Big(-4l^2\frac{\partial^2}{\partial z \partial \bar{z}} + \Big(z\frac{\partial}{\partial z} - \bar{z}\frac{\partial}{\partial \bar{z}}\Big) + \frac{|z|^2}{4l^2}\Big). \tag{15.29}$$

Also, since the state $\psi_0(\vec{r}) := |0\rangle$ is characterized by $a|0\rangle = 0$, $b|0\rangle = 0$ use of (15.28) gives

$$\frac{\partial}{\partial z}\psi_0 = -\frac{\bar{z}}{4l^2}\psi_0 \quad \text{and} \quad \frac{\partial}{\partial \bar{z}}\psi_0 = -\frac{z}{4l^2}\psi_0. \tag{15.30}$$

The first equation has solution $\psi_0 = f(\bar{z})e^{-z\bar{z}/4l^2}$ for any f analytic in $\bar{z}$, while the second equation has solution $\psi_0 = g(z)e^{-z\bar{z}/4l^2}$ for any function g analytic in z. These solutions are only compatible if f and g are constant functions, so after determining the normalization we have

$$\psi_0 = \frac{1}{\sqrt{2\pi l^2}}e^{-z\bar{z}/4l^2} = \frac{1}{\sqrt{2\pi l^2}}e^{-(x^2+y^2)/4l^2}.$$

Substituting this in (15.25) with the explicit form of $b^\dagger$ deduced from (15.28) gives

$$\psi_m(\vec{r}) = \frac{(\bar{z})^m e^{-(x^2+y^2)/4l^2}}{(2\pi l^2 2^m l^{2m} m!)^{1/2}}, \tag{15.31}$$

and this substituted in (15.26) together with the Vandermonde formula (1.173) gives

$$\psi(\vec{r}_1, \dots, \vec{r}_N) = \frac{1}{\sqrt{N!}} \prod_{j=1}^{N} \frac{e^{-(x_j^2+y_j^2)/4l^2}}{(2\pi l^2 2^{j-1} l^{2(j-1)}(j-1)!)^{1/2}} \prod_{1\le j<k\le N} (\bar{z}_k - \bar{z}_j). \tag{15.32}$$

Comparison with (15.1) shows that $|\psi|^2$ is indeed proportional to the Boltzmann factor for the one-component plasma at $\Gamma = 2$ in a disk with $\rho = 1/2\pi l^2$.

As in the case of random matrices, one immediate consequence of the plasma analogy is a prediction for the density of the quantum system: to leading order it will be uniform in a disk of radius $R = (2N)^{1/2} l$ with value $1/2\pi l^2$, and equal to zero outside this radius. This prediction is consistent with the fact that the maximum distance from the origin to the center of the cyclotron orbit of the single particle states is $\sqrt{2N-1}\,l$.

15.2.3 Zeros of a random single particle state

In Section 15.2.2 an N-body state has been constructed from the first N single-particle states (15.31). Here we will consider a single-particle state $\phi(\vec{r})$ constructed as a linear combination of the first $N+1$ states (15.31). From the form of these states and their orthogonality, we can write

$$\phi(\vec{r}) = e^{-(x^2+y^2)/2} p(\bar{z}), \qquad p(w) = \sum_{n=0}^{N} \frac{\alpha_n}{\sqrt{n!}} w^n \tag{15.33}$$

(for convenience we have set $2l^2 = 1$), where $\alpha_n = \langle p(z), z^n \rangle$ with the inner product defined by

$$\langle f, g \rangle = \frac{1}{\pi} \int_{-\infty}^{\infty} dx \int_{-\infty}^{\infty} dy\, e^{-(x^2+y^2)} \bar{f} g. \tag{15.34}$$

Suppose now $\phi(\vec{r})$ is random in the sense that the coefficients α_n in (15.33) are chosen at random from a complex Gaussian distribution with mean zero and variance unity. Then $p(w)$ is equivalent to a polynomial of the form (15.10) in which each a_j is chosen from a complex Gaussian distribution with mean zero and variance $1/j!$, and so the joint distribution of its zeros is given by (15.14) with $\sigma_j^2 = 1/j!$.

EXERCISES 15.2 1. For a classical particle of charge $-e$ in a magnetic field $\vec{B}$, the force is given by $-\frac{e}{c}\vec{v} \times \vec{B}$.

(i) Suppose the particle is confined to the xy-plane and the magnetic field is perpendicular to this plane, $\vec{B} = B\hat{\mathbf{z}}$, $B > 0$. Show that Newton's equation of motion implies

$$\dot{v}_x = -\omega_c v_y, \qquad \dot{v}_y = \omega_c v_x,$$

and thus with the initial condition $(v_x, v_y) = (v_0, 0)$,

$$v_x = v_0 \cos \omega_c t, \qquad v_y = v_0 \sin \omega_c t.$$

Note that $v_x^2 + v_y^2 = v_0^2$ independent of t.

(ii) Use the formulas for v_x and v_y to show

$$x = x_0 + \frac{v_0}{\omega_c} \sin \omega_c t, \qquad y = y_0 - \frac{v_0}{\omega_c} \cos \omega_c t.$$

Note that $((x - x_0)^2 + (y - y_0)^2)^{1/2} = v_0/\omega_c$, so the solution corresponds to a circular cyclotron orbit of radius v_0/ω_c with center at (x_0, y_0).

(iii) Make the correspondence $mv_x \leftrightarrow \Pi_x$, $mv_y \leftrightarrow \Pi_y$ to interpret $\Pi_x^2 + \Pi_y^2$ as proportional to the square of the cyclotron orbit radius.

2. (i) [143] Show from (15.29) that the Hamiltonian H in the symmetric gauge can be written

$$H = -\frac{\hbar\omega_c}{2}\Big(\Big(\frac{\partial^2 \Phi}{\partial z \partial \bar{z}}\Big)^{-1}\Big(\frac{\partial}{\partial \bar{z}} - \frac{\partial \Phi}{\partial \bar{z}}\Big)\Big(\frac{\partial}{\partial z} + \frac{\partial \Phi}{\partial z}\Big) - 1\Big) \tag{15.35}$$

with $\Phi = |z|^2/(2l)^2$. Also, with $A_z := A_x + iA_y$ and $A_{\bar{z}} := A_x - iA_y$, check that

$$A_z = \frac{iB}{2}(2l^2)\frac{\partial \Phi}{\partial \bar{z}}, \qquad A_{\bar{z}} = -\frac{iB}{2}(2l^2)\frac{\partial \Phi}{\partial z}$$

and note too that with dS denoting an element of surface area,

$$dS = (2l)^2 \frac{\partial^2 \Phi}{\partial z \partial \bar{z}} dx dy.$$

(ii) With H given by (15.35) and Φ general show that

$$e^{\Phi} H e^{-\Phi} = -\frac{\hbar\omega_c}{2}\Big(\Big(\frac{\partial^2 \Phi}{\partial z \partial \bar{z}}\Big)^{-1}\Big(\frac{\partial}{\partial \bar{z}} - 2\frac{\partial \Phi}{\partial \bar{z}}\Big)\frac{\partial}{\partial z} - 1\Big).$$

Note from this that

$$e^{-\Phi} f(\bar{z}) \tag{15.36}$$

is an eigenstate in the lowest Landau level for any f analytic in $\bar{z}$.

(iii) Verify that $J := z\frac{\partial}{\partial z} - \bar{z}\frac{\partial}{\partial \bar{z}}$ commutes with H as defined by (15.35) for general $\Phi = \Phi(|z|^2)$, and in these cases the states (15.36) which are simultaneous eigenstates of J are

$$e^{-\Phi}(\bar{z})^n, \qquad n \in \mathbb{Z}_{\ge 0}. \tag{15.37}$$

3. [205] Consider the wave function

$$\tilde{\psi}_0 = \frac{1}{\sqrt{2l^2\pi \cosh \mu}} \exp\Big(-\frac{1}{4l^2} z\bar{z} + \frac{1}{4l^2} \tanh \mu\, \bar{z}^2\Big),$$

which is of the form $e^{-z\bar{z}/4l^2} f(\bar{z})$ for f analytic in $\bar{z}$ and so belongs to the lowest Landau level. The objective here is to define operators $\tilde{b}$ and $\tilde{b}^\dagger$ such that $\tilde{b}\tilde{\psi}_0 = 0$ and $[\tilde{b}, \tilde{b}^\dagger] = 1$, and to use the operator $\tilde{b}^\dagger$ to construct states orthogonal to $\tilde{\psi}_0$ using the analogue of (15.25).

(i) Make a Bogoliubov transformation involving the operator b in (15.28) by writing $\tilde{b} = \alpha b + \beta b^\dagger$ with $\alpha^2 - \beta^2 = 1$ to verify that

$$\begin{aligned}\tilde{b}^\dagger &= (\cosh \mu) b^\dagger - (\sinh \mu) b \\ &= \frac{1}{\sqrt{2}} \cosh \mu \Big(\frac{1}{2l}\bar{z} - 2l\frac{\partial}{\partial z}\Big) - \frac{1}{\sqrt{2}} \sinh \mu \Big(\frac{1}{2l} z + 2l \frac{\partial}{\partial \bar{z}}\Big)\end{aligned}$$

has the sought properties.

(ii) By defining $\tilde{\psi}_m = \frac{1}{\sqrt{m!}}(\tilde{b}^\dagger)^m \tilde{\psi}_0$ analogous to (15.25) show that

$$\tilde{\psi}_m = P_m(\bar{z}) \exp\Big(- \frac{1}{4l^2} z\bar{z} + \frac{1}{4l^2} \tanh\mu\, \bar{z}^2 \Big),$$

where $P_m(z)$ is a polynomial of degree m defined by the recurrence

$$P_{m+1}(z) = \frac{1}{\sqrt{2}\sqrt{m+1}} \Big(\frac{1}{l\cosh\mu} z P_m(z) - 2l \sinh\mu \frac{dP_m(z)}{dz} \Big), \quad P_0(z) = (2\pi l^2 \cosh\mu)^{-1/2}.$$

Compare this with the Hermite polynomial three-term recurrence

$$2xH_n(x) = H_{n+1}(x) + 2nH_{n-1}(x)$$

and thus conclude that

$$P_m(z) = \frac{1}{\sqrt{m!}} \Big(\frac{1}{2} \tanh\mu \Big)^{m/2} \Big(2\pi l^2 \cosh\mu \Big)^{-1/2} H_m \Big(\frac{z}{\sqrt{2l^2 \sinh 2\mu}} \Big).$$

4. [110], [205] From q.3 (ii) it follows that the Slater determinant ψ constructed from the states $\tilde{\psi}_m, m = 0, 1, \dots, N-1$ is such that $|\psi|^2$ is proportional to

$$\exp\Big(- \frac{1}{2l^2} \sum_{j=1}^{N} (x_j^2 + y_j^2) + \frac{1}{2l^2} \tanh\mu \sum_{j=1}^{N} (x_j^2 - y_j^2) \Big) \prod_{1 \le j < k \le N} |\vec{r}_j - \vec{r}_k|^2. \tag{15.38}$$

This is proportional to the Boltzmann factor for a 2dOCP with a uniform background of a certain shape. The objective of this exercise is to deduce that the shape is an ellipse.

(i) Consider an ellipse Ω with semi-axes A and B such that $A = a\cosh\xi_b$ and $B = a\sinh\xi_b$. Introduce elliptic coordinates (ξ, η), $0 \le \xi \le \xi_b$, $0 \le \eta \le 2\pi$, related to the Cartesian coordinates by

$$z := x + iy = \frac{a}{2}\cosh(\xi + i\eta).$$

Show that the Jacobian J in the change of variables $dxdy = J d\xi d\eta$ is $J = (a^2/8)(\cosh 2\xi - \cos 2\eta)$.

(ii) Use this formula for J and the expansion [408, Eq. (10.132)]

$$- \log|z - z'| = -\Big(\xi' + \log\frac{a}{4} \Big) + \sum_{n=1}^{\infty} \frac{2}{n} \Big(\cosh n\xi \cos n\eta\, e^{-n\xi'} \cos n\eta' + \sinh n\xi \sin n\eta\, e^{-n\xi'} \sin n\eta' \Big) \tag{15.39}$$

valid for $\xi < \xi'$ (otherwise interchange ξ and ξ' on the r.h.s.) to show that

$$\begin{aligned} -\int_\Omega dx'dy' \log|z - z'| &= -\frac{\pi a^2}{16} \Big(\cosh 2\xi + \cos 2\eta - \cosh 2\xi \cos 2\eta\, e^{-2\xi_b} \Big) + \text{const.} \\ &= -b_1 x^2 - b_2 y^2 + \text{const.}, \end{aligned}$$

where $b_1 = \frac{\pi}{2}(1 - e^{-2\xi_b})$, $b_2 = \frac{\pi}{2}(1 + e^{-2\xi_b})$.

(iii) Use (ii) to interpret the exponent in (15.38) as due to a uniformly charged ellipse, charge density $-1/2\pi l^2$, with $B/A = e^{-2\mu}$ and $a = l\sqrt{2N}$.

(iv) Use (ii) to interpret the exponent in (15.16) as due to a uniformly charged ellipse, charge density $-1/(\pi(1-\tau^2))$, with $A = \sqrt{N}(1+\tau)$, $B = \sqrt{N}(1-\tau)$.

15.3 CORRELATION FUNCTIONS

15.3.1 The 2dOCP in a disk

Introducing polar coordinates (r, θ), and using the Vandermonde identity (1.173) we have

$$\prod_{1 \le j < k \le N} |r_k e^{i\theta_k} - r_j e^{i\theta_j}|^2 = \Big| \det[(r_j e^{i\theta_j})^{k-1}]_{j,k=1,\dots,N} \Big|^2. \tag{15.40}$$

Let Ω_R be a disk of radius R centered on the origin, and let $f(r)$ be a function of the polar coordinate r only. Observing the orthogonality

$$\int_{\Omega_R} f(r)(re^{i\theta})^k (re^{-i\theta})^j \, d\vec{r} = \delta_{j,k} \Big(2\pi \int_0^R r^{2j+1} f(r)\, dr \Big) \tag{15.41}$$

the method of proof of Propositions 5.1.1 and 5.1.2 allows the correlations to be computed for a class of Boltzmann factors extending that for the 2dOCP at $\Gamma = 2$ in a disk.

PROPOSITION 15.3.1 *For the 2dOCP at $\Gamma = 2$ subject to a rotationally invariant one-body potential with Boltzmann factor such that*

$$e^{-\beta U(\vec{r}_1,\dots,\vec{r}_N)} d\vec{r}_1 \cdots d\vec{r}_N = \prod_{j=1}^N f(r_j) \prod_{1 \le j < k \le N} |r_k e^{i\theta_k} - r_j e^{i\theta_j}|^2 \prod_{j=1}^N r_j dr_j d\theta_j, \tag{15.42}$$

and confined to a disk of radius R, the n-particle correlation function is given by

$$\rho_{(n)}(\vec{r}_1, \dots, \vec{r}_n) = \det \left[\frac{1}{2\pi} \sum_{j=1}^N \frac{(f(r_\mu) f(r_\gamma))^{1/2} (r_\mu r_\gamma)^{j-1} e^{i(\theta_\mu - \theta_\gamma)(j-1)}}{\int_0^R f(r) r^{2j-1} dr} \right]_{\mu,\gamma=1,\dots,n}. \tag{15.43}$$

It remains to take the thermodynamic limit. Now, for a uniform background in a disk of radius R we have from (1.72) that $f(r) = Ae^{-\pi\rho r^2}$ in (15.42), A constant. In taking the thermodynamic limit there are two distinct cases of interest: the first when $\vec{r}_1, \dots, \vec{r}_n$ are fixed in the bulk of the system, and the second when $\vec{r}_1, \dots, \vec{r}_n$ are some finite distance from the surface at $|\vec{r}| = R$. The first case is particularly simple [313].

PROPOSITION 15.3.2 *For the n-particle correlation (15.43) with $f(r) = Ae^{-\pi\rho r^2}$, corresponding to the 2dOCP in a disk with uniform background,*

$$\lim_{\substack{N\to\infty \\ \rho \text{ fixed}}} \rho_{(n)}(\vec{r}_1, \dots, \vec{r}_n) = \rho^n \det \left[e^{-\pi\rho(r_\mu^2 + r_\gamma^2)/2} e^{\pi\rho z_\mu \bar{z}_\gamma} \right]_{\mu,\gamma=1,\dots,n}. \tag{15.44}$$

In particular

$$\rho_{(1)}(\vec{r}) = \rho, \quad \rho_{(2)}(\vec{r}_1, \vec{r}_2) = \rho^2 \left(1 - e^{-\pi\rho|\vec{r}_1 - \vec{r}_2|^2} \right), \tag{15.45}$$

where it is understood $\rho_{(1)}$ and $\rho_{(2)}$ refer to the thermodynamic values.

Proof. In (15.43) we have

$$\begin{aligned} \frac{1}{2\pi\rho} \lim_{\substack{N\to\infty \\ \rho\text{ fixed}}} \sum_{j=1}^N \frac{e^{-\pi\rho(r_\mu^2+r_\gamma^2)/2} (r_\mu r_\gamma)^{j-1} e^{i(\theta_\mu - \theta_\gamma)(j-1)}}{\int_0^R r e^{-\pi\rho r^2} r^{2(j-1)}\, dr} &= e^{-\pi\rho(r_\mu^2+r_\gamma^2)/2} \sum_{j=1}^\infty \frac{(\pi\rho r_\mu r_\gamma)^{j-1} e^{i(\theta_\mu-\theta_\gamma)(j-1)}}{\Gamma(j)} \\ &= e^{-\pi\rho(r_\mu^2+r_\gamma^2)/2} e^{\pi\rho z_\mu \bar{z}_\gamma}, \end{aligned}$$

where $z = re^{i\theta}$, as required. □

We will see in Exercises 15.3 q.5 that a general property of $g_{(2)}(r) := \rho_{(2)}(\vec{0}, \vec{r})/\rho^2$ for the 2dOCP is the functional equation [479]

$$g_{(2)}(r) = e^{\pm \pi i \Gamma/2} e^{-\pi \rho \Gamma r^2/2} g_{(2)}(ir). \tag{15.46}$$

Note that this is indeed satisfied by the exact evaluation. Another property of the exact evaluation is that the truncated two-particle correlation has a Gaussian decay, which is in contrast to the algebraic decay of the same quantity in the one-dimensional log-gas systems. The exact result gives that at $\Gamma = 2$ the dimensionless structure function

$$\tilde{S}(k) := 1 + \frac{1}{\rho} \int_{\mathbb{R}^2} \rho_{(2)}^T(\vec{r}, \vec{0}) e^{i\rho^{1/2}\vec{k}\cdot\vec{r}}\, d\vec{r} \tag{15.47}$$

(cf. (7.3)) $k := |\vec{k}|$ is given by

$$\tilde{S}(k) = 1 - e^{-k^2/4\pi}. \tag{15.48}$$

The evaluation of the surface correlation functions requires a more refined analysis than that of Proposition 15.3.2 [493]. It is convenient to choose as the origin $(r, \theta) = (R, 3\pi/2)$, and to specify points in Cartesian coordinates from this origin, so that

$$r_\mu^2 = x_\mu^2 + (R - y_\mu)^2, \qquad \theta_\mu \sim -\frac{3\pi}{2} + \frac{x_\mu}{R}. \tag{15.49}$$

Note that in the limit $R \to \infty$ the background then occupies the upper half-plane $y > 0$.

PROPOSITION 15.3.3 *With (x_μ, y_μ) $(\mu = 1, \dots, n)$ as given in (15.49),*

$$\lim_{\substack{N\to\infty \\ \rho \text{ fixed}}} \rho_{(n)}(\vec{r}_1, \dots, \vec{r}_n) = \rho^n \det \left[e^{-\pi\rho(y_\mu^2 + y_\gamma^2)} h\Big(\frac{1}{2}(y_\mu + y_\gamma + i(x_\mu - x_\gamma))\Big) \right]_{\mu,\gamma=1,\dots,n}$$

where

$$h(z) := \Big(\frac{1}{\pi}\Big)^{1/2} \int_0^\infty \frac{e^{2(2\pi\rho)^{1/2} z t} e^{-t^2}}{\frac{1}{2}(1 + \operatorname{erf} t)}\, dt$$

with $\operatorname{erf} x := \frac{2}{\sqrt{\pi}} \int_0^x e^{-t^2} dt$ *denoting the error function.*

Proof. Introducing x and y according to (15.49), and replacing j by $N + 1 - j$ in the summation gives

$$\frac{1}{2\pi\rho} \sum_{j=1}^N \frac{e^{-\pi\rho(r_\mu^2 + r_\gamma^2)/2} (r_\mu r_\gamma)^{j-1} e^{i(\theta_\mu - \theta_\gamma)(j-1)}}{\int_0^R r e^{-\pi\rho r^2} r^{2(j-1)}\, dr}$$

$$\sim e^{-\pi\rho(x_\mu^2 + x_\gamma^2)/2} e^{-\pi\rho((R-y_\mu)^2 + (R-y_\gamma)^2)/2} \sum_{j=1}^N \frac{(\pi\rho)^{N-j} e^{(N-j)(\log(R-y_\mu) + \log(R-y_\gamma))} e^{i(x_\mu - x_\gamma)(N-j)/R}}{\gamma(N-j+1; N)},$$

where

$$\gamma(k; a) := \int_0^a t^{k-1} e^{-t} dt \tag{15.50}$$

denotes the incomplete gamma function. Next we make use of the uniform asymptotic expansion

$$\gamma(N - j + 1; N) \sim \frac{1}{2} \Gamma(N - j + 1) \left(1 + \operatorname{erf}\left(\frac{j}{\sqrt{2N}}\right)\right), \tag{15.51}$$

and in this use Stirling's formula to approximate the gamma function therein by

$$\Gamma(N - j + 1) \sim (2\pi N)^{1/2} N^{N-j} e^{-N} e^{-j^2/2N}.$$

An expression of the form $\frac{1}{\sqrt{N}}\sum_{j=1}^{N} f(j/\sqrt{N})$ results. Since this sum tends to $\int_0^\infty f(t)dt$ in the limit $N \to \infty$, the stated formula follows. □

In particular, Proposition 15.3.3 gives for the density profile as measured from the wall at $y = 0$,

$$\rho_{(1)}(y) = \frac{\rho}{\pi^{1/2}} e^{-2\pi\rho y^2} \int_0^\infty \frac{e^{2(2\pi\rho)^{1/2} y t} e^{-t^2}}{\frac{1}{2}(1 + \operatorname{erf} t)} dt. \tag{15.52}$$

At contact with the wall ($y = 0$), the fact that $\frac{d}{dt}\log(1 + \operatorname{erf} t) = \frac{2}{\sqrt{\pi}} e^{-t^2}/(1 + \operatorname{erf} t)$ reveals the closed form evaluation

$$\rho_{(1)}(0) = \rho \log 2. \tag{15.53}$$

We also draw attention to the asymptotic behavior of the truncated two-particle correlation at large distances $|x_1 - x_2|$ along the wall. Thus integration by parts of $h(z)$ gives

$$\rho_{(2)}^T(\vec{r}_1, \vec{r}_2) \underset{|x_1 - x_2| \to \infty}{\sim} -\frac{2\rho e^{-2\pi\rho(y_1^2 + y_2^2)}}{\pi^2 (x_1 - x_2)^2}. \tag{15.54}$$

15.3.2 Complex random matrices

According to Proposition 15.1.1 the only difference between the 2dOCP at $\Gamma = 2$, with $\rho = 1/\pi$, and complex random matrices from the viewpoint of (15.44) is that in the latter we require $R = \infty$. By the readily verified formula

$$\frac{\Gamma(N; x)}{\Gamma(N)} = e^{-x} \sum_{j=1}^{N} \frac{x^{j-1}}{(j-1)!}, \qquad \Gamma(j; x) := \int_x^\infty t^{j-1} e^{-t}\, dt = \Gamma(x) - \gamma(j; x) \tag{15.55}$$

the correlations are immediate.

Proposition 15.3.4 *For complex random matrices as specified in Proposition 15.1.1,*

$$\rho_{(n)}(\vec{r}_1, \dots, \vec{r}_n) = \pi^{-n} \det\left[e^{-(r_\mu^2 + r_\gamma^2)/2} e^{z_\mu \bar{z}_\gamma} \frac{\Gamma(N; z_\mu \bar{z}_\gamma)}{\Gamma(N)} \right]_{\mu,\gamma=1,\dots,n}.$$

In the limit $N \to \infty$ the result for the bulk correlations (15.44), with $\rho = 1/\pi$, is evident. To compute the edge correlation, with $\vec{r} = (x, y)$ we move the origin to $x = 0$, $y = -\sqrt{N}$, so that (15.49) applies. Making use then of (15.55) gives the following result.

Proposition 15.3.5 *For complex random matrices as specified in Proposition 15.1.1*

$$\lim_{N\to\infty} \rho_{(n)}((x_1, -N + y_1), \dots, (x_n, -N + y_n))$$
$$= \pi^{-n} \det\left[e^{-(\vec{r}_\mu - \vec{r}_\gamma)^2/2} H\Big(\frac{1}{2}(y_\mu + y_\gamma - i(x_\mu - x_\gamma))\Big) \right]_{\mu,\gamma=1,\dots,n},$$

where

$$H(z) := \frac{1}{2\pi}\Big(1 + \operatorname{erf}(\sqrt{2} z)\Big).$$

15.3.3 Correlation functions for the zeros of some complex random polynomials

For the p.d.f. (15.14) it is not known how to compute the correlation functions directly. However, by returning to the expression (15.12) for the joint distribution of the coefficients, a closed form expression for $\rho_{(n)}$ can be obtained [288]. Note that the distribution in (15.12) can be regarded as an $(N+1)$-dimensional complex Gaussian distribution with covariance matrix $\mathbf{L} := \mathrm{diag}(\sigma_1^{-2}, \dots, \sigma_{N+1}^{-2})$ (see Exercises 15.3 q.3).

The first step is to introduce complex numbers $z_1^{(0)}, \dots, z_k^{(0)}$ $(k < (N+1)/2)$ and to define $2k$ linear combinations of the coefficients $a_0, a_1, \dots, a_N$ by $p(z_l^{(0)}) =: p_l$ and $p'(z_l^{(0)}) =: p_l'$ $(l = 1, \dots, k)$, where the prime denotes differentiation. Then general properties of the Gaussian distribution as revised in Exercises 15.3 q.3 give that in terms of these $2k$ complex variables (with the other $N+1-2k$ complex variables integrated out), the probability measure (15.12) reduces to

$$\frac{1}{(2\pi)^{2k}} \frac{1}{\det \mathbf{M}} \exp\Big(-\frac{1}{2}(\vec{p}, \vec{p}\,')^\dagger \mathbf{M}^{-1} (\vec{p}, \vec{p}\,') \Big) \prod_{l=1}^k dp_l^{\mathrm{r}} dp_l^{\mathrm{i}} dp_l'^{\mathrm{r}} dp_l'^{\mathrm{i}},$$

where $\mathbf{M}$ is the covariance matrix

$$\mathbf{M} = \begin{bmatrix} \mathbf{A} & \mathbf{B} \\ \mathbf{B}^\dagger & \mathbf{C} \end{bmatrix}, \quad \mathbf{A} := [\langle p_j \bar{p}_l \rangle_{\mathbf{L}}]_{j,l=1,\dots,k}, \quad \mathbf{B} := [\langle p_j \bar{p}_l' \rangle_{\mathbf{L}}]_{j,l=1,\dots,k}, \quad \mathbf{C} := [\langle p_j' \bar{p}_l' \rangle_{\mathbf{L}}]_{j,l=1,\dots,k}.$$

For given points $z_1^{(0)}, \dots, z_k^{(0)}$ the above expression gives the p.d.f. for the corresponding values of $p(z_1^{(0)}), \dots, p(z_k^{(0)})$ and $p'(z_1^{(0)}), \dots, p'(z_k^{(0)})$. We want to now change variables so that the points $z_1^{(0)}, \dots, z_k^{(0)}$ replace the function values $p(z_1^{(0)}), \dots, p(z_k^{(0)})$ as the variables. The Jacobian for each such change of variables is simply $|p_l'|^2$ so we obtain

$$\frac{1}{(2\pi)^{2k}} \frac{1}{\det \mathbf{M}} \exp\Big(-\frac{1}{2}(\vec{p}, \vec{p}\,')^\dagger \mathbf{M}^{-1} (\vec{p}, \vec{p}\,') \Big) \prod_{l=1}^k |p_l'|^2 dz_l^{\mathrm{r}} dz_l^{\mathrm{i}} dp_l'^{\mathrm{r}} dp_l'^{\mathrm{i}}.$$

Of course the change of variables is only locally one to one, as there will be in general k points giving the same function value. This shows that if we set $\vec{p} = \vec{0}$ and integrate over each $p_l'^{\mathrm{r}}, p_l'^{\mathrm{i}}$ the k-point correlation function for the zeros will result, giving

$$\rho_{(k)}(z_1^{(0)}, \dots, z_k^{(0)}) = \frac{1}{(2\pi)^{2k}} \frac{1}{\det \mathbf{M}} \int_{(-\infty,\infty)^{2k}} \prod_{l=1}^k dp_l'^{\mathrm{r}} dp_l'^{\mathrm{i}} \, |p_l'|^2 \exp\Big(-\frac{1}{2}(\vec{0}, \vec{p}\,')^\dagger \mathbf{M}^{-1} (\vec{0}, \vec{p}\,') \Big). \quad (15.56)$$

PROPOSITION 15.3.6 *The integral formula (15.56) for $\rho_{(k)}$ can be evaluated with the result [288]*

$$\rho_{(k)}(z_1^{(0)}, \dots, z_k^{(0)}) = \frac{1}{\pi^k} \frac{\mathrm{per}(\mathbf{C} - \mathbf{B}^\dagger \mathbf{A}^{-1} \mathbf{B})}{\det \mathbf{A}} \quad (15.57)$$

(recall (4.35) for the definition of a permanent).

Proof. Introducing the auxiliary complex vector $\vec{\nu}$, (15.56) can be rewritten

$$\rho_{(k)}(z_1^{(0)}, \dots, z_k^{(0)}) = \frac{1}{\pi^{2k}} \frac{1}{\det \mathbf{M}} \frac{\partial^{2k}}{\partial \nu_1 \cdots \partial \nu_k \partial \bar{\nu}_1 \cdots \partial \bar{\nu}_k} \int_{(-\infty,\infty)^{2k}} \prod_{l=1}^k dp_l'^{\mathrm{r}} dp_l'^{\mathrm{i}}$$
$$\times \exp\Big(-\frac{1}{2}(\vec{0}, \vec{p}\,')^\dagger \mathbf{M}^{-1} (\vec{0}, \vec{p}\,') + \frac{1}{2} i \vec{\nu}^* \cdot \vec{p}\,' - \frac{1}{2} i \vec{\nu} \cdot \vec{p}\,'^* \Big) \Big|_{\vec{\nu} = \vec{0}},$$

where $\vec{t}^*$ denotes the complex conjugate of entries of $\vec{t}$. Now write $(\vec{0}, \vec{p}\,')^\dagger \mathbf{M}^{-1} (\vec{0}, \vec{p}\,') = \vec{p}\,'^{*T} \mathbf{N}^{-1} \vec{p}\,'$, where $\mathbf{N}^{-1}$ is the $k \times k$ matrix which is the lower-right $k \times k$ submatrix of $\mathbf{M}^{-1}$, and so related to the blocks of $\mathbf{M}$ by

$$\mathbf{N} = \mathbf{C} - \mathbf{B}^\dagger \mathbf{A}^{-1} \mathbf{B}. \quad (15.58)$$

The integral is then of the type evaluated by the formula (15.65) below, so we have

$$\rho_{(k)}(z_1^{(0)}, \dots, z_k^{(0)}) = \Big(\frac{2}{\pi}\Big)^k \frac{\det \mathbf{N}}{\det \mathbf{M}} \frac{\partial^{2k}}{\partial \nu_1 \cdots \partial \nu_k \partial \bar{\nu}_1 \cdots \partial \bar{\nu}_k} \exp\Big(\frac{1}{2} \vec{\nu}^* \cdot \mathbf{N} \vec{\nu} \Big) \Big|_{\vec{\nu} = \vec{0}}.$$

Carrying out the differentiation and noting $\det \mathbf{M} = \det \mathbf{A} \det(\mathbf{C} - \mathbf{B}^\dagger \mathbf{A}^{-1}\mathbf{B}) = \det \mathbf{A} \det \mathbf{N}$ gives the stated formula. □

In general the structure of the $\rho_{(k)}$ for the zeros of random complex polynomials implied by Proposition 15.3.6 is more complicated than the structure exhibited by the $\rho_{(k)}$ for the eigenvalues of complex random matrices as given in Proposition 15.3.2. In particular, in general there is no simple formula for $\rho^T_{(k)}$ in the random polynomial problem (for an exception see Exercises 15.7 q.3).

The simplest correlation is the one body density, when the formula of Proposition 15.3.6 simplifies to read

$$\rho_{(1)}(z) = \frac{1}{\pi} \frac{\langle p\bar{p}\rangle_{\mathbf{L}} \langle p'\bar{p}'\rangle_{\mathbf{L}} - \langle p\bar{p}'\rangle_{\mathbf{L}} \langle \bar{p}p'\rangle_{\mathbf{L}}}{(\langle p\bar{p}\rangle_{\mathbf{L}})^2} = \frac{1}{\pi} \frac{\partial^2}{\partial z \partial \bar{z}} \log \langle p\bar{p}\rangle_{\mathbf{L}}. \tag{15.59}$$

In the specific case that $\sigma_j^2 = 1/j!$, as occurred in the study of the random single particle state confined to the lowest Landau level (recall Section 15.1.2), we have

$$\langle p\bar{p}\rangle_{\mathbf{L}} = \sum_{l=0}^{N} \frac{|z|^{2l}}{l!} = e^{|z|^2} \frac{\Gamma(N+1; |z|^2)}{\Gamma(N+1)}, \tag{15.60}$$

where the second equality follows from (15.55). We know from (15.51) that the asymptotics of $\Gamma(N+1; |z|^2)$ switches behavior at $|z|^2 = N+1$. This indicates that $|z|^2 = N$ defines to leading order the boundary of the support of the density (which must therefore be equal to $1/\pi$ inside the disk, and zero outside). Indeed by writing $z = x + i(y - \sqrt{N})$ and using (15.51) we see from (15.59) and (15.60) that in the limit $N \to \infty$

$$\rho_{(1)}(y) = \frac{1}{\pi}\Big(1 + \frac{1}{4}\frac{\partial^2}{\partial y^2} \log(1 + \mathrm{erf}(\sqrt{2}y))\Big) \sim \begin{cases} 1/\pi, & y \to \infty, \\ 1/4y^2, & y \to -\infty. \end{cases}$$

The two-particle correlation function in the case $\sigma_j^2 = 1/j!$ can also readily be computed in the limit $N \to \infty$. For $z_1 = x_1 + iy_1$ and $z_2 = x_2 + iy_2$ fixed one finds [288]

$$\begin{aligned} \rho_{(2)}(z_1, z_2) &= \frac{1}{\pi^2} f(|z_1 - z_2|^2/2), \\ f(x) &:= \frac{(\sinh^2 x + x^2)\cosh x - 2x \sinh x}{\sinh^3 x} = \frac{1}{2}\frac{d^2}{dx^2}\Big(x^2 \coth x\Big). \end{aligned} \tag{15.61}$$

This result has been proved to be universal, in the sense that it persists for random polynomials with complex coefficients a_j drawn from a large class of distributions of mean zero and variance $1/j!$ [67].

EXERCISES 15.3 1. [320] The aim of this exercise is to compute the asymptotic expansion of the free energy for the 2dOCP in a disk at $\Gamma = 2$.

(i) Use (4.160) and (1.72) to show that for the 2dOCP in a disk at $\Gamma = 2$

$$\beta F = -\log \frac{Q_N}{N!} + N^2 \log R - \frac{3}{4} N^2,$$

where

$$Q_N := \int_\Omega d\vec{r}_1 \cdots \int_\Omega d\vec{r}_N e^{-\pi\rho \sum_{j=1}^N |\vec{r}_j|^2} \prod_{1 \le j < k \le N} |\vec{r}_k - \vec{r}_j|^2.$$

(ii) Make use of (15.40) and (15.41) to show that

$$Q_N = N! \pi^N (\pi\rho)^{-N(N+1)/2} \prod_{j=1}^N \gamma(j; N),$$

where $\gamma(k; a)$ is specified by (15.50).

(iii) Make use of the asymptotic expansions (15.51) and (4.186) (with $\beta = 2$) to deduce that for $N, R \to \infty$ with ρ fixed

$$\beta F \sim \pi R^2 \beta f_v + 2\pi R \beta\gamma + \frac{1}{6}\log\left((\pi\rho)^{1/2}R\right) + \mathrm{O}(1), \tag{15.62}$$

where

$$\beta f_v = \frac{\rho}{2}\log\left(\rho/2\pi^2\right), \quad \beta\gamma = -\left(\frac{\rho}{2\pi}\right)^{1/2}\int_0^\infty \log\left(\frac{1}{2}(1+\mathrm{erf}\, y)\right)dy \tag{15.63}$$

(f_v is the free energy per volume and γ the surface tension).

(iv) [313] Use an appropriate generalization of (4.179) together with (15.45) to show that in the thermodynamic limit the internal energy per particle is given by

$$u = -\frac{1}{4}(\log \pi\rho + C),$$

where C is Euler's constant.

2. [203] Calculate the leading two terms in the $X \to \infty$ asymptotic expansion of $\mathrm{erf}(Y + iX)$, and from this and Proposition 15.3.5 conclude that the edge correlations for complex random matrices has the property

$$\rho_{(2)}^T(y_1, y_2; x_1 - x_2) \underset{|x_1 - x_2|\to\infty}{\sim} -\frac{f(y_1)f(y_2)}{4\pi^2(x_1-x_2)^2}, \qquad f(y) := \pi\frac{d}{dy}\rho_{(1)}(y). \tag{15.64}$$

3. Consider the integral identity (1.93) with n replaced by $2n$ and $\mathbf{A}$ of the form

$$\mathbf{A} = \begin{bmatrix} \mathbf{a} & \mathbf{b} \\ -\mathbf{b} & \mathbf{a} \end{bmatrix},$$

where $\mathbf{a}$ and $\mathbf{b}$ are $n \times n$ matrices, and write $\vec{b}^T = (\vec{u}^T, \vec{v}^T)$ where $\vec{u}$ and $\vec{v}$ are n-component vectors, and let * denote complex conjugation.

(i) Show that (1.93) can be written as

$$\int_{\mathbb{R}^{2n}} dx_1 dy_1 \cdots dx_n dy_n \, \exp\left(-\frac{1}{2}\vec{z}^\dagger \mathbf{C}\vec{z} + \frac{1}{2}\vec{w}^*\cdot\vec{z} + \frac{1}{2}\vec{w}\cdot\vec{z}^*\right) = (2\pi)^n\left(\det \mathbf{C}\right)^{-1}\exp\left(\frac{1}{2}\vec{w}^*\cdot\mathbf{C}^{-1}\vec{w}\right), \tag{15.65}$$

where $\vec{z} = \vec{x} + i\vec{y}$ and $\vec{w} = \vec{u} + i\vec{v}$.

(ii) Let $\langle f \rangle_{\mathbf{C}}$ denote the normalized average with respect to $\exp(-(1/2)\vec{z}^\dagger\mathbf{C}\vec{z})$. From (15.65) deduce that

$$2^{-p}\langle z_{i_1}\bar{z}_{j_1}\cdots z_{i_p}\bar{z}_{j_p}\rangle_{\mathbf{C}} = \sum_{\substack{\text{all permutations} \\ P \text{ of } \{j_1,\dots,j_p\}}} C^{-1}_{i_1 P(j_1)}\cdots C^{-1}_{i_p P(j_p)}$$

while

$$\langle z_{i_1}\cdots z_{i_p}\bar{z}_{j_1}\cdots\bar{z}_{j_q}\rangle_{\mathbf{C}} = 0 \quad \text{for} \quad p \neq q.$$

Use the first of these results to deduce that

$$\mathbf{C}^{-1} = \frac{1}{2}[\langle z_j \bar{z}_k\rangle_{\mathbf{C}}]_{j,k=1,\dots,N}.$$

(iii) Regarding the r.h.s. of (15.65) as a p.d.f. for $\vec{w}$—the multivariable complex Gaussian distribution—show that the results of Exercises 1.5 q.1(iii)&(iv) (appropriately modified) hold.

4. [479] The objective of this exercise is to derive the functional formula (15.46).

(i) Check that

$$\prod_{j=3}^N |r_1 e^{i\theta_1} - r_j e^{i\theta_j}|^\Gamma |r_2 e^{i\theta_2} - r_j e^{i\theta_j}|^\Gamma = \prod_{j=3}^N \left((z_1 - z_j)(\bar{z}_1 - \bar{z}_j)(z_2 - z_j)(\bar{z}_2 - \bar{z}_j)\right)^{\Gamma/2}$$

is unchanged by the interchange $z_1 \leftrightarrow z_2$ with $\bar{z}_1$ and $\bar{z}_2$ fixed.

(ii) Use this property and the definition of $\rho_{(2)}(\vec{r}_1, \vec{r}_2) = \rho_{(2)}((z_1, \bar{z}_1); (\bar{z}_2, \bar{z}_2))$ to show that for a 2dOCP with Boltzmann factor

$$\prod_{j=1}^{N} f(\vec{r}_j) \prod_{1 \le j < k \le N} |\vec{r}_j - \vec{r}_k|^\Gamma = \prod_{j=1}^{N} f(z_j, \bar{z}_j) \prod_{1 \le j < k \le N} |z_j - z_k|^\Gamma$$

the functional property

$$\frac{\rho_{(2)}((z_1, \bar{z}_1); (z_2, \bar{z}_2))}{f(z_1, \bar{z}_1) f(z_2, \bar{z}_2)((z_1 - z_2)(\bar{z}_1 - \bar{z}_2))^\Gamma} = \frac{\rho_{(2)}((z_2, \bar{z}_1); (z_1, \bar{z}_2))}{f(z_2, \bar{z}_1) f(z_1, \bar{z}_2)((z_2 - z_1)(\bar{z}_1 - \bar{z}_2))^\Gamma}$$

holds. Put $f(\vec{r}) = e^{-\pi\rho\Gamma|\vec{r}|^2/2}$, set $z_1 = 0$ and assume a fluid state in the thermodynamic limit to deduce (15.46).

(iii) Show that (15.46) is equivalent to saying $g_{(2)}(r)$ possesses the functional form

$$g_{(2)}(r) = r^\Gamma e^{-\pi\rho\Gamma r^2/4} f(r^4)$$

for $f(x)$ analytic.

5. [172] In this exercise the eigenvalue density of a matrix with complex eigenvalues will be related to that in a related chiral ensemble.

(i) Note from the Poisson equation (1.40), written in terms of the operators $\frac{\partial}{\partial z}, \frac{\partial}{\partial \bar{z}}$ defined below (15.27), that for any two-dimensional statistical mechanical system with N particles at (x_j, y_j) $(j = 1, \dots, N)$,

$$\rho_{(1)}(\vec{r}) = \frac{1}{\pi} \frac{\partial^2}{\partial z \partial \bar{z}} \Big\langle \sum_{j=1}^{N} \log(z - z_j)(\bar{z} - \bar{z}_j) \Big\rangle = \frac{2}{\pi} \frac{\partial^2}{\partial z \partial \bar{z}} \int_{\mathbb{R}^2} \log|z - w| \rho_{(1)}((x', y'))\, dx' dy', \quad (15.66)$$

where $z := x + iy$, $w = x' + iy'$.

(ii) In the case of a random matrix $\mathbf{X}$ with complex eigenvalues $\{z_j\}_{j=1,\dots,N}$, note that

$$\sum_{j=1}^{N} \log(z - z_j)(\bar{z} - \bar{z}_j) = \mathrm{Tr}\, \log(z\mathbf{1}_N - \mathbf{X})(\bar{z}\mathbf{1}_N - \bar{\mathbf{X}}) = \mathrm{Tr}\, \log\Big((-1)^N \mathbf{H}\Big), \quad (15.67)$$

where

$$\mathbf{H} = \begin{bmatrix} \mathbf{0} & z\mathbf{1}_N - \mathbf{X} \\ \bar{z}\mathbf{1}_N - \bar{\mathbf{X}}^T & \mathbf{0} \end{bmatrix}.$$

Let $\rho_{(1)}^{\mathbf{H}}(x; z, \bar{z})$ denote the eigenvalue density of the positive eigenvalues for the class of Hermitian matrices $\mathbf{H}$, and $\rho_{(1)}^{\mathbf{X}}(\vec{r})$ the eigenvalue density for the random matrices $\mathbf{X}$. By substituting (15.67) in the first equality of (15.66) deduce the inter-relation

$$\rho_{(1)}^{\mathbf{X}}(\vec{r}) = \frac{2}{\pi} \frac{\partial^2}{\partial z \partial \bar{z}} \int_0^\infty (\log t) \rho_{(1)}^{\mathbf{H}}(t; z, \bar{z})\, dt.$$

6. (i) [242] Use the result of Exercises 15.2 q.3 to deduce that $\{C_n(z)\}_{n=0,1,\dots}$ with

$$C_n(z) := \Big(\frac{\tau}{2}\Big)^{n/2} H_n\Big(\frac{z}{\sqrt{2\tau}}\Big)$$

satisfies the orthogonality

$$\int_{-\infty}^{\infty} dx \int_{-\infty}^{\infty} dy\, e^{-x^2/(1+\tau) - y^2/(1-\tau)} C_m(z) C_n(\bar{z}) = \pi m! \sqrt{1 - \tau^2} \delta_{m,n}.$$

As in going from (15.41) to (15.43), conclude that the correlation kernel for the p.d.f. (15.16) is given by

$$(1-\tau^2)^{-n/2}\pi^{-n}\exp\Big(-\frac{1}{1-\tau^2}\Big(\sum_{l=1}^{n}|z_l|^2-\frac{\tau}{2}(z_l^2+\bar{z}_l^2)\Big)\Big)\sum_{l=0}^{N-1}\frac{C_l(z_j)C_l(\bar{z}_k)}{l!}. \tag{15.68}$$

(ii) [250] The weakly non-Hermitian limit is specified by writing $\tau=(1-\alpha^2/N)$, scaling the coordinates $z\mapsto \pi z/\sqrt{N}$ and taking the limit $N\to\infty$. This has the effect of concentrating the eigenvalues in an $O(1)$ neighborhood of the x-axis. Use the asymptotic expansion (7.1) and the result (15.68) to show that the limiting correlation kernel (the scale factor π^2/N must first be multiplied to account for the scale of $dxdy$) is equal to

$$\frac{(2\pi)^{1/2}}{\alpha}\exp\Big(-\frac{\pi^2}{\alpha^2}(y^2+(y')^2)\Big)\int_0^1 e^{-\alpha^2u^2}\cos\pi u(z-z')\,du.$$

15.4 GENERAL PROPERTIES OF THE CORRELATIONS AND FLUCTUATION FORMULAS

15.4.1 Dipole moment and Stillinger-Lovett sum rules

In Chapter 14 macroscopic physical characterizations of log-gases were used to make certain predictions relating to correlations for general β. These considerations carry over to the case of the 2dOCP, or more generally two-dimensional Coulomb systems in their conductive phase [386]. Here, for definiteness, only the case of the 2dOCP will be considered. One-component systems are special because only then is the microscopic particle density $n_{(1)}(\vec{r})$ proportional to the microscopic charge density $c_{(1)}(\vec{r})$.

We will begin by generalizing (14.12) (generalization is possible because, as distinct from the log-gas on a line, the correlations for the 2dOCP are expected to decay faster than any algebraic power). For this we use the linear response relation (14.1), with the perturbation due to an external charge δq at the origin, $\delta U=-\delta q\int_{\mathbb{R}^2}c_{(1)}(\vec{r})\log|\vec{r}|\,d\vec{r}$, and with the observable $A=\prod_{j=1}^{l}n_{(1)}(\vec{r}_j-\vec{a})$, which gives

$$\langle A\rangle_{\delta q}-\langle A\rangle_0=\beta\delta q\int_{\mathbb{R}^2}\log|\vec{r}+\vec{a}|\Big\langle c_{(1)}(\vec{r})\prod_{j=1}^{l}n_{(1)}(\vec{r}_j)\Big\rangle^T d\vec{r}. \tag{15.69}$$

With the assumption that the l.h.s. decays faster than any algebraic power for $|\vec{a}|\to\infty$, and use of the multipole expansion

$$\log|\vec{r}+\vec{a}|=\log|\vec{a}|+\sum_{p=1}^{\infty}\frac{(-1)^{p+1}}{p}\Big(\frac{r}{a}\Big)^p\cos p(\theta-\phi)$$

where in polar coordinates $\vec{r}=(r,\theta)$, $\vec{a}=(a,\phi)$ and it is assumed $r/a<1$, taking $|\vec{a}|\to\infty$ in (15.69) gives

$$\int_{\mathbb{R}^2}(x-iy)^p\Big\langle c_{(1)}(\vec{r})\prod_{j=1}^{l}n_{(1)}(\vec{r}_j)\Big\rangle^T d\vec{r}=0 \qquad (p=0,1,\dots).$$

Proceeding as in Exercises 14.1 q.1(i) this can be rewritten as

$$\int_{\mathbb{R}^2}(x-iy)^p\rho_{(l+1)}^T(\vec{r}_1,\dots,\vec{r}_l,\vec{r})\,d\vec{r}=-\Big(\sum_{j=1}^{l}(x_j-iy_j)^p\Big)\rho_{(l)}^T(\vec{r}_1,\dots,\vec{r}_l). \tag{15.70}$$

In the case $p=0$ this is formally the same as (14.12). For $p=1$ it is referred to as the dipole moment sum rule. Its validity for $\Gamma=2$ can be checked from Proposition 15.3.2 and (5.9), together with use of polar coordinates.

For the log-gas on a line the structure function $\hat{S}(k)$, related to the dimensionless structure function $\tilde{S}(k)$ by (14.7), exhibits the small $|k|$ behavior (14.8). Applying essentially the same linear response argument, but with a two-dimensional external charge density $\epsilon e^{\imath\vec{k}\cdot\vec{r}}$, and the two-dimensional generalized function result

$$-\int_{\mathbb{R}^2} \log|\vec{r}|e^{-\imath\vec{k}\cdot\vec{r}}\,d\vec{r} = \frac{2\pi}{k^2}, \quad k := |\vec{k}|, \tag{15.71}$$

one obtains the prediction that

$$\hat{S}(k) \underset{k\to 0}{\sim} \frac{k^2}{2\pi\beta}. \tag{15.72}$$

In terms of the notation (14.16), and assuming fast decay of the latter, equating powers of k^2 on both sides shows

$$\int_{\mathbb{R}^2} r^2 C_{(2)}(\vec{r},\vec{0})\,d\vec{r} = -\frac{2}{\pi\beta}. \tag{15.73}$$

This condition on the second moment of $C_{(2)}(\vec{r},\vec{0})$ is known as the *Stillinger-Lovett sum rule*. It can in fact be rigorously derived from the $n=2$ Born-Green-Yvon equation, together with the assumptions that (15.70) holds for $p=1$, $l=1,2$ and the three-particle correlations have a sufficiently fast decay [387].

Analogous to (14.7), in two dimensions $\hat{S}(k) = \rho\tilde{S}(k/\rho^{1/2})$, where $\tilde{S}(k/\rho^{1/2})$ is dimensionless. If we now expand $\tilde{S}(k/\rho^{1/2})$ as a power series in $k/\rho^{1/2}$, and then take $\rho\to\infty$, we see from (15.73) that

$$\lim_{\rho\to\infty} \hat{S}(k) = \frac{k^2}{2\pi\beta}. \tag{15.74}$$

This is equivalent to the statement that [316]

$$\lim_{\rho\to\infty} C_{(2)}(\vec{r},\vec{0}) = -\frac{1}{2\pi\beta}\nabla^2\delta(\vec{r}). \tag{15.75}$$

For the 2dOCP, $C_{(2)}(\vec{r},\vec{0}) = q^2\rho_{(2)}^T(\vec{r},\vec{0})$. In this case, the linear response argument leading to (14.15) can be generalized, implying the fourth moment condition

$$\rho\int_{\mathbb{R}^2} r^4\rho_{(2)}^T(\vec{r},\vec{0})\,d\vec{r} = -\frac{16}{(\pi\Gamma)^3}\Big(1-\frac{\Gamma}{4}\Big). \tag{15.76}$$

In fact the expansion of $\hat{S}(k)$ is known to $\mathrm{O}(k^6)$ [340]. It has been computed using exact diagrammatic expansion techniques, and implies the sixth moment sum rule

$$\rho^2\int_{\mathbb{R}^2} r^6\rho_{(2)}^T(\vec{r},\vec{0})\,d\vec{r} = -\frac{18}{(\pi\Gamma)^4}(\Gamma-6)\Big(\Gamma-\frac{8}{3}\Big). \tag{15.77}$$

The sum rules (15.73), (15.76) and (15.77) can be checked for $\Gamma=2$ on the exact result (15.45).

15.4.2 Contact theorem and related sum rules

The sum rule (15.73) is expected to be true of general two-dimensional Coulomb systems in their conductive phase. In the special case of the 2dOCP, a different viewpoint on (15.73) can be obtained by studying the consequence of scaling properties of the partition function.

On this theme, we begin by noting from (1.72) that with

$$Z_N = e^{-(\Gamma/2)N^2\log R + 3\Gamma N^2/8}\frac{1}{N!}\int_{|\vec{r}_1|<R} d\vec{r}_1\cdots\int_{|\vec{r}_N|<R} d\vec{r}_N \prod_{l=1}^N e^{-\pi\rho\Gamma|\vec{r}_l|^2/2}\prod_{1\le j<k\le N}|\vec{r}_k-\vec{r}_j|^\Gamma, \tag{15.78}$$

$\rho = N/\pi R^2$, changing variables $\vec{r}_j \mapsto R\vec{s}_j$, shows that the dependence on R can be scaled from the integral.

Hence, analogous to (4.169) one has

$$\beta P := \frac{\partial}{\partial(\pi R^2)} \log Z_N = \Big(1 - \frac{\Gamma}{4}\Big)\rho. \tag{15.79}$$

A simple consequence of this is a formula expressing the density at contact in terms of a moment of the density profile in the finite system [109].

PROPOSITION 15.4.1 *With the density measured inward from the boundary*

$$\rho_{(1)}(0) - \Big(1 - \frac{\Gamma}{4}\Big)\rho = -\frac{\Gamma\rho^2\pi^2}{N}\int_0^R r^3\Big(\rho_{(1)}(R-r) - \rho\Big)\, dr. \tag{15.80}$$

Proof. Direct differentiation of (15.78) and use of the general definition of the density (5.1) with $n = 1$ gives

$$\frac{\partial}{\partial(\pi R^2)} \log Z_N = \rho_{(1)}(0) + \frac{\pi N\Gamma}{2}\Big(\frac{1}{\pi R^2}\Big)^2 \int_0^R r^3\Big(\rho_{(1)}(R-r) - \rho\Big)\, dr. \tag{15.81}$$

Equating (15.79) and (15.81) gives (15.80). □

In the thermodynamic limit (15.80) reduces to a formula for the density at contact with the boundary and the potential drop.

PROPOSITION 15.4.2 *For the 2dOCP in the half plane $y \ge 0$, the potential drop $\Delta\phi$ is specified by*

$$\Delta\phi = 2\pi\int_0^\infty y(\rho_{(1)}(y) - \rho)\, dy. \tag{15.82}$$

In terms of this quantity, and with βP specified by (15.79), we have

$$\rho(0) = \beta P + \rho\Gamma\Delta\phi. \tag{15.83}$$

Proof. Neutrality implies

$$\int_0^R r(\rho_{(1)}(R-r) - \rho)\, dr = \int_0^R (R-r)(\rho_{(1)}(r) - \rho)\, dr = 0. \tag{15.84}$$

Changing variables in (15.81) $r \mapsto R - r$ and making use of the above equation implies

$$\begin{aligned} &-\frac{\Gamma\rho^2\pi^2}{N}\int_0^R r^3(\rho_{(1)}(R-r) - \rho)\, dr \\ &= 2\Gamma\rho\pi\int_0^R r(\rho_{(1)}(r) - \rho)\, dr - \frac{3\Gamma(\rho\pi)^{3/2}}{N^{1/2}}\int_0^R r^2(\rho_{(1)}(r) - \rho)\, dr + \frac{\Gamma\rho^2\pi^2}{N}\int_0^R r^3(\rho_{(1)}(r) - \rho)\, dr. \end{aligned}$$

For fast decay of $\rho_{(1)}(r) - \rho$, the second two terms on the r.h.s. tend to zero as $N, R \to \infty$ and so the stated result follows from (15.80). □

The so-called contact theorem (15.82) can be verified at $\Gamma = 2$ using the exact results (15.52) and (15.53).

The idea behind the derivation of Proposition 15.4.1 can be applied to give a formula for the second moment of $\rho_{(2)}^T$ in the 2dOCP in the finite system [511].

PROPOSITION 15.4.3 *We have*

$$\int_{|\vec{r}|<R} r^2\rho_{(2)}^T(\vec{0},\vec{r})\, d\vec{r} = -\frac{2}{\pi\Gamma}\Big(\rho_{(1)}(0)/\rho + N\rho_{(2)}^T(0,R)/\rho^2\Big), \tag{15.85}$$

where here $\rho_{(1)}(0)$ refers to the density at the center of the system.

Proof. Consider

$$\tilde{Z}_{N-1} := e^{-(\Gamma/2)N^2 \log R + 3\Gamma N^2/8}$$
$$\times \frac{1}{(N-1)!} \int_{|\vec{r}_1|<R} d\vec{r}_1 \cdots \int_{|\vec{r}_{N-1}|<R} d\vec{r}_{N-1} \prod_{l=1}^{N-1} |\vec{r}_l|^\Gamma e^{-\pi\rho\Gamma|\vec{r}_l|^2/2} \prod_{1\le j<k\le N-1} |\vec{r}_k - \vec{r}_j|^\Gamma,$$

which corresponds to the partition function for the 2dOCP consisting of N particles with one of the particles fixed at the origin. By scaling out the radius we deduce

$$\frac{1}{\tilde{Z}_N} \frac{\partial}{\partial(\pi R^2)} \tilde{Z}_{N-1} = \frac{N-1}{\pi R^2} \rho_{(1)}(0) - \frac{\Gamma}{4} \rho \rho_{(1)}(0), \tag{15.86}$$

and direct differentiation gives

$$\frac{1}{\tilde{Z}_N} \frac{\partial}{\partial(\pi R^2)} \tilde{Z}_{N-1} = \rho_{(2)}(0, R) + \frac{\Gamma\rho}{2R^2} \int_{|\vec{r}|<R} r^2 \rho_{(2)}(\vec{0}, \vec{r})\, d\vec{r} - \frac{\Gamma}{4} \rho\rho_{(1)}(0). \tag{15.87}$$

Now write

$$\int_{|\vec{r}|<R} r^2 \rho_{(2)}(\vec{0}, \vec{r})\, d\vec{r} = \int_{|\vec{r}|<R} r^2 \rho^T_{(2)}(\vec{0}, \vec{r})\, d\vec{r} + \rho_{(1)}(0) \Big(\frac{2N}{\Gamma\rho^2\pi}\Big) \Big(\rho_{(1)}(R) - (1 - \Gamma/4)\rho\Big),$$

where the second equality follows upon use of (15.80), paying attention to the fact that the origin is the boundary in that equation. Substituting in (15.87) shows

$$\frac{1}{\tilde{Z}_N} \frac{\partial}{\partial(\pi R^2)} \tilde{Z}_{N-1} = \rho_{(2)}(0, R) + \frac{\Gamma\rho}{2R^2} \int_{|\vec{r}|<R} r^2 \rho^T_{(2)}(\vec{0}, \vec{r})\, d\vec{r} - \rho\rho_{(1)}(0). \tag{15.88}$$

Equating (15.86) and (15.88) gives (15.85). □

In the thermodynamic limit, with the assumption of fast decay of correlations, it is expected that $\rho_{(1)}(0)/\rho \to 1$ while $N\rho^T_{(2)}(0, R)/\rho^2 \to 0$ at rates faster than any power law, so implying (15.73).

15.4.3 Surface correlations

Whereas the particle correlations in the bulk of a two-dimensional Coulomb system in the conductive phase are expected to decay exponentially fast, the same correlations parallel to a boundary will typically exhibit a slow algebraic decay [314]. This is evident from the exact result (15.54) for the asymptotic decay of the truncated two-particle correlation parallel to a hard wall (with a vacuum outside the wall). In fact a universal behavior can be identified if instead of the charge-charge correlation $C_{(2)}(\vec{r}, \vec{r}\,') := \langle c_{(1)}(\vec{r}) c_{(1)}(\vec{r}\,')\rangle^T$ one considers the asymptotic surface charge–surface charge correlation

$$\langle \sigma(x)\sigma(x')\rangle^T := \int_0^\infty dy \int_0^\infty dy'\, S_{\rm as}(\vec{r}, \vec{r}\,'),$$

where $S_{\rm as}(\vec{r}, \vec{r}\,')$ is the leading large $|x - x'|$ asymptotic form of $S(\vec{r}, \vec{r}\,')$. (In the case of the soft edge 2dOCP as realized by the eigenvalue PDF for complex Gaussian matrices, the terminals of integration should begin at $-\infty$.)

We see from (15.54) and (15.64) that for the 2dOCP at $\Gamma = 2$, with a hard wall or soft edge,

$$\beta\langle \sigma(x)\sigma(x')\rangle^T \underset{|x-x'|\to\infty}{\sim} -\frac{1}{2\pi^2(x - x')^2}. \tag{15.89}$$

This result bears a striking resemblance to the asymptotic behavior of the charge-charge correlation for the log-gas (recall (14.9)). In fact it is identical apart from an extra factor of $\frac{1}{2}$ on the r.h.s. of (15.89).

Before giving a derivation of this result, let us consider the heuristic reason behind the slow decay. When a charge is isolated near the surface, or any other point within the Coulomb system for that matter, a screening cloud of equal and opposite total charge will form (implying the charge sum rule must be obeyed). However, as distinct from isolating a charge in the bulk, the screening cloud about a charge isolated near the boundary cannot be spherically symmetric. As a result, it will carry a nonzero dipole moment, and it is this which is responsible for the slow decay of the correlation.

For the derivation of (15.89) we apply the method of Section 14.3 due to Jancovici [316]. This requires that the macroscopic potential $\phi(\vec{r}\,')$ due to the charge distribution in the Coulomb system when an external charge δq, regarded as a perturbation, is fixed at $\vec{r}$. There are four cases to consider, depending on the location of the points $\vec{r}$ and $\vec{r}\,'$ inside or outside the domain Ω, which for the time being will be taken to be an arbitrary planar region with its boundary forming a simple curve.

PROPOSITION 15.4.4 *The macroscopic potential $\phi(\vec{r}\,')$ defined above is given by*

$$\phi(\vec{r}\,') = \begin{cases} \delta q \log|\vec{r} - \vec{r}\,'| + \delta q/C, & \vec{r}, \vec{r}\,' \in \Omega, \\ \delta q \log|\vec{r} - \vec{r}\,'| + \delta q F(\vec{r}\,'), & \vec{r} \in \Omega, \ , \vec{r}\,' \notin \Omega, \\ \delta q \log|\vec{r} - \vec{r}\,'|, & \vec{r} \notin \Omega, \ , \vec{r}\,' \in \Omega, \\ \delta q \log|\vec{r} - \vec{r}\,'| + \delta q G(\vec{r}, \vec{r}\,'), & \vec{r}, \vec{r}\,' \notin \Omega \end{cases}$$

where C is the capacitance, $\delta q G(\vec{r}, \vec{r}\,')$ is the potential at the point $\vec{r}\,'$ due to the charge at $\vec{r}$ and the induced surface charge, and $\delta q F(\vec{r}\,')$ is the potential at $\vec{r}\,'$ due to the surface charge.

Derivation. For both $\vec{r}$ and $\vec{r}\,'$ in the system, a screening cloud of charge $-\delta q$ will surround the charge δq at $\vec{r}$. From large distances this appears as a point charge, and so creates a potential at $\vec{r}\,'$ equal to $\delta q \log|\vec{r} - \vec{r}\,'|$. Also, due to charge conservation, a charge δq will spread itself around the boundaries creating a constant potential $\delta q/C$, where C is the capacitance. Thus

$$\langle \Phi(\vec{r}\,') \rangle_\epsilon - \langle \Phi(\vec{r}\,') \rangle_0 = \delta q \log|\vec{r} - \vec{r}\,'| + \delta q/C. \tag{15.90}$$

For an infinite system in two dimensions $1/C$ will diverge (e.g., for a disk of radius R, $1/C = -\log R$). However, we will see below that only the derivatives $\partial^2/\partial r_\mu \partial r'_\nu$ are relevant to the derivation of (15.89).

Now suppose $\vec{r}$ is in the system, but $\vec{r}\,'$ is outside. Reasoning as above gives

$$\langle \Phi(\vec{r}\,') \rangle_\epsilon - \langle \Phi(\vec{r}\,') \rangle_0 = \delta q \log|\vec{r} - \vec{r}\,'| + \delta q F(\vec{r}\,'), \tag{15.91}$$

where $\delta q F(\vec{r}\,')$ is the potential at $\vec{r}\,'$ due to the surface charge. In the case $\vec{r}$ is outside and the observation point $\vec{r}\,'$ is in, there is no screening cloud but there will be an induced surface charge of zero total charge with a potential cancelling that due to the external charge, so that

$$\langle \Phi(\vec{r}\,') \rangle_\epsilon - \langle \Phi(\vec{r}\,') \rangle_0 = \delta q \log|\vec{r} - \vec{r}\,'|. \tag{15.92}$$

The final case, when both $\vec{r}$ and $\vec{r}\,'$ are outside the conductor, follows by definition of G.

Substituting the results of Proposition 15.4.4, via the linear response relation (14.1), in the formula (14.43) for the electric field–electric field correlation and proceeding as in the derivation of (14.45) we see that

$$\beta \langle \sigma(\vec{r}) \sigma(\vec{r}\,') \rangle^T = -\frac{1}{(2\pi)^2} \frac{\partial^2 G(\vec{r}, \vec{r}\,')}{\partial r_n \partial r'_n} \Big|_{\vec{r}, \vec{r}\,' \in \text{surface}}, \tag{15.93}$$

where r_n is the direction normal to the surface. It remains to specify $G(\vec{r}, \vec{r}\,')$. In the case of the half-plane boundary, the method of images gives

$$G(z, z') = -\log|z - z'| + \log|z - \bar{z}'|.$$

Substituting in (15.93) gives the formula (15.89). As another explicit example, consider the case of a disk of radius R. Then the method of images gives for G the expression (15.174) below, which when substituted in

(15.93) implies

$$\beta\langle\sigma(\vec{r})\sigma(\vec{r}\,')\rangle^T = -\frac{1}{2\pi^2(2R\sin\frac{\theta}{2})^2}, \tag{15.94}$$

where θ is the angle between $\vec{r}$ and $\vec{r}\,'$.

15.4.4 Distribution of a linear statistic

In general the Fourier transform of the p.d.f. for a linear statistic $A = \sum_{j=1}^N a(\vec{r}_j)$ is given by

$$\hat{P}(k) = \Big\langle \prod_{l=1}^N e^{ika(\vec{r}_l)} \Big\rangle \tag{15.95}$$

(recall Section 14.4). In the case of the OCP in a disk with a soft wall at $\Gamma = 2$, or equivalently the problem of Gaussian random matrices with complex entries, the quantity (15.95) can be evaluated exactly for rotationally symmetric linear statistics $A = \sum_{j=1}^N a(|\vec{r}_j|)$. Furthermore, by scaling the variables so that the support of the system is the unit disk, the limiting behavior of the distribution can be explicitly determined [194].

PROPOSITION 15.4.5 *Let $e^{-\beta W}$ be given by (15.1) with $\vec{r}_j \mapsto \vec{r}_j/\sqrt{N}$, and defined $\hat{P}(k)$ by (15.95) (generalized to a two-dimensional domain) with a linear statistic $A = \sum_{j=1}^N a(|\vec{r}_j|)$. Then for $N \to \infty$*

$$\hat{P}(k) \sim e^{ik\mu}e^{-k^2\sigma^2/2}$$

with

$$\mu = 2N\int_0^1 ra(r)\,dr, \quad \sigma^2 = \frac{1}{2}\int_0^1 r(a'(r))^2\,dr.$$

Proof. From (15.95)

$$\hat{P}(k) = \frac{\int_{\mathbf{R}^2} d\vec{r}_1 \cdots \int_{\mathbf{R}^2} d\vec{r}_N e^{-N|\vec{r}_l|^2 + ika(\vec{r}_l)} \prod_{1\le j<k\le N} |\vec{r}_k - \vec{r}_j|^2}{\int_{\mathbf{R}^2} d\vec{r}_1 \cdots \int_{\mathbf{R}^2} d\vec{r}_N e^{-N|\vec{r}_l|^2} \prod_{1\le j<k\le N} |\vec{r}_k - \vec{r}_j|^2}. \tag{15.96}$$

Introducing the Vandermonde determinant as in the proof of Proposition 15.3.1 allows the integrals in this expression to be evaluated, with the result

$$\hat{P}(k) = \prod_{l=1}^N \frac{\int_0^\infty e^{-s}s^{l-1}e^{ika(\sqrt{s/N})}\,ds}{\int_0^\infty e^{-s}s^{l-1}\,ds}.$$

To analyze the $N \to \infty$ limit, we change variables $s \to ls$ and expand the integrand about its large l maximum at $s = 1$. This gives

$$\hat{P}(k) \sim \prod_{l=1}^N e^{ika(\sqrt{l/N})}e^{-k^2(a'(\sqrt{l/N}))^2}.$$

Writing the product of exponentials as the exponential of a sum shows that in fact Riemann sums occur, and their leading behaviors written as integrals gives the stated formula. □

The result of Proposition 15.4.5 exhibits two key features in common with the fluctuation formula (14.68) for the log-gas. These are that the distribution is a Gaussian, and that the variance is O(1). Indeed the argument used to deduce that the distribution will be a Gaussian in the infinite density (macroscopic) limit applies immediately to the two-dimensional OCP, with the key formula (14.61) from macroscopic electrostatics giving an exact linear relationship between an external potential and the corresponding induced charge density. No assumption of a rotationally invariant statistic is required.

Let us now turn our attention to the computation of the variance in the case of general Γ and no assumed symmetry for the linear statistic. With $q = 1$ for convenience, according to the formulas (14.38) and (14.16) we have

$$\mathrm{Var}(A) = \int_\Omega d\vec{r}_1 \int_\Omega d\vec{r}_2\, a(\vec{r}_1)a(\vec{r}_2)C_{(2)}(\vec{r}_1,\vec{r}_2), \tag{15.97}$$

where Ω denotes the unit disk. The significance of this is that in the infinite density limit we know that $C_{(2)}$ exhibits universal behavior in the bulk and at the boundary.

In the bulk, from (15.75), we have

$$C_{(2)}(\vec{r}_1 - \vec{r}_2, \vec{0}) = \frac{1}{2\pi\beta}\Big(\frac{\partial}{\partial x^{(1)}} + i\frac{\partial}{\partial y^{(1)}}\Big)\Big(\frac{\partial}{\partial x^{(2)}} - i\frac{\partial}{\partial y^{(2)}}\Big)\delta(\vec{r}_1 - \vec{r}_2).$$

Substituting in (15.97), and integrating by parts (ignoring possible boundary terms, which are treated separately below) gives

$$\sigma^2_{\rm bulk} = \frac{1}{2\pi\beta}\int_\Omega \Big(\Big(\frac{\partial a(x,y)}{\partial x}\Big)^2 + \Big(\frac{\partial a(x,y)}{\partial y}\Big)^2\Big)\, dxdy. \tag{15.98}$$

In the special case $a(\vec{r}) = a(|\vec{r}|)$, $\beta = 2$, this reproduces the exact result of Proposition 15.4.5 for σ^2.

At the surface, in terms of polar coordinates, we have from (15.94) that

$$C_{(2)}(\vec{r}_1 - \vec{r}_2, \vec{0}) = -\frac{1}{2\pi^2\beta}\Big(\frac{\partial^2}{\partial\theta_1\partial\theta_2}\log\Big|\sin\frac{\theta_1-\theta_2}{2}\Big|\Big)\delta(r_1-1)\delta(r_2-1).$$

Substituting in (15.97), integrating by parts and introducing the Fourier series (14.68) gives

$$\sigma^2_{\rm surface} = \frac{2}{\beta}\sum_{n=1}^\infty n|a_n|^2, \qquad a(1,\theta) = \sum_{n=-\infty}^\infty a_n e^{in\theta}. \tag{15.99}$$

Note that $\sigma^2_{\rm surface}$ vanishes for $a(\vec{r}) = a(|\vec{r}|)$, but in general $\mathrm{Var}(A)$ is equal to the sum of (15.98) and (15.99), which are both O(1).

A simple check on these predictions is possible. Consider the definition (15.96) of $\hat{P}(k)$, suitably generalized to account for general β. We notice that with the special choice of linear statistic $a(x,y) = c_{10}x + c_{01}y$, the dependence on k can be factored out of the integral by completing the square, giving the exact formula

$$\hat{P}(k) = e^{-k^2(c_{10}^2+c_{01}^2)/2\beta}$$

independent of N. Thus the distribution is indeed Gaussian and the variance is given by $\sigma^2 = (c_{10}^2 + c_{01}^2)/\beta$. This agrees with the prediction from the general formulas (15.98) and (15.99), since they give $\sigma^2_{\rm bulk} = (c_{10}^2 + c_{01}^2)/2\beta$ and $\sigma^2_{\rm surface} = (c_{10}^2+c_{01}^2)/2\beta$ and thus sum to σ^2. For the case $\beta = 2$ with a soft wall (corresponding to the eigenvalues of complex Gaussian matrices considered in Section 15.1) the above predictions have been rigorously proved [468], [15].

The variance of a linear statistic $A = \sum_{j=1}^N a(\vec{r}_j/\alpha)$ in the infinite system, with $\alpha \to \infty$, is considered in Exercises 15.4 q.1.

EXERCISES 15.4 1. Consider the formula (15.97) for the variance of the linear statistic and suppose $\Omega = \mathbb{R}^2$. In the case that $a(\vec{r}) \mapsto a(\vec{r}/\alpha)$, use (15.75) to show that

$$\lim_{\alpha\to\infty}\mathrm{Var}(A) = \frac{1}{2\pi\beta}\int_{\mathbb{R}^2}\Big(\Big(\frac{\partial a(x,y)}{\partial x}\Big)^2 + \Big(\frac{\partial a(x,y)}{\partial y}\Big)^2\Big)\, dxdy.$$

By using the convolution theorem in (15.97) and using (15.71), obtain the alternative formula

$$\lim_{\alpha\to\infty}\mathrm{Var}(A) = \frac{1}{2\pi\beta}\frac{1}{(2\pi)^2}\int_{\mathbb{R}^2}|\hat{a}(k)|^2|\vec{k}|^2\, d\vec{k}$$

(cf. (14.54)).

2. Use (15.64) to verify (15.89) in the case of the 2dOCP with a soft edge.
3. Applying the expansion (7.5) to the exact result of Proposition 15.3.2 gives the expansion [313]

$$g_{(2)}(\vec{r},\vec{0}) = -e^{-\pi\rho r^2} + (\Gamma-2)\Big(\mathrm{Ei}(-\pi\rho r^2) - \frac{1}{2}\mathrm{Ei}(-\pi\rho r^2/2)$$
$$+e^{-\pi\rho r^2}\Big(\frac{1}{2}\mathrm{Ei}(\pi\rho r^2/2) - (\log\pi\rho r^2 + C)\Big) + \mathrm{O}((\Gamma-2)^2),$$

where C denotes Euler's constant and $\mathrm{Ei}(x) := \int_x^\infty (e^{-t}/t)\,dt$. Deduce from this that [340]

$$\frac{1}{\rho}\Big(\frac{\pi\Gamma\rho}{2}\Big)^j \int_{\mathbb{R}^2} r^{2j}\rho_{(2)}^T(\vec{r},\vec{0})\,d\vec{r} = -j! + (\Gamma-2)j!\Big(\sum_{k=0}^{j}\frac{2^k-1}{k+1} - \frac{j}{2}\Big) + \mathrm{O}((\Gamma-2)^2),$$

and so verify that the sum rules (15.70) (in the case $l=1, p=0$), (15.73), (15.76) and (15.77) are obeyed to the same order.

15.5 SPACING DISTRIBUTIONS

15.5.1 The bulk

The orthogonality (15.41) allows some two-dimensional analogues of spacing distributions to be computed explicitly in terms of the incomplete gamma functions. This contrasts with the one-dimensional case, where we saw in Chapters 8 and 9 that the calculation of spacing distributions requires Painlevé transcendents. Now, for one-dimensional systems we have defined $E_\beta(n;s)$ as the probability that an interval of length s contains exactly n particles, and $P_\beta(n;s)$ as the p.d.f. for the event that our randomly chosen particle is a distance s from another particle, and there are exactly n particles in between. For a two-dimensional system in a fluid state in the bulk, similar quantities can be defined.

DEFINITION 15.5.1 *In a two-dimensional statistical mechanical system at inverse temperature β, introduce the notation $E_\beta(n;\alpha)$ ($F_\beta(n;\alpha)$) denote for the probability that a randomly chosen point (particle) has exactly n particles within a radius α. Furthermore, let $P_\beta(n;\alpha)$ ($Q_\beta(n;\alpha)$) denote the probability density that a randomly chosen point (particle) is a distance α from another particle, and there are exactly n particles within this distance.*

In fact P_β can be deduced from E_β, and Q_β from F_β.

PROPOSITION 15.5.2 *We have*

$$P_\beta(n;\alpha) = -\frac{d}{d\alpha}E_\beta(n;\alpha) + P_\beta(n-1;\alpha), \qquad Q_\beta(n;\alpha) = -\frac{d}{d\alpha}F_\beta(n;\alpha) + Q_\beta(n-1;\alpha)$$

valid for $n = 0,1,2,\dots$ with $P_\beta(-1;\alpha) = Q_\beta(-1;\alpha) = 0$.

Proof. Let $\delta\alpha$ denote a small change in α. We see from the definitions that

$$\delta\alpha\, P_\beta(n-1;\alpha) + E_\beta(n;\alpha) - E_\beta(n;\alpha+\delta\alpha) = \delta\alpha P_\beta(n;\alpha),$$

which implies the first formula. The second result follows similarly. □

For a finite system in the canonical ensemble we have

$$E_\beta(n;\alpha) = \frac{1}{\hat{Z}_N}\binom{N}{n}\int_{\Omega_\alpha} d\vec{r}_1\cdots\int_{\Omega_\alpha} d\vec{r}_n\int_{\overline{\Omega}_\alpha} d\vec{r}_{n+1}\cdots\int_{\overline{\Omega}_\alpha} d\vec{r}_N e^{-\beta W(\vec{r}_1,\dots,\vec{r}_N)},$$

where Ω_α denotes the disk of radius α centered at the origin and $\overline{\Omega}_\alpha := \Omega - \Omega_\alpha$. Note that each $E_\beta(n;\alpha)$ can be calculated via the generating function

$$\sum_{n=0}^{N} u^n E_\beta(n;\alpha) = \frac{1}{\hat{Z}_N} \prod_{l=1}^{N} \Big(\int_{\overline{\Omega}_\alpha} + u \int_{\Omega_\alpha} \Big) d\vec{r}_1 \cdots \Big(\int_{\overline{\Omega}_\alpha} + u \int_{\Omega_\alpha} \Big) d\vec{r}_N e^{-\beta W(\vec{r}_1,\dots,\vec{r}_N)}. \tag{15.100}$$

For the 2dOCP in a disk at $\Gamma = 2$ we have $e^{-\beta W}$ given by (15.42) with $f(r)$ proportional to $e^{-\pi\rho r^2}$. Using the same integration technique which led to (15.43), and recalling the definitions (15.50) and (15.62), then allows (15.100) to be evaluated with the result

$$\sum_{n=0}^{N} u^n E_2(n;\alpha) = \prod_{j=1}^{N} \frac{(\Gamma(j;\pi\rho\alpha^2) - \Gamma(j;N) + u\gamma(j;\pi\rho\alpha^2))}{\gamma(j;\pi\rho R^2)}. \tag{15.101}$$

Taking the $N \to \infty$ limit, we obtain a closed form expression for the generating function [186].

PROPOSITION 15.5.3 *For the 2dOCP at $\Gamma = 2$ in the thermodynamic limit we have*

$$\sum_{n=0}^{\infty} u^n E_2(n;\alpha) = \prod_{j=1}^{\infty} \frac{\Gamma(j;\pi\rho\alpha^2) + u\gamma(j;\pi\rho\alpha^2)}{\Gamma(j)}.$$

Thus

$$E_2(0;\alpha) = \prod_{j=1}^{\infty} \frac{\Gamma(j;\pi\rho\alpha^2)}{\Gamma(j)}, \qquad \frac{E_2(n;\alpha)}{E_2(0;\alpha)} = \sum_{1 \le j_1 < \cdots < j_n < \infty} \prod_{l=1}^{n} \frac{\gamma(j_l;\pi\rho\alpha^2)}{\Gamma(j_l;\pi\rho\alpha^2)}.$$

Similar formulas can be derived for $F_2(0;\alpha)$ and $F_2(n;\alpha)/F_2(0;\alpha)$ (see Exercises 15.5 q.2). In particular

$$F_2(0;\alpha) = e^{\pi\rho\alpha^2} E_2(0;\alpha). \tag{15.102}$$

According to Proposition 15.5.2 we can use this formula to compute $Q_2(0;\alpha)$ which is the p.d.f. for the nearest neighbour spacing between the particles in the bulk.

The large α asymptotics of $E_2(0;\alpha)$ and $E_2(n;\alpha)/E_2(0;\alpha)$ (for fixed n) can be determined from the above explicit formulas. In the former case this requires taking the logarithm and breaking the resulting summation over j into three regions,

$$\text{(i) } 1 \le j \le J, \quad \text{(ii) } J+1 \le j \le [\alpha^2], \quad \text{(iii) } [\alpha^2]+1 \le j,$$

where $J := \alpha^2 - \alpha^{1+\epsilon}$, $0 < \epsilon \ll 1$. In each region we approximate $\Gamma(j+1;x)$ by a different asymptotic expansion [512],

$$\text{(i) } \Gamma(1+j;x) \sim \frac{e^{-x}x^{j+1}}{x-j}\left[1 - \frac{j}{(x-j)^2} + \frac{2j}{(x-j)^3} + \mathrm{O}\Big(\frac{j^2}{(x-j)^4}\Big)\right]$$

$$\text{(ii) } \Gamma(1+j;x) \sim \Gamma(1+j)\left[\frac{1-\operatorname{erf}\tau}{2} + \frac{e^{-\tau^2}}{\sqrt{2\pi j}}\Big(\frac{x}{x-j} - \frac{1}{\tau}\sqrt{\frac{j}{2}}\Big)\right]$$

where $\tau := (x - j - j\log(x/j))^{1/2}$, while in region (iii) the expansion (15.51) is used. After some further manipulation, the following result is obtained [186].

PROPOSITION 15.5.4 *With $\mu = (\pi\rho)^{1/2}\alpha$ and $\operatorname{erfc} x := 1 - \operatorname{erf} x$ denoting the complementary error* *function, we have*

$$\begin{aligned}\log E_2(0;\alpha) &\sim -\frac{1}{4}\mu^4 - \mu^2\log\mu + \Big(1 - \frac{1}{2}\log 2\pi\Big)\mu^2 \\ &\quad + 2^{1/2}\mu\left(\int_0^\infty \log\left(\pi^{1/2} t e^{t^2}\operatorname{erfc} t\right) dt + \int_0^\infty \log\left(\frac{1}{2}(1+\operatorname{erf} t)\right) dt\right) + \mathrm{o}(\mu) \\ &= -\frac{1}{4}\mu^4 - \mu^2\log\mu + 0.081\mu^2 - 2.294\mu + \mathrm{o}(\mu),\end{aligned}$$

where the decimal digits are accurate to three places, as $\mu \to \infty$.

A macroscopic argument of the type used in Section 14.6 can be used to predict the leading term in this expansion (see Exercises 15.5 q.3).

The asymptotic expansion of $E_2(n;\alpha)/E_2(0;\alpha)$ for fixed n is obtained from the term $j_1 = 1, j_2 = 2, \dots, j_n = n$ in the formula of Proposition 15.5.3. Using the formula

$$\Gamma(j+1;x) \sim x^j e^{-j}(1 + j/x + \mathrm{O}(1/x^2))$$

valid for j fixed and x large, which follows from (i) above, we deduce that

$$\frac{\gamma(j+1;\alpha^2)}{\Gamma(j+1;\alpha^2)} = \frac{\Gamma(j+1)}{\Gamma(j+1;\alpha^2)} - 1 \sim \frac{\Gamma(j+1)e^{\alpha^2}}{\alpha^{2j}}\Big(1 - \frac{j}{\alpha^2} + \mathrm{O}\Big(\frac{1}{\alpha^4}\Big)\Big)$$

and thus obtain the sought asymptotic expression [186].

PROPOSITION 15.5.5 *We have*

$$\frac{E_2(n;\alpha)}{E_2(0;\alpha)} \sim \prod_{j=0}^{n-1}\Gamma(j+1)\frac{e^{n\alpha^2}}{\alpha^{n(n-1)}}\left(1 - \frac{n(n-1)}{2\alpha^2} + \mathrm{O}\left(\frac{1}{\alpha^4}\right)\right),$$

valid for n fixed and α large.

15.5.2 The edge

We next turn our attention to the distribution of the largest eigenvalue (in modulus) for complex Gaussian matrices. Now, the probability that all eigenvalues are inside the disk Ω_α of radius α about the origin is given by

$$E_2(0;(\alpha,\infty)) = \frac{1}{\hat{Z}_N}\int_{\Omega_\alpha} d\vec{r}_1 \cdots \int_{\Omega_\alpha} d\vec{r}_N \prod_{l=1}^N e^{-|\vec{r}_l|^2} \prod_{1\le j<k\le N} |\vec{r}_k - \vec{r}_j|^2,$$

where

$$\hat{Z}_N := \int_{\mathbb{R}^2} d\vec{r}_1 \cdots \int_{\mathbb{R}^2} d\vec{r}_N \prod_{l=1}^N e^{-|\vec{r}_l|^2} \prod_{1\le j<k\le N} |\vec{r}_k - \vec{r}_j|^2.$$

The working required for Exercises 15.3 q.1(ii) shows that the integrals can be evaluated with the result

$$E_2(0;(\alpha,\infty)) = \prod_{j=1}^N \frac{\gamma(j;\alpha^2)}{\Gamma(j)}. \tag{15.103}$$

From this the following limit formula can be deduced [466].

PROPOSITION 15.5.6 *Let $\gamma_N = \log(N/2\pi) - 2\log\log N$, and $\alpha_N(x) = \sqrt{N} + \sqrt{\gamma_N}/2 + x/2\sqrt{\gamma_N}$. We* *have*

$$\lim_{N\to\infty} \log E_2(0;(\alpha_N(x),\infty)) = \exp(-\exp(-x)). \tag{15.104}$$

Proof. The 2dOCP interpretation tells us that to leading order the support of the eigenvalues is a disk of radius $\sqrt{N}$. Writing $\alpha = \sqrt{N} + f_N(x)/2$ with $f_N(x)$ linear in x, setting $N \mapsto N - j + 1$, then making use of the uniform expansion (15.51) shows

$$\log E_2\Big(0;\Big(\sqrt{N} + \frac{1}{2} f_N(x),\infty\Big)\Big) \sim \sum_{j=1}^{N} \log \frac{1}{2}\Big(1 + \mathrm{erf}\Big(\frac{j}{\sqrt{2N}} + \frac{f_N(x)}{\sqrt{2}}\Big)\Big)$$
$$\sim \sqrt{2N} \int_{f_N(x)}^{\infty} \log \frac{1}{2}(1 + \mathrm{erf}\, t)\, dt.$$

For this to be of order unity, we require $f_N(x) \to \infty$ as $N \to \infty$. Estimating the integrand in this limit gives

$$\log E_2\Big(0;\Big(\sqrt{N} + \frac{1}{2} f_N(x),\infty\Big)\Big) \sim -\sqrt{\frac{N}{2\pi}} \frac{e^{-(f_N(x))^2/2}}{(f_N(x))^2}.$$

With $\frac{1}{2} f_N(x) = \alpha_N(x) - \sqrt{N}$ the result (15.104) follows. □

Although not the setting of the present problem, the distribution in (15.104) is well known as the extremal distribution exhibited by the largest of N independent random variables, for $N \to \infty$ and after suitable scaling. In fact this coincidence persists for the distribution of the kth largest eigenvalue in magnitude [467] $(k = 1, 2, \dots)$ telling us that after appropriate scaling these magnitudes are in fact statistically independent.

EXERCISES 15.5 1. [275] Let $e(k;x) := \sum_{l=0}^{k} x^l/l!$. Use the formula (15.55) to show that the formula for $E_2(0;\alpha)$ in Proposition 15.5.3 can be written as

$$E_2(0;\alpha) = \prod_{j=1}^{\infty} \Big(e^{-\pi\rho\alpha^2} e(j;\pi\rho\alpha^2)\Big).$$

2. Follow the method used to derive Proposition 15.5.3 to show that $F_2(0;\alpha) = e^{\pi\rho\alpha^2} E_2(0;\alpha)$, and

$$\frac{F_2(n;\alpha)}{F_2(0;\alpha)} = \sum_{1 \le j_1 < \cdots < j_n < \infty} \prod_{l=1}^{n} \frac{\gamma(j_l + 1;\pi\rho\alpha^2)}{\Gamma(j_l + 1;\pi\rho\alpha^2)}.$$

3. [319] In this exercise the leading large α form of the probability $E_\beta(0;\alpha)$ of there being no eigenvalues in a disk of radius α—the hole probability—for the 2dOCP at general Γ will be deduced.

(i) Hypothesize that the creation of a hole of radius α, with $\alpha \gg 1$, causes at $r = \alpha$ a neutralizing uniform excess surface charge density $Q = \pi\rho\alpha^2$.

(ii) Calculate the electrostatic energy of the surface charge/surface charge, and surface charge/disk background interaction (for the latter use can be made of (1.71)), and substitute in (14.113) to deduce

$$E_\beta(0;\alpha) \underset{\alpha\to\infty}{\sim} e^{-\Gamma(\pi\rho\alpha^2)^2/8}.$$

15.6 THE SPHERE

15.6.1 Poisson equation

The pair potential Φ for two-dimensional unit charges on a sphere of radius R must obey the charge neutral Poisson equation on the sphere,

$$\frac{1}{R^2 \sin\theta} \frac{\partial}{\partial\theta}\left(\sin\theta \frac{\partial\Phi}{\partial\theta}\right) + \frac{1}{R^2 \sin^2\theta} \frac{\partial^2\Phi}{\partial\phi^2} = -2\pi\delta_S\left((\theta,\phi),(\theta',\phi')\right) + \frac{1}{2R^2}, \tag{15.105}$$

where θ refers to the usual azimuthal angle, ϕ to the polar angle and $\delta_S\left((\theta,\phi),(\theta',\phi')\right)$ is the delta function on the sphere, defined by the property

$$R^2 \int_0^\pi d\theta \sin\theta \int_0^{2\pi} d\phi f(\theta,\phi)\delta_S\left((\theta,\phi),(\theta',\phi')\right) = f(\theta',\phi'). \tag{15.106}$$

Note from (15.106) that the integral over the sphere of the r.h.s. of (15.105) gives 0. The interpretation is that as well as there being a charge at (θ',ϕ') there is a uniform neutralizing background contributing to the potential at (θ,ϕ); without charge neutrality, the Poisson equation on the sphere—a compact surface—does not have a solution.

With $\theta' = 0$ (source at the north pole) we see from the rotational invariance of the sphere about the azimuthal axis that Φ must be independent of ϕ (of course the solution must also be independent of ϕ'). Furthermore, for small θ we must reclaim the potential in a plane with free boundary conditions, and thus $\Phi \sim -\log|R\theta|$. It is easy to check by substitution in (15.105) that the required solution in this circumstance is

$$\Phi\left((\theta,\phi),(0,\phi')\right) = -\log\left(2R\sin(\theta/2)\right).$$

Geometrically $2R\sin(\theta/2)$ is the chord length between the points (θ,ϕ) and $(0,\phi')$ on the sphere. Hence, by rotational invariance of the sphere, for a general source point (θ',ϕ'), Φ must be given by minus the logarithm of the chord length. Writing $\vec{r}$ and $\vec{r}\,'$ for the vector from the center of the sphere to the points with spherical coordinates (θ,ϕ) and (θ',ϕ') we thus have

$$\Phi\left((\theta,\phi),(\theta',\phi')\right) = -\log|\vec{r}-\vec{r}\,'|.$$

Explicitly, if α denotes the relative angle between the vectors $\vec{r}$ and $\vec{r}\,'$, then

$$\Phi\left((\theta,\phi),(\theta',\phi')\right) = -\log\left(2R\sin(\alpha/2)\right). \tag{15.107}$$

By noting $\sin^2(\alpha/2) = \frac{1}{2}(1-\cos\alpha)$, $\cos\alpha = \vec{r}\cdot\vec{r}\,'/R^2$ and that in Cartesian coordinates

$$\vec{r} = (R\sin\theta\cos\phi, R\sin\theta\sin\phi, R\cos\theta),$$

a short calculation verifies that (15.107) can be rewritten as

$$\Phi\left((\theta,\phi),(\theta',\phi')\right) = -\log\left(2R|u'v - uv'|\right), \tag{15.108}$$

where u, v (and similarly u', v') are the *Cayley-Klein parameters*

$$u := \cos(\theta/2)e^{i\phi/2}, \quad v := -i\sin(\theta/2)e^{-i\phi/2}. \tag{15.109}$$

The total potential of an N-particle system is calculated from the pair potential according to the method of Section 1.4.1. Due to the rotational invariance of the sphere the potential energy for the particle-background interaction only contributes a constant. From the total potential we find that the Boltzmann factor for the system is

$$\left(\frac{1}{2R}\right)^{N\Gamma/2} e^{\Gamma N^2/4} \prod_{1\le j<k\le N} |u_k v_j - u_j v_k|^\Gamma. \tag{15.110}$$

15.6.2 Correlation functions

At $\Gamma = 2$ use of the Vandermonde determinant (1.173) shows that the Boltzmann factor is proportional to

$$\prod_{j=1}^N (\cos(\theta_j/2))^{2(N-1)} \prod_{1\le j<k\le N} \left|\det\left[\left(\frac{v_j}{u_j}\right)^{k-1}\right]_{j,k=1,\dots,N}\right|^2.$$

With Ω_R denoting the sphere of radius R, noting the orthogonality

$$\int_{\Omega_R} g(\theta)(v/u)^m(\bar{v}/\bar{u})^n\, d\Omega = 2\pi R^2 \delta_{m,n} \int_0^\pi g(\theta)(\tan\theta/2)^{m+n}\sin\theta\, d\theta, \tag{15.111}$$

the method of proof of Proposition 15.3.1 gives that at $\Gamma = 2$ the correlation function is given in terms of the Cayley-Klein parameters by [104]

$$\rho_{(n)}((\theta_1,\phi_1),\dots,(\theta_n,\phi_n)) = \rho^n \det\left[(\langle(u_k,v_k),(u_j,v_j)\rangle)^{N-1}\right]_{j,k=1,\dots,n}, \tag{15.112}$$

where $\langle(u_k,v_k),(u_j,v_j)\rangle := u_j\bar{u}_k + v_j\bar{v}_k$. A feature of (15.112) is that it is invariant with respect to a rotation of the sphere. This follows from the fact that a rotation of the sphere corresponds to the multiplication of the vectors (u, v) by a unitary matrix, and that the inner product $\langle\cdot,\cdot\rangle$ is invariant with respect to such a transformation.

For the particles in the neighborhood of the north pole, we can substitute (15.109) for the Cayley-Klein parameters in (15.112), set $\theta_j = r_j/R$ and take the limit $N, R \to \infty$ ($N/4\pi R^2 = \rho$). In this limit the neighborhood of the north pole becomes a plane, and (r_j, ϕ_j) specifies the position of the particles in polar coordinates. A straightforward calculation shows that the n-particle correlation (15.112) tends to the planar bulk n-particle correlation specified in Proposition 15.3.2, as expected.

15.6.3 Magnetic analogy

In the theory of a charged particle confined to a sphere of radius R with a radial magnetic field $\vec{B} = B\hat{r}$ at the surface (due to a magnetic monopole at the center of the sphere), the operator formalism of Section 15.2.1 is no longer applicable since there it is assumed xy-coordinates are used. The situation is further complicated by the fact (see, e.g., [549]) that it is not possible to define a singularity free vector potential $\vec{A}$ such that $\nabla\times\vec{A} = B\hat{r}$ over the entire sphere. Instead, the sphere is divided into overlapping regions

$$\begin{aligned}&\text{(a) } 0\le\theta<\pi/2+\delta,\ 0\le\phi<2\pi,\\ &\text{(b) } \pi/2-\delta<\theta\le\pi,\ 0\le\phi<2\pi,\end{aligned}$$

where $0\le\delta\le\pi/2$. Using the general formula

$$\nabla\times\vec{A} = \frac{1}{R\sin\theta}\left(\left(\frac{\partial}{\partial\theta}(A_\phi\sin\theta) - \frac{\partial A_\theta}{\partial\phi}\right)\hat{r} + \frac{\partial A_r}{\partial\phi}\hat{\theta} - \sin\theta\frac{\partial A_r}{\partial\theta}\hat{\phi}\right),$$

valid for $r = R$, we see that the equation $\nabla\times\vec{A} = B\hat{r}$ is satisfied in the region (a) and (b) by choosing

$$\vec{A}_a = \left(0, 0, \frac{BR}{\sin\theta}(1-\cos\theta)\right), \quad \vec{A}_b = \left(0, 0, -\frac{BR}{\sin\theta}(1+\cos\theta)\right) \tag{15.113}$$

respectively, where the coordinates in (15.113) refer to the spherical coordinates (r, θ, ϕ).

In the overlap region the wave functions implied by different vector potentials are related, and this implies a restriction on the allowed value of B. Now, using the formula

$$\nabla f = \frac{1}{R}\frac{\partial f}{\partial \theta}\hat{\theta} + \frac{1}{R\sin\theta}\frac{\partial f}{\partial \phi}\hat{\phi}$$

we see that $\vec{A}_a = \vec{A}_b + \nabla f$, where $f = 2BR^2\phi$. But if $\vec{A}_a$ and $\vec{A}_b$ are so related and the solution of the Schrödinger equation

$$\frac{1}{2m}\Big(-i\hbar\nabla + \frac{e}{c}\vec{A}\Big)^2 \psi = E\psi \tag{15.114}$$

with vector potential $\vec{A} = \vec{A}_b$ is given by ψ_b, then the solution of (15.114) with the same value of E but vector potential $\vec{A} = \vec{A}_a$ is given by $\psi_b e^{-ief/\hbar c}$. In the present situation this means

$$\psi_a(\theta,\phi) = \psi_b(\theta,\phi)e^{-i2BeR^2\phi/\hbar c}.$$

Furthermore ψ_a and ψ_b must both be periodic of period 2π in ϕ. This is only possible if

$$B = n\frac{\hbar c}{2eR^2}, \; n = 1,2,\dots, \tag{15.115}$$

which is the quantization condition for the strength of the magnetic monopole.

Our task is to determine the minimal energy states of (15.114) with $\vec{A} = \vec{A}_a$. An algebraic formalism is possible by the introduction of $\vec{\Lambda} := \vec{r} \times (-i\hbar\nabla + \frac{e}{c}\vec{A})$, which is a type of angular momentum operator. Direct calculation gives that the components of $\vec{\Lambda}$ satisfy the commutation relation

$$[\Lambda_x, \Lambda_y] = i\hbar\left(\Lambda_z - \frac{n\hbar}{2}\frac{z}{R}\right),$$

together with the relations obtained by cyclic permutation of the indices. This tells us that if we define

$$\vec{L} = \vec{\Lambda} + \frac{n\hbar}{2}\hat{r} \tag{15.116}$$

then we obtain the canonical commutation relations for angular momentum,

$$[L_x, L_y] = i\hbar L_z, \; [L_y, L_z] = i\hbar L_x, \; [L_z, L_x] = i\hbar L_y. \tag{15.117}$$

The Hamiltonian operator in (15.114) can be written in terms of $\vec{L}$ by noting

$$\vec{L}^2 = \Lambda^2 + \frac{n^2\hbar^2}{4} = R^2\Big(-i\hbar\nabla + \frac{e}{c}\vec{A}\Big)^2 + \frac{n^2\hbar^2}{4}, \tag{15.118}$$

where we have used the vector identity $(\vec{B}\times\vec{C})^2 = \vec{B}^2\vec{C}^2 - (\vec{B}\cdot\vec{C})^2$. Thus we have

$$H\psi(\theta,\phi) = \frac{1}{2mR^2}\left(\vec{L}^2 - \frac{\hbar^2 n^2}{4}\right)\psi(\theta,\phi) = E\psi(\theta,\phi). \tag{15.119}$$

The angular momentum algebra (see Exercises 15.6) gives that the allowed values of $\vec{L}^2$ are $l(l+1)\hbar^2$, $l = 0, \frac{1}{2}, 1, \frac{3}{2}, \dots$. However, the original Hamiltonian (15.114) must have non-negative eigenvalues, so the smallest allowed value of l is $n/2$, which from (15.119) and (15.115) is seen to correspond to $E = \frac{1}{2}\hbar\omega_c$.

As is well known, and revised in Exercises 15.6, for a given value of l the allowed values of L_z are $m\hbar$ with $m = -l, -l+1, \dots, l$. Now, from (15.116) we can deduce that

$$L_z = -\hbar\left(i\frac{\partial}{\partial\phi} - \frac{n}{2}\right),$$

which has corresponding eigenfunctions

$$\psi_m(\theta, \phi) = A_m(\theta) e^{i(m-n/2)\phi}. \tag{15.120}$$

The function $A_m(\theta)$ can be determined by writing out the explicit form of $\vec{L}^2$,

$$\vec{L}^2 = \hbar^2 \left[-\frac{1}{\sin\theta}\frac{\partial}{\partial\theta}\left(\sin\theta\frac{\partial}{\partial\theta}\right) + \frac{1}{\sin^2\theta}\left(i\frac{\partial}{\partial\phi} + \frac{n}{2}(\cos\theta - 1)\right)^2 + \frac{n^2}{4}\right] \tag{15.121}$$

(cf. (15.105)). Substituting (15.120) and (15.121) in the eigenvalue equation (15.119) with $E = \frac{1}{2}\hbar w_c$ gives that $A_m(\theta)$ must satisfy the equation

$$A_m''(\theta) + \cot\theta A_m'(\theta) + \left(\frac{n}{2} - \left(\frac{\frac{n}{2}\cos\theta - m}{\sin\theta}\right)^2\right) A_m(\theta) = 0. \tag{15.122}$$

With $m = -n/2$, direct substitution verifies that $A_{-n/2}(\theta) = C_{-n/2}\sin^n(\theta/2)$ is a solution of (15.122), where the constant $C_{-n/2}$ is a normalization. The solution for $m = -n/2+1, -n/2+2, \dots, n/2$ can be constructed from $A_{-n/2}$ by applying the raising operator

$$L_+ = L_x + iL_y = \hbar e^{i\phi}\left[\frac{\partial}{\partial\theta} + i\cot\theta\frac{\partial}{\partial\phi} + \frac{n}{2\sin\theta}(1-\cos\theta)\right].$$

This gives

$$A_m(\theta) = C_m \cos^{n/2+m}(\theta/2)\sin^{n/2-m}(\theta/2),$$

which when substituted in (15.120) shows that

$$\psi_m(\theta,\phi) = C_m u^{n/2+m} v^{n/2-m} e^{-in\phi/2}, \quad C_m = \left(\frac{n/2+1}{4\pi R^2}\binom{n}{n/2+m}\right)^{1/2}, \tag{15.123}$$

where u and v are the Cayley-Klein parameters (15.109) (the value of the normalization constant can be determined by making use of the trigonometric Euler integral of Exercises 4.1 q.1(i)).

Setting $N = n+1$, an N-particle state can be formed by constructing a Slater determinant from the states (15.120),

$$\psi\left((\theta_1,\phi_1),\dots,(\theta_N,\phi_N)\right) = Ce^{-i(N-1)\sum_{j=1}^N \phi_j/2}\prod_{j=1}^N v_j^{N-1}\det\left[(u_j/v_j)^{k-1}\right]_{j,k=1,\dots,N}. \tag{15.124}$$

Use of the Vandermonde formula and comparison with (15.110) shows that $|\psi|^2$ is proportional to the Boltzmann factor for the 2dOCP on a sphere at $\Gamma = 2$.

15.6.4 Stereographic projection

The 2dOCP on a sphere can be mapped onto a 2dOCP in the plane with a particular one-body potential. The mapping is achieved by making a stereographic projection from the south pole of the sphere to a plane tangent to the north pole (see Figure 15.2 and cf. Section 2.5). This is specified by the equation

$$z = 2Re^{i\phi}\tan\frac{\theta}{2}, \qquad z = x + iy. \tag{15.125}$$

Using this to rewrite the argument of the logarithm in (15.108) and comparing with (15.107) we see that

$$2R\sin\frac{\alpha}{2} = \cos\frac{\theta}{2}\,|z - z'|\cos\frac{\theta'}{2}, \tag{15.126}$$

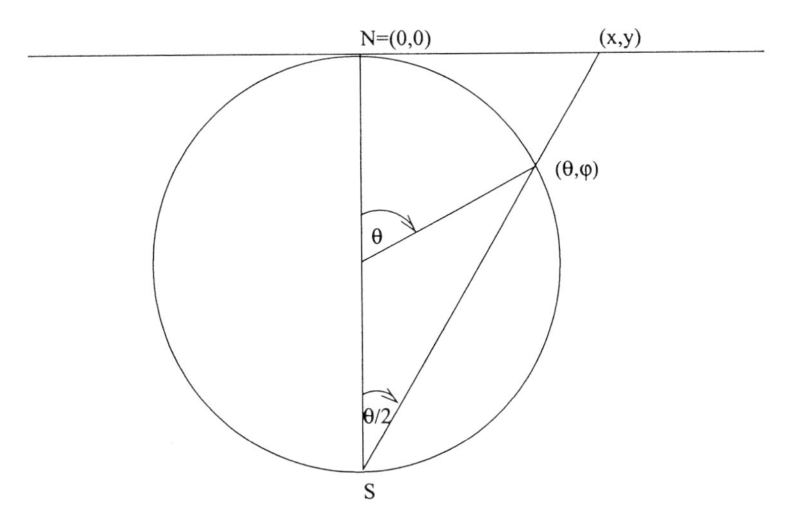

Figure 15.2 Cross-section of the stereographic projection from the south pole of the sphere to the plane tangent to the north pole

while calculation of the relevant Jacobian gives

$$dS := R^2 \sin\theta d\theta d\phi = \frac{1}{(1+|z|^2/4R^2)^2} dxdy. \tag{15.127}$$

The Boltzmann factor (15.110) therefore transforms according to

$$\begin{aligned}\prod_{1\le j<k\le N} |u_k v_j - u_j v_k|^\Gamma dS_1 \cdots dS_N \\ = \prod_{j=1}^N \frac{1}{(1+|z_j|^2/(4R^2))^{2+\Gamma(N-1)/2}} \prod_{1\le j<k\le N} |z_j - z_k|^\Gamma d\vec{r}_1 \cdots d\vec{r}_N.\end{aligned} \tag{15.128}$$

Writing the one-body terms on the r.h.s. of (15.128) in the form $\exp(-\beta\sum_{j=1}^N V(\vec{r}_j))$, we see that the effective one body potential is $V(\vec{r}) = (\frac{1}{2}(N-1) + \frac{2}{\Gamma})\log(1+|z|^2/(4R^2))$. Substituting this into the (flat space) Poisson equation (1.40) shows that the corresponding background density is

$$\frac{1}{2\pi R^2}\Big(\frac{1}{2}(N-1) + \frac{2}{\Gamma}\Big)\frac{1}{(1+|z|^2/(4R^2))^2}.$$

The Hamiltonian for the quantum particle on a sphere in the presence of a perpendicular magnetic field can also be transformed by the stereographic projection [143]. First, we note from (15.119) and (15.121) that the Hamiltonian with the vector potential $\vec{A}_a$ (15.113) (recall this has a singularity at the south pole, which will be projected to infinity) is given by

$$H = \frac{\hbar^2}{2mR^2}\Big(-\frac{1}{\sin\theta}\frac{\partial}{\partial\theta}\Big(\sin\theta\frac{\partial}{\partial\theta}\Big) + \frac{1}{\sin^2\theta}\Big(i\frac{\partial}{\partial\phi} + \Big(\frac{R}{l}\Big)^2(\cos\theta - 1)\Big)^2\Big).$$

The change of variables (15.125) gives the projected Hamiltonian as

$$H = \frac{\hbar^2}{2m}\Big(-4l^2\Big(1+\frac{|z|^2}{4R^2}\Big)^2\frac{\partial^2}{\partial z\partial\bar{z}} + \Big(1+\frac{|z|^2}{4R^2}\Big)\Big(z\frac{\partial}{\partial z} - \bar{z}\frac{\partial}{\partial\bar{z}}\Big) + \frac{|z|^2}{4l^2}\Big). \tag{15.129}$$

This Hamiltonian is of the form (15.35) with

$$\Phi = \Big(\frac{R}{l}\Big)^2 \log\Big(1 + \frac{|z|^2}{(2R)^2}\Big). \tag{15.130}$$

It therefore follows from Exercises 15.2 q.2 that the states

$$\Big(1 + \frac{|z|^2}{(2R)^2}\Big)^{-(R/l)^2} \bar{z}^p, \quad p \in \mathbb{Z}_{\ge 0} \tag{15.131}$$

are orthogonal states in the lowest Landau level. Choosing $n = N - 1$ in the quantization condition (15.115), and forming a Slater determinant out of the states (15.131) with $p = 0, 1, \dots, N-1$, shows that the absolute value squared of the corresponding N-particle free fermion state, when multiplied by the product of measures (15.127), is proportional to (15.128).

15.6.5 Random matrix analogy

The p.d.f. corresponding to (15.128) with $\Gamma = 2$, $R = \frac{1}{2}$ can be realized as an eigenvalue p.d.f. [366].

PROPOSITION 15.6.1 *Let* $\mathbf{A}$, $\mathbf{B}$ *be complex Gaussian matrices as specified by Proposition 15.1.1. The eigenvalue p.d.f. of* $\mathbf{A}^{-1}\mathbf{B}$ *is proportional to*

$$\prod_{l=1}^N \frac{1}{(1+|z_l|^2)^{N+1}} \prod_{1 \le j < k \le N} |z_k - z_j|^2. \tag{15.132}$$

Proof. We follow [298]. From (3.82) we have that the distribution of $\mathbf{Y} := \mathbf{A}^{-1}\mathbf{B}$ is proportional to

$$\frac{1}{\det(\mathbf{1} + \mathbf{Y}\mathbf{Y}^\dagger)^{2N}}.$$

Introduce now the Schur decomposition (15.3) for $\mathbf{Y}$ to conclude that the distribution of the upper triangular matrix $\mathbf{T} =: \mathbf{T}_N$ therein is proportional to

$$\frac{1}{\det(\mathbf{1}_N + \mathbf{T}_N\mathbf{T}_N^\dagger)^{2N}} \prod_{1 \le j < k \le N} |z_k - z_j|^2, \tag{15.133}$$

where $z_j = T_{jj}$ are the eigenvalues of $\mathbf{Y}$. The task then is to integrate out the strictly upper triangular elements of $\mathbf{T}_N$, $\tilde{\mathbf{T}}_N$, say, and thus compute

$$\prod_{1 \le j < k \le N} |z_k - z_j|^2 I_N(z_1, \dots, z_N), \qquad I_N(z_1, \dots, z_N) := \int \frac{1}{\det(\mathbf{1}_N + \mathbf{T}_N\mathbf{T}_N^\dagger)^{2N}} (d\tilde{\mathbf{T}}_N). \tag{15.134}$$

In fact it is convenient to consider the more general family of integrals

$$I_{n,p}(z_1, \dots, z_n) := \int \frac{1}{\det(\mathbf{1}_n + \mathbf{T}_n\mathbf{T}_n^\dagger)^p} (d\tilde{\mathbf{T}}_n), \tag{15.135}$$

where $p \ge n$ (this ensures convergence). For these integrals a recurrence can be obtained by proceeding in a way similar to the derivation of the result (b) in Proposition 2.5.1. Let $\vec{u}$ denote the last column of $\tilde{T}_n$. We can then write

$$\mathbf{1}_n + \mathbf{T}_n\mathbf{T}_n^\dagger = \begin{bmatrix} \mathbf{1}_{n-1} + \mathbf{T}_{n-1}\mathbf{T}_{n-1}^\dagger + \vec{u}\vec{u}^\dagger & z_n\vec{u} \\ \bar{z}_n\vec{u}^\dagger & 1 + |z_n|^2 \end{bmatrix}, \tag{15.136}$$

showing

$$\det(\mathbf{1}_n + \mathbf{T}_n\mathbf{T}_n^\dagger) = (1+|z_n|^2)\det\Big(\mathbf{1}_{n-1} + \mathbf{T}_{n-1}\mathbf{T}_{n-1}^\dagger + \frac{\vec{u}\vec{u}^\dagger}{1+|z_n|^2}\Big)$$
$$= (1+|z_n|^2)\det(\mathbf{1}_{n-1} + \mathbf{T}_{n-1}\mathbf{T}_{n-1}^\dagger)\Big(1 + \frac{1}{1+|z_n|^2}\vec{u}^\dagger(\mathbf{1}_{n-1} + \mathbf{T}_{n-1}\mathbf{T}_{n-1}^\dagger)^{-1}\vec{u}\Big),$$

where the last equality follows by noting that the final term in the determinant of the previous line is of rank 1.

Substituting this last equality in (15.135) shows

$$I_{n,p}(z_1,\dots,z_n) = \frac{1}{(1+|z_n|^2)^p}\int (d\tilde{\mathbf{T}}_{n-1})\,\frac{1}{\det(\mathbf{1}_{n-1} + \mathbf{T}_{n-1}\mathbf{T}_{n-1}^\dagger)^p}$$
$$\times \int (d\vec{u})\,\Big(1 + \frac{1}{1+|z_n|^2}\vec{u}^\dagger(\mathbf{1}_{n-1} + \mathbf{T}_{n-1}\mathbf{T}_{n-1}^\dagger)^{-1}\vec{u}\Big)^{-p}. \tag{15.137}$$

The change of variables

$$\vec{v} = (1+|z_n|^2)^{-1/2}(\mathbf{1}_{n-1} + \mathbf{T}_{n-1}\mathbf{T}_{n-1}^\dagger)^{-1/2}\vec{u}$$

then gives the recurrence

$$I_{n,p}(z_1,\dots,z_n) = \frac{C_{n,p}}{(1+|z_n|^2)^{p-n+1}},\, I_{n-1,p-1}(z_1,\dots,z_{n-1}), \tag{15.138}$$

where

$$C_{n,p} = \int \frac{(d\vec{v})}{(1+\vec{v}^\dagger\vec{v})^p}. \tag{15.139}$$

Iterating (15.138) with $n = 2N, p = N$ shows

$$I_N(z_1,\dots,z_N) \propto \prod_{l=1}^N \frac{1}{(1+|z_l|^2)^{N+1}},$$

which when substituted in (15.134) implies (15.132). □

15.6.6 Zeros of a random spin state

We have noted above that the eigenvalues of $\vec{L}^2$ are $l(l+1)\hbar^2$, $l = 0, \frac{1}{2}, 1, \frac{3}{2}, \dots$. In the quantum theory of angular momentum applied to spin, eigenstates with a value of $\vec{L}^2$ equal to $l(l+1)\hbar^2$ are said to have spin l. According to (15.123) a general state $\Phi(\theta,\phi)$ of spin $N/2$ (and thus $n = N$) is then of the form

$$\Phi(\theta,\phi) := \frac{1}{\mathcal{N}}\Big(\frac{N/2+1}{4\pi R^2}\Big)^{1/2}(-i\sin\theta/2)^N e^{-iN\phi}p(e^{i\phi}\cot\theta/2),$$
$$p(z) := \sum_{j=0}^N \binom{N}{j}^{1/2}\alpha_j z^j, \tag{15.140}$$

where $\mathcal{N} := (\sum_{j=0}^N |\alpha_j|^2)^{1/2}$.

Consider the function $p(e^{i\phi}\cot\theta/2)$. Now the mapping $z = e^{i\phi}\cot\theta/2$ can be interpreted geometrically as a stereographic projection from the north pole of a unit diameter sphere to a plane passing through the equator of the sphere. This means that if $p(z)$ is factorized with zeros at $z_1,\dots,z_N$, then $p(e^{i\phi}\cot\theta/2)$ vanishes at the value of (θ_j,ϕ_j) on the sphere corresponding to the stereographic projection of the z_j. Such a parametrization of a spin state is due to Majorana [382].

With all the coefficients of (15.140) chosen as independent Gaussian random variables with mean zero and unit variances, the density of zeros can be computed from (15.59) with $\sigma_j^2 = \binom{N}{j}$, while the two-point function follows by simplifying the general formula of Proposition 15.3.6. To compute the density we note that

$$\langle p\bar{p}\rangle_{\mathbf{L}} = \sum_{j=0}^{N} \binom{N}{j} |z|^{2j} = (1+|z|^2)^N$$

and thus

$$\rho_{(1)}(z) = \frac{N}{\pi} \frac{1}{(1+|z|^2)^2}.$$

But with $z = e^{i\phi} \cot \theta/2$,

$$dS = \frac{1}{(1+|z|^2)^2} dx\, dy$$

(recall (15.127)), so in terms of the spherical coordinates (θ, ϕ)

$$\rho_{(1)}(\theta, \phi) = \frac{N}{\pi},$$

and thus the zeros are uniformly distributed on the unit diameter sphere [288], [71].

15.6.7 Ground state

The ground state configuration for the 2dOCP in the infinite plane is expected to be the triangular lattice [12]. The question of the configuration with minimum energy can also be addressed in the finite system; indeed in the case of a disk with a uniform background of variable density it will be seen in Section 15.8.2 that this problem arises in studying the configurations of vortices in a rotating superfluid. On the unit sphere, the problem of minimizing the energy of a general repulsive pair potential of the form

$$v(\hat{r}_1, \hat{r}_2) = -\mathrm{sgn}(\alpha)|\hat{r}_1 - \hat{r}_2|^{-\alpha} = -\mathrm{sgn}(\alpha)|2(1 - \hat{r}_1 \cdot \hat{r}_2)|^{-\alpha/2} \tag{15.141}$$

has attracted much attention (see, e.g., [474]).

In general the configuration of minimum energy can only be determined numerically. Associated with the configuration is a tiling of *Dirichlet (Voronoi) cells* D_j, $(j = 1, \dots, N)$, which are defined as the boundary of the region of the sphere closest to the jth lattice point,

$$D_j := \{x \in S^2 : |x - x_j| = \min_{1 \le k \le N} |x - x_k|\}.$$

For general potentials (15.141) it is observed numerically that all Dirichlet cells are hexagons or pentagons, and furthermore that for N large enough there are precisely 12 pentagons. In fact this latter fact can be understood [474] as a consequence of Euler's relation (1.107).

PROPOSITION 15.6.2 *A tiling of the sphere consisting only of hexagons and pentagons, in which there are 3 edges from each vertex, must have exactly 12 pentagons.*

Proof. Let p denote the number of hexagons and q denote the number of pentagons so that $F = p+q$. Since each edge is shared by 2 cells, we have $E = \frac{1}{2}(6p+5q)$, while the fact that each vertex is shared by three cells gives $V = \frac{1}{3}(6p+5q)$. Furthermore, the Euler number for a sphere is 2. Substituting the above formulas for F, E and V in (1.107) with $g = 0$ gives $q = 12$. □

Note that the faces of a conventional soccer ball conform to Proposition 15.6.2.

The logarithmic potential

$$v(\hat{r}_1, \hat{r}_2) = -\log|\hat{r}_1 - \hat{r}_2| = -\frac{1}{2}\log|2(1 - \hat{r}_1 \cdot \hat{r}_2)| \tag{15.142}$$

has the special property that the dipole moment of the configuration of minimum energy must vanish [57].

PROPOSITION 15.6.3 *Let $\hat{r}_1, \dots, \hat{r}_N$ denote a configuration of minimum energy for the pair potential (15.142) on the unit sphere. Then $\sum_{j=1}^N \hat{r}_j = 0$.*

Proof. The force experienced by particle i due to particle j is given by

$$-\nabla_i \log |\vec{r}_i - \vec{r}_j| = \frac{\vec{r}_i - \vec{r}_j}{|\vec{r}_i - \vec{r}_j|^2}.$$

For an equilibrium configuration the total force must be directed radially in the direction of $\hat{r}_i$. Taking the dot product with $\hat{r}_i$ and noting $|\hat{r}_i - \hat{r}_j|^2 = 2(1 - \hat{r}_i \cdot \hat{r}_j)$ shows that the magnitude is $\frac{1}{2}(N-1)$ independent of i. Summing the resulting expression over i gives

$$\sum_{\substack{i,j=1 \\ i \neq j}}^N \frac{\hat{r}_i - \hat{r}_j}{|\hat{r}_i - \hat{r}_j|^2} = \frac{1}{2}(N-1) \sum_{i=1}^N \hat{r}_i.$$

But the left-hand side is antisymmetric with respect to interchanging labels i and j and so vanishes, implying the result. □

EXERCISES 15.6 1. (i) [104] Consider the Boltzmann factor (15.110) with $\Gamma = 2$. Use the orthogonality (15.111) to show

$$\hat{Z}_N = N! R^{2N} \left(\frac{1}{2R}\right)^N e^{N^2/2} \prod_{p=0}^{N-1} \left(4\pi \frac{\Gamma(N-p)\Gamma(p+1)}{\Gamma(N+1)}\right). \tag{15.143}$$

(ii) [320] Use Stirling's formula and the asymptotic expansion (4.186) with $\beta = 2$ to deduce that for $N, R \to \infty$ with ρ fixed

$$\beta F := -\frac{1}{N} \log \frac{\hat{Z}_N}{N!} = 4\pi R^2 \beta f_v + \frac{1}{3} \log \left((4\pi\rho)^{1/2} R\right) + \frac{1}{12} - 2\zeta'(-1) + \mathrm{o}(1), \tag{15.144}$$

where βf_v is given by (15.63).

2. (i) For Hermitian operators satisfying the canonical commutation relations for angular momentum (15.117), show that

$$[L^2, L_z] = 0, \quad [L^2, L_\pm] = 0, \quad [L_z, L_\pm] = \pm\hbar L_\pm, \quad L_\mp L_\pm = L^2 - L_z^2 \mp \hbar L_z, \tag{15.145}$$

where $L_\pm = L_x \pm i L_y$.

(ii) Suppose $L_z|\psi\rangle = \mu|\psi\rangle$, where μ is a simple (i.e., nondegenerate) eigenvalue. From the first commutation relation in (i) note that $L^2|\psi\rangle = \lambda|\psi\rangle$ for some $\lambda \geq \mu^2$, and from the second and third commutation relations show that

$$L^2(L_\pm|\psi\rangle) = \lambda L_\pm|\psi\rangle, \quad L_z(L_\pm|\psi\rangle) = (\mu \pm \hbar)(L_\pm|\psi\rangle),$$

with $\lambda \geq (\mu \pm \hbar)^2$. From this last inequality deduce that there exist highest $|\psi_+\rangle$ and lowest $|\psi_-\rangle$ eigenstates of L^2 and L_z for which $L_+|\psi_+\rangle = L_-|\psi_-\rangle = 0$.

(iii) Use the final relation in (i) to deduce that $(L^2 - L_z^2 \mp \hbar L_z)|\psi_\pm\rangle = 0$ and thus $(\lambda - \mu_\pm^2 \mp \hbar\mu_\pm) = 0$, where $L_z|\psi_+\rangle = \mu_\pm|\psi_+\rangle$. From this and the second displayed equation in (ii), deduce that $\mu_+ - \mu_- = 2l\hbar$ for some $l = 0, \frac{1}{2}, 1, \frac{3}{2}, \dots$, and so $\mu_+ = -\mu_- = l\hbar$.

(iv) From the results of (iii) conclude that the allowed values of the eigenvalues of L^2 are $\lambda = l(l+1)\hbar^2$, $l = 0, \frac{1}{2}, 1, \frac{3}{2}, \dots$ and that for each value of l there exist simultaneous eigenfunctions of L^2 and L_z which have eigenvalue of L_z given by $\mu = m\hbar$, where m is an integer $-l \leq m \leq l$.

15.7 THE PSEUDOSPHERE

15.7.1 Geometry

The sphere of radius R can be characterized by the fact that its *Gaussian curvature* is equal to $1/R^2$.

DEFINITION 15.7.1 *For a surface parametrized by the coordinates (τ, ϕ), and possessing a line element ds of the form $(ds)^2 = (d\tau)^2 + (f(\tau)d\phi)^2$, the Gaussian curvature κ is defined as*

$$\kappa = -\frac{1}{f}\frac{\partial^2 f}{\partial \tau^2}. \tag{15.146}$$

Parametrizing the sphere by $(\theta/R, \phi)$ gives

$$(ds)^2 = (d\theta)^2 + (R\sin(\theta/R)d\phi)^2 \tag{15.147}$$

and formula (15.146) gives $\kappa = 1/R^2$ as stated.

A closely related, but very different, two-dimensional surface is the pseudosphere (see, e.g., [42]). The pseudosphere is characterized by having constant negative Gaussian curvature $\kappa = -1/a^2$, and is of infinite extent. Furthermore, unlike the sphere, it cannot in its entirety be embedded in three-dimensional Euclidean space. Instead, the pseudosphere is naturally embedded in the three-dimensional Minkowski space with coordinates (y_0, y_1, y_2) in which the distance from the origin is defined as $-y_0^2 + y_1^2 + y_2^2$. The equation

$$-y_0^2 + y_1^2 + y_2^2 = -a^2 \tag{15.148}$$

(set of points at "distance" $-a^2$ from the origin in Minkowski space) defines a surface with two branches, intersecting the y_0-axis at $y_0 = \pm a$, respectively. Note that in Euclidean space (15.148) defines a hyperboloid. In Minkowski space the upper hyperboloid is called the pseudosphere.

The equation (15.148) can be parametrized by

$$y_0 = a\cosh\tau, \quad y_1 = a\sinh\tau\cos\phi, \quad y_2 = a\sinh\tau\sin\phi. \tag{15.149}$$

The Minkowski space line element $(ds)^2 = -(dy_0)^2 + (dy_1)^2 + (dy_2)^2$ then becomes

$$(ds)^2 = a^2(d\tau)^2 + a^2\sinh^2\tau(d\phi)^2. \tag{15.150}$$

Replacing τ by τ/a puts (15.150) into the form required in Definition 15.7.1, and (15.146) gives $\kappa = -1/a^2$ as remarked. Note also that making the replacements

$$a \mapsto iR, \qquad \tau \mapsto i\theta \tag{15.151}$$

in the modified form of (15.150) reclaims the formula (15.147) for the sphere.

Like the sphere, the pseudosphere is homogeneous, in the sense that any two points can be connected by generalized translations (isometries). These can be constructed from a rotation R_{ϕ_0} and boost T_{τ_0} specified by the orthogonal matrices (in Minkowski space)

$$R_{\phi_0} = \begin{pmatrix} 1 & 0 & 0 \\ 0 & \cos\phi_0 & -\sin\phi_0 \\ 0 & \sin\phi_0 & \cos\phi_0 \end{pmatrix}, \qquad T_{\tau_0} = \begin{pmatrix} \cosh\tau_0 & \sinh\tau_0 & 0 \\ \sinh\tau_0 & \cosh\tau_0 & 0 \\ 0 & 0 & 1 \end{pmatrix}. \tag{15.152}$$

15.7.2 The plasma system

After these preliminary remarks on the geometry of the pseudosphere, let us turn our attention to the 2dOCP confined to this surface [321]. First, we require the solution of the corresponding Poisson equation. Now, in terms of the coordinates (τ, ϕ), the Laplacian on the pseudosphere can be written down from the Laplacian on the boundary of the sphere (l.h.s. of (15.105)) by making the replacements (15.151). Thus the Poisson equation on the pseudosphere reads

$$\frac{1}{a^2 \sinh\tau}\frac{\partial}{\partial\tau}\Big(\sinh\tau\frac{\partial\Phi}{\partial\tau}\Big) + \frac{1}{a^2\sinh^2\tau}\frac{\partial^2\Phi}{\partial\phi^2} = -2\pi\delta((\tau,\phi),(\tau',\phi')), \tag{15.153}$$

where the delta function on the r.h.s. is defined by the property

$$a^2\int_0^\infty d\tau\,\sinh\tau\int_0^{2\pi} d\phi\, f(\tau,\phi)\delta((\tau,\phi),(\tau',\phi')) = f(\tau',\phi')$$

($a^2\sinh\tau$ is the Jacobian associated with the change of variables (15.149)). Note that, with the pseudosphere being an infinite surface, there is no need to have the r.h.s. of the Poisson equation (15.153) charge neutral, as distinct from (15.105).

For $\tau' = 0$, the solution of (15.153) must be independent of ϕ and ϕ'. With the boundary condition that $\Phi \to 0$ as $\tau \to \infty$, the solution is verified to be

$$\Phi((\tau,\phi),(0,\phi')) = -\log\Big(\tanh\frac{\tau}{2}\Big). \tag{15.154}$$

Noting that $a\tau$ is the geodesic distance from the point (τ,ϕ) to the point $(0,\phi')$, homogeneity of the pseudosphere implies that

$$\Phi((\tau,\phi),(\tau',\phi')) = -\log\Big(\tanh\frac{s}{2a}\Big), \tag{15.155}$$

where s is the geodesic distance between (τ,ϕ) and (τ',ϕ'). To obtain an explicit formula for s we note

$$\cosh\frac{s}{a} = \cosh\tau\cosh\tau' - \sinh\tau\sinh\tau'\cos(\phi-\phi'), \tag{15.156}$$

which can be deduced from the analogous formula for the sphere

$$\cos\frac{s}{a} = \frac{1}{R^2}\vec{r}\cdot\vec{r}\,' = \cos\theta\cos\theta' + \sin\theta\sin\theta'\cos(\phi-\phi')$$

(recall the formula below (15.107)) by making the replacements $s \mapsto is$ and $\theta \mapsto i\tau$, while $\sinh^2(s/2a) = \frac{1}{2}(1-\cosh(s/a))$. This implies

$$\tanh\frac{s_{jk}}{2a} = \left|\frac{\tanh\frac{\tau_j}{2}\,e^{i\phi_j} - \tanh\frac{\tau_k}{2}\,e^{i\phi_k}}{1-\tanh\frac{\tau_j}{2}\tanh\frac{\tau_k}{2}e^{i\phi_j - i\phi_k}}\right| = \left|\frac{U_jV_k - U_kV_j}{\bar{U}_jU_k - \bar{V}_jV_k}\right| = \left|\frac{(z_j - z_k)/2a}{1 - z_j\bar{z}_k/4a^2}\right|, \tag{15.157}$$

where U and V are the analogues of the Cayley-Klein parameters (15.109),

$$U = \cosh\frac{\tau}{2}\,e^{i\phi/2}, \qquad V = \sinh\frac{\tau}{2}\,e^{-i\phi/2} \tag{15.158}$$

while

$$z := 2a\tanh\frac{\tau}{2}\,e^{i\phi} \tag{15.159}$$

(cf. (15.125)). In Euclidean space the transformation (15.159) represents a stereographic projection from a hyperboloid into a disk (the Poincaré disk) of radius $R = 2a$. Substituting (15.157) in (15.155) shows that the pair potential is formally the same as the pair potential (15.174) below for a charge inside a disk of radius $R = 2a$ with metallic boundary conditions. However the systems are different because the pseudosphere is not a flat space.

The explicit formula (15.157) allows the total potential energy for a system of N equal charges confined to the portion of the pseudosphere $0 \le \tau \le a_0$, $0 \le \phi \le 2\pi$, in the presence of a uniform background of charge density $-\eta$, (the particle density is given by $\rho = N/|\Omega|$, $|\Omega| = 2\pi a^2(\cosh a_0 - 1)$, but because of the analogy with the metal wall problem, it is not required that $\rho = \eta$). However it is not possible to then calculate the free energy from the formula

$$\beta F = - \lim_{\substack{a_0 \to \infty \\ \rho \text{ fixed}}} \frac{1}{N} \log Z_N .$$

The problem is that the length of the boundary at $\tau = a_0$ is $2\pi a \sinh a_0$. For $a_0 \to \infty$ this is of the same order e^{a_0} as the volume, so, contrary to the disk and semiperiodic boundary conditions, the boundary contribution to βF cannot be distinguished from the bulk contribution. Alternative ways of accessing the bulk free energy and bulk energy per particle are discussed in [321] (see also [317] for computation of the pressure using the Maxwell stress tensor); they essentially rely on the observation that the pair potential (15.155) decays exponentially fast as a function of τ, so virial expansion techniques from the theory of fluids with short range potentials can be used. We will not address this problem, but focus attention on the calculation of the correlation functions.

15.7.3 The correlation functions

For the one-component system confined to the portion of the pseudosphere $0 \le \tau \le a_0$, $0 \le \phi \le 2\pi$, the general formulas of Section 1.1 give that the background-particle potential is given by $\sum_{j=1}^N V(\tau_j)$ with

$$\begin{aligned} V(\tau) &= \frac{1}{2} a^2 \eta \int_0^{a_0} d\tau' \, \sinh \tau' \int_0^{2\pi} d\phi' \, \log \left| \frac{\tanh \frac{\tau}{2} - \tanh \frac{\tau'}{2} e^{i\phi'}}{1 - \tanh \frac{\tau}{2} \tanh \frac{\tau'}{2} e^{-i\phi'}} \right|^2 \\ &= 2\pi a^2 \eta \Big(\cosh a_0 \, \log \tanh \frac{a_0}{2} - \log \sinh a_0 + \log \Big(2 \cosh^2 \frac{\tau}{2} \Big) \Big). \end{aligned}$$

The corresponding Boltzmann factor is thus

$$\begin{aligned} e^{-\beta U} &= e^{-\Gamma C_1} e^{-\Gamma N C_2} e^{-\Gamma 2\pi a^2 \eta \sum_{j=1}^N \log \cosh^2 \tau_j / 2} \\ &\quad \times \prod_{1 \le j < k \le N} \left| \frac{\tanh \frac{\tau_j}{2} e^{i\phi_j} - \tanh \frac{\tau_k}{2} e^{i\phi_k}}{1 - \tanh \frac{\tau_j}{2} \tanh \frac{\tau_k}{2} e^{i\phi_j - i\phi_k}} \right|^\Gamma , \end{aligned} \tag{15.160}$$

where C_1 and C_2 are independent of N. The explicit form of C_1 does not affect the correlations, and the fact that $C_2 \to -\infty$ as $a_0 \to \infty$ is all that is relevant regarding C_2. Note also that unlike the metal wall problem below, the self-energy term is not included.

In terms of the variable z defined in (15.159) the pseudosphere surface element dS is given by

$$dS = \frac{d\vec{r}}{(1 - |z|^2/4a^2)^2} \tag{15.161}$$

(cf. (15.127)). Hence the generalized grand partition function associated with the Boltzmann factor (15.160) can be written

$$\Xi_\Gamma[a;\zeta] = e^{-\Gamma C_1} \sum_{N=0}^\infty \frac{(e^{-\Gamma C_2} \zeta)^N}{N!} \int_{(\Omega_{2a \tanh a_0/2})^N} d\vec{r}_1 \cdots d\vec{r}_N \prod_{j=1}^N \Big(1 - |z_j|^2/4a^2 \Big)^{\Gamma 2\pi a^2 \eta - 1} a(\vec{r}_j) \, W_{N\Gamma},$$

where

$$W_{N\Gamma} := \prod_{j=1}^N \frac{1}{1 - |z_j|^2/(2a)^2} \prod_{1 \le j < k \le N} \left| \frac{(z_k - z_j)/2a}{1 - z_j \bar{z}_k / 4a^2} \right|^\Gamma$$

and Ω_α denotes the disk of radius α about the origin. At the special coupling $\Gamma = 2$, making use of the identity (4.34) and the Fredholm formula (5.32) shows that we have

$$\Xi_2[a;\zeta] = e^{-\Gamma C_1} \det(1 + \zeta K), \tag{15.162}$$

where K is the integral operator defined by the mapping rule

$$K[f(z')](z) := e^{-2C_2} \int_{\Omega_{2a \tanh a_0/2}} \Big(1 - |\vec{r}'|^2/4a^2\Big)^{4\pi a^2\eta - 1} \frac{a(\vec{r}')f(z')}{1 - z\bar{z}'/4a^2}\, d\vec{r}'.$$

Analogous to the situation in Section 15.8 below for the OCP in a disk with metallic boundary conditions, the form (15.162) implies

$$\rho_{(n)}(\vec{r}_1, \dots, \vec{r}_n) = \det \Big[G(\vec{r}_j, \vec{r}_k)\Big]_{j,k=1,\dots,n},$$

where $G(\vec{r}, \vec{r}')$ is given by (15.179). Thus the quantity

$$\tilde{G}(\vec{r}, \vec{r}') := (1 - r^2/4a^2)^{2\pi a^2\eta - 1/2} G(\vec{r}, \vec{r}')(1 - r'^2/4a^2)^{-2\pi a^2\eta + 1/2}$$

satisfies the integral equation

$$\tilde{G}(\vec{r}_1, \vec{r}_3)(1 - r_3^2/4a^2)^{-2\pi a^2\eta + 1/2} + \zeta e^{-2C_2} \int_{\Omega_{2a \tanh a_0/2}} (1 - r_2^2/4a^2)^{2\pi a^2\eta - 1/2} \frac{\tilde{G}(\vec{r}_1, \vec{r}_2)}{1 - z_2\bar{z}_3/4a^2}\, d\vec{r}_2$$
$$= \zeta e^{-2C_2} \frac{1}{1 - z_1\bar{z}_3/4a^2} (1 - r_1^2/4a^2)^{-2\pi a^2\eta + 1/2}.$$

Note that this implies

$$(1 - r_1^2/4a^2)^{-2\pi a^2\eta + 1/2} \tilde{G}(\vec{r}_1, \vec{r}_3)(1 - r_3^2/4a^2)^{2\pi a^2\eta - 1/2}$$

is analytic in z_1 and $\bar{z}_3$. Furthermore G itself has the symmetry (15.181).

Taking the limit $a_0 \to \infty$, using the fact that then $C_2 \to -\infty$, and setting $z_3 = 0$ shows

$$\int_{\Omega_{2a}} (1 - r_2^2/4a^2)^{2\pi a^2\eta - 1/2} \tilde{G}(\vec{r}_1, \vec{r}_2)\, d\vec{r}_2 = (1 - r_1^2/4a^2)^{2\pi a^2\eta - 1/2}. \tag{15.163}$$

This has a solution of the required type,

$$\tilde{G}(\vec{r}_1, \vec{r}_2) = (1 - r_1^2/4a^2)^{2\pi a^2\eta - 1/2}(1 - r_2^2/4a^2)^{2\pi a^2\eta - 1/2} f(z_1\bar{z}_2),$$

where f must be such that

$$f(0) = 1 \Big/ \int_{\Omega_{2a}} (1 - r_2^2/4a^2)^{4\pi a^2\eta - 1}\, d\vec{r}_2 = \eta.$$

The fact that (15.163) is independent of ζ indicates that the particle density must equal the background density. Hence $\tilde{G}(\vec{r}, \vec{r}) = \eta = \rho$ and f is uniquely specified with the result

$$\tilde{G}(\vec{r}_1, \vec{r}_2) = \rho \frac{(1 - r_1^2/4a^2)^{2\pi a^2\rho - 1/2}(1 - r_2^2/4a^2)^{2\pi a^2\rho - 1/2}}{(1 - z_1\bar{z}_2/4a^2)^{4\pi a^2\rho - 1}}.$$

Projecting back onto the pseudosphere and recalling (15.161), (15.158) and (15.159), we see that

$$\rho_{(n)}((\tau_1, \phi_1), \dots, (\tau_n, \phi_n)) = \rho^n \det \Big[(\langle (U_k, V_k), (U_j, V_j) \rangle)^{-4\pi a^2\rho - 1}\Big]_{j,k=1,\dots,n}, \tag{15.164}$$

where $\langle (U_k, V_k), (U_j, V_j) \rangle := U_j\bar{U}_k - V_j\bar{V}_k$. This expression is essentially identical to (15.112) for the sphere (write $N = 4\pi R^2\rho$ and $R \mapsto ia$).

15.7.4 Random matrix analogy

It is to be shown in Exercises 15.7 q.2 (ii) that for the one-component plasma with background density η on the pseudosphere, and with pair potential

$$\Phi((\tau,\phi),(\tau',\phi')) = -\log\Big(\sinh\frac{s}{2a}\Big), \tag{15.165}$$

the Boltzmann factor is proportional to

$$\prod_{j=1}^N \Big(1-|z_j|^2/(2a)^2\Big)^{\Gamma(4\pi\eta a^2+1)/2} \prod_{1\le j<k\le N} |z_j - z_k|^\Gamma, \tag{15.166}$$

$|z_j| < 2a$. For the coupling $\Gamma = 2$ this same Boltzmann factor, with $4\pi\eta a^2+1$ suitably chosen, is also proportional to the eigenvalue p.d.f. for a certain class of random matrices [559].

Proposition 15.7.2 *Let* $\mathbf{S} \in U(N+n)$ *be chosen to be Haar distributed. The eigenvalue p.d.f. of the top* $N\times N$ *sub-block of* $\mathbf{S}$, $\mathbf{S}_N$, *say, is proportional to (15.166) with* $\Gamma=2$, $(2a)^2=1$ *and* $4\pi\eta a^2+1 = n-1$.

Proof. Following [209], which requires $n \ge N$, we begin with the result of Proposition 3.8.2 with $N \mapsto N+n$ and $n_1 = n_2 = N$. This tells us that the distribution of $\mathbf{S}_N$ is proportional to

$$\det(\mathbf{1}_N - \mathbf{S}_N\mathbf{S}_N^\dagger)^{n-N}.$$

Introducing the Schur decomposition $\mathbf{S}_N = \mathbf{U}\mathbf{T}_N\mathbf{U}^{-1}$, with $\mathbf{U}$ and $\mathbf{T}_N$ as specified in (15.3), and recalling the change of variables formula (15.9), we see the eigenvalue p.d.f. for $\mathbf{S}_N$ is proportional to

$$\prod_{1\le j<k\le N} |z_k - z_j|^2 J_N(z_1,\dots,z_N), \qquad J_N(z_1,\dots,z_N) := \int \det(\mathbf{1}_N - \mathbf{T}_N\mathbf{T}_N^\dagger)^{n-N}(d\tilde{\mathbf{T}}_N). \tag{15.167}$$

The fact that $\mathbf{S}_N$ is a sub-block of a unitary matrix tells us that the p.d.f. is supported on $|z_j| < 1$ $(j=1,\dots,N)$. We will now proceed as in the proof of Proposition 15.6.1 and compute J_N by recurrence. For this purpose, for $m \ge p$ introduce

$$I_{m,p}(z_1,\dots,z_m) := \int \frac{1}{\det(\mathbf{1}_m - \mathbf{T}_m\mathbf{T}_m^\dagger)^p}(d\tilde{\mathbf{T}}_m) \tag{15.168}$$

(cf. (15.135)). Analogous to (15.136), with $\vec{u}$ denoting the last column of $\tilde{\mathbf{T}}_m$, we can write

$$\mathbf{1}_m - \mathbf{T}_m\mathbf{T}_m^\dagger = \begin{bmatrix} \mathbf{1}_{m-1} + \mathbf{T}_{m-1}\mathbf{T}_{m-1}^\dagger + \vec{u}\vec{u}^\dagger & z_m\vec{u} \\ \bar{z}_m\vec{u}^\dagger & 1-|z_m|^2 \end{bmatrix}, \tag{15.169}$$

which gives

$$\begin{aligned}&\det(\mathbf{1}_m - \mathbf{T}_m\mathbf{T}_m^\dagger)\\ &\quad = (1-|z_m|^2)\det(\mathbf{1}_{m-1} - \mathbf{T}_{m-1}\mathbf{T}_{m-1}^\dagger)\Big(1+\frac{1}{1+|z_m|^2}\vec{u}^\dagger(\mathbf{1}_{m-1} - \mathbf{T}_{m-1}\mathbf{T}_{m-1}^\dagger)^{-1}\vec{u}\Big).\end{aligned}$$

Substituting in (15.168) and proceeding as in the derivation of (15.138) shows

$$J_{m,p}(z_1,\dots,z_m) = \frac{\tilde{C}_{m,p}}{(1-|z_m|^2)^{p-m+1}} J_{m-1,p-1}(z_1,\dots,z_{m-1}), \tag{15.170}$$

for a suitable $\tilde{C}_{m,p}$. Iterating (15.170) with $m=N, p=N-n$ shows

$$J_N(z_1,\dots,z_N) \propto \prod_{l=1}^N (1-|z_l|^2)^{n-1}$$

and this substituted in (15.167) gives the stated result. □

EXERCISES 15.7 1. [321] In this exercise it will be argued that for a general Coulomb system on the pseudosphere, replacing the pair potential (15.155) by (15.165) does not affect the correlation functions in the thermodynamic limit. This helps explain the relationship between (15.164) and (15.112), as making the replacement $a \mapsto iR$ gives that the potential (15.165) maps to $-\log \sin(s/2R)$, which according to (15.107) is the Coulomb potential on a sphere.

(i) Use (15.157) to show that for the pair potential (15.165), the total potential energy of a system of charges q_i ($\sum_i q_i = 0$) can be written

$$\Big(-\frac{1}{2}\sum_{i\neq j} q_i q_j \log\Big|\frac{(z_i - z_j)/2a}{1 - z_i\bar{z}_j/4a^2}\Big| + \frac{1}{2}\sum_i q_i^2 \log(1-(r_i/2a)^2)\Big) - \frac{1}{2}\sum_i q_i^2 \log(1-(r_i/2a)^2),$$

which can be interpreted as a Coulomb system in a disk of radius $2a$ with metallic boundary conditions (the second term corresponds to the self energy), also coupled to an external potential $-\frac{1}{2}q_i^2\log(1-(r_i/2a)^2)$.

(ii) Use the formula

$$\sinh\frac{s_{jk}}{2a} = \frac{(|z_j - z_k|/2a)}{(1-(r_j/2a)^2)^{1/2}(1-(r_k/2a)^2)^{1/2}}$$

to show that for the pair potential (15.165) the total potential energy of the system in (i) is

$$-\frac{1}{2}\sum_{i\neq j} q_i q_j \log(|z_i - z_j|/2a) - \frac{1}{2}\sum_i q_i^2 \log(1-(r_i/2a)^2),$$

which is the same as in (i) apart from the image terms.

(iii) Recalling (15.161), argue that for $\beta q_i^2/2 - 2 \le 0$ at least, the density will be infinite at the boundary $|z| = 2a$ in the thermodynamic limit. Conclude that the image particles will be perfectly screened from the interior of the disk and so can be ignored.

2. (i) Show that for the one-component plasma with background density η and pair potential (15.165), the Boltzmann factor is proportional to (15.166).

(ii) Consider now a quantum particle on the pseudosphere in the presence of a perpendicular magnetic field. With the vector potential $\vec{A} = (0, 0, \frac{Ba}{\sinh\tau}(1-\cosh\tau))$, repeat the working which led to (15.129) to show that the projected Hamiltonian is given by (15.35) with

$$\Phi = -\Big(\frac{a}{l}\Big)^2 \log\Big(1 - \frac{|z|^2}{(2a)^2}\Big).$$

(iii) Use the result of Exercises 15.2 q.2 to conclude that the states

$$\Big(1 - \frac{|z|^2}{(2a)^2}\Big)^{(a/l)^2} \bar{z}^p, \quad p \in \mathbb{Z}_{\geq 0}, \tag{15.171}$$

are orthogonal states in the lowest Landau level. Form an N-particle state ψ as a Slater determinant from the single-particle states with $p = 0, 1 \dots, N-1$, and note that $|\psi|^2$ is proportional to the Boltzmann factor in (i) with $\Gamma = 2$ provided $4\pi\eta a^2 + 1 = 2(a/l)^2$.

(iv) Set $a = 1/2$, $(a/l)^2 = L/2 - 1$ in (15.171). Note that the normalized state is then equal to

$$\Big(\Big(\frac{L-1}{\pi}\Big)\binom{L+p-1}{p}\Big)^{1/2} (1-|z|^2)^{L/2-1}\bar{z}^p.$$

Analogous to (15.140), a general linear combination of these states with $p = 0, \dots, N+1$ is specified by

$$\phi(z) := \frac{1}{\mathcal{N}}\Big(\frac{L-1}{\pi}\Big)^{1/2}(1-|z|^2)^{L/2-1}p(z), \qquad p(z) := \sum_{j=0}^{N}\binom{L+j-1}{j}^{1/2}\alpha_j z^j. \tag{15.172}$$

With $N \to \infty$ and $L = 1$, the correlation function for the zeros of $p(z)$, which are given in general by the permanent formula (15.57), can be written as a determinant [443]

$$\rho_{(n)}(z_1, \dots, z_n) = \pi^{-n} \det \left[\frac{1}{(1 - z_j \bar{z}_k)^2} \right]_{j,k=1,\dots,n} \tag{15.173}$$

(this relies crucially on the Borchardt identity (4.36)). Use the general formula (15.43) to show that this coincides with the $N \to \infty$ correlation function for the 2dOCP in a unit disk without a background.

15.8 METALLIC BOUNDARY CONDITIONS

Metallic boundary conditions along some boundary $\partial\Omega$ specify that the electrostatic potential must vanish along $\partial\Omega$. Physically, such boundary conditions arise when a perfect conductor occupies the region outside $\partial\Omega$. Prescribing the value of the potential along a boundary is also referred to as Dirichlet boundary conditions.

With Ω a disk of radius R, and a perfect conductor occupying the region $r > R$, the electrostatic potential which satisfies the 2d Poisson equation (1.40) inside the disk, vanishes on the boundary of the disk and is symmetric in $\vec{r}$ and $\vec{r}\,'$ is

$$\Phi(\vec{r}, \vec{r}\,') = -\log \left| \frac{(z - z')/R}{1 - z\bar{z}'/R^2} \right| \tag{15.174}$$

(for $R = 1$ cf. (14.48)). The metal wall thus has the well-known property of creating an effective charge of opposite sign at the image point $R^2/\bar{z}'$.

Consider now the total energy of a one-component system, with a uniform background of density η in the region $r < R$. For hard wall boundary conditions, as considered in Exercises 1.4.1 q.3, it has been assumed that η is equal to the particle density ρ so that the system is overall charge neutral. However, as noted in Section 5.6.2, for metal boundary conditions, the image charges of opposite sign allow for a grand canonical formalism in which the background density is fixed independent of the number of particles, with the latter being summed over. For such a system with a fixed number N of particles of unit charge, the total potential energy consists of the particle-particle energy

$$U_1 = -\sum_{1 \le j < k \le N} \log \left| \frac{(z_j - z_k)/R}{1 - z_j z_k/R^2} \right|;$$

the self-energy U_1', which is defined by

$$U_1' := \frac{1}{2} \sum_{j=1}^{N} \lim_{r' \to r_j} \Big(\Phi(\vec{r}_j, \vec{r}\,') - \log |\vec{r}_j - \vec{r}\,'| \Big), \tag{15.175}$$

(recall Section 2.7.2) and so is equal to

$$U_1' = \frac{1}{2} \sum_{j=1}^{N} \log(1 - r_j^2/R^2) + \frac{N}{2} \log R;$$

the particle-background energy

$$U_2 = \sum_{j=1}^{N} V(r_j), \qquad V(r) = -\frac{\pi}{2}\eta(R^2 - r^2), \tag{15.176}$$

which can be calculated from the fact that $V(r)$ must satisfy the Poisson equation $\nabla^2 V(r) = 2\pi\eta$, together with the boundary condition $V(R) = 0$; and the background-background potential, which is calculated from the formula for $V(r)$ using (1.43) to be given by

$$U_3 = \frac{(\pi\eta R^2)^2}{8}.$$

Observe that the Boltzmann factor $e^{-\beta U_1'(r)} \propto 1/|1 - r^2/R^2|^{\Gamma/2}$ for a single particle and its image is not integrable at $|r| = R$ for $\Gamma \geq 2$. To prevent short distance collapse in these cases the particles can be confined to the region $|r| < R - \epsilon$, $\epsilon > 0$. The corresponding generalized grand partition function is then given by

$$\begin{aligned}\Xi_\Gamma[a;\zeta] = {} & e^{-\Gamma(\pi\eta R^2)^2/8} \\ & \times \sum_{N=0}^{\infty} \left(\frac{\zeta}{R^{\Gamma/2}}\right)^N \frac{1}{N!} \int_{\Omega_{R-\epsilon}} d\vec{r}_1\, a(\vec{r}_1) e^{\Gamma\pi\eta(R^2 - r_1^2)/2} \cdots \int_{\Omega_{R-\epsilon}} d\vec{r}_N\, a(\vec{r}_N) e^{\Gamma\pi\eta(R^2 - r_N^2)/2} \\ & \times \prod_{j=1}^{N} \frac{1}{(1 - |z_j|^2/R^2)^{\Gamma/2}} \prod_{1 \leq j < k \leq N} \left| \frac{(z_j - z_k)/R}{1 - z_j \bar{z}_k/R^2} \right|^\Gamma, \end{aligned} \tag{15.177}$$

where $\Omega_{R-\epsilon}$ denotes the disk of radius $R - \epsilon$ centered at the origin.

At the special coupling $\Gamma = 2$, it is possible to write the second line of (15.177) as a determinant, and to proceed to express $\Xi_2[a;\zeta]$ as a determinant.

PROPOSITION 15.8.1 *For $\Gamma = 2$ we have*

$$\Xi_2[a;\zeta] = e^{-(\pi\eta R^2)^2/4} \det[1 + \zeta K],$$

where K is the integral operator defined by the mapping rule

$$K[f(z')](z) := \frac{1}{R} \int_{\Omega_{R-\epsilon}} a(\vec{r}\,') e^{\pi\eta(R^2 - r'^2)} \frac{f(z')}{1 - z\bar{z}'/R^2}\, d\vec{r}\,'.$$

Proof. Setting $x_j = z_j/R$ and $y_j = \bar{z}_j/R$ in the Cauchy double alternant (4.34) shows that

$$\prod_{j=1}^{N} \frac{1}{(1 - |z_j|^2/R^2)} \prod_{1 \leq j < k \leq N} \frac{|(z_j - z_k)/R|^2}{(1 - z_j \bar{z}_k/R^2)^2} = \det\left[\frac{1}{1 - z_j \bar{z}_k/R^2}\right]_{j,k=1,\dots,N}. \tag{15.178}$$

Substituting this in (15.177) and using the Fredholm formula (5.32) gives the stated result. □

15.8.1 Correlation functions

Substituting the formula of Proposition 15.8.1 in (5.20) and using (5.31) gives

$$\rho_{(n)}(\vec{r}_1, \dots, \vec{r}_n) = \det[G(\vec{r}_j, \vec{r}_k)]_{j,k=1,\dots,n},$$

where

$$G(\vec{r}, \vec{r}\,') := \langle \vec{r} | \zeta K (1 + \zeta K)^{-1} |_{a=1} | \vec{r}\,' \rangle. \tag{15.179}$$

Thus G satisfies the integral equation

$$G(\vec{r}_1, \vec{r}_3) + \frac{\zeta}{R} e^{\pi\eta(R^2 - r_3^2)} \int_{\Omega_{R-\epsilon}} \frac{G(\vec{r}_1, \vec{r}_2)}{1 - z_2 \bar{z}_3/R^2}\, d\vec{r}_2 = \frac{\zeta}{R} \frac{e^{\pi\eta(R^2 - r_3^2)}}{1 - z_1 \bar{z}_3/R^2}. \tag{15.180}$$

A useful symmetry of G is that

$$G(\vec{r}, \vec{r}\,') = \bar{G}(\vec{r}\,', \vec{r}). \tag{15.181}$$

This follows by noting $\zeta K(1 + \zeta K)^{-1} = (1 + \zeta K)^{-1} \zeta K$ and writing out the corresponding integral equation for G. With $\vec{r}_1$ and $\vec{r}_3$ fixed, the limit $R \to \infty$ gives the correlation functions in the bulk. The integral equation then reduces to

$$\int_{\mathbb{R}^2} G(\vec{r}_1, \vec{r}_2)\, d\vec{r}_2 = 1,$$

which has solution satisfying (15.181) $G(\vec{r}_1, \vec{r}_2) = \eta e^{-\pi\eta(r_1^2+r_2^2)/2} e^{\pi\eta z_1 \bar{z}_2}$. Notice that there is no longer any dependence on the fugacity ζ. The corresponding distribution functions are identical to the distribution functions in the bulk with hard wall boundary conditions and $\eta = \rho$, as expected.

15.8.2 Vortex analogy

The theory of two-dimensional incompressible fluid flow (here two-dimensional means that the flow is the same through any plane in the fluid parallel to the reference plane), has close analogies with two-dimensional electrostatics, and hence two-dimensional Coulomb systems. This can be seen from the continuity equation $\nabla \cdot \vec{V}(x,y) = 0$ for the velocity field $\vec{V}(x,y)$, which expresses in a mathematical form the assumption that the fluid is incompressible. Because $\vec{V}(x,y)$ is divergence-free, it is possible to choose a function $\Psi(x,y)$ such that $\vec{V} = \nabla \times (\Psi \hat{\mathbf{z}})$ (Ψ is known as a stream function; we will see below that motion of the fluid particles occurs along constant values of the stream function). This means that

$$\nabla^2 \Psi(x,y) = -\xi(x,y), \qquad \xi(x,y)\hat{\mathbf{z}} = \nabla \times \vec{V}(x,y).$$

The function $\xi(x,y)$ is called the vorticity — only when $\xi(x,y) \neq 0$ is there a net circulation about the point (x,y). This equation is the same as the two-dimensional Poisson equation satisfied by an electrostatic potential $\Psi(x,y)$ due to a continuous charge distribution $\xi(x,y)/2\pi$. Thus we can write

$$\Psi(x,y) = \frac{1}{2\pi} \int_\Omega G(\vec{r}, \vec{r}\,')\xi(\vec{r}\,')\, d\vec{r}\,', \tag{15.182}$$

where $G(\vec{r}, \vec{r}\,')$ satisfies the 2d Poisson equation with delta function source term (1.40).

In free boundary conditions we would have $G(\vec{r}, \vec{r}\,') = -\log|\vec{r} - \vec{r}\,'|$. But for a fluid confined to a region Ω the boundary condition is $\vec{V}(x,y) \cdot \vec{n} = 0$, where $\vec{n}$ is the unit normal at the point (x,y) on the boundary. Now, in terms of $\xi(x,y)$,

$$\vec{V}(x,y) = \Big(\frac{\partial}{\partial y}, -\frac{\partial}{\partial x}\Big) \frac{1}{2\pi} \int_\Omega G(\vec{r}, \vec{r}\,')\xi(\vec{r}\,')\, d\vec{r}\,'$$

which shows that the boundary condition can be satisfied if

$$\vec{n} \cdot \Big(\frac{\partial}{\partial y} G(\vec{r}, \vec{r}\,'), -\frac{\partial}{\partial x} G(\vec{r}, \vec{r}\,')\Big) = 0$$

for $\vec{r} \in \partial\Omega$. But $(-n_y, n_x)$ is tangent to the boundary at (x,y), so this is equivalent to saying the directional derivative of $G(\vec{r}, \vec{r}\,')$ vanishes along the boundary, which in turn implies $G(\vec{r}, \vec{r}\,')$ must be constant for $\vec{r} \in \partial\Omega$. Choosing the constant to be zero shows that this is equivalent to metallic boundary conditions in the corresponding electrostatic problem. In particular, for a disk of radius R, $G(\vec{r}, \vec{r}\,')$ is therefore given by (15.174).

To further exploit the analogy (see, e.g., [111]), consider the total kinetic energy of the fluid

$$E = \frac{1}{2}\rho \int_\Omega \vec{V}^2\, d\vec{r},$$

where ρ denotes the mass density. Use of the vector identity

$$\nabla \cdot (\vec{A} \times \vec{B}) = \vec{B} \cdot (\nabla \times \vec{A}) - \vec{A} \cdot (\nabla \times \vec{B}),$$

with $\vec{A} = \Psi\hat{\mathbf{z}}$, $\vec{B} = \nabla \times \vec{A}$, shows we can write

$$E = \frac{1}{2}\rho \int_\Omega \Big(\Psi(\vec{r})\xi(\vec{r}) - \nabla \cdot (\vec{V} \times (\Psi\hat{\mathbf{z}})) \Big)\, d\vec{r}.$$

But from the divergence theorem the second integral can be rewritten as

$$\oint_{\partial\Omega} (\vec{V} \times (\Psi\hat{\mathbf{z}})) \cdot \vec{n}\, ds,$$

which vanishes since Ψ vanishes on the boundary. Hence, after making use of (15.182), we have

$$E = \frac{\rho}{4\pi} \int_\Omega d\vec{r} \int_\Omega d\vec{r}\,'\, \xi(\vec{r})\xi(\vec{r}\,')G(\vec{r}, \vec{r}\,'), \tag{15.183}$$

which is formally proportional to the expression for the potential energy of a continuous charge distribution $\xi(\vec{r})$ in Ω with metallic boundary conditions.

In the case that there are N vortices with a small support centered at $\vec{r}_j$ $(j = 1, \dots, N)$, one writes

$$\xi(\vec{r}) = \sum_{j=1}^{N} \xi_j(\vec{r} - \vec{r}_j) \approx \sum_{j=1}^{N} \Gamma_j \delta(\vec{r} - \vec{r}_j). \tag{15.184}$$

The quantities Γ_j represent the circulation about the point $\vec{r}_j$, since, with C denoting a small circle about $\vec{r}_j$ and D the corresponding disk,

$$\oint_C \vec{V} \cdot d\vec{r} = \int_D (\nabla \times V) \cdot \hat{\mathbf{z}}\, d\vec{r} = \int_D \xi(\vec{r})\, d\vec{r} = \Gamma_j.$$

Substituting (15.184) for ξ in the formula (15.183) for E gives

$$E = \frac{\rho}{2\pi} \sum_{j<k} \Gamma_j \Gamma_k G(\vec{r}_j, \vec{r}_k) + \frac{\rho}{4\pi} \sum_{j=1}^{N} \Gamma_j^2 G_{\text{finite}}(\vec{r}_j, \vec{r}_j), \tag{15.185}$$

where G_{finite} represents the finite part of the self-energy (the infinite part which results from the point approximation (15.184) is disregarded, as in electrostatics). This expression is proportional to that for the potential energy of a system of charges of strength Γ_j in Ω with metallic boundary conditions.

For Ω a disk of radius R, another case of interest arises when the whole fluid is rotated with angular velocity ω. Taking the rotating system as the frame of reference, we see that we should subtract ω from $\xi(\vec{r})$ in (15.183). This implies that in the formula (15.185) there is an additional one-body term

$$-\frac{\rho\omega}{4} \sum_{j=1}^{N} \Gamma_j (R^2 - r_j^2), \tag{15.186}$$

as well as a constant. Comparing with (15.176), we see that this additional term corresponds to a uniform background in the analogy with charges in a disk with metallic boundary conditions.

15.9 ANTIMETALLIC BOUNDARY CONDITIONS

The metallic boundary condition is the $\epsilon \to \infty$ boundary condition at the interface between a vacuum and a dielectric material with dielectric constant ϵ. To see this, with the dielectric material on the outside of some boundary $\partial\Omega$, and $\hat{n}$ denoting the (outward) normal, we recall from macroscopic electrostatics that the potential Φ must satisfy the boundary condition

$$\frac{\partial \Phi}{\partial \hat{n}}\Big|_{z\to\partial\Omega^-} = \epsilon \frac{\partial \Phi}{\partial \hat{n}}\Big|_{z\to\partial\Omega^+}. \tag{15.187}$$

In the limit $\epsilon \to \infty$ these equations can only be satisfied if $\Phi = C$ (C constant) for $z \in \partial\Omega^+$, and thus by continuity of Φ, $\Phi = C$ for $z \in \partial\Omega$. The other extreme limit is $\epsilon = 0$. Although this is unphysical since $\epsilon \geq 1$, it is mimicked by situations in which the dielectric constant of the material inside the system is much greater than that outside. With $\epsilon = 0$, (15.187) gives that the normal derivative of the potential must vanish at the boundary, a situation referred to as *Neumann boundary conditions*.

Suppose now that the dielectric material occupies the outside of a disk of radius R. Then for the source point $\vec{r}'$ inside the disk one can verify that the solution of the 2d Poisson equation (1.40) inside the disk, subject to the boundary condition (15.187), is

$$\Phi(\vec{r}, \vec{r}') = -\log\Big(\frac{|z - z'|}{|R - zz'/R|^{(\epsilon-1)/(\epsilon+1)}}\Big). \tag{15.188}$$

Here the arbitrary additive constant has been chosen so that (15.188) reduces to (1.41) for $\epsilon = 1$ and to (15.174) for $\epsilon \to \infty$. The formula (15.188) exhibits the well-known feature that the dielectric material creates an effective particle of charge $-(\epsilon - 1)/(\epsilon + 1)$ at the image point $R^2/\bar{z}'$. In the case that $\epsilon = 0$, this image particle has charge of the same sign and strength as the test particle—such a boundary condition will therefore be referred to as anti-metallic.

We will consider a one-component system formed by placing N particles of unit charge in a disk of radius R with the anti-metallic boundary condition. Inside the disk we assume that there is a uniform neutralizing background of density ρ. A short calculation using the formulas of Section 1.4.1, and taking into consideration the self energy as defined by (15.175), shows that the total energy of such a system is

$$U = \frac{N^2}{2}\log R - \frac{3}{8}N^2 + \frac{1}{2}\sum_{j=1}^{N}\Big(\pi\rho r_j^2 - \log(1 - r_j^2/R^2)\Big) - \sum_{1\leq j<k\leq N}\log\Big(|z_k - z_j||1 - z_j\bar{z}_k/R^2|\Big) \tag{15.189}$$

(note that the constant terms are the same as for hard wall boundary conditions calculated in Exercises 1.4 q.3). The corresponding Boltzmann factor is thus

$$Ae^{-\pi\Gamma\rho\sum_{j=1}^{N} r_j^2/2}\prod_{1\leq j<k\leq N}|z_k - z_j|^{\Gamma}|1 - z_j\bar{z}_k/R^2|^{\Gamma}\prod_{j=1}^{N}(1 - r_j^2/R^2)^{\Gamma/2}, \tag{15.190}$$

where $A = e^{-\Gamma N^2((1/2)\log R - 3/8)}$.

15.9.1 Random matrix analogy

In the case of a dielectric material with $\epsilon = 0$ in the region $y < 0$, the pair potential is

$$\Phi(\vec{r}, \vec{r}') = -\log\Big(|z - z'||z - \bar{z}'|\Big).$$

For a one-component system confined to the semi-disk $|z| < R$, $y > 0$, with a neutralizing background of density $\rho = N/(\frac{1}{2}\pi R^2)$, we see that the corresponding Boltzmann factor is proportional to

$$\prod_{j=1}^{N} e^{-\pi\Gamma\rho r_j^2/2}|z_j - \bar{z}_j|^{\Gamma/2}\prod_{1\leq j<k\leq N}|z_k - z_j|^{\Gamma}|z_k - \bar{z}_j|^{\Gamma}. \tag{15.191}$$

With $\Gamma = 2$, this same expression (up to a factor of $\prod_{j=1}^{N}|z_j - \bar{z}_j|$) occurs as the eigenvalue p.d.f. for a certain ensemble of random matrices [264], [395].

PROPOSITION 15.9.1 *Let $\mathbf{X}$ be an $N \times N$ random matrix with real quaternion elements, and suppose the four real numbers $u_{jk}^{(l)}$ $(l = 1, \dots, 4)$ in each such element are chosen independently with p.d.f.*

$$\frac{1}{\pi^2} e^{-\sum_{l=1}^{4} (u_{jk}^{(l)})^2}.$$

For $\mathbf{X}$ a $2N \times 2N$ matrix with complex elements, the eigenvalues λ_j of $\mathbf{X}$ come in complex conjugate pairs $(\lambda_{j+N} = \bar{\lambda}_j)$, and the p.d.f. of the N eigenvalues with $Im(\lambda_j) > 0$ is proportional to (15.191) with $\Gamma = 2$ and $\rho = 2/\pi$, multiplied by an additional self-energy type factor $\prod_{j=1}^{N} |z_j - \bar{z}_j|$.

Proof. The mechanism for the eigenvalues occurring in complex conjugate pairs is the same as that for the Kramer's degeneracy discussed in Exercises 1.3 q.1. Formally, the result for the p.d.f. follows from the p.d.f. for complex random matrices

$$\frac{1}{C} e^{-\sum_{j=1}^{N} |\lambda_j|^2} \prod_{1 \le j < k \le N} |\lambda_k - \lambda_j|^2,$$

by simply replacing N by $2N$ and putting $\lambda_{j+N} = \bar{\lambda}_j$ for $j = 1, \dots, N$. This prescription can be verified by starting with the Schur decomposition (15.3) where now $\mathbf{U}$ has real quaternion elements, and proceeding as in the proof of Proposition 15.1.1 (see [395]). □

15.9.2 Correlation functions

At the special coupling $\Gamma = 2$, the products in (15.190) can be written in terms of the Vandermonde determinant (1.173) of dimension $2N$ with $x_j = z_j/R$, $x_{j+N} = R/\bar{z}_j$ $(j = 1, \dots, N)$, and the free energy and correlation functions calculated exactly [493]. First, for the products in the Boltzmann factor we have

$$\prod_{1 \le j < k \le N} |z_k - z_j|^2 \prod_{j,k=1}^{N} (1 - z_j \bar{z}_k / R^2) = R^{-N^2} \prod_{j=1}^{N} (\bar{z}_j)^{2N-1} \det \begin{bmatrix} (z_j/R)^{k-1} \\ (R/\bar{z}_j)^{k-1} \end{bmatrix}_{\substack{j=1,\dots,N \\ k=1,\dots,2N}}. \tag{15.192}$$

The determinant has a similar structure to the one appearing in (6.14), which suggests we proceed using skew orthogonal polynomials.

The appropriate skew symmetric inner product for this purpose is

$$\langle f | g \rangle_s = \frac{1}{2} \int_0^R dr\, r^{2N} e^{-\pi \rho r^2} \int_0^{2\pi} d\theta\, e^{-i\theta(2N-1)} \Big(f(z/R) g(R/\bar{z}) - f(R/\bar{z}) g(z/R) \Big), \tag{15.193}$$

$z := re^{i\theta}$. With respect to this inner product we can check that

$$Q_{2n+1}(z) = z^{2N-1-n}, \quad Q_{2n}(z) = z^n \quad (n = 0, \dots, N-1) \tag{15.194}$$

are skew orthogonal with normalization

$$q_n = \frac{\pi}{2} R^{2N+1} \Big(N^{-n-1} \gamma(n+1; N) - N^{n-2N} \gamma(2N - n; N) \Big). \tag{15.195}$$

Here we have relaxed the condition that $Q_j(z)$ be of degree j, and replaced it by the requirement that

$$\det \begin{bmatrix} Q_{k-1}(z_j/R) \\ Q_{k-1}(R/\bar{z}_j) \end{bmatrix}_{\substack{j=1,\dots,N \\ k=1,\dots,2N}} \propto \det \begin{bmatrix} (z_j/R)^{k-1} \\ (R/\bar{z}_j)^{k-1} \end{bmatrix}_{\substack{j=1,\dots,N \\ k=1,\dots,2N}}.$$

The analogue of the formulas of Propositions 6.1.6 and 6.1.7 apply for the evaluation of the partition function and correlation functions (see Exercises 15.9 q.1).

In particular, for the density $\rho_{(1)}(z)$ in the finite system these formulas imply

$$\rho_{(1)}(z) = \frac{(\bar{z})^{2N-1} e^{-\pi \rho |z|^2}}{\pi R^{2N+1}} \sum_{n=0}^{N-1} \frac{(z/R)^n (R/\bar{z})^{2N-1-n} - (z/R)^{2N-1-n} (R/\bar{z})^n}{N^{-n-1} \gamma(n+1; N) - N^{n-2N} \gamma(2N - n; N)}. \tag{15.196}$$

Taking the limit $N, R \to \infty$ with z and ρ fixed gives for the bulk density $\rho(z) = \rho$ as expected. If instead the coordinates are chosen as in (15.49) and then the limit $N \to \infty$ taken (which requires using (15.51)), we obtain for the density profile as measured from the boundary at $y = 0$,

$$\rho_{(1)}(y) = \rho e^{-2\pi\rho y^2} g(2y), \quad g(y) := \frac{2}{\sqrt{\pi}} \int_{-\infty}^{\infty} e^{-t^2} \frac{\sinh(2\pi\rho)^{1/2} yt}{\operatorname{erf} t}\, dt. \tag{15.197}$$

With the particle coordinates fixed in the bulk of the system, we find that the general n-particle correlation deduced from the quaternion determinant formalism reduces to the bulk expression of Proposition 15.3.2 in the thermodynamic limit. For the n-particle correlation in the neighborhood of the boundary, using the coordinates (15.49) and the asymptotic expansion (15.51), we find

$$\rho_{(n)}(\vec{r}_1, \dots, \vec{r}_n) = \rho^n$$
$$\times \text{qdet} \begin{bmatrix} e^{-\pi\rho(y_j^2+y_k^2)} g(y_j + y_k + i(x_j - x_k)) & -e^{-\pi\rho(y_j^2+y_k^2)} g(y_j - y_k + i(x_j - x_k)) \\ e^{-\pi\rho(y_j^2+y_k^2)} g(y_j - y_k - i(x_j - x_k)) & e^{-\pi\rho(y_j^2+y_k^2)} g(y_j + y_k - i(x_j - x_k)) \end{bmatrix}_{j,k=1,\dots,n}, \tag{15.198}$$

where $g(u)$ is defined in (15.197).

EXERCISES 15.9 1. (i) Use the method of the proof of Proposition 6.1.6 (with $\mathbf{C}(x)$ replaced by the identity) to derive the formula

$$\prod_{l=1}^N e^{-\pi\rho|z_l|^2} \prod_{1 \le j < k \le N} |z_k - z_j|^2 \prod_{j,k=1}^N (1 - z_j \bar{z}_k / R^2) = \prod_{l=0}^{N-1} 2q_l \, \text{qdet} \begin{bmatrix} S(z_j, z_k) & I(z_j, z_k) \\ D(z_j, z_k) & S(z_k, z_j) \end{bmatrix}_{N \times N},$$

where with $\tilde{w}(z) := e^{-\pi\rho|z|^2/2} \bar{z}^{(N-1/2)}$

$$S(z,u) := \tilde{w}(z)\tilde{w}(u) \sum_{m=0}^{N-1} \frac{1}{2q_m} \Big(Q_{2m}(z/R) Q_{2m+1}(R/\bar{u}) - Q_{2m+1}(z/R) Q_{2m}(R/\bar{u}) \Big),$$

$$I(z,u) := -\tilde{w}(z)\tilde{w}(u) \sum_{m=0}^{N-1} \frac{1}{2q_m} \Big(Q_{2m}(z/R) Q_{2m+1}(u/R) - Q_{2m+1}(z/R) Q_{2m}(u/R) \Big),$$

$$D(z,u) := \tilde{w}(z)\tilde{w}(u) \sum_{m=0}^{N-1} \frac{1}{2q_m} \Big(Q_{2m}(R/\bar{z}) Q_{2m+1}(R/\bar{u}) - Q_{2m+1}(R/\bar{z}) Q_{2m}(R/\bar{u}) \Big).$$

(ii) With $\{Q_j(x)\}_{j=0,1,\dots,2N-1}$ skew orthogonal with respect to the skew inner product (15.193), note that the analogue of the formulas (6.17) hold and thus so does the analogue of Proposition 6.1.7.

2. From Proposition 15.9.1 we know that the eigenvalue p.d.f. for random matrices with real quaternion elements is, after scaling $z_j \mapsto z_j/\sqrt{2}$,

$$\frac{1}{Q_N} \prod_{j=1}^N e^{-|z_j|^2} |z_j - \bar{z}_j|^2 \prod_{1 \le j < k \le N} |z_k - z_j|^2 |z_k - \bar{z}_j|^2,$$

where $\text{Im}(z_j) \ge 0$ and

$$Q_N = 2^{-N} \int_{\mathbb{R}^2} dx_1 dy_1 \dots \int_{\mathbb{R}^2} dx_N dy_N \prod_{l=1}^N e^{-(x_l^2+y_l^2)} |z_l - \bar{z}_l|^2 \prod_{1 \le j < k \le N} |z_k - z_j|^2 |z_k - \bar{z}_j|^2.$$

(i) Show from the Vandermonde determinant (1.173)

$$\prod_{l=1}^{N}(\bar{z}_l - z_l) \prod_{1\le j<k\le N} |z_k - z_j|^2 |z_k - \bar{z}_j|^2 = \det \begin{bmatrix} z_j^{k-1} \\ (\bar{z}_j)^{k-1} \end{bmatrix}_{\substack{j=1,\dots,N \\ k=1,\dots,2N}}.$$

(ii) Define the skew symmetric inner product

$$\langle f|g\rangle_s = \frac{1}{2}\int_0^\infty dr\, re^{-r^2} \int_0^{2\pi} d\theta\, (z-\bar{z})\Big(f(z)g(\bar{z}) - f(\bar{z})g(z)\Big),$$

and verify that with respect to this inner product

$$Q_{2n}(z) = 2^n n! \sum_{l=0}^{n} \frac{z^{2l}}{2^l l!}, \quad Q_{2n+1}(z) = z^{2n+1} \quad (n = 0,\dots,N-1)$$

are skew orthogonal with normalization $q_n = \pi\Gamma(2n+2)$.

(iii) Use the results of (i) and (ii) to show

$$Q_N = \prod_{n=0}^{N-1} q_n = \pi^N \prod_{n=0}^{N-1} \Gamma(2n+2).$$

(iv) Use the formulas of q.1 with $\tilde{w}(z) = e^{-|z|^2/2}(z-\bar{z})^{1/2}$ and $Q_n(z/R)$, $Q_n(R/\bar{z})$, $2q_n$ replaced by $Q_n(z)$, $Q_n(\bar{z})$, q_n, respectively, as specified above to show that in the limit $N \to \infty$

$$\rho_{(n)}(\vec{r}_1,\dots,\vec{r}_n) = \prod_{j=1}^{n} e^{-|\vec{r}_j|^2} 2y_j \,\mathrm{qdet}\begin{bmatrix} f(z_j,\bar{z}_k) & -f(z_j,z_k) \\ f(\bar{z}_j,\bar{z}_k) & f(z_k,\bar{z}_j) \end{bmatrix}_{n\times n},$$

where

$$f(w,z) = \frac{i}{(2\pi)^{1/2}} \sum_{n=0}^{\infty} I_{n+1/2}(wz)\bigg(\Big(\frac{z}{w}\Big)^{n+1/2} - \Big(\frac{w}{z}\Big)^{n+1/2}\bigg),$$

with $I_a(z)$ denoting the Bessel function of pure imaginary argument, specified by the integral (8.67) for $a \in \mathbb{Z}$, and more generally by the series

$$I_{n+1/2}(x) := \sum_{k=0}^{\infty} \frac{(x/2)^{2k+n+1/2}}{k!\Gamma(k+n+3/2)}. \tag{15.199}$$

(v) [344] Verify that

$$\frac{\partial f}{\partial z} = zf + \frac{i}{\pi}e^{zw}, \qquad \frac{\partial f}{\partial w} = wf - \frac{i}{\pi}e^{zw}.$$

Add these two equations together to deduce

$$f(w,z) = e^{(w^2+z^2)/2}A(z-w), \qquad A(0) = 0,$$

for some $A(x)$, and substitute this back in either of the equations to deduce

$$\frac{\partial A(s)}{\partial s} = \frac{i}{\pi}e^{-s^2/2}.$$

Hence conclude

$$f(w,z) = \frac{i}{\sqrt{2\pi}} e^{(w^2+z^2)/2}\mathrm{erf}\Big(\frac{z-w}{\sqrt{2}}\Big).$$

15.10 EIGENVALUES OF REAL RANDOM MATRICES

For $N \times N$ random matrices with real entries there is a nonzero probability that there will be exactly k real eigenvalues for each $k = 0, \ldots, N$, provided N and k have the same parity (this latter constraint is required because the complex eigenvalues must occur in complex conjugate pairs). In the case that the entries of such a matrix, $\mathbf{A}$ say, are independent standard Gaussians (the matrix $\mathbf{A}$ is then said to belong to the real Ginibre ensemble) the joint p.d.f. for the k real and $N-k$ complex eigenvalues can be calculated exactly [371], [155]. For this purpose one uses the real Schur decomposition

$$\mathbf{A} = \mathbf{QRQ}^T, \tag{15.200}$$

where $\mathbf{Q}$ is an $N \times N$ orthogonal matrix with elements of the first row positive while

$$\mathbf{R} = \begin{bmatrix} \lambda_1 & \cdots & R_{1,k} & R_{1,k+1} & \cdots & R_{1,m} \\ & \ddots & \vdots & \vdots & \cdots & \vdots \\ & & \lambda_k & R_{k,k+1} & \cdots & R_{k,m} \\ & & & \mathbf{Z}_{k+1} & \cdots & R_{k+1,m} \\ & & & & \ddots & \vdots \\ & & & & & \mathbf{Z}_m \end{bmatrix}. \tag{15.201}$$

In (15.201) all elements not explicitly shown are zero, $m = (N+k)/2$, while R_{ij} is of size $p \times q$ with

$$p \times q = \begin{cases} 1 \times 1 & \text{if } i \le k,\ j \le k, \\ 1 \times 2 & \text{if } i \le k,\ j > k, \\ 2 \times 1 & \text{if } i > k,\ j \le k, \\ 2 \times 2 & \text{if } i > k,\ j > k. \end{cases}$$

Furthermore, the λ_j $(j \le k)$ are the real eigenvalues of $\mathbf{A}$, while the 2×2 matrices $\mathbf{Z}_j$ have the structure

$$\mathbf{Z}_j = \begin{bmatrix} x_j & b_j \\ -c_j & x_j \end{bmatrix}, \qquad b_j, c_j > 0, \tag{15.202}$$

such that the complex eigenvalues of $\mathbf{A}$ are $x_j \pm iy_j$, $y_j = \sqrt{b_j c_j}$ (see Exercises 15.10 q.2).

Let $\tilde{\mathbf{R}}$ denote the strictly upper triangular part of $\mathbf{R}$, involving only entries labeled $R_{j,k}$ in (15.201). We see that $\tilde{\mathbf{R}}$ has $N(N-1)/2 - (N-k)/2$ independent variables, while $\{\lambda_i\}$ has k independent variables, $\{\mathbf{Z}_i\}$ has $3(N-k)/2$ independent variables and $\mathbf{Q}$ has $N(N-1)/2$ independent variables. This adds up to the N^2 independent parameters of $\mathbf{A}$.

The change of variables (15.200) can be carried out according to the following result [155].

PROPOSITION 15.10.1 *We have*

$$(d\mathbf{A}) = 2^{(N-k)/2} \prod_{j<p} |\lambda(R_{pp}) - \lambda(R_{jj})| (d\tilde{\mathbf{R}})(\mathbf{Q}^T d\mathbf{Q}) \times \prod_{l=k+1}^{(N+k)/2} |b_l - c_l| d\lambda_1 \cdots d\lambda_k (d\mathbf{Z}), \tag{15.203}$$

where $\lambda(R_{ll}) = \lambda_l$ for $l \le k$, while $\lambda(R_{ll}) = x_l \pm iy_l$ for $l > k$, and

$$(d\mathbf{Z}) = \prod_{j=k+1}^{(N+k)/2} dx_j db_j dc_j.$$

Proof. Analogous to (15.4) one has

$$\mathbf{Q}^T d\mathbf{A}\,\mathbf{Q} = d\mathbf{R} + \mathbf{Q}^T d\mathbf{Q}\,\mathbf{R} - \mathbf{R}\mathbf{Q}^T d\mathbf{Q}.$$

With $d\mathbf{O} := \mathbf{Q}^T d\mathbf{Q}$ the ij (block) element on the r.h.s. is

$$dR_{ij} + dO_{ij}R_{jj} - R_{ii}dO_{ij} + \sum_{l<j} dO_{il}R_{lj} - \sum_{l>i} R_{il}dO_{lj} \tag{15.204}$$

(cf. (15.5)). Arguing now as in the derivation of (15.8) shows that the wedge product of the off-diagonal elements in (15.204) is

$$\prod_{j<k} |\lambda(R_{kk}) - \lambda(R_{jj})|(d\tilde{\mathbf{O}})(d\tilde{\mathbf{R}}), \tag{15.205}$$

where $\tilde{\mathbf{R}}$ is as defined below (15.202), while $(d\tilde{\mathbf{O}})$ refers to the product of (block) off-diagonal differentials in $d\mathbf{O}$.

Noting that

$$d\mathbf{O}_{ii} = \begin{bmatrix} 0 & do_i \\ -do_i & 0 \end{bmatrix}$$

for some do_i shows that with $i = j > k$ (15.204) has the explicit form

$$\begin{bmatrix} dx_i + (b_i - c_i)do_i & db_i \\ -dc_i & dx_i + (c_i - b_i)do_i \end{bmatrix}.$$

The wedge product of these entries is therefore equal to

$$2^{(N-k)/2} \prod_{l=k+1}^{(N+k)/2} |b_l - c_l|(d\mathbf{Z}) \prod_{i=k+1}^{(N+k)/2} do_i. \tag{15.206}$$

Multiplying together (15.206) and (15.205), and noting too that for $i = j < k$ (15.204) reduces to $d\lambda_i$, gives the result. □

Suppose now that

$$\Pr(\mathbf{A}) = \left(\frac{1}{2\pi}\right)^{N^2/2} e^{-\mathrm{Tr}(\mathbf{A}\mathbf{A}^T)/2} \tag{15.207}$$

so that the elements of $\mathbf{A}$ are independent real Gaussians with distribution $\mathrm{N}[0,1]$. Changing variables according to (15.200) the corresponding eigenvalue p.d.f. can be computed.

PROPOSITION 15.10.2 *For $N \times N$ real Gaussian matrices distributed according to (15.207), with k real eigenvalues $\lambda_1, \dots, \lambda_k$ and $(N-k)/2$ complex conjugate pairs of eigenvalues $x_j \pm iy_j$ $(j = 1, \dots, (N-k)/2)$ the eigenvalue p.d.f. is equal to*

$$\frac{1}{2^{N(N+1)/4}\prod_{l=1}^N \Gamma(l/2)} \frac{2^{(N-k)/2}}{k!((N-k)/2)!} \Big|\Delta(\{\lambda_l\}_{l=1,\dots,k} \cup \{x_j \pm iy_j\}_{j=1,\dots,(N-k)/2})\Big|$$
$$\times e^{-\sum_{j=1}^k \lambda_j^2/2} e^{\sum_{j=1}^{(N-k)/2}(y_j^2 - x_j^2)} \prod_{j=1}^{(N-k)/2} \mathrm{erfc}(\sqrt{2}y_j), \tag{15.208}$$

where $\Delta(\{z_p\}_{p=1,\dots,m}) := \prod_{j<l}^m (z_l - z_j)$. Here the earlier ordering constraint on the eigenvalues has been relaxed. Integrating this over $\lambda_j \in (-\infty, \infty)$, $(x_j, y_j) \in \mathbb{R}_+^2$, $\mathbb{R}_+^2 := \{(x,y) \in \mathbb{R} : y > 0\}$ gives the probability that a matrix from the ensemble (15.207) has exactly k real eigenvalues.

Proof. Substituting for $\mathbf{A}$ using (15.200) and making use of (15.201), (15.202) shows

$$e^{-\mathrm{Tr}(\mathbf{A}\mathbf{A}^T)/2} = e^{-\sum_{i<j} r_{ij}^2/2} e^{-\sum_{j=1}^k \lambda_j^2/2} e^{-\sum_{j=1}^{(N-k)/2}(x_j^2 + y_j^2 + \delta_j^2/2)}, \tag{15.209}$$

where

$$\delta_j := b_j - c_j \tag{15.210}$$

and $[r_{ij}] = \tilde{\mathbf{R}}$. Furthermore, a direct calculation gives

$$db_j dc_j = \frac{2y_j}{\sqrt{\delta_j^2 + 4y_j^2}} dy_j d\delta_j. \tag{15.211}$$

Multiplying (15.209) and (15.203), and further changing variables according to (15.210), making use of (15.211) we see that the eigenvalues separate from the other variables apart from the integration over δ_j. For this one uses

$$e^{-y^2} \int_{-\infty}^{\infty} \frac{|\delta| e^{-\delta^2/2}}{\sqrt{\delta^2 + 4y^2}} \, d\delta = \sqrt{2\pi} e^{y^2} \operatorname{erfc}(\sqrt{2}y),$$

which can be derived by appropriate changes of variables. To calculate the precise form of the constants, note from (1.38) in the case $\beta = 1$ that

$$\int (\mathbf{Q}^T d\mathbf{Q}) = \pi^{N(N+1)/4} \prod_{j=1}^{N} \frac{1}{\Gamma(j/2)},$$

while the integration over $\{r_{ij}\}$ are simple Gaussians. □

Setting $k = N$ in (15.208), and integrating over the λ_j using (1.163) in the case $\beta = 1$ shows that the probability of all eigenvalues being real is equal to $2^{-N(N-1)/4}$.

Generalized partition function

Let the p.d.f. (15.208) be denoted $p(\{\lambda_l\}_{l=1,\dots,k}; \{x_j \pm iy_j\}_{j=1,\dots,(N-k)/2})$, and specify the corresponding generalized partition function by

$$Z_{k,(N-k)/2}[u,v] = \int_{-\infty}^{\infty} d\lambda_1 \cdots \int_{-\infty}^{\infty} d\lambda_k \prod_{l=1}^{k} u(\lambda_l) \int_{\mathbb{R}_+^2} dx_1 dy_1 \cdots \int_{\mathbb{R}_+^2} dx_{(N-k)/2} dy_{(N-k)/2}$$
$$\times \prod_{l=1}^{(N-k)/2} v(x_l, y_l) p(\{\lambda_l\}_{l=1,\dots,k}; \{x_j \pm iy_j\}_{j=1,\dots,(N-k)/2}).$$

This can be evaluated in terms of the series expansion of a particular Pfaffian, analogous to the expression of Proposition 6.7.1.

PROPOSITION 15.10.3 *Let $\{p_{l-1}(x)\}_{l=1,\dots,N}$ be a set of monic polynomials, with $p_{l-1}(x)$ of degree $l-1$. Let*

$$\alpha_{j,k} = \int_{-\infty}^{\infty} dx \, u(x) \int_{-\infty}^{\infty} dy \, u(y) \, e^{-(x^2+y^2)/2} p_{j-1}(x) p_{k-1}(y) \operatorname{sgn}(y - x),$$
$$\beta_{j,k} = 2i \int_{\mathbb{R}_+^2} dx dy \, v(x,y) e^{y^2 - x^2} \operatorname{erfc}(\sqrt{2}y) \Big(p_{j-1}(x+iy) p_{k-1}(x-iy) - p_{k-1}(x+iy) p_{j-1}(x-iy) \Big). \tag{15.212}$$

For k, N even we have

$$Z_{k,(N-k)/2}[u,v] = \frac{1}{2^{N(N+1)/4} \prod_{l=1}^{N} \Gamma(l/2)} [\zeta^{k/2}] \mathrm{Pf}[\zeta \alpha_{j,l} + \beta_{j,l}]_{j,l=1,\dots,N}. \tag{15.213}$$

Proof. We have

$$\Delta\Big(\{\lambda_l\}_{l=1,\dots,k} \cup \{x_j \pm iy_j\}_{j=1,\dots,(N-k)/2}\Big) = \det\left[\begin{array}{c} [p_{l-1}(\lambda_j)]_{j=1,\dots,k} \\ \left[\begin{array}{c} p_{l-1}(x_j+iy_j) \\ p_{l-1}(x_j-iy_j) \end{array}\right]_{j=1,\dots,(N-k)/2} \end{array}\right]_{l=1,\dots,N}.$$

With the λs ordered $\lambda_k > \cdots > \lambda_1$, the absolute value signs about Δ in (15.208) can be removed, provided one multiplies by $i^{(N-k)/2}$. Substituting for Δ according to the above formula, we can apply the method of integration over alternate variables. With k assumed even, this gives

$$\begin{aligned} Z_{k,(N-k)/2}[u,v] = {} & i^{(N-k)/2} A_{k,N} \frac{k!}{(k/2)!} \int_{\mathbb{R}^2_+} dx_1 dy_1 \cdots \int_{\mathbb{R}^2_+} dx_{(N-k)/2} dy_{(N-k)/2} \prod_{l=1}^{(N-k)/2} v(x_l,y_l) e^{y_l^2 - x_l^2} \\ & \times \mathrm{erfc}(\sqrt{2}y_l) \int_{-\infty}^{\infty} d\lambda_2 \int_{-\infty}^{\infty} d\lambda_4 \cdots \int_{-\infty}^{\infty} d\lambda_k \det \left[\begin{array}{c} \left[\begin{array}{c} \int_{-\infty}^{\lambda_{2j}} u(\lambda) e^{-\lambda^2/2} p_{l-1}(\lambda)\, d\lambda \\ u(\lambda_{2j}) e^{-\lambda_{2j}^2/2} p_{l-1}(\lambda_{2j}) \end{array}\right]_{j=1,\dots,k} \\ \left[\begin{array}{c} p_{l-1}(x_j+iy_j) \\ p_{l-1}(x_j-iy_j) \end{array}\right]_{j=1,\dots,(N-k)/2} \end{array}\right]_{l=1,\dots,N}, \end{aligned}$$

where $A_{k,N}$ denotes the numerical prefactor in (15.208). Expanding the determinant according to its definition, then performing the remaining integrations shows

$$Z_{k,(N-k)/2}[u,v] = A_{k,N} \frac{k!}{(k/2)!} \sum_{P \in S_N} \varepsilon(P) \prod_{l=1}^{k/2} a_{P(2l-1),P(2l)} \prod_{l=k/2+1}^{N/2} b_{P(2l-1),P(2l)},$$

where

$$\begin{aligned} a_{j,k} &= \int_{-\infty}^{\infty} dx\, u(x) e^{-x^2/2} p_{k-1}(x) \int_{-\infty}^{x} dy\, u(y) e^{-y^2/2} p_{j-1}(y), \\ b_{j,k} &= \int_{\mathbb{R}^2_+} dx dy\, v(x,y) e^{y^2-x^2} \mathrm{erfc}(\sqrt{2}y) p_{j-1}(x+iy) p_{k-1}(x-iy). \end{aligned}$$

Imposing the restriction $P(2l) > P(2l-1)$ and using the notation of the statement of (15.213) this reads

$$\begin{aligned} Z_{k,(N-k)/2}[u,v] &= 2^{(k-N)/2} A_{k,N} \frac{k!}{(k/2)!} \sum_{\substack{P \in S_N \\ P(2l) > P(2l-1)}} \varepsilon(P) \prod_{l=1}^{k/2} \alpha_{P(2l-1),P(2l)} \prod_{l=k/2+1}^{N/2} \beta_{P(2l-1),P(2l)} \\ &= 2^{(k-N)/2} A_{k,N} k!((N-k)/2)! [\zeta^{k/2}] \mathrm{Pf}[\zeta \alpha_{j,l} + \beta_{j,l}]_{j,l=1,\dots,N} \end{aligned}$$

(cf. (6.159)), where the final equality, which immediately implies (15.208), can be verified directly from the definition of the Pfaffian. □

With N even and $Z_N[u,v] := \sum_{k=0}^{N/2} Z_{2k,(N-2k)/2}[u,v]$, we read off from (15.213) that [491]

$$Z_N[u,v] = \frac{1}{2^{N(N+1)/4} \prod_{l=1}^N \Gamma(l/2)} \mathrm{Pf}[\alpha_{j,k} + \beta_{j,k}]_{j,k=1,\dots,N}. \tag{15.214}$$

Similarly we see that $Z_N(\zeta) := \sum_{k=0}^{N/2} \zeta^k Z_{2k,(N-2k)/2}[1,1]$, which is the generating function for the probability $p_{2k,N} = Z_{2k,(N-2k)/2}[1,1]$ of there being exactly $2k$ real eigenvalues, has the Pfaffian form

$$Z_N(\zeta) = \frac{1}{2^{N(N+1)/4} \prod_{l=1}^N \Gamma(l/2)} \mathrm{Pf}[\zeta \alpha_{j,k} + \beta_{j,k}]_{j,k=1,\dots,N} \Big|_{u=v=1}. \tag{15.215}$$

If we choose, say, $p_j(x) = x^j$ in (15.212) all the integrals defining $\alpha_{j,k}, \beta_{j,k}$ can be computed explicitly

[155] (see also the proof of Proposition 15.10.4 below). Furthermore, we can check that by the symmetry of the integrand

$$\alpha_{2j,2k}|_{u=1} = \alpha_{2j-1,2k-1}|_{u=1} = 0, \qquad \beta_{2j,2k}|_{v=1} = \beta_{2j-1,2k-1}|_{v=1} = 0.$$

Thus the entries of the Pfaffian in (15.215) vanish in a checkerboard fashion. As in deducing (8.132) from (8.134) we can therefore write the Pfaffian as a determinant of half the size [10]

$$Z_N(\zeta) = \frac{1}{2^{N(N+1)/4}\prod_{l=1}^N \Gamma(l/2)} \det \Big[\zeta\alpha_{2j-1,2k}|_{u=1} + \beta_{2j-1,2k}|_{v=1} \Big]_{j,k=1,\dots,N/2}, \tag{15.216}$$

giving a computable formula for the $p_{2k,N}$. From this viewpoint the general N exact result $p_{N,N} = 2^{-N(N-1)/4}$ follows by noting from the coefficient of $\zeta^{N/2}$ in (15.215) that

$$p_{N,N} = \frac{1}{2^{N(N+1)/4}\prod_{l=1}^N \Gamma(l/2)} \det[\alpha_{2j-1,2k}|_{u=1}]_{j,k=1,\dots,N/2}.$$

Choosing $p_j(x) = R_j(x)$, where $\{R_j(x)\}$ are the skew orthogonal polynomials for the $\beta = 1$ Gaussian weight, as discussed in Section 6.3 the determinant becomes diagonal, implying the sought result.

Correlation functions

Since the number of real eigenvalues is a variable, the summed up generalized partition function (15.214) is the appropriate quantity to use in the functional derivative formula for the correlation functions. The latter may involve both real and complex eigenvalues. In the case that it involves only real eigenvalues (to be denoted $\rho^{\rm r}_{(n)}$) the functional differentiation formula reads

$$\rho^{\rm r}_{(n)}(x_1,\dots,x_n) = \frac{1}{Z_N[1,1]} \frac{\delta^n}{\delta u(x_1)\cdots\delta u(x_n)} Z_N[u,1]\Big|_{u=1}. \tag{15.217}$$

To compute (15.217), suppose furthermore that the polynomials $\{p_{l-1}(x)\}_{l=1,2,\dots}$ in Proposition 15.10.3 have been chosen so that the matrix $[(\alpha_{j,k}+\beta_{j,k})|_{u=v=1}]_{j,k=1,\dots,2N}$ evaluates to the block diagonal structure of (6.2). This is equivalent to supposing that $\{p_{l-1}(x)\}_{l=1,2,\dots}$ are skew orthogonal with respect to the skew product implied by the matrix elements. Now, the only term in the matrix elements dependent on u is $\alpha_{j,k}$, and this is proportional to the $\beta = 1$ inner product (6.61) with $e^{-V(x)} = e^{-x^2/2}$ (Gaussian case). As a consequence the functional differentiation formula (15.217) must give a formula for $\rho^{\rm r}_{(n)}$ which is structurally identical to that implied by Propositions 6.3.3 and 6.3.2 for $\rho^{\rm GOE}_{(n)}$. Hence [214], [89], [497]

$$\rho^{\rm r}_{(n)}(x_1,\dots,x_n) = \mathrm{qdet}\begin{bmatrix} S^{\rm r}(x_j,x_k) & \tilde{I}^{\rm r}(x_j,x_k) \\ D^{\rm r}(x_j,x_k) & S^{\rm r}(x_k,x_j) \end{bmatrix} \tag{15.218}$$

with

$$S^{\rm r}(x,y) = \sum_{k=0}^{N/2-1} \frac{e^{-y^2/2}}{u_k}\Big(\Phi_{2k}(x)p_{2k+1}(y) - \Phi_{2k+1}(x)p_{2k}(y)\Big), \tag{15.219}$$

$$D^{\rm r}(x,y) = \frac{\partial}{\partial x}S^{\rm r}(x,y), \qquad \tilde{I}^{\rm r}(x,y) = \frac{1}{2}\mathrm{sgn}(y-x) - \int_x^y S^{\rm r}(x,z)\,dz. \tag{15.220}$$

In (15.219)

$$u_{k-1} := (\alpha_{2k-1,2k} + \beta_{2k-1,2k})|_{u=v=1}, \quad \Phi_k(x) = \int_{-\infty}^{\infty} \mathrm{sgn}(x-y)p_k(y)e^{-y^2/2}\,dy.$$

The crucial difference between $\rho^{\rm r}_{(n)}$ and $\rho^{\rm GOE}_{(n)}$ is the explicit form of the skew orthogonal polynomials, which in the former case are given by the following result.

PROPOSITION 15.10.4 *Consider the matrix* $\mathbf{X} := [(\alpha_{j,k} + \beta_{j,k})|_{u=v=1}]_{j,k=1,\dots,2N}$. *The polynomials* $\{p_l(x)\}_{l=0,\dots}$ *with the property that* $\mathbf{X}$ *evaluates to the block diagonal structure (6.2) are specified by*

$$p_{2j}(x) = x^{2j}, \qquad p_{2j+1}(x) = x^{2j+1} - 2jx^{2j-1}, \tag{15.221}$$

and furthermore

$$(\alpha_{2k-1,2k} + \beta_{2k-1,2k})|_{u=v=1} = 2\sqrt{2\pi}\Gamma(2k-1). \tag{15.222}$$

Proof. For general $\{p_j\}_{j=0,1,\dots}$ set

$$\langle p_j, p_k \rangle := \alpha_{j,k}|_{u=1} + \beta_{j,k}|_{v=1}.$$

For any even p_{2j} and odd p_{2j+1}, by changing variables $x \mapsto -x$, $y \mapsto -y$ in the definition of $\alpha_{j,k}$, and changing variables $\theta \mapsto \pi - \theta$ in the definition of $\beta_{j,k}$ it is easy to see that

$$\langle p_{2j}, p_{2k} \rangle = \langle p_{2j+1}, p_{2k+1} \rangle = 0.$$

It remains to verify that with p_j as stated in (15.221),

$$\langle p_{2j-1}, p_{2k} \rangle = 0$$

for $j \ne k$, and that for $j = k$ the normalization (15.222) results. This is an immediate corollary of the explicit formula

$$\langle x^{2j+1}, x^{2k} \rangle = \begin{cases} -2^{j+k+3/2} j!\Gamma(k+1/2), & j \ge k, \\ 0, & j < k. \end{cases} \tag{15.223}$$

To derive (15.223), use can be made of recurrences satisfied by the corresponding integrals [215]. However the details are lengthy, and so will not be given in full. Briefly, using integration by parts one finds the recurrences

$$\begin{aligned}
\alpha_{2j+4,2k+1}|_{u=1} &= (2j+2)\alpha_{2j+2,2k+1}|_{u=1} - 2\Gamma\Big(j+k+\frac{3}{2}\Big), \\
\alpha_{2j+2,2k+3}|_{u=1} &= (2k+1)\alpha_{2j+2,2k+1}|_{u=1} + 2\Gamma\Big(j+k+\frac{3}{2}\Big), \\
\beta_{2j+4,2k+1}|_{v=1} &= (2j+2)\beta_{2j+2,2k+1}|_{v=1} + 2\Gamma\Big(j+k+\frac{3}{2}\Big) - 2\sqrt{2\pi}(j+k+1)!\delta_{j+1,k}, \\
\beta_{2j+2,2k+3}|_{v=1} &= (2k+1)\beta_{2j+2,2k+1}|_{v=1} - 2\Gamma\Big(j+k+\frac{3}{2}\Big) + 2\sqrt{2\pi}(j+k+1)!\delta_{j+1,k}.
\end{aligned}$$

With $\gamma_{j,k} = \langle x^j, x^k \rangle := \alpha_{j,k}|_{u=1} + \beta_{j,k}|_{v=1}$ it follows that

$$\begin{aligned}
\gamma_{2j+4,2k+1} &= (2j+2)\gamma_{2j+2,2k+1} - 2\sqrt{2\pi}(j+k+1)!\delta_{j+1,k} \quad (j \ge 0, k \ge 1), \\
\gamma_{2j+2,2k+3} &= (2k+1)\gamma_{2j+2,2k+1} + 2\sqrt{2\pi}(j+k+1)!\delta_{j+1,k} \quad (j \ge 0, k \ge 0).
\end{aligned}$$

With initial condition $\gamma_{2,1} = -2\sqrt{2\pi}$, these recurrences can be verified to have the solution (15.223). □

Substituting the result of Proposition 15.10.4 in (15.219) shows

$$S^{\mathrm{r}}(x,y) = \frac{e^{-(x^2+y^2)/2}}{\sqrt{2\pi}} \Big(\frac{1}{2} (\sqrt{2}y)^{N-1} e^{x^2/2} \mathrm{sgn}(x) \frac{\gamma(N/2-1/2; x^2/2)}{\Gamma(N-1)} + e^{xy} \frac{\Gamma(N-2; xy)}{\Gamma(N-1)} \Big), \tag{15.224}$$

where use has been made of (15.50) and (15.55). According to (15.218), setting $x = y$ in this gives the density of the real eigenvalues. Plots show that to leading order the support of the density is in the region $|x| < \sqrt{N}$ and indicate the boundary layer to be O(1). This suggests that for $N \to \infty$ the correlations are well defined in the neighborhood of this edge. Indeed use of (15.51) shows

$$\lim_{N\to\infty} S^{\mathrm{r}}(\sqrt{N}+X, \sqrt{N}+Y) = \frac{1}{\sqrt{2\pi}} \Big(\frac{1}{2} e^{-(X-Y)^2/2} \Big(1 - \mathrm{erf} \frac{X+Y}{\sqrt{2}} \Big) + \frac{e^{-Y^2}}{2\sqrt{2}} (1 + \mathrm{erf}\, X) \Big). \tag{15.225}$$

For $X, Y \to -\infty$ (15.225) reveals the bulk limiting form

$$\lim_{N\to\infty} S^{\rm r}(X,Y) = \frac{1}{\sqrt{2\pi}} e^{-(X-Y)^2/2}, \tag{15.226}$$

which with $X = Y$ implies that the density of real eigenvalues is the constant $1/\sqrt{2\pi}$. One consequence of this is that the expected number of real eigenvalues, $E_N := \int_{-\infty}^{\infty} \rho^{\rm r}(x)\, dx$, has the large N behavior $E_N \sim \sqrt{2N/\pi}$. In fact direct integration of (15.224) in the case $x = y$ leads to the evaluation [156]

$$E_N = \frac{1}{2} + \sqrt{\frac{2}{\pi}} \frac{\Gamma(N+1/2)}{\Gamma(N)} {}_2F_1(1, -1/2; N; 1/2), \tag{15.227}$$

and from this the asymptotic expansion of E_N can be systematically generated.

Application of Proposition 15.10.4 to the calculation of the correlation between complex eigenvalues is given in Exercises 15.10 q.3. Although not considered here, the mixed correlation involving both real and complex eigenvalues can be computed [89] by working similar to that in the proof of Proposition 6.7.2.

Consider the probability $\Pr(R_s)$, say, that either there is no real eigenvalue, or all real eigenvalues are less than s. This is given in terms of the generalized partition function (15.214) according to

$$\Pr(R_s) = Z_N[\chi_{x\in(-\infty,s)}, 1]. \tag{15.228}$$

With $\mathbf{f}^{\rm r}$ denoting the 2×2 matrix integral operator on $(-\infty, \infty)$ with kernel equal to the block matrix exhibited by (15.218), and $\chi_{(-\infty,s)} = \mathrm{diag}[\chi_{(-\infty,s)}, \chi_{(-\infty,s)}]$, the reasoning which led to (9.18) implies

$$\Big(\Pr(R_s)\Big)^2 = \det[\mathbf{1}_2 - \mathbf{f}^{\rm r}\chi_{(-\infty,s)}].$$

However, from the perspective of numerical computation it is easier to work directly with (15.228). A case of particular interest is $s = \sqrt{N}$, which corresponds to the probability that either no eigenvalue is real, or all real eigenvalues are inside the leading order support. The task of computing such probabilities first arose in the context of stability analysis of biological webs [374], [391] (see also Exercises 15.10 q.4).

EXERCISES 15.10

1. [170] A polynomial of the form

$$f(z) = z^N + a_1 z^{N-1} + \cdots + a_N, \qquad a_{N-j} = a_N \bar{a}_j, \quad a_N = e^{i\phi}$$

is said to be self-reciprocal. Such polynomials have a nonzero probability of having zeros on the unit circle, while zeros off the unit circle occur in pairs $\rho_j e^{i\theta_j}, (1/\rho_j)e^{i\theta_j}$. Suppose N is odd, and let there be L zeros $\{\alpha_j = e^{i\delta_j}\}_{j=1,\dots,L}$ on the unit circle (L odd) and $2M$ zeros $\{\beta_j = \rho_j e^{i\theta_j}, 1/\bar{\beta}_j = (1/\rho_j)e^{i\theta_j}\}_{j=1,\dots,M}$ off the unit circle. The Jacobian J for the change of variables from $\mathrm{Re}\, a_1, \mathrm{Im}\, a_1, \dots \mathrm{Re}\, a_{(N-1)/2}, \mathrm{Im}\, a_{(N-1)/2}, \phi$ specifying the coefficients, to $\{\delta_j\}_{j=1,\dots,L}, \{\rho_j, \theta_j\}_{j=1,\dots,M}$ specifying the zeros has been calculated to be equal to

$$2^{M-(N-1)/2} \Big| \prod_{m=1}^{M} \frac{1}{\rho_m} \Delta(\beta_1, 1/\bar{\beta}_1, \dots, \beta_M, 1/\bar{\beta}_M, \alpha_1, \dots, \alpha_L) \Big|,$$

where Δ is as in (15.208). Relate this to the Boltzmann factor for a certain one-component log-potential system is a unit disk with anti-metallic boundary conditions.

2. [155] Let $\mathbf{X}_N$ be a real $N \times N$ matrix. Let $\alpha \pm i\beta$, $\beta > 0$ be a pair of complex eigenvalues with corresponding eigenvectors $\vec{x} \pm ic\vec{y}$, where $|\vec{x}|^2 = |\vec{y}|^2 = 1$ and $c > 0$.

   (i) Show that $\vec{x} \cdot \vec{y} = \vec{0}$ and

$$\mathbf{X}_N[\vec{x} \;\; \vec{y}] = [\vec{x} \;\; \vec{y}] \begin{bmatrix} \alpha & \beta/c \\ -\beta c & \alpha \end{bmatrix}. \tag{15.229}$$

   Show too that the effect of interchanging $\vec{x}$ and $\vec{y}$ is to replace $c \mapsto -1/c$ in (15.229). Conclude that the decomposition (15.229) for $|\vec{x}|^2 = |\vec{y}|^2 = 1$, the first component of $\vec{x}, \vec{y}$ positive and $\vec{x} \cdot \vec{y} = \vec{0}$ is unique provided $c \geq 1$.

(ii) Show from the result of Exercises 1.9 q.3 that there is a Householder matrix $\mathbf{U}_1$ such that $\mathbf{U}_1\vec{x} = \vec{e}_1$, and use the fact that $\vec{x} \cdot \vec{y} = 0$ to show that $\mathbf{U}_1\vec{y} = \vec{y}\,'$, where $\vec{y}\,'$ is a unit vector with first component 0. Now construct a Householder transformation $\mathbf{U}_2$ of the form

$$\begin{bmatrix} 1 & \vec{0}_{N-1}^T \\ \vec{0}_{N-1} & \mathbf{V} \end{bmatrix},$$

where $\mathbf{V}$ is an $N-1 \times N-1$ Householder transformation with the property $\mathbf{U}_2\vec{y}\,' = \vec{e}_2$, to deduce that

$$\mathbf{U}_2\mathbf{U}_1[\vec{x} \quad \vec{y}] = [\vec{e}_1 \quad \vec{e}_2].$$

(iii) Conclude from the results of (i) and (ii) that for $\mathbf{Q} = \mathbf{U}_2\mathbf{U}_1$ an orthogonal matrix we have

$$\mathbf{X}_N = \mathbf{Q}\begin{bmatrix} \mathbf{Z} & \mathbf{Y}_{2\times N-2} \\ \mathbf{0}_{N-2\times 2} & \mathbf{X}_{N-2} \end{bmatrix}\mathbf{Q}^T, \qquad \mathbf{Z} := \begin{bmatrix} \alpha & b_1 \\ -b_2 & \alpha \end{bmatrix},$$

where $b_1, b_2 > 0$, $\sqrt{b_1 b_2} = \beta$.

3. [214], [155] In this exercise the explicit form of $\rho_{(n)}^{\rm c}$ is given.

(i) Let $\mathbf{f}^{\rm c}$ denote the 2×2 matrix integral operator on $(-\infty,\infty)$ with kernel equal to $\mathbf{f}^{\rm c}(w,z)$ and let $g = \mathrm{diag}\,[g(z), g(z)]$. With the polynomials $\{p_{l-1}(x)\}_{l=1,2,\dots}$ chosen as in Proposition 15.10.4, and $q_{2j-2}(z) := -p_{2j-1}(z)$, $q_{2j-1}(z) := p_{2j-2}(z)$, use the method of Proposition 6.1.9 to show

$$Z_n[1, 1+g] = \det(\mathbf{1}_2 + \mathbf{f}^{\rm c}\mathbf{g}),$$

where, with $z = x + iy$, $w = u + iv$, and u_k as in (15.219),

$$\mathbf{f}^{\rm c}(w,z) := 2ie^{v^2-u^2}\mathrm{erfc}(\sqrt{2}v)\begin{bmatrix} S^{\rm c}(\bar{w},z) & -S^{\rm c}(\bar{w},\bar{z}) \\ S^{\rm c}(w,z) & -S^{\rm c}(w,\bar{z}) \end{bmatrix}, \qquad S^{\rm c}(w,z) := \sum_{j=1}^{N}\frac{p_{j-1}(w)q_{j-1}(z)}{u_{[(j-1)/2]}}.$$

(ii) Use the theory of the paragraph including (6.31) to deduce from (i) that

$$\rho_{(n)}^{\rm c}((x_1,y_1),\dots,(x_n,y_n)) = \mathrm{qdet}[\mathbf{f}^{\rm c}((x_j,y_j),(x_k,y_k))]_{j,k=1,\dots,n}$$

and thus in particular

$$\rho_{(1)}^{\rm c}((x,y)) = \sqrt{\frac{2}{\pi}}\frac{\Gamma(N-1, x^2+y^2)}{\Gamma(N-1)}ye^{2y^2}\mathrm{erfc}(\sqrt{2}y). \tag{15.230}$$

4. Let $\vec{y}(t)$ be an $N\times 1$ column vector, and let $\mathbf{B}$ be an $N\times N$ random real Gaussian matrix in which the entries are chosen from $\mathrm{N}[0,\sigma]$. Consider the matrix linear differential equation

$$\frac{d\vec{y}(t)}{dt} = (-\mathbf{1}_N + \mathbf{B})\vec{y}(t).$$

(i) Show that for $\vec{y}(t) \to \vec{0}$ as $t\to\infty$, the real parts of all eigenvalues of $\mathbf{B}$ must be less than 1.

(ii) Deduce from (15.225) and (15.230) that for $\sigma = 1$ and N large the spectral radius is equal to $\sqrt{N} + \mathrm{O}(1)$. Use this to show that for $\sigma = \epsilon/\sqrt{N}$, and with $N\to\infty$, the condition of (i) holds provided $0 < \epsilon < 1$.

5. [371], [215] Let $\mathbf{S}$ be an element of the GOE. Let $\mathbf{A}$ be an antisymmetric real Gaussian matrix with joint distribution of its elements proportional to $\exp(-\mathrm{Tr}\,\mathbf{A}^2/2)$. With $0 < \tau < 1$ and $c := (1-\tau)/(1+\tau)$ define random matrices $\mathbf{X}$ according to

$$\mathbf{X} = \frac{1}{\sqrt{b}}(\mathbf{S} + \sqrt{c}\mathbf{A}). \tag{15.231}$$

(i) When $\tau = 0$, $b = 1$ observe that $\mathbf{X}$ is a real Gaussian matrix with distribution (15.207), while when $\tau = 1$, $b = 1$, $\mathbf{X}$ is a member of the GOE.

(ii) Write $\mathbf{S}$ and $\mathbf{A}$ in terms of $\mathbf{X}$ and $\mathbf{X}^T$, and also show from (15.231) that

$$(d\mathbf{X}) = 2^{N(N-1)/2}(\sqrt{c})^{N(N-1)/2}(\sqrt{b})^{-N^2}(d\mathbf{S})(d\mathbf{A}),$$

to deduce from the joint distribution of the elements of $\mathbf{S}$ and $\mathbf{A}$ that the joint distribution of the elements of $\mathbf{X}$ is equal to

$$A_{\tau,b}\exp\Big(-\frac{b}{2(1-\tau)}\Big(\mathrm{Tr}\,\mathbf{X}\mathbf{X}^T - \tau\mathrm{Tr}\,\mathbf{X}^2\Big)\Big) \tag{15.232}$$

(cf. (15.15)) where

$$A_{\tau,b} = (\sqrt{c})^{-N(N-1)/2}(\sqrt{b})^{N^2}(2\pi)^{-N^2/2}.$$

(iii) Denote (15.208) by $P_{k,(N-k)/2}(\{\lambda_j\}_{j=1,\dots,k};\{x_j \pm iy_j\}_{j=1,\dots,(N-k)/2})$, which corresponds to the eigenvalue p.d.f. in the case $\tau = 0$, $b = 1$, conditioned so that there are exactly k real eigenvalues. For these matrices scale $\mathbf{X} \mapsto \sqrt{b}\mathbf{X}/(1-\tau)^{1/2}$ to obtain that for matrices with p.d.f.

$$A_{0,1}b^{N^2/2}(1-\tau)^{-N^2/2}e^{-b\mathrm{Tr}\,\mathbf{X}\mathbf{X}^T/2(1-\tau)}, \tag{15.233}$$

the eigenvalue p.d.f. is equal to

$$b^{N/2}(1-\tau)^{-N/2}$$
$$\times P_{k,(N-k)/2}\Big(\{\sqrt{b}\lambda_j/(1-\tau)^{1/2}\}_{j=1,\dots,k};\{\sqrt{b}x_j/(1-\tau)^{1/2} \pm i\sqrt{b}y_j/(1-\tau)^{1/2}\}_{j=1,\dots,(N-k)/2}\Big).$$

By comparing (15.233) to (15.232) deduce that for matrices with p.d.f. (15.232) the eigenvalue p.d.f, conditioned so that there are exactly k eigenvalues, is equal to

$$\begin{aligned}
&\frac{A_{\tau,b}}{A_{0,1}}(1-\tau)^{N(N-1)/2}\exp\Big(\frac{\tau b}{2(1-\tau)}\Big(\sum_{j=1}^{k}\lambda_j^2 + 2\sum_{j=1}^{(N-k)/2}(x_j^2-y_j^2)\Big)\Big)\\
&\quad\times P_{k,(N-k)/2}(\{\sqrt{b}\lambda_j/(1-\tau)^{1/2}\}_{j=1,\dots,k};\{\sqrt{b}x_j/(1-\tau)^{1/2} \pm i\sqrt{b}y_j/(1-\tau)^{1/2}\}_{j=1,\dots,(N-k)/2})\\
&= \frac{(\sqrt{b})^{N(N+1)/2}(\sqrt{1+\tau})^{N(N-1)/2}}{2^{N(N+1)/4}\prod_{l=1}^{N}\Gamma(l/2)}\frac{2^{(N-k)/2}}{k!((N-k)/2)!}\Big|\Delta(\{\lambda_l\}_{l=1,\dots,k}\cup\{x_j \pm iy_j\}_{j=1,\dots,(N-k)/2})\Big|\\
&\quad\times e^{-b\sum_{j=1}^{k}\lambda_j^2/2}e^{b\sum_{j=1}^{(N-k)/2}(y_j^2-x_j^2)}\prod_{j=1}^{(N-k)/2}\mathrm{erfc}\Big(\sqrt{\frac{2b}{1-\tau}}y_j\Big).
\end{aligned}$$

15.11 CLASSIFICATION OF NON-HERMITIAN RANDOM MATRICES

In our studies of Hermitian random matrices in Chapters 1 and 3, a total of ten distinct ensembles were identified on the basis of symmetry constraints. Furthermore, these were shown to be identical to the Hermitian part of the ten infinite families of symmetric spaces. The task of classifying non-Hermitian random matrices according to symmetries has been undertaken in [59], [379].

DEFINITION 15.11.1 *Let $\mathbf{A}$ be a square matrix and let $\mathbf{p}$, $\mathbf{c}$, $\mathbf{q}$ and $\mathbf{k}$ be unitary matrices of the same size as $\mathbf{A}$ and with*

$$\mathbf{p}^2 = \mathbf{1}, \quad \mathbf{c}^T\mathbf{c}^\dagger = \pm\mathbf{1}, \quad \mathbf{q}^2 = \mathbf{1}, \quad \mathbf{k}\bar{\mathbf{k}} = \pm\mathbf{1}.$$

The matrix $\mathbf{A}$ is said to have a symmetry of P-type, C-type, Q-type and K-type respectively if

$$\mathbf{A} = -\mathbf{p}\mathbf{A}\mathbf{p}^\dagger, \quad \mathbf{A} = \pm\mathbf{c}\mathbf{A}^T\mathbf{c}^\dagger, \quad \mathbf{A} = -\mathbf{q}\mathbf{A}^\dagger\mathbf{q}^\dagger, \quad \mathbf{A} = \mathbf{k}\bar{\mathbf{A}}\mathbf{k}^\dagger,$$

respectively. For $\mathbf{A}$ to have two or more of these symmetries, the symmetries are required to commute.

In addition to the ten classes of Hermitian random matrices already catalogued, in [379] consideration of these symmetries gave rise to twenty types of non-Hermitian random matrices. Three of these are the real, complex and real quaternion Ginibre ensembles, corresponding to $\mathbf{k} = \mathbf{1}$, no symmetry and $\mathbf{k} = \mathbb{Z}_{2N}$, respectively. We will single out just three others, obtained by requiring that the Ginibre matrices anticommute with the P-type symmetry

$$\begin{bmatrix} \mathbf{1}_p & \mathbf{O}_{p\times q} \\ \mathbf{O}_{q\times p} & -\mathbf{1}_q \end{bmatrix}$$

(with the sizes doubled in the quaternion case). This gives the non-Hermitian matrices

$$\begin{bmatrix} \mathbf{O}_{p\times p} & \mathbf{A}_{p\times q} \\ \mathbf{B}_{q\times p} & \mathbf{O}_{q\times q} \end{bmatrix} \tag{15.234}$$

with the elements of $\mathbf{A} := \mathbf{A}_{p\times q}$ and $\mathbf{B} := \mathbf{B}_{q\times p}$ real, complex and real quaternion. With the matrices $\mathbf{A}$ and $\mathbf{B}$ chosen according to a Gaussian measure, this class of random matrices have been applied to studies of QCD [439], [9]. Following these references, we'll take up the problem of computing the eigenvalue p.d.f. in the case the elements of (15.234) are complex.

In numerical linear algebra, for square matrices $\mathbf{A}$, $\mathbf{B}$ the decomposition

$$\mathbf{A} = \mathbf{Q}\mathbf{T}_A\mathbf{Z}^\dagger, \qquad \mathbf{B} = \mathbf{Q}\mathbf{T}_B\mathbf{Z}^\dagger$$

is called the QZ decomposition. Here $\mathbf{Q}$, $\mathbf{Z}$ are unitary matrices, while $\mathbf{T}_A$, $\mathbf{T}_B$ are upper triangular matrices with diagonal elements $(T_A)_{jj}$, $(T_B)_{jj}$, such that $(T_A)_{jj}/(T_B)_{jj}$ are the eigenvalues of $\mathbf{B}^{-1}\mathbf{A}$. The matrices $\mathbf{Q}$ and $\mathbf{Z}$ are unique provided the entries of the first row are chosen to be positive.

A variation of this is the joint decomposition of $p\times q$, $q\times p$ $(p \ge q)$ matrices $\mathbf{A}$, $\mathbf{B}$ given by [439]

$$\mathbf{A} = \mathbf{U}\mathbf{T}_A\mathbf{V}, \qquad \mathbf{B} = \mathbf{V}^\dagger\mathbf{T}_B\mathbf{U}^\dagger, \tag{15.235}$$

where $\mathbf{T}_A$, $\mathbf{T}_B$ are $p\times q$, $q\times p$ upper triangular matrices and $\mathbf{U}$, $\mathbf{V}$ are unitary. With the diagonal entries of $\mathbf{T}_A$, $\mathbf{T}_B$ denoted $\{x_k\}$, $\{y_k\}$, respectively, and the nonzero eigenvalues of the product $\mathbf{AB}$ denoted $\{-z_k^2\}$ so that the eigenvalues of (15.234) are $\{\pm iz_k\}$, we have $-z_k^2 = x_k y_k$. The matrix $\mathbf{V}$ is unique provided the entries of the first row are chosen to be positive, while $\mathbf{U}$ requires this condition, but then is only unique up to multiplication on the right by

$$\begin{bmatrix} \mathbf{1}_q & \mathbf{O}_{q\times(p-q)} \\ \mathbf{O}_{(p-q)\times q} & \mathbf{S} \end{bmatrix}$$

for $\mathbf{S}$ a $(p-q)\times(p-q)$ unitary matrix.

For $\mathbf{A}$ and $\mathbf{B}$ having all entries independent standard complex Gaussians, we want to compute the distribution of the independent eigenvalues $\{iz_k\}$ of (15.234). The first step is to compute the Jacobian for the change of variables implied by (15.235).

PROPOSITION 15.11.2 *Let $\tilde{\mathbf{T}}_A, \tilde{\mathbf{T}}_B$ denote the strictly upper triangular part of $\mathbf{T}_A, \mathbf{T}_B$. We have*

$$(d\mathbf{A})(d\mathbf{B}) = \prod_{l=1}^q |x_l|^{2(p-q)} \prod_{1\le j<k\le q} |z_k^2 - z_j^2|^2 \prod_{l=1}^q dx_l^{\rm r} dx_l^{\rm i} dy_l^{\rm r} dy_l^{\rm i} (d\tilde{\mathbf{T}}_A)(d\tilde{\mathbf{T}}_B)(\mathbf{U}^\dagger d\mathbf{U})(\mathbf{V}^\dagger d\mathbf{V}). \tag{15.236}$$

Proof. From (15.235) we compute

$$(\mathbf{U}^\dagger\mathbf{A}\mathbf{V}^\dagger)_{jk} = (d\mathbf{T}_A)_{jk} + \sum_{l\le k}(\mathbf{U}^\dagger d\mathbf{U})_{jl}(\mathbf{T}_A)_{lk} - \sum_{j\le l}(\mathbf{T}_A)_{jl}(\mathbf{V}d\mathbf{V}^\dagger)_{lk},$$

$$(\mathbf{V}\mathbf{B}\mathbf{U})_{jk} = (d\mathbf{T}_B)_{jk} + \sum_{l\le k}(\mathbf{V}d\mathbf{V}^\dagger)_{jl}(\mathbf{T}_B)_{lk} - \sum_{j\le l}(\mathbf{T}_B)_{jl}(\mathbf{U}^\dagger d\mathbf{U})_{lk}$$

(cf. (15.5)). We want to take the wedge product of all the independent differentials in these equations. By choosing the order of doing this so that a triangular structure results, as done below (15.7), we see that the only contributing terms are

$$(\mathbf{T}_A)_{kk}(\mathbf{U}^\dagger d\mathbf{U})_{jk} - (\mathbf{T}_A)_{jj}(\mathbf{V}d\mathbf{V}^\dagger)_{jk}, \qquad (\mathbf{T}_B)_{kk}(\mathbf{V}d\mathbf{V}^\dagger)_{jk} - (\mathbf{T}_B)_{jj}(\mathbf{U}^\dagger d\mathbf{U})_{jk} \tag{15.237}$$

and

$$(d\mathbf{T}_A)_{jk}, \qquad (d\mathbf{T}_B)_{jk}$$

for $j > k$ and $j \le k$, respectively. Furthermore, for $j > q$, $k \le q$ only the term

$$(\mathbf{T}_A)_{kk}(\mathbf{U}^\dagger d\mathbf{U})_{jk}$$

of (15.237) contributes. Recalling that $(\mathbf{T}_A)_{jj} = x_j$, $(\mathbf{T}_B)_{jj} = y_j$, $x_j y_j = -z_j^2$, we see by taking the wedge product of the contributing terms that (15.236) results. □

In terms of the decomposition (15.235) the joint p.d.f. for $\mathbf{A}$ and $\mathbf{B}$ is proportional to

$$e^{-\sum_{j=1}^N (|x_j|^2 + |y_j|^2) - \sum_{j<k}(|(\tilde{T}_A)_{jk}|^2 + |(\tilde{T}_B)_{jk}|^2)}.$$

The integration of $(d\tilde{\mathbf{T}}_A)$ and $(d\tilde{\mathbf{T}}_B)$ in (15.236) therefore only contributes a constant, as does the integration over $(\mathbf{U}^\dagger d\mathbf{U})(\mathbf{V}^\dagger d\mathbf{V})$. The variable x_j is independent of y_j and so to deduce the distribution of the z_j we must substitute

$$y_j = -z_j^2/x_j, \qquad dy_j^{\mathrm{r}} dy_j^{\mathrm{i}} = \Big|\frac{dy}{dz}\Big|^2 dz_j^{\mathrm{r}} dz_j^{\mathrm{i}} = 4\Big|\frac{z_j}{x_j}\Big|^2 dz_j^{\mathrm{r}} dz_j^{\mathrm{i}}$$

and integrate over x_j. Our remaining task then is to compute

$$\int_{\mathbb{R}^2} |x|^{-2+2(p-q)} e^{-|x|^2 - |z|^2/|x|^2}\, dx^{\mathrm{r}} dx^{\mathrm{i}}. \tag{15.238}$$

Changing to polar coordinates gives an integral representation of the modified Bessel function of the second kind $K_\nu(w)$ results, showing that (15.238) is proportional to

$$|z|^{2(p-q)} K_{p-q}(2|z|^2),$$

and giving the sought eigenvalue p.d.f. [439].

PROPOSITION 15.11.3 *Consider matrices (15.234) in the case that the elements of* $\mathbf{A}_{p\times q}$ *and* $\mathbf{B}_{p\times q}$ *are independent standard complex Gaussians. The eigenvalues come in ± pairs, which we write as* $\{\pm i z_k\}_{k=1,\dots,q}$. *The p.d.f. for* $\{z_k\}$, *defined on* $z_k^{\mathrm{r}} \ge 0$ $(k = 1, \dots, q)$ *is proportional to*

$$\prod_{l=1}^q |z_l|^{2(p-q+1)} K_{p-q}(2|z_l|^2) \prod_{1\le j<k\le q} |z_k^2 - z_j^2|^2.$$

We remark that the method of derivation of Proposition 15.3.1 can be used to write down the corresponding n-point correlation functions as an $n \times n$ determinant. Consider in particular the case $p = q$, and change variables $w_j = -z_j^2$ to the eigenvalues of the matrix product $\mathbf{AB}$. According to Proposition 15.11.3 the latter have p.d.f. proportional to

$$\prod_{l=1}^p K_0(2|w_l|) \prod_{1\le j<k\le p} |w_k - w_j|^2. \tag{15.239}$$

Making use of the definite integral

$$\int_0^\infty r^{\nu+2j-1} K_\nu(2r)\, dr = \frac{1}{4}\Gamma(j)\Gamma(j+\nu)$$

and recalling (15.199) we see that the limiting bulk correlation kernel is given by

$$\frac{2}{\pi}\Big(K_0(2|u|)K_0(2|v|)\Big)^{1/2} I_0(2(u\bar{v})^{1/2}). \tag{15.240}$$

It follows from (15.240) and the large $|z|$ expansions [541]

$$I_0(z) \sim \frac{e^z}{(2\pi z)^{1/2}}, \qquad K_0(z) \sim \Big(\frac{\pi}{2z}\Big)^{1/2} e^{-z} \tag{15.241}$$

that the eigenvalue density $\rho_{(1)}(z)$ exhibits the large $|z|$ form $1/(2\pi|z|)$. This could have been anticipated, as it is generally true that for a product of M independent random complex $N \times N$ Gaussian matrices (N large), the decay of the bulk eigenvalue density will be proportional to $|z|^{-2+2/M}$ [103]. Furthermore, with $|r_0|$ large, writing $u \mapsto r_0 + (2\pi r_0)^{1/2}u$ and $v \mapsto (2\pi r_0)^{1/2}v$ in (15.240), forming the product with its complex conjugate, and making use of (15.241) we reclaim the bulk form of the two-point correlation function as given in (15.45). We would expect the scaled n-point correlations, $n \geq 3$, to similarly reduce to (15.43). Related to this is the fact that for the general one body term $f(r) = e^{-cr^\alpha}$, $\alpha \geq 2$, in (15.42) (asymptotically (15.239) corresponds to the case $\alpha = 1$), scaling of the general n-point correlation to (15.43) has been explicitly demonstrated [528].

Bibliography

[1] A.Y. Abdul-Magd and M.H. Simbel, *Wigner surmise for high order level spacing distributions of chaotic systems*, Phys. Rev. E **60** (1999), 5371–5374.

[2] P.A. Absil, A. Edelman, and P. Koev, *On the largest principal angle between random subspaces*, Linear Algebra Appl. **414** (2006), 288–294.

[3] M. Adler, P.J. Forrester, T. Nagao, and P. van Moerbeke, *Classical skew orthogonal polynomials and random matrices*, J. Stat. Phys. **99** (2000), 141–170.

[4] M. Adler, E. Horozov, and P. van Moerbeke, *The Pfaff lattice and skew-orthogonal polynomials*, Int. Math. Res. Notices **(11)** (1999), 569–588.

[5] M. Adler and P. van Moerbeke, *Matrix integrals, Toda symmetries, Virasora constraints and orthogonal polynomials*, Duke Math. Journal **80** (1995), 863–911.

[6] ______, *Toda versus Pfaff latice and related polynomials*, Duke Math. J. **112** (2002), 1–58.

[7] ______, *Recursion relations for unitary integrals, combinatorics and the Toeplitz lattice*, Commun. Math. Phys. **237** (2003), 397–440.

[8] A.C. Aitken, *Determinants and matrices*, 9th ed., Oliver and Boyd, Edinburgh & London, 1956.

[9] G. Akemann, *The complex Laguerre symplectic ensemble of non-Hermitian matrices*, Nucl. Phys. B **73** (2005), 253–299.

[10] G. Akemann and E. Kanzieper, *Integrable structure of Ginibre's ensemble of real random matrices and a Pfaffian integration theorem*, J. Stat. Phys. **129** (2007), 1159–1231.

[11] T. Akuzawa and M. Wadati, *Non-Hermitian random matrices and integrable quantum Hamiltonians*, J. Phys. Soc. Japan **65** (1996), 1583–1588.

[12] A. Alastuey and B. Jancovici, *On the two-dimensional one-component Coulomb plasma*, J. Physique **42** (1981), 1–12.

[13] James T. Albrecht, C. Chan, and A. Edelman, *Sturm sequences and random eigenvalue distributions*, Found. Comput. Math. **9** (2009), 461–483.

[14] A. Altland and M.R. Zirnbauer, *Nonstandard symmetry classes in mesoscopic normal-superconducting hybrid compounds*, Phys. Rev. B **55** (1997), 1142–1161.

[15] Y. Ameur, H. Hedenmalm, and N. Makarov, *Fluctuations of eigenvalues of random matrices*, arXiv:0807.0375, 2008.

[16] G.W. Anderson, *A short proof of Selberg's generalized beta formula*, Forum Math. **3** (1991), 415–417.

[17] G.E. Andrews, R. Askey, and R. Roy, *Special functions*, Cambridge University Press, New York, 1999.

[18] K. Aomoto, *The complex Selberg integral*, Quart. J. Math. Oxford **38** (1987), 385–399.

[19] ______, *Jacobi polynomials associated with Selberg's integral*, SIAM J. Math. Analysis **18** (1987), 545–549.

[20] ______, *2 conjectural formulae for symmetric A-type Jackson integrals*, unpublished, 1994.

[21] K. Aomoto and P.J. Forrester, *On the Jacobian identity associated with real hyperplane arrangements*, Compositio Math. **121** (2000), 263–295.

[22] A.I. Aptekarev, A. Branquinho, and W. Van Assche, *Multiple orthogonal polynomials for classical weights*, Trans. Amer. Math. Soc. **355** (2003), 3887–3914.

[23] R. Askey, *An excursion in classical mathematics*, unpublished, 1986.

[24] ______, *Integration and computers*, Computers in mathematics (D.V. Chudnovsky and R.D. Jenks, eds.), Lecture notes in pure and applied mathematics, vol. 125, CRC Press, Boca Raton, FL, 1990, pp. 35–82.

[25] H. Aslaksen, *Quaternion determinants*, Math. Intell. **18** (1996), 57–65.

[26] Z.D. Bai, *Circular law*, Ann. Prob. **25** (1997), 494–529.

[27] J. Baik, *Painlevé expressions for LOE, LSE, and interpolating ensembles*, Int. Math. Res. Notices **(33)** (2002), 1739–1789.

[28] ———, *Riemann-Hilbert problem problems for last passage percolation*, Recent developments in integrable systems and Riemann-Hilbert problems (K. McLaughlin and X. Zhou, eds.), Cont. Math., vol. 326, AMS, Providence, RI, 2003, pp. 1–22.

[29] J. Baik, G. Ben Arous, and S. Péché, *Phase transition of the largest eigenvalue for nonnull complex sample covariance matrices*, Annals of Prob. **33** (2005), 1643–1697.

[30] J. Baik, R. Buckingham, and J. DiFranco, *Asymptotics of Tracy-Widom distributions and the total integral of a Painlevé II function*, Commun. Math. Phys. **280** (2008), 463–497.

[31] J. Baik, P. Deift, and K. Johansson, *On the distribution of the length of the longest increasing subsequence of random permutations*, J. Amer. Math. Soc. **12** (1999), 1119–1178.

[32] J. Baik, P. Deift, and E. Strahov, *Products and ratios of characteristic polynomials of random Hermitian matrices*, J. Math. Phys. **44** (2003), 3657–3670.

[33] J. Baik and E.M. Rains, *Limiting distributions for a polynuclear growth model with external sources*, J. Stat. Phys. **100** (2000), 523–541.

[34] ———, *Algebraic aspects of increasing subsequences*, Duke Math. J. **109** (2001), 1–65.

[35] ———, *The asymptotics of monotone subsequences of involutions*, Duke Math. J. **109** (2001), 205–281.

[36] T.H. Baker and P.J. Forrester, *The Calogero-Sutherland model and generalized classical polynomials*, Commun. Math. Phys. **188** (1997), 175–216.

[37] ———, *Finite-n fluctuation formulas for random matrices*, J. Stat. Phys. **88** (1997), 1371–1386.

[38] ———, *Generalizations of the q-Morris constant term identity*, J. Comb. Theory. Ser. A **81** (1998), 69–87.

[39] ———, *Nonsymmetric jack polynomials and integral kernels*, Duke Math. J. **95** (1998), 1–50.

[40] ———, *Symmetric Jack polynomials from nonsymmetric theory*, Annals Comb. **3** (1999), 159–170.

[41] ———, *Random walks and random fixed point free involutions*, J. Phys. A **34** (2001), L381–L390.

[42] N.L. Balazs and A. Voros, *Chaos on the pseudosphere*, Phys. Rep. **143** (1986), 109–240.

[43] H.U. Baranger and P.A. Mello, *Mesoscopic transport through chaotic cavities: a random S-matrix theory approach*, Phys. Rev. Lett. **73** (1994), 142–145.

[44] E.W. Barnes, *The theory of the G-function*, Quart. J. Pure Appl. Math. **31** (1900), 264–313.

[45] Y. Baryshnikov, *GUEs and queues*, Probab. Theory Relat. Fields **119** (2001), 256–274.

[46] E. Basor and T. Ehrhardt, *Determinant computations for some classes of Toeplitz-Hankel matrices*, arXiv:0804.3073, 2008.

[47] E.L. Basor, *Toeplitz determinants, Fisher-Hartwig symbols, and random matrices*, Recent perspectives in random matrix theory and number theory (F. Mezzadri and N.C. Snaith, eds.), London Mathematical Society Lecture Note Series, vol. 322, Cambridge University Press, Cambridge, 2005, pp. 309–336.

[48] E.L. Basor and P.J. Forrester, *Formulas for the evaluation of Toeplitz determinants with rational generating functions*, Math. Nachr. **170** (1994), 5–18.

[49] E.L. Basor, C.A. Tracy, and H. Widom, *Asymptotics of level spacing distribtions for random matrices*, Phys. Rev. Lett. **69** (1992), 5–8.

[50] T. Bayes, *An essay towards solving a problem in the doctrine of chances*, Phil. Trans. Roy. Soc. London **53** (1763), 370–418.

[51] C.W.J. Beenakker, *Random-matrix theory of mesoscopic fluctuations in conductors and superconductors*, Phys. Rev. B **47** (1993), 15763–15775.

[52] C.W.J. Beenakker, *On the universality of Brezin's and Zee's spectral correlator*, Nucl. Phys. B **422** (1994), 515–520.

[53] ———, *Random-matrix theory of quantum transport*, Rev. Mod. Phys. **69** (1997), 731–808.

[54] C.W.J. Beenakker and B. Rejaei, *Random-matrix theory of parametric correlations in the spectra of disordered metals and chaotic billiards*, Physica A **203** (1994), 61–90.

[55] E.A. Bender, *Central and local limit theorems applied to asymptotic enumeration*, J. Combin. Theory Ser. A **15** (1973), 91–111.

[56] I. Bengtsson and K. Zyczkowski, *Geometry of quantum states*, Cambridge University Press, Cambridge, 2006.

[57] B. Bergersen, D. Boal, and P. Palffy-Muhoray, *Equilibrium configurations of particles on the sphere: the case of logarithmic interactions*, J. Phys. A **27** (1994), 2579–2586.

[58] D. Bernard, M. Gaudin, F.D.M. Haldane, and V. Pasquier, *Yang-Baxter equation in long-range interacting systems*, J. Phys. A **26** (1993), 5219–5236.

[59] D. Bernard and A. LeClair, *A classification of non-Hermitian random matrices*, Statistical field theories (A. Cappelli and G. Mussardo, eds.), NATO Science Series II, vol. 73, Springer, Berlin, 2002, pp. 207–214.

[60] D. Bernard, V. Pasquier, and D. Serban, *Exact solution of long-range interacting spin chains with boundaries*, Europhs. Lett. **30** (1995), 301–306.

[61] B.A. Bernevig and F.D.M. Haldane, *Generalized clustering conditions of jack polynomials at negative Jack parameter α*, Phys. Rev. B **77** (2008), 184502.

[62] M.V. Berry, *Riemann's zeta function: a model for quantum chaos?*, Quantum Chaos and Statistical Nuclear Physics (T.H. Seligman and H. Nishioka, eds.), Lecture Notes in Physics, vol. 263, Springer, Berlin, 1986, pp. 1–17.

[63] ______, *Some quantum to classical asymptotics*, Quantum Chaos (M.J. Giannoni, A. Voros, and J. Zinn-Justin, eds.), Les Houches Lecture Series, vol. 52, North-Holland, Amsterdam, 1991, pp. 251–304.

[64] M.V. Berry and J.P. Keating, *The Riemann zeros and eigenvalue asymptotics*, SIAM Rev. **41** (1999), 236–266.

[65] M.V. Berry and M. Robnik, *Semi-classical level spacings when regular and chaotic orbits coexist*, J. Phys. A **17** (1984), 2413–2421.

[66] P. Billingsley, *Probability and measure*, John Wiley, New York, 1979.

[67] P. Bleher and X. Di, *Correlations between zeros of non-Gaussian random polynomials*, Int. Math. Res. Notices **2004** (2004), 2443–2484.

[68] P.M. Bleher and A. Kuijlaars, *Random matrices with external source and multiple orthogonal polynomials*, Int. Math. Res. Notices **2004** (2004), 109–129.

[69] E. Bogomolny, *Quantum and arithmetic chaos*, Frontiers in number theory, physics and geometry I (P. Cartier, B. Julia, P. Moussa, and P. Vanhove, eds.), Springer, Berlin, 2005, pp. 3–106.

[70] E. Bogomolny, O. Bohigas, and P. Leboeuf, *Distribution of roots of random polynomials*, Phys. Rev. Lett. **68** (1992), 2726–2729.

[71] ______, *Quantum chaotic dynamics and random polynomials*, J. Stat. Phys. **85** (1996), 639–679.

[72] E. Bogomolny, O. Bohigas, P. Leboeuf, and A.C. Monastra, *On the spacing distribution of the Riemann zeros: corrections to the asymptotic result*, J. Phys. A **39** (2006), 10743–10754.

[73] E.B. Bogomolny and J.P. Keating, *Random matrix theory and the Riemann zeros II: n-point correlations*, Nonlinearity **9** (1996), 911–935.

[74] E.B. Bogomolny and P. Leboeuf, *Statistical properties of the zeros of zeta functions–beyond the Riemann case*, Nonlinearity **7** (1994), 1155–1167.

[75] O. Bohigas, *Compound nucleus resonances, random matrices, quantum chaos*, Recent perspectives in random matrix theory and number theory (F. Mezzadri and N.C. Snaith, eds.), London Mathematical Society Lecture Note Series, vol. 322, Cambridge University Press, Cambridge, 2005, pp. 147–183.

[76] O. Bohigas and M.P. Pato, *Missing levels in correlated spectra*, Phys. Lett. B **595** (2004), 171–176.

[77] G. Boole, *On the comparison of transcendents with certain applications to the theory of definite integrals*, Philos. Trans. Roy. Soc. **147** (1857), 745–803.

[78] F. Bornemann, *On the numerical evaluation of Fredholm determinants*, arXiv:0804.2543, 2008.

[79] ______, *On the numerical evaluation of distributions in random matrix theory: a review with an invitation to experimental mathematics*, arXiv:0904.1581, 2009.

[80] A. Borodin, *Biorthogonal ensembles*, Nucl. Phys. B **536** (1998), 704–732.

[81] ______, *Riemann-Hilbert problem and the discrete Bessel kernel*, Inter. Math. Res. Notices **(9)** (2000), 467–494.

[82] ______, *Discrete gap probabilities and discrete Painlevé equations*, Duke Math. J. **117** (2003), 489–542.

[83] A. Borodin and P. Deift, *Fredholm determinants, Jimbo-Miwa-Ueno tau-functions and representation theory*, Commun. Pur. Appl. Math. **55** (2002), 1160–1230.

[84] A. Borodin, P.L. Ferrari, M. Prähoffer, and T. Sasamoto, *Fluctuation properties of the TASEP with periodic initial configuration*, J. Stat. Phys. **129** (2006), 1055–1080.

[85] A. Borodin and P.J. Forrester, *Increasing subsequences and the hard-to-soft transition in matrix ensembles*, J.Phys. A **36** (2003), 2963–2981.

[86] A. Borodin and A. Okounkov, *A Fredholm determinant formula for Toeplitz determinants*, Integral Equations Operator Theory **37** (2000), 386–396.

[87] A. Borodin and S. Péché, *Airy kernel with two sets of parameters in directed percolation and random matrix theory*, J. Stat. Phys. **132** (2008), 275–290.

[88] A. Borodin and E.M. Rains, *Eynard-Mehta theorem, Schur process and their Pfaffian analogs*, J. Stat. Phys. **121** (2005), 291–317.

[89] A. Borodin and C.D. Sinclair, *The Ginibre ensemble of real random matrices and its scaling limit*, arXiv:0805.2986.

[90] A. Borodin and A. Soshnikov, *Janossy densities I. Determinantal ensembles*, J. Stat. Phys. **113** (2003), 595–610.

[91] A. Borodin and E. Strahov, *Averages of characteristic polynomials in random matrix theory*, Commun. Pure Appl. Math. **59** (2004), 161–253.

[92] A. Böttcher, *On the determinant formula by Borodin, Okounkov, Baik, Deift and Rains*, Operator Theory: Advances and Applications **135** (2002), 91–99.

[93] P. Bourgade, A. Nikeghbali, and A. Rouault, *Circular Jacobi ensembles and deformed Verblunsky coefficients*.

[94] C. Boutillier, *The bead model and limit behaviors of dimer models*, Ann. Prob. **37** (2009), 107–142.

[95] E. Brézin and S. Hikami, *Level spacing of random matrices in an external source*, Phys. Rev. E **58** (1998), 7176–7185.

[96] ———, *Characteristic polynomials of random matrices*, Commun. Math. Phys. **214** (2000), 111–135.

[97] ———, *Characteristic polynomials of real symmetric random matrices*, Commun. Math. Phys. **223** (2001), 363–382.

[98] E. Brézin, C. Itzykson, G. Parisi, and J.B. Zuber, *Planar diagrams*, Commun. Math. Phys. **59** (1978), 35–51.

[99] E. Brézin and A. Zee, *Universality of the correlations between eigenvalues of large random matrices*, Nucl. Phys. B **402** (1993), 613–627.

[100] T.A. Brody, J. Flores, J.B. French, P.A. Mello, A. Pandey, and S.S.M. Wong, *Random matrix theory*, Rev. Mod. Phys. **53** (1981), 329–351.

[101] P.W. Brouwer, *Generalized circular ensemble of scattering matrices for a chaotic cavity with nonideal leads*, Phys. Rev. B **51** (1995), 16878–16884.

[102] P.W. Brouwer, K. Frahm, and C.W.J. Beenakker, *Quantum mechanical time-delay matrix in chaotic scattering*, Phys. Rev. Lett. **78** (1997), 4737–4740.

[103] Z. Burda, R.A. Janik, and B. Waclaw, *Spectrum of the product of independent random Gaussian matrices*, arXiv:0912.3422, 2009.

[104] J.M. Caillol, *Exact results for a two-dimensional one-component plasma on a sphere*, J. Phys. Lett. (Paris) **42** (1981), L245–L247.

[105] F. Calogero, *Solution of the three-body problem in one dimension*, J. Math. Phys. **10** (1969), 2191–2196.

[106] J. Cardy, *Stochastic Loewner evolution and Dyson's circular ensembles*, J. Phys. A **36** (2003), L379–L386.

[107] Y. Chen, M.E.H. Ismail, and K.A. Muttalib, *A solvable random matrix model for disordered conductors*, J. Phys. Cond. Matter **4** (1992), L417–L413.

[108] I.V. Cherednik, *Double affine Hecke algebras and Macdonald's conjectures*, Ann. Math. **141** (1995), 191–216.

[109] Ph. Choquard, P. Favre, and Ch. Gruber, *On the equation of state of classical one-component systems with long-range forces*, J. Stat. Phys. **23** (1980), 405–442.

[110] Ph. Choquard, B. Piller, and R. Rentsch, *On the dielectric susceptibility of classical Coulomb systems II*, J. Stat. Phys. **46** (1987), 599–633.

[111] A.J. Chorin, *Vorticity and turbulence*, Applied Mathematical Sciences, vol. 103, Springer, New York, 1994, pp. 407–424.

[112] C. Cohen-Tannoudji, B. Diu, and F. Laloë, *Quantum mechanics*, Wiley, Paris, 1977.

[113] B. Collins, *Product of random projections, Jacobi ensembles and universality problems arising from free probability*, Prob. Theory Related Fields **133** (2005), 315–344.

[114] B. Collins and P. Sniady, *Integration with respect to Haar measure on unitary, orthogonal and symplectic groups*, Commun. Math. Phys. **264** (2006), 773–795.

[115] J.B. Conrey, *Notes on l-functions and random matrices*, Frontiers in number theory, physics and geometry I (P. Cartier, B. Julia, P. Moussa, and P. Vanhove, eds.), Springer, Berlin, 2006, pp. 107–162.

[116] J.B. Conrey, P.J. Forrester, and N.C. Snaith, *Averages of ratios of characteristic polynomials for the compact classical groups*, Int. Math. Res. Notices **2005** (2005), 397–431.

[117] J.B. Conrey, M.O. Rubinstein, and N.C. Snaith, *Moments of the derivative of characteristic polynomials with an application to the Riemann zeta function*, Commun. Math. Phys. **267** (2006), 611–629.

[118] A.G. Constantine, *Some noncentral distribution problems in multivariate analysis*, Ann. Math. Statist. **34** (1963), 1270–1285.

[119] C.M. Cosgrove, *Chazy classes IX–XI of third-order differential equations*, Stud. in Appl. Math. **104** (2000), 171–228.

[120] C.M. Cosgrove and G. Scoufis, *Painlevé classification of a class of differential equations of the second order and second degree*, Stud. in Appl. Math. **88** (1993), 25–87.

[121] O. Costin and J.L. Lebowitz, *Gaussian fluctuations in random matrices*, Phys. Rev. Lett. **75** (1995), 69–72.

[122] A.W. Davis, *On the marginal distributions of the latent roots of the multivariable beta matrix*, Ann. Math. Statist. **43** (1972), 1664–1669.

[123] N.G. de Bruijn, *On some multiple integrals involving determinants*, J. Indian Math. Soc. **19** (1955), 133–151.

[124] D.S. Dean and S.N. Majumdar, *Extreme value statistics of eigenvalues of Gaussian random matrices*, Phys. Rev. E **77** (2008), 041108.

[125] P. Deift and D. Gioev, *Universality in random matrix theory for orthogonal and symplectic ensembles*, Int. Math. Res. Papers **2007** (2007), 004–116.

[126] P. Deift, A. Its, and I. Krasovsky, *Asymptotics of Airy kernel determinant*, Commun. Math. Phys. **278** (2008), 643–678.

[127] P. Deift, A. Its, I. Krasovsky, and X. Zhou, *The Widom-Dyson constant for the gap probability in random matrix theory*, J. Comp. Applied Math. **202** (2006), 26–47.

[128] P.A. Deift, *Application of a commutation formula*, Duke Math. J. **45** (1978), 267–310.

[129] P.A. Deift, A.R. Its, and X. Zhou, *A Riemann-Hilbert approach to asymptotic problems arising in the theory of random matrices and also in the theory of integrable statistical mechanics*, Ann. Math. **146** (1997), 149–235.

[130] A.P. Dempster, *Elements of continuous multivariate analysis*, Addison-Wesley, Reading, MA, 1969.

[131] P. Desrosiers, *Duality in random matrix ensembles for all β*, Nucl. Phys. B **817** (2009), 224–251.

[132] P. Desrosiers and P.J. Forrester, *Hermite and Laguerre β-ensembles: asymptotic corrections to the eigenvalue density*, Nucl. Phys. B **743** (2006), 307–332.

[133] P. Diaconis, *Longest increasing subsequences: from patience sorting to the Baik-Deift-Johansson theorem*, Bull. Amer. Math. Soc. **36** (1999), 413–432.

[134] P. Diaconis and M. Shahshahani, *The subgroup algorithm for generating uniform random variables*, Prob. Eng. Inf. Sci. **1** (1987), 15–32.

[135] M. Dieng, *Distribution functions for edge eigenvalues in orthogonal and symplectic ensembles: Painlevé representations*, Int. Math. Res. Notices **2005** (2005), 2263–2287.

[136] R.B. Dingle, *Asymptotic expansions*, Academic Press, New York, 1973.

[137] A.L. Dixon, *Generalizations of Legendre's formula $ke' - (k - e)k' = \frac{1}{2}\pi$*, Proc. London Math. Soc. **3** (1905), 206–224.

[138] V.S. Dotsenko and V.A. Fateev, *Four-point correlation functions and the operator algebra in 2D conformal invariant theories with central charge $C \leq 1$*, Nucl. Phys. B **251** (1985), 691–734.

[139] E. Dueñez, *Random matrix ensembles associated to compact symmetric spaces*, Ph.D. thesis, Princeton University, 2001.

[140] I. Dumitriu and A. Edelman, *Matrix models for beta ensembles*, J. Math. Phys. **43** (2002), 5830–5847.

[141] C.F. Dunkl, *Difference-differential operators associated to reflection groups*, Trans. Amer. Math. Soc. **311** (1989), 167–183.

[142] ———, *Integral kernels with reflection group invariance*, Can. J. Math. **43** (1991), 1213–1227.

[143] ———, *Hankel transforms associated to finite reflection groups*, Contemp. Math. **138** (1992), 123–128.

[144] C.F. Dunkl and Y. Xu, *Orthogonal polynomials of several variables*, Cambridge University Press, Cambridge, 2001.

[145] F.J. Dyson, *A Brownian motion model for the eigenvalues of a random matrix*, J. Math. Phys. **3** (1962), 1191–1198.

[146] ———, *Statistical theory of energy levels of complex systems I*, J. Math. Phys. **3** (1962), 140–156.

[147] ———, *Statistical theory of energy levels of complex systems II*, J. Math. Phys. **3** (1962), 157–165.

[148] ———, *Statistical theory of energy levels of complex systems III*, J. Math. Phys. **3** (1962), 166–175.

[149] ———, *The three fold way. Algebraic structure of symmetry groups and ensembles in quantum mechanics*, J. Math. Phys. **3** (1962), 1199–1215.

[150] ———, *Correlations between the eigenvalues of a random matrix*, Commun. Math. Phys. **19** (1970), 235–250.

[151] ———, *Quaternion determinants*, Helv. Phys. Acta **45** (1972), 289–302.

[152] ———, *The Coulomb fluid and the fifth Painlevé transcendent*, Chen Ning Yang (S.-T. Yau, ed.), International Press, Cambridge MA, 1995, p. 131.

[153] F.J. Dyson and M.L. Mehta, *Statistical theory of the energy levels of complex systems. IV*, J. Math. Phys. **4** (1963), 701–712.

[154] A. Edelman, *Eigenvalues and condition numbers of random matrices*, SIAM J. Matrix Anal. Appl. **9** (1988), 543–560.

[155] ———, *The probability that a random real Gaussian matrix has k real eigenvalues, related distributions, and the circular law*, J. Multivariate. Anal. **60** (1997), 203–232.

[156] A. Edelman, E. Kostlan, and M. Shub, *How many eigenvalues of random matrix are real?*, J. Amer. Math. Soc. **7** (1994), 247–267.

[157] A. Edelman and N. Raj Rao, *Random matrix theory*, Acta Numerica (A. Iserles, ed.), vol. 14, Cambridge University Press, Cambridge, 2005.

[158] A. Edelman and B.D. Sutton, *From random matrices to stochastic operators*, J. Stat. Phys. **127** (2006), 1121–1165.

[159] ———, *The beta-Jacobi matrix model, the CS decomposition, and generalized singular value problems*, Found. Comput. Math. **8** (2008), 259–285.

[160] K.B. Efetov, *Supersymmetry and theory of disordered metals*, Adv. Phys. **32** (1983), 53–127.

[161] ———, *Supersymmetry in disorder and chaos*, Cambridge University Press, Cambridge, 1997.

[162] T. Ehrardt, *The asymptotics a Bessel kernel determinant which arises in random matrix theory*, arXiv:1001.2340, 2010.

[163] T. Ehrhardt, *Dyson's constant in the asymptotics of the Fredholm determinant of the sine kernel*, Commun. Math. Phys. **262** (2006), 317–341.

[164] N. Elkies, G. Kuperberg, M. Larsen, and J. Propp, *Alternating sign matrices and domino tilings I*, J. Algebraic Combin. **1** (1992), 111–132.

[165] A. Erdélyi et al.(ed.), *Higher transcendental functions*, vol. I and II, McGraw-Hill, New York, 1953.

[166] R. J. Evans, *Multidimensional beta and gamma integrals*, Contemp. Math. **166** (1994), 341–357.

[167] B. Eynard, *Asymptotics of skew-orthogonal polynomials*, J. Phys. A **34** (2001), 7591–7605.

[168] B. Eynard and M.L. Mehta, *Matrices coupled in a chain. I. Eigenvalue correlations*, J. Phys. A **31** (1998), 4449–4456.

[169] E. Faleiro, J.M.G. Gomez, R.A. Molina, A. Relano, and J. Retamosa, *Theoretical derivation of $1/f$ noise in quantum chaos*, Phys. Rev. Lett. **93** (2004), 244101.

[170] D.W. Farmer, F. Mezzadri, and N.C. Snaith, *Random polynomials, random matrices and L-functions*, Nonlinearity **19** (2006), 919–936.

[171] B. Feigin, M. Jimbo, T. Miwa, and E. Mukhin, *A differential ideal of symmetric polynomials spanned by Jack polynomials at* $\beta = -(r-1)/(k+1)$, Int. Math. Res. Not. **no. 23** (2002), 1223–1237.

[172] J. Feinberg and A. Zee, *Non-Hermitean random matrix theory: method of Hermitian reduction*, Nucl. Phys. B **504** (1997), 579–608.

[173] P.L. Ferrari and H. Spohn, *A determinantal formula for the GOE Tracy Widom distribution*, J. Phys. A **38** (2005), L557–L561.

[174] R.P. Feynman, *Forces in molecules*, Phys. Rev. **56** (1939), 340–343.

[175] M.E. Fisher, *Walks, walls, wetting, and melting*, J. Stat. Phys. **34** (1984), 667–729.

[176] M.E. Fisher and R.E. Hartwig, *Toeplitz determinants—some applications, theorems and conjectures*, Adv. Chem. Phys. **15** (1968), 333–353.

[177] B.J. Fleming, P.J. Forrester, and E. Nordenstam, *A finitization of the bead process*, arXiv:0902.0709, 2009.

[178] M. Fogler and B.I. Shklovskii, *The probability of an eigenvalue fluctuation in an interval of a random matrix spectrum*, Phys. Rev. Lett. **74** (1995), 3312–3315.

[179] P.J. Forrester, *An exactly solvable two-component classical Coulomb system*, J. Aust. Math. Soc. Series B **26** (1984), 119–128.

[180] ______, *Exact results for correlations in a two-component log-gas*, J. Stat. Phys. **59** (1989), 57–79.

[181] ______, *Exact solution of the lockstep model of vicious walkers*, J. Phys. A **23** (1990), 1259–1273.

[182] ______, *Theta function generalizations of some constant term identities in the theory of random matrices*, SIAM J. Math. Anal. **21** (1990), 270–280.

[183] ______, *Exact results for vicious walker models of domain walls*, J. Phys. A **24** (1991), 203–218.

[184] ______, *A constant term identity and its relationship to the log-gas and some quantum many body systems*, Phys. Lett. A **163** (1992), 121–126.

[185] ______, *Selberg correlation integrals and the* $1/r^2$ *quantum many body system*, Nucl. Phys. B **388** (1992), 671–699.

[186] ______, *Some statistical properties of the eigenvalues of complex random matrices*, Phys. Lett. A **169** (1992), 21–24.

[187] ______, *Exact results and universal asymptotics in the Laguerre random matrix ensemble*, J. Math. Phys. **35** (1993), 2539–2551.

[188] ______, *Recurrence equations for the computation of correlations in the* $1/r^2$ *quantum many body system*, J. Stat. Phys. **72** (1993), 39–50.

[189] ______, *The spectrum edge of random matrix ensembles*, Nucl. Phys. B **402** (1993), 709–728.

[190] ______, *Addendum to Selberg correlation integrals and the* $1/r^2$ *quantum many body system*, Nucl. Phys. B **416** (1994), 377–385.

[191] ______, *Properties of an exact crystalline many-body ground state*, J. Stat. Phys. **76** (1994), 331–346.

[192] ______, *Integration formulas and exact calculations in the Calogero-Sutherland model*, Mod. Phys. Lett B **9** (1995), 359–371.

[193] ______, *Normalization of the wave function for the Calogero-Sutherland model with internal degrees of freedom*, Int. J. Mod. Phys. B **9** (1995), 1243–1261.

[194] ______, *Fluctuation formula for complex random matrices*, J. Phys. A **32** (1999), L159–L163.

[195] ______, *Painlevé transcendent evaluation of the scaled distribution of the smallest eigenvalue in the Laguerre orthogonal and symplectic ensembles*, nlin.SI/0005064, 2000.

[196] ______, *Growth models, random matrices and Painlevé transcendents*, Nonlinearity **16** (2003), R27–R49.

[197] ______, *Evenness symmetry and inter-relationships between gap probabilities in random matrix theory*, Forum Math. **18** (2006), 711–743.

[198] ______, *Hard and soft edge spacing distributions for random matrix ensembles with orthogonal and symplectic symmetry*, Nonlinearity **19** (2006), 2989–3002.

[199] ———, *A random matrix decimation procedure relating $\beta = 2/(r+1)$ to $\beta = 2(r+1)$*, Commun. Math. Phys. **285** (2007), 653–672.

[200] ———, *Beta random matrix ensembles*, Random matrix theory and its applications (Z. Bai, Y. Chen, and Y.-C. Liang, eds.), Lecture Notes Series, IMS, NUS, vol. 18, World Scientific, Singapore, 2009, pp. 27–68.

[201] P.J. Forrester and N.E. Frankel, *Applications and generalizations of Fisher-Hartwig asymptotics*, J. Math. Phys. **45** (2003), 2003–2028.

[202] P.J. Forrester, N.E. Frankel, and T.M. Garoni, *Asymptotic form of the density profile for Gaussian and Laguerre random matrix ensembles with orthogonal and symplectic symmetry*, J. Math. Phys. **47** (2006), 023301.

[203] P.J. Forrester and G. Honner, *Exact statistical properties of the zeros of complex random polynomials*, J. Phys. A **32** (1999), 2961–2981.

[204] P.J. Forrester and T.D. Hughes, *Complex Wishart matrices and conductance in mesoscopic systems: exact results*, J. Math. Phys. **35** (1994), 6736–6747.

[205] P.J. Forrester and B. Jancovici, *Two-dimensional one-component plasma in a quadrupolar field*, Int. J. Mod. Phys. A **11** (1996), 941–949.

[206] ———, *Exact and asymptotic formulas for overdamped Brownian dynamics*, Physica A **238** (1997), 405–424.

[207] P.J. Forrester, B. Jancovici, and D.S. McAnally, *Analytic properties of the structure function for the one-dimensional one-component log-gas*, J. Stat. Phys. **102** (2000), 737–780.

[208] P.J. Forrester, B. Jancovici, and G. Téllez, *Universality in some classical Coulomb systems of restricted domain*, J. Stat. Phys. **84** (1996), 359–378.

[209] P.J. Forrester and M. Krishnapur, *Derivation of an eigenvalue probability density function relating to the Poincaré disk*, J. Phys. A **42** (2009), 385204 (10pp).

[210] P.J. Forrester and A. Mays, *A method to calculate correlation functions for $\beta = 1$ random matrices of odd size*, J. Stat. Phys. **134** (2009), 443–462.

[211] P.J. Forrester, D.S. McAnally, and Y. Nikoyalevsky, *On the evaluation formula for Jack polynomials with prescribed symmetry*, J. Phys. A **34** (2001), 8407–8424.

[212] P.J. Forrester and T. Nagao, *Correlations for the circular Dyson Brownian motion model with Poisson initial conditions*, Nucl. Phys. B **532** (1998), 733–752.

[213] ———, *Correlations for the Cauchy and generalized circular ensembles with orthogonal and symplectic symmetry*, J. Phys. A **34** (2001), 7917–7932.

[214] ———, *Eigenvalue statistics of the real Ginibre ensemble*, Phys. Rev. Lett. **99** (2007), 050603.

[215] ———, *Determinantal correlations for classical projection processes*, arXiv:0801.0100, 2008.

[216] P.J. Forrester, T. Nagao, and G. Honner, *Correlations for the orthogonal-unitary and symplectic-unitary transitions at the hard and soft edges*, Nucl. Phys. B **553** (1999), 601–643.

[217] P.J. Forrester, T. Nagao, and E.M. Rains, *Correlations for the orthogonal-unitary and symplectic-unitary transitions at the hard and soft edges*, Int. Math. Res. Notices **2006** (2006), 89796 (35 pages).

[218] P.J. Forrester and A.M. Odlyzko, *Gaussian unitary ensemble eigenvalues and Riemann ζ function zeros: a nonlinear equation for a new statistic*, Phys. Rev. E **54** (1996), R4493–R4495.

[219] P.J. Forrester and C. Pisani, *The hole probability in log-gas and random matrix systems*, Nucl. Phys. B **374** (1992), 720–740.

[220] P.J. Forrester and E.M. Rains, *Inter-relationships between orthogonal, unitary and symplectic matrix ensembles*, Random matrix models and their applications (P.M. Bleher and A.R. Its, eds.), Mathematical Sciences Research Institute Publications, vol. 40, Cambridge University Press, Cambridge, 2001, pp. 171–208.

[221] ———, *Correlations for superpositions and decimations of Laguerre and Jacobi orthogonal matrix ensembles with a parameter*, Prob. Theory Related Fields **130** (2004), 518–576.

[222] ———, *Interpretations of some parameter dependent generalizations of classical matrix ensembles*, Prob. Theory Related Fields **131** (2005), 1–61.

[223] ———, *Jacobians and rank 1 perturbations relating to unitary Hessenberg matrices*, Int. Math. Res. Not. **2006** (2006), 48306 (36 pages).

[224] ———, *Symmetrized models of last passage percolation and non-intersecting lattice paths*, J. Stat. Phys. **2007** (2007), 833–856.

[225] P.J. Forrester and J.B. Rogers, *Electrostatics and the zeros of the classical polynomials*, SIAM J. Math. Anal. **17** (1986), 461–468.

[226] P.J. Forrester, N.S. Snaith, and J.J.M. Verbaarschot, *Developments in random matrix theory*, J. Phys. A **36** (2003), R1–R10.

[227] P.J. Forrester and S.O. Warnaar, *The importance of the Selberg integral*, Bull. Am. Math. Soc. **45** (2008), 489–534.

[228] P.J. Forrester and N.S. Witte, *Application of the τ-function theory of Painlevé equations to random matrices: PIV, PII and the GUE*, Commun. Math. Phys. **219** (2000), 357–398.

[229] ———, *Exact Wigner surmise type evaluation of the spacing distribution in the bulk of the scaled random matrix ensembles*, Lett. Math. Phys. **53** (2000), 195–200.

[230] ———, *Application of the τ-function theory of Painlevé equations to random matrices: PV, PIII, the LUE, JUE and CUE*, Commun. Pure Appl. Math. **55** (2002), 679–727.

[231] ———, *τ-function evaluation of gap probabilities in orthogonal and symplectic matrix ensembles*, Nonlinearity **15** (2002), 937–954.

[232] ———, *Discrete Painlevé equations and random matrix averages*, Nonlinearity **16** (2003), 1919–1944.

[233] ———, *Application of the τ-function theory of Painlevé equations to random matrices: PVI, the JUE,CyUE, cJUE and scaled limits*, Nagoya Math. J. **174** (2004), 29–114.

[234] ———, *Discrete Painlevé equations, orthogonal polynomials on the unit circle and n-recurrences for averages over $U(N)$—$P_{\rm III'}$ and $P_{\rm V}$ τ-functions*, Int. Math. Res. Not. **(4)** (2004), 159–183.

[235] ———, *Boundary conditions associated with the Painlevé III′ and V evaluations of some random matrix averages*, J. Phys. A **39** (2006), 8983–8996.

[236] ———, *Discrete Painlevé equations for a class of $P_{\rm VI}$ τ-functions given as $U(N)$ averages*, Nonlinearity **16** (2006), 1919–1944.

[237] P.J. Forrester and J.A. Zuk, *Applications of the Dotsenko-Fateev integral in random-matrix models*, Nucl. Phys. B **473** (1996), 616–630.

[238] K. Frahm and J.L. Pichard, *Brownian motion ensembles and parametric correlations of the transmission eigenvalues: applications to coupled quantum billiards and to disordered systems*, J. Phys. I (France) **5** (1995), 877–906.

[239] P. Di Francesco, *Matrix model combinatorics: applications to folding and coloring*, Random matrix models and their applications (P.M. Bleher and A.R. Its, eds.), Mathematical Sciences Research Institute Publications, vol. 40, Cambridge University Press, Cambridge, 2001, pp. 111–170.

[240] ———, *Rectangular matrix models and combinatorics of colored graphs*, Nucl. Phys. B **648** (2003), 461–496.

[241] ———, *2d quantum gravity, matrix models and graph combinatorics*, Applications of random matrices to physics (E. Brezin et al., ed.), Springer, Dordrecht, 2006, pp. 33–88.

[242] P. Di Francesco, M. Gaudin, C. Itzykson, and F. Lesage, *Laughlin's wave functions, Coulomb gases and expansions of the discriminant*, Int. J. Mod. Phys. A **9** (1994), 4257–4351.

[243] J.B. French, P.A. Mello, and A. Pandey, *Statistical properties of many-particle spectra. II. Two-point correlations and fluctuations*, Ann. Phys. **113** (1978), 277–293.

[244] G. Freud, *On the coefficients in the recursion formulas of orthogonal polynomials*, Proc. Roy. Irish Acad. Sect. A **76** (1976), 1–6.

[245] W.H.J. Fuchs, *On the eigenvalues of an integral equation arising in the theory of band-limited signals*, J. Math. Anal. Appl. **9** (1964), 317–330.

[246] W. Fulton, *Young tableaux*, London Mathematical Society Student Texts, CUP, Cambridge, 1997.

[247] Y.V. Fyodorov, *Negative moments of characteristic polynomials of random matrices: Ingham-Siegel integral as an alternative to Hubbard-Stratonovich transformation*, Nucl. Phys. B **621** (2002), 643–674.

[248] ———, *Complexity of random energy landscapes, glass transition and absolute value of spectral determinant of random matrices*, Phys. Rev. Lett. **92** (2004), 240601, Erratum: Phys. Rev. Lett. 93 (2004), 149901.

[249] Y.V. Fyodorov and B.A. Khoruzhenko, *On absolute moments of characteristic polynomials of a certain class of complex random matrices*, Comm. Math. Phys. **273** (2007), 561–599.

[250] Y.V. Fyodorov, B.A. Khoruzhenko, and H.-J. Sommers, *Almost-Hermitian random matrices: crossover from Wigner-Dyson to Ginibre eigenvalue statistics*, Phys. Rev. Lett. **79** (1997), 557–560.

[251] Y.V. Fyodorov and H.-J. Sommers, *Universality of "level curvature" distribution for large random matrices: systematic analytical approaches*, Zeit. für Phys. B **99** (1995), 123–135.

[252] ———, *Random matrices close to hermitian or unitary: overview of methods and results*, J. Phys. A **36** (2003), 3303–3347.

[253] G. Gallavoti and C. Marchiori, *On the calculation of an integral*, J. Math. Ann. Appl. **44** (1973), 661–675.

[254] B. Gambier, *Sur les équations différentielles du second ordre et du premier degré dont l'intégrale générale est a points critiques fixes*, Acta Math. **33** (1909), 1–55.

[255] F.R. Gantmacher, *Matrix theory*, 2nd ed., Springer, Berlin, 1986.

[256] H. Gao and P.J. Smith, *A determinant representation for the distribution of quadratic forms in complex normal vectors*, J. Multivariate. Anal. **73** (2000), 155–165.

[257] T.M. Garoni, Unpublished notes, 2003.

[258] T.M. Garoni, P.J. Forrester, and N.E. Frankel, *Asymptotic corrections to the eigenvalue density of the GUE and LUE*, J. Math. Phys. **46** (2005), 103301.

[259] M. Gaudin, *Sur la loi limite de l'espacement des valeurs propres d'une matrice aléatoire*, Nucl. Phys. **25** (1961), 447–458.

[260] J.S. Geronimo and K.M. Case, *Scattering theory and polynomials orthogonal on the unit circle*, J. Math. Phys. **20** (1979), 299–310.

[261] Ya.L. Geronimus, *Orthogonal polynomials*, Consultants Bureau, New York, 1961.

[262] I.M. Gessel, *Symmetric functions and p-recursiveness*, J. Comb. Th. A **53** (1990), 257–285.

[263] S. Ghosh, *Skew-orthogonal polynomials and random matrix theory*, CRM Monograph Series, vol. 28, American Mathematical Society, Providence, RI, 2009.

[264] J. Ginibre, *Statistical ensembles of complex, quaternion, and real matrices*, J. Math. Phys. **6** (1965), 440–449.

[265] V. Girko, *Circular law*, Th. Prob. Appl. **29** (1984), 694–706.

[266] ———, *An introduction to statistical analysis of random arrays*, Walter de Gruyter, 1998.

[267] P. Glynn and W. Whitt, *Departures form many queues in series*, Ann. Appl. Probability **1** (1991), 546–572.

[268] I.J. Good, *Short proof of a conjecture of Dyson*, J. Math. Phys. **11** (1972), 1884.

[269] N.R. Goodman, *Statistical analysis based on a certain multivariate complex Gaussian distribution (an introduction)*, Ann. Math. Stat. **34** (1963), 152–176.

[270] I.S. Gradshteyn and I.M. Ryzhik, *Table of integrals, series, and products*, 4th ed., Academic Press, New York, 1980.

[271] W.B. Gragg, *Positive definite Toeplitz matrices, the Arnoldi process for isometric operators, and Gaussian quadrature on the unit circle*, J. Comput. Appl. Math. **46** (1993), 183–198.

[272] J. Gravner, C.A. Tracy, and H. Widom, *Limit theorems for height fluctuations in a class of discrete space and time fluctuations*, J. Stat. Phys. (2001), 1085–1132.

[273] C. Greene, *An extension of Schensted's theorem*, Adv. in Math. **14** (1974), 254–265.

[274] U. Grenander, *Probabilites on algebraic structures*, John Wiley and Sons, New York, 1963.

[275] R. Grobe, F. Haake, and H.-J. Sommers, *Quantum distinction of regular chaotic dissipative motion*, Phys. Rev. Lett. **61** (1988), 1899–1902.

[276] T. Guhr, A. Müller-Groeling, and H.A. Weidenmüller, *Random matrix theories in quantum physics: common concepts*, Phys. Rep. **299** (1998), 189–425.

[277] T. Guhr and T. Wettig, *An Itzykson-Zuber like integral and diffusion for complex ordinary and supermatrices*, J. Math. Phys. **37** (1996), 6395–6413.

[278] J. Gunson, *Proof of a conjecture of Dyson in the statistical theory of energy levels*, J. Math. Phys. **4** (1962), 752–753.

[279] A.K. Gupta and D.K. Nagar, *Matrix variate distributions*, Chapman & Hall/CRC, Boca Raton, FL, 1999.

[280] R.J. Gustafson, *A generalization of the Selberg beta-integral*, Bull. Am. Math. Soc. **22** (1990), 97–105.

[281] J. Gustavsson, *Gaussian fluctuations in the GUE*, Ann. l'Inst. Henri Poincaré (B) **41** (2005), 151–178.

[282] A.J. Guttmann, A.L. Owczarek, and X.G. Viennot, *Vicious walkers and Young tableaux I: without walls*, J. Phys. A **31** (1998), 8123–8135.

[283] Z.N.C. Ha, *Fractional statistics in one dimension: view from an exactly solvable model*, Nucl. Phys. B **435** (1995), 604–636.

[284] F. Haake, *Quantum signatures of chaos*, 2nd ed., Springer, Berlin, 2000.

[285] M. Hallnäs and E. Langmann, *Quantum Calogero-Sutherland type models and generalized classical polynomials*, arXiv:math-ph/0703090, 2007.

[286] J.M. Hammersley, *The zeros of random polynomials*, Proceedings of the Third Berkeley Symposium on Probability and Statistics (J. Neyman, ed.), vol. 2, Univ. California Press, Berekeley, CA, 1956, pp. 89–111.

[287] P.J. Hanlon, R.P. Stanley, and J.R. Stembridge, *Some combinatorial aspects of the spectra of normally distributed random matrices*, Contemp. Math. **138** (1992), 151–174.

[288] J. H. Hannay, *Chaotic analytic zero points: exact statistics for those of a random spin state*, J. Phys. A **29** (1996), L101–L105.

[289] J.P. Hansen and I.R. McDonald, *Theory of simple liquids*, 2nd ed., Academic Press, London, 1990.

[290] G.H. Hardy and J.E. Littlewood, *Some problems of "Partitio Numerorum" III. On the expression of a number as a sum of primes*, Acta Math. **44** (1923), 1–70.

[291] J. Harer and D. Zagier, *The Euler characteristic of the moduli space of curves*, Inven. Math. **85** (1986), 457–485.

[292] Harish-Chandra, *Differential operators on a semisimple Lie algebra*, Amer. J. Math. **79** (1957), 87–120.

[293] J. Harnad and A.Yu. Orlov, *Determiantal identity for multilevel systems and finite determinantal point processes*, arXiv:0712.3892, 2007.

[294] S.P. Hastings and J.B. McLeod, *A boundary value problem associated with the second Painlevé transcendent and the Korteweg-de Vries equation*, Arch. Rat. Mech. Anal. **73** (1980), 31–51.

[295] S. Helgason, *Differential geometry and symmetric spaces*, Academic, New York, 1962.

[296] E. Hellmann, *Einführung in die quantenchemie*, Deuticke, Vienna, 1937.

[297] K. Hikami, *Dunkl operator formalism for quantum many body problems associated with classical root systems*, J. Phys. Soc. Japan **65** (1996), 394–401.

[298] J.B. Hough, M. Krishnapur, Y. Peres, and B. Virág, *Zeros of Gaussian analytic functions and determinantal point processes*, American Mathematical Society, Providence, RI, 2009.

[299] P.L. Hsu, *On the distribution of the roots of certain determinantal equations*, Ann. Eugen. **9** (1939), 250–258.

[300] L.K. Hua, *Analysis of functions of several complex variables in the classical domains*, American Mathematical Society, Providence, RI, 1963.

[301] A. Hurwitz, *Über die Composition der quadratischen Formen von beliebig vielen Variabeln*, Nachr. Ges. Wiss. Göttingen (1898), 309–316.

[302] ———, *Mathematische werke*, vol. II, Birkhäuser, Basel, 1933.

[303] M. Ishikawa, S. Okada, and M. Wakayama, *Applications of minor summation formula. I Littlewood's formulas*, J. Algebra **183** (1996), 193–216.

[304] M. Ishikawa and M. Wakayama, *Minor summation formulas of Pfaffians*, Linear and Multilinear Algebra **39** (1995), 285–305.

[305] M.E. Ismail, *An electrostatics model for zeros of general orthogonal polynomials*, Pacific J. Math. **193** (1999), 153–169.

[306] M.E.H. Ismail and N.S. Witte, *Discriminants and functional equations for polynomials orthogonal on the unit circle*, J. Approx. Theory **110** (2001), 200–228.

[307] A.R. Its, A.G. Izergin, V.E. Korepin, and N.A. Slavnov, *Differential equations for quantum correlation functions*, Int. J. Mod. Phys B **4** (1990), 1003–1037.

[308] A.R. Its and A.V. Kitaev, *Mathematical aspects of the non-perturbative 2d quantum gravity*, Mod. Phys Lett. **A5** (1990), 2079–2083.

[309] C. Itzykson and J.B. Zuber, *Planar approximation 2*, J. Math. Phys. **21** (1980), 411–421.

[310] K. Iwasaki, H. Kimura, S. Shimomura, and M. Yoshida, *From Gauss to Painlevé. a modern theory of special functions*, Vieweg-Verlag, Braunschweig, 1991.

[311] A.D. Jackson, M.K. Şener, and J.J.M. Verbaarschot, *Finite volume partition functions and Itzyson-Zuber integrals*, Phys. Lett. B **387** (1997), 355–360.

[312] A.T. James, *Distributions of matrix variate and latent roots derived from normal samples*, Ann. Math. Statist. **35** (1964), 475–501.

[313] B. Jancovici, *Exact results for the two-dimensional one-component plasma*, Phys. Rev. Lett. **46** (1981), 386–388.

[314] ———, *Classical Coulomb systems near a plane wall. II*, J. Stat. Phys. **29** (1982), 263–280.

[315] ———, *Sum rules for the one-component plasma*, unpublished note, 1994.

[316] ———, *Classical Coulomb systems: screening and correlations revisited*, J. Stat. Phys. **80** (1995), 445–459.

[317] ———, *Pressure and Maxwell tensor in a Coulomb fluid*, cond-mat/99112246, 1999.

[318] B. Jancovici and P.J. Forrester, *Derivation of an asymptotic expression in Beenakker's general fluctuation formula for correlations near an edge*, Phys. Rev. B **50** (1994), 14599–14600.

[319] B. Jancovici, J.L. Lebowitz, and G. Manificat, *Large charge fluctuations in classical Coulomb systems*, J. Stat. Phys. **72** (1993), 773–787.

[320] B. Jancovici, G. Manificat, and C. Pisani, *Coulomb systems seen as critical systems: finite-size effects in two dimensions*, J. Stat. Phys. **76** (1994), 307–330.

[321] B. Jancovici and G. Téllez, *Two-dimensional Coulomb systems on a surface of constant negative curvature*, J. Stat. Phys. **91** (1998), 953–977.

[322] R.A. Janik and M.A. Nowak, *Wishart and anti-Wishart random matrices*, J. Phys. A **36** (2003), 3629–3637.

[323] M. Jimbo and T. Miwa, *Monodromony preserving deformations of linear ordinary differential equations with rational coefficients II*, Physica **2D** (1981), 407–448.

[324] M. Jimbo, T. Miwa, Y. Môri, and M. Sato, *Density matrix of an impenetrable Bose gas and the fifth Painlevé transcendent*, Physica **1D** (1980), 80–158.

[325] M. Jimbo, T. Miwa, and K. Ueno, *Monodromony preserving deformations of linear ordinary differential equations with rational coefficients I*, Physica **2D** (1981), 306–352.

[326] S. Jitomirskaya, H. Schulz-Blades, and G. Stolz, *Delocalization in random polymer models*, Commun. Math. Phys. **233** (2003), 27–48.

[327] K. Johansson, *On Szegö's asymptotic formula for Toeplitz determinants and generalizations*, Bull. Sci. Math., 2nd sér **112** (1988), 257–304.

[328] ———, *On fluctuation of eigenvalues of random Hermitian matrices*, Duke Math. J. **91** (1998), 151–204.

[329] ———, *Shape fluctuations and random matrices*, Commun. Math. Phys. **209** (2000), 437–476.

[330] ———, *Non-intersecting paths, random tilings and random matrices*, Prob. Theory Related Fields **123** (2002), 225–280.

[331] ———, *Discrete polynuclear growth and determinantal processes*, Commun. Math. Phys. **242** (2003), 277–329.

[332] ———, personal communication, 2005.

[333] K. Johansson and E. Nordenstam, *Eigenvalues of GUE minors*, Elect. J. Probability **11** (2006), 1342–1371.

[334] I.M. Johnstone, *On the distribution of the largest principal component*, Ann. Math. Stat. **29** (2001), 295–327.

[335] R.C. Jones, J.M. Kosterlitz, and D.J. Thouless, *The eigenvalue spectrum of a large symmetric random matrix with a finite mean*, J. Phys. A **11** (1978), L45–L48.

[336] K.W.J. Kadell, *A proof of Askey's conjectured q-analogue of Selberg's integral and a conjecture of Morris*, SIAM J. Math. Anal. **19** (1988), 969–986.

[337] ———, *A proof of the q-Macdonald-Morris conjecture for BC_n*, Mem. Amer. Math. Soc. **108** (1994).

[338] ———, *The Selberg-Jack symmetric functions*, Adv. Math. **130** (1997), 33–102.

[339] K. Kajiwara, T. Masuda, M. Noumi, Y. Ohta, and Y. Yamada, *Determinant formulas for the Toda and discrete Toda equations*, Funkcialaj Ekvacioj **44** (2001), 291–307.

[340] P. Kalinay, P. Markos, L. Samaj, and I. Travenec, *The sixth-moment sum rule for the pair correlations of the two-dimensional one-component plasma: exact results*, J. Stat. Phys. **98** (2000), 639–666.

[341] F. Kalisch and D. Braak, *Exact density of states for finite Gaussian random matrix ensembles via supersymmetry*, J. Phys. A **35** (2002), 9957–9969.

[342] J. Kaneko, *Selberg integrals and hypergeometric functions associated with Jack polynomials*, SIAM J. Math Anal. **24** (1993), 1086–1110.

[343] ———, *Constant term identities of Forrester-Zeilberger-Cooper*, Discrete Math. **173** (1997), 79–90.

[344] E. Kanzieper, *Eigenvalue correlations in non-Hermitian symplectic random matrices*, J. Phys. A **35** (2002), 6631–6644.

[345] E. Kanzieper and V. Freilikher, *Two band random matrices*, Phys. Rev. E **57** (1998), 6604–6611.

[346] A.A. Kapaev and E. Hubert, *A note on the Lax pairs for Painlevé equations*, J. Math. Phys. **32** (1999), 8145–8156.

[347] S. Karlin and L. McGregor, *Coincidence probabilities*, Pacific J. Math. **9** (1959), 1109–1140.

[348] M. Katori and N. Komatsuda, *Moments of vicious walkers and Möbius graph expansions*, Phys. Rev. E **67** (2003), 051110.

[349] M. Katori and H. Tanemura, *Scaling limit of vicious walks and two-matrix model*, Phys. Rev. E **66** (2002), 011105.

[350] ———, *Symmetry of matrix-valued stochastic processes and noncolliding diffusion particle systems*, J. Math. Phys. **45** (2004), 3058–3085.

[351] N.M. Katz and P. Sarnak, *Random matrices, Frobenius eigenvalues, and monodromy*, Colloquium publications, vol. 45, American Mathematical Society, Providence, RI, 1999.

[352] ———, *Zeroes of zeta functions and symmetry*, Bull. Amer. Math. Soc. **36** (1999), 1–26.

[353] J.P. Keating, *The Riemann zeta function and quantum chaology*, Quantum Chaos (I. Guarneri and U. Smilansky, eds.), North Holland, Amsterdam, 1993, pp. 145–185.

[354] J.P. Keating and N.C. Snaith, *Random matrix theory and* $\zeta(1/2 + it)$, Commun. Math. Phys. **214** (2001), 57–89.

[355] R. Killip, *Gaussian fluctuations for* β *ensembles*, Int. Math. Res. Not. **2008** (2008), rnn007.

[356] R. Killip and I. Nenciu, *Matrix models for circular ensembles*, Int. Math. Res. Not. **50** (2004), 2665–2701.

[357] R. Killip and M. Stoiciu, *Eigenvalue statistics for CMV matrices: from Poisson to clock via circular beta ensembles*, Duke Math. J. **146** (2009), 361–399.

[358] F. Knop, *Symmetric and non-symmetric quantum Capelli polynomials*, Comm. Math. Helv. **72** (1997), 84–100.

[359] F. Knop and S. Sahi, *A recursion and combinatorial formula for Jack polynomials*, Inv. Math. **128** (1997), 9–22.

[360] D.E. Knuth, *Permutations, matrices and generalized Young tableaux*, Pacific J. Math. **34** (1970), 709–727.

[361] H. Kohler, *Exact diagonalisation of 1-d interacting spinless fermions*, arXiv:0801.0132, 2008.

[362] I.K. Kostov, *Conformal field theory techniques in random matrix ensembles*, hep-th/9907060, 1999.

[363] I. Krasovsky, *Gap probability in the spectrum of random matrices and asymptotics of polynomials orthogonal on the unit circle*, Int. Math. Res. Not. **2004** (2004), 1249–1272.

[364] C. Krattenthaler, *Generating functions for plane partitions of a given shape*, Manuscripta Math. **69** (1990), 173–202.

[365] C. Krattenthaler, A.J. Guttmann, and X.G. Viennot, *Vicious walkers, friendly walkers and Young tableaux II: with a wall*, J. Phys. A **33** (2000), 8835–8866.

[366] M. Krishnapur, *Zeros of random analytic functions*, Ann. Prob. **37** (2009), 314–346.

[367] H. Kunz, *The one-dimensional classical electron gas*, Ann. Phys. (N.Y.) **85** (1974), 303–335.

[368] ———, *The probability distribution of the spectral form factor in random matrix theory*, J. Phys. A **32** (1999), 2171–2182.

[369] V.B. Kuznetsov, V.V. Mangazeev, and E.K. Sklyanin, *q-operator and factorized separation chain for Jack polynomials*, Indag. Math. **14** (2003), 451–482.

[370] L. Lapointe, A. Lascoux, and J. Morse, *Determinantal formula and recursion for Jack polynomials*, Electro. J. Comb. **7** (2001), 467.

[371] N. Lehmann and H.-J. Sommers, *Eigenvalue statistics of random real matrices*, Phys. Rev. Lett. **67** (1991), 941–944.

[372] A. Lenard, *Correlation functions and the uniqueness of the state in classical statitical mechanics*, Comm. Math. Phys. **30** (1973), 35–44.

[373] I.G. Macdonald, *Hypergeometric functions*, unpublished manuscript.

[374] ———, *Affine root systems and Dedekind's eta function*, Inven. Math. **15** (1972), 91–143.

[375] ———, *Some conjectures for root systems*, SIAM J. Math. Anal. **13** (1982), 998–1007.

[376] ———, *Hall polynomials and symmetric functions*, 2nd ed., Oxford University Press, Oxford, 1995.

[377] C.C. MacDuffee, *Vectors and matrices*, Mathematical Association of America, Buffalo, NY, 1943.

[378] A.M.S. Macêdo, *Universal parametric correlations at the soft edge of the spectrum of random matrices*, Europhys. Lett. **26** (1994), 641–646.

[379] U. Magnea, *Random matrices beyond the Cartan classification*, J. Phys. A **71** (2008), 045203(27pp).

[380] C. Mahaux and H.A. Weidenmüller, *Shell-model approach to nuclear reactions*, North Holland, New York, 1969.

[381] G. Mahoux and M.L. Mehta, *A method of integration over matrix variables IV*, J. Physique I (France) **1** (1991), 1093–1108.

[382] E. Majorana, *Atomi orientati in campo magnetico variablile*, Nuovo Cimento **9** (1932), 43–50.

[383] J. Malmquist, *Sur les équations différentialles du second ordre dont l'intégrale générale a ses points critiques fixes*, Arkiv Mat. Astron. Fys. **18** (1922), 1–89.

[384] F. Marcellan and R. Alvarez-Nordarse, *On the "Favard theorem" and its extensions*, J. Comput. Appl. Math. **127** (2001), 231–254.

[385] M. Marino, *Chen-Simon thoery, matrix integrals and perturbative three-manifold invariants*, Commun. Math. Phys. **253** (2005), 25–49.

[386] Ph. A. Martin, *Sum rules in charged fluids*, Rev. Mod. Phys. **60** (1988), 1075–1127.

[387] Ph. A. Martin and Ch. Gruber, *A new proof of the Stillinger-Lovett complete shielding condition*, J. Stat. Phys. **31** (1983), 691–710.

[388] V.A. Marčenko and L.A. Pastur, *Distributions of eigenvalues for some sets of random matrices*, Math. USSR-Sbornik **1** (1967), 457–483.

[389] A.M. Mathai, *Jacobians of matrix transformations and functions of matrix arguments*, World Scientific, Singapore, 1997.

[390] D.C. Mattis, *Statistical mechanics made simple*, 2nd ed., World Scientific, River Edge, NJ, 2008.

[391] R.E McMurtrie, *Determinants of stability of large randomly connected systems*, J. Theor. Biol. **50** (1975), 1–11.

[392] M.L. Mehta, *Random matrices and the statistical theory of energy levels*, Academic Press, New York, 1967.

[393] ———, *Matrix theory*, Éditions de Physique, Les Ulis, France, 1989.

[394] ———, *A non-linear differential equation and a Fredholm determinant*, J. de Physique I (France) **2** (1991), 1721–1729.

[395] ———, *Random matrices*, 2nd ed., Academic Press, New York, 1991.

[396] ———, *Power series for level spacing functions of random matrix ensembles*, Z. Phys. B **86** (1992), 285–290.

[397] M.L. Mehta and F.J. Dyson, *Statistical theory of the energy levels of complex systems. V*, J. Math. Phys. **4** (1963), 713–719.

[398] P.A. Mello, *Averages on the unitary group and applications to the problem of disordered conductors*, J. Phys. A **23** (1990), 4061–4080.

[399] P.A. Mello, A. Pereyra, and N. Kumar, *Macroscopic approach to multichannel disordered conductors*, Ann. Phys. **181** (1988), 290–317.

[400] P.A. Mello, P. Pereyra, and T.H. Seligman, *Information theory and statistical nuclear reactions I*, Ann. Phys. (N.Y.) **161** (1985), 254–275.

[401] A. Messiah, *Quantum mechanics*, vol. I and II, North Holland, Amsterdam, 1972.

[402] F. Mezzadri, *How to generate random matrices from the classical groups*, Notices AMS **54** (2007), 592–604.

[403] S.J. Miller and R. Takloo-Bighash, *An invitation to modern number theory*, Princeton University Press, Princeton, NJ, 2006.

[404] K. Mimachi, *A duality of Macdonald-Koornwinder polynomials and its application to integral representations*, Duke Math. J **107** (2001), 265–281.

[405] H.L. Montgomery, *The pair correlation of zeros of the zeta function*, Proc. Sympos. Pure Math., vol. 24, American Mathematical Society, Providence, RI, 1973, pp. 181–193.

[406] E.H. Moore, *On the determinant of a Hermitian matrix of quaternionic elements*, Bull. Am. Math. Soc. **28** (1922), 161–162.

[407] W. G. Morris, *Constant term identities for finite and affine root systems, conjectures and theorems*, Ph.D. thesis, University of Wisconsin, 1982.

[408] P.M. Morse and H. Feshbach, *Methods of theoretical physics*, vol. 2, McGraw-Hill, New York, 1953.

[409] M. Moshe, H. Neuberger, and B. Shapiro, *Generalized ensemble of random matrices*, Phys. Rev. Lett. **73** (1994), 1497–1500.

[410] R.J. Muirhead, *Aspects of multivariate statistical theory*, Wiley, New York, 1982.

[411] F.D. Murnaghan, *The unitary and rotation groups*, Spartan Books, Washington DC, 1962.

[412] N.I. Muskhelishvili, *Singular integral equations*, 2nd ed., Dover, New York, 1992.

[413] K.A. Muttalib, Y. Chen, M.E.H. Ismail, and V.N. Nicopolous, *New family of unitary random matrices*, Phys. Rev. Lett. **71** (1994), 471–474.

[414] T. Nagao, *Universal correlations near a singularity of random matrix spectrum*, J. Phys. Soc. Japan **64** (1995), 3675–3681.

[415] T. Nagao and P.J. Forrester, *Asymptotic correlations at the spectrum edge of random matrices*, Nucl. Phys. B **435** (1995), 401–420.

[416] ______, *Multilevel dynamical correlation functions for Dyson's Brownian motion model of random matrices*, Phys. Lett. A **247** (1998), 42–46.

[417] ______, *Quaternion determinant expressions for multilevel dynamical correlation functions of parametric random matrices*, Nucl. Phys. B **563** (1999), 547–572.

[418] T. Nagao and K. Slevin, *Laguerre ensembles of random matrices: nonuniversal correlation functions*, J. Math. Phys. **34** (1993), 2317–2330.

[419] T. Nagao and M. Wadati, *Correlation functions of random matrix ensembles related to classical orthogonal polynomials*, J. Phys. Soc. Japan **60** (1991), 3298–3322.

[420] ______, *Correlation functions of random matrix ensembles related to classical orthogonal polynomials II, III*, J. Phys. Soc. Japan **61** (1992), 78–88, 1910–1918.

[421] ______, *Correlation functions for Jastrow-product wave functions*, J. Phys. Soc. Japan **62** (1993), 480–488.

[422] F. Nijhoff, J. Satsuma, K. Kajiwara, B. Grammaticos, and A. Ramani, *A study of the alternate discrete painlevé II equation*, Inverse problems **12** (1996), 697–716.

[423] E.J.G. Nordenstam, *Erratum to eigenvalues of GUE minors*, Elect. J. Probability **12** (2007), 1048–1051.

[424] M. Noumi and Y. Yamada, *Higher order Painlevé equations of type* $A_l^{(1)}$, Funkcial. Ekvac **41** (1998), 483–503.

[425] A.M. Odlyzko, *On the distribution of spacings between zeros of the zeta function*, Math. Comput. **48** (1987), 273–308.

[426] ______, *The* 10^{20}*th zero of the Riemann zeta function and 70 million of its neighbours*, Preprint, 1989.

[427] K. Okamoto, *Polynomial Hamiltonians associated with Painlevé equations*, Proc. Japan Acad. Ser. A Math. Sci. **56** (1980), 264–268.

[428] ______, *Studies of the Painlevé equations I. Sixth Painlevé equation* P_{VI}, Ann. Math. Pura Appl. **146** (1987), 337–381.

[429] ______, *Studies of the Painlevé equations IV. Third Painlevé equation* P_{III}, Funkcialaj. Ekvacioj. **30** (1987), 305–332.

[430] A. Okounkov, *Infinite wedge and random partitions*, Selecta Math. (N.S.) **7** (2001), 57–81.

[431] A. Okounkov and G. Olshanski, *Shifted Jack polynomials, binomial formula and applications*, Math. Res. Lett. **4** (1997), 69–78.

[432] ______, *Shifted Schur polynomials, binomial formula and applications*, Math. Res. Lett. **4** (1997), 533–553.

[433] I. Olkin, *The 70th anniversary of the distribution of random matrices: a survey*, Linear Algebra Appl. **354** (2002), 231–243.

[434] M.A. Olshanetsky and A.M. Perelomov, *Quantum integrable systems related to Lie algebras*, Phys. Rep. **94** (1983), 313–404.

[435] F. Olver, *Asymptotics and special functions*, Academic Press, London, 1974.

[436] E.M. Opdam, *Harmonic analysis for certain representations of graded Hecke algebras*, Acta Math. **175** (1995), 75–121.

[437] ———, *Lectures on Dunkl operators*, Math. Soc. Japan Mem. **8** (1998), 1–62.

[438] A.Y. Orlov, *New solvable matrix integrals*, Int. J. Mod. Phys. A **19** (2004), 276–293.

[439] J.C. Osborn, *Universal results from an alternative random matrix model for QCD with a baryon chemical potential*, Phys. Rev. Lett. **93** (2004), 222001.

[440] A. Pandey, *Statistical properties of many-particle spectra III.*, Ann. Phys. **119** (1979), 170–191.

[441] A. Pandey and M.L. Mehta, *Gaussian ensembles of random Hermitian matrices intermediate between orthogonal and unitary ones*, Commun. Math. Phys. **87** (1983), 449–468.

[442] S. Péché, *The largest eigenvalue of small rank perturbations of Hermitian random matrices*, Prob. Theory Related Fields **134** (2006), 127–173.

[443] Y. Peres and B. Virág, *Zeros of the i.i.d. Gaussian power series: a conformally invariant determinantal process*, Acta. Math. **194** (2005), 1–35.

[444] N. Du Plessis, *An introduction to potential theory*, Oliver and Boyd, Edinburgh, 1970.

[445] H.D. Politzer, *Random-matrix description of the distribution of mesoscopic conductance*, Phys. Rev. B **40** (1989), 11917–11919.

[446] A.P. Polychronakos, *Exhange operator formalism for integrable systems of particles*, Phys. Rev. Lett. **69** (1992), 703.

[447] C.E. Porter, *Statistical theories of spectra: fluctuations*, Academic Press, New York, 1965.

[448] D. Porter and D.S.G. Stirling, *Integral equations: a practical treatment, from spectral theory to applications*, Cambridge University Press, Cambridge, 1990.

[449] M. Prähofer, *High precision moments corresponding to GUE/GOE/GSE*, Unpublished note, 2004.

[450] M. Prähofer and H. Spohn, *Universal distributions for growth processes in* $1+1$ *dimensions and random matrices*, Phys. Rev. Lett. **84** (2000), 4882–4885.

[451] ———, *Exact scaling functions for one-dimensional stationary KPZ growth*, J. Stat. Phys. **108** (2004), 1071–1106.

[452] R.A. Proctor, *Odd symplectic groups*, Inv. Math. **92** (1988), 307–332.

[453] ———, *Equivalence of the combinatorial and the classical definitions of Schur functions*, J. Combin. Theory A **51** (1989), 135–137.

[454] E.M. Rains, *Attack of the zonal polynomials*, Preprint, 1995.

[455] ———, *Increasing subsequences and the classical groups*, Elect. J. of Combinatorics **5** (1998), #R12.

[456] ———, *Correlations for symmetrized increasing subsequences*, math.CO/0006097, 2000.

[457] ———, Unpublished notes, 2001,04.

[458] ———, *Images of eigenvalue distributions under power maps*, Prob. Theory Related Fields **125** (2003), 522–538.

[459] ———, *Limits of elliptic hypergeometric integrals*, Ramanujan J. **18** (2009), 257–306.

[460] E.M. Rains and M. Vazirani, *Vanishing integrals of Macdonald and Koornwinder polynomials*, Transform. Groups **12** (2007), 725–759.

[461] A. Ramani, Y. Ohta, and B. Grammaticos, *Discrete integrable systems from continuous Painlevé equations through limiting procedures*, Nonlinearity **13** (2000), 1073–1085.

[462] J. Ramirez and B. Rider, *Diffusion at the random matrix hard edge*, arXiv:0803.2043, 2008.

[463] J. Ramirez, B. Rider, and B. Virag, *Beta ensembles, stochastic Airy spectrum, and a diffusion*, arXiv:math/060733, 2006.

[464] A. Raposo, H.J. Weber, D.E. Alvarez, and M. Kirchback, *Romanovski polynomials in selected physics problems*, Central Europ. J. Phys. **5** (2007), 253–284.

[465] B. Rider, *Deviations from the circular law*, Prob. Theory Related Fields **130** (2003), 337–367.

[466] ______, *A limit theorem at the edge of a non-Hermitian random matrix ensemble*, J. Phys. A **36** (2003), 3401–3410.

[467] ______, *Order statistics and Ginibre's ensemble*, J. Stat. Phys. **114** (2004), 1139–1148.

[468] B. Rider and B. Virág, *The noise in the circular law and the Gaussian free field*, Int. Math. Res. Not. **2007** (2007), rnm006(32 pp).

[469] H. Risken, *The Fokker-Planck equation*, Springer, Berlin, 1992.

[470] V. Romanovsky, *Sur quelques classes nouvelles de polynômes orthogonaux*, C.R. Acad. Sci. Paris **188** (1929), 1023–1025.

[471] M. Rubinstein, *Computational methods and experiments in analytic number theory*, Recent perspectives in random matrix theory and number theory (F. Mezzadri and N.C. Snaith, eds.), London Mathematical Society Lecture Note Series, vol. 322, Cambridge University Press, Cambridge, 2005, pp. 425–506.

[472] Z. Rudnick and P. Sarnak, *Zeros of principal L-functions and random matrix theory*, Duke Math. J. (1995), 269–322.

[473] E. Ryckman, *Linear statistics of point processes via orthogonal polynomials*, J. Stat. Phys. (2008), 473–486.

[474] E. Saff and A. Kuijlaars, *Distributing many points on a sphere*, Math. Intelligencer **10** (1997), 5–11.

[475] B.E. Sagan, *The symmetric group*, 2nd ed., Springer, New York, 2000.

[476] S. Sahi, *A new scalar product for nonsymmetric jack polynomials*, Int. Math. Res. Not. **(20)** (1996), 997–1004.

[477] ______, *The binomial formula for nonsymmetric Macdonald polynomials*, Duke Math. J. **1998** (1998), 465–477.

[478] H. Sakai, *Rational surfaces associated with affine root systems and geometry of the Painlevé equations*, Commun. Math. Phys. **220** (2001), 165–229.

[479] L. Samaj and J.K. Percus, *A functional relation among the pair correlations of the two-dimensional one-component plasma*, J. Stat. Phys. **80** (1995), 495–512.

[480] T. Sasamoto, *Spatial correlations of the 1D KPZ surface on a flat substrate*, J. Phys. A **38** (2005), L549–L556.

[481] C. Schensted, *Longest increasing and decreasing subsequences*, Canad. J. Math. **13** (1963), 179–191.

[482] B. Schlittgen and T. Wettig, *Generalizations of some integrals over the unitary group*, J. Phys. A **36** (2003), 3195–3202.

[483] A. Selberg, *Bemerkninger om et multipelt integral*, Norsk. Mat. Tidsskr. **24** (1944), 71–78.

[484] D. Serban, *Some properties of the Calogero-Sutherland model with reflections*, J. Phys. A **30** (1997), 4215–4225.

[485] T. Shifrin, *Multivariable mathematics*, Wiley, New York, 2005.

[486] J.W. Silverstein, *The smallest eigenvalues of a large dimensional Wishart matrix*, Ann. Probab. **13** (1985), 1364–1368.

[487] S.H. Simon and A. L. Moustakas, *Crosssover from conserving to lossy transport in circular random-matrix ensembles*, Phys. Rev. Lett. **96** (2006), 136805.

[488] S.H. Simon, A. L. Moustakas, and L. Marinelli, *Capacity and character expansions: moment generating function and other exact results for MIMO correlated channels*, IEEE Trans. Inform. Theory **52** (2006), 5336–5351.

[489] S.H. Simon, A. L. Moustakas, M. Stoytchev, and H. Safar, *Communication in a disordered world*, Physics Today **54** (2001), 38.

[490] B.D. Simons and B. L. Altshuler, *Universal velocity correlations in disordered and chaotic systems*, Phys. Rev. Lett. **70** (1993), 4063.

[491] C.D. Sinclair, *Averages over Ginibre's ensemble of random real matrices*, Int. Math. Res. Not. **2007** (2007), rnm015.

[492] K. Slevin and T. Nagao, *Impurity scattering in mesoscopic quantum wires and the Laguerre ensemble*, Phys. Rev. B **50** (1994), 2380–2392.

[493] E.R. Smith, *Effects of surface charge on the two-dimensional one-component plasma: I. Single double layer*, J. Phys. A **15** (1982), 3861–3868.

[494] A. Soshnikov, *Level spacings distribution for large random matrices: Gaussian fluctuations*, Ann. Math. **148** (1998), 573–617.

[495] ______, *Determinantal random point fields*, Russian Math. Surveys **55** (2000), 923–975.

[496] ———, *A note on the universality of the distribution of the largest eigenvalues in certain sample covariance matrices*, J. Stat. Phys. **108** (2002), 1033–1056.

[497] ———, *Statistics of extreme spacing in determinantal random point processes*, Mosc. Math. J. **5** (2007), 705–719.

[498] ———, *Gaussian fluctuation for the number of particles in Airy, Bessel, Sine, and other determinantal random point fields*, J. Stat. Phys. **100** (2000), 491–522.

[499] A. Sri Ranga and W. Van Assche, *Blumenthal's theorem for Laurent orthogonal polynomials*, J. Approx. Th. **117** (2002), 255–278.

[500] R.P. Stanley, *Some combinatorial properties of Jack symmetric functions*, Advan. Math. **77** (1989), 76–115.

[501] ———, *Enumerative combinatorics*, vol. 2, Cambridge University Press, Cambridge, 1999.

[502] J.R. Stembridge, *Non-intersecting paths, Pfaffians and plane partitions*, Adv. Math. **83** (1990), 96–131.

[503] T.J. Stieltjes, *Sur quelques théorèms d'algèbra*, Oeuvres complètes, vol. 1, Société Mathématique d'Amsterdam, Groningen, 1914, pp. 440–441.

[504] A.D. Stone, P.A. Mello, K.A. Muttalib, and J.-L. Pichard, *Random matrix theory and maximum entropy models for disordered conductors*, Mesoscopic phenomena in solids (P.A. Lee B.L. Altshuler and R.A. Webb, eds.), North Holland, Amsterdam, 1991, pp. 369–448.

[505] B. Sutherland, *Exact results for a quantum many body problem in one dimension*, Phys. Rev. A **4** (1971), 2019–2021.

[506] ———, *Quantum many-body problem in one dimension*, J. Math. Phys. **12** (1971), 246–250.

[507] ———, *Exact ground state wave function for a one-dimensional plasma*, Phys. Rev. Lett. **34** (1975), 1083–1085.

[508] G. Szegö, *Orthogonal polynomials*, 4th ed., American Mathematical Society, Providence RI, 1975.

[509] A. Takemura, *Zonal polynomials*, Institute of Mathematical Statistics, Hayward, CA, 1984.

[510] T. Tao and V. Vu, *Random matrices: universality of esds and the circular law*, arXiv:0807.4898, 2008.

[511] G. Telléz and P.J. Forrester, *Finite size study of the 2dOCP at* $\Gamma = 4$ *and* $\Gamma = 6$, J. Stat. Phys. **97** (1999), 489–521.

[512] N.M. Temme, *Uniform asymptotic expansions of the incomplete gamma functions and the incomplete beta function*, Math. Comp. **29** (1975), 1109.

[513] Y. Tian, *Matrix representations of octonions and their applications*, Adv. Appl. Clifford Algebras **10** (2000), 61–90.

[514] M. Tierz, *Soft matrix models and chern-simons partition functions*, Mod. Phys. Lett. A **19** (2004), 1365–1378.

[515] E.C. Titchmarsh, *Theory of functions*, Oxford University Press, London, 1939.

[516] ———, *The theory of the Riemann zeta function*, 2nd ed., Clarendon Press, Oxford, 1986.

[517] C.A. Tracy and H. Widom, *Introduction to random matrices*, Geometric and quantum aspects of integrable systems (G.F. Helminck, ed.), Lecture Notes in Physics, vol. 424, Springer, New York, 1993, pp. 407–424.

[518] ———, *Fredholm determinants, differential equations and matrix models*, Commun. Math. Phys. **163** (1994), 33–72.

[519] ———, *Level-spacing distributions and the Airy kernel*, Commun. Math. Phys. **159** (1994), 151–174.

[520] ———, *Level-spacing distributions and the Bessel kernel*, Commun. Math. Phys. **161** (1994), 289–309.

[521] ———, *On orthogonal and symplectic matrix ensembles*, Commun. Math. Phys. **177** (1996), 727–754.

[522] ———, *Correlation functions, cluster functions and spacing distributions in random matrices*, J. Stat. Phys. **92** (1998), 809–835.

[523] A.M. Tulino and S. Verdú, *Random matrix theory and wireless communications*, Foundations and Trends in Communcations and Information Theory, vol. 1, Now Publisher, 2004, pp. 1–182.

[524] B. Valkó and B. Virág, *Continuum limits of random matrices and the Brownian carousel*, Inv. Math. **177** (2008), 463–508.

[525] ———, *Large gaps between random eigenvalues*, arXiv:0811.0007, 2008.

[526] P. van Moerbeke, *Integrable lattices: random matrices and random permutations*, Random matrix models and their applications (P.M. Bleher and A.R. Its, eds.), Mathematical Sciences Research Institute Publications, vol. 40, Cambridge University Press, Cambridge, 2001, pp. 321–406.

[527] R. Vein and P. Dale, *Determinants and their applications in mathematical physics*, Springer, New York, 1999.

[528] A.M. Veneziani, T. Pereira, and D.H.U. Marchetti, *Conformal universality in normal matrix ensembles*, arXiv:0909.3418, 2009.

[529] J.J.M. Verbaarschot, *The spectrum of the Dirac operator near zero virtuality for $n_c = 2$ and chiral random matrix theory*, Nucl. Phys. B **426** (1994), 559–574.

[530] J.J.M. Verbaarschot and T. Wettig, *Random matrix theory and chiral symmetry in QCD*, Ann. Rev. Nucl. Part. Sci. **50** (2000), 343–410.

[531] J.M. Verbaarschot, H.A. Weidenmüller, and M.R. Zirnbauer, *Grassman integration in stochastic quantum physics: the case of compound nucleus scattering*, Phys. Rep. **129** (1985), 367–438.

[532] G. Viennot, *Une forme géométrique de la correspondance de Robinson-Schensted*, Combinatoire et représentation du groupe symétrique (D. Foata, ed.), Lecture Notes in Mathematics, vol. 579, Springer, Berlin, 1977, pp. 29–58.

[533] J.P. Vinson, *Closest spacing of eigenvalues*, Ph.D. thesis, Princeton University, 2001.

[534] F. von Oppen, *Exact distribution of eigenvalue curvatures of chaotic quantum systems*, Phys. Rev. Lett. **73** (1994), 798–801.

[535] K.W. Wachter, *The strong limits of random matrix spectra for sample matrices of independent elements*, Annal. Prob. **6** (1978), 1–18.

[536] ———, *The limiting measure of multiple disriminant ratios*, Ann. Statist. **8** (1980), 937–957.

[537] S.O. Warnaar, *On the generalized Selberg integral of Richards and Zheng*, Adv. Applied Math. **40** (2008), 212–218.

[538] H. Watanabe, *Defining variety and birational canonical transformations of the fifth Painlevé equation*, Analysis **18** (1998), 351–357.

[539] G.N. Watson, *Theory of Bessel functions*, 2nd ed., Cambridge University Press, Cambridge, 1966.

[540] H. Weyl, *The classical groups: Their invariants and representations*, Princeton University Press, Princeton, NJ, 1939.

[541] E.T. Whittaker and G.N. Watson, *A course of modern analysis*, 2nd ed., Cambridge University Press, Cambridge, 1965.

[542] H. Widom, *Strong Szegö theorem on circular arcs*, Indiana Univ. Math. J. **21** (1971), 277–283.

[543] ———, *The asymptotics of a continuous analogue of orthogonal polynomials*, J. Approx. Theory **77** (1996), 51–64.

[544] ———, *On the relation between orthogonal, symplectic and unitary random matrices*, J. Stat. Phys. **94** (1999), 347–364.

[545] H.S. Wilf, *Ascending subsequences of permutations and the shapes of tableaux*, J. Combin. Theory Ser. A **60** (1992), 155–157.

[546] J.H. Wilkinson, *The algebraic eigenvalue problem*, Clarendon Press, Oxford, 1965.

[547] J. Wishart, *The generalized product moment distribution in samples from a normal multivariate population*, Biometrika **20A** (1928), 32–43.

[548] N.S. Witte, *Gap probabilities for double intervals in Hermitian random matrix ensembles as τ-functions — spectrum singularity case*, Lett. Math. Phys. **68** (2004), 139–149.

[549] T.T. Wu and C.N. Yang, *Dirac monopole without string: monopole harmonics*, Nucl. Phys. B **107** (1976), 365–380.

[550] Z. Yan, *A class of generalized hypergeometric functions in several variables*, Canad. J. Math. **44** (1992), 1317–1338.

[551] I. Zakharevich, *A generalization of Wigner's law*, Comm. Math. Phys. **268** (2006), 403–414.

[552] D. Zeilberger, *A Stembridge-Stanton style elementary proof of the Habsieger-Kadell q Morris identity*, Discrete Math. **79** (1989), 313–322.

[553] A. Zhedanov, *Biorthogonal rational functions and the generalized eigenvalue problem*, J. Approx. Th. **101** (1999), 303–329.

[554] P. Zinn-Justin, *Universality of correlation functions of Hermitian random matrices in an external field*, Commun. Math. Phys. **194** (1998), 631–650.

[555] ———, *Some matrix integrals related to knots and links*, Random matrix models and their applications (P.M. Bleher and A.R. Its, eds.), Mathematical Sciences Research Institute Publications, vol. 40, Cambridge University Press, Cambridge, 2001, pp. 421–438.

[556] M.R. Zirnbauer, *Riemannian symmetric superspaces and their origin*, J. Math. Phys. **37** (1996), 4986–5018.

[557] A. Zvonkin, *Matrix integrals and map enumeration: an accessible introduction*, Math. Comp. Modelling **26** (1997), 281–304.

[558] K. Zyczkowski and M. Kus, *Unitary random matrices*, J. Phys. A **27** (1994), 4235–4245.

[559] K. Zyczkowski and H.-J. Sommers, *Truncations of random unitary matrices*, J. Phys. A **33** (2000), 2045–2057.

[560] ———, *Induced measures in the space of mixed quantum states*, J. Phys. A **34** (2001), 7111–7125.

Index